Tonicity changes (effects of hypertonicity and hypotonicity on osmotic movement of water in cells): Chapter 8 (Box 8A on pp. 200–1; Fig. 8A-1)

Voltage recording: Chapter 9 (pp. 236–37; Fig. 9-12)

Microscopy

Electron microscopy: Chapter 1 (pp. 6–8); *Guide to Microscopy* (pp. 17–20)

Autoradiography: *Guide to Microscopy* (p. 22; Fig. 24b)

Earliest use by biologists: Chapter 1 (p. 6)

Freeze etching: *Guide to Microscopy* (p. 25)

Freeze fracturing: Chapter 1 (p. 8); Chapter 7 (pp. 175–76; Figs. 7-16, 7-17, and 7-18; p. 188; Fig. 7-29): *Guide to Microscopy* (pp. 23–25; Figs. 34, 35, and 36)

High-voltage electron microscopy: Chapter 1 (p. 7); *Guide to Microscopy* (p. 19, Fig. 27)

Immunoelectron microscopy: *Guide to Microscopy* (p. 22; Fig. 30)

Negative staining: Chapter 1 (pp. 7–8); Chapter 14 (pp. 404–405; Fig. 14-6); *Guide to Microscopy* (p. 22; Fig. 31)

Sample preparation techniques in scanning electron microscopy: *Guide to Microscopy* (p. 26)

Sectioning: *Guide to Microscopy* (pp. 20–22; Fig. 29)

Shadowing: *Guide to Microscopy* (pp. 22–23; Figs. 32, 33)

Staining: *Guide to Microscopy* (p. 21)

Scanning electron microscopy: Chapter 1 (p. 7; Fig. 1-4); *Guide to Microscopy* (pp. 19–20; Figs. 26b and 28)

Stereo electron microscopy: *Guide to Microscopy* (pp. 25–26; Fig. 37)

Transmission electron microscopy: Chapter 1 (p. 7; Fig. 1-3b); *Guide to Microscopy* (pp. 18–19; Figs. 25 and 26a)

Light microscopy: Chapter 1 (pp. 5–6; Table 1-1); *Guide to Microscopy* (pp. 5–17; Fig. 6)

Autoradiography: *Guide to Microscopy* (pp. 16–17; Figs. 23 and 24a)

Brightfield microscopy: *Guide to Microscopy* (p. 6; Table 1)

Confocal microscopy: Chapter 1 (p. 6); *Guide to Microscopy* (pp. 11–12; Table 1, Figs. 16 and 18)

Digital deconvolution microscopy: *Guide to Microscopy* (p. 13; Fig. 20)

Digital video microscopy: Chapter 1 (p. 6); *Guide to Microscopy* (pp. 13–14; Fig. 21)

Differential interference contrast (DIC) microscopy: Chapter 1 (p. 6); *Guide to Microscopy* (p. 7; Table 1, Figs. 9, 10)

Earliest use by biologists: Chapter 1 (pp. 1–3)

Fixation: *Guide to Microscopy* (p. 16)

Fluorescence microscopy: Chapter 1 (p. 6); *Guide to Microscopy* (pp. 8–11; Figs. 12, 14)

Fluorescent probes: *Guide to Microscopy* (pp. 9–11)

Green fluorescent protein: *Guide to Microscopy* (p. 10; Fig. 15)

Immunofluorescence microscopy: *Guide to Microscopy* (pp. 9–10; Fig. 13)

Multiphoton excitation microscopy: *Guide to Microscopy* (p. 13; Fig. 19)

Phase-contrast microscopy: Chapter 1 (p. 6); *Guide to Microscopy* (pp. 6–7; Table 1, Figs. 7, 8)

Sectioning: *Guide to Microscopy* (p. 16; Fig. 22)

Staining: *Guide to Microscopy* (p. 16)

Optical principles of microscopy: *Guide to Microscopy* (pp. 1–5; Fig. 1)

Scanning probe microscopy: *Guide to Microscopy* (pp. 26–27)

Atomic force microscope: *Guide to Microscopy* (p. 27)

Scanning tunneling microscope: *Guide to Microscopy* (pp. 26–27; Fig. 38)

Nucleic Acids and Recombinant DNA

Cloning of genes: Chapter 18 (pp. 606–14; Figs. 18-26, 18-27, 18-28, 18-29)

Cloning of organisms: Chapter

Colony hybridization with nucleic acid probe: Chapter 18 (p. 611; Fig. 18-30)

cDNA preparation (reverse transcription): Chapter 18 (p. 612; Fig. 18-31)

cDNA molecules for transcription studies: Chapter 21 (pp. 717 and 736; Fig. 21-20)

DNA denaturation and renaturation: Chapter 16 (pp. 490–91 and 499–500; Figs. 16-10 and 16-15)

DNA fingerprinting: Chapter 16 (Box 16C on pp. 502–3)

DNA microarrays: Chapter 21 (pp. 716–17, Fig. 21-20)

DNA sequencing: Chapter 7 (Box 7A on p. 182); Chapter 16 (pp. 496–97; Fig. 16-14)

DNase sensitivity of active genes in chromatin: Chapter 21 (pp. 712–13; Fig. 21-17)

Electrophoresis of DNA: Chapter 16 (pp. 493–94; Fig. 16-12)

Equilibrium density centrifugation of DNA: Chapter 17 (pp. 525–27; Figs. 17-3, 17-4)

Footprinting technique for detecting protein-binding sites on DNA: Chapter 19 (Box 19B on p. 638)

Hybridization of nucleic acids: Chapter 16 (p. 491)

Northern blot: Chapter 21 (p. 737)

Nuclease digestion of chromatin (to isolate nucleosomes): Chapter 16 (pp. 505–6; Fig. 16-19)

Phage particle concentration: Chapter 16 (pp. 484–85; Fig. 16A-3)

Polymerase chain reaction (PCR): Chapter 17 (Box 17A on pp. 533–34)

Restriction enzymes: Chapter 16 (Box 16B on pp. 494–95)

Restriction mapping of DNA: Chapter 16 (pp. 495–96; Fig. 16-13)

Run-on transcription assay: Chapter 21 (p. 716; Fig. 21-19)

Southern blot: Chapter 16 (p. 503)

Temperature-sensitive mutants (DNA replication): Chapter 17 (p. 530)

Transgenic animals: Chapter 18 (Box 18A on p. 617)

Transgenic plants: Chapter 18 (pp. 614–16; Fig. 18-33)

Yeast artificial chromosomes (YACs): Chapter 18 (pp. 613–14; Fig. 18-32)

X-ray crystallography: Chapter 3 (Box 3A on p. 60); Chapter 16 (p. 487); *Guide to Microscopy* (pp. 27–28; Fig. 40)

Proteins

Antibodies against specific cell-surface molecules (to identify specific proteins involved in cell-cell adhesion): Chapter 11 (p. 303)

Affinity labeling of proteins: Chapter 7 (Box 7A on p. 182)

Hydropathic analysis of proteins (to identify possible transmembrane segments): Chapter 7 (pp. 179 and 193; Figs. 7-22 and 7-30)

Immunoblotting: Chapter 21 (p. 736)

Protein denaturation and renaturation: Chapter 2 (pp. 31–33; Fig. 2-18a)

Protein sequencing: Chapter 3 (p. 48)

SDS-polyacrylamide gel electrophoresis: Chapter 7 (pp. 180–81; Fig. 7-23)

X-ray crystallography: Chapter 6 (pp. 138–139; Fig. 6-7); Chapter 14 (p. 432); *Guide to Microscopy* (pp. 27–28; Fig. 40)

Scientific Method

Use of scientific method: Chapter 1 (Box 1B on pp. 11–12)

Separation of Cells, Organelles, and Molecules

Density gradient centrifugation: Chapter 12 (pp. 328–30; Figs. 12A-5 and 12A-6)

Differential centrifugation: Chapter 12 (pp. 327–28; Figs. 12A-2 and 12A-4); used to isolate lysosomes: Chapter 4 (Box 4A on p. 92)

Equilibrium density centrifugation: Chapter 12 (pp. 329–30 and 358–59; Figs. 12A-7 and 12-23); Chapter 17 (pp. 525–27; Figs. 17-3, 17-4)

Gel electrophoresis of DNA: Chapter 16 (pp. 493–94; Fig. 16-12)

SDS-polyacrylamide gel electrophoresis as a means of separating proteins: Chapter 7 (pp. 180–81; Fig. 7-23)

Subcellular fractionation: Chapter 12 (Box 12A on pp. 326–30)

Thin-layer chromatography (TLC) for analysis of membrane lipids: Chapter 7 (pp. 168–69; Fig. 7-9)

Ultracentrifugation, earliest use by biologists: Chapter 1 (pp. 9–10)

Free Student Aid.

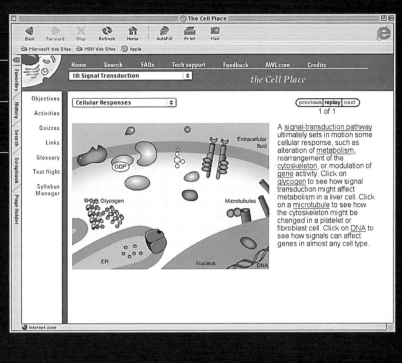

Log in.

Explore.

Succeed.

The Cell Place web site features over 30 animations and interactive activities that facilitate the learning of key concepts in the field of cell biology. Go to www.thecellplace.com.

Got technical questions?
For technical support, please visit www.aw.com/techsupport, send an email to online.support@pearsoned.com (for web site questions) or send an email to media.support@pearsoned.com (for CD-ROM questions) with a detailed description of your computer system and the technical problem. You can also call our tech support hotline at 1-800-677-6337 Monday-Friday, 8 a.m. to 5 p.m. CST.

What your system needs to use these media resources:

WINDOWS
250 MHz Windows-95/98/NT/2000
32 MB RAM installed
1024 X 768 screen resolution
Thousands of colors
Browser: Internet Explorer 5.0 or Netscape
    Navigator 4.7
Note: THIS SITE DOES NOT SUPPORT NETSCAPE
    NAV. 6.0
Plug-Ins: Flash Player 5, Shockwave

MACINTOSH
233 MHz PowerPC
OS 8.6 or higher
32 MB RAM minimum
1024 x 768 screen resolution, thousands of colors
Thousands of Colors
Browser: Internet Explorer 5.0 or Netscape
    Navigator 4.7
Note: THIS SITE DOES NOT SUPPORT NETSCAPE
    NAV. 6.0
Plug-Ins: Flash Player 5, Shockwave

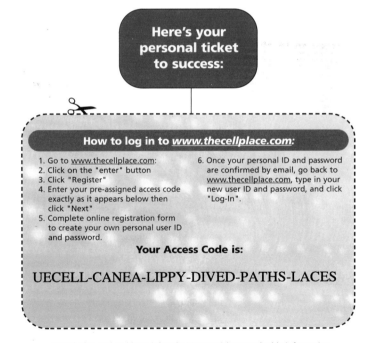

Here's your personal ticket to success:

How to log in to www.thecellplace.com:

1. Go to www.thecellplace.com:
2. Click on the "enter" button
3. Click "Register"
4. Enter your pre-assigned access code exactly as it appears below then click "Next"
5. Complete online registration form to create your own personal user ID and password.
6. Once your personal ID and password are confirmed by email, go back to www.thecellplace.com, type in your new user ID and password, and click "Log-In".

Your Access Code is:

UECELL-CANEA-LIPPY-DIVED-PATHS-LACES

Detach this card and keep it handy. It's your ticket to valuable information.

Important: Please read the License Agreement, located on the launch screen before using The Cell Place web site (www.thecellplace.com) or The Cell Place CD-ROM. By using the web site or CD-ROM, you indicate that you have read, understood and accepted the terms of this agreement.

0-8053-4547-7 / 0-8053-4854-9 / 0-8053-4548-5

The WORLD of the CELL

The WORLD of the CELL

5th edition

Wayne M. Becker
University of Wisconsin, Madison

Lewis J. Kleinsmith
University of Michigan, Ann Arbor

Jeff Hardin
University of Wisconsin, Madison

CONTRIBUTOR
John Raasch
University of Wisconsin, Madison
Chapters 12 and 15

Benjamin
Cummings

San Francisco Boston New York
Capetown Hong Kong London Madrid Mexico City
Montreal Munich Paris Singapore Sydney Tokyo Toronto

Editorial Director:	Frank Ruggirello
Sponsoring Editor:	Michele Sordi
Project Editor/ Production Manager:	Laura Kenney Editorial Services
Project Coordinator:	Electronic Publishing Services Inc., N.Y.C.
Designer, Text and Cover:	Jennifer Dunn
Art Editor:	Kelly Murphy
Illustrations:	Precision Graphics
Composition/Page Makeup:	Electronic Publishing Services Inc., N.Y.C.
Indexing:	Shane-Armstrong Information Systems
Prepress Supervisor:	Vivian McDougal
Marketing Manager:	Josh Frost
Publishing Assistant:	Michael McArdle

On the cover: "Disks of Newton, Study for Fugue" by Frank Kupka, 1912, Philadelphia Museum of Art/CORBIS.

ISBN 0-8053-4852-2

Library of Congress Cataloging-in-Publication Data
Becker, Wayne M.
    The world of the cell / Wayne M. Becker, Lewis J. Kleinsmith, Jeff Hardin;
contributor, John Raasch.—5th ed.
        p. cm.
    Includes bibliographical references and index.
    ISBN 0-8053-4547-7
    1. Cytology. 2. Molecular biology. I. Kleinsmith, Lewis J. II. Hardin, Jeff. III. Title.

QH581.2 .B43 2002
571.6—dc21   2002025613

Benjamin Cummings

Text and CD-ROM without solutions: 0-8053-4854-9

1 2 3 4 5 6 7 8 9 10—VH—05 04 03 02
www.aw.com/bc

About the Authors

WAYNE M. BECKER teaches cell biology at the University of Wisconsin, Madison. His interest in textbook writing grew out of notes, outlines, and problem sets that he assembled for his students, culminating in *Energy and the Living Cell*, a paperback text on bioenergetics published in 1977, and *The World of the Cell*, the first edition of which appeared in 1986. He earned all his degrees at the University of Wisconsin, Madison. All three degrees are in biochemistry, an orientation that is readily discernible in his textbooks. His research interests have been in plant molecular biology, focused specifically on the regulation of the expression of genes that encode enzymes of the photorespiratory pathway. His interests in teaching, learning, and research have taken him on sabbatical leaves at Harvard University, Edinburgh University, the University of Indonesia, the University of Puerto Rico, Canterbury University in Christchurch, New Zealand, and the Chinese University of Hong Kong. His honors include a Guggenheim Fellowship, a Chancellor's award For Distinguished Teaching, and a Visiting Scholar Award from the Royal Society of London.

LEWIS J. KLEINSMITH is a Professor of Molecular, Cellular, and Developmental biology at the University of Michigan, where he has served on the faculty since receiving his Ph.D. from Rockefeller University in 1968. His teaching experiences have involved courses in introductory biology, cell biology, and cancer biology, and his research interests have included studies of growth control in cancer cells, the role of pro-

tein phosphorylation in eukaryotic gene regulation, and the control of gene expression during development. Among his numerous publications, he is the author of *Principles of Cell and Molecular Biology*, first published in 1988, and several award-winning educational software programs. His honors include a Guggenheim Fellowship, the Henry Russell Award, a Michigan Distinguished Service Award, citations for outstanding teaching from the Michigan Students Association, a Thurnau Professorship, an NIH Plain Language Award, and a Best Curriculum Innovation Award from the EDUCOM Higher Education Software Awards Competition.

JEFF HARDIN received his Ph.D. in biophysics from the University of California, Berkeley, and pursued postdoctoral work at Duke University. In 1991 he joined the faculty of the Zoology Department at the University of Wisconsin, Madison, where he is currently an associate professor. His research interests center on how the cells move and change the shape of the embryo. Dr. Hardin's teaching is enhanced by his extensive use of video-microscopy and his Web-based teaching materials, which are used on many campuses in the United States and other countries. As part of his interest in teaching biology, Dr. Hardin has been involved in several teaching initiatives, including being a founding member of the University of Wisconsin Teaching Academy and a cofounder of a University of Wisconsin system-wide instructional technology initiative known as BioWeb. He is currently faculty director of the Biocore Curriculum, a four-semester honors biology sequence for undergraduates. His teaching awards include a Lily Teaching Fellowship and a National Science Foundation Young Investigator Award.

Preface

The past several decades have seen an explosive growth in our understanding of the properties and functions of living cells. As a consequence, the scientific literature is now growing so rapidly that it is almost a full-time job to keep abreast of the major developments relating to cellular organization and behavior. This enormous profusion of information presents a daunting challenge to authors as they confront the task of preparing an introductory textbook that is both modest in length and readily comprehensible to students encountering the field of cell and molecular biology for the first time. This fifth edition of *The World of the Cell* represents our most recent attempt to rise to that challenge. As with the previous edition, each of us has brought our own teaching and writing experience to the venture in ways that we have found mutually beneficial—a view that we hope our readers will share.

A major goal for this edition has been to update the artwork, including new and redesigned figures to communicate concepts more effectively, and more frequent and effective use of color throughout the text. Concomitant with the emphasis on improvement of the artwork, we have remained committed to the three central goals that have characterized every edition. As always, our primary goal is to introduce students to the fundamental principles that guide cellular organization and function. Second, we think it is important for students to understand some of the critical scientific evidence that has led to the formulation of these central concepts. And finally, we have sought to accomplish these goals in a book of manageable length that can be easily read and understood by a beginning cell biology student in the time allotted for a typical course—a quarter or semester, in most cases. To accomplish this third objective, we have necessarily been selective both in the types of examples chosen to illustrate key concepts and in the quantity of scientific evidence included. We have, in other words, attempted to remain faithful to the purpose of this text in each of its previous editions: To present the essential principles, processes, and methodology of molecular and cell biology as lucidly as possible.

Something Old and Something New

Like the proverbial bride, this edition has "something old and something new," in the sense that we have tried to retain the features of the first four editions that readers have identified as "user-friendly" while still reorganizing and updating the material. We have also added new features that we hope will make the text even more useful and accessible to introductory students.

Something Old . . . Features that we have been careful to retain from prior editions include the following:

- Organization of subject matter that is readily adaptable to a great variety of course plans.
- Subdivision of each chapter into a series of conceptual sections, each introduced by a sentence heading that summarizes the concept to be described.
- Frequent use of overview figures, which outline complicated structures or processes in broad strokes before the details are examined more closely in the text and figures that follow.
- Careful and selective use of micrographs, accompanied in most cases by size bars to indicate magnification.
- Problem sets that are intended to encourage thoughtful application of information, including in each case several very challenging problems, identified by red dots, that are meant to test the reasoning ability and problem-solving skills of especially able students.
- Boxed essays intended to explore selected topics in more depth, with each essay identified and colorcoded as focusing on historical perspectives, further insights, contemporary techniques, or clinical applications.
- Emphasis on the experimental evidence that underlies our understanding of cell structure and function, thereby reminding readers that advances in cell biology, as in all branches of science, come not from lecturers in their classrooms or textbook authors at their computers but from researchers in their laboratories.

- Careful attention to accuracy, consistency, vocabulary, and readability, hoping thereby to minimize confusion and maximize understanding for our readers.

. . . And Something New. New features that further enhance the usefulness of the text include the following:

- Substantial updating of many figures, with more color added throughout the text to facilitate an understanding of complex topics by the color coding of atoms, molecules, structures, pathways, and organelles, as appropriate.
- Chapter content significantly updated as necessary to reflect recent progress in rapidly advancing fields of molecular and cell biology.
- Updated list of Suggested Readings at the end of each chapter to reflect the most recent advances, often including references through 2000 and 2001.
- Inclusion of a Glossary containing definitions for all the bold-faced key terms in every chapter—more than 1500 terms in all, a veritable "dictionary of cell biology" in its own right.
- Discussion of principles and techniques of microscopy previously included as an Appendix now available as a free-standing *Guide to Microscopy*, which gives students more readily accessible information on microscopy without their having to locate and page through an appendix.

Techniques and Methods

Throughout the text, we have tried to explain not only *what* we know about cells but also *how* we know what we know. Toward that end, we have included descriptions of experimental techniques and findings in every chapter, almost always in the context of the questions they address and in anticipation of the answers they provide. For example, polyacrylamide gel electrophoresis is introduced not in a chapter that describes a variety of methods for studying cells but in Chapter 7, where it becomes important to our understanding of how membrane proteins can be separated from one another. Similarly, equilibrium density centrifugation is described in Chapter 12, where it is essential to our understanding of how lysosomes were originally distinguished from mitochondria and subsequently from peroxisomes as well.

To help readers locate techniques out of context, an alphabetical Guide to Techniques and Methods appears on the inside of the book's front cover, with references to chapters, pages, tables, figures, and boxed essays, as appropriate. To enhance its usefulness, the Guide to Techniques and Methods includes references not just to laboratory techniques but also to mathematical determination of values such as ΔG (free energy change) and $\Delta E_0'$ (standard reduction potential) and even to clinical procedures such as the determination of blood types and the treatment of methanol poisoning.

The only exception to the introduction of techniques in context is microscopy. The techniques of light and electron microscopy are so pervasively relevant to so much of contemporary cell biology that they warrant special consideration as a self-contained unit, now provided by the free-standing *Guide to Microscopy*, as noted above.

Pedagogical Features

To enhance the effectiveness of this text as a learning tool, each chapter includes the following basic features:

- **Boldface type** is used to highlight the most important terms in each chapter, all of which are defined in the Glossary. *Italics* are employed to identify additional technical terms that are less important than boldfaced terms but significant in their own right. Occasionally, italics are also used to highlight important phrases or sentences.
- A list of *Key Terms* at the end of each chapter includes all of the boldfaced terms in the chapter and provides the page number of the location at which each term appears in boldface and is defined or described.
- A *Suggested Reading* list is also included at the end of each chapter, with an emphasis on review articles and carefully selected research publications that motivated users are likely to find understandable. We have tried to avoid overwhelming readers with lengthy bibliographies of the original literature but have referenced articles that are especially relevant to the topics of the chapter. In most chapters, we have included a few citations of especially important historical publications, which are marked with red dots to alert the reader to the publications' historical significance.
- The inclusion of a *Problem Set* at the end of each chapter reflects our conviction that we learn science not just by reading or hearing about it, but by working with it. The problems are designed to emphasize understanding and application rather than rote recall. Many of the problems are class-tested, having been selected from problem sets and exams we have used in our own courses. To maximize the usefulness of the problem sets, detailed answers for all problems appear in the *Solutions Manual* described below. At the discretion of the instructor, this manual can be made available to students through the local bookstore or used by the instructor as a resource for homework and exam questions.
- Each chapter contains one or more *Boxed Essays* to aid students in their understanding of particularly important or intriguing aspects of cell biology. Some of the essays provide interesting historical perspectives on how science is done—the discovery of the double-helical structure of DNA as described in Box 3A, for example. Other essays are intended to help readers understand

potentially difficult principles, such as the essay that uses the analogy of monkeys shelling peanuts to explain enzyme kinetics (Box 6A). Still others provide insights into contemporary techniques used by cell biologists, as exemplified by the description of DNA fingerprinting in Box 16C. Yet another role of the boxed essays is to describe clinical applications of research findings in cell biology, as illustrated by the discussion of cystic fibrosis and the prospects for gene therapy in Box 8B.

Supplementary Learning Aids

Supplementary materials that are available with this text include the following:

- A *Solutions Manual* containing detailed answers to all problems in the text, available as ISBN 0-8053-4856-5.
- A *Guide to Microscopy* designed to move the discussion of microscopy out of the possible obscurity of an appendix and into a free-standing manual, available as ISBN 0-8053-4869-7.
- A set of 175 transparencies corresponding to selected figures from the text but with enlarged labels to enhance their usefulness in the classroom, available as ISBN 0-8053-4857-2.
- New *Instructor Art & PowerPoint CD-ROM* containing animations of key concepts and most of the line art from the text. A presentation program enables instructors to design a customized slide show of images, edit labels, import illustrations and photos from other sources, and export figures into other presentation software programs, including PowerPoint. ISBN: 0-8053-4861-1.
- *Cell Place Student CD-ROM with Free Solutions and Website*, packaged free with every new copy of the text, engages today's visually oriented students with over 30 animations and interactive activities that reinforce students' understanding of key concepts explained in the text. The CD-ROM also contains a pop-up, searchable glossary; practice quizzes; and hundreds of annotated web links. Solutions to all text problems come loaded free on the Cell Place CD-ROM, so students don't have to purchase a separate Solutions Manual. (An alternate version of the Cell Place CD-ROM without solutions is also available.)

- A *Printed Test Bank*, thoroughly revised and updated. Features multiple-choice, true/false, and short-answer questions. New inquiry-based application questions are also available. ISBN: 0-8053-4858-1.
- A *Computerized Test Bank* featuring multiple-choice, true/false, and short-answer questions. ISBN: 0-8053-4860-3.
- *Biology Labs On-Line: Cell Version* allows students to learn biological principles by designing and conducting simulated experiments on line. Explore HemoglobinLab, MitrochondriaLab, EnzymeLab, and TranslationLab at www.biologylabsonline.com.
- New booklets from the Benjamin Cummings Special Topics Series, Michael Palladino, Series Editor: *Stem Cells & Cloning, Biology of Cancer, Biological Terrorism,* and *Understanding the Human Genome Project.*

We Welcome Your Comments and Suggestions

The ultimate test of any textbook is how effectively it helps instructors teach and students learn. We welcome feedback and suggestions from readers and will try to acknowledge all correspondence. Please send your comments, criticisms, and suggestions to the appropriate authors, as follows:

Chapters 1–6, 12–15:   Wayne M. Becker
Department of Botany
University of Wisconsin
Madison, WI 53706
email: wbecker@facstaff.wisc.edu

Chapters 9–11, 22–23:   Jeff Hardin
Department of Zoology
University of Wisconsin
Madison, WI 53706
email: jdhardin@facstaff.wisc.edu

Chapters 7–8, 16–21   Lewis J. Kleinsmith
Department of Molecular, Cellular, and Developmental Biology
University of Michigan
Ann Arbor, MI 48109
email: lewisk@umich.edu

Acknowledgments

We want to acknowledge the contributions of the numerous people who have made this book possible. We are indebted especially to the many students whose words of encouragement catalyzed the writing of these chapters and whose thoughtful comments and criticisms have contributed much to whatever level of reader-friendliness the text may be judged to have. Each of us owes a special debt of gratitude to our colleagues, from whose insights and suggestions we have benefited greatly and borrowed freely. We are especially grateful to John Raasch for his very skillful and thorough revision of Chapters 12 and 15. We also acknowledge those who have contributed to previous editions of our textbooks, including David Deamer, Martin Poenie, Jane Reece, and Valerie Kish, as well as Peter Armstrong, John Carson, Ed Clark, Joel Goodman, David Gunn, Jeanette Natzle, Mary Jane Niles, Timothy Ryan, Beth Schaefer, Lisa Smit, David Spiegel, Akif Uzman, and Karen Valentine. In addition, we want to express our appreciation to the many colleagues who graciously consented to contribute micrographs to this endeavor, as well as to the authors and publishers who have kindly granted permission to reproduce copyrighted material.

The many reviewers listed below provided helpful criticisms and suggestions at various stages of manuscript development and revision. Their words of appraisal and counsel were gratefully received and greatly appreciated. Indeed, the extensive review process to which this and the prior editions of the book have been exposed should be considered a significant feature of the book. Nonetheless, the final responsibility for what you read here remains ours, and you may confidently attribute to us any errors of omission or commission encountered in these pages.

We are also deeply indebted to the many publishing professionals whose consistent encouragement, hard work, and careful attention to detail contributed much to the clarity of both the text and the art. Special recognition and sincere appreciation go to Laura Kenney; Jennifer Dunn; Michele Sordi, Kelly Murphy, and Michael McArdle at Benjamin Cummings; and Patty O'Connell at Electronic Publishing Services.

Finally, we are grateful beyond measure to our wives, families, graduate students, and postdoctoral associates, without whose patience, understanding, and forbearance this book would never have been written.

Reviewers of the Fifth Edition

James T. Bradley, *Auburn University*
Andrew Brittain, *Hawaii Pacific University*
Patrick J. Bryan, *Central Washington University*
Mitchell Chernin, *Bucknell University*
Guy E. Farish, *Adams State College*
Swapan K. Ghosh, *Indiana State University*
Dale W. Laird, *University of Western Ontario (London, Ontario, Canada)*
Esther M. Leise, *University of North Carolina, Greensboro*
Joel B. Piperberg, *Millersville University*
Sheldon S. Shen, *Iowa State University*
Bradley J. Stith, *University of Colorado, Denver*
Quinn Vega, *Montclair State University*
Linda Yasui, *Northern Illinois University*

Reviewers of Previous Editions

L. Rao Ayyagari, *Lindenwood College*
Margaret Beard, *Columbia University*
William Bement, *University of Wisconsin*
Paul Benko, *Sonoma State University*
Steve Benson, *California State University, Hayward*
Joseph J. Berger, *Springfield College*
Gerald Bergtrom, *University of Wisconsin, Milwaukee*
Frank L. Binder, *Marshall University*
Robert Blystone, *Trinity University*
R. B. Boley, *University of Texas at Arlington*

Edward M. Bonder, *Rutgers, the State University of New Jersey*

James T. Bradley, *Auburn University*

Suzanne Bradshaw, *University of Cincinnati*

J. D. Brammer, *North Dakota State University*

Chris Brinegar, *San Jose State University*

Alan H. Brush, *University of Connecticut, Storrs*

Brower R. Burchill, *University of Kansas*

Ann B. Burgess, *University of Wisconsin, Madison*

Thomas J. Byers, *Ohio State University*

P. Samuel Campbell, *University of Alabama, Huntsville*

George L. Card, *The University of Montana*

C. H. Chen, *South Dakota State University*

Edward A. Clark, *University of Washington*

Philippa Claude, *University of Wisconsin, Madison*

John M. Coffin, *Tufts University School of Medicine*

J. John Cohen, *University of Colorado Medical School*

Larry Cohen, *Pomona College*

Reid S. Compton, *University of Maryland*

Mark Condon, *Dutchess Community College*

Jonathan Copeland, *Georgia Southern University*

Bracey Dangerfield, *Salt Lake Community College*

David DeGroote, *St. Cloud State*

Douglas Dennis, *James Madison University*

Elizabeth D. Dolci, *Johnson State College*

Aris J. Domnas, *University of North Carolina, Chapel Hill*

Michael P. Donovan, *Southern Utah University*

Robert M. Dores, *University of Denver*

Diane D. Eardley, *University of California, Santa Barbara*

James E. Forbes, *Hampton University*

Carl S. Frankel, *Pennsylvania State University, Hazleton Campus*

David R. Fromson, *California State University, Fullerton*

David M. Gardner, *Roanoke College*

Carol V. Gay, *Pennsylvania State University*

Stephen A. George, *Amherst College*

Joseph Gindhart, *University of Massachusetts, Boston*

Michael L. Gleason, *Central Washington University*

T. T. Gleeson, *University of Colorado*

Ursula W. Goodenough, *Washington University*

Thomas A. Gorell, *Colorado State University*

Marion Greaser, *University of Wisconsin, Madison*

Karen F. Greif, *Bryn Mawr College*

Mark T. Groudine, *Fred Hutchinson Cancer Research Center, University of Washington School of Medicine*

Gary Gussin, *University of Iowa*

Leah T. Haimo, *University of California, Riverside*

Arnold Hampel, *Northern Illinois University*

Laszlo Hanzely, *Northern Illinois University*

Bettina Harrison, *University of Massachusetts, Boston*

Lawrence Hightower, *University of Connecticut, Storrs*

James P. Holland, *Indiana University, Bloomington*

Johns Hopkins III, *Washington University*

Betty A. Houck, *University of Portland*

William R. Jeffery, *University of Texas, Austin*

Kwang W. Jeon, *University of Tennessee*

Kenneth C. Jones, *California State University, Northridge*

Patricia P. Jones, *Stanford University*

Martin A. Kapper, *Central Connecticut State University*

Lon S. Kaufman, *University of Illinois, Chicago*

Steven J. Keller, *University of Cincinnati Main Campus*

Robert Koch, *California State University, Fullerton*

Joseph R. Koke, *Southwest Texas State University*

Hal Krider, *University of Connecticut, Storrs*

William B. Kristan, Jr., *University of California, San Diego*

David N. Kristie, *Acadia University (Nova Scotia)*

Frederic Kundig, *Towson State University*

Elias Lazarides, *California Institute of Technology*

John T. Lis, *Cornell University*

Robert Macey, *University of California, Berkeley*

Roderick MacLeod, *University of Illinois, Urbana-Champaign*

Shyamal K. Majumdar, *Lafayette College*

Gary G. Matthews, *State University of New York, Stony Brook*

Douglas McAbee, *California State University, Long Beach*

Thomas D. McKnight, *Texas A & M University*

Robert L. Metzenberg, *University of Wisconsin, Madison*

Hugh A. Miller III, *East Tennessee State University*

Tony K. Morris, *Fairmont State College*

Deborah B. Mowshowitz, *Columbia University*

Carl E. Nordahl, *University of Nebraska, Omaha*

Richard Nuccitelli, *University of California, Davis*

Donata Oertel, *University of Wisconsin, Madison*

Joanna Olmsted, *University of Rochester*

Alan Orr, *University of Northern Iowa*

Curtis L. Parker, *Morehouse School of Medicine*

Lee D. Peachey, *University of Colorado*

Debra Pearce, *Northern Kentucky University*

Howard Petty, *Wayne State University*

Susan Pierce, *Northwestern University*

Gilbert C. Pogany, *Northern Arizona University*

Ralph Quatrano, *Oregon State University*

Ralph E. Reiner, *College of the Redwoods*

Gary Reiness, *Lewis and Clark College*

Douglas Rhoads, *University of Arkansas*

Michael Robinson, *North Dakota State University*

Adrian Rodriguez, *San Jose State University*

Donald J. Roufa, *Kansas State University*

Donald H. Roush, *University of North Alabama*

Edmund Samuel, *Southern Connecticut State University*

Mary Jane Saunders, *University of South Florida*

John I. Scheide, *Central Michigan University*

Mary Schwanke, *University of Maine, Farmington*

David W. Scupham, *Valparaiso University*

Edna Seaman, *University of Massachusetts, Boston*

Diane C. Shakes, *University of Houston*

Sheldon S. Shen, *Iowa State University*

Randall D. Shortridge, *State University of New York, Buffalo*

Dwayne D. Simmons, *University of California, Los Angeles*

Robert D. Simoni, *Stanford University*

William R. Sistrom, *University of Oregon*

Donald Slish, *Plattsburgh State College*

Robert H. Smith, *Skyline College*

Mark Staves, *Grand Valley State University*

Barbara Y. Stewart, *Swarthmore College*
Richard D. Storey, *Colorado College*
Antony O. Stretton, *University of Wisconsin, Madison*
Philip Stukus, *Denison University*
Stephen Subtelny, *Rice University*
Millard Susman, *University of Wisconsin, Madison*
Elizabeth J. Taparowsky, *Purdue University*
Barbara J. Taylor, *Oregon State University*

Bruce R. Telzer, *Pomona College*
John J. Tyson, *Virginia Polytechnic University*
Akif Uzman, *University of Texas, Austin*
Fred D. Warner, *Syracuse University*
James Watrous, *St. Joseph's University*
Fred H. Wilt, *University Of California, Berkeley*
James Wise, *Hampton University*
David Worcester, *University of Missouri, Columbia*

Brief Contents

About the Authors   v

Preface   vii

Acknowledgments   x

Detailed Contents   xv

1   A Preview of the Cell   1

2   The Chemistry of the Cell   17

3   The Macromolecules of the Cell   41

4   Cells and Organelles   76

5   Bioenergetics: The Flow of Energy
     in the Cell   106

6   Enzymes: The Catalysts of Life   130

7   Membranes: Their Structure, Function,
     and Chemistry   158

8   Transport Across Membranes: Overcoming
     the Permeability Barrier   195

9   Signal Transduction Mechanisms: I. Electrical
     Signals in Nerve Cells   225

10   Signal Transduction Mechanisms: II. Messengers
     and Receptors   256

11   Beyond the Cell: Extracellular Structures,
     Cell Adhesion, and Cell Junctions   290

12   Intracellular Compartments: The Endoplasmic
     Reticulum, Golgi Complex, Endosomes,
     Lysosomes, and Peroxisomes   323

13   Chemotrophic Energy Metabolism: Glycolysis
     and Fermentation   368

14   Chemotrophic Energy Metabolism: Aerobic
     Respiration   398

15   Phototropic Energy Metabolism:
     Photosynthesis   445

16   The Structural Basis of Cellular Information:
     DNA, Chromosomes, and the Nucleus   479

17   The Cell Cycle: DNA Replication, Mitosis,
     and Cancer   523

18   Sexual Reproduction, Meiosis,
     and Genetic Recombination   577

19   Gene Expression: I. The Genetic Code
     and Transcription   623

20   Gene Expression: II. Protein Synthesis
     and Sorting   660

21   The Regulation of Gene Expression   692

22   Cytoskeletal Systems   742

23   Cellular Movement: Motility
     and Contractility   769

     Glossary   G-1

     Photo, Illustration, and Text Credits   C-1

     Index   I-1

Detailed Contents

About the Authors v

Preface vii

Acknowledgments x

Chapter 1
A Preview of the Cell 1

The Cell Theory: A Brief History 1

The Emergence of Modern Cell Biology 3
> The Cytological Strand Deals with Cellular Structure 5
> The Biochemical Strand Covers the Chemistry of Biological
>   Structure and Function 8
> The Genetic Strand Focuses on Information Flow 10
> "Facts" and the Scientific Method 11

Perspective 13

Key Terms for Self-Testing 13

Problem Set 13

Suggested Reading 15

Box 1A: *Units of Measurement in Cell Biology* 2

Box 1B: *Further Insights: Biology, "Facts," and the
Scientific Method* 11

Chapter 2
The Chemistry of the Cell 17

The Importance of Carbon 18
> Carbon-Containing Molecules Are Stable 18
> Carbon-Containing Molecules Are Diverse 19
> Carbon-Containing Molecules Can Form Stereoisomers 20

The Importance of Water 21
> Water Molecules Are Polar 22
> Water Molecules Are Cohesive 22

Water Has a High Temperature-Stabilizing Capacity 22
Water Is an Excellent Solvent 23

The Importance of Selectively Permeable Membranes 24
> A Membrane Is a Lipid Bilayer with Proteins
>   Embedded in It 24
> Membranes Are Selectively Permeable 25

The Importance of Synthesis by Polymerization 26
> Macromolecules Are Responsible for Most of the Form
>   and Function in Living Systems 26
> Cells Contain Three Different Kinds of Macromolecules 28
> Macromolecules Are Synthesized by Stepwise Polymerization
>   of Monomers 29

The Importance of Self-Assembly 31
> Many Proteins Self-Assemble 31
> Molecular Chaperones Assist the Assembly of Some Proteins 33
> Noncovalent Interactions Are Important in the Folding
>   of Macromolecules 34
> Self-Assembly Also Occurs in Other Cellular Structures 34
> The Tobacco Mosaic Virus Is a Case Study in Self-Assembly 35
> Self-Assembly Has Limits 35
> Hierarchical Assembly Provides Advantages for the Cell 36

Perspective 38

Key Terms for Self-Testing 38

Problem Set 38

Suggested Reading 40

Box 2A: *Further Insights: Tempus Fugit and the Fine Art
of Watchmaking* 37

Chapter 3
The Macromolecules of the Cell 41

Proteins 41
> The Monomers Are Amino Acids 41
> The Polymers Are Polypeptides and Proteins 44

Several Kinds of Bonds and Interactions Are Important in Protein Folding and Stability   45

Protein Structure Depends on Amino Acid Sequence and Interactions   47

Nucleic Acids   54

The Monomers Are Nucleotides   55

The Polymers Are DNA and RNA   56

A DNA Molecule Is a Double-Stranded Helix   58

Polysaccharides   60

The Monomers Are Monosaccharides   61

The Polymers Are Storage and Structural Polysaccharides   63

Polysaccharide Structure Depends on the Kinds of Glycosidic Bonds Involved   65

Lipids   66

Fatty Acids Are the Building Blocks of Several Classes of Lipids   68

Triacylglycerols Are Storage Lipids   68

Phospholipids Are Important in Membrane Structure   69

Glycolipids Are Specialized Membrane Components   70

Steroids Are Lipids with a Variety of Functions   70

Terpenes Are Formed from Isoprene   71

Perspective   71

Key Terms for Self-Testing   71

Problem Set   72

Suggested Reading   75

Box 3A: *Historical Perspectives: On the Trail of the Double Helix*   60

Chapter 4
Cells and Organelles   76

Properties and Strategies of Cells   76

All Cells Are Either Prokaryotic or Eukaryotic   76

Cells Come in Many Sizes and Shapes   76

Eukaryotic Cells Use Organelles to Compartmentalize Cellular Function   78

Prokaryotes and Eukaryotes Differ from Each Other in Many Ways   78

Cell Specialization Demonstrates the Unity and Diversity of Biology   82

The Eukaryotic Cell in Overview: Pictures at an Exhibition   83

The Plasma Membrane Defines Cell Boundaries and Retains Contents   83

The Nucleus Is the Cell's Information Center   84

Intracellular Membranes and Organelles Define Compartments   85

The Cytoplasm of Eukaryotic Cells Contains the Cytosol and Cytoskeleton   95

The Extracellular Matrix and the Cell Wall Are the "Outside" of the Cell   97

Viruses, Viroids, and Prions: Agents That Invade Cells   98

A Virus Consists of a DNA or RNA Core Surrounded by a Protein Coat   98

Viroids Are Small, Circular RNA Molecules   99

Prions Are "Proteinaceous Infective Particles"   99

Perspective   101

Key Terms for Self-Testing   102

Problem Set   102

Suggested Reading   104

Box 4A: *Historical Perspectives: Discovering Organelles: The Importance of Centrifuges and Chance Observations*   92

Chapter 5
Bioenergetics: The Flow of Energy in the Cell   106

The Importance of Energy   107

Cells Need Energy to Cause Six Different Kinds of Changes   107

Most Organisms Obtain Energy Either from Sunlight or from Organic Food Molecules   109

Energy Flows Through the Biosphere Continuously   109

The Flow of Energy Through the Biosphere Is Accompanied by Flow of Matter   111

Bioenergetics   112

To Understand Energy Flow, We Need to Understand Systems, Heat, and Work   112

The First Law of Thermodynamics Tells Us That Energy Is Conserved   113

The Second Law of Thermodynamics Tells Us That Reactions Have Directionality   114

Entropy and Free Energy Are Two Alternative Means of Assessing Thermodynamic Spontaneity   115

Understanding ΔG 120

The Equilibrium Constant Is a Measure of Directionality   120

ΔG Can Be Calculated Readily   121

The Standard Free Energy Change Is ΔG Measured Under Standard Conditions   122

Summing Up: The Meaning of $\Delta G'$ and $\Delta G^{\circ\prime}$ 123

Free Energy Change: Sample Calculations   124

Life and the Steady State: Reactions That Move Toward Equilibrium Without Ever Getting There   125

Perspective   125

Key Terms for Self-Testing   126

Problem Set   126

Suggested Reading   129

Box 5A: *Further Insights: Jumping Beans and Free Energy*   116

Chapter 6
Enzymes: The Catalysts of Life   130

Activation Energy and the Metastable State   130
Before a Chemical Reaction Can Occur, the Activation Energy Barrier Must Be Overcome   131
The Metastable State Is a Result of the Activation Barrier   131
Catalysts Overcome the Activation Energy Barrier   131

Enzymes as Biological Catalysts   132
Most Enzymes Are Proteins   132
Substrate Binding, Activation, and Reaction Occur at the Active Site   137

Enzyme Kinetics   140
Most Enzymes Display Michaelis-Menten Kinetics   140
What Is the Meaning of V_{max} and K_m?   142
Why are V_{max} and K_m Important to Cell Biologists?   143
The Double-Reciprocal Plot Is a Useful Means of Linearizing Kinetic Data   143
Determining K_m and V_{max}: An Example   144
Enzyme Inhibitors Act Irreversibly or Reversibly   145

Enzyme Regulation   147
Allosteric Enzymes Are Regulated by Molecules Other than Reactants and Products   147
Allosteric Enzymes Exhibit Cooperative Interactions Between Subunits   149
Enzymes Can Also Be Regulated by the Addition or Removal of Chemical Groups   149

RNA Molecules as Enzymes: Ribozymes   151

Perspective   153

Key Terms for Self-Testing   154

Problem Set   154

Suggested Reading   157

Box 6A: *Further Insights: Monkeys and Peanuts*   141

Chapter 7
Membranes: Their Structure, Function, and Chemistry   158

The Functions of Membranes   159
Membranes Define Boundaries and Serve as Permeability Barriers   159
Membranes Are Sites of Specific Functions   159
Membranes Regulate the Transport of Solutes   160
Membranes Detect and Transmit Electrical and Chemical Signals   160
Membranes Mediate Cell-to-Cell Communication   160

Models of Membrane Structure: An Experimental Perspective   160
Overton and Langmuir: Lipids Are Important Components of Membranes   161

Gorter and Grendel: The Basis of Membrane Structure Is a Lipid Bilayer   161
Davson and Danielli: Membranes Also Contain Proteins   162
Robertson: All Membranes Share a Common Underlying Structure   162
Further Research Revealed Major Shortcomings of the Davson-Danielli Model   163
Singer and Nicolson: A Membrane Consists of a Mosaic of Proteins in a Fluid Lipid Bilayer   164
Unwin and Henderson: Some Membrane Proteins Contain Transmembrane Segments   164
The Fluid Mosaic Model Is Now the Accepted View of Membrane Structure   166

Membrane Lipids: The "Fluid" Part of the Model   166
Membranes Contain Several Major Classes of Lipids   166
Thin-Layer Chromatography Is an Important Technique for Lipid Analysis   168
Fatty Acids Are Essential to Membrane Structure and Function   169
Membrane Asymmetry: Most Lipids Are Distributed Unequally Between the Two Monolayers   169
The Lipid Bilayer Is Fluid   171
Membranes Function Properly Only in the Fluid State   172
Most Organisms Can Regulate Membrane Fluidity   174

Membrane Proteins: The "Mosaic" Part of the Model   175
The Membrane Consists of a Mosaic of Proteins: Evidence from Freeze-Fracture Microscopy   175
Membranes Contain Integral, Peripheral, and Lipid-Anchored Proteins   177
Proteins Can Be Separated by SDS–Polyacrylamide Gel Electrophoresis   180
Molecular Biology Has Contributed Greatly to Our Understanding of Membrane Proteins   181
Membrane Proteins Have a Variety of Functions   184
Membrane Proteins Are Oriented Asymmetrically Across the Lipid Bilayer   184
Many Membrane Proteins Are Glycosylated   185
Membrane Proteins Vary in Their Mobility   186

Perspective   189

Key Terms for Self-Testing   190

Problem Set   190

Suggested Reading   193

Box 7A: *Contemporary Techniques: Revolutionizing the Study of Membrane Proteins: The Impact of Molecular Biology*   182

Chapter 8
Transport Across Membranes: Overcoming the Permeability Barrier   195

Cells and Transport Processes   195
Solutes Cross Membranes by Simple Diffusion, Facilitated Diffusion, and Active Transport   196

The Erythrocyte Plasma Membrane Provides Examples of
Transport Mechanisms 197

Simple Diffusion: Unassisted Movement Down the Gradient 197

Diffusion Always Moves Solutes Toward Equilibrium 198

Osmosis Is the Diffusion of Water Across a Differentially
Permeable Membrane 199

Simple Diffusion Is Limited to Small, Nonpolar
Molecules 199

The Rate of Simple Diffusion Is Directly Proportional
to the Concentration Gradient 201

Facilitated Diffusion: Protein-Mediated Movement Down the Gradient 203

Carrier Proteins and Channel Proteins Facilitate Diffusion
by Different Mechanisms 203

Carrier Proteins Alternate Between Two Conformational
States 203

Carrier Proteins Are Analogous to Enzymes in Their
Specificity and Kinetics 204

Carrier Proteins Transport Either One or Two Solutes 204

The Erythrocyte Glucose Transporter and Anion Exchange
Protein Are Examples of Carrier Proteins 204

Channel Proteins Facilitate Diffusion by Forming Hydrophilic
Transmembrane Channels 206

Active Transport: Protein-Mediated Movement Up the Gradient 207

The Coupling of Active Transport to an Energy Source May Be
Direct or Indirect 207

Direct Active Transport Depends on Four Types of Transport
ATPases 208

Indirect Active Transport Is Driven by Ion Gradients 210

Examples of Active Transport 211

Direct Active Transport: The Na^+/K^+ Pump Maintains
Electrochemical Ion Gradients 211

Indirect Active Transport: Sodium Symport Drives
the Uptake of Glucose 214

The Bacteriorhodopsin Proton Pump Uses Light Energy to
Transport Protons 215

The Energetics of Transport 218

For Uncharged Solutes, the ΔG of Transport Depends Only
on the Concentration Gradient 218

For Charged Solutes, the ΔG of Transport Depends on the
Electrochemical Gradient 218

On to Nerve Cells 220

Perspective 220

Key Terms for Self-Testing 221

Problem Set 221

Suggested Reading 224

Box 8A: *Further Insights: Osmosis: The Special Case of
Water Diffusion 200*

Box 8B: *Clinical Applications: Membrane Transport, Cystic
Fibrosis, and the Prospects For Gene Therapy 212*

Chapter 9
Signal Transduction Mechanisms: I. Electrical Signals in Nerve Cells 225

The Nervous System 225

Neurons Are Specially Adapted for the Transmission
of Electrical Signals 227

Understanding Membrane Potential 228

The Resting Membrane Potential Depends on
Differing Concentrations of Ions Inside and Outside
the Neuron 228

The Nernst Equation Describes the Relationship Between
Membrane Potential and Ion Concentration 230

Ions Trapped Inside the Cell Have Important Effects
on Resting Membrane Potential 230

Steady-State Concentrations of Common Ions Affect Resting
Membrane Potential 230

The Goldman Equation Describes the Combined Effects of
Ions on Membrane Potential 231

Electrical Excitability 233

Ion Channels Act Like Gates for the Movement of Ions
Through the Membrane 233

Patch Clamping and Molecular Biological Techniques Allow
the Activity of Single Ion Channels to Be Monitored 233

Specific Domains of Voltage-Gated Channels Act as Sensors
and Inactivators 234

The Action Potential 236

Action Potentials Propagate Electrical Signals Along
an Axon 236

Action Potentials Involve Rapid Changes in the Membrane
Potential of the Axon 236

Action Potentials Result from the Rapid Movement
of Ions Through Axonal Membrane Channels 237

Action Potentials Are Propagated Along the Axon Without
Losing Strength 240

The Myelin Sheath Acts Like an Electrical Insulator
Surrounding the Axon 241

Synaptic Transmission 243

Neurotransmitters Relay Signals Across Nerve
Synapses 243

Elevated Calcium Levels Stimulate Secretion
of Neurotransmitters from Presynaptic Neurons 245

Secretion of Neurotransmitters Requires the Docking
and Fusion of Vesicles with the Plasma Membrane 246

Neurotransmitters Are Detected by Specific Receptors
on Postsynaptic Neurons 247

Neurotransmitters Must Be Inactivated Shortly After Their
Release 249

Integration and Processing of Nerve Signals 250

Neurons Can Integrate Signals from Other Neurons Through
Both Temporal and Spatial Summation 251

Neurons Can Integrate Both Excitatory and Inhibitory Signals
from Other Neurons 251

Perspective 252

Key Terms for Self-Testing   253

Problem Set   254

Suggested Reading   255

Box 9A: *Clinical Applications: Poisoned Arrows, Snake Bites, and Nerve Gases*   250

Chapter 10
Signal Transduction Mechanisms: II. Messengers and Receptors   256

Chemical Signals and Cellular Receptors   256

Different Types of Chemical Signals Can Be Received by Cells   257

Receptor Binding Involves Specific Interactions Between Ligands and Their Receptors   257

Receptor Binding Activates a Sequence of Signal Transduction Events Within the Cell   258

G Protein–Linked Receptors   259

Seven-Membrane Spanning Receptors Act via G Proteins   259

Cyclic AMP Is a Second Messenger Used by One Class of G Proteins   261

Disruption of G Protein Signaling Causes Several Human Diseases   262

Many G Proteins Use Inositol Trisphosphate and Diacylglycerol as Second Messengers   263

The Release of Calcium Ions Is a Key Event in Many Signaling Processes   264

Nitric Oxide Couples G Protein-Linked Receptor Stimulation in Endothelial Cells to Relaxation of Smooth Muscle Cells in Blood Vessels   269

Protein Kinase–Associated Receptors   270

Receptor Tyrosine Kinases Aggregate and Undergo Autophosphorylation   270

Receptor Tyrosine Kinases Initiate a Signal Transduction Cascade Involving Ras and MAP Kinase   271

Receptor Tyrosine Kinases Activate a Variety of Other Signaling Pathways   272

Growth Factors as Messengers   272

Disruption of Growth Factor Signaling Through Receptor Tyrosine Kinases Can Have Dramatic Effects on Embryonic Development   273

Other Growth Factors Transduce Their Signals via Receptor Serine/Threonine Kinases   274

Growth Factor Receptor Pathways Share Common Themes   275

Disruption of Growth Factor Signaling Can Lead to Cancer   275

The Endocrine and Paracrine Hormone Systems   276

Hormonal Signals Can Be Classified by the Distance They Travel to Their Target Cells   276

Hormones Control Many Physiological Functions   276

Animal Hormones Can Be Classified by Their Chemical Properties   276

Adrenergic Hormones and Receptors Are a Good Example of Endocrine Regulation   278

Histamine and Prostaglandins Are Good Examples of Paracrine Regulation   280

Cell Signals and Apoptosis   282

Apoptosis Is Triggered by Death Signals or Withdrawal of Survival Factors   284

Perspective   286

Key Terms for Self-Testing   286

Problem Set   287

Suggested Reading   288

Chapter 11
Beyond the Cell: Extracellular Structures, Cell Adhesion, and Cell Junctions   290

The Extracellular Matrix of Animal Cells   290

Collagens Are Responsible for the Strength of the Extracellular Matrix   291

A Precursor Called Procollagen Forms Many Types of Tissue-Specific Collagens   292

Elastins Impart Elasticity and Flexibility to the Extracellular Matrix   293

Collagen and Elastin Fibers Are Embedded in a Matrix of Proteoglycans   294

Free Hyaluronate Lubricates Joints and Facilitates Cell Migration   294

Proteoglycans and Adhesive Glycoproteins Anchor Cells to the Extracellular Matrix   295

Fibronectins Bind Cells to the Matrix and Guide Cellular Movement   296

Laminins Bind Cells to the Basal Lamina   297

Integrins Are Cell Surface Receptors that Bind ECM Constituents   298

The Glycocalyx Is a Carbohydrate-Rich Zone at the Periphery of Animal Cells   302

Cell-Cell Recognition and Adhesion   302

Transmembrane Proteins Mediate Cell-Cell Adhesion   302

Carbohydrate Groups Are Important in Cell-Cell Recognition and Adhesion   304

Cell Junctions   306

Adhesive Junctions Link Adjoining Cells to Each Other   306

Tight Junctions Prevent the Movement of Molecules Across Cell Layers   310

Gap Junctions Allow Direct Electrical and Chemical Communication Between Cells   313

The Plant Cell Surface   314

Cell Walls Provide a Structural Framework and Serve as a Permeability Barrier   314

The Plant Cell Wall Is a Network of Cellulose Microfibrils,
    Polysaccharides, and Glycoproteins 315
Cell Walls Are Synthesized in Several Discrete Stages 316
Plasmodesmata Permit Direct Cell-Cell Communication
    Through the Cell Wall 318

Perspective 319

Key Terms for Self-Testing 320

Problem Set 320

Suggested Reading 322

Box 11A: *Clinical Applications: Anthrax, Food
Poisoning, and Other "Bad Bugs": The Cell
Surface Connection 308*

Chapter 12
Intracellular Compartments: The Endoplasmic Reticulum, Golgi Complex, Endosomes, Lysosomes, and Peroxisomes 323

The Endoplasmic Reticulum 323
The Two Basic Kinds of Endoplasmic Reticulum Differ in
    Structure and Function 324
Rough ER Is Involved in the Biosynthesis and Processing
    of Proteins 325
Smooth ER Is Involved in Drug Detoxification, Carbohydrate
    Metabolism, and Other Cellular Processes 330
The ER Plays a Central Role in the Biosynthesis
    of Membranes 332

The Golgi Complex 333
The Golgi Complex Consists of a Series of Membrane-
    Bounded Cisternae 333
Two Models Depict the Flow of Lipids and Proteins Through
    the Golgi Complex 335

**Roles of the ER and Golgi Complex in Protein
Glycosylation 336**

Roles of the ER and Golgi Complex in Protein Sorting 338
ER-Specific Proteins Contain Retrieval Tags 338
Golgi Complex Proteins May Be Sorted According to the
    Lengths of Their Membrane-Spanning Domains 339
Targeting of Soluble Lysosomal Proteins to Endosomes
    and Lysosomes Is a Model for Protein Sorting
    in the TGN 339
Secretory Pathways Transport Molecules to the Exterior of the
    Cell 341

**Exocytosis and Endocytosis: Transporting Material Across
the Plasma Membrane 342**
Exocytosis Releases Intracellular Molecules
    to the Extracellular Medium 342
Endocytosis Imports Extracellular Molecules by Forming
    Vesicles from the Plasma Membrane 343

Coated Vesicles in Cellular Transport Processes 349
Clathrin-Coated Vesicles Are Surrounded by Lattices
    Composed of Clathrin and Adaptor Protein 349
The Assembly of Clathrin Coats Drives the Formation
    of Vesicles from the Plasma Membrane and TGN 350
COPI- and COPII-Coated Vesicles Connect the ER
    and Golgi Complex Cisternae 351
The SNARE Hypothesis Connects Coated Vesicles
    and Target Membranes 352

Lysosomes and Cellular Digestion 353
Lysosomes Isolate Digestive Enzymes from the Rest
    of the Cell 353
Lysosomes Develop from Endosomes 354
Lysosomal Enzymes Are Important for Several Different
    Digestive Processes 355
Lysosomal Storage Diseases Are Usually Characterized
    by the Accumulation of Indigestible Material 357

The Plant Vacuole: A Multifunctional Organelle 357

Peroxisomes 358
The Discovery of Peroxisomes Depended on Innovations
    in Equilibrium Density Centrifugation 358
Most Peroxisomal Functions Are Linked to Hydrogen Peroxide
    Metabolism 359
Plant Cells Contain Types of Peroxisomes Not Found
    in Animal Cells 361
Peroxisome Biogenesis Occurs by Division of Preexisting
    Peroxisomes 362

Perspective 363

Key Terms for Self-Testing 364

Problem Set 365

Suggested Reading 367

Box 12A: *Contemporary Techniques: Centrifugation:
An Indispensable Technique of Cell Biology 326*

Box 12B: *Clinical Applications: Cholesterol, the LDL
Receptor, and Receptor-Mediated Endocytosis 346*

Chapter 13
Chemotrophic Energy Metabolism: Glycolysis and Fermentation 368

Metabolic Pathways 368

ATP: The Universal Energy Coupler 369
ATP Contains Two High-Energy Phosphoanhydride
    Bonds 369
ATP Hydrolysis Is Highly Exergonic Because of Charge
    Repulsion and Resonance Stabilization 370
ATP Is an Important Intermediate in Cellular Energy
    Transactions 371

Chemotrophic Energy Metabolism 373
Biological Oxidations Usually Involve the Removal of Both
    Electrons and Protons and Are Highly Exergonic 373
Coenzymes Such as NAD^+ Serve as Electron Acceptors
    in Biological Oxidations 374
Most Chemotrophs Meet Their Energy Needs
    by Oxidizing Organic Food Molecules 375

Glucose Is One of the Most Important Oxidizable Substrates
    in Energy Metabolism 375
The Oxidation of Glucose Is Highly Exergonic 375
Glucose Catabolism Yields Much More Energy
    in the Presence of Oxygen than in Its Absence 375
Based on Their Need for Oxygen, Organisms Are Aerobic,
    Anaerobic, or Facultative 378

**Glycolysis and Fermentation: ATP Generation Without
the Involvement of Oxygen 378**

Glycolysis Generates ATP by Catabolizing Glucose
    to Pyruvate 378
The Fate of Pyruvate Depends on Whether or Not Oxygen
    Is Available 383
In the Absence of Oxygen, Pyruvate Undergoes Fermentation
    to Regenerate NAD^+ 384
Fermentation Taps Only a Small Fraction of the Free Energy
    of the Substrate but Conserves That Energy Efficiently
    as ATP 385

Alternative Substrates for Glycolysis 386

Other Sugars Are Also Catabolized by the Glycolytic
    Pathway 386
Polysaccharides Are Cleaved to Form Sugar Phosphates That
    Also Enter the Glycolytic Pathway 386

Gluconeogenesis 386

The Regulation of Glycolysis and Gluconeogenesis 389

Key Enzymes in the Glycolytic and Gluconeogenic Pathways
    Are Subject to Allosteric Regulation 390
Fructose-2,6-Bisphosphate Is an Important Regulator
    of Glycolysis and Gluconeogenesis 391

Perspective 393

Key Terms for Self-Testing 393

Problem Set 394

Suggested Reading 397

Box 13A: *Further Insights: "What Happens to
the Sugar?"* 376

Chapter 14
Chemotrophic Energy Metabolism: Aerobic Respiration 398

Cellular Respiration: Maximizing ATP Yields 398

Aerobic Respiration Yields Much More Energy
    than Fermentation 400
Respiration Includes Glycolysis, the TCA Cycle, Electron
    Transport, and ATP Synthesis 400

The Mitochondrion: Where the Action Takes Place 400

Mitochondria Are Often Present Where the ATP Needs
    Are Greatest 401
Are Mitochondria Interconnected Networks Rather
    than Discrete Organelles? 401
The Outer and Inner Membranes Define Two Separate
    Compartments 402

Mitochondrial Functions Occur in or on Specific Membranes
    and Compartments 404
In Prokaryotes, Respiratory Functions Are Localized
    to the Plasma Membrane and the Cytoplasm 404

The Tricarboxylic Acid Cycle: Oxidation in the Round 405

Pyruvate Is Converted to Acetyl Coenzyme A
    by Oxidative Decarboxylation 406
The TCA Cycle Begins with the Entry of Acetate as
    Acetyl CoA 406
NADH Is Formed and CO_2 Is Released in Two Reactions of
    the TCA Cycle 406
Direct Generation of GTP (or ATP) Occurs at One Step in the
    TCA Cycle 409
The Final Oxidative Reactions of the TCA Cycle Generate
    $FADH_2$ and NADH 409
Summing Up: The Products of the TCA Cycle are CO_2, ATP,
    NADH, and $FADH_2$ 410
Several TCA Cycle Enzymes Are Subject to Allosteric
    Regulation 410
The TCA Cycle Also Plays a Central Role in the Catabolism of
    Fats and Proteins 412
The TCA Cycle Serves as a Source of Precursors for Anabolic
    Pathways 414
The Glyoxylate Cycle Converts Acetyl CoA to
    Carbohydrates 415

**Electron Transport: Electron Flow from Coenzymes
to Oxygen 415**

The Electron Transport System Conveys Electrons Stepwise
    from Reduced Coenzymes to Oxygen 418
The Electron Transport System Consists of Five Different
    Kinds of Carriers 418
The Electron Carriers Function in a Sequence Determined by
    Their Reduction Potentials 420
Most of the Carriers Are Organized into Four Large
    Respiratory Complexes 422
The Respiratory Complexes Move Freely Within
    the Inner Membrane 424

**The Electrochemical Proton Gradient: Key to
Energy Coupling 425**

Electron Transport and ATP Synthesis Are Coupled
    Events 426
The Chemiosmotic Model: The "Missing Link" Is
    a Proton Gradient 426
Coenzyme Oxidation Pumps Enough Electrons to Form
    3 ATP per NADH and 2 ATP per $FADH_2$ 427
The Chemiosmotic Model Is Affirmed by an Impressive Array
    of Evidence 427
1. Electron Transport Causes Protons to Be Pumped Out of
    the Mitochondrial Matrix 428
2. Components of the Electron Transport System Are
    Asymmetrically Oriented Within the Inner Mitochondrial
    Membrane 428
3. Membrane Vesicles Containing Complexes I, III,
    or IV Establish Proton Gradients 428
4. Oxidative Phosphorylation Requires a Membrane-Enclosed
    Compartment 429

5. Uncoupling Agents Abolish Both the Proton Gradient and ATP Synthesis 429
6. The Proton Gradient Has Enough Energy to Drive ATP Synthesis 429
7. Artificial Proton Gradients Can Drive ATP Synthesis in the Absence of Electron Transport 429

ATP Synthesis: Putting It All Together   430
F_1 Particles Have ATP Synthase Activity 430
The F_oF_1 Complex: Proton Translocation Through F_o Drives ATP Synthesis by F_1 431
ATP Synthesis by F_oF_1 Involves Physical Rotation of the Gamma Subunit 431
The Chemiosmotic Model Involves Dynamic Transmembrane Proton Traffic 434

Aerobic Respiration: Summing It All Up   434
The Maximum ATP Yield of Aerobic Respiration Is 36–38 ATPs per Glucose 435
Aerobic Respiration Is a Highly Efficient Process 438

Perspective   439

Key Terms for Self-Testing   439

Problem Set   440

Suggested Reading   443

Box 14A: *Further Insights: The Glyoxylate Cycle, Glyoxysomes, and Seed Germination* 416

Chapter 15
Phototrophic Energy Metabolism: Photosynthesis   445

An Overview of Photosynthesis   445

The Chloroplast: A Photosynthetic Organelle   447
Chloroplasts Are Composed of Three Membrane Systems 448

Photosynthetic Energy Transduction   452
Chlorophyll Is Life's Primary Link to Sunlight 452
Accessory Pigments Further Expand Access to Solar Energy 453
Light-Gathering Molecules Are Organized into Photosystems and Light-Harvesting Complexes 453
Oxygenic Phototrophs Have Two Types of Photosystems 454

Photoreduction (NADPH Synthesis) in Oxygenic Phototrophs   455
Photosystem II Transfers Electrons from Water to a Plastoquinone 456
The Cytochrome b_6/f Complex Transfers Electrons from a Plastoquinol to Plastocyanin 457
Photosystem I Transfers Electrons from Plastocyanin to Ferredoxin 458
Ferredoxin-NADP$^+$ Reductase Catalyzes the Reduction of NADP$^+$ 458

Photophosphorylation (ATP Synthesis) in Oxygenic Phototrophs   459

The ATP Synthase Complex Couples Transport of Protons Across the Thylakoid Membrane to ATP Synthesis 459
Cyclic Photophosphorylation Allows a Photosynthetic Cell to Balance NADPH and ATP Synthesis to Meet Its Precise Energy Needs 459
The Complete Energy Transduction System 460
Complexes of the Energy Transduction System Are Localized in Different Regions of the Thylakoids 461

A Photosynthetic Reaction Center from a Purple Bacterium   461

Photosynthetic Carbon Assimilation: The Calvin Cycle   462
Carbon Dioxide Enters the Calvin Cycle by Carboxylation of Ribulose-1,5-Bisphosphate 463
3-Phosphoglycerate Is Reduced to Form Glyceraldehyde-3-Phosphate 464
Regeneration of Ribulose-1,5-Bisphosphate Allows Continuous Carbon Assimilation 464
The Complete Calvin Cycle 465
The Calvin Cycle Operates Only When Light Is Available 465

Photosynthetic Energy Transduction and the Calvin Cycle   466

Carbohydrate Synthesis   467
Glyceraldehyde-3-Phosphate and Dihydroxyacetone Phosphate Are Combined to Form Glucose-1-Phosphate 467
The Biosynthesis of Sucrose Occurs in the Cytosol 468
The Biosynthesis of Starch Occurs in the Chloroplast Stroma 468

Other Photosynthetic Assimilation Pathways   468

Rubisco's Oxygenase Activity Decreases Photosynthetic Efficiency   468
The Glycolate Pathway Returns Reduced Carbon from Phosphoglycolate to the Calvin Cycle 469
C_4 Plants Minimize Photorespiration by Confining Rubisco to Cells Containing High Concentrations of CO_2 471
CAM Plants Minimize Photorespiration and Water Loss by Opening Their Stomata Only at Night 473

Perspective   474

Key Terms for Self-Testing   475

Problem Set   475

Suggested Reading   477

Box 15A: *Further Insights: The Endosymbiont Theory and the Evolution of Mitochondria and Chloroplasts from Ancient Bacteria* 450

Chapter 16
The Structural Basis of Cellular Information: DNA, Chromosomes, and the Nucleus   479

The Chemical Nature of the Genetic Material   479
Miescher's Discovery of DNA Led to Conflicting Proposals Concerning the Chemical Nature of Genes 480

Avery Showed That DNA Is the Genetic Material
of Bacteria 481
Hershey and Chase Showed That DNA Is the Genetic Material
of Viruses 482
Chargaff's Rules Reveal That A = T and G = C 486

DNA Structure 486
Watson and Crick Discovered That DNA Is a Double Helix 487
DNA Can Be Interconverted Between Relaxed and Supercoiled
Forms 489
The Two Strands of a DNA Double Helix Can Be Separated by
Denaturation and Rejoined by Renaturation 490

The Organization of DNA in Genomes 491
Genome Size Generally Increases with an Organism's
Complexity 492
Restriction Enzymes Cleave DNA Molecules at Specific
Sites 492
Rapid Procedures Exist for DNA Base Sequencing 496
The Genomes of Numerous Organisms Have Been
Sequenced 497
The New Field of Bioinformatics Has Emerged to Decipher
Genomes and Proteomes 498
Repeated DNA Sequences Partially Explain the Large Size
of Eukaryotic Genomes 499

DNA Packaging 501
Prokaryotes Package DNA in Bacterial Chromosomes and
Plasmids 501
Eukaryotes Package DNA in Chromatin and
Chromosomes 504
Nucleosomes Are the Basic Unit of Chromatin Structure 505
A Histone Octamer Forms the Nucleosome Core 506
Nucleosomes Are Packed Together to Form Chromatin Fibers
and Chromosomes 506
Eukaryotes Also Package Some of Their DNA in Mitochondria
and Chloroplasts 507

The Nucleus 510
A Double-Membrane Nuclear Envelope Surrounds
the Nucleus 510
Molecules Enter and Exit the Nucleus Through Nuclear
Pores 513
The Nuclear Matrix and Nuclear Lamina Are Supporting
Structures of the Nucleus 515
Chromatin Fibers Are Dispersed Within the Nucleus
in a Nonrandom Fashion 516
The Nucleolus Is Involved in Ribosome Formation 517

Perspective 518

Key Terms for Self-Testing 519

Problem Set 520

Suggested Reading 522

Box 16A: *Further Insights: Phages: Model Systems
for Studying Genes* 484

Box 16B: *Further Insights: A Closer Look at Restriction
Enzymes* 494

Box 16C: *Contemporary Techniques: DNA
Fingerprinting* 502

Chapter 17
The Cell Cycle: DNA Replication, Mitosis, and Cancer 523

An Overview of the Cell Cycle 523

DNA Replication 525
Equilibrium Density Centrifugation Shows That DNA
Replication Is Semiconservative 525
DNA Replication Is Usually Bidirectional 527
Eukaryotic DNA Replication Involves Multiple
Replicons 528
DNA Polymerases Catalyze the Elongation of DNA
Chains 530
DNA Is Synthesized as Discontinuous Segments That Are
Joined Together by DNA Ligase 532
Proofreading Is Performed by the 3′ ⟶ 5′ Exonuclease
Activity of DNA Polymerase 535
RNA Primers Initiate DNA Replication 535
Unwinding the DNA Double Helix Requires Helicases,
Topoisomerases, and Single-Strand Binding Proteins 536
Putting It All Together: DNA Replication in Summary 537
Telomeres Solve the DNA End-Replication Problem 539
Eukaryotic DNA Is Licensed for Replication 541

DNA Damage and Repair 541
DNA Damage Can Occur Spontaneously or in Response to
Mutagens 541
Translesion Synthesis and Excision Repair Correct Mutations
Involving Abnormal Nucleotides 543
Mismatch Repair Corrects Mutations That Involve
Noncomplementary Base Pairs 543
Damage Repair Helps Explain Why DNA Contains Thymine
Instead of Uracil 544

Nuclear and Cell Division 544
Mitosis Is Subdivided into Prophase, Prometaphase,
Metaphase, Anaphase, and Telophase 544
The Mitotic Spindle Is Responsible for Chromosome
Movements During Mitosis 549
Cytokinesis Divides the Cytoplasm 551

Regulation of the Cell Cycle 554
The Length of the Cell Cycle Varies Among Different
Cell Types 554
Cell Cycle Checkpoints Regulate Progression Through the
Cell Cycle 555
Cell Fusion Experiments Provide Evidence for Control
Molecules in the Cell Cycle 556
A Mitosis-Promoting Factor (MPF) Moves Cells Through
the G2 Checkpoint 557
Cell Cycle Mutants Facilitated the Identification
of Molecules that Control the Cell Cycle 557
The Cell Cycle Is Controlled by Cyclin-Dependent Kinase
(Cdk) Molecules 558
The Mitotic Cdk-Cyclin Complex (MPF) Controls
the G2 Checkpoint by Phosphorylating Proteins Involved
in the Early Stages of Mitosis 558
The G1 Cdk-Cyclin Complex Controls the G1 Checkpoint by
Phosphorylating the Rb Protein 559

The Mitotic Cdk-Cyclin Complex (MPF) Controls the Spindle Assembly Checkpoint by Activating the Anaphase–Promoting Complex   561
Putting It All Together: The Cell Cycle Regulation Machine   562

Growth Control and Cancer   562
Stimulatory Growth Factors Activate the Ras Pathway   562
Inhibitory Growth Factors Act Through Cdk Inhibitors   564
Cancer Involves Defective Cell Cycle Control Mechanisms   564
Oncogenes Can Trigger the Development of Cancer   565
Oncogenes Arise from Proto-oncogenes   565
Most Oncogenes Code for Components of Growth Factor Signaling Pathways   566
Cancer Can Arise from the Loss of Tumor Suppressor Genes That Normally Restrain Cell Proliferation   567
Genetic Instability Leads to the Accumulation of Multiple Mutations in Cancer Cells   569
Can Cancer Growth Be Stopped?   570

Perspective   571

Key Terms for Self-Testing   572

Problem Set   573

Suggested Reading   576

Box 17A: *Contemporary Techniques: The PCR Revolution*   533

Box 17B: *Clinical Applications: Attacking a Tumor's Blood Supply*   570

Chapter 18
Sexual Reproduction, Meiosis, and Genetic Recombination   577

Sexual Reproduction   577
Sexual Reproduction Generates Genetic Variety   577
The Diploid State Is an Essential Feature of Sexual Reproduction   578
Diploid Cells May Be Homozygous or Heterozygous for Each Gene   578
Gametes Are Haploid Cells Specialized for Sexual Reproduction   579

Meiosis   580
The Life Cycles of Sexual Organisms Have Diploid and Haploid Phases   580
Meiosis Converts One Diploid Cell into Four Haploid Cells   581
Meiosis I Produces Two Haploid Cells That Have Chromosomes Composed of Sister Chromatids   582
Meiosis II Resembles a Mitotic Division   588
Sperm and Egg Cells Are Generated by Meiosis Accompanied by Cell Differentiation   588

Meiosis Generates Genetic Diversity   590

Genetic Variability: Segregation and Assortment of Alleles   590
Information Specifying Recessive Traits Can Be Present Without Being Displayed   590
The Law of Segregation States That the Alleles of Each Gene Separate from Each Other During Gamete Formation   592
The Law of Independent Assortment States That the Alleles of Each Gene Separate Independently of the Alleles of Other Genes   592
Early Microscopic Evidence Suggested That Chromosomes Might Carry Genetic Information   593
Chromosome Behavior Explains the Laws of Segregation and Independent Assortment   593
The DNA Molecules of Homologous Chromosomes Have Similar Base Sequences   594

Genetic Variability: Recombination and Crossing Over   595
Chromosomes Contain Groups of Linked Genes That Are Usually Inherited Together   596
Homologous Chromosomes Exchange Segments During Crossing Over   597
Gene Locations Can Be Mapped by Measuring Recombination Frequencies   597

Genetic Recombination in Bacteria and Viruses   598
Co-infection of Bacterial Cells with Related Bacteriophages Can Lead to Genetic Recombination   598
Transformation and Transduction Involve Recombination with Free DNA or DNA Brought into Bacterial Cells by Bacteriophages   599
Conjugation Is a Modified Sexual Activity That Facilitates Genetic Recombination in Bacteria   600

Molecular Mechanism of Homologous Recombination   602
DNA Breakage-and-Exchange Underlies Homologous Recombination   602
Homologous Recombination Can Lead to Gene Conversion   603
Homologous Recombination Is Initiated by Single-Strand DNA Exchanges (Holliday Junctions)   604
The Synaptonemal Complex Facilitates Homologous Recombination During Meiosis   606

Recombinant DNA Technology and Gene Cloning   606
The Discovery of Restriction Enzymes Paved the Way for Recombinant DNA Technology   607
DNA Cloning Techniques Permit Individual Gene Sequences to Be Produced in Large Quantities   608
Genomic and cDNA Libraries Are Both Useful for DNA Cloning   612
Large DNA Segments Can Be Cloned in Cosmids and Yeast Artificial Chromosomes (YACs)   613

Genetic Engineering   614
Genetic Engineering Can Produce Valuable Proteins That Are Otherwise Difficult to Obtain   614
The Ti Plsmid Is a Useful Vector for Introducing Foreign Genes into Plants   614

Genetic Modification Can Improve the Traits of
Food Crops   615
Concerns Have Been Raised About the Safety and
Environmental Risks of GM Crops   616
Gene Therapies Are Being Developed for the Treatment of
Human Diseases   616

Perspective   618

Key Terms for Self-Testing   619

Problem Set   619

Suggested Reading   621

Box 18A: *Historical Perspectives: Supermouse, an Early*
Transgenic Triumph   617

Chapter 19
Gene Expression: I. The Genetic Code and Transcription   623

The Directional Flow of Genetic Information   623

The Genetic Code   624
Experiments on *Neurospora* Revealed That Genes Can Code
for Enzymes   624
The Base Sequence of a Gene Usually Codes for the Amino
Acid Sequence of a Polypeptide Chain   625
The Genetic Code Is a Triplet Code   628
The Genetic Code Is Degenerate and Nonoverlapping   631
Messenger RNA Guides the Synthesis of Polypeptide
Chains   632
The Codon Dictionary Was Established Using Synthetic RNA
Polymers and Triplets   633
Of the 64 Possible Codons in Messenger RNA, 61 Code for
Amino Acids   634
The Genetic Code Is (Nearly) Universal   634

Transcription in Prokaryotic Cells   635
Transcription Is Catalyzed by RNA Polymerase, Which
Synthesizes RNA Using DNA as a Template   635
Transcription Involves Four Stages: Binding, Initiation,
Elongation, and Termination   635

Transcription in Eukaryotic Cells   640
RNA Polymerases I, II, and III Carry Out Transcription in the
Eukaryotic Nucleus   640
Three Classes of Promoters Are Found in Eukaryotic Nuclear
Genes, One for Each Type of RNA Polymerase   641
General Transcription Factors Are Involved
in the Transcription of All Nuclear Genes   643
Elongation, Termination, and RNA Cleavage Are Involved
in Completing Eukaryotic RNA Synthesis   644

RNA Processing   644
Ribosomal RNA Processing Involves Cleavage of Multiple
rRNAs from a Common Precursor   644
Transfer RNA Processing Involves Removal, Addition, and
Chemical Modification of Nucleotides   646

Messenger RNA Processing in Eukaryotes Involves Capping,
Addition of Poly(A), and Removal of Introns   647
Spliceosomes Remove Introns from Pre-mRNA   650
Some Introns Are Self-Splicing   651
Why Do Eukaryotic Genes Have Introns?   652
RNA Editing Allows the Coding Sequence of mRNA
to Be Altered   653
RNA Proofreading and RNA Surveillance Help to Protect Cells
from Producing and Using Defective mRNAs   654

Key Aspects of mRNA Metabolism   654
Most mRNA Molecules Have a Relatively Short Life Span   655
The Existence of mRNA Allows Amplification of Genetic
Information   655

Perspective   655

Key Terms for Self-Testing   656

Problem Set   656

Suggested Reading   658

Box 19A: *Further Insights: Reverse Transcription,*
Retroviruses, and Retrotransposons   626

Box 19B: *Contemporary Techniques: The Footprinting*
Method for Identifying Protein-Binding Sites
on DNA   638

Chapter 20
Gene Expression: II. Protein Synthesis and Sorting   660

Translation: The Cast of Characters   660
The Ribosome Carries Out Polypeptide Synthesis   660
Transfer RNA Molecules Bring Amino Acids to the
Ribosome   662
Aminoacyl-tRNA Synthetases Link Amino Acids to the
Correct Transfer RNAs   664
Messenger RNA Brings Polypeptide-Coding Information to
the Ribosome   666
Protein Factors Are Required for the Initiation, Elongation,
and Termination of Polypeptide Chains   666

The Mechanism of Translation   666
The Initiation of Translation Requires Initiation Factors,
Ribosomal Subunits, mRNA, and Initiator tRNA   667
Chain Elongation Involves Sequential Cycles of
Aminoacyl tRNA Binding, Peptide Bond Formation,
and Translocation   669
Termination of Polypeptide Synthesis Is Triggered
by Release Factors That Recognize Stop Codons   671
Polypeptide Folding Is Facilitated by Molecular
Chaperones   672
Protein Synthesis Typically Utilizes a Substantial Fraction of a
Cell's Energy Budget   672
A Summary of Translation   674

Nonsense Mutations and Suppressor tRNA   674

Posttranslational Processing   675

Protein Targeting and Sorting   676

    Cotranslational Import Allows Some Polypeptides
      to Enter the ER as They Are Being Synthesized   678

    Posttranslational Import Allows Some Polypeptides to Enter
      Organelles After They Have Been Synthesized   683

Perspective   688

Key Terms for Self-Testing   688

Problem Set   689

Suggested Reading   691

Box 20A: *Clinical Applications: Protein-Folding*
Diseases   673

Box 20B: *Further Insights: A Mutation Primer   676*

Chapter 21
The Regulation of Gene Expression   692

Gene Regulation in Prokaryotes   692

    Catabolic and Anabolic Pathways Utilize Different Strategies
      for Adaptive Enzyme Synthesis   692

    The Genes Involved in Lactose Catabolism Are Organized
      into an Inducible Operon   694

    The *lac* Repressor Is an Allosteric Protein Whose Binding
      to DNA Is Controlled by Lactose   694

    The *lac* Operon Model Is Based Largely on Genetic
      Analysis   695

    The Genes Involved in Tryptophan Synthesis Are Organized
      into a Repressible Operon   698

    The *lac* and *trp* Operons Illustrate the Negative Control
      of Transcription   700

    Catabolite Repression Illustrates the Positive Control
      of Transcription   700

    Inducible Operons Are Often Under Dual Control   700

    Sigma Factors Can Regulate the Initiation of Transcription   701

    Attenuation Allows Transcription to Be Regulated After the
      Initiation Step   701

**Comparison of Gene Regulation in Prokaryotes
and Eukaryotes   704**

    Prokaryotes and Eukaryotes Exhibit Many Differences
      in Genome Organization and Expression   704

    Stem Cells Give Rise to Specialized Eukaryotic Cell Types
      By the Process of Cell Differentiation   706

    Eukaryotic Gene Expression Is Regulated at Multiple
      Levels   706

Eukaryotic Gene Regulation: Genomic Control   706

    As a General Rule, the Cells of a Multicellular Organism All
      Contain the Same Set of Genes   706

    Gene Amplification and Deletion Can Alter the Genome   707

    DNA Rearrangements Can Alter the Genome   709

    Chromosome Puffs Provide Visual Evidence That Chromatin
      Decondensation Is Involved in Genomic Control   710

    DNase I Sensitivity Provides Further Evidence for the Role of
      Chromatin Decondensation in Genomic Control   712

    DNA Methylation Is Associated with Inactive Regions
      of the Genome   714

    Changes in Histones, HMG Proteins, and the
      Nuclear Matrix Are Associated with Active Regions
      of the Genome   715

Eukaryotic Gene Regulation: Transcriptional Control   715

    Different Sets of Genes are Transcribed in Different
      Cell Types   715

    DNA Microarrays Allow the Expression of Thousands of
      Genes to Be Monitored Simultaneously   716

    Proximal Control Elements Lie Close to the Promoter   717

    Enhancers and Silencers Are Located at Variable Distances
      from the Promoter   718

    Coactivators Mediate the Interaction Between
      Regulatory Transcription Factors and the RNA
      Polymerase Complex   719

    Multiple DNA Control Elements and Transcription Factors
      Act in Combination   721

    Several Common Structural Motifs Allow Regulatory
      Transcription Factors to Bind to DNA and Activate
      Transcription   721

    DNA Response Elements Coordinate the Expression
      of Nonadjacent Genes   724

    Steroid Hormone Receptors Are Transcription Factors That
      Bind to Hormone Response Elements   724

    CREBs and STATs are Examples of Transcription Factors
      Activated by Phosphorylation   725

    The Heat-Shock Response Element Coordinates the Expression
      of Genes Activated by Elevated Temperatures   726

    Homeotic Genes Code for Transcription Factors That
      Regulate Embryonic Development   727

Eukaryotic Gene Regulation: Posttranscriptional Control   728

    Control of RNA Processing and Nuclear Export Follows
      Transcription   728

    Translation Rates Can Be Controlled by Initiation Factors and
      Translational Repressors   730

    Translation Can Also Be Controlled by Regulation
      of mRNA Half-Life   731

    Posttranslational Control Involves Modifications of Protein
      Structure, Function, and Degradation   732

    Ubiquitin Targets Proteins for Degradation by
      Proteasomes   733

    A Summary of Eukaryotic Gene Regulation   734

Perspective   736

Key Terms for Self-Testing   738

Problem Set   738

Suggested Reading   741

Box 21A: *Further Insights: Dolly: A Lamb with No*
Father   708

Box 21B: *Contemporary Techniques: Discriminating Among*
Multiple Levels of Control   735

Chapter 22
Cytoskeletal Systems 742

The Major Structural Elements of the Cytoskeleton 742

Techniques for Studying the Cytoskeleton 744
Modern Microscopy Techniques Have Revolutionized the Study of the Cytoskeleton 744
Drugs and Mutations Can Be Used to Disrupt Cytoskeletal Structures 744

Microtubules 744
Two Types of Microtubules Are Responsible for Many Functions in the Cell 744
Tubulin Heterodimers Are the Protein Building Blocks of Microtubules 746
Microtubules Form by the Addition of Tubulin Dimers at Their Ends 747
Addition of Tubulin Dimers Occurs More Quickly at the Plus Ends of Microtubules 747
GTP Hydrolysis Contributes to the Dynamic Instability of Microtubules 748
MTs Originate from Microtubule-Organizing Centers Within the Cell 749
MTOCs Organize and Polarize the Microtubules Within Cells 750
Microtubule Stability Within Cells Is Highly Regulated 751
Drugs Can Affect the Assembly of Microtubules 753
Microtubules Are Regulated by Microtubule-Associated Proteins (MAPs) 753

Microfilaments 754
Actin Is the Protein Building Block of Microfilaments 755
Different Types of Actin and Actin-Related Proteins Are Found in Cells 756
G-Actin Monomers Polymerize into F-Actin Microfilaments 756
Actin Polymerization Is Regulated by Small GTP-Binding Proteins 757
Specific Proteins and Drugs Affect Polymer Dynamics at the Ends of Microfilaments 758
Capping Proteins Stabilize the Ends of Microfilaments 758
Inositol Phospholipids Regulate Molecules That Affect Actin Polymerization 758
Actin-Binding Proteins Regulate Interactions Between Microfilaments 758
A Variety of Proteins Link Actin to Membranes 760

Intermediate Filaments 762
Intermediate Filament Proteins Are Tissue-Specific 762
Intermediate Filaments Assemble from Fibrous Subunits 763
Intermediate Filaments Confer Mechanical Strength on Tissues 764
The Cytoskeleton Is a Mechanically Integrated Structure 765

Perspective 765

Key Terms for Self-Testing 766

Problem Set 766

Suggested Reading 767

Chapter 23
Cellular Movement: Motility and Contractility 769

Motile Systems 769

Intracellular Microtubule-Based Movement: Kinesin and Dynein 770
Motor MAPs Move Organelles Along Microtubules During Axonal Transport 770
Kinesins Move Along Microtubules by Hydrolyzing ATP 771
Kinesins Are a Large Family of Proteins with Varying Structures and Functions 772
Dyneins Can Be Grouped in Two Major Classes: Axonemal and Cytoplasmic Dyneins 772
Motor MAPs Are Involved in the Transport of Intracellular Vesicles 773

Microtubule-Based Motility 773
Cilia and Flagella Are Common Motile Appendages of Eukaryotic Cells 773
Cilia and Flagella Consist of an Axoneme Connected to a Basal Body 774
Microtubule Sliding Within the Axoneme Causes Cilia and Flagella to Bend 776

Actin-Based Cell Movement: The Myosins 777
Myosins Have Diverse Roles in Cell Motility 777
Myosins Move Along Actin Filaments in Short Steps 779

Filament-Based Movement in Muscle 779
Skeletal Muscle Cells Are Made of Thin and Thick Filaments 779
Sarcomeres Contain Ordered Arrays of Actin, Myosin, and Accessory Proteins 780
The Sliding-Filament Model Explains Muscle Contraction 782
Cross-Bridges Hold Filaments Together and ATP Powers Their Movement 783
The Regulation of Muscle Contraction Depends on Calcium 786
ATP Generated Under Aerobic or Hypoxic Conditions Meets the Energy Needs Of Muscle Contraction 788
The Coordinated Contraction of Cardiac Muscle Cells Involves Electrical Coupling 789
Smooth Muscle Is More Similar to Nonmuscle Cells than to Skeletal Muscle 790

Actin-Based Motility in Nonmuscle Cells 792
Cell Migration via Lamellipodia Involves Cycles of Protrusion, Attachment, Translocation, and Detachment 792
Amoeboid Movement Involves Cycles of Gelation and Solation of the Actin Cytoskeleton 795

Cytoplasmic Streaming Moves Components Within
the Cytoplasm of Some Cells   795
Infectious Microorganisms Can Move Within Cells Using
Actin "Tails"   795
Rho, Rac, and Cdc42 Regulate the Actin Cytoskeleton   797
Chemotaxis Is a Directional Movement in Response to a
Graded Chemical Stimulus   798

Perspective   799

Key Terms for Self-Testing   799

Problem Set   800

Suggested Reading   801

Box 23A: *Clinical Applications: Cytoskeletal Motor Proteins
and Human Disease   778*

Glossary   G-I

Photo, Illustration, and Text Credits   C-1

Index   I-1

The WORLD of the CELL

1

A Preview of the Cell

The **cell** is the basic unit of biology. Every organism either consists of cells or is itself a single cell. Therefore, it is only as we understand the structure and function of cells that we can appreciate both the capabilities and the limitations of living organisms, whether animal, plant, or microorganism.

We are in the midst of a revolution in biology that has brought with it tremendous advances in our understanding of how cells are constructed and how they carry out all the intricate functions necessary for life. Particularly significant is the dynamic nature of the cell, as evidenced by its capacity to grow, reproduce, and become specialized and by its ability to respond to stimuli and to adapt to changes in its environment.

Cell biology itself is changing, as scientists from a variety of related disciplines focus their efforts on the common objective of understanding more adequately how cells work. The convergence of cytology, genetics, and biochemistry has made modern cell biology one of the most exciting and dynamic disciplines in contemporary biology.

In this chapter, we will look briefly at the beginnings of cell biology as a discipline. Then we will consider the three main historical strands that have given rise to our current understanding of what cells are and how they function.

The Cell Theory: A Brief History

The story of cell biology started to unfold more than 300 years ago, as European scientists began to focus their crude microscopes on a variety of biological material ranging from tree bark to human sperm. One such scientist was Robert Hooke, Curator of Instruments for the Royal Society of London. In 1665, Hooke used a microscope that he had built himself to examine thin slices of cork cut with a penknife. He saw a network of tiny boxlike compartments that reminded him of a honeycomb. Hooke called these little compartments *cellulae,* a Latin term meaning "little rooms." It is from this word that we get our present-day term, *cell.*

Actually, what Hooke observed were not cells at all but the empty cell walls of dead plant tissue, which is what tree bark really is. However, Hooke would not have thought of his *cellulae* as dead, because he did not understand that they could be alive! Although he noticed that cells in other plant tissues were filled with what he called "juices," he preferred to concentrate on the more prominent cell walls that he had first encountered.

One of the limitations inherent in Hooke's observations was that his microscope could only magnify objects 30-fold, making it difficult to learn much about the internal organization of cells. This obstacle was overcome a few years later by Antonie van Leeuwenhoek, a Dutch shopkeeper who devoted much of his spare time to the design of microscopes. Van Leeuwenhoek produced hand-polished lenses that could magnify objects almost 300-fold. Using these superior lenses, he became the first to observe living cells, including blood cells, sperm cells, and single-celled organisms found in pond water. He reported his observations to the Royal Society in a series of papers during the last quarter of the seventeenth century. His detailed reports attest to both the high quality of his lenses and his keen powers of observation.

Two factors restricted further understanding of the nature of cells. One was the limited resolution of the microscopes of the day, which even van Leeuwenhoek's

UNITS OF MEASUREMENT IN CELL BIOLOGY

The challenge of understanding cellular structure and organization is complicated by the problem of size. Most cells and their organelles are so small that they cannot be seen by the unaided eye. In addition, the units used to measure them are unfamiliar to many students and therefore often difficult to appreciate. The problem can be approached in two ways: by realizing that there

are really only two units necessary to express the dimensions of most structures of interest to us, and by illustrating a variety of structures that can be appropriately measured with each of these units.

The **micrometer** (µm) is the most useful unit for expressing the size of cells and larger organelles. A micrometer

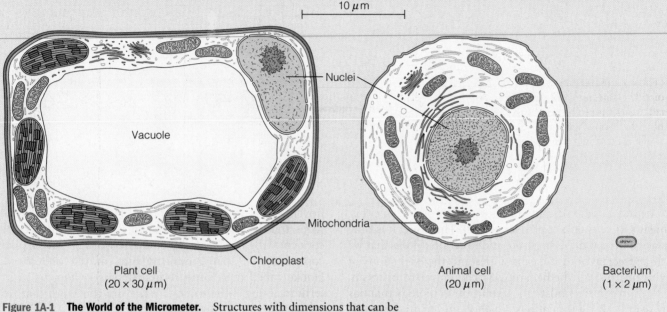

10 µm

Nuclei

Vacuole

Mitochondria

Chloroplast

Plant cell
(20 × 30 µm)

Animal cell
(20 µm)

Bacterium
(1 × 2 µm)

Figure 1A-1 The World of the Micrometer. Structures with dimensions that can be measured conveniently in micrometers include almost all cells and some of the larger organelles.

superior instruments could push just so far. The second and probably more fundamental factor was the essentially descriptive nature of seventeenth-century biology. It was basically an age of observation, with little thought given to explaining the intriguing architectural details of biological materials that were beginning to yield to the probing lens of the microscope.

More than a century passed before the combination of improved microscopes and more experimentally minded microscopists resulted in a series of developments that culminated in an understanding of the importance of cells in biological organization. By the 1830s, improved lenses led to higher magnification and better resolution, such that structures only 1 micrometer (µm) apart could be resolved. (A *micrometer* is 10^{-6} m, or one-millionth of a meter; see Box 1A for a discussion of the units of measurement appropriate to cell biology.)

Aided by such improved lenses, the English botanist Robert Brown found that every plant cell he looked at contained a rounded structure, which he called a *nucleus,* a term derived from the Latin word for "kernel." In 1838, his

German colleague Matthias Schleiden came to the important conclusion that all plant tissues are composed of cells and that an embryonic plant always arises from a single cell. Similar conclusions concerning animal tissue were reported only a year later by Theodor Schwann, thereby laying to rest earlier speculations that plants and animals might not resemble each other structurally. It is easy to understand how such speculations could have arisen. After all, plant cell walls provide conspicuous boundaries between cells that are readily visible even with a crude microscope, whereas individual animal cells, which lack cell walls, are much harder to distinguish in a tissue sample. It was only when Schwann examined animal cartilage cells that he became convinced of the fundamental similarity between plant and animal tissue, since cartilage cells, unlike most other animal cells, have boundaries that are well defined by thick deposits of collagen fibers. Schwann drew all these observations together into a single unified theory of cellular organization, which has stood the test of time and continues to provide the basis for our own understanding of the importance of cells and cell biology.

(sometimes also called a *micron*) corresponds to one-millionth of a meter (10^{-6} m). In general, bacterial cells are a few micrometers in diameter, and the cells of plants and animals are 10- to 20-fold larger in any single dimension. Organelles such as mitochondria and chloroplasts tend to have diameters or lengths of a few micrometers and are therefore comparable in size to whole bacterial cells. Smaller organelles are usually in the range of 0.2–1.0 µm. As a rule of thumb, if you can see it with a light microscope, you can probably express its dimensions conveniently in micrometers, since the resolution limit of the light microscope is about 0.20–0.35 µm. Figure 1A-1 illustrates a variety of structures that are usually measured in micrometers.

The **nanometer** (nm), on the other hand, is the unit of choice for molecules and subcellular structures that are too small or too thin to be seen with the light microscope. A nanometer is one-billionth of a meter (10^{-9} m). It takes 1000 nanometers to equal 1 micrometer. (An alternative to the term nanometer is therefore *millimicron, mµ.*) As a benchmark on the nanometer scale, a ribosome has a diameter of about 25–30 nm. Other structures that can be measured conveniently in nanometers are microtubules, microfilaments, membranes, and DNA molecules. The dimensions of these structures are indicated in Figure 1A-2.

Another unit frequently used in cell biology is the *angstrom* (Å) which corresponds to 10^{-10} m or 0.1 nm. Molecular dimensions, in particular, are often expressed in angstroms. However, because the angstrom differs from the nanometer by only a factor of ten, it adds little flexibility to the expression of dimensions at the cellular level and will therefore not be used in this text.

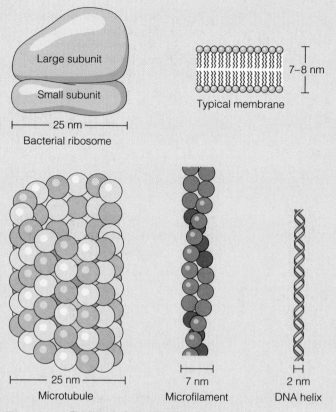

Figure 1A-2　The World of the Nanometer. Structures with dimensions that can be measured conveniently in nanometers include ribosomes, membranes, microtubules, microfilaments, and the DNA double helix.

As originally postulated by Schwann in 1839, the **cell theory** had two basic tenets:

1. All organisms consist of one or more cells.
2. The cell is the basic unit of structure for all organisms.

Less than 20 years later, a third tenet was added. This grew out of Brown's original description of nuclei, extended by Karl Nägeli to include observations on the nature of cell division. By 1855, Rudolf Virchow, a German physiologist, was able to conclude that cells arose in only one manner—by the division of other, preexisting cells. Virchow encapsulated this conclusion in the now-famous Latin phrase *omnis cellula e cellula,* which in translation becomes the third tenet of the modern cell theory:

3. All cells arise only from preexisting cells.

Thus, the cell is not only the basic unit of structure for all organisms but also the basic unit of reproduction. In other words, all of life has a cellular basis. No wonder, then, that an understanding of cells and their properties is so fundamental to a proper appreciation of all other aspects of biology.

The Emergence of Modern Cell Biology

Modern cell biology involves the weaving together of three distinctly different strands into a single cord. As the timeline of Figure 1-1 illustrates, each of the strands had its own historical origins, and most of the intertwining has occurred only within the last 70 years. Each strand should be appreciated in its own right, because each makes its own unique and significant contribution. The contemporary cell biologist must be adequately informed about all three strands, regardless of what his or her own immediate interests happen to be.

The first of these historical strands is **cytology,** which is concerned primarily with cellular structure. (The Greek prefix *cyto-* means "cell," as does the suffix *-cyte.*) As we have already seen, cytology had its origins more than three centuries ago and depended heavily on the light microscope for its initial impetus. The advent of electron microscopy and several related optical techniques has led to considerable additional cytological activity and understanding.

CELL BIOLOGY

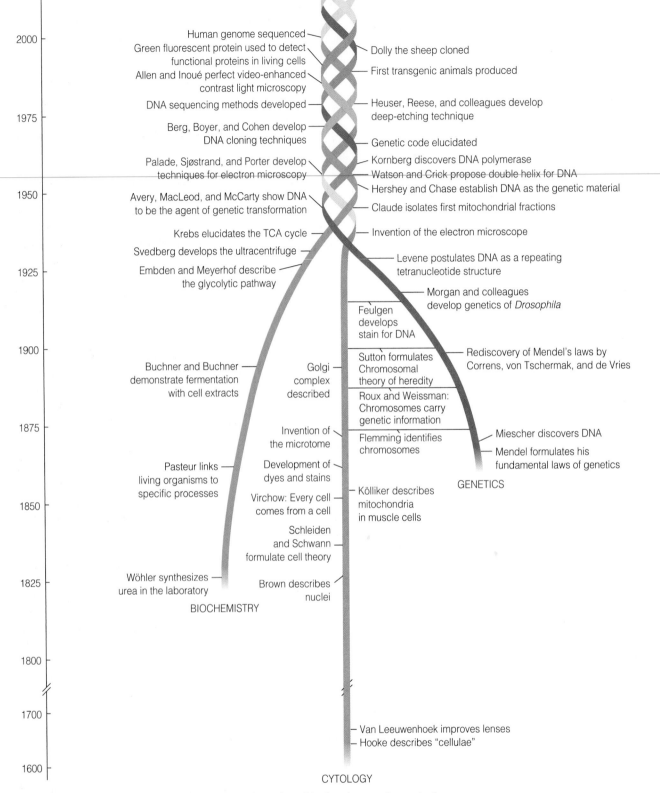

2000 — Human genome sequenced
Green fluorescent protein used to detect functional proteins in living cells
Allen and Inoué perfect video-enhanced contrast light microscopy
DNA sequencing methods developed
1975 — Berg, Boyer, and Cohen develop DNA cloning techniques
Palade, Sjøstrand, and Porter develop techniques for electron microscopy
Avery, MacLeod, and McCarty show DNA to be the agent of genetic transformation
1950 — Krebs elucidates the TCA cycle
Svedberg develops the ultracentrifuge
Embden and Meyerhof describe the glycolytic pathway
1925 —

Dolly the sheep cloned
First transgenic animals produced
Heuser, Reese, and colleagues develop deep-etching technique
Genetic code elucidated
Kornberg discovers DNA polymerase
Watson and Crick propose double helix for DNA
Hershey and Chase establish DNA as the genetic material
Claude isolates first mitochondrial fractions
Invention of the electron microscope
Levene postulates DNA as a repeating tetranucleotide structure
Morgan and colleagues develop genetics of *Drosophila*

Feulgen develops stain for DNA

1900 — Buchner and Buchner demonstrate fermentation with cell extracts
Golgi complex described
Sutton formulates Chromosomal theory of heredity
Rediscovery of Mendel's laws by Correns, von Tschermak, and de Vries
Roux and Weissman: Chromosomes carry genetic information
1875 — Invention of the microtome
Flemming identifies chromosomes
Miescher discovers DNA
Mendel formulates his fundamental laws of genetics
Pasteur links living organisms to specific processes
Development of dyes and stains
GENETICS
Virchow: Every cell comes from a cell
Kölliker describes mitochondria in muscle cells
1850 —
Schleiden and Schwann formulate cell theory
1825 — Wöhler synthesizes urea in the laboratory
Brown describes nuclei
BIOCHEMISTRY

1800 —

1700 — Van Leeuwenhoek improves lenses
Hooke describes "cellulae"

1600 —
CYTOLOGY

Figure 1-1 The Cell Biology Time Line.   Although cytology, biochemistry, and genetics began as separate disciplines, they have increasingly merged since about the second quarter of the twentieth century.

The second strand represents the contributions of **biochemistry** to our understanding of cellular function. Most of the developments in this field have occurred within the last 70 years, though again the roots go back much further. Especially important has been the development of techniques such as ultracentrifugation, chromatography, and electrophoresis for the separation of cellular components and molecules. The use of radioactively labeled compounds in the study of enzyme-catalyzed reactions and metabolic pathways is another very significant contribution of biochemistry to our understanding of how cells function. We will encounter these and other techniques in subsequent chapters as we explore various aspects of cellular structure and function and an understanding of relevant techniques becomes necessary. To locate discussions of specific techniques, see the Guide to Techniques and Methods inside the front cover.

The third strand is **genetics.** Here, the historical continuum stretches back more than a century to Gregor Mendel. Again, however, much of our present understanding has come within the last 70 years. An especially important landmark on the genetic strand came with the demonstration that DNA (deoxyribonucleic acid) is the bearer of genetic information in most life forms, specifying the order of subunits, and hence the properties, of the proteins that are responsible for most of the functional and structural features of cells. Recent accomplishments on the genetic strand include the sequencing of the entire *genomes* (all of the DNA) of humans and other species and the *cloning* (production of genetically identical organisms) of mammals, including sheep, cattle, and cats.

To understand present-day cell biology therefore means to appreciate its diverse roots and the important contributions that each of its component strands has made to our current understanding of what a cell is and what it can do. Each of the three historical strands of cell biology is discussed briefly here; a fuller appreciation of each will come as various aspects of cell structure, function, and genetics are explored in later chapters.

The Cytological Strand Deals with Cellular Structure

Strictly speaking, cytology is the study of cells. (Actually, the literal meaning of the Greek word *cytos* is "hollow vessel," which fits well with Hooke's initial impression of cells.) Historically, however, cytology has dealt primarily with cellular structure, mainly through the use of optical techniques. Here we describe briefly some of the microscopy that has been important in cell biology. For a more detailed discussion, see the *Guide to Microscopy* by Becker, Kleinsmith, and Hardin, which is intended as a supplement to this textbook.

The Light Microscope. The **light microscope** was the earliest tool of the cytologists and continues to play an important role in our elucidation of cellular structure. Light microscopy allowed cytologists to identify membrane-bounded structures such as *nuclei, mitochondria,* and *chloroplasts* within a variety of cell types. Such structures are called **organelles** ("little organs") and are prominent features of most plant and animal (but not bacterial) cells.

Other significant developments include the invention of the microtome in 1870 and the availability of various dyes and stains at about the same time. A *microtome* is an instrument for slicing thin sections of biological samples, usually after they have been dehydrated and embedded in paraffin or plastic. The technique enables rapid and efficient preparation of thin tissue slices of uniform thickness. The dyes that came to play so important a role in staining and identifying subcellular structures were developed primarily in the latter half of the nineteenth century by German industrial chemists working with coal tar derivatives.

Together with improved optics and more sophisticated lenses, these and related developments extended light microscopy as far as it could go—to the physical limits of resolution imposed by the wavelengths of visible light. As used in microscopy, the **limit of resolution** refers to how far apart adjacent objects must be in order to be distinguished as separate entities. For example, to say that the limit of resolution of a microscope is 400 nanometers (nm) means that objects need to be at least 400 nm apart to be recognizable as separate entities, whereas a resolution of 200 nm means that objects only 200 nm can be distinguished from each other. (A *nanometer* is 10^{-9} or one-billionth of a meter; 1 nm = 0.001 μm.) The smaller the limit of resolution, the greater the **resolving power** of the microscope. Expressed in terms of λ, the wavelength of the light used to illuminate the sample, the theoretical limit of resolution for the light microscope is $\lambda/2$. For visible light in the wavelength range of 400–700 nm, the limit of resolution is about 200–350 nm. Figure 1-2 illustrates the useful range of the light microscope and compares its resolving power with that of the human eye and the electron microscope.

Visualization of Living Cells. The type of microscopy described thus far is called *brightfield microscopy* because white light is passed directly through a specimen that is either stained or unstained, depending on the structural features to be examined. A significant limitation of this approach is that specimens must be fixed (preserved), dehydrated, and embedded in paraffin or plastic. The specimen is therefore no longer alive, which raises the possibility that features observed by this method could be artifacts or distortions due to the fixation, dehydration, and embedding processes. To overcome this disadvantage, a variety of special optical techniques have been developed that make it possible to observe living cells directly. These include fluorescence microscopy, phase-contrast microscopy, and differential interference contrast microscopy. Table 1-1 depicts the images seen with each of these techniques and compares them with the images seen with brightfield microscopy for both unstained and stained specimens. Each of these techniques is discussed in the *Guide to Microscopy* supplement mentioned earlier; here we will content ourselves with a brief description of each.

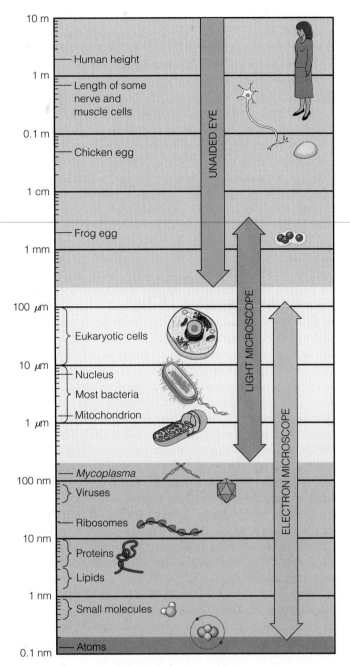

Figure 1-2 Resolving Power of the Human Eye, the Light Microscope, and the Electron Microscope.   Notice that the vertical axis is on a logarithmic scale to accommodate the range of sizes shown.

Fluorescence microscopy enables researchers to detect specific proteins or other molecules that are made fluorescent by coupling them to a fluorescent dye. By the simultaneous use of two or more such dyes, each coupled to a different kind of molecule, the distributions of different kinds of molecules can be followed in the same cell.

An inherent limitation of fluorescence microscopy is that the viewer can focus on only a single plane of the specimen at a given time, yet fluorescent light is emitted throughout the specimen. As a result, the visible image is blurred by light emitted from regions of the specimen above and below the focal plane, which historically limited the technique to flattened cells with minimal depth. This problem has been largely overcome by the recent development of *confocal scanning*, in which a laser beam is used to illuminate a single plane of the specimen at a time. This approach gives much better resolution than traditional fluorescence microscopy when used with thick specimens such as whole cells. Furthermore, the laser beam can be directed to successive focal planes sequentially, thereby generating a series of images that can be combined to provide a three-dimensional picture of the cell.

Phase-contrast and *differential interference contrast* (or *Nomarski*) *microscopy* make it possible to see living cells clearly (see Table 1-1). Both of these techniques enhance and amplify slight changes in the phase of transmitted light as it passes through a structure that has a different refractive index than the surrounding medium. Most modern light microscopes are equipped for phase-contrast and differential interference contrast in addition to the simple transmission of light, with conversion from one use to another accomplished by interchanging optical components.

Another recent development in light microscopy is *digital video microscopy*, which makes use of video cameras and computer storage, and allows computerized image processing to enhance and analyze images. Attachment of a highly light-sensitive video camera to a light microscope makes it possible to observe cells for extended periods of time using very low levels of light. This *image intensification* is particularly useful for visualizing fluorescent molecules in living cells with a fluorescence microscope. For *computerized image processing*, the electronic signal from the video camera is processed by a computer in a variety of ways to compensate for optical flaws in the microscope and to enhance the contrast between objects and background. The enhanced contrast makes it possible to see small transparent objects that would otherwise have been impossible to distinguish from the background.

The Electron Microscope.   Despite advances in optical techniques and contrast enhancement, light microscopy is inevitably subject to the limit of resolution imposed by the wavelength of the light used to view the sample. Even the use of ultraviolet radiation, with shorter wavelengths, increases the resolution only by a factor of two.

A major breakthrough in resolving power came with the development of the **electron microscope.** The electron microscope was invented in Germany in 1932 and came into widespread biological use in the early 1950s, with George Palade, Fritiof Sjøstrand, and Keith Porter among its most notable early users. In place of visible light and optical lenses, the electron microscope uses a beam of electrons that is deflected and focused by an electromagnetic field. Because the wavelength of electrons is so much shorter than that of photons of visible light, the limit of resolution of the electron microscope is much better than that of the light microscope: about 0.1–0.2 nm for the electron microscope compared with about 200–350 nm for the light microscope.

Table 1-1 Different Types of Light Microscopy: A Comparison

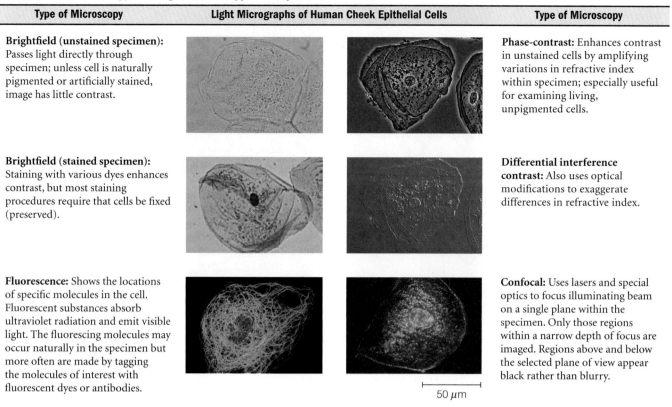

Type of Microscopy	Light Micrographs of Human Cheek Epithelial Cells		Type of Microscopy
Brightfield (unstained specimen): Passes light directly through specimen; unless cell is naturally pigmented or artificially stained, image has little contrast.			**Phase-contrast:** Enhances contrast in unstained cells by amplifying variations in refractive index within specimen; especially useful for examining living, unpigmented cells.
Brightfield (stained specimen): Staining with various dyes enhances contrast, but most staining procedures require that cells be fixed (preserved).			**Differential interference contrast:** Also uses optical modifications to exaggerate differences in refractive index.
Fluorescence: Shows the locations of specific molecules in the cell. Fluorescent substances absorb ultraviolet radiation and emit visible light. The fluorescing molecules may occur naturally in the specimen but more often are made by tagging the molecules of interest with fluorescent dyes or antibodies.			**Confocal:** Uses lasers and special optics to focus illuminating beam on a single plane within the specimen. Only those regions within a narrow depth of focus are imaged. Regions above and below the selected plane of view appear black rather than blurry.

50 μm

Source: From Campbell, Reece, and Mitchell, *Biology* 5th edition (Menlo Park, CA: Benjamin Cummings, 1999), p. 104.

However, for biological samples the practical limit of resolution is usually no better than 2 nm or more, because of problems with specimen preparation and contrast. Nevertheless, the electron microscope has about 100 times more resolving power than the light microscope (see Figure 1-2). As a result, the useful magnification is also greater: up to 100,000-fold for the electron microscope, compared with about 1000- to 1500-fold for the light microscope. Figure 1-3 shows how much more structural detail can be seen when a cell is examined with an electron microscope than with a light microscope.

Electron microscopes are of two basic designs: the **transmission electron microscope** (**TEM**) and the **scanning electron microscope** (**SEM**), both of which are described in detail in the *Guide to Microscopy* supplement, as are the several specialized techniques described below. Transmission and scanning electron microscopes are similar in that each employs a beam of electrons, but they use quite different mechanisms to form the image. As the name implies, a TEM forms an image from electrons that are transmitted through the specimen. An SEM, on the other hand, scans the surface of the specimen and forms an image by detecting electrons that are deflected from the outer surface of the specimen. Scanning electron microscopy is an especially spectacular technique because of the sense of depth it gives to biological structures (Figure 1-4). Most of the electron micrographs in this book were obtained by the use of either a TEM or an SEM and are identified as such by the appropriate three-letter abbreviation at the end of the figure legend.

Because of the low penetration power of electrons, samples prepared for electron microscopy must be exceedingly thin. The instrument used for this purpose is called an *ultramicrotome*. It is equipped with a diamond knife and can cut sections as thin as 20 nm. Substantially thicker samples can also be examined by electron microscopy, but a much higher accelerating voltage is then required to increase the penetration power of the electrons adequately. Such a *high-voltage electron microscope* uses accelerating voltages up to several thousand kilovolts (kV), compared with the range of 50–100 kV common to most conventional instruments. Sections up to 1 μm thick can be studied with such a high-voltage instrument. This thickness allows organelles and other cellular structures to be examined in more depth.

Specialized Techniques of Electron Microscopy. Several specialized techniques of electron microscopy have been developed, each of which is really just a different way of preparing samples for transmission electron microscopy. For the technique of *negative staining*, for example, samples are not cut into ultrathin sections but are instead simply suspended in an electron-dense stain, allowing the intact specimen to be visualized in relief against a darkly stained

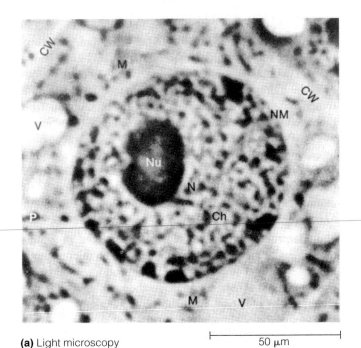

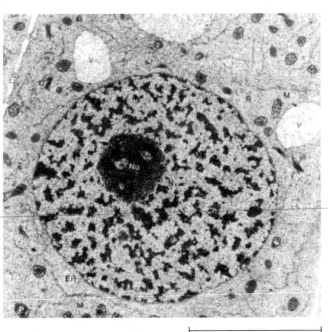

(a) Light microscopy   50 μm

(b) Electron microscopy   50 μm

Figure 1-3 Comparison of Resolving Power. Cells can be visualized by either **(a)** light microscopy or **(b)** electron microscopy, but much more detail is seen in the latter case because of the greater resolving power of the electron microscope. Both sections were cut from the same piece of onion root tissue. For light microscopy, a 1.5-μm section was cut and photographed with phase-contrast optics. For electron microscopy, a section of 0.03 μm was used. Labeled structures include the nucleus (N), nucleolus (Nu), chromatin (Ch), nuclear membranes (NM), plastids (P), mitochondria (M), endoplasmic reticulum (ER), vacuoles (V), and cell wall (CW). (750×. A magnification of 1000× is the upper limit for the light microscope but is a very low magnification for the electron microscope.)

background. This technique is obviously applicable only to very small objects such as viruses or isolated organelles, but it allows the shape and surface appearance of such objects to be examined with the objects still in intact form.

Freeze-fracturing involves a fundamentally different means of sample preparation. Instead of cutting uniform slices or examining unsectioned material, specimens are rapidly frozen—in liquid nitrogen, usually—then struck with a sharp knife edge. This causes the specimen to fracture along lines of natural weakness, which are the interiors of membranes in most cases. A thin layer of an electron-dense metal such as gold or platinum is sprayed on the sample surface to create a gold or platinum replica of the specimen, a technique called "shadowing." The replica is then examined with a TEM. Because the fracture line passes through the interiors of membranes wherever possible, the freeze-fracture replica is largely a view of the insides of membranes. Examination of freeze-fractured samples has contributed greatly to our understanding of membrane structure. Not surprisingly, we will encounter this technique in more detail when we get to Chapter 7, which is devoted to membrane structure and function (see pp. 175–176).

Electron microscopy has revolutionized our understanding of cellular architecture by making detailed ultrastructural investigations possible. Some organelles (such as nuclei or mitochondria) are large enough to be seen with a light microscope but can be studied in much greater detail with an electron microscope. In addition, electron microscopy has revealed cellular structures that are too small to be seen at all with a light microscope. These include ribosomes, membranes, microtubules and microfilaments (see Figure 1A-2 on p. 3).

The Biochemical Strand Covers the Chemistry of Biological Structure and Function

At about the time when cytologists were starting to explore cellular structure with their microscopes, other scientists were making observations that began to explain and clarify cellular function. Much of what is now called biochemistry dates from a discovery reported by the German chemist Friedrich Wöhler in 1828. Wöhler was a contemporary (as well as fellow countryman) of Schleiden and Schwann. He revolutionized our thinking about biology and chemistry by demonstrating that urea, an organic compound of biological origin, could be synthesized in the laboratory from an inorganic starting material, ammonium cyanate. Until then, it had been widely held that living organisms were a world unto themselves, not governed by the laws of chemistry and physics that apply to the nonliving world. By showing that a compound made by living organisms—a "bio-chemical"—could be synthesized in a laboratory just like any other chemical, Wöhler helped to break down the conceptual distinction between the living and nonliving worlds and to dispel the

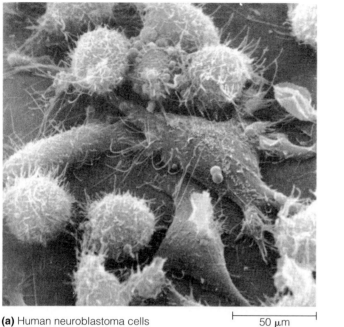
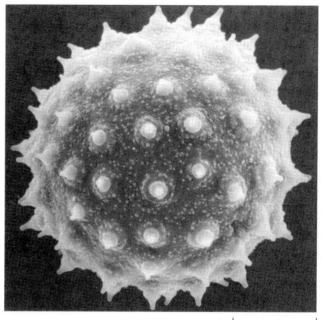

(a) Human neuroblastoma cells   50 μm

(b) Pollen grain   10 μm

Figure 1-4   **Scanning Electron Microscopy.** A scanning electron microscope was used to visualize **(a)** cultured human neuroblastoma cells and **(b)** a pollen grain.

notion that biochemical processes were somehow exempt from the laws of chemistry and physics.

Another major advance came about 40 years later, when Louis Pasteur linked the activity of living organisms to specific processes by showing that living yeast cells were needed to carry out the fermentation of sugar into alcohol. This observation was followed in 1897 by the finding of Eduard and Hans Buchner that fermentation could also take place with extracts from yeast cells—that is, the intact cells themselves were not required. Initially, such extracts were called "ferments," but gradually it became clear that the active agents in the extracts were specific biological catalysts that have since come to be called **enzymes.**

Significant progress in our understanding of cellular function came in the 1920s and 1930s as the biochemical pathways for fermentation and related cellular processes were elucidated. This was a period dominated by German biochemists such as Gustav Embden, Otto Meyerhof, Otto Warburg, and Hans Krebs. Several of these men have long since been immortalized by the pathways that bear their names. For example, the *Embden-Meyerhof pathway* for glycolysis was a major research triumph of the early 1930s. It was followed shortly by the *Krebs cycle* (also known as the TCA cycle). Both of these pathways are important because of their role in the process by which cells extract energy from foodstuffs. At about the same time, Fritz Lipmann, an American biochemist, showed that the high-energy compound *adenosine triphosphate (ATP)* is the principal energy storage compound in most cells.

An important advance in the study of biochemical reactions and pathways came as radioactive isotopes such as ^3H, ^{14}C, and ^{32}P began to be used to trace the metabolic fate of specific atoms and molecules. (As you may recall from chemistry, different atoms of a given chemical element may have the same atomic number and nearly identical properties but differ in the number of neutrons and hence in atomic weight; an *isotope* refers to the atoms with a specific number of neutrons and thus a particular atomic weight. A radioactive isotope, or *radioisotope,* is an isotope that is unstable, emitting subatomic particles [either alpha or beta particles] and, in some cases, gamma rays as it undergoes spontaneous conversion to a stable form.) Melvin Calvin and his colleagues at the University of California, Berkeley, were pioneers in this field as they traced the fate of ^{14}C-labeled carbon dioxide, $^{14}CO_2$, in illuminated algal cells that were actively photosynthesizing. Their work, carried out in the late 1940s and early 1950s, led to the elucidation of the *Calvin cycle,* as the most common pathway for photosynthetic carbon metabolism is called. The Calvin cycle was the first metabolic pathway to be elucidated using a radioisotope.

Biochemistry took another major step forward with the development of the **ultracentrifuge** as a means of separating subcellular structures and macromolecules on the basis of size, shape, and density. In many ways, the ultracentrifuge was as significant for biochemistry as the electron microscope was for cytology. In fact, both instruments were developed at about the same time, so the ability to see organelles and other subcellular structures with much greater resolution came almost simultaneously with the capacity to isolate and purify them. The ultracentrifuge was developed in Sweden by Theodor Svedberg during the period 1925–1930 and was used initially for determining the sedimentation rates of proteins.

Adaptation of centrifugation techniques for the isolation of subcellular fractions came in the 1940s and early 1950s, largely through the pioneering work of Albert Claude.

With an enhanced ability both to see subcellular structures and to isolate them, cytologists and biochemists began to realize the extent to which their respective observations on cellular structure and function could complement each other, thereby laying the foundations for modern cell biology.

The Genetic Strand Focuses on Information Flow

The third strand in the cord of cell biology is genetics. Like the other two, this strand also has important roots in the nineteenth century. In this case, the strand begins with Gregor Mendel, whose studies with the pea plants he grew in a monastery garden must surely rank among the most famous experiments in all of biology. His findings were published in 1866, laying out the principles of segregation and independent assortment of the "hereditary factors" that we know today as **genes.** These were singularly important principles, destined to provide the foundation for what would eventually be known as Mendelian genetics. But Mendel was clearly a man ahead of his time. His work went almost unnoticed when it was first published and was not fully appreciated until its rediscovery nearly 35 years later.

As a prelude to that rediscovery, the role of the nucleus in the genetic continuity of cells came to be appreciated in the decade following Mendel's work. In 1880, Walther Flemming identified **chromosomes,** threadlike bodies seen in dividing cells. Flemming called the division process *mitosis,* from the Greek word for thread. Chromosome number soon came to be recognized as a distinctive characteristic of a species and was shown to remain constant from generation to generation. That the chromosomes themselves might be the actual bearers of genetic information was suggested by Wilhelm Roux as early as 1883 and was expressed more formally by August Weissman shortly thereafter.

With the role of the nucleus and chromosomes established and appreciated, the stage was set for the rediscovery of Mendel's initial observations. This came in 1900, when his studies were cited almost simultaneously by three plant geneticists working independently: Carl Correns in Germany, Ernst von Tschermak in Austria, and Hugo de Vries in Holland. Within three years, the **chromosome theory of heredity** was formulated by Walter Sutton, who was the first to link the chromosomal "threads" of Flemming with the "hereditary factors" of Mendel.

Sutton's theory proposed that the hereditary factors responsible for Mendelian inheritance are located on the chromosomes within the nucleus. This hypothesis received its strongest confirmation from the work of Thomas Hunt Morgan and his students at Columbia University during the first two decades of the twentieth century. They chose as their experimental species *Drosophila melanogaster,* the common fruit fly. By identifying a variety of morphological mutants of *Drosophila,* Morgan and his co-workers were able to link specific traits to specific chromosomes.

Meanwhile, the foundation for our understanding of the chemical basis of inheritance was also slowly being laid. An important milestone was the discovery of DNA by Johann Friedrich Miescher in 1869. Using such unlikely sources as salmon sperm and human pus from surgical bandages, Miescher isolated and described what he called "nuclein." But, like Mendel, Miescher was ahead of his time. It was about 75 years before the role of his nuclein as the genetic information of the cell came to be fully appreciated.

As early as 1914, DNA was implicated as an important component of chromosomes by the staining technique of Robert Feulgen, a method that is still in use today. But little consideration was given to the possibility that DNA could be the bearer of genetic information. In fact, that was considered quite unlikely in light of the apparently uninteresting structure of the monomer constituents of DNA (called nucleotides) that were known by 1930. Until the middle of the twentieth century, it was widely held that genes were made up of proteins, since these were the only nuclear components that seemed to account for the obvious diversity of genes.

A landmark experiment that clearly pointed to DNA as the genetic material was reported in 1944 by Oswald Avery, Colin MacLeod, and Maclyn McCarty. Their work focused on the phenomenon of genetic transformation in bacteria, to be discussed in Chapter 16. Their evidence was compelling, but the scientific community remained largely unconvinced of the conclusion. Just eight years later, however, a considerably more favorable reception was accorded the report of Alfred Hershey and Martha Chase that DNA, and not protein, enters a bacterial cell when it is infected by a bacterial virus.

Meanwhile, George Beadle and Edward Tatum, working in the 1940s with the bread mold *Neurospora crassa,* formulated the "one gene—one enzyme" concept, asserting that the function of a gene is to control the production of a single, specific protein. Shortly thereafter, in 1953, James Watson and Francis Crick proposed their now-famous **double helix model** for DNA structure, with features that immediately suggested how replication and genetic mutations could occur. Thereafter, the features of DNA function fell rapidly into place, establishing that DNA specifies the order of monomers (amino acids) and hence the properties of proteins and that several different kinds of RNA (ribonucleic acid) molecules serve as intermediates in protein synthesis.

The 1960s brought especially significant developments, including the discovery of the enzymes that synthesize DNA and RNA (DNA polymerases and RNA polymerases, respectively) and the "cracking" of the genetic code, which specifies the relationship between the order of nucleotides in a DNA (or RNA) molecule and the order of amino acids in a protein. At about the same time, Jacques Monod and François Jacob deduced the mechanism responsible for the regulation of bacterial gene expression, thereby launching an era of molecular genetics that continues to revolutionize biology. In the process, the historical strand of genetics became inti-

Further Insights   BIOLOGY, "FACTS," AND THE SCIENTIFIC METHOD

If asked what they expect to get out of a science textbook, most readers would probably reply that they intend to learn the facts relevant to the particular scientific area the book is about—cell biology, in the case of the text you are reading right now. If pressed to explain what a fact is, most people would probably reply that a fact is "something that we know to be true." That sense of the word agrees with the dictionary, since one of the definitions of *fact* is "a piece of information presented as having objective reality."

To a scientist, however, a fact is a much more tenuous piece of information than such a definition might imply. The "facts" of science are really just attempts to state our current understanding of the natural world around us, based on observations that we make and experiments that we do. As such, a given "fact" is only as sound as the observations or experiments on which it is based and can be modified or superseded at any time by a better understanding based on more careful observations or more discriminating experiments. As one scientist so aptly put it, truth to a researcher "is not a citadel of certainty to be defended against error; it is a shady spot where one eats lunch before tramping on" (White, 1968, p. 3).

Cell biology is rich with examples of "facts" that were once widely held but have since been superseded as cell biologists have "tramped on" to a better understanding of the phenomena those "facts" attempted to explain. As recently as the early nineteenth century, for example, it was widely held (i.e., regarded as fact) that living matter consisted of substances quite different from those in nonliving matter. According to this view, called *vitalism,* the chemical reactions that occurred within living matter did not follow the known laws of chemistry and physics but were instead directed by a "vital force." Then came Friedrich

Wöhler's demonstration (in 1828) that the biological compound urea could be synthesized in the laboratory from an inorganic compound, thereby undermining one of the "facts" of vitalism. The other "fact" was refuted by the work of Eduard and Hans Buchner, who showed (in 1897) that nonliving extracts from yeast cells could ferment sugar into ethanol. Thus, a view held as "fact" by generations of scientists was eventually discredited and replaced by the new "fact" that the components and reactions of living matter are not a world unto themselves, but follow all the laws of chemistry and physics.

For a more contemporary example, consider what we know about the energy needed to support life. Until recently, it was regarded as a fact that the sun is the ultimate source of all energy in the biosphere, such that every organism either uses solar energy directly (i.e., green plants, algae, and certain bacteria) or is a part of a food chain that is sustained by such photosynthetic organisms. Then came the discovery of *deep-sea thermal vents* and the thriving communities of organisms that live around them, none of which depends on solar energy. Instead, these organisms depend on the bond energy of hydrogen sulfide (H_2S), which is extracted by bacteria that live around the thermal vents and used to synthesize organic compounds from carbon dioxide. These bacteria form the basis of food chains that include zooplankton (microscopic animals), worms, and other residents of the thermal vent environment.

Thus, the "facts" presented in biology textbooks such as this one are nothing more than our best current attempts to describe and explain the workings of the biological world around us. They are subject to change whenever we become aware of new or better information.

(continued)

---

mately entwined with those of cytology and biochemistry, and the discipline of cell biology as we know it today came into being.

"Facts" and the Scientific Method

To become familiar with an area of science such as cell biology means, at least in part, to learn the *facts* about that subject. Even in this short introductory chapter, we have already encountered a number of facts about cell biology. When we say, for example, that "all organisms consist of one or more cells" or that "DNA is the bearer of genetic information," we recognize these statements as facts of cell biology. But we also recognize that the first of these statements was initially regarded as part of a theory and the second statement actually replaced an earlier misconception that genes were made of proteins.

Clearly, then, a scientific "fact" is a much more tenuous piece of information than our everyday sense of the word might imply. To a scientist, a "fact" is simply an attempt to state our best current understanding of a specific phenomenon and is only valid until it is revised or

replaced by a better understanding. Box 1B explores the meaning of "facts" in biology and the **scientific method** by which new and better information becomes available.

As we consider the scientific method, we need to recognize several important terms that scientists use to indicate the degree of certainty with which a specific explanation or concept is regarded as true. Three terms are especially significant: *hypothesis, theory,* and *law.*

Of the three, **hypothesis** is the most tentative. A hypothesis is simply a statement or explanation that is consistent with most of the observational and experimental evidence to date. Suppose, for example, that you have experienced heartburn three times in the last month and that you had in each case eaten a pepperoni pizza shortly before experiencing the heartburn. A reasonable hypothesis might be that the heartburn is somehow linked to the consumption of pepperoni pizza. Often, a hypothesis takes the form of a *model* that appears to provide a reasonable explanation of the phenomenon in question.

To be useful to scientists, a hypothesis must be *testable*—that is, it must be possible to design a controlled experiment that will either confirm or discredit the

How does new and better information become available? Scientists usually follow a systematic approach to new information called the *scientific method.* As Figure 1B-1 indicates, the scientific method begins as a researcher *makes observations,* either in the field or in a research laboratory. Based on these observations and on knowledge gained in prior studies, the scientist *formulates a testable hypothesis,* a tentative explanation or model consistent with the observations and with prior knowledge that can be tested experimentally. Next, the investigator *designs a controlled experiment* to test the hypothesis by varying specific conditions while holding everything else as constant as possible. The scientist then *collects the data, interprets the results,* and *draws reasonable conclusions,* which obviously must be consistent not only with the results of this particular experiment but with prior knowledge as well.

To a practicing scientist, the scientific method is more a way of thinking than a set of procedures to be followed. Most likely, this is the way our ancestors explained and interpreted natural phenomena long before scientists were trained at universities—and long before students read essays about the scientific method!

For a good example of the scientific method in action, consider a report that points to a protein in cow's milk as a trigger of insulin-dependent diabetes. The research was conducted by Dr. Jutta Karjalainen and his colleagues at the Universities of Turku and Helsinki in Finland and at the Hospital for Sick Children in Toronto. As you read the next several paragraphs, refer to Figure 1B-1 and see if you can identify each of the elements of the scientific method in the work of Dr. Karjalainen and his colleagues.

In insulin-dependent diabetes, the body's immune system attacks and destroys the insulin-producing *beta cells* of the pancreas. The body is then unable to process sugars into energy because insulin is required for normal carbohydrate utilization, including the uptake of blood glucose into cells. Cow's milk was initially implicated as the possible culprit by experiments in which diabetes-prone rats did not get diabetes when fed a diet containing no cow's milk. Moreover, children who were breast-fed instead of receiving cow's milk showed a significantly lower risk of diabetes.

Armed with these facts and with preliminary data that suggested the milk protein *bovine serum albumin (BSA)* as the specific trigger molecule, the investigators formulated their hypothesis. They postulated that children with insulin-dependent diabetes produce antibodies against a specific segment of the BSA molecule and that these antibodies also recognize a specific protein, called *p69,* on the surface of the pancreatic beta cells, leading to the destruction of these cells. To test their hypothesis, the researchers used immunological techniques to look for antibodies against BSA in the blood of diabetic and nondiabetic children. They found that the diabetic children had a much higher level of anti-BSA antibodies in their blood than the nondiabetic children did. Furthermore, most of those antibodies were specific for the short segment of the BSA molecule that is almost identical to a portion of the p69 cell-surface protein.

Based on these results, the investigators concluded that diabetic children have antibodies capable of recognizing a protein on the surface of their own pancreatic beta cells, and suggested that these antibodies may be the connection between cow's milk and insulin-dependent diabetes. This link has not yet been proved, but it appears likely that eliminating cow's milk from

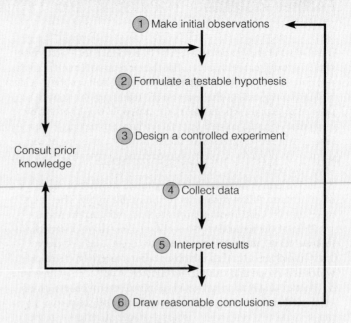

Figure 1B-1   The Scientific Method.

the diet of infants might dramatically reduce the incidence of insulin-dependent diabetes.

When illustrated by experiments such as this diabetes study, the scientific method sounds very neat and orderly. Not all scientific discoveries are made in this way, however. Many important advances in biology have come about more by accident than by plan. Alexander Fleming's discovery of penicillin in 1928 is a classic example. Fleming, a Scottish physician and bacteriologist, accidentally left a culture dish of *Staphylococcus* bacteria uncovered, such that it was inadvertently exposed to contamination by other microorganisms. Fleming was about to discard the contaminated culture when he happened to notice some clear patches where the bacteria were not growing. Reasoning that the bacterial growth may have been inhibited by some contaminant in the air and recognizing how important an inhibitor of bacterial growth might be, Fleming kept the culture dish and began attempts to isolate and characterize the substance. The actual identification of penicillin and the demonstration that it was the product of a mold was left to others, but Fleming is credited with the initial discovery.

Boxes in subsequent chapters will acquaint you with further examples of apparently accidental discoveries. Regardless of how accidental such discoveries may appear, however, it is almost always true that "chance favors the prepared mind." Behind the apparent "chance" of each such discovery is the "prepared mind" that has been trained to observe carefully and to think astutely.

As you proceed through this text, be on the outlook for applications of the scientific method. You will find that regardless of the approach, the conclusions from each experiment add to our knowledge of how biological systems work and usually lead to more questions as well, continuing the cycle of scientific inquiry. And that's good news if you aspire to a career in research, because it's your best insurance that there will still be questions to answer when you are ready to begin.

hypothesis. Based on initial observations and prior knowledge (from the work of other investigators, most likely), the scientist formulates a testable hypothesis and then designs a controlled experiment to determine whether or not the hypothesis will be supported by data or observations (see Figure 1B-1).

When a hypothesis or model has been tested critically under many different conditions—usually by many different investigators using a variety of approaches—and is consistently supported by the evidence, it gradually acquires the status of a **theory.** By the time an explanation or model comes to be regarded as a theory, it is generally and widely accepted by most scientists in the field. The *cell theory* described earlier in this chapter is an excellent example. There is little or no dissent or disagreement among biologists concerning its three tenets. Two more recent explanations that have acquired the status of theory are the *chemiosmotic model* that explains how mitochondrial ATP production is driven by an electrochemical proton gradient across the inner mitochondrial membrane (discussed in Chapter 14), and the *fluid*

mosaic model of membrane structure that we will encounter in Chapter 7.

When a theory has been so thoroughly tested and confirmed over a long period of time by a large number of investigators that virtually no doubt remains whatever, it may eventually come to be regarded as a **law.** The *law of gravity* comes readily to mind, as do the several *laws of thermodynamics* that we will encounter in Chapter 5. You may also be familiar with *Fick's law of diffusion*, the *ideal gas laws*, and other concepts from physics and chemistry that are generally regarded as laws. Some of the most notable biological examples are from genetics—Mendel's *laws of heredity* and the *Hardy-Weinberg law*, for instance. In general, however, biologists are quite conservative with the term. Even after more than 150 years, the cell theory is still regarded as just that—a theory. Perhaps our reluctance to label explanations of biological phenomena as laws is a reflection of the great diversity of life forms and the consequent difficulty we have convincing ourselves that we will never find organisms or cells that are exceptions to even our most well-documented theories.

Perspective

The biological world is a world of cells. All living organisms are made up of one or more cells, each of which came from a preexisting cell. Although the importance of cells in biological organization has been appreciated for about 150 years, the discipline of cell biology as we know it today is of much more recent origin. Modern cell biology has come about by the interweaving of three historically distinct strands—cytology, biochemistry, and genetics—which in their early development probably did not seem at all related. But the contemporary cell biologist must understand all three strands, because they complement one another in the quest to learn what cells are and how they function.

Key Terms for Self-Testing

cell (p. 1)

The Cell Theory: A Brief History
cell theory (p. 3)

The Emergence of Modern Cell Biology
cytology (p. 3)
biochemistry (p. 5)
genetics (p. 5)
light microscope (p. 5)
organelle (p. 5)
limit of resolution (p. 5)

resolving power (p. 5)
electron microscope (p. 6)
transmission electron microscope (TEM) (p. 7)
scanning electron microscope (SEM) (p. 7)
enzyme (p. 9)
ultracentrifuge (p. 9)
gene (p. 10)
chromosome (p. 10)
chromosome theory of heredity (p. 10)
double helix model (p. 10)

"Facts" and the Scientific Method
scientific method (p. 11)
hypothesis (p. 11)
theory (p. 13)
law (p. 13)

Box 1A: *Units of Measurement in Cell Biology*
micrometer (p. 2)
nanometer (p. 3)

Problem Set

More challenging problems are marked with a •.

1-1.   The Historical Strands of Cell Biology. For each of the following events, indicate whether it belongs mainly to the cytological (C), biochemical (B), or genetic (G) strand in the historical development of cell biology.

(a) Kölliker describes "sarcosomes" (now called mitochondria) in muscle cells (1857).

(b) Hoppe-Seyler isolates the protein hemoglobin in crystalline form (1864).

(c) Haeckel postulates that the nucleus is responsible for heredity (1868).

(d) Ostwald proves that enzymes are catalysts (1893).

(e) Muller discovers that X rays induce mutations (1927).

(f) Davson and Danielli postulate a model for the structure of cell membranes (1935).

(g) Beadle and Tatum formulate the one gene–one enzyme hypothesis (1940).

(h) Claude isolates the first mitochondrial fractions from rat liver (1940).

(i) Lipmann postulates the central importance of ATP in cellular energy transactions (1940).

(j) Avery, MacLeod, and McCarty demonstrate that bacterial transformation is attributable to DNA, not protein (1944).

(k) Palade, Porter, and Sjøstrand each develop techniques for fixing and sectioning biological tissue for electron microscopy (1952–1953).

(l) Lehninger demonstrates that oxidative phosphorylation depends for its immediate energy source on the transport of electrons in the mitochondrion (1957).

1-2.   Cell Sizes. To appreciate the differences in cell size illustrated in Figure 1A-1 on p. 2, consider the following specific examples. *Escherichia coli*, a typical bacterial cell, is cylindrical in shape, with a diameter of about 1 μm and a length of about 2 μm. As a typical animal cell, consider a human liver cell, which is roughly spherical in shape and has a diameter of about 20 μm. And for a typical plant cell, consider the columnar *palisade cells* located just beneath the upper surface of many plant leaves. These cells are cylindrical in shape, with a diameter of about 20 μm and a length of about 35 μm.

(a) Calculate the approximate volume of each of these three cell types in cubic micrometers. (Recall that $V = \pi r^2 h$ for a cylinder and that $V = 4\pi r^3/3$ for a sphere.)

(b) Approximately how many bacterial cells would fit in the internal volume of a human liver cell?

(c) Approximately how many liver cells would fit inside a palisade cell?

1-3.   Sizing Things Up. To appreciate the sizes of the subcellular structures shown in Figure 1A-2 on p. 3, consider the following calculations.

(a) All cells and many subcellular structures are surrounded by a membrane. Assuming a typical membrane to be about 8 nm wide, how many such membranes would have to be aligned side by side before the structure could be seen with the light microscope? How many with the electron microscope?

(b) Ribosomes are the structures in cells on which the process of protein synthesis takes place. A human ribosome is a roughly spherical structure with a diameter of about 30 nm. How many ribosomes would fit in the internal volume of the human liver cell described in Problem 1-2 if the entire volume of the cell were filled with ribosomes?

(c) The genetic material of the *Escherichia coli* cell described in Problem 1-2 consists of a DNA molecule with a diameter of 2 nm and a total length of 1.36 mm. (The molecule is actually circular, with 1.36 mm as its circumference.) To be accommodated in a cell that is only a few micrometers long, this large DNA molecule is tightly coiled and folded into a *nucleoid* that occupies a small proportion of the internal volume of the cell. Calculate the smallest possible volume into which the DNA molecule could fit, and express that as a percentage of the internal volume of the bacterial cell that you calculated in Problem 1-2a.

1-4.   Limits of Resolution Then and Now. Based on what you learned in this chapter about the limit of resolution of a light microscope, answer each of the following questions. Assume that the unaided human eye has a limit of resolution of about 0.25 mm and that a modern light microscope has a useful magnification of about 1000-fold.

(a) Define *limit of resolution* in your own words. What was the limit of resolution of Hooke's microscope? What about van Leeuwenhoek's microscope?

(b) What are the approximate dimensions of the smallest structure that Hooke would have been able to observe with his microscope? Would he have been able to see any of the structures shown in Figure 1A-1? If so, which ones? And if not, why not?

(c) What are the approximate dimensions of the smallest structure that van Leeuwenhoek would have been able to observe with his microscope? Would he have been able to see any of the structures shown in Figure 1A-1? If so, which ones? And if not, why not?

(d) What are the approximate dimensions of the smallest structure that a contemporary cell biologist should be able to observe with a modern light microscope?

(e) Of the seven intracellular structures in parts a–g of Problem 4-2 (p. 102), which, if any, would both Hooke and van Leeuwenhoek have been able to see with their respective microscopes? Which, if any, would van Leeuwenhoek have been able to see that Hooke could not? Explain your reasoning. Which, if any, would a contemporary cell biologist be able to see that neither Hooke nor van Leeuwenhoek could see?

1-5.   A Question of Resolution. The light microscope was the earliest tool of the cytologist and still plays a vital role in the study of cells. A drawback of the light microscope is its *limit of resolution*—how close together two points can be and still be distinguished from each other—and hence, its *useful magnification*. (To answer the following questions, you may find it useful to consult the first several pages of the *Guide to Microscopy*.)

(a) What is the major factor that determines the limit of resolution of a light microscope?

(b) Why does the use of ultraviolet light rather than visible (white) light increase the resolution of a light microscope by a factor of about two?

(c) What does it mean to say that the useful magnification of a light microscope is about 1000-fold?

(d) An image obtained with a light microscope can easily be magnified much more than 1000-fold by photographic enlargement. In what sense is any enlargement beyond 1000-fold called "empty magnification"?

1-6.   The "Facts" of Life. Each of the following statements was once regarded as a biological fact but is now understood to be untrue. In each case, indicate why the statement was once thought to be true and why it is no longer considered a fact.

(a) Plant and animal tissues are constructed quite differently because animal tissues do not have conspicuous boundaries that divide them into cells.

(b) Living organisms are not governed by the laws of chemistry and physics as is nonliving matter, but are subject to a "vital force" that is responsible for the formation of organic compounds.

(c) Genes most likely consist of proteins because the only other likely candidate, DNA, is a relatively uninteresting molecule consisting of only four kinds of monomers (nucleotides) arranged in a relatively invariant repeating tetranucleotide sequence.

(d) The fermentation of sugar to alcohol can take place only if living yeast cells are present.

• **1-7. More "Facts" of Life.** Each of the following statements was regarded as a biological fact until quite recently but is now either rejected or qualified to at least some extent. In each case, speculate on why the statement was once thought to be true, and then try to determine what evidence might have made it necessary to reject or at least to qualify the statement. (Note: This question requires a more intrepid sleuth than Problem 1-6 does, but chapter references are provided to aid in your sleuthing.)

(a) A biological membrane can be thought of as a protein-lipid "sandwich" consisting of an exclusively phospholipid interior coated on both sides with thin layers of protein. (Chapter 7)

(b) The enzymes required to catalyze the conversion of sugar into a compound called pyruvate invariably occur in the cytoplasm of the cell, rather than being compartmentalized in membrane-enclosed structures. (Chapter 13)

(c) The mechanism by which the oxidation of organic molecules such as sugars leads to the generation of ATP involves a high-energy phosphorylated molecule as an intermediate. (Chapter 14)

(d) When carbon dioxide from the air is "fixed" (covalently linked) into organic form by photosynthetic organisms such as green plants, the first form in which the carbon atom of the CO_2 molecule appears is always the three-carbon compound 3-phosphoglycerate. (Chapter 15)

(e) DNA always exists as a duplex of two strands wound together into a right-handed helix. (Chapter 16)

(f) The genetic code that specifies how the information present in the DNA molecule is used to make proteins is uni-versal in the sense that all organisms use the same code. (Chapter 19)

• **1-8. Cell Biology in 1875.** Friedrich Miescher (1844–1895), Gregor Mendel (1822–1884), Louis Pasteur (1822–1895), and Rudolf Virchow (1821–1902) were European scientists (from Switzerland, Austria, France, and Germany, respectively) whose most important discoveries were made in the 20-year period from 1855 to 1875. Assume that the four men met at a scientific conference in 1875 to discuss their respective contributions to biology.

(a) What might the four scientists have found they had in common?

(b) What do you think Pasteur might have found most intriguing or relevant about Virchow's work? Explain your answer.

(c) What do you think Virchow might have found most intriguing or relevant about Mendel's work? Explain your answer.

(d) In whose work do you think Miescher might have been the most interested? Explain your answer.

• **1-9. Facts and Truth.** An especially apt characterization of scientific fact was made by Lynn White, Jr. As quoted in Box 1B, White wrote that, to a scientist, truth "is not a citadel of certainty to be defended against error; it is a shady spot where one eats lunch before tramping on." Explain what White meant by this statement. In what sense does it aptly describe scientific facts? How does his statement relate to the scientific method?

• **1-10. Pizza, Heartburn, and the Scientific Method.** Although the scientific method may sound rather foreign when described in formal terms or when diagrammed as in Figure 1B-1 on p. 12, it is in reality not very different from the way most of us go about answering questions or solving problems. You probably use the scientific method frequently without even realizing it. Suppose, for example, that you have been experiencing heartburn quite often recently. By keeping track of your eating habits for a few weeks, you realize that the heartburn is most likely to occur on nights after you have eaten pizza for supper, especially if you have had your favorite pizza, with pepperoni, anchovies, and onions. You wonder if the heartburn is caused by eating pizza and, if so, which of the ingredients might be the culprit.

(a) Describe how you might go about determining whether the heartburn is due to the pizza and, if so, to which of the ingredients.

(b) Now compare your approach with the scientific method as shown in Figure 1B-1. How scientific was your method?

Suggested Reading

References of historical importance are marked with a •.

The Emergence of Modern Cell Biology
• Bonner, J. T. *Sixty Years of Biology.* New York: Princeton University Press, 1996.
• Bracegirdle, B. Microscopy and comprehension: The development of understanding of the nature of the cell. *Trends Biochem. Sci.* 14 (1989): 464.
• Bradbury, S. *The Evolution of the Microscope.* New York: Pergamon, 1967.
• Calvin, M. The path of carbon in photosynthesis. *Science* 135 (1962): 879.

• Claude, A. The coming of age of the cell. *Science* 189 (1975): 433.
• de Duve, C. Exploring cells with a centrifuge. *Science* 189 (1975): 186.
• de Duve, C., and H. Beaufay. A short history of tissue fractionation. *J. Cell Biol.* 91 (1981): 293s.
• Fruton, J. S. The emergence of biochemistry. *Science* 192 (1976): 327.
• Gall, J. G., K. R. Porter, and P. Siekevitz, eds. Discovery in cell biology. *J. Cell Biol.* 91, part 3 (1981).
• Judson, H. F. *The Eighth Day of Creation: Makers of the Revolution in Biology.* New York: Simon & Schuster, 1979.
• Kornberg, A. Centenary of the birth of modern biochemistry. *Trends Biochem. Sci.* 22 (1997): 282.

- Krebs, H. A. The history of the tricarboxylic acid cycle. *Perspect. Biol. Med.* 14 (1970): 154.
- Lipmann, F. Metabolic generation and utilization of phosphate bond energy. *Adv. Enzymol.* 18 (1941): 99.
- Mirsky, A. E. The discovery of DNA. *Sci. Amer.* 218 (June 1968): 78.
- Palade, G. E. Albert Claude and the beginning of biological electron microscopy. *J. Cell Biol.* 50 (1971): 5D.
- Peters, J. A. *Classic Papers in Genetics.* Englewood Cliffs, NJ: Prentice-Hall, 1959.
- Quastel, H. J. The development of biochemistry in the 20th century. *Canad. J. Cell Biol.* 62 (1984): 1103.
- Rasmussen, N. *Picture Control: The Electron Microscope and the Transformation of Biology in America.* Stanford, CA: Stanford University Press, 1997.
- Ruestow, E. G. *The Microscope in the Dutch Republic: The Shaping of Discovery.* Cambridge, England: Cambridge University Press, 1996.
- Schlenk, F. The ancestry, birth, and adolescence of ATP. *Trends Biochem. Sci.* 12 (1987): 367.
- Smith, M. P. Exploring molecular biology: An older surgeon looks at a new universe. *Arch. Surg.* 130 (1995): 811.
- Stent, G. C. That was the molecular biology that was. *Science* 160 (1968): 390.
- Stent, G. C., and R. Calendar. *Molecular Genetics: An Introductory Narrative,* 2d ed. New York: W. H. Freeman, 1978.
- Watson, J. D. *The Double Helix.* New York: Atheneum, 1968.

Methods in Modern Cell Biology

Antia, M. A microscope with an eye for detail. *Science* 285 (1999): 311.
Carlson, S. A kitchen centrifuge. *Sci. Amer.* 278 (January 1998): 102.

Flegler, S. L., J. W. Heckman, and K. L. Klomparens. *Scanning and Transmission Microscopy: An Introduction.* New York: W. H. Freeman, 1993.
Ford, T. C., and J. M. Graham. *An Introduction to Centrifugation.* Oxford, England: Bios Scientific.
Gupta, P. D. and H. Yamamoto, eds. *Electron Microscopy in Medicine and Biology.* Enfield, NH: Science Publishers, 2000.
Lichtman, J. W. Confocal microscopy. *Sci. Amer.* 271 (August 1994): 40.
Minsky, M. Memoir on inventing the confocal scanning microscope. *Scanning* 10 (1988): 128.
Olson, A. J., and D. S. Goodsell. Visualizing biological molecules. *Sci. Amer.* 267 (November 1992): 76.
Slayter, E. M., and H. S. Slayter. *Light and Electron Microscopy,* 3d ed. New York: Cambridge University Press, 1992.
Slayter, R. J., ed. *Radioisotopes in Biology: A Practical Approach.* New York: Oxford University Press.

The Scientific Method

Braben, D. *To Be a Scientist: The Spirit of Adventure in Science and Technology.* New York: Oxford University Press, 1994.
Karjalainen, J., J. M. Martin, M. Knip, et al. A bovine albumin peptide as a possible trigger of insulin-dependent diabetes mellitus. *New Engl. J. Med.* 327 (1992): 302.
Moore, J. A. *Science As a Way of Knowing.* Cambridge, MA: Harvard University Press, 1993.
- White, L., Jr. *Machina ex Deo: Essays in the Dynamism of Western Culture.* Cambridge, MA: MIT Press, 1968.

2

The Chemistry of the Cell

Students just beginning in cell biology are sometimes surprised—occasionally even dismayed—to find that almost all courses and textbooks dealing with cell biology involve a substantial amount of chemistry. Yet biology in general and cell biology in particular depend heavily on both chemistry and physics. After all, cells and organisms follow all the laws of the physical universe, so biology is really just the study of chemistry in systems that happen to be alive. In fact, everything cells are and do has a molecular and chemical basis. Therefore, we can truly understand and appreciate cellular structure and function only when we can describe that structure in molecular terms and express that function in terms of chemical reactions and events.

Trying to appreciate cellular biology without a knowledge of chemistry would be like trying to appreciate a translation of Goethe without a knowledge of German. Most of the meaning would probably get through, but much of the beauty and depth of appreciation would be lost in the translation. For this reason, we will consider the chemical background necessary for the cell biologist. Specifically, this chapter will focus on several principles that underlie much of cellular biology, preparing, in turn, for the next chapter, which focuses on the major classes of chemical constituents in cells.

The main points of this chapter can conveniently be structured around five principles:

1. *The importance of carbon.* The chemistry of cells is essentially the chemistry of carbon-containing compounds, because the carbon atom has several unique properties that make it especially suitable as the backbone of biologically important molecules.

2. *The importance of water.* The chemistry of cells is also the chemistry of water-soluble compounds, because the water molecule has several unique properties that make it especially suitable as the universal solvent of living systems.

3. *The importance of selectively permeable membranes.* Given that most biologically important molecules are water-soluble, membranes that do not dissolve in water and are differentially permeable to specific solutes are very important both in defining cellular spaces and compartments and in controlling the movements of molecules and ions into and out of such spaces and compartments.

4. *The importance of synthesis by polymerization of small molecules.* Most biologically important molecules are either small, water-soluble organic molecules that can be transported across membranes or large macromolecules that cannot in general be so transported. Biological macromolecules are polymers formed by the linking together of many similar or identical small molecules. The synthesis of macromolecules by polymerization of monomeric subunits is an important principle of cellular chemistry.

5. *The importance of self-assembly.* Proteins and other biological macromolecules made of repeating monomeric subunits are often capable of self-assembly into higher levels of structural organization. Self-assembly is possible because the information needed to specify the spatial configuration of the molecule is inherent in the linear array of monomers present in the polymer. Self-assembly is qualified, however, by the need in many cases for proteins called "molecular chaperones" that assist in the assembly process by inhibiting incorrect interactions that would lead to inactive structures.

Given these five principles, we can appreciate the main topics in cellular chemistry with which we need to be familiar before venturing further into our exploration of what it means to be a cell.

The Importance of Carbon

To study cellular molecules really means to study carbon-containing compounds. Almost without exception, molecules of importance to the cell biologist have a backbone, or skeleton, of carbon atoms linked together covalently. Actually, the study of carbon-containing compounds is the domain of **organic chemistry.** In its early days, organic chemistry was almost synonymous with biological chemistry because most of the carbon-containing compounds that chemists first investigated were obtained from biological sources (hence the word *organic*, acknowledging the organismal origins of the compounds). The terms have long since gone their separate ways, however, because organic chemists have now synthesized a bewildering variety of carbon-containing compounds that do not occur naturally (that is, in the biological world). Organic chemistry therefore includes all classes of carbon-containing compounds, whereas **biological chemistry** (*biochemistry* for short) deals specifically with the chemistry of living systems and is, as we have already seen, one of the several historical strands that form an integral part of modern cell biology.

The **carbon atom** (C) is the most important atom in biological molecules. The diversity and stability of carbon-containing compounds are due to specific properties of the carbon atom and especially to the nature of the interactions of carbon atoms with one another as well as with a limited number of other elements found in molecules of biological importance.

The single most fundamental property of the carbon atom is its **valence** of four, which means that the outermost electron orbital of the atom lacks four of the eight electrons needed to fill it completely (Figure 2-1a). Since a complete outer orbital is required for the most stable chemical state of an atom, carbon atoms tend to associate with one another or with other electron-deficient atoms, allowing adjacent atoms to share a pair of electrons. For each such pair, one electron comes from each of the atoms. Atoms that share each other's electrons in this way are said to be joined together by a **covalent bond.** Carbon atoms are most likely to form covalent bonds with one another and with atoms of oxygen (O), hydrogen (H), nitrogen (N), and sulfur (S).

The electronic configurations of several of these atoms are shown in Figure 2-1a. Notice that, in each case, one or more electrons are required to complete the outer orbital. The number of "missing" electrons corresponds in each case to the valence of the atom, which indicates, in turn, the number of covalent bonds the atom can form. Carbon, oxygen, hydrogen, and nitrogen are the lightest elements that form covalent bonds by sharing electron pairs. This lightness, or low atomic weight, makes the resulting compounds especially stable, because the strength of a covalent bond is inversely proportional to the atomic weights of the elements involved in the bond.

Because four electrons are required to fill the outer orbital of carbon, stable organic compounds have four

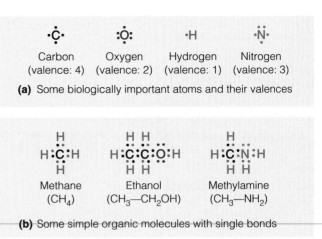

Figure 2-1   Electron Configurations of Some Biologically Important Atoms and Molecules.   Electronic configurations are shown for **(a)** atoms of carbon, oxygen, hydrogen, and nitrogen and for simple organic molecules with **(b)** single bonds, **(c)** double bonds, and **(d)** triple bonds. Only electrons in the outermost electron orbital are shown. In each case, the two electrons positioned between adjacent atoms represent a shared electron pair, with one electron provided by each of the two atoms. Electrons are color-coded; those from carbon, oxygen, hydrogen, and nitrogen are black, pink, blue, and gray, respectively. (All electrons are equivalent, of course; the color coding is simply to illustrate which electrons are contributed by each atom.)

covalent bonds for every carbon atom. Methane, ethanol, and methylamine are simple examples of such compounds, containing only **single bonds** between atoms (Figure 2-1b). Sometimes, two or even three pairs of electrons can be shared by two atoms, giving rise to **double bonds** or **triple bonds.** Ethylene and carbon dioxide are examples of double-bonded compounds (Figure 2-1c). Triple bonds are found in molecular nitrogen (N_2) and hydrogen cyanide (Figure 2-1d). Thus, both its valence and its low atomic weight confer on carbon unique properties that account for the diversity and stability of carbon-containing compounds and give carbon a preeminent role in biological molecules.

Carbon-Containing Molecules Are Stable

As already implied, the stability of organic molecules is a property of the favorable electronic configuration of each carbon atom in the molecule. This stability is expressed in terms of **bond energy**—the amount of energy required to break 1 mole (about 6×10^{23}) of such bonds. (The term *bond energy* is a frequent source of confusion. Be careful not to think of it as energy that is somehow "stored" in the bond but rather as the amount of energy that is needed to *break* the bond.) Bond energies are usually expressed in *calories per mole (cal/mol)*, where a **calorie** is the amount of energy needed to raise the temperature of one gram of water one degree centigrade.[1]

It takes a large amount of energy to break a covalent bond. For example, the carbon-carbon (C—C) bond has a bond energy of 83 kilocalories per mole (kcal/mol). The bond energies for carbon-nitrogen (C—N), carbon-oxygen (C—O), and carbon-hydrogen (C—H) bonds are all in the same range: 70, 84, and 99 kcal/mol, respectively. Even more energy is required to break a carbon-carbon double bond (C=C; 146 kcal/mol) or a carbon-carbon triple bond (C≡C; 212 kcal/mol), so these compounds are even more stable.

We can appreciate the significance of these bond energies by comparing them with other relevant energy values, as shown in Figure 2-2. Most noncovalent bonds in biologically important molecules have energies of only a few kilocalories per mole, and the energy of thermal vibration is even lower—about 0.6 kcal/mol. Covalent bonds are much higher in energy than noncovalent bonds and therefore very stable.

The fitness of the carbon-carbon bond for biological chemistry on Earth is especially clear when its energy is compared with that of solar radiation. As shown in Figure 2-3, there is an inverse relationship between the wavelength of electromagnetic radiation and its energy content. Specifically, the energy of electromagnetic radiation is related to the wavelength by the equation $E = 28,600/\lambda$, where λ is the wavelength in nm, E is the energy in kilocalories per einstein, and 28,600 is a constant with the units kcal-nm/einstein. (An *einstein* is equal to 1 mole of photons.) Using this equation, you can readily calculate that the visible portion of sunlight (wavelengths of 400–700 nm) is lower in energy than the carbon-carbon bond. For example, green light with a wavelength of 500 nm has an energy content of about 57.2 kcal/einstein. The energy of green light is therefore well below the energies of covalent bonds (see Figure 2-2). If this were not the case, visible light would break covalent bonds spontaneously, and life as we know it would not exist.

Figure 2-3 suggests another important point: the hazard that ultraviolet radiation poses to biological molecules. At a wavelength of 300 nm, for example, ultraviolet light has an energy content of about 95.3 kcal/einstein, clearly enough to break carbon-carbon bonds spontaneously. This threat underlies the current concern about pollutants that destroy the ozone layer in the upper

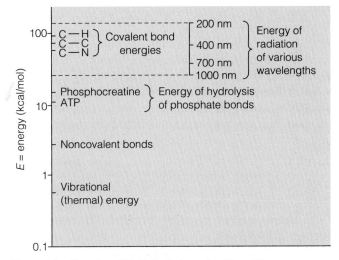

Figure 2-2 **Energies of Biologically Important Transitions, Bonds, and Wavelengths of Electromagnetic Radiation.** Note that energy is plotted on a logarithmic scale to accommodate the range of values shown.

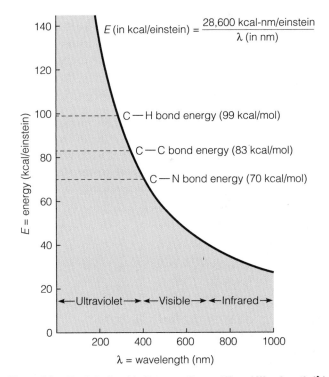

Figure 2-3 **The Relationship Between Energy (E) and Wavelength (λ) for Electromagnetic Radiation.** The dashed lines mark the bond energies of the C—H bond, the C—C bond, and the C—N bond.

atmosphere, because the ozone layer filters out much of the ultraviolet radiation that would otherwise reach the Earth's surface and wreak havoc with the covalent bonds that literally hold biological molecules together.

Carbon-Containing Molecules Are Diverse

In addition to their stability, carbon-containing compounds are characterized by the great diversity of molecules

---

[1] Energy, heat, and work can be expressed either in *calories* and *kilocalories* or in *joules* and *kilojoules*. The joule (J) is the unit of choice among physicists and some biochemists; the calorie (cal) continues to be used in most cell biology texts, including this one. Conversion is easy: 1 cal = 4.184 J, or 1 J = 0.239 cal. Similarly, 1 kcal = 4.184 kJ, or 1 kJ = 0.239 kcal.

that can be generated from relatively few different kinds of atoms. Again, this diversity is due to the tetravalent nature of the carbon atom and the resulting propensity of each carbon atom to form covalent bonds to four other atoms. Because one or more of these bonds can be to other carbon atoms, molecules consisting of long chains of carbon atoms can be built up. Ring compounds are also common. Further variety is possible by the introduction of branching and of double and single bonds into the carbon-carbon chains.

When only hydrogen atoms are used to complete the valence requirements of such linear or circular molecules, the resulting compounds are called **hydrocarbons** (Figure 2-4). Hydrocarbons are very important economically, because gasoline and other petroleum products are mixtures of short-chain hydrocarbons such as *octane*, an eight-carbon compound (C_8H_{18}).

In biology, on the other hand, hydrocarbons play only a very limited role because they are essentially insoluble in water, the universal solvent in biological systems. There is an important exception to this general rule, however: The interior of every biological membrane is a nonaqueous environment from which water and water-soluble compounds are excluded by the long hydrocarbon "tails" of phospholipid molecules that project into the interior of the membrane from either surface. This feature of membranes has important implications for their role as permeability barriers, as we will see shortly.

Most biological compounds contain, in addition to carbon and hydrogen, one or more atoms of oxygen and often nitrogen, phosphorus, or sulfur as well. These atoms are usually part of various **functional groups** that confer both water solubility and chemical reactivity on the molecules of which they are a part. (Even the phospholipid molecules whose hydrocarbon tails contribute so importantly to the nonaqueous nature of the membrane interior contain atoms other than hydrogen and carbon.)

Some of the more common functional groups present in biological molecules are shown in Figure 2-5. Several of these groups are ionized or protonated (i.e., have lost or gained a proton, respectively) at the near-neutral pH of most cells, including the negatively charged *carboxyl* and *phosphoryl* groups and the positively charged *amino* group. Other groups, such as the *hydroxyl, sulfhydryl, carbonyl,* and *aldehyde* groups, are uncharged at pH values near neutrality. However, they cause a significant redistribution of electrons within the molecules to which they are attached, thereby conferring on these molecules greater water solubility and chemical reactivity.

Carbon-Containing Molecules Can Form Stereoisomers

Carbon-containing molecules are capable of still greater diversity because the carbon atom is a **tetrahedral** structure with *geometric symmetry* (Figure 2-6). When four different atoms or groups of atoms are bonded to the four corners of such a tetrahedral structure, two different spatial configurations are possible. Although both forms have the same structural formula, they are not superimposable but are, in fact, mirror images of each other (see Figure 2-6). Such mirror-image forms of the same compound are called **stereoisomers.**

A carbon atom that has four different substituents is called an **asymmetric carbon atom.** Because two stereo-

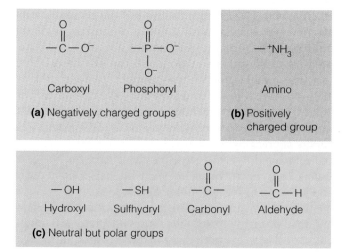

(a) Negatively charged groups

(b) Positively charged group

(c) Neutral but polar groups

Figure 2-5 Some Common Functional Groups Found in Biological Molecules. Each functional group is shown in the form that predominates at the near-neutral pH of most cells. **(a)** The carboxyl and phosphoryl groups are ionized and therefore negatively charged. **(b)** The amino group, on the other hand, is protonated and is therefore positively charged. **(c)** Hydroxyl, sulfhydryl, carbonyl, and aldehyde groups are uncharged at pH values near neutrality but are much more polar than hydrocarbons, thereby conferring greater polarity and hence greater water solubility on the organic molecules to which they are attached.

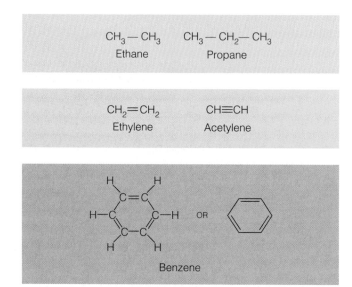

Figure 2-4 Some Simple Hydrocarbon Compounds. Compounds in the top row have single bonds only, whereas those in the second row have double or triple bonds. The condensed structure shown on the right for benzene is an example of the simplified structures that chemists frequently use for such compounds.

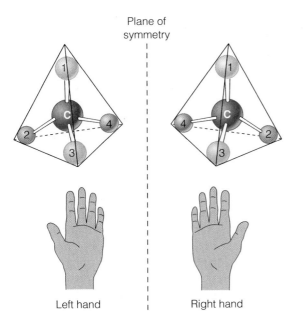

Figure 2-6 **Stereoisomers.** Stereoisomers of organic compounds occur when four different groups are attached to a tetrahedral carbon atom. Stereoisomers, like left and right hands, are mirror images of each other and cannot be superimposed on one another. (The dashed line down the center of the figure is the plane of the mirror.)

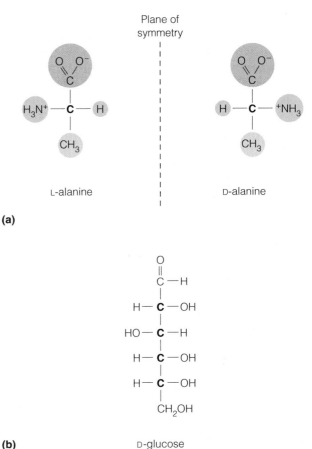

(a)

(b) D-glucose

Figure 2-7 **Stereoisomers of Biological Molecules.** **(a)** The amino acid alanine has a single asymmetric carbon atom (in the center) and can therefore exist in two spatially different forms, designated as L- and D-alanine. (The dashed line down the center of the figure is the plane of the mirror.) **(b)** The six-carbon sugar glucose has four asymmetric carbon atoms, so D-glucose is just one of 16 (2^4) possible stereoisomers of the $C_6H_{12}O_6$ molecule (though not nearly this many actually occur in nature).

isomers are possible for each asymmetric carbon atom, a compound with n asymmetric carbon atoms will have 2^n possible stereoisomers. As shown in Figure 2-7a, the three-carbon amino acid *alanine* has a single asymmetric carbon atom (in the center) and thus has two stereoisomers, called L-alanine and D-alanine. (Neither of the other two carbon atoms of alanine is an asymmetric carbon atom, because one has three identical substituents and the other has two bonds to a single oxygen atom.) Both stereoisomers of alanine occur in nature, but only L-alanine is present as a component of proteins.

As an example of a compound with multiple asymmetric carbon atoms, consider the six-carbon sugar *glucose* shown in Figure 2-7b. Of the six carbon atoms of glucose, the four shown in boldface are asymmetric. (Again, you should be able to figure out why the other two carbon atoms are not asymmetric.) With four asymmetric carbon atoms, the structure shown, D-glucose, is only one of 2^4, or 16, possible stereoisomers of the $C_6H_{12}O_6$ molecule. In this case, however, not all of the other possible stereoisomers exist in nature, mainly because some are energetically much less favorable than others.

The Importance of Water

Just as the carbon atom is uniquely important because of its role as the universal backbone of biologically important molecules, the water molecule commands special attention because of its indispensable role as the universal

solvent in biological systems. Water is, in fact, the single most abundant component of cells and organisms. Typically, about 75–85% of a cell by weight is water, and many cells depend on an extracellular environment that is essentially aqueous as well. In some cases this is the body of water—whether an ocean, lake, or river—in which the cell or organism lives, whereas in other cases it may be body fluids in which the cell is suspended or with which the cell is bathed.

Water is indispensable for life as we know it. True, there are life forms that can go into a "holding action" and survive periods of severe water scarcity. Seeds of plants and spores of bacteria and fungi are clearly in this category; the moisture content of a dry seed is frequently as low as 10–20%. Some lower plants and animals, notably certain mosses, lichens, nematodes, and rotifers, can also undergo physiological adaptations that allow them to dry out and survive in a highly desiccated form, sometimes for surprisingly long periods of time. Such adaptations are clearly an advantage in environments characterized by periods of drought. Yet all of these are at best holding actions, and resumption of normal activity always requires rehydration. Thus, an adequate water supply is needed to serve as the solvent system for the variety of activities that we associate with what it means to be alive.

To understand why water is so uniquely suitable for its role, we need to look at its chemical properties. The most critical attribute is clearly its *polarity,* because this property in turn accounts for its cohesiveness and its temperature-stabilizing capacity, both of which have important consequences for biological chemistry.

Water Molecules Are Polar

To understand the polar nature of water, we need to consider the shape of the molecule. As shown in Figure 2-8a, the water molecule is triangular rather than linear in shape, with the two hydrogen atoms bonded to the oxygen at an angle of 104.5° rather than 180°. It is not an overstatement to say that life as we know it depends critically on this angle, given the distinctive properties that the resulting asymmetry confers on the water molecule. Although the molecule as a whole is uncharged, the electrons tend to be unevenly distributed. The oxygen atom at the head of the molecule is **electronegative;** that is, it tends to draw elec-

trons toward it, giving that end of the molecule a partial negative charge and leaving the other end of the molecule with a partial positive charge around the hydrogen atoms. This charge separation gives the water molecule its **polarity,** which we can define as an uneven distribution of charge within a molecule. In the case of water, the polarity of the molecule has enormous consequences, accounting for the *cohesiveness,* the *temperature-stabilizing capacity,* and the *solvent properties* of water.

Water Molecules Are Cohesive

Because of their polarity, water molecules have an affinity for each other and tend to orient themselves spontaneously so that the electronegative oxygen atom of one molecule is associated with the electropositive hydrogen atoms of adjacent molecules. Each such association is called a *hydrogen bond* and is frequently represented by a dotted line, as in Figure 2-8b. Each oxygen atom can bond to two hydrogens, and both of the hydrogen atoms can associate in this way with the oxygen atoms of adjacent molecules. As a result, water is characterized by an extensive three-dimensional network of hydrogen-bonded molecules (Figure 2-8b). The hydrogen bonds between adjacent molecules are constantly being broken and reformed, with a typical bond having a half-life of a few microseconds. On the average, however, each molecule of water in the liquid state is hydrogen-bonded to about 3½ neighbor molecules at any given time. In ice, the hydrogen bonding is still more extensive, giving rise to a rigid, highly regular crystalline lattice with every oxygen hydrogen-bonded to hydrogens of two adjacent molecules and every water molecule therefore hydrogen-bonded to four neighboring molecules.

It is this tendency to form hydrogen bonds between adjacent molecules that makes water so highly *cohesive.* This cohesiveness accounts for the high *surface tension* of water, as well as for its high *boiling point,* high *specific heat,* and high *heat of vaporization.* The high surface tension of water causes the capillary action that enables water to move up the conducting tissues of plants and allows insects such as the water strider to move across the surface of a pond without breaking the surface (Figure 2-9).

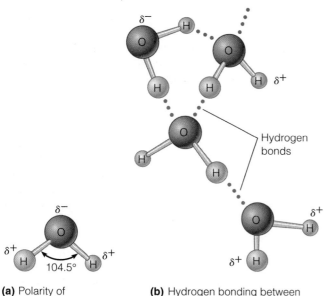

(a) Polarity of
water molecule

(b) Hydrogen bonding between
water molecules

Figure 2-8 Hydrogen Bonding Between Water Molecules.
(a) The water molecule is polar because it has an asymmetric charge distribution. The two hydrogen atoms are bonded to the oxygen at an angle of 104.5°. The oxygen atom bears a partial negative charge (δ^-; the Greek letter delta stands for "partial") and is thus the electronegative portion of the molecule. The two hydrogen atoms are electropositive; their end of the molecule has a partial positive charge (δ^+). **(b)** The extensive association of water molecules with one another in either the liquid or the solid state is due to hydrogen bonds (dotted lines) between the electronegative oxygen atom of one water molecule and the electropositive hydrogen atoms of adjacent molecules. In ice, the resulting crystal lattice is regular and complete; every oxygen is hydrogen-bonded to hydrogens of two adjacent molecules. In water, some of the structure is disrupted, but much is retained.

Water Has a High Temperature-Stabilizing Capacity

An important property of water that derives directly from the hydrogen bonding between adjacent molecules is the high specific heat that gives water its *temperature-stabilizing capacity.* **Specific heat** is the amount of heat a substance must absorb per gram to increase its temperature 1°C. The specific heat of water is 1.0 calorie per gram.

The specific heat of water is much higher than that of most other liquids because of its extensive hydrogen bonding. Much of the energy that in other liquids would contribute directly to an increase in the motion of solvent molecules and therefore to an elevation in temperature is used instead to break hydrogen bonds between neighboring

Figure 2-9   Walking on Water.   Insects such as this water strider are able to walk on the surface of a pond without breaking the surface because of the high surface tension of water that results from the collective strength of its hydrogen bonds.

water molecules. In effect, by absorbing heat that would otherwise increase the temperature of the water more rapidly, hydrogen bonds buffer aqueous solutions against large changes in temperature. This capability is an important consideration for the cell biologist, because cells release large amounts of energy during metabolic reactions. This release of energy would pose a serious overheating problem for cells if it were not for the extensive hydrogen bonding and the resulting high specific heat of water molecules.

Water also has a high **heat of vaporization,** which is defined as the amount of energy required to convert one gram of a liquid into vapor. This value is high for water because of the hydrogen bonds that must be disrupted in the process. This property makes water an excellent coolant and explains why people perspire, why dogs pant, and why plants lose water through transpiration. In each case, the heat required to evaporate water is drawn from the organism, which is therefore cooled in the process.

In light of the high temperature-stabilizing capacity that water has because of its extensive hydrogen bonding, it seems fair to conclude that life as we know it would not be possible were it not for the energy required to break the hydrogen bonds between adjacent water molecules.

Water Is an Excellent Solvent

Probably the single most important property of water from a biological perspective is its excellence as a general solvent. A **solvent** is a fluid in which another substance, called the **solute,** can be dissolved. Water is an especially good solvent for biological purposes because of its remarkable capacity to dissolve a great variety of solutes.

It is the polarity of water that makes it so useful as a solvent. Most of the molecules in cells are also polar and therefore interact electrostatically with water molecules, as do charged ions. Solutes that have an affinity for water and therefore dissolve readily in it are called **hydrophilic** ("water-loving"). Most small organic molecules found in cells are hydrophilic; examples are sugars, organic acids, and some of the amino acids. Molecules that are not very soluble in water are termed **hydrophobic** ("water-fearing"). Among the more important hydrophobic compounds found in cells are the lipids and most of the proteins of which membranes are made. In general, polar molecules tend to be hydrophilic and nonpolar molecules tend to be hydrophobic. Some biological macromolecules, notably proteins, have both hydrophobic and hydrophilic regions, so parts of the molecule have an affinity for an aqueous environment while other parts of the molecule do not.

To understand why polar substances dissolve so readily in water, first consider a salt such as sodium chloride (NaCl) (Figure 2-10). Because it is a salt, NaCl exists in crystalline form as a lattice of positively charged sodium ions (Na^+) and negatively charged chloride ions (Cl^-). For NaCl to dissolve in a liquid, solvent molecules must overcome the attraction of the oppositely charged Na^+ cations and Cl^- anions for each other and involve them instead in

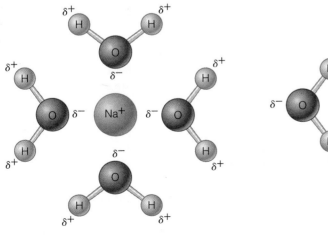

(a) Hydration of sodium ion

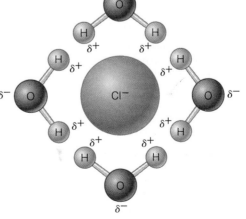

(b) Hydration of chloride ion

Figure 2-10   The Solubilization of Sodium Chloride.   Sodium chloride (NaCl) dissolves in water because of the formation of spheres of hydration around **(a)** the sodium ions and **(b)** the chloride ions. The oxygen atom and the sodium and chloride ions are drawn to scale.

electrostatic interactions with the solvent molecules themselves. Because of their polarity, water molecules can form **spheres of hydration** around both Na^+ and Cl^-, thereby neutralizing their attraction for each other and lessening their likelihood of reassociation. As Figure 2-10a shows, the sphere of hydration around a cation such as Na^+ involves water molecules clustered around the ion with their negative (oxygen) ends pointing toward it. For an anion such as Cl^-, the orientation of the water molecules is reversed, with the positive (hydrogen) ends of the solvent molecules pointing in toward the ion (Figure 2-10b).

Some biological compounds are soluble in water because they exist as ions at the near-neutral pH of the cell and are therefore solubilized and hydrated like the ions of Figure 2-10. Compounds containing carboxyl, phosphoryl, or amino groups are in this category (see Figure 2-5). Most organic acids, for example, are almost completely ionized at a pH near 7 and therefore exist as anions that are kept in solution by spheres of hydration, just as the chloride ion of Figure 2-10b is. Amines, on the other hand, are usually protonated at cellular pH and thus exist as hydrated cations, like the sodium ion of Figure 2-10a.

More frequently, organic molecules have no net charge—that is, they have neither lost nor gained protons and are therefore neutral molecules. However, many organic molecules are nonetheless hydrophilic because they have some regions that are positively charged and other regions that are negatively charged. Water molecules tend to cluster around such regions, and the resulting electrostatic interactions between solute and water molecules keep the solute molecules from associating with one another. Compounds containing the hydroxyl, sulfhydryl, carbonyl, or aldehyde groups shown in Figure 2-5 are usually in this category.

Hydrophobic molecules, on the other hand, have no such polar regions and therefore show no tendency to interact electrostatically with water molecules. In fact, they actually disrupt the hydrogen-bonded structure of water and, for this reason, tend to be excluded by the water molecules. Hydrophobic molecules therefore tend to coalesce in an aqueous medium, associating with one another rather than with the water. This association is driven not so much by any specific affinity of the hydrophobic molecules for one another as by the strong tendency of water molecules to form hydrogen bonds and to exclude molecules that disrupt hydrogen bonding. As we will see later in the chapter, such associations of hydrophobic molecules (or parts of molecules) are a major driving force in the folding of molecules, the assembly of cellular structures, and the organization of membranes.

The Importance of Selectively Permeable Membranes

Every cell and organelle needs some sort of physical barrier to keep its contents in and external materials out, as well as some means of controlling exchange between its internal environment and the extracellular environment. Ideally, such a barrier should be impermeable to most of the molecules and ions found in cells and their surroundings. Otherwise, substances could diffuse freely in and out, and the cell would not really have a defined content at all. On the other hand, the barrier cannot be completely impermeable, or else needful exchanges between the cell and its environment could not take place. Moreover, such a barrier must be insoluble in water, so that it will not be dissolved by the aqueous medium of the cell. At the same time, it must be readily permeable to water, because water is the basic solvent system of the cell and must be able to flow into and out of the cell as needed.

As you might expect, the membranes that surround cells and organelles satisfy these criteria admirably. A **membrane** is essentially a hydrophobic permeability barrier consisting of *phospholipids, glycolipids,* and *membrane proteins.* In most organisms other than bacteria, the membranes also contain *sterols—cholesterol* in the case of animal cells and related *phytosterols* in the membranes of plant cells. (Don't be concerned if you haven't encountered these kinds of molecules before; we'll meet them in Chapter 3.)

Most membrane lipids and proteins are not simply hydrophobic; they have both hydrophilic and hydrophobic regions and are therefore referred to as **amphipathic molecules** (the Greek prefix *amphi-* means "of both kinds"). The amphipathic nature of membrane phospholipids is illustrated in Figure 2-11, which shows the structure of *phosphatidyl ethanolamine,* a prominent phospholipid in many kinds of membranes. (Phosphatidyl ethanolamine is an example of a *phosphoglyceride,* the major class of membrane phospholipids in most cells. For other examples of phosphoglycerides and other classes of membrane phospholipids, see Figure 3-27.) The distinguishing feature of amphipathic phospholipids is that each molecule consists of a polar "head" and two nonpolar hydrocarbon "tails." The polarity of the hydrophilic head is due to the presence of a negatively charged phosphate group linked to a positively charged group—an amino group, in the case of phosphatidyl ethanolamine and most other phosphoglycerides.

A Membrane Is a Lipid Bilayer with Proteins Embedded in It

When exposed to an aqueous environment, amphipathic molecules undergo hydrophobic interactions. In a membrane, for example, phospholipids are organized into two layers with their polar heads facing outward toward the aqueous milieu on both sides and their hydrophobic tails hidden from the water by interacting with the tails of other molecules oriented in the opposite direction. The resulting structure is the **lipid bilayer,** shown in Figure 2-12. The heads of both layers face outward and the hydrocarbon tails extend inward, forming the continuous hydrophobic interior of the membrane.

Every known biological membrane has such a lipid bilayer as its basic structure. Each of the lipid layers is

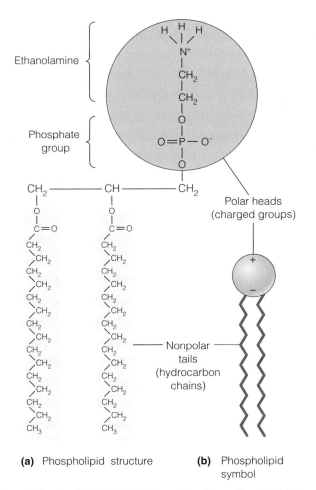

(a) Phospholipid structure   **(b)** Phospholipid symbol

Figure 2-11 **The Amphipathic Nature of Membrane Phospholipids.**
(a) A phospholipid molecule consists of two long nonpolar tails (yellow) and a polar head (orange). Shown here is phosphatidyl ethanolamine, an example of the phosphoglyceride class of membrane phospholipids. The polarity of the head of a phospholipid molecule results from a negatively charged phosphate group linked to a positively charged group—an amino group, in the case of phosphatidyl ethanolamine. Other common phosphoglycerides with both a phosphate group and an amino group include phosphatidyl serine and phosphatidyl choline.
(b) A phospholipid molecule is often represented schematically by a circle for the polar head (notice the plus and minus charges) and two zigzag lines for the nonpolar hydrocarbon chains.

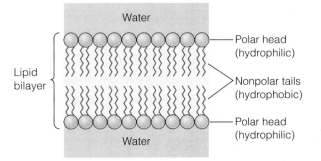

Figure 2-12 **The Lipid Bilayer as the Basis of Membrane Structure.** Due to their amphipathic nature, phospholipids in an aqueous environment orient themselves in a double layer, with the hydrophobic tails (gray) buried on the inside and the hydrophilic heads (orange) pointing toward the aqueous milieu on either side of the membrane.

phobic regions of the protein associate with the interior of the membrane, whereas hydrophilic regions protrude into the aqueous environment on either or both surfaces of the membrane. Some proteins of the plasma membrane have carbohydrate side chains attached to their outer surfaces.

Depending on the particular membrane, the membrane proteins may play any of a variety of roles. Some are *transport proteins,* responsible for moving specific substances across an otherwise impermeable membrane. Others are *enzymes* that catalyze reactions associated with the specific membrane. Still others are the *receptors* on the outer surface of the cell membrane, the *electron transport intermediates* of the mitochondrial membrane, or the *chlorophyll-binding proteins* of the chloroplast. We will encounter each of these kinds of membrane proteins in subsequent chapters.

Membranes Are Selectively Permeable

Because of its hydrophobic interior, a membrane is readily permeable to nonpolar molecules but is quite impermeable to most polar molecules and is highly impermeable to all ions. Because most cellular constituents are either polar or charged, they have little or no affinity for the membrane interior and are effectively prevented from entering or escaping from the cell or organelle. Very small molecules are an exception, however. Compounds with molecular weights below about 100 diffuse across membranes regardless of whether they are nonpolar (such as O_2 and CO_2) or polar (such as ethanol and urea). Water is an especially important example of a very small molecule that, although polar, diffuses very rapidly across membranes.

In contrast, even the smallest ions are very effectively excluded from the hydrophobic interior of the membrane. For example, a lipid bilayer is at least 10^8 times less permeable to such small cations as Na^+ or K^+ than it is to water. This striking difference is due to both the charge on an ion and the sphere of hydration that surrounds the ion.

about 3–4 nm thick, so the bilayer has a width of about 7–8 nm. It is the lipid bilayer that gives membranes their characteristic "railroad track" appearance when seen with the transmission electron microscope (TEM; Figure 2-13a). Apparently, the osmium used to prepare the tissue for electron microscopy reacts with the hydrophilic heads of the phospholipid molecules at both surfaces of the membrane but not with the hydrophobic tails in the interior of the membrane, which gives the membrane its *trilaminar* (three-layered) appearance.

The structure of biological membranes is illustrated in Figure 2-13b. Embedded within or associated with the membrane lipid bilayer are various membrane proteins. These proteins almost always are amphipathic, and they become oriented in the lipid bilayer accordingly: Hydro-

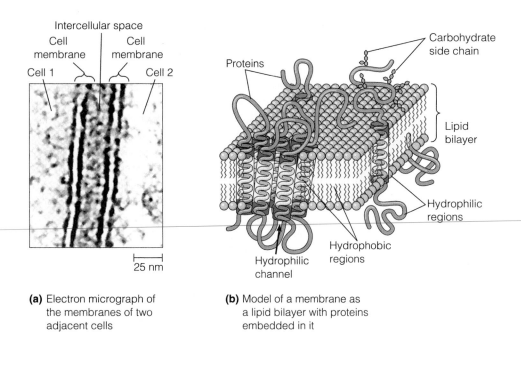

Intercellular space

Cell membrane
Cell 1

Cell membrane
Cell 2

25 nm

(a) Electron micrograph of the membranes of two adjacent cells

Proteins

Carbohydrate side chain

Lipid bilayer

Hydrophilic regions

Hydrophobic regions

Hydrophilic channel

Hydrophobic regions

(b) Model of a membrane as a lipid bilayer with proteins embedded in it

Figure 2-13 Membranes and Membrane Structure.   (a) With the electron microscope, the cell membranes surrounding two adjacent cells each appear as a pair of dark bands. This characteristic trilaminar, or "railroad track," appearance is thought to be due to the association of osmium with the hydrophilic heads of the phospholipid molecules but not with their hydrophobic tails, as seen in this transmission electron micrograph (TEM). **(b)** Biological membranes consist of amphipathic proteins embedded within a lipid bilayer. Proteins are positioned in the membrane so that their hydrophobic regions (light purple) are located in the hydrophobic interior of the phospholipid bilayer and their hydrophilic regions (dark purple) are exposed to the aqueous milieu on either side of the membrane. The short chains attached to the upper surface of one of the membrane proteins represent carbohydrate side chains.

Of course, it is essential that cells have ways of transferring ions such as Na^+ and K^+ as well as a wide variety of polar molecules across membranes that are not otherwise permeable to these substances. As already noted, membranes are equipped with transport proteins to serve this function. A transport protein is a specialized transmembrane protein that serves either as a *hydrophilic channel* through an otherwise hydrophobic membrane or as a *carrier* that binds a specific solute on one side of the membrane and then undergoes a conformational change to move the solute across the membrane.

Whether a channel or a carrier, each transport protein is specific for a particular molecule or ion (or, in some cases, for a class of closely related molecules or ions). Moreover, the activities of these proteins can be carefully regulated to meet cellular needs. As a result, biological membranes can best be described as *selectively permeable:* With the exception of very small molecules, the only molecules or ions that can move across a particular membrane are those for which the appropriate transport proteins are present in the membrane.

The Importance of Synthesis by Polymerization

For the most part, cellular structures such as ribosomes, chromosomes, membranes, flagella, and cell walls are made up of ordered arrays of linear polymers called **macromolecules.** (Some branching occurs, notably in the polysaccharides starch and glycogen, but the linearity of biological polymers is a good first approximation.) Exam-

ples of important macromolecules in cells include *proteins, nucleic acids* (both DNA and RNA), and *polysaccharides* such as starch, glycogen, and cellulose. (*Lipids* are sometimes regarded as macromolecules as well. However, they differ somewhat from the other classes of macromolecules in the way they are synthesized and will not be discussed further until Chapter 3.) Macromolecules are very important in both the function and the structure of cells. To understand the biochemical basis of cell biology, therefore, really means to understand macromolecules—how they are made, how they are assembled, and how they function.

Macromolecules Are Responsible for Most of the Form and Function in Living Systems

The importance of macromolecules in cell biology is emphasized by the cellular hierarchy shown in Figure 2-14. The compounds of which most cellular structures are made are small, water-soluble *organic molecules* (level 1) that cells either obtain from other cells or synthesize from simple nonbiological molecules such as carbon dioxide, ammonia, or phosphate ions available from the environment. These organic molecules polymerize to form *biological macromolecules* (level 2) such as nucleic acids, proteins, or carbohydrates. Macromolecules are then assembled into a variety of *supramolecular structures* (level 3), which in turn are components of organelles and other subcellular structures (level 4) and hence of the cell itself (level 5).

One of the examples in Figure 2-14 is that of cell wall biogenesis (panels on left). In plants, a major component of the primary cell wall (level 3) is the polysaccharide cellulose (level 2). Cellulose is in turn a repeating polymer of the

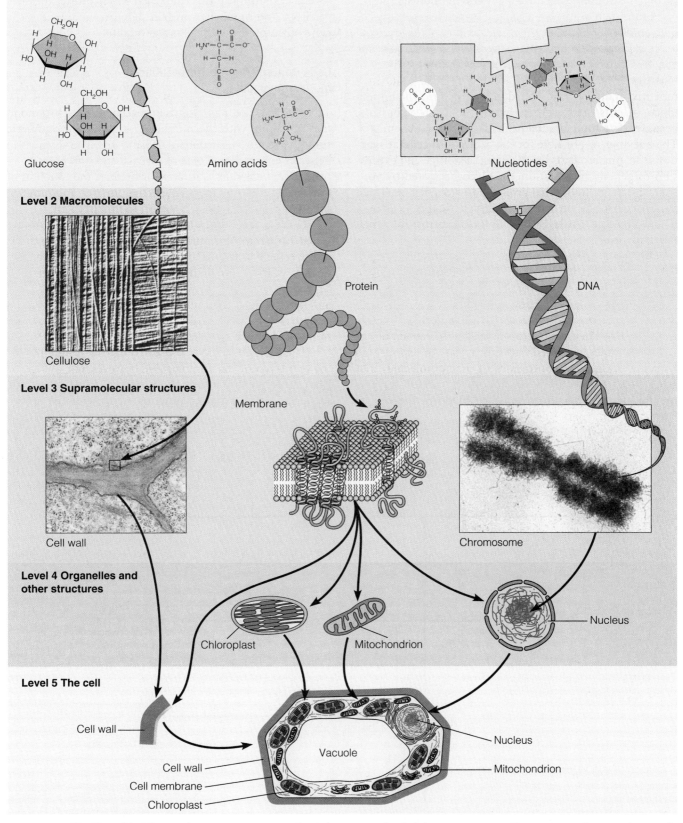

Level 1 Small organic molecules

Glucose

Amino acids

Nucleotides

Level 2 Macromolecules

Cellulose

Protein

DNA

Level 3 Supramolecular structures

Cell wall

Membrane

Chromosome

Level 4 Organelles and other structures

Chloroplast

Mitochondrion

Nucleus

Level 5 The cell

Cell wall

Cell wall

Cell membrane

Chloroplast

Vacuole

Nucleus

Mitochondrion

Figure 2-14   The Hierarchical Nature of Cellular Structures and Their Assembly.   Small organic molecules (level 1) are synthesized from simple inorganic substances and are polymerized to form macromolecules (level 2). The macromolecules then assemble into the supramolecular structures (level 3) that make up organelles and other subcellular structures (level 4) and, ultimately, the cell (level 5). (The supramolecular structures shown as level 3 are more complex in their chemical composition than the figure suggests. Chromosomes, for example, contain not only DNA but proteins as well—in about equal amounts, in fact. Similarly, membranes contain not only proteins but also a variety of lipids, and cell walls contain not just cellulose but also other carbohydrates and proteins.)

simple sugar glucose (level 1), formed by the plant cell from carbon dioxide and water in the process of photosynthesis.

From such examples, a general principle emerges: *The macromolecules that are responsible for most of the form and order characteristic of living systems are generated by the polymerization of small organic molecules.* This strategy of forming large molecules by joining smaller units in a repetitive manner is illustrated in Figure 2-15. The importance of this strategy can hardly be overemphasized, because it is a fundamental principle of cellular chemistry. The enzymes responsible for the catalysis of cellular reactions, the nucleic acids involved in the storage and expression of genetic information, the glycogen stored by your liver, and the cellulose that gives rigidity to a plant cell wall are all variations on the same design theme. Each is a macromolecule made by the linking together of small repeating units.

Examples of such repeating units, or **monomers,** are the *glucose* present in cellulose or glycogen, the *amino acids* needed to make proteins, and the *nucleotides* of which nucleic acids are made. In general, these are small, water-soluble organic molecules with molecular weights less than 350. They can be transported across most biological membranes, provided that appropriate transport proteins are present. By contrast, most macromolecules that are synthesized from these monomers cannot traverse membranes and therefore must be made in the cell or

compartment in which they are needed. (We will discover several exceptions to this rule when we encounter messenger RNA and organellar proteins later in the text, but the generalization is nonetheless a good one.)

Cells Contain Three Different Kinds of Macromolecules

Table 2-1 lists the major kinds of macromolecules found in the cell, along with the number and kind of monomeric subunits required for polymer formation. The distinction between *informational macromolecules* and *storage* or *structural macromolecules* is important because the function of these biological polymers affects the number and order of different monomers we can expect to find in them.

Nucleic acids (both DNA and RNA) and proteins are called **informational macromolecules** because the order of the several kinds of nonidentical subunits they contain is nonrandom and highly significant to their function. The order of the monomers (nucleotides in nucleic acids and amino acids in proteins) is genetically determined and carries important information that specifies the function of these macromolecules.

The term *information* is used somewhat differently for the two types of macromolecules, however. For a DNA or RNA molecule, the information inherent in its nucleotide sequence serves a *coding* function, specifying the amino

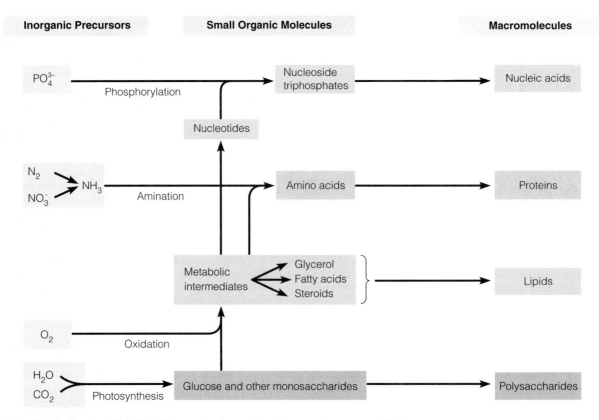

Figure 2-15 The Synthesis of Biological Macromolecules. Simple inorganic precursors (left) react to form small organic molecules (center), which are then used in the synthesis of the macromolecules (right) of which most cellular structures consist.

Table 2-1 Biologically Important Macromolecules and Their Repeating Units

	Biological Polymer			
	Proteins	**Nucleic Acids**	**Polysaccharides**	
Kind of macromolecule	Informational	Informational	Storage	Structural
Examples	Enzymes, hormones, antibodies	DNA, RNA	Starch, glycogen	Cellulose
Repeating monomers	Amino acids	Nucleotides	Monosaccharides	Monosaccharides
Number of kinds of repeating units	20	4 in DNA; 4 in RNA	One or a few	One or a few

acid sequence of a particular protein. For a protein, the information present in its amino acid sequence determines the three-dimensional structure of the protein, on which its biological activity depends. The sequence of amino acids in a protein is so important that any variation in that sequence is likely to have a deleterious effect on the ability of the protein to perform its function.

For informational macromolecules, polymerization of smaller subunits is not just an economical way of building large molecules; it is an essential part of the role these macromolecules play in the cell. For example, the synthesis of nucleic acids by linkage of small repeating units (nucleotides) is important because it is in the specific order of the monomeric units that these molecules store and transmit the genetic information of the cell.

Polysaccharides, on the other hand, are not informational macromolecules in the intrinsic sense that proteins and nucleic acids are. Most polysaccharides consist of either a single repeating subunit (a monosaccharide such as glucose, in most cases) or two subunits that occur in strict alternation. The order of monomers in a polysaccharide therefore carries no information and is not essential to the function of the polymer in the sense that it is for proteins or nucleic acids. Most polysaccharides are either **storage macromolecules** or **structural macromolecules.** The most familiar storage polysaccharides are the *starch* of plant cells and the *glycogen* found in animal cells. As

shown in Figure 2-16a, both of these storage polysaccharides consist of a single repeating monomer, the simple sugar glucose. (The main difference between glycogen and starch lies in the occurrence and extent of branching, a feature that is not shown in Figure 2-16a but will be encountered in Chapter 3.)

The best-known example of a structural polysaccharide is the *cellulose* present in plant cell walls. Like starch and glycogen, cellulose also consists of glucose units, but the units are linked together by a somewhat different bond, as we will see in Chapter 3. The cell walls of some bacterial cells contain a more complicated kind of structural polysaccharide. In this case, the molecule consists of two different kinds of monomers, *N*-acetylglucosamine (GlcNAc) and *N*-acetylmuramic acid (MurNAc), as illustrated in Figure 2-16b. The two monomers occur in strictly alternating sequence, however, and carry no information. A further example of a structural polysaccharide is *chitin,* found in insect exoskeletons and crustacean shells. Chitin consists of GlcNAc units only.

Macromolecules Are Synthesized by Stepwise Polymerization of Monomers

In Chapter 3, we will look at each of the major kinds of biologically important macromolecules. First, however, it will be useful to consider several important principles that

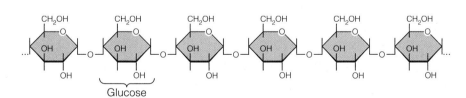

(a) A storage polysaccharide (segment of a starch or glycogen molecule)

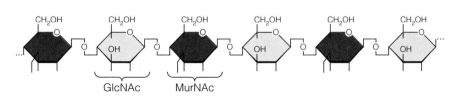

(b) A structural polysaccharide (segment of a bacterial cell wall component)

Figure 2-16 Storage and Structural Macromolecules. Storage and structural macromolecules contain one or a few kinds of repeating units in a strictly repetitious sequence. **(a)** A portion of the sequence of a linear segment of the storage polysaccharide starch or glycogen consisting of glucose units linked together by glycosidic bonds. **(b)** A portion of the sequence of a bacterial cell wall polysaccharide consisting of the sugar derivatives *N*-acetylglucosamine (GlcNAc) and *N*-acetylmuramic acid (MurNAc) in strict alternation. (For more detailed representations of the structures of glucose, starch, glycogen, GlcNAc, MurNAc, and a bacterial cell wall polysaccharide, see Figures 3-22, 3-24, and 3-26.)

underlie the polymerization processes by which these macromolecules arise. Although the chemistry of the monomeric units, and hence of the resulting polymers, differs markedly among such macromolecules as proteins, nucleic acids, and polysaccharides, the following basic principles apply in each case:

1. Macromolecules are always synthesized by the stepwise polymerization of similar or identical small molecules called monomers.

2. The addition of each monomeric unit occurs with the removal of a water molecule and is therefore termed a **condensation reaction.**

3. The monomeric units that are to be joined together must be present in an *activated form* before condensation can occur.

4. Activation usually involves coupling of the monomer to some sort of **carrier molecule** to form an **activated monomer.**

5. The energy to couple the monomer to the carrier molecule is provided by a molecule called *adenosine triphosphate* (*ATP*; see Figure 3-16 for its structure) or a related high-energy compound.

6. Because of the way in which they are synthesized, macromolecules have an inherent **directionality;** that is, the two ends of the polymer chain are chemically different from each other.

Because the elimination of water is essential in all biological polymerization reactions, each monomer must have a reactive hydrogen (H) on at least one functional group and a reactive hydroxyl group (—OH) elsewhere on the molecule. This structural feature is depicted schematically in Figure 2-17a, which represents the monomeric units (M) simply as boxes but indicates the reactive hydrogen and hydroxyl group on each. For a given kind of polymer, the monomers may differ from one another in other aspects of their structure—indeed, they *must* differ from one another if the polymer is an informational macromolecule—but each monomer possesses the same kind of reactive hydrogen and hydroxyl group.

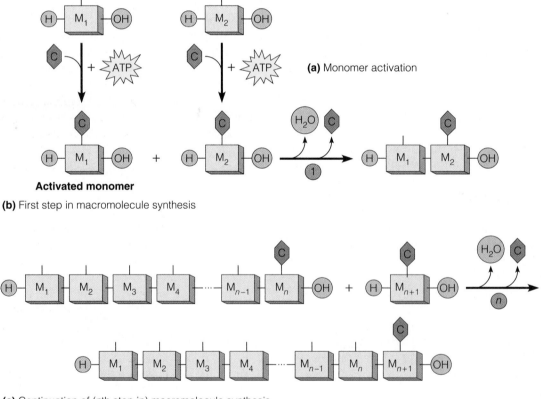

(a) Monomer activation

Activated monomer

(b) First step in macromolecule synthesis

(c) Continuation of (*n*th step in) macromolecule synthesis

Figure 2-17 The Synthesis of Macromolecules. Biological macromolecules are synthesized in a process that involves activation of monomeric units followed by their stepwise addition to the elongating polymer chain. Depending on the polymer, the monomers may all be identical (the glucose units in starch or glycogen, for example) or may differ from one another (the amino acids in proteins or the nucleotides in nucleic acids), but the polymerization process is conceptually the same in each case. **(a)** Monomers (M_1, M_2, etc.) with reactive H and OH groups (shown in blue) are activated by coupling to the appropriate carrier molecule (C, shown in purple), using energy provided by ATP or a similar high-energy compound. **(b)** The first step in polymer synthesis involves the condensation of two activated monomers, accompanied or followed by the release of one of the carrier molecules. **(c)** The *n*th step in the polymerization process involves the addition of the next activated monomer (M_{n+1}) to a polymer that already consists of *n* monomeric units, with release of the carrier molecule bound to the *n*th unit.

Figure 2-17a also shows the activation of monomers, an energy-requiring process. Regardless of the nature of the polymer, the addition of each monomer to a growing chain is always energetically unfavorable unless the incoming monomer is in an activated, energized form. Activation involves coupling of the monomer to some sort of carrier molecule (C), with the energy to drive this activation process provided by ATP or a closely related high-energy compound. A different kind of carrier molecule is used for each kind of polymer. For protein synthesis, amino acids are activated by linking them to carriers called *transfer RNA* (or *tRNA*) molecules, whereas polysaccharides are synthesized from sugar (often glucose) molecules that are activated by linking them to derivatives of nucleotides (*adenosine diphosphate* for starch, *uridine diphosphate* for glycogen).

Once activated, monomers are capable of reacting with each other in a condensation reaction that is followed or accompanied by the release of the carrier molecule from one of the two monomers (Figure 2-17b). Subsequent elongation of the polymer is a sequential, stepwise process, with one activated monomeric unit added at a time, thereby lengthening the elongating polymer by one unit. Figure 2-17c illustrates the *n*th step in such a process, in which the next monomer unit is added to an elongating polymer that already contains *n* monomeric units.

The chemical nature of the energized monomers, the carrier, and the actual activation process differ for each biological polymer, but the general principle is always the same: *Polymer synthesis always involves activated monomers, and ATP or a similar high-energy compound is always required to activate the monomers by linking them to an appropriate carrier molecule.*

The Importance of Self-Assembly

So far, we have seen that the macromolecules that characterize biological organization and function are polymers of small, hydrophilic organic molecules. The only requirements for polymerization are an adequate supply of the monomeric subunits, a source of energy, and, in the case of proteins and nucleic acids, sufficient information to specify the order in which the subunits (amino acids or nucleotides) are added. Still to be considered are the steps that lie beyond—the processes through which these macromolecules are organized into the supramolecular assemblies and organelles that are readily recognizable as cellular structures. Or, in terms of Figure 2-14, we need to ask how the macromolecules of level 2 are assembled into the higher-level structures of levels 3, 4, and 5.

Crucial to our understanding of these higher-level structures is the principle of **self-assembly,** which asserts that *the information required to specify the folding of macromolecules and their interactions to form more complicated structures with specific biological functions is inherent in the polymers themselves.* This principle says that once macro-molecules are synthesized in the cell, their assembly into more complex structures occurs spontaneously, without further input of energy or information. As we will see shortly, this principle has been qualified somewhat by the recognition that proteins called *molecular chaperones* are needed in some, maybe even many, cases of protein folding to prevent incorrect molecular interactions that would lead to inactive structures. Even in such cases, however, the chaperone molecules do not provide additional steric information; they simply assist the assembly process by inhibiting interactions that would produce incorrect structures.

Many Proteins Self-Assemble

A useful prototype for understanding self-assembly processes is the coiling and folding necessary to form a functional three-dimensional protein from one or more linear chains of amino acids. Although the distinction is not always properly made, the immediate product of amino acid polymerization is not a protein but a **polypeptide.** To become a functional protein, one or more such linear polypeptide chains must coil and fold in a very precise, predetermined manner to assume the unique three-dimensional structure necessary for biological activity.

The evidence for self-assembly of proteins comes largely from studies in which the "native," or natural, structure of a protein is disrupted by changing the environmental conditions. Such disruption, or unfolding, can be achieved by raising the temperature, by making the pH of the solution highly acidic or highly alkaline, or by adding certain chemical agents such as urea or any of several alcohols. The unfolding of a polypeptide under such conditions is called **denaturation,** because it results in the loss of the natural three-dimensional structure of the protein—and its function as well, such as the loss of catalytic activity, in the case of an enzyme.

When the denatured polypeptide is returned to conditions in which the native structure is stable, the polypeptide slowly undergoes **renaturation,** the return to its correct three-dimensional structure. In at least some cases, the renatured protein also regains its biological function—catalytic activity, in the case of an enzyme.

Figure 2-18a depicts the denaturation and subsequent renaturation of *ribonuclease,* the protein used by Christian Anfinsen and his colleagues in their classic studies on protein self-assembly. ① When a solution of ribonuclease is heated, the protein will denature, resulting in an unfolded, randomly coiled polypeptide with freedom of rotation about the covalent bonds in both the polypeptide chain and the functional groups of the various amino acids. In this form, ribonuclease has no fixed shape and no catalytic activity. ② If the solution of denatured molecules is then allowed to cool slowly, the ribonuclease molecules will regain their original form and will again have catalytic activity. Thus, all of the information necessary to specify the three-dimensional structure of the ribonuclease molecule is inherent in its amino acid sequence. Similar results have been obtained from denaturation/renaturation

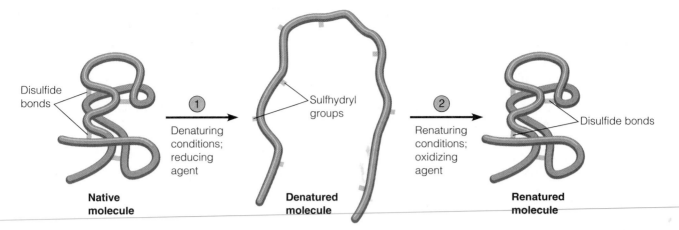

(a) Denaturation and renaturation of ribonuclease

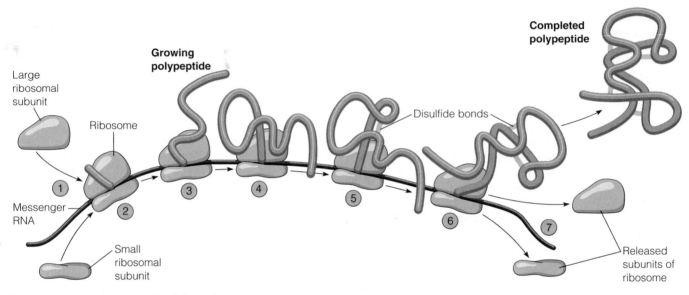

(b) Synthesis and self-assembly of ribonuclease

Figure 2-18   The Spontaneity of Polypeptide Folding. **(a)** To demonstrate that the information for the three-dimensional configuration of a polypeptide resides in its amino acid sequence, ① the intact polypeptide can be exposed to denaturing conditions. The protein depicted here is the enzyme ribonuclease, which consists of a single polypeptide chain. Heat was used as the denaturing agent in Anfinsen's original experiment with ribonuclease, but a variety of chemical agents can be used also. This treatment results in a denatured molecule that has no fixed shape and has lost its biological

function (enzyme activity, in the case of ribonuclease). ② Upon return to conditions that allow restoration of structure (gradual cooling or gradual removal of the denaturing agent), the polypeptide returns spontaneously to its original native configuration (renaturation) without the input of any additional information. (For polypeptides such as ribonuclease in which the native configuration of the molecule is stabilized by covalent disulfide bonds [—S—S—], complete denaturation requires the use of a reducing agent to reduce the disulfide bonds to their component sulfhydryl

[—SH] groups. An oxidizing agent is then used to restore the disulfide bonds as the polypeptide is renatured.) **(b)** The same interactions that ensure the return of a denatured polypeptide to its native form are thought to act on the elongating polypeptide during its synthesis on a ribosome to bring about the progressive coiling and folding of the polypeptide into its unique three-dimensional configuration. In this drawing, synthesis of the polypeptide ribonuclease is initiated on the left, such that each successive ribosome from left to right has a larger portion of the total polypeptide chain completed.

experiments with other proteins and protein-containing structures, although renaturation is much more difficult, sometimes impossible, to demonstrate with larger, more complex proteins.

Notice that the conditions for ribonuclease denaturation (Figure 2-18a, ①) include not only heat but also the

presence of a reducing agent. Similarly, renaturation (②) requires not just cooling but the presence of an oxidizing agent. The reducing agent is required because ribonuclease contains the amino acid cysteine at eight locations along the polypeptide chain, and each cysteine has a free sulfhydryl (—SH) group that is capable of forming a covalent *disulfide bond* (—S—S—) by reacting oxidatively with the sulfhydryl group of another cysteine elsewhere along the polypeptide chain. As shown in Figure 2-18, the native ribonuclease molecule has four such disulfide bonds, which confer additional stability upon the structure. To generate the fully denatured molecule, an agent is needed to reduce the disulfide bonds to free sulfhydryl groups. To renature the molecule, an oxidizing agent is needed to reconstitute the disulfide bonds. Remarkably, each disulfide bond in the renatured molecule is formed from the same two sulfhydryl groups constituting that particular disulfide bond in the original undenatured molecule, even though the cysteines of which those sulfhydryl groups are a part may be quite distant from one another along the polypeptide chain.

Molecular Chaperones Assist the Assembly of Some Proteins

Based on the ability of denatured proteins to return to their original configuration and to regain biological function as they do so, biologists have generally assumed that proteins and protein-containing structures self-assemble in cells as well. As illustrated in Figure 2-18b for ribonuclease, protein synthesis takes place on ribosomes. The polypeptide elongates as successive amino acids are added to one end. The self-assembly model envisions polypeptide chains coiling and folding spontaneously and progressively as polypeptides are synthesized in this way. By the time the fully elongated polypeptide is released from the ribosome, it is thought to have attained a stable, predictable three-dimensional structure without any input of energy or information beyond the basic polymerization process. Furthermore, folding is assumed to be unique, in the sense that each polypeptide with the same amino acid sequence will fold in an identical, reproducible manner under the same conditions. Thus, the self-assembly model assumes that interactions occurring within and between polypeptides are all that is necessary for the biogenesis of proteins in their functional forms.

However, this model for self-assembly in vivo (in the cell) is based entirely on studies with isolated proteins, and even under laboratory conditions, not all proteins regain their native structure. Based on their work with one such protein (*ribulose bisphosphate carboxylase/oxygenase,* the multimeric enzyme in chloroplasts that catalyzes the fixation of CO_2 in the process of photosynthesis), John Ellis and his colleagues at Warwick University concluded that the self-assembly model may not be adequate for all proteins, at least not in its simplest formulation. In many cases, the interactions that drive protein folding need to be assisted and controlled to reduce the probability of the formation of incorrect structures having no biological activity. For assembly processes with a low probability of incorrect interactions and nonfunctional structures (as is likely the case of the folding of a protein such as ribonuclease, which consists of a single polypeptide chain),

such control may not be needed. For more complex processes (such as the assembly of ribulose bisphosphate carboxylase/oxygenase from its 16 component subunits), control is essential to produce a sufficient number of correct structures for cellular needs.

This control of complex assembly processes is exerted by preexisting proteins called **molecular chaperones,** which facilitate the correct assembly of proteins and protein-containing structures but are not components of the assembled structures. The molecular chaperones that have been identified to date do not convey information either for polypeptide folding or for the assembly of multiple polypeptides into a single protein. Instead, they function by binding to specific structural features that are exposed only in the early stages of assembly, thereby inhibiting unproductive assembly pathways that would lead to incorrect structures. Commenting on the term *molecular chaperone,* Ellis and Van der Vies observe that "the term chaperone is appropriate for this family of proteins because the role of the human chaperone is to prevent incorrect interactions between people, not to provide steric information for those interactions" (Ellis and Van der Vies, 1991, p. 323).

The mode of action of molecular chaperones is best described as **assisted self-assembly.** We can therefore distinguish two types of self-assembly: *strict self-assembly,* for which no factors other than the structure of the polypeptide itself are required for proper folding; and *assisted self-assembly,* in which the appropriate molecular chaperone is required to ensure that correct assembly will predominate over incorrect assembly.

The list of known molecular chaperones has grown steadily since 1987, when Ellis and his colleagues first proposed the term. The first molecular chaperones studied were those of chloroplasts, mitochondria, and bacteria. Within a few years, however, chaperones were identified in other eukaryotic locations. Chaperone proteins are abundant under normal conditions and increase to still higher levels in response to stresses such as increased temperature—a condition called *heat shock*—or an increase in the cellular content of unfolded proteins. Many common chaperone proteins fall into two families, called *Hsp60* and *Hsp70*. ("Hsp" stands for *heat-shock protein,* and the numbers refer to the approximate molecular weights of the protein's polypeptide monomers—60,000 and 70,000, respectively.) The proteins within each Hsp family are evolutionarily related, and they are found throughout the biological world. Hsp70 proteins, for example, have been found in bacteria and in cells of a wide variety of eukaryotes, where they are present in several intracellular locations.

In addition to their role in the folding of newly made polypeptides, chaperone proteins have other functions. Some of these relate to protein activity in the cell, others to protein transport into organelles (see Chapter 20). One important class of chaperones, the *nucleoplasmins,* are nuclear proteins that perform a variety of functions during such nuclear processes as DNA replication, transcription and the processing of RNA, and the transport of molecules into and out of the nucleus.

Noncovalent Interactions Are Important in the Folding of Macromolecules

Whether assisted by molecular chaperones or not, polypeptides clearly fold and self-assemble without the input of further energy or information, and the three-dimensional structure of a protein is remarkably stable once it has been attained. To understand the self-assembly of proteins (and, as it turns out, of other biological molecules and structures as well), we need to consider both the covalent and the noncovalent bonds that hold polypeptides and other macromolecules together.

Covalent bonds are easy to understand. Every protein or other macromolecule in the cell is held together by strong covalent bonds such as those discussed earlier in the chapter. A covalent bond forms whenever two atoms *share* electrons rather than gaining or losing electrons completely. Electron sharing is an especially prominent feature of the carbon atom, which has four electrons in its outer orbital and is therefore at the midpoint between the tendency to gain or lose electrons.

Covalent bonds not only link the monomers of a polypeptide together, they also stabilize the three-dimensional structure of many proteins. Specifically, sulfur-sulfur covalent bonds (—S—S—) play an important role in protein structure. Consider, for example, the ribonuclease molecule in Figure 2-18. As noted earlier, this molecule contains the amino acid cysteine at eight locations along the polypeptide chain. Each cysteine has a free sulfhydryl group that can form a covalent disulfide bond by reacting oxidatively with the sulfhydryl group of another cysteine located elsewhere along the polypeptide chain. The native ribonuclease molecule has four disulfide bonds, which confer structural stability upon the molecule. Many proteins are stabilized in this way by disulfide bonds, both within a single polypeptide and between the monomeric polypeptides of a multimeric protein.

As important as covalent bonds are in cellular chemistry, the complexity of molecular structure cannot be described in terms of covalent bonds alone. Most of the structures in the cell are held together by much weaker forces—the **noncovalent interactions** within and between proteins and other macromolecules.

The most important noncovalent interactions in biological macromolecules are hydrogen bonds, ionic bonds, van der Waals interactions, and hydrophobic interactions. Each of these interactions is introduced here and discussed in more detail in Chapter 3 (see Figure 3-5). As we've already noted, *hydrogen bonds* involve weak, attractive interactions between a hydrogen atom that is covalently bonded to an electronegative atom and a second electronegative atom. *Ionic bonds* are noncovalent electrostatic interactions between two oppositely charged ions, which in the case of macromolecules are positively charged, and negatively charged functional groups such as hydroxyl groups, amino groups, carbonyl groups, and sulfhydryl groups. *Van der Waals interactions* (or *forces*) are weak attractive interactions between two atoms that are due to transient separations of charge (called *dipoles*) in both atoms. These interactions are seen only if the atoms are very close to one another and are oriented appropriately. The term *hydrophobic interactions* denotes the tendency of nonpolar groups within a macromolecule to associate with each other, thereby minimizing their contact with surrounding water molecules and with hydrophilic groups within the same (or even another) macromolecule.

Each of these bonds or interactions tends to draw parts of a macromolecule together, but when atoms or groups of atoms get too close, they begin to repulse each other due to overlap of their outer electron orbitals. The *van der Waals radius* of a specific atom or functional group specifies the "private space" around it that sets limits on how close other atoms or groups can come before being repulsed. Van der Waals radii for the most common atoms in biological molecules (H, O, N, C, S and P) are all in the range 0.12–0.19 nm. Van der Waals radii provide the basis for *space-filling models* of biological macromolecules, such as that for the protein insulin shown in Figure 2-19 and the DNA molecule shown in Figure 3-19b.

Self-Assembly Also Occurs in Other Cellular Structures

The same principle of self-assembly that accounts for the folding and interactions of polypeptides also applies to

Figure 2-19 A Space-Filling Model of a Protein. All of the atoms and chemical groups in this model of the protein insulin are represented as spheres with the appropriate van der Waals radii. Notice how tightly the atoms pack against one another. (Color key: C = red, H = white, O = blue, S = yellow.)

more complex cellular structures. Many of the character-istic structures of the cell are complexes of two or more different kinds of polymers and therefore clearly involve interactions chemically distinct from those of polypeptide folding and association. However, the principle of self-assembly may nonetheless apply. For example, *ribosomes* contain both RNA and proteins; *membranes,* as we have already seen, are made up of both phospholipids and pro-teins; and even the *primary cell wall* that surrounds a plant cell, although composed mainly of cellulose fibrils, also contains a variety of other carbohydrates as well as a small but apparently crucial protein component. Yet, despite the chemical differences among such polymers as proteins, nucleic acids, and polysaccharides, the interactions that drive these supramolecular assembly processes seem to be essentially the same as those that dictate the folding of individual protein molecules.

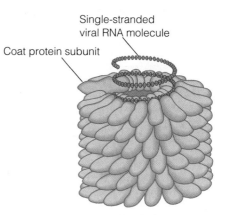

Figure 2-20   A Structural Model of Tobacco Mosaic Virus (TMV). A single-stranded RNA molecule is coiled into a helix, surrounded by a coat consisting of 2130 identical protein subunits. Only a portion of the entire TMV virion is shown here, and several layers of protein have been omitted from the upper end of the structure to reveal the helical RNA molecule inside.

The Tobacco Mosaic Virus Is a Case Study in Self-Assembly

Some of the most definitive findings concerning the self-assembly of complex biological structures have come from studies with viruses. As we will learn in Chapter 4, a *virus* is a complex of proteins and nucleic acid, either DNA or RNA. A virus is not itself alive, but it can invade and infect a living cell and subvert the synthetic machinery of the cell for the production of more viruses. Inherent in the pro-duction of more viruses is the synthesis of the viral nucleic acid and viral proteins and their subsequent assembly into the mature *virion,* or viral particle. Studies of this assem-bly process have provided a wealth of information on structure and assembly that exceeds what we know about any other self-assembly system.

An especially good example is tobacco mosaic virus (TMV), a plant virus that has long been popular with molecular biologists. TMV is a rodlike particle about 18 nm in diameter and 300 nm in length. It consists of a single strand of RNA with about 6000 nucleotides and about 2130 copies of a single kind of polypeptide, the *coat pro-tein,* each with 158 amino acids. The RNA molecule forms a helical core, with a cylinder of protein subunits clustered around it (Figure 2-20).

Heinz Fraenkel-Conrat and his colleagues carried out several important experiments that contributed signifi-cantly to our understanding of self-assembly. They sepa-rated TMV into its RNA and protein components and then allowed them to reassemble in vitro (i.e., in a test tube). Viral particles regenerated that were capable of infecting plant cells. This result was one of the first and most convincing demonstrations that the components of a complex biological structure can reassemble sponta-neously into functional entities without external informa-tion. Especially interesting was the finding that the RNA from one strain of virus could be mixed with the protein component from another strain to form a hybrid virus

that was also infective. As expected, the source of the RNA and not the protein determined the type of virus that was made by the infected cells.

The assembly process has since been studied in detail and is known to be surprisingly complex. The basic unit of assembly is a two-layered disk of coat protein, each layer consisting of 17 identical subunits arranged in a ring (Figure 2-21a). Each disk is initially a cylindrical structure but undergoes a conformational change that tightens it into a helical shape as it interacts with a short segment (about 102 nucleotides) of the RNA molecule (Figure 2-21b). This transition allows another disk to bind (Figure 2-21c), and each successive disk undergoes a conformational change from a cylinder to a helix and binds to another 102 bases of the RNA. This disk-by-disk elongation process contin-ues, the successively stacked disks creating a helical path for the RNA strand (Figure 2-21d). The process eventually gives rise to the mature virion, its RNA completely cov-ered with coat protein.

Self-Assembly Has Limits

In many cases, the information required to specify the exact configuration of a cellular structure seems to lie entirely within the polymers that contribute to the structure. Such self-assembling systems achieve stable three-dimensional configurations without additional information input because the information content of the component poly-mers is adequate to specify the complete assembly process. Even in assisted self-assembly, the molecular chaperones provide no additional information.

However, there also appear to be some assembly sys-tems that depend, in addition, on information supplied by a preexisting structure. In such cases, the ultimate structure arises not by assembling the components into a new struc-ture but rather by ordering the components into the matrix

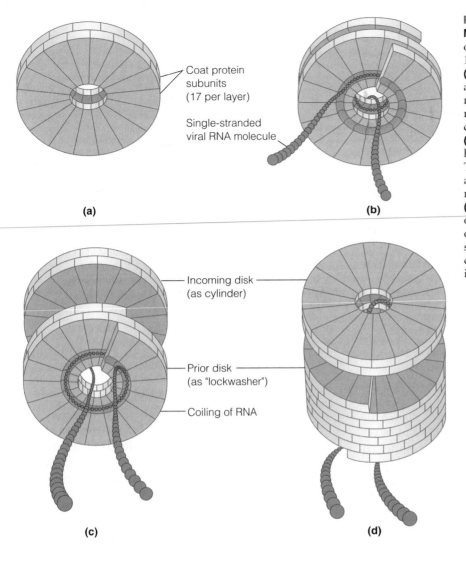

(a)

(b)

(c)

(d)

Coat protein
subunits
(17 per layer)

Single-stranded
viral RNA molecule

Incoming disk
(as cylinder)

Prior disk
(as "lockwasher")

Coiling of RNA

Figure 2-21 Self-Assembly of the Tobacco Mosaic Virus (TMV) Virion. **(a)** The unit of assembly is a double-layered disk with 17 protein subunits per layer. **(b)** Assembly begins as the viral RNA molecule associates with a disk, causing the RNA molecule to form a loop (involving 102 nucleotides) and the disk to change from a cylindrical to a helical conformation. **(c)** Another disk is then added, and another length of RNA becomes associated with it. The prior disk, meanwhile, has changed from a cylindrical to a helical conformation that makes it look somewhat like a lockwasher. **(d)** Each incoming disk adds two more layers of protein subunits and causes another length of RNA to become associated with the structure in a spiral loop. The process continues until the entire RNA molecule is involved and the virion is complete.

of an existing structure. Examples of cellular structures that are routinely built up by adding new material to existing structures are membranes, cell walls, and chromosomes.

On the other hand, such structures are not yet sufficiently characterized to determine whether the presence of a preexisting structure is obligatory or whether, under the right conditions, the components might be capable of self-assembly. Evidence from studies with artificial membranes and with chromatin (isolated chromosomal components), for example, suggests that a preexisting structure, though routinely present in vivo, may not be an indispensable requirement for the assembly process. Additional insight will be necessary before we can say with certainty whether, and to what extent, external information is required or exploited in cellular assembly processes.

Hierarchical Assembly Provides Advantages for the Cell

Each of the assembly processes we have looked at exemplifies the basic cellular strategy of **hierarchical assembly**

illustrated in Figure 2-14. Biological structures are almost always constructed in a hierarchical manner, with subassemblies acting as important intermediates en route from simple starting molecules to the end products of organelles, cells, and organisms. Consider how cellular structures are made. First, large numbers of similar, or even identical, monomeric subunits are assembled by condensation into polymers. These polymers then aggregate spontaneously but specifically into characteristic multimeric units. The multimeric units can, in turn, give rise to still more complex structures and eventually to assemblies that are recognizable as distinctive subcellular structures.

This hierarchical process has the double advantage of chemical simplicity and efficiency of assembly. To appreciate the chemical simplicity, we need only recognize that almost all structures found in cells and organisms are synthesized from about 30 small precursor molecules, which George Wald has called the "alphabet of biochemistry." This "alphabet" includes the 20 amino acids found in pro-

Further Insights TEMPUS FUGIT AND THE FINE ART OF WATCHMAKING

"Drat! Another defective watch!" With a look of disgust, Tempus Fugit the watchmaker tossed the faulty timepiece into the wastebasket and grumbled to himself, "That's two out of the last three watches I've had to throw away. What kind of a watchmaker am I, anyway?"

"A good question, Fugit," came a voice from the doorway. Tempus looked up to see Caveat Emptor entering the shop. "Maybe I can help you with it if you'll tell me a bit about how you make your watches."

"Nobody asked for your help, Emptor," growled Tempus testily, wishing fervently he hadn't been caught thinking out loud.

"Ah well, you'll get it just the same," continued Caveat, quite unperturbed. "Now tell me just how you make a watch and how long it takes you. I'd be especially interested in comparing your procedure with the way Pluribus Unum does it in his new shop down the street."

At the very mention of his competitor's name, Tempus groaned again. "Unum!" he blustered. "What does he know about the fine art of watchmaking?"

"A good deal, apparently; probably more than you do," replied Caveat. "But tell me exactly how you go about it. How many steps does it take you per watch, and how often do you make a mistake?"

"It takes exactly 100 operations to make a watch. Every step has to be done exactly right, or the watch won't work. It's tricky business, but I've got my error rate down to 1%," said Tempus with at least a trace of pride in his voice. "I can make a watch from start to finish in exactly one hour, but I only make 36 watches each week because I always take Tuesday afternoons off to play darts."

"Well now, let's see," said Caveat, as he pulled out a pocket calculator. "That means, my good Mr. Fugit, that you only make about 13 watches per week that actually work. All the rest you have to throw away just like the one you pitched as I entered your shop."

"How did you know that, you busybody?" Tempus asked defensively, wondering how this know-it-all with his calculator had guessed his carefully kept secret.

"Elementary, my dear Watson," returned Caveat gleefully. "Simple probability is all it takes. You told me that each watch requires 100 operations and that there's a 99% chance that you'll get a given step right. That's 0.99 times itself 100 times, which comes out to 0.366. So only about 37% of your watches will be put together right, and you can't even tell anything about a given watch until it's finished and you test it. Want to know how that compares with Unum's shop?"

Tempus was about to protest, but his tormentor hurried on with scarcely a pause for breath. "He can manage 100 operations per hour too, and his error rate is exactly the same as yours. He's also off every Tuesday afternoon, but he gets about 27 watches made every week. That's twice your output, Fugit! No wonder he's got so many watches in his window and so many customers at his door. I've heard that he's even thinking of expanding his shop. Want to know how he does it?"

Tempus was too depressed to even attempt a protest. And besides, although he wouldn't like to admit it, he really was dying to know how Unum managed it.

"Subunit assembly, Fugit, that's the answer! Subunit assembly! Instead of making each watch from scratch in 100 separate steps, Unum assembles the components into 10 pieces, each requiring 10 steps." Caveat began pushing calculator buttons again. "Let's see, he performs 10 operations with a 99% success rate for each, so I take 0.99 to the tenth power instead of the hundredth. That's 0.904, which means that about 90% of his subunits have no errors in them. So he spends about 33 hours each week making 330 subunits, throws the defective ones away, and still has about 300 to assemble into 30 watches during the last three hours on Friday afternoon. That takes 10 steps per watch, with the same error rate as before, so again he comes up with about 90% success. He has to throw away about 3 of his finished watches and ends up with about 27 watches to show for his efforts. Meanwhile, you've been working just as hard and just as accurately, yet you only make half as many watches. What do you have to say to that, my dear Fugit?"

Tempus managed a weary response. "Just one question, Emptor, and I'll probably hate myself for asking. But do you happen to know what Unum does on Tuesday afternoons?"

"I'm glad you asked," replied Caveat as he slipped his calculator back into his pocket and headed for the door. "But I don't think you'll be able to interest him in darts—he spends every Tuesday afternoon giving watchmaking lessons."

teins, the 5 aromatic bases present in nucleic acids, 2 sugars, and 3 lipid molecules. We will encounter each of these in Chapter 3 (see Table 3-1). Given these building blocks and the polymers that can be derived from them through just a few different kinds of condensation reactions, most of the structural complexity of life can be readily elaborated by hierarchical assembly into successively more complex structures.

The second advantage of hierarchical assembly lies in the "quality control" that can be exerted at each level of assembly, allowing defective components to be discarded at an early stage rather than being built into a more complex structure that would be more costly to reject and replace. Thus, if the wrong subunit has been inserted into a polymer at some critical point in the chain, that particular molecule may have to be discarded, but the cell will be spared the cost of synthesizing a more complicated supramolecular assembly or even a whole organelle before the defect is discovered. The story presented in Box 2A is intended to illustrate this basic principle.

Perspective

In summary, the basic molecular building blocks of the cell are small organic molecules that can be strung together by stepwise polymerization to form the macromolecules that are so important to cellular structure and function. About 30 different kinds of monomeric units account for most cellular structure. These monomeric units consist primarily of carbon, hydrogen, oxygen, nitrogen, and phosphorus, all readily available to the living world in the form of inorganic compounds such as carbon dioxide, water, ammonia, nitrate, and phosphate. Because of the presence of polar functional groups, most of these small organic molecules are quite water-soluble. The cell is able to retain these molecules within defined spaces because it has selectively permeable membranes with a hydrophobic interior that most polar molecules or ions cannot penetrate unless specific transport proteins are present in the membranes.

Polymerization into macromolecular form requires that the monomeric molecules be appropriately activated, usually at the expense of ATP. The resulting polymers can fold and coil spontaneously and then interact with one another in a unique, predictable manner to generate successively higher-order structures. The information needed for these assembly processes is inherent in the chemical nature of the monomeric units and the order in which chemically different monomers are strung together. In some (maybe even many) cases, proteins called molecular chaperones appear to mediate assembly, but they do so by inhibiting or preventing the formation of incorrect structure, not by providing additional steric information. This overall strategy of hierarchical assembly from subunits has the dual advantages of chemical simplicity and efficiency of assembly.

With these principles in mind, we are ready to proceed to Chapter 3. There, we will examine the major kinds of biological macromolecules and the chemical nature of the subunits from which they are synthesized.

Key Terms for Self-Testing

The Importance of Carbon
organic chemistry (p. 18)
biological chemistry (p. 18)
carbon atom (p. 18)
valence (p. 18)
covalent bond (p. 18)
single bond (p. 18
double bond (p. 18)
triple bond (p. 18)
bond energy (p. 19)
calorie (p. 19)
hydrocarbon (p. 20)
functional group (p. 20)
tetrahedral (carbon atom) (p. 20)
stereoisomer (p. 20)
asymmetric carbon atom (p. 20)

The Importance of Water
electronegative (p. 22)

polarity (p. 22)
specific heat (p. 22)
heat of vaporization (p. 23)
solvent (p. 23)
solute (p. 23)
hydrophilic (p. 23)
hydrophobic (p. 23)
sphere of hydration (p. 24)

The Importance of Selectively Permeable Membranes
membrane (p. 24)
amphipathic molecule (p. 24)
lipid bilayer (p. 24)

The Importance of Synthesis by Polymerization
macromolecule (p. 26)
monomer (p. 28)

informational macromolecule (p. 28)
storage macromolecule (p. 29)
structural macromolecule (p. 29)
condensation reaction (p. 30)
carrier molecule (p. 30)
activated monomer (p. 30)
directionality (p. 30)

The Importance of Self-Assembly
self-assembly (p. 31)
polypeptide (p. 31)
denaturation (p. 31)
renaturation (p. 31)
molecular chaperone (p. 33)
assisted self-assembly (p. 33)
noncovalent interaction (p. 34)
hierarchical assembly (p. 36)

Problem Set

More challenging problems are marked with a •.

2-1. The Fitness of Carbon. Carbon has a number of properties that make it especially fit to play a key role in biological molecules. One way to appreciate these properties is to compare and contrast them with those of silicon, the element immediately beneath carbon in the periodic table. Contrast each of the following properties of silicon with the comparable property of carbon, and indicate, if appropriate, why carbon is a fitter element than silicon for biological purposes in that particular regard.

(a) Silicon is a larger atom, with an atomic weight of 28.

(b) Silicon combines with oxygen to form insoluble silicates or network polymers of silicon dioxide (quartz).

(c) Silicon does not readily form double or triple bonds.

(d) Silicon has four valence electrons and is quite abundant in the Earth's crust (28% vs. 0.19% for carbon).

(e) Polymers of silicon are not stable in water.

2-2. **The Fitness of Water.** For each of the following statements about water, decide whether the statement is true and describes a property that makes water a desirable component of cells (T); is true but describes a property that has no bearing on water as a cellular constituent (X); or is false (F). For each true statement, indicate a possible benefit to living organisms.

(a) Water is a polar molecule and hence an excellent solvent for polar compounds.

(b) Water can be formed by the reduction of molecular oxygen (O_2).

(c) The density of water is less than the density of ice.

(d) The molecules of liquid water are extensively hydrogen-bonded to one another.

(e) Water does not absorb visible light.

(f) Water is odorless and tasteless.

(g) Water has a high specific heat.

(h) Water has a high heat of vaporization.

2-3. **True or False.** Identify each of the following statements as either true (T) or false (F). For each false statement, change the statement to make it true.

(a) Oxygen, nitrogen, and carbon are the elements that most readily form strong multiple bonds.

(b) Water has a higher heat of vaporization than most other liquids because of its high specific heat.

(c) Oil droplets in water coalesce to form a separate phase because of the strong attraction of hydrophobic molecules for each other.

(d) In a hydrogen bond, electrons are shared between a hydrogen atom and two neighboring electronegative atoms.

(e) Formation of internal disulfide bonds is the major interaction that drives protein folding.

2-4. **Bond Energies.** Of considerable importance in assessing the fitness of the bonds in organic molecules is the relationship between the amount of energy required to break these bonds and the amount of energy available in the solar radiation to which these bonds are exposed.

(a) Given that the carbon-carbon single bond is safe in green light (500 nm) but can be broken by the energy of ultraviolet light (300 nm, for example), where does the cutoff come? That is, what wavelength of light has just enough energy to break a carbon-carbon single bond?

(b) Would a carbon-nitrogen bond be more or less stable than a carbon-carbon single bond at this wavelength?

(c) What about a carbon-carbon double bond?

2-5. **Stereoisomers.** For each compound below, indicate how many stereoisomers exist and draw the structure of each.

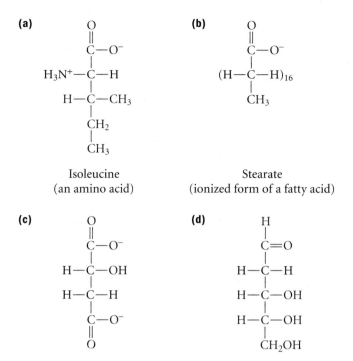

Isoleucine (an amino acid)

Stearate (ionized form of a fatty acid)

Malate (ionized form of a carboxylic acid)

Deoxyribose (pentose present in DNA)

2-6. **Solubility Properties of Biological Molecules.**

(a) Why is the organic compound hexadecane not a common chemical constituent of cells, whereas palmitate is?

$$CH_3-(CH_2)_{14}-CH_3$$

Hexadecane

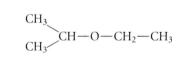

Palmitate

(b) Which of the following compounds is more likely to occur in cells? Explain your answer.

(i)

$$\begin{array}{c} CH_3 \\ {}^{\diagdown} \\ CH_3 {}^{\diagup} \end{array} CH-O-CH_2-CH_3$$

(ii)

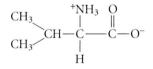

2-7. **The Polarity of Water.** Defend the assertion that all of life as we know it depends critically on the fact that the angle between the two hydrogen atoms in the water molecule is 104.5° and not 180°.

•2-8. **The Principle of Polymers.** Polymers clearly play an important role in the molecular economy of the cell. What advantage does each of the following features of biological polymers have for the cell?

(a) A particular kind of polymer is formed using the same kind of condensation reaction to add each successive monomeric unit.

(b) The bonds between monomers are formed by the removal of water and are broken or cleaved by the addition of water.

• 2-9.   TMV Assembly. Each of the following statements is an experimental observation concerning the reassembly of tobacco mosaic virus (TMV) virions from TMV RNA and coat protein subunits. In each case, state as carefully as possible a reasonable conclusion that can be drawn from the experimental finding.

(a) When RNA from a specific strain of TMV is mixed with coat protein from the same strain, infectious virions are formed.

(b) When RNA from strain A of TMV is mixed with coat protein from strain B, the reassembled virions are infectious, giving rise to strain A virus particles in the infected tobacco cells.

(c) Isolated coat protein monomers can polymerize into a viruslike helix in the absence of RNA.

(d) In infected plant cells, the TMV virions that form contain only TMV RNA and never any of the various kinds of cellular RNAs present in the host cell.

(e) Regardless of the ratio of RNA to coat protein in the starting mixture, the reassembled virions always contain RNA and coat protein in the ratio of three nucleotides of RNA per coat protein monomer.

• 2-10.   Here's the Answer; What's the Question? Each of the following statements is an *answer;* indicate in each case what the *question* is.

(a) This component of solar radiation is sufficiently energetic to break carbon-carbon bonds.

(b) This property of water makes it possible for land animals to cool themselves by surface evaporation with minimum loss of body fluid.

(c) The next monomeric unit to be added to a polymer must be in this form in order for the condensation reaction to be energetically favorable.

(d) This principle of macromolecular biosynthesis minimizes the amount of genetic information needed and allows imperfect components to be discarded at several stages.

(e) This property of a molecule makes the molecule likely to be found in a membrane.

Suggested Reading

References of historical importance are marked with a •.

General References and Reviews
• Herriott, J., G. Jacobson, J. Marmur, and W. Parsom. *Papers in Biochemistry.* Reading, MA: Addison-Wesley, 1984.
Lehninger, A. L., D. L. Nelson, and M. M. Cox. *Principles of Biochemistry,* 3d ed. New York: Worth, 1999.
Mathews, C. K., K. E. van Holde, and K. G. Ahern. *Biochemistry,* 3d ed. Menlo Park, CA: Benjamin/Cummings, 2000.
Thornton, R. M.. *The Chemistry of Life.* Menlo Park, CA: Benjamin/Cummings, 1998: 42a. (A CD-ROM that uses interactive exercises to help bridge the gap between chemistry and biology.)
Wald, G. The origins of life. *Proc. Natl. Acad. Sci. USA* 52 (1994): 595.

The Importance of Water
Bryant, R. G. The dynamics of water-protein interactions. *Annu. Rev. Biophys. Biolmolec. Struct.* 25 (1996): 29.
Mathews, R. Wacky water. *New Scientist* (June 21, 1997).
Pennisi, E. Water, water everywhere. *Science News* (February 20, 1993).
Westof, E. ed. *Water and Biological Macromolecules.* Boca Raton, FL: CRC Press, 1993.
Wiggins, P. M. Role of water in some biological processes. *Microbiol. Rev.* 54 (1990): 432.

The Importance of Membranes
Baldwin, S. A. *Membrane Transport: A Practical Approach.* Oxford: Oxford University Press, 2000.
Lipowsky, R., and E. Sackmann. *The Structure and Dynamics of Membranes.* Amsterdam: Elsevier Science, 1995. (Volume 1 in the series *Handbook of Biological Physics,* A. Hoff, ed.)
Petty, H. R. *Molecular Biology of Membranes: Structure and Function.* New York: Springer Verlag, 1993.
Tien, H. T., and A. Ottova-Leitmannova. *Membrane Biophysics: As Viewed from Experimental Bilayer Lipid Membranes.* New York: Elsevier, 2000.
Yeagle, P. L. *The Membranes of Cells,* 2d ed. New York: Academic Press, 1993.

The Importance of Macromolecules
Brandon, C., and J. Tooze. *Introduction to Protein Structure.* New York: Garland, 1991.
Creighton, T. E. *Proteins: Structure and Molecular Properties,* 2d ed. New York: W. H. Freeman, 1993.
Gesteland, R. F., and J. F. Atkins, eds. *The RNA World.* Cold Spring Harbor, NY: Cold Spring Harbor Laboratory Press, 1994.
Ingber, D. E. The architecture of life. *Sci. Amer.* 278 (January 1998): 48.
Schulz, G. E., and R. H. Schirmer. *Principles of Protein Structure.* New York: Springer Verlag, 1990.
• Watson, J. D., and J. Tooze. *The DNA Story.* San Francisco: Freeman, 1981.
Yon, J. M. Protein folding: Concepts and perspectives. *Cell. Mol. Life Sci.* 53 (1997): 557.

The Importance of Self-Assembly
• Anfinsen, C. B. Principles that govern the folding of protein chains. *Science* 181 (1973): 223.
Buchner, J. Supervising the fold: Functional principles of molecular chaperones. *FASEB J.* 10 (1996): 10.
• Butler, P. J. G., and A. Klug. The assembly of a virus. *Sci. Amer.* (November 1978): 62.
Craig, E. A. Chaperones: Helpers along the pathways to protein folding. *Science* 260 (1993): 1902.
Creighton, T. E. Protein folding: An unfolding story. *Curr. Biol.* 5 (1995): 353.
• Ellis, R. J. Discovery of molecular chaperones. *Cell Stress and Chaperones* 1 (1996): 155.
Ellis, R. J., ed. *The Chaperonins.* New York: Academic Press, 1996.
Ellis, R. J., and S. M. Van der Vies. Molecular chaperones. *Annu. Rev. Biochem.* 60 (1991): 321.
• Fraenkel-Conrat, H., and R. C. Williams. Reconstitution of active tobacco mosaic virus from its inactive protein and nucleic acid components. *Proc. Natl. Acad. Sci. USA* 41 (1955): 690.
Georgopoulos, C. The emergence of the chaperone machine. *Trends Biochem. Sci.* 17 (1992): 295.
Hendrick, J. P., and F.-U. Hartl. Molecular chaperone functions of heat-shock proteins. *Annu. Rev. Biochem.* 62 (1993): 349.
Richards, F. M. The protein folding problem. *Sci. Amer.* (January 1991): 54.
Strauss, E. How proteins take shape. *Science News* (September 6, 1997).

3

The Macromolecules of the Cell

In Chapter 2, we looked at some of the basic chemical principles of cellular organization. We saw that each of the major kinds of biological macromolecules—proteins, nucleic acids, and polysaccharides—consists of a relatively small number (from 1 to 20) of repeating monomeric units. These polymers are synthesized by condensation reactions in which activated monomers are linked together and water is removed. Once synthesized, the individual polymer molecules fold and coil spontaneously into stable, three-dimensional shapes. These folded molecules then associate with one another in a hierarchical manner to generate higher levels of structural complexity, usually without further input of energy or information.

We are now ready to examine the major kinds of biological macromolecules. In each case, we will focus first on the chemical nature of the monomeric components and then on the synthesis and properties of the polymer itself. Some of the most common monomers are included in Table 3-1, which lists and categorizes the 30 most common small molecules in cells. We begin our survey with proteins because they play such important and widespread roles in cellular structure and function. We then move on to nucleic acids and polysaccharides. The tour concludes with lipids, which do not quite fit the definition of a polymer but are important cellular components in their own right.

Proteins

Proteins are without a doubt the most important and ubiquitous macromolecules in the cell. Almost all enzymes are proteins, as are all antibodies and some hor-mones. In addition, proteins form the basis of most cellular structures. Connective tissue, muscle fibrils, cilia, flagella—all are made primarily or exclusively of proteins. Whether we are talking about carbon dioxide fixation in photosynthesis, oxygen transport in the blood, or the motility of a flagellated bacterium, we are dealing with processes that depend crucially on particular proteins with specific properties and functions.

Based on function, proteins fall into four major classes. Many proteins are *enzymes,* serving as catalysts that greatly increase the rates of the thousands of chemical reactions on which life depends. *Structural proteins,* on the other hand, provide support and shape to cells and organelles, giving cells their characteristic appearances. *Motility proteins* play key roles in the contraction and movement of cells and intracellular structures. *Regulatory proteins* are responsible for control and coordination of cellular functions, ensuring that cellular activities are regulated to meet cellular needs. Most proteins are *monofunctional,* which simply means that a specific protein has a single function, whether catalytic, structural, motile, or regulatory. However, some proteins are *bifunctional,* which means that an individual protein plays two distinctly different roles.

With virtually everything that a cell is or does dependent on the proteins it contains, it is clear we need to understand what proteins are and why they have the properties they do. We will begin our discussion by looking at the amino acids present in the proteins and will then consider some properties of proteins themselves.

The Monomers Are Amino Acids

Proteins are linear polymers of **amino acids.** More than 60 different kinds of amino acids are typically present in a

Table 3-1   The 30 Most Common Small Molecules in Cells*

Kind of Molecule	Number Present	Names of Molecules		Role in Cell	Figure Number for Structures
Amino acid	20	Alanine Arginine Asparagine Aspartate Cysteine Glutamate Glutamine Glycine Histidine Isoleucine	Leucine Lysine Methionine Phenylalanine Proline Serine Threonine Tryptophan Tyrosine Valine	Monomeric units of all proteins	3-2
Aromatic base (purines and pyrimidines)	5	Adenine Cytosine Guanine	Thymine Uracil	Components of nucleic acids (DNA and RNA)	3-15
Sugar (monosaccharides)	2	Ribose		Component of nucleic acids	3-15
		Glucose		Energy metabolism; component of starch and glycogen	3-21
Lipid	3	Choline Glycerol Palmitate		Components of phospholipids	 3-27b 3-28a

*Adapted from Wald (1994).

cell, but only 20 different kinds are used in protein synthesis, as indicated in Table 3-1. (Some proteins contain more than 20 different kinds of amino acids, but that is usually because of chemical modifications that occur after the protein has been synthesized.) Although most proteins contain all or most of the 20 amino acids, the proportions vary greatly among proteins, and no two proteins have the same amino acid sequence.

Every amino acid has the basic structure illustrated in Figure 3-1, with a carboxyl group, an amino group, a hydrogen atom, and a so-called R group all attached to a single carbon atom. Except for glycine, for which the R group is just a hydrogen atom, all amino acids have at least one asymmetric carbon atom and therefore exist in two stereoisomeric forms, called D- and L-amino acids. Both kinds exist in nature, but only L-amino acids occur in proteins.

Because the carboxyl and amino groups in Figure 3-1 are common features of all amino acids, the specific properties of a particular amino acid obviously depend on the chemical nature of the R groups, which range from a single hydrogen atom to relatively complex aromatic groups. Shown in Figure 3-2 are the structures of the 20 L-amino acids found in proteins. The three-letter abbreviations given in parentheses for each amino acid are widely used by biochemists and molecular biologists. Table 3-2 lists these three-letter abbreviations and the corresponding one-letter abbreviations that are also commonly used.

Nine of these amino acids have nonpolar R groups and are therefore *hydrophobic* (Group A). As you look at their structures, you will notice the hydrocarbon nature of the R groups, with oxygen and nitrogen conspicuous by

their absence. These are the amino acids that tend to become localized in the interior of the molecule as a polypeptide folds into its three-dimensional shape. If a protein (or a region of the molecule) has a preponderance of

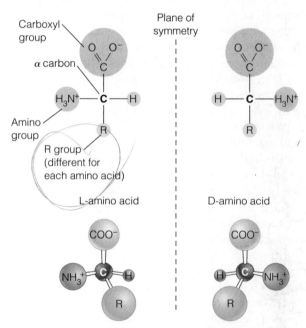

Figure 3-1   The Structure and Stereochemistry of an Amino Acid. Because the α carbon atom is asymmetric in all amino acids except glycine, most amino acids can exist in two isomeric forms, designated L and D and shown here as (top) conventional structural formulas and (bottom) ball-and-stick models. The L and D forms are stereoisomers, with the vertical dashed line as the plane of symmetry. Of the two forms, only L-amino acids are present in proteins.

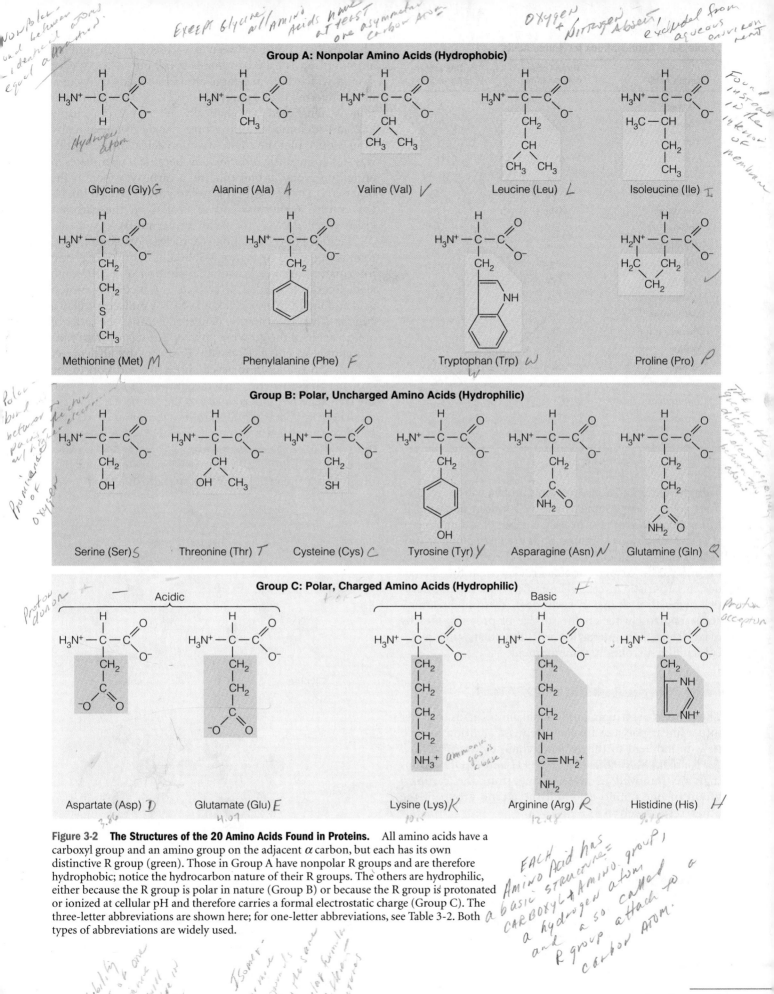

Figure 3-2 The Structures of the 20 Amino Acids Found in Proteins. All amino acids have a carboxyl group and an amino group on the adjacent α carbon, but each has its own distinctive R group (green). Those in Group A have nonpolar R groups and are therefore hydrophobic; notice the hydrocarbon nature of their R groups. The others are hydrophilic, either because the R group is polar in nature (Group B) or because the R group is protonated or ionized at cellular pH and therefore carries a formal electrostatic charge (Group C). The three-letter abbreviations are shown here; for one-letter abbreviations, see Table 3-2. Both types of abbreviations are widely used.

Table 3-2   Abbreviations for Amino Acids

Amino Acid	Three-Letter Abbreviation	One-Letter Abbreviation
Alanine	Ala	A
Arginine	Arg	R
Asparagine	Asn	N
Aspartate	Asp	D
Cysteine	Cys	C
Glutamate	Glu	E
Glutamine	Gln	Q
Glycine	Gly	G
Histidine	His	H
Isoleucine	Ile	I
Leucine	Leu	L
Lysine	Lys	K
Methionine	Met	M
Phenylalanine	Phe	F
Proline	Pro	P
Serine	Ser	S
Threonine	Thr	T
Tryptophan	Trp	W
Tyrosine	Tyr	Y
Valine	Val	V

The covalent bond between a carboxyl group and an amino group is called an *amide bond,* but in the special case in which both reactants are amino acids, it is commonly referred to as a **peptide bond.** Peptide bond formation is illustrated schematically in Figure 3-3 using ball-and-stick models of the amino acids glycine and alanine. Notice that the chain of amino acids formed in this way has an intrinsic *directionality,* because it terminates in an amino group at one end and a carboxyl group at the other end. The end with the amino group is called the **N-** (or **amino**) **terminus,** and the end with the carboxyl group is called the **C-** (or **carboxyl**) **terminus.**

Peptide bond formation is actually more complicated than is suggested by Figure 3-3, because both energy and information are needed for the addition of each amino acid. Energy is needed to activate the incoming amino acid and link it to a special kind of RNA molecule called a *transfer RNA.* Information is needed because the order of amino acids in the growing polypeptide chain is not random but is specified genetically. Ensuring that the correct amino acid is chosen for each successive addition to the growing chain depends on a recognition process between the transfer RNA that is linked to the amino acid and the *messenger RNA* that is bound to the ribosome. We will get to all of these details in due course. Here, we need only note that as each new peptide bond is formed, the growing chain of amino acids is lengthened by one, and that the process requires both energy and information.

hydrophobic amino acids, the whole protein will be excluded from aqueous environments and will instead be found in hydrophobic locations, such as the interior of a membrane.

The remaining 11 amino acids are *hydrophilic,* with R groups that are either distinctly *polar* (Group B; notice the prominence of oxygen) or actually *charged* at the pH values characteristic of cells (Group C; notice the net negative or positive charge in each case). Hydrophilic amino acids tend to cluster on the surface of proteins, thereby maximizing their interaction with the polar water molecules in the surrounding environment.

The Polymers Are Polypeptides and Proteins

The process of stringing individual amino acids together into a linear polymer involves stepwise addition of each new amino acid to the growing chain by a *dehydration* (or *condensation*) *reaction.* The —H and —OH groups that are removed as water come from the carboxyl group of one amino acid and the amino group of the other, respectively:

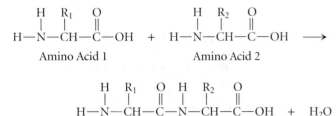

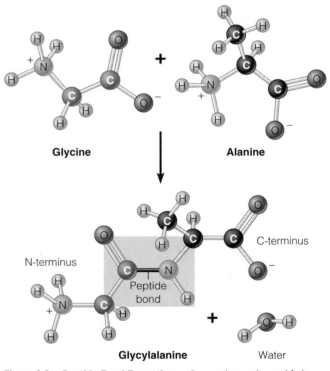

Figure 3-3   Peptide Bond Formation. Successive amino acids in a polypeptide are linked to one another by peptide bonds between the carboxyl group of one amino acid and the amino group of the next. Shown here is the formation of a peptide bond between the amino acids glycine and alanine.

Although this process of elongating a chain of amino acids is often called *protein synthesis,* the term is not entirely accurate because the immediate product of amino acid polymerization is not a protein but a **polypeptide.** A protein is a polypeptide chain (or several such chains) that has attained a unique, stable, three-dimensional shape and is biologically active as a result. Some proteins consist of a single polypeptide and therefore achieve their final shape as a consequence of the folding and coiling that occur spontaneously as the chain is being formed (see Figure 2-18b). Such proteins are called **monomeric proteins.** (Be careful with the terminology here. On the one hand, a polypeptide is a *polymer,* with amino acids as its monomeric repeating units; on the other hand, such a polypeptide becomes the *monomer* of which proteins are made.) The enzyme ribonuclease depicted in Figure 2-18 is an example of a monomeric protein. For such proteins, the term protein synthesis is appropriate for the polymerization because the functional protein forms spontaneously as the polypeptide is elongated.

Many other proteins are **multimeric proteins,** consisting of two or more polypeptide subunits. A multimeric protein is said to be *homomeric* if its subunits are identical, *heteromeric* if it consists of two or more different kinds of subunits. The *hemoglobin* that carries oxygen in your bloodstream is a multimeric protein because it contains four polypeptides, two called α *globin subunits,* or α chains, and two called β *globin subunits,* or β chains (Figure 3-4). In such cases, protein synthesis involves not only elongation and folding of the individual polypeptide chains but also their subsequent interaction and assembly.

Several Kinds of Bonds and Interactions Are Important in Protein Folding and Stability

As we noted in Chapter 2, the initial folding of a polypeptide into its proper shape, or **conformation,** depends on several different kinds of bonds and interactions, as does the subsequent maintenance and stability of that conformation. Depicted in Figure 3-5 are the most important of these bonds and interactions, including the covalent disulfide bond and several noncovalent interactions.

The most common covalent bond that contributes to the stabilization of protein conformation is the **disulfide bond** that forms when the sulfhydryl groups of two cysteine residues react oxidatively (Figure 3-5a). Once formed, a disulfide bond confers considerable stability because of its covalent nature. It can, in fact, be broken only by reducing it again and regenerating the two sulfhydryl groups. In many cases, the cysteine residues involved in a particular disulfide bond are a part of the same polypeptide and are usually quite distant from each other along the polypeptide but are juxtaposed by the folding process. Such *intramolecular disulfide bonds* stabilize the conformation of the polypeptide, as is the case for ribonuclease, the monomeric protein we encountered in Chapter 2 (see Figure 2-18). In the case of multimeric proteins, a disulfide bond may form between cysteine residues located in two different polypeptides, thereby linking the two polypeptides to one another covalently. The hormone insulin is a dimeric protein that has its two subunits linked in this manner (see Figure 3-7 on p. 48).

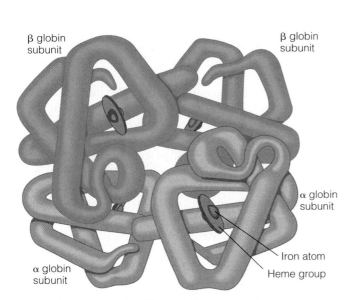

Figure 3-4 The Structure of Hemoglobin. Hemoglobin is a multimeric protein with four subunits (two α polypeptides and two β polypeptides). Each subunit contains a heme group with an iron atom. Each heme iron can bind a single oxygen molecule.

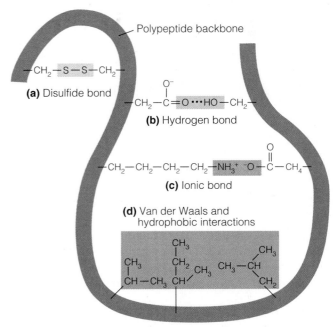

Figure 3-5 Bonds and Interactions Involved in Protein Folding and Stability. The initial folding and subsequent stability of a polypeptide depend on **(a)** covalent disulfide bonds and several kinds of noncovalent bonds and interactions, including **(b)** hydrogen bonds, **(c)** ionic bonds, **(d)** van der Waals interactions, and hydrophobic interactions.

As important as covalent disulfide bonds are in maintaining protein structure, **noncovalent bonds and interactions** are even more important, mainly because they are so diverse and numerous. As we noted briefly in Chapter 2, these include *hydrogen bonds, ionic bonds, van der Waals interactions,* and *hydrophobic interactions* (Figure 3-5).

Hydrogen Bonds. **Hydrogen bonds** are already familiar from our discussion of the properties of water in Chapter 2. In the case of water molecules, a hydrogen bond forms between a covalently bonded hydrogen atom on one water molecule and a pair of nonbonding electrons of the oxygen atom on another molecule (see Figure 2-8b). In the case of a polypeptide, hydrogen bonding is particularly important in stabilizing helical and sheet structures that are very prominent parts of many proteins, as we will see shortly. In addition, the R groups of many amino acids have functional groups that are either good hydrogen bond donors or good acceptors, allowing hydrogen bonds to form between amino acid residues that may be distant from one another along the amino acid sequence but brought in close proximity by the folding of the polypeptide (Figure 3-5b). Examples of good donors include the hydroxyl groups of several amino acids and the amino groups of others. The carbonyl and sulfhydryl groups of several other amino acids are examples of good acceptors. An individual hydrogen bond is quite weak (about 2–5 kcal/mol, compared to 70–100 kcal/mol for covalent bonds), but hydrogen bonds are very numerous in biological macromolecules such as proteins and DNA such that, in the aggregate, they become a force to be reckoned with.

Ionic Bonds. The role of **ionic bonds** (or *electrostatic interactions*) in protein structure is easy to understand. Because the R groups of some amino acids are positively charged and the R groups of others are negatively charged, polypeptide folding is dictated in part by the tendency of charged groups to repel groups with the same charge and to attract groups with the opposite charge (Figure 3-5c). Several features of ionic bonds are particularly significant. The strength of such interactions—about 3 kcal/mol—allows them to exert an attractive force over greater distances than some of the other noncovalent bonds. Moreover, the attractive force is nondirectional, so that ionic bonds are not limited to discrete angles, as is the case with covalent bonds. Because ionic bonds depend on both groups remaining charged, they will be disrupted if the pH value becomes so high or so low that either of the groups loses its charge. This loss of ionic bonds accounts in part for the denaturation that most proteins undergo at high or low pH.

Van der Waals Interactions. Interactions based on charge are not limited to ions that carry a discrete charge. Even molecules with nonpolar covalent bonds may have transient positively and negatively charged regions. Electrons are in constant motion and are not necessarily always symmetrically distributed about the molecule. At any

given instant, the density of electrons on one side of an atom may be greater than on the other side, even though the atom shares the electron equally with another atom. These momentary asymmetries in the distribution of electrons and hence in the separation of charge within a molecule are called *dipoles.* When two molecules that have such transient dipoles are very close to each other and are oriented appropriately with respect to each other, they are attracted to each other, though for only as long as the asymmetric electron distribution persists in both molecules. This transient attraction of two nonpolar molecules is called a **van der Waals interaction,** or *van der Waals force* (Figure 3-5d). A single such interaction is not only very transient; it is also very weak—0.1–0.2 kcal/mol, typically—and is only effective when the two molecules are very close together—within 0.2 nm of each other, in fact. Van der Waals interactions are nonetheless important in the structure of proteins and other biological macromolecules and also in the binding together of two molecules with complementary surfaces.

Hydrophobic Interactions. The fourth type of noncovalent interaction that plays a role in maintaining protein conformation is usually called a **hydrophobic interaction**, but it is not really a bond or interaction at all. Rather, it is the tendency of hydrophobic molecules or parts of molecules to be excluded from interactions with water molecules (Figure 3-5d). As already noted, the side chains of the 20 different amino acids vary greatly in their affinity for water. Some of them have R groups that are hydrophilic and are capable of forming hydrogen bonds not just with each other but also with the water molecules of the medium. Not surprisingly, such groups tend to be located near the surface of a folded polypeptide, where they can interact maximally with the surrounding water molecules. Other amino acids have hydrophobic R groups and are therefore essentially nonpolar. Such amino acids tend to be located on the inside of the polypeptide, where they interact with one another in an essentially nonaqueous milieu. This tendency of hydrophobic groups to be excluded from the aqueous surface of the polypeptide accounts for hydrophobic interactions.

Thus, protein structure is in part the result of a balance between the tendency of hydrophilic groups to seek an aqueous environment near the surface of the molecule and that of hydrophobic groups to minimize contact with water by associating with one another in the interior of the molecule. If all or even most of the amino acids in a protein were hydrophobic, the protein would be virtually insoluble in water and would be found instead in a nonpolar environment. Membrane proteins are localized in membranes for this very reason. Similarly, if all or most of the amino acids were hydrophilic, the polypeptide would most likely remain in a fairly distended, random shape, allowing maximum access of each amino acid to an aqueous environment. But precisely because most polypeptide chains contain both hydrophobic and hydrophilic amino acids (and in a genetically prescribed order), parts of the

Table 3-3   Levels of Organization in Protein Structure

Level of Structure	Basis of Structure	Kinds of Bonds and Interactions Involved*
Primary	Amino acid sequence	Covalent peptide bonds
Secondary	Folding into α helix, β sheet, or random coil	Hydrogen bonds
Tertiary	Three-dimensional folding of a single polypeptide chain	Disulfide bonds, hydrogen bonds, ionic bonds, van der Waals interactions, hydrophobic interactions
Quaternary	Association of two or more folded polypeptides to form a multimeric protein	Same as for tertiary structure

*For the structures of these bonds, see Figure 3-5.

molecule are drawn toward the surface whereas other parts are driven toward the interior.

Overall, then, the stability of the folded structure of a polypeptide depends on an interplay of covalent disulfide bonds and four noncovalent factors: hydrogen bonds between side groups that are good donors and good acceptors, ionic bonds between charged amino acid R groups, transient van der Waals interactions between nonpolar molecules with temporary electron asymmetries, and hydrophobic interactions that drive nonpolar groups to the interior of the molecule. The final conformation of the fully folded polypeptide is the net result of these forces and tendencies. Any one of these bonds is quite low in energy. A hydrogen bond, for example, is only about 3–5% as strong as a covalent bond. But because there are so many such bonds between the various side groups of the hundreds of amino acids that make up a typical polypeptide, their aggregate strength is sufficient

to confer great stability upon the conformation of the folded polypeptide.

Protein Structure Depends on Amino Acid Sequence and Interactions

The structure of a protein is usually described in terms of four hierarchical levels of organization: the *primary, secondary, tertiary,* and *quaternary* structures (Table 3-3). Primary structure refers to the amino acid sequence, while the higher levels of organization concern the interactions between the amino acid R groups that give the protein its characteristic conformation, or three-dimensional arrangement of atoms in space (Figure 3-6).

Secondary structure involves local interactions between contiguous amino acids along the chain; tertiary structure results from long-distance interactions between stretches of amino acids from different parts of the molecule; and

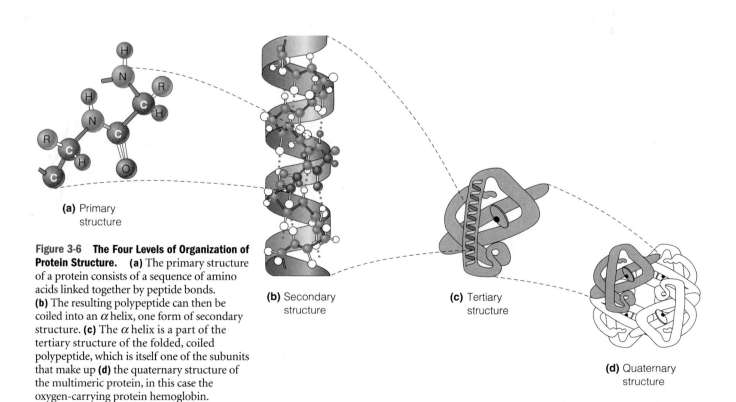

(a) Primary structure

Figure 3-6   The Four Levels of Organization of Protein Structure.   (a) The primary structure of a protein consists of a sequence of amino acids linked together by peptide bonds. **(b)** The resulting polypeptide can then be coiled into an α helix, one form of secondary structure. **(c)** The α helix is a part of the tertiary structure of the folded, coiled polypeptide, which is itself one of the subunits that make up **(d)** the quaternary structure of the multimeric protein, in this case the oxygen-carrying protein hemoglobin.

(b) Secondary structure

(c) Tertiary structure

(d) Quaternary structure

quaternary structure concerns the interaction of two or more individual polypeptides to form a single multimeric protein. All are dictated by the primary structure, but each is important in its own right in the overall structure of the protein. Secondary and tertiary structures are obviously involved in determining the conformation of the individual polypeptide, while quaternary structure is relevant only for proteins consisting of more than one polypeptide.

Primary Structure. As already noted, the **primary structure** of a protein is just a formal designation for the amino acid sequence of the constituent polypeptide(s) (Figure 3-6a). When we describe the primary structure of a protein, we are simply specifying the order in which its amino acids appear from one end of the molecule to the other. By convention, amino acid sequences are always written from the N-terminus to the C-terminus of the polypeptide, which is also the direction in which the polypeptide is synthesized. Once incorporated into a polypeptide chain, the individual amino acids are called *amino acid residues.*

The first protein to have its complete amino acid sequence determined was the hormone *insulin,* reported in 1956 by Frederick Sanger, who eventually received a Nobel Prize for this important technical advance. To determine the sequence of the insulin molecule, Sanger cleaved it into smaller fragments and analyzed the amino acid order within the individual fragments. Insulin consists of two polypeptides, called the *A subunit* and the *B subunit,* with 21 and 30 amino acid residues, respectively. Figure 3-7 shows the primary structure of insulin, with each of the chains written from its N-terminus to its C-terminus. Notice also the covalent disulfide (—S—S—) bond between two cysteine residues within the A chain and the two disulfide bonds linking the A and B chains. As we will see shortly, these bonds play an important role in stabilizing the tertiary structure of many proteins.

Sanger's techniques paved the way for the sequencing of hundreds of other proteins and led ultimately to the design of machines that can determine an amino acid sequence automatically. A more recent approach for determining protein sequences has emerged from the understanding that nucleotide sequences in DNA code for the amino acid sequences of protein molecules. It is now usu-

ally easier to determine a DNA nucleotide sequence than to purify and analyze the amino acid sequence of a protein molecule. Once a DNA nucleotide sequence has been determined, the amino acid sequence of the polypeptide encoded by that DNA segment can be inferred. Computerized data banks are now available that contain thousands of polypeptide sequences, making it easy to compare sequences and look for regions of similarity between polypeptides.

The primary structure of a protein is important both genetically and structurally. Genetically, it is significant because the amino acid sequence of the polypeptide derives directly from the order of nucleotides in the corresponding messenger RNA. The messenger RNA is in turn encoded by the DNA that represents the gene for this protein, so the primary structure of a protein is an inevitable result of the order of nucleotides in the DNA of the gene.

Of more immediate significance are the implications of the primary structure for higher levels of protein structure. In essence, all three higher levels of protein organization—secondary, tertiary, and quaternary structures— are direct consequences of the primary structure. This means that much (though for some proteins, not all) of the information necessary to specify how polypeptide chains will coil and fold and, if appropriate, how they will interact with one another is inherent in the amino acid sequence. Thus, if polypeptides are made that correspond in sequence to the α and β subunits of hemoglobin, they will assume the three-dimensional conformations appropriate to these respective subunits and will then interact spontaneously in just the right way to form the $\alpha_2\beta_2$ tetramer that we recognize as hemoglobin.

Secondary Structure. The **secondary structure** of a protein involves repetitive patterns of local structure. These patterns result from hydrogen bonding between atoms in the peptide bonds along the polypeptide backbone (Figure 3-6b). Such local interactions result in two major structural patterns, referred to as the α **helix** and β **sheet** conformations (Figure 3-8).

The α helix structure was proposed in 1951 by Linus Pauling and Robert Corey. As shown in Figure 3-8a, an α helix is spiral in shape, consisting of a backbone of pep-

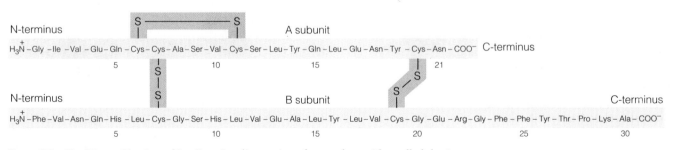

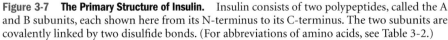

Figure 3-7   The Primary Structure of Insulin. Insulin consists of two polypeptides, called the A and B subunits, each shown here from its N-terminus to its C-terminus. The two subunits are covalently linked by two disulfide bonds. (For abbreviations of amino acids, see Table 3-2.)

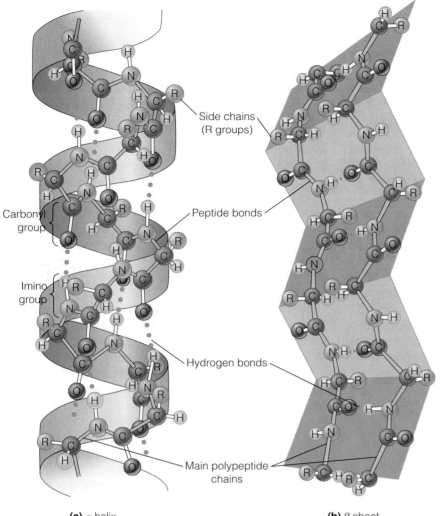

Side chains (R groups)

Carbonyl group

Peptide bonds

Imino group

Hydrogen bonds

Main polypeptide chains

(a) α helix

(b) β sheet

Figure 3-8 The α Helix and β Sheet. The α helix and β sheet are the two most important elements in the secondary structure of proteins. **(a)** The α helix resembles a coiled telephone cord, but with each turn of the coil stabilized by hydrogen bonds (blue dots) between the carbonyl and imino groups of one peptide bond and those of the peptide bonds just "below" and "above" it in the helix. The hydrogen bonds of an α helix are therefore within a single polypeptide chain and parallel to the polypeptide axis. **(b)** The β sheet resembles a pleated skirt, with successive atoms of each polypeptide chain located at folds of the pleats and with the R groups of the amino acids jutting out on alternating sides of the sheet. The structure is stabilized by hydrogen bonds (blue dots) between the carbonyl and imino groups of peptide bonds in the adjacent polypeptides (or adjacent segments of the same polypeptide). For the structure shown here, hydrogen bonds are between adjacent polypeptide chains, but β sheets can also form within the same polypeptide when segments of the molecule are folded back alongside one another.

tide bonds with the specific R groups of the individual amino acids jutting out from it. A helical shape is common to repeating polymers, as we will see when we get to the nucleic acids and the polysaccharides. For the α helix, there are 3.6 amino acids per turn of the helix, bringing the peptide bonds of every fourth amino acid in close proximity. The distance between these juxtaposed peptide bonds is, in fact, just right for the formation of a hydrogen bond between the *imino group* (—NH—) of one peptide bond and the *carbonyl group* (—CO—) of the other, as shown in Figure 3-8a.

As a result, every peptide bond in the helix is hydrogen-bonded through its carbonyl group to the peptide bond immediately "below" it in the spiral and through its imino group to the peptide bond just "above" it. In each case, however, the hydrogen-bonded peptide bonds are separated in linear sequence by the three amino acids required to advance the helix far enough to allow the two bonds to be juxtaposed. These hydrogen bonds are all nearly parallel to the main axis of the helix and therefore tend to stabilize the spiral structure by holding successive turns of the helix together.

A major alternative to the α helix is the β sheet, also initially proposed by Pauling and Corey. As shown in Figure 3-7b, this structure is an extended sheetlike conformation with successive atoms in the polypeptide chain located at the "peaks" and "troughs" of the pleats. The R groups of successive amino acids jut out on alternating sides of the sheet.

Like the α helix, the β sheet is characterized by a maximum of hydrogen bonding. In both cases, all of the imino groups and carbonyl groups of the peptide bonds are involved. However, hydrogen bonding in an α helix is invariably intramolecular, between peptide bonds within the same polypeptide, whereas hydrogen bonding in the β sheet is perpendicular to the plane of the sheet and can be either intramolecular (between two juxtaposed segments of the same polypeptide) or intermolecular (linking peptide bonds in two adjacent polypeptides). The protein regions that form β sheets can interact with each other in two different ways. If the two interacting regions run in the same N-terminus to C-terminus direction, the structure is called a *parallel β sheet;* if the two strands run in opposite N-terminus to C-terminus directions, the structure is called an *antiparallel β sheet*. Because the carbon atoms that make up the backbone of the polypeptide chain are successively located a little above and a little below the plane of the β sheet, such structures are sometimes called *pleated β sheets*.

To depict localized regions of structure within a protein, biochemists have adopted the conventions shown in Figure 3-9a. An α-helical region is represented as either a *spiral* (on left) or a *cylinder* (on right), whereas a β-sheet region is drawn as a *flat ribbon* or *arrow* with the arrowhead pointing in the direction of the C-terminus. A looped segment that connects α-helical and/or β-sheet regions is called a *random coil* and is depicted as a *narrow ribbon*.

Certain combinations of α helix and β sheet have been identified in many proteins. These units of secondary structure, called **motifs,** consist of small segments of α helix and/or β sheet connected to each other by looped regions of varying length. Among the most commonly encountered motifs are the β-α-β motif shown in Figure 3-9a and the hairpin loop and helix-turn-helix motifs depicted in Figure 3-9b and c, respectively. When the same motif is present in different proteins, it usually serves the same purpose in each. For example, the helix-turn-helix motif is one of several secondary structure

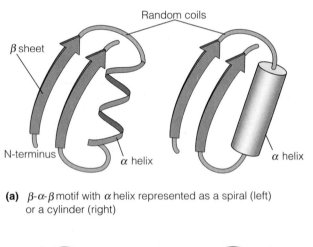

(a) β-α-β motif with α helix represented as a spiral (left) or a cylinder (right)

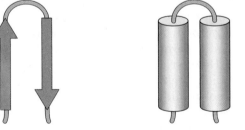

(b) Hairpin loop motif   **(c)** Helix-turn-helix motif

Figure 3-9 Common Structural Motifs. Three common units of secondary structure include the **(a)** β-α-β, **(b)** hairpin loop, and **(c)** helix-turn-helix motifs. Notice that an α helix can be represented as either a spiral (a, left) or a cylinder (a, right), whereas a β sheet is represented as a flat ribbon or arrow. The short segments that connect α helices and β sheets are called random coils. In each diagram, only a small portion of the entire protein is shown.

motifs that are characteristic of the DNA-binding proteins we will encounter when we consider the regulation of gene expression in Chapter 21.

Tertiary Structure. The **tertiary structure** of a protein can probably be best understood by contrasting it with secondary structure (Figure 3-6a). Secondary structure is a predictable, repeating conformational pattern that derives from the repetitive nature of the polypeptide because it involves hydrogen bonding between peptide bonds, the common structural elements along every polypeptide chain. If proteins contained only one or a few kinds of similar amino acids, virtually all aspects of protein conformation could probably be understood in terms of secondary structure, with only modest variations among proteins.

Tertiary structure comes about precisely because of the variety of amino acids present in proteins and the very different chemical properties of their R groups. In fact, tertiary structure depends almost entirely on interactions between the various R groups, regardless of where along the primary sequence they happen to come. Tertiary structure therefore reflects the nonrepetitive aspect of a polypeptide because it depends not on the carboxyl and amino groups common to all of the amino acids in the chain, but on the very feature that makes each amino acid distinctive—its R group.

Tertiary structure is neither repetitive nor readily predictable; it involves competing interactions between side groups with different properties. Hydrophobic R groups, for example, will tend to associate with one another and spontaneously seek out a nonaqueous environment in the interior of the molecule, whereas polar amino acids will be drawn to the surface, by both their affinity for one another and their attraction to the polar water molecules in the surrounding milieu. As a result, the polypeptide chain will be folded, coiled, and twisted into the **native conformation** that represents the most stable state for that particular sequence of amino acids. As we learned in Chapter 2, this folding occurs spontaneously for some polypeptides, whereas for others, folding is assisted by proteins called *molecular chaperones.* Once the tertiary structure of a polypeptide has been achieved, it is stabilized and maintained by the several types of noncovalent bonds and interactions shown in Figure 3-5 and by covalent disulfide bonds, if present.

The relative contributions of secondary and tertiary structures to the overall shape of a polypeptide vary from protein to protein and depend critically on the relative proportions and sequence of amino acids in the chain. Broadly speaking, proteins can be divided into two categories: *fibrous proteins* and *globular proteins.*

Fibrous proteins have extensive secondary structure (either α helix or β sheet) throughout the molecule, giving them a highly ordered, repetitive structure. In general, secondary structure is much more important than tertiary interactions in determining the shape of such proteins, which often have an extended, filamentous structure. Especially prominent examples of fibrous proteins include silk *fibroin* and the *keratins* of hair and wool, as well as *collagen* (found in tendons and skin) and *elastin* (present in ligaments and blood vessels).

The amino acid sequence of each of these proteins favors a particular kind of secondary structure, which in turn confers a specific set of desirable mechanical properties on the protein. Fibroin, for example, consists mainly of long stretches of β sheet, with the polypeptide chains running parallel to the axis of the silk fiber. The most prevalent amino acids in fibroin are glycine, alanine, and serine. These amino acids have small R groups that pack together well (see Figure 3-2). The result is a silk fiber that is strong and relatively inextensible because the covalently bonded chains are stretched to nearly their maximum possible length.

Hair and wool fibers, on the other hand, consist of the protein α-keratin, which is almost entirely helical. The individual keratin molecules are very long and lie with their helix axes nearly parallel to the fiber axis. As a result, hair is quite extensible because stretching of the fiber is opposed not by the covalent bonds of the polypeptide chain, as in β sheets, but by the hydrogen bonds that stabilize the α-helical structure. The individual α helices in a hair are wound together to form a ropelike structure, as shown in Figure 3-10. First, three helices are wrapped around each other to form a *protofibril*. Eleven protofibrils then aggregate laterally to form a *microfibril*. Microfibrils, in turn, are packed together to form *macrofibrils*, bundles of which pass through and around the cells in a hair.

As important as fibrous proteins may be, they represent only a small fraction of the kinds of proteins present in most cells. Most of the proteins involved in cellular structure are **globular proteins,** so named because their polypeptide chains are folded into compact structures rather than extended filaments. The polypeptide chain of a globular protein is often folded locally into regions with α-helical or β-sheet structures, and these regions are themselves folded on one another to give the protein its compact, globular shape. This folding is possible because regions of α helix or β sheet are interspersed with random coils, irregularly structured regions that allow the polypeptide chain to loop and fold (see Figure 3-9). Thus, every globular protein has its own unique tertiary structure, made up of secondary structural elements (helices and sheets) folded in a specific way that is especially suited to the protein's particular functional role.

Whether a specific segment of a polypeptide will form an α helix, a β sheet, or neither depends on the amino acids present in that segment. For example, leucine, methionine, and glutamate are strong "helix formers," whereas isoleucine, valine, and phenylalanine are strong "sheet formers." Both glycine and proline, the only cyclic amino acid, are "helix breakers" and are, in fact, largely responsible for the bends and turns in α helices. Such bends and turns usually occur at the surface of a polypeptide.

Figure 3-11 shows the tertiary structure of a typical globular protein, the enzyme ribonuclease. (Recall that ribonuclease was one of the first proteins to have its

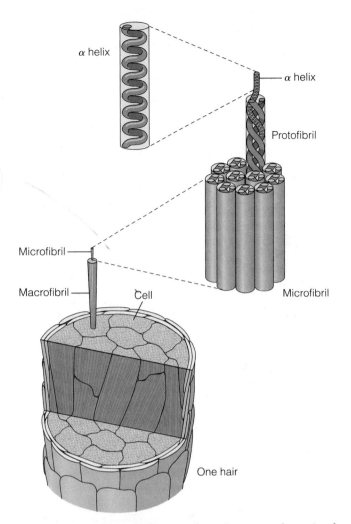

Figure 3-10   The Structure of Hair.   The main structural protein of hair is α-keratin, a fibrous protein with an α-helical shape. Three helices of α-keratin wrap into protofibrils, which then bond together to form microfibrils. Microfibrils have a characteristic "9 + 2" structure in which two central protofibrils are surrounded by nine others. Microfibrils, in turn, aggregate laterally to form macrofibrils. Bundles of macrofibrils pass through and around the cells in a hair.

sequence determined. We also used it in Figure 2-18 to illustrate the denaturation and renaturation of a polypeptide and the spontaneity of its folding.) Two different conventions are used to represent the structure of ribonuclease: the *ball-and-stick* model and the *spiral-and-ribbon* model we saw in Figure 3-9. For clarity, most of the side chains of ribonuclease have been omitted in both models. The gold groups in Figure 3-11a are the four disulfide bonds that help to stabilize the three-dimensional structure of the ribonuclease model (see Figure 2-18a).

Globular proteins can be mainly α helical, mainly β sheet, or a mixture of both structures. These categories are illustrated in Figure 3-12 by the coat protein of tobacco mosaic virus (TMV), a portion of an immunoglobulin molecule, and a portion of the enzyme hexokinase, respectively. Helical segments of globular proteins often consist of bundles of helices, as seen for the coat protein of TMV

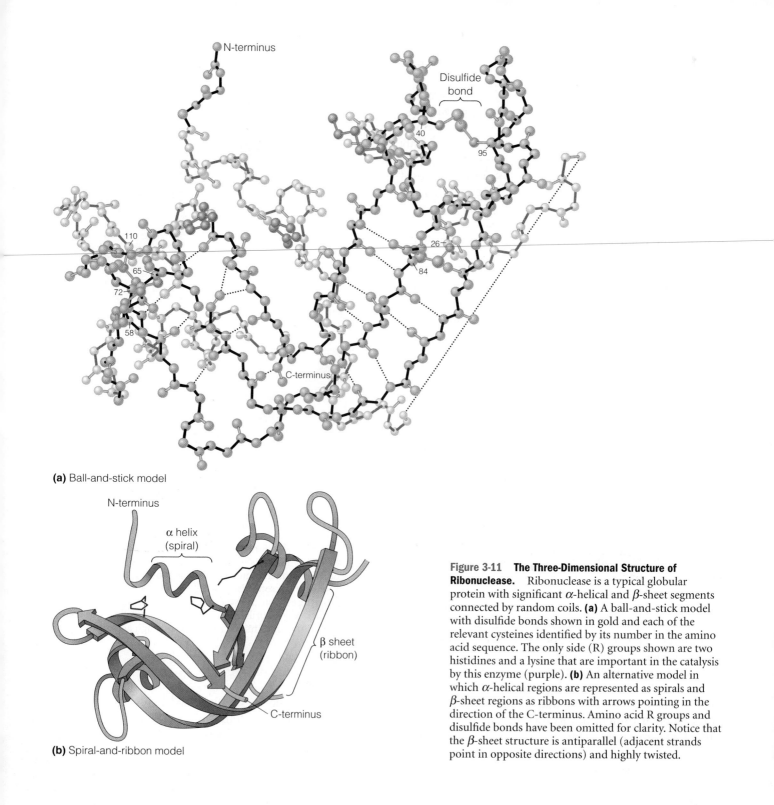

(a) Ball-and-stick model

N-terminus

Disulfide bond

40

95

110

65

72

58

26

84

C-terminus

N-terminus

α helix (spiral)

β sheet (ribbon)

C-terminus

(b) Spiral-and-ribbon model

Figure 3-11   The Three-Dimensional Structure of Ribonuclease.   Ribonuclease is a typical globular protein with significant α-helical and β-sheet segments connected by random coils. **(a)** A ball-and-stick model with disulfide bonds shown in gold and each of the relevant cysteines identified by its number in the amino acid sequence. The only side (R) groups shown are two histidines and a lysine that are important in the catalysis by this enzyme (purple). **(b)** An alternative model in which α-helical regions are represented as spirals and β-sheet regions as ribbons with arrows pointing in the direction of the C-terminus. Amino acid R groups and disulfide bonds have been omitted for clarity. Notice that the β-sheet structure is antiparallel (adjacent strands point in opposite directions) and highly twisted.

in Figure 3-12a. Segments with mainly β-sheet structure are usually characterized by a barrel-like configuration (Figure 3-12b) or by a twisted sheet (Figure 3-12c).

Most globular proteins consist of a number of segments called domains. A **domain** is a discrete, locally folded unit of tertiary structure, often containing regions of α helices and β sheets packed together compactly. A domain typically includes about 50–350 amino acids and usually has a specific function. Small globular proteins

tend to be folded into a single domain, as shown in Figure 3-11b for ribonuclease, a relatively small protein. Large globular proteins usually have multiple domains. The portions of the immunoglobulin and hexokinase molecules shown in Figure 3-12b and c are, in fact, specific domains of these proteins (the V_2 domain of immunoglobulin and domain 2 of hexokinase, respectively). Figure 3-13 is an example of a protein that consists of a single polypeptide folded into two functional domains.

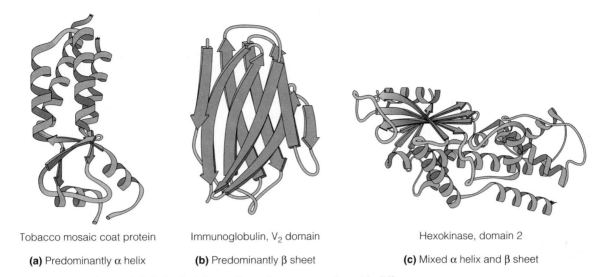

(a) Predominantly α helix   **(b)** Predominantly β sheet   **(c)** Mixed α helix and β sheet

Tobacco mosaic coat protein   Immunoglobulin, V₂ domain   Hexokinase, domain 2

Figure 3-12 Structures of Several Globular Proteins. Shown here are proteins with different tertiary structures: **(a)** a predominantly α-helical structure (blue spirals), the coat protein of TMV; **(b)** a mainly β-sheet structure (purple ribbons with arrows), the V_2 domain of immunoglobulin; and **(c)** a structure that mixes α helix and β sheet, domain 2 of hexokinase. The immunoglobulin V_2 domain is an example of an antiparallel β-barrel structure, whereas the hexokinase domain 2 illustrates a twisted β sheet. (Green segments are random coils.)

Proteins having a common function (such as binding a specific metal ion or recognizing a specific molecule) usually have a common domain. Moreover, proteins with multiple functions usually have a separate domain for each function. Thus, domains can be thought of as the modular units of function from which globular proteins are constructed.

Before leaving the topic of tertiary structure, we should emphasize again the dependence of these higher levels of organization on the primary structure of the polypeptide. The significance of primary structure is exemplified especially well by the inherited condition *sickle-cell anemia.* People with this trait have red blood cells that are distorted from their normal disk shape into a "sickle" shape, which causes the cells to clog blood vessels, thereby impeding blood flow. This condition is caused by a slight change in the hemoglobin within the red blood cells. The hemoglobin molecules have normal α chains, but their β chains have a single amino acid that is different. At a specific position in the chain (the sixth residue from the N-terminus), the glutamate normally present at that point is replaced by valine. The other 145 amino acids in the chain are correct, but this single substitution causes enough of a difference in the tertiary structure of the β chain that the hemoglobin molecules tend to crystallize, deforming the cell into a sickle shape. Not all amino acid substitutions cause such dramatic changes in structure and function, but the example serves to underscore the crucial relationship between the amino acid sequence of a polypeptide and the final shape (and often the biological activity) of the molecule.

Quaternary Structure. The **quaternary structure** of a protein is the level of organization concerned with subunit

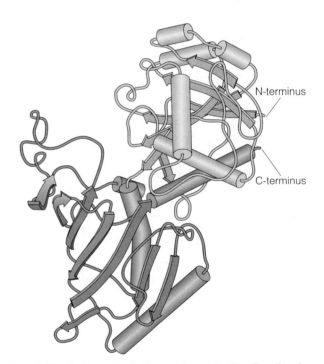

Figure 3-13 An Example of a Protein Containing Two Functional Domains. This model of the enzyme glyceraldehyde phosphate dehydrogenase shows a single polypeptide chain folded into two domains. One domain binds to the substance being metabolized, whereas the other domain binds to a chemical factor required for the reaction to occur. The two domains are indicated by different shadings.

interactions and assembly (Figure 3-6d). Quaternary structure therefore applies only to multimeric proteins. Many proteins are included in this category, particularly those with molecular weights above 50,000. Hemoglobin, for example, is a multimeric protein (see Figure 3-4). As noted earlier, some multimeric proteins contain identical polypeptide subunits; others, such as hemoglobin, contain several different kinds of polypeptides.

The bonds and forces that maintain quaternary structure are the same as those responsible for tertiary structure: hydrogen bonds, electrostatic interactions, hydrophobic interactions, and covalent disulfide bonds. As noted earlier, disulfide bonds may be either within a polypeptide chain or between chains. When they occur within a polypeptide, they stabilize tertiary structure; when they occur between polypeptides, they help maintain quaternary structure. As in the case of polypeptide folding, the process of subunit assembly is often, though not always, spontaneous, with most, if not all, of the requisite information provided by the amino acid sequence of the individual polypeptides.

In some cases, a still higher level of assembly is possible, in the sense that two or more proteins (enzymes, usually) are organized into a **multiprotein complex,** with each protein involved sequentially in a common multistep process. An example of such a complex is an enzyme system called the *pyruvate dehydrogenase complex.* This complex catalyzes the oxidative removal of a carbon atom (as CO_2) from the three-carbon compound pyruvate (or pyruvic acid), a reaction that will be of interest to us when we get to Chapter 14. Three individual enzymes and five kinds of molecules called coenzymes constitute a highly organized multienzyme complex that has a mass of 4.6 million daltons in bacterial cells and almost twice that size in the mitochondria of your liver cells. (A *dalton* is the mass of one hydrogen atom, or 1.66×10^{-24} g.) The pyruvate dehydrogenase complex is one of the best understood examples of how cells can achieve economy of function by ordering the enzymes that catalyze sequential reactions into a single multienzyme complex.

Nucleic Acids

Next we come to the **nucleic acids,** macromolecules of paramount importance to the cell because of their role in the storage, transmission, and expression of genetic information. Nucleic acids are linear polymers of nucleotides, strung together in a genetically determined order that is critical to their role as informational macromolecules. The two major types of nucleic acids are **DNA (deoxyribonucleic acid)** and **RNA (ribonucleic acid).** DNA and RNA differ in their chemistry and their role in the cell. As the names suggest, RNA contains the five-carbon sugar **ribose** in each of its nucleotides, whereas DNA contains the closely related sugar **deoxyribose.** Functionally, DNA serves primarily as the repository of genetic information, whereas RNA molecules play several different roles in the expression of that information—that is, during protein synthesis. Figure 3-14 illustrates these roles in a typical plant or animal cell. Most of the cell's DNA is located in the nucleus, which is therefore the major site of RNA synthesis in the cell. A specific

segment of a DNA molecule directs the synthesis of a particular *messenger RNA (mRNA)* in a process called *transcription.* The mRNA is processed within the nucleus and then moves through nuclear pores (tiny channels in the nuclear membrane) into the cytoplasm, where the nucleotide sequence of the mRNA is used to direct the amino acid sequence of a specific protein in a process called *translation.* Translation takes place on ribosomes, which are complexes of ribosomal proteins and *ribosomal RNA (rRNA)* molecules. The rRNA is also synthesized in the nucleus, as are the *transfer RNA (tRNA)* molecules that bring the correct amino acids to the ribosome as the growing polypeptide chain is lengthened by the successive addition of amino acids. Thus, mRNA, rRNA and tRNA are the three major species of RNA.

These roles of DNA and RNA in the storage, transmission, and expression of genetic information will be

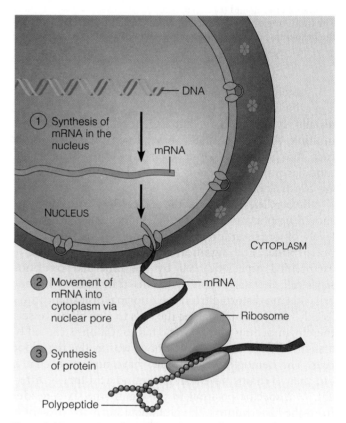

Figure 3-14   Genetic information is stored in the nucleotide sequences of DNA molecules.   In eukaryotes (organisms other than bacteria), most of the DNA in a cell is located in the nucleus. (Smaller amounts of DNA are present in organelles called mitochondria and plastids, the latter found only in plant cells.) Nuclear DNA programs protein synthesis in the cytoplasm by directing the synthesis of messenger RNA (mRNA) molecules that leave the nucleus via nuclear pores and bind to ribosomes in the cytoplasm. As a ribosome moves along a specific mRNA, the genetic message is translated into a polypeptide with a specific amino acid sequence. Though not shown in the diagram, DNA also directs the synthesis of ribosomal RNA (rRNA) molecules, which are structural components of the ribosome, and transfer RNA (tRNA) molecules, which bring the correct amino acids to the ribosome as polypeptide synthesis progresses.

considered in detail in Part Four. For now, we will focus on the chemistry of nucleic acids and the nucleotides of which they are composed.

The Monomers Are Nucleotides

Like proteins, nucleic acids are informational macromolecules and therefore contain nonidentical monomeric units in a specified sequence. The monomeric units of nucleic acids are called **nucleotides.** Nucleotides exhibit less variety than amino acids; DNA and RNA each contain only four different kinds. (Actually, there is more variety than this suggests, especially in some RNA molecules; but virtually all of the variant structures found in nucleic acids represent one of the four basic nucleotides that has been chemically modified after insertion into the chain.)

As shown in Figure 3-15, each nucleotide consists of a five-carbon sugar to which is attached a phosphate group and a nitrogen-containing aromatic base. The sugar is either D-ribose (for RNA) or D-deoxyribose (for DNA). The phosphate is joined by a phosphoester bond to the 5′ carbon of the sugar, and the base is attached at the 1′ carbon. The base may be either a **purine** or a **pyrimidine.** DNA contains the purines **adenine (A)** and **guanine (G)** and the pyrimidines **cytosine (C)** and **thymine (T).** RNA also has adenine, guanine, and cytosine but contains the pyrimidine

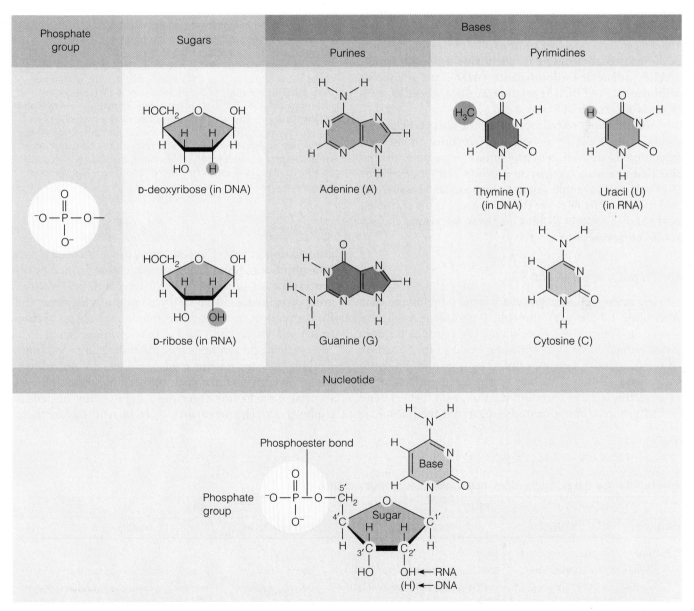

Figure 3-15   The Structure of a Nucleotide. In RNA, a nucleotide consists of the five-carbon sugar D-ribose with an aromatic nitrogen-containing base attached to the 1′ carbon and a phosphate group linked to the 5′ carbon by a phosphoester bond. (Carbon atoms in the sugar of a nucleotide are numbered from 1′ to 5′ to distinguish them from those in the base, which are numbered without the prime.) In DNA, the hydroxyl group on the 2′ carbon is replaced by a hydrogen atom, so the sugar is D-deoxyribose. The bases in DNA are the purines adenine (A) and guanine (G) and the pyrimidines thymine (T) and cytosine (C). In RNA, thymine is replaced by the pyrimidine uracil (U).

uracil (U) in place of thymine. Like the 20 amino acids present in proteins, these 5 aromatic bases are among the 30 most common small molecules in cells (see Table 3-1).

If the phosphate is removed from a nucleotide, the remaining base-sugar unit is called a **nucleoside.** Each pyrimidine and purine may therefore occur as the free base, the nucleoside, or the nucleotide. The appropriate names for these compounds are given in Table 3-4. Notice that nucleotides and nucleosides containing deoxyribose are specified by a lowercase *d* preceding the letter that identifies the base.

As the nomenclature indicates, a nucleotide can be thought of as a **nucleoside monophosphate** because it is a nucleoside with a single phosphate group attached to it. This terminology can be readily extended to molecules with two or three phosphate groups attached to the 5′ carbon. For example, the nucleoside adenosine (adenine plus ribose) can have one, two, or three phosphates attached and is designated accordingly as **adenosine monophosphate (AMP), adenosine diphosphate (ADP),** or **adenosine triphosphate (ATP).** The relationships among these compounds are shown in Figure 3-16.

You probably recognize ATP as the energy-rich compound used to drive a variety of reactions in the cell, including the activation of monomers for polymer formation that we encountered in the previous chapter (see Figure 2-17). As this example suggests, nucleotides actually play two roles in cells: They are the monomeric units of nucleic acids and they serve as intermediates in various energy-transferring reactions.

The Polymers Are DNA and RNA

Nucleic acids are linear polymers formed by linking each nucleotide to the next through a phosphate group, as shown in Figure 3-17. Specifically, the phosphate group already attached by a phosphoester bond to the 5′ carbon of one nucleotide becomes linked by a second phosphoester bond to the 3′ carbon of the next nucleotide. In essence, this is a condensation reaction, with the —H and —OH groups coming from the sugar and the phosphate

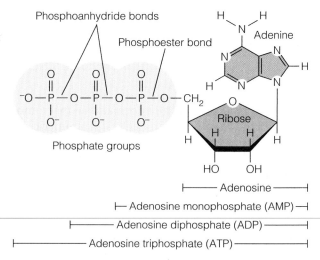

Figure 3-16   The Phosphorylated Forms of Adenosine.   Adenosine occurs as the free nucleoside, the monophosphate (AMP), the diphosphate (ADP), and the triphosphate (ATP). The bond that links the first phosphate to the ribose of adenosine is a low-energy phosphoester bond, whereas the bonds that link the second and third phosphate groups to the molecule are higher-energy phosphoanhydride bonds. (Recall that the terms "low-energy" and "higher-energy" refer not to the bonds themselves, which always require energy to break them, but to the amount of free, or useful, energy that is liberated upon the hydrolysis of the bond. As we will see in Chapter 13, the hydrolysis of a phosphoanhydride typically liberates 2–3 times as much free energy as does the hydrolysis of a phosphoester bond.)

groups, respectively. The resulting linkage is called a 3′, 5′ **phosphodiester bond.** The **polynucleotide** formed by this process has an intrinsic directionality, with a 5′ hydroxyl group at one end and a 3′ hydroxyl group at the other end. By convention, nucleotide sequences are always written from the 5′ end to the 3′ end of the polynucleotide.

Nucleic acid synthesis requires both energy and information, just as protein synthesis does. To provide the energy needed to form each phosphodiester bond, each successive nucleotide enters as a high-energy nucleoside triphosphate. The precursors for DNA synthesis are there-

Table 3-4   The Bases, Nucleosides, and Nucleotides of RNA and DNA

Bases	RNA		DNA	
	Nucleoside	Nucleotide	Deoxynucleoside	Deoxynucleotide
Purines				
Adenine (A)	Adenosine	Adenosine monophosphate (AMP)	Deoxyadenosine	Deoxyadenosine monophosphate (dAMP)
Guanine (G)	Guanosine	Guanosine monophosphate (GMP)	Deoxyguanosine	Deoxyguanosine monophosphate (dGMP)
Pyrimidines				
Cytosine (C)	Cytidine	Cytidine monophosphate (CMP)	Deoxycytidine	Deoxycytidine monophosphate (dCMP)
Uracil (U)	Uridine	Uridine monophosphate (UMP)	—	—
Thymine (T)	—	—	Thymidine	Thymidine monophosphate (dTMP)

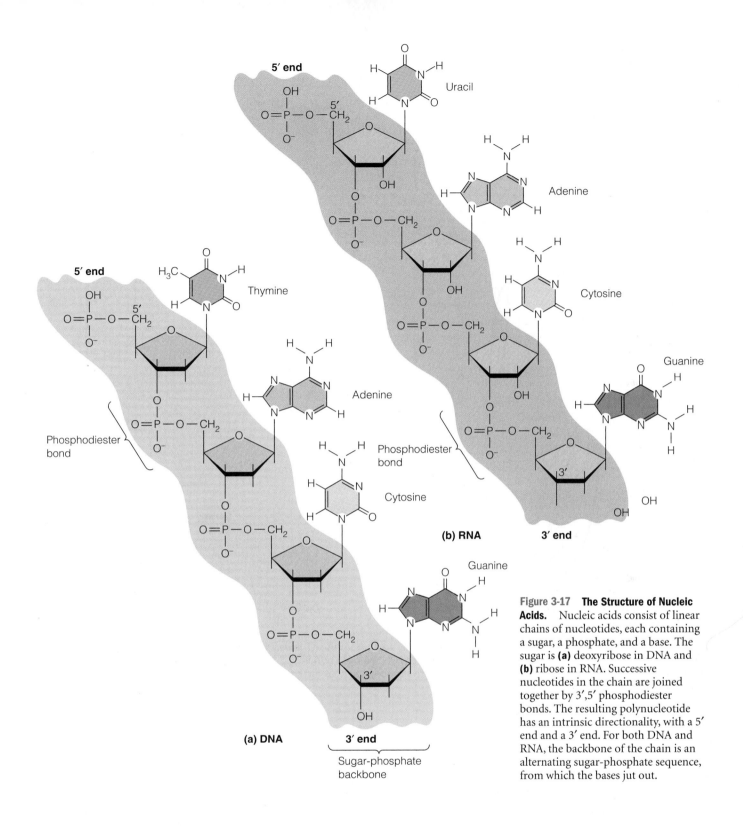

Figure 3-17 The Structure of Nucleic Acids. Nucleic acids consist of linear chains of nucleotides, each containing a sugar, a phosphate, and a base. The sugar is **(a)** deoxyribose in DNA and **(b)** ribose in RNA. Successive nucleotides in the chain are joined together by 3′,5′ phosphodiester bonds. The resulting polynucleotide has an intrinsic directionality, with a 5′ end and a 3′ end. For both DNA and RNA, the backbone of the chain is an alternating sugar-phosphate sequence, from which the bases jut out.

fore dATP, dCTP, dGTP, and dTTP. For RNA synthesis, ATP, CTP, GTP, and UTP are needed.

Information is required for nucleic acid synthesis because successive incoming nucleotides must be added in a specific, genetically determined sequence. For this purpose, a preexisting molecule is used as a **template** to specify nucleotide order. For both DNA and RNA synthesis, the template is usually DNA. The essence of template-directed nucleic acid synthesis is that each incoming nucleotide is selected because its base can be recognized by (will interact with) the base of the nucleotide already present at that position in the template.

This recognition process depends on an important chemical feature of the purine and pyrimidine bases

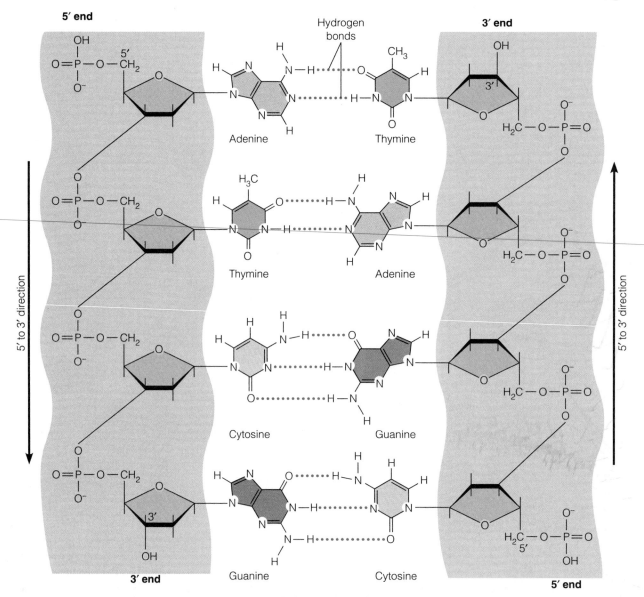

Figure 3-18 Hydrogen Bonding in Nucleic Acid Structure. The hydrogen bonds (blue dots) between adenine and thymine and between cytosine and guanine account for the AT and CG base pairing of DNA. Notice that the AT pair is held together by two hydrogen bonds, whereas the CG pair has three hydrogen bonds. If one or both strands were RNA instead, the pairing partner for adenine would be uracil (U).

shown in Figure 3-18. These bases have carbonyl groups and nitrogen atoms capable of hydrogen bond formation under appropriate conditions. Furthermore, complementary relationships between purines and pyrimidines allow A to form two hydrogen bonds with T (or U) and G to form three hydrogen bonds with C, as shown in Figure 3-18. This pairing of A with T (or U) and G with C is a fundamental property of nucleic acids. Genetically, this **base pairing** provides a mechanism for nucleic acids to recognize one another, as we will see in Part Four. For now, however, let us concentrate on the structural implications.

A DNA Molecule Is a Double-Stranded Helix

One of the most significant biological advances of the twentieth century came in 1953 in a two-page article in the scientific journal *Nature*. In the article, Francis Crick and James Watson postulated a double-stranded helical structure for DNA—the now-famous **double helix**—that not only accounted for the known physical and chemical properties of DNA but also suggested a mechanism for replication of the structure. Some highlights of this exciting chapter in the history of contemporary biology are related in Box 3A.

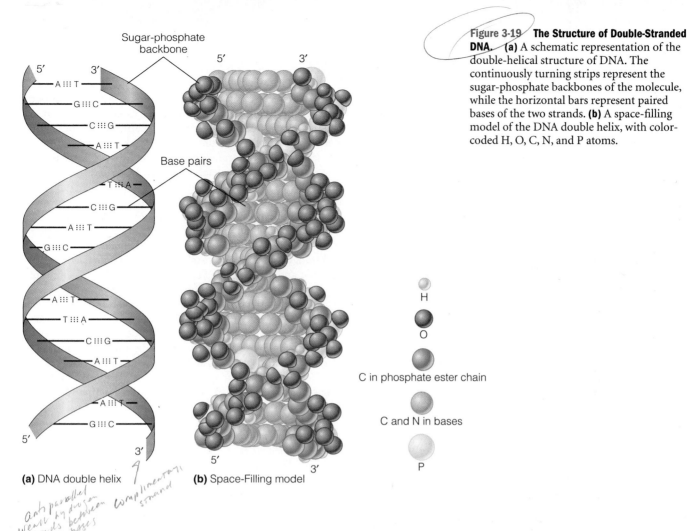

Figure 3-19 **The Structure of Double-Stranded DNA.** **(a)** A schematic representation of the double-helical structure of DNA. The continuously turning strips represent the sugar-phosphate backbones of the molecule, while the horizontal bars represent paired bases of the two strands. **(b)** A space-filling model of the DNA double helix, with color-coded H, O, C, N, and P atoms.

H

O

C in phosphate ester chain

C and N in bases

P

(a) DNA double helix

(b) Space-Filling model

[handwritten annotations: anti parallel, weak hydrogen bonds between bases, complimentary strand]

Essentially, the double helix consists of two complementary chains of DNA twisted together around a common axis to form a right-handed helical structure that resembles a circular staircase (Figure 3-19). The two chains are oriented in opposite directions along the helix, one running in the 5′ ⟶ 3′ direction and the other in the 3′ ⟶ 5′ direction. The backbone of each chain consists of sugar molecules alternating with phosphate groups. The phosphate groups are charged and the sugar molecules contain polar hydroxyl groups, so it is not surprising that the sugar-phosphate backbones of the two strands are on the outside of the DNA helix, where their interaction with the surrounding aqueous milieu can be maximized. The pyrimidine and purine bases, on the other hand, are aromatic compounds with less affinity for water. Accordingly, they are oriented inward, forming the base pairs that hold the two chains together.

To form a stable double helix, the two component strands must be not only *antiparallel* (running in opposite directions) but also *complementary*. By this we mean that each base in one strand can form specific hydrogen bonds with the base in the other strand directly across from it. From the pairing possibilities shown in Figure 3-18, this means that each A must be paired with a T, and each G with a C. In both cases, one member of the pair is a

pyrimidine (T or C) and the other is a purine (A or G). The distance between the two sugar-phosphate backbones in the double helix is just sufficient to accommodate one of each kind of base. If we envision the sugar-phosphate backbones of the two strands as the sides of a circular staircase, then each step or rung of the stairway corresponds to a pair of bases held in place by hydrogen bonding (see Figure 3-19).

The right-handed Watson-Crick helix shown in Figure 3-19 is actually an idealized version of what is called *B-DNA*. B-DNA is the main form of DNA in cells, but two other forms may also exist, perhaps in short segments interspersed within molecules that consist mainly of B-DNA. *A-DNA* has a right-handed helical configuration that is shorter and thicker than B-DNA. *Z-DNA*, on the other hand, is a left-handed double helix that derives its name from the zigzag pattern of its longer, thinner sugar-phosphate backbone. (For a comparison of the structures of B-DNA and Z-DNA, see Figure 16-5.)

RNA structure also depends in part on base pairing, but this pairing is usually between complementary regions within the same strand and is much less extensive than the interstrand pairing of the DNA duplex. Of the various RNA species, secondary and tertiary structures are well understood only for tRNA molecules, as we will

[handwritten annotations: A = T, G = C]

Historical Perspectives ON THE TRAIL OF THE DOUBLE HELIX

"I have never seen Francis Crick in a modest mood. Perhaps in other company he is that way, but I have never had reason so to judge him." With this typically irreverent observation as an introduction, James Watson goes on to describe in a very personal and highly entertaining way the events that eventually led to the discovery of the structure of DNA. The account, published in 1968 under the title *The Double Helix,* is still fascinating reading for the personal, unvarnished insights it provides into how an immense scientific discovery came about. Commenting on his reasons for writing the book, Watson observes in the preface that "there remains general ignorance about how science is 'done.' That is not to say that all science is done in the manner described here. This is far from the case, for styles of scientific research vary almost as much as human personalities. On the other hand, I do not believe that the way DNA came out constitutes an odd exception to a scientific world complicated by the contradictory pulls of ambition and the sense of fair play."

As portrayed in Watson's account, Crick and Watson are about as different from each other in nature and background as they could be. But there was one thing they shared, and that was an unconventional but highly productive way of "doing" science. They did little actual experimentation on DNA, choosing instead to draw heavily on the research findings of others and to bring their own considerable ingenuities to bear building models and exercising astute insights and hunches (Figure 3A-1). And out of it all emerged, in a relatively short time, an understanding of the double-helical structure of DNA that has come to rank as one of the major scientific events of this century.

To appreciate their findings and their brilliance, we must first understand the setting in which Watson and Crick worked. The early 1950s was an exciting time in biology. It had been only a few years since Avery, MacLeod, and McCarty had published evidence on the genetic transformation of bacteria, but the work of Hershey and Chase that confirmed DNA as the genetic material had not yet appeared in print. Meanwhile, at Columbia University, Erwin Chargaff's careful chemical analyses had revealed that although the relative proportions of the four bases—A, T, C, and G—varied greatly from one species to the next, it was always the same for all members of a single species. Even more puzzling and portentous was Chargaff's second finding: For a given species, A and T always occurred in the same proportions, and so did G and C (i.e., A = T and C = G).

The most important clues came from the work of Maurice Wilkins and Rosalind Franklin at King's College in London. Wilkins and Franklin were using the technique of X-ray diffraction to study DNA structure, and they took a rather dim view of Watson and Crick's strategy of model building. X-ray diffraction is a useful tool for detecting regularly occurring structural elements in a crystalline substance, because any structural feature that repeats at some fixed interval in the crystal contributes in a characteristic way to the diffraction pattern that is obtained. From Franklin's painstaking analysis of the diffraction patterns of DNA, it became clear that the molecule was long and thin, with some structural element being repeated every 0.34 nm and another being repeated every 3.4 nm. Even more intriguing, the molecule appeared to be some sort of helix.

This stirred the imaginations of Watson and Crick, because they had heard only recently of Pauling and Corey's α-helical structure for proteins. Working with models of the bases cut from stiff cardboard, Watson and Crick came to the momentous insight that DNA was also a helix, but with an all-important difference: It was a *double* helix, with hydrogen-bonded pairing of purines and pyrimidines. The actual discovery is best recounted in Watson's own words:

When I got to our still empty office the following morning, I quickly cleared away the papers from my desk top so that I would have a large, flat surface on which to form pairs of bases held together by hydrogen bonds. Though I initially went back to my like-with-like prejudices, I saw all too well that they led nowhere. When Jerry [Donohue, an American crystallographer working in the same laboratory] came in I looked up, saw that it was not Francis, and began shifting the bases in and out of various other pairing possibilities. Suddenly I became aware that an adenine-thymine pair held together by two hydrogen bonds was identical in shape to a guanine-cytosine pair held together by at least two hydrogen bonds. All the hydrogen bonds seemed to form naturally; no fudging was required to make the two types of base pairs identical in shape. Quickly I called Jerry over to ask him whether this time he had any objections to my new base pairs.

When he said no, my morale skyrocketed, for I suspected that we now had the answer to the riddle of why the number of purine residues exactly equaled the number of pyrimidine

see when we get to Chapter 19. (For the extent of internal complementarity and base pairing in tRNA molecules, see Figure 19–17.)

Polysaccharides

The next macromolecules we will consider are the **polysaccharides**, which are long-chain polymers of sugars and sugar derivatives. Polysaccharides usually consist of a single kind of repeating unit, or sometimes an alternating pattern of two kinds, and are not informational molecules. (As we will see in Chapter 7, however, shorter polymers called *oligosaccharides*, when attached to proteins on the cell surface, play very important roles in cellular recognition of extracellular signal molecules and of other cells.) As noted earlier, the major polysaccharides in higher organisms are the storage polysaccharides starch and glycogen and the structural polysaccharide cellulose. Each of these polymers contains the six-carbon sugar glucose as its single repeating unit, but they differ in both the nature of the bond between successive glucose units and the presence and extent of side branches on the chains.

Figure 3A-1   James Watson (left) and Francis Crick (right) at work with their model of DNA.

residues. Two irregular sequences of bases could be regularly packed in the center of a helix if a purine always hydrogen-bonded to a pyrimidine. Furthermore, the hydrogen-bonding requirement meant that adenine would always pair with thymine, while guanine could pair only with cytosine. Chargaff's rules then suddenly stood out as a consequence of a double-helical structure for DNA. Even more exciting, this type of double helix suggested a replication scheme much more satisfactory than my briefly considered like-with-like pairing. Always pairing adenine with thymine and guanine with cytosine meant that the base sequences of the two intertwined chains were complementary to each other. Given the base sequence of one chain, that of its partner was automatically determined. Conceptually, it was thus very easy to visualize how a single chain could be the template for the synthesis of a chain with the complementary sequence.

Upon his arrival Francis did not get more than halfway through the door before I let loose that the answer to everything was in our hands. Though as a matter of principle he maintained skepticism for a few moments, the similarly shaped AT and GC pairs had their expected impact. His quickly pushing the bases together in a number of different ways did not reveal any other way to satisfy Chargaff's rules. A few minutes later he spotted the fact that the two glycosidic bonds (joining base and sugar) of each base pair were systematically related by a diad axis perpendicular to the helical axis. Thus, both pairs could be flip-flopped over and still have their glycosidic bonds facing in the same direction. This had the important consequence that a given chain could contain both purines and pyrimidines. At the same time, it strongly suggested that the backbones of the two chains must run in opposite directions.

The question then became whether the AT and GC base pairs would easily fit the backbone configuration devised during the previous two weeks. At first glance this looked like a good bet, since I had left free in the center a large vacant area for the bases. However, we both knew that we would not be home until a complete model was built in which all the stereochemical contacts were satisfactory. There was also the obvious fact that the implications of its existence were far too important to risk crying wolf. Thus I felt slightly queasy when at lunch Francis winged into the Eagle to tell everyone within hearing distance that we had found the secret of life.*

The rest is history. Shortly thereafter, the prestigious journal *Nature* carried an unpretentious two-page article entitled simply "Molecular Structure of Nucleic Acids: A Structure for Deoxyribose Nucleic Acid," by James Watson and Francis Crick. Though modest in length, that paper has had far-reaching implications, for the double-stranded model that Watson and Crick worked out in 1953 has proved to be correct in all its essential details, unleashing a revolution in the field of biology.

*Excerpted from The Double Helix, pp. 194–197. Copyright © 1968 James D. Watson. Reprinted with the permission of the author and Atheneum Publishers, Inc.

The Monomers Are Monosaccharides

The repeating units of polysaccharides are simple sugars called **monosaccharides** (from the Greek *mono*, meaning "single," and *saccharide*, meaning "sugar"). A sugar can be defined as an aldehyde or ketone that has two or more hydroxyl groups. Thus, there are two categories of sugars: the *aldosugars*, with a terminal carbonyl group (Figure 3-20a); and the *ketosugars*, with an internal carbonyl group (Figure 3-20b). Within these categories, sugars are named generically according to the number of carbon atoms they contain. Most sugars have between three and seven carbon atoms and are therefore classified as a *triose* (three carbons), a *tetrose* (four carbons), a *pentose* (five carbons), a *hexose* (six carbons), or a *heptose* (seven carbons). We have already encountered two pentoses—the ribose of RNA and the deoxyribose of DNA.

The single most common monosaccharide in the biological world is the aldohexose D-glucose, represented by the formula $C_6H_{12}O_6$ and by the structure shown in Figure 3-21. The formula $C_nH_{2n}O_n$ is characteristic of sugars and gave rise to the general term **carbohydrate** because compounds of this sort were originally thought of as "hydrates of carbon"—$C_n(H_2O)_n$. The term persists, although in no

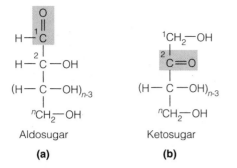

Aldosugar

(a)

Ketosugar

(b)

Figure 3-20 **Structures of Monosaccharides.** **(a)** Aldosugars have a carbonyl group on carbon atom 1. **(b)** Ketosugars have a carbonyl group on carbon atom 2. The number of carbon atoms in a monosaccharide (n) varies from three to seven.

sense can a carbohydrate be thought of as hydrated carbon atoms.

In keeping with the general rule for numbering carbon atoms in organic molecules, the carbons of glucose are numbered beginning with the more oxidized end of the molecule, the aldehyde group. Notice that glucose has four asymmetric carbon atoms (carbon atoms 2, 3, 4, and 5). There are therefore 2^4 different possible stereoisomers of the aldosugar $C_6H_{12}O_6$, but we will concern ourselves only with D-glucose, which is the most stable of the 16 isomers.

Figure 3-21a illustrates D-glucose as it appears in what chemists call a **Fischer projection,** with the —H and —OH groups intended to be projecting slightly out of the plane of the paper. This structure indicates that glucose is a linear molecule, and it is often a useful representation of glucose for pedagogic purposes. In reality, however, glucose exists in the cell in a dynamic equilibrium between the linear (or open-chain) configuration of Figure 3-21a and the ring form of Figure 3-21b. The ring form is the predominant structure, because it is energetically more stable and therefore favored. The ring form results from the addition of the hydroxyl group on carbon atom 5 across the carbonyl group of carbon atom 1. Although the juxtaposition of carbon atoms 1 and 5 required for ring formation seems

unlikely from the Fischer projection, it is actually favored by the tetrahedral nature of each carbon atom in the chain.

A more satisfactory representation of glucose is that shown in Figure 3-21c. The advantage of this **Haworth projection** is that it at least suggests the spatial relationship of different parts of the molecule and makes the spontaneous formation of a bond between carbon atoms 1 and 5 appear more likely. Any of the three representations of glucose shown in Figure 3-21 is valid, but the Haworth projection is preferred because it indicates both the ring form and the spatial relationship of the carbon atoms.

Notice that formation of the ring structure results in the generation of one of two alternative forms of the molecule, depending on the spatial orientation of the hydroxyl group on carbon atom 1. These alternative forms of glucose are designated α and β. As shown in Figure 3-22, α-D-glucose has the hydroxyl group on carbon atom 1 pointing downward in the Haworth projection, and β-D-glucose has the hydroxyl group on carbon atom 1 pointing upward. Starch and glycogen both have α-D-glucose as their repeating unit, whereas cellulose consists of strings of β-D-glucose.

In addition to the free monosaccharide and the long-chain polysaccharides, glucose also occurs in **disaccharides,** which consist of two monosaccharide units linked covalently. Three common disaccharides are shown in Figure 3-23. *Maltose* (malt sugar) consists of two glucose units linked together, whereas *lactose* (milk sugar) contains a glucose linked to a galactose and *sucrose* (common table sugar) has a glucose linked to a fructose. Both galactose and fructose will be discussed in more detail in Chapter 13, where the chemistry and metabolism of several sugars are considered.

Each of these disaccharides is formed by a condensation reaction in which two monosaccharides are linked together by the elimination of water. The resulting **glycosidic bond** is characteristic of linkages between sugars. In the case of maltose, both of the constituent glucose molecules are in the α form, and the glycosidic bond forms between carbon atom 1 of one glucose and carbon atom 4 of the other. This is called an *α glycosidic bond* because it involves

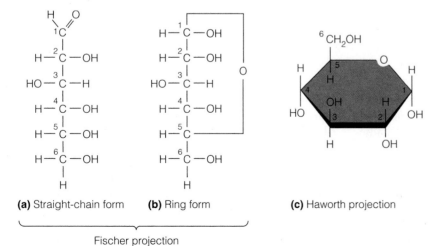

(a) Straight-chain form   **(b)** Ring form   **(c)** Haworth projection

Fischer projection

Figure 3-21 **The Structure of Glucose.** The glucose molecule can be represented by Fischer projections of **(a)** the straight-chain form or **(b)** the ring form of the molecule, as well as by **(c)** the Haworth projection of the ring form. In the Fischer projections, the —H and —OH groups are intended to be projecting slightly out of the plane of the paper. In the Haworth projection, carbon atoms 2 and 3 are intended to be jutting out of the plane of the paper, and carbon atoms 5 and 6 are behind the plane of the paper. The —H and —OH groups then project upward or downward, as indicated. Notice that the carbon atoms are numbered from the more oxidized end of the molecule.

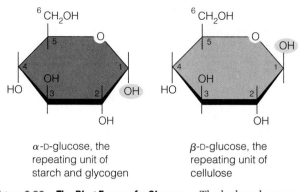

α-D-glucose, the repeating unit of starch and glycogen

β-D-glucose, the repeating unit of cellulose

Figure 3-22 The Ring Forms of D-Glucose. The hydroxyl group on carbon atom 1 points downward in the α form and upward in the β form.

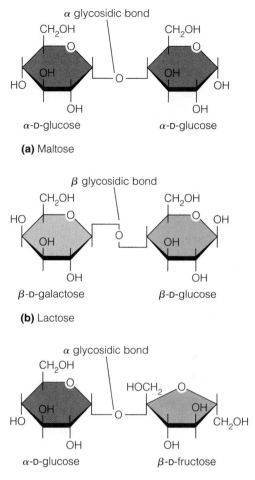

(a) Maltose

(b) Lactose

(c) Sucrose

Figure 3-23 Some Common Disaccharides. (a) Maltose (malt sugar) consists of two molecules of α-D-glucose linked by an α glycosidic bond. **(b)** Lactose (milk sugar) consists of a molecule of β-D-galactose linked to a molecule of β-D-glucose by a β glycosidic bond. **(c)** Sucrose (table sugar) consists of a molecule of α-D-glucose linked to a molecule of β-D-fructose by an α glycosidic bond.

a carbon atom 1 with its hydroxyl group in the α configuration. Lactose, on the other hand, is characterized by a β *glycosidic bond* because the hydroxyl group on carbon atom 1 of the galactose is in the β configuration. The distinction between α and β glycosidic bonds becomes critical when we get to the polysaccharides, because both the three-dimensional configuration and the biological role of the polymer depend on the nature of the bond between the repeating monosaccharide units.

The Polymers Are Storage and Structural Polysaccharides

Polysaccharides perform either storage or structural functions in cells. The most familiar *storage polysaccharides* are the **starch** of plant cells (Figure 3-24a) and the **glycogen** of animal cells (Figure 3-24b). Both of these polymers consist of α-D-glucose units linked together by α glycosidic bonds. In addition to $\alpha(1 \longrightarrow 4)$ bonds that link carbon atoms 1 and 4 of adjacent glucose units, these polysaccharides may contain occasional $\alpha(1 \longrightarrow 6)$ linkages along the backbone, giving rise to side chains (Figure 3-24c). Storage polysaccharides can therefore be branched or unbranched polymers, depending on the presence or absence of $\alpha(1 \longrightarrow 6)$ linkages.

Glycogen is highly branched, with $\alpha(1 \longrightarrow 6)$ linkages occurring every 8 to 10 glucose units along the backbone and giving rise to short side chains of about 8 to 12 glucose units (Figure 3-24b). Glycogen is stored mainly in the liver and in muscle tissue. In the liver it is used as a source of glucose to maintain blood sugar levels, whereas in muscle it serves as a fuel source to generate the ATP needed for muscle contraction.

Starch occurs both as unbranched **amylose** and as branched **amylopectin.** Like glycogen, amylopectin has $\alpha(1 \longrightarrow 6)$ branches, but these occur less frequently along the backbone (once every 12 to 25 glucose units) and give rise to longer side chains (lengths of 20 to 25 glucose units are common) (see Figure 3-24a). Starch deposits are usually about 10–30% amylose and 70–90% amylopectin. Starch is stored in plant cells as *starch grains* within the plastids—either within the *chloroplasts* that are the sites of

carbon fixation and sugar synthesis in photosynthetic tissue, or within the *amyloplasts* that are specialized plastids for starch storage. The potato tuber, for example, is filled with starch-laden amyloplasts.

The best-known example of a *structural polysaccharide* is the **cellulose** found in plant cell walls (Figure 3-25). Cellulose is an important polymer quantitatively; more than half of the carbon in higher plants is present in cellulose! Like starch and glycogen, cellulose is also a polymer of glucose, but the repeating monomer is β-D-glucose and the linkage is therefore $\beta(1 \longrightarrow 4)$. This linkage has structural consequences that we will get to shortly, but it also has nutritional implications. Mammals do not possess an enzyme that can hydrolyze a $\beta(1 \longrightarrow 4)$ bond; therefore, mammals cannot utilize cellulose as food. As a result, you can digest potatoes (starch) but not grass (cellulose). Animals such as cows and sheep might seem to be exceptions because they do eat grass and

(a) Starch

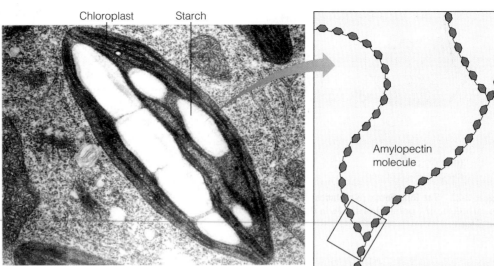

Chloroplast   Starch

Plant leaf cell with starch grains in chloroplast   ⊢ 1 μm ⊣

Amylopectin molecule

Glycogen granules   Mitochondrion

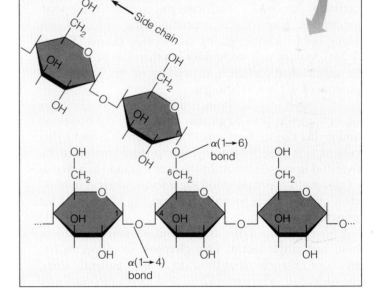

(b) Glycogen

Liver cell with glycogen granules in the cytosol   ⊢ 0.5 μm ⊣

Glycogen molecule

Figure 3-24   The Structure of Starch and Glycogen.
(a) The starch found in plant cells and **(b)** the glycogen found in animal cells are both storage polysaccharides composed of linear chains of α-D-glucose units, with or without occasional branch points (TEMs). Starch occurs in two forms: branched amylopectin, as shown in part a, and unbranched amylose (not shown). Glycogen occurs only as the branched form shown in part b. **(c)** The straight-chain portion of all three kinds of molecules consists of α-D-glucose units linked by α(1 ⟶ 4) glycosidic bonds. In the case of amylopectin and glycogen, branch chains originate at α(1 ⟶ 6) glycosidic bonds.

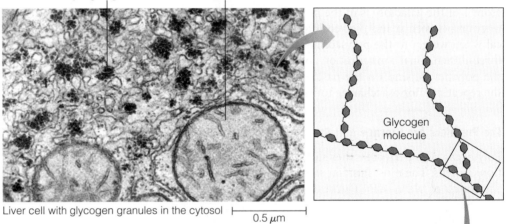

Side chain

α(1→6) bond

α(1→4) bond

(c) Glycogen or amylopectin structure

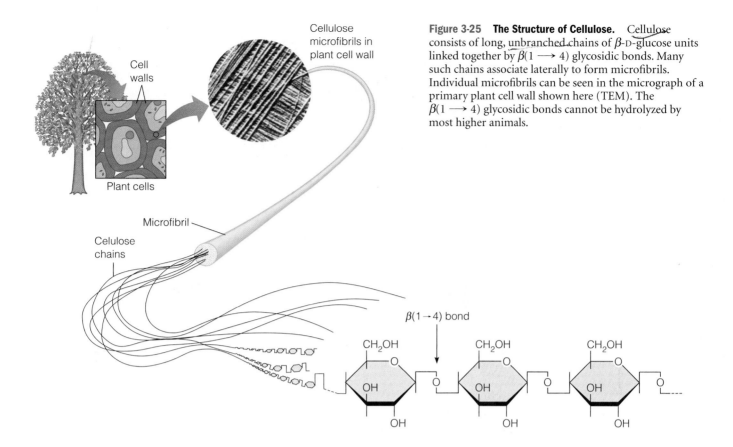

Cell walls

Cellulose microfibrils in plant cell wall

Plant cells

Microfibril

Celulose chains

$\beta(1 \rightarrow 4)$ bond

Figure 3-25 The Structure of Cellulose. Cellulose consists of long, unbranched chains of β-D-glucose units linked together by $\beta(1 \longrightarrow 4)$ glycosidic bonds. Many such chains associate laterally to form microfibrils. Individual microfibrils can be seen in the micrograph of a primary plant cell wall shown here (TEM). The $\beta(1 \longrightarrow 4)$ glycosidic bonds cannot be hydrolyzed by most higher animals.

similar plant products. But they cannot cleave β glycosidic bonds either; they depend on the population of bacteria and protozoa in their rumen (part of their compound stomach) to do this for them. The microorganisms digest the cellulose, and the host animal then obtains the end-products of microbial digestion, now in a form the animal can use.

Although $\beta(1 \longrightarrow 4)$-linked cellulose is quantitatively the most significant structural polysaccharide, others are also known. The celluloses of fungal cell walls, for example, contain either $\beta(1 \longrightarrow 4)$ or $\beta(1 \longrightarrow 3)$ linkages, depending on the species. The cell wall of many bacteria is somewhat more complex and contains two kinds of sugars, *N-acetylglucosamine (GlcNAc)* and *N-acetylmuramic acid (MurNAc)*. As shown in Figure 3-26a, GlcNAc and MurNAc are derivatives of *β-glucosamine,* a glucose molecule with the hydroxyl group on carbon atom 2 replaced by an amino group. GlcNAc is formed by acetylation of the amino group, and MurNAc requires the further addition of a three-carbon lactyl group to carbon atom 3. The cell wall polysaccharide is then formed by the linking of GlcNAc and MurNAc in a strictly alternating sequence with $\beta(1 \longrightarrow 4)$ bonds (Figure 3-26b). Figure 3-26c shows the structure of yet another structural polysaccharide, the **chitin** found in insect exoskeletons and crustacean shells. Chitin consists of GlcNAc units only, joined by $\beta(1 \longrightarrow 4)$ bonds.

Polysaccharide Structure Depends on the Kinds of Glycosidic Bonds Involved

The distinction between the α and β glycosidic bonds of storage and structural polysaccharides has more than just nutritional significance. Because of the difference in linkages and therefore in the spatial relationship between successive glucose units, the two classes of polysaccharides differ markedly in secondary structure. The helical shape already established as a characteristic of both proteins and nucleic acids is also found in polysaccharides. Both starch and glycogen coil spontaneously into loose helices, but often the structure is not highly ordered because of the numerous side chains of amylopectin and glycogen.

Cellulose, by contrast, forms rigid, linear rods. These, in turn, aggregate laterally into *microfibrils* (see Figure 3-25). Microfibrils are about 25 nm in diameter and are composed of about 2000 cellulose chains. Plant and fungal cell walls consist of these rigid microfibrils of cellulose embedded in a *noncellulosic matrix* containing a rather variable mixture of several other polymers (*hemicellulose* and *pectin,* mainly) and a protein called *extensin* that occurs only in the cell wall. Cell walls have been aptly compared to reinforced concrete, in which steel rods are embedded in the cement before it hardens to add strength. In cell walls, the cellulose microfibrils are the "rods" and the noncellulosic matrix is the "cement."

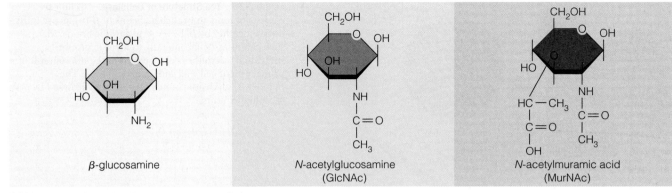

β-glucosamine   N-acetylglucosamine (GlcNAc)   N-acetylmuramic acid (MurNAc)

(a) Polysaccharide subunits

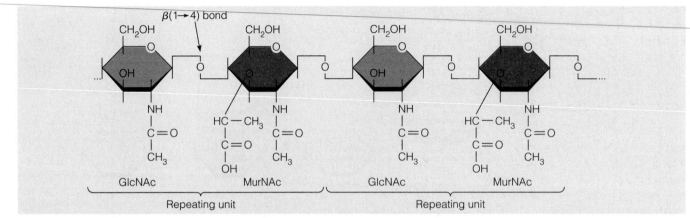

(b) Bacterial cell wall polysaccharide

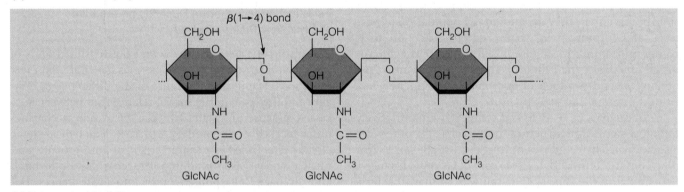

(c) Polysaccharide chitin

Figure 3-26 Polysaccharides of Bacterial Cell Walls and Insect Exoskeletons. **(a)** The subunits glucosamine, N-acetylglucosamine (GlcNAc), and N-acetylmuramic acid (MurNAc). **(b)** A bacterial cell wall polysaccharide, consisting of alternating GlcNAc and MurNAc units. **(c)** The polysaccharide chitin found in insect exoskeletons and crustacean shells, with GlcNAc as its single repeating unit.

Lipids

Strictly speaking, **lipids** do not qualify for inclusion in this chapter because they are not formed by the kind of stepwise polymerization that gives rise to proteins, nucleic acids, and polysaccharides. However, they are commonly regarded as macromolecules because of their molecular weights. Moreover, any discussion of cellular structure and chemical components would be incomplete without reference to this important group of molecules. Their inclusion here also seems reasonable in light of their fre-

quent association with the macromolecules we have already discussed, especially proteins.

Lipids constitute a rather heterogeneous category of cellular components that resemble one another more in their solubility properties than in their chemical structures. The distinguishing feature of lipids is their hydrophobic nature. They have little, if any, affinity for water but are readily soluble in nonpolar solvents such as chloroform or ether. Accordingly, we can expect to find that they are rich in nonpolar hydrocarbon regions and have relatively few polar groups. Some lipids, however, are

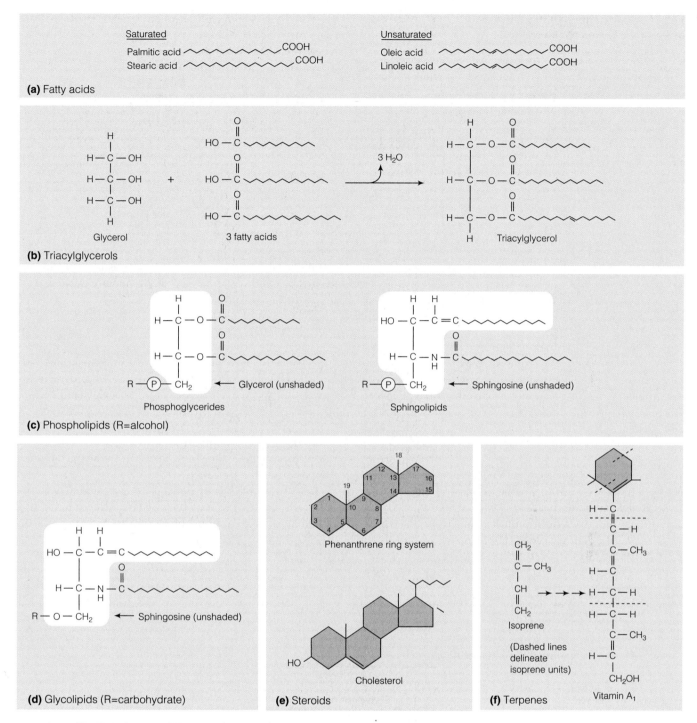

Figure 3-27 **The Main Classes of Lipids.** The zigzag lines in parts a–d represent the long hydrocarbon chains of fatty acids, with a CH$_3$ group on the end. Each vertex, or "zig," represents a methylene ($-CH_2-$) group.

amphipathic, having both a polar and a nonpolar region. As we have already seen, this characteristic has implications for membrane structure.

Because they are defined in terms of solubility characteristics rather than chemical structure, we should not be surprised to find that lipids as a group include molecules that are both functionally and chemically diverse. Functionally, lipids play at least three main roles in cells. Some serve as forms of *energy storage,* others are involved in

membrane structure, and still others have *specific biological functions* such as the transmission of chemical signals into and within the cell. In terms of chemical structure, the six main classes of lipids are *fatty acids, triacylglycerols, phospholipids, glycolipids, steroids,* and *terpenes.* Each of these classes is illustrated in Figure 3-27, which includes representative examples of each class. We will look briefly at each of these six kinds of lipids, pointing out their functional roles in the process.

Fatty Acids Are the Building Blocks of Several Classes of Lipids

We will begin our discussion with **fatty acids** because they are components of several other kinds of lipids. A fatty acid is a long, unbranched hydrocarbon chain with a carboxyl group at one end (Figure 3-27a). The fatty acid molecule is therefore amphipathic; the carboxyl group renders one end (often called the "head") polar, whereas the hydrocarbon "tail" is nonpolar. Fatty acids contain a variable, but usually even, number of carbon atoms. The usual range is from 12 to 24 carbon atoms per chain, with 16- and 18-carbon fatty acids especially common.

Table 3-5 summarizes the nomenclature of fatty acid chain length. Even numbers of carbon atoms are greatly favored because of the mode of fatty acid synthesis. Each molecule is generated by the stepwise addition of two-carbon units as acetyl coenzyme A, a carrier of acyl groups that we will consider in more detail in Chapter 14. Once added, each new two-carbon increment is then reduced to the hydrocarbon level.

Because they are highly reduced, fatty acids yield a great deal of energy upon oxidation and are therefore efficient forms of energy storage. We will see how to quantify the efficiency of storing energy as fat rather than as carbohydrate when we get to Part Three. For the present, simply note that a gram of fat contains more than twice as much usable energy as a gram of sugar or polysaccharide.

Table 3-5 also shows the variability in fatty acid structure due to the presence of double bonds between carbons. Fatty acids without double bonds are referred to as **saturated fatty acids** because every carbon atom in the chain has the maximum number of hydrogen atoms attached to it (Figure 3-28a). The general formula for a saturated fatty acid with n carbon atoms is $C_nH_{2n}O_2$. By contrast, **unsaturated fatty acids** contain one or more double bonds. The presence of such sites of unsaturation affects the shape of the molecule and therefore the kinds

of structures of which it can be a part. Saturated fatty acids have long straight tails that pack together well, whereas each double bond results in a bend or kink in the molecule that prevents tight packing (Figure 3-28b).

Triacylglycerols Are Storage Lipids

The **triacylglycerols,** also called *triglycerides,* consist of a glycerol molecule with three fatty acids linked to it. As shown in Figure 3-27b, **glycerol** is a three-carbon alcohol with a hydroxyl group on each carbon. Fatty acids are linked to glycerol by *ester bonds,* which are formed by the removal of water. Triacylglycerols are synthesized stepwise, with one fatty acid added at a time. *Monoacylglycerols* contain a single esterified fatty acid, *diacylglycerols* have two, and *triacylglycerols* have three. The three fatty acids of a given triacylglycerol need not be identical. They can, and generally do, vary in either chain length or degree of unsaturation, or both.

The main function of triacylglycerols is to store energy, which will be of special interest to our discussion of energy metabolism in Part Three. In some animals, triacylglycerols also provide insulation against cold temperatures. Animals such as walruses, seals, and penguins that live in very cold climates store triacylglycerols under their skin and depend on the insulating properties of this fat for survival in harsh environments.

Triacylglycerols containing a preponderance of saturated fatty acids are usually solid or semisolid at room temperature and are called *fats.* Fats are prominent in the bodies of animals, as evidenced by the fat that you buy with most cuts of meat, by the large quantity of lard that is obtained as a by-product of the meat-packing industry, and by the widespread concern people have that they are "getting fat." In plants, most triacylglycerols are liquid at room temperature, as the term *vegetable oil* suggests. Because the fatty acids of oils are predominantly unsaturated, their hydrocarbon chains have kinks that prevent an

Table 3-5   Nomenclature of the Fatty Acids*

Number of Carbons	Number of Double Bonds	Common Name	Systematic Name	Formula
12	0	Laurate	*n*-dodecanoate	$CH_3(CH_2)_{10}COO^-$
14	0	Myristate	*n*-tetradecanoate	$CH_3(CH_2)_{12}COO^-$
16	0	Palmitate	*n*-hexadecanoate	$CH_3(CH_2)_{14}COO^-$
18	0	Stearate	*n*-octadecanoate	$CH_3(CH_2)_{16}COO^-$
20	0	Arachidate	*n*-eicosanoate	$CH_3(CH_2)_{18}COO^-$
16	1	Palmitoleate	*cis*-Δ^9-hexadecenoate	$CH_3(CH_2)_5CH{=}CH(CH_2)_7COO^-$
18	1	Oleate	*cis*-Δ^9-octadecenoate	$CH_3(CH_2)_7CH{=}CH(CH_2)_7COO^-$
18	2	Linoleate	*cis, cis*-Δ^9, Δ^{12}-octadecadienoate	$CH_3(CH_2)_4(CH{=}CHCH_2)_2(CH2)_6COO^-$
18	3	Linolenate	All *cis*-Δ^9, Δ^{12}, Δ^{15}-octadecatrienoate	$CH_3CH_2(CH{=}CHCH_2)_3(CH_2)_6COO^-$
20	4	Arachidonate	All *cis*-Δ^5, Δ^8, Δ^{11}, Δ^{14}-eicosatetraenoate	$CH_3(CH_2)_4(CH{=}CHCH_2)_4(CH_2)_2COO^-$

*The common names, systematic names, and formulas are for the ionized (anionic) forms of the fatty acids, because fatty acids exist primarily in the anionic form at the near-neutral pH of most cells. For the names and structures of the free fatty acids, simply replace the "-ate" ending with "-ic acid" and substitute a hydrogen atom (H) for the negative charge in each case.

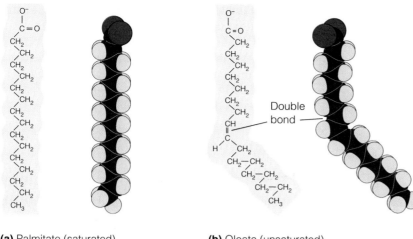

Figure 3-28 Structures of Saturated and Unsaturated Fatty Acids. (a) The saturated 16-carbon fatty acid palmitate. (b) The unsaturated 18-carbon fatty acid oleate. The space-filling models are intended to emphasize the overall shape of the molecules. Notice the kink that the double bond creates in the oleate molecule.

(a) Palmitate (saturated)   **(b)** Oleate (unsaturated)

orderly packing of the molecules. As a result, vegetable oils have lower melting temperatures than most animal fats. Soybean oil and corn oil are two familiar vegetable oils. Vegetable oils can be converted into solid products such as margarine and shortening by partial hydrogenation (saturation) of the double bonds, a process explored further in Problem 3-16 at the end of the chapter.

Phospholipids Are Important in Membrane Structure

Phospholipids make up a third class of lipids (Figure 3-27c). They are similar to triacylglycerols in some chemical details but differ strikingly in their properties and their role in the cell. First and foremost, phospholipids are important in membrane structure. In fact, as we saw in Chapter 2, they are critical to the bilayer structure found in all membranes (see Figure 2-12). In terms of chemistry, phospholipids are *phosphoglycerides* or *sphingolipids* (Figure 3-27c, left and right, respectively).

Phosphoglycerides are the predominant phospholipids present in most membranes. Like triacylglycerols, a phosphoglyceride consists of fatty acids esterified to a glycerol molecule. However, the basic component of a phosphoglyceride is **phosphatidic acid,** which has just two fatty acids and a phosphate group (Figure 3-29a). Phosphatidic acid is a key intermediate in the synthesis of other phosphoglycerides but is itself not at all prominent in membranes. Instead, membrane phosphoglycerides invariably have, in addition, a small hydrophilic alcohol linked to the phosphate by an ester bond and represented in Figure 3-29a as an R group. The alcohol is usually *serine, ethanolamine, choline,* or *inositol* (Figure 3-29b). Except

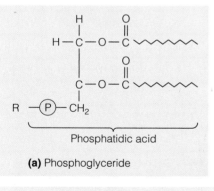

(a) Phosphoglyceride

Figure 3-29 Structures of Common Phosphoglycerides. (a) A phosphoglyceride consists of a molecule of phosphatidic acid (glycerol esterified to two fatty acids and a phosphate group) with a small polar alcohol, represented as R, also esterified to the phosphate group. (b) The four most common R groups found in phosphoglycerides are serine, ethanolamine, choline, and inositol, the first three of which contain a positively charged amino group.

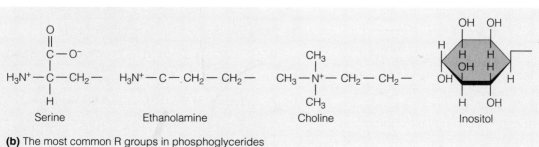

Serine   Ethanolamine   Choline   Inositol

(b) The most common R groups in phosphoglycerides

for inositol, these alcohols contain an amino group that is protonated and therefore charged at cellular pH. The presence of a negatively charged phosphate and a positively charged amine in juxtaposition makes these phosphoglycerides electrically neutral but gives them a highly polar head region.

The combination of a highly polar head and two long nonpolar chains gives the phosphoglycerides the characteristic amphipathic nature that is so critical to their role in membrane structure. As we saw earlier, the fatty acids can vary considerably in both length and the presence and position of sites of unsaturation. In membranes, 16- and 18-carbon fatty acids are most common, and a typical phosphoglyceride molecule is likely to have one saturated and one unsaturated fatty acid. The length and the degree of unsaturation of fatty acid chains in membrane phospholipids profoundly affect membrane fluidity and can, in fact, be regulated by the cells of some organisms to control this crucial membrane property.

In addition to the phosphoglycerides, some membranes contain another class of phospholipid called **sphingolipids.** As the name suggests, these lipids are based not on glycerol but on the amine alcohol **sphingosine.** As shown in Figure 3-27c, sphingosine has a long hydrocarbon chain with a single site of unsaturation near the polar end. Through its amino group, sphingosine can form an amide bond to a long-chain fatty acid. The resulting molecule is called a *ceramide* and consists of a polar region flanked by two long nonpolar tails. Because of their common nonpolar nature, the two tails tend to bend around and associate with each other, giving the molecule a hairpin bend and a shape that approximates that of the phospholipids.

The hydroxyl group on carbon atom 1 of sphingosine juts out from what is effectively the head of this hairpin molecule. A sphingolipid is formed when any of several polar groups becomes linked to this hydroxyl group. Actually, a whole family of sphingolipids exists, differing only in the chemical nature of the polar group attached to the hydroxyl group of the ceramide (i.e., the R group of Figure 3-27c). The sphingomyelins, for example, contain ethanolamine or choline and therefore closely resemble phosphoglycerides in overall shape and the chemical nature of the polar head.

Glycolipids Are Specialized Membrane Components

Glycolipids are derivatives of sphingosine (or sometimes glycerol) that contain a carbohydrate group instead of a phosphate group. Those containing sphingosine are called *glycosphingolipids*. The carbohydrate group attached to a glycolipid may contain one to six sugar units, which can be D-glucose, D-galactose, or N-acetyl-D-galactosamine. These carbohydrate groups, like phosphate groups, are water-soluble, giving the glycolipid an amphipathic nature. Glycolipids are specialized constituents of some membranes, especially those found in certain plant cells and in the cells of the nervous system. In such cells, glycolipids occur largely in the outer face of the plasma membrane.

When sphingolipids were first discovered by Johann Thudicum in the late nineteenth century, their biological role seemed as enigmatic as the Sphinx, after which he named them. We now know that sphingolipids in general, and glycosphingolipids in particular, are sites of biological recognition on the surface of the plasma membrane, a topic we will consider further in Chapter 10.

Steroids Are Lipids with a Variety of Functions

The **steroids** constitute yet another distinctive class of lipids. Steroids are derivatives of a four-membered ring compound called *phenanthrene* (Figure 3-27e), which makes them structurally distinct from other lipids. In fact, the only property that links them to the other classes of lipids is that they are relatively nonpolar and therefore hydrophobic. As Figure 3-30 illustrates, steroids differ from one another in the number and positions of double bonds and functional groups.

Steroids are found only in eukaryotic cells. The most common steroid in animal cells is **cholesterol,** the structure of which is shown in Figure 3-27e. Cholesterol is an amphipathic molecule, with a polar head group (the hydroxyl group at position 3) and a nonpolar hydrocarbon body and tail (the four-membered ring and the hydrocarbon side chain at position 17). Because most of the molecule is hydrophobic, cholesterol is found primarily in membranes. It occurs in the plasma membrane of animal cells and in most of the membranes of organelles,

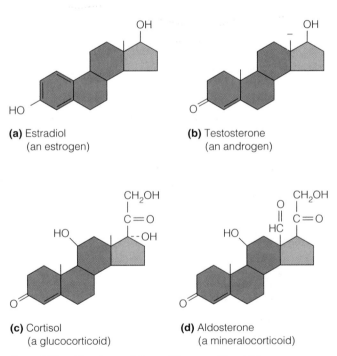

(a) Estradiol
(an estrogen)

(b) Testosterone
(an androgen)

(c) Cortisol
(a glucocorticoid)

(d) Aldosterone
(a mineralocorticoid)

Figure 3-30 Structures of Several Common Steroid Hormones. Among the many steroids that are synthesized from cholesterol are the hormones **(a)** estradiol, an estrogen; **(b)** testosterone, an androgen; **(c)** cortisol, a glucocorticoid; and **(d)** aldosterone, a mineralocorticoid.

except the inner membranes of mitochondria and chloroplasts. Similar membrane steroids occur in other eukaryotic cells also, including *stigmasterol* and *sitosterol* in plant cells and *ergosterol* in fungal cells.

Cholesterol is the starting point for the synthesis of all the **steroid hormones** (Figure 3-30), which include the male and female *sex hormones*, the *glucocorticoids*, and the *mineralocorticoids*. The sex hormones include the *estrogens* produced by the ovaries of females (*estradiol*, for example) and the *androgens* produced by the testes of males (*testosterone*, for example). The glucocorticoids (*cortisol*, for example) are a family of hormones that promote gluconeogenesis (synthesis of glucose) and suppress inflammation reactions. Mineralocorticoids such as *aldosterone* regulate ion balance by promoting the reabsorption of sodium, chloride, and bicarbonate ions in the kidney.

Terpenes Are Formed from Isoprene

The final class of lipids shown in Figure 3-27 consists of the **terpenes.** Terpenes, synthesized from the five-carbon compound *isoprene*, are also called *isoprenoids*. Isoprene and its derivatives are joined together in various combinations to produce such substances as *vitamin A_1* (Figure 3-27f), *carotenoid pigments*, compounds called *dolichols* that are involved in activating sugar derivatives, and electron carriers such as *coenzyme Q* and *plastoquinone*, which we will encounter in Part Three.

Perspective

Three kinds of macromolecules are very prominent in cells: proteins, nucleic acids, and polysaccharides. Proteins and nucleic acids are informational macromolecules that depend directly or indirectly on genetic information to determine the order of amino acids and nucleotides that is so critical to their roles in the cell. Polysaccharides need no such information; they usually contain one or a few kinds of repeating units and play storage or structural roles instead.

Proteins consist of linear chains of amino acids that differ markedly in the chemical properties of their R groups. The amino acid sequence, or primary structure, of a polypeptide is all-important because it contains most, if not all, of the information necessary to specify the local folding or orientation of the amino acid chain (secondary structure), the overall shape of the polypeptide (tertiary structure), and, in the case of multimeric proteins, the further association with other polypeptides (quaternary structure). A major force behind protein folding and polypeptide interaction is the tendency of hydrophobic amino acids to avoid an aqueous environment. In addition, hydrogen bonds, ionic bonds, van der Waals interactions, and covalent disulfide bonds are important in stabilizing protein structure.

Nucleic acids also attain a shape that is dictated by the chemical nature of their subunits. This is seen most strikingly in the DNA double helix, in which complementary base pairing (A with T, C with G) is stabilized by hydrogen bonding. The elucidation of the double-helical structure of DNA was one of the defining biological advances of the twentieth century.

Polysaccharides are chains of monosaccharides linked together by either α or β glycosidic bonds. The difference is critical because α linkages are readily digested by animals and are therefore suitable for storage polysaccharides such as starch or glycogen. By contrast, the β glycosidic bonds of structural polysaccharides such as cellulose and chitin are not digestible by animals and give the molecule a rigid shape suitable to its function.

Lipids are not macromolecules but are included in this chapter because of their general importance as constituents of cells (especially membranes) and their frequent association with macromolecules, particularly proteins. Lipids differ substantially in chemical structure, but they all share the common property of solubility in organic solvents but not in water. The major classes of lipids include the triacylglycerols that make up fats and oils, the phospholipids and sphingolipids found in membranes, the glycolipids involved in recognition phenomena, and the steroids and terpenes, which perform a variety of functions in eukaryotic cells.

Key Terms for Self-Testing

Proteins

protein (p. 41)	multimeric protein (p. 45)	primary structure (p. 48)
amino acid (p. 41)	conformation (p. 45)	secondary structure (p. 48)
peptide bond (p. 44)	disulfide bond (p. 45)	α helix (p. 48)
N- (amino) terminus (p. 44)	noncovalent bonds and interactions (p. 46)	β sheet (p. 48)
C- (carboxyl) terminus (p. 44)	hydrogen bond (p. 46)	motif (p. 50)
polypeptide (p. 45)	ionic bond (p. 46)	tertiary structure (p. 50)
monomeric protein (p. 45)	van der Waals interaction (p. 46)	native conformation (p. 50)
	hydrophobic interaction (p. 46)	fibrous protein (p. 50)

globular protein (p. 51)
domain (p. 52)
quaternary structure (p. 53)
multiprotein complex (p. 54)

Nucleic Acids
nucleic acid (p. 54)
DNA (deoxyribonucleic acid) (p. 54)
RNA (ribonucleic acid) (p. 54)
ribose (p. 54)
deoxyribose (p. 54)
nucleotide (p. 55)
purine (p. 55)
pyrimidine (p. 55)
adenine (A) (p. 55)
guanine (G) (p. 55)
cytosine (C) (p. 55)
thymine (T) (p. 55)
uracil (U) (p. 56)
nucleoside (p. 56)
nucleoside monophosphate (p. 56)

adenosine monophosphate (AMP) (p. 56)
adenosine diphosphate (ADP) (p. 56)
adenosine triphosphate (ATP) (p. 56)
phosphodiester bond (p. 56)
polynucleotide (p. 56)
template (p. 57)
base pairing (p. 58)
double helix (p. 58)

Polysaccharides
polysaccharide (p. 60)
monosaccharide (p. 61)
carbohydrate (p. 61)
Fischer projection (p. 62)
Haworth projection (p. 62)
disaccharide (p. 62)
glycosidic bond (p. 62)
starch (p. 63)
glycogen (p. 63)
amylose (p. 63)
amylopectin (p. 63)

cellulose (p. 63)
chitin (p. 65)

Lipids
lipid (p. 66)
fatty acid (p. 68)
saturated fatty acid (p. 68)
unsaturated fatty acid (p. 68)
triacylglycerol (p. 68)
glycerol (p. 68)
phospholipid (p. 69)
phosphoglyceride (p. 69)
phosphatidic acid (p. 69)
sphingolipid (p. 70)
sphingosine (p. 70)
glycolipid (p. 70)
steroid (p. 70)
cholesterol (p. 70)
steroid hormone (p. 71)
terpene (p. 71)

Problem Set

More challenging problems are marked with a •.

3-1. Polymers and Their Properties. For each of the six biological polymers listed below, indicate which of the properties apply. Each polymer has multiple properties, and a given property may be used more than once.

Polymers

(a) Cellulose   (d) Amylopectin

(b) Messenger RNA   (e) DNA

(c) Globular protein (f) Fibrous protein

Properties

1. Branched-chain polymer
2. Extracellular location
3. Glycosidic bonds
4. Informational macromolecule
5. Peptide bond
6. β linkage
7. Phosphodiester bond
8. Nucleoside triphosphates
9. Helical structure possible
10. Synthesis requires a template

3-2. Stability of Protein Structure. Several different kinds of bonds or interactions are involved in generating and maintaining the structure of proteins. List four or five such bonds or interactions, give an example of an amino acid that might be involved in each, and indicate which level(s) of protein structure might be generated or stabilized by that particular kind of bond or interaction.

3-3. Amino Acid Localization in Proteins. Amino acids tend to be localized either in the interior or on the exterior of a

globular protein molecule, depending on their relative affinities for water.

(a) Classify each of the following amino acids as likely to be found in the interior, on the exterior, or at either location, and explain.

valine   aspartate

alanine   glycine

phenylalanine   lysine

(b) For each of the following pairs of amino acids, choose the one that is more likely to be found in the interior of a protein molecule, and explain why.

alanine; glycine   glutamate; aspartate

tyrosine; phenylalanine   methionine; cysteine

(c) Explain why cysteines with free sulfhydryl groups tend to be localized on the exterior of a protein molecule, whereas those involved in disulfide bonds are more likely to be buried in the interior of the molecule.

3-4. Myoglobin Versus Hemoglobin. Myoglobin and hemoglobin are both oxygen-binding proteins. Myoglobin is a monomeric protein found in muscle cells, whereas hemoglobin is tetrameric and is found in red blood cells. The tertiary structure of myoglobin is strikingly similar to that of both the α and β subunits of hemoglobin; yet when the primary structures are compared, myoglobin can be shown to have hydrophilic amino acids at several positions that in the hemoglobin chains are occupied by hydrophobic amino acids. Given the extent to which tertiary structure is thought to depend on primary structure, how can a relatively hydrophobic polypeptide such as the α or β subunit of hemoglobin have a tertiary structure very much like that of the relatively hydrophilic myoglobin?

3-5. Sickle-Cell Anemia. Sickle-cell anemia is a striking example of the drastic effect that a single amino acid substitution can have on the structure and function of a protein.

(a) Given the chemical nature of glutamate and valine, can you suggest why the substitution of the latter for the former at position 6 of the β chain would be especially deleterious?

(b) Suggest several other amino acids that would be much less likely to cause impairment of hemoglobin function if substituted for the glutamate at position 6 of the β chain.

(c) Can you see why in some cases two proteins could differ at a number of points in their amino acid sequence and still be very similar in structure and function? Explain.

3-6.   Hair Versus Silk. The α-keratin of human hair is a good example of a fibrous protein with extensive α-helical structure. Silk fibroin is also a fibrous protein, but it consists primarily of β-sheet structure. Fibroin is essentially a polymer of alternating glycines and alanines, whereas α-keratin contains most of the common amino acids and has many disulfide bonds.

(a) If you were able to grab onto both ends of an α-keratin polypeptide and pull, you would find it to be both extensible (it can be stretched to about twice its length in moist heat) and elastic (when you let go, it will return to its normal length). In contrast, a fibroin polypeptide has essentially no extensibility, but it has great tensile strength. Explain these differences.

(b) Can you suggest why fibroin assumes a pleated sheet structure, whereas α-keratin exists as an α helix and even reverts spontaneously to a helical shape when it has been stretched artificially?

3-7.   The "Permanent" Wave That Isn't. The "permanent" wave that your local beauty parlor offers depends critically on rearrangements in the extensive disulfide bonds of keratin that give your hair its characteristic shape. To change the shape of your hair (to give it a wave or curl), the beautician first treats your hair with a sulfhydryl reducing agent, then uses curlers or rollers to impose the desired artificial shape, and follows this by treatment with an oxidizing agent.

(a) What is the chemical basis of a "permanent"? Be sure to include the use of a reducing agent and an oxidizing agent in your explanation.

(b) Why do you suppose a "permanent" isn't permanent? (Explain why the wave or curl is gradually lost during the weeks following your visit to the beautician.)

(c) Can you suggest an explanation for naturally curly hair?

3-8.   The Importance of Hydrogen Bonds. Hydrogen bonds play an important role in stabilizing the secondary structure of both proteins and nucleic acids. When a solution of either a protein or a nucleic acid is heated, one of the main effects of the thermal energy is to break hydrogen bonds; this is called *thermal denaturation*. For each statement below, decide for which, if any, of the following polymers the statement is true: the fibrous protein α-keratin (k); the silk protein fibroin (f); the DNA double helix (d).

(a) All of the hydrogen bonds involve one nitrogen atom.

(b) The hydrogen bonds are between units of the same polymer strand.

(c) The hydrogen bonds are perpendicular to the main axis of the polymer.

(d) The hydrogen bonds make the polymer more polar than it would otherwise be.

(e) The number of hydrogen bonds in a given length of the polymer does not depend on the particular amino acids or nucleotides present in that specific segment of the polymer.

(f) One of the effects of heating will be to separate polymer strands that are otherwise bonded to each other.

(g) The two ends of an individual heat-denatured polymer strand are likely to be much farther away from each other than they are in the fully hydrogen-bonded structure.

(h) Upon cooling under the appropriate conditions, the strands will return spontaneously to the original three-dimensional conformation.

(i) The renaturation process described in (h) will proceed much more slowly if the solution of denatured strands is diluted before being cooled.

3-9.   The Size of DNA Molecules. The DNA double helix contains exactly ten nucleotide pairs per turn, and each turn adds 3.4 nm to the overall length of the duplex. The circular DNA molecule of *E. coli* has about 4 million nucleotide pairs and a molecular weight of about 2.8×10^9.

(a) What is the circumference of the *E. coli* DNA molecule? What problems might this pose for a cell that is only about 2 µm long?

(b) Human mitochondrial DNA is also a circular duplex, with a diameter of about 1.8 µm. About how many nucleotide pairs does it contain? If it takes about 1000 nucleotides of DNA to encode (specify the amino acid sequence of) a typical polypeptide, how many polypeptides can mitochondrial DNA encode?

(c) What is the weight (in grams) of one nucleotide pair? How many nucleotide pairs are present in the 6 picograms (6×10^{-12} g) of double-stranded DNA present in the nucleus of most cells in your body?

(d) What is the total length of the 6 picograms of DNA present in a single nucleus?

3-10.   Storage Polysaccharides. The only common examples of branched-chain polymers in cells are the storage polysaccharides glycogen and amylopectin. Both are degraded exolytically, which means by stepwise removal of terminal glucose units.

(a) Why might it be advantageous for a storage polysaccharide to have a branched-chain structure instead of a linear structure?

(b) Can you foresee any metabolic complications in the process of glycogen degradation? How do you think the cell handles this?

(c) Can you see why cells that must degrade amylose instead of amylopectin have enzymes capable of endolytic (internal) as well as exolytic cleavage of glycosidic bonds?

(d) Why do you suppose the structural polysaccharide cellulose does not contain branches?

3-11.   Carbohydrate Structure. From the following descriptions of gentiobiose, raffinose, and a dextran, draw Haworth projections of each.

(a) *Gentiobiose* is a disaccharide found in gentians and other plants. It consists of two molecules of β-D-glucose linked to each other by a $\beta(1 \longrightarrow 6)$ glycosidic bond.

(b) *Raffinose* is a trisaccharide found in sugar beets. It consists of one molecule each of α-D-galactose, α-D-glucose, and β-D-fructose, with the galactose linked to the glucose by an α(1 ⟶ 6) glycosidic bond and the glucose linked to the fructose by an α(1 ⟶ 2) bond.

(c) *Dextrans* are polysaccharides produced by some bacteria. They are polymers of a-D-glucose linked by α(1 ⟶ 6) glycosidic bonds, with frequent α(1 ⟶ 3) branching. Draw a portion of a dextran, including one branch point.

3-12.   Reducing Sugars. A *reducing sugar* is one that will undergo the *Fehling reaction* shown in Figure 3-31 for the monosaccharide glucose. In this reaction, cupric ions (Cu^{2+}) are reduced in an alkaline solution to insoluble cuprous oxide (Cu_2O), with concomitant oxidation of the sugar to the corresponding acid—gluconic acid, in the case of glucose. (To understand the oxidation of glucose, draw the ring form of the sugar in its Fischer projection, convert it to the straight-chain form, and note the free aldehyde group on carbon atom 1 that can be oxidized to a carboxyl group.) The production of a red precipitate of Cu_2O indicates that the sugar being tested is a reducing sugar. The Fehling reaction was at one time used to test for excess sugar in the urine of people thought to have diabetes.

(a) Which of the disaccharides shown in Figure 3-23 are reducing sugars? Explain.

(b) Is either gentiobiose or raffinose a reducing sugar? Explain. (See Problem 3-11 for information on these sugars.)

•**3-13.   Cotton and Potatoes.** A cotton fiber consists almost exclusively of cellulose, whereas a potato tuber contains mainly starch. Cotton is tough, fibrous, and virtually insoluble in water. The starch present in a potato tuber, on the other hand, is neither tough nor fibrous and can be dispersed in hot water to form a turbid solution. Yet both the cotton fiber and the potato tuber consist primarily of polymers of D-glucose in (1 ⟶ 4) linkage.

(a) How can two polymers consisting of the same repeating subunit have such different properties?

(b) What is the advantage of the respective properties in each case?

•**3-14.   How Much Cellulose?** Assume you are working out the answer to this question on a sheet of paper that is 21.5 cm (8.5 inches) wide and 28 cm (11 inches) long and weighs 4.0 grams. Assume that the paper consists only of cellulose, the chemical formula for which can be written as $(C_6H_{10}O_5)_n$, where *n* is a large variable number, and that each glucose unit in a cellulose molecule is 0.45 nm long and 0.30 nm wide.

(a) Why is the chemical formula written as $(C_6H_{10}O_5)_n$ rather than as $(C_6H_{12}O_6)_n$, since cellulose is a polymer of glucose units?

(b) Based on the molecular weight of the $C_6H_{10}O_5$ unit, calculate the number of such units present in the sheet of paper.

(c) Assuming that all of the strands of cellulose in the sheet of paper are oriented vertically, how many glucose units does it take to reach from the top of the sheet to the bottom?

(d) How many such vertically oriented cellulose strands have to be packed side by side to span the width of the paper?

•**3-15.   Thinking About Lipids.** You should be able to answer each of the following questions based on the properties of lipids discussed in this chapter.

(a) How would you define a lipid? In what sense is the operational definition different from that of proteins, nucleic acids, or carbohydrates?

(b) Arrange the following lipids in order of decreasing polarity: cholesterol, estradiol, fatty acid, phosphatidyl choline, triacylglyceride. Explain your reasoning.

(c) Which would you expect to resemble a sphingomyelin molecule more closely, a molecule of phosphatidyl choline containing two molecules of palmitate acid as its fatty acid side chains, or a phosphatidyl choline molecule with one molecule of palmitate and one molecule of oleate as its fatty acid side chains? Explain your reasoning.

(d) Assume you and your lab partner Mort determined the melting temperature for each of the following fatty acids: arachidic, linoleic, linolenic, oleic, palmitic, and stearic acids. Mort recorded the melting points of each but neglected to note the specific fatty acid to which each value belongs. Assign each of the following melting temperatures (in °C) to the appropriate fatty acid and explain your reasoning: −11, +5, +16, +63, +70, and +76.5.

(e) For each of the following amphipathic molecules, indicate which part of the molecule is hydrophilic: phosphatidyl serine; sphingomyelin; cholesterol; triacylglycerol.

3-16.   Shortening. A popular brand of shortening has a label on the can that identifies the product as "partially hydrogenated soybean oil, palm oil, and cottonseed oil."

(a) What does the process of partial hydrogenation accomplish chemically?

(b) What did the product in the can look like before it was partially hydrogenated?

(c) What is the physical effect of partial hydrogenation?

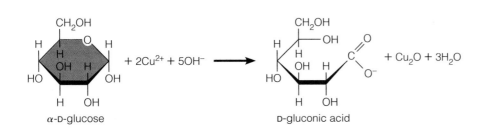

Figure 3-31   The Fehling Reaction: A Test for Reducing Sugars.

Suggested Reading

References of historical importance are marked with a •.

General References

• Fruton, J. S. *A Skeptical Biochemist.* Cambridge, MA: Harvard University Press, 1992.

Lehninger, A. L., D. L. Nelson, and M. M. Cox. *Principles of Biochemistry,* 3d ed. New York: Worth, 1999.

Mathews, C. K., K. E. van Holde, and K. G. Ahern. *Biochemistry,* 3d ed. Menlo Park, CA: Benjamin/Cummings, 2000.

Olson, A. J., and D. S. Goodsell. Visualizing biological molecules. *Sci. Amer.* 267 (November 1992): 76.

Wald, G. The origins of life. *Proc. Natl. Acad. Sci. USA* 52 (1994): 595.

Weinberg, R. A. The molecules of life. *Sci. Amer.* 253 (October 1985): 48.

Proteins

• Anfinsen, C. B. Principles that govern the folding of protein chains. *Science* 181 (1973): 223.

Branden, C., and J. Tooze. *Introduction to Protein Structure,* 2d ed. New York: Garland Press, 1999.

Danchin, A. From protein sequence to function. *Curr. Opin. Struct. Biol.* 9 (1999): 363.

Dinner, A. R., A. Sali, L. J. Smith, C. M. Dobson, and M. Karplus. Understanding protein folding via free-energy surfaces from theory and experiment. *Trends Biochem. Sci.* 25 (2000): 331.

Edgcomb, S. P. P. and K. P. Murphy. Structural energetics of protein folding and binding. *Curr. Opin. Biotech.* 11 (2000): 62.

Ellis, R. J., ed. *The Chaperonins.* New York: Academic Press, 1996.

Lesk, A.M. *Introduction to Protein Architecture: The Structural Biology of Proteins.* Oxford: Oxford University Press, 2001.

Moult, J., and E. Melamud. From fold to function. *Curr. Opin. Struct. Biol.* 10 (2000): 384.

Murphy, K.P., ed. *Protein Structure, Stability, and Folding.* Totowa, NJ: Humana Press, 2001.

Orengo, C.A., A. E. Todd, and J. M. Thornton. From protein structure to function. *Curr. Opin. Struct. Biol.* 9 (1999): 374.

Pennisi, E. Taking a structural approach to understanding proteins. *Science* 279 (1998): 978.

Nucleic Acids

Berg, P., and M. Singer. *Dealing with Genes. The Language of Heredity.* Mill Valley, CA: University Science Books, 1992.

• Crick, F. H. C. The structure of the hereditary material. *Sci. Amer.* (October 1954): 54.

• Dickerson, R. E. The DNA helix and how it is read. *Sci. Amer.* 249 (December 1983): 94.

Forsdye, D. R. Chargaff's legacy. *Gene* 261 (2000): 127.

Li, W. The study of correlation structures of DNA sequences: A critical review. *Computers & Chemistry* 21 (1997): 257.

Munoz, V., and L. Serrano. Helix design, prediction and stability. *Curr. Opin. Biotech.* 6 (1995): 382.

• Portugal, F. H., and J. S. Cohen. *The Century of DNA: A History of the Discovery of the Structure and Function of the Genetic Substance.* Cambridge, MA: MIT Press, 1977.

Skolnick, J., and J. S. Fetrow. From genes to protein structure and function: novel applications of computational approaches in the genomic era. *Trends Biotech.* 18 (2000): 34.

Carbohydrates and Lipids

Binkley, R. W. *Modern Carbohydrate Chemistry.* San Diego, CA: Marcel Dekker, 1988.

Flatt, J. P. Use and storage of carbohydrate and fat. *Amer. J. Clin. Nutr.* 61 (1995): 952S.

Gurr, M. I., and J. L. Harwood. *Lipid Biochemistry. An Introduction,* 4th ed. London: Chapman and Hall, 1990.

Hakamori, S. Glycosphingolipids. *Sci. Amer.* 254 (May 1986): 44.

Kobata, A. Structures and functions of the sugar chains of glycoproteins. *Europ. J. Biochem.* 209 (1992): 483.

Merrill, A. H., Jr., E. M. Schmelz, D. L. Dillehay, S. Spiegel, J. A. Shayman, J. J. Schroeder, R. T. Riley, K. A. Voss, and E. Wang. Sphingolipids—the enigmatic lipid class: Biochemistry, chemistry, physiology, and pathophysiology. *Toxicol. Appl. Pharmacol.* 142(1997): 208.

Sharon, N., and H. Lis. Carbohydrates in cell recognition. *Sci. Amer.* 268 (January 1993): 82.

Vance, D. E., and J. E. Vance, eds. *Biochemistry of Lipids, Lipoproteins, and Membranes.* New Comprehensive Biochemistry, Vol. 20. New York: Elsevier, 1991.

Cells and Organelles

In the previous two chapters, we encountered most of the major kinds of molecules found in cells, as well as some of the principles that govern the assembly of these molecules into the supramolecular structures of which cells and their organelles consist (see Figure 2-14). Now we are ready to focus our attention on cells and organelles directly.

Properties and Strategies of Cells

As we begin to consider what cells are and how they function, several general characteristics of cells quickly emerge. These include the classification of cells on the basis of their organizational complexity, the sizes and shapes of cells, and the specializations that cells undergo.

All Cells Are Either Prokaryotic or Eukaryotic

With the advent of electron microscopy, biologists came to recognize two fundamentally different plans of cellular organization, the simpler one characteristic of bacteria and the more complex one found in all other kinds of cells. Based on the structural differences of their cells, organisms can be divided into two broad groups, the **prokaryotes** (bacteria) and the **eukaryotes** (all other forms of life). The most fundamental distinction between the two groups is that eukaryotic cells have a true, membrane-bounded nucleus (*eu-* is Greek for "true" or "genuine"; *karyon* means "nucleus"), whereas prokaryotic cells do not (*pro-* means "before," suggesting an evolutionarily earlier form of life).

On the basis of molecular criteria, especially sequence analysis of ribosomal RNAs (rRNAs), we now recognize that the prokaryotes can be further differentiated as either **eubacteria** or **archaebacteria**. The eubacteria ("true bacteria") include most present-day **bacteria** and **cyanobacteria** (bacteria that have chlorophyll and can carry out photosynthesis)—most of the commonly encountered bacteria, in other words. The archaebacteria (also called *archaea*) are similar to eubacteria in cellular structure but are as different from eubacteria as they are from eukaryotes in terms of molecular and biochemical distinctions. Archaebacteria are regarded as modern descendants of an evolutionarily ancient form of prokaryote that differed fundamentally from the ancestors of present eubacteria. (*Archae-* is a Greek prefix meaning "ancient" or "original.") Present-day archaebacteria can be subdivided into three main groups: the *methanobacteria*, which obtain energy by converting carbon dioxide and hydrogen into methane; the *halobacteria*, which can grow in salty environments (up to 5.5 M NaCl); and the *sulfobacteria*, which obtain energy from sulfur-containing compounds. Some of these archaebacteria are also *thermacidophiles*, which thrive in acidic hot springs with pH as low as 2 and temperatures that can exceed 100°C.

The work of Carl Woese, C. Fred Fox, and their collaborators on rRNA sequences has led to the recognition of eukaryotes, eubacteria, and archaebacteria as three fundamentally different groups of organisms. In addition to characteristics of their rRNAs, the cells in each of the three groups have other molecular properties in common as well, including RNA polymerases (the enzymes that synthesize RNA) and sensitivity to inhibitors of protein and nucleic acid synthesis.

Cells Come in Many Sizes and Shapes

Cells come in a variety of sizes, shapes, and forms. Some of the smallest bacterial cells, for example, are only about

0.2–0.3 μm in diameter—so small that about 40,000 such cells could fit side by side across the head of a thumbtack! At the other extreme are highly elongated nerve cells, which may extend one or more meters. Those that run the length of a giraffe's neck or legs are especially dramatic examples. On the other hand, the often-cited examples of bird eggs, especially ostrich eggs, are rather misleading because although they are indeed single cells, most of their internal volume is occupied by yolk—large deposits of stored food intended as nourishment for the embryo that will develop if the egg is fertilized.

Despite these extremes, most cells fall into a rather narrow and predictable range of sizes. Bacterial cells, for example, are usually about 1–5 μm in diameter, and most cells of higher plants and animals have dimensions in the range of 10–50 μm. (Recall that Box 1A on pages 2–3 describes the units used to express cellular dimensions and illustrates the relative sizes of cells and cellular structures.) Of the factors that limit cell size, the three most important are the requirement for an adequate surface area/volume ratio, the rates at which molecules diffuse, and the need to maintain adequate local concentrations of the specific substances and enzymes involved in various cellular processes. We will look at each of these three factors in turn.

Surface Area/Volume Ratio. In most cases, the main constraint on cell size is that set by the need to maintain an adequate **surface area/volume ratio**. Surface area is important because it is at the cell surface that the needful exchanges between a cell and its environment take place. The internal volume of the cell determines the amount of nutrients that will have to be imported and the quantity of waste products that must be excreted, but the surface area effectively measures the amount of membrane available for such uptake and excretion.

The problem of maintaining adequate surface area arises because the volume of a cell increases with the cube of the cell's length or diameter, whereas its surface area only increases with the square. Consider, for example, the cube-shaped cells shown in Figure 4-1. The cell on the left is 20 μm on a side and has a volume of 8000 μm^3 ($V = s^3$, where $s = 20$ μm) and a surface area of 2400 μm^2 ($A = 6s^2$). The surface area/volume ratio is therefore 2400 μm^2/8000 μm^3, or 0.3 μm^{-1}. When this single large cell is divided into smaller cells, the total volume remains the same but the surface area increases. Thus, the surface area/volume ratio increases as the linear dimension of the cell decreases. The 1000 cells on the right, for example, still have a total volume of 8000 μm^3 (1000 × 2^3) but the total surface area is 24,000 μm^2 (1000 × 6 × 2^2), so the surface area/volume ratio is 24,000 μm^2/8000 μm^3, or 3.0 μm^{-1}.

This comparison illustrates a major constraint on cell size: As a cell increases in size, its surface area does not keep pace with its volume, and the necessary exchange of substances between the cell and its surroundings becomes more and more problematic. Cell size therefore can increase only over the range of values for which the membrane surface area is still adequate for the passage of materials into and out of the cell. Once the limiting surface area/volume ratio is reached, further increases in cell size would generate more cytoplasmic volume and therefore greater exchange requirements than could be met by the more modest increases in membrane surface area.

Some cells, particularly those that play a role in absorption, have characteristics that maximize their surface area. Effective surface area is most commonly increased by the inward folding or outward protrusion of the cell membrane.

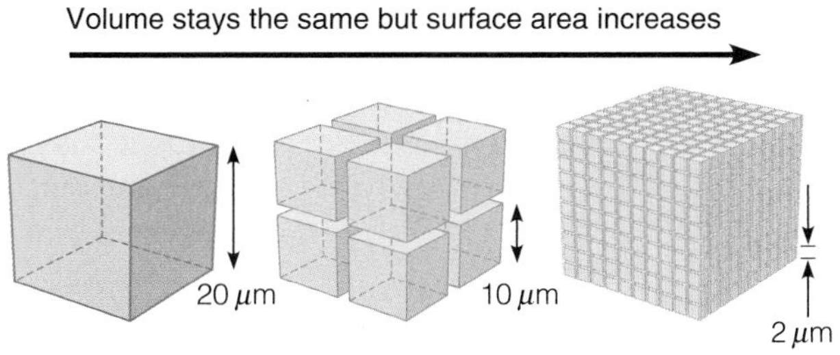

Volume stays the same but surface area increases →

Figure 4-1 The Effect of Cell Size on the Surface Area/Volume Ratio. The single large cell on the left, the eight smaller cells in the center, and the 1000 tiny cells on the right have the same total volume (8000 μm^3), but total surface area increases as the cell size decreases. The surface area/volume ratio increases from left to right as the linear dimension of the cell decreases. Thus, 1000 prokaryotic cells with a linear dimension of 2 μm have a total surface area five times that of the four small eukaryotic cells with a linear dimension of 10μm and ten times that of a single eukaryotic cell with a linear dimension of 20 μm.

Length of one side	20 μm	10 μm	2 μm
Total surface area (height x width x number of sides x number of cubes)	2400 μm^2	4800 μm^2	24,000 μm^2
Total volume (length x width x height x number of cubes)	8000 μm^3	8000 μm^3	8000 μm^3
Surface area to volume ratio (surface area ÷ volume)	0.3	0.6	3.0

The cells that line your small intestine, for example, contain many fingerlike projections called *microvilli* that greatly increase the effective membrane surface area and therefore the absorbing capacity of these cells (Figure 4-2).

Diffusion Rates of Molecules. Cell size is also limited by the rates at which molecules can move around in the cell to reach sites of specific cellular activities. In general, molecules move through the cytoplasm (or within membranes, in the case of membrane proteins and lipids) by **diffusion** from regions where they are more highly concentrated to regions where their concentrations are lower. Molecular movement is therefore limited by the diffusion rates for molecules of various sizes. Because the rate of diffusion varies inversely with the size of the molecule, this limitation is most significant for macromolecules such as proteins and nucleic acids. Some cells of higher organisms get around this limitation to some extent by **cytoplasmic streaming** (also called **cyclosis** in plant cells), a process that involves active movement and mixing of cytoplasmic contents rather than diffusion. In general, however, the size of a cell is quite strictly limited by the diffusion rates of the molecules it contains.

The Need for Adequate Concentrations of Reactants and Catalysts. A third limit on cell size is that imposed by the need to maintain adequate concentrations of the essential compounds and catalysts (enzymes) for the various processes that cells must carry out. For a chemical reaction to occur in a cell, the appropriate reactants must collide with and bind to the surface of a particular enzyme. The frequency of such random collisions will be greatly increased by the larger concentrations of the reactants and of the enzyme itself. To maintain appropriate levels of reactants and

enzymes as the size of a cell increases, the number of all such molecules must increase eightfold every time the three dimensions of the cell double. This increase obviously taxes the synthetic capabilities of the cell.

Eukaryotic Cells Use Organelles to Compartmentalize Cellular Function

One effective solution to the concentration problem is *compartmentalization of activities* within specific regions of the cell. If all the enzymes and compounds necessary for a particular process are localized within a specific region, high concentrations of those substances are needed only in that region rather than throughout the whole cell.

To compartmentalize activities, most eukaryotic cells have a variety of **organelles**, internal compartments that are highly specialized for specific functions. For example, the cells in a plant leaf have most of the enzymes, compounds, and pigments needed for photosynthesis compartmentalized together into structures called *chloroplasts*. Such a cell can therefore maintain appropriately high concentrations of everything that is essential for photosynthesis within its chloroplasts without having to maintain correspondingly high concentrations of these substances elsewhere in the cell. In a similar way, other functions are localized within other compartments. This internal compartmentalization of specific functions enables the large cells of plants and animals to maintain locally high concentrations of the specific enzymes and compounds needed for particular cellular processes. Such processes can therefore proceed efficiently, even though as a whole such cells are orders of magnitude larger than bacterial cells.

Prokaryotes and Eukaryotes Differ from Each Other in Many Ways

Returning to the basic distinction between prokaryotes and eukaryotes, we recognize many important structural, biochemical, and genetic differences between eubacteria and archaebacteria on the one hand and eukaryotic cells on the other. Some of these differences are summarized in Table 4-1 and are discussed here briefly.

Presence or Absence of a Membrane-Enveloped Nucleus. As already noted, the most fundamental distinction between eukaryotes and prokaryotes is reflected in the nomenclature itself: A eukaryotic cell has a true, membrane-bounded nucleus whereas a prokaryotic cell does not. Instead of being enveloped by a membrane, the genetic information of a prokaryotic cell is localized in a region of the cytoplasm called the *nucleoid* (Figure 4-3). Within a eukaryotic cell, on the other hand, most of the genetic information is localized to the nucleus, which is surrounded not by a single membrane, but by a *nuclear envelope* that consists of two membranes (Figure 4-4). Other structural features of the nucleus include the *nucleolus*, the site of ribosome synthesis, and the DNA-bearing *chromosomes*, which in Figure 4-4 are dispersed as chro-

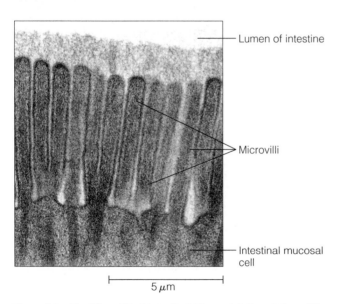

— Lumen of intestine

— Microvilli

— Intestinal mucosal cell

5 μm

Figure 4-2   The Microvilli of Intestinal Mucosal Cells. Microvilli are fingerlike projections of the cell membrane that greatly increase the absorptive surface area of intestinal mucosal cells such as those that line the inner surface of your small intestine (TEM).

Table 4-1   A Comparison of Some Properties of Prokaryotic and Eukaryotic Cells

Property	Prokaryotic Cells	Eukaryotic Cells
Size*	Small (a few micrometers in length or diameter, in most cases)	Large (10–50 times the length or diameter of prokaryotes, in most cases)
Membrane-bounded nucleus	No	Yes
Organelles	No	Yes
Microtubules	No	Yes
Microfilaments	No	Yes
Intermediate filaments	No	Yes
Exocytosis and endocytosis	No	Yes
Mode of cell division	Cell fission	Mitosis and meiosis
Genetic information	DNA molecule complexed with relatively few proteins	DNA complexed with proteins (notably histones) to form chromosomes
Processing of RNA	Little	Much
Ribosomes**	Small (70S); 3 RNA molecules and 55 proteins	Large (80S); 4 RNA molecules and about 78 proteins

*The disparity in size between prokaryotic and eukaryotic cells indicated here is generally valid, but cells of both types vary greatly in size, with some overlap in size ranges.
**Ribosomes are characterized in terms of their *sedimentation coefficients* or *S values*, a measure of their sedimentation rate based on size and shape. Sedimentation coefficients are normally expressed in Svedberg units (S), where $1S = 1 \times 10^{-13}$ sec. Unlike molecular weights, S values are not additive.

matin throughout the semifluid *nucleoplasm* that fills the internal volume of the nucleus.

Use of Internal Membranes to Segregate Function.   As Figure 4-3 illustrates, prokaryotic cells contain few internal membranes; most cellular functions occur either in the cytoplasm or on the plasma (cell) membrane. By contrast, eukaryotic cells make extensive use of internal membranes to compartmentalize specific functions (Figures 4-5 and 4-6).

Examples of internal membrane systems in eukaryotic cells include the *endoplasmic reticulum,* the *Golgi complex,* and the membranes that surround and delimit organelles such as *mitochondria, chloroplasts, lysosomes,* and *peroxisomes,* as well as various kinds of *vacuoles* and *vesicles.* Each of these organelles has its own characteristic membrane (or pair of membranes, in the case of mitochondria and chloroplasts), similar to other membranes

in basic structure but often with its own specific chemical composition and proteins. Localized within each such organelle is the molecular machinery needed to carry out the particular cellular functions for which the structure is specialized. We will meet each of these organelles later in this chapter and then return to each in its appropriate context in succeeding chapters.

Tubules and Filaments.   Also found in the cytoplasm of eukaryotic cells are several nonmembranous structures that are involved in cellular contraction and motility, and in the establishment and support of cellular architecture. These include the *microtubules* found in the cilia and flagella of many cell types, the *microfilaments* of actin found in muscle fibrils and other structures involved in motility, and the *intermediate filaments,* which are especially prominent in parts of the cell that are subject to stress. Microtubules,

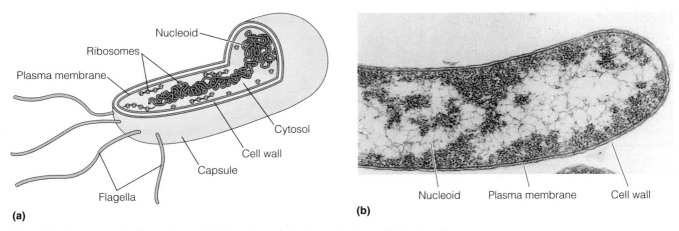

(a)

(b)

Figure 4-3   Structure of a Typical Bacterial Cell.   (a) A three-dimensional model showing the components of a typical bacterium. **(b)** An electron micrograph of a bacterial cell with several of the same components labeled. Notice that the nucleoid is simply a region within the cell, not a membrane-bounded compartment (TEM).

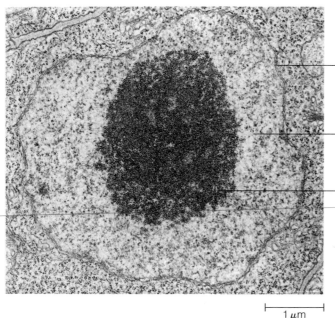

Nuclear envelope

Nucleoplasm with dispersed chromosomes

Nucleolus

Figure 4-4   The Nucleus of a Eukaryotic Cell.   The nucleus is enclosed by a pair of membranes called the nuclear envelope. Because the cell shown here is between divisions, the chromosomes are dispersed as chromatin in the semifluid nucleoplasm within the nucleus. The nucleolus is a structure within the nucleus that is involved in the synthesis of ribosomal components (TEM).

1 μm

microfilaments, and intermediate filaments are also involved in the *cytoskeleton* that imparts structure and elasticity to almost all eukaryotic cells, as we will learn shortly and explore in more detail in Chapter 22. Like ribosomes, these structures are examples of organelles without membranes.

Exocytosis and Endocytosis.   A further feature of eukaryotic cells is their ability to exchange materials between the membrane-bounded compartments within the cell and the exterior of the cell. This exchange is possible because of *exocytosis* and *endocytosis,* processes that are unique to eukaryotic cells. In endocytosis, portions of the plasma membrane invaginate and are pinched off to form membrane-bounded cytoplasmic vesicles containing substances that were previously on the outside of the cell. Exocytosis is essentially the reverse of this process: membrane-bounded vesicles inside the cell fuse with the plasma membrane and release their contents to the outside of the cell.

Organization of DNA.   Another distinction between prokaryotes and eukaryotes becomes apparent when we consider the amount and organization of the genetic material. Prokaryotes characteristically contain amounts of DNA that might be described as "reasonable"; that is, we can account for much of the DNA in terms of known proteins for which the DNA serves as a genetic blueprint. Prokaryotic DNA is usually present in the cell as one or more circular molecules with which relatively few proteins are associated.

Though of reasonable size in a genetic sense, the circular DNA molecule of a prokaryotic cell is much longer than the cell itself. It therefore has to be folded and packed together tightly to fit into the nucleoid of the cell. For example, the common intestinal bacterium *Escherichia coli* is only about a micrometer or two long, yet it has a circular DNA molecule that is about 1300 μm in circumference. Clearly, a great deal of folding and packing is necessary to fit

that much DNA into a small region of such a small cell. By way of analogy, it is roughly equivalent to packing about 60 feet (18 m) of very thin thread into a typical thimble.

But if DNA appears to pose a packaging problem for prokaryotic cells, consider the case of the eukaryotic cell! Although some of the lower eukaryotes (such as yeasts and fruit flies) contain only 10 to 50 times as much DNA as bacteria, most eukaryotic cells have at least 1000 times as much DNA as *E. coli.* It is tempting to label such amounts of DNA as "unreasonable" because we cannot at present assign any known function to much of it. But that is probably a more telling commentary on cell biologists than on cells.

Whatever the genetic function of such large amounts of DNA, the packaging problem is clearly acute. It is solved universally among eukaryotes by the organization of DNA into complex structures called **chromosomes** that contain at least as much protein as DNA. It is as chromosomes that the DNA of eukaryotic cells is packaged, segregated during cell division, transmitted to daughter cells, and transcribed as needed into the molecules of RNA that are involved in protein synthesis. Figure 4-7 shows a chromosome from an animal cell as seen by high-voltage electron microscopy.

Segregation of Genetic Information.   A further contrast between prokaryotes and eukaryotes is the way they allocate genetic information to daughter cells upon division. Prokaryotic cells merely replicate their DNA and divide by a relatively simple process called *cell fission,* with one molecule of DNA going to each daughter cell. Eukaryotic cells also replicate their DNA, but they then use the more complex processes of *mitosis* and *meiosis* to distribute chromosomes equitably to daughter cells. The chromosome in Figure 4-7 was prepared from a cell that was undergoing mitosis. If the process had been allowed to proceed, the chromosome would have divided into two daughter chromosomes, each destined for one of the two daughter cells.

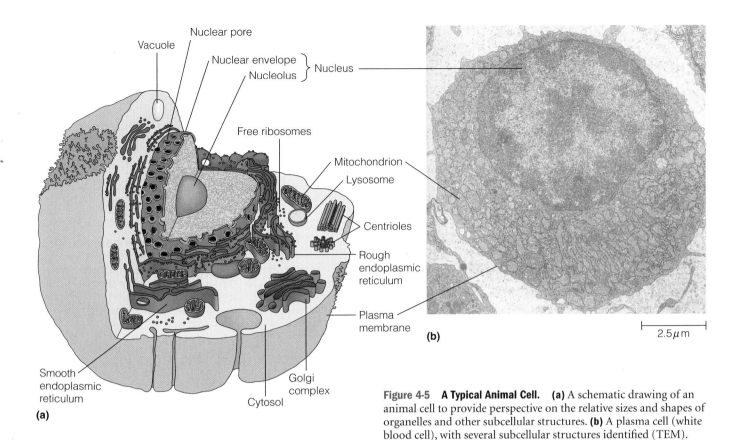

Vacuole
Nuclear pore
Nuclear envelope ⎫
Nucleolus ⎭ Nucleus
Free ribosomes
Mitochondrion
Lysosome
Centrioles
Rough
endoplasmic
reticulum
Plasma
membrane
Smooth
endoplasmic
reticulum
Golgi
complex
Cytosol
Nucleus
2.5 μm
(a)
(b)

Figure 4-5 A Typical Animal Cell. **(a)** A schematic drawing of an animal cell to provide perspective on the relative sizes and shapes of organelles and other subcellular structures. **(b)** A plasma cell (white blood cell), with several subcellular structures identified (TEM).

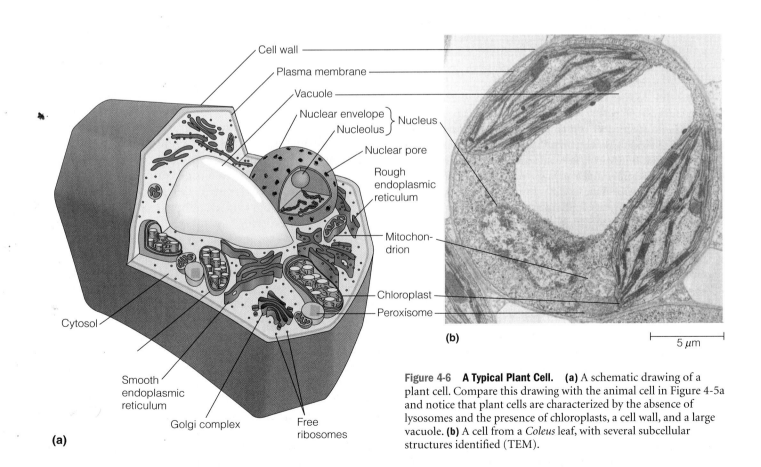

Cell wall
Plasma membrane
Vacuole
Nuclear envelope ⎫
Nucleolus ⎭ Nucleus
Nuclear pore
Rough
endoplasmic
reticulum
Mitochon-
drion
Chloroplast
Peroxisome
Cytosol
Smooth
endoplasmic
reticulum
Golgi complex
Free
ribosomes
5 μm
(a)
(b)

Figure 4-6 A Typical Plant Cell. **(a)** A schematic drawing of a plant cell. Compare this drawing with the animal cell in Figure 4-5a and notice that plant cells are characterized by the absence of lysosomes and the presence of chloroplasts, a cell wall, and a large vacuole. **(b)** A cell from a *Coleus* leaf, with several subcellular structures identified (TEM).

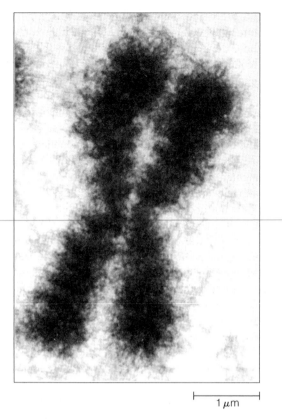

1 μm

Figure 4-7   A Eukaryotic Chromosome.   This chromosome was obtained from a cultured Chinese hamster cell and visualized by high-voltage electron microscopy (HVEM). The cell was undergoing mitosis, and the chromosome is therefore highly coiled and condensed. If mitosis had been allowed to proceed, the chromosome would have divided into two daughter chromosomes, one destined for each of the two daughter cells that result from subsequent cell division.

Expression of DNA.   The differences between prokaryotic and eukaryotic cells extend to the expression of genetic information. Eukaryotic cells tend to transcribe genetic information in the nucleus into large RNA molecules and depend on later processing and transport processes to deliver RNA molecules of the proper sizes to the cytoplasm for protein synthesis.

By contrast, prokaryotes transcribe very specific segments of genetic information into RNA messages, and little or no processing or selection appears to be either necessary or possible. In fact, the absence of a nuclear membrane makes it possible for new messenger RNA molecules to become involved in the process of protein synthesis even before they are themselves completely synthesized (Figure 4-8). Prokaryotes and eukaryotes also differ in the size and composition of the ribosomes used to synthesize proteins (see Table 4-1). We will return to this distinction in more detail later in the chapter.

Cell Specialization Demonstrates the Unity and Diversity of Biology

In terms of structure and function, cells are characterized by both unity and diversity, as you can see in the typical

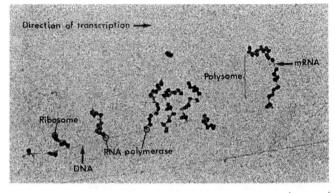

0.2 μm

Figure 4-8   The Expression of Genetic Information in a Prokaryotic Cell.   This electron micrograph shows small segments of DNA molecules from a bacterial cell, one of which is being transcribed into messenger RNA (mRNA) molecules by the enzyme RNA polymerase (TEM). Before transcription of an mRNA molecule is completed, ribosomes associate with it and begin synthesizing protein molecules. Multiple ribosomes associate with a single mRNA molecule, forming structures called polysomes (TEM).

animal and plant cells shown in Figures 4-5 and 4-6. By unity and diversity, we simply mean that all cells resemble one another in fundamental ways, yet differ from one another in other important ways. In upcoming chapters, we will concentrate on aspects of structure and function common to most cell types, as these are the features of cells that are of the greatest general interest. We will find, for example, that virtually all cells oxidize sugar molecules for energy, transport ions across membranes, transcribe DNA into RNA, and undergo division to generate daughter cells. These, then, become topics of legitimate interest to us.

Much the same is true in terms of structural features. All cells are surrounded by a selectively permeable membrane, all have ribosomes for the purpose of protein synthesis, and all contain double-stranded DNA as their genetic information. Clearly, we can be confident that we are dealing with fundamental aspects of cellular organization and function when we consider processes and structures common to most, if not all, cells.

But sometimes our understanding of cellular biology is enhanced by considering not just the unity but also the diversity of cells—not just features common to most cells, but also features that are especially prominent in a particular cell type. For example, to understand how the process of protein secretion works, it would obviously be an advantage to consider a cell that is highly specialized for that particular function. Cells from the human pancreas would be a good choice, for example, because they secrete large amounts of digestive enzymes, such as amylase and trypsin.

Similarly, to study functions known to occur in mitochondria, it would clearly be an advantage to select a cell type that is highly specialized in the energy-releasing processes that occur in the mitochondrion, since such a cell would probably have a lot of well-developed, highly active mitochondria. It was for this very reason, in fact, that Hans Krebs chose the flight muscle of the pigeon as the tissue with which to carry out the now-classic studies on the cyclic pathway of oxidative reactions that we know as the *tricarboxylic acid (TCA)*, or *Krebs, cycle.*

Whenever we exploit the specialized functions of specific cell types to study a particular function, we are acknowledging the diversity of cell structure and function that arises primarily because of cellular specialization. Although we may not often realize it, we are also taking advantage of the multicellularity of many organisms, because it is usually only as part of a multicellular organism that a cell can afford to commit itself to a specialized function. In general, the single cell of unicellular organisms such as bacteria, protozoa, and some algae must be capable of carrying out any and all of the functions necessary for survival, growth, and reproduction and cannot afford to overemphasize any single function at the expense of others. Multicellular organisms, on the other hand, are characterized by a division of labor among tissues and organs that not only allows for, but actually depends on, specialization of structure and function. Whole groups of cells become highly specialized for a particular task, which then becomes their specific role in the overall economy of the organism.

The Eukaryotic Cell in Overview: Pictures at an Exhibition

From the preceding discussion, it should be clear that all cells carry out many of the same basic functions and have some of the same basic structural features. However, the cells of eukaryotic organisms are far more complicated structurally than prokaryotic cells, primarily because of the organelles and other intracellular structures that eukaryotes use to compartmentalize various functions. The structural complexity of eukaryotic cells is illustrated by the typical animal and plant cells shown in Figures 4-5 and 4-6.

In reality, of course, there is no such thing as a truly "typical" cell; all eukaryotic cells have features that distinguish them from the generalized cells shown in Figures 4-5 and 4-6. Nonetheless, most eukaryotic cells are sufficiently similar to warrant a general overview of their structural features.

As we have seen, a typical eukaryotic cell has in essence at least four major structural features: a *plasma* (or *cell*) *membrane* to define its boundary and retain its contents, a *nucleus* to house the DNA that directs cellular activities, *membrane-bounded organelles* in which various cellular functions are localized, and the *cytosol* interlaced by a *cytoskeleton* of tubules and filaments. In addition, plant cells have a rigid *cell wall* external to the plasma membrane. Animal cells do not have a wall but are usually surrounded by an *extracellular matrix,* which consists primarily of *collagen* and a special class of proteins called *proteoglycans.*

Our intention here is to look at each of these structural features in overview, as an introduction to cellular architecture. We are not yet ready to consider any of these features in detail; that awaits us in later chapters, when we encounter the cellular processes in which the various organelles and other structures are involved. For the present, we will simply look at each structure as one might look at pictures at an exhibition. We will move through the gallery rather quickly, just to get a feel for the overall display, but with the intention of returning to each structure later for a careful and more detailed examination. (If you are interested in more information on a specific topic at this point, see Table 4-2 for a list of the chapters in which particular cellular structures or research techniques are considered in detail.)

The Plasma Membrane Defines Cell Boundaries and Retains Contents

Our tour begins with the **plasma membrane** that surrounds every cell (Figure 4-9a). The plasma membrane defines the boundaries of the cell and ensures that its contents are retained. Like all biological membranes, the plasma membrane consists of phospholipids, other lipids, and proteins organized into two layers (Figure 4-9b). Each phospholipid molecule consists of two hydrophobic "tails" and a hydrophilic "head" and is therefore an *amphipathic molecule* (see Figures 2-11 and 2-12).

The phospholipid molecules orient themselves in the two layers of the membrane such that the tails of each molecule (the nonpolar hydrocarbon chains of the fatty acids) face inward, toward the other layer. The hydrophilic head of each molecule (containing a negatively charged

Table 4-2 Chapter Cross-References for Cellular Structures and Techniques Used to Study Cells

For more detailed information about these cellular structures, see the following chapters:

Cell wall	Chapter 11
Chloroplast	Chapter 15
Cytoskeleton	Chapter 22
Endoplasmic reticulum	Chapter 12
Extracellular matrix	Chapter 11
Golgi complex	Chapter 12
Lysosome	Chapter 12
Mitochondrion	Chapter 14
Nucleus	Chapter 16
Peroxisome	Chapter 12
Plasma membrane	Chapter 7
Ribosome	Chapter 20

For more detailed information about techniques used to study cellular structures, see the following chapters and the Guide to Microscopy:

Autoradiography	Chapter 16 and Guide
Differential centrifugation	Chapter 12
Electron microscopy	Guide
Equilibrium density centrifugation	Chapter 12
Light microscopy	Guide

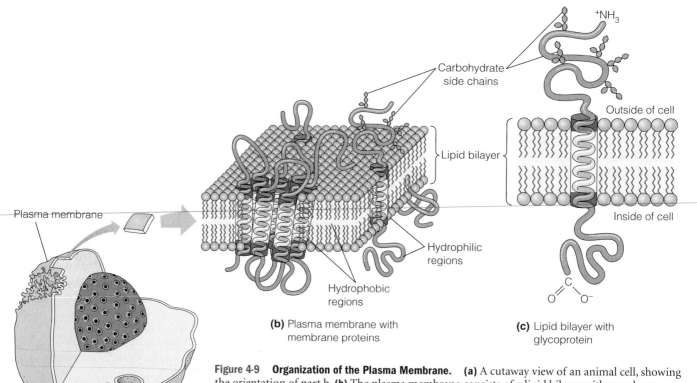

Figure 4-9 Organization of the Plasma Membrane. **(a)** A cutaway view of an animal cell, showing the orientation of part b. **(b)** The plasma membrane consists of a lipid bilayer with membrane proteins suspended in it such that their hydrophobic regions are associated with the interior of the bilayer and their hydrophilic regions protrude from the membrane on one or both sides of the bilayer. **(c)** Most membrane proteins have at least one hydrophobic membrane-spanning domain. Proteins in the plasma membrane are typically glycoproteins, with short carbohydrate side chains attached to the hydrophilic region(s) on the external, but not the internal, side of the membrane. (The membrane protein represented here is glycophorin, a prominent protein in the plasma membrane of the red blood cell.)

phosphate group and a positively charged amine) faces outward, either toward the inside or toward the outside of the cell, depending on the layer in which the lipid molecule is located (Figure 4-9c). The resulting **lipid bilayer** is the basic structural unit of all membranes and serves as a permeability barrier to most water-soluble substances. The lipid bilayer appears as a pair of dark bands when viewed with the electron microscope (see Figure 2-13a).

Membrane proteins are also amphipathic, with both hydrophobic and hydrophilic regions on their surface. They orient themselves in the membrane such that hydrophobic regions of the protein are located within the interior of the membrane, whereas hydrophilic regions protrude into the aqueous environment at the surface of the membrane. Most of the proteins with hydrophilic regions exposed on the external side of the plasma membrane have oligosaccharide (short carbohydrate) side chains attached to them and are therefore called *glycoproteins* (Figure 4-9c).

The proteins present in the plasma membrane play a variety of roles. Some are *enzymes,* which catalyze reactions known to be associated with the membrane. Others serve as *anchors* for structural elements of the cytoskeleton that we will encounter later in the chapter. Still others are *transport proteins,* responsible for moving specific sub-

stances (ions and hydrophilic solutes, usually) across the membrane. Membrane proteins are also important as *receptors* for specific chemical signals that impinge on the cell from the outside and trigger specific processes within the cell. Transport proteins and receptor proteins (and most other membrane proteins as well) are *transmembrane proteins,* with hydrophilic regions protruding on both sides of the membrane, connected by one or more hydrophobic membrane-spanning domains.

The Nucleus Is the Cell's Information Center

If we now enter the cell, one of the most prominent structures we encounter is the **nucleus** (Figure 4-10). The nucleus serves as the information center for the cell. Here, separated from the rest of the cell by a membrane boundary are the DNA-bearing *chromosomes* of the cell. Actually, the boundary around the nucleus consists of two membranes, called the *inner* and *outer nuclear membranes.* Taken together, the two membranes make up the **nuclear envelope.** Unique to the membranes of the nuclear envelope are numerous small openings called *pores* (Figure 4-10b and c). Each pore is a channel through which water-soluble molecules can move between the nucleus and cytoplasm. Ribosomes, messenger RNA molecules, chromo-

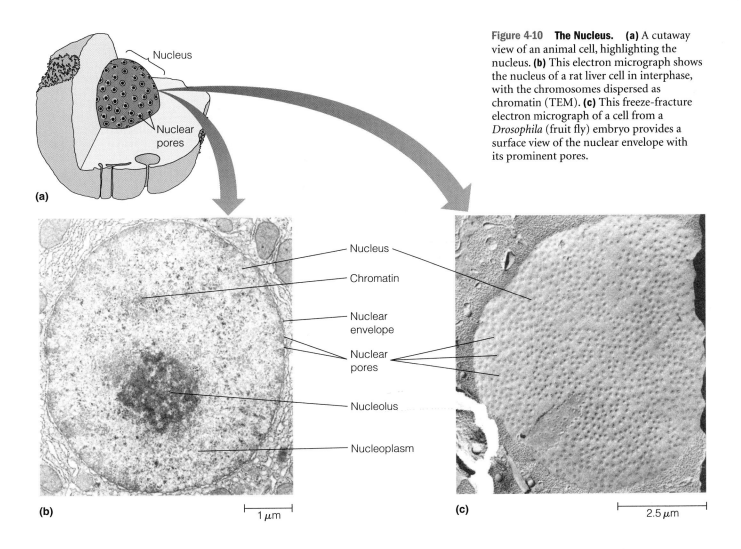

Figure 4-10 The Nucleus. (a) A cutaway view of an animal cell, highlighting the nucleus. **(b)** This electron micrograph shows the nucleus of a rat liver cell in interphase, with the chromosomes dispersed as chromatin (TEM). **(c)** This freeze-fracture electron micrograph of a cell from a *Drosophila* (fruit fly) embryo provides a surface view of the nuclear envelope with its prominent pores.

Labels in figure (a): Nucleus; Nuclear pores

Labels in figure (b): Nucleus; Chromatin; Nuclear envelope; Nuclear pores; Nucleolus; Nucleoplasm — 1 μm

Labels in figure (c): Nucleus; Nuclear pores — 2.5 μm

somal proteins, and enzymes needed for nuclear activities are also presumed to be transported across the nuclear envelope through its pores.

The number of chromosomes within the nucleus is characteristic of the species. It can be as low as two (in the sperm and egg cells of some grasshoppers, for example), or it can run into the hundreds. Chromosomes are most readily visualized during mitosis, when chromosomes in dividing cells are highly condensed and can easily be stained (see Figure 4-7). During the *interphase* between divisions, on the other hand, chromosomes are dispersed as DNA-protein fibers called **chromatin** and are not easy to visualize (see Figure 4-10b).

Also present in the nucleus are **nucleoli** (singular: **nucleolus**), structures responsible for the synthesis and assembly of most of the RNA and protein components needed to form the subunits that make up ribosomes. Nucleoli are usually associated with specific regions of particular chromosomes.

Intracellular Membranes and Organelles Define Compartments

The internal volume of the cell exclusive of the nucleus is called the *cytoplasm* and is occupied by *organelles* and by the

semifluid *cytosol* in which they are suspended. In this section, we will look at each of the major eukaryotic organelles. In a typical animal cell, these compartments make up almost half of the total internal volume of the cell.

The Mitochondrion. Our tour of the eukaryotic organelles begins with a very prominent organelle, the **mitochondrion** (plural: **mitochondria**). The structure of the mitochondrion is shown in Figure 4-11. Mitochondria are large by cellular standards—up to a micrometer across and usually a few micrometers long. A mitochondrion is therefore comparable in size to a whole bacterial cell. The fact that most eukaryotic cells contain hundreds of mitochondria, each approximately the size of an entire bacterial cell, emphasizes again the great difference in size between prokaryotic and eukaryotic cells. The mitochondrion is surrounded by two membranes, designated the *inner* and *outer mitochondrial membranes*.

Most of the chemical reactions involved in the oxidation of sugars and other cellular "fuel" molecules occur within the mitochondria. The purpose of these oxidative events is to extract energy from food and conserve as much of it as possible in the form of the high-energy compound *adenosine triphosphate (ATP)*. It is within the mitochondrion that the cell localizes most of

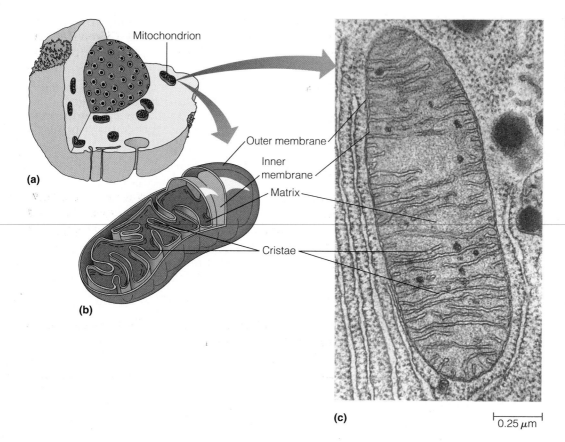

(a)

(b)

(c)

0.25 μm

the enzymes and intermediates involved in such important cellular processes as the tricarboxylic acid (TCA) cycle, fat oxidation, and ATP generation. Most of the intermediates involved in the transport of electrons from oxidizable food molecules to oxygen are located in or on the **cristae** (singular: **crista**), infoldings of the inner mitochondrial membrane. Other reaction sequences, particularly those of the TCA cycle and those involved in fat oxidation, occur in the semifluid **matrix** that fills the inside of the mitochondrion.

The number and location of mitochondria within a cell can often be related directly to their role in that cell. Tissues with an especially heavy demand for ATP as an energy source can be expected to have cells that are well endowed with mitochondria, and the organelles are usually located within the cell just where the energy need is greatest. This localization is illustrated by the sperm cell shown in Figure 4-12. As the drawing indicates, a sperm cell often has a single spiral mitochondrion wrapped around the central shaft, or *axoneme,* of the cell. Notice how tightly the mitochondrion coils around the axoneme, just where the ATP is actually needed to propel the sperm cell. Muscle cells and cells that specialize in the transport of ions also have numerous mitochondria located strategically to meet the special energy needs of such cells (Figure 4-13).

Although mitochondrial structure is usually illustrated by drawings and cross-sectional views such as those shown in Figure 4-11, fluorescent images of mitochondria in living cells show them to be very large and highly

branched. This type of structure is more widespread than just in sperm midpieces and may, in fact, be more representative of mitochondrial shape and size than most conventional diagrams indicate.

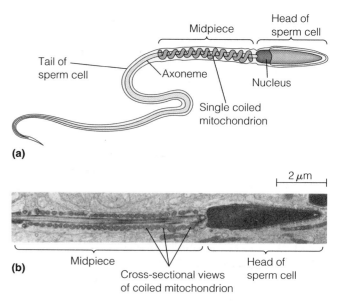

Figure 4-12 Localization of the Mitochondrion Within a Sperm Cell. The single mitochondrion present in a sperm cell is coiled tightly around the axoneme of the tail, reflecting the localized need of the sperm tail (flagellum) for energy. **(a)** A schematic drawing of a sperm. **(b)** An electron micrograph of the head and the midpiece of a sperm cell from a marmoset monkey (TEM).

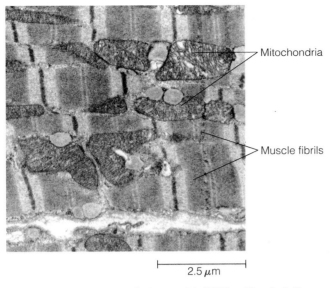

Figure 4-13 **Localization of Mitochondria Within a Muscle Cell.**
This electron micrograph of a muscle cell from a cat heart shows
the intimate association of mitochondria with the muscle fibrils
that are responsible for muscle contraction (TEM).

The Chloroplast. Next, our gallery tour takes in the
chloroplast, in many ways a close relative of the mitochondrion. A typical chloroplast is shown in Figure 4-14.

Chloroplasts are large organelles, typically a few micrometers in diameter and 5–10 μm long. Chloroplasts are
therefore substantially bigger than mitochondria and
larger than any other structure in the cell except the
nucleus. Like mitochondria, chloroplasts are surrounded
by both an inner and an outer membrane. In addition,
they have a third membrane system consisting of flattened
sacs called **thylakoids** and the membranes, called **stroma
thylakoids,** that interconnect them. Thylakoids are stacked
together to form the **grana** (singular: **granum**) that characterize most chloroplasts (Figure 4-14c and d).

Chloroplasts are the site of *photosynthesis,* the light-
driven process whereby carbon dioxide and water are used
to manufacture the sugars and other organic compounds
from which all life is ultimately fabricated. Chloroplasts are
found in leaves and other photosynthetic tissues of higher
plants, as well as in all of the eukaryotic algae. Located
within the organelle are most of the enzymes, intermediates,
and pigments (light-absorbing molecules) needed to "fix"
carbon from carbon dioxide into organic form and convert
it reductively into sugars. The reactions that depend directly
on solar energy are localized in or on the thylakoid membrane system. Reactions involved in the initial trapping of
carbon dioxide into organic form and its subsequent reduction and rearrangement into sugar molecules occur within
the semifluid **stroma** that fills the interior of the chloroplast.

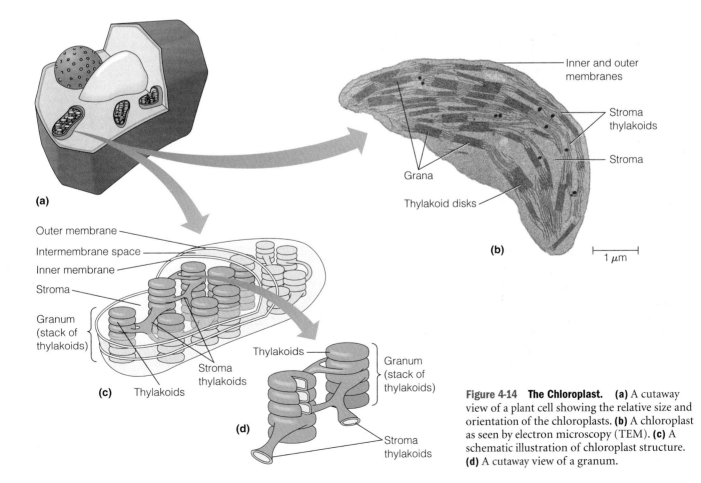

Figure 4-14 **The Chloroplast.** **(a)** A cutaway
view of a plant cell showing the relative size and
orientation of the chloroplasts. **(b)** A chloroplast
as seen by electron microscopy (TEM). **(c)** A
schematic illustration of chloroplast structure.
(d) A cutaway view of a granum.

Although known primarily for their role in photosynthesis, chloroplasts are involved in a variety of other processes as well. An important example involves the reduction of nitrogen from the oxidation level of the nitrate (NO_3^-) that plants obtain from the soil to the oxidation level of ammonia (NH_3), the form of nitrogen required for protein synthesis. Sulfur metabolism also occurs in this organelle. Furthermore, the chloroplast is only the most prominent example of a broader class of plant organelles, the **plastids.** Plastids perform a variety of functions in plant cells. *Chromoplasts,* for example, are pigment-containing plastids that are responsible for the characteristic coloration of flowers, fruits, and other plant parts. *Amyloplasts* are plastids that are specialized for starch storage.

The Endosymbiont Theory: Did Mitochondria and Chloroplasts Evolve from Ancient Bacteria?

Having just met the mitochondrion and chloroplast as eukaryotic organelles, we pause here for a brief digression concerning the possible evolutionary origins of these organelles. Both mitochondria and chloroplasts contain their own DNA and ribosomes, which enable them to carry out the synthesis of both RNA and proteins (although most of the proteins present in these organelles are in fact encoded by nuclear genes). As molecular biologists studied nucleic acid and protein synthesis in these organelles, they were struck by the many similarities between these processes in chloroplasts and mitochondria and the comparable processes in prokaryotic cells. Similarities are seen, for example, in DNA organization (circular molecules without associated histones), rRNA sequences, ribosome size, sensitivity to inhibitors of RNA and protein synthesis, and protein factors required for the initiation of protein synthesis. In addition to these molecular features, mitochondria and chloroplasts also resemble bacterial cells in size and shape.

These similarities led to the **endosymbiont theory** for the evolutionary origins of mitochondria and chloroplasts. This theory, developed most fully by Lynn Margulis, proposes that both of these organelles originated from prokaryotes that gained entry to, and established a symbiotic relationship with, the cytoplasm of ancient single-celled organisms called *protoeukaryotes.* (In biology, *symbiosis* involves the intimate living together of two organisms or cells for the mutual benefit of both). The endosymbiont theory assumes that protoeukaryotes developed two features that distinguished them from other cells existing at the time: large size and the ability to ingest nutrients from the environment by a process called *phagocytosis.* These features allowed a protoeukaryotic cell to ingest prokaryotes (bacteria and cyanobacteria) that later evolved into mitochondria and chloroplasts, respectively.

The Endoplasmic Reticulum.

Extending throughout the cytoplasm of almost every eukaryotic cell is a network of membranes called the **endoplasmic reticulum,** or **ER** (Figure 4-15). The name sounds complicated, but *endo-*

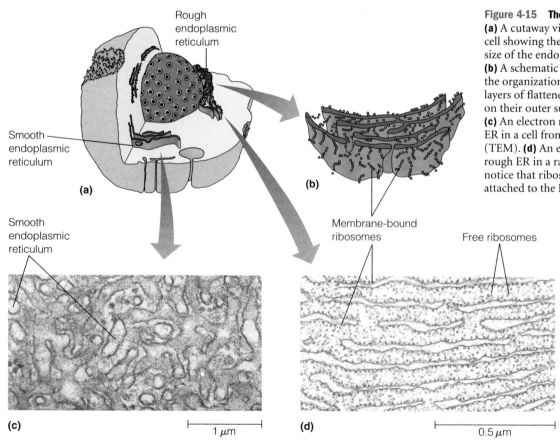

Smooth endoplasmic reticulum

Rough endoplasmic reticulum

Smooth endoplasmic reticulum

(a)

(b)

Membrane-bound ribosomes

Free ribosomes

(c)   1 μm

(d)   0.5 μm

Figure 4-15   The Endoplasmic Reticulum. (a) A cutaway view of a typical animal cell showing the location and relative size of the endoplasmic reticulum (ER). **(b)** A schematic illustration depicting the organization of the rough ER as layers of flattened membranes studded on their outer surface with ribosomes. **(c)** An electron micrograph of smooth ER in a cell from guinea pig testis (TEM). **(d)** An electron micrograph of rough ER in a rat pancreas cell (TEM); notice that ribosomes are either attached to the ER or free in the cytosol.

plasmic just means "within the plasm" (of the cell), and *reticulum* is simply a fancy word for "network." The endoplasmic reticulum consists of tubular membranes and flattened sacs, or **cisternae** (singular: **cisterna**), that appear to be interconnected. The internal space enclosed by the ER membranes is called the **lumen.** The ER is continuous with the outer membrane of the nuclear envelope (see Figure 4-15a). The space between the two nuclear membranes is therefore a part of the same compartment as the lumen of the ER.

The ER can be either *rough* or *smooth.* **Rough endoplasmic reticulum** (**rough ER**) appears "rough" in the electron microscope because it is studded with ribosomes on the side of the membrane that faces the cytosol (Figure 4-15b and d). These ribosomes are actively synthesizing proteins. Most of these proteins are transported into or across the membrane as they are synthesized, accumulating in completed form within either the membrane or the lumen of the ER. *Secretory proteins* (proteins destined to be exported from the cell) are synthesized in this way. They then make their way to the cell surface by a complex process that involves not only the rough ER but also the Golgi complex and secretory vesicles described below.

Not all proteins are synthesized on the rough ER, however. Much protein synthesis occurs on ribosomes that are not attached to the ER but are instead found free in the cytosol (Figure 4-15d). In general, secretory proteins and membrane proteins are made by ribosomes on the rough ER, whereas proteins intended for use within the cytosol are made on free ribosomes.

The **smooth endoplasmic reticulum** (**smooth ER**) has no role in protein synthesis and hence no ribosomes.

It therefore has a characteristic smooth appearance when viewed by electron microscopy (Figure 4-15c). Smooth ER is involved in the synthesis of lipids and steroids, such as cholesterol and the steroid hormones derived from it. In addition, smooth ER is responsible for the inactivation and detoxification of drugs and other compounds that might otherwise be toxic or harmful to the cell.

The Golgi Complex. Closely related to the smooth ER in both proximity and function is the **Golgi complex** (or *Golgi apparatus*), named after its Italian discoverer, Camillo Golgi. The Golgi complex, shown in Figure 4-16, consists of a stack of flattened vesicles. The Golgi complex plays an important role in the processing and packaging of secretory proteins and in the synthesis of complex polysaccharides. Vesicles that arise by budding off the ER are accepted by the Golgi complex. Here, the contents of the vesicles (proteins, for the most part) and sometimes the vesicle membranes are further modified and processed. The processed contents are then passed on to other components of the cell by means of vesicles that arise by budding off the Golgi complex (Figure 4-16c).

Most membrane proteins and secretory proteins are glycoproteins. The initial steps in *glycosylation* (sugar addition) take place within the lumen of the rough ER, but the process is usually completed within the Golgi complex. The Golgi complex should therefore be understood as a processing station, with vesicles both fusing with it and arising from it. Almost everything that goes into it comes back out, but in a modified, packaged form, often ready for export from the cell.

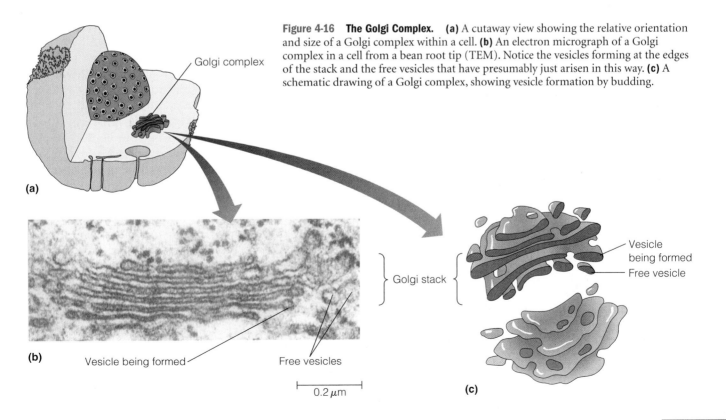

Figure 4-16 The Golgi Complex. **(a)** A cutaway view showing the relative orientation and size of a Golgi complex within a cell. **(b)** An electron micrograph of a Golgi complex in a cell from a bean root tip (TEM). Notice the vesicles forming at the edges of the stack and the free vesicles that have presumably just arisen in this way. **(c)** A schematic drawing of a Golgi complex, showing vesicle formation by budding.

Golgi complex

(a)

(b)

Vesicle being formed

Free vesicles

0.2 μm

Golgi stack

Vesicle being formed

Free vesicle

(c)

Secretory Vesicles. Once processed by the Golgi complex, secretory proteins and other substances intended for export from the cell are packaged into **secretory vesicles.** The cells of your pancreas, for example, are likely to contain many such vesicles because the pancreas is responsible for the synthesis of several important digestive enzymes. These enzymes are synthesized on the rough ER, packaged by the Golgi complex, and then released from the cell via secretory vesicles, as shown in Figure 4-17. These vesicles move from the Golgi region to the plasma membrane that surrounds the cell. The vesicles then fuse with the plasma membrane and discharge their contents to the exterior of the cell by the process of exocytosis. The whole process of protein synthesis, processing, and export via the ER, the Golgi complex, and secretory vesicles will be considered in more detail in Chapter 12.

The Lysosome. The next picture at our cellular exhibition is of the **lysosome,** an organelle about 0.5–1.0 μm in diameter and surrounded by a single membrane (Figure 4-18). Lysosomes were discovered in the early 1950s by Christian de Duve and his colleagues. The story of that discovery is recounted in Box 4A on page 92, both to underscore the significance of chance observations when they are made by astute investigators and to illustrate the importance of new techniques to the progress of science.

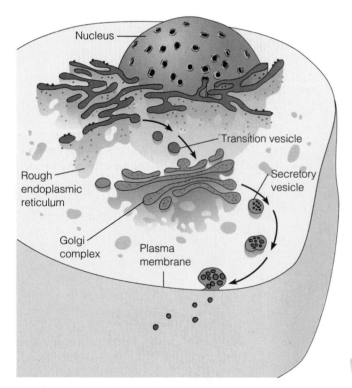

Figure 4-17   The Process of Secretion in Eukaryotic Cells.   Proteins to be packaged for export are synthesized on the rough ER, passed to the Golgi complex for processing, and eventually compartmentalized into secretory vesicles. These vesicles then make their way to the plasma membrane and fuse with it, releasing their contents to the exterior of the cell.

Lysosomes are used by the cell as a means of storing *hydrolases,* enzymes capable of digesting specific biological molecules such as proteins, carbohydrates, or fats. It is important for cells to possess such enzymes, both to digest food molecules that the cell may acquire from its environment and to break down cellular constituents that are no longer needed. But it is also essential that such enzymes be carefully sequestered until actually needed, lest they digest cellular components that were not scheduled for destruction.

We are considering the lysosome at this point in our overview because of its relationship to the ER and the Golgi complex. Lysosomal enzymes are somewhat similar to secretory proteins in their synthesis and packaging. They are synthesized on the rough ER and transported to the Golgi, where a mannose-6-phosphate tag is added to mark them as lysosomal enzymes. These enzymes are then packaged into vesicles that transport them to organelles called *early endosomes,* which subsequently mature into *late endosomes.* A late endosome is therefore a collection of newly synthesized digestive enzymes packaged in a way that protects the cell from their activities until they are required for specific hydrolytic functions.

A late endosome, in turn, either matures to form a new lysosome or delivers its enzyme complement to an active lysosome. Equipped with its repertoire of hydrolases, the lysosome is able to break down virtually any kind of biological molecule. Eventually, the digestion products are small enough to pass through the membrane into the cytosol of the cell where they can be utilized for the synthesis of macromolecules—recycling at the cellular level!

The Peroxisome. The next organelle at our exhibition is the **peroxisome.** Peroxisomes resemble lysosomes in size and general lack of obvious internal structure. Like lysosomes, they are surrounded by a single, rather than a double, membrane. Peroxisomes are found in both plant and animal cells, as well as in fungi, protozoa, and algae. They perform several distinctive functions that differ with cell type, but they have the common property of both generating and degrading hydrogen peroxide (H_2O_2). Hydrogen peroxide is highly toxic to cells but can be decomposed into water and oxygen by the enzyme *catalase.* Eukaryotic cells protect themselves from the detrimental effects of hydrogen peroxide by packaging peroxide-generating reactions together with catalase in a single compartment, the peroxisome.

In animals, peroxisomes are found in most cell types but are especially prominent in liver and kidney cells (Figure 4-19). Beyond their role in the detoxification of hydrogen peroxide, animal peroxisomes may have several other functions, including the detoxification of other harmful compounds (such as methanol, ethanol, formate, and formaldehyde) and the catabolism of unusual substances (such as D-amino acids). Some researchers speculate that peroxisomes may also be involved in the regulation of oxygen tension within the cell.

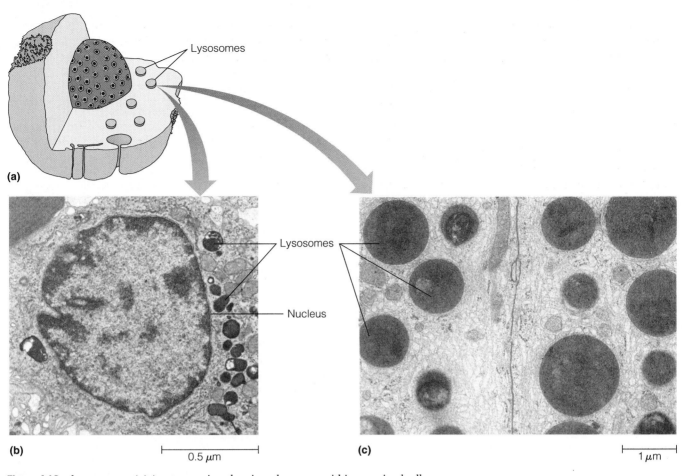

(a)

Lysosomes

(b)

0.5 μm

Lysosomes

Nucleus

(c)

1 μm

Figure 4-18 Lysosomes. (a) A cutaway view showing a lysosome within an animal cell. **(b)** Lysosomes in an animal cell stained cytochemically for acid phosphatase, a lysosomal enzyme. The cytochemical staining technique results in dense deposits of lead phosphate at the site of acid phosphatase activity (TEM). **(c)** An electron micrograph of lysosomes in two adjacent cells of human epididymal epithelium.

Animal peroxisomes also play a role in the oxidative breakdown of fatty acids, which are components of triacylglycerols, phospholipids, and glycolipids (see Table 3-5 and Figure 3-27). As we will see in Chapter 14, fatty acid breakdown occurs primarily in the mitochondrion. However, fatty acids with more than 12 carbon atoms are oxidized relatively slowly by mitochondria. In the peroxisomes, on the other hand, fatty acids up to 22 carbon atoms long are oxidized rapidly. The long chains are degraded two carbon units at a time until they get to a length (10–12 carbon atoms) that can be handled efficiently by the mitochondria.

The vital role that peroxisomes play in breaking down long-chain fatty acids is underscored by the serious human diseases that result when one or more peroxisomal enzymes involved in fatty acid degradation are defective or absent. One such disease is adrenoleukodystrophy (ALD), a sex-linked (male-only) disorder that leads to profound neurological debilitation and eventually to death.

The best-understood metabolic roles of peroxisomes occur in plant cells. During the germination of fat-storing seeds, specialized peroxisomes called **glyoxysomes** play a key role in the conversion of stored fat into carbohydrates. In photosynthetic tissue, **leaf peroxisomes** are prominent because of their role in *photorespiration,* a light-dependent pathway that detracts from the efficiency of photosynthesis in many plants by "unfixing" some of the carbon that is fixed by the chloroplasts (to be discussed further in Chapter 15). The photorespiratory pathway is an example of a cellular process that involves several organelles. Some of the enzymes that catalyze the reactions in this sequence occur in the peroxisome, whereas others are located in the chloroplast or the mitochondrion. This mutual involvement in a common cellular process is suggested by the intimate association of peroxisomes with mitochondria and chloroplasts in many leaf cells, as shown in Figure 4-20.

Vacuoles. Cells also contain a variety of other membrane-bounded organelles called **vacuoles.** In animal cells, vacuoles are frequently used for temporary storage or transport. Some protozoa, for example, take up food

DISCOVERING ORGANELLES: THE IMPORTANCE OF CENTRIFUGES AND CHANCE OBSERVATIONS

Have you ever wondered how the various organelles within eukaryotic cells were discovered? There are almost as many answers to that question as there are kinds of organelles. In general, they were described by microscopists before their role in the cell was understood. As a result, the names of organelles usually reflect structural features rather than physiological roles. Thus, *chloroplast* simply means "green particle" and *endoplasmic reticulum* just means "network within the plasm (of the cell)."

Such is not the case for the *lysosome*, however. This organelle was the first to have its biochemical properties described before it had ever been reported by microscopists. Only after fractionation data had predicted the existence and properties of such an organelle were lysosomes actually observed in cells. A suggestion of its function is even inherent in the name given to the organelle, because the Greek root *lys-* means "to digest." (The literal meaning is "to loosen," but that's essentially what digestion does to chemical bonds!)

The lysosome is something of a newcomer on the cellular biology scene; it was not discovered until the early 1950s. The story of that discovery is fascinating because it illustrates how important chance observations can be, especially when made by the right people at the right time. The account also illustrates how significant new techniques can be, since the discovery depended on subcellular fractionation, a technique that was then still in its infancy.

The story begins in 1949 in the laboratory of Christian de Duve, who later received a Nobel Prize for this work. Like so many scientific advances, the discovery of lysosomes depended on a chance observation made by an astute investigator. Because of an interest in the effect of insulin on carbohydrate metabolism, de Duve was attempting at the time to pinpoint the cellular location of *glucose-6-phosphatase*, the enzyme responsible for the release of free glucose in liver cells. As a control enzyme (that is, one not involved in carbohydrate metabolism), de Duve happened to choose *acid phosphatase*.

De Duve first homogenized liver tissue and resolved it into several fractions by the new technique of *differential centrifugation,* which separates cellular components on the basis of differences in size and density. In this way, he was able to show that the glucose-6-phosphatase activity could be recovered with the microsomal fraction. (*Microsomes* are small vesicles that form from ER fragments when tissue is homogenized.) This in itself was an important observation because it helped establish the identity of microsomes, which at the time tended to be dismissed as fragments of mitochondria.

But the acid phosphatase results turned out to be still more interesting, even though they were at first quite puzzling. When de Duve and his colleagues assayed their liver homogenates for this enzyme, they found only a fraction of the expected activity. When assayed again for the same enzyme a few days later, however, the same homogenates had about ten times as much activity.

Speculating that he was dealing with some sort of activation phenomenon, de Duve subjected the homogenates to differential centrifugation to see with what subcellular fraction the phenomenon was associated. He and his colleagues were able to demonstrate that much of the acid phosphatase activity could be recovered in the mitochondrial fraction, and that this fraction showed an even greater increase in activity after standing a few days than did the original homogenates.

To their surprise, they then discovered that upon recentrifugation, this elevated activity no longer sedimented with the mitochondria but stayed in the supernatant. They went on to show that the activity could be increased and the enzyme solubilized by a variety of treatments, including harsh grinding, freezing and thawing, and exposure to detergents or hypotonic conditions. From these results, de Duve concluded that the enzyme must be present in some sort of membrane-bounded particle that could easily be ruptured to release the enzyme. Apparently, the enzyme could not be detected within the particle, probably because the membrane was not permeable to the substrates used in the enzyme assay.

Assuming that particle to be the mitochondrion, they continued to isolate and study this fraction of their liver homogenates. At this point, another chance observation occurred, this time because of a broken centrifuge. The unexpected breakdown forced one of de Duve's students to use an older, slower centrifuge, and the result was a mitochondrial fraction that had little or no acid phosphatase in it. Based on this unexpected finding, de Duve speculated that the mitochondrial fraction as they usually prepared it might in fact contain two kinds of organelles: the actual mitochondria, which could be sedimented with either centrifuge, and some sort of more slowly sedimenting particle that came down only in the faster centrifuge.

This led them to devise a fractionation scheme that allowed the original mitochondrial fraction to be subdivided into a rapidly sedimenting component and a slowly sedimenting component. As you might guess, the rapidly sedimenting component contained the mitochondria, as evidenced by the presence of enzymes known to be mitochondrial markers. The acid phosphatase, on the other hand, was in the slowly sedimenting component, along with several other hydrolytic enzymes, including ribonuclease, deoxyribonuclease, β-glucuronidase, and a protease. Each of these enzymes showed the same characteristic of increased activity upon membrane rupture, a property that de Duve termed *latency.*

By 1955, de Duve was convinced that these hydrolytic enzymes were packaged together in a previously undescribed organelle. In keeping with his speculation that this organelle was involved in intracellular lysis (digestion), he called it a *lysosome.*

Thus, the lysosome became the first organelle to be identified entirely on biochemical criteria. At the time, no such particles had been described by microscopy. But when de Duve's lysosome-containing fractions were examined with the electron microscope, they were found to contain membrane-bounded vesicles that were clearly not mitochondria and were in fact absent from the mitochondrial fraction. Knowing what the isolated particles looked like, microscopists were then able to search for them in fixed tissue. As a result, lysosomes were soon identified and reported in a variety of animal tissues. Within six years, then, the organelle that began as a puzzling observation in an insulin experiment became established as a bona fide feature of most animal cells.

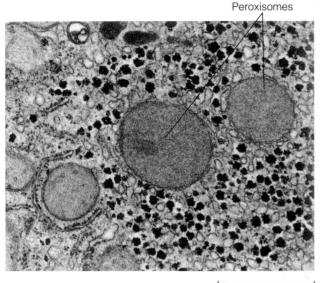

Peroxisomes

1 μm

Figure 4-19 **Animal Peroxisomes.** Several peroxisomes can be seen in this cross section of a liver cell (TEM). Peroxisomes are found in most animal cells but are especially prominent features of liver and kidney cells.

particles or other materials from their environment by a process called *phagocytosis* ("cell eating"). Phagocytosis is a form of endocytosis that involves an inpocketing of the plasma membrane around the desired substance, followed by a pinching-off process that internalizes the membrane-bounded particle as a vacuole.

Plant cells also contain vacuoles. In fact, a single large vacuole is a characteristic feature of most mature plant cells (Figure 4-21). This vacuole, sometimes called the **central vacuole,** may play a limited role in storage and is apparently also capable of a lysosome-like function in intracellular digestion. However, its real importance lies in the maintenance of the turgor pressure of the plant cell. As you already know, a plant cell is surrounded by a rigid cell wall. The vacuole has a high concentration of solutes and is surrounded by a differentially permeable membrane called the *tonoplast*. In response to the high solute concentration within the vacuole, water tends to move inward across the tonoplast, causing the vacuole to swell. As a result, the vacuole presses the rest of the cell constituents against the cell wall, thereby maintaining the *turgor pressure* characteristic of nonwilted plant tissue. The limp, flaccid appearance associated with wilting comes about when the central vacuole does not provide adequate pressure, allowing the cell (and hence the tissue) to go limp. We can easily demonstrate this by placing a piece of crisp celery in salt water. The high concentration of salt on the outside of the cells will cause water to move out of the cells; the turgor pressure will then decrease, and the tissue will quickly become flaccid and lose its crispness.

Ribosomes. The last portrait in our gallery is the **ribosome.** As noted earlier, the ribosome is an example of an organelle

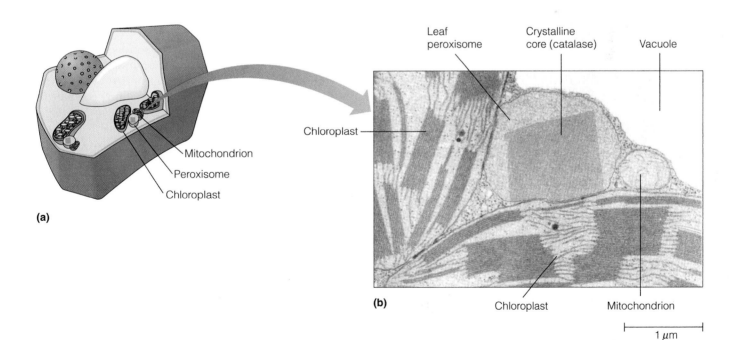

(a)

Mitochondrion
Peroxisome
Chloroplast

Leaf peroxisome | Crystalline core (catalase) | Vacuole
Chloroplast
(b)
Chloroplast | Mitochondrion

1 μm

Figure 4-20 **A Leaf Peroxisome and Its Relationship to Other Organelles in a Plant Cell.** **(a)** A cutaway view showing a peroxisome, a mitochondrion, and a chloroplast within a plant cell. **(b)** A peroxisome in close proximity to chloroplasts and to a mitochondrion within a tobacco leaf cell (TEM). This is probably a functional relationship because all three organelles participate in the process of photorespiration. The crystalline core frequently observed in leaf peroxisomes is the enzyme catalase, which catalyzes the decomposition of hydrogen peroxide into water and oxygen.

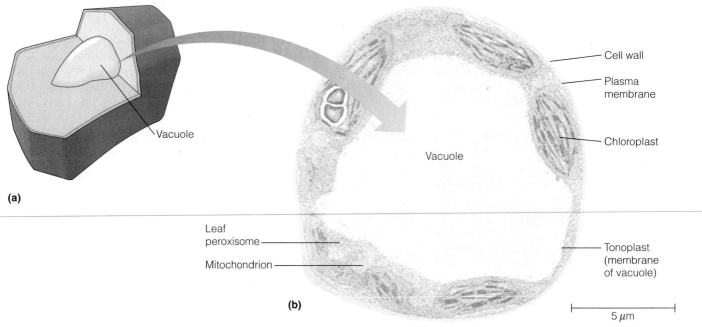

(a)

(b)

Cell wall

Plasma membrane

Chloroplast

Vacuole

Vacuole

Leaf peroxisome

Mitochondrion

Tonoplast (membrane of vacuole)

5 µm

Figure 4-21   The Vacuole in a Plant Cell.   (a) A cutaway view showing the vacuole in a plant cell. **(b)** An electron micrograph of a bean leaf cell with a large central vacuole (TEM). The vacuole occupies much of the internal volume of the cell, with the cytoplasm sandwiched into a thin sphere between the vacuole and the plasma membrane. The membrane of the vacuole is called the tonoplast.

that is not surrounded by a membrane. Like other organelles, ribosomes are the focal point for a specific cellular activity—in this case, protein synthesis. Unlike most other organelles, ribosomes are found in both eukaryotic and prokaryotic cells. Even here, however, the dichotomy between the two basic cell types manifests itself: prokaryotic and eukaryotic ribosomes differ characteristically in size and in the number and kinds of protein and RNA molecules they contain. Ribosomes also exemplify the molecular dichotomy between eubacteria and archaebacteria; in fact, as we discussed earlier, scientists first differentiated these two groups of organisms on the basis of characteristic differences in their ribosomal RNA sequences.

Compared with membrane-bounded organelles, ribosomes are tiny structures. The ribosomes of eukaryotic and prokaryotic cells have diameters of about 30 and 25 nm respectively. An electron microscope is therefore required to visualize ribosomes (see Figure 4-15d). To appreciate how small ribosomes are, consider that more than 350,000 ribosomes could fit inside a typical bacterial cell, with room to spare!

Another way to express the size of such a small particle is to refer to its **sedimentation coefficient.** The sedimentation coefficient of a particle or macromolecule is a measure of how rapidly the particle sediments in an ultracentrifuge and is expressed in *Svedberg units (S)*. Sedimentation coefficients are widely used to indicate relative size, especially for large macromolecules such as proteins and nucleic acids and small particles such as ribosomes. Ribosomes from eukaryotic cells have sedimentation coeffi-

cients of about 80S; those from prokaryotic cells are about 70S (see Table 4-1).

A ribosome consists of two subunits differing in size, shape, and composition (Figure 4-22). In eukaryotic cells, the **large** and **small ribosomal subunits** have sedimentation coefficients of about 60S and 40S, respectively. For prokaryotic ribosomes, the corresponding values are about 50S and 30S. (Note that the sedimentation coefficients of the subunits do not add up to that of the intact ribosome; this is because sedimentation coefficients

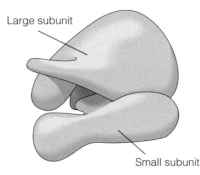

Large subunit

Small subunit

**Figure 4-22   Structure of a Eukaryotic Ribosome.   Each ribosome is made up of a large subunit and a small subunit that join together when they attach to a messenger RNA and begin to make a protein. The large subunit of a eukaryotic ribosome consists of about 45 different protein molecules and 3 molecules of RNA. The small subunit has about 33 proteins and 1 molecule of RNA. The fully assembled ribosome of a eukaryotic cell has a diameter of about 30 nm. The ribosomes and ribosomal subunits of prokaryotic cells are somewhat smaller than those of eukaryotic cells and consist of their own distinctive protein and RNA molecules.

depend critically on both size and shape and are therefore not linearly related to molecular weight.) In both eukaryotic and prokaryotic cells, ribosomal subunits are synthesized and assembled separately in the cell but come together for the purpose of making proteins.

Ribosomes are far more numerous than most other intracellular structures. Prokaryotic cells usually contain thousands of ribosomes, and eukaryotic cells may have hundreds of thousands or even millions of them. Ribosomes are also found in both chloroplasts and mitochondria, where they function in organelle-specific protein synthesis. Although the ribosomes of these eukaryotic organelles differ in size and composition from the ribosomes found in the cytoplasm of the same cell, they are strikingly similar to those found in bacteria and cyanobacteria. This similarity is particularly striking when the nucleotide sequences of ribosomal RNA (rRNA) from mitochondria and chloroplasts are compared with those of prokaryotic rRNAs. These similarities provide further support for the endosymbiotic origins of mitochondria and chloroplasts.

The Cytoplasm of Eukaryotic Cells Contains the Cytosol and Cytoskeleton

The **cytoplasm** of a eukaryotic cell consists of that portion of the interior of the cell not occupied by the nucleus. Thus, the cytoplasm includes organelles such as the mitochondria; it also includes the **cytosol,** the semifluid substance in which the organelles are suspended. In a typical animal cell, the cytosol occupies more than half of the total internal volume of the cell. Many cellular activities take place in the cytosol, including the synthesis of proteins, the synthesis of fats, and the initial steps in the release of energy from sugars.

In the early days of cell biology, the cytosol was regarded as a rather amorphous, gel-like substance, and its proteins were thought to be soluble and freely diffusible. However, several new techniques have done much to change this view greatly. We now know that the cytosol of eukaryotic cells, far from being a structureless fluid, is permeated by an intricate three-dimensional array of interconnected filaments and tubules called the **cytoskeleton** (Figure 4-23). Specifically, the cytoskeleton consists of a network of microtubules, microfilaments, and intermediate filaments, all of which are unique to eukaryotes (see Table 4-1).

As the name suggests, the cytoskeleton is an internal framework that gives a eukaryotic cell its distinctive shape and high level of internal organization. This elaborate array of filaments and tubules forms a highly structured yet very dynamic matrix that not only helps establish and maintain shape but also plays important roles in cell movement and cell division. As we will see in later chapters, the filaments and tubules that make up the cytoskeleton function in the contraction of muscle cells, the beating of cilia and flagella, the movement of chromosomes during cell division, and in some cases the locomotion of the cell itself.

In addition, the cytoskeleton serves as a framework for positioning and actively moving organelles within the

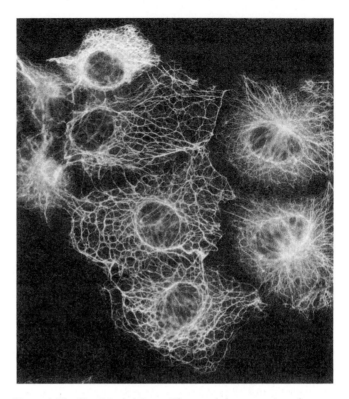

Figure 4-23   The Cytoskeleton.   The cytoskeleton consists of a network of tubules and filaments that gives the cell shape, anchors organelles and directs their movement, and enables the whole cell to move or to change its shape. This micrograph shows the cytoskeleton of human amnion epithelial cells, as revealed by the technique of immunofluorescence microscopy using antibodies against keratin proteins found in the intermediate filaments of the cytoskeleton. Notice that these filaments surround, but do not extend into, the space occupied by the nucleus.

cytosol. It may also function similarly in regard to ribosomes and enzymes. Some researchers estimate that up to 80% of the proteins of the cytosol are not freely diffusible but are instead associated with the cytoskeleton. Even water, which accounts for about 70% of the cell volume, may be influenced by the cytoskeleton. It has been estimated that as much as 20–40% of the water in the cytosol may be bound to the filaments and tubules of the cytoskeleton.

The three major structural elements of the cytoskeleton are *microtubules, microfilaments,* and *intermediate filaments.* They can be visualized by phase-contrast, immunofluorescence, and electron microscopy. Some of the structures in which they are found (such as cilia, flagella, or muscle fibrils) can even be seen by ordinary light microscopy. Microfilaments and microtubules are best known for their roles in contraction and motility. In fact, these roles were appreciated well before it became clear that the same structural elements are also integral parts of the pervasive network of filaments and tubules that gives cells their characteristic shape and structure.

Chapter 22 provides a detailed description of the cytoskeleton, followed in Chapter 23 by a discussion of microtubule- and microfilament-mediated contraction and motility. We will also encounter microtubules and microfilaments in Chapter 17 because of their roles in chromosome separation and cell division, respectively. Here, we will focus on the structural features of the three major components of the cytoskeleton: microtubules, microfilaments, and intermediate filaments.

Microtubules. Of the structural elements found in the cytoskeleton, **microtubules** (**MTs**) are the largest. A well-known MT-based cellular structure is the *axoneme* of cilia and flagella, the appendages responsible for motility of eukaryotic cells. We have already encountered an example of such a structure—the axoneme of the sperm tail in Figure 4-12 consists of microtubules. MTs also form the *spindle fibers* that separate chromosomes prior to cell division, as we will see in Chapter 17.

In addition to their involvement in motility and chromosome movement, MTs also play an important role in the organization of the cytoplasm. They contribute to the overall shape of the cell, the spatial disposition of its organelles, and the distribution of microfilaments and intermediate filaments. Examples of the diverse phenomena that are governed by MTs include the asymmetric shapes of animal cells, the plane of cell division in plant cells, the ordering of filaments during muscle development, and the positioning of mitochondria around the axoneme of motile appendages.

As shown in Figure 4-24a, MTs are hollow cylinders with an outer diameter of about 25 nm and an inner diameter of about 15 nm. Although microtubules are usually drawn as if they are straight and rigid, video microscopy reveals them to quite flexible in living cells. The wall of the microtubule consists of longitudinal arrays of *protofilaments*, usually 13 arranged side by side around the hollow center, or *lumen*. Each protofilament is a linear polymer of *tubulin* molecules. Tubulin is a dimeric protein, consisting of two similar but distinct polypeptide subunits,

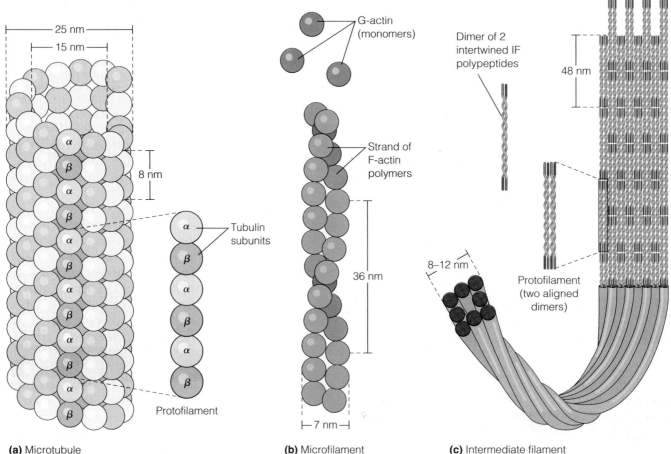

(a) Microtubule   **(b)** Microfilament   **(c)** Intermediate filament

Figure 4-24 Structures of Microtubules, Microfilaments, and Intermediate Filaments. **(a)** A schematic diagram of a microtubule, showing 13 protofilaments arrayed longitudinally to form a hollow cylinder with an outer diameter of about 25 nm and an inner diameter of about 15 nm. Each protofilament is a polymer of tubulin dimers, each about 8 nm long. All dimers are oriented in the same direction, thereby accounting for the polarity of the protofilament and hence of the whole microtubule. **(b)** A schematic diagram of a microfilament, showing a strand of F-actin twisted into a helical structure with a diameter of about 7 nm. A half-turn of the helix occurs every 36 nm. The F-actin polymer consists of monomers of G-actin, all oriented in the same direction to give the microfilament its inherent polarity. **(c)** A schematic diagram of an intermediate filament. The structural unit is the tetrameric protofilament, consisting of two pairs of coiled polypeptides, with a length of about 48 nm. Protofilaments assemble by end-to-end and side-to-side alignment, forming an intermediate filament having a diameter of 8–12 nm that is thought to be eight protofilaments thick at any point.

α-*tubulin* and β-*tubulin*. All of the tubulin dimers in each of the protofilaments are oriented in the same direction, such that all of the subunits face the same end of the microtubule. This uniform orientation gives the MT an inherent *polarity*. As we will see in Chapter 22, the polarity of microtubules has important implications for their assembly and for the directional movement of membrane-bounded organelles in which microtubules are involved.

Microfilaments. Microfilaments (**MFs**) are much thinner than microtubules. They have a diameter of about 7 nm, which makes them the smallest of the major cytoskeletal components. MFs are best known for their role in the contractile fibrils of muscle cells; we will meet them again in Chapter 23. However, microfilaments are involved in a variety of other cellular phenomena as well. They can form connections with the plasma membrane and thereby influence locomotion, amoeboid movement, and cytoplasmic streaming. MFs also produce the *cleavage furrows* that divide the cytoplasm of animal cells after chromosomes have been separated by the spindle fibers. In addition, MFs contribute importantly to the development and maintenance of cell shape.

MFs are polymers of the protein *actin* (Figure 4-24b). Actin is synthesized as a monomer called *G-actin* (G for globular). G-actin monomers polymerize reversibly into long, double-helical strands of *F-actin* (F for filamentous), each strand about 4 nm wide. A microfilament consists of a single chain of actin monomers that are assembled at angles to one another, creating a filament with a helical appearance and a diameter of about 7 nm (Figure 4-24b). Like microtubules, microfilaments are polar structures; all of the subunits are oriented in the same direction. This polarity influences the direction of MF elongation; assembly usually proceeds more readily at one end of the growing microfilament, whereas disassembly is favored at the other end.

Intermediate Filaments. Intermediate filaments (**IFs**) make up the third structural element of the cytoskeleton. IFs have a diameter of about 8–12 nm, larger than the diameter of microfilaments but smaller than that of microtubules. Intermediate filaments are the most stable and the least soluble constituents of the cytoskeleton. Because of this stability, some researchers regard IFs as a scaffold that supports the entire cytoskeletal framework. Intermediate filaments are also thought to have a tension-bearing role in some cells because they often occur in areas that are subject to mechanical stress.

In contrast to microtubules and microfilaments, intermediate filaments differ in their composition from tissue to tissue. Based on biochemical and immunological criteria, IFs from animal cells can be grouped into six classes. A specific cell type usually contains only one or sometimes two classes of IF proteins. Because of this tissue specificity, animal cells from different tissues can be distinguished on the basis of the IF proteins present. This *intermediate filament typing* serves as a diagnostic tool in medicine.

Despite their heterogeneity of size and chemical properties, all IF proteins share common structural features. They all have a central rodlike segment that is remarkably similar from one IF protein to the other. Flanking the central region of the protein are N-terminal and C-terminal segments that differ greatly in size and sequence, presumably accounting for the functional diversity of these proteins.

Several models have been proposed for IF structure. One possibility is shown in Figure 4-24c. The basic structural unit is a dimer of two intertwined IF polypeptides. Two such dimers align laterally to form a tetrameric *protofilament*. Protofilaments then interact with each other to form an intermediate filament that is thought to be eight protofilaments thick at any point, with protofilaments probably joined end to end in an overlapping manner.

The Extracellular Matrix and the Cell Wall Are the "Outside" of the Cell

So far, we have considered the plasma membrane that surrounds every cell, the nucleus and cytoplasm within, and the variety of organelles, membrane systems, tubules, and filaments found in the cytoplasm of most eukaryotic cells. While it may seem that our tour of the cell is complete, most cells are also characterized by extracellular structures formed from materials that the cells transport outward across the plasma membrane. For animal cells, these structures are called the **extracellular matrix** (**ECM**) and consist primarily of *collagen fibers* and *proteoglycans*. For plant cells, the extracellular structure is the rigid **cell wall,** which consists mainly of *cellulose microfibrils* embedded in a matrix of other polysaccharides and small amounts of protein.

This difference between plant and animal cells is in keeping with the lifestyles of these two broad groups of eukaryotes. Plants are generally *nonmotile,* a lifestyle that is compatible with the rigidity cell walls confer on an organism. Animals, on the other hand, are usually *motile,* an essential feature for an organism that needs not only to find food but to escape becoming food for other organisms. Accordingly, their cells are not encased in rigid walls but are instead surrounded by an elastic network of collagen fibers.

The extracellular matrix of animal cells forms a variety of structures, depending on the cell type. The primary function of the ECM is support, but the kinds of extracellular materials and the patterns in which they are deposited may regulate such diverse processes as cell motility and migration, cell division, cell recognition and adhesion, and cell differentiation during embryonic development. The main constituents of animal extracellular structures are the collagen fibers and the network of proteoglycans that surround them. The collagens are a group of generally insoluble glycoproteins that contain large amounts of glycine and the hydroxylated forms of lysine and proline. In vertebrates, collagen is such a prominent part of tendons, cartilage, and bone that it is the single most abundant protein of the animal body.

Figure 4-25 illustrates the prominence of the cell wall as a structural feature of a typical plant cell. Although the distinction is not made in the figure, plant cell walls are actually of two types. The wall that is laid down during cell division, called the *primary cell wall*, consists mainly of cellulose fibrils embedded in a gel-like polysaccharide matrix. Primary walls are quite flexible and extensible, which allows them to expand somewhat in response to cell enlargement and elongation. As a cell reaches its final size and shape, a much thicker and more rigid *secondary cell wall* may form by deposition of additional cell wall material on the inner surface of the primary wall. The secondary wall usually contains more cellulose than the primary wall and may have a high content of *lignin,* a major component of wood. Deposition of a secondary cell wall renders the cell very inextensible and therefore specifies the final size and shape of the cell definitively.

Neighboring plant cells, though separated by the wall between them, are actually connected by numerous cytoplasmic bridges, called **plasmodesmata** (singular: plasmodesma), which pass through the cell wall. The plasma membranes of adjacent cells are continuous through each plasmodesma, such that the channel is membrane-lined. The diameter of a typical plasmodesma is large enough to allow water and small solutes to pass freely from cell to cell. Most of the cells of the plant are interconnected in this way.

Animal cells can also communicate with each other. But instead of plasmodesmata they have intercellular connections called **gap junctions** that are specialized for the transfer of material between the cytoplasms of adjacent cells. Two other types of intercellular junctions are also characteristic of animal cells. **Tight junctions** hold cells together so tightly that the transport of substances through the spaces between the cells is effectively blocked. **Adhesive junctions** also link adjacent cells, but for the purpose of connecting them tightly into sturdy yet flexible sheets. Each of these types of junctions is discussed in detail in Chapter 11.

Most prokaryotes are also surrounded by an extracellular structure called a cell wall. However, bacterial cell walls consist not of cellulose but mainly of *peptidoglycans.* Peptidoglycans are built up from a backbone of repeating units of *N*-acetyglucosamine (GlcNAc) and *N*-acetylmuramic acid (MurNAc), amino sugars we encountered in Chapter 3 (see Figure 3-26). In addition, bacterial cell walls contain a variety of other constituents, some of which are unique to each of the major structural groups of bacteria. The presence or absence of such features is indicated by the cell's response to the *Gram stain,* named for Hans Christian Gram, the Danish bacteriologist who devised it. Bacterial cells are stained with a solution of crystal violet and iodine, then destained with alcohol or acetone. Gram-positive cells are resistant to destaining and remain deep blue in color, whereas gram-negative cells are rapidly decolorized.

Cell wall constituents unique to gram-negative bacteria include *lipopolysaccharides* and *lipoproteins.* Compounds found only in gram-positive bacteria include the *teichoic acids,* which consist of three- or five-carbon alcohols (glycerol and ribitol, respectively) linked together by phosphate groups.

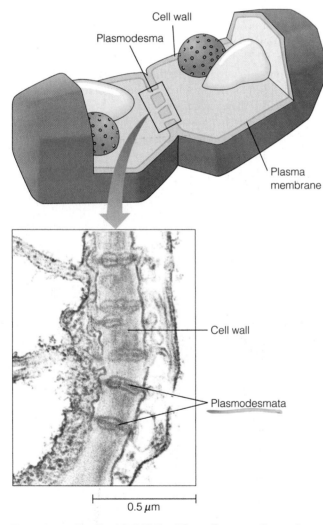

Cell wall
Plasmodesma
Plasma membrane
Cell wall
Plasmodesmata
0.5 μm

Figure 4-25 The Plant Cell Wall. The wall surrounding a plant cell consists of rigid microfibrils of cellulose embedded in a noncellulosic matrix of proteins and sugar polymers. Notice how the neighboring cells are connected by plasmodesmata (TEM).

Viruses, Viroids, and Prions: Agents That Invade Cells

Before concluding this preview of cellular biology, we need to look at several kinds of agents that invade cells, subverting normal cellular functions and often killing their unwilling hosts. These include the *viruses* and two other agents about which we know much less, the *viroids* and *prions*.

A Virus Consists of a DNA or RNA Core Surrounded by a Protein Coat

Viruses are subcellular parasites that are incapable of a free-living existence. Instead, they invade and infect cells and redirect their host's synthetic machinery toward the production of more viruses. Viruses cannot carry on all of the functions required for independent existence and must therefore depend for most of their needs on the cells they invade.

Viruses are responsible for many diseases in humans, animals, and plants and are therefore important in their own right. They are also very significant research tools for cell and molecular biologists because they are much less complicated than cells. The *tobacco mosaic virus (TMV)* that we encountered in Chapter 2 (see Figures 2-20 and 2-21) is a good example of a virus that is important both economically and scientifically—economically because of the threat it poses to tobacco and other crop plants that it can infect, and scientifically because it is so amenable to laboratory study. Fraenkel-Conrat's studies on TMV self-assembly described in Chapter 2 are good examples of the usefulness of viruses to cell biologists.

Viruses are sometimes named for the diseases they cause. Poliovirus, influenza virus, herpes simplex virus, and TMV are several common examples. Other viruses have more cryptic laboratory names such as T4, Qβ, λ, or Epstein-Barr virus. Viruses that infect bacterial cells are called **bacteriophages,** or often just **phages** for short. Bacteriophages and other viruses will figure prominently in our discussion of molecular genetics in Chapters 16–21 because of the striking genetic similarities between viruses and cells. In terms of the storage, expression, and transmission of genetic information, viruses seem to follow most of the same rules and use many of the same mechanisms that cells do, yet they are simpler than cells and much easier to manipulate. Viruses are therefore very useful in the study of genetics at the molecular level, a point to which we will return in Chapter 16.

Viruses range in size from about 25 nm to 300 nm. The smallest viruses are therefore about the size of a ribosome, whereas the largest ones are about one-quarter the diameter of a typical bacterial cell. Each virus has its own characteristic shape, as shown in Figure 4-26.

Despite their morphological diversity, viruses are chemically quite simple. Most viruses consist of little more than a *coat* (or *capsid*) of protein surrounding a *core* that contains one or more molecules of either RNA or DNA, depending on the type of virus. The simplest viruses, such as TMV, have a single nucleic acid molecule surrounded by a capsid consisting of proteins of a single type (see Figure 2-20). More complex viruses have cores that contain several nucleic acid molecules and capsids consisting of several (or even many) different kinds of proteins. Some viruses are surrounded by a membrane that is derived from the plasma membrane of the host cell in which the viral particles are made and assembled. Such viruses are called *enveloped viruses;* human immunodeficiency virus, the virus that causes AIDS, is an example.

The question is sometimes asked whether or not viruses are living. The answer depends crucially on what we mean by "living," and it is probably worth pondering only to the extent that it helps us more fully understand what viruses are—and what they are not. The most fundamental properties of living things are *metabolism* (cellular reactions organized into coherent pathways), *irritability* (perception of, and response to, environmental stimuli), and the *ability to reproduce.* Viruses clearly do not satisfy the first two criteria. Outside their host cells, viruses are inert and inactive. They can, in fact, be isolated and crystallized almost like a chemical compound. It is only in an appropriate host cell that a virus becomes functional, undergoing a cycle of synthesis and assembly that gives rise to more viruses.

Even the ability of viruses to reproduce has to be qualified carefully. A basic tenet of the cell theory is that cells arise only from preexisting cells, but this is not true of viruses. No virus can give rise to another virus by any sort of self-duplication process. Rather, the virus must subvert the metabolic and genetic machinery of the host cell, reprogramming it for synthesis of the proteins necessary to package the DNA or RNA molecules that arise by copying the genetic information of the parent virus.

It is only in a genetic sense that one can think of viruses as living at all. Another fundamental property of living things is the capability of specifying and directing the genetic composition of progeny—an ability that viruses clearly possess. It is probably most helpful to think of viruses as "quasi-living," satisfying part but not all of the basic definition of life.

Viroids Are Small, Circular RNA Molecules

As simple as viruses are, even simpler agents are known that can infect eukaryotic cells (though apparently not prokaryotic cells, as far as is presently known). The **viroids** found in some plant cells represent one class of such agents. Viroids are small, circular RNA molecules, rather like an RNA virus without its capsid. They are, in fact, the smallest known infectious agents. The RNA circles are only about 300–400 nucleotides long and are replicated in the host cell even though they do not code for any protein.

It is not yet known how viroids are transmitted from one host to another, since they apparently do not occur in a free form. Most likely, they pass from one plant cell to another only when the surfaces of adjacent cells are damaged so that there is no membrane barrier for the RNA molecules to cross.

Viroids are responsible for diseases of several crop plants, including potatoes and tobacco. A viroidal disease that has severe economic consequences is *cadang-cadang disease* of the coconut palm. It is not yet clear how viroids cause disease. They may enter the nucleus and interfere with the transcription of DNA into RNA. Alternatively, they may interfere with the subsequent processing required of most eukaryotic RNA transcripts.

Prions Are "Proteinaceous Infective Particles"

Prions represent another class of infectious agents about which little is yet known. The term was coined to describe *proteinaceous infective particles* that are thought to be responsible for neurological diseases such as scrapie in sheep and goats, kuru in humans, and "mad cow" disease in cattle. *Scrapie* is so named because infected animals rub

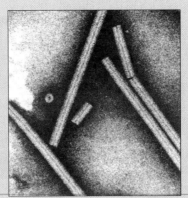

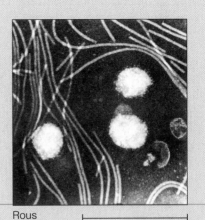

Polio 0.1 μm

Tobacco mosaic 0.1 μm

Rous sarcoma 0.5 μm

(a) RNA-containing viruses

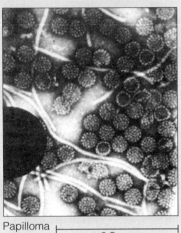

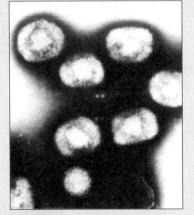

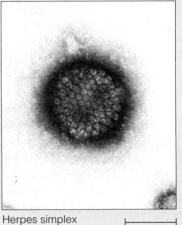

Papilloma 0.5 μm

Vaccinia 0.5 μm

Herpes simplex 0.1 μm

(b) DNA-containing viruses

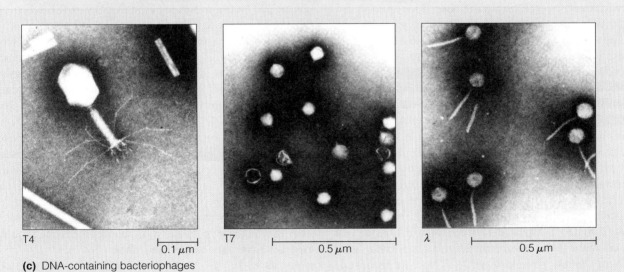

T4 0.1 μm

T7 0.5 μm

λ 0.5 μm

(c) DNA-containing bacteriophages

Figure 4-26 Sizes and Shapes of Viruses. These electron micrographs illustrate the morphological diversity of viruses. **(a)** RNA-containing viruses (left to right): polio, tobacco mosaic, and Rous Sarcoma viruses. **(b)** DNA-containing viruses: papilloma, vaccinia, and herpes simplex viruses. **(c)** DNA-containing bacteriophages: T4, T7, and λ (all TEMs.)

incessantly against trees or other objects, scraping off most of their wool in the process. *Kuru* is a degenerative disease of the central nervous system originally reported among native peoples in New Guinea. Patients with this or other prion-based diseases suffer initially from mild physical weakness and dementia, but the effects slowly become more severe, and the diseases are eventually fatal. For a discussion of *"mad cow" disease,* see Box 20A (p. 673).

The proteinaceous particles to which the name refers are the only agents that have thus far been isolated from infected individuals. The best guess at present is that prion proteins are abnormally folded versions of normal cellular proteins. Both the normal and variant forms of the prion protein are found on the surfaces of neurons, suggesting that the protein may somehow affect the receptors that detect nerve signals.

Perspective

A fundamental distinction in biology is the one between the prokaryotes (eubacteria and archaebacteria) and the eukaryotes (animals, plants, fungi, and protists). Equally fundamental is the distinction between the eubacteria and the archaebacteria. In fact, analyses of ribosomal RNA sequences and other molecular data suggest a tripartite view of organisms, with eukaryotes, eubacteria, and archaebacteria as the three main groups.

A plasma membrane and ribosomes are the only two structural features common to all three groups. All other organelles are found only in eukaryotic cells, where they play indispensable roles in the compartmentalization of function. Prokaryotic cells are relatively small and structurally less complex than eukaryotic cells, lacking most of the internal membrane systems and organelles of the latter.

Eukaryotic cells have at least four major structural features: a plasma membrane that defines the boundaries of the cell and retains its contents, a nucleus that houses most of the cell's DNA, a variety of organelles, and the cytosol with its cytoskeleton of tubules and filaments. In addition, plant cells almost always have a rigid cell wall, and animal cells are usually surrounded by an extracellular matrix of collagen and proteoglycans.

The nucleus is surrounded by a double membrane called the nuclear envelope. The chromosomes within the nucleus contain most of the DNA of the cell, complexed with proteins.

Mitochondria play an important role in the oxidation of food molecules to release energy, which is used to make ATP. Chloroplasts trap solar energy and use it to "fix" carbon from carbon dioxide into organic form and convert it to sugar. Mitochondria and chloroplasts are surrounded by a double membrane and have an extensive system of internal membranes in which most of the components involved in ATP generation are embedded.

The endoplasmic reticulum is an extensive network of membranes that are either rough (studded with ribosomes) or smooth. Rough ER is responsible for the synthesis of secretory and membrane proteins, whereas smooth ER is involved in lipid synthesis and drug detoxification. Proteins synthesized on the rough ER are further processed and packaged in the Golgi complex and are then transported to the surface of the cell by secretory vesicles.

Lysosomes contain hydrolytic enzymes and are involved in cellular digestion. They were the first organelles to be discovered on the basis of their function rather than their morphology.

Peroxisomes are often about the same size as lysosomes but function in the generation and degradation of hydrogen peroxide. Animal peroxisomes play an important role in the catabolism of long-chain fatty acids. In plants, specialized peroxisomes are involved in the conversion of stored fat into carbohydrate during seed germination and in the process of photorespiration.

The ribosome is an example of an organelle that is not membrane-bounded. Ribosomes serve as sites of protein synthesis in both prokaryotic and eukaryotic cells and also in mitochondria and chloroplasts. The striking similarities between mitochondrial and chloroplast ribosomes and those of bacteria and cyanobacteria, respectively, lend strong support to the endosymbiont theory that these organelles are of prokaryotic origin.

The cytoskeleton is an extensive network of microtubules, microfilaments, and intermediate filaments that gives eukaryotic cells their distinctive shapes. The cytoskeleton is also important in cellular motility and contractility, topics of later chapters.

Viruses satisfy some, though not all, of the basic criteria of living things. Viruses are important both as infectious agents that cause diseases in humans, animals, and plants and as laboratory tools, particularly for geneticists. Viroids and prions are infectious agents that are even smaller (and less well understood) than viruses. Viroids are small RNA molecules, whereas prions are thought to be abnormal products of normal cellular genes.

Key Terms for Self-Testing

Properties and Strategies of Cells
prokaryote (p. 76)
eukaryote (p. 76)
eubacterium (p. 76)
archaebacterium (p. 76)
bacterium (p. 76)
cyanobacterium (p. 76)
surface area/volume ratio (p. 77)
diffusion (p. 78)
cytoplasmic streaming (cyclosis) (p. 78)
organelle (p. 78)
chromosome (p. 80)

The Plasma Membrane
plasma membrane (p. 83)
lipid bilayer (p. 84)

The Nucleus
nucleus (p. 84)
nuclear envelope (p. 84)
chromatin (p. 85)
nucleolus (p. 85)

Intracellular Membranes and Organelles
mitochondrion (p. 85)
crista (p. 86)
matrix (p. 86)

chloroplast (p. 87)
thylakoid (p. 87)
stroma thykaloid (p. 87)
granum (p. 87)
stroma (p. 87)
plastid (p. 88)
endosymbiont theory (p. 88)
endoplasmic reticulum (ER) (p. 88)
cisterna (p. 89)
lumen (p. 89)
rough endoplasmic reticulum
   (rough ER) (p. 89)
smooth endoplasmic reticulum
   (smooth ER) (p. 89)
Golgi complex (p. 89)
secretory vesicle (p. 90)
lysosome (p. 90)
peroxisome (p. 90)
glyoxysome (p. 91)
leaf peroxisome (p. 91)
vacuole (p. 91)
central vacuole (p. 93)
ribosome (p. 93)
sedimentation coefficient (p. 94)
large ribosomal subunit (p. 94)
small ribosomal subunit (p. 94)

The Cytoplasm, Cytosol, and Cytoskeleton
cytoplasm (p. 95)
cytosol (p. 95)
cytoskeleton (p. 95)
microtubule (MT) (p. 96)
microfilament (MF) (p. 97)
intermediate filament (IF) (p. 97)

Outside the Cell: The Extracellular Matrix and the Cell Wall
extracellular matrix (ECM) (p. 97)
cell wall (p. 97)
plasmodesma (p. 98)
gap junction (p. 98)
tight junction (p. 98)
adhesive junction (p. 98)

Viruses, Viroids, and Prions
virus (p. 98)
bacteriophage (phage) (p. 99)
viroid (p. 99)
prion (p. 99)

Problem Set

More challenging problems are marked with a • .

4-1.   Prokaryotes and Eukaryotes. Indicate whether each of the following statements is true (T) or false (F). If false, reword the statement to make it true.

(a) Eukaryotic cells are in most cases larger than prokaryotic cells.

(b) Some cells are large enough to be seen with the unaided eye.

(c) Prokaryotic cells possess none of the following features: mitochondria, membrane-bounded nucleus, plasma membrane, microtubules.

(d) The surface area/volume ratio is generally greater for a prokaryotic cell than for a eukaryotic cell.

(e) The ribosomes found in the mitochondria of your muscle cells are more like those of the bacteria in your intestine than they are like the ribosomes in the cytosol of your muscle cells.

(f) Because prokaryotic cells have neither mitochondria nor chloroplasts, they cannot carry out either ATP synthesis or photosynthesis.

4-2.   That's About the Size of It. At the beginning of this chapter (p. 77), we noted that some of the smallest bacteria are only about 0.3 μm in diameter, such that more than 40,000 cells could fit side by side across the head of a thumbtack. For each of the following cells or subcellular structures, use the specified dimensions to determine (i) the minimum number of them required to span the head of a thumbtack with a diameter of 1.2

cm and (ii) approximately how many of them are needed to form a single layer on the head of the thumbtack.

(a) Mycoplasma cell (0.3 μm in diameter)

(b) Human liver cell (20 μm in diameter)

(c) Nucleus (6 μm in diameter)

(d) Chloroplast (2×8 μm)

(e) Mitochondrion (1×2 μm)

(f) Peroxisome (0.5 μm in diameter)

(g) Microtubule (1 μm $\times$ 25 nm)

(h) Microfilament (7×200 nm)

(i) Eukaryotic ribosome (30 nm in diameter)

4-3.   Cellular Specialization. Each of the cell types listed here is a good example of a cell that is specialized for a specific function. Match each cell type in list A with the appropriate function from list B, and explain why you matched each as you did.

List A	List B
(a) Pancreatic cell	Cell division
(b) Cell from flight muscle	Absorption
(c) Palisade cell from leaf	Motility
(d) Cell of intestinal lining	Photosynthesis
(e) Nerve cell	Secretion
(f) Bacterial cell	Transmission of electrical impulses

4-4. **Cellular Structure.** Indicate whether each of the following cellular structures or components is found in animal cells (A), bacterial cells (B), and/or plant cells (P).

(a) Chloroplasts

(b) Cell wall

(c) Microtubules

(d) DNA

(e) Nuclear envelope

(f) Nucleoli

(g) Golgi complex

(h) Central vacuole

(i) Thylakoids

(j) Ribosomes

(k) Lipid bilayers

(l) Actin

4-5. **Matching.** For each of the structures in list A, choose the single term from list B that matches best, and explain your choice.

List A

(a) Plant cell wall

(b) Cyanobacterium

(c) Lysosome

(d) Nuclear envelope

(e) Nucleolus

(f) Rough ER

(g) Smooth ER

(h) Plasma membrane

(i) Virus

(j) Viroid

List B

Bacteriophage	Pores
Hydrolases	Ribosome synthesis
RNA	Glycoprotein
Cellulose	Eukaryote
Photorespiration	Svedberg
Lipid synthesis	Glyoxysomes
Muscle cells	Granum
Peptidoglycan	Interphase
Prokaryote	Secretory proteins

4-6. **Sentence Completion.** Complete each of the following statements about cellular structure in ten words or less.

(a) If you were shown an electron micrograph of a section of a cell and were asked to identify the cell as plant or animal, one thing you might do is . . .

(b) A slice of raw apple placed in a concentrated sugar solution will . . .

(c) A cellular structure that is visible with an electron microscope but not with a light microscope is . . .

(d) Several environments in which you are more likely to find archaebacteria than eubacteria are . . .

(e) One reason that it might be difficult to separate lysosomes from peroxisomes by centrifugation techniques is that . . .

4-7. **Telling Them Apart.** Suggest a way to distinguish between the two elements in each of the following pairs.

(a) Bacterial cell vs. a protist of the same size

(b) Rough ER vs. smooth ER

(c) Animal peroxisome vs. leaf peroxisome

(d) Late endosome vs. lysosome

(e) Virus vs. viroid

(f) Microfilament vs. intermediate filament

(g) Polio virus vs. herpes simplex virus

(h) Eukaryotic ribosome vs. prokaryotic ribosome

(i) Eubacterial ribosome vs. archaebacterial ribosome

4-8. **Structural Relationships.** For each pair of structural elements, indicate with an A if the first element is a constituent part of the second, with a B if the second element is a constituent part of the first, and with an N if they are separate structures with no particular relationship to each other.

(a) Mitochondrion; crista

(b) Golgi complex; nucleus

(c) Cytoplasm; cytoskeleton

(d) Cell wall; extracellular matrix

(e) Nucleolus; nucleus

(f) Smooth ER; ribosome

(g) Lipid bilayer; plasma membrane

(h) Peroxisome; thylakoid

(i) Chloroplast; granum

4-9. **The Palisade Cell: A Look Inside.** Just under the upper surface of many plant leaves is a layer of columnar *palisade cells,* the site of much photosynthesis. A typical palisade cell is cylindrical in shape, with a diameter of 20 μm and a length of 35 μm. In round numbers, such a cell might contain 200 mitochondria, 40 chloroplasts, 100 peroxisomes, 2 million ribosomes, and 1 nucleus. The dimensions of each of these organelles are given in Problem 4-2. Except for the region occupied by the nucleus, the cytoplasm of the palisade cell is restricted to a 2.5-μm layer just beneath the plasma membrane because of the large vacuole, which, like the cell itself, is roughly cylindrical in shape.

(a) Calculate the proportion of the total internal volume of the cell that is occupied by each of these populations of organelles. Consider the cell, the vacuole, the chloroplasts, and the mitochondria to be cylindrical in shape ($V = \pi r^2 h$) and the nucleus, peroxisomes, and ribosomes to be approximately spherical ($V = 4\pi r^3/3$).

(b) What proportion of the total internal volume of the cell is not accounted for by the named organelles? What other major structural features must be accommodated in this remaining cytoplasmic volume?

•**4-10.** **Protein Synthesis and Secretion.** Although we will not encounter protein synthesis and secretion in detail until later chapters, you already have enough information about these processes to place in order the seven events that are now listed randomly. Order events 1–7 so that they represent the correct sequence of events corresponding to steps a–g, tracing a typical secretory protein from the initial transcription (readout) of the relevant genetic information in the nucleus to the eventual secretion of the protein from the cell by exocytosis.

Transcription $\longrightarrow$ (a) $\longrightarrow$ (b) $\longrightarrow$ (c) $\longrightarrow$
(d) $\longrightarrow$ (e) $\longrightarrow$ (f) $\longrightarrow$ (g) $\longrightarrow$ Secretion

(1) The protein is partially glycosylated within the lumen of the ER.

(2) The secretory vesicle arrives at and fuses with the plasma membrane.

(3) The RNA transcript is transported from the nucleus to the cytoplasm.

(4) The final sugar groups are added to the protein in the Golgi complex.

(5) As the protein is synthesized, it passes across the ER membrane into the lumen of a cisterna.

(6) The enzyme is packaged into a secretory vesicle and released from the Golgi complex.

(7) The RNA message associates with a ribosome and begins synthesis of the desired protein on the surface of the rough ER.

•4-11.   **Disorders at the Organelle Level.** Each of the following medical problems involves a disorder in the function of an organelle or other cell structure. In each case, identify the organelle or structure involved, and indicate whether it is likely to be underactive or overactive.

(a) A girl inadvertently consumes cyanide and dies almost immediately because ATP production ceases.

(b) A boy is diagnosed with adrenoleukodystrophy (ALD), characterized by an inability of his body to break down very long-chain fatty acids.

(c) A smoker develops lung cancer and is told that the cause of the problem is a population of cells in her lungs that are undergoing mitosis at a much greater rate than is normal for lung cells.

(d) A young man learns that he is infertile because his sperm are nonmotile.

(e) A young child dies of Tay-Sachs disease because her cells lack the hydrolase that normally breaks down a membrane component called ganglioside G_{M2}, which therefore accumulated in the membranes of her brain.

(f) A young child is placed on a milk-free diet because the mucosal cells that line his small intestine do not secrete the enzyme necessary to hydrolyze lactose, the disaccharide present in milk.

•4-12.   **The Smallest Bacteria.** The "smallest bacteria" mentioned on p. 77 and in Problem 4-2 are called *mycoplasma*. Most mycoplasma are only about 0.3 μm in diameter, which makes them the smallest known free-living organisms. To appreciate how small these cells are, answer the following. (The volume of a sphere and a cylinder are given by $4\pi r^3/3$ and $\pi r^2 h$, respectively.)

(a) What is the internal volume of a mycoplasma cell in milliliters? How does this compare with the internal volume of

Escherichia coli, a common intestinal bacterium, which is a cylinder with a diameter of 1 μm and a length of 2 μm?

(b) How many ribosomes can a mycoplasma cell accommodate if ribosomes can occupy no more than 20% of the internal volume of the cell? How does this compare with the thousands of ribosomes present in an *E. coli* cell?

(c) Assuming that the concentration of glucose in a typical mycoplasma cell is 1.0 mM, how many molecules of glucose are present in the cell?

(d) Most compounds are present in cells at much lower concentrations than glucose. The concentration of a substance called NAD (for nicotinamide adenine dinucleotide) in a bacterial cell is typically about 2 μM. How many molecules of NAD are present in a mycoplasma cell?

(e) A typical mycoplasma cell has as its genetic information a single DNA molecule with a molecular weight of about 2×10^7 and a circumference of about 100 nm. What fraction of the weight of a mycoplasma cell does its DNA molecule represent? (Assume the density of a mycoplasma cell to be 1.1 g/cm³.) What, if any, problem might the length of the DNA molecule pose for the cell?

(f) Based on your calculations, what do you think might be the factor(s) that set the lower limit on the size of a cell?

•4-13.   **The Giant Bacterium.** In 1993, E. R. Angert and colleagues published a paper with the title, "The largest bacterium" (*Nature* 362: 239–241), in which they reported the discovery of a bacterium, *Epulopiscium fishelsoni*, that measures up to 80 μm by 600 μm. In the abstract, the authors comment that to prove that *E. fishelsoni* is a prokaryote, they "isolated the genes encoding the small subunit ribosomal RNA . . . and used them in a phylogenetic analysis."

(a) What is the significance of the genes encoding ribosomal RNA in this kind of phylogenetic analysis?

(b) Why is a bacterium of this size such a surprise?

(c) Assuming that the "largest bacterium" is cylindrical in shape with a diameter of 80 μm and a length of 600 μm, how many cells of the common intestinal bacterium *Escherichia coli* (also a cylinder, with a diameter of 1 μm and a length of 2 μm) would fit inside one *E. fishelsoni* cell?

Suggested Reading

References of historical importance are marked with a •.

General References

Alberts, B., D. Bray, J. Lewis, M. Raff, K. Roberts, and J. D. Watson. *Molecular Biology of the Cell,* 3d ed. New York: Garland Press, 1994.

Lodish, H., A. Berk, S. L. Zipursky, P. Matsudaira, D. Baltimore, and J. Darnell. *Molecular Cell Biology,* 4th ed. New York: Scientific Books, 2000.

Eukaryotic and Prokaryotic Cells/Eubacteria and Archaebacteria

Angert, E. R., K. D. Clements, and N. R. Pace. The largest bacterium. *Nature* 362 (1993): 239.

Dixon, B. *Power Unseen: How Microbes Rule the World.* New York: W. H. Freeman, 1994.

Gray, M. W. The third form of life. *Nature* 383 (1996): 299.

Gupta, R. S., and C. B. Golding. The origin of the eukaryotic cell. *Trends Biochem. Sci.* 21 (1996): 166.

Madigan, M. T., and B. L. Marrs. Extremophiles. *Sci. Amer.* 276 (April 1997): 82.

Sogin, M. L. Giants among the prokaryotes. *Nature* 362 (1993): 207.

•Vidal, G. The oldest eukaryotic cells. *Sci. Amer.* 244 (February 1984): 48.

•Woese, C. R., and G. E. Fox. Phylogenetic structure of the prokaryotic domain: The primary kingdoms. *Proc. Nat. Acad. Sci. USA* 74 (1977): 5088.

Woese, C. R., O. Kandler, and M. L. Wheelis. Towards a natural system of organisms: Proposal for the domains Archaea, Bacteria, and Eucarya. *Proc. Nat. Acad. Sci. USA* 87 (1990): 4576.

The Nucleus

Hoffman, M. The cells' nucleus shapes up. *Science* 259 (1993): 1257.

Lamond, A. I., and W. C. Earnshaw. Structure and function in the nucleus. *Science* 280 (1998): 547.

Strouboulis, J., and A. P. Wolfe. Functional compartmentalization of the nucleus. *J. Cell Sci.* 109 (1996): 1991.

Intracellular Membranes and Organelles

Balch, W. E., ed. *Molecular Mechanisms of Intracellular Vesicular Traffic.* San Diego: Academic Press, 2000.

Berger, E. G., and J. Roth, eds. *The Golgi Apparatus.* Boston: Birkhauser Verlag, 1997.

Cuervo, A. M., and J. F. Dice. Lysosomes, a meeting point of proteins, chaperones, and proteases. *J. Molec. Med.* 76 (1998): 76.

• de Duve, C. The peroxisome in retrospect. *Ann. N.Y. Acad. Sci.* 804 (1996): 1.

Frey, T. G., and C. A. Mannella. The internal structure of mitochondria. *Trends Biochem. Sci.* 25 (2000): 319.

Garrett, R. A., et al., eds. *The Ribosome: Structure, Function, Antibiotics, and Cellular Interactions.* Washington, DC: ASM Press, 2000.

Granzier, H. L., and G. H. Pollack, eds. *Elastic Filaments of the Cell.* New York: Kluwer Academic/Plenum Publishers, 2000.

Green, R., and H. F. Noller. Ribosomes and translation. *Annu. Rev. Biochem.* 66 (1997): 679.

Leigh, R.A., and D. Sanders, eds. *The Plant Vacuole.* San Diego: Academic Press, 1997.

Moore, R. B. Ribosomes: protein synthesis in slow motion. *Curr. Biol.* 7 (1997): R179.

Pon, L. A., and E. A. Schon, eds. *Mitochondria.* San Diego: Academic Press, 2001.

Schettler, I. E. *Mitochondria.* New York: Wiley-Luss, 1999.

Spirin, A. S. *Ribosomes.* New York: Kluwer Academic/Plenum Publishers, 1999.

Subramani, S. Components involved in peroxisome import, biogenesis, proliferation, turnover, and movement. *Physiol. Rev.* 78 (1998): 171.

The Cytoplasm and the Cytoskeleton

Carraway, K. L., and C. A. Carraway, eds. *Cytoskeleton: Signaling and Cell Regulation: A Practical Approach.* Oxford: Oxford University Press, 2000.

Drubin, D., and N. Hirokawa, eds. The cytoskeleton. *Curr. Opin. Cell Biol.* vol. 10, number 1 (1998).

Fowler, V. M., and R. Vale. The cytoskeleton: a collection of reviews. *Curr. Opin. Cell Biol.,* vol. 8, number 1 (1996).

Fuchs, E., and D. W. Cleveland. A structural scaffolding of intermediate filaments in health and disease. *Science* 279 (1998): 514.

The Extracellular Matrix and the Cell Wall

Aumailley, M., and B. Gayraud. Structure and biological activity of the extracellular matrix. *J. Molec. Med.* 76 (1998): 253.

Cosgrove, D. J. Assembly and enlargement of the primary cell wall in plants. *Annu. Rev. Cell Develop. Biol.* 13 (1997): 171.

Ekblom, P. E., and R. Timpl, eds. Cell-to-cell contact and extracellular matrix. *Curr. Opin. Cell Biol.* 8 (1996): 599.

Kreis, T., and R. Vale, eds. *Guidebook to the Extracellular Matrix, Anchor, and Adhesion Proteins,* 2d ed. Oxford: Oxford University Press, 1999.

Sakai, L. Y. The extracellular matrix. *Sci. and Med.* 2 (May 1995): 58.

Scott, J. E. Extracellular matrix, supramolecular organisation, and shape. *J. Anat.* 187 (1995): 259.

Viruses, Viroids, and Prions

Aguzzi, A., and C. Weissmann. The prion's perplexing persistence. *Nature* 392 (1998): 763.

Balter, M. Viruses have many ways to become unwelcome guests. *Science* 280 (1998): 204.

Cann, A. J. *Principles of Molecular Virology,* 3d ed. San Diego: Academic Press, 2001

Diener, T. O. Origin and evolution of viroids and viroid-like satellite RNAs. *Virus Genetics* 11 (1995): 119.

Horwich, A. L., and J. S. Weissman. Deadly conformations: Protein misfolding in prion disease. *Cell* 89 (1997): 499.

Prusiner, S. B. The prion diseases. *Sci. Amer.* 272 (January 1995): 48.

Prusiner, S. B., and M. R. Scott. Genetics of prions. *Annu. Rev. Genetics* 31 (1997): 139.

• Reisner, D., and H. H. Gross. Viroids. *Annu. Rev. Biochem.* 54 (1985): 531.

5

Bioenergetics:
The Flow of Energy in the Cell

Broadly speaking, every cell has four essential needs. To remain alive and functional, every cell requires *molecular building blocks, chemical catalysts* called enzymes, *information* to guide all its activities, and *energy* to drive the various reactions and processes that are essential to life and biological function.

In Chapter 3 we considered the various molecules that cells need, including amino acids, nucleotides, sugars, and lipids. To these we can add other essential molecules and ions, such as water, inorganic salts, metal ions, oxygen, and carbon dioxide. Some of these materials are produced by cells; others must be obtained from the environment.

In the absence of catalysts, most of the chemical reactions that must take place in cells would occur much too slowly to maintain life as we know it. To overcome this limitation, cells use *enzymes,* which can speed up reactions by many orders of magnitude. We will consider enzymes and enzyme-catalyzed reactions in Chapter 6.

The third general requirement of cells is for *information* to guide and direct their activities. As we know from Chapter 3, information is encoded within the nucleotide sequences of DNA and RNA and expressed in the synthesis of specific proteins. The genetic information that is stored, transmitted, and expressed as DNA, RNA, and proteins determines what kinds of chemical reactions a cell can carry out, what kinds of structures it can form, and what kinds of functions it can carry out. We will discuss the flow and expression of genetic information in Chapters 16–20.

In addition to molecules, enzymes, and information, all cells also require **energy.** Energy is needed to drive the chemical reactions involved in the formation of cellular components and to power the many activities that these components carry out. The capacity to obtain, store, and use energy is, in fact, one of the obvious features of most living things (Figure 5-1). Like the flow of information, the flow of energy is a major theme of this text. Energy flow is introduced in this chapter and considered in detail in Chapters 13–15.

Figure 5-1   Energy and Life. The capacity to expend energy is one of the most obvious features of life at both the cellular and organismal levels.

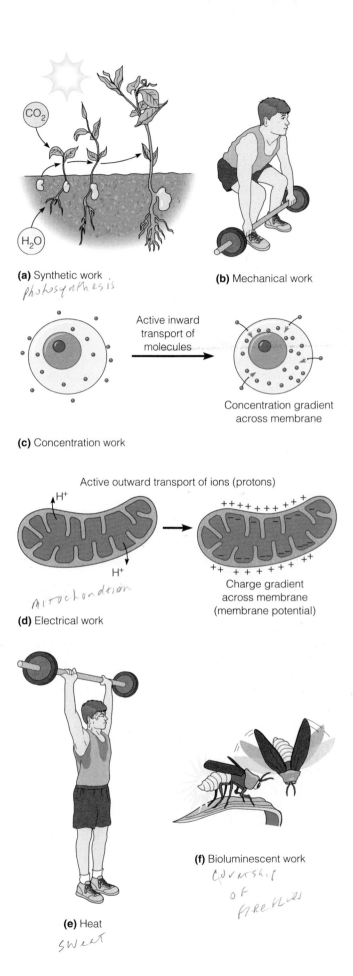

Figure 5-2 Several Kinds of Biological Work. The six major categories of biological work are shown here. **(a)** Synthetic work is illustrated by the process of photosynthesis, **(b)** mechanical work by the contraction of a weight lifter's muscles, and **(c)** concentration work by the uptake of molecules into a cell against a concentration gradient. **(d)** Electrical work is represented by the membrane potential of a mitochondrion (shown being generated by active proton transport), **(e)** heat production is illustrated by the sweat of the weight lifter, and **(f)** bioluminescence is depicted by the courtship of fireflies.

(a) Synthetic work
photosynthesis

(b) Mechanical work

Active inward transport of molecules

Concentration gradient across membrane

(c) Concentration work

Active outward transport of ions (protons)

H^+

H^+

mitochondrion

Charge gradient across membrane (membrane potential)

(d) Electrical work

(e) Heat
sweat

(f) Bioluminescent work
Courtship of fireflies

The Importance of Energy

All living systems require an ongoing supply of energy. Before discussing why cells need energy, it might be useful to consider what we mean by energy. Usually, energy is defined as the capacity to do work. But that turns out to be a somewhat circular definition because work is frequently defined in terms of energy changes. A more useful definition is that *energy is the ability to cause specific changes.* Since life is characterized first and foremost by change, this definition underscores the total dependence of all forms of life on the continuous availability of energy.

Cells Need Energy to Cause Six Different Kinds of Changes

Now that we have defined energy in this way, we recognize that asking about cellular needs for energy really means inquiring into the kinds of changes that cells must effect—that is, the cellular activities that give rise to change. Six categories of change come to mind, which in turn define six kinds of work: synthetic, mechanical, concentration, and electrical work, as well as the generation of heat and light (Figure 5-2).

Synthetic Work: Changes in Chemical Bonds. An important activity of virtually every cell at all times is the work of **biosynthesis**, which results in the formation of new bonds and the generation of new molecules. This activity is especially obvious in a population of growing cells, where it can be shown that additional molecules are being synthesized if the cells are increasing in size or number or both. But synthetic work is required to maintain structures just as surely as it is needed to generate them originally. Most existing structural components of the cell are in a state of constant turnover. The molecules that make up the structure are continuously being degraded and replaced.

In terms of the hierarchy of cellular structure shown in Figure 2-14, almost all of the energy that cells require for biosynthetic work is used to make energy-rich organic molecules from simpler starting materials and to activate such organic molecules for incorporation into macromolecules. As we already know from Chapters 2 and 3, higher levels of structural complexity usually occur by spontaneous self-assembly, without further energy input.

Mechanical Work: Changes in Location or Orientation of a Cell or a Subcellular Structure.

Mechanical work involves a physical change in the position or orientation of a cell or some part of it. An especially good example is the movement of a cell with respect to its environment. This movement requires the presence of some sort of motile appendage such as a flagellum or a cilium. Many prokaryotic cells propel themselves through the environment, as in the case of the flagellated bacterium in Figure 5-3. Sometimes, however, the environment is moved past the cell, as when the ciliated cells that line your trachea beat upward to sweep inhaled particles back to the mouth or nose, thus protecting the lungs. Muscle contraction is another good example of mechanical work, involving not just a single cell but a large number of muscle cells (Figure 5-4). Other examples of mechanical work that occur within the cell include the movement of chromosomes along the spindle fibers during mitosis, the streaming of cytoplasm, and the movement of a ribosome along a strand of messenger RNA.

Concentration Work: Movement of Molecules Across a Membrane Against a Concentration Gradient.

Less conspicuous than either of the previous two categories but every bit as important to the cell is the work of moving molecules or ions against a concentration gradient. The purpose of concentration work is either to accumulate substances within a cell or subcellular compartment, or to remove by-products of cellular activity that cannot be further utilized by the cell and indeed might be toxic if allowed to build up in the cell or compartment. Examples of concentration work include the pumping of sodium and potassium ions across plasma membranes and the light-driven accumulation of protons within the chloroplasts of a plant cell.

Electrical Work: Movement of Ions Across a Membrane Against an Electrochemical Gradient.

Electrical work is often considered a specialized case of concentration work because it also involves movement across membranes. In this case, however, ions are transported and the result is not just a change in concentration but also the establishment of an electrical potential across the membrane. In fact, about two-thirds of your energy consumption when at rest is used to maintain the electrical potential across the membranes of your cells.

Every membrane has some characteristic potential that is generated in this way. An electrochemical gradient of protons across the mitochondrial or chloroplast membrane is essential to the production of ATP in both respiration (Chapter 14) and photosynthesis (Chapter 15). Electrical work is also important in the mechanism whereby impulses are conducted in nerve and muscle cells (Chapter 9). An especially dramatic example of electrical work is found in *Electrophorus electricus*, the electric eel. The electric organ of *Electrophorus* consists of layers of cells called *electroplaxes*, each of which can generate a membrane potential of about 150 millivolts (mV). Because the electric organ contains thousands of such cells arranged in series, the eel can develop potentials of several hundred volts.

Heat: An Increase in Temperature that Is Useful to Warm-Blooded Animals.

It is easy to forget about **heat** because living organisms do not use heat as a form of energy in the same way that a steam engine does. But heat is, in fact, a major use of energy in all *homeotherms* (animals that regulate their body temperature independent of the environment). In fact, as you read these lines, about two-thirds of your metabolic energy is being used just to stay warm. How can two-thirds of our energy be used for heat when we noted earlier that about two-thirds of our energy is used for ion transport? The answer is straightforward: Some of the energy is used to transport ions, and some of it is released as heat. This is the heat that warm-blooded animals use to maintain body temperature near 37°C, where their metabolism is most efficient. The relationship between work and heat energy is demonstrated when you sweat while exercising or shiver when cold. You should now be able to understand the relationship between these seemingly unrelated processes and ATP hydrolysis, which powers muscle contraction.

Bioluminescence: The Production of Light.

To be complete, we must also include the production of light, or **bioluminescence,** as yet another way in which energy is

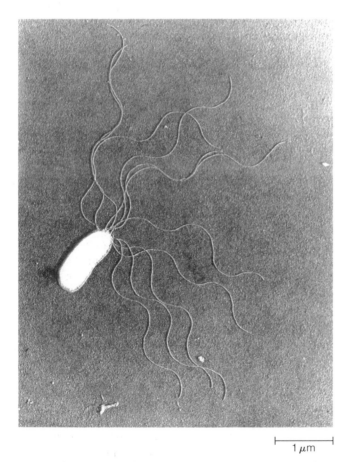

1 μm

Figure 5-3   A Flagellated Bacterium.   The whipping motion of bacterial flagella is an example of mechanical work, providing motility for some bacterial species (TEM).

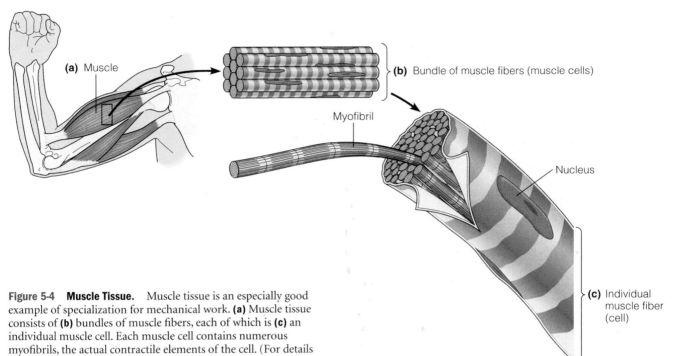

Figure 5-4 **Muscle Tissue.** Muscle tissue is an especially good example of specialization for mechanical work. **(a)** Muscle tissue consists of **(b)** bundles of muscle fibers, each of which is **(c)** an individual muscle cell. Each muscle cell contains numerous myofibrils, the actual contractile elements of the cell. (For details about muscle contraction, see Chapter 23.)

used by cells. The light produced by bioluminescent organisms is generated by the reaction of ATP with specific luminescent compounds and is usually pale blue in color. This is a much more specialized kind of energy use than the other five categories, and for present purposes we can leave it to the fireflies, luminous toadstools, dinoflagellates, deep-sea fish, and other creatures that live in its strange, cold light.

Most Organisms Obtain Energy Either from Sunlight or from Organic Food Molecules

Nearly all life on Earth is sustained, directly or indirectly, by the sunlight that continuously floods our planet with energy. This *solar radiation* is readily quantified: Solar energy arrives at the upper surface of the Earth's atmosphere at the rate of 1.94 cal/min per square centimeter of cross-sectional area, a value known as the *solar energy constant.* Not all organisms can obtain energy from sunlight directly, of course. In fact, organisms (and therefore cells) can be classified into one of two groups on the basis of their energy sources.

The first group consists of organisms capable of capturing light energy by means of photosynthetic pigment systems, then storing the energy in the form of organic molecules like glucose. Such organisms are called **phototrophs** (literally, "light-feeders") and include plants, algae, cyanobacteria, and certain groups of bacteria that are capable of photosynthesis.

The chemical bonds in organic molecules provide an energy source that can be used by a second group of organisms, called **chemotrophs** (literally, "chemical-feeders") because they require the intake of chemical compounds

such as carbohydrates, fats, and proteins. All animals, protists, and fungi and most bacteria are chemotrophs. The chemical energy is released by two methods. In the first, the chemicals are partially broken down, releasing some of the stored energy. The general term for this process is *fermentation.* The more specific term *glycolysis* is used to describe the breakdown of the glucose molecule during the fermentation process. The second method involves oxidation, in which electrons are transferred from the chemical bonds to molecular oxygen or other electron acceptor. This process is called *cellular respiration.* When oxygen serves as the electron acceptor, the process is called *aerobic respiration.* During respiration, essentially all the energy stored in chemical bonds is released.

A point that is often not appreciated about phototrophs is that although they can utilize solar energy when it is available, they are also capable of functioning as chemotrophs and, in fact, do so whenever they are not illuminated. Most higher plants are really a mixture of phototrophic and chemotrophic cells. A plant root cell, for example, though part of an obviously phototrophic organism, is in most cases incapable of carrying out photosynthesis and is every bit as chemotrophic as an animal cell.

Phototrophs can function as chemotrophs

Energy Flows Through the Biosphere Continuously

So far we have seen that both chemotrophs and phototrophs depend on their environment for the energy they need, but differ in the forms of energy they can use. Chemotrophs require organic molecules, whereas phototrophs trap solar radiation and transduce it into chemical bond energies.

The flow of energy through the biosphere is depicted in Figure 5-5. Solar energy is trapped by phototrophs and used to convert carbon dioxide and water into more complex (and more reduced) cellular materials in the process of photosynthesis. As we will see in Chapter 15, the immediate products of photosynthetic carbon fixation are sugars, but in a sense we can consider the entire phototrophic organism to be the "product" of photosynthesis because every carbon atom in every molecule of that organism is derived from carbon dioxide that is fixed into organic form by the photosynthetic process.

Chemotrophs, on the other hand, are unable to use solar energy directly and depend on the chemical energy of oxidizable molecules. The energy needs of the chemotrophs can be met either *anaerobically* (in the absence of oxygen) by a variety of fermentation processes or *aerobically* (in the presence of oxygen) by the complete oxidation of chemical compounds in the process of aerobic respiration. Chemotrophs therefore depend completely on energy that has been packaged into fermentable or oxidizable food molecules by phototrophs. A world composed only of chemotrophs would last only so long as food supplies held out, for even though we live on a planet that is flooded each day with solar energy, it is in a form that we cannot use to meet our energy needs.

Both phototrophs and chemotrophs use energy to carry out work—that is, to effect the various kinds of changes we have already catalogued. In the process, two kinds of losses occur. One of the principles of energy conversion is that no chemical or physical process occurs with 100% efficiency; some energy is lost as heat. In fact, most processes that involve the conversion of energy from one form to another actually dissipate more energy as heat than they succeed in converting to the desired form. An incandescent light bulb, for example, generates more heat than light.

As we will see in Part Three, biological processes are remarkably efficient in energy conversion. *Heat losses* are nonetheless inevitable in every biological energy transaction. Sometimes the heat that is liberated during cellular processes is put to good use. As discussed earlier, warm-blooded animals use heat to maintain the body temperature at some constant level, usually well above ambient. Some plants use metabolically generated heat to melt overlying snow or to attract pollinators (Figure 5-6). But in general, the heat is simply dissipated into the environment and lost.

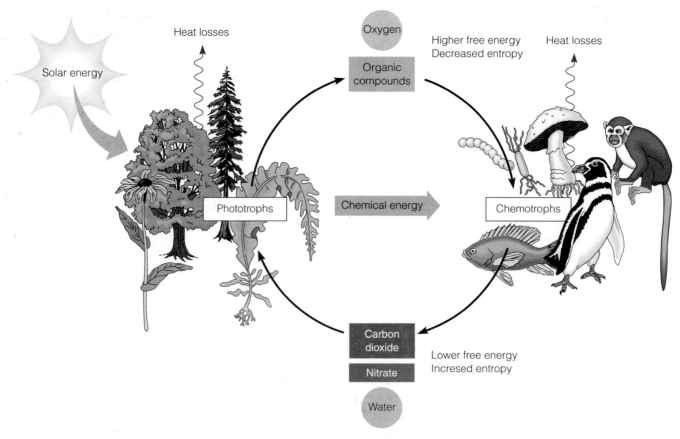

Figure 5-5 The Flow of Energy Through the Biosphere. Most of the energy in the biosphere originates in the sun and contributes eventually to the ever-increasing entropy of the universe. Accompanying the unidirectional flow of energy from phototrophs to chemotrophs is a cyclic flow of matter between the two groups of organisms. (As we will learn in Chapter 15, phototrophs are further divided into two groups, depending on their source of carbon. *Photoautotrophs* obtain their carbon from CO2, whereas *photoheterotrophs* depend on organic sources of reduced carbon. All of the phototrophs shown in this diagram are photoautotrophs; most photoheterotrophs are bacteria.)

Figure 5-6 Voodoo Lily, a Plant That Depends on Metabolically Generated Heat to Attract Pollinators. The voodoo lily (*Sauromatum guttatum*) warms certain parts of its flowers. The plant is pollinated by flies, which apparently mistake the flowers for dead meat. The flower emits odors that help attract the flies, and heating helps disperse the smelly gases.

Even more fundamental is the *increase in entropy* that accompanies cellular activities. We will get to that in more detail shortly; here, we can simply note that every process or reaction that occurs anywhere in the universe always does so in such a way that the total entropy, or disorder, in the universe is increased. This change in entropy occurs at the expense of energy that might otherwise have been available to do useful work and is therefore an inevitable "sink" into which energy is lost. Just as the ultimate source of nearly all energy in the biosphere is the sun, the ultimate fate of all energy in the biosphere is to become randomized in the universe as increased entropy.

Viewed on a cosmic scale, there is a continuous, massive, and unidirectional flow of energy from its source in the nuclear fusion reactions of the sun to its eventual sink, the entropy of the universe. We here in the biosphere are the transient custodians of an almost infinitesimally small portion of that energy, but it is precisely that small but critical fraction of energy and its flow through living systems that is of concern to us. The flow begins with green plants, which use light energy to drive electrons energetically "uphill" into new chemical bonds. This energy is then released by both plants and animals in "downhill" fermentative or oxidative reactions. This flux of energy through living matter—from the sun, to phototrophs, to chemotrophs, to heat—drives the molecular machinery of all life processes.

The Flow of Energy Through the Biosphere Is Accompanied by a Flow of Matter

Energy enters the biosphere unaccompanied by matter (that is, as photons of light) and leaves the biosphere similarly unaccompanied (as heat losses and increases in entropy). While it is passing through the biosphere, however, energy exists primarily in the form of chemical bond energies of oxidizable organic molecules in cells and organisms. As a result, the flow of energy in the biosphere is coupled to a correspondingly immense flow of matter.

Whereas energy flows unidirectionally from the sun through phototrophs to chemotrophs, matter flows in cyclic fashion between the two groups of organisms (see Figure 5-5). During respiration, aerobic chemotrophs take in organic nutrients from their surroundings, usually by ingesting phototrophs or other chemotrophs that have in turn eaten phototrophs. These nutrients are oxidized to carbon dioxide and water, low-energy molecules that are returned to the environment. Those molecules then become the raw materials that phototrophic organisms use to make new organic molecules photosynthetically, returning oxygen to the environment in the process.

In addition, there is an accompanying cycle of nitrogen. Phototrophs obtain nitrogen from the environment in inorganic form (often as nitrate from the soil, in some cases as N_2 from the atmosphere), convert it into ammonia, and use it in the synthesis of amino acids, proteins, nucleotides, and nucleic acids. Eventually, these molecules, like other components of phototrophic cells, are consumed by chemotrophs. The nitrogen is then converted back into ammonia and eventually to nitrate—by soil microorganisms, in the latter case.

Carbon dioxide, oxygen, nitrogen, and water thus cycle continuously between the phototrophic and chemotrophic worlds, always entering the chemotrophic sphere as energy-rich compounds and leaving again in an energy-poor form. The two great groups of organisms can therefore be thought of as living in symbiotic relationship with each other, with a cyclic flow of matter and a unidirectional flow of energy as components of that symbiosis.

When we deal with the overall macroscopic flux of energy and matter through living organisms, we find cellular biology interfacing with ecology. Ecologists are very much concerned with cycles of energy and nutrients, with the roles of various species in these cycles, and with environmental factors that affect the flow. At the cellular level, our ultimate concern is how the flux of energy and matter we have been considering on a macroscopic scale can be expressed and explained on a molecular scale in terms of energy transactions and the chemical processes that occur within cells. We therefore leave the macroscopic cycles to the ecologist and turn our attention to the reactions that occur within individual cells of bacteria, plants, and animals to account for those cycles. First, however, we must acquaint ourselves with the physical principles underlying energy transactions, and for that we turn to the topic of bioenergetics.

Bioenergetics

The principles that govern energy flow are incorporated in an area of science that the physical chemist calls **thermodynamics.** Although the prefix *thermo-* suggests that the term is limited to heat (and that is indeed its historical origin), thermodynamics also takes into account other forms of energy and processes that convert energy from one form to another. Specifically, thermodynamics concerns the laws governing the energy transactions that inevitably accompany most physical processes and all chemical reactions. **Bioenergetics,** in turn, can be thought of as *applied thermodynamics*—that is, it concerns the application of thermodynamic principles to reactions and processes in the biological world.

To Understand Energy Flow, We Need to Understand Systems, Heat, and Work

As we have already seen, it is useful to define energy not simply as the ability to do work but specifically as the ability to cause change. Without energy, all processes would be at a standstill, including those that we associate with living cells.

Energy exists in a variety of forms, many of them of interest to biologists. Think, for example, of the energy represented by a ray of sunlight, a teaspoon of sugar, a moving flagellum, an excited electron, or the concentration of ions or small molecules within a cell or an organelle. These phenomena are diverse, but they are all governed by certain basic principles of energetics.

Energy is distributed throughout the universe, and for some purposes it is necessary to consider the total energy of the universe, at least in a theoretical way. Usually, however, we are interested not in the whole universe but only in a small portion of it. We might, for example, be concerned with a reaction or process occurring in a beaker of chemicals, in a cell, or in a block of metal. By convention, the restricted portion of the universe that one wishes to consider at the moment is called the **system,** and all the rest of the universe is referred to as the **surroundings.** Sometimes the system has a natural boundary, such as a glass beaker or a cell membrane. In other cases, the boundary between the system and its surroundings is a hypothetical one used only for convenience of discussion, such as the imaginary boundary around one mole of glucose molecules in a solution.

Systems can be either open or closed, depending on whether or not they can exchange energy with their surroundings (Figure 5-7). A *closed system* is sealed from its environment and can neither take in nor release energy in any form. An *open system,* on the other hand, can have energy added to it or removed from it. As we will see later, the levels of organization that biological systems routinely display are possible only because cells and organisms are open systems, capable of both the uptake and the release of energy. Specifically, biological systems require a constant, large-scale influx of energy from their surroundings both to attain and to maintain the levels of complexity characteristic of them. That is why plants need sunlight and you need food.

Whenever we talk about a system, we have to be careful to specify the state of the system. A system is said to be in a specific **state** if each of its variable properties (such as temperature, pressure, and volume) is held constant at a specified value. In such a situation, the total energy content of the system, while not directly measurable, has some unique value. If such a system then changes from one state to another as a result of some interaction between the system and its surroundings, the change in its total energy is determined uniquely by the initial and final states of the

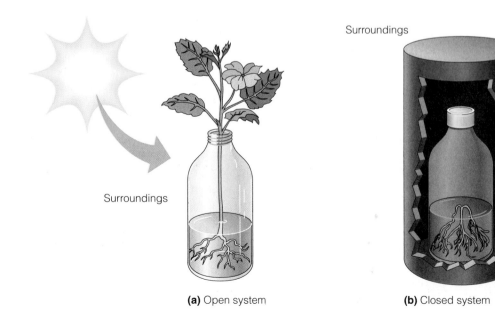

(a) Open system

(b) Closed system

Figure 5-7 Open and Closed Systems. A system is that portion of the universe under consideration. The rest of the universe is called the surroundings of the system. **(a)** An open system can exchange energy with its surroundings, whereas **(b)** a closed system cannot. The open system can use incoming energy to increase its orderliness, thereby decreasing its entropy. The closed system tends toward equilibrium and increases its entropy. All living organisms are open systems, exchanging energy freely with their surroundings.

system and is not affected at all by the mechanism by which the change occurs or the intermediate states through which the system may pass. This is a very useful property because it allows energy changes to be determined from a knowledge of the initial and final states only.

The problem of keeping track of system variables and their effect on energy changes can be simplified if one or more of the variables are held constant. Fortunately, this is the case with most biological reactions, because they usually occur in dilute solutions within cells that are at approximately the same temperature and pressure during the entire course of the reaction. These environmental conditions, as well as the cell volume, are generally slow to change compared with the speed of biological reactions. This means that three of the most important system variables that physical chemists usually concern themselves with—temperature, pressure, and volume—are essentially constant for most biological reactions.

The exchange of energy between a system and its surroundings occurs in two ways: as heat and as work. Heat is energy transfer from one place to another as a result of a temperature difference between the two places. Spontaneous transfer always occurs from the hotter place to the colder place. Heat is an exceedingly useful form of energy for many machines and other devices designed to accomplish mechanical work. However, it has only limited biological utility because most biological systems operate under conditions of either fixed or only minimally variable temperature. Such *isothermal* systems lack the temperature gradients required to convert heat into other forms of energy. As a result, heat is not a useful source of energy for cells—although it can be used for such purposes as maintaining body temperature or attracting pollinators, as we noted earlier.

In biological systems, **work** is the use of energy to drive any process other than heat flow. For example, work is performed when the muscles in your arm expend chemical energy to lift this book, when a corn leaf uses light energy to synthesize sugar, or when an electrical eel draws on the ion concentration gradients of its electroplax tissue to deliver a shock. It is the amount of useful energy available to do cellular work that we will be primarily interested in when we begin calculating energy changes associated with specific reactions that cells carry out.

To quantify energy changes during chemical reactions or physical processes, we need units in which energy can be expressed. In biological chemistry, energy changes are usually expressed in terms of the **calorie (cal),** which is defined as the amount of energy required to warm 1 gram of water 1 degree centigrade (specifically, from 14.5°C to 15.5°C) at a pressure of 1 atmosphere. (Again, note that the unit of energy measurement, like the very term *thermodynamics*, is based on heat but is applied generally to all forms of energy.) An alternative energy unit, the **joule (J),** is preferred by physicists and is used in some biochemistry texts. Conversion is easy: 1 cal = 4.184 J, or 1 J = 0.239 cal.

Energy changes are often measured on a per-mole basis, and the most common form in which we will

encounter energy units in biological chemistry will be as calories (or sometimes kilocalories) per mole (cal/mol or kcal/mol). (Be careful to distinguish between the *calorie* as defined here and the *nutritional Calorie* that is often used to express the energy content of foods. The nutritional Calorie is represented with a capital C and is really a *kilocalorie* as defined here.)

The First Law of Thermodynamics Tells Us That Energy Is Conserved

Much of what we understand about the principles governing energy flow can be summarized by the three laws of thermodynamics. Of these, only the first and second laws are of particular relevance to the cell biologist. The **first law of thermodynamics** is called the *law of conservation of energy.* Simply put, the first law states that *in every physical or chemical change, the total amount of energy in the universe remains constant, although the form of the energy may change.* Or, in other words, *energy can be converted from one form to another but can never be created or destroyed.* (If you are familiar with the conversion of mass to energy that occurs in nuclear reactions, you will recognize that a more accurate statement would take both mass and energy into account. For purposes of biological chemistry, however, the law is adequate as stated.)

Applied to the universe as a whole or to a closed system, the first law means that the total amount of energy present in all forms must be the same before and after any process or reaction occurs. Applied to an open system such as a cell, the first law says that during the course of any reaction or process, the total amount of energy that leaves the system must be exactly equal to the energy that enters the system minus any energy that remains behind and is therefore stored within the system:

$$\text{energy out} = \text{energy in} - \text{energy stored} \qquad (5\text{-}1)$$

Or, by simple rearrangement,

$$\text{energy stored} = \text{energy in} - \text{energy out} \qquad (5\text{-}2)$$

The total energy stored within a system is called the **internal energy** of the system, represented by the symbol E. We are not usually concerned with the actual value of E for a system because that value cannot be measured directly. However, it is possible to measure the *change in internal energy*, ΔE, that occurs during a given process. ΔE is the difference in internal energy of the system before the process (E_1) and after the process (E_2):

$$\Delta E = E_2 - E_1 \qquad (5\text{-}3)$$

Equation 5-3 is valid for all physical and chemical processes under any conditions. For a chemical reaction, we can write

$$\Delta E = E_{\text{products}} - E_{\text{reactants}} \qquad (5\text{-}4)$$

In the case of biological reactions and processes, we are usually more interested in the change in **enthalpy,** or *heat content*. Enthalpy is represented by the symbol H (for heat) and is related to the internal energy E by a term that combines both pressure (P) and volume (V):

$$H = E + PV \qquad (5\text{-}5)$$

Unlike many chemical reactions, biological reactions generally proceed with little or no change in either pressure (1 atmosphere, usually) or volume. So for biological reactions, both ΔP and ΔV are usually zero (or at least negligible), and we can write:

$$\Delta H = \Delta E + \Delta(PV) \cong \Delta E \qquad (5\text{-}6)$$

Thus, biologists routinely determine changes in heat content for reactions of interest, confident that the values are valid estimates of ΔE.

The enthalpy change that accompanies a specific reaction is simply the difference in the heat content between the reactants and the products of the reaction:

$$\Delta H = H_{products} - H_{reactants} \qquad (5\text{-}7)$$

The ΔH value for a specific reaction or process will be either negative or positive (Figure 5-8). If the heat content of the products is less than that of the reactants, ΔH will be negative and the reaction is said to be *exothermic* (Figure 5-8a). If the heat content of the products is greater than that of the reactants, ΔH will be positive and the reaction is *endothermic* (Figure 5-8b). Thus, the ΔH value for a reaction is simply a measure of the heat that is either liberated or taken up by that reaction (for negative and positive values of ΔH, respectively) as it occurs under conditions of constant temperature and pressure.

The Second Law of Thermodynamics Tells Us That Reactions Have Directionality

So far, all that thermodynamics has been able to tell us is that energy is conserved whenever a process or reaction occurs—that all of the energy going into a system must either be stored within the system or released again to the surroundings. We have seen the usefulness of ΔH as a measure of how much the total enthalpy of a system would change if a given process were to occur, but we have no way as yet of predicting whether and to what extent the process will in fact occur under the prevailing conditions.

We have, at least in some cases, an intuitive feeling that some reactions or processes are possible, whereas others are not. We are somehow very sure that if we set a match to a sheet of paper, it will burn. The oxidation of cellulose to carbon dioxide and water is, in other words, a possible reaction. Or, to use more precise terminology, it is a *thermodynamically spontaneous* reaction. In the context of thermodynamics, the term *spontaneous* has a specific, restricted meaning that is different from its commonplace usage. **Thermodynamic spontaneity** is a measure of

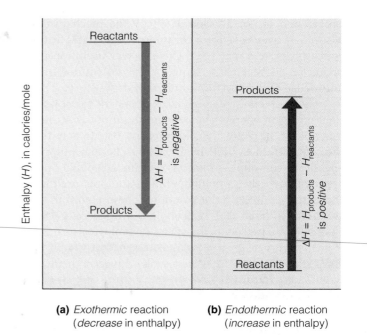

(a) *Exothermic* reaction (*decrease* in enthalpy) **(b)** *Endothermic* reaction (*increase* in enthalpy)

Figure 5-8 Changes in Enthalpy for Chemical Reactions. The change in enthalpy, or heat content, that accompanies a chemical reaction is the difference in the heat content between the reactions and the products. **(a)** If the products are lower in enthalpy than the reactants, the reaction involves a decrease in enthalpy ($\Delta H < 0$) and is therefore exothermic. **(b)** If the products are higher in enthalpy than the reactants, the reaction involves an increase in enthalpy ($\Delta H > 0$) and is therefore endothermic.

whether a reaction or process *can go* but says nothing about whether it *will go*. Our sheet of paper illustrates this point well. The oxidation of cellulose is clearly a *possible* reaction, but we know that it does not "just happen"; it needs some impetus—a match, in this particular case.

Not only are we convinced that a sheet of paper will burn if ignited but we also know, if only intuitively, that there is *directionality* to the property. We are, in other words, equally convinced that the reverse reaction will not occur—that if we were to stand around clutching the charred remains, the paper would not spontaneously reassemble in our hands. We have, in other words, a feeling for both the possibility and the directionality of cellulose oxidation.

You can probably think of other processes for which you can make such thermodynamic predictions with equal confidence. We know, for example, that drops of dye diffuse in water, that ice cubes melt at room temperature, and that sugar dissolves in water, and we can therefore label these as thermodynamically spontaneous events. But if we ask why we recognize them as such, the answer has to do with repeated prior experience. We have seen paper burn, ice cubes melt, and sugar dissolve often enough to know intuitively that these are processes that really occur, and with such predictability that we can label them as spontaneous, provided only that we know the conditions.

However, when we move from the world of familiar physical processes to the realm of chemical reactions in cells, we quickly find that we cannot depend on prior experience to guide us in our predictions. Consider, for example, the conversion of glucose-6-phosphate into fructose-6-phosphate:

$$\text{glucose-6-phosphate} \rightleftharpoons \text{fructose-6-phosphate} \quad \textbf{(5-8)}$$

Glucose is a six-carbon aldosugar and fructose is its keto equivalent (see Figure 3-21). Both can form a phosphoester bond between a phosphoric acid (phosphate) molecule and the hydroxyl group on carbon 6 of the sugar, giving rise to the phosphorylated compounds. Reaction 5-8 therefore involves the interconversion of a phosphorylated aldosugar and the corresponding phosphorylated ketosugar, as shown in Figure 5-9.

This particular interconversion is a significant reaction in all cells. It is, in fact, the second step in an important and universal reaction sequence called the *glycolytic pathway*. In addition to illustrating an important thermodynamic principle, therefore, reaction 5-8 introduces a bit of cellular chemistry that will come in handy later on. For now, however, just focus on the reaction from a thermodynamic point of view, and ask yourself what predictions you can make about the likelihood of glucose-6-phosphate being converted into fructose-6-phosphate. You will probably be at a loss to make any predictions at all. We know what will happen with burning paper and melting ice, but we lack the familiarity and prior experience with phosphorylated sugars even to make an intelligent guess. Clearly, what we need is a reliable means of determining whether a given physical or chemical change can occur under specific conditions without having to rely on prior experience, familiarity, or intuition.

Thermodynamics provides us with exactly such a measure of spontaneity in the **second law of thermodynamics,** or the *law of thermodynamic spontaneity.* As we will see shortly, the second law can be expressed in several different ways. Simply put, however, it tells us that *in every physical or chemical change, the universe always tends toward greater disorder or randomness.* The second law is useful for our purposes because it allows us to predict in what direction a reaction will proceed under specified conditions, how much energy the reaction will release as it proceeds, and how the energetics of the reaction will be affected by specific changes in the conditions.

An important point to note is that *no process or reaction disobeys the second law of thermodynamics.* Some processes may *seem* to do so because they result in more, rather than less, order. Think, for example, of the increase in order when a house is built, your room is cleaned, or a human being develops from a single egg cell. In each of these cases, however, the increase in order is limited to a specific system (the house, the room, or the embryo) and is possible only because it is an open system so that energy can be added from the outside, whether by means of an electric saw, your arm muscles, or nutrients supplied by the mother. The input of energy means, in turn, that greater disorder is occurring elsewhere in the universe—as water flows through the turbines of a hydroelectric plant to power the saw, for example, or as you consume and digest a bag of potato chips to power your arm muscles.

Entropy and Free Energy Are Two Alternative Means of Assessing Thermodynamic Spontaneity

Thermodynamic spontaneity—whether a reaction *can* go—can be measured by changes in either of two parameters: *entropy* or *free energy.* These concepts are abstract and can be somewhat difficult to understand. We will therefore limit our discussion here to their use in determining what kinds of changes can occur in biological systems. For further help, see Box 5A for an essay that uses jumping beans to introduce the concepts of internal energy, entropy, and free energy.

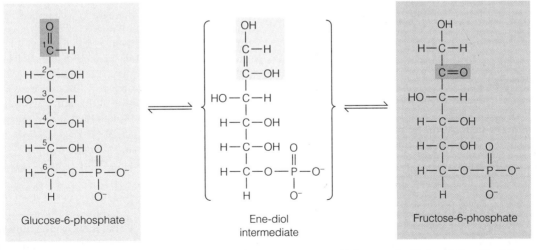

Figure 5-9   The Interconversion of Glucose-6-Phosphate and Fructose-6-Phosphate.   This reaction involves the interconversion of the phosphorylated forms of an aldosugar (glucose) and a ketosugar (fructose). The reaction is catalyzed by an enzyme called phosphoglucoisomerase and is readily reversible. The reaction proceeds via an enzyme-bound intermediate called an ene-diol because it has a carbon-carbon double bond ("ene") with two alcohol groups ("diol") attached. This reaction is part of the glycolytic pathway, as we will see in Chapter 13.

Further Insights   JUMPING BEANS AND FREE ENERGY

If you are finding the concepts of free energy, entropy, and equilibrium constants difficult to grasp, perhaps a simple analogy might help.* For this we will need an imaginary supply of jumping beans, which are really seeds of certain Mexican shrubs, with larvae of the moth *Laspeyresia saltitans* inside. Whenever the larvae inside the seed wiggle about, the seeds wiggle, too. The "jumping" action probably serves to get the larvae out of direct sunlight, which could heat them to lethal temperatures.

The Jumping Reaction

For purposes of illustration, imagine that we have some high-powered jumping beans in two chambers separated by a low partition, as shown below. Notice that the chambers have the same floor area and are at the same level, although we will want to vary both of these properties shortly. As soon as we place a handful of jumping beans in chamber 1, they begin jumping about randomly. Although most of the beans jump only to a modest height most of the time, occasionally one of them, in a burst of ambition, gives a more energetic leap, surmounting the barrier and falling into chamber 2. We can write this as the *jumping reaction*:

$$\text{Beans in chamber 1} \rightleftharpoons \text{Beans in chamber 2}$$

We will imagine this to be a completely random event, happening at irregular, infrequent intervals. Occasionally, one of the beans that has reached chamber 2 will happen to jump back into chamber 1, which is the *back reaction*. At first, of course, there will be more beans jumping from chamber 1 to chamber 2 because there are more beans in chamber 1, but things will eventually even out so that, on average, there will be the same number of beans in both compartments. The system will then be at *equilibrium*. Beans will still continue to jump between the two chambers, but the numbers jumping in both directions will be equal.

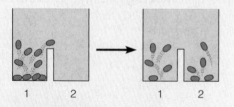

The Equilibrium Constant

Once our system is at equilibrium, we can count up the number of beans in each chamber and express the results as the ratio of the number of beans in chamber 2 to the number in chamber 1. This is simply the *equilibrium constant* K_{eq} for the jumping reaction:

$$K_{eq} = \frac{\text{number of beans in chamber 2 at equilibrium}}{\text{number of beans in chamber 1 at equilibrium}}$$

We are indebted to Princeton University Press for permission to use this analogy, which was first developed by Harold F. Blum in the book Time's Arrow and Evolution *(3rd ed, 1968), pp.17–26.*

For the specific case shown in the previous column, the numbers of beans in the two chambers are equal at equilibrium, so the equilibrium constant for the jumping reaction under these conditions is 1.0.

Enthalpy Change (ΔH)

Now suppose that the level of chamber 1 is somewhat higher than that of chamber 2, as shown in the next diagram. Jumping beans placed in chamber 1 will again tend to distribute themselves between chambers 1 and 2, but this time a higher jump is required to get from 2 to 1 than from 1 to 2, so the latter will occur more frequently. As a result, there will be more beans in chamber 2 than in chamber 1 at equilibrium, and the equilibrium constant will therefore be greater than 1.

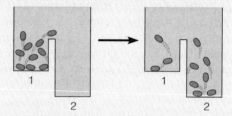

The relative heights of the two chambers can be thought of as measures of the *enthalpy*, or *heat content* (H), of the chambers, such that chamber 1 has a higher H value than chamber 2, and the difference between them is represented by ΔH. Since it is a "downhill" jump from chamber 1 to chamber 2, it makes sense that ΔH has a negative value for the jumping reaction from chamber 1 to chamber 2. Similarly, it seems reasonable that ΔH for the reverse reaction should have a positive value because that jump is "uphill."

Entropy Change (ΔS)

So far, it might seem as if the only thing that can affect the equilibrium distribution of beans between the two chambers is the difference in enthalpy, ΔH. But that is only because we have kept the floor area of the two chambers constant. Imagine instead the situation shown below, where the two chambers are again at the same height, but chamber 2 now has a greater floor area than chamber 1. The probability of a bean finding itself in chamber 2 is therefore correspondingly greater, so there will be more beans in chamber 2 than in chamber 1 at equilibrium, and the equilibrium constant will be greater than 1 in this case also. This means that the equilibrium position of the jumping reaction has been shifted to the right, even though there is no change in enthalpy.

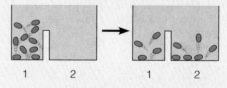

The floor area of the chambers can be thought of as a measure of the *entropy*, or randomness, of the system, S, and the *difference* between the two chambers can be represented by ΔS. Since chamber 2 has a greater floor area than chamber 1, the entropy change is positive for the jumping reaction as it proceeds from left to right under these conditions. Note that for ΔH, negative values are associated with favorable reactions, while for ΔS, favorable reactions are indicated by positive values.

Free Energy Change (ΔG)

So far, we have encountered two different factors that affect the distribution of beans: the difference in levels of the two chambers (ΔH) and the difference in floor area (ΔS). Moreover, it should be clear that neither of these factors by itself is an adequate indicator of how the beans will be distributed at equilibrium because a favorable (negative) ΔH could be more than offset by an unfavorable (negative) ΔS, and a favorable (positive) ΔS could be more than offset by an unfavorable (positive) ΔH. You should, in fact, be able to design chamber conditions that illustrate both of these situations, as well as situations in which ΔH and ΔS tend to reinforce rather than counteract each other.

Clearly, what we need is a way of summing these two effects algebraically to see what the net tendency will be. The new measure we come up with is called the *free energy change, ΔG,* which turns out to be the most important thermodynamic parameter for our purposes. ΔG is defined so that *negative* values correspond to favorable (i.e., thermodynamically spontaneous) reactions and *positive* values represent unfavorable reactions. Thus, ΔG should have the *same* sign as ΔH (since a negative ΔH is also favorable) but the *opposite* sign from ΔS (since for ΔS a positive value is favorable). In terms of real-life thermodynamics, the expression for ΔG in terms of ΔH and ΔS is

$$\Delta G = \Delta H - T\,\Delta S$$

(Notice that the temperature dependence of ΔS is the only feature of this relationship that cannot be readily explained by our model, unless we assume that the effect of changes in room size is somehow greater at higher temperatures.)

ΔG and the Capacity to Do Work

You should be able to appreciate the difficulty of suggesting a physical equivalent for ΔG because it represents an algebraic sum of entropy and energy changes, which may either reinforce or partially offset each other. But as long as ΔG is negative, beans will continue to jump from chamber 1 to chamber 2, whether driven primarily by changes in entropy, internal energy, or both. This means that if some sort of bean-powered "bean wheel" is placed between the two chambers as shown below,

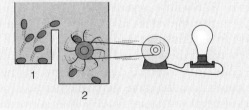

the movement of beans from one chamber to the other can be harnessed to do work until equilibrium is reached, at which point no further work is possible. Furthermore, the greater the difference in free energy between the two chambers (that is, the more highly negative ΔG is), the more work the system can do.

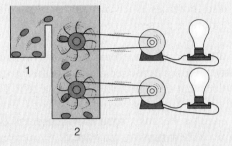

Thus, ΔG is first and foremost a measure of the capacity of a system to do work under specified conditions. You might, in fact, want to think of ΔG as free energy in the sense of *energy that is free or available to do useful work.* Moreover, if we contrive to keep ΔG negative by continuously adding beans to chamber 1 and removing them from chamber 2, we have a dynamic *steady state,* a condition that effectively harnesses the inexorable drive to achieve equilibrium. Work can then be performed continuously by beans that are forever jumping toward equilibrium but that never actually reach it.

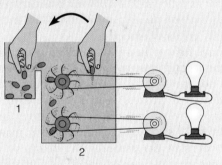

Looking Ahead

To anticipate the transition from the thermodynamics of this chapter to the kinetics of the next, begin thinking about the *rate* at which beans actually proceed from chamber 1 to chamber 2. Clearly, ΔG measures how much energy will be released if beans do jump, but it says nothing at all about the rate. That would appear to depend critically on how high the barrier between the two chambers is. Label this the *activation energy barrier,* and then contemplate means by which you might get the beans to move over the barrier more rapidly. One approach might be to heat the chambers; this would be effective because the larvae inside the seeds wiggle more vigorously if they are warmed. Cells, on the other hand, have a far more effective and specific means of speeding up reactions: They lower the activation barrier by using catalysts called *enzymes,* which we will meet in the next chapter.

Entropy. Although we cannot experience **entropy** directly, we can get some feel for it by considering it to be a measure of *randomness* or *disorder*. Entropy is represented by the symbol **S**. For any system, the *change in entropy, ΔS,* represents a change in the degree of randomness or disorder of the components of the system. For example, the combustion of paper involves an increase in entropy because the carbon, oxygen, and hydrogen atoms of cellulose are much more randomly distributed in space once they are converted to carbon dioxide and water. Entropy also increases as ice melts or as a volatile solvent such as gasoline is allowed to evaporate.

Entropy Change as a Measure of Thermodynamic Spontaneity.

How can the second law of thermodynamics help predict what changes will occur in a cell? There is a very important link between spontaneous events and entropy changes because, whenever a process occurs in nature, the randomness or disorder of the universe (that is, the entropy of the universe) invariably increases. This is one of two alternative ways to state the second law of thermodynamics. According to this formulation, *all processes or reactions that occur spontaneously result in an increase in the total entropy of the universe.* Or, in other words, *the value of $\Delta S_{universe}$ is positive for every real process or reaction.* Processes that would lead to a decrease in the entropy of the universe simply do not occur. Remembering that entropy is a measure of randomness, we can phrase the second law as follows: *The universe becomes more random with every reaction that occurs; reactions that would make it less random are never observed.*

We have to keep in mind, however, that this formulation of the second law pertains to the universe as a whole and may not apply to the specific system under consideration. Every real process, without exception, must be accompanied by an increase in the entropy of the universe, but for a given system the entropy may increase, decrease, or stay the same as the result of a specific process. For example, the oxidation of glucose to carbon dioxide and water (see reaction 5-11 on p. 119) is spontaneous under standard conditions (25°C, a pressure of 1 atm, and pH 7.0) and is accompanied by an *increase* in the system entropy (ΔS_{system} = +43.6 cal/mol-degree). On the other hand, the freezing of water at −0.1°C is also a spontaneous event, yet it involves a *decrease* in the system entropy (ΔS_{system} = −0.5 cal/mol-degree). This makes sense when you consider the greater ordering of water molecules in ice crystals. Thus, while the change in entropy of the universe is a valid measure of the spontaneity of a process, the change in entropy of the system is not.

To understand how the entropy of the universe can increase during a process while the entropy of the system decreases, we need only realize that the decrease in entropy of the system can be accompanied by an equal or even greater increase in the entropy of the surroundings. On the basis of the second law, such local increases in order (decreases in entropy) must be offset by an even

greater decrease in the order (increase in entropy) of the surroundings. This is, in fact, the situation that normally prevails within living cells, which by their growth and reproduction produce striking local increases in order.

This means, however, that the second law, stated in terms of entropy of the universe, is of limited value in predicting the spontaneity of biological processes, for it would require keeping track of changes that occur not only within the system but also in its surroundings. Far more convenient would be a parameter that would enable prediction of the spontaneity of reactions from a consideration of the system alone.

Free Energy. As you might guess, a measure of spontaneity for the system alone does in fact exist. It is called **free energy** and is represented by the symbol **G** after Willard Gibbs, who first developed the concept. Because of its predictive value and its ease of calculation, the free energy function is one of the most useful thermodynamic concepts in biology. One could even make the case that our entire discussion of thermodynamics so far has really been a way of getting us to free energy, because it is here that the usefulness of thermodynamics for cell biologists becomes apparent.

Like most other thermodynamic functions, free energy is defined only in terms of mathematical relationships. But for biological systems at constant pressure, volume, and temperature, the **free energy change, ΔG,** is related to the changes in enthalpy and entropy as follows:

$$\Delta H = \Delta G + T \Delta S \qquad (5\text{-}9)$$

or

$$\Delta G = \Delta H - T \Delta S \qquad (5\text{-}10)$$

where ΔH is the change in enthalpy, ΔG the change in free energy, ΔS the change in entropy, and T the temperature of the system in degrees Kelvin (K = °C + 273).

Notice that ΔG is the algebraic sum of two terms, ΔH and $-T\Delta S$. Like ΔH, ΔS for a specific reaction or process will be either positive (increase in entropy) or negative (decrease in entropy). Because of the minus sign, the term $-T\Delta S$ will be negative if entropy increases or positive if entropy decreases. If the entropy of the products is less than that of the reactants, ΔS will be negative and the term $-T\Delta S$ will be positive. If the entropy of the products is greater than that of the reactants, ΔS will be positive and the term $-T\Delta S$ will be negative.

Given that the values for ΔH and $-T\Delta S$ can be either positive or negative, the value of ΔG for a given reaction will depend on the signs and numerical values of the ΔH and $-T\Delta S$ terms (Figure 5-10). The terms will be additive if they both have the same sign, whether positive (Figure 5-10a) or negative (Figure 5-10b). Thus, a reaction that is exothermic (i.e., ΔH is negative) and results in an increase in entropy (i.e., ΔS is positive and $-T\Delta S$ is negative) has a

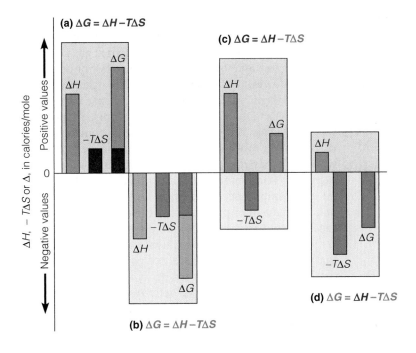

The ΔG value for a specific process or reaction is the algebraic sum of the change in enthalpy (ΔH) and the temperature-dependent entropy term (−TΔS). When both ΔH and −TΔS have the same sign, the numerical value of ΔG will be the *sum* of the two terms, whether both terms are **(a)** positive or **(b)** negative. When the two terms differ in sign, the numerical value of ΔG will be the *difference* between the two terms, whether that results in ΔG having **(c)** a positive value or **(d)** a negative value.

ΔG value that is the *sum* of two negative terms and is therefore more negative than either term (Figure 5–10b). However, if the ΔH and −TΔS terms differ in sign, the ΔG value will have the sign of the larger but its value will be the numerical *difference* between the two terms. Thus, a reaction that is endothermic (i.e., ΔH is positive) and results in an increase in entropy (i.e., ΔS is positive and −TΔS is negative) will have a ΔG value that is either positive (Figure 5-10c) or negative (Figure 5-10d), depending on the numerical values of ΔH and −TΔS.

Free Energy Change as a Measure of Thermodynamic Spontaneity. Free energy is an exceptionally useful concept as a readily measurable indicator of spontaneity. As we shall see shortly, ΔG for a reaction can be readily calculated from the equilibrium constant for the reaction and from easily measurable system variables, such as the concentrations of reactants and products. Once determined, ΔG provides exactly what we have been looking for: a measure of the spontaneity of a reaction based solely on the properties of the system in which the reaction is occurring.

Specifically, every spontaneous reaction is characterized by a *decrease* in the free energy of the system ($\Delta G_{system} < 0$) just as surely as it is characterized by an *increase* in the entropy of the universe ($\Delta S_{universe} > 0$). This is true because with the temperature and pressure held constant, ΔG for the system is related to ΔS for the universe in a simple but inverse way. This gives us a second, equally valid way of expressing the second law: *All processes or reactions that occur spontaneously result in a decrease in the free energy content of the system.* Or, in other words, *the value of ΔG_{system} is negative for every real process or reaction.*

Such processes or reactions are called **exergonic,** which means *energy-yielding.* Note carefully that the reference is specifically to the change in free energy and not to the changes either in enthalpy or in the entropy of the system; these values may be negative, positive, or zero for a given reaction and are therefore *not* valid measures of thermodynamic spontaneity. Conversely, any process or reaction that would result in an increase in the free energy of the system is called **endergonic** (*energy-requiring*) and cannot proceed under the conditions for which ΔG was calculated.

For a biological example of an exergonic reaction, consider the oxidation of glucose to carbon dioxide and water:

$$C_6H_{12}O_6 + 6O_2 \longrightarrow 6CO_2 + 6H_2O + \text{energy} \quad \textbf{(5-11)}$$

You may recognize this as the summary equation for the process of aerobic respiration, whereby chemotrophs obtain energy from glucose. (Most of the cells in your body are carrying out this process right now.) We'll encounter this equation again when we discuss aerobic respiration in Chapter 14; for now, our interest lies in the energetics of glucose oxidation. As Figure 5-11a illustrates (and as you already know if you've ever held a marshmallow over the campfire too long), the oxidation of glucose is a highly exergonic process, with a highly negative ΔG value. By combusting glucose under standard conditions of temperature, pressure, and concentration, we can show that 673 kcal of heat are liberated for every mole of glucose that is oxidized, which means that ΔH for the above reaction is −673 kcal/mol. The −TΔS term can also be determined experimentally and is known to be −13 kcal/mol at 25°C, so this is an example of a reaction in

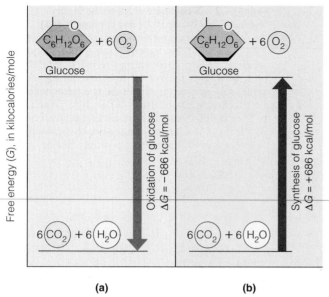

Figure 5-11 Changes in Free Energy for the Oxidation and Synthesis of Glucose. **(a)** The oxidation of glucose to carbon dioxide and water is a highly exergonic reaction, with a ΔG value of -686 kcal/mol under standard conditions. This value is calculated by summing the ΔH and $-T\Delta S$ terms, which under standard conditions have the values of -673 kcal/mol and -13 kcal/mol, respectively. The highly negative value for ΔH is due to the oxidation of the many C-O and C-C bonds in the glucose molecule. The negative value of the $-T\Delta S$ term means that ΔS must be positive, which accords well with the greater randomness to be expected when the carbon, oxygen, and hydrogen atoms of a single glucose molecule are converted into six molecules each of carbon dioxide and water. **(b)** The synthesis of glucose from carbon dioxide and water is as highly endergonic as its oxidation is exergonic. The ΔG value of $+686$ kcal/mol under standard conditions is the sum of the ΔH and $-T\Delta S$ terms, which have values of $+673$ kcal/mol and $+13$ kcal/mol, respectively.

which the ΔH and $-T\Delta S$ terms are additive, with a ΔG of -686 kcal/mol.

Now consider the reverse reaction, by which phototrophs synthesize sugars such as glucose from carbon dioxide and water, with the release of oxygen:

$$6CO_2 + 6\,H_2O + energy \longrightarrow C_6H_{12}O_6 + 6O_2 \quad \textbf{(5-12)}$$

As you might guess, the values of ΔH, ΔS, and ΔG for this reaction are identical in magnitude but opposite in sign when determined under standard conditions and compared to the corresponding values for reaction 5-11. Specifically, this reaction has a ΔG value of $+686$ kcal/mol, which makes it a highly endergonic reaction (Figure 5-11b). Phototrophs must therefore use large amounts of energy to drive this reaction in the direction of glucose synthesis—and that, of course, is where the energy of the sun comes in, as we'll see when we get to Chapter 15.

The Meaning of Spontaneity. Before looking at how we can actually calculate ΔG and use it as a measure of thermodynamic spontaneity, we need to look more closely at what is—and what is not—meant by the term *spontaneous*. As we noted earlier, spontaneity tells us only that a

reaction *can* go; it says nothing at all about whether it *will* go. A reaction can have a negative ΔG value and yet not actually proceed to any measurable extent at all. The cellulose of paper obviously burns spontaneously once ignited, consistent with a highly negative ΔG value of -686 kcal/mol of glucose units. Yet in the absence of a match, paper is reasonably stable and might require hundreds of years to oxidize. Thus, ΔG can really tell us only whether a reaction or process is thermodynamically feasible—whether it has the *potential* for occurring. Whether an exergonic reaction will in fact proceed depends not only on its favorable (negative) ΔG but also on the availability of a mechanism or pathway to get from the initial state to the final state. Usually, an initial input of activation energy is required as well, such as the heat energy from the match that was used to ignite the piece of paper.

Thermodynamic spontaneity is therefore a necessary but insufficient criterion for determining whether a reaction will actually occur. In Chapter 6, we will explore the subject of reaction rates in the context of enzyme-catalyzed reactions. For the moment, we need only note that when we designate a reaction as thermodynamically spontaneous, we simply mean that it is an energetically feasible event that will liberate free energy if and when it actually takes place.

Understanding ΔG

Our final task in this chapter will be to understand how ΔG is calculated and how it can then be used to assess the thermodynamic feasibility of reactions under specified conditions. For that, we come back to the reaction that converts glucose-6-phosphate into fructose-6-phosphate (reaction 5-8) and ask what we can learn about the spontaneity of the conversion in the direction written (from left to right). Prior experience and familiarity provide no clues here, nor is it obvious how the entropy of the universe would be affected if the reaction were to proceed. Clearly, we need to be able to calculate ΔG and to determine whether it is positive or negative under the particular conditions we specify for the reaction.

The Equilibrium Constant Is a Measure of Directionality

One means of assessing whether a reaction can proceed in a given direction under specified conditions involves the **equilibrium constant K_{eq}**, which is the ratio of product concentrations to reactant concentrations at equilibrium. For the general reaction in which A is converted reversibly into B, the equilibrium constant is simply the ratio of the equilibrium concentrations of A and B:

$$A \rightleftharpoons B \quad \textbf{(5-13)}$$

$$K_{eq} = \frac{[B]_{eq}}{[A]_{eq}} \quad \textbf{(5-14)}$$

where $[A]_{eq}$ and $[B]_{eq}$ are the concentrations of A and B, in moles per liter, when reaction 5-13 is at equilibrium at 25°C. Given the equilibrium constant for a reaction, you can easily tell whether a specific mixture of products and reactants is at equilibrium, and, if not, how far the reaction is away from equilibrium and in which direction it must proceed to reach equilibrium.

For example, the equilibrium constant for reaction 5-8 at 25°C is known to be 0.5. This means that at equilibrium there will be one-half as much fructose-6-phosphate as glucose-6-phosphate, regardless of the actual magnitudes of the concentrations:

$$K_{eq} = \frac{[\text{fructose–6–phosphate}]_{eq}}{[\text{glucose–6–phosphate}]_{eq}} = 0.5 \qquad \textbf{(5-15)}$$

If the two compounds are present in any other concentration ratio, the reaction will not be at equilibrium and will tend toward (be thermodynamically spontaneous in the direction of) equilibrium. Thus, a concentration ratio less than K_{eq} means that there is too little fructose-6-phosphate present, and the reaction will tend to proceed to the right to generate more fructose-6-phosphate at the expense of glucose-6-phosphate. Conversely, a concentration ratio greater than K_{eq} indicates that the relative concentration of fructose-6-phosphate is too high, and the reaction will tend to proceed to the left.

Figure 5-12 illustrates this concept for the interconversion of A and B (reaction 5-13), showing the relationship between the free energy of the reaction and how far the concentrations of A and B are from equilibrium. (Note that K_{eq} is assumed to be 1.0 in this illustration; for other values of K_{eq}, the curve would be the same but the numbers along the X axis would be different.) The point of Figure 5-12 is clear: The free energy is lowest at equilib-

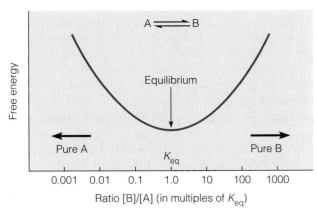

Figure 5-12 Free Energy and Chemical Equilibrium. The amount of free energy available from a chemical reaction depends on how far the components are from equilibrium. This is illustrated here for a reaction that interconverts A and B and has an equilibrium constant, K_{eq}. The free energy of the system increases as the [B]/[A] ratio changes on either side of the equilibrium point. Note that the x-axis is an exponential scale, with all [B]/[A] ratios expressed as multiples of K_{eq}.

rium and increases as the system is displaced from equilibrium in either direction. Moreover, if we know how the ratio of prevailing concentrations compares with the equilibrium concentration ratio, we can predict in which direction a reaction will tend to proceed *and* how much free energy will be released as it does so. Thus, the tendency toward equilibrium provides the driving force for every chemical reaction, and a comparison of prevailing and equilibrium concentration ratios provides one measure of that tendency.

ΔG Can Be Calculated Readily

It should come as no surprise that ΔG is really just a means of calculating how far from equilibrium a reaction lies under specified conditions and how much energy will be released as the reaction proceeds toward equilibrium. Nor should it be surprising that both the equilibrium constant and the prevailing concentrations of reactants and products are needed to calculate ΔG. For reaction 5-13, the equation relating these variables is as follows:

$$\begin{aligned} \Delta G &= RT \ln \frac{[B]_{pr}}{[A]_{pr}} - RT \ln \frac{[B]_{eq}}{[A]_{eq}} \\ &= RT \ln \frac{[B]_{pr}}{[A]_{pr}} - RT \ln K_{eq} \\ &= -RT \ln K_{eq} + RT \ln \frac{[B]_{pr}}{[A]_{pr}} \end{aligned} \qquad \textbf{(5-16)}$$

where ΔG is the free energy change, in cal/mol, under the specified conditions; R is the gas constant (1.987 cal/mol-K); T is the temperature in kelvins (use 25°C = 298 K unless otherwise specified); ln is the natural log (i.e., to the base e); $[A]_{pr}$ and $[B]_{pr}$ are the prevailing concentrations of A and B in moles per liter; $[A]_{eq}$ and $[B]_{eq}$ are the equilibrium concentrations of A and B in moles per liter; and K_{eq} is the equilibrium constant at the standard temperature of 298 K (25°C).

More generally, for a reaction in which a molecules of reactant A combine with b molecules of reactant B to form c and d molecules, respectively, of products C and D,

$$a\text{A} + b\text{B} \rightleftharpoons c\text{C} + d\text{D} \qquad \textbf{(5-17)}$$

ΔG is calculated as

$$\Delta G = -RT \ln K_{eq} + RT \ln \frac{[C]_{pr}^{c}[D]_{pr}^{d}}{[A]_{pr}^{a}[B]_{pr}^{b}} \qquad \textbf{(5-18)}$$

where all the constants and variables are as previously defined, and K_{eq} is the equilibrium constant for reaction 5-17.

Returning to reaction 5-8, assume that the prevailing concentrations of glucose-6-phosphate and fructose-6-phosphate in a cell are 10 μM (10 × 10⁻⁶ M) and 1 μM (1 × 10⁻⁶ M), respectively, at 25°C. Since the ratio of prevailing product concentrations to reactant concentrations is 0.1 and the equilibrium constant is 0.5, there is clearly too little fructose-6-phosphate present relative to glucose-6-phosphate for the reaction to be at equilibrium. The reaction should therefore tend toward the right (in the direction of fructose-6-phosphate generation). In other words, the reaction is thermodynamically favorable in the direction written. This, in turn, means that ΔG is negative under these conditions.

The actual value for ΔG is calculated as follows:

$$\Delta G = -(1.987\,cal/mol\text{-}K)(298K)\ln(0.5) +$$

$$(1.987\,cal/mol\text{-}K)(298K)\ln\frac{1\times10^{-6}M}{10\times10^{-6}M}$$

$$= -(592\,cal/mol)\ln(0.5) + (592\,cal/mol)\ln(0.1)$$

$$= -(592\,cal/mol)(-0.693) + (592\,cal/mol)(-2.303)$$

$$= +410\,cal/mol - 1364\,cal/mol$$

$$= -954\,cal/mol \qquad\qquad\qquad (5\text{-}19)$$

Notice that our expectation of a negative ΔG is confirmed, and we now know exactly how much free energy is liberated upon the spontaneous conversion of 1 mole of glucose-6-phosphate into 1 mole of fructose-6-phosphate under the specified conditions. (As we will see later, the free energy liberated in an exergonic reaction can be either "harnessed" to do work or lost as heat.)

It is important to understand exactly what this calculated value for ΔG means and under what conditions it is valid. Because it is a thermodynamic parameter, ΔG can tell us whether a reaction is thermodynamically possible as written, but it says nothing about the rate or mechanism. It simply says that if the reaction does occur, it will be to the right and will liberate 954 calories of free energy for every mole of glucose-6-phosphate that is converted to fructose-6-phosphate, *provided* that the concentrations of both the reactant and the product are *maintained at the initial values* (10 μM and 1 μM, respectively) throughout the course of the reaction.

More generally, *ΔG is a measure of thermodynamic spontaneity for a reaction in the direction in which it is written (from left to right), at the specified concentrations of reactants and products.* In a beaker or test tube, this requirement for constant reactant and product concentrations means that reactants must be added continuously and products must be removed continuously. In the cell, each reaction is part of some metabolic pathway, and its reactants and products are maintained at fairly constant, nonequilibrium concentrations by the reactions that precede and follow it in the sequence.

The Standard Free Energy Change Is ΔG Measured Under Standard Conditions

Because it is a thermodynamic parameter, ΔG is independent of the actual mechanism or pathway for a reaction, but it depends crucially on the conditions under which

the reaction occurs. A reaction characterized by a large decrease in free energy under one set of conditions may have a much smaller (but still negative) ΔG or may even have a positive ΔG under a different set of conditions. The melting of ice, for example, depends on temperature; it proceeds spontaneously above 0°C but goes in the opposite direction (freezing) below that temperature. It is therefore important to identify the conditions under which a given measurement of ΔG is made.

By convention, biochemists have agreed on certain arbitrary conditions to define the **standard state** of a system for convenience in reporting, comparing, and tabulating free energy changes in chemical reactions. For systems consisting of dilute aqueous solutions, these are usually a standard temperature of 25°C (298 K), a pressure of 1 atmosphere, and all products and reactants present in their most stable forms at a standard concentration of 1 mol/L (or a pressure of 1 atmosphere for gases). The only common exception to this standard concentration rule is water. The concentration of water in a dilute aqueous solution is approximately 55.5 M and does not change significantly during the course of reactions, even when water is itself a reactant or product. By convention, biochemists do not include the concentration of water in calculations of free energy changes, even though the reaction may indicate a net consumption or production of water.

In addition to standard conditions of temperature, pressure, and concentration, biochemists also frequently specify a standard pH of 7.0 because most biological reactions occur at or near neutrality. The concentration of hydrogen ions (and of hydroxyl ions as well) is therefore 10^{-7} M, so the standard concentration of 1.0 M does not apply to H^+ or OH^- ions when a pH of 7.0 is specified. Values of K_{eq}, ΔG, or other thermodynamic parameters determined or calculated at pH 7.0 are always written with a prime (as K'_{eq}, $\Delta G'$, and so on) to indicate this fact.

Energy changes for reactions are usually reported in standardized form as the change that would occur if the reactions were run under the standard conditions. More precisely, the standard change in any thermodynamic parameter refers to the conversion of a mole of a specified reactant to products or the formation of a mole of a specified product from the reactants under conditions where the temperature, pressure, pH, and concentrations of all relevant species are maintained at their standard values. (As for the more general case, the maintenance of concentrations at 1.0 M implies that reactants are added as they are used up and that products are removed as they are formed.)

The free energy change calculated under these conditions is called the **standard free energy change,** designated **$\Delta G^{o\prime}$,** where the superscript (°) refers to standard conditions of temperature, pressure, and concentration and the prime (′) emphasizes that the standard hydrogen ion concentration for biochemists is 10^{-7} M, not 1.0 M.

It turns out that $\Delta G^{o\prime}$ bears a simple relationship to the equilibrium constant K'_{eq} (Figure 5-13). This relationship can readily be seen by rewriting equation 5-18 with primes and then assuming standard concentrations for all

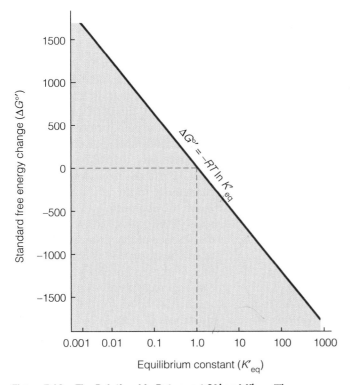

Figure 5-13   The Relationship Between $\Delta G^{\circ\prime}$ and K'_{eq}.   The standard free energy change and the equilibrium constant are related by the equation $\Delta G^{\circ\prime} = -RT \ln K'_{eq}$. Note that the equilibrium constant is plotted on an exponential scale. Note also that if the equilibrium constant is 1.0, the standard free energy change is zero.

reactants and products. All concentration terms are now 1.0 and the logarithm of 1.0 is zero, so the second term in the general expression for $\Delta G^{\circ\prime}$ is eliminated, and what remains is an equation for $\Delta G^{\circ\prime}$, the free energy change under standard conditions:

$$\Delta G^{\circ\prime} = -RT \ln K'_{eq} + RT \ln 1 = -RT \ln K'_{eq} \quad \textbf{(5-20)}$$

In other words, $\Delta G^{\circ\prime}$ can be calculated directly from the equilibrium constant, provided that the latter has also been determined under the same standard conditions of temperature, pressure, and pH. This, in turn, allows equation 5-18 to be expressed in somewhat simpler form as

$$\Delta G' = \Delta G^{\circ\prime} + RT \ln \frac{[C]^c_{pr}[D]^d_{pr}}{[A]^a_{pr}[B]^b_{pr}} \quad \textbf{(5-21)}$$

At the standard temperature of 25°C (298 K), the term RT becomes $(1.987)(298) = 592$ cal/mol, so equations 5-20 and 5-21 can be rewritten as follows, in what are the most useful formulas for our purposes:

$$\Delta G^{\circ\prime} = -592 \ln K'_{eq} \quad \textbf{(5-22)}$$

$$\Delta G' = \Delta G^{\circ\prime} + 592 \ln \frac{[C]^c_{pr}[D]^d_{pr}}{[A]^a_{pr}[B]^b_{pr}} \quad \textbf{(5-23)}$$

Summing Up: The Meaning of $\Delta G'$ and $\Delta G^{\circ\prime}$

Equations 5-22 and 5-23 represent the most important contribution of thermodynamics to biochemistry and cell biology—a means of assessing the feasibility of a chemical reaction based on the prevailing concentrations of products and reactants and a knowledge of the equilibrium constant. Equation 5-22 expresses the relationship between the standard free energy change $\Delta G^{\circ\prime}$ and the equilibrium constant K'_{eq} and enables us to calculate the free energy change that would be associated with any reaction of interest if all reactants and products were maintained at a standard concentration of 1.0 M.

If K'_{eq} is greater than 1.0, then $\ln K'_{eq}$ will be positive and $\Delta G^{\circ\prime}$ will be negative, and the reaction can proceed to the right under standard conditions. This makes sense because if K'_{eq} is greater than 1.0, products will predominate over reactants at equilibrium. A predominance of products can only be achieved from the standard state by the conversion of reactants to products, so the reaction will tend to proceed spontaneously to the right. Conversely, if K'_{eq} is less than 1.0, then $\Delta G^{\circ\prime}$ will be positive and the reaction cannot proceed to the right. Instead, it will tend toward the left because $\Delta G^{\circ\prime}$ for the reverse reaction will have the same absolute value but will be opposite in sign. This is in keeping with the small value for K'_{eq}, which specifies that reactants are favored over products (that is, the equilibrium lies to the left).

The $\Delta G^{\circ\prime}$ values are convenient both because of the ease with which they can be determined from the equilibrium constant and because they provide a uniform convention for reporting free energy changes. But bear in mind that a $\Delta G^{\circ\prime}$ value is an arbitrary standard in that it refers to an arbitrary state specifying conditions of concentration that cannot be achieved with most biologically important compounds. $\Delta G^{\circ\prime}$ is therefore useful for standardized reporting, but *it is not a valid measure of the thermodynamic spontaneity of reactions as they occur under real conditions.*

For that purpose we need $\Delta G'$, which provides a direct measure of how far from equilibrium a reaction is at the concentrations of reactants and products that actually prevail in the cell or other system of interest (equation 5-23). Therefore, $\Delta G'$ is the most useful measure of thermodynamic spontaneity. If it is negative, the reaction in question is thermodynamically spontaneous and can proceed as written under the conditions for which the calculations were made. Its magnitude serves as a measure of how much free energy will be liberated as the reaction occurs. This, in turn, determines the maximum amount of work that can be performed on the surroundings, provided a mechanism is available to conserve and use the energy as it is liberated.

A positive $\Delta G'$, on the other hand, indicates that the reaction cannot occur in the direction written under the conditions for which the calculations were made. Such reactions can sometimes be rendered spontaneous, however, by changes in the concentrations of products or reactants. For the special case where $\Delta G' = 0$, the reaction is clearly at equilibrium, and no net energy change accompanies the conversion of reactant molecules into product molecules or vice versa. These features of K'_{eq}, $\Delta G^{\circ\prime}$, and $\Delta G'$ are summarized in Table 5-1.

Free Energy Change: Sample Calculations

To illustrate the calculation and utility of $\Delta G'$ and $\Delta G^{\circ\prime}$, we return once more to the interconversion of glucose-6-phosphate and fructose-6-phosphate (reaction 5-8). We already know that the equilibrium constant for this reaction under standard conditions of temperature, pH, and pressure is 0.5 (equation 5-15). This means that if the enzyme that catalyzes this reaction in cells is added to a solution of glucose-6-phosphate at 25°C, 1 atmosphere, and pH 7.0 and the solution is incubated until no further reaction occurs, fructose-6-phosphate and glucose-6-phosphate will be present in an equilibrium ratio of 0.5. (Note that this ratio is independent of the actual starting concentration of glucose-6-phosphate and could have been achieved equally well by starting with any concentration of fructose-6-phosphate or any mixture of both, in any starting concentrations.)

The standard free energy change $\Delta G^{\circ\prime}$ can be calculated from K'_{eq} as follows:

$$\begin{aligned} \Delta G^{\circ\prime} &= -RT \ln K_{eq} = -592 \ln K'_{eq} \\ &= -592 \ln 0.5 = -592(-0.693) \\ &= +410 \text{ cal/mol} \end{aligned} \quad \textbf{(5-24)}$$

The positive value for $\Delta G^{\circ\prime}$ is therefore another way of expressing the fact that the reactant (glucose-6-phosphate) is the predominant species at equilibrium. A positive $\Delta G^{\circ\prime}$ value also means that under standard conditions of concentration, the reaction is nonspontaneous (thermodynamically impossible) in the direction written. In other words, if we begin with both glucose-6-phosphate and fructose-6-phosphate present at concentrations of 1.0 M, no net conversion of glucose-6-phosphate to fructose-6-phosphate can occur.

As a matter of fact, given an appropriate catalyst, the reaction will proceed to the *left* under standard conditions because the $\Delta G^{\circ\prime}$ for the reaction in that direction is −410 cal/mol. Fructose-6-phosphate will therefore be converted into glucose-6-phosphate until the equilibrium ratio of 0.5 was reached. Alternatively, if both species were added or removed continuously as necessary to maintain the concentrations of both at 1.0 M, the reaction would proceed continuously and spontaneously to the left (assuming the presence of a catalyst), with the liberation of 410 cal of free energy per mole of fructose-6-phosphate converted to glucose-6-phosphate. In the absence of any provision for conserving this energy, it would be dissipated as heat.

In a real cell, neither of these phosphorylated sugars would ever be present at a concentration even approaching 1.0 M. In fact, experimental values for the actual concentrations of these substances in human red blood cells are as follows:

[glucose-6-phosphate]: 83 μM (83 × 10^{-6} M)
[fructose-6-phosphate]: 14 μM (14 × 10^{-6} M)

Using these values, we can calculate the actual $\Delta G'$ for the interconversion of these sugars in red blood cells as follows:

$$\begin{aligned} \Delta G' &= \Delta G^{\circ\prime} + 592 \ln \frac{[\text{fructose}-6-\text{phosphate}]_{pr}}{[\text{glucose}-6-\text{phosphate}]_{pr}} \\ &= +410 + 592 \ln \frac{14 \times 10^{-6}}{83 \times 10^{-6}} \\ &= +410 + 592 \ln 0.169 \\ &= +410 + 592(-1.78) = +410 - 1054 \\ &= -644 \text{ cal/mol} \end{aligned} \quad \textbf{(5-25)}$$

Table 5-1 The Meaning of $\Delta G^{\circ\prime}$ and $\Delta G'$

The Meaning of $\Delta G^{\circ\prime}$		
$\Delta G^{\circ\prime}$ Negative ($K'_{eq} > 1.0$)	**$\Delta G^{\circ\prime}$ Positive ($K'_{eq} < 1.0$)**	**$\Delta G^{\circ\prime} = 0$ ($K'_{eq} = 1.0$)**
Products predominate over reactants at equilibrium at standard temperature, pressure, and pH.	Reactants predominate over products at equilibrium at standard temperature, pressure, and pH.	Products and reactants are present equally at equilibrium at standard temperature, pressure, and pH.
Reaction goes spontaneously to the right under standard conditions.	Reaction goes spontaneously to the left under standard conditions.	Reaction is at equilibrium under standard conditions.

The Meaning of $\Delta G'$		
$\Delta G'$ Negative	**$\Delta G'$ Positive**	**$\Delta G' = 0$**
Reaction is thermodynamically feasible as written under conditions for which $\Delta G'$ was calculated.	Reaction is not feasible as written under the conditions for which $\Delta G'$ was calculated.	Reaction is at equilibrium under the conditions for which $\Delta G'$ was calculated.
Work can be done by the reaction under conditions for which $\Delta G'$ was calculated.	Energy must be supplied to drive the reaction under the conditions for which $\Delta G'$ was calculated.	No work can be done nor is energy required by the reaction under the conditions for which $\Delta G'$ was calculated.

The negative value for $\Delta G'$ means that the conversion of glucose-6-phosphate into fructose-6-phosphate is thermodynamically possible under the conditions of concentration actually prevailing in red blood cells and that the reaction will yield 644 cal of free energy per mole of reactant converted to product. Thus, the conversion of reactant to product is thermodynamically impossible under standard conditions, but the red blood cell maintains these two phosphorylated sugars at concentrations adequate to offset the positive $\Delta G^{\circ\prime}$, thereby rendering the reaction possible. This adaptation, of course, is essential if the red blood cell is to be successful in carrying out the glucose-degrading process of glycolysis of which this reaction is a part.

Life and the Steady State: Reactions That Move Toward Equilibrium Without Ever Getting There

As this chapter has emphasized, the driving force in all reactions is their tendency to move toward equilibrium. Indeed, $\Delta G^{\circ\prime}$ and $\Delta G'$ are really nothing more than convenient means of quantifying how far and in what direction from equilibrium a reaction lies under the specific conditions dictated by standard or prevailing concentrations of products and reactants. But to understand how cells really function, we must appreciate the importance of reactions that move toward equilibrium without ever achieving it. At equilibrium, the forward and backward rates are the same for a reaction and there is therefore no net flow of matter in either direction. Most importantly, no further energy can be extracted from the reaction because $\Delta G'$ is zero for a reaction at equilibrium.

For all practical purposes, then, a reaction at equilibrium is a reaction that has stopped. But a living cell is characterized by reactions that are continuous, not stopped. A cell at equilibrium would be a dead cell. We might, in fact, define life as a continual struggle to maintain myriad cellular reactions in positions far from equilibrium because at equilibrium no net reactions are possible, no energy can be released, no work can be done, and the order of the living state cannot be maintained.

Thus, life is possible only because living cells maintain themselves in a **steady state,** with most of their reactions far from thermodynamic equilibrium. The levels of glucose-6-phosphate and fructose-6-phosphate found in red blood cells illustrate this point. As we have seen, these compounds are maintained in the cell at steady-state concentrations far from the equilibrium condition predicted by the K'_{eq} value of 0.5. In fact, the levels are so far from equilibrium concentrations that the conversion of glucose-6-phosphate to fructose-6-phosphate can occur continuously in the cell, even though the equilibrium state has a positive $\Delta G^{\circ\prime}$ and actually favors glucose-6-phosphate. The same is true of most reactions and pathways in the cell. They can proceed and can be harnessed to perform various kinds of cellular work because reactants, products, and intermediates are maintained at steady-state concentrations far from the thermodynamic equilibrium.

This state, in turn, is possible only because a cell is an open system and receives large amounts of energy from its environment. If the cell were a closed system, all its reactions would gradually run to equilibrium and the cell would come inexorably to a state of minimum free energy, after which no further changes could occur, no work could be accomplished, and life would cease. The steady state so vital to life is possible only because the cell is able to take up energy continuously from its environment, whether in the form of light or preformed organic food molecules. This continuous uptake of energy and the accompanying flow of matter make possible the maintenance of a steady state in which all the reactants and products of cellular chemistry are kept far enough from equilibrium to ensure that the thermodynamic drive toward equilibrium can be harnessed by the cell to perform useful work, thereby maintaining and extending its activities and structural complexity.

We will focus on how this is accomplished in later chapters. In the next chapter, we will look at principles of enzyme catalysis that determine the rates of cellular reactions, i.e., that translate the "can go" of thermodynamics into the "will go" of kinetics. Then we will be ready to move on to subsequent chapters, where we will encounter functional metabolic pathways that result from series of such reactions acting in concert.

Perspective

The high level of order that exists in cells is possible only because of the availability of energy from the environment. Cells require energy to carry out various kinds of change, including synthesis, movement, concentration, charge separation, the generation of heat, and bioluminescence. The energy needed for these processes comes either from the sun or from the bonds of oxidizable organic molecules such as carbohydrates, fats, and proteins. Since chemotrophs feed directly or indirectly on phototrophs, there is a unidirectional flow of energy through the biosphere, with the sun as the ultimate source and entropy and heat losses as the eventual fate of all the energy that moves through living systems.

The flow of energy through cells is governed by the laws of thermodynamics. The first law specifies that energy can change form but must always be conserved. The second law provides a measure of thermodynamic spontaneity, although this means only that a reaction can occur and says nothing about whether it will actually occur, or at what rate. Spontaneous processes are always accompanied by an *increase* in the entropy of the universe and by a *decrease* in the free energy of the system. The latter is a far more practical indicator of spontaneity because it can be calculated readily from the equilibrium constant, the prevailing concentrations of reactants and products, and the temperature.

Cells obtain the energy they need to carry out their activities by maintaining the many reactants and products of the various reaction sequences at steady-state concentrations far from equilibrium, thereby allowing the reactions to move exergonically toward equilibrium without ever actually reaching it. A negative $\Delta G'$ is a necessary prerequisite for a reaction to proceed, but it does not guarantee that the reaction will actually occur at a reasonable rate. To assess that, we must know more about the reaction than just its thermodynamic status. We need to know whether an appropriate catalyst is on hand and at what rate the reaction can occur in the presence of the catalyst. In other words, we need the enzymes that will be discussed in Chapter 6.

Key Terms for Self-Testing

The Importance of Energy
energy (p. 106)
biosynthesis (p. 107)
mechanical work (p. 108)
concentration work (p. 108)
electrical work (p. 108)
heat (p. 108)
bioluminescence (p. 108)
phototroph (p. 109)
chemotroph (p. 109)

Bioenergetics
thermodynamics (p. 112)

bioenergetics (p. 112)
system (p. 112)
surroundings (p. 112)
state (p. 112)
work (p. 113)
calorie (cal) (p. 113)
joule (J) (p. 113)
first law of thermodynamics (p. 113)
internal energy (E) (p. 113)
enthalpy (H) (p. 114)
thermodynamic spontaneity (p. 114)
second law of thermodynamics (p. 115)

entropy (S) (p. 118)
free energy (G) (p. 118)
free energy change (ΔG) (p. 118)
exergonic (p. 119)
endergonic (p. 119)

Understanding ΔG
equilibrium constant (K_{eq}) (p. 120)
standard state (p. 122)
standard free energy change ($\Delta G^{o\prime}$) (p. 122)

Life and the Steady State
steady state (p. 125)

Problem Set

More challenging problems are marked with a •.

5-1.   Solar Energy. Although we sometimes hear concerns about a global energy crisis, we actually live on a planet that is flooded continuously with an extravagant amount of energy in the form of solar radiation. Every day, year in and year out, solar energy arrives at the upper surface of the Earth's atmosphere at the rate of 1.94 cal/min per square centimeter of cross-sectional area (the *solar energy constant*).

(a) Assuming the cross-sectional area of the Earth to be about 1.28×10^{18} cm², what is the total annual amount of incoming energy?

(b) A substantial portion of that energy, particularly in the wavelength ranges below 300 nm and above 800 nm, never reaches the Earth's surface. Can you suggest what happens to it?

(c) Of the radiation that reaches the Earth's surface, only a small proportion is actually trapped photosynthetically by phototrophs. (You can calculate the actual value in Problem 5-2.) Why do you think the efficiency of utilization is so low?

5-2.   Photosynthetic Energy Transduction. The amount of energy trapped and the volume of carbon converted to organic form by photosynthetic energy transducers are mind-boggling: about 5×10^{16} g of carbon per year over the entire Earth's surface.

(a) Assuming that the average organic molecule in a cell has about the same proportion of carbon as glucose does, how many grams of organic matter are produced annually by carbon-fixing phototrophs?

(b) Assuming that all the organic matter in part a is glucose (or any molecule with an energy content equivalent to that of glucose), how much energy is represented by that quantity of organic matter? Assume that glucose has a free energy content (free energy of combustion) of 3.8 kcal/g.

(c) Refer to the answer for Problem 5-1a. What is the average efficiency with which the radiant energy incident on the upper atmosphere is trapped photosynthetically on the Earth's surface?

(d) What proportion of the net annual phototrophic production of organic matter calculated in part a do you think is consumed by chemotrophs each year?

5-3.   Energy Conversion. Most cellular activities involve the conversion of energy from one form to another. For each of the following cases, give a biological example and explain the significance of the conversion.

(a) Chemical energy into mechanical energy

(b) Chemical energy into radiant energy

(c) Solar (light) energy into chemical energy

(d) Chemical energy into electrical energy

(e) Chemical energy into the potential energy of a concentration gradient

5-4. Enthalpy, Entropy, and Free Energy. The oxidation of glucose to carbon dioxide and water is represented by the following reaction whether the oxidation occurs by combustion in the laboratory or by biological oxidation in living cells:

$$C_6H_{12}O_6 + 6O_2 \rightleftharpoons 6CO_2 + 6H_2O \qquad (5-26)$$

When combustion is carried out under controlled conditions in the laboratory, the reaction is highly exothermic, with an enthalpy change (ΔH) of −673 kcal/mol. As you will discover in Chapter 13, ΔG for this reaction at 25°C is −686 kcal/mol, so the reaction is also highly exergonic.

(a) Explain in your own words what the ΔH and ΔG values mean. What do the negative signs mean in each case?

(b) What does it mean to say that the difference between the ΔG and ΔH values is due to entropy?

(c) Without doing any calculations, would you expect ΔS (entropy change) for this reaction to be positive or negative? Explain your answer.

(d) Now calculate ΔS for this reaction at 25°C. Does the calculated value agree in sign with your prediction in part c?

(e) What are the values of ΔG, ΔH, and ΔS for the reverse of the above reaction as carried out by a photosynthetic algal cell that is using CO_2 and H_2O to make $C_6H_{12}O_6$?

5-5. The Equilibrium Constant. The following reaction is one of the steps in the glycolytic pathway, which we will encounter again in Chapter 13. You should recognize it already, however, because we used it as an example earlier (reaction 5-8):

$$\text{glucose-6-phosphate} \rightleftharpoons \text{fructose-6-phosphate} \qquad (5-27)$$

The equilibrium constant K_{eq} for this reaction at 25°C is 0.5.

(a) Assume that you incubate a solution containing 0.15 M glucose-6-phosphate (G6P) overnight at 25°C with the enzyme phosphoglucomutase that catalyzes the above reaction. How many millimoles of fructose-6-phosphate (F6P) will you recover from 10 mL of the incubation mixture the next morning, assuming you have an appropriate chromatographic procedure to separate F6P from G6P?

(b) What answer would you get for part a if you had started with a solution containing 0.15 M F6P instead?

(c) What answer would you expect for part a if you had started with a solution containing 0.15 M G6P but forgot to add phosphoglucomutase to the incubation mixture?

(d) Would you be able to answer the question in part a if you had used 15°C as your incubation temperature instead of 25°C? Why or why not?

5-6. Calculating $\Delta G°'$ and $\Delta G'$. Like the reaction in Problem 5-5, the conversion of 3-phosphoglycerate (3PG) to 2-phosphoglycerate (2PG) is an important cellular reaction because it is one of the steps in the glycolytic pathway (see Chapter 13):

$$\text{3-phosphoglycerate} \rightleftharpoons \text{2-phosphoglycerate} \qquad (5-28)$$

If the enzyme that catalyzes this reaction is added to a solution of 3PG at 25°C and pH 7.0, the equilibrium ratio between the two species will be 0.165:

$$K'_{eq} = \frac{[2-\text{phosphoglycerate}]_{eq}}{[3-\text{phosphoglycerate}]_{eq}} = 0.165 \qquad (5-29)$$

Experimental values for the actual steady-state concentrations of these compounds in human red blood cells are 61 μM for 3PG and 4.3 μM for 2PG.

(a) Calculate $\Delta G°'$. Explain in your own words what this value means.

(b) Calculate $\Delta G'$. Explain in your own words what this value means. Why are $\Delta G'$ and $\Delta G°'$ different?

(c) If conditions in the cell change such that the concentration of 3PG remains fixed at 61 μM but the concentration of 2PG begins to rise, how high can the 2PG concentration get before reaction 5-28 will cease because it is no longer thermodynamically feasible?

5-7. Backward or Forward? The interconversion of dihydroxyacetone phosphate (DHAP) and glyceraldehyde-3-phosphate (G3P) is a part of both the glycolytic pathway (see Chapter 13) and the Calvin cycle for photosynthetic carbon fixation (see Chapter 15):

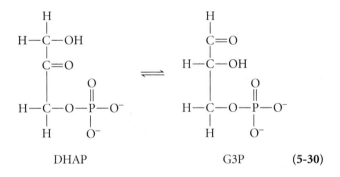

DHAP   G3P   (5-30)

The value of $\Delta G°'$ for this reaction is +1.8 kcal/mol at 25°C. In the glycolytic pathway, this reaction goes to the *right,* converting DHAP to G3P. In the Calvin cycle, this reaction proceeds to the *left,* converting G3P to DHAP.

(a) In which direction does the equilibrium lie? What is the equilibrium constant at 25°C?

(b) In which direction does this reaction tend to proceed under standard conditions? What is $\Delta G'$ for the reaction *in that direction?*

(c) In the glycolytic pathway, this reaction is driven to the right because G3P is consumed by the next reaction in the sequence, thereby maintaining a low G3P concentration. What will $\Delta G'$ be (at 25°C) if the concentration of G3P is maintained at 1% of the DHAP concentration (i.e., if [G3P]/[DHAP] = 0.01)?

(d) In the Calvin cycle, this reaction proceeds to the left. How high must the [G3P]/[DHAP] ratio be to ensure that the reaction is exergonic by at least −3.0 kcal/mol (at 25°C)?

5-8. Synthesis of Citrate. The enzyme *citrate synthase* catalyzes the first reaction of the tricarboxylic acid (TCA) cycle, an important metabolic pathway that we will encounter in Chapter 14. In this reaction, a two-carbon acetate unit is transferred from an acyl carrier called *coenzyme A (CoA)* to oxaloacetate (OAA), a four-carbon compound, releasing coenzyme A and forming citrate, a six-carbon compound:

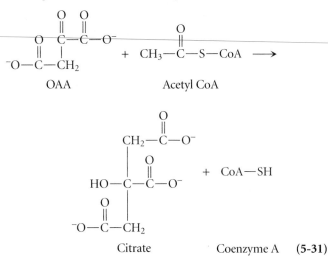

$\Delta G^{\circ\prime}$ for this reaction at 25°C is −7.7 kcal/mol.

(a) What is K_{eq}' for this reaction?

(b) If the [acetyl CoA]/[CoA] ratio is 3, at what [OAA]/ [citrate] ratio will this reaction be at equilibrium?

(c) If the [acetyl CoA]/[CoA] ratio is 3 and the concentration of citrate in a cell is 0.4 m*M*, at what OAA concentration will $\Delta G'$ be −2.0 kcal/mol?

(d) With the citrate concentration at 0.4 m*M* and the OAA concentration as calculated in part c, how low could the [acetyl CoA]/[CoA] ratio drop before the reaction would no longer proceed to the right?

5-9. Succinate Oxidation. Like the reaction in Problem 5-8, the oxidation of succinate to fumarate is an important cellular reaction because it is one of the steps in the tricarboxylic acid (TCA) cycle (see Chapter 14). The two hydrogen atoms that are removed from succinate are accepted by a coenzyme molecule called flavin adenine dinucleotide (FAD), which is thereby reduced to FADH$_2$:

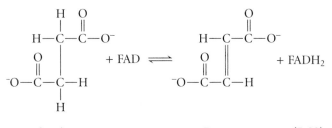

$\Delta G^{\circ\prime}$ for this reaction is 0 cal/mol.

(a) If you start with a solution containing 0.01 *M* each of succinate and FAD and add an appropriate amount of the

enzyme that catalyzes this reaction, will any fumarate be formed? If so, calculate the resulting equilibrium concentrations of all four species. If not, explain why not.

(b) Answer part a assuming that 0.01 *M* FADH$_2$ is also present initially.

(c) Assuming that the steady-state conditions in a cell are such that the FADH$_2$/FAD ratio is 5 and the fumarate concentration is 2.5 μ*M*, what steady-state concentration of succinate would be necessary to maintain $\Delta G'$ for succinate oxidation at −1.5 kcal/mol?

•5-10. Proof of Additivity. A useful property of thermodynamic parameters such as $\Delta G'$ or $\Delta G^{\circ\prime}$ is that they are additive for sequential reactions. Assume that K_{AB}', K_{BC}', and K_{CD}' are the respective equilibrium constants for reactions 1, 2, and 3 of the following sequence:

$$A \underset{\text{Reaction 1}}{\rightleftharpoons} B \underset{\text{Reaction 2}}{\rightleftharpoons} C \underset{\text{Reaction 3}}{\rightleftharpoons} D \qquad (5\text{-}33)$$

(a) Prove that the equilibrium constant K_{AD}' for the overall conversion of A to D is the *product* of the three component equilibrium constants:

$$K_{AD}' = K_{AB}' K_{BC}' K_{CD}' \qquad (5\text{-}34)$$

(b) Prove that the $\Delta G^{\circ\prime}$ for the overall conversion of A to D is the *sum* of the three component $\Delta G^{\circ\prime}$ values:

$$\Delta G_{AD}^{\circ\prime} = \Delta G_{AB}^{\circ\prime} + \Delta G_{BC}^{\circ\prime} + \Delta G_{CD}^{\circ\prime} \qquad (5\text{-}35)$$

(c) Prove that the $\Delta G'$ values are similarly additive.

•5-11. Utilizing Additivity. The additivity of thermodynamic parameters discussed in Problem 5-10 applies not just to sequential reactions in a pathway, but to *any* reactions or processes. Moreover, it also applies to subtraction of reactions. Use this information to answer the following questions.

(a) The phosphorylation of glucose using inorganic phosphate (abbreviated P$_i$) is endergonic ($\Delta G^{\circ\prime}$ = +3.3 kcal/mol), whereas the dephosphorylation (hydrolysis) of ATP is exergonic ($\Delta G^{\circ\prime}$ = −7.3 kcal/mol):

$$\text{glucose} + \text{P}_i \rightleftharpoons \text{glucose-6-phosphate} + \text{H}_2\text{O} \quad (5\text{-}36)$$

$$\text{ATP} + \text{H}_2\text{O} \rightleftharpoons \text{ADP} + \text{P}_i \qquad (5\text{-}37)$$

Write a reaction for the phosphorylation of glucose by the transfer of a phosphate group from ATP, and calculate $\Delta G^{\circ\prime}$ for the reaction.

(b) Phosphocreatine is used by your muscle cells to store energy. The dephosphorylation of phosphocreatine, like that of ATP (reaction 5-37), is a highly exergonic reaction with $\Delta G^{\circ\prime}$ = −10.3 kcal/mol:

$$\text{phosphocreatine} + \text{H}_2\text{O} \rightleftharpoons \text{creatine} + \text{P}_i \quad (5\text{-}38)$$

Write a reaction for the transfer of phosphate from phosphocreatine to ADP to generate creatine and ATP, and calculate $\Delta G^{\circ\prime}$ for the reaction.

5-12. The Energetics of Glucose Phosphorylation. The enzyme *hexokinase* catalyzes the first reaction in the glycolytic pathway, which we will encounter in Chapter 13. In this reaction, glucose is phosphorylated to form glucose-6-phosphate, with ATP as the phosphate donor and energy source:

$$\text{glucose} + \text{ATP} \xrightarrow[\text{hexokinase}]{} \text{glucose-6-phosphate} + \text{ADP} \qquad \text{(5-39)}$$

$\Delta G°'$ for this reaction at 25°C is −4.0 kcal/mol. The concentrations of ATP, ADP, and glucose-6-phosphate (G6P) in a typical bacterial cell are as follows:

$$[\text{ATP}] = 2.0 \text{ m}M \qquad [\text{ADP}] = 0.15 \text{ m}M \qquad [\text{G6P}] = 0.05 \text{ m}M$$

(a) What is K_{eq} for reaction 5-39 at 25°C? Show your calculations.

(b) In a bacterial cell with ATP, ADP, and glucose-6-phosphate present at the above concentrations, at what concentration of glucose would this reaction be in equilibrium? Would you expect this reaction to be at equilibrium in a typical bacterial cell? Explain your reasoning.

(c) Assuming that the concentration of glucose in the bacterial cell is 5.0 m*M*, what is $\Delta G'$ for this reaction at 25°C? Show your calculations.

(d) Knowing that $\Delta G°'$ for the hydrolysis of ATP is −7.3 kcal/mol, what is $\Delta G°'$ for the direct phosphorylation of glucose by inorganic phosphate? Is this reaction ever likely to be exergonic in a bacterial cell? Explain your reasoning.

(e) What does the $\Delta G'$ value that you calculated in part c tell you about the rate at which the ATP-dependent phosphorylation of glucose will occur in a typical bacterial cell? Explain your reasoning.

5-13. Protein Folding. In Chapter 2, you learned that a polypeptide in solution usually folds into its proper three-dimensional shape spontaneously. The driving force for this folding is the tendency to achieve the thermodynamically most favored conformation. A folded polypeptide can be induced to unfold (i.e., will undergo denaturation) if the solution is heated or made acidic or alkaline. The denatured polypeptide is a random structure, with many possible conformations.

(a) What is the sign of ΔG for the folding process? What about for the unfolding (denaturation) process?

(b) What is the sign of ΔS for the folding process? What about for the unfolding (denaturation) process?

(c) Will the contribution of ΔS to the free energy change be positive or negative?

Suggested Reading

References of historical importance are marked with a • .

General References
Lehninger, A. L., D. L. Nelson, and M. M. Cox. *Principles of Biochemistry*, 3[d] ed. New York: Worth, 1999.
Mathews, C. K., K. E. van Holde, and K G. Ahern. *Biochemistry*, 3[d] ed. San Francisco: Benjamin/Cummings, 2000.

Historical References
• Blum, H. F. *Time's Arrow and Evolution*, 3d. ed. Princeton, NJ: Princeton University Press, 1968.
• Gates, D. M. The flow of energy in the biosphere. *Sci. Amer.* 224 (September 1971): 88.
• Racker, E. From Pasteur to Mitchell: A hundred years of bioenergetics. *Fed. Proc.* 39 (1980): 210.

Bioenergetics
Cooper, A. Thermodynamic analysis of biomolecular interactions. *Curr. Opin. Chem. Biol.* 3 (1999): 557.
Demetrius, L. Thermodynamics and evolution. *J. Theor. Biol.* 206 (2000): 1.
Harris, D. A. *Bioenergetics at a Glance.* London: Blackwell Press, 1995.
Haynie, D. T. *Biological Thermodynamics.* Cambridge: Cambridge University Press, 2001.
Papa, S., F. Guerrieri, and J. M. Tager. *Frontiers of Cellular Bioenergetics: Molecular Biology, Biochemistry, and Physiopathology.* New York: Kluwer Academic/Plenum Publishers, 1999.
Peschek, G. A., W. Loffelhardt, and G. Schmetterer, eds. *Phototrophic Prokaryotes.* New York: Kluwer Academic/Plenum Publishers, 1999.
Rees, D. C., and J. B. Howard. Structural bioenergetics and energy transduction mechanisms. *J. Molec. Med.* 293 (1999): 343.

6 Enzymes: The Catalysts of Life

In Chapter 5, we encountered $\Delta G'$, the change in free energy, and saw its importance as an indicator of thermodynamic spontaneity. Specifically, the *sign* of $\Delta G'$ tells us whether a reaction is possible in the indicated direction, and the *magnitude* of $\Delta G'$ indicates how much energy will be released (or must be provided) as the reaction proceeds in that direction under the conditions for which $\Delta G'$ was calculated. At the same time, we were careful to note that because it is a thermodynamic parameter, $\Delta G'$ can provide us with no clue as to whether a feasible reaction will actually take place, or at what rate. In other words, $\Delta G'$ tells us only whether a reaction *can* go but says nothing at all about whether it actually *will* go. For that distinction we need to know not just the direction and energetics of the reaction, but something about the mechanism and rate as well.

This brings us to the topic of **enzyme catalysis,** because virtually all cellular reactions or processes are mediated by protein (or, in certain cases, RNA) catalysts called **enzymes.** The only reactions that occur at any appreciable rate in a cell are those for which the appropriate enzymes are present and active. Thus, enzymes almost always spell the difference between "can go" and "will go" for cellular reactions. It is only as we explore the nature of enzymes and their catalytic properties that we begin to understand how reactions that are energetically feasible actually take place in cells and how the rates of such reactions are controlled.

In this chapter, we will first examine why thermodynamically spontaneous reactions do not usually occur at appreciable rates without a catalyst, and then we will look at the role of enzymes as specific biological catalysts. We will also see how the rate of an enzyme-catalyzed reaction is affected by the concentration of substrate available to it, as well as some of the ways in which reaction rates are regulated to meet the needs of the cell.

Activation Energy and the Metastable State

If you stop to think about it, you are already familiar with many reactions that are thermodynamically feasible yet do not occur to any appreciable extent. An obvious example from Chapter 5 is the oxidation of glucose (see reaction 5-11). This reaction (or series of reactions, really) is highly exergonic ($\Delta G^{\circ\prime} = -686$ kcal/mol) and yet does not take place on its own. In fact, glucose crystals or a glucose solution can be exposed to the oxygen in the air indefinitely, and little or no oxidation occurs. The cellulose in the paper on which these words are printed is another example—and so, for that matter, are *you*, consisting as you do of a complex collection of thermodynamically unstable molecules.

Not nearly as familiar but equally important to cellular chemistry are the many thermodynamically feasible reactions in cells that could go, but do not proceed at an appreciable rate on their own. As an example, consider the high-energy molecule adenosine triphosphate (ATP), which has a highly favorable $\Delta G^{\circ\prime}$ (-7.3 kcal/mol) for the hydrolysis of its terminal phosphate group to form the corresponding diphosphate (ADP) and inorganic phosphate (P_i):

$$\text{ATP} + \text{H}_2\text{O} \rightleftharpoons \text{ADP} + \text{P}_i \qquad \textbf{(6-1)}$$

This reaction is very exergonic under standard conditions and is even more so under the conditions that prevail in cells. Yet despite the highly favorable free energy change, this reaction occurs only slowly on its own, so that ATP remains stable for several days when dissolved in pure water. This property turns out to be shared by many biologically important molecules and reactions, and it is important to understand why.

Before a Chemical Reaction Can Occur, the Activation Energy Barrier Must Be Overcome

Molecules that should react with one another often do not because they lack sufficient energy. For every reaction, there is a specific **activation energy** (E_A), which is the minimum amount of energy that reactants must have before collisions between them will be successful in giving rise to products. More specifically, reactants need to reach an intermediate chemical stage called the **transition state**, the free energy of which is higher than that of the initial reactants. Figure 6-1a shows the activation energy required for molecules of ATP and H_2O to reach their transition state. $\Delta G^{o\prime}$ measures the difference in free energy between reactants and products (−7.3 kcal/mol for this particular reaction), whereas E_A indicates the minimum energy required for the reactants to reach the transition state and hence to be capable of giving rise to products.

The actual rate of a reaction is always proportional to the fraction of molecules that have an energy content equal to or greater than E_A. When in solution at room temperature, molecules of ATP and water move about readily, each possessing a certain amount of energy at any instant. As Figure 6-1b shows, the energy distribution among molecules will be bell-shaped; some molecules will have very little energy, some will have a lot, and most will be somewhere near the average. The important point is that only those with enough energy to exceed the *activation energy barrier* are capable of reacting at a given instant.

The Metastable State Is a Result of the Activation Barrier

For most biologically important reactions at normal cellular temperatures, the activation energy is sufficiently high that the proportion of molecules possessing that much energy at any instant is extremely small. Accordingly, the rates of uncatalyzed reactions in cells are very low, and most molecules appear to be stable even though they are potential reactants in thermodynamically favored reactions. They are, in other words, thermodynamically unstable but they do not have enough energy to exceed the activation energy barrier.

Such seemingly stable molecules are said to be in a **metastable state.** For cells and cell biologists, high activation energies and the resulting metastable state of cellular constituents are crucial, because life by its very nature is a system maintained in a steady state a long way from equilibrium. Were it not for the metastable state, all reactions would proceed quickly to equilibrium, and life as we know it would be impossible. Life, then, depends critically on the high activation energies that prevent most cellular reactions from occurring at appreciable rates in the absence of a suitable catalyst.

Catalysts Overcome the Activation Energy Barrier

Important as it is to the maintenance of the metastable state, the activation energy requirement is a barrier that must be overcome if desirable reactions are to proceed at

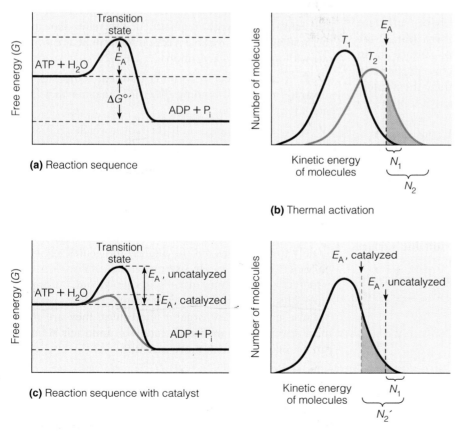

(a) Reaction sequence

(b) Thermal activation

(c) Reaction sequence with catalyst

(d) Catalytic activation

Figure 6-1 The Effect of Catalysis on Activation Energy and Number of Molecules Capable of Reaction. (a) The activation energy E_A is the minimum amount of kinetic energy that reactant molecules (here ATP and H_2O) must possess to permit collisions leading to product formation. After reactants overcome the activation energy barrier and enter into a reaction, the products have less free energy by the amount $\Delta G^{o\prime}$. **(b)** The number of molecules N_1 that have sufficient energy to exceed the activation energy barrier (E_A) and collide successfully can be increased to N_2 by raising the temperature from T_1 to T_2. **(c)** Alternatively, the activation energy can be lowered by a catalyst, thereby **(d)** increasing the number of molecules from N_1 to $N_2{}^\prime$.

reasonable rates. Since the energy content of a given molecule must exceed E_A before that molecule is capable of undergoing reaction, the only way a reaction involving metastable reactants can be made to proceed at an appreciable rate is to increase the proportion of molecules with sufficient energy. This can be achieved either by increasing the average energy content of all molecules or by lowering the activation energy requirement.

One way to increase the energy content of the system is by the input of heat. As Figure 6-1b illustrates, simply increasing the temperature of the system from T_1 to T_2 will increase the kinetic energy of the average molecule, thereby ensuring a greater number of reactive molecules (N_2 instead of N_1). Thus, the hydrolysis of ATP could be facilitated by heating the solution, giving each ATP and water molecule more energy.

The problem with using an elevated temperature is that such an approach is incompatible with life, because biological systems require a relatively constant temperature. Cells are basically *isothermal* (constant-temperature) systems and require isothermal methods to solve the activation problem. Moreover, we can anticipate a later topic if we note that this approach would also suffer from a lack of specificity because the input of heat would result in many activation energy barriers being indiscriminately overcome. Yet the essence of successful regulation of cellular activities lies in the ability to facilitate specific reactions under specific conditions while leaving other metastable molecules undisturbed.

The alternative to thermal activation is to lower the activation energy requirement, thereby ensuring that a greater proportion of molecules will have sufficient energy to collide successfully and undergo reaction. If the reactants can be bound on some sort of surface in an arrangement that brings potentially reactive portions of adjacent molecules into close juxtaposition, their interaction will be greatly favored and the activation energy effectively reduced.

Providing such a reactive surface is the task of a **catalyst,** an agent that enhances the rate of a reaction by lowering the energy of activation (Figure 6-1c), thereby ensuring that a higher proportion of the molecules are energetic enough to undergo reaction without the input of heat (Figure 6-1d). A primary feature of a catalyst is that it is not permanently changed or consumed as the reaction proceeds. It simply provides a suitable surface and environment to facilitate the reaction.

For a specific example of catalysis, consider the decomposition of hydrogen peroxide (H_2O_2) into water and oxygen:

$$2H_2O_2 \rightleftharpoons 2H_2O + O_2 \qquad \text{(6-2)}$$

This is a thermodynamically favored reaction, yet hydrogen peroxide exists in a metastable state because of the high activation energy of the reaction. However, if we add a small number of ferric ions (Fe^{3+}) to a hydrogen per-

oxide solution, the decomposition reaction proceeds about 30,000 times faster than without the ferric ions. Clearly, Fe^{3+} is a catalyst for this reaction, lowering the activation energy (as shown in Figure 6-1c) and thereby ensuring that a significantly greater proportion (30,000-fold more) of the hydrogen peroxide molecules possess adequate energy to decompose at the existing temperature without the input of added energy.

In cells, the solution to hydrogen peroxide breakdown is not the addition of ferric ions but the enzyme *catalase,* an iron-containing protein. In the presence of catalase, the reaction proceeds about 100,000,000 times faster than the uncatalyzed reaction. Catalase contains iron atoms bound in chemical structures called *porphyrins,* thus taking advantage of inorganic catalysis within the context of a protein molecule. This combination is obviously a much more effective catalyst for hydrogen peroxide decomposition than ferric ions by themselves. The rate enhancement of about 10^8 for catalase is not at all an atypical value; the rate enhancements of enzyme-catalyzed reactions range from 10^7 to as high as 10^{14} compared with the uncatalyzed reaction. These values underscore the extraordinary importance of enzymes as catalysts and bring us to the main theme of this chapter.

Enzymes as Biological Catalysts

Regardless of their chemical nature, all catalysts share the following three basic properties.

1. A catalyst increases the rate of a reaction by lowering the activation energy requirement, thereby allowing a thermodynamically feasible reaction to occur at a reasonable rate without thermal activation.

2. A catalyst acts by forming transient complexes with substrate molecules, binding them in a manner that facilitates their interaction.

3. A catalyst changes only the *rate* at which equilibrium is achieved; it has no effect on the *position* of the equilibrium. This means that a catalyst can enhance the rate of exergonic reactions but cannot somehow drive an endergonic reaction. Catalysts, in other words, are not thermodynamic genies.

These properties are common to all catalysts, organic and inorganic alike. In terms of our example, they apply equally to ferric ions and to catalase molecules. However, biological systems rarely use inorganic catalysts. Instead, essentially all catalysis in cells is carried out by organic molecules (proteins, in most cases) called *enzymes.* Because enzymes are organic molecules, they are much more specific than inorganic catalysts, and their activities can be regulated much more carefully.

Most Enzymes Are Proteins

The capacity of cellular extracts to catalyze chemical reactions has been known since the fermentation studies of

Eduard and Hans Buchner in 1897. In fact, one of the first terms for what we now call enzymes was *ferments*. However, it was not until 1926 that a specific enzyme, *urease*, was crystallized (from jack beans, by James B. Sumner) and shown to be a protein. (Urease catalyzes the decomposition of urea, NH_2—CO—NH_2, to ammonia and carbon dioxide. Its crystallization from a common plant not only established the protein nature of enzymes but also drove the final nail in the coffin of *vitalism*, the belief that biochemical reactions could only occur in living cells through the agency of a "vital force.")

Even then, it took a while for biochemists and enzymologists to appreciate that the "ferments" they were studying were, in fact, protein catalysts, and that to understand them as catalysts really meant understanding their structure and function as proteins. (Since the early 1980s, biologists have recognized that, in addition to proteins, certain RNA molecules, called *ribozymes*, also have catalytic activity. Ribozymes will be discussed in a subsequent section. Here, we will consider enzymes as proteins—which, in fact, most are.)

The Active Site. One of the most important concepts to emerge from our understanding of enzymes as proteins is the active site. Every enzyme, regardless of the reaction it catalyzes or the details of its structure, contains somewhere within its tertiary configuration a characteristic cluster of amino acids forming the **active site** where the catalytic event occurs for which that enzyme is responsible. Usually, the active site is an actual groove or pocket with chemical and structural properties that accommodate the intended substrate with high specificity. Figure 6-2 shows computer graphic models of the enzymes *lysozyme* and *carboxypeptidase A* that illustrate well the precise fit of the substrate molecule into pockets produced by the characteristic folding of polypeptide chains. Lysozyme is an enzyme that breaks the glycosidic bond between *N*-acetylglucosamine (GlcNAc) and *N*-acetylmuramic acid (MurNAc) groups in the peptidoglycan of bacterial cell walls, thereby causing the bacterial cells to lyse (break open) and die. (For the structure of the peptidoglycan, see Figure 3-26b.) Carboxypeptidase A is an enzyme that breaks down polypeptides by removing one amino acid at a time from the C-terminus of the polypeptide.

The amino acids that make up the active site of an enzyme are not usually contiguous to one another along the primary sequence of the protein. Instead, they are brought together in just the right conformation by the specific three-dimensional folding of the polypeptide chain. For the carboxypeptidase molecule, for example, the active site consists of a tightly bound zinc ion bonded to the side chains of the histidine residue at position 69, the glutamate residue at position 72, and another histidine residue at position 196 of the polypeptide chain. This configuration is shown in Figure 6-3a. When the substrate for the enzyme binds to the active site, the structure of the enzyme changes, bringing the arginine residue at position 145, the glutamate residue at position 270, and the tyrosine residue at position 248 into close proximity with the substrate as well (Figure 6-3b). Each of these amino acids plays a critical role in the actual catalytic process.

For carboxypeptidase, then, 6 amino acids out of a total chain length of 307 are actually involved at the active site. This is typical of most enzymes; the active site usually involves about 5% of the surface area of the enzyme. Also typical is the involvement of amino acids from distant positions along the primary structure of the polypeptide. This involvement underscores the importance of the overall tertiary structure of the protein; only as an enzyme molecule attains its stable three-dimensional conformation are the specific amino acids brought together to constitute the active site.

Of the 20 different amino acids that make up proteins, only a few are actually involved in the active sites of the many proteins that have been studied. In most cases, these are the amino acids cysteine, histidine, serine, aspartate, glutamate, and lysine. All of these can participate in binding the substrate to the active site during the catalytic process, and several (histidine, aspartate, and glutamate) also serve as donors or acceptors of protons.

Some enzymes consist not only of one or more polypeptide chains but of specific nonprotein components

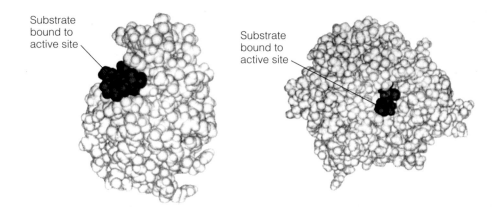

Substrate bound to active site

Substrate bound to active site

(a) Lysozome

(b) Carboxypeptidase

Figure 6-2 Molecular Structures of Lysozyme and Carboxypeptidase A. The enzymes **(a)** lysozyme and **(b)** carboxypeptidase A are shown here as computer-generated space-filling models with a substrate molecule bound to the active site in each case—a short segment of a bacterial peptidoglycan in the case of lysozyme and an artificial peptide in the case of carboxypeptidase A. For a detailed view of the carboxypeptidase active site, see Figure 6-3.

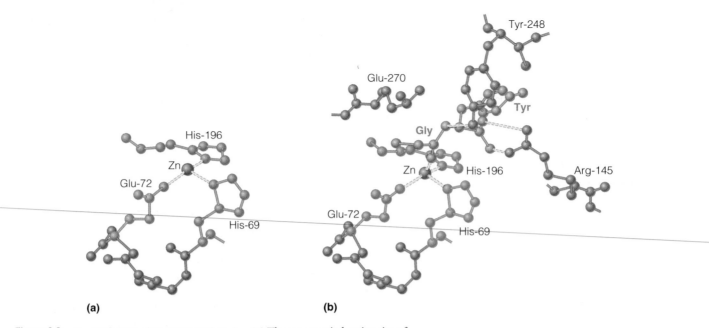

Figure 6-3 The Active Site of Carboxypeptidase A. **(a)** The unoccupied active site of carboxypeptidase A consists of a zinc ion bonded coordinately to two histidine residues (His-69 and His-196) and a glutamate residue (Glu-72). **(b)** Upon substrate binding (using an artificial peptide, shown in red), the active site is induced to close in on the substrate, bringing an arginine residue (Arg-145), a glutamate residue (Glu-270), and a tyrosine residue (Tyr-248) into close proximity with the substrate. Like many other enzymes, carboxypeptidase A contains a prosthetic group at its active site—a zinc ion, in this case.

as well. These components, called **prosthetic groups,** are usually either small organic molecules or metal ions, such as the iron in catalase. Frequently, they function as electron acceptors, because none of the amino acid side chains is a good electron acceptor. Where present, prosthetic groups are located at the active site and are indispensable for the catalytic activity of the enzyme. Carboxypeptidase A is an example of an enzyme with a prosthetic group, containing a zinc atom at its active site (see Figure 6-3a).

Enzyme Specificity. A consequence of the structure of the active site is that enzymes display a high degree of **substrate specificity,** evidenced by an ability to discriminate between very similar molecules. Specificity is probably one of the most characteristic properties of living systems, and enzymes are especially dramatic examples of biological specificity.

We can illustrate their specificity by comparing enzymes with inorganic catalysts. Most inorganic catalysts are quite nonspecific in that they will act on a variety of compounds that share some general chemical feature. Consider, for example the *hydrogenation* of (addition of hydrogen to) an unsaturated C=C bond:

$$
\begin{array}{c}
\quad\quad\;\; H \;\; H \quad\quad\quad\quad\quad\quad\quad H \;\; H \\
\quad\quad\;\; | \;\;\; | \quad\quad\quad\quad\quad\quad\quad\; | \;\;\; | \\
R{-}C{=}C{-}R' + H_2 \xrightarrow{\;\text{Pt or Ni}\;} R{-}C{-}C{-}R' \quad \textbf{(6-3)}\\
\quad\quad\quad\quad\quad\quad\quad\quad\quad\quad\quad\quad\quad | \;\;\; | \\
\quad\quad\quad\quad\quad\quad\quad\quad\quad\quad\quad\quad\quad H \;\; H
\end{array}
$$

This reaction can be carried out in the laboratory using a platinum (Pt) or nickel (Ni) catalyst, as indicated. These inorganic catalysts are very nonspecific, however; they can catalyze the hydrogenation of a wide variety of unsaturated compounds. In fact, nickel or platinum is used commercially to hydrogenate polyunsaturated vegetable oils in the manufacture of solid cooking fats or shortenings. Regardless of the exact structure of the unsaturated compound, it can be effectively hydrogenated in the presence of nickel or platinum.

By way of contrast, consider the biological example of hydrogenation involved in the conversion of fumarate to succinate, a reaction we will encounter again in Chapter 14:

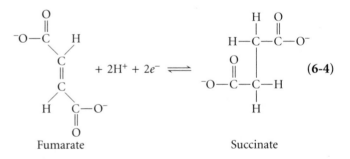

$$+ 2H^+ + 2e^- \rightleftharpoons \quad\quad\quad\quad\quad\quad \textbf{(6-4)}$$

Fumarate   Succinate

This particular reaction is catalyzed in cells by the enzyme *succinate dehydrogenase* (so named because it normally functions in the opposite direction during energy metabolism). This dehydrogenase, like most enzymes, is highly specific. It will not add or subtract hydrogens from any compounds except those shown in reaction 6-4. In fact, this particular enzyme is so specific that it will not even recognize maleate, a stereoisomer of fumarate (Figure 6-4).

Not all enzymes are quite this specific; some accept a number of closely related substrates, and others accept any of a whole group of substrates as long as they possess some common structural feature. Such **group specificity** is seen most often with enzymes involved in the synthesis or degradation of polymers. The purpose of carboxypeptidase A is to degrade polypeptide chains from the carboxyl end; thus, it makes sense for the enzyme to accept any of a wide variety of polypeptides as substrate, since it would be needlessly extravagant of the cell to require a separate enzyme for every different peptide bond that has to be hydrolyzed in polypeptide degradation.

In general, however, enzymes are highly specific with respect to substrate, such that a cell must possess almost as many different kinds of enzymes as it has reactions to catalyze. For a typical cell, this means that thousands of different enzymes are necessary to carry out its full metabolic program. At first, that may seem wasteful in terms of proteins to be synthesized, genetic information to be stored and read out, and enzyme molecules to have on hand in the cell. But you should also be able to see the tremendous regulatory possibilities this suggests, a point we will return to later.

Enzyme Diversity and Nomenclature. Given the specificity of enzymes and the large number of reactions that occur within a cell, it is not surprising that thousands of different enzymes have been identified. This enormous diversity of enzymes and enzymatic functions led initially to a variety of schemes for naming enzymes as they were discovered and characterized. Some were given names indicative of the substrate; *ribonuclease, protease,* and *amylase* are examples. Others, such as *succinate dehydrogenase* and *phosphoglucoisomerase,* were named to describe their function. Still other enzymes have names that provide little indication of either substrate or function. *Trypsin, catalase,* and *lysozyme* are enzymes in this category.

The proliferation of common names for enzymes and the resulting confusion eventually prompted the International Union of Biochemistry to appoint an Enzyme Commission (EC) charged with devising a rational system for naming enzymes. The EC system is represented in

Table 6-1. Enzymes are divided into six major classes based on their general functions, with subgroups used to define their functions more precisely. The six major classes are *oxidoreductases, transferases, hydrolases, lyases, isomerases,* and *ligases.* Table 6-1 provides an example of each class, using enzymes that catalyze reactions to be discussed later in the text.

The EC system assigns every known enzyme a four-part number. For example, EC 3.4.17.1 is the number for carboxypeptidase A. The first three numbers define the major class, subclass, and sub-subclass, and the final number is the serial number assigned to the enzyme when it was added to the list. Thus, carboxypeptidase A is the first entry in the 17th sub-subclass of the fourth subclass of class 3 (hydrolases). The list of enzymes that have been classified in this way grows continuously as new enzymes are discovered and characterized.

Sensitivity to Temperature. In addition to their specificity and diversity, enzymes are also characterized by their sensitivity to temperature. This temperature dependence is not usually a practical concern for enzymes in the cells of mammals or birds because these organisms are *homeotherms,* capable of regulating body temperature independent of the environment. However, organisms such as lower animals, plants, protists, and bacteria function at the temperature of their environment, which can vary greatly. For these organisms, the dependence of enzyme activity on temperature is very significant.

Within limits, the rate of an enzyme-catalyzed reaction increases with temperature because the greater kinetic energy of both enzyme and substrate molecules ensures more frequent collisions, thereby increasing the likelihood of correct substrate binding. At some point, however, further increases in temperature are counterproductive because the enzyme molecule begins to denature. Hydrogen bonds break, hydrophobic interactions change, and the structural integrity of the active site is disrupted, causing a loss of activity.

The temperature range over which an enzyme denatures varies greatly from enzyme to enzyme and especially from organism to organism. Figure 6-5a contrasts the temperature dependence of a typical enzyme from the human body with that of a typical enzyme from a thermophilic bacterium. Not surprisingly, the reaction rate of a human enzyme is maximum at about 37°C, which is normal body temperature. The sharp decrease in activity thereafter reflects the progressive denaturation of the enzyme molecules. Most enzymes of homeotherms are inactivated by temperatures above about 50–55°C. For such enzymes, heat inactivation is clearly a laboratory phenomenon only. However, some enzymes are remarkably sensitive to heat and are denatured and inactivated at lower temperatures—in some cases, even by body temperatures encountered in people with high fevers.

On the other end of the spectrum, some enzymes retain activity at unusually high temperatures. The right-hand

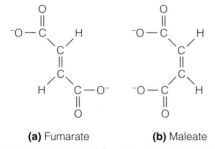

(a) Fumarate   (b) Maleate

Figure 6-4   The Stereoisomers (a) fumarate and (b) maleate.

Table 6-1 The Major Classes of Enzymes with an Example of Each*

Class	Reaction Type	Enzyme Name	Example — Reaction Catalyzed
1. Oxidoreductases	Oxidation-reduction reactions	Alcohol dehydrogenase (EC 1.1.1.1) (oxidation with NAD^+)	CH_3-CH_2-OH (Ethanol) $\; \overset{NAD^+ \;\; NADH + H^+}{\rightleftharpoons} \;$ $CH_3-\overset{O}{\overset{\|}{C}}-H$ (Acetaldehyde)
2. Transferases	Transfer of functional groups from one molecule to another	Glycerokinase (EC 2.4.3.2) (phosphorylation)	$HO-CH_2-\overset{OH}{\overset{\|}{CH}}-CH_2-OH$ (Glycerol) $\; \xrightarrow{ATP \;\; ADP} \;$ $HO-CH_2-\overset{OH}{\overset{\|}{CH}}-CH_2-O-PO_3^{2-}$ (Glycerol phosphate)
3. Hydrolases	Hydrolytic cleavage of one molecule into two molecules	Carboxypeptidase A (EC 3.4.17.1) (peptide bond cleavage)	R_{n-1} $\overset{O}{\overset{\|}{}}$ R_n $\overset{O}{\overset{\|}{}}$ $-NH-\overset{}{\underset{}{CH}}-\overset{}{C}-NH-\overset{}{\underset{}{CH}}-\overset{}{C}-O^-$ (C-terminus of polypeptide) $\; \xrightarrow{H_2O} \;$ $-NH-\overset{R_{n-1}}{\underset{}{CH}}-\overset{O}{\overset{\|}{C}}-O^-$ (Shortened polypeptide) $+\; H_3N^+-\overset{R_n}{\underset{}{CH}}-\overset{O}{\overset{\|}{C}}-O^-$ (C-terminal amino acid)
4. Lyases	Removal of a group from, or addition of a group to, a molecule with rearrangement of electrons	Pyruvate decarboxylase (EC 4.1.1.1) (decarboxylation)	$CH_3-\overset{O}{\overset{\|}{C}}-\overset{O}{\overset{\|}{C}}-O^- + H^+$ (Pyruvate) $\; \longrightarrow \;$ $CH_3-\overset{O}{\overset{\|}{C}}-H \; + \; CO_2$ (Acetaldehyde)
5. Isomerases	Movement of a functional group within a molecule	Maleate isomerase (EC 5.2.1.1) (cis-trans isomerization)	Maleate $\; \rightleftharpoons \;$ Fumarate
6. Ligases	Joining of two molecules to form a single molecule	Pyruvate carboxylase (EC 6.4.1.1) (carboxylation)	$CH_3-\overset{O}{\overset{\|}{C}}-\overset{O}{\overset{\|}{C}}-O^- + CO_2$ (Pyruvate) $\; \xrightarrow{ATP \;\; ADP + P_i} \;$ $^-O-\overset{O}{\overset{\|}{C}}-CH_2-\overset{O}{\overset{\|}{C}}-\overset{O}{\overset{\|}{C}}-O^-$ (Oxaloacetate)

*This system for classifying, naming, and assigning numbers to enzymes was devised by the Enzyme Commission of the International Union of Biochemistry.

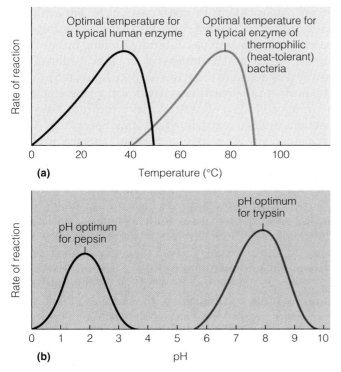

(a)

(b)

Figure 6-5 **The Effect of Temperature and pH on the Reaction Rate of Enzyme-Catalyzed Reactions.** **(a)** The dependence of reaction rate on temperature for a typical human enzyme (black) and a typical enzyme from a thermophilic bacterium (green). The reaction rate is maximized at the optimal temperature, which is about 37°C (body temperature) for the human enzyme and about 75°C (the temperature of a typical hot spring) for the bacterial enzyme. The initial increase in activity with temperature is due to the greater kinetic energy of enzyme and substrate molecules at higher temperatures. Eventually, however, an inactivating effect is seen as further increases in temperature result in thermal denaturation of the enzyme protein. **(b)** The dependence of reaction rate on pH for the gastric enzyme pepsin (black) and the intestinal enzyme trypsin (red). The reaction rate is maximized at the optimal pH, which is about 2.0 for pepsin and about 8.0 for trypsin. The pH optimum for an enzyme corresponds to the proton concentration at which ionizable groups on both the enzyme and the substrate molecules are in the most favorable form for maximum reactivity. Changes in pH away from the optimum usually reflect titration of charged groups on the enzyme, the substrate, or both. The pH optimum of an enzyme usually reflects the pH of the intracellular compartment or other environment in which the enzyme is active.

curve in Figure 6-5a depicts the temperature dependence of an enzyme from one of the thermophilic archaebacteria mentioned in Chapter 4. These organisms thrive in acidic hot springs at temperatures as high as 80°C! Clearly, each of the thousands of different kinds of enzymes such organisms need for normal cellular activity must be capable of functioning at temperatures that would denature most of the enzymes in the cells of other organisms, your own included.

Sensitivity to pH. Enzymes are also sensitive to pH, with many enzymes active only within a pH range of about 3–4 pH units. This pH dependence is usually due to the presence of one or more charged amino acids at the active site and/or on the substrate itself, with activity usually dependent on having such groups present in a specific form, either charged or uncharged. For example, the active site of carboxypeptidase A involves the carboxyl groups from each of two glutamate residues, located at positions 72 and 270 of the protein (see Figure 6-3). These carboxyl groups must be present in the charged (ionized) form, so the enzyme becomes inactive if the pH is decreased to the point at which the glutamate carboxyl groups on most of the enzyme molecules are protonated and therefore uncharged.

As you might expect, the pH dependence of an enzyme usually reflects the environment in which that enzyme is normally active. Figure 6-5b shows the pH dependence of two protein-degrading enzymes found in the human digestive tract. Pepsin is present in the stomach, where the pH is usually about 2, whereas trypsin is secreted into the small intestine, which has a pH of about 8. Both enzymes are active over a range of almost 4 pH units but differ greatly in their pH optima, consistent with the conditions in their respective locations within the body.

Sensitivity to Other Factors. In addition to temperature and pH, enzymes are also sensitive to other factors, including substances that act as inhibitors or activators of the enzyme, a topic to which we will return later in the chapter. In the case of enzymes with group specificity, activity can also be affected by the presence of alternative substrates. Most enzymes are also sensitive to the ionic environment, which is likely to affect the overall conformation of the enzyme because of its dependence on hydrogen bonds. The ionic environment may also affect binding of the substrate because ionic interactions are often involved in the interaction between the substrate and the active site.

Substrate Binding, Activation, and Reaction Occur at the Active Site

Because of the precise chemical fit between the active site of an enzyme and its substrates, enzymes are highly effective as catalysts. This effectiveness can be seen in the much greater extent to which reaction rates are catalyzed by enzymes compared with inorganic catalysts. As we noted previously, enzyme-catalyzed reactions proceed 10^7 to 10^{14} times faster than uncatalyzed reactions, versus 10^3 to 10^4 times faster for inorganic catalysts. As you might guess, most of the interest in enzymes focuses on the active site, where binding, activation, and chemical transformation of the substrate occur.

Substrate Binding. Initial contact between the active site of an enzyme and a potential substrate molecule depends on random collision. Once in the groove, or pocket, of the active site, however, the substrate molecules are bound temporarily to the enzyme surface in just the right orientation to one another and to specific catalytic groups on the enzyme to facilitate reaction. Substrate binding usually involves hydrogen bonds or ionic bonds (or both) to charged amino acids. These are generally weak bonds, but several bonds may hold a single molecule in place. The strength of the bonds between an enzyme and a substrate molecule is often in the range of 3–12 kcal/mol, less than one-tenth the strength of a single covalent bond (see Figure 2-2). Substrate binding is therefore readily reversible.

For many years, enzymologists regarded the active site as a rigid structure. They likened the fit of a substrate into the active site to that of a key into a lock, an analogy first suggested in 1894 by the German biochemist Emil Fischer. This *lock-and-key model*, shown in Figure 6-6a, explained enzyme specificity but did little to enhance our understanding of the catalytic event. A more helpful view of enzyme-substrate interaction is provided by the **induced-fit model,** first proposed in 1958 by Daniel Koshland. As Figure 6-6b illustrates, this model assumes that the initial binding of the substrate molecule(s) at the active site distorts both the enzyme and the substrate. Distortion of the substrate puts one or more covalent bonds under stress, thereby making those bond(s) more susceptible to catalytic attack and favoring the reaction.

Distortion of the enzyme involves a conformational change in the shape of the enzyme molecule and hence in the configuration of the active site. This change positions the proper reactive groups of the enzyme optimally for the catalytic reaction in which they are involved, thereby enhancing the likelihood of that reaction. Specifically, the conformational change brings into the active site amino acid side chains that are critical to the catalytic process but are not in the immediate area of the active site in the unin-

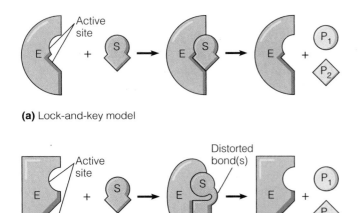

(a) Lock-and-key model

(b) Induced-fit model

Figure 6-6 Two Models for Enzyme-Substrate Interaction. (a) The lock-and-key model. In this early model, the enzyme molecule (E) was regarded as a rigid structure with an active site that fits the substrate (S) as a lock does a key. Reaction at the active site then converts molecules of substrate into molecules of products (P_1 and P_2). **(b)** The induced-fit model. According to this model, binding of a substrate molecule to the active site of an enzyme induces a conformational change in the enzyme molecule that positions the proper reactive groups at the active site optimally for the catalytic reaction. In addition, the tighter enzyme-substrate fit distorts one or more of the bonds of the substrate, thereby making these bonds more susceptible to catalytic attack.

duced conformation. In many cases, these are acidic or basic groups that promote catalysis. In the case of carboxypeptidase A, for example, substrate binding brings three critical amino acid residues—an arginine, a glutamate, and a tyrosine—into the active site (see Figure 6-3).

Evidence that such conformational changes actually occur upon binding of substrate has come from X-ray diffraction studies of crystallized proteins. Specifically, X-ray crystallography is used to determine the shape of an enzyme molecule with and without substrate bound to the active site. Figure 6-7 illustrates the conformational changes that take place upon substrate binding for *lysozyme,* the enzyme shown in Figure 6-2a, and for hexokinase, an enzyme that we will encounter again later in this chapter. For both enzymes, a noticeable change takes place in the conformation of the protein molecule in response to substrate binding. In the case of hexokinase, for example, the binding of substrate (a molecule of D-glucose) causes two domains of the enzyme to fold toward each other, closing the binding site cleft about the substrate (Figure 6-7b).

The conformational change upon substrate binding can result in remarkably large displacements of amino acid groups within the protein molecule. In the case of carboxypeptidase A, for example, substrate binding causes the tyrosine residue at position 248 (see Figure 6-3) to move 1.2 nm, a distance equal to about a quarter of the diameter of the enzyme molecule!

Substrate Activation. The role of the active site is not just to recognize and bind the appropriate substrate but also to *activate* it by subjecting it to the right chemical environment for catalysis. A given enzyme-catalyzed reaction may involve one or more means of **substrate activation.** Three of the most common mechanisms are as follows:

1. The change in enzyme conformation induced by initial substrate binding to the active site not only causes better complementarity and a tighter enzyme-substrate fit but also distorts one or more of its bonds, thereby weakening the bond and making it more susceptible to catalytic attack.

2. The enzyme may also accept or donate protons, thereby increasing the chemical reactivity of the substrate. This accounts for the importance of charged amino acids in active-site chemistry, which in turn explains why enzyme activity is so often dependent on pH.

3. As a further means of substrate activation, enzymes may also accept or donate electrons, thereby forming temporary covalent bonds between the enzyme and its substrate. This mechanism requires that the substrate have a region that is either electropositive (electron-deficient) or electronegative (electron-rich) and that the active site of the enzyme have one or more groups of opposite polarity. In one case, the electronegative side group of an appropriate amino acid at the active site donates electrons to the electropositive region of the substrate, in what is called a **nucleophilic substitution** reaction (Figure 6-8a). The hydroxyl group of serine, the sulfhydryl group of cysteine, and the indole group of histi-

more precisely, consider the special case where [S] is exactly equal to K_m. Under these conditions, the Michaelis-Menten equation can be written as follows:

$$v = \frac{V_{max}[S]}{K_m + [S]} = \frac{V_{max}[S]}{2[S]} = \frac{V_{max}}{2} \qquad \textbf{(6-10)}$$

This equation provides us with the definition we have been looking for: K_m is that specific substrate concentration at which the reaction proceeds at one-half of its maximum (upper-limiting) velocity. This specific concentration is a fixed value for a given enzyme catalyzing a specific reaction under specified conditions and is called the **Michaelis constant** (hence the designation K_m) after the enzymologist who first elucidated its meaning. Figure 6-9 illustrates the meaning of both V_{max} and K_m.

Why Are V_{max} and K_m Important to Cell Biologists?

Now that we understand what V_{max} and K_m mean, it is fair to ask why these kinetic parameters are important to cell biologists. The K_m value is useful because it allows us to estimate where along the Michaelis-Menten plot of Figure 6-9 an enzyme is functioning in a cell (providing, of course, that the normal substrate concentration in the cell is known). We can then estimate at what fraction of the maximum velocity the enzyme-catalyzed reaction is likely to be proceeding in the cell. Moreover, K_m can be regarded as an estimate of how "good" an enzyme is, in the sense that the lower the K_m value, the lower the substrate concentration range in which the enzyme is effective.* K_m values for several enzymes are given in Table 6-2.

The V_{max} for a particular reaction is important because it provides a measure of the rate of the reaction. Few enzymes actually encounter saturating substrate concentrations in vivo, so enzymes are not likely to be functioning at their maximum rate under cellular conditions. However, by knowing the V_{max} value, the K_m value, and the substrate concentration in vivo, we can at least estimate the likely rate of the reaction under cellular conditions.

V_{max} can also be used to determine another useful parameter called the **turnover number** (k_{cat}), which expresses the rate at which substrate molecules are converted to product by a single enzyme molecule when the enzyme is operating at its maximum velocity. The constant k_{cat} has the units of reciprocal time (s^{-1}, for example) and is calculated as the quotient of V_{max} over $[E_t]$, the concentration of the enzyme in moles/liter:

$$k_{cat} = \frac{V_{max}}{[E_t]} \qquad \textbf{(6-11)}$$

Turnover numbers vary greatly among enzymes, as is clear from the several examples given in Table 6-2.

The Double-Reciprocal Plot Is a Useful Means of Linearizing Kinetic Data

A classic Michaelis-Menten plot of v versus [S] is a faithful representation of the dependence of velocity on substrate concentration, but it is not an especially useful tool for the quantitative determination of the key kinetic parameters K_m and V_{max}. Its hyperbolic shape makes it difficult to extrapolate accurately to infinite substrate concentration, as would be required to determine the critical parameter V_{max}, and if V_{max} is not known accurately, K_m cannot be determined. This problem is readily apparent from Figure 6-9, since it would be difficult to estimate V_{max} accurately if it were not already sketched in, and without V_{max}, K_m cannot easily be estimated either.

To circumvent this problem and provide a more useful graphic approach, Hans Lineweaver and Dean Burk converted the hyperbolic relationship of the Michaelis-Menten equation into a linear function by inverting both sides of equation 6-7 and simplifying the resulting expression into the form of an equation for a straight line:

$$\frac{1}{v} = \frac{K_m + [S]}{V_{max}[S]} = \frac{K_m}{V_{max}[S]} + \frac{[S]}{V_{max}[S]}$$

$$= \frac{K_m}{V_{max}}\left(\frac{1}{[S]}\right) + \frac{1}{V_{max}} \qquad \textbf{(6-12)}$$

Equation 6-12 is the **Lineweaver-Burk equation.** If it is plotted as $1/v$ versus $1/[S]$, as in Figure 6-11, the resulting **double-reciprocal plot** is linear, with a y-intercept of $1/V_{max}$, an x-intercept of $-1/K_m$, and a slope of K_m/V_{max}. (You should be able to convince yourself of these intercept values by setting first $1/[S]$ and then $1/v$ equal to zero in

---

*K_m is often considered a measure of an enzyme's affinity for its substrates or, in effect, the dissociation constant for ES. In fact, however, it is a ratio of rate constants: $K_m = (k_2 + k_3)/k_1$. For the derivation of this relationship, see Problem 6-13 at the end of the chapter.

Table 6-2 K_m and k_{cat} Values for Some Enzymes

Enzyme Name	Substrate	K_m (M)	k_{cat} (s^{-1})
Acetylcholinesterase	Acetylcholine	9×10^{-5}	1.4×10^4
Carbonic anhydrase	CO_2	1×10^{-2}	1×10^6
Fumarase	Fumarate	5×10^{-6}	8×10^2
Triose phosphate isomerase	Glyceraldehyde-3-phosphate	5×10^{-4}	4.3×10^3
β-lactamase	Benzylpenicillin	2×10^{-5}	2×10^3

Source of data: R. Fersht (1985).

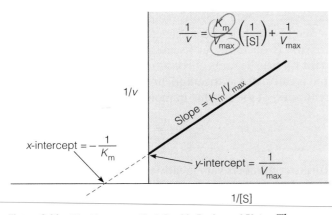

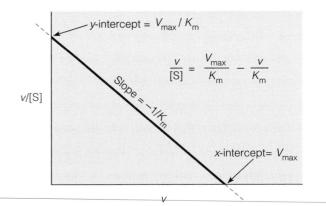

Figure 6-11 The Lineweaver-Burk Double-Reciprocal Plot. The reciprocal of the initial velocity, $1/v$, is plotted as a function of the reciprocal of the substrate concentration, $1/[S]$. K_m can be calculated from the x-intercept and V_{max} from the y-intercept.

Figure 6-12 The Eadie-Hofstee Plot. The ratio $v/[S]$ is plotted as a function of v. K_m can be determined from the slope and V_{max} from the x-intercept.

equation 6-12 and solving for the other value.) Therefore, once the double-reciprocal plot has been constructed, V_{max} can be determined directly from the reciprocal of the y-intercept and K_m from the negative reciprocal of the x-intercept. Furthermore, the slope can be used to check both values.

Thus, the Lineweaver-Burk plot is useful because it confirms by its linearity that the reaction in question is following Michaelis-Menten kinetics, and it allows determination of the parameters V_{max} and K_m without the complication of a hyperbolic shape. It also serves as a useful diagnostic in the analysis of enzyme inhibition, because the several different kinds of reversible inhibitors affect the shape of the plot in characteristic ways.

The Lineweaver-Burk equation has some limitations, however. The main problem is that a long extrapolation is often necessary to determine K_m, with resulting uncertainty in the result. Moreover, the data points that are often the most crucial in determining the slope of the curve are the farthest from the y-axis, and those points represent the samples with the lowest substrate concentrations, the lowest levels of enzyme activity, and hence the most uncertain values.

To circumvent these disadvantages, several alternatives to the Lineweaver-Burk equation have come into use to linearize kinetic data. One such alternative is the **Eadie-Hofstee equation,** which is represented graphically as a plot of $v/[S]$ versus v. As Figure 6-12 illustrates, V_{max} is determined from the x-intercept and K_m from the slope of this plot. (To explore the Eadie-Hofstee plot and another alternative to the Lineweaver-Burk plot further, see Problem 6-14 at the end of this chapter.)

Determining K_m and V_{max}: An Example

To illustrate the value of the double-reciprocal plot in determining V_{max} and K_m, consider a specific example involving the enzyme hexokinase, as illustrated in Figures 6-13 and 6-14. Hexokinase is an important enzyme in cel-

lular energy metabolism because it catalyzes the first reaction in the glycolytic pathway, which is discussed in detail in Chapter 13. Using ATP as a source of both the phosphate group and the energy needed for the reaction, hexokinase catalyzes the phosphorylation of glucose on carbon atom 6:

$$\text{glucose} + \text{ATP} \xrightarrow{\text{hexokinase}} \text{glucose-6-phosphate} + \text{ADP} \quad \textbf{(6-13)}$$

To analyze this reaction kinetically, we must determine the initial velocity at each of several substrate concentrations. When an enzyme has two substrates, the usual approach is to vary the concentration of one substrate at a time, holding that of the other one constant at a sufficiently high (near-saturating) level to ensure that it does not become rate-limiting. We must also take care to ensure that the velocity determination is made before product accumulates to the point that the back reaction becomes significant.

In the experimental approach shown in Figure 6-13, glucose is the variable substrate, with ATP present at a saturating concentration in each tube. Of the nine reaction mixtures set up for this experiment, one is designated the reagent blank (B) because it contains no glucose. The other eight tubes contain graded concentrations of glucose ranging from 0.05 to 0.40 mM. With all tubes prepared and maintained at some favorable temperature (25°C is often used), the reaction in each is initiated by the addition of a fixed amount of hexokinase.

The rate of product formation can then be determined either by continuous spectrophotometric monitoring of the reaction mixture (provided that one of the reactants or products absorbs light of a specific wavelength) or by allowing each reaction mixture to incubate for some short, fixed period of time, followed by chemical assay for either substrate depletion or product accumulation. In the case of the hexokinase reaction, the latter procedure is used because there is no direct photometric means of detecting any of the products or reactants.

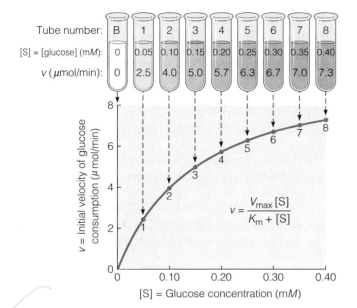

Figure 6-13 Experimental Procedure for Studying the Kinetics of the Hexokinase Reaction. Test tubes containing graded concentrations of glucose and a saturating concentration of ATP were incubated with a standard amount of hexokinase, and the initial rate of product appearance, v, was then plotted as a function of the substrate concentration [S]. The curve is hyperbolic, approaching V_{max} as the substrate concentration gets higher and higher. For the double-reciprocal plot derived from these data, see Figure 6-14.

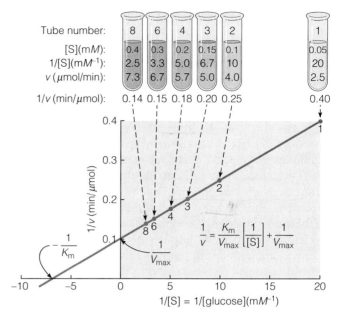

Figure 6-14 Double-Reciprocal Plot for the Hexokinase Data of Figure 6-13. For each test tube, $1/v$ and $1/[S]$ were calculated from the data of Figure 6-13, and $1/v$ was then plotted as a function of $1/[S]$. The y-intercept of 0.1 corresponds to $1/V_{max}$, so V_{max} is 10 μmol/min. The x-intercept of -6.7 corresponds to $-1/K_m$, so K_m is 0.15 mM. (Note that some of the tubes depicted in Figure 6-13 are not shown here due to lack of space.)

As Figure 6-13 indicates, the initial velocity for tubes 1–8 ranged from 2.5 to 7.3 μmol of glucose consumed per minute, with no detectable reaction in the blank. (If any glucose consumption were noted in the blank, the values of tubes 1–8 would have to be corrected for that amount of noncatalytic reaction.) When these reaction velocities are plotted as a function of glucose concentration, the eight data points generate the hyperbolic curve shown in Figure 6-13. Although the data of Figure 6-13 are idealized for illustrative purposes, most kinetic data generated by this approach do, in fact, fit a hyperbolic curve unless the enzyme has some special properties that cause departure from Michaelis-Menten kinetics.

The hyperbolic curve of Figure 6-13 illustrates the need for some means of linearizing the analysis because neither V_{max} nor K_m can be determined from the values as plotted, even though the data are known to be idealized. This need is met by the linear double-reciprocal plot of Figure 6-14. To obtain the data plotted here, reciprocals were calculated for each value of [S] and v from Figure 6-13. Thus, the [S] values of 0.05–0.40 mM generate reciprocals of 20–2.5 mM^{-1}, and the v values of 2.5–7.3 μmol/min give rise to reciprocals of 0.4–0.14 min/μmol. Because these are reciprocals, the data point for tube 1 is farthest from the origin, and each successive tube is represented by a point closer to the origin.

When these data points are connected by a straight line, the y-intercept is found to be 0.1 min/μmol and the x-intercept is -6.7 mM^{-1}. From these intercepts, we can calculate that $V_{max} = 1/0.1 = 10$ μmol/min and $K_m = -(1/-6.7) = 0.15$ mM. If we now go back to the Michaelis-Menten plot of Figure 6-13, we can see that both of these values are eminently reasonable because we can readily imagine that the plot is rising hyperbolically to a maximum of 10 μmol/min. Moreover, the graph reaches one-half of this value at a substrate concentration of 0.15 mM, which turns out to be the data point for tube 3. This, of course, is the K_m of hexokinase for glucose, often written $K_{m,glucose}$.

The enzyme also has a K_m value for the other substrate, $K_{m,ATP}$, but that would have to be determined by varying the ATP concentration while holding the glucose concentration constant. Interestingly, hexokinase phosphorylates not only glucose but also other hexoses, and has a distinctive K_m value for each. The K_m for fructose, for example, is 1.5 mM, which means that it takes ten times more fructose than glucose to sustain the reaction at one-half of its maximum velocity.

Though somewhat simplified and idealized, this is the approach that enzymologists take in studying the kinetics of enzyme-catalyzed reactions. Their analyses are often more complicated than this, and they almost always use a computer to calculate and plot double-reciprocal data and to determine K_m and V_{max} values, but the basic approach is the same.

Enzyme Inhibitors Act Irreversibly or Reversibly

Thus far, we have assumed that the only substances in cells that affect the activities of enzymes are their substrates. However, enzymes are also influenced by products, alternative substrates, substrate analogues, drugs, toxins, and an especially important class of regulators called *allosteric effectors*. Most of these substances have an inhibitory effect on enzyme activity, reducing (or sometimes even abolishing) the reaction rate with the desired substrate.

This **inhibition** of enzyme activity is important for several reasons. First and foremost, enzyme inhibition plays a vital role as a control mechanism in cells. As discussed in the next section, many enzymes are subject to regulation by specific small molecules other than their substrates. Enzyme inhibition is also important in the action of drugs and poisons, which frequently exert their effects by inhibiting specific enzymes. Inhibitors are also useful to enzymologists as tools in their studies of reaction mechanisms. Especially important in this latter case are *substrate analogues*, compounds that resemble the real substrate closely enough to bind to the active site but are then chemically unable to undergo reaction.

Inhibitors may be either *reversible* or *irreversible*. An **irreversible inhibitor** binds to the enzyme covalently, causing an irrevocable loss of catalytic activity. Not surprisingly, irreversible inhibitors are usually toxic to cells. Ions of heavy metals are often irreversible inhibitors, as are alkylating agents and nerve gas poisons. This is, in fact, the reason many insecticides and nerve gases are so toxic. These substances bind irreversibly to *acetylcholinesterase*, an enzyme that is vital to the transmission of nerve impulses (see Chapter 9). Inhibition of acetylcholinesterase activity leads to rapid paralysis of vital functions and therefore to death. One such nerve gas is *diisopropyl fluorophosphate*, which binds covalently to the hydroxyl group of a critical serine at the active site of the enzyme, thereby rendering the enzyme molecule permanently inactive.

Many natural toxins are also irreversible inhibitors of enzymes. For example, the alkaloid *physostigmine*, a natural constituent of calabar beans, is toxic to animals because it is a potent inhibitor of acetylcholinesterase. The antibiotic *penicillin* is an irreversible inhibitor of serine-containing enzymes involved in bacterial cell wall synthesis. Penicillin is therefore effective in treating bacterial infections because it prevents the bacterial cells from forming cell walls.

In contrast, a **reversible inhibitor** binds to an enzyme in a noncovalent, dissociable manner, such that the free and bound forms of the inhibitor exist in equilibrium with each other. We can represent such binding as follows, with E as the free, active enzyme, I as the inhibitor, and EI as the inactive enzyme-inhibitor complex:

$$E + I \rightleftharpoons EI \qquad (6\text{-}14)$$

Clearly, the fraction of the enzyme that is available to the cell in active form depends on the concentration of the inhibitor and the strength of the enzyme-inhibitor complex.

The two most common forms of reversible inhibitors are called *competitive inhibitors* and *noncompetitive inhibitors*. A competitive inhibitor binds to the active site of the enzyme and therefore competes directly with substrate molecules for the same site on the enzyme (Figure 6-15a) and reduces enzyme activity to the extent that the active sites of the enzyme molecules have inhibitor molecules rather than substrate molecules bound to them at any point in time.

A noncompetitive inhibitor, on the other hand, binds to the enzyme surface at a location *other* than the active site. It therefore does not block substrate binding but

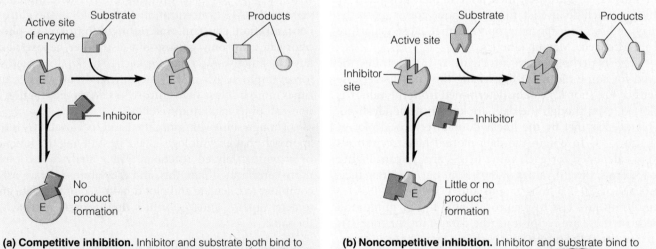

(a) Competitive inhibition. Inhibitor and substrate both bind to the active site of the enzyme. Binding of an inhibitor prevents substrate binding, thereby inhibiting enzyme activity.

(b) Noncompetitive inhibition. Inhibitor and substrate bind to different sites. Binding of an inhibitor distorts the enzyme, thereby decreasing the likelihood of substrate binding.

Figure 6-15   Modes of Action of Competitive and Noncompetitive Inhibitors.   Both **(a)** competitive and **(b)** noncompetitive inhibitors (red) bind reversibly to the enzyme (E), thereby inhibiting its activity. The two kinds of inhibitors differ in the site on the enzyme to which they bind, however.

nonetheless inhibits enzyme activity because the enzyme-catalyzed reaction does not take place at the active site when the inhibitor is bound to its site (Figure 6-15b).

Enzyme Regulation

To understand the role of enzymes in cellular function, it is important to recognize that it is rarely in the cell's best interest to allow an enzyme to function at an indiscriminately high rate. Instead, the rates of enzyme-catalyzed reactions and the biochemical sequences of which they are a part must be continuously adjusted to keep them finely tuned to the needs of the cell. An important aspect of that adjustment lies in the cell's ability to control enzyme activities with specificity and precision.

We have already encountered a variety of regulatory mechanisms, including changes in substrate (and product) concentrations, alterations in pH, and the presence and concentration of inhibitors. Regulation that depends directly on the interactions of substrates and products with the enzyme is called **substrate-level regulation.** As the Michaelis-Menten equation makes clear, increases in substrate concentration result in higher reaction rates (see Figure 6-9). Conversely, increases in product concentration reduce the rate at which substrate is converted to product. (This inhibitory effect of product concentration is why v needs to be identified as the *initial* reaction velocity in the Michaelis-Menten equation, as given by equation 6-7. If a significant amount of product is already present, or accumulates during the course of the reaction, the equation becomes more complex than the simple form we have considered.)

As an example of substrate-level regulation, consider the phosphorylation of glucose to generate glucose-6-phosphate (reaction 6-13). The enzyme hexokinase that catalyzes this reaction is inhibited by its own product, glucose-6-phosphate. If utilization of glucose-6-phosphate is blocked for any reason, it will accumulate, inhibiting the hexokinase reaction and slowing down the further entry of glucose into the pathway.

Substrate-level regulation is an important control mechanism in cells, but it is not sufficient for the regulation of most reactions or reaction sequences. For most pathways, enzymes are regulated by other mechanisms as well. Two of the most important mechanisms are *allosteric regulation* and *covalent modification.* These mechanisms allow cells to turn enzymes on or off or to fine-tune their reaction rates by modulating enzyme activities appropriately.

Almost invariably, an enzyme that is regulated by such a mechanism catalyzes a reaction that represents the first step of a multistep sequence. By increasing or reducing the rate at which the first step functions, the whole sequence is effectively controlled. Pathways that are regulated in this way include those required to break down large molecules (such as sugars, fats, or amino acids) to extract energy from them, as well as pathways that lead to

the synthesis of substances needed by the cell (such as amino acids and nucleotides).

We will discuss allosteric regulation and covalent modification in an introductory manner here, with the intention of returning to these mechanisms as we encounter specific examples in later chapters.

Allosteric Enzymes Are Regulated by Molecules Other than Reactants and Products

The single most important control mechanism whereby the rates of enzyme-catalyzed reactions are adjusted to meet cellular needs is *allosteric regulation.* To understand this mode of regulation, consider the pathway shown below, by which a cell converts some precursor A into some final product P via a series of intermediates B, C, and D, in a sequence of reactions catalyzed respectively by enzymes E_1, E_2, E_3, and E_4:

$$A \xrightarrow{E_1} B \xrightarrow{E_2} C \xrightarrow{E_3} D \xrightarrow{E_4} P \qquad \text{(6-15)}$$

Product P could, for example, be an amino acid needed by the cell for protein synthesis, and A could be some common cellular component that serves as the starting point for the specific reaction sequence leading to P.

Feedback Inhibition. If allowed to proceed at a constant, unrestrained rate, the pathway 6-15 has the capacity to convert large amounts of A to P, with possible adverse effects resulting from a depletion of A or an excessive accumulation of P (or both). Clearly, the best interests of the cell are served when the pathway is functioning not at its maximum rate or even some constant rate, but at a rate that is carefully tuned to the cellular need for P. Somehow, the enzymes of this pathway must be responsive to the cellular level of the product P in somewhat the same way that a furnace needs to be responsive to the temperature of the rooms it is intended to heat.

In the latter case, a thermostat provides the necessary regulatory link between the furnace and its "product," heat. In our enzyme example, the desired regulation is possible because the product P is a specific inhibitor of E_1, the enzyme that catalyzes the first reaction in the sequence. This phenomenon is called **feedback** (or **end-product**) **inhibition** and is represented by the dashed arrow that connects the product P to enzyme E_1 in the following pathway:

$$A \xrightarrow{E_1} B \xrightarrow{E_2} C \xrightarrow{E_3} D \xrightarrow{E_4} P \qquad \text{(6-16)}$$

Feedback inhibition of E_1 by P

More generally, feedback inhibition occurs whenever a metabolic product inhibits one of the enzymes involved in the pathway by which that product is synthesized. Feedback inhibition is one of the most common mechanisms used by cells to ensure that the activities of reaction sequences are adjusted to cellular needs.

Figure 6-16 provides a specific example of such a pathway—the five-step sequence whereby the amino acid *isoleucine* is synthesized from *threonine*, another amino acid. In this case, the first enzyme in the pathway, *threonine deaminase*, is regulated by the concentration of isoleucine within the cell. If isoleucine is being used by the cell (in the synthesis of proteins, most likely), the isoleucine concentration will be low. Under these conditions, threonine deaminase is active and the pathway functions to produce more isoleucine, thereby meeting the ongoing need for this amino acid. If the need for isoleucine decreases, isoleucine will begin to accumulate in the cell, and the increase in its concentration will lead to a decrease in the activity of threonine deaminase and hence to a reduced rate of isoleucine synthesis.

Allosteric Regulation. How can the first enzyme in a pathway (e.g., enzyme E_1 in reaction sequence 6-16) be sensitive to the concentration of a substance P that is neither its substrate nor its immediate product? Or, in terms of Figure 6-16, how can the activity of threonine deaminase be responsive to the concentration of isoleucine, when isoleucine differs sufficiently from threonine in structure that it is unlikely to be recognized by the active site of threonine deaminase?

The answer to this question was first proposed in 1963 by Jacques Monod, Jean-Pierre Changeux, and François Jacob. Although based initially on inconclusive data, their model was quickly substantiated and went on to become the foundation for our understanding of **allosteric regulation.** The term *allosteric* derives from the Greek for "another shape (or state)," thereby indicating that all enzymes capable of allosteric regulation can exist in two different states. In one of the two forms, the enzyme has a high affinity for its substrate(s), whereas in the other form, it has little or no affinity for its substrate. Enzymes with this property are called **allosteric enzymes.** The two different forms of an allosteric enzyme are readily interconvertible and are, in fact, in equilibrium with each other. Obviously, the reaction rate is high when the enzyme is in its high-affinity form and low or even zero when the enzyme is in its low-affinity form.

Whether the active or inactive form of an allosteric enzyme is favored depends on the cellular concentration of the appropriate regulatory substance, called an **allosteric effector.** In the case of isoleucine synthesis, the allosteric enzyme is threonine deaminase and the allosteric effector is isoleucine. More generally, an allosteric effector is a small organic molecule that regulates the activity of an enzyme for which it is neither the substrate nor the immediate product.

An allosteric effector influences enzyme activity by binding to one of the two interconvertible forms of the enzyme, thereby stabilizing it in that state. In other words, an allosteric enzyme can exist in either a complexed or uncomplexed form, depending on whether it has an effector molecule bound to it or not. The effector binds to the enzyme because of the presence on the enzyme surface of an **allosteric** (or **regulatory**) **site** that is distinct from the active site at which the catalytic event occurs. Thus, a distinguishing feature of all allosteric enzymes (and other allosteric proteins, as well) is the presence on the enzyme surface of an *active site* to which the substrate binds and an *allosteric site* to which the effector binds. In fact, some allosteric enzymes have multiple allosteric sites, each capable of recognizing a different effector.

An effector may be either an **allosteric inhibitor** or an **allosteric activator,** depending on the effect it has when bound to the allosteric site on the enzyme—that is, depending on whether the complexed form is the low-affinity or high-affinity state of the enzyme (Figure 6-17). The binding

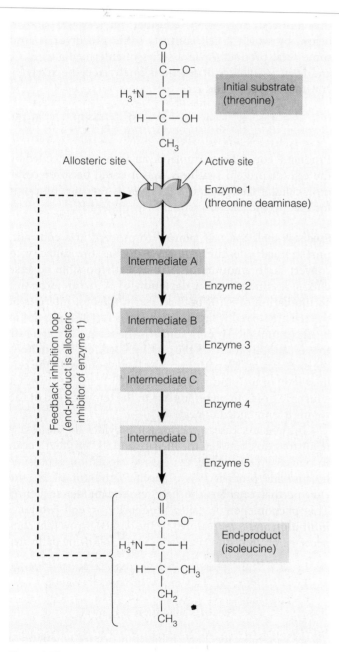

Figure 6-16 Allosteric Regulation of Enzyme Activity. A specific example of feedback inhibition is seen in the pathway by which the amino acid isoleucine is synthesized from threonine, another amino acid. The first enzyme in the sequence, threonine deaminase, is allosterically inhibited by isoleucine, which binds to the enzyme at a site *other* than the active site.

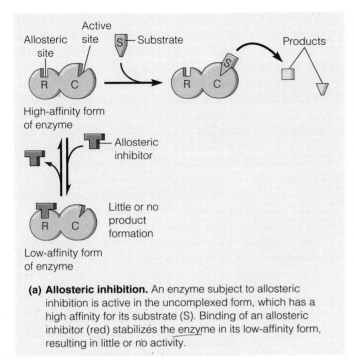

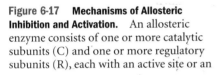

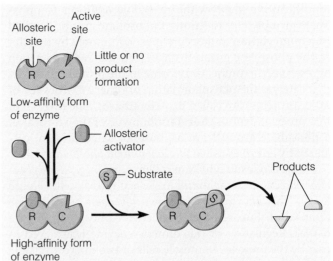

(a) **Allosteric inhibition.** An enzyme subject to allosteric inhibition is active in the uncomplexed form, which has a high affinity for its substrate (S). Binding of an allosteric inhibitor (red) stabilizes the enzyme in its low-affinity form, resulting in little or no activity.

(b) **Allosteric activation.** An enzyme subject to allosteric activation is inactive in its uncomplexed form, which has a low affinity for its substrate. Binding of an allosteric activator (green) stabilizes the enzyme in its high-affinity form, resulting in enzyme activity.

Figure 6-17 Mechanisms of Allosteric Inhibition and Activation. An allosteric enzyme consists of one or more catalytic subunits (C) and one or more regulatory subunits (R), each with an active site or an allosteric site, respectively. The enzyme exists in two forms, one with a high affinity for its substrate (and therefore a high likelihood of product formation) and the other with a low affinity (and a correspondingly low likelihood of product formation). Which form an enzyme is in depends on the concentration of the allosteric effector(s) for that enzyme.

of an allosteric inhibitor shifts the equilibrium between the two forms of the enzyme to favor the low-affinity state (Figure 6-17a). The binding of an allosteric activator, on the other hand, shifts the equilibrium in favor of the high-affinity state (Figure 6-17b). In either case, binding of the effector to the allosteric site stabilizes the enzyme in one of its two interconvertible forms, thereby either decreasing or increasing the likelihood of substrate binding.

Most allosteric enzymes are large, multisubunit proteins with an active site or an allosteric site on each subunit. In fact, the active sites and allosteric sites are usually on different subunits of the protein, referred to as **catalytic subunits** and **regulatory subunits,** respectively (see Figure 6-17). This means, in turn, that the binding of effector molecules to the allosteric sites affects not just the shape of the regulatory subunits but that of the catalytic subunits as well.

Allosteric Enzymes Exhibit Cooperative Interactions Between Subunits

Many allosteric enzymes exhibit a property known as **cooperativity.** This means that as the multiple catalytic sites on the enzyme bind substrate molecules, the enzyme undergoes conformational changes that affect the affinity of the remaining sites for substrate. Some enzymes show *positive cooperativity,* in which the binding of a substrate molecule to one catalytic subunit increases the affinity of other catalytic subunits for substrate. Other enzymes show *negative cooperativity,* in which the substrate binding to one catalytic subunit reduces the affinity of the other catalytic sites for substrate.

The cooperativity effect enables cells to produce enzymes that are more sensitive or less sensitive to changes in substrate concentration than would otherwise be predicted by Michaelis-Menten kinetics. Positive cooperativity causes an enzyme's catalytic activity to increase faster than normal as the substrate concentration is increased, whereas negative cooperativity means that enzyme activity increases more slowly than expected.

Enzymes Can Also Be Regulated by the Addition or Removal of Chemical Groups

In addition to allosteric regulation, many enzymes are also subject to control by **covalent modification.** In this form of regulation, the activity of an enzyme is affected by the addition or removal of specific chemical groups. Common modifications include the addition of phosphate groups, methyl groups, acetyl groups, or derivatives of nucleotides. Some of these modifications can be reversed, whereas others cannot. In each case, the effect of the modification is to activate or to inactivate the enzyme, or at least to adjust its activity upward or downward.

Phosphorylation/Dephosphorylation. One of the most frequently encountered and best understood covalent modifications involves the reversible addition of phosphate groups. The addition of phosphate groups is called **phosphorylation** and occurs most commonly by transfer of the phosphate group from ATP to the hydroxyl group of a serine, threonine, or tyrosine in the protein. Enzymes that catalyze the phosphorylation of other enzymes (or of other proteins) are called **protein kinases.** The reversal of this process, **dephosphorylation,** involves the removal of a phosphate group from a phosphorylated protein, catalyzed by enzymes called *phosphoprotein phosphatases.*

This mode of regulation is illustrated by *glycogen phosphorylase,* an enzyme found in skeletal muscle cells (Figure 6-18). This enzyme breaks down glycogen by successive removal of glucose units as glucose-1-phosphate (Figure 6-18a). Regulation of this dimeric enzyme is achieved in part by the presence in muscle cells of two interconvertible forms of the enzyme, an active form called *phosphorylase a*

and an inactive form called *phosphorylase b* (Figure 6-18b). When glycogen breakdown is required in the muscle cell, the inactive *b* form of the enzyme is converted into the active *a* form by the addition of a phosphate group to a particular serine on each of the two subunits of the phosphorylase molecule. The reaction is catalyzed by *phosphorylase kinase* and results in a conformational change of phosphorylase to the active form. When glycogen breakdown is no longer needed, the phosphate groups are removed from phosphorylase *a* by the enzyme *phosphorylase phosphatase. Glycogen synthase,* the enzyme that adds glucose units onto the glycogen chain, responds in the opposite manner: It is inactive in the phosphorylated form and is activated by dephosphorylation.

In addition to regulation by the phosphorylation/dephosphorylation mechanism shown in Figure 6-18, glycogen phosphorylase is also an allosteric enzyme,

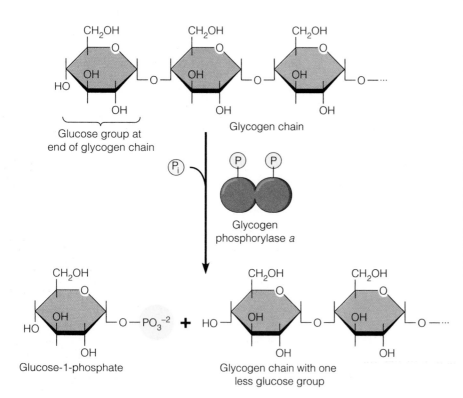

(a) The reaction catalyzed by glycogen phosphorylase

Figure 6-18   The Regulation of Glycogen Phosphorylase by Phosphorylation.
(a) Glycogen phosphorylase is a dimeric enzyme in muscle cells that releases glucose units from glycogen molecules as glucose-1-phosphate, which can then be used by the muscle cell as an energy source.
(b) Glycogen phosphorylase is regulated in part by a phosphorylation/dephosphorylation mechanism. The inactive form of the enzyme, phosphorylase *b*, can be converted to the active form, phosphorylase *a*, by the transfer of phosphate groups from ATP to a particular serine on each of the two subunits of the enzyme. This phosphorylation reaction is catalyzed by the enzyme phosphorylase kinase. Removal of the phosphate groups by phosphorylase phosphatase returns the phosphorylase molecule to the inactive *b* form.

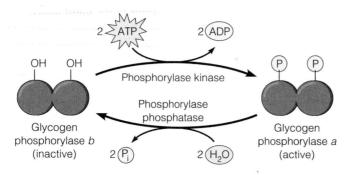

(b) Regulation of glycogen phosphorylase

inhibited by glucose and ATP and activated by AMP. If a hormonal signal triggers the phosphorylation and hence the activation of the phosphorylase in a muscle cell that still has an adequate supply of glucose, the glucose will inhibit the enzyme allosterically until it is actually needed. On the other hand, muscle cells with a low level of glucose will benefit immediately from the conversion of phosphorylase to the active form.

The existence of two levels of regulation for glycogen phosphorylase illustrates an important aspect of enzyme regulation. Many enzymes are controlled by two or more regulatory mechanisms, thereby enabling the cell to make appropriate responses to a variety of situations.

Proteolytic Cleavage. A different kind of covalent activation of enzymes involves the one-time, irreversible removal of a portion of the polypeptide chain by an appropriate proteolytic (protein-degrading) enzyme. This kind of modification, called **proteolytic cleavage,** is exemplified especially well by the proteolytic enzymes of the pancreas, which include trypsin, chymotrypsin, and carboxypeptidase. Synthesized in the pancreas, these enzymes are secreted into the duodenum of the small intestine in response to a hormonal signal. These proteases, along with the pepsin of the stomach and other proteolytic enzymes secreted by the cells of the intestine, can digest almost all ingested proteins into free amino acids, which can then be absorbed by the intestinal epithelial cells.

Pancreatic proteases are not synthesized in their active form; that would likely cause problems for the cells of the pancreas, which must protect themselves against their own proteolytic enzymes. Instead, each of these enzymes is synthesized as a slightly larger, catalytically inactive molecule called a *zymogen.* Zymogens must themselves be cleaved proteolytically to yield active enzymes. Several such events are shown in Figure 6-19. For example, trypsin is synthesized initially as a zymogen called *trypsinogen.* When trypsinogen reaches the duodenum, it is activated by the removal of a hexapeptide (a string of six amino acids) from its N-terminus by the action of *enterokinase,* a membrane-bound protease produced by the duodenal cells. The active trypsin then activates the other zymogens by specific proteolytic cleavages.

RNA Molecules as Enzymes: Ribozymes

Prior to the early 1980s, it was thought that all enzymes were proteins. Indeed, that statement was regarded as one of the fundamental truths of cellular biology and was found in every textbook. Cell biologists became convinced that all enzymes were proteins because every enzyme isolated in the 55 years following Sumner's purification of urease in 1926 turned out to be a protein. But biology is full of surprises, and we now know that the statement needs to be revised to include RNA catalysts called **ribozymes**.

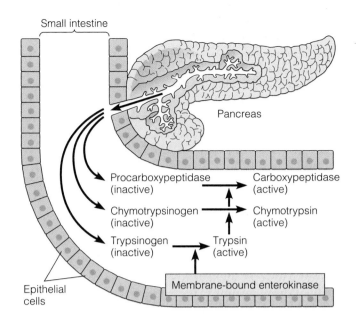

Figure 6-19 Activation of Pancreatic Zymogens by Proteolytic Cleavage. Pancreatic proteases are synthesized and secreted into the small intestine as inactive precursors called zymogens. Procarboxypeptidase, trypsinogen, and chymotrypsinogen are zymogens. Activation of trypsinogen to trypsin requires removal of a hexapeptide segment by enterokinase, a membrane-bound duodenal enzyme. Trypsin then activates other zymogens by proteolytic cleavage. Procarboxypeptidase is activated by a single cleavage event, whereas the activation of chymotrypsinogen is a somewhat more complicated two-step process, the details of which are not shown here.

The first evidence came in 1981, when Thomas Cech and his colleagues discovered an apparent exception to the "all enzymes are proteins" rule. They were studying the splicing of an internal segment from a specific ribosomal RNA precursor (pre-rRNA) in *Tetrahymena thermophila*, a single-celled eukaryote. (As you will learn in Chapter 19, many eukaryotic RNA molecules require the removal of one or more internal segments called *introns* before they become functional in the cell. The removal process involves the excision of the intron and a splicing together of the two pieces of the original molecule at the excision site.) In the course of their work, the researchers made the remarkable observation that the process apparently proceeded without the presence of proteins! Describing their attempt to study intron excision in vitro, Cech later wrote:

It turned out that some of the small molecules—notably magnesium ion and any of several forms of the nucleotide guanosine—were required for the reaction to proceed. To our great surprise, however, the nuclear extract containing the enzymes was not. We were forced to conclude either that the enzymatic activity came from a protein bound so tightly to the RNA that we were unable to strip it off or that the RNA was catalyzing its own splicing. Given the deeply rooted nature of the idea that all biological catalysts are proteins, the hypothesis of catalysis by RNA was not easy to accept. (Cech, 1987, p. 1533.)

But further experimentation bore out their initial conclusion: The removal of a 413-nucleotide intron from the *Tetrahymena* pre-rRNA is catalyzed by the pre-rRNA molecule itself.

Figure 6-20 shows the excision process, which involves ① folding of the unspliced pre-rRNA molecule to form a loop, ② attack by a hydroxyl group of a free guanosine nucleotide that acts as a cofactor, ③ cleavage and splicing of the rRNA molecule with release of the intron, and ④ further autocatalytic cleavage of the intron to remove 19 more nucleotides. The fully processed intron is itself a ribozyme, capable of shortening or elongating small oligonucleotides.

One might argue that the rRNA molecule in Figure 6-20 fails to satisfy the definition of a catalyst, which requires that the catalyst itself is not altered in the reaction process. However, another RNA-based catalyst discovered at about the same time in the laboratory of Sidney Altman overcomes this objection. The enzyme, called *ribonuclease P*, cleaves transfer RNA precursors (pre-tRNAs) to yield functional RNA molecules. (In this case, a terminal seg-

ment of the RNA molecule is removed rather than an intron as for pre-rRNA processing.)

It had been known for some time that ribonuclease P consisted of a protein component and an RNA component, and it was generally assumed that the active site was on the protein component. By isolating the components and studying them separately, however, Altman and his colleagues showed unequivocally that the isolated protein component was completely inactive, whereas the RNA component was capable of catalyzing the specific cleavage of tRNA precursors on its own and was not itself altered in the process. Furthermore, the RNA-catalyzed reaction followed Michaelis-Menten kinetics, further evidence that the RNA component was acting like a true enzyme. (The protein component enhances activity but is not required for either substrate binding or cleavage.)

The significance of these findings was recognized by the Nobel Prize that Cech and Altman shared in 1989 for their discovery of ribozymes. Since these initial discoveries, additional examples of ribozymes have been reported. Especially significant are recent reports that ribosomes

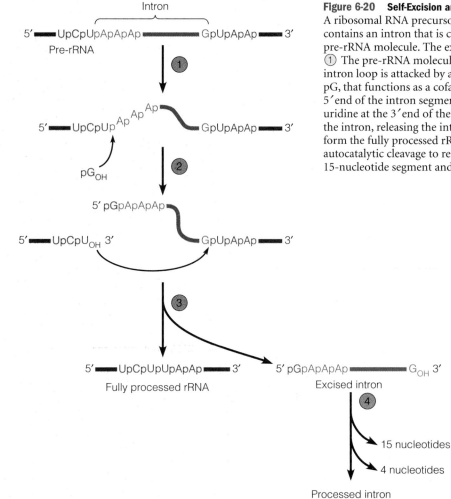

Figure 6-20 Self-Excision and Splicing of the Intron from *Tetrahymena* Pre-rRNA. A ribosomal RNA precursor (pre-rRNA) molecule from *Tetrahymena* contains an intron that is capable of catalyzing its own excision from the pre-rRNA molecule. The excision and splicing process occurs in four steps. ① The pre-rRNA molecule forms a loop at the intron region. ② The intron loop is attacked by a hydroxyl group of a free guanosine nucleotide, pG, that functions as a cofactor. ③ The pre-rRNA molecule is cleaved at the 5′ end of the intron segment, with the addition of G to the intron. The uridine at the 3′ end of the other rRNA fragment then attacks the 3′ end of the intron, releasing the intron and splicing the two rRNA pieces together to form the fully processed rRNA. ④ The intron then undergoes further autocatalytic cleavage to remove 19 more nucleotides by cleaving first a 15-nucleotide segment and then a 4-nucleotide piece.

may have active sites involving RNA. Ribosomes can be thought of as very large enzymes because they catalyze the formation of the peptide bonds that add successive amino acids to a growing polypeptide chain (see Figure 4-8). Ribosomes consist of both protein and RNA molecules, and it had long been assumed that the active sites involved in the polymerization process were on the protein molecules. However, evidence from the laboratory of Harry Noller indicates that the proteins can be extracted from ribosomes without destroying the ability of the ribosomes to catalyze peptide bond formation.

Although most biologists were initially astonished by the discovery of ribozymes, in retrospect there is no reason that RNA molecules should not be able to function as enzymes. Like proteins, RNA molecules can take on com-plex tertiary structure, and that is the prerequisite for catalytic function in both cases.

The discovery of ribozymes has markedly changed the way we think about the origin of life on Earth. For many years, scientists had speculated that the first catalytic macromolecules must have been amino acid polymers resembling proteins. But this concept immediately ran into difficulty because there was no obvious way for a primitive protein to carry information or to replicate itself, two primary attributes of life. However, if the first catalysts were RNA rather than protein molecules, it becomes conceptually easier to imagine a system of RNA molecules acting both as catalysts and as replicating systems capable of transferring information between generations.

Perspective

We come full circle and return to the theme raised in the introduction to this chapter—namely, that thermodynamics allows us to assess the feasibility of a reaction but says nothing about the likelihood that the reaction will actually occur at a reasonable rate in the cell. To ensure that the activation energy requirement is met and the transition state is achieved, a catalyst is required, which in biological systems is always an enzyme. All protein enzymes are chains of amino acids in a genetically programmed sequence that are sensitive to temperature and pH. They are also exquisitely specific, either for a single specific substrate or for a class of closely related compounds. The actual catalytic process takes place at the active site, a critical cluster of amino acids responsible for substrate binding and activation and for the actual chemical reaction. Binding of the appropriate substrate at the active site induces a more stringent fit between enzyme and substrate, thereby facilitating substrate activation.

An enzyme-catalyzed reaction proceeds via an enzyme-substrate intermediate. Most enzymes follow Michaelis-Menten kinetics, characterized by a hyperbolic relationship between the initial reaction velocity and the substrate concentration. The upper limit on velocity is called V_{max} and the substrate concentration needed to reach one-half of the maximum velocity is termed the Michaelis constant, K_m. The hyperbolic relationship between v and [S] can be linearized by a double-reciprocal equation and plot, from which V_{max} and K_m can be determined graphically or by computer analysis.

Enzyme activity is influenced not only by substrate availability but also by products, alternative substrates, substrate analogues, drugs, and toxins, most of which have an inhibitory effect. Inhibition may be either reversible or irreversible, with the latter category involving covalent bonding of the inhibitor to the enzyme surface. A reversible inhibitor, on the other hand, binds to an enzyme in a reversible manner, either at the active site (competitive inhibition) or elsewhere on the enzyme surface (noncompetitive inhibition).

Enzymes must be regulated to adjust their activity levels to cellular needs. Substrate-level regulation involves the effects of substrate and product concentrations on the reaction rate. Additional control mechanisms include allosteric regulation and covalent modification. Most allosterically regulated enzymes catalyze the first step in a reaction sequence and are multisubunit proteins with multiple catalytic subunits and multiple regulatory subunits. Each of the catalytic subunits has an active site that recognizes substrates and products, whereas each regulatory subunit has one or more allosteric sites that recognize specific effector molecules. A given effector may either inhibit or activate the enzyme, depending on which form of the enzyme is favored by effector binding. The most common covalent modifications include phosphorylation and dephosphorylation, as exemplified by the enzyme glycogen phosphorylase, and proteolytic cleavage, as occurs in the activation of the zymogen forms of proteolytic enzymes secreted by the pancreas.

Although it was long thought that all enzymes were proteins, we now recognize the catalytic properties of certain RNA molecules called ribozymes. These include some rRNA molecules that are able to catalyze the removal of their own introns, RNA components of enzymes that also contain protein components, and perhaps even the RNA components of assembled ribosomes. The discovery of ribozymes has changed the way we think about the origin of life on Earth because RNA molecules, unlike proteins, are capable of replicating themselves.

Key Terms for Self-Testing

enzyme catalysis (p. 130)
enzyme (p. 130)

**Activation Energy
and the Metastable State**
activation energy (E_A) (p. 131)
transition state (p. 131)
metastable state (p. 131)
catalyst (p. 132)

Enzymes as Biological Catalysts
active site (p. 133)
prosthetic group (p. 134)
substrate specificity (p. 134)
group specificity (p. 135)
induced-fit model (p. 138)
substrate activation (p. 138)
nucleophilic substitution (p. 138)
electrophilic substitution (p. 139)

Enzyme Kinetics
enzyme kinetics (p. 140)
initial reaction velocity (v) (p. 140)
saturation (p. 140)
Michaelis-Menten equation (p. 142)
maximum velocity (V_{max}) (p. 142)
Michaelis constant (K_m) (p. 143)
turnover number (k_{cat}) (p. 143)
Lineweaver-Burk equation (p. 143)
double-reciprocal plot (p. 143)
Eadie-Hofstee equation (p. 144)

Enzyme Inhibition
inhibition (p. 146)
irreversible inhibitor (p. 146)
reversible inhibitor (p. 146)

Enzyme Regulation
substrate-level regulation (p. 147)

feedback (end-product) inhibition
  (p. 147)
allosteric regulation (p. 148)
allosteric enzyme (p. 148)
allosteric effector (p. 148)
allosteric (regulatory) site (p. 148)
allosteric inhibitor (p. 148)
allosteric activator (p. 148)
catalytic subunit (p. 149)
regulatory subunit (p. 149)
cooperativity (p. 149)
covalent modification (p. 149)
phosphorylation (p. 150)
protein kinase (p. 150)
dephosphorylation (p. 150)
proteolytic cleavage (p. 151)

RNA Molecules as Enzymes: Ribozymes
ribozyme (p. 151)

Problem Set

More challenging problems are marked with a • .

6-1. The Need for Enzymes. You should now be in a position to appreciate the difference between the thermodynamic feasibility of a reaction and the likelihood that it will actually proceed.

(a) Many reactions that are thermodynamically possible do not occur at an appreciable rate because of the activation energy required for the reactants to achieve the transition state. In molecular terms, what does this mean?

(b) One way to meet this requirement is by an input of heat, which in some cases need only be an initial, transient input. Give an example, and explain what this accomplishes in molecular terms.

(c) An alternative solution is to lower the activation energy. What does it mean in molecular terms to say that a catalyst lowers the activation energy of a reaction?

(d) Organic chemists often use inorganic catalysts such as nickel, platinum, or cations in their reactions, whereas cells use proteins called enzymes. What advantages can you see to the use of enzymes? Can you think of any disadvantages?

6-2. Activation Energy. As shown in reaction 6-2, hydrogen peroxide, H_2O_2, decomposes to H_2O and O_2. The activation energy, E_A, for the uncatalyzed reaction at 20°C is 18 kcal/mol. The reaction can be catalyzed by colloidal platinum ($E_A = 13$ kcal/mol) or by the enzyme *catalase* ($E_A = 7$ kcal/mol).

(a) Draw an activation energy diagram for this reaction under catalyzed and uncatalyzed conditions, and explain what it means for the activation energy to be lowered from 18 to 13 kcal/mol by platinum but from 18 to 7 kcal/mol by catalase.

(b) Suggest two properties of catalase that make it a more suitable intracellular catalyst than platinum.

(c) Suggest yet another way in which the rate of hydrogen peroxide decomposition can be accelerated. Is this a suitable means of increasing reaction rates within cells? Why or why not?

(d) Recall from Chapter 4 that the catalase present in eukaryotic cells is localized within the peroxisomes, along with any of several H_2O_2-generating enzymes. Given the toxicity of hydrogen peroxide to the cell, explain in terms of enzyme kinetics why it is advantageous to have H_2O_2-generating enzymes and catalase compartmentalized together within a membrane-bounded organelle.

6-3. Rate Enhancement by Catalysts. The decomposition of H_2O_2 to H_2O and O_2 shown in reaction 6-2 can be catalyzed either by an inorganic catalyst (ferric ions) or by an enzyme (catalase). Compared with the uncatalyzed rate, this reaction proceeds about 30,000 times faster in the presence of ferric ions but about 100,000,000 times faster in the presence of catalase, an iron-containing enzyme. Assume that 1 μg of catalase can decompose a given quantity of H_2O_2 in 1 minute at 25°C and that all reactions are carried out under sterile conditions.

(a) How long would it take for the same quantity of H_2O_2 to be decomposed in the presence of an amount of ferric ions equivalent to the iron content of 1 μg of catalase?

(b) How long would it take for the same quantity of H_2O_2 to decompose in the absence of a catalyst?

(c) Explain how these calculations illustrate the indispensability of catalysts and the superiority of enzymes over inorganic catalysts.

6-4. Temperature and pH Effects. Figure 6-5 illustrates enzyme activities as functions of temperature and pH. In general, the activity of a specific enzyme is highest at the temperature and pH that are characteristic of the environment in which the enzyme normally functions.

(a) Explain the shapes of the curves in Figure 6-5 in terms of the major chemical or physical factors that affect enzyme activity.

(b) For each enzyme in Figure 6-5, suggest the adaptive advantage of having the enzyme activity profile shown in the figure.

(c) Some enzymes have a very flat pH profile—that is, they have essentially the same activity over a broad pH range. How might you explain this observation?

6-5. Enzyme Specificity. All enzymes are highly specific in both the reactions they catalyze and their choice of substrates. Substrate specificity can be for a particular molecule or for a class of closely related molecules.

(a) A proteolytic enzyme such as trypsin can usually degrade a variety of polypeptide chains, whereas a dehydrogenase is usually absolutely specific for a particular substrate. Explain.

(b) Subtilisin is a bacterial protease that can cleave any peptide bond, regardless of the specific amino acids involved. Trypsin, on the other hand, splits peptide bonds only on the carboxyl side of lysine and arginine groups. What differences might you expect in the active sites of these two enzymes?

(c) Compounds that are sufficiently similar in structure to a bona fide substrate to allow them to bind to the active site of an enzyme but that cannot then undergo the reaction catalyzed by that enzyme are usually highly effective competitive inhibitors of enzyme activity. Explain.

6-6. Michaelis-Menten Kinetics. Figure 6-21 represents a Michaelis-Menten plot for a typical enzyme, with initial reaction velocity plotted as a function of substrate concentration. Three regions of the curve are identified by the letters A, B, and C. For each of the statements that follow, indicate with a single letter which one of the three regions of the curve fits the statement best. A given letter can be used more than once.

(a) The active site of an enzyme molecule is occupied by substrate most of the time.

(b) The active site of an enzyme molecule is free most of the time.

(c) The range of substrate concentration in which most enzymes usually function in normal cells.

(d) Includes the point $(K_m, V_{max}/2)$.

(e) Reaction velocity is limited mainly by the number of enzyme molecules present.

(f) Reaction velocity is limited mainly by the number of substrate molecules present.

6-7. Enzyme Kinetics. The enzyme β-galactosidase catalyzes the hydrolysis of the disaccharide lactose into its component monosaccharides:

$$\text{lactose} + H_2O \xrightarrow{\beta\text{-galactosidase}} \text{glucose} + \text{galactose} \quad \textbf{(6-17)}$$

To determine V_{max} and K_m of β-galactosidase for lactose, the same amount of enzyme (1 μg per tube) was incubated with a series of lactose concentrations under conditions where product concentrations remained negligible. At each lactose concentration, the initial reaction velocity was determined by assaying for the amount of lactose remaining at the end of the assay. The following data were obtained:

Lactose concentration (mM)	Rate of lactose consumption (μmol/min)	
1	10.0	5
2	16.7	8.3
4	25.0	12.5
8	33.3	16.6
16	40.0	20
32	44.4	22.2

(handwritten at top of table: 1.0   0.5)

(a) Why is it necessary to specify that product concentrations remained negligible during the course of the reaction?

(b) Plot v (rate of lactose consumption) versus [S] (lactose concentration). Why is it that when the lactose concentration is doubled, the increase in velocity is always less than twofold?

(c) Calculate $1/v$ and $1/[S]$ for each entry on the data table and plot $1/v$ versus $1/[S]$.

(d) Determine K_m and V_{max} from your double-reciprocal plot.

(e) On the same graph as part b, plot the results you would expect if each tube contained only 0.5 μg of enzyme. Explain.

6-8. More Enzyme Kinetics. The galactose formed in reaction 6-17 can be phosphorylated by the transfer of a phosphate group from ATP, a reaction catalyzed by the enzyme galactokinase:

$$\text{galactose} + \text{ATP} \xrightarrow{\text{galactokinase}} \text{galactose-1-phosphate} + \text{ADP}$$

$$\textbf{(6-18)}$$

Assume that you have isolated the galactokinase enzyme and have determined its kinetic parameters by varying the concentration of galactose in the presence of a constant, high (i.e., saturating) concentration of ATP. The double-reciprocal (Lineweaver-Burk) plot of the data is shown as Figure 6-22.

(a) What is the K_m of galactokinase for galactose under these assay conditions? What does K_m tell us about the enzyme?

(b) What is the V_{max} of the enzyme under these assay conditions? What does V_{max} tell us about the enzyme?

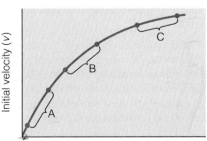

Figure 6-21 Analysis of the Michealis-Menten Plot. See Problem 6-6.

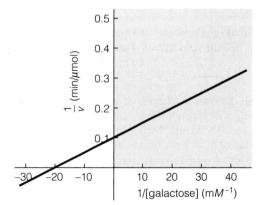

Figure 6-22 Double-Reciprocal Plot for the Enzyme Galactokinase. See Problem 6-8.

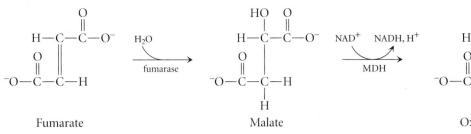 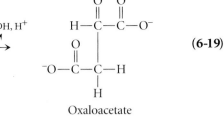

Fumarate **Malate** **Oxaloacetate**

(6-19)

(c) Assume that you now repeat the experiment, but with the ATP concentration varied and galactose present at a constant, high concentration. Assuming that all other conditions are maintained as before, would you expect to get the same V_{max} value as in part b? Why or why not?

(d) In the experiment described in part c, the K_m value turned out to be very different from the value determined in part b. Can you explain why?

6-9. Still More Enzyme Kinetics. Reaction sequence 6-19 is part of the tricarboxylic acid (TCA) cycle, a metabolic pathway that you will encounter in Chapter 14. The enzyme fumarase converts the four-carbon compound fumarate into malate by the addition of water across the carbon-carbon double bond. The enzyme malate dehydrogenase (MDH) then catalyzes the oxidation of malate to oxaloacetate, with a coenzyme called nicotinamide adenine dinucleotide (NAD) as the electron acceptor. The K_m value of fumarase for fumarate is $5 \times 10^{-6} M$ ($5\,\mu M$).

(a) At what fraction of its maximum velocity will fumarase proceed in a test tube containing a negligible concentration of malate with fumarate present at a concentration of (i) $0.5\,\mu M$, (ii) $5\,\mu M$, or (iii) $50\,\mu M$?

(b) In each of two test tubes (A and B), MDH is incubated with malate in the presence of excess NAD^+ (i.e., the coenzyme concentration is nonlimiting). The starting concentrations of malate are 0.1 and 0.2 mM for test tubes A and B, respectively, and the concentrations of NADH and oxaloacetate remain negligible during the experiment. The reaction in test tube B is found to have an initial reaction velocity (v_B) that is 1.5 times the initial velocity in tube A (v_A). What is the K_m of MDH for malate?

(c) Why is it necessary to specify in part b that v_A and v_B are *initial* reaction velocities?

(d) How would you proceed if you wished to determine the K_m value of MDH for its other substrate, NAD^+?

6-10. Turnover Number. Carbonic anhydrase catalyzes the reversible hydration of carbon dioxide to form bicarbonate ion:

$$CO_2 + H_2O \rightleftharpoons HCO_3^- + H^+ \qquad (6\text{-}20)$$

As we will discover in Chapter 8, this reaction is important in the transport of carbon dioxide from body tissues to the lungs by red blood cells. Carbonic anhydrase has a molecular weight of 30,000 and a turnover number (k_{cat} value) of 1×10^6 sec^{-1}. Assume that you are given a 1-mL solution containing 2.0 μg of pure carbonic anhydrase.

(a) At what rate (in millimoles of CO_2 consumed per second) will you expect this reaction to proceed under optimal conditions?

(b) Assuming standard temperature and pressure, how much CO_2 is that in mL per second?

• 6-11. Competitive Inhibition. A distinguishing characteristic of a competitive inhibitor is that it binds reversibly to the active site of the enzyme, thereby competing with the substrate for the active site and reducing the rate at which the substrate is processed. To understand competitive inhibition, consider that the monkeys in Box 6A have access not only to real peanuts but also to plastic peanuts that look just like real peanuts. A monkey will therefore pick up plastic peanuts and real peanuts indiscriminately, but only the latter can be shelled. The initial reaction velocity v is expressed in terms of real peanuts shelled per minute. Decide whether each of the following statements is true (T) or false (F) of a roomful of monkeys confronted with a fixed concentration of plastic peanuts and variable concentrations of real peanuts. If false, reword the statement to make it true.

(a) At a fixed concentration of plastic peanuts, the reaction velocity achieved for a specific concentration of real peanuts will be less than would be observed in the absence of the plastic peanuts.

(b) If the concentration of plastic peanuts is increased, the concentration of real peanuts necessary to reach a specific reaction velocity will be higher also.

(c) The K_m value of the monkeys for real peanuts will be higher when determined in the presence of a fixed concentration of plastic peanuts than in their absence.

(d) The V_{max} value for a specific number of monkeys will be the same in the presence or absence of plastic peanuts, because any finite concentration of plastic peanuts would have no effect on reaction velocity if the concentration of real peanuts were infinite.

(e) The Lineweaver-Burk double-reciprocal plot for data obtained in the presence of a fixed concentration of plastic peanuts will have the same y-intercept but a steeper slope than it would if the data had been obtained in the absence of the plastic peanuts.

• 6-12. Biological Relevance. Explain the biological relevance of each of the following observations concerning enzyme regulation.

(a) When you need a burst of energy, the hormones epinephrine and glucagon are secreted into your bloodstream and circulated to your muscle cells, where they initiate a cascade of reactions that leads to the phosphorylation of the inactive (*b*) form of glycogen phosphorylase, thereby converting the enzyme into the active (*a*) form.

(b) Even in the *a* form, glycogen phosphorylase is allosterically inhibited by a high concentration of glucose or ATP within a specific muscle cell.

(c) Your pancreas synthesizes and secretes the proteolytic enzyme carboxypeptidase in the form of an inactive precursor called procarboxypeptidase, which is activated as a result of proteolytic cleavage by the enzyme trypsin in the duodenum of your small intestine.

•6-13.   Derivation of the Michaelis-Menten Equation. For the enzyme-catalyzed reaction in which a substrate S is converted into a product P (see reaction 6-5), velocity can be defined as the disappearance of substrate or the appearance of product per unit time:

$$v = -\frac{d[S]}{dt} = +\frac{d[P]}{dt} \qquad (6\text{-}21)$$

Beginning with this definition and restricting your consideration to the initial stage of the reaction when [P] is essentially zero, derive the Michaelis-Menten equation (see equation 6-7). The following points may help you in your derivation:

(a) Begin by expressing the rate equations for $d[S]/dt$, $d[P]/dt$, and $d[ES]/dt$ in terms of concentrations and rate constants.

(b) Assume a steady state at which the enzyme-substrate complex of reaction 6-6 is being broken down at the same rate at which it is being formed, such that the net rate of change, $d[ES]/dt$, is zero.

(c) Note that the total amount of enzyme present, E_t, is the sum of the free form E_f plus the amount of complexed enzyme ES: $E_t = E_f + ES$.

(d) When you get that far, note that V_{max} and K_m can be defined as follows:

$$V_{max} = k_3[E_t] \qquad K_m = \frac{k_2 + k_3}{k_1} \qquad (6\text{-}22)$$

•6-14.   Linearizing Michaelis and Menten. In addition to the Lineweaver-Burk plot (see Figure 6-11), two other straight-line forms of the Michaelis-Menten equation are sometimes used. The Eadie-Hofstee plot is a graph of $v/[S]$ versus v (see Figure 6-12), and the Hanes-Woolf plot graphs $[S]/v$ versus $[S]$.

(a) In both cases, show that the equation being graphed can be derived from the Michaelis-Menten equation by simple arithmetic manipulation.

(b) In both cases, indicate how K_m and V_{max} can be determined from the resulting graph.

(c) Make a Hanes-Woolf plot, with the intercepts and slope labeled as for the Lineweaver-Burk and Eadie-Hofstee plots (see Figures 6-11 and 6-12). Can you suggest why the Hanes-Woolf plot is the most statistically satisfactory of the three?

Suggested Reading

References of historical importance are marked with a •.

General References
Colowick, S. P., and N. O. Kaplan, eds. *Methods in Enzymology.* New York: Academic Press, 1970–present (ongoing series).
Lehninger, A. L., D. L. Nelson, and M. M. Cox. *Principles of Biochemistry*, 3ᵈ ed. New York: Worth, 1999.
Mathews, C. K., K. E. van Holde, and K G. Ahern. *Biochemistry*, 3ᵈ ed. San Francisco: Benjamin/Cummings, 2000.
Webb, E. *Enzyme Nomenclature.* Orlando, FL: Academic Press, 1992.

Historical References
• Changeux, J. P. The control of biochemical reactions. *Sci. Amer.* 212 (April 1965): 36.
• Cori, C. F. James B. Sumner and the chemical nature of enzymes. *Trends Biochem. Sci.* 6 (1981): 194.
• Friedmann, H., ed. *Benchmark Papers in Biochemistry. Vol. 1: Enzymes.* Stroudsburg, PA: Hutchinson Ross, 1981.
• Monod, J., J. P. Changeux, and F. Jacob. Allosteric proteins and cellular control systems. *J. Mol. Biol.* 6 (1963): 306.
• Phillips, D. C. The three-dimensional structure of an enzyme molecule. *Sci. Amer.* 215 (November 1966): 78.

Structure and Function of Enzymes
DeSantis, G., and J. B. Jones. Chemical modification of enzymes for enhanced functionality. *Curr. Opin. Biotech.* 10 (1999): 324.

Erlandsen, H., E. E. Abola, and R. C. Stevens. Combining structural genomics and enzymology: completing the picture in metabolic pathways and enzyme active sites. *Curr. Opin. Struct. Biol.* 10 (2000): 719.
Fersht, A. *Structure and Mechanism in Protein Science: A Guide to Enzyme Catalysis and Protein Folding.* New York: W.H. Freeman, 1999.
Suckling, C. J., C. L. Gibson, and A. R. Pitt, eds. *Enzyme Chemistry: Impact and Applications*, 3d ed. New York: Blackie Academic & Professional, 1998.
Walsh, C. Enabling the chemistry of life. *Nature* 409 (2001): 226.

Mechanisms of Enzyme Catalysis
Frey, P.A., and D. B. Northrop, eds. *Enzymatic Mechanisms.* Washington, DC: IOS Press, 1999.
Kraut, J. How do enzymes work? *Science* 242 (1988): 533.
Tsou, C.L. Active site flexibility in enzyme catalysis. *Ann. N. Y. Acad. of Sci.* 864 (1998): 1.

Ribozymes: RNA as an Enzyme
• Cech, T. R. The chemistry of self-splicing RNA and RNA enzymes. *Science* 236 (1987): 1532.
Jaeger, L. The new world of ribozymes. *Curr. Opin. Struct. Biol.* 7 (1997): 324.
Joyce, G. L. Building the RNA world: Ribozymes. *Curr. Biol.* 6 (1996): 965.
Krupp, G., and R. K. Gaur. *Ribozymes: Biochemistry and Biotechnology.* Natick, MA: Eaton Publishing, 2000.

Membranes: Their Structure, Function, and Chemistry

An essential feature of every cell is the presence of **membranes** that define the boundaries of the cell and its various internal compartments. Even the casual observer of electron micrographs is likely to be struck by the prominence of membranes around and within cells, especially those of eukaryotic organisms (Figure 7-1). We encountered membranes and membrane-bounded organelles in an introductory manner in Chapter 4; now we are ready to look at membrane structure and function in greater detail. In this chapter, we will examine the molecular structure of membranes and explore the multiple roles that membranes play in the life of the cell. In Chapter 8, we will discuss the

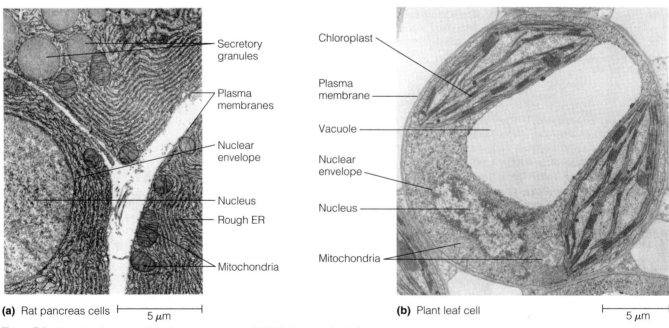

(a) Rat pancreas cells ├────── 5 μm ──────┤

(b) Plant leaf cell ├──── 5 μm ────┤

Figure 7-1   The Prominence of Membranes Around and Within Eukaryotic Cells.   Among the structures of eukaryotic cells that involve membranes are the plasma membrane, nucleus, chloroplasts, mitochondria, endoplasmic reticulum (ER), secretory granules, and vacuoles. These structures are shown here in **(a)** portions of three cells from a rat pancreas and **(b)** a plant leaf cell (TEMs).

transport of solutes across membranes. We will then go on to consider the mechanisms whereby signals that arrive at the surface of a cell are detected and transmitted into the cell, focusing first on electrical signals (Chapter 9) and then on chemical signals (Chapter 10). Membranes also play prominent roles in cell-cell recognition and adhesion (Chapter 11), and in the structure and function of the endoplasmic reticulum and Golgi complex (Chapter 12).

The Functions of Membranes

We begin our discussion by noting that biological membranes play five related yet distinct roles. They ① define the boundaries of the cell and delineate its compartments, ② serve as loci of specific functions, and ③ possess transport proteins that control the movement of substances into and out of the cell and its compartments. In addition, membranes ④ contain the receptors required for the detection of external signals and ⑤ provide mechanisms for cell-to-cell communication. Figure 7-2 summarizes these five main membrane functions. In the following five sections, which are keyed by numbers to the figure, we briefly introduce each of these functions.

① Membranes Define Boundaries and Serve as Permeability Barriers

One of the most obvious functions of membranes is to define the boundaries of the cell and its compartments and to serve as permeability barriers. The interior of the cell must be physically separated from the surrounding environment not only to keep desirable substances in the cell but also to keep undesirable substances out. Membranes serve this purpose well because the hydrophobic interior of the membrane is an effective barrier for hydrophilic molecules and ions.

The permeability barrier for the cell as a whole is the **plasma** (or **cell**) **membrane,** a membrane that surrounds the cell and regulates the passage of materials both into and out of cells. (The term *plasma* was originally used to describe the contents of a cell; we now use the term cytoplasm instead, but the outer limiting membrane is still called the plasma membrane.) In addition to the plasma membrane, various **intracellular membranes** serve to compartmentalize functions within eukaryotic cells. For example, the mitochondrion is the specific organelle that contains the enzymes, substrates, and other molecules required for the process of aerobic respiration, whereas the chloroplast contains the enzymes, pigments, and intermediates needed for photosynthesis. Without such compartments, eukaryotic life as we know it would not exist.

② Membranes Are Sites of Specific Functions

Membranes have specific functions associated with them because the molecules and structures responsible for those functions—proteins, in most cases—are either embedded in or localized on membranes. One of the most useful ways to characterize a specific membrane, in fact, is to describe the particular enzymes, transport proteins, receptors, and other molecules associated with it.

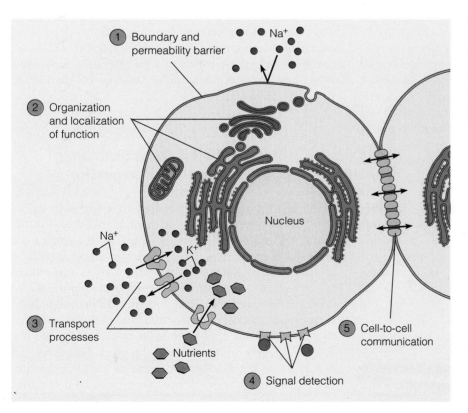

Figure 7-2 Functions of Membranes.
Membranes ① define the boundaries of the cell and its organelles, ② serve as loci for specific functions, ③ provide for and regulate transport processes, ④ contain the receptors needed to detect external signals, and ⑤ provide mechanisms for cell-to-cell contact, communication, and adhesion.

For example, many distinctive enzymes are present in or on the membranes of organelles such as the endoplasmic reticulum (ER), Golgi complex, lysosomes, and peroxisomes, as we will learn in Chapter 12. Such enzymes are often useful as *markers* during the isolation of organelles from suspensions of disrupted cells. For example, *glucose phosphatase* is a membrane-bound enzyme found in the endoplasmic reticulum. When ER membranes are isolated and purified (as tiny vesicles called *microsomes*), glucose phosphatase can be used as a marker enzyme, enabling the investigator to determine both the distribution of microsomes among the various fractions and the purity of the final preparation.

③ Membranes Regulate the Transport of Solutes

Another function of membranes is to carry out and regulate the **transport** of substances into and out of cells and their organelles. Nutrients, ions, gases, water, and other substances are taken up into various compartments, and various products and wastes must be removed.

As we will see in Chapter 8, the modes of transport for various substances differ. Many substances move in the direction dictated by their concentration gradients (or, in the case of ions, by the combined effect of the concentration and charge gradients across the membrane). This process, which does not require energy because movement is "down" the gradient, occurs by two different modes. Some molecules such as water, oxygen, and ethanol can cross membranes by *simple diffusion*. Larger, more polar molecules such as sugars and amino acids move across membranes aided by specific *transport proteins,* a process called *facilitated diffusion.*

Alternatively, substances can be transported against, or "up," a concentration gradient and/or a charge gradient, with ATP as the usual energy source. This process, called *active transport,* is carried out by membrane-bound enzyme systems that are commonly referred to as *pumps.* Nutrient molecules such as sugars and amino acids that are present in low concentrations outside the cell are transported by this mechanism, as are ions such as potassium, calcium, and hydrogen (protons).

Even molecules as large as proteins can be transported across membranes. In some cases, intracellular vesicles facilitate the movement of such molecules either into the cell (*endocytosis*) or out of the cell (*exocytosis*). In other cases, proteins that are synthesized on the endoplasmic reticulum or in the cytosol can be imported into specific membrane-bounded organelles such as lysosomes, peroxisomes, or mitochondria. We will discuss both of these processes in later chapters.

④ Membranes Detect and Transmit Electrical and Chemical Signals

Cells receive information from their environment, usually in the form of electrical or chemical signals that impinge on the outer surface of the cell. The nerve impulses being sent from your eyes to your brain as you read these words are examples of such signals, as are the various hormones present in your circulatory system. **Signal transduction** is the term used to describe both the detection of specific signals at the outer surface of cells and the specific mechanisms used to transmit such signals to the cell interior.

In the case of chemical signal transduction, some signal molecules enter directly into cells and act internally. The hormone estrogen is an example. Because estrogen is a steroid (see Figure 3-30a), it is nonpolar and can therefore cross membranes readily. As a result, estrogen enters into its target cells and interacts with regulatory proteins inside the cell. In most cases, however, the impinging signal molecules do not enter the cell but instead bind to specific **receptors** on the outer surface of the plasma membrane. Binding of such substances, called *ligands,* is followed by specific chemical events on the inner surface of the membrane, thereby generating internal signals called *second messengers.* Membrane receptors therefore allow cells to recognize, transmit, and respond to a variety of specific chemical signals, a topic we will explore in Chapter 10.

⑤ Membranes Mediate Cell-to-Cell Communication

Membranes also provide a means of communication between adjacent cells. Although textbooks often depict cells as separate, isolated entities, most cells in multicellular organisms are in contact with other cells via direct cytoplasmic connections that allow the exchange of at least some cellular components. This **intercellular communication** is provided by *gap junctions* in animal cells and by *plasmodesmata* in plant cells.

All the functions we have just considered—compartmentalization, localization of function, transport, signal detection, and intercellular communication—depend on the chemical composition and structural features of membranes. It is to these topics that we now turn as we consider how our present understanding of membrane structure developed.

Models of Membrane Structure: An Experimental Perspective

Until electron microscopy was applied to the study of cell structure in the early 1950s, no one had ever seen a membrane. Yet indirect evidence led biologists to postulate the existence of membranes long before they could actually be seen. In fact, researchers have been trying to understand the molecular organization of membranes for more than a century. Because cells contain many different kinds of membranes, finding the structural features that are common to all membranes has been a special challenge. The intense research effort paid off, however, because it led eventually to the *fluid mosaic model* of membrane structure. This model, which is now thought to be descriptive of all biological membranes, envisions a membrane as a fluid "sea" of phospholipids and other lipids with proteins "floating" in and on it.

Before looking at the model in detail, we will describe some of the central experiments that have led to this view of membrane structure and function. As we do so, you may also gain some insight into how such developments come about, as well as a greater respect for the diversity of approaches and techniques that are often important in advancing our understanding of biological phenomena. Figure 7-3 presents a chronology of membrane studies that began over a century ago with the understanding that lipid layers are a part of membrane structure and led eventually to our current understanding of membranes as fluid mosaics. Refer to Figure 7-3 as you read the sections that follow.

Overton and Langmuir: Lipids Are Important Components of Membranes

A good starting point for our experimental overview is the pioneering work of Charles Overton in the 1890s. Overton was aware that cells appear to be enveloped by some sort of selectively permeable layer that allows different substances to enter and leave cells at significantly different rates. Working with cells of plant root hairs, he observed that lipid-soluble substances penetrate readily into cells, whereas water-soluble substances do not. In fact, he found a good correlation between the *lipophilic* ("lipid-loving") nature of a substance and the ease with which it can enter cells. From his studies, Overton concluded that lipids are present on the cell surface as some sort of "coat" (Figure 7-3a). He even suggested that cell coats are probably mixtures of cholesterol and lecithin, an insight that proved to be remarkably foresighted in light of what we now know about the prominence of sterols and phospholipids as membrane components.

A second important advance came about a decade later through the work of Irving Langmuir, who studied the behavior of purified phospholipids by dissolving them in benzene and layering samples of the benzene-lipid solution onto a water surface. As the benzene evaporated, the molecules were left as a lipid film one molecule thick, i.e., a "monolayer." Langmuir knew that phospholipids are *amphipathic* molecules, meaning that they possess both hydrophilic and hydrophobic regions (see Figure 2-11 for an illustration of the polar "head" groups and nonpolar "tails" of phospholipid molecules). He reasoned that phospholipids orient themselves on water such that their hydrophilic heads face the water and their hydrophobic tails protrude away from the water (Figure 7-3b). Langmuir's lipid monolayer became the basis for further thought about membrane structure in the early years of the twentieth century.

Gorter and Grendel: The Basis of Membrane Structure Is a Lipid Bilayer

The next major advance came in 1925 when two Dutch physiologists, E. Gorter and F. Grendel, read Langmuir's

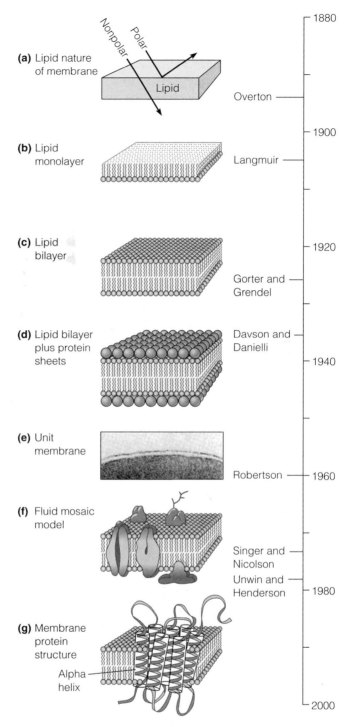

Figure 7-3   Timeline for the Development of the Fluid Mosaic Model. The fluid mosaic model of membrane structure that Singer and Nicholson proposed in 1972 was the culmination of studies that date back to the 1890s.

papers and thought that his approach might help answer a question regarding the surface of the red blood cells, or *erythrocytes*, with which they worked. Overton's earlier research had shown the presence of a lipid coat on the cell surface. But how many lipid layers are present in the coat? To answer this question, Gorter and Grendel extracted the lipids from a known number of erythrocytes and used

Langmuir's method to spread the lipids out as a monolayer on a water surface. Finding that the area of the lipid film on the water was about twice the estimated total surface area of the erythrocyte, they concluded that the erythrocyte plasma membrane consists of not one, but *two* layers of lipids. (As it turned out later, Grendel and Gorter made two errors, underestimating by about one-third both the surface area of the red blood cell and the amount of lipid present in its plasma membrane. Fortunately, these errors canceled each other out, so Gorter and Grendel's conclusion was correct even though their data were not. For a chance to repeat their calculations and discover the source of their errors, see Problem 7-3 at the end of the chapter.)

Hypothesizing a bilayer structure, Grendel and Gorter reasoned that it would be thermodynamically favorable for the nonpolar hydrocarbon chains of each layer to face inward, away from the aqueous milieu on either side of the membrane. The polar hydrophilic groups of each layer would then face outward, toward the aqueous environment on either side of the membrane (Figure 7-3c). Their experiment and their conclusions were momentous because this work represented the first attempt to understand membranes at the molecular level. Moreover, the **lipid bilayer** that they envisioned became the basic underlying assumption for each successive refinement in our understanding of membrane structure.

Davson and Danielli: Membranes Also Contain Proteins

Shortly after Gorter and Grendel proposed their bilayer model in 1925, it became clear that a simple lipid bilayer, though an important feature of membrane structure, could not explain all the properties of membranes, particularly those related to *surface tension, solute permeability,* and *electrical resistance*. For example, the surface tension of a lipid film was shown to be significantly higher than that of cellular membranes but could be lowered by adding protein to the lipid film. Moreover, sugars, ions, and other hydrophilic solutes moved into and out of cells much more readily than could be explained by the permeability of pure lipid bilayers to water-soluble substances. For example, the leakage rate of potassium ions across artificial lipid bilayers is measured in days, whereas the same amount of leakage across erythrocyte membranes occurs in an hour or so.

To explain such differences, Hugh Davson and James Danielli invoked the presence of proteins in membranes, proposing in 1935 that biological membranes consist of lipid bilayers that are coated on both sides with thin sheets of protein (Figure 7-3d). Thus, the original *Davson-Danielli model* was in essence a protein-lipid-protein "sandwich." Their model was the first detailed representation of membrane organization and dominated the thinking of cell biologists for the next several decades.

The original model was subsequently modified to accommodate additional findings. Particularly notable was the suggestion, made in 1954, that hydrophilic proteins might penetrate into the membrane in places to provide *polar pores* through an otherwise hydrophobic bilayer. These proteins could then account for permeability and resistivity properties of membranes that were not easily explained in terms of the lipid bilayer alone. Specifically, the lipid interior accounted for the hydrophobic properties of membranes and the protein components explained their hydrophilic properties.

The real significance of the Davson-Danielli model, however, was its recognition of the importance of proteins in membrane structure. This feature, more than any other, made the Davson-Danielli sandwich the basis for much subsequent research on membrane structure.

Robertson: All Membranes Share a Common Underlying Structure

All of the membrane models discussed so far were developed before anyone had ever seen a biological membrane, and each was intended specifically as a model of the plasma membrane. With the advent of electron microscopy in the 1950s, cell biologists could finally verify the presence of a plasma membrane around each cell. In addition, they observed that most subcellular organelles are bounded by similar membranes. Furthermore, when membranes were stained with osmium, a heavy metal, and then examined closely at high magnification, they were found to have extensive regions of "railroad track" structure that appeared as two dark lines separated by a lightly stained central zone, with an overall thickness of 6–8 nm. This pattern is seen in Figure 7-4 for the plasma membranes of two adjacent cells that are separated from each other by a thin intercellular space. The fact that this same

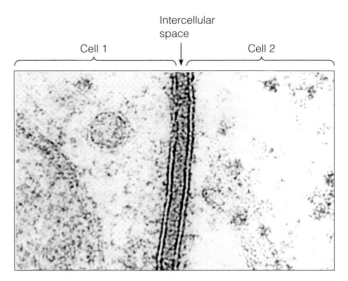

Figure 7-4 Trilaminar Appearance of Cellular Membranes. This electron micrograph shows two adjacent cells with their plasma membranes separated by a small intercellular space. Each membrane appears as two dark lines separated by a lightly stained central zone, a staining pattern that gives each membrane a trilaminar, or "railroad track," appearance (TEM).

trilaminar, or three-layered, staining pattern was observed with many different kinds of membranes led J. David Robertson to suggest that all cellular membranes share a common underlying structure, which he called the *unit membrane* (Figure 7-3e).

When first proposed, the unit membrane structure appeared to agree remarkably well with the Davson-Danielli model. Robertson suggested that the lightly stained space (between the two dark lines of the trilaminar pattern) contains the hydrophobic region of the lipid molecules, which do not readily stain. Conversely, the two dark lines were thought to represent phospholipid head groups and the thin sheets of protein bound to the membrane surfaces, which appear dark because of their affinity for heavy metal stains. This interpretation appeared to provide strong support for the Davson-Danielli view that a membrane consists of a lipid bilayer coated on both surfaces with thin sheets of protein.

Further Research Revealed Major Shortcomings of the Davson-Danielli Model

In spite of its apparent confirmation by electron microscopy and its extension to all membranes by Robertson, the Davson-Danielli model encountered difficulties as more and more data emerged in the 1960s that could not be reconciled with it. Consider, for example, the problem of membrane dimensions: Based on electron microscopy, most membranes were reported to be about 6–8 nm in thickness and of this, the lipid bilayer accounted for about 4–5 nm. That left only about 1–2 nm of space on either surface of the bilayer for the membrane protein, a space that could at best accommodate a thin monolayer of protein consisting primarily of extended regions of β-sheet structure. Yet as membrane proteins were isolated and studied, it became apparent that most of them were globular proteins with extensive regions of α helix.

Such proteins have sizes and shapes that are inconsistent with the concept of thin sheets of protein on the two surfaces of the membrane, suggesting that they must protrude into the interior of the membrane.

As a further complication, the Davson-Danielli model did not readily account for the distinctiveness of different kinds of membranes. Depending on their source, membranes vary considerably in chemical composition and especially in the ratio of protein to lipid (Table 7-1). The *protein/lipid ratio* can be as high as 3 or more in some bacterial cells and as low as 0.23 for the *myelin sheath* that serves as a membranous electrical insulation around nerve axons. Even the two membranes of the mitochondrion differ significantly: The protein/lipid ratio is about 1.2 for the outer membrane and about 3.5 for the inner membrane, which contains all the enzymes and proteins related to electron transport and ATP synthesis. Yet all of these membranes look essentially the same when visualized for electron microscopy by the osmium staining technique of Robertson and others. As more membranes were studied, it became increasingly difficult to reconcile such enormous variations in protein content with the unit membrane model, because the width and appearance of the "rails" simply did not vary correspondingly. (We now understand that the osmium used in this staining procedure reacts primarily with the hydrophilic head groups of phospholipids, so that the "rails" of the trilaminar pattern seen in electron micrographs such as Figure 7-4 represent mainly the head groups on the two sides of the membrane and do not include thin sheets of protein, as initially supposed.)

The Davson-Danielli model was also called into question by studies in which membranes were exposed to *phospholipases*, enzymes that degrade phospholipids by removing their head groups. According to the model, the hydrophilic head groups of membrane lipids should be covered by a layer of protein and therefore protected from phospholipase digestion. Yet a significant proportion (up

Table 7-1 Protein, Lipid, and Carbohydrate Content of Biological Membranes

Membrane	Approximate Percent by Weight			Protein/Lipid Ratio
	Protein	Lipid	Carbohydrate	
Plasma membrane				
Human erythrocyte	49	43	8	1.14
Mammalian liver cell	54	36	10	1.50
Amoeba	54	42	4	1.29
Myelin sheath of nerve axon	18	79	3	0.23
Nuclear envelope	66	32	2	2.06
Endoplasmic reticulum	63	27	10	2.33
Golgi complex	64	26	10	2.46
Chloroplast thylakoids	70	30	0	2.33
Mitochondrial outer membrane	55	45	0	1.22
Mitochondrial inner membrane	78	22	0	3.54
Gram-positive bacterium	75	25	0	3.00

to 75%, depending on the cell type) of the membrane phospholipid can be degraded when the membrane is exposed to phospholipases. This susceptibility to enzymatic digestion suggests that many of the phospholipid head groups are exposed at the membrane surface rather than being covered by a layer of protein.

Moreover, the surface localization of membrane proteins specified by the Davson-Danielli model was not supported by the experience of scientists who tried to isolate such proteins. Most membrane proteins turn out to be quite insoluble in water and can be extracted only by the use of organic solvents or detergents. These observations indicated that many membrane proteins are hydrophobic (or at least amphipathic) and suggested that they are located, at least in part, within the hydrophobic interior of the membrane rather than on either of its surfaces.

The Davson-Danielli model was further discredited by experimental evidence, to be discussed later, indicating that membranes are fluid structures in which most of the lipids and many of the proteins are free to move within the plane of the membrane. Mobility of membrane proteins and lipids is not easy to reconcile with a model that envisions sheets of surface proteins linked by ionic bonds to the underlying lipid bilayer.

Singer and Nicolson: A Membrane Consists of a Mosaic of Proteins in a Fluid Lipid Bilayer

The preceding problems with the Davson-Danielli model stimulated considerable interest in the development of new ideas about membrane organization, culminating in 1972 with the **fluid mosaic model** proposed by S. Jonathan Singer and Garth Nicolson. This model, which now dominates our view of membrane organization, has two key features, both implied by its name. Simply put, the model envisions a membrane as a *mosaic* of proteins discontinuously embedded in, or at least attached to, a *fluid* lipid bilayer (Figure 7-3f). In other words, the Singer-Nicolson model retained the basic lipid bilayer structure of earlier models but viewed membrane proteins in an entirely different way—not as thin sheets on the membrane surface, but as discrete globular entities that associate with the membrane on the basis of their affinity for the hydrophobic interior of the lipid bilayer (Figure 7-5a).

This way of thinking about membrane proteins was revolutionary when first proposed by Singer and Nicolson, but it turned out to fit the data quite well. Based on differences in the nature of their linkage to the bilayer, three broad classes of membrane proteins are recognized. *Integral membrane proteins* are embedded within the lipid bilayer, where they are held in place by the affinity of hydrophobic segments of the protein for the hydrophobic interior of the lipid bilayer. *Peripheral proteins,* on the other hand, are much more hydrophilic and are therefore located on the surface of the membrane, where they are linked noncovalently to the polar head groups of phospholipids and/or to the hydrophilic parts of other membrane proteins. *Lipid-anchored proteins,* though not a part of the original fluid mosaic model, are now recognized as a third class of membrane proteins. These are essentially hydrophilic proteins and therefore reside on membrane surfaces, but they are covalently attached to lipid molecules that are embedded within the bilayer.

The fluid nature of the membrane is the second critical feature of the Singer-Nicholson model. Rather than being rigidly locked in place, most of the lipid components of a membrane are in constant motion, capable of lateral mobility (i.e., movement parallel to the membrane surface). Many membrane proteins are also able to move laterally within the membrane, although some proteins are anchored to structural elements on one side of the membrane or the other and are therefore restricted in their mobility.

The major strength of the fluid mosaic model is that it provides ready explanations for most of the criticisms of the Davson-Danielli model. For example, the concept of proteins partially embedded within the lipid bilayer accords well with the hydrophobic nature and globular structure of most membrane proteins and eliminates the need to accommodate membrane proteins in thin surface layers of invariant thickness. Moreover, the variability in the protein/lipid ratios of different membranes simply means that some membranes have relatively few proteins embedded within the lipid bilayer whereas other membranes have more such proteins. The exposure of lipid head groups at the membrane surface is obviously compatible with their susceptibility to phospholipase digestion, while the intermingling of lipids and proteins within the membrane makes it easy to envision the mobility of both lipids and proteins.

Unwin and Henderson: Some Membrane Proteins Contain Transmembrane Segments

The final illustration in the timeline (Figure 7-3g) depicts an important property of integral membrane proteins that cell biologists began to understand in the 1970s: Most such proteins have in their primary structure one or more hydrophobic sequences that span the lipid bilayer (Figure 7-5b and c). These *transmembrane segments* anchor the protein to the membrane and hold it in proper alignment within the lipid bilayer.

The example in Figure 7-3g is *bacteriorhodopsin*, the first membrane protein shown to possess this structural feature. Bacteriorhodopsin is a plasma membrane protein found in archaebacteria of the genus *Halobacterium*, where its presence allows cells to obtain energy directly from sunlight. To capture this solar energy, bacteriorhodopsin has as part of its structure a molecule of *retinal*, the same light-absorbing molecule used by the human eye to detect light. Upon absorbing light energy, retinal triggers a conformational change in bacteriorhodopsin that causes the protein to pump protons out of the cell. The resulting proton gradient across the plasma membrane can be used as a source of energy.

Bacteriorhodopsin molecules are arranged in the membrane in a crystalline lattice, a property that enabled

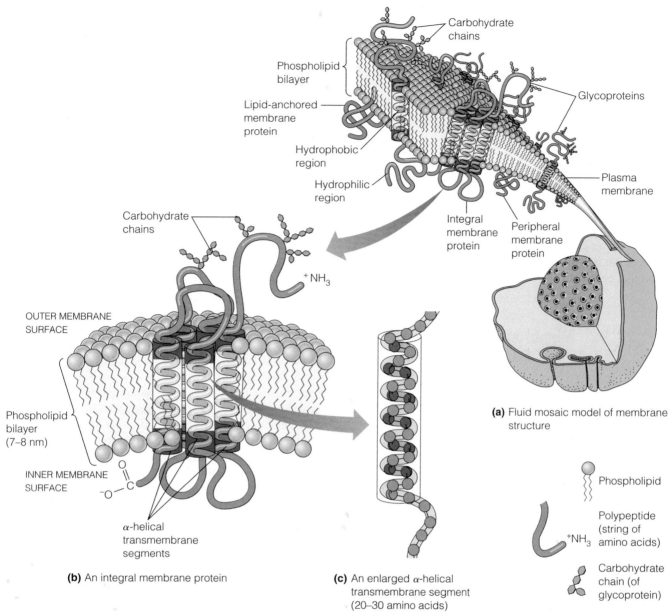

Figure 7-5 **The Fluid Mosaic Model of Membrane Structure.**   **(a)** Singer and Nicolson's model envisions the membrane as a fluid bilayer of lipids with a mosaic of associated proteins that are either integral or peripheral parts of the membrane. Integral membrane proteins are anchored to the hydrophobic interior of the membrane by one or more hydrophobic transmembrane segments (light purple), usually with an α-helical conformation. Hydrophilic segments (dark purple) extend outward on one or both sides of the membrane. Peripheral membrane proteins are associated with the membrane surface by weak electrostatic forces that bind them to hydrophilic regions of adjacent integral membrane proteins or to the polar head groups of phospholipids. **(b)** An integral membrane protein with multiple α-helical transmembrane segments, as revealed by the later work of Unwin and Henderson. Many integral membrane proteins of the plasma membrane have carbohydrate side chains attached to their hydrophilic segments on the outer membrane surface. **(c)** An α-helical transmembrane segment with amino acids represented by circles within the helix.

Nigel Unwin and Richard Henderson to determine the three-dimensional structure of the protein and its orientation in the membrane. Their remarkable finding, reported in 1975, was that bacteriorhodopsin consists of a single peptide chain folded back and forth across the lipid bilayer a total of seven times. Each of the seven transmembrane segments of the protein is a closely packed α helix composed mainly of hydrophobic amino acids. Successive transmembrane segments are linked to each other by short loops of hydrophilic amino acids that extend into or protrude from the polar surfaces of the membrane. Based on subsequent work in many laboratories, membrane biologists now believe that all transmembrane proteins are anchored in the lipid bilayer by one or more transmembrane segments, a topic to which we will return later in the chapter.

The Fluid Mosaic Model Is Now the Accepted View of Membrane Structure

Almost from the moment Singer and Nicolson proposed it, the fluid mosaic model has revolutionized the way scientists think about membrane structure. The model launched a new era in membrane research that has not only confirmed the basic model but has augmented and extended it. As a result, the fluid mosaic model is now the universally accepted view of membrane structure. It is therefore important for us to examine the essential features of the model in detail. These features include the chemistry, the asymmetric distribution, and the fluidity of membrane lipids, the relationship of membrane proteins to the bilayer, and the mobility of proteins within the bilayer. We will discuss each of these features in turn, focusing on both the supporting evidence and the implications of each feature for membrane function.

Membrane Lipids: The "Fluid" Part of the Model

We will begin our detailed look at membranes by considering lipids, which are the "fluid" part of the fluid mosaic model.

Membranes Contain Several Major Classes of Lipids

An obvious property of Singer and Nicolson's model is its retention of the lipid bilayer initially proposed by Gorter and Grendel, though with a greater diversity and fluidity of lipid components than early investigators recognized. The main classes of membrane lipids are *phospholipids, glycolipids,* and *sterols.* Figure 7-6 lists the main lipids in each of these categories and depicts the structures of several.

Phospholipids.   As we already know from Chapter 3, the most abundant lipids found in membranes are the **phospholipids** (Figure 7-6a). Membranes contain many different kinds of phospholipids, including both the glycerol-based **phosphoglycerides** and the sphingosine-based **sphingolipids.** The most common phosphoglycerides are *phosphatidylcholine, phosphatidylethanolamine, phosphatidylserine,* and *phosphatidylinositol;* the most common sphingolipid is *sphingomyelin,* the structure of which is shown in Figure 7-6a. The kinds and relative proportions of phospholipids present vary significantly among membranes from different sources (Figure 7-7). Sphingomyelin, for example, is one of the main phospholipids of animal plasma membranes, but it is absent from the plasma membranes of plants and bacteria and from the membranes of mitochondria and chloroplasts.

Glycolipids.   As their name indicates, **glycolipids** are formed by adding carbohydrate groups to lipids. Some glycolipids are glycerol-based, but most are derivatives of sphingosine and are therefore called *glycosphingolipids.* The most common examples are **cerebrosides** and **gangliosides.** Cerebrosides are called *neutral glycolipids* because each molecule has a single uncharged sugar as its head group—galactose, in the case of the galactocerebroside shown in Figure 7-6b. A ganglioside, on the other hand, always has an oligosaccharide head group that contains one or more negatively charged sialic acid residues, giving the molecule a net negative charge. Cerebrosides and gangliosides are especially prominent in the membranes of brain and nerve cells. Gangliosides exposed on the surface of the plasma membrane also function as antigens recognized by antibodies in immune reactions, including those responsible for blood group interactions. The human ABO blood groups, for example, involve glycosphingolipids that serve as cell surface markers of red blood cells.

Several serious human diseases are known to result from impaired metabolism of glycosphingolipids. The best-known example is *Tay-Sachs disease,* which is caused by the absence of a lysosomal enzyme, β-N-acetylhexosaminidase, that is responsible for one of the steps in ganglioside degradation. As a result of the genetic defect, gangliosides accumulate in the brain and other nervous tissue, leading to impaired nerve and brain function and eventually to paralysis, severe mental deterioration, and death.

Sterols.   In addition to phospholipids and glycolipids, the membranes of most eukaryotic cells also contain significant amounts of **sterols** (Figure 7-6c). The main sterol in animal cell membranes is **cholesterol**, whereas the membranes of plant cells contain small amounts of cholesterol and larger amounts of **phytosterols**, including campesterol, sitosterol, and stigmasterol. To the extent that they have been studied, the membranes of fungi and protists also contain sterols similar in structure to cholesterol.

Sterols are not found in the membranes of prokaryotic cells and are also absent from the inner membranes of both mitochondria and chloroplasts, which are believed to be derived evolutionarily from the plasma membranes of prokaryotic cells. However, the plasma membranes of at least some prokaryotes, including both bacteria and cyanobacteria, contain sterol-like molecules called **hopanoids** that appear to substitute for sterols in membrane

Figure 7-6   The Three Major Classes of Membrane Lipids.   (a) Phospholipids contain phosphate groups covalently attached to a lipid backbone, forming either phosphoglycerides (glycerol-based) or sphingolipids (sphingosine-based). **(b)** Glycolipids are also either glycerol- or sphingosine- based, with the latter more common. A cerebroside has a neutral sugar as its head group, whereas a ganglioside has an oligosaccharide chain containing one or more sialic acid residues and therefore carries a negative charge. **(c)** The most common membrane sterols are cholesterol in animals and several related phytosterols in plants.

(a) PHOSPHOLIPIDS

Phosphatidylcholine (shown)
Phosphatidylethanolamine
Phosphatidylserine
Phosphatidylthreonine
Phosphatidylinositol
Phosphatidylglycerol
Diphosphatidylglycerol (cardiolipin)

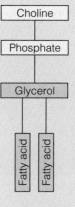

Sphingomyelin (a sphingolipid)

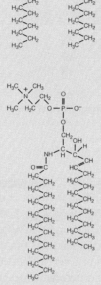

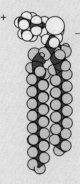

(b) GLYCOLIPIDS

Cerebrosides
 (galactocerebroside shown)
Gangliosides

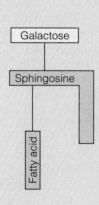

(c) STEROLS

Cholesterol (shown)
Campesterol
Sitosterol   } Phytosterols
Stigmasterol

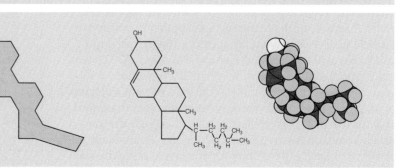

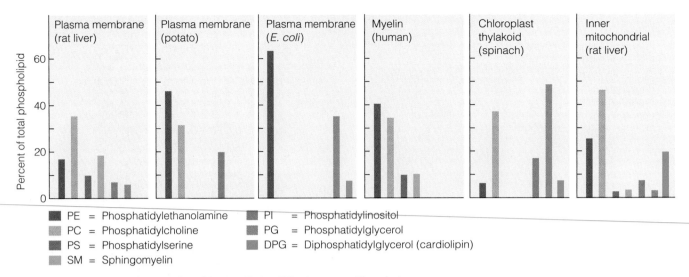

Figure 7-7 Phospholipid Composition of Several Kinds of Membranes. The relative abundance of different kinds of phospholipids in biological membranes varies greatly with the source of the membrane.

structure. The hopanoid molecule is rigid and strongly hydrophobic, with a short hydrophilic side chain extending from one side (Figure 7-8). Hopanoids are abundant in petroleum deposits, suggesting that they might have been membrane components of the ancient prokaryotes that presumably contributed to the formation of fossil fuels. If they are in fact as abundant in underground petroleum reserves as some experts believe, hopanoids may well be the single most abundant class of organic molecules in the world!

Thin-Layer Chromatography Is an Important Technique for Lipid Analysis

How do we know so much about the lipid components of membranes? The answer, of course, is that biologists and biochemists have been isolating, separating, and studying membrane lipids for more than a century. One important technique for the analysis of lipids is **thin-layer chromatography (TLC)**, depicted schematically in Figure 7-9. This technique separates different kinds of lipids based on their relative affinities for a hydrophilic *stationary phase* and a hydrophobic *mobile phase*. The stationary phase is usually a thin layer of silicic acid on a glass or metal plate, while the mobile phase is a mixture of appropriate solvents.

In this procedure, the lipids are first extracted from a membrane preparation using a mixture of organic solvents. A sample of the extract is then applied to one end of a silicic acid–coated plate by spotting the extract onto a small area called the *origin* (Figure 7-9a). After the solvent has evaporated, the edge of the plate is dipped into a solvent system that typically consists of chloroform, methanol, and water. As the solvent moves past the origin and up the plate by capillary action, the lipids are separated based on their polarity. Nonpolar lipids such as cholesterol have little affinity for the silicic acid (the stationary phase) and therefore move up the plate with the solvent system (the mobile phase). Lipids that are more

polar, such as phospholipids, interact more strongly with the silicic acid, which slows their movement. In this way, the various lipids are separated progressively as the leading edge of the mobile phase continues to move up the plate. When the leading edge, or *solvent front*, approaches the top, the plate is removed from the solvent system and dried, and the separated lipids are then recovered from the plate (by dissolving them in a solvent such as chloroform) for identification and further study.

Figure 7-9b shows the TLC pattern seen for the lipids of the erythrocyte plasma membrane. The main components of this membrane are phospholipids (55%) and cholesterol (25%), with phosphatidylethanolamine (PE),

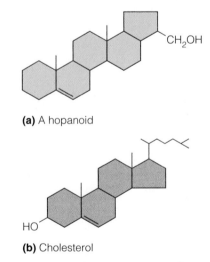

(a) A hopanoid

(b) Cholesterol

Figure 7-8 The Structure of Hopanoids. **(a)** A hopanoid, one of a class of sterol-like molecules that appear to function in the plasma membranes of at least some prokaryotes as sterols do in the membranes of eukaryotic cells. **(b)** The structure of cholesterol, for comparison. A weakly hydrophilic side chain (−CH$_2$OH or −OH) protrudes from each molecule.

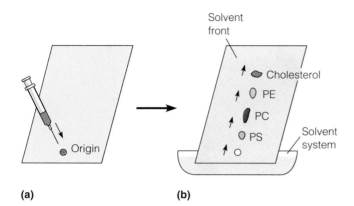

Figure 7-9 Use of Thin-Layer Chromatography in the Analysis of Membrane Lipids. Thin-layer chromatography (TLC) is a useful technique for the analysis of membrane lipids. Lipids are extracted from a membrane preparation with a mixture of organic solvents, and **(a)** a sample is spotted onto a small area of a glass or metal plate coated with a thin layer of silicic acid. When the sample has dried at this origin point, the plate is dipped into a solvent system, usually a mixture of chloroform, methanol, and water. As the solvent moves up the plate by capillary action, the lipids are separated according to their polarity: Nonpolar lipids such as cholesterol do not adhere strongly to the silicic acid and move further up the plate, while more polar lipids remain closer to the origin. **(b)** When the solvent front nears the top, the plate is removed from the solvent system and the lipids are eluted and identified. The pattern shown is that for lipids of the erythrocyte plasma membrane. The main components are cholesterol, phosphatidylethanolamine (PE), phosphatidylcholine (PC), and phosphatidylserine (PS).

phosphatidylcholine (PC), and phosphatidylserine (PS) as the most prominent phospholipids. The erythrocyte plasma membrane also contains phosphatidylinositol and sphingolipids, but these components are present in smaller amounts and are less likely to be detected by TLC. These lipids are nonetheless important membrane components, more prominent in other types of membranes. Sphingolipids, for example, are especially prevalent in nervous tissue, as already noted.

Fatty Acids Are Essential to Membrane Structure and Function

Fatty acids are components of all membrane lipids except the sterols. They are essential to membrane structure because their long hydrocarbon tails form an effective hydrophobic barrier to the diffusion of polar solutes. Most fatty acids in membranes are between 12 and 20 carbon atoms in length, with 16- and 18-carbon fatty acids especially common. This size range appears to be optimal for bilayer formation, because chains with fewer than 12 or more than 20 carbons are unable to form a stable bilayer. Thus, the thickness of membranes (about 6–8 nm, depending on the source) is dictated primarily by the chain length of the fatty acids required for bilayer stability.

In addition to differences in length, the fatty acids found in membrane lipids also vary considerably in the presence and number of double bonds. Table 7-2 shows the structures of several fatty acids that are especially com-

mon in membrane lipids. *Palmitate* and *stearate* are saturated fatty acids with 16 and 18 carbon atoms, respectively. *Oleate* and *linoleate* are 18-carbon unsaturated fatty acids with one and two double bonds, respectively. Other common unsaturated fatty acids in membranes are *linolenate* with 18 carbons and three double bonds, and *arachidonate* with 20 carbons and four double bonds. All unsaturated fatty acids in membranes are in the *cis* configuration, resulting in a sharp bend, or kink, in the hydrocarbon chain at every double bond. Because of the nonlinearity of their side chains, fatty acids with double bonds do not pack tightly in the membrane, a feature with considerable implications for membrane fluidity, as we will see shortly.

Membrane Asymmetry: Most Lipids Are Distributed Unequally Between the Two Monolayers

Since membranes contain many different kinds of lipids, the question arises as to whether the various lipids are randomly distributed between the two *monolayers* of lipid that together constitute a lipid bilayer. Chemical investigations involving membranes derived from a variety of different cell types have revealed that most lipids are unequally distributed between the two monolayers. This **membrane asymmetry** includes differences both in the kinds of lipids present and in the degree of unsaturation of the fatty acids in the phospholipid molecules. For example, most of the glycolipids present in the plasma membrane of an animal cell are restricted to the outer of the two monolayers. As a result, their carbohydrate groups protrude from the outer membrane surface, where they are involved in signaling and recognition events. Phosphatidylethanolamine, phosphatidylinositol, and phosphatidylserine, on the other hand, are more prominent in the inner monolayer, where they are involved in transmitting various kinds of signals from the plasma membrane to the interior of the cell. Further details of signal detection and transduction await us in Chapter 10.

Membrane asymmetry is established during membrane biogenesis by the insertion of different lipids, or different proportions of the various lipids, into each of the two monolayers. Once established, asymmetry tends to be maintained because the movement of lipids from one monolayer to the other requires their hydrophilic head groups to pass through the hydrophobic interior of the membrane, an event that is thermodynamically unfavorable. While such "flip-flop," or **transverse diffusion**, of membrane lipids does occur occasionally, it is relatively rare. For instance, a typical phospholipid molecule undergoes flip-flop less than once a week in a pure phospholipid bilayer. This contrasts strikingly with the **rotation** of phospholipid molecules about their long axis and with the **lateral diffusion** of phospholipids in the plane of the membrane, both of which occur freely and rapidly. Figure 7-10 illustrates these three types of lipid movements.

While phospholipid flip-flop is relatively rare, its frequency is greater in natural membranes than in artificial lipid bilayers because some membranes—the smooth

Table 7-2 Structures of Some Common Fatty Acids

Name of Fatty Acid	Number of Carbon Atoms	Number of Double Bonds	Structural Formula	Space-Filling Model
Saturated				
Palmitate	16	0		
Stearate	18	0		
Unsaturated				
Oleate	18	1		
Linoleate	18	2		

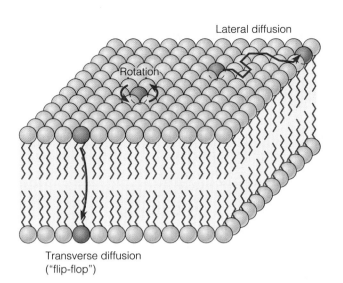

Lateral diffusion

Rotation

Transverse diffusion
("flip-flop")

Figure 7-10   Movements of Phospholipid Molecules Within Membranes. A phospholipid molecule is capable of three kinds of movement in a membrane: rotation about its long axis; lateral diffusion by exchanging places with neighboring molecules in the same monolayer; and transverse diffusion, or "flip-flop," from one monolayer to the other. In a pure phospholipid bilayer at 37°C, a typical lipid molecule exchanges places with neighboring molecules about 10 million times per second and can move laterally at a rate of about several micrometers per second. By contrast, the frequency with which an individual phospholipid molecule flip-flops from one layer to the other ranges from less than once a week in a pure phospholipid bilayer to once every few hours in some natural membranes. (The latter difference is due to the presence in some membranes of enzymes called phospholipid translocators, or flippases, that catalyze the transverse diffusion of phospholipid molecules from one monolayer to the other.)

endoplasmic reticulum, in particular—have proteins called **phospholipid translocators,** or **flippases,** that catalyze the flip-flop of membrane lipids from one monolayer to the other. Such proteins act only on specific kinds of lipids. For example, one of these proteins catalyzes the translocation of phosphatidylcholine from one side of the ER membrane to the other but does not recognize other phospholipids. This ability to move lipid molecules selectively from one side of the bilayer to the other contributes further to the asymmetric distribution of phospholipids across the membrane. The role of smooth ER in the synthesis and selective flip-flop of membrane phospholipids is a topic to which we will return in Chapter 12.

The Lipid Bilayer Is Fluid

One of the most striking properties of membrane lipids is that rather than being fixed in place within the mem-

brane, they form a fluid bilayer that permits lateral diffusion of membrane lipids as well as proteins. Lipid molecules move especially fast because they are much smaller than proteins. A typical phospholipid molecule, for example, has a molecular weight of about 800 and can travel the length of a bacterial cell (a few micrometers, in most cases) in a second or less! Proteins move much more slowly than lipids, mainly because they are much larger molecules, with molecular weights many times greater than those of phospholipids.

The lateral diffusion of membrane lipids can be demonstrated experimentally by a technique called **fluorescence recovery after photobleaching** (Figure 7-11). The investigator *tags,* or labels, lipid molecules in the membrane of a living cell by covalently linking molecules of a fluorescent dye to them. A high-intensity laser beam is then used to bleach the dye in a tiny spot (a few square micrometers) on the cell surface. If the cell surface is examined immediately thereafter with a fluorescence microscope, a dark, nonfluorescent spot is seen on the membrane. Within seconds, however, the edges of the spot

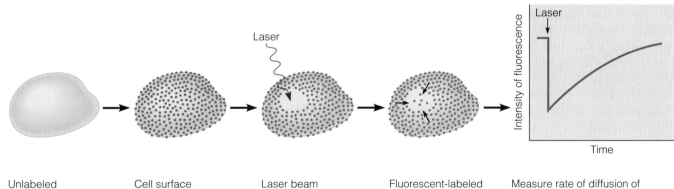

| Unlabeled cell surface | Cell surface molecules labeled with fluorescent dye | Laser beam bleaches an area of the cell surface | Fluorescent-labeled molecules diffuse into bleached area | Measure rate of diffusion of fluorescence into bleached area |

Figure 7-11   Measuring Lipid Mobility Within Membranes by Fluorescence Recovery After Photobleaching. Membrane lipids are labeled with a fluorescent compound, and the fluorescence in a local area is then bleached by irradiating the cell with a laser beam. Membrane fluidity is measured by determining the rate at which the diffusion of fluorescent molecules from surrounding regions into the bleached area causes fluorescence to reappear in the laser-bleached spot. Similar experiments can be carried out with membrane proteins.

become fluorescent as bleached lipid molecules diffuse out of the laser-treated area and fluorescent lipid molecules from adjoining regions of the membrane diffuse in. Eventually, the spot is indistinguishable from the rest of the cell surface. Not only does this technique demonstrate that membrane lipids are in a fluid rather than a static state, it also provides a direct means of measuring the lateral movement of specific molecules.

Membranes Function Properly Only in the Fluid State

As you might guess, membrane fluidity changes with temperature, decreasing as the temperature drops and increasing as it rises. In fact, we know from studies with artificial lipid bilayers that every lipid bilayer has a characteristic **transition temperature** (T_m) at which it gels ("freezes") when cooled and becomes fluid again ("melts") when warmed. This change in the state of the membrane is called a **phase transition** and is somewhat similar to the change that butter or margarine undergoes upon heating or cooling. To function properly, a membrane must be maintained in the fluid state—that is, at a temperature above its T_m value. At a temperature below the T_m value, all functions that depend on the mobility or conformational changes of membrane proteins will be impaired or disrupted, including such vital processes as transport of solutes across the membrane, detection and transmission of signals, and cell-to-cell communication (recall Figure 7-2).

The technique of **differential scanning calorimetry** is one means of determining the transition temperature of a given membrane. This procedure monitors the uptake of heat that occurs during the transition from one physical state to another—the gel-to-fluid transition, in the case of membranes. The membrane of interest is placed in a sealed chamber, the *calorimeter,* and the uptake of heat is measured as the temperature is slowly increased. The point of maximum heat absorption corresponds to the transition temperature (Figure 7-12a).

The Effects of Fatty Acid Composition on Membrane Fluidity. The fluidity of a membrane depends primarily on the kinds of lipids it contains. Two aspects of a membrane's lipid makeup are especially important in determining fluidity: the length of the fatty acid side chains and their degree of unsaturation (that is, the number of double bonds present). Long-chain fatty acids have higher transition temperatures than shorter-chain fatty acids, which means that membranes enriched in long-chain fatty acids tend to be less fluid. For example, as the chain length of saturated fatty acids increases from 10 to 20 carbon atoms, the transition temperature rises from 32°C to 76°C and hence the membrane becomes progressively less fluid (Figure 7-13a). The presence of unsaturation affects melting temperature even more markedly. For fatty acids with 18 carbon atoms, the melting points are 70, 16, 5, and −11°C for zero, one, two, and three double bonds, respectively (Figure 7-13b). As a result, membranes containing many unsaturated fatty acids tend to have lower transition temperatures and thus are more fluid than

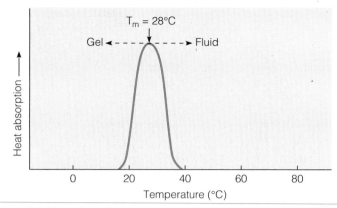

(a) Normal membrane

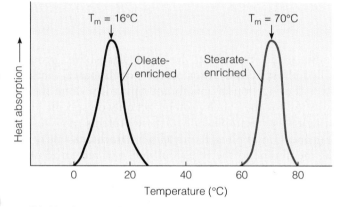

(b) Membrane enriched in either oleate or stearate

Figure 7-12 **Determination of Membrane Transition Temperature by Differential Scanning Calorimetry.** **(a)** When the temperature of a membrane preparation is increased slowly in a calorimeter chamber, a peak of heat absorption marks the gel-to-fluid transition temperature, T_m. **(b)** Membranes from cells grown in media enriched in oleate, an unsaturated fatty acid, are more fluid than normal membranes and therefore have a lower transition temperature. Membranes from cells grown in media enriched in stearate, a saturated fatty acid, are less fluid than normal membranes and therefore have a higher transition temperature.

membranes with many saturated fatty acids. Figure 7-12b illustrates this increased fluidity for membranes enriched in oleate (18 carbons, one double bond) versus membranes enriched in stearate (18 carbons, saturated).

The effect of unsaturation on membrane fluidity is so dramatic because the kinks that double bonds introduce into fatty acids prevent the hydrocarbon chains from fitting together snugly. Membrane lipids with saturated fatty acids pack together tightly, thereby optimizing their van der Waals interactions (Figure 7-14a), whereas lipids with unsaturated fatty acids do not (Figure 7-14b). The lipids of most plasma membranes contain fatty acids that vary in both chain length and degree of unsaturation. In fact, the variability is often intramolecular because membrane lipids commonly contain one saturated and one unsaturated fatty acid. This property helps to ensure that membranes are in the fluid state at physiological temperatures.

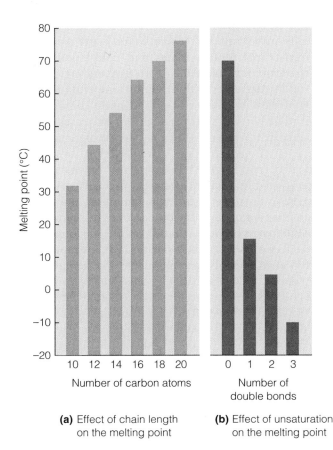

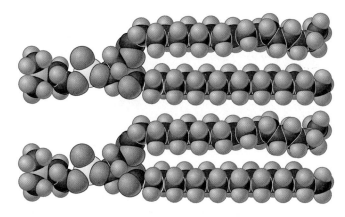

(a) Lipids with saturated fatty acids pack together well in the membrane

Figure 7-13 The Effect of Chain Length and the Number of Double Bonds on the Melting Point of Fatty Acids. The melting point of fatty acids **(a)** increases with chain length for saturated fatty acids and **(b)** decreases dramatically with the number of double bonds for fatty acids with a fixed chain length. Part **(b)** shows data for the 18-carbon fatty acids stearate, oleate, linoleate, and linolenate, with 0, 1, 2, and 3 double bonds, repectively.

(a) Effect of chain length on the melting point

(b) Effect of unsaturation on the melting point

The Effects of Sterols on Membrane Fluidity. For eukaryotic cells, membrane fluidity is also affected by the presence of sterols—mainly cholesterol in animal cell membranes and phytosterols in plant cell membranes. Sterols are prominent components in the membranes of many cell types. A typical animal cell, for example, contains large amounts of cholesterol—up to 50% of the total membrane lipid on a molar basis. Cholesterol molecules are usually found in both layers of the plasma membrane, but a given molecule is localized to one of the two layers (Figure 7-15a). The molecule orients itself in the layer with its single hydroxyl group—the only polar part of an otherwise hydrophobic molecule—close to the polar head group of a neighboring phospholipid molecule, where it can form a hydrogen bond with the oxygen of the ester bond between the glycerol backbone and a fatty acid of the phospholipid molecule (Figure 7-15b). The rigid hydrophobic steroid rings and the hydrocarbon side chain of the cholesterol molecule interact with the portions of adjacent hydrocarbon chains that are closest to the phospholipid head groups.

This intercalation of rigid cholesterol molecules into the membrane of an animal cell makes the membrane less fluid at higher temperatures than it would otherwise be.

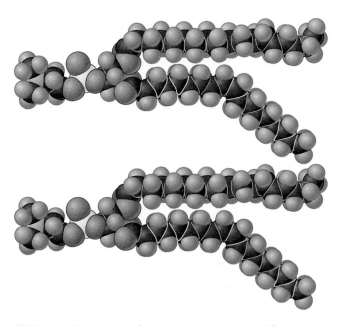

(b) Lipids with a mixture of saturated and unsaturated fatty acids do not pack together well in the membrane

Figure 7-14 The Effect of Unsaturated Fatty Acids on the Packing of Membrane Lipids. **(a)** Membrane phospholipids with no unsaturated fatty acids fit together tightly because the fatty acid chains are parallel to each other. **(b)** Membrane lipids with one or more unsaturated fatty acids do not fit together as tightly because the *cis* double bonds cause bends in the chains, which interfere with packing. Each of the structures shown is a phosphatidylcholine molecule, with either two 18-carbon saturated fatty acids (stearate) (part a) or two 18-carbon fatty acids, one saturated (stearate) and the other with one double bond (oleate) (part b).

However, cholesterol also effectively prevents the hydrocarbon chains of phospholipids from fitting snugly together as the temperature is decreased, thereby reducing the tendency of membranes to gel upon cooling. Thus, cholesterol has the paradoxical effect of *decreasing* membrane fluidity at high temperatures and *increasing* it at low temperatures. The sterols in the membranes of other eukaryotes presumably function in the same way.

In addition to their effects on membrane fluidity, sterols also decrease the permeability of a lipid bilayer to

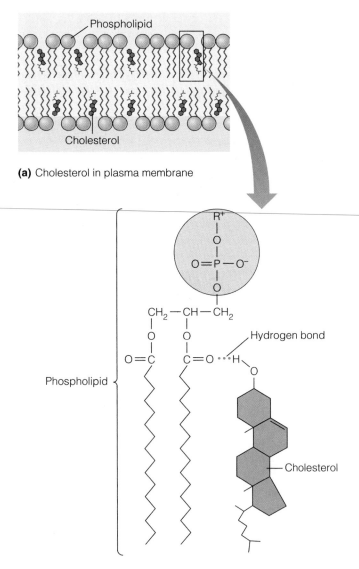

(a) Cholesterol in plasma membrane

(b) Bonding of cholesterol to phospholipid

Figure 7-15 Orientation of Cholesterol Molecules in a Lipid Bilayer.
(a) Cholesterol molecules are present in both lipid layers in the plasma membranes of most animal cells, but each molecule is localized to one of the two layers. **(b)** Each molecule orients itself in the lipid layer so that its single hydroxyl group is close to the polar head group of a neighboring phospholipid molecule, where it forms a hydrogen bond with the oxygen of the ester bond between the glycerol backbone and a fatty acid. The steroid rings and hydrocarbon side group of the cholesterol molecule interact with adjacent hydrocarbon chains of the membrane phospholipids.

ions and small polar molecules. They probably do so by filling in spaces between hydrocarbon chains of membrane phospholipids, thereby plugging small channels through which ions and small molecules might otherwise pass. In general, a lipid bilayer containing sterols is less permeable to ions and small molecules than is a bilayer lacking sterols.

Most Organisms Can Regulate Membrane Fluidity

Most organisms, whether prokaryotic or eukaryotic, are able to regulate membrane fluidity, primarily by changing the lipid composition of the membranes. This ability is especially important for *poikilotherms*—organisms such as bacteria, fungi, protists, plants, and "cold-blooded" animals that cannot regulate their own temperature. Because lipid fluidity decreases with dropping temperature, membranes of these organisms would gel upon cooling if the organism had no way to compensate for decreases in environmental temperature. (You may have experienced this effect even though you are a *homeotherm,* or "warm-blooded" organism: On chilly days, your fingers and toes can get so cold that the membranes of sensory nerve endings cease to function, resulting in temporary numbness.) At high temperatures, on the other hand, the lipid bilayers of poikilotherms become so fluid that they no longer serve as an effective permeability barrier. For example, most cold-blooded animals are paralyzed by temperatures much above 45°C, probably because nerve cell membranes become so leaky to ions that ion gradients cannot be maintained and overall nervous function is disabled.

Fortunately, most poikilotherms can compensate for temperature changes by altering the lipid composition of their membranes, thereby regulating membrane fluidity. This capability is called **homeoviscous adaptation** because the main effect of such regulation is to keep the viscosity of the membrane approximately the same despite changes in temperature. Consider, for example, what happens when bacterial cells are transferred from a warmer to a cooler environment. In some species of the genus *Micrococcus,* a drop in temperature triggers an increase in the proportion of 16-carbon versus 18-carbon fatty acids in the plasma membrane, which helps the cell maintain membrane fluidity. (Remember that the presence of shorter fatty acid chains increases the fluidity of membranes by decreasing the melting temperature.) In this particular case, the desired increase in membrane fluidity is accomplished by activating an enzyme that removes two terminal carbons from 18-carbon hydrocarbon tails. In other bacterial species, adaptation to environmental temperature involves an alteration in the extent of unsaturation of membrane fatty acids rather than in their length. In the common intestinal bacterium *Escherichia coli,* for example, a decrease in environmental temperature triggers the synthesis of a *desaturase* enzyme that introduces double bonds into the hydrocarbon chains of fatty acids. As these unsaturated fatty acids are incorporated into membrane phospholipids, they decrease the transition temperature of the membrane, thereby ensuring that the membrane remains fluid at the lower temperature.

Homeoviscous adaptation also occurs in yeasts and plants. In these organisms, temperature-related changes in membrane fluidity appear to depend on the increased solubility of oxygen in the cytoplasm at lower temperatures. Oxygen is a substrate for the desaturase enzyme system involved in the generation of unsaturated fatty acids; with more oxygen available at lower temperatures, unsaturated fatty acids are synthesized at a greater rate and membrane fluidity increases, thereby offsetting the temperature effect. This capability has great agricultural significance because plants that can adapt in this way are cold-hardy (resistant to chilling) and can therefore be grown in colder environments. Poikilothermic animals such as amphibians and reptiles also adapt to lower temperatures by increasing the proportion of unsaturated fatty acids in their membranes. In addition, these organisms can increase the proportion of cholesterol in the membrane, thereby decreasing the interaction between hydrocarbon chains and reducing the tendency of the membrane to gel.

Although homeoviscous adaptation is of greatest general relevance to poikilothermic organisms, it is also important to mammals that hibernate because body temperature often drops substantially as an animal enters hibernation— a decrease of more than 30°C for some rodents. An animal entering hibernation adapts to this change by incorporating a greater proportion of unsaturated fatty acids into membrane phospholipids as its body temperature falls.

Membrane Proteins: The "Mosaic" Part of the Model

Having looked in some detail at the "fluid" aspect of the fluid mosaic model, we come now to the "mosaic" part. We come, in other words, to the *proteins* present in mem-

branes, recalling that the novel feature of the Singer-Nicolson model was to envision the membrane as a mosaic of proteins floating in and on a fluid lipid bilayer. We will look first at the confirming evidence that microscopists provided for the membrane as a mosaic of proteins and then consider the major classes of membrane proteins.

The Membrane Consists of a Mosaic of Proteins: Evidence from Freeze-Fracture Microscopy

Strong support for the fluid mosaic model came from studies in which artificial bilayers and natural membranes were prepared for electron microscopy by **freeze-fracturing.** In this technique, a lipid bilayer or a membrane (or a cell containing membranes) is frozen quickly and then subjected to a sharp blow from a diamond knife. The resulting fracture often follows the plane between the two layers of membrane lipid, because the nonpolar interior of the bilayer is the path of least resistance through the frozen specimen. As a result, the bilayer is split into its inner and outer monolayers, revealing the inner surface of each (Figure 7-16a).

Electron micrographs of membranes prepared in this way provide striking evidence that proteins are actually suspended within membranes. Whenever a fracture plane splits the membrane into its two layers, particles having the

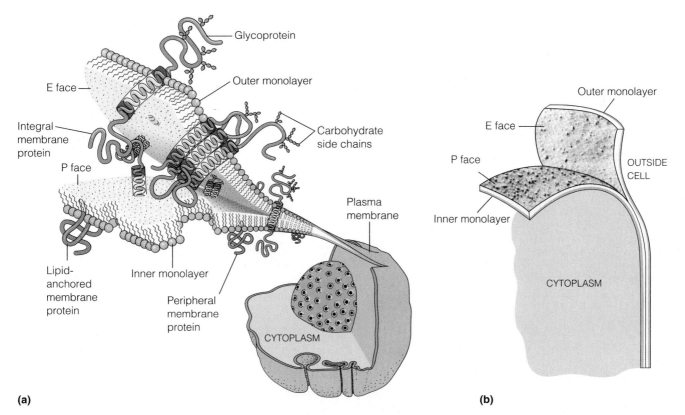

(a)

(b)

Figure 7-16 Freeze-Fracture Analysis of a Membrane. **(a)** Sketch of a freeze-fractured membrane in which the fracture plane has passed through the hydrophobic interior of the membrane, revealing the inner surfaces of the two monolayers. Hydrophobic segments of proteins are shown in light purple, hydrophilic segments in dark purple. Integral membrane proteins that remain with the outer monolayer are seen on the E (exterior) face, whereas those that remain with the inner monolayer are seen on the P (protoplasmic) face. **(b)** Sketch of a freeze-fractured membrane, with electron micrographs of the E and P faces from the plasma membrane of a mouse kidney tubule cell superimposed on the drawing.

size and shape of globular proteins can be seen adhering to one or the other of the inner membrane surfaces, called the *E* (for *exterior*) and *P* (for *protoplasmic*) *faces* (Figure 7-16b). Moreover, the abundance of such particles correlates well with the known protein content of the particular membrane under investigation. The electron micrographs in Figure 7-17 illustrate this well: The erythrocyte plasma membrane has a rather low protein/lipid ratio (1.14; see Table 7-1) and a rather low density of particles when subjected to freeze-fracture (Figure 7-17a), whereas a chloroplast membrane has a higher protein/lipid ratio (2.33) and

a correspondingly higher density of intramembranous particles, especially on the inner lipid layer (Figure 7-17b).

Confirmation that the particles seen in this way really *are* proteins came from work by David Deamer and Daniel Branton, who used the freeze-fracture technique to examine artificial bilayers with and without added protein. Bilayers formed from pure phospholipids showed no evidence of particles on their interior surfaces (Figure 7-18a). When proteins were added to the artificial bilayers, however, particles similar to those seen in natural membranes were readily visible (Figure 7-18b).

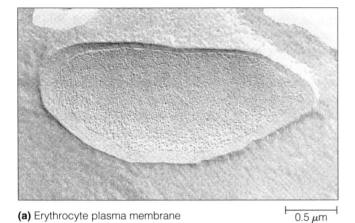

(a) Erythrocyte plasma membrane ⊢ 0.5 μm ⊣

(b) Chloroplast membrane ⊢ 0.2 μm ⊣

Figure 7-17 Membrane Proteins Visualized by Freeze-Fracture Electron Microscopy. Membrane proteins appear as discrete particles embedded within the lipid bilayer. The lower density of particles in **(a)** the erythrocyte membrane compared with **(b)** the chloroplast membrane agrees well with the protein/lipid ratios of the two membranes (1.14 and 2.33, respectively) (TEMs).

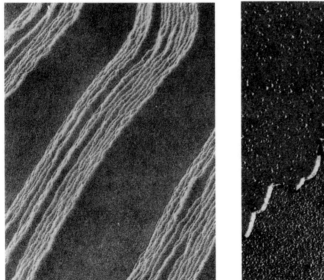

(a) Artificial bilayers without proteins

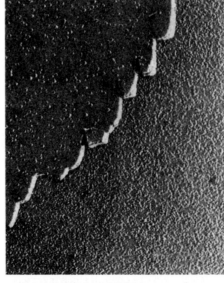

(b) Artificial bilayers with proteins ⊢ 0.1 μm ⊣

Figure 7-18 Freeze-Fracture Comparison of Lipid Bilayers With and Without Added Proteins. This figure compares the appearance by freeze-fracture electron microscopy of **(a)** artificial lipid bilayers with **(b)** artificial bilayers to which proteins were added. The white lines in the artificial membranes of part a represent individual lipid bilayers in a multilayered specimen, and the gray regions show where single bilayers have split to reveal smooth surfaces. In contrast, the artificial membrane of part b shows large numbers of globular particles in the fracture surface. These are the proteins that were added to the membrane preparation (TEMs).

Membranes Contain Integral, Peripheral, and Lipid-Anchored Proteins

Membrane proteins differ in their affinity for the hydrophobic interior of the membrane and therefore in the extent to which they interact with the lipid bilayer. That, in turn, determines how easy or difficult it is to extract a given protein from the membrane. Based on the conditions required to extract them and thus, by extension, on the nature of their association with the lipid bilayer, membrane proteins fall into one of three categories: integral, peripheral, or lipid-anchored. We will consider each of these in turn, referring in each case to the schematic diagrams shown in Figure 7-19.

The plasma membrane of the human erythrocyte, shown in Figure 7-20, provides a cellular context for our discussion. This has been one of the most widely studied membranes since the time of Gorter and Grendel, mainly because of the ready availability of red blood cells and the ease with which pure plasma membrane preparations can be made from them. We will refer back to the erythrocyte plasma membrane at several points in the following discussion to note examples of different types of proteins and their roles within the membrane.

Integral Membrane Proteins. Most membrane proteins are amphipathic molecules possessing one or more hydrophobic regions that exhibit an affinity for the hydrophobic interior of the lipid bilayer. These proteins are called **integral membrane proteins** because their hydrophobic regions are embedded within the membrane interior and so these molecules cannot be easily removed from membranes. However, such proteins also have one or more hydrophilic regions that extend outward from the membrane into the aqueous phase on one or both sides of the membrane. Because of their affinity for the lipid bilayer, integral membrane proteins are difficult to isolate and study by standard protein purification techniques, most of which are designed for water-soluble proteins. Treatment with a detergent that disrupts the lipid bilayer is usually necessary to solubilize and extract integral membrane proteins.

A few integral membrane proteins are known to be embedded in, and therefore to protrude from, only one side of the bilayer; these are called **integral monotopic proteins** (Figure 7-19a). However, most integral membrane proteins are <u>**transmembrane proteins,**</u> which means that they span the membrane and have hydrophilic regions protruding from the membrane on both sides. Such proteins cross the membrane either once (*singlepass proteins;* Figure 7-19b) or several times (*multipass proteins;* Figure 7-19c). Some multipass proteins consist of a single polypeptide (Figure 7-19c), whereas others have two or more polypeptides (*multisubunit proteins;* Figure 7-19d).

Most transmembrane proteins are anchored to the lipid bilayer by one or more hydrophobic **transmembrane segments,** one for each time the protein crosses the bilayer. In most cases, the polypeptide chain appears to span the membrane in an α-helical conformation consisting of about 20–30 amino acid residues, most—sometimes even *all*—of which have hydrophobic R groups. In some multipass

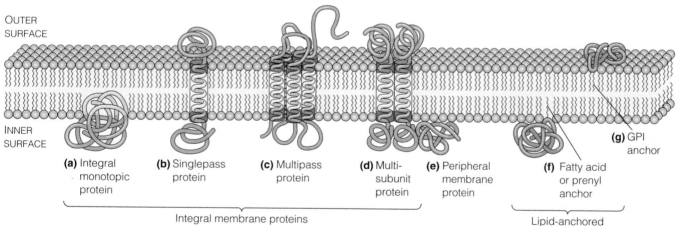

OUTER SURFACE

INNER SURFACE

(a) Integral monotopic protein   **(b)** Singlepass protein   **(c)** Multipass protein   **(d)** Multi-subunit protein   **(e)** Peripheral membrane protein   **(f)** Fatty acid or prenyl anchor   **(g)** GPI anchor

Integral membrane proteins

Lipid-anchored membrane proteins

Figure 7-19 The Main Classes of Membrane Proteins. Membrane proteins are classified according to their mode of attachment to the membrane. Integral membrane proteins **(a-d)** contain one or more hydrophobic regions that are embedded within the lipid bilayer. **(a)** A few integral proteins appear to be embedded in the membrane on only one side of the bilayer (integral monotopic proteins.) However, most integral proteins are transmembrane proteins that span the lipid bilayer either **(b)** once (singlepass proteins) or **(c)** multiple times (multipass proteins). Multipass proteins may consist of either a single polypeptide, as in part c, or **(d)** several associated polypeptides (multisubunit proteins). **(e)** Peripheral membrane proteins are too hydrophilic to penetrate into the membrane but are attached to the membrane by electrostatic and hydrogen bonds that link them to adjacent membrane proteins or to phospholipid head groups. Lipid-anchored proteins **(f-g)** are hydrophilic and do not penetrate into the membrane; they are covalently bound to lipid molecules that are embedded in the lipid bilayer. **(f)** Proteins on the inner surface of the membrane are usually anchored by either a fatty acid or a prenyl group. **(g)** On the outer membrane surface, the most common lipid anchor is glycosylphosphatidylinositol (GPI).

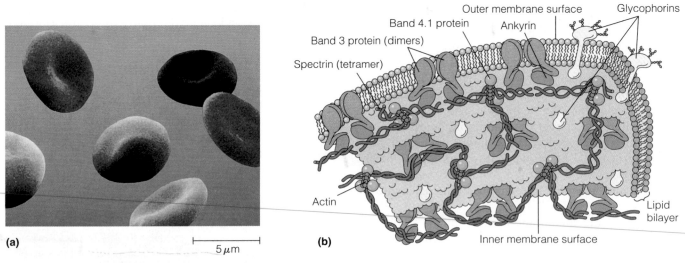

Figure 7-20 Structural Features of the Erythrocyte Plasma Membrane. (a) An erythrocyte is a small disk-shaped cell with a diameter of about 7 μm. A mammalian erythrocyte contains no nucleus or other organelles, which makes it easy to obtain very pure plasma membrane preparations without contamination by organellar membranes, as often occurs with plasma membrane preparations from other cell types. (b) The erythrocyte plasma membrane as seen from the inside of the cell. The membrane has a relatively simple protein composition. The two major integral membrane proteins are glycophorin and an anion exchange protein known from its electrophoretic mobility as band 3. Glycophorin has a single transmembrane segment, whereas each of the polypeptides of the band 3 protein spans the membrane at least six times. The membrane is anchored to the underlying cytoskeleton by long, slender strands of tetrameric spectrin, $(\alpha\beta)_2$, that are linked to glycophorin molecules by band 4.1 protein and to band 3 protein by ankyrin, another peripheral membrane protein. The free ends of adjacent spectrin tetramers are held together by short chains of actin and band 4.1 Not shown is another protein, band 4.2, that assists ankyrin in linking spectrin to band 3 protein. For the fractionation of these proteins by SDS–polyacrylamide gel electrophoresis, see Figure 7-23.

proteins, however, several transmembrane segments are arranged as a β sheet in the form of a closed β sheet—the so-called β barrel. This structure is especially prominent in a group of pore-forming transmembrane proteins called *porins* that are found in the outer membrane of many bacteria and of chloroplasts and mitochondria as well. Regardless of their conformation, transmembrane segments are usually separated along the primary structure of the protein by hydrophilic sequences that protrude or loop out on the two sides of the membrane.

Singlepass membrane proteins have just one transmembrane segment, with a hydrophilic carboxyl (C-) terminus extending out of the membrane on one side and a hydrophilic amino (N-) terminus protruding on the other side. Depending on the particular protein, the C-terminus may protrude on either side of the membrane. An example of a singlepass protein is *glycophorin*, a prominent protein in the erythrocyte plasma membrane (Figure 7-20b). Glycophorin is oriented in the membrane so that its C-terminus is on the inner surface of the membrane and its N-terminus is on the outer surface (Figure 7-21a).

Multipass membrane proteins have several transmembrane segments, ranging from 2 or 3 to 20 or more such segments. An example of a multipass protein in the erythrocyte plasma membrane is a dimeric transport protein called *band 3 protein* (also known as the *anion exchange protein*). Each of its two polypeptides span the lipid bilayer at least six times, with both the C-terminus and the N-terminus on the same side of the membrane.

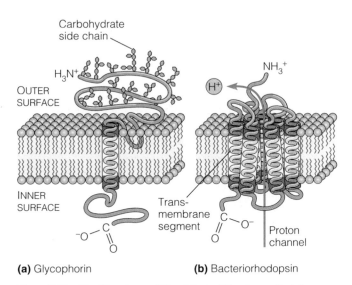

(a) Glycophorin   **(b)** Bacteriorhodopsin

Figure 7-21 The Structures of Two Integral Membrane Proteins. (a) Glycophorin is a singlepass integral membrane protein in the erythrocyte plasma membrane. Its α-helical transmembrane segment consists entirely of hydrophobic amino acids. The N-terminus protrudes on the outer surface, the C-terminus on the cytoplasmic surface. Glycophorin is a glycoprotein, with 16 carbohydrate chains attached to its outer surface. (b) Bacteriorhodopsin is a multipass integral membrane protein in the plasma membrane of *Halobacterium*. Its seven transmembrane segments, which account for about 70% of its 248 amino acids, are organized into a proton channel. The C- and N-termini of the protein have short hydrophilic segments that protrude on the inner and outer surfaces of the plasma membrane, respectively. Short hydrophilic segments also link each of the transmembrane segments.

One of the best-studied examples of a multipass protein is bacteriorhodopsin, the first integral membrane protein to have its detailed three-dimensional structure determined by X-ray crystallography (the work of Unwin and Henderson in 1975; recall Figure 7-3g). Bacteriorhodopsin turned out to have seven α-helical membrane-spanning segments, each of which corresponds to a sequence of about 20 hydrophobic amino acids in the primary structure of the protein (Figure 7-21b). The seven transmembrane segments are positioned in the membrane to form a channel and thereby to facilitate the light-activated pumping, or active transport, of protons across the membrane, a topic to which we will return in Chapter 8.

Most integral membrane proteins are hard to isolate and even harder to crystallize, so few others have as yet been subjected to X-ray crystallography, which is the usual means of determining protein structure. However, an alternative approach is possible if an integral protein can at least be isolated and sequenced. Once the amino acid sequence of a membrane protein is known, the number and positions of transmembrane segments can be inferred from a **hydropathy** (or **hydrophobicity**) **plot,** as shown in Figure 7-22.

Such a plot is constructed by using a computer program to identify clusters of hydrophobic amino acids. The amino acid sequence of the protein is scanned through a series of "windows" of about 10 amino acids at a time, with each successive window one amino acid further along the sequence. Based on the known hydrophobicity values for the various amino acids, a **hydropathy index** is calculated for each successive window by averaging the hydrophobicity values of the amino acids in the window. (By convention, hydrophobic amino acids have positive hydropathy values and hydrophilic residues have negative values.) The hydropathy index is then plotted against the positions of the windows along the sequence of the protein. The resulting hydropathy plot predicts how many membrane-spanning regions are present in the protein, based on the number of positive peaks. The hydropathy plot shown in Figure 7-22a is for a plasma membrane protein called *connexin*. The plot shows four positive peaks and therefore predicts that connexin has four stretches of hydrophobic amino acids and hence four transmembrane segments, as shown in Figure 7-22b.

Peripheral Membrane Proteins. In contrast to integral membrane proteins, some membrane-associated proteins lack discrete hydrophobic sequences and therefore do not penetrate into the lipid bilayer. Instead, these **peripheral membrane proteins** are bound to membrane surfaces through weak electrostatic forces and hydrogen bonding with the hydrophilic portions of integral proteins and perhaps with the polar head groups of membrane lipids (Figure 7-19e). Peripheral proteins are more readily removed from membranes than integral proteins and can usually be extracted by changing the pH or ionic strength.

The main peripheral proteins of the erythrocyte plasma membrane are *spectrin, ankyrin,* and a protein called *band*

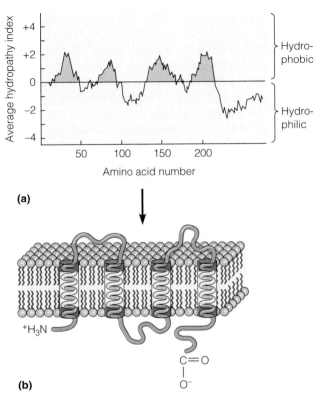

(a)

(b)

Figure 7-22 Hydropathy Analysis of an Integral Membrane Protein. A hydropathy plot is a means of representing hydrophobic regions (positive values) and hydrophilic regions (negative values) along the length of a protein—the plasma membrane protein *connexin,* in this case. **(a)** The hydropathy index on the vertical axis is a numerical measure of the relative hydrophobicity of successive segments of the polypeptide chain based on its amino acid sequence. **(b)** Connexin has four distinct hydrophobic regions, which correspond to four α-helical segments that span the plasma membrane. (The hydrophobicity value for a given amino acid is the standard free-energy change, $\Delta G°$, for the transfer of one mole of that amino acid from a hydrophobic environment into an aqueous environment, expressed in kilojoules/mole. For a hydrophilic amino acid, that process is exergonic and hence has a negative $\Delta G°$ value; for a hydrophobic amino acid, the process is endergonic and the $\Delta G°$ value is positive.)

4.1 (see Figure 7-20b). These proteins are bound to the inner surface of the plasma membrane, where they form a skeletal meshwork that supports the plasma membrane and helps maintain the shape of the erythrocyte (see Figure 7-20a).

Lipid-Anchored Membrane Proteins. When it was initially proposed by Singer and Nicholson, the fluid mosaic model regarded all membrane proteins as either peripheral or integral membrane proteins. However, we now recognize a third class of proteins that are neither peripheral nor integral but have some of the characteristics of both. The polypeptide chains of these **lipid-anchored membrane proteins** sit on one of the surfaces of the lipid bilayer but are covalently bound to lipid molecules embedded within the bilayer (Figure 7-19f and g).

Several mechanisms are employed for attaching lipid-anchored proteins to membranes. Proteins bound to the inner surface of the plasma membrane are attached by covalent linkage either to a fatty acid or to an isoprene derivative called a *prenyl* group (Figure 7-19f). In the case of **fatty acid-anchored membrane proteins**, the protein is synthesized in the cytosol and then covalently attached to a saturated fatty acid embedded within the membrane bilayer, usually *myristic acid* (14 carbons) or *palmitic acid* (16 carbons). **Prenylated membrane proteins**, on the other hand, are synthesized as soluble cytosol proteins before being modified by addition of a prenyl group, usually a 15-carbon *farnesyl* group or 20-carbon *geranylgeranyl* group. After attachment, the farnesyl or geranylgeranyl group is inserted into the lipid bilayer of the membrane.

Many lipid-anchored proteins attached to the external surface of the plasma membrane are covalently linked to *glycosylphosphatidylinositol (GPI)*, a glycolipid found in the external monolayer of the plasma membrane (Figure 7-19g). These **GPI-anchored membrane proteins** are made in the endoplasmic reticulum as singlepass transmembrane proteins, which subsequently have their transmembrane segments cleaved off and replaced by GPI anchors. The proteins are then transported from the ER to the exterior of the plasma membrane by a pathway we will encounter in Chapter 12. Once at the cell surface, GPI-anchored proteins can be released from the membrane by the enzyme *phospholipase C*, which is specific for phosphatidylinositol linkages.

Proteins Can Be Separated by SDS–Polyacrylamide Gel Electrophoresis

Before continuing our discussion of membrane proteins, it is useful to consider briefly how membrane proteins are isolated and studied. We will look first at the general problem of solubilizing and extracting proteins from membranes, and we will then learn about an electrophoretic technique that is very useful in the fractionation and characterization of proteins.

Isolation of Membrane Proteins.

A major challenge to protein chemists has been the difficulty of isolating and studying membrane proteins, many of which are hydrophobic. Peripheral membrane proteins are in general quite amenable to isolation. As noted earlier, they are bound to the membrane by weak electrostatic interactions and hydrogen bonding with either the hydrophilic portions of integral membrane proteins or the polar head groups of membrane lipids. Peripheral proteins can therefore be extracted from the membrane by changes in pH or ionic strength; in fact, peripheral membrane proteins were originally defined as those that could be extracted from membranes with an alkaline carbonate solution of a specific ionic strength. Peripheral membrane proteins can also be solubilized by the use of a chelating (cation-binding) agent to remove calcium or by addition of urea, which breaks hydrogen bonds. Lipid-anchored proteins are similarly amenable to isolation, though with the requirement that the covalent bond to the lipid must first be cleaved. Once extracted from the membrane, most peripheral and lipid-anchored proteins are sufficiently hydrophilic to be purified and studied with techniques commonly used by protein chemists.

Integral membrane proteins, on the other hand, are difficult to isolate from membranes, especially in a manner that preserves their biological activity. In most cases, they can be solubilized only by the use of detergents that disrupt hydrophobic interactions and dissolve the lipid bilayer. As we shall now see, the use of strong ionic detergents such as *sodium dodecyl sulfate (SDS)* allows integral membrane proteins not just to be isolated, but also to be fractionated and analyzed by the technique of electrophoresis.

SDS–Polyacrylamide Gel Electrophoresis.

Cells contain thousands of different macromolecules that must be separated from one another before the properties of individual components can be investigated. One of the most common approaches for separating molecules from each other is **electrophoresis**, a group of related techniques that utilize an electrical field to separate electrically charged molecules. The rate at which any given molecule moves during electrophoresis depends upon its charge as well as its size. Electrophoresis can be carried out using a variety of support media, such as paper, cellulose acetate, starch, polyacrylamide, or agarose (a polysaccharide obtained from seaweed). Of these media, gels made of polyacrylamide or agarose provide the best resolution and are most commonly employed for the electrophoresis of nucleic acids and proteins.

When using electrophoresis to study membrane proteins, membrane fragments are first solubilized with the detergent SDS, which disrupts most protein-protein and protein-lipid associations. The proteins denature, unfolding into stiff polypeptide rods that cannot refold because their surfaces are coated with negatively charged detergent molecules. The solubilized, SDS-coated polypeptides are then applied to the top of a polyacrylamide gel and an electrical potential is applied across the gel, such that the bottom of the gel is the positively charged anode (Figure 7-23). Because the polypeptides are coated with negatively charged SDS molecules, they migrate down the gel toward the anode. The polyacrylamide gel can be thought of as a fine meshwork that impedes the movement of large molecules more than that of small molecules. As a result, polypeptides move down the gel at a rate that is inversely related to their size.

When the smallest polypeptides approach the bottom of the gel, the process is terminated and the gel is stained with a dye that binds to polypeptides and makes them visible. (*Coomassie brilliant blue* is commonly used for this purpose.) The particular polypeptide profile shown in Figure 7-23 is for the membrane proteins of human erythrocytes, most of which we have already encountered (recall Figure 7-20b).

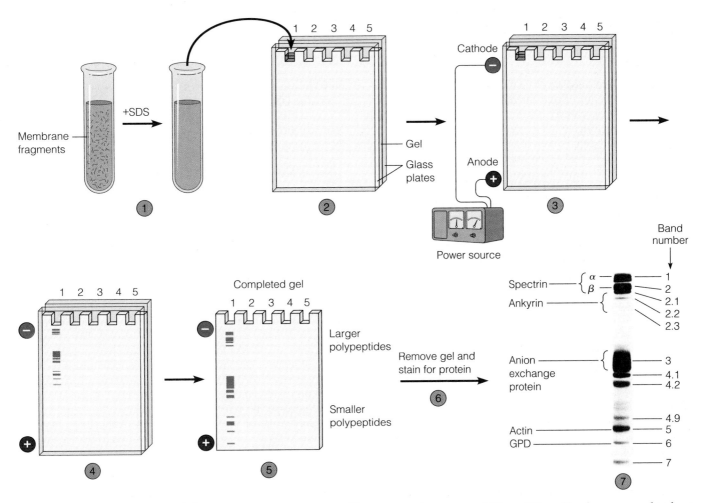

Figure 7-23 SDS–Polyacrylamide Gel Electrophoresis of Membrane Proteins. Membrane fragments are suspended in buffer and ① treated with sodium dodecyl sulfate (SDS). ② A small sample of the solubilized polypeptides is then applied to one of the wells at the top of a gel of polyacrylamide that has been polymerized and crosslinked between two glass plates. (Only one of the five numbered wells are used in this exmple, but in practice other samples would be placed in the additional wells.) ③ An electrical potential is applied across the gel, ④ causing the polypeptide molecules to migrate toward the far end of the gel. ⑤ Each polypeptide moves down the gel at a rate that is inversely related to its size. ⑥ When the smallest polypeptides are near the bottom, the gel is removed from between the plates and stained with a dye that binds to polypeptides and makes them visible. (The bands shown in color in steps 4 and 5 are not actually visible until the gel has been removed and stained.) ⑦ The polypeptide profile shown here is for the main membrane proteins of the human erythrocyte. The separated bands of protein were initially named using sequential numbers, e.g., "band 1," "band 2," etc. We now know the identity of some of these separated proteins, e.g., band 3 is the anion exchange protein and band 6 is GPD (glyceraldehyde-3-phosphate dehydrogenase).

Molecular Biology Has Contributed Greatly to Our Understanding of Membrane Proteins

Membrane proteins have not yielded as well as other proteins to biochemical techniques, mainly because of the problems involved in isolating and purifying hydrophobic proteins in physiologically active form. Procedures such as SDS-polyacrylamide electrophoresis and hydropathy analysis have certainly been useful, as have labeling techniques involving radioisotopes or fluorescent antibodies, which we will discuss later in the chapter. Within the past two decades, however, the study of membrane proteins has been revolutionized by the techniques of molecular biology, especially DNA sequencing and recombinant DNA technology. DNA sequencing makes it possible to deduce the amino acid sequence of a protein, a prerequisite for hydropathy analysis to identify transmembrane segments. Moreover, sequence comparisons between proteins often reveal evolutionary and functional relationships that might not otherwise have been appreciated. DNA pieces can also be used as probes to identify and isolate sequences that encode related proteins. In addition, the DNA sequence for a particular protein can be altered at specific nucleotide positions to determine the effects of selected changes in the sequence on the activity of the mutant protein for which it codes. Box 7A describes these exciting developments in more detail.

Membrane proteins mediate a remarkable variety of cellular functions and are therefore of great interest to cell biologists. Only within recent years, however, has the study of these proteins begun to yield definitive insights and answers. Some of these answers have come from the application of biochemical techniques to membrane proteins. We have encountered several such applications in this chapter, including hydropathy analysis, SDS–polyacrylamide gel electrophoresis, and procedures for labeling membrane proteins with radioactivity or fluorescent antibodies (see Figures 7-22, 7-23, 7-24, and 7-28, respectively). Two other biochemical approaches that can be used to study membrane proteins are affinity labeling and membrane reconstitution.

Affinity labeling utilizes radioactive molecules that bind to specific membrane proteins because of known functions of the proteins. For example, a compound called *cytochalasin B* is known to be a potent inhibitor of glucose transport. Membranes that have been exposed to radioactive cytochalasin B are therefore likely to contain radioactivity bound specifically to the protein(s) involved in glucose transport.

Membrane reconstitution involves the formation of artificial membranes from specific purified components. In this approach, proteins are extracted from membranes with detergent solutions and separated into their individual protein components. The purified proteins are then mixed together with phospholipids under conditions known to promote the formation of membrane vesicles called *liposomes.* These reconstituted vesicles can then be tested for their ability to carry out specific functions that are known, or thought, to be mediated by membrane proteins.

In spite of some success with these and similar approaches, membrane biologists have often found themselves stymied in their attempts to isolate, purify, and study membrane proteins. Biochemical techniques that work well with soluble proteins are not often useful with proteins that are hydrophobic. Within the past two decades, however, the study of membrane proteins has been revolutionized by the techniques of molecular biology, especially *DNA sequencing* and *recombinant DNA technology.* We will consider these techniques in detail in Chapters 16 and 18, but we need not wait until then to appreciate the enormous impact that molecular biology has had on the study of membranes and membrane proteins. Figure 7A-1 summarizes several approaches that have proven especially powerful.

Vital to these approaches is the isolation of a gene, or at least a fragment of a gene, that encodes a specific membrane protein (Figure 7A-1, top). With a DNA molecule in hand, the first priority of the molecular biologist is almost always to determine its nucleotide sequence ①. *DNA sequencing* is in fact one of the triumphs of molecular biology; it is now far easier to determine the nucleotide sequence of a DNA molecule than to determine the amino acid sequence of the protein for which it codes. Moreover, most of the sequencing procedure is carried out

quickly and automatically by DNA sequencing machines. Once the DNA for a particular protein has been sequenced, the *putative,* or predicted, *amino acid sequence* of the protein can be deduced using the genetic code that equates every possible sequence of three nucleotides with a particular amino acid (see Figure 19-8)②. The amino acid sequence can then be subjected to *hydropathy analysis* (see Figure 7-22) to identify likely transmembrane segments of the protein ③.

Knowing the amino acid sequence of the protein also allows the investigator to prepare synthetic peptides that correspond to specific segments of the protein ④. Antibodies made against these peptides can then be radioactively labeled and used to determine which segments are exposed on one side of the membrane or the other. The information gained in this way combined with the hydropathy data often provides compelling evidence for the likely structure of the protein and its orientation within the membrane and possibly for its mode of action as well. The structure of the CFTR protein that is defective in people with cystic fibrosis was determined in this way, for example (see Box 8B on p. 212).

Another powerful molecular technique, called **site-specific mutagenesis,** is used to examine the effect of specific changes in the amino acid sequence of a protein ⑤. The DNA sequence encoding a specific segment of the protein can be altered by changing particular nucleotides. The mRNA transcribed from the mutant DNA is then injected into living cells (either cultured mammalian cells or amphibian oocytes). The cells use the mRNA to direct the synthesis of a mutant protein, the functional properties of which can then be readily determined. In this way, functionally important amino acids can be identified.

Yet another way to use an isolated gene or gene segment is as a *DNA probe* to isolate other DNA sequences that are similar to the probe ⑥. DNA identified in this way is likely to encode proteins that are structurally similar to the protein for which the probe DNA codes. Such proteins are likely to be related to each other both in evolutionary origin and also possibly in their mechanisms of action.

From such studies, we know that transport of a great variety of solutes across membranes is mediated by a remarkably small number of protein families, although the families often contain many different proteins. Even a single member of such a family may be present in a variety of forms that differ in such properties as time of expression during development, tissue distribution, or location within the cell.

Most of these molecular approaches are indirect, in the sense that they allow scientists to deduce properties and functions of proteins rather than to prove them directly. Nevertheless, these techniques are powerful tools that have already had a great impact on our understanding of membrane proteins. Almost certainly, they will continue to revolutionize the study of membranes and their proteins.

Figure 7A-1 (opposite page)   Application of Techniques of Molecular Biology to the Study of Membrane Proteins.   Most molecular approaches begin with an isolated piece of DNA corresponding to all or part of a gene for a membrane protein, obtained by methods to be discussed in Chapter 18. ① The nucleotide sequence of the DNA is determined, usually by a rapid, automated sequencing machine. ② The amino acid sequence of the protein is then readily deduced using the genetic code. The deduced sequence can be used ③ in hydropathy analysis to determine the frequency and location of likely transmembrane segments and/or ④ in the preparation of synthetic peptides and then of antibodies, which can be radioactively

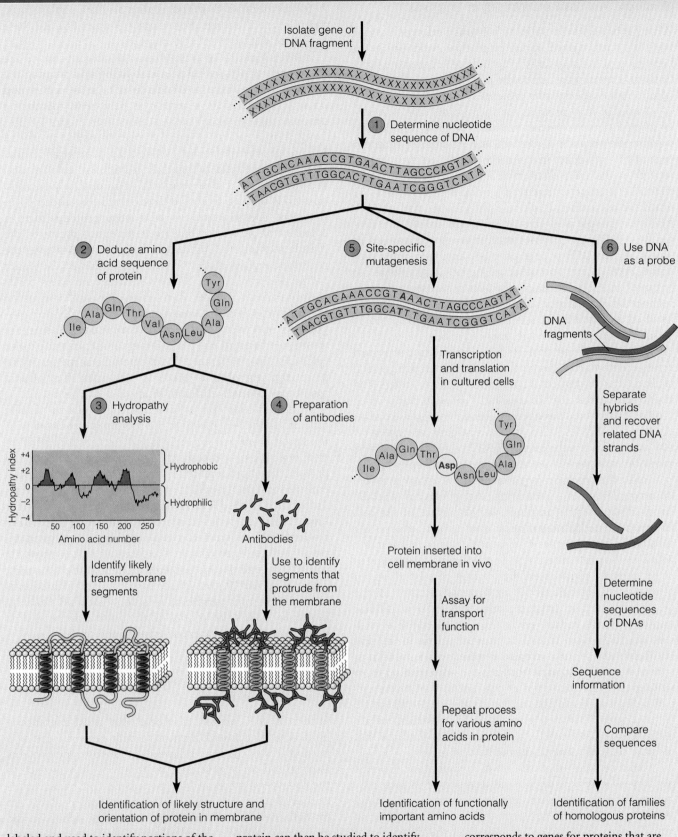

Isolate gene or
DNA fragment

1 Determine nucleotide
sequence of DNA

...ATTGCACAAACCGTGAACTTAGCCCAGTAT...
...TAACGTGTTTGGCACTTGAATCGGGTCATA...

2 Deduce amino
acid sequence
of protein

Ile — Ala — Gln — Thr — Val — Asn — Leu — Ala — Gln — Tyr

5 Site-specific
mutagenesis

...ATTGCACAAACCGT**AA**ACTTAGCCCAGTAT...
...TAACGTGTTTGGCA**TT**TGAATCGGGTCATA...

6 Use DNA
as a probe

DNA
fragments

3 Hydropathy
analysis

4 Preparation
of antibodies

Transcription
and translation
in cultured cells

Separate
hybrids
and recover
related DNA
strands

Hydropathy index
+4
+2
0
−2
−4

50 100 150 200 250
Amino acid number

Hydrophobic

Hydrophilic

Antibodies

Ile — Ala — Gln — Thr — **Asp** — Asn — Leu — Ala — Gln — Tyr

Identify likely
transmembrane
segments

Use to identify
segments that
protrude from
the membrane

Protein inserted into
cell membrane in vivo

Assay for
transport
function

Repeat process
for various amino
acids in protein

Determine
nucleotide
sequences
of DNAs

Sequence
information

Compare
sequences

Identification of likely structure and
orientation of protein in membrane

Identification of functionally
important amino acids

Identification of families
of homologous proteins

labeled and used to identify portions of the protein that protrude from the membrane. ⑤ Site-specific mutagenesis can be used to make specific changes in the DNA sequence and hence in the amino acid sequence of the protein; properties of the resulting mutant protein can then be studied to identify functionally important amino acids or protein segments. ⑥ The isolated DNA can also be used as a probe to identify related sequences of DNA in the same or other genomes. Such DNA presumably corresponds to genes for proteins that are related structurally and perhaps also functionally and evolutionarily.

Membrane Proteins Have a Variety of Functions

What functions do membrane proteins perform? Most of what we summarized about membrane function at the beginning of the chapter is relevant here because the functions of a membrane are really just those of its chemical components, especially its proteins.

Some of the proteins in membranes are *enzymes*, which accounts for the localization of specific functions to specific membranes. As we will see in coming chapters, each of the organelles in a eukaryotic cell is in fact characterized by its own distinctive set of membrane-bound enzymes. We encountered an example earlier when we noted the association of glucose phosphatase with the ER. Another example is glyceraldehyde-3-phosphate dehydrogenase (GPD), an enzyme involved in the catabolism of blood glucose. GPD is a peripheral plasma membrane protein in erythrocytes and other cell types. Closely related to enzymes in their function are *electron transport proteins* such as the cytochromes and iron-sulfur proteins that are involved in oxidative processes in mitochondria, chloroplasts, and the plasma membranes of prokaryotic cells.

Other membrane proteins function in solute transport across membranes. These include *transport proteins,* which facilitate the movement of nutrients such as sugars and amino acids across membranes, and *channel proteins,* which provide hydrophilic passageways through otherwise hydrophobic membranes. Also in this category are *transport ATPases,* which use the energy of ATP to pump ions across membranes, as we will see in more detail in Chapter 8.

Still other membrane proteins are *receptors* involved in recognizing and mediating the effects of specific chemical signals that impinge on the surface of the cell. Hormones, neurotransmitters, and growth-promoting substances are examples of chemical signals that interact with specific protein receptors in the plasma membrane of target cells. In most cases, the binding of a hormone or other signal molecule to the appropriate receptor on the membrane surface triggers some sort of intracellular response, which in turn elicits the desired effect; we will consider such signal-transduction mechanisms in Chapter 10. Membrane proteins are also involved with intercellular communication. Examples include the proteins that form structures called *connexons* at *gap junctions* between animal cells, and those that make up the *desmotubules* of *plasmodesmata* between plant cells. These structures are discussed in Chapter 11, as are membrane proteins that mediate cell-cell recognition and adhesion.

Other cellular functions in which membrane proteins play key roles include uptake and secretion of various substances by endocytosis and exocytosis; targeting, sorting, and modification of proteins within the endoplasmic reticulum and the Golgi complex; and the detection of light, whether by the human eye or by a seedling as it emerges from the soil. In addition, membrane proteins are vital components of various structures, including the links between the plasma membrane and the extracellular matrix located outside of the cell; the pores found in the outer membranes of mitochondria and chloroplasts; and the pores of the nuclear envelope (Chapter 16). All these topics are discussed in later chapters.

A final group of membrane-associated proteins are those with structural roles in stabilizing and shaping the cell membrane. Examples include ankyrin, spectrin, and band 4.1 protein, the erythrocyte peripheral membrane proteins that we encountered earlier (see Figure 7-20b). Long, thin tetramers of α and β spectrin, $(\alpha\beta)_2$, are linked to glycophorin molecules by band 4.1 protein and short actin filaments and to band 3 proteins by ankyrin and another protein, called band 4.2. (Appropriately enough, *ankyrin* is derived from the Greek word for "anchor.") In this way, spectrin and its associated proteins form a cytoskeletal network that underlies and supports the plasma membrane. This spectrin-based network gives the red blood cell its distinctive biconcave shape (see Figure 7-20a) and enables the cell to withstand the stress on its membrane as it is forced through narrow capillaries in the circulatory system. Proteins that are structurally homologous to spectrin and spectrin-associated proteins are found just beneath the plasma membrane in many other cell types also, indicating that a cytoskeletal meshwork of peripheral membrane proteins underlies the plasma membrane of many different kinds of cells.

The function of a membrane protein is usually reflected in the way in which the protein is associated with the lipid bilayer. For example, a protein that functions on only one side of a membrane is likely to be a peripheral protein or lipid-anchored protein. Membrane-bound enzymes that catalyze reactions on only one side of a membrane, such as the ER enzyme glucose phosphatase, are in this category. In contrast, the tasks of transporting solutes or transmitting signals across a membrane clearly require transmembrane proteins. In Chapter 8 we will examine the role of transmembrane proteins in membrane transport, and in Chapter 10, we will see how signal transduction is mediated by transmembrane receptors that bind signaling molecules such as hormones on the outside of the plasma membrane and generate signals inside the cell.

Membrane Proteins Are Oriented Asymmetrically Across the Lipid Bilayer

Earlier in the chapter, we noted that most membrane lipids are distributed asymmetrically between the two monolayers of the lipid bilayer. Most membrane proteins also exhibit an asymmetric orientation with respect to the bilayer. For example, peripheral proteins, lipid-anchored proteins, and integral monotopic proteins are by definition associated with one or the other of the membrane surfaces (recall Figure 7-19); once in place, they cannot move across the membrane from one surface to the other. Integral membrane proteins that span the membrane are embedded in both monolayers, but they are asymmetrically arranged because all of the molecules of a given protein are oriented the same way—that is, the regions of the protein molecule that are exposed on one side of the membrane are structurally and chemically different from the regions of the protein exposed on the other side.

To determine how proteins are oriented in a membrane, radioactive labeling procedures have been devised that distinguish between proteins exposed on the inner and outer surfaces of membrane vesicles. One such approach makes use of the enzyme *lactoperoxidase (LP)*, which catalyzes the covalent binding of iodine to proteins. When the reaction is carried out in the presence of ^{125}I, a radioactive isotope of iodine, LP labels the protein. Because LP is too large to pass through membranes, only proteins exposed on the outer surface of intact membrane vesicles are labeled (Figure 7-24a). To label only those proteins that are exposed on the inner membrane surface, vesicles are first exposed to a hypotonic (low ionic strength) solution to make them more permeable to large molecules. Under these conditions, LP can enter the vesicles. When the vesicles are subsequently incubated with ^{125}I, the enzyme labels the proteins exposed on the inner surface of the vesicle membrane (Figure 7-24b and c). In this way, it is possible to determine whether a given membrane protein is exposed on the inner membrane surface only, on the outer surface only, or on both surfaces.

In a similar approach, the enzyme *galactose oxidase (GO)* can be used to label carbohydrate side chains that are attached to membrane proteins or lipids. Vesicles are first treated with GO to oxidize galactose residues in carbohydrate side chains. The vesicles are then exposed to tritiated borohydride ($^3H\text{-}BH_4$), which reduces the galactose groups, introducing labeled hydrogen atoms in the process. To explore these techniques further, try Problems 7-12 and 7-13 at the end of the chapter.

The orientation of membrane proteins is created when each protein is inserted into the membrane during membrane biogenesis, and it is then preserved by the thermodynamic restrictions on transverse diffusion. The result is a membrane in which the lipid and protein molecules present on one side of the membrane are different from those on the other side. As a result, membranes are *functionally asymmetric*, in the sense that the reactions and processes that occur on one side of the membrane are usually quite different from the activities on the other side.

Many Membrane Proteins Are Glycosylated

In addition to lipids and proteins, most membranes contain small but significant amounts of carbohydrates. The plasma membrane of the human erythrocyte, for example, contains about 52% protein, 40% lipid, and 8% carbohydrate by weight. The glycolipids that we encountered earlier account for a small portion of membrane carbohydrate but most of it is in the form of **glycoproteins,** membrane proteins with carbohydrate chains linked covalently to amino acid side chains.

The addition of a carbohydrate side chain to a protein is called **glycosylation.** As Figure 7-25 shows, glycosylation involves linkage of the carbohydrate either to the nitrogen atom of an amino group (*N-linked glycosylation*) or to the oxygen atom of a hydroxyl group (*O-linked glycosylation*). N-linked carbohydrates are attached to the amino group on the side chain of asparagine (Figure 7-25a), whereas O-linked carbohydrates are usually bound to the hydroxyl groups of either serine or threonine (Figure 7-25b). In some cases, O-linked carbohydrates are attached to the hydroxyl group of hydroxylysine or hydroxyproline, which are derivatives of the amino acids lysine and proline, respectively (Figure 7-25c).

The carbohydrate chains attached to glycoproteins can be either straight or branched and range in length from 2 to about 60 sugar units. The most common sugars used in constructing these chains are *galactose, mannose, N-acetylglucosamine,* and *sialic acid* (Figure 7-26a). Figure 7-26b shows the carbohydrate chain found in the erythrocyte plasma membrane protein, glycophorin. This integral membrane protein has 16 such carbohydrate chains attached to the portion of the molecule (the N-terminus) that extends outward from the erythrocyte membrane; of these, 1 is N-linked and 15 are O-linked. Notice that both branches of the carbohydrate chain terminate in a negatively charged sialic acid. Because of these anionic groups on their surfaces, erythrocytes repel each other, thereby reducing blood viscosity.

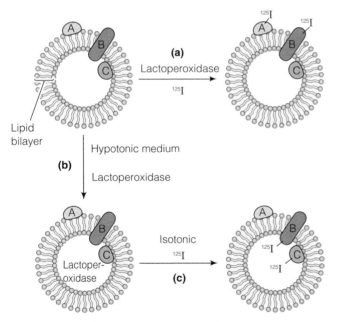

Figure 7-24   A Method For Labeling Proteins Exposed on One or Both Surfaces of a Membrane Vesicle.   (a) When lactoperoxidase (LP) and ^{125}I are present in the solution outside a membrane vesicle, LP catalyzes the labeling of membrane proteins exposed on the outer membrane surface (i.e., proteins A and B). If membrane vesicles are **(b)** first incubated in a hypotonic medium to make them permeable to LP and **(c)** then transferred to an isotonic solution containing ^{125}I but no external LP, proteins exposed on the inner membrane surface (i.e., proteins B and C) become labeled.

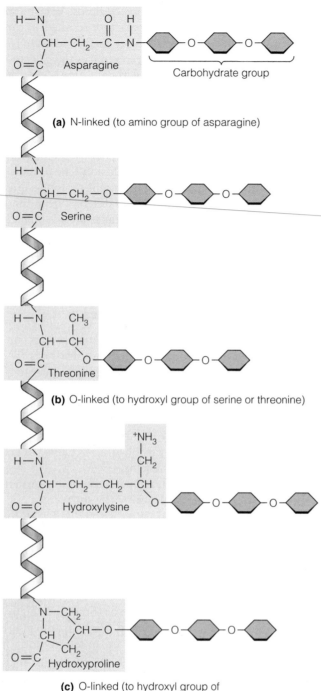

(a) N-linked (to amino group of asparagine)

(b) O-linked (to hydroxyl group of serine or threonine)

(c) O-linked (to hydroxyl group of hydroxylysine or hydroxyproline)

Figure 7-25   N-Linked and O-Linked Glycosylation of Membrane Proteins. Carbohydrate groups are linked to membrane proteins in two different ways. In each case, the amino acid shown is part of the primary structure of a membrane protein. **(a)** N-linked carbohydrates are attached to the amino group of asparagine side chains by amide bonds. **(b)** O-linked carbohydrates are attached to the hydroxyl group of serine or threonine side chains by ester bonds. **(c)** In some cases, O-linked carbohydrates are linked to the hydroxyl group of the modified amino acids hydroxylysine and hydroxyproline.

terminate in *N*-acetylglucosamine, whereas *concanavalin A,* a lectin from jack beans, recognizes mannose groups in internal positions. Investigators visualize these lectins by linking them to *ferritin,* an iron-containing protein that shows up as a very electron-dense spot when viewed with an electron microscope. When such ferritin-linked lectins are used as probes to localize the oligosaccharide chains of membrane glycoproteins, binding is always specifically to the outer surface of the membrane.

In many animal cells, the carbohydrate groups of plasma membrane glycoproteins and glycolipids protrude from the cell surface and form a surface coat called the **glycocalyx** (meaning "sugar coat"). Figure 7-27 shows the prominent glycocalyx of an intestinal epithelial cell. The carbohydrate groups on the cell surface are important components of the recognition sites of membrane receptors such as those involved in binding extracellular signal molecules, in antibody-antigen reactions, and in intercellular adhesion to form tissues.

Membrane Proteins Vary in Their Mobility

Earlier in the chapter, we noted that lipid molecules can diffuse laterally within the plane of a membrane (see Figure 7-11). Now we can ask the same question about membrane proteins: Are they also free to move within the membrane? In fact, membrane proteins are much more variable than lipids in their mobility. Some proteins appear to move freely within the lipid bilayer whereas others are constrained, often because they are anchored to protein complexes located adjacent to one side of the membrane or the other.

Experimental Evidence for Protein Mobility. Particularly convincing evidence for the mobility of at least some membrane proteins has come from *cell fusion experiments* such as those summarized in Figure 7-28. In these studies, David Frye and Michael Edidin took advantage of two powerful techniques, one that enabled them to fuse cells from two different species and another that made it possible for them to label, or tag, the surfaces of cells with fluorescent dye molecules. Their approach was to tag mouse and human cells with **fluorescent antibodies** against plasma membrane proteins specific to the two species. They then fused the two types of cells and observed the hybrid human/mouse cells with a micro-

Glycoproteins are most prominent in plasma membranes, where they play an important role in cell-cell recognition. Consistent with this role, glycoproteins are always positioned so that the carbohydrate groups protrude on the external surface of the cell membrane. This arrangement, which contributes to membrane asymmetry, has been shown experimentally using *lectins,* plant proteins that bind specific sugar groups very tightly. For example, *wheat germ agglutinin,* a lectin found in wheat embryos, binds very specifically to oligosaccharides that

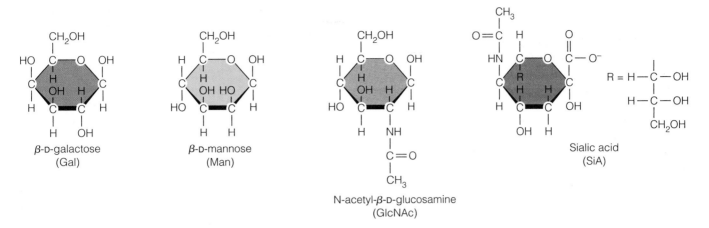

(a) Common sugars found in glycoproteins

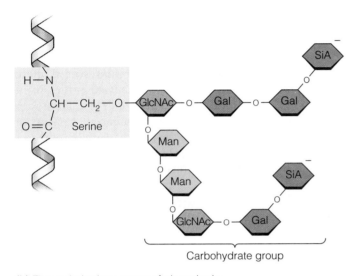

(b) The carbohydrate group of glycophorin

Figure 7-26 Examples of Carbohydrates Found in Glycoproteins. **(a)** The four most common carbohydrates in glycoproteins are galactose (Gal), mannose (Man), *N*-acetylglucosamine (GlcNAc), and sialic acid (SiA). **(b)** For example, 15 of the 16 carbohydrate chains of glycophorin are linked to the amino acid serine in the hydrophilic portion of the protein on the outer surface of the erythrocyte plasma membrane. Each of these carbohydrate chains consists of three units of Gal and two units each of GlcNAc, Man, and SiA. The sialic acid groups are negatively charged.

scope to see what happened to the plasma membrane proteins from the two parent cells.

The antibodies needed for this kind of experiment are prepared by injecting small quantities of purified membrane proteins from human and mouse cells into separate experimental animals. (Rabbits or goats are commonly used for this purpose.) The animals respond immunologically to the foreign protein by producing *antibodies* (blood proteins called *immunoglobulins*) that react specifically with the membrane proteins used in the injections. After the antibodies have been isolated from the blood of the animal, fluorescent dyes are covalently linked to them so that the protein/antibody/dye complexes can be visualized with a microscope.

Frye and Edidin tagged human and mouse plasma membrane proteins with antibodies that had two different dyes linked to them so that the two types of proteins could be distinguished by the color of their fluorescence. The mouse cells were reacted with mouse-specific antibodies linked to a green fluorescent dye called *fluorescein*, whereas the human cells were reacted with human-specific antibodies linked to a red fluorescent dye, *rhodamine*. They then exposed the tagged cells to Sendai virus, an agent known to cause fusion of eukaryotic cells, even those from different species. When the cells were first fused, the green fluorescent membrane proteins from the mouse cell were localized on one half of the hybrid cell surface, whereas the red fluorescent membrane proteins derived from the human cell were restricted to the other half (see Figure 7-28). In a few minutes, however, the proteins from the two parent cells began to intermix, and after 40 minutes, the separate regions of green and red fluorescence were completely intermingled. If the fluidity of the membrane was depressed by lowering the temperature below the transition temperature of the lipid bilayer, this intermixing could be prevented. Frye and Edidin therefore concluded that the intermingling of the fluorescent proteins had been caused by lateral diffusion of the human and mouse proteins through the fluid lipid bilayer of the plasma membrane.

If proteins are completely free to diffuse within the plane of the membrane, then they should eventually become randomly distributed. Support for the idea that at least some membrane proteins behave in this way has emerged from freeze-fracture microscopy, which directly visualizes proteins embedded within the lipid bilayer. When plasma membranes are examined in freeze-fracture micrographs, their embedded protein particles often tend to be randomly distributed. Such evidence for protein mobility is not restricted to the plasma membrane. It has also been found, for example, that the protein particles of the inner mitochondrial membranes are randomly arranged (Figure 7-29a). If isolated mitochondrial membrane vesicles are exposed to an electrical potential, the protein particles, which bear a net negative charge, all move to one end of the vesicle (Figure 7-29b). Removing the electrical potential causes the particles to become randomly distributed again, indicating that these proteins are free to move within the lipid bilayer. We will return to this evidence in Chapter 14, where we will learn that the transfer of electrons within the inner mitochondrial membrane depends on random collisions between mobile complexes of membrane proteins.

Experimental Evidence for Restricted Mobility. Although many types of membrane proteins have been shown to diffuse through the lipid bilayer, their rates of movement vary. A widely used approach for quantifying the rates at which membrane proteins diffuse is fluorescence photobleaching recovery, a technique discussed earlier in the context of lipid mobility (see Figure 7-11). The rate at which unbleached molecules from adjacent parts of the membrane move back into the bleached area can be used to calculate the diffusion rates of various kinds of fluorescent lipid or protein molecules.

Membrane proteins are much more variable in their diffusion rates than lipids. A few membrane proteins diffuse almost as rapidly as lipids, but most diffuse more slowly than would be expected if they were completely free to move within the lipid bilayer. Moreover, the diffusion of many membrane proteins is restricted to a limited area of the membrane, indicating that at least some membranes consist of a series of separate *membrane domains* that differ in their protein compositions and hence in their functions. For example, the cells lining your small intestine have membrane proteins that transport solutes such as sugars and amino acids out of the intestine and into the body. These transport proteins are restricted to the side of the cell where the corresponding type of transport is required (see Figure 11-22).

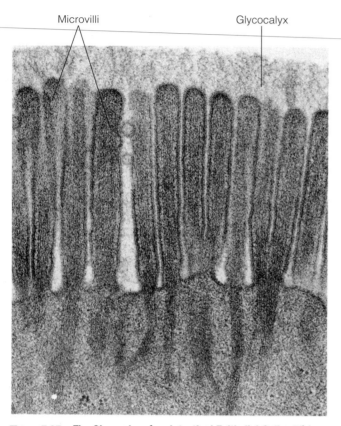

Figure 7-27 The Glycocalyx of an Intestinal Epithelial Cell. This electron micrograph of a cat intestinal epithelial cell shows the microvilli (fingerlike projections) that are involved in absorption and the glycocalyx on the cell surface. The glycocalyx on this cell is about 150 nm thick and consists primarily of oligosaccharide chains about 1.2–1.5 nm in diameter (TEM).

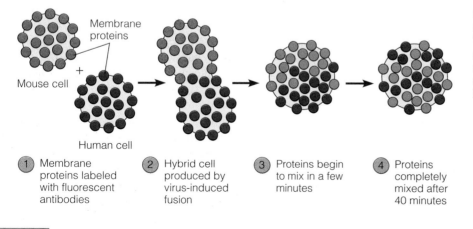

① Membrane proteins labeled with fluorescent antibodies

② Hybrid cell produced by virus-induced fusion

③ Proteins begin to mix in a few minutes

④ Proteins completely mixed after 40 minutes

Figure 7-28 Demonstration of the Mobility of Membrane Proteins by Cell Fusion. The mobility of membrane proteins can be shown experimentally by the mixing of membrane proteins that occurs when two cells are tagged with different fluorescent labels and then induced to fuse.

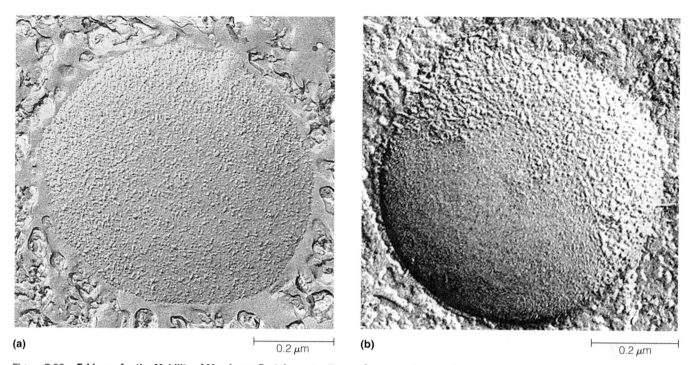

(a) 0.2 μm   **(b)** 0.2 μm

Figure 7-29 Evidence for the Mobility of Membrane Proteins. **(a)** Freeze-fracture micrograph showing the random distribution of protein particles in vesicles prepared from the inner mitochondrial membrane. **(b)** Exposure of the vesicles to an electrical field causes the membrane particles to migrate to one end of the vesicle (upper right of micrograph).

Mechanisms for Restricting Protein Mobility. Several different mechanisms account for restricted protein mobility. In some cases, membrane proteins aggregate within the membrane, forming large complexes that move only sluggishly, if at all. In other cases, membrane proteins form structures that become barriers to the diffusion of other membrane proteins, thereby effectively creating specific membrane domains; the *tight junctions* to be discussed in Chapter 11 are one example. The most common restraint on the mobility of membrane proteins, however, is imposed by the binding, or *anchoring*, of such proteins to structures located adjacent to one side of the membrane or the other. For example, many proteins of the plasma membrane are anchored either to elements of the cytoskeleton on the inner surface of the membrane or to extracellular structures such as the extracellular matrix of animal cells. We will encounter both of these anchoring mechanisms in Chapter 11.

Perspective

Cells need membranes to define and compartmentalize space, regulate the flow of materials, detect external signals, and mediate interactions between cells. Our current understanding of membrane structure represents the culmination of more than a century of studies, beginning with the recognition that lipids are an important membrane component and moving progressively from a lipid monolayer to a phospholipid bilayer. Once proteins were recognized as important components, Davson and Danielli proposed their "sandwich" model, with the lipid bilayer surrounded on both sides by layers of proteins. This model stimulated much research and was presumed to apply to the variety of membranes revealed by electron microscopy beginning in the 1950s. As membranes and membrane proteins were examined in more detail, however, the model was called increasingly into question and was eventually discredited.

In its place, Singer and Nicholson's fluid mosaic model emerged and is now the universally accepted description of membrane structure. According to this model, proteins with varying affinities for the hydrophobic membrane interior float in and on a fluid lipid bilayer. Prominent lipids in most membranes include phospholipids (both phosphoglycerides and phosphosphingolipids) and glycolipids such as cerebrosides and gangliosides. In eukaryotic cells, sterols are also important membrane components; these include cholesterol in animal cells and several related phytosterols in plant cells. Sterols are not found in the membranes of prokaryotes, but at least some bacterial cells contain somewhat similar compounds called hopanoids.

In terms of their association with the lipid bilayer, membrane proteins are classified as integral, peripheral, or lipid-anchored. Integral membrane proteins have one or more short segments of predominantly hydrophobic amino acids that anchor the protein to the membrane. In many cases, these are transmembrane segments, such that portions of the protein are exposed on both sides. Most of these transmembrane segments are α-helical sequences of about 20–30 amino

acids. Peripheral membrane proteins are hydrophilic and remain on the membrane surface. Lipid-anchored proteins are also hydrophilic in nature but are covalently linked to the membrane by any of several lipid anchors embedded in the lipid bilayer.

Most membrane phospholipids and proteins are free to move within the plane of the membrane unless they are specifically anchored to structures on the inner or outer membrane surface. However, transverse diffusion, or "flip-flop," between monolayers is not generally possible, except for phospholipids when catalyzed by enzymes called phospholipid translocators or flippases. As a result, most membranes are characterized by an asymmetric distribution of lipids between the two monolayers and an asymmetric orientation of proteins within the membranes, so that the two sides of the membrane are structurally and functionally dissimilar.

Membrane proteins function as enzymes, electron carriers, transport molecules, and receptor sites for chemical signals such as neurotransmitters and hormones. Membrane proteins also stabilize and shape the membrane and mediate intercellular communication and cell-cell adhesion. Many proteins in the plasma membrane are glycoproteins, with carbohydrate side chains that protrude from the membrane on the external side, where they play important roles as recognition markers on the cell surface.

With the understanding of membrane structure and function that we have acquired in this chapter, we are now ready to explore the mechanisms by which solutes and signals move across membranes. The transport of small molecules and ions across membranes will be discussed in Chapter 8, followed in Chapters 9 and 10 by a consideration of the mechanisms involved in the transduction of electrical and chemical signals, respectively.

Key Terms for Self-Testing

membrane (p. 158)

The Functions of Membranes
plasma (cell) membrane (p. 159)
intracellular membrane (p. 159)
transport (p. 160)
signal transduction (p. 160)
receptor (p. 160)
intercellular communication (p. 160)

Models of Membrane Structure:
An Experimental Perspective
lipid bilayer (p. 162)
fluid mosaic model (p. 164)

Membrane Lipids:
The "Fluid" Part of the Model
phospholipid (p. 166)
phosphoglycerides (p. 166)
sphingolipid (p. 166)
glycolipid (p. 166)
cerebroside (p. 166)
ganglioside (p. 166)

sterol (p. 166)
cholesterol (p. 166)
phytosterol (p. 166)
hopanoid (p. 166)
thin-layer chromatography (TLC) (p. 168)
fatty acid (p. 169)
membrane asymmetry (p. 169)
transverse diffusion (p. 169)
rotation (of phospholipid molecules) (p. 169)
lateral diffusion (p. 169)
phospholipid translocator (flippase) (p. 171)
fluorescence recovery after photobleaching (p. 171)
transition temperature (T_m) (p. 172)
phase transition (p. 172)
differential scanning calorimetry (p. 172)
homeoviscous adaptation (p. 174)

Membrane Proteins:
The "Mosaic" Part of the Model
freeze-fracturing (p. 175)

integral membrane protein (p. 177)
integral monotopic protein (p. 177)
transmembrane protein (p. 177)
transmembrane segments (p. 177)
hydropathy (hydrophobicity) plot (p. 179)
hydropathy index (p. 179)
peripheral membrane protein (p. 179)
lipid-anchored membrane protein (p. 179)
fatty acid-anchored membrane protein (p. 180)
prenylated membrane protein (p. 180)
GPI-anchored membrane protein (p. 180)
electrophoresis (p. 180)
glycoprotein (p. 185)
glycosylation (p. 185)
glycocalyx (p. 186)
fluorescent antibody (p. 186)

Box 7A: *Revolutionizing the Study of Membrane Proteins: The Impact of Molecular Biology*
site-specific mutagenesis (p. 182)

Problem Set

More challenging problems are marked with a •.

7-1.   Functions of Membranes. For each of the following statements, specify which one of the five general membrane functions (permeability barrier, localization of function, regulation of transport, detection of signals, or intercellular communication) the statement illustrates.

(a) When cells are disrupted and fractionated into subcellular components, the enzyme cytochrome *c* reductase is recovered with the endoplasmic reticulum fraction.

(b) Cells of multicellular organisms carry tissue-specific glycoproteins on their outer surface that are responsible for cell-cell adhesion.

(c) The interior of a membrane consists primarily of the hydrophobic portions of phospholipids and amphipathic proteins.

(d) Photosystems I and II are embedded in the thylakoid membrane of the chloroplast.

(e) All of the acid phosphatase in a mammalian cell is found within the lysosomes.

(f) The membrane of a plant root cell has an ion pump that exchanges phosphate inward for bicarbonate outward.

(g) The inner mitochondrial membrane contains an ATP-ADP carrier protein that couples outward ATP movement to inward ADP movement.

(h) Insulin does not enter a target cell, but instead binds to a specific membrane receptor on the external surface of the membrane, thereby activating the enzyme adenylyl cyclase on the inner membrane surface.

(i) Adjacent plant cells frequently exchange cytoplasmic components through membrane-lined channels called plasmodesmata.

7-2. Elucidation of Membrane Structure. Each of the following observations played an important role in enhancing our understanding of membrane structure. Explain the significance of each, and indicate in what decade of the timeline shown in Figure 7-3 the observation was most likely made.

(a) When a membrane is observed in the electron microscope, both of the thin, electron-dense lines are about 2 nm thick, but the two lines are often distinctly different from each other in appearance.

(b) Ethylurea penetrates much more readily into a membrane than does urea, and diethylurea penetrates still more readily.

(c) The addition of phospholipase to living cells causes rapid digestion of the lipid bilayers of the membranes, which suggests that the enzyme has access to the membrane phospholipids.

(d) When artificial lipid bilayers are subjected to freeze-fracture, no particles are seen on either face.

(e) The electrical resistivity of artificial lipid bilayers is several orders of magnitude greater than that of real membranes.

(f) Some membrane proteins can be readily extracted with 1 M NaCl, whereas others require the use of an organic solvent or a detergent.

(g) When halobacteria are grown in the absence of oxygen, they produce a purple pigment that is embedded in their plasma membranes and has the ability to pump protons outward when illuminated. If the purple membranes are isolated and viewed by freeze-fracture electron microscopy, they are found to contain patches of crystalline particles.

7-3. Gorter and Grendel Revisited. Gorter and Grendel's classic conclusion that the plasma membrane of the human erythrocyte consists of a lipid bilayer was based on the following observations: (i) the lipids that they extracted with acetone from 4.74×10^9 erythrocytes formed a monolayer 0.89 m² in area when spread out on a water surface; and

(ii) the surface area of one erythrocyte was about 100 μm², according to their measurements.

(a) Show from these data how they came to the conclusion that the erythrocyte membrane is a bilayer.

(b) We now know that the surface area of a human erythrocyte is about 145 μm². Explain how Gorter and Grendel could have come to the right conclusion when one of their measurements was only about two-thirds of the correct value.

7-4. Lipid, Protein, and Bilayers in Erythrocytes. The membrane of a human erythrocyte has a surface area of about 145 μm² and is about 8 nm thick. It contains about 0.52 picogram of lipid and 0.60 picogram of protein. (A picogram is 1×10^{-12} g.) Assume equimolar amounts of cholesterol and phospholipids, with molecular weights of 386 and about 800, respectively. In a monolayer, each phospholipid molecule occupies a surface area of 0.55 nm² per molecule, and each cholesterol molecule occupies 0.38 nm² per molecule.

(a) Assuming an average molecular weight of 50,000 for membrane proteins, how many protein molecules are there in a single erythrocyte membrane?

(b) What is the ratio of lipid molecules to protein molecules in the erythrocyte membrane?

(c) What proportion of the total surface area of the membrane is occupied by the lipids in the bilayer?

7-5. That's About the Size of It. From chemistry, we know that each methylene ($-CH_2-$) group in a straight-chain hydrocarbon advances the chain length by about 0.13 nm. And from studies of protein structure, we know that one turn of an α helix includes 3.6 amino acid residues and extends the long axis of the helix by about 0.56 nm. Use this information to answer the following:

(a) How long is a single molecule of palmitate (16 carbon atoms) in its fully extended form? What about molecules of laurate (12C) and arachidate (20C)?

(b) How does the thickness of the hydrophobic interior of a typical membrane compare with the length of two palmitate molecules laid end to end? What about two molecules of laurate or arachidate?

(c) Approximately how many amino acids must a helical transmembrane segment of an integral membrane protein have if the segment is to span the lipid bilayer defined by two palmitate molecules laid end to end?

(d) The protein bacteriorhodopsin has 248 amino acids and seven transmembrane segments. Approximately what portion of the amino acids are part of the transmembrane segments? Assuming that most of the remaining amino acids are present in the hydrophilic loops that link the transmembrane segments together, approximately how many amino acids are present in each of these loops, on the average?

7-6. Temperature and Membrane Composition. Which of the following responses are *not* likely to be seen when a bacterial culture growing at 37°C is transferred to a culture room maintained at 25°C?

(a) Initial decrease in membrane fluidity

(b) Gradual replacement of shorter fatty acids by longer fatty acids in the membrane phospholipids

(c) Gradual replacement of stearate by oleate in the membrane phospholipids

(d) Enhanced rate of synthesis of unsaturated fatty acids.

(e) Incorporation of more cholesterol into the membrane

7-7.   Glycophorin: The "Sugar Carrier." The word *glycophorin* is derived from Greek roots meaning "to carry sugar." To understand why one of the major integral membrane proteins of the erythrocyte membrane is so named, consider the following:

(a) The protein consists of a single polypeptide with 131 amino acids. Assuming that amino acids have an average molecular weight of 110, what is the molecular weight of the glycophorin polypeptide?

(b) Glycophorin contains 16 carbohydrate side chains, each with three galactose units ($C_6H_{12}O_6$), two mannose units ($C_6H_{12}O_6$), two glucosamine units ($C_6H_{13}O_5N$), and two sialic acid units ($C_{11}H_{19}O_5N$). (Note: One molecule of water is lost as each of these units is linked to the carbohydrate side chain and as the chain is linked to the protein.) Determine the molecular weight of each unit in one carbohydrate side chain, and then calculate the total molecular weight of the 16 carbohydrate side chains.

(c) What is the total molecular weight of the glycophorin polypeptide plus its 16 carbohydrate side chains? What percentage of the total molecular weight do the 16 carbohydrate side chains represent?

(d) In what sense is glycophorin an aptly named protein?

7-8.   Membrane Fluidity and Temperature. The effects of temperature and lipid composition on membrane fluidity are often studied by using artificial membranes containing only one or a few kinds of lipids and no proteins. Assume that you and your lab partner have made the following artificial membranes:

Membrane 1: Made entirely from phosphatidylcholine with saturated 16-carbon fatty acids.

Membrane 2: Same as membrane 1, except that each of the 16-carbon fatty acids has a single *cis* double bond.

Membrane 3: Same as membrane 1, except that each of the saturated fatty acids has only 14 carbon atoms.

After determining the transition temperatures of samples representing each of the membranes, you discover that your lab partner failed to record the membranes to which the samples correspond. The three values you determined are $-36°C$, $23°C$, and $41°C$. Can you assign each of these transition temperatures to the correct artificial membrane? Explain your reasoning.

7-9.   Membrane Potpourri. Explain each of the following observations as succinctly as possible.

(a) Proteins A and B both have a molecular weight of 33,000 and consist of a single polypeptide chain with 240 hydrophilic amino acids and 60 hydrophobic amino acids, yet protein A is an integral membrane protein and protein B is a soluble protein.

(b) When a photobleaching experiment similar to that described in Figure 7-11 is carried out using plasma membranes from myoblasts (precursors to muscle cells) with fluorescently tagged surface glycoproteins, it takes much longer for the dark, nonfluorescent spot to become fluorescent again than when phospholipids are labeled with a fluorescent dye. Moreover, only 55% of the protein fluorescence eventually returns to the laser-bleached spot.

(c) In the legs of a reindeer, the membranes of cells near the hoof have a higher proportion of unsaturated fatty acids than the membranes of cells in the rest of the leg.

(d) Most of the agents used as anesthetics in medicine are non-polar and act by increasing the fluidity of the lipid bilayers of cellular membranes to the extent that the transmission of nerve impulses is disrupted and sensations of pain are eliminated.

7-10.   The Little Bacterium That Can't. *Acholeplasma laidlawii* is a small bacterium that cannot synthesize its own fatty acids and must therefore construct its plasma membrane from whatever fatty acids are available in the environment. As a result, the *Acholeplasma* membrane takes on the physical characteristics of the fatty acids available at the time.

(a) If you give *Acholeplasma* cells access to a mixture of saturated and unsaturated fatty acids, they will thrive at room temperature. Can you explain why?

(b) If you transfer the bacteria of part a to a medium containing only saturated fatty acids but make no other changes in culture conditions, they will stop growing shortly after the change in medium. Explain why.

(c) What is one way you could get the bacteria of part b growing again without changing the medium? Explain your reasoning.

(d) If you were to maintain the *Acholeplasma* culture of part b under the conditions described there for an extended period of time, what do you predict will happen to the bacterial cells? Explain your reasoning.

(e) What result would you predict if you were to transfer the bacteria of part a to a medium containing only unsaturated fatty acids without making any other changes in the culture conditions? Explain your reasoning.

• 7-11.   Hydropathy: The Plot Thickens. A hydropathy plot can be used to predict the structure of a membrane protein based on its amino acid sequence and the hydrophobicity values of the amino acids. Hydrophobicity is measured as the standard free-energy change, $\Delta G^{o\prime}$, for the transfer of a given amino acid residue from a hydrophobic solvent into water, in kilojoules/mole (kJ/mol). The hydropathy index is calculated by averaging the hydrophobicity values for a series of short segments of the polypeptide, with each segment displaced one amino acid further from the N-terminus. The hydropathy index of each successive segment is then plotted as a function of the location of that segment in the amino acid sequence and the plot is examined for regions of high hydropathy index.

(a) Why do scientists try to predict the structure of a membrane protein by this indirect means when the technique of X-ray crystallography would reveal the structure directly?

(b) Given the way it is defined, would you expect the hydrophobicity index of a hydrophobic residue such as valine or isoleucine to be positive or negative? What about a hydrophilic residue such as aspartic acid or arginine?

(c) Listed below are four amino acids and five hydrophobicity values, respectively. Match the hydrophobicity values with the correct amino acids and explain your reasoning.

Amino acids: alanine; arginine; isoleucine; serine

Hydrophobicity (in kJ/mol): +3.1; +1.0; −1.1; −7.5

(d) Shown in Figure 7-30 is a hydropathy plot for a specific integral membrane protein. Draw a horizontal bar over each transmembrane segment as identified by the plot. How long is the average transmembrane segment? How well does that value compare with the number you calculated in Problem 7-5c? How many transmembrane segments do you think the protein has? Can you guess which protein this might be?

•**7-12. Inside or Outside?** From Figure 7-24, we know that exposed regions of membrane proteins can be labeled with ^{125}I by the lactoperoxidase (LP) reaction. Similarly, carbohydrate side chains of membrane glycoproteins can be labeled with ^{3}H by oxidation of galactose groups with galactose oxidase (GO), followed by reduction with tritiated borohydride (^{3}H-BH[4]). Noting that both LP and GO are too large to penetrate into the interior of an intact cell, explain each of the following observations made with intact erythrocytes.

(a) When intact cells are incubated with LP in the presence of ^{125}I and peroxide and the membrane proteins are then extracted and analyzed on SDS-polyacrylamide gels, several of the bands on the gel are found to be radioactive.

(b) When intact cells are incubated with GO and then reduced with ^{3}H-BH[4], several of the bands on the gel are found to be radioactive.

(c) All of the proteins of the plasma membrane that are known to contain carbohydrates are labeled by the GO/^{3}H-BH[4] method.

(d) None of the proteins of the erythrocyte plasma membrane that are known to be devoid of carbohydrate is labeled by the LP/^{125}I method.

(e) If the erythrocytes are ruptured before the labeling procedure, the LP procedure labels virtually all of the major membrane proteins.

•**7-13. Inside-Out Membranes.** It is technically possible to prepare sealed vesicles from erythrocyte membranes in which the original orientation of the membrane is inverted. Such vesicles have what was originally the cytoplasmic side of the membrane facing outward.

(a) What results would you expect if such inside-out vesicles were subjected to the GO/^{3}H-BH[4] procedure described in Problem 7-12?

(b) What results would you expect if such inside-out vesicles were subjected to the LP/^{125}I procedure of Problem 7-12?

(c) What conclusion would you draw if some of the proteins that become labeled by the LP/^{125}I method of part b were among those that had been labeled when intact cells were treated in the same way in Problem 7-12a?

(d) Knowing that it is possible to prepare inside-out vesicles from erythrocyte plasma membranes, can you think of a way to label a transmembrane protein with ^{3}H on one side of the membrane and with ^{125}I on the other side?

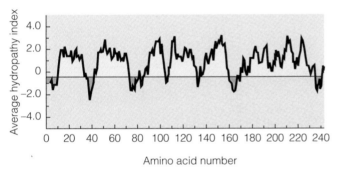

Figure 7-30 Hydropathy Plot for an Integral Membrane Protein. See Problem 7-11d.

Suggested Reading

References of historical importance are marked with a •.

General References

Lee, A. G., ed. *Biomembranes, Volume 1: General Principles.* Greenwich, CT: JAI Press, Inc., 1995.

Mathews, C. K., K. E. van Holde, and K. G. Ahern. *Biochemistry,* 3d ed. Menlo Park, CA: Benjamin/Cummings, 2000.

Vance, D. E., and J. E. Vance, eds. *Biochemistry of Lipids, Lipoproteins, and Membranes.* New York: Elsevier Science, 1996. (Volume 31 in the series *New Comprehensive Biochemistry*)

Yeagle, P. L. *The Membranes of Cells,* 2d ed. New York: Academic Press, 1993.

Membrane Structure and Function

Bennett, V., and D. M. Gilligan. The spectrin-based membrane structure and micron-scale organization of the plasma membrane. *Annu. Rev. Cell Biol.* 9 (1993): 27.

•Branton, D., and R. Park. *Papers on Biological Membrane Structure.* Boston: Little, Brown, 1968.

Cossins, A. R. *Temperature Adaptations of Biological Membranes.* London: Portland Press, 1994.

•Davson, H., and J. F. Danielli, eds. *The Permeability of Natural Membranes.* Cambridge, England: Cambridge University Press, 1943.

•Gorter, E., and F. Grendel. On bimolecular layers of lipids on the chromocyte of the blood. *J. Exptl. Med.* 41 (1925): 439.

Jacobson, K., E. D. Sheets, and R. Simson. Revisiting the fluid mosaic model of membranes. *Science* 268 (1995): 1441.

Lipowsky, R., and E. Sackmann. *The Structure and Dynamics of Membranes.* Amsterdam: Elsevier Science, 1995. (Volume 1 in the series *Handbook of Biological Physics,* A. Hoff, ed.)

Packer, L., and S. Fleischer, eds. *Biomembranes.* San Diego: Academic Press, 1997.

Petty, H. R. *Molecular Biology of Membranes: Structure and Function.* New York: Plenum, 1993.

- Robertson, J. D. The ultrastructure of cell membranes and their derivatives. *Biochem. Soc. Symp.* 16 (1959): 3.
- Robertson, J. D. Membrane structure. *J. Cell Biol.* 91 (1981): 189s.
- Rothstein, A. The cell membrane—a short historic perspective. *Curr. Topics Membr. Transport* 11 (1978): 1.
- Singer, S. J. The molecular organization of membranes. *Annu. Rev. Biochem.* 43 (1974): 805.
- Singer, S. J., and G. L. Nicolson. The fluid mosaic model of the structure of cell membranes. *Science* 175 (1972): 720.
- Stoeckenius, W. The purple membrane of salt-loving bacteria. *Sci. Amer.* 234 (June 1976): 38.
- Unwin, N., and R. Henderson. The structure of proteins in biological membranes. *Sci. Amer.* 250 (February 1984): 78.

Membrane Lipids

Dowhan, W. Molecular basis for membrane phospholipid diversity: Why are there so many lipids? *Annu. Rev. Biochem.* 66 (1997): 199.

Higgins, C. F. Flip-flop: The transmembrane translocation of lipids. *Cell* 79 (1994): 393.

Katsaros, J., and T. Gutberlet, eds. *Lipid Bilayers: Structure and Interactions.* New York: Springer-Verlag, 2000.

Menon, A. Flippases. *Trends Cell Biol.* 5 (1995): 355.

Quinn, P. J., and R. J. Cherry. *Structural and Dynamic Properties of Lipids and Membranes.* London: Portland Press, 1992.

Thompson, T. E. Lipids. *Curr. Opin. Struct. Biol.* 5 (1995): 529.

Zachowski, A. Phospholipids in animal eukaryotic membranes: Transverse asymmetry and movement. *Biochem. J.* 294 (1993): 1.

Membrane Proteins

- Frye, L. D., and M. Edidin. The rapid intermixing of cell surface antigens after formation of mouse-human heterokaryons. *J. Cell Sci.* 7 (1970): 319.

Henderson, R., J. M. Baldwin, T. A. Ceska, F. Zemlin, D. Beckmann, and K. H. Downing. Model for the structure of bacteriorhodopsin based on high-resolution electron cryomicroscopy. *J. Mol. Biol.* 213 (1990): 899.

Popot, J.-L. Integral membrane protein structure: Transmembrane alpha helices as autonomous folding domains. *Curr. Opin. Struct. Biol.* 3 (1993): 512.

Reits, E. A. J., and J. J. Neefjes. From fixed to FRAP: Measuring protein mobility and activity in living cells. *Nature Cell Biol.* 3 (2001): E145.

Singer, S. J. The structure and insertion of integral proteins in membranes. *Annu. Rev. Cell Biol.* 6 (1990): 247.

Sussman, M. R. Molecular analysis of proteins in the plant plasma membrane. *Annu. Rev. Plant Physiol. Plant Mol. Biol.* 45 (1994): 211.

von Heijne, G. Membrane proteins: From sequence to structure. *Annu. Rev. Biophys. Biomol. Structure* 23 (1994): 167.

8 Transport Across Membranes: Overcoming the Permeability Barrier

In Chapter 7, we focused on the structure and chemistry of membranes. We noted that its hydrophobic interior makes a membrane an effective barrier to the passage of most molecules and ions, thereby keeping some substances inside the cell and others out. Within eukaryotic cells, membranes also delineate organelles by retaining the appropriate molecules, ions, enzymes, and other factors needed for specific functions.

However, it is not enough to think of membranes simply as permeability barriers. Crucial to the proper functioning of a cell or organelle is the ability to overcome the permeability barriers for specific molecules and ions so that they can be moved into and out of the cell or organelle selectively. In other words, not only are membranes barriers to the indiscriminate movement of substances into and out of cells and organelles, but they are also *selectively permeable*, allowing the controlled passage of specific molecules and ions from one side of the membrane to the other. In this chapter, we will look at the ways in which substances are moved selectively across membranes and consider the significance of such transport processes to the life of the cell.

Cells and Transport Processes

An essential feature of every cell and subcellular compartment is its ability to accumulate a variety of substances at concentrations that are often strikingly different from those in the surrounding milieu. Some of these substances are proteins and other macromolecules, which are moved into and out of cells and organelles by mechanisms we will

consider in later chapters. Specifically, Chapter 12 includes a discussion of *endocytosis* and *exocytosis,* bulk transfer processes whereby substances are moved into and out of cells as part of volumes of fluid enclosed within membrane-bounded vesicles. Mechanisms for the secretion of proteins from cells and for the import of proteins into organelles are discussed in Chapter 20.

As important as these topics are, most of the substances that move across membranes are not macromolecules or fluids but dissolved ions and small organic molecules—*solutes,* in other words. These solutes cross membranes not by bulk flow but single file, one ion or molecule at a time. Some of the more common ions transported across membranes are sodium (Na^+), potassium (K^+), calcium (Ca^{2+}), chloride (Cl^-), and hydrogen (H^+). Most of the small organic molecules are *metabolites*—substrates, intermediates, and products in the various metabolic pathways that take place within cells or specific organelles. Sugars, amino acids, and nucleotides are some common examples. For most such solutes, their concentration inside the cell or organelle is markedly higher than that outside. Very few cellular reactions or processes could occur at reasonable rates if they had to depend on the concentrations at which essential substrates are present in the cell's surroundings.

A central aspect of cell function, then, is **transport:** the ability to move ions and organic molecules across membranes selectively. The importance of membrane transport is evidenced by the fact that about 20% of the genes that have been identified in the bacterium *Escherichia coli* are involved in some aspect of transport. Figure 8-1 summarizes a few of the many transport processes that occur within eukaryotic cells.

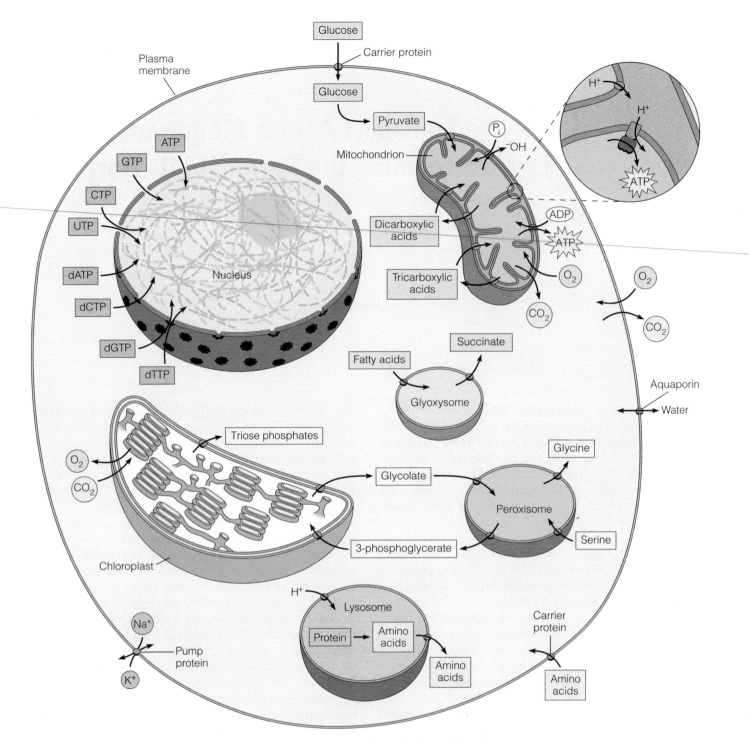

Figure 8-1 Transport Processes Within a Composite Eukaryotic Cell. The molecules and ions shown in this composite plant/animal cell are some of the many kinds of solutes that are transported across the membranes of eukaryotic cells. Note that the nucleoside triphosphate precursors to DNA and RNA enter the nucleus through nuclear pores. The enlargement at the upper right depicts a small portion of a mitochondrion, illustrating the outward pumping of protons and the use of the resulting electrochemical gradient to drive ATP synthesis.

Solutes Cross Membranes by Simple Diffusion, Facilitated Diffusion, and Active Transport

Three different mechanisms are involved in the movement of solutes across membranes. Certain small, nonpolar molecules such as oxygen, carbon dioxide, and ethanol move across membranes by *simple diffusion*—direct, unaided movement of solute molecules into and through the lipid bilayer in the direction dictated by the concentrations of the solute on the two sides of the membrane. For most solutes, however, movement across biological membranes at a significant rate is possible only because of the presence of transport proteins—membrane proteins that recognize substances with great specificity and speed their movement across the membrane. In some cases, transport proteins are involved in *facilitated diffusion* of solutes, moving them down their free-energy gradient (a gradient of concentration, charge, or most commonly, both) in the direction of thermodynamic equilibrium. In other cases, transport proteins mediate the *active transport* of solutes, moving them against their respective free-energy gradients, with transport driven by an energy-yielding process such as the hydrolysis of ATP or the concomitant transport of another solute, usually an ion such as H^+ or Na^+, down its free-energy gradient.

Active transport of ions is an especially important process. Recall from Chapter 5 that more than two-thirds of the energy your body expends in the resting state is used to maintain gradients of ions such as H^+, K^+, Na^+, and Ca^{2+} across the membranes of your cells. Because ions have an electrical charge, these ion gradients can create an electrical voltage, or **membrane potential**, across the membrane that makes one side of the membrane negative and the other side positive. The combined effect of the concentration gradient and membrane potential for an ion is referred to as the **electrochemical gradient** for that ion. The energy stored in electrochemical gradients is used to drive a variety of processes in cells, including the uptake of other solutes and the synthesis of ATP during the processes of cellular respiration and photosynthesis. (We will discuss these processes in Chapters 14 and 15, respectively.) In nerve cells, gradients of K^+ and Na^+ ions are responsible for the transmission of nerve impulses, as we will see in Chapter 9.

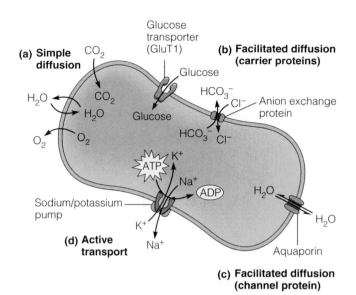

Figure 8-2 Important Transport Processes of the Erythrocyte. Depicted here are several types of transport that are vital to erythrocyte function. **(a)** *Simple diffusion:* Oxygen and carbon dioxide diffuse across the plasma membrane in response to their concentrations inside and outside the cell. **(b)** *Facilitated diffusion mediated by carrier proteins:* The movement of glucose across the plasma membrane is facilitated by a glucose transporter (GluT1) that recognizes only D-glucose and the D-isomers of a few closely related monosaccharides. An anion exchange protein facilitates the reciprocal transport of chloride (Cl^-) and bicarbonate (HCO_3^-) ions on a 1 : 1 basis. **(c)** *Facilitated diffusion mediated by channel proteins:* Aquaporin channel proteins facilitate the rapid inward or outward movement of water molecules, although water can also cross the membrane by simple diffusion. **(d)** *Active transport:* Driven by the hydrolysis of ATP, the Na^+/K^+ pump moves sodium ions outward and potassium ions inward.

The Erythrocyte Plasma Membrane Provides Examples of Transport Mechanisms

In our discussion of membrane transport processes, we will use as examples the erythrocyte transport proteins that we introduced in Chapter 7. These are some of the most extensively studied transport proteins and are therefore among the best understood. Vital to the erythrocyte's role in providing oxygen to body tissues is the movement across its plasma membrane of O_2, CO_2, and bicarbonate ion (HCO_3^-), as well as glucose, which serves as the cell's main energy source. Also important is the membrane potential maintained across the plasma membrane by the active transport of potassium ions inward and sodium ions outward. In addition, special pores, or channels, allow water and ions to enter and leave the cell rapidly in response to cellular needs. These transport activities are summarized in Figure 8-2 and will be used as examples in the following sections.

Simple Diffusion: Unassisted Movement Down the Gradient

The most straightforward way for a solute to get from one side of a membrane to the other is **simple diffusion,** which is the unassisted net movement of a solute from a region where its concentration is higher to a region where its concentration is lower (Figure 8-2a). Because membranes have a hydrophobic interior, simple diffusion is a relevant means of transport only for small, relatively nonpolar molecules. Such molecules simply permeate, or move into, the phospholipid bilayer on one side, diffuse passively across the bilayer, and emerge on the other side, again in aqueous solution.

Oxygen is an example of a molecule that moves across membranes by simple diffusion. This behavior enables erythrocytes in the circulatory system to take up oxygen in the lungs and release it again in body tissues. In the capillaries of body tissues, where oxygen concentration is low, oxygen is released from hemoglobin and diffuses passively from the cytoplasm of the erythrocyte into the blood plasma and from there into the cells that line the capillaries (Figure 8-3a). In the capillaries of the lungs, the opposite occurs: Oxygen diffuses from the inhaled air in the lungs, where its concentration is higher, into the cytoplasm of the erythrocytes, where its concentration is lower (Figure 8-3b). Carbon dioxide is also able to cross membranes by simple diffusion; however, most CO_2 is actually transported in the form of bicarbonate ion (HCO_3^-), as we will see later in the chapter. Not surprisingly, carbon dioxide

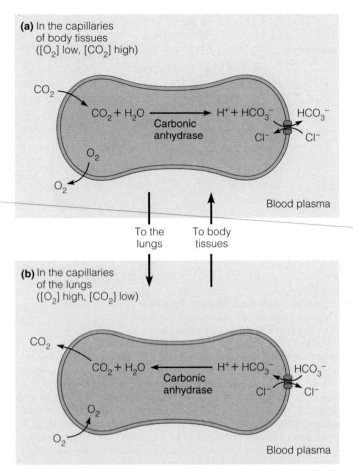

(a) In the capillaries of body tissues ([O_2] low, [CO_2] high)

CO_2

$CO_2 + H_2O \longrightarrow H^+ + HCO_3^-$
Carbonic anhydrase

HCO_3^-

Cl^- Cl^-

O_2

O_2

Blood plasma

To the lungs

To body tissues

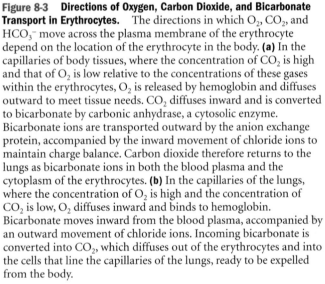

(b) In the capillaries of the lungs ([O_2] high, [CO_2] low)

CO_2

$CO_2 + H_2O \longleftarrow H^+ + HCO_3^-$
Carbonic anhydrase

HCO_3^-

Cl^- Cl^-

O_2

O_2

Blood plasma

Figure 8-3 Directions of Oxygen, Carbon Dioxide, and Bicarbonate Transport in Erythrocytes. The directions in which O_2, CO_2, and HCO_3^- move across the plasma membrane of the erythrocyte depend on the location of the erythrocyte in the body. **(a)** In the capillaries of body tissues, where the concentration of CO_2 is high and that of O_2 is low relative to the concentrations of these gases within the erythrocytes, O_2 is released by hemoglobin and diffuses outward to meet tissue needs. CO_2 diffuses inward and is converted to bicarbonate by carbonic anhydrase, a cytosolic enzyme. Bicarbonate ions are transported outward by the anion exchange protein, accompanied by the inward movement of chloride ions to maintain charge balance. Carbon dioxide therefore returns to the lungs as bicarbonate ions in both the blood plasma and the cytoplasm of the erythrocytes. **(b)** In the capillaries of the lungs, where the concentration of O_2 is high and the concentration of CO_2 is low, O_2 diffuses inward and binds to hemoglobin. Bicarbonate moves inward from the blood plasma, accompanied by an outward movement of chloride ions. Incoming bicarbonate is converted into CO_2, which diffuses out of the erythrocytes and into the cells that line the capillaries of the lungs, ready to be expelled from the body.

and oxygen move across the erythrocyte membrane in opposite directions, with carbon dioxide diffusing inward in body tissues and outward in the lungs.

Diffusion Always Moves Solutes Toward Equilibrium

No matter how a population of molecules is distributed initially, diffusion always tends to create a random solution in which the concentration is the same everywhere.

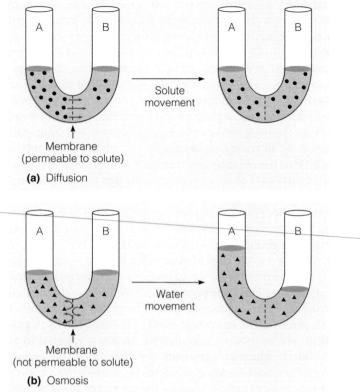

A B → A B
Solute movement
Membrane (permeable to solute)
(a) Diffusion

A B → A B
Water movement
Membrane (not permeable to solute)
(b) Osmosis

Figure 8-4 Comparison of Simple Diffusion and Osmosis. **(a)** Simple diffusion takes place when the membrane that separates chambers A and B is permeable to molecules of dissolved solute, represented by the black dots. Solute molecules cross the membrane in both directions, but with net movement from chamber A to B. Equilibrium is reached when the solute concentration is the same in both chambers. **(b)** To understand osmosis, assume that the membrane between the two chambers is not permeable to the dissolved solute, represented by the black triangles. Instead, water diffuses from the chamber where the solute concentration is lower to the chamber where the solute concentration is higher. At equilibrium, the pressure that builds up in the chamber containing the excess water exactly counterbalances the tendency of water to continue diffusing into that chamber.

To illustrate this point, consider the apparatus shown in Figure 8-4a, consisting of two chambers separated by a membrane that is freely permeable to molecules of S, an uncharged solute represented by the black dots. Initially, chamber A has a higher concentration of S than does chamber B. Other conditions being equal, random movements of solute molecules back and forth through the membrane will lead to a net movement of solute S from chamber A to chamber B. When the concentration of S is equal on both sides of the membrane, the system is at equilibrium; random back-and-forth movement of individual molecules continues, but no further net change in concentration occurs. Thus, *diffusion is always movement toward equilibrium.*

Another way to express this is to say that diffusion always tends toward minimum free energy. As we learned in Chapter 5, chemical reactions and physical processes always proceed in the direction of decreasing free energy, in accord with the second law of thermodynamics. Diffusion through membranes is no exception: Free energy is minimized as molecules flow down their concentration gradient and as ions flow down their electrochemical gradient. We will apply this principle later in the chapter when we calculate ΔG, the free energy change that accompanies the transport of molecules and ions across membranes. In most membrane transport processes, the free energy change depends on the concentration or electrochemical gradient alone, but factors such as heat, pressure, and entropy also contribute to the total free energy and may in some cases need to be taken into consideration as well. Strictly speaking, therefore, *diffusion always proceeds from regions of higher to lower free energy.* At thermodynamic equilibrium, no further net movement occurs because the free energy of the system is at a minimum.

Osmosis Is the Diffusion of Water Across a Differentially Permeable Membrane

Several properties of water cause it to behave in a special way. First of all, water molecules are not charged, so they are not affected by the membrane potential. Moreover, the concentration of water is not appreciably different on opposite sides of a membrane. What, then, determines the direction in which water molecules diffuse? When a solute is dissolved in water, the solute molecules disrupt the ordered three-dimensional interactions that normally occur between individual molecules of water, thereby increasing the entropy and decreasing the free energy of the solution. Water, like other substances, tends to diffuse from areas where its free energy is higher to areas where its free energy is lower. Thus, water tends to move from regions of lower solute concentration (higher free energy) to regions of higher solute concentration (lower free energy).

This principle is illustrated by Figure 8-4b. Solutions of differing solute concentrations are placed in chambers A and B as in Figure 8-4a, but with the two chambers now separated by a *differentially permeable membrane* that is permeable to water but *not* to the dissolved solute. Under these conditions, water moves, or diffuses, across the membrane from chamber B to chamber A. Such movement of water in response to differences in solute concentration is called **osmosis.** Osmotic movement of water across a mem-

brane is always from the side with the higher free energy to the side with the lower free energy. For most cells, this means that water will tend to move inward because the concentration of solutes is almost always higher inside a cell than outside. If not controlled, the inward movement of water would cause cells to swell and perhaps to burst. For further discussion of osmotic water movement and the means whereby organisms from different kingdoms control water content and movement, see Box 8A.

Simple Diffusion Is Limited to Small, Nonpolar Molecules

To investigate the factors that influence the diffusion of solutes through membranes, scientists frequently use membrane models. An important advance in developing such models was provided by Alec Bangham and his colleagues, who found that when lipids are extracted from cell membranes and dispersed in water, they form *liposomes,* which are small vesicles about 0.1 μm in diameter, each consisting of a closed spherical lipid bilayer that is devoid of membrane proteins. Bangham showed that it is possible to trap solutes such as potassium ions in the liposomes as they form, and then measure the rate at which the solute escapes by diffusion across the liposome bilayer.

The results were remarkable: Ions such as potassium and sodium were trapped in the vesicles for days, whereas small uncharged molecules such as oxygen exchanged so rapidly that the rates could hardly be measured. The inescapable conclusion was that the lipid bilayer represents the primary permeability barrier of a membrane. Small uncharged molecules can pass through the barrier by simple diffusion, whereas sodium and potassium ions can hardly pass at all. Based on subsequent experiments by many investigators using a variety of lipid bilayer systems and thousands of different solutes, we can predict with considerable confidence how readily a solute will diffuse across a lipid bilayer. As Table 8-1 indicates, the three main factors affecting diffusion of solutes are size, polarity, and whether the solute is an ion. We will consider each of these factors in turn.

Solute Size.   Generally speaking, lipid bilayers are more permeable to smaller molecules than to larger molecules. The smallest molecules relevant to cell function are water,

Table 8-1   Factors Governing Diffusion Across Lipid Bilayers

| Factor | Examples | | Permeability Ratio* |
	More Permeable	Less Permeable	
1. Size: bilayer more permeable to smaller molecules	H_2O (Water)	$H_2N-CO-NH_2$ (Urea)	$10^2 : 1$
2. Polarity: bilayer more permeable to nonpolar molecules	$CH_3-CH_2-CH_2-OH$ (Propanol)	$HO-CH_2-CHOH-CH_2-OH$ (Glycerol)	$10^3 : 1$
3. Ionic: bilayer highly impermeable to ions	O_2 (Oxygen)	OH^- (Hydroxide ion)	$10^9 : 1$

*Ratio of diffusion rate for the more permeable solute to the less permeable solute.

Further Insights OSMOSIS: THE SPECIAL CASE OF WATER DIFFUSION

Most of the discussion in this chapter focuses on the transport of *solutes*—ions and small molecules that are dissolved in the aqueous milieu of cells, their organelles, and their surroundings. That emphasis is quite appropriate because most of the traffic across membranes involves ions such as K^+, Na^+, and H^+, and hydrophilic molecules such as sugars, amino acids and a variety of metabolic intermediates. But to fully understand solute transport, we need to understand the forces that act on water, thereby determining its movement within and between cells. As we do, we will recognize that water, the universal solvent of the biological world, represents a special case in several ways.

The movement of most substances across a membrane can usually be understood in terms of transmembrane concentration gradients and, for charged solutes, the membrane potential as well. Not so for water, however; its concentration is essentially the same on the two sides of a membrane, and as an uncharged molecule it is unaffected by membrane potential. Instead, water tends to move across membranes in response to differences in solute concentrations. Specifically, water tends to diffuse from the side of the membrane with the lower solute concentration to the side with the higher solute concentration. This diffusion of water, called *osmosis*, is readily observed when a differentially permeable membrane separates two compartments, one of which contains a solute to which the membrane is not permeable (see Figure 8-4b).

Osmotic movement of water into and out of a cell is related to the **osmolarity,** or relative solute concentration, of the solution in which the cell finds itself. A solution with a higher solute concentration than that inside a cell is called a *hypertonic solution,* whereas one with a solute concentration lower than that inside a cell is referred to as a *hypotonic solution.* Hypertonic solutions cause water molecules to diffuse out of the cell, whereas hypotonic solutions cause water to diffuse into the cell. Osmotic movement of water, in other words, is always from a hypotonic

to a hypertonic solution. A solution that triggers no net flow of water is called an *isotonic* solution.

Osmosis accounts for a well-known observation: Cells tend to shrink or swell as the solute concentration of the extracellular medium changes. Consider, for example, the scenario of Figure 8A-1. An animal cell that starts out in an isotonic solution (0.25 *M* sucrose is isotonic or nearly so for many cells) will shrink and shrivel if it is transferred to a hypertonic solution (Figure 8A-1a). On the other hand, the cell will swell if placed in a hypotonic solution and will actually lyse, or burst, if placed in a very hypotonic solution, such as water containing no solutes (Figure 8A-1b).

Osmolarity: A Common Problem With Different Solutions

The osmotic movements of water shown in Figure 8A-1 occur because of differences in the osmolarity of the cytoplasm on the inside of the membrane and the solution on the outside. Under most conditions, the concentration of solutes is greater inside a cell than outside. This is due in part to the high concentrations of ions and small organic molecules required for normal metabolic processes and other cellular functions. In addition, most of these metabolites are charged, as are most biological macromolecules, and the *counterions* required to balance these charges also contribute significantly to the intracellular osmolarity. As a result, most cells are hypertonic compared with their surroundings, which means that water will tend to move inward across the plasma membrane. Left unchecked, this tendency would cause cells to swell and possibly lyse.

How cells cope with the problem of high osmolarity and the resulting osmotic influx of water depends mainly on the kingdom to which the organism belongs. Cells of plants, algae, fungi, and many bacteria are surrounded by cell walls that are sufficiently rigid and sturdy to keep the cells from swelling and bursting in a hypotonic solution (Figure 8A-1d). Instead, the cells—and the tissues, in the case of a multicellular organism such

oxygen, and carbon dioxide. Membranes are quite permeable to these molecules; no specialized transport processes are required to move them into and out of cells. But even such small molecules do not move across membranes freely. Water molecules, for example, diffuse across a bilayer 10,000 times slower than they move when allowed to diffuse freely in the absence of a membrane! The best way to think about the passive movement of small molecules across membranes is that their diffusion is strongly hindered by the presence of the lipid bilayer, but they occasionally permeate the bilayer and diffuse randomly to the other side, where they leave the membrane again.

The size rule holds for molecules up to about the size of glucose. Ethanol (CH_3CH_2OH; 46 daltons) and glycerol ($C_3H_8O_3$; 92 daltons) are able to diffuse across membranes at reasonable rates, but glucose ($C_6H_{12}O_6$; 180 daltons) cannot. Most cells therefore need specialized proteins in

their plasma membranes to facilitate the entry of glucose and other large nutrient molecules.

Solute Polarity. In general, lipid bilayers are relatively permeable to nonpolar molecules and less permeable to polar molecules. This is because nonpolar molecules dissolve more readily in the hydrophobic phase of the lipid bilayer and can therefore cross the membrane much more rapidly than polar molecules of similar size. As you may recall from Chapter 7, the relationship that Charles Overton discovered between the lipid solubility of various solutes and their membrane permeability was one of the first indications that a nonpolar hydrophobic phase might be present in membranes.

Figure 8-5 illustrates the relationship between the hydrophobicity of a solute as measured by its relative solubility in a nonpolar solvent (olive oil) and its rate of diffusion across a membrane. In general, the more hydro-

as a plant—become very firm as a result of the **turgor pressure** that builds up due to the inward movement of water. At equilibrium, the turgor pressure forces out as much water as the high solute concentration draws in. The resulting turgidity accounts for the firmness, or *turgor,* of fully hydrated plant tissue.

In a hypertonic solution, on the other hand, the outward movement of water causes the plasma membrane to pull away from the cell wall, a process called **plasmolysis** (Figure 8A-1c). The wilting of a plant or a plant part under conditions of water deprivation is due to the plasmolysis of its cells. You can demonstrate plasmolysis readily by dropping a piece of celery into a solution with a high concentration of salt or sugar. Plasmolysis can be a practical problem when plants are grown under conditions of high salinity, as is sometimes the case in locations near an ocean.

Animal cells, and other cells without walls, solve the osmolarity problem by continuously and actively pumping out inorganic ions, thereby reducing the intracellular osmolarity and thus minimizing the difference in solute concentration between the cell and its surroundings. Animal cells continuously remove sodium ions; this is, in fact, the main purpose of the Na^+/K^+ *pump* (see Figures 8-11 and 8-12). The importance of the Na^+/K^+ pump in regulating cell volume is indicated by the observation that animal cells swell and sometimes even lyse when treated with *ouabain,* an inhibitor of the Na^+/K^+ pump. Thus, cells without walls must expend significant amounts of energy to ensure that the intracellular osmolarity remains low and the cell does not swell and burst, whereas cells with walls can tolerate considerable osmotic pressure without danger of rupture.

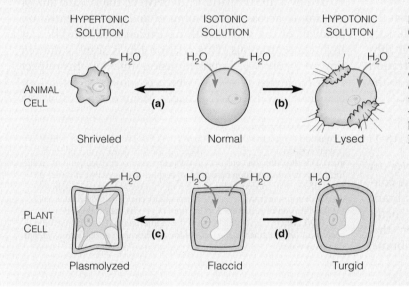

Figure 8A-1 Responses of Animal and Plant Cells to Changes in Osmolarity. (a) When an animal cell (or other cell not surrounded by a cell wall) is transferred from an isotonic solution to a hypertonic solution, water leaves the cell and the cell shrivels. **(b)** In a hypotonic solution, water enters the cell and the cell swells, sometimes to the point that it bursts. **(c)** Plant cells (or other cells with rigid cell walls) also shrivel (plasmolyze) in a hypertonic solution, but **(d)** they will only become turgid—not burst—in a hypotonic solution.

phobic, or nonpolar, a substance is, the more readily and rapidly it can move across a membrane. Note, for example, that a polar compound like urea has a relatively low membrane permeability. However, if it is made less polar by adding methyl groups to form dimethylurea, both its solubility in lipid and its membrane permeability increase.

Ion Permeability. As a general rule, lipid bilayers are very impermeable to ions. This is because a great deal of energy (about 40 kcal/mol) is required to move ions from an aqueous environment into a nonpolar environment. The impermeability of membranes to ions is very important to cell activity, because every cell must maintain an ion gradient across its plasma membrane in order to function. In most cases, this is a gradient either of sodium ions (animal cells) or protons (most other cells). Similarly, mitochondria and chloroplasts depend on proton gradients for their function. On the other hand, membranes must also allow

ions to cross the barrier in a controlled manner. As we will see later in the chapter, the proteins that facilitate ion transport provide hydrophilic channels that shield the ions from the hydrophobic interior of the membrane.

The Rate of Simple Diffusion Is Directly Proportional to the Concentration Gradient

So far, we have focused on qualitative aspects of simple diffusion. We can be more quantitative by considering the thermodynamic and kinetic properties of the process (Table 8-2). Thermodynamically, simple diffusion is always an exergonic process, requiring no input of metabolic energy. Individual molecules simply diffuse randomly in both directions, but net flux will always be in the direction of minimum free energy—which in the case of uncharged molecules means down the concentration gradient.

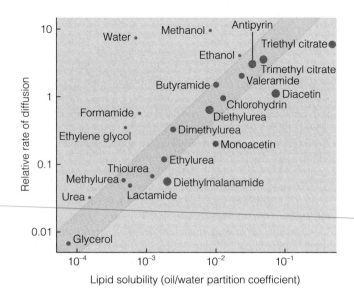

Figure 8-5 **Relationship Between Lipid Solubility of a Solute and Its Rate of Diffusion Across a Membrane.** The rates at which various solutes cross the plasma membrane are plotted against their lipid solubility. A logarithmic scale is used for both axes to accommodate the wide (~10,000-fold) range of values. The diameter of each point is roughly proportional to the size of the molecule it represents. Note that in spite of the general correlation between diffusion rate and lipid solubility, small deviations occur. Those molecules penetrating faster than expected (above the shaded area) are relatively small, whereas those penetrating more slowly than expected (below the shaded area) are relatively large.

Kinetically, a key feature of simple diffusion is that the net rate of transport for a specific substance is directly proportional to the concentration difference for that substance across the membrane over a broad concentration range. For the diffusion of solute S from the outside to the inside of a cell, the expression for the rate, or velocity of inward diffusion through the membrane, v_{inward}, is

$$v_{inward} = P \, \Delta S \qquad (8\text{-}1)$$

where v_{inward} is the rate of inward diffusion (in moles per second per cm² of membrane surface) and ΔS is the concentration gradient of the solute across the membrane ($\Delta S = [S]_{outside} - [S]_{inside}$). P is the permeability coefficient, an experimentally determined parameter that depends on the thickness and viscosity of the membrane;

the size, shape, and polarity of S; and the solubility of S in the membrane. As defined, v_{inward} is actually the inward *flux* of the solute because flux is a measure of the amount of solute moving per unit area per unit time.*

As equation 8-1 indicates, simple diffusion is characterized by a linear relationship between the inward flux of the solute across the membrane and the concentration gradient of the solute, with no evidence of saturation at high concentrations. This relationship is seen as the red line on Figure 8-6. Simple diffusion differs in this respect from facilitated diffusion, which is subject to saturation and generally follows Michaelis-Menten kinetics, as we

---

* Equation 8-1 is a modified version of *Fick's first law of diffusion*, which is usually written as J = −DΔC/Δx, where J is the flux density (flux per unit area), D is the diffusion constant (usually expressed as cm²/sec), and ΔC is the difference in concentration between two regions separated by a distance Δx (the difference in concentration across a membrane of thickness Δx, in the case of membrane transport). The permeability coefficient P is related to D for the diffusion of substance S in the lipid bilayer by the expression P = KD/L, where L is the thickness of the membrane and K is the partition coefficient for S, defined as the ratio of the equilibrium concentrations of S in oil versus water.

Table 8-2 **Comparison of Simple Diffusion, Facilitated Diffusion, and Active Transport**

Properties		Simple Diffusion	Facilitated Diffusion	Active Transport
Solutes transported	**Examples**			
Small nonpolar	Oxygen	Yes	No	No
Large nonpolar	Fatty acids	Yes	No	No
Small polar	Water	Yes	No	No
Large polar	Glucose	No	Yes	Yes
Ions	N⁺, K⁺, Ca²⁺	No	Yes	Yes
Thermodynamic properties				
Direction relative to electrochemical gradient		Down	Down	Up
Effect on entropy		Increased	Increased	Decreased
Metabolic energy required		No	No	Yes
Intrinsic directionality		No	No	Yes
Kinetic properties				
Membrane protein-mediated		No	Yes	Yes
Saturation kinetics		No	Yes*	Yes
Competitive inhibition		No	Yes	Yes

*For channel proteins, saturation occurs at much higher (nonphysiological) concentrations of solute than it does for carrier proteins.

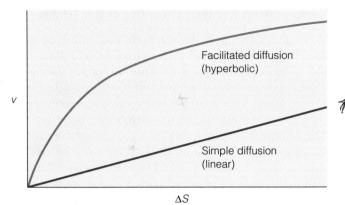

Figure 8-6 Comparison of the Kinetics of Simple Diffusion and Facilitated Diffusion.   For simple diffusion across a membrane, the relationship between v, the rate of diffusion, and ΔS, the solute concentration gradient, is linear over a broad concentration range (red line). For facilitated diffusion, the relationship exhibits saturation kinetics and is therefore hyperbolic (green line). For simplicity, the initial solute concentration is assumed to be [S] on one side of the membrane and zero on the other side.

will see shortly. Simple diffusion can therefore be distinguished from facilitated diffusion by its kinetic properties, as indicated in Table 8-2.

We can summarize simple diffusion by noting that it is relevant only to molecules, such as ethanol and O_2, that are small enough and/or nonpolar enough to cross membranes at a reasonable rate without the aid of transport proteins. Simple diffusion proceeds exergonically in the direction dictated by the concentration gradient, with a linear, nonsaturating relationship between the diffusion rate and the concentration gradient.

Facilitated Diffusion: Protein-Mediated Movement Down the Gradient

Most substances in cells are too large or too polar to cross membranes at reasonable rates by simple diffusion even if the process is exergonic. Such solutes can move into and out of cells and organelles at appreciable rates only with the assistance of transport proteins that mediate the movement of solute molecules across the membrane. If such a process is exergonic, it is called **facilitated diffusion** because the solute still diffuses down the concentration or electrochemical gradient, but with no input of energy needed. The role of the transport protein is simply to facilitate the diffusion of a polar or charged solute across an otherwise impermeable barrier.

As an example of facilitated diffusion, consider the movement of glucose across the plasma membrane of an erythrocyte (or almost any other cell in your body, for that matter). The concentration of glucose is higher in the blood than in the erythrocyte, so the transport of glucose

across the plasma membrane of the cell is passive. However, glucose is too large and too polar to diffuse across the membrane unaided; a transport protein is required to facilitate its inward movement (see Figure 8-2b).

Carrier Proteins and Channel Proteins Facilitate Diffusion by Different Mechanisms

Transport proteins involved in facilitated diffusion of small molecules and ions are integral membrane proteins that contain several, or even many, transmembrane segments and therefore traverse the membrane multiple times. Functionally, these proteins fall into two main classes that transport solutes in quite different ways. Carrier proteins (also called *transporters* or *permeases*) bind one or more solute molecules on one side of the membrane and then undergo a conformational change that transfers the solute to the other side of the membrane. In so doing, a carrier protein presumably binds the solute molecules in such a way as to shield the polar or charged groups of the solute from the nonpolar interior of the membrane.

Channel proteins, on the other hand, form hydrophilic *channels* through the membrane that allow the passage of solutes without any change in the conformation of the protein. Some of these channels are relatively large and nonspecific, such as the *pores* found in the outer membranes of bacteria, mitochondria, and chloroplasts. Pores are formed by transmembrane proteins called *porins* and allow selected hydrophilic solutes with molecular weights up to about 600 Da to diffuse across the membrane. However, most channels are small and highly selective. Most of these smaller channels are involved in the transport of ions rather than molecules, and are therefore referred to as *ion channels*. The movement of solutes through ion channels is much more rapid than transport by carrier proteins, presumably because complex conformational changes are not necessary.

Carrier Proteins Alternate Between Two Conformational States

An important topic of contemporary membrane research concerns the mechanisms whereby transport proteins facilitate the movement of solutes across membranes. For carrier proteins, the most likely explanation is the **alternating conformation model** proposed by S. Jonathan Singer and others. According to this model, a carrier protein is an allosteric protein that alternates between two conformational states, such that the solute-binding site of the protein is open or accessible first to one side of the membrane and then to the other. Binding of a solute molecule or ion to the protein on one side of the membrane triggers a conformational change in the protein that opens the binding site to the other side of the membrane, thereby transferring the solute across the membrane. We will encounter an example of this mechanism shortly when we discuss the facilitated diffusion of glucose into erythrocytes.

Carrier Proteins Are Analogous to Enzymes in Their Specificity and Kinetics

As we noted earlier, carrier proteins are sometimes called *permeases*. This term is apt because the suffix -*ase* suggests a similarity between carrier proteins and enzymes that is in fact valid. Like an enzyme-catalyzed reaction, facilitated diffusion involves an initial binding of the solute to a specific site on a protein surface, the subsequent release of product, and a reduction in the activation energy of the reaction due to the involvement of the protein catalyst.

Specificity of Carrier Proteins. Another property that carrier proteins share with enzymes is *specificity*. Like enzymes, transport proteins are highly specific, often for a single compound or a small group of closely related compounds and sometimes even for a specific stereoisomer. A good example is the carrier protein that facilitates the diffusion of glucose into erythrocytes (see Figure 8-2b). This protein recognizes only glucose and a few closely related monosaccharides, such as galactose and mannose. Moreover, the protein is *stereospecific:* It accepts the D- but not the L-isomer of these sugars. This specificity is presumably a result of the precise stereochemical fit between the solute and its binding site on the carrier protein.

Kinetics of Carrier Protein Function. As you might expect from the analogy with enzymes, carrier proteins become saturated as the concentration of the transportable solute is raised. This is because the number of transport proteins is limited and each has some finite maximum velocity at which it can function. As a result, carrier-facilitated transport, like enzyme catalysis, exhibits *saturation kinetics* (p. 140), with an upper limiting velocity V_{max} and a constant K_m corresponding to the concentration of transportable solute needed to achieve one-half of the maximum rate of transport (see equation 6-7 on p. 142). This means that the rate of solute transport v can be described mathematically by an equation similar to that used for enzyme kinetics:

$$v = \frac{V_{max} \Delta S}{K_m + \Delta S} \qquad (8\text{-}2)$$

where ΔS is the concentration gradient across the membrane. A plot of transport rate versus the solute concentration gradient is therefore hyperbolic for facilitated diffusion instead of linear as for simple diffusion (see Figure 8-6, green line). As we have already noted, this difference is an important means of distinguishing between simple and facilitated diffusion (see Table 8-2).

A further similarity with enzymes is that carrier proteins are often subject to *competitive inhibition* by molecules or ions that are structurally related to the intended "substrate." For example, the transport of glucose by a glucose carrier protein is competitively inhibited by the other monosaccharides that the protein also accepts—that is, the rate of glucose transport is reduced in the presence of other transportable sugars.

Carrier Proteins Transport Either One or Two Solutes

Although carrier proteins resemble each other in their kinetics and their presumed mechanism of action, they also differ in significant ways. One important difference concerns the number of solutes transported and the direction in which they move. When a carrier protein transports a single solute across the membrane, the process is called **uniport** (Figure 8-7a). The glucose carrier protein shown in Figure 8-8 is a uniporter. When two solutes are transported simultaneously and their transport is coupled such that transport of either stops if the other is absent, the process is referred to as **cotransport** or *coupled transport* (Figure 8-7b). In cotransport, the process is called **symport** if the two solutes are moved in the same direction or **antiport** if the two solutes are moved in opposite directions. As we will see later, these same terms apply whether the mode of transport is facilitated diffusion or active transport; the term does not indicate anything about the energetics of the process.

The Erythrocyte Glucose Transporter and Anion Exchange Protein Are Examples of Carrier Proteins

Now that we have described the general properties of carrier proteins, let us briefly consider two specific examples: the uniport carrier for glucose and the antiport anion carrier.

The Glucose Transporter: A Uniport Carrier. As we noted earlier, the movement of glucose into an erythrocyte is an example of facilitated diffusion mediated by a uniport carrier protein (Figure 8-2b). The concentration of glucose in the blood plasma is usually in the range of 65–90 mg/100 mL, or about 3.6–5.0 m*M*. The erythrocyte

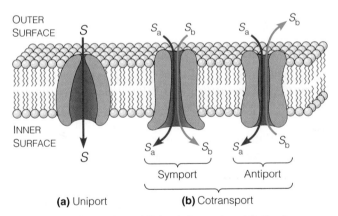

Figure 8-7 A Comparison of Uniport, Symport, and Antiport Transport. **(a)** In uniport, a membrane transport protein moves a single solute across a membrane. **(b)** Cotransport involves the concomitant transport of two solutes, S_a and S_b. Cotransport may be either symport (both solutes moved in the same direction) or antiport (the two solutes moved in opposite directions). These terms are used for both facilitated diffusion and active transport; they therefore tell us nothing about the energetics of the process.

(or almost any other cell in contact with the blood, for that matter) is capable of glucose uptake by facilitated diffusion because of its low intracellular glucose concentration and the presence in its plasma membrane of a glucose carrier protein, or **glucose transporter (GluT).** The GluT of erythrocytes is called GluT1 to distinguish it from related GluTs in other mammalian tissues. GluT1 allows glucose to enter the cell about 50,000 times faster than it would enter by free diffusion through a lipid bilayer.

GluT1-mediated uptake of glucose displays all of the classic features of facilitated diffusion: It is specific for glucose (and a few related sugars, such as galactose and mannose), exhibits saturation kinetics, and is susceptible to competitive inhibition by related monosaccharides. GluT1 is an integral membrane protein with 12 hydrophobic transmembrane segments. These are presumably folded and assembled in the membrane to form a cavity lined with hydrophilic side chains that form hydrogen bonds with glucose molecules as they move through the membrane.

GluT1 is thought to transport glucose by an alternating conformation mechanism, as illustrated in Figure 8-8. The two conformational states are called T_1, which has the binding site for glucose open to the outside of the cell, and T_2, with the site open to the interior of the cell. The process begins when a molecule of glucose collides with a GluT1 molecule in the T_1 conformation ①. Binding of glucose causes the protein to shift to the T_2 conformation, with the binding site now facing the inner surface of the membrane ②. The conformational change facilitates the release of glucose and the GluT1 molecule returns to its original conformation, with the binding site again facing outward ③.

The example shown in Figure 8-8 is for inward transport, but the process is readily reversible because carrier proteins function equally well in either direction. A carrier protein is really just a gate in an otherwise impenetrable wall, and, like most gates, it facilitates traffic in either direction. Individual solute molecules may be transported either inward or outward, with the net direction of solute movement determined by the relative concentrations of the solute on the two sides of the membrane. If the concentration is higher outside, net flow will be inward; if the higher concentration occurs inside, net flow will be outward.

GluT1 is just one of several glucose transporters in mammals. Each of these proteins is encoded by a separate gene and each has physical and kinetic characteristics that suit it especially for the specific tissues in which it is found. For example, GluT2, the glucose transporter present in liver cells, has kinetic properties that adapt it well to its role in transporting glucose out of the cells when liver glycogen is broken down to replenish the supply of glucose in the blood.

The low intracellular glucose concentration that makes facilitated diffusion possible for most animal cells exists because incoming glucose is quickly phosphorylated to glucose-6-phosphate by the enzyme hexokinase, with ATP as the phosphate donor and energy source:

$$\text{glucose} \xrightarrow[\text{hexokinase}]{\text{ATP} \quad \text{ADP}} \text{glucose-6-phosphate} \qquad \textbf{(8-3)}$$

This hexokinase reaction is the first step in glucose metabolism, which we will discuss in Chapter 13. The low K_m of hexokinase for glucose (1.5 mM) and the highly exergonic nature of the reaction ($\Delta G^{\circ\prime} = -4.0$ kcal/mol) ensure that the concentration of glucose within the cell is kept low. For many mammalian cells, the glucose concentration inside the cell ranges from 0.5 to 1.0 mM, about 15–20% of the glucose level in the blood plasma outside the cell.

The phosphorylation of glucose also has the effect of locking glucose in the cell, because the plasma membrane of the erythrocyte does not have a transport protein for glucose-6-phosphate. (GluT1, like most sugar transporters, does not recognize the phosphorylated form of the sugar.) This is, in fact, a general strategy for retaining molecules within the cell because most cells do not have membrane proteins capable of transporting phosphorylated compounds. For example, all of the intermediates beyond glucose-6-phosphate in the pathway for glucose

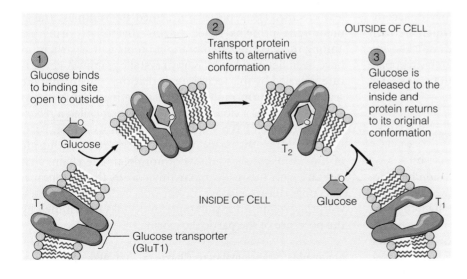

Figure 8-8 The Alternating Conformation Model for Facilitated Diffusion of Glucose by the Glucose Transporter GluT1. GluT1, the glucose transporter present in the erythrocyte plasma membrane, is a transmembrane protein that moves glucose molecules across the membrane by alternating between two conformations, called T_1 and T_2. The transport process is shown here in three steps, arranged around the periphery of a cell. ① With GluT1 in its T_1 conformation, a molecule of D-glucose collides with and binds to the binding site on the protein. ② Binding of glucose causes the transporter to shift to its alternate (T_2) conformation, with the binding site now open to the inside of the cell. ③ As glucose is released from the binding site to the inside, the GluT1 protein reverts to its original (T_1) conformation, ready for a further transport cycle.

metabolism are also phosphorylated compounds and are therefore effectively trapped inside the cell.

The Erythrocyte Anion Exchange Protein: An Antiport Carrier. Another well-studied example of facilitated diffusion involves the transport of chloride and bicarbonate ions by the **anion exchange protein** of the erythrocyte plasma membrane (see Figure 8-2b). This antiport protein, also called the *chloride-bicarbonate exchanger* (or *band 3 protein*; see Figure 7-20b), facilitates the reciprocal exchange of chloride (Cl^-) and bicarbonate (HCO_3^-) ions across the plasma membrane. The coupling of chloride and bicarbonate transport is obligatory. Moreover, the anion exchange protein is very selective; it exchanges bicarbonate for chloride in a strict 1 : 1 ratio and it accepts no other anions.

Like the glucose transporter, the anion exchange protein is thought to function by alternating between two conformational states. The major difference is that the solute binding site of the anion exchange protein interacts with different ions on opposite sides of the membrane. In cells where the bicarbonate concentration is high, the binding site of the anion exchange protein binds to bicarbonate at the interior membrane surface and chloride at the exterior surface. In cells where the bicarbonate concentration is low, the reciprocal process occurs: bicarbonate ions bind at the exterior membrane surface and chloride ions bind at the interior surface.

The rapid bidirectional movement of bicarbonate across the plasma membrane made possible by the anion exchange protein plays a central role in the process by which CO_2 produced in metabolically active tissues is delivered to the lung (see Figure 8-3). In gaseous form, CO_2 is not very soluble in aqueous solutions such as cytoplasm or blood plasma. Instead, CO_2 diffuses from tissues into erythrocytes, where the cytosolic enzyme *carbonic anhydrase* converts it to bicarbonate ions. As the concentration of bicarbonate in the erythrocyte rises, it diffuses out of the cell. To prevent a net charge imbalance, the expulsion of each negatively charged bicarbonate ion is accompanied by the uptake of one negatively charged chloride ion. In the lungs, this entire process is reversed; chloride ions are transported out of erythrocytes accompanied by the uptake of bicarbonate ions, which are then converted back to CO_2 by carbonic anhydrase. The net result is the movement of CO_2 (in the form of bicarbonate ions) from tissues to the lung, where the CO_2 is exhaled from the body.

Channel Proteins Facilitate Diffusion by Forming Hydrophilic Transmembrane Channels

While some transport proteins facilitate diffusion by functioning as carrier proteins that alternate between different conformational states, others do so by forming hydrophilic *transmembrane channels* that allow specific solutes—mainly ions—to move across the membrane directly. We will consider three kinds of transmembrane protein channels: *ion channels*, *porins*, and *aquaporins*.

Ion Channels: Transmembrane Proteins That Allow Rapid Passage of Specific Ions. Despite their apparently simple design—just a tiny pore lined with hydrophilic amino acid side chains—**ion channels** are remarkably selective. Most allow passage of only one kind of ion, so separate channels are needed for transporting such ions as Na^+, K^+, Ca^{2+}, and Cl^-. This selectivity is remarkable given the small differences in size and charge among these ions. The underlying mechanism is not yet well understood; a reasonable hypothesis is that selectivity may involve both ion-specific binding sites and a constricted center that serves as a size filter. The rate of transport is equally remarkable: In some cases, a single channel can conduct almost a million ions per second!

Most ion channels are *gated*, which means that they can be opened and closed by conformational changes in the protein, thereby regulating the flow of ions through the channel. (Notice that this regulatory role is quite different from the role of conformational changes in carrier proteins, where they are an inherent part of the transport mechanism.) In animal cells, three different kinds of stimuli control the opening and closing of gated channels: *Voltage-gated channels* open and close in response to changes in the membrane potential; *ligand-gated channels* are triggered by the binding of specific substances to the channel protein; and *mechanosensitive channels* respond to mechanical forces that act on the membrane.

Regulation of ion movement across membranes plays an important role in many types of cellular communication. For example, as we will see in the next chapter, the transmission of electrical signals by nerve cells depends critically on rapid, controlled changes in the movement of Na^+ and K^+ ions through their respective channels. These changes are so rapid that they are measured in milliseconds. In addition to such short-term regulation, most ion channels are also subject to longer-term regulation, usually in response to external stimuli such as hormones.

Porins: Transmembrane Proteins That Allow Rapid Passage of Various Solutes. Compared with ion channels, the pores found in the outer membranes of mitochondria, chloroplasts, and many bacteria are somewhat larger and much less specific. These pores are formed by multipass transmembrane proteins called **porins.** Bacterial porins are among the few transmembrane proteins whose structures have been determined by X-ray crystallography. A key feature revealed by this technique is that the transmembrane segments of porin molecules cross the membrane not as an α helix but as a closed cylindrical *β sheet* called a *β barrel*. The β barrel has a water-filled pore at its center. Polar side chains line the inside of the pore, whereas the outside of the barrel consists mainly of nonpolar side chains that interact with the hydrophobic interior of the membrane. The pore allows passage of various hydrophilic solutes, with the size limit for the solute molecules determined by the pore size of the particular porin.

Aquaporins: Transmembrane Channels That Allow Rapid Passage of Water. Water crosses both natural and artificial membranes at rates much higher than would be expected for such a polar molecule. The reason for this behavior is not well understood. One proposal is that membranes contain pores that allow the passage of water molecules but are too small for any other polar substance. An alternative suggestion is that in the continual movements of membrane lipids, transient "holes" are created in the lipid monolayers that allow water molecules to move first through one monolayer and then through the other. There is little experimental evidence to support either of these hypotheses, however, and water movement across most membranes remains an enigma.

For cells in at least some tissues, a specific kind of water movement is now better understood, thanks to the recent discovery of a family of channel proteins called **aquaporins (AQPs).** Aquaporins do not account for all water movement across membranes (see Figure 8-2a and c). Instead, they facilitate the rapid movement of water molecules into or out of cells in specific tissues that require this capability. For example, the proximal tubules of your kidneys reabsorb water as part of urine formation and cells in this tissue have a high density of AQPs in their plasma membrane. The same is true of erythrocytes, which must be able to expand or shrink rapidly in response to sudden changes in osmotic pressure as they move through the kidney or other arterial passages (note the aquaporin in Figure 8-2c). In plants, AQPs are a prominent feature of the vacuolar membrane, reflecting the rapid movement of water that is required to develop turgor, as discussed in Box 8A. Aquaporins may well be responsible for rapid transport of water in other cell types as well, but these are some of the better-characterized examples at present.

All aquaporins described to date are integral membrane proteins with six helical transmembrane segments. In the case of AQP-1, the aquaporin found in proximal kidney tubules, the functional unit is a tetramer of four identical monomers. The four monomers appear to associate side by side in the membrane with their 24 transmembrane segments oriented to form a central channel lined with hydrophilic side chains. The diameter of the channel is about 0.3 nm, just large enough for water molecules to pass through one at a time. Although constrained to single-file passage, water molecules flow through a tetrameric AQP-1 channel at the rate of several billion per second.

Active Transport: Protein-Mediated Movement Up the Gradient

Facilitated diffusion is an important mechanism for speeding up the movement of substances across cellular membranes but it only accounts for the transport of molecules *toward* equilibrium, which means *down* a concentration or electrochemical gradient. What happens when a substance needs to be transported *against* a gradient? Such situations require **active transport,** a process that differs from facilitated diffusion in one crucial aspect: Active transport always moves solutes *away* from thermodynamic equilibrium (that is, *up* a concentration or electrochemical gradient), and therefore it always requires an input of energy. In other words, active transport is thermodynamically unfavorable (i.e., endergonic) and occurs only when coupled to an exergonic process. As a result, membrane proteins involved in active transport must provide mechanisms not only for moving desired solute molecules across the membrane but also for coupling such movements to energy-yielding reactions.

Active transport performs three major functions in cells and organelles. It makes possible the uptake of essential nutrients from the environment or surrounding fluid, even when their concentrations in the environment are much lower than inside the cell. In addition, it allows various substances, such as secretory products and waste materials, to be removed from the cell or organelle, even when the concentration outside is greater than that inside. Third, it enables the cell to maintain constant, nonequilibrium intracellular concentrations of specific inorganic ions, notably K^+, Na^+, Ca^{2+}, and H^+.

This ability to create an internal cellular environment whose solute concentrations are far removed from equilibrium is a crucial feature of active transport in terms of its impact on cells. In contrast to simple or facilitated diffusion, which tends to create conditions that are the same on opposite sides of a membrane, active transport is a means of establishing differences in solute concentration and/or electrical potential across membranes. The end result is a nonequilibrium steady state, without which life as we know it would be impossible.

The membrane proteins involved in active transport are often called *pumps,* both in the scientific literature and in textbooks, though no functional analogy with mechanical pumps is intended. On the contrary, mechanical pumps invariably effect a mass flow of fluid from one location to another, whereas membrane pumps selectively transport specific components—molecules or ions—from one fluid mass to another.

An important distinction between active transport and simple or facilitated diffusion concerns the direction of transport. Simple and facilitated diffusion are both *nondirectional* with respect to the membrane; solute can move in either direction, depending entirely on the prevailing concentration or electrochemical gradient. Active transport, on the other hand, has **directionality.** An active transport system that transports a solute across a membrane in one direction will not transport that solute actively in the other direction. Active transport is therefore said to be a *unidirectional,* or *vectorial,* process.

The Coupling of Active Transport to an Energy Source May Be Direct or Indirect

Active transport mechanisms can be divided into several categories that differ primarily in the source of energy and whether or not two solutes are transported concomitantly.

Depending on the energy source, active transport is regarded as being either direct or indirect (Figure 8-9). In **direct active transport**, the accumulation of solute molecules or ions on one side of the membrane is coupled *directly* to an exergonic chemical reaction, most commonly the hydrolysis of ATP (Figure 8-9a). Transport proteins driven directly by ATP hydrolysis are called *transport ATPases* or *ATPase pumps*.

Indirect active transport, on the other hand, depends on the cotransport of two solutes, with the movement of one solute *down* its gradient driving the movement of the other solute *up* its gradient. In most cases, one of the two solutes is an ion (usually Na^+ or H^+) that moves exergonically down its electrochemical gradient, driving the concomitant transport of the second solute (such as a monosaccharide or an amino acid) against its concentration gradient (Figure 8-9b). The transport process is either symport or antiport, depending on whether the two solutes move in the same or opposite directions. For animal cells, sodium symport is the usual option, with the uptake of specific solute molecules driven by the steep sodium ion gradient maintained across the plasma membrane of most animal cells. Bacteria, fungi, and plants, on the other hand, depend on an electrochemical proton gradient as the driving force for indirect active transport. We will consider direct active transport first; then we will discuss indirect active transport.

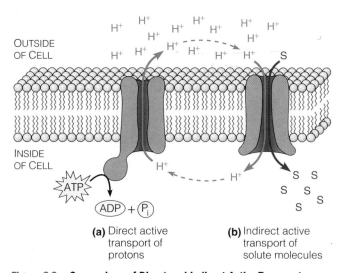

(a) Direct active transport of protons

(b) Indirect active transport of solute molecules

Figure 8-9   Comparison of Direct and Indirect Active Transport.
(a) Direct active transport involves a transport system coupled to an exergonic chemical reaction, most commonly the hydrolysis of ATP. As shown here, ATP drives the outward transport of protons, thereby establishing an electrochemical proton gradient across the membrane. **(b)** Indirect active transport involves a transport system that is driven by the cotransport of ions—protons, in this case—down the electrochemical gradient. The exergonic inward movement of protons provides the energy to move the transported solute, S, against its concentration or electrochemical gradient. Notice the circulation of protons across the membrane that results from this coupling of direct and indirect mechanisms of active transport.

Direct Active Transport Depends on Four Types of Transport ATPases

The most common mechanism employed for direct active transport involves transport ATPases that link active transport to the hydrolysis of ATP. Four main types of transport ATPases have been identified, known as P-type, V-type, F-type, and ABC-type ATPases (Table 8-3). These four types of transport proteins differ in structure, mechanism, localization, and role; most, however, move sodium ions or protons outward across the plasma membrane of the cell.

P-type ATPases.   P-type ATPases (P for "phosphorylation") are reversibly phosphorylated by ATP as part of the transport mechanism, with an aspartic acid residue phosphorylated in each case. P-type ATPases also share several other properties: They have 8–10 transmembrane segments in a single polypeptide that zigzags back and forth across the membrane; they are all cation transporters; and they are all sensitive to inhibition by vanadate, VO_4^{3-}. Most P-type pumps are located in the plasma membrane, where they are responsible for maintaining an ion gradient across the membrane. The best known example is the Na^+/K^+ pump found in almost all animal cells. We will consider this ATPase in more detail shortly (see Figure 8-11). Other examples include the H^+ ATPase responsible for acidification of the gastric juice in your stomach, the H^+ ATPase that pumps protons outward across the plasma membrane of most plant and fungal cells, and the Ca^{2+} ATPase that moves calcium ions out of cells or into the endoplasmic reticulum against their electrochemical gradient. Most P-type ATPases are found in eukaryotes, although at least one bacterial K^+ ATPase is also a member of this family of transporters.

V-type ATPases.   V-type ATPases (V for "vesicle") pump protons into such organelles as vesicles, vacuoles, lysosomes, endosomes, and the Golgi complex. Typically, the proton gradient across the membranes of these organelles ranges from 10-fold to over 10,000-fold. V-type pumps are not inhibited by vanadate and do not undergo phosphorylation as part of the transport process. They have two multisubunit components: an integral component embedded within the membrane, and a peripheral component that juts out from the membrane surface. The peripheral component contains the ATP-binding site and hence the ATPase activity.

F-type ATPases.   F-type ATPases (F for "factor") are found in bacteria, mitochondria, and chloroplasts, where they are an integral part of the mechanism that conserves the energy of solar radiation or of substrate oxidation as ATP. F-type ATPases are involved in proton transport and have two components, both of which are multisubunit complexes. The integral component, called F_o, serves as a transmembrane pore for protons. The peripheral component, called F_1, includes the ATP-binding site. F-type ATPases

Table 8-3   Main Types of Transport ATPases (Pumps)

Solutes Transported	Kind of Membrane	Kind of Organisms	Function of ATPase
P-type ATPases (*P* for "phosphorylation")			
Na^+ and K^+	Plasma membrane	Animals	Keeps $[Na^+]$ low and $[K^+]$ high within cell; maintains membrane potential
H^+	Plasma membrane	Plants, fungi	Pumps protons out of cell; generates membrane potential
Ca^{2+}	Plasma membrane	Eukaryotes	Pumps Ca^{2+} out of cell; keeps $[Ca^{2+}]$ low in cytosol
Ca^{2+}	Sarcoplasmic reticulum (SR; specialized ER)	Animals	Pumps Ca^{2+} into SR; keeps $[Ca^{2+}]$ low in cytosol
V-type ATPases (*V* for "vesicle")			
H^+	Lysosomes, secretory vesicles	Animals	Keeps pH in organelle low, which activates hydrolytic enzymes
H^+	Vacuolar membrane	Plants, fungi	Keeps pH in vacuole low, which activates hydrolytic enzymes
F-type ATPases I (*F* for "factor"); also called ATP synthases			
H^+	Inner mitochondrial membrane	Eukaryotes	Generates H^+ gradient that drives ATP synthesis
H^+	Thylakoid membrane	Plants	Generates H^+ gradient that drives ATP synthesis
H^+	Plasma membrane	Prokaryotes	Generates H^+ gradient that drives ATP synthesis
ABC ATPases (*ABC* for "ATP-binding cassette")			
A variety of solutes*	Plasma membrane, organellar membranes	Prokaryotes, eukaryotes	Nutrient uptake; protein export; possibly also transport into and out of organelles
Antitumor drugs**	Plasma membrane	Animal tumor cells	Removes hydrophobic drugs (and hydrophobic natural products) from cell

*Solutes include ions, sugars, amino acids, carbohydrates, peptides, and proteins.
**Drugs include colchicine, taxol, vinblastine, actinomycin D, and puromycin.

can use the energy of ATP hydrolysis to pump protons against their electrochemical gradient. They can also catalyze the reverse process, in which the exergonic flow of protons down their electrochemical gradient is used to drive ATP synthesis. When they function in this latter mode, these proteins are more appropriately called **ATP synthases.** It is in this role that these proteins function in cellular energy metabolism, in which the energy of solar radiation or of exergonic oxidative reactions is used to maintain a transmembrane proton gradient that then drives ATP synthesis. We will return to these ATP synthases when we consider the processes of cellular respiration and photosynthesis in Chapters 14 and 15, respectively.

F-type ATPases illustrate an important principle: *Not only can ATP be used as an energy source to generate and maintain electrochemical ion gradients, but such gradients can be used as an energy source to synthesize ATP.* This principle, which was discovered in studies of F-type pumps, is the basis of ATP-synthesizing mechanisms in all eukaryotic organisms and in most prokaryotes as well.

ABC-type ATPases. The fourth major class of ATP-driven pumps is the **ABC-type ATPases,** also called **ABC transporters.** The ABC designation is for "*ATP-b*inding *c*assette," with the term *cassette* used to describe catalytic domains of the protein that bind ATP as an integral part of the transport process. The ABC transporters comprise a large superfamily of transport proteins that are related to each other in sequence and probably also in molecular mechanism. Most of the ABC-type ATPases are from prokaryotic species, but increasing numbers are being reported in eukaryotes as well, some of great clinical importance. The typical ABC transporter has four domains, two of which are highly hydrophobic and are embedded in the membrane, while the other two are peripherally located on the cytoplasmic side of the membrane. Each of the two embedded domains consists of six membrane-spanning segments that presumably form the structure through which solute molecules pass. The two peripheral domains are the cassettes that bind ATP and couple its hydrolysis to the transport process. These four domains are separate polypeptides in most cases, especially in prokaryotic cells. However, examples are also known in which the four domains are part of a large multifunctional polypeptide.

Whereas the other three classes of ATPases transport only cations, the ABC-type ATPases handle a remarkable variety of solutes. Most are specific for a particular solute or class of closely related solutes, but the variety of solutes transported by the numerous members of this superfamily is great, including ions, sugars, amino acids, and even peptides and polysaccharides. ABC transporters are of considerable medical interest because some of them pump antibiotics or other drugs out of the cell, thereby making the cell resistant to the drug. For example, some human tumors are remarkably resistant to a variety of drugs that are normally quite effective at arresting tumor growth. Cells of such tumors have unusually high concentrations of a large protein called the **multidrug resistance (MDR) transport protein,** which was in fact the first ABC-type ATPase to be identified in humans. The MDR transport protein uses the energy of ATP hydrolysis to pump hydrophobic drugs out of cells, thereby reducing the cytoplasmic concentration of the drugs and hence their effectiveness as therapeutic agents. Unlike most ABC transporters, the MDR protein has a remarkably broad specificity: It can export a wide range of chemically dissimilar drugs commonly used in cancer chemotherapy, so that a cell with the MDR protein in its plasma membrane becomes resistant to a wide variety of therapeutic agents.

Medical interest in this class of transport proteins was heightened when cystic fibrosis was shown to be caused by a genetic defect in a plasma membrane protein that is structurally related to the ABC transporters. We have long known that people with cystic fibrosis accumulate unusually thick mucus in their lungs, a condition that often leads to pneumonia and other lung disorders. Now we understand that the underlying problem is an inability to secrete chloride ions and that the genetic defect is in a protein that functions as a chloride ion channel. The protein, called the *cystic fibrosis transmembrane conductance regulator (CFTR)*, has two transmembrane domains, two peripheral domains with ATP-binding sites, and a highly hydrophilic *regulatory domain*. The transmembrane and peripheral domains are similar in sequence and likely topology to the core domains of ABC transporters. However, CFTR is an ion channel and does not use ATP to drive transport, as most of the ABC-type ATPases do. Instead, ATP hydrolysis appears to be involved in opening the channel. Recent developments in our understanding of cystic fibrosis at the molecular level are reported in Box 8B, along with prospects for using gene therapy as a treatment, which may soon become a realistic therapeutic option.

Indirect Active Transport Is Driven by Ion Gradients

In contrast to direct active transport, which is powered by energy released from a chemical reaction such as ATP hydrolysis, indirect active transport is driven by the movement of an ion down its electrochemical gradient. This principle has emerged from studies of the active uptake of sugars, amino acids, and other organic molecules into cells: The inward transport of such molecules *up* their concentration gradients is often coupled to, and driven by, the simultaneous inward movement of either sodium ions (for animal cells) or protons (for most plant, fungal, and bacterial cells) *down* their respective electrochemical gradients.

The widespread existence of such *symport* mechanisms explains why most cells continuously pump either sodium ions or protons out of the cell. In animals, for example, the relatively high extracellular concentration of sodium ions

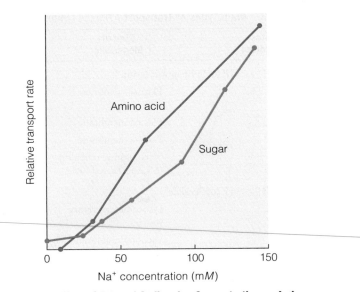

Figure 8-10 Effect of External Sodium Ion Concentration on Amino Acid and Sugar Transport. In this experiment, investigators varied the extracellular concentration of sodium ions and measured the transport rate of the amino acid glycine into erythrocytes or the sugar 7-deoxy-D-glucoheptose into intestinal lining cells. Studies of this type provided the first evidence that transport of amino acids and sugars into cells is stimulated by sodium ions present in the extracellular medium.

maintained by the Na^+/K^+ pump (discussed in the following section) serves as the driving force for the uptake of a variety of sugars and amino acids (Figure 8-10). The uptake of such compounds is regarded as indirect active transport because it is not directly driven by the hydrolysis of ATP, although it still depends on ATP because the Na^+/K^+ pump that maintains the sodium ion gradient is itself driven by ATP hydrolysis. The continuous outward pumping of Na^+ by the ATP-driven Na^+/K^+ pump and inward movement of Na^+ by symport (coupled to the uptake of another solute) establishes a circulation of sodium ions across the plasma membrane of every animal cell.

While animal cells use sodium ions to drive indirect active transport, most other organisms rely on a proton gradient instead. For example, fungi and plants utilize proton symport for the uptake of organic solutes, with an ATP-driven proton pump responsible for the generation and maintenance of the electrochemical proton gradient. Bacterial cells also make extensive use of proton cotransport to drive the uptake of organic solutes, but in this case the proton gradient is maintained by an electron transfer process that accompanies cellular respiration, as we will see in Chapter 14.

In addition to the symport *uptake* of organic molecules such as sugars and amino acids, sodium ion or proton gradients can also be used to drive the *export* of other ions, including Ca^{2+} and K^+. This type of indirect active transport is always antiport, involving the exchange of potassium ions for protons or of calcium ions for sodium ions, for example.

Examples of Active Transport

Having considered some general features of active transport, we are now ready to look at three specific examples, including one example each of direct and indirect active transport from animal cells, plus an unusual type of light-driven transport in a bacterium. In each case, we will note what kinds of solutes are transported, what the driving force is, and how the energy source is coupled to the transport mechanism. We will look first at the *Na$^+$/K$^+$ ATPase* (or *pump*) present in all animal cells, a well-understood example of direct active transport by a P-type ATPase. Then we will consider a second example from animal cells: the indirect active transport of glucose by a *Na$^+$/glucose symporter*, using the energy of the sodium ion gradient established by the Na$^+$/K$^+$ ATPase. Finally, we will explore light-driven *proton transport* in certain bacteria.

Direct Active Transport: The Na$^+$/K$^+$ Pump Maintains Electrochemical Ion Gradients

A characteristic feature of most animal cells is a high intracellular level of potassium ions and a low intracellular level of sodium ions, such that the $[K^+]_{inside}/[K^+]_{outside}$ ratio is much greater than 1 (about 35:1 in a typical animal cell) and the $[Na^+]_{inside}/[Na^+]_{outside}$ ratio is much less than 1 (about 0.08:1 in a typical animal cell). These differences are required to ensure equal concentrations of solutes on both sides of the membrane, thereby maintaining osmotic equilibrium and protecting the cell from swelling and lysis. In addition, the resulting gradients of potassium and sodium ions are essential as the driving force for cotransport as well as for the transmission of nerve impulses (see Chapter 9).

The potassium ion concentration is usually maintained at about 100–150 m*M* inside most animal cells, whereas the external K$^+$ concentration is generally much lower than that and may fluctuate widely. Conversely, the intracellular concentration of sodium ions is about 10–15 m*M*, which is considerably less than that in the surrounding medium. Both the inward pumping of potassium ions and the outward pumping of sodium ions are therefore energy-requiring processes.

The pump responsible for this process was discovered in the mid-1950s, which made it the first documented case of active transport. The pump uses ATP as its energy source and is therefore an example of a transport ATPase. Specifically, the **Na$^+$/K$^+$ ATPase**, or **Na$^+$/K$^+$ pump**, as this P-type ATPase is usually called, couples the exergonic hydrolysis of ATP to the inward transport of potassium ions and the outward transport of sodium ions. The Na$^+$/K$^+$ pump is present in the plasma membrane of virtually all animal cells but has been studied in the greatest detail in red blood cells (see Figure 8-2d). Like other active transport systems, this pump has inherent directionality: Potassium ions are always pumped inward and sodium ions are pumped only outward. In fact, sodium and potassium ions activate the ATPase only on the side of the membrane from which they are transported—sodium ions from the inside, potassium ions from the outside.

Because the gradients against which sodium and potassium ions are pumped by the Na$^+$/K$^+$ ATPase seldom exceed 50:1, the energy requirement per ion moved is relatively low, usually less than 2 kcal/mol. This suggests that the hydrolysis of one ATP molecule ($\Delta G^{\circ\prime} = -7.3$ kcal/mol) can drive the transport of several ions at a time, which is in fact what happens in animal cells. The stoichiometry apparently varies somewhat with cell type, but for red blood cells, three sodium ions are moved out and two potassium ions are moved in per molecule of ATP hydrolyzed.

Figure 8-11 is a schematic illustration of the Na$^+$/K$^+$ pump consistent with available evidence. The pump is a tetrameric transmembrane protein, with two α and two β subunits. The α subunits contain binding sites for sodium ions and ATP on the cytoplasmic side and for potassium ions on the external side of the membrane. We know that the β subunits are located on the extracellular side and are glycosylated, but their function is not yet clear.

The Na$^+$/K$^+$ pump is an allosteric protein exhibiting two alternative conformational states, referred to as E$_1$ and E$_2$. The E$_1$ conformation is open to the inside of the cell and has a high affinity for sodium ions, whereas E$_2$ is open to the outside, with a high affinity for potassium ions. Phosphorylation of the enzyme, a sodium-triggered event, stabilizes it in the E$_2$ form. Dephosphorylation, on the other hand, is triggered by K$^+$ and stabilizes the enzyme in the E$_1$ form.

As illustrated schematically in Figure 8-12, the actual transport mechanism probably involves an initial binding of three sodium ions to E$_1$ on the inner side of the membrane ①, upper right. The binding of sodium ions

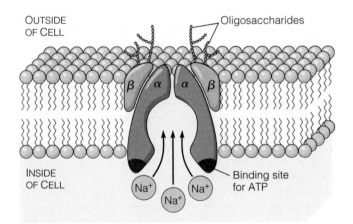

Figure 8-11 The Na$^+$/K$^+$ Pump. The Na$^+$/K$^+$ pump found in most animal cells consists of two α and two β subunits. The α subunits are transmembrane proteins, with binding sites for ATP on the cytoplasmic side. The β subunits are located on the outer side of the membrane and are glycosylated. The pump is shown in the E$_1$ conformation, which is open to the inside of the cell. Binding of sodium ions causes a conformational change to the E$_2$ form, which opens to the outside.

Clinical Applications — MEMBRANE TRANSPORT, CYSTIC FIBROSIS, AND THE PROSPECTS FOR GENE THERAPY

Transport proteins located in the plasma membrane play critical roles in speeding up and controlling the movement of molecules and ions into and out of cells. To remain healthy, our bodies depend on the proper functioning of many such membrane proteins. If any of these proteins is defective, the movement of a particular ion or molecule across cell membranes is likely to be impaired, and disease may result.

An example that has attracted the attention of researchers and doctors alike is **cystic fibrosis (CF),** a disease caused by genetic defects in a transport protein in the plasma membrane. Cystic fibrosis is a common genetic disease that typically affects children and young adults. The parts of the body that are most noticeably affected are the lungs, pancreas, and sweat glands. Complications in the lungs are the most severe medical problems because they are difficult to treat and can become life-threatening. The airways of a CF patient are often obstructed with abnormally thick mucus and are vulnerable to chronic bacterial infections, especially by *Pseudomonas aeruginosa.*

Using the tools of molecular and cellular biology, researchers have achieved a better understanding of this disease. During the 1980s, cells from CF patients were shown to be defective in the secretion of chloride ions (Cl⁻). The cells that line unaffected lungs secrete chloride ions in response to a substance called cyclic AMP, whereas cells from CF patients do not. (Cyclic AMP is a form of AMP that is involved in a variety of regulatory roles in cells; for its structure, see Figure 10-5.) Experiments with tissue from CF patients suggested that this difference might be due to a defect in a membrane protein that normally serves as a channel for the movement of chloride ions across the membrane.

Many symptoms of CF can be explained by faulty Cl⁻ secretion, as Figure 8B-1a illustrates. In the lungs of an unaffected person (top panel), chloride ions are secreted from the cells that line the airways and enter the lumen of the passage normally. (A lumen is the space inside a passage or duct.) The movement of Cl⁻ out of the cell and into the lumen provides the driving force for the concurrent movement of sodium ions into the lumen. Osmotic pressure causes water to follow the sodium and chloride ions, resulting in the secretion of a dilute salt solution. The water that moves into the lumen in this way provides vital hydration to the mucus lining of the air passages. In the cells of a person with cystic fibrosis, Cl⁻ ions cannot exit into the lumen, so sodium ions and water do not either (bottom panel). As a result, the mucus is insufficiently hydrated, a condition that favors bacterial growth.

An exciting breakthrough in CF research came in 1989 when investigators in the laboratories of Francis Collins at the University of Michigan and of Lap-Chee Tsui and John Riordan at the University of Toronto isolated the gene that is defective in

CF patients. The gene encodes a protein called the **cystic fibrosis transmembrane conductance regulator (CFTR).** The sequence of nucleotide bases in the gene was determined by using methods that are described in Chapter 16 (see Figure 16-14). Knowing the base sequence of the gene, scientists were able to predict the amino acid sequence and the structure of the CFTR protein. As shown in Figure 8B-1b, the protein is thought to have two sets of *transmembrane domains* that anchor the protein in the plasma membrane and two *nucleotide-binding folds* that serve as binding sites for ATP, which provides the energy to drive transport of chloride ions across the membrane. In addition, the protein has a large cytoplasmic domain called the *regulatory domain,* which has several serine hydroxyl groups that can be reversibly phosphorylated. The CFTR protein has since been shown to function as a chloride channel in cells, and channel function is known to be affected when the phosphorylation sites in the regulatory domain are changed as a result of a mutation in the CFTR gene.

By sequencing the *CFTR* genes from CF patients, investigators have identified more than 600 mutations in the gene. The most common of these mutations causes the deletion of a single amino acid in the first nucleotide-binding domain. The question of how this mutation causes CF remained unanswered until researchers examined the location of CFTR in cells with and without the mutation. Normal CFTR was found in the cell membrane, as predicted. In contrast, mutant CFTR was not detected in the plasma membrane.

The most likely explanation at present is that normal CFTR is synthesized on the rough endoplasmic reticulum (ER), moves through the Golgi complex, and is eventually inserted in the plasma membrane by a route that is explained in Chapter 12. Mutant CFTR, on the other hand, is apparently trapped in the ER, perhaps because it is folded improperly. It is therefore recognized as a defective protein and degraded. Consequently, CFTR is not present in the plasma membrane of CF cells, chloride ion secretion cannot take place, and disease results.

Armed with information about the gene and the protein associated with CF, researchers are now trying to develop new treatments or perhaps even a cure for the disease. A promising approach is *gene therapy,* in which a normal copy of a gene is introduced into affected cells of the body. Investigators would like to direct normal copies of the *CFTR* gene into the cells that line the airways of CF patients. These cells should then be able to synthesize a correct CFTR protein, which, unlike mutant CFTR, would be located in the plasma membrane, thereby allowing proper Cl⁻ secretion and correcting the disease.

Two kinds of practical problems must be overcome if gene therapy is to work: The *CFTR* gene must be delivered efficiently to the affected tissue and its expression must be regulated to achieve and maintain normal production of the CFTR protein.

triggers phosphorylation of the enzyme by ATP ②, resulting in a conformational change from E_1 to E_2. As a result, the bound sodium ions are moved through the membrane to the external surface, where they are released to the outside ③. Then potassium ions from the outside

bind to the α subunits ④, triggering dephosphorylation and a return to the original conformation ⑤. During this process, the potassium ions are moved to the inner surface, where they dissociate, leaving the carrier ready to accept more sodium ions ⑥.

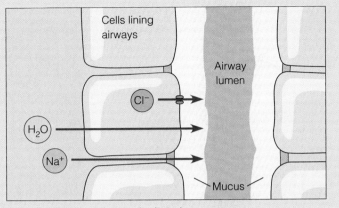

Normal cells lining airways; hydrated mucus

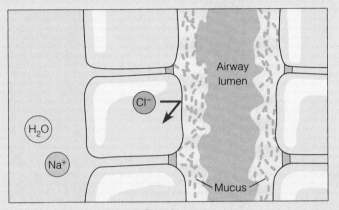

Cells of a person with cystic fibrosis; dehydrated mucus infected with bacteria

(a) Normal and faulty chloride ion secretion

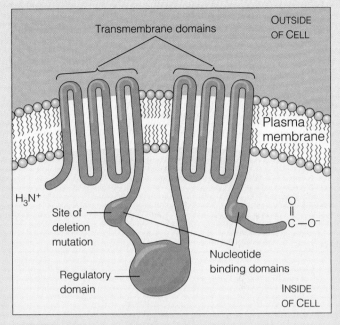

(b) CFTR protein

Figure 8B-1 Cystic Fibrosis and Chloride Ion Secretion. **(a)** Cystic fibrosis is caused by a defect in the secretion of chloride ions in cells lining the lungs, leading to insufficient hydration and the promotion of bacterial growth. **(b)** The cystic fibrosis transmembrane conductance regulator (CFTR) is an integral membrane protein that functions as a chloride ion channel. The most common mutation found in cystic fibrosis patients causes the deletion of a single amino acid in the first nucleotide-binding domain of the CFTR protein.

In most clinical studies to date, the normal *CFTR* gene has been introduced into CF patients by one of two means: Either the *CFTR* is incorporated into the DNA of a virus called *adenovirus* or it is mixed with fat droplets called *liposomes*. The viral or liposomal preparation is sprayed as an aerosol into the nose or lungs of CF patients, who are then monitored for the correction of the chloride transport abnormality.

In initial experiments with CF mutant mice, several groups of British investigators showed that the chloride ion channel defect could be corrected by spraying the respiratory tract with liposomes containing the normal human *CFTR* gene. The same approach is now being used in clinical trials with human patients. In one such study, a "gene spray" of the *CFTR* gene mixed with liposomes was administered nasally. Gene delivery and expression was demonstrated in most of the patients, but expression was short-lived and the ion channel defect only partially corrected. More recently, Eric Alton and his colleagues in London reported a well-controlled experiment in which they administered liposomes with or without the *CFTR* gene in the form of an aerosol to the lungs and the nose. A short-term improvement in chloride transport was seen in the patients who received the *CFTR* gene, but not in the patients who received the placebo (liposome-only) treatment. They also reported reduced bacterial adherence to respiratory epithelial cells in the *CFTR*-treated patients, an observation that may have significant clinical relevance. "Gene therapy for cystic fibrosis continues to make steady progress towards becoming a realistic therapeutic option for this disease," concluded Alton and his colleagues. And so may we.

This text was originally written by Dr. Lisa S. Smit, University of Michigan.

The Na⁺/K⁺ pump is not only one of the best-understood transport systems but also one of the most important for animal cells. In addition to maintaining the appropriate intracellular concentrations of both potassium and sodium ions, it is responsible for maintaining the membrane potential that exists across the plasma membrane. The Na⁺/K⁺ pump assumes still more significance when we take into account the vital role that sodium ions play in the inward transport of organic substrates, a topic we now come to as we consider sodium symport.

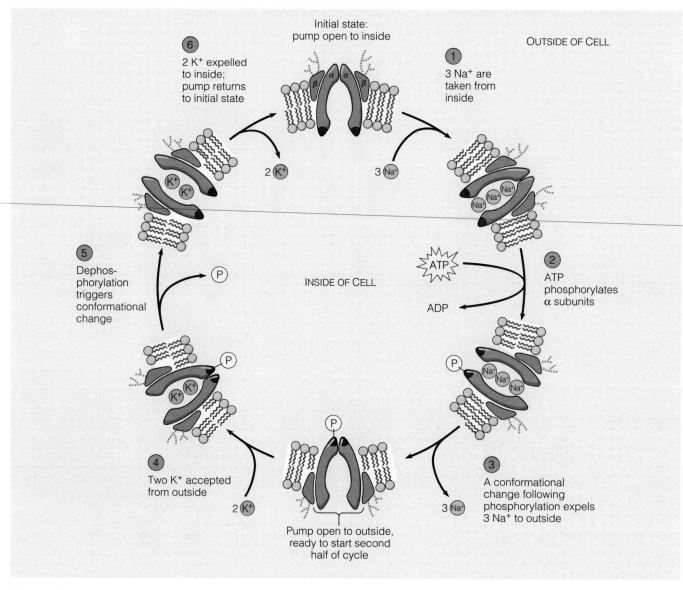

Figure 8-12   A Model Mechanism for the Na⁺/K⁺ Pump.

Labels in figure:

Initial state: pump open to inside

OUTSIDE OF CELL

⑥ 2 K⁺ expelled to inside; pump returns to initial state

① 3 Na⁺ are taken from inside

⑤ Dephos-phorylation triggers conformational change

INSIDE OF CELL

② ATP phosphorylates α subunits

ATP

ADP

④ Two K⁺ accepted from outside

③ A conformational change following phosphorylation expels 3 Na⁺ to outside

Pump open to outside, ready to start second half of cycle

Figure 8-12   A Model Mechanism for the Na⁺/K⁺ Pump.   The transport process is shown here in six steps arranged around the periphery of a cell. The outward transport of sodium ions is coupled to the inward transport of potassium ions, both against their respective electrochemical gradients. The driving force is provided by ATP hydrolysis, which is required for phosphorylation of the α subunit of the pump. Selective transport of sodium ions and potassium ions in opposite directions is possible because sodium ions activate the ATPase only on the inner surface of the membrane, whereas potassium ion activation of the dephosphorylation reaction occurs only on the outer surface. E_1 and E_2 are the conformation states of the protein with the channel open to the inside (top of figure) and to the outside (bottom of figure) of the cell, respectively. As depicted here, the binding of sodium ions precedes phosphorylation of the α subunit, but these steps may occur in the reverse order.

Indirect Active Transport: Sodium Symport Drives the Uptake of Glucose

As an example of indirect active transport, consider the uptake of glucose by the **Na⁺/glucose symporter.** Although most glucose transport in your body occurs by facilitated diffusion as shown in Figure 8-8, the epithelial cells that line your intestine have transport proteins that are able to take up glucose and certain amino acids from the intestine even when their concentrations there are much lower than in the epithelial cells. Although the uptake of glucose under these conditions is endergonic, the process is driven by the concomitant uptake of sodium ions, which is exergonic because of the steep electrochemical sodium ion gradient maintained across the plasma membrane of the epithelial cells.

Figure 8-13 depicts the transport mechanism of the Na⁺/glucose symporter. Transport is initiated by the binding of an external sodium ion to its binding site on the symporter. As a consequence, the conformation of the symporter changes, giving its glucose binding site a high affinity for glucose ①, upper right. Binding of glucose ② presumably causes a further conformational change in the protein that results in the movement of both sodium ions and glucose to the inner surface of the membrane ③.

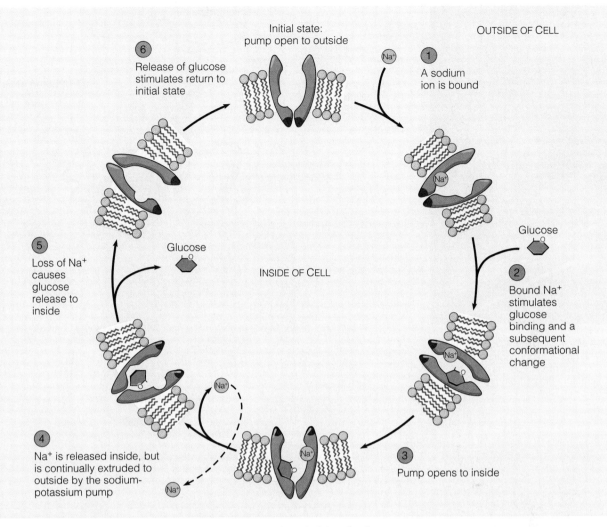

Figure 8-13   A Model Mechanism for the Na⁺/Glucose Symporter. The transport process is shown here in six steps arranged around the periphery of a cell. The inward transport of an organic solute such as glucose against its concentration gradient is driven by the concomitant inward transport of sodium ions down their electrochemical gradient. The sodium ion gradient is in turn maintained by the continuous outward extrusion of sodium ions (dashed arrow) by the Na⁺/K⁺ pump of Figure 8-12, such that sodium ions circulate across the plasma membrane, pumped outward by the Na⁺/K⁺ pump and flowing back into the cell as the driving force for sodium symport of molecules such as glucose.

There, the sodium ions dissociate in response to the low intracellular sodium ion concentration ④. This causes glucose to be ejected from its site regardless of the cellular glucose level ⑤, probably because the affinity of that site for glucose is greatly reduced upon dissociation of the sodium ions. The protein then reverts to its original conformation, returning the empty binding sites to the outer surface of the membrane ⑥.

Similar mechanisms are involved in the uptake of amino acids and other organic substrates by sodium symport in animal cells and by proton symport in plant, fungal, and bacterial cells.

The Bacteriorhodopsin Proton Pump Uses Light Energy to Transport Protons

The final active transport system we will consider is the simplest; nothing more is involved than a small integral membrane protein called **bacteriorhodopsin**. This protein, briefly introduced in Chapter 7, is a proton pump found in the plasma membrane of halophilic (salt-loving) archaebacteria belonging to the genus *Halobacterium*. In contrast to transport proteins that utilize direct or indirect active transport mechanisms based on energy derived from ATP or ion gradients, respectively, bacteriorhodopsin uses energy derived from photons of light to drive active transport. This ability is important when Halobacteria are deprived of oxygen, thereby blocking the oxygen-driven metabolic pathways that represent their normal source of energy. Under such conditions, these cells synthesize bacteriorhodopsin as an alternative means of capturing energy. Bacteriorhodopsin traps light energy and uses it to drive the active transport of protons outward across the plasma membrane, thereby creating an electrochemical proton gradient that powers the synthesis of ATP.

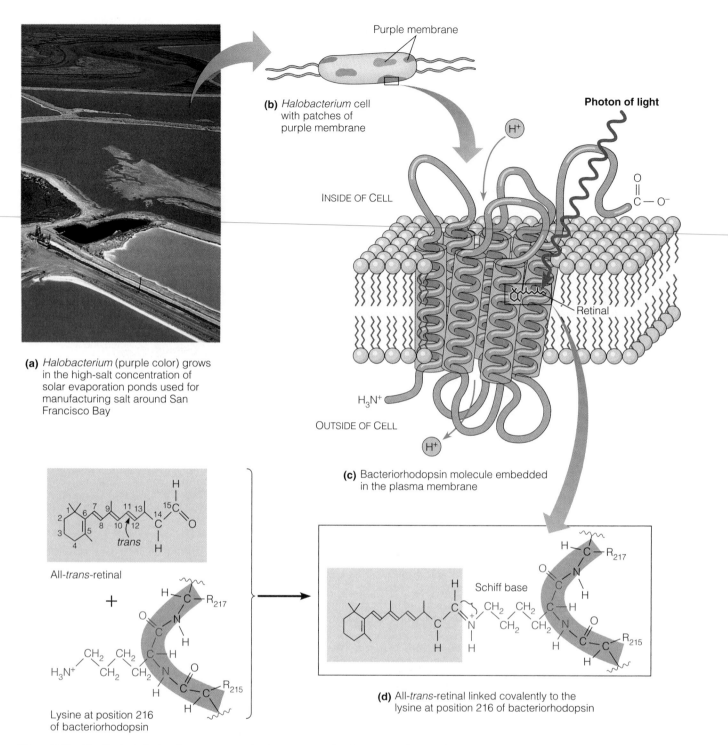

(b) *Halobacterium* cell with patches of purple membrane

INSIDE OF CELL

Photon of light

Retinal

OUTSIDE OF CELL

(c) Bacteriorhodopsin molecule embedded in the plasma membrane

(a) *Halobacterium* (purple color) grows in the high-salt concentration of solar evaporation ponds used for manufacturing salt around San Francisco Bay

All-*trans*-retinal

trans

+

Lysine at position 216 of bacteriorhodopsin

Schiff base

(d) All-*trans*-retinal linked covalently to the lysine at position 216 of bacteriorhodopsin

Figure 8-14   The Bacteriorhodopsin Proton Pump of Halobacteria. **(a)** Bacteria belonging to the genus *Halobacterium* are characterized by a purple color that is due to the protein bacteriorhodopsin. **(b)** Bacteriorhodopsin is a light-activated proton pump that is present in the plasma membrane of *Halobacterium* cells as bright purple patches known as purple membrane. **(c)** The seven α-helical transmembrane segments of bacteriorhodopsin are separated by short nonhelical segments and are oriented in the membrane to form an overall cylindrical shape. **(d)** The chromophore, all-*trans*-retinal, is linked as a Schiff base to the lysine at position 216 in the seventh transmembrane segment of the protein.

The light-absorbing pigment, or *chromophore,* of bacteriorhodopsin is *retinal,* a carotenoid derivative related to vitamin A. (Retinal also serves as the visual pigment in the retina of your eyes.) Retinal makes bacteriorhodopsin bright purple in color, which is the reason that halobacteria are also called *purple photosynthetic bacteria* (Figure 8-14a). Bacteriorhodopsin appears in the plasma membrane of the *Halobacterium* cell as colored patches called *purple membrane* (Figure 8-14b).

Bacteriorhodopsin is an integral membrane protein with seven α-helical membrane-spanning segments that are oriented in the membrane to form an overall cylindri-

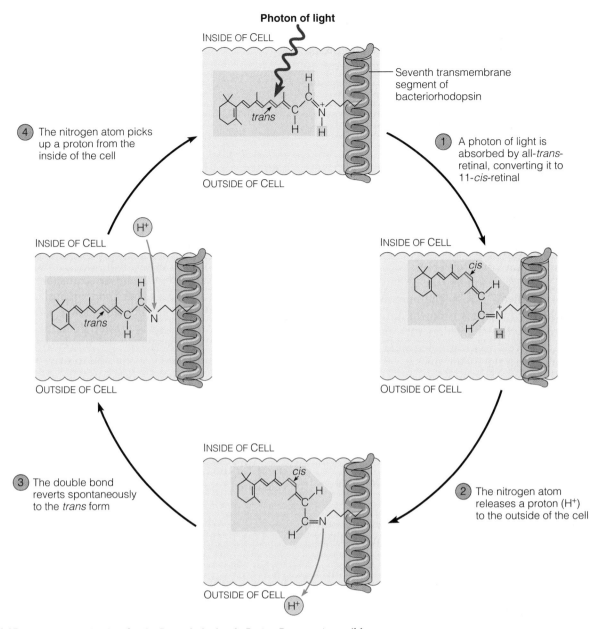

Figure 8-15   A Model Mechanism for the Bacteriorhodopsin Proton Pump.   A possible mechanism for light-activated proton pumping by bacteriorhodopsin is shown here in four steps. Protons are always released to the outside of the cell but taken from the inside of the cell, thereby accounting for the outward pumping of protons that generates an electrochemical proton gradient across the plasma membrane of illuminated cells. The proton gradient formed in this way is then used to drive ATP synthesis.

cal shape (Figure 8-14c). The retinal molecule is present in the all-*trans* form and is covalently linked to the side chain of the lysine at position 216, forming a positively charged *Schiff base* (Figure 8-14d). When the retinal absorbs a photon of light, the photoactivated bacteriorhodopsin molecule is capable of transferring one or two protons from the inside to the outside of the cell. The details of the proton transfer process are not yet known, even though bacteriorhodopsin has been successfully isolated and incorporated into artificial membranes in functional form.

One possible mechanism for proton pumping is shown in Figure 8-15. According to this model, the absorption of an incoming photon of light causes one of the double bonds of all-*trans* retinal to isomerize to the *cis* conformation ① (upper right). This change decreases the ability of the nitrogen atom to bind protons and a proton is released to the outside of the cell ②, permitting the double bond to relax to the *trans* form ③. The nitrogen again binds a proton, but always from the inside of the cell ④. Because protons are always drawn from the inside and released to the outside of the cell, the process leads to a vectorial pumping of protons from inside to outside. This pumping results in an electrochemical proton gradient across the plasma membrane that is then used to drive ATP synthesis.

Energy-dependent proton pumping is one of the most basic concepts in cellular energetics. Proton pumping occurs in all bacteria, mitochondria, and chloroplasts and represents the driving energy of life on Earth because it is an absolute requirement for the efficient synthesis of ATP. We will discuss the mechanisms underlying the generation of proton gradients and the use of the energy of such gradients in more detail in Chapters 13–15.

The Energetics of Transport

Every transport event is an energy transaction; energy is either released as transport occurs or is required to drive transport. To understand the energetics of transport, we must recognize that two different factors may be involved. For uncharged solutes, the only variable is the concentration gradient across the membrane, which determines whether transport is downhill (exergonic) or uphill (endergonic). For charged solutes, however, there may be both a concentration gradient and an electrical charge gradient across the membrane; the two may either reinforce each other or oppose each other, depending on the charge on the ion and the direction of transport. We will first look at the transport of uncharged substances and then consider the additional complication that arises when charged substances are moved across membranes.

For Uncharged Solutes, the ΔG of Transport Depends Only on the Concentration Gradient

For solutes with no net charge—molecules, in other words—we are concerned only with the concentration gradient across the membrane (assuming no appreciable changes in pressure, temperature, or volume). We can therefore treat the transport process as a simple chemical reaction and calculate ΔG as we would for any other reaction.

Calculating ΔG for the Transport of Molecules. The general "reaction" for the transport of molecules of solute S from the outside of a membrane-bounded compartment to the inside can be represented as

$$S_{outside} \longrightarrow S_{inside} \qquad (8\text{-}4)$$

From Chapter 5, we know that the free energy change for this reaction can be written as

$$\Delta G = \Delta G° + RT \ln \frac{[S]_{inside}}{[S]_{outside}} \qquad (8\text{-}5)$$

where ΔG is the free energy change, $\Delta G°$ is the standard free energy change, R is the gas constant (1.987 cal/mol-K), T is the absolute temperature, and $[S]_{inside}$ and $[S]_{outside}$ are the prevailing concentrations of S on the inside and outside of the membrane, respectively. However, the equilibrium constant K_{eq} for the transport of an uncharged solute

is always 1, because at equilibrium the solute concentrations on the two sides of the membrane will be the same:

$$K_{eq}' = \frac{[S]_{inside}}{[S]_{outside}} = 1.0 \qquad (8\text{-}6)$$

This means that $\Delta G°$ is always zero:

$$\Delta G° = -RT \ln K_{eq}' = -RT \ln 1 = 0 \qquad (8\text{-}7)$$

So the expression for ΔG of inward transport of an uncharged solute simplifies to

$$\Delta G_{inward} = +RT \ln \frac{[S]_{inside}}{[S]_{outside}} \qquad (8\text{-}8)$$

Notice that if $[S]_{inside}$ is less than $[S]_{outside}$, then ΔG will be negative, indicating that the inward transport of substance S is exergonic and may occur spontaneously, as would be expected for facilitated diffusion down a concentration gradient. But if $[S]_{inside}$ is greater than $[S]_{outside}$, inward transport of S will be against the concentration gradient and the amount of energy required to drive the transport is indicated by the positive value of ΔG.

An Example: The Uptake of Lactose. As an example, suppose that the concentration of lactose within a bacterial cell is to be maintained at 10 mM, while the external lactose concentration is only 0.20 mM. The energy requirement for the inward transport of lactose at 25°C can be calculated from equation 8-8 as

$$\begin{aligned} \Delta G_{inward} &= +RT \ln \frac{[lactose]_{inside}}{[lactose]_{outside}} \\ &= +(1.987)(273+25) \ln \frac{0.010}{0.0002} \\ &= +592 \ln 50 = +2316 \, cal/mol \\ &= +2.32 \, kcal/mol \qquad (8\text{-}9) \end{aligned}$$

Clearly, this is an energy requirement that can be met easily by coupling lactate transport to ATP hydrolysis ($\Delta G° = -7.3$ kcal/mol), which is one of the several sources of energy that bacterial cells can use for lactose uptake.

As written, equation 8-8 applies to inward transport. For outward transport, the positions of S_{inside} and $S_{outside}$ are simply interchanged within the logarithm. As a result, the absolute value of ΔG remains the same, but the sign is changed. As for any other process, a transport reaction that is exergonic in one direction will be endergonic to the same degree in the opposite direction. The equations for calculating ΔG of inward and outward transport of uncharged solutes are summarized in Table 8-4.

For Charged Solutes, the ΔG of Transport Depends on the Electrochemical Gradient

For charged solutes—ions, in other words—we need to take into account both the concentration gradient and the

Table 8-4 Calculation of ΔG for the Transport of Charged and Uncharged Solutes

Reaction:

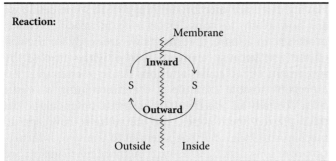

ΔG for Transport of Uncharged Solutes:

$$\Delta G_{\text{inward}} = +RT \ln \frac{[S]_{\text{inside}}}{[S]_{\text{outside}}}$$

$R = 1.987 \text{ cal/mol-K}$
$T = K = \degree C + 273$

$$\Delta G_{\text{outward}} = +RT \ln \frac{[S]_{\text{outside}}}{[S]_{\text{inside}}}$$

ΔG for Transport of Charged Solutes:

$$\Delta G_{\text{inward}} = +RT \ln \frac{[S]_{\text{inside}}}{[S]_{\text{outside}}} + zFV_m$$

$z = $ charge on ion
$F = 23{,}062 \text{ cal/mol-V}$

$V_m = $ membrane potential (in volts)

$$\Delta G_{\text{outward}} = +RT \ln \frac{[S]_{\text{outside}}}{[S]_{\text{inside}}} - zFV_m$$

---

electrical charge gradient. A charge gradient exists across the plasma membrane of almost all cells because active transport causes the concentrations of various electrically charged solutes inside the cell to differ from those outside the cell. This unequal distribution of charge creates a voltage across the plasma membrane, referred to earlier in the chapter as the *membrane potential*. The magnitude of the membrane potential, or V_m, is expressed either in volts (V) or millivolts (mV). For animal cells, V_m usually falls in the range of −60 to −90 mV, while in bacterial and plant cells it is significantly more negative, often about −150 mV in bacteria and between −200 and −300 mV in plants. By convention, the minus sign indicates that the negative charge is on the inside of the cell. Thus, the V_m value indicates how negative (or positive, in the case of a plus sign) the *inside* of the cell is compared with the *outside*.

The membrane potential obviously has no effect on uncharged solutes, but it affects the energetics of ion transport significantly. Because it is almost always negative, the membrane potential *favors* the inward movement of cations and *opposes* their outward movement. Conversely, anions move *against* the membrane potential when they are transported inward but *with* it when they are transported outward. As we mentioned earlier, the net effect of both the concentration gradient and the potential gradient for an ion is called the *electrochemical gradient* for that ion.

Calculating ΔG for the Transport of Ions. Both components of the electrochemical gradient must be considered when determining the energetics of ion transport. To calculate ΔG for the transport of ions therefore requires an equation with two terms, one to express the effect of the concentration gradient across the membrane and the other to take into account the membrane potential.

If we let S^z represent a solute with a charge z, then we can calculate ΔG for the inward transport of S^z as

$$\Delta G_{\text{inward}} = +RT \ln \frac{[S]_{\text{inside}}}{[S]_{\text{outside}}} + zFV_m \qquad \text{(8-10)}$$

where R, T, and $[S]$ are defined as before, z is the charge on S (such as +1, +2, −1, or −2), F is the Faraday constant (23,062 cal/mol-V), and V_m is the membrane potential (in volts).

For the outward transport of S, ΔG has the same value as for inward transport but is opposite in sign, so we can write

$$\Delta G_{\text{outward}} = -\Delta G_{\text{inward}}$$

$$= -RT \ln \frac{[S]_{\text{inside}}}{[S]_{\text{outside}}} - zFV_m \qquad \text{(8-11)}$$

Or, by interchanging terms within the logarithm,

$$\Delta G_{\text{outward}} = +RT \ln \frac{[S]_{\text{outside}}}{[S]_{\text{inside}}} - zFV_m \qquad \text{(8-12)}$$

An Example: The Uptake of Chloride Ions. To illustrate the use of equation 8-12—and to point out that intuition does not always serve us well in predicting the direction of ion transport—consider what happens when nerve cells with an intracellular chloride ion concentration of 50 mM are placed in a solution containing 100 mM Cl⁻. Since the Cl⁻ concentration is twice as high outside the cell as inside, you might expect chloride ions to diffuse passively into the cell without the need for active transport. However, this expectation ignores the membrane potential of about −60 mV (−0.06 V) that exists across the plasma membrane of nerve cells. The minus sign reminds us that the inside of the cell is negative with respect to the outside, which means that the inward movement of negatively charged anions such as Cl⁻ will be *against* the charge gradient. Thus, the inward movement of chloride ions is *down* the concentration gradient but *up* the charge gradient.

To quantify the relative magnitudes of these two opposing forces at 25°C, we can use equation 8-12 with the relevant values inserted:

$$\Delta G_{\text{inward}} = +RT \ln \frac{[S]_{\text{inside}}}{[S]_{\text{outside}}} + zFV_m$$

$$= +(1.987)(273 + 25) \ln\left(\frac{0.05}{0.10}\right) + (-1)(23{,}062)(-0.06)$$

$$= +592 \ln(0.5) + (23{,}062)(0.06)$$

$$= -410 + 1384 = 974 \text{ cal/mol}$$

$$= +0.97 \text{ kcal/mol} \qquad \text{(8-13)}$$

The fact that ΔG is positive rather than negative means that even though the chloride ion concentration is twice as high outside the cell as inside, energy will still be required to drive the movement of chloride ions into the cell. This is because the movement of a mole of chloride ions up the charge gradient represented by the membrane potential requires more energy (+1384 calories) than is released by the movement of the mole of chloride ions down their concentration gradient (−410 calories).

To guide you in such calculations, the equations for inward and outward transport of charged solutes are included along with those for the transport of uncharged solutes in Table 8-4, which summarizes the thermodynamic properties of each of these processes.

On to Nerve Cells

Having discussed membrane transport in some detail, we are now ready to consider its role in specific cellular processes. In a sense, that means we are ready for the rest of the text because many cellular activities involve the movement of molecules and ions across membranes. An especially notable example is the functioning of nerve cells. Virtually all cells maintain electrical potentials across their plasma membranes, but nerve cells have special mechanisms for using this potential to transmit information over long distances. To explore those mechanisms, we take our understanding of membrane transport phenomena and move on to Chapter 9.

Perspective

The selective transport of molecules and ions across membrane barriers ensures that the necessary substances are moved into and out of cells and cell compartments at the appropriate time and at useful rates. Small, nonpolar molecules such as O_2 and ethanol cross the membrane by simple diffusion. Transport of all other solutes, including all ions and most molecules of biological relevance, is mediated by specific transport proteins that provide solute-specific mechanisms for passage through an otherwise impermeable membrane. Each such protein has at least one, and frequently many, hydrophobic membrane-spanning sequences that embed the protein within the membrane and determine its molecular mechanism of action.

Transport can either be downhill or uphill in relation to a solute's concentration (or electrochemical) gradient. Downhill transport, called facilitated diffusion, is mediated by carrier proteins and channel proteins. Carrier proteins function by alternating between two conformational states; examples include the glucose transporter and the anion exchange protein found in the plasma membrane of the erythrocyte. Transport of a single kind of molecule or ion is called uniport. The coupled transport of two or more molecules or ions at a time is called cotransport, which may involve movement of both solutes in the same direction (symport) or in opposite directions (antiport). Channel proteins facilitate diffusion by forming hydrophilic transmembrane channels. Three important categories of channel proteins are ion channels (which are used mainly for transport of Na^+, K^+, Ca^{2+}, Cl^-, and H^+), and porins and aquaporins (which facilitate the rapid movement of various solutes and water, respectively).

Uphill transport, called active transport, requires energy and may be powered by ATP hydrolysis, an ion gradient, or light. Active transport powered by ATP hydrolysis utilizes four major classes of transport proteins called P-type, V-type, F-type, and ABC-type ATPases. One widely encountered example is the ATP-powered Na^+/K^+ pump (a P-type ATPase), which maintains electrochemical gradients of sodium and potassium ions across the plasma membrane of animal cells. Transport driven by ion gradients is usually powered by gradients of either sodium ions (animal cells) or protons (plant, fungal, and many prokaryotic cells). Examples include the transport of organic molecules across the plasma membrane driven by symport of Na^+. When this transport mechanism is operating at the same time as the Na^+/K^+ pump, sodium ions cycle back and forth across the plasma membrane, being pumped outward by the Na^+/K^+ pump and flowing back into the cell as they drive inward transport of sugars, amino acids, and other organic molecules.

The ΔG for transport can be readily calculated. For uncharged solutes, ΔG depends only on the concentration gradient, whereas for charged solutes, both the concentration gradient and the membrane potential must be taken into account.

An understanding and appreciation of membrane proteins and membrane transport will prove vital in many of the coming chapters in this text. That is especially true for the next chapter, which expands our discussion of ion transport to explain the transmission of electrical signals by nerve cells.

Key Terms for Self-Testing

Cells and Transport Processes
transport (p. 195)
transport protein (p. 197)
membrane potential (p. 197)
electrochemical gradient (p. 197)

Simple Diffusion
simple diffusion (p. 197)
osmosis (p. 199)

Facilitated Diffusion
facilitated diffusion (p. 203)
carrier protein (p. 203)
channel protein (p. 203)
alternating conformation model
    (p. 203)
uniport (p. 204)
cotransport (p. 204)
symport (p. 204)

antiport (p. 204)
glucose transporter (GluT) (p. 205)
anion exchange protein (p. 206)
ion channel (p. 206)
porin (p. 206)
aquaporin (AQP) (p. 207)

Active Transport
active transport (p. 207)
directionality (p. 207)
direct active transport (p. 208)
indirect active transport (p. 208)
P-type ATPase (p. 208)
V-type ATPase (p. 208)
F-type ATPase (p. 208)
ATP synthase (p. 209)
ABC-type ATPase (ABC transporter)
    (p. 209)

multidrug resistance (MDR) transport
    protein (p. 210)

Examples of Active Transport
Na^+/K^+ ATPase (Na^+/K^+ pump) (p. 211)
Na^+/glucose symporter (p. 214)
bacteriorhodopsin (p. 215)

Box 8A: *Osmosis: The Special Case
of Water Diffusion*
osmolarity (p. 200)
turgor pressure (p. 201)
plasmolysis (p. 201)

Box 8B: *Membrane Transport, Cystic Fibrosis,
and the Prospects for Gene Therapy*
cystic fibrosis (CF) (p. 212)
cystic fibrosis transmembrane conductance
    regulator (CFTR) (p. 212)

Problem Set

More challenging problems are marked with a • .

8-1.   True or False? Indicate whether each of the following
statements about membrane transport is true (T) or false (F).
If false, reword the statement to make it true.

(a) Facilitated diffusion of a cation occurs only from a compartment of higher concentration to a compartment of
lower concentration.

(b) Active transport is always driven by the hydrolysis of high-
energy phosphate bonds.

(c) The K_{eq} value for the diffusion of polar molecules out of
the cell is less than one because membranes are essentially
impermeable to such molecules.

(d) A 0.25 M sucrose solution would not be isotonic for a
mammalian cell if the cell had sucrose carrier proteins in
its plasma membrane.

(e) The permeability coefficient for a particular solute is likely
to be orders of magnitude lower if a transport protein for
that solute is present in the membrane.

(f) Plasma membranes have few, if any, transport proteins that
are specific for phosphorylated compounds.

(g) Carbon dioxide and bicarbonate anions usually move in the
same direction across the plasma membrane of an erythrocyte.

(h) Treatment of an animal cell with an inhibitor that is specific for the Na^+/K^+ pump is not likely to affect the uptake
of glucose by sodium cotransport.

8-2.   Telling Them Apart. From the list of properties below,
indicate which one(s) can be used to distinguish between each
of the following pairs of transport mechanisms.

(a) Simple diffusion; facilitated diffusion

(b) Facilitated diffusion; active transport

(c) Simple diffusion; active transport

(d) Direct active transport; indirect active transport

(e) Symport; antiport

(f) Uniport; cotransport

(g) P-type ATPase; V-type ATPase

Properties
1. Directions in which two transported solutes move
2. Direction in which solute moves relative to its concentration or electrochemical gradient
3. Kinetics of solute transport
4. Requirement for metabolic energy
5. Requirement for simultaneous transport of two solutes
6. Intrinsic directionality
7. Competitive inhibition
8. Sensitivity to the inhibitor vanadate

8-3.   Mechanisms of Transport. For each of the following
statements, answer with a D if the statement is true of simple
diffusion, with an F if it is true of facilitated diffusion, and with
an A if it is true of active transport. Any, all, or none (N) of the
choices may be appropriate for a given statement.

(a) Requires the presence of an integral membrane protein.

(b) Depends primarily on solubility properties of the solute.

(c) Doubling the concentration gradient of the molecule to be
transported will double the rate of transport over a broad
range of concentrations.

(d) Involves proteins called ATPases.

(e) Work is done during the transport process.

(f) Applies only to small, nonpolar solutes.

(g) Applies only to ions.

(h) Transport can occur in either direction across the membrane, depending on the prevailing concentration gradient.

(i) $\Delta G° = 0$.

(j) A Michaelis constant (K_m) can be calculated.

8-4.   Discounting the Transverse Carrier Model. At one time, membrane biologists thought that transport proteins might act by binding a solute molecule or ion on one side of the membrane and then diffusing across the membrane to release the solute molecule on the other side. We now know that this transverse carrier model is almost certainly wrong. Suggest two reasons that argue against such a model. One of your reasons should be based on our current understanding of membrane structure and the other on thermodynamic considerations.

8-5.   Potassium Ion Transport. Most of the cells of your body pump potassium ions inward to maintain an internal K⁺ concentration that is 35 times the external concentration.

(a)   What is ΔG (at 37°C) for the transport of potassium ions into a cell that maintains no membrane potential across its plasma membrane?

(b)   For a nerve cell with a membrane potential of −60 mV, what is ΔG for the inward transport of potassium ions at 37°C?

(c)   For the nerve cell in part b, what is the maximum number of potassium ions that can be pumped inward by the hydrolysis of one ATP molecule if the ATP/ADP ratio in the cell is 5 : 1 and the inorganic phosphate concentration is 10 mM? (Assume $\Delta G°' = -7.3$ kcal/mol for ATP hydrolysis.)

8-6.   Ion Gradients and ATP Synthesis. The ion gradients maintained across the plasma membranes of most cells play a very significant role in cellular energetics. Ion gradients are often either generated by the hydrolysis of ATP or used to make ATP by the phosphorylation of ADP.

(a)   Cite an example in which ATP is used to generate and maintain an ion gradient. What is another way that an ion gradient can be generated and maintained?

(b)   Cite an example in which an ion gradient is used to make ATP. What is another use to which ion gradients are put?

(c)   Assume that the sodium ion concentration is 12 mM inside a cell and 145 mM outside the cell, and that the membrane potential is −90 mV. Can a cell use ATP hydrolysis to drive the outward transport of sodium ions on a 2 : 1 basis (two sodium ions transported per ATP hydrolyzed) if the ATP/ADP ratio is 5, the inorganic phosphate concentration is 50 mM, and the temperature is 37°C? What about on a 3 : 1 basis? Explain your answers.

(d)   Assume that a bacterial cell maintains a proton gradient across its plasma membrane such that the pH inside the cell is 8.0 when the outside pH is 7.0. Can the cell use the proton gradient to drive ATP synthesis on a 1 : 1 basis (one ATP synthesized per proton transported) if the membrane potential is +180 mV, the temperature is 25°C, and the ATP, ADP, and inorganic phosphate concentrations are as in part c? What about on a 1 : 2 basis? Explain your answers.

8-7.   Sodium Ion Transport. A marine protozoan is known to pump sodium ions outward by a simple ATP-driven Na⁺ pump that operates independently of potassium ions. The intracellular concentrations of ATP, ADP, and P$_i$ are 20, 2, and 1 mM, respectively, and the membrane potential is −75 mV.

(a)   Assuming that the pump transports three sodium ions outward per molecule of ATP hydrolyzed, what is the lowest internal sodium ion concentration that can be maintained at 25°C when the external sodium ion concentration is 150 mM?

(b)   If you were dealing with an uncharged molecule rather than an ion, would your answer for part a be higher or lower, assuming all other conditions remained the same? Explain.

8-8.   The Case of the Acid Stomach. The gastric juice in your stomach has a pH of 2.0. This acidity is due to the secretion of protons (hydrogen ions) into the stomach by the epithelial cells of the gastric mucosa. Epithelial cells have an internal pH of 7.0 and a membrane potential of −70 mV (inside negative) and function at body temperature (37°C).

(a)   What is the concentration gradient of protons across the epithelial membrane?

(b)   Calculate the free energy change associated with the secretion of 1 mole of protons into gastric juice at 37°C.

(c)   Do you think that proton transport can be driven by ATP hydrolysis at the ratio of one molecule of ATP per proton transported?

(d)   If protons were free to move back into the cell, calculate the membrane potential that would be required to prevent them from doing so.

8-9.   Sodium Cotransport as a Mechanism of Amino Acid Uptake. The epithelial cells that line your small intestine are thought to take up amino acids from the gut by a process of cotransport, in which the electrochemical gradient of sodium ions across the plasma membrane drives the uptake of amino acids against their concentration gradients. The following values were obtained at 37°C:

pH of epithelial cells = 7.0

V_m = membrane potential = − 65 mV (inside negative)

[Na⁺]$_{inside}$ = 7.5 mM [Na⁺]$_{outside}$ = 105 mM

[glycine]$_{inside}$ = 15.0 mM [glycine]$_{outside}$ = 0.10 mM

[aspartate]$_{inside}$ = 22.5 mM [aspartate]$_{outside}$ = 0.15 mM

The structures of glycine and aspartate at pH 7.0 are as follows:

Glycine

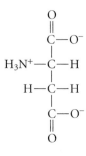

Aspartate

(a) Calculate ΔG for the movement of 1 mole-equivalent of sodium ions into an epithelial cell.

(b) Calculate ΔG for the movement of 1 mole of glycine into an epithelial cell. What is the minimum number of sodium ions that must be cotransported with one glycine molecule to drive the uptake?

(c) Calculate ΔG for the movement of 1 mole of aspartate into an epithelial cell. What is the minimum number of sodium ions that must be cotransported with one aspartate molecule to drive the uptake?

(d) Explain why the number of sodium ions required for amino acid cotransport is different for glycine (part b) and aspartate (part c).

(e) Do you think the diffusion of aspartate back out through the epithelial cell membrane would be a problem? Why or why not? What about that of glycine?

• 8-10.   The Calcium Pump of the Sarcoplasmic Reticulum.
Muscle cells use calcium ions to regulate the contractile process. Calcium is both released and taken up by the sarcoplasmic reticulum (SR). The release of calcium from the SR activates muscle contraction, and ATP-driven calcium uptake causes the muscle cell to relax afterward. When muscle tissue is disrupted by homogenization, the SR forms small vesicles called *microsomes* that maintain their ability to take up calcium. In the experiment shown in Figure 8-16, a reaction medium was prepared to contain 5 m*M* Mg²⁺-ATP and 0.1 *M* KCl at pH 7.5. An aliquot of SR microsomes containing 1.0 mg protein was added to 1 mL of the reaction mixture, followed by 0.4 mmol of calcium. Two minutes later, a calcium ionophore was added. (An ionophore is a substance that facilitates the movement of an ion across a membrane.) ATPase activity was monitored during the additions, with the results shown in the figure.

(a) What is the ATPase activity, calculated as micromoles of ATP hydrolyzed per milligram of protein per minute?

(b) The ATPase is calcium-activated, as shown by the increase in ATP hydrolysis when the calcium was added and the decrease in hydrolysis when all the added calcium was taken up into the vesicles 1 minute after it was added. How many calcium ions are taken up for each ATP hydrolyzed?

(c) The final addition is an ionophore that carries calcium ions across membranes. Why does ATP hydrolysis begin again?

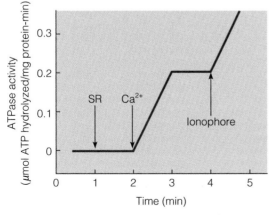

Figure 8-16   Calcium Uptake by the Sarcoplasmic Reticulum. See Problem 8-10.

• 8-11.   Inverted Vesicles. An important advance in transport research was the development of methods for making closed membrane vesicles that retain the activity of certain transport systems. One such system uses resealed vesicles from red blood cell membranes, in which the orientation of membrane proteins may be either the same as in the intact cell (right-side-out) or inverted (inside-out). Such vesicles have demonstrable ATP-driven Na⁺/K⁺ pump activity. By resealing the vesicles in one medium and then placing them in another, it is possible to have ATP, sodium ions, and potassium ions present inside the vesicle, present outside the vesicle, or not present at all.

(a) Suggest one or two advantages that such vesicles might have compared to intact red blood cells for studying the Na⁺/K⁺ pump. Can you think of any possible disadvantages?

(b) For the inverted vesicles, indicate whether each of the following should be present inside the vesicle (I), outside the vesicle (O), or not present at all (N) in order to demonstrate ATP hydrolysis: Na⁺, K⁺, ATP.

(c) If you were to plot the rate of ATP hydrolysis as a function of time after initiating transport in such inverted vesicles, what sort of a curve would you expect to obtain?

• 8-12.   Ouabain Inhibition. Ouabain is a very specific inhibitor of the active transport of sodium ions out of the cell and is therefore a valuable tool in studies of membrane transport mechanisms. Which of the following processes in your own body would you expect to be sensitive to inhibition by ouabain? Explain your answer in each case.

(a) Facilitated diffusion of glucose into a muscle cell

(b) Active transport of dietary phenylalanine across the intestinal mucosa

(c) Uptake of potassium ions by red blood cells

(d) Active uptake of lactose by the bacteria in your intestine

• 8-13.   Deducing the Structure. As is the case for most membrane proteins, the three-dimensional structure of the glucose transporter GluT1 has not yet been determined by X-ray crystallography because of difficulties encountered in obtaining crystals. However, membrane biologists have deduced its likely structure based on its amino acid sequence and a variety of topological data. Using the following information, propose a structure of GluT1 and indicate how the protein might be organized within the membrane.

1. GluT1 consists of a single polypeptide with 492 amino acids.

2. The protein has 12 hydrophobic segments, five of which (#3, 5, 7, 8 and 11) contain three or four polar or charged amino acid residues, often separated by several hydrophobic residues. (Note: There are about 3.5 residues per turn in an α-helical protein.)

3. When a helix with a polar residue in every third or fourth position is viewed end-on, the hydrophilic residues cluster along one side of the helix.

4. Side-by-side association of five such helices, each with its hydrophilic side oriented inward, can generate a transmembrane channel lined with polar and charged residues, which can form transient hydrogen bonds with glucose as it moves through the channel.

Suggested Reading

General References

Assmann, S. M., and L. L. Haubrick. Transport proteins of the plant plasma membrane. *Curr. Opin. Cell Biol.* 8 (1996): 458.

Baldwin, S. A., ed. *Membrane Transport: A Practical Approach.* New York: Oxford University Press, 2000.

Hoffman, J. F., and M. J. Welsh, eds. *Molecular Biology of Membrane Transport Disorders.* New York: Plenum, 1996.

Stein, W. D. *Channels, Carriers, and Pumps: An Introduction to Membrane Transport.* New York: Academic Press, 1990.

Van Winkle, L. J. *Biomembrane Transport.* San Diego: Academic Press, 1999.

Facilitated Diffusion

Chrispeels, M. J., and C. Maurel. Aquaporins: The molecular basis of facilitated water movement through living plant cells? *Plant Physiol.* 105 (1994): 9.

Henderson, P. J. F. The 12-transmembrane helix transporters. *Curr. Opin. Cell Biol.* 5 (1993): 708.

King, L. S., and P. Agre. Pathophysiology of the aquaporin water channels. *Annu. Rev. Physiol.* 58 (1996): 619.

Lienhard, G. E., J. W. Slot, D. E. James, and M. M. Mueckler. How cells absorb glucose. *Sci. Amer.* 266 (January 1992): 86.

Malandro, M. S., and M. S. Kilberg. Molecular biology of mammalian amino acid transporters. *Annu. Rev. Biochem.* 65 (1996): 305.

Silverman, M. Structure and function of glucose transporters. *Annu. Rev. Biochem.* 60 (1994): 757.

Zeuthen, T. How water molecules pass through aquaporins. *Trends Biochem. Sci.* 26 (2001): 77.

Active Transport

Higgins, C. F., and K. J. Linton. The xyz of ABC transporters. *Science* 293 (2001): 1782.

Horisberger, J. D., V. Lemas, J. P. Krahenbuhl, and B. C. Rossier. Structure-function relationships of the Na,K-ATPase. *Annu. Rev. Physiol.* 53 (1991): 565.

Kaback, H. R., M. Sahin-Tóth, and A. B. Weinglass. The kamikaze approach to membrane transport. *Nature Reviews Mol. Cell Biol.* 2 (2001): 610.

Kühlbrandt, W. Bacteriorhodopsin—the movie. *Nature* 406 (2000): 569.

Läuger, P. *Electrogenic Ion Pumps.* Sunderland, MA: Sinauer, 1991.

Mercer, R. W. Structure of the Na,K-ATPase. *Internat. Rev. Cytol.* 137C (1993): 139.

Moller, J. M., B. Juul, and M. le Maire. Structural organization, ion transport, and energy transduction of P-type ATPases. *Curr. Opin. Cell Biol.* 4 (1992): 1

Nelson, N. Organellar proton-ATPases. *Curr. Opin. Cell Biol.* 4 (1992): 654.

Stevens, T., and M. Forgac. Structure, function, and regulation of the vacuolar (H⁺)-ATPase. *Annu. Rev. Cell Biol.* 13 (1997): 779.

Wright, E. M., D. D. F. Loo, E. Turk, and B. A. Hirayama. Sodium cotransporters. *Curr. Opin. Cell Biol.* 8 (1996): 468.

Box 8B: *Membrane Transport, Cystic Fibrosis, and the Prospects for Gene Therapy*

Alton, E. W. F. W., M. Stern, R. Farley et al. Cationic lipid-mediated *CFTR* gene transfer to the lungs and nose of patients with cystic fibrosis: A double-blind placebo-controlled trial. *Lancet* 353 (1999): 947.

Collins, F. S. Cystic fibrosis: Molecular biology and therapeutic implications. *Science* 256 (1992): 774.

Porteous, D. J., J. R. Dorin, G. McLachlan et al. Evidence for the safety and efficacy of DOTAP cationic liposome mediated *CFTR* gene transfer to nasal epithelium of patients with cystic fibrosis. *Gene Therapy* 4 (1997): 210.

Riordan, J.R. The cystic fibrosis transmembrane conductance regulator. *Annu. Rev. Physiol.* 55 (1993): 609.

Welsh, M. J., and A. E. Smith. Cystic fibrosis. *Sci. Amer.* 273 (December 1995): 52.

9

Signal Transduction Mechanisms: I. Electrical Signals in Nerve Cells

In the previous chapter we saw that a key property of cell membranes is the ability to regulate the flow of ions between the interior of the cell and the outside environment. Since ions are charged solutes, cells can, in effect, regulate the flow of electric current through their membranes, as well as the electrical potentials across them. The most dramatic example of the regulation of the electrical properties of cells is the functioning of the main cellular component of the nervous system of animals: the nerve cell, or *neuron.* By studying neurons, we will gain important insights into how cells in general regulate their electrical properties, and how cells can send signals to one another. This again underscores an important principle in biological research: Cellular functions are often best studied in cells that are highly specialized for the function of interest.

In this chapter, we will examine the cellular mechanisms that allow neurons to act as the "wiring" of the nervous system, the means by which they can transmit signals at remarkably high speeds throughout the body, and the ways these cellular "wires" connect with one another to form circuits. In the first part of this chapter, we will see that although virtually all cells maintain electrical potentials across their plasma membranes, nerve cells have special mechanisms for using this potential to transmit information over long distances. In the second part of the chapter, we will focus on the processes by which the information is passed between the *senders*—nerve cells or sensory cells— and the *receivers*—other nerve cells, glands, or muscle cells.

The Nervous System

Almost all animals have a **nervous system,** in which electrical impulses are transmitted along the specialized plasma membranes of nerve cells. The nervous system performs three functions: It *collects* information from the environment ("the light just turned green"), it *processes* that information ("green means go"), and it *elicits responses* to that information by triggering specific *effectors,* usually muscle tissue or glands ("push on the accelerator").

To accomplish these functions, the nervous system has special components for sensing and processing information and for triggering the appropriate response (Figure 9-1). We will briefly introduce these components here for vertebrates, and discuss them in more detail throughout the chapter. The nervous system is divided into two components, the central nervous system and the peripheral nervous system. The **central nervous system (CNS)** consists of the brain and the spinal cord, including both sensory and motor cells; the **peripheral nervous system (PNS)** comprises all other sensory and motor components, including the somatic nervous system and the autonomic nervous system. The **somatic nervous system** controls voluntary movements of skeletal muscles, whereas the **autonomic nervous system** controls the involuntary activities of cardiac muscle, the smooth muscles of the gastrointestinal tract and blood vessels, and a variety of secretory glands.

Cells that make up the nervous system can be broadly divided into two groups: neurons and glial cells. **Neurons** can be subdivided into three basic types based on function: sensory neurons, motor neurons, and interneurons. **Sensory neurons** are a diverse group of cells specialized for the detection of various types of stimuli; they provide a continuous stream of information to the brain about the state of the body and its environment. Examples of sensory neurons include the photoreceptors of the retina (used in vision), olfactory neurons (smell), and the various touch,

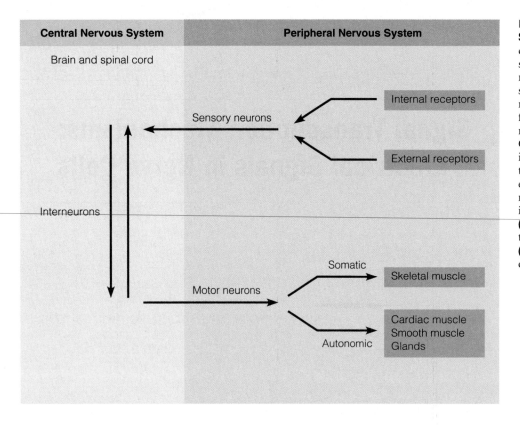

Central Nervous System	Peripheral Nervous System

Brain and spinal cord

Internal receptors

Sensory neurons

External receptors

Interneurons

Somatic

Motor neurons → Skeletal muscle

Autonomic → Cardiac muscle / Smooth muscle / Glands

Figure 9-1   The Vertebrate Nervous System. The nervous system consists of the central nervous system (tan) and the peripheral nervous system (yellow). The sensory neurons of the peripheral nervous system receive information from **(a, b)** external and **(c)** internal receptors and transmit it to the CNS. Interneurons in the CNS integrate and coordinate response to the sensory input, a primary example being **(d)** cerebral cortical neurons. Motor responses originate in the CNS and are transmitted to **(e)** skeletal muscles by neurons of the somatic nervous system and to **(f, g)** involuntary muscles by means of the autonomic nervous system.

pressure, pain, and temperature-sensitive neurons located in the skin and joints. **Motor neurons** transmit signals from the CNS to the muscles or glands they **innervate**—that is, the tissues to which they send signals. **Interneurons** process signals received from other neurons and relay the information to other parts of the nervous system.

The term **glial cell** (from *glia,* the Greek word for "glue") encompasses a variety of different cell types, including microglia, oligodendrocytes, Schwann cells, and astrocytes. *Microglia* are phagocytic cells that fight infections and remove debris. *Oligodendrocytes* and *Schwann cells* form the insulating *myelin sheath* around neurons of the CNS and those of the peripheral nerves, respectively. *Astrocytes* control access of blood vessels to the extracellular fluid surrounding nerve cells, thereby forming the blood-brain barrier.

Intricate networks of neurons make up the complex tissues of the brain that are responsible for coordinating nervous function—in humans, about 10 billion nerve cells. Each neuron can receive input from thousands of other neurons, so the brain's connections number well into the trillions. Computers, by comparison, do not even approach this number of connections, despite their seeming complexity. Figure 9-2 gives you a sense of how an actual neuron compares in size with a computer microprocessor.

Like data in a computer, however, neural information consists of individual "bits," or *signals*. Rather than discussing the overall functioning of the nervous system, our purpose here is to examine the cellular mechanisms by

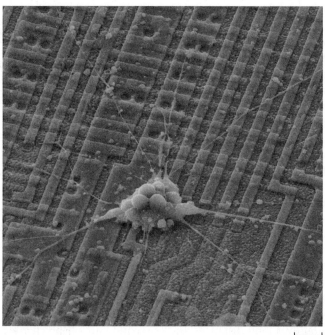

10 μm

Figure 9-2   A Neuron on a Microprocessor. This micrograph shows a neuron growing on top of a Motorola 68000 microprocessor chip. The neuron is the fundamental information-processing unit of the brain, which might be compared to the transistor as the fundamental processing unit in the computer. However, the brain has 15 billion neurons, whereas microprocessors may have up to a few million transistors (SEM).

which these electrical signals, called *nerve impulses,* are spread. In addition to understanding the functions of nerve cells specifically, we will also acquire a better appreciation for several general aspects of membrane function, of which nerve cells present a specialized, highly developed example.

Neurons Are Specially Adapted for the Transmission of Electrical Signals

The structure of a typical motor neuron is shown schematically in Figure 9-3. The **cell body** of most neurons is similar to that of other cells, consisting of a nucleus and most of the same organelles. Neurons also contain extensions, or branches, called **processes.** There are two types of processes: Those that receive signals and combine them with signals received from other neurons are called **dendrites,** and those that conduct signals, sometimes over long distances, are called **axons.** The cytoplasm within an axon is commonly referred to as **axoplasm.** Most vertebrate axons are surrounded by a discontinuous **myelin sheath,** which insulates the segments of axon separating

the **nodes of Ranvier.** Axons can be very long—up to several thousand times longer than the diameter of the cell body. For example, a motor neuron that innervates your foot has its cell body in your spinal cord, and its axon extends approximately a meter down your leg! A **nerve** is simply a tissue composed of bundles of axons.

As Figure 9-3 illustrates, a motor neuron has multiple, branched dendrites and a single axon leading away from the cell body. The axon of a typical neuron is much longer than the dendrites and forms multiple branches. Each branch terminates in structures called *terminal bulbs.* The terminal bulbs are responsible for transmitting the signal to the next cell, which may be another neuron or a muscle or gland cell. In each case, the junction is called a **synapse.** For neuron-to-neuron junctions, synapses usually occur between an axon and a dendrite, but they can occur between two dendrites. Typically, neurons have synapses with many other neurons.

Neurons display more structural variability than Figure 9-3 suggests. Some sensory neurons have only one process, which conducts signals both toward and away from the cell body. Moreover, the structure of the

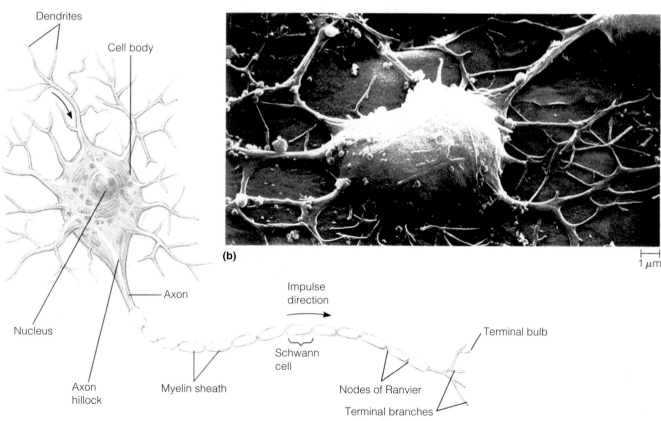

(b)

(a)

Figure 9-3   The Structure of a Typical Motor Neuron.   (a) A diagram of a typical motor neuron. The cell body contains the nucleus and most of the usual organelles. Dendrites conduct signals passively inward to the cell body; starting near the axon hillock the axon transmits signals outward (the direction of transmission is shown by black arrows). At the end of the axon are numerous terminal bulbs. Some, though not all, neurons have a discontinuous myelin sheath around their axons to insulate them electrically. Each segment of the sheath consists of a concentric layer of membranes wrapped around the axon by a Schwann cell (or an oligodendrocyte, in the CNS). The breaks in the myelin sheath, called nodes of Ranvier, are concentrated regions of electrical activity. **(b)** A scanning electron micrograph of the cell body and dendrites of a motor neuron.

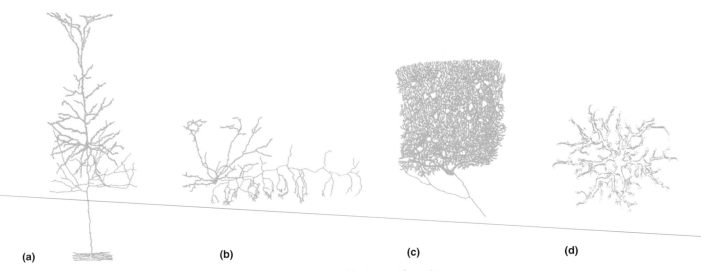

Figure 9-4 Neuron Shapes.   Neurons of the central nervous system display a wide variety of characteristic shapes. Axons are shown in yellow and dendrites in black. **(a)** A pyramidal neuron from the cerebral cortex. **(b)** Several short-axon cells in the cerebral cortex. **(c)** A Purkinje cell in the cerebellum. **(d)** An axonless horizontal cell in the retina of the eye.

processes is not random; many different classes of neurons in the central nervous system can be identified by structure alone (Figure 9-4).

Understanding Membrane Potential

Recall from Chapter 8 that **membrane potential** is a fundamental property of all cells. It results from an excess of negative charge on one side of the plasma membrane and an excess of positive charge on the other side. Cells at rest normally have an excess of negative charge inside and an excess of positive charge outside the cell; the resulting electrical potential is called the **resting membrane potential,** denoted V_m. The membrane potential can be measured by placing one tiny electrode inside the cell and another outside the cell; its value is generally given in millivolts (mV). (Do not confuse the potential V_m and the unit mV!) The electrodes compare the ratio of negative to positive charge inside the cell and outside the cell. Because the inside of a cell typically has an excess of negative charge, we say that the cell has a *negative resting membrane potential.* For example, the resting membrane potential is approximately −60 mV for the squid giant axon (see Figure 9-11).

Nerve, muscle, and certain other cell types such as the islet cells of the pancreas of vertebrates exhibit a special property called **electrical excitability.** In electrically excitable cells, certain types of stimuli trigger a rapid sequence of changes in membrane potential known as an *action potential.* During an action potential, the membrane potential changes from negative to positive values and then back to negative values again, all in little over a millisecond. In nerve cells, the action potential has the specific function of transmitting an electrical signal along the axon. To understand how nerve cells use action potentials to transmit signals, we must first examine how cells generate a resting membrane potential and how the membrane potential changes during an action potential.

The Resting Membrane Potential Depends on Differing Concentrations of Ions Inside and Outside the Neuron

The resting membrane potential develops because the cytosol of the cell and the extracellular fluid contain different compositions of cations and anions. Extracellular fluid is a watery solution of salts, including sodium chloride and lesser amounts of potassium chloride. The cytosol contains potassium rather than sodium as its main cation because of the action of the Na^+/K^+ pump (described in Chapter 8). The anions in the cytosol consist largely of macromolecules such as proteins, RNA, and a variety of other molecules that are not present outside the cell. These negatively charged macromolecules cannot pass through the plasma membrane and, therefore, remain inside the cell. The presence of these impermeable anions in the cytosol is the main reason that cells develop a resting membrane potential. The concentrations of important ions for two types of well-studied neurons, the squid axon and one type of mammalian axon, are shown in Table 9-1.

Table 9-1   Ionic Concentrations Inside and Outside Axons and Neurons

Ion	Squid Axon		Mammalian Neuron (cat motor neuron)	
	Outside (mM)	Inside (mM)	Outside (mM)	Inside (mM)
Na^+	440	50	145	10
K^+	20	400	5	140
Cl^-	560	50	125	10

To understand how a membrane potential forms, we need to recall a few basic physical principles. First, all substances tend to diffuse from an area where they are more highly concentrated to an area of their lower concentration. Cells normally have a high concentration of potassium ions inside and a low concentration of potassium ions outside. We refer to this uneven distribution of potassium ions as a *potassium ion gradient*. By convention, the potassium ion gradient is expressed as the molar concentration of potassium ions outside the cell ($[K^+]_{outside}$) divided by the molar concentration of potassium in the cytoplasm ($[K^+]_{inside}$)—that is, $[K^+]_{outside}/[K^+]_{inside}$. Given the large potassium concentration gradient, potassium ions will tend to diffuse out of the cell.

The second basic principle is that of *electroneutrality*. When ions are in solution, they are always present in pairs, one positive ion for each negative ion, so that there is no net charge imbalance. For any given ion, which we will call A, there must be an oppositely charged ion B in the solution; therefore, we refer to B as the *counterion* for A. In the cytosol, potassium ions (K^+) serve as the counterions for the trapped anions. Outside the cell, sodium (Na^+) is the main cation and chloride (Cl^-) is its counterion.

Although a solution must have an equal number of positive and negative charges overall, these charges can be locally separated so that one region has more positive charges while another region has more negative charges. It takes work to separate charges; consequently, once they have been separated, they tend to move back toward each other. The tendency of oppositely charged ions to flow back toward each other is called a **potential** or **voltage.** When negative or positive ions are actually moving, one toward the other, we say that **current** is flowing; this current is measured in amperes (A). Given these principles, we can understand how a resting membrane potential will form as a result of the ionic compositions of the cytosol and the extracellular fluid, and the characteristics of the plasma membrane.

The plasma membrane is normally permeable to potassium because of the leakiness of some types of potassium channels, which permits potassium ions to diffuse out of the cell. There are no channels for negatively charged macromolecules, however. As potassium leaves the cytosol, increasing numbers of trapped anions are left behind without counterions. Excess negative charge therefore accumulates in the cytosol and excess positive charge accumulates on the outside of the cell, resulting in a membrane potential.

Figure 9-5 illustrates the formation of a membrane potential in a container subdivided into two compartments by a semipermeable membrane. The left compartment represents the cytosol of the cell and the right side the extracellular fluid. The semipermeable membrane is permeable to potassium ions but not to negatively charged macromolecules (represented as M^-). The cytosolic compartment contains a mixture of potassium ions and negatively charged macromolecules. For simplicity, we will assume that the extracellular compartment starts with nothing but water. As a result of the concentration gradient, potassium ions diffuse across the membrane from the left side to the right side. However, the negatively charged macromolecules are not free to follow. The result is an

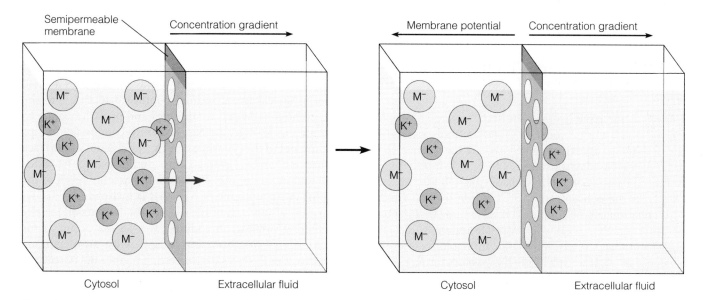

Figure 9-5   Development of the Equilibrium Membrane Potential.   A two-compartment container is used to represent the cytosol of a cell and the extracellular space, with a semipermeable membrane separating the compartments. In each container, the left-hand compartment represents the cytosol, and the right-hand compartment represents the extracellular fluid. The cytosol contains a high concentration of potassium ions (K^+) and impermeable anions (M^-) relative to their concentrations in the extracellular fluid. As potassium ions diffuse out of the cell (from left to right), the impermeable anions are left behind, creating a membrane potential. The magnitude of the membrane potential increases until an equilibrium is reached in which the electrical attraction of the anions for potassium ions prevents any further net diffusion of potassium ions out of the cell. At this point, the membrane potential is in equilibrium with the potassium ion gradient.

accumulation of anions on the left side and of cations on the right side. A membrane potential is created by a separation of negative charge from positive charge. The source of work (defined here as force × distance) needed to produce this charge separation is the potassium ion gradient.

As potassium ions diffuse from left to right in Figure 9-5, the left compartment becomes increasingly negative. This potential ultimately builds to a point at which the positively charged potassium ions are pulled back into the left compartment as fast as they leave. In this way, an equilibrium is reached in which the force of attraction due to the membrane potential balances the tendency of potassium to diffuse down its concentration gradient. This type of equilibrium, in which a chemical gradient is balanced with an electrical potential, is referred to as an **electrochemical equilibrium.** The membrane potential at the point of equilibrium is known as an **equilibrium membrane potential.** Since the potassium ion gradient does the work needed to separate negative and positive charges, resulting in an equilibrium membrane potential, the magnitude of the potassium ion gradient determines the magnitude of the equilibrium membrane potential. In the next section, we will present a mathematical description of this relationship that enables us to estimate the membrane potential.

The Nernst Equation Describes the Relationship Between Membrane Potential and Ion Concentration

The **Nernst equation** is named for the German physical chemist and Nobel laureate Walther Nernst, who first formulated it in the late 1880s in the context of his work on electrochemical cells (forerunners of modern batteries). The Nernst equation describes the mathematical relationship between an ion gradient and the equilibrium membrane potential that will form when the membrane is permeable only to that ion:

$$E_X = \frac{RT}{zF} \ln \frac{[X]_{outside}}{[X]_{inside}} \qquad (9\text{-}1)$$

Here E_X is the equilibrium membrane potential for ion X (in volts), R is the gas constant (1.987 cal/mol-degree), T is the absolute temperature in kelvins, z is the valence of the ion, F is the Faraday constant (23,062 cal/V-mol), $[X]_{outside}$ is the concentration of X outside the cell (in M), and $[X]_{inside}$ is the concentration of X inside the cell (in M). This equation can be simplified if we assume that the temperature is 20°C (a value appropriate for the squid giant axon) and that X is a monovalent cation and therefore has a valence of +1. If we substitute these values for R, T, F, and z in the Nernst equation and convert from natural logs to $\log_{10}$ ($\log_{10} = \ln/2.303$), equation 9-1 reduces to

$$E_X = 0.058 \log_{10} \frac{[X]_{outside}}{[X]_{inside}} \qquad (9\text{-}2)$$

In this simplified form we can see that for every tenfold increase in the cation gradient, the membrane potential changes by −0.058 V, or −58 mV.

Ions Trapped Inside the Cell Have Important Effects on Resting Membrane Potential

Though Figure 9-5 explains the formation of the membrane potential, it is incomplete, mainly because it does not take into account the effects of anions. The cell membrane is relatively permeable to some anions, but impermeable to others. The main anion in the extracellular fluid is chloride, which moves relatively freely across cell membranes. In addition to chloride, the cell contains impermeable anions (M^-), and to maintain electrical neutrality, it must also contain a matching number of cations. We can express this idea mathematically as

$$[C^+]_{inside} = [M^-]_{inside} + [A^-]_{inside} \qquad (9\text{-}3)$$

where C^+ and A^- represent cations and diffusible anions, respectively. Outside the cell, there will be only diffusible anions with a matching number of cations. Electrical neutrality is achieved when

$$[C^+]_{outside} = [A^-]_{outside} \qquad (9\text{-}4)$$

If we were to add up the concentrations of solutes inside the cell, we would find that there are more dissolved substances inside the cell. Therefore, water would tend to enter the cytosol, causing the cell to swell. If this process were to continue unchecked, the cell would eventually burst.

To prevent swelling and rupture, an impermeable solute is needed outside the cell to balance the osmolarity of the cytosol. The high concentration of sodium ions present in the extracellular fluid fulfills this requirement because of the poor permeability of the membrane to sodium ions. In addition, when sodium ions do leak through the membrane, they are removed from the cytosol by an ATPase known as the Na^+/K^+ pump.

The Na^+/K^+ Pump. Although the plasma membrane is relatively impermeable to sodium ions, there is always a small amount of leakage. If sodium continued to leak into the cell, eventually the cell would swell and burst. To compensate for this leakage, the **Na^+/K^+ pump** continually pumps sodium out of the cell while carrying potassium inward (see Figure 8-12). On average, the pump transports three sodium ions out of the cell and two potassium ions into the cell for every molecule of ATP that is hydrolyzed. This net transport of ions out of the cell dilutes the cytosol, thereby preventing swelling. In addition, the Na^+/K^+ pump maintains the large potassium ion gradient across the membrane that provides the basis for the resting membrane potential.

Steady-State Concentrations of Common Ions Affect Resting Membrane Potential

As we have seen, sodium, potassium, and chloride ions are the major ionic components present in both the cytosol and the extracellular fluid. Due to their unequal distributions across the cell membrane, each ion has a different impact on the membrane potential. The magnitude of the concentration gradient for each ion for a mammalian neuron is illustrated in Figure 9-6. Each ion will tend to diffuse down its electrochemical gradient (shown as red arrows in Figure 9-6) and thereby produce a change in the membrane potential.

Potassium ions tend to diffuse out of the cell, which makes the membrane potential more negative. Sodium ions tend to flow into the cell, driving the membrane potential in the positive direction and thereby causing a **depolarization** of the membrane (that is, causing the membrane potential to be less negative). Chloride ions tend to diffuse into the cell, which should, in principle, make the membrane potential more negative. However, chloride ions are also repelled by the negative membrane potential, so that chloride ions usually enter the cell in association with positively charged ions such as sodium. This paired movement nullifies the depolarizing effect of sodium entry. Increasing the permeability of cells to chloride can have two effects, both of which decrease neuronal excitability. First, the net entry of chloride ions (chloride entry without a matching cation) causes *hyperpolarization* of the membrane (that is, the membrane potential becomes more highly negative than usual). Second, when the membrane becomes permeable to sodium ions, some chloride will tend to enter the cell along with sodium. This effect of chloride entry will be prominent later when we discuss inhibitory neurotransmitters.

The Goldman Equation Describes the Combined Effects of Ions on Membrane Potential

The relative contributions of the major ions are important to the resting membrane potential because, even in its resting state, the cell has some permeability to sodium and chloride ions as well as to potassium ions. To take into account the leakage of sodium and chloride ions into the cell, we cannot use the Nernst equation because it deals with only one type of ion at a time, and it assumes that this ion is in electrochemical equilibrium. We must move from the more static concept of an equilibrium membrane potential to a consideration of **steady-state ion movements** across the membrane.

We can illustrate the concept of steady-state ion movements by returning once more to our model of a cell in electrochemical equilibrium (Figure 9-7). Remember, a cell that is permeable only to potassium will have a membrane potential equal to the equilibrium potential for potassium ions. Under these conditions, there will be no net movement of potassium out of the cell. If we assume that the membrane is also slightly permeable to sodium ions, what will happen? We know that the cell will have both a large sodium gradient across the membrane and a negative membrane potential corresponding to the potassium equilibrium potential. These forces tend to drive sodium ions into the cell. As sodium ions leak inward, the membrane is partially depolarized. At the same time, as the membrane potential is neutralized, there is now less restraining force preventing potassium from leaving the cell, so potassium ions diffuse outward, balancing the inward movement of sodium. The inward movement of sodium ions shifts the membrane potential in the positive direction, while the outward movement of potassium ions shifts the membrane potential back in the negative direction.

Movements of sodium and potassium ions across the membrane, therefore, have essentially opposite effects on the membrane potential. For one type of squid axon, the sodium ion gradient tends toward a cell membrane potential of about +55 mV, while the potassium ion gradient

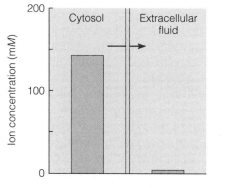
(a) Potassium ions (K⁺)

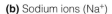
(b) Sodium ions (Na⁺)

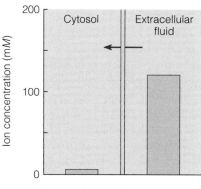
(c) Chloride ions (Cl⁻)

Figure 9-6 Relative Concentrations of Potassium, Sodium, and Chloride Ions Across the Plasma Membrane of a Mammalian Neuron. Each of the major ions in the cytosol is found in a concentration gradient across the plasma membrane, and each is capable of having an effect on membrane potential. The red arrow indicates the direction the ion would tend to flow in response to its electrochemical gradient. **(a)** Potassium ions are more concentrated in the cytosol than in the extracellular fluid. As a result, potassium ions have a tendency to move out of the cell, leaving behind trapped anions. Therefore, the loss of potassium causes the membrane potential to become more negative. **(b)** Sodium ions are much more concentrated outside the cell than inside; therefore, sodium ions tend to enter the cell. As sodium ions enter, they neutralize some of the excess negative charge in the cytosol. As a result, the membrane potential becomes more positive. **(c)** Chloride ions usually cross the membrane together with a permeable cation (normally a potassium ion). Although chloride ions are concentrated outside the cell, the extracellular potassium ion concentration is low, which limits the rate of chloride entry. In contrast, the potassium concentration in the cytosol is high, allowing chloride to leave the cell even at low cytosolic chloride concentrations.

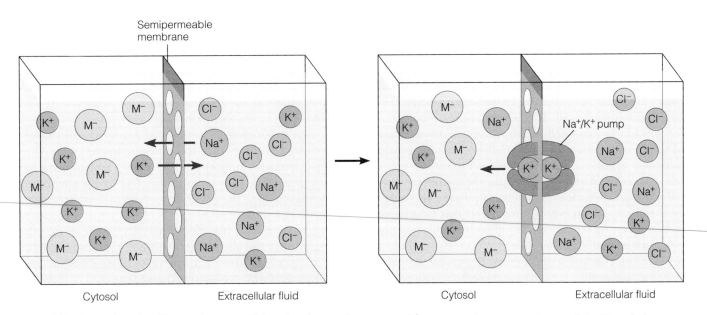

Semipermeable membrane

Cytosol   Extracellular fluid   Cytosol   Extracellular fluid

Figure 9-7   Steady-State Ion Movements.
The actual membrane potential of a cell depends on the permeability of the plasma membrane to various ions and the steady-state movements of ions across the membrane. As illustrated here with our two-compartment model, a small number of sodium ions continually leak into the cell.

This makes the membrane potential more positive, weakening the electrical restraint on the movement of potassium ions. A small number of ions now leak out of the cell. As sodium ions accumulate in the cytosol, they are pumped outward in exchange for potassium ions by the Na^+/K^+ pump. (In this figure, the Na^+ have already

been pumped out and the K^+ are being pumped in.) The end result is a low concentration of sodium ions inside the cell. The presence of a low concentration of sodium ions in the cytosol causes the membrane potential to be more positive than the equilibrium membrane potential for potassium ions.

tends toward a membrane potential of about −75 mV. At what value will the membrane potential come to rest? In principle, it could be anywhere between these points. The membrane potential now becomes a function not only of electrochemical gradients but also of the *rate* at which ions can flow through the membrane. At any given time, the membrane potential will depend on the outward rate of movement of potassium ions relative to the inward movement of sodium ions. Changes in the permeability of the cell to either ion cause corresponding changes in the membrane potential. In a living cell, sodium ions continually leak into the cell and potassium ions leak out, but steady-state concentrations of the two ions are maintained because the Na^+/K^+ pump acts to move sodium ions outward and potassium ions inward.

The pioneering neurobiologists David E. Goldman, Alan Lloyd Hodgkin, and Bernard Katz were the first to describe how gradients of several different ions each contribute to the membrane potential as a function of relative ionic permeabilities. The Goldman-Hodgkin-Katz equation, more commonly known as the **Goldman equation,** is as follows:

$$V_m = \frac{RT}{F} \ln \frac{(P_K)[K^+]_{out} + (P_{Na})[Na^+]_{out} + (P_{Cl})[Cl^-]_{in}}{(P_K)[K^+]_{in} + (P_{Na})[Na^+]_{in} + (P_{Cl})[Cl^-]_{out}} \quad \text{(9-5)}$$

Because chloride ions have a negative valence, $[Cl^-]_{inside}$ appears in the numerator and $[Cl^-]_{outside}$ in the denominator.

One of the key differences between the Nernst equation and the Goldman equation is the incorporation of terms for permeability. Here P_K, P_{Na}, and P_{Cl} are the *relative permeabilities* of the membrane for the respective ions. The use of relative permeabilities circumvents the complicated task of determining the absolute permeability of each ion. While the equation shown here takes into account only the contributions of potassium, sodium, and chloride ions, other ions could be added as well. Except under special circumstances, however, the permeability of the plasma membrane to other ions is usually so low that their contributions are negligible.

We can use a squid axon to illustrate how we can accurately estimate the resting membrane potential from the known steady-state concentrations and relative permeabilities of sodium, potassium, and chloride ions. To do so, we assign K^+ a permeability value of 1.0, and the permeability values of all other ions are determined relative to that of K^+. For squid axons, the permeability of sodium ions is only about 4% of that for potassium ions, and for chloride ions, the value is 45%. Relative values of P_K, P_{Na}, and P_{Cl} are therefore 1.0, 0.04, and 0.45, respectively. Using these values, a temperature of 20°C, and the intracellular and extracellular concentrations of Na^+, K^+, and Cl^- for the squid axon from Table 9-1, you should be able to estimate the resting membrane potential of a squid axon as −60.3 mV. Typical measured values for the resting membrane potential of a squid axon are about −60 mV, which is remarkably close to our calculated potential.

Electrical Excitability

The establishment of a resting membrane potential and its dependence on ion gradients and ion permeability are properties of almost all cells. The unique feature of electrically excitable cells is their response to membrane depolarization. Whereas a nonexcitable cell that has been temporarily and slightly depolarized will simply return to its original resting membrane potential, an electrically excitable cell that is depolarized to the same degree will respond with an action potential.

Electrically excitable cells produce an action potential because of the presence of particular types of ion channels in the plasma membrane. To understand how nerve cells communicate signals electrically, we need to know the characteristics of the ion channels in the nerve cell membrane, a subject to which we now turn.

Ion Channels Act Like Gates for the Movement of Ions Through the Membrane

In Chapter 8 we introduced **ion channels,** integral membrane proteins that form ion-conducting pores through the lipid bilayer. Recall that **voltage-gated ion channels,** as the name suggests, respond to changes in the voltage across a membrane. Voltage-gated sodium and potassium channels are responsible for the action potential. **Ligand-gated ion channels,** by contrast, open when a particular molecule binds to the channel. (*Ligand* comes from the Latin root *ligare,* meaning to bind.) Some channels appear to be leaky, and are not completely closed at rest. For example, some potassium channels are leaky, making resting cells somewhat permeable to potassium ions. Detailed knowledge of the structure and function of voltage-gated sodium and potassium channels is fundamental to understanding the events of the action potential. We begin our exploration by looking at how ion channels are studied experimentally.

Patch Clamping and Molecular Biological Techniques Allow the Activity of Single Ion Channels to Be Monitored

Our clearest picture of how channels operate comes from using a technique that permits the recording of ion currents passing through individual channels. This technique, known as *single-channel recording,* or more commonly as **patch clamping** (Figure 9-8), was developed by Erwin Neher and Bert Sackman, who earned a Nobel Prize in 1991 for their discovery.

To record single-channel currents, a fire-polished glass micropipette with a tip diameter of approximately 1 μm is carefully pressed against the surface of a cell such as a neuron (Figure 9-8a). Gentle suction is then applied, so that a tight seal forms between the pipette and the plasma membrane. There is now a patch of membrane under the mouth of the micropipette that is sealed off from the surrounding medium (Figure 9-8b). This patch is small enough that it will usually contain only one or perhaps a few ion channels. Current can enter and leave the pipette only through these

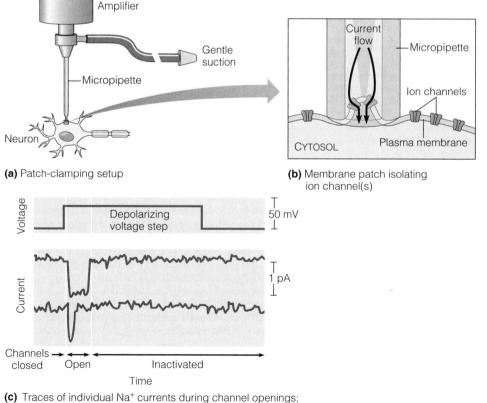

(a) Patch-clamping setup

(b) Membrane patch isolating ion channel(s)

(c) Traces of individual Na⁺ currents during channel openings; two separate traces are shown

Figure 9-8   Patch Clamping.
(a) A fire-polished micropipette with a diameter of about 1 μm is carefully placed against a cell, such as the neuron shown here, and **(b)** gentle suction is applied to form a tight seal between the pipette and the plasma membrane. The density of channels is usually low enough that only one or a few channels will be in the membrane under the mouth of the pipette. **(c)** When current flow (flow of ions) is recorded while the membrane is subjected to a depolarizing step in voltage, a burst of channel opening occurs. When each channel opens, the amount of current that flows through the channel is always the same (1 pA for the sodium channels shown here). Following the burst of channel opening, a quiescent period occurs because of channel inactivation.

channels, thereby enabling an experimenter to study various properties of the individual channels. The channels can be studied in the intact cell, or the patch can be pulled away from the cell so that the researcher has access to the cytosolic side of the membrane.

During the experimental process, an amplifier maintains voltage across the membrane with the addition of a sophisticated electronic feedback circuit called a voltage clamp (hence the term *patch clamp*). The voltage clamp keeps the cell at a fixed membrane potential, regardless of changes in the electrical properties of the plasma membrane, by injecting current as needed to hold the voltage constant. The voltage clamp then measures tiny changes in current flow—actual ionic currents through individual channels—from the patch pipette.

The patch-clamp method has been used to show that when a sodium channel opens, it always conducts the same amount of *current*—that is, the same number of ions per unit of time. There are, in other words, no partially open or closed states in which the channel conducts more or less current. Therefore, we can characterize a particular channel in terms of its conductance. *Conductance* is an indirect measure of the permeability of a channel when a specified voltage is applied across the membrane. In electrical terms, conductance is the inverse of resistance. For voltage-gated sodium channels, when a voltage of 50 mV is applied across the membrane, a current of approximately 1 picoampere (1 pA = 1×10^{-12} A) is generated. This current corresponds to about 6 million sodium ions flowing through the channel per second. This can be seen in the traces shown in Figure 9-8c.

In single-channel recording, membrane depolarization (triggered by changing the applied voltage to a more positive potential) increases the probability that a channel will open. Even before the membrane is depolarized, a sodium channel will occasionally flicker open and closed. Once the channel has opened, it cannot reopen unless the membrane potential is restored to a more negative level. The cessation of channel activity that occurs while the membrane is still depolarized is due to channel inactivation, which we will discuss later in this chapter.

Much of the present research on ion channels combines patch clamping with molecular biology to study how ion channels of excitable membranes sense and respond to changes in membrane potential. Isolation and cloning of the genes that encode channel proteins has made it possible to synthesize large amounts of channel proteins and to study their functions in lipid bilayers or in frog eggs. Specific molecular modifications or mutations of the channel can be used to determine how various regions of the channel protein are involved in channel function. This approach has been used to study the domains of the sodium and potassium channels responsible for voltage-gating.

Specific Domains of Voltage-Gated Channels Act as Sensors and Inactivators

The structure of voltage-gated ion channels falls into two different, though similar, categories. *Voltage-gated potassium channels* are *multimeric* proteins—that is, they consist of several separate protein subunits that associate with one another to form the functional channel. In the case of potassium channels, four separate protein subunits come together in the membrane, forming a central pore through which ions can pass (Figure 9-9). *Voltage-gated sodium channels*, by contrast, are large, *monomeric* proteins (in other words, they consist of a single polypeptide) with four separate domains, each of which is similar to one of

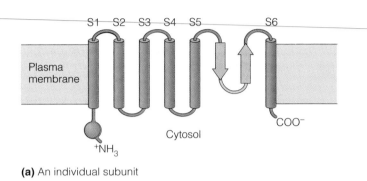

(a) An individual subunit

(b) Arrangement of two of four subunits in a channel

Figure 9-9 The General Structure of Voltage-Gated Ion Channels. The voltage-gated channels for sodium, potassium, and calcium ions all share the same basic structural themes. The channel is essentially a rectangular tube whose four walls are formed from either four subunits (e.g., potassium channels), or four domains of a single polypeptide (e.g., sodium channels). **(a)** Each subunit or domain contains six transmembrane α helices, labeled S1–S6. The fourth transmembrane α helix, S4, contains many positively charged residues, which make it a good candidate for a voltage sensor and part of the gating mechanism. For voltage-gated sodium channels and some types of potassium channels, a region near the N-terminus protrudes into the cytosol and forms an inactivating particle. **(b)** Two of the four subunits of a voltage-gated potassium channel are brought together to show how the pore forms in the middle. The inactivating particle (when present) causes channel inactivation by extending over the mouth of the channel to block the passage of ions.

the subunits of the voltage-gated potassium channel. In both kinds of channels, each subunit or domain contains six transmembrane α helices (called subunits S1–S6). One of these transmembrane α helices, S4, has positively charged amino acid groups in the middle of its transmembrane segment. When these positively charged amino acids are replaced with neutral amino acids, the channel does not open, suggesting that the transmembrane helix S4 functions as the **voltage sensor** that makes these channels responsive to changes in potential. Changes in voltage across the membrane appear to act on the charges of these amino acids, either rotating the S4 subunit upward, resulting in the opening of the channel, or downward, closing it.

The size of the central pore and, more importantly, the way it interacts with an ion, give a channel its ion selectivity (Figure 9-10a). In part, channels select for ions of the right charge through electrostatic attraction to, or repulsion from, charged amino acids at the opening of the pore. Ultimately, however, ions must bind to the channel before they can pass through. When an ion binds to a channel, most of the water molecules bound to the ion are released. A channel thus selects for ions that bind strongly enough to displace the water molecules surrounding the ion. Once this happens, the ion can pass through the pore.

Voltage-gated sodium channels have the ability to open in response to some stimulus and then to close again. This open or closed state is an all-or-none phenomenon—that is, gates do not appear to remain partially open. However, a sodium channel can go into either of two different closed states. When a sodium channel closes as a result of a shift in the S_4 subunit, a phenomenon known as **channel gating,** the channel closes but remains capable of opening again in response to a depolarizing signal (Figure 9-10b). The other closed state is referred to as **channel inactivation,** which is an important feature of voltage-gated sodium channels (Figure 9-10c). When a channel is inactivated, it is closed in such a way that it cannot reopen immediately, even if stimulated to do so. To extend our gate analogy, channel inactivation is like placing a padlock on the closed gate; only when the padlock is unlocked can the gate be opened again. Inactivation is caused by a portion of the channel protein (called the *inactivating particle*) that protrudes into the cytosol, somewhat like a ball on a chain (Figure 9-10c). During inactivation, this particle covers the opening of the

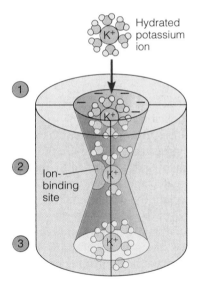

(a) Ion selectivity of channels

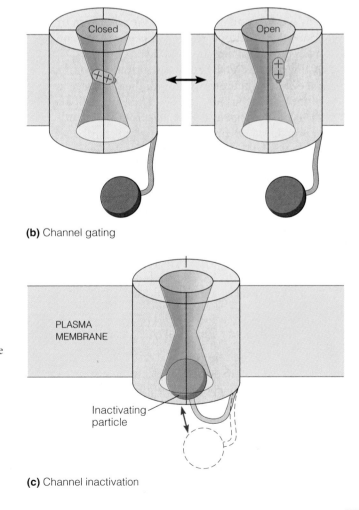

(b) Channel gating

(c) Channel inactivation

Figure 9-10   The Function of a Voltage-Gated Ion Channel.
(a) Several ways that a channel can select for different ions are shown here. ① Negative charges at the opening of the channel repel anions and attract cations. ② The pore diameter restricts the size of ions that can pass. ③ Ion-selective binding strips off water molecules so that ions can pass through the pore. **(b)** Channel gating occurs because a portion of the channel changes conformation when the membrane potential changes. Here, one of the inactivating particles is depicted as swinging open and shut. **(c)** Inactivation of sodium channels occurs when an inactivating particle blocks the opening of the pore.

channel pore. For such a channel to reactivate and open in response to a stimulus, the inactivating particle must move away from the pore. When the cytosolic side of the channel is treated with a protease or with antibodies prepared against the fragment of the channel thought to be responsible for inactivation, the inactivating particle can no longer function, and channels can no longer be inactivated.

As we will see in the next few sections of this chapter, the regulation of ion channels is crucial for the proper functioning of neurons. Recently, defects in several voltage-gated ion channels have been linked to human neurological diseases (such defects have been termed "channelopathies"). For example, humans carrying mutations in certain K^+ channels suffer from ataxia (a defect in muscle coordination), and one form of epilepsy is caused by a mutation in one type of voltage-gated Na^+ channnel.

The Action Potential

We have seen how an ion gradient across a selectively permeable membrane can generate a membrane potential, and how, according to the Goldman equation, the membrane potential will change in response to changes in ion permeability. We have also examined the nature of membrane ion channels, which regulate the permeability of the membrane to the different ions. Now we are ready to explore how the coordinated opening and closing of ion channels can lead to an action potential. We will begin by examining how the membrane potential is measured and how it changes during an action potential.

One of the great technical advances in understanding the electrical events in neurons came with the discovery in the 1930s of the giant axons of the squid. These nerve fibers stimulate the explosive expulsion of water from the mantle cavity of the squid, enabling it to propel itself quickly backward to escape predators (Figure 9-11). The **squid giant axon** has a diameter of about 0.5–1.0 mm, allowing the easy insertion of microelectrodes to measure and control electrical potentials and ionic currents across the axonal membrane. Figure 9-12 illustrates the method for doing this. Alan Hodgkin and Andrew Huxley used the squid giant axon to learn how action potentials are generated. Since its discovery, the squid giant axon has become one of the most extensively studied and well characterized of all axons, and we will use it as the model system for our discussion.

Action Potentials Propagate Electrical Signals Along an Axon

A resting neuron is a system poised for electrical action. As we have seen, the membrane potential of the cell is set by a delicate balance of ion gradients and ion permeability. Depolarization of the membrane upsets this balance. If the level of depolarization is small—less than about +20 mV—the membrane potential will normally drop back to resting levels without further consequences. Further depolarization brings the membrane to the **threshold**

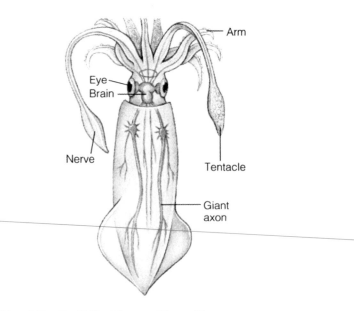

Figure 9-11 Squid Giant Axons. The squid nervous system includes motor nerves that control swimming movements. The nerves contain giant axons (fibers) with diameters ranging up to 1 mm, providing a convenient system for studying resting and action potentials in a biological membrane.

potential. Above the threshold potential, the nerve cell membrane undergoes rapid and dramatic alterations in its electrical properties and permeability to ions, and an action potential is initiated.

An **action potential** is a brief but large electrical depolarization and repolarization of the neuronal plasma membrane caused by the inward movement of sodium ions and the subsequent outward movement of potassium ions. These ion movements are controlled by the opening and closing of voltage-gated sodium and potassium channels. In fact, we can explain the development of an action potential solely in terms of the behavior of these channels. Once an action potential is initiated in one region of the membrane, it will travel along the membrane away from the site of origin by a process called **propagation.**

Action Potentials Involve Rapid Changes in the Membrane Potential of the Axon

The development and propagation of an action potential can be readily studied in large axons such as those of the squid using an apparatus similar to that shown in Figure 9-12a. An electrode called the *stimulating electrode* is connected to a power source and inserted into the axon some distance from the recording electrode (Figure 9-12b). A brief impulse from this stimulating electrode depolarizes the membrane by about 20 mV (i.e., from −60 to about −40 mV), bringing the neuron beyond the threshold potential. This triggers an action potential that propagates away from the stimulating electrode. As the action potential passes the recording electrode, the voltmeter or oscilloscope will display the characteristic pattern of membrane potential changes shown in Figure 9-13.

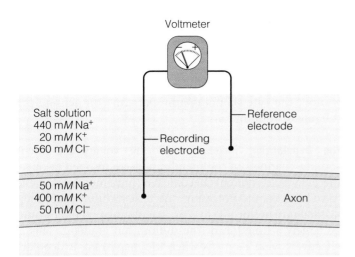

(a) Measuring the resting membrane potential in a squid axon

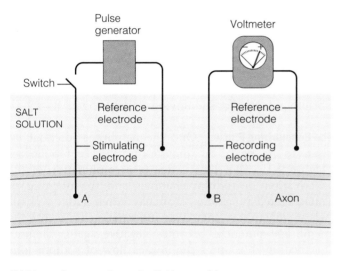

(b) Measuring an action potential in a squid axon

Figure 9-12   An Apparatus for Measuring Membrane Potentials.
(a) Measurement of the resting membrane potential requires two electrodes, one inserted inside the axon (the recording electrode) and one placed in the fluid surrounding the cell (the reference electrode). Differences in potential between the recording and reference electrodes are amplified by a voltage amplifier and displayed on a voltmeter, an oscilloscope, or a computer monitor. **(b)** Measurement of an action potential requires four electrodes, one in the axon for stimulation, another in the axon for recording, and two in the fluid surrounding the cell for reference. The stimulating electrode is connected to a pulse generator, which delivers a pulse of current to the axon when the switch is momentarily closed. The nerve impulse this generates is propagated down the axon and can be detected a few milliseconds later by the recording electrode. The impulse is detected as a transient change in transmembrane potential, measured with respect to the reference electrode.

In less than a millisecond, the membrane potential rises dramatically from the resting membrane potential to about +40 mV—the interior of the membrane actually becomes positive for a brief period. The potential then falls somewhat more slowly, dropping to about −75 mV (*undershoot* or *hyperpolarization*) before stabilizing again

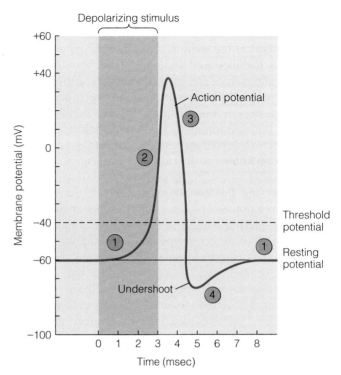

Figure 9-13   The Action Potential of the Squid Axon. ① The resting potential prior to the start of the action potential is approximately −60 mV. ② An action potential begins when the neuron is depolarized by about 20 mV to a point known as the threshold potential. Once the action potential is initiated, the potential swings rapidly in the positive direction. At the peak of the positive swing, the membrane potential reaches a value of about +40 mV. ③ Once the cell reaches the peak positive potential, it begins to repolarize, returning to a negative membrane potential. ④ Repolarization often leads to a membrane potential that is actually hyperpolarized, or more negative than the resting potential. This is referred to as undershoot. The membrane potential then returns to its resting state.

at the resting potential of about −60 mV. As Figure 9-13 indicates, the complete sequence of events during an action potential takes place within a few milliseconds.

The apparatus shown in Figure 9-12 can also be used to measure the ion currents that flow through the membrane at different phases of an action potential. To do so, an additional electrode known as the *holding electrode* is inserted into the cell and connected to a voltage clamp, thereby enabling the investigator to set and hold the membrane at a particular potential, regardless of changes in the membrane's electrical properties. Using the voltage-clamp apparatus, a researcher can measure the current flowing through the membrane at any given membrane potential. Such experiments have contributed fundamentally to our present understanding of the mechanism that causes an action potential.

Action Potentials Result from the Rapid Movement of Ions Through Axonal Membrane Channels

In a resting neuron, the voltage-dependent sodium and potassium channels are usually closed. Because of the leakiness of potassium channels, the cell is roughly 100 times more permeable to potassium than to sodium ions at this point. When a region of the nerve cell is slightly depolarized, a fraction of the sodium channels respond and open. As they do, the increased sodium current acts to further depolarize the membrane. Increasing depolarization causes an even larger sodium current to flow, which depolarizes the membrane even more. This relationship between depolarization, the opening of voltage-gated sodium channels, and an increased sodium current constitutes a positive feedback loop known as the *Hodgkin cycle* (Figure 9-14).

Subthreshold and Threshold Depolarization. If the Hodgkin cycle were not opposed by other forces, even a small amount of sodium entering the cell would always lead to complete depolarization of the cell membrane. However, the outward movement of potassium ions restores the resting membrane potential. As mentioned earlier, when the membrane is depolarized by a small amount, the membrane potential recovers and no action potential is generated. Levels of depolarization that are too small to produce an action potential are referred to as *subthreshold depolarizations.*

To understand why subthreshold depolarizations lead to recovery whereas larger depolarizations lead to an action potential, we need to consider how potassium ions respond to depolarization. When sodium ions enter a cell and begin to depolarize the membrane, the electrical restraint that keeps potassium ions in the cell is weakened, potassium ions diffuse outward, and the membrane potential becomes more negative again.

This outward movement of potassium ions can effectively oppose the Hodgkin cycle as long as the rate of outflow of potassium ions is greater than or equal to the rate of inward movement of sodium ions. The rate of

movement of sodium ions into the cell varies with the degree of depolarization: The greater the depolarization, the faster sodium enters. Up to a point, increasing rates of sodium entry are matched by increasing rates of potassium exit. For subthreshold levels of depolarization, the rate of potassium exit can compensate for the rate of sodium entry.

The Depolarizing Phase. If all the voltage-dependent sodium channels in the membrane were to open at once, the cell would suddenly become ten times more permeable to sodium than to potassium. Because sodium would then be the more permeable ion, the membrane potential would be largely a function of the sodium ion gradient. This is effectively what happens when the membrane is depolarized past the threshold potential (Figure 9-15, ① and ②). Once the threshold potential is reached, potassium exit (efflux) can no longer compensate for the rate of sodium entry (influx). At this point, the membrane potential shoots rapidly upward. When the rate of sodium entry slightly exceeds the maximum rate of potassium exit, an action potential is triggered. When the membrane potential peaks, at approximately +40 mV, the action potential approaches, although it does not actually reach, the equilibrium potential for sodium ions (about +55 mV).

The Repolarizing Phase. Once the membrane potential has risen to its peak, the membrane quickly repolarizes (Figure 9-15, ③). This is due to a combination of the inactivation of sodium channels and the opening of voltage-gated potassium channels. When sodium channels are inactivated, they close and remain closed until the membrane potential becomes negative again. Channel inactivation thus stops the inward flow of sodium ions and temporarily blocks the Hodgkin cycle. The cell will now automatically repolarize as potassium ions leak out.

The difference in response speed between voltage-gated potassium channels and voltage-gated sodium channels plays an important role in the generation of the action potential. When the cell is depolarized, the potassium channels open more slowly. As a result, an action potential begins with an increase in the membrane's permeability to sodium, which depolarizes the membrane, followed by an increased permeability to potassium, which repolarizes the membrane.

The Hyperpolarizing Phase (Undershoot). At the end of an action potential, most neurons show a transient **hyperpolarization,** or **undershoot,** in which the membrane potential briefly becomes even more negative than it normally is at rest (Figure 9-15, ④). The undershoot occurs because of the increased potassium permeability that exists while voltage-gated potassium channels remain open. Note that the potential of the undershoot closely approximates the equilibrium potential for potassium ions (about −75 mV for the squid axon). As the voltage-gated potassium channels close, the membrane potential returns to its original resting state. Note that the restora-

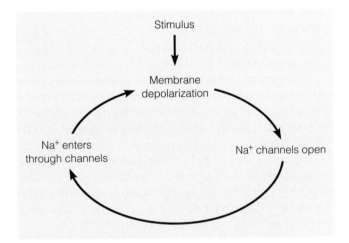

Figure 9-14 The Hodgkin Cycle. This diagram illustrates the positive feedback relationship between depolarization, the opening of sodium channels, and the corresponding increase in sodium current. As long as the membrane's permeability to potassium ions is greater than its permeability to sodium ions, diffusion of potassium ions out of the cell will counteract the Hodgkin cycle. When the membrane's permeability to sodium ions approaches its permeability to potassium ions, the Hodgkin cycle will operate unopposed, leading to maximum sodium permeability.

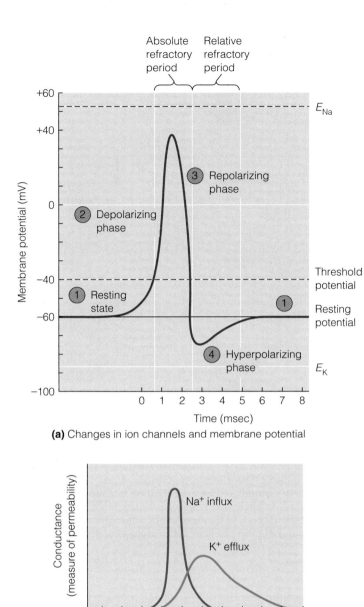

(a) Changes in ion channels and membrane potential

(b) Change in membrane conductance

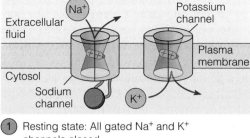

① Resting state: All gated Na$^+$ and K$^+$ channels closed

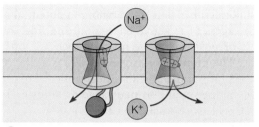

② Depolarizing phase: Na$^+$ channels open

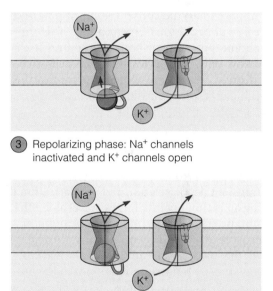

③ Repolarizing phase: Na$^+$ channels inactivated and K$^+$ channels open

④ Hyperpolarizing phase (undershoot): K$^+$ channels remain open and Na$^+$ channels inactivated

Figure 9-15 Changes in Ion Channels and Currents in the Membrane of a Squid Axon During an Action Potential. **(a)** The change in membrane potential caused by movement of Na$^+$ and K$^+$ through their voltage-gated channels, which are shown at each step of the action potential at right. The absolute refractory period is caused by sodium channel inactivation. Notice that, at the peak of the action potential, the membrane potential approaches the E_{Na} (sodium equilibrium potential) value of about +55 mV; similarly, the potential undershoots nearly to the E_K (potassium equilibrium potential) value of about −75 mV. **(b)** The change in membrane conductance (permeability of the membrane to specific ions). The depolarized membrane initially becomes very permeable to sodium ions, facilitating a large inward rush of sodium. Thereafter, as permeability to sodium declines, the permeability of the membrane to potassium increases transiently, causing the membrane to hyperpolarize.

tion of the resting potential following an action potential does not use the Na$^+$/K$^+$ pump, but involves the passive movements of ions. In cells that that have been treated with a metabolic inhibitor so that they cannot produce ATP (and hence their Na$^+$/K$^+$ pumps cannot use it), action potentials can still be generated.

The Refractory Periods. For a few milliseconds after an action potential, it is impossible to trigger a new action potential. During this interval, known as the **absolute refractory period,** sodium channels are inactivated and cannot be opened by depolarization. During the period of the undershoot, when the sodium channels have reactivated and are capable of opening again, it is possible but difficult to trigger an action potential, because sodium currents are opposed by larger potassium currents. This interval is known as the **relative refractory period.**

Changes in Ion Concentrations Due to an Action Potential. Our discussion of ion movements might give the impression that an action potential involves large changes in the cytosolic concentrations of sodium and potassium ions. In fact, during a single action potential, the cellular concentrations of sodium and potassium ions hardly change at all. Remember that the membrane potential is due to a slight excess of negative charge on one side and a slight excess of positive charge on the other side of the membrane. The number of excess charges is a tiny fraction of the total ions in the cell, and the number of ions that must cross the membrane to neutralize or alter the balance of charge is likewise small.

Nevertheless, intense neuronal activity can lead to significant changes in overall ion concentrations. For example, as a neuron continues to generate large numbers of action potentials, the concentration of potassium outside the cell will begin to rise perceptibly. This can affect the membrane potential of both the neuron itself and surrounding cells. Astrocytes, the glial cells that form the blood-brain barrier, are thought to control this problem by taking up excess potassium ions.

Action Potentials Are Propagated Along the Axon Without Losing Strength

In order for neurons to transmit signals to one another, the transient depolarization and repolarization that occur during an action potential must travel along the neuronal membrane. The depolarization at one point on the membrane spreads to adjacent regions through a process called the **passive spread of depolarization,** in which cations (mostly potassium ions) move away from the site of depolarization to regions of membrane where the potential is more negative. As a wave of depolarization spreads passively away from the site of origin, it also decreases in magnitude. This fading of the depolarization with distance from the source makes it difficult for signals to travel very far by passive means only. For signals to travel longer distances, an action potential must be *propagated*, or actively generated, from point to point along the membrane.

To understand the difference between the passive spread of depolarization and the propagation of an action potential, consider how a signal travels along the neuron from the site of origin at the dendrites to the end of the axon (Figure 9-16). Incoming signals are transmitted to a neuron at synapses that form points of contact between the terminal bulbs of the transmitting neuron and the dendrites of the receiving neuron. When these incoming signals depolarize the dendrites of the receiving neuron, the depolarization spreads passively over the membrane from the dendrites to the base of the axon, the **axon hillock.** The

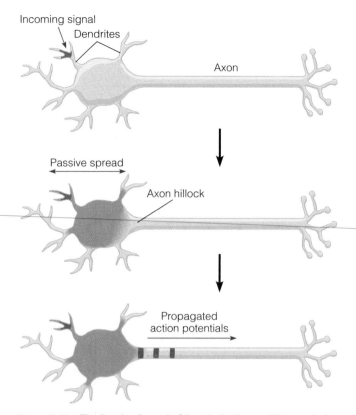

Figure 9-16 The Passive Spread of Depolarization and Propagated Action Potentials in a Neuron.   The transmission of a nerve impulse along a neuron depends on both the passive spread of depolarization and the propagation of action potentials. A neuron is stimulated when its dendrites receive a depolarizing stimulus from other neurons. A depolarization starting at a dendrite spreads passively over the cell body to the axon hillock, where an action potential forms. The action potential is then propagated down the axon.

axon hillock is the region where action potentials are initiated most easily. This is because sodium channels are distributed sparsely over the dendrites and cell body but are concentrated at the axon hillock and nodes of Ranvier; a given amount of depolarization will produce the greatest amount of sodium entry at sites where sodium channels are abundant. The action potentials initiated at the axon hillock are then propagated along the axon.

The mechanism for propagating an action potential in nonmyelinated nerve cells is illustrated in Figure 9-17. Stimulation of a resting membrane at point *P* results in a depolarization of the membrane and a sudden rush of sodium ions into the axon at that location. Membrane polarity is temporarily reversed at that point, and this depolarization then spreads to an adjacent point *Q*. Depolarization at point *Q* is sufficient to bring it above the threshold potential, triggering the inward rush of sodium ions. By this time, the membrane at point *P* has become highly permeable to potassium ions. As potassium ions rush out of the cell, negative polarity is restored, and that portion of the membrane returns to its resting state.

Meanwhile, the events at *Q* have stimulated the membrane in the neighboring region at *R*, initiating the same

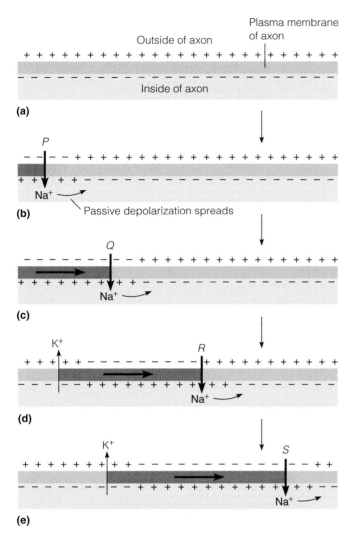

(a)

(b)
Na⁺
Passive depolarization spreads

(c)
Q
Na⁺

(d)
K⁺
R
Na⁺

(e)
K⁺
S
Na⁺

Figure 9-17   The Transmission of an Action Potential Along a Nonmyelinated Axon. A nonmyelinated axon might be viewed as an infinite string of points, each capable of undergoing an action potential. For simplicity, we will consider only the four points *P*, *Q*, *R*, and *S*, which represent adjacent regions along the plasma membrane of the axon. **(a)** At the start, the membrane is completely polarized. **(b)** When an action potential is initiated at point *P*, this region of the membrane depolarizes and briefly has a positive potential. As a result, the adjacent regions become depolarized. **(c)** When the adjacent point *Q* is depolarized to its threshold, an action potential starts there. **(d)** Meanwhile, point *P* is recovering from its action potential and has repolarized because of the outward flow of potassium ions. The action potential at point *Q* continues to propagate in the direction of point *R*. (It cannot propagate backward toward *P* because sodium channels are in a refractory, or inactivated, state, and the membrane in this region has become hyperpolarized.) The depolarization spreading from point *Q* will trigger an action potential at *R*. **(e)** Likewise, the depolarization at point *R* will eventually trigger an action potential at point *S*.

sequence of events there, and the depolarization then moves on to point *S*. In this way, the signal moves along the membrane as a ripple of depolarization-repolarization events, with the membrane polarity reversed in the immediate vicinity of the signal, but returned to normal again as the signal travels down the axon. The propagation of this cycle of events along the nerve fiber is called a *propa-*

gated action potential, or **nerve impulse.** The nerve impulse can only move *away* from the initial site of depolarization because the sodium channels that have just been depolarized are in the inactivated state and cannot respond immediately to further stimulation.

We can establish that an action potential is actively propagated by showing that it does not fade as it travels. This can be demonstrated by measuring changes in membrane potential with two recording electrodes, each inserted at a different distance from a stimulating electrode. The stimulating electrode will trigger an action potential, which will then travel along the axon, first passing by recording electrode 1 and then by recording electrode 2. The magnitude of response detected by the two electrodes will be the same, even though the signal has had to travel further along the membrane to reach the second electrode. A nerve impulse can be propagated along the membrane with no reduction in strength because it is generated anew as an all-or-none event at each successive point along the membrane. Thus, a nerve impulse can be transmitted over essentially any distance with no decrease in strength.

The Myelin Sheath Acts Like an Electrical Insulator Surrounding the Axon

Most axons in vertebrates have an additional specialization: they are surrounded by a discontinuous *myelin sheath* consisting of many concentric layers of membrane. The myelin sheath is a reasonably effective electrical insulation for the segments of the axon that it envelops. The myelin sheath of neurons in the CNS is formed by **oligodendrocytes;** in the PNS the myelin sheath is formed by **Schwann cells** (see Figure 9-3), each of which wraps layer after layer of its own plasma membrane around the axon in a tight spiral (Figure 9-18). Since each Schwann cell surrounds a short segment (about 1 mm) of a single axon, numerous Schwann cells are required to encase a PNS axon with a discontinuous sheath of myelin. Myelination decreases the ability of the neuronal membrane to retain electric charge (i.e. myelination decreases an axon's *capacitance*), permitting a depolarization event to spread farther and faster than it would along a nonmyelinated axon.

Myelination does not eliminate the need for propagation, however. For depolarization to spread from one site to the rest of the neuron, the action potential must still be renewed periodically down the axon. This happens at the *nodes of Ranvier,* interruptions in the myelin layer that are spaced just close enough together (1–2 mm) to ensure that the depolarization spreading out from an action potential at one node is still strong enough to bring an adjacent node above its threshold potential (Figure 9-19). The nodes of Ranvier are the only places on a myelinated axon that an action potential can be generated, because they are where voltage-gated sodium channels are concentrated. Thus, action potentials jump from node to node along myelinated axons, rather than moving as a steady ripple along the membrane. This so-called **saltatory propagation** is much more rapid than the

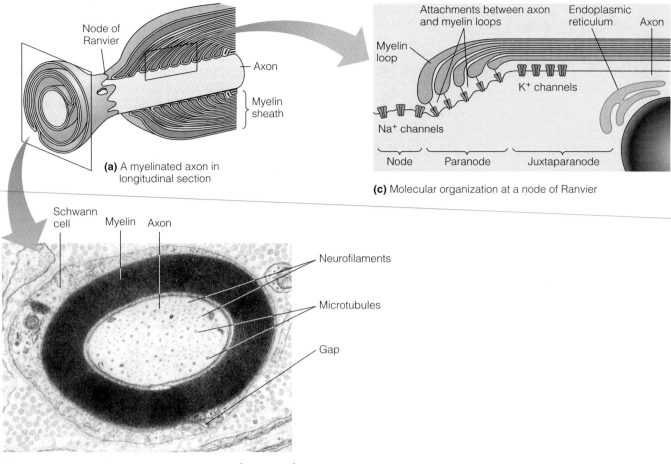

(a) A myelinated axon in longitudinal section

(c) Molecular organization at a node of Ranvier

(b) A myelinated axon in cross section

1 μm

Figure 9-18 Myelination of Axons. (a) This cross-sectional view of a myelinated axon from the nervous system of a cat shows the concentric layers of membrane that have been wrapped around the axon by the Schwann cell that envelops it. The gap in the plasma membrane of the Schwann cell is the point at which the membrane initially invaginated to begin enveloping the axon. Neurofilaments and microtubules will be discussed in Chapter 22 (TEM). **(b)** An axon of the peripheral nervous system that has been myelinated by a Schwann cell. Each Schwann cell gives rise to one segment of myelin sheath by wrapping its own plasma membrane concentrically around the axon. The layers of myelin terminate one by one as a node of Ranvier is approached, exposing the membrane of the axon. **(c)** Organization of a typical node of Ranvier in the peripheral nervous system. Sodium channels (red) are concentrated in the node. Myelin loops attach to the regions next to the node ("Paranodal" regions) via proteins on the axonal membrane and on the myelin loops (green). Potassium channels (light geen) cluster next to the paranodal regions.

continuous propagation that occurs in nonmyelinated axons (*saltatory* is derived from the Latin word for "dancing."). Myelination is a crucial feature of mammalian axons. The debilitating human disease multiple sclerosis results when a patient's immune system attacks his or her own myelinated nerve fibers, disrupting their conductive properties. The result is defective propagation of nerve impulses. If the affected nerves innervate muscles, the patient's capacity for movement can be severely compromised.

Nodes of Ranvier are highly organized structures that involve close contact between the loops of glial or Schwann cell membrane and the plasma membrane of the axon(s) they myelinate. It is now clear that there are three distinct regions associated with these specialized sites of contact. In the node of Ranvier itself, voltage-sen-

sitive sodium channels are highly concentrated. In the adjacent regions, called paranodal regions ("para" means "alongside"), the axonal and glial cell membranes contain specialized transmembrane proteins. Finally, in the region next to the paranodal areas, called juxtaparanodal regions ("juxta" means "next to"), potassium channels are highly concentrated (Figure 9-18). In the paranodal areas, the axons produce two proteins that interact with one another within the axonal plasma membrane (one is called *contactin*; the other is called *Caspr*, for "contactin-associated protein"). Schwann cell membranes contain a specific form of a protein called *neurexin*, which is thought to interact with the contactin protruding from the axonal cell membranes. This interaction prevents free movement of the sodium and potassium channels within

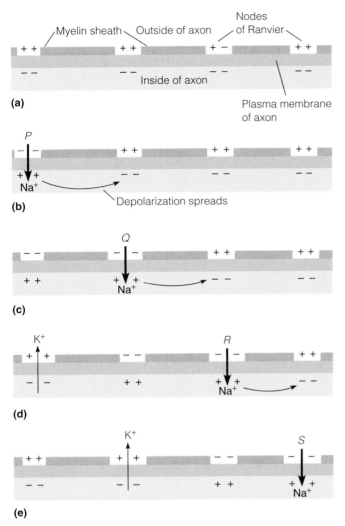

(a)

(b)

Depolarization spreads

(c)

(d)

(e)

Figure 9-19 **The Transmission of an Action Potential Along a Myelinated Axon.** Myelination reduces membrane capacitance, thereby allowing a given amount of sodium current, entering at one point of the membrane, to spread much farther along the membrane than it would in the absence of myelin. **(a)** In myelinated neurons, an action potential is usually triggered at the axon hillock, just before the start of the myelin sheath. The depolarization then spreads along the axon. Action potentials can be generated only at nodes of Ranvier, here represented by points *P*, *Q*, *R*, and *S*. **(b)** With myelination, the depolarization at point *P* spreads passively all the way to point *Q* and **(c)** brings *Q* to its threshold. Here a new action potential is generated that **(d)** triggers point *R* to undergo an action potential and **(e)** so on to point *S*. A nerve impulse consists of a wave of depolarization-repolarization events that is propagated along the axon from node to node.

Synaptic Transmission

Nerve cells communicate with one another and with glands and muscles at synapses. There are two structurally distinct types of synapses, electrical and chemical. In an **electrical synapse,** the axon of one neuron, called the **presynaptic neuron,** is connected to the dendrite(s) of another cell, the **postsynaptic neuron,** by gap junctions (Figure 9-20; we will encounter gap junctions in more detail in Chapter 11). As ions move back and forth between the two cells, the depolarization in one cell spreads passively to the connected cell. Electrical synapses provide for transmission with virtually no delay and occur in places in the nervous system where speed of transmission is critical.

In a **chemical synapse,** the presynaptic and postsynaptic neurons are close to each other but not directly connected (Figure 9-21). Typically, the presynaptic membrane is separated from the postsynaptic membrane by a gap of about 20–50 nm, known as the **synaptic cleft.** A nerve signal arriving at the terminals of the presynaptic neuron cannot bridge the synaptic cleft as an electrical impulse. For synaptic transmission to take place, the electrical signal must be converted at the presynaptic neuron to a chemical signal carried by a neurotransmitter. Neurotransmitter molecules are stored in the **terminal bulbs** (also called *synaptic knobs*) of the presynaptic neuron. An action potential arriving at the terminal causes the neurotransmitter to be secreted into and diffuse across the synaptic cleft. The neurotransmitter molecules then bind to specific proteins embedded within the plasma membrane of the postsynaptic neuron and are converted back to electrical signals, setting in motion a sequence of events that either stimulates or inhibits the production of an action potential in the postsynaptic neuron, depending on the kind of synapse.

Understanding how a nerve signal is transmitted across a synapse requires insight into the nature of neurotransmitters and their respective receptors. Specifically, we need to know how the binding of neurotransmitter molecules to their receptors can alter the electrical activity of the postsynaptic cell, either to excite it to the threshold point or to inhibit its electrical activity. Finally, we need to understand how an action potential regulates the secretion of neurotransmitter into the synaptic cleft and the processes that terminate the signal by removing the neurotransmitter from the synaptic cleft.

Neurotransmitters Relay Signals Across Nerve Synapses

A **neurotransmitter** is a small molecule whose function is to bind to a receptor within the membrane of a postsynaptic neuron. Many kinds of molecules act as neurotransmitters, each with at least one specific type of receptor; some neurotransmitters have more than one type of receptor. When a neurotransmitter molecule binds to its receptor, the properties of the receptor are altered.

the axon's plasma membrane in regions around the nodes. When this organization is disrupted, the rate and frequency of action potentials transmitted along an axon are impaired. For example, in mutant mice lacking contactin, the refractory period is longer (apparently due to effects on potassium channels); this slows the rate of propagation of successive action potentials along the axon. Similarly, mice lacking Caspr have much slower rates of action potential transmission.

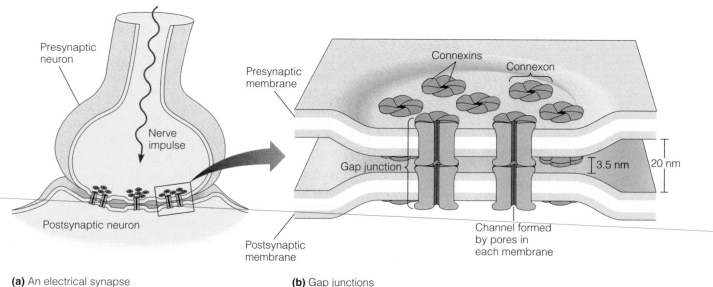

(a) An electrical synapse

(b) Gap junctions

Figure 9-20   An Electrical Synapse.   (a) In electrical synapses, the presynaptic and postsynaptic neurons are coupled by gap junctions. Gap junctions allow small molecules and ions to pass freely from the cytosol of one cell to the next. Therefore, when an action potential arrives at the presynaptic side of an electrical synapse, the depolarization spreads passively, due to the flow of positively charged ions, across the gap junction **(b)** The gap junction is composed of sets of channels. Each channel is made up of six protein subunits called connexins. The entire set of six subunits together is called a connexon. Two connexons, one in the presynaptic membrane and one in the postsynaptic membrane, make up a gap junction.

We might compare receptors and neurotransmitters to a lock and key. The neurotransmitter, by virtue of its unique chemical structure, is the key that fits into the receptor lock. Its function is typically to turn the receptor from an "off" state to an "on" state. For a particular receptor, however, the specific meanings of "off" and "on" lie in the properties of that receptor. An *excitatory neurotransmitter* causes depolarization of the postsynaptic neuron, whereas an *inhibitory neurotransmitter* typically causes the postsynaptic cell to hyperpolarize.

To qualify as a neurotransmitter, a compound must satisfy the following three criteria: (1) It must elicit the appropriate response when introduced into the synaptic cleft, (2) it must occur naturally in the presynaptic neuron, and (3) it must be released at the right time when the presynaptic neuron is stimulated. At present, molecules known to meet these criteria include acetylcholine, a group of biogenic amines called the catecholamines, certain amino acids and their derivatives, and some neuropeptides (see below). Figure 9-22 shows several common neurotransmitters; we will discuss a number of them here.

Acetylcholine.   In vertebrates, **acetylcholine** (Figure 9-22a) is the most common neurotransmitter for synapses between neurons outside the central nervous system, as well as for neuromuscular junctions (see Chapter 23). Acetylcholine is an excitatory neurotransmitter. Bernard Katz and his collaborators were the first to make the important observation that acetylcholine increases the permeability of the postsynaptic membrane to sodium within 0.1 msec of binding to its receptor. Synapses that use acetylcholine as their neurotransmitter are called **cholinergic synapses.**

Catecholamines.   The **catecholamines** include *dopamine* and the hormones *norepinephrine* and *epinephrine,* all derivatives of the amino acid tyrosine (Figure 9-22b). Because the catecholamines are also synthesized in the adrenal gland, synapses that use them as neurotransmitters are termed **adrenergic synapses.** Adrenergic synapses are found at the junctions between nerves and smooth muscles in internal organs such as the intestines, as well as at nerve-nerve junctions in the brain. The mode of action of the adrenergic hormones will be considered in Chapter 10.

Other Amino Acids and Derivatives.   Other neurotransmitters that consist of amino acids and derivatives include *histamine, serotonin,* and *γ-aminobutyric acid (GABA)* (Figure 9-22), as well as *glycine* and *glutamate.* Serotonin functions in the central nervous system. It is considered an excitatory neurotransmitter because it indirectly causes potassium channels to close, which has an effect similar to opening sodium channels in that the postsynaptic cell is depolarized. However, its effect is exerted much more slowly than that of sodium channels. GABA and glycine are inhibitory neurotransmitters, whereas glutamate has an excitatory effect.

Neuropeptides.   Short chains of amino acids called **neuropeptides** are formed by proteolytic cleavage of precursor

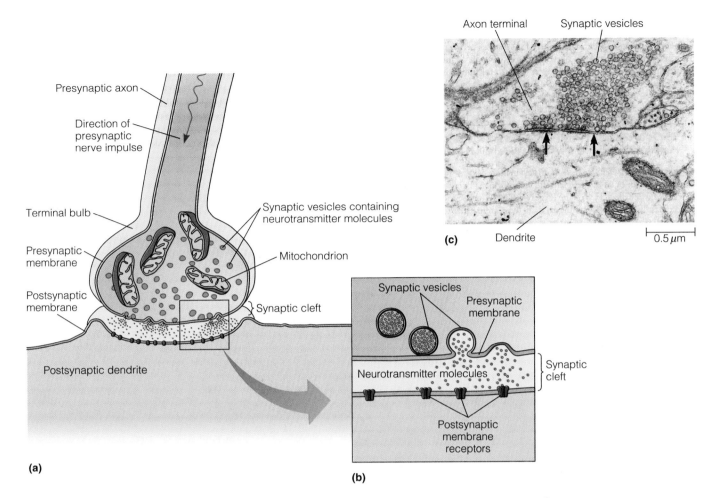

Figure 9-21 A Chemical Synapse. **(a)** When a nerve impulse from the presynaptic axon arrives at the synapse (red arrow), it causes synaptic vesicles containing neurotransmitter in the terminal bulb to fuse with the presynaptic membrane, releasing their contents into the synaptic cleft. **(b)** Neurotransmitter molecules diffuse across the cleft from the presynaptic (axonal) membrane to the postsynaptic (dendritic) membrane, where they bind to specific membrane receptor sites and change the polarization of the membrane, either exciting or inhibiting the next cell. **(c)** Electron micrograph of a chemical synapse (TEM).

proteins and can be stored in secretory vesicles similar to those that hold neurotransmitters. At present, more than 50 different neuropeptides have been identified. Some neuropeptides exhibit characteristics similar to neurotransmitters in that they excite, inhibit, or modify the activity of other neurons in the brain. However, they differ from typical neurotransmitters in that they act on groups of neurons and have long-lasting effects.

Examples of neuropeptides include the *enkephalins,* which are naturally produced in the mammalian brain and inhibit the activity of neurons in regions of the brain involved in the perception of pain. The modification of neural activity by these neuropeptides appears to be responsible for the insensitivity to pain experienced by individuals under conditions of great stress or shock. The analgesic (i.e., pain-killing) effectiveness of drugs such as morphine, codeine, Demerol, and heroin derives from their ability to bind to the same sites within the brain that are normally targeted by enkephalins.

Elevated Calcium Levels Stimulate Secretion of Neurotransmitters from Presynaptic Neurons

The secretion of neurotransmitters by the presynaptic cell is directly controlled by the concentration of calcium ions in the terminal bulb (Figure 9-23). Each time an action potential arrives, the depolarization causes the calcium concentration in the terminal bulb to increase temporarily due to the opening of voltage-gated calcium channels in the terminal bulbs. Normally, the cell is relatively impermeable to calcium ions, so that the cytosolic calcium concentration remains low (about 10^{-4} mM). However, there is a very large concentration gradient of calcium across the membrane, because the calcium concentration outside the cell is about 10,000 times higher than that of the cytosol. As a result, calcium ions will rush into the cell when the calcium channels open.

Before they are released, neurotransmitter molecules are stored in small membrane-bounded **neurosecretory vesicles**

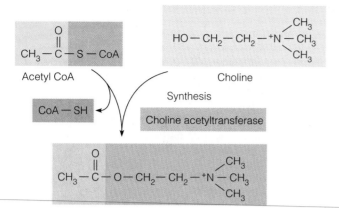

(a) Acetylcholine

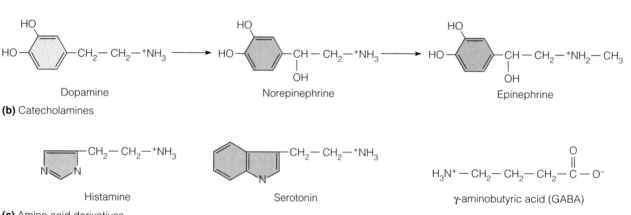

(b) Catecholamines

Histamine

Serotonin

γ-aminobutyric acid (GABA)

(c) Amino acid derivatives

Figure 9-22   The Structure and Synthesis of Neurotransmitters. **(a)** Acetylcholine is synthesized from acetyl CoA and choline by choline acetyltransferase. **(b)** The catecholamines dopamine, norepinephrine, and epinephrine are synthesized from the amino acid tyrosine and are inactivated by the enzyme monoamine oxidase. Dopamine can be converted to norepinephrine, and norepinephrine to epinephrine, as indicated by the arrows. **(c)** Other amino acid derivatives are histamine, serotonin, and γ-aminobutyric acid (GABA). The amino acids glycine and glutamate (not shown) are also neurotransmitters.

in the terminal bulbs (see Figure 9-23). The release of calcium within the terminal bulb has two main effects on neurosecretory vesicles. First, vesicles held in storage are mobilized for rapid release. For any given action potential, only a tiny fraction of the total number of vesicles stored in the terminal release their contents. Neurons hold vesicles in reserve by linking them to the cytoskeleton so that they cannot move close to the synaptic membrane for secretion. For neurotransmitter vesicles to become available for secretion, they must become disengaged from the cytoskeleton. This process is stimulated by calcium release, and involves the addition of phosphate groups to *synapsin*, an integral membrane protein found in the membrane of neurosecretory vesicles. In its unphosphorylated form, synapsin binds to the cytoskeleton. When phosphorylated, synapsin no longer binds to the cytoskeleton, freeing the vesicles so that they can proceed to the next step in neurotransmitter release.

The second calcium-sensitive effect on neurotransmitter release is the rapid docking and fusion of neurosecretory vesicles with the plasma membrane in the terminal bulb region. During this process, the membrane of a vesicle moves into close contact with the plasma membrane of the axon terminal and then fuses with it to release the contents of the vesicle. We will now examine this process in more detail.

Secretion of Neurotransmitters Requires the Docking and Fusion of Vesicles with the Plasma Membrane

For the neurotransmitter to act on a postsynaptic cell, it must be secreted by the process of exocytosis (see Chapter 12 for details of exocytosis). The key event in secretion is the fusion of neurosecretory vesicles with the plasma membrane, which discharges the vesicle contents into the synaptic cleft. The process requires ATP and proceeds through several steps, one of which is calcium-dependent.

The release of neurotransmitters from vesicles requires that the neurosecretory vesicles dock with the plasma membrane and then fuse with it to release their contents into the synaptic cleft. When an action potential arrives at an axon terminal and triggers the opening of voltage-gated calcium channels, calcium enters the terminal bulb. In the presence of high levels of calcium, a sequence of docking events is initiated. Once synaptic vesicles are released from the cytoskeleton, they are not quite ready to fuse with the plasma membrane. Such vesicles cluster near, but not in contact with, the plasma membrane, and are associated with a meshwork of fine filaments (Figure 9-24). In order to be ready to fuse with the plasma membrane, such released vesicles must be "primed" by proteins that form part of this meshwork. Once such priming has occured, the freed vesicles are now capable of docking and fusing with the plasma

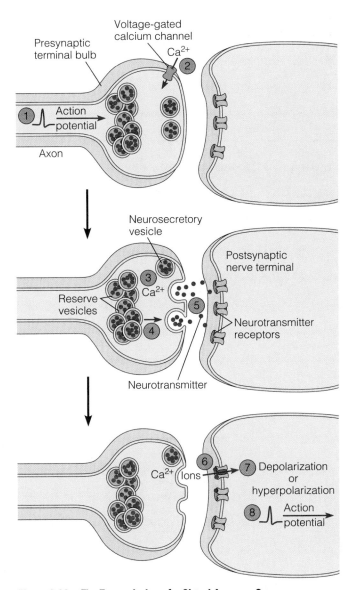

Figure 9-23 The Transmission of a Signal Across a Synapse.
① An action potential arrives at the presynaptic terminal bulb, resulting in a transient depolarization. ② Depolarization opens voltage-gated calcium channels, allowing calcium ions to rush into the terminal. ③ This increase in the calcium concentration in the terminal bulb induces the secretion of a fraction of the neurosecretory vesicles. ④ Calcium also causes reserve vesicles to be released from the actin cytoskeleton so that they are ready for secretion. ⑤ Secreted neurotransmitter molecules diffuse across the synaptic cleft to receptors on the postsynaptic cell. ⑥ Binding of neurotransmitter to the receptor alters the receptor properties. ⑦ For receptors that are ligand-gated channels, the channel opens, letting ions flow into the postsynaptic cell. Depending on the ion, channel opening leads to either depolarization or hyperpolarization of the postsynaptic cell membrane. ⑧ If depolarization results, a sufficient amount of excitatory neurotransmitter will result in an action potential in the postsynaptic cell.

membrane of the presynaptic neuron in response to elevated calcium.

Docking takes place at a specialized site, called the **active zone,** within the membrane of the presynaptic neuron. The active zone is a highly organized structure. Synap-

tic vesicles, the proteins to which the vesicles attach, and the calcium channels that elicit their release are all clustered in to ordered rows poised for secretion (Figure 9-24). The close proximity of calcium channels to the vesicles helps to explain the extremely rapid fusion of the primed, releasable population of vesicles with the presynaptic neuron's plasma membrane when that neuron is stimulated.

Docking and fusion of neurosecretory vesicle are mediated by docking proteins within the vesicle and within the plasma membrane of the active zone. Within the vesicle membrane *synaptotagmin, synaptobrevin,* and other proteins dock with proteins, such as *syntaxin* at the plasma membrane of the active zone (Figure 9-24; the specific mechanisms involved in the docking steps of exocytosis are discussed in more detail in Chapter 12). Once the initial docking occurs, additional proteins from the cytosol stabilize the attachment, forming a multiprotein docking complex. Although the next step, fusion, is not well understood, it depends on the successful docking of vesicles at the active zone and the formation of a complete docking complex. Synaptotagmin probably acts as a "calcium sensor" during neurotransmitter release. When the calcium binding region of synaptoagmin is mutated in mice, their neurons require much more calcium for neurotransmitter release to occur.

Two familiar and potentially deadly illnesses result from interference with the docking events we have just described. Both tetanus and botulism result from interference by **neurotoxins** with vesicle docking and release. *Tetanus toxin* prevents the release of neurotransmitter from inhibitory neurons in the spinal cord, resulting in uncontrolled muscle contraction (which is why tetanus has been colloquially referred to as "lockjaw"). *Botulinum toxin* prevents release of neurotransmitter from motor neurons, resulting in muscle weakness and paralysis.

Neurotransmitters Are Detected by Specific Receptors on Postsynaptic Neurons

Neurotransmitter receptors fall into two broad groups: ligand-gated ion channels, in which activation has a direct effect on the cell, and receptors that exert their effects indirectly through a system of intracellular messengers. We will discuss the latter category of receptors in Chapter 10; here we focus on *ligand-gated channels*. These membrane ion channels open in response to the binding of a neurotransmitter, and they can mediate either excitatory or inhibitory responses in the postsynaptic cell.

The Acetylcholine Receptor. Acetylcholine binds to a ligand-gated sodium channel known as the *acetylcholine receptor*. When two molecules of acetylcholine bind, the channel opens and lets sodium ions rush into the postsynaptic neuron, causing a depolarization.

Our understanding of synaptic transmission has been greatly aided by the ease with which membranes rich in acetylcholine receptors can be isolated from the electric organs of the electric ray (*Torpedo californica*). The electric

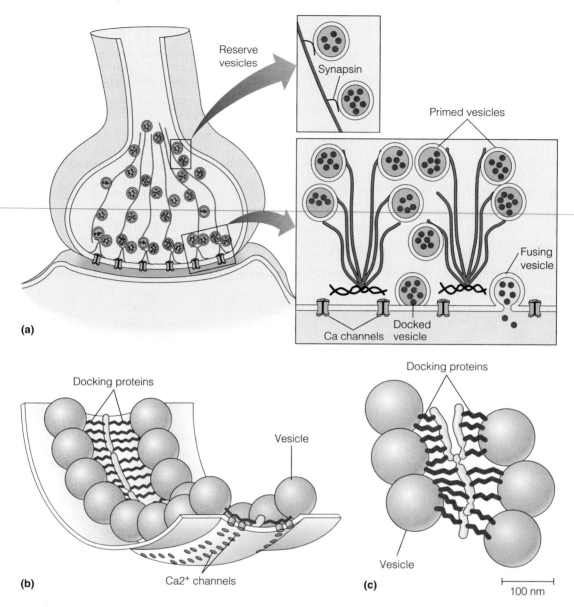

Figure 9-24 Docking of Synaptic Vesicles with the Plasma Membrane of the Presynaptic Neuron. (a) In response to local elevation of calcium in the presynaptic neuron, synaptic vesicles are released from the cytoskeleton, allowing them to fuse with the plasma membrane of the terminal bulb with the active zone. Vesicles bound to the cytoskeleton via synapsin form a reserve pool of vesicles. Once released form the cytoskeleton, vesicles move near the plasma membrane and associate with a fine meshwork of proteins; such vesicles have not been "primed." Vesicles that have been primed are tightly associated with the plasma membrane, or "docked." When nearby calcium channels open, such docked vesicles fuse with the plasma membrane, releasing their contents. **(b,c)** A drawing **(b)** and actual TEM reconstruction **(c)** of the active zone of a motor neuron from a frog. Docked vesicles (blue structures in c) are arranged in rows and connected by a complex of proteins associated with docking (c, yellow). Calcium channels lie immediately beneath the vesicles.

organ consists of *electroplaxes*—stacks of cells that are innervated on one side but not on the other. The innervated side of the stack can undergo a potential change from about −90 mV to about +60 mV upon excitation, whereas the noninnervated side stays at −90 mV. Therefore, at the peak of an action potential, a potential difference of about 150 mV can be built up across a single electroplax. Because the electric organ contains thousands of electroplaxes arranged in series, their voltages are additive, allowing the ray to deliver a jolt of several hundred volts.

When electroplax membranes are examined under the electron microscope, they are found to be rich in rosette-like particles about 8 nm in diameter (Figure 9-25a). Each particle consists of five subunits arranged around a central axis, which is assumed to be the ion channel. Their size and reactions with antibodies indicate that these particles are the acetylcholine receptors.

Biochemical purification of the acetylcholine receptor was greatly aided by the availability of several substances from snake venom, including *α-bungarotoxin* and *cobra-*

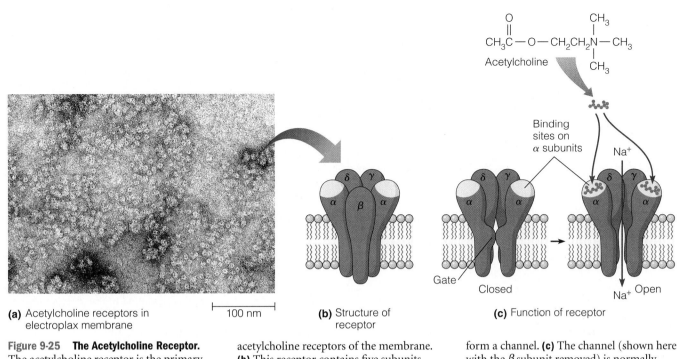

(a) Acetylcholine receptors in electroplax membrane

(b) Structure of receptor

(c) Function of receptor

Figure 9-25 The Acetylcholine Receptor. The acetylcholine receptor is the primary excitatory receptor of the central nervous system. **(a)** This TEM micrograph of an electroplax postsynaptic membrane shows the rosettelike particles thought to be the acetylcholine receptors of the membrane. **(b)** This receptor contains five subunits, including two α subunits with binding sites for acetylcholine and one each of β, γ, and δ. The subunits aggregate in the lipid bilayer in such a way that the transmembrane portions form a channel. **(c)** The channel (shown here with the β subunit removed) is normally closed, but when acetylcholine binds to the two sites on the α subunits, the subunits are altered in such a way that the channel opens to allow sodium ions to cross.

toxin. These neurotoxins serve as a highly specific means of locating and quantifying acetylcholine receptors, because they can be radiolabeled easily and they bind to the receptor protein very tightly and specifically. Other neurotoxins act on other ion channels, such as calcium channels (Box 9A).

The purified acetylcholine receptor has a molecular weight of about 300,000 and consists of four kinds of subunits—α, β, γ, and δ—each containing about 500 amino acids. The transmembrane segment of each subunit includes sequences of relatively hydrophobic amino acids, which probably form α helices grouped together in the plane of the bilayer. The intact receptor contains the subunits in the ratio $2:1:1:1$ (Figure 9-25b and c), so the simplest empirical formula for the receptor protein is $\alpha_2\beta\gamma\delta$. Acetylcholine receptors play a key role in the transmission of nerve impulses to muscle (for more information see Box 9A). In some cases, human patients develop an autoimmune response to their own acetylcholine receptors (in other words, their immune system attacks their own receptors as if they were foreign invaders). When this happens, a condition known as *myasthenia gravis* can develop, in which degenerative muscle weakness occurs.

The GABA Receptor. The γ-aminobutyric acid (GABA) receptor is also a ligand-gated channel, but when open it conducts chloride ions rather than sodium ions. Since chloride ions are normally found at a higher concentration in the medium surrounding a neuron (see Table 9-1), opening GABA receptors produces an influx of chloride into the postsynaptic neuron. The main effect of chloride influx is to inhibit the depolarization of postsynaptic neurons. To understand how the GABA receptor accomplishes this, recall that sodium ions depolarize the membrane as long as they enter the cell unaccompanied by a negatively charged ion. However, if a chloride ion entered at the same time as a sodium ion, there would be no net effect on the membrane potential. By opening chloride channels, the activated GABA receptor helps to neutralize the effects of sodium influx on membrane potential, thereby reducing the amount of depolarization that occurs and decreasing the chance that an action potential will be initiated in the postsynaptic neuron. Benzodiazepine drugs, such as Valium and Librium, can enhance the effects of GABA on its receptor. Presumably, this produces the tranquilizing effects of these drugs.

Neurotransmitters Must Be Inactivated Shortly After Their Release

For neurons to transmit signals effectively, it is just as important to turn the stimulus off as it is to turn it on. Whether excitatory or inhibitory, once the neurotransmitter has been secreted, it must be rapidly removed from the synaptic cleft. If it were not, stimulation or inhibition of a postsynaptic neuron would be abnormally prolonged, even in the absence of further signals from presynaptic neurons. In fact, the persistence of an excitatory neurotransmitter such as acetylcholine renders muscles unable to relax, ultimately leading to death.

Clinical Applications | POISONED ARROWS, SNAKE BITES, AND NERVE GASES

Because the coherent functioning of the human body depends so critically on the nervous system, anything that disrupts the transmission of nerve impulses is likely to be very harmful. And because of the importance of acetylcholine as a neurotransmitter, any substance that interferes with its function is almost certain to be lethal. Various toxins are known that disrupt nerve and muscle function by specific effects on cholinergic synapses. We will consider several of these substances, not only to underscore the serious threat they pose to human health, but also to illustrate how clearly their modes of action can be explained once the physiology of synaptic transmission is understood.

Once acetylcholine has been released into the synaptic cleft and depolarization of the postsynaptic membrane has occurred, excess acetylcholine must be rapidly hydrolyzed. If it is not, the membrane cannot be restored to its polarized state, and further transmission will not be possible.

The enzyme acetylcholinesterase is therefore essential, and substances that inhibit its activity are usually very toxic.

One such family of acetylcholinesterase inhibitors consists of *carbamoyl esters*. These compounds inhibit acetylcholinesterase by covalently blocking the active site of the enzyme, effectively preventing the breakdown of acetylcholine. An example of such an inhibitor is *physostigmine* (sometimes also called *eserine*), a naturally occurring alkaloid produced by the Calabar bean.

Many synthetic organic phosphates form even more stable covalent complexes with the active site of acetylcholinesterase and are therefore still more potent inhibitors. Included in this class of compounds are the widely used insecticides *parathion* and *malathion,* as well as nerve gases such as *tabun* and *sarin.* The primary effect of these compounds is muscle paralysis,

caused by an inability of the postsynaptic membrane to regain its polarized state.

Nerve transmission at cholinergic synapses can be blocked not only by inhibitors of acetylcholinesterase, but also by substances that compete with acetylcholine for binding to its receptor on the postsynaptic membrane. A notorious example of such a poison is *curare,* a plant extract once used by native South Americans to poison arrows. One of the active factors in curare is *d-tubocurarine.* Snake venoms act in the same way. Both α-bungarotoxin (from snakes of the genus *Bungarus*) and *cobratoxin* (from cobra snakes) are small, basic proteins that bind covalently to the acetylcholine receptor, thereby blocking depolarization of the postsynaptic membrane.

Substances that function in this way are called *antagonists* of cholinergic systems. Other compounds, called *agonists*, have just the opposite effect. Agonists also bind to the acetylcholine receptor, but in doing so they mimic acetylcholine, causing depolarization of the postsynaptic membrane. Unlike acetylcholine, however, they cannot be rapidly inactivated, so the membrane does not regain its polarized state.

Though of disparate origins and uses, poisoned arrows, snake venom, nerve gases, and surgical muscle relaxants all turn out to have some features in common. Each interferes in some way with the normal functioning of acetylcholine, and each is therefore a neurotoxin because it disrupts the transmission of nerve impulses, potentially with lethal consequences. Each has also turned out to be useful as an investigative tool, illustrating again the strange but powerful arsenal of exotic tools upon which biologists and biochemists are able to draw in their continued probings into the intricacies of cellular function.

Neurotransmitters are removed from the synaptic cleft by two specific mechanisms: degradation into inactive molecules or reuptake into the presynaptic terminals. An example of the first mechanism is found in the case of acetylcholine. The enzyme *acetylcholinesterase* hydrolyzes acetylcholine into acetic acid (or acetate ion) and choline, neither of which stimulates the acetylcholine receptor. Reuptake is a much more common method of terminating synaptic transmission. Specific transporter proteins in the membrane of the presynaptic terminals pump the neurotransmitter back into the presynaptic axon terminals. The rate of neurotransmitter uptake can be very rapid; for some neurons, the synapse maybe cleared of stray neurotransmitter within as little as a millisecond. Since the resorbed neurotransmitter can be recycled and used again, less new neurotransmitter must be synthesized by the presynaptic neuron. Reuptake is carried out by a process known as *endocytosis;* we will examine this very important process in detail in Chapter 12. Some antidepressant drugs act by blocking the reuptake of specific neurotransmitters. For example, Prozac blocks the reuptake of serotonin, leading to a local increase in the level of serotonin available to postsynaptic neurons.

Integration and Processing of Nerve Signals

Sending a signal across a synapse does not automatically generate an action potential in the postsynaptic cell. There is not necessarily a one-to-one relationship between an action potential arriving at the presynaptic neuron and one initiated in the postsynaptic neuron. A single action potential can cause the secretion of enough neurotransmitter to produce a detectable depolarization in the postsynaptic neuron, but usually not enough to cause the firing of an action potential in the postsynaptic cell. These small incremental changes in potential due to the binding of neurotransmitter are referred to as *postsynaptic potentials (PSPs)*. If a neurotransmitter is excitatory, it will cause a small amount of depolarization known as an **excitatory postsynaptic potential** (EPSP) (Figure 9-26a). Likewise, if the neurotransmitter is inhibitory, it will hyperpolarize the postsynaptic neuron by a small amount; this is called an **inhibitory postsynaptic potential (IPSP).**

For a presynaptic neuron to stimulate the formation of an action potential in the postsynaptic neuron, the EPSP must build up to a point at which the postsynaptic membrane reaches its threshold potential. EPSPs can do this in two different ways, which we will examine next.

Neurons Can Integrate Signals from Other Neurons Through Both Temporal and Spatial Summation

As we have noted, an action potential is an all-or-none event: neurons do not produce larger or smaller action potentials. Yet neurons can detect whether a stimulus is weak or strong, and respond by firing action potentials at different rates. If a neuron is maintained in a strongly depolarized state, it will fire a train of action potentials in rapid succession.

An individual action potential will produce only a temporary EPSP. However, if two action potentials fire in rapid succession at the presynaptic neuron, the postsynaptic neuron will not have time to recover from the first EPSP before experiencing a second EPSP. The result is that the postsynaptic neuron will be more depolarized. A rapid sequence of action potentials effectively sums EPSPs over time and brings the postsynaptic neu-

ron to its threshold. This is called **temporal summation** (Figure 9-26b).

The amount of neurotransmitter released at a single synapse following an action potential is usually not sufficient to produce an action potential in the postsynaptic cell. When many action potentials cause the release of neurotransmitter simultaneously, their effects combine; sometimes this results in a large depolarization of the postsynaptic cell. This is known as **spatial summation** because the postsynaptic neuron integrates the numerous small depolarizations that occur over its surface into one large depolarization (Figure 9-26c).

Neurons Can Integrate Both Excitatory and Inhibitory Signals from Other Neurons

In addition to receiving stimuli of varying strength, postsynaptic neurons can receive inputs from different kinds of neurons. These synapses do not necessarily all use the same neurotransmitter. In addition, a postsynaptic neuron can receive synapses from both excitatory and inhibitory neurons (Figure 9-26d). Neurons can receive literally thousands of synaptic inputs from other neurons. When these different neurons fire at the same time, they

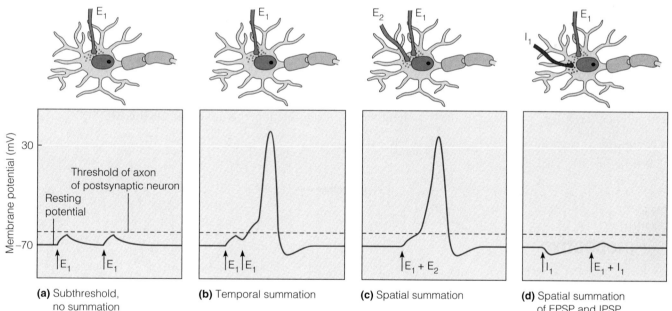

(a) Subthreshold, no summation

(b) Temporal summation

(c) Spatial summation

(d) Spatial summation of EPSP and IPSP

Figure 9-26   Summation of EPSPs and IPSPs.   Neurons detect the strength of incoming signals and the strength of excitatory versus inhibitory inputs. **(a)** A single presynaptic action potential does not cause the release of enough neurotransmitter to produce an action potential in the postsynaptic neuron. **(b)** In temporal summation, the response of the postsynaptic neuron is determined by the rate of action potentials arriving at the presynaptic terminal bulb. The strength of a nerve signal is often expressed as the frequency of action potentials—a weak stimulus produces infrequent action potentials and a strong stimulus produces frequent action potentials. When two action potentials arrive at close intervals, the effects of the two overlap and sum to produce an action potential in the postsynaptic cell. **(c)** In spatial summation, even infrequent action potentials can cause an action potential when many neurons form synapses with a single postsynaptic cell. If two or more neurons send an action potential at the same time, these action potentials sum. **(d)** One neuron can receive many synaptic inputs, either excitatory or inhibitory. The stimulation of inhibitory inputs makes it more difficult for excitatory transmissions to cause an action potential in the postsynaptic cell.

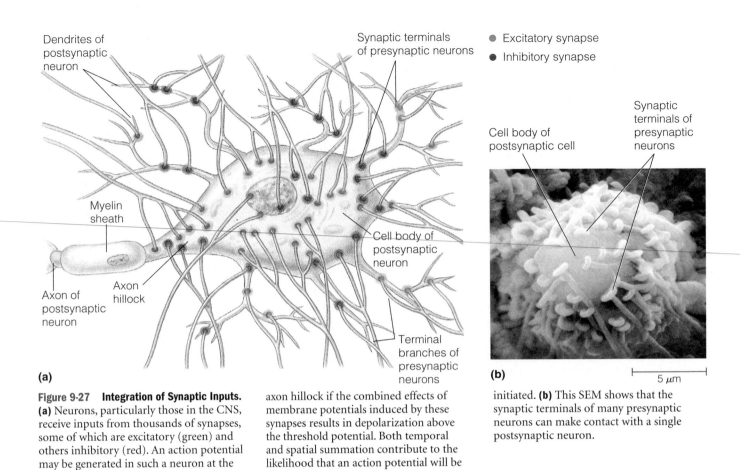

Dendrites of postsynaptic neuron

Synaptic terminals of presynaptic neurons

● Excitatory synapse
● Inhibitory synapse

Myelin sheath

Axon of postsynaptic neuron

Axon hillock

Cell body of postsynaptic neuron

Terminal branches of presynaptic neurons

(a)

Cell body of postsynaptic cell

Synaptic terminals of presynaptic neurons

(b)

5 μm

Figure 9-27   Integration of Synaptic Inputs.
(a) Neurons, particularly those in the CNS, receive inputs from thousands of synapses, some of which are excitatory (green) and others inhibitory (red). An action potential may be generated in such a neuron at the axon hillock if the combined effects of membrane potentials induced by these synapses results in depolarization above the threshold potential. Both temporal and spatial summation contribute to the likelihood that an action potential will be initiated. **(b)** This SEM shows that the synaptic terminals of many presynaptic neurons can make contact with a single postsynaptic neuron.

exert combined effects on the membrane potential of the postsynaptic neuron. (Figure 9-27) Thus, by physically summing EPSPs and IPSPs, an individual neuron effectively integrates incoming signals (excitatory or inhibitory). Yet another level of complexity involves long-term changes in the synaptic connections between neurons that are involved in such complex processes as learning and memory. The changes within neurons that allow for such long-term changes require an understanding of signal transduction, the subject of Chapter 10.

Perspective

All cells maintain an electrical potential across their membranes; neurons, however, are specialized to use membrane potentials as a means of transmitting signals from one part of an organism to another. For this function they possess slender processes (dendrites and axons) that either receive transmitted impulses or conduct them to the next cell. The membrane of an axon may or may not be encased in a myelin sheath.

Cells develop a membrane potential due to the separation of positive and negative charges across the plasma membrane. This potential develops as each ion to which the membrane is permeable moves down its electrochemical gradient. The maximum membrane potential that an ion gradient can produce is the equilibrium potential for that ion—a theoretical condition that is not achieved in cells because it requires that the membrane be permeable only to that ion. To calculate the resting membrane potential of a cell, the Goldman equation is used. The resting potential for the plasma membrane of most animal cells is usually in the range −60 to −75 mV. These values are quite near the equilibrium potential for potassium ion, but very far from that for sodium ion (about +55 mV), reflecting the greater permeability of the resting membrane for potassium.

The action potential of a neuron represents a transient depolarization and repolarization of its membrane, due to the sequential opening and closing of sodium and potassium ion channels. These channels have been characterized structurally by molecular techniques and functionally by patch clamping. They are voltage-gated ion channels whose probability of opening, and consequently their conductance, depends on membrane potential.

An action potential is initiated when the membrane is depolarized to its threshold, a point at which the rate of sodium influx exceeds the maximum rate of potassium efflux under resting conditions. The entry of sodium ions drives the membrane potential to approximately +40 mV before voltage-gated sodium channels inactivate. Depolarization also stimulates the opening of slower voltage-gated potassium channels, which leads to repolarization of the membrane, including a short period of hyperpolarization. This sequence of channel opening and closing generally takes a few milliseconds.

The depolarization of the membrane due to an action potential spreads by passive conductance to adjacent regions of the membrane, which in turns generates an action potential. In this way, an action potential is propagated along the membrane, eventually reaching a synapse, or junction, between a nerve cell and another cell with which it communicates. Such synapses may be either electrical or chemical. In an electrical synapse, depolarization is transmitted from the presynaptic cell to the postsynaptic cell by a direct gap junctional connection. In a chemical synapse, the electrical impulse increases the permeability of the membrane to calcium. As calcium ions cross the presynaptic membrane, they cause synaptic vesicles to fuse with the membrane. The synaptic vesicles contain neurotransmitter molecules, which are released into the synaptic cleft by the fusion event. Neurotransmitter molecules diffuse across the cleft to the postsynaptic membrane, where they bind to specific receptors, often ligand-gated ion channels.

The best-understood receptor is the acetylcholine receptor of the neuromuscular junction. Binding of acetylcholine stimulates this receptor channel to open, permitting sodium to enter. The resulting sodium influx produces a local depolarization of the postsynaptic membrane, which in turn can initiate an action potential in the postsynaptic cell. Following depolarization, the enzyme acetylcholinesterase hydrolyzes the acetylcholine, thereby returning the synapse to its resting state. An example of an inhibitory receptor is the GABA (γ-aminobutyric acid) receptor, which is a voltage-gated chloride channel. Upon binding GABA, this receptor allows increased chloride influx, leading to hyperpolarization of the postsynaptic membrane and reduced neuronal excitability.

Specific neurotransmitters can produce either excitatory or inhibitory postsynaptic potentials. Therefore, transmission of an action potential from one neuron to another requires that the cell body of the postsynaptic neuron integrate the excitatory and inhibitory activity of thousands of synaptic inputs. Through temporal and/or spatial summation, incoming signals can depolarize the nerve cell body sufficiently to initiate a new action potential at the axon hillock.

Key Terms for Self-Testing

The Nervous System
nervous system (p. 225)
central nervous system (CNS) (p. 225)
peripheral nervous system (PNS) (p. 225)
somatic nervous system (p. 225)
autonomic nervous system (p. 225)
neuron (p. 225)
sensory neuron (p. 225)
motor neuron (p. 226)
innervate (p. 226)
interneuron (p. 226)
glial cell (p. 226)
cell body (p. 227)
process (p. 227)
dendrite (p. 227)
axon (p. 227)
axoplasm (p. 227)
myelin sheath (p. 227)
node of Ranvier (p. 227)
nerve (p. 227)
synapse (p. 227)

Understanding Membrane Potential
membrane potential (p. 228)
resting membrane potential (V_m)
    (p. 228)
electrical excitability (p. 228)
potential (voltage) (p. 229)

current (p. 229)
electrochemical equilibrium (p. 230)
equilibrium membrane potential (p. 230)
Nernst equation (p. 230)
Na^+/K^+ pump (p. 230)
depolarization (p. 231)
steady-state ion movement (p. 231)
Goldman equation (p. 232)

Electrical Excitability
ion channel (p. 233)
voltage-gated ion channel (p. 233)
ligand-gated ion channel (p. 233)
patch clamping (p. 233)
voltage sensor (p. 235)
channel gating (p. 235)
channel inactivation (p. 235)

The Action Potential
squid giant axon (p. 236)
threshold potential (p. 236)
action potential (p. 236)
propagation (p. 236)
hyperpolarization (undershoot)
    (p. 238)
absolute refractory period (p. 239)
relative refractory period (p. 240)
passive spread of depolarization (p. 240)
axon hillock (p. 240)

nerve impulse (p. 241)
oligodendrocyte (p. 241)
Schwann cell (p. 241)
saltatory propagation (p. 241)

Synaptic Transmission
electrical synapse (p. 243)
presynaptic neuron (p. 243)
postsynaptic neuron (p. 243)
chemical synapse (p. 243)
synaptic cleft (p. 243)
terminal bulb (p. 243)
neurotransmitter (p. 243)
acetylcholine (p. 244)
cholinergic synapse (p. 244)
catecholamine (p. 244)
adrenergic synapse (p. 244)
neuropeptide (p. 244)
neurosecretory vesicle (p. 245)
active zone (p. 247)
neurotoxin (p. 247)

Integration and Processing of Nerve Signals
excitatory postsynaptic potential (EPSP)
    (p. 250)
inhibitory postsynaptic potential (IPSP)
    (p. 250)
temporal summation (p. 251)
spatial summation (p. 251)

Problem Set

More challenging problems are marked with a • .

9-1. The Truth About Nerve Cells. For each of the following statements, indicate whether it is true of all nerve cells (A), of some nerve cells (S), or of no nerve cells (N).

(a) The axonal endings are separated from the cells they innervate by a small cleft.

(b) The resting membrane potential of the axonal membrane is negative.

(c) Nodes of Ranvier are found at regular intervals along the axon.

(d) The resting potential of the membrane is much closer to the equilibrium potential for potassium ions than to that for sodium ions because the permeability of the axonal membrane is much greater for potassium than for sodium.

(e) Excitation of the membrane results in a permanent increase in its permeability to sodium ions.

(f) The electrical potential across the membrane of the axon can be easily measured using electrodes.

(g) Calcium elevation stimulates release of neurosecretory vesicles containing serotonin.

•9-2. The Resting Membrane Potential. The Goldman equation is used to calculate V_m, the resting potential of a biological membrane. As presented in the chapter, this equation contains terms for sodium, potassium, and chloride ions only.

(a) Why do only these three ions appear in the Goldman equation as it applies to nerve impulse transmission?

(b) Suggest a more general formulation for the Goldman equation that would be applicable to membranes that might be selectively permeable to other monovalent ions as well.

(c) How much would the resting potential of the membrane change if the relative permeability for sodium ions were 1.0 instead of 0.01?

(d) Would you expect a plot of V_m versus the relative permeability of the membrane to sodium to be linear? Why or why not?

9-3. Patch Clamping. Patch-clamp instruments enable researchers to measure the opening and closing of a single channel in a membrane. A typical acetylcholine receptor channel passes about 5 pA (picoamperes) of ionic current (1 picoampere $= 10^{-12}$ ampere) over a period of about 5 msec at −60 mV.

(a) Given that an electrical current of 1 A is about 6.2×10^{18} electrical charges per second, how many ions (potassium or sodium) pass through the channel during the time it is open?

(b) Do you think the opening of a single receptor channel would be sufficient to depolarize a postsynaptic membrane? Why or why not?

9-4. The Equilibrium Membrane Potential. Answer each of the following questions with respect to E_{Cl}, the equilibrium membrane potential for chloride ions. The chloride ion concentration inside the squid giant axon can vary from 50 to 150 mM.

(a) Before doing any calculations, predict whether E_{Cl} will be positive or negative. Explain.

(b) Now calculate E_{Cl}, assuming an internal chloride concentration of 50 mM.

(c) How much difference would it make in the value of E_{Cl} if the internal chloride concentration were 150 mM instead?

(d) Why do you suppose the chloride concentration inside the axon is so variable?

•9-5. Heart Throbs. An understanding of muscle cell stimulation involves some of the same principles as nerve cell stimulation, except that calcium ions play an important role in the former. The following ion concentrations are typical of those in human heart muscle and in the serum that bathes the muscles:

[K+]: 150 mM in cell, 4.6 mM in serum

[Na+]: 10 mM in cell, 145 mM in serum

[Ca2+]: 0.001 mM in cell, 6 mM in serum

Figure 9-28 depicts the change in membrane potential with time upon stimulation of a cardiac muscle cell.

(a) Calculate the equilibrium membrane potential for each of the three ions, given the concentrations listed.

(b) Why is the resting membrane potential significantly more negative than that of the squid axon (−75 mV versus −60 mV)?

(c) The increase in membrane potential in region Ⓐ of the graph could in theory be due to the movement across the membrane of one or both of two cations. Which cations are they, and in what direction would you expect each to move across the membrane?

(d) How might you distinguish between the possibilities suggested in part c?

(e) The rapid decrease in membrane potential that is occurring in the region Ⓑ is caused by the outward movement of potassium ions. What are the driving forces that cause potassium to leave the cell at this point? Why aren't the same forces operative in region Ⓐ of the curve?

(f) People with heart disease often take drugs that can double or triple their serum potassium levels to about 10 mM without altering intracellular potassium levels. What effect should this increase have on the rate of potassium ion movement across the heart cell membrane during muscle stimulation? What effect should it have on the resting potential of the muscle?

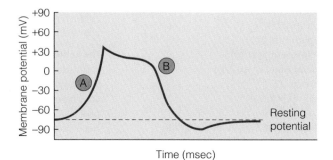

Figure 9-28 The Action Potential of a Muscle Cell of the Human Heart. See Problem 9-5.

9-6. The All-or-None Response of Membrane Excitation. A nerve cell membrane exhibits an all-or-none response to excitation; that is, the magnitude of the response is independent of the magnitude of the stimulus, once a threshold value is exceeded.

(a) Explain in your own words why this is so.

(b) Why is it necessary that the stimulus exceed a threshold value?

(c) If every neuron exhibits an all-or-none response, how do you suppose the nervous system of an animal can distinguish different intensities of stimulation? How do you think your own nervous system can tell the difference between a warm iron and a hot iron, or between a chamber orchestra and a rock band?

9-7. Multiple Sclerosis and Action Potential Propagation. Multiple sclerosis (MS) is an autoimmune disease that attacks myelinated nerves and degrades the myelin sheath around them. How do you think the propagation of action potentials would be affected in a nerve cell that has been damaged in a patient with MS?

9-8. One-Way Propagation. Why does the action potential move in only one direction down the axon?

9-9. Inhibitory Neurotransmitters. Some inhibitory neurotransmitters cause chloride channels to open, whereas others stimulate the opening of potassium channels. Explain why increasing the permeability of the neuronal membrane to either chloride or potassium would make it more difficult to stimulate an action potential. What generalization can you make about inhibitory neurotransmitters?

9-10. Trouble at the Synapse. Transmission of a nerve impulse across a cholinergic synapse is subject to inhibition by a variety of neurotoxins. Indicate, as specifically as possible, what effect each of the following poisons or drugs has on synaptic transmission and what effect each has on the polarization of the postsynaptic membrane (see Box 9A).

(a) The snake poison α-bungarotoxin

(b) The insecticide malathion

(c) Succinylcholine

(d) The carbamoyl ester neostigmine

•**9-11. More Trouble at the Synapse.** Drugs that affect the reuptake of neurotransmitters are in widespread use for the treatment of attention-deficit/hyperactivity disorder and clinical depression. In molecular terms, explain what effect(s) such reuptake inhibitors have on a postsynaptic neuron, assuming that the neurotransmitter in question is excitatory for that postsynaptic neuron.

Suggested Reading

References of historical importance are marked with a • .

General References

Albright, T.D., T.M. Jessell, E.R. Kandel, and M.I. Posner. Neural Science: a century of progress and the mysteries that remain. *Cell* 100 (2000): S1.

Delcomyn, F. *Foundations of Neurobiology.* New York: W. H. Freeman, 1998.

Hille, B. *Ionic Channels of Excitable Membranes.* Sunderland, MA: Sinauer, 1992.

Levitan, I. B., and L. K. Kaczmarek. *The Neuron: Cell and Molecular Biology,* 2d ed. New York: Oxford University Press, 1997.

Nicholls, J.G., A.R. Martin, B.G. Wallace, and P.A. Fuchs. *From Neuron to Brain,* 4th ed. Sunderlan, MA: Sinauer, 2001.

The Na⁺/K⁺ Pump

Mercer, R. W. Structure of the Na, K-ATPase. *Internat. Rev. Cytol.* 137C (1993): 139.

Patch Clamping

Neher, E., and B. Sakmann. The patch-clamp technique. *Sci. Amer.* 266 (March, 1992): 28.

Ion Channels and Membrane Excitation

Caterall, W.A. From ionic currents to molecular mechanisms: The structure and function of voltage-gated sodium channels. *Neuron* 26 (2000): 13.

Caterall, W.A. Structure and regulation of voltage-gated Ca⁺ channels. *Ann. Rev. Cell Dev. Biol.* 16 (2000): 521.

•Hodgkin, A.L., and A.F. Huxley. A quantitative description of membrane current and its application to conduction and excitation in nerve. *J. Physiol.* 117 (1952): 500.

•Karlin, A. The structure of nicotinic acetylcholine receptors. *Curr. Opin. Neurobiol.* 3 (1993): 299.

•Unwin, N. Neurotransmitter action: Opening of ligand-gated ion channels. *Cell* 72 (1993): 31.

Weinreich, F., and T. J. Jentsch. Neurological diseases caused by ion-channel mutations. *Curr. Opin. Neurobiol.* 10 (2000): 409.

Yi, B.A., and L.Y. Jan. Taking apart the gating of voltage-gated K+ channels. *Neuron* 27 (2000): 423.

Myelin

Pedraza, L., J.K. Huang, and D. R. Colman. Organizing principles of the axoglial apparatus. *Neuron* 30 (2001): 335.

Steinman, L. Multiple sclerosis: A coordinated immunological attack against myelin in the central nervous system. *Cell* 85 (1996): 299.

Neurotransmitters and Neuropeptides

Hokfelt, T. Neuropeptides in perspective: The last ten years. *Neuron* 7 (1991): 867.

Nemeroff, C.B. The neurobiology of depression. *Sci. Amer.* 278 (June 1998): 42

Vincent, A., D. Beesonn, and B. Lang. Molecular targets for autoimmune and genetic disorders of neuromuscular transmittion. *Eur. J. Biochem.* 267 (2000) 6717.

Synaptic Transmission

Augustine, G.J. How does calcium trigger neurotransmitter release? *Curr. Opinion Neurobiol.* 11 (2001): 320.

Garner, C.G, S. Kindler, and E.D. Gundelfinger. Molecular determinants of presynaptic active zones. *Curr. Opinion Neurobiol.* 10 (2000): 321.

Geppert, M., and T.C. Sudhof. RAB3 and synaptotagmin: The yin and yang of synaptic membrane fusion. *Annu. Rev. Neurosci.* 211 (1998): 75.

Harlow, M.L., D. Ress, A. Stoschek, R.M. Marshall, and U.J. McMahan. The architecture of the active zone material at the frog's neuromuscular junction. *Nature* 409 (2001): 479.

10

Signal Transduction Mechanisms: II. Messengers and Receptors

In the previous chapter, we learned how nerve cells communicate with one another and with other types of cells by means of electrochemical changes in their membranes. We saw how, in most cases, the arrival of an action potential at a synapse causes the release of neurotransmitters, which in turn bind to receptors on the adjacent postsynaptic cell membrane, thereby passing on the signal. Now we are ready to explore a second major means of intercellular communication, which likewise involves interactions between chemicals and receptors. In this case, however, the signal is transmitted by regulatory chemical messengers, and the receptors are located on the surfaces of cells that may be quite distant from the secreting cells. Thus, animals have two different but complementary systems of communication and control, and receptors play a crucial role in both systems. In this chapter, we will encounter several general mechanisms by which receptors detect and pass on chemical signals to which a cell is exposed. We focus on humans, but consider several other organisms as well.

Chemical Signals and Cellular Receptors

All cells have some ability to sense and respond to specific aspects of their environment. Characteristics of the environment to which cells can respond include physical factors, such as light, heat, or gravity; and chemical factors, such as the presence and concentration of particular extracellular molecules. Prokaryotes, for instance, have membrane-bound receptor molecules on the cell surface that enable them to respond to chemicals in their environ-

ment. The human body has receptors for light (the rods and cone cells of the retina), sound (the hair cells of the inner ear), and various other physical stimuli. Receptors on the tongue and in the nose detect chemicals in our food and in the environment. As we will see, even the cells of the early animal embryo possess sophisticated machinery for detecting changes in their environment.

Cells also communicate with one another. One way they do this is by displaying molecules on their surfaces that are recognized by receptors on the surfaces of other cells. This kind of cell-to-cell communication requires that cells come into direct contact with each other. Alternatively, one cell can release chemical signals that are recognized by another cell, either nearby or at a distant location. For example, simple eukaryotic organisms such as amoeboid cells of the slime mold *Dictyostelium* secrete a compound called cyclic AMP (cAMP) at one stage in their life cycle. The compound binds to receptors on the surfaces of neighboring cells, triggering a process whereby thousands of individual amoeboid cells aggregate and differentiate into a multicellular organism.

In more complex multicellular organisms, the problem of regulating and coordinating the various activities of cells or tissues is particularly important. Here, the whole organism is organized into different tissues made up of specialized cells. Furthermore, the specific functions of these cells may be critical only for certain occasions, or one tissue may need to perform different functions in different circumstances. The nervous system regulates tissue function very quickly, but it is confined to tissues that are innervated. Another form of regulation is through the release of *chemical messengers,* the topic of this chapter.

Different Types of Chemical Signals Can Be Received by Cells

A variety of compounds can function as **chemical messengers** that serve as signals between cells. Some messengers, such as hormones, are produced at great distances from their target tissues and are carried in the circulatory system to various sites in the body. Others, such as growth factors, are released locally, acting only on nearby tissues (Figure 10-1). Once a messenger reaches its target tissue, it binds to receptors on the surface of the target cells, initiating the signaling process. The logic and general flow of information involved in such chemical signaling is shown in Figure 10-2. A molecule coming from either a long or a short distance functions as a **ligand** by binding to a receptor. A ligand, as we saw in Chapter 9, often binds to a receptor embedded within the plasma membrane of the cell receiving the signal. In other cases, such as steroid hormones, the ligand binds to a receptor inside the cell. In either case, the ligand is a **primary messenger**. The binding of ligand to receptor often results in the production of additional molecules within the cell receiving the signal. Such **second messengers** are small molecules or ions that relay the signals from one location in the cell, such as the plasma membrane, to the interior of the cell, initiating a cascade of changes within the receiving cell. Often these events affect the expression of specific genes within the receiving cell. The ultimate result is a change in the identity or function of the cell. The ability of a cell to translate a receptor-ligand interaction to changes in its behavior or gene expression is known as **signal transduction.**

Messenger molecules can be chemically characterized as amino acids or their derivatives, peptides, proteins, fatty acids, lipids, nucleosides, or nucleotides. Many messengers are hydrophilic compounds whose function lies entirely in their ability to bind to one or more specific receptors on a target cell. In this case, the chemical com-position of the messenger does not have any direct bearing on the kind of message delivered to the target cell; it is important mainly in that it provides the correct molecular shape and bonding characteristics to fit the receptor.

Hydrophobic messengers, on the other hand, act on receptors in the nucleus or cytosol whose function is to regulate the transcription of particular genes. Among the hydrophobic messengers that bind to intracellular receptors are the *steroid hormones,* which are derived from the compound cholesterol, and the *retinoids,* derived from vitamin A. Because such receptor-ligand complexes act in the nucleus, we will postpone a discussion of them until Chapter 21, when we have gained a better understanding of how the transcription of specific genes is regulated within cells. Here we will focus on hydrophilic ligands and their receptors on the cell surface.

Receptor Binding Involves Specific Interactions Between Ligands and Their Receptors

How do cells distinguish messengers from the multitude of other chemicals in the environment or from messengers intended for other cells? The answer lies in the highly specific way the messenger molecule binds to the receptor. A messenger forms noncovalent chemical bonds with the receptor protein. Individual noncovalent bonds are generally weak; therefore, several bonds must form to achieve strong binding. For a receptor to make numerous bonds with its ligand, the receptor must have a *binding site* (or *binding pocket*) that fits the messenger molecule closely, like a hand in a glove. Furthermore, within the ligand-binding site on the receptor, appropriate amino acid side chains must

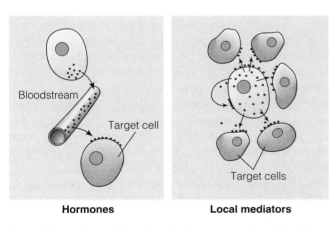

Figure 10-1 Cell-to-Cell Signaling by Hormones and Local Mediators. The main distinction between these classes of signaling molecules involves the distance the molecule travels before encountering its target cell. Although most signaling molecules act on the cell surface, certain hydrophobic hormones and local mediators enter their target cells.

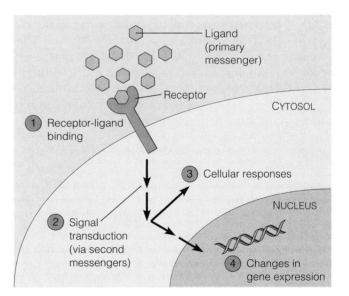

Figure 10-2 The Overall Flow of Information During Cell Signaling. Binding of ligand by a receptor activates a series of events known as signal transduction, which relays the signal to the interior of the cell, resulting in specific cellular responses and/or changes in gene expression.

be positioned so that they can form chemical bonds with the messenger molecule. This combination of binding site shape and the strategic positioning of amino acid side chains within the binding site is what enables the receptor to distinguish its specific ligand from thousands of other chemicals.

Receptor Affinity. In most cases, the binding reaction between a ligand and the receptor specific for it, known as its **cognate receptor,** is similar to the binding of an enzyme to its substrate. Ligands in the fluid surrounding the cell can collide with receptors, and some of these collisions lead to binding. When a receptor binds its ligand, the receptor is said to be *occupied*. Similarly, the ligand can be either bound (to a receptor) or free in the solution. The amount of receptor that is occupied by ligand is proportional to the concentration of free ligand in solution. As the ligand concentration increases, an increasingly greater proportion of its cognate receptors will become occupied, until all the receptors are occupied. Further increases in ligand concentration will, in principle, have no further effect on the target cell.

The relationship between the concentration of ligand in solution and the number of receptors occupied can be described qualitatively in terms of **receptor affinity.** When almost all of the receptors are occupied at low concentrations of free ligand, we say that the receptor has a high affinity for its ligand. Conversely, when it takes a relatively high concentration of ligand for most receptors to be occupied, we say that the receptor has a low affinity for its ligand. Receptor affinity can be described quantitatively in terms of the **dissociation constant, K_d,** the concentration of free ligand needed to produce a state in which half the receptors are occupied. Values for K_d range from roughly 10^{-4} to 10^{-9} mM. As with the Michaelis constant (K_m; see Chapter 6) in enzyme kinetics, the importance of the value is that it tells us at what concentration a particular ligand will be effective in producing a cellular response. Thus receptors with a high affinity for their ligands have a very small dissociation constant, and conversely, low-affinity receptors have a high K_d. Typically, the ligand concentration must be in the range of the K_d value of the receptor in order for the ligand to have an effect on the target tissue.

Receptor Down-Regulation. Although receptors have a characteristic affinity for their ligands, cells are geared to sense *changes* in ligand concentration rather than fixed ligand concentrations. When a ligand is present and receptors are occupied for prolonged periods of time, the cell adapts, or becomes desensitized, so that it no longer responds to the ligand. To further stimulate the cell, the ligand concentration must be increased.

Desensitization is due mainly to changes in the properties or cellular location of the receptor, a phenomenon known as *receptor down-regulation*. The down-regulation of receptors occurs in several ways, including (1) removal of the receptor from the cell surface, (2) alterations to the receptor that lower its affinity for ligand, or (3) alterations to the receptor that render it unable to initiate changes in

cellular function. The removal of receptors from the cell surface takes place through the process of *receptor-mediated endocytosis*, in which small portions of the plasma membrane containing receptors invaginate and are internalized. (Receptor-mediated endocytosis will be discussed in detail in Chapter 12.) The reduced number of receptors on the cell surface results in a diminished cellular response to ligand. Receptors that are altered chemically, often by phosphorylation, can remain on the surface yet not respond to the presence of ligand, either because they have reduced affinity for ligand or they are unable to initiate changes in cellular function.

Desensitization provides a way for cells to adapt to permanent differences in levels of messenger concentration. However, it also leads to the phenomenon of *tolerance*. An example of this is seen in nasal sprays, which contain neosynephrine or other compounds that stimulate particular receptors called α-adrenergic receptors that will be discussed later in this chapter. The drug causes blood vessels in the nose to constrict, thereby inhibiting mucous secretions. However, prolonged use leads to a loss of effectiveness due to receptor down-regulation.

Understanding the nature of receptor-ligand binding has provided great opportunities for researchers and pharmaceutical companies. Although receptors have binding sites that fit the messenger molecule quite closely, it is possible to make similar synthetic ligands that bind even more tightly or selectively. This is especially important when more than one type of receptor exists for the same ligand. Many synthetic compounds have been developed that selectively affect only one type of receptor. In addition, whereas normal messengers cause a change in the receptor when they bind, both synthetic and natural compounds have been discovered that can bind to receptors without triggering such a change. These compounds inhibit the receptor by preventing the naturally occurring messenger from binding and activating the receptor.

Drugs that selectively activate or inhibit particular kinds of receptors have become central to the treatment of many medical problems. For example, *isoproterenol* and *propranolol* activate or inhibit β-adrenergic receptors, which will be discussed later in the chapter. Isoproterenol is used to treat asthma or to stimulate the heart, whereas propranolol is used to reduce blood pressure and the strength of cardiac contractions and to control anxiety attacks. Another example is *famotidine*, a compound that selectively binds and inhibits a particular type of histamine receptor found on cells in the stomach. Famotidine is sold as a stomach "acid controller" under the trade name Pepcid AC. Another drug, *cimetidine*, sold under the trade name Tagamet, acts in a similar manner.

Receptor Binding Activates a Sequence of Signal Transduction Events Within the Cell

As we have noted, when a ligand binds to its cognate receptor, the receptor is altered in a way that causes changes in cellular activities. In general, the binding of a ligand either induces a change in receptor conformation or causes receptors to cluster together. Once one of these changes takes place, the receptor initiates a preprogrammed sequence of events inside the cell. By *preprogrammed,* we mean that cells have a greater repertoire of functions than are in use at any particular time. Some of these cellular processes remain unused until particular signals are received that trigger them. We might compare the activation effect of a messenger to the stimulation of a reflex. In the knee-jerk reflex, for example, the nerve connections are already in place that can cause the leg to extend in response to a tap just below the knee, but the tap is needed to activate the reflex. In a similar manner, a messenger activates a series of cellular functions that are already in place but are not presently used.

Signal transduction events are initiated or altered within a cell when a ligand binds to its cognate receptor on the cell surface or within the cell. To understand these pathways and the ways particular hormones or other messengers cause changes in cellular activities, we need to know what a receptor does when it binds to its ligand. Depending on their mode of action, receptors can be classified into several basic categories. We discussed *ligand-gated channels* in Chapter 9, and we will consider *intracellular receptors* in Chapter 21. A third category, *plasma membrane receptors,* is the focus of our discussion in this chapter. Many of these receptors can be classified into two families: those linked to G proteins and those linked to protein kinases. We begin by looking at the former.

G Protein–Linked Receptors

Seven-Membrane Spanning Receptors Act via G Proteins

The **G protein–linked receptor family** is so named because ligand binding causes a change in receptor conformation that activates a particular **G protein** (an abbreviation for *guanine-nucleotide binding protein*). The activated G protein in turn binds to a target protein such as an enzyme or a channel protein, thereby altering the target's activity. All G protein–linked receptors initiate signal transduction inside the cell in this way.

The Structure of G Protein–Linked Receptors. The G protein–linked receptors are remarkable in that they all have a similar structure yet differ significantly in their amino acid sequences. In each case, the receptor protein forms seven transmembrane α helices connected by alternating cytosolic or extracellular loops. The N-terminus of the protein is exposed to the extracellular fluid, while the C-terminus resides in the cytosol (Figure 10-3). The extracellular portion of each G protein–linked receptor has a unique messenger-binding site, and a cytosolic loop connecting the fifth and sixth transmembrane α helices is specific for a particular G protein. G protein–linked receptors therefore provide a versatile method for linking different messengers to different signal transduction pathways.

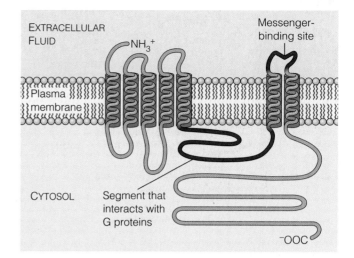

Figure 10-3 The Structure of G Protein–Linked Receptors. Each G protein–linked receptor has seven transmembrane α helices. The primary messenger binds to the extracellular portion of the receptor. This binding causes an intracellular portion of the receptor to activate an adjacent G protein.

The Structure and Activation of G Proteins. We might describe G proteins as a type of molecular switch whose "on" or "off" state depends on whether the G protein is bound to GTP or GDP. There are two distinct classes of G proteins: the *large heterotrimeric G proteins* and the *small monomeric G proteins.* The large heterotrimeric G proteins contain three different subunits, called G alpha (G_α), G beta (G_β), and G gamma (G_γ). These heterotrimeric G proteins mediate signal transduction through G protein–linked receptors. The small monomeric G proteins include *Ras,* which will be discussed later in the chapter in the context of tyrosine kinase receptors. Other small monomeric G proteins similar in structure to Ras that regulate the cytoskeleton will be discussed in Chapter 23. Though the term *G protein* can apply to both trimeric and monomeric types, we will use it only to refer to the trimeric type.

All G proteins have the same basic structure and mode of activation. Of the three subunits in the $G_{\alpha\beta\gamma}$ heterotrimer, G_α, the largest, binds to a guanine nucleotide (GDP or GTP). When G_α binds to GTP, it also detaches from the $G_{\beta\gamma}$ complex. The G_β and G_γ subunits, on the other hand, are permanently bound together. Some G proteins, such as **Gs,** act as stimulators of signal transduction (hence s, for "stimulatory"); others, such as **Gi,** act to inhibit signal transduction (hence i, for "inhibitory").

When a messenger binds to a G protein–linked receptor on the exterior surface of the cell, the change in conformation of the receptor causes the G protein to associate with the receptor, which in turn causes the G_α subunit to release its bound GDP, acquire a GTP, and then detach from the complex (Figure 10-4). Depending on the G protein and the cell type, either the free GTP-G_α subunit or the $G_{\beta\gamma}$ complex can then initiate signal transduction events in the cell. Each portion of the G protein exerts its effect by binding to a particular enzyme or other protein

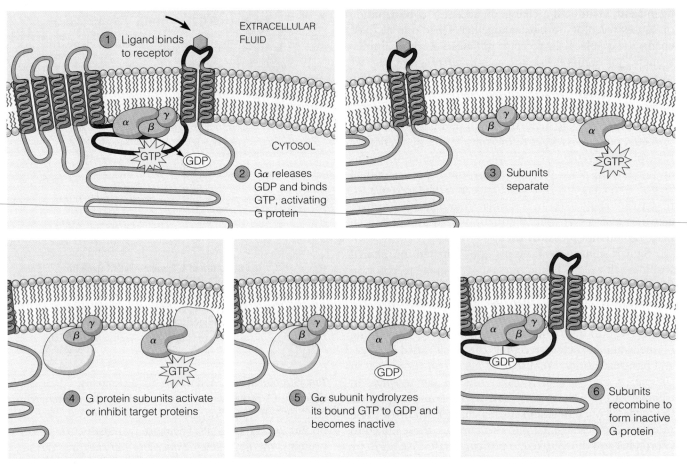

Figure 10-4 The G Protein Activation/ Inactivation Cycle. G protein–linked receptors contain seven transmembrane segments that form a ligand-binding site on the outside of the cell and a G protein– binding site on the inside. ① When the ligand binds, ② the receptor activates a G protein by causing the G$_\alpha$ subunit to release GDP and acquire GTP. ③ The G$_\alpha$ and G$_{\beta\gamma}$ subunits then separate and ④ initiate signal transduction events. ⑤ The GTP-G$_\alpha$ subunit eventually hydrolyzes its bound GTP, converting the subunit back to its inactive GDP-G$_\alpha$ form. ⑥ The inactive GDP-G$_\alpha$ subunit then recombines with G$_{\beta\gamma}$ to form the inactive G heterotrimer.

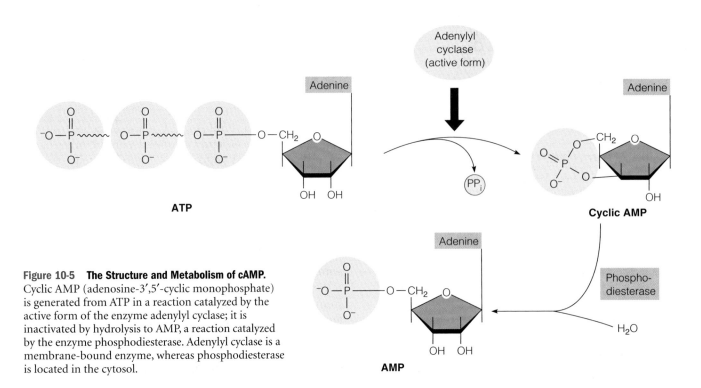

Figure 10-5 The Structure and Metabolism of cAMP. Cyclic AMP (adenosine-3′,5′-cyclic monophosphate) is generated from ATP in a reaction catalyzed by the active form of the enzyme adenylyl cyclase; it is inactivated by hydrolysis to AMP, a reaction catalyzed by the enzyme phosphodiesterase. Adenylyl cyclase is a membrane-bound enzyme, whereas phosphodiesterase is located in the cytosol.

Figure 10-6 The Roles of G Proteins and Cyclic AMP in Signal Transduction. G proteins mediate signal transduction through G protein–linked receptors. When a messenger binds to its receptor, the Gs protein is activated. In the inactive state, the α, β, and γ subunits are present as a complex, with GDP bound to the α subunit. **(a)** When a receptor is activated by binding of its specific ligand on the outer surface of the plasma membrane, the receptor-messenger complex associates with the Gs protein, causing the displacement of GDP by GTP and the dissociation of the Gs$_\alpha$-GTP complex. **(b)** The GTP-Gs$_\alpha$ complex then binds tightly to a molecule of membrane-bound adenylyl cyclase, activating it for synthesis of cAMP. **(c)** Activation ends when the ligand leaves the receptor, the GTP is hydrolyzed to GDP by the GTPase activity of the Gs$_\alpha$ subunit, and the Gs$_\alpha$ dissociates from the adenylyl cyclase. **(d)** Adenylyl cyclase then reverts to the inactive form, the Gs$_\alpha$ reassociates with the Gs$_{\beta\gamma}$ complex, and cAMP molecules in the cytosol are hydrolyzed to AMP by the enzyme phosphodiesterase.

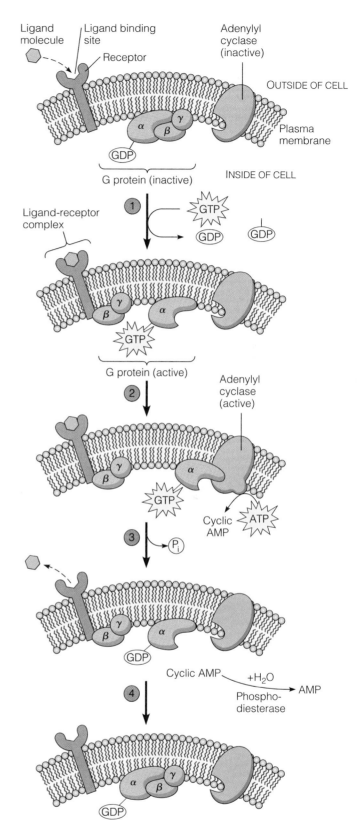

in the cell. In some cases, both the GTP-G$_\alpha$ and GTP-G$_{\beta\gamma}$ subunits simultaneously regulate different processes in the cytosol. However, the activity of the G protein persists only as long as the G$_\alpha$ is bound to GTP and the subunits remain separated. Because the G$_\alpha$ subunit hydrolyzes GTP, it will only remain active for a short time before reverting to the GDP-bound state and reassociating with G$_{\beta\gamma}$. This feature allows the signal transduction pathway to shut down rapidly when the messenger is removed.

The large number of different G proteins provides for a diversity of G protein–mediated signal transduction events. Some G proteins interact directly with certain types of potassium or calcium ion channels to mediate the actions of specific neurotransmitters. In some organisms, G proteins activate kinases. Perhaps the most important and widespread G protein–mediated signal transduction events are the release or formation of second messengers. The two most widely used second messengers are cyclic AMP and calcium ions, which stimulate the activity of target enzymes when their cytosolic concentrations are elevated. As we will see next, G protein–linked receptors can stimulate the elevation of either cyclic AMP or calcium, depending on which G protein is activated.

Cyclic AMP Is a Second Messenger Used by One Class of G Proteins

Cyclic AMP (cAMP) is formed from cytosolic ATP by the enzyme **adenylyl cyclase** (Figure 10-5). Adenylyl cyclase is anchored in the plasma membrane with its catalytic portion protruding into the cytosol. Normally, the enzyme is inactive until it binds to the α subunit of a specific G protein, Gs. When a G protein–linked receptor is coupled to Gs, the binding of ligand stimulates the Gs$_\alpha$ subunit to release GDP and acquire a GTP (Figure 10-6). This in turn causes GTP-Gs$_\alpha$ to detach from the Gs$_{\beta\gamma}$ subunits and bind to adenylate cyclase. When GTP-Gs$_\alpha$ binds to adenylyl cyclase, the enzyme becomes active and converts ATP to cAMP.

As we have mentioned, G proteins respond quickly to changes in ligand concentration because they remain

active for only a short period of time before the G$_\alpha$ subunit hydrolyzes its bound GTP and converts to the inactive state. Once the G protein becomes inactive, the adenylyl cyclase ceases to make cAMP. However, cAMP levels would still remain elevated in the cell if not for the enzyme **phosphodiesterase,** which degrades the cAMP. This further ensures that the signal transduction pathway will shut down promptly when the concentration of the ligand outside the cell declines.

Cyclic AMP appears to have one main intracellular target, an enzyme known as cAMP-dependent kinase, or **protein kinase A (PKA).** Protein kinase A phosphorylates a wide variety of cellular proteins, each of which contains a similar short sequence of amino acids that is recognized by the kinase as the phosphorylation site. The kinase transfers a phosphate from ATP to a serine or threonine found within this particular stretch of amino acids. Cyclic AMP regulates the activity of PKA by causing the detachment of its two regulatory subunits from its two catalytic subunits (Figure 10-7). Once the catalytic subunits are free, PKA can catalyze the phosphorylation of various proteins in the cell.

An increase in cAMP concentration can produce many different effects. When cAMP is elevated in skeletal muscle and liver cells, the breakdown of glycogen is stimulated. In cardiac muscle the elevation of cAMP strengthens heart contraction, whereas in smooth muscle contraction is inhibited. In blood platelets the elevation of cAMP inhibits their mobilization during blood clotting, and in intestinal epithelial cells it causes the secretion of salts and water into the lumen of the gut. Each of these reactions is an example of the preprogrammed response discussed earlier. In fact, if the concentration of cAMP is artificially raised in these different types of cells, these same cellular responses can be triggered even in the absence of a ligand. This can be done in two different ways: either by stimulating cAMP production directly or by inhibiting the enzyme phosphodiesterase that degrades cAMP. The latter leads to elevated cAMP levels because there is always a low level of cAMP production in cells, and the inhibition of phosphodiesterase thus allows the cAMP to accumulate without being degraded. Examples of phosphodiesterase inhibitors are the methylxanthines, compounds such as caffeine and theophylline, found in coffee, tea, and soft drinks. (Theophylline is often used to treat asthma because it relaxes bronchial smooth muscle.)

To summarize, the cAMP pathway transduces extracellular signals that affect numerous cellular functions. Cyclic AMP is produced when a ligand binds to its membrane receptor, activating adenylyl cyclase for cAMP production. Cyclic AMP then activates PKA. The events through the activation of PKA take place in all cells that use cAMP as a second messenger. The events that follow the activation of PKA are usually specific for each cell type.

Disruption of G Protein Signaling Causes Several Human Diseases

What would happen if the G protein–adenylyl cyclase system could not be shut off? This question can be answered by examining what happens in two human diseases caused by bacteria. The bacteria *Vibrio cholerae* (which causes cholera) and *Bordetella pertussis* (which causes whooping cough) both cause disease through their effects on heterotrimeric G proteins.

Cholera results from the secretion of *cholera toxin* when *V. cholerae* colonizes the gut. The toxin alters secretion of salts (sodium chloride and sodium bicarbonate) and fluid in the intestine, which is normally regulated by hormones that act through the G protein G_s to alter intracellular levels of cAMP. A portion of the cholera toxin is an enzyme that chemically modifies G_s so that it can no longer hydrolyze GTP to GDP. As a result, G_s cannot be shut off, cAMP levels remain high, and the intestinal cells secrete large amounts of salt and water. If left untreated, this condition can result in death by dehydration. The *pertussis toxin* secreted by *B. pertussis* acts in a similar manner, but on the inhibitory G protein, G_i. This protein normally shuts off adenylyl cyclase. When inactivated by pertussis toxin, G_i no longer inhibits adenylyl cyclase. However, it is not yet clear why inactivation of adenylyl cyclase leads to whooping cough.

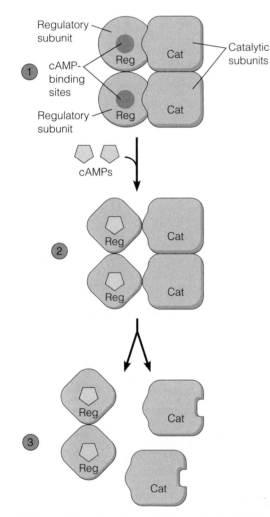

Figure 10-7 The Activation of Protein Kinase A by Cyclic AMP. ① Protein kinase A is composed of four subunits, two catalytic and two regulatory. The regulatory subunits inhibit the catalytic subunits in the absence of cAMP. ② Cyclic AMP activates protein kinase A by binding to the regulatory subunits. When cAMP binds, the regulatory subunits change conformation and ③ detach, freeing the catalytic subunits. Once the catalytic subunits are free, they are activated and capable of phosphorylating target proteins in the cell.

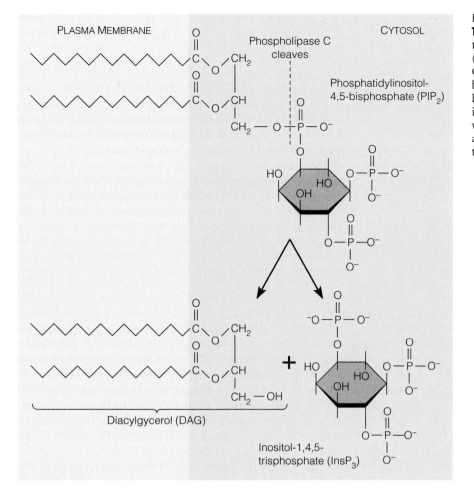

PLASMA MEMBRANE

Phospholipase C cleaves

CYTOSOL

Phosphatidylinositol-4,5-bisphosphate (PIP$_2$)

Diacylgycerol (DAG)

+

Inositol-1,4,5-trisphosphate (InsP$_3$)

Figure 10-8   The Formation of Inositol Trisphosphate and Diacylglycerol.   Inositol trisphosphate (InsP$_3$) and diacylglycerol (DAG) are formed when phospholipase C cleaves phosphatidylinositol-4,5-bisphosphate (PIP$_2$), one of the phospholipids present in membranes. InsP$_3$ is released into the cytosol, whereas DAG remains within the membrane. Both InsP$_3$ and DAG are second messengers in a variety of signal transduction pathways.

The discovery of these toxins and their mode of action has provided not only advances in medical treatment, but also powerful tools for studying G protein–mediated signal transduction. Because these two toxins act on different G proteins, researchers have been able to use the purified toxins to associate a signaling pathway with a particular G protein.

Many G Proteins Use Inositol Trisphosphate and Diacylglycerol as Second Messengers

The importance of inositol phospholipids in cell signaling was first brought to light in the pioneering studies of Robert Michell and Michael Berridge. In the early 1980s, Berridge noticed that when the salivary glands of certain insect larvae were stimulated to secrete, changes occurred in membrane inositol phospholipids. We now know that **inositol-1,4,5-trisphosphate (InsP$_3$)**, one of the breakdown products of inositol phospholipids, also functions as a second messenger. InsP$_3$ is generated from phosphatidylinositol-4,5-bisphosphate (PIP$_2$), a relatively uncommon membrane phospholipid, when the enzyme **phospholipase C** is activated. Phospholipase C cleaves PIP$_2$ into two molecules: inositol trisphosphate and **diacylglycerol (DAG)** (Figure 10-8). If the sequence of events that Berridge observed occurred only in insect sali-

vary glands, his findings would have warranted no more than a modest note in the story of signal transduction. However, InsP$_3$ and DAG were quickly shown to be second messengers in a variety of regulated cell functions, some of which are indicated in Table 10-1. Many different cell-surface receptors are known to activate this pathway, with new examples being reported each year.

The roles of InsP$_3$ and DAG as second messengers in the inositol-phospholipid-calcium pathway are shown in Figure 10-9. The sequence begins with the binding of a ligand to its membrane receptor, leading to the activation of a specific G protein called G_p (the "p" stands for activation of phospholipase C). G_p then activates a form of

Table 10-1   Examples of Cell Functions Regulated by Inositol Trisphosphate and Diacylglycerol

Regulated Function	Target Tissue	Messenger
Platelet activation	Blood platelets	Thrombin
Muscle contraction	Smooth muscle	Acetylcholine
Insulin secretion	Pancreas, endocrine	Acetylcholine
Amylase secretion	Pancreas, exocrine	Acetylcholine
Glycogen degradation	Liver	Antidiuretic hormone
Antibody production	B lymphocytes	Foreign antigens

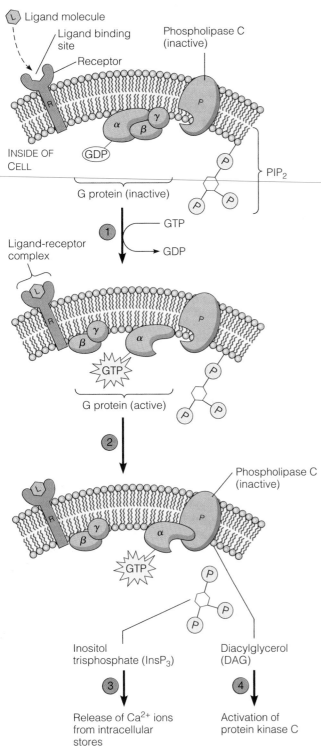

OUTSIDE OF CELL

Ligand molecule

Ligand binding site

Phospholipase C (inactive)

Receptor

INSIDE OF CELL

GDP

G protein (inactive)

PIP_2

① GTP → GDP

Ligand-receptor complex

GTP

G protein (active)

②

Phospholipase C (inactive)

GTP

Inositol trisphosphate ($InsP_3$)

Diacylglycerol (DAG)

③ Release of Ca^{2+} ions from intracellular stores

④ Activation of protein kinase C

Figure 10-9   The Role of InsP₃ and DAG in Signal Transduction. ① When a receptor (R) is activated by the binding of its ligand (L) on the outer surface of the plasma membrane, the receptor-ligand complex associates with the G protein G_p, causing the displacement of GDP by GTP and the dissociation of the GTP-G_α complex. ② The GTP-G_α complex then binds to phospholipase C (P), activating it and causing cleavage of PIP_2 into one molecule each of $InsP_3$ and DAG. ③ The $InsP_3$ is released into the cytosol, where it triggers the release of calcium. ④ The DAG remains in the membrane, where it activates protein kinase C.

phospholipase C known as C_β, thereby generating both $InsP_3$ and DAG. Inositol trisphosphate is water-soluble and quickly diffuses through the cytosol, binding to a ligand-gated calcium channel known as the **InsP₃ receptor** channel in the endoplasmic reticulum. When $InsP_3$ binds, the channel opens, releasing calcium ions into the cytosol. Calcium then binds to a protein known as calmodulin (to be discussed shortly), and the calcium-calmodulin complex activates the desired physiological process.

The DAG generated by phospholipase C activity remains in the membrane, where it activates the enzyme **protein kinase C (PKC).** This enzyme can then phosphorylate specific serine and threonine groups on a variety of target proteins, depending on the cell type. PKC is in fact a family of six or more related enzymes. Most members of the PKC family require membrane phospholipids and DAG for activity, and some require calcium as well. The role of DAG was established experimentally by showing that its effects can be mimicked by *phorbol esters,* plant metabolites that bind to PKC, activating it directly. Using a variety of pharmacological agents, researchers showed that $InsP_3$-stimulated calcium release and DAG-mediated kinase activity are both required to produce a full response in target cells.

A wide variety of cellular effects have been linked to the activation of protein kinase C, including the stimulation of cell growth, the regulation of ion channels, changes in the cytoskeleton, increases in cellular pH, and effects on secretion of proteins and other substances. The stimulation of cell growth by PKC is probably due to its ability to phosphorylate and activate proteins called *MAP kinases,* which we will discuss later in the chapter.

The Release of Calcium Ions Is a Key Event in Many Signaling Processes

How do we know that the sequence of events shown in Figure 10-9, especially the link between $InsP_3$ signaling and the release of calcium within cells, actually occurs in cells? The answer begins with the observation that a specific physiological phenomenon—salivary secretion, in this case—could be activated by the products of phospholipase action: $InsP_3$ and DAG. Evidence for the role of calcium was provided by an experimental approach involving the injection of a calcium-dependent fluorescent dye, such as fura-2, into a target cell. Because the fluorescence of the dye varies with the calcium concentration, the dye is a sensitive indicator of the intracellular calcium concentration. Such dyes are typically referred to as *calcium indicators* (Figure 10-10). By measuring the increase in fluorescence in response to activation by a ligand or by $InsP_3$ directly, investigators were able to establish that ligand binding leads to an increase in $InsP_3$ concentration, which in turns triggers an increase in cytosolic calcium concentration. More recently, genetically engineered proteins called "cameleons," which increase their fluorescence in response to elevated calcium, have been used to monitor cytosolic calcium levels.

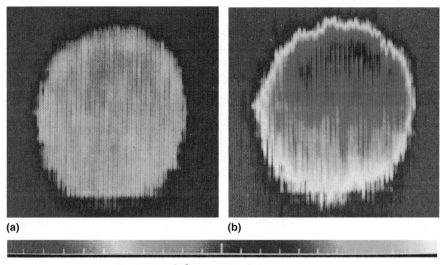

Figure 10-10   Increase in Free Cytosolic Ca²⁺ Concentration Triggered by a Hormone That Stimulates the Formation of Inositol Trisphosphate. Cells from a bovine adrenal gland were loaded with fura-2, a dye that fluoresces when bound to Ca²⁺. The color of the fluorescence changes with alterations in Ca²⁺ concentration. **(a)** Fluorescence micrograph of an unstimulated adrenal cell color-coded to reflect different levels of fluorescence in response to free Ca²⁺. The yellow color indicates a relatively low concentration of free Ca²⁺. **(b)** Fluorescence in the adrenal cell stimulated by angiotensin, a hormone that triggers the formation of inositol trisphosphate. The green and blue color indicates an increased concentration of free Ca²⁺.

Increasing Ca²⁺ concentration ⟶

To complete the sequence of events, the increase in calcium concentration had to be linked to the actual physiological response of the target cell. This link was established by treating target cells with a **calcium ionophore** (such as the drugs ionomycin or A23187) in the absence of extracellular calcium. The ionophore renders membranes permeable to calcium, thereby releasing intracellular stores of calcium in the absence of a physiological stimulus. Treatment with the calcium ionophore mimicked the effect of InsP₃, thus implicating calcium as an intermediary in the InsP₃ signal transduction pathway.

Calcium ions (Ca²⁺) play an essential role in regulating a variety of cellular functions. Regulation is achieved by changes in the calcium concentration of the cytosol in response to external signals. Figure 10-11 provides an overview of the various mechanisms of calcium regulation. The concentration of calcium is normally maintained at very low levels in the cytosol because of the presence of **calcium pumps** in the plasma membrane and the endoplasmic reticulum. The calcium pump in the plasma membrane transports calcium out of the cell, whereas the calcium pump in the ER sequesters calcium ions in the lumen of the ER. In addition, some cells have sodium-calcium exchangers that further reduce the cytosolic calcium concentration. Finally, mitochondria can transport calcium into the mitochondrial matrix. For most cells in their resting state, the action of calcium pumps maintains the calcium concentration in the cytosol at about 1×10^{-4} mM.

Just as a number of pumps keep calcium concentration low in the cytosol, there are a number of different ways that various stimuli can cause cytosolic calcium concentrations to increase. One way, discussed in Chapter 9 in relation to neurons, is by the opening of calcium channels in the plasma membrane. The calcium concentration in the extracellular fluid and the blood is about 1.2 mM, more than 10,000 times as high as that of the cytosol. As a result, when calcium channels open, calcium ions rush into the cell.

Calcium levels can also be elevated by the release of calcium from intracellular stores. Calcium ions sequestered in the ER can be released through the InsP₃ receptor channel, discussed earlier, and through the ryanodine receptor channel. The **ryanodine receptor channel,** so named because it is sensitive to the plant alkaloid *ryanodine,* is particularly important for calcium release from the sarcoplasmic reticulum of cardiac and skeletal muscle (see Chapter 23), but nonmuscle cells such as neurons also have ryanodine calcium-release channels. Surprisingly, the ligand that opens the ryanodine receptor channel appears to be calcium itself. When a neuron is depolarized, calcium channels in the plasma membrane open and allow some calcium to enter the cytosol. Upon exposure to a rapid increase in calcium ions, the ryanodine receptor channel opens, allowing calcium to escape from the ER into the cytosol. This phenomenon has been aptly named *calcium-induced calcium release.*

The Calcium-Calmodulin Complex and Its Intracellular Effects. Calcium can bind directly to some proteins and thereby alter their activity, but more often it works through the protein **calmodulin.** This protein plays a pivotal role in the regulation of a variety of cellular processes.

How does calmodulin mediate calcium-activated events in the cell? The calmodulin molecule has been compared to a flexible "arm" with a "hand" at each end (Figure 10-12a). Two calcium ions bind at each of two hand regions, causing the calmodulin to undergo a change in shape that forms the active **calcium-calmodulin complex** (Figure 10-12b, steps 1 and 2). When a protein is present that contains a calmodulin-binding site, the hands and arm bind to it by wrapping around the binding site (Figure 10-12b, step 3).

One of the important features of calmodulin is its affinity for calcium, which must be such that calmodulin will bind to calcium only when the cytosolic calcium concentration rises in response to a specific stimulus. Thus,

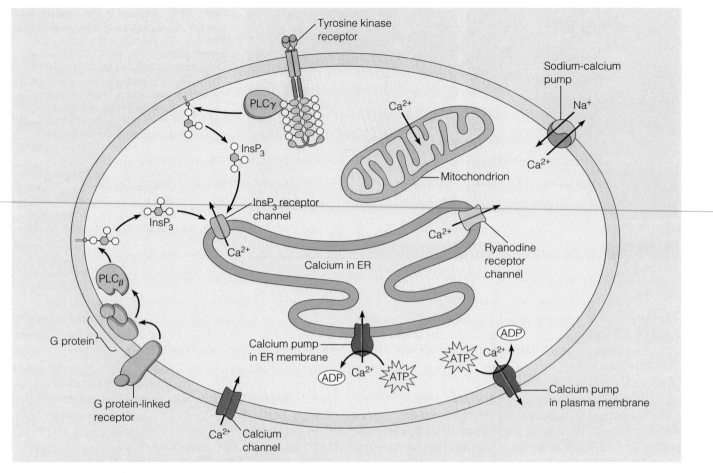

Figure 10-11   An Overview of Calcium Regulation in Cells.   Cytosolic calcium concentration is lowered by the actions of the ER calcium pump, the plasma membrane calcium pump, sodium-calcium exchangers, and mitochondria. Calcium concentration increases in the cytosol because of the opening of calcium channels in the plasma membrane and the release of calcium through the InsP₃ or ryanodine receptor channels in the ER membrane.

calmodulin binds to calcium when the cytosolic calcium concentration increases to about 10^{-3} mM, but it releases calcium when cytosolic calcium levels decline back to the resting level of 10^{-4} mM. For calmodulin to respond to such low calcium concentrations, calcium ions must bind tightly—but not too tightly. Calmodulin is uniquely suited to operate within the typical range of cytosolic calcium concentrations.

Most calmodulin-binding proteins are enzymes such as protein kinases and protein phosphatases. The response of a target cell to an increase in calcium concentration depends on the particular calmodulin-binding proteins that are present in the cell. This means that the same change in calcium concentration can produce markedly different effects in two target cells if each possesses different calmodulin-sensitive enzyme systems.

Calcium Release Following Fertilization of Animal Eggs.   Fertilization of animal eggs is a striking example of the importance of calcium-mediated signal transduction following a receptor-ligand interaction. In many animals, when sperm that have undergone several prior steps of

activation bind to the surface of mature eggs and unite with them at fertilization, a striking sequence of events ensues. One of the early responses of the egg—within 30 seconds to several minutes after fertilization—is the release of calcium from internal stores. Calcium release occurs initially at the site where the sperm penetrates the egg surface, and then spreads like a wave across the egg, much as a ripple on the surface of a pond spreads away from the site where a pebble strikes the water. The wavelike propagation of calcium release can be visualized using calcium indicators (Figure 10-13).

The calcium release is necessary for two crucial events. First, it stimulates the fusion of vesicles, known as *cortical granules,* with the egg plasma membrane, resulting in the release of the granule contents outside of the cell (Figure 10-14). Cortical granules contain several proteins and enzymes, the release of which results in alterations of the protein coat surrounding most eggs (typically known as the *vitelline envelope*). These alterations render the egg unable to bind additional sperm, thereby preventing more than one sperm from fertilizing the egg. This process is known as the *slow block to polyspermy.* (An earlier *fast*

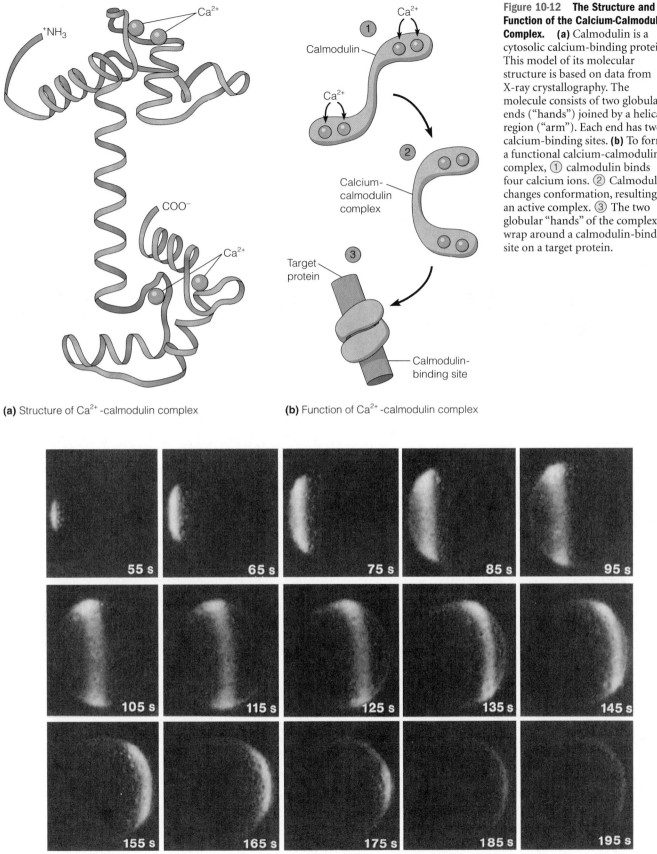

Figure 10-12 **The Structure and Function of the Calcium-Calmodulin Complex.** **(a)** Calmodulin is a cytosolic calcium-binding protein. This model of its molecular structure is based on data from X-ray crystallography. The molecule consists of two globular ends ("hands") joined by a helical region ("arm"). Each end has two calcium-binding sites. **(b)** To form a functional calcium-calmodulin complex, ① calmodulin binds four calcium ions. ② Calmodulin changes conformation, resulting in an active complex. ③ The two globular "hands" of the complex wrap around a calmodulin-binding site on a target protein.

(a) Structure of Ca^{2+}-calmodulin complex

(b) Function of Ca^{2+}-calmodulin complex

Figure 10-13 **Transient Increase in Free Ca^{2+} Concentration That Occurs in an Egg Cell Immediately After Fertilization.** In this classic experiment a fish egg has been injected with aequorin, a dye that emits light when bound to free Ca^{2+} in the cytosol. The amount of time that has elapsed after fertilization is indicated in seconds in each photograph. These pictures show that in the first few minutes after fertilization, a transient wave of increased Ca^{2+} concentration passes across the egg, starting from the point of sperm entry on the left.

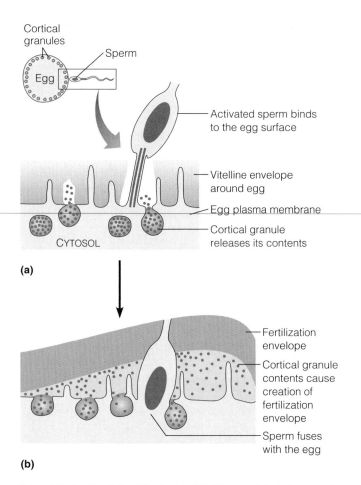

(a)

Cortical granules

Sperm

Egg

Activated sperm binds to the egg surface

Vitelline envelope around egg

Egg plasma membrane

Cortical granule releases its contents

CYTOSOL

(b)

Fertilization envelope

Cortical granule contents cause creation of fertilization envelope

Sperm fuses with the egg

Figure 10-14   The Role of Calcium in the Slow Block to Polyspermy in Sea Urchins.   **(a)** A sperm cell that has been activated binds to the egg surface, resulting in local calcium release and cortical granule exocytosis. **(b)** The result is the creation of the fertilization envelope, which prevents additional sperm from penetrating the egg.

block to polyspermy involves a transient depolarization of the egg plasma membrane.)

The second major function of calcium is *egg activation*. Egg activation involves the resumption of many metabolic processes, the reorganization of the internal contents of the egg, and other events that initiate the process of embryonic development. Many features of the slow block to polyspermy and egg activation can be initiated by treating unfertilized eggs with calcium ionophore in the absence of sperm, demonstrating the key role elevated calcium levels within the egg play in its activation.

Calcium Oscillations and the Opening and Closing of Guard Cells.   Egg activation is a good example of how a dynamic change in calcium concentration can result in dramatic cellular responses. In other cases, it is the oscillation of calcium concentration over time that elicits a cellular response. Calcium oscillations occur in neurons and in fertilized mammalian eggs and may contribute to stable changes in the state of these cells. Calcium oscillations are also important in regulating the opening and closing of *stomata* in plants, the openings in leaves through which plants carry out respi-

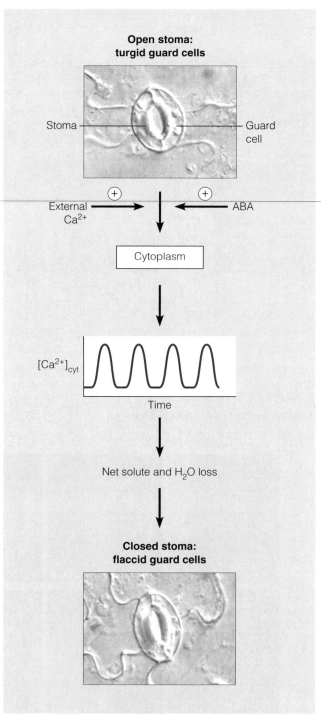

Open stoma: turgid guard cells

Stoma

Guard cell

External Ca²⁺   (+)   (+)   ABA

Cytoplasm

$[Ca^{2+}]_{cyt}$

Time

Net solute and H_2O loss

Closed stoma: flaccid guard cells

Figure 10-15   Calcium Oscillations and Guard Cells.   Guard cells control the opening of stomata, which allows the entry and exit of gases and water vapor from a plant's leaves. When guard cells are turgid (swollen with water), stomata are open. Treatment of guard cells with external calcium or the hormone abscisic acid (ABA) results in calcium oscillations within the guard cells, leading to loss of turgor and opening of stomata.

ration (you will learn more about these specialized structures in Chapter 15). Cells known as *guard cells* regulate the opening and closing of stomata by changing their turgor pressure (turgor pressure was discussed in Chapter 8, p. 201). Under drought conditions, guard cells become flaccid (i.e., they lose turgor), and their stomata close to prevent water loss. Changes in guard cell turgor pressure are in turn regulated by signal transduction events.

One signal that initiates closing of stomata is the hormone *abscisic acid (ABA)*. An increase in ABA concentration or the addition of extracellular calcium results in an influx of extracellular Ca^{2+} into guard cells, which in turn causes release of Ca^{2+} from internal stores. Ultimately, such signals result in steady-state oscillation of Ca^{2+} levels inside guard cells that are necessary for the loss in turgor pressure that results in stomatal closure (Figure 10-15). The changes in turgor pressure only take place when the oscillations are allowed to occur; when they are prevented, no change in the turgor pressure of the guard cells occurs.

Nitric Oxide Couples G Protein-Linked Receptor Stimulation in Endothelial Cells to Relaxation of Smooth Muscle Cells in Blood Vessels

An important signaling molecule in the cardiovascular system is **nitric oxide (NO),** a toxic, short-lived gas molecule produced by the enzyme *NO synthase,* which converts the amino acid arginine to NO and citrulline. Nitric oxide's mode of action on blood vessels involves G proteins.

It has been known for many years that acetylcholine dilates blood vessels by causing their smooth muscles to relax. In 1980, Robert Furchgott demonstrated that acetylcholine dilated blood vessels only if the *endothelium* (the inner lining of the blood vessel) was intact. He concluded that blood vessels are dilated because the endothelial cells produce a signal molecule (or *vasodilator*) that makes vascular smooth muscle cells relax. In 1986 work by Furchgott and parallel work by Louis Ignarro identified NO as the signal released by endothelial cells that causes relaxation of the vascular smooth muscle.

Figure 10-16 illustrates how the binding of acetylcholine to the surface of vascular endothelial cells results in release of NO. There are six steps to this process. ① Acetylcholine binds to G protein–linked receptors that activate the phosphoinositide signaling pathway, causing InsP₃ to be produced by the endothelial cells. ② InsP₃ causes the release of calcium from the endoplasmic reticulum. ③ The calcium ions bind to calmodulin, forming a complex that stimulates NO synthase to produce nitric oxide. ④ Nitric oxide is a gas that readily diffuses through plasma membranes, allowing it to pass from the endothelial cell into the adjacent smooth muscle cells. ⑤ Once inside the smooth muscle cell, NO activates the enzyme *guanylyl cyclase,* which catalyzes the formation of *cyclic GMP (cGMP)*. Cyclic GMP is derived from GTP in a manner analogous to the production of cAMP from ATP, and, like cAMP, cGMP can act as a second messenger. ⑥ The

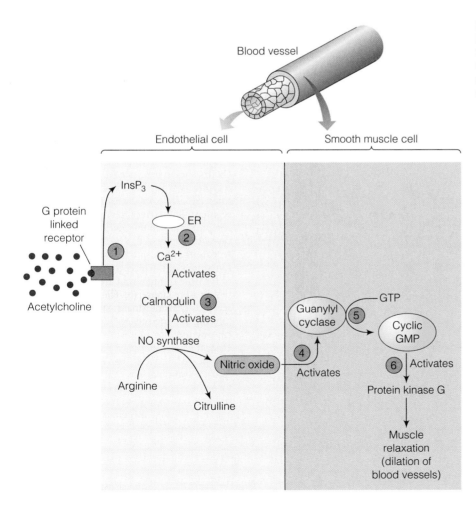

Figure 10-16 The Action of Nitric Oxide on Blood Vessels. The binding of acetylcholine to endothelial cells triggers the production of nitric oxide, which diffuses into the adjacent smooth muscle cells and stimulates guanylyl cyclase, thereby leading to muscle relaxation.

increase in cGMP concentration activates a protein known as *protein kinase G,* which induces muscle relaxation by catalyzing the phosphorylation of the appropriate muscle proteins.

The mechanism by which acetylcholine stimulation of the endothelial cells leads to smooth muscle relaxation also explains the mechanism of action of the chemical *nitroglycerin.* Nitroglycerin is often taken by patients with angina (chest pain due to inadequate blood flow to the heart) to relieve constriction of coronary arteries. In 1977 Ferid Murad found that nitroglycerin and similar vasodilators elicit release of nitric oxide, which relaxes arterial smooth muscle cells. In 1998 Furchgott, Ignarro, and Murad received a Nobel Prize for their elucidation of NO's effects on the cardiovascular system.

Nitric oxide is also used by neurons to signal nearby cells. For example, nitric oxide released by the neurons in the penis results in the blood vessel dilation responsible for penile erection. The drug *sildenafil,* sold under the trade name Viagra, is an inhibitor of a cyclic GMP–specific phosphodiesterase that normally catalyzes the breakdown of cyclic GMP. By maintaining elevated levels of cyclic GMP in erectile tissue, this pathway is stimulated for a longer time period following NO release.

Protein Kinase–Associated Receptors

We have seen that G protein–linked receptors transmit their signals to the interior of the cell by causing changes in a G protein, which in turn initiates a cascade of signal transduction events. These events can include production of cyclic AMP, production of InsP$_3$, and release of calcium. Another large family of proteins uses a different strategy to transmit their signals. They not only function as receptors, but are also protein kinases. When they bind to the appropriate ligand, their kinase activity is stimulated, and they transmit signals through a cascade of phosphorylation events within the cell. We will now examine these **protein kinase–associated receptors** in some detail, using as our main example one especially well-studied class of such receptors, the receptor tyrosine kinases.

Receptor Tyrosine Kinases Aggregate and Undergo Autophosphorylation

Recall that kinases add phosphate groups to particular amino acids within substrate proteins (see Chapter 6). Most kinases involved in regulating cellular enzymes phosphorylate a serine or threonine residue on the target enzyme and are thus known as serine/threonine kinases; we discuss them later in this chapter. Many **receptor tyrosine kinases** trigger a chain of signal transduction events inside the cell that ultimately leads to cell growth, proliferation, or the specialization of cells in a process known as differentiation. These processes are tightly controlled, so that only specific cells respond when the appropriate ligand is available. Figure 10-17 summarizes the structure and activation of a typical receptor tyrosine kinase, the epidermal growth factor (EGF) receptor.

The Structure of Receptor Tyrosine Kinases. Receptor tyrosine kinases differ structurally from G protein–linked

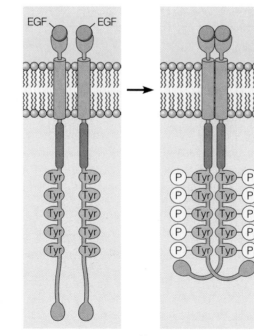

Figure 10-17 The Structure and Activation of a Receptor Tyrosine Kinase. (a) The receptor for epidermal growth factor (EGF), shown here, is typical of many receptor tyrosine kinases. These receptors often have only one transmembrane segment. The extracellular portion of the receptor binds to the ligand (EGF in this case). Inside the cell, a portion of the receptor has tyrosine kinase activity. The remainder of the receptor contains a series of tyrosine residues that are substrates for the tyrosine kinase. **(b)** The activation of receptor tyrosine kinases starts with the binding of a messenger (EGF, in this case), causing receptor aggregation or clustering. Once the receptors aggregate, they cross-phosphorylate each other at a number of tyrosine amino acid residues. The formation of tyrosine phosphate (Tyr-P) residues on the receptor creates binding sites for cytosolic proteins that contain SH2 domains.

(a) Structure of the epidermal growth factor (EGF) receptor

(b) Activation of the EGF receptor

receptors in many ways. These receptors often consist of a single polypeptide chain with only one transmembrane segment. Within this polypeptide chain are several distinct domains (Figure 10-17a). One end of the polypeptide chain projects out into the extracellular fluid, forming the *ligand-binding domain.* The other end of the peptide protrudes through the plasma membrane into the cytosol. On the cytosolic side, a portion of the receptor forms the tyrosine kinase while the remainder constitutes a cytosolic tail. The cytosolic portion of the receptor contains tyrosine residues that are in fact substrates or targets for the tyrosine kinase portion of the receptor.

The tyrosine kinase is frequently an integral part of the receptor protein. In some cases, however, the receptor and the tyrosine kinase are two separate proteins, and the tyrosine kinase is then referred to as a **nonreceptor tyrosine kinase.** However, it can bind to the receptor, and be activated when the receptor binds its ligand, so the net effect is quite similar to the activation of a typical receptor tyrosine kinase.

Nonreceptor tyrosine kinases were, in fact, the first tyrosine kinases to be discovered. The first nonreceptor tyrosine kinase identified was the Src protein, which is encoded by the *src* gene of the avian sarcoma virus. Since then, many nonreceptor tyrosine kinases have been discovered, most of which are similar in structure to the Src protein.

The Activation of Receptor Tyrosine Kinases. Signal transduction is initiated when a ligand binds, causing the receptor tyrosine kinases to aggregate (Figure 10-17b). In many of the best-understood cases, two receptor molecules cluster together within the plasma membrane when they bind to ligand; in other words, receptor tyrosine kinases often act as dimers. Once the receptors cluster in this way, the tyrosine kinase associated with each receptor phosphorylates the tyrosines of neighboring receptors. Since the receptors phosphorylate other receptors of the same type, this process is referred to as **autophosphorylation.**

Once the cytosolic portion of a receptor becomes phosphorylated, the receptor recruits a number of cytosolic adapter proteins to interact with itself. Each of these proteins binds to the receptor at a phosphorylated tyrosine residue. To bind to the receptor, each cytosolic protein must contain a stretch of amino acids that recognizes the phosphotyrosine and a few neighboring amino acids on the receptor. The portion of a protein that recognizes one of these phosphorylated tyrosines is called an **SH2 domain.** The term *SH2,* for Src homology (domain) 2, was originally used because proteins with SH2 domains have sequences of amino acids that are strikingly similar to a portion of the Src protein. Several proteins with SH2 domains have been identified and shown to bind to tyrosine phosphate residues, either on a receptor or on other tyrosine-phosphorylated proteins.

Receptor tyrosine kinases can activate several different signal transduction pathways at the same time. These include the inositol-phospholipid-calcium second messenger pathway, which we have already discussed, and the

Ras pathway, which ultimately activates the expression of genes involved in growth or development. We next turn to the Ras pathway.

Receptor Tyrosine Kinases Initiate a Signal Transduction Cascade Involving Ras and MAP Kinase

Receptor tyrosine kinases use a different method of initiating signal transduction from that seen with G protein–linked receptors. The key initial event is the cross-phosphorylation of tyrosine residues on the cytosolic portion of the receptor that occurs when receptors aggregate in response to ligand binding (① and ②, Figure 10-18). These phosphorylated tyrosine residues become binding sites for other proteins in the cytosol that contain SH2 domains. At present, several important signaling pathways are known to be initiated by receptor tyrosine kinases. One of these involves the activation of the small monomeric G protein called **Ras.**

Ras is important in regulating the growth of cells, as we will see in Chapter 17. Like other types of G proteins, Ras can be bound to either GDP or GTP, but it is active only when bound to GTP. In the absence of receptor stimulation, Ras is normally in the GDP-bound state. For Ras to become active, it must release GDP and acquire a molecule of GTP. For this to take place, Ras needs the help of another type of protein called a **guanine-nucleotide exchange factor (GEF).**

The GEF that activates Ras is **Sos** (so called because it was originally identified from a genetic mutation in fruit flies called Son of Sevenless that results in the failure of cells in the compound eye to develop properly). For Sos to become active, it must bind indirectly to the receptor tyrosine kinase through another protein, called *GRB2* (③, Figure 10-18). This protein contains an SH2 domain and binds a tyrosine phosphate on the receptor. Thus, to activate Ras, the receptor becomes tyrosine phosphorylated, and GRB2 and Sos form a complex that binds to the receptor, activating Sos. Sos then stimulates Ras to release GDP and acquire GTP, which converts Ras to its active state.

Once Ras is active, it triggers a cascade of cellular events collectively referred to as the *Ras pathway* (⑤, Figure 10-18). One important event in the pathway is the activation of **mitogen-activated protein kinases,** or **MAP kinases (MAPKs).** MAPKs are activated when cells receive a stimulus to grow and divide (such a signal is sometimes called a *mitogen,* hence the name of the kinase). One of the functions of MAPKs is to phosphorylate a nuclear protein called *Jun.* When Jun is phosphorylated, it assembles along with other proteins into a transcription factor called *AP-1,* which appears to stimulate the production of proteins that are needed for cells to grow and divide. (We will see exactly how transcription factors work in Chapters 19 and 21.)

Once Ras is in its active state, it must be inactivated by hydrolysis of the GTP bound to it to avoid continued stimulation of the Ras/MAPK pathway. GTP hydrolysis is facilitated by a **GTPase activating protein (GAP).** GAPs can accelerate inactivation of Ras a hundred fold.

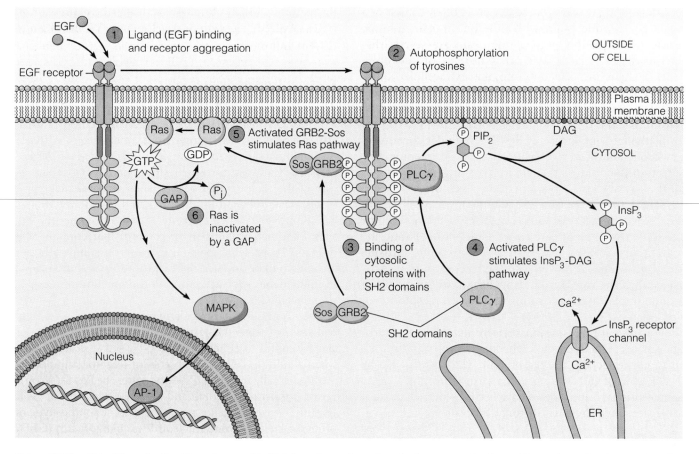

Figure 10-18 Signal Transduction Through Receptor Tyrosine Kinases. ① Upon ligand binding, receptor tyrosine kinases, such as that shown here for the epidermal growth factor (EGF), aggregate and ② undergo autophosphorylation. Once a receptor is phosphorylated at tyrosine residues in its cytosolic tail, ③ proteins with SH2 domains such as phospholipase C (PLC) and GRB2 bind to the receptor. ④ The binding of phospholipase C results in its activation and the cleavage of PIP₂ into InsP₃ and DAG. ⑤ The binding of GRB2 causes the activation of Sos, to which it is bound. Sos (a GEF) then causes the activation of the Ras protein by helping it release GDP and acquire GTP. Activated Ras initiates a cascade of events that ultimately results in the formation of AP-1, a transcription factor in the nucleus that stimulates the expression of genes needed for cell growth. ⑥ Ras is inactivated by hydrolysis of its bound GTP, a step facilitated by a GAP.

Receptor Tyrosine Kinases Activate a Variety of Other Signaling Pathways

Receptor tyrosine kinases, like G protein-linked receptors, can also activate phospholipase C (④ , Figure 10-18). We have already seen that the activation of phospholipase C leads to production of InsP₃ and DAG, and that InsP₃ releases calcium from intracellular stores. However, the phospholipase C_γ (activated by receptor tyrosine kinases) is different from phospholipase C_β (activated by the G protein–linked receptors) in that it contains an SH2 domain and must bind to the receptor. Once it binds to the receptor, phospholipase C_γ is phosphorylated by the receptor tyrosine kinase and becomes active.

In addition, receptor tyrosine kinases can activate other enzymes, such as phosphatidylinositol-3-kinase, which phosphorylates the plasma membrane phospholipid phosphatidylinositol. This enzyme is important in regulating cell growth and cell movement via its action on various phosphatidylinositides. The roles of this kinase are diverse, and the network of signaling events mediated by these other derivatives of inositol within the cell are complex.

Growth Factors as Messengers

The pathways we have examined are important in many cellular processes. One well-studied case in which they come into play is cell growth.

In order for a cell to grow, it must have all the nutrients needed for synthesis of its component parts, but the availability of nutrients is usually not itself sufficient for growth. Cells often also need messengers that act on specific receptors to stimulate cell growth. Biologists encountered the requirements for cell growth when they first tried to culture cells in vitro. Although provided with a growth medium rich in nutrients, including the presence of blood plasma, cells would not grow. A turning point came when blood serum was used instead of plasma: Serum was able to support the growth of cells, whereas plasma would not. Many of the messengers present within the serum have now been purified, and they are members of various classes of proteins known as **growth factors.**

The difference between blood serum and blood plasma held an important clue about growth factors. *Plasma* is whole blood, including unreacted *platelets* (which contain clotting components) but without the red and white blood cells. *Serum* is the clear fluid remaining after blood has clotted. During clotting, platelets secrete growth factors into the blood that stimulate the growth of cells called *fibroblasts,* which form the new connective tissue that makes up a scar. After clotting, the resulting serum is full of *platelet-derived growth factor (PDGF)*. Plasma does not contain this factor because the clotting reaction has not taken place.

We now know that the receptor for PDGF is a receptor tyrosine kinase. In fact, several growth factors act by stimulating receptor tyrosine kinases, including *insulin, insulin-like growth factor-1, fibroblast growth factor, epidermal growth factor,* and *nerve growth factor.* We will examine their mechanism of action in some detail below. Many other types of growth factors have also been isolated. A small sampling is shown in Table 10-2, which includes some of the types of cells affected by each factor, and the general class of molecule that serves as a receptor for each factor.

Although collectively known as growth factors, the proteins that activate tyrosine kinase and other types of receptors function in many diverse events. These include not only growth and cell division, but also crucial events during the development of embryos, responses to tissue injury, and many other activities. We will examine the effects of growth factors on cell division in Chapter 17. For now, all we need to recognize is that growth factors are secreted molecules that act at short range and have specific effects on cells possessing the appropriate receptor to sense the presence of the growth factor.

Disruption of Growth Factor Signaling Through Receptor Tyrosine Kinases Can Have Dramatic Effects on Embryonic Development

Many receptors for growth factors are receptor tyrosine kinases, falling into families based on which growth factor(s) they can bind. One well-studied class of growth factors and their receptors are the **fibroblast growth factors (FGFs)** and their receptor tyrosine kinases, the **fibroblast growth factor receptors (FGFRs).** FGFs and FGFRs are used in signaling events in both adult animals and embryos. We will focus here on the role of FGFRs during embryonic development.

FGFRs have been shown to play an important role in the development of cells derived from the middle embryonic cell layer of early embryos, known as the *mesoderm.* The mesoderm forms many cell types, including muscle, cartilage, bone, and blood cells, as well as the forerunner of the vertebral column. When specific FGFRs fail to function properly, the development of particular mesodermal tissues is affected. In one class of FGFR defects, a mutation in the receptor results in dominant effects on the developing embryo. In other words, even though the embryo makes a substantial quantity of normal, functional receptor, the presence of the mutant receptor within the embryo's cells prevents the normal receptors from functioning properly. Normal function is inhibited because FGFRs must act together as dimers to bind FGFs. If a normal receptor dimerizes with a mutant receptor, then the phosphorylation events that normally occur within the tyrosine kinase portion of the receptor fail to occur, blocking signal transduction (Figure 10-19). Such a mutation that overrides the function of the normal receptor is sometimes called a **dominant negative mutation.**

Dominant negative mutations can have dramatic effects on cells in developing embryos. For example, when genetically engineered dominant negative FGFRs are expressed in frog eggs and the eggs are fertilized, the embryos fail to develop mesodermal tissues in the trunk and tail, resulting in "tadpoles" with heads, but no bodies (Figure 10-20). In humans, dominant mutations in the transmembrane portion of the *FGFR-3* gene result in the most common form of dwarfism, known as *achondroplasia.* Heterozygous individuals have abnormal bone growth, in which the long bones suffer from abnormally slow ossification (i.e., the process by which cartilage is converted to bone during childhood). Human fetuses homozygous for the same mutation appear similar to fetuses that have a slightly different condition known as *thanatophoric dysplasia,* which often results from a single amino acid

Table 10-2 **Examples of Growth Factor Families**

Growth Factor	Target Cells	Type of Receptor Complex
Epidermal growth factor (EGF)	Wide variety of epithelial and mesenchymal cells	Tyrosine kinase
Transforming growth factor-α (TGF-α)	Same as EGF	Tyrosine kinase
Platelet-derived growth factor (PDGF)	Mesenchyme, smooth muscle, trophoblast	Tyrosine kinase
Transforming growth factor-β (TGF-β)	Fibroblastic cells	Serine-threonine kinase
Fibroblast growth factor (FGF)	Mesenchyme, fibroblasts, many other cell types	Tyrosine kinase
Interleukin-2 (IL-2)	Cytotoxic T lymphocytes	Complex of three subunits
Colony stimulating factor-1 (CSF-1)	Macrophage precursors	Tyrosine kinase
Wnts	Many types of embryonic cells	Seven-pass protein

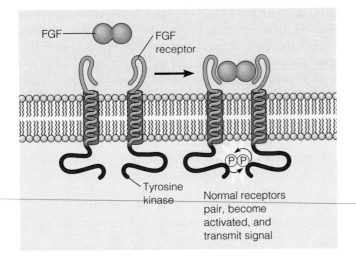

(a) Normal FGFR: FGF binds; FGFRs form a dimer

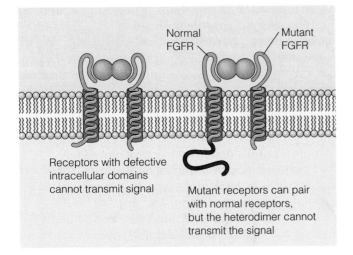

(b) Dominant negative mutation in FGFR

Figure 10-19 **Dominant Negative Disruption of FGF Receptor (FGFR) Function.** **(a)** Normal receptors form dimers after binding FGF, and transmit the appropriate signal via Ras and MAPK. **(b)** When a cell makes mutant FGF receptors, normal receptors can dimerize as in **(a)**, or the defective receptors can bind to FGF and dimerize with normal receptors. In this case no signal is transmitted. When sufficient quantities of the mutant receptor are present, most of the normal receptors are paired with mutant receptors, resulting in overall disruption in signaling.

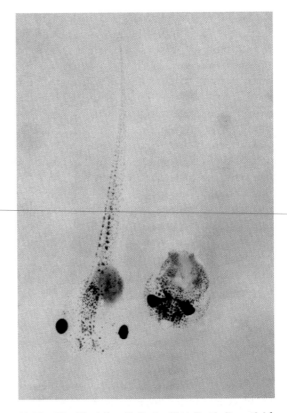

Figure 10-20 **Fibroblast Growth Factor Signaling Is Essential for Mesoderm Production in Embryos.** Left: Normal frog embryo. Right: Embryo derived from an egg injected with mRNA encoding a dominant negative mutant receptor for fibroblast growth factor (FGF). As a result, mesoderm fails to form properly, and the embryo lacks a trunk and tail.

change in the cytosolic portion of the FGFR-3 protein. In this case, more severe bone abnormalities result, and affected individuals die soon after birth. The dramatic effect of a single amino acid change in a single protein demonstrates the key role that growth factor receptors play during human development.

Other Growth Factors Transduce Their Signals via Receptor Serine/Threonine Kinases

We have seen that when receptor tyrosine kinases bind ligand, they activate a set of signal transduction events within the cell that results in changes within the cell receiving such a signal. Another major class of transmembrane receptors uses a very different set of signal transduction pathways to elicit changes within the cell. These receptors are also protein kinases, but they phosphorylate serine and threonine residues rather than tyrosines. One major class of **serine/threonine kinases receptor** comprises a family of proteins that bind members of the **transforming growth factor β (TGFβ)** family of growth factors. This growth factor family regulates a wide range of cellular functions in both embryos and adult animals, including cell proliferation, programmed cell death, the specialization of cells, and key events in embryonic development. Some of the TGFβ family members form dimers with one another prior to binding to the appropriate receptors.

The first step in TGFβ signaling is the binding of growth factor by the transmembrane receptor (Figure 10-21). In general, TGFβ family members bind to two types of receptors within the receiving cell: the *type I and type II receptors.* When ligand is bound, the type II receptor phosphorylates the type I receptor. The type I receptor then initiates a signal transduction cascade within the cell receiving the growth factor signal. The targets of the type I receptor are a class of proteins known as **Smads** (a name coined from two of the founding members of this class of

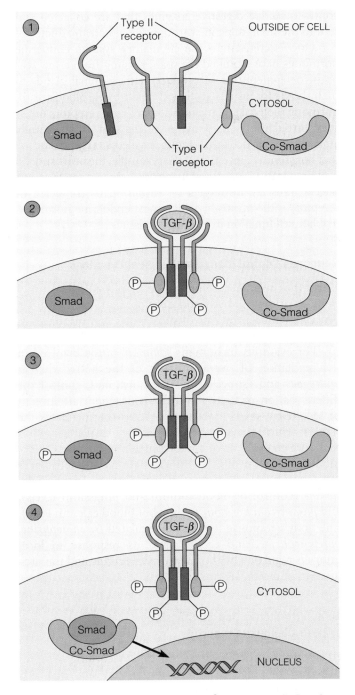

Figure 10-21   Signal Transduction by TGFβ Receptor Family Proteins.
① Type I and type II receptors for TGFβ in a cell prior to binding of the growth factor. ② Binding of growth factor results in clustering of type I and type II receptors, and phosphorylation of type I receptors by type II receptors. ③ The activated type I receptors then phosphorylate particular receptor-mediated Smads. ④ These Smads then bind to other Smads (co-Smads), and together they enter the nucleus.

proteins). In response to the appropriate signal, the phosphorylated receptor–activated Smad forms a complex with an additional Smad (a "co-Smad"), and the two Smads move into the nucleus. Once inside the nucleus, the Smad complex can associate with DNA-binding proteins to regulate gene expression. Several mutations in

Smad proteins in human cells appear to result in specific forms of cancer.

Growth Factor Receptor Pathways Share Common Themes

We have seen that a common theme in cell signaling involving growth factors is the phosphorylation of a transmembrane receptor, followed by a complex chain reaction of events, often leading to changes in proteins that can alter the expression of genes within a target cell. However, not all growth factor receptors that rely on such phosphorylation events have their own kinase activity. For example, you will learn in Chapter 21 that the binding of some growth factors to their receptors causes the receptor to activate a separate intracellular protein kinase called the *Janus activated kinase* (p. ◊◊◊).

Although the details are different in each case, it has become clear that there are several common themes involved in all growth factor signaling events. First, ligand binding often results in the activation and/or clustering of receptors. Receptor activation then leads to a cascade of events, in which various intermediary proteins often lose or gain phosphate groups. Finally, this cascade leads to changes in one or more proteins that enter the nucleus to alter the expression of genes within the cell receiving the signal. The specific response of the cell receiving such signals depends on the history of the cell and on the aggregate of signals the cell is being exposed to at any given time. In the next section, we will see that a similar logic operates during the long-range signaling mediated by hormones.

Disruption of Growth Factor Signaling Can Lead to Cancer

We have seen that growth factors regulate events such as cell proliferation, cell movement, and gene expression. We have also seen that embryonic development requires growth factors to tightly regulate cell biological events. If embryonic development is a prime example of the processes that are correctly regulated by growth factors, many types of cancer are exactly the opposite. We now know that a number of cancers result from loss of regulation of growth factor signaling. In the case of receptor tyrosine kinases, for example, mutations in the epidermal growth factor receptor can result in breast cancer, glioblastoma (a cancer of glial cells in the brain), and fibrosarcoma (a type of cancer of long bones). Similar discoveries have been made for receptor serine threonine kinases. Mutations in the TGF-β type I receptor occur in one-third of ovarian cancers, and mutations in the type II receptor occur in many colorectal cancers. Mutations in one of the Smads (Smad4) occurs in one-half of all pancreatic cancers. Indeed, as cell and developmental biologists have elucidated these and other signaling pathways, cancer biologists have discovered that there are fundamental links between how cancers arise and the basic biology of signal transduction.

The Endocrine and Paracrine Hormone Systems

To regulate the function of various cells and tissues, both plants and animals use chemical signals called **hormones.** Hormones are chemical messengers secreted by one tissue that regulate the function of other cells or tissues in the same organism. In contrast to growth factors, hormones often act over large distances. In plants and animals, hormones are often transported via the vasculature. Hundreds of different hormones regulate a wide variety of functions, many critical for maintaining the physiological steady state of an organism.

Though we can consider them as a group based on their regulatory functions, hormones differ in many ways. Some hormones are steroids or other hyprdrophobic molecules that are targeted to intracellular receptors. Other hormones, such as the adrenergic hormones discussed later in this section, are targeted to a wide variety of different G protein–linked receptors. Still others, such as insulin, are ligands for receptor tyrosine kinases.

Both plants and animals produce a wide array of hormones. For example, plants produce steroid hormones called *brassinosteroids* that regulate leaf growth. Similarly, the organic molecule *ethylene* regulates ripening of fruit, and we have seen that *abscicic acid* causes the closing of stomata during drought conditions. However, because animal hormones are so much better understood, we will focus on them for the remainder of our discussion.

Hormonal Signals Can Be Classified by the Distance They Travel to Their Target Cells

Every animal hormone can be placed in one of two categories depending on the distance over which it operates. An **endocrine hormone** is a chemical that travels by means of the circulatory system from the cells where it is released to other cells, where it regulates one or more specific functions. (The word *endocrine* comes from the Greek, meaning "to secrete into.") A **paracrine hormone** is a more local signal that is taken up, destroyed, or immobilized so rapidly that it can act only on cells in the immediate environment. (The word *paracrine* means "to secrete around.") There is not a strong functional distinction between growth factors and paracrine hormones; which term is used often reflects how or when a particular signaling molecule was discovered.

Endocrine hormones are synthesized by the **endocrine tissues** of the body and are secreted directly into the bloodstream. This mode of secretion distinguishes endocrine tissues from **exocrine tissues,** which secrete their products into ducts that then transport the secretions to other parts of the body. Some organs of the body have both endocrine and exocrine tissues. The pancreas is a good example of such an organ. Cells of the pancreatic endocrine tissue secrete two hormones, glucagon and insulin, that regulate the concentration of glucose circulating in the blood. Cells of the pancreatic exocrine tissue, by contrast, secrete digestive enzymes that are collected and sent to the small intestine through the pancreatic duct.

Once secreted into the circulatory system, endocrine hormones have a limited life span, ranging from a few seconds for epinephrine (a product of the adrenal gland) to many hours for insulin. As they circulate in the bloodstream, hormone molecules come into contact with receptors in tissues throughout the body. A tissue that is specifically affected by a particular hormone is called a **target tissue** for that hormone (Figure 10-22). For example, the heart and the liver are target tissues for epinephrine, whereas the liver and skeletal muscles are targets for insulin. This contact with receptors is the means by which most endocrine hormones regulate cell function in a variety of tissues.

Hormones Control Many Physiological Functions

Hormones regulate a wide range of physiological functions, including growth and development, rates of body processes, concentrations of substances, and responses to stress and injury (Table 10-3).

For example, in humans *somatotropin* is involved in the regulation of overall growth of the body, whereas *androgens* and *estrogens,* the sex hormones, control the differentiation of tissues and the consequent attainment of secondary sex characteristics. *Thyroxine* regulates the rate at which the body makes energy available and is therefore an example of a rate-controlling hormone. Hormones that control the concentrations of substances include *insulin* (control of blood glucose level), *aldosterone* (control of blood sodium and potassium levels), and *parathyroid hormone* (control of blood calcium level). The body's response to stress is regulated by *epinephrine, norepinephrine,* and *cortisol,* and its response to local injury is regulated by the release of *histamine* and the production of *prostaglandins.*

Animal Hormones Can Be Classified by Their Chemical Properties

Hormones can be classified not only according to function and the distances over which they act, but also according to their chemical properties (Table 10-4). Chemically, the endocrine hormones fall into four categories: amino acid derivatives, peptides, proteins, and lipid-like hormones such as the steroids. Examples of hormones derived from amino acids, peptides, or proteins are shown in Figure 10-23. An example of an amino acid derivative is epinephrine, derived from tyrosine. *Antidiuretic hormone* (also called *vasopressin*) is an example of a peptide hormone, whereas insulin is a protein. *Testosterone* is an example of a steroid hormone. The steroid hormones are derivatives of cholesterol (see Figure 3-30) that are synthesized either in the gonads (the *sex hormones*) or in the adrenal cortex (the *corticosteroids*). The mechanism of action of steroid hormones will be discussed in Chapter 21; in this chapter, we will focus on water-soluble (nonlipid) hormones.

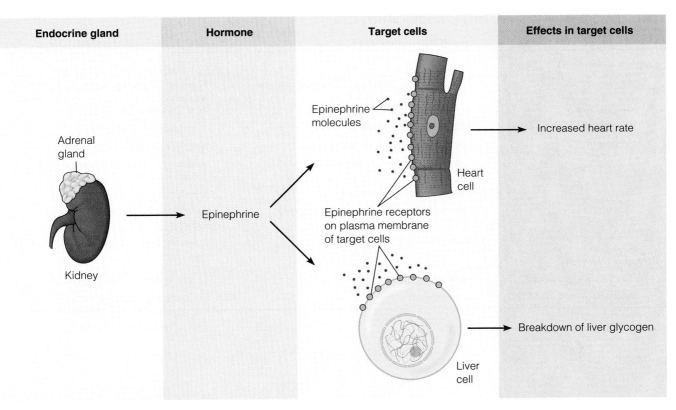

Figure 10-22 Target Tissues for Endocrine Hormones. Cells in a target tissue have hormone-specific receptors embedded in their plasma membranes (or, in the case of the steroid hormones, present in the nucleus or cytosol). Heart and liver cells can respond to epinephrine synthesized by the adrenal glands because these cells have epinephrine-specific receptors on their outer surfaces. A specific hormone may elicit different responses in different target cells. Epinephrine causes an increase in the heart rate but stimulates glycogen breakdown in the liver.

Table 10-3   Physiological Functions of Hormones

Function Under Hormonal Control	Hormone	Source (Endocrine Tissue)
Growth and Development		
Body size	Somatotropin (growth hormone)	Anterior pituitary
Sexual development	Androgens (males)	Testes
	Estrogens (females)	Ovaries
Reproductive cycle	Luteinizing hormone	Anterior pituitary
	Follicle-stimulating hormone	Anterior pituitary
	Chorionic gonadotropin	Follicle
Rates of Body Processes		
Hormone secretion	Tropic hormones	Anterior pituitary
Basal metabolism	Thyroxine	Thyroid
Glucose uptake	Insulin	Pancreas
Kidney filtration	Antidiuretic hormone (vasopressin)	Posterior pituitary
Uterine contraction	Oxytocin	Posterior pituitary
Concentrations of Substances		
Blood glucose	Glucagon, insulin	Pancreas
Mineral balance	Corticosteroids	Adrenal cortex
Blood calcium	Parathyroid hormone	Parathyroid
Responses to Stress and Injury		
Heart rate	Epinephrine	Adrenal medulla
Blood pressure	Epinephrine	Adrenal medulla
Inflammation	Histamine	Mast cells
	Prostaglandins	All tissues
	Corticosteroids (cortisol)	Adrenal cortex

Table 10-4 Chemical Classification and Function of Hormones

Chemical Classification	Examples	Regulated Function
Endocrine Hormones		
Amino acid derivatives	Epinephrine (adrenaline) and norepinephrine (both derived from tyrosine)	Stress responses: regulation of heart rate and blood pressure; release of glucose and fatty acids from storage sites
	Thyroxine (derived from tyrosine)	Regulation of metabolic rate
Peptides	Antidiuretic hormone (vasopressin)	Regulation of body water and blood pressure
	Hypothalamic hormones (releasing factors)	Regulation of tropic hormone release from pituitary gland
Proteins	Anterior pituitary hormones	Regulation of other endocrine systems
Steroids	Sex hormones (androgens and estrogens)	Development and control of reproductive capacity
	Corticosteroids	Stress responses; control of blood electrolytes
Paracrine Hormones		
Amino acid derivative	Histamine	Local responses to stress and injury
Arachidonic acid derivatives	Prostaglandins	Local responses to stress and injury

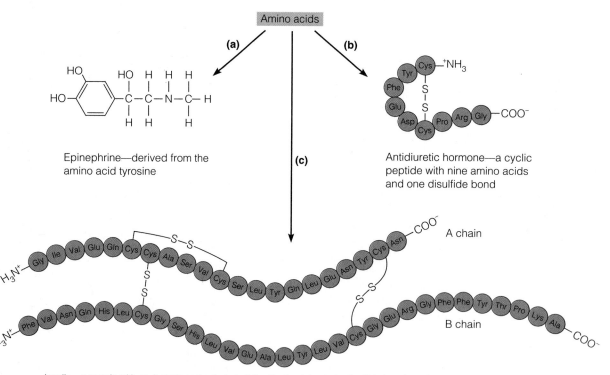

Epinephrine—derived from the amino acid tyrosine

Antidiuretic hormone—a cyclic peptide with nine amino acids and one disulfide bond

Insulin—a protein with an A chain and a B chain linked by two interchain disulfide bonds and one intrachain disulfide bond

Figure 10-23 The Chemistry of Animal Hormones. Many endocrine hormones are **(a)** derivatives of amino acids, **(b)** peptides, or **(c)** proteins. Note that insulin is actually a globular protein and is shown here in extended form only for purposes of illustration. (Steroid hormones are illustrated in Figure 3-30.)

Examples of paracrine hormones are histamine and the prostaglandins. **Histamine** is produced by decarboxylation of the amino acid histidine and is responsible for local inflammatory responses. The **prostaglandins,** so named because they were first identified in human semen as a secretion of the prostate gland, are derived from arachidonic acid and are important in smooth muscle function.

Adrenergic Hormones and Receptors Are a Good Example of Endocrine Regulation

Since hormones modulate the function of particular target tissues, an important aspect of studying hormones is understanding the specific functions of those target tissues. To illustrate how endocrine hormones act, we will look more closely at the **adrenergic hormones** epinephrine and norepinephrine. (Epinephrine is also called adrenaline; the two words are of Greek and Latin derivation, respectively, and mean "above, or near, the kidney," referring to the location in the body of the *adrenal glands,* which synthesize this hormone.) When secreted into the bloodstream, epinephrine and norepinephrine stimulate changes in many different tissues or organs, all aimed at preparing the body for dangerous or stressful situations (the so-called "fight-or-flight response"). Overall, the adrenergic hormones trigger increased cardiac output, shunting blood from the visceral organs to the muscles and the heart, as well as dilation of arterioles to facilitate oxygenation of the blood. In addition, these hormones stimulate the breakdown of glycogen to supply glucose to the muscles.

Adrenergic hormones bind to a family of G protein–linked receptors known as **adrenergic receptors.** The individual members of this family differ mainly in their preference for epinephrine or norepinephrine and in which G protein is linked to the receptor. They can be broadly classified into α- and β-adrenergic receptors. The α-adrenergic receptors bind both epinephrine and norep-

inephrine. These receptors are located on the smooth muscles that regulate blood flow to visceral organs. The β-adrenergic receptors bind epinephrine much better than norepinephrine. These receptors are found on smooth muscles associated with arterioles that feed the heart, smooth muscles of the bronchioles in the lungs, and skeletal muscles.

The α- and β-adrenergic receptors stimulate different signal transduction pathways, because they are linked to different G proteins (Figure 10-24). For example, the G proteins activated by one type of α-adrenergic receptor, the α_1-adrenergic receptors, are G_p proteins, whereas the β-adrenergic receptors activate G_s. As we discussed earlier, activation of G_s stimulates the cAMP signal transduction pathway, leading to relaxation of certain smooth muscles. Activation of G_p stimulates phospholipase C, leading to the production of InsP$_3$ and DAG, which in turn elevates intracellular calcium levels. Table 10-5 summarizes some of the major physiological functions that are regulated by cAMP. As a specific example of cAMP-mediated regulation, we will consider the control by the hormone epinephrine of glycogen degradation in liver or muscle cells.

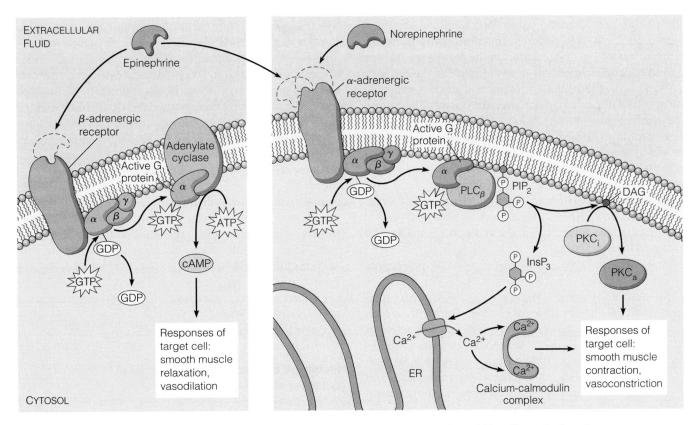

(a) cAMP pathway initiated by activation of β-adrenergic receptor

(b) Inositol-phospholipid-calcium pathway initiated by activation of α-adrenergic receptor

Figure 10-24   The Stimulation of G Protein–Linked Signal Transduction Pathways by α- and β-Adrenergic Receptors.   The α- and β-adrenergic receptors are a closely related family of G protein–linked receptors. Each receptor binds to epinephrine or norepinephrine. While both types of receptors can bind the same hormone (epinephrine, for example), they trigger different signal transduction pathways because they activate different G proteins. **(a)** The β receptors activate Gs and stimulate the cAMP signal transduction pathway. **(b)** The α receptors activate Gp and stimulate the inositol-phospholipid-calcium signal transduction pathway.

Table 10-5   Examples of Cell Functions Regulated by cAMP

Regulated Function	Target Tissue	Hormone
Glycogen degradation	Muscle, liver	Epinephrine
Fatty acid production	Adipose	Epinephrine
Heart rate, blood pressure	Cardiovascular	Epinephrine
Water reabsorption	Kidney	Antidiuretic hormone
Bone resorption	Bone	Parathyroid hormone

Intracellular Effects of the cAMP Pathway: Control of Glycogen Degradation.

One of the actions of the adrenergic hormones is to stimulate the breakdown of glycogen to provide muscle cells with an adequate supply of glucose. The breakdown of glycogen is catalyzed by the enzyme *glycogen phosphorylase,* which cleaves glucose units from glycogen as glucose-1-phosphate by the addition of inorganic phosphate (P_i). The glycogen phosphorylase system was the first cAMP-mediated regulatory sequence to be elucidated. The original work was published in 1956 by Earl Sutherland, who received a Nobel Prize in 1971 for this discovery.

The sequence of events that leads from hormonal stimulation to enhanced glycogen degradation is shown in Figure 10-25. It begins when an epinephrine molecule binds to a β-adrenergic receptor on the plasma membrane of a liver or muscle cell. As described earlier, the receptor activates a neighboring G_s protein, and the G_s protein in turn stimulates adenylyl cyclase, the membrane-bound enzyme that generates cAMP from ATP (see Figure 10-24a). The resulting transient increase in the concentration of cAMP in the cytosol activates protein kinase A. PKA then activates another cascade of events that begins with the phosphorylation of the enzyme *phosphorylase kinase.* This leads to the conversion of *glycogen phosphorylase* from phosphorylase *b,* the less active form, to phosphorylase *a,* the more active form, and thus to an increased rate of glycogen breakdown.

cAMP also stimulates the inactivation of the enzyme system responsible for glycogen synthesis. In this case, cAMP activates PKA as we have seen, which in turn phosphorylates the enzyme *glycogen synthase.* Rather than activating this enzyme, however, phosphorylation inactivates it. Thus the overall effect of cAMP involves both an increase in glycogen breakdown and a decrease in its synthesis. The examples of glycogen phosphorylase and glycogen synthase also demonstrate that phosphorylation can have very different effects on enzyme activity, depending on which enzyme is being considered.

α-Adrenergic Receptors and the Inositol-Phospholipid-Calcium Pathway.

Another important adrenergic pathway is represented by the α_1-adrenergic receptors, which stimulate the formation of inositol trisphosphate (InsP$_3$) and diacylglycerol (DAG). The α_1-adrenergic receptors are found mainly on smooth muscles in blood vessels, including those controlling blood flow to the intestines. When α_1-adrenergic receptors are stimulated, the formation of InsP$_3$ causes an increase in the intracellular calcium concentration. The elevated level of calcium causes smooth muscle contraction, resulting in constriction of the blood vessels and diminished blood flow. Thus, the activation of α_1-adrenergic receptors affects smooth muscle cells in a manner opposite to that of β-adrenergic activation, which causes smooth muscles to relax.

Coordination of the Responses to Adrenergic Hormones.

We can now consider how the secretion of epinephrine and norepinephrine leads to the coordinated changes in various tissues that are required to prepare the body for stressful situations. The overall strategy of adrenergic hormone actions is to put many of the normal bodily functions on hold and to deliver vital resources to the heart and skeletal muscles instead, as well as to produce a heightened state of alertness. This is effected by the adrenergic hormones and the system of α- and β-adrenergic receptors that produce essentially opposite effects on a particular tissue such as smooth muscle. The predominant receptors present in a particular tissue determine how that tissue will respond to adrenergic hormones.

Using smooth muscle as our example, we can see how the elevation of cAMP level causes smooth muscle to relax, and the elevation of intracellular calcium level causes smooth muscle to contract (cAMP and calcium coordinate many of the needed changes in blood flow). In the smooth muscles surrounding the blood vessels of the heart and the bronchioles (narrow air passageways) of the lungs, stimulation of β-adrenergic receptors results in muscle relaxation. As a result, blood flow to the heart increases and airway passages widen, facilitating better oxygenation of the blood. At the same time, stimulation of α_1-adrenergic receptors in the smooth muscles surrounding peripheral veins causes these muscles to contract, constricting the smaller veins. This results in diminished blood flow to the skin, kidneys, and digestive tract, increasing blood pressure and shunting the blood to the tissues where it is needed most: the heart, lungs, and skeletal muscles. These are just a few effects of adrenergic stimulation, but they give a sense of the coordinated effort required to mobilize the body for stress or strenuous activity.

Histamine and Prostaglandins Are Good Examples of Paracrine Regulation

Recall that paracrine hormones are substances secreted by cells that affect other cells a short distance away. In this case, the messenger clearly has a limited range of action. One example of a messenger commonly classified as a paracrine hormone is histamine, which is important in allergic responses. In allergies, certain types of foreign substances called **allergens** induce the body to make an antibody protein of the class IgE. Once a person is

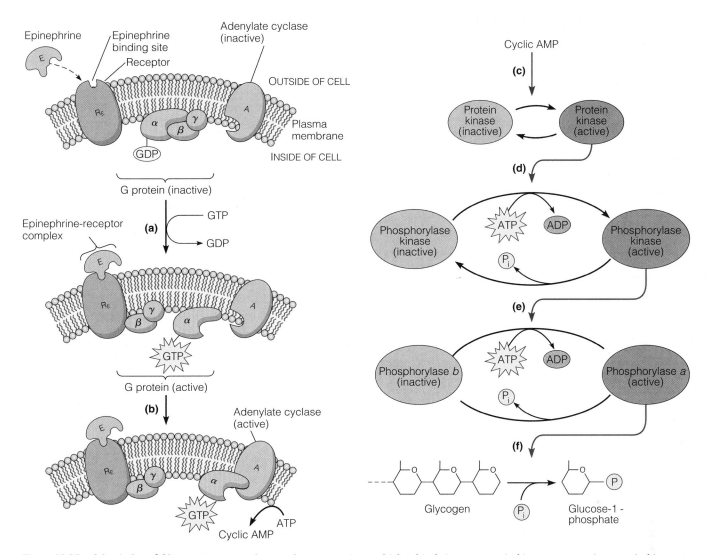

Figure 10-25 Stimulation of Glycogen Breakdown by Epinephrine. Muscle and liver cells respond to an increased concentration of epinephrine in the blood by increasing their rate of glycogen breakdown. The stimulatory effect of extracellular epinephrine on intracellular glycogen catabolism is mediated by a G protein–cyclic AMP regulatory cascade. In this case, the ligand is epinephrine (E) and the membrane protein to which it binds is the β-adrenergic receptor (R_E). **(a)** The binding of epinephrine to its receptor activates the Gs protein, causing dissociation of the GTP-Gα complex. **(b)** The GTP-Gα complex then binds to and activates adenylyl cyclase (A). **(c)** As the intracellular concentration of cyclic AMP rises, cAMP initiates a further cascade of regulatory events that begins with the activation of protein kinase A. **(d)** Active protein kinase A then converts inactive phosphorylase kinase to the active form by ATP-dependent phosphorylation. **(e)** Active phosphorylase kinase, in turn, phosphorylates phosphorylase b, converting it to phosphorylase a, the active form of the enzyme. **(f)** Phosphorylase a then catalyzes the phosphorolytic cleavage of glycogen into molecules of glucose-1-phosphate.

exposed to the allergen and IgE protein has been formed, it binds receptors on membranes of cells known as **mast cells,** which store histamine inside secretory granules. The next time a person is exposed to the allergen, it binds to the antibody and stimulates an increase in calcium in the mast cell, which induces the secretion of histamine. The release of histamine generally acts locally on neighboring tissues and causes many of the symptoms of allergy.

Other examples of paracrine hormones include the prostaglandins. Typically, the prostaglandins act on G protein-linked receptors to stimulate either the cAMP or the inositol-trisphosphate-calcium second messenger pathway.

Prostaglandins have a variety of effects, many of which involve smooth muscle. For example, prostaglandins contained in semen stimulate uterine smooth muscle contraction, which helps transport sperm to the egg. Prostaglandins also help to initiate smooth muscle contraction during labor and can be used to induce labor clinically. Some prostaglandins cause smooth muscle relaxation and can, for example, cause bronchiole dilation or lowering of blood pressure.

An example of how prostaglandins function in paracrine signaling is seen in the activation of **blood platelets.** Platelets are essential components of the blood-clotting mechanism that plug sites where blood vessels are ruptured. Platelets are not true cells; they have no nucleus, and are produced by the fragmentation of certain bone marrow cells.

Clotting is a complex cascade of enzymatic events that is triggered by injury and leads rapidly to the formation of a *clot*, a tangled mass of red blood cells, platelets, and fibers of a protein called *fibrin*. The clotting mechanism must be carefully regulated, because inappropriate blood clots lead quickly to life-threatening situations. The coronary arteries of the heart are only 1 millimeter or so in diameter, and most heart attacks result from a small blood clot that blocks coronary circulation. Similarly, clots in the venous system can produce *phlebitis* (inflammation of the veins) in the lower limbs and *embolisms* (obstructions) in the lungs. Thus, clot formation is essential, but it must be confined to the area where the damage occurs.

Platelets have many different receptors that sense extracellular signals produced when a tissue is damaged. Most relevant to our discussion of paracrine regulation, platelets can be activated by members of the prostaglandin family such as thromboxane A2, as well as by extracellular adenosine diphosphate. These substances are released by activated platelets and then function as paracrine hormones to activate other platelets.

The generally accepted platelet activation sequence is shown in Figure 10-26. Most of the known platelet activators act through G protein–linked receptors to activate phospholipase C, which triggers the release of calcium from a calcium- storing tubular network known as the *dense tubular system;* this results in the release of *arachidonic acid*. An enzymatic pathway starting with *cyclooxygenase* then converts the arachidonic acid to thromboxane A2. Thromboxane A2 diffuses out of the platelet and acts on G protein–linked thromboxane receptors of neighboring platelets. The thromboxane receptors in turn activate phospholipase C, triggering the activation of nearby platelets. As a result, platelets in the area are recruited to the site of injury.

Knowledge of the regulatory pathways involved in prostaglandin formation and platelet activation has enabled us to understand the mechanism of action of certain drugs. It has long been known, for example, that aspirin significantly reduces the incidence of heart attacks in older individuals, apparently by decreasing the likelihood of small internal blood clots that might block blood flow through the coronary arteries in the heart. We now understand that aspirin specifically blocks cyclooxygenase activity, thereby inhibiting synthesis of thromboxane A2 and other prostaglandins from arachidonic acid. This inhibition slows platelet activation and recruitment, reducing the probability of small blood clot formation. Interestingly, prostaglandins also appear to increase body heat production, causing fever, by their effect on cells of the hypothalamus. By inhibiting the formation of these prostaglandins, aspirin causes the fever to subside.

Cell Signals and Apoptosis

Signal transduction is not only important during hormonal regulation and in growth factor signaling. Cell signaling also regulates a kind of programmed cell death, or "cellular suicide," known as **apoptosis.** Apoptosis is a key event in many biological processes. In embryos, apoptosis occurs in a variety of circumstances. Examples include removal of the webbing between the digits (fingers and toes) during the development of hands and feet, the resorption of the tail of tadpoles when they undergo metamorphosis, and the "pruning" of neurons that occurs in human infants during the first few months of life as connections mature within the developing brain. In adult humans, apoptosis occurs continually; when cells become infected by pathogens or when white blood cells reach the end of their lifespan, they are eliminated through apoptosis. As a result, millions of cells die every minute in the human body. When cells that should die via apoptosis do not, the consequences can be dire. We now know that mutations in some of the proteins that participate in apoptosis can lead to cancer; melanoma frequently results from a mutation in Apaf-1, a protein that participates in apoptosis.

Apoptosis is very different from another type of cell death, known as *necrosis*, which sometimes follows massive tissue injury. Whereas necrosis involves the swelling and rupture of the injured cells, apoptosis involves a specific series of events that leads to the dismantling of the internal contents of the cell (Figure 10-27). During the early phases of apoptosis the cell's DNA segregates near the periphery of the nucleus and the volume of the cytoplasm decreases. Next, the cell begins to produce small bubble-like cytoplasmic extensions ("blebs"), and the nucleus and organelles begin to fragment. The cell's DNA is cleaved by an apoptosis-specific DNA endonuclease, or *DNAse* (an enzyme that digests DNA), clipping it at sites between nucleosomes. (You will learn about nucleosomes in Chapter 16.) Nucleosomes are found at regular intervals along the DNA. As a result, the DNA fragments, which are multiples of 200 base pairs in length, form a diagnostic "ladder" of fragments. Eventually the cell is dismantled into small pieces called **apoptotic bodies.** Ultimately the remnants of the affected cell are engulfed by other nearby cells (typically macrophages) via phagocytosis. (You will learn more about phagocytosis in Chapter 12.) The macrophages act as scavengers to remove the resulting cellular debris.

That cells have a "death program" was first conclusively demonstrated in the nematode, *Caenorhabditis elegans*, where key genes that control apoptosis were first identified. Subsequent research showed that many other organisms, including mammals, use similar proteins to commit cellular suicide. As a result of these studies, many of the molecular events that underlie apoptosis are now known. A key event in apoptosis is the activation of a series of enzymes called **caspases** (caspases get their name

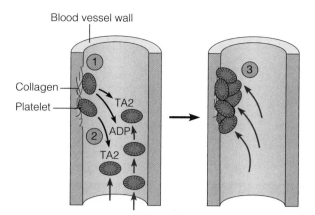

(a) Platelet aggregation

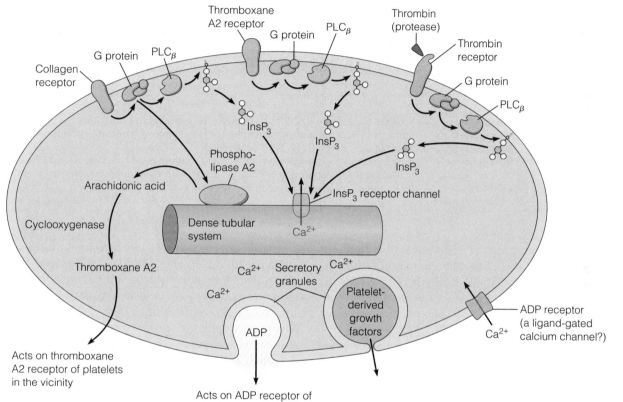

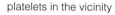

(b) Platelet activation

Figure 10-26   The Process of Platelet Activation.   (a) Platelets are activated by several substances and aggregate upon activation, forming a blood clot. ① Collagen exposed in a damaged blood vessel wall stimulates platelet activation and adhesion. ② Activated platelets release the platelet factors ADP and thromboxane A2 (TA2), which ③ stimulate the activation and aggregation of more platelets in the vicinity. The release of TA2 and ADP is an example of paracrine signaling. **(b)** Paracrine signaling by platelets uses some of the same classes of receptors and signal transduction pathways described for the endocrine hormones. The receptors for thrombin, TA2, and collagen all appear to be G protein–linked receptors that activate phospholipase C. However, the activation of the thrombin receptor is unusual in that it depends on the protease thrombin removing a portion of the receptor protein. When newly exposed collagen in a damaged blood vessel wall binds to a G protein–linked collagen receptor, membrane phospholipase C is activated, causing the production of InsP$_3$ and DAG, and ultimately the release of calcium from the dense tubular system. Increased calcium concentration inside the platelet stimulates the secretion of granules that contain growth factors and ADP. Phospholipase A is thought to be simultaneously activated by a G protein. The activation of phospholipase A2 releases arachidonic acid, which is converted to TA2. TA2 binds to the TA2 receptors on nearby platelets, causing elevated calcium levels, whereas ADP binds to what is probably a ligand-gated calcium channel. Both messengers stimulate platelets to aggregate.

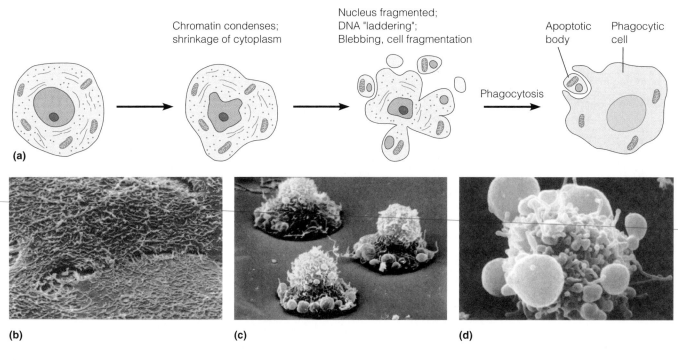

Nucleus fragmented;
DNA "laddering";
Blebbing, cell fragmentation

Chromatin condenses;
shrinkage of cytoplasm

Apoptotic   Phagocytic
body   cell

Phagocytosis

(a)

(b)   **(c)**   **(d)**

Figure 10-27   Major Steps in Apoptosis.
(a) As a cell begins to undergo apoptosis, its chromatin condenses and the cytoplasm shrinks. Eventually the nucleus becomes fragmented, its DNA is digested at regular intervals ("laddering"), the cytoplasm becomes fragmented, and the cell extends numerous blebs. Ultimately the remnants of the dead cell (apoptotic bodies) are ingested by phagocytic cells. **(b)–(d)** SEMs of epithelial cells undergoing apoptosis. **(b)** Epithelial cells in contact with one another in culture form flat sheets. **(c)** As apoptosis ensues, the cells round up, withdraw their connections with one another, and bleb. **(d)** A single dead cell with many apoptotic bodies.

because they contain a *c*ysteine at their active site, and they cleave proteins at sites that contain an *aspar*tic acid residue followed by four amino acids that are specific to each caspase). Caspases are produced as inactive precursors known as **procaspases**, which are subsequently cleaved to create active enzymes, often by other caspases, in a proteolytic cascade. Once they are activated, caspases cleave other proteins within cells, resulting in efficient and precise killing of the cell in which they are activated. The ladder DNAse is a good example; it is bound to an inhibitory protein that is cleaved by a caspase.

Apoptosis Is Triggered by Death Signals or Withdrawal of Survival Factors

There are two main routes by which cells can activate caspases and enter the apoptotic pathway (Figure 10-28). In some cases, cells receive *cell death signals* that initiate the apoptotic cascade. Sometimes these signals are positive, i.e., the cell that dies comes in contact with a "cell death inducer." Two well-known death signals are received by the *tumor necrosis factor receptor* and the *CD95/Fas receptor*. In other cases, cells require *trophic,* or *survival, factors* to stay alive. When these trophic factors are withdrawn, the cell enters apoptosis. *Cytokines*, which foster survival of cells in the bloodstream, are well-studied examples of such survival factors. Although similar, there are some dif-

ferences between the ways in which apoptosis is triggered by cell death signals and withdrawal of survival factors; for the remainder of our discussion, we will focus on one example of a cell death inducer, shown in Figure 10-28.

When cells in the human body are infected by certain viruses, a population of killer lymphocytes are activated and induce the infected cells to initiate apoptosis. One way in which lymphocytes do this involves the binding of the protein *CD95 ligand* (also called *Fas ligand*), on the surface of the lymphocyte, to *CD95/Fas* receptors on the surface of the infected cell (Figure 10-28, ①). CD95 receptor aggregation results in the attachment of adaptor proteins to the cluster (one of these is called FADD), which in turn causes a particular procaspase (*procaspase-8*) to assemble at the site of receptor clustering. The clustered procaspases are thought to cleave one another into active caspase-8 protein (②). Procaspase-8 is a protease that can cleave other proteins involved in apoptosis. One of these is a second procaspase (*procaspase-3*; ③), which is cleaved to produce *caspase-3*. The action of active caspases -3 and -8 together is important for activating many steps in apoptosis. Caspase-8 also activates a protein that promotes cell death (*Bid*) by interacting with proteins at the surface of mitochondria (④).

The involvement of mitochondria in apoptosis is shared by a second pathway that leads to cell death. In this case, apoptosis occurs in the absence of cell death signals.

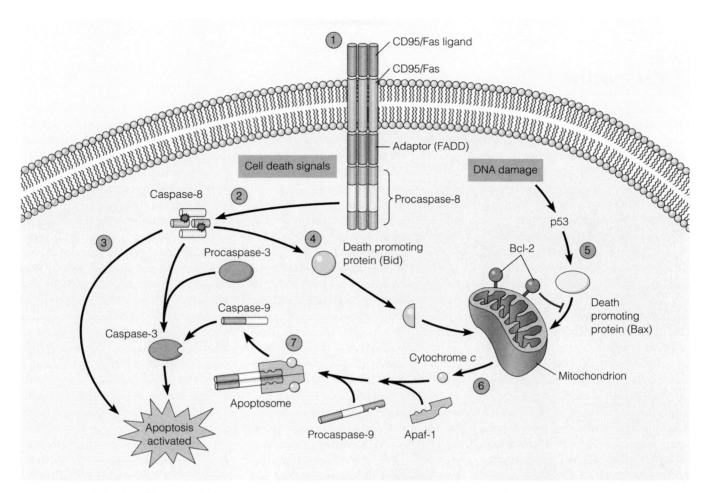

Figure 10-28 Induction of Apoptosis by Cell Death Signals or by DNA Damage. Cell death signals, e.g., CD95/Fas ligand on the surface of a killer lymphocyte, can lead to apoptosis. ① CD95/Fas ligand binds to CD95/Fas receptors on the surface of a target cell. Receptor binding causes clustering of receptors and recruitment of adaptor proteins in the target cell, resulting in clustering of procaspase-8 molecules. ② Caspase-8 molecules become activated and contribute to the initiation of apoptosis. ③ One of the targets of caspase-8 is procaspase-3, which is cleaved and activated to become caspase-3, a key initiator of apoptosis. ④ In addition, caspase-8 activates death-promoting proteins (such as Bid) of the Bcl-2 family. DNA damage can also lead to apoptosis through p53 ⑤, which results in activation of other death-promoting proteins (e.g., Bax). The balance between death-promoting and death-preventing proteins (e.g., Bcl-2) at the mitochondrial outer membrane determines whether the mitochondrion releases cytochrome c (⑥). Cytochrome c forms a complex with Apaf-1 and procaspase-9 called the apoptosome ⑦ that produces activated caspase-9, which can also activate more caspase-3 and hence initiate the events of apoptosis.

When a cell's DNA is damaged (for example, by radiation or ultraviolet light), it can enter apoptosis via the activity of a protein known as *p53* (p53 is an important regulator of the cell cycle; you will learn more about p53 in Chapter 17). The p53 pathway can activate another death-promoting protein (*Bax*), which, as in the case of cell death signals, acts at the surface of mitochondria (⑤).

At the mitochondrion, the balance between death-promoting and death-preventing proteins appears to determine whether the rest of the apoptotic pathway will be activated. These proteins are structurally related to one another and to an anti-apoptotic protein within the outer membrane of the mitochondrion called **Bcl-2**. When the balance shifts toward death-promoting members of this protein family, a cell is more likely to undergo apoptosis. The connection between mitochondria and cell death may be surprising, but it is clear that in addition to their role in energy production, mitochondria are important in apoptosis. If a cell death signal or DNA damage is the sentence of execution, then the executioners are mitochondria.

How does the mitochondrion hasten cell death? In the presence of cell death-promoting proteins, **cytochrome c**, normally within the mitochondria, is released (⑥). Although cytochrome c is normally involved in electron transport (see Chapter 14), the released cytochrome c recruits proteins, including *Apaf-1* and another procaspase, *procaspase-9*, which assemble into a complex sometimes called an **apoptosome**. The procaspase-9 in the apoptosome is cleaved into activated caspase-9, which in turn stimulates the activation of caspase-3 (⑦). Thus in the end, both cell death mechanisms lead to the activation of a common caspase that sets apoptosis in motion.

Perspective

Most cells respond to hormones, growth factors, and other substances present in the extracellular fluid. Such responses are mediated by receptor proteins, either at the cell surface or inside the cell. Each receptor protein has a binding site that is specific for its particular ligand. In the case of membrane receptors, ligand binding is followed by transmission of the signal to the interior of the cell, thereby regulating specific intracellular events.

Several different mechanisms for signal transduction are known. One important signal transduction pathway involves receptors linked to G proteins. G proteins are activated when the binding of a ligand to a neighboring receptor causes a conformational change in the G protein, resulting in the displacement of GDP by GTP. The G protein then activates an enzyme system that produces intracellular chemical signals called second messengers. One of the most common second messengers is cyclic AMP, synthesized when the enzyme adenylyl cyclase is activated by a G protein. In an alternative pathway, the second messengers are inositol trisphosphate and diacylglycerol, which are produced from phosphatidylinositol bisphosphate when a G protein activates the enzyme phospholipase C. Regardless of the pathway, second messengers mediate specific intracellular responses by activating specific enzymes or enzyme cascades.

Calcium ions play a central role in cellular activation processes and can be considered second messengers as well. Calcium effects are often mediated by calmodulin, a protein that is activated when calcium ions bind to it. Depending on the target cell, the calcium-calmodulin complex can activate any of a variety of enzymes, thereby modulating enzyme activity in response to the cytosolic calcium concentration. Release of calcium from internal stores within the endoplasmic reticulum is an important step in many signaling processes, including fertilization of animal eggs.

Receptor protein kinases, such as receptor tyrosine kinases, act via a third category of signal transduction mechanisms. Upon binding of the appropriate ligand, such receptors become phosphorylated on specific tyrosines. The phosphorylated receptor becomes a binding site for other proteins, which contain SH2 domains. When these proteins bind to the receptor, they are activated either by the binding itself or by subsequent phosphorylation. SH2 domain-containing proteins activate major signal transduction pathways, including the Ras and phospholipase C pathways.

Growth factors are messengers that play a specific role in regulating cell growth and behavior. Many growth factors bind receptor tyrosine kinases; other growth factor receptors have serine/threonine kinase activity. In these and other cases, signal transduction events result in alterations in phosphorylation of cytosolic proteins, ultimately resulting in changes in the function of proteins that enter the nucleus to affect the expression of genes. The signal transduction pathways initiated by these different types of receptors provide an important link to our understanding of embryonic development. Mutations can lead to unregulated activity or lack of function of the receptor or of components of the signal transduction pathway. As a result, cells may experience changes in the stimuli they perceive, which can lead to abnormal function.

Hormones are messengers that regulate the activities of body tissues distant from the tissues that secrete them. Hormones can be endocrine or paracrine, depending on the mode of delivery. The adrenergic hormones secreted by the adrenal medulla are examples of endocrine hormones. Adrenergic hormone receptors can be classified as α-and β-adrenergic receptors. The β-adrenergic receptors are linked to a G protein that stimulates the formation of cAMP, whereas the α-adrenergic receptors stimulate phospholipase C, resulting in elevation of the intracellular calcium concentration. The different receptors mediate opposite effects in different smooth muscles but work together to coordinate changes in blood flow and other activities, thereby preparing the body for stressful situations.

Apoptosis is a form of cellular suicide that can be triggered by external signals or by internal changes in a cell, such as DNA damage. Apoptosis involves the orderly dismantling of a dying cell's contents, including the fragmentation of its DNA and cytoplasm, and eventually results in the dead cell's remnants being engulfed by phagocytic cells. Specific proteases called caspases are key mediators of apoptosis. Caspases can be activated in several ways; one major route by which caspases are activated is through the release of cytochrome c from mitochondria.

Key Terms for Self-Testing

Chemical Signals and Cellular Receptors
chemical messenger (p. 257)
ligand (p. 257)
primary messenger (p. 257)
second messenger (p. 257)
signal transduction (p. 257)
cognate receptor (p. 258)

receptor affinity (p. 258)
dissociation constant (K_d) (p. 258)

G-Protein Linked Receptors
G protein–linked receptor family (p. 259)
G protein (p. 259)
Gs (p. 259)

Gi (p. 259)
cyclic AMP (cAMP) (p. 261)
adenylyl cyclase (p. 261)
phosphodiesterase (p. 262)
protein kinase A (PKA) (p. 262)
inositol-1,4,5-trisphosphate (InsP$_3$) (p. 263)
phospholipase C (p. 263)

diacylglycerol (DAG) (p. 263)
InsP$_3$ receptor (p. 264)
protein kinase C (PKC) (p. 264)
calcium ionophore (p. 265)
calcium pump (p. 265)
ryanodine receptor channel (p. 265)
calmodulin (p. 265)
calcium-calmodulin complex (p. 265)
nitric oxide (NO) (p. 269)

Protein Kinase–Associated Receptors
protein kinase–associated receptor (p. 270)
receptor tyrosine kinase (p. 270)
nonreceptor tyrosine kinase (p. 271)
autophosphorylation (p. 271)
SH2 domain (p. 271)
Ras (protein) (p. 271)
guanine-nucleotide exchange factor (GEF)
    (p. 271)
Sos (protein) (p. 271)

mitogen-activated protein kinase (MAP
    kinase, MAPK) (p. 271)
GTPase activating protein (GAP) (p. 271)

Growth Factors as Messengers
growth factor (p. 273)
fibroblast growth factor (FGF) (p. 273)
fibroblast growth factor receptor (FGFR)
    (p. 273)
dominant negative mutation (p. 273)
serine/threonine kinases receptor
    (p. 274)
transforming growth factor β (TGFβ)
    (p. 274)
Smad (protein) (p. 274)

**The Endocrine and Paracrine
Hormone Systems**
hormone (p. 276)
endocrine hormone (p. 276)

paracrine hormone (p. 276)
endocrine tissue (p. 276)
exocrine tissue (p. 276)
target tissue (p. 276)
histamine (p. 278)
prostaglandin (p. 278)
adrenergic hormone (p. 279)
adrenergic receptor (p. 279)
allergen (p. 280)
mast cell (p. 281)
blood platelet (p. 281)

Cell Signals and Apoptosis
apoptosis (p. 282)
apoptotic body (p. 282)
caspase (p. 282)
procaspase (p. 284)
Bcl-2 (p. 285)
cytochrome c (p. 285)
apoptosome (p. 285)

Problem Set

More challenging problems are marked with a • .

10-1.   Chemical Signals and Second Messengers. Fill in the blanks with the appropriate terms:

(a) _____ is an intracellular protein that binds calcium and activates enzymes.

(b) The _____ is a gland that regulates other endocrine tissues of the body.

(c) A substance that fits into a specific binding site on the surface of a protein molecule is called a _____.

(d) Two products of phospholipase C activity that serve as second messengers are _____ and _____.

(e) Cyclic AMP is produced by the enzyme _____ and degraded by the enzyme _____.

10-2.   Heterotrimeric and Monomeric G Proteins. G protein–linked receptors activate heterotrimeric G proteins through interactions between the receptor and the G$_{\beta\gamma}$ subunit. Upon binding to the receptor, the G$_{\beta\gamma}$ subunit then catalyzes GDP/GTP exchange by the G$_\alpha$ subunit. How is this similar to the activation of Ras by a receptor tyrosine kinase?

• 10-3.   Why Calcium? Compare the use of sodium ions (see Chapter 9) and calcium ions as intracellular signals.

(a) What happens when sodium ions rush into a cell? What happens when the cytosolic calcium concentration rises? What is the difference?

(b) Why would it be better to use calcium ions as a second messenger rather than sodium ions, if the ion must bind to a protein as part of a signal transduction pathway?

10-4.   Calcium Chelators and Ionophores. Two important tools that have aided studies of the role of calcium in triggering different cellular events are *calcium chelators* and *calcium ionophores*. Chelators are compounds such as EGTA and EDTA that bind very tightly to calcium ions, thereby effectively reducing the free (or available) calcium ion concentration outside of the cell nearly to zero. Ionophores are compounds that shuttle ions across lipid bilayers and membranes, including the plasma membrane and internal membranes such as the ER. For calcium ions, two of the commonly used ionophores are A23187 and ionomycin. Using these tools, describe how you could demonstrate that a hormone exerts its effect by (1) causing calcium to enter the cell through channels or (2) releasing calcium from intracellular stores such as the ER.

10-5.   Anti-inflammatory vs. Inflammatory Drugs. Anti-inflammatory drugs can generally be classified as nonsteroidal and steroidal. Aspirin is a nonsteroidal anti-inflammatory drug, whereas cortisone is a steroid. Cortisone induces the synthesis of a protein that inhibits phospholipase A.

(a) Are the actions of these two kinds of compounds related? Explain your answer.

(b) Based on how these drugs act, what kind of compound would be a good candidate for stimulating inflammation?

10-6.   Membrane Receptors and Medicine. *Hypertension,* or high blood pressure, is often seen in elderly people. A typical prescription to reduce a patient's blood pressure includes compounds called *beta-blockers,* which block β-adrenergic receptors throughout the body. These receptors bind epinephrine, thereby activating a cellular response. Why do you think beta-blockers are effective in reducing blood pressure?

• 10-7.   Double Negative. In frog embryos, mRNA encoding dominant negative fibroblast growth factor receptors are injected into one or more cells of the early embryo. We saw that the receptors produced in this experiment interfere with the function of normal receptors (see Figure 10-19). In some experiments, an excess of normal, wild-type receptor mRNA was injected as a control. Why was this done? Would injecting excess wild-type receptor mRNA cancel out the effects of the mutant receptors? Explain your answer.

10-8.   Chemoattractant Receptors on Neutrophils. *Neutrophils* are blood cells that are normally responsible for killing bacteria at sites of infection. Neutrophils are able to find their way toward sites of infection by the process of *chemotaxis*. In this process, neutrophils sense the presence of bacterial protein with receptors on their plasma membranes and then follow the trail of these proteins toward the site of infection. Suppose you find that chemotaxis is inhibited by pertussis toxin. What kind of receptor is likely to be involved in responding to bacterial proteins?

10-9.   Once Is Enough. A person who smokes for the first time tends to experience much more severe effects from a cigarette than someone who smokes frequently. Given that the smoke contains nicotine, provide a reasonable explanation for this difference.

• 10-10.   Scrambled Eggs. G protein–stimulated calcium release has been proposed to be involved in egg activation in starfish eggs. Starfish eggs can be induced to produce foreign proteins by injection with mRNA encoding the protein of choice. What do you predict would happen in the following cases, and why?

(a)   When mRNA encoding a phospholipase β that is always activated is injected into an unfertilized egg

(b)   When mRNA for $InsP_3$ receptors that do not allow calcium release are injected, and then the egg is fertilized

(c)   When aequorin, a jellyfish protein that fluoresces in response to elevated calcium levels, is injected into normal eggs, which are then fertilized

• 10-11.   Apoptosis and Medicine. A current focus of molecular medicine is to trigger or prevent apoptosis in specific cells. Several components of the apoptic pathway are being targeted in clinical trials. For each of the following approaches, state specifically how the treatment would be expected to stimulate or inhibit apoptosis:

(a)   Treatment of cells with antisense oligonucleotides targeted against the Bcl-2 gene (antisense oligonucleotides bind to mRNAs, preventing their translation into protein)

(b)   Exposing cells to recombinant TRAIL protein, a ligand for the tumor necrosis factor family of receptors

(c)   Treatment of cells with organic compounds that enter the cell and bind with high affinity to the active site of caspase-3

Suggested Reading

References of historical importance are marked with a •.

Characteristics and Functions of Receptors

• Berridge, M. J. The molecular basis of communication within the cell. *Sci. Amer.* 253 (October 1985): 142.

Blume-Jensen, P., and T. Hunter. Oncogenic kinase signaling. *Nature* (2001) 411: 355.

Clapham, D. E. Mutations in G protein-linked receptors: Novel insights on disease. *Cell* 75 (1993): 1237.

Hepler, J., and A. Gilman. G proteins. *Trends Biochem. Sci.* 17 (1992): 383.

Karin, M., and T. Hunter. Transciptional control by protein phosphorylation: Signal transduction from the cell surface to the nucleus. *Curr. Biol.* 5 (1997): 747.

Levitsky, A. *Receptors: A Quantitative Approach.* Menlo Park, CA: Benjamin/Cummings, 1984.

Pawson, T. Protein modules and signalling networks. *Nature* 373 (1995): 573.

Scott, J.D., and Pawson, T. Cell communication: the inside story. *Sci. Amer.* (2000) 282: 72.

Signal Transduction Pathway: G Proteins and Second Messenger Systems

Berridge, M. J. Inositol trisphosphate and calcium signaling. *Nature* 361 (1993): 315.

Berridge, M. J., M. D. Bootman, and P. Lipp. Calcium—a life and death signal. *Nature* 395 (1998): 645.

Berridge, M.J., P. Lipp, and M.D. Bootman. The versatility and universality of calcium signalling. *Nat. Rev. Mol. Cell Biol.* (2000) 1: 11.

Brownlee, C. Cellular calcium imaging: So, what's new? *Trends Cell Biol.* (2000) 10: 451.

Evans, N.H., and A.M. Hetherington. The ups and downs of guard cell signalling. *Curr Biol.* (2001) 11: R92.

Head, J. F. A better grip on calmodulin. *Curr. Biol.* 2 (1992): 609.

Lai, C.-Y. The chemistry and biology of cholera toxin. *CRC Crit. Rev. Biochem.* 9 (1987): 171.

• Levitsky, A. From epinephrine to cyclic AMP. *Science* 241 (1988): 800.

Linder, M. E., and A. G. Gilman. G proteins. *Sci. Amer.* 267 (July 1992): 56.

Michell, R. Inositol lipids in cellular signaling mechanisms. *Trends Biochem. Sci.* 17 (1992): 274.

Mikoshiba, K. Inositol 1,4,5-trisphosphate receptor. *Trends Pharmacol. Sci.* 14 (1993): 86.

Newton, A. C. Protein kinase C. Seeing two domains. *Curr. Biol.* 5 (1995): 973.

Strader, C. D., T. M. Fong, M. R. Tota, D. Underwood, and R. A. F. Dixon. Structure and function of G protein–coupled receptors. *Ann. Rev. Biochem.* 63 (1994): 101.

• Sutherland, E. W. Studies on the mechanism of hormone action. *Science* 177 (1972): 401.

Tang, W. J., and A. G. Gilman. Adenylyl cyclases. *Cell* 71 (1992): 1069.

Growth Factor Signaling

• Amaya, E., T. J. Musci, and M. W. Kirschner. Expression of a dominant negative mutant of the FGF receptor disrupts mesoderm formation in *Xenopus* embryos. *Cell* 66 (1991): 257.

De Moerlooze, L., and C. Dickson. Skeletal disorders associated with fibroblast growth factor receptor mutations. *Curr. Opinion Genetics & Dev.* 7 (1997): 378.

Gilbert, S. F. *Developmental Biology,* 6th ed. Sunderland, MA: Sinauer Associates, 2000.

Massague, J.S., W. Blain, and R. S. Lo. TGFβ signaling in growth control, cancer, and heritable disorders. *Cell* (2000) 103: 295.

Mochly-Rosen, D. Localization of protein kinases by anchoring proteins: A theme in signal transduction. *Science* 268 (1995): 247.

Porter, A. C., and R. R. Vaillancourt. Tyrosine kinase receptor–activated signal transduction pathways which lead to oncogenesis. *Oncogene* 17 (1998): 1343.

• Rozakis-Adcock, M., R. Fernley, J. Wade, T. Pawson, and D. Bowtell. The SH2 and SH3 domains of mammalian Grb 2 couple the EGF receptor to the Ras activator mSos1. *Nature* 363 (1993): 83.

Schlessinger, J. How receptor tyrosine kinases activate Ras. *Trends Biochem. Sci.* 18 (1993): 273.

Nitric Oxide and Cell Signaling

Mayer, B., and B. Hemmens. (1997). Biosynthesis and action of nitric oxide in mammalian cells. *Trends Biochem. Sci.* 22 (1997): 477.

Snyder, S. H., and D. S. Bredt. Biological roles for nitric oxide. *Sci. Amer.* 266 (May, 1992): 68.

The Endocrine and Paracrine Hormone Systems

Baulieu, E. E., and P. A. Kelly. *Hormones: From Molecules to Disease.* London: Chapman & Hall, 1990.

Bjorkman, D. J. The effect of aspirin and nonsteroidal anti-inflammatory drugs on prostaglandins. Am. J. Med. 105 (1998): 8S.

Brass, L. F., J. A. Hoxie, and D. R. Manning. Signaling through G proteins and G protein–coupled receptors during platelet activation. *Thromb. Haemost.* 70 (1993): 217.

Norman, A. W., and G. Litwack. *Hormones.* San Diego: Academic Press, 1987.

• Weissmann, G. Aspirin. *Sci. Amer.* 264 (January 1991): 84.

Cell Signals and Apoptosis

Adrain, C., and S.J. Martin. The mitochondrial apoptosome: A killer unleashed by the cytochrome seas. *Trends Biochem.* (2001) 26: 390.

Hengartner, M.O. The biochemistry of apoptosis. *Nature* (2000) 407: 770.

Nicholson, D.W. From bench to clinic with apoptosis-based therapeutic agents. *Nature* (2000) 407: 810.

Raff, M. Cell suicide for beginners. *Nature* 396 (1998): 119.

Vaux, D., and S. J. Korsmeyer. Cell death in development. *Cell* 96 (1999): 245.

11

Beyond the Cell: Extracellular Structures, Cell Adhesion, and Cell Junctions

In the preceding chapters, we have treated cells as if they exist in isolation and as if they "end" at the plasma membrane. However, many kinds of cells—including most of those in your own body—spend all their lives linked to neighboring cells. In addition, almost all cells have some sort of structure that is external to the plasma membrane but is nonetheless an integral part of the cell, both structurally and functionally. These **extracellular structures** consist mainly of macromolecules that are secreted by the cell. The chemical nature of the macromolecules differs among organisms, but the extracellular structures of most eukaryotes have a common theme: They contain long, flexible fibers embedded in an amorphous, hydrated matrix of branched molecules that are usually glycoproteins or polysaccharides.

Animal cells have an *extracellular matrix* that takes on a variety of forms and plays important roles in cellular processes as diverse as division, motility, differentiation, and adhesion. In plants, fungi, algae, and prokaryotes, the extracellular structure is called a *cell wall*—although its chemical composition differs considerably among these several kinds of organisms. Cell walls confer rigidity on the cells they encase, serve as permeability barriers, and protect the cells from physical damage and from attack by viruses and infectious organisms.

Most of our attention in this chapter is devoted to the extracellular structures of animal cells. We will consider the extracellular matrix, the adhesion of cells to each other and to the extracellular matrix, and the several kinds of junctions that link cells together into multicellular tissues. Then we will turn to the walls that surround plant cells and the specialized structures that allow direct cell-to-cell communication despite the presence of a wall between neighboring cells. In each case, we

will focus on the molecules involved and the contributions they make to the structural and functional properties of cells.

The Extracellular Matrix of Animal Cells

The **extracellular matrix (ECM)** of animal cells takes on a remarkable variety of forms in different tissues. Figure 11-1 illustrates just three examples. *Bone* consists largely of a rigid extracellular matrix that contains a tiny number of interspersed cells. *Cartilage* is another tissue constructed almost entirely of matrix materials, although the matrix is much more flexible than in bone. In contrast to bone and cartilage, the *connective tissue* surrounding glands and blood vessels has a relatively gelatinous extracellular matrix containing numerous interspersed fibroblast cells.

These examples illustrate the roles that the ECM plays in determining the shape and mechanical properties of organs and tissues. The matrix does more than just provide structural support, however; it also influences properties such as tissue extensibility, cell shape and movement, and development of specialized cellular characteristics.

Despite this diversity of function, the ECM of animal cells almost always consists of the same three classes of molecules: (1) structural proteins such as *collagens* and *elastins,* which give the ECM its strength and flexibility; (2) protein-polysaccharide complexes called *proteoglycans* that provide the matrix in which the structural molecules are embedded; and (3) adhesive glycoproteins such as *fibronectins* and *laminins,* which attach cells to the matrix (Table 11-1). The considerable variety in the properties of the ECM in different tissues results not only from differ-

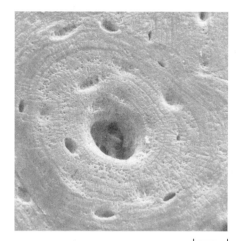

(a) Bone |20 μm

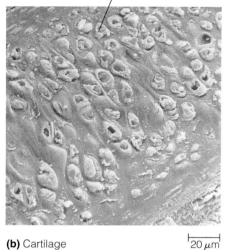

Cartilage cell

(b) Cartilage |20 μm

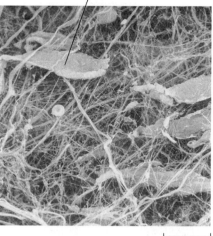

Fibroblast

(c) Connective tissue |20 μm

Figure 11-1 Different Kinds of Extracellular Matrix. The ECM takes on different forms in different tissues, as these SEMs illustrate. **(a)** In bone tissue, a hard, calcified ECM is laid down in concentric rings around central canals. The small elliptical depressions are regions in which bone cells are found. **(b)** In cartilage, the cells are embedded in a flexible matrix that contains large amounts of proteoglycans. **(c)** In the connective tissue found under skin, fibroblasts are surrounded by an ECM that contains large numbers of collagen fibers.

ences in the types of structural proteins and the kinds of proteoglycans present but also from variations in the ratio of structural proteins—collagen, most commonly—to proteoglycans and in the kinds and amounts of adhesive glycoproteins present. We will consider each of these classes of ECM constituents in turn.

Collagens Are Responsible for the Strength of the Extracellular Matrix

The most abundant component of the ECM in animals is a large family of closely related proteins called **collagens,** which form fibers with high tensile strength and thus account for much of the strength of the ECM. Considered collectively, collagen is the most abundant protein in vertebrates, accounting for as much as 25–30% of total body protein. Collagen is secreted by several types of connective-tissue cells, including *fibroblasts.* Collagen is especially prominent in connective tissue such as tendons and ligaments. Without collagen, cells in these and other tissues would not have sufficient adhesive strength to maintain a given form.

Two defining characteristics are shared by all collagens: their occurrence as a rigid *triple helix* of three inter-twined polypeptide chains and their unusual amino acid composition. Specifically, collagens are high in both the common amino acid glycine and the unusual amino acids hydroxylysine and hydroxyproline, which rarely occur in other proteins. (For the structures of hydroxylysine and hydroxyproline, see Figure 7-25c.) The high glycine content makes the triple helix possible because the spacing of the glycine residues places them in the axis of the helix, and glycine is the only amino acid small enough to fit in the interior of a triple helix.

In most animal tissues, **collagen fibers** can be seen in bundles throughout the extracellular matrix when viewed by scanning electron microscopy (Figure 11-2a). One of the most striking features of collagen fibers is their enormous physical strength. For example, it takes a load of more than 20 pounds (about 9 kg) to tear a collagen fiber just 1 millimeter in diameter! Not surprisingly, collagen is largely responsible for the mechanical strength of protective and supporting tissues such as skin, bone, tendon, and cartilage. As illustrated in Figure 11-2b, each collagen fiber is composed of numerous *fibrils.* A fibril, in turn, is made up of many collagen molecules, each of which consists of three polypeptides

Table 11-1 Extracellular Structures of Eukaryotic Cells

Kind of Organism	Extracellular Structure	Structural Fiber	Components of Hydrated Matrix	Adhesive Molecules
Animals	Extracellular matrix (ECM)	Collagens and elastins	Proteoglycans	Fibronectins and laminins
Plants	Cell wall	Cellulose	Hemicelluloses and extensins	Pectins

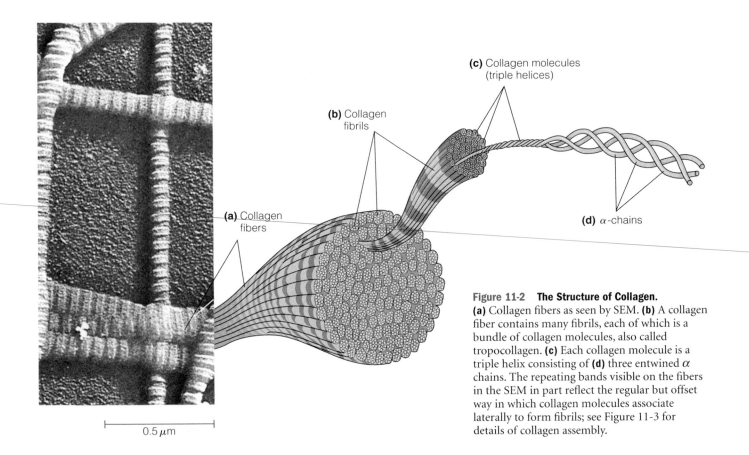

(c) Collagen molecules
(triple helices)

(b) Collagen
fibrils

(a) Collagen
fibers

(d) α-chains

0.5 μm

Figure 11-2   The Structure of Collagen.
(a) Collagen fibers as seen by SEM. **(b)** A collagen fiber contains many fibrils, each of which is a bundle of collagen molecules, also called tropocollagen. **(c)** Each collagen molecule is a triple helix consisting of **(d)** three entwined α chains. The repeating bands visible on the fibers in the SEM in part reflect the regular but offset way in which collagen molecules associate laterally to form fibrils; see Figure 11-3 for details of collagen assembly.

called α *chains* that are twisted together into a rigid, right-handed triple helix (Figure 11-2c and d). Collagen molecules are about 270 nm in length and 1.5 nm in diameter and are aligned both laterally and end-to-end within the fibrils. A typical collagen fiber has about 270 collagen molecules in its cross section.

A Precursor Called Procollagen Forms Many Types of Tissue-Specific Collagens

Given the complexity of a collagen fiber, it is important to ask how such an elaborate structure is generated. Figure 11-3 illustrates our current understanding of the process. In the lumen of the endoplasmic reticulum (ER), three α chains assemble to form a triple helix called **procollagen.** At both ends of the triple-helical structure, short nonhelical sequences of amino acids prevent the formation of collagen fibrils as long as the procollagen remains within the cell. Once procollagen is secreted from the cell into the intercellular space, it is converted to collagen by *procollagen peptidase,* an enzyme that removes the extra amino acids from both the N- and C-terminal ends of the triple helix. The resulting collagen molecules spontaneously associate and polymerize to form mature collagen fibrils, which then assemble laterally into fibers.

The stability of the collagen fibril is reinforced by hydrogen bonds that involve the hydroxyl groups of hydroxyproline and hydroxylysine residues in the α

chains. These hydrogen bonds form crosslinks both within and between the individual collagen molecules in a fibril. In addition, specialized types of collagen are often present on the surface of collagen fibrils. The triple-helical structure of these specialized collagens is interrupted at intervals, allowing the molecules to bend and hence to serve as flexible bridges between adjacent collagen fibrils or between collagen fibrils and other matrix components.

Vertebrates have about 25 different kinds of α chains, each of which is encoded by its own gene and has its own unique amino acid sequence. These different α chains combine in various ways to form at least 15 different types of collagen molecules, most of which are found in specific tissues. Table 11-2 lists the different types and the tissues where they are found. Types I, II, and III are the most abundant forms; type I alone makes up about 90% of the collagen in the human body.

When fibrils containing type I, II, III, or V collagen molecules are examined with an electron microscope, they exhibit a characteristic pattern of dark crossbands, or *striations,* that repeat at intervals of about 67 nm (see Figure 11-2a). These bands reflect the regular but offset manner in which triple helices associate laterally to form fibrils (see Figure 11-3, ③). The helices are aligned in parallel, staggered rows. Molecules in adjacent rows overlap by about one-quarter of their lengths, such that the repeat distance of about 67 nm corresponds to one-quarter of the length of an individual helix. Type IV collagen

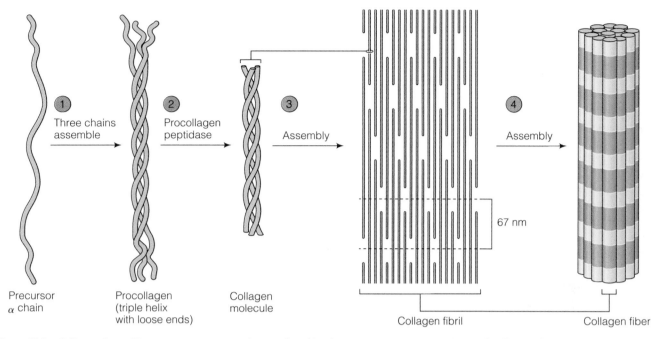

Occurs in ER lumen   Occurs after secretion from cell

① Three chains assemble   ② Procollagen peptidase   ③ Assembly   ④ Assembly

67 nm

Precursor α chain   Procollagen (triple helix with loose ends)   Collagen molecule   Collagen fibril   Collagen fiber

Figure 11-3 Collagen Assembly.
① Collagen precursor chains are assembled in the ER lumen to form triple-helical procollagen molecules. ② After secretion from the cell, procollagen is converted to collagen in a peptide-cleaving reaction catalyzed by the enzyme procollagen peptidase. ③ The molecules of collagen molecules, also called tropocollagen, then bind to each other and self-assemble into collagen fibrils. ④ The fibrils assemble laterally into collagen fibers. In striated collagen, the 67-nm repeat distance is created by packing together rows of collagen molecules in which each row is displaced by one-fourth the length of a single molecule.

forms very fine, unstriated fibrils. The structures of other collagen types are less well characterized.

Elastins Impart Elasticity and Flexibility to the Extracellular Matrix

Although collagen fibers give the ECM great tensile strength, their rigid, rodlike structure is not particularly suited to the elasticity and flexibility required by some tissues. For example, lung tissue must expand and contract as an organism breathes, and arteries must be able to dilate and constrict as the heart pumps blood through them. Other tissues that require a flexible ECM include skin and the intestines, which change shape continuously. Elasticity is provided by stretchable elastic fibers present in the ECM. Like a rubber band, *elastic fibers* can be stretched to several times their normal length and will snap back to their original size when the tension is released. The relative degree of elasticity of any given tissue is controlled by the ratio of nonstretchable collagen fibers to the stretchable elastic fibers.

The principle constituent of elastic fibers is a family of ECM proteins called **elastins.** Like the collagens, elastins

Table 11-2 Types of Collagens, Their Occurrence, and Their Structure

Type of Structure	Collagen Type(s)	Representative Tissues
Long fibrils	I, II, III, V, XI	Skin, bone, tendon, cartilage, muscle
Fibril-associated, with interrupted triple helices	IX, XII, XIV	Cartilage, embryonic skin and tendon
Fibril-associated, forms beaded filaments	VI	Interstitial tissues
Sheets	IV, VIII, X	Basal laminae, cartilage growth plates
Anchoring fibrils	VII	Epithelia
Transmembrane	XVII	Skin (XVII)
Other	XIII, XV, XVI, XVIII, XIX	Basement membranes, assorted tissues

are rich in the amino acids glycine and proline. However, the proline residues are not hydroxylated and no hydroxylysine is present. Elastin molecules are crosslinked to one another by covalent bonds between lysine residues. The elasticity of the resulting crosslinked elastin network is based on the ability of individual elastin molecules to adopt a variety of alternative conformations. Tension on the network causes the individual molecules to adopt extended conformations that permit the overall network to stretch (Figure 11-4a). When the tension is released, the individual molecules relax, returning to their normal, less extended conformations. The crosslinks between molecules then cause the network to recoil to its original shape (Figure 11-4b).

The important roles of collagens and elastins are clearly demonstrated as people age. Over time, collagens become increasingly crosslinked and inflexible, and elastins are lost from tissues like skin. As a result, older people often find that their bones and joints are less flexible, and their skin becomes wrinkled and returns to its original shape only slowly after being deformed (by gentle pinching, for instance).

Collagen and Elastin Fibers Are Embedded in a Matrix of Proteoglycans

The hydrated, gel-like network in which the collagen and elastin fibrils of the ECM are enmeshed consists primarily of *proteoglycans*, glycoproteins in which a large number of glycosaminoglycans are attached to a single protein molecule. *Glycosaminoglycans*, in turn, are long polysaccharide chains, usually consisting of two monosaccharides or monosaccharide derivatives in strictly alternating order.

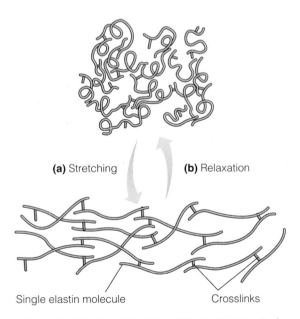

(a) Stretching   **(b)** Relaxation

Single elastin molecule   Crosslinks

Figure 11-4   Stretching and Recoiling of Elastin Fibers. Each elastin molecule in the crosslinked network can assume either an extended (bottom) or compact (top) configuration. The fiber **(a)** stretches to its extended form when tension is exerted on it and **(b)** recoils to its compact form when the tension is released.

We will begin our discussion of proteoglycans by considering glycosaminoglycans.

Glycosaminoglycans.   The main carbohydrate components of proteoglycans are the **glycosaminoglycans (GAGs),** also called *mucopolysaccharides*. GAGs are characterized by repeating disaccharide units, as illustrated in Figure 11-5a for the three most common types, *chondroitin sulfate, keratan sulfate*, and *hyaluronate*. In each case, one of the two sugars in the disaccharide repeating unit is an amino sugar, either *N-acetylglucosamine* (*GlcNAc*) or *N-acetylgalactosamine* (*GalNAc*). The other sugar in the disaccharide repeating unit is usually a sugar or a sugar acid, commonly *galactose* (*Gal*) or *glucuronate* (*GlcUA*). In most cases, the amino sugar has one or more sulfate groups attached; of the GAG repeating units shown in Figure 11-5a, only hyaluronate contains no sulfate groups. Because GAGs are hydrophilic molecules with many negatively charged sulfate and carboxyl groups, they attract both water and cations, thereby forming the hydrated, gelatinous matrix in which collagen and elastin fibrils become embedded.

Proteoglycans.   Most glycosaminoglycans in the ECM are covalently bound to protein molecules to form **proteoglycans,** also called *mucoproteins*. Each proteoglycan consists of numerous GAG chains attached along the length of a **core protein,** as shown in Figure 11-5b. Many different kinds of proteoglycans can be formed by the combination of different core proteins and GAGs of varying types and lengths. Proteoglycans vary greatly in size, depending on the molecular weight of the core protein (which ranges from about 10,000 to over 500,000) and the number and length of the carbohydrate chains (1–200 per molecule, with an average length of about 800 monosaccharide units). Most proteoglycans have molecular weights in the range of 0.25–3 million and a high carbohydrate content—up to 95%. Proteoglycans are linked directly to collagen fibers to make up the fiber/network structure of the ECM.

In many tissues, proteoglycans are present as individual molecules. In cartilage, however, numerous proteoglycans become attached to long molecules of hyaluronate, forming large complexes as shown in Figure 11-6. A single such complex can have a molecular weight of many millions and may exceed several micrometers in length. The remarkable resilience and pliability of cartilage are due mainly to the properties of these complexes.

An important role of proteoglycans is to trap water molecules, thereby slowing their flow. In effect, proteoglycan networks are like extracellular sponges that can hold remarkable quantities of water—up to 50 times their weight, in fact! Because of their high water content, proteoglycan networks are quite resistant to compression and regain their shape quickly if distorted.

Free Hyaluronate Lubricates Joints and Facilitates Cell Migration

Although most of the GAGs found in the extracellular matrix exist only as components of proteoglycans and not

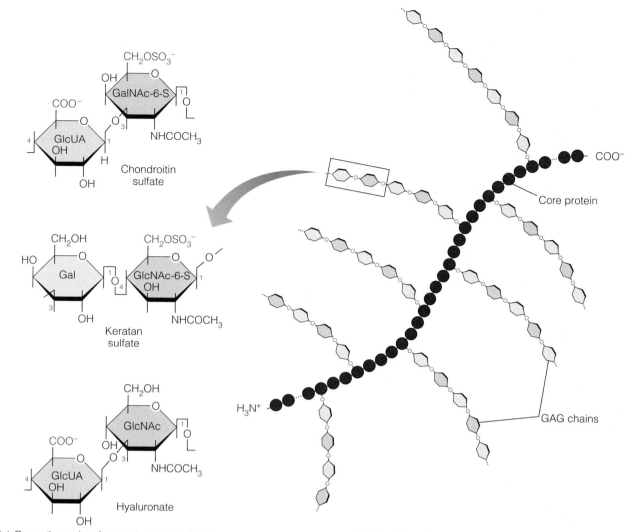

(a) Repeating units of several common GAGs

(b) Structure of a proteoglycan

Figure 11-5 The Structures of Glycosaminoglycans and Proteoglycans.
(a) Structures of the disaccharide repeating units in three common extracellular glycosaminoglycans (GAGs) found in the extracellular matrix of animal cells. The repeating unit of chondroitin sulfate (top) consists of glucuronate, the ionized form of glucuronic acid (GlcUA), and *N*-acetylgalactosamine (GalNAc). Keratan sulfate (middle) has as its repeating unit galactose (Gal) and *N*-acetylglucosamine (GlcNAc). Hyaluronate (bottom) has a repeating unit consisting of GlcUA and GlcNAc. In chondroitin sulfate and keratan sulfate, the GalNAc and GlcNAc residues are shown with a sulfate group (SO_4^{2-}) on carbon atom 6 and are therefore identified as GalNAc-6-S and GlcNAc-6-S, respectively. Other hydroxyl groups can also carry sulfate groups. **(b)** A proteoglycan molecule consists of a core protein with numerous short GAG chains attached along its length and radiating outward. The gray circles represent the amino acids of the core protein, and the colored hexagons represent the sugars or sugar acids of the GAG chains.

as free glycosaminoglycans, **hyaluronate** is an exception. In addition to its role as the backbone of the proteoglycan complex in cartilage, hyaluronate also occurs as a free molecule that consists of hundreds or even thousands of repeating disaccharide units. Hyaluronate molecules have lubricating properties and are most abundant in places where friction must be reduced, such as the joints between movable bones.

Hyaluronate is especially prevalent in the ECM of tissues where cells are actively proliferating or migrating. It is also found on the surface of migrating cells but disappears from the cell surface when migration ceases and cell-to-cell contacts are established. Such observations suggest that hyaluronate facilitates cell migration, perhaps by keeping cells separate from one another. Many cells bind hyaluronate via a 34kDa cell surface protein called *CD44*. Some cancer cells express a variant form of CD44 on their surfaces, suggesting that appropriate interaction with hyaluronate regulates invasive behavior.

Proteoglycans and Adhesive Glycoproteins Anchor Cells to the Extracellular Matrix

Cells are anchored to the ECM by proteoglycan linkages. In some cases, the proteoglycans are themselves integral components of the plasma membrane, with their core

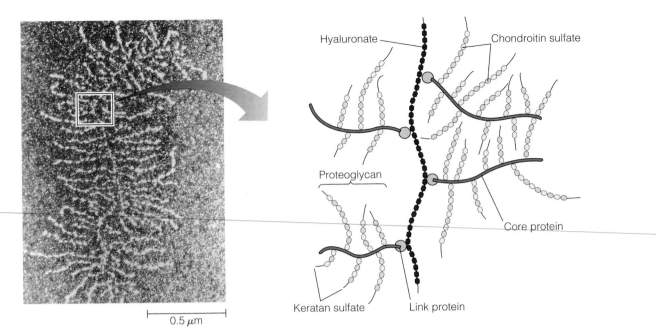

Figure 11-6 Proteoglycan Structure in Cartilage. In cartilage, many proteoglycans such as the one shown in Figure 11-5 associate with a hyaluronate backbone to form a complex that is readily visible with an electron microscope. **(Left)** A hyaluronate-proteoglycan complex isolated from bovine cartilage (TEM). **(Right)** A diagram of a small portion of the structure showing core proteins of proteoglycans attached by means of linker proteins to a long hyaluronate molecule. Short keratan sulfate and chondroitin sulfate chains are linked covalently along the length of the core proteins, as in Figure 11-5b. Proteoglycans have a carbohydrate content of about 95%.

polypeptides embedded within the membrane. In other cases, proteoglycans are linked covalently to membrane phospholipids. Alternatively, either proteoglycans or collagen may bind to specific receptor proteins on the outer surface of the plasma membrane.

Direct links between the ECM and the plasma membrane are reinforced by a family of *adhesive glycoproteins* that bind proteoglycans and collagen molecules to each other and to receptors on the membrane surface. These proteins typically have multiple domains, some with binding sites for macromolecules in the ECM and others with binding sites for membrane receptors. The two most common kinds of adhesive glycoproteins are the *fibronectins* and *laminins*. Many of the membrane receptors to which these glycoproteins bind belong to a family of transmembrane proteins called *integrins*. In the following sections, we discuss fibronectins, laminins, and integrins.

Fibronectins Bind Cells to the Matrix and Guide Cellular Movement

Fibronectin is the most common adhesive glycoprotein in the ECM and is widely distributed throughout the animal kingdom. Fibronectin is actually a collective term for a family of closely related proteins, each containing about 5% carbohydrate. Fibronectins occur in soluble form in blood and other body fluids, as insoluble fibrils in the extracellular matrix, and as an intermediate form loosely associated with cell surfaces. Despite these differences in solubility and location, the various forms of fibronectin are all encoded by the same gene. The different forms of the protein are generated because the RNA transcribed from the fibronectin gene is processed in various ways to generate mRNAs that code for the several different proteins.

A fibronectin molecule consists of two very large polypeptide subunits that are linked near their carboxyl ends by a pair of disulfide bonds (Figure 11-7). Each subunit has about 2500 amino acids and is folded into a series of rodlike domains connected by short, flexible segments of the polypeptide chain. To determine the function of the globular domains, researchers cleaved the polypeptide chains of fibronectin at specific sites, isolated individual domains, and analyzed each for binding activity. Such studies revealed that several of the domains recognize and bind to one or more specific kinds of macromolecules located in the ECM or on cell surfaces, including several types of collagen (I, II, and IV), heparin, and the blood-clotting protein *fibrin*. Other domains recognize and bind to cell-surface receptors. The receptor binding activity of these domains has been localized to a specific tripeptide sequence, RGD (arginine-glycine-aspartate). This *RGD sequence* is a common motif among extracellular adhesive proteins and is recognized by various integrins on the cell surface, including the fibronectin receptor to be discussed shortly.

Effects of Fibronectin on Cell Shape and Cell Movement. Fibronectin binds to cell-surface receptors as well as to ECM components such as collagen and heparin, and thus func-

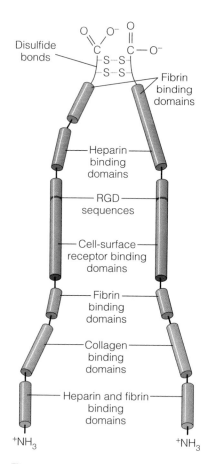

Figure 11-7 Fibronectin Structure. A fibronectin molecule consists of two nearly identical polypeptide chains joined by two disulfide bonds near their carboxyl ends. Each polypeptide chain is folded into a series of globular domains linked by short, flexible segments. The globular domains have binding sites for ECM components or for specific receptors on the cell surface. The receptor-binding domain contains the tripeptide sequence RGD (arginine-glycine-aspartate), which is recognized by fibronectin receptors. In addition to the binding activities noted, fibronectin also has binding sites for heparan sulfate, hyaluronate, and gangliosides (glycosphingolipids that contain sialic acid groups).

tions as a bridging molecule that attaches cells to the ECM, forming stable networks of ECM components. This anchoring role can be demonstrated experimentally by placing cells in a culture dish coated on the inside with fibronectin. Under these conditions, the cells attach to the surface of the dish more efficiently than they do in the absence of fibronectin. After attaching, the cells flatten out and components of the intracellular cytoskeleton become aligned with the fibronectin located outside the cell (Figure 11-8a). This alignment can be seen by staining the same cultured cells with fluorescent antibodies specific for either actin microfilaments, which are an integral part of the cytoskeleton, (Figure 11-8b) or fibronectin (Figure 11-8c). Because the orientation and organization of the cytoskeletal network are important in determining the shape of the cell, fibronectin is thought to be significant in the maintenance of cell shape.

Fibronectin is also involved in cellular movement, such as the cell migrations that occur during early embry-

onic development, when specific groups of cells move from one region of the embryo to another. When migratory embryonic cells are grown on fibronectin, they adhere readily to it (see Figure 11-8a). The pathways followed by migrating cells are rich in fibronectin, suggesting that such cells are guided by binding to fibronectin molecules along the way. Experimental support for this idea comes from studies in which developing amphibian embryos were injected with antibodies directed against fibronectin, thereby blocking the binding of cells to fibronectin. Normal cellular migration was disrupted, leading to the development of abnormal embryos.

More direct evidence for the importance of fibronectin in embryos comes from studying genetically engineered mice that cannot produce fibronectin. Such mice have severe defects in the cells that make the musculature along the length of the body and in the vasculature. These defects highlight the crucial importance of fibronectin during embryonic development.

A possible involvement of fibronectin in cancer is suggested by the observation that many kinds of cancer cells are unable to synthesize fibronectins, with an accompanying loss of normal cell shape and detachment from the ECM. If such cells are supplied with fibronectin, they often return to a more normal shape, recover their ability to bind to the ECM, and are no longer appear malignant.

Effects of Fibronectin on Blood Clotting. The soluble form of fibronectin present in the blood, called *plasma fibronectin,* is involved in blood clotting. Fibronectin promotes blood clotting because it has several binding domains that recognize fibrin, the blood-clotting protein, and it can attach blood platelets to fibrin as the blood clot forms.

Laminins Bind Cells to the Basal Lamina

Another major adhesive glycoprotein present in the ECM is a family of proteins called **laminins.** Unlike fibronectins, which occur widely throughout supporting tissues and body fluids, laminins are found mainly in the **basal lamina,** the thin sheet of specialized extracellular material, typically about 50 nm thick, that underlies epithelial cells, thereby separating them from connective tissues (Figure 11-9). In addition, basal laminae surround muscle cells, fat cells, and the Schwann cells that form myelin sheaths around nerve cells.

Properties of the Basal Lamina. The basal lamina serves as a structural support that maintains tissue organization and as a permeability barrier that regulates the movement of molecules as well as cells. In the kidney, for example, the basal lamina functions as a filter that allows small molecules but not blood proteins to move from the blood into the urine. The basal lamina beneath epithelial cells prevents the passage of underlying connective tissue cells into the epithelium but permits the migration of the white blood cells needed to fight infections. The effect of the basal lamina on cell migration is of special interest

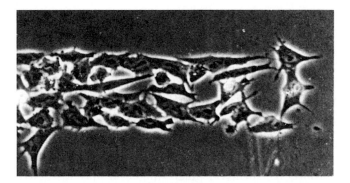

(a) Cells from the neural crest

Figure 11-8 **Influence of Fibronectin on Cell Shape and Cell Migration.** **(a)** Cells from the neural crest (a type of migratory embryonic cell) preferentially migrate along a strip of fibronectin in vitro. **(b, c)** Fluorescence micrographs show the same cultured cells stained with fluorescent antibodies specific for either **(b)** actin or **(c)** fibronectin. Note that the extracellular fibronectin network and the intracellular actin microfilaments (part of the cytoskeleton) are aligned in a similar pattern. (LMs)

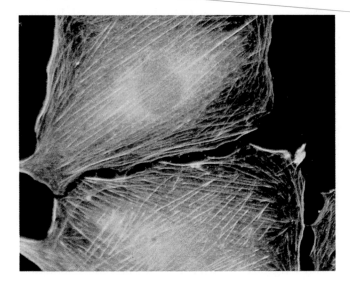

(b) Actin

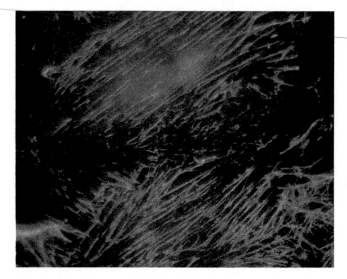

(c) Fibronectin

because some cancer cells show increased binding to the basal lamina. The resulting increase in the ability of cancer cells to bind to the basal lamina may facilitate their movement through the structure and allow them to migrate from one region of the body to another. The basal lamina also organizes proteins in the membranes of adjacent cells and is involved in the induction of cellular differentiation and the regeneration of injured tissue.

Despite differences in function and specific molecular composition from tissue to tissue, all forms of basal laminae contain type IV collagen, proteoglycans, laminins, and another glycoprotein called *entactin,* or *nidogen.* Fibronectins may be present, but laminins are the most abundant adhesive glycoproteins in basal laminae. Laminins are thought to be localized mainly on the surface of the lamina that faces the overlying epithelial cells, where they help bind the cells to the lamina. Fibronectins, on the other hand, are located on the other side of the lamina, where they help anchor cells of the connective tissue.

Properties of Laminin. Laminin, a very large protein with a molecular weight of about 850,000, consists of three long polypeptides, denoted α, β, and γ. There are several different types of each of the three subunits, which can combine

to form numerous different types of laminin. Disulfide bonds hold the polypeptide chains together in the shape of a cross, with part of the long arm wound into a three-stranded coil (Figure 11-10). Like fibronectin, laminin consists of several domains, which represent binding sites for type IV collagen, heparin, heparan sulfate, and entactin, as well as for laminin receptor proteins on the surface of the plasma membranes of the overlying cells. Its binding sites allow laminin to serve as a bridging molecule that attaches cells to the basal lamina. Entactin molecules have binding sites for both laminin and type IV collagen and are therefore thought to reinforce the binding of type IV collagen and laminin networks in basal laminae.

Integrins Are Cell Surface Receptors That Bind ECM Constituents

Fibronectins and laminins can bind to animal cells because the plasma membranes of most cells have specific *receptor glycoproteins* on their surfaces that recognize and bind to specific regions of the fibronectin or laminin molecule. These receptors—and those for a variety of other ECM constituents as well—belong to a large family of transmembrane proteins that are called **integrins** because

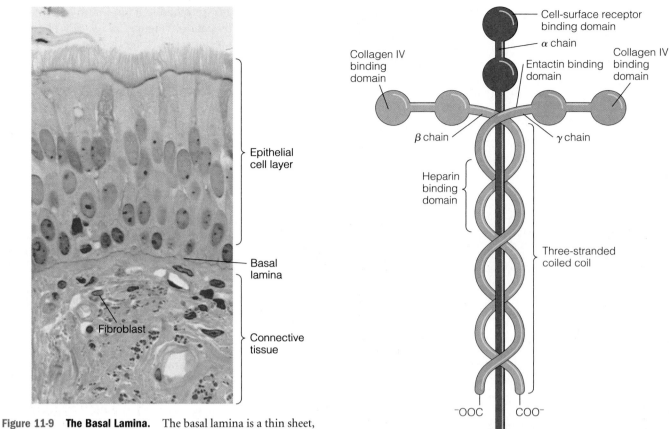

Figure 11-9 The Basal Lamina. The basal lamina is a thin sheet, typically about 50 nm, of matrix material that separates an epithelial cell layer from underlying connective tissue (TEM).

Figure 11-10 Laminin Structure. A laminin molecule consists of three large polypeptides—α, β, and γ—joined by disulfide bonds into a crosslinked structure. A portion of the long arm consists of a three-stranded coil. The functional domains on the ends of the α chain bind to organ-specific cell-surface receptors, whereas those at the ends of the two arms of the cross are specific for type IV collagen. The cross-arms also contain laminin-laminin binding sites, thereby enabling laminin molecules to bind to each other and form large aggregates. Laminin also contains binding sites for heparin and heparan sulfate, as well as for entactin (not shown).

of their role in *integrating* the cytoskeleton with the extracellular matrix. Integrins are very important receptors, because they are the primary means by which ECM proteins such as collagen, fibronectin, and laminin bind to cells and influence cellular functions and structure.

Structure of Integrins. An integrin consists of two large transmembrane polypeptides, the α and β subunits, that associate with each other noncovalently (Figure 11-11). Integrins differ from one another in their binding specificities and in the sizes of their subunits (molecular weight ranges: 110,000–140,000 for the α subunit; 85,000–91,000 for the β subunit). For a specific integrin, the α and β subunits interact on the outer membrane surface to form the binding site for a particular adhesive glycoprotein, with most of the binding specificity apparently depending on the α subunit. On the cytoplasmic side of the membrane, integrins have binding sites for specific molecules of the cytoskeleton, thereby enabling the cytoskeleton and the ECM to communicate across the plasma membrane.

The presence of multiple types of both α and β subunits results in a large number of different integrin heterodimers, which vary in their binding specificities for ECM proteins and also in the cell types on which they are found. For example, integrins containing a β_1 subunit are found on the surfaces of most vertebrate cells and mediate mainly cell-ECM interactions, whereas those containing a β_2 subunit are restricted to the surface of white blood cells and are involved primarily in cell-cell interactions. Specific α chains confer additional specificity on integrins. For example, the most common integrin that binds fibronectin is $\alpha_5\beta_1$, whereas $\alpha_6\beta_1$ binds laminin. Many of the integrins recognize the RGD sequence in the specific ECM glycoproteins that they bind. However, the binding site must recognize other parts of the glycoprotein molecule as well, because integrins display a greater specificity of glycoprotein binding than can be accounted for by the RGD sequence alone.

Integrins and the Cytoskeleton. Although integrins link the extracellular matrix and the cytoskeleton, they do not do so directly. Instead, the tails of integrins interact with proteins in the cytosol that link integrins to cytoskeletal

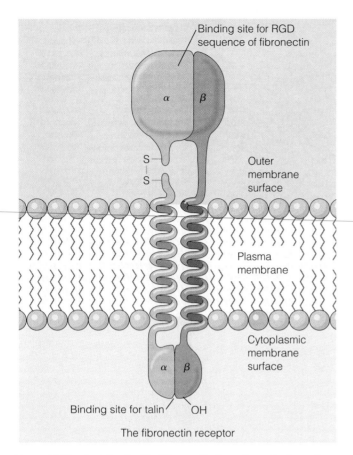

Figure 11-11 An Integrin: The Fibronectin Receptor. An integrin consists of α and β subunits, transmembrane polypeptides that associate with each other noncovalently to form a binding site for the ligand on the outer membrane surface and a binding site for a specific cytoskeletal protein on the inner membrane surface. Shown here is the fibronectin receptor ($\alpha_5\beta_1$), which has a binding site for fibronectin on the outer surface and a binding site for talin on the cytoplasmic side of the membrane. In this and several other integrins, the α subunit is split into two segments held together by a disulfide bond. Both the α and β subunits are glycosylated on the exterior side, although the sugar side chains are not shown here.

Within the figure: "Binding site for RGD sequence of fibronectin", "α", "β", "S", "S", "Outer membrane surface", "Plasma membrane", "Cytoplasmic membrane surface", "α", "β", "Binding site for talin", "OH", "The fibronectin receptor"

proteins. There are two main types of connections that integrins make to the cytoskeleton (Figure 11-12). Migratory and non-epithelial cells, such as fibroblasts, attach to extracellular matrix molecules via **focal adhesions**. Focal adhesions contain clustered integrins (for example, in the case of fibronectin, $\alpha_5\beta_1$ integrin) that interact with bundles of actin microfilaments via several linker proteins (Figure 11-12a–b). These include *talin*, which can bind to the actin-binding protein *vinculin*, and α-*actinin*, which can bind to actin microfilaments directly. Many other cytosolic proteins are recruited to focal adhesions; the interactions among them are complex and are the subject of intensive investigation.

The other major type of integrin-mediated attachment is found in epithelial cells. Epithelial cells attach to laminin in the basal lamina via **hemidesmosomes** (so called because they resemble "half desmosomes"; desmosomes are discussed later in this chapter). The integrin found in hemidesmosomes is $\alpha_6\beta_4$ integrin; in this case,

the integrins are not attached to actin but to intermediate filaments, typically *keratin* (Figure 11-12c–d). The linker proteins in hemidesmosomes form a dense **plaque** that connects clustered integrins to the cytoskeleton. Among the linker proteins, members of the **plakin** family of proteins are very prominent. A plakin known as *plectin* attaches keratin filaments to the integrins. In addition, another transmembrane protein called *BPAG2* and its associated plakin, *BPAG1*, can serve as a bridge between keratin and laminin. These proteins were first identified using antibodies derived from human patients who develop an autoimmune reaction against antigens in their own hemidesmosomes. The resulting disease, called *bullous pemphigoid*, affects predominantly elderly people, and involves blistering of the skin. This disease also gives rise to the names of these proteins (*BPAG* stands for *bullous pemphigoid antigen*).

Integrin Function. Because they bind to components of the ECM on the outside of the cell and to actin microfilaments on the inside, integrins play a crucial role in mediating interactions between the cytoskeleton and the ECM. These interactions are reciprocal: The ECM can influence the organization of the cytoskeleton (recall Figure 11-8), and the cytoskeleton can affect the orientation of ECM components. For example, filaments of fibronectin produced by cultured fibroblast cells normally assemble on the surface of the cells in alignment with the organization of actin microfilament bundles in the cytoskeleton. However, if the actin microfilaments are disrupted experimentally (by treatment with a drug called *cytochalasin B*, for example), the fibronectin filaments dissociate from the cell surface.

Integrins play important roles in regulating cell movement and cell attachment, both in embryos and adult tissues. For example, mice lacking the α_V integrin subunit cannot make blood vessels properly. Similarly, mice lacking a particular integrin subunit (α_7) required for attachment to laminin develop a distinctive form of progressive muscular dystrophy. Humans that carry the same mutation develop a unique form of progressive muscle degeneration. Recall that hemidesmosomes use $\alpha_6\beta_4$ integrins to attach to laminin. Humans that carry mutations in the β_4 subunit develop *junctional epidermolysis bullosa*, a severe blistering disease of the skin. Human patients who cannot produce (or produce very little) laminin 5 develop similar symptoms. In either case, the defects result from failure of epithelial cells in the skin to attach to the basal lamina.

Integrins and Signaling. Although a key role of integrins is their ability to link the cytoskeleton and the extracellular matrix, it is now clear that they also interact with intracellular signaling pathways. For example, signals such as binding of growth factors that lead to MAP kinase activation (see Chapter 10) can result in activation of integrin clustering. (Such effects are often called "inside out" signaling, because internal changes in the cell result in effects on integrins at the surface.) It has also become clear that

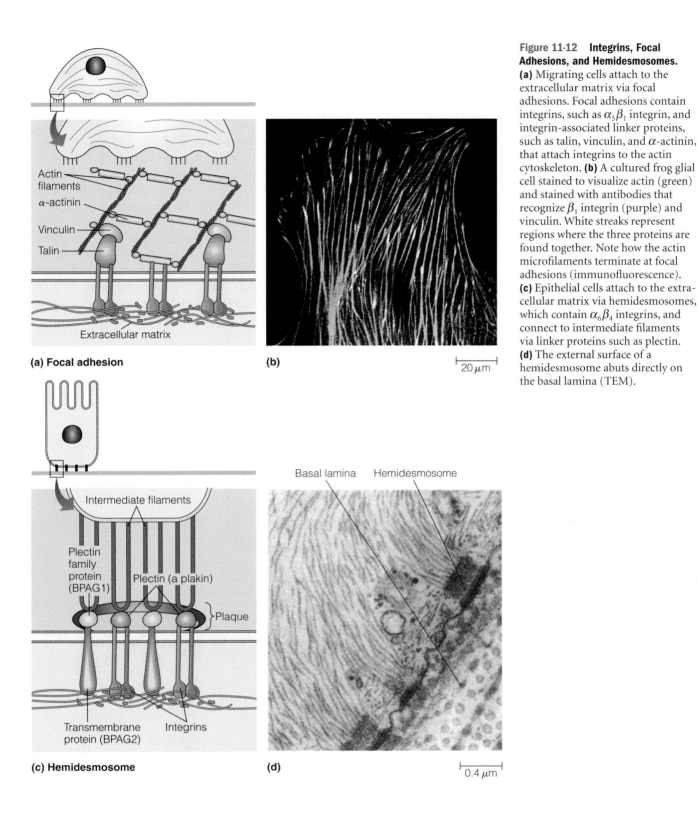

(a) Focal adhesion

Actin filaments

α-actinin

Vinculin

Talin

Extracellular matrix

(b)

20 μm

(c) Hemidesmosome

Intermediate filaments

Plectin family protein (BPAG1)

Plectin (a plakin)

Plaque

Transmembrane protein (BPAG2)

Integrins

(d)

Basal lamina   Hemidesmosome

0.4 μm

Figure 11-12 Integrins, Focal Adhesions, and Hemidesmosomes. **(a)** Migrating cells attach to the extracellular matrix via focal adhesions. Focal adhesions contain integrins, such as $\alpha_5\beta_1$ integrin, and integrin-associated linker proteins, such as talin, vinculin, and α-actinin, that attach integrins to the actin cytoskeleton. **(b)** A cultured frog glial cell stained to visualize actin (green) and stained with antibodies that recognize β_1 integrin (purple) and vinculin. White streaks represent regions where the three proteins are found together. Note how the actin microfilaments terminate at focal adhesions (immunofluorescence). **(c)** Epithelial cells attach to the extracellular matrix via hemidesmosomes, which contain $\alpha_6\beta_4$ integrins, and connect to intermediate filaments via linker proteins such as plectin. **(d)** The external surface of a hemidesmosome abuts directly on the basal lamina (TEM).

integrins can act as receptors that activate intracellular signaling themselves (sometimes called "outside in" signaling). The original observations that suggested involvement of integrins in signaling came from studies of cancer cells. In order for most normal cells to grow in culture, they must be attached to a substratum. Even in the presence of growth factors that normally stimulate their proliferation, if such cells are prevented from attaching to an extracellular matrix layer they will cease dividing and undergo apoptosis. Such behavior is referred to as **anchorage-dependent growth**. In contrast, cancer cells will continue to grow even when they are not firmly attached to an extracellular matrix layer, apparently because they no longer need to transduce signals that result from attachment to the extracellular matrix.

Anchorage-dependent growth appears to involve the activation of intracellular pathways following integrin clustering. Several kinases are activated at focal adhesions following integrin clustering. They are recruited to focal adhesions by adapter proteins such as *paxillin*. These include *focal adhesion kinase (FAK)* and *integrin-linked kinase (ILK)*. FAK appears to be important in regulating anchorage-dependent growth. Cancer cells contain activated FAK even when they are not attached, and cells can be transformed into cancer-like cells by the expression of a mutant, activated form of FAK. These experiments underscore the importance of cell adhesion for understanding cancer.

The Glycocalyx Is a Carbohydrate-Rich Zone at the Periphery of Animal Cells

The boundary between the extracellular matrix and the cell surface is rich in carbohydrates. In plants, algae, fungi, and bacteria, this carbohydrate-rich zone is organized into a rigid cell wall that provides protection and structural support for the cell. Although animal cells do not have a rigid enclosure like a cell wall, a carbohydrate-rich zone called the **glycocalyx** is often present immediately external to the plasma membrane (see Figure 7-27). Roles of the glycocalyx include cell recognition and adhesion, protection of the cell surface, and the creation of permeability barriers.

The glycocalyx has two components, called the *attached glycocalyx* and the *unattached glycocalyx*. The attached glycocalyx consists of inherent constituents of the cell surface that cannot be removed by mechanical means without simultaneously removing a portion of the plasma membrane itself. The major components of this layer are the carbohydrate chains attached to glycoproteins and to glycolipids that are integral components of the plasma membrane. In contrast, the unattached glycocalyx consists of material located external to the plasma membrane that can be readily removed without disrupting the membrane or affecting the viability of the cell. Specialized structures such as the outer coat of amoebas and the embryonic membranes that surround most animal eggs are examples of unattached glycocalyx. Because the unattached glycocalyx usually consists of glycoproteins and proteoglycans that are also components of the extracellular matrix, it is often difficult to specify precisely where the cell surface ends and the extracellular matrix begins.

Cell-Cell Recognition and Adhesion

Thus far, we have looked in some detail at the interactions between animal cells and the extracellular matrix in which most cells are enmeshed, noting the ways in which the ECM influences cells and vice versa. Now we turn to interactions between cells, which also involve components of the ECM. The integrity of multicellular organisms depends on the ability of individual cells to associate in precise patterns to form tissues, organs, and organ systems. Such ordered interactions require, in turn, that individual cells be able to recognize, adhere to, and communicate with each other. We discuss **cell-cell recognition**

and **cell-cell adhesion** in this section and consider intercellular communication in the context of cell junctions later in the chapter.

Transmembrane Proteins Mediate Cell-Cell Adhesion

The ability of cells to recognize and adhere to one another has been studied in many organisms and cell types. From such studies, a picture is beginning to emerge of the cell surface molecules that are involved and the ways in which they interact. Just as integrins serve as transmembrane receptors that bind cells to the extracellular matrix, adhesion receptors serve to bind cells to one another. We now know that such adhesion receptors fall into a relatively small number of classes. They include *immunoglobulin superfamily* proteins, *cadherins, selectins,* and, in a few cases, *integrins* (Figure 11-13). In each case, the adhesion

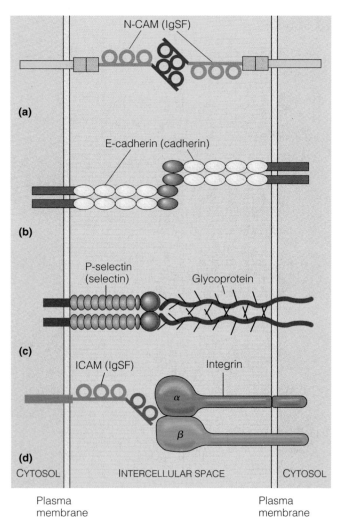

Figure 11-13 Different Types of Cell Adhesion Receptors. Cells adhere to other cells using transmembrane receptor proteins that fall into a few main classes. These include **(a)** immunoglobulin superfamily (IgSF) proteins, such as N-CAM, **(b)** cadherins, such as E-cadherin, **(c)** selectins, which bind to the carbohydrates of glycoproteins on other cells, and **(d)** in a few cases such as leukocytes, integrins, which bind to IgSF proteins such as ICAM on the surface of endothelial cells.

receptor on the surface of one cell binds to the appropriate ligand on the surface of a neighboring cell. In some cases, such as many cadherins and immunoglobulin superfamily members known as CAMs, cells interact with identical molecules on the surface of the cell to which they adhere. Such interactions are said to be **homophilic interactions** (from the Greek *homo,* "like," and *philia,* "friendship"). In other cases, such as the selectins, a cell adhesion receptor on one cell interacts with a different molecule on the surface of the cell to which it attaches. Such interactions are said to be **heterophilic interactions** (from the Greek *hetero,* "different"). As with the integrins, many transmembrane adhesion receptors attach to the cytoskeleton via linker proteins, which differ depending on the class of molecule and its location within the cell. In the next few sections, we consider examples of each of these major classes of cell-cell adhesion molecules.

CAMs. One way to identify molecules involved in cell-cell adhesion is to develop antibodies against specific types of cell membranes or cell-surface molecules. If such an antibody specifically perturbs cell-cell adhesion, the protein to which this antibody binds is likely involved in the adhesion process. This approach was first used successfully in the late 1970s by Gerald Edelman and his colleagues to identify a membrane glycoprotein from nerve tissue that they called **neural cell adhesion molecule (N-CAM).** When embryonic cells were exposed to antibodies directed against N-CAM, the cells no longer bound to one another and the orderly formation of neural tissue was disrupted. To study the binding mechanism, experiments were conducted in which purified N-CAM was incorporated into liposomes (artificial phospholipid vesicles) that do not otherwise adhere to one another. With the N-CAM present, the vesicles adhered to one another unless antibodies directed against N-CAM were also present. These and similar results led to the conclusion that cell adhesion is mediated by the binding of N-CAMs located on one cell to N-CAMs located on another cell.

CAMs are members of the **immunoglobulin superfamily (IgSF).** Proteins in this large superfamily are so named because they contain domains, characterized by well-organized loops, that are similar to those in the immunoglobulin subunits that constitute antibodies. CAMs such as N-CAM on one cell interact homophilically with CAMs on an adjacent cell via these domains. Other IgSF members, such as guidance receptors on the surface of embryonic neurons, interact heterophilically with their ligands. In contrast to cadherins, CAMs interact in a calcium-independent manner. IgSF members participate in a wide range of adhesion events, including interactions of lymphocytes with cells that mediate immune responses. In the embryonic nervous system, molecules from this superfamily, such as N-CAM and L1-CAM, are involved in the outgrowth and bundling of neurons. Humans with mutations in the L1-CAM gene show defects in the corpus callosum (a region that interconnects the two hemispheres of the brain), mental retardation, and other defects.

Cadherins. The use of antibodies that block cell adhesion also led to the discovery of the **cadherins,** a very important group of adhesive glycoproteins found in the plasma membranes of most animal cells. Like CAMs, cadherins play a crucial role in cell-cell recognition and adhesion. The two groups of proteins can be distinguished from each other because cadherins, but not CAMs, require calcium to function. It has been known for many years that removal of Ca^{2+} from the medium in which intact animal tissue is cultured causes the cells to dissociate from one another. We now understand that calcium is necessary for tissue integrity in part because Ca^{2+} induces a conformational change in cadherins that allows them to mediate cell-cell adhesion.

Cadherins are characterized by a series of structurally similar subunits (or "repeats") in their extracellular domain. In addition, the best studied cadherins have binding sites for calcium ions in the extracellular domain, consistent with the demonstrated requirement for calcium. Members of the cadherin superfamily have widely varying numbers of these repeats, and vary in the structure of their cytosolic ends. The best characterized cadherin, E-cadherin, has five such repeat domains. E-cadherin molecules associate in pairs in the plasma membrane. Based on X-ray crystallography, their extracellular domains have a structure that allows them to "zip" together in a homophilic fashion, as cadherins from one cell interlock with those from a neighboring cell (Figure 11-14). At their cytosolic ends, cadherins are connected to the cytoskeleton. As with integrins, these attachments link the cell surface to the cytoskeleton. We will examine these linkages in more detail when we discuss the structure of cell-cell junctions.

Different cadherins are expressed in specific tissues; their regulated expression is a particularly striking feature of embryonic development. The best-characterized members of the cadherin family are *E-cadherin* (prevalent in epithelial tissue), *N-cadherin* (in nervous tissue), and *P-cadherin* (in placental tissue). Each type of cadherin occurs in a different spectrum of cell types, and the timing of their production during embryonic development correlates with the association of specific cell types to form tissues and organs.

The role played by cadherins in cell-cell adhesion has been investigated in cultured fibroblasts called *L cells,* which bind poorly to one another and contain little cadherin. When purified DNA encoding E-cadherin or P-cadherin are introduced into L cells, the cells begin to produce cadherins and to bind more tightly to one another. Moreover, L cells that produce E-cadherin bind preferentially to other cells producing E-cadherin. Similarly, cells that produce P-cadherin bind selectively to other cells that are also producing P-cadherin (Figure 11-15). Such observations suggest that, as with CAMs, cadherin molecules on one cell bind to cadherin molecules of the same type on another cell, helping to segregate cells into specific tissues.

Cadherins have especially important roles during embryonic development. For example, mouse embryos

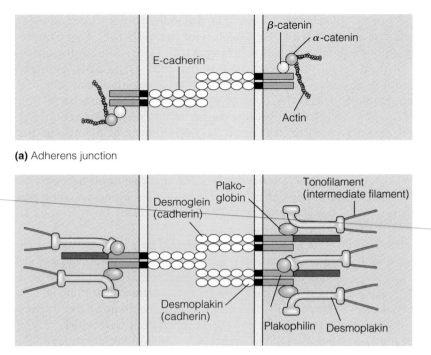

(a) Adherens junction

(b) Desmosome

Figure 11-14 Cadherin Structure. Cadherins are found at sites of cell-cell adhesion. **(a)** "Classical" cadherins, such as E-cadherin, mediate cell-cell adhesion at adherens junctions. E-cadherin molecules associate as pairs (homodimers) in the plasma membrane and interact with homodimers in the neighboring cell through their extracellular domains. Their cytosolic tail binds to the linker protein, β-catenin. β-catenin in turn binds α-catenin, which attaches to actin microfilaments. **(b)** The cadherins in desmosomes, called desmocollins and desmogleins, probably interact in pairs; neighboring cells probably attach to one another via their desmogleins. In the cytosol, desmosomal cadherins interact with proteins in the desmosomal plaque (plakoglobin, plakophilin, and desmoplakin), that anchor desmosomes to intermediate filaments.

lacking N-cadherin have severe heart defects. During the early development of vertebrates, the cells that result from early cell divisions must adhere to one another as they organize into tissues. When early mammalian embryos are treated with antibodies that interfere with E-cadherin, or when mutant embryos cannot produce E-cadherin, their cells lose their tight adhesion and the embryos fail to develop. Frog embryos have a similar requirement for cadherin; when they are depleted of mRNA for the main type of cadherin found in the early embryo, they lose their normal organization (Figure 11-16). Changes in cadherin expression also occur in cancer cells. Between 80 and 90% of cancers are derived from epithelial cells, and such cancers exhibit changes in cadherin expression. When well-defined tumors begin to metastasize and spread throughout the body, the metastatic cells lose cadherin from their surfaces; loss of adhesion is thought to contribute to the ability of such cells to detach and migrate throughout the body.

Carbohydrate Groups Are Important in Cell-Cell Recognition and Adhesion

The carbohydrate side chains of CAMs and cadherins affect both the strength and the specificity of cell-cell interactions. N-CAM molecules, for example, contain long repeating chains of *sialic acid*, a negatively charged carbohydrate (see Figure 7-26a). The amount of sialic acid bound to N-CAM changes significantly during development, suggesting a possible role in regulating cellular adhesion.

Lectins. A role for carbohydrate groups in cell adhesion is also suggested by the fact that many animal and plant cells secrete carbohydrate-binding proteins called **lectins,** which promote cell-cell adhesion by binding to a specific sugar or sequence of sugars exposed at the outer cell surface. Because a lectin molecule usually has more than one carbohydrate-binding site, it can bind to carbohydrate groups on two different cells, thereby linking the cells together.

Carbohydrates and the Survival of Erythrocytes. Although the phenomena of cell-cell recognition and adhesion are often discussed together, recognition events are also important in contexts that do not involve cell-cell adhesion. One especially well-known example is the determination of the human blood types A, B, AB, and O by a specific carbohydrate side chain present on a glycolipid of the erythrocyte plasma membrane.

The ABO blood group, as it is called, involves differences in carbohydrate side chains on the surface of the red blood cell that can be detected by antibodies present in the blood, leading to clumping of the red blood cells and likely to the death of the patient if the wrong type is used in transfusion.

The distinctions between the four blood types in the ABO system depend on modest genetically determined differences in the structure of a branched-chain carbohydrate attached to a specific glycolipid in the erythrocyte plasma membrane. Individuals with blood type A have the amino sugar *N*-acetylgalactosamine (GalNAc) at the ends of this carbohydrate, whereas individuals with blood type B have galactose instead. Individuals with blood type AB have both *N*-acetylgalactosamine and galactose present, and in individuals with type O blood, these terminal sugars are missing entirely.

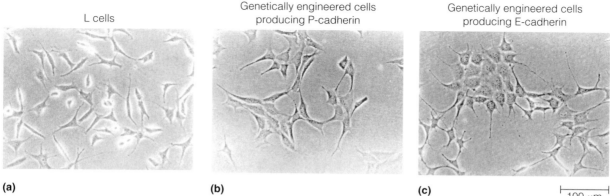

L cells

Genetically engineered cells producing P-cadherin

Genetically engineered cells producing E-cadherin

(a)

(b)

(c)

100 μm

(d)

Stained for P-cadherin

200 μm

(e)

Stained for E-cadherin

200 μm

Figure 11-15 The Effect of Cadherins on Cell Adhesions. **(a)** A light micrograph of cultured L cells shows that these cells do not normally adhere to one another. **(b–c)** When DNA coding for E-cadherin or P-cadherin are introduced into these cells, cadherin is produced and cell-cell adhesion occurs.

(d–e) Fluorescence micrographs of cell aggregates that form when cells producing P-cadherin are mixed with cells producing E-cadherin. The same field is stained with fluorescent antibodies specific for P-cadherin (red) or E-cadherin (green). The staining pattern shows that cells producing P-cadherin

are located in different areas from the cells producing E-cadherin. This indicates that cells making E-cadherin bind preferentially to other cells producing E-cadherin, and similarly, cells making P-cadherin bind preferentially to other cells producing P-cadherin.

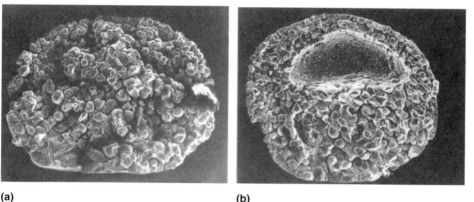

(a)

(b)

Figure 11-16 Cadherins During Embryonic Development. After many rounds of cell division, frog embryos form a blastula, which contains a fluid-filled space lined by cells. When embryos are depleted of mRNA for a cadherin known as EP-cadherin, they fail to make EP-cadherin protein, and they lose their normal organization **(a)** compared with control embryos **(b)**.

These minor differences have major effects on the compatibility of blood transfusions, because people with blood types A, B, or O have antibodies in their bloodstream that recognize and bind to the respective terminal sugars of this specific glycolipid. An individual whose blood contains antibodies against one or both sugars

(GalNAc or Gal) cannot accept blood containing the glycolipid with that terminal sugar because the antibodies present in the blood would bind to the carbohydrate groups on the erythrocyte surface and cause the erythrocytes to coagulate, or clump. Individuals with type A blood have antibodies against carbohydrate chains ending in galactose, which occur in type B and type AB blood. They therefore cannot be transfused with type B or AB blood but can accept blood from B or O donors. Conversely, individuals with type B blood have antibodies against carbohydrate chains ending in GalNAc, which occur in blood of types A and AB. They therefore cannot be transfused with type A or AB blood but can accept blood from A or O donors. Individuals with type O blood are called universal donors, because their erythrocytes do not generate an immune response when transfused into individuals of any blood type.

Carbohydrate chains are also thought to be involved in the means by which aging erythrocytes are recognized and targeted for destruction. (Erythrocytes have an average life span of 3–4 months, after which they are destroyed in the spleen or liver.) In this case, however, the important feature is not simply the presence of carbohydrate chains, but the progressive loss of sialic acid groups from them. The relevant carbohydrate chains are attached to *glycophorin*, an integral membrane protein present in the erythrocyte plasma membrane (see Figure 7-21a). Like N-CAM, glycophorin has a high sialic acid content. In the case of glycophorin, however, the sialic acid groups are located at the ends of oligosaccharide chains, where they are susceptible to removal by the enzyme *neuraminidase*. Loss of these sialic acid groups appears to target an erythrocyte for destruction, probably by exposing the underlying galactose residues in the carbohydrate chains.

Selectins and Leukocyte Adhesion. Carbohydrate recognition also plays an important role during the interactions of leukocytes with endothelial cells lining blood vessels or with platelets. Cell surface glycoproteins called **selectins** mediate these interactions; a different selectin is expressed by each cell type (*L-selectin* on leukocytes, *E-selectin* on the endothelial cells of blood vessels, and *P-selectin* on platelets and endothelial cells). Leukocytes roll along the walls of blood vessels. During inflammation, they must attach to the wall of a blood vessel in the vicinity of the inflammation and then migrate through the blood vessel to the inflammation site. The initial attachment of leukocytes is mediated by binding of selectins on the leukocyte to carbohydrates on the surface of the endothelial cells and by selectins on the endothelial cells binding to carbohydrates on the surface of the leukocyte. This is followed by more stable adhesions, which are mediated by a specific integrin on the surface of the leukocyte and immunoglobulin superfamily proteins called *ICAMs* on the surface of the endothelial cells (Figure 11-17). This sort of interaction is a good example of a heterophilic adhesive interaction.

Cell Junctions

By definition, unicellular organisms have no permanent associations between cells; each cell is an entity unto itself. Multicellular organisms, on the other hand, have specific

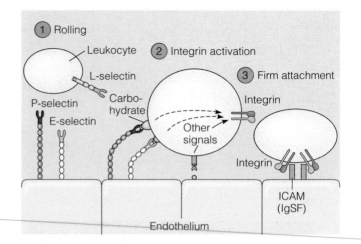

Figure 11-17 Leukocyte Adhesion and Selectins. The initial attachment of leukocytes to endothelial cells that line blood vessels is mediated by selectins. ① Attached leukocytes roll along the blood vessel wall, where they sense activating factors that are deposited on the surface of endothelial cells. ② This leads to activation of leukocyte integrins, which ③ bind to IgSF proteins (such as ICAMs), allowing leukocytes to adhere firmly to endothelial cells. Such adhesion allows leukocytes to stop and to pass through blood vessels to sites of inflammation.

means of joining cells in long-term associations to form tissues and organs. Such associations usually involve specialized modifications of the plasma membrane at the point where two cells come together. These specialized structures are called **cell junctions.** In animals, the three most common kinds of cell junctions are *adhesive junctions, tight junctions* and *gap junctions*. Figure 11-18 illustrates each of these kinds of junctions, which we are about to consider.

In plants, the presence of a cell wall between the plasma membranes of adjacent cells precludes the kinds of cell junctions that link animal cells. However, the cell wall and special structures called *plasmodesmata* carry out similar functions, as we will see a bit later in the chapter.

Adhesive Junctions Link Adjoining Cells to Each Other

One of the three main kinds of junctions in animal cells is the **adhesive** (or **anchoring**) **junction** (Table 11-3). Adhesive junctions link cells together into tissues, thereby enabling the cells to function as a unit. Like focal adhesions and hemidesmosomes, all junctions in this category anchor the cytoskeleton of one cell either to the cytoskeletons of neighboring cells or to the extracellular matrix that surrounds the cell. The resulting interconnected cytoskeletal network helps to maintain tissue integrity and to withstand mechanical stress. Adhesive junctions occur widely in animal tissue but are especially prominent in tissues such as heart muscle and skin epithelium that are subject to mechanical stress and stretching.

The two main kinds of cell-cell adhesive junctions are *adherens junctions* and *desmosomes* (Figure 11-18). Despite structural and functional differences among them, adhe-

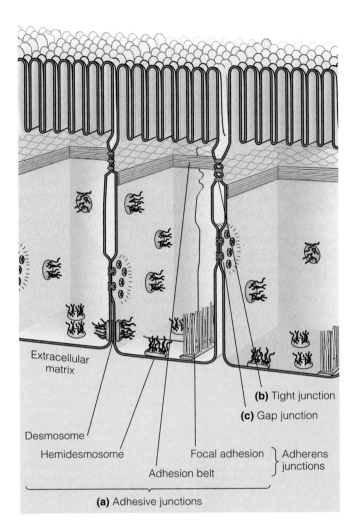

Extracellular
matrix

(b) Tight junction

(c) Gap junction

Desmosome

Hemidesmosome

Focal adhesion ⎤
Adhesion belt   ⎦ Adherens junctions

(a) Adhesive junctions

Figure 11-18   Major Types of Cell Junctions in Animal Cells.
(a) Adhesive junctions are specialized for cell-cell and cell-ECM adhesion. Desmosomes and hemidesmosomes bind to intermediate filaments within the cell, whereas adhesion belts and focal adhesions, both adherens junctions, bind to actin microfilaments. **(b)** Tight junctions create an impermeable seal between cells, thereby preventing fluids, molecules, or ions from crossing a cell layer via the intercellular space. **(c)** Gap junctions provide direct chemical and electrical communication between cells by allowing the passage of small molecules and ions from one cell to another.

or to the extracellular matrix. In the case of desmosomes, intracellular attachment proteins form a fibrous *plaque* on the cytoplasmic side of the plasma membrane. As we have already seen, the cytoskeletal elements to which these linker proteins attach are different for the two types of junction: microfilaments in the case of adherens junctions or intermediate filaments in the case of desmosomes.

Adherens Junctions. Cadherin-mediated adhesive junctions that are connected to the cytoskeleton by actin microfilaments are called **adherens junctions** (see Table 11-3). At adherens junctions the space between the adjacent membranes is about 20–25 nm. Adherens junctions are especially prominent in heart muscle and in the thin layers of tissue that line body cavities and cover body organs. In epithelial cells, adherens junctions typically form a continuous *adhesion belt*, an extensive zone that completely encompasses entire cells in a sheet of tissue (see Figure 11-18a). In nonepithelial cells, such as neurons, adherens-like junctions form smaller points of attachment. At neuronal synapses, for example, regions where neurons are in close contact contain many of the same proteins found in the adherens junctions of epithelial cells.

As we saw with focal adhesions, adherens junctions are points of attachment between the cell surface and the cytoskeleton. And just as integrins at focal adhesions are connected to the actin cytoskeleton by linker proteins, so

sive junctions contain two distinct kinds of proteins: *intracellular attachment proteins,* which link the junction to the appropriate cytoskeletal filaments on the inside of the plasma membrane, and *cadherins,* which protrude on the outer surface of the membrane and bind cells to each other

Table 11-3   Junctions Between Animal Cells

Type of Junction	Function	Intermembrane Features	Space	Associated Structures
Adhesive junctions				
Focal adhesion	Cell-ECM adhesion	Localized points of attachment	20–25 nm	Actin microfilaments
Hemidesmosome	Cell-basal lamina adhesion	Localized points of attachment	25–35 nm	Intermediate filaments (tonofilaments)
Adherens junctions	Cell-cell adhesion	Continuous zones of attachment	20–25 nm	Actin microfilaments
Desmosome	Cell-cell adhesion	Localized points of attachment	25–35 nm	Intermediate filaments (tonofilaments)
Tight junction	Sealing spaces between cells	Membranes joined along ridges	None	Transmembrane junctional proteins
Gap junction	Exchange of ions and molecules between cells	Connexons (transmembrane proteins with 3nm pores)	2–3 nm	Connexons in one membrane align with those in another to form channels between cells

Clinical Applications

ANTHRAX, FOOD POISONING, AND OTHER "BAD BUGS": THE CELL SURFACE CONNECTION

We have seen that cell adhesion and cell recognition play important roles during the construction of normal tissues in the body, and in the normal functioning of cells in the adult human body, such as blood cells. Surprisingly, foreign invaders that attack the human body can use the very same proteins that healthy cells require for cell adhesion to gain entry into the body. Here we consider two examples of how bacteria attach to and infect human cells: enteropathogenic bacteria and anthrax.

Enteropathogenic Bacteria Use Normal Cell Adhesion Proteins to Infect Host Cells

One good example of this sort of "molecular hijacking" of normal cell adhesion processes occurs when pathogenic bacteria enter the digestive tract (such bacteria are called *enteropathogenic* bacteria; *entero-* comes from the Greek word for "intestine"). Such bacteria are responsible for several types of food poisoning, and their combined effects have a substantial impact on public health. Although some pathogenic bacteria can use multiple methods for gaining entry into the gut, in several well-studied cases they attach to cell adhesion molecules such as integrins or cadherins.

One of the best-studied examples of such subversion of cell adhesion is the enteric pathogen *Yersinia pseudotuberculosis*. Infection by bacteria of the genus *Yersinia* typically results in gastroenteritis with diarrhea and vomiting 24–48 hours after exposure, which usually occurs via contaminated water and food. *Y. pseudotuberculosis* uses a 986 amino acid outer membrane protein, called *invasin*, to penetrate mammalian cells. Surprisingly, the cellular receptors for invasin are integrins that contain β_1 subunits on the surface of cells lining the gut (Figure 11A-1). The identification of the cellular receptor for *Yersinia* was an important discovery, because it showed that bacteria could invade cells by targeting common mammalian cell surface proteins.

As the molecular pathways used by other bacteria to invade host cells have been identified, such co-opting of normal cell surface proteins has emerged as a common theme in pathogenesis. A second enteropathogenic bacterium, *Shigella flexnerii*, also attaches to an integrin ($\alpha_5\beta_1$) via proteins known as IpaA, IpaB, and IpaC on its surface. *Shigella* infection results in dysentery and usually results from contamination of raw foods by food handlers. Approximately 300,000 cases of shigellosis occur annually in the United States, making it a significant health problem.

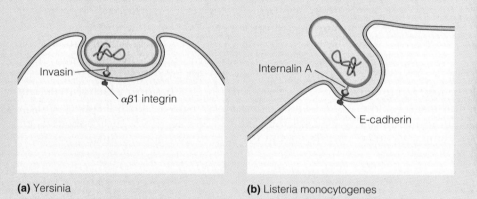

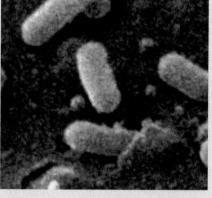

(a) Yersinia   (b) Listeria monocytogenes   (c) Listeria   1 μm

Figure 11A-1 Bacterial Pathogens and Cell Adhesion Proteins. Bacteria that invade the lining of the human digestive tract attach to cell adhesion receptors on the surface of intestinal cells. **(a)** Species in the genus *Yersinia* express a protein called invasin that attaches to integrins on gut cells that have β_1 subunits. **(b)** *Listeria monocytogenes* expresses a protein called internalin A that binds to E-cadherin on intestinal cells. **(c)** *Listeria* on the surface of cultured epithelial cells (SEM).

are the cadherens at adherens junctions (Figure 11-14a). In this case, a protein known as *β-catenin* (or in some cases a related protein known as *plakoglobin*) binds to the cytosolic tail of the cadherin; *β-catenin* in turn is bound by a second protein called *α-catenin*, which can bind to actin to complete the linkage. In some cases such linkages

are very strong. In heart muscle cells, for example, actin microfilaments at adherens junctions are continuous with the microfilaments responsible for muscle contraction.

We have seen that cell-cell and cell-extracellular matrix adhesion are important for the normal functions of cells and tissues in the body. A surprising finding of

Another well-studied example of subversion of cell adhesion is *Listeria monocytogenes*. *Listeria* infection can occur through exposure to improperly prepared raw foods, such as raw milk and hamburger. Although the incidence of listeriosis in the United States is not high (about 2000 people annually develop symptoms), once infection occurs, it is very serious. Approximately 25% of those infected die; many of these contract bacterial meningitis. *Listeria* expresses a protein on its surface called *internalin A* that can bind to E-cadherin on the surface of cells in the gut (Figure 11A-1). Binding of pathogenic bacteria to cells in the intestine results in dramatic changes in the cytoskeleton of the infected cells. We now know that just as these bacteria "hijack" the normal cell adhesion machinery of gut cells, they can do the same with the cytoskeleton once they are inside an infected cell. In Chapter 23, we will examine this process in more detail in the case of *Listeria*.

Anthrax Toxins Act via a Multistep Process

Enteropathogenic bacteria can use well-studied cell adhesion molecules to infect human cells. For other bacteria, the host cell surface molecules involved are less well characterized. One highly publicized example is the anthrax bacterium, *Bacillus anthracis*. Although anthrax has received much publicity in connection with its use in bioterrorism, the disease has been known since at the least the time of ancient Egypt. Historically, humans have typically acquired anthrax from direct contact with infected livestock, or in occupations involving processing of animal products.

Humans can contract anthrax through one of several routes. Anthrax can be contracted by contact of the skin with the bacteria (the mild form of the disease), by ingesting the bacteria, or by inhalation. Unless treated immediately after infection with antibiotics, inhalation anthrax is difficult to treat and typically results in the death of infected patients.

B. anthracis targets macrophages. Ironically, macrophages are the cells of the immune system that normally ingest bacteria and other foreign invaders. Although the details are still being worked out, anthrax kills macrophages in a multistep process (Figure 11A-2). Like many bacteria, *B. anthracis* kills cells by producing toxins. *B. anthracis* produces three proteins that result in the death of affected macrophages. One of these proteins, known as *protective antigen (PA)*, binds to the surface of macrophages. The receptor to which PA binds is a novel 368 amino acid protein that has only recently been identified. After binding to the receptor, PA is cleaved by a cell-surface protease called furin, resulting in an active, 63 kDa protein. Once acti-

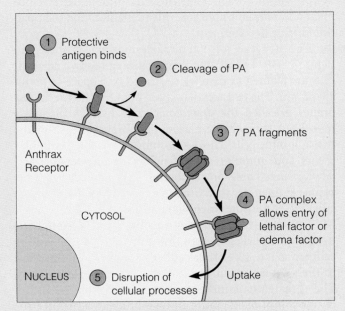

Figure 11A-2 Mechanisms of Anthrax Toxicity. *B. anthracis* produces three proteins that are primarily responsible for its toxic effects on macrophages. ① Protective antigen (PA) binds to receptors on the surface of the macrophage, which ② stimulates the cleavage of PA into an active fragment by a cell surface protease. ③ Seven activated PA fragments form a complex, which ④ allows lethal factor (LF) or edema factor (EF) to enter the cell. LF and EF disrupt cellular processes that are probably related to intracellular signaling, resulting in the death of the affected cell ⑤.

vated, seven activated PA fragments associate and insert into the plasma membrane of the macrophage. The resulting protein complex allows two other proteins, *edema factor (EF)* and *lethal factor (LF)*, to enter the cytosol. EF is an adenylyl cyclase (i.e., an enzyme that catalyzes the formation of cyclic AMP; see Chapter 10) that impairs host defenses through a variety of mechanisms. In particular, EF appears to inhibit phagocytosis of bacteria by macrophages, leading to further spread of the infection. LF is a protease that appears to act by cleaving kinases (called *MAPK kinases*, or MAPKKs). Recall from Chapter 10 that MAPK is a protein that is activated as a result of signal transduction events within cells. LF may help to kill macrophages by disrupting their own normal intracellular signaling, once again illustrating the opportunistic means by which infectious agents attack cells in the human body.

modern microbiology is that many pathogens, such as those responsible for several types of food poisoning, infect the body by using these very same adhesion systems to gain entry into healthy cells. In other cases, such as the bacterium that causes anthrax, toxins produced by the pathogen attach to cell surface proteins to initiate their

toxic effects. These mechanisms of infection are discussed in more detail in Box 11A.

Desmosomes. **Desmosomes** are buttonlike points of strong adhesion between adjacent cells in a body tissue. This cell-cell adhesion gives the tissue structural integrity, enabling the cells to function as a unit and to resist stress. Desmosomes are found in many tissues but are especially abundant in skin, heart muscle, and the neck of the uterus. Desmosomes form early in embryonic development and play an important role in maintaining cell position during development.

The structure of a typical desmosome is shown in Figure 11-19. The plasma membranes of the two adjacent cells are aligned in parallel, separated by a space of about 25–35 nm. The extracellular space between the two membranes is called the *desmosome core*. A thick plaque is found just beneath the plasma membrane of each of the two adjoining cells. The desmosome core is filled with filaments and granules of proteins called *desmocollins* and *desmogleins*. These are cadherins that interact with the plaque at the inner membrane surface and mediate cell-cell adhesion at the outer membrane surface.

Unlike E-cadherin, desmocollins and desmogleins probably interact heterophilically across the intercellular space. Like other cadherins, linker proteins bind to their cytosolic tail and link them to the cytoskeleton. The β-catenin family protein *plakoglobin* binds to desmocollin; plakoglobin in turn binds to a plakin protein called *desmoplakin* (Figure 11-14b). Desmoplakin in turn attaches to **tonofilaments**, which, as we saw in the case of hemidesmosomes, are composed of intermediate filaments such as *vimentin, desmin,* or *keratin*.

Tonofilaments extend inward from the plaque, anchoring the desmosome in the cytoplasm. The main protein present in the tonofilaments is the intermediate filament keratin, desmin, or vimentin, depending on the cell type. Each desmosome has anchoring sites for intermediate filaments in the cells on both sides of the junction, thereby linking the intermediate filaments of adjacent cells and forming what is, in effect, a continuous cytoskeletal network throughout the tissue.

Human patients that develop autoimmune reactions against components of their desmosomes develop blistering diseases of the skin known as *pemphigus*. Some patients develop antibodies against desmogleins, while others generate antibodies against linker proteins, such as desmoplakin.

Tight Junctions Prevent the Movement of Molecules Across Cell Layers

As the name implies, a **tight junction** leaves no space at all between the plasma membranes of adjacent cells (Figure 11-20). Tight junctions serve as seals, preventing the flow of fluids—and hence the movement of molecules and ions—between the cells of a cell layer that separates two body compartments. The properties of tight junctions have been studied by incubating tissues in the presence of electron-opaque tracer molecules and then using electron microscopy to observe the movement of the tracer through the extracellular space. As Figure 11-21 illustrates, tracer molecules diffuse into the narrow spaces between adjacent cells until they encounter a tight junction, which blocks further movement. In other words,

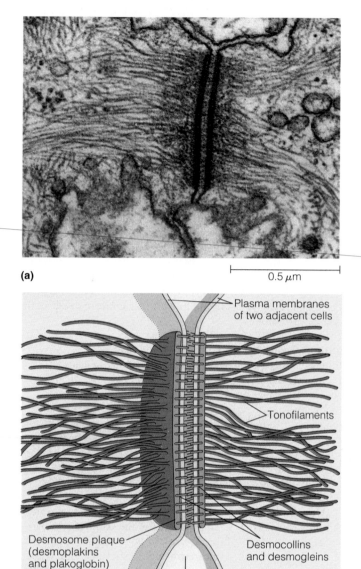

(a)

|———————| 0.5 μm

(b)

Cell 1 — Intercellular space — Cell 2

Labels: Plasma membranes of two adjacent cells; Tonofilaments; Desmosome plaque (desmoplakins and plakoglobin); Desmocollins and desmogleins

Figure 11-19 Desmosome Structure. (a) An electron micrograph of a desmosome joining two cells in the skin of a newt (TEM). (b) A schematic diagram of a desmosome. The distance between cells in the desmosome region is 25–35 nm, about that for a nonjunction region. The desmosome core between the two membranes is filled with cadherins (desmocollins and desmogleins). The plaque on the cytoplasmic side of the membrane contains desmoplakins and plakoglobin and is linked to tonofilaments, which are intermediate filaments that consist of keratin, desmin, or vimentin, depending on the cell type.

tight junctions ensure that substances in the extracellular fluid do not pass from one side of a cell layer to another.

The tight junctions between adjoining cells form a continuous belt around the lining of an organ or body cavity such that the spaces between cells are tightly sealed (Figure 11-20a). As a result, the space on one side of the junction plane is effectively separated from the space on the other side, preventing flow of fluids from one side of the cell layer to the other. The only way for molecules to

Figure 11-20 **Tight Junction Structure.** **(a)** A schematic representation of several adjoining epithelial cells fused to each other by tight junctions. **(b)** Transmembrane junctional proteins in the plasma membranes of two adjacent cells are clustered along the points of contact, forming ridges of protein particles that join the two plasma membranes together tightly. Tight junctions prevent the passage of extracellular molecules through the spaces between cells (red arrows) and also block lateral movement of transmembrane proteins, as shown in Figure 11-22. **(c)** This electron micrograph illustrates tight junctions between cells in a frog bladder, as revealed by the freeze-fracture technique. Tight junctions appear as raised ridges on the protoplasmic (P) face of the membrane. The lumen is the cavity of the bladder (TEM).

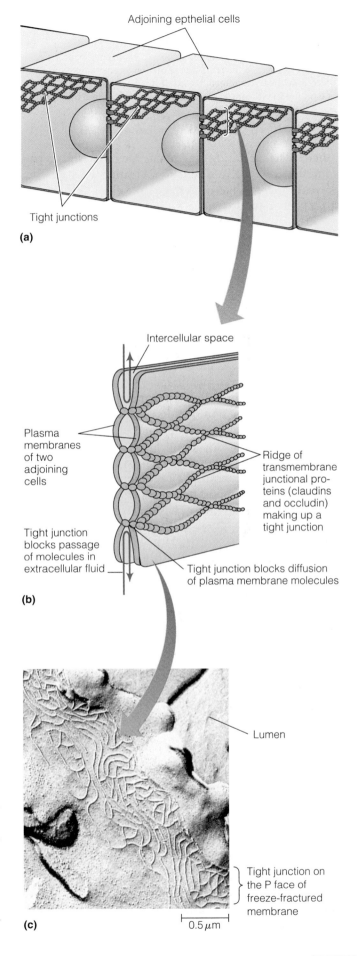

Adjoining epthelial cells

Tight junctions

(a)

Intercellular space

Plasma membranes of two adjoining cells

Tight junction blocks passage of molecules in extracellular fluid

Ridge of transmembrane junctional proteins (claudins and occludin) making up a tight junction

Tight junction blocks diffusion of plasma membrane molecules

(b)

Lumen

Tight junction on the P face of freeze-fractured membrane

0.5 μm

(c)

cross the cell layer is by passing through the cells themselves, a process that is regulated by plasma membrane transport proteins.

Tight junctions are especially prominent in intestinal epithelial cells, which must form an effective barrier so that liquid from the intestine cannot cross the epithelial layer. Tight junctions are also abundant in the ducts and cavities of glands, such as the liver and pancreas, that connect with the digestive tract, as well as in the urinary bladder, where they ensure that the urine stored in the bladder does not seep out between cells.

Structure of Tight Junctions. Although tight junctions seal the membranes of adjacent cells together very effectively, the membranes are not actually in close contact over broad areas. Rather, they are connected along sharply defined ridges (Figure 11-20b). Tight junctions can be seen especially well by freeze-fracture microscopy, which reveals the inner faces of membranes. Each junction appears as a series of ridges that form an interconnected network extending across the junction (Figure 11-20c). Each such ridge consists of a continuous row of tightly packed transmembrane junctional proteins about 3–4 nm in diameter. Recently, two different types of transmembrane proteins, *claudins* and *occludin*, have been identified as the major protein components of tight junctions. Each has four membrane-spanning domains, and the occludin and claudins in the plasma membrane of adjacent cells are thought to interlock to form a tight seal. **Claudins** seem to be the more important of the two proteins for the formation of a tight junction. When cells that do not make tight junctions are forced to express claudins, they form junctions that look very similar to tight junctions. Claudins are a large family of proteins; different claudins are expressed in different epithelial tissues, and it is thought that different claudins may influence the tightness or leakiness of tight junctions in particular tissues.

Claudins and occludin in the plasma membranes of adjoining cells make contact across the intercellular space and bind tightly to each other, thereby connecting the cells together. The result is rather like placing two pieces of corrugated metal together so that their ridges are aligned and then fusing the two pieces lengthwise along each ridge of

Figure 11-21 Experimental Evidence Demonstrating That Tight Junctions Create a Permeability Barrier. **(a)** When an electron-opaque tracer is added to the extracellular space on one side of an epithelial cell layer, tracer molecules penetrate into the space between adjacent cells only to the point where they encounter a tight junction. **(b)** Because the tracer molecules are electron-opaque, their penetration into the intercellular space can be visualized by electron microscopy (TEM).

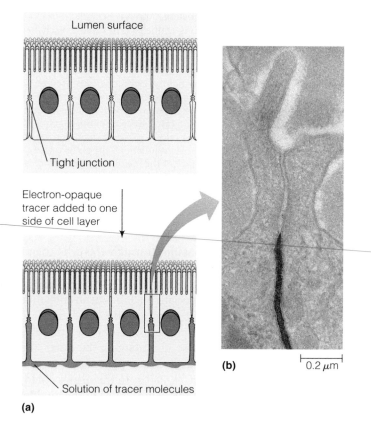

(b)   0.2 μm

contact. The fused ridges eliminate the intercellular space and effectively seal the junction, creating a barrier that prevents the passage of extracellular fluid through the spaces between adjacent cells. Not surprisingly, the number of such ridges across a junction correlates well with the tightness of the seal that the junction makes.

Role of Tight Junctions in Blocking Lateral Movement of Membrane Proteins.   In addition to preventing the movement of fluids, ions, and molecules between cells, tight junctions also block the lateral movement of lipids and proteins within the membrane. Lipid movement is blocked in the outer monolayer only, but the movement of integral membrane proteins is blocked entirely. As a result, different kinds of integral membrane proteins can be maintained in the portions of a plasma membrane on opposite sides of a tight junction belt. In this way, tight junctions help to keep the plasma membrane of an epithelial cell *polarized*—that is, organized into discrete functional domains at opposite ends of the cell.

To understand why the localization of integral membrane proteins is important, consider the epithelial cells that line the small intestine. These cells carry out the *trans-*

cellular transport of glucose and other nutrients from the intestinal lumen on one side of the cell into the bloodstream on the other side (Figure 11-22). The concentration of glucose is almost always lower in the intestinal lumen than in the blood, so energy is required to drive the trans-

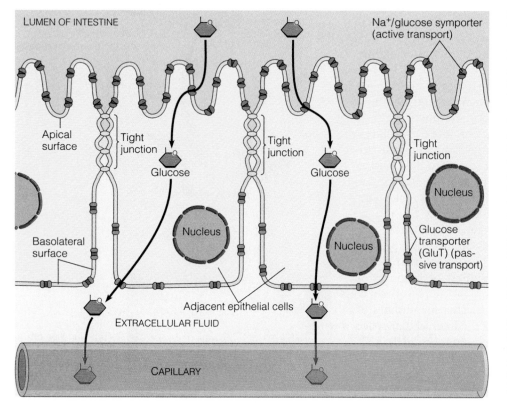

Figure 11-22 The Transcellular Transport of Glucose Across the Intestinal Epithelium.   Illustrated here is the transcellular transport of glucose from the lumen of the intestine (low glucose content) into the cytoplasm of epithelial cells (high glucose content) and then via the extracellular fluid into a capillary of the bloodstream. Glucose is actively transported into the epithelial cell by Na+/glucose symporters that are localized to the apical surface of the cell. Movement of glucose out of the cell into the extracellular fluid is facilitated by diffusion, mediated by glucose transporter (GluT) proteins that are localized to the basolateral surface of the cell. This transcellular transport of glucose is possible because the tight junctions between adjacent cells not only prevent seepage of the extracellular fluid into the lumen of the intestine but also restrict Na+/glucose symporters to the apical surface and glucose transporters (GluT) to the basolateral surface.

cellular movement of glucose across intestinal epithelial cells. In these cells, the Na+/glucose symport proteins that are responsible for the indirect active uptake of glucose from the intestinal tract are restricted to the *apical surface* (the cell surface that faces the lumen of the intestine), whereas the glucose transport (GluT) proteins that facilitate passive movement of glucose from the cell into the extracellular fluid are present only on the *basolateral surface* (the cell surface that faces the circulatory system). This polarity, together with the inherent directionality of Na+-driven glucose uptake from the intestine, ensures that glucose is moved unidirectionally across the epithelial cells from the intestinal tract into the bloodstream.

Thus, tight junctions not only seal the spaces between adjacent epithelial cells so that transported molecules such as glucose cannot diffuse back into the lumen; they also prevent transport proteins on the apical surface from moving laterally in the membrane to the basolateral surface, and vice versa.

Gap Junctions Allow Direct Electrical and Chemical Communication Between Cells

A **gap junction** is a region at which the plasma membranes of two cells are aligned and brought into intimate contact, with a gap of only 2–3 nm in between. The gap junction provides a point of cytoplasmic contact between two adjacent cells through which ions and small molecules can pass. Adjacent cells are thus in direct electrical and chemical communication with each other.

The structure of gap junctions is illustrated in Figure 11-23. At a gap junction, the two plasma membranes from adjacent cells are joined by tightly packed, hollow cylinders called **connexons.** A single gap junction

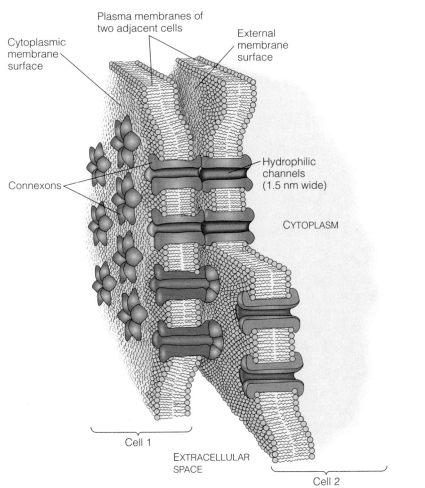

(a) Gap junction diagram

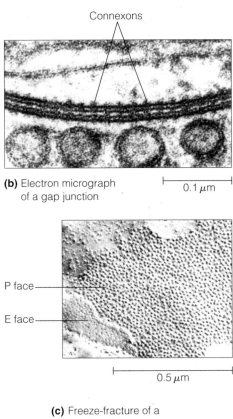

(b) Electron micrograph of a gap junction   0.1 μm

(c) Freeze-fracture of a gap junction   0.5 μm

Figure 11-23   Gap Junction Structure.
(a) A schematic representation of a gap junction. A gap junction consists of a large number of hydrophilic channels formed by the alignment of connexons in the plasma membranes of two adjoining cells. **(b)** An electron micrograph of a gap junction between two adjacent nerve cells. The connexons that extend through the membranes are visible here as beadlike projections spaced about 17 nm apart on either side of the membrane-membrane junction (TEM). **(c)** A gap junction as revealed by the freeze-fracture technique. The junction appears as an aggregation of intramembranous particles on the protoplasmic (P) face and as a series of pits on the exterior (E) face of the membrane (TEM).

may consist of just a few or as many as thousands of clustered connexons. In vertebrates, each connexon is a circular assembly of six subunits of the protein *connexin*. There are many different connexins (more than a dozen) that are found in different tissues, but each functions similarly in forming connexons. The assembly spans the membrane and protrudes into the space, or gap, between the two cells (Figure 11-23a). Each connexon has a diameter of about 7 nm and a hollow center that forms a very thin hydrophilic channel through the membrane. The channel is about 3 nm in diameter at its narrowest point, just large enough to allow the passage of ions and small molecules. Invertebrates do not have connexins. Instead, they produce proteins called *innexins* that appear to serve the same function in gap junctions.

When connexons in the plasma membranes of two adjacent cells are aligned, the cylinders in the two membranes meet end to end, forming direct channels of communication between the two cells that can be seen with an electron microscope (Figure 11-23b and c). The idea that such channels permit the direct movement of molecules and ions between adjoining cells owes much to the pioneering work of Werner Loewenstein and his colleagues, who monitored the flow of electric current in insect salivary glands. When they applied a small voltage to microelectrodes placed in neighboring cells, they found that current flowed readily from cell to cell, but not from cells to the extracellular fluid. Since current in living cells is carried by small ions, Loewenstein and his co-workers concluded that cells contain channels that permit ions to pass directly from the cytoplasm of one cell to the cytoplasm of an adjoining cell without passing through the extracellular space.

Similarly, Loewenstein and his co-workers injected fluorescent molecules that passed from cell to cell much more rapidly than would be expected if the molecules had to move into and out of the cells across their respective plasma membranes. By using fluorescent molecules of different sizes, Loewenstein and his collaborators showed that gap junctions allow the passage of solutes with molecular weights up to about 1200. Included in this range are such common substances as monosaccharides, amino acids, and nucleotides—most of the molecules involved in cellular metabolism, in other words. Problem 11-9 at the end of the chapter gives you an opportunity to consider this experimental approach in more detail.

Loewenstein and his colleagues went on to show that gap junction permeability can be changed rapidly by experimental manipulation of the Ca^{2+} concentration of the coupled cells. Gap junctions tend to be open at low intracellular Ca^{2+} concentrations and to close as the Ca^{2+} concentration is raised. Subsequent studies showed that the permeability of gap junctions can also be altered by changes in intracellular pH or membrane potential.

Gap junctions occur in most vertebrate and invertebrate cell types. They are especially abundant in tissues such as muscle and nerve, where extremely rapid communication between cells is required (recall that gap junc-

tions are important in electrical synapses; see Chapter 9). In heart tissue, gap junctions facilitate the flow of electrical current that causes the heart to beat; in the brain, they are concentrated in the cerebellum, which is involved in coordinating rapid muscular activities.

These roles are confirmed by analyzing mutations in connexins and innexins. Mutation of one type of connexin results in deafness in humans, probably because support cells in the cochlea do not function properly. Mice lacking another connexin have defects in conducting electrical impulses in the heart.

The Plant Cell Surface

Our discussion so far has focused on animal cell surfaces. The surfaces of plant, algal, fungal, and bacterial cells exhibit some of the same properties, but they also have several unique features of their own. In the remainder of this chapter we consider some of the distinctive features of the plant cell surface.

Cell Walls Provide a Structural Framework and Serve as a Permeability Barrier

One of the most remarkable features of plants is that they have no bones or related skeletal structures and yet exhibit remarkable strength. This strength is provided by the rigid **cell walls** that surround all plant cells except sperm and some eggs. The presence of the cell wall is one of the major factors that causes the properties of plants to differ so much from those of animals. Specifically, the rigidity of the wall makes cell movements virtually impossible, which presumably explains why plants have not developed the kind of neuromuscular system that defines many of the unique properties of animals. At the same time, the sturdy cell walls enable plant cells to withstand the considerable turgor pressure that is exerted by the uptake of water. Turgor pressure is vital to plants because it accounts for much of the turgidity, or firmness, of plant tissues and provides the driving force behind cell expansion.

Although the plant cell wall was once viewed as an inert structure that simply encloses and contains the cell it surrounds, we now recognize the wall as a dynamic structure that serves a variety of functions. For example, enzymes associated with the cell wall degrade extracellular nutrients, thereby generating smaller compounds that can pass through the plasma membrane and into the cell. The cell wall also plays a role in certain types of secretory and metabolic events. In addition, it protects individual cells from osmotic rupture, mechanical injury, and invasion by microorganisms, especially fungi and bacteria.

The wall that surrounds a plant cell is a permeability barrier for large molecules. For water, gases, ions, and small water-soluble molecules such as sugars or amino acids, the cell wall is not a significant obstacle; these substances diffuse through the wall readily. In fact, the cell wall typically allows the passage of globular molecules with molecular weights up to about 20,000. However, the passage of molecules larger than this size range is greatly or even completely impeded. Not surprisingly, the hormones that serve as intercellular signals in plants are without exception small molecules, with molecular weights well below 1000.

The Plant Cell Wall Is a Network of Cellulose Microfibrils, Polysaccharides, and Glycoproteins

Like the extracellular matrix of animal cells, plant cell walls consist predominantly of long fibers embedded in a network of branched molecules. Instead of collagen and proteoglycans, however, plant cell walls contain *cellulose microfibrils* enmeshed in a complex network of branched polysaccharides and glycoproteins called *extensins* (see Table 11-1). The two main types of polysaccharides are *hemicelluloses* and *pectins*. On a dry-weight basis, cellulose typically makes up about 40% of the cell wall, hemicelluloses account for another 20%, pectins represent about 30% and glycoproteins make up about 10%. Figure 11-24 illustrates the relationships among these cell wall components. Cellulose, hemicelluloses, and glycoproteins are linked together to form a rigid interconnected network that is embedded in a pectin matrix. Not shown are

lignins, which in woody tissues are localized between the cellulose fibrils and make the wall especially strong and rigid. We will consider each of these components in turn.

Cellulose. The predominant polysaccharide of the plant cell wall is **cellulose,** which is the single most abundant organic macromolecule on Earth. As we saw in Chapter 3, cellulose is an unbranched polymer that consists of thousands of β-D-glucose units linked together by $\beta(1 \longrightarrow 4)$ bonds (see Figure 3-25). Cellulose molecules are long, ribbonlike structures that are stabilized by intramolecular hydrogen bonds. Many such molecules (50–60, typically) associate laterally to form the **microfibrils** found in cell walls. The cellulose molecules in a microfibril all have the same polarity and are tightly crosslinked by hydrogen bonds. Microfibrils are as much as 25 nm in diameter, which means that they can be seen readily with an electron microscope (Figure 11-25). Cellulose microfibrils are often twisted together in a ropelike fashion to generate even larger structures, called *macrofibrils* (see Figure 11-24). Cellulose macrofibrils are as strong as an equivalent-sized piece of steel!

Hemicelluloses. Despite the name, **hemicelluloses** are chemically and structurally distinct from cellulose. The hemicelluloses are a heterogeneous group of polysaccharides, each consisting of a long, linear chain of a single kind of sugar (glucose or xylose) with short side chains. The side chains usually contain several different kinds of sugars, including the hexoses glucose, galactose, and mannose and the pentoses xylose and arabinose. The sugar units in the hemicellulose backbone form hydrogen bonds along the surface of the cellulose, creating a coating that helps to bond the fibrils together into a rigid interconnected network (see Figure 11-24).

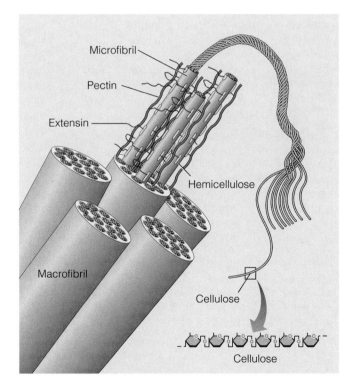

Figure 11-24 Structural Components of the Plant Cell Walls. Cellulose microfibrils are linked by hemicelluloses and glycoproteins called extensins to form a rigid interconnected network embedded in a matrix of pectins. Cellulose microfibrils are often twisted together to form larger structures called macrofibrils. (Not shown are the lignins that are localized between the cellulose microfibrils in woody tissues.)

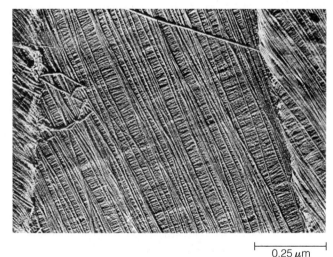

Figure 11-25 The Structure of Cellulose Microfibrils. This electron micrograph shows individual cellulose microfibrils in the cell wall of a green alga. Each microfibril consists of many cellulose molecules aligned laterally (TEM).

Pectins. **Pectins** are also branched polysaccharides, but with backbones called *rhamnogalacturonans* that consist mainly of negatively charged galacturonic acid and rhamnose. The side chains attached to the backbone contain some of the same monosaccharides found in hemicelluloses, including glucose, galactose, xylose, arabinose, and fucose. The rhamnose units that interrupt the strings of galacturonic acid units introduce kinks into the molecule and serve as attachment sites for neutral pectins that crosslink rhamnogalacturonan molecules to hemicelluloses. Pectin molecules form the matrix in which cellulose microfibrils are embedded (see Figure 11-24). In addition, they bind adjacent cell walls together.

Because of their highly branched structure and their negative charge, pectins trap and bind water molecules. As a result, pectins have a gel-like consistency that ranges from very fluid to very rigid, depending on the chemical structure and physical properties of the specific pectin molecules that are present. (It is because of their gel-forming capacity that pectins are added to fruit juice in the process of making fruit jams and jellies.) Negatively charged pectins also bind cations avidly. Calcium is an especially important cation because it crosslinks pectin molecules to other components of the cell wall.

Extensins. In addition to hemicelluloses and pectins, cell walls also contain a group of related glycoproteins called **extensins.** The polypeptide backbone of the extensins is rich in the amino acids serine, hydroxyproline, and lysine. Because of the high lysine content, extensin molecules have a net positive charge and therefore a high affinity for the negatively charged pectin molecules. Extensins contain numerous short oligosaccharides, most of which are attached to the hydroxyl groups of serine and hydroxyproline. Although short, the oligosaccharide side chains are so numerous that they account for about two-thirds of the extensin molecule by weight.

Despite the name, extensins are not very extensible; they are actually rigid, rodlike molecules that are tightly woven into the complex polysaccharide network of the cell wall (see Figure 11-24). In fact, extensins are so integral a part of the cell wall matrix that attempts to extract them chemically usually result in the loss of cell wall structure. Extensins are initially deposited in the cell wall in a soluble form. Once deposited, however, extensin molecules become covalently crosslinked to one another (primarily through tyrosine side groups) and to cellulose, generating a reinforced protein-polysaccharide complex. Extensins are least abundant in the cell walls of actively growing tissues and most abundant in the cell walls of tissues that provide mechanical support to the plant.

Lignins. **Lignins** are very insoluble polymers of aromatic alcohols that occur mainly in woody tissues. (The Latin word for "wood" is *lignum.*) These aromatic alcohols are deposited in the cell wall and become linked covalently by the action of the enzyme peroxidase, forming large crosslinked networks of insoluble polymers that contribute to the hardening of the cell wall and to the structural strength we associate with wood. Lignin molecules are localized mainly between the cellulose fibrils, where they function to resist compression forces. Lignin accounts for as much as 25% of the dry weight of woody plants, making it second only to cellulose as the most abundant organic compound on Earth.

Cell Walls Are Synthesized in Several Discrete Stages

The plant cell wall components are secreted from the cell stepwise, creating a series of layers in which the first layer to be synthesized ends up farthest away from the plasma membrane. The first structure to be laid down is called the **middle lamella;** it is shared by neighboring cell walls and holds adjacent cells together (Figure 11-26). The next structure to be formed is called the **primary cell wall,** which forms when the cells are still growing. Primary walls are about 100–200 nm thick, only several times the thickness of the basal lamina of animal cells. The primary cell wall consists of a loosely organized network of cellulose microfibrils associated with hemicelluloses, pectins, and glycoproteins (Figure 11-27a). Pectins are especially important in imparting flexibility to the primary cell wall, which enables the wall to expand during cell growth. The cellulose microfibrils are generated by cellulose-synthesizing enzyme complexes called *rosettes* that are localized within the plasma membrane. Because the microfibrils are anchored to other wall components, the rosettes must

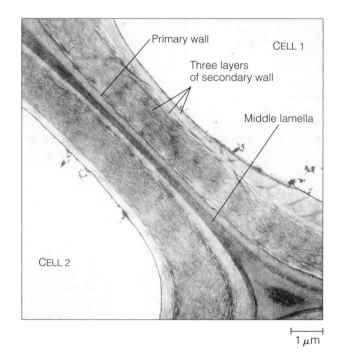

Figure 11-26 The Middle Lamella. The middle lamella is a layer of cell wall material that is shared by two adjacent plant cells. Consisting mainly of sticky pectins, the middle lamella binds the cells together tightly.

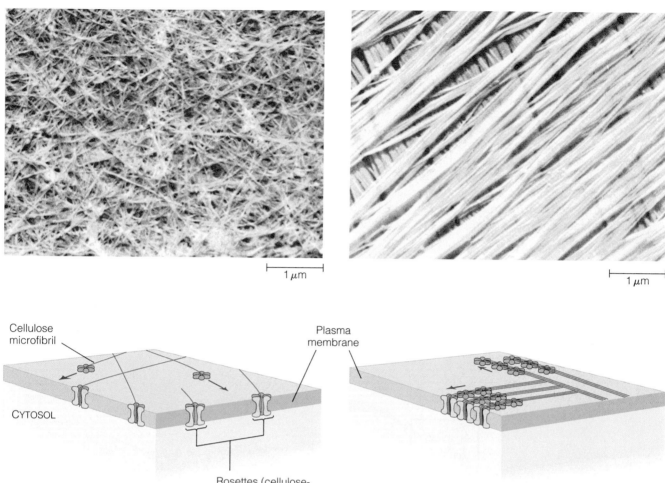

(a) Primary cell wall

Cellulose microfibril

Plasma membrane

CYTOSOL

Rosettes (cellulose-synthesizing enzymes)

(b) Secondary cell wall

Figure 11-27 Cellulose Microfibrils of Primary and Secondary Plant Cell Walls. (a) *Primary cell wall:* The electron micrograph shows the loosely organized cellular microfibrils of a primary cell wall and the diagram depicts how cellular microfibrils are synthesized by rosettes, each of which is a cluster of cellulose-synthesizing enzymes embedded in the plasma membrane. As a rosette synthesizes a bundle of cellulose molecules, it moves through the plasma membrane in the direction indicated by the arrows. **(b)** *Secondary cell wall:* The electron micrograph shows densely packed cellulose macrofibrils oriented in parallel and the diagram depicts how rosettes form dense aggregates that synthesize large numbers of microfibrils in parallel, generating cellular macrofibrils. (TEMs)

move in the plane of the membrane as they lengthen the growing cellulose microfibrils.

When the shoots or roots of a plant are growing, cell walls must be remodeled. Existing cells must change shape, and those cells that undergo rapid division in the shoot or root (regions called *meristems*) must undergo such divisions in a particular orientation, which is influenced by their cell walls. A family of proteins called **expansins** help cell walls to retain their pliability. Although expansins are a very minor component of cell walls (about 0.02% by weight), they appear to be key regulators of cell wall expansion. When added to denatured cell wall components maintained at an acidic pH, expansins allow those cell walls to extend much more readily when they are pulled on. When expansin-soaked agarose beads are placed on tomato shoots, new areas of growth are induced. One way in which expansins may act is by disrupting the normal hydrogen bonding of glycans within the microfibrils of the cell wall to allow the rearrangement of the microfibrils.

The loosely textured organization of the primary cell wall creates a relatively thin, flexible structure that can expand during cell growth. In some plant cells, development of the cell wall does not proceed beyond this point. However, many cells that have stopped growing add a thicker, more rigid set of layers that are referred to collectively as the **secondary cell wall** (Figure 11-27b). The components of the multilayered secondary wall are added to the inner surface of the primary wall after cell growth has ceased. Cellulose and lignins are the primary constituents of the secondary wall, making this structure significantly stronger, harder, and more rigid than the primary wall. Each layer of the secondary wall consists of densely packed bundles of cellulose microfibrils arranged in parallel and oriented so that they lie at an angle to the microfibrils of adjacent layers. This organization imparts great mechanical strength and rigidity to the secondary cell wall and is, along with the presence of lignin, responsible for the characteristic strength of woody tissues.

The orderly arrangement of cellulose microfibrils in secondary walls arises during the formation of each successive layer of the secondary wall. At this stage, microtubules located beneath the plasma membrane are oriented in the same direction as the newly formed cellulose microfibrils, and the cellulose-synthesizing rosettes are aggregated into large arrays containing dozens of rosettes. The underlying microtubules are thought to guide the movement of these rosette aggregates, which synthesize large parallel bundles of cellulose microfibrils as they move.

Plasmodesmata Permit Direct Cell-Cell Communication Through the Cell Wall

Recognizing that every plant cell is surrounded by a plasma membrane and a cell wall, you may wonder whether plant cells are capable of intercellular communication such as that afforded by the gap junctions of animal cells. Plasmodesmata accomplish this very purpose. As shown in Figure 11-28, **plasmodesmata** (singular: **plasmodesma**) are cytoplasmic channels through relatively large openings

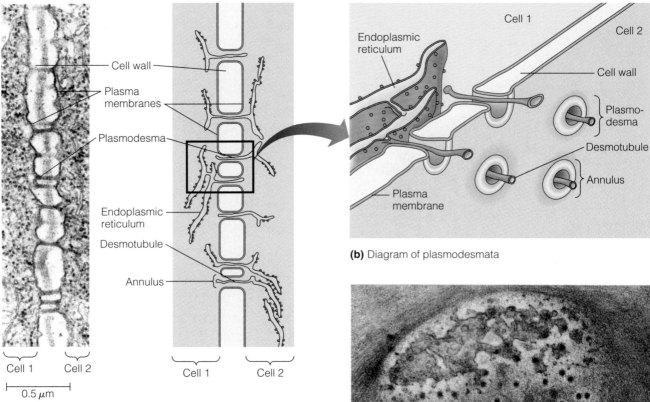

(a) Electron micrograph and diagram of plasmodesmata in longitudinal section

(b) Diagram of plasmodesmata

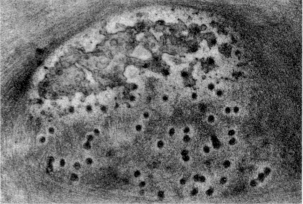

(c) Electron micrograph of plasmodesmata in cross section

Figure 11-28 Plasmodesmata. A plasmodesma is a channel through the cell wall between two adjacent plant cells, allowing cytoplasmic exchange between the cells. The plasma membrane of one cell is continuous with that of the other cell at each plasmodesma. Most plasmodesmata have a narrow cylindrical desmotubule at the center that is derived from the ER and appears to be continuous with the ER of both cells. Between the desmotubule and the plasma membrane that lines the plasmodesma is a narrow ring of cytoplasm called the annulus. **(a)** This electron micrograph and diagram show the cell wall between two adjacent root cells of timothy grass, with numerous plasmodesmata (TEM). **(b)** A diagrammatic view of a cell wall with numerous plasmodesmata, illustrating the continuity of the ER and cytoplasm between adjacent cells. **(c)** This electron micrograph shows many plasmodesmata in cross section (TEM).

in the cell wall, allowing continuity of the plasma membranes from two adjacent cells. Each plasmodesma is therefore lined with plasma membrane common to the two connected cells. A plasmodesma is cylindrical in shape, with the cylinder narrower in diameter at both ends. The channel diameter varies from about 20 to about 200 nm. A single tubular structure, the **desmotubule**, usually lies in the central channel of the plasmodesma. Endoplasmic reticulum (ER) cisternae are often seen near the plasmodesmata on either side of the cell wall. ER membranes from adjoining cells are continuous with the desmotubule and with the ER of the other cells, as depicted in Figure 11-28b.

The ring of cytoplasm between the desmotubule and the membrane that lines the plasmodesma is called the **annulus.** The annulus is thought to provide cytoplasmic continuity between adjacent cells, thereby allowing molecules to pass freely from one cell to the next without the stringent size limits that are otherwise set by the cell wall and the plasma membrane. The plasmodesmata therefore provide for continuity of the plasma membrane, the ER, and the cytoplasm between adjacent cells. Even after cell division and deposition of new cell walls between the two daughter cells, cytoplasmic continuities are maintained between the daughter cells by plasmodesmata that pass through the newly formed walls. In fact, most plasmodesmata are formed at the time of cell division, when the new cell wall is being formed. Minor changes may occur later, but the number and location of plasmodesmata are largely fixed at the time of division.

In many respects, plasmodesmata appear to be similar to gap junctions in function. They reduce electrical resistance between adjacent cells by about 50-fold compared with cells that are completely separated by plasma membranes. In fact, the movement of ions between adjacent cells (measured as current flow) is proportional to the number of plasmodesmata that connect the cells. As noted earlier for gap junctions, calcium appears to regulate movement through the plasmodesmata, because traffic between cells can be decreased significantly by the injection of Ca^{2+} into cells.

Perspective

Both plant and animal cells have extracellular structures that consist of long, rigid fibers embedded in an amorphous, hydrated matrix of branched molecules. In the case of animal cells, the extracellular matrix (ECM) consists of collagen (and, in some tissues, elastin) fibers embedded in a network of glycosaminoglycans and proteoglycans. Collagen is responsible for the strength of the ECM and elastin imparts elasticity. The ECM is held in place by adhesive glycoproteins such as fibronectins, which link cells to the ECM, and laminins, which attach cells to the basal lamina. These adhesive glycoproteins bind to cell-surface receptor glycoproteins called integrins. Cell-cell recognition and adhesion is mediated by plasma membrane glycoproteins such as IgSF proteins and cadherins. The carbohydrate groups on cell sufaces are also important in cell-cell recognition and adhesion; an example is the role of selectins in recognizing carbohydrates during inflammatory responses.

Most of the cells in multicellular organisms are in close and ongoing association with neighboring cells. The junctions that link animal cells together are of three general types. Adhesive junctions hold cells together in fixed positions within tissues or bind cells to the ECM. On the cytoplasmic side of the membrane, adhesive junctions are anchored to the cytoskeleton by linker proteins that attach to actin microfilaments (adherens junctions) or intermediate filaments (desmosomes and hemidesmosomes). Desmosomes are particularly prominent in tissues that must withstand considerable mechanical stress, such as skin, heart muscle, and the tissue at the neck of the uterus.

Tight junctions serve mainly as seals between two compartments, preventing the leakage of fluid such as gastric juice, urine, or blood. Tight junctions also prevent the lateral movement of membrane proteins, thereby partitioning the membrane into discrete functional domains. Gap junctions form open channels between cells, allowing direct chemical and electrical communication between cells. Gap junction permeability is limited to ions and small molecules and depends on Ca^{2+} and proton concentrations.

The primary cell wall of a plant cell consists mainly of cellulose fibers embedded in a complex network of hemicelluloses, pectins, and extensins. The secondary cell wall that forms as a cell reaches its final size and shape is reinforced with lignins, a major component of wood. Plasmodesmata are membrane-lined cytoplasmic channels between adjacent plant cells that allow chemical and electrical communication rather like that facilitated by the gap junctions between animal cells. The flow of molecules and ions between cells appears to be regulated in both plasmodesmata and gap junctions.

Key Terms for Self-Testing

The Extracellular Matrix of Animal Cells

extracellular structure (p. 290)
extracellular matrix (ECM) (p. 290)
collagen (p. 291)
collagen fiber (p. 291)
procollagen (p. 292)
elastin (p. 293)
glycosaminoglycan (GAG) (p. 294)
proteoglycan (p. 294)
core protein (p. 294)
hyaluronate (p. 295)
fibronectin (p. 296)
laminin (p. 297)
basal lamina (p. 297)
integrin (p. 298)
focal adhesion (p. 300)
hemidesmosome (p. 300)
plaque (p. 300)
plakin (p. 300)
anchorage-dependent growth (p. 301)
glycocalyx (p. 302)

Cell-Cell Recognition and Adhesion

cell-cell recognition (p. 302)
cell-cell adhesion (p. 302)
homophilic interaction (p. 303)
heterophilic interaction (p. 303)
neural cell adhesion molecule (N-CAM) (p. 303)
immunoglobulin superfamily (IgSF) protein (p. 303)
cadherin (p. 303)
lectin (p. 304)
selectin (p. 306)

Cell Junctions

cell junction (p. 306)
adhesive (anchoring) junction (p. 306)
adherens junction (p. 307)
desmosome (p. 309)
tonofilament (p. 310)
tight junction (p. 310)
claudin (p. 311)

gap junction (p. 313)
connexon (p. 313)

The Plant Cell Surface

cell wall (p. 314)
cellulose (p. 315)
microfibril (p. 315)
hemicellulose (p. 315)
pectin (p. 316)
extensin (p. 316)
lignin (p. 316)
middle lamella (p. 316)
primary cell wall (p. 316)
expansin (p. 317)
secondary cell wall (p. 317)
plasmodesma (p. 318)
desmotubule (p. 319)
annulus (p. 319)

Problem Set

More challenging problems are marked with a •.

11-1. Beyond the Membrane: ECM and Cell Walls. Compare and contrast the extracellular matrix (ECM) of animal cells with the walls around plant cells.

(a) What basic organizational principle underlies both ECM and cell walls?

(b) What are the chemical constituents in each case?

(c) What functional roles do the ECM and cell wall share in common?

(d) What functional roles are unique to the ECM? To the cell wall?

11-2. Anchoring Cells to the ECM. According to our current understanding, animal cells are anchored to the ECM by several different kinds of protein-mediated linkages.

(a) What are the two main kinds of linkages? How could you distinguish between them experimentally, based on the proteins involved?

(b) What is the role of the adhesive glycoproteins that are involved?

(c) Briefly explain how the various domains of the fibronectin molecule (see Figure 11-7) or the laminin molecule (see Figure 11-10) were identified experimentally.

11-3. Compare and Contrast. For each of the terms in list A, choose a related term in list B and explain the relationship between the two terms by comparing or contrasting them structurally or functionally.

	List A	List B
(a)	Collagen	Basolateral surface
(b)	Fibronectin	Focal adhesion
(c)	Integrin	Elastin
(d)	IgSF	Laminin

(e)	ECM	Cadherin
(f)	Hemidesmosome	Glycocalyx
(g)	Apical surface	Selectin

11-4. Compaction. In mammalian embryos such as the mouse, the fertilized egg divides three times to form eight loosely packed cells, which become tightly adherent in a process known as *compaction*. In the late 1970s, several laboratories made antibodies against mouse cell surface proteins. The antibodies prevented compaction, as did removal of Ca^{2+} from the medium. What sort of protein do the antibodies probably recognize, and why?

11-5. Cellular Junctions and Plasmodesmata. Indicate whether each of the following statements is true of adhesive junctions (A), tight junctions (T), gap junctions (G), and/or plasmodesmata (P). A given statement may be true of any, all, or none (N) of these structures.

(a) Associated with filaments that confer either contractile or tensile properties.

(b) Sites of membrane fusion are limited to abutting ridges of adjacent membranes.

(c) Require the alignment of connexons in the plasma membranes of two adjacent cells.

(d) Seal membranes of two adjacent cells tightly together.

(e) Allow the exchange of metabolites between the cytoplasms of two adjacent cells.

11-6. Junction Proteins. Indicate whether each of the following proteins or structures is a component of adhesive junctions (A), tight junctions (T), gap junctions (G), or plasmodesmata (P), and describe briefly the role the protein plays in the junction.

(a) connexin

(b) vinculin

(c) desmocollins

(d) desmotubule

(e) desmoplakin

(f) annulus

(g) cadherins

(h) claudins

11-7. Plant Cell Walls. Distinguish between the terms in each of the following pairs with respect to the structure of the plant cell wall, and indicate the significance of each.

(a) Primary wall; secondary wall

(b) Cellulose; hemicellulose

(c) Xylan; xyloglucan

(d) Extensin; lignin

(e) Desmotubule; annulus

(f) Plasmodesma; gap junction

• **11-8. Scurvy and Collagen.** Scurvy is a disease that until the last century was common among sailors and others whose diets were deficient in vitamin C (ascorbic acid). Individuals with scurvy suffer from a variety of disorders, including extensive bruising, hemorrhages, and breakdown of supporting tissues. Ascorbic acid serves as a reducing agent responsible for maintaining the activity of prolyl hydroxylase, the enzyme that catalyzes hydroxylation of proline residues within the collagen triple helix, which is required for helix stability.

(a) Based on this information, postulate a role for hydroxyproline in collagen triple helices and explain the sequence of events that leads from a dietary vitamin C deficiency to symptoms such as bruising and breakdown of supporting tissues.

(b) What does your answer to part a imply about the degradation and replacement of collagen in at least some tissues?

(c) Can you guess why sailors are no longer susceptible to scurvy? And why do you think British sailors are called "limeys" to this day?

• **11-9. Experimental Evidence for Gap Junctions.** Our understanding of gap junctions owes much to the pioneering work of Werner Loewenstein and his colleagues, who performed the experiments described below, and depicted in Figure 11-29. In each case, indicate (i) what hypothesis the experiment was designed to test, (ii) what you think these workers concluded from the experiment, (iii) how their findings support that conclusion, and (iv) a further experiment that might be done with the same methodology to extend their findings.

(a) Microelectrodes were inserted into individual cells of insect salivary glands. When a small voltage was applied to electrodes in two neighboring cells, the current that flowed between the cells was found to be several orders of magnitude higher than the current that could be measured when one electrode was placed in a cell and the second in the external medium.

(b) Cells were injected with fluorescent molecules of different molecular weights and a fluorescence microscope was then used to observe the movement of the molecules into adjacent cells. Figure 11-29a illustrates the results obtained with fluorescent molecules with molecular weights of 1158 (top) and 1926 (bottom).

(c) A fluorescent dye was injected into one cell (number 3) in a sequence of five cells and fluorescence microscopy was used to monitor the movement of the dye into adjacent cells in Ca^{2+}-containing medium. Figure 11-29b illustrates the results obtained with intact cells that are impermeable to Ca^{2+} (left) and with holes punched into cells 2 and 4 to allow the Ca^{2+}-containing solution to enter, thereby raising the intracellular Ca^{2+} concentration (right).

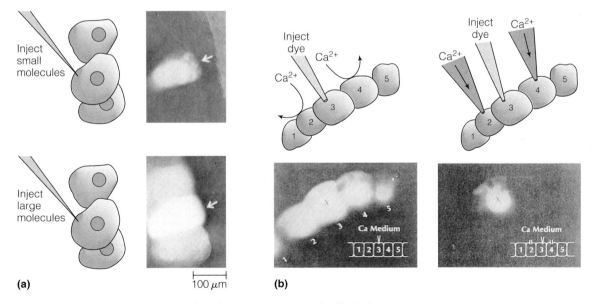

Figure 11-29 Experimental Evidence for Properties of Gap Junctions. (a) Size limitations of gap junction permeability as determined by fluorescence microscopy following microinjection of fluorescent molecules of two different sizes: 1158 Da (top) and 1926 Da (bottom). (b) Use of a fluorescent dye to determine how changes in the intracellular Ca^{2+} concentration affect the permeability of gap junctions. See Problem 11-9.

Suggested Reading

References of historical importance are marked with a •.

General References

Gumbiner, B.M. Cell adhesion: The molecular basis of tissue architecture and morphogenesis. *Cell* 84 (1996): 345.

Horwitz, A.F. Integrins and health. *Sci. Am.* 276 (1997): 68.

Hynes, R.O. Targeted mutations in cell adhesion genes: What have we learned from them? *Dev. Biol.* 180 (1996): 402.

Hynes, R.O. Cell adhesion: Old and new questions. *Trends Cell Biol.* 9 (1999): M33.

The Extracellular Matrix of Animal Cells

Colognato, H., and P.D. Yurchenco. Form and function: The laminin family of heterotrimers. *Dev. Dyn.* 218 (2000): 213.

Gorski, J.P., and B.R. Olsen. Mutations in extracellular matrix molecules. *Curr. Opin. Cell Biol.* 10 (1998): 586.

Sherman, L., J. Sleeman, P. Herrlich, and H. Ponta. Hyaluronate receptors: Key players in growth, differentiation, migration and tumor progression. *Curr. Opin. Cell Biol.* 6 (1994): 726.

Integrins, Focal Adhesions, and Hemidesmosomes

Boudreau, N.J., and P.L. Jones. Extracellular matrix and integrin signalling: The shape of things to come. *Biochem. J.* 339 (1999): 481.

De Arcangelis A., and E. Georges-Labouesse. Integrin and ECM functions: Roles in vertebrate development. *Trends Genet.* 16 (2000): 389.

Giancotti, F.G., and E. Ruoslahti. Integrin signaling. *Science* 285 (1999): 1028.

Jones, J.C., S.B. Hopkinson, and L.E. Goldfinger. Structure and assembly of hemidesmosomes. *Bioessays* 20 (1998): 488.

Kühn, K., and J. Eble. The structural bases of integrin-ligand interactions. *Trends Cell Biol.* 4 (1994): 256.

Sheppard, D. Epithelial integrins. *Bioessays* 18 (1996): 655.

Cell-Cell Adhesion

Angst, B.D., C. Marcozzi, and A.I. Magee. The cadherin superfamily: Diversity in form and function. *J. Cell Sci.* 114 (2001): 629.

Baillie, L., and T.D. Read. *Bacillus anthracis,* a bug with attitude! *Curr. Opin. Microbiol.* 4 (2001): 78.

Bradley, K.A, J. Mogridge, M. Mourez, R.J. Collier, and J.A. Young. Identification of the cellular receptor for anthrax toxin. *Nature* 41 (2001): 225.

Dramsi, S., and P. Cossart. Intracellular pathogens and the actin cytoskeleton. *Annu. Rev. Cell Dev. Biol.* 14 (1998): 137.

Ireton, K., and P. Cossart. Interaction of invasive bacteria with host signaling pathways. *Curr. Opin. Cell Biol.* 10 (1998): 276.

• Kemler, R. Classical cadherins. *Seminar Cell Biol.* 3 (1992): 149.

Leckband, D., and S. Sivasankar. Mechanism of homophilic cadherin adhesion. *Curr. Opin. Cell Biol.* 12 (2000): 587.

Sharon, N., and H. Lis. Carbohydrates in cell recognition. *Sci. Amer.* 268 (January 1993): 82.

Vestweber, D. and J.E. Blanks. Mechanisms that regulate the function of the selectins and their ligands. *Physiol. Rev.* 79 (1999): 181.

Young, J.A.T., and R.J. Collier. Attacking anthrax. *Sci. Amer.* 286 (March 2002): 48.

Cell Junctions

Kumar, N.M., and N.B. Gilula. The gap junction communication channel. *Cell* 84 (1996): 381.

Mitic, L.L., and J.M. Anderson. Molecular architecture of tight junctions. *Annu. Rev. Physiol.* 60 (1998): 121.

Mitic, L.L., C.M. Van Itallie, and J.M. Anderson. Molecular physiology and pathophysiology of tight junctions I. Tight junction structure and function: Lessons from mutant animals and proteins. *Am. J. Physiol. Gastrointest. Liver Physiol.* 279 (2000): G250.

North, A.J., W.G. Bardsley, J. Hyam, E.A. Bornslaeger, H.C. Cordingley, B. Trinnaman, M. Hatzfeld, K.J. Green, A.I. Magee, and D.R. Garrod. Molecular map of the desmosomal plaque. *J. Cell Sci.* 112 (1999): 4325.

Phelan, P., and T.A. Starich. Innexins get into the gap. *Bioessays* 23 (2001): 388.

Simon, A.M., and D.A. Goodenough. Diverse functions of vertebrate gap junctions. *Trends Cell Biol.* 8 (1998): 477.

• Stauffer, K.A., and N. Unwin. Structure of gap junction channels. *Semin. Cell Biol.* 3 (1992): 17.

Tsukita, S., and M. Furuse. Occludin and claudins in tight-junction strands: Leading or supporting players? *Trends Cell Biol.* 9 (1999): 268.

Yap, A.S., W.M. Brieher, and B.M. Gumbiner. Molecular and functional analysis of cadherin-based adherens junctions. *Annu. Rev. Cell Develop. Biol.* 13 (1997): 119.

The Plant Cell Wall

Cosgrove, D.J. Assembly and enlargement of the primary cell wall in plants. *Annu. Rev. Cell Develop. Biol.* 13 (1997): 171.

Cosgrove, D.J. Loosening of plant cell walls by expansins. *Nature* 407 (2000): 321.

Kieliszewski, M.J., and D.T. Lamport. Extensin: Repetitive motifs, functional sites, posttranslational codes, and phylogeny. *Plant J.* 5 (1994): 157.

Nicol, F., and H. Hofte. Plant cell expansion: Scaling the wall. *Curr. Opin. Plant Biol.* 1 (1998): 12.

Showalter, A.M. Structure and function of plant cell wall proteins. *Plant Cell* 5 (1993): 9.

Wyatt, S.E., and N.C. Carpita. The plant cytoskeleton-cell-wall continuum. *Trends Cell Biol.* 3 (1992): 413.

Plasmodesmata

Lucas, W.J. Plasmodesmata: Intercellular channels for macro-molecular transport in plants, *Curr. Opin. Cell Biol.* 7 (1995): 673.

Zambryski, P. Plasmodesmata: Plant channels for molecules on the move. *Science* 270 (1995): 1943.

12

Intracellular Compartments: The Endoplasmic Reticulum, Golgi Complex, Endosomes, Lysosomes, and Peroxisomes

The study of eukaryotic cells is essentially the exploration of an elaborate array of intracellular membrane-bounded compartments—called *organelles*—that house various cellular activities. Whether we consider the storage and transcription of genetic information, the biosynthesis of secretory proteins, the breakdown of long-chain fatty acids, or any of the myriad other metabolic processes occurring within eukaryotic cells, several if not all of the reactions of a pathway occur within a distinct membrane-bounded organelle. Therefore, a full appreciation of eukaryotic cells depends on an understanding of the prominent role of intracellular membranes and the compartmentalization of function such membranes make possible.

In Chapter 4, we briefly encountered the major organelles found in eukaryotic cells. We are now ready to consider individual organelles in more detail. In this chapter, we will focus on the five organelles highlighted in Figure 12-1. We will begin with the *endoplasmic reticulum* and *Golgi complex,* which are sites for protein synthesis, processing, and sorting. Next, we will look at endosomes and lysosomes. *Early endosomes* are important for sorting material brought into the cell by endocytosis, whereas *late endosomes*—which mature to form *lysosomes*—contain enzymes capable of breaking down intracellular and extracellular molecules. We will conclude with a discussion of the diverse functions of *peroxisomes,* which house peroxide-generating reactions and have an essential role in the oxidation of fatty acids.

As you study the role of each organelle, keep in mind that the endoplasmic reticulum, the Golgi complex, endosomes, and lysosomes (but not peroxisomes) are all components of the eukaryotic cell's **endomembrane system.** The *nuclear envelope,* which we will study in Chapter 16, is also a part of this system. As illustrated in Figure 12-1, the

outer membrane of the nuclear envelope is continuous with the surrounding membrane of the endoplasmic reticulum, and the *perinuclear space* between the two membranes of the nuclear envelope is continuous with the interior of the endoplasmic reticulum. As a result, molecules can diffuse freely from either one of these compartments to the other. Material can also flow from the endoplasmic reticulum to the Golgi complex, endosomes, and lysosomes by means of *transport vesicles* that shuttle between the various organelles. Such vesicles convey membrane lipids and membrane-bound proteins, as well as soluble cargo. Thus, these organelles and the vesicles connecting them make up a single system of surrounding membranes and internal spaces. This intimate association has important consequences for cellular metabolism, as we will soon see.

The Endoplasmic Reticulum

The **endoplasmic reticulum** (ER) is a continuous network of flattened sacs, tubules, and associated vesicles that stretches throughout the eukaryotic cell's cytoplasm. The sacs are called **ER cisternae** (singular: **ER cisterna**), and the space enclosed by the sacs and tubules is called the **ER lumen** (or *ER cisternal space*). Of a cell's total complement of membrane, 50–90% is devoted to surrounding the ER lumen. Unlike more prominent organelles, such as the mitochondrion or chloroplast, however, the ER is not visible by light microscopy unless one or more of its components are stained with a dye or labeled with a fluorescent molecule.

The ER was first observed in the late nineteenth century, when it was noted that some eukaryotic cells, particularly

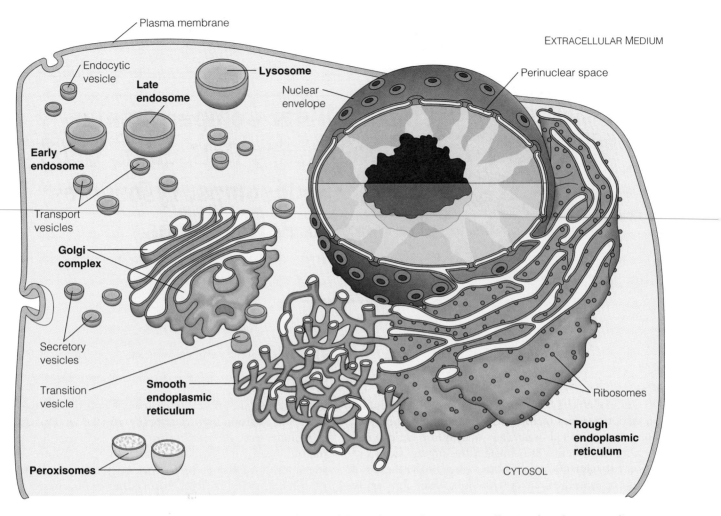

Plasma membrane

EXTRACELLULAR MEDIUM

Endocytic
vesicle

Lysosome

**Late
endosome**

Nuclear
envelope

Perinuclear space

**Early
endosome**

Transport
vesicles

**Golgi
complex**

Secretory
vesicles

Transition
vesicle

**Smooth
endoplasmic
reticulum**

Ribosomes

**Rough
endoplasmic
reticulum**

Peroxisomes

CYTOSOL

Figure 12-1   Cellular Structures Discussed in This Chapter.   This chapter focuses on the endoplasmic reticulum, the Golgi complex, endosomes, lysosomes, and peroxisomes. Also relevant to the discussion are the nuclear envelope and plasma membrane. The perinuclear space between the two membranes of the nuclear envelope is continuous with the interior of the endoplasmic reticulum. The interior of the endoplasmic reticulum, in turn, is linked to the interiors of the Golgi complex, endosomes, and lysosomes by transport vesicles that shuttle from organelle to organelle. Together, the surrounding membranes and internal spaces of these structures define the endomembrane system of the eukaryotic cell; the remaining portion of the cytoplasm is called the cytosol.

those involved in secretion, contained regions that stained intensely with basic dyes. These regions were called *ergastoplasm,* but their significance remained in doubt until the 1950s, when the enormous increase in resolving power provided by electron microscopy allowed cell biologists to visualize for the first time the ER's elaborate network of intracellular membranes and to investigate the role of the ER in cellular processes.

We now know that enzymes associated with the ER are responsible for the biosynthesis of proteins that are incorporated into the ER, the Golgi complex, endosomes, lysosomes, and the plasma membrane, as well as the biosynthesis of proteins secreted by the cell. Equally important is the ER's central role in the biosynthesis of lipids, including triacylglycerols, cholesterol, and related compounds. The ER is the source of most of the lipids that are assembled to form intracellular membranes and the plasma membrane.

Although the name sounds formidable, it is actually quite descriptive. *Endoplasmic* simply means "within the plasm" (of the cell), and *reticulum* is a Latin word meaning "network." The term was originally coined because of the extent to which the ER resembled a *reticule,* a netted handbag that was popular in the late nineteenth century.

The Two Basic Kinds of Endoplasmic Reticulum Differ in Structure and Function

The two basic kinds of endoplasmic reticulum found in eukaryotic cells are generally distinguished from one another by the presence or absence of ribosomes attached to the surrounding membrane (Figure 12-2). **Rough endoplasmic reticulum (rough ER)** is characterized by ribosomes attached to the cytosolic side (outer surface) of the membrane (Figure 12-2a). Ribosomes contain RNA, and it was this RNA that reacted strongly with the basic

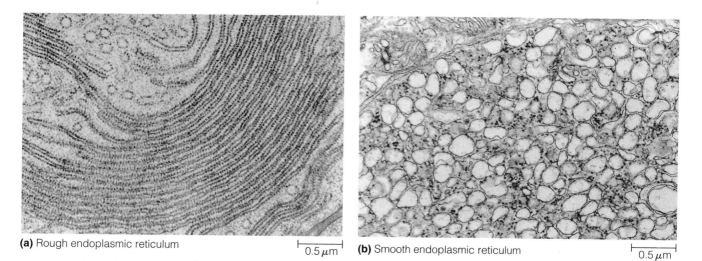

(a) Rough endoplasmic reticulum
0.5 μm

(b) Smooth endoplasmic reticulum
0.5 μm

Figure 12-2   Rough and Smooth Endoplasmic Reticulum.   (a) Electron micrograph of endoplasmic reticulum. The rough ER is studded with ribosomes (TEM). **(b)** Electron micrograph of smooth endoplasmic reticulum (TEM).

dyes originally used to identify ergastoplasm. In other words, ergastoplasm turned out to be rough ER. A subdomain of rough ER, the **transitional elements (TEs),** play an important role in the formation of **transition vesicles** that shuttle lipids and proteins from the ER to the Golgi complex. **Smooth endoplasmic reticulum (smooth ER)** is smooth because of the absence of ribosomes attached to the membrane (Figure 12-2b).

Rough and smooth ER can also be distinguished morphologically. As illustrated in Figure 12-1, rough ER membranes usually form large flattened sheets, whereas smooth ER membranes generally form tubular structures. The transitional elements of the rough ER are an exception to this rule; they often resemble smooth ER. The presence or absence of ribosomes, as well as the morphological differences, reflect functional differences between the types of ER in a eukaryotic cell. Ultimately, however, the rough and smooth ER are not separate organelles; electron micrographs reveal continuity of ER lumenal spaces. Thus, material can travel between the rough and smooth ER without the aid of vesicles.

Both types of ER are present in most eukaryotic cells, but there is considerable variation in relative amounts, depending on the activities of the cell. Cells characterized by the biosynthesis of secretory proteins, such as liver cells, tend to have very prominent rough ER networks. On the other hand, cells producing steroid hormones, such as the Leydig cells of the testes, contain extensive smooth ER networks. This reflects the importance of the smooth ER in the biosynthesis of steroids and related compounds.

When tissue is homogenized for subcellular fractionation, the ER membranes break into smaller fragments that spontaneously close to form sealed vesicles known as **microsomes.** When microsomes are prepared by differential centrifugation, fractions can be isolated with and without attached ribosomes, depending on whether the membrane originated from rough or smooth ER, respec-

tively. Such preparations are tremendously useful for exploring both types of ER. Keep in mind, however, that microsomes do not exist in the cell; they are simply an artifact of the fractionation process. Box 12A presents more detailed information about subcellular fractionation by differential centrifugation.

Rough ER Is Involved in the Biosynthesis and Processing of Proteins

The ribosomes attached to the cytosolic side of the rough ER membrane are responsible for synthesizing both membrane-bound and soluble proteins. These proteins are usually destined for incorporation into organelles of the endomembrane system, incorporation into the plasma membrane, or export from the cell as secretory products.

Most proteins entering the endomembrane system are inserted into the rough ER lumen *cotranslationally*—that is, as the polypeptide is synthesized on the ribosome. After biosynthesis, membrane-bound proteins remain anchored to the ER membrane by hydrophobic regions of the polypeptide or by covalent attachment to membrane lipids, whereas soluble proteins and most secretory proteins are released into the ER lumen. Only a small number of proteins are imported into the ER *posttranslationally*—that is, after synthesis—which is the general scheme observed for peroxisomes, mitochondria, and chloroplasts. We will discuss the mechanisms for the biosynthesis and targeting of proteins to the rough ER in detail in Chapter 20.

In addition to biosynthesis of polypeptide chains, the rough ER is also the site for the initial steps of addition of carbohydrate groups to glycoproteins, the folding of polypeptides, and the assembly of multimeric proteins. Thus, ER-specific proteins include a host of enzymes that catalyze cotranslational and posttranslational modifications. The glycosylation of polypeptides within the ER lumen

(text continues on p. 330)

Centrifugation is an indispensable procedure for the isolation and purification of organelles and macromolecules. This method is based on the fact that when a particle is subjected to centrifugal force its rate of movement through a specific solution depends on the particle's *size* and *density,* as well as the solution's density and viscosity. The larger or more dense a particle is, the higher its **sedimentation rate,** or rate of movement through the solution. Because most organelles and macromolecules differ significantly from one another in size and/or density, centrifuging a mixture of cellular components will separate, or resolve, the faster-moving components from the slower-moving ones. This procedure enables researchers to isolate and purify specific organelles and smaller cellular structures—a process called **subcellular fractionation**—which can then be manipulated and studied in vitro.

Albert Claude, George Palade, and Christian de Duve shared a Nobel Prize in 1974 for their pioneering work in centrifugation and subcellular fractionation. Claude played a key role in developing differential centrifugation as a method of isolating organelles. Palade was quick to use this technique in studies of the endoplasmic reticulum and the Golgi complex, establishing the roles of these organelles in the biosynthesis, processing, and secretion of proteins. De Duve, in turn, discovered both lysosomes and peroxisomes. In each case, subcellular fractionation was vital to the detection and characterization of a new class of organelles. De Duve's discovery of lysosomes depended on *differential centrifugation* (see Box 4A), whereas his discovery of peroxisomes, described in this chapter, depended on *equilibrium density centrifugation*. In addition to differential centrifugation and equilibrium density centrifugation, *density gradient centrifugation* is also commonly employed for resolving organelles. The latter two techniques are also useful for resolving the components of mixtures of macromolecules, such as nucleic acids or proteins. We will discuss each of these techniques, beginning with a general consideration of centrifuges and sample preparation.

Centrifuges

In essence, a **centrifuge** consists of a rotor—often housed in a refrigerated chamber—spun by an electric motor. The rotor holds tubes that contain solutions or suspensions of particles for fractionation. There are two basic types of rotors: A *fixed-angle rotor* holds the tubes at a specific angle (Figure 12A-1a), whereas a *swinging-bucket rotor* has hinges that allow the tubes to swing out as the rotor spins (Figure 12A-1b). Centrifugation at very high speeds—above 20,000 revolutions per minute (rpm)—requires an **ultracentrifuge** equipped with a vacuum system to reduce friction (between the rotor and air) and armor plating around the chamber to contain the rotor in the event of an accident. Some ultracentrifuges reach speeds over 100,000 rpm, subjecting samples to forces exceeding 500,000 times the force of gravity (*g*).

Sample Preparation

Tissues must undergo **homogenization,** or disruption, before the cellular components can be separated by centrifugation. Homogenization is usually done in a cold isotonic solution to

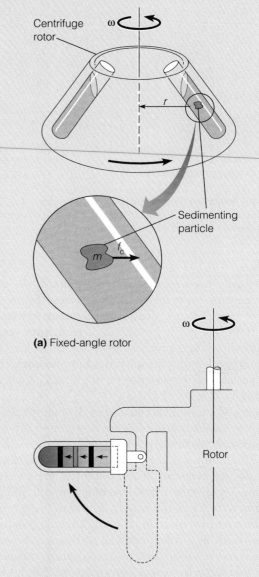

(a) Fixed-angle rotor

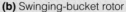

(b) Swinging-bucket rotor

Figure 12A-1   Centrifuge Rotors.   A centrifuge rotor either **(a)** holds the tubes at a fixed angle or **(b)** has hinged buckets to allow the tubes to swing out parallel to the centrifugal force during centrifugation. The centrifugal force f_c acting on a particle is given by the equation $f_c = m\omega^2 r$, where m is the mass of the particle (in grams), ω is the angular velocity of the rotor (in radians/sec), and r is the distance from particle to the axis of rotation (in centimeters). Therefore, f_c is expressed as g-cm/sec^2. The term f_r is the resistance that the particle encounters as it moves outward, so the net force acting on the particle is $f_c - f_r$.

preserve the integrity of organelles. A common homogenization solution is 0.25 *M* sucrose. Disruption can be achieved by forcing cells through a narrow orifice, by subjecting tissue to ultrasonic vibration, by osmotic shock, or by grinding the material

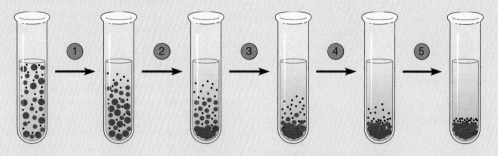

- ● Particles with large sedimentation coefficients
- ● Particles with intermediate sedimentation coefficients
- · Particles with small sedimentation coefficients

Figure 12A-2 Differential Centrifugation. Differential centrifugation separates particles based on differences in sedimentation rate, which reflect differences in size and/or density. The technique is illustrated here for three particles that differ significantly in size. The particles, which are initially part of a homogeneous mixture, are subjected to a fixed centrifugal force for five successive time intervals (circled numbers). Particles that are large or dense (purple spheres) sediment rapidly, those that are intermediate in size or density (blue spheres) sediment less rapidly, and the smallest or least dense particles (black spheres) sediment very slowly. Eventually, all of the particles reach the bottom of the tube, unless the process is interrupted after a specific length of time.

with a mortar and pestle. The resulting **homogenate** is a suspension of organelles, smaller cellular components, and molecules. If tissue is homogenized gently enough, most organelles and other structures remain intact and retain their original biochemical properties.

Differential Centrifugation

Differential centrifugation separates organelles on the basis of size and/or density differences. As illustrated in Figure 12A-2, particles that are large or dense (purple spheres) sediment rapidly, those that are intermediate in size or density (blue

spheres) sediment less rapidly, and the smallest or least dense particles (black spheres) sediment very slowly.

One way to express the relative size of an organelle or macromolecule is in terms of its **sedimentation coefficient,** a measure of how rapidly the particle sediments when subjected to centrifugation. Sedimentation coefficients are expressed in **Svedberg units** (S), in honor of Theodor Svedberg, the Swedish chemist who developed the ultracentrifuge between 1920 and 1940. (One Svedberg unit = 1×10^{-13} sec.) The sedimentation coefficients of some organelles, macromolecules, and viruses are shown in Figure 12A-3.

(continued)

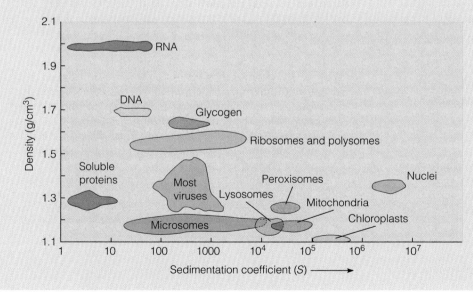

Figure 12A-3 Sedimentation Coefficients and Densities of Organelles, Macromolecules, and Viruses. A particle's sedimentation coefficient, expressed in Svedberg units (S), indicates how rapidly it sediments when subjected to a centrifugal force. A particle's density (expressed in g/cm^3) determines how far it will move during equilibrium density centrifugation.

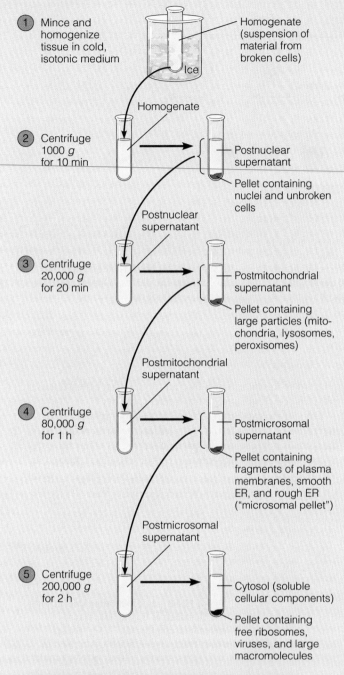

(1) Mince and homogenize tissue in cold, isotonic medium

Homogenate (suspension of material from broken cells)

Ice

Homogenate

(2) Centrifuge 1000 *g* for 10 min

Postnuclear supernatant

Pellet containing nuclei and unbroken cells

Postnuclear supernatant

(3) Centrifuge 20,000 *g* for 20 min

Postmitochondrial supernatant

Pellet containing large particles (mitochondria, lysosomes, peroxisomes)

Postmitochondrial supernatant

(4) Centrifuge 80,000 *g* for 1 h

Postmicrosomal supernatant

Pellet containing fragments of plasma membranes, smooth ER, and rough ER ("microsomal pellet")

Postmicrosomal supernatant

(5) Centrifuge 200,000 *g* for 2 h

Cytosol (soluble cellular components)

Pellet containing free ribosomes, viruses, and large macromolecules

Figure 12A-4   Differential Centrifugation and the Isolation of Organelles.

An example of differential centrifugation is illustrated in Figure 12A-4. The tissue of interest is first homogenized (①). Subcellular fractions are then isolated by subjecting the homogenate and subsequent supernatant fractions to succes-

sively higher centrifugal forces or longer centrifugation times (②–⑤). The **supernatant** is the clarified homogenate that remains after particles of a given size and density are removed as a **pellet** by the centrifugation process. In each case, the supernatant from one step is decanted, or poured off, into a new centrifuge tube and then returned to the centrifuge and subjected to greater centrifugal force to obtain the next pellet. In successive steps, the pellets are enriched in: unbroken cells and debris(②); mitochondria, lysosomes, and peroxisomes (③); ER and other membrane fragments (④); free ribosomes and large macromolecules (⑤). The material in each pellet can be resuspended and used for electron microscopy or biochemical studies. The final supernatant consists mainly of soluble cellular components and is called the *cytosol.*

Each fraction obtained in this way is enriched for the respective organelles, but is also likely contaminated with other organelles and cellular components. Often, most of the contaminants in a pellet can be removed by resuspending the pellet in an isotonic solution and repeating the centrifugation procedure.

Density Gradient Centrifugation

In the previous example of differential centrifugation, the particles about to be separated were uniformly distributed throughout the solution prior to centrifugation. **Density gradient** (or **rate-zonal**) **centrifugation** is a variation of differential centrifugation in which the sample for fractionation is placed as a thin layer on top of a *gradient of solute.* The gradient consists of an increasing concentration of solute—and therefore density—from the top of the tube to the bottom. The sample for fractionation and the top portion of the gradient are often isotonic. When subjected to a centrifugal force, particles differing in size and/or density move downward as discrete *zones,* or bands, that migrate at different rates. Because of the gradient of solute in the tube, the particles at the leading edge of each zone continually encounter a slightly denser solution and are therefore slightly impeded. As a result, each zone remains very compact, maximizing the resolution of different particles.

The process is illustrated in Figure 12A-5. Particles that are large or dense (purple) move into the gradient as a rapidly sedimenting band, particles that are intermediate in size or density (blue) sediment less rapidly, and the smallest particles (black) move quite slowly. The centrifuge is stopped after the bands of interest have moved sufficiently far into the gradient to be resolved from each other, but *before* any of the bands reach the bottom of the tube. Stopping at this point is essential because all of the particles are more dense than the solution in the tube, even at the very bottom of the tube. If centrifugation continues too long, the bands will reach the bottom of the tube one after another, piling up and negating the very purpose of the process.

Density gradient centrifugation is commonly used for separating both organelles and macromolecules. Figure 12A-6 illustrates the separation of mitochondria and lysosomes by this

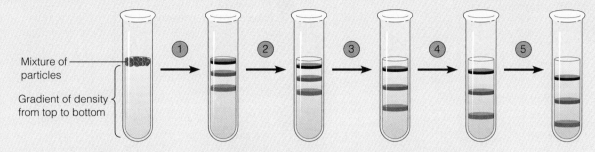

Mixture of particles

Gradient of density from top to bottom

● Particles with large sedimentation coefficients
● Particles with intermediate sedimentation coefficients
· Particles with small sedimentation coefficients

Figure 12A-5   Density Gradient Centrifugation.   Density gradient centrifugation, like differential centrifugation (see Figure 12A-2), is a technique for separating particles based on differences in sedimentation rate. For this centrifugation method, however, the sample for fractionation is placed as a thin layer on top of a gradient of solute that increases in density from the top of the tube to the bottom. The effect is illustrated here for three particles that differ significantly in size. Subjected to a fixed centrifugal force for five successive time intervals (circled numbers), the particles migrate through the gradient as distinct bands. Furthermore, the bands are sharpened as the leading edge of each one is impeded when it encounters increasingly denser solution.

technique. (For simplicity, we will assume that the tissue contains few peroxisomes.) The tissue of interest is first homogenized and subjected to differential centrifugation to obtain a pellet enriched in mitochondria and lysosomes (①). The pellet is then resuspended in a small volume of isotonic solution and layered over a gradient of solute that increases in concentration and density from the top of a plastic tube to the bottom (②). Notice that we cannot use a glass tube because the bottom must be punctured to collect individual bands.

During centrifugation, the mitochondria, which are both larger and denser than lysosomes, move into the gradient as a rapidly sedimenting band (③). After a suitable length of time, the centrifuge is stopped, the bottom of the tube is punctured, and *fractions* are collected (④). The mitochondria, which move faster than the lysosomes, are closest to the bottom of the tube and are collected first. Lysosomes are collected in later fractions. By assaying each fraction for marker enzymes that are unique to either mitochondria or lysosomes, the fractions containing these organelles can be identified and the extent of cross-contamination determined (⑤).

Equilibrium Density Centrifugation

Equilibrium density (or **buoyant density**) **centrifugation** is a powerful method for resolving organelles and macromolecules based on density differences (see Figure 12A-3). Like density gradient centrifugation, this procedure includes a gradient of solute that increases in concentration and density, but in this case the solute is concentrated so that the density gradient spans the range of densities of the organelles or macromolecules about to be separated. For organelles, a gradient of sucrose is often used, and the density range is 1.10–1.30 g/cm³ (0.75–2.3 M sucrose). For macromolecules, such as RNA and DNA, which have higher densities, a heavy metal salt, such as cesium chloride (CsCl), is commonly used.

Figure 12A-7 illustrates the use of equilibrium density centrifugation for the separation of organelles. The tissue of interest is first homogenized and subjected to differential centrifugation to obtain a pellet enriched in mitochondria, lysosomes, and peroxisomes (①). The pellet is then resuspended in a 0.25 M sucrose solution and layered over a gradient of sucrose that spans the densities of the three organelles (②). During centrifugation, the organelles move into the gradient until each reaches its *equilibrium,* or *buoyant, density*—the point in the gradient at which the density of the organelle is exactly equal to the density of the sucrose (③). At its buoyant density, an organelle has no net force acting on it, so it moves no further. Given enough time, all the organelles will reach their characteristic buoyant density positions in the gradient and will remain there. When the organelles are at their equilibrium positions, the centrifuge is stopped, the bottom of the tube is punctured, and the fractions are collected (④). Because of their different densities, the three classes of organelles are collected in different fractions (⑤).

For a classic experiment in which CsCl density gradients were used to resolve double-stranded DNA molecules containing nitrogen isotopes ^{14}N versus ^{15}N, as well as to detect hybrid DNA molecules containing both isotopes, see Figure 17-4.

(continued)

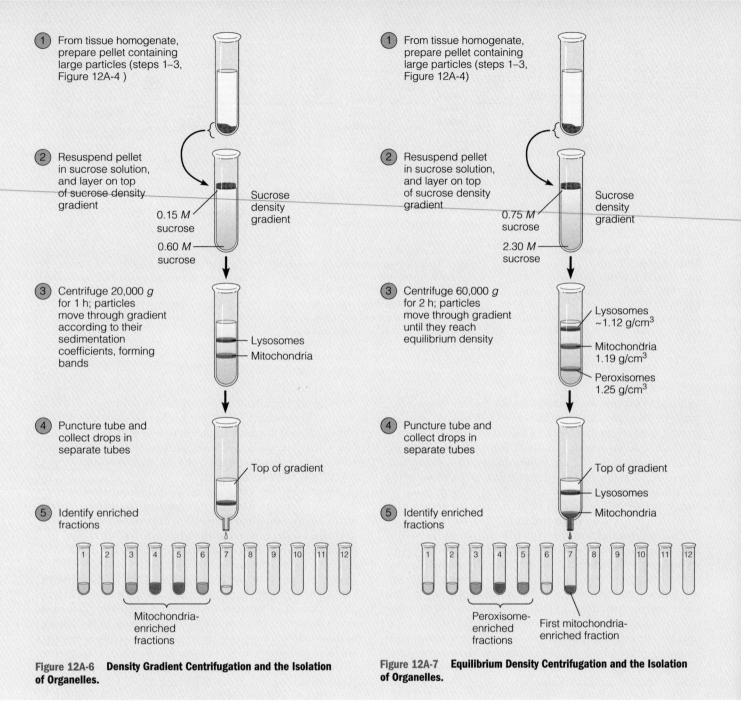

Figure 12A-6   Density Gradient Centrifugation and the Isolation of Organelles.

Figure 12A-7   Equilibrium Density Centrifugation and the Isolation of Organelles.

will be discussed in detail later in this chapter. An example of an enzyme involved in folding polypeptides is *protein disulfide isomerase,* which catalyzes the formation and breakage of disulfide bonds between cysteine residues of a polypeptide chain. As we learned in Chapter 3, formation of such bonds is often essential for proper folding of polypeptides (see Figures 3-5 and 3-8). We will return to the topic of protein folding in Chapter 20.

Along with folding and assembly of proteins, the ER is a site for quality control. Proteins that are not properly modified, folded, or assembled are exported from the ER

for degradation before they have an opportunity to move on to the Golgi complex.

Smooth ER Is Involved in Drug Detoxification, Carbohydrate Metabolism, and Other Cellular Processes

Smooth ER is involved in several different cellular processes, including drug detoxification, carbohydrate metabolism, calcium storage, and steroid biosynthesis. Especially important in the latter category are the male and female sex hormones and the steroid hormones of the adrenal cortex.

Drug Detoxification. A reaction common to most pathways for drug detoxification and steroid biosynthesis is **hydroxylation,** the addition of hydroxyl groups to organic acceptor molecules. In each of these cases, hydroxylation depends on a cytochrome P-450 protein. Members of this group of proteins are especially prevalent in the smooth ER of hepatocytes (liver cells) but are also found in lung and intestinal cells. An electron transport system in the smooth ER transfers electrons stepwise from the reduced coenzyme nicotinamide adenine dinucleotide phosphate (NADPH) to cytochrome P-450. The reduced form of P-450 can then donate an electron to molecular oxygen (O_2), activating the molecule for hydroxylation. Omitting the electron transport system, the net reaction is

$$RH + NADPH + H^+ + O_2 \longrightarrow ROH + NADP^+ + H_2O \quad \textbf{(12-1)}$$

Hydroxylation reactions catalyzed by ER enzymes can also use nicotinamide adenine dinucleotide (NADH) as an electron donor, depending on the substrate for hydroxylation. While NADPH is the electron donor for drug detoxification and steroid biosynthesis, NADH serves as an electron donor during the hydroxylation of fatty acids. In addition to NADPH or NADH, the second essential molecule for hydroxylation is oxygen. As indicated by reaction 12-1, one atom of the oxygen molecule is used to hydroxylate the substrate and the other is reduced to water. Enzymes that catalyze such hydroxylation reactions are called *mixed-function oxidases* or *monooxygenases.*

During drug detoxification, hydroxylation increases the solubility of hydrophobic drugs in water. This alteration is critical because most hydrophobic compounds are soluble in the lipid layers of membranes and are therefore retained by the body, whereas water-soluble compounds are more easily flushed away by the blood and subsequently excreted from the body.

The elimination of barbiturate drugs, for example, is enhanced by hydroxylation enzymes associated with the smooth ER. This can be demonstrated by injecting the sedative phenobarbital into a rat. One of the most striking effects is a rapid increase in the level of barbiturate-detoxifying enzymes in the liver, accompanied by a dramatic proliferation of smooth ER. A concomitant effect, however, is that increasingly higher doses of the drug are necessary to achieve the same sedative effect in habitual users of phenobarbital. Furthermore, the mixed-function oxidase induced by phenobarbital is of such broad specificity that it can hydroxylate and therefore solubilize a variety of other drugs, including such useful agents as antibiotics, anticoagulants, and steroids. As a result, the chronic use of barbiturates decreases the effectiveness of a host of other clinically useful drugs.

Another prominent protein found in the smooth ER and involved in hydroxylation is cytochrome P-448, which is part of an enzyme complex called *aryl hydrocarbon hydroxylase.* This complex is involved in metabolizing *polycyclic hydrocarbons,* organic molecules composed of two or more linked benzene rings (see Figure 2-4). While hydroxylation of such molecules can be important for increasing the solubility of toxic substances in water, the oxidized products are often more toxic than the original compounds. Research has revealed that aryl hydrocarbon hydroxylase can convert potential carcinogens into their chemically active forms. Mice synthesizing high levels of this hydroxylase exhibit a higher incidence of spontaneous cancer than normal mice, whereas mice treated with an inhibitor of aryl hydrocarbon hydrolase develop few tumors. Significantly, cigarette smoke is a potent inducer of aryl hydrocarbon hydroxylase.

Carbohydrate Metabolism. The smooth ER of hepatocytes is also involved in the catabolism (breakdown) of stored glycogen, as evidenced by the presence of the enzyme *glucose-6-phosphatase.* Glucose-6-phosphatase is one of several prominent membrane-bound enzymes that are unique to the ER, and it is often used as a marker to identify and follow the ER during subcellular fractionation. This enzyme catalyzes the removal of the phosphate group from glucose-6-phosphate to form free glucose and inorganic phosphate (P_i):

$$\text{glucose-6-phosphate} + H_2O \longrightarrow \text{glucose} + P_i \quad \textbf{(12-2)}$$

To understand the importance of this phosphatase, we need to appreciate both the way in which liver cells store glycogen and the reason and mechanism for its subsequent breakdown.

A major role of the liver is to keep the level of glucose in the blood relatively constant. The liver stores glucose as glycogen and releases it as needed by the body, especially between meals and in response to muscular activity. Liver glycogen is stored as granules associated with smooth ER (Figure 12-3a). The mobilization of liver glycogen is hormonally mediated (see Figure 10-25). This process involves the stimulation of a membrane-bound enzyme called *adenylyl cyclase,* resulting in a transient elevation in the intracellular level of *cyclic AMP (cAMP).* This form of AMP is involved in the regulation of a variety of cellular processes (see Figure 10-5). The cAMP elevation in turn triggers a complex cascade of events, leading eventually to the activation of *glycogen phosphorylase,* an enzyme that cleaves glycogen into glucose-1-phosphate units (Figure 12-3b). Glucose-1-phosphate can be readily converted to glucose-6-phosphate in the cytosol by an enzyme called *phosphoglucomutase.*

Membranes, however, are generally impermeable to phosphorylated sugars. To leave the liver cell and enter the bloodstream, the glucose-6-phosphate must be converted to free glucose. The glucose-6-phosphatase bound to the smooth ER membrane removes the phosphate group from glucose-6-phosphate, allowing the glucose to be moved out of the liver cell by a permease (the glucose transporter encountered in Chapter 8) and into the blood for transport to other cells that need energy. Significantly, glucose-6-phosphatase activity is present in liver, kidney, and intestinal cells, but not in muscle or brain cells.

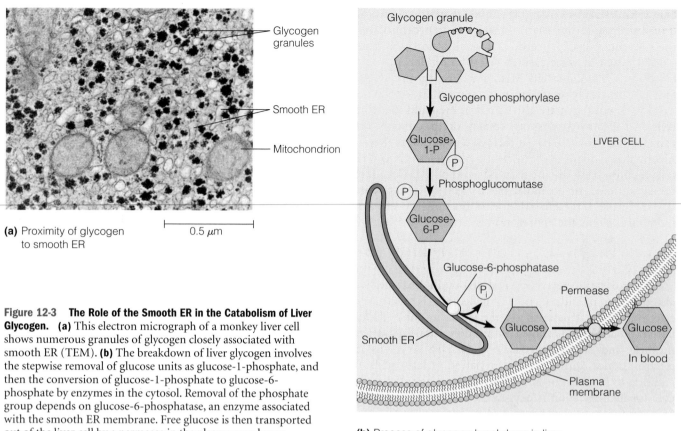

(a) Proximity of glycogen to smooth ER

0.5 μm

Figure 12-3   The Role of the Smooth ER in the Catabolism of Liver Glycogen. **(a)** This electron micrograph of a monkey liver cell shows numerous granules of glycogen closely associated with smooth ER (TEM). **(b)** The breakdown of liver glycogen involves the stepwise removal of glucose units as glucose-1-phosphate, and then the conversion of glucose-1-phosphate to glucose-6-phosphate by enzymes in the cytosol. Removal of the phosphate group depends on glucose-6-phosphatase, an enzyme associated with the smooth ER membrane. Free glucose is then transported out of the liver cell by a permease in the plasma membrane.

(b) Process of glycogen breakdown in liver

Muscle and brain cells retain glucose-6-phosphate and use it to meet their own substantial energy needs.

Calcium Storage.   Depending on the type of cell, specialized subdomains of smooth ER often serve as sites for the storage of calcium ions. In such cells, the ER lumen contains high concentrations of calcium-binding proteins. The calcium ions are pumped into the ER by *ATP-dependent calcium pumps* and are released in response to extracellular signals (see Figure 10-11). The *sarcoplasmic reticulum* found in muscle cells is an example of smooth ER that specializes in the storage of calcium. ATP-dependent calcium pumps continuously move calcium ions into the sarcoplasmic reticulum lumen. Binding of neurotransmitter molecules to receptors on the surface of the muscle cell triggers a signal cascade that leads to the release of calcium and the contraction of muscle fibers.

The ER Plays a Central Role in the Biosynthesis of Membranes

Studies of the biosynthesis of lipids and their fates within eukaryotic cells reveal that the ER is the primary source of membrane lipids, including phospholipids and cholesterol. Indeed, most of the enzymes required for the biosynthesis of the various membrane phospholipids are found nowhere else in the cell. There are, however, important exceptions. For example, while mitochondria import from the ER all of the phosphatidylcholine, phosphatidylinositol, phosphatidylserine, and cholesterol found in their exterior and interior membranes, they acquire phosphatidylethanolamine indirectly by decarboxylating imported phosphatidylserine. Another significant exception is the biosynthesis of certain lipids by peroxisomal enzymes. Peroxisomes can contribute as much as 20% of the cholesterol and 50% of the dolichol found in some cells, especially in liver tissue.

Biosynthesis of phospholipid molecules is restricted to one monolayer of the ER membrane. Specifically, the active sites of the enzymes involved are exposed to the cytosol, and newly synthesized lipids are incorporated into the monolayer of the membrane that faces the cytosol. Cellular membranes, of course, are phospholipid *bilayers*, with phospholipids distributed to both sides. Thus, there must be a mechanism for transferring phospholipids from one layer of the membrane to the other. Because it is thermodynamically unfavorable for phospholipids to spontaneously flip at a significant rate from one side of a bilayer to the other, transfer depends on **phospholipid translocators,** or **flippases,** which catalyze the translocation of phospholipids through ER membranes.

Phospholipid translocators, like other enzymes, are quite specific and affect only the rate of a process. As a result, the precise phospholipid molecules transferred across a membrane depend on the complement of translocators available. Moreover, transfer occurs only until the forward and reverse processes reach equilibrium. This translocator specificity contributes to the *membrane asymmetry* described in Chapter 7. For example, the ER membrane contains a translocator for phosphatidylcholine, but not for phosphatidylethanolamine, phosphatidylinositol, or phosphatidylserine. Consequently, phosphatidylcholine is found in both the cytosolic and lumenal layers of the ER membrane whereas the latter three phospholipids are confined to the cytosolic layer. When vesicles form from the ER membrane and fuse with other organelles of the endomembrane system, the distinct compositions of the cytosolic and lumenal layers established in the ER are transferred to other cellular membranes.

Movement of phospholipids from the ER to a mitochondrion or chloroplast poses a unique problem. Unlike organelles of the endomembrane system, mitochondria and chloroplasts do not grow by fusion with ER-derived vesicles. Instead, **phospholipid exchange proteins** (or *phospholipid transfer proteins*) found in the cytosol convey phospholipid molecules from the ER membrane to the outer mitochondrial and chloroplast membranes. Each exchange protein recognizes a specific kind of phospholipid, removes it from one membrane, and carries it through the cytosol to another membrane. Such transfer proteins also contribute to the movement of phospholipids from the ER to other cellular membranes, including the plasma membrane.

Although the ER is the source of most membrane lipids, the compositions of other cellular membranes vary significantly from the composition of the ER membrane (Table 12-1). For example, a striking feature of the plasma membrane of hepatocytes is the relatively low amount of phosphoglycerides and high amounts of cholesterol, sphingomyelin, and glycolipids. Researchers have observed an increasing gradient of cholesterol content from the ER through the compartments of the endomembrane system to the plasma membrane. This corre-

lates with an increasing gradient of membrane thickness. ER membranes are about 5 nm thick, whereas plasma membranes are about 8 nm thick. Some biologists believe that the observed change in membrane thickness has implications for sorting and targeting of integral membrane proteins, which we will discuss after we look at the Golgi complex and its role in protein processing.

The Golgi Complex

We now turn our attention to the Golgi complex, a component of the endomembrane system that is closely linked, both physically and functionally, to the ER. The **Golgi complex** (or *Golgi apparatus*) derives its name from Camillo Golgi, the Italian biologist who first described this structure in 1898. He reported that nerve cells soaked in osmium tetroxide showed deposits of heavy metal in a threadlike network surrounding the nucleus. The same staining reaction was demonstrated with a variety of cell types and other heavy metals. The results were inconsistent, however, and no cellular structure could be identified that explained the staining. As a result, the nature—actually, the very existence—of the Golgi complex remained highly controversial until the 1950s, when its existence was finally confirmed by electron microscopy. Since then, we have come to understand much about the Golgi complex, its close relationship with the ER, and its role in the chemical modification, sorting, and packaging of the proteins that it receives from the ER.

The Golgi Complex Consists of a Series of Membrane-Bounded Cisternae

The Golgi complex consists of a series of flattened membrane-bounded *cisternae*, disk-shaped sacs that are stacked together as illustrated in Figure 12-4a. A series of such cisternae is called a **Golgi stack** (also called a *dictyosome* in plant cells). These structures can be readily visualized by electron microscopy (Figure 12-4b). Usually, there are 3–8 cisternae per stack, though the Golgi stacks of some organisms can include several dozen cisternae. The number and size of Golgi stacks vary with cell type and with the metabolic activity of the cell. Some cells have one large stack, whereas others—especially those that are highly active in secretion—have hundreds or even thousands of Golgi stacks. The Golgi complex lumen, or *intracisternal space*, is part of the endomembrane system's network of internal spaces (see Figure 12-1).

Transport Vesicles. The static view of the endoplasmic reticulum and Golgi complex presented by electron micrographs such as Figure 12-4b is misleading; these organelles are actually very dynamic structures. Both the ER and Golgi complex are typically surrounded by numerous **transport vesicles,** which bud off membranes in one region of the cell and fuse with other membranes.

Table 12-1 Composition of the ER and Plasma Membranes of Rat Liver Cells

Membrane Components	ER Membrane	Plasma Membrane
Membrane components as % of membrane by weight		
Carbohydrate	10	10
Protein	62	54
Total lipid	27	36
Membrane lipids as % of total lipids by weight		
Phosphatidylcholine	40	24
Phosphatidylethanolamine	17	7
Phosphatidylserine	5	4
Cholesterol	6	17
Sphingomyelin	5	19
Glycolipids	trace	7
Other lipids	27	22

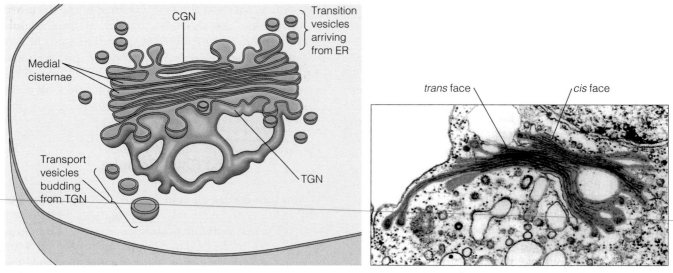

(a) A Golgi stack in an animal cell

(b) A dictyosome in an algal cell

0.5 µm

Figure 12-4 Golgi Structure. A Golgi stack, or dictyosome, usually consists of a small number of flattened cisternae. **(a)** At the *cis* face, transition vesicles arriving from the ER fuse with membranes of the *cis*-Golgi network (CGN). At the *trans* face, transport vesicles arise by budding from the *trans*-Golgi network (TGN). The transport vesicles convey lipids and proteins to other components of the endomembrane system or form secretory vesicles. **(b)** This electron micrograph shows a dictyosome (highlighted blue) lying next to the nuclear envelope of an algal cell (TEM).

Such vesicles convey lipids and proteins from the transitional elements of the ER to the Golgi complex, between the Golgi stack cisternae, and from the Golgi complex to various destinations in the cell, including secretory granules, endosomes, and lysosomes.

Most of the vesicles involved in lipid and protein transfer are also referred to as *coated vesicles* because of the characteristic coats, or layers, of proteins covering their surfaces as they form. The coat's precise composition depends on the role of the vesicle in the cell. The most studied coat proteins are *clathrin, COPI,* and a pair of protein complexes called *COPII*. (*COP* is an abbreviation for "cytosolic coat protein.") Coated vesicles are a common feature of most cellular processes that involve the transfer or exchange of substances between specific membrane-bounded compartments of eukaryotic cells. We will discuss coated vesicles and their roles in lipid and protein transport in more detail later in the chapter.

The Two Faces of the Golgi Stack. Each Golgi stack has two distinct sides, or *faces* (see Figure 12-4a). The *cis* (or **forming**) **face** is oriented toward the transitional elements of the ER. The Golgi compartment closest to the transitional elements is a network of membrane-bounded tubules referred to as the *cis*–**Golgi network (CGN)**. Coated transition vesicles containing newly synthesized lipids and proteins from transitional elements of the ER continuously arrive at the CGN, where they fuse with the CGN membrane. The opposite side of the Golgi complex is called the

trans (or **maturing**) **face.** The compartment on this side of the Golgi complex, like the CGN, is a network of membrane-bounded tubules. This structure is referred to as the *trans*–**Golgi network (TGN)**. Coated transport vesicles continuously bud from the tips of TGN cisternae, carrying processed proteins from the Golgi complex to secretory granules, endosomes, lysosomes, and the plasma membrane. The sacs between the CGN and TGN comprise the **medial cisternae** of the Golgi stack, in which much of the processing of proteins occurs.

The CGN, TGN, and medial cisternae of the Golgi complex are biochemically and functionally distinct, each compartment containing enzymes necessary for specific steps in protein processing. Immunological and cytochemical staining techniques have shown that specific receptor proteins and enzymes are concentrated within the CGN, whereas other proteins are found primarily in medial cisternae or in the TGN. This biochemical *polarity* is illustrated in Figure 12-5, which shows a Golgi stack in a rabbit kidney cell. Staining to detect *N-acetylglucosamine transferase I,* an enzyme that adds *N*-acetylglucosamine to carbohydrate side chains previously attached to proteins in the ER or CGN, shows that the enzyme is concentrated in medial cisternae of the Golgi complex.

The CGN, TGN, and medial cisternae of the Golgi complex also differ in the kinds of coated vesicles associated with each compartment. Vesicles that bud from the ER are coated with either COPI or COPII proteins, while vesicles that bud from the CGN or medial cisternae of the

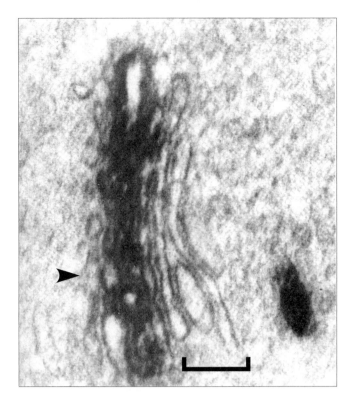

Figure 12-5   Immunochemical Staining of a Golgi Complex. This electron micrograph shows a Golgi stack in a rabbit kidney cell. The cell section has been stained to detect the enzyme *N-acetylglucos-amine transferase I,* which plays a role in terminal glycosylation of proteins. The arrow and bracket indicate unlabeled cisternae, showing that the enzyme is concentrated in a few medial cisternae close to the *cis* face of the Golgi stack. This indicates that *N*-acetyl-glucosamine is added to existing oligosaccharides on glycoproteins shortly after the proteins enter the Golgi stack (TEM).

Golgi stack are coated with COPI proteins only. Vesicles that bud from the TGN are usually coated with clathrin.

Two Models Depict the Flow of Lipids and Proteins Through the Golgi Complex

There are at least two models to explain the movement of lipids and proteins through the medial cisternae of the Golgi complex. According to the **stationary cisternae model,** each compartment of the Golgi stack is a stable structure. Traffic between successive cisternae is mediated by **shuttle vesicles** that bud from one cisterna and fuse with another, usually the next cisterna in the *cis*-to-*trans* sequence. Proteins destined for the TGN—where they will be sorted and packaged into vesicles targeted for various destinations—are not necessarily selected and concentrated but are simply carried forward by budding shuttle vesicles, while molecules that belong in the ER and successive Golgi compartments are actively retained or retrieved.

According to the second model, known as the **cisternal maturation model,** the Golgi cisternae are transient compartments that gradually change from CGN cisternae through medial cisternae to TGN cisternae. In this model, transition vesicles from the ER converge to form the CGN. This CGN accumulates specific enzymes for the early steps of protein processing. Step by step, the *cis* cisterna is transformed into an intermediate medial cisterna, and then into a TGN cisterna, as it acquires additional enzymes. Enzymes that are no longer needed in late compartments return in vesicles to early compartments. Finally, the TGN forms transport vesicles or secretory granules containing sorted cargo targeted for various destinations beyond the Golgi complex.

The two models are not mutually exclusive; it is likely that both apply to some degree, depending on the organism and the role of the cell. While the first model is supported by substantial evidence, some cellular components observed in medial compartments of the Golgi complex are clearly too large to travel by the small shuttle vesicles found in cells. For example, polysaccharide scales produced by several algal species first appear in early Golgi compartments. Too large to fit inside transport vesicles, the scales nevertheless reach late Golgi compartments, which eventually fuse with the plasma membrane and release the finished product for incorporation into the cell wall. A similar process of cisternal maturation apparently occurs as procollagen in fibroblasts and casein in lactating mammary gland cells travel through the Golgi complex.

Anterograde and Retrograde Transport.   The movement of material from the ER through the Golgi complex toward the plasma membrane is called **anterograde transport** (*antero* is derived from a Latin word meaning "front" and *grade* is related to a word meaning "to go"). Included among the lipids and proteins carried by vesicles are molecules essential only for packaging cargo and directing vesicles to their destinations. For example, every time a secretory granule fuses with the plasma membrane and discharges its contents by exocytosis, a bit of membrane that originated from the ER becomes a part of the plasma membrane. To balance the flow of lipids toward the plasma membrane and to ensure a supply of components for forming new vesicles, the cell recycles lipids and proteins no longer needed during the late stages of anterograde transport. This is accomplished by **retrograde transport** (*retro* is a Latin word meaning "back"), the flow of vesicles from Golgi cisternae back toward the ER.

In the stationary cisternae model, retrograde flow facilitates the recovery of ER-specific lipids and proteins that are passed from the ER to the *cis* face of a nearby Golgi stack, or the return of compartment-specific proteins back to distinct medial cisternae of the Golgi stack. Material destined for the TGN continues forward. Whether such retrograde traffic occurs directly from all medial cisternae of the Golgi stack back to the ER or by reverse flow through successive cisternae is not clear. In the cisternal maturation model, retrograde flow carries material back toward newly forming compartments after receptors and enzymes are no longer needed in the more mature compartments.

Roles of the ER and Golgi Complex in Protein Glycosylation

Much of the protein processing carried out within the ER and Golgi complex involves **glycosylation,** the addition of carbohydrate side chains to specific amino acid residues of proteins, forming **glycoproteins.** Subsequent enzyme-catalyzed reactions then modify the oligosaccharide side chain that was attached to the protein. Two general kinds of glycosylation are observed in cells (see Figure 7-25). **N-linked glycosylation** (or **N-glycosylation**) involves the addition of a specific oligosaccharide unit to the terminal amino group of asparagine residues, whereas **O-linked glycosylation** involves addition of oligosaccharides to hydroxyl groups of serine or threonine residues. We will focus here on N-glycosylation. Each step of glycosylation is strictly dependent on preceding modifications. An error at one step, perhaps due to a defective enzyme, will probably block further modification of a carbohydrate side chain and can lead to severe consequences for cellular metabolism. In fact, aberrant enzymes involved in glycosylation have been implicated in carcinogenesis.

Conceptually, we can divide N-glycosylation into two stages—the initial glycosylation event and the subsequent modification of the carbohydrate side chain. As illustrated in Figure 12-6, the enzymes that catalyze various steps of glycosylation and subsequent modifications are present in different compartments, or groups of compartments, of the ER and Golgi complex.

The first stage of N-glycosylation, called **core glycosylation,** takes place in the ER. Invariably, the carbohydrate directly linked to asparagine is *N-acetylglucosamine (GlcNAc),* a modified carbohydrate we encountered in Chapter 3. Despite the variety of oligosaccharides found in glycoproteins, all the carbohydrate groups added to proteins in the ER have a common high-mannose core structure consisting of two GlcNAc units, nine mannose units, and three glucose units. This **core oligosaccharide** is built up by the sequential addition of monosaccharide units to an activated lipid carrier called *dolichol phosphate.* Assembly of the core oligosaccharide is illustrated in Figure 12-7. Notice that early steps of the process occur on the cytosolic side of the ER membrane and later steps occur in the ER lumen. Translocation from the cytosol to the ER lumen is catalyzed by a specific flippase. The completed core oligosaccharide is then transferred as a single unit from dolichol to an asparagine residue of the protein, a reaction catalyzed by a membrane-bound *oligosaccharyl transferase.* In some cases, the core oligosaccharide is added to the protein as the polypeptide is synthesized by a ribosome bound to the ER membrane.

During the second stage of N-glycosylation, the core oligosaccharide is trimmed and modified. As in core glycosylation, the initial steps of processing are confined to the ER and often occur before biosynthesis of the polypeptide chain is completed. Specifically, the three glucose units and one of the mannose units are usually removed by *glucosidases* and *mannosidases.* Further processing of

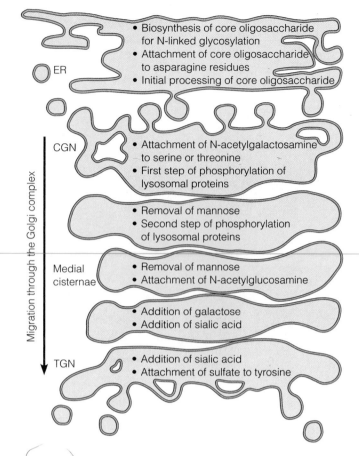

Figure 12-6 **Compartmentalization of the Steps of Glycosylation and Subsequent Modification of Proteins.** Enzymes that catalyze specific steps of glycosylation and further modification of proteins reside in different compartments, or groups of compartments, of the ER and Golgi complex. Processing occurs sequentially as proteins travel from compartment to compartment.

N-glycosylated proteins is carried out in the Golgi complex as the glycoproteins move from the *cis* face through the medial cisternae to the *trans* face of the Golgi stack. This sequence of events is called **terminal glycosylation** to distinguish it from the initial core glycosylation that occurs in the ER.

Terminal glycosylation always includes the removal of a few of the carbohydrate units that were added as part of the core oligosaccharide in the ER. In some cases, no further processing occurs when the glycoprotein reaches the Golgi complex. The resulting *high-mannose oligosaccharides* retain the two GlcNAc units and six to eight of the mannose groups present in the core oligosaccharide. In other cases, *complex oligosaccharides* are generated by the further addition of N-acetylglucosamine and other carbohydrates, including galactose, sialic acid, and fucose. (The structures of galactose and sialic acid are illustrated in Figure 7-26a.) Depending on the carbohydrate, further additions occur in a medial compartment near the *trans* face of the Golgi complex or in the TGN. As glycoproteins exit the Golgi complex, they may contain only high-mannose oligosaccharides, only complex oligosaccharides, or both.

Isopreie group

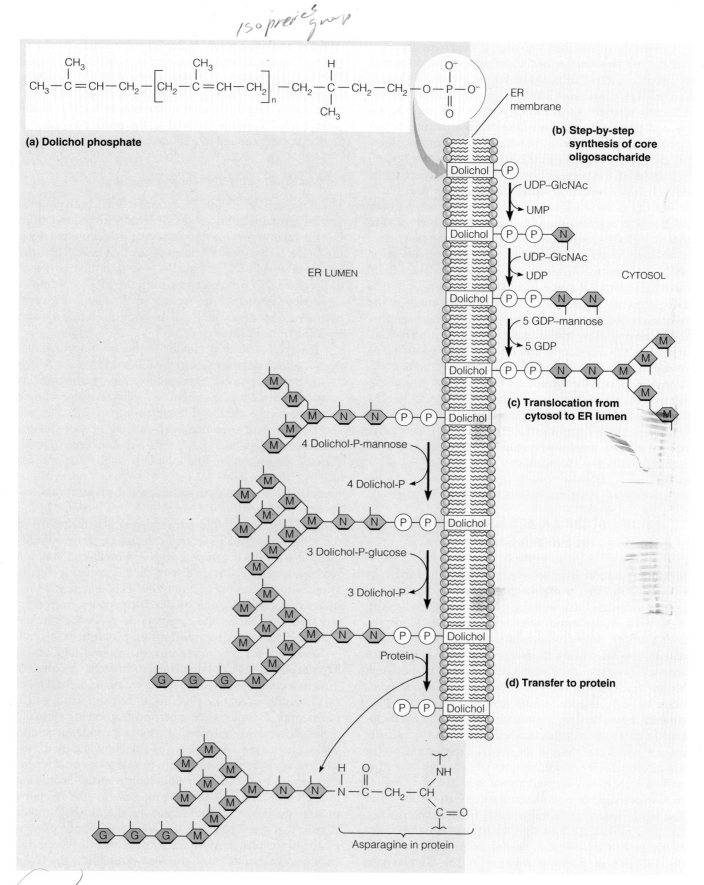

Figure 12-7 Assembly of Core Oligosaccharide and Transfer to Protein. (a) Dolichol phosphate is the carrier of oligosaccharide units that are transferred to asparagine during core glycosylation. (b) Step-by-step synthesis of the high- mannose core oligosaccharide by successive addition of *N*-acetylglucosamine (N), mannose (M), and glucose (G) units to the growing chain. (c) Translocation of the incomplete core oligosaccharide from the cystol to the ER lumen is catalyzed by a phospholipid translocator (flippase). (d) Transfer of the completed core oligosaccharide from dolichol phosphate to an asparagine residue of the protein.

Given the role of the Golgi complex in glycosylation, it is not surprising that two of the most important categories of enzymes present in Golgi stacks are *glucan synthetases* and *glycosyl transferases*. The synthetases catalyze the formation of oligosaccharides from monosaccharides, and the glycosyl transferases are instrumental in attaching carbohydrate groups to proteins. There are more than 200 different glycosyl transferases in the ER and Golgi complex, an indication of the potential complexity of oligosaccharide side chains. These and other enzymes are distributed within the Golgi stack in a manner consistent with their role in the glycosylation process, such that each cisterna in the stack contains a distinctive set of processing enzymes.

Notice in the preceding discussion that glycosylation occurs only on the lumenal and not the cytosolic side of the ER and Golgi complex membranes. The glycoproteins and glycolipids assembled in these organelles are therefore additional factors that contribute to membrane asymmetry. Following biosynthesis, the glycosylated proteins and lipids face the interiors of the ER and Golgi complex. As vesicles bud from these organelles and fuse with other cellular membranes, the asymmetry established during glycosylation is carried over to the other membranes. Because the lumenal side of the ER membrane is topologically equivalent to the exterior surface of the cell, it is easy to see why all plasma membrane glycoproteins are found on the extracellular side of the membrane.

Roles of the ER and Golgi Complex in Protein Sorting

Membrane-bound and soluble proteins synthesized by ribosomes attached to the rough ER must be directed to a variety of intracellular locations, including the ER itself, the Golgi complex, endosomes, and lysosomes. Moreover, once a protein reaches an organelle where it is supposed to remain, there must be a mechanism for preventing it from leaving. Other groups of proteins synthesized in the rough ER are destined for incorporation into the plasma membrane or for release to the outside of the cell. Each protein contains a specific "tag" that targets the protein for inclusion in transport vesicles that will convey material from one specific cellular location to another. Depending on the protein and its destination, the tag may be a specific amino acid sequence, an oligosaccharide side chain, a hydrophobic domain, or another structural feature. Tags may also be involved in excluding material from certain vesicles.

The traffic associated with the ER and Golgi complex is outlined in Figure 12-8. Sorting of proteins begins in the ER and early compartments of the Golgi stack, which contain mechanisms for retrieving or retaining compartment-specific proteins. This is an important step in the sorting process, for it preserves the compartment-specific functions needed to maintain the integrity of glyco-

sylation and processing pathways. The final sorting of material that will leave the Golgi complex occurs in the TGN, where lipids and proteins are selectively packaged into distinct populations of transport vesicles, each destined for a different location in the cell. In some cells, the Golgi complex is also involved in the processing of proteins that enter the cell by endocytosis.

ER-Specific Proteins Contain Retrieval Tags

The composition of the ER is maintained both by preventing proteins from escaping when vesicles bud from the ER membrane and by retrieving proteins that reach the CGN. It is not clear how proteins that never leave the ER are retained or what signal is built into their structure to prevent them from escaping from the ER. According to one theory, proteins that remain in the rough ER form extensive complexes that are physically excluded from budding transition vesicles.

Retrieval—which occurs by retrograde flow of transport vesicles from the CGN to the ER (see Figure 12-8, ⑨)—is better understood. Many soluble ER-specific proteins contain retrieval tags that bind to specific transmembrane receptors facing the Golgi complex lumen. The tags are short amino acid sequences such as KDEL (Lys-Asp-Glu-Leu) in mammals and HDEL (His-Asp-Glu-Leu) in yeast. (These sequences use the one-letter abbreviations for amino acids; see Table 3-2 on p. 44.) When a protein containing such a tag binds to its receptor, the receptor undergoes a conformational change and the receptor-ligand complex is packaged into a transport vesicle for return to the ER. An ER-specific protein can also be retrieved indirectly if it forms a complex with a protein containing a retrieval sequence. The acidity of the ER lumen is slightly lower than the acidity of the CGN. This apparently promotes dissociation of the ER-specific proteins from their receptors and allows the receptors to cycle back to the Golgi complex to recover more errant ER proteins.

Evidence for the importance of retrieval tags comes from experiments involving *chimeric proteins.* A chimeric protein is composed of polypeptide segments derived from two naturally occurring proteins, linked to form a single polypeptide. For this set of experiments, proteins normally secreted from the cell were altered to include an amino acid sequence thought to serve as a retrieval tag for a protein known to remain in the ER. In many cases, such proteins return to the ER after undergoing partial processing in the Golgi complex. Partial processing of the chimeric protein by enzymes found only in the Golgi complex reveals that the suspected tag does not simply prevent escape of the protein from the ER; retrieval of ER-specific enzymes is possible from cisternae throughout the Golgi stack. What is not clear, however, is whether retrograde transport occurs directly from a given Golgi cisterna back to the ER or by movement of the protein stepwise through each preceding cisterna until it reaches the ER.

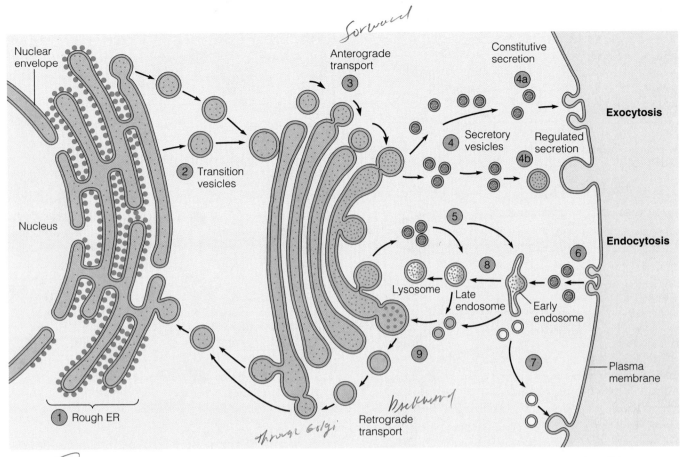

Forward

Backward

Through Golgi

Figure 12-8 Traffic Through the Endomembrane System. Vesicles carry lipids and proteins along several routes from the ER through the Golgi complex to various destinations, including secretory vesicles, endosomes, and lysosomes. ① Proteins are synthesized by ribosomes attached to the cytosolic surface of the rough ER. Initial glycosylation steps occur within the ER lumen. ② Transition vesicles carry newly synthesized lipids and glycosylated proteins to the CGN. ③ Lipids and proteins move through the cisternae of the Golgi stack via shuttle vesicles or as cisternae mature. At the TGN, vesicles bud off to form ④ secretory vesicles or ⑤ endosomes, depending on their protein content. Secretory vesicles move to the plasma membrane, where they release their contents by exocytosis, either ④a constitutively or ④b in response to an appropriate signal. ⑥ Proteins and other materials are taken into the cell by endocytosis, forming endocytic vesicles that fuse with early endosomes. ⑦ Cellular components not destined for digestion following endocytosis are recycled to the plasma membrane. ⑧ Early endosomes containing material for digestion mature to form late endosomes and then lysosomes. ⑨ Retrograde traffic returns compartment-specific proteins to earlier compartments.

Golgi Complex Proteins May Be Sorted According to the Lengths of Their Membrane-Spanning Domains

Like resident proteins of the ER, some resident proteins of the Golgi complex probably contain amino acid sequences that serve as tags for retention or retrieval. Moreover, as suggested for ER-specific proteins, the formation of large complexes that are excluded from transport vesicles may play a role in maintaining the composition of the Golgi complex. In addition, a third, distinctly different mechanism is also likely at work in this organelle.

All known Golgi complex proteins are membrane-bound proteins. Thus, each one has one or more hydrophobic membrane-spanning domains that anchor it to the membrane of the Golgi complex. The length of the hydrophobic domains may determine which cisterna of the Golgi complex each membrane-bound protein settles in as it moves through the organelle. Recall that the thickness of cellular membranes increases progressively from the ER (about 5 nm) to the plasma membrane (about 8 nm). Among Golgi-specific proteins, there is a corresponding increase in the lengths of hydrophobic membrane-spanning domains going from the CGN to the TGN. Such proteins tend to move from compartment to compartment until the thickness of the membrane exceeds the length of their membrane-spanning domains and blocks further migration.

Targeting of Soluble Lysosomal Proteins to Endosomes and Lysosomes Is a Model for Protein Sorting in the TGN

During their journey through the ER and early compartments of the Golgi complex, soluble lysosomal enzymes, like other glycoproteins, undergo N-glycosylation followed by removal of glucose and mannose units. Within the Golgi

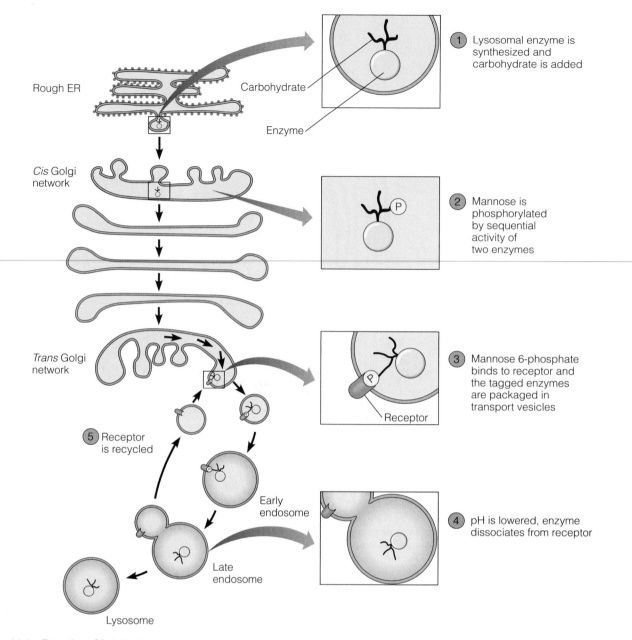

Rough ER

Cis Golgi network

Trans Golgi network

⑤ Receptor is recycled

Early endosome

Late endosome

Lysosome

① Lysosomal enzyme is synthesized and carbohydrate is added

Carbohydrate

Enzyme

② Mannose is phosphorylated by sequential activity of two enzymes

③ Mannose 6-phosphate binds to receptor and the tagged enzymes are packaged in transport vesicles

Receptor

④ pH is lowered, enzyme dissociates from receptor

Figure 12-9 Targeting of Soluble Lysosomal Enzymes to Endosomes and Lysosomes by a Mannose-6-Phosphate Tag. ① In the ER, soluble lysosomal enzymes undergo N-glycosylation followed by removal of glucose and mannose units. ② Within the Golgi complex, mannose residues on the lysosomal enzymes are phosphorylated by two enzymes. The first adds *N*-acetylglucos- amine-1-phosphate to carbon atom 6 of mannose and the second removes *N*-acetylglucosamine, leaving behind the phosphorylated mannose residue. ③ The tagged lysosomal enzymes bind to mannose-6-phosphate receptors in the TGN and are packaged into coated transport vesicles that convey the enzymes to an endosome. ④ When the receptor-ligand complexes reach a late endosome, the acidity of the endosomal lumen causes the enzymes to dissociate from their receptors. ⑤ The receptors are recycled in vesicles that return to the TGN. The late endosome either matures to form a lysosome or transfers its contents to an active lysosome.

complex, however, mannose residues on the carbohydrate side chain of lysosomal enzymes are phosphorylated, form- ing an unusual oligosaccharide containing mannose- 6-phosphate. This oligosaccharide tag distinguishes soluble lysosomal proteins from other glycoproteins and ensures their delivery to lysosomes (Figure 12-9).

The phosphorylation of mannose residues is cata- lyzed by two Golgi-specific enzymes. The first, located in an early compartment of the Golgi stack, is a phospho-

transferase that adds GlcNAc-1-phosphate to carbon atom 6 of mannose. The second, located in a mid-Golgi compartment, removes GlcNAc, leaving behind the man- nose-6-phosphate residue.

The interior surface of the TGN membrane, in turn, displays mannose-6-phosphate receptors (MPRs). The pH of the TGN is around 6.4, which favors binding of soluble lysosomal enzymes to the receptors. Following binding of tagged lysosomal proteins to the MPRs, the receptor-ligand complexes are packaged into clathrin-coated transport vesicles and conveyed to an endosome. In animal cells, most lysosomal enzymes are transported from the TGN to *early endosomes*—organelles formed by the coalescence of vesicles from the TGN and plasma membrane. As an early endosome matures to form a *late endosome*, the pH of the lumen decreases to about 5.5. This acidity causes lysosomal enzymes to dissociate from the MPRs. The receptors are then recycled in vesicles that return to the TGN, and the late endosome either matures to form a new lysosome or delivers its contents to an active lysosome. A few lysosomal proteins follow a shorter route, traveling directly from the TGN to a late endosome. Others follow a longer route, traveling to the plasma membrane and reentering the cell by endocytosis and transport to an early endosome.

Strong support for this concept of lysosomal enzyme targeting came from studies of a human genetic disorder called *I-cell disease.* Fibroblast cells from patients with this disease synthesize all the expected lysosomal enzymes in culture, but then release most of the soluble proteins to the extracellular medium instead of incorporating them into lysosomes. The distinguishing feature of the misdirected proteins is the absence of mannose-6-phosphate residues on their oligosaccharide side chains. The modified oligosaccharide tag is clearly essential for diverting this set of soluble glycoproteins from a secretory pathway. I-cell disease is now known to result from a defective phosphotransferase, the first of the two enzymes essential for adding mannose-6-phosphate to oligosaccharide chains on lysosomal enzymes.

While mannose-6-phosphate residues are important for directing traffic from the ER to the lysosome, other pathways may be involved as well. For example, in patients with I-cell disease, lysosomal acid hydrolases still reach the lysosomes in liver cells. Furthermore, in contrast to soluble lysosomal proteins, membrane-bound lysosomal proteins follow a pathway that does not depend on the mannose-6-phosphate receptor.

Secretory Pathways Transport Molecules to the Exterior of the Cell

Integral to the vesicular traffic shown in Figure 12-8 are **secretory pathways** by which proteins move from the ER through the Golgi complex to **secretory vesicles** and **secretory granules,** which then discharge their contents to the exterior of the cell (④). The concerted roles of the ER and the Golgi complex in secretion were first demonstrated by George Palade and his colleagues, who used autoradiography to trace the fate of radioactively labeled protein through cells in the parotid salivary gland.

The results of one of Palade's experiments are presented in Figure 12-10. For the first few minutes after injection of a radioactive amino acid into rabbits, newly synthesized protein containing the radioactive amino acid is found only in the rough ER of the secretory tissue. Shortly thereafter, radioactivity begins to appear in the

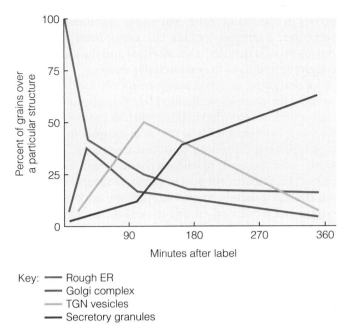

Key: ―― Rough ER
―― Golgi complex
―― TGN vesicles
―― Secretory granules

Figure 12-10 Autoradiographic Evidence of a Secretory Pathway. To trace the path of newly synthesized proteins through a cell, George Palade and his colleagues injected rabbits with a radioactive amino acid and used autoradiography to determine the distribution of radioactively labeled protein in parotid gland cells at several points following injection. Most of the radioactivity (measured as silver grains appearing over different regions of a cell) was initially confined to the rough ER. Then it moved through the Golgi complex to vesicles budding from the TGN, which Palade called condensing vacuoles. Within 6 hours, over 50% of the radioactively labeled protein was concentrated in mature secretory granules ready for secretion by exocytosis. For a description of microscopic autoradiography, see the *Guide to Microscopy* that accompanies this text.

Golgi complex, peaking at about 30–40 minutes. By 30 minutes, radioactively labeled protein can also be detected in vesicles budding from the TGN, which Palade called *condensing vacuoles.* After 90 minutes, radioactivity begins to accumulate in *secretory granules,* the vesicles that discharge secretory proteins to the exterior of the cell. Although not shown in the figure, radioactively labeled protein eventually appears in the extracellular medium, demonstrating that the secretory granules actually release their contents beyond the plasma membrane.

Based on Palade's initial experiments and numerous similar studies since then, the secretory pathways shown in Figure 12-8 are now understood in considerable detail. Moreover, we now distinguish two different modes of secretion by eukaryotic cells. *Constitutive secretion* involves the continuous discharge of vesicles at the plasma membrane surface, whereas *regulated secretion* entails controlled rapid releases, usually in response to an extracellular signal.

Constitutive Secretion. After budding from the TGN, some secretory vesicles move directly to the cell surface, where they immediately fuse with the plasma membrane and release their contents. This process, which is continuous and independent of specific extracellular signals, occurs in most eukaryotic cells and is called **constitutive secretion.** Examples include the continuous release of mucus by cells that line your intestine and the secretion of the glycoproteins of the extracellular matrix.

Constitutive secretion was once assumed to be a *default pathway* for proteins synthesized by ribosomes attached to the rough ER. According to this model, all proteins destined for transport from the ER to the Golgi complex must have a tag that diverts them from constitutive secretion. Unless an amino acid sequence or oligosaccharide chain identifies a protein for retention in or transport to a specific organelle, the protein passively moves through the endomembrane system and is released outside the cell by default. Support for this model came from studies in which the retrieval tags on ER-specific proteins were removed and the fates of the proteins were traced. Removal of the KDEL sequence from resident proteins of the ER generally led to secretion of the protein. More recent evidence, however, points to a variety of short amino acid sequences that identify proteins for constitutive secretion. Moreover, there may be several distinct pathways for this process.

Regulated Secretion. While vesicles containing constitutively secreted proteins move continuously and directly from the TGN to the plasma membrane, secretory vesicles involved in **regulated secretion** accumulate in the cell and then fuse with the plasma membrane only in response to specific extracellular signals. An important example is the release of neurotransmitters, which was described in Chapter 9 (see Figure 9-21). Two additional examples of regulated secretion are the release of insulin from the β cells of the pancreatic islets of Langerhans and the release of digestive enzymes from acinar cells of the pancreas .

Regulated secretory vesicles form by budding from the TGN as immature secretory vesicles, which lose their clathrin coats and undergo a maturation process. Maturation involves concentration of the proteins—referred to as *condensation*—and frequently also the proteolytic processing of secretory proteins. The mature secretory vesicles then move close to the site of secretion and wait near the plasma membrane for the signal that triggers release of their content by exocytosis.

Mature regulated secretory vesicles are usually quite large and contain much more highly concentrated proteins than do constitutive secretory vesicles. Such large dense vesicles are often called *secretory* or *zymogen granules* to distinguish them from other secretory vesicles. As shown in Figure 12-11, zymogen granules (ZG) are concentrated in the region of the cell between the Golgi stacks from which they arise and the portion of the plasma membrane bordering the lumen into which the contents of the granules are eventually discharged.

The information needed to direct a protein to a regulated secretory vesicle is presumably inherent in the amino acid sequence of the protein, though the precise signals and mechanisms for sorting proteins to secretory vesicles are not clear. Current evidence suggests that high concentrations of secretory proteins in secretory granules promote

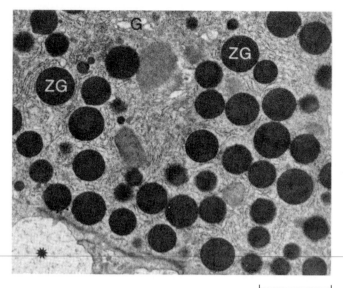

Figure 12-11   Zymogen Granules.   This electron micrograph of an acinar (secretory) cell from the exocrine pancreas of a rat illustrates the prominence of zymogen granules (ZG). Zymogen granules are usually concentrated in the region of the cell between the Golgi complex (G) from which they arise and the portion of the plasma membrane bordering the acinar lumen (marked by an * on the micrograph) into which the contents of the granules are eventually discharged by exocytosis (TEM).

the formation of large *protein aggregates* that exclude nonsecretory proteins. This could occur in the TGN, where only aggregates would be packaged in vesicles destined for secretory granules, or it could occur in the secretory granule itself. The pH of the TGN and the secretory granule lumens may serve as a trigger favoring aggregation just as material leaves the TGN. The soluble proteins that do not become part of an aggregate in the TGN or a secretory granule would be carried by transport vesicles to other locations.

Exocytosis and Endocytosis: Transporting Material Across the Plasma Membrane

Exocytosis, by which secretory granules release their contents to the exterior of the cell, and *endocytosis*, by which cells internalize materials that were previously outside the cell, are means of transporting macromolecules and other substances across the plasma membrane. Both processes are unique to eukaryotic cells. We will first consider exocytosis, because it is the final step in a secretory pathway that began with the ER and the Golgi complex.

Exocytosis Releases Intracellular Molecules to the Extracellular Medium

In **exocytosis,** proteins sequestered within a vesicle are released to the exterior of the cell as the membrane of the vesicle fuses with the plasma membrane. A variety of proteins are exported from both animal and plant cells by exocytosis. Animal cells secrete peptide and protein hormones, mucus, milk proteins, and digestive enzymes in this manner. Plant cells secrete proteins associated with the cell wall, including both enzymes and structural proteins.

The process of exocytosis is illustrated schematically in Figure 12-12. Vesicles containing cellular products destined for secretion move to the cell surface (①), where the membrane of the vesicle fuses with the plasma membrane (②). The plasma membrane ruptures, discharging the vesicle contents to the exterior of the cell (③). In the process, the membrane of the vesicle becomes integrated into the plasma membrane, with the *inner* (lumenal) surface of the vesicle becoming the *outer* (extracellular) surface of the plasma membrane (④). Glycoproteins and glycolipids that remain anchored to the plasma membrane will face the extracellular space.

The mechanism underlying the movement of exocytic vesicles to the cell surface is not yet clear. Current evidence points to the involvement of microtubules in both exocytic and endocytic vesicle movement. For example, in some cells vesicles appear to move from the Golgi complex to the plasma membrane along "tracks" of microtubules that are oriented parallel to the direction of vesicle movement. Moreover, vesicle movement stops when the cells are treated with *colchicine,* a plant alkaloid that binds to tubulin monomers and prevents their assembly into microtubules. We will discuss the movement of vesicles along microtubules in Chapter 23.

The Role of Calcium in Triggering Exocytosis. Fusion of regulated secretory vesicles with the plasma membrane is generally triggered by a specific extracellular signal. In most cases, the signal is a hormone or a neurotransmitter that binds to specific receptors on the cell surface and triggers the synthesis or release of a *second messenger* within the cell (Chapter 10). During regulated secretion, a transient elevation of the intracellular concentration of calcium ions often appears to be an essential step in the signal cascade leading from the receptor on the cell surface to exocytosis. For example, microinjection of calcium into pancreatic cells induces mature secretory granules to discharge their contents to the extracellular medium. The specific role of calcium, however, is not yet clear. It may be that an elevation in the intracellular calcium concentration leads to the activation of protein kinases whose target proteins are components of either the vesicle membrane or the plasma membrane.

Polarized Secretion. In many cases, exocytosis of specific proteins is limited to a specific surface of the cell. For example, the secretory cells that line your intestine release digestive enzymes only on the side of the cell that faces the interior of the intestine. On the opposite side of the cell, a completely different set of proteins is secreted. This phenomenon, called **polarized secretion,** is also seen in nerve cells, which secrete neurotransmitter molecules only at junctions with other nerve cells (see Figure 9-21). Proteins destined for polarized secretion, as well as lipid and protein components of the two different membrane domains, are

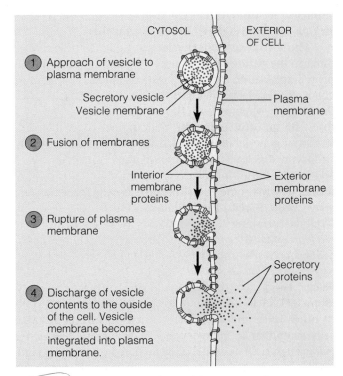

Figure 12-12 Exocytosis. Release of the contents of a secretory vesicle or granule to the exterior of the cell. For clarity, the proteins that mediate docking and fusion of a vesicle with a membrane have been omitted. See Figure 12-19 for more detail.

sorted into vesicles that bind to localized recognition sites on subdomains of the plasma membrane.

Endocytosis Imports Extracellular Molecules by Forming Vesicles from the Plasma Membrane

Most eukaryotic cells carry out one or more forms of **endocytosis.** A small segment of the plasma membrane progressively folds inward (Figure 12-13, ①), and then it pinches off to form an **endocytic vesicle** containing ingested substances or particles (② and ③). By this means, materials that were previously outside the cell are brought into the cell (④). Endocytosis is important for several cellular processes, including ingestion of essential nutrients and defense against microorganisms.

In terms of membrane flow, exocytosis and endocytosis clearly have opposite effects. Whereas exocytosis adds lipids and proteins to the plasma membrane as vesicles fuse with it, endocytosis internalizes portions of the plasma membrane. Through endocytosis and retrograde transport, the cell can recycle molecules essential for exocytosis by recovering lipids and proteins deposited in the plasma membrane by secretory vesicles. The magnitude of the resulting membrane exchange can be impressive. For example, the secretory cells in your pancreas recycle an amount of membrane equal to the whole surface area of the cell within about 90 minutes. Cultured macrophages (large white blood cells) are faster, replacing an amount of membrane equivalent to the plasma membrane within about 30 minutes. And cells of the slime mold *Dictyostelium* accomplish the same feat within about 20 minutes.

The term *endocytosis* encompasses several processes that differ in the nature of the material ingested and the mechanism employed. In each case, however, the membrane of an endocytic vesicle isolates the internalized substances from the cytosol. Moreover, most endocytic vesicles eventually fuse with an early endosome, adding their contents to the stream of material moving toward a lysosome. Alternatively, endocytic vesicles may form temporary connections with endosomes, acquiring digestive enzymes and maturing to form a new lysosome. A distinction is usually made between *phagocytosis* (Greek for "cellular eating") and *pinocytosis* ("cellular drinking"). Pinocytosis is then subdivided into *receptor-mediated endocytosis*—also called *clathrin-dependent endocytosis*—and *clathrin-independent endocytosis*.

Phagocytosis. The ingestion of large particles (>0.5 μm diameter), including aggregates of macromolecules, parts of other cells, and even whole microorganisms or other cells is known as **phagocytosis.** For many single-celled eukaryotes, such as amoebas and ciliated protozoa, phagocytosis is a routine means for acquiring food. Phagocytosis is also used by some primitive animals, notably flatworms, coelenterates, and sponges, as a means of obtaining nutrients. In more complex organisms, however, phagocytosis is usually restricted to specialized cells called **phagocytes.** For example, your body contains two classes of white blood cells that routinely function as phagocytes: *neutrophils* and *macrophages*. Both of these types of cells use phagocytosis for defense rather than nutrition, engulfing and digesting foreign material or invasive microorganisms found in the bloodstream. Macrophages have an additional role as scavengers, ingesting cellular debris and whole damaged cells. Under certain conditions, other mammalian cells engage in phagocytosis. For example, fibroblasts found in connective tissue can take up collagen to allow remodeling of the tissue, and dendritic cells in the mammalian spleen can ingest bacteria as part of an immune response.

Phagocytosis has been studied most extensively in the amoeba. The cell surface of most amoebas is covered with either fine hairs or a coat of glycosaminoglycans, which absorb and trap food particles or smaller organisms. White blood cells, on the other hand, display cell surface receptors that recognize specific ligands on the surfaces of particles and microorganisms. In each case, contact with an appropriate target triggers the onset of phagocytosis, which is illustrated in Figure 12-14. In a process involving poly-

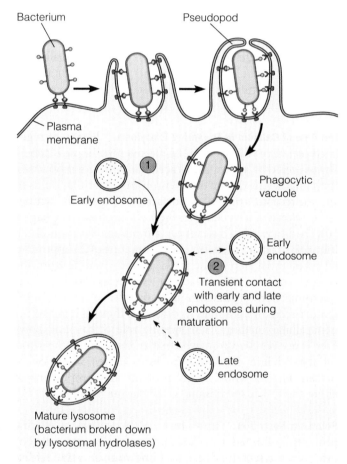

Figure 12-14 Phagocytosis. Particles or microorganisms bind to receptors on the cell surface, triggering the onset of phagocytosis. In a process involving polymerization of actin, folds of membrane called *pseudopods* gradually surround the particle. Eventually, the pseudopods meet and engulf the particle, forming a *phagocytic vacuole.* The vacuole then ① fuses with an early endosome or ② forms transient connections (indicated by dashed lines) with early and late endosomes and matures into a lysosome where digestion of the internalized material occurs.

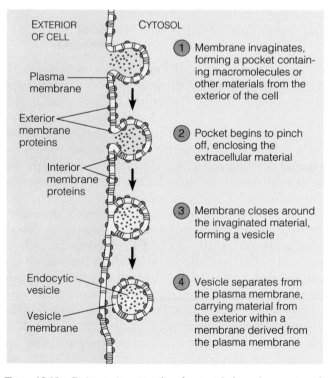

Figure 12-13 Endocytosis. Uptake of materials from the exterior of the cell. For clarity, the coat proteins at the site of invagination and around the endocytic vesicle have been omitted from this diagram. See Figure 12-15 for a description of clathrin-dependent endocytosis.

merization of actin, folds of membrane called *pseudopods* gradually surround the object. Eventually, the pseudopods meet and engulf the particle, forming an intracellular **phagocytic vacuole.** This endocytic vesicle, also called a *phagosome,* then fuses with a late endosome or matures directly into a lysosome, forming a large vesicle in which the ingested material is digested. As part of their role in the immune system, phagocytes generate toxic concentrations of hydrogen peroxide, hypochlorous acid, and other oxidants in the phagocytic vacuole to kill microorganisms before the vacuole matures into a lysosome.

Receptor-Mediated Endocytosis. **Receptor-mediated endocytosis** (also called **clathrin-dependent endocytosis**) is a pathway for concentrating and ingesting extracellular molecules by means of specific receptors on the outer surface of the plasma membrane. (Although phagocytosis is also receptor-mediated, the process does not concentrate macromolecules, nor is it clathrin-dependent.) Receptor-mediated endocytosis is the primary mechanism for the specific internalization of most macromolecules by eukaryotic cells. Depending on the cell type, mammalian cells can ingest hormones, growth factors, enzymes, serum proteins, antibodies, iron, and even some viruses and bacterial toxins by this mechanism.

The discovery of receptor-mediated endocytosis and its role in the internalization of *low-density lipoproteins (LDL)* is highlighted in Box 12B. Receptor-mediated endocytosis of LDL carries cholesterol into mammalian cells. An interest in familial hypercholesterolemia, a hereditary predisposition to high blood cholesterol levels and hence to atherosclerosis and heart disease, led Michael Brown and Joseph Goldstein to the discovery of receptor-mediated endocytosis, for which they shared a Nobel Prize in 1986.

Receptor-mediated endocytosis is illustrated in Figure 12-15. The process begins with the binding of ligand molecules to their respective receptors on the outer surface of the plasma membrane (①). As the receptor-ligand complexes diffuse laterally in the membrane, they encounter specialized membrane regions—called *coated pits*—that serve as sites for the collection and internalization of such complexes (②). In a typical mammalian cell, the coated pits occupy about 20% of the total surface area of the plasma membrane. Accumulation of receptor-ligand complexes within the coated pits triggers accumulation of additional proteins—including *adaptor protein, clathrin,* and *dynamin*—that are required for invagination of the pit (③). These proteins are found on the inner (cytosolic) surface of the plasma membrane. Invagination continues until the pit pinches off from the plasma membrane, forming a **coated vesicle** (④). The clathrin coat is released, leaving an uncoated vesicle (⑤). The coat proteins and dynamin are now available for forming new vesicles, while the endocytic vesicle is free to fuse with an early endosome

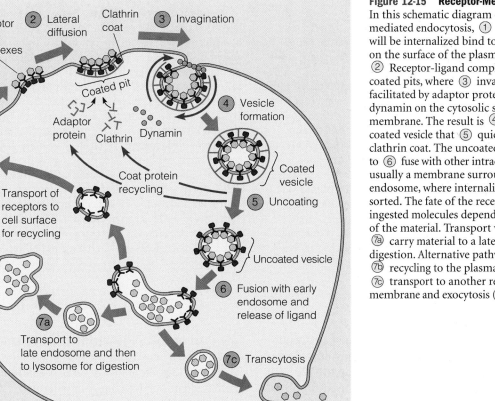

Figure 12-15 Receptor-Mediated Endocytosis. In this schematic diagram of receptor-mediated endocytosis, ① the molecules that will be internalized bind to specific receptors on the surface of the plasma membrane. ② Receptor-ligand complexes accumulate in coated pits, where ③ invagination is facilitated by adaptor protein, clathrin, and dynamin on the cytosolic surface of the membrane. The result is ④ an internalized coated vesicle that ⑤ quickly loses its clathrin coat. The uncoated vesicle is now free to ⑥ fuse with other intracellular membranes, usually a membrane surrounding an early endosome, where internalized material is sorted. The fate of the receptors and the ingested molecules depends on the nature of the material. Transport vesicles often ⑦a carry material to a late endosome for digestion. Alternative pathways include ⑦b recycling to the plasma membrane or ⑦c transport to another region of the plasma membrane and exocytosis (called transcytosis).

Clinical Applications CHOLESTEROL, THE LDL RECEPTOR, AND RECEPTOR-MEDIATED ENDOCYTOSIS

Receptor-mediated endocytosis (RME) is a highly efficient pathway for the uptake of specific macromolecules by eukaryotic cells. We now know of several dozen different kinds of macromolecules that can be taken up by this means, each recognized by its own specific receptor on the plasma membrane of the appropriate cell types. If we consider the discovery of RME, however, we focus on a specific receptor—and, as it turns out, on a health issue that many of us are concerned about: the level of cholesterol in our blood.

As you may know, one of the primary factors that predisposes a person to heart attacks is an abnormally high level of cholesterol in the blood serum, a condition called *hypercholesterolemia*. Because of its insolubility in the aqueous serum, cholesterol tends to be deposited on the inside walls of blood vessels. These deposits build up over time, forming the *atherosclerotic plaques* that cause *atherosclerosis*, commonly known as hardening of the arteries. Ultimately, the plaques may block the flow of blood through the vessels, causing strokes and heart attacks.

Although a high blood cholesterol level is often linked to dietary intake of cholesterol and fatty acids, some people are genetically predisposed to high blood cholesterol levels and hence to atherosclerosis and heart disease. This hereditary predisposition, called **familial hypercholesterolemia (FH)**, is especially debilitating to homozygous individuals—that is, to those who inherited a defective FH gene from both parents. People with FH have grossly elevated levels of serum cholesterol (about 650–1000 mg/100 mL of blood serum, compared with the normal range of about 130–200 mg/100 mL), and develop atherosclerosis early in life, often leading to death from heart disease before the age of 20. People who are heterozygous—those with one defective and one normal copy of the gene—are affected less severely, but they nonetheless have elevated serum cholesterol levels (about 250–500 mg/100 mL) and are at high risk for heart attacks in their thirties and forties.

The link between FH and RME came about because of a study of FH by Michael Brown and Joseph Goldstein that was begun in 1972 and led not only to the discovery of RME but also to Nobel Prizes for both scientists in 1986. Brown and Goldstein began by culturing fibroblast cells from FH patients in the laboratory and showing that such cells synthesized cholesterol at abnormally high rates compared with normal cells. Their next key observation was that normal cells also synthesized cholesterol at abnormally high rates when they were deprived of the **low-density lipoproteins (LDL)** that were usually present in the culture medium. LDL is one form in which cholesterol is transported in the blood and taken up into cells.

LDL is one of several classes of *blood lipoproteins,* which are classified according to their density. Basically, a lipoprotein consists of a monolayer of phospholipid and cholesterol molecules and one or more protein molecules, with the lipids oriented so that their polar head groups face the aqueous medium on the outside and their nonpolar tails extend into the interior of the particle (Figure 12B-1). LDL is the class of lipoproteins with the highest cholesterol content: Free cholesterol and cholesterol esters make up more than half of the LDL particle by weight. The esterified form of cholesterol has a long-chain fatty acid linked to it, which makes it highly hydrophobic. Esterified cholesterol molecules (about 1500 per particle) therefore cluster in the interior of the particle, whereas the free, or unesterified, cholesterol molecules (about 500 per particle) are found mainly in the lipid monolayer (see Figure 12B-1).

In addition to phospholipids and cholesterol, each LDL particle has a single molecule of a large protein called *apoprotein B-100* embedded in its lipid monolayer. This protein is crucial to our understanding of the difference in the response of FH cells and normal cells to the level of LDL in the medium. The ability of normal fibroblasts to maintain an appropriately low rate of cholesterol synthesis in the presence of LDL suggested to Brown and Goldstein that LDL was involved in the transport of cholesterol into the cell, where the cholesterol then regulated its own synthesis by *allosteric* (or *feedback*) *inhibition* (see Figure 6-17). FH fibroblasts, on the other hand, synthesized cholesterol at a high rate regardless of whether LDL was present in the medium, suggesting that these cells might be defective in LDL-dependent cholesterol uptake.

Based on their observations, Brown and Goldstein postulated that the uptake of cholesterol into cells requires the action of a specific receptor on the cell surface and that this receptor is absent or defective in FH patients. In a brilliant series of experiments, these investigators and their colleagues demonstrated the existence of an LDL-specific membrane protein, called the **LDL receptor,** and showed that it recognizes the apoprotein B-100 molecule that is present in every LDL particle. They also showed that the cells from FH patients either lacked this protein entirely or had receptor molecules that were defective in any of several ways.

To visualize the LDL particles, these scientists conjugated, or linked, them to molecules of ferritin, a protein that binds iron atoms. Because iron atoms are electron-dense, they appear as dark dots in the electron microscope (Figure 12B-2). Using this technique, Brown and Goldstein showed that the ferritin-conjugated LDL particles bound to the surface of the cell and clustered at specific locations (Figure 12B-2a). We now recognize these sites

(⑥). The speed and scope of receptor-mediated endocytosis are impressive. A coated pit usually invaginates within a minute or so of being formed, and up to 2500 such coated pits invaginate per minute in a cultured fibroblast cell. Figure 12-16 shows progressive formation of a coated vesicle from a coated pit as particles of yolk protein are taken up by the maturing oocyte (egg cell) of a chicken.

The roles of coat proteins in vesicle formation and transport will be discussed in the next section.

There are actually several variations of receptor-mediated endocytosis. Epidermal growth factor (EGF), for example, undergoes endocytosis by the mechanism shown in Figure 12-15. The EGF receptors are concentrated in coated pits only after the formation of receptor-ligand

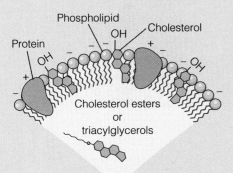

Phospholipid
Cholesterol
OH
Protein
OH
+
−
− − + +
−
− +
OH
Cholesterol esters
or
triacylglycerols

Figure 12B-1 Lipoprotein Structure. A typical lipoprotein consists of a monolayer of phospholipid and free (unesterified) cholesterol molecules surrounding a hydrophobic interior. One or more protein molecules are imbedded in the monolayer. Cholesterol molecules esterified to long-chain fatty acids are highly hydrophobic and tend to cluster within the lipoprotein's interior. Lipoproteins differ from one another in density, depending on the relative amounts of lipid and proteins present. Because lipids are less dense than proteins, the density of the particle is inversely related to the abundance of lipids. The lipoprotein shown here is a low-density lipoprotein (LDL), with a density of 1.02–1.06 g/mL. A typical LDL particle contains about 800 phospholipid molecules and 500 free cholesterol molecules in the lipid monolayer and about 1500 esterified cholesterol molecules in the interior. The single protein molecule, called apoprotein B-100, has a molecular weight of about 500,000 and is embedded in the lipid monolayer. The protein provides structural organization to the particle and mediates the binding of the LDL to the LDL receptors on the surfaces of cells.

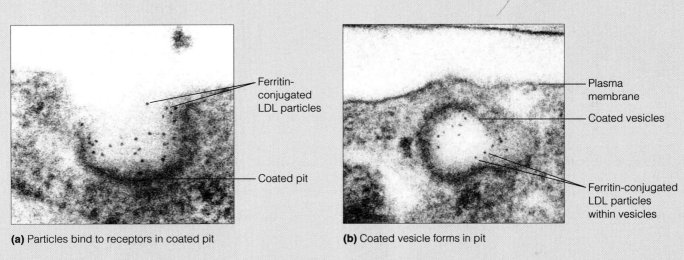

(a) Particles bind to receptors in coated pit

Ferritin-conjugated LDL particles

Coated pit

Plasma membrane

Coated vesicles

Ferritin-conjugated LDL particles within vesicles

(b) Coated vesicle forms in pit

Figure 12B-2 Visualization of LDL Binding. Conjugation of LDL particles with ferritin, an iron-binding protein, allows visualization of the LDL-ferritin complexes by electron microscopy because of the density of the iron atoms bound to the ferritin. **(a)** LDL-ferritin conjugates, visible as dark dots, bind to receptors concentrated in a coated pit on the surface of a cultured human fibroblast cell. **(b)** The LDL-ferritin conjugates are internalized when a coated vesicle forms from the coated pit region following invagination of the plasma membrane (TEMs).

as *coated pits,* localized regions of the plasma membrane characterized by the presence of *clathrin* on the cytoplasmic side of the membrane and by the accumulation of membrane-bound receptor-ligand complexes on the exterior of the membrane.

Dark dots were also seen on the inside of vesicles that formed by invagination and pinching off of coated pits (Figure 12B-2b). The receptors, in other words, not only bound the LDL on the cell surface but were also apparently involved in the internalization of LDL within vesicles. In short, these workers had discovered a new mechanism by which cells can take up macromolecules from their environment. And since it was an endocytic process involving specific receptors, Brown and Goldstein gave it the name by which we know it today—receptor-mediated endocytosis.

complexes. In another variation of receptor-mediated endocytosis, receptors are constitutively concentrated in coated pits independent of formation of receptor-ligand complexes. Binding of ligands to receptors simply triggers internalization. In yet another variation, the receptors are not only constitutively concentrated, they are also constitutively internalized regardless of whether ligands have bound to the receptors. For example, the LDL receptors described in Box 12B are constitutively internalized.

Following receptor-mediated endocytosis, the uncoated vesicles fuse with **early endosomes** found in peripheral regions of the cell. Early endosomes form from vesicles budding off the TGN and are sites for the sorting and recycling of extracellular material brought into the cell

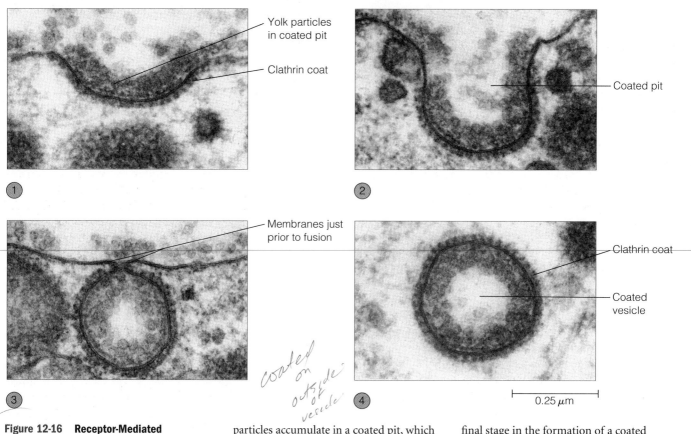

1 Yolk particles in coated pit

Clathrin coat

2 Coated pit

3 Membranes just prior to fusion

coated on outside of vesicle

4

Clathrin coat

Coated vesicle

0.25 μm

Figure 12-16 Receptor-Mediated Endocytosis of Yolk Protein by a Chicken Oocyte. This series of electron micrographs illustrates the formation of a coated vesicle from a coated pit during receptor-mediated endocytosis. ① Yolk particles accumulate in a coated pit, which initially appears as a shallow invagination of the plasma membrane with a clathrin coat on its inner surface. ② A deeper coated pit containing several free particles in addition to those adhering to the membrane. ③ The final stage in the formation of a coated vesicle, just prior to constriction of the membrane at the neck of the budding vesicle. ④ A coated vesicle that has just formed below the plasma membrane and still has an intact clathrin coat (All TEMs).

by endocytosis. Protein molecules essential for new rounds of endocytosis are often—but not always—recycled after separation from the material fated for digestion. The early endosome continues to acquire lysosomal proteins from the TGN and matures to form a late endosome. The roles of endosomes in digestion will be discussed in more detail when we examine lysosomes later in the chapter.

Recycling of receptor molecules is facilitated by acidification of the early endosome. The interior of an endocytic vesicle has a pH of about 7.0, whereas the interior of an early endosome has a pH of 5.9–6.5. The lower pH is maintained by an *ATP-dependent proton pump* in the endosomal membrane. The slightly acidic environment of the early endosome decreases the affinity of most receptor-ligand complexes (for example, LDL and its receptor), thereby freeing receptors to be recycled to the plasma membrane while newly ingested material is diverted to other locations.

Sorting receptors and ligands is not always as simple as sending the receptors to the plasma membrane and retaining the ligands in the endosome. Depending on the ligand, some receptor-ligand complexes do not dissociate in the early endosome. While dissociated ligands are swept along to their fate in a lysosome, intact receptor-ligand com-

plexes are still subject to sorting and packaging into transport vesicles. There are at least three alternative fates for these complexes: (1) Some receptor-ligand complexes (for example, epidermal growth factor and its receptor) are carried to a lysosome for degradation. (2) Others are carried to the TGN, where they enter a variety of pathways transporting material throughout the endomembrane system. (3) Receptor-ligand complexes can also travel by transport vesicles to a different region of the plasma membrane, where they are secreted as part of a process called **transcytosis.** This pathway accommodates the transfer of extracellular material from one side of the cell, where endocytosis occurs, through the cytoplasm to the opposite side, where exocytosis occurs. For example, immunoglobulins are transported across epithelial cells from maternal blood to fetal blood by transcytosis.

Clathrin-Independent Endocytosis. An example of a clathrin-independent endocytic pathway is **fluid-phase endocytosis,** a process for nonspecific internalization of extracellular fluid. Fluid-phase endocytosis, unlike receptor-mediated endocytosis, does not concentrate ingested material. Because the cell engulfs fluid without a mechanism for collecting or excluding particular molecules, the concentration of material trapped in vesicles reflects its concentration in the extracellular environment. Components of the plasma membrane, such as receptors, can also be indiscriminately swept up in fluid-phase endocytic vesicles. In contrast to other forms of endocytosis, fluid-phase uptake apparently proceeds at a relatively constant rate in most eukaryotic cells and may be the primary mechanism whereby some types of cells compensate for the membrane segments that are continuously added to the plasma membrane by exocytosis. Thus, it is a means for controlling a cell's volume and surface area. Once inside the cell, fluid-phase endocytic vesicles, like clathrin-dependent endocytic vesicles, are routed to early endosomes.

Coated Vesicles in Cellular Transport Processes

Coated vesicles were first reported in 1964 by Thomas Roth and Keith Porter, who described their involvement in the selective uptake of yolk protein by developing mosquito oocytes. Since then, coated vesicles have been shown to play vital roles in diverse cellular processes. We have seen that coated vesicles and their precursors, coated pits, are involved in vesicular traffic throughout the endomembrane system, as well as transport during exocytosis and endocytosis. It is probable that such vesicles participate in most, if not all, vesicular traffic connecting the various membrane-bounded compartments and plasma membrane of the eukaryotic cell.

A common feature of coated vesicles is the presence of a layer, or coat, of protein on the cytosolic side of the membrane surrounding the vesicle. As we noted earlier, the most studied coat proteins are clathrin, COPI, and a pair of protein complexes collectively called COPII. A fourth, more mysterious coat protein is caveolin. Coat proteins participate in several steps of the formation of transport vesicles. The variety of coat protein complexes reflects their participation in the sorting of molecules that are fated for different destinations into specific vesicles. More general roles may include forcing nearly flat membranes to form spherical buds, preventing premature fusion of a budding vesicle with nearby membranes, and regulating the inter-

actions between budding vesicles and microtubules that are important for moving vesicles through the cell.

The specific set of proteins covering the exterior of a vesicle is an indicator of the origin and destination of the vesicle within the cell (Table 12-2). As we have already seen, clathrin-coated vesicles are involved in the selective transport of proteins from the TGN to endosomes, and in the endocytosis of receptor-ligand complexes from the plasma membrane. COPI-coated vesicles, on the other hand, facilitate bidirectional transport of proteins between the ER and Golgi complex, as well as between cisternae of the Golgi complex. COPII-coated vesicles are involved in the transport of material from the ER to the Golgi. The precise role of caveolin-coated vesicles, called caveolae, is still controversial. Transient caveolae form from the plasma membrane and internalized material appears in endosomes, but the pathway connecting the plasma membrane and endosomes is not yet clear.

Clathrin-Coated Vesicles Are Surrounded by Lattices Composed of Clathrin and Adaptor Protein

Clathrin-coated vesicles are surrounded by coats composed of two multimeric proteins, **clathrin** and **adaptor protein (AP)**. The term clathrin comes from clathratus, the Latin word for "lattice," and as you can see in Figure 12-17, clathrin and AP assemble to form polygonal protein lattices. Flat clathrin lattices are composed entirely of hexagons, whereas curved lattices, which form under coated pits and surround vesicles, are composed of hexagons and pentagons.

Clathrin-coated vesicles readily dissociate into soluble clathrin complexes, adaptor protein complexes, and uncoated vesicles. These complexes and vesicles, in turn, will spontaneously reassemble under appropriate conditions. In a slightly acidic solution containing calcium ions, clathrin complexes will even reassemble independent of adaptor protein and membrane-bounded vesicles, resulting in empty shells called clathrin cages. Assembly occurs remarkably fast—within seconds under favorable conditions. Ease of assembly and disassembly is an important feature of the clathrin coat, because fusion of the underlying membrane with the membrane of another structure appears to require partial or complete uncoating of the vesicle.

Table 12-2   Coated Vesicles Found Within Eukaryotic Cells

Coated Vesicle	Coat Proteins*	Origin	Destination
Clathrin	Clathrin, AP1, ARF	TGN	Endosomes
Clathrin	Clathrin, AP2	Plasma membrane	Endosomes
COPI	COPI, ARF	Involved in bidirectional transport between the ER and Golgi complex, and between Golgi complex cisternae	
COPII	COPII (Sec13/31 and Sec23/24), Sar1	ER	Golgi complex
Caveolin	Caveolin	Plasma membrane	?

*ARF designates ADP ribosylation factor 1; AP1 and AP2 designate different adaptor protein complexes (also called assembly protein complexes).

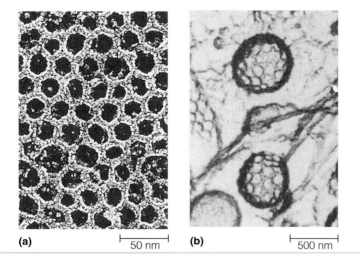

(a) ├── 50 nm **(b)** ├── 500 nm

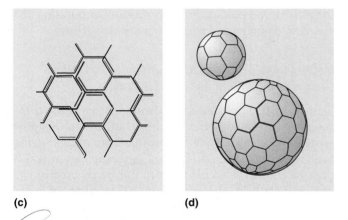

(c) **(d)**

Figure 12-17 Clathrin Lattices. Clathrin-coated vesicles are involved in a variety of transport processes in eukaryotic cells. Each vesicle is surrounded by a cage of overlapping clathrin complexes. **(a)** Freeze-etch electron micrograph of a clathrin lattice in a human carcinoma cell. This flat lattice is composed primarily of hexagonal units. **(b)** Electron micrograph of clathrin cages isolated from calf brain. Cages include both pentagonal and hexagonal units (TEMs). **(c)** and **(d)** Interpretive drawings of the lattice and clathrin cages.

Components of Clathrin Lattices. In 1981, Ernst Ungewickell and Daniel Branton visualized the basic structural units of clathrin lattices, three-legged structures called **triskelions** (Figure 12-18a). Each triskelion is a multimeric protein composed of three large polypeptides and three small polypeptides radiating from a central vertex, as illustrated in Figure 12-18b. The large polypeptides, each with a molecular weight of 192,000, are *clathrin heavy chains;* they form the legs of the triskelion. Each leg is slightly curved near the middle, or knee, and has a globular domain at its outer tip. The small polypeptides, with molecular weights in the range of 30,000–36,000, are *clathrin light chains.* Antibodies against clathrin light chains bind to the legs of the triskelion near the central vertex, suggesting that the light chains are associated with the inner half of each leg.

By combining information gathered from electron microscopy and X-ray crystallography, researchers have assembled a model for the organization of triskelions into the characteristic hexagons and pentagons of clathrin-coated pits and vesicles (Figure 12-18c). According to their model, one clathrin triskelion is located at each vertex of the polygonal lattice. Each polypeptide leg extends along two edges of the lattice, with the knee of the clathrin heavy chain located at an adjacent vertex. Because the knees of the heavy chains are located at adjacent vertices, each edge of the clathrin lattice is composed of overlapping legs from three triskelions. This arrangement of triskelions into overlapping networks ensures extensive longitudinal contact between clathrin polypeptides. Such contact may confer the mechanical strength needed when a coated vesicle forms from a membrane. The knees, however, do not interact with other polypeptides. This may provide flexibility for forming both hexagons and pentagons and also may accommodate vesicles of different sizes.

The second major component of clathrin coats—adaptor protein—was originally identified simply by its ability to promote the assembly of clathrin coats around vesicles; thus, AP is also called *assembly protein.* We now know that eukaryotic cells contain a variety of AP complexes, each composed of four polypeptides—two adaptin subunits, one medium chain, and one small chain. The four polypeptides, which are slightly different in each type of AP complex, bind to different transmembrane receptor proteins and confer specificity during vesicle budding and targeting. In addition to ensuring that appropriate macromolecules will be concentrated in coated pits, AP complexes also mediate the attachment of clathrin to the plasma membrane. Considering the central role of APs, it is not surprising that AP complexes are sites for regulation of clathrin assembly and disassembly; the ability of AP complexes to bind to clathrin is affected by pH, phosphorylation, and dephosphorylation.

The Assembly of Clathrin Coats Drives the Formation of Vesicles from the Plasma Membrane and TGN

(trans Golgi Network)

The binding of AP complexes to the plasma membrane and the concentration of receptors or receptor-ligand complexes in coated pits require ATP and GTP—though perhaps only for regulation of the process. The assembly of clathrin coats around budding vesicles, however, can occur without additional expenditure of ATP and GTP. An exception is the assembly of clathrin coats around vesicles forming from the TGN, which depends on the hydrolysis of nucleotide triphosphates.

The accumulation of clathrin triskelions and the assembly of a clathrin coat on the cytosolic side of a membrane appear to provide part of the driving force for formation of a vesicle at the site. In the case of receptor-mediated endocytosis, the assembly of the clathrin coat on the inner side of the plasma membrane causes the membrane to fold inward. As more clathrin triskelions are incorporated into the growing lattice, a combination of hexagonal and pentagonal units allows the new clathrin coat to curve around the budding vesicle.

As clathrin accumulates around the budding vesicle, at least one more protein—**dynamin**—participates in the

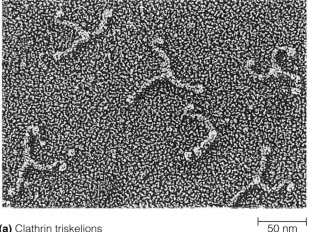

(a) Clathrin triskelions

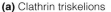
50 nm

(b) Structure of clathrin triskelion

Terminal
globular
domain

Clathrin
heavy chains

Vertex

Clathrin
light chains

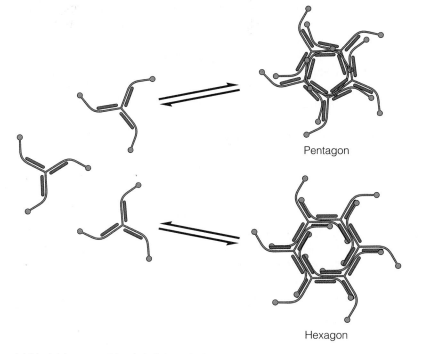

Pentagon

Hexagon

(c) Model for assembly of clathrin triskelions

Figure 12-18 Clathrin Triskelions. **(a)** This micrograph shows individual triskelions of clathrin (TEM). **(b)** Each triskelion consists of three clathrin heavy chains radiating from a central vertex, with a terminal globular domain at the tip of each triskelion leg and a clathrin light chain bound to the inner half of each leg. (Not shown are the adaptor protein complexes that are also found in clathrin coats.) **(c)** Under appropriate conditions, clathrin triskelions assemble into the pentagonal and hexagonal structures characteristic of coated pits and vesicles. According to the model presented here, one clathrin triskelion is located at each polyhedral vertex, with each leg extending along one of the legs of each of two neighboring triskelions.

process. Dynamin is a cytosolic GTPase required for coated pit constriction and closing of the budding vesicle. This essential protein was first identified in *Drosophila.* Flies expressing a temperature-sensitive form of dynamin were instantly paralyzed after a temperature shift. Further investigation revealed an accumulation of coated pits in the presynaptic membranes of neuromuscular junctions in the affected flies. Binding of GTP probably allows dynamin to form helical rings around the neck of the coated pit. As GTP is hydrolyzed, the dynamin rings tighten and separate the fully sealed endocytic vesicle from the plasma membrane.

Some mechanism is also required to *uncoat* clathrin-coated membranes. Moreover, uncoating must be done in a regulated manner because, in most cases, the clathrin coat remains intact as long as the membrane is part of a coated pit or budding vesicle but dissociates rapidly once

the vesicle is fully formed. Like assembly, dissociation of the clathrin coat is an energy-consuming process, accompanied by the hydrolysis of about three ATP molecules per triskelion. At least one protein, an *uncoating ATPase,* is essential for this process, though the uncoating ATPase releases only the clathrin triskelions from the APs; the factors responsible for releasing APs from the membrane have not yet been identified.

COPI- and COPII-Coated Vesicles Connect the ER and Golgi Complex Cisternae

COPI-coated vesicles have been found in all eukaryotic cells examined, including mammalian, insect, and yeast cells. Such vesicles are surrounded by coats composed of **COPI** (or **coatomer**) and **ADP ribosylation factor (ARF).** Viewed by electron microscopy, COPI-coated vesicles do not display polyhedral lattices like those surrounding clathrin-coated vesicles. Instead, they have dense "fuzzy" coats. The major component of the coat, COPI, is a protein multimer composed of seven subunits.

COPI-coated vesicles are likely involved in bidirectional transport between the ER and Golgi complex as well as between Golgi complex cisternae, though their precise role is controversial. Originally, evidence strongly suggested that COPI-coated vesicles formed from both ER and Golgi complex membranes. Recent investigations, however, suggest COPI-coated vesicles do not bud from the ER. Moreover, COPI-coated vesicles may not even move between compartments but may be involved in *lateral* sorting of proteins and lipids from one region of a compartment to another.

Assembly of a COPI coat is mediated by ARF, which is a small GTP-binding protein. In the cytosol, ARF occurs as part of an ARF-GTP complex. When ARF encounters a specific *guanine nucleotide exchange factor* associated with the membrane from which a new coated vesicle is about to form, however, the GDP is exchanged for GTP. ARF is then able to bind to the membrane by inserting a hydrophobic tail into the lipid bilayer. Once firmly anchored, ARF binds to COPI multimers, and assembly of the coat drives the formation and budding of a new vesicle. After the formation of a free vesicle, a protein in the donor membrane triggers hydrolysis of GTP, and ARF releases the coat proteins for another cycle of vesicle budding.

COPII-coated vesicles were first discovered in yeast, where they have a role in transport from the ER to the Golgi complex. Mammalian homologues of some of the components of COPII coats have been identified, but COPII-coated vesicles have not been found in mammals. The COPII coat found in yeast is assembled from two protein complexes—called **Sec13/31** and **Sec23/24**—and a small GTP-binding protein called **SarI**, which is similar to ARF. By a mechanism resembling formation of a COPI coat, a SarI molecule with GDP bound to it approaches the membrane from which a vesicle is about to form. A peripheral membrane protein then triggers exchange of GTP for GDP, enabling SarI to bind to Sec13/31 and Sec23/24. After the formation of a free vesicle, a component of the COPII coat triggers GTP hydrolysis and SarI releases Sec13/31 and Sec23/24.

The SNARE Hypothesis Connects Coated Vesicles and Target Membranes

Much of the intracellular traffic mediated by coated vesicles is highly specific. As we have seen, the final sorting of proteins synthesized in the ER occurs in the TGN, when lipids and proteins are packaged into vesicles for transport to various destinations. Recall that when clathrin-coated vesicles form from the TGN, the adaptor complexes include two adaptin subunits. The two adaptin subunits are apparently responsible for the specificity displayed when receptors are concentrated for inclusion in a budding vesicle. Once the vesicle forms, however, additional proteins must ensure delivery of the vesicle to the appropriate destination. The problem of finding the appropriate target membrane to fuse with can be quite complex in cells in which vesicles continually pass back and forth between the ER and the Golgi complex, between various Golgi cisternae, and between the Golgi complex and the plasma membrane. There must be a means for preventing these various migrating vesicles from inadvertently fusing with the wrong membrane. The **SNARE hypothesis** provides a model for this important sorting and targeting step in intracellular transport (Figure 12-19).

According to the SNARE hypothesis, the molecular components that facilitate sorting and targeting of vesicles in eukaryotic cells include two families of **SNARE (SNAP receptor) proteins**: the **v-SNAREs (vesicle-SNAP receptors)** found on transport vesicles and the **t-SNAREs (target-SNAP receptors)** found on target membranes. Apparently, v-SNAREs and t-SNAREs are complementary molecules that, along with additional proteins, allow a vesicle to recognize a target organelle. Both v-SNAREs and t-SNAREs were originally investigated because of their role in neuronal exocytosis. Since their discovery in brain tissue, both families of proteins have also been implicated in transport from the ER to the Golgi complex in yeast and other organisms.

When a vesicle reaches its destination, a third family of proteins, the **Rab GTPases,** comes into play. Rab GTPases are also specific; vesicles fated for different destinations have distinct members of the Rab family associated with them. As illustrated in Figure 12-19, the affinity of complementary v-SNAREs and t-SNAREs for one another ensures that, when they collide, they will remain in contact long enough for a Rab protein associated with the vesicle to hydrolyze a GTP molecule and lock the complementary t- and v-SNAREs together.

At this point, another set of proteins, including **N-ethylmaleimide-sensitive factor (NSF)** and a group of **SNAPs (soluble NSF attachment proteins)**, mediates release of the v- and t-SNAREs and fusion of the donor and target membranes, accompanied by the hydrolysis of ATP. NSF and SNAPs are involved in fusion between a variety of cellular membranes, indicating they are not responsible for specificity during targeting.

Manipulation of the structure and expression of Rab proteins has allowed investigators to localize members of this protein family to specific transport routes throughout the endomembrane system. For example, overexpression of Rab5 in a cell enhances uptake of a receptor for transferrin—a key molecule in the transport of iron across the plasma membrane—but does not affect the rate of recycling of the receptor. Rab5 is apparently confined to a role in receptor-mediated endocytosis and transport to the early endosome. A closer look at cells containing excessive amounts of Rab5 reveals accumulation of large vesicles due to the failure of endocytic vesicles to fuse with the early endosome. On the other hand, overexpression of Rab4 leads to accumulation of receptors for transferrin in the plasma membrane. Moreover, transferrin is returned

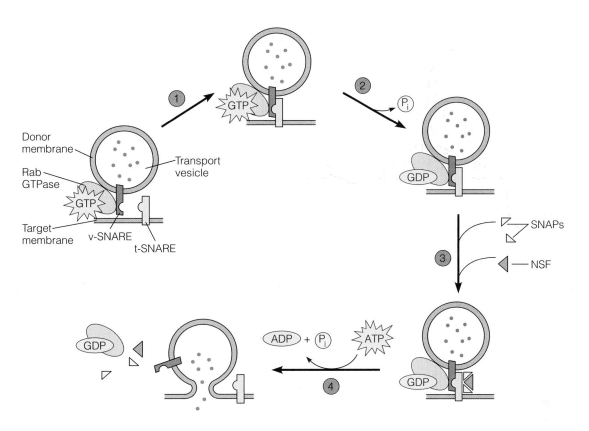

Figure 12-19   The SNARE Hypothesis for Transport Vesicle Targeting and Fusion.   The basic molecular components that mediate sorting and targeting of vesicles in eukaryotic cells include *v-SNAREs* on transport vesicles, *t-SNAREs* on target membranes, *Rab GTPase, NSF,* and several *SNAPs*. ① When complementary v-SNAREs and t-SNAREs meet, ② a Rab GTPase associated with the vesicle hydrolyzes GTP and locks the complementary v- and t-SNAREs together. ③ NSF and several SNAPs then bind to the complex and ④ mediate release of the v- and t-SNAREs and fusion of the donor and target membranes, accompanied by hydrolysis of ATP.

to the surface of the cell before it releases iron in the early endosome. Rab4 is apparently essential for recycling receptors to the plasma membrane.

Lysosomes and Cellular Digestion

The **lysosome** is an organelle containing digestive enzymes capable of degrading all the major classes of biological macromolecules, including lipids, carbohydrates, nucleic acids, and proteins. The digestive enzymes are needed to degrade extracellular materials brought into the cell by endocytosis and to digest intracellular structures and macromolecules that are damaged or no longer needed. We will first look at the organelle itself and then consider the digestive processes the lysosome is involved in, as well as some of the diseases that result from lysosomal malfunction.

Lysosomes Isolate Digestive Enzymes from the Rest of the Cell

As we saw in Chapter 4, lysosomes were discovered in the early 1950s by Christian de Duve and his colleagues (see Box 4A on p. 92). Differential centrifugation led the researchers to realize that an acid phosphatase initially thought to be located in the mitochondrion was in fact associated with a class of particles that had never been reported before. Along with the acid phosphatase, the new organelle contained several other hydrolytic enzymes, including β-glucuronidase, a deoxyribonuclease, a ribonuclease, and a protease. Because of its apparent role in cellular lysis, de Duve called the organelle a *lysosome.*

Only after the lysosome's existence had been predicted, its properties described, and its enzyme content specified was the organelle actually observed by electron microscopy and recognized as a normal constituent of most animal cells. Final confirmation came from cytochemical staining reactions capable of localizing the acid phosphatase and other lysosomal enzymes to specific structures that can be seen by electron microscopy (Figure 12-20).

Lysosomes vary considerably in size and shape, but are generally about 0.5 μm in diameter. Like the ER and Golgi complex, the lysosome is bounded by a single membrane. This membrane is crucial for protecting the rest of the cell from the hydrolytic enzymes present within the lysosomal lumen. ATP-dependent proton pumps maintain an acidic environment (pH 4.0–5.0) within the lysosome, which favors enzymatic digestion of macromolecules

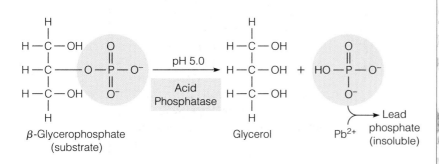

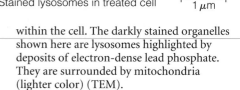

β-Glycerophosphate (substrate) → Acid Phosphatase, pH 5.0 → Glycerol + Lead phosphate (insoluble), Pb²⁺

(a) Cytochemical process

(b) Stained lysosomes in treated cell

Lysosome

1 μm

Figure 12-20 Cytochemical Localization of Acid Phosphatase, a Lysosomal Enzyme. **(a)** Thin sections of tissue are fixed in glutaraldehyde and incubated in a medium containing β-glycerophosphate (a substrate for the enzyme acid phosphatase) and a soluble lead salt (commonly lead nitrate). Acid phosphatase cleaves the β-glycerophosphate substrate, leaving free glycerol and phosphate anions. The phosphate anions react with lead ions to form lead phosphate, which precipitates at the site of enzyme activity. **(b)** Electron microscopy indirectly reveals the location of acid phosphatase within the cell. The darkly stained organelles shown here are lysosomes highlighted by deposits of electron-dense lead phosphate. They are surrounded by mitochondria (lighter color) (TEM).

by partially denaturing the macromolecules targeted for degradation. The products of digestion are then transported—either passively or actively—across the membrane to the cytosol, where they enter various synthetic pathways or are exported from the cell.

The list of lysosomal enzymes has expanded considerably since de Duve's original work, but all have the common property of being *acid hydrolases*—hydrolytic enzymes with a pH optimum around 5.0. The list includes at least 5 phosphatases, 14 proteases and peptidases, 2 nucleases, 6 lipases, 13 glycosidases, and 7 sulfatases. Taken together, these lysosomal enzymes can digest all the major classes of biological molecules. No wonder, then, that they are sequestered from the rest of the cell, where they would quickly wreak havoc. We might ask why hydrolases capable of digesting entire organelles do not damage the lysosomal membrane itself. Apparently, extensive glycosylation of membrane components exposed to the interior of the mature lysosome protects the lysosomal membrane from degradation.

Lysosomes Develop from Endosomes

Lysosomal enzymes are synthesized by ribosomes attached to the rough ER and are threaded into the ER lumen before transport to the Golgi complex. After modification and processing in the ER and Golgi complex compartments by some of the same enzymes that modify and process secretory and plasma membrane proteins, the lysosomal enzymes are sorted from other proteins in the TGN. Earlier in the chapter, we described the addition of a unique mannose-6-phosphate tag to soluble lysosomal enzymes. Distinctive sorting signals are also present on membrane-bound lysosomal proteins. The lysosomal enzymes are packaged in clathrin-coated vesicles that bud from the TGN, lose their protein coats, and travel to one of the endosomal compartments (see Figure 12-9).

Most lysosomal enzymes are delivered from the TGN to an early endosome. You will recall that early endosomes are formed by the coalescence of vesicles from the TGN and plasma membrane. Over time, the early endosome matures to form a **late endosome**, an organelle with a full complement of acid hydrolases but not engaged in digestive activity. During this process, ATP-dependent proton pumps lower the pH of the early endosomal lumen from about 6.0 to 5.5, and the organelle loses its capacity to fuse with endocytic vesicles. The late endosome is essentially a collection of newly synthesized digestive enzymes as well as extracellular and intracellular material fated for digestion, packaged in a way that protects the cell from hydrolytic enzymes as the late endosome matures into a lysosome or delivers its contents to an existing lysosome.

The final step in lysosome development is the activation of the acid hydrolases. This activation depends on moving the enzymes and their substrates to a more acidic environment. There are two ways eukaryotic cells accomplish this step. ATP-dependent proton pumps may lower the pH of the late endosomal lumen to 4.0–5.0, transforming the late endosome into a lysosome and generating a new organelle. Alternatively, the late endosome may transfer material to an existing lysosome. There are at least two models for the transfer of material from late endosomes to lysosomes. In the **transient fusion model** (sometimes also called the "kiss-and-run" model), the late endosome forms temporary connections with the lysosome. Only new lysosomal proteins and lipids, as well as material fated for digestion, are transferred to the lysosome. This is followed by dissociation of the organelles. In the **hybrid model**, the late endosome and lysosome fuse to form a temporary hybrid organelle, with proteins and lipids from the two organelles not clearly segregated. This step is followed by sorting and recycling of endosomal components to earlier compartments and digestion of the remaining material.

Lysosomal Enzymes Are Important for Several Different Digestive Processes

Lysosomes are important for cellular activities as diverse as nutrition, defense, recycling of cellular components, and differentiation. We can distinguish the digestive processes that depend on lysosomal enzymes by both the site of their activity and the origin of the material that is digested, as shown in Figure 12-21. Usually, the site of activity is intracellular; in some cases, though, lysosomes may release their enzymes to the outside of the cell by exocytosis. The materials to be digested are often of extracellular origin, although there are important processes known to involve lysosomal digestion of cellular components. To distinguish between mature lysosomes of different origins, we refer to those containing substances of extracellular origin as **heterophagic lysosomes,** whereas those with materials of intracellular origin are called **autophagic lysosomes.** The specific processes in which lysosomal enzymes are involved are *phagocytosis, receptor-mediated endocytosis, autophagy,* and *extracellular digestion.* These are illustrated in Figure 12-21 as pathways Ⓐ, Ⓑ, Ⓒ, and Ⓓ, respectively.

Phagocytosis and Receptor-Mediated Endocytosis: Lysosomes in Defense and Nutrition.
One of the most important functions of lysosomal enzymes is the degradation of foreign material brought into eukaryotic cells by *phagocytosis* and *receptor-mediated endocytosis* (see Figures 12-14 and 12-15). Phagocytic vacuoles are transformed into lysosomes by at least two processes (pathway Ⓐ of Figure 12-21). Such endocytic vesicles can either (1) accumulate lysosomal proteins by forming temporary connections with early and late endosomes, never permanently fusing with an endosomal compartment, or (2) fuse directly with early endosomes. As a result, the lysosomes generated by phagocytosis vary considerably in size, appearance, content of digestible material, and stage of digestion. Most of the material brought into a cell by receptor-mediated endocytosis follows a single path. After internalized macromolecules are sorted by early endosomes, material destined for digestion is carried along as early endosomes mature to form late endosomes and lysosomes, bringing it in contact with acid hydrolases (pathway Ⓑ of Figure 12-21).

Soluble products of digestion, such as sugars, amino acids, and nucleotides, are then transported across the lysosomal membrane and are used as a source of nutrients by the cell. Some may cross by facilitated diffusion, whereas others undergo active transport. The acidity of the lysosomal lumen contributes to an electrochemical proton gradient across the lysosomal membrane, which can provide energy for driving transport to and from the cytosol.

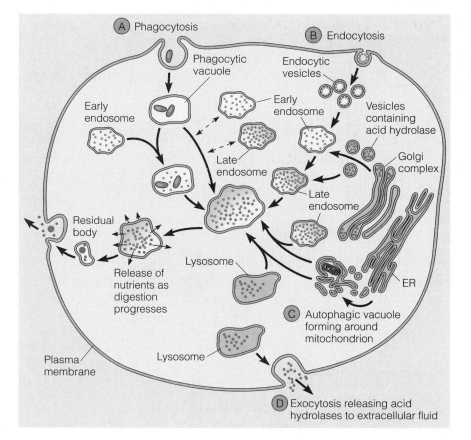

Figure 12-21 **The Formation of Lysosomes and Their Roles in Cellular Digestive Processes.** Illustrated in this composite cell are the major processes in which lysosomes are involved. The pathways depicted are Ⓐ phagocytosis, Ⓑ receptor-mediated endocytosis, Ⓒ autophagy, and Ⓓ extracellular digestion.

Eventually, however, only indigestible material remains in the lysosome, which becomes a **residual body** as digestion ceases. In protozoa, residual bodies routinely fuse with the plasma membrane and expel their contents to the outside by exocytosis, as illustrated in Figure 12-21. In vertebrates, there appears to be no such mechanism, so residual bodies accumulate in the cytoplasm. This accumulation of debris is thought to contribute to cellular aging, particularly in long-lived cells such as those of the nervous system.

Autophagy: The Original Recycling System. A second important task for lysosomes is the breakdown of cellular structures and components that are damaged or no longer needed. Most cellular organelles are in a state of dynamic flux, with new organelles continuously synthesized while old organelles are destroyed. The digestion of old or unwanted organelles or other cell structures is called **autophagy,** which is Greek for "self-eating." Autophagy is illustrated as pathway Ⓒ in Figure 12-21.

There are two types of autophagy—*macrophagy* and *microphagy*. **Macrophagy** begins when an organelle or other structure becomes wrapped in a double membrane derived from the ER. The resulting vesicle is called an **autophagic vacuole** (or *autophagosome*). It is often possible to see identifiable remains of cellular structures in these vacuoles, as shown in Figure 12-22. **Microphagy** involves formation of a much smaller autophagic vacuole, surrounded by a single phospholipid bilayer that encloses small bits of cytoplasm

rather than whole organelles. The fate of an autophagic vacuole is apparently slightly different from that of a phagocytic vacuole. Rather than gradually accumulating lysosomal enzymes, autophagic vacuoles tend to fuse with late endosomes or directly with active lysosomes.

Autophagy occurs at varying rates in most cells under most conditions, but it is especially prominent in certain developmental situations. For example, during maturation of a red blood cell virtually all of the intracellular content is destroyed, including all of the mitochondria. This is accomplished by autophagic digestion. A marked increase in autophagy is also noted in cells stressed by starvation. Presumably, the process represents a desperate attempt on the part of the cell to continue providing for its energy needs, even if it has to consume its own structures to do so.

Extracellular Digestion. Most of the digestive processes involving lysosomal enzymes occur intracellularly, following either endocytosis or formation of an autophagic vacuole. In rare cases, however, lysosomes discharge their enzymes to the outside of the cell by exocytosis, resulting in **extracellular digestion** (Figure 12-21, pathway Ⓓ). One example of extracellular digestion occurs during fertilization of animal eggs. The head of the sperm releases lysosomal enzymes capable of degrading chemical barriers that would otherwise keep the sperm from penetrating the egg surface. Certain inflammatory diseases, such as rheumatoid arthritis, apparently result from inadvertent release of lysosomal enzymes

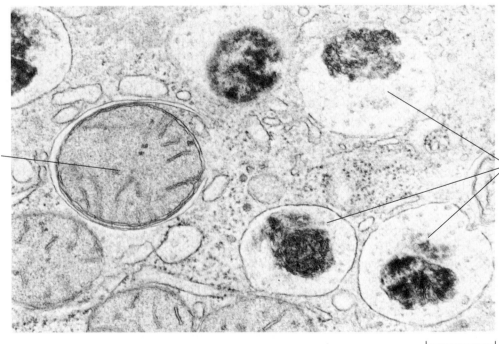

Mitochondrion being sequestered by membrane of the smooth ER

Autophagic vacuoles with remnants of mitochondria

0.5 μm

Figure 12-22 Autophagic Digestion. Early and late stages of autophagic digestion are shown here in a rat liver cell. On the left an autophagic vacuole is formed as a mitochondrion is sequestered by a membrane derived from the ER. On the right are several autophagic vacuoles containing remnants of mitochondria (TEM). Autophagic digestion is the means by which most old organelles are eliminated.

into the joints. The steroid hormones cortisone and hydrocortisone are thought to be effective antiinflammatory agents because of their role in stabilizing lysosomal membranes and thereby inhibiting enzyme release.

Lysosomal Storage Diseases Are Usually Characterized by the Accumulation of Indigestible Material

The essential role of lysosomes in the recycling of cellular components is clearly seen in disorders caused by deficiencies of specific lysosomal proteins. Over 40 such **lysosomal storage diseases** are known, each characterized by the harmful accumulation of specific substances, usually polysaccharides or lipids. In most cases, the substances accumulate because digestive enzymes are defective or missing, but sometimes because the proteins that transport degradation products from the lysosomal lumen to the cytosol are defective. In either case, the cells in which material accumulates are severely impaired, if not destroyed. Skeletal deformities, muscle weakness, and mental retardation commonly result, often with a fatal outcome. Unfortunately, most lysosomal storage diseases are not yet treatable.

The first storage disease to be understood was *type II glycogenosis,* in which young children accumulate excessive amounts of glycogen in the liver, heart, and skeletal muscles, and die at an early age. The problem turned out to be a defective form of the lysosomal enzyme *α-1,4-glucosidase,* which catalyzes glycogen hydrolysis in normal cells. Glycogen metabolism occurs predominantly in the cytosol, but a small amount of glycogen can enter the lysosome through autophagy and will accumulate to a damaging level if not broken down to glucose.

Two of the best-known lysosomal storage diseases are *Hurler syndrome* and *Hunter syndrome.* Both arise from defects in the degradation of glycosaminoglycans, which are the major carbohydrate components of the proteoglycans found in the extracellular matrix (see Chapter 11). The defective enzyme in a patient with Hurler syndrome is *α-L-iduronidase,* which is required for the degradation of glycosaminoglycans. Electron microscopic observation of sweat gland cells from a patient with Hurler syndrome reveals large numbers of atypical vacuoles that stain for both acid phosphatase and undigested glycosaminoglycans. These vacuoles are apparently aberrant late endosomes filled with indigestible material.

Mental retardation is a common feature of lysosomal storage diseases because of the impaired metabolism of glycolipids, which are important components of brain tissue and the sheaths of nerve cell axons. One particularly well-known example is *Tay-Sachs disease,* a condition inherited as a recessive trait. Afflicted children show rapid mental deterioration after about six months of age, followed by paralysis and death within three years. The disease results from the accumulation in nervous tissue of a particular glycolipid called a *ganglioside.* The missing lysosomal enzyme in this case is *β-N-acetylhexosaminidase,* which is responsible for cleaving the terminal *N*-acetylgalactosamine from the carbohydrate portion of the ganglioside. Lysosomes from children afflicted with Tay-Sachs disease are filled with membrane fragments containing undigested gangliosides.

The Plant Vacuole: A Multifunctional Organelle

Plant cells contain acidic membrane-enclosed compartments called **vacuoles** that resemble the lysosomes found in most animal cells, but generally serve additional roles. The biogenesis of a vacuole parallels that of a lysosome. Most of the components are synthesized in the ER and transferred to a **dictyosome** (as a Golgi stack in a plant cell is usually called), where proteins undergo further processing. Coated vesicles then convey lipids and proteins destined for the vacuole from the dictyosome to a **provacuole,** which is analogous to an endosome. Alternatively, a provacuole can form by autophagy. The provacuole eventually matures to form a functional vacuole that can fill as much as 90 percent of the volume of a plant cell.

In addition to confining hydrolytic enzymes, plant vacuoles have a variety of other functions essential to the health of plant cells. Most functions reflect the plant's lack of mobility and consequent susceptibility to changes in the surrounding environment. As mentioned in Chapter 4, a major role of the vacuole lies in maintenance of *turgor pressure,* the osmotic pressure that prevents plant cells from collapsing. Not only does turgor pressure prevent a plant from wilting, it can also drive the expansion of cells. During development, softening of the cell wall—accompanied by higher turgor pressure—allows the cell to expand. The direction of expansion can be controlled by selective softening of specific segments of the cell wall. Maintenance of turgor pressure is closely connected to another role of plant vacuoles, regulation of the concentrations of various solutes in the cytoplasm. An important example is the control of cytosolic pH. ATP-dependent proton pumps in the vacuolar membrane can compensate for a decline in cytosolic pH (perhaps due to a change in the extracellular environment) by transferring protons from the cytosol to the lumen of the vacuole.

The vacuole also serves as a storage compartment. Seed storage proteins are generally synthesized by ribosomes attached to the rough ER and cotranslationally inserted into the ER lumen. Some of the storage proteins remain in the ER while others are transferred to vacuoles, either by autophagy of vesicles budding from the ER or by way of dictyosomes. When the seeds germinate, the storage proteins are available for hydrolysis by vacuolar proteases, thereby releasing amino acids for the biosynthesis of new proteins. Other substances stored in vacuoles include the anthocyanins that impart color to flowers and attract pollinating insects and birds, toxic substances that deter predators, inorganic and organic nutrients, compounds that shield cells from ultraviolet light, and residual indigestible waste. Storage of soluble as well as insoluble waste is an important function of plant vacuoles. Unlike animals, most plants do not have a mechanism for excreting soluble waste from the organism. The large vacuoles found in plant cells enable the cells to accumulate solutes to a degree that would inhibit or restrict metabolic processes if the material were to remain in the cytosol.

Peroxisomes

Peroxisomes, like the Golgi complex and lysosomes, are bounded by single membranes; however, they are not derived from the endoplasmic reticulum and are therefore not part of the endomembrane system that includes all the other organelles studied in this chapter. This organelle is found in all eukaryotic cells but is especially prominent in mammalian kidney and liver cells, in algae and photosynthetic cells of plants, and in germinating seedlings of plant species that store fat in their seeds. Peroxisomes are somewhat smaller than mitochondria, though there is considerable variation in size, depending on the tissue in which they are found.

Regardless of location or size, the defining characteristic of a peroxisome is the presence of *catalase,* an enzyme essential for the degradation of hydrogen peroxide (H_2O_2). Hydrogen peroxide is a potentially toxic compound that is formed by a variety of oxidative reactions catalyzed by *oxidases.* Both

catalase and the oxidases are confined to peroxisomes. Thus, the generation and degradation of H_2O_2 occur within the same organelle, thereby protecting other parts of the cell from exposure to this harmful compound. Before discussing the functions of peroxisomes further, let us look at how peroxisomes were discovered and how they are distinguished from other organelles when viewed by electron microscopy.

The Discovery of Peroxisomes Depended on Innovations in Equilibrium Density Centrifugation

Christian de Duve and his colleagues discovered not only lysosomes, but also peroxisomes. During the course of their early studies on lysosomes, the researchers encountered at least one enzyme, *urate oxidase,* that appeared to be associated with lysosomal fractions but was not an acid hydrolase. Moreover, when they subjected the organelles to differential centrifugation, the urate oxidase behaved somewhat differently than known lysosomal enzymes.

The clue that eventually persuaded de Duve and his colleagues that urate oxidase was associated with a new class of organelles was provided by the technique of equilibrium density centrifugation (Figure 12-23; see also Box 12A). By using a gradient of sucrose concentration, the

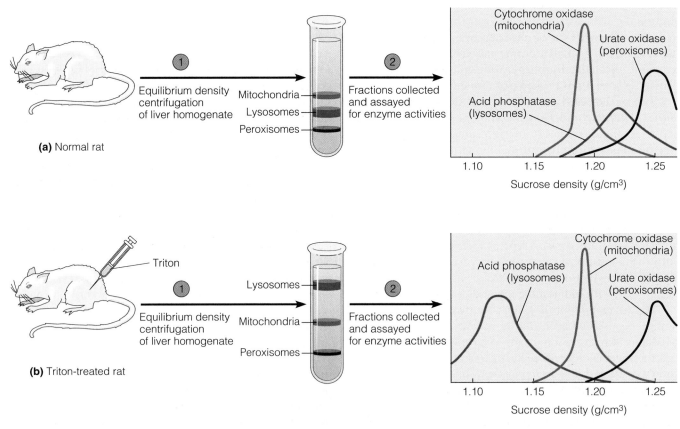

Figure 12-23 Separation of Lysosomes from Peroxisomes by Equilibrium Density Centrifugation. (a) For organelles obtained from the liver of a normal rat and centrifuged to equilibrium on a sucrose density gradient, lysosomes (marker enzyme: *acid phosphatase*) have a range of densities between those of mitochondria (marker enzyme: *cytochrome oxidase*) and peroxisomes (marker enzyme: *urate oxidase*), making resolution of the three organelles difficult. **(b)** For organelles obtained from the liver of a rat that was treated with the detergent Triton WR-1339, the lysosomes are concentrated at a much lower density, permitting separation of lysosomes from peroxisomes.

researchers found that urate oxidase from rat liver was recovered in a region of the gradient having a slightly higher density (1.25 g/cm³) than that of other organelles, particularly lysosomes (about 1.20–1.24 g/cm³) and mitochondria (about 1.19 g/cm³). However, the density differences were too small and the range of the lysosome densities was too broad to allow the new class of organelles to be separated adequately from lysosomes under normal conditions (Figure 12-23a). By using an experimental trick, however, de Duve was able to achieve good separation. The trick was based on the chance observation that animals injected with the detergent Triton WR-1339 accumulate the detergent preferentially in their lysosomes, giving these organelles a much lower buoyant density than normal (about 1.10–1.14 g/cm³). By using liver homogenates from detergent-treated rats, de Duve was able to separate urate oxidase cleanly from both lysosomal and mitochondrial marker enzymes (Figure 12-23b).

Once separation was achieved, additional enzymes were identified in the fractions containing urate oxidase, including catalase and D–amino acid oxidase. Catalase, as we have seen, degrades H_2O_2; and, like urate oxidase, D–amino acid oxidase generates H_2O_2. Because of its apparent involvement in hydrogen peroxide metabolism, the new organelle became known as a *peroxisome*. Other peroxisomal enzymes have since been identified and it is now clear that the enzyme complement of the organelle varies significantly from species to species and sometimes from organ to organ, or from one developmental stage to another within the same organ. However, the presence of catalase and one or more hydrogen peroxide–generating oxidases remains a distinguishing characteristic of all peroxisomes.

Once peroxisomes had been identified and isolated biochemically, the existence of organelles with the expected properties—first in isolated peroxisomal fractions from density gradients and then in intact cells—was confirmed by electron microscopy. Peroxisomes turned out to be the functional equivalents of organelles that had been seen earlier in electron micrographs of both animal and plant cells. Because their function was not known at the time, these organelles were simply called **microbodies.** In both plant and animal cells, a microbody is usually about 0.2–2.0 μm in diameter, is surrounded by a single membrane, and generally has a finely granular *matrix* (interior of the organelle). Figure 12-24 shows the appearance of microbodies, or peroxisomes, in a rat liver cell.

As seen in Figure 12-24, animal peroxisomes often contain a distinct crystalline core, which usually consists of a crystalline form of urate oxidase. Crystalline cores are also often present in the peroxisomes of plant leaves, but these usually consist of catalase instead (see Figure 4-20). When such cores are present, it is easy to identify microbodies as peroxisomes, since urate oxidase and catalase are two of the enzymes by which peroxisomes are defined. In the absence of a crystalline core, however, it is not always easy to spot peroxisomes ultrastructurally.

A useful technique in such cases is a cytochemical test for catalase called the *diaminobenzidine (DAB) reaction*. This

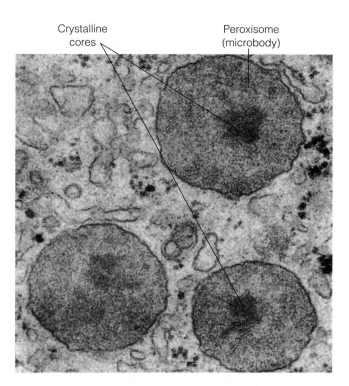

Figure 12-24 Peroxisomes in Animal Cells. This electron micrograph shows several peroxisomes (microbodies) in the cytoplasm of a rat liver cell. A crystalline core is readily visible in each microbody. In animal microbodies, the cores are almost always crystalline urate oxidase (TEM).

assay depends on the ability of catalase to oxidize DAB to a polymeric form that causes deposition of electron-dense osmium atoms when the tissue is treated with osmium tetroxide (OsO_4). The resulting electron-dense deposits can be readily seen in cells from stained tissue. In animal peroxisomes, the entire internal space often stains intensely with DAB, indicating that catalase exists as a soluble enzyme uniformly distributed throughout the matrix of the organelle. In plant leaf cells, DAB treatment preferentially stains the crystalline cores of the peroxisomes (Figure 12-25), thereby definitively identifying the cores as crystalline catalase. Because catalase is the single enzyme present in all peroxisomes and does not routinely occur in any other organelle, the DAB reaction is a very reliable and highly specific means of identifying organelles unambiguously as peroxisomes.

Most Peroxisomal Functions Are Linked to Hydrogen Peroxide Metabolism

Peroxisomes occur widely in animals, plants, algae, and some fungi. In animals, peroxisomes are most prominent in liver and kidney tissue. The essential roles of peroxisomes in eukaryotic cells have become more apparent in recent years, stimulating new research into the metabolic pathways and the disorders arising from defective components of the pathways found in these organelles. There are at least five general categories of peroxisomal functions: hydrogen peroxide metabolism, detoxification of harmful compounds,

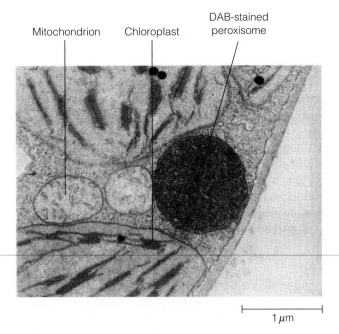

Figure 12-25 Cytochemical Localization of Catalase in Plant Peroxisomes. Shown here is a tobacco leaf cell similar to the one shown in Figure 4-20, but stained by diaminobenzidine (DAB). The principle of this assay is similar to that of the cytochemical test for acid phosphatase described in Figure 12-20. Catalase oxidizes DAB to a polymeric form that causes the deposition of electron-dense osmium atoms in tissue treated with osmium tetroxide (OsO_4). The DAB technique reveals that the deposition of osmium is confined to peroxisomes and, therefore, that catalase is a component of the crystalline core (TEM).

oxidation of fatty acids, metabolism of nitrogen-containing compounds, and catabolism of unusual substances.

Hydrogen Peroxide Metabolism. The most obvious role of peroxisomes in eukaryotic cells is the detoxification of H_2O_2, which is accomplished by the coexistence in the same organelle of catalase and the oxidases that generate H_2O_2. The oxidases in peroxisomes vary considerably in the specific reactions they catalyze, but they all share the property of transferring electrons from their respective substrates directly to molecules of oxygen (O_2), forming H_2O_2. Using RH_2 to represent an oxidizable substrate, the general reaction catalyzed by oxidases can be written as

$$RH_2 + O_2 \longrightarrow R + H_2O_2 \qquad \text{(12-3)}$$

The hydrogen peroxide formed in this manner is broken down by catalase in one of two ways. Usually, catalase functions in what is called its *catalic mode,* in which one molecule of H_2O_2 is oxidized to oxygen and a second is reduced to water:

$$2H_2O_2 \longrightarrow O_2 + 2H_2O \qquad \text{(12-4)}$$

Dividing reaction 12-4 by 2 and adding it to reaction 12-3 yields a summary reaction for the two-step process:

$$RH_2 + \tfrac{1}{2}O_2 \longrightarrow R + H_2O \qquad \text{(12-5)}$$

Alternatively, catalase can function in its *peroxidatic mode,* in which electrons derived from an organic donor are used to reduce hydrogen peroxide to water:

$$R'H_2 + H_2O_2 \longrightarrow R' + 2H_2O \qquad \text{(12-6)}$$

(The prime on the R group simply indicates that this substrate is likely to be different from the substrate in reaction 12-3.) The corresponding summary reaction in this case is

$$RH_2 + R'H_2 + O_2 \longrightarrow R + R' + 2H_2O \qquad \text{(12-7)}$$

The result is the same in either case: Hydrogen peroxide is degraded without ever leaving the peroxisome. Given the toxicity of hydrogen peroxide (which is the main active ingredient in a variety of disinfectants), it makes good sense for the enzymes responsible for peroxide generation to be compartmentalized together with the catalase that catalyzes its degradation. Indeed, catalase is the most abundant protein in most peroxisomes, representing up to 15% of the total protein content of this organelle. In this way, virtually every molecule of hydrogen peroxide generated by the oxidases will encounter a molecule of catalase almost immediately and thereby be promptly degraded.

Detoxification of Harmful Compounds. In its peroxidatic mode (reaction 12-6), catalase can use a variety of substances as electron donors, including methanol, ethanol, formic acid, formaldehyde, nitrites, and phenols. Because all of these compounds are harmful to cells, their oxidative detoxification by catalase may be a vital peroxisomal function. The prominent peroxisomes of liver and kidney cells are thought to be important in such detoxification reactions.

Oxidation of Fatty Acids. Peroxisomes found in animal, plant, and fungal cells contain enzymes necessary for oxidizing fatty acids. This process, called β **oxidation,** also occurs in the mitochondrion and will be encountered again in Chapter 14. About 25–50% of fatty acid oxidation in animal tissues occurs in peroxisomes, with the remainder localized in mitochondria. In plant and yeast cells, on the other hand, all β oxidation is confined to peroxisomes.

In animal cells, peroxisomal β oxidation appears to be especially important for the catabolism of long-chain (16–20 carbons), very long-chain (24–26 carbons), and branched fatty acids. The primary product of β oxidation, acetyl-CoA, is then exported to the cytosol and enters biosynthetic pathways. Once fatty acids are shortened to fewer than 16 carbons, further oxidation usually occurs in the mitochondria. Thus, in animal cells, the peroxisome is important for shortening fatty acids in preparation for subsequent metabolism in the mitochondrion rather than completely breaking them down to acetyl-CoA. In plants and yeast, on the other hand, peroxisomes are essential for the complete catabolism of fatty acids to acetyl-CoA.

As we learned in Chapter 10, prostaglandins and related compounds are important paracrine and autocrine hormones. Peroxisomal enzymes, by shortening the fatty acid side chains on prostaglandins, play an important role in terminating the activity of such compounds.

Metabolism of Nitrogen-Containing Compounds. Except for primates, most animals require urate oxidase (also called *uricase*) to oxidize urate, a purine that is formed during the catabolism of nucleic acids and some proteins. Like other oxidases, urate oxidase catalyzes the direct transfer of electrons from the substrate to molecular oxygen, generating H_2O_2:

$$\text{urate} + O_2 \longrightarrow \text{allantoin} + H_2O_2 \qquad \textbf{(12-8)}$$

As noted earlier, the H_2O_2 is immediately degraded in the peroxisome by catalase. The allantoin is further metabolized and excreted by the organism, either as allantoic acid or, in the case of crustaceans, fish, and amphibians, as urea.

Additional peroxisomal enzymes involved in nitrogen metabolism include *aminotransferases*. Members of this collection of enzymes catalyze the transfer of amino groups ($-NH_3^+$) from amino acids to α-keto acids:

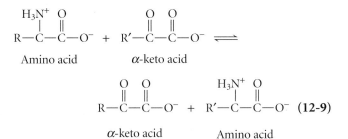

Such enzymes play important roles in the biosynthesis and degradation of amino acids by moving amino groups from one molecule to another.

Catabolism of Unusual Substances. Some of the substrates for peroxisomal oxidases are rare compounds for which the cell has no other degradative pathways. Such compounds include D–amino acids, which are not recognized by enzymes capable of degrading the far more common L–amino acids that constitute polypeptides. In some fungi, the peroxisomes also contain enzymes that break down unusual substances called **xenobiotics,** chemical compounds foreign to biological organisms. This category includes *alkanes,* short-chain hydrocarbon compounds found in oil and other petroleum products. Fungi containing enzymes capable of metabolizing such xenobiotics may turn out to be useful for cleaning up oil spills that would otherwise contaminate the environment.

Peroxisomal Disorders. Considering the variety of metabolic pathways found in peroxisomes, it is not surprising that a large number of disorders arise from defective peroxisomal proteins. The most common peroxisomal disorder is *X-linked adrenoleukodystrophy.* The defective protein causing this disorder is an integral membrane protein that may be responsible for transporting very long-chain fatty acids into the peroxisome for β oxidation.

Accumulation of these long-chain fatty acids in body fluids destroys the myelin sheath in nervous tissues.

Plant Cells Contain Types of Peroxisomes Not Found in Animal Cells

In plants and algae, peroxisomes are involved in several specific aspects of energy metabolism and will therefore be discussed in more detail in Chapters 14 and 15. Here, we will simply introduce several plant-specific peroxisomes and briefly describe their functions.

Leaf Peroxisomes. Cells of leaves and other photosynthetic plant tissues contain characteristic large, prominent **leaf peroxisomes,** which often appear in close contact with chloroplasts and mitochondria (see Figure 12-25; see also Figure 4-20). The spatial proximity of the three organelles probably reflects their mutual involvement in the *glycolate pathway,* which is also called the *photorespiratory pathway* because it involves the light-dependent uptake of O_2 and release of CO_2. Several enzymes of this pathway, including a peroxide-generating oxidase and two aminotransferases, are confined to leaf peroxisomes. Because of the link between peroxisomes, photosynthesis, and the glycolate pathway, we will return to leaf peroxisomes in Chapter 15.

Glyoxysomes. Another functionally distinct type of plant peroxisome occurs transiently in seedlings of plant species that store carbon and energy reserves in the seed as fat (primarily triacylglycerols). In such species, stored triacylglycerols are mobilized and converted to sucrose during early postgerminative development by a sequence of events that includes β oxidation of fatty acids as well as a pathway known as the *glyoxylate cycle.* All of the enzymes needed for these processes are localized to specialized peroxisomes called **glyoxysomes.** Glyoxysomes are found only in the tissues in which the fat is stored (endosperm or cotyledons, depending on the species) and are present only for the relatively short period of time (a week or two, in most cases) required for the seedling to deplete its supply of stored fat. Once they fulfill their role in the seedling, the glyoxysomes are converted to peroxisomes. Glyoxysomes have been reported to appear again in the senescing (aging) tissues of some plant species, presumably to degrade lipids derived from the membranes of the senescent cells. However, it is not yet clear how important their involvement in senescence is. Because of their role in the oxidative breakdown of fat, we will encounter glyoxysomes again in Chapter 14.

Other Kinds of Plant Peroxisomes. In addition to their presence in tissues that carry out either photorespiration or β oxidation of fatty acids, peroxisomes are also found in other plant tissues. For example, another kind of specialized peroxisome is present in *nodules,* the structures on plant roots in which plant cells and certain bacteria cooperate in the fixation of atmospheric nitrogen (that is, the conversion of N_2 into organic form). The peroxisomes in these cells are involved in the processing of fixed nitrogen.

Peroxisome Biogenesis Occurs by Division of Preexisting Peroxisomes

Like other organelles, peroxisomes increase in number as cells grow and divide; this proliferation of organelles is called *biogenesis*. In the case of endosomes and lysosomes, biogenesis occurs by fusion of vesicles budding from the Golgi complex. Peroxisomes were once thought to form from vesicles, but investigators now agree that their biogenesis occurs by the division of preexisting peroxisomes, a mode similar to that of mitochondria and chloroplasts. This mode of biogenesis raises two questions.

First, where do the lipids that make up the newly synthesized peroxisomal membrane come from? The answer is that a portion of the lipids are synthesized by peroxisomal enzymes and the rest, especially phosphatidylcholine and phosphatidylethanolamine, are probably acquired from the ER. Rather than arriving by vesicles, however, lipids are conveyed to peroxisomes by the phospholipid exchange proteins mentioned earlier in the chapter.

Second, where are the new enzymes and other proteins that are present in the peroxisomal membrane and matrix synthesized? Proteins destined for peroxisomes are synthesized not on ER-associated ribosomes, but on free cytosolic ribosomes. Thus, the new proteins—full-length polypeptides—are incorporated into preexisting peroxisomes posttranslationally. This passage of polypeptides across the peroxisomal membrane is an ATP-dependent process mediated by specific membrane proteins.

Figure 12-26 illustrates the incorporation of membrane components and matrix enzymes from the cytosol into a peroxisome, and the formation of new peroxisomes by division of a preexisting organelle. The protein depicted in the figure is catalase, a tetrameric protein with a heme group bound to each subunit. The subunits are synthesized individually on cytosolic ribosomes, imported into the peroxisome, refolded, and assembled, along with heme, into the active tetrameric enzyme.

For posttranslational import to work, each protein destined for a specific organelle must have some sort of signal that directs, or targets, the protein to the correct organelle. Such a signal functions by recognizing specific receptors or other features on the surface of the appropriate membrane. The signal in each case is a sequence of amino acids, which differs in sequence, length, and location for proteins targeted to different organelles. The signal that targets at least some peroxisomal proteins to their destination consists of just three amino acids and is found at or near the carboxyl terminus of the molecule. The most common sequence is serine-lysine-leucine (SKL), though a limited number of other amino acids are possible at each of the three locations. Thus, we expect each of the four subunits of catalase shown in Figure 12-26 to

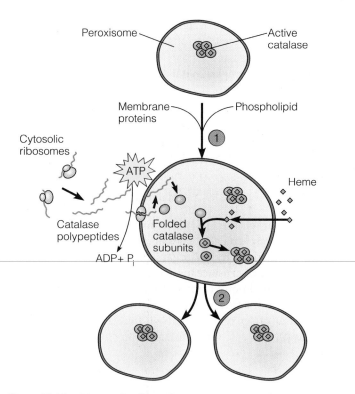

Figure 12-26 Biogenesis of Peroxisomes. New peroxisomes arise by the division of existing peroxisomes rather than by fusion of vesicles from the Golgi complex. ① Lipids, membrane proteins, and matrix enzymes are added to existing peroxisomes from cytosolic sources. The enzyme shown here is catalase, a tetrameric protein. The polypeptides are synthesized on cytosolic ribosomes and are threaded through the membrane posttranslationally. Heme, a cofactor, enters the peroxisomal lumen via a separate pathway. The polypeptides are then folded and assembled to form the active tetrameric protein. ② After lipids and protein are added, new peroxisomes are formed by the division of the existing organelle.

have the SKL sequence or one of the acceptable variants near its carboxyl terminus.

Suresh Subramani and others have shown that a protein normally targeted to the peroxisome will remain in the cytosol if its SKL sequence is removed. Conversely, the addition of an SKL sequence to a protein normally found in the cytosol will direct that protein to the peroxisome. Thus, the SKL sequence, or one of the acceptable variants, is both *necessary* and *sufficient* to direct proteins to peroxisomes. Moreover, a peroxisomal-targeting sequence identified in one species often functions in other species, even when the organisms are as evolutionarily diverse from one another as plants, yeast, insects, and animals. In one striking example, the gene for *luciferase,* a peroxisomal enzyme that enables fireflies to emit flashes of light, was transferred into plant cells from which whole plants were then grown. When cells of the genetically transformed plant were carefully examined, the luciferase was found in peroxisomes, the same organelle in which it is located in fireflies.

A full appreciation of eukaryotic cells depends on an understanding of intracellular membranes and the compartmentalization of function such membranes make possible. Especially prevalent within most eukaryotic cells is the endomembrane system, an elaborate array of membrane-bounded organelles derived either directly or indirectly from the endoplasmic reticulum (ER). The ER itself is a network of sacs, tubules, and vesicles that separates the lumen (interior space of the ER and related compartments) from the surrounding cytosol. The rough ER is studded with ribosomes and is the site of the synthesis of proteins destined for incorporation into various organelles—the nuclear envelope, Golgi complex, endosomes, and lysosomes—or for secretion from the cell. Both the rough and smooth ER synthesize most lipids for cellular membranes, and the smooth ER is the site of other processes, including hydroxylation reactions, drug detoxification, carbohydrate metabolism, calcium storage, and steroid biosynthesis. Before proteins leave the ER—by means of transition vesicles—they undergo the first few steps of protein modification. A host of ER-specific proteins catalyze glycosylation and folding of polypeptides, as well as the assembly of multimeric proteins.

The Golgi complex plays an equally important role in the glycosylation of proteins, and it has the additional role of sorting proteins for transport to other organelles and the plasma membrane. Transition vesicles that bud from the ER fuse with the membrane surrounding the *cis*-Golgi network, delivering lipids and proteins to the Golgi complex. Proteins then move through the Golgi stack toward the *trans*-Golgi network. During their journey, proteins are further modified as oligosaccharide side chains are trimmed or glycosylated. Transport vesicles that bud from the *trans*-Golgi network convey the processed proteins to their final destinations within the cell.

Among the proteins processed by enzymes of the Golgi complex are those intended for secretion from the cell by exocytosis, which may be constitutive (continuous) or regulated (in response to a specific extracellular signal). Secretory vesicles and secretory granules release their contents to the extracellular medium by fusing with the plasma membrane.

Exocytosis, which adds lipids and proteins to the plasma membrane, is balanced by endocytosis, which involves the ingestion of extracellular substances through invagination of the plasma membrane. Endocytic processes include phagocytosis, receptor-mediated endocytosis (also called clathrin-dependent endocytosis), and clathrin-independent endocytosis. Receptor-mediated endocytosis is highly specific because it depends on the binding of ligands to receptors on the cell surface. The receptor-ligand complexes are concentrated in clathrin-coated pits and brought into the cell by clathrin-coated vesicles. Once inside the cell, vesicles promptly lose their protein coats and fuse with intracellular membranes, typically membranes surrounding early endosomes, which are sites where ingested material is sorted. Receptors and membrane lipids are often recycled to the plasma membrane for another round of endocytosis, while material destined for degradation is carried along as early endosomes mature to form late endosomes and lysosomes.

Late endosomes are organelles containing inactive acid hydrolases. ATP-dependent proton pumps lower the pH of the endosomal lumen and transform the late endosome into a lysosome. Alternatively, a late endosome may deliver its contents to an existing lysosome. In both cases, latent acid hydrolases capable of degrading most biological molecules become active. Phagocytic vacuoles acquire lysosomal enzymes by fusing with early endosomes or forming temporary connections with endosomes. Autophagic

vacuoles generally fuse with late endosomes or active lysosomes. Autophagy is important for recycling cellular structures that are damaged or no longer needed.

Transport vesicles carry material throughout the endomembrane system. Coat proteins—which include clathrin, COPI, COPII, and caveolin—participate in the sorting of molecules fated for different destinations, as well as in the formation of vesicles. The specific set of proteins covering the exterior of a vesicle is an indicator of its origin and its destination within the cell. Once a transport vesicle reaches its destination, several additional proteins catalyze fusion with the target membrane. Two families of receptor proteins, the v-SNAREs and t-SNAREs, ensure that the vesicle has reached its target. A third family of proteins, Rab GTPases, locks complementary v- and t-SNAREs together. And finally, NSF and a group of SNAPs catalyze fusion of the vesicle membrane with the target membrane.

Peroxisomes, which are not part of the endomembrane system, increase in number by the division of preexisting organelles rather than by the coalescence of vesicles. Some peroxisomal membrane lipids are synthesized by peroxisomal enzymes; the rest are conveyed from the ER to peroxisomes by phospholipid exchange proteins. Peroxisomal proteins are synthesized by cytosolic ribosomes and are imported posttranslationally. The defining enzyme of a peroxisome is catalase. This enzyme degrades hydrogen peroxide that is generated by various oxidases before the chemical can harm cellular components. In animal cells, the reactions occurring in peroxisomes are important for detoxification of harmful substances, oxidation of fatty acids, and metabolism of nitrogen-containing compounds. In plants, peroxisomes play distinctive roles in the conversion of stored lipids into carbohydrate (glyoxysomes), and in photorespiration (leaf peroxisomes).

In this chapter, we have covered several major membranous structures found in eukaryotic cells, including the ER, the Golgi complex, endosomes, lysosomes, exocytic and endocytic vesicles, and peroxisomes. In each case, we recognized that each structure is characterized by a unique set of proteins and is essential for specific functions in the cell. In the next few chapters, we will discuss other important cellular structures, including the mitochondrion (Chapter 14), the chloroplast (Chapter 15), and the nucleus (Chapter 16).

Key Terms for Self-Testing

endomembrane system (p. 323)

The Endoplasmic Reticulum
endoplasmic reticulum (ER) (p. 323)
ER cisterna (p. 323)
ER lumen (p. 323)
rough endoplasmic reticulum (rough ER) (p. 324)
transitional element (TE) (p. 325)
transition vesicle (p. 325)
smooth endoplasmic reticulum (smooth ER) (p. 325)
microsome (p. 325)
hydroxylation (p. 331)
phospholipid translocator (flippase) (p. 332)
phospholipid exchange protein (p. 333)

The Golgi Complex
Golgi complex (p. 333)
Golgi stack (p. 333)
transport vesicle (p. 333)
cis (forming) face (p. 334)
cis-Golgi network (CGN) (p. 334)
trans (maturing) face (p. 334)
trans-Golgi network (TGN) (p. 334)
medial cisterna (p. 334)
stationary cisternae model (p. 335)
shuttle vesicle (p. 335)
cisternal maturation model (p. 335)
anterograde transport (p. 335)
retrograde transport (p. 335)

Roles of the ER and Golgi Complex in Protein Glycosylation
glycosylation (p. 336)
glycoprotein (p. 336)
N-linked glycosylation (N-glycosylation) (p. 336)
O-linked glycosylation (p. 336)
core glycosylation (p. 336)
core oligosaccharide (p. 336)
terminal glycosylation (p. 336)

Roles of the ER and Golgi Complex in Protein Sorting
secretory pathway (p. 341)
secretory vesicle (p. 341)
secretory granule (p.341)

constitutive secretion (p. 342)
regulated secretion (p. 342)

Exocytosis and Endocytosis: Transporting Material Across the Plasma Membrane
exocytosis (p. 342)
polarized secretion (p. 343)
endocytosis (p. 343)
endocytic vesicle (p. 343)
phagocytosis (p. 344)
phagocyte (p. 344)
phagocytic vacuole (p. 345)
receptor-mediated (clathrin-dependent) endocytosis (p. 345)
coated vesicle (p. 345)
early endosome (p. 347)
transcytosis (p. 348)
fluid-phase endocytosis (p. 348)

Coated Vesicles in Cellular Transport Processes
clathrin (p. 349)
adaptor protein (AP) (p. 349)
triskelion (p. 350)
dynamin (p. 350)
COPI (coatomer) (p. 351)
ADP ribosylation factor (ARF) (p. 351)
Sec13/31 (p. 352)
Sec23/24 (p. 352)
SarI (p. 352)
SNARE hypothesis (p. 352)
SNARE (SNAP receptor) protein (p. 352)
vesicle-SNAP receptor (v-SNARE) (p. 352)
target-SNAP receptor (t-SNARE) (p. 352)
Rab GTPase (p. 352)
N-ethylmaleimide-sensitive factor (NSF) (p. 352)
soluble NSF attachment protein (SNAP) (p. 352)

Lysosomes and Cellular Digestion
lysosome (p. 353)
late endosome (p. 354)
transient fusion model (p. 354)
hybrid model (p. 354)
heterophagic lysosome (p. 355)
autophagic lysosome (p. 355)
residual body (p. 356)

autophagy (p. 356)
macrophagy (p. 356)
autophagic vacuole (p. 356)
microphagy (p. 356)
extracellular digestion (p. 356)
lysosomal storage disease (p. 357)

The Plant Vacuole: A Multifunctional Organelle
vacuole (p. 357)
dictyosome (p. 357)
provacuole (p. 357)

Peroxisomes
peroxisome (p. 358)
microbody (p. 359)
β oxidation (p. 360)
xenobiotic (p. 361)
leaf peroxisome (p. 361)
glyoxysome (p. 361)

Box 12A: *Centrifugation: An Indispensable Technique of Cell Biology*
centrifugation (p. 326)
sedimentation rate (p. 326)
subcellular fractionation (p. 326)
centrifuge (p. 326)
ultracentrifuge (p. 326)
homogenization (p. 326)
homogenate (p. 327)
differential centrifugation (p. 327)
sedimentation coefficient (p. 327)
Svedberg unit (p. 327)
supernatant (p. 328)
pellet (p. 328)
density gradient (rate-zonal) centrifugation (p. 328)
equilibrium density (buoyant density) centrifugation (p. 329)

Box 12B: *Cholesterol, the LDL Receptor, and Receptor-Mediated Endocytosis*
familial hypercholesterolemia (FH) (p. 346)
low-density lipoprotein (LDL) (p. 346)
LDL receptor (p. 346)

Problem Set

More challenging problems are marked with a • .

12-1. Compartmentalization of Function. Each of the following processes is associated with one or more specific eukaryotic organelles. In each case, identify the organelle or organelles, and suggest one advantage of confining the process to the organelle or organelles.

(a) β oxidation of long-chain fatty acids

(b) Biosynthesis of cholesterol

(c) Biosynthesis of insulin

(d) Biosynthesis of testosterone (a male sex hormone)

(e) Degradation of damaged or no longer needed organelles

(f) Glycosylation of membrane proteins

(g) Hydroxylation of phenobarbital

(h) Sorting of lysosomal proteins from secretory proteins

12-2. Endoplasmic Reticulum. For each of the following statements, indicate if it is true of the rough ER only (R), of the smooth ER only (S), of both rough and smooth ER (RS), or of neither (N).

(a) Contains less cholesterol than does the plasma membrane.

(b) Has ribosomes attached to its outer (cytosolic) surface.

(c) Is involved in detoxification of drugs.

(d) Is involved in the breakdown of glycogen.

(e) Is the site for biosynthesis of secretory proteins.

(f) Is the site for the folding of membrane-bound proteins.

(g) Tends to form tubular structures.

(h) Usually consists of flattened sacs.

(i) Visible only by electron microscopy.

12-3. Biosynthesis of Integral Membrane Proteins. In addition to their role in cellular secretion, the rough ER and the Golgi complex are also responsible for the biosynthesis of integral membrane proteins. More specifically, these organelles are the source of glycoproteins commonly found in the outer phospholipid monolayer of the plasma membrane.

(a) In a series of diagrams, depict the synthesis and glycosylation of glycoproteins of the plasma membrane.

(b) Explain why the carbohydrate groups of membrane glycoproteins are always found on the outer surface of the plasma membrane.

(c) What assumptions did you make about biological membranes in order to draw the diagrams in part a and answer the question in part b?

12-4. Coated Vesicles in Intracellular Transport. For each of the following statements, indicate for which coated vesicle the statement is true: clathrin- (C), COPI- (I), or COPII-coated (II). Each statement may be true for one, several, or none (N) of the coated vesicles discussed in this chapter.

(a) Binding of the coat protein to an LDL receptor is mediated by an adaptor protein complex.

(b) Fusion of the vesicle (after dissociation of the coat) with the Golgi membrane is facilitated by specific t-SNARE and Rab proteins.

(c) Has a role in bidirectional transport between the ER and Golgi complex.

(d) Has a role in sorting proteins for intracellular transport to specific destinations.

(e) Has a role in transport of acid hydrolases to late endosomes.

(f) Is essential for all endocytic processes.

(g) Is important for retrograde traffic through the Golgi complex.

(h) Is involved in the movement of membrane lipids from the TGN to the plasma membrane.

(i) The basic structural component of the coat is called a triskelion.

(j) The protein coat dissociates shortly after formation of the vesicle.

(k) The protein coat always includes a specific small GTP-binding protein.

•12-5. Interpreting Data. Each of the following statements summarizes the results of an experiment related to exocytosis or endocytosis. In each case, explain the relevance of the experiment and its result to our understanding of these processes.

(a) Addition of the drug colchicine to cultured fibroblast cells inhibits movement of transport vesicles.

(b) Certain pituitary gland cells secrete laminin continuously, but secrete adrenocorticotropic hormone only in response to specific signals.

(c) Certain adrenal gland cells can be induced to secrete epinephrine when their intracellular calcium concentration is experimentally increased.

(d) Cells expressing a temperature-sensitive form of dynamin do not display receptor-mediated endocytosis after a temperature shift, yet they continue to ingest extracellular fluid (at a reduced level initially, and then at a normal level within 30–60 minutes).

12-6. Cellular Digestion. For each of the following statements, indicate the specific digestion process or processes of which the statement is true: phagocytosis (P), receptor-mediated endocytosis (R), autophagy (A), or extracellular digestion (E). Each statement may be true of one, several, or none (N) of these processes.

(a) Can involve exocytosis.

(b) Can involve fusion of vesicles or vacuoles with a lysosome.

(c) Digested material is of extracellular origin.

(d) Digested material is of intracellular origin.

(e) Essential for sperm penetration of the egg during fertilization.

(f) Important for certain developmental processes.

(g) Involves acid hydrolases.

(h) Involves fusion of endocytic vesicles with an early endosome.

(i) Involves fusion of lysosomes with the plasma membrane.

(j) Occurs within lysosomes.

(k) Serves as a source of nutrients within the cell.

12-7. Peroxisomal Properties. For each of the following statements, indicate whether it is true of all (A), some (S), or none (N) of the various kinds of peroxisomes described in this chapter, and explain your answer.

(a) Acquires proteins from the ER and Golgi complex.

(b) Capable of catabolizing fatty acids.

(c) Contains acid hydrolases.

(d) Contains catalase.

(e) Contains peroxide-generating chemical reactions.

(f) Contains the genes coding for luciferase.

(g) Contains urate oxidase.

(h) Is a source of dolichol.

(i) Is surrounded by a lipid bilayer.

12-8. Lysosomal Storage Diseases. Despite a bewildering variety of symptoms, lysosomal storage diseases have a number of properties in common. For each of the following statements, indicate if you would expect the property to be common to most lysosomal storage diseases (M), to be true of a specific lysosomal storage disease (S), or not true of any lysosomal storage disease (N).

(a) Impaired metabolism of glycolipids causes mental deterioration.

(b) Leads to accumulation of degradation products in the lysosome.

(c) Leads to accumulation of excessive amounts of glycogen in the lysosome.

(d) Results from an inability to regulate synthesis of glycos-aminoglycans.

(e) Results from an absence of functional acid hydrolases.

(f) Results in accumulation of lysosomes in the cell.

(g) Symptoms include muscle weakness and mental retardation.

(h) Triggers proliferation of organelles containing catalase.

12-9. Sorting Proteins. Specific structural features tag proteins for transport to various intracellular and extracellular destinations. Three examples were described in this chapter: (1) the short peptide Lys-Asp-Glu-Leu, (2) characteristic hydrophobic membrane-spanning domains, and (3) mannose-6-phosphate residues attached to oligosaccharide side chains. For each structural feature, answer the following questions.

(a) Where in the cell is the tag incorporated into the protein?

(b) How does the tag ensure that the protein reaches its destination?

(c) Where would the protein likely go if you were to remove the tag?

• 12-10. Silicosis and Asbestosis. *Silicosis* is a debilitating miner's disease that results from the ingestion of silica particles (such as sand or glass) by macrophages in the lungs. *Asbestosis* is a similarly serious disease caused by inhalation of asbestos fibers. In both cases, the particles or fibers are found in lysosomes, and fibroblasts, which secrete collagen, are stimulated to deposit

nodules of collagen fibers in the lungs, leading to reduced lung capacity, impaired breathing, and eventually death.

(a) How do you think the fibers get into the lysosomes?

(b) What effect do you think fiber or particle accumulation has on the lysosomes?

(c) How might you explain the death of silica-containing or asbestos-containing cells?

(d) What do you think happens to the silica particles or asbestos fibers when such cells die? How can cell death continue almost indefinitely, even after prevention of further exposure to silica dust or asbestos fibers?

(e) Cultured fibroblast cells will secrete collagen and produce connective tissue fibers after the addition of material from a culture of lung macrophages that have been exposed to silica particles. What does this tell you about the deposition of collagen nodules in the lungs of silicosis patients?

• 12-11. Subcellular Fractionation. Figure 12A-3 shows a two-dimensional diagram depicting sedimentation coefficients (in Svedberg units) and the densities (in g/cm³) of various organelles, molecules, and viruses. In each case, the approximate oval-shaped area delimits the usual range of S values and densities for a given particle. Use Figure 12A-3 and your knowledge of the centrifugation techniques described in Box 12A to answer the following questions.

(a) Which organelles or molecules have the lowest S values? The highest S values? The lowest densities? The highest densities?

(b) Compare the density of a ribosome with the densities of RNA and protein molecules. Given what you know about ribosomes, how can you explain your observation?

(c) Given that most viruses consist of DNA or RNA bound to protein molecules, why is the sedimentation coefficient of a typical virus greater than that of an individual DNA, RNA, or protein molecule?

(d) Can differential centrifugation be used to effectively separate nuclei from mitochondria? Why or why not? Can this technique be used to separate peroxisomes from mitochondria? Why or why not?

(e) Can equilibrium density centrifugation be used to effectively separate microsomes from mitochondria? Why or why not? Can this technique be used to separate RNA from DNA? Why or why not?

(f) Devise a purification scheme by which you can separate lysosomes, peroxisomes, and mitochondria from other components of a homogenate, and then separate the mixture of organelles into three distinct fractions.

• 12-12. Enzymes and Organelles. Most organelles are identified and described in terms of the enzymes they contain. Both the lysosome and the peroxisome were discovered because specific enzymes displayed properties not easily explained by their presence in already-known structures.

(a) What criteria would you invoke to identify a new organelle unequivocally, based on enzymes contained within that organelle?

(b) Would your criteria have been adequate to identify lysosomes in a mitochondrial preparation? Explain your answer.

(c) Would your criteria have been adequate to identify peroxisomes in a lysosomal preparation? Explain you answer.

Suggested Reading

References of historical importance are marked with a •.

General References

Lee, A. G., ed. *Biomembranes, Volume 1: General Principles.* Greenwich, CT: JAI, 1995.

Nunnari, J., and P. Walter. Regulation of organelle biogenesis. *Cell* 84 (1996): 389.

The Endoplasmic Reticulum and the Golgi Complex

Berger, E. G., and J. Roth, eds. *The Golgi Apparatus.* Basel, Switzerland: Birkhäuser Verlag, 1997.

Glick, B. S. Organization of the Golgi apparatus. *Curr. Opin. Cell Biol.* 12 (2000): 450.

Harter, C., and F. Wieland. The secretory pathway: Mechanisms of protein sorting and transport. *Biochim. Biophys. Acta* 1286 (1996): 75.

Helenius, A., and M. Aebi. Intracellular functions of N-linked glycans. *Science* 291 (2001): 2364.

Keller, P., and K. Simons. Post-Golgi biosynthetic trafficking. *J. Cell. Sci.* 110 (1997): 3001.

Klumperman, J. Transport between ER and Golgi. *Curr. Opin. Cell Biol.* 12 (2000): 445.

Lippincott-Schwartz, J., T. H. Roberts, and K. Hirschberg. Secretory protein trafficking and organelle dynamics in living cells. *Annu. Rev. Cell Dev. Biol.* 16 (2000): 557.

Staehelin, L. A., and I. Moore. The plant Golgi apparatus: Structure, functional organization, and trafficking mechanisms. *Annu. Rev. Plant Physiol. Plant Mol. Biol.* 46 (1995): 261.

Exocytosis and Endocytosis

Blázquez, M., and K. I. J. Shennan. Basic mechanisms of secretion: Sorting into the regulated secretory pathway. *Biochem. Cell Biol.* 78 (2000): 181.

Lee, A. G., ed. *Biomembranes, Volume 4: Endocytosis and Exocytosis.* Greenwich, CT: JAI, 1996.

Mostov, K. E., M. Verges, and Y. Altschuler. Membrane traffic in polarized epithelial cells. *Curr. Opin. Cell Biol.* 12 (2000): 483.

Mukherjee, S., R. N. Ghosh, and F. R. Maxfield. Endocytosis. *Physiol. Rev.* 77 (1997): 759.

Robinson, M. S., C. Watts, and M. Zerial. Membrane dynamics in endocytosis. *Cell* 84 (1996): 13.

Tjelle, T. E., T. Lùvdal, and T. Berg. Phagosome dynamics and function. *BioEssays* 22 (2000): 255.

Transport Vesicles

• Crowther, R. A., and B. M. F. Pearse. Assembly and packing of clathrin into coats. *J. Cell. Biol.* 91 (1981): 790.

Hinshaw, J. E. Dynamin and its role in membrane fission. *Annu. Rev. Cell Dev. Biol.* 16 (2000): 483.

Kirchhausen, T. Clathrin. *Annu. Rev. Biochem.* 69 (2000): 699.

Nickel, W., and F. T. Wieland. Biogenesis of COPI-coated transport vesicles. *FEBS Lett.* 413 (1997): 395.

Rothman, J. E., and F. T. Wieland. Protein sorting by transport vesicles. *Science* 272 (1996): 227.

Schekman, R., and L. Orci. Coat proteins and vesicle budding. *Science* 271 (1996): 1526.

Schlegel, A., and M. P. Lisanti. Caveolae and their coat proteins, the caveolins: From electron microscopic novelty to biological launching pad. *J. Cell. Physiol.* 186 (2001): 329.

Storrie, B., R. Pepperkok, and T. Nilsson. Breaking the COPI monopoly on Golgi recycling. *Trends Cell Biol.* 10 (2000): 385.

Urbé, S., S. A. Tooze, and F. A. Barr. Formation of secretory vesicles in the biosynthetic pathway. *Biochim. Biophys. Acta* 1358 (1997): 6.

Lysosomes and Vacuoles

• Bainton, D. The discovery of lysosomes. *J. Cell. Biol.* 91 (1981): 66s.

De, D. N. *Plant Cell Vacuoles: An Introduction.* Collingwood, Victoria: CSIRO Publishing, 2000.

• de Duve, C. The lysosome. *Sci. Amer.* 208 (May 1963): 64.

Hunziker, W., and H. J. Geuze. Intracellular trafficking of lysosomal membrane proteins. *BioEssays* 18 (1995): 379.

Lloyd, J. B., and R. W. Mason, eds. *Subcellular Biochemistry, Volume 27: Biology of the Lysosome.* New York: Plenum, 1996.

Luzio, J. P., B. A. Rous, N. A. Bright, P. R. Pryor, B. M. Mullock, and R. C. Piper. Lysosome-endosome fusion and lysosome biogenesis. *J. Cell. Sci.* 113 (2000): 1515.

Winchester, B., A. Vellodi, and E. Young. The molecular basis of lysosomal storage diseases and their treatment. *Biochem. Soc. Trans.* 28 (2000): 150.

Peroxisomes

• de Duve, C. The peroxisome: A new cytoplasmic organelle. *Proc. Roy. Soc. London Ser. B* 173 (1969): 71.

Masters, C., and D. Crane. *The Peroxisome: A Vital Organelle.* Cambridge, England: Cambridge University Press, 1995.

Reumann, S. The structural properties of plant peroxisomes and their metabolic significance. *Biol. Chem.* 381 (2000): 639.

Sacksteder, K. A., and S. J. Gould. The genetics of peroxisome biogenesis. *Annu. Rev. Genetics* 34 (2000): 623.

Wanders, R. J. A., E. G. van Grunsven, and G. A. Jansen. Lipid metabolism in peroxisomes: Enzymology, functions and dysfunctions of the fatty acid α- and β-oxidation systems in humans. *Biochem. Soc. Trans.* 28 (2000): 141.

Waterham, H. R., and J. M. Craig. Peroxisome biogenesis. *BioEssays* 19 (1997): 57.

13 Chemotrophic Energy Metabolism: Glycolysis and Fermentation

As we learned in earlier chapters, cells cannot survive without a source of energy and a source of chemical "building blocks"—the small molecules from which macromolecules such as proteins, nucleic acids, and polysaccharides are synthesized. These two requirements are related in that the desired energy and small molecules are both present in the food molecules that organisms produce or ingest, and both are simultaneously transformed every time a chemical reaction takes place. In this and the following two chapters, we will deal mainly with the question of how cells obtain energy. Keep in mind, however, that the reactions whereby cells obtain energy also provide the various small molecules that cells need for synthesis of macromolecules and other cellular constituents.

As we consider cellular energy flow in these several chapters, we will focus especially on the sources of energy available to the biological world and the processes of energy conversion and utilization in cells. In the process, we will draw on foundations that have been laid in previous chapters. In Chapter 5, for example, we considered the thermodynamic principles that govern all reactions and processes in cells. We saw that $\Delta G'$, the free energy change, determines in which direction a reaction will proceed and how much free energy can be derived from (or must be put into) the system during the reaction. In Chapter 6, we learned that almost all reactions in cells are catalyzed by enzymes and that the rate of an enzyme-catalyzed reaction depends on a variety of factors, including temperature, pH, and the concentrations of substrates and products. In Chapters 7 and 8, we encountered the membranes that define cell boundaries and compartments and we learned how energy is used to transport various solute molecules across those boundaries. And in Chapters 9–12, we considered a variety of cellular processes and reactions that also require energy.

Still unanswered, however, are the questions of where all this energy comes from and how cells gain access to it. We will answer these questions by looking at several different ways in which cells obtain the energy they need to drive cellular reactions and processes. In this chapter and the next, we will consider how *chemotrophs* such as animals and most microorganisms obtain energy from the food they engulf or ingest, focusing especially on the oxidative breakdown of sugar molecules. Then, in Chapter 15, we will discuss the process by which *phototrophs* such as green plants, algae and some bacteria tap the solar radiation that is the ultimate energy source for almost all living organisms.

Metabolic Pathways

When we encountered enzymes in Chapter 6, we considered individual chemical reactions catalyzed by individual enzymes functioning in isolation. But that is not the way cells really operate. To accomplish any major task, a cell requires a series of reactions occurring in an ordered sequence. This, in turn, requires many different enzymes because most enzymes catalyze only a single reaction, and many such reactions are usually needed to accomplish a major biochemical operation.

When we consider all the chemical reactions that occur within a cell, we are talking about **metabolism** (from the Greek word *metaballein,* meaning "to change"). The overall metabolism of a cell consists, in turn, of many specific **metabolic pathways,** each of which accomplishes a particular task. From a biochemist's perspective, *life at the cellular level can be defined as a network of integrated*

anabolic = increase entropy
decrease entropy
endergonic (energy requiring))   Catabolic =
decrease
increase entropy
exergonic =
energy
liberating

and carefully regulated metabolic pathways, each contributing to the sum of activities that a cell must carry out.

Metabolic pathways are of two general types. Pathways concerned with the synthesis of cellular components are called **anabolic pathways** (from the Greek prefix *ana-*, meaning "up"), whereas those involved in the breakdown of cellular constituents are called **catabolic pathways** (from the Greek prefix *kata-*, meaning "down"). Anabolic pathways usually involve a substantial increase in molecular order (and therefore a local decrease in entropy) and are *endergonic* (energy-requiring). Catabolic pathways, by contrast, are *exergonic* (energy-liberating), in part because they involve a decrease in molecular order (increase in entropy). Catabolic pathways play two roles in cells: They release the free energy needed to drive cellular functions and they give rise to the small organic molecules, or *metabolites*, that are the building blocks for biosynthesis. These two roles are complementary because some of the free energy liberated by catabolic reactions is used to drive anabolic pathways in which macromolecules and other cellular components are synthesized from metabolites such as sugars, amino acids, and nucleotides.

As we will see shortly, catabolism can be carried out either in the presence or absence of oxygen (i.e., under either *aerobic* or *anaerobic* conditions). The energy yield is much greater in the presence of oxygen, which undoubtedly explains the preponderance of aerobic organisms in the world. However, anaerobic catabolism is also important, not only for organisms in environments that are always devoid of oxygen but also for organisms and cells that are temporarily deprived of oxygen.

ATP: The Universal Energy Coupler

The anabolic reactions of cells are responsible for the growth and repair processes characteristic of all living systems, whereas catabolic reactions release the energy needed to drive the anabolic reactions and to carry out other kinds of cellular work. The efficient linking, or *coupling,* of energy-yielding processes to energy-requiring processes is therefore crucial to cell function. This coupling is made possible by specific kinds of molecules that conserve the energy derived from exergonic reactions and release it again whenever and wherever energy is needed. In virtually all cells, the molecule most commonly used as an energy intermediate is the phosphorylated compound **adenosine triphosphate (ATP).** ATP is, in other words, the energy "currency" of the biological realm. Because ATP is involved in most cellular energy transactions, it is essential that we understand its structure and function and appreciate the properties that make this molecule so suitable for its role as the universal energy coupler.

ATP Contains Two High-Energy Phosphoanhydride Bonds

As we learned in Chapter 3, ATP is a complex molecule containing the aromatic base adenine, the five-carbon sugar ribose, and a chain of three phosphate groups. The phosphate groups are linked to each other by **phosphoanhydride bonds** and to the ribose by a **phosphoester bond,** as shown for the ATP molecule in Figure 13-1. The compound formed by linking adenine and ribose is called *adenosine*. Adenosine may occur in the

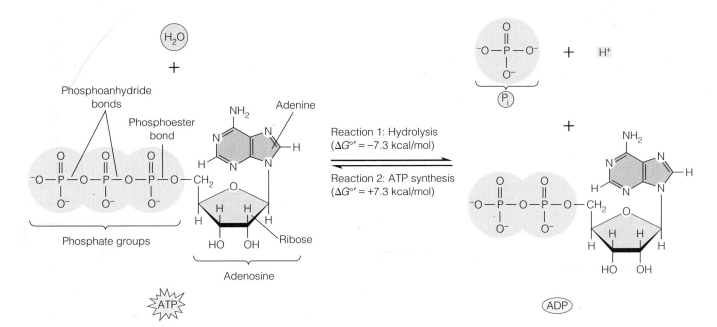

Figure 13-1 Structure and Function of ATP. ATP consists of adenosine (adenine + ribose) with a chain of three phosphate groups attached to carbon atom 5 of the ribose. *Reaction 1:* The hydrolysis of ATP to ADP and inorganic phosphate (P_i) is a highly exergonic reaction, with a standard free energy change of about −7.3 kcal/mol. (The two protons shown as products of the reaction result from the ionization of the ADP and P_i after hydrolysis.) *Reaction 2:* The synthesis of ATP by phosphorylation of ADP is a highly endergonic reaction, with a standard free energy change of about +7.3 kcal/mol.

cell in the unphosphorylated form or with one, two, or three phosphates attached to carbon atom 5 of the ribose, forming *adenosine monophosphate (AMP), diphosphate (ADP),* and *triphosphate (ATP),* respectively.

The ATP molecule serves well as an intermediate in cellular energy metabolism because energy is released when the phosphoanhydride bond that links the third (outermost) phosphate to the second undergoes hydrolysis. (*Hydrolysis* is the general term for any reaction whereby a bond in a molecule is broken by reaction with water. Two products are formed; one receives an —H from the water molecule and the other gets an —OH group.) Thus, the hydrolysis of ATP—that is, its reaction with water to form ADP and inorganic phosphate (PO_4^{3-}, often written as P_i)—is exergonic, with a standard free energy change ($\Delta G^{\circ\prime}$) of -7.3 kcal/mol (Figure 13-1, reaction 1). The reverse reaction, whereby ATP is synthesized from ADP and P_i with the loss of a water molecule, is correspondingly endergonic, with a $\Delta G^{\circ\prime}$ of $+7.3$ kcal/mol (Figure 13-1, reaction 2). In other words, energy is required to drive ATP synthesis from ADP and is released again upon hydrolysis. (The phosphoanhydride bond between the first and second phosphates has the same properties and is important in some reactions—particularly those involved in the synthesis of macromolecules—but at present we are mainly interested in the terminal phosphoanhydride bond.)

Biochemists sometimes refer to bonds such as the phosphoanhydride bonds of ATP as "high-energy" or "energy-rich" bonds, a very useful convention introduced in 1941 by Fritz Lipmann, a leading bioenergetics researcher of the time. However, the term *high-energy bond* needs to be understood correctly, lest it convey the erroneous impression that the bond somehow contains energy that can be released. In fact, of course, all chemical bonds *require* energy to be broken and *release* energy when they form. What we really mean by "high-energy bond" is that free energy is released when the bond is hydrolyzed. And this, in turn, means that more energy is released as the new bonds are formed (between the —H and —OH groups of the water molecule and the two products of the reaction) than is required to break the so-called high-energy bond. The energy is therefore a characteristic of the reaction in which the molecule is involved and not of a particular bond within that molecule. Thus, to call ATP or any other molecule a **"high-energy compound"** should always be understood as a shorthand way of saying that the molecule possesses one or more bonds—the phosphoanhydride bonds, in the case of ATP—the hydrolysis of which is highly exergonic.

ATP Hydrolysis Is Highly Exergonic Because of Charge Repulsion and Resonance Stabilization

What is it about the ATP molecule that makes the hydrolysis of its phosphoanhydride bonds so exergonic? The answer to this question has two parts: Hydrolysis of ATP to ADP and P_i is exergonic because of *charge repulsion* between the adjacent negatively charged phosphate groups and because of *resonance stabilization* of both products of hydrolysis (ADP and inorganic phosphate).

Charge repulsion is easy to understand. By way of analogy, imagine holding two magnets together with like poles touching. The like poles repel each other, and you need to make an effort (i.e., you need to put energy into the system) to force them together. If you let go, the magnets spring apart, releasing the energy. Now consider the three phosphate groups of ATP. As Figure 13-1 shows, each group bears at least one negative charge due to its ionization at the near-neutral pH of the cell. These negative charges tend to repel one another, thereby straining the covalent bond linking the phosphate groups together. As a result, less energy is needed to break the bond than would otherwise be required.

A second, even more important contribution to ATP bond energy is **resonance stabilization.** To understand this phenomenon, we need to realize that functional groups like the carboxylate or phosphate groups, though formally written with one double bond and one or more single bonds to oxygen, have in reality an unshared electron pair that is delocalized over all of the bonds to oxygen. The true structure of the carboxylate and phosphate groups is actually an average of the contributing structures called a **resonance hybrid,** in which the extra electrons are delocalized over all possible bonds.

When its electrons are maximally delocalized in this way, a molecule is in its most stable (lowest-energy) configuration and is said to be *resonance-stabilized.* Consider now what happens if one (or more) of the oxygen atoms of the carboxylate or phosphate group becomes involved in a covalent bond to an organic compound: The oxygen atom is no longer available for electron delocalization, so the extra electrons are delocalized over fewer than the maximum number of oxygen atoms and the molecule is consequently locked into a higher-energy (less delocalized) configuration (Figure 13-2; note that the Greek letter δ [delta] is used in resonance hybrids to indicate a partial charge on an atom). Hence any chemical bond to a carboxylate or phosphate group will result in a compound that is higher in free energy than the free carboxylate or phosphate group itself. Specifically, esters, phosphoesters, anhydrides, and phosphoanhydrides are higher-energy compounds than the alcohol, carboxylate, and phosphate compounds they yield upon hydrolysis, and they will therefore liberate energy if they are hydrolyzed to these products.

For esters, only a moderate amount of energy is liberated upon hydrolysis, because only one of the two products—the acid—can undergo resonance stabilization. For anhydrides, however, both products can be resonance-stabilized, so the hydrolysis reaction is highly exergonic. In addition, both products of anhydride hydrolysis are charged and therefore repel each other, which is not the case with ester hydrolysis. In general, ester and phosphoester bonds have free energies of hydrolysis of about -3 to -4 kcal/mol, whereas the hydrolysis of anhydride and phosphoanhydride bonds releases roughly twice that much free energy. ATP illustrates this difference well: Hydrolysis of either of the phosphoanhydride bonds that link the second and third phosphate groups to the rest of the molecule has a standard free energy change of about −7.3 kcal/mol, whereas hydrolysis of the phosphoester bond that links the first (innermost) phosphate group to the ribose has a $\Delta G^{\circ\prime}$ of about −3.6 kcal/mol:

$$ATP + H_2O \longrightarrow ADP + P_i$$
$$\Delta G^{\circ\prime} = -7.3 \text{ kcal/mol} \qquad \textbf{(13-1)}$$

$$ADP + H_2O \longrightarrow AMP + P_i$$
$$\Delta G^{\circ\prime} = -7.3 \text{ kcal/mol} \qquad \textbf{(13-2)}$$

$$AMP + H_2O \longrightarrow \text{adenosine} + P_i$$
$$\Delta G^{\circ\prime} = -3.6 \text{ kcal/mol} \qquad \textbf{(13-3)}$$

Thus, ATP and ADP are both "higher-energy compounds" than AMP, to use the shorthand of biochemists.

If anything, the $\Delta G^{\circ\prime}$ value of −7.3 kcal/mol underestimates the free energy change associated with the hydrolysis of ATP to ADP under biological conditions. As

we learned in Chapter 5, the actual free energy change, $\Delta G'$, depends on the prevailing concentrations of reactants and products (see equation 5-21). For the hydrolysis of ATP (reaction 13-1), $\Delta G'$ is calculated as

$$\Delta G' = \Delta G^{\circ\prime} + RT \ln \frac{[ADP][P_i]}{[ATP]} \qquad \textbf{(13-4)}$$

In most cells, the ATP/ADP ratio is significantly greater than 1 : 1, often in the range of about 5 : 1. As a result, the term $\ln([ADP][P_i]/[ATP])$ is negative and $\Delta G'$ is therefore more negative than −7.3 kcal/mol, usually in the range of −10 to −14 kcal/mol. (Can you use equation 13-4 to convince yourself that $\Delta G'$ is in this range when the ATP/ADP ratio is about 5 : 1? What must you assume about $[P_i]$?)

ATP Is an Important Intermediate in Cellular Energy Transactions

Although scientists often refer to ATP as a high-energy compound, we really should think of it as an *intermediate-energy compound* because it actually occupies an intermediate position in the overall spectrum of phosphorylated compounds in the cell. This point becomes clear when some of the more common phosphorylated intermediates in cellular energy metabolism are ranked according to their $\Delta G^{\circ\prime}$ values, as in Table 13-1. The values are negative, so the compounds closest to the top of the table release the most energy upon hydrolysis of the phosphate group.

This means that, under standard conditions, any compound is capable of exergonically phosphorylating any compound below it (i.e., one that has a less negative $\Delta G^{\circ\prime}$ value) but none of the compounds above it. Thus, ATP can be formed from ADP by the transfer of a phosphate group from phosphoenolpyruvate (PEP) but not

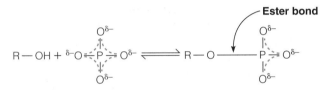

(a) Ester bond formation

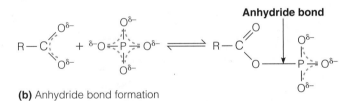

(b) Anhydride bond formation

Figure 13-2 Decreased Resonance Stabilization of the Phosphate and Carboxyl Groups Due to Bond Formation. Resonance stabilization is an important feature of both the phosphate ion and the carboxylate group. In each case, the extra electron pair that is usually shown as formally localized to one of the oxygen bonds of the contributing structures is in fact delocalized over all available C—O or P—O bonds, as illustrated in the resonance hybrid structures on the left. (Delocalized electrons are shown by dashed lines on the resonance hybrids, with the Greek letter δ used to indicate a partial negative charge on each of the oxygen atoms.) Involvement of a phosphate ion in the formation of either **(a)** an ester bond or **(b)** an anhydride bond decreases the opportunity for electron delocalization. As a result, the ester or anhydride product is a higher-energy compound than the phosphate group and the alcohol or carboxylic acid that are required for its formation. An ester bond has a lower free energy of hydrolysis because hydrolysis (reversal of the reaction in part a) results in increased electron delocalization for only one of the two products. An anhydride bond has a higher energy of hydrolysis because hydrolysis (reversal of the reaction in part b) leads to increased electron delocalization for *both* of the products and relieves charge repulsion as well. (The R refers to the rest of the molecule; only the hydroxyl or carboxyl group is shown.)

Table 13-1 Standard Free Energies of Hydrolysis for Phosphorylated Compounds Involved in Energy Metabolism

Phosphorylated Compound and Its Hydrolysis Reaction	$\Delta G^{\circ\prime}$
Phosphoenolpyruvate (PEP)	(Kcal/mol)
$+ H_2O \longrightarrow$ pyruvate $+ P_i$	−14.8
1,3-bisphosphoglycerate	
$+ H_2O \longrightarrow$ 3-phosphoglycerate $+ P_i$	−11.8[1]
Phosphocreatine	
$+ H_2O \longrightarrow$ creatine $+ P_i$	−10.3
Adenosine triphosphate (ATP)	
$+ H_2O \longrightarrow$ adenosine diphosphate $+ P_i$	−7.3
Glucose-1-phosphate	
$+ H_2O \longrightarrow$ glucose $+ P_i$	−5.0
Glucose-6-phosphate	
$+ H_2O \longrightarrow$ glucose $+ P_i$	−3.3
Glycerol phosphate	
$+ H_2O \longrightarrow$ glycerol $+ P_i$	−2.2

[1]The $\Delta G^{\circ\prime}$ value for 1, 3-bisphosphoglycerate is for the hydrolysis of the phosphoanhydride bond on carbon atom 1.

from glucose-6-phosphate, as Figure 13-3 illustrates. Similarly, ATP can phosphorylate glucose, but it cannot phosphorylate pyruvate. The following reactions, in other words, have negative $\Delta G^{\circ\prime}$ values, which we will designate as $\Delta G^{\circ\prime}_{\text{transfer}}$ values to emphasize that they represent the standard free energy change that accompanies the transfer of a phosphate group from a donor to an acceptor molecule. $\Delta G^{\circ\prime}_{\text{transfer}}$ values can be predicted from Table 13-1 and calculated as shown:

$$\text{glucose} + \text{ATP} \longrightarrow \text{glucose-6-phosphate} + \text{ADP}$$
$$\Delta G^{\circ\prime}_{\text{transfer}} = \Delta G^{\circ\prime}_{\text{donor}} - \Delta G^{\circ\prime}_{\text{acceptor}}$$
$$= -7.3 - (-3.3) = -4.0 \text{ kcal/mol} \qquad \textbf{(13-5)}$$

$$\text{PEP} + \text{ADP} \longrightarrow \text{pyruvate} + \text{ATP}$$
$$\Delta G^{\circ\prime}_{\text{transfer}} = -14.8 - (-7.3) = -7.5 \text{ kcal/mol} \qquad \textbf{(13-6)}$$

As we will discover shortly, reactions 13-5 and 13-6 are the first and final reactions, respectively, in an important catabolic sequence called the *glycolytic pathway,* which is the

starting point for chemotrophic energy metabolism in virtually all forms of life.

Reactions such as 13-5 and 13-6 that involve the movement of a chemical group from one molecule to another are called **group transfer reactions.** Group transfer reactions represent one of the most common processes in cellular metabolism, and the phosphate group is one of the most frequently transferred groups, especially in energy metabolism.

The most important point to understand from Table 13-1 and Figure 13-3 is that the ATP/ADP pair occupies a crucial intermediate position in terms of bond energies. This means that ATP can serve as a phosphate *donor* in some biological reactions and that its dephosphorylated form, ADP, can serve as a phosphate *acceptor* in other reactions because there are compounds both above and below the ATP/ADP pair in energy. Thus, the important role of ATP in energy metabolism really depends on the hydrolysis of its terminal phosphate group being not a "high-energy" but an "intermediate-energy" reaction.

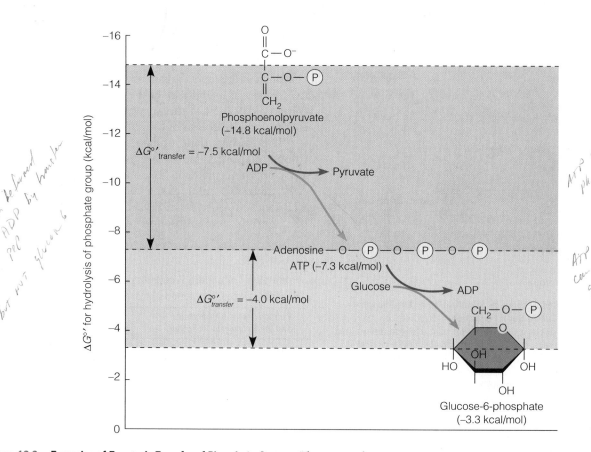

Figure 13-3   Examples of Exergonic Transfer of Phosphate Groups. These examples are based on the $\Delta G^{\circ\prime}$ values shown in Table 13-1. For the hydrolysis of phosphate groups, phosphenolpyruvate (PEP; −14.8 kcal/mol) is considered a high-energy compound, ATP (−7.3 kcal/mol) an intermediate-energy compound and glucose-6-phosphate (−3.3 kcal/mol) a low-energy compound. PEP can therefore transfer its phosphate group exergonically onto ADP to form ATP [$\Delta G^{\circ\prime}_{\text{transfer}} = \Delta G^{\circ\prime}_{\text{donor}} - \Delta G^{\circ\prime}_{\text{receptor}} = -14.8 - (-7.3) = -7.5$ kcal/mol] and ATP can be used to phosphorylate glucose exergonically [$\Delta G^{\circ\prime}_{\text{transfer}} = -7.3 - (-3.3) = -4.0$ kcal/mol], but the reverse reactions are not possible under standard conditions. In fact, the $\Delta G^{\circ\prime}$ values for these transfers are so highly negative that both reactions are irreversible under all cellular conditions.

In summary, the ATP/ADP pair represents a reversible means of conserving, transferring, and releasing energy within the cell (Figure 13-4). ATP is the "charged," or higher-energy, form, whereas ADP is the "discharged," or lower-energy, form. As catabolic processes occur in the cell, whether anaerobically (Figure 13-4a) or aerobically (Figure 13-4b), the energy-liberating reactions in the sequence are coupled to the ATP/ADP system such that the available free energy drives the formation of ATP from ADP. The free energy released upon hydrolysis of ATP then provides the driving force for the many processes (such as biosynthesis, active transport, charge separation, and muscle contraction) that are essential to life and require the input of energy. Thus, we are dealing with an energy-conserving system that is charged during the catabolism of nutrients (or during the photosynthetic trapping of solar energy) and is discharged as it is used to provide the energy that drives cellular work.

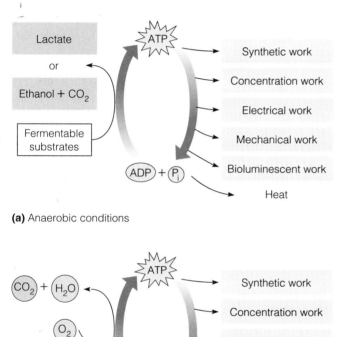

(a) Anaerobic conditions

(b) Aerobic conditions

Figure 13-4 The ATP/ADP System as a Means of Conserving and Releasing Energy Within the Cell. ATP is generated during the oxidative catabolism of nutrients (left side) and is used to do cellular work (right side). **(a)** Under anaerobic or hypoxic (oxygen-deficient) conditions, ATP is generated by fermentation, with lactate as the most common end-product in some organisms and ethanol plus carbon dioxide as the most common end-products in other organisms. **(b)** Under aerobic conditions, ATP is generated by respiration, as oxidizable nutrients are catabolized completely to carbon dioxide and water.

Chemotrophic Energy Metabolism

Now we have the essential concepts in hand to take up the main theme of this chapter and the next. We are ready, in other words, to discuss **chemotrophic energy metabolism**— the reactions and pathways by which cells catabolize nutrients such as carbohydrates, fats, and proteins, conserving as ATP some of the free energy that is released in the process. Or, to put it more personally, we will be looking at the specific metabolic processes by which the cells of your own body make use of the food you eat to meet your energy needs. We begin our discussion by considering *oxidation,* because much of chemotrophic energy metabolism involves energy-yielding oxidative reactions.

Biological Oxidations Usually Involve the Removal of Both Electrons and Protons and Are Highly Exergonic

To say that nutrients such as carbohydrates, fats, or proteins are sources of energy for cells means that these are oxidizable organic compounds and that their oxidation is highly exergonic. Recall from chemistry that *oxidation is the removal of electrons.* Thus, for example, a ferrous ion (Fe^{2+}) is oxidizable because it readily gives up an electron as it is converted to a ferric ion (Fe^{3+}):

$$Fe^{2+} \longrightarrow Fe^{3+} + e- \qquad (13\text{-}7)$$

The only difference in biological chemistry is that the oxidation of organic molecules frequently involves the removal not just of electrons but of hydrogen ions (protons) as well, so that the process is also one of **dehydrogenation.** Consider, for example, the oxidation of ethanol to the corresponding aldehyde:

$$CH_3-CH_2-OH \xrightarrow{\text{oxidation}} CH_3-\overset{\displaystyle H}{\underset{}{C}}=O + 2e^- + 2H^+$$
$$\text{Ethanol} \qquad\qquad \text{Acetaldehyde}$$
$$(13\text{-}8)$$

Electrons are removed, so this is clearly an oxidation. But protons are liberated as well, and an electron plus a proton is the equivalent of a hydrogen atom. Therefore, what happens, in effect, is the removal of the equivalent of two hydrogen atoms:

$$CH_3-CH_2-OH \xrightarrow[\text{(dehydrogenation)}]{\text{oxidation}} CH_3-\overset{\displaystyle H}{\underset{}{C}}=O + [2H]$$
$$\text{Ethanol} \qquad\qquad \text{Acetaldehyde}$$
$$(13\text{-}9)$$

Thus, for cellular reactions involving organic molecules, oxidation is almost always manifested as a dehydrogenation reaction. Many of the enzymes that catalyze oxidative reactions in cells are in fact called *dehydrogenases.* (The brackets around the hydrogen atoms in this and the following reaction are to remind us that hydrogen atoms are not released or taken up as such. Rather, they are picked up and released by carriers called *coenzymes,* as we will see shortly.)

None of the preceding reactions can take place in isolation, of course; the electrons must be transferred to another molecule, which is *reduced* in the process. **Reduction** is defined as the addition of electrons, but in biological reactions the electrons are frequently accompanied by protons and the overall reaction is therefore a **hydrogenation:**

$$CH_3-\overset{\overset{\displaystyle H}{|}}{C}=O \ + \ [2H] \ \xrightarrow[\text{(hydrogenation)}]{\text{reduction}} \ CH_3-CH_2-OH$$
Acetaldehyde   Ethanol

$$(13\text{-}10)$$

Reactions 13-9 and 13-10 both illustrate the further general feature that biological oxidation-reduction reactions almost always involve two-electron (and therefore two-proton) transfers.

As written, reactions 13-9 and 13-10 are only *half-reactions,* representing an oxidation and a reduction event, respectively. In real reactions, however, oxidation and reduction always take place simultaneously. Any time an oxidation occurs, a reduction must occur as well because the electrons (and protons) removed from one molecule must be added to another molecule. Despite the way reactions 13-9 and 13-10 are written, hydrogen atoms are never actually released into solution as such; they are transferred to another molecule instead.

Coenzymes Such as NAD⁺ Serve as Electron Acceptors in Biological Oxidations

In chemotrophic energy metabolism, the ultimate acceptor of electrons is usually, though not always, oxygen. (We will encounter some alternative acceptors in Chapter 14, including H_2 and H_2S; here, we will regard oxygen as the standard electron acceptor.) Rarely, however, are electrons passed directly from an oxidizable substrate to oxygen. Instead, the immediate electron acceptor in most biological oxidations (and the immediate electron donor in most biological reductions) is any of several **coenzymes.** In general, coenzymes are small molecules that function along with enzymes (hence the name), usually by serving as carriers of electrons or small functional groups. As we will see shortly, coenzymes actually function as substrates in the reactions in which they participate. However, they are not consumed stoichiometrically. Rather, they are continuously recycled within the cell, so the relatively low intracellular concentration of a given coenzyme is adequate to meet the needs of the cell for that particular coenzyme.

The most common coenzyme involved in energy metabolism is **nicotinamide adenine dinucleotide (NAD⁺).** Its structure is shown in Figure 13-5. Despite its formidable name and structure, its function is very straightforward: NAD⁺ serves as an electron acceptor by adding two electrons and one proton to a carbon atom on its aromatic ring, thereby generating the reduced form, NADH, with the concomitant release of a proton into the medium:

$$NAD^+ + 2[H] \longrightarrow NADH + H^+ \qquad (13\text{-}11)$$

As a nutritional note, the nicotinamide of NAD⁺ is a derivative of *niacin,* which we recognize as a *B vitamin*—one of a family of water-soluble compounds essential to the diet of humans and other vertebrates unable to synthesize these compounds for themselves. In Chapter 14, we will meet two other coenzymes that also have B vitamin derivatives as part of their structures. It is precisely because these vitamins are components of indispensable coenzymes that they are essential in the diet of any organism that cannot manufacture its own supply of them.

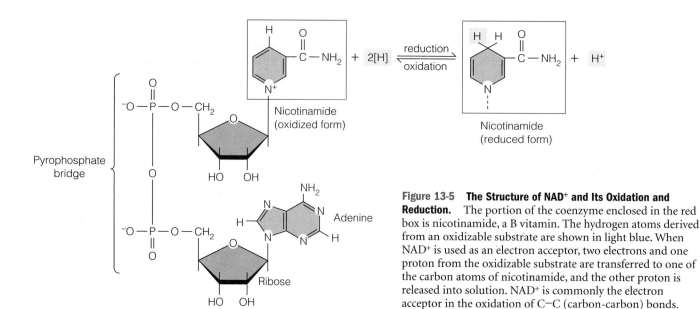

Figure 13-5 The Structure of NAD⁺ and Its Oxidation and Reduction. The portion of the coenzyme enclosed in the red box is nicotinamide, a B vitamin. The hydrogen atoms derived from an oxidizable substrate are shown in light blue. When NAD⁺ is used as an electron acceptor, two electrons and one proton from the oxidizable substrate are transferred to one of the carbon atoms of nicotinamide, and the other proton is released into solution. NAD⁺ is commonly the electron acceptor in the oxidation of C–C (carbon-carbon) bonds.

Most Chemotrophs Meet Their Energy Needs by Oxidizing Organic Food Molecules

We are interested in oxidation because it is the means by which chemotrophs meet most of their energy needs. Many different kinds of substances serve as substrates for biological oxidation. For example, some microorganisms can use inorganic compounds such as reduced forms of iron, sulfur, or nitrogen as their energy sources. These organisms play important roles in the inorganic economy of the biosphere and utilize rather specialized oxidative reactions. However, most chemotrophs depend on organic food molecules as oxidizable substrates, with carbohydrates, fats, and proteins as the three major categories.

Glucose Is One of the Most Important Oxidizable Substrates in Energy Metabolism

To simplify our discussion initially and to provide a unifying metabolic theme, we will concentrate on the biological oxidation of the six-carbon sugar **glucose** ($C_6H_{12}O_6$). Glucose is a good choice for several reasons. In many vertebrates, including humans, glucose is the main sugar in the blood, and hence the main energy source for most of the cells in the body. Blood glucose comes primarily from dietary carbohydrates such as sucrose or starch and from the breakdown of stored glycogen. Glucose is therefore an especially important molecule for you personally, a point that is underscored in Box 13A. Glucose is also important to plants because it is the monosaccharide released upon starch breakdown. In addition, glucose makes up one-half of the disaccharide sucrose, the major sugar in the vascular system of most plants. Moreover, the catabolism of most other energy-rich substances in plants, animals, and microorganisms alike begins with their conversion into one or another of the intermediates in the pathway for glucose catabolism. Rather than looking at the fate of a single compound, then, we are considering a metabolic pathway that is at the very heart of chemotrophic energy metabolism.

Chemically, glucose is an *aldohexose,* a six-carbon sugar that has a terminal carbonyl group when the molecule is shown in its straight-chain form. Recall from Chapter 3 that the glucose molecule can be represented by any of the three structures shown in Figure 3-21. Overall, the *Haworth projection* is the most satisfactory of the three because it indicates both the ring form and the spatial relationship of the carbon atoms. However, we will use the straight-chain *Fischer projection* because the chemistry of the initial steps of glucose catabolism is somewhat easier to understand with this model.

The Oxidation of Glucose Is Highly Exergonic

Glucose is a good potential source of energy because its oxidation is a highly exergonic process, with a $\Delta G^{o\prime}$ of −686 kcal/mol for the complete combustion of glucose to carbon dioxide and water:

$$C_6H_{12}O_6 + 6O_2 \longrightarrow 6CO_2 + 6H_2O \qquad \textbf{(13-12)}$$

As a thermodynamic parameter, $\Delta G^{o\prime}$ is unaffected by route and will therefore have the same value whether the oxidation is by direct combustion, with all of the energy released as heat, or by biological oxidation, with some of the energy conserved as ATP. Thus, oxidation of the sugar molecules in a marshmallow will release the same amount of free energy whether you burn the marshmallow over a campfire or eat it and catabolize the sugar molecules in your body. Biologically, however, the distinction is critical: *Uncontrolled combustion occurs at temperatures that are incompatible with life, whereas biological oxidation is mediated by a series of enzyme-catalyzed reactions that are coupled to ATP generation and occur without significant temperature changes.*

In this and the following chapter, we will consider the biological processes whereby the energy of the oxidizable bonds of glucose is released under conditions compatible with life and in ways that ensure the conservation of as much energy as possible in the form of ATP. We will also discuss the means by which cells continuously adjust the rate of glucose oxidation and synthesis to meet actual cellular needs for ATP.

Glucose Catabolism Yields Much More Energy in the Presence of Oxygen than in Its Absence

To speak of glucose (or any other compound) as an *oxidizable substrate* assumes the availability of some sort of electron acceptor, without which there can be no oxidation. As noted earlier, the electron acceptor for most organisms is molecular oxygen (O_2). Notice, for example, that reaction 13-12 assumes the availability of oxygen. For most organisms, access to the full 686 kcal/mol of free energy in glucose is possible only with oxygen as the final electron acceptor and hence requires aerobic conditions. The complete oxidation of glucose to carbon dioxide and water in the presence of oxygen is called **aerobic respiration.** Aerobic respiration, or *respiration* for short, is a complex, multistep process, to be discussed in detail in Chapter 14.

Even in the absence or scarcity of oxygen, most organisms can still extract limited amounts of energy from glucose and related substrates but with no net oxidation involved and with much lower energy yields. They do so by means of *glycolysis,* a pathway that does not require oxygen. Instead, electrons are removed from an intermediate at one step in the pathway but are returned to another intermediate later in the same pathway. Such an anaerobic process is called **fermentation** and is identified in terms of the principal end-product. In some animal cells and many bacteria, the end-product is lactate, so the process of anaerobic glucose catabolism is called *lactate fermentation.* In most plant cells and in microorganisms such as yeast, the process is termed *alcoholic fermentation* because the end-product is ethanol, an alcohol. (As we will learn in Chapter 14, some organisms can carry out *anaerobic respiration,* using electron acceptors other than oxygen. Unlike fermentation, however, anaerobic respiration involves a net transfer of electrons to an acceptor that is not an intermediate in the pathway and therefore results in the net oxidation of the substrate.)

Further Insights "WHAT HAPPENS TO THE SUGAR?"

While it's important to acquire an academic understanding of processes like glycolysis and gluconeogenesis, it is also important to appreciate what that knowledge means for you as a person—or as an organism, to use a biological term. Will you, for example, be able to use the information in this chapter and the next to understand more adequately how your body meets its energy needs and what it does with the nutrients you eat?

To put it more specifically, can you relate what you are learning in these chapters to what the cells in your body are doing with the food you had for breakfast this morning? As you add sugar to your coffee or cereal, can you answer the question, "What happens to the sugar?" This brief essay will address these questions and might help you appreciate the relevance of these chapters by letting you see more immediately how the information they contain applies to your daily life.

To keep the topic manageable, let's focus on your bowl of cereal and ask what your body does with the sugar you sprinkle on it, as well as with the sugar in the milk and the carbohydrate in the cereal itself (Figure 13A-1). We are, in other words, considering the disaccharides sucrose (from the sugar bowl) and lactose (in the milk) and the polysaccharide starch (in the cereal). First we'll follow the sugars and starch through your digestive tract; then we'll consider the glucose in your blood and the several ways the cells in different parts of your body use it.

Let's start with a spoonful of cereal you've just eaten. The sucrose and lactose remain intact until they reach your small intestine, but the digestion of starch begins in your mouth because saliva contains salivary amylase, an enzyme that splits starch into smaller polysaccharides. (Theoretically, amylase can hydrolyze starch to the disaccharide maltose, but food is not usually in the mouth long enough for digestion to proceed that far.) Further digestion occurs in your small intestine, where pancreatic amylase completes the breakdown of starch to maltose. (Note, by the way, that the intestinal digestion of starch involves simple hydrolysis of the glycosidic bonds between glucose units. This contrasts with the phosphorolytic cleavage that occurs during glycogen breakdown in liver or muscle cells and during starch breakdown in plants.)

The maltose generated from starch is hydrolyzed to glucose in your intestine by the enzyme maltase. Maltase is one of a family of intestinal disaccharidases, each specific for a different disaccharide. The lactose from the milk and the sucrose that you sprinkled on the cereal are hydrolyzed by other members of this family—lactase and sucrase, respectively. Lactose yields one molecule each of glucose and galactose, whereas sucrose is hydrolyzed to one molecule each of glucose and fructose (see Figure 13-10). (In some people, intestinal lactase disappears gradually after age 4 or so, when milk drinking usually decreases. If such people ingest milk or other dairy products, they are likely to experience cramps and diarrhea, a condition called *lactose intolerance.*)

Glucose, galactose, and fructose molecules are absorbed by intestinal epithelial cells. These cells are well suited to the task because of the numerous microvilli that project into the lumen of the intestine, thereby greatly increasing the absorptive surface of the cell (see Figure 4-2). Moreover, only two layers of epithelial cells separate nutrients in the lumen of your intestine from the blood in your capillaries. Some sugars, such as fructose, move across the plasma membrane of an epithelial cell by facili-

Figure 13A-1 What Happens to the Sugar? Is this a question you ever ponder as you sprinkle sugar on your breakfast cereal?

tated diffusion, because the concentrations of these sugars are lower in capillary blood than in the intestinal lumen. Glucose, however, is moved by active transport because of its high concentration in blood. (Other, less abundant nutrients such as vitamins and amino acids are also transported actively from the intestine into the bloodstream.)

Sugars such as fructose and galactose are transported by your bloodstream to the various tissues of your body. These sugars are eventually absorbed by body cells and converted to intermediates in the glycolytic (and therefore also the gluconeogenic) pathway, as shown in Figures 13-10 and 13-12. The pathway for galactose utilization is more complex than that of most other simple sugars, with five reactions required to convert a molecule of galactose into glucose-6-phosphate. (A genetic defect in this pathway may result in an inability to metabolize galactose, resulting in high levels of galactose in the blood and high levels of galactose-1-phosphate in tissues. This disorder, called *galactosemia,* has serious consequences, including mental retardation. Not surprisingly, it occurs most commonly in infants because the major dietary source of galactose is milk. Provided the condition is detected early, the symptoms can be prevented or alleviated by removing milk and dairy products from the diet.)

The main sugar in the blood, of course, is glucose. Its concentration in your blood a few hours after a meal is probably about 80 mg% (80 mg per 100 mL of blood, or about 4.4 mM). The level may rise to 120 mg% (6.6 mM) shortly after you've eaten, and it decreases somewhat if you wait longer than usual between meals. In general, however, the blood glucose level is maintained within rather narrow limits. In fact, maintenance of the blood glucose level is one of the most important regulatory functions in your body, particularly for proper functioning of your nervous system, including your brain. Your blood glucose level is under the control of several hormones, including insulin, glucagon, epinephrine, and norepinephrine (see Chapter 10).

Once in your bloodstream, glucose is transported to cells in all parts of your body, where it has four main fates: It can be catabolized to CO_2 and H_2O by aerobic respiration, converted to lactate, used to synthesize the storage polysaccharide glycogen, or converted via glycolysis and pyruvate oxidation to acetyl CoA, from which body fat is made. Aerobic respiration is the most common fate of blood glucose because most of the tissues in your body function aerobically most of the time. Your brain is particularly noteworthy as an aerobic organ. It needs large amounts of energy to maintain the membrane potentials essential for the transmission of nerve impulses, and it normally depends solely on glucose to meet this need. In fact, your brain needs about 120 g of glucose per day, which is about 15% of your total energy consumption. When you are at rest, your brain accounts for about 60% of your glucose usage. The brain is also a heavy user of oxygen—about 20% of your total oxygen consumption. Moreover, the brain has no significant stores of glycogen, so the supply of both oxygen and glucose must be continuous. Even a short interruption of either has dire consequences. Your heart has similar requirements because it is also a completely aerobic organ and has little or no energy reserves. The supply of oxygen and fuel molecules must therefore be constant, though the heart, unlike the brain, can use a variety of fuel molecules, including glucose, lactate, and fatty acids. In fact, heart muscle cells are so richly supplied with blood—and therefore with oxygen—that they can continue aerobic respiration under conditions of strenuous exertion, even when skeletal muscle cells have already switched to anaerobic catabolism, as described below. Heart muscle cells can even carry out aerobic catabolism of lactate that is being generated anaerobically by skeletal muscle cells at the same time and hence under the same conditions.

In addition to aerobic respiration in a wide variety of tissues, glucose can also be catabolized anaerobically, especially in red blood cells and in skeletal muscle cells. Red blood cells have no mitochondria and cannot carry out aerobic respiration. They depend exclusively on glycolysis to meet their energy needs. Unlike heart muscle, skeletal muscle can function in either the presence or the absence of oxygen. When you exert yourself strenuously, the ATP demand of your muscle cells temporarily exceeds the capacity of your circulatory system to supply oxygen, so the rate of glycolysis exceeds that of aerobic respiration, and excess pyruvate is converted to lactate. Lactate generated in this way is released into the blood and taken up not only by your heart for use as fuel but also by gluconeogenic tissues, especially your liver. When lactate molecules enter a liver cell, they are reoxidized to pyruvate, which is then used to make glucose by gluconeogenesis (see Figure 13-12). The glucose is returned to the bloodstream, where it can be taken up by muscle (or any other) cells again.

Skeletal muscle is the main source of blood lactate, and the liver is the primary site of gluconeogenesis, so a cycle is set up, as shown in Figure 13A-2. Lactate produced by glycolysis in hypoxic (oxygen-deficient) muscle cells is transported via the blood to the liver. There, gluconeogenesis converts the lactate to glucose, which is released into the blood. This process is called the *Cori cycle*, for Carl and Gerti Cori, whose studies in the 1930s and

(continued)

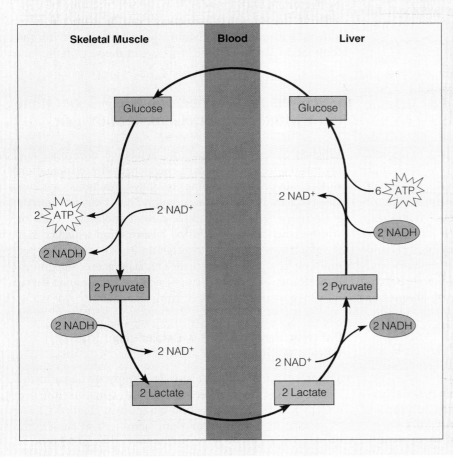

Figure 13A-2   The Cori Cycle: The Link Between Glycolysis in Muscle Cells and Gluconeogenesis in the Liver.   Skeletal muscle cells derive much of their energy from glycolysis, especially during periods of strenuous exercise. The lactate produced in this way is transported by the bloodstream to the liver, where it is reoxidized to pyruvate. Pyruvate is used as substrate for gluconeogenesis within the liver, generating glucose that is then returned to the blood. This cyclic process is called the *Cori cycle*, for Carl and Gerti Cori, the team of scientists who first described this process.

1940s led to an understanding of it. The next time you are resting after strenuous exercise, think of what is happening: The lactate your muscle cells have just released into the bloodstream is being taken up by liver cells and converted back to glucose. The reason you are breathing heavily is to provide the oxygen your body needs to return your muscle cells to aerobic conditions, and to generate all the ATP and GTP needed for gluconeogenesis in your liver and for rebuilding body glycogen stores.

Glycogen storage is, in fact, the third significant fate of blood glucose. Glycogen is stored primarily in the cells of your liver and skeletal muscle. Muscle glycogen is used to supply glucose during times of strenuous exertion, when the glycolytic rate of the muscle cells temporarily exceeds the capability of the circulatory system to deliver blood glucose. Liver glycogen, on the other hand, is used as a source of glucose when the liver is stimulated hormonally to release glucose into the bloodstream to maintain the blood glucose level.

The fourth possible fate of blood glucose is its use for the synthesis of body fat. The route to fat is via pyruvate to acetyl CoA, just as in the initial phase of aerobic respiration. Whenever you eat more food than your body needs for energy and for the biosynthesis of other molecules, excess glucose is oxidized to acetyl CoA and used for the synthesis

of triacylglycerols. These are then stored as body fat, especially in adipose tissue that is specialized for this purpose. Thus, your body has three sources of energy at all times: the glucose in your blood, the glycogen in your liver and skeletal muscle cells, and the triacylglycerols stored in adipose tissue.

To conclude, let's come back to our original question: "What happens to the sugar?" Hopefully, you're in a better position to answer that question now. All the glucose (and other sugars) in your body come originally from the food you eat—either as monosaccharides directly, or from the breakdown of disaccharides and polysaccharides in your intestinal tract. The ultimate fate of that glucose is oxidation to CO_2 and water, which you then exhale and excrete. But in the meantime, glucose molecules can circulate in your bloodstream or be stored as glycogen in liver or muscle cells. In its circulating form, glucose can be oxidized immediately by aerobic tissues such as the brain, it can be converted to lactate and become a part of the Cori cycle, or it can be used to synthesize glycogen or fat for storage.

It may look like just a modest spoonful of sugar as you sprinkle it on your cereal or add it to your coffee, but treat it with respect—it plays an important role in the energy metabolism of all of the cells in your body!

Based on Their Need for Oxygen, Organisms Are Aerobic, Anaerobic, or Facultative

Organisms can be classified in terms of their need for and use of oxygen as an electron acceptor in energy metabolism. Most organisms have an absolute requirement for oxygen and are called **obligate aerobes.** You look at such an organism every morning in the mirror. **Obligate anaerobes,** on the other hand, cannot use oxygen as an electron acceptor under any conditions. In fact, oxygen is usually toxic to these organisms. Not surprisingly, such organisms occupy environments from which oxygen is generally excluded, such as deep puncture wounds or the sludge at the bottoms of ponds. Most strict anaerobes are bacteria, including those responsible for gangrene, food poisoning, and methane production.

Facultative organisms can function under either aerobic or anaerobic conditions. Given the availability of oxygen, most facultative organisms carry out the full respiratory process, but they can switch to a fermentative mode if oxygen is limiting or absent. Many bacteria and fungi are facultative organisms, as are most molluscs and annelids (worms). Some cells or tissues of otherwise aerobic organisms can function in the temporary absence or scarcity of oxygen if required to do so. Your skeletal muscle cells are an example; they normally function aerobically but switch to lactate fermentation whenever the oxygen supply becomes limiting—during periods of prolonged or strenuous exercise, for example.

The rest of this chapter is devoted mainly to the anaerobic generation of ATP by the fermentation of glucose, with lactate and alcohol as the main products of interest.

Aerobic energy metabolism then becomes the focus of the next chapter. It seems appropriate to consider fermentation processes first, because the glycolytic pathway is common to both fermentation and aerobic respiration. Thus, by beginning with fermentation, we will not only be considering the ways in which energy can be extracted from glucose without net oxidation; we will also be laying the foundation for the aerobic processes of the next chapter.

Glycolysis and Fermentation: ATP Generation Without the Involvement of Oxygen

Whether it is an obligate or facultative anaerobe, any organism or cell that meets its energy needs by fermentation carries out energy-yielding oxidative reactions without using oxygen as an electron acceptor. The six-carbon glucose molecule is split into 2 three-carbon molecules, each of which is then partially oxidized by a reaction sequence that is sufficiently exergonic to generate 2 ATP molecules per molecule of glucose fermented. This is in fact the maximum possible energy yield that can be achieved without access to oxygen or to an alternative electron acceptor.

Glycolysis Generates ATP by Catabolizing Glucose to Pyruvate

The process of **glycolysis,** also called the **glycolytic pathway,** is a ten-step reaction sequence that converts glucose into pyruvate, a three-carbon compound (Figure 13-6). Glycolysis is common to both aerobic and anaerobic

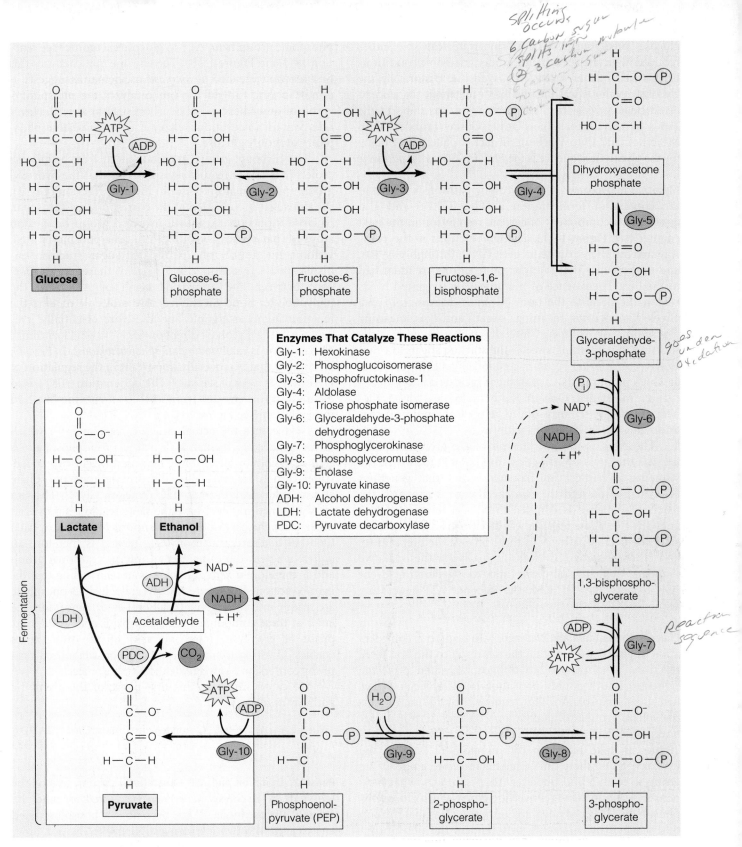

Handwritten annotations (top): Splitting occurs 6 carbon sugar splits into 2 3-carbon sugar

Handwritten annotations (right): goes under oxidation; Reaction sequence

Enzymes That Catalyze These Reactions

Gly-1: Hexokinase
Gly-2: Phosphoglucoisomerase
Gly-3: Phosphofructokinase-1
Gly-4: Aldolase
Gly-5: Triose phosphate isomerase
Gly-6: Glyceraldehyde-3-phosphate dehydrogenase
Gly-7: Phosphoglycerokinase
Gly-8: Phosphoglyceromutase
Gly-9: Enolase
Gly-10: Pyruvate kinase
ADH: Alcohol dehydrogenase
LDH: Lactate dehydrogenase
PDC: Pyruvate decarboxylase

Figure 13-6 The Glycolytic Pathway from Glucose to Pyruvate, with Two Fermentation Alternatives. Glycolysis is a sequence of ten reactions in which glucose (or any of several related sugars) is catabolized to pyruvate, with a single oxidative reaction (Gly-6) and two ATP-generating steps (Gly-7 and Gly-10). In the absence of oxygen or another electron acceptor, the NADH generated by reaction Gly-6 is reoxidized by transferring its electrons to pyruvate. The most common products of glucose fermentation are lactate or ethanol plus carbon dioxide. The enzymes that catalyze these reactions are identified in the box. In most cells, these enzymes occur in the cytosol. In some parasitic protozoans called trypanosomes, however, the first seven enzymes are compartmentalized in membrane-bounded organelles called *glycosomes*.

glucose metabolism because pyruvate can be either reduced to lactate (or ethanol) under anaerobic conditions or oxidized further if oxygen is available. Historically, the glycolytic pathway was the first major metabolic sequence to be elucidated. Most of the decisive work was done in the 1930s by the German biochemists Gustav Embden, Otto Meyerhof, and Otto Warburg. In fact, an alternative name for the glycolytic sequence is the *Embden-Meyerhof pathway*.

Glycolysis in Overview.   The glycolytic pathway is shown in the context of fermentation in Figure 13-6 and also appears as a component of aerobic respiration in the next chapter (see Figure 14-1). The ten reactions in the pathway are numbered in sequence (Gly-1 through Gly-10), and the enzymes that catalyze the reactions are identified in the box. The essence of the process is suggested by its very name because the term *glycolysis* derives from two Greek roots: *glykos*, meaning "sweet," and *lysis*, meaning "loosening" or "splitting." Literally, then, glycolysis is the splitting of something sweet—the starting sugar, in other words. The splitting occurs at reaction Gly-4 in Figure 13-6, at which point the six-carbon sugar is cleaved into 2 three-carbon molecules. One of these molecules, glyceraldehyde-3-phosphate, turns out to be the only molecule that undergoes oxidation in this pathway.

The most important features of the glycolytic pathway are the sugar-splitting reaction (Gly-4) for which the sequence is named, the oxidative event that generates NADH (Gly-6), and the two specific steps at which the reaction sequence is coupled to ATP generation (Gly-7 and Gly-10). These features will be emphasized as we consider the overall pathway in three phases: the preparatory and cleavage steps (Gly-1 through Gly-5); the oxidative sequence, which is also the first ATP-generating event (Gly-6 and Gly-7); and the second ATP-generating sequence (Gly-8 through Gly-10).

Phase 1: Preparation and Cleavage.   To begin our consideration of glycolysis, note that the net result of the first three reactions is to convert an unphosphorylated molecule (glucose) into a doubly phosphorylated molecule (fructose-1, 6-biphosphate). This requires the addition of two phosphate groups to glucose, one on each terminal carbon. Looking at glucose, it is easy to see how phosphorylation can take place on carbon atom 6, because the hydroxyl group there can be readily linked to a phosphate group to form a phosphoester. That is, in fact, what happens in reaction Gly-1, converting glucose into glucose-6-phosphate. ATP provides not only the phosphate group but also the driving force that renders the phosphorylation reaction strongly exergonic ($\Delta G^{\circ\prime} = -4.0$ kcal/mol), making it essentially irreversible in the direction of glucose phosphorylation. Notice, by the way, that the bond formed when glucose is phosphorylated is a *phosphoester bond,* whereas the bond by which the terminal phosphate is linked to ATP is a *phosphoanhydride bond.* This difference is what makes the transfer of the

phosphate group from ATP to glucose exergonic (see reaction 13-5 and Figure 13-3). The enzyme that catalyzes this first reaction is called *hexokinase;* as the name suggests, it is not specific for glucose but catalyzes the phosphorylation of other hexoses (six-carbon sugars) as well. (Liver cells contain an additional enzyme, *glucokinase,* that phosphorylates only glucose.)

The carbonyl group on carbon atom 1 of the glucose molecule is not as readily phosphorylated as the hydroxyl group on carbon atom 6. But in the next reaction (Gly-2), the aldosugar is converted to the corresponding ketosugar, fructose-6-phosphate, with a hydroxyl group on carbon atom 1. That hydroxyl group can then be phosphorylated, yielding the doubly phosphorylated sugar, fructose-1,6-bisphosphate (reaction Gly-3). Again, the energy difference between the anhydride bond of ATP and the phosphoester bond on the fructose molecule renders the reaction highly exergonic and therefore essentially irreversible in the glycolytic direction ($\Delta G^{\circ\prime} = -3.4$ kcal/mol). This reaction is catalyzed by *phosphofructokinase-1 (PFK-1),* an enzyme that is especially important in the regulation of glycolysis, as we will see later. The designation PFK-1 is to distinguish this enzyme from PFK-2, an enzyme involved in regulation (see Figure 13-14).

Next comes the actual cleavage reaction from which glycolysis derives its name. Fructose-1,6-bisphosphate is split reversibly by the enzyme *aldolase* to yield two trioses (three-carbon sugars) called dihydroxyacetone phosphate and glyceraldehyde-3-phosphate (reaction Gly-4). The two trioses formed in Gly-4 have the same relationship to each other as do glucose-6-phosphate and fructose-6-phosphate: One has a terminal carbonyl group and is therefore an *aldose,* whereas the other has an internal carbonyl group and is therefore a *ketose.* Thus it is not surprising that dihydroxyacetone phosphate and glyceraldehyde-3-phosphate are readily interconvertible (reaction Gly-5). Since only the latter of these compounds is directly oxidizable in the next phase of glycolysis, interconversion of the two trioses enables dihydroxyacetone phosphate to be catabolized simply by conversion to glyceraldehyde-3-phosphate.

We can summarize this first phase of the glycolytic pathway as follows:

$$\text{glucose} + 2\text{ATP} \longrightarrow 2 \text{ glyceraldehyde-3-phosphate} + 2\text{ADP}$$

$$(13\text{-}13)$$

Phase 2: Oxidation and ATP Generation.   So far, five of the ten steps of glycolysis have been accounted for, and the original glucose molecule has been doubly phosphorylated and cleaved into two interconvertible trioses. Notice, however, that the energy yield is negative thus far: Two molecules of ATP have been *consumed* per molecule of glucose up to this point. But the ATP debt is about to be repaid with interest as we encounter the two energy-yielding phases of glycolysis. In the first sequence (reactions Gly-6 and Gly-7), ATP production is linked directly to an oxidative event, and then in the second phase (Gly-8 through

Gly-10), a highly unstable form of the pyruvate molecule serves as the driving force behind ATP generation.

The oxidation of glyceraldehyde-3-phosphate to the corresponding acid, 3-phosphoglycerate, is highly exergonic—sufficiently so, in fact, to drive both the reduction of the coenzyme NAD^+ (Gly-6) and the phosphorylation of ADP with inorganic phosphate, P_i (Gly-7). Historically, this was the first example of a reaction sequence in which the coupling of ATP generation to an oxidative event was understood. Because of its usefulness as a prototype for understanding such coupling, this two-reaction sequence is illustrated in detail in Figure 13-7. The figure focuses especially on the mechanism of action of *glyceraldehyde-3-phosphate dehydrogenase*, the enzyme that catalyzes the actual oxidative reaction (Gly-6).

The important features of this highly exergonic reaction are the involvement of NAD^+ as the electron acceptor and the coupling of the oxidation to the formation of a high-energy, doubly phosphorylated intermediate, 1,3-bisphosphoglycerate. The phosphoanhydride bond on carbon atom 1 of this intermediate has such a highly negative $\Delta G^{\circ\prime}$ of hydrolysis (-11.8 kcal/mol; see Table 13-1) that the transfer of the phosphate to ADP, catalyzed by the enzyme *phosphoglycerate kinase*, is a highly exergonic, irreversible reaction. ATP generation by the direct transfer to ADP of a high-energy phosphate group from a phosphorylated substrate such as 1,3-bisphosphoglycerate mediated by a water-soluble enzyme is called **substrate-level phosphorylation.** This mode of ATP synthesis is distinct from *oxidative phosphorylation*, in which phosphorylation of ADP is driven by the exergonic transfer of electrons from reduced coenzymes to oxygen, as we will see in Chapter 14.

To summarize the substrate-level phosphorylation of reactions Gly-6 and Gly-7, we can write an overall reaction with a stoichiometry that accounts for one of the two glyceraldehyde-3-phosphate molecules generated from each glucose molecule in the first phase of glycolysis:

$$\text{glyceraldehyde-3-phosphate} + NAD^+ + ADP + P_i \longrightarrow$$

$$\text{3-phosphoglycerate} + NADH + H^+ + ATP$$

$$(13\text{-}14)$$

Keep in mind that each reaction in the glycolytic pathway beyond glyceraldehyde-3-phosphate occurs twice per starting molecule of glucose, thereby accounting for both molecules of triose phosphate generated in the cleavage reaction (Gly-4). This means that, on a per-glucose basis, two molecules of NADH need to be reoxidized in order to regenerate the NAD^+ that is needed for continual oxidation of glyceraldehyde-3-phosphate. It also means that the initial investment of two ATP molecules in phase 1 is recovered here in phase 2, so the net ATP yield is now zero.

Thus far, then, seven of the ten reactions of glycolysis have been used to convert one molecule of glucose into two molecules of 3-phosphoglycerate, but there is as yet no net ATP generation. That comes in the final phase of the pathway, to which we now turn.

Phase 3: Pyruvate Formation and ATP Generation. Generating another molecule of ATP at the expense of 3-phosphoglycerate depends on the phosphate group on carbon atom 3. At this stage, the phosphate group is linked to the carbon atom by a phosphoester bond with an unpromisingly low free energy of hydrolysis ($\Delta G^{\circ\prime} = -3.3$ kcal/mol). In the final phase of the glycolytic pathway, this phosphoester bond is converted to a *phosphoenol bond*, the hydrolysis of which is highly exergonic ($\Delta G^{\circ\prime} = -14.8$ kcal/mol; see Table 13-1). This increase in the amount of free energy released upon hydrolysis involves a rearrangement of internal energy within the molecule. To accomplish this, the phosphate group of 3-phosphoglycerate is moved to the adjacent carbon atom, forming 2-phosphoglycerate (reaction Gly-8). Water is then removed from 2-phosphoglycerate by the enzyme enolase (reaction Gly-9), thereby generating phosphoenolpyruvate (PEP).

If you look carefully at the structure of PEP (see Figure 13-6, reaction Gly-9), you will notice that, unlike the phosphoester bonds of either 3- or 2-phosphoglycerate, the phosphoenol bond of PEP has what we might define as a distinguishing characteristic of a high-energy phosphate bond: a phosphate group on a carbon atom that is linked by a double bond to another atom, almost always an atom of either carbon or oxygen. In fact, hydrolysis of the phosphoenol bond of PEP is one of the most exergonic hydrolytic reactions known in biological systems.

To understand why PEP is a higher-energy compound than most other phosphorylated intermediates in the cell, consider Figure 13-8. Like many keto compounds, pyruvate can exist in either of two interconvertible forms, called the *enol* and the *keto* forms. These two forms differ only in the location of the double bond (Figure 13-8a). However, the equilibrium of the interconversion greatly favors the keto form. This means that the keto form of pyruvate is much more stable, whereas the enol form is a highly unstable, thermodynamically unlikely configuration. Thus, when pyruvate is locked in the enol form by chemical constraints that do not allow its transition to the keto form, the molecule will be highly unstable.

Such is the case with the PEP generated in reaction Gly-9. When water is removed from 2-phosphoglycerate, the product is pyruvate locked in the enol form by the presence of a phosphate group on carbon atom 2 that prevents transition (or *tautomerization*) to the more stable keto form (Figure 13-8b). PEP is therefore higher in energy than most other phosphorylated compounds; in addition to the usual decrease in free energy when the extra electron pair becomes maximally delocalized over the P=O bonds of the phosphate group, reversion of the pyruvate from the enol form to the keto form is also highly exergonic.

To say that the phosphate bond of PEP has a highly negative free energy of hydrolysis is to make the last step in the sequence, reaction Gly-10, entirely reasonable because this reaction involves the transfer of that phosphate to ADP, generating another molecule (or, on a per-glucose basis, two more molecules) of ATP. That transfer,

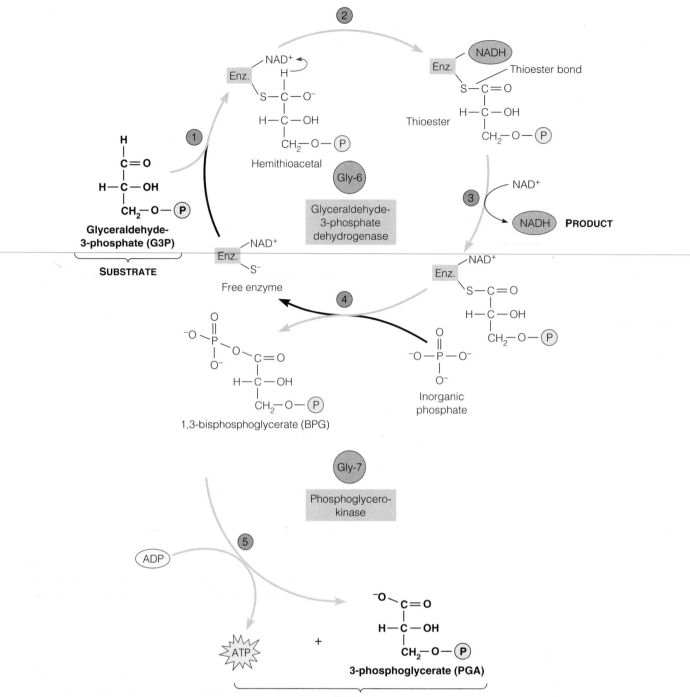

Figure 13-7   Substrate-Level Phosphorylation: A Detailed Mechanism for Reactions Gly-6 and Gly-7 of the Glycolytic Pathway.   ATP is generated by substrate-level phosphorylation in reactions Gly-6 and Gly-7, a two-reaction sequence in which glyceraldehyde-3-phosphate (G3P) is oxidized to 3-phosphoglycerate (PGA) with 1,3-bisphosphoglycerate (BPG) as an intermediate. ① The reaction sequence begins with the covalent binding of G3P to a sulfhydryl group at the active site of the enzyme glyceraldehyde-3-phosphate dehydrogenase, forming a *hemithioacetal.* ② The hemithioacetal is oxidized to a thioester by an enzyme-bound molecule of NAD+. ③ The reduced NADH is displaced from the enzyme surface by a molecule of NAD+. ④ Inorganic phosphate attacks the thioester bond, forming BPG and displacing it from the enzyme surface. ⑤ BPG, a high-energy phosphoanhydride, then binds to the active site of phosphoglycero-kinase, where the high energy of hydrolysis of the phosphoanhydride bond is used to drive the synthesis of ATP, releasing PGA as the product. The essential feature of the sequence is that a thermodynamically unfavorable reaction, the formation of an anhydride between a carboxylic acid and inorganic phosphate, is driven by a thermodynamically favorable reaction, the oxidation of an aldehyde. The two reactions are coupled by an enzyme-bound thioester intermediate, which preserves much of the free energy that would otherwise have been released as heat in the oxidation reaction.

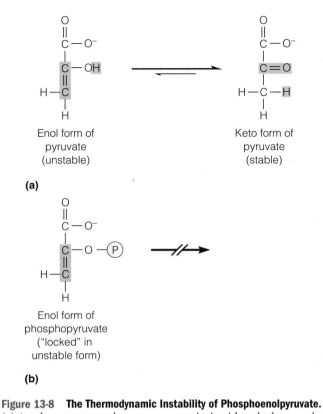

(a)

Enol form of
pyruvate
(unstable)

Keto form of
pyruvate
(stable)

(b)

Enol form of
phosphopyruvate
("locked" in
unstable form)

Figure 13-8   The Thermodynamic Instability of Phosphoenolpyruvate.
(a) As a keto compound, pyruvate can exist in either the keto or the enol form, but the equilibrium greatly favors the keto form. Conversion of the enol form to the keto form is therefore thermodynamically very favorable. **(b)** Phosphoenolpyruvate, which is the form generated in reaction Gly-9, is in the thermodynamically unfavorable enol configuration but cannot undergo conversion to the more stable keto form because of the phosphate group covalently linked to carbon atom 2. (The two diagonal lines across the arrow indicate that the reaction does not occur.) Release of this phosphate group is highly exergonic because the pyruvate molecule is then free to assume the more stable (low-energy) keto form.

catalyzed by the enzyme *pyruvate kinase,* is highly exergonic ($\Delta G°' = -7.5$ kcal/mol; see Figure 13-3 and reaction 13-6) and is therefore essentially irreversible in the direction of pyruvate and ATP formation.

To summarize the third phase of glycolysis, we can write an overall reaction for pyruvate formation, again using stoichiometry that accounts for one of the two trioses derived from glucose:

$$\text{3-phosphoglycerate} + \text{ADP} \longrightarrow \text{pyruvate} + \text{H}_2\text{O} + \text{ATP}$$
(13-15)

Summary of Glycolysis.   The two molecules of ATP initially invested in reactions Gly-1 and Gly-3 were recouped in the first phosphorylation event (Gly-7), so the two molecules of ATP formed per molecule of glucose by the second phosphorylation event (Gly-10) represent the net

ATP yield of the glycolytic pathway. This becomes clear when we add up the three reactions that summarize the three phases of the pathway (reactions 13-13, 13-14, and 13-15), with the latter two reactions multiplied by two to account for both triose molecules generated in reaction 13-13. The result is an overall expression for the pathway from glucose to pyruvate:

$$\text{glucose} + 2\text{NAD}^+ + 2\text{ADP} + 2\text{P}_i \xrightarrow{\text{reactions Gly-1 through Gly-10}}$$
$$\text{2 pyruvate} + 2\text{NADH}^+ + 2\text{H}^+ + 2\text{ATP} + 2\text{H}_2\text{O}$$
(13-16)

This sequence is exergonic in the direction of pyruvate formation. Under typical intracellular conditions in your body, for example, $\Delta G'$ for the overall pathway from glucose to pyruvate with the concomitant generation of two molecules each of ATP and NADH is about -20 kcal/mol.

The glycolytic pathway is one of the most universal metabolic pathways known. Virtually all cells possess the ability to convert glucose to pyruvate, with some of the energy of the oxidative event in the pathway conserved in the form of two molecules of ATP per molecule of glucose. What happens next, however, usually depends on the availability of oxygen, because catabolism beyond pyruvate is different under aerobic conditions than under the anaerobic conditions we are presuming at present.

The Fate of Pyruvate Depends on Whether or Not Oxygen Is Available

Pyruvate occupies a key position as a branching point in chemotrophic energy metabolism (Figure 13-9). Its fate depends on the kind of organism involved and whether oxygen is available. In the presence of oxygen, pyruvate undergoes further oxidation to a molecule called acetyl coenzyme A (acetyl CoA), and the glycolytic pathway becomes the first of several major components of aerobic respiration (Figure 13-9a). As we will see in Chapter 14, this results in the complete oxidation of pyruvate to carbon dioxide, with the generation of much more ATP than is otherwise possible.

An important feature of glycolysis, however, is that it can also take place in the absence of oxygen. Under these conditions, no further oxidation of pyruvate is possible, no acetyl coenzyme A is formed, and no additional ATP can be generated. Instead, the energy needs of the cell are met by the modest ATP yield of the glycolytic pathway. Rather than being oxidized, pyruvate is reduced by accepting the electrons (and protons) that must continually be removed from NADH to provide for the continued regeneration of NAD^+, without which glycolysis could not continue for very long. As Figure 13-9 illustrates, the most common products of pyruvate reduction are lactate (part b) or ethanol and carbon dioxide (part c).

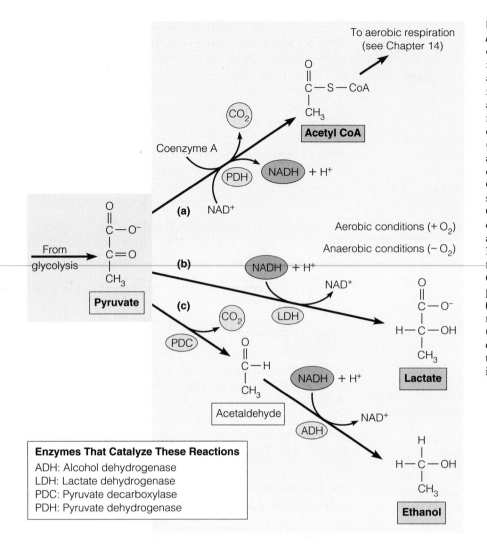

Figure 13-9 The Fate of Pyruvate Under Aerobic and Anaerobic Conditions. The fate of pyruvate depends on the organism involved and on whether oxygen is available. **(a)** Under aerobic conditions, most organisms convert pyruvate to an activated form of acetate in a reaction that involves both oxidation (with NAD^+ as the electron acceptor) and decarboxylation (liberation of a carbon atom as CO_2). The activated acetate is bound to the carrier coenzyme A as acetyl coenzyme A (acetyl CoA). Acetyl CoA then becomes the substrate for aerobic respiration (see Chapter 14). Under anaerobic or hypoxic conditions, pyruvate serves as the electron acceptor for the oxidation of NADH to NAD^+, thereby regenerating the oxidized form of the coenzyme required in reaction Gly-6 of glycolysis. The most common products of pyruvate reduction are **(b)** lactate (in most animal cells and many bacteria) or **(c)** ethanol and CO_2 (in many plant cells and in yeasts and other microorganisms). The enzymes that catalyze these reactions are identified in the box.

Enzymes That Catalyze These Reactions
ADH: Alcohol dehydrogenase
LDH: Lactate dehydrogenase
PDC: Pyruvate decarboxylase
PDH: Pyruvate dehydrogenase

In the Absence of Oxygen, Pyruvate Undergoes Fermentation to Regenerate NAD^+

As usually defined, the glycolytic pathway ends with pyruvate. Fermentative processes cannot end there, however, because of the need to regenerate NAD^+, the oxidized form of the coenzyme. As reaction 13-16 indicates, the conversion of glucose to pyruvate involves the stoichiometric reduction of NAD^+: One molecule of NAD^+ is reduced to NADH per molecule of pyruvate generated. Coenzymes are present in cells at only modest concentrations, however, so the conversion of NAD^+ to NADH during glycolysis would cause cells to run out of NAD^+ very quickly if there were not some mechanism for regenerating NAD^+. Thus, the reduction of NAD^+ to NADH that occurs during glycolysis must be accompanied by the concurrent reoxidation of NADH to NAD^+, or glycolysis would quickly come to a halt.

In the presence of oxygen, NADH is reoxidized by the transfer of its electrons (and protons) to oxygen, as we will see in Chapter 14. Under anaerobic conditions, however, the electrons are transferred to pyruvate, which has a carbonyl group that can be readily reduced to a hydroxyl group (see Figure 13-9b and c). The two most common pathways for fermentation use pyruvate as the electron acceptor, converting it either to lactate or to CO_2 and ethanol. We will consider both of these alternatives briefly.

Lactate Fermentation. The anaerobic process that terminates in lactate is called **lactate fermentation.** As Figure 13-9b indicates, lactate is generated by the direct transfer of electrons from NADH to the carbonyl group of pyruvate, reducing it to the hydroxyl group of lactate. On a per-glucose basis, this can be represented as follows:

$$2 \text{ pyruvate} + 2NADH + 2H^+ \longrightarrow 2 \text{ lactate} + 2NAD^+ \quad \textbf{(13-17)}$$

This reaction is readily reversible; in fact, the enzyme that catalyzes it is called *lactate dehydrogenase* because of its ability to catalyze the oxidation, or dehydrogenation, of lactate to pyruvate.

By adding reactions 13-16 and 13-17, we can write an overall reaction for the metabolism of glucose to lactate under anaerobic conditions:

$$\text{glucose} + 2ADP + 2P_i \longrightarrow 2\text{lactate} + 2ATP + 2H_2O \quad \textbf{(13-18)}$$

Lactate fermentation is the major energy-yielding pathway in many anaerobic bacteria, as well as in animal cells operating under anaerobic or hypoxic (oxygen-deficient) conditions. Lactate fermentation is very important commercially because the production of cheese, yogurt, and other dairy products depends on microbial fermentation of lactose, the main sugar found in milk.

A more personal example of lactate fermentation involves your own muscles during periods of strenuous exertion. Whenever muscle cells use oxygen faster than it can be supplied by the circulatory system, the cells become temporarily hypoxic. Under these conditions, some, or even most, of the pyruvate is reduced to lactate instead of being further oxidized, as it is under aerobic conditions. (The resulting local buildup of lactate is what causes the muscle pain that sometimes accompanies vigorous exercise.) The lactate produced in this way is transported by the circulatory system from the muscle to the liver. There it is converted to glucose again by the process of *gluconeogenesis*. As we will see later in this chapter, gluconeogenesis is essentially the reverse of lactate fermentation but with several critical differences that enable the pathway to proceed exergonically in the direction of glucose formation. The relationship between glycolysis and gluconeogenesis in your own body is explored in Box 13A on p. 376; see especially Figure 13A-2.

Alcoholic Fermentation. **Alcoholic fermentation** also involves both NADH and pyruvate, but with different end-products. Plant cells carry out alcoholic fermentation under anaerobic conditions (in water-logged roots, for example), as do yeasts and other microorganisms. In this case, the pyruvate first undergoes decarboxylation (liberation of a carbon atom as CO_2) to form the two-carbon compound, acetaldehyde, which accepts electrons from NADH. Acetaldehyde reduction gives rise to ethanol, the alcohol for which the process is named. This reductive sequence actually involves two separate events (pyruvate decarboxylation and subsequent acetaldehyde reduction) catalyzed by two separate enzymes, *pyruvate decarboxylase* and *alcohol dehydrogenase*, respectively (Figure 13-9c). The overall reaction can be summarized as follows:

$$2 \text{ pyruvate} + 2\text{NADH} + 2\text{H}^+ \longrightarrow$$
$$2 \text{ ethanol} + 2CO_2 + 2\text{NAD}^+ \qquad \textbf{(13-19)}$$

By adding this reductive step to the overall equation for glycolysis (reaction 13-16), we arrive at the following summary equation for alcoholic fermentation:

$$\text{glucose} + 2\text{ADP} + 2\text{P}_i \longrightarrow$$
$$2 \text{ ethanol} + 2CO_2 + 2\text{ATP} + 2\text{H}_2\text{O} \qquad \textbf{(13-20)}$$

Alcoholic fermentation has considerable economic significance because fermentation as carried out by yeast cells is a key process in the baking, brewing, and winemaking industries. For the baker, carbon dioxide is the important end-product. The yeast cells that are added to bread dough function anaerobically, generating both CO_2 and ethanol. Carbon dioxide becomes entrapped within the mass of dough, causing it to rise, and the alcohol is driven off harmlessly during the subsequent baking process, contributing to the pleasant aroma of baking bread. For the brewer, both CO_2 and ethanol are essential; ethanol makes the product an alcoholic beverage and CO_2 accounts for the carbonation.

Other Fermentation Pathways. Although lactate and ethanol are the fermentation products of greatest physiological or economic significance, they by no means exhaust the microbial repertoire with respect to fermentation. In *propionate fermentation*, for example, bacteria convert pyruvate reductively to propionate (CH_3—CH_2—COO^-), an important reaction in the production of Swiss cheese. Many bacteria that cause food spoilage do so by *butylene glycol fermentation*. Other fermentation processes yield acetone, isopropyl alcohol, or butyrate, which is responsible for the rotten smell of rancid food. All these reactions, however, are just metabolic variations on the common theme of reoxidizing NADH by the transfer of electrons to an organic acceptor.

Fermentation Taps Only a Small Fraction of the Free Energy of the Substrate but Conserves That Energy Efficiently as ATP

An essential feature of every fermentative process is that no external electron acceptor is involved and no net oxidation occurs. In both lactate and alcoholic fermentation, for example, the NADH generated by the single oxidative step of glycolysis (reaction Gly-6) is reoxidized in the final reaction of the sequence (reactions 13-17 and 13-19, respectively). Because no net oxidation occurs, fermentation is characterized by a modest ATP yield—two molecules of ATP per molecule of glucose, in the case of either lactate or alcoholic fermentation.

Most of the free energy of the glucose molecule is still present in the two lactate or ethanol molecules. In the case of lactate fermentation, for example, the two lactate molecules produced from every glucose molecule account for most of the 686 kcal of free energy present per mole of glucose, because the complete aerobic oxidation of lactate has a $\Delta G^{\circ\prime}$ of −319.5 kcal/mol. In other words, about 93% ($2 \times 319.5/686 \times 100\%$) of the original free energy of glucose is still present in the two lactate molecules to which the glucose is converted. Lactate fermentation is therefore able to tap only about 7% (47 kcal/mol) of the free energy potentially available from glucose. The calculated values for alcoholic fermentation are very similar to these.

Although the energy yield from lactate fermentation is low, the available free energy is conserved efficiently as ATP. The ATP yield, of course, is two molecules of ATP per molecule of glucose. Based on standard free energy changes, these two molecules of ATP represent $2 \times 7.3 = 14.6$ kcal/mol, corresponding to an efficiency of energy conservation of about 30% ($14.6/47 \times 100\%$). If anything, this underestimates the efficiency of fermentation, because actual $\Delta G^\prime$ values for

ATP hydrolysis under cellular conditions are usually substantially more negative than −7.3 kcal/mol, often in the range of −10 to −14 kcal/mol. Based on these values, two molecules of ATP represent at least 20 kcal/mol, which means that the efficiency of energy conservation probably exceeds 40%.

Alternative Substrates for Glycolysis

Thus far, we have assumed glucose to be the starting point for glycolysis and therefore, by implication, for all of cellular energy metabolism. Glucose is certainly a major substrate for both fermentation and respiration in a variety of organisms and tissues. It is not the only such substrate, however; there are in fact many organisms and tissues within organisms for which glucose is not very significant at all. So it is important to ask what some of the major alternatives to glucose are and how they are handled by cells.

One principle quickly emerges: *Regardless of the chemical nature of the alternative substrate, it is converted in the fewest possible steps into an intermediate in the main pathway for glucose catabolism.* Most carbohydrates, for example, are converted to intermediates in the glycolytic pathway. To emphasize this point, we will briefly consider two classes of alternative substrates—other sugars and storage carbohydrates.

Other Sugars Are Also Catabolized by the Glycolytic Pathway

Many sugars other than glucose are available to cells, depending on the food sources of the organism in question. Most of them are either monosaccharides (usually hexoses or pentoses), or disaccharides that can be readily hydrolyzed into their component monosaccharides. Ordinary table sugar (sucrose), for example, is a disaccharide consisting of the hexoses glucose and fructose, and milk sugar (lactose) contains glucose and galactose (see Figure 3-23). In addition to glucose, fructose, and galactose, mannose is another relatively common dietary hexose.

Figure 13-10 illustrates the reaction sequences that bring various carbohydrates into the glycolytic pathway. In general, disaccharides are hydrolyzed into their component monosaccharides, and each monosaccharide is converted to a glycolytic intermediate in one or a few steps. Glucose and fructose enter most directly because their conversion requires only phosphorylation on carbon atom 6, catalyzed by the enzyme hexokinase. Hexokinase can also phosphorylate mannose, and the resulting mannose-6-phosphate undergoes conversion (by the enzyme phosphomannoisomerase) to fructose-6-phosphate, a glycolytic intermediate. The entry of galactose requires a somewhat more complex reaction sequence, the details of which are explored in Problem 13-9 at the end of this chapter.

Phosphorylated pentoses can also be channeled into the glycolytic pathway, but only after being converted to hexose phosphates. That conversion is accomplished by a metabolic sequence called the *phosphogluconate pathway,* which shuffles carbon atoms between 3-, 4-, 5-, 6-, and 7-carbon sugars. Thus, the typical cell has metabolic capabilities to convert most naturally occurring sugars (and a variety of other compounds as well) to one or another of the glycolytic intermediates for further catabolism under anaerobic or aerobic conditions.

Polysaccharides Are Cleaved to Form Sugar Phosphates That Also Enter the Glycolytic Pathway

Although glucose is the immediate substrate for both fermentation and respiration in many cells and tissues, it is not present in cells to any large extent as the free monosaccharide. Instead, it occurs primarily in the form of storage polysaccharides, most commonly starch in plants and glycogen in animals. As indicated in Figure 13-11, these storage polysaccharides are mobilized by a process called *phosphorolytic cleavage.* Inorganic phosphate is used to break the $\alpha(1 \longrightarrow 4)$ bond between successive glucose units, liberating the glucose monomers as glucose-1-phosphate. Both glycogen and starch are cleaved in this manner, primarily by the enzymes *glycogen phosphorylase* and *starch phosphorylase,* respectively. The glucose-1-phosphate that is formed in this way can be converted by the enzyme *phosphoglucomutase* to glucose-6-phosphate, which is then catabolized by the glycolytic pathway.

Notice that glucose stored in polymerized form enters the glycolytic pathway as glucose-6-phosphate, without the input of the ATP that would be required for the initial phosphorylation of the free sugar. Consequently, the overall energy yield for glucose is greater by one molecule of ATP when it is catabolized from the polysaccharide level than when it is catabolized with the free sugar as the starting substrate. (This is not a case of getting something for nothing, however, because energy is required in the polymerization process whereby glucose units are added to the growing starch or glycogen chain during polysaccharide synthesis.)

Gluconeogenesis

Cells are not only able to catabolize glucose and other carbohydrates to meet their energy needs but they can also synthesize sugars and polysaccharides to meet cellular or organismal needs. The process of glucose synthesis is called **gluconeogenesis,** which literally means the genesis, or formation, of new glucose. More specifically, gluconeogenesis is defined as the process by which animal (and other) cells synthesize glucose (and other carbohydrates) from three-carbon and four-carbon precursors that are usually noncarbohydrate in nature. The most common

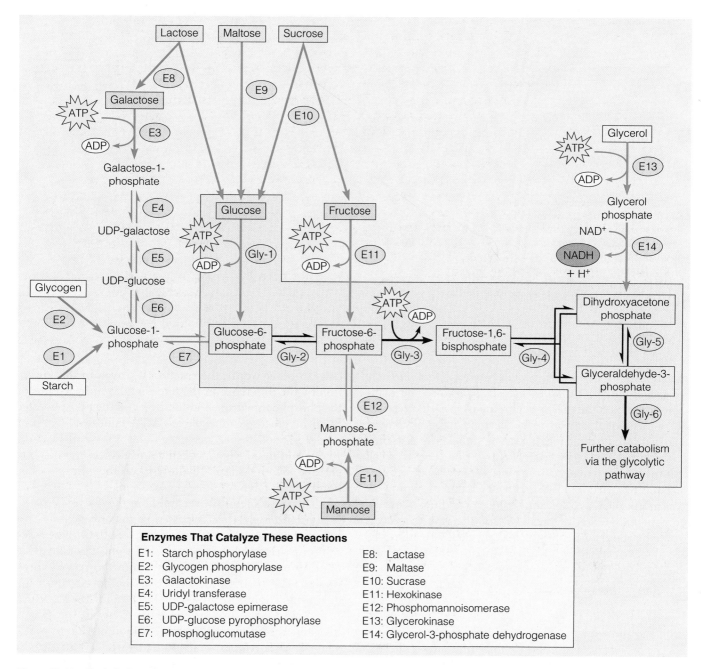

Enzymes That Catalyze These Reactions

E1: Starch phosphorylase	E8: Lactase
E2: Glycogen phosphorylase	E9: Maltase
E3: Galactokinase	E10: Sucrase
E4: Uridyl transferase	E11: Hexokinase
E5: UDP-galactose epimerase	E12: Phosphomannoisomerase
E6: UDP-glucose pyrophosphorylase	E13: Glycerokinase
E7: Phosphoglucomutase	E14: Glycerol-3-phosphate dehydrogenase

Figure 13-10 Carbohydrate Catabolism by the Glycolytic Pathway. Carbohydrate substrates that can be metabolized by conversion to an intermediate in the glycolytic pathway are enclosed in colored boxes. These include the hexoses galactose, glucose, fructose, and mannose; the disaccharides lactose, maltose, and sucrose; the polysaccharides glycogen and starch; and the three-carbon compound glycerol. The conversion reactions are shown by blue arrows. The enzymes that catalyze these reactions are identified in the box. The first six reactions of the glycolytic pathway are highlighted in green; for the names of the enzymes that catalyze these reactions, see Figure 13-6. In some cases, other enzymes or reaction sequences may be involved, depending on the organism and tissue.

starting material is pyruvate, or the lactate into which pyruvate is converted under anaerobic conditions.

Figure 13-12 compares gluconeogenesis and glycolysis. The two pathways share much in common; in fact, seven of the reactions in the gluconeogenic pathway occur by simple reversal of the corresponding reactions in glycolysis (reactions Gly-2 and Gly-4 through Gly-9). In each case, the same enzyme is used in both directions. Not all of the steps of the gluconeogenic pathway are simply the reversal of glycolytic reactions, however. As Figure 13-12 illustrates, three of the reactions of the glycolytic pathway—the first, third, and tenth—are accomplished by other means in the direction of gluconeogenesis. In fact, these differences illustrate well an important principle of cellular

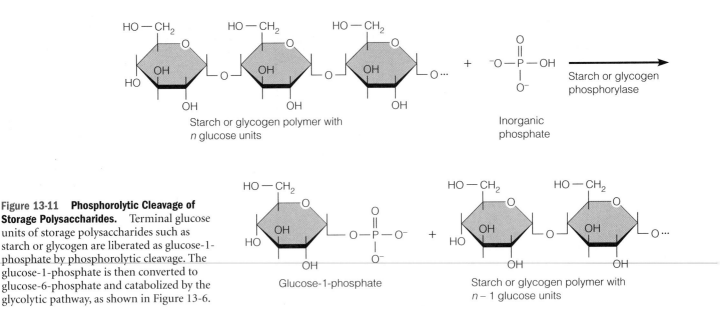

Figure 13-11 Phosphorolytic Cleavage of Storage Polysaccharides. Terminal glucose units of storage polysaccharides such as starch or glycogen are liberated as glucose-1-phosphate by phosphorolytic cleavage. The glucose-1-phosphate is then converted to glucose-6-phosphate and catabolized by the glycolytic pathway, as shown in Figure 13-6.

Starch or glycogen polymer with *n* glucose units

Inorganic phosphate

Starch or glycogen phosphorylase

Glucose-1-phosphate

Starch or glycogen polymer with *n* – 1 glucose units

metabolism: *Biosynthetic pathways are seldom just the reversal of the corresponding catabolic pathways.* This principle is based on the energy requirements in each direction. For a metabolic pathway to be thermodynamically favorable in a specific direction, it must be sufficiently exergonic in that direction. That certainly is true of glycolysis; recall that the overall sequence from glucose to pyruvate as summarized by reaction 13-16 has a $\Delta G'$ value of about −20 kcal/mol under typical intracellular conditions in the human body. Clearly, then, $\Delta G'$ for the reverse process would be about +20 kcal/mol, making glucose synthesis by the direct reversal of glycolysis highly endergonic and therefore thermodynamically impossible.

Gluconeogenesis is possible because the three most exergonic reactions in the glycolytic pathway (Gly-1, Gly-3, and Gly-10) do not simply "run in reverse" in the gluconeogenic direction. Instead, the gluconeogenic pathway has *bypass reactions* at each of those three sites—alternative reactions that effectively circumvent the three glycolytic reactions that would be the most difficult to drive in the reverse direction. In fact, the three reactions of the glycolytic pathway that are bypassed in gluconeogenesis in Figure 13-12 are the only three that are shown as unidirectional in Figure 13-6. In each of these three instances, the bypass reactions of gluconeogenesis circumvent the irreversibility of the glycolytic step.

In the case of both Gly-1 and Gly-3, the requirement for ATP synthesis in the reverse direction is bypassed by a simple hydrolytic reaction, catalyzed by glucose-6-phosphatase and fructose-1,6-bisphosphatase, respectively (see Figure 13-12). Notice how effectively this simple metabolic ploy overcomes the thermodynamic hurdle. In the case of the interconversion of glucose and glucose-6-phosphate, for example, the reaction is exergonic in the glycolytic direction because of the input of an ATP molecule. In other words, the reaction is driven in that direction by the difference in the free energies of hydrolysis of the phosphoanhydride bond of ATP and the phosphoester bond of glucose-6-phosphate ($\Delta G^{\circ\prime}$ values of −7.3 and −3.3 kcal/mol, respectively; see Table 13-1 and reaction 13-5). And in the gluconeogenic direction, exergonicity is ensured by the simple hydrolysis of the phosphoester bond, which has a $\Delta G^{\circ\prime}$ of −3.3 kcal/mol.

The third site of irreversibility in the glycolytic pathway, reaction Gly-10, is bypassed in gluconeogenesis by a two-reaction sequence (Figure 13-12). Both of these reactions are driven by the hydrolysis of a phosphoanhydride bond, from ATP in one case and from the related compound GTP in the other. (GTP is the abbreviation for guanosine triphosphate; see Figure 3-15 for the structure of guanine.) The first of these two steps in the gluconeogenic direction involves the addition of CO_2 to pyruvate—a *carboxylation* reaction—to form a four-carbon compound called oxaloacetate, which we will meet again in the next chapter. In the second step, the carboxyl group is removed—a *decarboxylation* reaction—to form phosphoenolpyruvate (PEP). In this case, both the phosphate group and the energy are provided by GTP, which is energetically the equivalent of ATP. (GTP and ATP have identical $\Delta G^{\circ\prime}$ values for hydrolysis of their respective terminal phosphate groups.) The enzymes that catalyze these reactions are called *pyruvate carboxylase* and *PEP carboxykinase*, respectively.

What these bypass reactions accomplish becomes clear when the glycolytic and gluconeogenic pathways are compared directly (see Figure 13-12). Glycolysis uses two ATPs but generates four ATPs, for a net yield of two ATP molecules formed per molecule of glucose catabolized. Gluconeogenesis, on the other hand, requires four ATPs and two GTPs per glucose, or the equivalent of six ATP molecules

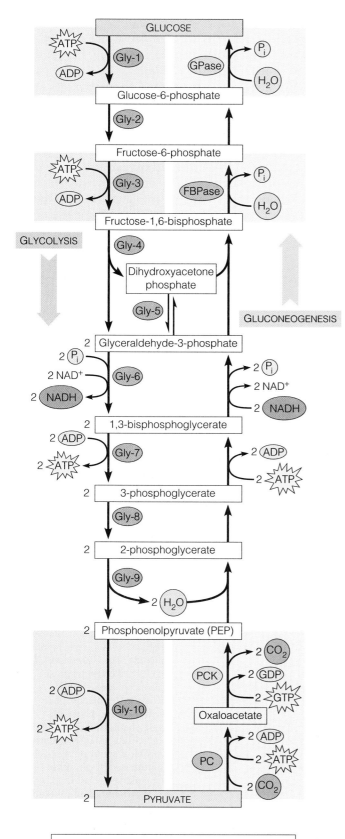

Figure 13-12 Pathways for Glycolysis and Gluconeogenesis Compared. The pathways for glycolysis (left) and gluconeogenesis (right) have nine intermediates and seven enzyme-catalyzed reactions in common. The three essentially irreversible reactions of the glycolytic pathway (in green shading) are circumvented in gluconeogenesis by four bypass reactions (in yellow shading). As a catabolic pathway, glycolysis is inherently exergonic, capable of yielding two ATPs per glucose. Gluconeogenesis, on the other hand, is an anabolic pathway, requiring the coupled hydrolysis of six phosphoanhydride bonds (four from ATP, two from GTP) to drive it in the direction of glucose formation. The enzymes that catalyze the bypass reactions are shown in gold and are identified in the box. (For the names of the glycolytic enzymes, see Figure 13-6.) Glycolysis occurs in muscle and various other tissues, whereas gluconeogenesis occurs mainly in the liver.

Enzymes That Catalyze the Bypass Reactions of Gluconeogenesis

PC: Pyruvate carboxylase
PCK: Phosphoenolpyruvate carboxykinase
FBPase: Fructose-1,6-bisphosphatase
GPase: Glucose-6-phosphatase

consumed per molecule of glucose synthesized. The difference of four ATP molecules per glucose represents enough energy to ensure that gluconeogenesis proceeds at least as exergonically in the direction of glucose synthesis as glycolysis does in the direction of glucose breakdown.

The Regulation of Glycolysis and Gluconeogenesis

If cells have enzymes to catalyze the reactions of both the glycolytic and the gluconeogenic pathways, it seems reasonable to ask what keeps both pathways from proceeding at the same time in the same cell, which would obviously be a futile metabolic exercise. How, we may ask, do cells regulate the synthesis and breakdown of glucose? The answer to that question has two parts, at least for humans and other mammals. Part of the answer lies in the differences in metabolic capabilities of cells from different parts of the body. Box 13A takes up this topic by tracing the fate of sugar molecules in the human body, including glycolysis in muscle (and other) cells and gluconeogenesis in liver and kidney cells. Here, we will consider the other part of the answer by asking how glycolysis and gluconeogenesis are regulated within the same cell.

Like all metabolic pathways, glycolysis and gluconeogenesis are regulated to function at rates that are responsive to cellular and organismal needs for their products, which are ATP and glucose, respectively. Not surprisingly, glycolysis and gluconeogenesis are regulated in a reciprocal, or inverse, manner: Intracellular conditions known to stimulate one pathway usually have an inhibitory effect on the other. In addition, glycolysis is closely coordinated with other major pathways of energy generation and utilization in the cell, especially the pathways involved in aerobic respiration that we will be considering in Chapter 14.

Most of the regulation of glycolysis and gluconeogenesis in animal cells involves one or both of two major control mechanisms—*allosteric regulation* and *hormonal regulation*. Allosteric regulation affects enzyme activity and is an intracellular mechanism. Hormonal regulation, on the other hand, is an intrinsically organismal mechanism because the initiating signal is a hormone produced in another, often distant, part of the body. We will consider allosteric regulation here, focusing specifically on liver cells, which possess the enzymes for both glycolysis and gluconeogenesis. Hormonal regulation will also be mentioned, but only briefly because it has already been discussed in the context of signal transduction in Chapter 10 (see Figures 10-22 and 10-25).

Key Enzymes in the Glycolytic and Gluconeogenic Pathways Are Subject to Allosteric Regulation

Recall from Chapter 6 that **allosteric regulation** of enzyme activity involves the interconversion of an enzyme between two forms, one of which is catalytically active (or more active), whereas the other is inactive (or less active). Whether an enzyme molecule is in its active or inactive form depends on whether a specific allosteric effector is bound to the allosteric site, and whether that effector is an allosteric activator or an allosteric inhibitor (see Figure 6-17). Every allosterically regulated enzyme has one or more catalytic subunits with an active site on each and one or more regulatory subunits containing the allosteric site(s) to which the effector(s) bind.

Figure 13-13 shows the key regulatory enzymes of the glycolytic and gluconeogenic pathways and the allosteric effectors that regulate each enzyme. For glycolysis, the key enzymes are hexokinase, phosphofructokinase-1 (PFK-1), and pyruvate kinase. For gluconeogenesis, fructose-1,6-bisphosphatase and pyruvate carboxylase are the key regulatory enzymes. Based primarily on studies with liver cells, each allosteric effector shown in Figure 13-13 is identified as either an activator (+) or an inhibitor (−) of the enzyme(s) to which it binds. Several points quickly become apparent from the figure. Notice, for instance, that each of the regulatory enzymes is unique to its pathway, thereby making it possible for each pathway to be regulated independently of the other. Notice also the reciprocal nature of the regulation of the two pathways: AMP and acetyl CoA, the two effectors to which both pathways are sensitive, have opposite effects in the two directions. AMP, for example, activates glycolysis but

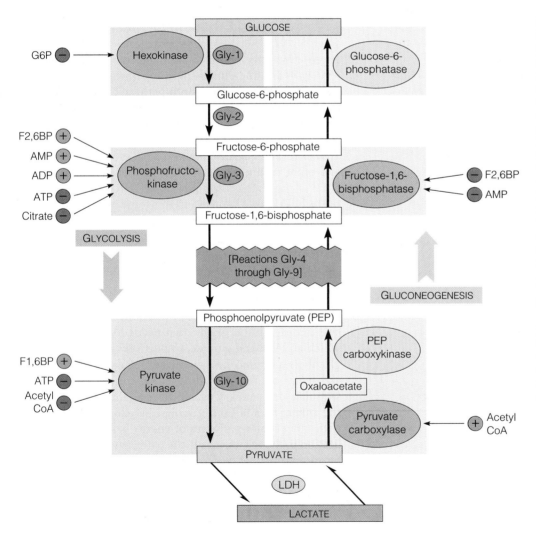

Figure 13-13 The Regulation of Glycolysis and Gluconeogenesis. Glycolysis and gluconeogenesis are regulated in a reciprocal manner. In both cases, regulation involves allosteric activation (+) or inhibition (−) of enzymes that catalyze reactions unique to the pathway. For glycolysis, the key regulatory enzymes are those that catalyze the three irreversible reactions unique to this pathway (green). For gluconeogenesis, two of the four bypass enzymes (gold) that are unique to this pathway are the main sites of allosteric regulation. Allosteric regulators include acetyl CoA, AMP, ADP, ATP, citrate, fructose-1,6-bisphosphate (F1,6BP), fructose-2,6-bisphosphate (F2,6BP), and glucose-6-phosphate (G6P). Acetyl CoA and citrate are intermediates in aerobic respiration. F2,6BP is synthesized by phosphofructokinase-2 (PFK-2), as shown in Figure 13-14a.

inhibits gluconeogenesis, whereas acetyl CoA activates gluconeogenesis but inhibits glycolysis.

Moreover, the effects of the regulatory agents make sense—that is, they are invariably in the direction one would predict, based on an understanding of the role each pathway plays in the cell. Consider, for example, the effects of ATP, ADP, and AMP, which can be thought of as the charged, discharged, and very discharged forms of the adenosine phosphates, with two, one, and zero phospho-anhydride bonds, respectively. When the concentration of ATP is low and the ADP and/or AMP concentrations are high, the cell is clearly low on energy, so it is reasonable for ADP and AMP to activate glycolysis. Conversely, as the ATP concentration increases and the ADP and/or AMP concentrations decrease, the stimulatory effects of AMP and ADP on glycolysis lessen, and the inhibitory effect of ATP on both PFK-1 and pyruvate kinase comes into play, reducing the rate of glycolysis appropriately.

You may be surprised to learn that ATP is an allosteric inhibitor of PFK-1 because this enzyme uses ATP as a substrate. This seems contradictory because increases in substrate concentration should *increase* the rate of an enzyme-catalyzed reaction, yet increasing the concentration of an allosteric inhibitor should *decrease* enzyme activity by converting the enzyme to its less active form. This apparent contradiction is readily explained, however. As an allosteric enzyme, PFK-1 has both an active site and at least one kind of effector site. The active site of PFK-1 has a high affinity (i.e., a low K_m) for ATP, whereas the affinity of the effector site for ATP is lower. Thus, at low ATP concentrations, binding occurs at the catalytic site but not to any appreciable extent at the allosteric site, so most of the PFK-1 molecules remain in the active form and glycolysis proceeds. As the ATP concentration increases, however, binding is enhanced at the effector site, converting more and more of the PFK-1 molecules to the inactive form and thereby serving effectively as a throttle for the whole glycolytic sequence.

Both pathways shown in Figure 13-13 are also subject to allosteric regulation by compounds involved in respiration. As we will learn in the next chapter, acetyl CoA and citrate are key players in an aerobic pathway called the *tricarboxylic acid cycle*. (They are, in fact, a substrate and a product, respectively, of the first reaction of that sequence.) High levels of acetyl CoA and citrate indicate that the cell is well supplied with substrate for the next phase of respiratory metabolism beyond pyruvate. Thus, it is not surprising to find that acetyl CoA and citrate both have inhibitory effects on a glycolytic enzyme (pyruvate kinase and PFK-1, respectively), thereby decreasing the rate at which pyruvate is formed. Similarly, the stimulatory effect of acetyl CoA on a gluconeogenic enzyme is consistent with the availability of pyruvate for conversion to glucose.

Fructose-2,6-Bisphosphate Is an Important Regulator of Glycolysis and Gluconeogenesis

Although each of the preceding mechanisms plays a significant role in the regulation of glycolysis and gluconeogenesis, the most important regulator of both pathways is **fructose-2,6-bisphosphate (F2,6BP)**. F2,6BP is synthesized by ATP-dependent phosphorylation of fructose-6-phosphate (Figure 13-14a), the same reaction that gives rise to fructose-1,6-bisphosphate at reaction Gly-3 in the

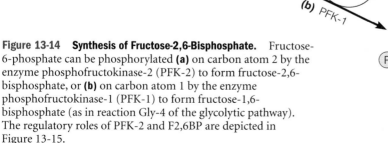
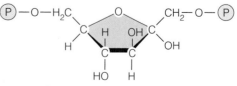

Figure 13-14 Synthesis of Fructose-2,6-Bisphosphate. Fructose-6-phosphate can be phosphorylated **(a)** on carbon atom 2 by the enzyme phosphofructokinase-2 (PFK-2) to form fructose-2,6-bisphosphate, or **(b)** on carbon atom 1 by the enzyme phosphofructokinase-1 (PFK-1) to form fructose-1,6-bisphosphate (as in reaction Gly-4 of the glycolytic pathway). The regulatory roles of PFK-2 and F2,6BP are depicted in Figure 13-15.

glycolytic pathway (Figure 13-14b). However, synthesis of F2,6BP is catalyzed by a separate form of phosphofructokinase, which is called **phosphofructokinase-2 (PFK-2)** to distinguish it from PFK-1, the glycolytic enzyme. As Figure 13-13 indicates, F2,6BP activates the glycolytic enzyme (PFK-1) that phosphorylates fructose-6-phosphate, and it inhibits the gluconeogenic enzyme (FBPase) that catalyzes the reverse reaction.

Figure 13-15 depicts these regulatory roles of PFK-2 and F2,6BP in more detail. The activity of PFK-2 depends on the phosphorylation status of one of its subunits. The kinase activity of the enzyme is high when that subunit is in the unphosphorylated form and low when it is phosphorylated. The phosphorylation of PFK-2 by ATP is catalyzed by a protein kinase (Figure 13-15a). The activity of this enzyme depends, in turn, on cyclic AMP (cAMP), which is a key intermediate in many cellular signal transduction pathways (see Figure 10-5).

In addition to the PFK activity responsible for phosphorylation of F2,6BP, another enzyme activity, initially called *fructose-2,6-bisphosphatase,* was found to cleave the phosphate group from F2,6BP, converting the compound back to fructose-6-phosphate (see Figure 13-15). This activity is also regulated by cAMP-stimulated phosphorylation, but in this case phosphorylation increases the activity of the enzyme. We now understand that both of these activities—the kinase that phosphorylates fructose-6-phosphate and the phosphatase that removes the phosphate group—are in fact properties of the same enzyme, PFK-2. Because it has two separate catalytic activities, PFK-2 is called a *bifunctional enzyme.* When the subunit with the phosphorylation site is in the nonphosphorylated form, the bifunctional enzyme acts as a kinase, forming F2,6BP from fructose-6-phosphate (Figure 13-15b). When the subunit is phosphorylated, the enzyme is active as a phosphatase, cleaving F2,6BP to fructose-6-phosphate (Figure 13-15c).

As noted earlier, F2,6BP activates the glycolytic enzyme PFK-1 (Figure 13-15d) and inhibits the gluconeogenic enzyme FBPase (Figure 13-15e). cAMP, in turn, affects the F2,6BP concentration in two ways: It inactivates the kinase activity and stimulates the phosphatase activity. Both of these effects tend to decrease the concentration of F2,6BP in the cell. This change leads, in turn, to less stimulation of PFK-1 and less inhibition of fructose-1,6-bisphosphatase, thereby decreasing the glycolytic flux and increasing the gluconeogenic flux.

The effects of cAMP shown in Figure 13-15 provide a link to the topic of hormonal regulation because the cAMP level in liver cells is controlled primarily by the hormones glucagon and epinephrine (adrenalin). These hormones cause an increase in cAMP concentration, thereby stimulating gluconeogenesis. Moreover, the increase in cAMP concentration also stimulates a regulatory cascade that increases the rate of glycogen breakdown (see Figure 10-25). Not surprisingly, the effect of cAMP on glycogen synthesis is just the opposite: Whether triggered by glucagon or epinephrine, an increase in liver

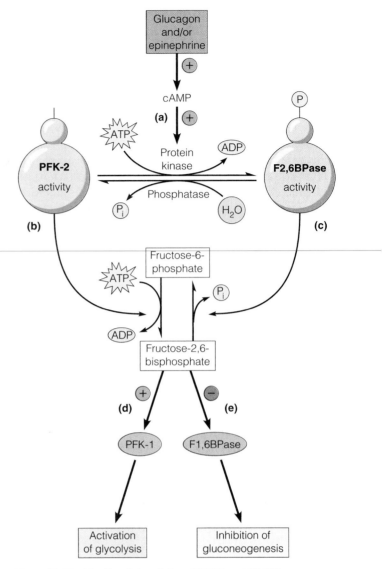

Figure 13-15 The Regulatory Roles of PFK-2 and F2,6BP. Phosphofructokinase-2 (PFK-2) is a bifunctional enzyme, with two different catalytic activities. **(a)** PFK-2 is phosphorylated by a cAMP-activated protein kinase and is dephosphorylated by a phosphatase. **(b)** In its unphosphorylated form, PFK-2 has phosphofructokinase activity, catalyzing the phosphorylation of fructose-6-phosphate to form fructose-2,6-bisphosphate (F2,6BP). **(c)** In its phosphorylated form, the enzyme acts as a phosphatase, catalyzing the hydrolysis of the phosphate group from carbon atom 2 of F2,6BP. F2,6BP is **(d)** an allosteric activator of the glycolytic enzyme PFK-1 and **(e)** an allosteric inhibitor of the gluconeogenic enzyme fructose-1,6-bisphosphatase. Hormonal regulation of cAMP by glucagon and epinephrine is shown at the top of the figure. The effect of these hormones is to increase the cAMP level, thereby stimulating the phosphorylation of PFK-2 and converting it to the form that has fructose-2,6-bisphosphatase activity. The resulting decrease in the concentration of F2,6BP eliminates the activation of glycolysis and alleviates the inhibition of gluconeogenesis.

cAMP concentration leads to a decrease in the rate of glycogen formation. For further details on hormonal regulation and the role of cAMP in mediating hormonal effects, review the discussion on hormonal signal transduction in Chapter 10.

Perspective

Metabolic pathways in cells are either anabolic (synthetic) or catabolic (degradative). The latter reactions provide the energy necessary to drive the former. ATP is a useful intermediate for this purpose because its terminal anhydride bond has a free energy of hydrolysis that allows ATP to serve as a donor, and ADP to serve as an acceptor, of phosphate groups. Most chemotrophs derive the energy needed for ATP generation from the catabolism of organic nutrients such as carbohydrates, fats, and proteins. They do so either by fermentative processes in the absence of oxygen or by respiratory metabolism, which is usually, though not always, an aerobic process.

Using glucose as a prototype substrate, catabolism under both anaerobic and aerobic conditions begins with glycolysis, a ten-step pathway that converts glucose into pyruvate with the net production of two molecules of ATP per molecule of glucose. In the absence of oxygen, the reduced coenzyme NADH generated during glycolysis must be reoxidized at the expense of pyruvate, leading to fermentation end-products such as lactate or ethanol and carbon dioxide. This severely limits the extent to which the free energy content of the glucose molecule can be released, but the 7% or so that is available is conserved as ATP quite efficiently. Although usually written with glucose as the starting substrate, the glycolytic sequence is also the mainstream pathway for the catabolism of a variety of related sugars, as well as for the utilization of the glucose-1-phosphate derived by phosphorolytic cleavage of storage polysaccharides such as starch or glycogen.

Gluconeogenesis is, in a sense, the opposite of glycolysis because it is the pathway whereby animal cells synthesize glucose (and other carbohydrates) from three- and four-carbon noncarbohydrate starting materials such as pyruvate. However, the gluconeogenic pathway is not just glycolysis in reverse. The two pathways share seven enzyme-catalyzed reactions in common, but the three most exergonic reactions of glycolysis are bypassed in gluconeogenesis by reactions that render the pathway exergonic in the gluconeogenic direction by the input of energy from ATP and GTP.

Like other metabolic pathways, glycolysis and gluconeogenesis are regulated to ensure that the rate of product formation (ATP and glucose, respectively) is carefully tuned to cellular needs. Both allosteric and hormonal regulation are involved. The enzymes that are subject to allosteric regulation catalyze reactions unique to the respective pathways. These enzymes are regulated by one or more effectors, which include ATP, ADP, and AMP, as well as acetyl CoA and citrate, key intermediates in aerobic respiration. In animal cells, the most important allosteric regulator of both glycolysis and gluconeogenesis is fructose-2,6-bisphosphate, the concentration of which depends on the relative kinase and phosphatase activities of the bifunctional enzyme PFK-2. The function of PFK-2 is regulated, in turn, by the hormones glucagon and epinephrine, mediated by the intracellular concentration of cyclic AMP.

As complex as glycolysis may seem upon first encounter, it represents the simplest mechanism by which glucose can be degraded in dilute solution at temperatures compatible with life and with a large portion of the free energy yield conserved as ATP. Coupled to an appropriate reductive sequence to regenerate the coenzyme NAD$^+$, glycolysis serves the cell well under anaerobic conditions, meeting energy needs despite the absence of oxygen. All we have seen so far, however, pales in comparison with the potential for energy release and conservation in the presence of oxygen, for aerobic respiration is the capstone of bioenergetics and the mainspring of cellular energy metabolism for most chemotrophic forms of life. For that, we take the NADH and the pyruvate provided by glycolysis and proceed to aerobic respiration, the subject of Chapter 14.

Key Terms for Self-Testing

Metabolic Pathways
metabolism (p. 368)
metabolic pathway (p. 368)
anabolic pathway (p. 369)
catabolic pathway (p. 369)

ATP: The Universal Energy Coupler
adenosine triphosphate (ATP) (p. 369)
phosphoanhydride bond (p. 369)
phosphoester bond (p. 369)
high-energy compound (p. 370)
charge repulsion (p. 370)
resonance stabilization (p. 370)
resonance hybrid (p. 370)
group transfer reaction (p. 372)

Chemotrophic Energy Metabolism
chemotrophic energy metabolism (p. 373)
oxidation (p. 373)
dehydrogenation (p. 373)
reduction (p. 374)
hydrogenation (p. 374)
coenzyme (p. 374)
nicotinamide adenine dinucleotide (NAD$^+$)
  (p. 374)
glucose (p. 375)
aerobic respiration (p. 375)
fermentation (p. 375)
obligate aerobe (p. 378)
obligate anaerobe (p. 378)
facultative organism (p. 378)

Glycolysis and Fermentation
glycolysis (glycolytic pathway) (p. 378)
substrate-level phosphorylation (p. 381)
lactate fermentation (p. 384)
alcoholic fermentation (p. 385)

Gluconeogenesis
gluconeogenesis (p. 386)

The Regulation of Glycolysis and Gluconeogenesis
allosteric regulation (p. 390)
fructose-2,6-bisphosphate (F2,6BP)
  (p. 391)
phosphofructokinase-2 (PFK-2) (p. 392)

Problem Set

More challenging problems are marked with a •.

13-1. **"High-Energy Bonds."** When first introduced by Fritz Lipmann in 1941, the term "high-energy bond" was considered a useful concept for describing the energetics of biochemical molecules and reactions. However, the term can lead to confusion when relating ideas about cellular energy metabolism to those of physical chemistry. To check out your own understanding, indicate whether each of the following statements is true (T) or false (F). If false, reword the statement to make it true.

(a) Energy is stored in special high-energy bonds in molecules such as ATP and is released when these bonds are broken.

(b) Energy is always released whenever a covalent bond is formed and is always required to break a covalent bond.

(c) To a physical chemist, high-energy bond means a very stable bond that requires a lot of energy to break, whereas to a biochemist, the term is likely to mean a bond that releases a lot of energy upon hydrolysis.

(d) The terminal phosphate of the ATP is a high-energy phosphate that takes its high energy with it when it is hydrolyzed.

(e) Phosphoester bonds are low-energy bonds because they require less energy to break than the high-energy bonds of phosphoanhydrides.

(f) The term high-energy molecule should be thought of as a characteristic of the reaction in which the molecule is involved and not as an intrinsic property of a particular bond within that molecule.

13-2. **The History of Glycolysis.** Following are several historical observations that led to the elucidation of the glycolytic pathway. In each case, suggest a metabolic basis for the observed effect, and explain the significance of the observation for the elucidation of the pathway.

(a) Alcoholic fermentation in yeast extracts requires a heat-labile fraction originally called *zymase* and a heat-stable fraction (*cozymase*) that is necessary for the activity of zymase.

(b) Alcoholic fermentation does not take place in the absence of inorganic phosphate.

(c) In the presence of iodoacetate, a known inhibitor of glycolysis, fermenting yeast extracts accumulate a doubly phosphorylated hexose.

(d) In the presence of fluoride ion, another known glycolytic inhibitor, fermenting yeast extracts accumulate two phosphorylated three-carbon acids.

13-3. **What's Your Reaction?** Table 13-1 indicates the free energy of hydrolysis for a variety of phosphorylated compounds commonly found in cells. Use the information in the table and your familiarity with $\Delta G°'$ and $\Delta G'$ from Chapter 5 to answer the following questions.

(a) For each of the following compounds, indicate whether the transfer of its phosphate group to ADP to form ATP would be exergonic or endergonic, assuming a temperature of 25°C and conditions such that $\Delta G' = \Delta G°'$: phospho-

enolpyruvate; glucose-1-phosphate; phosphocreatine; and glycerol phosphate.

(b) If equimolar concentrations of ADP, ATP, 1,3-bisphosphoglycerate, and 3-phosphoglycerate are mixed together at 25°C with an appropriate amount of the enzyme phosphoglycerokinase, what reaction do you predict will occur, and how exergonic will it be in the direction written? Explain your answer.

(c) If equimolar concentrations of ADP, phosphoenolpyruvate, and glucose are mixed together at 25°C with appropriate amounts of the enzymes pyruvate kinase and hexokinase, what reaction(s) do you predict will occur? Explain your answer.

(d) If the mixture in part c is allowed to come to equilibrium, what compounds will be present in the equilibrium mixture? Explain your answer.

(e) The conversion of glycerol to glycerol phosphate requires the transfer of a phosphate group from a compound with a suitable $\Delta G'$ for its hydrolysis. In most cells, the donor is ATP. Would 1,3-bisphosphoglycerate (BPG) be a possible donor also? If you answer no, explain why not. If you answer yes, suggest a reason that this reaction does not actually occur in cells. (Base your reason on bioenergetics in either case.)

(f) Under conditions such that $\Delta G' = \Delta G°'$, would the transfer of a phosphate group from 1,3-bisphosphoglycerate (BPG) to ADP be able to drive the synthesis of glycerol phosphate from free glycerol and inorganic phosphate, assuming the existence of an appropriate coupling mechanism? Making the same assumptions, would phosphate transfer from BPG to ADP be able to drive the synthesis of glucose-1-phosphate from free glucose and inorganic phosphate? Explain your answer in both cases.

13-4. **Glycolysis in 25 Words or Less.** Complete each of the following statements about the glycolytic pathway in 25 words or less.

(a) Although the brain is an obligately aerobic organ, it still depends on glycolysis because…

(b) Although one of its reactions is an oxidation, glycolysis can proceed in the absence of oxygen because…

(c) What happens to the pyruvate generated by the glycolytic pathway depends on…

(d) If you bake bread or brew beer, you depend on glycolysis for…

(e) Two organs in your body that can use lactate are…

(f) The synthesis of glucose from lactate in a liver cell requires more molecules of nucleoside triphosphates (ATP and GTP) than are formed during the catabolism of glucose to lactate in a muscle cell because…

13-5. **Energetics of Carbohydrate Utilization.** The anaerobic fermentation of free glucose has an ATP yield of 2 ATP molecules per molecule of glucose. For glucose units in a glycogen molecule, the yield is 3 ATP molecules per molecule of glucose. The corresponding value for the disaccharide sucrose is 2 ATP molecules per molecule of monosaccharide if the sucrose is eaten by an animal, but 2.5 ATP molecules per molecule of monosaccharide if the sucrose is metabolized by a bacterium.

(a) Explain why the glucose units present in glycogen have a higher ATP yield than free glucose molecules.

(b) What is the likely mechanism for sucrose breakdown in the gut of an animal to explain the energy yield of 2 ATP molecules per molecule of monosaccharide?

(c) Based on what you know about the process of glycogen breakdown, suggest a mechanism for bacterial sucrose metabolism that is consistent with an energy yield of 2.5 ATP molecules per molecule of monosaccharide.

(d) What energy yield (in molecules of ATP per molecule of monosaccharide) would you predict for the bacterial catabolism of raffinose, a trisaccharide?

13-6.   Glucose Phosphorylation. The direct phosphorylation of glucose by inorganic phosphate is a thermodynamically unfavorable reaction:

$$\text{glucose} + \text{P}_i \longrightarrow \text{glucose-6-phosphate} + \text{H}_2\text{O}$$
$$\Delta G^{\circ\prime} = +3.3 \text{ kcal/mol} \qquad \text{(13-21)}$$

In the cell, glucose phosphorylation is accomplished by coupling the reaction to the hydrolysis of ATP, a highly exergonic reaction:

$$\text{ATP} + \text{H}_2\text{O} \longrightarrow \text{ADP} + \text{P}_i$$
$$\Delta G^{\circ\prime} = -7.3 \text{ kcal/mol} \qquad \text{(13-22)}$$

Typical concentrations of these intermediates in yeast cells are as follows:

$[\text{glucose-6-phosphate}] = 0.08 \text{ m}M$ $[\text{ATP}] = 1.8 \text{ m}M$
$[\text{P}_i] = 1.0 \text{ m}M$ $[\text{ADP}] = 0.15 \text{ m}M$

Assume a temperature of 25°C for all calculations.

(a) What minimum concentration of glucose would have to be maintained in a yeast cell for direct phosphorylation (reaction 13-21) to be thermodynamically spontaneous? Is this physiologically reasonable? Explain your reasoning.

(b) What is the overall equation for the coupled (ATP-driven) phosphorylation of glucose? What is its $\Delta G^{\circ\prime}$ value?

(c) What minimum concentration of glucose would have to be maintained in a yeast cell for the coupled reaction to be thermodynamically spontaneous? Is this physiologically reasonable?

(d) By about how many orders of magnitude is the minimum required glucose concentration reduced when the phosphorylation of glucose is coupled to the hydrolysis of ATP?

(e) Assuming a yeast cell to have a glucose concentration of 5.0 mM, what is $\Delta G^{\prime}$ for the coupled phosphorylation reaction?

13-7.   Ethanol Intoxication and Methanol Toxicity. The enzyme alcohol dehydrogenase was mentioned in this chapter because of its role in the final step of alcoholic fermentation. However, the enzyme also occurs commonly in aerobic organisms, including humans. The ability of the human body to catabolize the ethanol in alcoholic beverages depends on the presence of alcohol dehydrogenase in the liver. One of the

effects of ethanol intoxication is a dramatic decrease in the NAD^+ concentration in liver cells, which decreases the aerobic utilization of glucose. Methanol, on the other hand, is not just an intoxicant; it is a deadly poison because of the toxic effect of the formaldehyde to which it is converted in the liver.

(a) Why does ethanol consumption lead to a reduction in NAD^+ concentration and to a decrease in aerobic respiration?

(b) Most of the unpleasant effects of hangovers result from an accumulation of acetaldehyde and its metabolites. Where does the acetaldehyde come from?

(c) The medical treatment for methanol poisoning usually involves administration of large doses of ethanol. Why is this treatment effective?

• 13-8.   Propionate Fermentation. Although lactate and ethanol are the best-known products of fermentation, other pathways are also known, some with important commercial applications. Swiss cheese production, for example, depends on the bacterium *Propionibacterium freudenreichii*, which converts pyruvate to propionate ($\text{CH}_3\text{—CH}_2\text{—COO}^-$). Fermentation of glucose to propionate always generates at least one other product as well.

(a) Why is it not possible to devise a scheme for the fermentation of glucose with propionate as the sole end-product?

(b) Suggest an overall scheme for propionate production that generates only one additional product, and indicate what that product might be.

(c) If you know that Swiss cheese production actually requires both propionate and carbon dioxide and that both are produced by *Propionibacterium* fermentation, what else can you now say about the fermentation process that this bacterium carries out?

13-9.   Galactose Metabolism. The glycolytic pathway is usually written with glucose as the starting substrate because it is the single most important sugar for most organisms. Cells can utilize a variety of sugars, however. The diet of a young mammal, for example, consists almost entirely of milk, which contains as its principal carbohydrate the disaccharide lactose. When lactose is hydrolyzed in the intestine, it yields one molecule each of the hexoses glucose and galactose, so the cells of these animals (and of human babies) must metabolize just as much galactose as glucose. Galactose is metabolized by phosphorylation and conversion to glucose. The reaction sequence is a bit complicated, though, because the conversion to glucose (an epimerization reaction on carbon atom 4) occurs while the sugar is attached to the carrier uridine diphosphate (UDP), a close relative of ADP. The reactions are as follows:

$$\text{galactose} + \text{ATP} \longrightarrow \text{galactose-1-phosphate} + \text{ADP} \quad \textbf{(13-23)}$$

$$\text{galactose-1-phosphate} + \text{UDP-glucose} \longrightarrow$$
$$\text{glucose-1-phosphate} + \text{UDP-galactose} \quad \textbf{(13-24)}$$

$$\text{UDP-galactose} \longrightarrow \text{UDP-glucose} \quad \textbf{(13-25)}$$

$$\text{glucose-1-phosphate} \longrightarrow \text{glucose-6-phosphate} \quad \textbf{(13-26)}$$

(a) Write a reaction for the overall conversion of galactose to glucose-6-phosphate.

(b) How do you think the $\Delta G^{\circ\prime}$ value for the overall reaction of part a compares with that for the hexokinase reaction (reaction Gly-1)?

(c) If you know that the epimerase reaction (reaction 13-25) has an absolute requirement for the coenzyme NAD^+ and involves 4-ketoglucose as an enzyme-bound intermediate, can you suggest a reaction sequence to explain the conversion of galactose to glucose? (Use Haworth projections in this case.)

(d) One form of the congenital disease galactosemia is caused by a genetic absence of the enzyme that catalyzes reaction 13-24. The symptoms of galactosemia, which include mental disorders and cataracts of the eye, are thought to result from high levels of galactose in the blood and from an intracellular accumulation of galactose-1-phosphate. Why does this seem a reasonable hypothesis?

(e) At least one other form of galactosemia is known, also caused by the genetic absence of a specific enzyme. Which enzyme do you think is missing? Explain your reasoning.

13-10. Glycolysis and Gluconeogenesis. As Figure 13-12 indicates, gluconeogenesis is accomplished by what is essentially the reverse of the glycolytic pathway but with bypass reactions in place of the first, third, and tenth reactions in glycolysis.

(a) Explain why it is not possible to accomplish gluconeogenesis by a simple reversal of all the reactions in glycolysis.

(b) Write an overall reaction for gluconeogenesis that is comparable to reaction 13-16 for glycolysis.

(c) Explain why gluconeogenesis requires the input of six molecules of nucleoside triphosphates (four ATPs and two GTPs) per molecule of glucose synthesized, whereas glycolysis only yields two molecules of ATP per molecule of glucose.

(d) Assuming concentrations of ATP, ADP, and P_i are such that $\Delta G^\prime$ for the hydrolysis of ATP is about -10 kcal/mol, what is the approximate $\Delta G^\prime$ value for the overall reaction for gluconeogenesis that you wrote in part b?

(e) With all of the enzymes for glycolysis and gluconeogenesis present in a liver cell, how does the cell "know" whether it should be synthesizing or catabolizing glucose at any given time?

• 13-11. Trypanosomes, Glycosomes, and the Compartmentalization of Glycolysis. In most organisms, all of the glycolytic enzymes occur in the cytosol. However, in certain parasitic protozoa known as *trypanosomes,* the enzymes that catalyze the first seven steps of glycolysis are compartmentalized in membrane-bounded organelles called *glycosomes.*

(a) What experimental evidence most likely led to the discovery that seven of the ten glycolytic enzymes are localized in an organelle rather than in the cytosol?

(b) What benefit do you think trypanosomes derive from this compartmentalization of glycolysis?

(c) What specific transport proteins do you predict are present in the glycosomal membrane? Explain.

(d) Glycosomes are usually regarded as a specialized kind of peroxisome. What other enzymes would you therefore expect to find in this organelle? Explain.

• 13-12. You've Got Some Explaining to Do. Explain each of the following observations:

(a) In his classic studies of glucose fermentation by yeast cells, Louis Pasteur observed that the rate of glucose consumption by yeast cells was much higher under anaerobic conditions than under aerobic conditions.

(b) In 1905, Arthur Harden and William Young found that addition of inorganic phosphate to a yeast extract stimulated and prolonged the fermentation of glucose.

(c) An alligator is normally very sluggish but is capable of very rapid movements of its legs, jaws and tail if provoked. However, such bursts of activity must be followed by long periods of recovery.

(d) Fermentation of glucose to lactate is an energy-yielding process despite the fact that it involves no net oxidation (i.e., even though the oxidation of glyceraldehyde-3-phosphate to glycerate is accompanied by the reduction of pyruvate to lactate and no net accumulation of NADH occurs).

• 13-13. Arsenate Poisoning. Arsenate ($HAsO_4^{2-}$) is a potent poison to almost all living systems. Among other effects, arsenate is known to uncouple the phosphorylation event from the oxidation of glyceraldehyde-3-phosphate. This uncoupling occurs because the enzyme involved, glyceraldehyde-3-phosphate dehydrogenase, can utilize arsenate instead of inorganic phosphate, forming glycerate-1-arseno-3-phosphate. This product is a highly unstable compound that immediately undergoes nonenzymatic hydrolysis into glycerate-3-phosphate and free arsenate.

(a) In what sense might arsenate be called an *uncoupler* of substrate-level phosphorylation?

(b) Why is arsenate such a toxic substance for an organism that depends critically on glycolysis to meet its energy needs?

(c) Can you think of other reactions that are likely to be uncoupled by arsenate in the same way as the glyceraldehyde-3-phosphate dehydrogenase reaction?

• 13-14. Regulation of Phosphofructokinase-1. Shown in Figure 13-16 are plots of initial reaction velocity (expressed as % of V_{max}) vs. fructose-6-phosphate concentration for liver phosphofructokinase PFK-1 in the presence and absence of fructose-2,6-bisphosphate (F2,6BP) (Figure 13-16a) and in the presence of a low or high concentration of ATP (Figure 13-16b).

(a) Explain the effect of F2,6BP on enzyme activity as shown in Figure 13-16a.

(b) Explain the effect of the ATP concentration on the data shown in Figure 13-16b.

(c) What assumptions do you have to make about the concentration of ATP in Figure 13-16a and about the concentration of F2,6BP in Figure 13-16b? Explain.

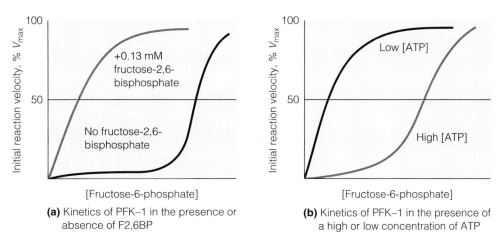

(a) Kinetics of PFK–1 in the presence or absence of F2,6BP

(b) Kinetics of PFK–1 in the presence of a high or low concentration of ATP

Figure 13-16   Allosteric Regulation of Phosphofructokinase-1.   Shown here are Michaelis-Menten plots of liver phosphofructokinase (PFK-1) activity, depicting **(a)** the dependence of initial reaction velocity on concentration of the substrate fructose-6-phosphate in the presence (red line) or absence (black line) of fructose-2,6-bisphosphate, and **(b)** the dependence of initial reaction velocity on fructose-6-phosphate concentration at high (red line) or low (black line) ATP concentrations. In both cases, initial reaction velocity is expressed as a percentage of V_{max}, the maximum velocity. See Problem 13-14.

Suggested Reading

References of historical importance are marked with a •.

General References

Lehninger, A. L., D. L. Nelson, and M. M. Cox. *Principles of Biochemistry*, 3d ed. New York: Worth, 1999.

Mathews, C. K., and K. E. van Holde. *Biochemistry*, 2d ed. Menlo Park, CA: Benjamin/Cummings, 1996.

Metzler, D.E. *Biochemistry: The Chemical Reactions of Living Cells.* 2d ed. San Diego: Academic Press, 2001.

ATP and ATP Generation

deMeis, L. The concept of energy-rich phosphate compounds: Water, transport ATPases, and entropic energy. *Arch. Biochem. Biophys.* 306 (1993): 287.

• Hinkle, P. C., and R. E. McCarty. How cells make ATP. *Sci. Amer.* 238 (March 1978): 104.

• Lipmann, F. *Wanderings of a Biochemist.* New York: Wiley, 1971.

Westheimer, F. Why nature chose phosphates. *Science* 235 (1987): 1173.

Glycolysis and Fermentation

Bittar, E. E., and K. M. Brindle. *Enzymology in Vivo.* Greenwich, CT: JAI Press, 1995.

Flores, C. L., C. Rodriguez, T. Petit, and C. Gancedo. Carbohydrate and energy-yielding metabolism in non-conventional yeasts. *FEMS Microbiol. Rev.* 24 (2000): 507.

Goncalves, P., and R. J. Planta. Starting up yeast glycolysis. *Trends Microbiol.* 6 (1998): 314.

Van Schaftingen, E. Glycolysis revisited. *Diabetologia* 36 (1993): 581.

Gluconeogenesis

Brosnan, J. T. Comments on metabolic needs for glucose and the role of gluconeogenesis. *Europ. J. Clin. Nutr.* 53 (1999): 107.

Previs, S. F., and H. Brunengraber. Methods for measuring gluconeogenesis in vivo. *Curr. Opin. Clin. Nutr. Metabol. Care* 1 (1998): 461.

Regulation of Glycolysis and Gluconeogenesis

Corssmit, E. P., J. A. Romijn, and H. P. Sauerwein. Regulation of glucose production with special attention to nonclassical regulatory mechanisms. *Metabolism: Clin. Exper.* 50 (2001): 742.

Nordlie, R .C., J. D. Foster, and A. J. Lange. Regulation of glucose production by the liver. *Annu. Rev. Nutr.* 19 (1999): 379.

Plaxton, W. C. The organization and regulation of plant glycolysis. *Annu. Rev. Plant Physiol. Mol. Struct.* 47 (1996): 185.

Rousseau, G. G., and L. Hue. Mammalian 6-phosphofructo-2-kinase/fructose-2,6-bisphosphatase: A bifunctional enzyme that controls glycolysis. *Progr. Nucleic Acid Mol. Biol.* 45 (1993): 99.

14

Chemotrophic Energy Metabolism: Aerobic Respiration

In the previous chapter, we learned that some cells meet their energy needs by anaerobic fermentation, either because they are strict anaerobes or because they are facultative cells functioning temporarily in the absence or scarcity of oxygen. However, we also noted that fermentation yields only modest amounts of energy. In the absence of an external electron acceptor—one that is not itself a part of the glycolytic pathway—the electrons that are removed from one three-carbon organic compound (glyceraldehyde-3-phosphate) are eventually transferred to another three-carbon compound (pyruvate), and the difference in energy is such that only two molecules of ATP can be generated per molecule of glucose.

In short, fermentation is a feasible way of meeting energy needs, but the ATP yield is low because the cell has access to only a limited portion of the total free energy potentially available from the oxidizable molecules it uses as substrates. In addition, fermentation always results in the accumulation of waste products such as ethanol or lactate that are toxic to the cells as they accumulate—unless, of course, they are removed and metabolized elsewhere in the environment (or in the organism, as in the Cori cycle, shown in Figure 13A-2). Furthermore, fermentation products cannot be used as starting materials by phototrophs, thereby failing to complete the chemotrophic portion of the cyclic flow of matter shown in Figure 5-5.

Cellular Respiration: Maximizing ATP Yields

All of this changes dramatically when we come to **cellular respiration,** or *respiration* for short. With an external electron acceptor available, complete substrate oxidation becomes possible, and ATP yields are much higher. As a formal definition, *cellular respiration is the oxidation-driven flow of electrons, through or within a membrane, from reduced coenzymes to an electron acceptor, usually accompanied by the generation of ATP.* We will get to the "membrane" and "ATP generation" parts of the definition later in the chapter; for now, let's focus on the reduced coenzymes and electron acceptors. We have already encountered NADH as the reduced coenzyme generated by the glycolytic catabolism of sugars or related compounds. As we will see shortly, two other coenzymes, *FAD* (for *flavin adenine dinucleotide*) and *coenzyme Q* (or *ubiquinone*), also collect the electrons that are removed from oxidizable organic substrates and pass them to the ultimate, or terminal, electron acceptor via a series of electron carriers.

For many organisms, including the authors and readers of this textbook, the ultimate electron acceptor is *oxygen,* the reduced form of the acceptor is *water,* and the overall process is called **aerobic respiration.** The vast majority of chemotrophs on Earth carry out aerobic respiration, which is hardly a surprise, given the availability of oxygen in the air and water of our planet. However, a variety of other acceptors are also used by other organisms, especially bacteria. Examples of alternative acceptors and their reduced forms include elemental sulfur (S/H_2S), protons (H^+/H_2O), and ferric ions (Fe^{3+}/Fe^{2+}). Respiratory processes that involve electron acceptors such as these require no oxygen and are therefore examples of **anaerobic respiration.** Anaerobic respiratory processes play important roles in the cycling of elements such as sulfur, hydrogen, and iron in the environment and contribute significantly to the overall energy economy of the biosphere. Here, however, we will focus on aerobic respira-

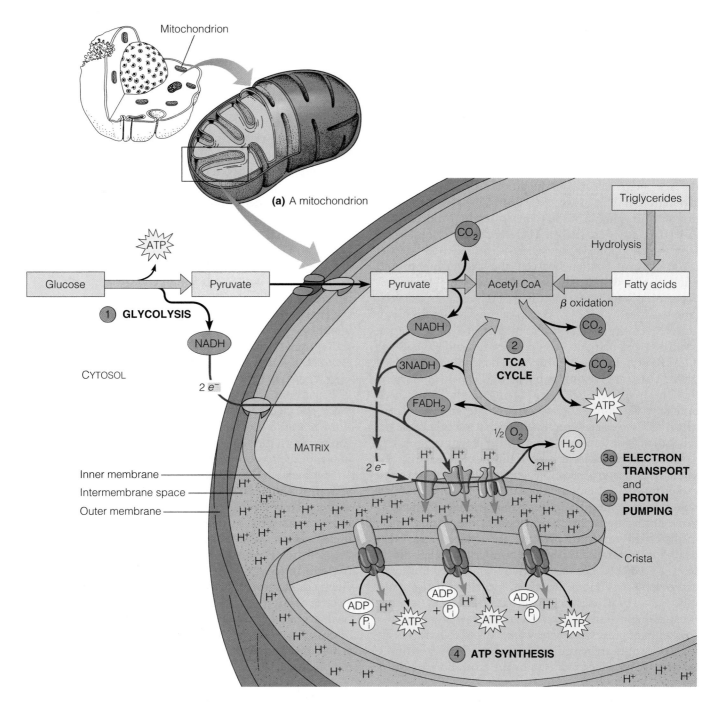

(a) A mitochondrion

(b) Localization of aerobic respiration within the mitochondrion

Figure 14-1 The Role of the Mitochondrion in Aerobic Respiration. **(a)** The mitochondrion plays a central role in aerobic respiration; most respiratory ATP production in eukaryotic cells occurs in this organelle. **(b)** Oxidation of glucose and other sugars begins in the cytosol with glycolysis (stage 1), producing pyruvate. Pyruvate is transported across the inner mitochondrial membrane and is oxidized within the matrix to acetyl CoA, the primary substrate of the TCA cycle (stage 2). Acetyl CoA can also be formed by β oxidation of fatty acids. Electron transport (stage 3a) is coupled to proton pumping (stage 3b), with the energy of electron transport conserved as an electrochemical proton gradient across the inner membrane of the mitochondrion (or across the plasma membrane, in the case of prokaryotes). The energy of the proton gradient is used in part to drive the synthesis of ATP from ADP and inorganic phosphate (stage 4).

tion because it is the mainspring of energy metabolism in the aerobic world of which we and all other higher organisms are a part.

At this point, attention focuses on the *mitochondrion* (Figure 14-1) because most aerobic ATP production in eukaryotic cells takes place within this organelle. For example, the complete aerobic oxidation of glucose by a muscle cell releases enough free energy to drive the synthesis of 36 molecules of ATP per molecule of glucose, and most of these ATPs (all but the two produced by glycolysis) are generated within the mitochondrion. Except for glycolysis, in fact, all of aerobic energy metabolism in eukaryotic cells occurs within the mitochondrion.

Aerobic Respiration Yields Much More Energy than Fermentation

With oxygen available as the ultimate electron acceptor, we can begin with the pyruvate generated by glycolysis and ask how pyruvate molecules are catabolized and how the free energy released in the process is conserved by the generation of ATP. As we know from Chapter 13, the end-products of aerobic respiration are carbon dioxide and water (see reaction 13-12). These are the very molecules with which photosynthesis begins, as required for the cyclic flow of matter between the chemotrophic and phototrophic worlds. Moreover, the energy yield for aerobic respiration is remarkably higher than that for fermentation. Instead of just 2 ATP molecules per molecule of glucose, aerobic respiration has the potential of generating up to 38 ATPs per glucose in prokaryotes and either 36 or 38 ATPs per glucose in eukaryotes, depending on the cell type. (These are maximum possible ATP yields, assuming that the free energy released during respiration is used only for ATP synthesis. As we will see later in this chapter, actual ATP yields are typically somewhat lower because the cell uses at least some of the energy for other purposes.)

The oxygen that makes all of this possible serves as the ultimate, or terminal, electron acceptor, thereby providing a means for the continuous reoxidation of NADH and other reduced coenzymes. These coenzyme molecules accept electrons during the stepwise oxidation of the organic intermediates derived from pyruvate (as well as from other oxidizable substrates). They then transfer these electrons to oxygen via a sequence of membrane-bound electron carriers. Aerobic respiration therefore involves oxidative pathways in which electrons are removed from organic substrates and transferred to coenzyme carriers, as well as concomitant processes whereby the reduced (electron-bearing) coenzymes are reoxidized by the transfer of electrons to oxygen, accompanied indirectly by the generation of ATP.

Respiration Includes Glycolysis, the TCA Cycle, Electron Transport, and ATP Synthesis

Aerobic respiration can be considered in four stages, two concerned with coenzyme-mediated oxidative processes and two involving coenzyme reoxidation and the generation of ATP—driven, as it turns out, by the energy stored in a transmembrane gradient of protons. These stages are shown in Figure 14-1b. In cells, of course, all these processes occur continuously and simultaneously. Their division into four arbitrary stages may be a useful framework for our discussion, but keep in mind that none of these stages functions in isolation; each is an integral part of the overall respiratory process.

Stage 1 is the *glycolytic pathway* that we encountered in Chapter 13. The function of glycolysis is the same under aerobic and anaerobic conditions: the conversion of glucose to pyruvate. But the fate of the pyruvate is different in the presence of oxygen (see Figure 13-9). Instead of serving as an electron acceptor as in fermentation, pyruvate is further oxidized to a compound called *acetyl coenzyme A (acetyl CoA)*, which enters stage 2, the *tricarboxylic acid (TCA) cycle*. This cyclic pathway oxidizes incoming carbon atoms to CO_2 and conserves the energy as reduced coenzyme molecules, which are high-energy compounds in their own right.

Stage 3 involves (a) *electron transport*, or transfer, from reduced coenzymes to oxygen, coupled to (b) the *active transport*, or *pumping*, of protons across a membrane. The transfer of electrons from coenzymes to oxygen is exergonic and occurs stepwise via a sequence of membrane-bound electron carriers called the *electron transport system*. The exergonic transfer of electrons between carrier molecules within the membrane provides the energy that drives the pumping of protons across the membrane in which the carriers are embedded. This active transport of protons generates and maintains an *electrochemical proton gradient* across the membrane. In stage 4, the energy of the proton gradient is used to drive ATP synthesis. This mode of oxygen-dependent ATP synthesis is called *oxidative phosphorylation* to distinguish it from *substrate-level phosphorylation*, which takes place at two points in the glycolytic pathway—and, as it turns out, in one of the reactions of the TCA cycle as well.

Our goal in this chapter is to understand the processes identified as stages 2, 3, and 4 in Figure 14-1b. Specifically, we want to consider (1) what happens to pyruvate (and other oxidizable substrates such as fats and amino acids) under aerobic conditions; (2) how coenzymes and other electron carriers mediate the exergonic transfer of electrons from oxidizable substrates to oxygen; (3) how the energy of electron-transfer events is used to maintain the electrochemical proton gradient; and (4) how the energy of that gradient drives ATP synthesis. We will begin our discussion of aerobic energy metabolism with a close look at the mitochondrion because of the prominent role this organelle plays in eukaryotic energy metabolism.

The Mitochondrion: Where the Action Takes Place

Beginning our discussion of aerobic respiration with the **mitochondrion** is appropriate because most of aerobic energy metabolism in eukaryotic cells takes place within this organelle. Thus, most of the ATP generation of most animal cells occurs in the mitochondrion, as does much of the nonphotosynthetic energy metabolism of plant and algal cells. No wonder, then, that the mitochondrion is called the "energy powerhouse" of the eukaryotic cell.

Biologists have known about mitochondria and have studied them for more than 150 years. As early as 1850, the German biologist Rudolph Kölliker described the presence of what he called "ordered arrays of particles" in muscle cells. When these particles were isolated, they were found to swell in water, leading Kölliker to conclude that each particle was surrounded by a semipermeable membrane. Various names were given to such particles in early work, but the term *mitochondrion* (meaning "threadlike granule"), first introduced in 1898, gradually replaced other names and is now the universally recognized term.

Evidence suggesting a role for the organelle in oxidative events began to accumulate almost a century ago. In 1913, for example, Otto Warburg showed that granules obtained by filtering tissue homogenates could consume oxygen. However, most of our understanding of the role of mitochondria in energy metabolism dates from the development of differential centrifugation, pioneered by Albert Claude (see Figure 12A-2). Intact, functionally active mitochondria were first isolated by this technique in 1948 and were subsequently shown by Eugene Kennedy, Albert Lehninger, and others to be capable of carrying out all the reactions of the TCA cycle, electron transport, and oxidative phosphorylation.

Mitochondria Are Often Present Where the ATP Needs Are Greatest

Mitochondria are found in virtually all aerobic cells of eukaryotes and are prominent features of many cell types when examined by electron microscopy (Figure 14-2). Mitochondria are present in both chemotrophic and phototrophic cells and are therefore common not only to animals but also to plants. Their occurrence in phototrophic cells reminds us that photosynthetic organisms are also capable of respiration, an ability on which they depend to meet energy and carbon needs during periods of darkness—and at all times in nonphotosynthetic tissue, such as the roots of plants.

The crucial role of the mitochondrion in meeting cellular ATP needs is often reflected in the localization of mitochondria within the cell. Frequently, mitochondria are clustered in regions of cells with the most intense metabolic activity, where the ATP need is greatest. An especially good example is in muscle cells (see Figure 4-13). Just as Kölliker originally observed, the mitochondria in such cells are organized in rows along the fibrils responsible for contraction. Presumably, this association minimizes the distance that ATP molecules must diffuse to get from the site of ATP generation in the mitochondrion to the site of ATP utilization in the contracting fibrils. A similar strategic localization of mitochondria occurs in flagella and cilia as well as in sperm tails (see Figure 4-12).

Are Mitochondria Interconnected Networks Rather than Discrete Organelles?

When seen in electron micrographs such as those in Figures 14-2, 4-12, or 4-13, mitochondria typically appear as oval, sausage-shaped structures measuring several micrometers in length and 0.5–1.0 μm across. This appearance has fostered the widely accepted view that mitochondria are discrete entities and as such are large and often very numerous organelles. In fact, a mitochondrion with these dimensions is similar in size to an entire bacterial cell and is the largest

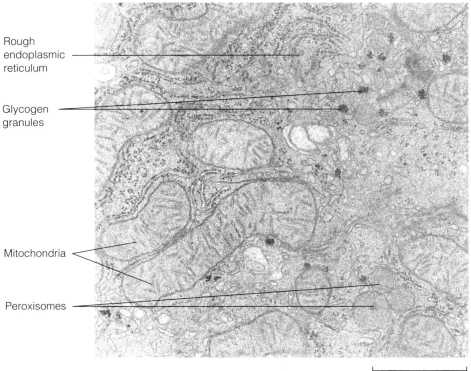

Rough endoplasmic reticulum

Glycogen granules

Mitochondria

Peroxisomes

| 0.5 μm |

Figure 14-2 The Prominence of Mitochondria. Mitochondria are prominent features in this electron micrograph of a rat liver cell. Other cellular components shown include the rough endoplasmic reticulum, glycogen granules, and peroxisomes (TEM).

organelle in most animal cells other than the nucleus (see Figure 1A-1). (In plant cells, the chloroplasts and some of the vacuoles are also typically larger than the mitochondria.)

Calculations based on cross-sectional views of mitochondria in electron micrographs indicate that the number of mitochondria per cell is highly variable, ranging from one or a few per cell in many protists, fungi, and algae (and in some mammalian cells as well) up to several hundred or even a few thousand per cell in some tissues of higher plants and animals. Mammalian liver cells, for example, are thought to contain about 500–1000 mitochondria each, though only a small fraction of these are seen in a typical thin section prepared for electron microscopy, such as that shown in Figure 14-2.

The notion that the mitochondrial profiles seen in electron micrographs represent discrete organelles of known size and abundance has been challenged by the work of Hans-Peter Hoffman and Charlotte Avers, who showed that the three-dimensional shape of mitochondria within intact cells cannot be determined by looking at a few thin-section micrographs. After examining a complete series of thin sections through an entire yeast cell, these investigators concluded that the oval profiles observed in individual micrographs all represent slices through a single large, extensively branched mitochondrion. This view is supported by similar analyses of other eukaryotic cells, in which mitochondrial profiles seen in thin-section micrographs appear to represent portions of larger, interconnected mitochondrial networks (Figure 14-3). Such results suggest that the number of mitochondria present in a cell may be considerably smaller than generally believed—and that the size of a mitochondrion may be much larger than can be calculated from individual cross-sectional profiles seen in thin-section electron micrographs.

Further support for the concept of mitochondria as interconnected networks rather than numerous separate organelles comes from studies in which intact living cells were examined by phase-contrast microscopy. Such investigations revealed that living cells contain large branched mitochondria in a dynamic state of flux, with segments of one mitochondrion frequently pinching off and fusing with another mitochondrion. These dynamic interactions suggest that the concept of the number of mitochondria present in any given cell may in fact be meaningless.

Most of the discussion and illustrations in this chapter will presume the conventional view of mitochondria as discrete entities, but keep in mind that what we regard as individual mitochondria may well be parts of a large, dynamic network instead, at least in some types of cells.

The Outer and Inner Membranes Define Two Separate Compartments

Figure 14-4 depicts a typical mitochondrion—or perhaps a small segment of a much larger network. In either case, a distinctive feature is the presence of two membranes, called the outer and inner membranes. The **outer membrane** is not a significant permeability barrier for ions and small molecules, because it contains transmembrane channel proteins called **porins** that permit the passage of solutes with molecular weights up to about 5000. Similar proteins are found in the outer chloroplast membrane and in the outer membrane of gram-negative bacteria. Because porins allow the free movement of small molecules and ions across the outer membrane, the **intermembrane space** between the inner and outer membranes of the mitochondrion or chloroplast is essentially continuous with the cytosol with respect to the solutes rel-

Figure 14-3 Model of Interconnected Mitochondrial Networks. This model was created by examining a complete series of thin sections through a skin cell. It indicates that individual mitochondrial profiles observed in thin-section micrographs may represent portions of larger, interconnected mitochondrial networks. Further support for this model comes from studies of living cells using phase-contrast microscopy.

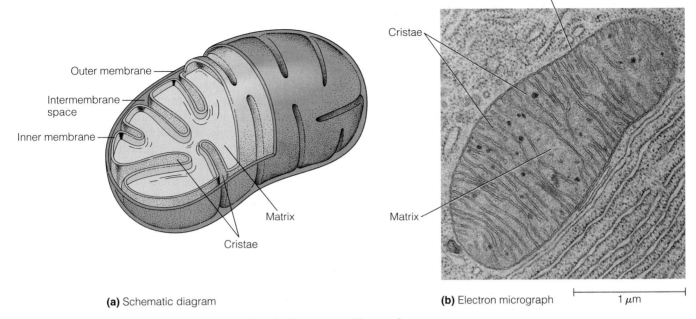

(a) Schematic diagram

Inner and outer membranes

Cristae

Matrix

(b) Electron micrograph   1 μm

Figure 14-4   Mitochondrial Structure.   (a) Mitochondrial structure is illustrated schematically in this cutaway view. **(b)** A mitochondrion of a bat pancreas cell as seen by electron microscopy (TEM). The cristae are infoldings of the inner membrane.

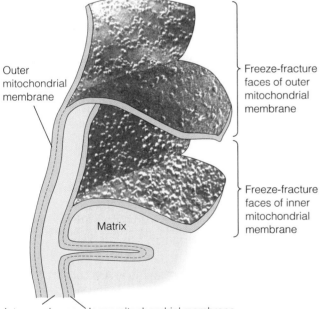

Freeze-fracture faces of outer mitochondrial membrane

Freeze-fracture faces of inner mitochondrial membrane

Outer mitochondrial membrane

Matrix

Intermembrane space   Inner mitochondrial membrane

Figure 14-5   Structure of the Inner and Outer Mitochondrial Membranes.   When the inner and outer mitochondrial membranes are subjected to freeze fracturing, each of the membranes splits along its hydrophobic interior, separating each membrane into two fracture faces. Segments of electron micrographs of the two fracture faces for both inner and outer membranes are superimposed here on a schematic diagram of the freeze-fractured membranes to illustrate the density of protein particles in each membrane.

most solutes, thereby partitioning the intermembrane space and the interior of the organelle into two separate compartments. Not surprisingly, the inner membrane contains the transport proteins needed to move solutes such as pyruvate, fatty acids, ATP, ADP, and inorganic phosphate across the membrane. The inner membrane is also the locale of the protein complexes involved in electron transport and ATP synthesis.

The inner membrane of most mitochondria has many distinctive infoldings called **cristae** (singular: **crista**) that greatly increase its surface area. In a typical liver mitochondrion, for example, the area of the inner membrane is about five times greater than that of the outer membrane. Because of its large surface area, the inner membrane can accommodate large numbers of the protein complexes needed for electron transport and ATP synthesis, thereby enhancing the capacity of the mitochondrion for ATP generation. Figure 14-5 illustrates structural details of the inner and outer mitochondrial membranes as revealed by the freeze-fracture technique described in

evant to the functions of these organelles. However, enzymes targeted to the intermembrane space are effectively confined there because enzymes and other soluble proteins are too large to pass through the porin channels.

In contrast to the outer membrane, the **inner membrane** of the mitochondrion presents a permeability barrier to

the *Guide to Microscopy*. Notice especially the high density of protein particles associated with both fracture faces of the inner membrane. Proteins account for about 75% of the inner membrane by weight, which is a higher proportion than in any other cellular membrane. These proteins include the transmembrane portions of proteins involved in solute transport, electron transport, and ATP synthesis.

The relative prominence of cristae within the mitochondrion frequently reflects the relative metabolic activity of the cell or tissue in which the organelle is located. Heart, kidney, and muscle cells have high respiratory activities, and their mitochondria have correspondingly large numbers of prominent cristae. The flight muscles of birds are especially high in respiratory activity and have mitochondria that are exceptionally well endowed with cristae. Plant cells, by contrast, have lower rates of respiratory activity than most animal cells and have correspondingly fewer cristae within their mitochondria.

The interior of the mitochondrion is filled with a semifluid **matrix.** Within the matrix are many of the enzymes involved in mitochondrial function as well as the DNA molecules and ribosomes that give the organelle its genetic competence. In most mammals, the mitochondrial genome consists of a circular DNA molecule of about 15,000–20,000 base pairs that codes for ribosomal RNAs, transfer RNAs, and about a dozen polypeptide subunits of inner-membrane proteins.

Mitochondrial Functions Occur in or on Specific Membranes and Compartments

Specific functions and pathways have been localized within the mitochondrion by disruption of the organelle and fractionation of the various components. Table 14-1 lists some of the main functions localized to each of the compartments of the mitochondrion.

Enzymes and other proteins that are readily solubilized upon disruption are assumed to be either in the matrix or only loosely associated with the inner membrane. By this criterion, most of the mitochondrial enzymes involved in pyruvate oxidation, in the TCA cycle, and in the catabolism of fatty acids and amino acids are matrix enzymes. In fact, when mitochondria are disrupted very gently, either by osmotic lysis or ultrasonic vibration, six of the eight enzymes of the TCA cycle are released as a single large multiprotein complex, suggesting that the product of one enzyme can pass directly to the next enzyme without having to diffuse through the matrix.

On the other hand, most of the intermediates in the electron transport system are integral components of the inner membrane, where they are organized into large complexes. Protruding from the inner membrane into the matrix are knoblike spheres called F_1 **complexes** (Figure 14-6). Each complex is an assembly of six polypeptides. Individual F_1 complexes can be seen in Figure 14-6a, an electron micrograph taken at high magnification using a technique called *negative staining*. (As described in the *Guide to Microscopy*, negative staining results in a light image against a dark background and is the preferred method for examining very small objects.) F_1 complexes are about 9 nm in diameter and are especially abundant along the cristae (Figure 14-6b).

Each F_1 complex is attached by a short protein stalk to an F_o **complex,** an assembly of hydrophobic polypeptides that are embedded within the inner membrane (Figure 14-6c). The combination of an F_1 complex linked to an F_o complex is called an F_oF_1 **complex** (or F_1F_o complex; both forms are used in the literature). As you may recall from Chapter 8, this structural feature is the defining characteristic of F-type ATPases (see Table 8-3 on p. 209) and indeed the mitochondrial F_1 complex has ATPase activity under certain conditions. Functionally, however, the mitochondrial F_oF_1 complex is best regarded as an *ATP synthase* because that is its normal role in energy metabolism. The F_oF_1 complex is in fact responsible for most of the ATP generation that occurs in the mitochondrion—and, as it turns out, in prokaryotic cells and chloroplasts as well. In each of these settings, ATP generation by F_oF_1 complexes is driven by an electrochemical gradient of protons across the membrane in which the F_oF_1 complexes are anchored, a topic we will explore in detail later in this chapter.

In Prokaryotes, Respiratory Functions Are Localized to the Plasma Membrane and the Cytoplasm

Prokaryotes do not have mitochondria, yet most prokaryotic cells are capable of aerobic respiration. Where in the prokaryotic cell are the various components of respiratory metabolism localized? Essentially, the cytoplasm and plasma membrane of a prokaryotic cell perform the same functions as the mitochondrial matrix and inner membrane, respectively. In the prokaryotic cell, therefore, most of the enzymes of the TCA cycle and of fatty acid and amino acid catabolism are found in the cytoplasm, whereas the electron transport proteins are located in the plasma membrane. The F_oF_1 complex is also localized to the plasma membrane in prokaryotes, with the F_o compo-

Table 14-1   Localization of Metabolic Functions Within the Mitochondrion

Membrane or Compartment	Metabolic Functions
Outer membrane	Phospholipid synthesis Fatty acid desaturation Fatty acid elongation
Inner membrane	Electron transport Oxidative phosphorylation Metabolite transport
Intermembrane space	Nucleotide phosphorylation
Matrix	Pyruvate oxidation TCA cycle β oxidation of fats DNA replication RNA synthesis (transcription) Protein synthesis (translation)

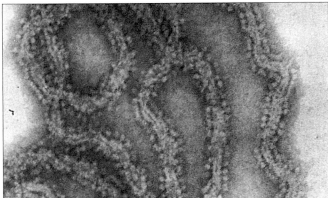

(a) Mitochondrial inner membrane

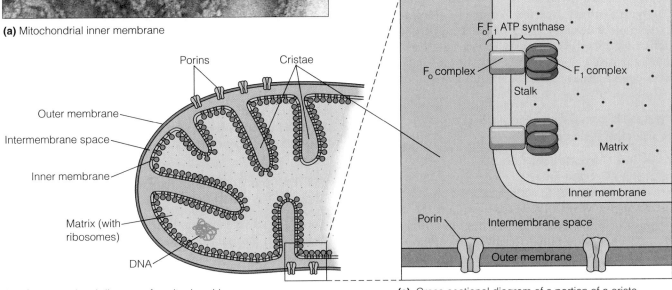

Figure 14-6   The F₁ and F₀ Complexes of the Inner Mitochondrial Membrane.   (a) This electron micrograph was prepared by negative staining to show the spherical F_1 complexes that line the matrix side of the inner membrane of a bovine heart mitochondrion (TEM). **(b)** Cross section of a mitochondrion, showing major structural features. **(c)** An enlargement of a small portion of a crista, showing the F_1 complexes that project from the inner membrane on the matrix side and the F_0 complexes embedded in the inner membrane. Each F_1 complex is attached to an F_0 complex by a short protein stalk. Together, an F_0F_1 pair constitutes a functional ATP synthase.

(b) Cross-sectional diagram of a mitochondrion

(c) Cross-sectional diagram of a portion of a crista showing F_0F_1 complexes

nent embedded in the membrane and the F_1 component protruding from the membrane into the cytoplasm.

In the absence of mitochondria, glycolysis and the TCA cycle are localized to the same compartment, the cytoplasm, in prokaryotes rather than to two separate compartments, the cytosol and the mitochondrial matrix, as in eukaryotes. As we will see later in this chapter, this difference explains why the maximum ATP yield per glucose is somewhat higher in prokaryotes than in at least some eukaryotic cells.

The Tricarboxylic Acid Cycle: Oxidation in the Round

Having considered localization of respiratory functions in mitochondria and in prokaryotic cells, we will now return to the eukaryotic context and follow a molecule of pyruvate across the inner membrane of the mitochondrion to see what fate awaits it inside.

In the presence of oxygen, pyruvate is oxidized fully to carbon dioxide and the released energy is used to drive

ATP synthesis. The first stage in this process is a cyclic pathway that is a central feature of energy metabolism in almost all aerobic chemotrophs. An important intermediate in this cyclic series of reactions is citrate, which has three carboxylic acid groups and is therefore a tricarboxylic acid. For this reason, this pathway is usually called the **tricarboxylic acid (TCA) cycle.** It is also commonly referred to as the *Krebs cycle* in honor of Hans Krebs, whose laboratory played a key role in elucidating this metabolic sequence in the 1930s.

As we noted earlier, the TCA cycle begins with acetyl coenzyme A (acetyl CoA), which consists of a two-carbon acetate group linked to a carrier called *coenzyme A*. (Coenzyme A was discovered by Fritz Lipmann, who shared a Nobel Prize with Krebs in 1953 for their work on aerobic respiration.) Acetyl CoA arises either by oxidative decarboxylation of pyruvate or by the stepwise oxidative breakdown of fatty acids (see Figure 14-1b). Regardless of its origin, acetyl CoA transfers its acetate group to a four-carbon acceptor called oxaloacetate, thereby generating citrate, the tricarboxylic acid for which the cycle is named. In a cyclic series of reactions, citrate is subjected to two successive decarboxylations and several oxidations, leaving a four-carbon compound from which the starting oxaloacetate is regenerated.

Each round of TCA cycle activity involves the entry of two carbons (as the acetate from acetyl CoA), the release of two carbons as carbon dioxide, and the regeneration of oxaloacetate. Oxidation occurs at five steps: four in the cycle itself and one in the reaction that converts pyruvate to acetyl CoA. In each case, electrons are accepted by coenzyme molecules. The substrate for the TCA cycle is therefore acetyl CoA, and the products are carbon dioxide, reduced coenzymes, and a molecule of ATP (or GTP, a closely related nucleotide).

With this brief overview in mind, let's look at the TCA cycle in more detail, focusing on what happens to the carbon molecules that enter as acetyl CoA and how the energy released by each of the oxidations is conserved as high-energy electrons of reduced coenzymes.

Pyruvate Is Converted to Acetyl Coenzyme A by Oxidative Decarboxylation

As we just noted, carbon enters the TCA cycle in the form of acetyl CoA, but from Chapter 13 we know that the glycolytic pathway ends with pyruvate, not acetyl CoA. To get from pyruvate to acetyl CoA requires the activity of *pyruvate dehydrogenase (PDH),* a huge multiprotein complex that has a molecular weight of about 4.6×10^6 and consists of three different enzymes, five coenzymes, and two regulatory proteins. These components work together to catalyze the *oxidative decarboxylation* of pyruvate:

$$CoA-SH \ + \ ^3CH_3-^2\overset{O}{\overset{\|}{C}}-^1\overset{O}{\overset{\|}{C}}-O^- \ \xrightarrow{\ NAD^+ \quad NADH + H^+\ }$$

$$CoA-S-^1\overset{O}{\overset{\|}{C}}-^2CH_3 \ + \ CO_2 \quad \textbf{(14-1)}$$

Coenzyme A   Pyruvate   Acetyl CoA

This reaction is a *decarboxylation* because one of the carbons of pyruvate (carbon atom 1) is liberated as carbon dioxide. As a result, carbon atoms 2 and 3 of pyruvate become, respectively, carbon atoms 1 and 2 of acetate. In addition, this reaction is an *oxidation* because two electrons (and a proton as well) are transferred from the substrate to the coenzyme NAD^+. The electrons carried by NADH represent potential energy that is tapped when the NADH is reoxidized by the electron transport system.

The oxidation of pyruvate occurs on carbon atom 2, which is oxidized from an α-keto to a carboxylic acid group. This oxidation is possible because of the concomitant elimination of carbon atom 1 as carbon dioxide. The oxidation is highly exergonic ($\Delta G^{\circ\prime} = -7.5$ kcal/mol), with the free energy used to energize, or activate, the acetate molecule for further metabolism by linking it to the sulfhydryl group of one of the enzymes in the PDH complex, followed by transfer of the acetyl group from the enzyme to **coenzyme A (CoA),** forming **acetyl CoA.**

As shown in Figure 14-7, CoA is a complicated molecule containing the B vitamin *pantothenic acid.* (Like the nicotinamide of NAD^+, pantothenic acid is classified as a vitamin because humans and other vertebrates need it as a part of an essential coenzyme but cannot synthesize it themselves.) To understand the role of coenzyme A in cellular metabolism, focus on the free sulfhydryl, or *thiol,* group at the end of the molecule. It is this thiol group that can form a thioester bond with organic acids such as acetate. The thiol group is sufficiently important to the function of the molecule that coenzyme A is sometimes abbreviated not just as CoA but as CoA—SH. Compared with an ester bond, a thioester bond is a higher-energy bond because significantly more free energy is released upon its hydrolysis.* Just as NAD^+ is adapted for transfer of electrons, so coenzyme A is well suited as a carrier of acetate and other acyl groups. (The "A" in the name comes, in fact, from its role as a carrier of *acyl* groups.) Thus, the acetyl group that is transferred to coenzyme A from one of the enzymes of the PDH complex (represented in Figure 14-7 as E—SH) is in a higher-energy, or activated, form.

The TCA Cycle Begins with the Entry of Acetate as Acetyl CoA

The TCA cycle is shown in Figure 14-8. As already noted, it begins with the entry of acetate in the form of acetyl CoA. With each round of TCA cycle activity, two carbon atoms enter in organic form (as acetate) and two carbon atoms leave in inorganic form (as carbon dioxide). In the first reaction (TCA-1), the acetate group of acetyl CoA is added onto oxaloacetate to form citrate, with the condensation driven by the free energy of hydrolysis of the thioester bond. The reaction is catalyzed by the enzyme *citrate synthase.* Acetate has two carbon atoms and oxaloacetate has four, so citrate is a six-carbon compound, with three of the carbons present as carboxylic acid groups. (Notice that the two incoming carbon atoms are shown in pink in Figure 14-8 to help you keep track of them in subsequent reactions.)

NADH Is Formed and CO_2 Is Released in Two Reactions of the TCA Cycle

Since the function of the TCA cycle is to oxidize the carbon atoms of the incoming acetate group completely, it should not be too surprising to find that four out of the eight steps

---

* Having learned in Chapter 13 that esters are relatively low-energy compounds compared with anhydrides, you may wonder why a thioester is a high-energy compound that has an even more negative standard free energy change ($\Delta G^{\circ\prime} = -7.5$ kcal/mol) than do the acid anhydride bonds of ATP. Resonance stabilization provides the answer here, as it did for the instability of anhydrides in Chapter 13. Although not mentioned there, the C=O bond of an ester has a partial double-bond character because of the partial overlap of π electrons, thereby allowing the ester to resonate between two forms. By contrast, the larger atomic size of S compared with O reduces π-electron overlap between C and S, thereby eliminating any significant contribution of the C=S structure to resonance stabilization. A thioester is therefore thermodynamically less stable than an ester, which results in a more highly negative $\Delta G^{\circ\prime}$ for hydrolysis.

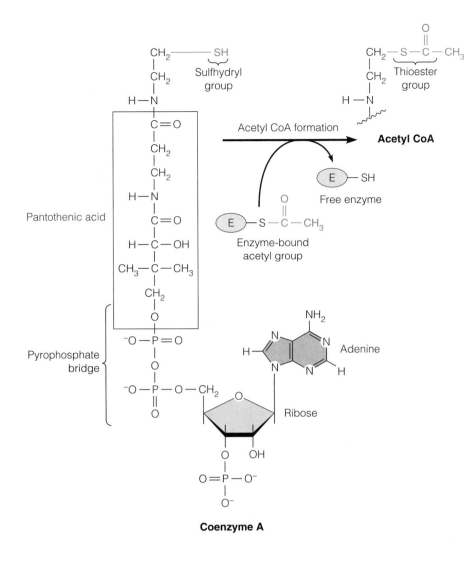

Figure 14-7 Structure of Coenzyme A and Acetyl CoA Formation. The portion of the coenzyme enclosed in the red box is pantothenic acid, a B vitamin. Formation of a thioester bond between CoA and an acetyl group generates acetyl coenzyme A. The acetyl group is formed by the oxidative decarboxylation of pyruvate (reaction PDH in Figure 14-8), or by β oxidation of fatty acids (Figure 14-11). It is transferred to CoA from one of the enzymes of the pyruvate dehydrogenase complex, to which it is bound as a thioester after the oxidation of pyruvate.

in the cycle are oxidations. This is evident in Figure 14-8 because four steps (TCA-3, TCA-4, TCA-6, and TCA-8) involve coenzymes that enter in the oxidized form and leave in the reduced form. Each of these reactions is catalyzed by a dehydrogenase that is specific for the particular substrate. The first two of these reactions, TCA-3 and TCA-4, are also decarboxylation steps; one molecule of carbon dioxide is eliminated in each step, reducing the number of carbons first from six to five and then from five to four.

Prior to these events, however, reaction TCA-2 converts citrate to the related compound isocitrate. If you look closely at the structures of these two compounds in Figure 14-8, you will see why this conversion is essential. The carbon atom of citrate that carries the hydroxyl group has no hydrogen atoms attached to it. In the chemist's terminology, therefore, citrate is a *tertiary alcohol,* and the hydroxyl group of a tertiary alcohol is not easily oxidizable. Isocitrate, on the other hand, has a hydrogen atom and its hydroxyl group attached to the same carbon atom and is therefore a *secondary alcohol,* with a hydroxyl group that can be quite easily oxidized. The enzyme that catalyzes the interconversion of citrate and isocitrate is called *aconitase.* Aconitase actually carries out successive dehydration and rehydration reac-

tions, in which the elements of water are removed from citrate to generate an unsaturated intermediate called *aconitate* (not shown), to which the —H and —OH of water are added back again, but in reversed positions.

The hydroxyl group of isocitrate is now the target of the first oxidation, or dehydrogenation, of the cycle. Isocitrate is oxidized by the enzyme *isocitrate dehydrogenase* to a six-carbon α-keto compound called *oxalosuccinate* (not shown), with NAD$^+$ as the usual electron acceptor. Oxalosuccinate never leaves the active site of the enzyme, however; it is unstable and immediately undergoes decarboxylation to the five-carbon compound α-ketoglutarate. This reaction, TCA-3, is therefore one of the two decarboxy-lation steps of the cycle.

The second decarboxylation event occurs in the next step, reaction TCA-4. This reaction is also an oxidation, with NAD$^+$ as the electron acceptor. To understand this reaction, compare the structure of α-ketoglutarate with that of pyruvate. Both compounds are α-keto acids, so it should not be surprising that the mechanism of oxidation is the same for both, complete with decarboxylation (from pyruvate to acetate in one case, and from α-ketoglutarate to succinate in the other) and linkage of the oxidized

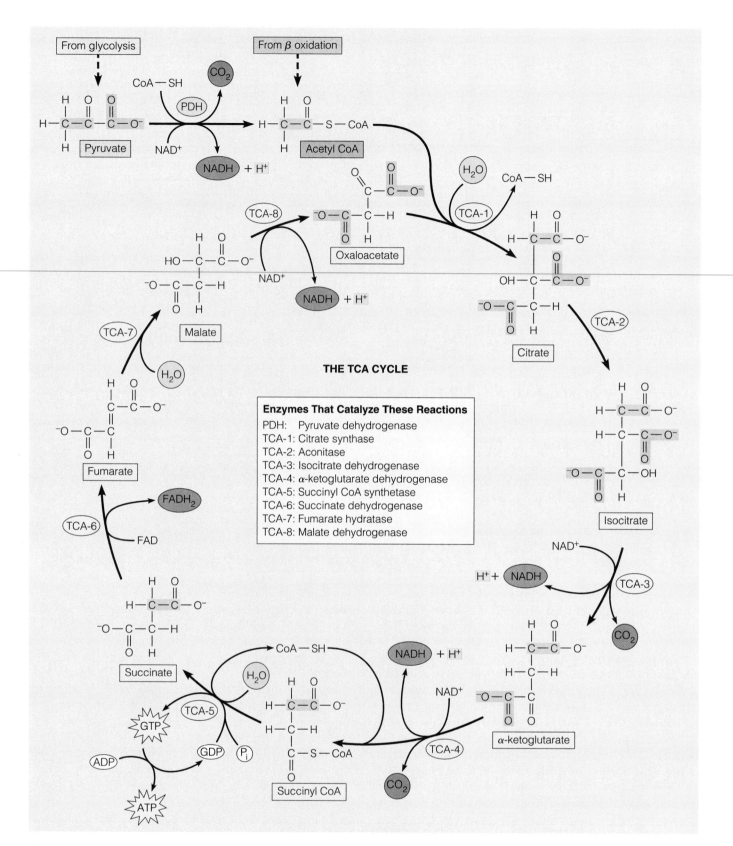

Figure 14-8 The Tricarboxylic Acid (TCA) Cycle. The two carbon atoms of pyruvate that enter the cycle via acetyl CoA are shown in pink in citrate and subsequent molecules until they are randomized by the symmetry of the fumarate molecule. The carbon atom of pyruvate that is lost as CO_2 is shown in gray, as are the two carboxyl groups of oxaloacetate that give rise to CO_2 in steps TCA-3 and TCA-4. Five of the reactions are oxidations, with NAD^+ as the electron acceptor in four reactions (PDH, TCA-3, TCA-4, and TCA-8) and FAD as the electron acceptor in one case (TCA-6). The reduced form of the coenzyme is shown in purple in each case. The generation of GTP shown in reaction TCA-5 is characteristic of animal mitochondria. In bacterial cells and plant mitochondria, ATP is formed directly, but the reactions are energetically equivalent because GTP and ATP are identical in free energy of hydrolysis and are readily interconvertible, as shown.

product to coenzyme A as a thioester. Thus, α-ketoglutarate is oxidized to succinyl CoA, a four-carbon compound with a thioester bond. The enzyme that catalyzes this reaction is called *α-ketoglutarate dehydrogenase*.

Direct Generation of GTP (or ATP) Occurs at One Step in the TCA Cycle

We are already halfway around the cycle, so let's pause a moment to take stock. Note that the carbon balance of the cycle is already satisfied: Two carbon atoms entered as acetyl CoA, and two carbon atoms have now been released as carbon dioxide. (But notice carefully from Figure 14-8 that the two carbon atoms released in a given cycle are not the *same* two that entered in reaction TCA-1 of that cycle. Instead, the two CO_2 molecules arise from what were initially the two carboxyl carbons of oxaloacetate, which are color-coded gray in the figure.) We have also encountered two of the four oxidation reactions of the TCA cycle and therefore have two molecules of NADH that will require reoxidation via the electron transport system. In addition, we recognize succinyl CoA as a compound that, like acetyl CoA, has a thioester bond, the hydrolysis of which is highly exergonic.

Unlike acetyl CoA, however, succinyl CoA is not subsequently transferred to another molecule, so the energy of the thioester bond is not needed for that purpose. Instead, the energy is used to generate a molecule of either ATP (in bacterial cells and plant mitochondria) or GTP (in animal mitochondria). In fact, it is the formation of succinyl CoA rather than free succinate in the preceding reaction (TCA-4) that conserves the energy of α-keto-

glutarate oxidation in a form suitable for subsequent generation of ATP or GTP in reaction TCA-5. Although animal mitochondria form GTP rather than ATP, the two are energetically equivalent because the terminal phosphoanhydride bonds of GTP and ATP have identical free energies of hydrolysis. Moreover, the terminal phosphate group of GTP can be readily transferred to ADP by a mitochondrial enzyme that exchanges phosphate groups among nucleotides. Thus the net result of succinyl CoA hydrolysis is the generation of one molecule of ATP, whether directly or via GTP, as depicted in Figure 14-8.

The Final Oxidative Reactions of the TCA Cycle Generate FADH₂ and NADH

Of the remaining three steps in the TCA cycle, two are oxidations. In reaction TCA-6, the succinate formed in the previous step is oxidized to fumarate. This reaction is unique in that both electrons come from adjacent carbon atoms, generating a C=C double bond. (All the other oxidations we have encountered thus far involve the removal of electrons from adjacent carbon and oxygen atoms, generating a C=O double bond.) The oxidation of a carbon-carbon bond releases less energy than the oxidation of a carbon-oxygen bond—not enough energy to transfer electrons exergonically to NAD$^+$, in fact. Accordingly, the electron acceptor for this dehydrogenation is not NAD$^+$ but a lower-energy coenzyme, **flavin adenine dinucleotide (FAD)**. Like NAD$^+$ and coenzyme A, FAD contains a B vitamin as part of its structure—riboflavin, in this case (Figure 14-9). FAD accepts two protons and two electrons,

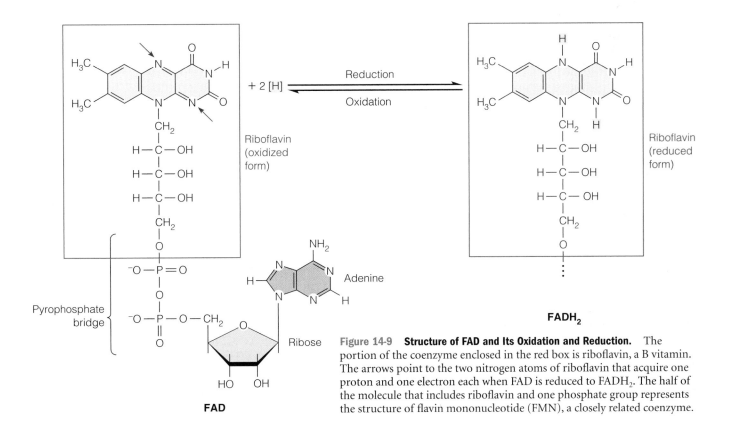

Figure 14-9   Structure of FAD and Its Oxidation and Reduction. The portion of the coenzyme enclosed in the red box is riboflavin, a B vitamin. The arrows point to the two nitrogen atoms of riboflavin that acquire one proton and one electron each when FAD is reduced to FADH₂. The half of the molecule that includes riboflavin and one phosphate group represents the structure of flavin mononucleotide (FMN), a closely related coenzyme.

so the reduced form is written as $FADH_2$. Note that FAD accepts electrons on nitrogen atoms (Figure 14-9, arrows), whereas NAD^+ accepts electrons on carbon atoms (see Figure 13-5). This feature has implications for the ATP yield when the two coenzymes are oxidized. As we will see later, the maximum ATP yield upon coenzyme oxidation is about three for NADH but only about two for $FADH_2$.

Reaction TCA-6 has a further unique feature: The enzyme that catalyzes it, succinate dehydrogenase, does not occur as a soluble protein in the matrix like the other enzymes of the TCA cycle. Instead, it is an integral membrane protein, embedded in the inner membrane of the mitochondrion (or in the plasma membrane, in the case of prokaryotes). Furthermore, the FAD coenzyme is tightly bound to the enzyme, passing its electrons to coenzyme Q, a membrane-soluble quinone that we will meet later in our discussion of the electron transport system. (Some scientists point out that FAD is really a prosthetic group and that coenzyme Q ought to be regarded as the actual electron acceptor for the succinate dehydrogenase reaction. The oxidized and reduced forms of the coenzyme for reaction TCA-6 would then be represented as CoQ and $CoQH_2$, respectively. However, most textbooks continue to represent FAD as the electron acceptor for this reaction, a convention observed in Figure 14-8 also.)

In the next step of the cycle, the double bond of fumarate is hydrated, producing malate (reaction TCA-7, catalyzed by the enzyme *fumarate hydratase*). Because fumarate is a symmetric molecule, the hydroxyl group of water has an equal chance of adding to either of the internal carbon atoms. As a result, the carbon atoms of Figure 14-8 that were color-coded pink to keep track of the most recent acetate group to enter the cycle, are randomized at this step between the "upper" and "lower" two carbon atoms of malate and are therefore not color-coded from this point on. In reaction TCA-8, the hydroxyl group of malate becomes the target of the final oxidation in the cycle. Again, NAD^+ serves as the electron acceptor, and the product is the corresponding keto compound, oxaloacetate.

Summing Up: The Products of the TCA Cycle are CO_2, ATP, NADH, and $FADH_2$

With the regeneration of oxaloacetate, one cycle is complete. We can summarize what has been accomplished by noting the following properties of the TCA cycle:

1. Acetate enters the cycle as acetyl CoA and is joined to a four-carbon acceptor molecule to form citrate, a six-carbon compound.
2. Decarboxylation occurs at two steps in the cycle so that the input of two carbons as acetate is balanced by the loss of two carbons as carbon dioxide.
3. Oxidation occurs at four steps, with NAD^+ as the electron acceptor in three cases and enzyme-bound FAD as the electron acceptor in one case.
4. ATP is generated at one point, with GTP as an intermediate in animal cells.

5. The cycle is completed upon regeneration of the original four-carbon acceptor—namely, oxaloacetate.

By summing the eight component reactions of the TCA cycle as shown in Figure 14-8, we arrive at the following overall reaction:

$$\text{acetyl CoA} + 3H_2O + 3NAD^+ + FAD + ADP + P_i \longrightarrow$$
$$2CO_2 + 3NADH + 3H^+ + FADH_2 + CoA\text{—}SH + ATP + H_2O$$
$$(14\text{-}2)$$

Because the cycle must, in effect, occur twice to metabolize both of the acetyl CoA molecules derived from a single molecule of glucose, the summary reaction on a per-glucose basis can be obtained by doubling all the coefficients in reaction 14-2. If we then add to this reaction the summary reactions for glycolysis through pyruvate (reaction 13-16) and for the oxidative decarboxylation of pyruvate to acetyl CoA (reaction 14-1, also multiplied by 2), we arrive at the following overall reaction for the entire sequence from glucose through the TCA cycle:

$$\text{glucose} + 6H_2O + 10NAD^+ + 2FAD + 4ADP + 4P_i \longrightarrow$$
$$6CO_2 + 10NADH + 10H^+ + 2FADH_2 + 4ATP + 4H_2O$$
$$(14\text{-}3)$$

As you consider this summary reaction, two points may strike you: how modest the ATP yield is thus far, and how many coenzyme molecules are reduced during the oxidation of glucose. With only four ATP molecules generated per glucose molecule, we have as yet little evidence for the substantially greater ATP yield that is supposed to be characteristic of respiratory metabolism. However, we also recognize the reduced coenzymes NADH and $FADH_2$ as high-energy compounds in their own right because the transfer of electrons from these coenzymes to oxygen is highly exergonic. Thus the reoxidation of the 12 reduced coenzyme molecules shown on the right side of reaction 14-3 is a very exergonic process that provides the energy needed to drive the synthesis of the majority of the ATP molecules produced during the complete oxidation of glucose.

For the release of that energy, we must look to the remaining stages of respiratory metabolism—electron transport and oxidative phosphorylation. Before doing so, however, we will consider several additional features of the TCA cycle: its regulation, its centrality in energy metabolism, and its role in other metabolic pathways.

Several TCA Cycle Enzymes Are Subject to Allosteric Regulation

Like all metabolic pathways, the TCA cycle must be carefully regulated to ensure that its level of activity reflects cellular needs for its products. The main regulatory sites are shown in Figure 14-10. Most of the control involves **allosteric regulation** of four key enzymes by the specific *effector molecules* that bind reversibly to them. As you may recall from Chapter 6, effector molecules can be either inhibitors or activators, which are indicated in Figure 14-10 as red minus

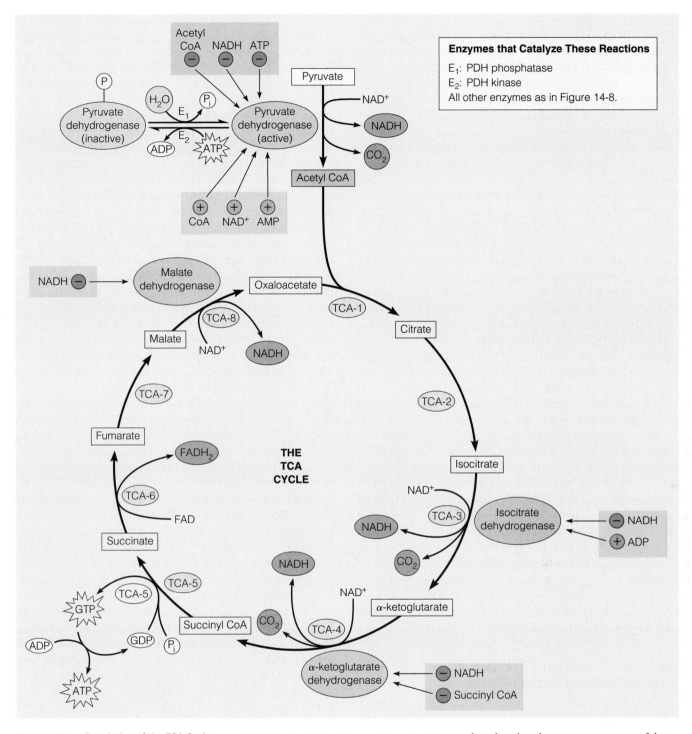

Figure 14-10 Regulation of the TCA Cycle. The TCA cycle and the prefatory pyruvate dehydrogenase reaction are shown here in outline form, with regulatory enzymes highlighted in blue. Major regulatory effects are indicated as either activation (+) or inhibition (−). Allosteric regulators include CoA, NAD$^+$, AMP, and ADP as activators and acetyl CoA, NADH, ATP, and succinyl CoA as inhibitors. In addition to its allosteric effect on pyruvate dehydrogenase activity, ATP activates PDH kinase (E$_2$), the enzyme that phosphorylates one component of the PDH complex, thereby converting it to an inactive form. The enzyme PDH phosphatase (E$_1$) removes the phosphate group, returning the enzyme to its active form.

signs and green plus signs, respectively. In addition to allosteric regulation, the pyruvate dehydrogenase complex is controlled by reversible phosphorylation and dephosphorylation of one of its protein components, thereby regulating the production of acetyl CoA from pyruvate.

To appreciate the logic of its regulation, remember that the TCA cycle uses acetyl CoA as its substrate and generates as its products NADH, FADH$_2$, CO$_2$, and ATP (see reaction 14-3). FADH$_2$ is membrane-bound and CO$_2$ is a gas, so it seems reasonable to exclude these as likely regulatory molecules. However, the other three products—NADH, ATP, and acetyl CoA—are important allosteric effectors of one or more enzymes, as shown in Figure 14-10. In addition, NAD^+, ADP, and AMP each activate at least one of the regulatory enzymes. In this way, the cycle is highly sensitive to the energy status of the cell, as assessed by both the $NADH/NAD^+$ ratio and the relative concentrations of ATP, ADP, and AMP.

All four of the NADH-generating dehydrogenases shown in Figure 14-10 are allosterically inhibited by NADH. An increase in the NADH concentration of the mitochondrion therefore decreases the activities of these dehydrogenases, leading to a reduction in TCA cycle activity. In addition, pyruvate dehydrogenase (PDH) is inhibited by ATP, and both PDH and isocitrate dehydrogenase are activated by "discharged" forms of ATP (AMP and ADP, respectively).

The overall availability of acetyl CoA is determined primarily by the activity of the PDH complex (see reaction 14-1), which is allosterically inhibited by NADH, ATP, and acetyl CoA and is activated by NAD^+, AMP, and free CoA (Figure 14-10). In addition, this enzyme complex is inactivated by phosphorylation of one of its protein components when the [ATP]/[ADP] ratio in the mitochondrion is high and is activated by removal of the phosphate group when the [ATP]/[ADP] ratio is low. These phosphorylation and dephosphorylation reactions are catalyzed by *PDH kinase* and *PDH phosphatase*, respectively. Not surprisingly, ATP is an activator of the kinase and an inhibitor of the phosphatase. As a result of these multiple control mechanisms, the generation of acetyl CoA is sensitive to the [acetyl CoA]/[CoA] and [NADH]/[NAD^+] ratios within the mitochondrion and to the mitochondrial ATP status as well.

In addition to these regulatory effects on reactions of the TCA cycle, feedback control from the cycle to the glycolytic pathway is provided by the inhibitory effects of citrate and acetyl CoA on phosphofructokinase and pyruvate kinase, respectively (see Figures 13-13 and 14-10).

The TCA Cycle Also Plays a Central Role in the Catabolism of Fats and Proteins

It is important to understand the central role of the TCA cycle in all of aerobic energy metabolism. Thus far, we have regarded glucose as the main substrate for cellular respiration. True, a variety of alternative carbohydrate substrates were considered in Chapter 13 (see Figure 13-10), but most summary reactions written for chemotrophic energy metabolism assume glucose as the starting compound, as we have done in reaction 14-3. In a sense, the assumption is very reasonable, given the importance of glucose as an energy source for the chemotrophic world. However, we must also note the roles of other substrates in cellular energy metabolism and the centrality of the TCA cycle in the catabolism of a variety of alternative fuel molecules, especially fats and proteins. Far from being a minor pathway for the catabolism of a single sugar, the TCA cycle represents the main conduit of aerobic energy metabolism. Moreover, it plays that role across a broad spectrum of organisms from microbes to higher plants and animals, which indicates that this process arose early in evolution.

Fat as a Source of Energy. When we first encountered fats in Chapter 3, we noted their role in energy storage and observed that they are highly reduced compounds that liberate more energy per gram upon oxidation than do carbohydrates. For this reason, fats are an important long-term energy storage form for many organisms. Storage polysaccharides such as starch and glycogen are important as localized, mobilizable energy reserves, but for long-term energy stores, fats are often used. Fat reserves are especially important in hibernating animals and migrating birds and also represent a common (though by no means the only) form in which energy and carbon are stored by plants in their seeds. Fats are well suited for this storage function because they allow a maximum number of calories to be stored with a minimum of volume and weight, a feature of obvious significance for both animal motility and seed dispersal.

Most fat is stored as deposits of **triacylglycerols** (also called *triglycerides*), which are neutral triesters of *glycerol* and long-chain *fatty acids* (see Figure 3-27b). Catabolism of triacylglycerols begins with their hydrolysis to glycerol and free fatty acids, usually in response to a hormone such as epinephrine. The glycerol is channeled into the glycolytic pathway by oxidative conversion to dihydroxyacetone phosphate, as shown in Figure 13-10. The fatty acids are linked to coenzyme A to form fatty acyl CoAs, which are then oxidatively degraded by a sequential, stepwise process that involves the successive removal of two-carbon units as acetyl CoA. Thus, the fatty acids derived from fats, like the pyruvate derived from carbohydrates, are oxidatively converted into acetyl CoA, which is then further catabolized by the TCA cycle. Moreover, the enzymes of fatty acid oxidation are localized to the mitochondrion in many (though not all) eukaryotic cells, so the acetyl CoA derived from fats is usually generated and catabolized within the same cellular compartment (see Figure 14-1b).

The sequential process of fatty acid catabolism to acetyl CoA is called β **oxidation** because the initial oxidative event in each successive cycle occurs on the carbon atom in the β position of the fatty acid (i.e., the second carbon in from the carboxylic acid group). The process involves successive cycles of oxidative attack on the fatty acid, as shown in Figure 14-11. Each cycle begins with oxidation of the β carbon, results in the release of two carbon atoms as acetyl CoA, and leaves the fatty acid shortened by two carbon atoms but ready for another round of oxidation.

In brief, β oxidation of a fatty acid starts with an activation step in which the energy of ATP is used to form the *fatty acyl CoA* derivative (reaction FA-1 in Figure 14-11). This activated form of the fatty acid is then partially oxidized in a series of four reactions, three of which (FA-2,

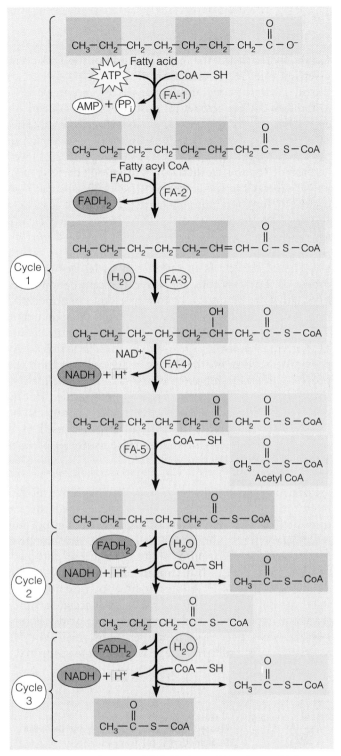

Figure 14-11 The Process of β Oxidation. Most fatty acids have an even number of carbon atoms ranging from $n = 10$ to $n = 24$ and must undergo $(n − 2)/2$ cycles of β oxidation to be catabolized to $n/2$ molecules of acetyl CoA, yielding $(n − 2)/2$ molecules each of NADH and FADH$_2$ in the process. The specific fatty acid shown here is octanoic acid (octanoate), a saturated fatty acid that has only eight carbon atoms and therefore undergoes just three cycles of β oxidation, yielding three molecules of NADH, three molecules of FADH$_2$, and four molecules of acetyl CoA. (Reaction numbers and chemical details are shown for the first cycle of β oxidation only.) The acetyl CoA molecules are then subject to further catabolism by the TCA cycle.

FA-3, and FA-4) parallel exactly the sequence whereby succinate is converted into oxaloacetate in the TCA cycle (see reactions TCA-6, TCA-7, and TCA-8 in Figure 14-8). Just as that series of reactions begins with the FAD-mediated dehydrogenation of succinate, here fatty acyl CoA is dehydrogenated to the corresponding *α,β-unsaturated acyl CoA*, with FAD as the electron acceptor (reaction FA-2). The enzyme that catalyzes this reaction, *fatty acyl dehydrogenase*, is an integral membrane protein with a molecule of FAD bound tightly to it.

In the next step (reaction FA-3), water is added across the double bond, generating *β-hydroxyl fatty acyl CoA* in the same way that malate is formed from fumarate. The hydroxyl group is then oxidized in an NAD$^+$-dependent reaction to form the corresponding *β-keto acid* (reaction FA-4). This compound is not very stable; in the final step of the sequence, another molecule of coenzyme A is taken up and the β-keto acid is split into acetyl CoA and an acyl CoA compound that is two carbon atoms shorter than the original molecule (reaction FA-5).

The newly formed acyl CoA compound is identical to the activated starting compound that served as the substrate for reaction FA-2, except that it is two carbons shorter. This new acyl CoA molecule then undergoes the same series of reactions represented by FA-2 through FA-5, with another two-carbon unit removed as acetyl CoA. Thus, β oxidation is a repetitive process in which oxidation continues down the fatty acid backbone, removing two carbon atoms at a time. The repetitive nature of β oxidation is illustrated in Figure 14-11 for a fatty acid with eight carbon atoms, which requires 3 cycles of β oxidation.

Protein as a Source of Energy. Proteins are not regarded primarily as energy sources because they have more fundamental roles in the cell—as enzymes, transport proteins, hormones, and receptors, for example. But proteins, too, can be catabolized to generate ATP if necessary. In animals, protein catabolism is prominent under conditions of fasting or starvation, or when the dietary intake of proteins exceeds the need for amino acids. In plants, catabolism of proteins is especially important during the germination of protein-storing seeds and in the senescence, or aging, of leaves. In addition, all cells undergo metabolic turnover of most proteins and protein-containing structures, and the amino acids to which the proteins are degraded can either be recycled into proteins or degraded oxidatively to yield energy.

Protein catabolism begins with hydrolysis of the peptide bonds that link amino acids together in the polypeptide chain. The process is called **proteolysis,** and the enzymes responsible for it are called *proteases*. The products of proteolytic digestion are small peptides and free amino acids. Further digestion of peptides is catalyzed by peptidases, which either hydrolyze internal peptide bonds (*endopeptidases*) or remove successive amino acids from the end of the peptide (*exopeptidases*). Exopeptidases are in turn either *aminopeptidases* or *carboxypeptidases,* depending on the end of the peptide from which digestion proceeds.

Free amino acids, whether ingested as such or obtained by the digestion of proteins, can be catabolized for energy. The pathways by which they are degraded illustrate the general principle that alternative substrates are converted to intermediates of mainstream catabolism in as few steps as possible. In spite of their number and chemical diversity, all these pathways eventually lead to a few key intermediates in the TCA cycle, notably acetyl CoA, α-ketoglutarate, oxaloacetate, fumarate, and succinyl CoA.

Most pathways for amino acid catabolism begin with removal of the amino group, either by transamination or by direct oxidative deamination (Figure 14-12). **Transamination** involves the transfer of an amino group from an amino acid to an α-keto acid acceptor (Figure 14-12a). The deamination of the amino acid is therefore accompanied by the amination of the α-keto acid. Transamination is a common means of shifting amino groups between carbon skeletons and is used by the cell not only in degradative processes but also in synthetic pathways. Transamination does not accomplish the net removal of nitrogen from the pool of organic molecules, as must happen if net catabolism of amino acids is to occur. It does, however, allow the transfer of amino groups to several common carbon skeletons, especially glutamate. The amino group can then be liberated as free ammonia by a process called **oxidative deamination** (Figure 14-12b). All amino acids can undergo transamination, but only a few can be oxidatively deaminated.

Of the 20 amino acids found in proteins, three give rise to TCA cycle intermediates or precursors directly. These are the amino acids alanine, aspartate, and glutamate, which can be transaminated to form pyruvate, oxaloacetate, and α-ketoglutarate, respectively (Figure 14-13). All the other amino acids require more complicated pathways, often with many intermediates. You will probably encounter these pathways at some future point, most likely in a biochemistry course. When you do, be sure to note how many of them have end-products that are TCA cycle intermediates, because that will further emphasize the centrality of the TCA cycle for all of cellular energy metabolism, regardless of the starting substrate.

The TCA Cycle Serves as a Source of Precursors for Anabolic Pathways

Catabolism is clearly the major function of the TCA cycle, given its role in the oxidation of acetyl CoA to carbon dioxide, with the concomitant conservation of free energy as reduced coenzymes. Yet in most cells, there is a considerable flow of four-, five-, and six-carbon intermediates into and out of the cycle. These side reactions replenish the supply of intermediates in the cycle as needed as well as provide for the synthesis of compounds derived from any of several intermediates in the cycle. Because the TCA cycle can function both in a catabolic mode and as a source of precursors for anabolic pathways, it is called an **amphibolic pathway** (from the Greek prefix *amphi-*, meaning "both").

The TCA cycle is involved in a variety of anabolic processes. For example, the three transamination reactions shown in Figure 14-13 convert α-keto intermediates into the amino acids alanine, aspartate, and glutamate. These amino acids are constituents of proteins, so the TCA cycle is indirectly involved in protein synthesis by providing several of the amino acids required for the process. Other amphibolic precursors in the cycle include succinyl CoA and citrate. Succinyl CoA is the starting point for the biosynthesis of heme, whereas citrate can be transported out of the mitochondrion and used as a source of acetyl CoA for the stepwise synthesis of fatty acids in the cytosol.

In each of these cases, intermediates of the TCA cycle are drawn off for biosynthetic purposes. This, in turn, dictates the need for mechanisms that can replenish the intermediates in the cycle as needed. The most important replenishment reactions are those in which oxaloacetate is formed by carboxylation of either pyruvate or phosphoenolpyruvate (PEP). Transamination reactions can also serve as a source of TCA intermediates by functioning in the direction of α-ketoglutarate and oxaloacetate synthesis at the expense of the analogous amino acids (Figure 14-13). By a variety of such side reactions, the TCA cycle acquires a metabolic versatility well beyond its primary catabolic role.

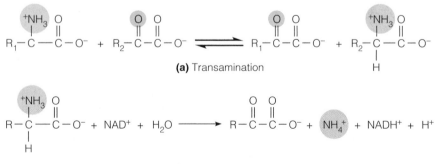

(a) Transamination

(b) Oxidative deamination

Figure 14-12 Transamination and Oxidative Deamination. **(a)** Transamination of an amino acid involves transfer of the amino group to an α-keto acid acceptor. **(b)** In oxidative deamination, the amino group is liberated as NH_4^+ or NH_3, depending on the pH. All amino acids can undergo transamination, but only a few can be oxidatively deaminated.

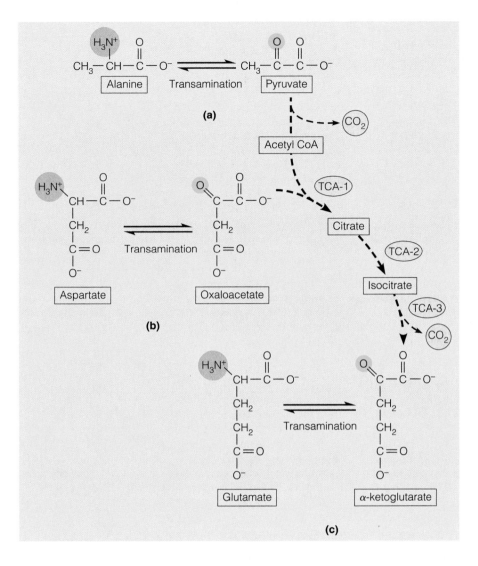

Figure 14-13 Interconversion of Several Amino Acids and Their Cognate Keto Acids in the TCA Cycle. The amino acids **(a)** alanine, **(b)** aspartate, and **(c)** glutamate can be reversibly transaminated into the corresponding α-keto acids: pyruvate, oxaloacetate, and α-ketoglutarate, respectively. Each of these keto acids is an intermediate in the TCA cycle, a portion of which is shown to provide the metabolic context for these transamination reactions. In each case, the amino group is shown in blue and the keto group is shown in yellow. These transamination reactions are readily reversible and perform the catabolic function of converting amino acids to TCA-cycle intermediates for oxidation to CO_2 and H_2O and the anabolic function of converting TCA-cycle intermediates to amino acids for use in protein synthesis. The degradative and synthetic pathways for other amino acids are less direct but also lead to and from intermediates in respiratory metabolism.

The Glyoxylate Cycle Converts Acetyl CoA to Carbohydrates

A pathway that is related to the TCA cycle but that performs a specialized anabolic function is the *glyoxylate cycle* described in Box 14A. The glyoxylate cycle shares several reactions with the TCA cycle but lacks the two decarboxylating reactions of the latter cycle. Instead, two acetate molecules (which enter the pathway as acetyl CoA) are used to generate succinate, a four-carbon compound (see Figure 14A-2). The succinate is then converted to pyruvate, from which sugars can be synthesized by gluconeogenesis.

Organisms that possess this pathway can synthesize sugars from two-carbon compounds such as acetate. The glyoxylate cycle also makes possible the conversion of stored fat to carbohydrate, with acetyl CoA as the intermediate. This capability is vital to seed germination in those plant species that store significant amounts of carbon reserves in their seeds as fats. In the seedlings of such species, both β-oxidation and the glyoxylate cycle take place in organelles called *glyoxysomes* (see Figure 14A-2).

Electron Transport: Electron Flow from Coenzymes to Oxygen

Having considered the first two stages of aerobic respiration—glycolysis and the TCA cycle—let's pause briefly to ask what has been achieved thus far. As reaction 14-3 indicates, chemotrophic energy metabolism through the TCA cycle accounts for the synthesis of four ATP molecules per glucose, two from glycolysis and two from the TCA cycle. This modest enhancement in ATP yield hardly seems to justify the metabolic jungle we have already come through. Where, one might ask, is all the rest of the free energy? And when will we get to the substantially greater ATP yield that is supposed to be characteristic of respiration?

The answer is straightforward: The free energy is right there in reaction 14-3, represented by the reduced coenzyme molecules NADH and $FADH_2$. As we will see shortly, large amounts of free energy are released when these reduced coenzymes are reoxidized by transfer of their electrons to molecular oxygen. In fact, about 90% of

Further Insights THE GLYOXYLATE CYCLE, GLYOXYSOMES, AND SEED GERMINATION

Plant species that store substantial carbon and energy reserves in their seeds as fats face a special metabolic challenge when their seeds germinate: They must convert the stored fat to sucrose, which is the immediate source of carbon and energy for most cells in the seedling. Many plant species are in this category, including such well-known oil-bearing species as soybeans, peanuts, sunflowers, castor beans, and maize. The fat consists mainly of triacylglycerols (triglycerides) and is stored as *lipid bodies,* either in the cotyledons of the plant embryo or in the endosperm, the nutritive tissue that surrounds and nourishes the developing embryo. The electron micrograph in Figure 14A-1 shows the prominence of lipid bodies in the cotyledon of a cucumber seedling.

The advantage of storing fat rather than carbohydrate is clear when you consider that one gram of triacylglycerol contains more than twice as much energy as one gram of carbohydrate. This difference enables fat-storing species to pack the greatest amount of carbon and calories into the least amount of space. However, it also means that such species must be able to convert the stored fat into sugar when the seeds germinate.

The conversion of fat to sugar is not possible for most organisms. Many organisms readily convert sugars and other carbohydrate to stored fat—some of us all too readily, in fact! But most eukaryotic organisms cannot carry out the reverse process. For the seedlings of fat-storing plant species, however, the conversion of storage triacylglycerols to sucrose is essential because sucrose is the form in which carbon and energy are transported to the growing shoot and root tips of the developing seedling.

The metabolic pathways that make this conversion possible are β oxidation, with which you should already be familiar, and the **glyoxylate cycle,** with which you probably are not. The function of β oxidation is to degrade the stored fat to acetyl CoA (see

Figure 14-11). The acetyl CoA then enters the glyoxylate cycle (Figure 14A-2), a five-step cyclic pathway that is named for one of its intermediates, the two-carbon keto acid called *glyoxylate.* The glyoxylate cycle is related to the TCA cycle, with which it has three reactions in common. There is a critical difference, however: The glyoxylate cycle bypasses the two decarboxylation reactions of the TCA cycle at which carbon dioxide would otherwise be evolved. Instead, the glyoxylate cycle has a strategic two-reaction sequence that enables it to take in not one but two molecules of acetyl CoA per cycle, generating succinate, a four-carbon compound. Thus, the glyoxylate cycle is anabolic (carbon enters as two-carbon molecules and leaves as a four-carbon molecule), whereas the TCA cycle is catabolic (carbon enters as a two-carbon molecule but leaves as two CO_2 molecules).

In the seedlings of fat-storing species, the enzymes of β oxidation and the glyoxylate cycle are localized in organelles called **glyoxysomes.** Recall from Chapter 12 that a glyoxysome is a specialized kind of plant peroxisome, found only in the seedlings of fat-storing species (and sometimes in senescing leaves). Glyoxysomes are visible in the electron micrograph of Figure 14A-1. The intimate association of glyoxysomes with lipid bodies presumably facilitates the transfer of fatty acids from the latter to the former.

Figure 14A-2 brings all the relevant metabolism together in an intracellular context. Storage triacylglycerols are hydrolyzed in the lipid bodies, releasing fatty acids. These are transported into the glyoxysome, where they are degraded by β oxidation to acetyl CoA. The acetyl CoA is converted to succinate by the enzymes of the glyoxylate cycle. The succinate moves to the mitochondrion, where it is converted via fumarate to malate by a reaction sequence that is a part of the TCA cycle. (Notice that mitochondria are also present in the cucumber cotyledon in

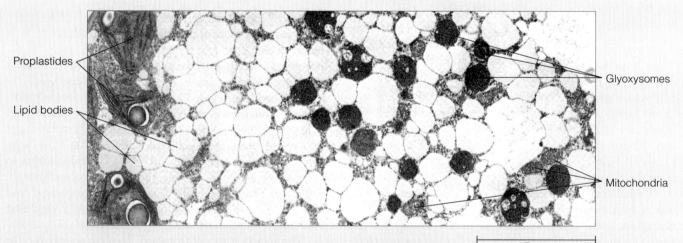

Proplastides

Lipid bodies

Glyoxysomes

Mitochondria

5 μm

Figure 14A-1 The Association of Glyoxysomes and Lipid Bodies in Fat-Storing Seedlings.   Shown here is a cell from the cotyledon of a cucumber seedling during early postgerminative development. The glyoxysomes and mitochondria involved in fat mobilization and gluconeogenesis (or "sucroneogenesis," to coin a term) are intimately associated with the lipid bodies in which the fat is stored. The fat is present primarily as triacylglycerols (also called triglycerides). Evidence that the cotyledon is not yet photosynthetically active and is therefore still heterotrophic in its nutritional mode can be seen in the presence of proplastids instead of mature chloroplasts (TEM).

Figure 14A-1.) Malate is then transported to the cytosol and oxidized to oxaloacetate, which is decarboxylated to form phosphoenolpyruvate (PEP). PEP serves as the starting point for gluconeogenesis, also a cytosolic pathway. Gluconeogenesis leads to the formation of the phosphorylated monosaccharides glucose-6-phosphate (G-6-P) and fructose-6-phosphate (F-6-P), from which sucrose can be synthesized. The route from stored triacylglycerols to sucrose is obviously quite complex, involving enzymes located in lipid bodies, glyoxysomes, mitochondria, and the cytosol. But it is the metabolic lifeline on which the seedlings of all fat-storing plant species depend. And it consolidates for us much of the metabolism we've been learning in Chapters 13 and 14, including gluconeogenesis, the TCA cycle, β oxidation—and now the glyoxylate cycle as well.

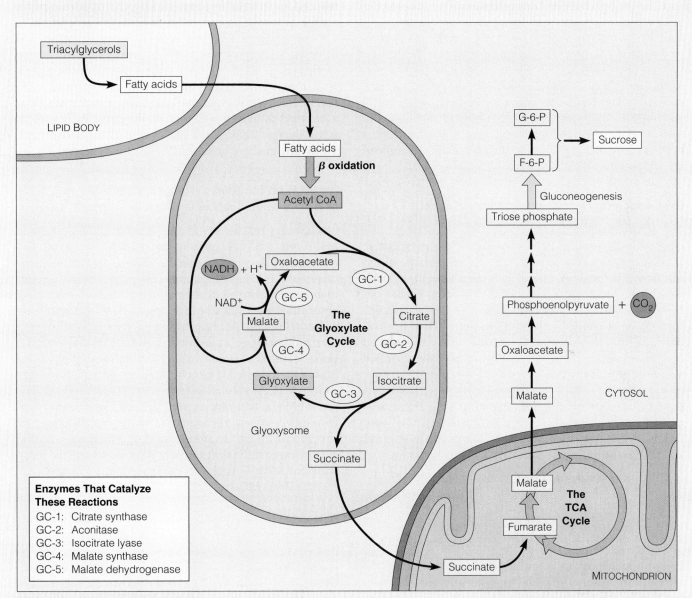

Figure 14A-2   The Glyoxylate Cycle and Gluconeogenesis in Fat-Storing Seedlings. Seedlings of fat-storing plant species can convert stored fat into sugar. Fatty acids derived from the hydrolysis of storage triacylglycerols are oxidized to acetyl CoA by the process of β oxidation. Acetyl CoA is then converted into succinate by the glyoxylate cycle, a five-reaction anabolic sequence named for glyoxylate, the molecule that is generated and consumed by reactions GC-3 and GC-4, respectively. All the enzymes of β oxidation and the glyoxylate cycle are located in the glyoxysome. Conversion of succinate to malate occurs within the mitochondrion, whereas the further metabolism of malate via phosphoenolpyruvate to hexoses and hence to sucrose takes place in the cytosol. In seedlings of fat-storing plant species such as soybean, peanut, maize, and castor bean, acetyl CoA is derived from β oxidation of fatty acids. In bacteria and eukaryotic microorganisms capable of growing on two-carbon substrates such as ethanol or acetate, acetyl CoA is generated from acetyl phosphate, which is formed by ATP-dependent phosphorylation of free acetate.

the potential free energy present in the chemical bonds of glucose is conserved in the 10 molecules of NADH and the 2 molecules of $FADH_2$ formed when a molecule of glucose is oxidized to CO_2 (see reaction 14-3). Such coenzymes are therefore high-energy compounds in their own right. Put more rigorously, *reduced coenzymes carry high-energy electrons with potential energy that can be tapped to drive ATP synthesis.* But to be tapped for ATP synthesis, the energy of NADH and $FADH_2$ must first be converted into the potential energy of a *transmembrane electrochemical proton gradient.*

The process of coenzyme reoxidation by the transfer of electrons to oxygen is called **electron transport.** Electron transport is the third stage of respiratory metabolism (see Figure 14-1b). The accompanying process of ATP synthesis (stage 4) will be discussed later in this chapter. Keep in mind, however, that electron transport and ATP synthesis are not isolated processes. They are both integral parts of cellular respiration and are functionally linked to each other by the electrochemical proton gradient that is both the "product" of electron transport and the source of the energy that drives ATP synthesis.

The Electron Transport System Conveys Electrons Stepwise from Reduced Coenzymes to Oxygen

Clearly, much of the energy released by the oxidative catabolism of carbohydrates, fats, or proteins is stored transiently in the high-energy reduced coenzymes generated in the process. Moreover, these oxidative events can be sustained only if oxidized coenzyme molecules continue to be available as electron acceptors. That, in turn, depends on the continuous reoxidation of reduced coenzymes by the transfer of electrons to the terminal acceptor—O_2, in the case of aerobic respiration.

Electron Transport and Coenzyme Oxidation. Electron transport involves oxidation of the coenzymes NADH and $FADH_2$ with molecular oxygen as the ultimate electron acceptor, so we can write summary reactions as follows:

$$NADH + H^+ + \tfrac{1}{2}O_2 \longrightarrow NAD^+ + H_2O$$
$$\Delta G^{o\prime} = -52.4 \text{ kcal/mol} \quad \textbf{(14-4)}$$

$$FADH_2 + \tfrac{1}{2}O_2 \longrightarrow FAD + H_2O$$
$$\Delta G^{o\prime} = -45.9 \text{ kcal/mol} \quad \textbf{(14-5)}$$

Electron transport therefore accounts not only for the reoxidation of coenzymes and the consumption of oxygen but also for the formation of water, which we recognize as the reduced form of oxygen and one of the two end-products of aerobic energy metabolism. (The other end-product, of course, is the carbon dioxide generated in the TCA cycle; see Figure 14-8 and reaction 14-19.)

The Electron Transport System. The most important aspect of reactions 14-4 and 14-5 is the large amount of free energy released upon oxidation of NADH and $FADH_2$ by the transfer of electrons to oxygen. The $\Delta G^{o\prime}$ values for

these reactions make it clear that the oxidation of a coenzyme is an extraordinarily exergonic process. It should therefore come as no surprise that electrons are not passed directly from reduced coenzymes to oxygen; such direct transfers would liberate excessive amounts of energy as heat. Rather, the transfer is accomplished as a multistep process that involves a series of reversibly oxidizable electron carriers functioning together in what is called the **electron transport system (ETS).** In this way, the total free energy difference between reduced coenzymes and oxygen is parceled out among a series of electron transfers and is released in increments, maximizing the opportunity for energy conservation and, indirectly, for ATP generation.

Our discussion of the electron transport system will focus on three questions: (1) What are the major electron carriers in the ETS? (2) What is the sequence of these carriers in the system? (3) How are these carriers organized in the membrane to ensure that the flow of electrons from reduced coenzymes to oxygen is coupled to the pumping of protons across the membrane, thereby maintaining the electrochemical proton gradient on which ATP synthesis depends?

The Electron Transport System Consists of Five Different Kinds of Carriers

The carriers that make up the ETS include *flavoproteins, iron-sulfur proteins, cytochromes, copper-containing cytochromes,* and a quinone called *coenzyme Q.* Most of these carriers absorb light of a particular wavelength, allowing them to be identified experimentally by their characteristic absorption spectra. Except for coenzyme Q, all the carriers are proteins with specific prosthetic groups capable of being reversibly oxidized and reduced. Almost all the events of electron transport occur within membranes, so it is not surprising that, except for cytochrome c, all of these carriers are hydrophobic molecules. In fact, most of these intermediates occur in the membrane as parts of large assemblies of proteins called *respiratory complexes.* We will look briefly at the chemistry of these electron carriers and then see how they are organized into respiratory complexes and ordered into a sequence of carriers that transfer electrons from reduced coenzymes to oxygen.

Flavoproteins. Several membrane-bound **flavoproteins** participate in electron transport, using either *flavin adenine dinucleotide (FAD)* or *flavin mononucleotide (FMN)* as the prosthetic group. FMN is essentially the flavin-containing half of the FAD molecule shown in Figure 14-9. An example of a flavoprotein is *NADH dehydrogenase,* an FMN-containing protein that is part of the protein complex responsible for the transfer of electrons from NADH in the mitochondrial matrix to coenzyme Q in the inner membrane. Another example, already familiar to us from the TCA cycle, is the enzyme succinate dehydrogenase, which has FAD as its prosthetic group. Unlike the other enzymes of the TCA cycle, succinate dehydrogenase is an integral membrane protein. It is, in fact, the respiratory complex that transfers electrons from succinate via its

FAD prosthetic group to coenzyme Q. As we will see shortly, both NADH dehydrogenase and succinate dehydrogenase are major structural components of the ETS as it functions in the membrane. An important characteristic of the flavoproteins (and of the coenzyme NADH as well) is that they transfer both electrons and protons as they are reversibly oxidized and reduced.

Iron-Sulfur Proteins. **Iron-sulfur proteins,** also called *nonheme iron proteins,* are a family of proteins, each with *iron-sulfur (Fe-S) center* that consists of iron and sulfur atoms complexed with cysteine groups of the protein. Iron-sulfur proteins are the most numerous intermediates in the mitochondrial ETS. At least a dozen different Fe-S centers are known to be involved in the mitochondrial transport system. The iron atoms of the iron-sulfur centers are the actual electron acceptors and donors of the iron-sulfur proteins. Each iron atom alternates between the Fe^{3+} (ferric) and Fe^{2+} (ferrous) states as the centers are oxidized and reduced. In contrast to NADH and flavoproteins, oxidation and reduction of an iron-sulfur center involves the transfer of a single electron rather than an electron pair. Moreover, iron-sulfur proteins do not pick up and release protons as they cycle between the oxidized and reduced states, a point whose importance will shortly become clear.

Cytochromes. Like the iron-sulfur proteins, **cytochromes** also contain iron, but as part of a porphyrin prosthetic group called *heme,* which you probably recognize as a component of hemoglobin (see Figure 3-4). There are at least five different kinds of cytochromes in the electron transport system, designated as cytochromes *b, c, c₁, a,* and a_3, each with its own characteristic absorption spectrum. Cytochromes *b, c,* and c_1 all contain a form of heme called *iron-protoporphyrin IX* (Figure 14-14), whereas cytochromes *a* and a_3 contain a modified prosthetic group called *heme A.*

The iron atom of the heme prosthetic group, like that of the iron-sulfur center, is reversibly oxidizable and serves as the electron acceptor for the cytochromes. Both cytochromes and iron-sulfur proteins are therefore one-electron carriers that transfer electrons but not protons. Cytochromes *b, c₁, a,* and a_3 are integral membrane proteins, the latter two occurring together at the end of the transport sequence as part of a respiratory complex called *cytochrome c oxidase.* Cytochrome *c,* on the other hand, is a peripheral membrane protein that is loosely associated with the outer surface of the membrane. Moreover, cytochrome *c* is not a part of a large complex and can therefore diffuse much more rapidly, a key property in its role in transferring electrons between protein complexes.

Copper-Containing Cytochromes. In addition to their iron atoms, cytochromes *a* and a_3 also contain copper atoms. In each case, a single copper atom is bound to the heme group of the cytochrome, where it associates with an iron atom to form a **bimetallic iron-copper (Fe-Cu) center.**

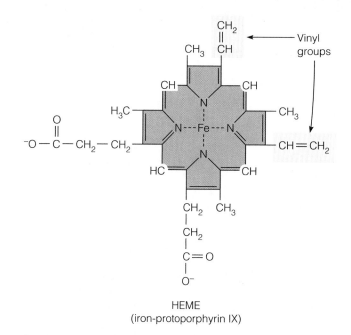

HEME
(iron-protoporphyrin IX)

Figure 14-14 The Structure of Heme. Heme, also called iron-protoporphyrin IX, is the prosthetic group in cytochromes *b, c,* and c_1. A similar molecule, called heme A, is present in cytochromes a_1 and a_3. The heme of cytochromes *c* and c_1 is covalently attached to the protein by thioether bonds between the sulfhydryl groups of two cysteines in the protein and the vinyl ($-CH=CH_2$) groups of the heme (highlighted in yellow). In other cytochromes, the heme prosthetic group is linked noncovalently to the protein.

Like iron atoms, copper ions can be reversibly converted from the oxidized to the reduced form—that is, from the Cu^{2+} (cupric) to the Cu^+ (cuprous) form—by accepting or donating single electrons. The iron-copper center plays a critical role in keeping an O_2 molecule bound to the cytochrome oxidase complex until it has picked up the requisite four electrons and four protons, at which point the oxygen atoms are released as two molecules of water. (A nutritional note: If you have ever wondered why you require the elements iron and copper in your diet, their roles in electron transport are part of the reason.)

Coenzyme Q. **Coenzyme Q (CoQ)** is the only nonprotein component of the ETS. Coenzyme Q is a quinone with the structure shown in Figure 14-15. Because of its ubiquitous occurrence in nature, coenzyme Q is also known as *ubiquinone.* Figure 14-15 also illustrates the reversible reduction, in two successive one-electron steps, from the quinone form (CoQ) via the *semiquinone* form (CoQH) to the *dihydroquinone* form (CoQH₂). Unlike most of the proteins of the ETS, coenzyme Q is not part of a respiratory complex. Instead, the nonpolar interior of the inner mitochondrial membrane (or of the plasma membrane, in the case of prokaryotes) contains a large pool of CoQ molecules that transfer electrons between other carriers. CoQ molecules are the most abundant electron carriers in the membrane. As we will see shortly, coenzyme Q occupies a central position in the ETS, serving as a collection

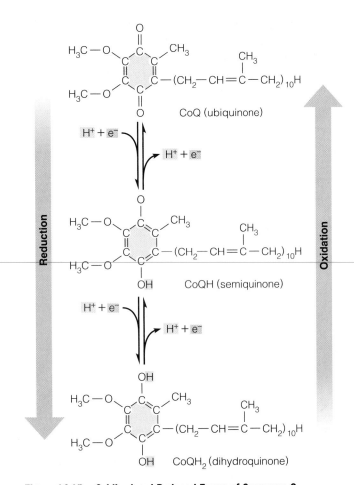

Figure 14-15 **Oxidized and Reduced Forms of Coenzyme Q.**
Coenzyme Q (also called ubiquinone) accepts both electrons and protons as it is reversibly reduced in two successive one-electron steps to form first CoQH (the semiquinone form) and then CoQH$_2$ (the dihydroquinone form).

point for electrons from the reduced prosthetic groups of FMN- and FAD-linked dehydrogenases in the membrane.

Note that coenzyme Q accepts not only electrons but also protons when it is reduced and that it releases both electrons and protons when it is oxidized. This property is vital to the role of coenzyme Q in the active transport, or pumping, of protons across the inner mitochondrial membrane. To understand why, imagine a closed membrane vesicle with coenzyme Q embedded in the membrane. Assume that whenever CoQ is reduced to CoQH$_2$, it always accepts protons from *inside* the vesicle. The reduced form, CoQH$_2$, then diffuses across the membrane to the outer surface, where it is oxidized to CoQ, with the protons ejected to the *outside* of the vesicle. Under these conditions, protons would be picked up by coenzyme Q on the inside and delivered to the outside, thereby providing a proton pump coupled to electron transport. As we will see, this mode of proton pumping is thought to be one of the mechanisms whereby mitochondria, chloroplasts, and prokaryotes establish and maintain the electrochemical proton gradients that are used to store the energy of electron transport.

The Electron Carriers Function in a Sequence Determined by Their Reduction Potentials

Now that we are acquainted with the kinds of electron carriers that make up the ETS, the next question concerns the sequence in which these carriers operate. To answer that question, we need to understand the **reduction potential, E',** which is a measure, in volts, of the *electron transfer potential,* or the relative reducing power, of a particular redox pair. A **redox (reduction-oxidation) pair** consists of two molecules or ions that are interconvertible by the loss or gain of electrons. For example, NAD$^+$ and NADH constitute a redox pair, as do the ferric and ferrous forms of iron. By convention, redox pairs are specified with the oxidized form given first, separated from the reduced form by a slash. Thus, NAD$^+$/NADH is a redox pair, as are Fe^{3+}/Fe^{2+} and O$_2$/H$_2$O.

Understanding Reduction Potentials. For any redox pair, the reduction potential refers to the *half-cell reaction,* which is defined as the reduction of the oxidized form of the pair. Thus, the respective half-cell reactions for the Fe^{3+}/Fe^{2+} and the NAD$^+$/NADH redox pairs are

$$Fe^{3+} + e^- \longrightarrow Fe^{2+} \qquad \textbf{(14-6)}$$
$$NAD^+ + H^+ + 2e^- \longrightarrow NADH \qquad \textbf{(14-7)}$$

Half-cell reactions such as these do not occur in isolation, of course; every reduction must be accompanied by an oxidation. However, the concept of half-cell reactions is useful because it provides us with a means of thinking about the tendency of a reduction to occur and enables us to quantify that tendency as an E' value. The E' values, in turn, allow us to compare redox pairs and to predict in which direction electrons will tend to flow when several redox pairs are present in the same system, as is clearly the case for the electron transport system.

Essentially, a reduction potential is a measure of the affinity that the oxidized form of a redox pair has for electrons or, in other words, how strongly the half-reaction tends toward reduction. For a redox pair to have a positive E' means that the oxidized form has a high affinity for electrons and is therefore a good electron *acceptor.* For example, the E' value for the Fe^{3+}/Fe^{2+} couple is highly positive, meaning that Fe^{3+} is a good electron acceptor and that the reaction 14-6 tends to proceed in the direction written under most conditions. On the other hand, the NAD$^+$/NADH couple has a highly negative E' value, meaning that NAD$^+$ is a poor electron acceptor and that reaction 14-7 will *not* tend to proceed in the direction written.

Alternatively, a negative E' value can be thought of as a measure of how good an electron *donor* the reduced form of a redox pair is. Thus, the highly negative E' value for the NADH pair means that NADH is a good electron donor and that reaction 14-7 will tend in the direction not of NAD$^+$ reduction, but of NADH oxidation.

Understanding Standard Reduction Potentials. To standardize calculations and comparisons of reduction potentials for various redox pairs, we clearly need values determined under specified conditions. For that purpose, we turn to the **standard reduction potential (E_0')**, which is the reduction potential for a redox pair under standard conditions (25°C, 1 M concentration, 1 atmosphere pressure, and pH 7.0). The standard reduction potentials of redox pairs relevant to energy metabolism are given in Table 14-2. For a description of how these values are determined experimentally, see Figure 14-16.

By convention, the $2H^+/H_2$ redox pair is used as a standard and is assigned the value 0.00 V (Table 14-2, boldface line). For a redox pair to have a positive standard reduction potential means that, under standard conditions, the oxidized form of the pair is a better electron acceptor, and therefore a better oxidizing agent, than H^+ (or, alternatively, that the reduced form of the pair is not

as good an electron donor, and hence a poorer reducing agent, than H_2). Conversely, a negative reduction potential means that the oxidized form of the pair has less affinity for electrons than H^+ (or, alternatively, that the reduced form is a better reducing agent than H_2).

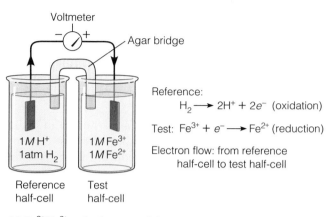

(a) Fe^{3+}/Fe^{2+} reduction potential

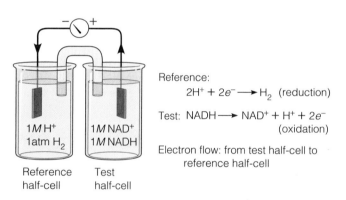

(b) $NAD^+/NADH$ reduction potential

Table 14-2 **Standard Reduction Potentials for Redox Pairs of Biological Relevance***

Redox Half-Reaction (oxidized form ⟶ reduced form)	No. of Electrons	E_0' (V)
Acetate ⟶ pyruvate	2	−0.70
Succinate ⟶ α-ketoglutarate	2	−0.67
Acetate ⟶ acetaldehyde	2	−0.60
3-phosphoglycerate ⟶ glyceraldehyde-3-P	2	−0.55
α-ketoglutarate ⟶ isocitrate	2	−0.38
NAD^+ ⟶ NADH	2	−0.32
FMN ⟶ $FMNH_2$	2	−0.30
1,3-bisphosphoglycerate ⟶ glyceraldehyde-3-P	2	−0.29
Acetaldehyde ⟶ ethanol	2	−0.20
Pyruvate ⟶ lactate	2	−0.19
FAD ⟶ $FADH_2$	2	−0.18**
Oxaloacetate ⟶ malate	2	−0.17
Fumarate ⟶ succinate	2	−0.03
$2H^+$ ⟶ H_2	**2**	**0.00*****
CoQ ⟶ $CoQH_2$	2	+0.04
Cytochrome b (Fe^{3+} ⟶ Fe^{2+})	1	+0.07
Cytochrome c (Fe^{3+} ⟶ Fe^{2+})	1	+0.25
Cytochrome a (Fe^{3+} ⟶ Fe^{2+})	1	+0.29
Cytochrome a_3 (Fe^{3+} ⟶ Fe^{2+})	1	+0.55
Fe^{3+} ⟶ Fe^{2+} (inorganic iron)	1	+0.77
O_2 ⟶ H_2O	2	+0.816

*Each $\Delta E_0'$ value is for the following half-reaction, where n is the number of electrons transferred:

oxidized form + nH^+ + ne^- ⟶ reduced form

$\Delta E_0'$ values are determined at pH 7.0 and 25°C relative to the standard hydrogen half-cell. For a two-electron reaction, a difference in $\Delta E_0'$ of 0.10 V corresponds to a $\Delta G^{\circ\prime}$ value of −4.6 kcal/mol.

**The value for the $FAD/FADH_2$ pair assumes the free coenzyme; when bound to a flavoprotein, the coenzyme has an $\Delta E_0'$ value in the range of from 0.0 to +0.3V, depending on the specific protein.

***This is the standard hydrogen half-cell; it requires that $[H^+]$ = 1.0M and therefore specifies pH 0.0. At pH 7.0, the value for the $2H^+/H_2$ pair is −0.42 V.

Figure 14-16 Determination of Standard Reduction Potentials. Standard reduction potentials are determined using an electrochemical cell that consists of two half-cells, each containing the oxidized and reduced forms of a redox pair under standard conditions. The *reference half-cell* (on the left) contains protons at a concentration of 1 M and hydrogen gas (H_2) at a pressure of 1 atmosphere. The *test half-cell* (on the right) contains the oxidized and reduced forms (both at 1 M) of the redox pair under study. An agar bridge connects the two half-cells to maintain electrical neutrality. The voltmeter that links the two half-cells measures the electromotive force (emf), which is the difference between the reduction potentials in the reference and test half-cells. Electrons may flow either away from or toward the reference half-cell, depending on whether H_2 has a greater or lesser tendency to lose electrons than the reduced form of the redox pair in the test half-cell. **(a)** When the Fe^{3+}/Fe^{2+} redox pair is in the test half-cell, as shown here, electrons flow from the reference half-cell to the test half-cell because H_2 loses electrons more readily than does Fe^{3+}. As a result, H_2 will be oxidized, Fe^{3+} ions will be reduced, and the voltmeter will record a positive emf: +0.77 V. The Fe^{3+}/Fe^{2+} redox pair therefore has a positive standard reduction potential. **(b)** When the test half-cell contains the $NAD^+/NADH$ redox pair, electron flow will be from the test half-cell to the reference half-cell because NADH is a better electron donor than H_2. Under these conditions, NADH will be oxidized, H^+ will be reduced, and the voltmeter will record a negative emf: −0.32 V. The $NAD^+/NADH$ redox pair therefore has a negative standard reduction potential.

Calculating E′ and ΔE′ Values. The E_0' values in Table 14-2 are valid only under standard conditions. To calculate E', the actual reduction potential under nonstandard conditions, an additional term is needed that takes into account the actual concentrations of the oxidized and reduced forms of the redox pair. The equation is

$$E' = E_0' + \frac{RT}{nF} \ln \frac{[\text{oxidized form}]}{[\text{reduced form}]} \qquad \textbf{(14-8)}$$

where E_0' is the standard reduction potential, R is the gas constant (1.987 cal/mol-K), T is the temperature in kelvins, n is the number of electrons transferred in the half-cell reaction (1 or 2; see Table 14-2), and F is the Faraday constant (23,062 cal/mol-V).

The redox pairs of Table 14-2 are arranged in order, with the most negative E_0' values (i.e., the best electron donors and hence the strongest reducing agents) at the top. Any redox pair shown in Table 14-2 can undergo a redox reaction with any other pair. The direction of such a reaction under standard conditions can be predicted by inspection, because, under standard conditions, *the reduced form of any redox pair will spontaneously reduce the oxidized form of any pair below it on the table.* Thus, NADH can reduce pyruvate to lactate but cannot reduce α-ketoglutarate to isocitrate.

The tendency of the reduced form of one pair to reduce the oxidized form of another pair can be quantified by determining $\Delta E_0'$, the difference in the E_0' values between the two pairs:

$$\Delta E_0' = E_{0,\text{acceptor}}' - E_{0,\text{donor}}' \qquad \textbf{(14-9)}$$

As you may have already guessed, $\Delta E_0'$ is a measure of thermodynamic spontaneity for the redox reaction between any two redox pairs under standard conditions. The spontaneity of a redox reaction under standard conditions can therefore be expressed as either $\Delta G^{\circ\prime}$ or $\Delta E_0'$. The sign convention for $\Delta E_0'$ is the opposite of that for $\Delta G^{\circ\prime}$, however: A thermodynamically feasible reaction is one with a negative $\Delta G^{\circ\prime}$ but a positive $\Delta E_0'$. For example, $\Delta E_0'$ for the transfer of electrons from NADH to O_2 (reaction 14-4) is calculated as follows, with NADH as the donor and O_2 as the acceptor:

$$\Delta E_0' = E_{0,\text{acceptor}}' - E_{0,\text{donor}}' = +0.816 - (-0.32) = +1.136 \text{ V}$$
$$\textbf{(14-10)}$$

The $\Delta E_0'$ value for the reaction is positive, so the transfer of electrons from NADH to O_2 is thermodynamically spontaneous under standard conditions. It is, in fact, this difference in reduction potentials between the NAD$^+$/NADH and O_2/H_2O redox pairs that drives the ETS and maintains the electrochemical proton gradient.

The Relationship Between $\Delta G^{\circ\prime}$ and $\Delta E_0'$. Because both $\Delta G^{\circ\prime}$ and $\Delta E_0'$ are measures of thermodynamic spontaneity, we should expect some sort of mathematical relation-

ship between them. The relationship is linear but with a change of sign because of the difference in conventions for $\Delta G^{\circ\prime}$ and $\Delta E_0'$. For any oxidation-reduction reaction, $\Delta G^{\circ\prime}$ is related to $\Delta E_0'$ by the equation

$$\Delta G^{\circ\prime} = -nF \, \Delta E_0' \qquad \textbf{(14-11)}$$

where n is the number of electrons transferred and F is the Faraday constant as previously defined. For example, the reaction of NADH with oxygen (reaction 14-4) involves the transfer of two electrons, so $\Delta G^{\circ\prime}$ for the reaction can be calculated as

$$\Delta G^{\circ\prime} = -2F \, \Delta E_0' = -2(23,062)(+1.136)$$
$$= -52,400 \text{ cal/mol} = -52.4 \text{ kcal/mol} \qquad \textbf{(14-12)}$$

Putting the Electron Transport System Together. We now have the information needed to put the pieces of the ETS together. As we already know, the respiratory ETS consists of several FMN- and FAD-linked dehydrogenases, iron-sulfur proteins with a total of 12 or more Fe/S centers, five cytochromes, including two with Fe/Cu centers, and a pool of coenzyme Q molecules that exist in the oxidized (CoQ), partially reduced (CoQH), or reduced (CoQH$_2$) states. Shown in Figure 14-17 are the major ETS components from free NADH ($E_0' = -0.32$ V) and the FADH$_2$ of succinate dehydrogenase ($E_0' = -0.18$ V) to oxygen ($E_0' = +0.816$ V), arranged according to their standard reduction potentials (E_0' values as obtained from Table 14-2). The E_0' scale is shown on the left; note that, by convention, the E_0' scale becomes increasingly negative (i.e., decreasingly positive) as it ascends. The scales on the right are for $\Delta E_0'$ and $\Delta G^{\circ\prime}$ calculated according to equations 14-9 and 14-11, respectively, for a two-electron reaction with O_2 as the electron acceptor.

In terms of energy, the key point of Figure 14-17 is that *the position of each carrier is determined by its standard reduction potential.* The electron transport system, in other words, consists of a series of chemically diverse electron carriers, with their order of participation in electron transfer determined by their relative reduction potentials. This means, in turn, that electron transfer from NADH or FADH$_2$ at the top of the figure to O_2 at the bottom is spontaneous and exergonic, with some of the transfers between successive carriers characterized by quite large differences in E_0' values and hence by large changes in free energy.

Most of the Carriers Are Organized into Four Large Respiratory Complexes

Although there are many electron carriers involved in the ETS, most are not present in the membrane as separate entities. Instead, the behavior of these carriers in mitochondrial extraction experiments indicates that they are organized into *multiprotein complexes.* For example, when mitochondrial membranes are extracted under gentle conditions, only coenzyme Q and cytochrome c are read-

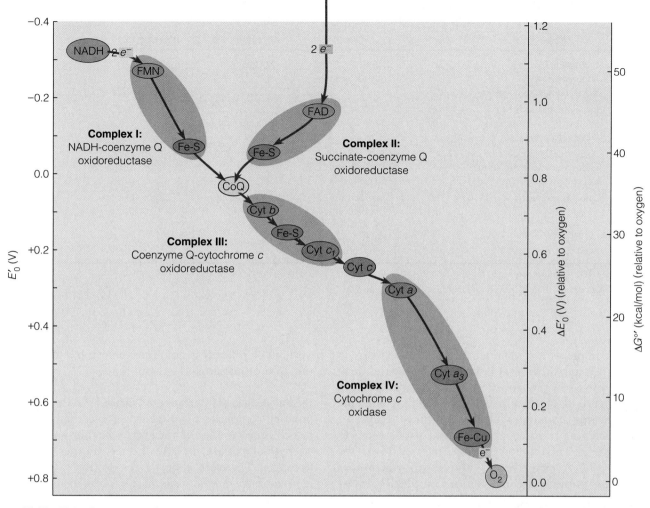

Figure 14-17 Major Components of the Respiratory Complexes and Their Energetics. The major intermediates in the transport of electrons from NADH (−0.32 V) and FADH$_2$ (−0.18 V) to oxygen (+0.816 V) are positioned vertically according to their energy levels, as measured by their standard reduction potentials (E_0', left axis). The four respiratory complexes are shown as brown ovals, with the major electron carriers in each complex enclosed within inset ovals. Coenzyme Q and cytochrome c are small, mobile intermediates that transfer electrons between the several complexes as they collide with them as a result of random diffusion, either within the membrane (coenzyme Q) or at the outer membrane surface (cytochrome c). The pink lines trace the exergonic flow of electrons through the system. On the right axes are the $\Delta E_0'$ and $\Delta G^{\circ\prime}$ values relative to oxygen (i.e., the changes in the standard reduction potential and the standard free energy for the transfer of electrons to O$_2$).

ily removed. The other electron carriers remain bound to the inner membrane and are not released unless the membrane is exposed to detergents or concentrated salt solutions. Such treatments then extract the remaining electron carriers from the membrane not as individual molecules but as four large complexes.

Based on these and similar findings, most of the electron carriers in the ETS are thought to be organized within the inner mitochondrial membrane into four different kinds of **respiratory complexes.** These complexes are identified in Figure 14-17 by the brown ovals and Roman numerals. Figure 14-17, in other words, indicates not only the ordering of the various electron carriers on the basis of their E_0' values but also their organization within the membrane.

Properties of the Respiratory Complexes. Each of the respiratory complexes consists of a distinctive assembly of polypeptides and prosthetic groups, and each has a unique role to play in the electron transport process. Table 14-3 summarizes some of the properties of these complexes.

Complex I transfers electrons from NADH to coenzyme Q and is called the **NADH-coenzyme Q oxidoreductase complex** (or *NADH dehydrogenase complex*). **Complex II** transfers to CoQ the electrons derived from succinate in reaction TCA-6. This complex is properly called the **succinate–coenzyme Q oxidoreductase complex,** although it is often also referred to by its more common name, *succinate dehydrogenase.* (Note that succinate dehydrogenase can be referred to as either an enzyme in the TCA cycle or a respiratory complex; in reality, it is both.) Similar but

Table 14-3 Properties of the Mitochondrial Respiratory Complexes

Respiratory Complex		Number of Polypeptides*	Prosthetic Groups	Electron Flow		Protons translocated (per electron pair)
Number	Name			Accepted from	Passed to	
I	NADH–coenzyme Q oxidoreductase (NADH dehydrogenase)	42 (7)	1 FMN 6–9 Fe-S centers	NADH	Coenzyme Q	4
II	Succinate–coenzyme Q oxidoreductase (succinate dehydrogenase)	4 (0)	1 FAD 3 Fe-S centers	Succinate (via enzyme-bound FAD)	Coenzyme Q	0
III	Coenzyme Q–cytochrome c oxidoreductase (cytochrome b-c_1 complex)	11 (1)	2 cytochrome b 1 cytochrome c_1 1 Fe-S center	Coenzyme Q	Cytochrome c	4**
IV	Cytochrome c oxidase	13 (3)	1 cytochrome a 1 cytochrome a_3 2 Cu centers (as Fe-Cu centers with cytochrome a_3)	Cytochrome c	Oxygen (O_2)	2

*The number of polypeptides encoded by the mitochondrial genome is indicated in parentheses for each complex.

**The value for complex III includes 2 protons translocated by coenzyme Q.

separate complexes are required to transfer electrons to coenzyme Q from other FAD-linked dehydrogenases, such as that involved in the oxidation of fatty acids (reaction FA-2 of Figure 14-11). **Complex III** is called the **coenzyme Q–cytochrome c oxidoreductase complex** because it accepts electrons from coenzyme Q and passes them to cytochrome c. This complex is also referred to as the *cytochrome b/c_1 complex* because those two cytochromes are its most prominent components. **Complex IV** transfers electrons from cytochrome c to oxygen and is called **cytochrome c oxidase.**

Each of these complexes consists of multiple polypeptide subunits and has as its prosthetic groups flavin nucleotides (complexes I and II), cytochromes (complexes III and IV), iron-sulfur centers (complexes I, II, and III), and/or iron-copper centers (complex IV). Complex I, for example, has a mass of about 850 kDa and contains about 42 polypeptides, an FMN prosthetic group, and 6–9 Fe/S centers. The respiratory complexes are very numerous in the inner mitochondrial membrane, as are the F_oF_1 complexes responsible for ATP synthesis. One estimate puts the numbers of complexes I, II, III, and IV in the inner membrane of a single liver mitochondrion at about 2600, 5500, 5500, and 15,600, respectively, in addition to about 15,000 F_oF_1 complexes.

Figure 14-18 places the three complexes needed for NADH oxidation (complexes I, III, and IV) in their proper eukaryotic context, the inner mitochondrial membrane. (Complex II is not included in the figure because it is not involved in NADH oxidation.) Also shown in the figure is the outward pumping of protons across the membrane that accompanies electron transport within the membrane. Each of these complexes represents one site of proton pumping. Complex II, on the other hand, is not a site of coupled proton pumping. The passage of electrons through complex II is not accompanied by the translocation of protons across the membrane.

The Unique Role of Cytochrome c Oxidase. Of the several respiratory complexes involved in aerobic respiration, only cytochrome c oxidase, at the end of the transport system, is a **terminal oxidase,** capable of direct transfer of electrons to oxygen. Therefore, almost every electron extracted from any oxidizable organic molecule anywhere in an aerobic cell passes eventually through cytochrome c oxidase. This complex is therefore the critical link between aerobic respiration and the oxygen that makes it all possible. Cyanide and azide ions are highly toxic to aerobic cells because they bind tightly to the iron-copper center of cytochrome c oxidase, thereby blocking all electron transport.

The Respiratory Complexes Move Freely Within the Inner Membrane

Unlike some integral proteins of the plasma membrane, the protein complexes of the mitochondrial inner membrane are mobile, free to diffuse within the plane of the membrane. This diffusion can be demonstrated experimentally. In one study, for example, inner mitochondrial membranes were placed in an electric field and the distribution of protein complexes was assessed subsequently by freeze-fracture analysis. Particles corresponding to protein complexes were found to accumulate at one end of the membrane. When the electrical field was turned off, the particles returned to a random distribution within seconds, clearly demonstrating their freedom to diffuse in a fluid lipid bilayer.

The results of this and similar studies make it clear that the respiratory complexes are not lined up in the membrane in the orderly fashion often seen in textbook

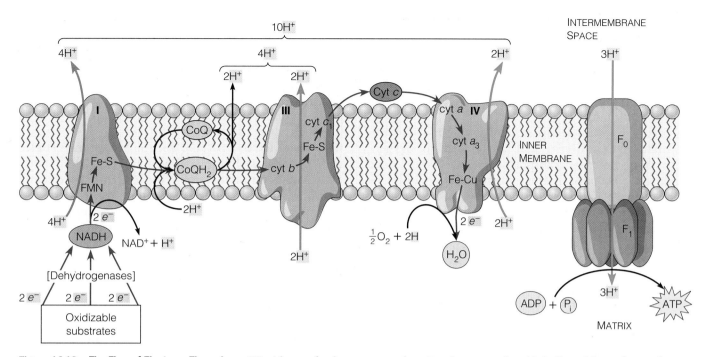

Figure 14-18 The Flow of Electrons Through Respiratory Complexes I, III, and IV with Concomitant Directional Proton Pumping. Electrons derived from oxidizable substrates in the mitochondrial matrix and carried to the surface of the inner membrane by NADH flow exergonically from NADH to oxygen via respiratory complexes I, III, and IV, with transfers between complexes I and III and between III and IV mediated by coenzyme Q (CoQ) and cytochrome c (cyt c), respectively. The main electron carriers of each complex are indicated; most are either Fe-S proteins or cytochromes. The number of protons pumped from the matrix into the intermembrane space per pair of electrons transferred is indicated for each complex. These numbers represent the best present consensus values in each case. The number of electrons required to drive the synthesis of one molecule by the F_oF_1 ATP synthase is most likely 3 or 4; when needed for energy calculations in this chapter, the value is assumed to be 3.

diagrams, but exist in the membrane as random arrays of mobile complexes. In fact, the inner mitochondrial membrane has a high ratio of unsaturated to saturated phospholipids and virtually no cholesterol, so its fluidity is very high and the mobility of the respiratory complexes is correspondingly high.

The electron transport system is sometimes referred to as a "chain." This is an apt description if we understand it to mean that the net flow of electrons is from NADH to cytochrome oxidase because each of the carriers in the sequence has a higher affinity for electrons (i.e., a more positive reduction potential) than the previous carrier. The transport process is also a chain in the sense that each carrier in the sequence can only interact with the component from which it receives electrons and the component to which it donates electrons. The word "chain" is *not* appropriate, however, if it suggests any sort of fixed location or ordered sequence of carriers within the membrane. Instead, each of the complexes diffuses randomly within the membrane, as do the coenzyme Q and cytochrome c molecules that shuttle electrons between them.

As Figure 14-17 indicates, NADH, coenzyme Q, and cytochrome c are key intermediates in the electron-transfer process. NADH links the ETS to the dehydrogenase (oxidation) reactions of the TCA cycle and to most other oxidation reactions in the matrix of the mitochondrion (or in the cytoplasm, in the case of the prokaryotic cell),

whereas coenzyme Q and cytochrome c transfer electrons between the respiratory complexes. Coenzyme Q accepts electrons from both complexes I and II, and is, in fact, the "funnel" that collects electrons from virtually every oxidation reaction in the cell. Coenzyme Q and cytochrome c are both relatively small molecules that can diffuse rapidly, either within the membrane (coenzyme Q) or on the membrane surface (cytochrome c). They are also quite numerous—about 10 molecules of cytochrome c and 50 molecules of ubiquinone for every complex I, according to one estimate. Because of their abundance and mobility, these carriers collide with the major complexes frequently enough to account for the observed rates of electron transfer in actively respiring mitochondria.

The Electrochemical Proton Gradient: Key to Energy Coupling

So far, we have learned that coenzymes are reduced during the oxidative events of the first two stages of aerobic respiration (stages 1 and 2 of Figure 14-1b). We have also seen that reduced coenzymes are reoxidized by the exergonic transfer of electrons to oxygen via a system of reversibly oxidizable intermediates located within the inner mitochondrial membrane (stage 3a). Now we are ready to tackle the final stages of respiration: the processes by which the free energy released during electron transport is used to generate an electrochemical proton gradient and the energy of the gradient is then used to drive ATP synthesis. In terms of Figure 14-1b, we are about to consider stage 3b (proton pumping) and stage 4 (ATP synthesis) of aerobic respiration.

Because this means of ATP synthesis involves phosphorylation events that are linked to oxygen-dependent electron transport, the process is called **oxidative phosphorylation,** thereby distinguishing it from *substrate-level phosphorylation,* which occurs as an integral part of a specific reaction in a metabolic pathway. (Reactions Gly-7 and Gly-10 of the glycolytic pathway and reaction TCA-5 of the TCA cycle are examples of substrate-level phosphorylation; see Figures 13-6 and 14-8.)

Electron Transport and ATP Synthesis Are Coupled Events

Mechanistically, oxidative phosphorylation is more complex than substrate-level phosphorylation. It has, in fact, been a confusing and highly controversial topic for much of its 60-year history, prompting Efraim Racker, a respected researcher in the field, to comment at one point, "Anyone who is not confused about oxidative phosphorylation just doesn't understand the situation." Much of the confusion and controversy is focused on the mechanism responsible for the actual coupling of electron transport to ATP generation. Now, however, there is good general agreement that the crucial link between electron transport and ATP production is an **electrochemical proton gradient** that is established by the directional pumping of protons across the membrane in which electron transport is occurring. We will come to the details of this model shortly, but first we need to consider the evidence that ATP synthesis is really coupled to electron transport and to understand the implications of that coupling.

Coupling of Electron Transport and ATP Generation. When we use the term **coupling,** we mean that two processes are mutually dependent on each other. To say that electron transport and ATP synthesis are coupled means not only that ATP synthesis depends critically on electron flow but also that electron flow is possible only when ATP can be synthesized. Coupling is an important regulatory mechanism because it ensures that the ETS—and therefore all of aerobic respiration—is dependent on, and responsive to, the ADP and ATP concentrations in the mitochondrion and hence to the energy status of the cell.

If two processes can be coupled, one might imagine that they could also be uncoupled. In fact, chemical compounds called *uncouplers* exist and have proved to be useful tools in the study of oxidative phosphorylation mechanisms. A classic uncoupler is *2,4-dinitrophenol (DNP).* If DNP is added to respiring mitochondria that are actively synthesizing ATP, the rates of coenzyme oxidation and oxygen consumption increase suddenly, and ATP

synthesis drops precipitously. The DNP has *uncoupled* electron transport from ATP synthesis. In a sense, this is like putting your car's automatic transmission in neutral: The engine speeds up somewhat, yet forward motion ceases. Any mechanism that attempts to explain the coupling between electron transport and ATP synthesis must also account for the action of uncoupling agents.

Respiratory Control. Because electron transport is coupled to ATP synthesis, the availability of ADP regulates the rate of oxidative phosphorylation and therefore of electron transport. This is called **respiratory control,** and its physiological significance is easy to appreciate: Electron transport and ATP generation will be favored when the ADP concentration is high (i.e., when the ATP concentration is low) and inhibited when the ADP concentration is low (when the ATP concentration is high). Oxidative phosphorylation is therefore regulated by cellular ATP needs, such that electron flow from organic fuel molecules to oxygen is adjusted to the energy needs of the cell. This regulatory mechanism becomes apparent during exercise, when the accumulation of ADP in muscle tissue causes an increase in electron transport rates, followed by a dramatic rise in the need for oxygen.

The Chemiosmotic Model: The "Missing Link" Is a Proton Gradient

How can ATP synthesis, a dehydration reaction, be tightly coupled to electron transport, which involves the sequential oxidation and reduction of various protein complexes in a lipid bilayer? Scientists puzzled over this question but the answer eluded their best efforts for several decades. Most were convinced that high-energy intermediates must be involved in oxidative phosphorylation just as surely as they are in substrate-level phosphorylation such as the ATP- generating steps of the glycolytic pathway and the TCA cycle (see reaction Gly-6 in Figure 13-7 and reaction TCA-4 in Figure 14-8).

But while others continued their feverish search for high-energy intermediates, the British biochemist Peter Mitchell made the revolutionary suggestion that such intermediates, like the emperor's new clothes, might not exist at all. In 1961 Mitchell proposed an alternative explanation, which he called the **chemiosmotic coupling model.** According to this model, the exergonic transfer of electrons between and within the respiratory complexes is accompanied by the unidirectional pumping of protons across the membrane in which the transport system is localized, either the inner mitochondrial membrane or the prokaryotic plasma membrane. The electrochemical proton gradient that is generated and maintained in this way then provides the driving force for ATP synthesis—by the F_oF_1-ATP complex, we now know. In other words, the "missing link" between electron transport and ATP synthesis was not a high-energy chemical intermediate but an electrochemical gradient. This mechanism is illustrated in

Figure 14-18, which depicts both the outward pumping of protons that accompanies electron transport through complexes I, III, and IV and the proton-driven synthesis of ATP by the F_oF_1 complex.

As you might expect, Mitchell's theory met with considerable initial skepticism and resistance. Not only was it a radical departure from conventional wisdom on coupling, but Mitchell proposed it in the virtual absence of experimental data. Over the years, however, a large body of evidence has been amassed in its support, and in 1978 Mitchell was awarded a Nobel Prize for his pioneering work. Today the chemiosmotic coupling model is a well-verified concept that provides a unifying framework for understanding energy transformations not just in mitochondrial membranes, but in chloroplast and bacterial membranes as well.

The essential feature of the chemiosmotic model is that *the link between electron transport and ATP formation is an electrochemical potential across a membrane.* It is, in fact, this feature that gives the model its name: The "chemi" part of the term refers to the chemical processes of oxidation and electron transfer and the "osmotic" part comes from the Greek word *osmos,* which means "to push"—to push protons across the membrane, in this case. The chemiosmotic model has turned out to be exceptionally useful not only because of the very plausible explanation it provides for coupled ATP generation but also because of its pervasive influence on the way we now think about energy conservation in biological systems.

Coenzyme Oxidation Pumps Enough Electrons to Form 3 ATP per NADH and 2 ATP per $FADH_2$

Before considering the experimental evidence that has confirmed the chemiosmotic model, notice that Figure 14-18 provides us with estimates of the numbers of protons pumped outward by the three respiratory complexes involved in NADH oxidation, as well as the number of protons required to drive the synthesis of ATP by the F_oF_1 complex. The word *estimate* is important here because investigators still disagree on some of these numbers. However, the values shown in the figure are generally accepted and are in any case not likely to be very far from the actual values.

As we have already seen, most of the dehydrogenases in the mitochondrial matrix transfer electrons from oxidizable substrates to NAD^+, generating NADH. NADH, in turn, transfers electrons to the FMN component of complex I, thereby initiating the electron transport system. As Figure 14-18 shows, transfer of two electrons from NADH down the respiratory chain to oxygen is accompanied by the transmembrane pumping of 4 protons by complex I, 4 protons by complex III, and 2 protons by complex IV, for a total of 10 protons per NADH. (The four protons attributed to complex III include two that are transferred by the reduction and subsequent oxidation of coenzyme Q but are included in the complex III total because of the close physical association of coenzyme Q with complex III.)

The number of protons required to drive the synthesis of one molecule of ATP by the F_oF_1 ATP synthase is thought to be 3 or 4, with 3 generally regarded as more likely. If we assume that 10 protons are pumped per NADH oxidized and that 3 protons are required per ATP molecule, then we can conclude that oxidative phosphorylation yields about 3 molecules of ATP synthesized per molecule of NADH oxidized. This agrees well with evidence dating back to the 1940s that a fixed relationship usually exists between the number of molecules of ATP generated and the number of oxygen atoms consumed in respiration. This relationship was expressed as the **P/O ratio,** defined as the number of molecules of ATP generated as a pair of electrons passes along the sequence of carriers from reduced coenzyme to oxygen.

Early investigators estimated the P/O ratios to be about 3 for NADH and about 2 for $FADH_2$. Racker and his colleagues, for example, incorporated each of the respiratory complexes into synthetic phospholipid vesicles, along with mitochondrial F_oF_1 complexes. Upon addition of the appropriate oxidizable substrates, they showed that vesicles containing complex I, complex III, or complex IV could generate 1 molecule of ATP per pair of electrons that passed through the complex. These results are consistent with our current understanding that each of these three complexes (but not complex II) serves as a proton pump, thereby contributing to the proton gradient that drives ATP synthesis.

For purposes of our discussion, we will assume the ATP yields to be 3 for NADH and 2 for $FADH_2$, recognizing that the values on which these yields are based are still imprecisely known. These are in any case reasonable numbers, considering the energetics of the system. As noted earlier (see reaction 14-4), the $\Delta G^{o\prime}$ for the oxidation of NADH by molecular oxygen is -52.4 kcal/mol. Assuming conditions such that the $\Delta G'$ for this reaction is similar to the $\Delta G^{o\prime}$ value and recognizing that 10 kcal/mol is a reasonable value to assume for the $\Delta G'$ of ATP synthesis from ADP and P_i under cellular conditions, a P/O ratio of 3 would mean that NADH oxidation drives ATP synthesis with an efficiency of about 57% ($3 \times 10/52.4 = 0.573$). Similar calculations for $FADH_2$, which sends electrons through complexes II, III, and IV but not I, indicate an efficiency of about 44% ($2 \times 10/45.9 = 0.436$).

The Chemiosmotic Model Is Affirmed by an Impressive Array of Evidence

Since its initial formulation in 1961, the chemiosmotic model has come to be universally accepted as the link between electron transport and ATP synthesis. To understand why, we will consider briefly several of the most important lines of experimental evidence that support this model. In the process, we will also learn more about the mechanism of chemiosmotic coupling.

1. Electron Transport Causes Protons to Be Pumped Out of the Mitochondrial Matrix

Shortly after proposing the chemiosmotic model, Mitchell and his colleague Jennifer Moyle demonstrated experimentally that the flow of electrons through the ETS is accompanied by the movement of protons across the inner mitochondrial membrane. They first suspended mitochondria in a medium in which electron transfer could not occur because oxygen was lacking. The proton concentration (i.e., the pH) of the medium was then monitored as electron transfer was stimulated by the addition of oxygen. Under these conditions, the pH of the medium declined rapidly with either NADH or succinate as the electron source (Figure 14-19).

Because a decline in pH reflects an increase in proton concentration, Mitchell and Moyle concluded that electron transfer within the mitochondrial inner membrane is accompanied by **unidirectional pumping of protons** from the mitochondrial matrix into the external medium. (Recall that the outer mitochondrial membrane is freely permeable to small molecules and ions and can therefore be ignored in experiments such as this.) Similar observations were soon made with chloroplasts, with light as the energy source rather than respiration.

The mechanism whereby the transfer of electrons from one carrier to another is coupled to the directional transport of protons is still not well understood. There may, in fact, be more than one such mechanism. An attractive possibility is that discussed earlier for coenzyme Q. As we already know, coenzyme Q picks up a proton from the medium whenever it accepts an electron and releases the proton again when it passes the electron on (see Figure

14-15). Coenzyme Q can move freely within the membrane, so it could easily accept electrons from complex I or II on the inner surface of the membrane and pass them to complex III at the outer surface, thereby moving one proton across the membrane for every electron transferred. Because coenzyme Q is known to associate with complex III, this is thought to explain at least a part of the proton pumping by this complex (see Figure 14-18).

In the case of the other proton-pumping complexes (I and IV), current evidence suggests that the pumping mechanism may involve allosteric changes in protein conformation. Suppose, for example, that a specific polypeptide in one of these complexes is an allosteric protein with two different conformational states. In one conformational state, the polypeptide might spontaneously bind a proton at the inner surface of the membrane. Transfer of one or more electrons through the complex might then provide the energy needed for the polypeptide to assume an alternative conformational state, capable of releasing the proton at the outer surface of the membrane. The polypeptide could then return to its initial conformation, ready to bind another proton at the inner surface. Thus, the pumping of protons may be achieved either by the physical movement of a carrier molecule such as coenzyme Q or by a conformational change in a polypeptide within a respiratory complex.

2. Components of the Electron Transport System Are Asymmetrically Oriented Within the Inner Mitochondrial Membrane

The unidirectional pumping demonstrated by Mitchell and Moyle suggests strongly that the electron carriers making up the ETS must be asymmetrically oriented within the membrane. If such were not the case, protons would presumably be pumped randomly in both directions. To study the topographical organization of the ETS, investigators have exposed inner membrane vesicles to a variety of antibodies, enzymes, and labeling agents designed to interact with various membrane components (see Figure 7-24). Such studies have shown clearly that some constituents of the respiratory complexes face the matrix side of the inner membrane, others are exposed on the opposite side, and some protrude on both sides and are therefore transmembrane proteins. These findings confirm the prediction that ETS components are asymmetrically distributed across the inner mitochondrial membrane, as suggested schematically in Figure 14-18.

3. Membrane Vesicles Containing Complexes I, III, or IV Establish Proton Gradients

Further support for the chemiosmotic model has come from experiments involving the reconstitution of membrane vesicles from mixtures of isolated components. Since each of respiratory complexes I, III, and IV has a coupling site for ATP synthesis (Figure 14-18), the chemiosmotic model predicts that each of these com-

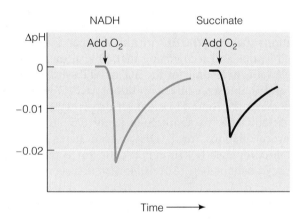

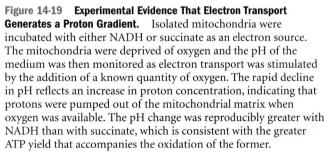

Figure 14-19 Experimental Evidence That Electron Transport Generates a Proton Gradient. Isolated mitochondria were incubated with either NADH or succinate as an electron source. The mitochondria were deprived of oxygen and the pH of the medium was then monitored as electron transport was stimulated by the addition of a known quantity of oxygen. The rapid decline in pH reflects an increase in proton concentration, indicating that protons were pumped out of the mitochondrial matrix when oxygen was available. The pH change was reproducibly greater with NADH than with succinate, which is consistent with the greater ATP yield that accompanies the oxidation of the former.

plexes should be capable of pumping protons across the inner mitochondrial membrane. This prediction has been confirmed experimentally by reconstituting artificial phospholipid vesicles containing complex I, III, or IV. When provided with appropriate oxidizable substrates, each of these three complexes is able to pump protons across the membrane of the vesicle.

4. Oxidative Phosphorylation Requires a Membrane-Enclosed Compartment

An obvious prediction of the chemiosmotic model is that oxidative phosphorylation occurs only within a compartment enclosed by an intact mitochondrial membrane. Otherwise, the proton gradient that drives ATP synthesis could not be maintained. This prediction has been verified by experimental demonstration that electron transfer carried out by isolated complexes cannot be coupled to ATP synthesis unless the complexes are incorporated into membranes that form enclosed, intact vesicles.

5. Uncoupling Agents Abolish Both the Proton Gradient and ATP Synthesis

Additional evidence for the role played by proton gradients in ATP formation has come from studies using agents such as dinitrophenol (DNP) that are known to uncouple ATP synthesis from electron transport. For example, Mitchell showed in 1963 that membranes become freely permeable to protons in the presence of DNP. Dinitrophenol, in other words, abolished both ATP synthesis and the capability of a membrane to maintain a proton gradient, a finding that is clearly consistent with the concept of ATP synthesis driven by an electrochemical proton gradient.

6. The Proton Gradient Has Enough Energy to Drive ATP Synthesis

For the chemiosmotic model to be viable, the proton gradient generated by electron transport must store enough energy to drive ATP synthesis. We can address this point with a few thermodynamic calculations. The electrochemical proton gradient across the inner membrane of a metabolically active mitochondrion involves both a membrane potential (the "electro" component of the gradient) and a concentration gradient (the "chemical" component). As you may recall from Chapter 8, the equation used to quantify this electrochemical gradient has two terms, one for the membrane potential, V_m, and the other for the concentration gradient, which in the case of protons is a pH gradient (see Table 8-4).

A mitochondrion actively involved in aerobic respiration typically has a membrane potential of about 0.16 V (positive on the side that faces the intermembrane space) and a pH gradient of about 1.0 pH unit (higher on the matrix side). This electrochemical gradient exerts a **proton motive force (pmf)** that tends to drive protons back down their concentration gradient—back into the matrix of the

mitochondrion, that is. The pmf can be calculated by summing the contributions of the membrane potential and the pH gradient using the following equation:

$$pmf = V_m + 2.303 \ RT \ \Delta pH/F \qquad (14\text{-}13)$$

where pmf is the proton motive force in volts, V_m is the membrane potential in volts, ΔpH is the difference in pH across the membrane ($\Delta pH = pH_{matrix} - pH_{cytosol}$), and other terms are as previously defined (see equation 14-8).

For a mitochondrion at 37°C with a V_m of 0.16 V and a pH gradient of 1.0 unit, the pmf can be calculated as follows:

$$pmf = 0.16 + \left(\frac{2.303(1.987)(37 + 273)(1.0)}{23{,}062} \right)$$
$$= 0.16 + 0.062 = 0.222 \ V \qquad (14\text{-}14)$$

Notice that the membrane potential accounts for more than 70% of the mitochondrial pmf. In the next chapter we will find that for the chloroplast, most of the pmf is due to a steep transmembrane pH difference.

Like the redox potential, pmf is an electrical force expressed in volts. And as with $\Delta E_0'$, pmf can be used to calculate $\Delta G^{\circ\prime}$, the standard free energy change for the movement of protons across the membrane, using the following equation:

$$\Delta G^{\circ\prime} = -nF(pmf) = -(23{,}062)(0.222)$$
$$= -5120 \ cal/mol = -5.1 \ kcal/mol \qquad (14\text{-}15)$$

Thus, a proton motive force of 0.222 V across the inner mitochondrial membrane corresponds to a free energy change of about −5.1 kcal per mole of protons, which is the amount of energy that will be released as protons return to the matrix.

Is this enough to drive ATP synthesis? Not surprisingly, the answer is yes, though it depends on how many protons are required to drive the synthesis of one ATP molecule by the F_oF_1 complex. Mitchell's original model assumed two protons per ATP, which would provide about 10.2 kcal/mol, barely enough to drive ATP formation, assuming the $\Delta G'$ for phosphorylation of ADP to be about 10 kcal/mol under mitochondrial conditions. Differences of opinion still exist concerning the number of protons per ATP but the real number is probably closer to three or four, which would provide about 15–20 kcal of energy per ATP. That is a comfortable excess, almost certainly enough to ensure that the reaction is driven strongly in the direction of ATP formation.

7. Artificial Proton Gradients Can Drive ATP Synthesis in the Absence of Electron Transport

Just because a proton gradient is established during electron transport and contains sufficient energy to drive ATP formation does not prove that it actually does so. However, direct evidence that a proton gradient can in fact drive ATP synthesis has been obtained by exposing mitochondria or inner membrane vesicles to artificial pH gradients. When mitochondria are suspended in a solution in which the external proton concentration is suddenly increased by the addition of acid, ATP is generated in response to the artificially created proton gradient. Because such artificial pH gradients induce ATP formation even in the absence of oxidizable substrates that would otherwise pass electrons to the ETS, it is evident that ATP synthesis can be induced by a proton gradient even in the absence of electron transport.

ATP Synthesis: Putting It All Together

We are now ready to take up the fourth, and final, stage of aerobic respiration: ATP synthesis. We have seen how the energy of an oxidizable substrate such as glucose is transferred to reduced coenzymes during the oxidation reactions of glycolysis and the TCA cycle (stages 1 and 2 of Figure 14-1b) and then is used to generate an electrochemical proton gradient across the inner membrane of the mitochondrion (stage 3). Now we can ask how the

pmf of that gradient is harnessed to drive ATP synthesis. For that, we return to the F_1 complexes that can be seen along the inner surfaces of cristae (see Figure 14-6a) and ask about the evidence that these spheres are capable of synthesizing ATP.

F_1 Particles Have ATP Synthase Activity

Key evidence concerning the role of the F_1 particles came from studies that Efraim Racker and his colleagues conducted to test the prediction of the chemiosmotic hypothesis that a reversible, proton-translocating ATPase is present in membranes capable of coupled ATP synthesis. Beginning with intact mitochondria (Figure 14-20a), Racker and his colleagues found that they could disrupt the mitochondria in such a way that fragments of the inner membrane formed small vesicles, which they called *submitochondrial particles* (Figure 14-20b). Like the intact mitochondrial membrane from which they were derived, these submitochondrial particles were capable of carrying out electron transport and ATP synthesis. By subjecting these submitochondrial particles to mechanical agitation

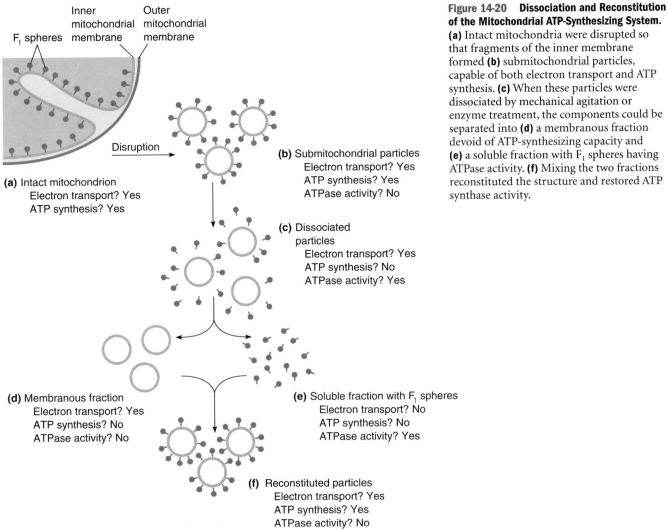

Figure 14-20 Dissociation and Reconstitution of the Mitochondrial ATP-Synthesizing System. (a) Intact mitochondria were disrupted so that fragments of the inner membrane formed **(b)** submitochondrial particles, capable of both electron transport and ATP synthesis. **(c)** When these particles were dissociated by mechanical agitation or enzyme treatment, the components could be separated into **(d)** a membranous fraction devoid of ATP-synthesizing capacity and **(e)** a soluble fraction with F_1 spheres having ATPase activity. **(f)** Mixing the two fractions reconstituted the structure and restored ATP synthase activity.

F_1 spheres
Inner mitochondrial membrane
Outer mitochondrial membrane

Disruption

(a) Intact mitochondrion
Electron transport? Yes
ATP synthesis? Yes

(b) Submitochondrial particles
Electron transport? Yes
ATP synthesis? Yes
ATPase activity? No

(c) Dissociated particles
Electron transport? Yes
ATP synthesis? No
ATPase activity? Yes

(d) Membranous fraction
Electron transport? Yes
ATP synthesis? No
ATPase activity? No

(e) Soluble fraction with F_1 spheres
Electron transport? No
ATP synthesis? No
ATPase activity? Yes

(f) Reconstituted particles
Electron transport? Yes
ATP synthesis? Yes
ATPase activity? No

or enzyme treatment, Racker was able to dislodge the F_1 structures from the membranous vesicles (Figure 14-20c).

When the F_1 particles and the membranous vesicles were separated from each other by centrifugation, the membranous fraction could still carry out electron transport but could no longer synthesize ATP; the two functions had become uncoupled (Figure 14-20d). The isolated F_1 particles, on the other hand, were not capable of either electron transport or ATP synthesis but had ATPase activity (Figure 14-20e), a property consistent with the F-type ATPase that we now recognize the mitochondrial F_0F_1 complex to be (see Table 8-3 for a description of F-type ATPases). The ATP-generating capability of the membranous fraction was restored by adding the F_1 particles back to the membranes, suggesting that the spherical projections seen on the inner surface of the inner mitochondrial membrane are an important part of the ATP-generating complex of the membrane. These particles were therefore referred to as *coupling factors* (from which we get the "F" in F_1) and are now known to be the structures responsible for the ATP-synthesizing activity of the inner mitochondrial membrane or the bacterial plasma membrane.

The F_0F_1 Complex: Proton Translocation Through F_0 Drives ATP Synthesis by F_1

As we saw in Figure 14-18, the F_1 complex is only part of the ATP synthase complex. F_1 is attached by a short stalk to complex F_0, which is embedded in the inner mitochondrial membrane at the base of the stalk (see Figure 14-6c). We now know that the F_0 complex serves as the **proton translocator,** the channel through which protons flow when the electrochemical gradient across the membrane is being used to drive the ATP-synthesizing activity of the F_1 complex. Thus, the F_0F_1 complex is the functional **ATP synthase.** The F_0 component provides a channel for the exergonic flow of protons from the outside to the inside of the membrane, thereby tapping into the pmf, or driving force, of the electrochemical proton gradient, and the F_1

component carries out the actual synthesis of ATP, driven by the energy of the proton gradient.

Table 14-4 presents the polypeptide composition of the F_0F_1 complex from *Escherichia coli,* and Figure 14-21 illustrates their assembly into the functional complex. (We know the most about the *E. coli* F_0F_1 complex; the mitochondrial F_0F_1 complex is thought to be similar, though with greater subunit complexity.) The bacterial F_1 headpiece consists of three α and three β polypeptides, organized as three $\alpha\beta$ complexes that form a *catalytic hexagon.* The catalytic site for ATP synthesis and hydrolysis is located on the β subunit; the α subunit serves as an ATP/ADP-binding site, thereby promoting the catalytic activity of the β subunit. The stalk consists of three polypeptides: gamma (γ), delta (δ), and epsilon (ε). These subunits extend into both the F_1 and the F_0 structures, as Figure 14-21 indicates. The δ and ε subunits are required for assembly of the F_0F_1 complex, and the γ subunit appears to rotate as protons move through the channel in the F_0 structure, an intriguing mechanism that is described further in the next section.

The F_0 complex consists of polypeptides *a, b,* and *c,* with 1 *a* subunit, 2 *b* subunits, and 9 to 12 *c* subunits present in the functional complex. The *c* subunits are thought to be organized in a circle, forming the *proton channel* through the membrane. The *a* and *b* subunits apparently stabilize the proton channel. In addition, subunit *b* binds to the stalk, thereby anchoring the F_1/stalk structure to the F_0 base.

ATP Synthesis by F_0F_1 Involves Physical Rotation of the Gamma Subunit

Once the link between electron transport within the membrane and proton pumping across the membrane had been established, the next piece of the puzzle must have seemed almost as daunting: How does the exergonic flux of electrons through the F_0 component of the F_0F_1 ATP synthase drive the otherwise-endergonic synthesis of ATP by the catalytic sites on the three β subunits of the F_1

Table 14-4   Polypeptide Composition of the E. coli F_0F_1-ATP Synthase (ATPase)*

Structure	Polypeptide	Molecular Weight**	Number Present	Function
F_1	α	52,000	3	ATP/ADP binding site; promotes activity of β subunit
	β	55,000	3	Catalytic site for ATP hydrolysis and synthesis
Stalk	γ	31,000	1	Rotates to transmit energy from F_0 to F_1
	δ	19,000	1	Main component of stalk; required for F_0F_1 assembly
	ε	15,000	1	Binds to δ subunit; required for F_0F_1 assembly
F_0	*a*	30,000	1	Stabilizes proton channel
	b	17,000	2	Stabilizes proton channel
	c	8,000	10***	Forms proton channel

*The mitochondrial F_0F_1 complex is similar to the bacterial complex but with one additional polypeptide in F_1 and seven additional polypeptides in F_0.

**The molecular weights of the three components of the *E. coli* F_0F_1 are about 321,000 for F_1 ($\alpha_3\beta_3$), 65,000 for the stalk ($\gamma\delta\varepsilon$), and 144,000 for F_0 (ab_2c_{10}). The total molecular weight for the assembled complex ($\alpha_3\beta_3\gamma\delta\varepsilon ab_2c_{10}$) is therefore about 530,000.

***Estimates of the number of *c* subunits in the functional F_0F_1 complex from *E. coli* range from 9 to 12, with 10 regarded as the most likely number.

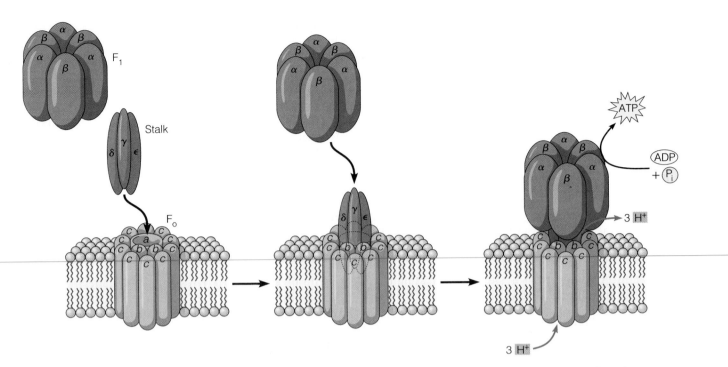

(a) F_1, stalk, and F_o components   **(b)** Assembly of F_oF_1 complex   **(c)** Functional ATP synthase

Figure 14-21 F_1 and F_o Components of the Bacterial F_oF_1 ATP Synthase. The F_1 complex is a spherical projection on the inner surface of the plasma membrane of a prokaryotic cell or the inner mitochondrial membrane of a eukaryotic cell that is responsible for ATP generation. The F_o complex is embedded in the membrane and provides a channel for exergonic translocation of protons from the outside of the membrane to the inside. **(a)** In a bacterial cell the F_1 particle is a complex of 3 α and 3 β subunits that occur as 3 $\alpha\beta$ assemblies organized into a catalytic hexagon. The F_o complex consists of 1 a subunit, 2 b subunits, and 9 to 12 c subunits, the latter arranged in a circle to form a cylindrical channel through the membrane. **(b)** F_1 and F_o are attached to each other through a stalk that consists of γ, δ, and ε subunits, which extend into both the F_1 and F_o complexes. **(c)** The assembled F_oF_1 complex is capable of using the energy of proton flux through the F_o complex to drive ATP synthesis by the β subunits of the F_1 particle. The mechanism that couples proton translocation to ATP synthesis is thought to involve physical rotation of the γ subunit, as illustrated in Figure 14-22.

complex? A novel answer to this question was suggested in 1979 by Paul Boyer, who proposed the **binding change model** shown in Figure 14-22. Boyer's model envisioned the catalytic site on each of the three β subunits of the F_1 complex progressing through three distinctly different conformations with quite different affinities for the substrates (ADP and P_i) and the product (ATP). Boyer identified these as the L (for loose) conformation, which binds ADP and P_i loosely, the T (for tight) conformation, which binds ADP and P_i tightly thereby catalyzing their spontaneous condensation into ATP, which also binds tightly and the O (for open) conformation, which has a very low affinity for substrates or product and is therefore unoccupied most of the time. Based on his own research findings, Boyer proposed that, at any point in time, each of the three active sites is in a different conformation and that the hexagonal ring of α and β subunits actually *rotates* with respect to the central stalk (or the stalk with respect to the ring of α and β subunits, as we now recognize to be the case), driven by the exergonic flux of protons through the F_o complex within the membrane.

Though not widely accepted at first, Boyer's model was substantially confirmed in 1994 by John Walker and co-workers, who used X-ray crystallography to provide the first detailed atomic model of the F_1 complex. Specifically, their model identified structures in the active sites of the β subunits that corresponded to Boyer's O, L, and T conformations, and they confirmed that each of the three active sites is in a different conformation at any given time. Even more remarkably, the γ subunit within the stalk that connects the F_1 and F_o complexes (see Figure 14-21) was shown to extend into the center of the F_1 structure in an asymmetric manner, such that its contact with and effect upon each of the β subunits is different at any given instant.

The Binding Change Model in Action. The sequence of events shown in Figure 14-22 is a current model of how the exergonic flux of protons through the F_o complex drives the synthesis of ATP by the catalytic sites on the three β subunits of the F_1 complex. As protons flow through a channel in the a subunit of F_o, they drive the rotation of the ring of c subunits in the F_o structure—and therefore also rotation of the γ subunit of the stalk, which is attached to the c-ring. (The ε subunit of the stalk is thought to rotate as well, though it is not shown in the figure.) The asymmetry of the γ subunit not only results in characteristically differ-

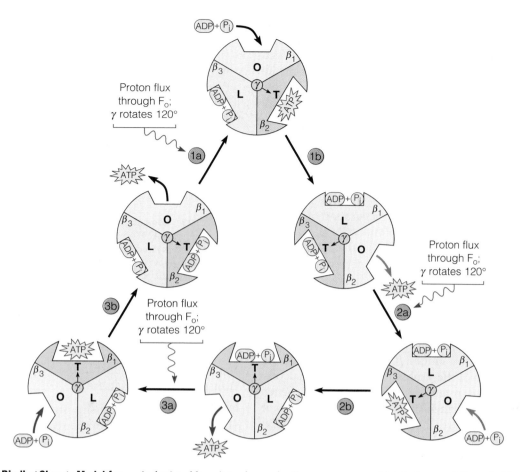

Figure 14-22 The Binding-Change Model for ATP synthesis by the β Subunits of the F_oF_1 Complex. According to this model, each of the β subunits of the F_1 complex is in a different conformation at any instant and each undergoes a sequence of conformational changes from the O (open) conformation through the L (loose) conformation to the T (tight) conformation, driven by the rotation of the γ subunit, the asymmetry of which causes it to affect each β subunit differently at any point in its rotation. (The ε subunit of the stalk also rotates—and probably the ring of c subunits within the F_o complex as well—but only the γ subunit is depicted here.) As shown for the subunit at the top of the diagram (which is arbitrarily identified as β_1; the three subunits are actually identical in amino acid sequence and structure), the process of ATP synthesis begins with the β subunit in its O conformation and involves ① a 120° rotation of the γ subunit (indicated schematically by the change in the orientation of the arrow projecting outward from the γ in the center), which converts the β_1 subunit into its L conformation, thereby causing loose binding of ADP and P_i; ① generation of ATP by one of the other subunits (β_3); ② a second 120° rotation of the γ subunit, which induces a shift to the T conformation that causes the substrates to be tightly bound to the catalytic site, resulting in ② their condensation into a molecule of ATP; ③ a third 120° rotation of the γ subunit, which returns the β_1 subunit to the O conformation and results in release of the newly formed ATP molecule; followed by ③ generation of ATP by the β_2 site, thereby completing one full cycle of catalysis. Note that the same sequence of events has occurred at each of the other sites, but it is offset temporally, such that 1 molecule of ATP is released after each 120° rotation of the γ subunit, for a total of 3 ATP molecules per full revolution.

ent interactions with the three β subunits at any point in time but also causes each β subunit to pass successively through the O, L, and T conformations as the γ subunit rotates 360 degrees.

To understand the model in Figure 14-22, imagine that you are looking up at the F_1 complex from a position on the membrane surface right next to the stalk. From that vantage point, you would see the three α and β subunits arranged hexagonally around the F_1 sphere (though only the β subunits are shown in Figure 14-22 because they alone have the catalytic sites at which ATP is synthesized). Keep your eye on one of the three subunits—let's pick the β_1 subunit shown in buff at the top of the diagram—and follow that subunit through one complete rotation of the γ subunit, the orientation of which is shown by the small arrow at the center of each diagram.

The β_1 subunit starts out in the O conformation, with the substrates ADP and P_i free to enter the catalytic site, though the catalytic site has little affinity for them in this conformation. As protons (3 or 4, most likely) flow through the F_o subunit in the membrane, the c-ring and the attached γ subunit rotate 120° degrees (step 1a), inducing a shift in the β_1 site from the O to the L conformation (and shifts in the conformation of the other two β subunits as well: from T to O for the β_2 subunit and from L to T for the β_3 subunit). The somewhat greater affinity of the L conformation for ADP and P_i results in the binding of these substrates to the catalytic site.

Following the generation of an ATP molecule by one of the other subunits (the β_3 subunit; step 1b), the γ subunit rotates another 120°, driven again by the proton flux through the F_o channel (step 2a). This induces a shift in the β_1 subunit to the T conformation, which increases the affinity of the site for the substrate molecules, such that they are now tightly bound to the catalytic site in an orientation that is maximally conducive to interaction. Under these conditions, the ADP and P_i molecules condense spontaneously to form ATP (step 2b). The γ subunit now rotates a further 120°, returning the β_1 catalytic site to the O conformation, thereby releasing the ATP from the catalytic site (step 3a). Following the generation of an ATP molecule by one of the other subunits (the β_2 subunit; step 3b), one full cycle is completed and the β_1 subunit is available for substrate binding again, thereby initiating another round of ATP synthesis.

Although we've focused specifically on just one of the three β subunits, the same sequence of events has occurred at the other two subunits as well, but offset temporally, such that newly synthesized ATP is released from subunit β_2 after step 1a and from subunit β_3 after step 2a. Thus, one complete revolution of the γ subunit involves three successive energy-requiring rotations of 120° each, which sends each β subunit through their O, L, and T conformations in an offset manner and results in the synthesis of 3 ATP molecules, one by each of the three β subunits.

Spontaneous Synthesis of ATP? Close scrutiny of Figure 14-22 reveals that each of the steps at which ATP is actually formed from ADP and P_i (i.e., steps 1b, 2b, and 3b) occurs without an accompanying input of energy. Yet we know that the synthesis of ATP is a highly endergonic reaction, with a $\Delta G^{o\prime}$ of +7.3 kcal/mol. How, one might then ask, does ATP synthesis proceed spontaneously at the catalytic site of a β subunit in its T conformation? Actually, what we really know is that the synthesis of ATP is a highly endergonic reaction *in dilute aqueous solution*, which is the only context in which we have encountered ATP synthesis before. Under those conditions, ATP synthesis is indeed a highly endergonic reaction (Figure 14-23a). However, the reaction *at the catalytic site of a β subunit* involves enzyme-bound intermediates in a drastically different environment—so different, in fact, that the reaction under these conditions has a $\Delta G^{o\prime}$ close to zero (Figure 14-23b) and can therefore occur spontaneously, without any immediate energy requirement.

This doesn't mean that ATP synthesis proceeds without thermodynamic cost, though; it simply means that the input of energy is required at another point in the cycle. In fact, energy is required both before the actual formation of ATP and thereafter: Prior energy input is needed to drive the transition of the catalytic site from the L to the T conformation, thereby packing the ADP and P_i together tightly and facilitating their interaction (as in step 2a of Figure 14-22), while subsequent energy input is required for release of the tightly bound ATP product from the catalytic site (as in step 3a).

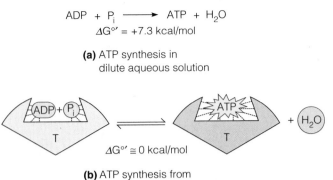

$$ADP + P_i \longrightarrow ATP + H_2O$$
$$\Delta G^{o\prime} = +7.3 \text{ kcal/mol}$$

(a) ATP synthesis in dilute aqueous solution

$$\Delta G^{o\prime} \cong 0 \text{ kcal/mol}$$

(b) ATP synthesis from protein-bound ADP and P_i

Figure 14-23 Comparative Energetics of ATP Synthesis. The energy requirement for ATP synthesis differs greatly, depending on the environment. **(a)** In dilute aqueous solution, ATP synthesis from soluble ADP and P_i is highly endergonic, with a $\Delta G^{o\prime}$ of +7.3 kcal/mol. **(b)** At the catalytic site of a β subunit in its T conformation, however, the environment is drastically different and the reaction has a $\Delta G^{o\prime}$ close to 0 (i.e., the K_{eq} is close to 1) and can therefore proceed spontaneously without any immediate energy requirement.

The Chemiosmotic Model Involves Dynamic Transmembrane Proton Traffic

The dynamics of the chemiosmotic model are summarized in Figure 14-24. Complexes I, III, and IV of the ETS pump protons outward across the inner membrane of the mitochondrion, and the resulting electrochemical gradient then drives ATP generation by means of the F_oF_1 complexes associated with the same membrane. There is, in other words, continuous, dynamic proton traffic across the inner membrane (or the plasma membrane, in the case of prokaryotes). Presuming that the number of protons extruded by each of the respiratory assemblies is as shown and that 3 protons are required to drive the synthesis of 1 ATP molecule, then the approximate P/O ratios of 3 for NADH and 2 for $FADH_2$ are understandable.

Aerobic Respiration: Summing It All Up

To summarize aerobic respiration, let's return to Figure 14-1 one last time and review the role of each of the components. As the glycolytic pathway and the TCA cycle (or other catabolic pathways, such as β oxidation) proceed, coenzymes are continuously reduced. These reduced coenzymes represent a storage form of much of the free energy of substrate oxidation—energy that can be tapped and released as the coenzymes are themselves reoxidized by the ETS. As electrons are transported from NADH or $FADH_2$ to oxygen, they pass through several respiratory complexes. Three of the four complexes are sites at which electron transport is coupled to the directional pumping of protons across the membrane. The resulting electrochemical gradient exerts a pmf that serves as the driving force for ATP synthesis. Under most conditions, a steady-state pmf will be maintained across the membrane, with

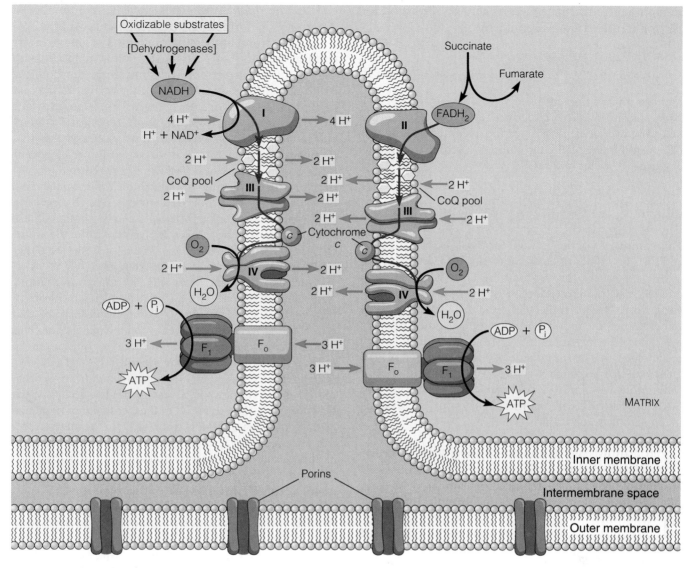

Figure 14-24 Dynamics of the Electrochemical Proton Gradient. The respiratory complexes I, II, III, and IV are integral components of the inner mitochondrial membrane. Complexes I, III, and IV (but not complex II) couple the exergonic flow of electrons (pink lines) through the complex with the outward pumping of protons (blue) across the membrane. The proton motive force of the resulting electrochemical proton gradient drives ATP synthesis by F_1 as protons are translocated back across the membrane by the F_o complex, which is also embedded in the inner membrane.

the transfer of electrons from coenzymes to oxygen carefully and continuously adjusted so that the outward pumping of protons balances the inward proton flux necessary to synthesize ATP at the desired rate.

The Maximum ATP Yield of Aerobic Respiration Is 36–38 ATPs per Glucose

Now we can return to the question of the **maximum ATP yield** per molecule of glucose under aerobic conditions. Recall from reaction 14-3 that the complete oxidation of glucose to carbon dioxide by glycolysis and the TCA cycle results in the generation of 4 molecules of ATP by substrate-level phosphorylation, with most of the remaining free energy of glucose oxidation stored in the 12 coenzyme molecules—10 of NADH and 2 of $FADH_2$. In prokaryotes and some eukaryotic cells, electrons from all the NADH molecules pass through all three ATP-generating complexes of the ETS, yielding 3 ATP molecules per molecule of coenzyme. Electrons from $FADH_2$, on the other hand, traverse only two of the three complexes, yielding only 2 ATP molecules per molecule of coenzyme. The maximum theoretical ATP yield obtainable upon reoxidation of the 12 coenzyme molecules formed per glucose can therefore be represented as follows:

$$10NADH + 10H^+ + 5O_2 + 30ADP + 30P_i \longrightarrow$$
$$10NAD^+ + 10H_2O + 30ATP + 30H_2O \quad (14\text{-}16)$$

$$2FADH_2 + O_2 + 4ADP + 4P_i \longrightarrow$$
$$2FAD + 2H_2O + 4ATP + 4H_2O \quad (14\text{-}17)$$

Summing these reactions gives us an overall reaction for electron transport and ATP synthesis:

$$10NADH + 10H^+ + 2FADH_2 + 6O_2 + 34ADP + 34P_i \longrightarrow$$
$$10NAD^+ + 2FAD + 12H_2O + 34ATP + 34H_2O \quad \textbf{(14-18)}$$

Addition of reaction 14-18 to the summary reaction for glycolysis and the TCA cycle (reaction 14-3) leads to the following overall expression for the maximum theoretical ATP yield obtainable by the complete aerobic respiration of glucose or other hexoses:

$$C_6H_{12}O_6 + 6O_2 + 38ADP + 38P_i \longrightarrow$$
$$6CO_2 + 38ATP + 44H_2O \quad \textbf{(14-19)}$$

This summary reaction is valid for most prokaryotic cells and for some types of eukaryotic cells. Depending on the cell type, however, the maximum ATP yield for a eukaryotic cell may be 36 instead of 38 because of the lesser ATP yield from NADH molecules generated in the cytosol rather than in the mitochondrion (see question 2 below).

Before leaving reaction 14-19 and the aerobic energy metabolism that it summarizes, we will consider several questions that are often asked concerning the number of water molecules and ATP molecules in the reaction, with the hope that each will further enhance your understanding of aerobic respiration.

1. Why Does Reaction 14-19 Have So Many Water Molecules on the Right?   Summary reactions such as reaction 14-19 are commonly written with only 6 H_2O on the right-hand side. However, doing so ignores the 38 water molecules that result when ADP and P_i condense to form ATP. Keep in mind that for every glucose molecule that is catabolized by aerobic respiration, electron transport generates 12 water molecules, the TCA cycle consumes 6 water molecules, and ATP synthesis produces another 38 water molecules. The net result is the production of 44 molecules of H_2O per molecule of glucose, as shown in reaction 14-19. To keep better track of the water molecules, we can factor out those that arise from ATP formation by rewriting reaction 14-19 as

$$C_6H_{12}O_6 + 6O_2 \xrightarrow[\quad 38ADP + 38P_i \quad]{\quad 38ATP + 38H_2O \quad} 6CO_2 + 6H_2O \quad \textbf{(14-19a)}$$

2. Why Does the Maximum ATP Yield in Eukaryotic Cells Vary Between 36 and 38 ATPs Per Glucose?   Recall that when glucose is catabolized aerobically in a eukaryotic cell, glycolysis gives rise to two molecules of NADH per glucose in the cytosol, while the catabolism of pyruvate generates another eight molecules of NADH in the matrix of the mitochondrion. This spatial distinction is important because the inner membrane of the mitochondrion does not have a carrier protein for NADH or NAD$^+$, so NADH generated in the cytosol cannot enter the mitochondrion to deliver its electrons to complex I of the ETS. Instead, the electrons and H$^+$ ions are passed inward by one of sev-

eral *electron shuttle systems* that differ in the number of ATP molecules formed per NADH molecule oxidized.

An **electron shuttle system** consists of one or more electron carriers that can be reversibly reduced, with transport proteins present in the membrane for both the oxidized and the reduced forms of the carrier. In the case of electron movement into the mitochondrion, cytosolic NADH passes its electrons and proton to a carrier molecule in the cytosol, thereby oxidizing the coenzyme so that it can again be used as an electron acceptor in glycolysis or other cytosolic processes. Meanwhile, the reduced form of the carrier is transported into the mitochondrion, where it is oxidized by a mitochondrial enzyme. The difference in ATP yield among eukaryotic cells comes about because the oxidation of the carrier molecule within the mitochondrion may involve the transfer of electrons to either NAD$^+$ or FAD, depending on the cell type.

In liver, kidney, and heart cells, electrons from cytosolic NADH are transferred into the mitochondrion by means of the *malate-aspartate shuttle*, a complex mechanism that uses NAD$^+$ as the electron acceptor on the matrix side. This shuttle is freely reversible and is therefore only useful in cells that maintain a significantly higher [NADH]/[NAD$^+$] ratio in the cytosol than in the mitochondrion, so that inward electron transport is exergonic. In such cells, the oxidation of NADH to NAD$^+$ in the cytosol is accompanied by the generation of NADH from NAD$^+$ in the matrix. Electrons derived from cytosolic NADH therefore pass through all three proton-pumping complexes of the transport system and are therefore capable of generating up to three molecules of ATP per molecule of coenzyme.

In skeletal muscle, brain, and other tissues, the [NADH]/[NAD$^+$] ratio is sometimes lower in the cytosol than in the mitochondrion. Cells in these tissues deliver electrons from cytosolic NADH to the mitochondrial respiratory complexes by means of a shuttle mechanism that uses FAD rather than NAD$^+$ as the mitochondrial electron acceptor. This mechanism, called the **glycerol phosphate shuttle** (or *glycerol-3-phosphate/dihydroxyacetone phosphate shuttle*, to use its formal name) is shown in Figure 14-25. The glycerol phosphate shuttle involves cytosolic and mitochondrial forms of the enzyme glycerol-3-phosphate dehydrogenase (G3PDH), which reversibly oxidizes glycerol-3-phosphate (glycerol-3-P) to dihydroxyacetone phosphate (DHAP). The cytosolic form of the enzyme transfers electrons from NADH to DHAP, reducing it to glycerol-3-P.

The glycerol-3-P then passes into the mitochondrion, where it is reoxidized to DHAP by an FAD-dependent G3PDH located on the outer face of the inner membrane. Because this membrane-bound enzyme uses FAD instead of NAD$^+$ as its electron acceptor, the electrons are transferred directly to coenzyme Q. As a result, these electrons bypass the first energy-conserving site of the ETS, generating only two molecules of ATP instead of three, and reducing the maximum theoretical yield by one ATP per cytosolic NADH and therefore by two ATP molecules per molecule of glucose.

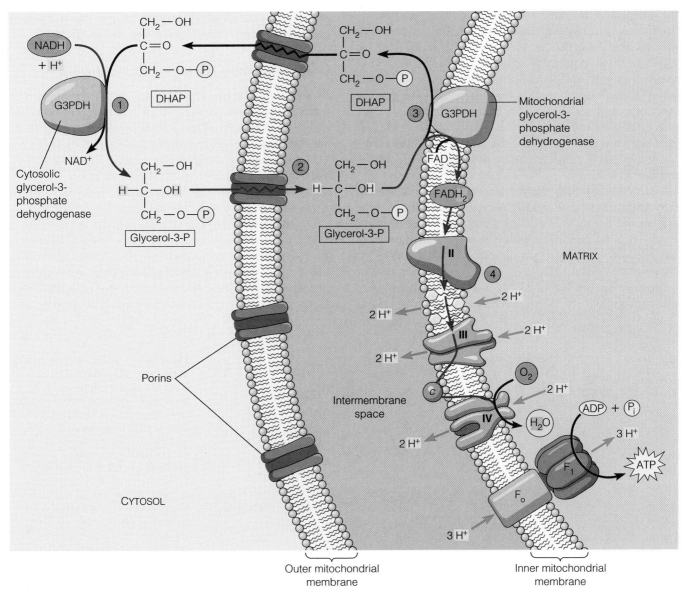

Figure 14-25 The Glycerol Phosphate Shuttle. The inner mitochondrial membrane is impermeable to NADH, so electrons from cytosolic NADH are moved into the mitochondrion by one of two shuttle mechanisms. Cells of skeletal muscle, brain, and other tissues use the glycerol phosphate shuttle for this purpose. The pink line traces the path of electrons from cytosolic NADH to oxygen. ① The cytosolic enzyme glycerol-3-phosphate dehydrogenase (G3PDH) uses electrons (and protons) from NADH to reduce dihydroxyacetone phosphate (DHAP) to glycerol-3-phosphate (glycerol-3-P), which ② can cross the outer mitochondrial membrane by diffusing through a porin channel. ③ Glycerol-3-P is then reoxidized to DHAP by an FAD-linked G3PDH in the inner membrane, with concomitant reduction of FAD to $FADH_2$. Because NADH is a more energy-rich coenzyme than $FADH_2$, the inward transport of electrons is exergonic, driven by the difference in the reduction potentials of the two coenzymes, even when the cytosolic NADH concentration is lower than the NADH concentration in the mitochondrial matrix. The cost of this inward transport is a decreased ATP yield because ④ electrons from the $FADH_2$ generated by the mitochondrial G3PDH bypass complex I, thereby reducing the number of protons pumped across the inner membrane.

This shuttle may seem inherently wasteful because the oxidation of NADH to NAD^+ in the cytosol leads to the generation of $FADH_2$ rather than NADH in the mitochondrion, with a consequent reduction in the maximum theoretical ATP yield. However, the difference in E_0' between NADH and $FADH_2$ provides the driving force necessary for inward electron transport when the $[NADH]/[NAD^+]$ ratio is lower in the cytosol than in the mitochondrion. In fact, the differ-ence in $\Delta E_0'$ values is sufficiently large that the shuttle is essentially irreversible and functions effectively even when the $[NADH]/[NAD^+]$ ratio in the cytosol is very low.

3. Why Is the ATP Yield of Aerobic Respiration Referred to as the "Maximum Theoretical ATP Yield"? This wording is a reminder that yields of 36 or 38 ATP molecules per molecule of glucose are possible only if we assume that the energy of the electrochemical proton gradient is used solely to drive ATP synthesis. Such an assumption may be useful for the calculation of maximum possible yields, but it is not otherwise realistic because the pmf of the proton gradient provides the driving force not only for ATP synthesis but also for other energy-requiring reactions and processes. For example, some of the energy of the proton gradient is used to drive the transport of various metabolites and ions across the membrane. Several of these transport processes are illustrated in Figure 14-26.

Depending on the relative concentrations of pyruvate, fatty acids, amino acids, and TCA-cycle intermediates in the cytosol and the mitochondrial matrix, variable amounts of energy may be needed to ensure that the mitochondrion has adequate supplies of oxidizable substrates and TCA-cycle intermediates. Moreover, the inward transport of phosphate ions needed for ATP synthesis is accompanied by the concomitant outward movement of hydroxyl ions,

which are neutralized by protons in the intermembrane space, thereby also drawing on the proton gradient.

Aerobic Respiration Is a Highly Efficient Process

To determine the overall efficiency of ATP production by aerobic respiration, we need to ask what proportion of the energy of glucose oxidation is preserved in the 36 or 38 molecules of ATP generated per molecule of glucose. The $\Delta G^{\circ\prime}$ value for the complete oxidation of glucose to CO_2 and H_2O is -686 kcal/mol. ATP hydrolysis has a $\Delta G^{\circ\prime}$ of about -7.3 kcal/mol but the actual $\Delta G^{\prime}$ is typically in the range of -10 to -14 kcal/mol. Assuming a value of 10 kcal/mol as we did earlier in the chapter, the 36–38 moles of ATP generated by aerobic respiration of 1 mole of glucose in an aerobic cell correspond to about 360–380 kcal of energy conserved per mole of glucose oxidized. This is an efficiency of about 52–55%, well above that obtainable with the most efficient machines we are capable of creating.

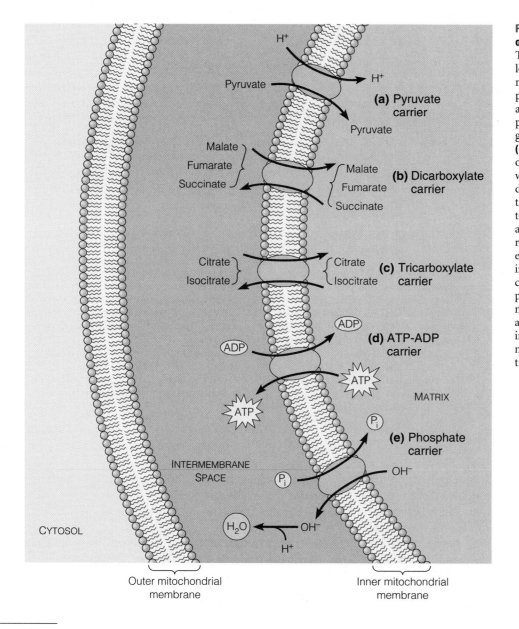

Figure 14-26   Major Transport Systems of the Inner Mitochondrial Membrane. The major transport proteins localized in the inner mitochondrial membrane are shown here. **(a)** The pyruvate carrier cotransports pyruvate and protons inward, driven by the pmf of the electrochemical proton gradient. The **(b)** dicarboxylate and **(c)** tricarboxylate carriers exchange organic acids across the membrane, with the direction of transport depending on the relative concentrations of dicarboxylic and tricarboxylic acids on the inside and outside of the inner membrane, respectively. **(d)** The ATP-ADP carrier exchanges ATP outward for ADP inward, and **(e)** the phosphate carrier couples the inward movement of phosphate with the outward movement of hydroxyl ions, which are neutralized by protons in the intermembrane space. (For the mechanism of inward electron transport, see Figure 14-25.)

Perspective

Compared with fermentation, aerobic respiration gives the cell access to much more of the free energy that is available from organic substrates such as sugars, fats, and proteins. The complete catabolism of carbohydrates begins with the glycolytic pathway, but the pyruvate that is formed is then passed into the mitochondrion, where it is oxidatively decarboxylated to acetyl CoA. The acetyl CoA is then oxidized fully by enzymes of the TCA cycle.

Fatty acids are alternative substrates for energy metabolism in many cells. Their catabolism occurs in the mitochondrial matrix and begins with β oxidation to acetyl CoA, which then enters the TCA cycle. Proteins can also be used as energy sources, particularly under conditions of fasting or starvation. In such cases, proteins are degraded to amino acids, each of which is then catabolized to one or more end-products that enter either the glycolytic pathway or the TCA cycle.

Reduced coenzymes (NADH and $FADH_2$) are reoxidized by an electron transport system that consists of respiratory complexes, large multiprotein assemblies embedded in the inner mitochondrial membrane (or, in the case of prokaryotes, in the plasma membrane). The respiratory complexes are free to move laterally within the membrane. Key intermediates in the electron transport system are coenzyme Q and cytochrome c, which transfer electrons between the complexes. In aerobic organisms, oxygen is the ultimate electron acceptor and water is the final product.

Of the four main respiratory complexes, three (complexes I, III, and IV) couple the transfer of electrons to the outward pumping of protons. This establishes an electrochemical proton gradient that is the driving force for ATP generation. The ATP-synthesizing system consists of a proton translocator, F_o, embedded in the membrane and an ATP synthase, F_1, a knoblike structure that projects from the inner membrane on the matrix side (or on the cytoplasmic side of the plasma membrane in prokaryotic cells). As we learned in Chapter 8, an F_oF_1 complex can function either as an ATP synthase driven by a proton gradient or as an ATPase that uses the energy of ATP to create and maintain a proton gradient. Thus, the electrochemical proton gradient and ATP are, in effect, interconvertible forms of stored energy.

Mitochondria are the site of respiratory metabolism in eukaryotic cells and are prominent organelles in both size and numbers. Mitochondria may form large, interconnected networks in some cell types but are regarded here as discrete organelles. They are usually several micrometers long and range in abundance from one or a few up to hundreds or even a few thousand per cell. A mitochondrion is surrounded by two membranes. The inner membrane has many infoldings called cristae, which greatly increase the surface area of the membrane and hence its ability to accommodate the numerous respiratory complexes, F_oF_1 complexes, and transport proteins needed for respiratory function.

The outer membrane of the mitochondrion is freely permeable to ions and small molecules due to the presence of porins. However, specific carriers are required for the inward transport of pyruvate, fatty acids, and other organic molecules across the inner membrane of the organelle. ATP transport outward is coupled to the inward movement of ADP, and the concurrent inward movement of phosphate ions is coupled to the outward movement of hydroxyl ions, driven by the proton gradient. The electrons of coenzyme molecules that undergo reduction in the cytosol must be passed inward to the electron transport system by specific shuttle mechanisms because the inner membrane is not permeable to the coenzymes themselves.

This, then, is aerobic energy metabolism. No transistors, no mechanical parts, no noise, no pollution—and all done in units of organization that require an electron microscope to visualize. Yet the process goes on continuously in living cells with a degree of integration, efficiency, fidelity, and control that we can scarcely understand well enough to appreciate fully, let alone aspire to reproduce in our test tubes.

Key Terms for Self-Testing

Cellular Respiration: Maximizing ATP Yields
cellular respiration (p. 398)
aerobic respiration (p. 398)
anaerobic respiration (p. 398)

**The Mitochondrion: Where
the Action Takes Place**
mitochondrion (p. 400)
outer membrane (p. 402)
porin (p. 402)
intermembrane space (p. 402)
inner membrane (p. 403)

crista (p. 403)
matrix (p. 404)
F_1 complex (p. 404)
F_o complex (p. 404)
F_oF_1 complex (p. 404)

**The Tricarboxylic Acid Cycle: Oxidation
in the Round**
tricarboxylic acid (TCA) cycle (p. 405)
coenzyme A (CoA) (p. 406)
acetyl CoA (p. 406)
flavin adenine dinucleotide (FAD) (p. 409)

allosteric regulation (p. 410)
triacylglycerols (p. 412)
β oxidation (p. 412)
proteolysis (p. 413)
transamination (p. 414)
oxidative deamination (p. 414)
amphibolic pathway (p. 414)

**Electron Transport: Electron Flow
from Coenzymes to Oxygen**
electron transport (p. 418)
electron transport system (ETS) (p. 418)

flavoprotein (p. 418)
iron-sulfur (Fe-S) protein (p. 419)
cytochrome (p. 419)
bimetallic iron-copper (Fe-Cu) center (p. 419)
coenzyme Q (CoQ) (p.419)
reduction potential (E′) (p. 420)
redox (reduction-oxidation) pair (p. 420)
standard reduction potential (E_0') (p. 421)
respiratory complex (p. 423)
complex I (NADH-coenzyme Q
oxidoreductase) (p. 423)
complex II (succinate–coenzyme Q
oxidoreductase) (p. 423)
complex III (coenzyme Q–cytochrome c
oxidoreductase) (p. 424)

complex IV (cytochrome c oxidase) (p. 424)
terminal oxidase (p. 424)

**The Electrochemical Gradient:
Key To Energy Coupling**
oxidative phosphorylation (p. 426)
electrochemical proton gradient (p. 426)
coupling (p. 426)
respiratory control (p. 426)
chemiosmotic coupling model (p. 426)
P/O ratio (p. 427)
unidirectional pumping of protons (p. 428)
proton motive force (pmf) (p. 429)

ATP Synthesis: Putting It All Together
proton translocator (p. 431)

ATP synthase (F_oF_1 complex) (p. 431)
binding change model (p. 432)

Aerobic Respiration: Summing It All Up
maximum ATP yield (p. 435)
electron shuttle system (p. 436)
glycerol phosphate shuttle (p. 436)

Box 14A: *The Glyoxylate Cycle, Glyoxysomes,
and Seed Germination*
glyoxylate cycle (p. 416)
glyoxysome (p. 416)

Problem Set

More challenging problems are marked with a •.

**14-1.   Localization of Molecules and Functions Within the
Mitochondrion.** Indicate whether you would expect to find each
of the following molecules or functions in the matrix (MA), the
inner membrane (IM), the outer membrane (OM), the intermembrane space (IS), or not in the mitochondrion at all (NO).

(a) Coenzyme A

(b) Coenzyme Q

(c) Nucleotide phosphorylation

(d) Succinate dehydrogenase

(e) Malate dehydrogenase

(f) Fatty acid elongation

(g) Dicarboxylate carrier

(h) Conversion of lactate into pyruvate

(i) ATP synthase

(j) Accumulation of a high proton concentration

**14-2.   Localization of Molecules and Functions Within the
Prokaryotic Cell.** Repeat Problem 14-1, but indicate where in a
prokaryotic cell you would expect to find each of the listed
molecules or functions. Choices: cytoplasm (CY), plasma membrane (PM), exterior of cell (EX), or not present at all (NO).

14-3.   True or False. Indicate whether each of the following
statements is true (T) or false (F). If false, reword the statement
to make it true.

(a) The orderly flow of carbon through the TCA cycle is possible because each of the enzymes of the cycle is embedded
in the inner mitochondrial membrane in such a manner
that their order in the membrane is the same as their
sequence in the cycle.

(b) Thermodynamically, acetyl CoA should be capable of driving the phosphorylation of ADP (or GDP), just as succinyl
CoA does, assuming availability of the appropriate enzyme.

(c) Respiration is an aerobic process in all organisms because
oxygen is the only known electron acceptor for the reoxidation of coenzymes.

(d) We can predict that the flow of electrons through the electron transport system is exergonic because the NAD^+/NADH
redox pair has a highly negative $\Delta E_0'$ and the O_2/H_2O redox
pair has a highly positive $\Delta E_0'$.

(e) Unlike NAD^+, the coenzyme FAD tends to be tightly associated with the dehydrogenase enzymes that use it as an electron acceptor.

(f) Nine cycles of β oxidation are required to degrade the 18-
carbon fatty acid oleate completely to acetyl CoA.

14-4.   Mitochondrial Transport. For aerobic respiration, a
variety of substances must be in a state of flux across the inner
mitochondrial membrane. Assuming a brain cell in which
glucose is the sole energy source, indicate for each of the following substances whether you would expect a net flux across the
membrane and, if so, how many molecules will move in which
direction per molecule of glucose catabolized.

(a) Pyruvate

(b) Oxygen

(c) ATP

(d) ADP

(e) Acetyl CoA

(f) Glycerol-3-phosphate

(g) NADH

(h) $FADH_2$

(i) Oxaloacetate

(j) Water

(k) Electrons

(l) Protons

14-5.   Completing the Pathway. In each of the following cases,
complete the pathway by indicating the structures and order of
the intermediates.

(a) The conversion of citrate to α-ketoglutarate by reactions
TCA-2 and TCA-3 involves as intermediates not only iso-
citrate but also molecules identified (but not shown in
Figure 14-8) as aconitate and oxalosuccinate. Illustrate the

pathway from citrate to α-ketoglutarate by showing the structures and order of all three intermediates.

(b) Synthesis of the amino acid glutamate can be effected from pyruvate and alanine by a metabolic sequence that illustrates the amphibolic role of the TCA cycle. Devise such a pathway, assuming the availability of whatever additional enzymes may be needed.

(c) If pyruvate-2-^{14}C (pyruvate with the middle carbon atom radioactively labeled) is provided to actively respiring mitochondria, most of the radioactivity will be incorporated into citrate. Trace the route whereby radioactively labeled carbon atoms are incorporated into citrate, and indicate where in the citrate molecule the label will first appear.

14-6. The Calculating Cell Biologist. Use Table 14-2 as the basis for the calculations needed to answer the following questions.

(a) Without doing any calculations initially, predict whether isocitrate can pass electrons exergonically to NAD^+ under standard conditions. How do you know that?

(b) What is the $\Delta E_0'$ for the oxidation of isocitrate by NAD^+ under standard conditions? Does this calculation support the prediction you made in part a? Explain your answer.

(c) Calculate the $\Delta G^{\circ\prime}$ for the oxidation of isocitrate by NAD^+ under standard conditions. In what way is this calculation relevant to aerobic energy metabolism?

(d) Repeat parts a–c for the oxidation of lactate to pyruvate by NAD^+. In what way is this calculation relevant to aerobic energy metabolism?

(e) Now repeat parts a–c for the oxidation of succinate to fumarate with NAD^+ as the electron acceptor. You should find that the $\Delta G^{\circ\prime}$ is highly positive. What does this tell us about the likelihood that NAD^+ could serve as the electron acceptor for the succinate dehydrogenase reaction of the TCA cycle?

(f) Finally, repeat parts a–c for the oxidation of succinate to fumarate with coenzyme Q as the electron acceptor. Why does it make sense to regard coenzyme Q as the electron acceptor when succinate dehydrogenase is shown in Figure 14-8 with FAD as the immediate electron acceptor?

14-7. Calculating Maximum ATP Yields. Table 14-5 is intended as a means of summarizing the ATP yield during the aerobic oxidation of glucose.

(a) Complete Table 14-5 for an aerobic prokaryote. What is the maximum ATP yield?

(b) Indicate on Table 14-5 the changes that are necessary to calculate the maximum ATP yield for a eukaryotic cell that uses the glycerol phosphate shuttle to move electrons from the cytosol into the matrix of the mitochondrion.

14-8. Regulation of Catabolism. Explain the advantage to the cell of each of the following regulatory mechanisms.

(a) Isocitrate dehydrogenase (reaction TCA-3) is allosterically activated by ADP.

(b) The dehydrogenases that oxidize isocitrate, α-ketoglutarate, and malate (reactions TCA-3, TCA-4, and TCA-8) are all allosterically inhibited by NADH.

(c) Pyruvate dehydrogenase (reaction 14-1) is allosterically inhibited by ATP.

(d) Phosphofructokinase (Figure 13-6, reaction Gly-3) is allosterically inhibited by citrate.

(e) Pyruvate dehydrogenase kinase (Figure 14-10, enzyme E_1) is allosterically activated by NADH.

(f) α-ketoglutarate dehydrogenase (reaction TCA-4) is allosterically inhibited by succinyl CoA.

14-9. Lethal Synthesis. The leaves of *Dichapetalum cymosum*, a South African plant, are very poisonous. Animals that eat the

Table 14-5   Calculation of the Maximum ATP Yield from Aerobic Oxidation of Glucose

Stage of Respiration	Glycolysis (glucose ⟶ 2 pyruvate)	Pyruvate Oxidation (2 pyruvate ⟶ 2 acetyl CoA)	TCA Cycle (2 turns)
Yield of CO_2			
Yield of NADH			
ATP per NADH			
Yield of $FADH_2$			
ATP per $FADH_2$			
ATP from substrate-level phosphorylation			
ADP from oxidative phosphorylation			
Maximum ATP yield			

leaves have convulsions and usually die shortly thereafter. One of the most pronounced effects of poisoning is a marked elevation in citrate concentration and a blockage of the TCA cycle in many organs of the affected animal. The toxic agent in the leaves of the plant is fluoroacetate, but the actual poison in the tissues of the animal is fluorocitrate. If fluoroacetate is incubated with purified enzymes of the TCA cycle, it has no inhibitory effect on enzyme activity.

(a) Why might you expect fluorocitrate to have an inhibitory effect on one or more of the TCA cycle enzymes if incubated with the purified enzymes in vitro, even though fluoroacetate has no such effect?

(b) Which enzyme in the TCA cycle do you suspect is affected by fluorocitrate? Give two reasons for your answer.

(c) How do you suppose fluoroacetate gets converted to fluorocitrate?

(d) Why is this phenomenon referred to as *lethal synthesis*?

14-10. Oxidation of Cytosolic NADH. In some eukaryotic cells, the NADH generated by glycolysis in the cytosol is reoxidized by the glycerol phosphate shuttle shown in Figure 14-25.

(a) Write balanced reactions for the reduction of dihydroxyacetone phosphate (DHAP) to glycerol-3-phosphate (glycerol-3-P) by cytosolic NADH and for the oxidation of glycerol-3-P to DHAP by the FAD-linked glycerol-3-P dehydrogenase in the inner membrane of the mitochondrion.

(b) Add the two reactions in part a to obtain a summary reaction for the transfer of electrons from cytosolic NADH to mitochondrial FAD. Calculate $\Delta E_0'$ and $\Delta G^{\circ\prime}$ for this reaction. Is the inward movement of electrons thermodynamically feasible under standard conditions?

(c) Write a balanced reaction for the reoxidation of $FADH_2$ by coenzyme Q within the inner membrane, assuming that CoQ is reduced to $CoQH_2$. Calculate $\Delta E_0'$ and $\Delta G^{\circ\prime}$ for this reaction. Is this transfer thermodynamically feasible under standard conditions?

(d) Write a balanced reaction for the transfer of electrons from cytosolic NADH to mitochondrial CoQ, and calculate $\Delta E_0'$ and $\Delta G^{\circ\prime}$ for this reaction. Is this transfer thermodynamically feasible under standard conditions?

(e) Assume that the $[NADH]/[NAD^+]$ ratio in the cytosol is 5.0 and that the $[CoQH_2]/[CoQ]$ ratio in the inner membrane is 2.0. What is $\Delta G'$ for the reaction in part d at 25°C and pH 7.0?

(f) Is $\Delta G'$ for the inward transfer of electrons from NADH to CoQ affected by the ratio of the reduced to the oxidized forms of the enzyme-bound FAD in the inner membrane? Why or why not?

• **14-11. Brown Fat and Thermogenin.** Most newborn mammals, including human infants, have a special type of fat tissue called *brown fat*, in which a naturally occurring uncoupling protein called *thermogenin* is present in the inner mitochondrial membrane. This protein uncouples ATP synthesis from electron transport so that the energy released as electrons flow through the electron transport chain is lost as heat instead.

(a) What happens to the energy that is released as electron transport continues but ATP synthesis ceases? Why might

it be advantageous for a baby to have thermogenin present in the inner membrane of the mitochondria that are present in brown fat tissue?

(b) Some adult mammals also have brown fat. Would you expect to find more brown fat tissue and more thermogenin in a hibernating bear or in a physically active bear? Explain your reasoning.

(c) Given its location in the cell, suggest a mode of action for thermogenin. What kind of an experiment can you suggest to test your hypothesis?

(d) What would happen to a mammal if all of its mitochondria were equipped with uncoupling protein, rather than just those in brown fat tissue?

• **14-12. Multiple Uses of the Electrochemical Proton Gradient.** Most of our discussion in this chapter focused on the use of the electrochemical proton gradient to drive the synthesis of ATP. In particular, the conclusion that the electrochemical proton gradient across the inner membrane of the mitochondrion can generate 34 molecules of ATP per molecule of glucose (reaction 14-18) assumes that the energy of the proton gradient is not tapped for any other purpose. Actually, however, the energy of the proton gradient is also used to drive a variety of other processes, notably the transport of ions and metabolites across the inner membrane. Assume (1) that two protons are required to drive the symport of each molecule of pyruvate into the mitochondrion; (2) that the coupled outward transport of ATP and inward transport of ADP is exergonic; and (3) that the inward movement of phosphate anions (P_i) is coupled on an equimolar basis to the outward transport of hydroxyl ions, each of which is neutralized by reaction with a proton on the outer surface of the inner membrane. Assume further (4) that each of the respiratory complexes pumps three protons outward across the inner membrane per pair of electrons transported, and that (5) ATP synthesis by the F_oF_1 complex is driven by the inward movement of three protons per ATP.

(a) Under these conditions, what is the total number of protons pumped outward by the electron transport that accompanies the aerobic catabolism of one molecule of glucose? How many ATP molecules could be synthesized per glucose if the energy of the proton gradient were used for no other purpose?

(b) How many pyruvate molecules must be transported into the mitochondrion per molecule of glucose catabolized? How many protons are required for this purpose?

(c) How many ATP molecules can be synthesized per glucose by the remaining protons, recalling that the inward transport of phosphate is coupled to the outward movement of hydroxyl ions, which are then neutralized by protons?

(d) What is the total ATP yield per molecule of glucose under these conditions? Write a balanced reaction to show the aerobic catabolism of glucose accompanied by the generation of the appropriate number of ATP molecules.

• **14-13. Glyoxysomal Function.** The glyoxysomes of fat-storing seedlings contain all the enzymes necessary to degrade fatty acids completely to acetyl CoA and to synthesize succinate from the resulting acetyl CoA. Fatty acid oxidation (Figure 14-11) begins with ATP-dependent formation of fatty acyl CoA thioester (reaction FA-1). Each cycle of β oxidation generates one molecule each of $FADH_2$ (FA-2) and NADH (FA-4), culminating in the release of two carbons as acetyl CoA (FA-5). In the glyoxysomes, $FADH_2$, but not NADH, is reoxidized by the direct, oxidase-mediated transfer of electrons to oxygen, generating hydrogen peroxide and no ATP. Synthesis of succinate from the acetyl CoA occurs by means of the glyoxylate cycle (see Box 14A). The further steps required to convert succinate to hexoses (and then to sucrose) occur elsewhere in the cell (Figure 14A-2).

(a) What are the main products of β oxidation of a fatty acid in the glyoxysome? What happens to each?

(b) How many molecules of succinate are produced from a single molecule of palmitate ($C_{16}H_{32}O_2$)? How many molecules of $FADH_2$ and of NADH are generated in the process?

(c) How many hexose molecules can be formed from the succinate of part b? Where in the cell does this process occur? What other products are formed in the conversion of succinate to hexose?

(d) How many molecules of sucrose can be produced from a single molecule of palmitate? How many of the original 16 carbon atoms of palmitate eventually appear in sucrose? What happens to the others?

(e) Assuming a quantitative conversion of fat to carbohydrate, how many grams of sucrose can a fat-storing seedling produce from 1 gram of stored palmitate?

(f) In addition to fatty acids and succinate, what other substances would you predict have to be transported across the glyoxysomal membrane? In which direction would you expect each of these substances to move?

• 14-14. **Dissecting the Electron Transport System.** To determine which segments of the electron transport system (ETS) are responsible for proton pumping and hence for ATP synthesis, investigators usually incubate isolated mitochondria under conditions such that only a portion of the ETS is functional. One approach is to supply the mitochondria with an electron donor and an electron acceptor that are known to tap into the ETS at specific points. In addition, inhibitors of known specificity are often added. In one such experiment, mitochondria were incubated with β-hydroxybutyrate, oxidized cytochrome c, ADP, P_i, and cyanide. (Mitochondria have an NAD^+-dependent dehydrogenase that is capable of oxidizing β-hydroxybutyrate to β-ketobutyrate.)

(a) What is the electron donor in this system? What is the electron acceptor? What is the most likely pathway of electron transport in this system?

(b) Based on what you already know about the ETS, how many moles of ATP would you expect to be formed per mole of β-hydroxybutyrate oxidized? Write a balanced equation for the reaction that occurs in this system.

(c) Cyanide is the only reagent added to the system that is not a part of the balanced equation for the reaction. What was the purpose of adding cyanide to the system? What result would you expect if the cyanide had not been added?

(d) Would you expect the enzymes of the TCA cycle to be active in this assay system? Explain your reasoning.

(e) Why is it important that β-ketobutyrate cannot be further metabolized in this system? Lactate is quite similar in structure to β-ketobutyrate; what effect would it have had if the investigators had used lactate as the oxidizable substrate instead of β-hydroxybutyrate?

Suggested Reading

References of historical importance are marked with a • .

General References

Mathews, C. K., and K. E. van Holde. *Biochemistry*, 2ᵈ ed., Menlo Park, CA.: Benjamin/Cummings, 1996.

Metzler, D. E. *Biochemistry: The Chemical Reactions of Living Cells.* 2ᵈ ed. San Diego: Academic Press, 2001.

• Nicholls, D. G. *Bioenergetics: An Introduction to the Chemiosmotic Theory.* New York: Academic Press, 1982.

Papa, S., F. Guerrieri, and J. M. Tager, eds. *Frontiers of Cellular Bioenergetics: Molecular Biology, Biochemistry, and Physiopathology.* New York: Kluwer Academic/Plenum Publishers, 1999.

• Racker, E. From Pasteur to Mitchell: A hundred years of bioenergetics. *Fed. Proc.* 39 (1980): 210.

Saks, V. A., R. Ventura-Clapier, X. Leverve, M. Rigoulet, and A. Rossi, eds. *Bioenergetics of the Cell: Quantitative Aspects.* Dordrecht, the Netherlands: Kluwer Academic Publishers, 1998.

Mitochondrial Structure and Function

Andre, J. Mitochondria. *Biol. Cell* 80 (1994): 103.

Bereiter-Hahn, J. Behavior of mitochondria in the living cell. *Internat. Rev. Cytol.* 122 (1990): 1.

• Ernster, L., and G. Schatz. Mitochondria: An historical overview. *J. Cell Biol.* 91 (1981): 227s.

Manella, C. A. The "ins" and "outs" of mitochondrial membrane channels. *Trends Biochem. Sci.* 17 (1992): 315.

Tyler, D. D. *The Mitochondrion in Health and Disease.* New York: VCH Publishers, 1992.

The Tricarboxylic Acid Cycle

Cammack, R. $FADH_2$ as a "product" of the citric acid cycle. *Trends Biochem. Sci.* 12 (1987): 377.

• Kornberg, H. L. Tricarboxylic acid cycles. *Bioessays* 7 (1987): 236.

• Krebs, H. A. The history of the tricarboxylic acid cycle. *Perspect. Biol. Med.* 14 (1970): 154.

Electron Transport

Beinert, H., R. Holm, and E. Münck. Iron-sulfur clusters: Nature's modular, multipurpose structures. *Science* 277 (1997): 653.

Calhoun, M. W., J. W. Thomas, and R. B. Gennis. The cytochrome oxidase superfamily of redox-driven proton pumps. *Trends Biochem. Sci.* 19 (1994): 325.

Canters, G. W., and E. Vijgenboom, eds. *Biological Electron Transfer Chains: Genetics, Composition, and Mode of Operation.* Dordrecht, the Netherlands: Kluwer Academic Publishers, 1998.

Crane, F. L., and P. Navas. The diversity of coenzyme Q function. *Mol. Aspects Med.* 18 (1997): S1.

Kiel, J. L. *Type-B Cytochromes: Sensors and Switches.* Boca Raton, FL: CRC Press, 1995.

Malatesta, F., G. Antonini, P. Sarti, and M. Brunori. Structure and function of a molecular machine: Cytochrome c oxidase. *Biophys. Chem.* 54 (1995): 1.

• Mitchell, P. Keilin's respiratory chain concept and its chemiosmotic consequences. *Science* 206 (1979): 1148.

Ostermeier, C., S. Iwata, and H. Michel. Cytochrome c oxidase. *Curr. Opin. Struct. Biol.* 6 (1996): 460.

Ramirez, B. E., B. G. Malmstrom, J. R. Winkler, and H. B. Gray. The currents of life: The terminal electron-transfer complex of respiration. *Proc. Nat. Acad. Sci. U.S.* 92 (1995): 11949.

Rauchova, H., Z. Drahota, and G. Lenaz. Function of coenzyme Q in the cell: Some biochemical and physiological properties. *Physiol. Res.* 44 (1995): 209.

Reed, J. C. Cytochrome *c*: Can't live with it—can't live without it. *Cell* 91 (1997): 559.

Scott, R. A., and A. G. Mauk, eds. *Cytochrome c: A Multidisciplinary Approach.* Sausalito, CA: University Science Books, 1996.

Smith, J. 1998. The secret life of cytochrome bc_1. *Science* 281 (1998): 58.

Soole, K. L., and R. I. Menz. Functional molecular aspects of the NADH dehydrogenases of plant mitochondria. *J. Bioenerg. Biomembr.* 27 (1995): 397.

Williams, R. Bioenergetics: Purpose of proton pathways. *Nature* 376 (1996): 643.

Oxidative Phosphorylation and ATP Synthesis

Boyer, P. D. The ATP synthase—A splendid molecular machine. *Annu. Rev. Biochem.* 66 (1997): 717.

Elston, T., H. Wang, and G. Oster. Energy transduction in ATP synthase. *Nature* 391 (1998): 510.

Engelbrecht, S., and W. Junge. ATP synthase: A tentative structural model. *FEBS Letters* 414 (1997): 485.

Fillingame, R. H. Coupling H^+ transport and ATP synthesis in F_oF_1-ATP synthases: Glimpses of interacting parts in a dynamic molecular machine. *J. Exp. Biol.* 200 (1997): 217.

Kinosita, K., R. Yasuda, H. Noji, S. Ishiwata, and M. Yoshida. F_1-ATPase: A rotary motor made of a single molecule. *Cell* 93 (1998): 21.

• Mitchell, P. Coupling of phosphorylation to electron and hydrogen transfer by a chemiosmotic type of mechanism. *Nature* 191 (1961): 144.

Nakamoto, R. K., C. J. Ketchum, and M. K. Shawi. Rotational coupling in the F_oF_1 ATP synthase. *Annu. Rev. Biophys. Biomol. Struct.* 28 (1999): 205.

Oster, G., and H. Wang. ATP synthase: Two motors, two fuels. *Structure with Folding & Design.* 7 (1999): 67.

Saraste, M. Oxidative phosphorylation at the *fin de siecle. Science* 283 (1999): 1488.

Zhou, Y., T. Duncan, and R. Cross. Subunit rotation in *Escherichia coli* F_oF_1-ATP synthase during oxidative phosphorylation. *Proc. Nat. Acad. Sci.* 94 (1997): 10583.

Regulation of Respiratory Metabolism

Brown, G. C. Control of respiration and ATP synthesis in mammalian mitochondria and cells. *Biochem. J.* 284 (1992): 1.

Kell, D. B. The protonmotive force as an intermediate in electron transport-linked phosphorylation: Problems and prospects. *Curr. Topics Cell. Regul.* 33 (1992): 279.

Moore, A. L., G. Leach, and D. G. Whitehouse. The regulation of oxidative phosphorylation in plant mitochondria: The roles of the quinone-oxidizing and -reducing pathways. *Biochem. Soc. Transactions* 21 (1993): 765.

15

Phototrophic Energy Metabolism: Photosynthesis

In the two preceding chapters, we studied fermentation and respiration, two chemotrophic solutions to the universal challenge of meeting the energy and carbon needs of a living cell. Although a few prokaryotic chemotrophs can acquire energy, and even reduced carbon, from inorganic molecules, most chemotrophs depend on organic substrates for survival. This creates a secondary problem: Left alone, most chemotrophs would perish shortly after they oxidized the Earth's limited supply of reduced carbon to carbon dioxide and water. Moreover, the first organisms affected would be those dependent on aerobic respiration, because molecular oxygen (O_2), like reduced carbon, is an exhaustible resource.

Nature's ultimate solution to the steady drain of energy and organic carbon from the biosphere is **photosynthesis**— the conversion of light energy to chemical energy, and its subsequent use in synthesizing organic molecules. Nearly all life on Earth is sustained by the cascade of energy that arrives at the planet as sunlight. **Phototrophs** are organisms that convert solar energy to chemical energy in the form of ATP. For one group of phototrophs, the halobacteria described in Chapter 7, this is the extent of their photosynthetic ability. The halobacteria are **photoheterotrophs**, organisms that acquire energy from sunlight, but depend on organic sources of reduced carbon. Most other phototrophs—including plants, algae, and most photosynthetic bacteria—are photoautotrophs. **Photoautotrophs** use solar energy to drive the biosynthesis of energy-rich organic molecules from simple inorganic starting materials: carbon dioxide and water.* Moreover, members of one group

of photoautotrophs, appropriately called *oxygenic phototrophs,* release molecular oxygen as a by-product of photosynthesis. Thus, phototrophs replenish reduced carbon in the biosphere and molecular oxygen in the atmosphere, completing the cyclic flow of energy and carbon introduced in Chapter 5.

In this chapter, we will study two general aspects of photosynthesis: how photoautotrophs capture solar energy and convert it to chemical energy, and how this energy is used to transform energy-poor carbon dioxide and water into energy-rich organic molecules, such as carbohydrates.

An Overview of Photosynthesis

In photoautotrophs, photosynthesis may be divided into two major processes: energy transduction and carbon assimilation (Figure 15-1). During the **energy transduction reactions,** light energy is converted to chemical energy in the form of ATP and the coenzyme NADPH. To emphasize the importance of light, the energy transduction reactions are also called the *light reactions* of photosynthesis. ATP and NADPH generated by the energy transduction reactions subsequently provide energy and reducing power for the **carbon assimilation reactions.** During the carbon assimilation reactions, fully oxidized carbon atoms from carbon dioxide are *fixed* (covalently attached) to organic acceptor molecules and then reduced and rearranged to form carbohydrates and other organic compounds required for building a living cell. The primary pathway for carbon dioxide fixation in most phototrophs is the *Calvin cycle.* Unfortunately, because the Calvin cycle can operate in the dark as long as ATP and NADPH are

---

*Some phototrophs are *facultative* phototrophs, which can function as either photoautotrophs or photoheterotrophs, depending on the availability of nutrients. Plants are *obligate* photoautotrophs; they must acquire carbon from carbon dioxide.

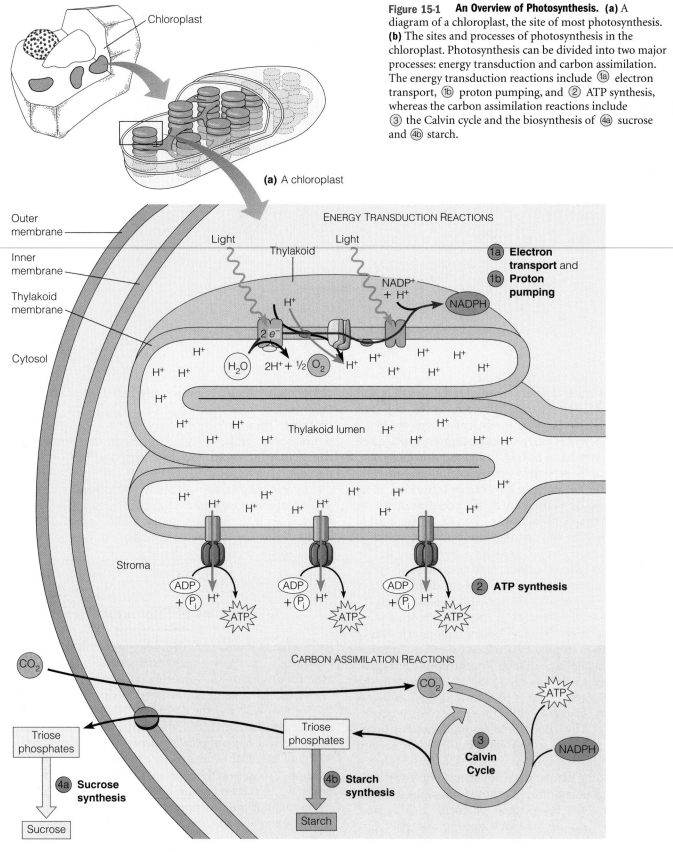

Figure 15-1 An Overview of Photosynthesis. (a) A diagram of a chloroplast, the site of most photosynthesis. **(b)** The sites and processes of photosynthesis in the chloroplast. Photosynthesis can be divided into two major processes: energy transduction and carbon assimilation. The energy transduction reactions include ①a electron transport, ①b proton pumping, and ② ATP synthesis, whereas the carbon assimilation reactions include ③ the Calvin cycle and the biosynthesis of ④a sucrose and ④b starch.

(a) A chloroplast

(b) Localization of photosynthesis within the chloroplast

provided, the carbon assimilation reactions are traditionally referred to as the *dark reactions* of photosynthesis. This term is not accurate, however—in nature both energy transduction and the Calvin cycle operate only when light is available.

A family of green pigment molecules called *chlorophyll* has a key role in every photoautotroph's energy transduction pathway. While a variety of pigments found in photoautotrophs absorb light energy, only one specific form of chlorophyll converts solar energy to chemical energy as it donates photoexcited electrons to organic acceptor molecules. From chlorophyll, photoexcited electrons flow energetically downhill through an electron transport system (ETS). Like electron transport in mitochondria, this flow of electrons is coupled to unidirectional proton pumping, which stores energy in an electrochemical proton gradient that drives an ATP synthase. The light-dependent synthesis of ATP by this process is called **photophosphorylation.**

To incorporate fully oxidized carbon atoms from carbon dioxide into organic molecules, photoautotrophs also need NADPH, a reduced form of the coenzyme nicotinamide adenine dinucleotide phosphate ($NADP^+$). In **oxygenic phototrophs**—plants, algae, and cyanobacteria—light energy absorbed by chlorophyll and other pigment molecules drives the movement of electrons from water, which has a very positive reduction potential, to *ferredoxin,* which has a very negative reduction potential. From ferredoxin, electrons then travel exergonically to $NADP^+$, thereby generating NADPH. In **anoxygenic phototrophs**—green and purple bacteria—compounds with less positive reduction potentials than that of water, such as sulfide (SH^-), thiosulfate ($S_2O_3^{2-}$), or succinate, serve as electron donors. In this case, the light-dependent generation of reductant is often less direct, the ATP generated by photophosphorylation driving the movement of electrons from donor molecules to $NADP^+$. In both oxygenic and anoxygenic phototrophs, the light-dependent generation of NADPH is called **photoreduction.**

Most of the energy accumulated within photosynthetic cells by the light-dependent generation of ATP and NADPH is rapidly consumed by carbon assimilation pathways, where it drives carbon dioxide fixation and reduction. A general reaction for the complete process may be written as

$$\text{light} + CO_2 + 2H_2A \longrightarrow [CH_2O] + 2A + H_2O \quad \text{(15-1)}$$

where H_2A is a suitable electron donor, $[CH_2O]$ represents an organic molecule with carbon at the oxidation level of an aldehyde, and A is the oxidized form of the electron donor. By expressing photosynthesis in this way, we avoid perpetuating the incorrect notion that all phototrophs use water as an electron donor.

When we focus on oxygenic phototrophs, which do use water as an electron donor, we can rewrite reaction 15-1 in the following more specific and familiar form:

$$\text{light} + 3CO_2 + 6H_2O \longrightarrow C_3H_6O_3 + 3O_2 + 3H_2O \quad \text{(15-2)}$$

As indicated by reaction 15-2, the immediate product of photosynthesis is a three-carbon carbohydrate. Later, we will see that the primary products of photosynthetic carbon assimilation are two specific three-carbon carbohydrates, the triose phosphates *glyceraldehyde-3-phosphate (G-3-P)* and *dihydroxyacetone phosphate (DHAP).* Carbon assimilation generally diverges at this point, as the triose phosphates enter a variety of biosynthetic pathways. The most important pathways for our consideration are the biosynthesis of sucrose and starch. Sucrose conveys energy and reduced carbon from photosynthetic cells to nonphotosynthetic cells and is therefore the major transport carbohydrate in most plant species. Starch, on the other hand, accumulates when photosynthetic carbon assimilation exceeds the energy and carbon demands of a photoautotroph and is therefore the major storage carbohydrate.

Our model organisms for studying photosynthesis in this chapter are the most familiar oxygenic phototrophs—green plants.* Before we study photosynthetic energy transduction and carbon assimilation in more detail, however, we will look at the structure and function of the *chloroplast.* In every eukaryotic phototroph, most of the events of photosynthesis are localized in this organelle.

The Chloroplast: A Photosynthetic Organelle

In plants and algae, the primary events of photosynthetic energy transduction and carbon assimilation are confined to specialized organelles, called **chloroplasts.** Because chloroplasts are usually large (1–5 μm wide and 5–10 μm long) and opaque, they were described and studied early in the history of cell biology. Antonie van Leeuwenhoek and Nehemiah Grew described these organelles in the seventeenth century, at the very dawn of microscopic observation. The prominence of chloroplasts within a photosynthetic cell is clearly illustrated in the electron micrograph of a *Coleus* leaf cell shown in Figure 15-2a. A mature leaf cell usually contains 20–100 chloroplasts, the precise number depending on the plant species and habitat, whereas an algal cell typically contains only one or a few chloroplasts. The shapes of these organelles vary from the simple flattened spheres common in plants to the more elaborate forms found in green

---

*Several plants, including Indian Pipe (*Monotropa uniflora*) and Spotted Coral Root (*Corallorhiza maculata*), lack chlorophyll, the green pigment essential for photosynthesis. Such plants derive all of their energy and reduced carbon from green plants or from organic compounds in the soil.

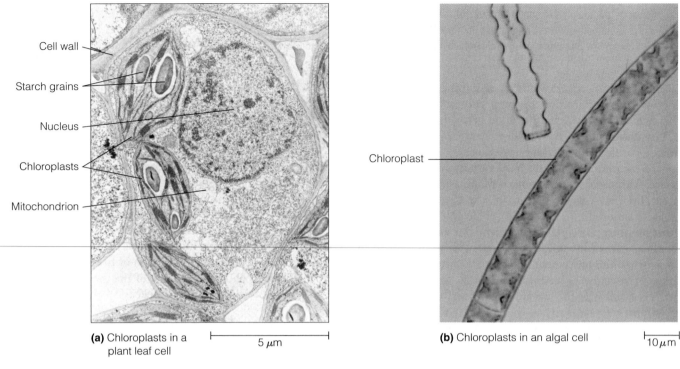

Cell wall	
Starch grains	
Nucleus	
Chloroplasts	
Mitochondrion	

(a) Chloroplasts in a plant leaf cell   5 μm

Chloroplast

(b) Chloroplasts in an algal cell   10 μm

Figure 15-2 Chloroplasts. (a) The prominence of chloroplasts in a leaf cell of a plant is demonstrated by this electron micrograph of a parenchyma cell from a *Coleus* leaf. The cell contains many chloroplasts, three of which are seen in this particular cross section. The presence of large starch grains in the chloroplasts indicates that the cell was photosynthetically active just prior to fixation for electron microscopy (TEM). **(b)** This light micrograph reveals the unusual ribbon-shaped chloroplasts of the filamentous green alga *Spirogyra*.

algae. A cell of the filamentous green alga *Spirogyra,* for example, contains one or more ribbon-shaped chloroplasts, as shown in Figure 15-2b.

Not all plant cells contain chloroplasts. The various types of cells found in a plant all arise from rapidly dividing, undifferentiated tissue called *meristem.* Meristem cells lack chloroplasts, but they have smaller organelles called **proplastids.** Depending on where they occur in the plant and how much light they receive, proplastids develop into any of several kinds of **plastids** equipped to serve different functions. Chloroplasts are only one example of a plastid. Some proplastids differentiate into *amyloplasts,* which are sites for storing starch. Other proplastids acquire red, orange, or yellow pigments, forming the *chromoplasts* that give flowers and fruits their distinctive colors. Proplastids can also develop into organelles for storing protein (*proteinoplasts*) and lipids (*elaioplasts*).

Chloroplasts Are Composed of Three Membrane Systems

A closer view of a typical chloroplast is shown in the electron micrograph of Figure 15-3a. A chloroplast, like a mitochondrion, has both an **outer membrane** and an **inner membrane,** often separated by a narrow **intermembrane space** (Figure 15-3b). Enclosed by the inner membrane is the **stroma,** a gel-like matrix teeming with enzymes for carbon, nitrogen, and sulfur assimilation. The outer membrane contains **porins,** transmembrane proteins that permit the passage of solutes with molecular weights up to about 5000, and is therefore freely permeable to most small organic molecules and ions, but the inner membrane forms a significant permeability barrier. Transport proteins control the flow of most metabolites between the intermembrane space and the stroma. Three important metabolites that are able to diffuse freely across both the outer and the inner membranes, however, are water, carbon dioxide, and oxygen.

Unlike mitochondria, chloroplasts have a third membrane system, the **thylakoids,** seen in Figure 15-3c and d. Thylakoids are flat, saclike structures suspended in the stroma and usually arranged in stacks called **grana** (singular: **granum**). Grana, which resemble stacks of coins, are interconnected by a network of longer thylakoids called **stroma thylakoids.** Essentially all of the photosynthetic pigments, the enzymes required for the photoreactions, the carriers involved in electron transport, and the proteins that couple electron transport to proton pumping and ATP synthesis are localized on or in the thylakoid membranes. Regions where one thylakoid membrane within a granum touches another are called *appressed regions.* Two of the photosynthetic complexes described later, the *photosystem I complex* and the *ATP synthase complex,* are excluded from these appressed regions.

The thylakoid membranes might arise from invaginations of the inner membrane during chloroplast develop-

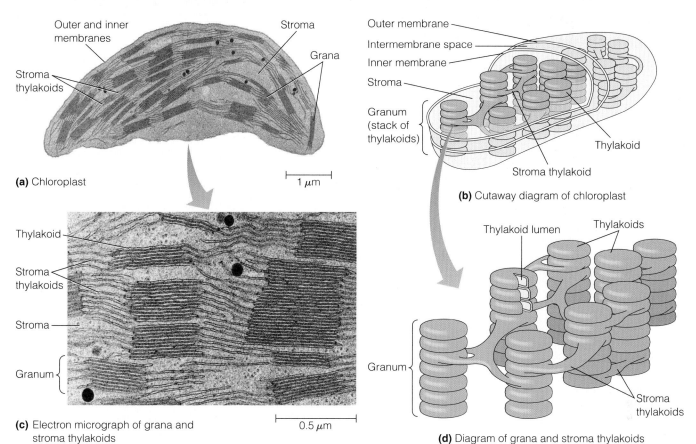

(a) Chloroplast

Outer and inner membranes
Stroma
Grana
Stroma thylakoids

1 μm

(b) Cutaway diagram of chloroplast

Outer membrane
Intermembrane space
Inner membrane
Stroma
Granum (stack of thylakoids)
Thylakoid
Stroma thylakoid

(c) Electron micrograph of grana and stroma thylakoids

Thylakoid
Stroma thylakoids
Stroma
Granum

0.5 μm

(d) Diagram of grana and stroma thylakoids

Thylakoid lumen
Thylakoids
Granum
Stroma thylakoids

Figure 15-3   Structural Features of a Chloroplast. **(a)** An electron micrograph of a chloroplast from a leaf of timothy grass (*Phleum pratense*) (TEM). **(b)** A diagram showing the three-dimensional structure of a typical chloroplast. **(c)** A more magnified electron micrograph of the chloroplast in part a, showing the arrangement of grana and stroma thylakoids (TEM). **(d)** A diagram depicting the continuity of the thylakoid membranes, the arrangement of thylakoids into stacks called grana, and the stroma thylakoids that interconnect the grana. The thylakoid membranes enclose a space called the thylakoid lumen.

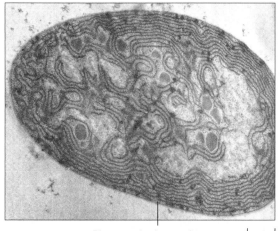

Photosynthetic membranes

1 μm

Figure 15-4   The Photosynthetic Membranes of a Cyanobacterium. This electron micrograph of a thin section of *Anabaena azollae* reveals the extensive folded membranes of cyanobacterial cells that resemble the thylakoids of chloroplasts (TEM).

thylakoid lumen. Separation of the lumen from the stroma by the thylakoid membrane plays an important role in the generation of an electrochemical proton gradient and the synthesis of ATP, because protons pumped into the lumen during light-driven electron transport drive ATP synthesis as they return to the stroma.

Photosynthetic prokaryotes do not have chloroplasts. In some prokaryotes, however, such as the cyanobacteria, the plasma membrane is folded inward and forms *photosynthetic membranes.* Such structures, shown in Figure 15-4, are analogous to the thylakoids. Indeed, to some extent cyanobacteria appear to be free-living chloroplasts. Similarities between mitochondria, chloroplasts, and bacterial cells have led biologists to formulate the *endosymbiont theory,* which suggests mitochondria and chloroplasts evolved from bacteria that were engulfed by primitive cells 1 to 2 billion years ago. A summary of this theory is presented in Box 15A.

ment. They are not, however, physically contiguous with the inner membrane in mature chloroplasts. Moreover, their lipid and protein content is clearly different from that of either the outer or inner membrane. Electron micrographs of serial sections suggest that the grana and stroma thylakoids enclose a single compartment, the

The Chloroplast: A Photosynthetic Organelle   **449**

Further Insights

THE ENDOSYMBIONT THEORY AND THE EVOLUTION OF MITOCHONDRIA AND CHLOROPLASTS FROM ANCIENT BACTERIA

The debate on the evolutionary origins of mitochondria and chloroplasts has a long history. As early as 1883, Andreas F. W. Schimper suggested that chloroplasts arose from a symbiotic relationship between photosynthetic bacteria and nonphotosynthetic cells. By the mid-1920s, other investigators had extended Schimper's idea by proposing a symbiotic origin for mitochondria. Such ideas encountered ridicule and neglect for decades, however, until the 1960s, when it was discovered that mitochondria and chloroplasts contain their own DNA. Further research revealed that mitochondria and chloroplasts are **semiautonomous organelles,** containing not only DNA, but also mRNA, tRNAs, and ribosomes.

The gradual realization that DNA, RNA, and protein synthesis in mitochondria and chloroplasts display more similarities to the cognate processes in prokaryotic cells than to those in the nucleus or cytoplasm of eukaryotic cells led biologists to formulate the **endosymbiont theory.** This theory, developed most fully by Lynn Margulis, proposes that mitochondria and chloroplasts evolved from ancient prokaryotes that established a **symbiotic relationship** (a mutually beneficial association) with primitive nucleated cells 1 to 2 billion years ago. The proposed sequence of events leading to mitochondria and chloroplasts is outlined in Figure 15A-1.

A preliminary assumption of the endosymbiont theory is that the absence of molecular oxygen in Earth's primitive atmosphere limited early cells to *anaerobic* mechanisms for acquiring energy. A few anaerobic cells subsequently developed pigments capable of converting light energy to chemical energy, allowing them to use sunlight as a source of energy. The first photosynthetic organisms probably used hydrogen sulfide or molecular hydrogen as electron donors, but some of their early descendants developed mechanisms for using water as an electron donor. As a result, oxygen was released as a by-product and the composition of Earth's atmophere was dramatically altered.

As oxygen accumulated in Earth's atmosphere, some anaerobic bacteria evolved into *aerobic* organisms by developing oxygen-dependent electron transport and oxidative phosphorylation pathways. At this point, the stage was set for the emergence of eukaryotic cells. The endosymbiont theory suggests that the ancestor of eukaryotic cells (called a **protoeukaryote**) developed at least one important feature distinguishing it from other primitive cells: the ability to ingest nutrients from the environment by phagocytosis. This characteristic enabled protoeukaryotes to establish endosymbiotic relationships with primitive bacteria.

Mitochondria Apparently Evolved from Ancient Purple Bacteria

The first step toward the evolution of mitochondria may have occurred when a protoeukaryote ingested smaller aerobic bacteria by phagocytosis. Some scientists believe that prior to this event, protoeukaryotes were anaerobic and depended entirely on glycolysis for energy. The ingested aerobic bacteria, with their electron transport and oxidative phosphorylation pathways, would have provided larger amounts of useful energy

than the protoeukaryotic cell could produce by glycolysis alone. In turn, the host cell provided protection and nutrients to the bacteria residing in its cytoplasm. The ingested bacteria and the protoeukaryote, which both benefited from the association, established a stable symbiotic relationship. As the cytoplasmic bacteria and host cell adapted to living together over hundreds of millions of years, the bacteria gradually lost functions that were not essential in their new cytoplasmic environment and developed into mitochondria. To determine what kind of bacterium might have been drawn into this scenario, the base sequences of contemporary mitochondrial ribosomal RNAs (rRNAs) have been compared with the base sequences of various bacterial rRNAs. The closest matches occur among *purple bacteria,* suggesting that the ingested ancestor of mitochondria was an ancient member of this group.

Critics of the preceding scenario believe that viewing the first eukaryotic cell as a primitive anaerobic organism containing aerobic mitochondria is not accurate. In fact, a eukaryotic cell's cytoplasm contains oxygen-dependent enzymes and proteins, such as superoxide dismutase and cytochrome P450, that reflect a long history of aerobic metabolism. If the ancestral protoeukaryote were already capable of aerobic metabolism, what would be the advantage of acquiring an aerobic endosymbiont? According to one variation of the endosymbiont theory, the primitive protoeukaryote's aerobic metabolism was less efficient for trapping energy than the ingested purple bacterium's electron transport and oxidative phosphorylation pathways. Thus, even if the host protoeukaryote were already aerobic, the ingestion of aerobic bacteria capable of coupling electron transport to oxidative phosphorylation could have been advantageous.

Chloroplasts Apparently Evolved from Ancient Cyanobacteria

According to the endosymbiont theory, the first step toward the evolution of chloroplasts occurred when members of a subgroup of early eukaryotes, already equipped with aerobic bacteria or primitive mitochondria, ingested primitive photosynthetic cells. As just described for the evolution of mitochondria, the ingested organisms probably provided a useful pathway for meeting the energy needs of the host cell in exchange for shelter and nutrients. The photosynthetic cells gradually lost functions that were not essential in their new environment and evolved into an integral component of the eukaryotic host. To determine what kind of photosynthetic bacterium might have been involved in this scenario, the base sequences of chloroplast rRNAs, like mitochondrial rRNAs, have been compared with the base sequences of various bacterial rRNAs. In this case, the closest matches occur among *cyanobacteria,* suggesting that the ingested ancestor of chloroplasts was an ancient member of this group.

The chloroplasts found in different phototrophic eukaryotes, however, display biochemical and structural variations that lead some evolutionary biologists to wonder if the ingestion of photosynthetic cells and their subsequent development into chloro-

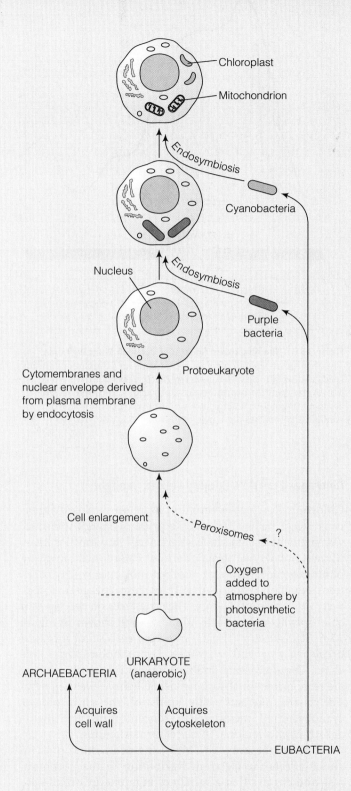

Major Events That Might Have Occurred During the Evolution of Eukaryotic Cells. Considerable evidence exists for an endosymbiotic origin for mitochondria and chloroplasts. Most biologists agree that the primitive cells that were ingested by protoeukaryotes and then evolved into mitochondria and chloroplasts were, respectively, purple bacteria and cyanobacteria. Earlier endosymbiotic events might have led to the evolution of peroxisomes and other organelles. The ancestral cells that did not have internal membrane systems are called urkaryotes.

dinoflagellates, and diatoms. Does this mean three different kinds of ancient photosynthetic bacteria, each carrying a distinct combination of pigments, entered into symbiotic relationships with early eukaryotes? Chlorophyll *a* and phycobilins are used in existing cyanobacteria, while chlorophylls *a* and *b* are used in a closely related group of bacteria called prochlorophytes. Although the base sequences of rRNAs from cyanobacteria display the closest match to rRNAs of chloroplasts from higher plants and green algae, the rRNAs of prochlorophytes are also very similar to the rRNAs of contemporary chloroplasts, not ruling out ancient members of either group as candidates for the symbiont, or symbionts, that evolved into chloroplasts.

The endosymbiont theory is based primarily on biochemical similarities observed among mitochondria, chloroplasts, and bacteria, but support is also provided by contemporary symbiotic relationships that resemble what might have occurred in the distant past. Algae, dinoflagellates, diatoms, and photosynthetic prokaryotes live as endosymbionts in the cytoplasm of cells occurring in more than 150 different kinds of existing protists and invertebrates. The cell wall of the ingested organism is often no longer present, and in a few instances the cell structure is even further reduced, with only the chloroplasts of the endosymbiont remaining.

A striking example of endosymbiosis occurs in certain marine slugs and related mollusks, where the cells lining the animal's digestive tract contain clearly identifiable chloroplasts. These chloroplasts originate from the green algae the mollusks feed on, and continue to carry out photosynthesis long after being incorporated into the animal's cells. The carbohydrates produced by the photosynthetic process are even distributed as a source of nutrients to the rest of the organism. Although the chloroplasts do not grow and divide, the organelles continue to function in the cytoplasm of the animal cell for several months. The discovery that chloroplasts derived from algae are ingested by animal cells and assume a stable symbiotic relationship provides support for the idea that chloroplasts could have had their evolutionary origins in endosymbiotic associations between ancient photosynthetic bacteria and ancient eukaryotes.

In the final analysis, our ideas about how eukaryotic cells acquired their complex array of organelles over billions of years of evolution must remain speculative because the events under consideration are inaccessible to direct laboratory experimentation. However, one of the strengths of the proposed role of endosymbiosis in the evolutionary origins of eukaryotic organelles is that it involves interactions and events that are observed in contemporary cells.

plasts occurred more than once during the evolution of eukaryotic cells. Consider, for example, the light-absorbing pigments found in the chloroplasts of different phototrophic eukaryotes. The major pigments used for photosynthesis are chlorophylls *a* and *b* in higher plants and green algae, chlorophyll *a* and phycobilins in red algae, and chlorophylls *a* and *c* in brown algae,

Photosynthetic Energy Transduction

To understand how light energy is converted into chemical energy within a chloroplast or bacterium, we require some knowledge of the nature of light (electromagnetic radiation) and its interaction with molecules. Light is often regarded as a wave; the visible portion of the electromagnetic spectrum, for example, consists of light having wavelengths ranging from about 380 to 750 nm. However, light also behaves as a stream of discrete particles called **photons,** each photon carrying a **quantum** (indivisible packet) of energy. The wavelength of a photon and the precise quantum of energy carried are inversely related (see Figure 2-3). A photon of ultraviolet or blue light, for example, has a shorter wavelength and carries a larger quantum of energy than a photon of red or infrared light.

When a photon is absorbed by a **pigment** (light-absorbing molecule), such as chlorophyll, the energy of the photon is transferred to an electron, which is energized from its *ground state* in a low-energy orbital to an *excited state* in a high-energy orbital. This event, called **photoexcitation,** is the first step in photosynthesis. A critical feature of photoexcitation is that absorption can occur only if the quantum of energy carried by the photon is *identical* to the difference in energy between the ground-state orbital and the excited-state orbital of an electron in the pigment. Because each pigment has a different configuration of electrons, and electron orbitals are characterized by discrete energy levels, pigments display characteristic **absorption spectra.** The wavelengths of light absorbed by a pigment correspond to the orbital transitions that electrons in the molecule can undergo. The absorption spectra of several common pigments found in photosynthetic organisms, along with the spectrum of solar radiation reaching the Earth's surface, are shown in Figure 15-5.

A photoexcited electron in a pigment molecule is not stable and must either return to its ground state in a low-energy orbital or undergo transfer to a relatively stable high-energy orbital, usually in a different molecule. When the electron returns to a low-energy orbital in the pigment molecule, the absorbed energy has two potential fates. Frequently, the excitation energy is lost when the pigment molecule releases a combination of heat and a photon carrying a smaller quantum of energy (fluorescence). Alternatively, most or all of the excitation energy is transferred from the photoexcited electron to an electron in an adjacent pigment molecule, exciting the second electron to a high-energy orbital. This process, called **resonance energy transfer,** is important for moving captured energy from light-absorbing molecules to molecules capable of passing an excited electron to an organic acceptor molecule. The transfer of the photoexcited electron itself to a high-energy orbital in another molecule is called **photochemical reduction.** This transfer, we will soon see, is essential for converting light energy to chemical energy.

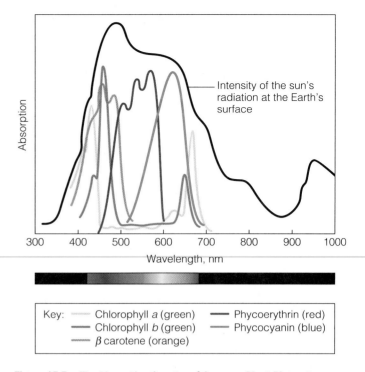

Figure 15-5   The Absorption Spectra of Common Plant Pigments. The absorption spectra of various pigments are compared with the spectral distribution of solar energy reaching the Earth's surface. The overlapping absorption spectra of various chlorophylls and accessory pigments effectively cover almost the entire spectrum of sunlight that reaches the Earth's surface.

Chlorophyll Is Life's Primary Link to Sunlight

Chlorophyll, which is found in all photosynthetic organisms except halobacteria, is the primary energy-transduction pigment that channels solar energy into the biosphere. Moreover, only one specific form of chlorophyll can pass a photoexcited electron to another molecule in an oxidation-reduction reaction that traps light energy and converts it to chemical energy. The structures of two types of chlorophyll—chlorophylls *a* and *b*—are shown in Figure 15-6. The skeleton of each molecule consists of a central *porphyrin ring* and a strongly hydrophobic *phytol* side chain. The alternating double bonds in the porphyrin ring are responsible for absorbing visible light, while the phytol side chain interacts with lipids of the thylakoid or cyanobacterial membranes, concentrating the light-absorbing molecules in these cellular structures.

The magnesium ion (Mg^{2+}) found in chlorophylls *a* and *b* affects the electron distribution in the porphyrin ring and ensures that a variety of high-energy orbitals are available. As a result, several specific wavelengths of light can be absorbed. Chlorophyll *a*, for example, has a broad absorption spectrum, with maxima at about 420 and 660 nm (see Figure 15-5). The precise absorption spectrum of each form of chlorophyll is determined by characteristic structural features and by the presence of specific *chlorophyll-binding proteins.* Chlorophyll *b* is distinguished from

Figure 15-6 The Structures of Chlorophylls a and b. The skeleton of each chlorophyll molecule consists of a central porphyrin ring (highlighted) and a hydrophobic side chain called phytol. Chlorophyll *a* has a methyl (—CH₃) group at the indicated position, whereas chlorophyll *b* has a formyl (—CHO) group. The bacteriochlorophyll found in anoxygenic phototrophs has a saturated carbon-carbon bond at the location indicated by the arrow.

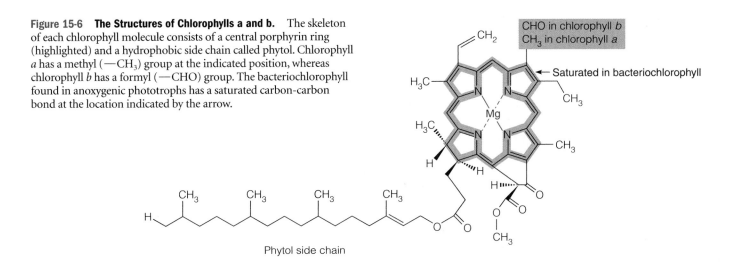

Phytol side chain

chlorophyll *a* by a *formyl* (—CHO) group in place of one of the *methyl* (—CH₃) groups on the porphyrin ring. This minor structural alteration shifts the absorption maxima toward the center of the visible spectrum (see Figure 15-5).

All plants and green algae contain both chlorophyll *a* and *b*. The combined absorption spectra of the two forms of chlorophyll provide access to a broader range of wavelengths of sunlight and enable such organisms to collect more photons than either pigment alone would. Other photosynthetic eukaryotes supplement chlorophyll *a* with either chlorophyll *c* (brown algae, diatoms, and dinoflagellates) or *d* (red algae).

Bacteriochlorophyll is a subfamily of chlorophyll molecules restricted to anoxygenic phototrophs (photosynthetic bacteria) and characterized by a saturated site not found in other chlorophyll molecules (indicated by an arrow in Figure 15-6). In this case, structural alterations shift the absorption maxima of bacteriochlorophylls toward the near-ultraviolet and far-red regions of the spectrum. While this enables such organisms to avoid competition with plants, green algae, and cyanobacteria, it also prevents them from exploiting water as an electron donor for photosynthesis; the reduction potentials of bacteriochlorophylls, unlike other chlorophylls, are not sufficiently positive to accept electrons from water.

Accessory Pigments Further Expand Access to Solar Energy

Most photosynthetic organisms also contain **accessory pigments,** which absorb photons that cannot be captured by chlorophyll. This feature enables organisms to collect energy from a much larger portion of the sunlight reaching the Earth's surface, as shown in Figure 15-5. Two types of accessory pigments are **carotenoids** and **phycobilins.** Two carotenoids that are abundant in the thylakoid membranes of most plants and green algae are β-*carotene* and *lutein*. When sufficiently abundant and not masked by chlorophyll, these pigments confer an orange or yellow tint to leaves. With absorption maxima between 420 and 480 nm, carotenoids absorb photons from a broad range of the blue region of the spectrum.

Phycobilins are found only in red algae and cyanobacteria. Two common examples are *phycoerythrin* and *phycocyanin*. Phycoerythrin absorbs photons from the blue, green, and yellow regions of the spectrum, enabling red algae to utilize the dim light that penetrates an ocean's surface water. Phycocyanin, on the other hand, absorbs photons from the orange region of the spectrum and is characteristic of cyanobacteria living close to the surface of a lake or on land. Thus, variations in the amounts and properties of accessory pigments often reflect a phototroph's adaptation to a specific environment.

Light-Gathering Molecules Are Organized into Photosystems and Light-Harvesting Complexes

Chlorophyll molecules, accessory pigments, and associated proteins are organized into functional units called **photosystems,** which are localized to thylakoid or photosynthetic bacterial membranes. The chlorophyll molecules are anchored to the membranes by long hydrophobic side chains. **Chlorophyll-binding proteins** stabilize the arrangement of the chlorophyll within a photosystem and modify the absorption spectra of specific chlorophyll molecules; other proteins bind components of electron transport systems or catalyze oxidation-reduction reactions.

Most pigments of a photosystem serve solely as light-gathering **antenna pigments,** absorbing photons and passing the energy to a neighboring chlorophyll molecule or accessory pigment by resonance energy transfer. The photochemical events that drive electron flow and proton pumping do not begin until energy reaches the **reaction center** of a photosystem, where two distinct chlorophyll *a* molecules reside. Surrounded by other components of the reaction center, this special pair of chlorophyll molecules catalyzes the conversion of solar energy into chemical energy.

The absorption of a photon and the transfer of energy to the reaction center of a photosystem is illustrated in Figure 15-7. Each chlorophyll molecule and accessory

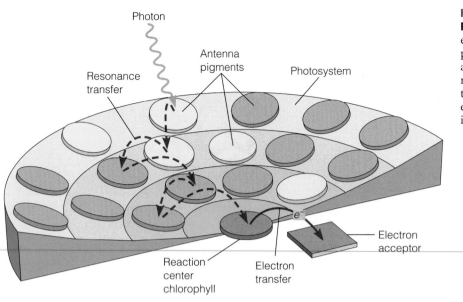

Figure 15-7 The Transfer of Energy to the Reaction Center of a Photosystem. Light energy absorbed by antenna pigments is passed from one antenna pigment to another by resonance energy transfer until it reaches a specific chlorophyll molecule at the reaction center of the photosystem. The energy is captured when an excited electron is transferred to an organic acceptor molecule.

pigment within a photosystem has a characteristic absorption maximum that depends on its structure and immediate chemical environment. Following the laws of thermodynamics, electrons always flow exergonically from antenna pigments that absorb light of the shortest wavelengths (those with electrons that can be excited by the largest quanta of energy) to pigments that absorb light of the longest wavelengths (those with electrons that can be excited by the smallest quanta of energy). Thus, pigments are arranged to funnel energy toward the reaction center of a photosystem. The chlorophyll molecules at the reaction center absorb light of the longest wavelength (smallest quantum of energy) and therefore act as a *sink*. Within a nanosecond after absorption of a photon by any pigment within a photosystem, an electron at the reaction center is excited and passed to an acceptor molecule positioned at the "top" of an electron transport system. The oxidized chlorophyll, with a positive charge referred to as an *electron hole,* is then a strong oxidant, ready to accept an electron from a suitable donor.

Each photosystem is generally associated with a **light-harvesting complex (LHC),** which, like a photosystem, collects light energy. The LHC does not, however, contain a reaction center; instead, it passes the collected energy to a nearby photosystem by resonance energy transfer. Plants and green algae have LHCs composed of about 80–250 chlorophyll *a* and *b* molecules, along with carotenoids and pigment-binding proteins. Red algae and cyanobacteria have a different type of LHC, called a **phycobilisome,** which contains phycobilins rather than chlorophyll and carotenoids. Together, a photosystem and the associated LHCs are referred to as a **photosystem complex.**

Oxygenic Phototrophs Have Two Types of Photosystems

In the 1940s, Robert Emerson and his colleagues at the University of Illinois discovered that two separate photo-reactions are involved in photoreduction in oxygenic phototrophs. Initially, they observed a dramatic drop in the **action spectrum,** at a wavelength of about 690 nm, for photosynthesis by the green alga *Chlorella* (Figure 15-8); an action spectrum for photosynthesis depicts the relationship between wavelength of light and photosynthetic efficiency. Emerson's group considered this *red drop* odd because *Chlorella* actually contains chlorophyll molecules that absorb a significant amount of light at wavelengths above 690 nm. When they supplemented the longer wavelengths of light with a shorter wavelength (about 650 nm), the red drop was less severe, as shown in Figure 15-8. Indeed, photosynthesis driven by a combination of long and short wavelengths of red light exceeded the sum of activities obtained with either wavelength alone. This syn-

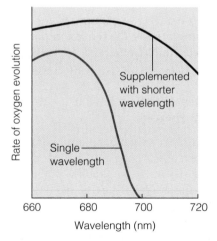

Figure 15-8 The Emerson Enhancement Effect. When cells of the green alga *Chlorella* are illuminated by light of a single wavelength, there is a dramatic drop in the action spectrum (measured by oxygen evolution) above a wavelength of about 690 nm. This red drop is less severe when long wavelengths of light are supplemented with a shorter wavelength (650 nm).

ergistic phenomenon became known as the **Emerson enhancement effect.**

We now know that the Emerson enhancement effect in oxygenic phototrophs is the result of two distinct photosystems working in concert. **Photosystem I (PSI),** with an absorption maximum of 700 nm, absorbs both long and short wavelengths of red light, whereas **photosystem II (PSII),** with an absorption maximum of 680 nm, absorbs only short wavelengths of red light. As we are about to see, each electron that passes from water to $NADP^+$ must be photoexcited twice, once by PSI and once by PSII. When illumination is restricted to wavelengths above 690 nm, PSII is not active and photosynthesis is severely impaired.

The distinct pairs of chlorophyll molecules within the reaction centers of PSI and PSII are designated, respectively, **P700** and **P680** to reflect their specific absorption maxima.

Photoreduction (NADPH Synthesis) in Oxygenic Phototrophs

In oxygenic phototrophs, two distinct photosystems are apparently essential for efficient conservation of energy as photoexcited electrons, donated by water, are transferred to $NADP^+$ to form NADPH. As shown in Figure 15-9, the

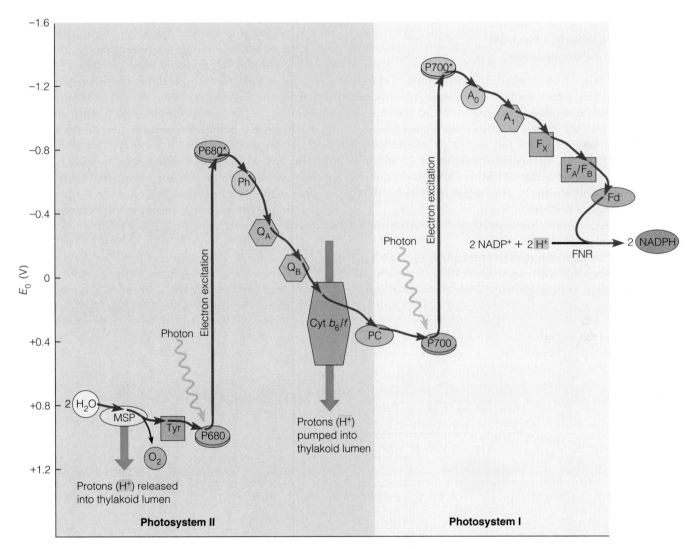

Figure 15-9 Noncyclic Electron Flow in Oxygenic Phototrophs. Components of the electron transport system that are part of PSII include manganese-stabilizing protein (MSP), a tyrosine residue (Tyr) on protein D1 (see Figure 15-11), a special pair of chlorophyll a molecules (P680), pheophytin (Ph), and two plastoquinones (Q_A and Q_B). Components of PSI include another special pair of chlorophyll a molecules (P700), a modified chlorophyll a molecule (A_0),

phylloquinone (A_1), and three iron-sulfur centers (F_X, F_A, and F_B). The photosystems are linked by a cytochrome b_6/f complex, which couples electron transport (the path highlighted in red) to pumping of protons into the thylakoid lumen, and by plastocyanin (PC). P680 and P700 are unusual because absorption of photons alters their reduction potentials (vertical axis), enabling them to accept electrons from a donor with a highly positive

reduction potential and donate electrons to an acceptor with a highly negative reduction potential, thereby capturing energy. Ferredoxin-$NADP^+$ reductase (FNR) catalyzes the transfer of electrons from ferredoxin (Fd) to $NADP^+$. See Figure 15-11 for a model showing the orientation of the various components described here within a thylakoid membrane.

complete photoreduction pathway includes several components. Absorption of light energy by each photosystem boosts electrons to the "top" of an electron transport system (ETS). As the electrons flow exergonically from PSII to PSI, a portion of their energy is conserved in an electrochemical proton gradient across the thylakoid (or cyanobacterial) membrane. From PSI, electrons are passed to *ferredoxin (Fd)* and then to **nicotinamide adenine dinucleotide phosphate (NADP$^+$)**, conserving reducing power in the form of NADPH. As indicated in Figure 15-10, NADP$^+$ differs from NAD$^+$ by having an additional phosphate group attached to its adenosine. NADP$^+$ is the coenzyme of choice for a large number of anabolic pathways, whereas NAD$^+$ tends to be involved in catabolic pathways.

Photosystem II Transfers Electrons from Water to a Plastoquinone

Our detailed tour of the photoreduction pathway begins at photosystem II, which must use electrons from water to reduce a *plastoquinone* to a *plastoquinol*. The PSII reaction center, part of which is shown in Figure 15-11, includes two large proteins, D1 and D2, that bind not only chlorophyll, but also components of an ETS. Surrounding the reaction center of PSII are 40–50 additional chlorophyll *a* and about 10 β-carotene molecules that enhance photon collection.

Photoreduction does not depend on direct absorption of a photon at the reaction center of a photosystem, or even by a nearby pigment. In plants and green algae, PSII is generally associated with **light-harvesting complex II (LHCII)**, which contains about 250 chlorophyll and numerous carotenoid molecules. Energy captured by antenna pigments of PSII or of an associated LHCII is funneled to the reaction center by resonance energy transfer (see Figure 15-7). When the energy reaches the reaction center, it lowers the reduction potential of a pair of chlorophyll *a* molecules, designated P680, to about −0.80 V (see Figure 15-9). A photoexcited electron is then passed to *pheophytin (Ph)* in an oxidation-reduction reaction. Pheophytin is a modified chlorophyll *a* molecule with two protons in place of the magnesium ion. The charge separation between oxidized P680 and reduced pheophytin traps the electron, preventing it from returning to its ground state in P680.

Next the electron is passed to Q_A, a **plastoquinone** that is tightly bound to protein D2 (see Figure 15-11). Plastoquinone is similar to *coenzyme Q*, which is a component of the mitochondrial electron transport system (see Figure 14-15). Another plastoquinone, Q_B, is tightly bound to protein D1, but only until it receives two electrons, one at time, from Q_A. Reduced Q_B picks up two protons from the stroma and enters a pool of **plastoquinol**, Q_BH_2, in the lipid phase of the photosynthetic membrane. Q_BH_2, which is a mobile electron carrier, can then convey two electrons and two protons to the *cytochrome b_6/f complex*. Because a chlorophyll molecule catalyzes the transfer of only one electron per photon absorbed, the formation of one mobile plastoquinol molecule depends on two sequential photoreactions at the same reaction center:

$$2 \text{ photons} + Q_B + 2H^+_{stroma} \longrightarrow Q_BH_2 \qquad \textbf{(15-3)}$$

To absorb photons repeatedly, oxidized P680 must be reduced each time an electron is lost to plastoquinone. To

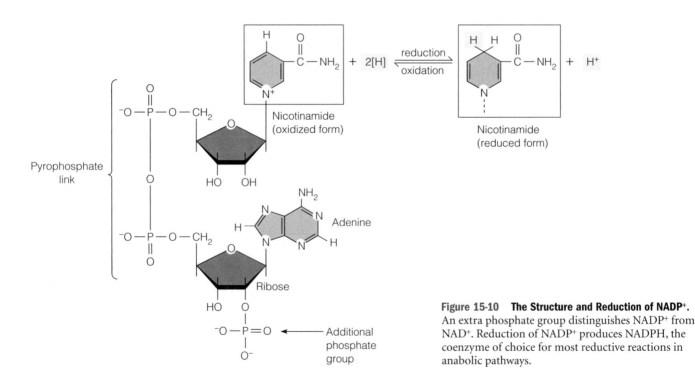

Figure 15-10 The Structure and Reduction of NADP$^+$. An extra phosphate group distinguishes NADP$^+$ from NAD$^+$. Reduction of NADP$^+$ produces NADPH, the coenzyme of choice for most reductive reactions in anabolic pathways.

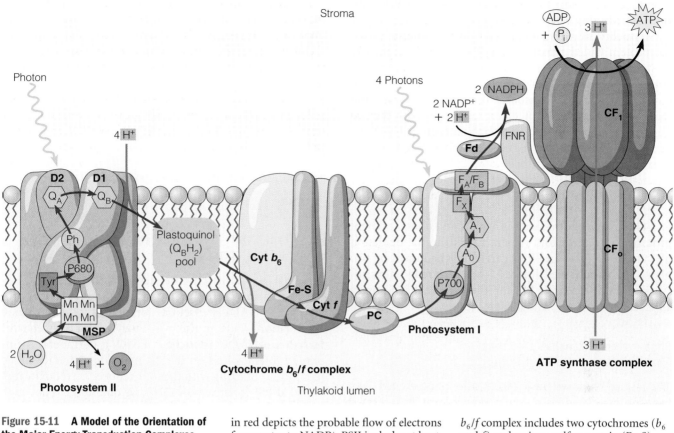

Stroma

Photon

4 H⁺

$4\ H^+$

D2   D1

Q_A Q_B

Ph

P680

Tyr

Mn Mn
Mn Mn

MSP

$2\ H_2O$

$4\ H^+ + O_2$

Photosystem II

Plastoquinol
(Q_BH_2)
pool

Cyt b_6

Fe-S

Cyt f

PC

$4\ H^+$

Cytochrome b_6/f complex

Thylakoid lumen

4 Photons

2 NADPH

2 NADP⁺
+ 2 H⁺

Fd

FNR

F_A/F_B

F_X

A_1

A_0

P700

Photosystem I

ADP
+ Pᵢ

3 H⁺

ATP

CF₁

CF₀

3 H⁺

ATP synthase complex

Figure 15-11   A Model of the Orientation of the Major Energy Transduction Complexes Within the Thylakoid Membrane.   This diagram shows the possible arrangement of PSII, the cytochrome b_6/f complex, PSI, and the ATP synthase complex within the thylakoid membrane. The path highlighted in red depicts the probable flow of electrons from water to NADP⁺. PSII includes at least two major proteins, D1 and D2, that bind components of the electron transport system. The cluster of four manganese atoms at the base of PSII is part of the oxygen-evolving complex. The cytochrome b_6/f complex includes two cytochromes (b_6 and f) and an iron-sulfur protein (Fe-S). The stoichiometry shown is for the passage of four electrons from H_2O to NADPH. See Figure 15-9 for a general key to symbols.

replenish electrons, PSII includes an **oxygen-evolving complex (OEC),** an assembly of proteins and manganese ions that catalyzes the oxidation of water to molecular oxygen (O_2). Two water molecules donate four electrons, one at a time by way of a *tyrosine* on protein D1, to oxidized P680 (see Figures 15-9 and 15-11). In the process, four protons and one oxygen molecule are released within the thylakoid lumen. The protons contribute to an electrochemical proton gradient across the membrane and oxygen diffuses out of the chloroplast (or cyanobacterium). Because the complete oxidation of two water molecules to molecular oxygen depends on four photoreactions, the net reaction catalyzed by PSII can be summarized as

$$4 \text{ photons} + 2H_2O + 2Q_B + 4H^+_{stroma} \longrightarrow$$
$$4H^+_{lumen} + O_2 + 2Q_BH_2 \quad \textbf{(15-4)}$$

The protons removed from the stroma are actually still in transit to the lumen as part of Q_BH_2, while the protons added to the lumen at this point are derived from oxidation of water. The light-dependent oxidation of water to protons and molecular oxygen, called *water photolysis,* probably appeared in cyanobacteria between 2 and 3

billion years ago, thereby permitting exploitation of water as an abundant electron donor. The oxygen released by the process dramatically changed the Earth's early atmosphere, which did not originally contain free oxygen, and allowed the development of aerobic respiration.

The Cytochrome b_6/f Complex Transfers Electrons from a Plastoquinol to Plastocyanin

Before the electrons carried by Q_BH_2 are delivered to PSI for another boost in energy, they flow through an ETS coupled to unidirectional proton pumping across the thylakoid membrane. This happens by way of the **cytochrome b_6/f complex,** which is analogous to the mitochondrial *respiratory complex III* described in Chapter 14 (see Figure 14-17). The cytochrome b_6/f complex is composed of four distinct integral membrane proteins, including two cytochromes and an iron-sulfur protein.

Q_BH_2 diffuses within the lipid phase of the thylakoid membrane from PSII to a cytochrome b_6/f complex. There it donates two electrons, one at a time via *cytochrome b_6* and the iron-sulfur protein, to *cytochrome f*. Oxidation of Q_BH_2 releases two protons into the thylakoid lumen and enables Q_B to return to the pool of plastoquinone available for accepting electrons from PSII.*

Cytochrome f donates electrons to a copper-containing protein called **plastocyanin (PC).** Plastocyanin, like plastoquinol, is a mobile electron carrier. Unlike plastoquinol, however, it is a peripheral membrane protein found on the lumenal side of the thylakoid membrane and it carries only one electron at a time. Starting with the two Q_BH_2 molecules generated at PSII (see reaction 15-4), the net reaction catalyzed by the cytochrome b_6/f complex can be summarized as

$$2Q_BH_2 + 4PC(Cu^{2+}) \longrightarrow 2Q_B + 4H^+_{lumen} + 4PC(Cu^+) \quad \textbf{(15-5)}$$

Thus, four photoreactions at PSII add a total of eight protons to the thylakoid lumen: four from the oxidation of water to oxygen, and four carried to the lumen by two plastoquinol molecules (Figure 15-11). Each reduced plastocyanin now diffuses to a PSI complex, where it can donate an electron to an oxidized P700 molecule.

Photosystem I Transfers Electrons from Plastocyanin to Ferredoxin

The task of photosystem I is to transfer photoexcited electrons from reduced plastocyanin to *ferredoxin*. As indicated in Figure 15-11, the reaction center of PSI includes a third chlorophyll a molecule, called A_0, instead of a pheophytin molecule. Other components of the reaction center include a *phylloquinone* (A_1) and three *iron-sulfur centers* (F_X, F_A, and F_B) that form an ETS linking A_0 to ferredoxin. Surrounding the reaction center of PSI, as in PSII, are additional pigment molecules—about 100 chlorophyll a and several β-carotene molecules—that enhance photon collection.

PSI in plants and green algae is associated with **light-harvesting complex I (LHCI),** which contains fewer antennae molecules than LHCII does—between 80 and 120 chlorophyll and a few carotenoid molecules. As in the PSII complex, energy captured by antenna pigments in the photosystem or in an associated light-harvesting complex is funneled to a reaction center containing a special pair of chlorophyll a molecules. In PSI, however, the energy lowers the reduction potential of the pair of chlorophylls, designated P700, to about -1.30 V (see Figure 15-9). A photoexcited electron is then rapidly passed to A_0, and charge separation between oxidized P700 and the reduced A_0 prevents the electron from returning to its original ground state. This oxidation-reduction reaction and the analogous one in PSII are the true *photoreactions* of photosynthetic energy transduction.

From A_0, electrons flow exergonically through the ETS to ferredoxin, the final electron acceptor for PSI.

*According to several models, the oxidation of Q_BH_2 is actually accompanied by the transfer of four protons into the lumen by a mechanism called *Q-cycling*. The evidence is equivocal, however, so we leave this mechanism out of our discussion.

Ferredoxin (Fd) is a mobile iron-sulfur protein found in the chloroplast stroma. As we will see later, Fd is an important reductant in several metabolic pathways, including those for nitrogen and sulfur assimilation. To transfer electrons repeatedly to ferredoxin, oxidized P700 must be reduced each time an electron is lost. The donor, of course, is reduced plastocyanin. Starting with the four reduced plastocyanin molecules generated at cytochrome b_6/f complex (reaction 15-5), the net reaction catalyzed by PSI can be summarized as

$$4 \text{ photons} + 4PC(Cu^+) + 4Fd(Fe^{3+}) \longrightarrow \\ 4PC(Cu^{2+}) + 4Fd(Fe^{2+}) \quad \textbf{(15-6)}$$

Ferredoxin-NADP+ Reductase Catalyzes the Reduction of NADP+

The final step in the photoreduction pathway is the transfer of electrons from ferredoxin to $NADP^+$, thereby providing the NADPH essential for photosynthetic carbon assimilation. This transfer is catalyzed by the enzyme **ferredoxin-NADP+ reductase (FNR),** a peripheral membrane protein found on the stromal side of the thylakoid membrane. Starting with the four reduced ferredoxin generated at PSI (see reaction 15-6), we obtain the following:

$$4Fd(Fe^{2+}) + 2NADP^+ + 2H^+_{stroma} \longrightarrow \\ 4Fd(Fe^{3+}) + 2NADPH \quad \textbf{(15-7)}$$

Notice that the reduction of one $NADP^+$ molecule consumes two electrons, each from a single reduced ferredoxin. While not actually moving protons from the stroma to the lumen, this reaction nonetheless raises the pH of the stroma and contributes to the electrochemical proton gradient across the thylakoid membrane.

Functioning together within the chloroplast, the various components of the ETS provide a continuous, unidirectional flow of electrons from water to $NADP^+$, as indicated in Figures 15-9 and 15-11. This is referred to as **noncyclic electron flow,** primarily to distinguish it from the *cyclic electron flow* we will encounter shortly. The net result of the complete light-dependent oxidation of two water molecules to molecular oxygen can be obtained by summing reactions 15-4 through 15-7, assuming ferredoxin does not donate electrons to other pathways:

$$4 \text{ photons at PSII} + 4 \text{ photons at PSI} + \\ 2H_2O + 6H^+_{stroma} + 2NADP^+ \longrightarrow \\ 8H^+_{lumen} + O_2 + 2NADPH \quad \textbf{(15-8)}$$

For every eight photons absorbed, two NADPH molecules are generated. Furthermore, the ETS for photoreduction is coupled to a mechanism for unidirectional proton pumping across the thylakoid membrane from the stroma to the lumen. Thus, solar energy is captured and stored in two forms: the reductant NADPH and an electrochemical proton gradient.

Photophosphorylation (ATP Synthesis) in Oxygenic Phototrophs

Because the thylakoid membrane is virtually impermeable to protons, a substantial electrochemical proton gradient can develop across the membrane in illuminated chloroplasts. It is not unusual, for example, for the thylakoid lumen to have a pH of 4.5, while the stroma pH is in the range 8.0–8.5. As described in Chapter 14 for mitochondria, we can calculate the total proton motive force (pmf) across the thylakoid membrane, or the equivalent free energy change ($\Delta G'$), for the movement of protons from the lumen to the stroma by summing a concentration (pH) term and a membrane potential (V_m) term (see equation 14-13). In mitochondria, the pH difference across the inner membrane is only about 1.0–1.5 units, so the membrane potential term is usually more significant than the concentration term. In chloroplasts, however, magnesium ions readily diffuse from the lumen to the stroma, maintaining charge neutrality across the thylakoid membrane. Thus, nearly all of the pmf is due to the light-induced pH gradient, which can be as steep as 3.5–4.0 units. This difference corresponds to a $\Delta G'$ value of about −5.1 kcal/mol of protons moving from the lumen to the stroma.

The ATP Synthase Complex Couples Transport of Protons Across the Thylakoid Membrane to ATP Synthesis

In chloroplasts as in mitochondria, the movement of protons from an acidic to a basic compartment drives the synthesis of ATP by an **ATP synthase** complex. The ATP synthase complexes found in chloroplasts, designated **$CF_o CF_1$ complexes,** are remarkably similar to the $F_o F_1$ complexes of mitochondria described in Chapter 14 (see Figure 14-21). The **CF_1** component is a hydrophilic assembly of polypeptides that protrudes from the stromal side of the thylakoid membrane and contains the catalytic site for ATP synthesis. Like the mitochondrial F_1 and its stalk, CF_1 is composed of five distinct polypeptides with a stoichiometry of $\alpha_3 \beta_3 \gamma \delta \varepsilon$.

The other half of the chloroplast ATP synthase complex, the **CF_o** component, is an assembly of polypeptides that is embedded in the thylakoid membrane, where it forms a hydrophilic channel between the lumen and the stroma. CF_o is the **proton translocator** through which protons flow under the pressure of the pmf. This movement is probably coupled to ATP synthesis by the CF_1 component as described in Chapter 14 for the mitochondrial ATP synthase (see Figure 14-22), with one ATP generated for every three protons that pass through the channel. The reaction catalyzed by the ATP synthase complex can therefore be summarized as

$$3H^+_{lumen} + ADP + P_i \longrightarrow 3H^+_{stroma} + ATP + H_2O \quad \textbf{(15-9)}$$

This is quite reasonable in terms of thermodynamics. The $\Delta G'$ value for the synthesis of ATP within the chloroplast stroma is usually 10–14 kcal/mol, whereas the energy available is about 15 kilocalories per 3 moles of protons passing through the ATP synthase complex (5.1 kcal/mol × 3 mol = 15.3 kcal). Thus, the flow of four electrons through the noncyclic pathway shown in Figures 15-9 and 15-11 not only generates two NADPH molecules, but leads indirectly to the formation of almost three ATP molecules from ADP and P_i (4 electrons × 2 protons/electron × 1 ATP/3 protons = 2.7 ATP molecules).

Recall from Chapter 6 that enzymes change only the *rate* at which equilibrium between the reactants and the products of a chemical reaction is achieved; they have no effect on the *position* of the equilibrium. Within an illuminated chloroplast, the net free energy change favors ATP synthesis: The net $\Delta G'$ is between −1 and −5 kcal/mol. But what happens in the dark, when the light-induced proton gradient is no longer maintained? If the $CF_o CF_1$ complex remained active, it would become an *ATPase,* consuming ATP as it actively transported protons into the lumen. Not surprisingly, chloroplasts have systems for regulating ATP synthase.

In the dark, deterioration of the light-induced proton gradient across the thylakoid membrane substantially reduces enzyme activity, but not simply because there are no protons available for transport from the thylakoid lumen to the stroma. The chloroplast ATP synthase does not become an ATPase that consumes ATP and transports protons back into the lumen. Apparently, the enzyme's catalytic activity is affected by the acidity of the thylakoid lumen. When the concentration of protons in the lumen drops, both ATP synthesis and ATP hydrolysis activity are reduced. A second mechanism, to be described later, links the activation of chloroplast ATP synthase to the presence of reduced ferredoxin in the stroma. In the absence of photoreduction, there is not enough reduced ferredoxin to maintain the ATP synthase in an active state.

Cyclic Photophosphorylation Allows a Photosynthetic Cell to Balance NADPH and ATP Synthesis to Meet Its Precise Energy Needs

Having seen how NADPH and ATP are generated by light-driven electron flow, we are near the end of our discussion of the photosynthetic energy transduction reactions. But before we move on to carbon assimilation, we must consider how phototrophs might balance NADPH and ATP synthesis to meet the precise energy needs of a living cell. Notice that noncyclic electron flow leads to the generation of slightly less than three ATP for every two NADPH molecules. Because both compounds are consumed by a variety of metabolic pathways, which may be more or less active at any given time, it is very unlikely that a photosynthetic cell will always require ATP and NADPH in the precise ratio generated by noncyclic electron flow.

When NADPH consumption is low, an optional pathway may divert the reducing power generated at PSI into ATP synthesis rather than NADP+ reduction. As shown in Figure 15-12, the reduced ferredoxin generated

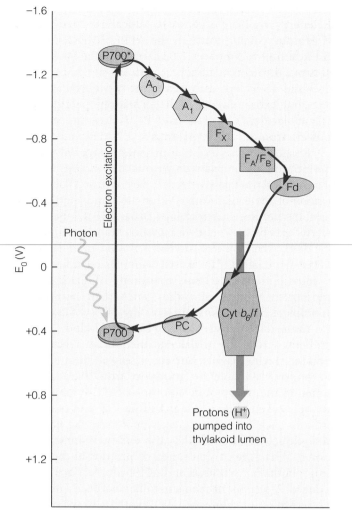

Figure 15-12   Cyclic Electron Flow.   Cyclic electron flow through PSI enables oxygenic phototrophs to regulate the ratio of NADPH to ATP within photosynthetic cells. When the concentration of NADP⁺ is low (i.e., when the concentration of NADPH is high), ferredoxin (Fd) donates electrons to the cytochrome b_6/f complex. Electrons then return to P700 via plastocyanin (PC). Because this cyclic electron flow is coupled to unidirectional proton pumping across the thylakoid membrane, excess reducing power is channeled into ATP synthesis. See Figure 15-9 for a general key to symbols.

by PSI can pass electrons to a cytochrome b_6/f complex instead of donating them to NADP⁺. By a mechanism that probably involves plastoquinol, the exergonic flow of electrons from reduced ferredoxin to plastocyanin is coupled to proton pumping, thereby contributing to the pmf across the thylakoid membrane. At least three protons are moved from the stroma to the lumen for every two electrons, but the precise number is not known. From plastocyanin, electrons return to an oxidized P700 molecule in PSI, completing a closed circuit and allowing P700 to absorb another photon. This is referred to as **cyclic electron flow,** and the ATP synthesis that it supports is called *cyclic photophosphorylation.* No water is oxidized and no oxygen is released, since the flow of electrons from PSII is bypassed.

The Complete Energy Transduction System

The model shown in Figure 15-11 represents the entire photosynthetic energy transduction system within a thylakoid membrane. The essential features of the complete system for the transfer of electrons from water to NADP⁺ concomitant with ATP synthesis can be summarized in terms of the following component parts:

1. *Photosystem II complex:* An assembly of chlorophyll molecules, accessory pigments, and proteins, with a sufficiently positive reduction potential to allow a reaction center chlorophyll to accept electrons exergonically from water (E_0 = +0.816 V). Photoexcited from about +0.9 to −0.8 V by a single photon, PSII has a sufficiently negative reduction potential to donate electrons to plastoquinone, a mobile electron carrier.

2. *Cytochrome b_6/f complex:* An electron transport system linking plastoquinol, the electron shuttle for PSII, with plastocyanin, the electron shuttle for PSI. The flow of electrons through the cytochrome b_6/f complex is coupled to unidirectional proton pumping across the thylakoid membrane, thereby establishing and maintaining an electrochemical proton gradient. An option for cyclic electron flow allows a cytochrome b_6/f complex to accept electrons from ferredoxin as well as from plastoquinol, providing for the generation of additional ATP as needed.

3. *Photosystem I complex:* A second assembly of chlorophyll molecules, accessory pigments, and proteins, with a sufficiently positive reduction potential to allow a reaction center chlorophyll to accept electrons exergonically from plastocyanin (E_0 = +0.36 V). Photoexcited from about +0.48 to −1.3 V by a single photon, PSI has a sufficiently negative reduction potential to donate electrons to ferredoxin (E_0 = −0.42 V), a stromal protein.

4. *Ferredoxin-NADP⁺ reductase:* An enzyme on the stromal side of the thylakoid membrane that catalyzes the transfer of electrons from two reduced ferredoxin proteins to a single NADP⁺ molecule (E_0 = −0.32 V). The NADPH generated is an essential reducing agent in anabolic pathways, particularly carbon assimilation.

5. *ATP synthase complex (also called the CF_0CF_1 complex):* A proton channel and ATP synthase that couples the exergonic flow of protons, from the thylakoid lumen to the stroma, to the synthesis of ATP from ADP and P_i. Like NADPH, the ATP accumulates in the stroma, where it provides energy for carbon assimilation.

Within a chloroplast, both noncyclic and cyclic pathways of electron flow are operative, thereby providing flexibility in the relative amounts of ATP and NADPH generated. ATP can be produced on a close to equimolar basis with respect to NADPH if the noncyclic pathway is operating alone, or it can be generated in any desired excess simply by shunting more and more of the electrons from ferredoxin to the cyclic rather than the noncyclic side of PSI.

Complexes of the Energy Transduction System Are Localized in Different Regions of the Thylakoids

The complexes we just described are not necessarily contiguous within a membrane. Instead, components appear to be segregated to different regions of the thylakoids, as illustrated in Figure 15-13. PSII complexes are localized primarily in *appressed* regions of stacked thylakoids (grana), whereas PSI complexes and ATP synthase complexes are localized exclusively in *nonappressed* regions of stacked thylakoids and in stroma thylakoids. This allows NADPH and ATP to be generated on the stromal side of the membrane. Cytochrome b_6/f complexes are present in roughly equal concentration in both stacked and stroma thylakoids. Plastoquinol and plastocyanin, of course, are mobile electron carriers that can diffuse over long distances, linking the noncontiguous components into a complete system. As already noted, plastoquinol is confined to the lipid phase of the thylakoid membrane, whereas plastocyanin is a peripheral membrane protein on the lumenal side.

A Photosynthetic Reaction Center from a Purple Bacterium

Much of our knowledge of photosynthetic reaction centers and the chemistry of capturing light energy has come from studies of reaction center complexes isolated from photosynthetic bacteria. The stage was set in the late 1960s when Roderick Clayton successfully purified the reaction center complex of a purple bacterium, *Rhodopseudomonas sphaeroides*. More recently, Hartmut Michel, Johann Deisenhofer, and Robert Huber crystallized a reaction center complex from a different purple bacterium, *Rhodopseudomonas viridis*, and determined its molecular structure by X-ray crystallography. Michel and his colleagues not only provided the first detailed look at how pigment molecules are arranged to capture light energy, they were the first group to crystallize any membrane protein complex at all. For their exciting contributions, Michel, Deisenhofer, and Huber shared a Nobel Prize in 1988.

As shown in Figure 15-14, the reaction center of *R. viridis* includes four protein subunits. The first subunit is a cytochrome *c* molecule bound to the periplasmic surface of the bacterial membrane. (The periplasm is the space between the plasma membrane and cell wall of a bacterium.) The second and third subunits, called L and M, span the membrane and stabilize a total of four bacteriochlorophyll *b* molecules, two bacteriopheophytin molecules, and two quinones. Subunits L and M are homologous to proteins D1 and D2, respectively, of PSII in oxygenic phototrophs (see Figure 15-11). A fourth subunit, called H, is bound to the cytoplasmic surface of the membrane and is homologous to a PSII protein called CP43. Electron flow through this bacterial photosystem and a second membrane protein complex, the *cytochrome b/c_1 complex*, resembles the flow of electrons through PSII and the cytochrome b_6/f complex of oxygenic phototrophs, with one major difference: The bacterial photosystem does not include an oxygen-evolving complex.

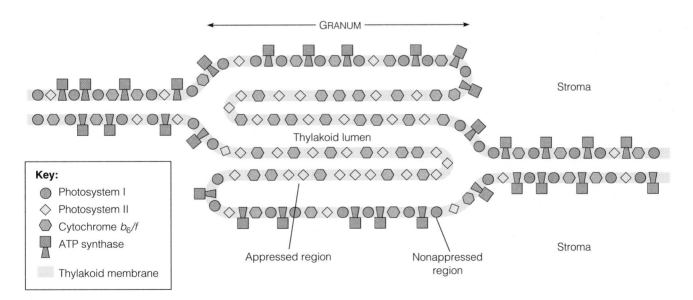

Figure 15-13 Localization of the Four Major Complexes of Photosynthetic Energy Transduction.
The complexes are localized to different regions of the stacked thylakoids (grana) and stroma thylakoids. Each PSI and PSII complex includes one photosystem and one light-harvesting complex. The noncontiguous components are linked by plastoquinol, which is soluble in the lipid phase of the thylakoid membrane, and plastocyanin, which is a peripheral membrane protein on the lumenal side.

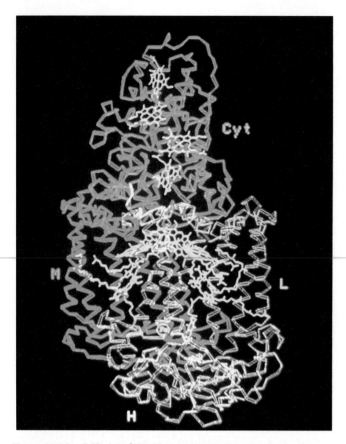

Figure 15-14   A Model of the Photosynthetic Reaction Center from a Purple Bacterium.   The structure for the reaction center from *Rhodopseudomonas viridis* was deduced from X-ray diffraction measurements. The cytochrome (blue-green) with four heme groups (yellow) lies in the periplasm outside the bacterial plasma membrane. Subunits L and M span the membrane and bind components for light harvesting and electron transfer. Subunit H lies mostly on the cytosolic side of the membrane.

Two of the bacteriochlorophyll *b* molecules, designated *P960* to indicate their absorption maxima, have a direct role in catalyzing the light-dependent transfer of an electron to an ETS. Absorption of a photon, either directly or by accessory pigments, lowers the reduction potential of P960 from about +0.5 to −0.7 V. The photoexcited electron is immediately transferred to bacteriopheophytin, thereby stabilizing the charge separation. From bacteriopheophytin, the electron flows exergonically to a tightly bound quinone, and then to a second quinone. After accepting two electrons, the reduced quinone diffuses to a cytochrome b/c_1 complex, which then passes electrons to cytochrome *c*. This event is coupled to unidirectional proton pumping across the bacterial membrane from the cytoplasm to the periplasm. Cytochrome *c* then returns the electron to oxidized P960. As in chloroplasts and mitochondria, the electrochemical proton gradient across the membrane drives an ATP synthase, or CF_oCF_1 complex.

Notice that the flow of electrons just described is cyclic, with no net gain of reducing power. How, then, does this particular phototroph generate reductant? In *R. viridis*, the cytochrome b/c_1 complex and cytochrome *c* accept electrons from donors such as hydrogen sulfide, thiosulfate, or succinate. The ATP generated by cyclic photophosphorylation is then used to push electrons energetically uphill from the cytochrome b/c_1 complex or cytochrome *c* to NAD^+. The cyclic flow of electrons described above is therefore very different than the cyclic flow through PSI in oxygenic phototrophs. The bacterial photosystem is most accurately described as a somewhat simplified version of PSII in plants, green algae, and cyanobacteria, the major product of the reaction center being a reduced quinone.

Photosynthetic Carbon Assimilation: The Calvin Cycle

With information about chloroplast structure and photosynthetic energy transduction in mind, we are now prepared to look closely at photosynthetic carbon assimilation. More specifically, we will look at the primary events of carbon assimilation: the initial fixation and reduction of carbon dioxide to form simple three-carbon carbohydrates. The fundamental pathway for the movement of inorganic carbon into the biosphere is the **Calvin cycle,** which is found in all oxygenic and most anoxygenic phototrophs. This pathway is named after Melvin Calvin, who received a Nobel Prize in 1961 for the work he and his colleagues Andrew Benson and James Bassham did to elucidate the process. As mentioned earlier, the primary products of carbon dioxide fixation enter a variety of metabolic pathways, the most important of these being sucrose and starch biosynthesis.

In plants and algae, the Calvin cycle is confined to the chloroplast stroma, where the ATP and NADPH generated by photosynthetic energy transduction reactions accumulate. In plants, the carbon dioxide about to be assimilated generally enters a leaf through special pores called **stomata** (singular: **stoma**), which permit gas exchange between the atmosphere and the interior of a leaf. Once inside the leaf, carbon dioxide molecules diffuse into **mesophyll cells** and, in over 95% of plant species, travel unhindered into the chloroplast stroma, where the inorganic carbon enters the Calvin cycle. In a relatively small number of species, however, the Calvin cycle is preceded by one of two types of preliminary carboxylation/decarboxylation pathways that enhance photosynthetic efficiency under certain conditions. After we consider the Calvin cycle in detail, we will look at how it may be augmented by such additional pathways.

For convenience, we will divide the Calvin Cycle into three stages (Figure 15-15):

1. The carboxylation of the organic acceptor molecule *ribulose-1,5-bisphosphate,* followed by hydrolysis to generate two molecules of *3-phosphoglycerate.*
2. The reduction of 3-phosphoglycerate from the oxidation level of a carboxylic acid to the level of an aldehyde, forming the triose phosphate *glyceraldehyde-3-phosphate.*
3. The regeneration of the initial acceptor molecule, thereby allowing continuous carbon assimilation.

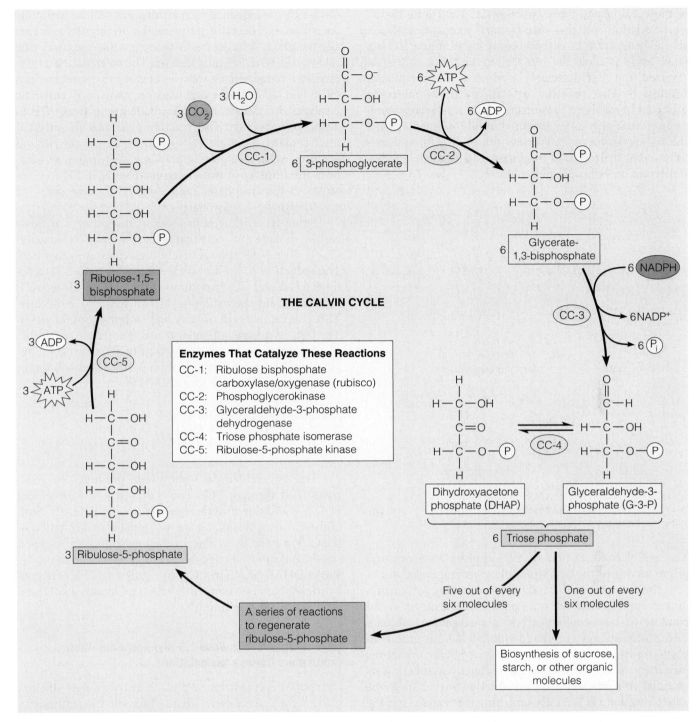

THE CALVIN CYCLE

Enzymes That Catalyze These Reactions

CC-1: Ribulose bisphosphate carboxylase/oxygenase (rubisco)
CC-2: Phosphoglycerokinase
CC-3: Glyceraldehyde-3-phosphate dehydrogenase
CC-4: Triose phosphate isomerase
CC-5: Ribulose-5-phosphate kinase

Figure 15-15   The Calvin Cycle for Photosynthetic Carbon Assimilation. The stoichiometry shown is for the net synthesis of one triose phosphate molecule. Reactions CC-1, CC-2, and CC-3 provide for the initial fixation and reduction of carbon dioxide, generating the interconvertible trioses glyceraldehyde-3-phosphate (G-3-P) and dihydroxyacetone phosphate (DHAP; formed by reaction CC-4). On the average, one out of every six triose phosphate molecules—the product of the Calvin cycle—is used for the biosynthesis of sucrose, starch, or other organic molecules. Five out of every six triose phosphate molecules must be used to form ribulose-5-phosphate. The ribulose-5-phosphate is then phosphorylated at the expense of ATP in reaction CC-5 to regenerate the acceptor molecule for reaction CC-1, thus completing the cycle.

When looking at individual stages, keep in mind that each is an integral part of the complete pathway shown in Figure 15-15 and that all three processes occur simultaneously in illuminated chloroplasts.

Carbon Dioxide Enters the Calvin Cycle by Carboxylation of Ribulose-1,5-Bisphosphate

Our tour of the Calvin cycle begins with the covalent attachment of carbon dioxide to the carbonyl carbon of ribulose-1,5-bisphosphate (reaction CC-1 in Figure 15-15). Carboxylation of this five-carbon acceptor molecule should generate a six-carbon product, but no one has isolated such a molecule. As Calvin and his colleagues revealed, the first detectable product of carbon dioxide fixation by this pathway is a three-carbon molecule, 3-phosphoglycerate. Presumably, the six-carbon compound exists only as an enzyme-bound intermediate, and the carboxylation of the acceptor molecule is immediately followed by hydrolysis of this intermediate to generate two molecules of 3-phosphoglycerate:

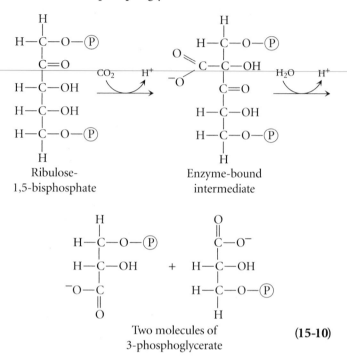

(15-10)

The newly fixed carbon dioxide appears as a carboxyl group on one of the two 3-phosphoglycerate molecules.

The enzyme that catalyzes the capture of carbon dioxide and formation of 3-phosphoglycerate is called **ribulose-1,5-bisphosphate carboxylase/oxygenase (rubisco).** This relatively large enzyme (about 560 kDa) is unique to phototrophs and is found in all photosynthetic organisms except for a few photosynthetic bacteria. Considering its essential role in carbon dioxide fixation for virtually the entire biosphere, it is hardly surprising that rubisco has the distinction of being the most abundant protein on the planet. About 10–25% of soluble leaf protein is rubisco, and one estimate puts the total amount of rubisco on the Earth at 40 million tons, or almost 15 lb (about 7kg) for each living person. After we look at the next two stages of the Calvin cycle, we will return to rubisco and consider regulation of the enzyme's *carboxylase* activity, as well as problems created by its *oxygenase* activity.

3-Phosphoglycerate Is Reduced to Form Glyceraldehyde-3-Phosphate

The 3-phosphoglycerate molecules formed during carbon dioxide fixation are reduced to glyceraldehyde-3-phosphate

(G-3-P) by a sequence of reactions that is essentially the reverse of the oxidative sequence of glycolysis (reactions Gly-6 and Gly-7 in Figure 13-6), except that the coenzyme involved is NADPH, not NADH. The steps for NADPH-mediated reduction are shown in Figure 15-15 as reactions CC-2 and CC-3. In the first reaction, *phosphoglycerokinase* catalyzes the transfer of a phosphate group from ATP to 3-phosphoglycerate. This reaction generates an activated intermediate, glycerate-1,3-bisphosphate. In the second reaction, *glyceraldehyde-3-phosphate dehydrogenase* catalyzes the transfer of two electrons from NADPH to glycerate-1,3-bisphosphate, reducing the molecule to glyceraldehyde-3-phosphate.

Some accounting is in order at this point. For every carbon dioxide molecule that is fixed by rubisco (reaction CC-1), two 3-phosphoglycerate molecules are generated. The reduction of both of these molecules to G-3-P (reactions CC-2 and CC-3) requires the hydrolysis of two ATP molecules and the oxidation of two NADPH molecules. This is for a net gain of only one carbon atom, however. The net synthesis of one triose phosphate molecule requires the fixation and reduction of *three* carbon dioxide molecules to maintain carbon balance and will therefore consume six ATP and six NADPH:

$$3 \text{ ribulose-1,5-bisphosphate} + 3CO_2 +$$
$$3H_2O + 6ATP + 6NADPH \longrightarrow$$
$$6 \text{ G-3-P} + 6ADP + 6P_i + 6NADP^+ \quad \textbf{(15-11)}$$

The glyceraldehyde-3-phosphate produced by reactions CC-1 through CC-3 enters an interconvertible pool of G-3-P and dihydroxyacetone phosphate (DHAP). Both of these triose phosphates are required for continuation of the Calvin cycle, as well as for the biosynthesis of sucrose, starch, and other organic molecules. The conversion of G-3-P to DHAP (reaction CC-4 in Figure 15-15) is catalyzed by *triose phosphate isomerase*, which maintains a balanced pool of the two compounds.

Regeneration of Ribulose-1,5-Bisphosphate Allows Continuous Carbon Assimilation

One out of every six triose phosphate molecules—the net gain of the Calvin cycle—is used for the biosynthesis of sucrose, starch, or other organic molecules. The five remaining triose phosphates are required within the Calvin cycle for regeneration of the initial acceptor molecule, ribulose-1,5-bisphosphate (reaction CC-5). For each three-carbon compound that leaves the cycle, a total of 3 five-carbon acceptor molecules must be regenerated. Four basic reactions—catalyzed by *aldolases, transketolases, phosphatases,* and *isomerases*—transform two molecules of DHAP and three molecules of G-3-P to three molecules of *ribulose-5-phosphate*. Then, *ribulose-5-phosphate kinase* phosphorylates each ribulose-5-phosphate to form the ribulose-1,5-bisphosphate that accepts carbon dioxide in reaction CC-1. Thus, regeneration of ribulose-1,5-bisphosphate from DHAP and G-3-P consumes three ATP molecules. The series of events may be summarized by the following net reaction:

$$2DHAP + 3\text{ G-3-P} + 3ATP + 2H_2O \longrightarrow$$
$$3\text{ ribulose-1,5-bisphosphate} + 3ADP + 2P_i \quad \textbf{(15-12)}$$

The Complete Calvin Cycle

Primary carbon assimilation by the Calvin cycle may be summarized by simply combining reactions 15-11 and 15-12, which encompass all the chemistry of the cycle from carbon dioxide fixation and reduction through triose phosphate synthesis and regeneration of the organic acceptor molecule. The resulting net reaction is

$$3CO_2 + 9ATP + 6NADPH + 5H_2O \longrightarrow$$
$$\text{G-3-P} + 9ADP + 6NADP^+ + 8P_i \quad \textbf{(15-13)}$$

Note the requirements for ATP and NADPH: The Calvin cycle consumes nine ATP molecules and six NADPH molecules for every three-carbon carbohydrate synthesized, or three ATP and two NADPH for each carbon dioxide molecule fixed. As noted earlier, the product of the Calvin cycle is actually an interconvertible pool of glyceraldehyde-3-phosphate and dihydroxyacetone phosphate that may be diverted to various metabolic pathways.

Because triose phosphates generated by the Calvin cycle are consumed by metabolic pathways in the cytosol as well as the chloroplast stroma, there must be a mechanism for transporting them across the chloroplast inner membrane. The most abundant protein in the chloroplast inner membrane is a *phosphate translocator* that catalyzes the exchange of DHAP, G-3-P, or 3-phosphoglycerate in the stroma for P_i in the cytosol. This *antiport* system (see Figure 8-7b) ensures that triose phosphates will not be exported unless P_i—which is released when triose phosphates are used in the cytosol and is required for synthesizing new triose phosphates—returns to the stroma. Moreover, the phosphate translocator is sufficiently specific to prevent other intermediates of the Calvin cycle from leaving the stroma. Within the cytosol, triose phosphates may be used for sucrose synthesis, as we will describe below, or they may enter the glycolytic pathway to provide ATP and NADH in the cytosol. The triose phosphates remaining in the stroma are generally used for starch synthesis, which we also will describe shortly.

The Calvin Cycle Operates Only When Light Is Available

In the dark, all phototrophs essentially become chemotrophs. To meet the steady demand for energy and carbon, such organisms must tap the surplus of carbohydrates that accumulates when light is available. This activity would be futile, however, if the Calvin cycle were not rendered inoperable while the glycolytic and other pathways consume storage carbohydrates. Remember that all biological pathways run at less than 100% thermodynamic efficiency—some energy is inevitably lost as heat and

entropy. Channeling energy from carbohydrates into ATP or NADH synthesis and then back into carbohydrate synthesis would eventually drain all energy and carbon reserves. It is not surprising, then, that phototrophs have several regulatory systems to ensure that the Calvin cycle does not operate unless light is available.

The first level of control, appropriate for enzymes unique to a specific pathway, is the regulation of gene expression (see Chapter 21). Enzymes that are important only for light-dependent carbon dioxide fixation and reduction are generally not present in plant tissues that are not exposed to light. Plastids in root cells, for example, contain less than 1% of the amount of rubisco activity found in the chloroplasts of leaf cells.

Enzymes of the Calvin cycle are also regulated by metabolites. Because the photosynthetic energy transduction reactions and the Calvin cycle are localized in the same compartment—the cytoplasm of a photosynthetic bacterium or the chloroplast stroma of a plant or algal cell—there is ample opportunity for metabolites of one process to affect the other. Consider the changes that occur in the chloroplast stroma as light drives the movement of electrons from water to ferredoxin. As protons are pumped from the stroma to the lumen, the pH of the stroma rises from about 7.2 (a typical value in the dark) to about 8.5. Simultaneously, magnesium ions diffuse from the lumen to the stroma (to maintain charge neutrality across the thylakoid membrane), and the concentration of magnesium ions in the stroma rises about tenfold. Eventually, high levels of reduced ferredoxin, NADPH, and ATP also accumulate. Each of these factors serves as a signal that can affect enzymes of the Calvin cycle as well as enzymes of other metabolic pathways.

Three enzymes unique to the Calvin cycle are logical points for metabolic control. Rubisco is an obvious candidate, since it catalyzes the first reaction of the Calvin cycle. The others are *sedoheptulose bisphosphatase* and *ribulose-5-phosphate kinase*, which have roles in regenerating the organic acceptor molecule ribulose-1,5-bisphosphate. Rubisco, sedoheptulose bisphosphatase, and ribulose-5-phosphate kinase are all stimulated by a high pH and a high concentration of magnesium ions. In the dark, pH and magnesium ion levels in the stroma decline, rendering these enzymes less active. Each of these enzymes is not only unique to the Calvin cycle but catalyzes an essentially irreversible reaction. Recall from Chapter 13 that the glycolytic pathway is also regulated at sites of unique and irreversible reactions. This is a common theme in metabolism.

Another system for coordinating photosynthetic energy transduction and carbon assimilation, shown in Figure 15-16, depends on ferredoxin. During illumination, electrons donated by water are used to reduce ferredoxin. An enzyme called *ferredoxin-thioredoxin reductase* then catalyzes the transfer of electrons from ferredoxin to *thioredoxin*, another mobile electron carrier. Thioredoxin affects enzyme activity by reducing disulfide (S—S) bonds to sulfhydryl (—SH) groups, which can cause a protein to undergo a conformational change. Glyceraldehyde-3-phosphate dehydrogenase

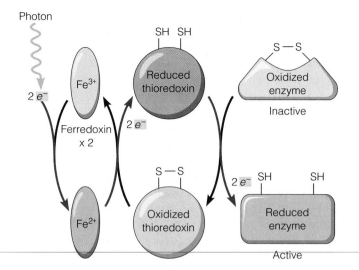

Figure 15-16   Thioredoxin-Mediated Activation of a Calvin Cycle Enzyme. Two reduced ferredoxin molecules generated by the photoreduction pathway donate electrons to thioredoxin, an intermediate electron carrier. Thioredoxin subsequently activates a target enzyme by reducing a disulfide bond to two sulfhydryl groups. In other cases, thioredoxin may inactivate enzymes.

(reaction CC-3 in Figure 15-15), sedoheptulose bisphosphatase, and ribulose-5-phosphate kinase contain disulfide bonds that can be reduced by thioredoxin. For these enzymes, the resulting conformational change stimulates activity. In the dark, no reduced ferredoxin is available, and the sulfhydryl groups spontaneously reoxidize to disulfide bonds, thereby inactivating the enzymes. The CF_oCF_1 ATP synthase described earlier is also affected by the thioredoxin system.

Because the Calvin cycle produces three-carbon molecules that can directly enter the glycolytic pathway, the breakdown of complex carbohydrates is not necessary in the light. Not surprisingly, the same mechanisms that activate enzymes of the Calvin cycle inactivate enzymes of degradative pathways. One example is *phosphofructokinase*, the most important control point in glycolysis. While the Calvin cycle is operating in the light, inhibition of phosphofructokinase ensures that the early steps of the glycolytic pathway are bypassed, thereby preventing the development of another potentially futile cycle.

Photosynthetic Energy Transduction and the Calvin Cycle

We are now ready to write an overall reaction for photosynthesis that takes both energy transduction and carbon assimilation into account. Considering the needs of the Calvin cycle and assuming no diversion of ATP or NADPH to other pathways, the requirements are clear.

According to reaction 15-13, the synthesis of one molecule of glyceraldehyde-3-phosphate (or of dihydroxyacetone phosphate) requires nine ATP and six NADPH molecules. This demand can be met by the flow of 12 electrons through the noncyclic pathway, which provides all six of the NADPH molecules ($3 \times$ reaction 15-8) and eight of the nine ATP molecules ($8 \times$ reaction 15-9), plus the flow of 2 electrons through the cyclic pathway, which will provide at least one more ATP molecule ($1 \times$ reaction 15-9). Thus, the net reaction for energy transduction is

$$26 \text{ photons} + 9ADP + 9P_i + 6NADP^+ \longrightarrow \\ O_2 + 9ATP + 6NADPH + 3H_2O \qquad \textbf{(15-14)}$$

The absorption of 26 photons is accounted for by 12 photoexcitation events at PSI and 12 at PSII during noncyclic electron flow, and 2 photoexcitation events at PSI during cyclic electron flow.

By adding the summary reaction for the Calvin cycle (reaction 15-13) to the net reaction for energy transduction (reaction 15-14), we obtain the following overall expression:

$$26 \text{ photons} + 3CO_2 + 5H_2O + P_i \longrightarrow \\ \text{G-3-P} + 3O_2 + 3H_2O \qquad \textbf{(15-15)}$$

Because the phosphate group is generally removed by hydrolysis when G-3-P is incorporated into more complex organic compounds, we may rewrite reaction 15-15 as

$$26 \text{ photons} + 3CO_2 + 6H_2O \longrightarrow \\ \text{glyceraldehyde} + 3O_2 + 3H_2O \qquad \textbf{(15-16)}$$

This reaction is almost identical to the net photosynthetic reaction we introduced near the beginning of this chapter (reaction 15-2). We are now, however, in a much better position to understand some of the photochemical and metabolic complexity behind what might otherwise appear to be a simple reaction. Moreover, we have replaced the vague terms "light" and "$C_3H_6O_3$" with a specific number of photons and a primary product of carbon dioxide fixation.

This information enables us to calculate the maximum efficiency of photosynthetic energy transduction and carbon assimilation. For red light, assuming a wavelength of 670 nm, 26 photons represent 26×43 kcal/einstein (see Figure 2-3 for the relationship between wavelength and energy; an einstein equals one mole of photons), or a total of 1118 kcal of energy. Since glyceraldehyde differs in free energy from carbon dioxide and water by 343 kcal/mol, the maximum efficiency of photosynthetic energy transduction is about 31% (343/1118 = 0.31). Although an efficiency of this order might be observed in a laboratory with isolated chloroplasts or algal cells, photosynthesis in a natural environment is far less efficient. After we look at carbohydrate synthesis, we will consider a few of the factors that limit the overall productivity of a photosynthetic organism.

Carbohydrate Synthesis

As we have already learned, the Calvin cycle generates an interconvertible pool of glyceraldehyde-3-phosphate and dihydroxyacetone phosphate, both of which can be transported out of the chloroplast by the phosphate translocator located in the chloroplast inner membrane. These triose phosphates are, in turn, the starting points for starch synthesis within the chloroplast and for sucrose synthesis in the cytosol (Figure 15-17). We will consider both pathways, each of which begins with the formation of hexose phosphates.

Glyceraldehyde-3-Phosphate and Dihydroxyacetone Phosphate Are Combined to Form Glucose-1-Phosphate

The key hexose phosphate required for both starch and sucrose synthesis is glucose-1-phosphate, which is formed from triose phosphates as shown in Figure 15-17, reactions S-1 through S-3. This set of reactions occurs both in the cytosol (c) and in the chloroplast stroma (s). Hexoses and hexose phosphates must be synthesized in each compartment because they cannot cross the chloroplast inner membrane. The initial steps in the sequence resemble reactions we encountered in our discussion of gluconeogenesis in Chapter 13 (see Figure 13-12). The first step in triose phosphate utilization is a condensation reaction that generates fructose-1,6-bisphosphate; this is catalyzed by aldolase. As in gluconeogenesis, the pathway is rendered exergonic by the hydrolytic removal of a phosphate group from fructose-1,6-bisphosphate to form fructose-6-phosphate. The enzyme that catalyzes this reaction, fructose-1,6-bisphosphatase, exists in two forms, one in the cytosol and the other in the chloroplast stroma. The multiple forms of such an enzyme are called **isoenzymes**—physically distinct proteins that catalyze the same reaction. The fructose-6-phosphate can then be converted, via glucose-6-phosphate, to glucose-1-phosphate (reactions S-2 and S-3). Again, separate isoenzymes catalyze each of these reactions in the cytosol and in the chloroplast stroma.

Textbooks often depict glucose as the end-product of photosynthetic carbon assimilation. But this portrayal is more a definition of convenience for writing summary reactions than a metabolic fact, as very little free glucose actually accumulates in photosynthetic cells. Most glucose is converted into either transport carbohydrates such as sucrose or storage carbohydrates such as starch. In reality, the designation of an end-product for photosynthetic carbon assimilation becomes rather arbitrary once you get past the pool of triose phosphates produced by the Calvin cycle. Indeed, we could regard the whole phototrophic organism—plant, algal cell, or bacterium—as the "end-product" of photosynthesis.

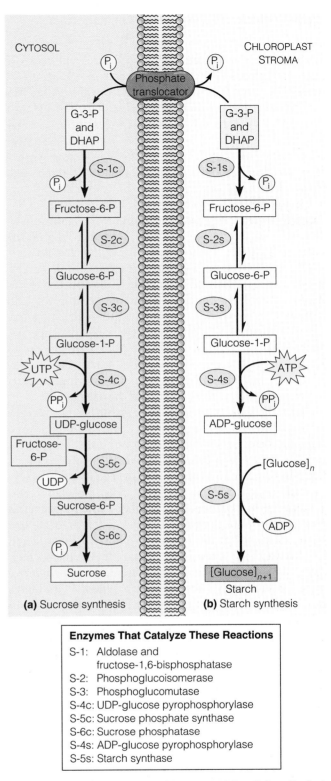

Enzymes That Catalyze These Reactions

S-1: Aldolase and fructose-1,6-bisphosphatase
S-2: Phosphoglucoisomerase
S-3: Phosphoglucomutase
S-4c: UDP-glucose pyrophosphorylase
S-5c: Sucrose phosphate synthase
S-6c: Sucrose phosphatase
S-4s: ADP-glucose pyrophosphorylase
S-5s: Starch synthase

Figure 15-17 The Biosynthesis of Sucrose and Starch from Products of the Calvin Cycle.   The triose phosphates glyceraldehyde-3-phosphate (G-3-P) and dihydroxyacetone phosphate (DHAP) from the chloroplast stroma are exchanged for inorganic phosphate from the cytosol via a phosphate translocator. **(a)** Sucrose synthesis is confined to the cytosol, whereas **(b)** starch synthesis occurs only in the chloroplast stroma. The enzymes and isoenzymes that catalyze these reactions are restricted to either the cytosol (c) or the stroma (s).

The Biosynthesis of Sucrose Occurs in the Cytosol

Recall from Chapter 3 that sucrose is a disaccharide consisting of one molecule each of glucose and fructose linked by a glycosidic bond (see Figure 3-23). Sucrose is of interest because it is the major carbohydrate used for transporting stored energy and reduced carbon in most plant species. Moreover, in some species, such as sugar beets and sugar cane, sucrose also serves as a storage carbohydrate. As shown in Figure 15-17a, sucrose synthesis is localized in the cytosol of a photosynthetic cell. Triose phosphates exported from the chloroplast stroma and not consumed by other metabolic pathways are converted to fructose-6-phosphate and glucose-1-phosphate. Glucose from glucose-1-phosphate is then activated by the reaction of glucose-1-phosphate with *uridine triphosphate (UTP)*, generating *UDP-glucose* (reaction S-4c). Finally, the glucose is transferred to fructose-6-phosphate to form the phosphorylated disaccharide *sucrose-6-phosphate*, and hydrolytic removal of the phosphate group generates free sucrose (reactions S-5c and S-6c). (In some plant species, glucose may be transferred directly from UDP-glucose to free fructose.) The sucrose is exported from leaves by way of vascular bundles and conveys energy and reduced carbon to nonphotosynthetic tissues of the organism.

Like the Calvin cycle, sucrose synthesis is precisely controlled to prevent conflict with degradation pathways. *Cytosolic* fructose-1,6-bisphosphatase, for example, is inhibited by *fructose-2,6-bisphosphate*, a metabolite introduced in Chapter 13 as an important regulator of glycolysis and gluconeogenesis in liver cells. In plant cells, fructose-2,6-bisphosphate accumulates when high levels of fructose-6-phosphate and P_i (signals that indicate low sucrose demand) or low levels of 3-phosphoglycerate and dihydroxyacetone phosphate (signals that indicate high triose phosphate demand) are present. Another control point for sucrose synthesis is *sucrose phosphate synthase*, the enzyme that catalyzes the transfer of glucose from UDP-glucose to fructose-6-phosphate (reaction S-5c). This enzyme is stimulated by glucose-6-phosphate and inhibited by sucrose-6-phosphate, UDP, and P_i.

The Biosynthesis of Starch Occurs in the Chloroplast Stroma

Starch synthesis in plant cells is confined to plastids. In photosynthetic plant cells, starch synthesis is generally restricted to chloroplasts, which are essentially photosynthetic plastids. When sufficient energy and carbon are available to meet the metabolic needs of a plant, the excess triose phosphates within the chloroplast stroma are converted to glucose-1-phosphate, which is then used for starch synthesis. As shown in Figure 15-17b, glucose is activated by the reaction of glucose-1-phosphate with ATP, generating *ADP-glucose* (reaction S-4s). The activated glucose is then transferred directly to a growing starch chain by *starch synthase*, leading to elongation of the polysaccharide (reaction S-5s). The two forms of starch commonly found in plants are amylose and amylopectin, as described in Chapter 3. As shown in Figure 15-2a, starch may accumulate in large storage granules within the chloroplast stroma. When photosynthesis is limited by darkness or other factors, starch is degraded to triose phosphates, which can then enter glycolysis or be converted to sucrose in the cytosol and exported from the cell.

Like the Calvin cycle and sucrose synthesis, starch synthesis is precisely controlled. *Chloroplast* fructose bisphosphatase, which channels fructose-1,6-bisphosphate into glucose and starch biosynthesis in the stroma, is activated by the same thioredoxin system that affects enzymes of the Calvin cycle. This regulation ensures that starch synthesis occurs only when there is sufficient illumination for photoreduction. The key enzyme for regulation, though, is *ADP-glucose pyrophosphorylase*, which catalyzes the activation of glucose and commits it to starch synthesis (reaction S-4s). ADP-glucose pyrophosphorylase is stimulated by glyceraldehyde-3-phosphate and inhibited by P_i. Thus, when triose phosphates are diverted to the cytosol and ATP is hydrolyzed to ADP and P_i (signals that indicate high energy demand), starch synthesis is blocked.

Other Photosynthetic Assimilation Pathways

Photosynthesis encompasses more than carbon dioxide fixation and carbohydrate synthesis. In plants and algae, the ATP and NADPH generated by photosynthetic energy transduction reactions are consumed by a variety of other anabolic pathways found in chloroplasts. Carbohydrate synthesis is only one example of carbon metabolism; the synthesis of fatty acids, chlorophyll, and carotenoids also occurs in chloroplasts. Moving beyond carbon metabolism, several key steps of nitrogen and sulfur assimilation are localized in chloroplasts. The reduction of nitrite (NO_2^-) to ammonia (NH_3), for example, is catalyzed by a reductase enzyme in the chloroplast stroma, with reduced ferredoxin serving as an electron donor. The ammonia is then channeled into amino acid and nucleotide synthesis, portions of which also occur in chloroplasts. Furthermore, much of the reduction of sulfate (SO_4^{2-}) to sulfide (S^{2-}) is catalyzed by enzymes in the chloroplast stroma. In this case, both ATP and reduced ferredoxin provide energy and reducing power. The sulfide, like ammonia, may then be used for amino acid synthesis.

Rubisco's Oxygenase Activity Decreases Photosynthetic Efficiency

The primary reaction catalyzed by rubisco—as a *carboxylase*—is the addition of carbon dioxide and water to ribulose-1,5-bisphosphate, forming two molecules of 3-phosphoglycerate. However, rubisco can also function as an *oxygenase*. Through this activity, rubisco catalyzes the addition of molecular oxygen, rather than carbon dioxide, to ribulose-1,5-bisphosphate:

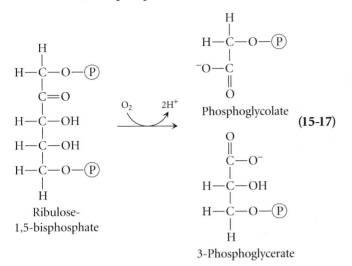

Ribulose-1,5-bisphosphate

Phosphoglycolate

3-Phosphoglycerate

(15-17)

The result of rubisco's oxygenase activity is a single three-carbon product—3-phosphoglycerate—and one two-carbon product, **phosphoglycolate.** Because phosphoglycolate cannot be used in the next step of the Calvin cycle, it appears to be a wasteful diversion of material from carbon assimilation. Furthermore, alternative functions for rubisco's oxygenase activity have not been clearly demonstrated. Indeed, not only does the production of phosphoglycolate appear to waste energy and carbon, the accumulation of phosphoglycolate may kill a plant by inhibiting the triose phosphate isomerase that maintains a balance of glyceraldehyde-3-phosphate and dihydroxyacetone phosphate in the chloroplast stroma.

Considering that rubisco has a much lower affinity for oxygen than for carbon dioxide and that the oxygenase reaction proceeds much more slowly than the carboxylase reaction, one might not expect the oxygenase reaction to be a serious problem. However, the low carbon dioxide and high oxygen concentrations in the Earth's atmosphere—about 0.035 and 21% respectively—lead to significant competition between carbon dioxide and oxygen for binding to rubisco's catalytic site. Within a typical leaf or algal cell exposed to the atmosphere, the low $CO_2:O_2$ ratio leads to one oxygenation of ribulose-1,5-bisphosphate for every two or three carboxylation events, seriously reducing photosynthetic efficiency and generating large amounts of phosphoglycolate.

Plants living in hot arid environments under intense illumination are particularly affected by rubisco's oxygenase activity, since such conditions may further lower the $CO_2:O_2$ ratio in the chloroplast stroma. Although the solubilities of both carbon dioxide and oxygen decrease as temperature increases, the solubility of carbon dioxide declines more rapidly, thereby lowering the $CO_2:O_2$ ratio in solution. Another problem occurs when plants respond to drought by closing their stomata during the day to reduce water loss. When the stomata are closed, carbon dioxide cannot enter the leaf. Without a steady supply of carbon dioxide for assimilation, the concentration of carbon dioxide in leaf cells may decline. Moreover, water photolysis continues to generate oxygen, which accumulates because it cannot diffuse out of the leaf when the stomata are closed. Intense sunlight exacerbates this problem by increasing the rate of water photolysis, which depends on the absorption of light that drives noncyclic electron flow and photoreduction.

All efforts to reduce rubisco's oxygenase activity through alteration of the amino acid sequence of the enzyme have failed. According to one theory, the oxygenase activity is an evolutionary relic from a time when oxygen did not make up a large part of the Earth's atmosphere, and it cannot be eliminated without seriously compromising the carboxylase function. Not even natural selection appears to be up to the task of altering this enzyme. Instead, phototrophs that depend on rubisco have developed three alternative strategies for coping with the enzyme's apparently wasteful oxygenase activity. We will briefly consider each strategy.

The Glycolate Pathway Returns Reduced Carbon from Phosphoglycolate to the Calvin Cycle

In all photosynthetic plant cells, phosphoglycolate generated by rubisco's oxygenase activity is channeled into the **glycolate pathway.** This *salvage pathway* disposes of phosphoglycolate and returns about 75% of the reduced carbon (three out of every four carbon atoms) present in phosphoglycolate to the Calvin cycle as 3-phosphoglycerate. Because the glycolate pathway is characterized by light-dependent uptake of oxygen and evolution of carbon dioxide, it is also referred to as **photorespiration.**

Several steps of the glycolate pathway are localized in a specific type of peroxisome called a **leaf peroxisome.** Recall from Chapter 12 that peroxisomes are membrane-bounded microbodies containing oxidase enzymes that generate hydrogen peroxide. The potentially destructive hydrogen peroxide is then eliminated by another peroxisomal enzyme, catalase, which degrades it to water and oxygen. Because of their essential role in the glycolate pathway, leaf peroxisomes are found not only in leaf cells but in all photosynthetic plant tissues.

A typical leaf peroxisome from a mesophyll cell is shown in the electron micrograph of Figure 4-20b. The crystalline core within the matrix of the organelle is composed of catalase, a common feature of peroxisomes. Notice the close proximity of the leaf peroxisome to a chloroplast and a mitochondrion. This association is frequently found in photosynthetic plant cells and may reflect the functional relationship among the three organelles, because two or more steps of the glycolate pathway occur in each organelle (Figure 15-18).

In the chloroplast, the phosphoglycolate generated by rubisco is rapidly dephosphorylated by a phosphatase attached to the stromal side of the chloroplast inner membrane (reaction GP-1). The product is *glycolate,* which diffuses to a nearby leaf peroxisome, where an *oxidase*

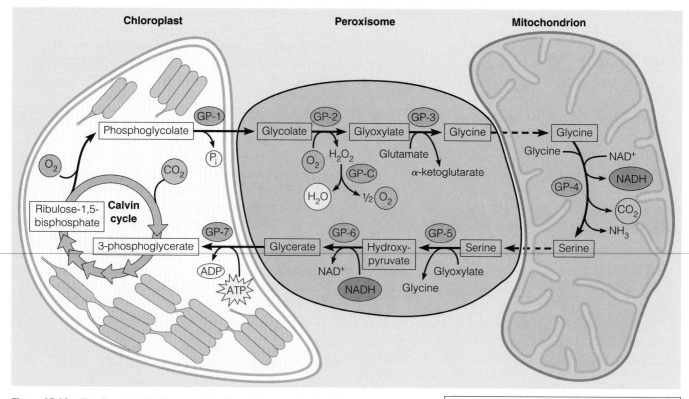

Figure 15-18   The Glycolate Pathway.   Glycolate arises as a result of the oxygenase activity of ribulose-1,5-bisphosphate carboxylase/oxygenase (rubisco). The immediate product is phosphoglycolate, which is converted to free glycolate by a phosphatase localized in the chloroplast membrane (reaction GP-1). Free glycolate diffuses out of the chloroplast stroma and is metabolized by a five-step pathway (GP-2 through GP-6) that occurs partially in the peroxisome and partially in the mitochondrion. Glycerate then diffuses into the chloroplast and is phosphorylated to form 3-phosphoglycerate (reaction GP-7), which enters the Calvin cycle. The oxygen uptake and carbon dioxide evolution characteristic of photorespiration occur in the peroxisome (reaction GP-2) and mitochondrion (reaction GP-4), respectively.

Enzymes That Catalyze These Reactions

GP-1: Phosphoglycolate phosphatase
GP-2: Glycolate oxidase
GP-3: Glutamate: glyoxylate aminotransferase
GP-4: Glycine decarboxylase and serine hydroxymethyl transferase
GP-5: Serine: glyoxylate aminotransferase
GP-6: Hydroxypyruvate reductase
GP-7: Glycerate kinase
GP-C: Catalase

converts it to glyoxylate (reaction GP-2). Presumably, the close juxtaposition of chloroplasts and peroxisomes in photosynthetic cells contributes to the efficient transfer of metabolites from one organelle to the other. As you might expect of a peroxisomal process, the oxidation of glycolate is accompanied by the uptake of oxygen and the generation of hydrogen peroxide, which is immediately degraded to oxygen and water by catalase (reaction GP-C). During the next reaction in the peroxisome, an *aminotransferase* catalyzes the transfer of an amino group from glutamate to glyoxylate, forming glycine (reaction GP-3).

Glycine diffuses from the leaf peroxisome to a mitochondrion, where two enzyme activities working in series—a *decarboxylase* and an *hydroxymethyl transferase*—convert two glycine molecules to a single *serine*, concomitant with the generation of NADH and the release of carbon dioxide and ammonia (reaction GP-4). Rubisco's oxygenase activity therefore leads not only to a loss of carbon, but to a potential loss of nitrogen. To prevent depletion of nitrogen reserves, the ammonia must be reassimilated at the expense of ATP and reductant.

Back in the peroxisome, another aminotransferase removes the amino group from serine, generating *hydroxypyruvate* (reaction GP-5). A *reductase,* using NADH as an electron donor, then reduces hydroxypyruvate to *glycerate* (reaction GP-6). Finally, glycerate diffuses to the chloroplast, where it is phosphorylated by *glycerate kinase* to generate 3-phosphoglycerate (reaction GP-7), a key intermediate of the Calvin cycle.

What is the benefit of this long salvage pathway, winding through several organelles? Three out every four carbon atoms that exit the Calvin cycle as part of phosphoglycolate are recovered as 3-phosphoglycerate. Without this pathway, phosphoglycolate would not only accumulate to toxic levels, but triose phosphates essential for the regeneration of ribulose-1,5-bisphosphate and the continuation of the Calvin cycle would be depleted. In terms of energy and reduced carbon, however, phosphoglycolate metabolism is expensive. For every three carbon atoms salvaged, an ammonia molecule must be reassimilated at the expense of one ATP and two reduced ferredoxin molecules, and the glycerate generated by reaction GP-5 must be phosphorylated at the expense of one ATP molecule. With rubisco's apparently unavoidable oxygenase activity, however, photorespiration is a net gain for the plant. Just consider the value of the three carbon atoms salvaged: nine ATP and six NADPH were consumed when they were originally fixed and reduced.

C₄ Plants Minimize Photorespiration by Confining Rubisco to Cells Containing High Concentrations of CO₂

As mentioned earlier, plants in hot arid environments under intense illumination are particularly affected by rubisco's oxygenase activity. In some cases, the potential for energy and carbon drain through photorespiration is so overwhelming that plants must depend on adaptive strategies for solving the problem. One general approach is to confine rubisco to cells that contain a high concentration of carbon dioxide, thereby minimizing the enzyme's inherent oxygenase activity.

In many tropical grasses, including economically important plants such as maize, sorghum, and sugar cane, the isolation of rubisco is accomplished by a short carboxylation/decarboxylation pathway referred to as the **Hatch-Slack cycle,** after Marshall D. Hatch and C. Roger Slack, two plant physiologists who played key roles in the elucidation of the pathway. Plants containing this pathway are referred to as **C₄ plants** because the immediate product of carbon dioxide fixation by the Hatch-Slack cycle is the four-carbon organic acid oxaloacetate. This term distinguishes such plants from **C₃ plants,** in which the first detectable product of carbon dioxide fixation is the three-

carbon compound 3-phosphoglycerate; in C₃ plants, the Calvin cycle is not preceded by an additional carbon assimilation pathway.

To appreciate the advantage of the Hatch-Slack cycle, we must first consider the arrangement of the Hatch-Slack and Calvin cycles within the leaf of a C₄ plant. As shown in Figure 15-19, C₄ plants, unlike C₃ plants, have in their leaves two distinct types of photosynthetic cells—*mesophyll cells* and *bundle sheath cells*—that differ in their enzyme composition and hence their metabolic activities. The first steps of carbon dioxide fixation within a C₄ plant are accomplished by the Hatch-Slack cycle in mesophyll cells, which are exposed to the carbon dioxide and oxygen that enter a leaf through its stomata. The carbon dioxide that is fixed in mesophyll cells is subsequently released in **bundle sheath cells,** which are relatively isolated from the atmosphere. The entire Calvin cycle, including rubisco, is confined to chloroplasts in the bundle sheath cells. Because of the activity of the Hatch-Slack cycle, the carbon dioxide concentration in bundle sheath cells may be as much as ten times the level in the atmosphere, strongly favoring rubisco's carboxylase activity and minimizing its oxygenase activity.

As detailed in Figure 15-20, the Hatch-Slack cycle begins with the carboxylation of *phosphoenolpyruvate (PEP)* to form oxaloacetate (reaction HS-1). Both PEP and oxaloacetate should be familiar to you, because the same carboxylation reaction was encountered earlier as one of the means of replenishing oxaloacetate for the TCA cycle (see Figure 14-15). Carboxylation is catalyzed by a specific cytosolic form of *PEP carboxylase,* which is particularly

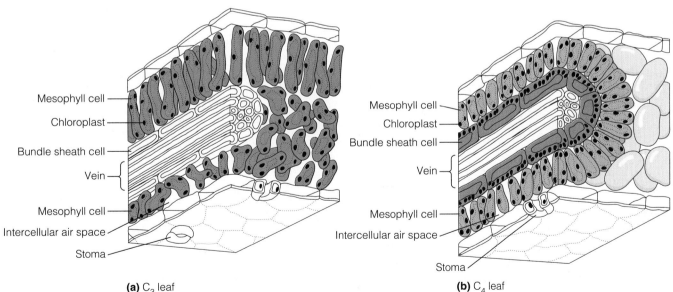

(a) C₃ leaf

(b) C₄ leaf

Figure 15-19 Structural Differences Between Leaves of C₃ and C₄ Plants.
(a) In C₃ plants, the Calvin cycle occurs in mesophyll cells. **(b)** In C₄ plants, the Calvin cycle is confined to bundle sheath cells, which are relatively isolated from

atmospheric carbon dioxide and oxygen. C₄ plants utilize the Hatch-Slack cycle for collecting carbon dioxide in mesophyll cells and concentrating it in bundle sheath cells. The bundle sheath cells surround the vascular bundles (veins) of the leaf, which

carry carbohydrates to other parts of the plant. This concentric arrangement is called Kranz (German for "halo" or "wreath") anatomy and is essential to the photosynthetic efficiency of C₄ plants.

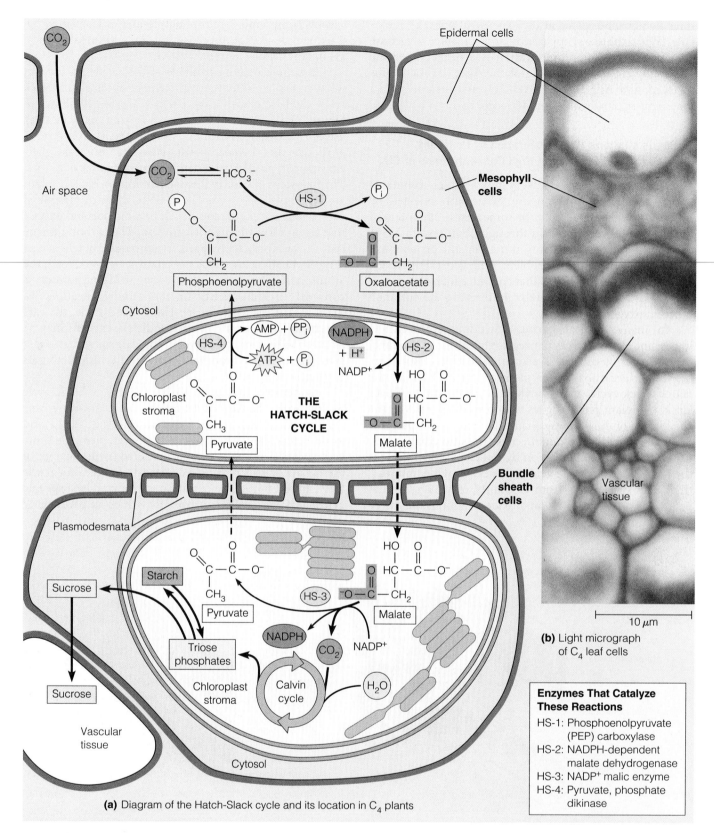

(a) Diagram of the Hatch-Slack cycle and its location in C$_4$ plants

(b) Light micrograph of C$_4$ leaf cells

Enzymes That Catalyze These Reactions

HS-1: Phosphoenolpyruvate (PEP) carboxylase
HS-2: NADPH-dependent malate dehydrogenase
HS-3: NADP$^+$ malic enzyme
HS-4: Pyruvate, phosphate dikinase

Figure 15-20 Localization of the Hatch-Slack Cycle Within Different Cells of a C$_4$ Leaf. **(a)** Carbon dioxide fixation in C$_4$ plants initially occurs by the Hatch-Slack cycle within mesophyll cells. (The path of incoming carbon is indicated by heavy black arrows.) Malate is then passed inward to the bundle sheath cells, where it is decarboxylated. The carbon dioxide is refixed by the Calvin cycle, eventually yielding sucrose, which passes into the adjacent vascular tissue for transport to other parts of the plant. The enzymes that catalyze the Hatch-Slack reactions are listed in the box at the lower right. The enzyme pyruvate, phosphate dikinase, that catalyzes phosphorylation of pyruvate is unique to Hatch-Slack cycle. **(b)** The particular C$_4$ leaf shown is that of maize, *Zea mays* (LM).

abundant in mesophyll cells of C_4 plants. Not only does this carboxylase lack rubisco's oxygenase activity, it is an excellent scavenger for carbon dioxide. In other words, it has a high affinity (a low K_m) for its substrate, *bicarbonate* (HCO_3^-), and operates very efficiently even when the concentration of bicarbonate is quite low. (Bicarbonate forms when carbon dioxide dissolves in water; its concentration therefore reflects the availability of carbon dioxide gas.)

In one version of the Hatch-Slack pathway, the oxaloacetate generated by PEP carboxylase is rapidly converted to malate by an *NADPH-dependent malate dehydrogenase* (reaction HS-2 in Figure 15-20). Malate is a stable four-carbon acid that carries carbon from mesophyll cells to chloroplasts of bundle sheath cells, where decarboxylation occurs (reaction HS-3). The liberated carbon dioxide is then refixed and reduced by the Calvin cycle. Because decarboxylation of malate is accompanied by the generation of NADPH, the Hatch-Slack cycle also conveys reducing power from mesophyll to bundle sheath cells. This might limit the demand for noncyclic electron flow from water to $NADP^+$ in the bundle sheath cells, thereby minimizing the formation of oxygen by PSII complexes and further favoring rubisco's carboxylase activity.

The pyruvate generated by decarboxylation of malate diffuses into a mesophyll cell, where it is phosphorylated at the expense of ATP to regenerate PEP (reaction HS-4), the original carbon dioxide acceptor of the Hatch-Slack cycle. Thus, the overall process is cyclic, and the net result is a feeder system that captures carbon dioxide in mesophyll cells and passes it to the Calvin cycle in bundle sheath cells. The Hatch-Slack cycle is not a substitute for the Calvin cycle, but is simply a preliminary carboxylation/decarboxylation sequence that concentrates CO_2 in the bundle sheath cells.

Because ATP is hydrolyzed to AMP in reaction HS-4, the actual cost of moving carbon from mesophyll to bundle sheath cells is equivalent to two ATP molecules per carbon dioxide molecule. Carbon assimilation within a C_4 plant therefore consumes a total of five ATP molecules per carbon atom, rather than the three required in C_3 plants. In an environment that enhances rubisco's oxygenase activity, however, the energy required to prevent the formation of phosphoglycolate may be far less than the energy that would otherwise be lost through photorespiration.

When temperatures exceed about 30°C, the photosynthetic efficiency of a C_4 plant exposed to intense sunlight may be twice that of a C_3 plant. While the higher efficiency is largely due to the reduced photorespiration in the C_4 plant and enhanced photorespiration in the C_3 plant, other factors are also important. In a C_3 plant, photosynthesis is often limited by the low atmospheric concentration of carbon dioxide, not by the availability of sunlight. In a C_4 plant, however, the Hatch-Slack cycle actively concentrates carbon dioxide in bundle sheath cells, where the Calvin cycle is localized, enabling the plant to take advantage of higher levels of illumination.

Enrichment of carbon dioxide in the vicinity of rubisco by the Hatch-Slack cycle confers an additional advantage on C_4 plants. Because PEP carboxylase is an efficient scavenger of carbon dioxide, gas exchange through the stomata of C_4 plants can be substantially reduced to conserve water without adversely affecting photosynthetic efficiency. As a result, C_4 plants are able to assimilate over twice as much carbon as C_3 plants for each unit of water transpired. This adaptation makes C_4 plants suitable for regions of periodic drought, such as tropical savannas.

Although fewer than 1% of plant species investigated depend on the Hatch-Slack cycle, the pathway is of particular interest because several economically important species are in this group. Moreover, C_4 plants such as maize and sugar cane are characterized by net photosynthetic rates that are often two or three times those of C_3 plants such as cereal grains. Little wonder, then, that crop physiologists and plant breeders have devoted so much attention to C_4 species and to the question of whether it is possible to improve on the relatively inefficient carbon dioxide fixation pathway of the C_3 plants. Some scenarios of genetic engineering even envision the genetic conversion of C_3 plants into C_4 plants.

CAM Plants Minimize Photorespiration and Water Loss by Opening Their Stomata Only at Night

Some plant species that live in deserts, salt marshes, and other environments where access to water is severely limited contain a preliminary carbon dioxide fixation pathway closely related to the Hatch-Slack cycle. The sequence of reactions is similar, but these plants segregate the carboxylation and decarboxylation reactions by *time* rather than by *space*. Because the pathway was first recognized in the family Crassulaceae, it is called **crassulacean acid metabolism (CAM),** and plants that take advantage of CAM photosynthesis are called **CAM plants.** CAM photosynthesis has been found in about 4% of plant species investigated, including many succulents, cacti, and orchids.

CAM plants, unlike most C_3 and C_4 plants, generally open their stomata only at night, when the atmosphere is relatively cool and moist. As carbon dioxide diffuses into mesophyll cells, it is assimilated by the first two steps of a pathway similar to the Hatch-Slack cycle, and accumulates as malate. Instead of being exported from mesophyll cells, however, the malate is stored in large vacuoles, which become very acidic. The process of moving malate into vacuoles consumes ATP, but is necessary to protect cytosolic enzymes from a large drop in pH at night.

During the day, CAM plants close their stomata to conserve water. The malate then diffuses from vacuoles to the cytosol, where the Hatch-Slack cycle continues. Carbon dioxide released by decarboxylation of malate diffuses into the chloroplast stroma, where it is refixed and reduced by the Calvin cycle. The high carbon dioxide and low oxygen concentrations established when light is available for generating ATP and NADPH strongly favor rubisco's carboxylase activity and minimize the loss of carbon through photorespiration. Notice that the carboxylation of PEP and decarboxylation of malate occur in the same compartment. Because of this, the activity of PEP carboxylase in CAM plants must be strictly inhibited during the day to prevent a futile cycle from developing.

With their remarkable ability to conserve water, CAM plants may assimilate over 25 times as much carbon as a C_3 plant for each unit of water transpired. Moreover, some CAM plants display a process called *CAM idling,* whereby the plant keeps its stomata closed night *and* day. Carbon dioxide is simply recycled between photosynthesis and respiration, with virtually no loss of water. Such plants will not, of course, display a net gain of carbohydrate. This ability, however, may enable them to survive droughts lasting up to several months.

Perspective

Photosynthesis is the single most vital metabolic process for virtually all forms of life on Earth, because all of us, whatever our immediate sources of energy, ultimately depend on the energy radiating from the sun. Photosynthesis involves both energy transduction reactions and carbon assimilation reactions. During the energy transduction reactions, photons of light are absorbed by chlorophyll or accessory pigment molecules within the thylakoid or photosynthetic bacterial membranes, and the energy is rapidly passed to a pair of special chlorophyll molecules at the reaction center of a photosystem. There, the energy is used to excite and eject an electron. In the case of photosystem I of oxygenic phototrophs, this electron is passed via ferredoxin to $NADP^+$, generating the NADPH required for carbon dioxide fixation and reduction.

The source of electrons in oxygenic phototrophs is water. Electron transfer from water to $NADP^+$ depends on two photosystems acting in series, with photosystem II responsible for the oxidation of water and photosystem I responsible for the reduction of $NADP^+$. In plants, electron flow between the two photosystems (or in cyclic fashion around photosystem I) passes through a cytochrome b_6/f complex, where it drives the pumping of protons into the thylakoid lumen. The resulting proton motive force across the thylakoid membrane is due largely to the pH differ-

ential and is used to drive ATP synthesis by the CF_1 particles that protrude outward from the thylakoid membranes into the stroma of the chloroplast.

In the stroma, ATP and NADPH are used for the fixation and reduction of carbon dioxide into organic form by enzymes of the Calvin cycle. In C_3 plants, carbon dioxide is directly attached to ribulose-1,5-bisphosphate by rubisco, generating 3-phosphoglycerate. In C_4 and CAM plants, however, carbon dioxide is fixed by a preliminary carboxylation/decarboxylation pathway that concentrates it within a photosynthetic cell—either a different cell or at a different time of day—for subsequent assimilation by the Calvin cycle. The eventual product of carbon dioxide fixation in each case is glyceraldehyde-3-phosphate, which can be converted to a second triose phosphate called dihydroxyacetone phosphate. Some of these triose phosphate molecules are used for the biosynthesis of more complex carbohydrates, such as sucrose or starch, or as sources of energy for other metabolic pathways. The remainder must be used for regenerating the acceptor molecule with which the Calvin cycle began. The net synthesis of one triose phosphate molecule requires the fixation of three carbon dioxide molecules and uses nine ATP and six NADPH molecules. A combination of noncyclic and cyclic electron flow ensures that the ratio of ATP to NADPH within a

photosynthetic cell meets the metabolic demands imposed not only by carbon assimilation but also by other pathways, including those involved in nitrogen and sulfur assimilation.

This transduction of solar energy into chemical energy is crucial to the continued existence of the biological world. All the energy stored in organic molecules on which chemotrophs depend represents the energy of sunlight, originally trapped within the molecules of organic compounds during photosynthesis. We have not yet discovered anything particularly unique about the carbon metabolism of photoautotrophs, since carbon dioxide fixation (carboxylation) occurs in animals and other nonphotosynthetic organisms. What is remarkable about photosynthetic organisms is their ability to carry out sustained *net* fixation and reduction of carbon dioxide using solar energy to drive what would otherwise be a highly endergonic process. Only phototrophs can utilize sunlight to extract electrons from such poor (i.e., electropositive) donors as water and use them to reduce carbon atoms in carbon dioxide to the level of an organic compound. And they can do so because of the photochemical events that are initiated whenever light of the appropriate wavelength is absorbed by chlorophyll, a remarkable molecule that has transformed the biosphere of an entire planet—Earth.

Key Terms for Self-Testing

photosynthesis (p. 445)
phototroph (p. 445)
photoheterotroph (p. 445)
photoautotroph (p. 445)

An Overview of Photosynthesis
energy transduction reactions (p. 445)
carbon assimilation reactions (p. 445)
photophosphorylation (p. 447)
oxygenic phototroph (p. 447)
anoxygenic phototroph (p. 447)
photoreduction (p. 447)

The Chloroplast: A Photosynthetic Organelle
chloroplast (p. 447)
proplastid (p. 448)
plastid (p. 448)
outer membrane (p. 448)
inner membrane (p. 448)
intermembrane space (p. 448)
stroma (p. 448)
porin (p. 448)
thylakoid (p. 448)
granum (p. 448)
stroma thylakoid (p. 448)
thylakoid lumen (p. 449)

Photosynthetic Energy Transduction
photon (p. 452)
quantum (p. 452)
pigment (p. 452)
photoexcitation (p. 452)
absorption spectrum (p. 452)
resonance energy transfer (p. 452)
photochemical reduction (p. 452)
chlorophyll (p. 452)
bacteriochlorophyll (p. 453)
accessory pigment (p. 453)

carotenoid (p. 453)
phycobilin (p. 453)
photosystem (p. 453)
chlorophyll-binding protein (p. 453)
antenna pigment (p. 453)
reaction center (p. 453)
light-harvesting complex (LHC) (p. 454)
phycobilisome (p. 454)
photosystem complex (p. 454)
action spectrum (p. 454)
Emerson enhancement effect (p. 455)
photosystem I (PSI) (p. 455)
photosystem II (PSII) (p. 455)
P700 (p. 455)
P680 (p. 455)

**Photoreduction (NADPH Synthesis)
in Oxygenic Phototrophs**
nicotinamide adenine dinucleotide
    phosphate ($NADP^+$) (p. 456)
light-harvesting complex II
    (LHCII) (p. 456)
plastoquinone (p. 456)
plastoquinol (p. 456)
oxygen-evolving complex (OEC) (p. 457)
cytochrome b_6/f complex (p. 457)
plastocyanin (PC) (p. 458)
light-harvesting complex I
    (LHCI) (p. 458)
ferredoxin (Fd) (p. 458)
ferredoxin-$NADP^+$ reductase
    (FNR) (p. 458)
noncyclic electron flow (p. 458)

**Photophosphorylation (ATP Synthesis)
in Oxygenic Phototrophs**
ATP synthase (CF_oCF_1 complex) (p. 459)
CF_1 (p. 459)

CF_o (p. 459)
proton translocator (p. 459)
cyclic electron flow (p. 460)

**Photosynthetic Carbon Assimilation:
The Calvin Cycle**
Calvin cycle (p. 462)
stoma (p. 462)
mesophyll cell (p. 462)
ribulose-1, 5-bisphosphate carboxylase/
    oxygenase (rubisco) (p. 464)

Carbohydrate Synthesis
isoenzyme (p. 467)

**Rubisco's Oxygenase Activity Decreases
Photosynthetic Efficiency**
phosphoglycolate (p. 469)
glycolate pathway (p. 469)
photorespiration (p. 469)
leaf peroxisome (p. 469)
Hatch-Slack cycle (p. 471)
C_4 plant (p. 471)
C_3 plant (p. 471)
bundle sheath cell (p. 471)
crassulacean acid metabolism
    (CAM) (p. 473)
CAM plant (p. 473)

Box 15A: *The Endosymbiont Theory*
semiautonomous organelle (p. 450)
endosymbiont theory (p. 450)
symbiotic relationship (p. 450)
protoeukaryote (p. 450)

Problem Set

More challenging problems are marked with a • .

15-1. True, False, or Insufficient Information. Indicate whether each of the following statements is true (T), false (F), or does not provide enough information for you to make a decision (I).

(a) Although traditionally referred to as the dark reactions of photosynthesis, photosynthetic carbon assimilation actually depends indirectly on light and does not continue long in the dark.

(b) Sucrose is synthesized in the chloroplast stroma and exported from photosynthetic cells to provide energy and reduced carbon for nonphotosynthetic plant cells.

(c) 90% of the solar energy collected by a photosystem complex is absorbed when photons strike a special pair of chlorophyll molecules at the reaction center of the complex.

(d) The energy requirement expressed as ATP consumed per molecule of carbon dioxide fixed is higher for a C_3 plant than for a C_4 plant.

(e) The ultimate electron donor for the photosynthetic generation of NADPH is always water.

(f) The enzyme rubisco is unusual in that, depending on conditions, it exhibits two different enzymatic activities.

15-2. The Advantage of the Hatch-Slack Cycle. A C_4 plant can be more efficient than a C_3 plant at fixing carbon dioxide, an advantage that becomes more evident as the atmospheric carbon dioxide concentration decreases.

(a) Explain in your own words why a C_4 plant can be inherently more efficient than a C_3 plant at fixing carbon dioxide.

(b) Why would this advantage be more apparent at an atmospheric carbon dioxide concentration of 0.0035% than at the normal level of about 0.035%?

(c) If a C_4 plant and a C_3 plant are grown under constant illumination in a sealed container with an initial carbon dioxide concentration of 0.04%, the C_4 plant will eventually kill the C_3 plant. Explain why.

• **15-3.   The Role of Sucrose.** A plant uses solar energy to make ATP and NADPH, which then drive the synthesis of carbohydrates in the leaves. At least one carbohydrate, sucrose, is translocated to nonphotosynthetic parts of the plant (stems, roots, flowers, and fruits) for use as a source of energy. Thus, ATP is used to make sucrose, and the sucrose is then used to make ATP. It would seem simpler for the plant just to make ATP and translocate the ATP itself directly to other parts of the plant, thereby completely eliminating the need for a Calvin cycle, a glycolytic pathway, and a TCA cycle (and making life a lot easier for cell biology students). Suggest at least two major reasons that plants do not manage their energy economies in this way.

15-4.   Effects on Photosynthesis. Assume that you have an illuminated suspension of *Chlorella* cells carrying out photosynthesis in the presence of 0.1% carbon dioxide and 20% oxygen. What will be the short-term effects of the following changes in conditions on the levels of 3-phosphoglycerate and ribulose-1,5-bisphosphate? Explain your answer in each case.

(a) Carbon dioxide concentration is suddenly reduced 1000-fold.

(b) Light is restricted to green wavelengths (510–550 nm).

(c) An inhibitor of photosystem II is added.

(d) Oxygen concentration is reduced from 20% to 1%.

15-5.   Alternative Electron Donors. Eukaryotic phototrophs use water as an electron donor, but most prokaryotic phototrophs depend on a variety of inorganic and organic electron donors. Purple bacteria, for example, often use H_2S as their source of electrons, generating elemental sulfur instead of molecular oxygen as a by-product.

(a) Reaction 15-1 is intended as a general reaction to cover all cases. Write a balanced overall reaction for photosynthesis based on H_2S as the electron donor and a triose phosphate as the end-product.

(b) Why are photosynthetic bacteria able to use H_2S but not water as an electron donor?

(c) Under what circumstances would H_2S, rather than water, be a more suitable electron donor?

(d) The E_0 value for the Fe^{3+}/Fe^{2+} redox pair is +0.77 V. Do you think photosynthetic bacteria are able to generate reduced coenzyme (either NADH or NADPH) by oxidizing ferrous ions to ferric ions?

15-6.   Energy Flow in Photosynthesis. A portion of the solar energy that arrives at the surface of a leaf is eventually converted to chemical energy and appears in the chemical bonds of carbohydrates that are generally regarded as the end-products of photosynthesis. In between the photons and the carbohydrate molecules, however, the energy exists in a variety of forms. Trace the flow of energy from photon through ATP to starch molecule, assuming the wavelength of light to be in the absorption range of one of the accessory pigments rather than that of chlorophyll.

15-7.   The Mint and the Mouse. Joseph Priestley, a British clergyman, was a prominent figure in the early history of research in photosynthesis. In 1771, Priestley wrote these words:

One might have imagined that since common air is necessary to vegetable as well as to animal life, both plants and animals had affected it in the same manner; and I own that I had that expectation when I first put a sprig of mint into a glass jar standing inverted in a vessel of water; but when it had continued growing there for some months, I found that the air would neither extinguish a candle, nor was it at all inconvenient to a mouse which I put into it.

Explain the basis of Priestley's observations, and indicate their relevance to the early understanding of the nature of photosynthesis.

15-8.   The Hill Reaction. A highly significant advance in our understanding of photosynthesis occurred in 1937 when Robert Hill showed that isolated chloroplasts, though not capable of fixing carbon dioxide, were able to produce molecular oxygen. This only happened if the chloroplasts were illuminated and provided with an artificial electron acceptor. (Ferricyanide was the acceptor of choice.) Which of the following statements are valid conclusions from Hill's experiment? All, some, or none may be correct.

(a) The oxygen generated during photosynthesis apparently does not come from carbon dioxide, as was believed earlier.

(b) These findings are in accord with a proposal made six years earlier by Cornelius B. van Niel that the oxygen evolved during photosynthesis comes from water.

(c) Reaction 15-2 would be less indicative of what happens during photosynthesis in eukaryotic phototrophs if three molecules of water were subtracted from each side of the reaction.

15-9.   Photophosphorylation. ATP generation in higher plants can occur as the result of either cyclic or noncyclic electron flow. To sustain the ratio of three ATP to two NADPH molecules required by the Calvin cycle, the ratio of cyclic to noncyclic electron flow would have to be about 1 : 6. But other processes also consume ATP and NADPH. For each of the following conditions, indicate whether you would expect the ratio of cyclic to noncyclic flow to be higher than 1 : 6 (H), about 1 : 6 (S), or lower than 1 : 6 (L).

(a) The reduction of nitrite to ammonia (NH_3), which occurs in the chloroplast stroma, consumes NADPH but not ATP.

(b) Extensive active transport across the chloroplast inner membrane consumes ATP but not NADPH.

(c) In C_4 plants, the Hatch-Slack cycle conveys carbon dioxide from mesophyll cells to bundle sheath cells.

(d) A plant is treated with DCMU, a herbicide that blocks electron transfer from Q_A to Q_B in photosystem II.

(e) A plant is treated with ferricyanide, which accepts electrons from the "bottom" of the electron transport system, becoming reduced to ferrocyanide. (The E_0 for the ferricyanide/ferrocyanide redox pair is $+ 0.4$ V.)

(f) High temperatures favor rubisco's oxygenase activity, leading to the production of additional phosphoglycolate that must be processed by the glycolate pathway.

• 15-10. Photosynthetic Efficiency. On p. 466 we estimated the maximum photosynthetic efficiency for the conversion of red light, carbon dioxide, and water to glyceraldehyde. Under laboratory conditions, a photosynthetic organism *might* convert 31% of the light energy striking it to chemical bond energy of organic molecules. In reality, however, photosynthetic efficiency is far lower, closer to 5% or less. Considering a plant growing in a natural environment, suggest four reasons for this discrepancy.

15-11. Chloroplast Structure. Where in a chloroplast are the following substances or processes localized? Be as specific as possible.

(a) Ferredoxin-NADP$^+$ reductase

(b) Cyclic electron flow

(c) Starch synthase

(d) Light-harvesting complex I

(e) Plastoquinol

(f) Proton pumping

(g) P700

(h) Reduction of 3-phosphoglycerate

(i) Carotenoid molecules

(j) Oxygen-evolving complex

15-12. Metabolite Transport Across Membranes. For each of the following metabolites, indicate whether you would expect it to be in steady-state flux across one or more membranes in a photosynthetically active chloroplast and, if so, indicate which membrane(s) the metabolite must cross.

(a) CO_2

(b) P_i

(c) Electrons

(d) Starch

(e) Glyceraldehyde-3-phosphate

(f) NADPH

(g) ATP

(h) O_2

(i) Protons

(j) Pyruvate

15-13. Crassulacean Acid Metabolism. A CAM plant uses a pathway very similar to the Hatch-Slack cycle for preliminary carbon dioxide fixation. Trace the flow of carbon from the atmosphere to glyceraldehyde-3-phosphate within a CAM plant. How does this minimize water loss by such plants?

• 15-14. The Endosymbiont Theory. The endosymbiont theory suggests that mitochondria and chloroplasts evolved from ancient bacteria that were ingested by primitive nucleated cells. Biologists have proposed that endosymbiosis led to the evolution of other cellular structures, such as flagella and peroxisomes, as well. Over hundreds of millions of years, the ingested bacteria lost features not essential for survival inside the host cell.

(a) Based on what you have learned about phagocytosis and the membrane systems of organelles, for each of the membrane systems found in mitochondria (two systems) and chloroplasts (three systems) indicate whether it arose from an ingested bacterium or from the nucleated host. Explain your reasoning.

(b) Describe one structure or metabolic process that purple bacteria might have dispensed with once they became endosymbionts in a eukaryotic cell. How would loss of this feature prevent the bacteria from living outside the host?

(c) Describe one structure or metabolic process that cyanobacteria might have dispensed with once they became endosymbionts in a eukaryotic cell. How would loss of this feature prevent the bacteria from living outside the host?

(d) Peroxisomes, unlike mitochondria or chloroplasts, scarcely resemble free-living organisms. Assuming peroxisomes evolved from ingested bacteria, describe three features mitochondria retain but peroxisomes apparently lost over hundreds of millions of years. Describe one advantage peroxisomes might have conferred on ancient nucleated cells.

Suggested Reading

References of historical importance are marked with a **•** .

General References

Foyer, C. H., and W. P. Quick, eds. *A Molecular Approach to Primary Metabolism in Higher Plants.* London: Taylor & Francis, 1997.

Hall, D.O., and K.K. Rao. *Photosynthesis,* 6th ed. New York: Cambridge University Press, 1999.

Leegood, R. C., T. D. Sharkey, and S. von Caemmerer, eds. *Photosynthesis: Physiology and Metabolism. Advances in Photosynthesis,* Vol. 9. Dordrecht: Kluwer Academic Publishers, 2000.

Mathews, C. K., and K. E. van Holde. *Biochemistry,* 2d ed. Menlo Park, CA: Benjamin/Cummings, 1996.

Salisbury, F. B., and C. W. Ross. *Plant Physiology,* 4th ed. Belmont, CA: Wadsworth, 1992.

Singhal, G. S., G. Renger, S. K. Sopory, K-D. Irrgang, and Govindjee, eds. *Concepts in Photobiology: Photosynthesis and Photomorpho-genesis.* Dordrecht: Kluwer Academic Publishers; Delhi: Narosa Pub. House, 1999.

Tobin, A. K., ed. *Plant Organelles: Compartmentation of Metabolism in Photosynthetic Cells.* New York: Cambridge University Press, 1992.

The Chloroplast

Douce, R., and J. Joyard. Biochemistry and function of the plastid envelope. *Annu Rev. Cell Biol.* 6 (1990): 173.

Flügge, U. I., and H. W. Heldt. Metabolite translocators of the chloroplast envelope. *Annu. Rev. Plant Physiol. Plant Mol. Biol.* 42 (1991): 129.

Trissl, H. W., and C. Wilhelm. Why do thylakoid membranes from higher plants form grana stacks? *Trends Biochem. Sci.* 18 (1993): 415

Photosynthetic Energy Transduction

Chitnis, P. R. Photosystem I. *Plant Physiol.* 111 (1996): 661.

Cramer, W. A., G. M. Soriano, M. Ponomarev, D. Huang, H. Zhang, S. E. Martinez, and J. L. Smith. Some new structural aspects and old controversies concerning the cytochrome b_6f complex of oxygenic photosynthesis. *Annu. Rev. Plant Physiol. Plant Mol. Biol.* 47 (1996): 477.

• Deisenhofer, J., and H. Michel. The photosynthetic reaction center from the purple bacterium *Rhodopseudomonas viridis. Science* 245 (1989): 1463.

Govindjee and W. J. Coleman. How plants make oxygen. *Sci. Am.* 262 (1990):42.

Hankamer, B., J. Barber, and E. Boekema. Structure and membrane organization of photosystem II in green plants. *Annu. Rev. Plant Physiol. Plant Mol. Biol.* 48 (1997): 641.

Nugent J. H. Oxygenic photosynthesis. Electron transfer in photosystem I and photosystem II. *Eur. J. Biochem.* 237 (1996): 519

Ort, D. R., and C. F. Yocum, eds. *Oxygenic Photosynthesis: The Light Reactions. Advances in Photosynthesis,* Vol. 4. Dordrecht: Kluwer Academic Publishers, 1996.

Rögner, M., E. J. Boekema, and J. Barber. How does photosystem 2 split water? The structural basis of efficient energy conversion. *Trends Biochem. Sci.* 21 (1996): 44.

Szalai, V., and G. W. Brudvig. How plants produce dioxygen. *Amer. Sci.* 86 (1998): 542.

The Calvin Cycle and Carbohydrate Synthesis

Buchanan, B. B. Carbon dioxide assimilation in oxygenic and anoxygenic photosynthesis. *Photosynth. Res.* 33 (1992): 147.

• Calvin, M. The path of carbon in photosynthesis. *Science* 135 (1962): 879.

Geiger, D. R., and J. C. Servaites. Diurnal regulation of photosynthetic carbon metabolism in C_3 plants. *Annu. Rev. Plant Physiol. Plant Mol. Biol.* 45 (1994): 235.

Hartman, F. C., and M. R. Harpel. Structure, function, regulation, and assembly of D-ribulose-1,5-bisphosphate carboxylase/oxygenase. *Annu. Rev. Biochem.* 63 (1994): 197.

Nevins, D. J. Sugars: Their origin in photosynthesis and subsequent biological interconversions. *Am. J. Clin. Nutr.* 61(1995): 915S.

Portis, A. R., Jr. Regulation of ribulose-1,5-bisphosphate carboxylase/oxygenase activity. *Annu. Rev. Plant Physiol. Plant Mol. Biol.* 43 (1992): 415.

• Schnarrenberger, C., and W. Martin. The Calvin cycle: A historical perspective. *Photosynthetica* 33 (1997): 331.

Photorespiration, C4 Plants, and CAM Plants

Bazzaz, F. A., and E. D. Fajer. Plant life in a CO_2-rich world. *Sci. Amer.* 266 (January 1992): 68.

Drake, B. G., and M. A. Gonzàlez-Meler. More efficient plants: A consequence of rising atmospheric CO_2? *Annu. Rev. Plant Physiol. Plant Mol. Biol.* 48 (1997): 609.

Hatch, M. D. C_4 photosynthesis: An unlikely process full of surprises. *Plant Cell Physiol.* 33 (1992): 333.

Sage, R. F., and R. K. Monson. C_4 *Plant Biology.* San Diego: Academic Press, 1998.

Wingler, A., P. J. Lea, W. P. Quick, and R. C. Leegood. Photorespiration: Metabolic pathways and their role in stress protection. *Phil. Trans. R. Soc. Lond. B* 355 (2000): 1517.

Winter, K., and J. A. C. Smith, eds. *Crassulacean Acid Metabolism: Biochemistry, Ecophysiology, and Evolution. Ecological Studies,* Vol. 114. Berlin: Springer-Verlag, 1996.

Box 15A: *The Endosymbiont Theory*

Gray, M. W. The endosymbiont hypothesis revisited. *Internat. Rev. Cytol.* 141 (1992): 233.

Margulis, L. *Symbiosis in Cell Evolution,* 2nd ed. New York: Freeman, 1993.

Schenk, H. E. A., R. G. Herrmann, K. W. Jeon, N. E. Müller, and W. Schwemmler, eds. *Eukaryotism and Symbiosis.* Berlin: Springer-Verlag, 1997.

Tibor, V., K. Takács, and G. Vida. A new aspect to the origin and evolution of eukaryotes. *J. Mol. Evol.* 46 (1998): 499.

16

The Structural Basis of Cellular Information: DNA, Chromosomes, and the Nucleus

Implicit in our earlier discussions of cellular structure and function has been a sense of predictability, order, and control. We have come to expect that organelles and other cellular structures will have a predictable appearance and function, that metabolic pathways will proceed in an orderly fashion in specific intracellular locations, and that all of a cell's activities will be carried out in a carefully controlled, highly efficient, and heritable manner.

Such expectations express our confidence that cells possess a set of "instructions" that specify their structure, dictate their functions, and regulate their activities, and that these instructions can be passed on faithfully to daughter cells. More than a hundred years ago, the Augustinian monk Gregor Mendel worked out rules accounting for the inheritance patterns he observed in pea plants, although he had little inkling of the cellular or molecular basis for these rules. These studies led Mendel to conclude that hereditary information is transmitted in the form of distinct units that we now call *genes*. We also now know that genes are made of DNA, and we can tell a coherent genetic story starting with this molecule.

But first let's step back and use Figure 16-1 to preview how DNA carries out its instructional role in cells and, at the same time, examine how this set of chapters on information flow is organized. The information carried by DNA flows both *between* generations of cells and *within* each individual cell. During the first of these two processes (Figure 16-1a), the information stored in a cell's DNA molecules undergoes replication, generating two DNA copies that are distributed to the daughter cells when the cell divides. The initial three chapters in this section focus on the structures and events associated with this aspect of information flow. The present chapter covers the structural organization of DNA and the chromosomes in

which it is packaged; it also discusses the nucleus, which is the organelle that houses the chromosomes of eukaryotic cells. Chapter 17 then discusses DNA replication and cell division, while Chapter 18 considers the cellular and molecular events associated with information flow between generations of sexually reproducing organisms (including Mendel's work and its chromosomal basis).

Figure 16-1b summarizes how information residing in DNA is used *within* a cell. Instructions stored in DNA are utilized in a two-stage process called *transcription* and *translation*. During transcription, RNA is synthesized in an enzymatic reaction that copies information from DNA. During translation, the base sequences of the resulting messenger RNA molecules are used to determine the amino acid sequences of proteins. *Thus, the information initially stored in DNA base sequences is ultimately used to code for the synthesis of specific protein molecules.* It is the particular proteins synthesized by a cell that ultimately determine most of a cell's structural features as well as the functions it performs. Transcription and translation, which together constitute the expression of genetic information, are the subjects of Chapters 19–21.

We open this chapter by describing the discovery of DNA, the molecule whose informational role lies at the heart of this group of six chapters.

The Chemical Nature of the Genetic Material

When Mendel first postulated the existence of genes, he did not know the identity of the molecule that allows them to store and transmit inherited information. But a few years later, this molecule was unwittingly discovered

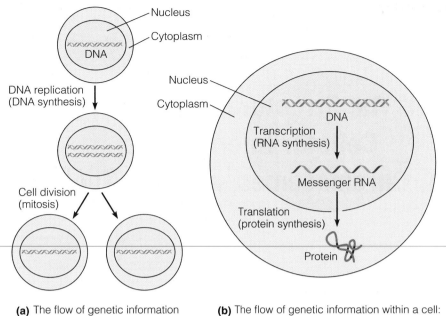

Figure 16-1 The Flow of Information in Cells. The diagrams here feature eukaryotic cells, but DNA replication, cell division, transcription, and translation are processes that occur in prokaryotic cells as well. **(a)** Genetic information encoded in DNA molecules is passed on to successive generations of cells by DNA replication and cell division (in eukaryotic cells, by means of mitosis). The DNA is first duplicated and then divided equally between the two daughter cells. In this way, each daughter cell is assured of having the same genetic information as the cell from which it arose. **(b)** Within each cell, genetic information encoded in the DNA is expressed through the processes of transcription (RNA synthesis) and translation (protein synthesis). Transcription involves the use of selected segments of DNA as templates for the synthesis of messenger RNA and other RNA molecules. Translation is the process whereby amino acids are joined together in a sequence dictated by the sequence of nucleotides in messenger RNA.

(a) The flow of genetic information between generations of cells

(b) The flow of genetic information within a cell: the expression of genetic information

by Johann Friedrich Miescher, a Swiss physician. Miescher reported the discovery of the substance now known as DNA in 1869, just a few years before the cell biologist Walther Flemming first observed chromosomes as he studied dividing cells under the microscope.

Miescher's Discovery of DNA Led to Conflicting Proposals Concerning the Chemical Nature of Genes

Miescher was interested in studying the chemistry of the nucleus, which most scientists guessed was the site of the cell's genetic material. In his initial experiments, Miescher isolated nuclei from white blood cells obtained from pus found on surgical bandages. Upon extracting these nuclei with alkali, he discovered an unusual substance, which he called "nuclein" and which we now know to have been largely DNA. Miescher then went on to study DNA from a more pleasant source, salmon sperm. Fish sperm may seem a somewhat unusual source material, until we realize that the nucleus accounts for more than 90% of the mass of a typical sperm cell and therefore DNA accounts for most of the mass of sperm cells. For this reason, Miescher initially believed that DNA is involved in the transmission of hereditary information. He soon rejected this idea, however, because his crude measuring techniques incorrectly suggested that egg cells contain much more DNA than sperm. Reasoning that sperm and egg must contribute roughly equal amounts of hereditary information to the offspring, it seemed to him that DNA could not be carrying hereditary information.

Although Miescher was led astray concerning the role of DNA, in the early 1880s a botanist named Eduard Zacharias reported that extracting DNA from cells causes the staining of the chromosomes to disappear. Since evidence was already beginning to suggest a role for chromosomes in transmitting hereditary information, Zacharias and others inferred that DNA is the genetic material. This view prevailed until the early 1900s, when incorrectly interpreted staining experiments led to the false conclusion that the amount of DNA changes dramatically within cells. Because cells would be expected to maintain a constant amount of the substance that stores their hereditary instructions, these mistaken observations led to a repudiation of the idea that DNA carries genetic information.

As a result, from around 1910 to the 1940s, most scientists believed that genes were made of protein rather than DNA. The chemical building blocks of both proteins and nucleic acids had been identified by the early 1900s, and proteins were perceived to be more complex and hence more likely to store genetic information. It was argued that proteins are constructed from 20 different amino acids that can be assembled in a vast number of combinations, thereby generating the sequence diversity and complexity expected of a molecule that stores and transmits genetic information. In contrast, DNA was widely perceived to be a simple polymer consisting of the same sequence of four bases (e.g., the tetranucleotide –ATCG–) repeated over and over, thereby lacking the variability expected of a genetic molecule. Such a simple polymer was thought to serve merely as a structural support for the genes, which were in turn made of protein. This view prevailed until two lines of evidence resolved the matter in favor of DNA as the genetic material, as we describe next.

Avery Showed That DNA Is the Genetic Material of Bacteria

A great surprise was in store for biologists who were studying protein molecules to determine how genetic information is stored and transmitted. The background was provided in 1928 by the British physician Frederick Griffith, who was studying a pathogenic strain of a bacterium, then called "pneumococcus," that causes a fatal pneumonia in animals. Griffith discovered that this bacterium (now called *Streptococcus pneumoniae*) exists in two forms called the *S strain* and the *R strain*. When grown on a solid agar medium, the S strain produces colonies that are smooth and shiny because of the mucous, polysaccharide coat each cell secretes, whereas the R strain lacks the ability to manufacture a mucous coat and therefore produces colonies exhibiting a rough boundary.

When injected into mice, S-strain but not R-strain bacteria trigger a fatal pneumonia. The ability to cause disease is directly related to the presence of the S strain's polysaccharide coat, which protects the bacterial cell from attack by the mouse's immune system. One of the most intriguing discoveries made by Griffith, however, was that pneumonia can also be induced by injecting animals with a mixture of live R-strain bacteria and dead S-strain bacteria (Figure 16-2). This finding was surprising because neither live R-strain nor dead S-strain organisms cause pneumonia if injected alone. When Griffith autopsied the animals that had been injected with the mixture of live R-strain and dead S-strain bacteria, he found them teeming with live S-strain bacteria. Since the animals had not been injected with any live S-strain cells, he concluded that the nonpathogenic R bacteria were somehow converted into pathogenic S bacteria by a substance present in the heat-killed S bacteria that had been co-injected. He called this phenomenon **genetic transformation** and referred to the active (though still unknown) substance in the S cells as the "transforming principle."

Griffith's discoveries set the stage for 14 years of work by Oswald Avery and his colleagues at the Rockefeller Institute in New York. These researchers pursued the investigation of bacterial transformation to its logical conclusion by asking which component of the heat-killed S bacteria was actually responsible for the transforming activity. They fractionated cell-free extracts of S-strain bacteria and found that only the nucleic acid fraction was capable of causing transformation. Moreover, the activity was specifically eliminated by treatment with deoxyribonuclease, an enzyme that degrades DNA. This and other evidence convinced them the transforming substance of pneumococcus was DNA, a conclusion published by Avery, Colin MacLeod, and Maclyn McCarty in 1944.

This was the first rigorously documented assertion that DNA can carry genetic information. Some of the excitement of that discovery and a glimpse into Avery's appreciation of its implications can be found in a letter that Avery wrote to his brother Roy in May 1943. Here is an excerpt from that letter:

> *For the past two years, first with MacLeod and now with Dr. McCarty, I have been trying to find out what is the chemical nature of the substance in the bacterial extract which induces this specific change. The crude*

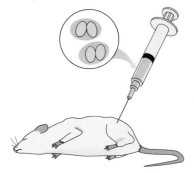  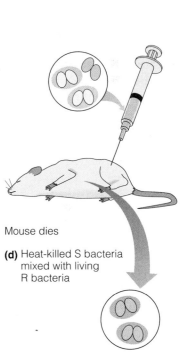

Mouse dies	Mouse remains healthy	Mouse remains healthy	Mouse dies
(a) Living S (smooth) bacteria	**(b)** Living R (rough) bacteria	**(c)** Heat-killed S bacteria	**(d)** Heat-killed S bacteria mixed with living R bacteria

(e) Living S bacteria in blood from dead mouse

Figure 16-2 Griffith's Experiment on Genetic Transformation in Pneumococcus. S (smooth) cells of the pneumococcus bacterium *(Streptococcus pneumoniae)* are pathogenic in mice; R (rough) cells are not. **(a)** Injection of living S bacteria into a mouse results in pneumonia and death. **(b)** Injection of living R bacteria leaves the mouse healthy. **(c)** Heat-killed S bacteria have no effect when injected alone. **(d)** When a mixture of living R bacteria and heat-killed S bacteria is injected, the result is pneumonia and death. **(e)** The finding that living pathogenic S-strain bacteria could be recovered from the blood of the mouse in part d suggested to Griffith that some chemical factor from the heat-killed S cells was able to cause a heritable change (transformation) of nonpathogenic R bacteria into pathogenic S bacteria. The chemical factor was later identified as DNA.

extract of Type III [the pathogenic, S strain of bacteria] is full of capsular polysaccharide, … carbohydrate, nucleoproteins, free nucleic acids of both the yeast [RNA] and thymus [DNA] type, lipids, and other cell constituents. Try to find in the complex mixtures the active principle! Try to isolate and chemically identify the particular substance that will by itself, when brought into contact with the R cell … cause it to elaborate Type III capsular polysaccharide and to acquire all the aristocratic distinctions of the same specific type of cells as that from which the extract was prepared! Some job, full of headaches and heartbreaks. But at last perhaps we have it.

*If we prove to be right—and of course that is a big if—then it means that both the chemical nature of the inducing stimulus is known and the chemical structure of the substance produced is also known, the former being thymus nucleic acid [DNA], the latter Type III polysaccharide, and both are thereafter reduplicated in the daughter cells …. Of course the problem bristles with implications. It touches the biochemistry of … [DNA molecules,] which are known to constitute the major part of chromosomes but have been thought to be alike regardless of origin and species…. But today it takes a lot of well documented evidence to convince anyone that the sodium salt of deoxyribose nucleic acid, protein free, could possibly be endowed with such biologically active and specific properties and that is the evidence we are now trying to get. It is lots of fun to blow bubbles but it is wiser to prick them yourself before someone else tries to.**

Though the experiments of Avery and his colleagues were rigorous, the assignment of a genetic role to DNA did not meet with immediate acceptance. Skepticism was due in part to the persistent, widespread conviction that DNA lacked the necessary complexity for such a role. In addition, many scientists questioned whether genetic information in bacteria had anything to do with heredity in other organisms. However, most remaining doubts were alleviated eight years later when DNA was also shown to be the genetic material of a virus, the bacteriophage T2.

Hershey and Chase Showed That DNA Is the Genetic Material of Viruses

Bacteriophages—or **phages**, for short—are viruses that infect bacteria. They have been objects of scientific study since the 1930s, and much of our early understanding of molecular genetics came from experiments involving these viruses. Box 16A describes the anatomy and replica-

tion cycle of some phages and highlights their advantages for genetic studies.

One of the most thoroughly studied of the phages that infect the bacterium *Escherichia coli* is bacteriophage T2. During infection, this virus attaches to the bacterial cell surface and injects material into the cell. Shortly thereafter, the bacterial cell begins to produce thousands of new copies of the virus. This scenario suggests that material injected into the bacterial cell carries the genetic information that guides the production of the virus. What is the chemical nature of the injected material? In 1952 Alfred Hershey and Martha Chase designed an experiment to address this question. Only two possibilities exist because the T2 virus is constructed from only two kinds of molecules: DNA and protein. To distinguish between these two alternatives, Hershey and Chase took advantage of the fact that the proteins of the T2 virus, like most proteins, contain the element sulfur (in the amino acids methionine and cysteine) but not phosphorus, while the viral DNA contains phosphorus (in its sugar-phosphate backbone) but not sulfur. Hershey and Chase therefore prepared two batches of T2 phage particles (as intact phages are called) with different kinds of radioactive labeling. In one batch, the phage proteins were labeled with the radioactive isotope ^{35}S; in the other batch, the phage DNA was labeled with the isotope ^{32}P.

By using radioactive isotopes in this way, Hershey and Chase were able to trace the fates of both protein and DNA during the infection process (Figure 16-3a). They began the experiment by mixing radioactive phage with intact bacterial cells and allowing the phage particles to attach to the bacterial cell surface and inject their genetic material into the cells. At this point, Hershey and Chase found that the empty protein coats (or phage "ghosts") could be effectively removed from the surface of the bacterial cells by agitating the suspension in an ordinary kitchen blender and recovering the bacterial cells by centrifugation. They then measured the radioactivity in the supernatant liquid and in the pellet of bacteria at the bottom of the tube.

The data revealed that most (65%) of the ^{32}P remained with the bacterial cells, while the bulk (80%) of the ^{35}S was released into the surrounding medium (Figure 16-3b). Since the ^{32}P labeled the viral DNA and the ^{35}S labeled the viral protein, Hershey and Chase concluded that DNA, not protein, had been injected into the bacterial cells and hence must function as the genetic material of phage T2. This conclusion received further support from the following observation: When the infected, radioactive bacteria were resuspended in fresh liquid and incubated longer, the ^{32}P was transferred to some of the offspring phage particles, but the ^{35}S was not.

As a result of the experiments we have described, by the early 1950s most biologists came to accept the view that genes are made of DNA, not protein. Unfortunately Oswald Avery, the visionary most responsible for the complete turnabout in views concerning the function of DNA, never received the credit he so richly deserved. The Nobel

*DNA as the Transforming Principle of Pneumococcus. Excerpt from a letter written by Oswald Avery to his brother Roy, capturing some of the excitement and implications of Avery's findings on the transforming principle in "Pneumococcus." Reproduced by R. D. Hotchkiss in Phage and the Origins of Molecular Biology, J. Cairns, G. S. Stent, and J. D. Watson, eds., Cold Spring Harbor, New York: Cold Spring Harbor Laboratory, 1966. Reprinted by permission.

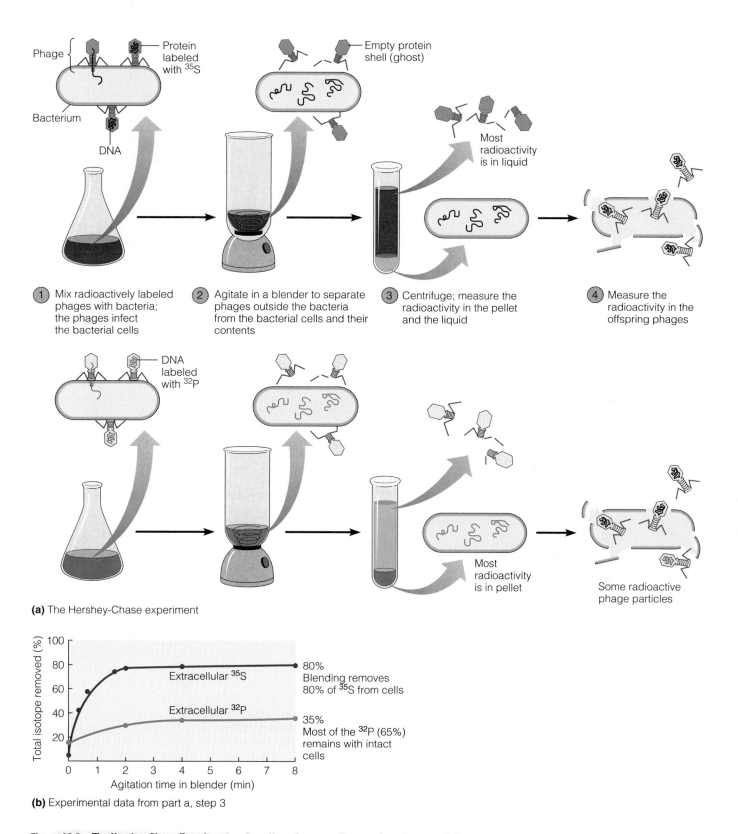

1. Mix radioactively labeled phages with bacteria; the phages infect the bacterial cells

2. Agitate in a blender to separate phages outside the bacteria from the bacterial cells and their contents

3. Centrifuge; measure the radioactivity in the pellet and the liquid

4. Measure the radioactivity in the offspring phages

Phage

Protein labeled with ^{35}S

Bacterium

DNA

Empty protein shell (ghost)

Most radioactivity is in liquid

DNA labeled with ^{32}P

Most radioactivity is in pellet

Some radioactive phage particles

(a) The Hershey-Chase experiment

Total isotope removed (%)

Extracellular ^{35}S

Extracellular ^{32}P

80% Blending removes 80% of ^{35}S from cells

35% Most of the ^{32}P (65%) remains with intact cells

Agitation time in blender (min)

(b) Experimental data from part a, step 3

Figure 16-3 The Hershey-Chase Experiment: DNA as the Genetic Material of Phage T2.
(a) ① T2 labeled with either ^{35}S (to label protein) or ^{32}P (to label DNA) is used to infect bacteria. The phages adsorb to the cell surface and inject their DNA. ② Agitation of the infected cells in a blender dislodges most of the ^{35}S from the cells, whereas most of the ^{32}P remains. ③ Centrifugation causes the cells to form a pellet; any free phage particles, including ghosts, remain in the supernatant liquid. ④ When the cells in each pellet are incubated further, the phage DNA within them dictates the synthesis and eventual release of new phage particles. Some of these phages contain ^{32}P in their DNA (because the old, labeled phage DNA is packaged into some of the new particles), but none contain ^{35}S in their coat proteins. **(b)** The graph shows the extent to which ^{35}S and ^{32}P are removed from the intact cells at step ③, as a function of time in the blender. A few minutes of blending is enough to remove most (80%) of the ^{35}S, while leaving most (65%) of the ^{32}P with the cells.

Further Insights PHAGES: MODEL SYSTEMS FOR STUDYING GENES

From its inception in the mid-nineteenth century, genetics has drawn upon a wide variety of organisms for its experimental materials. Initially, attention focused on plants and animals, such as Mendel's peas and the fruit flies popularized by later investigators. Around 1940, however, bacteria and viruses came into their own, providing biologists with experimental systems that literally revolutionized the science of genetics by bringing it to the molecular level.

Bacteriophages have been especially important. Bacteriophages, or phages for short, are viruses that infect bacterial cells. It is easy to obtain huge numbers of phage particles in a brief period of time, which greatly facilitates screening for mutants—phages with heritable variations—thereby making it possible for geneticists to identify particular genes. Some of the most thoroughly studied phages are the T2, T4, and T6 (the so-called T-even) bacteriophages, which infect the bacterium *E. coli*. The three T-even phages have similar structures, which are quite elaborate. T4 is shown in Figure 16A-1. The *head* of the phage is a protein capsule that is shaped like a hollow icosahedron (a 20-sided object) and filled with DNA. The head is attached to a protein *tail*, which consists of a hollow *tail core* surrounded by a contractile *tail sheath* and terminating in a hexagonal *baseplate*, to which six *tail fibers* are attached.

Figure 16A-2 depicts the main events in the replication cycle of the T4 phage. The drawings are not to scale; the bacterium is proportionately larger, as the electron micrograph indicates. The process begins with the adsorption of a phage particle to the wall of a bacterial cell. When the phage collides with the cell, it "squats" so that its baseplate attaches to a specific receptor protein in the wall (Figure 16A-2a, ①). Next, the tail sheath contracts, driving the hollow tail core through the cell wall. The core forms a needle through which the bacteriophage DNA is injected into the bacterium (②). Once this DNA has gained entry to the bacterial cell, the genetic information of the phage is transcribed and translated (③). This gives rise to a few key proteins that subvert the metabolic machinery of the host cell for the phage's benefit, which is usually its own rapid multiplication. Since the phage consists simply of a DNA molecule surrounded by a protein coat (its *capsid*), most of the metabolic activity in the infected cell is channeled toward the replication of phage DNA and the synthesis of capsid proteins. The phage DNA and capsid proteins then self-assemble into hundreds of new phage particles (④). Within about half an hour, the infected cell lyses (breaks open), releasing the new phage particles into the medium (⑤). Each new phage can now infect another bacterial cell, making it possible to obtain enormous populations of phage—as many as 10^{11} phage particles per milliliter in infected bacterial cultures.

To determine the number of phage particles in a sample, a measured volume is mixed with bacterial cells growing in liquid medium to allow adsorption of the phages to the bacteria. The mixture is then spread onto solid (agar-containing) nutrient

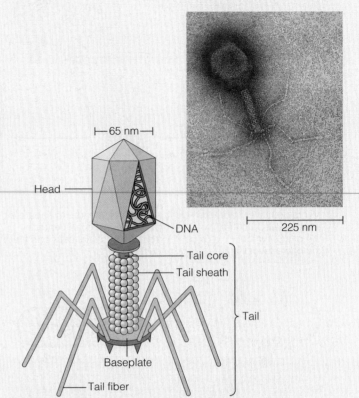

Figure 16A-1 The Structure of Bacteriophage T4. The drawing identifies the main structural components of this phage, not all of which are visible in the micrograph (TEM).

medium in a Petri dish. Upon incubation, the bacteria multiply to produce a dense "lawn" of cells on the surface of the nutrient medium. But wherever a virus particle has infected a bacterial cell, a clear spot appears in the lawn because the bacterial cells there have been killed by the multiplying phage population. Such clear spots are called *plaques*. The number of plaques appearing in the bacterial lawn represents the number of phage particles in the original phage-bacterium mixture, provided only that the initial number of phages was small enough to ensure that each gives rise to a separate plaque. Figure 16A-3 shows plaques formed by T4 bacteriophage on a lawn of *E. coli* cells.

The course of events shown in Figure 16A-2a is called *lytic growth* and is characteristic of a *virulent phage*. Lytic growth results in lysis of the host cell and the production of many progeny phage particles. In contrast, a *temperate phage* can either produce lytic growth, as a virulent phage does, or integrate its DNA into the bacterial chromosome without causing any immediate harm to the host cell. An especially well-studied example of a temperate

Prize Committee discussed Avery's work but decided he had not done enough. Perhaps Avery's modest and unassuming nature was responsible for this lack of recognition. After Avery died in 1955, the biochemist Erwin Chargaff

wrote in tribute: "He was a quiet man; and it would have honored the world more, had it honored him more."

Why did the Hershey-Chase experiments receive a warmer welcome than Avery's earlier work on bacterial

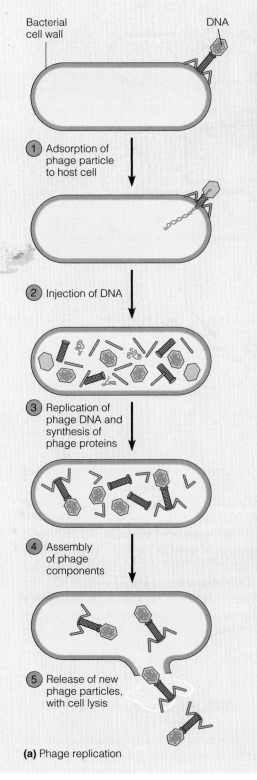

Bacterial cell wall

DNA

① Adsorption of phage particle to host cell

② Injection of DNA

③ Replication of phage DNA and synthesis of phage proteins

④ Assembly of phage components

⑤ Release of new phage particles, with cell lysis

(a) Phage replication

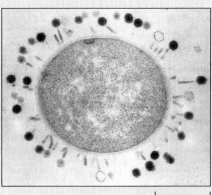

1 µm

(b) Bacterium with phage particles attached

Figure 16A-2   Replication of a T-Even Phage.   (a) The replication cycle of a T-even phage begins when a phage particle ① becomes adsorbed to the surface of a bacterial cell and ② injects its DNA into the cell. ③ The phage DNA replicates in the host cell and codes for the production of phage proteins. ④ These components assemble into new phage particles. ⑤ Eventually, the host cell lyses, releasing offspring phage particles that can infect additional bacteria. This replication process is typical of the lytic growth of many phages. **(b)** Electron micrograph of a bacterium with phage particles attached to its surface (TEM).

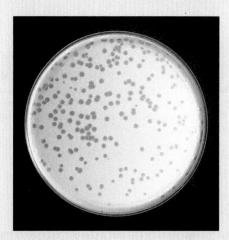

Figure 16A-3   Phage Plaques on a Lawn of Bacteria.   Phage plaques have formed on a lawn of *E. coli* infected with phage T4. Each plaque arises from the reproduction of a single phage particle in the original mixture.

phage is bacteriophage λ (*lambda*), which, like the T-even phages, infects *E. coli* cells. In the integrated or *lysogenic state,* the DNA of the temperate phage is called a *prophage.* The prophage is replicated along with the bacterial DNA, often through many generations of host cells (Figure 16A-4). During this time, the phage genes, though potentially lethal to the host, are inactive, or *repressed.* Under certain conditions, however, the prophage DNA is excised from the bacterial chromosome and again enters a lytic cycle, producing progeny phage particles and lysing the host cell.

(continued)

transformation, even though both led to the same conclusion? The main reason seems to have been simply the passage of time and the accumulation of additional, circumstantial evidence after Avery's 1944 publication.

Perhaps most important was evidence that DNA is indeed variable enough in structure to serve as the genetic material. This evidence came from studies of DNA base composition, as we see next.

One reason bacteriophages are so attractive to geneticists is that the small size of their genomes (which may be either DNA or RNA) makes it relatively easy to identify and study their genes. The genome of bacteriophage λ, for example, is a single DNA molecule containing fewer than 60 genes, compared with several thousand genes in a bacterium such as *E. coli*. Other phages are still smaller. For example, a phage called ϕX174 (ϕ is the Greek letter *phi*) has only 9 genes. This phage is noteworthy because its DNA genome is single-stranded, rather than double-stranded as in all cells and many other viruses. The ϕX174 genome is a circular DNA molecule 5375 nucleotides long and was the first complete genome to be sequenced, a feat accomplished in 1977 by Frederick Sanger and his colleagues.

Because of their simple genomes, their rapidity of multiplication, and the enormous numbers of progeny that can be produced in a small volume of culture medium, bacteriophages are among the best understood of all "organisms." They have proven exceedingly useful as model systems in our continuing quest to understand the much more complex genomes of true organisms.

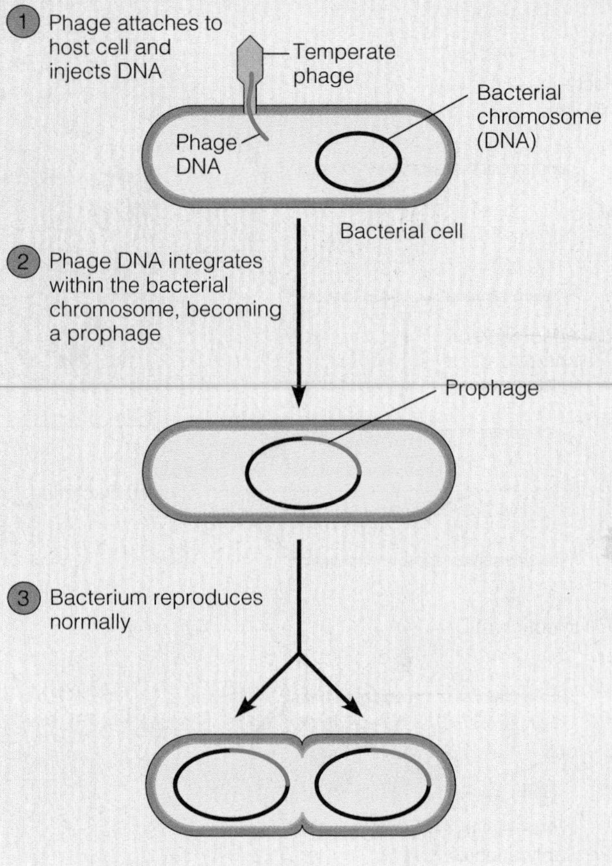

Figure 16A-4 Propagation of a Prophage Within a Bacterial Chromosome.   The DNA injected by a temperate phage can become integrated into the DNA of the bacterial chromosome. The integrated phage DNA, called a prophage, is replicated along with the bacterial DNA each time the bacterium reproduces.

Chargaff's Rules Reveal That A = T and G = C

Despite the lukewarm reaction initially received by Avery's work, it was an important influence on several other scientists. Among them was Erwin Chargaff, who was interested in the base composition of DNA. Between 1944 and 1952, Chargaff used chromatographic methods to separate and quantify the amounts of the four bases—adenine (A), guanine (G), cytosine (C), and thymine (T)—found in DNA. From his analyses came several important discoveries. First, he showed that DNA isolated from different cells of a given species have the same percentage of each of the four bases (Table 16-1, lines 1–4), and that this percentage does not vary with individual, tissue, age, nutritional state, or environment. This is exactly what would be expected of the chemical substance that stores genetic information, because the cells of a given species would be expected to have similar genetic information. However, Chargaff did find that DNA base composition varies from species to species. This can be seen by examining the last column of Table 16-1, which shows the relative amounts of the bases A and T versus G and C in the DNAs of various organisms. Comparison of such data revealed to Chargaff that DNA preparations from closely related species have similar base compositions, whereas those from very different species tend to exhibit quite different

base compositions. Again, this is what would be expected of a molecule that stores genetic information.

But Chargaff's most striking observation was his discovery that for all DNA samples examined, the number of adenines is equal to the number of thymines (A = T), and the number of guanines is equal to the number of cytosines (G = C). This meant that the number of purines is equal to the number of pyrimidines (A + G = C + T). The significance of these equivalencies, known as **Chargaff's rules,** was an enigma and remained so until the double-helical model of DNA was established by Watson and Crick in 1953.

DNA Structure

As the scientific community gradually came to accept the conclusion that DNA stores genetic information, a new set of questions began to emerge concerning the way in which DNA performs its genetic function. One of the first issues to arise was the question of how DNA is accurately replicated so that duplicate copies of the genetic information can be passed on from cell to cell during cell division, and from parent to offspring. Answering this question required an understanding of the three-dimensional

Table 16-1 Base Composition of DNA from Various Sources

Source of DNA	Number of Each Type of Nucleotide*				Nucleotide Ratios**		
	A	T	G	C	A/T	G/C	(A+T)/(G+C)
Bovine thymus	28.4	28.4	21.1	22.1	1.00	0.95	1.31
Bovine liver	28.1	28.4	22.5	21.0	0.99	1.07	1.30
Bovine kidney	28.3	28.2	22.6	20.9	1.00	1.08	1.30
Bovine brain	28.0	28.1	22.3	21.6	1.00	1.03	1.28
Human liver	30.3	30.3	19.5	19.9	1.00	0.98	1.53
Locust	29.3	29.3	20.5	20.7	1.00	1.00	1.41
Sea urchin	32.8	32.1	17.7	17.3	1.02	1.02	1.85
Wheat germ	27.3	27.1	22.7	22.8	1.01	1.00	1.19
Marine crab	47.3	47.3	2.7	2.7	1.00	1.00	7.50
Aspergillus (mold)	25.0	24.9	25.1	25.0	1.00	1.00	1.00
Saccharomyces cerevisiae (yeast)	31.3	32.9	18.7	17.1	0.95	1.09	1.79
Clostridium (bacterium)	36.9	36.3	14.0	12.8	1.02	1.09	2.73

*The values in these four columns are the average number of each type of nucleotide found per 100 nucleotides in DNA.

**The A/T and G/C ratios are not all exactly 1.00 because of experimental error.

structure of DNA, which was provided in 1953 when Watson and Crick formulated their double-helical model of DNA. We described the structure of the double helix in Chapter 3 and its discovery in Box 3A, but we return to it now for review and some further details.

Watson and Crick Discovered That DNA Is a Double Helix

In 1952, James Watson and Francis Crick were among a small number of scientists who were convinced that DNA was the genetic material and that knowledge of its three-dimensional structure would provide valuable clues as to how it functioned. Working at Cambridge University in England, they approached the puzzle by building wire models of possible structures. DNA had been known for years to be a long polymer with a backbone of repeating deoxyribose and phosphate units, and a nitrogenous base attached to each sugar. Watson and Crick were aided in their model building by knowing that the particular forms in which the bases A, G, C, and T exist at physiological pH permit specific hydrogen bonds to form between pairs of them. The crucial experimental evidence, however, came from an X-ray diffraction pattern of DNA determined by Rosalind Franklin, working in the laboratory of Maurice Wilkins at King's College in London. Franklin's picture told Watson and Crick that DNA was a helical structure composed of two strands—a **double helix**. Watson and Crick then put this information together with what they already knew to arrive at the model shown in Figure 3A-1 (p. 61). In their model, the sugar-phosphate backbones are on the outside of the helix, and the bases face inward toward the center of the helix, forming the "steps" of the "circular staircase" that the structure resembles.

Figure 16-4a illustrates several structural features of the Watson-Crick double helix. The helix is right-handed,

meaning that it curves "upward" to the right (notice that this is true even if you turn the diagram upside down). It contains ten nucleotide pairs per turn and advances 0.34 nm per nucleotide pair. Consequently, each complete turn of the helix adds 3.4 nm to the length of the molecule. The diameter of the helix is 2 nm. This distance turns out to be too small for two purines and too great for two pyrimidines, but it accommodates a purine and a pyrimidine well, consistent with Chargaff's rules. Pyrimidine-purine pairing, in other words, was necessitated by steric considerations. The two strands are held together by hydrogen bonding between the bases in opposite strands. Moreover, the hydrogen bonds holding together the two strands of the double helix *only fit when they form between the base adenine (A) in one chain and thymine (T) in the other, or between the base guanine (G) in one chain and cytosine (C) in the other*. This means that the base sequence of one chain determines the base sequence of the opposing chain; the two chains of the DNA double helix are therefore said to be **complementary** to each other. Such a model explains why Chargaff had observed that DNA molecules contain equal amounts of the bases A and T and equal amounts of the bases G and C.

The most profound implication of the Watson-Crick model was that it suggested a mechanism by which cells can replicate their genetic information: The two strands of the DNA double helix could simply separate from each other prior to cell division so that each strand could function as a *template* dictating the synthesis of a new complementary DNA strand using the base-pairing rules. In other words, the base A in the template strand would specify insertion of the base T in the newly forming strand, the base G would specify insertion of the base C, the base T would specify insertion of the base A, and the base C would specify insertion of the base G. In the next chapter, we will discuss the experimental evidence for this

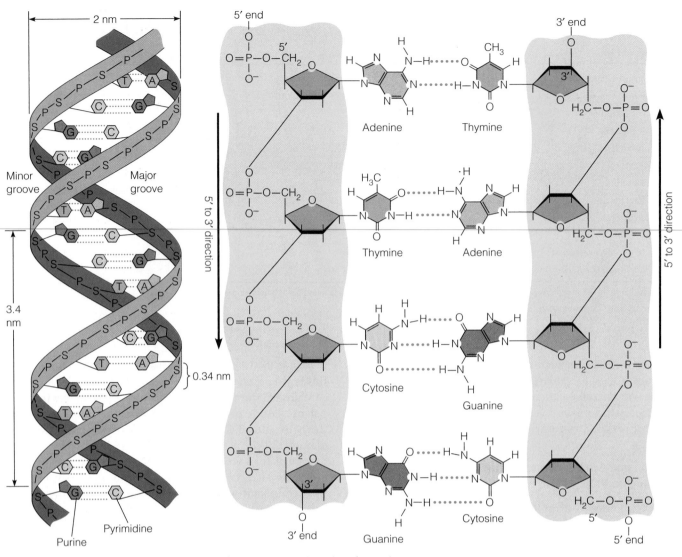

(a) Double helix

(b) Antiparallel orientation of strands

Figure 16-4 The DNA Double Helix.
(a) This schematic illustration shows the sugar-phosphate chains of the DNA backbone, the complementary base pairs, the major and minor grooves, and several important dimensions. A = adenine, G = guanine, C = cytosine, T = thymine, P = phosphate, and S = sugar (deoxyribose). **(b)** One of the strands of a DNA molecule is oriented 5′ ⟶ 3′ in one direction, whereas its complement has a 5′ ⟶ 3′ orientation in the opposite direction. This diagram also shows the hydrogen bonds that connect the bases in AT and GC pairs.

proposed mechanism and describe the molecular basis of DNA replication in detail.

Several other important features of the DNA double helix are illustrated in Figure 16-4. For example, the way in which the two strands are twisted around each other creates a *major groove* and a *minor groove*. These grooves play significant roles in the interactions of a variety of molecules with DNA. Another important feature is the *antiparallel* orientation of the two DNA strands, which is illustrated in Figure 16-4b. This diagram shows that as you move along one of the strands in a given direction, successive nucleotides are linked together by phosphodiester bonds that join the 5′ carbon of one nucleotide to the 3′ carbon of the next nucleotide; such a chain is said to have a *5′ ⟶ 3′ orientation*. But if you move along the opposing strand in the same direction, the order of the bonds exhibits a 3′ ⟶ 5′ orientation. In other words, the phosphodiester bonds of the two strands are oriented in *opposite* directions. The differing orientations of the two strands is a feature that has important implications for both DNA replication and DNA transcription, as we will see in Chapters 17 and 19.

The right-handed Watson-Crick helix is an idealized version of what is called *B-DNA* (Figure 16-5a). But naturally occurring B-DNA double helices are flexible molecules that often deviate from this ideal, with exact shapes and dimensions that depend on the local nucleotide sequence. Furthermore, although B-DNA is undoubtedly the main form of DNA in cells, other forms may also exist, perhaps in short segments interspersed in molecules that

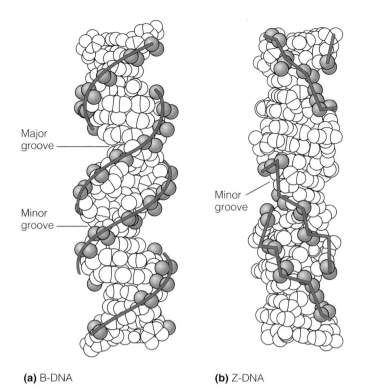

Major groove

Minor groove

Minor groove

(a) B-DNA   **(b)** Z-DNA

Figure 16-5   **Alternative Forms of DNA.**   **(a)** In the normal B-form of DNA, the sugar-phosphate backbone forms a smooth right-handed double helix. **(b)** In Z-DNA, the backbone forms a zigzag left-handed helix. Color is used to highlight the DNA backbone.

are mostly B-DNA. The most important of these alternative forms are A-DNA and Z-DNA. *A-DNA* has a right-handed helical configuration that is shorter and thicker than B-DNA. A-DNA can be created artificially by dehydrating B-DNA, but it is not known to exist under normal cellular conditions. As Figure 16-5b shows, *Z-DNA* is a *left-handed* double helix. Its name derives from the zigzag pattern of its sugar-phosphate backbone, and it is longer and thinner than B-DNA. The Z form arises most readily in DNA regions that contain either alternating purines and pyrimidines, or cytosines with extra methyl groups (which do occur in chromosomal DNA; see Chapter 21). But despite great excitement about Z-DNA immediately following its discovery in 1979, the amount of Z-DNA in cells and its biological significance, if any, are still unknown.

DNA Can Be Interconverted Between Relaxed and Supercoiled Forms

In many situations, the DNA double helix can be twisted upon itself to form **supercoiled DNA.** Although now known to be a widespread property of DNA, supercoiling was first identified in the DNA of certain small viruses containing circular DNA molecules that exist as closed loops. Circular DNA molecules are also found in bacteria, mitochondria, and chloroplasts. Although supercoiling is not restricted to circular DNA, it is easiest to study in such molecules.

A DNA molecule can go back and forth between the supercoiled state and the nonsupercoiled, or *relaxed,* state.

To understand the basic idea, you might perform the following exercise. Start with a length of rope consisting of two strands twisted together into a right-handed coil; this is the equivalent of a relaxed, linear DNA molecule. Just joining the ends of the rope together changes nothing; the rope is now circular but still in a relaxed state. But if before sealing the ends you first give the rope an extra right-handed twist (i.e., another twist in the direction in which the strands are already entwined around each other), the rope becomes overwound and is thrown into what is called a *positive supercoil.* Conversely, if the rope is given a left-handed twist before sealing (i.e., twisted in the direction opposite to that in which it is wound), it becomes underwound and is thrown into a *negative supercoil.* In the same manner, a relaxed DNA molecule can be converted to a positive supercoil by a right-handed twist and to a negative supercoil by a left-handed twist (Figure 16-6a). Circular DNA molecules found in nature, including those of bacteria, viruses, and eukaryotic organelles, are invariably negatively supercoiled. (Figure 16-6b shows a relaxed circular DNA molecule from a phage and a similar molecule with negative supercoils.)

Supercoiling also occurs in linear DNA molecules when regions of the molecule are anchored to some cell structure and so cannot freely rotate. At any given time, significant portions of the linear DNA in the nucleus of eukaryotic cells may be supercoiled, and when DNA is packaged into chromosomes at the time of cell division, extensive supercoiling helps to make the DNA more compact.

By influencing both the spatial organization and the energy state of DNA, supercoiling affects the ability of a DNA molecule to interact with other molecules. Positive supercoiling favors tighter winding of the double helix and therefore reduces opportunities for interaction. In contrast, negative supercoiling tends to unwind the double helix, thereby increasing access of its strands to proteins involved in DNA replication or transcription.

The interconversion between relaxed and supercoiled forms of DNA is catalyzed by enzymes known as **topoisomerases.** Most topoisomerases are classified as either *type I* or *type II.* Both catalyze the stepwise relaxation of supercoiled DNA, but type I enzymes relax one coil at a time, whereas type II enzymes relax two coils at a time. Topoisomerases function by attaching to a supercoiled duplex and producing a transient break, or *nick,* in either one strand (type I) or both strands (type II). The two strands can then rotate around each other, after which the nicks are resealed. This relaxation reaction requires no addition of energy, suggesting that the phosphodiester bond energy that is liberated when a strand is nicked must somehow be conserved and used to form the new bond.

DNA gyrase is a type II topoisomerase that can induce as well as relax supercoiling. As we will learn in Chapter 17, DNA gyrase is one of several enzymes involved in DNA replication. It can relax the positive supercoiling that results from partial unwinding of a double helix, or it can actively introduce negative supercoils that promote strand separation, thereby facilitating access

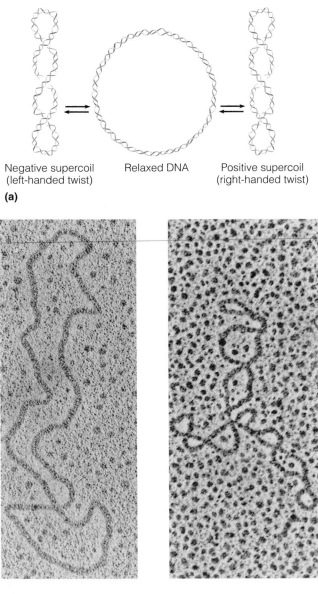

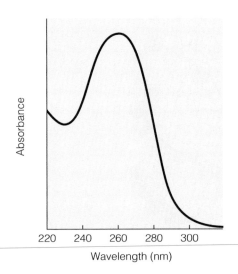

Figure 16-7   Ultraviolet Absorption Spectrum of DNA.   Nucleic acids absorb ultraviolet light with an absorption maximum at a wavelength of about 260 nm because of the absorption by the pyrimidine and purine bases.

(a)

Negative supercoil (left-handed twist)   Relaxed DNA   Positive supercoil (right-handed twist)

(b)

Figure 16-6   Interconversion of Relaxed and Supercoiled DNA.
(a) A schematic diagram of a relaxed circular DNA molecule and its conversion to supercoiled forms with a left-handed twist (negative supercoiling) or a right-handed twist (positive supercoiling).
(b) Electron micrographs of circular molecules of DNA from a bacteriophage called PM2, with a relaxed molecule on the left and a molecule with negative supercoils on the right (TEMs).

of other proteins involved in DNA replication. DNA gyrase requires ATP to generate supercoiling, but not to relax an already supercoiled molecule.

The Two Strands of a DNA Double Helix Can Be Separated by Denaturation and Rejoined by Renaturation

Because the two strands of the DNA double helix are bound together by relatively weak, noncovalent bonds, the two strands can be readily separated from each other under appropriate conditions. As we will see in coming chapters,

strand separation is an integral part of both DNA replication and RNA synthesis. Strand separation can also be induced experimentally, resulting in **DNA denaturation;** the reverse process, which reestablishes a double helix from separated DNA strands, is called **DNA renaturation.**

One way to denature DNA in the laboratory is to raise its temperature. If this is done slowly, the DNA retains its double-stranded, or native, state until a critical temperature is reached, at which point the duplex rapidly denatures, or "melts," into its component strands. The melting process is easy to monitor because double-stranded and single-stranded DNA differ in their light-absorbing properties. All DNA absorbs ultraviolet light, with an absorption maximum around 260 nm (Figure 16-7). When the temperature of a DNA solution is slowly raised, the absorbance at 260 nm remains constant until the double helix begins to melt into its component strands. As the strands separate, the absorbance of the solution increases rapidly because of the higher intrinsic absorption of single-stranded DNA (Figure 16-8).

The temperature at which one-half of the absorbance change has been achieved is called the **DNA melting temperature (T_m).** The value of the melting temperature reflects how tightly the DNA double helix is held together. For example, GC base pairs, held together by three hydrogen bonds, are more resistant to separation than AT base pairs, which have only two (see Figure 16-4b). The melting temperature therefore increases in direct proportion to the relative number of GC base pairs in the DNA (Figure 16-9). Likewise, DNA molecules in which the two strands of the double helix are properly base-paired at each position will melt at higher temperatures than DNA in which the two strands are not perfectly complementary.

Denatured DNA can be renatured by lowering the temperature to permit hydrogen bonds between the two

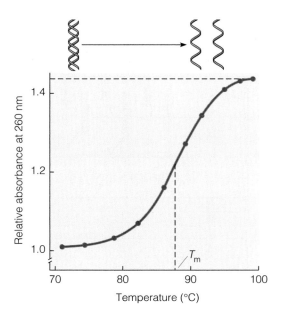

Figure 16-8 **A Thermal Denaturation Profile for DNA.** As the temperature of a solution of double-stranded (native) DNA is raised, a point is reached at which the heat energy causes the DNA to denature rapidly. The conversion to single strands is accompanied by a characteristic increase in the absorbance of light at 260 nm. The temperature at which the midpoint of this increase occurs is called the melting temperature, T_m. For the sample shown, the value of T_m is about 87°C.

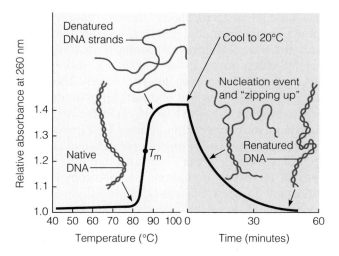

Figure 16-10 **DNA Denaturation and Renaturation.** If a solution of native (double-stranded) DNA is heated slowly under carefully controlled conditions, the DNA "melts" over a narrow temperature range, with an increase in absorbance at 260 nm. When the solution is allowed to cool, the separated DNA strands reassociate with kinetics that depend on the initial concentration. Complementary strands collide randomly in the nucleation event, followed by a rapid "zipping up" of adjacent nucleotide pairs. The reassociation requires varying amounts of time, depending on both the DNA concentration in the solution and the length of the DNA strands.

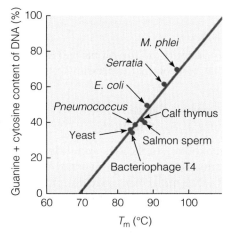

Figure 16-9 **The Dependence of Melting Temperature on DNA Base Composition.** The melting temperature of DNA increases linearly with its G + C content, as illustrated by the relationship between T_m and G + C content for DNA samples from a variety of organisms.

strands to reestablish (Figure 16-10). The ability to renature nucleic acids has a variety of important scientific applications. Most importantly, it forms the basis for **nucleic acid hybridization,** a family of procedures for identifying nucleic acids based on the ability of single-stranded chains with complementary base sequences to bind, or *hybridize,* to each other. Nucleic acid hybridization can be applied to DNA-DNA, DNA-RNA, and even RNA-RNA interactions. In DNA-DNA hybridization, for example, the DNA being examined is denatured and then

incubated with a purified, single-stranded radioactive DNA fragment, called a **probe,** whose sequence is complementary to the base sequence one is trying to detect.

Nucleic acid sequences do not need to be perfectly complementary to be able to hybridize. Changing the temperature, salt concentration, and pH used during hybridization can permit pairing to take place between sequences exhibiting numerous mismatched bases. Under such conditions, it is possible to detect DNA sequences that are related to one another but not identical. This approach is useful for identifying families of related genes, both within a given type of organism and among different kinds of organisms.

The Organization of DNA in Genomes

So far, we have considered several chemical and physical properties of DNA. But as cell biologists, we are primarily interested in its importance to the cell. We therefore want to know how much DNA cells have, how and where they store it, and how they utilize the genetic information it contains. We begin by inquiring about the amount of DNA present, because that determines the maximum amount of information that a cell can possibly possess.

The **genome** of an organism or virus consists of the DNA (or for some viruses, RNA) that contains one complete copy of all the genetic information of that organism or virus. For many viruses and prokaryotes, the genome resides in a single linear or circular DNA molecule, or a small number of them. Eukaryotic cells have a nuclear genome, a mitochondrial genome, and, in the case of plants and algae, a chloroplast genome as well. Mitochondrial and chloroplast genomes are single, usually circular DNA molecules, resembling those of bacteria. The nuclear genome generally consists of multiple DNA molecules dispersed among a haploid set of chromosomes. (As we will explore in more detail in Chapter 18, a *haploid* set of chromosomes consists of one representative of each type of chromosome, whereas a *diploid* set consists of two copies of each type of chromosome, one copy from the mother and one from the father. Sperm and egg cells each have a haploid set of chromosomes, whereas most other types of eukaryotic cells are diploid.)

Genome Size Generally Increases with an Organism's Complexity

Genome size is usually expressed as the total number of base-paired nucleotides, or **base pairs (bp).** For example, the circular DNA molecule that constitutes the genome of an *E. coli* cell has 4,639,221 bp. Since such numbers tend to be rather large, the abbreviations **Kb** (kilobases), **Mb** (megabases), and **Gb** (gigabases) are used to refer to a thousand, or million, or billion base pairs, respectively. Thus, the size of the *E. coli* genome can be expressed simply as 4.6 Mb. The range of genome sizes observed for various groups of organisms is summarized in Figure 16-11. These data reveal a spread of almost eight orders of magnitude in genome size, from a few thousand base pairs for the simplest viruses to more than 100 billion base pairs for certain plants and amphibians. In terms of the total length of DNA, this corresponds to a range of less than 2 μm of DNA for a small virus, such as SV40, to roughly 34 meters of DNA (more than 100 feet!) for certain plants, such as the wildflower *Trillium*.

Broadly speaking, genome size increases with the complexity of the organism. Viruses contain enough nucleic acid to code for only a few or a few dozen proteins, bacteria can specify a few thousand proteins, and eukaryotic cells have enough DNA (at least in theory) to encode hundreds of thousands of proteins. But closer examination of such data reveals some puzzling features. Most notably, the genome sizes of eukaryotes exhibit great variations that do not appear to correlate with any known differences in organismal complexity. Some amphibians and plants, for example, have gigantic genomes that are tens or even hundreds of times larger than those of other amphibians or plants, or of mammalian species. *Trillium*, for example, is a member of the lily family that has no obvious need for exceptional amounts of genetic information. Yet its genome size is more than 20 times that of pea plants and 30 times that of humans. We have no idea what *Trillium* does with all that DNA. Its presence highlights the fact that most eukaryotic genomes carry large amounts of DNA of no known function, a phenomenon we will discuss shortly. In the final analysis, genome size is less important than the

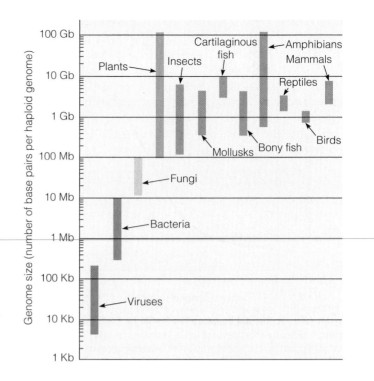

Figure 16-11   The Relationship Between Genome Size and Type of Organism.   For each group of organisms shown, the bar represents the approximate range in genome size measured as the number of nucleotide pairs per haploid genome. The same color (blue) is used for the various groups that involve members of the animal kingdom.

number and identity of functional genes and the DNA sequences that control their expression.

Restriction Enzymes Cleave DNA Molecules at Specific Sites

Since the hereditary similarities and differences observed among organisms derive from their DNA, we can expect the study of DNA molecules to yield important biological insights. Clues to a myriad of mysteries—from the control of gene expression within a cell to the evolution of new species—are to be found in the nucleotide sequences of genomic DNA. Most DNA molecules, however, are far too large to be studied intact. In fact, until the early 1970s, DNA was the most difficult biological molecule to analyze biochemically. Eukaryotic DNA seemed especially intimidating, given the size of most eukaryotic genomes, and no method was known for cutting DNA at specific sites to yield reproducible fragments. The prospect of ever being able to identify, isolate, sequence, or manipulate specific eukaryotic genes seemed unlikely. Yet in less than a decade, DNA became one of the easiest biological molecules to work with.

This breakthrough was made possible by the discovery of **restriction enzymes**, which are enzymes isolated from bacteria that cut foreign DNA molecules at specific sites. (Box 16B describes the biological role of restriction enzymes and some details about the sites they cut; here we focus on their use as analytical tools.) The cutting action of a restriction enzyme generates a specific set of DNA pieces called *restriction fragments*. Each restriction enzyme cleaves double-stranded DNA only in places where it encounters a specific recognition sequence, called a **restriction site,** that is usually four or six (but may be eight or more) nucleotides long. For example, here is the restriction site recognized by the widely used *E. coli* restriction enzyme called *Eco*RI:

$$5' \text{ G} \overset{\downarrow}{-}\text{A}-\text{A}-\text{T}-\text{T}-\text{C } 3'$$
$$3' \text{ C}-\text{T}-\text{T}-\text{A}-\text{A}\underset{\uparrow}{-}\text{G } 5'$$

The arrows indicate where *Eco*RI cuts the DNA. This restriction enzyme, like many others, makes a *staggered* cut in the double-stranded DNA molecule, as you can see in Figure 16B-1b (Box 16B).

The frequency with which restriction sites occur in DNA is such that a given restriction enzyme will typically cleave DNA into fragments ranging from a few hundred to a few thousand base pairs in length. Fragments of these sizes are far more amenable to further manipulation than the enormously long DNA molecules from which they are generated.

Separation of Restriction Fragments by Gel Electrophoresis.

Incubating a DNA sample with a specific restriction enzyme yields a collection of restriction fragments of different sizes. To determine the number and lengths of such fragments and to isolate individual fragments for further study, one must be able to separate the fragments from each other. The technique of choice for this purpose is **gel electrophoresis,** essentially the same method used for the separation of proteins and polypeptides (see Figure 7-23). In fact, the procedure for DNA is even simpler than for proteins, because DNA molecules have an inherent negative charge (due to their phosphate groups) and therefore do not need to be treated with a negatively charged detergent to make them move toward the anode. Small DNA fragments are usually separated in *polyacrylamide* gels, whereas larger DNA fragments are separated in more porous gels made of *agarose*, a polysaccharide.

Figure 16-12 illustrates the separation of restriction fragments of different sizes by gel electrophoresis. DNA samples are first incubated with the desired restriction enzyme; in the figure, three different restriction enzymes are used. The samples are then placed in separate compartments at one end of the gel. Next an electrical potential is applied across the gel, with the anode situated at the opposite end of the gel from the samples. Because of the negative charge of their phosphate groups, DNA fragments migrate toward the anode. Smaller fragments (i.e., those with lower molecular weight) move through the gel with relative ease and therefore migrate rapidly, while larger fragments move more slowly. The current is left on until the fragments are well spaced out on the gel. The final result is a series of DNA fragments that have been separated based on differences in size.

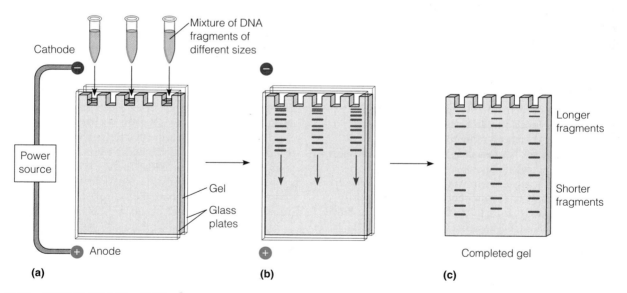

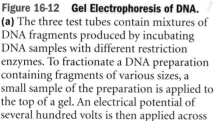

Figure 16-12 Gel Electrophoresis of DNA.
(a) The three test tubes contain mixtures of DNA fragments produced by incubating DNA samples with different restriction enzymes. To fractionate a DNA preparation containing fragments of various sizes, a small sample of the preparation is applied to the top of a gel. An electrical potential of several hundred volts is then applied across the gel, such that the anode (the positive electrode) is at the bottom of the gel and the cathode (the negative electrode) is at the top. **(b)** The DNA fragments in the applied sample migrate toward the anode at a rate that is inversely related to their size; the smaller fragments migrate more rapidly down the gel than do the larger ones. **(c)** After enough time is allowed for separating fragments, the gel is removed and stained with a dye such as ethidium bromide, which binds to the DNA fragments and causes them to fluoresce under ultraviolet light. Alternatively, autoradiography can be used to locate the DNA bands in the gel, provided that the DNA is radioactively labeled.

Further Insights   A CLOSER LOOK AT RESTRICTION ENZYMES

Restriction enzymes are a type of endonuclease (an enzyme that cuts DNA internally) found in most bacteria. These enzymes help bacteria protect themselves against invasion by foreign DNA molecules, particularly the DNA of bacterio- phages. In fact, the name "restriction" enzyme came from the discovery that these enzymes *restrict* the ability of foreign DNA to take over the transcription and translation machinery of the bacterial cell.

To protect its own DNA from being degraded, the bacterial cell has enzymes that add methyl groups ($-CH_3$) to specific nucleotides that its own restriction enzymes would otherwise recognize. Once methylated, the nucleotides are no longer recognized by the restriction enzymes, and so the bacterial DNA is not attacked by the cell's own restriction enzymes. Restriction enzymes are therefore said to be part of the cell's **restriction/methylation system:** Foreign DNA is cleaved by the restriction enzymes, while the bacterial genome is protected by prior methylation.

Restriction enzymes are named after the bacteria from which they are obtained. Each enzyme name is derived by combining the first letter of the bacterial genus with the first two letters of the species. The strain of the bacterium may also be indicated, and if two or more enzymes have been isolated from the same species, the enzymes are numbered (using Roman numerals) in order of discovery. Thus, the first restriction enzyme isolated from *E. coli* strain R is designated *Eco*RI, whereas the third enzyme isolated from *Hemophilus aegyptius* is called *Hae*III.

Restriction enzymes are specific for double-stranded DNA and always cleave both strands. Each restriction enzyme recog- nizes a specific DNA sequence that is usually four or six (but may be eight or more) nucleotide pairs long. For example, the enzyme *Hae*III recognizes the tetranucleotide sequence GGCC and cleaves the DNA double helix as shown in Figure 16B-la. The restriction sites for several other restriction enzymes are summarized in Table 16B-1. Some restriction enzymes, such as *Hae*III, cut both strands at the same point, generating restric- tion fragments with *blunt ends.* Many other restriction enzymes cleave the two strands in an offset (staggered) manner, generat- ing short, single-stranded tails on both fragments. *Eco*RI is an example of such an enzyme; it recognizes the sequence GAATTC and cuts the DNA molecule in an offset manner, leav- ing an AATT tail on both fragments (Figure 16B-1b). The restriction fragments generated by enzymes with this staggered cleavage pattern always have **sticky ends** (also called *cohesive ends*). These terms derive from the fact that the single-stranded tail at the end of each such fragment can base-pair with the tail at either end of any other fragment generated by the same

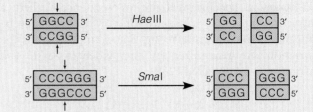

(a) Cleavage by enzymes producing blunt ends

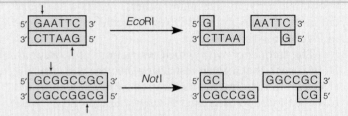

(b) Cleavage by enzymes producing sticky ends

Figure 16B-1   Cleavage of DNA by Restriction Enzymes.   (a) *Hae*III and *Sma*I are examples of restriction enzymes that cut both DNA strands in the same location, generating fragments with blunt ends. **(b)** *Eco*RI and *Not*I are examples of enzymes that cut DNA in a staggered fashion, generating fragments with sticky ends. A genetic "engineer" can use such sticky ends for joining DNA fragments from different sources, as we will explain in Chapter 18.

enzyme, causing the fragments to stick to one another by hydrogen bonding. Enzymes that generate such fragments are particularly useful because they can be employed experimen- tally to create recombinant DNA molecules, as we will see in Chapter 18.

The restriction sites for most restriction enzymes are *palin- dromes,* which means that the sequence reads the same in either direction. (The English word *radar* is a palindrome, for exam- ple.) The palindromic nature of a restriction site is due to its twofold rotational symmetry, which means that rotating the double-stranded sequence 180° in the plane of the paper yields a sequence that reads the same as it did before rotation. Palin- dromic restriction sites have the same base sequence on both strands when each strand is read in the 5' $\longrightarrow$ 3' direction.

The frequency with which any particular restriction site is likely to occur within a DNA molecule can be predicted statisti-

DNA fragments in the gel can be visualized either by staining or by using radioactively labeled DNA. A com- mon staining technique involves soaking the gel in the dye *ethidium bromide,* which binds to DNA and fluoresces pink when exposed to ultraviolet light. If the DNA frag- ments are radioactive, their locations can be determined

by **autoradiography,** a technique for detecting radioactive molecules by overlaying a sample with photographic film. When the film is developed, it yields an *autoradiogram* that is darkened wherever radioactivity has interacted with the film. After locating individual DNA fragments this way, they can be removed from the gel for futher study.

Table 16-B1 Some Common Restriction Enzymes and Their Recognition Sequences

Enzyme	Source Organism	Recognition Sequence*
*Ava*I	*Anabena variabilis*	5′ C—Py—C—G—Pu—G 3′ 3′ G—Pu—G—C—Py—C 5′
*Bam*HI	*Bacillus amyloliquefaciens*	5′ G—G—A—T—C^m—C 3′ 3′ C—C^m—T—A—G—G 5′
*Eco*RI	*Escherichia coli*	5′ G—A—A—T—T—C 3′ 3′ C—T—T—A—A—G 5′
*Hae*III	*Hemophilus aegyptius*	5′ G—G—C—C 3′ 3′ C—C—G—G 5′
*Hind*III	*Hemophilus influenzae*	5′ A^m—A—G—C—T—T 3′ 3′ T—T—C—G—A—A^m 5′
*Pst*I	*Providencia stuartii* 164	5′ C—T—G—C—A—G 3′ 3′ G—A—C—G—T—C 5′
*Pvu*I	*Proteus vulgaris*	5′ C—G—A—T—C—G 3′ 3′ G—C—T—A—G—C 5′
*Pvu*II	*Proteus vulgaris*	5′ C—A—G—C—T—G 3′ 3′ G—T—C—G—A—C 5′
*Sal*I	*Streptomyces albus* G	5′ G—T—C—G—A—C 3′ 3′ C—A—G—C—T—G 5′

*The arrows within the recognition sequence indicate the points at which the restriction enzyme cuts the two strands of the DNA molecule. Py = Pyrimidine (C or T), Pu = Purine (G or A), Am = N6-methyladenosine, Cm = 5-methylcytosine.

cally. For example, in a DNA molecule containing equal amounts of the four bases (A, T, C, and G), we can predict that, on average, a recognition site with four nucleotide pairs will occur once every 256 (i.e., 4^4) nucleotide pairs, whereas the likely frequency of a six-nucleotide sequence is once every 4096 (i.e., 4^6) nucleotide pairs. Restriction enzymes therefore tend to cleave DNA into fragments that typically vary in length from several hundred to a few thousand nucleotide pairs—gene-sized pieces, essentially. Such pieces are called *restriction fragments*. Because each restriction enzyme cleaves only a single, specific nucleotide sequence, it will always cut a given DNA molecule in the same predictable manner, generating a reproducible set of restriction fragments. This property makes restriction enzymes very powerful tools for generating manageable sized pieces of DNA for further study.

Restriction Mapping. How does a researcher determine the order in which a set of restriction fragments is arranged in a DNA molecule? One approach involves treating the DNA with two or more restriction enzymes, alone and in combination, followed by gel electrophoresis to determine the size of the resulting DNA fragments. Figure 16-13 shows how it would work for a simple DNA molecule cleaved with the restriction enzymes *Eco*RI and *Hae*III. In this example, each individual restriction enzyme cleaves the DNA into two fragments, indicating that the DNA contains one restriction site for each enzyme. Based on such information alone, two possible

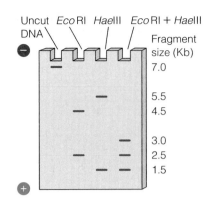

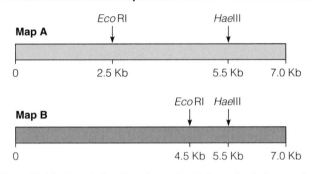

Possible restriction maps

Map A

EcoRI   HaeIII

| 0 | 2.5 Kb | 5.5 Kb | 7.0 Kb |

Map B

EcoRI HaeIII

| 0 | 4.5 Kb 5.5 Kb | 7.0 Kb |

Figure 16-13   Restriction Mapping. In this hypothetical example, the location of restriction sites for *Eco*RI and *Hae*III is determined in a DNA fragment 7.0 Kb long. Treatment with *Eco*RI cleaves the DNA into two fragments measuring 2.5 Kb and 4.5 Kb, indicating that DNA has been cleaved at a single point located 2.5 Kb from one end. Treatment with *Hae*III cleaves the DNA into two fragments measuring 1.5 Kb and 5.5 Kb, indicating that DNA has been cleaved at a single point located 1.5 Kb from one end. Based on this information alone, two possible restriction maps can be proposed. If map A were correct, simultaneous digestion of the DNA with *Eco*RI and *Hae*III should yield three fragments measuring 3.0 Kb, 2.5 Kb, and 1.5 Kb. If map B were correct, simultaneous digestion of the DNA with *Eco*RI and *Hae*III should yield three fragments measuring 4.5 Kb, 1.5 Kb, and 1.0 Kb. The experimental data reveal that map A must be correct.

restriction maps can be proposed (see maps A and B at the bottom of Figure 16-13). To determine which of the two maps is correct, an experiment must be done in which the starting DNA molecule is cleaved *simultaneously* with *Eco*RI and *Hae*III. The size of the fragments produced by simultaneous digestion with both enzymes reveals that map A must be the correct one.

In practice, restriction mapping usually involves data that are considerably more complex than in our simple example. In such situations, the DNA fragments produced by each restriction enzyme can be physically isolated—for example, by cutting the gel into slices and extracting the DNA from each slice. The isolated fragments are then individually cleaved with the second restriction enzyme, allowing the cleavage sites in each fragment to be analyzed separately. Problem 16-4 provides an example of how this approach can be used to

construct a **restriction map** depicting the location of all the restriction sites in the original DNA.

Rapid Procedures Exist for DNA Base Sequencing

Now that we know how to prepare reproducible fragments of manageable length from very long DNA molecules, it becomes feasible to determine DNA base sequences. At about the same time that techniques for preparing restriction fragments were developed, two methods were devised for rapidly determining the base sequences of DNA fragments. One method was devised by Allan Maxam and Walter Gilbert, the other by Frederick Sanger and his colleagues. The Maxam-Gilbert method, called the *chemical method,* is based on the use of (nonprotein) chemicals that cleave DNA preferentially at specific bases, whereas the Sanger procedure, called the *chain termination method,* utilizes *dideoxynucleotides* (nucleotides lacking a 3′ hydroxyl group) to interfere with the normal enzymatic synthesis of DNA. We will focus on Sanger's method because it has been adapted for use in automated machines that are now employed for most DNA sequencing tasks.

In this procedure, a single-stranded DNA fragment is employed as a template to guide the synthesis of new complementary DNA strands. DNA synthesis is carried out in the presence of the *deoxynucleotides* dATP, dCTP, dTTP, and dGTP, which are the normal substrates that provide the bases A, C, T, and G to growing DNA chains. But also included are four dye-labeled *dideoxynucleotides* (ddATP, ddCTP, ddTTP, and ddGTP), which lack the hydroxyl group attached to the 3′ carbon of normal deoxynucleotides. When a dideoxynucleotide is incorporated into a growing DNA chain in place of the normal deoxynucleotide, *DNA synthesis is prematurely halted* because the absence of the 3′ hydroxyl group makes it impossible to form a bond with the next nucleotide. Hence a series of incomplete DNA fragments are produced whose sizes provide information concerning the linear sequence of bases in the DNA.

Figure 16-14 illustrates how this procedure works. In step ①, a reaction mixture is assembled using five main ingredients: (1) a single-stranded DNA molecule that is to be sequenced; (2) the deoxynucleotides dATP, dCTP, dTTP, and dGTP, which are the normal precursors for DNA synthesis; (3) the dideoxynucleotides ddATP, ddCTP, ddTTP, and ddGTP, each labeled with a fluorescent dye of a different color, e.g., ddATP = red, ddCTP = blue, ddTTP = orange, and ddGTP = green; (4) DNA polymerase, which is the enzyme that copies DNA; and (5) a short, single-stranded DNA *primer* that is complementary in sequence to the 3′ end of the DNA strand being sequenced.

When these ingredients are mixed together and incubated, DNA polymerase catalyzes the attachment of nucleotides, one by one, to the 3′ end of the primer, producing a growing DNA strand that is complementary to the template whose sequence is being determined. Wherever the template has an A, the DNA polymerase inserts a

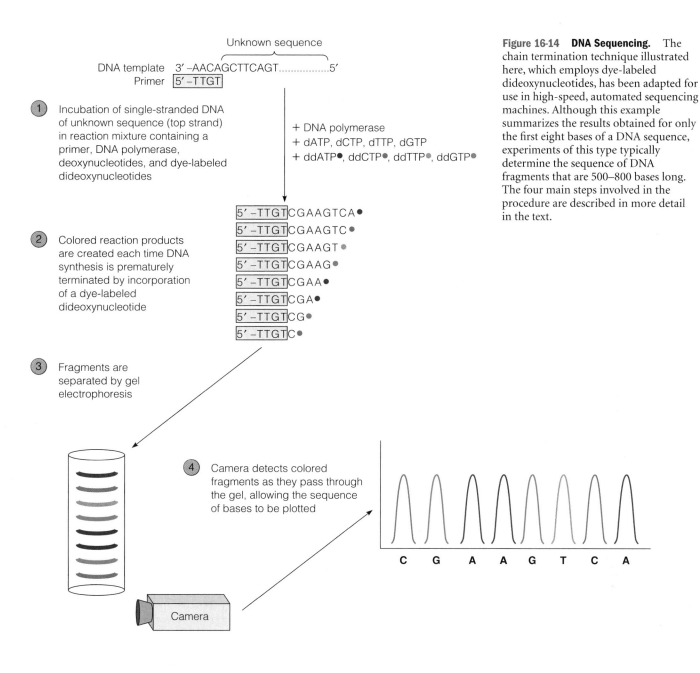

Unknown sequence

DNA template   3′ –AACAGCTTCAGT.................5′
Primer   5′ –TTGT

① Incubation of single-stranded DNA of unknown sequence (top strand) in reaction mixture containing a primer, DNA polymerase, deoxynucleotides, and dye-labeled dideoxynucleotides

+ DNA polymerase
+ dATP, dCTP, dTTP, dGTP
+ ddATP●, ddCTP●, ddTTP●, ddGTP●

5′ –TTGT CGAAGTCA●
5′ –TTGT CGAAGTC●
5′ –TTGT CGAAGT●
5′ –TTGT CGAAG●
5′ –TTGT CGAA●
5′ –TTGT CGA●
5′ –TTGT CG●
5′ –TTGT C●

② Colored reaction products are created each time DNA synthesis is prematurely terminated by incorporation of a dye-labeled dideoxynucleotide

③ Fragments are separated by gel electrophoresis

④ Camera detects colored fragments as they pass through the gel, allowing the sequence of bases to be plotted

C G A A G T C A

Camera

Figure 16-14   DNA Sequencing. The chain termination technique illustrated here, which employs dye-labeled dideoxynucleotides, has been adapted for use in high-speed, automated sequencing machines. Although this example summarizes the results obtained for only the first eight bases of a DNA sequence, experiments of this type typically determine the sequence of DNA fragments that are 500–800 bases long. The four main steps involved in the procedure are described in more detail in the text.

T in the new strand, and so forth, according to the base-pairing rules. Most of the nucleotides inserted are the normal deoxynucleotides because they are the preferred substrates for DNA polymerase. However, every so often, at random, a dye-labeled dideoxynucleotide is inserted instead of its normal equivalent. Each time a dye-labeled dideoxynucleotide is incorporated, it halts further DNA synthesis for that particular strand. Hence a mixture of strands of varying lengths is generated, each containing a colored base at the end where DNA synthesis was prematurely terminated by incorporation of a dye-labeled dideoxynucleotide (step ②).

Next, the sample is subjected to electrophoresis in a polyacrylamide gel, which allows the newly synthesized DNA fragments to be separated from each other because the shorter fragments migrate through the gel more quickly than the longer fragments (step ③). As they move through the gel, a special camera detects the colors of the various fragments as they pass by. Step ④ shows how such information allows the DNA base sequence to be determined. In this particular example, the shortest DNA fragment is blue and the next shortest fragment is green. Since blue and green are the colors of ddCTP and ddGTP, respectively, the first two bases added to the primer must have been C followed by G. In automatic sequencing machines, such information is collected for hundreds of bases in a row and fed into a computer, allowing the complete sequence of the initial DNA fragment to be quickly determined.

The Genomes of Numerous Organisms Have Been Sequenced

The significance of the technique we have just been describing can scarcely be overestimated. DNA sequencing is now so commonplace and automated that it is routinely applied not just to individual genes but also to entire genomes. Although DNA sequencing machines can only determine the sequence of short pieces of DNA, usually 500–800 bases long, one at a time, computer programs search for overlapping sequences between such fragments and thereby allow data from hundreds or thousands of DNA pieces to be assembled into longer stretches that can reach millions of bases in length.

Many of the initial successes in genome sequencing involved bacteria because they have relatively small genomes, typically a few million bases. Complete DNA sequences are now available for dozens of different bacteria, including those that cause a variety of human diseases. In fact, sequencing machines are so efficient that one research institute recently reported that it had been able to sequence 15 different bacterial genomes in a single month! But DNA sequencing has also been successfully applied to much larger genomes, including those from organisms that are most important in biological research (Table 16-2). For example, scientists have largely completed the genome sequences of the yeast *Saccharomyces cerevisiae* (12.1 million bases), the roundworm *Caenorhabditis elegans* (97 million bases), the mustard plant *Arabidopsis thaliana* (125 million bases), and the fruit fly *Drosophila melanogaster* (180 million bases).

To us as human beings, of course, the ultimate challenge of DNA sequencing is the human genome. How awesome a challenge is that? To answer this question, we need to realize that the human nuclear genome contains about 3.2 billion bases, which is roughly a thousand-fold more DNA than is present in an *E. coli* cell. One way to comprehend the magnitude of such a challenge is to note that in the early 1990s, when genome sequencing efforts began in earnest, it required almost 6 years for the laboratory of Frederick Blattner to determine the complete base sequence of the *E. coli* genome. At this rate, it would have taken a single lab almost 6000 years to sequence the entire human genome! As a consequence, scientists came together in 1990 to establish the *Human Genome Project*, a cooperative international effort involving hundreds of scientists who shared their data in an attempt to determine the entire sequence of the human genome. In the late 1990s a commercial company, Celera Genomics, tackled the job as well. As a result of these efforts, a working draft of the human genome that is roughly 90% complete was published in 2001, and most of the remaining sequence is expected by 2003.

The New Field of Bioinformatics Has Emerged to Decipher Genomes and Proteomes

Without doubt, sequencing the human genome has been one of the crowning achievements of modern biology, not only because of its sheer scale but also because of its potential impact on our understanding of human evolution, physiology, and disease. And yet, unraveling the sequence of bases was the "easy" part. Now comes the hard part of figuring out the meaning of this sequence of 3 billion A's, G's, C's, and T's. For example, which stretches of DNA correspond to genes, when and in what tissues are these genes expressed, what kinds of proteins do they code for, and how do all these proteins interact with each other and function?

The prospect of analyzing such a vast torrent of data has led to the identification of a new discipline, called **bioinformatics**, which merges computer science and biology in an attempt to make sense of it all. For example, computer programs that analyze DNA for stretches that could code for amino acid sequences are used to estimate the number of protein-coding genes. Such analyses suggest the existence of a minimum of 35,000 protein-coding genes in the human genome, roughly half of which were not known to exist prior to genome sequencing. The fascinating thing about this estimate is that it means humans may have only about twice the number of genes as do worms or flies! Computer analyses have also revealed that only about 1–2% of the human genome actually represents coding sequences. Although the remaining DNA contains some important regulatory elements, most of it appears to consist of "junk" DNA with no apparent function (for examples, see the discussions of repeated DNA in the following section and introns in Chapter 19).

Because the function of most genes is to produce proteins, scientists are beginning to look beyond the genome to study the **proteome**—the structure and properties of every protein produced by a genome. A big boost has come from recent advances in *mass spectrometry*, a technique that can measure the mass of peptides derived from proteins that have been separated by gel electrophoresis and then digested with specific proteases, such as trypsin. By comparing the resulting data to the predicted masses of peptides that would be produced by DNA sequences present in genomic databases, the proteins produced by newly discovered genes can be identified. Several new techniques also make it feasible to study interactions between the multitude of proteins found in the proteome. For example, it is possible to immobilize thousands of different proteins as

Table 16-2   Examples of Some Sequenced Genomes That Are Nearly or Entirely Complete

Organism	Genome Size	Estimated Gene Number
Bacteria		
Mycoplasma genitalium	0.6 Mb	470
Haemophilus influenza	1.8 Mb	1,740
Streptococcus pneumoniae	2.2 Mb	2,240
Escherichia coli	4.6 Mb	4,290
Yeast (*S. cerevisiae*)	12.1 Mb	6,000
Roundworm (*C. elegans*)	97 Mb	19,100
Mustard plant (*A. thaliana*)	125 Mb	25,500
Fruit fly (*D. melanogaster*)	180 Mb	13,600
Human (*H. sapiens*)	3200 Mb	35,000

tiny spots on a piece of glass smaller than a microscope slide. The resulting *protein microarrays* can then be used to study a variety of protein properties, such as the ability of each individual spot of protein to bind to other molecules added to the surrounding solution.

In addition to the proteins that it encodes, another important feature of the human genome is the way in which its base sequence differs from person to person. The published human genome sequence is actually a mosaic obtained from the DNA isolated from 10 anonymous individuals. In practice, roughly 99.9% of the bases in your genome will match perfectly with this published sequence, or with the DNA base sequence of your next-door neighbor. But the remaining 0.1% of the bases vary from person to person, creating features that make us unique individuals. These base sequence variations between individuals are called **single nucleotide polymorphisms,** or **SNPs** (pronounced "snips"). Though 0.1% might not sound like very much, 0.1% multiplied by the 3.2 billion bases in the human genome yields a total of 3.2 million SNPs. Scientists are busy creating a database of these SNPs (more than 2 million have already been identified), because these tiny genetic variations may influence how likely you are to become afflicted with a particular disease or how well you might respond to a particular treatment.

The impact of this growing body of genetic data is already becoming apparent as discoveries regarding the genetic basis of many human diseases—from breast cancer and colon cancer to diabetes and Alzheimer's disease—are being reported at a rapidly increasing pace. Such discoveries promise to revolutionize the future practice of medicine, because the ability to identify disease genes and investigate their function makes it possible to devise medical interventions for alleviating and even preventing disease.

But the ability to identify potentially harmful genes also raises ethical concerns, because all of us are likely to carry a few dozen genes that place us at risk for something. The possibility exists that such information could be misused—for example, genetic discrimination against individuals or groups of people by insurance companies, employers, or even government agencies. Moreover, detailed knowledge of the human genome increases the potential for using recombinant DNA techniques (see Chapter 18) to *alter* people's genes, not only to correct diseases in malfunctioning body tissues but also to change genes in sperm and eggs, and hence to alter the genetic makeup of future generations. What use to make of these abilities and how they should be regulated are clearly questions that concern not only the scientific community but human society as a whole.

Repeated DNA Sequences Partially Explain the Large Size of Eukaryotic Genomes

Besides the difficulties it creates for DNA sequencing studies, the enormous size of the human genome raises a more fundamental question: Is the large amount of DNA found in the cells of humans and other higher eukaryotes simply a reflection of the need for thousands of times more genes than are present in bacterial cells, or are other factors at play as well? A major breakthrough in answering this question occurred in the late 1960s, when DNA renaturation studies carried out by Roy Britten and David Kohne led to the discovery of repeated sequences in DNA.

In the Britten and Kohne experiments, DNA was broken into small fragments and dissociated into single strands by heating. The temperature was then lowered to permit the single-stranded fragments to renature. The rate of renaturation depends on the concentration of each individual kind of DNA sequence; the higher the concentration of any given kind of DNA sequence in the sample, the greater the probability that it will randomly collide with a complementary strand with which it can reassociate. How would the renaturation properties of different kinds of DNA be expected to compare? As an example, let us consider DNA derived from a bacterial cell and from a typical mammalian cell containing a thousand-fold more DNA. If this difference in DNA content reflects a thousand-fold difference in the kinds of DNA sequences present, then bacterial DNA should renature 1000 times faster than mammalian DNA. The rationale underlying this prediction is that any particular DNA sequence should be present in a thousand-fold lower concentration in the mammalian DNA sample because there are a thousand times more kinds of sequences present, and so each individual sequence represents a smaller fraction of the total population of sequences.

When studies comparing the renaturation rates of mammalian and bacterial DNA were actually performed by Britten and Kohne, the results were not exactly as expected. In Figure 16-15, which summarizes the data obtained for calf and *E. coli* DNA, renaturation is plotted as a function of the starting DNA concentration multiplied by the elapsed time. This parameter of DNA

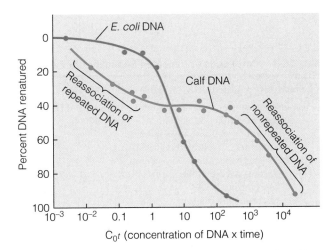

Figure 16-15 Renaturation of Calf and *E. coli* DNAs. The calf DNA that reassociates more rapidly than the bacterial DNA consists of repeated sequences.

concentration × time, or C_0t, is employed in place of time alone because it permits direct comparison of data obtained from reactions run at different DNA concentrations. When graphed this way, the data reveal that calf DNA consists of two classes of sequences that renature at markedly different rates. One type of sequence, which accounts for about 40% of the calf DNA, renatures more rapidly (i.e., at a lower C_0t value) than bacterial DNA. The most straightforward explanation for this unexpected result is that calf DNA contains **repeated DNA** sequences that are present in multiple copies. The existence of multiple copies increases the relative concentration of such sequences, thereby generating more collisions and a faster rate of reassociation than would be expected if each sequence were present in only a single copy.

The remaining 60% of the calf DNA renatures about a thousand times more slowly than *E. coli* DNA, which is the behavior expected of sequences present as single copies. This fraction is therefore called **nonrepeated DNA** to distinguish it from the repeated sequences that renature more quickly. Nonrepeated DNA sequences are each present in one copy per genome. Most protein-coding genes consist of nonrepeated DNA, although this does not mean that all nonrepeated DNA codes for proteins.

In bacterial cells virtually all of the DNA is nonrepeated, whereas eukaryotes exhibit a large variation in the relative proportion of repeated and nonrepeated sequences. This provides an explanation, at least in part, for the mystery of the seemingly excess amount of DNA in species such as *Trillium*: This organism contains a relatively high proportion of repeated DNA sequences. Using the sequencing techniques described earlier, researchers have been able to determine the base sequences of various types of repeated DNAs and to classify them into two main categories: *tandemly repeated DNA* and *interspersed repeated DNA* (Table 16-3).

Tandemly Repeated DNA. One major category of repeated DNA is referred to as **tandemly repeated DNA** because the multiple copies are arranged next to each other in a row—that is, tandemly. Tandemly repeated DNA accounts for 10–15% of a typical mammalian genome and consists of many different types of DNA sequences that vary in both the length of the basic repeat unit and the number of times this unit is repeated in succession. The length of the repeated unit can measure anywhere from 1 to 2000 bp or so. Most of the time, however, the repeated unit is shorter than 10 bp; consequently, this subcategory is called *simple-sequence repeated DNA*. The following is an example (showing one strand only) of a simple-sequence repeated DNA built from the five-base unit, GTTAC:

$$...GTTACGTTACGTTACGTTACGTTAC...$$

The number of sequential repetitions of the GTTAC unit can be as high as several hundred thousand at selected sites in the genome.

Tandemly repeated DNA of the simple-sequence type was originally called *satellite DNA* because, in many cases, its distinctive base composition causes it to appear in a "satellite" band that separates from the rest of the genomic DNA during centrifugation procedures designed to separate molecules by density. This difference in density arises because adenine and guanine differ slightly in molecular weight, as do cytosine and thymine; hence the densities of DNAs with differing base compositions will differ. In the procedures that reveal satellite bands, the genomic DNA is cleaved to short lengths and hence DNA segments of differing densities are free to migrate to different positions during centrifugation.

What is the function of simple-sequence repeated DNA (satellite DNA)? Because such sequences are not usually transcribed, it has been proposed that they may instead be responsible for imparting special physical properties to certain regions of the chromosome. In most eukaryotes, chromosomal regions called **centromeres,** which play an important role in chromosome distribution during cell division (see Chapter 17), are particularly rich in simple-sequence repeats, and it is possible that these sequences impart specialized structural properties to the centromere. **Telomeres,** which are DNA sequences located at the ends of

Table 16-3 Categories of Repeated Sequences in Eukaryotic DNA

I. Tandemly repeated DNA, including simple-sequence repeated DNA (satellite DNA)	
10–15% of most mammalian genomes is this type of DNA	
Length of each repeated unit:	1–2000 bp; typically 5–10 bp for simple-sequence repeated DNA
Number of repetitions per genome:	10^2–10^5
Arrangement of repeated units:	Tandem
Total length of satellite DNA at each site:	
Regular satellite DNA:	10^5–10^7 bp
Minisatellite DNA:	10^2–10^5 bp
Microsatellite DNA:	10^1–10^2 bp
II. Interspersed repeated DNA	
25–40% of most mammalian genomes is this type of DNA	
Length of each repeated unit:	10^2–10^4 bp
Arrangement of repeated units:	Scattered throughout the genome
Number of repetitions per genome:	10^1–10^6; "copies" not identical

chromosomes, also have simple-sequence repeats. In the next chapter we will learn how telomeres protect chromosomes from degradation at their vulnerable ends during each round of replication (see Figure 17-15). Human telomeres contain 250–1500 copies of the sequence TTAGGG, which has been highly conserved over hundreds of millions of years of evolution. All vertebrates studied so far have this same identical sequence, and even unicellular eukaryotes possess similar sequences. Apparently, such sequences are critical to the survival of these organisms.

The total length of satellite DNA found at any given site can vary enormously. Typical satellite DNAs usually fall in the range of 10^5 to 10^7 bp in total length. The term *minisatellite DNA* is used to refer to satellite DNAs occurring in shorter stretches, ranging from 10^2 to 10^5 bp at each site. And *microsatellite DNAs*, in which the repeated unit is only 1–4 bp, are even shorter (about 10–100 bp per site), although numerous sites in the genome may exhibit the same sequence. The short repeated sequences found in microsatellite and minisatellite DNAs are extremely useful in the laboratory for **DNA fingerprinting.** This procedure, described more fully in Box 16C, uses gel electrophoresis of DNA restriction fragments to compare various sites in the genomes of two or more individuals. It is a means of identifying individuals that is as accurate as conventional fingerprinting.

Within the past few years, medical researchers have made the surprising discovery that several genetic diseases affecting the nervous system involve changes in microsatellite DNAs. More specifically, these diseases are traceable to excessive numbers of repeated trinucleotide sequences within an otherwise normal gene. An example of this phenomenon, called *triplet repeat amplification,* is found in *Huntington's disease,* a devastating neurological disease that strikes in middle age and is invariably fatal. The normal version of the Huntington's gene contains the trinucleotide CAG tandemly repeated 11–34 times. The genes of afflicted individuals, however, possess up to 100 copies of the repeated unit. Neurological diseases resulting from the amplification of other triplet repeat sequences include *fragile X syndrome,* which is a major cause of mental retardation, and *myotonic dystrophy,* which affects the muscles. For some of these diseases, the repeated sequence is in a region of the affected gene that is not translated; for others, it is translated into a polypeptide segment consisting of a long string of the same amino acid. In both situations, the normal function of the repeated sequences is not known, nor is it known how the extra copies lead to neurological dysfunction.

Interspersed Repeated DNA. The second main category of repeated sequence DNA is **interspersed repeated DNA.** Rather than being clustered in tandem arrangements, the repeated units of this type of DNA are scattered around the genome. A single unit tends to be hundreds or even thousands of base pairs long, and its dispersed "copies," which may number in the hundreds of thousands, are similar but usually not identical to each other. Interspersed repeated DNA makes up 25–40% of most mammalian genomes.

In humans and other primates, a large portion of the interspersed repeated DNA consists of a family of related sequences called the *Alu family,* so named because the first ones identified all contained a restriction site for the restriction enzyme *Alu*I. A single Alu unit is about 300 bp long, and the hundreds of thousands of Alu units in the human genome add up to at least 5% of the DNA. The Alu sequence is similar in sequence to a special type of RNA involved in protein synthesis on the rough ER (as a component of the signal-recognition particle, discussed in Chapter 20). Although Alu units can be transcribed into RNA in cells, this Alu RNA is quickly degraded and it is not clear that it actually performs any function. Hence the role played by Alu sequences, if any, remains to be determined.

Although the functions of interspersed repeated sequences are largely a mystery, we do know something about how the numerous copies of these sequences may have arisen. Most interspersed repeated sequences, at least in mammals, seem to be a kind of molecular parasite resembling *transposable elements,* which are DNA segments that can readily move around the genome, leaving copies of themselves wherever they stop. (We will learn more about transposable elements in Chapter 19.)

We conclude this section on DNA organization by noting that we will defer discussion of some of the most interesting discoveries to come from molecular genetics and DNA sequencing—namely, the organization of the DNA that constitutes *genes*—until Chapter 19, where we discuss the structure of genes in detail.

DNA Packaging

The amount of DNA that a cell must accommodate is awesome, even for organisms with genomes of modest size. For example, the typical *E. coli* cell measures about 1 μm in diameter and 2 μm in length, yet it must accommodate a (circular) DNA molecule with a length of about 1600 μm—enough DNA to encircle the cell more than 400 times! Eukaryotic cells face an even greater challenge. A human cell of average size contains enough DNA to wrap around the cell more than 15,000 times. Somehow all this DNA must be efficiently packaged into cells and yet still remain accessible to the cellular machinery for both DNA replication and the transcription of specific genes. Clearly, DNA packaging is a challenging problem for all forms of life. We will look first at how prokaryotes accomplish this task of organizing their DNA and then consider how eukaryotes address the same problem.

Prokaryotes Package DNA in Bacterial Chromosomes and Plasmids

The genome of prokaryotes such as *E. coli* was once thought to be a "naked" DNA molecule lacking any elaborate

Contemporary Techniques | DNA FINGERPRINTING

Analysis of the restriction fragment patterns produced when DNA is digested with restriction enzymes has been exploited for a variety of purposes, ranging from research into the organization of genomes to practical applications such as diagnosing genetic diseases and solving violent crimes. The practical applications of restriction fragment analysis are based on the fact that no two people other than identical twins have the same exact set of DNA base sequences. Although the differences in DNA sequence between any two people are quite small, they alter the lengths of some of the DNA fragments produced by

restriction enzymes. These differences in fragment length, called **restriction fragment length polymorphisms (RFLPs),** can be analyzed by gel electrophoresis. The resulting pattern of fragments serves as a "fingerprint" that identifies the individual from whom the DNA was obtained.

In practical usage, the technique of *DNA fingerprinting* is performed in such a way that only a small, selected subset of restriction fragment bands is examined. To illustrate this point, Figure 16C-1 summarizes how DNA fingerprinting might be used to determine whether an individual carries a

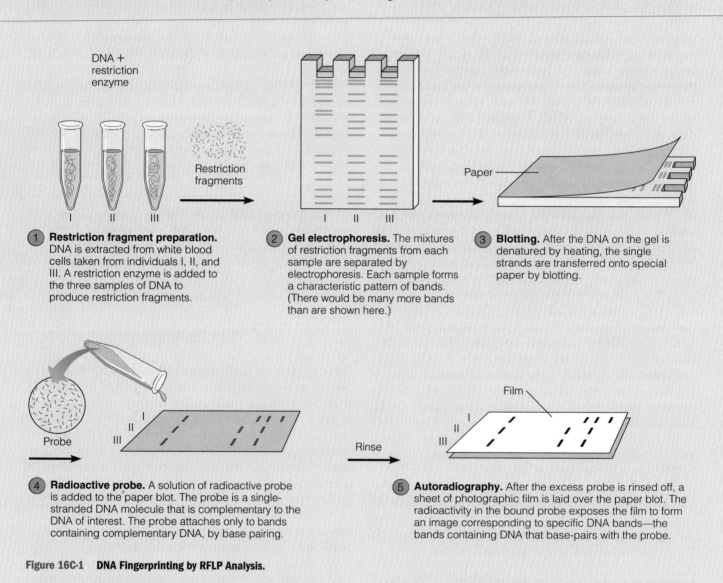

1 **Restriction fragment preparation.** DNA is extracted from white blood cells taken from individuals I, II, and III. A restriction enzyme is added to the three samples of DNA to produce restriction fragments.

2 **Gel electrophoresis.** The mixtures of restriction fragments from each sample are separated by electrophoresis. Each sample forms a characteristic pattern of bands. (There would be many more bands than are shown here.)

3 **Blotting.** After the DNA on the gel is denatured by heating, the single strands are transferred onto special paper by blotting.

4 **Radioactive probe.** A solution of radioactive probe is added to the paper blot. The probe is a single-stranded DNA molecule that is complementary to the DNA of interest. The probe attaches only to bands containing complementary DNA, by base pairing.

5 **Autoradiography.** After the excess probe is rinsed off, a sheet of photographic film is laid over the paper blot. The radioactivity in the bound probe exposes the film to form an image corresponding to specific DNA bands—the bands containing DNA that base-pairs with the probe.

Figure 16C-1   DNA Fingerprinting by RFLP Analysis.

organization and with only trivial amounts of protein associated with it. We now know that the organization of the bacterial genome is more like the chromosomes of eukaryotes than previously realized. Bacterial geneticists therefore refer to the structure that contains the main bacterial genome as the **bacterial chromosome.**

Bacterial Chromosomes.   The bacterial chromosome is typically a circular DNA molecule, containing some bound protein, that is localized in a special region of the cell called the **nucleoid** (Figure 16-16). Although the nucleoid is not membrane-enclosed, the bacterial chromosomal DNA residing in this region forms a threadlike mass of fibers

particular disease-causing gene even though he or she may not yet exhibit symptoms of the disease. In this example, imagine that individuals I, II, and III are members of a family in which the disease-causing gene is common. It is known that individual I carries the defective gene but individual II does not, and we want to determine the genetic status of their child, individual III.

The key step in DNA fingerprinting is called **Southern blotting** (after E. M. Southern, who developed it in 1975), but the fingerprinting procedure actually involves several distinct steps. In ① of Figure 16C-1, DNA obtained from the three individuals is digested with a restriction enzyme. In ②, the resulting fragments are then separated from each other by gel electrophoresis. Because each person's DNA represents an entire genome, hundreds of thousands of bands would appear at this stage if all were made visible. Here is where Southern's stroke of brilliance comes in, with a technique that enables us to single out the bands of interest. In the Southern blotting process (③), a special kind of "blotter" paper (nitro-cellulose or nylon) is pressed against the completed gel, allowing the pattern of DNA fragments to be transferred to the paper. In ④, a radioactive *probe* is added to the blot. The probe is simply a radioactive single-stranded DNA (or RNA) molecule whose base sequence is complementary to the DNA of interest—in this case, the DNA of the disease-causing gene. The probe binds to complementary DNA sequences by base pairing (the same process that occurs when denatured DNA renatures). In ⑤, the bands that bind the radioactive probe are made visible by autoradiography. The results indicate that the child's version of the gene matches that of the *healthy* parent, individual II.

The autoradiogram in Figure 16C-2 illustrates the use of DNA fingerprinting in a murder case. Here RFLP analysis makes it clear that blood found on the defendant's clothes came from the victim, strongly implicating the defendant in the murder. Another practical application of DNA fingerprinting in the legal area is the determination of paternity or maternity.

For many DNA fingerprinting applications, scientists now analyze the length of short repeated sequences (minisatellite DNAs) known as **variable number tandem repeats,** or **VNTRs.** The VNTR sites chosen for DNA fingerprinting vary greatly in length from person to person because of differences in the number of times the basic repeat unit is sequentially repeated. As a result, the pattern of VNTRs in a person's DNA can be used to identify that individual uniquely. For example, the DNA of three suspects in a murder case might all have different numbers of copies of the dinucleotide sequence TG at a particular place in the genome; one might have 19 copies, another 21 copies, and the third 32 copies. The differences in restriction

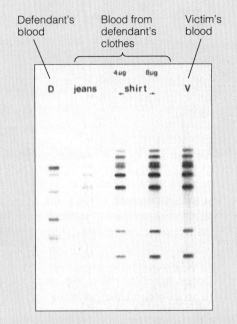

Figure 16C-2   DNA Fingerprints from a Murder Case.   DNA was isolated from bloodstains on the defendant's clothes and compared, by RFLP analysis, with DNA from the defendant and DNA from the victim. The band pattern for the bloodstain DNA matches that for the victim, showing that the blood on the defendant's clothes came from the victim. (Courtesy of Cellmark Diagnostics, Inc., Germantown, MD.)

fragment patterns would come from the differences in the number of repeated (TG) units between two restriction cutting sites, rather than from the absence or presence of a cutting site as for traditional RFLPs. In criminal cases, the FBI now routinely examines restriction fragment lengths at 13 VNTR sites in the genome. The chance that any two unrelated individuals would exhibit the same exact profile at all 13 sites is roughly one in a million billion.

The usefulness of DNA fingerprinting can be further enhanced by the *polymerase chain reaction (PCR)*, a technique for making multiple DNA copies that will be described in Chapter 17. Starting with DNA isolated from a single cell, the PCR reaction can be used to selectively synthesize millions of copies of any given DNA sequence within a few hours, easily producing enough DNA for fingerprinting analysis of the 13 VNTR sites. In this way, a few skin cells left on a pen or car keys touched by a person may yield enough DNA to uniquely identify that individual.

packed together in a way that maintains a distinct boundary between the nucleoid and the rest of the cell. The DNA of the bacterial chromosome is negatively supercoiled and folded into an extensive series of loops, averaging 50,000–100,000 bp in length, that are about the same size as similar looped domains observed in eukaryotic nuclei.

Because the two ends of each loop are anchored to structural components found within the nucleoid, the super-coiling of individual loops can be altered without influencing the supercoiling of adjacent loops.

The loops are thought to be held in place by RNA and protein molecules. Evidence for a structural role for RNA

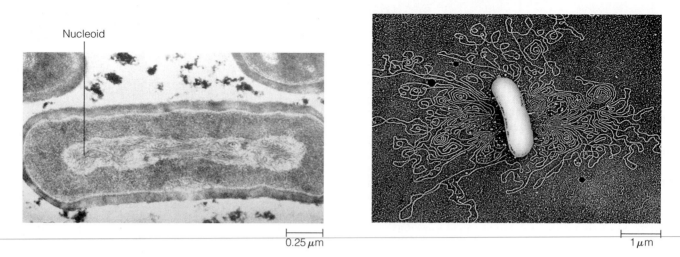

0.25 µm   1 µm

Figure 16-16   The Bacterial Nucleoid.   The electron micrograph on the left shows a bacterial cell with a distinct nucleoid, which is the region where the bacterial chromosome resides. When bacterial cells are ruptured, their chromosomal DNA is released from the cell. The micrograph on the right shows that the released DNA forms a series of loops that remain attached to a structural framework within the nucleoid (TEMs).

in the bacterial chromosome has come from studies showing that treatment with ribonuclease, an enzyme that degrades RNA, releases some of the loops, although it does not relax the supercoiling. Nicking the DNA with a topoisomerase, on the other hand, relaxes the supercoiling but does not disrupt the loops. The supercoiled DNA that forms each loop is organized into beadlike packets containing small, basic protein molecules, analogous to the histones of eukaryotic cells (discussed in the next section). Current evidence suggests that the DNA molecule is wrapped around particles of the basic protein. Thus, from what we know so far, the bacterial chromosome consists of supercoiled DNA that is bound to small, basic proteins and then folded into looped domains.

Bacterial Plasmids.   In addition to its chromosome, a bacterial cell may contain one or more plasmids. **Plasmids** are relatively small, circular molecules of DNA that carry genes both for their own replication and, often, for one or more cellular functions (usually nonessential ones). Most plasmids are supercoiled, giving them a condensed, compact form. Although plasmids replicate autonomously, the replication is usually in sufficient synchrony with the replication of the bacterial chromosome to ensure a roughly comparable number of plasmids from one generation to the next. In *E. coli* cells, three classes of plasmids are recognized. *F* (*fertility*) *factors* are involved in the process of conjugation, a sexual process we will discuss in Chapter 18. *R* (*resistance*) *factors* carry genes that confer drug resistance on the bacterial cell. And *col* (*colicinogenic*) *factors* allow the bacterium to secrete *colicins,* compounds that kill other bacteria lacking the col factor. In addition, some strains of *E. coli* contain *cryptic plasmids,* which have no known function.

Eukaryotes Package DNA in Chromatin and Chromosomes

When we turn from prokaryotic cells to eukaryotic cells, DNA packaging becomes more complicated. First, substantially larger amounts of DNA are involved. Each eukaryotic chromosome contains a single, linear DNA molecule of enormous size. In human cells, for example, just *one* of these DNA molecules may be 10 cm or more in length, roughly a hundred times the size of the DNA molecule found in a typical bacterial chromosome. Second, greater structural complexity is introduced by the association of eukaryotic DNA with greater amounts and numbers of proteins. When bound to these proteins, the DNA is converted into **chromatin** fibers measuring 10 to 30 nm in diameter, which are normally dispersed throughout the nucleus. At the time of cell division (and in a few other special situations), these fibers condense and fold into much larger, compact structures that become recognizable as individual **chromosomes.** Rap up DNA

The proteins with the most important role in chromatin structure are the **histones,** a group of relatively small proteins whose high content of the amino acids lysine and arginine gives them a strong positive charge. The binding of histones to DNA, which is negatively charged, is therefore stabilized by ionic bonds. In most cells, the mass of histones in chromatin is approximately equal to the mass of DNA. Histones are divided into five main types, designated H1, H2A, H2B, H3, and H4. Chromatin contains roughly equal numbers of H2A, H2B, H3, and H4 molecules, and about half that number of H1 molecules. These proportions are remarkably constant among different kinds of eukaryotic cells, regardless of the type of cell or its physiological state. In addition to histones, chromatin also contains a diverse group of *nonhistone proteins* that play a variety of enzymatic, structural, and regulatory roles.

Nucleosomes Are the Basic Unit of Chromatin Structure

The DNA contained within a typical nucleus would measure a meter or more in length if it were completely extended, whereas the nucleus itself is usually no more than 5–10 μm in diameter. The folding of such an enormous length of DNA into a nucleus that is almost a million times smaller presents a significant topological problem. One of the first insights into the folding process emerged in the late 1960s, when X-ray diffraction studies carried out by Maurice Wilkins revealed that purified chromatin fibers have a repeating structural subunit that is seen in neither DNA nor histones alone. Wilkins therefore concluded that histones impose a repeating structural organization upon DNA. A clue to the nature of this structure was provided in 1974, when Ada Olins and Donald Olins published electron micrographs of chromatin fibers isolated from cells in a way that avoided the harsh solvents used in earlier procedures for preparing chromatin for microscopic examination. Chromatin fibers viewed in this way appear as a series of tiny particles attached to one another by thin filaments. This "beads-on-a-string" appearance led to the suggestion that the beads are made of proteins (presumably histones) and the thin filaments connecting the beads correspond to DNA. We now refer to each bead, along with its associated short stretch of DNA, as a **nucleosome** (Figure 16-17).

On the basis of electron microscopy alone, it would have been difficult to determine whether nucleosomes are a normal component of chromatin or an artifact generated during sample preparation. Fortunately, independent evidence for the existence of a repeating structure in chromatin was reported at about the same time by Dean Hewish and Leigh Burgoyne, who discovered that rat liver nuclei contain a nuclease that is capable of cleaving the DNA in chromatin fibers. In one crucial set of experiments, these investigators exposed chromatin to this nuclease and then purified the partially degraded DNA to remove chromatin proteins. When they examined the purified DNA by gel electrophoresis, they found a distinctive pattern of fragments in which the smallest piece of DNA measured about 200 bp in length, and the remaining fragments were exact multiples of 200 bp (Figure 16-18). Since nuclease digestion of protein-free DNA does not generate this fragment pattern, they concluded that (1) chromatin proteins are clustered along the DNA molecule in a regular pattern that repeats at intervals of roughly 200 bp, and (2) the DNA located between these protein clusters is susceptible to nuclease digestion, yielding fragments that are multiples of 200 bp in length.

These observations raised the question of whether the protein clusters postulated to occur at 200-bp intervals correspond to the spherical particles observed in electron micrographs of chromatin fibers. Answering this question required a combination of the nuclease digestion and electron microscopic approaches. In these studies, chromatin was briefly exposed to *micrococcal nuclease,* a bacterial enzyme that, like the rat liver nuclease, cleaves chromatin DNA at intervals of 200 bp. The fragmented chromatin was then separated into fractions of varying sizes by centrifugation and examined by electron microscopy. The smallest fraction was found to contain single spherical particles, the next fraction contained clusters of two particles, the succeeding fraction contained clusters of three particles, and so forth (Figure 16-19). When DNA was isolated from these fractions and analyzed by gel electrophoresis, the DNA from the fraction containing single particles measured 200 bp in length, the DNA from the fraction containing clusters of two particles measured

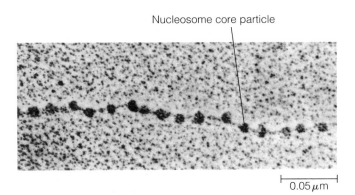

Nucleosome core particle

0.05 μm

Figure 16-17 Nucleosomes. The core particles of nucleosomes appear as beadlike structures spaced at regular intervals along eukaryotic chromatin fibers. A nucleosome is defined to include both a core particle and the stretch of DNA that connects to the next core particle (TEM).

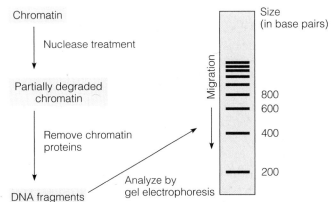

Figure 16-18 Evidence That Proteins Are Clustered at 200 Base-Pair Intervals Along the DNA Molecule in Chromatin Fibers. In these experiments, DNA fragments generated by nuclease digestion of rat liver chromatin were analyzed by gel electrophoresis. The discovery that the DNA fragments are multiples of 200 base pairs suggests that histones are clustered at 200 base-pair intervals along the DNA, thereby conferring a regular pattern of protection against nuclease digestion.

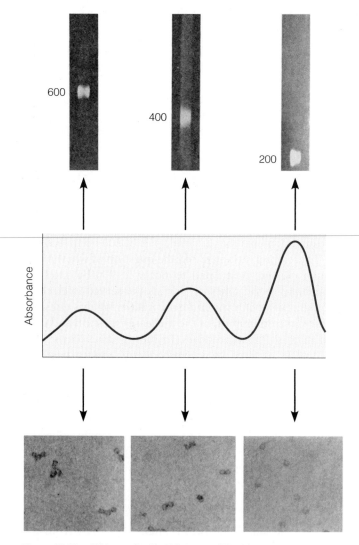

Figure 16-19 Evidence for the Existence of Nucleosomes.
Chromatin that had been partially degraded by treatment with micrococcal nuclease was fractionated by density gradient centrifugation (center graph). The individual peaks were then analyzed both by electron microscopy (bottom) and by gel electrophoresis after removal of chromatin proteins (top). The peak on the right consists of single protein particles associated with 200 base pairs of DNA, the middle peak consists of clusters of two particles associated with 400 base pairs of DNA, and the peak on the left consists of clusters of three particles associated with 600 base pairs of DNA. This indicates that the basic repeat unit in chromatin is a protein particle associated with 200 base pairs of DNA.

assembling nucleosomes from purified mixtures of DNA and protein. They found that chromatin fibers composed of nucleosomes can be generated by mixing together DNA and all five histones. However, when they attempted to use individually purified histones, they discovered that nucleosomes could be assembled only when the histones had been purified by gentle techniques that left histone H2A bound to histone H2B, and histone H3 bound to histone H4. When these H3-H4 and H2A-H2B complexes were mixed with DNA, chromatin fibers exhibiting normal nucleosomal structure were reconstituted. Kornberg therefore concluded that histone H3-H4 and H2A-H2B complexes are an integral part of the nucleosome.

To investigate the nature of these histone interactions in more depth, Kornberg and his colleague, Jean Thomas, treated isolated chromatin with a chemical reagent that forms covalent crosslinks between protein molecules that are located next to each other. After treatment with this reagent, the chemically crosslinked proteins were isolated and analyzed by polyacrylamide gel electrophoresis. Protein complexes the size of eight histone molecules were prominent in such gels, suggesting that the nucleosomal particle contains an *octamer* of eight histones. Given the knowledge that histones H3-H4 and histones H2A-H2B each form tight complexes, and that these four histones are present in roughly equivalent amounts in chromatin, Kornberg and Thomas proposed that histone octamers are created by joining together two H2A-H2B dimers and two H3-H4 dimers, and that the DNA double helix is then wrapped around the resulting octamer (Figure 16-20).

One issue not addressed by the preceding model concerns the significance of histone H1, which is not part of the octamer. If individual nucleosomes are isolated by briefly digesting chromatin with micrococcal nuclease, histone H1 is still present (along with the four other histones and 200 bp of DNA). When digestion is carried out for longer periods, the DNA fragment is further degraded until it reaches a length of about 146 bp; during the final stages of the digestion process, histone H1 is released. The remaining particle, consisting of a histone octamer associated with 146 bp of DNA, is referred to as a *core particle*. The DNA that is degraded during digestion from 200 to 146 bp in length is referred to as *linker DNA* because it joins one nucleosome to the next (Figure 16-20). Since histone H1 is released upon degradation of the linker DNA, histone H1 molecules are thought to be associated with the linker region. The length of the linker DNA varies somewhat among organisms, but the DNA associated with the core particle always measures close to 146 bp, which is enough to wrap around the core particle roughly 1.7 times.

Nucleosomes Are Packed Together to Form Chromatin Fibers and Chromosomes

The formation of nucleosomes is only the first step in the packaging of nuclear DNA (Figure 16-21). Isolated chromatin fibers exhibiting the "beads-on-a-string"

400 bp in length, and so on. It was therefore concluded that the spherical particles observed in electron micrographs are each associated with 200 bp of DNA. This basic repeat unit, containing an average of 200 bp of DNA associated with a protein particle, is the nucleosome.

A Histone Octamer Forms the Nucleosome Core

The first insights into the underlying molecular organization of the nucleosome emerged from the work of Roger Kornberg and his associates, who developed techniques for

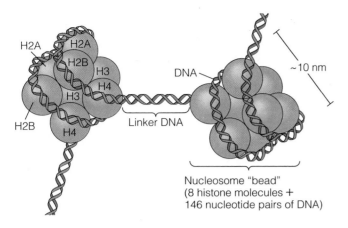

Figure 16-20   A Closer Look at Nucleosome Structure.   Each nucleosome consists of eight histone molecules (two each of histones H2A, H2B, H3, and H4) associated with 146 nucleotide pairs of DNA and a stretch of linker DNA about 50 nucleotide pairs in length. The diameter of the nucleosome "bead," or core particle, is about 10 nm. Histone H1 (not shown) is thought to bind to the linker DNA and facilitate the packing of nucleosomes into 30-nm fibers.

(Labels in figure: H2A, H2A, H2B, H3, H4, H3, H2B, H4, DNA, ~10 nm, Linker DNA, Nucleosome "bead" (8 histone molecules + 146 nucleotide pairs of DNA))

appearance measure about 10 nm in diameter, but the chromatin of intact cells often forms a slightly thicker fiber about 30 nm in thickness called the **30-nm chromatin fiber**. In preparations of isolated chromatin, the 10-nm and 30-nm forms of the chromatin fiber can be interconverted by changing the salt concentration of the solution. However, the 30-nm fiber does not form in chromatin preparations whose histone H1 molecules have been removed, suggesting that histone H1 facilitates the packing of nucleosomes into the 30-nm fiber. Several models have been proposed to explain how individual nucleosomes are packed together to form a 30-nm fiber. Most early models postulated that the chain of nucleosomes is twisted upon itself to form some type of coiled structure. However, recent studies suggest that the structure of the 30-nm fiber is much less uniform than a coiled model suggests. Instead, the nucleosomes of the 30-nm fiber seem to be packed together to form an irregular, three-dimensional zigzag structure that can interdigitate with its neighboring fibers.

The next level of chromatin packaging is the folding of the 30-nm fibers into **looped domains** averaging 50,000–100,000 bp in length. This looped arrangement is maintained by the periodic attachment of DNA to an insoluble network of nonhistone proteins, which form a chromosomal *scaffold* to which the long loops of DNA are attached. The looped domains can be most clearly seen in electron micrographs of chromosomes isolated from dividing cells and treated to remove all the histones and most of the nonhistone proteins (Figure 16-22). Loops can also be seen in specialized types of chromosomes that are not associated with the process of cell division (see the discussion of polytene chromosomes in Chapter 21). In these cases, the chromatin loops turn out to contain

"active" regions of DNA—that is, DNA that is being transcribed. It makes sense that active DNA would be less tightly packed than inactive DNA because it would allow easier access by proteins involved in gene transcription.

Even in cells where genes are being actively transcribed, significant amounts of chromatin may be further compacted (Figure 16-21d). The degree of folding in such cells varies over a continuum. Segments of chromatin so highly compacted that they show up as dark spots in micrographs are called **heterochromatin,** whereas the more loosely packed, diffuse form of chromatin is called **euchromatin** (see Figure 16-26a). The tightly packed heterochromatin contains DNA that is transcriptionally inactive, while the more loosely packed euchromatin is associated with DNA that is being actively transcribed. Much of the chromatin in metabolically active cells is loosely packed as euchromatin, but as a cell prepares to divide, *all* of its chromatin becomes highly compacted, generating a group of microscopically distinguishable chromosomes. Because the chromosomal DNA has been recently duplicated, each chromosome is composed of two duplicate units called *chromatids* (Figure 16-21e).

The extent to which a DNA molecule has been folded in chromatin and chromosomes can be quantified using the **DNA packing ratio,** which is calculated by determining the total extended length of a DNA molecule and dividing it by the length of the chromatin fiber or chromosome into which it has been packaged. The initial coiling of the DNA around the histone cores of the nucleosomes reduces the length by a factor of about seven, and formation of the 30-nm fiber results in a further sixfold condensation. The packing ratio of the 30-nm fiber is therefore about $7 \times 6 = 42$. Further folding and coiling brings the overall packing ratio of typical euchromatin to about 750. For heterochromatin and the chromosomes of dividing cells, the packing ratio is still higher. At the time of cell division, for example, a typical human chromosome measures about 4–5 μm in length, yet contains a DNA molecule that would measure almost 75 mm if completely extended. The packing ratio for such a chromosome therefore falls in the range of 15,000–20,000.

Eukaryotes Also Package Some of Their DNA in Mitochondria and Chloroplasts

Not all the DNA of a eukaryotic cell is contained in the nucleus. Though nuclear DNA accounts for the bulk of a cell's genetic information, mitochondria and chloroplasts also contain some DNA of their own, along with the machinery needed to replicate, transcribe, and translate the information encoded by this DNA. The DNA molecules that reside in mitochondria and chloroplasts are devoid of histones and are usually circular (Figure 16-23). In other words, they resemble the genomes of bacteria, as we might expect from the likely endosymbiotic origin of these organelles (discussed in Box 15A). Mitochondrial

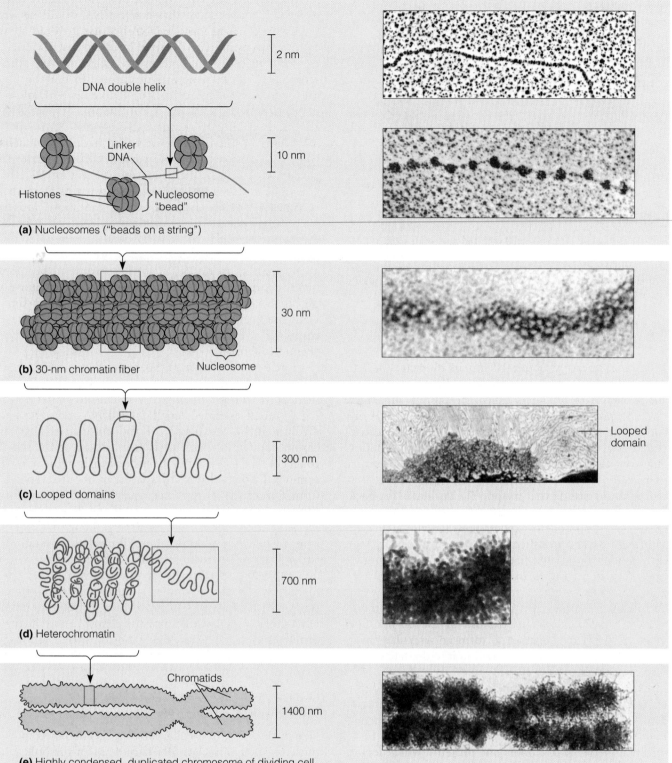

2 nm — DNA double helix

10 nm — Linker DNA, Histones, Nucleosome "bead"

(a) Nucleosomes ("beads on a string")

30 nm — Nucleosome

(b) 30-nm chromatin fiber

300 nm

(c) Looped domains — Looped domain

700 nm

(d) Heterochromatin

1400 nm — Chromatids

(e) Highly condensed, duplicated chromosome of dividing cell

Figure 16-21 Levels of Chromatin Packing.
These diagrams and TEMs show a current model for progressive stages of DNA coiling and folding, culminating in the highly compacted chromosome of a dividing cell. **(a)** "Beads on a string," an extended configuration of nucleosomes formed by the association of DNA with four types of histones. **(b)** The 30-nm chromatin fiber, shown here as a tightly packed collection of nucleosomes. The fifth histone, H1, may be located in the interior of the fiber. **(c)** Looped domains of 30-nm fibers, visible in the TEM here because a mitotic chromosome has been experimentally unraveled. **(d)** Heterochromatin, highly folded chromatin that is visible as discrete spots even in interphase cells. **(e)** A replicated chromosome (two attached chromatids) from a dividing cell, with all the DNA of the chromosome in the form of very highly compacted heterochromatin.

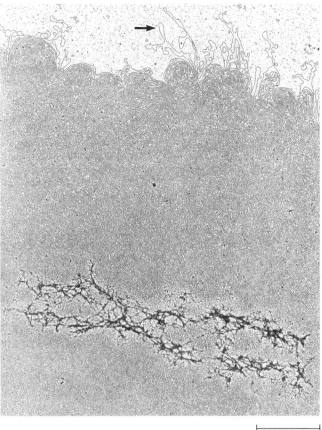

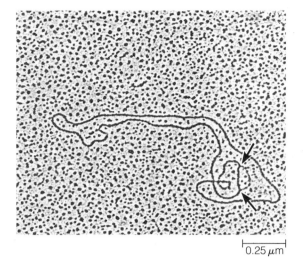

Figure 16-23 Mitochondrial DNA. Mitochondrial DNA from most organisms is circular, as seen in this electron micrograph. This molecule was caught in the act of replication; the arrows indicate the points at which replication was proceeding when the molecule was fixed for electron microscopy (TEM).

Figure 16-22 Electron Micrograph Showing the Protein Scaffold That Remains After Removing Histones from Human Chromosomes. The chromosomal DNA remains attached to the scaffold as a series of long loops. The arrow points to a region where a loop of the DNA molecule can be clearly seen (TEM).

and chloroplast genomes tend to be relatively small, comparable in size to a viral genome. Both organelles are therefore semiautonomous, able to code for some of their polypeptides but dependent on the nuclear genome to encode most of them.

The genome of the human mitochondrion, for example, consists of a circular DNA molecule containing 16,569 base pairs and measuring about 5 μm in length. It has been completely sequenced, and all of its 37 genes are known. The RNA and polypeptides encoded by this DNA are just a small fraction (about 5%) of the number of RNA molecules and proteins needed by the mitochondrion. This is nonetheless a vital genetic contribution, for these products include the RNA molecules present in mitochondrial ribosomes, all of the transfer RNA molecules required for mitochondrial protein synthesis, and 13 of the polypeptide subunits of the electron transport system. These include subunits of NADH dehydrogenase, cytochrome *b*, cytochrome *c* oxidase, and ATP synthase (Figure 16-24).

The size of the mitochondrial genome varies considerably among organisms. Mammalian mitochondria typically have about 16,500 bp of DNA, whereas yeast mitochondrial DNA is roughly five times larger and plant mitochondrial DNA is even bigger than that. It is not clear, however, that larger mitochondrial genomes necessarily code for correspondingly more polypeptides. A comparison of yeast and human mitochondrial DNA, for example, suggests that most of the additional DNA present in the yeast mitochondrion consists of noncoding sequences.

Chloroplasts typically possess circular DNA molecules around 120,000 bp in length containing about 120 genes. In addition to ribosomal and transfer RNAs and polypeptides involved in protein synthesis, the chloroplast genome also codes for a number of polypeptides specifically involved in photosynthesis. These include several polypeptide components of photosystems I and II and one of the two subunits of ribulose-1,5-bisphosphate carboxylase, the carbon-fixing enzyme of the Calvin cycle.

Interestingly, most polypeptides encoded by mitochondrial or chloroplast genomes are components of multimeric proteins that also contain subunits encoded by the nuclear genome. In other words, organelle proteins that contain subunits encoded within the organelle are typically hybrid protein complexes containing polypeptides encoded and synthesized within the organelle plus polypeptides encoded by the nuclear genome and synthesized by cytoplasmic ribosomes. This raises intriguing questions as to how the polypeptides synthesized in the cytoplasm enter the organelle, a topic to which we will return in Chapter 20.

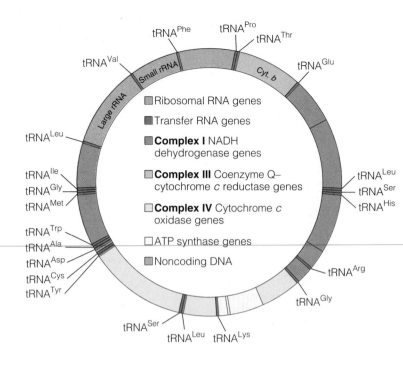

Figure 16-24   The Genome of the Human Mitochondrion.
The double-stranded DNA molecule of the human mitochondrion is circular and contains 16,569 base pairs. This genome codes for large and small ribosomal RNA molecules, transfer RNA (tRNA) molecules (each identified by a superscript with the three-letter abbreviation for the amino acid it carries), and subunits of a number of the proteins that make up the mitochondrial electron transport system complexes. The tRNA genes are very short because the RNA molecules they encode each contain only about 75 nucleotides. Notice that there are two tRNA genes for leucine and two for serine; they code for slightly different versions of tRNAs for these amino acids. The mitochondrial genome is extremely compact, with little noncoding DNA between genes.

The Nucleus

So far in this chapter, we have discussed DNA as the genetic material of the cell, the genome as a complete set of DNA instructions for the cells of a particular species, and the chromosome as the physical means of packaging DNA within cells. Now we come to the **nucleus,** the site within the eukaryotic cell where the chromosomes are localized and replicated and where the DNA they contain is transcribed. The nucleus is therefore both the repository of most of the cell's genetic information and the control center for the expression of that information.

The nucleus is one of the most prominent and characteristic features of eukaryotic cells (Figure 16-25). Recall that the term *eukaryon* means "true nucleus." The very essence of a eukaryotic cell is its membrane-bounded nucleus, which compartmentalizes the activities of the genome—both replication and transcription—from the rest of cellular metabolism. In the following discussion, we focus first on the membrane envelope that forms the boundary of the nucleus. Then we turn our attention to the pores that perforate the envelope, the structural matrix inside the nucleus, the arrangement of the chromatin, and finally, the organization of the nucleolus. Figure 16-26 provides an overview of some of these nuclear structures.

A Double-Membrane Nuclear Envelope Surrounds the Nucleus

The existence of a membrane around the nucleus was first suggested in the late nineteenth century, based primarily on the osmotic properties of the nucleus. Since light microscopy reveals only a narrow fuzzy border at the outer surface of the nucleus, little was known about the structure of this membranous boundary prior to the advent of electron microscopy. Transmission electron microscopy revealed that the nucleus is bounded by a **nuclear envelope** composed of two membranes—the inner and outer nuclear membranes—separated by a **perinuclear space** measuring about 20–40 nm across (see Figure 16-26b). Each membrane is about 7–8 nm thick and exhibits the same trilamellar appearance as most other cellular membranes. The inner nuclear membrane rests on a network of supporting fibers called the *nuclear lamina* (discussed later in this chapter). The outer nuclear membrane is continuous with the endoplasmic reticulum, making the perinuclear space continuous with the lumen of the ER. Like membranes of the rough ER, the outer membrane is often studded on its outer surface with ribosomes engaged in protein synthesis. In some cells, intermediate filaments of the cell's cytoskeleton extend outward from the outer membrane into the cytoplasm, anchoring the nucleus to other organelles or the plasma membrane.

One of the most distinctive features of the nuclear envelope is the presence of specialized openings, called **nuclear pores,** which are especially easy to see when the nuclear envelope is examined by freeze-fracture microscopy (Figure 16-27). Each pore is a small cylindrical channel that extends through both membranes of the nuclear envelope, thereby providing direct continuity between the cytosol and the **nucleoplasm,** the interior space of the nucleus (other than the region of the nucleolus). The density of pores (i.e., the number per unit surface area of the nuclear envelope) varies greatly with cell type and activity. A typical mammalian nucleus has about 3000–4000 pores, or about 10–20 pores per square micrometer.

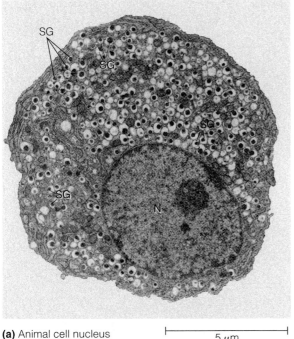

(a) Animal cell nucleus

5 μm

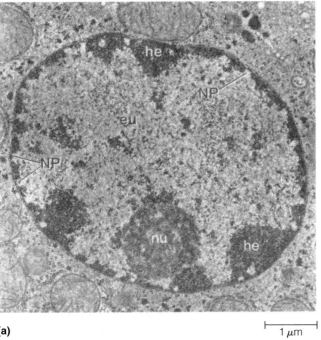

(a)

1 μm

(b) Plant cell nucleus

5 μm

Figure 16-25   The Nucleus.   The nucleus is a prominent structural feature in most eukaryotic cells. **(a)** The nucleus (N) of an animal cell. This is an insulin-producing cell from a rat pancreas; hence, the prominence of secretory granules (SG) in the cytoplasm. **(b)** The nucleus (N) of a plant cell. This is a cell from a soybean root nodule. The prominence of plastids (P) reflects their role in the storage of starch granules (TEMs).

Figure 16-26   The Structural Organization of the Nucleus and Nuclear Envelope.   **(a)** An electron micrograph of the nucleus from a mouse liver cell, with prominent structural features labeled (TEM). The nuclear envelope is a double membrane perforated by nuclear pores (NP). Internal structures include the nucleolus (nu), euchromatin (eu), and heterochromatin (he). **(b)** A drawing of a typical nucleus. Structural features included here but not visible in the micrograph include the nuclear lamina, ribosomes on the outer nuclear membrane, and the continuity between the outer nuclear membrane and the rough ER.

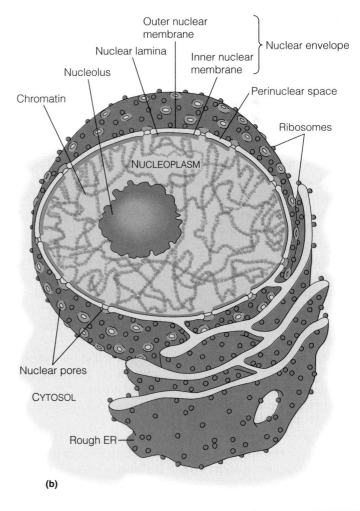

(b)

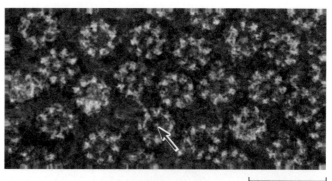

(a) Nuclear pores in the envelope

0.25 µm

Figure 16-27　**Nuclear Pores.**　Numerous nuclear pores (NP) are visible in this freeze-fracture micrograph of the nuclear envelope of an epithelial cell from a rat kidney. The fracture plane reveals faces of both the inner membrane (A) and the outer membrane (B). The ridges to which the arrows point represent the perinuclear space delimited by the two membranes (TEM).

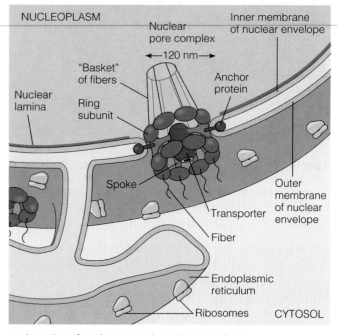

(b) Location of nuclear pores in nuclear membrane

Figure 16-28　**The Structure of the Nuclear Pore.**　**(a)** Negative staining of an oocyte nuclear envelope reveals the octagonal pattern of the nuclear pore complexes. The arrow shows a central granule. This nuclear envelope is from an oocyte of the newt *Taricha granulosa* (TEM). **(b)** A nuclear pore is formed by the fusion of the inner and outer nuclear membranes and is lined by an intricate protein structure called the nuclear pore complex. The structure is roughly wheel-shaped and has octagonal symmetry. Two parallel rings, each consisting of eight subunits (dark purple), outline the rim of the wheel. Eight spokes (green) connect the two rings (two of the spokes have been omitted from the drawing) and extend to the central transporter (dark pink) at the hub of the wheel; the transporter may be a diaphragm-like structure that opens and closes to allow the passage of particles of different sizes. Proteins extending from the rim-and-spoke assembly into the perinuclear space presumably help anchor the complex within the nuclear envelope. Fibers extend above and below the complex, with the ones on the nucleoplasmic side forming a basket of unknown function. The nuclear pore complex may consist of as many as 100 distinct polypeptides.

At each pore, the inner and outer membranes of the nuclear envelope are fused together, forming a channel that is lined with an intricate protein structure called the **nuclear pore complex (NPC).** The diameter of the entire pore complex is about 120 nm. It has an overall mass of some 120 million daltons and consists of dozens of different kinds of polypeptide subunits. In electron micrographs, the most striking feature of the pore complex is the octagonal arrangement of its subunits. Micrographs such as the one in Figure 16-28a show rings of eight subunits arranged in an octagonal pattern. In other views, the eight subunits are seen to protrude on both the cytoplasmic and nucleoplasmic sides of the envelope. Notice that central granules can be seen in some of the nuclear pore complexes in Figure 16-28a. Although these granules were once thought to be particles in transit through the pores, recent evidence suggests that they are an integral part of the pore complex. Apparently, they are easily lost during the preparation of samples for microscopy.

Figure 16-28b illustrates the main components of the nuclear pore complex. Examination of this diagram reveals that the pore complex as a whole is shaped somewhat like a wheel lying on its side within the nuclear envelope. Two

parallel rings, outlining the rim of the wheel, each consist of the eight subunits seen in electron micrographs. Eight spokes (shown in green) extend from the rings to the wheel's hub (dark pink), which is the "central granule"

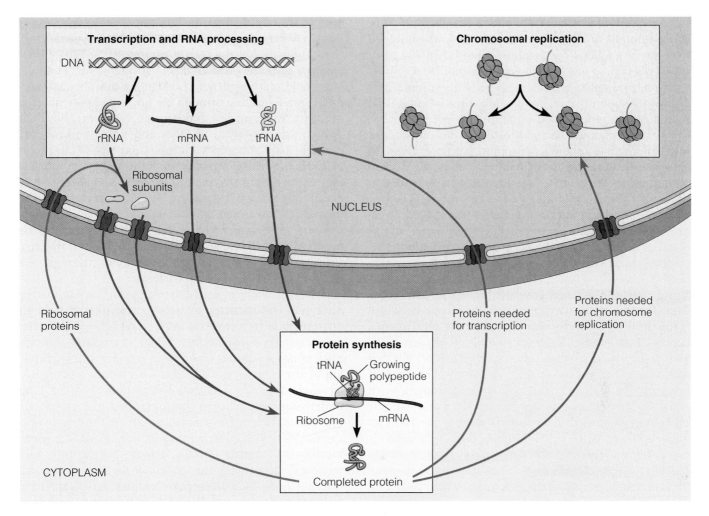

Figure 16-29 Macromolecular Transport into and out of the Nucleus. Because eukaryotic cells store their genetic information in the nucleus but synthesize proteins in the cytoplasm, all the proteins needed in the nucleus must be transported inward from the cytoplasm (purple arrows), and all the RNA molecules and ribosomal subunits needed for protein synthesis in the cytoplasm must be transported outward from the nucleus (red arrows). The three kinds of RNA molecules required for protein synthesis are ribosomal RNA (rRNA), messenger RNA (mRNA), and transfer RNA (tRNA).

seen in many EMs. This granule is now usually called the **transporter** because it is thought to move macromolecules across the nuclear envelope. Proteins extending from the rim into the perinuclear space are thought to help anchor the pore complex to the envelope. In addition, fibers extend from the rings into the cytosol and nucleoplasm, the ones on the nucleoplasm side forming a basket (sometimes called a "cage" or "fishtrap").

Molecules Enter and Exit the Nucleus Through Nuclear Pores

The nuclear envelope is both a solution to one problem and the source of another. As a means of localizing chromosomes and their activities to one part of the cell, it is an example of the general eukaryotic strategy of compartmentalization. Presumably it is advantageous for a nucleus to possess a barrier that keeps the chromosomes in and organelles such as ribosomes, mitochondria, lysosomes, and microtubules out. For example, the nuclear envelope protects newly synthesized RNA from being acted upon by cytoplasmic organelles or enzymes before it has been fully processed.

At the same time that it protects the chromosomes and immature RNA molecules from exposure to the cytoplasm, however, the nuclear envelope creates for the eukaryotic cell several formidable transport problems that are unknown in prokaryotes. All the enzymes and other proteins required for chromosome replication and transcription of DNA in the nucleus must be imported from the cytoplasm, and all the RNA molecules and partially assembled ribosomes needed for protein synthesis in the cytoplasm must be obtained from the nucleus (Figure 16-29). In response to these transport problems, specialized pores have evolved that mediate virtually all transport into and out of the nucleus.

To get some idea of how much traffic must travel through the nuclear pores, consider the flow of ribosomal subunits from the nucleus to the cytoplasm. Ribosomes are partially assembled in the nucleus as two classes of subunits, each of which is a complex of RNA and proteins. These subunits move to the cytoplasm and, when needed for protein synthesis, are combined into functional ribosomes containing one of each type of subunit. An actively growing mammalian cell can easily be synthesizing 20,000 ribosomal subunits per minute. We already know that such a cell has about 3000–4000 nuclear pores, so ribosomal subunits must be transported to the cytosol at a rate of about 5–6 subunits per minute per pore. Traffic in the opposite direction is, if anything, even heavier. When chromosomes are being replicated, histones are needed at the rate of about 300,000 molecules per minute. The rate of inward movement must therefore be about 100 histone molecules per minute per pore! And, in addition to all this macromolecular traffic, the pores also mediate the transport of smaller particles, molecules, and ions.

Passive Diffusion of Small Molecules Through Nuclear Pores.

The idea that small particles can diffuse freely back and forth through nuclear pores first received experimental support from studies in which colloidal gold particles of various sizes were injected into the cytoplasm of cells that were then examined by electron microscopy. Shortly after injection, the gold particles could be seen passing through the nuclear pores and into the nucleus. The rate of particle entry into the nucleus is inversely related to the particle's diameter—that is, the larger the gold particle, the slower it enters the nucleus. Particles larger than about 10 nm in diameter are excluded entirely. Since the overall diameter of a nuclear pore complex is much larger than such gold particles, it was concluded that the pore complex contains tiny, *aqueous diffusion channels* through which small particles and molecules can freely move.

To determine the diameter of these channels, investigators have injected radioactive proteins of various sizes into the cytoplasm of cells and observed how long it takes for the proteins to appear in the nucleus. A globular protein with a molecular weight of 20,000 takes only a few minutes to equilibrate between the cytoplasm and nucleus, but most proteins of 60,000 daltons or more are barely able to penetrate into the nucleus at all. These and other transport measurements indicate that the aqueous diffusion channels are about 9 nm in diameter, a size that creates a permeability barrier for molecules significantly larger than 20,000 in molecular weight. Until recently, researchers generally assumed that each nuclear pore complex has one such channel. However, it now appears there may be eight separate 9-nm channels located at the periphery of the pore complex, between the spokes, and perhaps an additional 9-nm channel at the center of the transporter. These aqueous channels are thought to be freely permeable to ions and small molecules (including small proteins) because such substances cross the nuclear envelope very quickly after being injected into cells. Thus, the nucleoside triphosphates required for DNA and RNA synthesis probably diffuse freely through the pores, as do other small molecules needed for metabolic pathways that function within the nucleus.

Active Transport of Large Proteins and RNA Through Nuclear Pores.

Many of the proteins involved in DNA packaging, replication, and transcription are small enough to pass through a 9-nm-wide channel. Histones, for example, have molecular weights of 21,000 or less and should therefore passively diffuse through the nuclear pores with little problem. Some nuclear proteins are very large, however. The enzymes involved in DNA and RNA synthesis, for example, have subunits with molecular weights in excess of 100,000, which is too large to fit through a 9-nm opening. Messenger RNA molecules pose a challenge, too, because they leave the nucleus bound to proteins in the form of RNA-protein ("ribonucleoprotein") complexes that are quite large. Ribosomal subunits must also be exported to the cytoplasm after assembly in the nucleus. Clearly, transporting all these particles through the nuclear pores is a significant challenge.

A large body of evidence suggests that such large molecules and particles are actively transported through nuclear pores by a selective process. Like active transport across single membranes, active transport through nuclear pores requires energy and involves specific binding of the transported substance to membrane proteins, which in this case are part of the pore complex. The underlying molecular mechanism is best understood for proteins that are actively transported from the cytosol into the nucleus. Such proteins possess one or more **nuclear localization signals (NLS),** which are amino acid sequences that enable the protein to be recognized and transported by the nuclear pore complex. A typical NLS is 8–30 amino acids in length and usually contains proline as well as the positively charged (basic) amino acids lysine and arginine.

The role played by NLS sequences in targeting proteins for the nucleus has been established by experiments in which gold particles larger than 9 nm in diameter—too large to pass through the aqueous diffusion channels of the nuclear pores—were coated with NLS-containing polypeptides and then injected into the cytoplasm of frog oocytes. After such treatment, gold particles as large as 26 nm in diameter are rapidly transported through the nuclear pore complexes and into the nucleus. Thus, the maximum diameter for *active* transport across the nuclear envelope seems to be 26 nm (versus 9 nm for *passive* diffusion). This is only one of numerous examples in which a short stretch of amino acids has been found to target a molecule or particle to a specific cellular site.

The process of transporting cytoplasmic proteins into the nucleus through nuclear pores is illustrated in Figure 16-30. In step ①, a cytoplasmic protein containing an NLS is recognized by a special type of receptor protein called an **importin,** which binds to the NLS and mediates the movement of the NLS-containing protein to a nuclear pore, perhaps with the pore's cytoplasmic fibers somehow serving as tracks. In step ②, the importin-NLS protein complex is transported into the nucleus by the transporter at the center of the nuclear pore complex (NPC). After arriving in the nucleus, the importin molecule associates

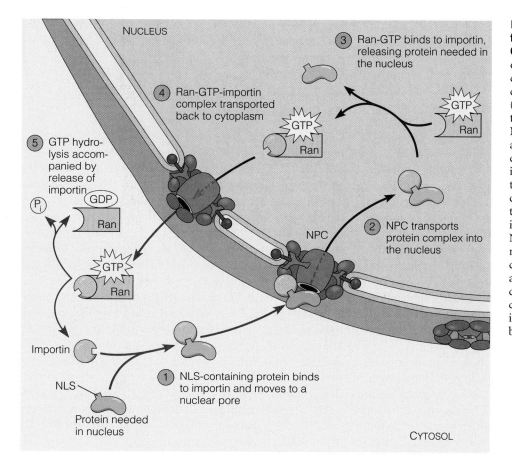

Figure 16-30 Proposed Mechanism for the Transport of Proteins from Cytoplasm to Nucleus. Proteins destined for use in the nucleus contain short stretches of amino acids called nuclear localization sequences (NLS) that target them for transport through the nuclear pores. ① The NLS-containing protein first binds to a receptor protein (importin) in the cytosol and moves to a pore. ② The importin-NLS protein complex is transported through the nuclear pore complex (NPC) and discharged into the nucleus. ③ Ran-GTP binds to importin, leading to the release of the NLS-containing protein in the nucleus. ④ The Ran-GTP-importin complex is transported back through a nuclear pore complex into the cytosol. ⑤ Upon arriving in the cytosol, the importin is released from its association with Ran, accompanied by the hydrolysis of GTP to GDP.

with a GTP-binding protein called *Ran.* This interaction between importin and Ran causes the NLS-containing protein to be released for use in the nucleus (step ③). The Ran-GTP-importin complex is then transported back through a nuclear pore to the cytoplasm (step ④), where the importin is released for reuse accompanied by hydrolysis of the Ran-bound GTP (step ⑤). Evidence that this GTP hydrolysis step provides the energy for nuclear import has come from experiments showing that nuclear transport can be inhibited by exposing cells to nonhydrolyzable analogs of GTP, but not to nonhydrolyzable analogs of ATP.

For the export of material out of the nucleus, comparable mechanisms operate. The main difference is that transport out of the nucleus is used mainly for RNA molecules that are synthesized in the nucleus but function in the cytoplasm, whereas nuclear import is devoted largely to the import of proteins that are synthesized in the cytoplasm but function in the nucleus. Although the main cargo for nuclear export is RNA rather than protein, most RNAs are transported out of the nucleus in the form of RNA-protein complexes. In several cases the protein components of these complexes have been shown to contain **nuclear export signals (NES),** which are amino acid sequences that target the protein, and hence its bound RNA, for export through the nuclear pores. These signal sequences are recognized by nuclear transport receptor proteins called **exportins,** which bind to molecules con-

taining NES sequences and mediate their transport out through the nuclear pores via a mechanism involving Ran-mediated GTP hydrolysis, just as importins mediate the transport of cytoplasmic molecules into the nucleus. The *direction* of transport through the nuclear pores is therefore specific to the type of targeting sequence a molecule contains, which in turn determines whether importins or exportins will bind to it. This mechanism means that any given pore can carry out transport in either direction, depending on whether importins or exportins are involved.

The Nuclear Matrix and Nuclear Lamina Are Supporting Structures of the Nucleus

Roughly 80–90% of the nuclear mass is accounted for by chromatin fibers, so you might expect that removing the chromatin would cause the nucleus to collapse into a relatively structureless mass. However, in the early 1970s researchers discovered that an insoluble fibrous network retaining the overall shape of the nucleus remains behind after more than 95% of the chromatin has been removed by a combination of nuclease and detergent treatments. This network, which was named the **nuclear matrix,** is thought to help maintain the shape of the nucleus and provide an organizing skeleton for the chromatin fibers. The existence of the nuclear matrix has not been accepted by all cell biologists, however. The fibers are visible only in certain micrographs (Figure 16-31), leading skeptics to question whether they are artifacts introduced in sample preparation.

In recent years, additional evidence has bolstered the case for the presence of a structural matrix that organizes nuclear activities. For example, a close connection between the matrix and chromatin fibers is suggested by the discovery that isolated nuclear matrix preparations always contain small amounts of tightly bound DNA and RNA. Nucleic acid hybridization techniques have revealed that the tightly bound DNA is enriched in sequences that are actively being transcribed into RNA. Moreover, when cells are incubated with ³H-thymidine, a radioactive precursor for DNA synthesis, the newly synthesized radioactive DNA is found to be preferentially associated with the nuclear matrix. These observations suggest that the nuclear matrix may be involved in anchoring chromatin fibers at locations where DNA or RNA is being synthesized, thereby organizing the DNA for orderly replication and transcription and perhaps even providing tracks that guide and propel newly formed messenger RNA to the nuclear pores for transport to the cytoplasm.

While the exact nature and functional significance of the nuclear matrix remains to be elucidated, the nucleus contains another fibrous structure whose role has been more clearly defined. This structure, called the **nuclear lamina,** is a thin, dense meshwork of fibers that lines the inner surface of the inner nuclear membrane and helps support the nuclear envelope. The nuclear lamina is about 10–40 nm in thickness and is constructed from intermediate filaments made of proteins called *lamins* (discussed in more detail in Chapter 22). At least some of these filaments seem to be attached to proteins of the inner nuclear membrane. In addition to providing structural support

for the nuclear envelope, the nuclear lamina may also provide attachment sites for chromatin, a topic to be addressed in the next section.

Chromatin Fibers Are Dispersed Within the Nucleus in a Nonrandom Fashion

Other than at the time of cell division, a cell's chromatin fibers tend to be highly extended and dispersed throughout the nucleus. You might therefore guess that the chromatin threads corresponding to each individual chromosome are randomly distributed and highly intertwined within the nucleus. Perhaps surprisingly, this seems not to be the case. Instead, the chromatin of each chromosome apparently has its own discrete location. This idea was first proposed in 1885, but evidence that it is true in a variety of cells awaited the techniques of modern molecular biology. Recently, using nucleic acid probes that hybridize to the DNA of specific chromosomes, several research groups have demonstrated that the chromatin fibers corresponding to individual chromosomes occupy discrete compartments within the nucleus, referred to as "chromosome territories." The positions of these territories do not seem to be fixed, however. They vary from cell to cell of the same organism and seem to change during a cell's life cycle, perhaps reflecting changes in gene activity of the different chromosomes.

The nuclear envelope helps organize the chromatin by binding certain segments of it to specific sites on the inner surface of the envelope, closely associated with the nuclear pores. The segments of chromatin that bind in this way are highly compacted—that is, they are heterochromatin. In electron micrographs, this material appears as a dark

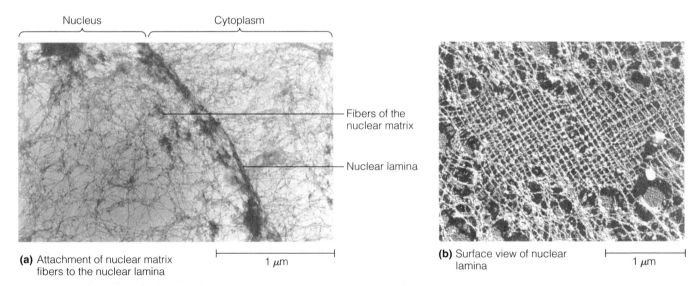

(a) Attachment of nuclear matrix fibers to the nuclear lamina

Nucleus | Cytoplasm

Fibers of the nuclear matrix

Nuclear lamina

1 μm

(b) Surface view of nuclear lamina

1 μm

Figure 16-31 **The Nuclear Matrix and the Nuclear Lamina.** **(a)** This electron micrograph of part of a mammalian cell nucleus shows a branched network of nuclear matrix filaments traversing the nucleus. These filaments seem attached to the nuclear lamina, the dense layer of filaments that lines the nucleoplasm side of the nuclear envelope. **(b)** A surface view of the nuclear lamina of a frog oocyte (TEMs).

irregular layer around the nuclear periphery, as you saw in Figure 16-26a. Most of it seems to be the type called **constitutive heterochromatin,** which exists in a highly condensed form at virtually all times in all cells of the organism. The DNA of constitutive heterochromatin consists of simple-sequence repeated DNA (recall that these are short sequences that repeat tandemly and are not transcribed). Two major chromosomal regions composed of constitutive heterochromatin are the centromere and the telomere. In many cases it is chromosomal telomeres—the highly repeated DNA sequences located at the ends of chromosomes—that are attached to the nuclear envelope at times other than during cell division.

In contrast to constitutive heterochromatin, **facultative heterochromatin** varies with the particular activities the cell is carrying out. Thus, it differs from tissue to tissue and can even vary from time to time within a given cell. Facultative heterochromatin appears to represent chromosomal regions that have become specifically inactivated in a specific cell type. The amount of facultative heterochromatin is usually low in embryonic cells but can be substantial in highly differentiated cells. The formation of facultative heterochromatin may therefore be an important means of inactivating entire blocks of genetic information during development.

The Nucleolus Is Involved in Ribosome Formation

The remaining structural component of the eukaryotic nucleus is the **nucleolus,** the ribosome factory of the cell. Typical eukaryotic cells contain one or two nucleoli, but the occurrence of several more is not uncommon and in certain situations hundreds or even thousands may be present. The nucleolus is usually a spherical structure measuring several micrometers in diameter, but wide variations in size and shape are observed. Because of their relatively large size, nucleoli are easily seen with the light microscope and were first observed more than 200 years ago. However, it was not until the advent of electron microscopy in the 1950s that the structural components of the nucleolus were clearly identified. In thin-section electron micrographs, each nucleolus appears as a membrane-free organelle consisting of fibrils and granules (Figure 16-32). The fibrils contain DNA that is being transcribed into *ribosomal RNA (rRNA)*, the RNA component of ribosomes. The granules are rRNA molecules being packaged with proteins (imported from the cytoplasm) to form ribosomal subunits. As we saw earlier, the ribosomal subunits are subsequently exported through the nuclear pores to the cytoplasm. Because of their role in synthesizing RNA, nucleoli become heavily radiolabeled when the cell is exposed to radioactive precursors of RNA (Figure 16-33).

The earliest evidence associating the nucleolus with ribosome formation was provided in the early 1960s by Robert Perry, who employed a microbeam of ultraviolet light to destroy the nucleoli of living cells. Such cells lost their ability to synthesize rRNA, suggesting that the

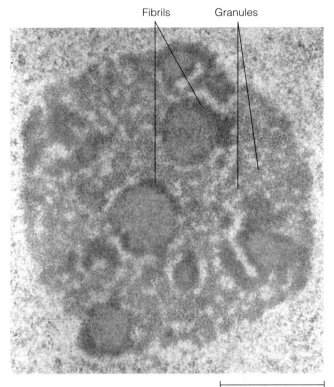

Figure 16-32 The Nucleolus. The nucleolus is a prominent intranuclear structure. It is a mass of fibrils and granules. The fibrils are DNA and rRNA; the granules are newly forming ribosomal subunits. Shown here is a nucleolus of a spermatogonium, a cell that gives rise to sperm cells (TEM).

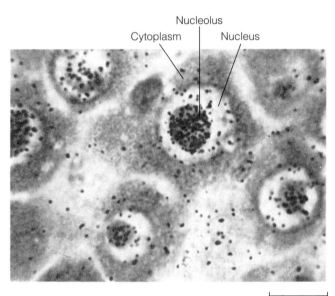

Figure 16-33 The Nucleolus as a Site of RNA Synthesis. To demonstrate the role of the nucleolus in RNA synthesis, a rat was injected with ^{3}H-cytidine, a radioactively labeled RNA precursor. Five hours later, liver tissue was removed and subjected to autoradiography. The black spots over the nucleoli in this autoradiograph indicate that the ^{3}H is concentrated in the nucleoli (TEM).

nucleolus is involved in manufacturing ribosomes. Additional evidence emerged from studies carried out by Donald Brown and John Gurdon on the African clawed frog, *Xenopus laevis*. Through genetic crosses, it is possible to produce *Xenopus* embryos whose cells lack nucleoli. Brown and Gurdon discovered that such embryos, termed *anucleolar mutants,* cannot synthesize rRNA and therefore die during early development, again implicating the nucleolus in ribosome formation.

If rRNA is synthesized in the nucleolus, then the DNA sequences coding for this RNA must reside in the nucleolus as well. This prediction has been verified by showing that isolated nucleoli contain the **nucleolus organizer region (NOR)**—a stretch of DNA carrying multiple copies of rRNA genes. These multiple rRNA genes occur in all genomes and are thus an important example of repeated DNA that carries genetic information. The number of copies of the rRNA genes varies greatly from species to species, but animal cells often contain hundreds of copies and plant cells usually contain thousands. The multiple copies are grouped into one or more NORs, which may reside on more than one chromosome; in each NOR, the multiple gene copies are tandemly arranged. A single nucleolus may contain rRNA genes derived from more than one NOR. For example, the human genome has five NORs per haploid chromosome set, or ten per diploid nucleus, each located near the tip of a different chromosome. But instead of ten separate nucleoli, the typical human nucleus has a single large nucleolus containing loops of chromatin derived from ten separate chromosomes.

The size of the nucleolus is correlated with its level of activity. In cells having a high rate of protein synthesis and hence a need for many ribosomes, nucleoli tend to be large and can account for 20–25% of the total volume of the nucleus. In less active cells, nucleoli are much smaller. The main difference is the amount of granular component present. Cells that are producing many ribosomes transcribe, process, and package large quantities of rRNA and have higher steady-state levels of partially complete ribosomal subunits on hand in the nucleolus, thus accounting for the prominent granular component.

The nucleolus disappears during mitosis, at least in the cells of higher plants and animals. As the cell approaches division, chromatin condenses into compact chromosomes accompanied by the shrinkage and then disappearance of the nucleoli. With our current knowledge of the nucleolus's composition and function, this makes perfect sense: The extended chromatin loops of the nucleolus cease being transcribed as they are coiled and folded, and any remaining rRNA and ribosomal protein molecules disperse or are degraded. Then, as mitosis is ending, the chromatin uncoils, the NORs loop out again, and rRNA synthesis resumes. In human cells, this is the only time when the ten NORs of the diploid nucleus are apparent; as rRNA synthesis begins again, ten tiny nucleoli become visible, one near the tip of each of ten chromosomes. As these nucleoli enlarge, they quickly fuse into the single large nucleolus found in human cells that are not in the process of dividing.

Although its primary function is clearly related to the production of ribosomes, the nucleolus contains some molecules whose presence suggests a role in additional activities, such as nuclear export, the chemical modification of small RNAs, and even the control of cell division. Microscopists have also identified several other types of small *nuclear bodies* that, like the nucleolus, are non-membrane-bounded structures composed of distinctive configurations of small fibers and/or granules. Several types of nuclear bodies have been characterized, each containing a different set of resident proteins. Although the details are not well understood, nuclear bodies are thought to play a variety of roles related to the processing and handling of messenger RNA molecules produced in the nucleus.

Perspective

The discovery of DNA dates back to the early studies of Miescher, but it was not until the mid-twentieth century that experiments with pneumococcal bacteria and bacteriophage T2 clearly revealed DNA to be the genetic material. This work was followed by Watson and Crick's elucidation of the double-helical structure of DNA, one of the landmarks in twentieth-century biology. The hydrogen bonds holding together the two strands of the double helix only fit when the base A is paired with T, and the base G is paired with C. Because the two strands are held together by relatively weak, non-covalent bonds, the two strands can be readily separated during DNA replication and RNA synthesis.

Molecular biologists have developed a number of powerful tools for studying DNA. For example, restriction enzymes isolated from bacteria can be used to cut DNA into reproducible fragments that are short enough to be easily manipulated in the laboratory. In this chapter, we described how the use of restriction enzymes makes it easier to study the structure of very long DNA molecules. Later, in Chapter 18, we will see how restriction enzymes are also the key to making recombinant DNA.

The DNA (or RNA for some viruses) that makes up one complete set of an organism's genetic information is called its genome. For most viruses and prokaryotes, the genome consists of a single DNA molecule or a small number of them. Eukaryotes have a nuclear genome divided among multiple chromosomes, each possessing one very long DNA molecule. Eukaryotes also have a mitochondrial genome and, in the case of plants, a chloroplast genome. Molecular biologists have developed a number of powerful tools for studying genomes. For example, restriction enzymes isolated from bacteria can be used to cut DNA into reproducible fragments that are short enough to be easily manipulated in the laboratory. And automated DNA sequencing techniques have allowed scientists to determine the entire base sequence of the genomes of numerous organisms, from bacteria to humans. One of the most striking features of the nuclear genomes of eukaryotes, especially multicellular organisms, is the large fraction of DNA that does not code for RNA or protein synthesis. Much of this noncoding DNA consists of repeated sequences. Little is known about the functions of this repeated DNA, but at least some of it is thought to play a structural role for the chromosome.

The enormous length of the DNA molecules present in cells (and even viruses) necessitates considerable packaging. In both prokaryotes and the nuclei of eukaryotic cells, the DNA is complexed with proteins, but this packaging is more elaborate in eukaryotes. The basic structural unit of the eukaryotic chromosome is the nucleosome, which consists of a short length of DNA wrapped around a protein particle constructed from eight histone molecules. Stretches of nucleosomes ("beads on a string") are packed together to form a 30-nm chromatin fiber, which can then loop and fold further. The more highly compacted the DNA, the less likely it is to be transcriptionally active in the cell. In nondividing cells that are actively transcribing DNA, much of the chromatin is in a relatively extended, highly diffuse form called euchromatin. However, other portions of chromatin are in a highly condensed state called heterochromatin. During cell division, all the chromatin becomes highly compacted, forming discrete chromosomes visible with the light microscope.

Eukaryotic chromosomes are localized within the nucleus. The double-membraned nuclear envelope is perforated with nuclear pores that mediate two-way transport of materials between the nucleoplasm and the cytosol. Within each pore is an elaborate protein structure called the nuclear pore complex. Ions and small molecules up to about 9 nm in diameter diffuse passively through aqueous channels in the pore complex; larger particles are actively transported through it. Nuclear pores thus control the inward movement of proteins used in the nucleus and the outward movement of RNA and ribosomal subunits. The nucleolus is a large, specialized nuclear structure that carries out ribosomal RNA synthesis and the assembly of ribosomal subunits. A structural network of fibers called the nuclear matrix is thought to be involved in localizing other nuclear activities, such as DNA replication and the production of messenger RNA, to discrete regions of the nucleus.

Key Terms for Self-Testing

The Chemical Nature of the Genetic Material
genetic transformation (p. 481)
bacteriophage (phage) (p. 482)
Chargaff's rules (p. 486)

DNA Structure
double helix (p. 487)
complementary (p. 487)
supercoiled DNA (p. 489)
topoisomerase (p. 489)
DNA gyrase (p. 489)
DNA denaturation (p. 490)
DNA renaturation (p. 490)
DNA melting temperature (T_m)(p. 490)
nucleic acid hybridization (p. 491)
probe (p. 491)

The Organization of DNA in Genomes
genome (p. 491)
base pairs (bp) (p. 492)
Kb, Mb, Gb (p. 492)
restriction enzyme (p. 492)
restriction site (p. 493)
gel electrophoresis (p. 493)
autoradiography (p. 494)
restriction map (p. 496)
bioinformatics (p. 498)
proteome (p. 498)

single nucleotide polymorphisms (SNPs) (p. 499)
repeated DNA (p. 500)
nonrepeated DNA (p. 500)
tandemly repeated DNA (p. 500)
centromere (p. 500)
telomere (p. 500)
DNA fingerprinting (p. 501)
interspersed repeated DNA (p. 501)

DNA Packaging
bacterial chromosome (p. 502)
nucleoid (p. 502)
plasmid (p. 504)
chromatin (p. 504)
chromosome (p. 504)
histone (p. 504)
nucleosome (p. 505)
30-nm chromatin fiber (p. 507)
looped domain (p. 507)
heterochromatin (p. 507)
euchromatin (p. 507)
DNA packing ratio (p. 507)

The Nucleus
nucleus (p. 510)
nuclear envelope (p. 510)
perinuclear space (p. 510)

nuclear pore (p. 510)
nucleoplasm (p. 510)
nuclear pore complex (NPC) (p. 512)
transporter (in the nuclear pore complex) (p. 513)
nuclear localization signal (NLS) (p. 514)
importin (p. 514)
nuclear export signal (NES)(p. 515)
exportin (p. 515)
nuclear matrix (p. 515)
nuclear lamina (p. 516)
constitutive heterochromatin (p. 517)
facultative heterochromatin (p. 517)
nucleolus (p. 517)
nucleolus organizer region (NOR) (p. 518)

Box 16B: *A Closer Look at Restriction Enzymes*
restriction/methylation system (p. 494)
sticky end (p. 494)

Box 16C: *DNA Fingerprinting*
restriction fragment length polymorphism (RFLP) (p. 502)
Southern blotting (p. 503)
variable number tandem repeat (VNTR) (p. 503)

Problem Set

More challenging problems are marked with a •.

16-1. The Genetic Material. Label each of the following statements concerning the chemical nature of the genetic material with a B if the statement is most appropriately dated to the several decades *before* 1944, with an I if it belongs to the *interim* period 1944–1952, with an A if it is most appropriate to the period *after* 1952, or with an N if it was *never* a widely held concept.

(a) Double-stranded DNA contains equal amounts of the bases A and T, and equal amounts of the bases G and C, but the significance of these equivalencies is a mystery.

(b) The genetic material in higher organisms is most likely protein.

(c) DNA is the genetic information in both bacteria and their phages.

(d) DNA is chemically too simple to be considered the genetic information of any cell.

(e) Nuclein is an important component of the cytoplasm.

(f) DNA may well be the genetic material of bacterial cells, but it is still an open question what that means for higher organisms.

(g) Smooth (S) strains of pneumococcus are capable of converting nonpathogenic rough (R) strains into pathogenic S strains, but we do not yet know which component of the S cells effects this transformation.

(h) Once adsorbed onto a host cell, a bacteriophage injects its proteins into the bacterium.

16-2. Prior Knowledge. Virtually every experiment performed by biologists builds on knowledge that has resulted from prior experiments.

(a) Of what significance to Avery and his colleagues was the finding (made in 1932 by J. L. Alloway) that the same kind of transformation of R cells into S cells that Griffith observed to occur in mice could also be demonstrated in culture with isolated pneumococcus cells?

(b) Of what significance to Hershey and Chase was the following suggestion (made in 1951 by R. M. Herriott)? "A virus may act like a little hypodermic needle full of transforming principles; the virus as such never enters the cell; only the tail contacts the host and perhaps enzymatically cuts a small hole through the outer membrane and then the nucleic acid of the virus head flows into the cell."

(c) Of what significance to Watson and Crick were the data of their colleagues at Cambridge that the particular forms in which A, G, C, and T exist at physiologic pH permit the formation of specific hydrogen bonds?

(d) How did the findings of Hershey and Chase help explain an earlier report (by T. F. Anderson and R. M. Herriott) that bacteriophage T2 loses its ability to reproduce when it is burst open osmotically by suspending the viral particles in distilled water prior to their addition to a bacterial culture?

16-3. DNA Structure. Carefully inspect the double-stranded DNA molecule shown here, and notice that it has twofold rotational symmetry:

3′ A—G—C—G—C—T—A—T—A—G—C—G—C—T 5′
5′ T—C—G—C—G—A—T—A—T—C—G—C—G—A 3′

Label each of the following statements as T if true or F if false.

(a) There is no way to distinguish the right end of the double helix from the left end.

(b) If a solution of these molecules were heated to denature them, every single-stranded molecule in the solution would be capable of hybridizing with every other molecule.

(c) If the molecule were cut at its midpoint into two halves, it would be possible to distinguish the left half from the right half.

(d) If the two single strands were separated from each other, it would not be possible to distinguish one strand from the other.

(e) In a single strand from this molecule, it would be impossible to determine which is the 3′ end and which is the 5′ end.

16-4. Restriction Mapping of DNA. The genome of a newly discovered bacteriophage is a linear DNA molecule 10,500 nucleotide pairs in length. One sample of this DNA has been incubated with restriction enzyme X and another sample with restriction enzyme Y. The lengths (in thousands of base pairs) of the restriction fragments produced by the two enzymes have been determined by gel electrophoresis to be as follows:

Enzyme X: Fragment X-1 = 4.5; X-2 = 3.6; X-3 = 2.4

Enzyme Y: Fragment Y-1 = 5.2; Y-2 = 3.8; Y-3 = 1.5

Next, the fragments from the enzyme X reaction are isolated and treated with enzyme Y, and the fragments from the enzyme Y reaction are treated with enzyme X. The results are as follows:

X fragments treated with Y : X-1 ⟶ 4.5 (unchanged)
X-2 ⟶ 2.1 + 1.5
X-3 ⟶ 1.7 + 0.7

Y fragments treated with X : Y-1 ⟶ 4.5 + 0.7
Y-2 ⟶ 2.1 + 1.7
Y-3 ⟶ 1.5 (unchanged)

Draw a restriction map of the phage DNA, indicating the positions of all the enzyme X and enzyme Y restriction sites and the lengths of DNA between them.

16-5. DNA Sequencing. You have isolated the DNA fragment shown in Problem 16-3 but do not know its complete sequence. From knowledge of the specificity of the restriction enzyme used to prepare it, you know the first four bases at the left end and have prepared a single-stranded DNA primer of sequence 5′ T–C–G–C 3′. Explain how you would determine the rest of the sequence using dye-labeled dideoxynucleotides. Draw the gel pattern that would be observed, indicating the base sequence of the DNA in each band and the color pattern that would be detected by the camera in a DNA sequencing machine.

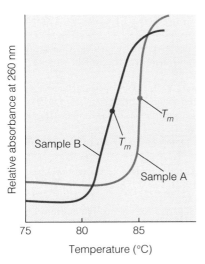

Figure 16-34   Thermal Denaturation of Two DNA Samples.
See Problem 16-6.

• **16-6.   DNA Melting.** Shown in Figure 16-34 are the melting curves for two DNA samples that were thermally denatured under the same conditions.

(a)   What conclusion can you draw concerning the base compositions of the two samples? Explain.

(b)   How might you explain the steeper slope of the melting curve for sample A?

(c)   Formamide and urea are agents known to form hydrogen bonds with pyrimidines and purines. What effect, if any, would the inclusion of a small amount of formamide or urea in the incubation mixture have on the melting curves?

• **16-7.   DNA Renaturation.** You are given two samples of DNA, each of which melts at 92°C when thermal denaturation is carried out. After denaturing the DNA, you mix the two samples together and then cool the mixture to allow the DNA strands to reassociate. When the newly reassociated DNA is denatured a second time, the sample now melts at 85°C.

(a)   How might you explain the lowering of the melting temperature from 92°C to 85°C?

(b)   What kind of experiment could be carried out to test your hypothesis?

(c)   If the newly reassociated DNA had melted at 92°C instead of 85°C, what conclusions might you have drawn concerning the base sequences of the two initial DNA samples?

• **16-8.   Nucleosomes.** You perform an experiment in which chromatin is isolated from sea urchin sperm cells and briefly digested with micrococcal nuclease. When the chromatin proteins are removed and the resulting purified DNA is analyzed by gel electrophoresis, you observe a series of DNA fragments that are multiples of 260 base pairs in length (i.e., 260 bp, 520 bp, 780 bp, and so forth).

(a)   Although these results differ somewhat from the typical results discussed in the chapter, explain why they still point to the likely existence of nucleosomes in this cell type.

(b)   What can you conclude about the amount of DNA that is associated with each nucleosome?

(c)   If the chromatin had been analyzed by density gradient centrifugation immediately after digestion with micrococcal nuclease, describe what you would expect to see.

(d)   Suppose you perform an experiment in which the chromatin is digested for a much longer period of time with micrococcal nuclease prior to removal of chromatin proteins. When the resulting DNA preparation is analyzed by electrophoresis, all of the DNA appears as fragments 146 bp in length. What does this suggest to you about the length of the linker DNA in this cell type?

16-9.   Nuclear Structure and Function. Indicate the implications for nuclear structure or function of each of the following experimental observations.

(a)   Sucrose crosses the nuclear envelope so rapidly that its rate of movement cannot be accurately measured.

(b)   Colloidal gold particles with a diameter of 5.5 nm equilibrate rapidly between the nucleus and cytoplasm when injected into an amoeba, but gold particles with a diameter of 15 nm do not.

(c)   Nuclear pore complexes sometimes stain heavily for ribonucleoprotein.

(d)   If gold particles up to 26 nm in diameter are coated with a polypeptide containing a nuclear localization signal (NLS) and are then injected into the cytoplasm of a living cell, they are transported into the nucleus. If they are injected into the nucleus, however, they remain there.

(e)   Many of the proteins of the nuclear envelope appear from electrophoretic analysis to be the same as those found in the endoplasmic reticulum.

(f)   Ribosomal proteins are synthesized in the cytoplasm but are packaged with rRNA into ribosomal subunits in the nucleus.

(g)   If nucleoli are irradiated with a microbeam of ultraviolet light, synthesis of ribosomal RNA is inhibited.

(h)   Treatment of nuclei with the nonionic detergent Triton X-100 dissolves away the nuclear envelope but leaves an otherwise intact nucleus.

16-10.   Nucleoli. Indicate whether each of the following statements is true (T) or false (F). If false, reword the statement to make it true.

(a)   Nucleoli are membrane-bounded structures present in the eukaryotic nucleus.

(b)   The fibrils seen in electron micrographs of nucleoli contain DNA and RNA.

(c)   The DNA of nucleoli carries the cell's tRNA genes, which are present in clusters of multiple copies.

(d)   A single nucleolus always corresponds to a single nucleolus organizer region (NOR).

(e)   Nucleoli become heavily radiolabeled when radioactive ribonucleotides are provided to the cell.

(f)   In animals and plants, the disappearance of nucleoli during mitosis correlates with cessation of ribosome synthesis.

Suggested Reading

References of historical importance are marked with a •.

The Chemical Nature of the Genetic Material

• Avery, O. T., C. M. MacLeod, and M. McCarty. Studies on the chemical nature of the substance inducing transformation of pneumococcal types. Induction of transformation by a desoxyribonucleic acid fraction isolated from *Pneumococcus* Type III. *J. Exp. Med.* 79 (1944): 137.

• Chargaff, E. Preface to a grammar of biology: A hundred years of nucleic acid research. *Science* 172 (1971): 637.

• Hershey, A. D., and M. Chase. Independent functions of viral protein and nucleic acid in growth of bacteriophage. *J. Gen. Physiol.* 36 (1952): 39.

• Portugal, F. H., and J. S. Cohen. *The Century of DNA: A History of the Discovery of the Structure and Function of the Genetic Substance.* Cambridge, MA: MIT Press, 1977.

DNA Structure

• Bauer, W. R., F. H. C. Crick, and J. H. White. Supercoiled DNA. *Sci. Amer.* 243 (July 1980): 118.

Marmur, J. DNA strand separation, renaturation and hybridization. *Trends Biochem. Sci.* 19 (1994): 343.

• Sanger, F. Determination of nucleotide sequences in DNA. *Science* 214 (1981): 1205.

• Watson, J. D. *The Double Helix.* New York: Atheneum, 1968.

• Watson, J. D., and F. H. C. Crick. Molecular structure of nucleic acids: A structure for deoxyribose nucleic acid. *Nature* 171 (1953): 737.

The Organization of DNA in Genomes

Arabidopsis Genome Initiative. Analysis of the genome sequence of the flowering plant *Arabidopsis thaliana. Nature* 408 (2000): 796.

Blattner, F. R., G. Plunkett III, C. A. Bloch, *et al.* The complete genome sequence of *Escherichia coli* K-12. *Science* 277 (1997): 1453.

• Britten, R. J., and D. E. Kohne. Repeated sequences of DNA. *Science* 161 (1968): 529.

Findlay, I., A. Taylor, P. Quirke, R. Frazier, and A. Urquhart. DNA fingerprinting from single cells. *Nature* 389 (1997): 555.

International Human Genome Sequencing Consortium. Initial sequencing and analysis of the human genome. *Nature* 409 (2001): 860.

Moxon, E. R., and C. Willis. DNA microsatellites: Agents of evolution? *Sci. Amer.* 280 (January 1999): 94.

Venter, J. C., et al. The sequence of the human genome. *Science* 291 (2001): 1304.

DNA Packaging

• Kornberg, R. D., and A. Klug. The nucleosome. *Sci. Amer.* 244 (February 1981): 52.

Kornberg, R. D., and Y. Lorch. Twenty-five years of the nucleosome, fundamental particle of the eukaryote chromosome. *Cell* 98 (1999): 285.

van Holde, K., and J. Zlatanova. Chromatin higher order structure: Chasing a mirage? *J. Biol. Chem.* 270 (1995): 8373.

Wallace, D. C. Mitochondrial DNA in aging and disease. *Sci. Amer.* 277 (August 1997): 40.

Wolffe, A. *Chromatin: Structure and Function,* 3d ed. San Diego: Academic Press, 1998.

The Nucleus

Berezney, R., and K. W. Jeon, eds. *Nuclear Matrix: Structural and Functional Organization.* San Diego: Academic Press, 1997.

Carmo-Fonseca, M., L. Mendes-Soares, and I. Campos. To be or not to be in the nucleolus. *Nature Cell Biol.* 2 (2000): E107.

Daneholt, B. A look at messenger RNP moving through the nuclear pore. *Cell* 88 (1997): 585.

Dingwall, C., and R. Laskey. The nuclear membrane. *Science* 258 (1992): 942.

Görlich, D., and U. Kutay. Transport between the cell nucleus and the cytoplasm. *Annu. Rev. Cell Dev. Biol.* 15 (1999): 607.

Matera, A. G. Nuclear bodies: Subdomains of the interchromatin space. *Trends Cell Biol.* 9 (1999): 302.

Shaw, P. J., and E. G. Jordan. The nucleolus. *Annu. Rev. Cell Dev. Biol.* 11 (1995): 93.

Weis, K. Importins and exportins: How to get in and out of the nucleus. *Trends Biochem. Sci.* 23 (1998): 185.

Wente, S. R. Gatekeepers of the nucleus. *Science* 288 (2000): 1374.

17

The Cell Cycle: DNA Replication, Mitosis, and Cancer

The ability to grow and reproduce is a fundamental property of living organisms. Whether an organism is composed of a single cell or trillions of cells, individual cells must be able to grow and divide. Cell growth is accomplished through the synthesis of new molecules of proteins, nucleic acids, carbohydrates, and lipids. As the accumulation of these molecules causes the volume of a cell to increase, the plasma membrane expands to prevent the cell from bursting. But cells cannot continue to enlarge indefinitely; as a cell grows larger, there is an accompanying decrease in its surface area/volume ratio and hence in its capacity for effective exchange with the environment. For this reason, cell growth must be accompanied by **cell division,** whereby one cell gives rise to two new daughter cells. (The term *daughter* is used by convention and does not indicate that cells have gender.) For single-celled organisms, cell division increases the total number of individuals in a population. In multicellular organisms, cell division either increases the number of cells, leading to growth of the organism, or replaces cells that have died. In an adult human, for example, about 2 million stem cells in bone marrow divide every second to maintain a constant number of red blood cells in the body.

When cells grow and divide, the newly formed daughter cells are usually genetic duplicates of the parent cell, containing the same (or virtually the same) DNA sequences. Therefore, all the genetic information in the nucleus of the parent cell must be duplicated and carefully distributed to the daughter cells during the division process. In accomplishing this task a cell passes through a series of discrete stages, collectively known as the cell cycle. In the present chapter we will examine the events associated with the cell cycle, paying special attention to the mechanisms that ensure each newly forming daughter

cell receives a complete set of genetic instructions. We then explore how the cell cycle is regulated to fit the needs of the organism, and finally end the chapter by discussing how abnormalities in cell cycle regulation contribute to the development of cancer.

An Overview of the Cell Cycle

The **cell cycle** begins when two new cells are formed by the division of a single parental cell and ends when one of these cells divides again into two cells (Figure 17-1). To early cell biologists studying eukaryotic cells with the microscope, the most dramatic events in the life of a cell were those associated with the point in the cycle when the cell actually divides. This division process, called the **M phase,** involves two overlapping events in which the nucleus divides first and the cytoplasm second. Nuclear division is called **mitosis,** and the division of the cytoplasm to produce two daughter cells is termed **cytokinesis.**

The stars of the mitotic drama are the chromosomes. As you can see in Figure 17-1a, the beginning of mitosis is marked by condensation (coiling and folding) of the cell's chromatin, which generates chromosomes that are thick enough to be individually discernible under the microscope. Because DNA replication has already taken place, each chromosome actually consists of two chromosome copies that remain attached to each other until the cell divides. As long as they remain attached, the two new chromosomes are referred to as **sister chromatids,** and the attachment site that holds each pair of chromatids together is called the *centromere*. As the chromosomes become visible, the nuclear envelope breaks into fragments.

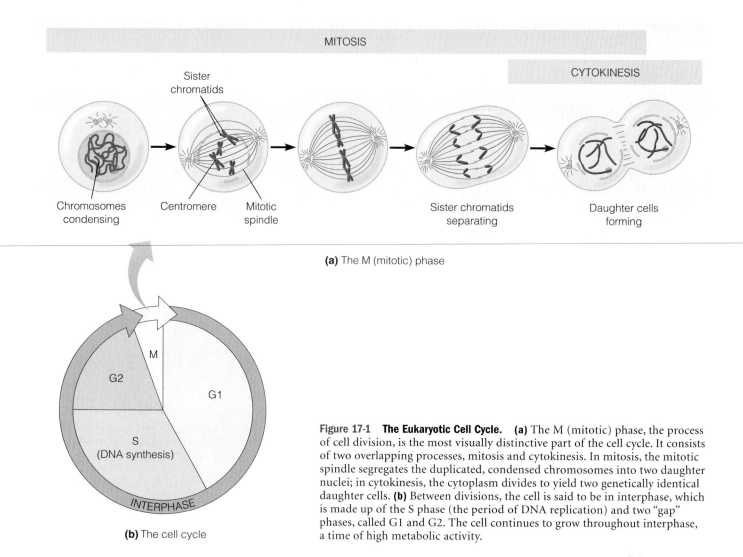

Sister chromatids

Chromosomes condensing

Centromere

Mitotic spindle

Sister chromatids separating

Daughter cells forming

(a) The M (mitotic) phase

M

G2

G1

S
(DNA synthesis)

INTERPHASE

(b) The cell cycle

Figure 17-1   The Eukaryotic Cell Cycle.   (a) The M (mitotic) phase, the process of cell division, is the most visually distinctive part of the cell cycle. It consists of two overlapping processes, mitosis and cytokinesis. In mitosis, the mitotic spindle segregates the duplicated, condensed chromosomes into two daughter nuclei; in cytokinesis, the cytoplasm divides to yield two genetically identical daughter cells. **(b)** Between divisions, the cell is said to be in interphase, which is made up of the S phase (the period of DNA replication) and two "gap" phases, called G1 and G2. The cell continues to grow throughout interphase, a time of high metabolic activity.

Then, in a stately ballet guided by the microtubules of the *mitotic spindle,* the sister chromatids separate and—each now a full-fledged chromosome—move to opposite ends of the cell. By this time cytokinesis has usually begun, and new nuclear membranes envelop the two groups of daughter chromosomes as cell division is completed.

While visually striking, the events of the mitotic phase account for a relatively small portion of the total cell cycle; for a typical mammalian cell, the mitotic phase usually lasts less than an hour. Cells spend the majority of their time in the growth phase between divisions, called **interphase** (Figure 17-1b). Most cellular contents are synthesized continuously during interphase, so cell mass gradually increases as the cell approaches division. Because the microscopic appearance of the chromatin does not change much during interphase, cell biologists originally believed that chromosome replication must take place during M phase. However, this idea was dispelled when DNA staining techniques showed that the amount of DNA doubles during interphase rather than M phase. Subsequent experiments using radioactive DNA precursors revealed that DNA is synthesized during a defined period of interphase, which was named the **S phase** (S for synthesis). A time gap called **G1 phase** separates the S phase from the preceding M phase, and a second gap, the **G2 phase,** separates the end of S phase from the beginning of the next M phase.

Although the cells of a multicellular organism divide at varying rates, most studies of the cell cycle involve cells growing in culture, where the length of the cycle tends to be similar for different cell types. We can easily determine the overall length of the cell cycle—the *generation time*—for cultured cells by counting the cells under a microscope and determining how long it takes for the cell population to double. In cultured mammalian cells, for example, the total cycle usually takes about 18–24 hours. Once we know the total length of the cycle, it is possible to determine the length of specific phases. To determine the length of the S phase, we can expose cells to a radioactively labeled DNA precursor (usually ^{3}H-thymidine) for a short period of time and then examine the cells by autoradiography. The fraction of cells with silver grains over their nuclei represents the fraction of cells that were somewhere in S phase when the radioactive compound was available. When we multiply this fraction by the total length of the cell cycle, the result is an estimate of the average length of the S phase. For mammalian cells in culture, this fraction is often around 0.33, which indicates that S phase is about 6–8 hours in length. Similarly, we can estimate the length of the M phase by multiplying the generation time by the percentage of the cells that are actually in mitosis at any given time. This percentage is called the **mitotic index.** The mitotic index for cultured mammalian cells is often about 3–5%, which means that M phase lasts less than an hour (usually 30–45 minutes).

In contrast to the S and M phases, whose lengths tend to be quite similar for different mammalian cells, the length of G1 is quite variable, depending on the cell type. Although a typical G1 phase lasts 8–10 hours, some cells spend only minutes or hours in G1, whereas others spend weeks, months, or years. During G1, a major "decision" is made as to whether and when the cell is to divide again. Cells that are arrested in G1 for long periods are often said to be in a **G0 state.** Some cells in G0 are destined never to divide again; most of the nerve cells in your body are in this state. In some cells, a similar kind of arrest also occurs in G2. In general, however, G2 is shorter than G1 and is more uniform in duration among different cell types, usually lasting 4–6 hours.

Now that we have looked at the entire cell cycle in overview, we will consider its workings in detail. Because DNA replication is, in a sense, the purpose of the cell cycle, we will start by examining the S phase.

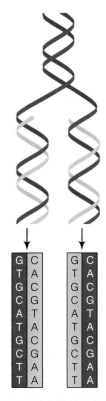

Figure 17-2   The Watson-Crick Model of DNA Replication.   In 1953, Watson and Crick proposed that the DNA double helix replicates semiconservatively, using a model like this one to illustrate the principle. The double-stranded helix unwinds, and each parent strand serves as a template for the synthesis of a complementary daughter strand, assembled according to the base-pairing rules. A, T, C, and G stand for the adenine, thymine, cytosine, and guanine nucleotides. A pairs with T, and G with C.

DNA Replication

One of the most significant features of the double-helical model of DNA is that it immediately suggests a mechanism for DNA replication. In fact, a month after Watson and Crick published their now-classic paper postulating a double helix for DNA, they followed it with an equally important paper suggesting how such a base-paired structure might duplicate itself. Here, in their own words, is the basis of that suggestion:

> Now our model for deoxyribonucleic acid is, in effect, a pair of templates, each of which is complementary to the other. We imagine that prior to duplication the hydrogen bonds are broken, and the two chains unwind and separate. Each chain then acts as a template for the formation onto itself of a new companion chain, so that eventually we shall have two pairs of chains, where we only had one before. Moreover, the sequence of the pairs of bases will have been duplicated exactly (Watson and Crick, 1953, p. 966).

The model Watson and Crick proposed for DNA replication is shown in Figure 17-2. The essence of their suggestion is that one of the two strands of every newly formed DNA molecule is derived from the parent molecule, whereas the other strand is newly synthesized. This is called **semiconservative replication,** because half of the parent molecule is retained by each daughter molecule.

Equilibrium Density Centrifugation Shows That DNA Replication Is Semiconservative

Within five years of its publication, the Watson-Crick model of semiconservative DNA replication was tested and proved correct by Matthew Meselson and Franklin Stahl. The ingenuity of their contribution lay in the method they devised, in collaboration with Jerome Vinograd, for distinguishing semiconservative replication from other possibilities. Their studies utilized two isotopic forms of nitrogen, ^{14}N and ^{15}N, to distinguish newly synthesized strands of DNA from old strands. Bacterial cells were first grown for many generations in a medium containing ^{15}N-labeled ammonium chloride to incorporate this *heavy* (but nonradioactive) isotope of nitrogen into their DNA molecules. Cells containing ^{15}N-labeled DNA were then transferred to a growth medium containing the normal *light* isotope of nitrogen, ^{14}N. Any new strands of DNA synthesized after this transfer would therefore incorporate ^{14}N rather than ^{15}N.

Since ¹⁵N-labeled DNA is significantly denser than ¹⁴N-labeled DNA, the old and new DNA strands can be distinguished from each other by **equilibrium density centrifugation,** a technique we encountered earlier when discussing the separation of organelles (see Figure 12A-7 on p. 330). Briefly, this technique allows organelles or macromolecules with differing densities to be separated from each other by centrifugation in a solution containing a gradient of increasing density from the top of the tube to the bottom. In response to centrifugal force, the particles migrate "down" the tube (actually, they move *outward,* away from the axis of rotation) until they reach a density equal to their own. They then remain at this equilibrium density and can be recovered as a band at that position in the tube after centrifugation.

For DNA analysis, equilibrium density centrifugation often uses cesium chloride (CsCl), a heavy metal salt that forms solutions of very high density. The DNA to be analyzed is simply mixed with cesium chloride and the solution is centrifuged at high speed for a relatively long time (with a modern centrifuge, 80,000 rpm for 8 hours, for example). As a density gradient of cesium chloride is established by the centrifugal force, the DNA molecules float "up" or sink "down" within the gradient to reach their equilibrium density positions. The difference in density between heavy (¹⁵N-containing) DNA and light (¹⁴N-containing) DNA causes them to come to rest at different positions in the gradient (Figure 17-3).

Using this approach, Meselson and Stahl analyzed the DNA obtained from bacterial cells that were first grown for many generations in ¹⁵N and then transferred to ¹⁴N for one or more additional cycles of replication (Figure 17-4). What results would be predicted for a semiconservative mechanism of DNA replication? After one replication cycle in ¹⁴N, each DNA molecule should consist of one ¹⁵N strand (the old strand) and one ¹⁴N strand (the new strand), and so the overall density would be intermediate between heavy DNA and light DNA. The experimental results clearly supported this model. After one replication cycle in the ¹⁴N medium, centrifugation in cesium chloride revealed a single band of DNA whose density was *exactly halfway* between that of ¹⁵N-DNA and ¹⁴N-DNA (Figure 17-4b). Because they saw no band at the density expected for heavy DNA, Meselson and Stahl concluded that the original, double-stranded parental DNA was not preserved intact in the replication process. Similarly, the absence of a band at the density expected for light DNA indicated that no daughter DNA molecules consisted exclusively of newly synthesized nucleotides. Instead, it appeared that a part of every daughter DNA molecule was newly synthesized, while another part was derived from the parent molecule. In fact, the density halfway between that of ¹⁴N-DNA and ¹⁵N-DNA meant that the hybrid DNA molecules were one-half parental and one-half newly synthesized, just as predicted by the semiconservative model of replication.

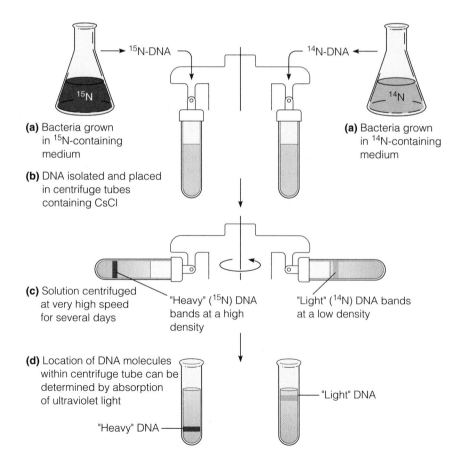

(a) Bacteria grown in ¹⁵N-containing medium

(b) DNA isolated and placed in centrifuge tubes containing CsCl

(a) Bacteria grown in ¹⁴N-containing medium

(c) Solution centrifuged at very high speed for several days

"Heavy" (¹⁵N) DNA bands at a high density

"Light" (¹⁴N) DNA bands at a low density

(d) Location of DNA molecules within centrifuge tube can be determined by absorption of ultraviolet light

"Light" DNA

"Heavy" DNA

Figure 17-3 Equilibrium Density Centrifugation in DNA Analysis. Equilibrium density centrifugation can be used to distinguish between heavy (¹⁵N-containing) and light (¹⁴N-containing) DNA. **(a)** If bacterial cells are grown on either ¹⁵N or ¹⁴N medium for many generations, we can distinguish the DNA from the two cultures by **(b)** placing it in tubes containing cesium chloride at the appropriate concentration and **(c)** centrifuging the tubes at a very high speed. Under these conditions, the CsCl forms a gradient of increasing density and the DNA molecules move to a position in the gradient corresponding to their equilibrium, or buoyant, density. Hence DNA containing atoms of ¹⁵N will form a band of higher density than will DNA containing atoms of ¹⁴N. **(d)** After centrifugation, the DNA bands can be visualized by their absorption of ultraviolet light. The density difference (about 1%) is sufficient not only to resolve the two bands, but also to detect hybrid DNA molecules of an intermediate density, as in Figure 17-4c.

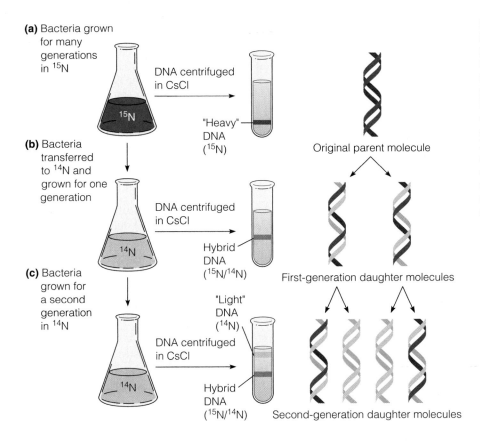

(a) Bacteria grown for many generations in ^{15}N

DNA centrifuged in CsCl

"Heavy" DNA (^{15}N)

(b) Bacteria transferred to ^{14}N and grown for one generation

DNA centrifuged in CsCl

Hybrid DNA ($^{15}N/^{14}N$)

(c) Bacteria grown for a second generation in ^{14}N

DNA centrifuged in CsCl

"Light" DNA (^{14}N)

Hybrid DNA ($^{15}N/^{14}N$)

Original parent molecule

First-generation daughter molecules

Second-generation daughter molecules

Figure 17-4 Semiconservative Replication of Density-Labeled DNA. Meselson and Stahl **(a)** grew bacteria for many generations on an ^{15}N-containing medium and then transferred the cells to an ^{14}N-containing medium for **(b)** one or **(c)** two further cycles of replication. In each case, DNA was extracted from the cells and centrifuged to equilibrium in cesium chloride, as described in Figure 17-3. Bacterial cultures appear on the left, cesium chloride gradients in the center, and schematic illustrations of the DNA molecules on the right. Dark blue strands contain ^{15}N, whereas light blue strands are synthesized with ^{14}N.

The data from cells allowed to grow in the presence of ^{14}N for additional generations provided further confirmation. After the second cycle of DNA replication, for example, Meselson and Stahl saw two equal bands, one at the hybrid density of the previous cycle and one at the density of purely ^{14}N-DNA (Figure 17-4c). As the figure illustrates, this is also consistent with a semiconservative mode of replication.

From these findings, Meselson and Stahl concluded "that the nitrogen of a DNA molecule is divided equally between two physically continuous subunits; that, following duplication, each daughter molecule receives one of these; and that the subunits are conserved through many duplications" (1958, p. 682). Further experimentation was then required to prove that the "physically continuous subunits" into which DNA is partitioned are indeed separate DNA strands. Meselson and Stahl obtained this proof by heating the $^{14}N/^{15}N$ hybrid DNA to separate its two strands and then showing that one strand exhibited the density of a ^{15}N-containing strand and the other exhibited the density of a ^{14}N-containing strand. Meanwhile, other researchers had used radioactive labeling and autoradiography to look at the process of DNA replication in eukaryotic chromosomes. In the end, Watson and Crick were proven right—DNA is replicated semiconservatively, and in all organisms.

DNA Replication Is Usually Bidirectional

The Meselson-Stahl experiments provided strong verification for the idea that during DNA replication, each strand

of the DNA double helix serves as a template for the synthesis of a new complementary strand. As biologists proceeded to unravel the molecular details of this process, it gradually became clear that DNA replication is a complex event involving numerous enzymes and other proteins, and even the participation of RNA. We will first examine the general features of this replication mechanism and then focus on some of its molecular details. In doing so, we will frequently refer to *E. coli*, where DNA replication is especially well understood. However, recent studies of mammalian viruses such as SV40 and of the yeast *Saccharomyces cerevisiae* have begun to reveal the details of eukaryotic DNA replication as well. DNA replication seems to be a drama whose plot and molecular actors are basically similar in prokaryotic and eukaryotic cells. This is perhaps not surprising for such a fundamental process—one that must have arisen very early in the evolution of life.

The first experiments to directly visualize DNA replication were carried out by John Cairns. Cairns grew *E. coli* cells for varying lengths of time in a medium containing the DNA precursor ^3H-thymidine and then used autoradiography to examine DNA molecules caught in the act of replication. One such molecule is shown in Figure 17-5a. The two Y-shaped structures indicated by the arrows represent the sites at which the DNA duplex is being replicated. These **replication forks** are created by a replication process that begins at a specific point and moves along the DNA, unwinding the helix and copying

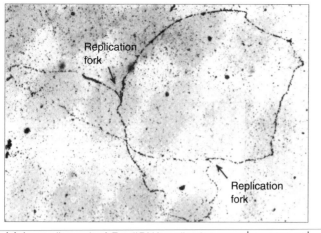

(a) Autoradiograph of *E.coli* DNA replication

|— 0.25 μm —|

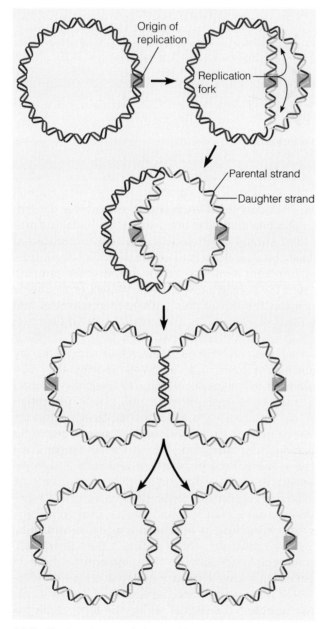

Origin of replication

Replication fork

Parental strand

Daughter strand

(b) Replication process of circular DNA

Figure 17-5 <u>**Replication of Circular DNA.**</u> **(a)** This autoradiograph shows an *E. coli* DNA molecule caught in the act of replication. The bacterium from which the molecule came had been grown in a medium containing ^{3}H-thymidine, thereby ensuring that the DNA molecule could be visualized by autoradiography. **(b)** Replication of a circular DNA molecule begins at a single origin and proceeds bidirectionally around the circle, with the two replication forks moving in opposite directions. The new strands are shown in light blue. The replication process generates intermediates that resemble the Greek letter theta (θ), from which this type of replication derives its name.

both strands as it goes. The specific point where replication is initiated contains a special DNA sequence called an **origin of replication.** To start the replication process, a specific group of *initiator proteins* must bind to the origin and, using energy derived from ATP, unwind the double helix to give the rest of the replication machinery access to single-stranded DNA.

In most cases, DNA replication proceeds from the origin in a *bidirectional* fashion—that is, two replication forks are formed that move in opposite directions away from the origin. For circular DNA, this process is called *theta replication* because it generates intermediates that look like the Greek letter theta (θ), as you can see in Figure 17-5b. Most circular DNA molecules are replicated in this way, starting from a single origin. Theta replication occurs not only in bacterial genomes such as that of *E. coli,* but also in the circular DNAs of plasmids, mitochondria, chloroplasts, and some viruses.

Eukaryotic DNA Replication Involves Multiple Replicons

In contrast to circular bacterial chromosomes—where DNA replication is initiated at a single origin—replication of the linear DNA molecules found in eukaryotic chromosomes is initiated at multiple sites, creating multiple replication units called **replicons** (Figure 17-6). The DNA of a typical large eukaryotic chromosome may contain several thousand replicons, each about 50,000–300,000 base pairs in length. At the center of each replicon is an origin of replication where DNA synthesis is initiated. The initiation of replication at each replicon requires that a small group of initiator proteins, called an *origin recognition complex (ORC)*, bind to the origin of replication, followed by the binding of several other proteins that are needed before DNA synthesis can be initiated. Following initiation, two replication forks begin to synthesize DNA in opposite directions away from the origin, creating a "replication bubble" that grows in size as replication proceeds in both directions. When the growing replication bubble of one replicon encounters the replication bubble of an adjacent replicon, the DNA synthesized by the two replicons is joined together. In this way, DNA synthesized at numerous replication sites is ultimately linked together to form two double-stranded daughter molecules, each composed of one parental strand and one new strand.

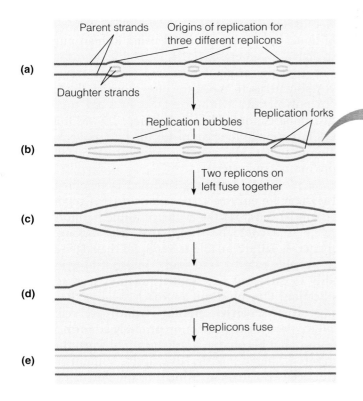

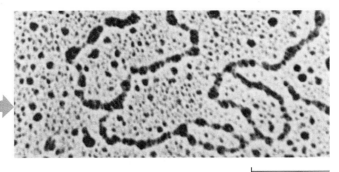

0.25 μm

Figure 17-6 **Multiple Replicons in Eukaryotic DNA.** Replication of linear eukaryotic DNA molecules is initiated at numerous origins along the DNA; the timing of initiation is specific for each cluster of origins. **(a)** Replication bubbles form at origins. **(b)** The bubbles grow as the replication forks move along the DNA in both directions from each origin. The micrograph shows three replication bubbles in DNA from cultured Chinese hamster cells (TEM). **(c)** Eventually, individual bubbles meet and fuse. **(d)** A Y-shaped structure forms as a replication fork reaches the end of a DNA molecule. **(e)** When all bubbles have fused, replication is complete, and the two daughter molecules separate.

The DNA sequences that act as replication origins for each replicon have been identified in yeast cells by isolating various naturally occurring DNA fragments and inserting them into DNA molecules that lack the ability to replicate. If the inserted DNA fragment gives the DNA molecule the ability to replicate within the yeast cell, it is called an *autonomously replicating sequence,* or *ARS element.* The number of ARS elements detected in normal yeast chromosomes is similar to the total number of replicons, suggesting that ARS sequences function as replication origins. ARS elements contain an 11-nucleotide sequence consisting largely of AT base pairs. Since the DNA double helix must be unwound when replication is initiated, the presence of AT base pairs serves a useful purpose at replication origins because AT base pairs, held together by two hydrogen bonds, are easier to disrupt than GC base pairs, which have three hydrogen bonds.

Why does DNA replication involve multiple replicons in eukaryotes but not in prokaryotes? Since eukaryotic chromosomes contain more DNA than bacterial chromosomes, it would take eukaryotes much longer to replicate their chromosomes if DNA synthesis were initiated from only a single replication origin. Moreover, the rate at which each replication fork synthesizes DNA is slower in eukaryotes than in bacteria (presumably because the presence of nucleosomes slows down the replication process). Measurements of the length of radioactive DNA synthesized by cells exposed to ^{3}H-thymidine for varying periods of time have revealed that eukaryotic replication forks synthesize DNA at a rate of about 2000 base pairs/minute, compared

with 50,000 base pairs/minute in bacteria. Since the average human chromosome contains about 10^8 base pairs of DNA, it would take more than a month to duplicate a chromosome if there were only a single replication origin!

The relationship between the speed of chromosome replication and the number of replicons is nicely illustrated by comparing the rates of DNA synthesis in embryonic and adult cells of the fruit fly *Drosophila.* In the developing embryo, where cell division must proceed very rapidly, embryonic cells employ a large number of simultaneously active replicons measuring only a few thousand base pairs in length. As a result, DNA replicates very rapidly and S phase takes only a few minutes. Adult cells, on the other hand, employ fewer replicons spaced at intervals of tens or hundreds of thousands of base pairs, generating an S phase that requires almost 10 hours to complete. Since the rate of DNA synthesis at any given replication fork is about the same in embryonic and adult cells, it is clear that the length of the S phase is determined by the number of replicons and the rate at which they are activated, and not by the rate at which each replicon synthesizes DNA.

During the S phase of a typical eukaryotic cell cycle, replicons are not all activated at the same time. Instead, certain clusters of replicons tend to replicate early during S phase whereas others replicate later. Information concerning the order in which replicons are activated has been obtained by incubating cells at various points during S phase with *5-bromodeoxyuridine (BrdU),* a substance that is incorporated into DNA in place of thymidine.

Because DNA that contains BrdU is denser than normal DNA, it can be separated from the remainder of the DNA by equilibrium density centrifugation. The BrdU-labeled DNA is then analyzed by hybridization with a series of DNA probes that are specific for individual genes. Such studies have revealed that genes that are being actively expressed in a given tissue are replicated early during S phase, whereas inactive genes are replicated later during S phase. If the same gene is analyzed in two different cell types, one in which the gene is active and one in which it is inactive, early replication is observed only in the cell type where the gene is being transcribed.

DNA Polymerases Catalyze the Elongation of DNA Chains

When the semiconservative model of DNA replication was first proposed in the early 1950s, biologists thought that DNA replication was so complex it could be carried out only by intact cells. A few years later, however, Arthur Kornberg found that an enzyme he had isolated from bacterial cells could copy DNA molecules in a test tube. This enzyme, which he named **DNA polymerase,** required that a small amount of DNA be initially present to act as a template. In the presence of such a template, DNA polymerase catalyzes the elongation of DNA chains using as substrates the triphosphate deoxynucleoside derivatives of the four bases found in DNA (dATP, dTTP, dGTP, and dCTP). As each of these substrates is incorporated into a newly forming DNA chain, its two terminal phosphate groups are released. Since deoxynucleoside triphosphates are high-energy compounds whose free energy of hydrolysis is comparable to that of ATP, the energy released as these phosphate bonds are broken drives what would otherwise be a thermodynamically unfavorable polymerization reaction.

In the DNA polymerase reaction, incoming nucleotides are covalently bonded to the 3′ hydroxyl end of the growing DNA chain. Each successive nucleotide is linked to the growing chain by a phosphoester bond between the phosphate group on its 5′ carbon and the hydroxyl group on the 3′ carbon of the nucleotide added in the previous step (Figure 17-7). In other words, chain elongation occurs at the 3′ end of a DNA strand and the strand is therefore said to grow in the 5′ → 3′ direction.

Soon after Kornberg's initial discovery, several other forms of DNA polymerase were detected in prokaryotic and eukaryotic cells. (Table 17-1 lists the main DNA polymerases used in DNA replication, along with other key proteins involved in the process.) In *E. coli*, the enzyme discovered by Kornberg has turned out not to be responsible for DNA replication in intact cells. This fact first became apparent when Peter DeLucia and John Cairns reported that mutant strains of bacteria lacking the Kornberg enzyme can still replicate their DNA and reproduce normally. With the Kornberg enzyme missing, it was possible to detect the presence of several other bacterial enzymes that synthesize DNA. These additional enzymes are named using Roman numerals (e.g., DNA polymerases II, III, IV, and V) to distinguish them from the original Kornberg enzyme, now called DNA polymerase I. When the rates at which the various DNA polymerases synthesize DNA in a test tube were first compared, only DNA polymerase III was found to work fast enough to account for the rate of DNA replication in intact cells, which averages about 50,000 base pairs/minute in bacteria.

Such observations suggested that DNA polymerase III is the main enzyme responsible for DNA replication in bacterial cells, but the evidence would be more convincing if it could be shown that cells lacking DNA polymerase III are unable to replicate their DNA. How is it possible to grow and study cells that have lost the ability to carry out an essential function such as DNA replication? One powerful approach involves the use of **temperature-sensitive mutants,** which are cells that produce proteins that function properly at normal temperatures but become seriously impaired when the temperature is altered slightly. For example, mutant bacteria have been isolated in which DNA polymerase III behaves normally at 37°C, but loses its function when the temperature is raised to 42°C. Such bacteria grow normally at 37°C, but lose the ability to replicate their DNA when the temperature is elevated to 42°C, indicating that DNA polymerase III plays an essential role in the process of normal DNA replication.

Though the preceding observations indicate that DNA polymerase III is central to bacterial DNA replication, it is not the only enzyme involved. As we will see shortly, a variety of other proteins are required for DNA replication, including DNA polymerase I. The other main types of bacterial DNA polymerase (II, IV, and V) play more specialized roles in events associated with DNA repair, a process to be described in detail later in the chapter.

Like bacteria, eukaryotic cells contain several types of DNA polymerase. At last count, more than a dozen different enzymes have been identified, each named with a different Greek letter. From among this group, DNA polymerases α (alpha), δ (delta), and perhaps ε (epsilon) are involved in nuclear DNA replication. DNA polymerase γ (gamma) is present only in mitochondria and is the main polymerase used in mitochondrial DNA replication. Most of the remaining eukaryotic DNA polymerases are involved either in DNA repair or in replication across regions of DNA damage.

In addition to their biological functions inside cells, DNA polymerases have found important practical applications in the field of biotechnology. Box 17A describes a technique called the **polymerase chain reaction (PCR),** in which a special type of DNA polymerase is the prime tool. Used for the rapid amplification of tiny samples of DNA, PCR is a powerful adjunct to the DNA fingerprinting method discussed in Chapter 16.

Figure 17-7 The Directionality of DNA Synthesis. Addition of the next nucleotide to a growing DNA strand is catalyzed by DNA polymerase and always occurs at the 3′ end of the strand. A phosphoester bond is formed between the 3′ hydroxyl group of the terminal nucleotide and the 5′ phosphate of the incoming deoxynucleoside triphosphate (here dTTP), extending the growing chain by one nucleotide, liberating pyrophosphate (PP$_i$), and leaving the 3′ end of the strand with a free hydroxyl group to accept the next nucleotide.

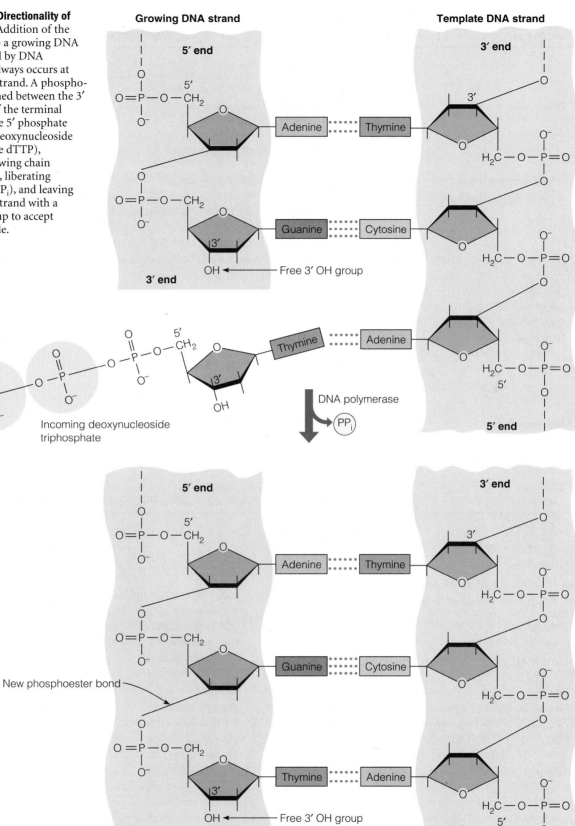

Table 17-1   Important DNA Replication Proteins

Protein	Cell Type	Main Activities and/or Functions
DNA polymerase I	Prokaryotic	DNA synthesis; 3′→5′ exonuclease (for proofreading); 5′→3′ exonuclease; removes and replaces RNA primers used in DNA replication (also functions in excision repair of damaged DNA)
DNA polymerase III	Prokaryotic	DNA synthesis; 3′→5′ exonuclease (for proofreading); used in synthesis of both DNA strands
DNA polymerase α (alpha)	Eukaryotic	Nuclear DNA synthesis; forms complex with primase and begins DNA synthesis at the 3′ end of RNA primers for both leading and lagging strands (also functions in DNA repair)
DNA polymerase γ (gamma)	Eukaryotic	Mitochondrial DNA synthesis
DNA polymerase δ (delta)	Eukaryotic	Nuclear DNA synthesis; 3′→5′ exonuclease (for proofreading); involved mainly in leading-strand synthesis (also functions in DNA repair)
DNA polymerase ε (epsilon)	Eukaryotic	Nuclear DNA synthesis; 3′→5′ exonuclease (for proofreading); carries out lagging-strand synthesis (also functions in DNA repair)
Primase	Both	RNA synthesis; makes RNA oligonucleotides that are used as primers for DNA synthesis
Helicase	Both	Unwinds double-stranded DNA
Single-strand binding protein (SSB)	Both	Binds to single-stranded DNA; stabilizes strands of unwound DNA in an extended configuration that facilitates access by other proteins
DNA topoisomerase (type I and type II)	Both	Makes single-strand cuts (type I) or double-strand cuts (type II) in DNA; induces and/or relaxes DNA supercoiling; can serve as swivel to prevent overwinding ahead of the DNA replication fork; can separate linked DNA circles at the end of DNA replication
DNA gyrase	Prokaryotic	Type II DNA topoisomerase that serves as a swivel to relax supercoiling ahead of the DNA replication fork in *E. coli*
DNA ligase	Both	Makes covalent bonds to join together adjacent DNA strands, including the Okazaki fragments in lagging-strand DNA synthesis and the new and old DNA segments in excision repair of DNA
Initiator proteins	Both	Bind to origin of replication and initiate unwinding of DNA double helix
Telomerase	Eukaryotic	Using an integral RNA molecule as template, synthesizes DNA for extension of telomeres (sequences at ends of chromosomal DNA)

DNA Is Synthesized as Discontinuous Segments That Are Joined Together by DNA Ligase

The discovery of DNA polymerase was merely the first step in unraveling the mechanism of DNA replication. An early conceptual problem arose from the finding that *DNA polymerases can only catalyze the addition of nucleotides to the 3′ end of an existing DNA chain;* in other words, DNA polymerases synthesize DNA exclusively in the 5′ → 3′ direction. Yet the two strands of the DNA double helix run in opposite directions. So how does an enzyme that functions solely in the 5′ → 3′ direction manage to replicate a DNA molecule that contains one chain running in the 5′ → 3′ direction and one chain running in the 3′ → 5′ direction?

An answer to this question was first proposed in 1968 by Reiji Okazaki, whose experiments suggested that DNA is synthesized as small fragments that are later joined together. Okazaki isolated DNA from bacterial cells that had been briefly exposed to a radioactive substrate that is incorporated into newly synthesized DNA. Analysis of this DNA revealed that much of the radioactivity was located in small DNA fragments measuring about a thousand nucleotides in length (Figure 17-8a). After longer labeling periods the radioactivity became associated with larger DNA molecules. These findings suggested to Okazaki that the smaller DNA pieces, now known as **Okazaki fragments,** are precursors of newly forming larger DNA molecules. Later research revealed that the conversion of Okazaki fragments into larger DNA molecules fails to take place in mutant bacteria that lack the enzyme **DNA ligase,** which joins DNA fragments together by catalyzing the ATP-dependent formation of a phosphoester bond between the 3′ end of one nucleotide chain and the 5′ end of another (Figure 17-8b).

The preceding observations suggest a model of DNA replication that is consistent with the fact that DNA polymerase only synthesizes DNA in the 5′ → 3′ direction. According to this model, DNA synthesis at each replication fork is *continuous* in the direction of fork movement for one strand but *discontinuous* in the opposite direction for the other strand (Figure 17-9). The two daughter strands can therefore be distinguished on the basis of their mode of growth. One of the two new strands, called the **leading strand,** is synthesized as a continuous chain

Contemporary Techniques THE PCR REVOLUTION

The ability to work with minuscule amounts of DNA is proving valuable in a wide range of endeavors, from paleontology to criminology. In Chapter 16, we described how DNA fingerprinting analysis can be used to identify and characterize particular sequences contained in as little as 1 µg of DNA, the amount in a small drop of blood (see Box 16C). But sometimes even that amount of DNA may not be available. In such cases another method, called the *polymerase chain reaction (PCR)*, can come to the rescue. With PCR, it is possible to rapidly replicate, or *amplify*, selected DNA segments that are initially present in extremely small amounts. In only a few hours, PCR can make millions or even billions of copies of a particular DNA sequence, thereby producing enough material for DNA fingerprinting, DNA sequencing, or other uses. Like DNA fingerprinting, PCR is often in the news in connection with the solving of violent crimes.

The complicated, multiprotein system that cells use for DNA replication is not required for the PCR method; neither origins of replication, nor DNA unwinding proteins, nor the apparatus for lagging-strand synthesis are involved. The keys to the simplicity of PCR are an unusual DNA polymerase and the fact that synthetic primers can set up a chain reaction that produces an exponentially growing population of specific DNA molecules. For this insight, biochemist Kary Mullis received a Nobel Prize.

To carry out PCR, it is usually necessary to know part of the base sequence of the DNA segment that one wishes to amplify. Based on this information, short single-stranded *DNA primers* are chemically synthesized; these primers are generally 15–20 nucleotides long and consist of sequences that are complementary to sequences located at the two ends of the DNA segment being amplified. (If sequences that naturally flank the sequence of interest are not known, artificial ones can be attached prior to running the polymerase chain reaction.) DNA polymerase is then added to catalyze the synthesis of complementary DNA strands using the two primers as starting points. The DNA polymerase routinely used for this purpose was first isolated from the bacterium *Thermus aquaticus,* an inhabitant of thermal hot springs where the waters are normally 70–80°C. The optimal temperature for this enzyme, called *Taq* polymerase, is 72°C, and it is stable at even higher temperatures—a property that made possible the automation of PCR.

The way in which the PCR procedure operates is summarized in Figure 17A-1. The ingredients of the initial reaction mixture include the DNA containing the sequence targeted for amplification, *Taq* DNA polymerase, the synthetic DNA primers, and the four deoxynucleoside triphosphates (dATP, dTTP, dCTP, and dGTP). Each reaction cycle begins with a short period of heating to near boiling (95°C) to denature the DNA double helix into its

(continued)

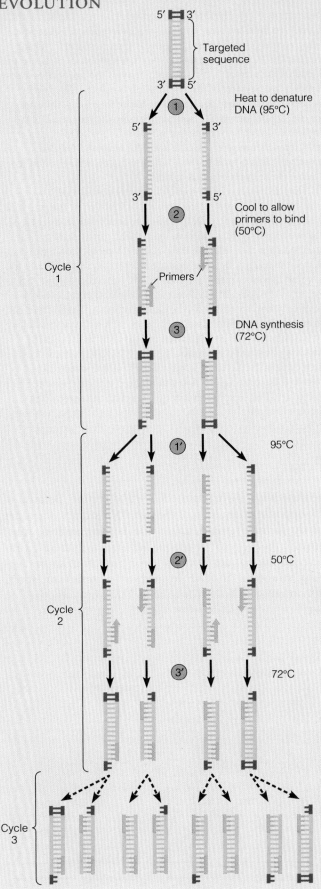

Figure 17A-1   DNA Amplification Using the Polymerase Chain Reaction.   See the description in the box text. PCR works best when the DNA segment to be amplified—the region flanked by the two primers—is 50–2000 nucleotides long. The cooling temperature used to promote primer binding, designated here as 50°C, may vary somewhat, depending on the melting temperature of the particular primer being used.

two strands (①). The DNA solution is then cooled to allow the primers to bind to complementary regions on the DNA strands being copied (②). The temperature is then raised to 72°C and the *Taq* DNA polymerase goes to work, adding nucleotides to the 3′ end of the primer (③). The specificity of the primers ensures the selective copying of the stretches of template DNA downstream from the primers. It takes no more than a few minutes for the *Taq* polymerase to completely copy the targeted DNA sequence, thereby doubling the amount of DNA. The reaction mixture is then heated again to melt the new double helices, more primer is bound to the DNA, and the cycle is repeated to double the amount of DNA again (①–③).

This reaction cycle is repeated as many times as necessary, with each cycle doubling the amount of DNA from the previous cycle. After the third cycle, more and more of the product DNA molecules will be of a uniform length that consists only of the targeted sequence (like the third and sixth molecules in the last line of the figure). Because heating to 95°C does not destroy the *Taq* polymerase, there is no need to add fresh enzyme for each round of the cycle. In most cases, 20–30 reaction cycles are sufficient to produce the desired quantity of DNA. The theoretical amplification accomplished by n cycles is 2^n, so 20 cycles yields an amplification of a millionfold or more ($2^{20} = 1,048,576$), and 30 cycles over a billionfold ($2^{30} = 1,073,741,824$). Since each cycle takes less than 5 minutes, several hundred billion copies of the original DNA sequence can be produced within a few hours. This is considerably quicker than the several days required for amplifying DNA by cloning it in bacteria, a method to be discussed in Chapter 18. Furthermore, PCR can be used

with as little as one molecule of DNA, and it does not require that the starting DNA sample be purified because the primers select the DNA region that will be amplified.

PCR therefore makes it possible to identify a person from the minuscule amount of DNA that is left behind when that person touches an object, inadvertently leaving a few skin cells behind. By using PCR to amplify the tiny amount of DNA in such a sample and then performing a DNA fingerprinting analysis on the amplified DNA, it is possible to obtain a DNA fingerprint from a person's actual fingerprints! Although such techniques have enormous potential in helping to solve crimes, this extraordinary sensitivity can also cause problems. A few contaminating DNA molecules (such as from skin cells shed by a lab technician) might be amplified along with the DNA of interest, yielding misleading results. In a legal case, such an error could lead to grave injustice, and for this reason courts are proceeding cautiously in allowing the introduction of PCR evidence.

Nevertheless, with proper precautions and controls, PCR is proving extremely valuable. As an aid in evolution research, it has been used to amplify DNA fragments recovered from ancient Egyptian mummies, a 40,000-year-old woolly mammoth frozen in a glacier, and a 30-million-year-old plant fossil. In medical diagnosis, PCR has been used to amplify DNA from single embryonic cells for rapid prenatal diagnosis, and it has made possible the detection of viral genes in cells infected with HIV or other viruses. Perhaps most importantly, PCR has revolutionized basic research in molecular genetics by allowing easy amplification of particular genes or sequences from among the thousands of genes in mammalian genomes.

---

because it is growing in the 5′ ⟶ 3′ direction. Because of the opposite orientation of the two DNA strands, the other newly forming strand, called the **lagging strand,** must grow in the 3′ ⟶ 5′ direction. But DNA polymerase cannot add nucleotides in the 3′ ⟶ 5′ direction, so the lagging strand is instead formed as a series of short, discontinuous Okazaki fragments that are synthesized in

the 5′ ⟶ 3′ direction. These fragments are then joined together by DNA ligase to make a continuous new 3′ ⟶ 5′ DNA strand. Okazaki fragments are generally about 1000–2000 nucleotides long in viral and bacterial systems, but only about one-tenth this length in eukaryotic cells. In *E. coli* the same DNA polymerase, polymerase III, is used for synthesizing the Okazaki fragments of the lagging

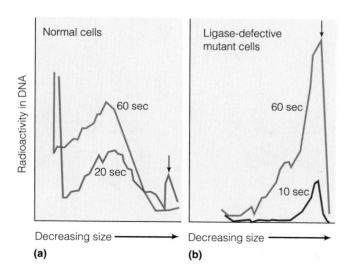

Figure 17-8 Summary of Okazaki's Experiments on the Mechanism of DNA Replication in Bacterial Cells. **(a)** Bacteria were incubated for brief periods with radioactive thymidine to label newly synthesized DNA. The DNA was then isolated, dissociated into its individual strands, and fractionated by centrifugation into molecules of differing size. In normal bacteria incubated with ³H-thymidine for 20 seconds, a significant amount of radioactivity is present in small DNA fragments (arrow). By 60 seconds the radioactivity present in small DNA fragments has all shifted to larger DNA molecules. **(b)** In bacterial mutants deficient in the enzyme DNA ligase, radioactivity remains in small DNA fragments even after 60 seconds of incubation. It was therefore concluded that DNA ligase normally functions to join small DNA fragments together into longer DNA chains.

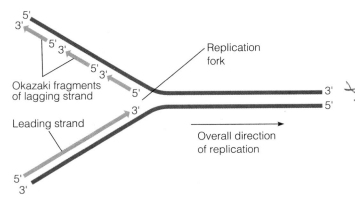

Figure 17-9   Directions of DNA Synthesis at a Replication Fork. Because DNA polymerases function only in the 5′ ⟶ 3′ direction, synthesis at each replication fork is continuous in the direction of fork movement for the leading strand but discontinuous in the opposite direction for the lagging strand. Discontinuous synthesis involves short intermediates called Okazaki fragments, which are 1000–2000 nucleotides long in bacteria and about 100–200 nucleotides long in eukaryotic cells. The fragments are later joined together by the enzyme DNA ligase. Parent DNA is shown in dark blue, newly synthesized DNA in lighter blue. Here and in subsequent figures, arrowheads indicate the direction in which the nucleic acid chain is being elongated.

strand and the continuous DNA chain of the leading strand. In eukaryotes, polymerase ε seems to be the main DNA polymerase for synthesis of the lagging strand, while polymerase δ works on the leading strand.

Proofreading Is Performed by the 3′ ⟶ 5′ Exonuclease Activity of DNA Polymerase

Given the complexity of the preceding model, you might wonder why cells have not simply evolved an enzyme that synthesizes DNA in the 3′ ⟶ 5′ direction. One possible answer is related to the need for error correction during DNA replication. About one out of every 100,000 nucleotides incorporated during DNA replication is incorrectly base-paired with the template DNA strand. Such mis-

takes are usually corrected by a **proofreading** mechanism that utilizes the same DNA polymerase molecules that catalyze DNA synthesis. Proofreading is made possible by the fact that in addition to catalyzing DNA synthesis, almost all DNA polymerases exhibit 3′ ⟶ 5′ exonuclease activity. An **exonuclease** is an enzyme that degrades nucleic acid (usually DNA) from one end, rather than making internal cuts, as *endo*nucleases do. A 3′ ⟶ 5′ exonuclease is one that clips off nucleotides from the 3′ end of a nucleotide chain. Hence the 3′ ⟶ 5′ exonuclease activity of DNA polymerase allows it to remove improperly base-paired nucleotides from the 3′ end of a growing DNA chain (Figure 17-10). This ability to remove incorrect nucleotides can improve the fidelity of DNA replication to an average of only a few errors for every billion base pairs replicated.

If cells did happen to contain an enzyme capable of synthesizing DNA in the 3′ ⟶ 5′ direction, proofreading could not work because a DNA chain growing in the 3′ ⟶ 5′ direction would have a nucleotide triphosphate at its growing 5′ end. If the 5′ nucleotide were an incorrect base that needed to be removed during proofreading, its removal would eliminate the triphosphate group that provides the free energy that allows DNA polymerase to add nucleotides to a growing DNA chain, and hence the chain could not elongate further.

RNA Primers Initiate DNA Replication

Since DNA polymerase can only add nucleotides to an existing nucleotide chain, how is replication of a DNA double helix initiated? Shortly after Okazaki fragments were first discovered, researchers implicated RNA in the initiation process through the following observations: (1) Okazaki fragments often have short stretches of RNA, usually 3–10 nucleotides in length, at their 5′ ends; (2) DNA polymerase can catalyze the addition of nucleotides to the 3′ end of RNA chains as well as to DNA chains; (3) cells contain an enzyme called **primase** that synthesizes RNA fragments about ten bases long using DNA as a template; and (4) unlike DNA polymerase, which adds nucleotides only to the ends of existing chains, primase can initiate RNA synthesis from scratch by joining two nucleotides together.

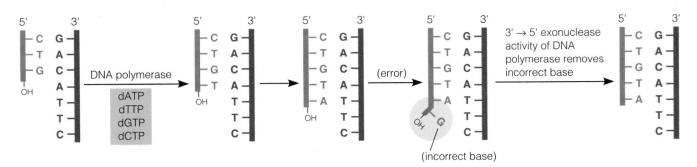

Figure 17-10   Proofreading by 3′ ⟶ 5′ Exonuclease.   If an incorrect base is inserted during DNA replication, the 3′ ⟶ 5′ exonuclease activity that is part of the DNA polymerase molecule catalyzes its removal so that the correct base can be inserted.

These observations led to the conclusion that DNA synthesis is initiated by the formation of short **RNA primers.** RNA primers are synthesized by primase, which uses a single DNA strand as a template to guide the synthesis of a complementary stretch of RNA (Figure 17-11, ①). Primase is a specific kind of RNA polymerase that is involved only in the process of DNA replication. Like all other RNA polymerases, and unlike all DNA polymerases, primases can *initiate* a new polynucleotide strand complementary to a template strand; they do not themselves require a primer.

In *E. coli*, primase is relatively inactive unless it is accompanied by six other proteins, forming a complex called a **primosome.** The other primosome proteins function in unwinding the parental DNA and recognizing target DNA sequences where replication is to be initiated. The situation in eukaryotic cells is slightly different, so the term *primosome* is not used. The eukaryotic primase is not as closely associated with unwinding proteins, but it is very tightly bound to DNA polymerase α, the main DNA polymerase involved in initiating DNA replication.

Once an RNA primer has been created, DNA synthesis can proceed, with DNA polymerase III (or DNA polymerase δ or ε in eukaryotes) adding successive deoxynucleotides to the 3′ end of the primer (Figure 17-11, ②). For the leading strand, initiation using a RNA primer only needs to occur once when a replication fork first forms; DNA polymerase can then add nucleotides to the chain continuously in the 5′ ⟶ 3′ direction. In contrast, the lagging strand is synthesized as a series of discontinuous Okazaki fragments, each of which must be initiated with a separate RNA primer. For each primer, DNA nucleotides are added by DNA polymerase III until the growing fragment reaches the adjacent Okazaki fragment. No longer needed at that point, the RNA segment is removed and DNA nucleotides are polymerized to fill its place. In *E. coli*, the RNA primers are removed by a 5′ ⟶ 3′ exonuclease activity inherent to the DNA polymerase I molecule (distinct from the 3′ ⟶ 5′ exonuclease activity involved in proofreading). At the same time, the DNA polymerase I molecule synthesizes DNA in the normal 5′ ⟶ 3′ direction to fill in the resulting gaps (Figure 17-11, ③). Adjacent fragments are subsequently joined together by DNA ligase.

Why do cells employ RNA primers that must later be removed rather than simply using a DNA primer in the first place? Again, the answer may be related to the need for error correction. We have already seen that DNA polymerase possesses a 3′ ⟶ 5′ exonuclease activity that allows it to remove incorrect nucleotides from the 3′ end of a DNA chain. In fact, DNA polymerase will only elongate an existing DNA chain if the nucleotide present at the 3′ end is properly base-paired. But an enzyme that *initiates* the synthesis of a new chain cannot perform such a proofreading function because it is not adding a nucleotide to an existing base-paired end. As a result, enzymes that initiate nucleic acid synthesis are not very good at correcting errors. By using RNA rather than DNA to initiate DNA synthesis, cells ensure that any incorrect bases inserted during initiation are restricted to RNA sequences destined to be removed by DNA polymerase I.

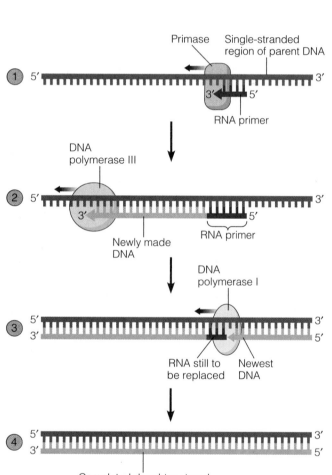

Figure 17-11 The Role of RNA Primers in DNA Replication. DNA synthesis is initiated with a short RNA primer in both prokaryotes and eukaryotes. This figure shows the process as it occurs for the lagging strand in *E. coli*. ① The primer (red) is synthesized by primase, an RNA polymerase that uses a single strand of DNA as its template. In *E. coli*. the primase is part of a protein complex called a primosome (not shown here). ② Once the short stretch of RNA is available, DNA polymerase III uses it as a primer to initiate DNA synthesis, which proceeds in the 5′ ⟶ 3′ direction. ③ The RNA primer is eventually removed by the 5′ ⟶ 3′ exonuclease activity of DNA polymerase I, which replaces ribonucleotides with deoxynucleotides as it proceeds. ④ The daughter strand is completed via joining by DNA ligase.

Unwinding the DNA Double Helix Requires Helicases, Topoisomerases, and Single-Strand Binding Proteins

During DNA replication, the two strands of the double helix must unwind at each replication fork to expose the single strands to the enzymes responsible for copying them. Three classes of proteins with distinct functions facilitate this unwinding process: *helicases, topoisomerases,* and *single-strand binding proteins* (Figure 17-12).

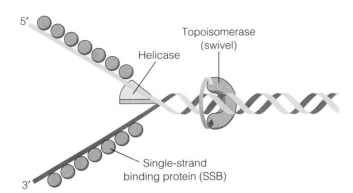

Figure 17-12 Proteins Involved in Unwinding the DNA at the Replication Fork. Three types of proteins are required for this aspect of DNA replication. The actual unwinding proteins are the helicases; the principal one in *E. coli*, which is part of the primosome, operates 5′ → 3′ along the template for the lagging strand, as shown here. Single-strand binding proteins (SSB) stabilize the unwound DNA in an extended position. A topoisomerase forms a swivel ahead of the replication fork; in *E. coli*, this topoisomerase is DNA gyrase.

The proteins directly responsible for unwinding DNA are the **helicases.** Using energy derived from the hydrolysis of ATP, helicases unwind the DNA in advance of the replication fork, breaking the hydrogen bonds as they go. In *E. coli,* at least two different helicases are involved in DNA replication; one attaches to the lagging-strand template and moves in a 5′ → 3′ direction, and the other attaches to the leading-strand template and moves 3′ → 5′. Both are part of the primosome, but the 5′ → 3′ helicase is more important for unwinding DNA at the replication fork.

The unwinding associated with DNA replication would create an intolerable amount of supercoiling and possibly tangling in the rest of the DNA were it not for the actions of **topoisomerases,** which we discussed in Chapter 16. These enzymes create swivel points in the DNA molecule by making and then quickly resealing single- or double-stranded breaks in the double helix. Of the ten or so topoisomerases found in *E. coli,* the one that appears to be most important for DNA replication is *DNA gyrase,* a type II topoisomerase (an enzyme that cuts both DNA strands). Using energy derived from ATP, gyrase introduces negative supercoils and thereby relaxes positive ones. DNA gyrase serves as the main swivel that prevents overwinding (positive supercoiling) of the DNA ahead of the replication fork. In addition, this enzyme has a role in both initiating and completing DNA replication in *E. coli*—in opening up the double helix at the origin of replication and in separating the linked circles of daughter DNA at the end. The situation in eukaryotic cells is not as well understood, although topoisomerases of both types have been isolated.

Once strand separation has begun, molecules of **single-strand binding protein (SSB)** quickly attach to the exposed single strands to keep the DNA unwound and therefore accessible to the DNA replication machinery. After a particular segment of DNA has been replicated, the SSB molecules fall off and are recycled, attaching to the next single-stranded segment.

Putting It All Together: DNA Replication in Summary

Figure 17-13 reviews the highlights of what we currently understand about the mechanics of DNA replication in *E. coli.* Starting at the origin of replication, the machinery at the replication fork gradually builds up until seven different proteins are involved. It is important to keep in mind that these and a number of additional proteins are all closely associated in one large complex. This complex, called a **replisome,** is about the size of a ribosome. The replisome is powered by the hydrolysis of nucleoside triphosphates. These include both the nucleoside triphosphates used by DNA polymerases and primase as building blocks for DNA (and RNA) synthesis and the ATP hydrolyzed by several other DNA-replication proteins: initiator protein, helicase, gyrase, and ligase.

Much remains to be learned about DNA replication, especially in eukaryotic cells. The great length and elaborate folding of eukaryotic DNA molecules pose special challenges for DNA replication. For example, how are the many replication origins coordinated and how is their activation linked to other key events in the cell cycle? Answering such questions requires a better understanding of the spatial organization of DNA replication within the nucleus. When cells are briefly incubated with DNA precursors that make the most recently formed DNA fluorescent, microscopic examination reveals that the newly made, fluorescent DNA is located in a series of discrete spots scattered throughout the nucleus. Such observations suggest the existence of apparently immobile structures, so-called "factories," where chromatin fibers are fed for DNA replication. These sites are closely associated with the inner surface of the nuclear envelope, although it is not clear whether they are anchored in the nuclear membrane or to some other nuclear support.

When a chromatin fiber is fed through one of these replication factories, how are the histones and other chromosomal proteins removed, so the DNA can be replicated, and then added back after the two new DNA strands are formed? Researchers have learned that, as the replication fork passes through a nucleosome, the old histone octamers appear to slide from the parental double helix ahead of the fork to the leading strand behind the fork. At the same time, new histone octamers are assembled mainly on the lagging strand. But when and how are other chromosomal proteins added, what controls the higher levels of chromatin packing, and how is all this accomplished without tangling the chromatin? Research is under way on these and other difficult questions, with varying rates of progress. Meanwhile, an answer has emerged to one of the most baffling questions related to DNA replication in eukaryotes, the end-replication problem.

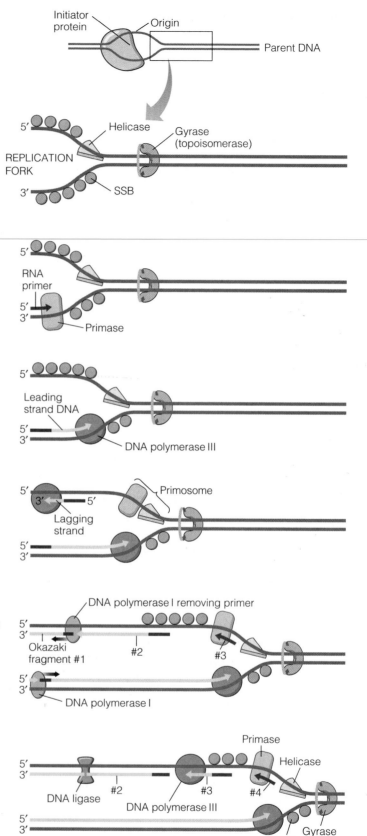

1. The initiator protein binds to the double-stranded DNA at the origin of replication and, using ATP energy, slightly unwinds the DNA.

2. Helicase attaches to the unwound DNA and continues the unwinding. Meanwhile, a topoisomerase, DNA gyrase, promotes strand separation by inducing negative supercoiling in advance of the helicase. Strand separation is maintained by molecules of single-strand binding protein (SSB), which stabilize the unwound region and allow the separated strands to serve as templates. A replication fork is now in evidence.

3. Primase binds to the first priming sequence on the leading-strand template and synthesizes a short RNA primer that is complementary to the DNA template.

4. DNA polymerase III (polymerase δ in eukaryotes) uses the primer to initiate DNA synthesis by adding deoxyribonucleotides to its 3′ end. The leading strand requires only one priming event, because DNA synthesis is continuous thereafter, in the 5′→3′ direction.

5. An RNA primer is now made for the lagging strand, and DNA polymerase III (polymerase ε in eukaryotes) extends the strand. In *E. coli,* the primase functions as part of a primosome, a complex of proteins that includes the helicase.

6. For the lagging strand, DNA synthesis is discontinuous and requires a series of RNA primers. DNA is synthesized at the 3′ end of each primer, generating an Okazaki fragment that grows until it meets the adjacent fragment. The RNA primer is then removed by the 5′→3′ exonuclease activity of DNA polymerase I and replaced with DNA by the polymerase acitivity of the same enzyme.

7. DNA ligase links together adjacent Okazaki fragments with covalent, phosphoester bonds. Hereafter, DNA ligase, polymerase I, polymerase III, primase, helicase, and gyrase will be working simultaneously in the vicinity of the replication fork.

Figure 17-13   A Summary of DNA Replication in Bacteria.   Starting with the initiation event at the replication origin, this figure depicts DNA replication in *E. coli* in seven steps. Two replication forks move in opposite directions from the origin, but only one fork is illustrated for steps 2–7. The various proteins shown here as separate entities are actually closely associated (along with others) in a single large complex called a replisome. The primase and helicase are particularly closely bound and, together with other proteins, form a primosome. Parental DNA is shown in dark blue, newly synthesized DNA in light blue, and RNA in red. This series of diagrams does not show the topological arrangement of the DNA strands.

Telomeres Solve the DNA End-Replication Problem

If we continue the process summarized in Figure 17-13 for the circular genome of *E. coli* (or any other circular DNA), we will eventually complete the circle. The leading strand can simply continue to grow 5′ → 3′ until its 3′ end is joined to the 5′ end of the lagging strand coming around in the other direction. And for the lagging strand, the very last bit of DNA to be synthesized—the replacement for the RNA primer of the last Okazaki fragment—can be added to the free 3′ OH end of the leading strand coming around in the opposite direction.

For linear DNA molecules, however, the fact that DNA polymerases can add nucleotides only to the 3′ OH end of a preexisting DNA chain creates a serious problem, which is depicted in Figure 17-14. When a growing lagging strand (lightest blue) reaches the end of the DNA molecule and the last RNA primer is removed by a 5′ → 3′ exonuclease, the final gap cannot be filled because there is no 3′ OH end to which deoxynucleotides can be added. As a result, linear DNA molecules are in danger of yielding shorter and shorter daughter DNA molecules each time they replicate. Clearly, if this trend continued indefinitely, we would not be here today!

Viruses with linear DNA genomes solve this problem in various ways. In some cases, such as bacteriophage λ, the linear DNA forms a closed circle before it replicates. In other cases, the viruses use more exotic reproduction strategies, although the DNA polymerases always progress 5′ to 3′, adding nucleotides to a 3′ end.

Eukaryotes have solved the end-replication problem by locating highly repeated DNA sequences at the terminal ends, or **telomeres,** of each linear chromosome. These special telomeric elements consist of short repeating sequences enriched in the base G in the 5′ → 3′ strand. The sequence TTAGGG, located at the ends of human chromosomes, is an example of such a telomeric or *TEL sequence.* Human telomeres typically contain between 100 and 1500 copies of the TTAGGG sequence repeated in tandem. Such noncoding sequences at the ends of each chromosome ensure that the cell will not lose any important genetic information if a DNA molecule is shortened slightly during the process of DNA replication. Moreover, a special DNA polymerase called **telomerase** can catalyze the formation of additional copies of the telomeric repeat sequence, thereby compensating for the gradual shortening that occurs at both ends of the chromosome during DNA replication.

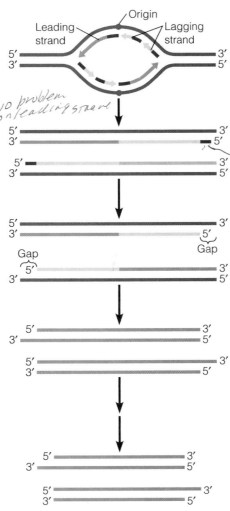

1. DNA replication is initiated at the origin; the replication bubble grows as the two replication forks move in opposite directions

2. Finally only one primer (red) remains on each daughter DNA molecule

3. The last primers are removed by a 5′→3′ exonuclease, but no DNA polymerase can fill the resulting gaps because there is no 3′–OH available to which a nucleotide can be added

4. Each round of replication generates shorter and shorter DNA molecules

Figure 17-14 The End-Replication Problem. For a linear DNA molecule, such as that of a eukaryotic chromosome, the usual DNA replication machinery is unable to replicate the ends. As a result, with each round of replication, the DNA molecules will get shorter, with potentially disastrous consequences for the cell. In this diagram, the initial parent DNA strands are dark blue, daughter DNA strands are lighter blue, and RNA primers are red. For simplicity, we show only one origin of replication and, in the last two steps, only the shortest of the progeny molecules. (The lagging-strand daughter DNA is shown in the lightest blue in the first three steps to make a point unrelated to the end-replication problem—that each daughter strand is leading at one end and lagging at the other. This is apparent in this figure because it shows the entire replicating molecule, with both replication forks.)

Telomerase is an unusual enzyme in that it is composed of RNA as well as protein. In the protozoan *Tetrahymena*, whose telomerase was the first to be isolated, the RNA component of the enzyme contains the sequence 3'-AACCCC-5', which is complementary to the 5'-TTGGGG-3' repeat sequence that makes up *Tetrahymena* telomeres. As shown in Figure 17-15, this enzyme-bound RNA serves as a template for creating the DNA repeat sequence that is added to the ends of the telomeres. After they have been lengthened, telomeres are protected by *telomere capping proteins* that bind to the exposed 3' end of the DNA. In addition, in many eukaryotes the 3' end of the DNA can loop back and base-pair with the opposite DNA strand, generating a closed loop that likewise protects the end of the telomere (see step ⑤ in Figure 17-15).

In multicellular organisms, telomerase resides mainly in the *germ cells* that give rise to sperm and eggs, and in a few other types of actively proliferating cells. The presence of telomerase allows these cells to divide indefinitely without telomere shortening. Because telomerase is not found in most cells, their chromosomal telomeres get shorter and shorter with each cell division. As a result, telomere length is a counting device that reveals how many times a cell has divided. If a cell divides enough times, the telomeres are in danger of disappearing entirely and the cell would then be at risk of eroding its coding DNA. This potentially hazardous scenario can be prevented by destroying such cells using a mechanism involving the telomeric DNA loop often found at chromosome ends. When telomeric DNA becomes too short to generate a loop, it exposes a bare, double-stranded DNA end whose presence is recognized by a signaling system that triggers *apoptosis*, an orchestrated type of cell death described in Chapter 10.

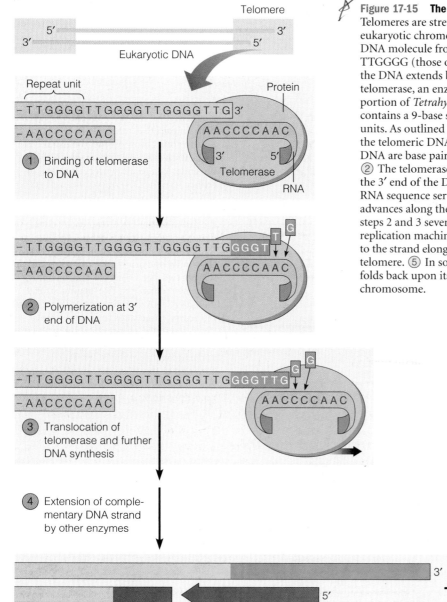

Figure 17-15 The Extension of Telomeres by Telomerase. Telomeres are stretches of repeated DNA located at the ends of eukaryotic chromosomes. This figure focuses on one end of a DNA molecule from *Tetrahymena*, whose telomeric repeat unit is TTGGGG (those of other species are very similar). The 3' end of the DNA extends beyond the 5' end and is the substrate for telomerase, an enzyme composed of protein and RNA. The RNA portion of *Tetrahymena* telomerase is 159 nucleotides long and contains a 9-base sequence complementary to 1½ telomeric repeat units. As outlined here, the telomerase ① binds to the 3' end of the telomeric DNA, positioning itself so the last few bases of the DNA are base paired with part of the 9-base RNA sequence. ② The telomerase then catalyzes the addition of nucleotides to the 3' end of the DNA strand, with the remainder of the 9-base RNA sequence serving as a template. ③ Next the telomerase advances along the DNA strand in the 3' direction and repeats steps 2 and 3 several times. ④ Meanwhile, the standard DNA replication machinery synthesizes a lagging strand complementary to the strand elongated by telomerase. The net result is a lengthened telomere. ⑤ In some eukaryotes the lengthened telomeric DNA folds back upon itself, creating a loop that caps the end of the chromosome.

It is possible that the telomeres at birth, about 10,000 base pairs in humans, may be long enough to endure more than a lifetime of replicative erosion without triggering cell death. But it is also possible that telomeres are a limiting factor in determining an organism's life span. In this regard, it is interesting to note that the introduction of telomerase into cultured human cells lengthens the amount of time they can survive in culture.

Besides being present in germ cells and a few other kinds of proliferating normal cells, telomerase has also been detected in human cancers. Because cancer cells undergo an abnormally large number of cell divisions, their telomeres become unusually short. This progressive shortening of the telomeres would presumably lead to self-destruction of the cancer, unless telomerase were produced to stabilize telomere length. This is exactly what seems to happen. A survey of a large number of human cell samples, including more than 100 tumors, found telomerase activity in almost all the cancer cells but in none of the samples from normal tissues. If telomerase is indeed an important factor in the development of cancer, it may eventually provide a useful target for anticancer drug therapy.

Eukaryotic DNA Is Licensed for Replication

In addition to needing telomeres to solve the end-replication problem, eukaryotic cells also require a mechanism for ensuring that they replicate their nuclear DNA molecules once, and only once, prior to cell division. To enforce this general rule, a process known as **licensing** is invoked to make certain that after DNA is replicated during S phase, chromatin does not become competent (or "licensed") for a further round of DNA replication until it has first passed through mitosis.

A group of proteins known as the *MCM complex* plays a central role in the licensing process. Before DNA can be replicated, the MCM complex must bind to DNA near each replication origin and "license" the DNA for replication, presumably by making the chromatin accessible to enzymes involved in DNA synthesis. When replication begins at each origin, the MCM proteins are displaced from the DNA. Before another round of DNA synthesis can be re-initiated at the same site, the MCM proteins must once again bind to the DNA. Several mechanisms prevent this re-binding, including the presence of *geminin*, a protein produced during S phase that blocks the binding of MCM proteins to DNA. Geminin continues to prevent MCM binding as the cell completes S phase, passes through G2 phase, and enters mitosis. When the cell completes mitosis, geminin is then degraded so that MCM proteins can once again license DNA for replication during the next cell cycle.

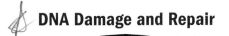

DNA Damage and Repair

The faithfulness with which the information stored in DNA is maintained from one generation of cells to the next requires not only that DNA be replicated accurately, but also that provision be made for repairing the DNA damage that can arise both spontaneously and from exposure to environmental agents. Some DNA alterations are actually desirable because DNA base-sequence changes, or **mutations,** provide the genetic variability that is the raw material of evolution. The net rate at which organisms accumulate mutations is quite low; by some estimates, an average gene retains only one mutation every 200,000 years. The underlying rate of DNA mutation, however, is far greater than this number suggests, because most damage is repaired shortly after it occurs and therefore does not affect future generations. Moreover, most mutations occur in cells other than sperm and eggs, and hence are never passed on to offspring.

DNA Damage Can Occur Spontaneously or in Response to Mutagens

During the normal process of DNA replication, several types of mutations occur spontaneously. The most common involve depurination and deamination reactions, which are spontaneous hydrolysis reactions caused by random interactions between DNA and the water molecules around it. *Depurination* refers to the loss of a purine base (either adenine or guanine) by spontaneous hydrolysis of the glycosidic bond that links it to deoxyribose (Figure 17-16a). This glycosidic bond is intrinsically unstable and is in fact so susceptible to hydrolysis that the DNA in a human cell may lose thousands of purine bases every day. *Deamination* is the removal of a base's amino group ($-NH_2$). This type of alteration, which can involve cytosine, adenine, or guanine, changes the base-pairing properties of the affected base. Of the three bases, cytosine is most susceptible to deamination, giving rise to uracil (Figure 17-16b). Like depurination, deamination is a hydrolytic reaction, usually caused by random collision of a water molecule with the bond that links the amino group of the base to the pyrimidine or purine ring. In a typical human cell, the rate of DNA damage by this means is about 100 deaminations per day.

If a DNA strand with missing purines or deaminated bases is not repaired, an erroneous base sequence may be propagated when the strand serves as a template in the next round of DNA replication. For example, where a cytosine has been converted to a uracil by deamination, the uracil behaves like thymine in its base-pairing properties, directing the insertion of an adenine on the opposite strand, rather than guanine, the correct base. The ultimate effect of this change in base sequence may be a change in the amino acid sequence and function of a protein encoded by the affected gene.

In addition to spontaneous mutations, DNA damage can also be caused by mutation-causing agents, or *mutagens,* in our environment. Environmental mutagens fall into two major categories: chemicals and radiation. Mutagenic chemicals alter DNA structure by a variety of mechanisms. *Base analogs* resemble nitrogenous bases in structure and are incorporated into DNA;

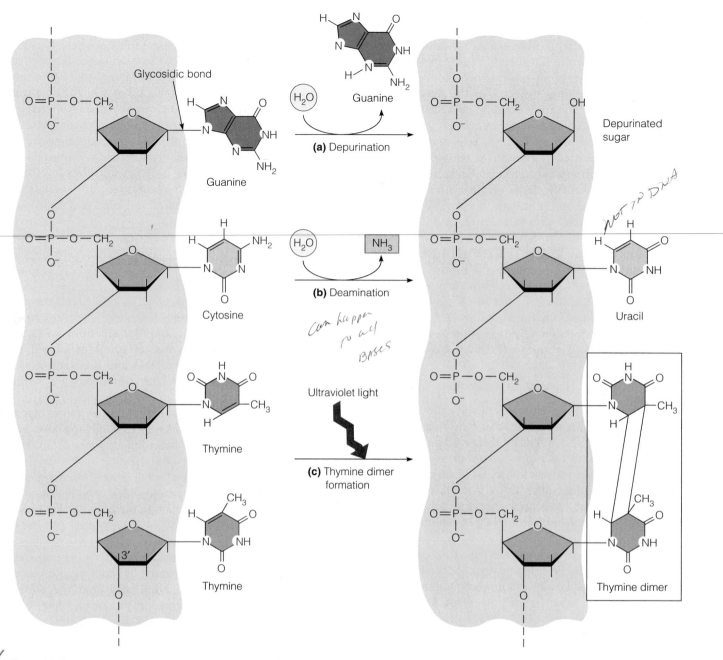

Figure 17-16 **Some Common Types of DNA Damage.** The most common kinds of chemical changes that can damage DNA are **(a)** depurination, **(b)** deamination, and **(c)** pyrimidine dimer formation (shown here are thymine dimers). Depurination and deamination are spontaneous hydrolytic reactions, whereas dimers result from covalent bonds induced to form by ultraviolet light.

base-modifying agents react chemically with DNA bases to alter their structures; and *intercalating agents* insert themselves between adjacent bases of the double helix, thereby distorting DNA structure and increasing the chance that a base will be deleted or inserted during DNA replication.

DNA mutations can also be caused by several types of radiation. Sunlight is a strong source of ultraviolet radiation, which alters DNA by triggering *pyrimidine dimer formation*—that is, the formation of covalent bonds between adjacent pyrimidine bases, usually two thymines (Figure 17-16c). Both replication and transcription are blocked by such dimers, presumably because the enzymes carrying out these functions cannot cope with the resulting bulge in the DNA double helix. Mutations can also be caused by X rays and related forms of radiation emitted by radioactive substances. This type of radiation is called ionizing radiation because it removes electrons from biological molecules, thereby generating highly reactive intermediates that cause various types of DNA damage.

Translesion Synthesis and Excision Repair Correct Mutations Involving Abnormal Nucleotides

As is perhaps not surprising for a molecule so important to an organism's health and survival, a variety of mechanisms have evolved for repairing damaged DNA. In some cases, repair is performed during the process of DNA replication using specialized DNA polymerases that carry out **translesion synthesis**—that is, the synthesis of new DNA across regions in which the DNA template is damaged. While this type of DNA synthesis is sometimes prone to error, it is also capable of correcting certain types of damage. For example, eukaryotic DNA polymerase η (eta) can catalyze DNA synthesis across a region containing a thymine dimer, correctly inserting two new adenines opposite the dimer.

Once left behind by the DNA replication machinery, errors that still remain (or that subsequently arise) become the province of a different group of enzymes and proteins. In *E. coli* alone, almost 100 genes code for proteins involved in removing and replacing abnormal nucleotides. These proteins are components of **excision repair** pathways, which correct DNA defects using a basic three-step process (Figure 17-17). In the first step, the defective nucleotides are cut out from one strand of the DNA double helix. This process is carried out by special enzymes called *repair endonucleases*, which recognize the defect and nick the phosphodiester backbone of the DNA near it. Additional enzymes then facilitate removal of the defective nucleotide(s). For example, a helicase might unwind the DNA located between two nicks to release the damaged DNA from the double helix; alternatively, an exonuclease might attach to an end created by a single nick and chew away the damaged strand one nucleotide at a time. The second step in excision repair is the replacement of the missing nucleotides with the correct ones by a DNA polymerase—in *E. coli*, usually DNA polymerase I. The nucleotide sequence of the complementary strand serves as a template to ensure correct base insertion, just as it does in DNA replication. Finally, in the third step, DNA ligase seals the remaining nick in the repaired strand by forming the missing phosphoester bond.

Excision repair pathways are classified into two main types, *base excision repair* and *nucleotide excision repair*. The first of these pathways, **base excision repair,** corrects single damaged bases in DNA. For example, deaminated bases are detected by specific *DNA glycosylases*, which recognize a specific deaminated base and remove it from the DNA molecule by cleaving the bond between the base and the sugar to which it is attached. The sugar with the missing base is then recognized by a repair endonuclease that detects depurination. This repair endonuclease breaks the phosphodiester backbone on one side of the sugar lacking a base, and a second enzyme then completes the removal of the sugar-phosphate unit.

For removing pyrimidine dimers and other bulky lesions in DNA, cells employ the second type of excision repair, namely **nucleotide excision repair (NER).** This repair system recognizes major distortions in the DNA double helix. The NER endonuclease (often called an excision nuclease, or *excinuclease*) makes two nicks, one on either side of the distortion. Then a helicase binds to the stretch of DNA between the nicks (12 nucleotides long in *E. coli*, 29 in humans) and unwinds it, freeing it from the rest of the DNA. Finally, the resulting gap is filled in by DNA polymerase and sealed by DNA ligase. The NER system is the most versatile of a cell's DNA repair systems, recognizing and correcting many types of damage that cannot otherwise be repaired. The importance of NER is underscored by the plight of people who have mutations affecting this pathway. Individuals with the disease *xeroderma pigmentosum*, for example, carry a mutation in any of seven genes coding for components of the NER system. As a result, such individuals cannot repair the DNA damage caused by the ultraviolet radiation in sunlight and therefore have a high risk of developing skin cancer.

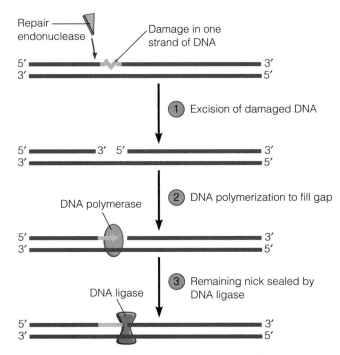

Figure 17-17 General Scheme for Excision Repair of DNA Damage. The three steps shown here are common to all types of excision repair of DNA damage in one strand. ① The damaged part of the DNA strand, and possibly some DNA on either side of it, is cut out (excised) from the double helix. An endonuclease, which nicks the DNA near the damage, is crucial for this step; a helicase and/or exonuclease may help remove the damaged segment. ② A DNA polymerase fills in the gap by adding nucleotides to the strand's 3′ end; DNA polymerase I plays this role in *E. coli*. ③ The remaining nick is sealed by DNA ligase.

Mismatch Repair Corrects Mutations That Involve Noncomplementary Base Pairs

Excision repair is a powerful mechanism for correcting damage involving the presence of abnormal nucleotides. This is not the only type of DNA error that cells can repair, however. An alternative repair pathway, called

mismatch repair, targets errors made during DNA replication, when improperly base-paired nucleotides sometimes escape the normal proofreading mechanisms. Because mismatched base pairs do not hydrogen bond properly, their presence can be detected and corrected by the mismatch repair system. But to operate properly, this repair system must solve a problem that puzzled biologists for many years: How is the *incorrect* member of an abnormal base pair distinguished from the *correct* member? Unlike the situation in excision repair, neither of the bases in a mismatched pair exhibits any structural alteration that would allow it to be recognized as an abnormal base. The pair is simply composed of two normal bases that are inappropriately paired with each other, such as the base A paired with C, or G paired with T. If the incorrect member of an AC base pair were the base C and the repair system instead removed the base A, the repair system would create a permanent mutation instead of correcting a mismatched base pair!

To solve this problem, the mismatch repair system must be able to recognize which of the two DNA strands was newly synthesized during the previous round of DNA replication (the new strand would be the one that contains the incorrectly inserted base). The bacterium *E. coli* employs a detection system that is based on the fact that a methyl group is normally added to the base adenine (A) wherever it appears in the sequence GATC in DNA. This process of *DNA methylation* does not take place until a short time after a new DNA strand has been synthesized; hence the bacterial mismatch repair system can detect the new DNA strand by its unmethylated state. A repair endonuclease then introduces a single nick in the unmethylated strand and an exonuclease removes the incorrect nucleotides from the nicked strand.

Although evidence for the existence of mismatch repair has come largely from studies of bacterial cells, comparable repair systems have been detected in eukaryotes, although they differ from bacterial systems in that they use mechanisms other than DNA methylation for distinguishing the newly synthesized DNA strand. The importance of mismatch repair for eukaryotes is highlighted by the discovery that one of the most common hereditary cancers, *hereditary nonpolyposis colon cancer (HNPCC)*, results from mutations in genes coding for proteins involved in mismatch repair.

Now that we have distinguished among the main types of DNA repair, it is important to emphasize that some of the same proteins are used in more than one repair system. Moreover, many of these "repair" proteins play additional roles in other important activities, including DNA replication, gene transcription, genetic recombination, and control of the cell cycle. In the words of the journal *Science,* which selected DNA repair enzymes as "molecule of the year" for 1994, this area of research has opened our eyes to "a more integrated picture of the cell, in which all processes relating to DNA are coordinated and the same molecular toolkit is used for different tasks." (*Science* 266 [1994]: 1927)

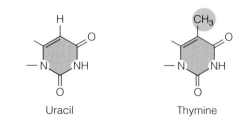

Figure 17-18 **Uracil and Thymine Compared.**

Damage Repair Helps Explain Why DNA Contains Thymine Instead of Uracil

For many years, it was not clear why DNA contains thymine instead of the uracil found in RNA. Both bases pair with adenine, but thymine contains a methyl group not present on uracil (Figure 17-18). In terms of energy, it would make more sense for DNA to contain uracil because the methylation step required during the synthesis of thymine is energetically expensive. But now that we understand how deamination damage is repaired, we also understand why thymine, rather than uracil, is present in DNA.

When DNA is damaged through deamination reactions, cytosine is converted to uracil (see Figure 17-16b), which is then detected and removed by the DNA repair enzyme, *uracil-DNA glycosylase*. But if uracil were present as a normal component of DNA, DNA repair would not work because these normal uracils would not be distinguishable from the uracils generated by the accidental deamination of cytosine. By using thymine in DNA in place of uracil, cells ensure that DNA damage caused by the deamination of cytosine can be recognized and repaired without causing other changes in the DNA molecule.

Nuclear and Cell Division

Having examined the mechanisms involved in DNA replication and repair, we can now turn to the question of how the two copies of each chromosomal DNA molecule created during the S phase of the cell cycle are subsequently separated from each other and partitioned into daughter cells. These events occur during the M phase, which encompasses both nuclear division (mitosis) and cytoplasmic division (cytokinesis).

Mitosis Is Subdivided into Prophase, Prometaphase, Metaphase, Anaphase, and Telophase

Mitosis has been studied for more than a century, but only in the past few decades has significant progress been made toward understanding the mitotic process at the molecular level. We will begin by surveying the morphological changes that occur in a cell as it undergoes mitosis, and we will then examine the underlying molecular mechanisms.

Morphologically, mitosis can be subdivided into five sequential stages, based primarily on the changing appearance and behavior of the chromosomes. These five phases are *prophase, prometaphase, metaphase, anaphase,* and *telophase.* (An alternative term for prometaphase is simply *late prophase.*) The micrographs and schematic diagrams of Figure 17-19 illustrate the phases in a typical animal cell; Figure 17-20 depicts the comparable stages in plant cells. As you follow the events of each phase, keep in mind that the purpose of mitosis is to ensure that each of the two daughter nuclei receives one copy of each duplicated chromosome.

Prophase. After completing DNA replication, cells exit from S phase and enter into G2 phase (see Figure 17-1b), where final preparations are made for the onset of mitosis. Toward the end of G2, the chromosomes start to condense from the extended, highly diffuse form of interphase chromatin fibers into the compact, extensively folded structures that are typical of mitosis. Chromosome condensation is an important event because interphase chromatin fibers are so long and intertwined that in an uncompacted form, they would become impossibly tangled during distribution of the chromosomal DNA at the time of cell division. Although the transition from G2 to prophase is not sharply defined, a cell is considered to be in **prophase** when individual chromosomes have condensed to the point of being visible as discrete objects in the light microscope. Because the chromosomal DNA molecules have replicated during S phase, each prophase chromosome is composed of two sister chromatids tightly attached to each other at a constricted region, the **centromere** (Figure 17-19a). In animal cells the nucleoli usually disperse as the chromosomes condense; plant cell nucleoli may either remain as discrete entities, undergo partial disruption, or disappear entirely.

Meanwhile, another important organelle has sprung into action. This is the **centrosome,** a small zone of granular material located adjacent to the nucleus. The centrosome is the cell's main organizing center for the assembly of microtubules. The centrosome is duplicated prior to mitosis (usually during S phase), and at the beginning of prophase, the two centrosomes separate from each other and begin to move toward opposite ends of the cell. As they move apart, the region between the two centrosomes begins to fill with growing microtubules that will form the **mitotic spindle,** the structure that distributes the chromosomes to the daughter cells later in mitosis. During this process, cytoskeletal microtubules disassemble and their tubulin subunits (see Figure 4-24a) are added to the growing mitotic spindle. At the same time, a dense starburst of microtubules called an *aster* forms in the immediate vicinity of each centrosome.

Embedded within the centrosome of animal cells is a pair of small, cylindrical, microtubule-containing structures called **centrioles,** which are often oriented at right angles to each other. Because centrioles are absent in certain cell types, including most higher plant cells, they cannot be essential to the process of mitosis. They do, however, play a role in the formation of cilia and flagella and so will be discussed in more detail when we cover these cellular appendages in Chapter 23.

Prometaphase. The onset of **prometaphase** is marked by the breakdown of the nuclear envelope, which is reduced to a collection of small membranous vesicles. As the centrosomes complete their movement to opposite poles of the cell (Figure 17-19b), the breakdown of the nuclear envelope allows the spindle microtubules to enter the nuclear area and make contact with the chromosomes. At this point, *CEN sequences*—a special type of DNA sequence located at the centromere of each chromosome—bind to a specialized set of proteins to form a protein-DNA complex called a **kinetochore.** Each chromosome develops two kinetochores located on opposite sides of the centromere, one associated with each of the two chromatids (Figure 17-21a). Once they are formed, the kinetochores bind to the free ends of spindle microtubules, thereby attaching the chromosomes to the spindle. Forces exerted by the spindle microtubules then throw the chromosomes into agitated motion and gradually move them toward the center of the cell (Figure 17-21b).

While all the spindle microtubules are essentially identical in composition, they can be classified into three types, based on the structures with which their tips interact. The ones attached to chromosomal kinetochores are called **kinetochore microtubules;** those that interact with microtubules from the opposite pole of the cell are called **polar microtubules;** and the shorter ones that form the asters at each pole are called **astral microtubules.** At least some of the astral microtubules appear to interact with proteins lining the plasma membrane.

Metaphase. The cell is said to be in **metaphase** when the chromosomes, which are now maximally condensed, all become aligned at the *metaphase plate,* the plane equidistant between the two poles of the mitotic spindle (Figure 17-19c). The cell seems to pause at metaphase, which occupies about 20 minutes out of the hour or so required for mitosis. Agents that interfere with the functioning of the spindle, such as the drug *colchicine,* can be used to generate metaphase-arrested cells. Microscopic examination of such cells allows individual chromosomes to be identified and classified on the basis of differences in size and shape, generating an analysis known as a **karyotype** (Figure 17-22).

At metaphase the chromosomes appear to be relatively stationary, but this appearance is misleading. Actually, the two sister chromatids of each chromosome are already being actively tugged toward opposite poles. They appear stationary because the forces acting on them are equal in magnitude and opposite in direction; the chromatids are the prizes in a tug-of-war between two equally strong opponents. (We will discuss the source of these opposing forces shortly.)

Anaphase. Usually the shortest phase of mitosis, **anaphase** typically lasts only a few minutes. At the beginning of anaphase, the two sister chromatids of each chromosome abruptly separate and begin moving toward opposite spindle poles at a rate of about 1 μm/min (Figure 17-19d).

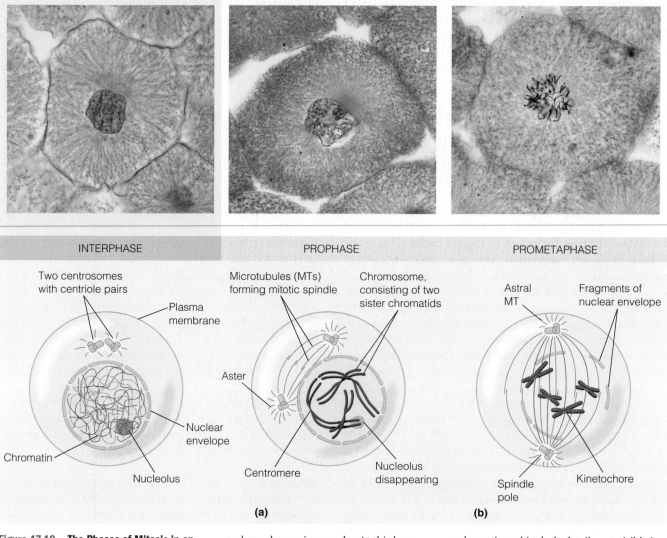

INTERPHASE	PROPHASE	PROMETAPHASE

Two centrosomes with centriole pairs
Plasma membrane
Chromatin
Nucleolus
Nuclear envelope

Microtubules (MTs) forming mitotic spindle
Chromosome, consisting of two sister chromatids
Aster
Centromere
Nucleolus disappearing

Astral MT
Fragments of nuclear envelope
Spindle pole
Kinetochore

(a)

(b)

Figure 17-19 The Phases of Mitosis in an Animal Cell. The micrographs show mitosis in cells from a fish embryo viewed by light microscopy. The mitotic spindle, including asters, is visible in the metaphase and anaphase micrographs. At this low magnification (about 600×), however, we see spindle "fibers," rather than individual microtubules; each fiber consists of a number of microtubules. The drawings are schematic and include details not visible in the micrographs; for simplicity, only four chromosomes are drawn.
MT = microtubule.

Figure 17-20 The Phases of Mitosis in a Plant Cell. These micrographs show mitosis in cells of an onion root viewed by light microscopy.

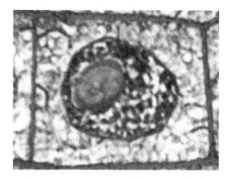

(a) Prophase

(b) Prometaphase

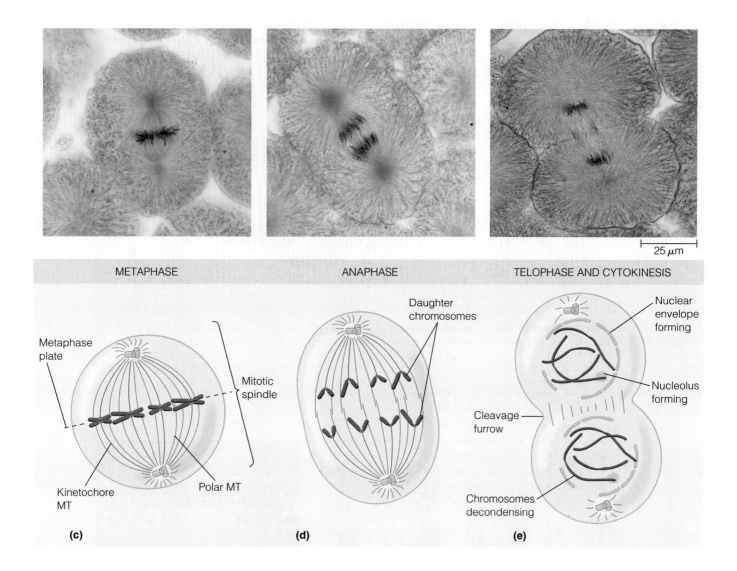

25 μm

| METAPHASE | ANAPHASE | TELOPHASE AND CYTOKINESIS |

Metaphase plate

Mitotic spindle

Kinetochore MT

Polar MT

(c)

Daughter chromosomes

(d)

Nuclear envelope forming

Nucleolus forming

Cleavage furrow

Chromosomes decondensing

(e)

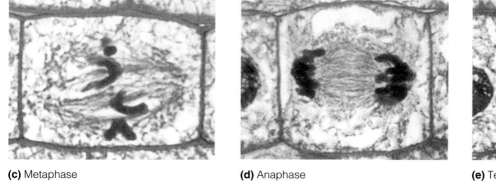

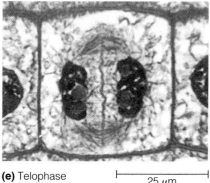

(c) Metaphase

(d) Anaphase

(e) Telophase

25 μm

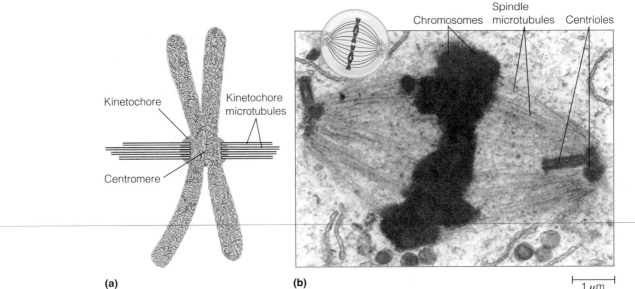

(a)

(b)

├─┤ 1 μm

Figure 17-21 Attachment of Chromosomes to the Mitotic Spindle. (a) A schematic model that summarizes the relationship between the centromere, kinetochores, and kinetochore microtubules of the spindle. **(b)** This electron micrograph shows the mitotic spindle of a metaphase cell from a rooster. The centrioles at the two poles of the spindle and the spindle between the poles are clearly visible. The chromosomes appear as a single mass aligned at the spindle equator. Although individual chromosomes cannot be distinguished in this type of micrograph, the individual chromosomes remain distinct from one another at this stage of mitosis (TEM).

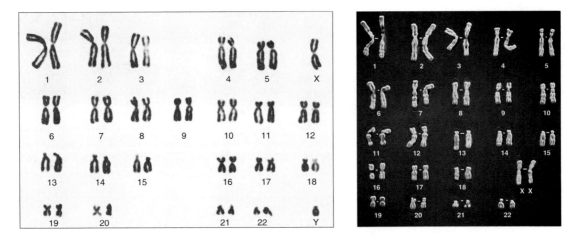

Figure 17-22 A Mitotic Karyotype of Human Chromosomes from Metaphase-Arrested Cells. (Left) This set of chromosomes obtained from the cells of a human male has been stained with a dye that reacts uniformly with the entire body of the chromosome. Human males contain 22 pairs of chromosomes, plus one X and one Y chromosome. The chromosomes in the karyotype have been arranged according to size and centromere position. (Right) This set of human female chromosomes has been stained with dyes that selectively react with certain chromosome regions, creating a unique banding pattern for each type of chromosome.

Anaphase is characterized by two kinds of movements, called anaphase A and anaphase B (Figure 17-23). In **anaphase A,** the chromosomes are pulled, centromere first, toward the spindle poles as the kinetochore microtubules get shorter and shorter. In **anaphase B,** the poles themselves move away from each other as the polar microtubules lengthen. Depending on the cell type involved, anaphase A and B may take place at the same time, or anaphase B may follow anaphase A.

Telophase. By the beginning of **telophase,** the daughter chromosomes have arrived at the poles of the spindle (Figure 17-19e). At this point, the chromosomes uncoil and revert to the extended fibers and homogeneous appearance of interphase chromatin. At the same time, nucleoli develop at the nucleolar organizing sites on the DNA, the spindle disassembles, and nuclear envelopes form around the two groups of daughter chromosomes, thereby completing the mitotic process. During this period

Figure 17-23 **The Two Types of Movement Involved in Chromosome Separation During Anaphase.** Anaphase A involves the movement of chromosomes toward the spindle pole to which they are attached. Anaphase B is the movement of the two spindle poles away from each other. Anaphase A and anaphase B may occur simultaneously.

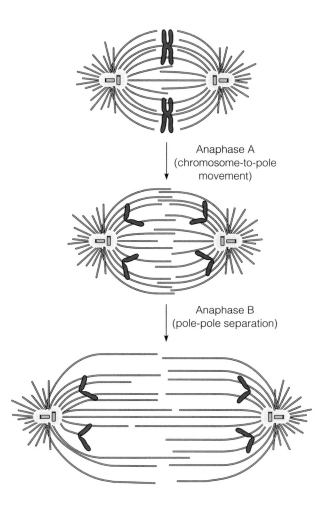

Anaphase A
(chromosome-to-pole movement)

Anaphase B
(pole-pole separation)

the cell usually undergoes cytokinesis, which divides the cell into two daughter cells.

The Mitotic Spindle Is Responsible for Chromosome Movements During Mitosis

The central purpose of mitosis is to separate the two sets of daughter chromosomes and partition them into the two, newly forming daughter cells. To understand the mechanisms that allow this to be accomplished, we need to take a closer look at the microtubule-containing apparatus responsible for these events, the mitotic spindle.

Spindle Assembly and Chromosome Attachment. As we mentioned in Chapter 4 (and will describe in more detail in Chapter 22), the fact that the tubulin subunits of a microtubule all face in the same direction gives microtubules an inherent *polarity;* that is, the two ends of each microtubule are chemically different (Figure 17-24). The end where microtubule assembly is initiated—located at the centrosome for spindle microtubules—is the minus (−) end, while the end where most growth occurs, located away from the centrosome, is the plus (+) end. Microtubules are dynamic structures, in that tubulin subunits are continually being added and subtracted from both ends. When more subunits are being added than removed, the microtubule gets longer. In general, the plus end is the site favored for the addition of tubulin subunits and the minus end is favored for subunit removal, so increases in

microtubule length come mainly from increased addition of subunits to the plus end.

During late prophase, microtubule-forming activity speeds up dramatically and the initiation of new microtubules at the centrosomes increases. Once the nuclear envelope disintegrates at the beginning of prometaphase,

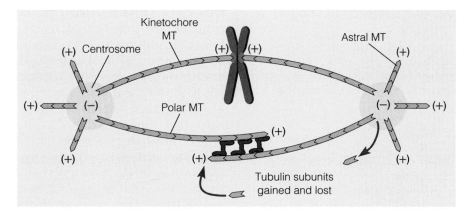

Figure 17-24 **Microtubule Polarity in the Mitotic Spindle.** This diagram shows only a few representatives of the many microtubules making up a spindle. The orientation of the tubulin subunits constituting a microtubule (MT) make the two ends of the MT different. The minus end is at the initiating centrosome; the plus end points away from the centrosome. MTs lengthen by adding tubulin subunits and shrink by losing subunits. In general, lengthening is due to addition at the plus ends and shortening to loss at the minus ends, but subunits can also be removed from the plus ends. The red structures between the plus ends of the polar MTs shown here represent proteins that crosslink them (see Figure 17-26a).

contact between microtubules and chromosomal kineto-chores becomes possible. When contact is made between a kinetochore and the plus end of a microtubule, they bind to each other and the microtubule becomes known as a *kinetochore microtubule*. This binding slows down the depolymerization rate at the plus end of the microtubule, although polymerization and depolymerization can still occur there.

Figure 17-25 is an electron micrograph of a metaphase chromosome with two sets of attached kinetochore microtubules. The plus ends of the microtubules are embedded in the two kinetochores. Each kinetochore is a platelike, three-layered structure made of proteins attached to CEN sequences located in the centromere's DNA. Kinetochores of different species vary in size. In yeast, for example, they are small and bind only one spindle microtubule each, whereas the kinetochores of mammalian cells are much larger, each binding 30–40 microtubules.

Because the two kinetochores are located on opposite sides of a chromosome, they usually attach to microtubules coming from opposite poles of the cell. (The orientation of each chromosome is random; either kinetochore can end up facing either pole.) Meanwhile, the other main group of microtubules—the polar microtubules—are making direct contact with polar microtubules coming from the opposite centrosome. When the plus-end regions of two microtubules of opposite polarity start to overlap, crosslinking proteins bind them to each other (see Figure 17-24). Like the crosslinking between kinetochores and kinetochore microtubules, this crosslinking stabilizes the polar microtubules. Thus we can picture a barrage of microtubules rapidly shooting out from each centrosome during late prophase and prometaphase. The ones that successfully hit a kineto-chore or a microtubule of opposite polarity are stabilized; the others retreat by disassembling.

Chromosome Alignment and Separation. When spindle microtubules first become attached to chromosomal kinetochores during early prometaphase, the chromosomes are randomly distributed throughout the spindle. The chromosomes then migrate toward the central region of the spindle through a series of agitated, back-and-forth motions generated by at least two different kinds of forces. First, the kinetochore microtubules exert a "pulling" force that moves the chromosomes toward the pole to which the microtubules are attached. This force can be demonstrated experimentally by using glass microneedles to tear individual chromosomes away from the spindle. A chromosome that has been removed from the spindle remains motionless until new microtubules attach to its kinetochore, at which time the chromosome is drawn back into the spindle.

The second force tends to "push" chromosomes away if they approach either spindle pole. The existence of this pushing force has been demonstrated by studies in which a laser microbeam is used to break off one end of a chromosome. Once the broken chromosome fragment has been cut free from its associated centromere and kinetochore, the fragment tends to move away from the nearest spindle pole, even though it is no longer attached to the spindle by microtubules. The nature of the pushing force that propels chromosomes in the absence of microtubule attachments has not yet been clearly identified.

The combination of pulling and pushing forces exerted on the chromosomes drives them to the metaphase plate, their most stable location, where they line up in random order. Although the chromosomes appear to stop moving at this point, careful microscopic study of living cells reveals that they continue to make small jerking movements, indicating that the chromosomes are under constant tension in both directions. If the kinetochore located on one side of a metaphase chromosome is severed using a laser microbeam, the chromosome promptly migrates toward the opposite spindle pole. Hence metaphase chromosomes remain at the center of the spindle because the forces pulling them toward opposite poles are precisely balanced.

At the beginning of anaphase the centromere region of each chromosome splits, allowing the paired chromatids to separate and move toward opposite spindle poles. The mechanism responsible for centromere splitting is not yet well understood, but several components appear to be involved. One is the enzyme topoisomerase II, which is concentrated near the centromere and which catalyzes changes in DNA supercoiling that are required for chromatid separation. In mutant cells lacking topoisomerase II, the paired chromatids still attempt to separate at the beginning of anaphase, but they are torn apart and damaged instead of separating properly. Chromatid separation also involves a group of proteins localized at the center of the centromere where the two chromatids are in close physical contact. Recent evidence suggests that these proteins function as an adhesive that holds paired

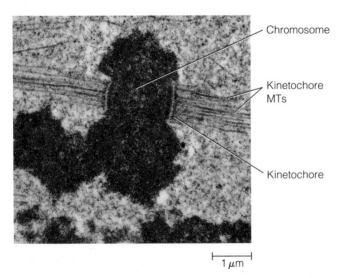

Chromosome

Kinetochore MTs

Kinetochore

1 μm

Figure 17-25 Kinetochores and Their Microtubules. The striped structures on either side of this metaphase chromosome are its kinetochores, each associated with one of the two sister chromatids. Numerous kinetochore MTs are attached to each kinetochore. The two sets of microtubules come from opposite poles of the cell (TEM).

chromatids together prior to the onset of anaphase, and that the proteins are degraded at the beginning of anaphase, allowing sister chromatids to separate.

Motor Proteins and Chromosome Movement. The splitting of the centromeres allows the two chromatids of each metaphase chromosome to separate, thereby generating two independent chromosomes that migrate to opposite spindle poles. Studies of the mechanisms underlying this movement have led to the discovery of a number of **motor proteins** that play active roles in mitosis. Like motor proteins involved in cellular motility and contractility (which we will discuss in Chapter 23), these proteins use energy derived from ATP to change shape in such a way that they exert force and cause attached structures to move. Motor proteins play at least three distinct roles in the movement of anaphase chromosomes.

The first of these roles underlies the mechanism by which chromosomes are pulled, centromere first, toward the spindle poles during anaphase A, accompanied by shortening of the kinetochore microtubules. As shown in Figure 17-26a, ①, this type of chromosomal movement is driven by a kinetochore motor protein that moves the kinetochore, and hence its attached chromosome, along the microtubule in the minus direction, thereby moving the chromosome toward the spindle pole. Several types of evidence suggest that microtubule depolymerization accompanies this motor-protein driven movement of chromosomes. For example, if cells are exposed to the drug taxol, which inhibits microtubule depolymerization, the shortening of the kinetochore microtubules is blocked and the chromosomes do not move toward the spindle poles. Conversely, exposing cells to increased atmospheric pressure, which increases the rate of microtubule depolymerization, causes the chromosomes to move more quickly toward the poles.

To investigate which end of the kinetochore microtubule is being depolymerized, investigators have injected metaphase cells with tubulin molecules linked to a substance that makes them visible with fluorescence microscopy. Initially the fluorescent tubulin is incorporated into the plus end of the kinetochore microtubules, where they attach to chromosomal kinetochores. But these same fluorescent tubulin molecules are quickly released from the kinetochore microtubules when anaphase begins, indicating that the microtubules are being disassembled at their plus ends adjacent to the kinetochore. Such observations suggest that during anaphase A, each kinetochore pulls its attached chromosome toward the spindle pole by advancing along a stationary "track" of kinetochore microtubules, depolymerizing the microtubules at the plus end as it proceeds (Figure 17-26b). In this model, the kinetochore motor proteins create the driving force, and the shortening of the microtubules is a secondary event.

The second role played by motor proteins during anaphase is associated with the movement of the spindle poles away from each other during anaphase B. In this case, motors associated with the polar microtubules (Figure 17-26a, ②) cause the polar microtubules emerging from opposite spindle poles to slide apart, thereby forcing the spindle poles away from each other. As the microtubules slide apart, they are lengthened by the addition of tubulin subunits to their plus ends near the center of the spindle where microtubules coming from opposite spindle poles overlap. Microtubule sliding can be experimentally induced by exposing isolated spindles to ATP, indicating that the motor proteins use energy derived from ATP hydrolysis to cause the overlapping microtubules to slide away from one another. The proteins that stabilize antiparallel pairs of polar microtubules by crosslinking them are thought to act as the molecular motors that push the overlapping microtubules in opposite directions, thereby driving the poles of the spindle apart. During anaphase B, this motor activity may be the primary force that elongates the spindle, while the lengthening of the polar microtubules is secondary. (This situation is analogous to the preceding model for movement at the chromosomal kinetochore, although there it is the *shortening* of microtubules that is secondary.) Figure 17-26c and d provides electron microscopic evidence for the sliding of overlapping polar microtubules during anaphase B.

The third type of motor-produced force detected during anaphase involves a group of motor proteins associated with astral microtubules (Figure 17-26a, ③). These motor proteins attach the plus ends of the astral microtubules to the *cell cortex,* a layer of actin microfilaments lining the inner surface of the plasma membrane. The astral motor proteins are thought to exert an outward pull on the spindle that—in addition to the outward push generated by the motor proteins that crosslink the overlapping polar microtubules—helps to separate the spindle poles during anaphase B.

Mitosis therefore involves at least three separate groups of motor proteins, operating at the plus ends of kinetochore microtubules, polar microtubules, and astral microtubules, respectively (Figure 17-26a). The relative contributions of the pushing and pulling forces generated by these three sets of motor proteins differ among organisms. For example, in diatoms and yeast the pushing (sliding) of microtubules against adjacent ones of opposite polarity is particularly important in anaphase B. In contrast, pulling at the asters is the main force operating in the cells of certain other fungi. In vertebrate cells, both mechanisms are probably operative, although astral pulling may play a greater role, especially during spindle formation.

Cytokinesis Divides the Cytoplasm

After the two sets of chromosomes have separated during anaphase, cytokinesis divides the cytoplasm in two, thereby completing the process of cell division. Cytokinesis usually starts during late anaphase or early telophase, as the nuclear envelope and nucleoli are re-forming and the chromosomes are decondensing. Cytokinesis is not inextricably linked to mitosis, however. In some cases, a significant time lag may

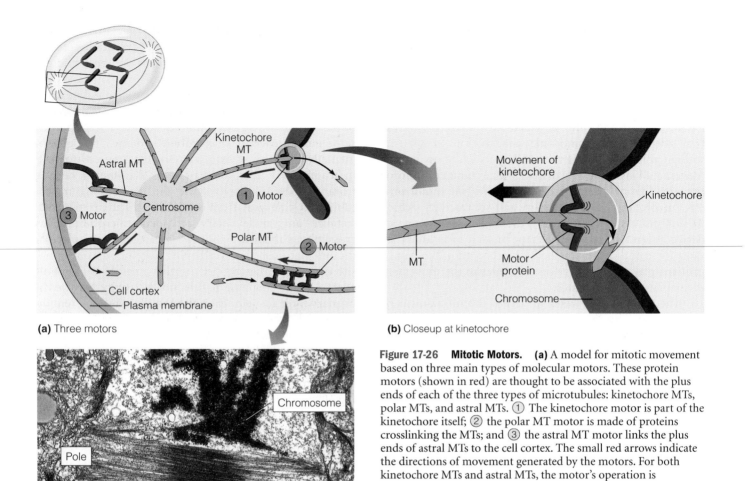

(a) Three motors

Astral MT

Kinetochore MT

Centrosome

③ Motor

① Motor

Polar MT

② Motor

Cell cortex

Plasma membrane

(b) Closeup at kinetochore

Movement of kinetochore

Kinetochore

MT

Motor protein

Chromosome

Figure 17-26   Mitotic Motors.   (a) A model for mitotic movement based on three main types of molecular motors. These protein motors (shown in red) are thought to be associated with the plus ends of each of the three types of microtubules: kinetochore MTs, polar MTs, and astral MTs. ① The kinetochore motor is part of the kinetochore itself; ② the polar MT motor is made of proteins crosslinking the MTs; and ③ the astral MT motor links the plus ends of astral MTs to the cell cortex. The small red arrows indicate the directions of movement generated by the motors. For both kinetochore MTs and astral MTs, the motor's operation is associated with loss of tubulin subunits at the plus ends; for polar MTs there is, in anaphase, a net gain of subunits, which leads to the lengthening of the spindle as the poles of the cell move farther apart. **(b)** A closer view of the kinetochore motor, showing a single MT. Energized by ATP, the motor proteins move in such a way that they "walk" toward the minus end of the MT, dragging the rest of the kinetochore and its attached chromosome along with them. Simultaneously, the MT, while remaining attached to the kinetochore, loses tubulin subunits from its tip. As a result, the kinetochore MT shortens and the chromosome is pulled toward one pole of the cell. **(c)** and **(d)** The two electron micrographs provide evidence for the sliding of polar MTs driven by polar MT motors. During metaphase, the polar MTs from opposite ends of the cell overlap significantly. In anaphase, the polar MT motors cause these two groups of MTs to slide away from each other, thereby resulting in a reduced region of overlap (TEMs).

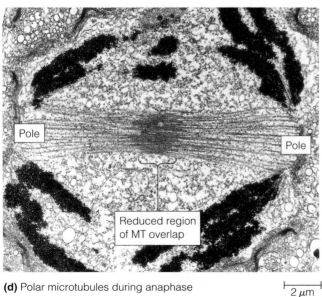

Chromosome

Pole

Pole

Polar MTs

Region of MT overlap

(c) Polar microtubules during metaphase

⊢——⊣ 2 μm

Pole

Pole

Reduced region of MT overlap

(d) Polar microtubules during anaphase

⊢——⊣ 2 μm

occur between nuclear division (mitosis) and cytokinesis, indicating that the two processes are not tightly coupled. Moreover, some cell types can undergo many rounds of chromosome replication and nuclear division in the absence of cytokinesis, thereby producing large multinucleated cells. In some cases, the multinucleate condition is permanent, while in other situations, the multinucleate state is only a temporary phase in the organism's development. This is the case, for example, in the development of a plant seed tissue called *endosperm* in cereal grains. Here nuclear division occurs for a time unaccompanied by cytokinesis, generating many nuclei in a common cytoplasm. Successive rounds of cytokinesis then occur without mitosis, walling off the many nuclei into separate endosperm cells. A similar process occurs in developing insect eggs. The fertilized egg undergoes mitosis but not cytokinesis and soon consists of hundreds of nuclei in the same cytoplasm; later, cytokinesis catches up.

Despite these examples, in most cases cytokinesis does accompany or closely follow mitosis, thereby ensuring that each of the daughter nuclei acquires its own cytoplasm and becomes a separate cell.

Cytokinesis in Animal Cells. The mechanism of cytokinesis is quite different in animals and plants. In animal cells, cytoplasmic division is called **cleavage.** The process begins as a slight indentation or puckering of the cell surface, which deepens into a **cleavage furrow** that encircles the cell, as shown in Figure 17-27 for a fertilized frog egg. (Cleavage has been studied most extensively in fertilized eggs of frogs and sea urchins.) The furrow continues to deepen until opposite surfaces make contact and the cell is split in two. The cleavage furrow divides the cell along a plane that passes through the central region of the spindle (the *spindle equator*), suggesting that the location of the spindle determines where the cytoplasm will be divided. This idea has been investigated experimentally by moving the mitotic spindle using either tiny glass needles or gravitational forces generated by centrifugation. If the experimenter moves the spindle before the end of metaphase, the orientation of the cleavage plane changes so that it passes through the new location of the spindle equator. However, if the spindle is not moved until metaphase has been completed, the cleavage plane then passes through the area originally occupied by the spindle equator. Hence the site of cytoplasmic division must be programmed by the end of metaphase.

Cleavage depends on a beltlike bundle of actin microfilaments called the **contractile ring,** which forms just beneath the plasma membrane during early anaphase. Examination of the contractile ring with an electron microscope reveals large numbers of actin filaments oriented with their long axes parallel to the furrow. As cleavage progresses, this ring of microfilaments tightens around the cytoplasm, like a belt around the waist, eventually pinching the cell in two. The force needed to tighten the contractile ring and divide the cytoplasm appears to be generated by interaction of the actin microfilaments with molecules of the protein *myosin.* (As we will see in Chapter 23, myosin is the protein that interacts with actin to produce the molecular movements that underlie muscle contraction.) In dividing cells, the band of microfilaments in the contractile ring apparently constricts progressively in a manner that involves sliding of actin filaments driven by interaction with myosin, with the necessary energy provided by ATP.

The contractile ring provides a dramatic example of the rapidity with which actin-myosin complexes can be assembled and disassembled in nonmuscle cells. Polymerization of actin monomers into microfilaments takes place just before initial indentation of the cleavage furrow, and the entire structure is dismantled again shortly after cytokinesis is complete. The actin monomers used in assembling the microfilaments of the contractile ring are obtained by disassembly of the actin filaments of the cytoskeleton, just as the tubulin needed for spindle microtubules is derived from cytoskeletal microtubules.

Cytokinesis in Plant Cells. Cytokinesis in higher plants differs in a fundamental way from the corresponding process in animal cells. Because plants cells are surrounded by a rigid cell wall, they cannot create a contractile ring at the cell surface that pinches the cell in two. Instead, they divide by assembling a plasma membrane and a cell wall between the two daughter nuclei (Figure 17-28). In other words, rather than pinching the cytoplasm in half with a contractile ring that moves from the outside of the cell toward the interior, the plant cell cytoplasm is divided by a process that begins in the cell interior and works toward the periphery.

Cytokinesis in plants is typically initiated during late anaphase or early telophase, when a group of small membranous vesicles derived from the Golgi complex align themselves across the equatorial region of the spindle. These vesicles, which contain polysaccharides and glycoproteins required for cell wall formation, are guided to the spindle equator by the **phragmoplast,** a parallel array of microtubules derived from polar microtubules and oriented perpendicular to the direction in which the new cell wall is being formed. After they arrive at the equator, the Golgi-derived vesicles fuse together to produce a large

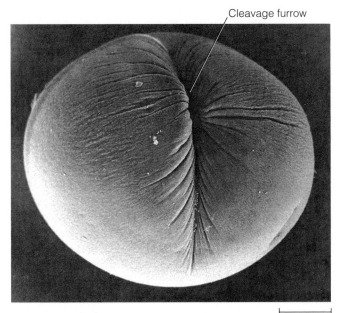

Cleavage furrow

100 μm

Figure 17-27 Cytokinesis in an Animal Cell. An electron micrograph of a fertilized frog egg caught in the act of dividing. The cleavage furrow is clearly visible as an inward constriction of the plasma membrane. Within the cell, mitosis is nearly complete, so the cleavage furrow will separate the two sets of chromosomes as it continues to constrict the membrane (SEM).

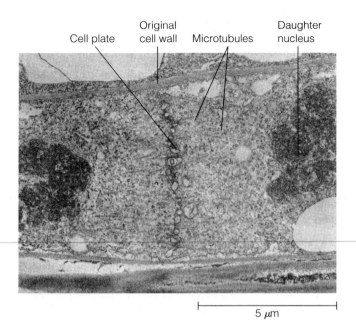

Cell plate | Original cell wall | Microtubules | Daughter nucleus

5 μm

Figure 17-28 Cytokinesis and Cell Plate Formation in a Plant Cell. This electron micrograph shows a plant cell at late telophase. The daughter nuclei with their sets of chromosomes are partially visible as the dark material on the far right and far left of the micrograph, and the developing cell plate is seen as a line of vesicles in the midregion of the cell. The microtubules of the phragmoplast are oriented perpendicular to the cell plate. The cell is from *Acer saccharinum,* the sugar maple (TEM).

flattened sac, called the **cell plate,** which represents the cell wall in the process of formation. The contents of the sac assemble to form the noncellulose components of the primary cell wall, which expands outward as clusters of microtubules and vesicles form at the lateral edges of the advancing cell plate. Eventually, the expanding cell plate makes contact with the original cell wall, separating the two daughter cells from each other. The new cell wall is then completed by deposition of cellulose microfibrils. The plasmodesmata that provide channels of continuity between the cytoplasms of adjacent plant cells are also present in the cell plate and the new wall as it forms.

Because the division plane passes through the spindle equator in both animal and plant cells, the two sets of chromosomes situated at the spindle poles are separated into the two daughter cells at the time of division. Cytokinesis does not always ensure that the cytoplasm is equally divided, however. If the spindle forms symmetrically across the cell, as is typically the case, then cytokinesis will also be symmetrical and the smaller organelles will be equally divided between the two daughter cells. Larger organelles, such as the endoplasmic reticulum and Golgi complex, tend to fragment into small vesicles early in mitosis and then reassemble in the daughter cells, thereby ensuring a roughly even division of cell components. But in some cases, cytokinesis is asymmetric. For example, in the budding yeast *Saccharomyces cerevisiae,* the interphase cell produces a small bud of cytoplasm that protrudes from the cell prior to entering mitosis. The mitotic spindle then

forms in the area of the cytoplasmic bud and cytokinesis divides the cytoplasm in a highly asymmetric fashion, creating one large cell and one tiny cell.

Regulation of the Cell Cycle

Earlier in the chapter, we described a typical eukaryotic cell cycle in which G1, S, G2, and M phases follow one another in orderly progression. Such a pattern is often the case, particularly in growing organisms or in cultured cells that have not run out of nutrients or space. But many variations are also possible, especially in the overall length of the cycle, the relative length of time spent in various phases of the cycle, and the immediacy with which mitosis and cytokinesis are coupled. This variability tells us that the cell cycle must somehow be regulated to meet the needs of a particular cell type or species. The molecular basis of this regulation is a subject of intense interest, not only for understanding the life cycles of normal cells but also for understanding how cancer cells manage to escape normal control mechanisms. Currently one of the most active areas of biological research, cell cycle regulation is beginning to reveal its underlying molecular mechanisms.

The Length of the Cell Cycle Varies Among Different Cell Types

Some of the most commonly encountered variations in the cell cycle involve differences in generation time. In multicellular organisms, generation times vary markedly among cell types, depending on their role in the organism. Some cells divide rapidly and continuously throughout the life of the organism as a means of replacing cells that are continually being lost or destroyed. Included in this category are cells involved in sperm formation and the *stem cells* that give rise to blood cells, skin cells, and the epithelial cells that form the inner lining of body organs such as the lungs and intestines. Human stem cells may have generation times as short as 8 hours.

In contrast, cells of slow-growing tissues may have generation times of several days or more, and some cells, such as those of mature nerve or muscle tissue, do not divide at all. Still other cell types do not divide under normal conditions but can be induced to begin dividing again by an appropriate stimulus. Liver cells are in this category; they do not normally proliferate in the mature liver but can be induced to do so if a portion of the liver is removed surgically. Lymphocytes (white blood cells) are another example; when exposed to a foreign protein, they begin dividing as part of the immune response.

Most of these variations in generation time are based on differences in G1, although S and G2 can also vary somewhat. Cells that divide very slowly may spend days, months, or even years in the offshoot of G1 called G0, whereas cells that divide very rapidly have a short G1 phase or even eliminate G1 entirely. The embryonic cells of insects, amphibians, and several other nonmammalian animals are dramatic examples of cells that have very short cell cycles, with no G1 phase and a very short S phase. For example, during early embryonic development of amphibians such as the frog *Xenopus laevis,* the cell cycle takes less than 30 minutes, even though the normal length of the cell cycle in adult tissues is about 20 hours. Under these conditions, the S phase is completed in less than 3 minutes, at least 100 times faster than in adult tissues. The rapid rate of DNA synthesis needed to sustain such a quick cell cycle is possible because all replicons are activated at the same time, in contrast to the sequential activation in adult tissues. The total number of replicons also increases, thereby decreasing the amount of DNA that each replicon must synthesize.

In addition, these embryonic cells have little or no need to synthesize components other than DNA because the fertilized egg is a very large cell with enough cytoplasm to sustain many rounds of cell division. Each round of division subdivides the initial cytoplasm into smaller and smaller cells, until the cell size characteristic of adult tissues is reached (Figure 17-29). This means that cell growth need not be part of the cell cycle. As a result, not

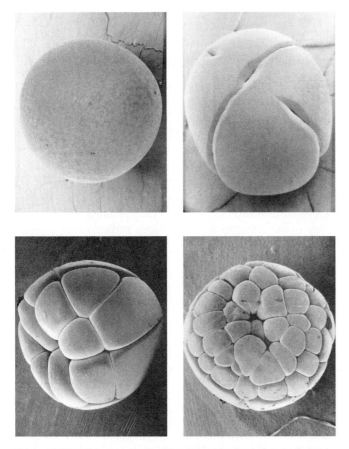

Figure 17-29   Cleavage of a Fertilized Egg into Progressively Smaller Cells. Amphibian eggs are very large, with enough cytoplasm to sustain many rounds of cell division after fertilization. Each round of division during early development parcels the cytoplasm into smaller cells.

only is G1 lacking but G2 is also quite short, allowing cells to go almost directly from DNA synthesis to mitosis and back to DNA synthesis. In fact, S phase begins even before mitosis is complete. From such examples, we know that cell growth during G1 and G2 is not an absolute prerequisite for cell division, even though growth and division are normally coupled events.

Cell Cycle Checkpoints Regulate Progression Through the Cell Cycle

The control system that regulates progression through the cell cycle must accomplish several tasks. First, it must ensure that the events associated with each phase of the cell cycle are carried out at the appropriate time and in the appropriate sequence. Second, it must make sure that each phase of the cycle has been properly completed before the next phase is initiated. And finally, the control system must be able to respond to external conditions that indicate the need for cell growth and division (e.g., the quantity of nutrients available or the presence of growth-signaling molecules).

A series of control points in the cell cycle known as *checkpoints* accomplish these objectives (Figure 17-30). At each checkpoint, conditions within the cell determine whether or not the cell will proceed to the next stage of the cycle. The first checkpoint occurs late during the G1 phase. We have already seen that G1 is the phase that varies most among cell types, and mammalian cells that have stopped dividing are almost always arrested during G1. For example, in cultured cells we can stop or slow down the process of cell division by allowing the cells to run out of either nutrients or space or by adding inhibitors of vital processes such as protein synthesis. In all such cases, the cells are arrested in G1. These findings suggest that passing from G1 into S phase is a critical control point in the cell cycle. In yeast, this **G1 checkpoint** is called *Start;* yeast must have sufficient nutrients and must reach a certain size before they can pass through Start. In animal cells, the G1 checkpoint is called the *restriction point.* The ability to pass through the restriction point is controlled to a large extent by extracellular, growth-signaling proteins called *growth factors* (p. 562), which multicellular organisms use to stimulate or inhibit cell proliferation. Cells that have successfully passed through the restriction point are committed to S phase, whereas those that have not passed this point can remain in G1 indefinitely, in the resting state called G0.

In addition to the G1 checkpoint, two other cell cycle checkpoints have been well characterized. At the **G2 checkpoint,** located at the boundary between G2 and M phase, proper completion of DNA synthesis is required before the cell can initiate mitosis. In certain cell types, the cell cycle can be indefinitely arrested at this point if cell division is not necessary; under such conditions, the cells enter a resting state analogous to G0. The relative importance of the G1 and G2 checkpoints in regulating the rate of cell division varies with the organism and cell type. For

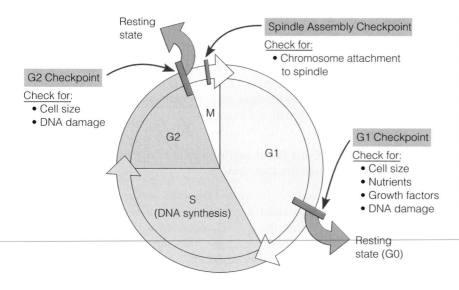

Figure 17-30 **Cell Cycle Checkpoints.** The red rectangles mark three of the main checkpoints in the eukaryotic cell cycle, points that determine whether or not the cell proceeds through the rest of the cycle. The determination is based on chemical signals reflecting both the cell's internal state and its external environment. Two main checkpoints are near the end of G1 (also called the restriction point in mammalian cells and Start in yeast cells) and at the end of G2. If conditions are not satisfactory at these checkpoints, the cell exits the cycle and goes into a resting state (called G0 for the G1 checkpoint). There is also a spindle assembly checkpoint during M phase, near the end of metaphase.

example, the G1 checkpoint is more important in the budding yeast *Saccharomyces cerevisiae*, as it is in most cells of multicellular organisms. On the other hand, the G2 checkpoint is more important in the mitotic divisions of fertilized frog eggs and in the yeast *Schizosaccharomyces pombe* (called a *fission yeast* because it reproduces by dividing evenly in two, rather than by budding).

The third cell cycle checkpoint, the **spindle assembly checkpoint,** is at the junction between metaphase and anaphase. Before cells can pass through the spindle assembly checkpoint and begin anaphase, all the chromosomes must be properly attached to the spindle. If the two chromatids that make up each chromosome are not properly attached to opposite spindle poles, the cell cycle is temporarily arrested at this point. In the absence of such a control mechanism, there would be no guarantee that each of the newly forming daughter cells would receive a complete set of chromosomes.

Cell behavior at the various checkpoints is influenced both by successful completion of preceding events in the cycle (such as DNA replication or chromosome attach-

ment to the spindle) and by factors in the cell's environment (such as nutrients and growth factors). But whatever the particular influences may be, their effects on cell cycle checkpoints are mediated by cellular proteins that activate or inhibit one another in chains of interactions that can be quite elaborate. As we will see shortly, there is an underlying unity in the types of molecules involved in this process.

Cell Fusion Experiments Provide Evidence for Control Molecules in the Cell Cycle

The first hints concerning the identity of the molecules involved in checkpoint control came from cell fusion experiments carried out in the early 1970s. In some of the earliest studies, two cultured mammalian cells in different phases of the cell cycle were fused to form a single cell with two nuclei, a **heterokaryon**. As Figure 17-31a indicates, if one of the original cells is in S phase and the other is in G1, the G1 nucleus in the heterokaryon quickly initiates DNA synthesis, even if it would not normally have reached S phase until many hours later. Such observations indicate that S phase

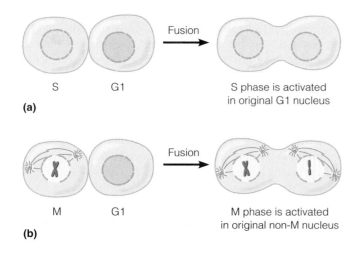

Figure 17-31 **Cell Fusion Evidence for the Role of Cytoplasmic Chemical Signals in Cell Cycle Regulation.** Important information can be obtained from experiments in which cells at two different points in the cell cycle are induced to fuse, forming a single cell with two nuclei, a heterokaryon. Cell fusion can be brought about by any of several methods, including the addition of certain viruses or polyethylene glycol, or the application of a brief electrical pulse, which causes plasma membranes to destabilize momentarily (electroporation). **(a)** If cells in S phase and G1 phase are fused, DNA synthesis begins in the original G1 nucleus, suggesting that a substance that activates S phase is present in the S phase cell. **(b)** If a cell in M phase is fused with one in any other phase, the latter cell immediately enters mitosis. If the cell was in G1, the condensed chromosomes that appear have not replicated and therefore are analogous to a single chromatid.

cells contain one or more molecules that trigger progression through the G1 checkpoint and into S phase. The controlling molecules are not simply the enzymes involved in DNA replication, since these enzymes can be present in high concentration in cells that do not enter S phase.

In contrast, when S phase cells are fused with cells in G2 (instead of G1), the G2 nucleus does not initiate DNA synthesis. This finding suggests that the G2 cell contains some factor that prevents it from carrying out an unwanted, second round of DNA replication prior to mitosis.

Cell fusion experiments have also been carried out in which cells undergoing mitosis were fused with interphase cells in either G1, S, or G2. After fusion, the nucleus of such interphase cells is immediately driven into the preparatory steps for mitosis, including chromatin condensation into visible chromosomes, spindle formation, and fragmentation of the nuclear envelope. If the interphase cell had been in G1, the condensed chromosomes will be unduplicated (Figure 17-31b).

A Mitosis-Promoting Factor (MPF) Moves Cells Through the G2 Checkpoint

The preceding cell fusion experiments suggested that specific molecules present in the cytoplasm are responsible for moving cells through the G1 and G2 checkpoints—that is, for triggering the onset of DNA replication (S phase) and mitosis (M phase). Even better evidence regarding the mitosis-triggering signal has come from experiments involving frog eggs. During development of the frog *oocyte* (an egg cell precursor), the cell cycle is arrested in G2 until hormones stimulate meiosis. (Meiosis, discussed in Chapter 18, is a special type of cell division that halves the number of chromosomes during egg or sperm production.) The oocyte then proceeds through most of the phases of meiosis but is arrested during metaphase of the second of two meiotic divisions. It is now a "mature" egg cell, capable of being fertilized. A crucial experiment demonstrated that if cytoplasm taken from a mature egg cell is injected into the cytoplasm of an immature oocyte, the oocyte immediately begins meiosis (Figure 17-32). The researchers hypothesized that a cytoplasmic chemical, which they named *maturation-promoting factor (MPF)*, induces this oocyte "maturation."

Subsequent experiments revealed that in addition to inducing meiosis, MPF can also trigger mitosis when injected into fertilized frog eggs. Similar MPF molecules were soon discovered in the cytoplasms of a broad range of dividing cell types, including yeasts, marine invertebrates, and mammals. The MPF activities in all these organisms have proven to be very similar to each other. For example, in yeast cells with a defective or missing MPF gene, the human version of the gene can substitute perfectly well, even though the last ancestor common to yeasts and humans probably lived about 3 billion years ago! Because of the general role played by MPF in triggering passage through the G2 checkpoint and into mitosis, the acronym **MPF,** which originally stood for "maturation promoting factor," is now understood to mean **mitosis-promoting factor** (or *M-phase promoting factor*), which more accurately describes this molecule's role.

Cell Cycle Mutants Facilitated the Identification of Molecules That Control the Cell Cycle

In the late 1980s, the study of cell cycle regulation entered a molecular era as the identity of MPF and a related family of control molecules in the cell cycle began to emerge. The identification of these molecules was greatly facilitated by genetic studies of yeasts. Because they are single-celled

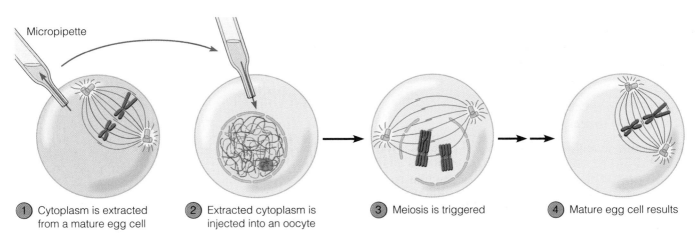

① Cytoplasm is extracted from a mature egg cell
② Extracted cytoplasm is injected into an oocyte
③ Meiosis is triggered
④ Mature egg cell results

Figure 17-32 Evidence for the Existence of MPF. In nature, hormones act on frog oocytes to trigger entry into meiosis and development into mature frog eggs, which are arrested (until fertilization) in metaphase of the second meiotic division. The experiment shown here, performed by Y. Masui and C. L. Markert in 1971, established the existence of a cytoplasmic substance involved in this process; they called it maturation-promoting factor (MPF). ① In their experiment, they used a micropipette to remove cytoplasm from a mature egg cell, arrested in metaphase of the second meiotic division, and ② inject it into an oocyte arrested in G2 of interphase. ③ The oocyte then entered meiosis and ④ became a mature egg cell. This experimental procedure could now be used as an assay for the detection and eventual isolation of MPF. The hormones that trigger oocyte maturation in the frog were presumed to act by stimulating the synthesis or activation of MPF.

organisms that can be readily grown and studied under defined laboratory conditions, yeasts are particularly convenient model organisms for investigating the genes involved in the control of the eukaryotic cell cycle.

Working with *S. cerevisiae*, geneticist Leland Hartwell undertook a search for yeast mutants that are "stuck" at some point in the cell cycle. It might be expected that most such mutants would be difficult or impossible to work with, because their blocked cell cycle would prevent them from reproducing. But Hartwell overcame this potential obstacle with a powerful strategy of microbial genetics, the use of temperature-sensitive mutants. As we saw earlier (p. 530), these are mutants whose defect is apparent only at temperatures above the normal range for the organism. Therefore, yeast carrying a temperature-sensitive mutation can be successfully grown at a lower ("permissive") temperature, even though their cell cycles would be blocked at higher temperatures. Presumably the protein encoded by the mutated cell cycle gene is close enough to the normal gene product to function at the lower temperature, while the increased thermal energy at higher temperatures disrupts its active conformation (the molecular shape needed for function) more readily than that of the normal protein.

Using this approach, Hartwell and his colleagues identified many genes involved in the cell cycle of *S. cerevisiae* and established the points in the cell cycle at which their products operate. Predictably, some of these genes turned out to encode DNA replication proteins, but others seemed to function in cell cycle regulation. A breakthrough discovery was made by Paul Nurse and his colleagues, who carried out similar research with the fission yeast *Schizosaccharomyces pombe*. They identified a gene called *cdc2*, whose activity is essential for the initiation of mitosis—that is, for passing through the G2 checkpoint. (The acronym *cdc* stands for cell division cycle.) The *cdc2* gene turned out to have counterparts in all eukaryotic cells studied. The yeast research came together with the frog egg research when it was established that the protein encoded by the *cdc2* gene is one of two proteins making up MPF. Researchers were now primed to unravel the mysteries of the cell cycle checkpoints.

The Cell Cycle Is Controlled by Cyclin-Dependent Kinase (Cdk) Molecules

Biochemical examination has revealed that the protein encoded by the yeast *cdc2* gene functions as a *protein kinase*—a type of enzyme that catalyzes the transfer of a phosphate group from ATP to certain other proteins. The phosphorylation of proteins by kinases, and their dephosphorylation by enzymes called *phosphatases*, is a common mechanism for regulating protein activity (e.g., see Figure 6-18b). And it is a mechanism that is used many times over in regulating the cell cycle.

Though the protein produced by the *cdc2* gene functions as a protein kinase, it is active only when bound to a member of another group of proteins known as **cyclins**. The protein product of the *cdc2* gene is therefore a

cyclin-dependent kinase (Cdk). Subsequent studies have revealed that control of the eukaryotic cell cycle involves several kinds of Cdk molecules and their interaction with multiple forms of cyclin, thereby creating a variety of different Cdk-cyclin complexes. As the name suggests, cyclins are proteins whose level in the cell oscillates, thereby allowing them to control the activity of the various Cdk molecules at different points in the cell cycle. Cyclins involved in regulating passage through the G2 checkpoint into M phase are called *mitotic cyclins*, and the Cdk molecules to which they bind are known as *mitotic Cdk's*. Likewise, cyclins involved in regulating passage through the G1 checkpoint into S phase are called *G1 cyclins*, and the Cdk molecules to which they bind are known as *G1 Cdk's*.

If progression through critical points in the cell cycle is controlled by Cdk molecules, how is the activity of these protein kinases regulated? One level of control is exerted by the availability of cyclins, which are required for Cdk activation, and a second type of regulation involves phosphorylation of Cdk molecules. We will illustrate both types of control by taking a closer look at mitotic Cdk molecules, which control progression through the G2 checkpoint.

The Mitotic Cdk-Cyclin Complex (MPF) Controls the G2 Checkpoint by Phosphorylating Proteins Involved in the Early Stages of Mitosis

The amount of mitotic Cdk present in a cell remains relatively constant throughout the cycle. However, mitotic Cdk is active as a protein kinase only after it becomes bound to a mitotic cyclin. The concentration of mitotic cyclin gradually increases during the G1, S, and G2 phases of the cycle, eventually reaching a critical threshold that permits it to activate mitotic Cdk at the end of G2 (Figure 17-33). The mitotic Cdk-cyclin complex is in turn responsible for trig-

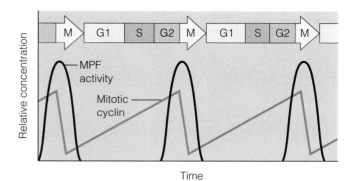

Figure 17-33   Fluctuating Levels of Mitotic Cyclin and MPF Activity During the Cell Cycle.   Cellular levels of mitotic cyclin rise during interphase (G1, S, and G2), then fall abruptly during M phase. The peaks of MPF activity (assayed by testing for its ability to stimulate mitosis) and cyclin concentration correspond, although the rise in MPF activity is not significant until a threshold concentration of cyclin is reached. Active MPF has been found to consist of a combination of mitotic cyclin and mitotic Cdk. The mitotic Cdk itself is present at a constant concentration (not shown on the graph), because the amount of mitotic Cdk increases at a rate corresponding to the overall growth of the cell.

gering passage through the G2 checkpoint and into mitosis. Earlier in the chapter, we stated that the G2 checkpoint was controlled by MPF. It is now clear that MPF is the same molecule as the mitotic Cdk-cyclin complex. In fact, in the scientific literature, "MPF" or "active MPF" often continues to be used to refer to this protein complex.

Although the activation of mitotic Cdk requires its binding to cyclin, phosphorylation and dephosphorylation of the Cdk protein also play key roles in the activation mechanism. When mitotic cyclin initially binds to mitotic Cdk, the resulting complex is inactive (Figure 17-34, ①). To trigger mitosis, the complex requires the addition of an activating phosphate group to a particular amino acid of the Cdk molecule. Before this phosphate is added, however, an *inhibiting* kinase phosphorylates the Cdk molecule at two other locations, causing the active site to be blocked (②). The activating phosphate group, highlighted with yellow in ③, is then added by a specific *activating* kinase. The last step in the activation sequence is the removal of the inhibiting phosphates by a specific *phosphatase* enzyme (④). Once the phosphatase begins removing the inhibiting phosphates, a positive feedback loop is set up: The activated Cdk-cyclin complex generated by this reaction stimulates the phosphatase, thereby causing the activation process to proceed more rapidly.

After the mitotic Cdk-cyclin complex has been activated, it functions as an active MPF whose protein kinase activity triggers the onset of mitosis (Figure 17-35). We have already learned that chromosome condensation, assembly of the mitotic spindle, and breakdown of the nuclear envelope are key events in the early stages of mitosis. How are these events triggered by the protein kinase activity of the activated MPF? In the case of nuclear envelope breakdown, the activated MPF phosphorylates (and stimulates other kinases to phosphorylate) the *lamin* proteins of the *nuclear lamina,* to which the inner nuclear membrane is attached (see Figure 16-31). This phosphorylation causes the lamins to depolymerize,

resulting in breakdown of the nuclear lamina. Disappearance of the nuclear lamina destabilizes the nuclear envelope, causing it to disperse into tiny membrane vesicles. In a similar fashion, phosphorylation of other proteins by MPF triggers other mitotic events. For example, phosphorylation of several chromosome-associated proteins, including histones and a multiprotein complex called the *condensin complex,* is thought to trigger chromosome condensation. Likewise, phosphorylation of microtubule-associated proteins by MPF may facilitate formation of the mitotic spindle.

It would obviously be disastrous for a cell to enter mitosis before all its chromosomal DNA had been replicated. The MPF pathway therefore includes mechanisms that ensure DNA has been replicated prior to permitting the cell to pass through the G2 checkpoint and into mitosis. For example, if cells are treated with an inhibitor that prevents DNA replication from being completed, the final dephosphorylation step in the activation of MPF (④ in Figure 17-34) does not take place and the cell is halted at the G2 checkpoint.

The G1 Cdk-Cyclin Complex Controls the G1 Checkpoint by Phosphorylating the Rb Protein

Although detailed information about the role of Cdk-cyclin complexes in cell cycle regulation first came from experiments involving the G2 checkpoint, subsequent studies revealed that other points in the cell cycle are also controlled by Cdk-cyclin complexes. While the components vary somewhat among organisms, eukaryotes contain a variety of different Cdk and cyclin molecules that interact in different combinations at various stages of the cell cycle. Two common themes underlie the way in which these various Cdk-cyclin complexes are regulated. First, different kinds of cyclins are synthesized and degraded during different phases of the cell cycle, and second, the activity of Cdk-cyclin complexes is controlled

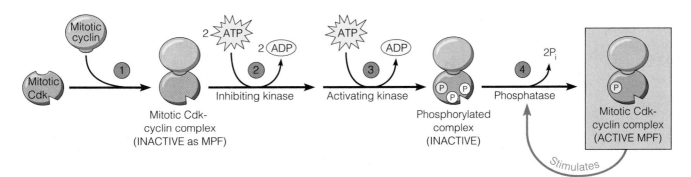

Figure 17-34 Phosphorylation and Dephosphorylation in the Activation of a Cdk-Cyclin Complex.   The series of reactions shown here for the formation of active MPF (mitotic Cdk-cyclin complex) was worked out using a combination of data from yeasts and frog eggs. ① The mitotic Cdk and cyclin proteins form an inactive complex. ② An inhibiting kinase adds two phosphate groups (white) to the complex, which block its active site. ③ An activating kinase phosphorylates a third site on the complex (yellow phosphate). ④ A phosphatase removes the inhibiting phosphate groups, converting the complex to a singly phosphorylated form, which is active as MPF. This active MPF in turn stimulates the phosphatase to produce additional active MPF. The result is a burst of MPF activity.

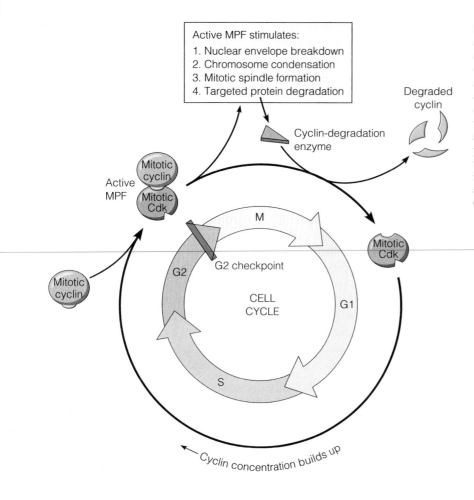

Active MPF stimulates:
1. Nuclear envelope breakdown
2. Chromosome condensation
3. Mitotic spindle formation
4. Targeted protein degradation

Degraded cyclin

Cyclin-degradation enzyme

Mitotic cyclin

Active MPF

Mitotic Cdk

Mitotic Cdk

M

G2 checkpoint

G2

CELL CYCLE

G1

Mitotic cyclin

S

Cyclin concentration builds up

Figure 17-35 The Mitotic Cdk Cycle. This diagram illustrates the activation and inactivation of the mitotic Cdk protein during the cell cycle. In G1, S, and G2, mitotic Cdk is made at a steady rate as the cell grows, while the mitotic cyclin concentration gradually increases. Mitotic Cdk and cyclin form a complex, "active MPF," that drives the cell cycle past the G2 checkpoint and stimulates the mitotic events listed. By activating a protein-degradation pathway that degrades cyclin, the mitotic Cdk-cyclin complex also brings about its own demise, allowing the completion of mitosis and entry into G1 of the next cell cycle. Mitotic Cdk, sometimes called "inactive MPF," is recycled.

by phosphorylation and dephosphorylation reactions catalyzed by protein kinases and phosphatases.

Having examined how these principles apply to the G2 checkpoint, let us briefly look at the operation of the G1 checkpoint (also known as the restriction point in mammals or Start in yeast, as mentioned earlier). Passing through the G1 checkpoint and into S phase is the main step that commits a cell to the process of cell division; it is therefore subject to control by factors such as cell size, the availability of nutrients, and the presence of external growth factors that signal the need for cell proliferation. These various types of signals function by activating Cdk-cyclin complexes that trigger entry into S phase by phosphorylating several target proteins.

The main target is the **Rb protein,** a molecule that controls the expression of genes coding for products needed for passage through the G1 checkpoint and into S phase. As shown in Figure 17-36, the Rb protein exerts this control by binding to the **E2F transcription factor,** a protein that, in the absence of bound Rb, activates the transcription of genes coding for enzymes and other proteins required for initiating DNA replication. As long as the Rb protein remains bound to E2F, the E2F molecule is inactive and these genes remain silent, thereby preventing the cell from entering into S phase. But in cells that have been stimu-

lated to divide (e.g., by the addition of growth factors), the growth signaling pathway leads to the production and activation of Cdk-cyclin complexes that in turn catalyze the phosphorylation of the Rb protein. Phosphorylated Rb molecules lose the ability to bind to E2F, allowing the E2F molecule to activate the transcription of genes whose products are required for entry into S phase.

Since the Rb protein regulates a key transition in the cell cycle, it is not surprising that defects in this protein can have disastrous consequences. For example, we will describe later in the chapter how mutations in the gene coding for the Rb protein play an important role in both hereditary and environmentally induced forms of cancer. Even when the Rb protein is normal, there are situations in which it would not be appropriate for this protein to trigger passage through the G1 checkpoint into S phase. For example, it would be dangerous for a cell with damaged DNA to enter S phase prior to repairing the DNA damage. The protein encoded by the *p53* gene plays a central role in stopping cells with damaged DNA from proceeding through the G1 checkpoint, in part by inhibiting the normal pathway for phosphorylating Rb. Because abnormalities in the *p53* gene are commonly encountered in cancer, we will delay further discussion of its role until we consider cancer cells later in the chapter.

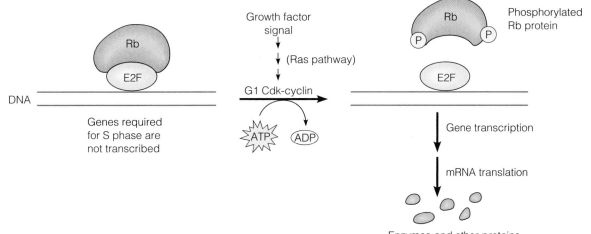

Figure 17-36   Role of the Rb Protein in Cell Cycle Control. In its normal, dephosphorylated state, the Rb protein binds to the E2F transcription factor. This binding prevents E2F from activating the transcription of genes coding for proteins required for DNA replication, which are needed before the cell can pass through the G1 checkpoint into S phase. In cells that have been stimulated by growth factors, the Ras pathway is activated (see Figure 17-39), which leads to the production and activation of a G1 Cdk-cyclin complex that catalyzes the phosphorylation of the Rb protein. Phosphorylated Rb can no longer bind to E2F, thereby allowing E2F to activate gene transcription and trigger the onset of S phase. During the subsequent M phase (not shown), the Rb protein is dephosphorylated so that it can once again inhibit E2F.

The Mitotic Cdk-Cyclin Complex (MPF) Controls the Spindle Assembly Checkpoint by Activating the Anaphase-Promoting Complex

In addition to acting at the G1 and G2 checkpoints, Cdk-cyclin complexes are also involved in the spindle assembly checkpoint, where the decision is made to separate the sister chromatids and thus initiate anaphase. But here, neither a new cyclin nor a new Cdk appears to be involved. Instead, the onset of anaphase is triggered by a protein degradation pathway activated near the end of metaphase by MPF (the mitotic Cdk-cyclin complex that also acts at the G2 checkpoint). MPF triggers passage through the spindle assembly checkpoint by catalyzing one or more protein phosphorylation reactions that lead to the activation of the **anaphase-promoting complex,** a large protein complex that controls many events associated with the final phases of mitosis. This complex exerts its effects by targeting selected proteins for degradation by joining them to *ubiquitin,* a molecule that causes the attached protein to be destroyed by a special mechanism to be described in Chapter 21.

As shown in Figure 17-37, the anaphase-promoting complex exerts control over the onset of anaphase by causing the destruction of proteins called **cohesins,** which hold sister chromatids together prior to the beginning of anaphase. Rather than acting directly on cohesins, the anaphase-promoting complex triggers the breakdown of *securin,* a protein that normally binds to and inhibits a special, protein-degrading enzyme called *separin.* It is the released separin molecule that cleaves cohesins, thereby freeing sister chromatids so they can separate from each other and begin the process of anaphase.

In addition to initiating anaphase, the anaphase-promoting complex also controls a series of events associated with the completion of mitosis. It exerts this control by targeting mitotic cyclin for destruction. Because mitotic cyclin is a component of MPF, the activity of MPF quickly falls. Evidence suggests that many of the events associated with the exit from mitosis—such as chromosome decondensation and re-assembly of the nuclear envelope—depend on this cyclin degradation step and the associated loss of MPF activity. For example, it has been shown that adding a nondegradable form of mitotic cyclin to frog egg extracts blocks both chromosome decondensation and re-assembly of the nuclear envelope, thereby preventing mitosis from being completed.

Since the anaphase-promoting complex triggers both the onset of anaphase and the subsequent completion of mitosis, it is crucial that these events do not begin until all the chromosomes have been properly attached to the mitotic spindle. This timing is controlled by the chromosomal kinetochores, which release a signaling protein called *Mad2* as long as they remain *unattached* to spindle microtubules. The Mad2 protein in turn inhibits the anaphase-promoting complex, thereby preventing the initiation of anaphase. After all kinetochores have become attached to the spindle, Mad2 is no longer released from the kinetochores; thus the anaphase-promoting complex is no longer inhibited and it can initiate the onset of anaphase.

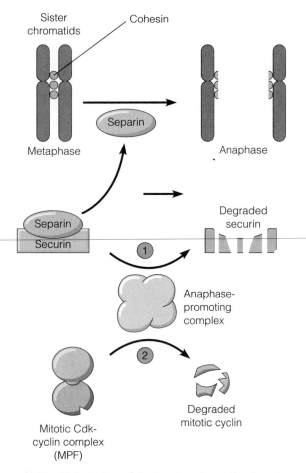

Sister chromatids

Cohesin

Separin

Metaphase

Anaphase

Separin

Securin

Degraded securin

①

Anaphase-promoting complex

②

Degraded mitotic cyclin

Mitotic Cdk-cyclin complex (MPF)

Figure 17-37 Main Actions of the Anaphase-Promoting Complex. The anaphase-promoting complex controls the final stages of mitosis by targeting selected proteins for degradation. Two important proteins targeted for destruction are ① securin and ② mitotic cyclin. The degradation of securin leads to the release of separin, a protease that triggers the onset of anaphase by cleaving the cohesin proteins that hold sister chromatids together. The degradation of mitotic cyclin leads to the loss of MPF activity, which in turn triggers chromosome decondensation and re-assembly of the nuclear envelope.

Putting It All Together: The Cell Cycle Regulation Machine

Figure 17-38 is a generalized and simplified summary of the main features of the molecular "machine" that regulates the eukaryotic cell cycle. The operation of this machine can be described in terms of two fundamental, interacting mechanisms. One mechanism is an autonomous clock that goes through a fixed cycle over and over again. The molecular basis of this clock is the synthesis and degradation of cyclins, alternating in a rhythmic fashion. These cyclins in turn bind to Cdk molecules, creating various Cdk-cyclin complexes that trigger the passage of cells through the three major cell cycle checkpoints. The second mechanism adjusts the clock as needed, by providing feedback from the cell's internal and external environments. This mechanism makes use of additional proteins that, directly or indi-

rectly, influence the activity of Cdk's and cyclins. Many of these additional proteins are protein kinases or phosphatases. It is this part of the cell cycle machine that relays information about the state of the cell's metabolism—including DNA replication—and about conditions outside the cell, thereby influencing whether or not the cell should commit to the process of cell division. As we will see next, growth-promoting and growth-inhibiting signaling molecules that come from outside the cell are prominent components of this regulatory mechanism.

Growth Control and Cancer

Simple unicellular organisms, such as bacteria and yeast, live under conditions in which the presence of sufficient nutrients in the external environment is the primary factor that determines whether cells grow and divide. In multicellular organisms, the situation is usually reversed; cells are typically surrounded by nutrient-rich extracellular fluids, but the organism as a whole would be quickly destroyed if every cell were to continually grow and divide just because it had access to adequate nutrients. Cancer is a potentially lethal reminder of what happens when cell proliferation continues unabated without being coordinated with the needs of the organism as a whole. To overcome this problem, multicellular organisms utilize extracellular signaling proteins called **growth factors** to control the rate of cell growth and division (see Table 10-2). Most growth factors are *mitogens,* which means that they stimulate cells to enter the S phase of the cell cycle, followed by G2 and then mitosis.

Stimulatory Growth Factors Activate the Ras Pathway

If mammalian cells are placed in a culture medium containing nutrients and vitamins but lacking growth factors, they normally become arrested in G1 despite the presence of adequate nutrients. Growth and division can be triggered by adding small amounts of blood serum, which contains several stimulatory growth factors. Among them is **platelet-derived growth factor (PDGF),** a protein produced by blood platelets that stimulates the proliferation of connective tissue cells and smooth muscle cells. Another important growth factor, called **epidermal growth factor (EGF),** is widely distributed in embryonic tissue. EGF was initially isolated from the salivary glands of mice by Stanley Cohen, who received a Nobel Prize in 1987 for his pioneering investigations of growth factors. Certain growth factors, such as EGF, stimulate the growth of a wide variety of cell types, whereas others act more selectively on particular target cells. Growth factors stimulate tissue growth during embryonic development and during wound repair and tissue regeneration in the adult. For example, release of the growth factor PDGF from blood platelets at wound sites is instrumental in stimulating the growth of tissue required for wound healing.

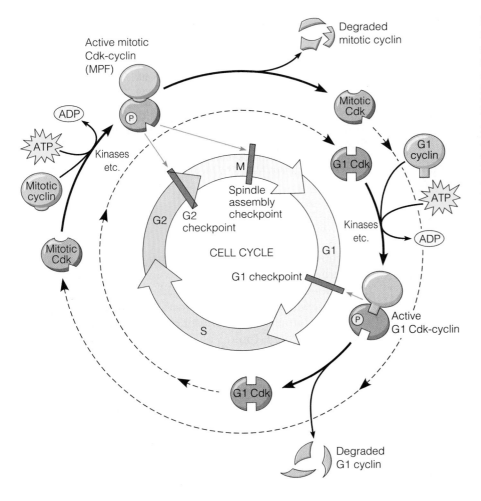

Figure 17-38   A General Model for Cell Cycle Regulation.   According to this model, passage through the three main checkpoints is triggered by protein complexes made of cyclin and Cdk, whose phosphorylation of other proteins induces progression through the cycle. Activation of the Cdk-cyclin complexes themselves requires their phosphorylation by other kinases, as well as appropriate dephosphorylation by phosphatases. Different cyclins and, in most eukaryotes, different Cdk proteins are used at different points in the cell cycle.

Growth factors such as PDGF and EGF act by binding to plasma membrane receptors located on the surface of target cells. Different kinds of cells have different receptors and hence differ in the growth factors to which they respond. In Chapter 10, we learned that growth factor receptors exhibit tyrosine kinase activity. The binding of a growth factor to its receptor activates this tyrosine kinase activity, leading to phosphorylation of tyrosine residues located in the region of the receptor protruding into the cytosol. Phosphorylation of these tyrosines in turn triggers a complex cascade of events that culminates in the cell passing through the G1 checkpoint and entering into S phase. The *Ras pathway* introduced in Chapter 10 plays a central role in these events, as shown by studies involving cells that have stopped dividing because growth factor is not present. When mutant, hyperactive forms of the Ras protein are injected into such cells, the cells enter S phase and begin dividing, even in the absence of growth factor. Conversely, injecting cells with antibodies that inactivate the Ras protein prevents cells from entering S phase and dividing in response to growth factor stimulation.

As shown in Figure 17-39, the mechanism by which the Ras pathway induces cells to pass through the G1 checkpoint and into S phase involves six steps. ① The stimulating growth factor first binds to its plasma membrane receptor. ② This binding activates the receptor's tyrosine kinase activity, thereby leading to phosphorylation of tyrosine residues. ③ The phosphorylated tyrosines serve as binding sites for a series of adaptor proteins that in turn activate the plasma membrane G protein, Ras. As is generally the case for G proteins (p. 259), activation of the Ras protein is accompanied by GTP binding and the release of GDP.

④ The activated Ras molecule then triggers a cascade of protein phosphorylation reactions, beginning with phosphorylation of a protein kinase called *Raf*. Activated Raf phosphorylates serine and threonine residues in a protein kinase called *MEK*, which in turn phosphorylates threonine and tyrosine residues in a group of protein kinases called *MAP kinases* (*mitogen-activated protein kinases; MAPK*). ⑤ The activated MAP kinases enter the nucleus and phosphorylate several regulatory **transcription factors,** which are proteins that activate the transcription of specific genes. Among the proteins activated by this phosphorylation mechanism are Jun (a component of the AP-1 transcription factor) and proteins that are members of the *Ets family* of transcription factors. These activated transcription factors turn on the transcription of "early genes" that code for the production of other transcription factors, including Myc, Fos, and Jun, which then activate the transcription of a family of "delayed genes." One of these latter genes codes for the E2F transcription factor

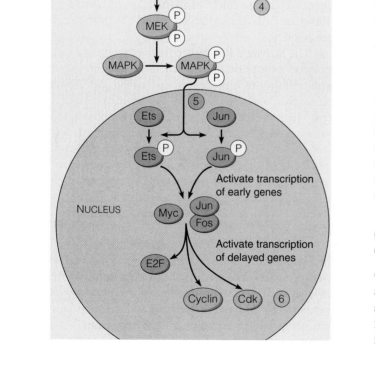

Figure 17-39   Growth Factor Signaling via the Ras Pathway. Scientists have pieced together much of the complex molecular chain of events that is triggered when the binding of a growth factor to its cell-surface receptor leads to a stimulation of cell growth and division. The Ras protein is a key intermediate in this pathway, which consists of six main steps: ① binding of a growth factor to its receptor; ② phosphorylation of the receptor; ③ activation of Ras; ④ activation of a cascade of cytoplasmic protein kinases (Raf, MEK, and MAPK); ⑤ activation or production of nuclear transcription factors (Ets, Jun, Fos, Myc, E2F); and ⑥ synthesis of cyclin and Cdk molecules. The resulting Cdk-cyclin complexes catalyze the phosphorylation of Rb and hence trigger passage from G1 into S phase. MAPK = Map kinases.

whose role in controlling entry into S phase was described earlier in the chapter. ⑥ Also included in the delayed genes are several coding for either Cdk or cyclin molecules, whose production leads to the formation of Cdk-cyclin complexes that phosphorylate Rb and hence trigger passage from G1 into S phase.

Thus, in summary, the Ras pathway is a complex signaling cascade in which the binding of a growth factor to a receptor on the cell surface ultimately causes the cell to pass through the G1 checkpoint and into S phase, thereby starting the cell on the road to cell division.

Inhibitory Growth Factors Act Through Cdk Inhibitors

Although we usually think of growth factors as being growth-stimulating molecules, the function of some growth factors is actually to inhibit cell proliferation. One example is **transforming growth factor β (TGFβ),** a protein that can exhibit either growth-stimulating or growth-inhibiting properties, depending on the target cell type. The antiproliferative actions of TGFβ include antagonism of the growth-stimulating effects of other growth factors, such as EGF and PDGF.

When acting as a growth inhibitor, the binding of TGFβ to its cell-surface receptor triggers a series of events that increases the activity of a protein called p15, which in turn suppresses the activity of Cdk-cyclin complexes and hence blocks progression through the cell cycle. The p15 protein is therefore said to function as a **Cdk inhibitor.** Several other Cdk inhibitors are also involved in pathways that inhibit cell growth and division. An especially important example is a protein called p21, which, as we will see shortly, plays a key role in preventing cells containing damaged DNA from passing through the G1 checkpoint.

Cancer Involves Defective Cell Cycle Control Mechanisms

Given the complexity of the cell cycle and its control mechanisms, it is perhaps not surprising that malfunctions occasionally occur. When normal growth control mechanisms fail, uncontrolled cell proliferation can produce a growing mass of cells called a **tumor.** Tumors are classified as benign or malignant based on their likelihood of spreading. **Benign tumors** are localized masses that do not spread, whereas **malignant tumors** can invade neighboring tissues and even other parts of the body, and thus are potentially life-threatening. The general term for a malignant tumor is **cancer.** The term *cancer,* derived from a Greek word meaning "crab," was coined by Hippocrates in the fifth century B.C. to describe diseases in which cells grow and spread unrestrained throughout the body, eventually choking off life. Roughly 35% of the people living in the United States are now expected to develop cancer and roughly half of this group will die from the disease, making it the most frequent cause of death after cardiovascular disease.

Although impaired control of cell proliferation is not the only defect observed in cancer cells, it is a fundamental, underlying feature. What causes cells to lose control of the mechanisms that normally regulate growth and division? A vast body of evidence points to the role played by DNA mutations. Some cancer-causing mutations are triggered by DNA-damaging chemicals and radiation, whereas others arise from spontaneous DNA mutations and replication errors. Mutant genes that contribute to the development of cancer can also be inherited, or in some cases, may be introduced into cells by viruses. In spite of such differences in origin, the end result is often the same: a mutation in one or more genes affecting cell cycle control. In the following sections, we will see that cancer-inducing mutations involve three main classes of genes: *oncogenes, tumor suppressor genes,* and *DNA repair genes.*

Oncogenes Can Trigger the Development of Cancer

An **oncogene** is a gene whose *presence* can trigger the development of cancer. Some oncogenes arise by mutation from normal cellular genes, whereas others are introduced into cells by cancer-causing viruses. In either case, oncogenes code for proteins that stimulate excessive cell proliferation.

The first oncogene to be discovered is in the genome of the *Rous sarcoma virus,* which causes cancer in chickens. Because the Rous virus has only four genes, it was relatively easy to determine which gene triggers the development of cancer. Mutational studies revealed that defects in one particular gene, now known as the *src* oncogene, yields mutant viruses that have lost the ability to cause cancer and yet can still infect cells and reproduce normally. In other words, a functional copy of the *src* gene must be present before the virus can cause sarcomas (a type of cancer). Similar studies involving other cancer viruses have led to the identification of several dozen additional oncogenes.

Although viruses cause a variety of different kinds of cancer in animals, only a few types of human cancer are associated with viruses. Among these are cervical cancer, which is associated with infection by the human papillomavirus (HPV), and Burkitt's lymphoma, which is linked to infection with the Epstein-Barr virus (EBV). The vast majority of human cancers are not associated with viral infections.

How do scientists identify the oncogenes involved in such nonviral cancers? The most common technique, called the *oncogene transfection assay,* analyzes the ability of DNA isolated from cancer cells to cause cancer. The assay involves three basic steps. First, the DNA from the tumor is isolated, cleaved with a restriction endonuclease, and presented to normal cultured cells, such as mouse fibroblasts, under conditions that favor incorporation of the foreign DNA fragments into the cells' chromosomes. The uptake of foreign DNA by cells under such artificial laboratory conditions is called *transfection.* Next, cells that grow excessively in culture are injected into animals. If a tumor results from this step (*transformation*), it confirms that the cells carry an oncogene. Finally, the DNA is isolated from these cells and analyzed. For example, by using the presence of human-specific repeated sequences to distinguish human DNA from the bulk of the surrounding mouse DNA, the oncogene in the human DNA can be identified.

Oncogenes Arise from Proto-oncogenes

These two independent means for identifying oncogenes—genetic analysis of cancer viruses and transfection with DNA isolated from human tumors—have led to the discovery of more than 50 different genes that can function as oncogenes. An unexpected finding to emerge from this type of analysis is that *most oncogenes are mutant forms of normal cellular genes.* The term **proto-oncogene** is used to refer to such normal cellular genes that closely resemble oncogenes. Oncogenes are typically denoted using italicized names, such as *src, myc,* or *ras,* while the corresponding normal version of each gene (i.e., the proto-oncogene), is denoted using the same name preceded by the letter "c-": c-*src,* c-*myc,* c-*ras* (the "c" stands for "cellular").

What types of mutations transform normal genes (i.e., proto-oncogenes) into genes that contribute to the development of cancer (i.e., oncogenes)? Recent research suggests that the following four mechanisms are involved.

Point Mutation. The simplest mechanism by which a proto-oncogene is converted into an oncogene is a *point mutation,* a single base-pair substitution in DNA that causes a single amino acid substitution in the protein encoded by the normal proto-oncogene. This phenomenon is frequently encountered in *ras* oncogenes.

DNA Rearrangement. The second mechanism for creating oncogenes is based on local DNA rearrangements that cause either deletions or base-sequence exchanges between proto-oncogenes and surrounding genes. The *trk* oncogene, for example, is generated by a local DNA rearrangement that disrupts one end of the c-*trk* proto-oncogene, thereby leading to the production of a hybrid protein in which the N-terminal end of the protein produced by the *trk* oncogene is replaced by an amino acid sequence encoded by a neighboring gene. The discovery of a somewhat different type of DNA rearrangement has come from the surprising finding that certain cancer viruses lack oncogenes. Such viruses cause cancer by integrating a DNA copy of their genetic information into a host chromosome in a region where a proto-oncogene is located, rearranging the structure of the normal host proto-oncogene and thereby converting it into an oncogene.

Gene Amplification.   The third way of creating oncogenes utilizes *gene amplification* (which we will discuss in Chapter 21) to increase the number of copies of a particular proto-oncogene. The increased number of gene copies causes overproduction of the normal protein product of the proto-oncogene rather than formation of an abnormal protein product. For example, certain forms of lung cancer are caused by amplification of the c-*myc* proto-oncogene, leading to excessive production of the Myc protein. (Note that the names of proteins encoded by oncogenes or proto-oncogenes are usually written without italics, starting with a capital letter. Hence the *myc* and c-*myc* genes are said to code for the Myc protein.)

Chromosomal Translocation.   In *chromosomal translocation*, a portion of one chromosome is physically removed and joined to another chromosome. A classic example occurs in *Burkitt's lymphoma*, a cancer of human lymphocytes in which a segment of chromosome 8 containing the c-*myc* proto-oncogene is translocated to chromosome 2, 14, or 22. The sites of translocation are close to genes coding for immunoglobulins (antibody molecules). These immunoglobulin genes are very active in lymphocytes, where approximately half of the total protein produced by the cell is immunoglobulin. The proximity of an active immunoglobulin gene causes the translocated *myc* gene to be overexpressed, resulting in overproduction of the Myc protein.

Most Oncogenes Code for Components of Growth Factor Signaling Pathways

We have just seen that alterations in proto-oncogenes can convert these normal genes into oncogenes, which in turn code for proteins that either are structurally abnormal or are produced in excessive amounts. But how do these oncogene-encoded proteins cause cancer? Examination of their functional activities has revealed that most oncogene proteins fit into one of six basic categories: *growth factors, receptors, plasma membrane G proteins, protein kinases, transcription factors,* or *Cdk-cyclins* (Table 17-2). These six functional groups correspond to the six main steps of the growth factor signaling pathway summarized in Figure 17-39. The following sections provide examples of oncogenes involved in each of the six steps in the growth signaling pathway.

Growth Factors.   Normal cells require an external supply of an appropriate growth factor and will fail to divide in its absence. If a cell acquires an oncogene that produces this growth factor, however, then the cell will divide even in the absence of an external supply. One oncogene that functions in this way is *sis,* an oncogene carried by the *simian sarcoma virus.* The *sis* oncogene codes for an altered form of a polypeptide subunit of the growth factor PDGF. Introducing the *sis* oncogene into a cell whose growth is controlled by PDGF causes that cell to produce

Table 17-2   Examples of Oncogenes Categorized by the Nature of Their Protein Products

Nature of Gene Product	Oncogene	Mechanism of Origin	Cancer Type
1. Growth factors	*sis*	Viral	Monkey sarcomas
	hst	Gene amplification	Human stomach cancer
2. Receptors	*erb-B*	Viral	Avian erythroblastosis
	fms	Viral	Feline sarcomas
	neu	Gene amplification	Human breast and ovarian cancers
	ret	DNA rearrangement	Human thyroid cancer
	trk	DNA rearrangement	Human colon and rectal cancer
3. Plasma membrane G proteins	*N-ras*	Point mutation	Human myeloid leukemia
	H-ras	Point mutation	Human bladder cancer
	K-ras	Point mutation	Human pancreatic, colon, rectal, and lung cancer, leukemia, others
4. Protein kinases	*raf*	Viral	Mouse sarcoma, avian carcinoma
	mos	Viral	Mouse sarcoma
5. Transcription factors	*myc*	Chromosomal translocation	Burkitt's lymphoma (human)
	L-myc	Gene amplification	Human small cell lung cancer
	N-myc	Gene amplification	Human neuroblastoma
	fos	Viral	Mouse bone cancer
	jun	Viral	Chicken sarcoma
6. Cdk-cyclins	*Cdk4*	Gene amplification	Human sarcomas
	CYCD1 (cyclin D1)	Gene amplification, chromosomal translocation	Human breast cancer, B cell lymphoma

its own supply of active PDGF. This PDGF in turn stimulates the cell's own growth, in contrast to the normal situation, in which cells are exposed to PDGF only if blood platelets in their immediate vicinity release PDGF into the surrounding environment. Hence the abnormal PDGF molecules produced by the *sis* oncogene cause cells to continually stimulate their own proliferation.

Receptors. A growth factor receptor may be altered so that its tyrosine kinase activity is permanently activated, even when growth factor is not bound to the receptor. An altered receptor of this type is produced by the *erb*-B oncogene, a gene found in the *erythroblastosis virus* that causes cancer of red blood cells in chickens. The Erb-B protein is an altered version of the receptor for the growth factor EGF that lacks the growth factor binding site, but retains tyrosine kinase activity. As a result, the Erb-B protein acts as a tyrosine kinase that is *constitutively active* (active in the absence of growth factor). In contrast, the normal EGF receptor exhibits tyrosine kinase activity only when the proper growth factor, EGF, is bound to the receptor's EGF binding site. The ability of Erb-B to function as a tyrosine kinase in the absence of EGF binding allows it to permanently stimulate the pathway leading to cell proliferation.

Plasma Membrane G Proteins. In most growth factor signaling pathways, the binding of a growth factor to its receptor leads to activation of the plasma membrane G protein Ras. Oncogenes coding for mutant Ras proteins are the most common type of oncogene in human cancers. *Ras* oncogenes produce abnormal Ras proteins that retain bound GTP instead of hydrolyzing it to GDP, thereby keeping the Ras protein in a permanently activated state. In this hyperactive state, the Ras protein constantly stimulates the rest of the pathway leading to cell proliferation, regardless of whether or not growth factor is bound to the cell's growth factor receptors.

Protein Kinases. Upon activation, Ras normally stimulates a cascade of protein kinases that phosphorylate the amino acids serine, threonine, and tyrosine in various combinations in different proteins. Ras initiates this protein kinase cascade by binding to and activating a protein kinase called Raf. Several cancer viruses have *raf* oncogenes that code for abnormal versions of the Raf protein. These mutant Raf proteins are permanently locked in an active configuration, causing them to continually stimulate the rest of the protein kinase cascade.

Transcription Factors. The protein kinase cascade triggers the phosphorylation of nuclear transcription factors, thereby turning on the expression of genes required for cell proliferation. Oncogenes called *myc*, *fos*, and *jun* all code for transcription factors that are part of this process. We have already described how chromosomal translocations associated with Burkitt's lymphoma cause the *myc* gene to be overexpressed, causing too much Myc protein to be produced. In several other types of human cancer

the c-*myc* gene is amplified, again creating oncogenes that produce too much Myc protein.

Cdk-Cyclins. In the final step of the signaling pathway, transcription factors activate genes coding for Cdk and cyclin molecules that govern passage through G1 and the G1 checkpoint. At least two oncogenes implicated in human cancers are known to produce such components. One is a Cdk oncogene called *Cdk4*, which is amplified in certain sarcomas, and the other is the cyclin oncogene *CYCD1*, which is overexpressed in several types of cancer, including some breast cancers. Such oncogenes cause the overproduction of Cdk-cyclin complexes that can stimulate progression through the cell cycle, even in the absence of growth factors.

Most oncogenes code for a protein that falls into one of the six preceding categories. Some oncogenes code for abnormal, hyperactive versions of such proteins, whereas others cause the overproduction of an otherwise normal protein. In either case, the result is excessive stimulation of the growth factor signaling pathway and hence excessive cell proliferation.

While these generalizations apply to most oncogenes, there are some exceptions. For example, an oncogene called *bcl*-2 codes for a protein that prevents *apoptosis,* a special type of cell death that will be discussed shortly. One function of apoptosis is to destroy cells that contain damaged DNA, thereby preventing the proliferation of cells possessing large numbers of mutations. When a *bcl*-2 oncogene causes too much Bcl-2 protein to be produced, the excess Bcl-2 blocks apoptosis and thus contributes to the development of cancer by prolonging the survival of genetically damaged cells rather than by stimulating cell proliferation.

Another group of oncogenes that do not fit into the six categories are those whose products interfere with the normal functions of tumor suppressor genes, which we discuss next.

Cancer Can Arise from the Loss of Tumor Suppressor Genes That Normally Restrain Cell Proliferation

In contrast to oncogenes, whose *presence* can induce cancer formation, the *absence* or *inactivation* of **tumor suppressor genes** can also lead to cancer. As its name indicates, the normal function of this type of gene is to restrain cell proliferation. In other words, tumor suppressor genes act as brakes on the process of cell proliferation, whereas oncogenes function more like an accelerator that speeds up cell proliferation.

The *Rb* Gene. The first tumor suppressor gene to be discovered was identified in *hereditary retinoblastoma,* a rare type of eye cancer that develops in young children who have a family history of the disease. Such children inherit a deletion in a specific region of one copy of chromosome 13. By itself, this deletion does not cause cancer. But during the many rounds of cell division that take place during development of the normal retina, cells occasionally acquire a deletion or mutation involving the same region in the second copy of chromosome 13. Cancer arises from these cells containing defects in both copies of chromosome 13. This pattern suggests that chromosome 13 contains a gene that normally inhibits cell division, and that deletion or disruption of both copies of the gene must occur before cancer develops. Comparison of DNA fragments isolated from chromosome 13 in both normal and retinoblastoma cells has led to the identification of the *Rb* gene as the missing gene.

The *Rb* gene codes for the Rb protein, whose role in controlling the transition from G1 into S phase was described earlier in the chapter (see Figure 17-36). The Rb protein functions as part of the braking mechanism that normally inhibits cells from entering into S phase in the absence of an appropriate signal from a growth factor. An absence of functional Rb protein, caused by the loss or disruption of both copies of the *Rb* gene, removes this restraining mechanism and allows uncontrolled proliferation. This effect of *Rb* mutations is not limited to the rare, inherited form of retinoblastoma from which the gene was originally isolated. Environmentally induced mutations leading to a loss or inactivation of the Rb protein have subsequently been detected in several common adult cancers, including some forms of lung, breast, and bladder cancer.

Inactivation of the Rb protein is also involved in the action of certain cancer viruses. One example involves the *human papillomavirus (HPV)*, which is associated with the development of human cervical cancer. HPV contains an oncogene that codes for a protein called E7, which binds to and inactivates the Rb protein, thereby interfering with the ability of Rb to restrain cell proliferation. In sum, cancers triggered by a loss of functional Rb protein can be induced in at least two distinct ways: through mutations that delete or disrupt both copies of the *Rb* gene, and through viral-encoded proteins that bind to and inactivate the Rb protein.

The *p53* Gene. Since the initial isolation of the *Rb* gene in 1986, more than a dozen other tumor suppressor genes have been discovered (Table 17-3). One of the more intriguing is the ***p53* gene,** which turns out to be the most frequently mutated gene in human cancer. More than half of the roughly 7 million people worldwide who will be diagnosed with some form of cancer this year will have *p53* mutations. Sometimes called the "guardian of the genome," the protein produced by the *p53* gene is responsible for protecting cells from the effects of DNA damage.

Figure 17-40 illustrates how this works. If normal cells are subjected to treatments that cause extensive DNA damage, such as exposure to ionizing radiation, they respond by increasing the amount of p53 protein. The mechanism responsible for this increase is related to the ability of damaged DNA to trigger activation of a protein kinase called *ATM*, which in turn leads to the phosphorylation of p53. Phosphorylation stabilizes p53 by preventing its interaction with *Mdm2*, a protein that normally promotes the degradation of p53 by linking it to *ubiquitin* (a

Table 17-3 Examples of Human Tumor Suppressor Genes

Gene	Inherited Syndrome	Cancer Type
APC	Familial adenomatous polyposis	Colon
BRCA1	Familial breast cancer	Breast, ovary
BRCA2	Familial breast cancer	Breast
DCC*	Colorectal cancer	Colon, rectal
NF-1	Neurofibromatosis type 1	Neurofibromas
NF-2	Neurofibromatosis type 2	Schwann cells, meninges
p16**	Familial melanoma	Melanoma, others
p53	Li-Fraumeni	Sarcomas, breast, brain, others
Rb	Hereditary retinoblastoma	Retina, bone, others
VHL	von Hippel-Lindau	Kidney, retina, brain
WT-1	Wilms' tumor	Kidney

*Recent evidence suggests that *SMAD4*, a gene located near *DCC*, may be the actual tumor suppressor gene.

**The *p16* gene codes for a CdK inhibitor that is closely related to the p15 protein.

molecule whose role in targeting proteins for destruction by proteasomes is described in Chapter 21).

Once it has been stabilized, the p53 protein activates two kinds of events: cell cycle arrest and cell death. In the first of these events, the p53 protein functions as a transcription factor that activates the gene coding for p21, a Cdk inhibitor that suppresses the activity of Cdk-cyclin complexes and hence blocks passage through the G1 checkpoint. Arresting the cell cycle in G1 gives cells time to repair DNA damage. To assist in the repair process, p53 also stimulates the production of enzymes involved in DNA repair.

If the DNA damage cannot be successfully repaired, p53 then activates the transcription of genes that trigger a cell death process called *apoptosis* (previously described in Chapter 10). Apoptosis is a carefully orchestrated death program whose early stages are usually characterized by changes in mitochondrial membrane permeability. This leads to the release of cytochrome *c* and other mitochondrial proteins into the cytosol. These proteins in turn help to activate a family of proteases called *caspases*. Through their proteolytic actions, caspases degrade the cell's key structural macromolecules, thereby leading to an orderly disassembly of the dying cell. This highly organized suicide program is characterized by cell shrinkage, collapse of the cytoskeleton, nuclear envelope breakdown, chromatin condensation, and degradation of DNA.

The ability of p53 to trigger cell cycle arrest and/or cell death allows it to function as a molecular stop light that responds to DNA damage by preventing cells from proliferating and passing the damage on to daughter cells. Loss of p53 function can thus contribute to the development of cancer by allowing the survival and reproduction of cells containing damaged DNA. It is therefore not surprising that, as was the case for the *Rb*

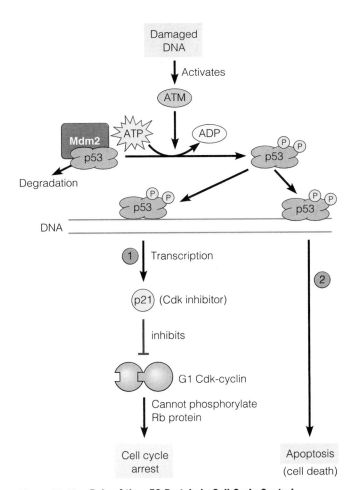

Figure 17-40 Role of the p53 Protein in Cell Cycle Control.
Damaged DNA activates the protein kinase ATM, leading to
phosphorylation of the p53 protein. Phosphorylation stabilizes
p53 by blocking its interaction with Mdm2, a protein that
normally promotes p53 degradation. As a result, p53 accumulates
and triggers two events. ① The p53 protein binds to DNA and
functions as a transcription factor that activates transcription of
the gene coding for the p21 protein, a Cdk inhibitor. The resulting
inhibition of Cdk-cyclin prevents phosphorylation of the Rb
protein, leading to cell cycle arrest at the G1 checkpoint. ② If the
DNA damage cannot be repaired, p53 activates the transcription
of other genes that trigger cell death via apoptosis.

gene, individuals who inherit only one functional copy of
the *p53* gene exhibit an increased risk of developing cancer. In this inherited condition, called the *Li-Fraumeni
syndrome*, cancers tend to arise by early adulthood as a
result of mutations that inactivate the second, normal
copy of the *p53* gene.

Although inherited defects in the *p53* gene can
increase a person's risk of developing cancer, most *p53*
mutations are not inherited but are instead caused by
environmental exposure to DNA-damaging chemicals and
radiation. For example, cancer-causing chemicals in
tobacco smoke are known to cause several kinds of point
mutations in the *p53* gene. In such cases, mutation in one
copy of the *p53* gene may be enough to interfere with
functioning of the p53 protein even when the other copy
of the *p53* gene is normal. The reason for this unexpected

finding is that functional p53 molecules actually consist of
four p53 chains bound together to form a tetramer. The
presence of even one mutant chain in such a tetramer may
be enough to keep the p53 protein complex from performing its normal function.

Like the Rb protein, the p53 protein is a target for certain cancer viruses. For example, the human papillomavirus
(HPV), in addition to producing the E7 protein that
inactivates Rb, has a second oncogene that codes for a
protein called E6, which binds to the p53 protein and
targets it for destruction. Thus HPV induces the
development of cancer by producing proteins that block
the actions of the proteins encoded by both the *Rb* and
p53 tumor suppressor genes.

Genetic Instability Leads to the Accumulation of Multiple Mutations in Cancer Cells

Although abnormalities in the *p53* tumor suppressor gene
have been detected in a majority of all human cancers, this
does not mean that the loss of p53 by itself causes cancer.
The most frequent types of human cancer, including
colon, lung, and breast cancer, are produced by multiple
mutations that involve the inactivation of tumor suppressor genes as well as the conversion of proto-oncogenes
into oncogenes. In other words, creating a cancer cell usually requires that the brakes on cell growth (tumor suppressor genes) be released and the accelerators for cell
growth (oncogenes) be activated.

Investigators first came to this conclusion by studying
the genetic abnormalities exhibited by tumors removed
from colon cancer patients. The results showed that
human colon cancer is commonly associated with formation of a *ras* oncogene as well as inactivation or loss of
three different tumor suppressor genes—*p53*, *APC*, and
DCC (or perhaps a neighboring gene called *SMAD4*). The
most rapidly growing colon cancers tend to exhibit all
four genetic alterations, whereas benign tumors have
acquired only one or two of the changes. This pattern suggests that cancers develop by a stepwise accumulation of
mutations affecting both proto-oncogenes and tumor
suppressor genes.

Because the number of mutations accumulated by
cancer cells is usually more than can be accounted for by
normal mutation rates, scientists have concluded that cancer cells are genetically unstable. One possible explanation
for this genetic instability is that many cancer cells exhibit
defects in the *p53* gene. As a result, the p53 protein cannot
prevent the accumulation of genetically damaged cells by
triggering apoptosis or arresting the cell cycle to allow
time for DNA repair.

Genetic instability also arises from defects in DNA
repair itself, which can create cells whose mutation rates
are hundreds or even thousands of times higher than normal. We noted earlier in the chapter that mutations in
genes coding for proteins involved in mismatch repair
occur in hereditary nonpolyposis colon cancer (HNPCC),
an inherited condition that increases a person's risk of

Clinical Applications ATTACKING A TUMOR'S BLOOD SUPPLY

The idea of trying to stop cancer growth by attacking a tumor's blood supply is based largely on the work of Judah Folkman, who first showed in the 1970s that cancers cannot grow beyond a few millimeters in diameter unless they first trigger the formation of new blood capillaries. This proliferation of new capillaries, called *angiogenesis*, provides tumor cells with fresh nutrients and allows them to get rid of their waste products. In a dramatic group of studies, Folkman placed cancer cells in the liquid-filled anterior chamber of a rabbit's eye and showed that the mass of suspended tumor cells could not grow much beyond a millimeter in diameter. The tiny tumors were nourished by the fluid in the anterior chamber, but blood vessels from nearby tissues could not reach them and so they did not obtain nutrients at a fast enough rate to sustain further growth. But when Folkman implanted these tiny tumors directly into the tissue of the eye, blood vessels quickly grew into the tumor mass, allowing it to grow rapidly to thousands of times its original size.

Folkman also performed some experiments in which rat tumor cells were placed in chambers surrounded by an artificial membrane possessing tiny pores that allow molecules, but not cells, to pass through the membrane. When Folkman implanted the chambers under the skin of a rat, he found that angiogenesis was induced in the surrounding normal tissue. From this observation he postulated that the tumors had produced molecules that diffused through the artificial membrane and stimulated the growth of blood vessels in the surrounding normal tissue. In recent years, several of these angiogenesis-stimulating molecules have been identified. Two of them—proteins called *vascular endothelial growth factor (VEGF)* and *basic fibroblast growth factor (bFGF)*—are now known to be produced by a variety of human cancers. VEGF and bFGF trigger angiogenesis by binding to tyrosine kinase receptors on the cells lining the inner surface of blood vessels, thereby stimulating the proliferation of these cells. In addition to secreting molecules that *stimulate* angiogenesis, cancer cells also produce substances that *inhibit* angiogenesis. Among these inhibitors are proteins called *angiostatin, thrombospondin,* and *endostatin*. The relative balance between the concentration of angiogenesis inhibitors and angiogenesis stimulators determines whether a tumor can induce the growth of new blood capillaries.

Since tumors must trigger angiogenesis before they can grow beyond a few millimeters in size, the question arises as to whether angiogenesis inhibitors might be used therapeutically to inhibit the growth of cancer cells. In one striking study, it was shown that injecting mice with angiostatin causes a large decrease in the number of metastases in animals with lung cancer. In another experiment, treating several different kinds of cancer with endostatin was found to virtually eliminate large primary tumors in mice after a few cycles of treatment (Figure 17B-1). Angiostatin and endostatin are just two of more than a dozen molecules now known to be capable of inhibiting angiogenesis. Several of these inhibitors are currently being tested in humans to determine whether we might eventually be able to prevent the growth and spread of cancer by using angiogenesis inhibitors to attack a tumor's blood supply.

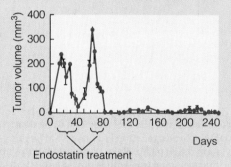

Figure 17B-1 **Treating Cancer by Blocking Angiogenesis.** In this experiment, cancer cells were allowed to grow in mice for about ten days to form a large tumor. The mice were then injected with endostatin until the tumor regressed. After allowing the tumor to grow again in the absence of endostatin, a second treatment cycle was begun. By the end of the second treatment cycle the tumor had permanently stopped growing, even after endostatin treatment was halted. (Data from Boehm, T., J. Folkman, T. Browder, and M. S. O'Reilly. *Nature* 390 [1997]: 404.)

---

developing colon cancer. Individuals with HNPCC have mutations in one of several mismatch repair genes, allowing their DNA to accumulate more mutations than normal. Another example of defective DNA repair is observed in *xeroderma pigmentosum,* a disease involving mutations in genes coding for components of the excision repair pathway. Because the skin cells of such persons are less able to repair DNA mutations caused by exposure to sunlight, the risk of developing skin cancer is increased.

In addition to its arising from faulty DNA repair, genetic instability can also be produced by defects in mitosis that cause chromosomes to separate improperly during cell division. The result is cells that are **aneuploid**—that is, they possess an abnormal number of chromosomes.

Aneuploidy is very common in cancer cells and may result in the loss of tumor suppressor genes or the gain or activation of oncogenes. Although the mechanisms that create aneuploidy are not well understood, recent observations suggest that cancer cells often possess an abnormal number of centrosomes, which might produce mitotic spindle abnormalities that interfere with chromosome sorting. Defects in proteins involved in attaching chromosomes to the mitotic spindle have also been detected in some cancers.

Can Cancer Growth Be Stopped?

If current trends continue, roughly one out of every three Americans will eventually develop cancer, so it is not surprising that we continually hear news reports about the latest attempts to find a cure for this dreaded disease. When a doctor treats a person with cancer, the first step is usually surgery to remove the primary tumor. If the tumor is confined to its initial location, surgery almost always results in a cure. But cancers are capable of spreading throughout the body. The direct spread of cancer cells to neighboring tissues is called **invasion,** while the spread to distant organs via the bloodstream and other body fluids is termed **metastasis.** The tumor nodules that develop from cells that have implanted at sites distant from the parent tumor are referred to as *metastases.*

If cancer cells have already spread to other parts of the body, surgical removal of the primary tumor is insufficient treatment. One way of attacking cancer cells that may have already metastasized to distant locations is to treat cancer patients with *radiation,* which selectively kills cells that are in the S or M phase of the cell cycle. Hence radiation destroys cells that are actively proliferating. *Chemotherapy* uses drugs that, like radiation, selectively destroy dividing cells. The drugs used in chemotherapy enter the bloodstream and travel throughout the body, killing cancer cells wherever they may reside. The problem with both radiation and chemotherapy is that they kill normal dividing cells as well as cancer cells, resulting in toxic side effects (such as susceptibility to infections caused by destruction of developing blood cells in the bone marrow).

Because of this shortcoming, scientists are currently looking for other, less toxic ways to treat metastatic cancer. One possible approach, *immunotherapy,* utilizes techniques that stimulate a person's immune system to seek out and destroy cancer cells. Immunotherapy sometimes involves treatment with molecules such as interferon, interleukin, and tumor necrosis factor, whose normal function is to stimulate the immune system. An alternative tactic utilizes purified antibodies targeted against proteins found on the surface of cancer cells.

A fundamentally different approach to cancer therapy is based on the discovery that before tumors can grow beyond a few millimeters in size, cancer cells must trigger the development of a blood supply by a process called **angiogenesis.** Box 17B describes evidence for this phenomenon and highlights recent experimental attempts to treat cancer by interfering with the development of a tumor's blood supply. Another strategy still in its early experimental stages involves the use of *gene therapy* techniques to repair the genetic defects that cause cancer. For example, experiments are being carried out in which a normal copy of the *p53* gene is inserted into a virus, which is then used to deliver the gene to cancer cells. Since more than half of all human cancers have *p53* mutations, designing an effective strategy for repairing this gene might lead to a cure for many types of cancer.

The preceding ideas represent just a few of the alternative strategies that are currently being investigated in the hopes of finding more effective ways of treating cancer after it has spread throughout the body. Although it is too soon to know which strategy will be the most effective, it is exciting to see alternatives to chemotherapy and radiation that are based on logical rationales for selectively destroying cancer cells.

Perspective

The eukaryotic cell cycle is divided into four main phases, called G1, S, G2, and M. Chromosomal DNA is replicated during S phase, and cell division (mitosis and cytokinesis) takes place during M phase. Interphase, consisting of G1, S, and G2, is a time of cell growth and metabolism that typically occupies about 95% of the cycle time. Cultured mammalian cells usually divide once every 18–24 hours, but cells in multicellular organisms differ greatly in generation time, ranging from stem cells that divide rapidly and continuously to differentiated cells that normally do not divide at all.

DNA is replicated by a semiconservative mechanism in which the two strands of the double helix unwind and each serves as a template for the synthesis of a complementary strand. Bacterial chromosome replication is initiated at a single point and moves in both directions around the circular DNA molecule. In contrast, eukaryotes initiate DNA replication at multiple replicons, with replication proceeding bidirectionally in each replicon. DNA synthesis is catalyzed by DNA polymerases, which add nucleotides to DNA chains in the $5' \longrightarrow 3'$ direction. DNA synthesis is continuous along the leading strand, but discontinuous along the lagging strand, generating small Okazaki fragments that are later joined together by DNA ligase. DNA replication is initiated by an enzyme called primase, which synthesizes short RNA primers that are later removed and replaced with DNA. During DNA replication, the double helix is unwound through the action of helicases, topoisomerases, and single-strand binding proteins. As replication proceeds, a proofreading mechanism based on the $3' \longrightarrow 5'$ exonuclease activity of DNA polymerase allows incorrectly base-paired nucleotides to be removed and replaced.

In eukaryotes, the problem of replicating the ends of linear chromosomal DNA molecules is solved by telomerase, an RNA-containing enzyme that uses its RNA as a template for creating short repeated DNA sequences at the ends of each chromosomal DNA molecule. A mechanism known as licensing also allows eukaryotes to ensure that DNA molecules are replicated only once prior to mitosis.

DNA damage arises both spontaneously and through the action of mutation-causing chemicals and radiation. Some types of DNA damage are corrected by DNA polymerases that carry out translesion synthesis of new DNA across regions where the template DNA is damaged. Alternatively, damaged regions can be repaired by nucleases that remove damaged stretches of DNA followed by replacement of the missing nucleotides by DNA polymerase. Excision repair pathways are used to correct mutations involving abnormal bases, while mismatch repair removes and replaces improperly base-paired nucleotides that have escaped the proofreading mechanism.

Mitosis, the process that distributes the two sets of duplicated chromosomes into two daughter nuclei, consists of five phases: prophase, prometaphase, metaphase, anaphase, and telophase. During prophase, replicated chromosomes condense as sister chromatids that are joined at the centromere. Meanwhile, the cell's two centrosomes move apart and initiate the assembly of the microtubules (MTs) of the mitotic spindle. In prometaphase, the nuclear envelope breaks down and the chromosomes then become attached to kinetochore MTs and move toward the spindle equator. At metaphase, the chromosomes line up at the metaphase plate. Anaphase begins with the separation of the sister chromatids and continues with their movement, as daughter chromosomes, toward the spindle poles. During this process the kinetochore MTs shorten, the polar MTs lengthen, and the cell starts to elongate. At telophase, the separated chromosomes decondense and a nuclear envelope is re-formed around each daughter nucleus.

Three groups of motor proteins are involved in the chromosomal movements that take place during mitosis. Motor proteins at the kinetochore move chromosomes along their kinetochore MTs toward the spindle poles, accompanied by disassembly of the MTs at their kinetochore ("plus") ends. At the same time, motor proteins that crosslink the polar microtubules push overlapping microtubules in opposite directions, thereby pushing the spindle poles away from each other. And finally, a third set of motor proteins pull astral MTs toward the plasma membrane at the cell poles, thereby pulling the spindle poles apart.

Cytokinesis usually begins before mitosis is complete. In animal cells, actin filaments form a cleavage furrow that progressively constricts the cell at the midline and eventually separates the cytoplasm into two daughter cells. In plant cells, a cell wall forms through the middle of the parent cell.

Progression through the eukaryotic cell cycle is regulated by cyclin-dependent kinases (Cdk's), which bind to cyclins to form Cdk-cyclin complexes. At the G1 checkpoint, a Cdk-cyclin complex catalyzes the phosphorylation of the Rb protein to trigger passage into S phase. At the G2 checkpoint, a different Cdk-cyclin complex called MPF triggers entry into mitosis by catalyzing the phosphorylation of various proteins, thereby promoting nuclear envelope breakdown, chromosome condensation, and spindle formation. And at the spindle assembly checkpoint, MPF activates the anaphase-promoting complex, triggering a protein degradation pathway that initiates chromatid separation. This protein degradation pathway also targets mitotic cyclin for breakdown, leading to an inactivation of MPF that in turn triggers events associated with the exit from mitosis, including chromatin decondensation and reassembly of the nuclear envelope.

The cells of multicellular organisms do not normally proliferate unless they are stimulated by an appropriate growth factor. Binding of a growth factor to its plasma membrane receptor triggers the Ras pathway, a complex cascade of events that culminates in the cell passing through the G1 checkpoint and into S phase. Cancer can be caused by the malfunctioning of this growth factor signaling pathway. Oncogenes, which are mutant genes whose presence can cause cancer, code for abnormal versions or excessive quantities of the main components of the growth factor signaling pathway; these components include growth factors, receptors, plasma membrane G proteins, protein kinases, transcription factors, and Cdk-cyclins. Some oncogenes are brought into cells by cancer viruses, but most oncogenes detected in human cancers arise by mutation from normal genes called proto-oncogenes.

Cancer is also associated with the loss or inactivation of tumor suppressor genes, whose normal function is to inhibit cell proliferation. Two proteins produced by tumor suppressor genes are the Rb protein, which normally stops cells from passing through the G1 checkpoint, and the p53 protein, which responds to DNA damage by arresting the cell cycle at the G1 checkpoint or triggering cell death by apoptosis. Genetic instability allows for the accumulation of multiple mutations in tumor suppressor genes and oncogenes that together lead to the development of cancer.

Key Terms for Self-Testing

cell division (p. 523)

An Overview of the Cell Cycle
cell cycle (p. 523)
M phase (p. 523)
mitosis (p. 523)
cytokinesis (p. 523)
sister chromatid (p. 523)
interphase (p. 524)

S phase (p. 524)
G1 phase (p. 524)
G2 phase (p. 524)
mitotic index (p. 525)
G0 state (p. 525)

DNA Replication
semiconservative replication (p. 525)
equilibrium density centrifugation (p. 526)

replication fork (p. 527)
origin of replication (p. 528)
replicon (p. 528)
DNA polymerase (p. 530)
temperature-sensitive mutant (p. 530)
polymerase chain reaction (PCR) (p. 530)
Okazaki fragment (p. 532)
DNA ligase (p. 532)
leading strand (p. 532)

lagging strand (p. 534)
proofreading (p. 535)
exonuclease (p. 535)
primase (p. 535)
RNA primer (p. 536)
primosome (p. 536)
helicase (p. 537)
topoisomerase (p. 537)
single-strand binding protein (SSB) (p. 537)
replisome (p. 537)
telomere (p. 539)
telomerase (p. 539)

DNA Damage and Repair
licensing (p. 541)
mutation (p. 541)
translesion synthesis (p. 543)
excision repair (p. 543)
base excision repair (p.543)
nucleotide excision repair (NER) (p. 543)
mismatch repair (p. 544)

Nuclear and Cell Division
prophase (p. 545)
centromere (p. 545)
centrosome (p. 545)
mitotic spindle (p. 545)

centriole (p. 545)
prometaphase (p. 545)
kinetochore (p. 545)
kinetochore microtubule (p. 545)
polar microtubule (p. 545)
astral microtubule (p. 545)
metaphase (p. 545)
karyotype (p. 545)
anaphase (p. 545)
anaphase A and B (p. 548)
telophase (p. 548)
motor protein (p. 551)
cleavage (p. 553)
cleavage furrow (p. 553)
contractile ring (p. 553)
phragmoplast (p. 553)
cell plate (p. 554)

Regulation of the Cell Cycle
G1 checkpoint (p. 555)
G2 checkpoint (p. 555)
spindle assembly checkpoint (p. 556)
heterokaryon (p. 556)
MPF (mitosis-promoting factor) (p. 557)
cyclin (p. 558)
cyclin-dependent kinase (Cdk) (p. 558)

Rb protein (p. 560)
E2F transcription factor (p. 560)
anaphase-promoting complex (p. 561)
cohesin (p. 561)

Growth Control and Cancer
growth factor (p. 562)
platelet-derived growth factor (PDGF)
   (p. 562)
epidermal growth factor (EGF) (p. 562)
transcription factor (p.563)
transforming growth factor β (TGFβ)
   (p. 564)
Cdk inhibitor (p. 564)
tumor (p. 564)
benign tumor (p. 565)
malignant tumor (p. 565)
cancer (p. 565)
oncogene (p. 565)
proto-oncogene (p. 565)
tumor suppressor gene (p. 567)
p53 gene (p. 568)
aneuploid (p. 570)
invasion (p. 571)
metastasis (p. 571)
angiogenesis (p. 571)

Problem Set

More challenging problems are marked with a • .

17-1.   Cell Cycle Phases. Indicate whether each of the following statements is true of the G1 phase of the cell cycle, the S phase, the G2 phase, or the M phase. A given statement may be true of any, all, or none of the phases.

(a)   The amount of nuclear DNA in the cell doubles.

(b)   The nuclear envelope breaks into fragments.

(c)   Sister chromatids separate from each other.

(d)   Cells that will never divide again are likely to be arrested in this phase.

(e)   The primary cell wall of a plant cell forms.

(f)   Chromosomes are present as diffuse, extended chromatin.

(g)   This phase is part of interphase.

(h)   Mitotic cyclin is at its lowest level.

(i)   A Cdk protein is present in the cell.

(j)   A cell cycle checkpoint has been identified in this phase.

17-2.   The Mitotic Index and the Cell Cycle. The mitotic index is a measure of the amount of mitotic activity in a population of cells. It is calculated as the percentage of cells in mitosis at any one time. Assume that upon examination of a sample of 1000 cells, you find 30 cells in prophase, 20 in prometaphase, 20 in metaphase, 10 in anaphase, 20 in telophase, and 900 in interphase. Of those in interphase, 400 are found (by microspectrophotometric analysis after staining the cells with a DNA-specific stain) to have *X* amount of DNA, 200 to have 2*X*, and 300 cells to be somewhere in between. Autoradiographic analysis indicates that the G2 phase lasted 4 hours.

(a)   What is the mitotic index for this population of cells?

(b)   Specify the proportion of the cell cycle spent in each of the following phases: prophase, prometaphase, metaphase, anaphase, telophase, G1, S, G2.

(c)   What is the total length of the cell cycle?

(d)   What is the actual amount of time (in hours) spent in each of the phases of part b?

(e)   To measure the G2 phase, radioactive thymidine (a DNA precursor) is added to the culture at some time *t*, and samples of the culture are analyzed autoradiographically for labeled nuclei at regular intervals thereafter. What specific observation would have to be made to assess the length of the G2 phase?

(f)   What proportion of the interphase cells would you expect to exhibit labeled nuclei in autoradiographs prepared shortly after exposure to the labeled thymidine? (Assume a labeling period just long enough to allow the thymidine to get into the cells and begin to be incorporated into DNA.)

17-3.   Meselson and Stahl Revisited. Although the Watson-Crick structure for DNA suggested a semiconservative model for DNA replication, at least two other models are conceivable. In a *conservative model*, the parental DNA double helix remains intact and a second, all-new copy is made. In a *dispersive model*, each strand of both daughter molecules contains a mixture of old and newly synthesized segments.

(a)   Starting with one parental double helix, sketch the progeny molecules for two rounds of replication according to each of these alternative models. Use one color for the original parent strands and another color for all the DNA synthesized thereafter (as is done in Figure 17-4 for the semiconservative model).

(b) For each of the alternative models, indicate the distribution of DNA bands that Meselson and Stahl would have found in their cesium chloride gradients after one and two rounds of replication.

17-4. DNA Replication. Sketch a replication fork of bacterial DNA in which one strand is being replicated discontinuously and the other is being replicated continuously. List six different enzyme activities associated with the replication process, identify the function of each, and indicate on your sketch where each would be located on the replication fork. In addition, identify the following features on your sketch: DNA template, RNA primer, Okazaki fragments, and single-strand binding protein.

• 17-5. More DNA Replication. The following are observations from five experiments carried out to determine the mechanism of DNA replication in the hypothetical organism *Fungus mungus*. For each experiment, indicate whether the results support (S), refute (R), or have no bearing (NB) on the hypothesis that this fungus replicates its DNA by the same mechanism as that known for *E. coli*. Explain your reasoning in each case.

(a) Neither of the two DNA polymerases of *F. mungus* appears to have an exonuclease activity.

(b) Replicating DNA from *F. mungus* shows discontinuous synthesis on both strands of the replication fork.

(c) Some of the DNA sequences from *F. mungus* are present in multiple copies per genome, whereas other sequences are unique.

(d) Short fragments of *F. mungus* DNA isolated during replication contain both ribose and deoxyribose.

(e) *F. mungus* cells are grown in the presence of the heavy isotopes ^{15}N and ^{13}C for several generations and then grown for one generation in normal (^{14}N, ^{12}C) medium; then DNA is isolated from these cells and denatured. The single strands yield a single band in a cesium chloride density gradient.

17-6. Still More DNA Replication. DNA replication seems an extremely complicated process. Perhaps it evolved from some simpler process in which there were not distinct mechanisms for leading-strand and lagging-strand synthesis. In this primitive process, there may only have been DNA replication of the leading-strand type, which is less complicated than the lagging-strand mechanism. For example, DNA could have been replicated by a mechanism that involved the synthesis of one daughter strand at a time:

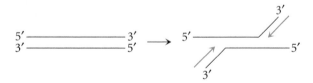

There is something wrong with this picture, however. All DNA synthesis requires a primer. Devise a simple model for DNA synthesis that would allow complete daughter molecules of DNA to be produced by a leading-strand replication mechanism. By simple model, we mean one that does not require more than one monofunctional enzyme in addition to the primordial DNA polymerase. You may assume that the primordial slime in which the DNA is replicating contains an ade-

quate supply of deoxynucleoside triphosphates (dNTPs). (There are numerous possible solutions to this problem.)

17-7. The Minimal Chromosome. To enable it to be transmitted intact from one cell generation to the next, the linear DNA molecule of a eukaryotic chromosome must have appropriate nucleotide sequences making up three special kinds of regions: origins of replication (at least one), a centromere, and two telomeres. What would happen if such a chromosomal DNA molecule somehow lost

(a) all of its origins of replication?

(b) all of the DNA constituting its centromere?

(c) one of its telomeres?

17-8. DNA Damage and Repair. Indicate whether each of the following statements is true of depurination (DP), deamination (DA), or pyrimidine dimer formation (DF). A given statement may be true of any, all, or none of these processes.

(a) This process is caused by spontaneous hydrolysis of a glycosidic bond.

(b) This process is induced by ultraviolet light.

(c) This can happen to guanine but not to cytosine.

(d) This can happen to thymine but not to adenine.

(e) This can happen to thymine but not to cytosine.

(f) Repair involves a DNA glycosylase.

(g) Repair involves an endonuclease.

(h) Repair involves DNA ligase.

(i) Repair depends on the existence of separate copies of the genetic information in the two strands of the double helix.

(j) Repair depends on cleavage of both strands of the double helix.

17-9. Nonstandard Purines and Pyrimidines. Shown in Figure 17-41a are three nonstandard nitrogenous bases that are formed by the deamination of naturally occurring bases in DNA.

(a) Indicate which base in DNA must be deaminated to form each of these bases.

(b) Why are there only three bases shown, when DNA contains four bases?

(c) Why is it important that none of the bases shown in Figure 17-41a occurs naturally in DNA?

(d) Figure 17-41b shows 5-methylcytosine, a pyrimidine that arises naturally in DNA when cellular enzymes methylate cytosine. Why is the presence of this base in the DNA sequence likely to increase the probability of a mutation at that site?

• 17-10. Chromosome Movement in Mitosis. It is possible to mark the microtubules of a spindle by photobleaching with a laser microbeam (Figure 17-42). When this is done, chromosomes move *toward* the bleached area during anaphase. Are the following statements consistent (C) or inconsistent (I) with this experimental result?

(a) Microtubules move chromosomes by disassembling at the spindle poles.

(b) Chromosomes move by disassembling microtubules at their kinetochore ends.

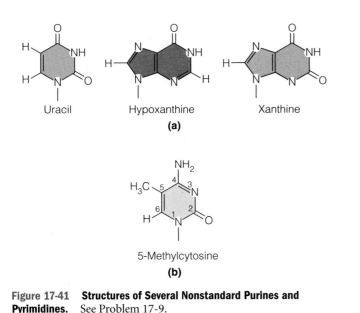

5-Methylcytosine

(b)

Figure 17-41   Structures of Several Nonstandard Purines and Pyrimidines.   See Problem 17-9.

(c)   Chromosomes are moved by a springlike elastic property of microtubules.

(d)   Chromosomes are moved along microtubules by a motor protein.

17-11.   More on Cell Cycle Phases. For each of the following pairs of phases from the cell cycle, indicate how you could tell which of the two phases a specific cell is in.

(a)   G1 and G2

(b)   G1 and S

(c)   G2 and M

(d)   G1 and M

• 17-12.   Cell Cycle Regulation. One approach to the study of cell cycle regulation has been to fuse cultured cells that are at different stages of the cell cycle and observe the effect of the fusion on the nuclei of the fused cells (heterokaryons). When cells in G1 are fused with cells in S, the nuclei from the G1 cells begin DNA replication earlier than they would have if they had not been fused. In fusions of cells in G2 and S, however, nuclei continue their previous activities, apparently uninfluenced by the fusion. Fusions between mitotic cells and interphase cells always lead to chromatin condensation in the nonmitotic nuclei. Based on these results, identify each of the following statements about cell cycle regulation as probably true (T), probably false (F), or not possible to conclude from the data (NP).

(a)   The activation of DNA synthesis may result from the stimulatory activity of one or more cytoplasmic factors.

(b)   The transition from S to G2 may result from the presence of a cytoplasmic factor that inhibits DNA synthesis.

(c)   The transition from G2 to mitosis may result from the presence in the G2 cytoplasm of one or more factors that induce chromatin condensation.

(d)   G1 is not an obligatory phase of all cell cycles.

(e)   The transition from mitosis to G1 appears to result from the disappearance or inactivation of a cytoplasmic factor present during M phase.

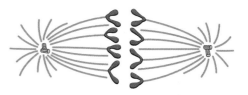

Microtubules are stained with a fluorescent antibody during anaphase.

A laser microbeam is used to mark two areas by bleaching the fluorescent dye.

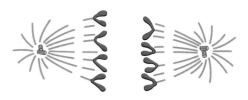

The chromosomes are observed to move toward the bleached areas.

Figure 17-42   The Use of Laser Photobleaching to Study Chromosome Movement During Mitosis.   See Problem 17-10.

• 17-13.   Role of Cyclin-Dependent Protein Kinases. Based on your understanding of the regulation of the eukaryotic cell cycle, how could you explain each of the following experimental observations?

(a)   When MPF is injected into cells that have just emerged from S phase, chromosome condensation and nuclear envelope breakdown occur immediately, rather that after the normal G2 delay of several hours.

(b)   When an abnormal, indestructible form of mitotic cyclin is introduced into cells, they enter into mitosis but cannot emerge from it and reenter G1 phase.

(c)   Mutations that inactivate the main protein phosphatase used to catalyze protein dephosphorylations cause a long delay in the reconstruction of the nuclear envelope that normally takes place at the end of mitosis.

• 17-14.   Growth Factor Signaling Pathway. Abnormalities in the Ras pathway are commonly observed in human cancers. The following questions focus on several experimental observations relating to this pathway.

(a)   When epidermal growth factor (EGF) is added to epithelial cells, the Ras and Raf proteins are both activated. From this observation *by itself,* what conclusion can you draw about the relationship between Ras and Raf?

(b)   In mutant cells containing a permanently inactivated Ras protein, addition of EGF does not activate the Raf protein as much as it does in normal cells. What conclusion can you draw from this observation?

(c) Suppose you discover a new type of cancer in which the enzymatic activity of the Raf protein is much higher than normal. Describe several distinct mechanisms that might have caused this, and describe some experiments that might be carried out to distinguish among your hypotheses.

17-15. Oncogenes and Tumor Suppressor Genes. Indicate whether each of the following descriptions applies to an oncogene (OG), a proto-oncogene (PO), or a tumor suppressor gene (TS). Some descriptions may apply to more than one of these gene types. Explain your answers.

(a) A type of gene found in normal cells

(b) A gene that could code for a normal growth factor

(c) A type of gene found in cancer cells

(d) A type of gene found only in cancer cells

(e) A gene whose presence can cause cancer

(f) A gene whose absence can cause cancer

(g) A type of gene that can be found in both normal cells and cancer cells

Suggested Reading

References of historical importance are marked with a • .

DNA Replication

de Lange, T., and R. A. DePinho. Unlimited mileage from telomerase? *Science* 283 (1999): 947.

DePamphilis, M. L., ed. *Concepts in Eukaryotic DNA Replication.* Cold Spring Harbor, NY: Cold Spring Harbor Laboratory Press, 1998.

Greider, C. W., and E. H. Blackburn. Telomeres, telomerase, and cancer. *Sci. Amer.* 274 (February 1996): 92.

Hübscher, U., H.-P. Nasheuer, and J. E. Syväoja. Eukaryotic DNA polymerases, a growing family. *Trends Biochem. Sci.* 25 (2000): 143.

Kelly, T. J., and G. W. Brown. Regulation of chromosome replication. *Annu. Rev. Biochem.* 69 (2000): 829.

Madine, M., and R. Laskey. Geminin bans replication license. *Nature Cell Biol.* 3 (2001): E49.

• Meselson, M., and F. W. Stahl. The replication of DNA in *E. coli. Proc. Natl. Acad. Sci. USA* 44 (1958): 671.

Mullis, K. B. The unusual origin of the polymerase chain reaction. *Sci. Amer.* 262 (April 1990): 56.

• Ogawa, T., and R. Okazaki. Discontinuous DNA replication. *Annu. Rev. Biochem.* 49 (1980): 421.

• Watson, J. D., and F. H. C. Crick. Genetical implications of the structure of deoxyribonucleic acid. *Nature* 171 (1953): 964.

DNA Repair

Cleaver, J. E. Xeroderma pigmentosum: The first of the cellular caretakers. *Trends Biochem. Sci.* 26 (2001): 398.

Culotta, E., and D. E. Koshland, Jr. DNA repair works its way to the top (Molecule of the Year). *Science* 266 (1994): 1926. (In the same issue, see Perspectives articles by A. Sancar [p. 1954], P. C. Hanawalt [p. 1957], and P. Modrich [p. 1959].)

Friedberg, E. C. *Correcting the Blueprint of Life. An Historical Account of the Discovery of DNA Repair Mechanisms.* Cold Spring Harbor, NY: Cold Spring Harbor Laboratory Press, 1997.

Lindahl, T., and R. D. Wood. Quality control by DNA repair. *Science* 286 (1999): 1897.

Modrich, P., and R. Lahue. Mismatch repair in replication fidelity, genetic recombination, and cancer biology. *Annu. Rev. Biochem.* 65 (1996): 101.

Sancar, A. DNA excision repair. *Annu. Rev. Biochem.* 65 (1996): 43.

Nuclear and Cell Division

Endow, S. A., and D. M. Glover, eds. *The Dynamics of Cell Division.* New York: Oxford University Press, 1998.

Glover, D. M., C. Gonzalez, and J. W. Raff. The centrosome. *Sci. Amer.* 268 (June 1993): 62.

Mitchison, T. J., and E. D. Salmon. Mitosis: A history of division. *Nature Cell Biol.* 3 (2001): E17.

Nasmyth, K., J.-M. Peters, and F. Uhlmann. Splitting the chromosomes: Cutting the ties that bind sister chromatids. *Science* 288 (2000): 1379.

Rappaport, R. *Cytokinesis in Animal Cells.* New York: Cambridge University Press, 1996.

Robinson, D. N., and J. A. Spudich. Towards a molecular understanding of cytokinesis. *Trends Cell Biol.* 10 (2000): 228.

Wittman, T., A. Hyman, and A. Desai. The spindle: A dynamic assembly of microtubules and motors. *Nature Cell Biol.* 3 (2001): E28.

Regulation of the Cell Cycle

den Boer, B. G. W., and J. A. H. Murray. Triggering the cell cycle in plants. *Trends Cell Biol.* 10 (2000): 245.

Hutchison, C. J., and D. Glover, eds. *Cell Cycle Control.* New York: IRL (Oxford University Press), 1995.

King, R. W., R. J. Deshaies, J. M. Peters, and M. W. Kirschner. How proteolysis drives the cell cycle. *Science* 274 (1996): 1652.

Maller, J. L. Maturation-promoting factor in the early days. *Trends Biochem. Sci.* 20 (1995): 524.

Murray, A. W., and T. Hunt. *The Cell Cycle: An Introduction.* New York: W. H. Freeman, 1993.

Nigg, E. A. Mitotic kinases as regulators of cell division and its checkpoints. *Nature Reviews Mol. Cell Biol.* 2 (2001): 21.

Nurse, P. A long twentieth century of the cell cycle and beyond. *Cell* 100 (2000): 71.

Weinberg, R. A. The retinoblastoma protein and cell cycle control. *Cell* 81 (1995): 323.

Growth Control and Cancer

Bishop, J. M., and R. A. Weinberg, eds. *Scientific American Molecular Oncology.* New York: Scientific American, 1996.

Cahill, D. P., K. W. Kinzler, B. Vogelstein, and C. Lengauer. Genetic instability and darwinian selection in tumors. *Trends Cell Biol.* 9 (1999): M57.

Cooper, G. M. *Oncogenes,* 2d ed. Sudbury, MA: Jones and Bartlett, 1995.

Culotta, E., and D. E. Koshland, Jr. *p53* sweeps through cancer research. *Science* 262 (1993): 1958.

Folkman, J. Fighting cancer by attacking its blood supply. *Sci. Amer.* 275 (September 1996): 150.

McCormick, F. Signalling networks that cause cancer. *Trends Cell Biol.* 9 (1999): M53.

18 Sexual Reproduction, Meiosis, and Genetic Recombination

Mitotic cell division, which we discussed in the preceding chapter, is used for the proliferation of most eukaryotic cells, leading to the production of more organisms or more cells per organism. Since mitosis involves one round of DNA replication followed by the segregation of identical chromatids into two daughter cells, mitotic division produces cells that are genetically identical, or very nearly so. This ability to perpetuate genetic traits faithfully allows mitotic division to form the basis of **asexual reproduction** in eukaryotes. During asexual reproduction, new individuals are generated by mitotic division of cells in a single parent organism, either unicellular or multicellular. Although the details vary among organisms, asexual reproduction is widespread in nature. Examples include *mitotic division* of unicellular organisms, *budding* of offspring from a multicellular parent's body, and *regeneration* of whole organisms from pieces of a parent organism. In plants, entire new organisms can even be regenerated from single cells taken from an adult plant.

Asexual reproduction can be an efficient and evolutionarily successful mode of perpetuating a species. As long as environmental conditions remain essentially constant, the genetic predictability of asexual reproduction is perfectly suited for maintaining the survival of a population. But if the environment changes, a population that reproduces asexually may not be able to adapt to the new conditions. Under such conditions, organisms that reproduce sexually rather than asexually will usually have an advantage, as we now discuss.

Sexual Reproduction

In contrast to asexual reproduction, in which progeny are genetically identical to the single parent from which they arise, **sexual reproduction** allows genetic information from two parents to be mixed together, thereby producing offspring that are genetically dissimilar, both from each other and from the parents. Moreover, the offspring are unpredictably dissimilar; that is, we cannot anticipate exactly which combination of genes a particular offspring will receive from its two parents. Since most plants and animals—and even many microorganisms—reproduce sexually, this type of reproduction must provide some distinct advantages.

Sexual Reproduction Generates Genetic Variety

The main advantage of sexual reproduction is that it allows desirable genetic traits found in different individuals to be combined in various ways in newly developing offspring, thereby generating enormous variety among the individuals that make up a population. Genetic variation ultimately depends on the occurrence of *mutations*, which are unpredictable alterations in DNA base sequence. Mutations are rare events, and beneficial mutations are even rarer. But when a beneficial mutation does arise, it is clearly advantageous to preserve the mutation in the population. It can be even more beneficial to combine several desirable mutations in a single individual—and therein lies the fundamental advantage of sexual reproduction. Although mutations occur in both sexual and asexual species, only sexual reproduction can bring together beneficial mutations that originally arose in two separate individuals.

Because sexual reproduction brings about a reshuffling of genetic information in each new generation, it generates individuals exhibiting a wide range of genetic combinations, including new combinations of mutations.

It follows that at least some of these genetic variations will confer advantages to particular members of the population. Of course, other members of the population will be less suited for survival. For the species as a whole, however, the large amount of genetic variation makes it more likely that some members will be able to survive if the environment changes.

The Diploid State Is an Essential Feature of Sexual Reproduction

During sexual reproduction, genetic information derived from two different parents is combined in an individual offspring. Therefore at some point in its life cycle, every sexually reproducing organism has cells that contain two copies of each chromosome, one inherited from each parent. The two members of each chromosome pair are called **homologous chromosomes.** Two homologous chromosomes carry the same lineup of genes, although for any given gene, the two versions may differ slightly in base sequence. Not surprisingly, homologous chromosomes usually look alike when viewed with a microscope (see Figure 17-22). An exception to this rule is the **sex chromosomes,** which determine whether an individual is male or female. The two kinds of sex chromosomes, generally called X and Y chromosomes, differ significantly in genetic makeup and appearance. In mammals, for example, females have two X chromosomes of the same size, whereas males have one X chromosome and a Y chromosome that is much smaller. Nonetheless, parts of the X and Y chromosomes are actually homologous, and during sexual reproduction, the X and Y chromosomes behave as homologues.

A cell or organism with two sets of chromosomes is said to be **diploid** (from the Greek word *diplous,* meaning "double") and contains two copies of its genome. A cell or organism with a single set of chromosomes, and therefore a single copy of its genome, is **haploid** (from the Greek word *haplous,* meaning "single"). By convention, the haploid chromosome number for a species is designated n (or $1n$) and the diploid number $2n$. For example, in humans $n = 23$, which means that most human cells contain two sets of 23 chromosomes, yielding a diploid total of 46. The diploid state is an essential feature of the life cycle of sexually reproducing species. Its main advantage over the haploid state lies in the greater genetic flexibility it allows. In a sense, a diploid cell contains an extra set of genes that is available for mutation and genetic innovation. Changes in a second copy of a gene usually will not threaten the survival of an organism, even if the mutation is deleterious to the original function of that particular gene.

Diploid Cells May Be Homozygous or Heterozygous for Each Gene

To further explore the genetic consequences of the diploid state, let's now focus on the behavior of an individual **gene locus** (plural: **loci**), which is the place on a chromosome that contains the DNA sequence for a particular gene. For simplicity, we will assume that the gene controls a single, clear-cut characteristic—or *character,* as geneticists usually say—in the organism. Let's also assume that only one copy of this gene is present per haploid genome, so that a diploid organism will have two copies of the gene, which may be either identical or slightly different. The two versions of the gene are called **alleles,** and the combination of alleles determines how an organism will express the character controlled by the gene. In garden peas, for example, the alleles at one particular locus determine seed color, which may be green or yellow (Figure 18-1). An organism with two identical alleles for a given gene is said to be **homozygous** for that gene or character. Thus, a pea plant that inherited the same allele for yellow seed color from both of its parents is said to be homozygous for seed color. An organism with two different alleles for a gene is said to be **heterozygous** for that gene or for the character it determines. A pea plant with one allele that specifies yellow seed color and a second allele that specifies green seed color is therefore heterozygous for seed color.

In a heterozygous individual, one of the two alleles is often **dominant** and the other **recessive.** These terms convey the idea that the dominant allele determines how the trait will appear in a heterozygous individual. (The word *trait* refers to a particular variant of a character, such as green or yellow seeds, where seed color is the character.) For seed color in peas, the yellow trait is dominant over green; this means that pea plants heterozygous for seed color have yellow seeds. A dominant allele is usually designated by an uppercase letter that stands for the trait, whereas a corresponding recessive allele is represented by

Genotype:

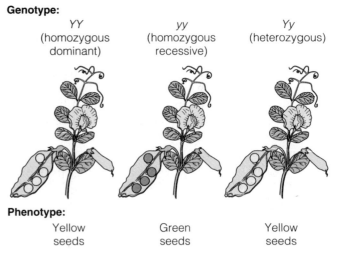

YY	*yy*	*Yy*
(homozygous dominant)	(homozygous recessive)	(heterozygous)

Phenotype:

Yellow seeds	Green seeds	Yellow seeds

Figure 18-1 **Genotype and Phenotype.** In garden peas, seed color (phenotype) can be either yellow or green. The seed-color alleles are *Y* (yellow, dominant) or *y* (green, recessive). Because the pea plant is a diploid organism, its genetic makeup (genotype) for seed color may be homozygous dominant (*YY*), homozygous recessive (*yy*), or heterozygous (*Yy*).

the same letter in lowercase. In both cases, italics are used. Thus, alleles for seed color in peas are represented by *Y* for yellow and *y* for green, because the yellow trait is dominant. As Figure 18-1 illustrates, a pea plant can be homozygous for the dominant allele (*YY*), homozygous for the recessive allele (*yy*), or heterozygous (*Yy*).

It is important to distinguish between the **genotype,** or genetic makeup of an organism, and its **phenotype,** or the physical expression of its genotype. The phenotype of an organism can usually be determined by inspection (e.g., by looking to see whether seeds are green or yellow). Genotype, on the other hand, can be *directly* determined only by studying an organism's DNA, although it can be deduced from indirect evidence, such as the organism's phenotype and information about the phenotypes of its parents and/or offspring. Organisms exhibiting the same phenotype do not necessarily have identical genotypes. In the example of Figure 18-1, pea plants with yellow seeds (phenotype) can be either *YY* or *Yy* (genotype). Figure 18-2 summarizes the genetic terminology introduced so far.

Gametes Are Haploid Cells Specialized for Sexual Reproduction

The hallmark of sexual reproduction is that genetic information contributed by two parents is brought together in a single individual. Because the offspring of sexual reproduction are diploid, the contribution from each parent must be haploid. The haploid cells produced by each parent that fuse together to form the diploid offspring are called **gametes,** and the process that produces them is **gametogenesis.** Biologists distinguish between male and female individuals on the basis of the gametes they produce. Gametes produced by males, called **sperm** (or *spermatozoa*), are usually quite small and may be inherently motile. Female gametes, called **eggs** or **ova** (singular: **ovum**), are specialized for the storage of nutrients and tend to be quite large and nonmotile. For example, in sea urchins the volume of an egg cell is more than 10,000 times greater than that of a sperm cell; in birds and amphibians, which have massive yolky eggs, the size difference is even greater. But in spite of their differing sizes, sperm and egg bring equal amounts of chromosomal DNA to the offspring.

The union of sperm and egg during sexual reproduction is called **fertilization.** The resulting fertilized egg, or **zygote,** is diploid, having received one chromosome set from the sperm and a homologous set from the egg. In the life cycles of multicellular organisms, fertilization is followed by **development**—a series of mitotic divisions and progressive specialization of various groups of cells to form a multicellular embryo, and eventually an adult.

The distinction between male and female parents is not universal, however. For example, the gametes produced by certain fungi or unicellular eukaryotes are identical in appearance, but differ slightly at the molecular level. Such gametes are said to differ in **mating type.** The union of two gametes requires that they be of different mating types, but the number of possible mating types in a species may be greater than two—in some cases, more than ten!

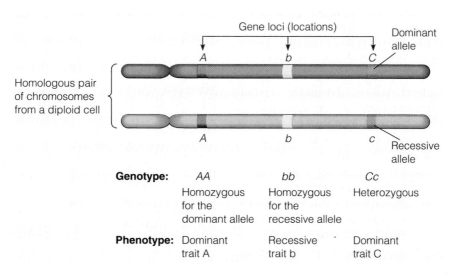

Figure 18-2 Some Genetic Terminology. This diagram shows a homologous pair of chromosomes from a diploid cell; the chromosomes are the same size and shape and carry genes for the same characters (characteristics), in the same order. The site of a gene on a chromosome is called a gene locus. The particular versions of a gene— alleles—found at comparable gene loci on homologous chromosomes may be identical (giving the organism a genotype that is homozygous for that gene) or different (making the genotype heterozygous). If one allele of a gene is dominant and the other recessive, the heterozygous organism exhibits a dominant phenotype with respect to the character in question. A recessive phenotypic trait is observed only if the genotype is homozygous for the recessive allele. Red and blue are used here and in most later figures to distinguish the different parental origins of the two chromosomes.

Meiosis

Since gametes are haploid, they cannot be produced from diploid cells by mitosis because mitosis creates daughter cells that are genetically identical to the original parent cell. In other words, if gametes were formed by mitotic division of diploid cells, both sperm and egg would have a diploid chromosome number, just like the parent diploid cells. The hypothetical zygote created by the fusion of such diploid gametes would be *tetraploid* (i.e., possess *four* homologous sets of chromosomes). Moreover, the chromosome number would continue to double for each succeeding generation—an impossible scenario. Thus, for the chromosome number to remain constant from generation to generation, a different type of cell division must occur during the formation of gametes. That special type of division, called **meiosis,** reduces the chromosome number from diploid to haploid.

Meiosis involves one round of chromosomal DNA replication followed by two successive nuclear divisions. This results in the formation of four daughter nuclei (usually in separate daughter cells) containing one haploid set of chromosomes per nucleus. Figure 18-3 outlines the principle of meiosis starting with a diploid cell containing four chromosomes ($2n = 4$). A single round of DNA replication is followed by two cell divisions, meiosis I and meiosis II, leading to the formation of four haploid cells.

The Life Cycles of Sexual Organisms Have Diploid and Haploid Phases

Meiosis and fertilization are indispensable components of the life cycle of every sexually reproducing organism, because the doubling of chromosome number that takes place at fertilization is balanced by the halving that occurs during meiosis. As a result, the life cycle of sexually reproducing organisms is divided into two phases: a diploid ($2n$) phase and a haploid ($1n$) phase. The diploid phase begins at fertilization and extends until meiosis, whereas the haploid phase is initiated at meiosis and ends with fertilization.

Organisms vary greatly in the relative prominence of the haploid and diploid phases of their life cycles, as shown for some representative groups in Figure 18-4. Fungi are examples of sexually reproducing organisms whose life cycles are primarily haploid but include a brief diploid phase that begins with gamete fusion (the fungal equivalent of fertilization) and ends with meiosis (Figure 18-4b). Meiosis usually takes place almost immediately after gamete fusion, so the diploid phase is very short, and, accordingly, only a very small fraction of fungal nuclei are diploid at any one time. Fungal gametes develop, without meiosis, from cells that are already haploid.

Mosses and ferns are probably the best examples of organisms in which both the haploid and diploid phases are prominent features of the life cycle. Every species of these plants has two alternative, morphologically distinct, multicellular forms, one haploid and the other diploid (Figure 18-4c). For mosses, the haploid form of the

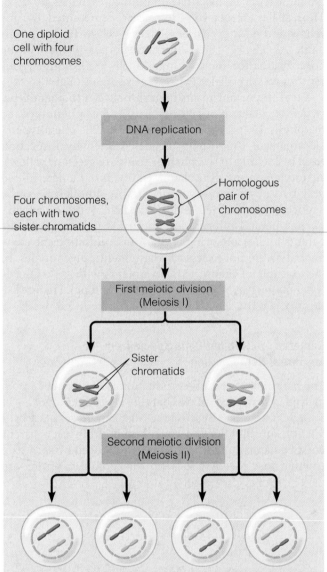

Figure 18-3   The Principle of Meiosis.   Meiosis involves a single round of DNA replication (chromosome duplication) in a diploid cell followed by two successive cell division events. In this example, the diploid cell has only four chromosomes, which can be grouped into two homologous pairs. After DNA replication, each chromosome consists of two sister chromatids. In the first meiotic division (meiosis I), homologous chromosomes separate, but sister chromatids remain attached. In the second meiotic division (meiosis II), sister chromatids separate, resulting in four haploid daughter cells with two chromosomes each. Notice that each haploid cell has one chromosome from each homologous pair that was present in the diploid cell. For simplicity, the effects of crossing over and genetic recombination are not shown in this diagram.

organism is larger and more prominent, and the diploid form is a modest, rather inconspicuous structure. For ferns, it is the other way around. In both cases, gametes develop from preexisting haploid cells.

Organisms that alternate between haploid and diploid multicellular forms in this way are said to display an

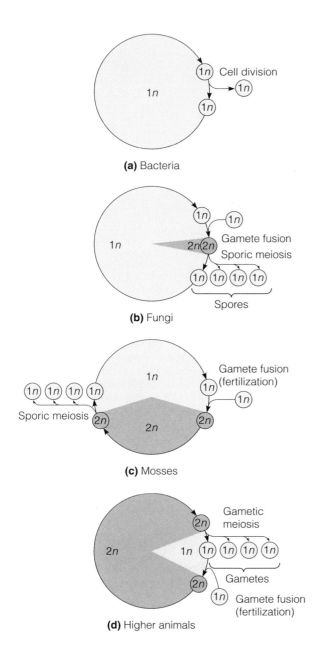

(a) Bacteria

(b) Fungi

(c) Mosses

(d) Higher animals

Figure 18-4 Types of Life Cycles. The relative prominence of the haploid ($1n$) and diploid ($2n$) phases of the life cycle differ greatly, depending on the organism. **(a)** Bacteria exist exclusively in the haploid state. **(b)** Most fungi exemplify a life form that is predominantly haploid but has a brief diploid phase. Because the products of meiosis in fungi are haploid spores, this type of meiosis is called sporic meiosis. The spores give rise to haploid cells, some of which later become gametes (without meiosis). **(c)** Mosses (and ferns as well) alternate between haploid and diploid forms, both of which are significant components in the life cycles of these organisms. Sporic meiosis produces haploid spores, which in this case grow into haploid plants. Eventually, some of the haploid plant's cells differentiate into gametes. (In seed plants, such as conifers and flowering plants, the haploid forms of the organism are vestigial, each consisting of only a small number of cells.) **(d)** Higher animals are the best examples of organisms that are predominantly diploid, with only the gametes representing the haploid phase of the life cycle. Animals are said to have a gametic meiosis, since the immediate products of meiosis are haploid gametes.

alternation of generations in their life cycles. In addition to mosses and ferns, eukaryotic algae and seed plants also exhibit an alternation of diploid and haploid generations. In all such organisms, the products of meiosis are **haploid spores,** which, after germination, give rise by mitotic cell division to the haploid form of the plant or alga. The haploid form in turn produces the gametes by specialization of cells that are already haploid. The gametes, upon fertilization, give rise to the diploid form. Because the diploid form produces spores, it is called a **sporophyte** ("spore-producing plant"). The haploid form produces gametes and is therefore called a **gametophyte.** While all plants exhibit an alternation of generations, in most cases the sporophyte generation predominates. In flowering plants, for example, the gametophyte generation is an almost vestigial structure located in the flower (female gametophyte in the *carpel,* male gametophytes in the flower's *anthers*).

The best examples of life cycles dominated by the diploid phase are found in animals (Figure 18-4d). In such organisms, including humans, meiosis gives rise not to spores but to gametes directly, so the haploid phase of the life cycle is represented only by the gametes. Meiosis in such species is called *gametic meiosis* to distinguish it from the *sporic meiosis* observed in spore-producing organisms exhibiting an alternation of generations. Meiosis is thus gametic in animals and sporic in plants.

Meiosis Converts One Diploid Cell into Four Haploid Cells

Wherever it occurs in an organism's life cycle, meiosis always involves chromosome duplication in a diploid cell followed by two successive divisions that convert the diploid nucleus into four haploid nuclei. Figure 18-5 illustrates the various phases of meiosis; refer to it as you read the following discussion.

During the first meiotic division, or **meiosis I,** the two chromosomes of each homologous pair come together during prophase to exchange genetic information (using a mechanism to be discussed shortly). This pairing of homologous chromosomes, called **synapsis,** is unique to meiosis; at all other times, including mitosis, the chromosomes of a homologous pair behave independently. The two chromosomes of each homologous pair bind together so tightly during the first meiotic prophase that they behave as a single unit, called a **bivalent** (or *tetrad,* which emphasizes that each of the two homologous chromosomes consists of two sister chromatids, yielding a total of four chromatids). After aligning at the spindle equator, each bivalent splits apart in such a way that its two homologous chromosomes move to opposite spindle poles. As a result, each daughter nucleus produced by the first meiotic division is haploid because it contains only one of the two chromosomes of each bivalent. During the second meiotic division (meiosis II), which closely resembles a mitotic division, the two sister chromatids of each chromosome

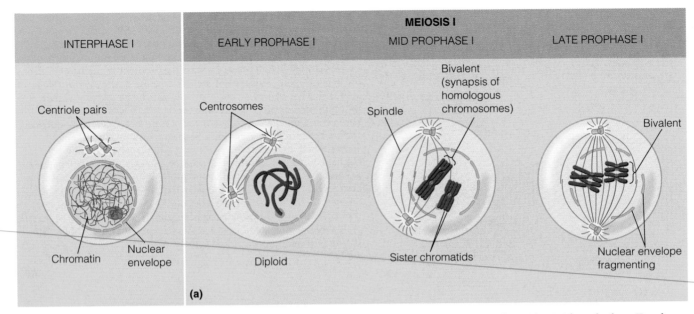

| INTERPHASE I | EARLY PROPHASE I | MID PROPHASE I | LATE PROPHASE I |

Interphase I: Centriole pairs, Chromatin, Nuclear envelope

Early Prophase I: Centrosomes, Diploid

Mid Prophase I: Spindle, Bivalent (synapsis of homologous chromosomes), Sister chromatids

Late Prophase I: Bivalent, Nuclear envelope fragmenting

(a)

Figure 18-5   Meiosis in an Animal Cell. Meiosis consists of two successive divisions, called meiosis I and II, with no intervening DNA synthesis or chromosome duplication. **(a)** During prophase I, the chromosomes (duplicated during the previous S phase) condense and the two centrosomes migrate to opposite poles of the cell. Each chromosome (four, in this example) consists of two sister chromatids. Homologous chromosomes pair to form bivalents. **(b)** Bivalents become aligned at the spindle equator (metaphase I). **(c)** Homologous chromosomes separate during anaphase I, but sister chromatids remain attached at the centromere. **(d)** Telophase and cytokinesis follow. Although not illustrated here, there may then be a short interphase (interphase II). In meiosis II, **(e)** chromosomes recondense (prophase II), **(f)** chromosomes align at the spindle equator (metaphase II), and **(g)** sister chromatids at last separate (anaphase II). **(h)** After telophase II and cytokinesis, the result is four haploid daughter cells, each containing one chromosome of each homologous pair. Prophase I is a complicated process shown in more detail in Figure 18-7. Meiosis in plants is similar, except for the absence of centrioles and the mechanism of cytokinesis, which involves formation of a cell plate.

separate into two daughter cells. Hence the events unique to meiosis happen during the first meiotic division: the synapsis of homologous chromosomes and their subsequent segregation into different daughter nuclei.

Each meiotic division involves the same basic stages as mitosis, although cell biologists do not usually distinguish prometaphase as a separate phase. Thus the meiotic phases are *prophase, metaphase, anaphase,* and *telophase.* Prophase I is much longer and more complicated than mitotic prophase, whereas prophase II tends to be quite short. Another important difference from mitosis is that a normal interphase does not intervene between the two meiotic divisions. If an interphase does take place, it is usually very short and—most important—does *not* include DNA replication because each chromosome already consists of a pair of replicated, sister chromatids that had been generated prior to the first meiotic division. The purpose of the second meiotic division, like that of a typical mitotic division, is to parcel these sister chromatids into two daughter nuclei.

The micrographs in Figure 18-6 show how chromosomes look at various points in meiosis. As you can see, there are several distinct stages of prophase I; we will discuss these as we examine meiosis I in more detail in the next section.

Meiosis I Produces Two Haploid Cells That Have Chromosomes Composed of Sister Chromatids

The first meiotic division segregates homologous chromosomes into different daughter cells. This feature of meiosis is of special genetic significance because it is at this stage in the life cycle of the organism that the two alleles for each gene part company. And it is this separation of alleles that makes possible the eventual remixing of different pairs of alleles at fertilization. Also of great significance during the first meiotic division are events involving the physical exchange of parts of DNA molecules. Such an exchange of DNA segments between two different sources is called **genetic recombination** by molecular biologists. As we will discuss shortly, this type of DNA exchange between homologous chromosomes takes place when the chromosomes are synapsed during prophase I.

Prophase I: Homologous Chromosomes Become Paired and Exchange DNA.   Prophase I is a particularly long and complex phase. Based on light microscopic observations, early cytologists divided prophase I into five stages called *leptotene, zygotene, pachytene, diplotene,* and *diakinesis* (Figure 18-7).

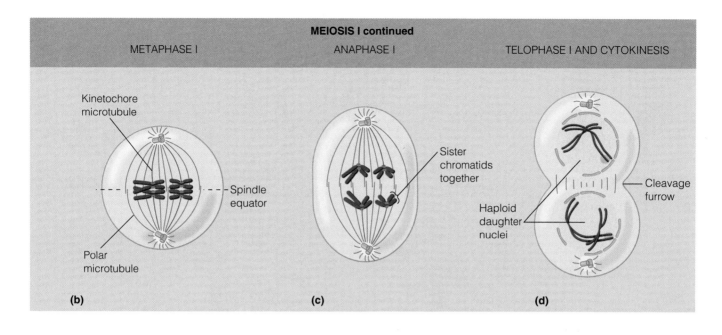

MEIOSIS I continued

METAPHASE I	ANAPHASE I	TELOPHASE I AND CYTOKINESIS

Kinetochore microtubule

Spindle equator

Polar microtubule

(b)

Sister chromatids together

(c)

Cleavage furrow

Haploid daughter nuclei

(d)

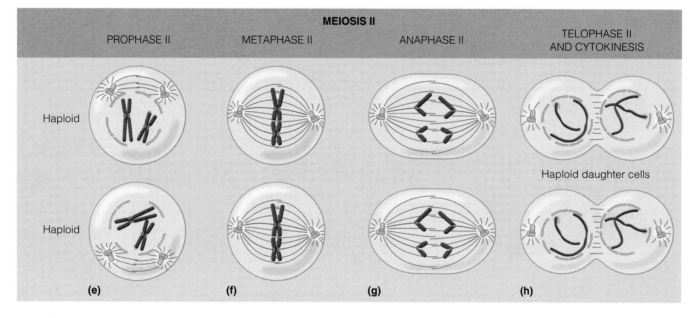

MEIOSIS II

PROPHASE II	METAPHASE II	ANAPHASE II	TELOPHASE II AND CYTOKINESIS

Haploid

Haploid

Haploid daughter cells

(e)

(f)

(g)

(h)

The **leptotene** stage begins with the condensation of chromatin fibers into long, threadlike structures, similar to what occurs at the beginning of mitosis. At **zygotene,** individual chromosomes become distinguishable and homologous chromosomes become closely paired with each other via the process of synapsis, forming bivalents. Keep in mind that each bivalent has four chromatids, two derived from each chromosome. Bivalent formation is of considerable genetic significance because the close proximity between homologous chromosomes allows DNA segments to be exchanged by a process called **crossing over,** which occurs during the **pachytene** stage. It is this physical exchange of genetic information between corresponding regions of homologous chromosomes during pachytene that accounts for genetic recombination.

Pachytene is marked by a dramatic compacting process that reduces each chromosome to less than a quarter of its previous length.

At the **diplotene** stage, the homologous chromosomes of each bivalent begin to separate from each other, particularly near the centromere. However, the two chromosomes of each homologous pair remain attached by connections known as **chiasmata** (singular: **chiasma**). Such connections are situated in regions where homologous chromosomes have exchanged DNA segments and hence provide visual evidence that crossing over has occurred between two chromatids, one derived from each chromosome.

In some organisms—female mammals, for instance—the chromosomes decondense during diplotene, transcription resumes, and the cells "take a break" from

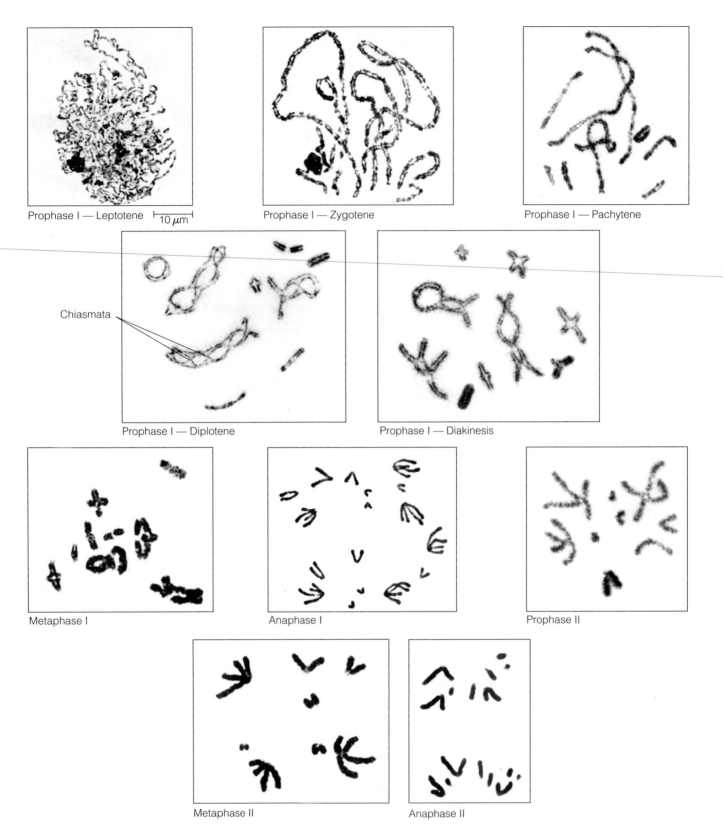

Prophase I — Leptotene |— 10 μm —|

Prophase I — Zygotene

Prophase I — Pachytene

Chiasmata

Prophase I — Diplotene

Prophase I — Diakinesis

Metaphase I

Anaphase I

Prophase II

Metaphase II

Anaphase II

Figure 18-6   Chromosome Appearance During the Various Stages of Meiosis.   Homologous chromosomes become paired during zygotene, and shorten and thicken at the pachytene stage. During diplotene the four chromatids present in each bivalent can be distinguished; at this stage, the chiasmata are visible. During anaphase I, each newly forming cell receives one member of each pair of homologous chromosomes. At metaphase II, each chromosome consists of a pair of chromatids held together by a centromere. During anaphase II the centromeres split, allowing the chromatids to separate and migrate in opposite directions.

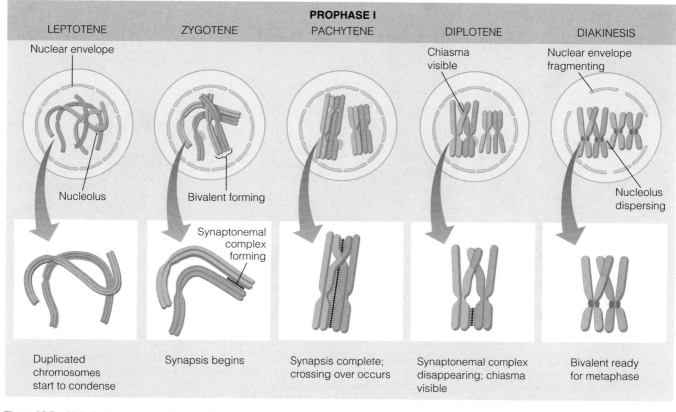

PROPHASE I

| LEPTOTENE | ZYGOTENE | PACHYTENE | DIPLOTENE | DIAKINESIS |

Nuclear envelope

Nucleolus

Bivalent forming

Synaptonemal complex forming

Chiasma visible

Nuclear envelope fragmenting

Nucleolus dispersing

Duplicated chromosomes start to condense

Synapsis begins

Synapsis complete; crossing over occurs

Synaptonemal complex disappearing; chiasma visible

Bivalent ready for metaphase

Figure 18-7 Meiotic Prophase I. Prophase I is a complicated process that is subdivided into five stages. The top diagrams depict the cell nucleus at each stage, for a diploid cell with a total of four chromosomes (two homologous pairs). The bottom diagrams focus in on a single homologous pair in greater detail, revealing the formation and subsequent disappearance of the synaptonemal complex, a protein structure that holds homologous chromosomes in close lateral apposition during pachytene. Red and blue distinguish the paternal and maternal chromosomes of each homologous pair; the synaptonemal complex is shown in shades of purple.

meiosis for a prolonged period of growth, sometimes lasting for years. (We will consider this situation at the end of our discussion of meiosis.) With the onset of **diakinesis,** the final stage of prophase I, the chromosomes recondense to their maximally compacted state. Now the centromeres of the homologous chromosomes separate further, and the chiasmata eventually become the only remaining attachments between the homologues. At this stage the nucleoli disappear, the spindle forms, and the nuclear envelope breaks down, marking the end of prophase I.

With the advent of modern tools, especially the electron microscope, cytologists have been able to refine our picture of what happens during prophase I. They have found that what holds homologous chromosomes in tight apposition during synapsis is the **synaptonemal complex,** an elaborate protein structure resembling a zipper (Figure 18-8). The *lateral elements* of the synaptonemal complex start to attach to individual chromosomes during leptotene, but the *central element,* which actually joins homologous chromosomes together, does not form until zygotene (see Figure 18-8b). How do the members of each pair of homologous chromosomes find each other so they can be joined by a synaptonemal complex? During early

zygotene, the ends (telomeres) of each chromosome become clustered on one side of the nucleus and attach to the nuclear envelope, with the body of each chromosome looping out into the nucleus. To picture this, imagine holding all the ends of four ropes (two long and two short) together. If you give the ropes a strong shake, they will settle into four loops, arranged according to length. This type of chromosome configuration, called a *bouquet,* is thought to promote chromosome alignment.

The alignment of similar-sized chromosomes facilitates formation of synaptonemal complexes, which become fully developed during pachytene. Formation of synaptonemal complexes is closely associated with the process of crossing over in higher eukaryotes, and some electron micrographs reveal additional protein complexes, called *recombination nodules,* that may mediate the crossing over process. The synaptonemal complexes then disassemble during diplotene, allowing the homologous chromosomes to separate (except where they are joined by chiasmata).

Metaphase I: Bivalents Align at the Spindle Equator. During metaphase I, the bivalents attach via their kinetochores to spindle microtubules and migrate to the spindle equator.

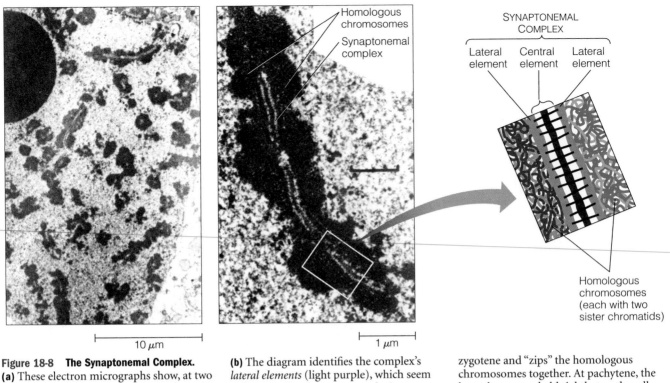

Figure 18-8 The Synaptonemal Complex.
(a) These electron micrographs show, at two magnifications, synaptonemal complexes in the nuclei of cells from a lily. The cells are at the pachytene stage of prophase I (TEMs).

(b) The diagram identifies the complex's *lateral elements* (light purple), which seem to form on the chromosomes during leptotene, and its *central* (or *axial*) *element* (dark purple), which starts to appear during

zygotene and "zips" the homologous chromosomes together. At pachytene, the homologues are held tightly together all along their lengths.

The presence of *paired* homologous chromosomes (i.e., bivalents) at the spindle equator during metaphase I is a crucial difference between meiosis I and a typical mitotic division, where such pairing is not observed (Figure 18-9). Because each bivalent contains four chromatids (two sister chromatids from each chromosome), four kinetochores are also present. The kinetochores of sister chromatids lie side by side—in many species appearing as a single mass—and face the same pole of the cell. Such an arrangement allows the kinetochores derived from the sister chromatids of one homologous chromosome to attach to microtubules emanating from one spindle pole, and the kinetochores derived from the sister chromatids of the other homologous chromosome to attach to microtubules emanating from the opposite spindle pole. This orientation sets the stage for separation of the homologous chromosomes during anaphase. The bivalents are randomly oriented at this point, in the sense that, for each bivalent, either the maternal or paternal homologue may face a given pole of the cell. As a result, each spindle pole (and hence each daughter cell) will receive a random mixture of maternal and paternal chromosomes when the two members of each chromosome pair move toward opposite spindle poles during anaphase.

At this stage, homologous chromosomes are held together solely by chiasmata. If, for some reason, prophase I

had occurred without crossing over, and hence without chiasma formation, the chromosomes might not pair properly at the spindle equator, and homologous chromosomes might not separate properly during anaphase I. This is exactly what happens during meiosis in mutant yeast cells that exhibit deficiencies in genetic recombination.

Anaphase I: Homologous Chromosomes Move to Opposite Spindle Poles.
At the beginning of anaphase I, the members of each pair of homologous chromosomes separate from each other and start migrating toward opposite spindle poles, pulled by their respective kinetochore microtubules. Again, note the fundamental difference between meiosis and mitosis (see Figure 18-9). During mitotic anaphase, *sister chromatids separate* and move to opposite spindle poles, whereas in anaphase I of meiosis, *homologous chromosomes separate* and move to opposite spindle poles while sister chromatids remain together. Because the two members of each pair of homologous chromosomes move to opposite spindle poles during anaphase I, each pole receives a haploid set of chromosomes.

Telophase I and Cytokinesis: Two Haploid Cells Are Produced.
The onset of telophase I is marked by the arrival of a haploid set of chromosomes at each spindle pole. Which

MEIOSIS I

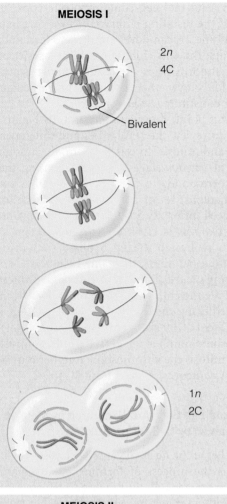

2n
4C

Bivalent

1n
2C

MEIOSIS II

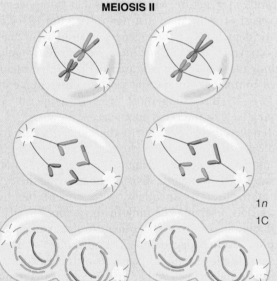

1n
1C

Result of meiosis: four haploid cells, each with half as many chromosomes as the original cell. Each haploid cell contains a random mixture of maternal and paternal chromosomes.

Prophase

Each condensing chromosome has two chromatids. In meiosis I, homologous chromosomes synapse, forming a bivalent. Crossing over occurs between nonsister chromatids, producing chiasmata. In mitosis, each chromosome acts independently.

Metaphase

In meiosis I, the bivalents align at the metaphase plate. In mitosis, individual chromosomes align at the metaphase plate.

Anaphase

In meiosis I, chromosomes (not chromatids) separate. In mitosis, chromatids separate.

Telophase and Cytokinesis

In meiosis II, sister chromatids separate.

MITOSIS

2n
4C

2n
2C

Result of mitosis: two cells, each with the same number of chromosomes as the original cell.

Figure 18-9 Meiosis and Mitosis Compared. Meiosis and mitosis are both preceded by DNA replication, resulting in two sister chromatids per chromosome at prophase. Meiosis, which occurs only in sex cells, includes two nuclear (and cell) divisions, halving the chromosome number to the haploid level. Moreover, during the elaborate prophase of the first meiotic division, homologous chromosomes synapse, and crossing over occurs between nonsister chromatids. Mitosis involves only a single division, producing two diploid nuclei (and, usually, cells), each with the same number of chromosomes as the original cell. In mitosis, the homologous chromosomes behave independently; at no point do they come together in pairs. The meaning of the C values in this figure (4C, 2C, 1C) is described on page 588.

member of a homologous pair ends up at which pole is determined entirely by how the chromosomes happened to be oriented at the spindle equator during metaphase I. After the chromosomes have arrived at the spindle poles, nuclear envelopes form around the chromosomes and cytokinesis ensues, generating two haploid cells whose chromosomes consist of sister chromatids. In most cases, the chromosomes do not decondense before meiosis II begins.

Meiosis II Resembles a Mitotic Division

After meiosis I has been completed, a brief interphase may intervene before **meiosis II** begins. However, this interphase is not accompanied by DNA replication because each chromosome already consists of a pair of replicated, sister chromatids that had been generated by DNA synthesis during the interphase preceding meiosis I. *In other words, DNA is replicated only once during meiosis, prior to the first meiotic division.* The purpose of meiosis II, like that of a typical mitotic division, is to parcel the sister chromatids created by this initial round of DNA replication into two newly forming cells.

Prophase II is very brief. If detectable at all, it is much like a mitotic prophase. Metaphase II also resembles the equivalent stage in mitosis, except that only half as many chromosomes are present at the spindle equator. The kinetochores of sister chromatids now face in opposite directions, allowing the sister chromatids to separate and move (as new daughter chromosomes) to opposite spindle poles during anaphase II. The remaining phases of the second meiotic division resemble the comparable stages of mitosis. The final result is the formation of four daughter cells, each containing a haploid set of chromosomes. Because the two members of each homologous chromosome pair were randomly distributed to the two cells produced by meiosis I, each of the haploid daughter cells produced by meiosis II contains a random mixture of maternal and paternal chromosomes. Moreover, each of these chromosomes is composed of a mixture of maternal and paternal DNA sequences created by crossing over during prophase I.

While each of the cells produced by meiosis normally contains a complete, haploid set of chromosomes, a rare malfunction called **nondisjunction** can produce cells that either lack a particular chromosome or contain an extra chromosome. Nondisjunction refers to the failure of the two members of a homologous chromosome pair to separate during anaphase I of meiosis. Instead, both chromosomes remain together and move into one of the two daughter cells, generating one cell containing both copies of the chromosome and one cell containing neither copy. The gametes resulting from such an abnormal meiosis have an incorrect number of chromosomes and tend to produce defective embryos that die before birth. However, a few such gametes can participate in the formation of embryos that do survive. For example, if an abnormal human sperm containing two copies of chromosome 21 fertilizes a normal egg containing one copy of chromosome 21, the resulting embryo, possessing three copies of chromosome 21, can develop fully and lead to the birth of a live child. But this child will exhibit a series of developmental abnormalities—including short stature, broad hands, folds over the eyes, and low intelligence—that together constitute *Down syndrome.*

At this point, you may want to study Figure 18-9 in its entirety to review the similarities and differences between meiosis and mitosis. In this diagram, the amount of DNA present at various stages is indicated using the **C value,** which corresponds to the amount of DNA present in a single (haploid) set of chromosomes. Hence in a normal diploid cell prior to S phase, the DNA content is 2C because two sets of chromosomes are present. After DNA has been replicated during S phase (but prior to sister chromatid separation), the value is increased to 4C. Note that during meiosis I, the DNA content is reduced from 4C to 2C while the chromosome number goes from 2n to 1n, and that during meiosis II, the DNA content is reduced from 2C to 1C (by sister chromatid separation) while the chromosome number remains at 1n. In contrast, during a normal mitosis the chromosome number remains at 2n as the DNA content is reduced from 4C to 2C.

Sperm and Egg Cells Are Generated by Meiosis Accompanied by Cell Differentiation

Meiosis lies at the heart of gametogenesis, which, as we saw earlier, is the process of forming haploid gametes from diploid precursor cells. But male and female gametes differ significantly in structure, which means that gametogenesis must consist of more than just meiosis. Figure 18-10 is a schematic depiction of gametogenesis in animals. In males, meiosis converts a diploid *spermatocyte* into four haploid *spermatids* of similar size (Figure 18-10a). After meiosis has been completed, the spermatids then differentiate into sperm cells by discarding most of their cytoplasm and developing flagella and other specialized structures.

In females, meiosis converts a diploid *oocyte* into four haploid cells, but only one of the four survives and gives rise to a functional egg cell (Figure 18-10b). This outcome is generated by two meiotic divisions that divide the cytoplasm of the oocyte unequally, with one of the four daughter cells receiving the bulk of the cytoplasm of the original diploid oocyte. The other three, smaller cells, called **polar bodies,** degenerate. The advantage of having only one of the four haploid products of meiosis develop into a functional egg cell is that the cytoplasm that would otherwise have been distributed among four cells is instead concentrated into one egg cell, maximizing the content of stored nutrients in each egg.

An important difference between sperm and egg formation concerns the stage at which the cells acquire the specialized characteristics that make them functionally mature gametes. During sperm cell development, meiosis creates haploid spermatids that must then discard most of their cytoplasm and develop flagella before they are func-

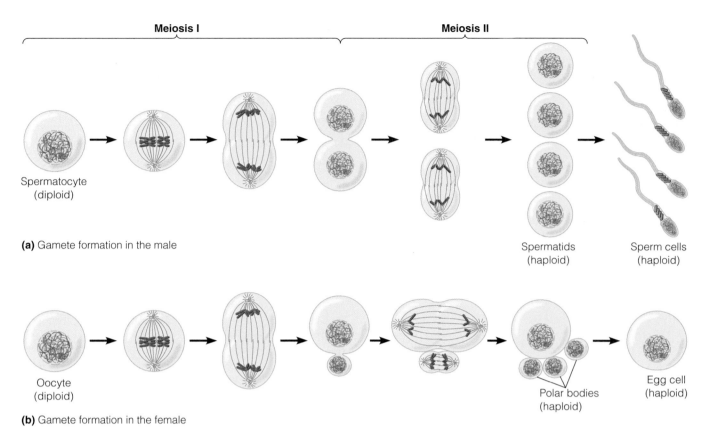

Meiosis I

Meiosis II

Spermatocyte
(diploid)

Spermatids
(haploid)

Sperm cells
(haploid)

(a) Gamete formation in the male

Oocyte
(diploid)

Polar bodies
(haploid)

Egg cell
(haploid)

(b) Gamete formation in the female

Figure 18-10 Gamete Formation. **(a)** In the male, all four haploid products of meiosis are retained and differentiate into sperm. **(b)** In the female, both meiotic divisions are asymmetric, forming one large egg cell and three (in some cases, only two) small cells called polar bodies that do not give rise to functional gametes. Although not indicated here, the mature egg cell has usually grown much larger than the oocyte from which it arose.

tionally mature. In contrast, developing egg cells acquire their specialized features *during* the process of meiosis. Many of the specialized features of the egg cell are acquired during prophase I, when meiosis is temporarily halted to allow time for extensive cell growth. During this *growth phase,* the cell also develops various types of external coatings designed to protect the egg from chemical and physical injury. The amount of growth that takes place during this phase can be quite extensive. A human egg cell, for example, has a diameter of about 100 μm, giving it a volume more than a hundred times as large as that of the diploid oocyte from which it arose. And consider the gigantic size of a bird egg!

After the growth phase has been completed, the developing oocyte remains arrested in prophase I until resumption of meiosis is triggered by an appropriate stimulus. In amphibians, the resumption of meiosis is triggered by the steroid hormone progesterone, whose presence leads to an increase in the activity of the protein kinase, *MPF.* In Chapter 17, we learned that MPF is a Cdk-cyclin complex that controls mitotic cell division by triggering passage through the G2 checkpoint. MPF also controls meiosis by triggering the transition from prophase I to metaphase I. Progesterone exerts its control over MPF by stimulating the production of *Mos,* a protein kinase that activates a

cascade of other protein kinases; this cascade in turn leads to the activation of MPF.

In response to the activation of MPF, the first meiotic division is completed. In some organisms, the second meiotic division then proceeds rapidly to completion; in others, it halts at an intermediate stage and is not completed until after fertilization. In vertebrate eggs, for example, the second meiotic division is often arrested at metaphase II until fertilization takes place. Recent evidence suggests that the molecule responsible for triggering this metaphase arrest is Mos, the same protein kinase that regulates the transition from prophase I to metaphase I.

By the time meiosis is completed, the egg cell is fully "mature" and may even have been fertilized. The mature egg is a highly differentiated cell that is specialized for the task of producing a new organism in much the same sense that a muscle cell is specialized to contract, or a red blood cell is specialized to transport oxygen. This inherent specialization of the egg is vividly demonstrated by the observation that even in the absence of fertilization by a sperm cell, many kinds of animal eggs can be stimulated to develop into a complete embryo by artificial treatments as simple as a pinprick. Hence everything needed for programming the early stages of development must already be present in the egg.

Normally, this developmental sequence is activated by interactions between sperm and egg that involve specialized features of sperm and egg cell structure. In animal sperm, enzymes released from a membrane-bounded sac called the *acrosome*, located at the front end of the sperm, dissolve the egg cell's outer coats, and species-specific proteins on the sperm's surface bind to receptors on the egg cell. After fusion of the plasma membranes of sperm and egg, the sperm nucleus enters the egg and fuses with the egg nucleus, thereby generating a diploid cell, the zygote. Union with the sperm triggers important biochemical changes in the egg. For example, the fertilized egg quickly develops a barrier to the entry of additional sperm and simultaneously undergoes a tremendous burst of metabolic activity in preparation for embryonic development. Although not described here, fertilization in plants involves similarly complex cellular and biochemical events.

Meiosis Generates Genetic Diversity

As we have pointed out, one of the main functions of meiosis is to preserve the chromosome number in sexually reproducing organisms. If it were not for meiosis, gametes would have as many chromosomes as other cells in the body, and the chromosome number would double each time gametes fused to form a new organism.

Equally important, however, is the role meiosis plays in generating genetic diversity in sexually reproducing populations. Meiosis is the point in the flow of genetic information where various combinations of chromosomes (and the alleles they carry) are assembled in gametes for potential passage to offspring. Although every gamete is haploid, and hence possesses one member of each pair of homologous chromosomes, the particular combination of paternal and maternal chromosomes in any given gamete is random. This randomness is generated by the random orientation of bivalents at metaphase I, where the paternal and maternal homologues of each pair can face either pole, independent of the orientations of the other bivalents. Consequently in human gametes, which contain 23 chromosomes, more than 8 million different combinations of maternal and paternal chromosomes are possible.

Moreover, the crossing over that takes place between homologous chromosomes during prophase I generates additional genetic diversity among the gametes. By allowing the exchange of DNA segments between homologous chromosomes, crossing over generates more combinations of alleles than the random assortment of maternal and paternal chromosomes would create by itself. We will return to crossing over (recombination) later in the chapter. First, however, we turn to the historic experiments that revealed the genetic consequences of chromosome segregation and random assortment during meiosis. These experiments were carried out by Gregor Mendel before chromosomes were even known to exist.

Genetic Variability: Segregation and Assortment of Alleles

Most students of biology have heard of Gregor Mendel and the classic genetics experiments he conducted in a monastery garden. Mendel's findings, first published in 1865, laid the foundation for what we now know as *Mendelian genetics*. Working with the common garden pea, Mendel chose seven readily identifiable characters of pea plants and selected in each case two varieties of plants that displayed different forms of the character. For example, seed color was one character Mendel chose, because he had one strain of peas with yellow seeds and another with green seeds (see Figure 18-1). He first established that each of the plant strains was **true-breeding** upon self-fertilization, which means that plants grown from his yellow seeds produced only yellow seeds, and plants grown from his green seeds produced only green seeds (Figure 18-11a). Once this had been established, Mendel was ready to investigate the principles that govern the inheritance of such traits.

Information Specifying Recessive Traits Can Be Present Without Being Displayed

In his first set of experiments, Mendel *cross*-fertilized the true-breeding parental plants (the **P₁ generation**) to produce **hybrid** strains. The outcome of this experiment must have seemed mystifying at first: In every case, the resulting offspring—called the **F₁ generation**—exhibited one or the other of the parental traits, but never both. In other words, one parental trait was always dominant and the other was always recessive. In the case of seed color, for example, all the F₁ plants had yellow seeds (Figure 18-11b), indicating that yellow seed color is dominant.

During the next summer, Mendel allowed all the F₁ hybrids to self-fertilize. For each of the seven characters under study, he made the same surprising observation: The recessive trait that had seemingly disappeared in the F₁ generation reappeared among the progeny in the next generation, the **F₂ generation.** Moreover, for each of the seven characters, the ratio of dominant to recessive phenotypes in the progeny was always about 3 : 1. In the case of seed color, for example, plants grown from the yellow F₁ seeds produced about 75% yellow seeds and 25% green (Figure 18-11c).

This outcome was quite different from the behavior of the true-breeding yellow seeds of the parent strain, which had produced only yellow seeds when self-fertilized. Clearly, there was an important difference between the yellow seeds of the P₁ stock and the yellow seeds of the F₁ generation. They looked alike, but the former bred true, whereas the latter did not.

Next, Mendel investigated the F₂ plants through self-fertilization (Figure 18-12). The F₂ plants that showed the recessive trait (green, in the case of seed color) always bred true (Figure 18-12c), suggesting they were genetically iden-

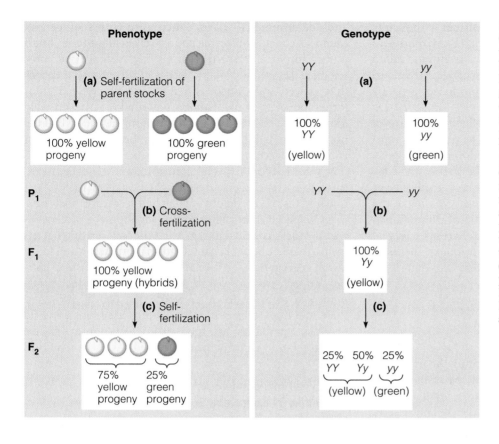

Figure 18-11   Genetic Analysis of Seed Color in Pea Plants.   Genetic crosses were performed starting with two true-breeding strains of pea plants, one having yellow seeds and one having green seeds. The yellow-seed trait is dominant (allele Y); the green-seed trait is recessive (allele y). The resulting phenotypes of the progeny are shown on the left and their genotypes on the right. (The genotypes were deduced later.) **(a)** The parent stocks are homozygous for either the dominant (YY) or recessive (yy) trait and breed true upon self-fertilization. **(b)** When crossed, the parent stocks yield F_1 plants (hybrids) that are all heterozygous (Yy) and therefore show the dominant trait. **(c)** Upon self-fertilization, the F_1 plants produce yellow and green seeds in the F_2 generation in a ratio of 3 : 1. See Figures 18-12 and 18-13 for further analyses.

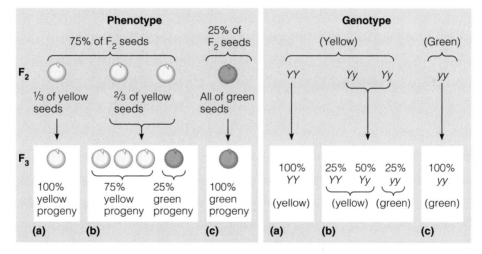

Figure 18-12   Analysis of F_2 Pea Plants by Self-Fertilization.   The F_2 plants of Figure 18-11 were analyzed through self-fertilization. The phenotypic results are shown on the left and the genotypes (deduced later) on the right. **(a)** One-third of the yellow F_2 progeny of Figure 18-11 (25% of the total F_2 progeny) breed true for yellow seed color upon self-fertilization, because they are genotypically YY. **(b)** Two-thirds of the yellow F_2 progeny (50% of the total F_2 progeny) yield yellow and green seeds upon self-fertilization, in a ratio of 3 : 1 (just as the F_1 plants of Figure 18-11 did upon self-fertilization). **(c)** All the green F_2 seeds (25% of the total F_2 progeny) breed true for green seed color upon self-fertilization because they are genotypically yy.

tical to the green-seeded P_1 strain with which Mendel had begun. F_2 plants with the dominant trait yielded a more complex pattern, however. One-third bred true for the dominant trait (Figure 18-12a) and therefore seemed to be identical to the P_1 plants with the dominant (yellow-seed) trait. However, the other two-thirds of the yellow-seeded F_2

plants produced progeny with both dominant and recessive phenotypes, in a 3 : 1 ratio (Figure 18-12b)—the same ratio that had arisen from the F_1 self-fertilization.

These results led Mendel to conclude that genetic information specifying the recessive trait must be present in the F_1 hybrid plants and seeds, even though the trait is not displayed. This conclusion was consistent with results from another set of experiments, in which F₁ hybrids were crossed with the original parent strains, a technique called **backcrossing** (Figure 18-13). Backcrossing F₁ hybrids to the dominant parent strain always produced progeny exhibiting the dominant trait (Figure 18-13a), whereas backcrossing to the recessive parent yielded a mixture of plants exhibiting dominant and recessive traits in a ratio of 1 : 1 (Figure 18-13b). Moreover, the dominant progeny from the latter cross behaved just like the F₁ hybrids: Upon self-fertilization, they gave rise to a 3 : 1 mixture of phenotypes (Figure 18-13c), and upon backcrossing to the recessive parent, they yielded a 1 : 1 ratio of dominant to recessive progeny (Figure 18-13d). (An alternative way of diagramming crosses, called the *Punnett square,* is shown in Problem 18-7 at the end of the chapter.)

The Law of Segregation States That the Alleles of Each Gene Separate from Each Other During Gamete Formation

After a decade of careful work documenting the preceding patterns of inheritance, Mendel formulated several principles—now known as **Mendel's laws of inheritance**—that explained the results he had observed. The first of these principles was that phenotypic traits are determined by discrete "factors" that are present in most organisms as pairs of "determinants." Today, we call these "factors" *genes*

and "determinants" *alleles* (alternative forms of genes). Mendel's conclusion seems almost self-evident to us, but it was an important assertion in his day. At that time, most scientists favored a *blending theory of inheritance,* a theory that viewed traits such as yellow and green seed color rather like cans of paint that are poured together to yield intermediate results. Other investigators had described the nonblending nature of inheritance before Mendel, but without the accompanying data and mathematical analysis that were Mendel's special contribution. Mendel's breakthrough is especially impressive when we recall that he formulated his theory before anyone had seen chromosomes.

Of special importance to the development of genetics was Mendel's conclusion regarding the way in which genes are parceled out during gamete formation. According to his **law of segregation,** *the two alleles of a gene are distinct entities that segregate, or separate from each other, during the formation of gametes.* In other words, the two alleles retain their identities even when both are present in a hybrid organism, and they are then parceled out into separate gametes so they can emerge unchanged in later generations.

The Law of Independent Assortment States That the Alleles of Each Gene Separate Independently of the Alleles of Other Genes

In addition to the crosses already described, each of which focused on a single pair of alleles, Mendel also studied

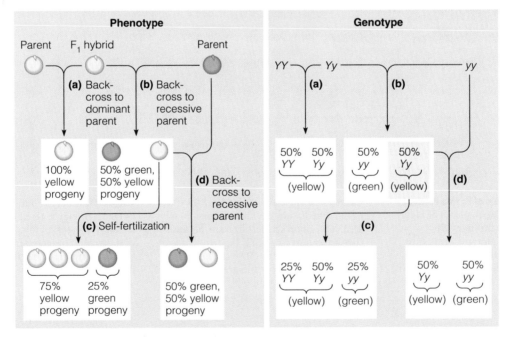

Figure 18-13 Analysis of F₁ Hybrids by Backcrossing. The F₁ hybrids of Figure 18-11 were analyzed by backcrossing to the parent (P₁) strains. The phenotypic results are shown on the left and the genotypes (deduced later) on the right. **(a)** Upon backcrossing of the F₁ hybrid (*Yy*) to the

dominant parent (*YY*), all the progeny have yellow seeds, because the genotype is either *YY* or *Yy*. **(b)** Backcrossing to the recessive parent (*yy*) yields yellow (*Yy*) or green (*yy*) seeds in equal proportions. **(c)** These yellow-seeded progeny will give rise, upon self-fertilization, to a 3 : 1 mixture of yellow and

green seeds (just as with the F₁ hybrids in Figure 18-11). **(d)** Backcrossing of the yellow-seeded progeny to the homozygous recessive parent again yields a 1 : 1 mixture of yellow and green seeds, as in the backcross of part b.

multifactor crosses between plants that differed in *several* characters. In addition to differing in seed color, for example, the plants he crossed might differ in seed shape and flower position. As in his single-factor crosses, he used parent plants that were true-breeding (homozygous) for the characters he was testing and generated F_1 hybrids heterozygous for each character. He then self-fertilized these hybrids and determined the frequency with which the dominant and recessive forms of the various characters appeared among the progeny.

Mendel found that all possible combinations of traits appeared in the F_2 progeny, and he concluded that all possible combinations of the different alleles must therefore have been present among the F_1 gametes. Furthermore, from the proportions in which the various phenotypes were found in the F_2 generation, Mendel deduced that all possible combinations of alleles occurred in the gametes with equal frequency. In other words, *the two alleles of each gene segregate independently of the alleles of other genes.* This is the **law of independent assortment,** another cornerstone of genetics. (Later it would be shown that this law only applies to genes on different chromosomes or very far apart on the same chromosome.)

Early Microscopic Evidence Suggested That Chromosomes Might Carry Genetic Information

Mendel's findings lay dormant in the scientific literature until 1900, when his paper was rediscovered almost simultaneously by three other European biologists. In the meantime, much had been learned about the cellular basis of inheritance. By 1875, for example, microscopists had identified chromosomes with the help of stains produced by the developing aniline dye industry. At about the same time, fertilization was shown to involve the fusion of sperm and egg nuclei, suggesting that the nucleus carries genetic information.

The first suggestion that chromosomes might be the bearers of this genetic information was made in 1883. Within ten years, chromosomes had been studied in dividing cells and had been seen to split longitudinally into two apparently identical daughter chromosomes. This led to the realization that the number of chromosomes per cell remains constant during the development of an organism. With the invention of better optical systems, more detailed analysis of chromosomes became possible. Mitotic cell division was shown to involve the movement of identical daughter chromosomes to opposite poles, thereby ensuring that daughter cells would have exactly the same complement of chromosomes as their parent cell.

Against this backdrop came the rediscovery of Mendel's paper, followed almost immediately by three crucial studies that established chromosomes as the carriers of Mendel's factors. The investigators were Edward Montgomery, Theodor Boveri, and Walter Sutton. Montgomery's contribution was to recognize the existence of homologous chromosomes. From careful observations of insect chromosomes, he concluded that the chromosomes

of most cells could be grouped into pairs, with one member of the pair of maternal origin and the other of paternal origin. He also noted that the two chromosomes of each type come together during synapsis in the "reduction division" (now called meiosis) of gamete formation, a process that had been reported a decade or so earlier.

Boveri then added the crucial observation that each chromosome plays a unique genetic role. This idea came from studies in which sea urchin eggs were fertilized in the presence of a large excess of sperm, causing some eggs to be fertilized by two sperm. The presence of two sperm nuclei in a single egg leads to the formation of an abnormal mitotic spindle that does not distribute the chromosomes equally to the newly forming embryonic cells. The resulting embryos exhibit various types of developmental defects, depending on which particular chromosomes they are missing. Boveri therefore concluded that each chromosome plays a unique role in development.

Sutton, meanwhile, was studying meiosis in the grasshopper. In 1902, he made the important observation that the orientation of each pair of homologous chromosomes (bivalent) at the spindle equator during metaphase I is purely a matter of chance. Any homologous pair, in other words, may lie with the maternal or paternal chromosome toward either pole, regardless of the positions of other pairs. Many different combinations of maternal and paternal chromosomes are therefore possible in the gametes produced by any given individual (note the similarity to Mendel's observation that all possible combinations of different alleles are present among the F_1 generation of gametes).

Chromosome Behavior Explains the Laws of Segregation and Independent Assortment

During 1902–1903, Sutton put the preceding observations together into a coherent theory describing the role of chromosomes in inheritance. This *chromosomal theory of inheritance* can be summarized as five main points:

1. Nuclei of all cells except those of the *germ line* (sperm and eggs) contain two sets of homologous chromosomes, one set of maternal origin and the other of paternal origin.

2. Chromosomes retain their individuality and are genetically continuous throughout an organism's life cycle.

3. The two sets of homologous chromosomes in a diploid cell are functionally equivalent, each carrying a similar set of genes.

4. Maternal and paternal homologues synapse during meiosis and then move to opposite poles of the division spindle, thereby becoming segregated into different cells.

5. The maternal and paternal members of different homologous pairs segregate independently during meiosis.

The chromosomal theory of inheritance provided a physical basis for understanding how Mendel's genetic factors could be carried, transmitted, and segregated. For example, the presence of two sets of homologous chromosomes in each cell parallels Mendel's suggestion of two determinants for each phenotypic trait. Likewise, the segregation of homologous chromosomes during the meiotic divisions of gamete formation provides an explanation for Mendel's law of segregation, and the random orientation of homologous pairs at metaphase I accounts for his law of independent assortment.

Figures 18-14 and 18-15 illustrate the chromosomal basis of Mendel's laws, using examples from his pea plants. The basis for the law of segregation is shown in Figure 18-14 for the alleles governing seed color in a heterozygous pea plant with genotype *Yy*. (Peas have seven pairs of chromosomes, but only the pair bearing the alleles for seed color is shown.) During meiosis the two homologous chromosomes, each with two sister chromatids, synapse during prophase I, align together at the spindle equator at metaphase I, and then segregate into separate daughter cells. The second meiotic division then separates sister chromatids, so that each haploid cell ends up with only one allele for seed color, either *Y* or *y*.

The basis for the law of independent assortment is illustrated in Figure 18-15 for the chromosomes carrying the genes for seed color (alleles *Y* and *y*) and seed shape (alleles *R* and *r*, where *R* stands for round seeds, a dominant trait, and *r* for wrinkled seeds, a recessive trait). The explanation for independent assortment is that there are two possible, equally likely arrangements of the chromosome pairs at metaphase I. Of the *YyRr* cells that undergo meiosis, half will produce gametes like the four at the bottom left of the figure, and half will produce gametes like the four at the bottom right. Therefore, the *YyRr* plant will produce equal numbers of gametes of the eight types.

As we will discuss shortly, the law of independent assortment is *guaranteed* to hold only for genes on *different* chromosomes. It is remarkable that Mendel happened to choose seven independently assorting characters in an organism that has only seven pairs of chromosomes. Mendel was also fortunate that each of the seven characters he chose to study turned out to be controlled by a single gene (pair of alleles). Perhaps he concentrated on pea strains that, in preliminary experiments, gave him the most consistent and comprehensible results.

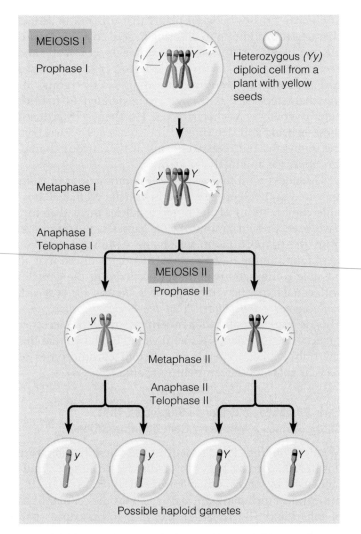

Figure 18-14 **The Meiotic Basis for Mendel's Law of Segregation.** Segregation of seed color alleles during meiosis is illustrated for the case of a pea plant heterozygous for this character. Peas have seven pairs of chromosomes, but only the homologous pair bearing the seed color alleles is shown here. During the first meiotic division (meiosis I), homologous chromosomes (each consisting of two sister chromatids) pair during prophase I, allowing crossing over to take place. The homologous chromosomes then align as a pair at the spindle equator during metaphase I, and segregate into separate cells at anaphase I and telophase I. In meiosis II, sister chromatids segregate to different daughter cells. The result is four haploid daughter cells, each of which has one allele for seed color.

The DNA Molecules of Homologous Chromosomes Have Similar Base Sequences

One of the five main points in the chromosomal theory of inheritance is the idea that homologous chromosomes are functionally equivalent, each carrying a similar set of genes. What does this mean in terms of our current understanding of the molecular organization of chromosomes? Simply put, it means that homologous chromosomes have DNA molecules whose base sequences are almost, but not entirely, identical. Thus, homologous chromosomes typically carry the same genes in exactly the same order. However, minor differences in base sequence along the chromosomal DNA molecule can create different alleles of the same gene. Such differences arise by mutation, and the different alleles for a gene that we find in a population—whether of pea plants or people—arise from mutations that have gradually occurred in an ancestral gene. Alleles are usually "expressed" by transcription into RNA and translation into proteins, and it is the behavior of these proteins that ultimately creates an organism's phenotype. As we will see in Chapter 19, a change as small as a single DNA base pair can create an allele that codes for an altered protein that is different enough to cause an observable

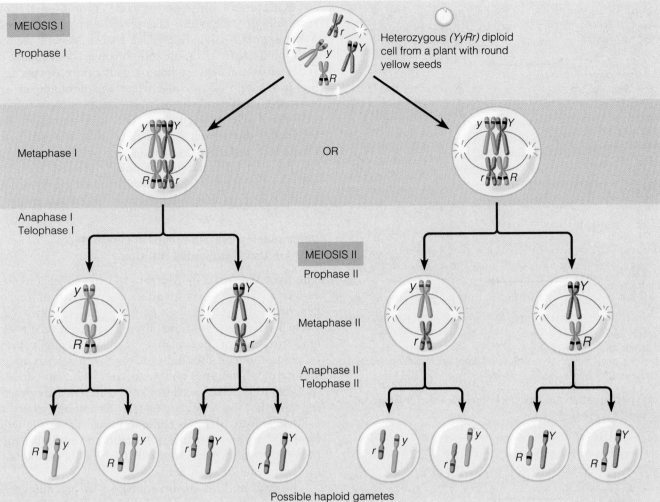

MEIOSIS I

Prophase I

Heterozygous (YyRr) diploid cell from a plant with round yellow seeds

Metaphase I

OR

Anaphase I
Telophase I

MEIOSIS II

Prophase II

Metaphase II

Anaphase II
Telophase II

Possible haploid gametes

Figure 18-15   The Meiotic Basis for Mendel's Law of Independent Assortment.
Independent assortment of the alleles of two genes on different chromosomes is illustrated by meiosis in a pea plant heterozygous for seed color (*Yy*) and seed shape (*Rr*). Pea plants have seven pairs of chromosomes, but only the two pairs bearing the seed color alleles and the seed shape alleles are shown. During meiosis, segregation of the seed color alleles occurs independently of segregation of the seed shape alleles. The basis of this independent assortment is found at metaphase I, when the homologous pairs (bivalents) align at the spindle equator. Each bivalent can face in either direction. Consequently, there are two alternative situations: Either the paternal homologues (blue) can both face the same pole of the cell, with the maternal homologues (red) facing the other pole, or they can face different poles. The arrangement at metaphase I determines which homologues subsequently go to which daughter cells. Since the alternative arrangements occur with equal probability, all the possible combinations of the alleles therefore occur with equal probability in the gametes. For simplicity, chiasmata and crossing over are not shown in this diagram.

change in an organism's phenotype—in fact, such a DNA alteration can be lethal.

The underlying similarity in the base sequences of their DNA molecules—*DNA homology*—probably explains the ability of homologous chromosomes to undergo synapsis (close pairing) during meiosis and is essential for normal crossing over. One popular model for synapsis holds that the correct alignment of homologous chromosomes is brought about before completion of the synaptonemal complex by some sort of base-pairing interaction between matching regions of DNA in the two chromosomes. Then, after the synaptonemal complex is fully formed, DNA recombination is completed at these (and perhaps other) sites.

Genetic Variability: Recombination and Crossing Over

Segregation and independent assortment of homologous chromosomes during the first meiotic division lead to random assortment of the alleles carried by different chromosomes, as was shown in Figure 18-15 for seed shape and color in pea plants. Figure 18-16a summarizes the outcome of meiosis for a generalized version of the same situation. Here we have a diploid organism heterozygous for two genes on nonhomologous chromosomes, with the allele pairs called *Aa* and *Bb*. Meiosis in such an organism will produce gametes in which allele *A* is just as

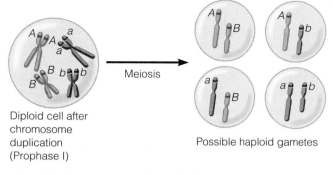

Diploid cell after
chromosome
duplication
(Prophase I)

Meiosis

Possible haploid gametes

(a) Unlinked genes assort independently

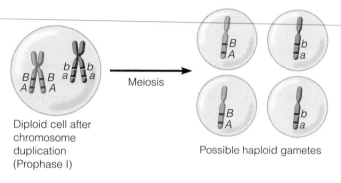

Diploid cell after
chromosome
duplication
(Prophase I)

Meiosis

Possible haploid gametes

(b) Linked genes end up together in the absence of crossing over

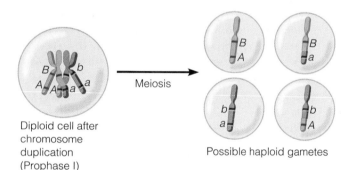

Diploid cell after
chromosome
duplication
(Prophase I)

Meiosis

Possible haploid gametes

(c) Linked genes do not always end up together if crossing over occurs

Figure 18-16 **Segregation and Assortment of Linked and Unlinked Genes.** In this figure, *A* and *a* are alleles of one gene, and *B* and *b* are alleles of another gene. **(a)** Alleles of genes on different chromosomes segregate and assort independently during meiosis; allele *A* is as likely to occur in a gamete with allele *B* as it is with allele *b*. **(b)** Alleles of genes on the same chromosome remain linked during meiosis in the absence of crossing over; in this case, allele *A* will occur routinely in the same gamete with allele *B*, but not with allele *b*. **(c)** Alleles of genes on the same chromosome can become interchanged when crossing over takes place, so that allele *A* occurs not only with allele *B* but also with allele *b*, and so forth.

likely to occur with allele *B* as it is with allele *b*, and *B* is just as likely to occur with *A* as it is with *a*. But what happens if genes *A* and *B* reside on the same chromosome? In that case, alleles *A* and *B* will routinely be linked together in the same gamete, as will alleles *a* and *b* (Figure 18-16b).

But even for genes on the same chromosome, some scrambling of alleles can take place because of the phenomenon of crossing over, which leads to genetic recombination. Genetic recombination involves the exchange of genetic material between homologous chromosomes during prophase I of meiosis when the homologues are synapsed, creating chromosomes exhibiting new combinations of alleles (Figure 18-16c). Recombination was originally discovered in studies with the fruit fly *Drosophila melanogaster,* conducted by Thomas Hunt Morgan and his colleagues beginning around 1910. We will therefore turn to Morgan's work to investigate recombination, beginning with his discovery of linkage groups.

Chromosomes Contain Groups of Linked Genes That Are Usually Inherited Together

The fruit flies used by Morgan and his colleagues had certain advantages over Mendel's pea plants as objects of genetic study, not the least of which was the fly's relatively brief generation time (about two weeks versus several months for pea plants). Unlike Mendel's peas, however, the fruit flies did not come with a ready-made variety of phenotypes and genotypes. Whereas Mendel was able to purchase seed stocks of different true-breeding varieties, the only type of fruit fly initially available to Morgan was what has come to be known as the **wild type,** or "normal" organism. Morgan and his colleagues therefore had to generate variants—that is, mutants— for their genetic experiments.

Morgan and his coworkers began by breeding large numbers of flies and then selecting mutant individuals having phenotypic modifications that were heritable. (Later, X-irradiation was used to enhance the mutation rate, but in their early work Morgan's group depended entirely on spontaneous mutations.) Within five years, they were able to identify about 85 different mutants, each carrying a mutation in a different gene. Each of these mutants could be propagated as a laboratory stock and used for matings as needed.

One of the first things Morgan and his colleagues realized as they began analyzing their mutants was that, unlike the genes Mendel had studied in peas, the mutant fruit fly genes did not all assort independently. Instead, some genes behaved as if they were linked together, and for such genes, the new combinations of alleles predicted by Mendel were infrequent or even nonexistent. In fact, it was soon recognized that fruit fly genes can be classified into four **linkage groups,** each group consisting of a collection of **linked genes** that are usually inherited together. Morgan quickly realized that the number of linkage groups was the same as the number of different chromosomes in the organism, since the haploid chromosome number for *Drosophila* is four. The conclusion was obvious: Each chromosome is the physical basis for a specific linkage group. Mendel had not observed linkage because the genes he studied all happened to reside on different chromosomes and so were not linked to one another.

Homologous Chromosomes Exchange Segments During Crossing Over

Although the genes they discovered could all be organized into linkage groups, Morgan and his colleagues found that linkage within such groups was incomplete. Most of the time, genes known to be linked (and therefore on the same chromosome) assorted together, as would be expected. Sometimes, however, two or more such traits would appear in the offspring in nonparental combinations. This phenomenon of less-than-complete linkage was called *recombination* because the different alleles appeared in new and unexpected associations in the offspring.

To explain such recombinant offspring, Morgan proposed that homologous chromosomes can exchange segments, presumably by some sort of breakage-and-fusion event, as illustrated in Figure 18-17a. By this process, which Morgan termed *crossing over,* a particular allele or group of alleles initially present on one member of a homologous pair of chromosomes could be transferred to the other chromosome in a reciprocal manner.

In the example of Figure 18-17a, two homologous chromosomes, one with alleles *A* and *B* and the other with alleles *a* and *b*, lie side by side at synapsis. Portions of an *AB* chromatid and a nonsister *ab* chromatid then exchange DNA segments, producing two recombinant chromatids, one with the alleles *A* and *b* and the other with the alleles *a* and *B*. Each of the four chromatids ends up in a different gamete at the end of the second meiotic division, so the products of meiosis will include two

parental gametes and two *recombinant* gametes, assuming a single crossover event (Figure 18-17b).

We now know that crossing over takes place during the pachytene stage of meiotic prophase I (see Figure 18-7), at a time when sister chromatids are packed tightly together and it is difficult to observe what is happening. As the chromatids begin to separate at diplotene, each of the four chromatids in a bivalent can be identified as belonging to one or the other of the two homologues. Wherever crossing over has taken place between nonsister chromatids, the two homologues remain attached to each other, forming a chiasma.

At the first meiotic metaphase, homologous chromosomes are almost always held together by at least one chiasma; if not, they may not segregate properly. Many bivalents contain multiple chiasmata. Human bivalents, for example, typically have two or three chiasmata because multiple crossover events routinely occur between paired homologues. To be of genetic significance, crossing over must involve nonsister chromatids. In some species, exchanges between sister chromatids are also observed, but such exchanges are of no genetic consequence because sister chromatids are genetically identical.

Gene Locations Can Be Mapped by Measuring Recombination Frequencies

Eventually, it became clear to Morgan and others that the frequency of recombinant progeny differed for different pairs of genes within the various linkage groups, but that

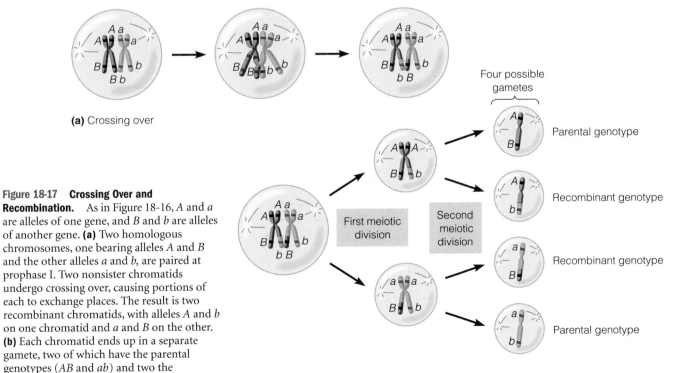

Figure 18-17 Crossing Over and Recombination. As in Figure 18-16, *A* and *a* are alleles of one gene, and *B* and *b* are alleles of another gene. **(a)** Two homologous chromosomes, one bearing alleles *A* and *B* and the other alleles *a* and *b*, are paired at prophase I. Two nonsister chromatids undergo crossing over, causing portions of each to exchange places. The result is two recombinant chromatids, with alleles *A* and *b* on one chromatid and *a* and *B* on the other. **(b)** Each chromatid ends up in a separate gamete, two of which have the parental genotypes (*AB* and *ab*) and two the recombinant genotypes (*Ab* and *aB*).

(a) Crossing over

(b) Results of crossing over

Four possible gametes

Parental genotype

Recombinant genotype

Recombinant genotype

Parental genotype

First meiotic division

Second meiotic division

for a specific pair of genes it was remarkably constant. This suggested that the frequency of observed recombination between two genes might be a measure of how far the genes are located from each other along the chromosome. If the probability of crossing over is the same at every point along a chromatid (an assumption that appears to be justified much of the time), then genes located very close to each other would be less likely to become separated by a crossover event between them than would genes that are far apart.

It was, in fact, quickly realized that the frequency of recombination, expressed as the percentage of progeny that are recombinant, was a useful means of quantifying the distance between different genes. This insight led Alfred Sturtevant, an undergraduate student in Morgan's lab, to suggest in 1911 that recombination data could be used to determine where genes are located along the chromosomes of *Drosophila*. Thus, a chromosome had come to be viewed as a linear string of genes whose positions can be determined on the basis of recombination data.

Determining the sequential order and spacing of genes on a chromosome based on recombinant frequencies is called **genetic mapping.** In the construction of such maps, the recombinant frequency is the map distance expressed in *map units*. If, for example, the alleles in Figure 18-17b appear among the progeny in their parental combination (*AB* and *ab*) 85% of the time and in the recombinant combinations (*Ab* and *aB*) 15% of the time, we conclude that the two genes are linked (are on the same chromosome) and are 15 map units apart. This approach has been used to map the chromosomes of many species of plants and animals, as well as bacteria and viruses. However, because bacteria and viruses do not reproduce sexually, the methods used to generate recombinants are somewhat different.

Genetic Recombination in Bacteria and Viruses

From what we have said so far, you might expect genetic recombination to be restricted to sexually reproducing organisms because recombination depends on crossing over between homologous chromosomes. As we have seen, sexual reproduction provides an opportunity for recombination once every generation when homologous chromosomes become closely juxtaposed during the meiotic divisions that produce gametes. In contrast, bacteria and viruses might seem to be poor candidates for recombination because they have haploid genomes and reproduce asexually, with no obvious mechanism for regularly bringing genomes from two different parents together.

Nonetheless, viruses and bacteria are still capable of genetic recombination. In fact, recombination data permitted the extensive mapping of viral and bacterial genomes well before the advent of modern DNA technology. To understand how recombination can take place despite a haploid genome, we need to examine the mecha-

nisms that allow two haploid genomes, or portions of genomes, to be brought together within the same cell.

Co-infection of Bacterial Cells with Related Bacteriophages Can Lead to Genetic Recombination

Much of our early understanding of genes and recombination at the molecular level, as well as the vocabulary in which we express that information, came from experiments involving bacteriophages, particularly T-even phages and phage λ (see Box 16A). Although phages are haploid and do not reproduce sexually, genetic recombination between related phages can take place when individual bacterial cells are simultaneously infected by different versions of the same phage. Figure 18-18 illus-

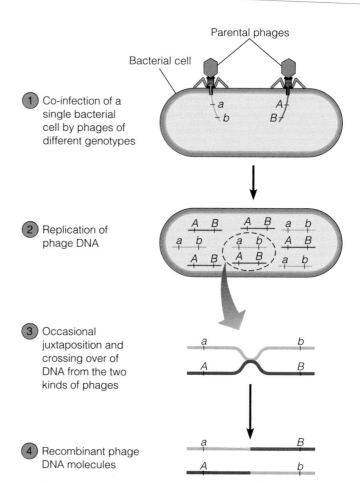

Figure 18-18   Genetic Recombination of Bacteriophages. ① Two parental phage populations with different genotypes are used to co-infect a bacterial cell, thereby ensuring the simultaneous presence of both kinds of phage genomes in a single bacterial cell. One of the parental phages carries the mutant alleles *a* and *b*, while the other has alleles *A* and *B*. ② As the phage DNA molecules replicate, ③ they occasionally become juxtaposed in a way that allows crossing over and recombination to occur, ④ giving rise to recombinant phage DNA molecules. The frequency of occurrence of recombinant genotypes provides a measure of the distance between genes. (As in other cases, biologists determine the genotypes of the progeny phage particles from the phenotypes they display. An example of a phage phenotype is the size of the plaques it forms on lawns of bacteria.)

trates an experiment in which bacterial cells are co-infected by two related types of T4 phage. As the phages replicate in the bacterial cell, their DNA molecules occasionally become juxtaposed in ways that allow DNA segments to be exchanged between homologous regions. The resulting recombinant phage arises at a frequency that depends on the distance between the genes under study, just as in diploid organisms. In general, the farther apart the genes, the greater the likelihood of recombination between them.

In phage recombination, crossing over involves relatively short, naked DNA molecules, rather than chromatids. The simplicity of this situation has facilitated research on the molecular mechanisms of recombination and the proteins that catalyze the process. It is clear that phage recombination involves the precise alignment of homologous DNA molecules at the region of crossing over—a requirement that presumably holds for the recombination of eukaryotic and prokaryotic DNA as well.

Transformation and Transduction Involve Recombination with Free DNA or DNA Brought into Bacterial Cells by Bacteriophages

In bacteria, several mechanisms exist for recombining genetic information. One such mechanism has already been mentioned in Chapter 16, where we discussed the experiments with rough and smooth strains of pneumo-coccal bacteria that led Oswald Avery to conclude that bacterial cells can be transformed from one genetic type to another by exposing them to purified DNA. This ability of a bacterial cell to take up DNA molecules and to incorporate some of that DNA into its own genome is called **transformation** (Figure 18-19a). Although initially described as a laboratory technique for artificially introducing DNA into bacterial cells, transformation is now recognized as a natural mechanism by which some (though by no means all) kinds of bacteria acquire genetic information when they have access to DNA from other cells.

A second mechanism for genetic recombination in bacteria, called **transduction,** involves DNA that has been brought into a bacterial cell by a bacteriophage. Most phages contain only their own DNA, but occasionally a phage will incorporate some bacterial host cell DNA sequences into its progeny particles. Such a phage particle can then infect another bacterium, acting like a syringe carrying DNA from one bacterial cell to the next (Figure 18-19b). Phages capable of carrying host cell DNA from one cell to another are called *transducing phages.*

The transducing phage known as P1, which infects *E. coli,* has been especially useful for gene mapping. The amount of DNA that will fit into a phage particle is small compared with the size of the bacterial genome. Two genes—or, more generally, genetic markers (specific base sequences)—must therefore be close together for both of them to be carried into a bacterial cell by a single phage

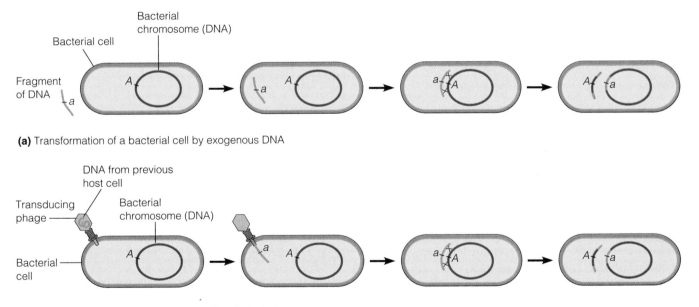

(a) Transformation of a bacterial cell by exogenous DNA

(b) Transduction of a bacterial cell by a transducing phage

Figure 18-19   Transformation and Transduction in Bacterial Cells.
(a) Transformation involves uptake by the bacterial cell of exogenous DNA, which occasionally becomes integrated into the bacterial genome by two crossover events (indicated by X). The exogenous DNA will be detectable in progeny cells only if integrated into the bacterial chromosome, because the fragment of DNA initially taken up does not normally have the capacity to replicate itself autonomously in the cell. (The main exception is an intact plasmid.) **(b)** Transduction involves the introduction of exogenous DNA into a bacterial cell by a phage. Once injected into the host cell, the DNA can become integrated into the bacterial genome in the same manner as in transformation. In both cases, linear fragments of DNA that end up outside the bacterial chromosome are eventually degraded by nucleases. In this figure, the letters A and a represent alleles of the same gene.

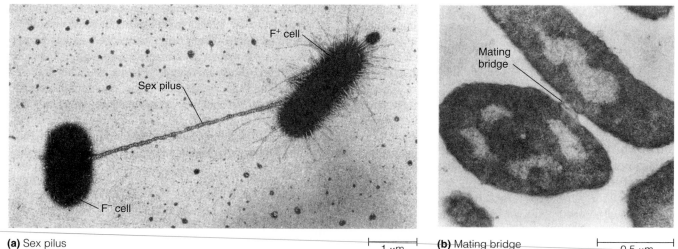

(a) Sex pilus
⊢————⊣
1 μm

(b) Mating bridge
⊢————⊣
0.5 μm

Figure 18-20   The Cellular Apparatus for Bacterial Conjugation.   (a) The donor bacterial cell on the right, an F⁺ cell, has numerous slender appendages, called pili, on its surface. Some of these pili are sex pili, including the very long pilus leading to the other cell, an F⁻ cell. Made of protein encoded by a gene on the F factor, sex pili enable a donor cell to attach to a recipient cell. **(b)** Subsequently, a cytoplasmic mating bridge forms, through which DNA is passed from the donor cell to the recipient cell (TEMs).

particle. This is the basis of *cotransductional mapping*, in which the proximity of one marker to another is determined by quantifying the frequency with which the markers accompany each other in a transducing phage particle. The closer two markers are, the more likely they are to be cotransduced into a bacterial cell. Studies using the transducing phage P1 have revealed that markers cannot be cotransduced if they are separated in the bacterial DNA by more than about 10^5 base pairs. This finding agrees with the observation that the P1 phage has a genome of about that size.

Conjugation Is a Modified Sexual Activity That Facilitates Genetic Recombination in Bacteria

In addition to transformation and transduction, some bacteria also transfer DNA from one cell to another by **conjugation.** As the name suggests, conjugation resembles a mating in that one bacterium is clearly identifiable as the donor (often called a "male") and another as the recipient ("female"). Although conjugation resembles a sexual process, this mode of DNA transfer is not an inherent part of the bacterial life cycle, and it usually involves only a portion of the genome; therefore, it does not qualify as true sexual reproduction. The existence of conjugation was postulated in 1946 by Joshua Lederberg and Edward L. Tatum, who were the first to show that genetic recombination occurs in bacteria. They also established that physical contact between two cells is necessary for conjugation to take place. We now understand that conjugation involves the directional transfer of DNA from the donor bacterium to the recipient bacterium. Let us look at some details of how the process works.

The F Factor.   The presence of a DNA sequence called the **F factor** (F for fertility) enables an *E. coli* cell to act as a donor during conjugation. The F factor can take the form of either an independent, replicating plasmid (p. 504) or a segment of DNA within the bacterial chromosome. Donor bacteria containing the F factor in its plasmid form are designated F⁺, whereas recipient cells, which usually lack the F factor completely, are designated F⁻. Donor cells develop long, hairlike projections called **sex pili** (singular: **pilus**) that emerge from the cell surface (Figure 18-20a). The end of each sex pilus contains molecules that selectively bind to the surface of recipient cells, thereby leading to the formation of a transient cytoplasmic **mating bridge** through which DNA is transferred from donor cell to recipient cell (Figure 18-20b).

When a donor cell contains an F factor in its plasmid form, a copy of the plasmid is quickly transferred to the recipient cell during conjugation, converting the recipient cell from F⁻ to F⁺ (Figure 18-21a). Transfer always begins at a point on the plasmid called its **origin of transfer,** represented in the figure by an arrowhead. During transfer of an F factor to an F⁻ cell, the donor cell does not lose its F⁺ status because the F factor is replicated in close association with the transfer process, allowing a copy of the F plasmid to remain behind in the donor cell. As a result, mixing an F⁺ population of bacteria with F⁻ cells will eventually lead to a population of cells that is entirely F⁺. "Maleness" is in a sense infectious, and the F factor is the infectious agent.

Hfr Cells and Bacterial Chromosome Transfer.   Thus far we have seen that donor and recipient cells are defined by the presence or absence of the F factor, which is in turn transmitted by conjugation. But how do recombinant bacteria arise by this means? The answer is that the F factor—while usually present as a plasmid—can sometimes become integrated into the bacterial chromosome, as shown in Figure 18-21b. (Integration results from crossing over between short DNA sequences in the chromosome and similar sequences in the F factor.) Chromosomal integra-

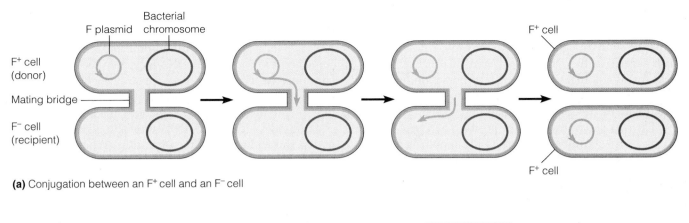

(a) Conjugation between an F⁺ cell and an F⁻ cell

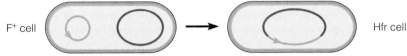

(b) Conversion of an F⁺ cell into an Hfr cell by integration of the F factor into the bacterial chromosome

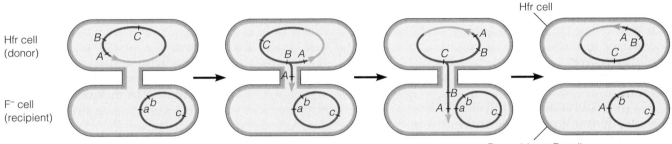

(c) Conjugation between an Hfr cell and an F⁻ cell

Figure 18-21   DNA Transfer by Bacterial Conjugation.   (a) Conjugation between an F⁺ donor bacterium and an F⁻ recipient involves the transfer of a copy of the F factor plasmid from the donor to the recipient, thereby converting the F⁻ cell to an F⁺ cell. Transfer of the F plasmid occurs through a mating bridge and always begins at the F factor's origin of transfer, indicated by the arrowhead. **(b)** Conversion of an F⁺ cell into an Hfr cell by integration of the F factor into the bacterial chromosome. **(c)** Conjugation between an Hfr donor cell and an F⁻ cell involves the transfer of a copy of the Hfr genome through a mating bridge into the F⁻ cell, beginning with the origin of transfer on the integrated F factor. Transfer is usually incomplete, because cells rarely remain in mating contact long enough for the entire bacterial chromosome to be transferred. Once inside the F⁻ cell, parts of the Hfr DNA can undergo recombination with the DNA of the F⁻ cell, just as in transformation or transduction (see Figure 18-19). Here, uppercase letters represent alleles carried by the Hfr; lowercase letters represent the corresponding alleles carried by the F⁻ cell. In the last step, allele A from the Hfr becomes recombined into the F⁻ cell's DNA in place of its a allele.

tion of the F factor converts an F⁺ donor cell into an **Hfr cell,** which is capable of producing a *h*igh *f*requency of *r*ecombination in further matings because it can now transfer *genomic DNA* during conjugation.

When an Hfr bacterium is mated to an F⁻ recipient, DNA is transferred into the recipient cell (Figure 18-21c). But instead of transferring just the F factor itself, the Hfr cell transfers at least part (and occasionally all) of its chromosomal DNA, retaining a copy, as in F⁺ DNA transfer. Transfer begins at the origin of transfer within the integrated F factor and proceeds in a direction dictated by the orientation of the F factor within the chromosome. Note that the chromosomal DNA is transferred in a linear form, with a small part of the F factor at the leading end and the remainder at the trailing end. Because the F factor is split in this way, only recipient cells that receive a complete bacterial chromosome from the Hfr donor actually become Hfr cells themselves. Transfer of the whole chromosome is extremely rare, however, because it takes about 90 minutes. Usually, mating contact is spontaneously disrupted before transfer is complete, leaving the recipient cell with only a portion of the Hfr chromosome, as shown in Figure 18-21c. As a result, genes located close to the origin of transfer on the Hfr chromosome are the most likely to be transmitted to the recipient cell.

Once a portion of the Hfr chromosome has been introduced into a recipient cell by conjugation, it can recombine with regions of the recipient cell's chromosomal DNA that are homologous (similar) in sequence. The recombinant bacterial chromosomes generated by this process contain some genetic information derived from the donor cell and some from the recipient. Only donor DNA sequences that are successfully integrated by this recombination mechanism survive in the recipient cell and its progeny. Donor DNA that is not integrated during recombination, as well as DNA removed from the recipient chromosome during the recombination process, is eventually degraded by nucleases.

The correlation between the position of a gene within the bacterial chromosome and its likelihood of transfer can be used to map genes with respect to the origin of transfer and therefore with respect to one another. For example, if gene A of an Hfr cell is transferred in conjugation 95% of the time, gene B 70% of the time, and gene C 55% of the time, then the sequence of genes is A-B-C, with gene A closest to the origin of transfer. Moreover, since the daughter cells of the recipient bacterium are recombinants, they can be used for genetic analysis. Typically, a cross is made between Hfr and F⁻ strains that differ in two or more genetic properties. After conjugation has taken place, the cells are plated on a nutrient medium on which recombinants can grow but "parent" strains cannot, thereby allowing the recombinants to be detected and recombinant frequencies to be calculated.

Molecular Mechanism of Homologous Recombination

We have now described five different situations in which genetic information can be exchanged between homologous DNA molecules: (1) prophase I of meiosis associated with gametogenesis in eukaryotes, (2) co-infection of bacteria with related bacteriophages, (3) transformation of bacteria with DNA, (4) transduction of bacteria by transducing phages, and (5) bacterial conjugation. In spite of their obvious differences, all five situations share a fundamental feature in common: Each involves **homologous recombination** in which genetic information is exchanged between DNA molecules exhibiting extensive sequence similarity. We are now ready to discuss the molecular mechanisms that underlie this type of recombination. Since the principles involved appear to be quite similar in prokaryotes and eukaryotes, we will use examples from both types of organisms.

DNA Breakage-and-Exchange Underlies Homologous Recombination

Shortly after it was first discovered that genetic information is exchanged between chromosomes during meiosis, two theories were proposed to explain how this might occur. The *breakage-and-exchange model* postulated that breaks occur in the DNA molecules of two adjoining chromosomes, followed by exchange and rejoining of the broken segments. In contrast, the *copy-choice model* proposed that genetic recombination occurs while DNA is being replicated. According to the latter view, DNA replication begins by copying a DNA molecule located in one chromosome and then switches at some point to copying the DNA located in the homologous chromosome. The net result would be a new DNA molecule containing information derived from both chromosomes. One of the more obvious predictions made by the copy-choice model is that DNA replication and genetic recombination should happen at the same time. When subsequent studies revealed that DNA replication takes place during S phase while recombination typically occurs during prophase I, the copy-choice idea had to be rejected as a general model of meiotic recombination.

The first experimental evidence providing support for the breakage-and-exchange model was obtained in 1961 by Matthew Meselson and Jean Weigle, who employed phages of the same genetic type labeled with either the heavy (^{15}N) or light (^{14}N) isotope of nitrogen. Simultaneous infection of bacterial cells with these two labeled strains of the same phage resulted in the production of recombinant phage particles containing genes derived from both phages. When the DNA from these recombinant phages was examined, it was found to contain a mixture of ^{15}N and ^{14}N (Figure 18-22). Since these experiments were performed under conditions that prevented any new DNA from being synthesized, the recombinant DNA molecules must have been produced by breaking and rejoining DNA molecules derived from the two original phages.

Subsequent experiments involving bacteria whose chromosomes had been labeled with either ^{15}N or ^{14}N revealed that DNA containing a mixture of both isotopes is also produced during genetic recombination between bacterial chromosomes. Moreover, when such recombinant DNA molecules are heated to dissociate them into single strands, a mixture of ^{15}N and ^{14}N is detected in each DNA strand; hence the DNA double helix must be broken and rejoined during recombination.

A similar conclusion emerged from experiments performed shortly thereafter on eukaryotic cells by J. Herbert Taylor. In these studies, cells were briefly exposed to ^3H-thymidine during the S phase preceding the last mitosis prior to meiosis, producing chromatids containing one radioactive DNA strand per double helix. During the following S phase, DNA replication in the absence of ^3H-thymidine generated chromosomes containing one labeled chromatid and one unlabeled chromatid (Figure 18-23). But during the subsequent meiosis, individual chromatids exhibited a mixture of radioactive and nonradioactive segments, as would be predicted by the breakage-and-exchange model. Moreover, the frequency of such exchanges was directly proportional to the frequency with which the genes located in these regions underwent genetic recombination. Such observations provided strong support for the notion that genetic recombination in eukaryotic cells, as in prokaryotes, involves DNA breakage and exchange. These experiments also showed that most DNA exchanges arise

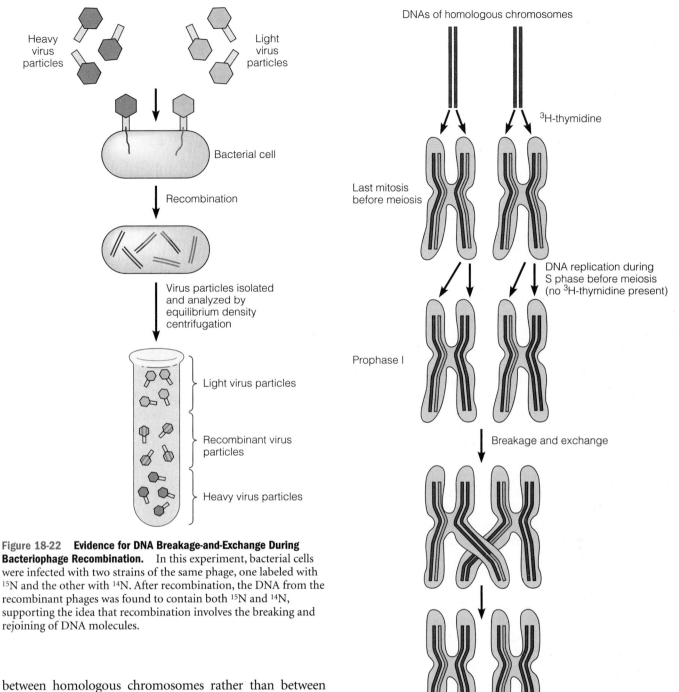

Figure 18-22   Evidence for DNA Breakage-and-Exchange During Bacteriophage Recombination.   In this experiment, bacterial cells were infected with two strains of the same phage, one labeled with ^{15}N and the other with ^{14}N. After recombination, the DNA from the recombinant phages was found to contain both ^{15}N and ^{14}N, supporting the idea that recombination involves the breaking and rejoining of DNA molecules.

between homologous chromosomes rather than between the two sister chromatids of a given chromosome. This selectivity is important because it ensures that genes are exchanged between paternal and maternal chromosomes.

Homologous Recombination Can Lead to Gene Conversion

The conclusion that homologous recombination is based on DNA breakage-and-exchange does not in itself provide much information concerning the underlying molecular mechanisms. One of the simplest breakage-and-exchange models that might be envisioned would involve the cleavage of two homologous, double-stranded DNA molecules at

Figure 18-23   Experimental Demonstration of Breakage-and-Exchange During Eukaryotic Recombination.   In this experiment, DNA was radioactively labeled by briefly exposing eukaryotic cells to ^{3}H-thymidine before the last mitosis prior to meiosis. When autoradiography was employed to examine the chromatids during meiosis, some were found to contain a mixture of labeled (orange) and unlabeled (dark blue) segments as predicted by the breakage-and-exchange model.

comparable locations, followed by exchange and rejoining of the cut ends. This model implies that genetic recombination should be completely reciprocal; that is, any genes exchanged from one chromosome should appear in the other chromosome, and vice versa. For example, consider a hypothetical situation involving two genes designated *P* and *Q*. If one chromosome contains forms of these genes called *P1* and *Q1,* and the other chromosome has alternative forms designated *P2* and *Q2,* reciprocal exchange would be expected to generate one chromosome with genes *P1* and *Q2,* and a second chromosome with genes *P2* and *Q1.*

Although this reciprocal pattern is usually observed, in some situations recombination has been found to be nonreciprocal. For example, recombination might generate one chromosome with genes *P1* and *Q2,* and a second chromosome with genes *P2* and *Q2.* In this particular example, the *Q1* gene expected on the second chromosome appears to have been converted to a *Q2* gene. For this reason, nonreciprocal recombination is often referred to as **gene conversion.** Gene conversion is most commonly observed when the recombining genes are located very close to one another. Because recombination between closely spaced genes is a rare event, gene conversion is most readily detectable in organisms that reproduce rapidly and generate large numbers of offspring, such as yeast and the common bread mold, *Neurospora.*

Neurospora is an especially convenient organism in which to study gene conversion, because its meiotic cells are enclosed in a small sac, called an **ascus,** which prevents the cells from moving around and thus allows the lineage of each cell to be easily followed (Figure 18-24). Initially each ascus contains a single diploid cell. Meiotic division of this cell produces four haploid cells that subsequently divide by mitosis, yielding a final total of eight cells. Since the final division is mitotic, it should produce two identical progeny cells for each cell that divides. Yet in a significant number of cases this mitosis produces two cells that are genetically different. Such unexpected results are most often observed with genes that are close to a site of genetic recombination.

How can mitosis produce two cells that are genetically different? The most straightforward explanation is that a chromosome contains one or more genes in which the two strands of the DNA double helix are not entirely complementary. When the two DNA strands in such a noncomplementary region split apart and serve as templates for DNA replication, the two newly forming DNA molecules will have slightly different base sequences in this region and thus will represent slightly different genes.

Homologous Recombination Is Initiated by Single-Strand DNA Exchanges (Holliday Junctions)

The preceding observations suggest that homologous recombination is more complicated than can be explained by a simple breakage-and-exchange model in which crossing over is accomplished by cleaving two double-stranded DNA molecules and then exchanging and rejoining the cut ends. Robin Holliday was the first to propose the alterna-

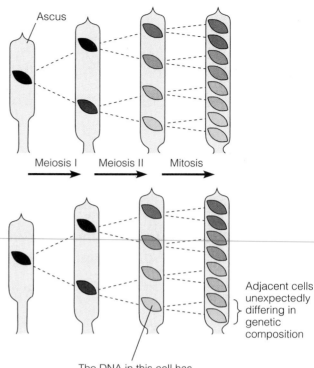

Figure 18-24   Meiosis in *Neurospora.* In the bread mold *Neurospora,* cells undergoing meiotic division are contained within a sac called an ascus. The ascus keeps the individual cells lined up in a row, making it easy to trace each cell's lineage. The separation of homologous chromosomes during meiosis I generates two cells exhibiting different genetic traits. Cells produced during meiosis II may also differ from each other genetically because of the crossing over that occurred during meiosis I. The third division is a simple mitosis, and hence is expected to produce a pair of genetically identical cells for each cell that divides (top). Occasionally, however, this terminal mitosis generates a pair of nonidentical cells (bottom).

tive idea that recombination is based on the exchange of single DNA strands between two double-stranded DNA molecules. A subsequent adaptation of Holliday's single-strand exchange model, developed by Matthew Meselson and Charles Radding, is illustrated in Figure 18-25. According to this model, the initial step (①) in recombination is the cleavage of one (or both) strands of the DNA double helix. In either case, a single broken DNA strand derived from one DNA molecule "invades" a complementary region of a homologous DNA double helix, displacing one of the two strands (② in Figure 18-25). The process of DNA repair (③) then generates a crossed structure, called a **Holliday junction,** in which a single strand from each DNA double helix crosses over and joins the opposite double helix (④). Electron microscopy has provided direct support for the existence of Holliday junctions, revealing the presence of DNA double helices joined by single-strand crossovers at sites of genetic recombination.

Once a Holliday junction has been formed, unwinding and rewinding of the DNA double helices causes the crossover point to move back and forth along the chromosomal DNA (⑤). This phenomenon, called *branch*

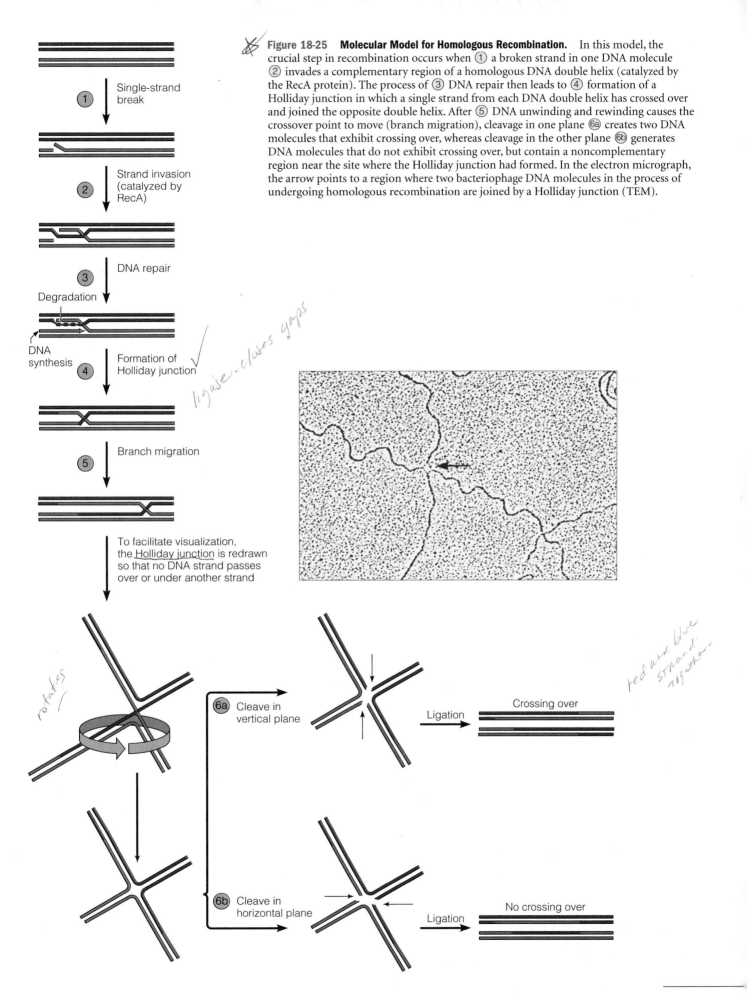

Figure 18-25 Molecular Model for Homologous Recombination. In this model, the crucial step in recombination occurs when ① a broken strand in one DNA molecule ② invades a complementary region of a homologous DNA double helix (catalyzed by the RecA protein). The process of ③ DNA repair then leads to ④ formation of a Holliday junction in which a single strand from each DNA double helix has crossed over and joined the opposite double helix. After ⑤ DNA unwinding and rewinding causes the crossover point to move (branch migration), cleavage in one plane ⑥ⓐ creates two DNA molecules that exhibit crossing over, whereas cleavage in the other plane ⑥ⓑ generates DNA molecules that do not exhibit crossing over, but contain a noncomplementary region near the site where the Holliday junction had formed. In the electron micrograph, the arrow points to a region where two bacteriophage DNA molecules in the process of undergoing homologous recombination are joined by a Holliday junction (TEM).

① Single-strand break

② Strand invasion (catalyzed by RecA)

③ DNA repair

Degradation

DNA synthesis

④ Formation of Holliday junction

⑤ Branch migration

To facilitate visualization, the Holliday junction is redrawn so that no DNA strand passes over or under another strand

⑥ⓐ Cleave in vertical plane

Ligation

Crossing over

⑥ⓑ Cleave in horizontal plane

Ligation

No crossing over

migration, can rapidly increase the length of single-stranded DNA that is exchanged between two DNA molecules. After branch migration has occurred, the Holliday junction is cleaved and the broken DNA strands are joined back together to produce two separate DNA molecules. There are two ways in which a Holliday junction can be cleaved and rejoined. If it is cleaved in one plane, the two DNA molecules that are produced will exhibit crossing over; that is, the chromosomal DNA beyond the point where recombination occurred will have been completely exchanged between the two chromosomes (⑥ⓐ). If the Holliday junction is cut in the other plane, crossing over does not occur but the DNA molecules exhibit a noncomplementary region near the site where the Holliday junction had formed (⑥ⓑ).

What is the fate of such noncomplementary regions? If they remain intact, an ensuing mitotic division will separate the mismatched DNA strands and each will serve as a template for the synthesis of a new complementary strand. The net result will be two new DNA molecules with differing base sequences, and hence two cells containing slightly different gene sequences in the affected region. This is the situation occasionally observed in *Neurospora,* where two genetically different cells can arise during the mitosis following meiosis (see Figure 18-24). Alternatively, a noncomplementary DNA region may be corrected by excision and repair. The net effect of DNA repair would be to convert genes from one form to another—in other words, gene conversion.

A key enzyme involved in homologous recombination has been identified using bacterial extracts that catalyze the formation of Holliday junctions. Such extracts contain a protein, called **RecA,** whose presence is required for recombination. Studies of the purified RecA protein have revealed that it catalyzes a "strand invasion" reaction in which a single-stranded DNA segment displaces one of the two strands of a DNA double helix; in other words, it catalyzes step ② in Figure 18-25. In catalyzing this reaction, the RecA protein first coats the single-stranded DNA region. The coated, single-stranded DNA then interacts with a DNA double helix, moving along the target DNA until it reaches a complementary sequence. Here RecA denatures the target DNA molecule and promotes base pairing between the single-stranded DNA and its complementary strand. Mutant bacteria that produce a defective RecA protein cannot carry out genetic recombination, nor are extracts prepared from such cells capable of creating Holliday junctions from homologous DNA molecules.

The Synaptonemal Complex Facilitates Homologous Recombination During Meiosis

Earlier in the chapter we learned that during prophase I of meiosis, homologous chromosomes are joined together by a zipperlike, protein-containing structure called the synaptonemal complex. A number of observations suggest that this structure plays an important role in genetic recombination. First, the synaptonemal complex appears at the time when recombination takes place. Second, its location between the opposed homologous chromosomes corresponds to the region where crossing over occurs. And finally, synaptonemal complexes are absent in organisms that fail to carry out meiotic recombination, such as male fruit flies.

Presumably, the synaptonemal complex facilitates recombination by maintaining a close pairing between adjacent homologous chromosomes along their entire length. If the synaptonemal complex functions to facilitate recombination, how do cells ensure that such structures form only between homologous chromosomes? Recent observations suggest the existence of a process called *homology searching,* in which a single-strand break in one DNA molecule produces a free strand that "invades" another DNA double helix and checks for the presence of complementary sequences (see ② in Figure 18-25). If extensive homology is not found, the free DNA strand invades another DNA molecule and checks for complementarity, repeating the process until a homologous DNA molecule is detected. Only then does a synaptonemal complex develop, bringing the homologous chromosomes together throughout their length to facilitate the recombination process.

Recombinant DNA Technology and Gene Cloning

In nature, genetic recombination usually takes place between two DNA molecules derived from organisms of the same species. In animals and plants, for example, an individual's two parents are the original sources of the DNA that recombines during meiosis. A naturally arising recombinant DNA molecule usually differs from the parental DNA molecules only in the combination of alleles it contains; the fundamental identities and sequences of its genes remain the same.

In the laboratory, such limitations do not exist. Since the development of **recombinant DNA technology** in the 1970s, scientists have had at their disposal a collection of techniques for making recombinant DNA in the laboratory. Any segment of DNA can now be excised from any genome and spliced together with any other piece of DNA. Initially derived from basic research on the molecular biology of bacteria, these techniques have enabled researchers to isolate and study genes from both prokaryotes and eukaryotes with greater ease and precision than was earlier thought possible.

A central feature of recombinant DNA technology is the ability to replicate—or *clone*—specific pieces of DNA in order to prepare large enough quantities for research and other uses. Cloning is accomplished by splicing the DNA of interest to the DNA of a genetic element, called a **cloning vector,** that can replicate autonomously when introduced into a cell grown in culture—in most cases, a bacterium such as *E. coli.* The cloning vector can be a plasmid or the DNA of a virus, usually a bacteriophage; in either case, the vector's DNA "passenger" is copied every time it replicates.

In this way, it is possible to generate large quantities of specific genes or other DNA segments—and of their protein products as well, if the passenger genes are transcribed and translated in proliferating cells that carry the vector.

To appreciate the importance of recombinant DNA technology, we need to grasp the magnitude of the problem that biologists faced as they tried to study the genomes of eukaryotic organisms. Much of our early understanding of information flow in cells came from studies with bacteria and viruses, whose genomes were mapped and analyzed in great detail using genetic methods that were not easily applied to eukaryotes. Until a few decades ago, investigators despaired of ever being able to understand and manipulate eukaryotic genomes to the same extent because the typical eukaryote has at least 10,000 times as much DNA as the best-studied phages—truly an awesome haystack in which to find a gene-sized needle. But the advent of recombinant DNA technology has made it possible to isolate individual eukaryotic genes in quantities large enough to permit them to be thoroughly studied, ushering in a new era in biology.

The Discovery of Restriction Enzymes Paved the Way for Recombinant DNA Technology

Much of what we call recombinant DNA technology was made possible by the discovery of *restriction enzymes* (see Box 16B). The ability of restriction enzymes to cleave DNA molecules at specific sequences called *restriction sites* makes them powerful tools for cutting large DNA molecules into smaller fragments that can be recombined in various ways. Restriction enzymes that make staggered cuts in DNA are especially useful because they generate single-stranded *sticky ends* (also called *cohesive ends*) that provide a simple means for joining DNA fragments obtained from different sources. In essence, any two DNA fragments generated by the same restriction enzyme can be joined together by complementary base pairing between their single-stranded, sticky ends.

Figure 18-26 illustrates how this general approach works. DNA molecules from two sources are first treated with a restriction enzyme known to generate fragments with sticky ends (①), and the fragments are then mixed together under conditions that favor base pairing between these sticky ends (②). Once joined in this way, the DNA fragments are covalently sealed together by DNA ligase (③), an enzyme normally involved in DNA replication and repair (see Chapter 17). The final product is a **recombinant DNA molecule** containing DNA sequences derived from two different sources.

The combined use of restriction enzymes and DNA ligase allows any two (or more) pieces of DNA to be spliced together, regardless of their origins. A piece of human DNA, for example, can be joined to bacterial or phage DNA just as easily as it can be linked to another piece of human DNA. In other words, it is possible to form recombinant DNA

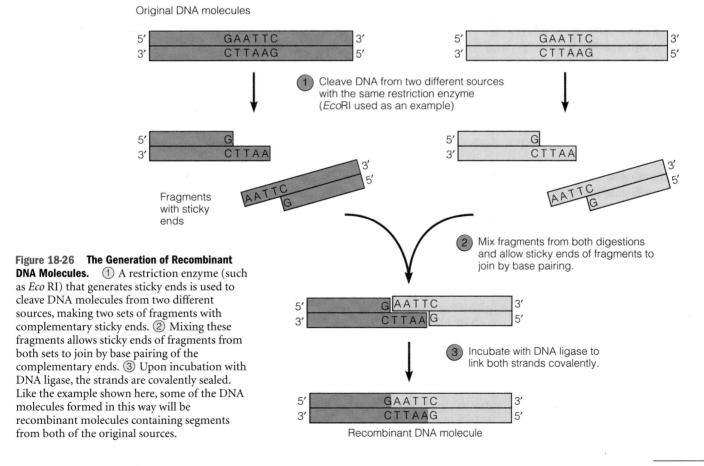

Figure 18-26 The Generation of Recombinant DNA Molecules. ① A restriction enzyme (such as *Eco* RI) that generates sticky ends is used to cleave DNA molecules from two different sources, making two sets of fragments with complementary sticky ends. ② Mixing these fragments allows sticky ends of fragments from both sets to join by base pairing of the complementary ends. ③ Upon incubation with DNA ligase, the strands are covalently sealed. Like the example shown here, some of the DNA molecules formed in this way will be recombinant molecules containing segments from both of the original sources.

molecules that never existed in nature, without any regard for the natural barriers that otherwise limit genetic recombination to genomes of the same or closely related species. Therein lies the power of (and, for some, the concern about) recombinant DNA technology.

DNA Cloning Techniques Permit Individual Gene Sequences to Be Produced in Large Quantities

The power of restriction enzymes is the ease with which they allow a desired segment of DNA, usually a segment containing a specific gene, to be inserted into a cloning vector that can replicate itself when introduced into bacterial cells. Suppose, for example, you wanted to isolate a gene that codes for a medically useful product, such as insulin needed for the treatment of diabetics or blood clotting factors needed by patients with hemophilia. By using restriction enzymes to insert DNA containing such genes into a cloning vector in bacterial cells and then identifying bacteria that contain the DNA of interest, it is possible to grow large masses of such cells and thereby obtain large quantities of the desired DNA (and its protein product). This process of generating many copies of specific DNA fragments is called **DNA cloning.** (In biology, a *clone* is a population of organisms that is derived from a single ancestor and hence is genetically homogeneous, and a *cell clone* is a population of cells derived from the division of a single cell. By analogy, a *DNA clone* is a population of DNA molecules that are derived from the replication of a single molecule and hence are identical to one another.)

Although the specific details of DNA cloning procedures vary, the following five steps are typically involved: (1) insertion of DNA into a cloning vector; (2) introduction of the recombinant vector into cultured cells, usually bacteria; (3) amplification of the recombinant vector in the bacteria; (4) selection of cells containing recombinant DNA; and (5) identification of clones containing the DNA of interest. Figure 18-27 provides an overview of these events using a bacterial cloning vector. You should refer to this figure as we consider each step in turn.

1. Insertion of DNA into a Cloning Vector.
The first step in cloning a desired piece of DNA is its insertion into an appropriate cloning vector, usually a bacteriophage or a plasmid. Most vectors used for DNA cloning are themselves recombinant DNA molecules, designed specifically for this purpose. For example, when bacteriophage λ DNA is used as a cloning vector, the phage DNA has had some of its nonessential genes removed to make room in the phage head for spliced-in DNA. Plasmids used as cloning vectors usually have a variety of restriction sites and often carry genes that confer antibiotic resistance on their host cells. The antibiotic-resistance genes facilitate the selection stage (④), while the presence of multiple kinds of restriction sites allows the plasmid to incorporate DNA fragments prepared with a variety of different restriction enzymes.

An example of a commonly used plasmid vector is *pUC19* ("puck-19"), shown schematically in Figure 18-28a.

Because this plasmid carries a gene that confers resistance to the antibiotic ampicillin (*amp^R*), bacteria containing the plasmid can be easily identified by their ability to grow in the presence of ampicillin. The pUC19 plasmid also has 11 different restriction sites clustered in a region of the plasmid containing the *lacZ* gene, which codes for the enzyme β-galactosidase. Integration of foreign DNA at any of these restriction sites will disrupt the *lacZ* gene, thereby blocking the production of β-galactosidase. As we will see shortly, this disruption in β-galactosidase production can be used later in the cloning process to detect the presence of plasmids containing foreign DNA.

Figure 18-28b illustrates how a specific gene of interest residing in a foreign DNA source is inserted into a plasmid cloning vector, using pUC19 as the vector and a restriction enzyme that cleaves pUC19 at a single site within the *lacZ* gene. Incubation with the restriction enzyme cuts the plasmid at that site (①), making the DNA linear (opening the circle). The same restriction enzyme is used to cleave the DNA molecule containing the gene to be cloned (②). The sticky-ended fragments of foreign DNA are then incubated with the linearized vector molecules under conditions that favor base pairing (③), followed by treatment with DNA ligase to link the molecules covalently (④). To keep the diagram simple, Figure 18-28b shows only the recombinant plasmid containing the desired fragment of foreign DNA. In practice, however, a variety of DNA products will be present, including nonrecombinant plasmids and recombinant plasmids containing other fragments generated by the action of the restriction enzyme.

2. Introduction of the Recombinant Vector into Bacterial Cells.
Once foreign DNA has been inserted into a cloning vector, the resulting recombinant vector is replicated by introducing it into an appropriate host cell, usually the bacterium *E. coli*. Cloning vectors are introduced into bacteria in one of two ways. If the cloning vector is a phage, it is allowed to infect an appropriate cell population. Plasmids, on the other hand, are simply introduced into the medium surrounding the target cells. Both prokaryotic and eukaryotic cells will take up plasmid DNA from the external medium, although special treatments are usually necessary to enhance the efficiency of the process. The addition of calcium ions, for example, markedly increases the rate at which cells take up DNA from the external environment.

3. Amplification of the Recombinant Vector in Bacteria.
After they have taken up the recombinant cloning vector, the host bacteria are plated out on a nutrient medium so that the recombinant DNA vector can be replicated, or *amplified*. In the case of a plasmid vector, the bacteria proliferate and form colonies, each derived from a single cell. Under favorable conditions *E. coli* will divide every 22 minutes, giving rise to a billion cells in less than 11 hours. As the bacteria multiply, the recombinant plasmids also replicate, producing an enormous number of vector

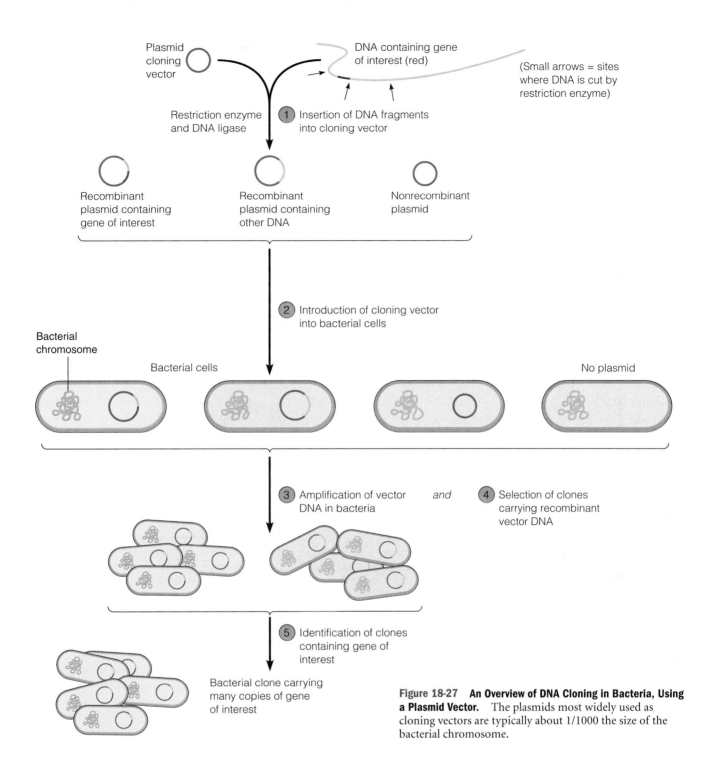

Figure 18-27 An Overview of DNA Cloning in Bacteria, Using a Plasmid Vector. The plasmids most widely used as cloning vectors are typically about 1/1000 the size of the bacterial chromosome.

molecules containing foreign DNA fragments. Under such conditions, a single recombinant plasmid introduced into one cell will be amplified several hundred billionfold in less than half a day.

In the case of phage vectors such as phage λ, a slightly different procedure is used. Phage particles containing recombinant DNA are mixed with bacterial cells and the mixture is then placed on a culture medium under conditions that produce a continuous "lawn" of bacteria across the plate. Each time a phage particle infects a cell, the phage is replicated and eventually causes the cell to rup-

ture and die. The released phage particles can then infect neighboring cells, repeating the process again. This cycle eventually produces a clear zone of dead bacteria called a **plaque**, which contains large numbers of replicated phage particles derived by replication from a single type of recombinant phage (see Figure 16A-3 on p. 485). The millions of phage particles in each plaque contain identical molecules of recombinant phage DNA.

4. Selection of Cells Containing Recombinant DNA. The next step in the cloning process is to select those cells that

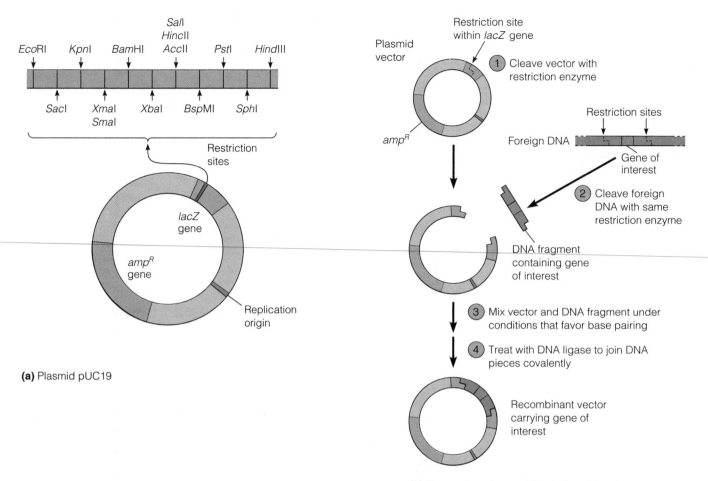

(a) Plasmid pUC19

(b) Preparation of recombinant plasmid vector

Figure 18-28   Cloning a Gene in the Plasmid Vector pUC19. **(a)** pUC19 is an *E. coli* plasmid of 2686 base pairs that contains a replication origin, an ampicillin resistance gene (*amp^R*), and a *lacZ* gene coding for β-galactosidase. Eleven different restriction sites are clustered within the *lacZ* gene. **(b)** Insertion of foreign DNA into the plasmid. ① The plasmid is cleaved with a restriction enzyme known to recognize a single site, in this case one that is within the *lacZ* gene.

② The same enzyme is used to cleave a foreign DNA molecule containing a gene of interest, thereby generating fragments that have the same sticky ends as the linearized plasmid DNA. ③ The fragments of foreign DNA are then incubated with the linearized plasmid DNA under conditions that favor base pairing between sticky ends. Among the expected products will be plasmid molecules recircularized by base pairing with a single fragment of foreign DNA, and

some of these will contain the gene of interest. ④ Incubation with DNA ligase covalently seals such a recombinant molecule, generating a recombinant plasmid vector carrying the gene of interest. Cells carrying such plasmids will be resistant to ampicillin and will fail to produce β-galactosidase because of the foreign DNA inserted within the *lacZ* gene.

have successfully taken up the cloning vector. For plasmid vectors such as pUC19, the method of selection depends on the plasmid's antibiotic-resistance genes. For example, bacteria carrying the recombinant plasmids generated in Figure 18-28b will all be resistant to the antibiotic ampicillin, since all plasmids have an intact ampicillin-resistance gene. The *amp^R* gene is a **selectable marker,** which allows only the cells carrying plasmids to grow on culture medium containing ampicillin (the medium "selects for" the growth of the ampicillin-resistant cells).

However, not all the ampicillin-resistant bacteria will carry *recombinant* plasmids—that is, plasmids containing spliced-in DNA. But those bacteria that do contain recombinant plasmids can be readily identified because the *lacZ* gene has been disrupted by the foreign DNA and,

as a result, β-galactosidase will no longer be produced. The lack of β-galactosidase can be detected by a simple color test in which bacteria are exposed to a substrate that is normally cleaved by β-galactosidase into a blue-staining compound. Bacterial colonies containing the normal pUC19 plasmid will therefore stain blue, whereas colonies containing recombinant plasmids with inserted DNA fragments will appear colorless.

A different approach is used with phage cloning vectors, which are usually derived from phage λ DNA molecules that are only about 70% as long as normal phage DNA. As a result, these DNA molecules are too small to be packaged into functional phage particles. But if an additional fragment of DNA is inserted into the middle of such a cloning vector, it creates a larger DNA molecule

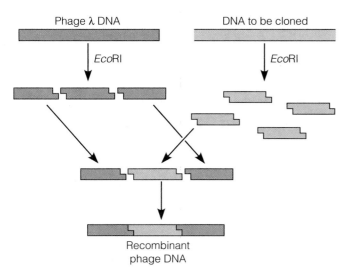

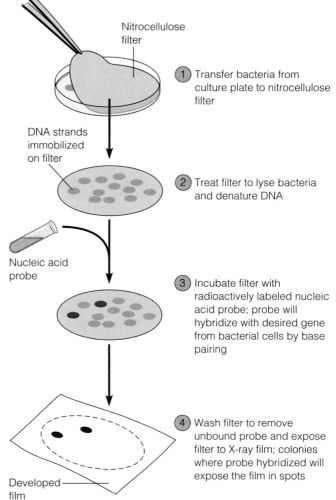

Figure 18-29 **Bacteriophage λ as a Cloning Vector.** The middle segment of the phage λ DNA molecule is removed by *Eco*RI cleavage and then replaced by the DNA fragment to be cloned. The inserted DNA fragment is necessary to make the phage λ DNA molecule large enough to be packaged into a functional phage particle.

that is again capable of being assembled into a functional phage (Figure 18-29). Hence when phage cloning vectors are employed, the only phage particles that can successfully infect bacterial cells are those that contain an inserted foreign DNA sequence.

5. Identification of Clones Containing the DNA of Interest.
The preceding steps typically generate vast numbers of bacteria producing many different kinds of recombinant DNA, only one or a few of which are relevant to the desired application. In Figure 18-27, for example, the bacterial colonies present on the Petri dishes at the end of step ④ are likely to contain at least as many different kinds of fragments as there are restriction sites in the DNA used in step ①. The final stage in any recombinant DNA procedure is therefore screening the bacterial colonies (or phage plaques) to identify those that contain the specific DNA fragment of interest. This is frequently the most difficult step in DNA cloning.

A number of techniques for screening colonies of bacteria exist. The particular technique used depends on what the researcher knows about the gene being cloned. If something is known about the base sequence of the gene of interest, the researcher can employ a **nucleic acid probe,** a radioactively labeled, single-stranded molecule of DNA or RNA that can identify a desired DNA sequence by base-pairing with it. (In Box 16C, we saw such a probe used to identify restriction fragment bands in Southern blotting.) The researcher prepares a labeled DNA or RNA probe containing all or part of the nucleotide sequence of interest and uses it to tag the colonies that contain complementary DNA. Figure 18-30 outlines how this *colony hybridization technique* can be used to screen for colonies carrying the desired DNA. Once the appropriate colonies

Figure 18-30 **The Colony Hybridization Technique.** This technique is used to screen bacterial colonies for the presence of DNA that is complementary in sequence to a nucleic acid probe. ① Bacterial colonies are transferred from the surface of an agar culture plate onto a nitrocellulose filter. ② The filter is treated with detergent to lyse the bacteria and alakali (NaOH) to denature their DNA. ③ The filters are then incubated with molecules of the nucleic acid probe—radioactively labeled, single-stranded DNA or RNA— which attach by base pairing to any complementary DNA present on the filter. ④ The filter is rinsed and subjected to autoradiography, which will make visible only those colonies containing DNA that is complementary in sequence to the probe. The base pairing between strands of nucleic acid from two sources, such as the probe nucleic acid and the target DNA here, is called *nucleic acid hybridization.*

have been identified, cloned DNA is recovered from these colonies by isolating the vector DNA from the bacterial cells and digesting it with the same restriction enzyme used initially.

Another approach focuses on the protein encoded by a gene of interest. If this protein is known and has been purified, labeled antibodies against it can be prepared and used as probes to check bacterial colonies for the presence of the protein. Alternatively, the *function* of the protein can be assayed in some way; for example, an enzyme could be tested for its catalytic activity. Protein-screening methods obviously depend on the ability of the bacterial cells to produce a foreign protein encoded by a cloned gene and will fail to detect cloned genes that are not expressed in the host cell. However, special *expression vectors* can be used to increase the likelihood that bacteria will transcribe eukaryotic genes properly and in large amounts. Expression vectors contain special DNA sequences that signal the bacterial cell to perform these processes.

Genomic and cDNA Libraries Are Both Useful for DNA Cloning

Cloning foreign DNA in bacterial cells is now a routine procedure. In practice, obtaining a good source of DNA to serve as starting material is often one of the most difficult steps. Two different approaches are commonly used for obtaining DNA starting material. In the "shotgun" approach, an organism's entire genome (or some substantial portion thereof) is cleaved into a large number of restriction fragments, which are then inserted into cloning vectors for introduction into bacterial cells (or phage particles). The resulting group of clones is called a **genomic library** because it contains cloned fragments representing most, if not all, of the genome. Genomic libraries of eukaryotic DNA are valuable resources from which specific genes can be isolated provided that a sufficiently sensitive identification technique is available. Once a rare bacterial colony that contains the desired DNA fragment has been identified, it can be grown on a nutrient medium to generate as many copies of the fragment as may be needed. Of course, the DNA cuts made by a restriction enzyme do not respect gene boundaries, and some genes may be divided among two or more restriction fragments. This problem can be circumvented by carrying out a *partial DNA digestion* in which a small quantity of restriction enzyme is used for a brief period of time. Under such conditions, some restriction sites remain uncut, increasing the probability that at least one intact copy of each gene will be present in the genomic library.

The alternative DNA source for cloning experiments is DNA that has been generated by copying messenger RNA (mRNA) with the enzyme *reverse transcriptase* (p. 626). This reaction generates a population of **complementary DNA (cDNA)** molecules that are complementary in sequence to the mRNA employed as template (Figure 18-31). If the entire mRNA population of a cell is isolated and copied into cDNA for cloning, the resulting group of clones is called a **cDNA library.** The advantage of a cDNA library is that it contains only those DNA sequences that are transcribed into RNA—presumably, the active genes in the cells or tissue from which the mRNA was prepared.

In addition to being limited to transcribed genes, a cDNA library has another important advantage as a starting point for the cloning of eukaryotic genes. Using mRNA to make cDNA guarantees that the cloned genes will contain only gene-coding sequences, without the noncoding interruptions called *introns* that are common

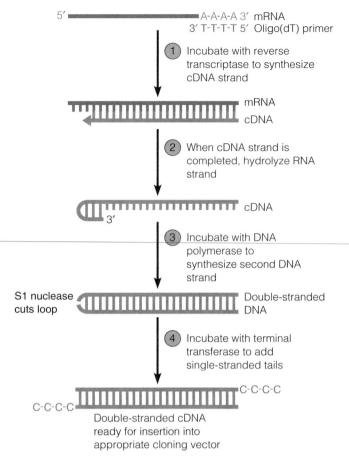

Figure 18-31   Preparation of Complementary DNA (cDNA) for Cloning. ① Messenger RNA is incubated with reverse transcriptase, which uses the mRNA as a template for synthesis of a complementary DNA (cDNA) strand. Oligo(dT), a short chain of thymine deoxynucleotides, can be used as a primer, because eukaryotic mRNA always has a stretch of adenine nucleotides at its 3′ end. ② The resulting mRNA-cDNA hybrid is treated with alkali or an enzyme to hydrolyze the RNA, leaving the single-stranded cDNA. ③ DNA polymerase can now synthesize the complementary DNA strand. The looped-around 3′ end of the first DNA strand can often be used as a primer. An enzyme called S1 nuclease is then used to cleave the loop. ④ For efficient insertion in a cloning vector, the double-stranded DNA must have single-stranded tails that are complementary to those of the vector. These can be added by incubation with terminal transferase, an enzyme that adds nucleotides one at a time to the ends of the molecule. If short stretches of cytosine (C) nucleotides, for example, are added to the cDNA and short stretches of guanine (G) nucleotides are added in the same way to a linearized cloning vector, recombinant molecules can be generated by allowing the single-stranded C tails in the cDNA to hybridize to the single-stranded G tails in the vector. (As an alternative to step 4, short synthetic "linker" molecules containing a variety of restriction sites can be ligated to the ends of both the cDNA and a blunt-ended cloning vector. The linkers are then cleaved with a restriction enzyme that generates sticky ends.)

in eukaryotic genes (see Chapter 19). Introns can be so extensive that the overall length of a eukaryotic gene becomes too unwieldy for recombinant DNA manipulation. Using cDNA eliminates this problem. In addition, bacteria cannot synthesize the correct protein product of

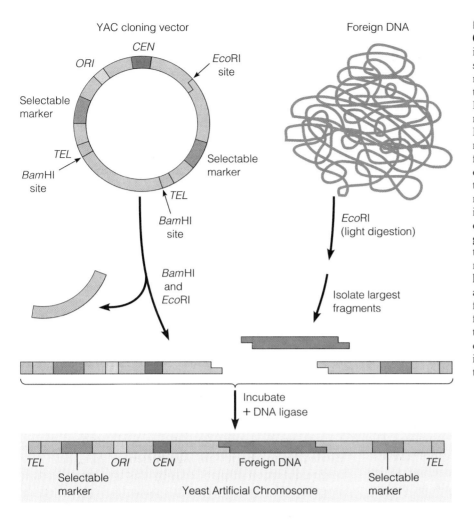

YAC cloning vector

Foreign DNA

Figure 18-32 Construction of a Yeast Artificial Chromosome (YAC). The YAC cloning vector is a circular DNA molecule with nucleotide sequences specifying an origin of DNA replication (*ORI*), a centromere (*CEN*), two telomeres (*TEL*), and two selectable markers. It has two recognition sequences for the restriction enzyme *Bam* HI and one for *Eco* RI. Digestion of the YAC vector with both restriction enzymes produces two linear DNA fragments that together contain all the essential sequences, as well as the fragment that connected the *Bam* HI sites, which is of no further use. The fragment mixture is incubated with fragments from light digestion of foreign DNA with *Eco* RI (light digestion generates large DNA fragments because not all the restriction sites are cut), and the resulting recombinant strands are sealed with DNA ligase. Among the products will be yeast artificial chromosomes—YACs—carrying foreign DNA, as shown at the bottom of the figure. After yeast cells are transformed with the products of the procedure, the colonies of cells that have received complete YACs can be identified by the properties conferred by the two selectable markers.

an intron-containing eukaryotic gene unless the introns have been removed—as they are in cDNA.

Large DNA Segments Can Be Cloned in Cosmids and Yeast Artificial Chromosomes (YACs)

DNA cloning using the vectors mentioned so far is a very powerful methodology, but it has an important limitation: the relatively small size of the foreign DNA fragments that can be successfully cloned in a single vector molecule. Certain phage λ vectors accommodate DNA fragments as large as 15,000 base pairs (bp), but eukaryotic genes are often bigger than this. Somewhat larger fragments can be cloned in **cosmids,** which are DNA molecules that share some of the features of both plasmids and phage cloning vectors. DNA fragments up to 30,000–40,000 bp in length can be inserted into cosmids, which are then packaged into phage particles. Although the resulting phage does not contain the genes required for replication, it can infect bacterial cells and release its cosmid DNA inside the cell. The released cosmid DNA then replicates like a plasmid.

For genome-mapping projects, the availability of clones containing even longer DNA segments is extremely valuable, since the more DNA per clone, the fewer the number of clones needed to cover the entire genome. Very long DNA segments can now be readily cloned using a vector called a **yeast artificial chromosome (YAC).** Such vectors are "minimalist" eukaryotic chromosomes that contain all the DNA sequences needed for normal chromosome replication and segregation to daughter cells, and very little else. As you might guess from your knowledge of chromosome replication and segregation (see Chapter 17), a eukaryotic chromosome requires three kinds of DNA sequences: (1) an origin of DNA replication; (2) two telomeres to allow periodic extension of the shrinking ends by telomerase; and (3) a centromere to ensure proper attachment, via a kinetochore, to spindle microtubules during cell division. If yeast versions of these three kinds of DNA sequences are combined with a segment of foreign DNA, the resulting YAC will replicate in yeast and segregate into daughter cells with each round of cell division, just like a natural chromosome. And under appropriate conditions, its foreign genes may be expressed.

Figure 18-32 outlines the construction of a typical YAC. The YAC cloning vector is a small circular DNA molecule carrying an origin of DNA replication (*ORI*), a centromere sequence (*CEN*), and two telomeres (*TEL*). In addition, the vector illustrated here carries two genes that function as selectable markers and three restriction sites, one for *Eco*RI and two for *Bam*HI. In cloning experiments, the vector DNA is cleaved with *Eco*RI and *Bam*HI and then incubated with DNA fragments that have been generated by digesting foreign DNA with *Eco*RI. After covalent sealing by DNA ligase, the products include a variety of YACs carrying different fragments of foreign DNA. The YACs are introduced by transformation into yeast cells whose cell walls have been removed. The presence of two selectable markers makes it easy to select for yeast cells containing YACs with both chromosomal "arms." The diagram in the figure is not to scale: The YAC vector alone is only about 10,000 bp, but the inserted foreign DNA usually ranges from 300,000 to 1.5 million bp in length. In fact, YACs must carry at least 50,000 bp to be reliably replicated and segregated.

Genetic Engineering

Recombinant DNA technology has had an enormous impact on the field of cell biology, leading to many new insights into the organization, behavior, and regulation of genes and their protein products. Many of these discoveries would have been virtually inconceivable in the absence of such powerful techniques for isolating gene sequences. The rapid advances in our ability to manipulate genes have also opened up a new field, called **genetic engineering,** which involves the application of recombinant DNA technology to practical problems, primarily in medicine and agriculture. In concluding the chapter, we will briefly examine some of the areas in which these practical benefits of recombinant DNA technology are beginning to be seen.

Genetic Engineering Can Produce Valuable Proteins That Are Otherwise Difficult to Obtain

One practical benefit to emerge from recombinant DNA technology is the ability to clone genes coding for medically useful proteins that are difficult to obtain by conventional means. Among the first proteins to be produced by genetic engineering was human *insulin,* which is required by roughly 2 million diabetics in the United States to treat their disease. Because supplies of insulin purified from human blood or pancreatic tissue are extremely scarce, for many years diabetic patients were treated with insulin obtained from pigs and cattle, which can cause toxic reactions. Now there are several ways of producing human insulin from genetically engineered bacteria containing the human insulin gene. As a result, diabetics can be treated with insulin molecules that are identical to the insulin produced by the human pancreas.

Like insulin, a variety of other medically important proteins that were once difficult to obtain in adequate amounts are now produced using recombinant DNA technology. Included in this category are the *blood clotting factors* needed for the treatment of hemophilia, *growth*

hormone utilized for treating pituitary dwarfism, *tissue plasminogen activator (TPA)* used for dissolving blood clots in heart attack patients, *erythropoietin* employed for stimulating the production of red blood cells in patients with anemia, and *tumor necrosis factor, interferon,* and *interleukin,* which are used in treating certain kinds of cancer. Traditional methods for isolating and purifying such proteins from natural sources are quite cumbersome and tend to yield only tiny amounts of protein; hence prior to the advent of recombinant DNA technology, the supplies of such substances were inadequate and their cost was extremely high. But now that the genes for these proteins have been cloned in bacteria and yeast, large quantities of protein can be produced in the laboratory at reasonable cost.

The Ti Plasmid Is a Useful Vector for Introducing Foreign Genes into Plants

Recombinant DNA technology is also being used to modify agriculturally important plants by inserting genes designed to introduce traits such as resistance to insects, herbicides, or viral disease, or to improve a plant's nitrogen-fixing ability, photosynthetic efficiency, nutritional value, or ability to grow under adverse conditions. Cloned genes are transferred into plants by inserting them first into the **Ti plasmid,** a naturally occurring DNA molecule carried by the bacterium *Agrobacterium tumefaciens.* In nature, infection of plant cells by this bacterium leads to insertion of a small part of the plasmid DNA, called the *T DNA region,* into the plant cell chromosomal DNA; expression of the inserted DNA then triggers the formation of an uncontrolled growth of tissue called a *crown gall tumor.* In the laboratory, the DNA sequences that trigger tumor formation can be removed from the Ti plasmid without stopping the transfer of DNA from the plasmid to the host cell chromosome. Inserting genes of interest into such modified plasmids produces vectors that will transfer foreign genes into plant cells.

This general approach for transferring genes into plants is summarized in Figure 18-33. The Ti plasmid is first isolated from *Agrobacterium* cells and the desired foreign gene is inserted into the T DNA region of the plasmid using standard recombinant DNA techniques. (This step disrupts the tumor-inducing genes.) The plasmid is then put back into *Agrobacterium,* and the genetically engineered bacterium is used to infect plant cells growing in culture. When the recombinant plasmid enters the plant cell, its T DNA becomes stably integrated into the plant genome and is passed on to both daughter cells at every cell division. Such cells are subsequently used to regenerate plants that contain the recombinant T DNA—and therefore the desired foreign gene—in all of their cells. The foreign gene will now be inherited by progeny plants just like any other gene. Such plants are said to be **transgenic,** a general term that refers to any type of organism, plant or animal, that carries one or more genes from another organism in all of its cells, including its

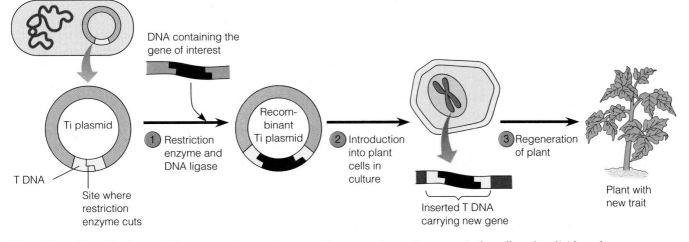

Agrobacterium tumefaciens

DNA containing the gene of interest

Ti plasmid

T DNA

Site where restriction enzyme cuts

① Restriction enzyme and DNA ligase

Recom- binant Ti plasmid

② Introduction into plant cells in culture

③ Regeneration of plant

Inserted T DNA carrying new gene

Plant with new trait

Figure 18-33 Using the Ti Plasmid to Transfer Genes into Plants. Most genetic engineering in plants uses the Ti plasmid as a vector. ① The Ti plasmid is isolated from the bacterium *Agrobacterium tumefaciens,* and a DNA fragment containing a gene of interest is inserted into a restriction site located in the T DNA region of the plasmid. ② When the recombinant plasmid is introduced into plant cells, the T DNA region becomes integrated into the plant cell's chromosomal DNA. ③ The plant cell is then allowed to divide and regenerate a new plant containing the recombinant T DNA stably incorporated into the genome of every cell.

reproductive cells. Transgenic plants are also commonly referred to as *GM (genetically modified)* plants.

Genetic Modification Can Improve the Traits of Food Crops

The ability to insert new genes into plants using the Ti plasmid has allowed scientists to create GM crops exhibiting a variety of new traits. For example, plants can be made more resistant to insect damage by the introduction of a gene cloned from the soil bacterium *Bacillus thuringiensis* (Bt). This Bt gene codes for a protein that is toxic to certain insects—especially caterpillars and beetles that cause crop damage by chewing on plant leaves. Putting the Bt gene into plants such as cotton and corn has allowed farmers to limit their use of more hazardous pesticides for controlling insects, in some cases allowing a substantial return of wildlife to crop fields. A similar rationale has led to the introduction of genes that enable crops to resist weed-killing herbicides, thereby allowing farmers to achieve higher crop yields using fewer toxic chemicals.

Genetic modification can also be used to improve the nutritional value of food. Consider rice, for example, which is the most common food source in the world. Currently eaten by more than 3 billion people daily, at least half the population of the world is expected to depend on rice for food by the year 2020, especially in poor, developing countries. Many of these same people suffer from deficiencies in essential nutrients and vitamins. To illustrate how genetic engineering might be used to improve this situation, the genes required for the synthesis of β-carotene, a precursor of vitamin A, have been genetically engineered into rice. The resulting product, called "golden rice" because of the color imparted by β-carotene, could help to alleviate a global vitamin A deficiency that now causes blindness and disease in millions of children. Of course, adding a single vitamin to a single crop is not enough to prevent malnutrition. But if similar techniques can be used to enhance the content of various vitamins and nutrients in staple crops such as rice, wheat, and corn, it could go a long way toward improving the overall nutritional status of the world's population.

Food spoilage is another area that has been tackled by genetic engineering. An interesting example involves the problems associated with storing and transporting tomatoes. To make them less susceptible to damage and to give them a longer shelf life, commercially grown tomatoes are generally picked before they have fully ripened. But such tomatoes do not taste as good as tomatoes that have ripened on the vine. Scientists have tried to overcome this problem by using genetic engineering techniques to create the so-called "Flavr Savr" tomato. Their strategy was to inhibit the production of polygalacturonase (PG), an enzyme that catalyzes cell wall breakdown and thereby causes tomatoes to rot. PG was inhibited by inserting a copy of the PG gene into cells in a backward orientation, causing the affected cells to produce an abnormal mRNA that is complementary to the mRNA produced by the normal PG gene. The abnormal mRNA, called an *antisense* mRNA, binds to the normal mRNA by complementary base pairing and prevents it from functioning, thereby leading to decreased production of the PG enzyme. Because they produce less PG, Flavr Savr tomatoes are less susceptible to rotting and can therefore be allowed to ripen on the vine rather than being picked green and hard. Although the resulting tomato tastes better and has a longer shelf life without spoilage, it has not been a commercial success because conventional tomato-handling equipment tends to damage the soft, ripened tomatoes.

Concerns Have Been Raised about the Safety and Environmental Risks of GM Crops

As the prevalence of GM crops has begun to increase, some people have expressed concerns about the possible risks associated with this technology. For consumers, the main focus has been on safety because it is well known that toxic and allergic reactions to things we eat can be serious and even life-threatening. Thus far, it seems clear that conventional foods already on the market, such as peanuts and Brazil nuts, pose greater allergy risks than have been demonstrated for any GM foods. Although it is certainly possible that new allergies or toxic reactions might be triggered by genes inserted into GM crops, the same problem also arises when new crop strains are produced by conventional breeding techniques. In fact, conventional foods have already undergone massive changes in genetic makeup by traditional plant-breeding methods, and it can be argued that it is safer to insert carefully selected genes into plants one at a time, as is done in genetic engineering, than it is to alter thousands of genes at a time, as is done with traditional cross-breeding of plants. When a single gene is inserted into a crop, the effects of that single gene and its protein product can be more easily assessed for safety hazards.

The possibility that GM crops may pose environmental risks has also raised concerns. In a 1999 laboratory experiment, scientists reported that monarch butterfly larvae died after being fed leaves dusted with pollen from GM corn containing the Bt gene. This observation led to fears that the Bt toxin produced by GM plants can harm friendly insects. However, the lab bench is not the farm field, and the initial studies were carried out under artificial laboratory conditions in which butterfly larvae consumed far higher doses of Bt toxin than they would in the real world. Subsequent data collected from farm fields containing GM crops suggest that the amount of Bt corn pollen encountered under real-life conditions does not pose a significant hazard to monarch butterflies.

Despite legitimate concerns about potential hazards, the GM experience has thus far revealed little evidence of significant risks to either human health or the environment. To the contrary, the introduction of certain GM crops has even been associated with reductions in pesticide use, suggesting the possibility of positive effects on the environment. Of course, no new technology is entirely without risk, and so the safety and environmental impact of GM crops must be continually assessed. At the same time, it is important to realize that GM crops can make a unique contribution to the fight against hunger and disease in developing countries, helping to ensure that the world will be able to feed the additional 2 billion people expected to inhabit the earth over the next quarter-century.

Gene Therapies Are Being Developed for the Treatment of Human Diseases

A large number of human diseases are caused by inherited defects in specific genes, such as the gene responsible for cystic fibrosis (see Box 8B). Because gene cloning techniques make it relatively straightforward to purify the normal versions of such genes, scientists are currently working on the development of *gene therapy* techniques that would allow genetic diseases to be treated by transplanting normal, functional copies of genes into people who possess defective, disease-causing genes.

One of the first successful attempts at gene transplantation was reported in animals by Richard Palmiter, who transferred the gene for *growth hormone* into a mouse egg, thereby creating a transgenic mouse that carries genes from another organism in all its cells, including its germ line (Box 18A). The success of this type of experiment raises the question of whether gene transplantation techniques can be used to replace defective genes in humans. Humans certainly suffer from many genetic diseases that might in theory be cured by replacing the defective gene with a normal copy. Potential candidates include cystic fibrosis, hemophilia, hypercholesterolemia, hemoglobin disorders, muscular dystrophy, lysosomal storage diseases, and an immune disorder called *severe combined immunodeficiency (SCID)*.

The first person to be treated using gene therapy was a 4-year-old girl with a form of SCID caused by an inherited defect in the enzyme *adenosine deaminase (ADA)*. The loss of ADA activity resulting from this disorder causes a potentially lethal immune deficiency in which children are unable to fight infections because they lack sufficient numbers of immune cells called *T lymphocytes*. Prior to treatment, the girl suffered from frequent infections and was generally lethargic. Starting in 1990, she underwent a two-year series of treatments in which a normal copy of the cloned ADA gene was inserted into a virus, the virus was used to infect T lymphocytes that had been isolated from the girl's blood, and the lymphocytes were then injected back into her bloodstream. These treatments led to a significant improvement in her immune function, although it has diminished somewhat in the years since she was treated.

Unfortunately, follow-up studies revealed that this type of treatment does not help most SCID patients. Nonetheless, the results with the initial patient were good enough to encourage attempts to improve the effectiveness of gene therapy, and considerable progress has been made in increasing the efficiency of techniques for delivering cloned genes into target cells and getting the genes to insert into the chromosomal DNA and function properly. Two studies reported in the year 2000 indicate that these efforts are finally beginning to pay off. The first involved two children in France suffering from an especially severe form of SCID caused by a defective receptor gene rather than a defective ADA gene. These experiments incorporated two major improvements over the initial SCID

A DNA fragment containing the gene of rat growth hormone was microinjected into the pronuclei of fertilized mouse eggs. Of 21 mice that developed from these eggs, seven carried the gene and six of these grew significantly larger than their littermates. (Palmiter et al., 1982, p. 611)

With these words, a team of investigators led by Richard Palmiter and Ralph Brinster reported how a genetic trait can be introduced experimentally into mice without going through the usual procedure of breeding—that is, sexual reproduction followed by the selection of desired traits. The researchers injected rat growth hormone genes into fertilized mouse eggs, and from one of these eggs developed a "supermouse" weighing almost twice as much as its littermates (Figure 18A-1). The accomplishment was heralded as a significant breakthrough because it proved the feasibility of applying genetic engineering to animals, with all the scientific and practical consequences such engineering is likely to have.

What Palmiter, Brinster, and their colleagues did to create the supermouse is an intriguing story that begins with the isolation of the gene for growth hormone (GH) from a library of rat DNA, using techniques similar to those described in this chapter. The cloned GH gene from which the regulatory region had been deleted was then fused to the regulatory portion of a mouse gene, the gene that codes for *metallothionein (MT)*. MT is a small metal-binding protein that is normally present in most mouse tissues and appears to be involved in regulating the level of zinc in the animal. The advantage of fusing the MT gene to the GH gene was that the expression of the MT gene could then be specifically induced (turned on) by zinc.

To make multiple copies of the MT-GH hybrid gene, it was cloned in *E. coli* using a plasmid as a cloning vector. After isolating the recombinant plasmid DNA from the bacterial cells, the MT-GH region was excised by digesting the DNA with two restriction enzymes, each of which cleaved a restriction site located at one end of the desired DNA fragment. About 600 copies of the excised DNA fragment were then microinjected into fertilized mouse eggs, in a volume of about 2 picoliters (0.000002 µL!). The DNA was injected into the male pronucleus, the haploid sperm nucleus that has not yet fused with the haploid egg nucleus. (The success rate for integration and retention of the MT-GH gene had been found to be higher when the male pronucleus was used than when the DNA was injected into either the female pronucleus or the cytoplasm.) From the 170 fertilized eggs that were injected and implanted back into the reproductive tracts of foster mothers, 21 animals developed. Seven of them turned out to be transgenic mice with MT-GH genes present in their cells. In at least one case, a transgenic mouse transmitted the MT-GH gene faithfully to about half of its offspring, suggesting that the gene had become stably integrated into one of its chromosomes.

Because the GH gene had been linked to an MT gene regulatory region, it was predicted that the hybrid gene could be turned on by giving the mice zinc in their drinking water. Three kinds of evidence confirmed that exposing mice to zinc caused the rat GH gene to be expressed. First, when mouse liver tissue was assayed for the presence of messenger RNA for GH, the results indicated about 800–3000 mRNA molecules per liver cell. Moreover, elevated levels of growth hormone were found in the blood: Four of the transgenic mice had blood GH levels that were 100–800 times higher than those of their nonengineered littermates! But the most dramatic evidence for expression of the rat GH genes was that the transgenic mice grew faster and weighed about twice as much as normal mice. During the period of maximum sensitivity to growth hormone (3 weeks to 3 months of age), the transgenic animals grew three to four times as fast as their normal littermates.

This dramatic experiment proved that it is possible to introduce cloned genes into the cells of higher organisms and that such genes can become stably integrated into the genome, where they are expressed and passed on to offspring. As the authors pointed out when the report was published, "this approach has implications for studying the biological effects of growth hormone, as a way to accelerate animal growth, as a model for gigantism (a human growth abnormality caused by growth hormone), as a means of correcting genetic diseases, and as a method of farming valuable products." Whether (and how fast) all these possibilities become realities remains, of course, to be seen. But in the years since supermouse's creation, rapid progress has been made in most of these areas. Supermouse, it seems, was just the beginning.

Figure 18A-1   Genetic Engineering in Mice.   "Supermouse" (on the left) is significantly larger than its littermate because it was engineered to carry and express at high levels the gene for rat growth hormone.

studies: The virus used for inserting a normal copy of the cloned receptor gene into the children's cells was more efficient at transferring cloned genes into human cells than the virus used in the original SCID studies, and improved conditions were employed for culturing the children's cells during the gene transfer process. As a consequence, the overall efficiency of gene transfer was much higher and the immune function of the two patients appeared to be restored to normal. The improvement was so dramatic that for the first time, the children were able to leave the protective isolation "bubble" that had been used in the hospital to protect them from infections. Although it is too soon to know whether this dramatic improvement will last, the initial results suggest that we may have seen the first clearcut success of gene therapy.

The other encouraging result involved the treatment of *hemophilia*, a rare, inherited disease characterized by life-threatening episodes of uncontrollable bleeding. Hemophilia is caused by genetic defects in proteins called *blood-clotting factors*, which catalyze steps involved in the formation of blood clots. In recent gene therapy trials, several hemophilia patients were injected in the leg with a virus containing a gene coding for the blood-clotting factor they require. A few months after the injections, their leg muscles were producing blood-clotting factor in sufficient quantities to enter the bloodstream and ameliorate, although not cure, the disease. This was an especially exciting discovery because unlike the SCID trials, where gene therapy was carried out on cells that had to be removed from patients for gene transfer and then reinjected into the body, the hemophilia studies employed genes carried by a virus that was injected directly into people. (Another example of gene therapy targeting cells that still reside within the body was covered in Box 8B, which discusses the potential use of nasal sprays containing the CFTR gene to treat cystic fibrosis.)

In the years since the enormous potential of gene therapy was first widely publicized in the early 1980s, the field has been criticized for promising too much and delivering too little. But most new technologies take time to be perfected and encounter disappointments along the way. Gene therapy is no exception, and it will take time to develop. But it now seems virtually certain that using normal genes to treat genetic diseases is a realistic prospect that will become common practice, at least for a few genetic diseases, within the next two decades, and that will eventually revolutionize the practice of medicine. Of course, the ability to alter people's genes raises a series of important ethical, safety, and legal concerns, and the ultimate question of how society will control this growing power to change the human genome is an issue that needs to be thoroughly discussed not just by scientists and physicians but by society as a whole.

Perspective

Asexual reproduction is based on mitosis and produces offspring that are genetically identical (or nearly so) to the single parent. Sexual reproduction, on the other hand, involves two parents and leads to a mixture of parental traits in the offspring. Sexual reproduction allows populations to adapt to environmental changes, enables desirable mutations to be combined in a single individual, and promotes genetic flexibility by maintaining a diploid genome.

The life cycle of every sexually reproducing, eukaryotic species includes both haploid and diploid phases. Haploid gametes are generated by meiosis and fuse at fertilization to restore the diploid chromosome number. Meiosis consists of two successive cell divisions without an intervening duplication of chromosomes. During the first meiotic division, homologous chromosomes separate and segregate into the two daughter cells. During the second meiotic division, sister chromatids separate and four haploid daughter cells are produced. In addition to reducing the chromosome number from diploid to haploid, meiosis differs from mitosis in that homologous chromosomes synapse during prophase of the first meiotic division, thereby allowing crossing over and genetic recombination between nonsister chromatids.

Mendel's laws of inheritance describe the genetic consequences of chromosome behavior during meiosis, even though chromosomes had not yet been discovered at the time of Mendel's experiments. According to these laws, maternal and paternal alleles segregate into different gametes during meiosis, and the alleles of genes located on separate chromosomes assort independently of one another. The enormous genetic variability among an organism's gametes arises in part from the independent assortment of chromosomes during anaphase I and in part from the recombination that occurs during prophase I. The frequency of recombination between genes located on the same chromosome is a measure of the distance between the two genes and can therefore be used to map their chromosomal locations.

In addition to occurring during meiosis in eukaryotes, homologous recombination is also observed in viruses (during co-infection) and when DNA is transferred into prokaryotic cells by transformation, transduction, or conjugation. The mechanism of recombination involves breakage-and-exchange between DNA molecules that exhibit extensive sequence homology. Recombination is sometimes accompanied by gene conversion or the formation of DNA molecules whose two strands are not completely complementary to one another. These phenomena can be explained by recombination models involving the formation of Holliday junctions, which are regions of single-strand exchange between double-stranded DNA molecules.

The development of recombinant DNA technology has made it possible to combine DNA from any two (or more) sources into a single molecule of recombinant DNA. Combining a gene of interest with a plasmid or phage cloning vector allows the gene to be cloned (amplified) in bacterial cells. In this way, large amounts of specific genes or their protein products can be prepared for research or practical purposes. Recombinant DNA technology has made possible the detailed analysis and manipulation of eukaryotic genomes, including the human genome. At the same time, practical applications of this technology have the potential to revolutionize modern medicine and agriculture.

Key Terms for Self-Testing

asexual reproduction (p. 577)

Sexual Reproduction
sexual reproduction (p. 577)
homologous chromosomes (p. 578)
sex chromosome (p. 578)
diploid (p. 578)
haploid (p. 578)
gene locus (p. 578)
allele (p. 578)
homozygous (p. 578)
heterozygous (p. 578)
dominant (allele) (p. 578)
recessive (allele) (p. 578)
genotype (p. 579)
phenotype (p. 579)
gamete (p. 579)
gametogenesis (p. 579)
sperm (p. 579)
egg (ovum) (p. 579)
fertilization (p. 579)
zygote (p. 579)
development (of an organism) (p. 579)
mating type (p. 579)

Meiosis
meiosis (p. 580)
alternation of generations (p. 581)
haploid spore (p. 581)
sporophyte (p. 581)
gametophyte (p. 581)
meiosis I (p. 581)
synapsis (p. 581)
bivalent (p. 581)
genetic recombination (p. 582)

leptotene (p. 583)
zygotene (p. 583)
crossing over (p. 583)
pachytene (p. 583)
diplotene (p. 583)
chiasma (p. 583)
diakinesis (p. 585)
synaptonemal complex (p. 585)
meiosis II (p. 588)
nondisjunction (p. 588)
C value (p. 588)
polar body (p. 588)

Genetic Variability: Segregation and Assortment of Alleles
true-breeding (plant strain) (p. 590)
P_1 generation (p. 590)
hybrid (p. 590)
F_1 generation (p. 590)
F_2 generation (p. 590)
backcrossing (p. 592)
Mendel's laws of inheritance (p. 592)
law of segregation (p. 592)
law of independent assortment (p. 593)

Genetic Variability: Recombination and Crossing Over
wild type (p. 596)
linkage group (p. 596)
linked genes (p. 596)
genetic mapping (p. 598)

Genetic Recombination in Bacteria and Viruses
transformation (p. 599)

transduction (p. 599)
conjugation (p. 600)
F factor (p. 600)
sex pilus (p. 600)
mating bridge (p. 600)
origin of transfer (p. 600)
Hfr cell (p. 601)

Molecular Mechanism of Homologous Recombination
homologous recombination (p. 602)
gene conversion (p. 604)
ascus (p. 604)
Holliday junction (p. 604)
RecA (p. 606)

Recombinant DNA Technology and Gene Cloning
recombinant DNA technology (p. 606)
cloning vector (p. 606)
recombinant DNA molecule (p. 607)
DNA cloning (p. 608)
plaque (p. 609)
selectable marker (p. 610)
nucleic acid probe (p. 611)
genomic library (p. 612)
complementary DNA (cDNA) (p. 612)
cDNA library (p. 612)
cosmid (p. 613)
yeast artificial chromosome (YAC) (p. 613)

Genetic Engineering
genetic engineering (p. 614)
Ti plasmid (p. 614)
transgenic (p. 614)

Problem Set

More challenging problems are marked with a •.

18-1.   The Truth About Sex. For each of the following statements, indicate with an S if it is true of sexual reproduction, with an A if it is true of asexual reproduction, with a B if it is true of both, and with an N if it is true of neither.

(a) Traits from two different parents can be combined in a single offspring.

(b) Each generation of offspring is virtually identical to the previous generation.

(c) Mutations are propagated to the next generation.

(d) Some offspring in every generation will be less suited for survival than the parents, but others may be better suited.

(e) Mitosis is involved in the life cycle.

18-2.   Ordering the Phases of Meiosis. Shown in Figure 18-34 are drawings of several phases of meiosis in an organism, labeled A through F.

(a) What is the diploid chromosome number in this species?

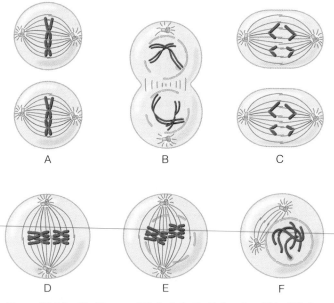

A   B   C

D   E   F

Figure 18-34 Six Phases of Meiosis to Be Ordered and Identified. See Problem 18-2.

(b) Place the six phases in chronological order, and name each one.

(c) Between which two phases do homologous centromeres separate?

(d) Between which two phases does recombination occur?

18-3. Telling Them Apart. Briefly describe how you might distinguish between each of the following pairs of phases in the same organism:

(a) Metaphase of mitosis and metaphase I of meiosis.

(b) Metaphase of mitosis and metaphase II of meiosis.

(c) Metaphase I and metaphase II of meiosis.

(d) Telophase of mitosis and telophase II of meiosis.

(e) Pachytene and diplotene stages of meiotic prophase I.

18-4. Your Centromere Is Showing. Suppose you have a diploid organism in which all the chromosomes contributed by the sperm have cytological markers on their centromeres that allow you to distinguish them visually from the chromosomes contributed by the egg.

(a) Would you expect all the somatic cells (cells other than gametes) to have equal numbers of maternal and paternal centromeres in this organism? Explain.

(b) Would you expect equal numbers of maternal and paternal centromeres in each gamete produced by that individual? Explain.

18-5. How Much DNA? Let X be the amount of DNA present in the gamete of an organism that has a diploid chromosome number of 4. Assuming all chromosomes to be of approxi-

mately the same size, how much DNA (X, $2X$, $\frac{1}{2}X$, and so on) would you expect in each of the following?

(a) A zygote immediately after fertilization

(b) A single sister chromatid

(c) A daughter cell following mitosis

(d) A single chromosome following mitosis

(e) A nucleus in mitotic prophase

(f) The cell during metaphase II of meiosis

(g) One bivalent

18-6. Meiotic Mistakes. Infants born with Patau syndrome have an extra copy of chromosome 13, which leads to developmental abnormalities such as cleft lip and palate, small eyes, and extra fingers and toes. Another type of genetic disorder, called Turner syndrome, results from the presence of only one sex chromosome—an X chromosome. Individuals born with one X chromosome are females exhibiting few noticeable defects until puberty, when they fail to develop normal breasts and internal sexual organs. Describe the meiotic events that could lead to the birth of an individual with either Patau syndrome or Turner syndrome.

18-7. Punnett Squares as Genetic Tools. A *Punnett square* is a diagram representing all possible outcomes of a genetic cross. The genotypes of all possible gametes from the male and female parents are arranged along two adjacent sides of a square, and each box in the matrix is then used to represent the genotype resulting from the union of the two gametes at the heads of the intersecting rows. By the law of independent assortment, all possible combinations are equally likely, so the frequency of a given genotype among the boxes represents the frequency of that genotype among the progeny of the genetic cross represented by the Punnett square.

Figure 18-35 on the following page shows the Punnett squares for two crosses of pea plants. The genetic characters involved are seed color (where Y is the allele for yellow seeds and y for green seeds) and seed shape (where R is the allele for round seeds and r for wrinkled seeds). The Punnett square in Figure 18-35a represents a one-factor cross between parent plants that are both heterozygous for seed color ($Yy \times Yy$). The Punnett square in Figure 18-35b is a two-factor cross between plants heterozygous for both seed color (Yy) and seed shape (Rr).

(a) Using the Punnett square of Figure 18-35a, explain the 3 : 1 phenotypic ratio Mendel observed for the offspring of such a cross.

(b) Explain why the Punnett square of Figure 18-35b is a 4 × 4 matrix with 16 genotypes. In general, what is the mathematical relationship between the number of heterozygous allelic pairs being considered and the number of different kinds of gametes?

(c) How does the Punnett square of Figure 18-35b reflect Mendel's law of independent assortment?

(d) Complete the Punnett square of Figure 18-35b by writing in each of the possible progeny genotypes. How many different genotypes will be found in the progeny? In what ratios?

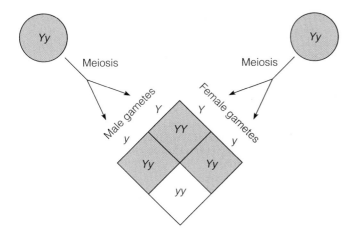

(a) One-factor cross

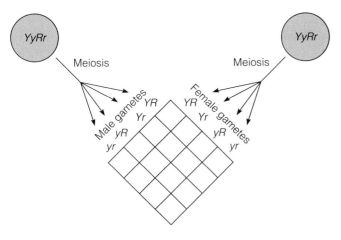

(b) Two-factor cross

Figure 18-35   Punnett Squares.   See Problem 18-7.

Genes	Recombination Frequency
w and *x*	25%
w and *y*	29%
w and *z*	17%
x and *y*	50%
x and *z*	9%
y and *z*	44%

(a) Construct a genetic map indicating the order in which these four genes occur and the number of map units that separate the genes from each other.

(b) In constructing this map, you may have noticed that the map distances are not exactly additive. Can you provide an explanation for this apparent discrepancy?

• **18-9.   Homologous Recombination.** Bacterial cells use at least three different pathways for carrying out genetic recombination. All three pathways require the RecA protein, but in each case a different set of steps precedes the action of RecA in catalyzing strand invasion. One of these three pathways utilizes an enzyme complex called RecBCD, which binds to double-strand breaks in DNA and exhibits both helicase and single-strand nuclease activities.

(a) Briefly describe a model showing how the RecBCD enzyme complex might set the stage for genetic recombination.

(b) When bacterial cells are co-infected with two different strains of bacteriophage λ, genes located near certain regions of the phage DNA, called *CHI sites,* recombine at much higher frequencies than other genes. However, in mutant bacteria lacking the RecBCD protein, genes located near CHI sites do not recombine any more frequently than other genes. How can you modify your model to accommodate this additional information?

• **18-10.   Recombinant DNA Technology.** A researcher wants to study an important protein involved in human gametogenesis. Because the protein is very difficult to prepare in sufficient quantities from human cells, she decides to clone its gene so that, if all goes well, she can use bacteria to make large batches of the protein. The amino acid sequence of the protein's single polypeptide chain has already been established. Briefly explain how she might clone the desired gene.

(e) For the case of Figure 18-35b, how many different phenotypes will be found in the progeny? In what ratios?

• **18-8.   Genetic Mapping.** The following table provides data concerning the frequency with which four genes (*w, x, y,* and *z*) located on the same chromosome recombine with each other.

Suggested Reading

References of historical importance are marked with a • .

Meiosis
de Lange, T. Ending up with the right partner. *Nature* 392 (1998): 753.
Haber, J. E. Meiosis: Searching for a partner. *Science* 279 (1998): 823.
John, B. *Meiosis.* New York: Cambridge University Press, 1990.
Masui, Y. The elusive cytostatic factor in the animal egg. *Nature Reviews Molecular Cell Biol.* 1 (2000): 228.
Murray, A. W. MAP kinases in meiosis. *Cell* 92 (1998): 157.

Scherthan, H. A bouquet makes ends meet. *Nature Reviews Molecular Cell Biol.* 2 (2001): 621.

Mendel's Experiments
Fincham, J. R. S. Mendel—now down to the molecular level (News and Views). *Nature* 343 (1990): 208.
• Mendel, G., H. de Vries, C. Correns, and E. Tschermak. The birth of genetics. *Genetics* 35 (1950, Suppl.): 1. (Original papers in English translation.)

• Sturtevant, A. H. *A History of Genetics.* New York: Harper & Row, 1965.

Mechanism of Recombination

Clark, A. J. *recA* mutants of *E. coli* K12: A personal turning point. *BioEssays* 18 (1996): 767.

Leach, D. *Genetic Recombination.* Cambridge, MA: Blackwell Science, 1996.

Lilley, D. M. J., and M. F. White. The junction-resolving enzymes. *Nature Reviews Molecular Cell Biol.* 2 (2001): 433.

• Meselson, M. S., and C. M. Radding. A general model for genetic recombination, *Proc. Natl. Acad. Sci. USA* 72 (1975): 358.

Shinagawa, H., and H. Iwasaki. Processing the Holliday junction in homologous recombination. *Trends Biochem. Sci.* 12 (1996): 107.

Stahl, F. Meiotic recombination in yeast: Coronation of the double-strand-break repair model. *Cell* 87 (1996): 965.

Tang, R. S. The return of copy-choice in DNA recombination. *BioEssays* 11 (1994): 785.

Recombinant DNA Technology and Genetic Engineering

Anderson, W. F. The best of times, the worst of times. *Science* 288 (2000): 627.

Brown, K., K. Hopkin, and S. Nemecek. Genetically modified foods: Are they safe? *Sci. Amer.* 284 (April 2001): 51.

• Chilton, M. D. A vector for introducing new genes into plants. *Sci. Amer.* 248 (June 1983): 50.

Friedman, T. Overcoming the obstacles to gene therapy. *Sci. Amer.* 276 (June 1997): 96.

Friedman, T., ed. *The Development of Human Gene Therapy.* Plainview, NY: Cold Spring Harbor Laboratory Press, 1999.

Glick, B. R., and J. J. Pasternak. *Molecular Biotechnology: Principles and Applications of Recombinant DNA.* Herndon, VA: ASM Press, 1994.

Kmiec, E. B. Gene therapy. *American Scientist* 87 (1999): 240.

• Palmiter, R. D., R. L. Brinster, R. E. Hammer, M. E. Trumbauer, M. G. Rosenfeld, N. C. Birnberg, and R. M. Evans. Dramatic growth of mice that develop from eggs microinjected with metallothionein-growth hormone fusion genes. *Nature* 300 (1982): 611.

Verma, I. M., and N. Somia. Gene therapy—promises, problems, and prospects. *Nature* 389 (1997): 239.

Watson, J. D., M. Gilman, J. Witkowski, and M. Zoller. *Recombinant DNA: A Short Course,* 2d ed. New York: Scientific American Books, 1992.

• Watson, J. D., and J. Tooze. *The DNA Story: A Documentary History of Gene Cloning.* New York: W. H. Freeman, 1981.

Gene Expression: I. The Genetic Code and Transcription

So far, we have described DNA as the genetic material of cells and organisms. We have come to understand its structure, chemistry, and replication, as well as the way it is packaged into chromosomes and parceled out to daughter cells during mitotic and meiotic cell divisions. Now we are ready to explore how DNA is expressed—that is, how the coded information it contains is used to guide the production of RNA and protein molecules. Our discussion of this important subject is divided among three chapters. The present chapter deals with the nature of the genetic code and the way in which information stored in DNA guides the synthesis of RNA molecules in the process we call *transcription*. Chapter 20 describes how RNA molecules are then used to guide the synthesis of specific proteins in the process known as *translation*. Finally, Chapter 21 elaborates on the various mechanisms used by cells to control transcription and translation, thereby leading to the *regulation* of gene expression. To put these topics in context, we start here with an overview of the roles played by DNA, RNA, and proteins in gene expression.

The Directional Flow of Genetic Information

As we mentioned at the beginning of this unit, the flow of genetic information in cells generally proceeds from DNA to RNA to protein (see Figure 16-1). DNA (more precisely, a segment of one DNA strand) first serves as a template for the synthesis of an RNA molecule, which in most cases then directs the synthesis of a particular protein. (In a few cases, the RNA is the final product of gene expression and functions as such within the cell.) The principle of directional information flow from DNA to RNA to protein is

known as the *central dogma of molecular biology,* a term coined by Francis Crick soon after the double-helical model of DNA was first proposed. This principle is summarized as follows:

Thus, the flow of genetic information involves replication of DNA, transcription of information carried by DNA into the form of RNA, and translation of this information from RNA into protein. The term **transcription** is used when referring to RNA synthesis using DNA as a template to emphasize that this phase of gene expression is simply a transfer of information from one nucleic acid to another, so the basic "language" remains the same. In contrast, protein synthesis is called **translation** because it involves a language change—from the nucleotide sequence of an RNA molecule to the amino acid sequence of a polypeptide chain.

RNA that is translated into protein is called **messenger RNA (mRNA)** because it carries a genetic message from DNA to the ribosomes, where protein synthesis actually takes place. In addition to mRNA, two other types of RNA are involved in protein synthesis: **ribosomal RNA (rRNA)** molecules, which are integral components of the ribosome, and **transfer RNA (tRNA)** molecules, which serve as intermediaries that translate the coded base sequence of messenger RNA and bring the appropriate amino acids to the ribosome. Note that ribosomal and transfer RNAs do not themselves code for proteins; thus genes coding for these two types of RNA are examples of genes whose final products are RNA molecules rather than protein chains. The involvement of all three major classes of RNA in the

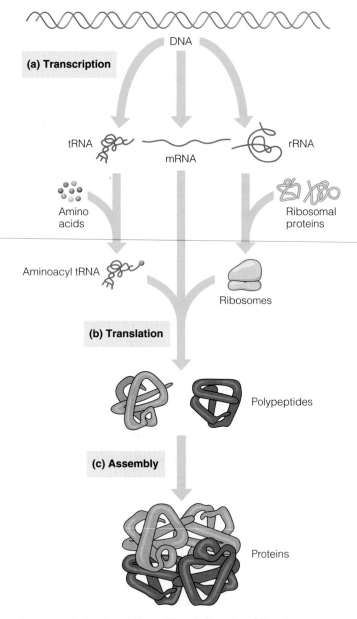

(a) Transcription

DNA

tRNA   mRNA   rRNA

Amino acids   Ribosomal proteins

Aminoacyl tRNA   Ribosomes

(b) Translation

Polypeptides

(c) Assembly

Proteins

Figure 19-1 RNAs as Intermediates in the Flow of Genetic Information. All three major classes of RNA—tRNA, mRNA, and rRNA—are **(a)** synthesized by transcription of the appropriate DNA sequences (genes) and **(b)** involved in the subsequent process of translation (polypeptide synthesis). The appropriate amino acids are brought to the mRNA and ribosome by tRNA. A tRNA molecule carrying an amino acid is called an aminoacyl tRNA. Polypeptides then fold and **(c)** assemble into functional proteins. The specific polypeptides shown here are the globin chains of the protein hemoglobin. For simplicity, this figure omits many details that will be described in this chapter and the next.

overall flow of information from DNA to protein is outlined in Figure 19-1.

During the years since Crick's first statement of the central dogma, it has been refined in various ways. For example, many viruses with RNA genomes have been found to synthesize RNA molecules using RNA as a template. Other RNA viruses, such as HIV, carry out *reverse transcription,* whereby the viral RNA is used as a template

for DNA synthesis—a "backward" flow of genetic information. (Box 19A discusses these viruses and the role of reverse transcription in rearranging DNA sequences.) But in spite of these variations on the original model, the principle that information flows from DNA to RNA to protein remains the main operating principle by which all cells express their genetic information.

The Genetic Code

The essence of gene expression lies in the relationship between the nucleotide base sequence of DNA molecules and the linear order of amino acids in protein molecules. This relationship is based on a set of rules known as the **genetic code.** The cracking of that code, which tells us how DNA can code for proteins, is one of the major landmarks of twentieth-century biology.

During the flow of information from DNA to RNA to protein, it is easy to envision how information residing in a DNA base sequence could be passed to mRNA through the mechanism of complementary base pairing. But how does mRNA pass its "message" to protein? To the uninitiated, the message shown in Figure 19-2a is just a series of nucleotides in an mRNA molecule, with no obvious meaning. But given access to the genetic code (Figure 19-2b; see also Figure 19-8), anyone can convert the sequence of purine and pyrimidine bases in mRNA into a string of amino acids, and the message becomes recognizable as a polypeptide (Figure 19-2c). Thus, a nucleotide base sequence can contain information for guiding protein synthesis, but it must be translated into an amino acid sequence to make sense to the cell.

What is needed, of course, is a knowledge of the appropriate code—the set of rules that determines which nucleotides in mRNA correspond to which amino acids. The encoded message can then be translated into proteins the cell can use. Until it was cracked in the early 1960s, the genetic code was a secret code in a double sense: Before scientists could figure out the exact coding relationship between the base sequence of a DNA molecule and the amino acid sequence of a protein, they first had to become aware that such a relationship existed at all. That awareness arose from the discovery that mutations in DNA can lead to changes in proteins.

Experiments on *Neurospora* Revealed That Genes Can Code for Enzymes

The link between gene mutations and proteins was first detected experimentally by George Beadle and Edward Tatum in the early 1940s using the common bread mold, *Neurospora crassa. Neurospora* is a relatively self-sufficient organism that can grow in a *minimal medium* containing only sugar, inorganic salts, and the vitamin biotin. From these few ingredients, *Neurospora*'s metabolic pathways produce everything else the organism requires. To investi-

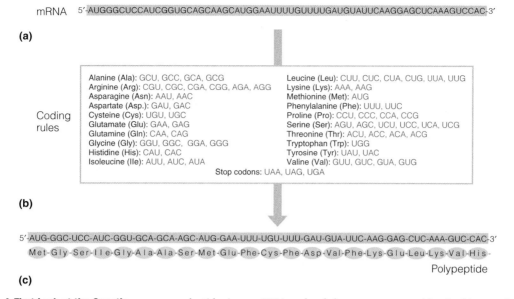

mRNA 5′-AUGGGCUCCAUCGGUGCAGCAAGCAUGGAAUUUUGUUUUGAUGUAUUCAAGGAGCUCAAAGUCCAC-3′

(a)

Coding rules

Alanine (Ala): GCU, GCC, GCA, GCG
Arginine (Arg): CGU, CGC, CGA, CGG, AGA, AGG
Asparagine (Asn): AAU, AAC
Aspartate (Asp.): GAU, GAC
Cysteine (Cys): UGU, UGC
Glutamate (Glu): GAA, GAG
Glutamine (Gln): CAA, CAG
Glycine (Gly): GGU, GGC, GGA, GGG
Histidine (His): CAU, CAC
Isoleucine (Ile): AUU, AUC, AUA

Leucine (Leu): CUU, CUC, CUA, CUG, UUA, UUG
Lysine (Lys): AAA, AAG
Methionine (Met): AUG
Phenylalanine (Phe): UUU, UUC
Proline (Pro): CCU, CCC, CCA, CCG
Serine (Ser): AGU, AGC, UCU, UCC, UCA, UCG
Threonine (Thr): ACU, ACC, ACA, ACG
Tryptophan (Trp): UGG
Tyrosine (Tyr): UAU, UAC
Valine (Val): GUU, GUC, GUA, GUG

Stop codons: UAA, UAG, UGA

(b)

5′-AUG-GGC-UCC-AUC-GGU-GCA-GCA-AGC-AUG-GAA-UUU-UGU-UUU-GAU-GUA-UUC-AAG-GAG-CUC-AAA-GUC-CAC-3′
Met-Gly-Ser-Ile-Gly-Ala-Ala-Ser-Met-Glu-Phe-Cys-Phe-Asp-Val-Phe-Lys-Glu-Leu-Lys-Val-His-

Polypeptide

(c)

Figure 19-2 A First Look at the Genetic Code. The genetic code is a system of purines and pyrimidines used to send messages from the genome to the ribosomes. **(a)** A message written as a sequence of nucleotides in an mRNA molecule has no obvious meaning, until **(b)** a set of equivalency rules for the genetic code is used to convert the sequence into **(c)** the amino acid sequence of a recognizable poly- peptide—in this case, the first 22 of the 385 amino acids in ovalbumin, the major protein of egg white. (The genetic code is shown in more conventional form in Figure 19-8.)

gate the influence of genes on these metabolic pathways, Beadle and Tatum treated a *Neurospora* culture with X rays to induce genetic mutations. Such treatments generated mutant strains that had lost the ability to survive in the minimal culture medium, although they could be grown on a *complete medium* supplemented with a variety of amino acids, nucleosides, and vitamins.

Such observations suggested that the *Neurospora* mutants had lost the ability to synthesize certain amino acids or vitamins and could survive only when these nutrients were added to the growth medium. To determine exactly which nutrients were required, Beadle and Tatum transferred the mutant organisms to a variety of different growth media, each containing a single amino acid or vitamin added as a supplement to the minimal medium. This approach led to the discovery that one mutant strain would grow only in a medium supplemented with vitamin B_6, a second mutant would grow only when the medium was supplemented with the amino acid arginine, and so forth. A large number of different mutants were eventually characterized, each impaired in its ability to synthesize a particular amino acid or vitamin.

Because amino acids and vitamins are synthesized by metabolic pathways involving multiple steps, Beadle and Tatum set out to identify the step in each pathway that had become defective. They approached this task by supplementing the minimal medium with metabolic precursors of a given amino acid or vitamin rather than with the amino acid or vitamin itself. By finding out which precursors supported the growth of a particular mutant strain, they were able to infer that in each mutation, a single enzyme-catalyzed step leading to the synthesis of a specific com-

pound was disabled. There was, in other words, a one-to-one correspondence between each genetic mutation and the lack of a specific enzyme required in a biochemical pathway. From these findings, Beadle and Tatum formulated the *one gene–one enzyme hypothesis,* which stated that each gene controls the production of a single enzyme molecule.

The Base Sequence of a Gene Usually Codes for the Amino Acid Sequence of a Polypeptide Chain

The theory that genes direct the production of enzyme molecules represented a major advance in our understanding of gene action, but it provided little insight into the question of how genes accomplish this task. The first clue to the underlying mechanism emerged a few years later in the laboratory of Linus Pauling, who was studying the inherited disease *sickle-cell anemia.* The red blood cells of individuals suffering from sickle-cell anemia exhibit an abnormal, "sickle" shape that causes the cells to become trapped and damaged when they pass through small blood vessels (Figure 19-3). In trying to identify the reason for this behavior, Pauling decided to analyze the properties of *hemoglobin,* the major protein of red blood cells. Because hemoglobin is a charged molecule, he used the technique of *electrophoresis,* which separates charged molecules from one another by placing them in an electric field. Pauling found that hemoglobin from sickle cells migrated at a different rate than normal hemoglobin, suggesting that the two proteins differ in electric charge. Since some amino acids have charged side chains, Pauling proposed that the difference between normal and sickle-cell hemoglobin lay in their amino acid compositions.

Further Insights REVERSE TRANSCRIPTION, RETROVIRUSES, AND RETROTRANSPOSONS

Transcription generally proceeds in the direction described in the central dogma, with DNA serving as a template for RNA synthesis. In certain cases, however, the process can be reversed and RNA serves as a template for DNA synthesis. This process of *reverse transcription* is catalyzed by the enzyme **reverse transcriptase,** first discovered by Howard Temin and David Baltimore in certain viruses with RNA genomes. Viruses that carry out reverse transcription are called *retroviruses.* Examples of retroviruses include some important pathogens, such as the *human immunodeficiency virus (HIV),* which causes AIDS, and a number of viruses that cause cancers in animals.

Retroviruses

Figure 19A-1 depicts the reproductive cycle of a typical retrovirus. In the virus particle, two copies of the RNA genome are enclosed within a protein capsid that is surrounded by a membranous envelope. Each RNA copy has a molecule of reverse transcriptase attached to it. The virus first (①) binds to the surface of the host cell and its envelope fuses with the plasma membrane, releasing the capsid and its contents into the cytoplasm. Once inside the cell, the viral reverse transcriptase (②) catalyzes the synthesis of a DNA strand that is complementary to the viral RNA and then (③) catalyzes the formation of a second DNA strand complementary to the first. The result is a double-stranded DNA version of the viral genome. (④) This double-stranded DNA then enters the nucleus and integrates into the host cell's chromosomal DNA, much as the DNA genome of a lysogenic phage integrates into the DNA of the bacterial chromosome (see Box 16A). Like a prophage, the integrated viral genome, called a *provirus,* is replicated every time the cell replicates its own DNA. (⑤) Transcription of the proviral DNA (by cellular enzymes) produces RNA transcripts that function in two ways. First, they serve as (⑥) mRNA molecules that direct the synthesis of viral proteins (capsid protein, envelope protein, and reverse transcriptase). And second, (⑦) some of these same RNA transcripts are packaged with the viral proteins into new virus particles. (⑧) The new viruses then "bud" from the plasma membrane without necessarily killing the cell.

The ability of a retroviral genome to integrate into host cell DNA helps explain how some retroviruses can cause cancer. These viruses, called RNA tumor viruses, are of two types. Viruses of the first type carry a cancer-causing *oncogene* in their genomes, along with the genes coding for viral proteins. As we learned in Chapter 17, an oncogene is a mutated version of a normal cellular gene (a proto-oncogene) that codes for proteins used to regulate cell growth and division. For example, the oncogene carried by the Rous sarcoma virus (a chicken virus

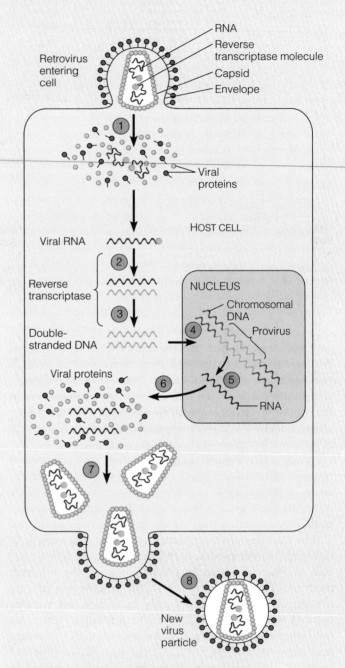

Figure 19A-1 The Reproductive Cycle of a Retrovirus.

To test this hypothesis, it seemed necessary to know the entire amino acid sequence of hemoglobin. At the time of Pauling's discovery in the early 1950s, the largest protein to have been sequenced was less than one-tenth the size of hemoglobin, so determining the complete amino acid sequence of hemoglobin would have been a monumental undertaking. Fortunately, an ingenious shortcut devised by Vernon Ingram made it possible to identify the amino acid abnormality in sickle-cell hemoglobin without determining the protein's complete amino acid sequence. Ingram used the protease *trypsin* to cleave hemoglobin into peptide fragments, which were then sep-

that was the first RNA tumor virus to be discovered) is a modified version of a cellular gene for a protein kinase. The protein product of the viral gene is hyperactive and the cell cannot control it in the normal way. As a result, cells expressing this gene proliferate wildly, producing cancerous tumors called *sarcomas*. RNA tumor viruses of the second type do not themselves carry oncogenes, but integration of their genomes into the host chromosome alters the cellular DNA in such a way that a normal proto-oncogene is converted into an oncogene.

Retrotransposons

Reverse transcription also takes place in normal eukaryotic cells in the absence of viral infection. Much of it involves DNA elements called *retrotransposons*. In Chapter 16 we learned that transposable elements are DNA segments that can move, or *transpose*, themselves from one site to another within the genome. Retrotransposons are a special class of transposable elements that use reverse transcription to carry out this movement. The mechanism is essentially identical to part of the retrovirus life cycle, with the reverse transcriptase often encoded in the retrotransposon DNA. As outlined in Figure 19A-2, the transposition process starts with (①) transcription of the retrotrans-poson DNA by the usual cellular machinery. (②) Translation of the RNA yields reverse transcriptase, which (③) catalyzes the synthesis of double-stranded DNA from the retrotransposon RNA. (④) The free retrotransposon DNA then inserts itself into the genome at some other location. In this way, retrotransposons can spread to many places in the genome. (Steps 3 and 4 in Figure 19A-2 correspond to steps 2–4 in Figure 19A-1.)

Retrotransposons are probably present in all eukaryotic organisms. Although they transpose themselves only rarely, they can attain very high numbers of copies within a genome. Over 10% of the human genome, for example, consists of retrotransposons. We encountered one type in Chapter 16—the *Alu family* of sequences. Alu sequences are only 300 base pairs long and do not encode a reverse transcriptase, but, using a reverse transcriptase encoded elsewhere in the genome, they have sent copies of themselves throughout the genomes of humans and other primates. The haploid genome of humans contains about half a million Alu sequences, making up 5% of human DNA! Apparently molecular "parasites," such sequences are not known to have any function other than their own propagation. The question of what keeps retrotransposons from completely overrunning the genome still awaits an answer.

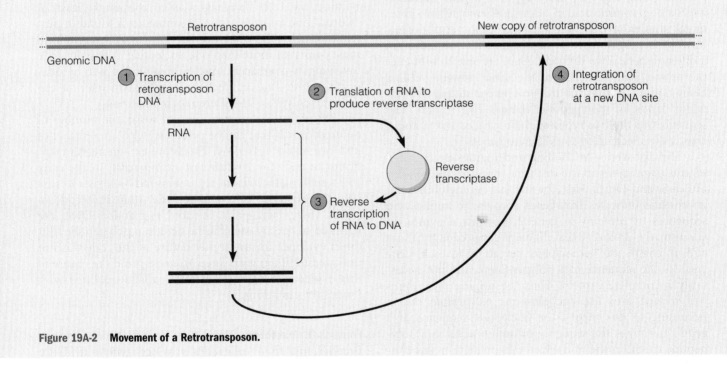

Figure 19A-2 Movement of a Retrotransposon.

arated from each other as shown in Figure 19-4. When Ingram examined the peptide patterns of normal and sickle-cell hemoglobin, he discovered that only one peptide differed between the two proteins. Analysis of the altered peptide revealed that a glutamic acid in normal hemoglobin had been replaced by a valine in sickle-cell hemoglobin. Since glutamic acid is negatively charged and valine is neutral, this substitution explains the difference in electrophoretic behavior between normal and sickle-cell hemoglobin originally observed by Pauling.

This single change from a glutamic acid to valine (caused by a single base-pair change in DNA) is enough to

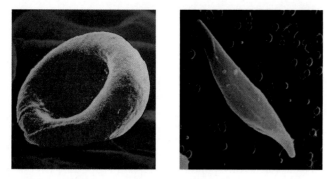

Figure 19-3   Normal and Sickled Red Blood Cells. The micrograph on the right reveals the abnormal shape of a sickled cell. This distorted shape, which is caused by a mutated form of hemoglobin, allows sickled cells to become trapped and damaged when passing through small blood vessels (SEMs).

kinds of genes that code for RNA molecules such as transfer RNAs (p. 662), ribosomal RNAs (p. 644), and small nuclear RNAs (p. 651). A direct linear correspondence between the complete base sequence of a gene and the amino acid sequence of the polypeptide chain it encodes is typical for bacterial genes, as shown by Yanofsky for tryptophan synthetase, but the situation in eukaryotes is usually more complex. As we will describe later in the chapter, most eukaryotic genes contain noncoding sequences interspersed among the coding regions of the gene, and hence do not show a complete linear correspondence with their polypeptide product. We will see that this complex type of gene organization allows more than one type of mRNA to be produced from a single gene, which means that an individual gene can actually code for more than one polypeptide.

alter the way in which hemoglobin molecules pack into red blood cells. Normal hemoglobin is jellylike in consistency, but sickle-cell hemoglobin tends to form a kind of crystal when it delivers oxygen to and picks up carbon dioxide from tissues. The crystalline array deforms the red blood cell into a sickled shape that blocks blood flow in capillaries and leads to a debilitating, potentially fatal, disease.

Following Ingram's discovery that a gene mutation alters a single amino acid in sickle-cell hemoglobin, subsequent studies revealed the existence of other abnormal forms of hemoglobin, some of which involve mutations in a different gene. Two different genes are able to influence the amino acid sequence of the same protein because hemoglobin is a multisubunit protein containing two different kinds of polypeptide chains. The amino acid sequences of the two types of chains, called the α and β chains, are specified by two different genes.

The discoveries by Pauling and Ingram necessitated several refinements in the one gene–one enzyme concept of Beadle and Tatum. First, the fact that hemoglobin is not an enzyme indicates that genes encode the amino acid sequences of proteins in general, not just enzymes. In addition, the discovery that different genes code for the α and β chains of hemoglobin reveals that each gene encodes the sequence of a polypeptide chain, not necessarily a complete protein. Thus the original hypothesis was refined into the *one gene–one polypeptide theory.* According to this theory, the nucleotide sequence of a **gene** determines the sequence of amino acids in a polypeptide chain. In the mid-1960s this prediction was confirmed in the laboratory of Charles Yanofsky, where the locations of dozens of mutations in the bacterial gene coding for a subunit of the enzyme tryptophan synthetase were determined. As predicted, the positions of the mutations within the gene correlated with the positions of the resulting amino acid substitutions in the tryptophan synthetase polypeptide chain (Figure 19-5).

The principle that the nucleotide sequence of a gene codes for the amino acid sequence of a polypeptide chain applies to virtually all genes, with the exception of a few

The Genetic Code Is a Triplet Code

Given a sequence relationship between DNA and proteins, the next question is how many nucleotides in DNA are needed to specify each amino acid in a protein? We know that the information in DNA must reside in the sequence of the four nucleotides that constitute the DNA: A, T, G, and C. These are the only "letters" of the DNA alphabet. Because the DNA language has to contain at least 20 "words," one for each of the 20 amino acids found in protein molecules, the DNA word coding for each amino acid must consist of more than one nucleotide. A doublet code involving two adjacent nucleotides would not be adequate, as four kinds of nucleotides taken two at a time can generate only $4^2 = 16$ different combinations.

But with three nucleotides per word, the number of different words that can be produced with an alphabet of just four letters is $4^3 = 64$. This number is more than sufficient to code for 20 different amino acids. In the early 1950s such mathematical arguments led biologists to suspect the existence of a **triplet code**—that is, a code in which three base pairs in double-stranded DNA are required to specify each amino acid in a polypeptide. But direct evidence for the triplet nature of the code was not provided until ten years later. To understand the nature of that evidence, we first need to become acquainted with frameshift mutations.

Frameshift Mutations. In 1961, Francis Crick, Sydney Brenner, and their colleagues provided genetic evidence for the triplet nature of the code by studying the mutagenic effects of the chemical *proflavin* on bacteriophage T4. Their work is well worth considering, not just because of the critical evidence it provided concerning the nature of the code but also because of the ingenuity of the deductive reasoning that was needed to understand the significance of their observations.

Proflavin is one of several *acridine dyes* commonly used as **mutagens** (mutation-inducing agents) in genetic research. Acridines are interesting mutagens because

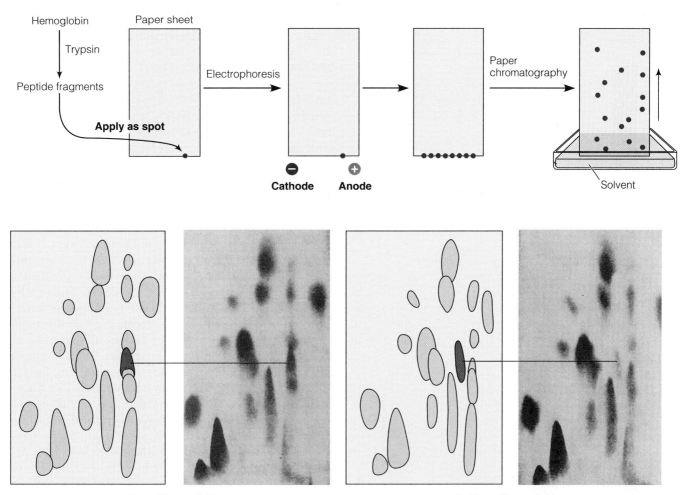

Figure 19-4 Peptide Patterns of Normal and Sickle-Cell Hemoglobin. In these experiments carried out by Vernon Ingram, hemoglobin was digested into peptide fragments using the protease trypsin. The peptide fragments were then separated from each other first by electrophoresis, and then by paper chromatography (a technique in which a solvent is allowed to move up the sheet of paper by capillary action). The photographs show the pattern of peptide fragments obtained from normal and sickle-cell hemoglobin. The colored spots in the drawings next to each photograph indicate the regions where the two patterns differ. The altered fragment in sickle-cell hemoglobin contains one altered amino acid (valine instead of glutamic acid).

they act by causing the addition or deletion of single base pairs in DNA. Sometimes, mutants generated by acridine treatment of a wild-type ("normal") virus or organism appear to revert to the wild type when treated with more of the same type of mutagen. Closer examination often reveals, however, that the reversion is not a true reversal of the original mutation, but the acquisition of a second mutation that maps very close to the first. In fact, the two mutations are in the same gene, and if they are separated by genetic recombination, each still gives rise to a mutant phenotype.

These mutations display an interesting kind of arithmetic. If the first alteration is called a plus (+) mutation, then the second can be called a minus (−) mutation. By itself, each creates a mutant phenotype, but when they occur close together as a double mutation, they cancel each other out and the virus or organism exhibits the normal,

wild-type phenotype. (Properly speaking, the phenotype is said to be *pseudo wild-type* because in spite of its wild-type appearance, two mutations are present.) Such behavior can be explained using the analogy in Figure 19-6. Suppose that line 1 represents a wild-type "gene" written in a language that uses three-letter words. When we "translate" the line by starting at the beginning and reading three letters at a time, the message of the gene is readily comprehensible. A plus mutation is the addition of a single letter within the message (line 2). That change may seem minor, but since the message is always read three letters at a time, the insertion of an extra letter early in the sequence means that all the remaining letters are read out of phase. There is, in other words, a shift in the *reading frame,* and the result is a garbled message from the point of the insertion onward. A minus mutation can be explained in a similar way because the deletion of a single letter also causes the reading frame

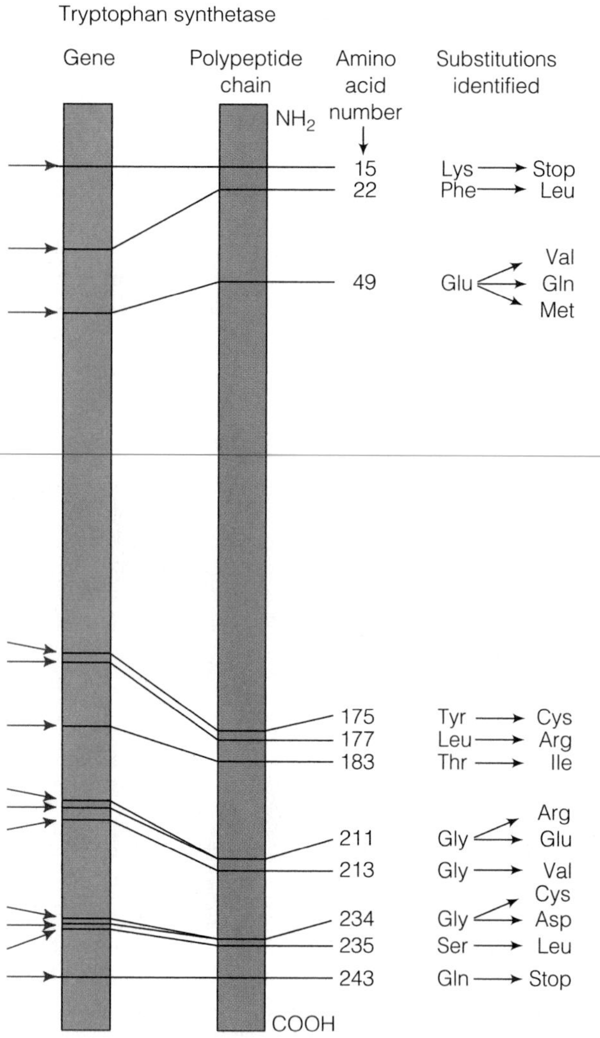

Figure 19-5 Relationship Between Mutation Sites in a Gene and Amino Acid Substitutions in Its Protein Product. In these data for the bacterial tryptophan synthetase gene, the linear order of mutation sites in the gene (red arrows) corresponds to the linear sequence of amino acid alterations in the polypeptide chain. The mutation sites in the gene and polypeptide do not line up precisely because the gene mutation sites were located using genetic recombination frequencies, which does not permit the exact localization of sites within a DNA base sequence.

to shift, resulting in another garbled message (line 3). Such **frameshift mutations** are typical effects of acridine dyes and other mutagens that cause the insertion or deletion of individual base pairs.

Individually, plus and minus mutations always change the reading frame and garble the message. But when a plus and a minus mutation occur in the same gene, they can largely cancel out each other's effect, particularly when they are located in close proximity. The insertion caused by the plus mutation compensates for the deletion caused by the minus mutation, and the message is intelligible from that point on (line 4). Notice, however, that double mutations with either two additions (+/+; line 5) or two

deletions (−/−; line 6) do not cancel in this way. They remain out of phase for the remainder of the message.

Evidence for a Triplet Code. When Crick and Brenner generated T4 phage mutants with proflavin, they obtained results similar to those in the hypothetical example illustrated in Figure 19-6 involving a language that uses three-letter words. They found that minus mutants, which exhibited an abnormal phenotype, could acquire a second mutation that caused them to revert to the wild-type (or more properly, pseudo wild-type) phenotype. The second mutation was always a plus mutation located at a site different from, but close to, the original minus mutation. In other words, phage reverting back to the wild-type phenotype exhibited a −/+ pattern of mutations. Crick and Brenner observed many examples of −/+ (or +/−) mutants exhibiting the wild-type phenotype in their experiments. But when they generated +/+ or −/− double mutants by recombination, no wild-type phenotypes were ever seen.

Crick and Brenner also constructed triple mutants of the same types (+/+/+ or −/−/−) and found that many of these did revert to wild-type phenotypes. This finding, of course, can be readily understood by consulting lines 7 and 8 of Figure 19-6: The reading frame (based on three-letter words) at the beginning and end of that hypothetical message remains the same when three letters are either added or removed. The portion of the message between the first and third mutations is garbled, but provided these are sufficiently close to each other, enough of the sentence may remain to convey an intelligible message.

Applying this concept of a three-letter code to DNA, Crick and Brenner concluded that adding or deleting a single base pair will shift the reading frame of the gene from that point onward, and a second, similar change shifts the reading frame yet again. Therefore from the site of the first mutation onward, the message is garbled. But after a third change of the same type, the original reading frame is restored, and the only segment of the gene translated incorrectly is the segment between the first and third mutations. Such errors can often be tolerated when the genetic message is translated into the amino acid sequence of a protein, provided the affected region is short and the changes in amino acid sequence do not destroy protein function. This is why the individual mutations in a triple mutant with wild-type phenotype map so closely together. Subsequent sequencing of "wild-type" polypeptides from such triple mutants confirmed the slightly altered sequences of amino acids that one would predict.

Based on their finding that wild-type phenotypes are often maintained in the presence of three base-pair additions (or deletions) but not in the presence of one or two, Crick and Brenner concluded that the nucleotides making up a DNA strand are read in groups of three. In other words, the genetic code is a triplet code in which the reading of a message begins at a specific starting place (to ensure the proper reading frame) and then proceeds three nucleotides at a time, with each such triplet translated into

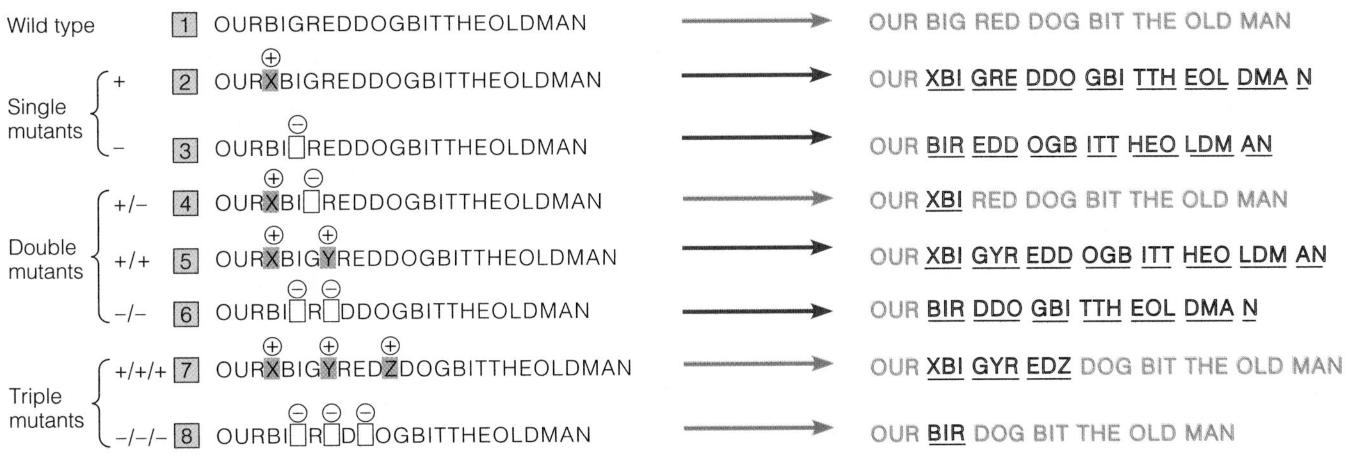

Wild type	☐1 OURBIGREDDOGBITTHEOLDMAN	→	OUR BIG RED DOG BIT THE OLD MAN
Single mutants { +	☐2 OUR⊕X̄BIGREDDOGBITTHEOLDMAN	→	OUR XBI GRE DDO GBI TTH EOL DMA N
−	☐3 OURBI⊖☐REDDOGBITTHEOLDMAN	→	OUR BIR EDD OGB ITT HEO LDM AN
Double mutants { +/−	☐4 OUR⊕X̄BI⊖☐REDDOGBITTHEOLDMAN	→	OUR XBI RED DOG BIT THE OLD MAN
+/+	☐5 OUR⊕X̄BIG⊕ȲREDDOGBITTHEOLDMAN	→	OUR XBI GYR EDD OGB ITT HEO LDM AN
−/−	☐6 OURBI⊖☐R⊖☐DDOGBITTHEOLDMAN	→	OUR BIR DDO GBI TTH EOL DMA N
Triple mutants { +/+/+	☐7 OUR⊕X̄BIG⊕ȲRED⊕Z̄DOGBITTHEOLDMAN	→	OUR XBI GYR EDZ DOG BIT THE OLD MAN
−/−/−	☐8 OURBI⊖☐R⊖☐D⊖☐OGBITTHEOLDMAN	→	OUR BIR DOG BIT THE OLD MAN

Figure 19-6 Frameshift Mutations. The effect of frameshift mutations can be illustrated with an English sentence. The wild-type sentence (line 1) consists of three-letter words. When read in the correct frame, it is fully comprehensible. The insertion (line 2) or deletion (line 3) of a single letter shifts the reading frame and garbles the message from that point onward. (Garbled words due to shifts in the reading frame are underscored.) Double mutants containing a deletion that "cancels" a prior insertion have a restored reading frame from the point of the second mutation onward (line 4). However, double insertions (line 5) or double deletions (line 6) produce garbled messages. Triple insertions (line 7) or deletions (line 8) garble part of the message but restore the reading frame with the net addition or deletion of a single word.

the appropriate amino acid, until the end of the message is reached. Keep in mind that in establishing the triplet nature of the code, Crick and Brenner did not have Figure 19-6 to assist them. Their ability to deduce the correct explanation from their analysis of proflavin-induced mutations is an especially inspiring example of the careful, often ingenious reasoning that almost always accompanies significant advances in science.

The Genetic Code Is Degenerate and Nonoverlapping

From the fact that so many of their triple mutants were viable, Crick and Brenner drew an additional conclusion: Most of the 64 possible nucleotide triplets must specify amino acids, even though proteins have only 20 different kinds of amino acids. If only 20 of the 64 possible combinations of nucleotides "made sense" to the cell, the chances of a meaningless triplet appearing in the out-of-phase stretches would be high. Such triplets would surely interfere with protein synthesis, and frameshift mutants would revert to wild-type behavior only rarely.

But Crick and Brenner detected reversion to wild-type behavior frequently, so they reasoned that most of the 64 possible triplets must code for amino acids. Since there are only 20 amino acids, this told them the genetic code is a **degenerate code**—that is, a given amino acid can be specified by more than one nucleotide triplet. In spite of the somewhat derogatory connotation of this word, degeneracy in no way implies malfunction. Quite the contrary, degeneracy serves a useful function in enhancing the adaptability of the coding system. If only 20 triplets were assigned a coding function (one for each of the 20 amino acids), any mutation in DNA that led to the formation of any of the other 44 possible triplets would interrupt the genetic message at that point. Therefore, the susceptibility of such a coding system to disruption would be very great.

A further conclusion from Crick and Brenner's work—which we have implicitly assumed—is that the genetic code is *nonoverlapping*. In an *overlapping* code, the reading frame would advance only one or two nucleotides at a time along a DNA strand, so that each nucleotide would be read two or three times. Figure 19-7 compares a nonoverlapping code with an overlapping code in which the reading frame advances one nucleotide at a time. With such an overlapping code, the insertion or deletion of a single base pair in the gene would lead to the insertion or deletion of one amino acid at one point in the polypeptide and would change several adjacent amino acids, but it would not affect the reading frame of the remainder of the gene. This means that if the genetic code were overlapping, Crick and Brenner would not have observed their frameshift mutations. Thus, their results clearly indicated the nonoverlapping nature of the code: Each nucleotide is a part of one, and only one, triplet.

Interestingly, although the genetic code is always translated in a nonoverlapping way, there are cases where a particular segment of DNA is translated in more than one reading frame. For example, certain viruses with very small genomes have overlapping genes, as was first discovered in 1977 for phage ϕX174. In this phage's DNA, one gene is completely embedded within another gene, and, to complicate matters further, a third gene overlaps them both! The three genes are translated in different reading frames. Presumably, it is highly advantageous for the phage to have its genome as small as possible. Other instances of overlapping genes are found in bacteria. As we will see in

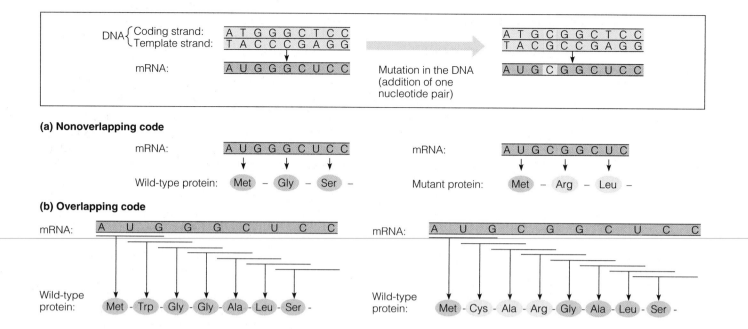

(a) Nonoverlapping code

(b) Overlapping code

Figure 19-7 Effect of the Insertion of a Single Nucleotide Pair on Proteins Encoded by Overlapping and Nonoverlapping Genetic Codes.   One strand of the DNA duplex at the top, called the *template strand,* is transcribed into the nine-nucleotide segment of mRNA shown, according to the same base-pairing rules used in DNA replication, except the base U is used in RNA in place of T. (The complementary DNA strand, with a sequence essentially identical to that of the mRNA, is called the *coding strand.*) **(a)** With a nonoverlapping code, the reading frame advances three nucleotides at a time, and this mRNA segment is therefore read as three successive triplets, coding for the amino acids

methionine, glycine, and serine. (See Figure 19-2 or 19-8 for amino acid coding rules.) If the DNA duplex is mutated by insertion of a single nucleotide pair (the yellow-shaded CG pair in the top box), the mRNA will have an additional nucleotide. This insertion alters the reading frame beyond that point, so the remainder of the mRNA is read incorrectly and all amino acids are wrong. In the example shown, the insertion occurs near the beginning of the message, and the only similarity between the wild-type protein and the mutant protein is the first amino acid (methionine). **(b)** In one type of overlapping code, the reading frame advances only one nucleotide at a time. The wild-type protein will therefore contain

three times as many amino acids as would a protein generated from the same mRNA using a nonoverlapping code. Insertion of a single nucleotide pair in the DNA again results in an mRNA molecule with one extra nucleotide. However, in this case the effect of the insertion on the protein is modest; two amino acids in the wild-type protein are replaced by three different amino acids in the mutant protein, but the remainder of the protein is normal. The frameshift mutations that Crick and Brenner found in their studies with the mutagen proflavin would not have been observed if the genetic code were overlapping. Accordingly, their data indicated the code to be nonoverlapping.

Chapter 21, some prokaryotic mRNAs encode more than one polypeptide; in such cases, genes sometimes overlap by a few nucleotides at their boundaries.

Messenger RNA Guides the Synthesis of Polypeptide Chains

After the publication of Crick and Brenner's historic findings in 1961, it took only five years for the meaning of each of the 64 triplets in the genetic code to be elucidated. Before we look at how that was done, let us first describe the role of RNA in the coding system. As we usually describe it, the genetic code refers not to the order of nucleotides in double-stranded DNA but to their order in the single-stranded mRNA molecules that actually direct protein synthesis. As indicated at the top of Figure 19-7, mRNA molecules are transcribed from DNA using a base-pairing mechanism similar to DNA replication, with two significant differences.

(1) In contrast to DNA replication, where both DNA strands are copied, only one of the two DNA strands—the **template strand**—serves as a template for mRNA formation during transcription. The nontemplate DNA strand, although not directly involved in transcription, is by convention called the **coding strand** because it is similar in sequence to the single-stranded mRNA molecules that carry the coded message.

(2) The mechanism used to copy sequence information from a DNA template strand to a complementary molecule of RNA utilizes the same base-pairing rules as apply to DNA replication, with the single exception that the base uracil (U) is employed in RNA where the base thymine (T) would have been incorporated into DNA. This substitution is permitted because U and T can both form hydrogen bonds with the base A. During DNA replication the base A pairs with T, whereas in transcription the base A pairs with U. Hence the sequence of an mRNA molecule is not exactly the same as the DNA coding

strand, in that mRNA contains the base U anywhere the coding DNA strand has the base T.

How do we know that mRNA molecules, produced by this transcription process, are responsible for directing the order in which amino acids are linked together during protein synthesis? This relationship was first demonstrated experimentally in 1961 by Marshall Nirenberg and J. Heinrich Matthei, who pioneered the use of *cell-free systems* for studying protein synthesis. In such systems, protein synthesis can be studied outside living cells by mixing together isolated ribosomes, amino acids, an energy source, and an extract containing soluble components of the cytoplasm. Nirenberg and Matthei found that adding RNA to cell-free systems increased the rate of protein synthesis, raising the question of whether the added RNA molecules were functioning as messages that determined the amino acid sequences of the proteins being manufactured. To address this question, they decided to add synthetic RNA molecules of known base composition to the cell-free system to see if such RNA molecules would influence the type of protein being made.

Their initial experiments took advantage of an enzyme called *polynucleotide phosphorylase,* which can be used to make synthetic RNA molecules of predictable base composition. Unlike the enzymes involved in cellular transcription, polynucleotide phosphorylase does not require a template but simply assembles available nucleotides randomly into a linear chain. If only one or two of the four ribonucleotides (ATP, GTP, CTP, and UTP) are provided, the enzyme will synthesize RNA molecules with a restricted base composition. The simplest RNA molecule results when a single kind of nucleotide is used, because the only possible product is an RNA *homopolymer* consisting of a single repeating nucleotide. For example, when polynucleotide phosphorylase is incubated with UTP as the sole substrate, the product is a homopolymer of uracil, called poly(U). When Nirenberg and Matthei added poly(U) to a cell-free protein-synthesizing system, they observed a marked increase in the incorporation of one particular amino acid, phenylalanine, into polypeptide chains. Synthetic RNA molecules containing bases other than uracil did not stimulate phenylalanine incorporation, whereas poly(U) enhanced the incorporation of only phenylalanine.

From these observations, Nirenberg and Matthei concluded that poly(U) directs the synthesis of polypeptide chains that consist solely of phenylalanine. This observation represented a crucial milestone in the development of the messenger RNA concept, for it was the first demonstration that the base sequence of an RNA molecule determines the order in which amino acids are linked together during protein synthesis.

The Codon Dictionary Was Established Using Synthetic RNA Polymers and Triplets

Once it had been shown that RNA functions as a messenger that guides the process of protein synthesis, the exact nature of the triplet coding system could be elucidated. Nucleotide triplets in mRNA, called **codons,** are the actual coding units read by the translational machinery during protein synthesis. The four bases present in RNA are the purines adenine (A) and guanine (G) and the pyrimidines cytosine (C) and uracil (U), so the 64 triplet codons consist of all 64 possible combinations of these four "letters" taken three at a time. And since mRNA molecules are synthesized in the $5' \longrightarrow 3'$ direction (like DNA) and are translated starting at the 5' end, the 64 codons by convention are always written in the $5' \longrightarrow 3'$ order.

These triplet codons in mRNA determine the amino acids that will be incorporated during protein synthesis, but which amino acid does each of the 64 triplets code for? The discovery that poly(U) directs the incorporation of phenylalanine during protein synthesis allowed Nirenberg and Matthei to make the first codon assignment: The triplet UUU in mRNA must code for the amino acid phenylalanine. Subsequent studies on the coding properties of other synthetic homopolymers, such as poly(A) and poly(C), quickly revealed that AAA codes for lysine and CCC codes for proline. (Because of unexpected structural complications, poly(G) is not a good messenger and was not tested.)

After the homopolymers had been tested, polynucleotide phosphorylase was used to create *copolymers* containing a mixture of two nucleotides, but the results were more difficult to interpret. For example, the copolymer synthesized by incubating polynucleotide phosphorylase with the precursors CTP and ATP contains C's and A's, but in no predictable order. Such a copolymer contains eight different codons: CCC, CCA, CAC, ACC, AAC, ACA, CAA, and AAA. When this copolymer was used to direct protein synthesis, the resulting polypeptides contained six of the 20 possible amino acids. The codons for two of these (CCC and AAA) were known from the homopolymer studies, but the other four could not be unambiguously assigned.

Further progress depended on an alternative means of codon assignment devised by Nirenberg's group. Instead of using long polymers, they synthesized 64 very short RNA molecules, each only three nucleotides long. They then conducted studies to see which amino acid bound to the ribosome in response to each of these triplets. (In such experiments, tRNA molecules actually carry the amino acids to the ribosome.) With this approach, they were able to determine most of the codon assignments.

Meanwhile, a refined method of polymer synthesis had been devised in the laboratory of H. Gobind Khorana. Khorana's approach was similar to that of Nirenberg and Matthei, but with the important difference that the polymers he synthesized had defined sequences. Thus, he could produce a synthetic mRNA molecule with the strictly alternating sequence UAUA.... Such an RNA copolymer has only two codons, UAU and AUA, and they alternate in strict sequence. When Khorana added this particular RNA to a cell-free protein-synthesizing system, a polypeptide containing only tyrosine and isoleucine was produced. Khorana was therefore able to narrow the possible codon assignments for UAU and AUA to these two particular amino acids. When the results obtained with such synthetic polymers were combined with the findings of Nirenberg's binding studies, most of the codons could be assigned unambiguously.

Of the 64 Possible Codons in Messenger RNA, 61 Code for Amino Acids

By 1966, just five years after the first codon was identified, the approaches we have just described allowed all 64 codons to be assigned—that is, the entire genetic code had been worked out, as shown in Figure 19-8. The elucidation of the code confirmed several properties that had been deduced earlier from indirect evidence. All 64 codons are in fact used in the translation of mRNA; 61 of the codons specify the addition of specific amino acids to the growing polypeptide, and one of these (AUG) also plays a prominent role as an **initiation codon** that starts the process of protein synthesis. The remaining three codons (UAA, UAG, and UGA) are **stop codons** that instruct the cell to terminate synthesis of the polypeptide chain.

It is clear from examining Figure 19-8 that the genetic code is *unambiguous:* Every codon has one and only one meaning. The figure also shows the *degenerate* nature of the code—that is, many of the amino acids are specified by more than one codon. There are, for example, two codons for histidine (His), four for threonine (Thr), and six for leucine (Leu). Although degeneracy may sound wasteful, it serves a useful function in enhancing the adaptability of the coding system. As we noted earlier, if there were only one codon for each of the 20 amino acids incorporated into proteins, then any mutation in DNA that led to the formation of one of the remaining codons would stop the synthesis of the growing polypeptide chain at that point. But with a degenerate code, most mutations simply cause codon changes that alter the specified amino acid. The change in a protein's behavior that results from a single amino acid alteration is often quite small, and in some cases may even be advantageous. Moreover, mutations in the third base of a codon frequently do not change the specified amino acid at all, as you can see in Figure 19-8. For example, a mutation that changes the codon ACU to ACC, ACA, or ACG does not alter the corresponding amino acid, which is threonine (Thr) in all four cases.

The validity of the codon assignments summarized in Figure 19-8 has been confirmed by analyzing the amino acid sequences of mutant proteins. For example, we learned earlier in the chapter that sickle-cell hemoglobin differs from normal hemoglobin at a single amino acid position, where valine is substituted for glutamic acid. The genetic code table reveals that glutamic acid may be encoded by either GAA or GAG. Whichever triplet is employed, a single base change could create a codon for valine. For example, GAA might have been changed to GUA, or GAG might have been changed to GUG. In either case, a glutamic acid codon would be converted into a valine codon. Many other mutant proteins have been examined in a similar way, and in nearly all cases the amino acid substitutions are consistent with a single base change in a triplet codon.

First position	Second position: U	C	A	G	Third position
U	UUU Phe	UCU Ser	UAU Tyr	UGU Cys	U
U	UUC Phe	UCC Ser	UAC Tyr	UGC Cys	C
U	UUA Leu	UCA Ser	UAA Stop	UGA Stop	A
U	UUG Leu	UCG Ser	UAG Stop	UGG Trp	G
C	CUU Leu	CCU Pro	CAU His	CGU Arg	U
C	CUC Leu	CCC Pro	CAC His	CGC Arg	C
C	CUA Leu	CCA Pro	CAA Gln	CGA Arg	A
C	CUG Leu	CCG Pro	CAG Gln	CGG Arg	G
A	AUU Ile	ACU Thr	AAU Asn	AGU Ser	U
A	AUC Ile	ACC Thr	AAC Asn	AGC Ser	C
A	AUA Ile	ACA Thr	AAA Lys	AGA Arg	A
A	AUG Met/start	ACG Thr	AAG Lys	AGG Arg	G
G	GUU Val	GCU Ala	GAU Asp	GGU Gly	U
G	GUC Val	GCC Ala	GAC Asp	GGC Gly	C
G	GUA Val	GCA Ala	GAA Glu	GGA Gly	A
G	GUG Val	GCG Ala	GAG Glu	GGG Gly	G

Figure 19-8 The Genetic Code. The code "words" are three-letter codons present in the nucleotide sequence of mRNA, as read in the $5' \longrightarrow 3'$ direction. Letters represent the nucleotide bases uracil (U), cytosine (C), adenine (A), and guanine (G). Each codon specifies either an amino acid or a stop signal. To decode a codon, read down the left edge for the first letter, then across the grid for the second letter, and then down the right edge for the third letter. For example, the codon AUG represents methionine. (As we shall see in Chapter 20, AUG is also a start signal.)

The Genetic Code Is (Nearly) Universal

A final property of the genetic code worth noting is its near universality. Except for a few cases, all organisms studied so far—prokaryotes as well as eukaryotes—use the same basic genetic code. Even viruses, though they are nonliving entities, employ this same code. In other words, the 64 codons almost always stand for the same amino acids or stop signals specified in Figure 19-8, suggesting that this coding system was established early in the history of life on earth and has remained largely unchanged over billions of years of evolution.

However, exceptions to the standard genetic code have been observed in a few situations, most notably in mitochondria and in a few bacteria and unicellular organisms. In the case of mitochondria, which contain their own DNA and carry out both transcription and translation, the genetic code can differ in several ways from the standard code. One difference involves the codon UGA, which is a stop codon in the standard code but is translated as tryptophan in mammalian and yeast mitochondria. Conversely, AGA is a stop codon in mammalian mitochondria, even though in most other systems (including yeast mitochondria) it codes for arginine. Such anomalies result from alterations in the properties of transfer RNA (tRNA) molecules found in mitochondria. As you will learn in the next chapter, tRNA molecules play a key role in the genetic code because they recognize codons in mRNA and bring the appropriate amino acid to each codon during the process of protein synthesis. Differences in the types of tRNA molecules present in mitochondria appear to underlie the ability to read codons such as UAG and AGA differently than in the standard genetic code.

In addition to the codon changes observed in mitochondria, bacteria called *mycoplasmas* also employ a few codons in a nonstandard way, as do the nuclear genomes of certain protozoa and fungi. For example, the fungus *Candida* produces an unusual tRNA that brings the amino acid serine to mRNAs containing the codon CUG rather than bringing the normally expected amino acid, leucine. And in an especially interesting case, observed in organisms as diverse as bacteria and mammals, a variation of the genetic code allows it to specify the incorporation of a twenty-first amino acid, *selenocysteine*, in which the sulfur atom of cysteine is replaced by an atom of selenium. In mRNAs coding for the few rare proteins that contain selenocysteine, the meaning of a UGA codon is changed from a stop codon to a codon specifying selenocysteine. In such cases, the folding of the mRNA molecule causes specific UGA codons to be recognized by a special tRNA carrying selenocysteine, rather than to function as stop codons that trigger the termination of protein synthesis.

Transcription in Prokaryotic Cells

Now that you have been introduced to the genetic code that governs the relationship between the nucleotide sequence of DNA and the amino acid sequence of protein molecules, we can discuss the specific steps involved in the flow of genetic information from DNA to protein. The first stage in this process is the transcription of a nucleotide sequence in DNA into a sequence of nucleotides in RNA. RNA is chemically similar to DNA, except it contains ribose instead of deoxyribose as its sugar, has the base uracil (U) in place of thymine (T), and is usually single-stranded (see Figure 3-17). As in other areas of molecular genetics, the fundamental principles of RNA synthesis were first elucidated in bacteria, where the molecules and mechanisms are relatively simple. For that reason, we will start with transcription in prokaryotes.

Transcription Is Catalyzed by RNA Polymerase, Which Synthesizes RNA Using DNA as a Template

Transcription of DNA is carried out by the enzyme **RNA polymerase,** which catalyzes the synthesis of RNA using DNA as a template. Bacterial cells have a single kind of RNA polymerase that synthesizes all three major classes of RNA—mRNA, tRNA, and rRNA. The enzymes from different bacteria are quite similar, and the RNA polymerase from *Escherichia coli* has been especially well characterized. It is a large protein consisting of two α subunits, two β subunits that differ enough to be identified as β and β', and a dissociable subunit called the **sigma (σ) factor.** Although the *core enzyme* lacking the sigma subunit is competent to carry out RNA synthesis, the complete *holoenzyme* ($\alpha_2\beta\beta'\sigma$) is required to ensure initiation at the proper sites within a DNA molecule. The sigma subunit plays a critical role in this process by promoting the binding of RNA polymerase to specific DNA sequences, called *promoters,* found at the beginnings of genes. Bacteria contain a variety of different sigma factors that selectively initiate the transcription of specific categories of genes, as we will discuss in Chapter 21.

Transcription Involves Four Stages: Binding, Initiation, Elongation, and Termination

Transcription is the synthesis of an RNA molecule whose base sequence is complementary to the base sequence of a template DNA strand. Figure 19-9 provides an overview of transcription that applies to both prokaryotes and eukaryotes. It illustrates RNA synthesis from a single **transcription unit,** a segment of DNA whose transcription gives rise to a single, continuous RNA molecule. The process of RNA synthesis begins with ① the binding of RNA polymerase to a DNA promoter sequence, which triggers local unwinding of the DNA double helix. Using one of the two DNA strands as a template, RNA polymerase then ② initiates the synthesis of an RNA chain. After initiation has taken place, the RNA polymerase molecule moves along the DNA template, unwinding the double helix and ③ elongating the RNA chain as it goes. During this process, the enzyme catalyzes the polymerization of nucleotides in an order determined by their base pairing with the DNA template strand. Eventually the enzyme transcribes a special base sequence called a *termination signal,* which ④ terminates RNA synthesis and causes the completed RNA molecule to be released and RNA polymerase to dissociate from the DNA template.

Although transcription is a complicated process, it can be thought of in four distinct stages: binding, initiation, elongation, and termination. We will now look at each stage in detail, as it occurs in *E. coli.* You can refer back to Figure 19-9 throughout this discussion to see how each steps fits into the overall process.

Binding of RNA Polymerase to a Promoter Sequence. The first step in RNA synthesis is the *binding* of RNA polymerase to

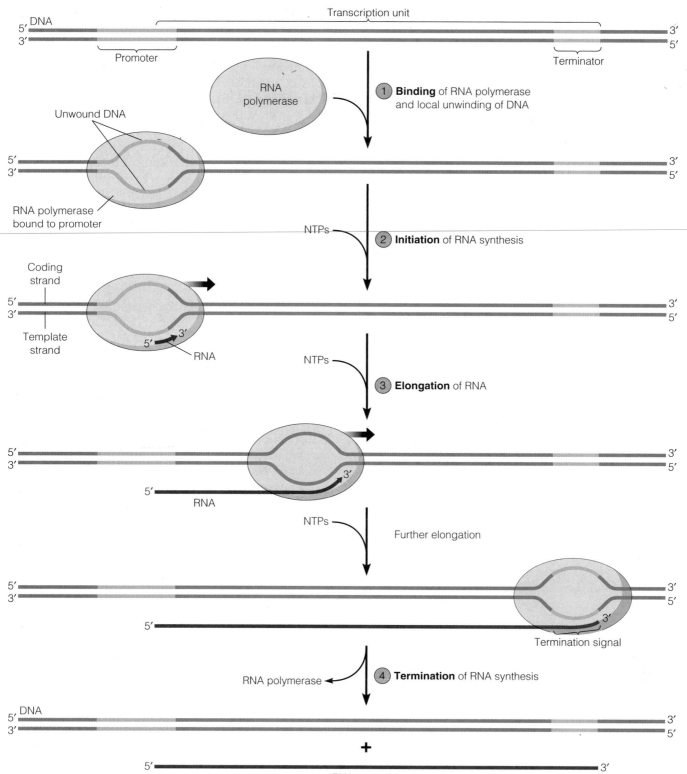

Figure 19-9 An Overview of Transcription.
Transcription of DNA occurs in four main stages: ① binding of RNA polymerase to DNA at a promoter, ② initiation of transcription on the template DNA strand, ③ subsequent elongation of the RNA chain, and ④ eventual termination of transcription, accompanied by the release of RNA polymerase and the completed RNA product from the DNA template. RNA polymerase moves along the template strand of the DNA in the 3′ ⟶ 5′ direction, and the RNA molecule grows in the 5′ ⟶ 3′ direction. The general scheme shown in this figure holds for both prokaryotic and eukaryotic transcription. NTPs (ribonucleoside triphosphate molecules) = ATP, GTP, CTP, and UTP.

a DNA **promoter site**—a specific sequence of several dozen base pairs that determines where RNA synthesis starts and which DNA strand is to serve as the template strand. Each transcription unit has a promoter site located near the beginning of the DNA sequence to be transcribed. By convention, promoter sequences are described in the $5' \rightarrow 3'$ direction on the coding strand, which is the strand that lies opposite the template strand. The terms **upstream** and **downstream** are used to refer to DNA sequences located toward the 5' or 3' end of the coding strand, respectively. Therefore the region where the promoter is located is said to be upstream of the transcribed sequence. In *E. coli,* the sigma subunit of RNA polymerase is required for promoter recognition and hence for the specificity of binding. Binding of RNA polymerase to the promoter leads to unwinding of the DNA double helix in the promoter region. Approximately 15–18 bp of DNA are unwound in the area surrounding the point where transcription will begin, exposing the two separate strands of the DNA double helix.

Promoter sequences were initially identified by **DNA footprinting,** a general technique for locating the DNA region to which a DNA-binding protein has become bound (Box 19B). When used for identifying promoter sequences, footprinting takes advantage of the fact that such sequences are specifically recognized and bound by RNA polymerase. Binding of RNA polymerase to DNA protects the nucleotide sequences at promoter sites from enzymatic (or chemical) attack, thereby allowing these sites to be identified. In addition to DNA footprinting, recombinant DNA techniques have also been used for studying promoter sequences. Essential sequences within the promoter region are identified by deleting or adding specific base sequences to cloned genes and then testing the ability of the altered DNA to be transcribed by RNA polymerase.

Studies using DNA footprinting and recombinant DNA techniques have revealed that DNA promoter sites differ significantly among bacterial transcription units. How, then, does a single kind of RNA polymerase recognize them all? Enzyme recognition and binding, it turns out, depend only on several very short sequences located at specific positions within each promoter site; the identities of the nucleotides making up the rest of the promoter are irrelevant for this purpose.

Figure 19-10 highlights the essential sequences in a typical prokaryotic promoter. The point where transcription will begin, called the *startpoint,* is almost always a purine and often an adenine. Approximately 10 bases upstream of the startpoint is the six-nucleotide sequence TATAAT, called the *−10 sequence* or the *Pribnow box,* after its discoverer. By convention, the nucleotides are numbered from the startpoint (+1), with positive numbers to the right (downstream) and negative ones to the left (upstream). The −1 nucleotide is immediately upstream of the startpoint (there is no "0"). At or near the −35 position is the six-nucleotide sequence TTGACA, called the *−35 sequence.*

The −10 and the −35 sequences (and their positions relative to the startpoint) have been conserved during evolution, but they are not identical in all bacterial promoters, or even in all the promoters in a single genome. For example, the −10 sequence in the promoter for one of the *E. coli* tRNA genes is TATGAT, whereas the −10 sequence for a group of genes needed for lactose breakdown in the same organism is TATGTT, and the sequence given in Figure 19-10 is TATAAT. The particular promoter sequences shown in the figure are **consensus sequences,** which consist of the most common nucleotides at each position within a given sequence. Mutations that cause significant deviations from the consensus sequences tend

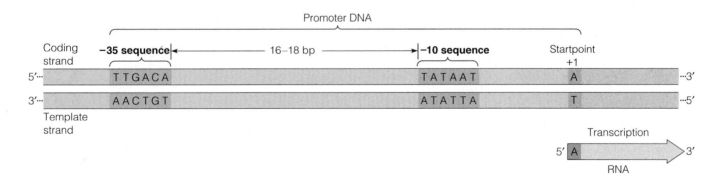

Figure 19-10   A Typical Prokaryotic Promoter.   The DNA promoter region in prokaryotes is a stretch of about 40 bp adjacent to and including the transcription startpoint. By convention, the critical DNA sequences are given as they appear on the coding strand (the nontemplate strand, which corresponds in sequence to the RNA transcript). The essential features of the promoter are the startpoint (designated +1 and usually an A), the six-nucleotide −10 sequence, and the six-nucleotide −35 sequence. As their names imply, the two key sequences are located approximately 10 nucleotides and 35 nucleotides upstream from the startpoint. The sequences shown here are consensus sequences, meaning that they have the most commonly found base at each position. The numbers of nucleotides separating the consensus sequences from each other and from the startpoint are important for promoter function, but the identity of these nucleotides is not.

Contemporary Techniques   THE FOOTPRINTING METHOD FOR IDENTIFYING PROTEIN-BINDING SITES ON DNA

The initiation of transcription depends on the interactions of proteins with specific DNA sequences. In prokaryotic cells, the RNA polymerase holoenzyme directly recognizes and binds to the promoter sequence at the start of a gene. In eukaryotic cells, certain transcription factors must bind to the promoter before an RNA polymerase can bind. And in all cells, the *regulation* of transcription depends on the interaction of still other proteins with specific sites on the DNA. Thus, the researcher seeking to understand transcription needs to know about transcriptional proteins and the DNA sequences to which some of them bind.

A technique called *DNA footprinting* can be used to locate the DNA regions where specific DNA-binding proteins attach. Footprinting can identify any DNA site that binds a protein specifically, as long as the protein binds tightly enough. The underlying principle is that the binding of a protein to a particular DNA sequence should protect that sequence from degradation by enzymes or chemicals. A widely used version of footprinting, outlined in Figure 19B-1, employs an endonuclease called DNAse I as the degradative agent. Unlike restriction endonucleases, this enzyme is not sequence-specific but attacks the bonds between nucleotides more or less at random.

In the figure here, the starting material is a piece of DNA known to contain a binding site for a particular DNA-binding protein (in a real case, the DNA might be several hundred base pairs long). Radioactive phosphate (indicated by the stars) has been added to the 5′ ends of the DNA by reacting the DNA with ^{32}P-labeled ATP and the enzyme polynucleotide kinase. (①) A sample of the end-labeled DNA is mixed with the DNA-binding protein under study; another sample, without the added protein, serves as the control. (②) Both samples are incubated for a short time with a low concentration of DNAse I—conditions ensuring that most of the DNA molecules will be cleaved only once. The arrowheads indicate possible cleavage sites in the DNA starting material. (③) The two incubation mixtures are submitted to electrophoresis in adjacent lanes of a polyacrylamide gel and visualized by autoradiography. The fragments resulting from the DNAse reactions are revealed as two series of bands. The control lane has nine bands spaced uniformly, because every possible cleavage site has been cut. However, the other lane is missing some of the bands, because the protein that was bound to the DNA protected some of the cleavage sites during DNAse treatment. The blank region in the left-hand lane is the "footprint" that identifies the location and length of the DNA sequence in contact with the DNA-binding protein. The exact sequence of the DNA

protein-binding site can be determined by subjecting a third sample of the starting DNA to DNA sequencing procedures.

Footprinting experiments like the one described here have revealed, for example, that when bacterial RNA polymerase binds to a promoter, it protects a segment of DNA extending from about 40 nucleotides upstream of the transcriptional startpoint to almost 20 nucleotides past that point.

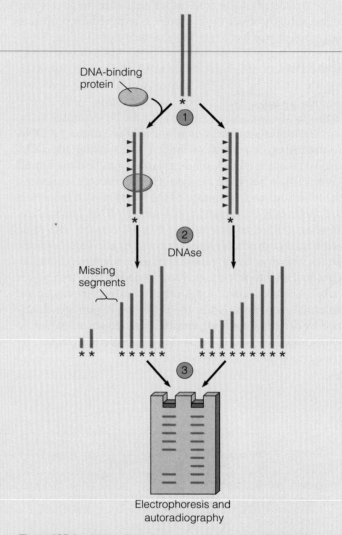

Figure 19B-1   **DNAse Footprinting as a Tool to Identify DNA Sites That Bind Specific Proteins.**

to interfere with promoter function and may even eliminate promoter activity entirely.

Initiation of RNA Synthesis. Once an RNA polymerase molecule has bound to a promoter site and locally unwound the DNA double helix, *initiation* of RNA synthesis can take place. One of the two exposed segments of single-stranded DNA serves as the template for the syn-

thesis of RNA, using incoming ribonucleoside triphosphate molecules (NTPs) as substrates. The DNA strand that carries the promoter sequence determines which way the RNA polymerase faces, and the enzyme's orientation in turn determines which DNA strand it transcribes (see Figure 19-9). As soon as the first two incoming NTPs are hydrogen bonded to the complementary bases of the DNA template strand at the startpoint, RNA polymerase cata-

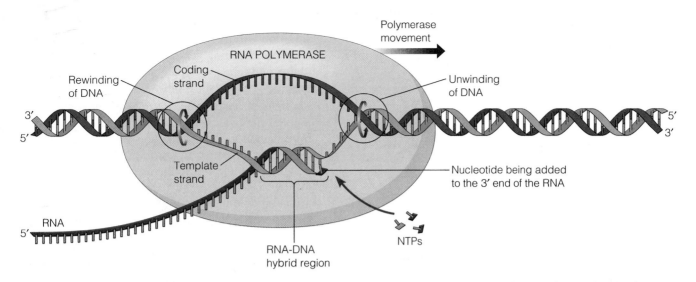

Figure 19-11   A Closeup of the Prokaryotic Elongation Complex. During elongation, RNA polymerase binds to about 30 bp of DNA (recall that each complete turn of the DNA double helix is about 10 bp). At any given moment, about 18 bp of DNA are unwound, and the most recently synthesized RNA is still hydrogen-bonded to the DNA, forming a short RNA-DNA hybrid. This hybrid is probably about 12 bp long, but it may be shorter. The total length of growing RNA bound to the enzyme and/or DNA is about 25 nucleotides. This diagram is drawn roughly to scale.

lyzes the formation of a phosphodiester bond between the 3′-hydroxyl group of the first NTP and the 5′-phosphate of the second, accompanied by the release of pyrophosphate (PP_i). The polymerase then moves along the promoter as additional nucleotides are added one by one, the 5′-phosphate of each new nucleotide joining to the 3′-hydroxyl group of the growing RNA chain, until the chain is about nine nucleotides long. At this point, the sigma factor generally detaches from the RNA polymerase molecule and the initiation stage is complete.

Elongation of the RNA Chain.   Chain *elongation* (Figure 19-11) now continues as RNA polymerase moves along the DNA molecule, untwisting the helix bit by bit and adding one complementary nucleotide at a time to the growing RNA chain. The enzyme moves along the template DNA strand from the 3′ toward the 5′ end. Because complementary base pairing between the DNA template strand and the newly forming RNA chain is antiparallel, *the RNA strand is elongated in the 5′ → 3′ direction* as each successive nucleotide is added to the 3′ end of the growing chain. (This is the same direction in which DNA strands are synthesized during DNA replication.) As the RNA chain grows, the most recently added nucleotides remain base-paired with the DNA template strand, forming a short RNA-DNA hybrid. Most evidence suggests that this hybrid is about 12 bp long, although it may be shorter. (Figure 19-11 is drawn roughly to scale.) In any case, the polymerase continues to move on, rewinding the DNA behind and unwinding the DNA ahead as it goes.

Unlike DNA polymerase (p. 535), some RNA polymerases lack 3′ → 5′ exonuclease activity that would allow them to correct mistakes by removing mismatched base pairs as RNA elongation proceeds. For this reason, RNA synthesis is subject to a higher error rate than DNA synthesis. This does not appear to create a problem, however, because multiple RNA molecules are usually transcribed from each gene, and so a small number of inaccurate copies can be tolerated. In contrast, only one copy of each DNA molecule is made when DNA is replicated prior to cell division. Since each newly forming cell receives only one set of DNA molecules, it is crucial that the copying mechanism used in DNA replication be extremely accurate.

Termination of RNA Synthesis.   Elongation of the growing RNA chain proceeds until RNA polymerase copies a special sequence, called a **termination signal,** that triggers the end of transcription. In bacteria, two classes of termination signals can be distinguished that differ in whether or not they require the participation of a protein called **rho (ρ) factor.** RNA molecules terminated without the aid of the rho factor contain a short GC-rich sequence followed by several U residues near their 3′ end (Figure 19-12). Since GC base pairs are held together by three hydrogen bonds, whereas AU base pairs are joined by only two hydrogen bonds, this configuration promotes termination in the following way: First, the GC region contains sequences that are complementary to each other, causing the RNA to spontaneously fold into a **hairpin loop** that tends to pull the RNA molecule away from the DNA. Then the weaker bonds between the sequence of U residues and the DNA template are broken, releasing the newly formed RNA molecule.

In contrast, RNA molecules that do not form a GC-rich hairpin loop require participation of the rho factor for termination. Genes coding for such RNAs were first discovered in experiments in which purified DNA obtained from bacteriophage λ was transcribed with

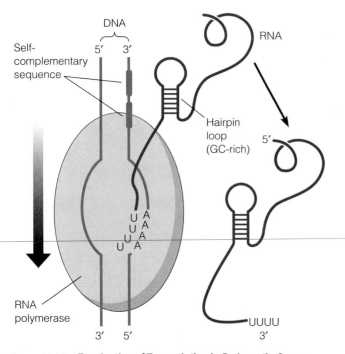

Figure 19-12   Termination of Transcription in Prokaryotic Genes That Do Not Require the Rho Termination Factor.   A short self-complementary sequence near the end of the gene allows the newly formed RNA molecule to form a hairpin loop structure that helps to dissociate the RNA from the DNA template.

purified RNA polymerase. Some genes were found to be transcribed into RNA molecules that are longer than the RNAs produced in living cells, suggesting that transcription was not terminating properly. This problem could be corrected by adding rho factor, which binds to specific termination sequences 50–90 bases long located near the 3′ end of newly forming RNA molecules. The rho factor acts as an ATP-dependent unwinding enzyme, moving along the newly forming RNA molecule toward its 3′ end and unwinding it from the DNA template as it proceeds.

Whether termination is rho-dependent or depends on the formation of a hairpin loop, it results in the release of the completed RNA molecule and of the core RNA polymerase. The core polymerase can then bind sigma factor again and reinitiate RNA synthesis at another promoter.

Transcription in Eukaryotic Cells

Transcription in eukaryotic cells involves the same four stages described in Figure 19-9, but the process in eukaryotes is more complicated than that in prokaryotes. The main differences are as follows:

- *Three different RNA polymerases* transcribe the nuclear DNA of eukaryotes. Each synthesizes one or more classes of RNA.

- *Eukaryotic promoters* are more varied than prokaryotic promoters. Not only are different types of promoters

employed for the three polymerases, but there is great variation within each type, especially among the ones for protein-coding genes. Furthermore, some eukaryotic promoters are actually located *downstream* from the transcription startpoint.

- Binding of eukaryotic RNA polymerases to DNA requires the participation of additional proteins, called *transcription factors.* Unlike the bacterial sigma factor, eukaryotic transcription factors are not part of the RNA polymerase molecule. Rather, some of them must bind to DNA *before* RNA polymerase can bind to the promoter and initiate transcription. Thus, transcription factors, rather than RNA polymerase itself, determine the specificity of transcription in eukaryotes. In this chapter, we limit our discussion to the class of factors that are essential for the transcription of all genes transcribed by an RNA polymerase. We defer discussion of the regulatory class of transcription factors, which selectively act on specific genes, until Chapter 21.

- *Protein-protein interactions* are very important in the first stage of eukaryotic transcription. Although some transcription factors bind directly to DNA, many attach to other proteins—either to other transcription factors or to RNA polymerase itself.

- *RNA cleavage* is more important than the site where transcription is terminated in determining the location of the 3′ end of the RNA product.

- Newly forming eukaryotic RNA molecules typically undergo extensive *RNA processing* (chemical modification) both during and, to a larger extent, after transcription.

We will now examine these various aspects of eukaryotic transcription, starting with the existence of multiple forms of RNA polymerase.

RNA Polymerases I, II, and III Carry Out Transcription in the Eukaryotic Nucleus

Table 19-1 summarizes some of the properties of the three RNA polymerases that function in the nucleus of the eukaryotic cell, as well as two others found in mitochondria and chloroplasts. The nuclear enzymes are designated RNA polymerases I, II, and III. As the table indicates, these enzymes differ in their location within the nucleus and in the kinds of RNA they synthesize. The nuclear RNA polymerases also differ in their sensitivity to various inhibitors, such as α-*amanitin,* a deadly toxin produced by the mushroom *Amanita phalloides* (known commonly as the "death cap" or "destroying angel").

RNA polymerase I resides in the nucleolus and is responsible for synthesizing an RNA molecule that serves as a precursor for three of the four types of rRNA found in eukaryotic ribosomes (28S rRNA, 18S rRNA, and 5.8S rRNA). This enzyme is not sensitive to α-amanitin. Its association with the nucleolus is understandable, as the nucleolus is the site of ribosomal RNA synthesis and ribosomal subunit assembly (p. 517). Because cells need large

Table 19-1   Properties of Eukaryotic RNA Polymerases

RNA Polymerase	Location	Main Products	α-Amanitin Sensitivity		
I	Nucleolus	Precursor for 28S rRNA, 18S rRNA, and 5.8S rRNA	Resistant	*40%*	*RNA synthesis*
II	Nucleoplasm	Pre-mRNA and most snRNA	Very sensitive	*60%*	
III	Nucleoplasm	Pre-tRNA, 5S rRNA, and other small RNAs	Moderately sensitive*	*3*	
Mitochondrial	Mitochondrion	Mitochondrial RNA	Resistant		
Chloroplast	Chloroplast	Chloroplast RNA	Resistant		

*In mammals.

amounts of rRNA, RNA polymerase I actually accounts for most cellular RNA synthesis.

RNA polymerase II is found in the nucleoplasm and synthesizes precursors to mRNA, the class of RNA molecules that code for proteins. In addition, RNA polymerase II synthesizes most of the *snRNAs*—small nuclear RNAs involved in posttranscriptional RNA processing. Thus, polymerase II is responsible for the synthesis of the greatest variety of RNA molecules. The enzyme is extremely sensitive to α-amanitin, which explains the toxicity of this compound to humans and other animals.

RNA polymerase III is also a nucleoplasmic enzyme, but it synthesizes a variety of small RNAs, including tRNA precursors and the smallest type of ribosomal RNA, 5S rRNA. Mammalian RNA polymerase III is sensitive to α-amanitin, but only at higher levels of the toxin than are required to inhibit RNA polymerase II. (The comparable enzymes of some other eukaryotes, such as insects and yeasts, are insensitive to α-amanitin.)

Structurally, RNA polymerases I, II, and III are somewhat similar to each other as well as to prokaryotic core RNA polymerase. The three enzymes are all quite large, with multiple polypeptide subunits and molecular weights around 500,000. RNA polymerase II, for example, has more than ten subunits of at least eight different types. The three biggest subunits are evolutionarily related to the prokaryotic RNA polymerase subunits α, β, and β'. Three of the smaller subunits lack that relationship but are also found in RNA polymerases II and III. The RNA polymerases of mitochondria and chloroplasts resemble their prokaryotic counterparts closely, as you might expect from the probable origins of these organelles as endosymbiotic bacteria (see Box 15A). Like bacterial RNA polymerase, they are resistant to α-amanitin.

Three Classes of Promoters Are Found in Eukaryotic Nuclear Genes, One for Each Type of RNA Polymerase

The promoters to which eukaryotic RNA polymerases bind are even more varied than prokaryotic promoters, but they can be grouped into three main categories, one for each type of polymerase. Figure 19-13 shows characteristic features of the three types of promoters.

The promoter used by RNA polymerase I—that is, the promoter of the transcription unit that produces the precursor for the three largest rRNAs—has two parts (Figure

19-13a). The part called the **core promoter**—defined as the minimal set of DNA sequences sufficient to direct the accurate initiation of transcription by RNA polymerase—actually extends into the nucleotide sequence to be transcribed. It is sufficient for proper initiation of transcription, but transcription is made more efficient by the presence of an *upstream control* element, which for RNA polymerase I is a fairly long sequence similar (though not identical) to the core promoter. Attachment of transcription factors to both parts of the promoter facilitates the binding of RNA polymerase I to the core promoter and enables it to initiate transcription at the startpoint.

In the case of RNA polymerase II, at least four types of DNA sequences are involved in core promoter function (Figure 19-13b). These four elements are (1) a short **initiator (Inr)** sequence surrounding the transcription startpoint (which is often an A, as in prokaryotes); (2) the **TATA box,** which consists of a consensus sequence of TATA followed by two or three more A's, usually located about 25 nucleotides upstream from the startpoint; (3) the **TFIIB recognition element** (BRE) located immediately upstream of the TATA box; and (4) the **downstream promoter element** (DPE) located about 30 nucleotides downstream from the startpoint. These four elements are organized into two general types of core promoters: *TATA-driven promoters*, which contain an Inr sequence and a TATA box with or without an associated BRE, and *DPE-driven promoters*, which contain DPE and Inr sequences, but no TATA box or BRE.

By itself, either type of core promoter is only capable of supporting a *basal* (low) level of transcription. However, most protein-coding genes have additional short sequences further upstream—upstream control elements—that improve the promoter's efficiency. Some of these upstream elements are common to many different genes; examples include the *CAAT box* (consensus sequence GCCCAATCT in animals and yeasts) and the *GC box* (consensus sequence GGGCGG). The locations of these elements relative to a gene's startpoint vary from gene to gene. The elements within 100–200 nucleotides of the startpoint are often called *proximal control elements* to distinguish them from *enhancer* elements, which tend to be farther away and can even be located downstream of the gene. We will return to proximal control elements and enhancers in Chapter 21.

The sequences important in promoter activity are often identified by deleting specific sequences from a cloned DNA molecule, which is then tested for its ability to

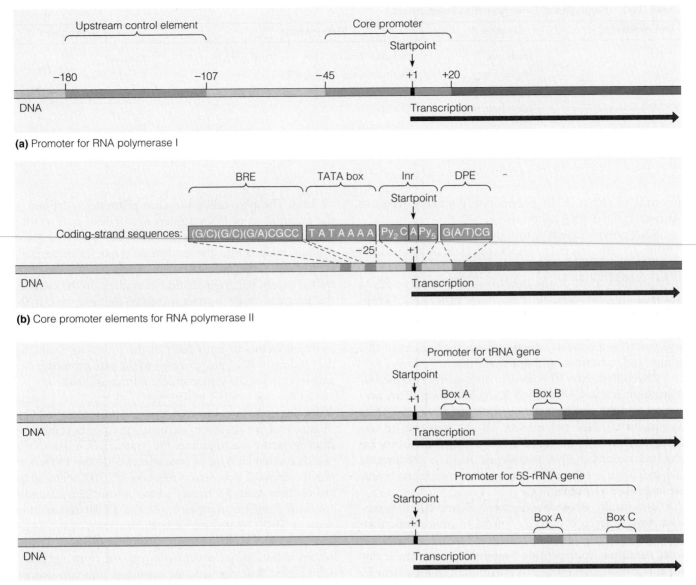

(a) Promoter for RNA polymerase I

(b) Core promoter elements for RNA polymerase II

(c) Two types of promoters for RNA polymerase III

Figure 19-13   Typical Eukaryotic Promoters Used by RNA Polymerases I, II, and III.
(a) The promoter for RNA polymerase I has two parts, a core promoter surrounding the startpoint and an upstream control element. After the binding of appropriate transcription factors to both parts, the RNA polymerase binds to the core promoter. **(b)** The typical promoter for RNA polymerase II has a short initiator (Inr) sequence, consisting mostly of pyrimidines (Py), combined with either a TATA box or a downstream promoter element (DPE). Promoters containing a TATA box may also include a TFIIB recognition element (BRE) as part of the core promoter. **(c)** The promoters for RNA polymerase III vary in structure, but the ones for tRNA genes and 5S rRNA genes are located entirely downstream of the startpoint, within the transcribed sequence. Boxes A, B, and C are DNA consensus sequences, each about 10 bp long. In tRNA genes, about 30–60 bp of DNA separate boxes A and B; in 5S rRNA genes, about 10–30 bp separate boxes A and C.

serve as a template for gene transcription, either in a test tube or after introduction of the DNA into cultured cells. For example, when transcription of the gene for β-globin (the β chain of hemoglobin) is investigated in this way, deletion of either the TATA box or an upstream CAAT box reduces the rate of transcription at least tenfold.

In contrast to RNA polymerases I and II, the RNA polymerase III molecule uses promoters that are entirely *downstream* of the transcription unit's startpoint when transcribing genes for tRNAs and 5S RNA. The promoters used by tRNA and 5S rRNA genes are different, but in both cases the consensus sequences fall into two blocks of about 10 bp each (Figure 19-13c). The tRNA promoter has consensus sequences called *box A* and *box B*. The promoters for 5S rRNA genes have box A (positioned farther from the startpoint than in tRNA-gene promoters) and another critical sequence, called *box C*. (Not shown in the figure is a third type of RNA polymerase III promoter, an upstream promoter that is used for the synthesis of other kinds of small RNA molecules.)

The promoters used by all the eukaryotic RNA polymerases must be recognized and bound by transcription

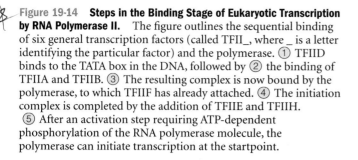

Figure 19-14 **Steps in the Binding Stage of Eukaryotic Transcription by RNA Polymerase II.** The figure outlines the sequential binding of six general transcription factors (called TFII_, where _ is a letter identifying the particular factor) and the polymerase. ① TFIID binds to the TATA box in the DNA, followed by ② the binding of TFIIA and TFIIB. ③ The resulting complex is now bound by the polymerase, to which TFIIF has already attached. ④ The initiation complex is completed by the addition of TFIIE and TFIIH. ⑤ After an activation step requiring ATP-dependent phosphorylation of the RNA polymerase molecule, the polymerase can initiate transcription at the startpoint.

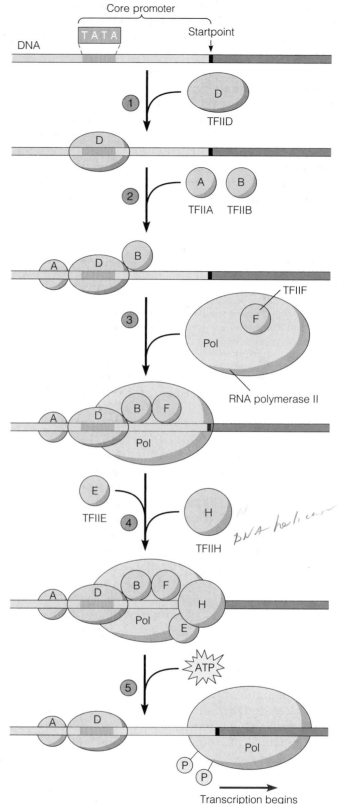

factors before the RNA polymerase molecule can bind to DNA. We turn now to the transcription factors.

General Transcription Factors Are Involved in the Transcription of All Nuclear Genes

We now take a closer look at the important role played by transcription factors in eukaryotes, using RNA polymerase II as our example. A **general transcription factor** (also called a *basal transcription factor*) is a protein that is always required for an RNA polymerase molecule to bind to its promoter and initiate RNA synthesis, regardless of the identity of the specific gene involved. Eukaryotes have many such transcription factors; their names usually include "TF" (for transcription factor), a roman numeral identifying the polymerase they aid, and a capital letter that identifies each individual factor (e.g., TFIIA, TFIIB, and so forth).

Figure 19-14 illustrates the involvement of general transcription factors in the binding of RNA polymerase II to a TATA-containing promoter site in DNA. General transcription factors bind to promoters in a defined order, starting with TFIID. Note that while TFIID binds directly to a DNA sequence (the TATA box in this example or the DPE sequence in the case of DPE-driven promoters), the other transcription factors interact primarily with each other. Hence *protein-protein interactions* play a crucial role in the binding stage of eukaryotic transcription. RNA polymerase II does not bind to the DNA until several steps into the process. Eventually, a large complex of proteins, including RNA polymerase, becomes bound to the promoter region to form a *preinitiation complex*.

Before RNA polymerase II can actually initiate RNA synthesis, it must be released from the preinitiation complex. A key role in this process is played by the general transcription factor TFIIH, which possesses both a helicase activity that unwinds DNA and a protein kinase activity that catalyzes the phosphorylation of RNA polymerase. Phosphorylation changes the shape of RNA polymerase, thereby releasing it from the transcription factors so that it can initiate RNA synthesis at the startpoint. At the same time, the helicase activity of TFIIH is thought to unwind the DNA so that the RNA polymerase molecule can begin to move.

TFIID, the initial transcription factor to bind to the promoter, is worthy of special note. Its ability to recognize

and bind to DNA promoter sequences is conferred by one of its subunits, the **TATA-binding protein (TBP)**, which combines with a variable number of additional protein subunits to form TFIID. In spite of its name, the ability of TBP to bind to DNA, illustrated in Figure 19-15, is not

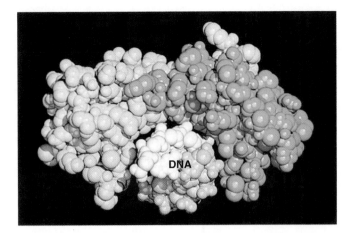

Figure 19-15 TATA-Binding Protein (TBP) Bound to DNA. In this computer graphic model, TBP is shown bound to DNA (white and gray, viewed looking down its axis). TBP differs from most DNA-binding proteins in that it interacts with the minor groove of DNA, rather than the major groove, and imparts a sharp bend to the DNA. The TBP molecule shown here is from the plant *Arabidopsis thaliana*, but TBP has been highly conserved during evolution. Dark and light blue differentiate the two, symmetrical domains of the polypeptide; light green is used for its nonconserved N-terminal segment. When TBP is bound to DNA, other transcription-factor proteins can interact with the convex surface of the TBP "saddle." TBP is involved in transcription initiation for all types of eukaryotic promoters.

restricted to TATA-containing promoters. TBP can also bind to promoters lacking a TATA box, including promoters used by RNA polymerases I and III. Depending on the type of promoter, TBP associates with different proteins, and for promoters lacking a TATA box, much of TBP's specificity is probably derived from its interaction with these associated proteins.

Because they play an essential role in the binding of RNA polymerase to DNA, the general transcription factors are all potential targets for the regulation of transcription. The recognition of upstream control elements in DNA by other regulatory proteins provides additional levels of potential control. Finally, the packing of eukaryotic DNA into nucleosomes represents yet another possible site of regulation: Nucleosomes can interfere with the ability of transcription factors and RNA polymerase to gain access to DNA, so mechanisms that modify nucleosome packing are a further way of exerting control over the initiation of transcription. These issues will be addressed in more detail in Chapter 21, where we discuss the regulation of transcription.

Elongation, Termination, and RNA Cleavage Are Involved in Completing Eukaryotic RNA Synthesis

After initiating transcription, RNA polymerase moves along the DNA and makes a complementary RNA copy of the DNA template strand. For elongation of the RNA chain to proceed, RNA polymerase must recruit proteins that unfold the higher-order nucleosomal packaging of the chromatin fibers that would otherwise block the enzyme from advancing along the DNA template.

Termination of transcription is governed by a variety of signals that vary with the enzyme involved. For example, transcription by RNA polymerase I is terminated by a protein factor that recognizes an 18-nucleotide termination signal in the growing RNA chain. Termination signals for RNA polymerase III are also known; the signals always include a short run of U's (as in prokaryotic termination signals), and no ancillary protein factors are needed for their recognition. Hairpin structures do not appear to be involved in termination by either polymerase I or polymerase III.

For RNA polymerase II, most transcripts destined to become mRNA are cleaved at a specific site before transcription is actually terminated. The cleavage site is 10–35 nucleotides downstream from an AAUAAA sequence in the growing RNA chain. The polymerase may continue transcription for hundreds or even thousands of nucleotides beyond the cleavage site, but this additional RNA is quickly degraded. The cleavage site is also the site for the addition of a *poly(A) tail*, a string of adenine nucleotides found at the 3′ end of almost all eukaryotic mRNAs. Addition of the poly(A) tail is part of RNA processing, our next topic.

RNA Processing

An RNA molecule newly produced by transcription, called a **primary transcript,** frequently must undergo chemical changes before it can function in the cell. We use the term **RNA processing** to mean all the chemical modifications necessary to generate a final RNA product from the primary transcript that serves as its precursor. Processing typically involves removal of portions of the primary transcript, and it may also include the addition or chemical modification of specific nucleotides. For example, methylation of bases or ribose groups is a common modification of individual nucleotides. In addition to chemical modifications, other posttranscriptional events, such as association with specific proteins or (in eukaryotes) passage from the nucleus to the cytoplasm, are often necessary before the RNA can function.

In this section, we examine the most important processing steps involved in the production of rRNA, tRNA, and mRNA from their respective primary transcripts. Although the term *RNA processing* is most often associated with eukaryotic systems, prokaryotes process some of their RNA as well. We therefore include examples involving both eukaryotic and prokaryotic RNAs.

Ribosomal RNA Processing Involves Cleavage of Multiple rRNAs from a Common Precursor

Ribosomal RNA (rRNA) is by far the most abundant and most stable form of RNA found in cells. Typically, rRNA

Table 19-2 RNA Components of Cytoplasmic Ribosomes

| | | rRNA | |
Source	Ribosomal Subunit	Sedimentation Coefficient	Nucleotides
Prokaryotic cells	Large (50S)	23S	2900
		5S	120
	Small (30S)	16S	1540
Eukaryotic cells	Large (60S)	25–28S	≤ 4700
		5.8S	160
		5S	120
	Small (40S)	18S	1900

represents about 70–80% of the total cellular RNA, tRNA represents about 10–20%, and mRNA accounts for less than 10%. In eukaryotes, cytoplasmic ribosomes contain four types of rRNA, usually identified by their differing sedimentation rates during centrifugation (p. 94). Table 19-2 lists the sedimentation coefficients (S values) of these different types of rRNA. The smaller of the two ribosomal subunits has a single 18S rRNA molecule; the larger subunit contains three rRNA molecules, one of about 28S (as low as 25S in some species) and the other two of about 5.8S and 5S. In prokaryotic ribosomes, only three species of rRNA are present: a 16S molecule associated with the small subunit and molecules of 23S and 5S associated with the large subunit.

Of the four kinds of rRNA in eukaryotic ribosomes, the three larger ones (28S, 18S, and 5.8S) are encoded by a single transcription unit that is transcribed by RNA polymerase I in the nucleolus to produce a single primary transcript called **pre-rRNA** (Figure 19-16b and c). The DNA sequences that code for these three rRNAs are separated within the transcription unit by segments of DNA called *transcribed spacers*. The presence of three different rRNA genes within a single transcription unit ensures that the cell makes these three rRNAs in equal quantities. Most eukaryotic genomes have multiple copies of the pre-rRNA transcription unit arranged in one or more tandem arrays (Figure 19-16a), facilitating production of the large amounts of ribosomal RNA typically needed by cells. The human haploid genome, for example, has 150–200 copies of the pre-rRNA transcription unit, distributed among five chromosomes. *Nontranscribed spacers* separate the transcription units within each cluster.

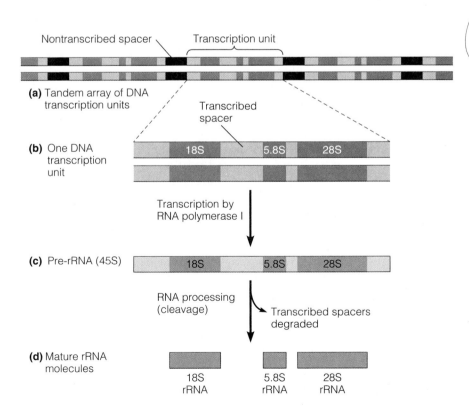

(a) Tandem array of DNA transcription units

Nontranscribed spacer   Transcription unit

Transcribed spacer

(b) One DNA transcription unit

18S   5.8S   28S

Transcription by RNA polymerase I

(c) Pre-rRNA (45S)

18S   5.8S   28S

RNA processing (cleavage)

Transcribed spacers degraded

(d) Mature rRNA molecules

18S rRNA   5.8S rRNA   28S rRNA

Figure 19-16 Eukaryotic rRNA Genes and the Processing of the Primary Transcript. (a) The eukaryotic transcription unit that includes the genes for the three largest rRNAs occurs in multiple copies, arranged in tandem arrays. Nontranscribed spacers (black) separate the units. (b) Each transcription unit includes the genes for the three rRNAs (darker blue) and four transcribed spacers (lighter blue). (c) The transcription unit is transcribed by RNA polymerase I into a single long transcript (pre-rRNA) with a sedimentation coefficient of about 45S. (d) RNA processing yields mature 18S, 5.8S, and 28S rRNA molecules. RNA cleavage actually occurs in a series of steps. The order of steps varies with the species and cell type, but the final products are always the same three types of rRNA molecules.

After RNA polymerase I has transcribed the pre-rRNA transcription unit, the resulting pre-rRNA molecule is processed by a series of cleavage reactions that remove the transcribed spacers and release the mature rRNAs (Figure 19-16c and d). The transcribed spacer sequences are then degraded. The pre-rRNA is also processed by the addition of methyl groups. The main site of methylation is the 2′-hydroxyl group of the sugar ribose, although a few bases are methylated as well. The methylation process, as well as pre-rRNA cleavage, are guided by a special group of RNA molecules, called **snoRNAs** (small nucleolar RNAs), which bind to complementary regions of the pre-rRNA molecule and target specific sites for methylation or cleavage.

Pre-rRNA methylation has been studied by incubating cells with radioactive *S-adenosyl methionine*, which is the methyl group donor for cellular methylation reactions. When human cells are incubated with radioactive S-adenosyl methionine, all of the radioactive methyl groups initially incorporated into the pre-rRNA molecule are eventually found in the finished 28S, 18S, and 5.8S rRNA products, indicating that the methylated segments are selectively conserved during rRNA processing. Methylation may help to guide RNA processing by protecting specific regions of the pre-rRNA molecule from cleavage. In support of this hypothesis, it has been shown that depriving cells of one of the essential components required for the addition of methyl groups leads to disruption of pre-rRNA cleavage.

In mammalian cells, the pre-rRNA molecule has about 13,000 nucleotides and a sedimentation coefficient of 45S. The three mature rRNA molecules generated by cleavage of this precursor contain only about 52% of the original RNA. The remaining 48% (about 6200 nucleotides) consists of transcribed spacer sequences that are removed and degraded during the cleavage steps. The rRNA precursors of some other eukaryotes contain smaller amounts of spacer sequences, but in all cases the pre-rRNA is larger than the aggregate size of the three rRNA molecules made from it. Thus, some processing is always required.

Processing of pre-rRNA in the nucleolus is accompanied by assembly of the RNA with proteins to form ribosomal subunits. In addition to the 28S, 18S, and 5.8S rRNAs generated by pre-rRNA processing, the ribosome assembly process also requires 5S rRNA. The gene for 5S rRNA constitutes a separate transcription unit that is transcribed by RNA polymerase III rather than RNA polymerase I. It, too, occurs in multiple copies arranged in long, tandem arrays. However, 5S rRNA genes are not usually located near the genes for the larger rRNAs and so do not tend to be associated with the nucleolus. Unlike pre-rRNA, the RNA molecules generated during transcription of 5S rRNA genes require little or no processing.

As in eukaryotes, ribosome formation in prokaryotic cells involves processing of multiple rRNAs from a larger precursor. *E. coli*, for example, has seven rRNA transcription units scattered about its genome. Each contains genes for all three prokaryotic rRNAs—23S rRNA, 16S rRNA, and 5S rRNA—plus several tRNA genes. Processing of the primary transcripts produced from these transcription units involves two sets of enzymes, one for the rRNAs and one for the tRNAs.

Transfer RNA Processing Involves Removal, Addition, and Chemical Modification of Nucleotides

Cells synthesize several dozen kinds of tRNA molecules, each designed to bring a particular amino acid to one or more codons in mRNA. However, all tRNA molecules share a common general structure, as illustrated in Figure 19-17. A mature tRNA molecule contains only 70–90 nucleotides, some of which are chemically modified. Base pairing between complementary sequences located in different regions causes each tRNA molecule to fold into a secondary structure containing several *hairpin loops*, illustrated in the figure. Most tRNAs have four base-paired regions, indicated by the light blue dots in part b of the figure; in some, a fifth such region is present at the *variable loop*. Each of these base-paired regions is a short stretch of RNA double helix. Molecular biologists call the tRNA secondary structure a *cloverleaf* structure because it resembles a cloverleaf when drawn in two dimensions. However, in its normal three-dimensional tertiary structure, the molecule is folded so that the overall shape actually resembles a letter "L" (see Figure 20-3b).

Like ribosomal RNA, transfer RNA is synthesized in a precursor form in both eukaryotic and prokaryotic cells. Processing of these **pre-tRNA** molecules involves several different events, as shown in Figure 19-17a for yeast tyrosine tRNA. ① At the 5′ end, a short *leader sequence* of 16 nucleotides is removed from the pre-tRNA. ② At the 3′ end, the two terminal nucleotides of the pre-tRNA are removed and replaced with the trinucleotide CCA, which is a common structural feature of all tRNA molecules. (Some tRNAs already have CCA in their primary transcripts and therefore do not require modification at the 3′ end.) ③ In a typical tRNA molecule, about 10–15% of the nucleotides are chemically modified during pre-tRNA processing. The principal modifications include methylation of bases and sugars, and creation of unusual bases such as dihydrouracil, ribothymine, psuedouridine, and inosine.

The processing of yeast tyrosine tRNA is also characterized by the removal of an internal 14-nucleotide sequence (④), although the transcripts for most tRNAs do not require this kind of excision. An internal segment of an RNA transcript that must be removed to create a mature RNA product is called an RNA *intron*. We will consider introns in more detail during our discussion of mRNA processing, because they are a nearly universal feature of mRNA precursors in eukaryotic cells. For the present, we simply note that some eukaryotic tRNA precursors contain introns that must be eliminated by a precise mechanism that cuts and splices the precursor molecule at exactly the same location every time. The

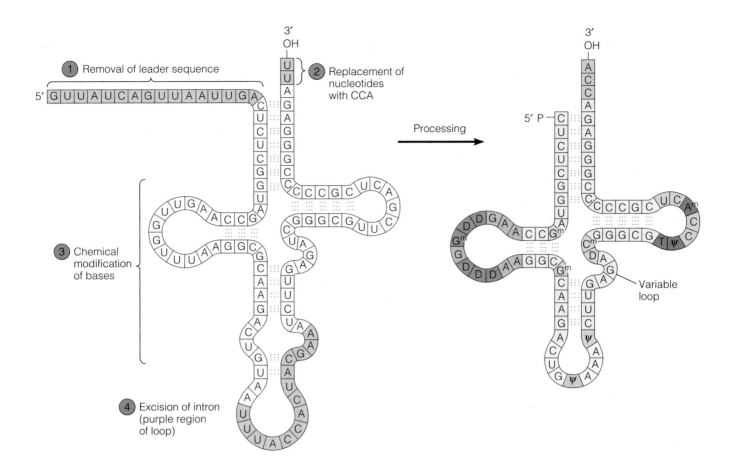

(a) Primary transcript (precursor) for yeast tyrosine tRNA

(b) Mature tRNA, secondary structure

Figure 19-17 Processing and Secondary Structure of Transfer RNA. (a) Every tRNA gene is transcribed as a precursor that must be processed into a mature tRNA molecule. In this primary transcript for yeast tyrosine tRNA, all regions highlighted in purple are removed during processing. Processing for this tRNA involves ① removal of the leader sequence at the 5′ end, ② replacement of two nucleotides at the 3′ end by the sequence CCA (with which all mature tRNA molecules terminate), ③ chemical modification of certain bases, and ④ excision of an intron. **(b)** The mature tRNA in a flattened, cloverleaf representation, which clearly shows the base pairing between self-complementary stretches in the molecule. Modified bases (darker colors) are abbreviated as A^m for methyladenine, G^m for methylguanine, C^m for methylcytosine, D for dihydrouracil, T for ribothymine, and ψ for pseudouridine.

cutting-splicing mechanism involves two separate enzymes, an RNA endonuclease and an RNA ligase, which are similar from species to species, even among organisms that are evolutionarily distant from one another. In an experiment that demonstrates this point vividly, cloned genes for the yeast tyrosine tRNA shown in Figure 19-17 were microinjected into eggs of *Xenopus laevis*, the African clawed frog. Despite the long evolutionary divergence between fungi and amphibians, the yeast genes placed in the frog eggs were transcribed and processed properly, including removal of the 14-nucleotide intron.

Messenger RNA Processing in Eukaryotes Involves Capping, Addition of Poly(A), and Removal of Introns

Prokaryotic mRNA is, in almost all cases, an exception to the generalization that RNA requires processing before it can be used by the cell. Most bacterial mRNA is synthesized in a form that is ready for translation, even before the entire RNA molecule has been completed. Moreover, transcription in bacteria is not separated by a membrane barrier from the ribosomes responsible for translation, so bacterial mRNA molecules in the process of being synthesized by RNA polymerase often have ribosomes already associated with them. The electron micrograph in Figure 19-18 shows this coupling of transcription and translation in a bacterial cell.

In eukaryotes, on the other hand, transcription and translation are separated in both time and space: Transcription takes place in the nucleus, whereas translation occurs mainly in the cytoplasm. Substantial processing is required in the nucleus to convert primary transcripts into mature mRNA molecules that are ready to be transported to the cytoplasm and translated. Although each

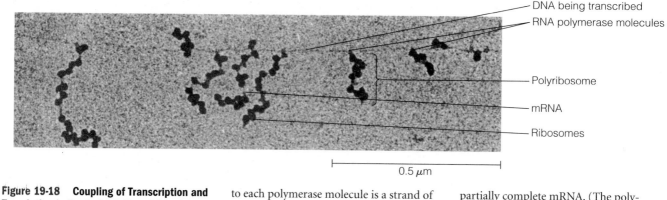

Figure 19-18 **Coupling of Transcription and Translation in Bacterial Cells.** This electron micrograph shows *E. coli* DNA being transcribed by RNA polymerase molecules that are moving from right to left. Attached to each polymerase molecule is a strand of mRNA still in the process of being transcribed. The large dark particles attached to each growing mRNA strand are ribosomes that are actively translating the partially complete mRNA. (The polypeptides being synthesized are not visible.) A cluster of ribosomes attached to a single mRNA strand is called a polyribosome (TEM).

eukaryotic mRNA transcription unit encodes only one polypeptide, the primary transcripts are often very long, typically ranging from 2000 to 20,000 nucleotides. This size heterogeneity is reflected in the term *heterogeneous nuclear RNA (hnRNA)*, which has been applied to the nonribosomal, nontransfer RNA found in eukaryotic nuclei. HnRNA consists of a mixture of mRNA molecules and their precursors, **pre-mRNA.** Conversion of pre-mRNA molecules into functional mRNAs usually requires the removal of nucleotide sequences as well as the addition of 5′ caps and 3′ tails, as we describe in the following sections.

5′ Caps and 3′ Poly(A) Tails. Most eukaryotic mRNA molecules bear distinctive modifications at both ends. At the 5′ end they all possess a modified nucleotide called a **5′ cap,** and at the 3′ end they usually have a long stretch of adenine ribonucleotides known as a **poly(A) tail.**

A 5′ cap is simply a guanosine nucleotide that has been methylated at position 7 of the purine ring and is "backward"; that is, the bond joining it to the 5′ end of the RNA molecule is a 5′ ⟶ 5′ linkage rather than the usual 3′ ⟶ 5′ bond (Figure 19-19). This distinctive feature of eukaryotic mRNA is added to the primary transcript shortly after initiation of RNA synthesis. As part of the capping process, the ribose rings of the first, and often the second, nucleotides of the RNA chain can also become methylated, as shown in Figure 19-19. The 5′ cap contributes to mRNA stability by protecting the molecule from degradation by nucleases that attack RNA at the 5′ end. The cap also plays an important role in positioning mRNA on the ribosome for the initiation of translation.

In addition to the 5′ cap, a poly(A) tail ranging from 50–250 nucleotides in length is present at the 3′ end of most eukaryotic mRNA molecules. In animal cells, only the mRNAs for the major histones lack them. It is clear that poly(A) must be added after transcription because genes do not contain long stretches of thymine (T) nucleotides that could serve as a template for the addition of poly(A). Direct support for this conclusion has come from the isolation of the enzyme *poly(A) polymerase,* which catalyzes the addition of poly(A) sequences to RNA without requiring a DNA template.

The addition of poly(A) is part of the process that creates the 3′ end of eukaryotic mRNA molecules. Unlike bacteria, where specific termination sequences halt transcription at the 3′ end of newly forming mRNAs, the transcription of eukaryotic pre-mRNAs often proceeds hundreds or even thousands of nucleotides beyond the site

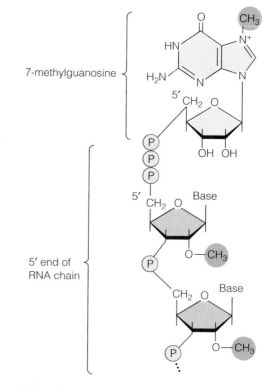

Figure 19-19 **Cap Structure Located at the 5′ End of Eukaryotic Pre-mRNA and mRNA Molecules.** The methyl groups attached to the first two riboses of the RNA chain are not always present.

destined to become the 3′ end of the final mRNA molecule. A special *AAUAAA signal sequence* located slightly upstream from this site in the growing pre-mRNA chain then signals where the poly(A) tail should be added. As shown in Figure 19-20, this signal sequence triggers cleavage of the primary transcript 10–35 nucleotides downstream from the AAUAAA sequence and poly(A) polymerase catalyzes formation of the poly(A) tail. In addition to creating the poly(A) tail, the processing events associated with the AAUAAA signal may also help to trigger the termination of transcription.

The poly(A) tail seems to have several functions. Like the 5′ cap, it protects mRNA from nuclease attack and, as a result, the length of the poly(A) influences mRNA stability (the longer the tail, the longer the life span of the mRNA in the cytoplasm). In addition, poly(A) is recognized by specific proteins involved in exporting mRNA from the nucleus to the cytoplasm, and it may also help ribosomes recognize mRNA as a molecule to be translated. In the laboratory, poly(A) tails can be used to isolate mRNA from the more prevalent rRNA and tRNA. RNA extracted from cells is simply passed through a column packed with particles coated with poly(dT), which are single strands of DNA consisting solely of thymine nucleotides. Molecules with poly(A) tails bind to the poly(dT) via complementary base pairing, while other RNA molecules pass through. The poly(A)-containing mRNA can then be removed from the column by changing the ionic conditions.

The Discovery of Introns. In eukaryotic cells, the precursors for most mRNAs (and for some tRNAs and rRNAs) contain **introns,** which are *sequences within the primary transcript that do not appear in the mature, functional RNA.* The dis-

covery of introns was a great surprise to biologists. It had been known for many years that in bacteria, the sequence of amino acids in a polypeptide chain correlates exactly with a sequence of contiguous nucleotides in DNA. This relationship was demonstrated in the colinearity experiments of Charles Yanofsky in the early 1960s and, as rapid methods for sequencing DNA and proteins became available, was confirmed by direct comparison of nucleotide and amino acid sequences. Biologists naturally assumed the same would turn out to be true for eukaryotes.

It was therefore a shock when several research groups reported in 1977 that eukaryotic genes do not follow this pattern, but are in fact interrupted by sequences of nucleotides—introns—that are not represented in either the functional mRNA or its protein product. The existence of introns was first shown by *R looping,* a technique in which single-stranded RNA is hybridized to double-stranded DNA under conditions that favor the formation of hybrids between complementary regions of RNA and DNA. To understand how this technique led to the discovery of introns, let us first examine what happens with a gene *lacking* introns. Figure 19-21a depicts the expected result when a prokaryotic mRNA is allowed to hybridize

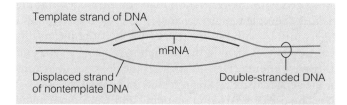

(a)

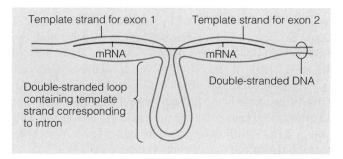

(b)

Figure 19-21 Demonstration of Introns in Protein-Coding Genes. Molecules of mature mRNA were allowed to hydrogen-bond with the DNA (genes) from which they had been transcribed. The resulting hybrid molecules were then examined under an electron microscope. The diagrams here show the results. **(a)** Hybridization of a prokaryotic mRNA molecule with DNA from its gene, which lacks introns. The displaced strand of DNA forms a wide loop. **(b)** Hybridization of a eukaryotic mRNA molecule with its gene, which has one intron. Three loops are observed in this case—two *single-stranded* DNA loops where the mRNA has hybridized to the DNA template strand, plus one prominent *double-stranded* DNA loop. The double-stranded DNA loop represents the intron, which contains sequences that do not appear in the final mRNA.

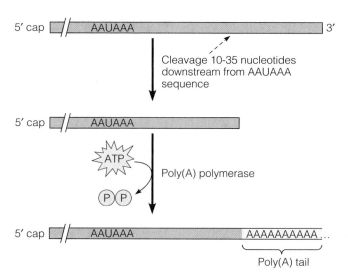

Figure 19-20 Addition of a Poly(A) Tail to Pre-mRNA. Transcription of eukaryotic pre-mRNAs often proceeds beyond the 3′ end of the mature mRNA. An AAUAAA sequence located slightly upstream from the proper 3′ end then signals that the RNA chain should be cleaved about 10–35 nucleotides downstream from the signal site, followed by addition of a poly(A) tail catalyzed by poly(A) polymerase.

with double-stranded DNA corresponding to the gene from which the mRNA was transcribed. The mRNA hybridizes to the template strand of the DNA, leaving the other, displaced strand as a single-stranded DNA loop that can be easily identified using electron microscopy.

If eukaryotic genes were constructed from contiguous linear sequences of coding nucleotides, their mRNAs would be expected to generate a similar looping pattern. However, in actual R-looping experiments with eukaryotic mRNAs coding for such proteins as human β-globin and chick ovalbumin, the surprising result was that *multiple* loops were seen (Figure 19-21b). This unexpected result indicated that the DNA sequences coding for a typical eukaryotic mRNA are not continuous with each other, but instead are separated by intervening sequences that do not appear in the final mRNA. The intervening sequences that disrupt the linear continuity of the message-encoding regions of a gene are the introns (*intervening* sequences), and the sequences destined to appear in the final mRNA are referred to as **exons** (because they are *ex*pressed).

Once they had been reported for a few genes, introns began popping up everywhere, especially as restriction enzyme techniques came to be applied to a wide variety of eukaryotic genes. Figure 19-22 illustrates how restriction mapping can be used to identify the presence of introns. In such studies, restriction maps of chromosomal genes are compared with the restriction maps of the corresponding cDNAs (complementary DNAs) made by transcribing a gene's mRNA with reverse transcriptase. If significant differences are observed in the restriction maps of a gene and its corresponding cDNA, it suggests the presence of introns in the gene that are not represented in the final mRNA.

The use of restriction mapping and DNA sequencing techniques has led to the conclusion that introns are present in most protein-coding genes of multicellular eukaryotes, although the size and number of the introns can vary considerably (Table 19-3). The human β-globin gene, for example, has only two introns, one of 120 bp and the other of 550 bp. Together, these account for about 40% of the total length of the gene. For many mammalian genes, an even larger fraction of the gene consists of introns. An extreme example is the human dystrophin gene, a mutant form of which causes Duchenne muscular dystrophy. This gene is over 2 million bp long and has at least 78 introns, representing more than 99% of the gene's DNA!

The discovery of introns that do not appear in mature mRNA molecules raises the question of whether the introns present in DNA are actually transcribed into the primary transcript (pre-mRNA). This question has been addressed by experiments in which pre-mRNA and DNA were mixed together and the resulting hybrids examined by electron microscopy. In contrast to the appearance of hybrids between mRNA and DNA, which exhibit multiple R loops where the DNA molecule contains sequences that are not present in the mRNA (see Figure 19-21b), pre-mRNA hybridizes in one continuous stretch to the DNA molecule, forming a single R loop (as in Figure 19-21a). Scientists have therefore concluded that pre-mRNA mole-

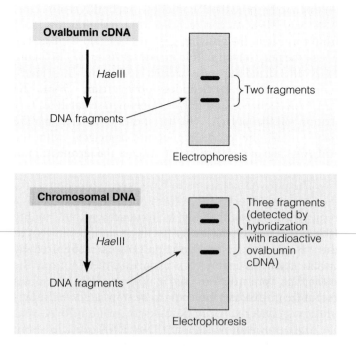

Figure 19-22 Detection of Introns Using Restriction Enzymes. Cleaving purified chicken ovalbumin cDNA with the restriction enzyme *Hae*III yields two fragments, indicating the presence of one *Hae*III site in the cDNA molecule. In contrast, cleaving chicken chromosomal DNA with the same enzyme, *Hae*III, generates three fragments containing ovalbumin sequences, thereby indicating the presence of two *Hae*III sites in the ovalbumin gene. The extra *Hae*III site is situated within an intron and therefore does not appear in the final ovalbumin mRNA molecule from which the ovalbumin cDNA is derived.

cules represent continuous copies of their corresponding genes, containing introns as well as sequences destined to become part of the final mRNA. This means that converting pre-mRNA into mRNA requires specific mechanisms for removing introns, as we now describe.

Spliceosomes Remove Introns from Pre-mRNA

To produce a functional molecule of mRNA from a pre-mRNA that contains introns, eukaryotes must somehow

Table 19-3 **Some Examples of Genes with Introns**

Gene	Organism	Number of Introns	Number of Exons
Actin	*Drosophilia*	1	2
β-Globin	Human	2	3
Insulin	Human	2	3
Actin	Chicken	3	4
Albumin	Human	14	15
Thyroglobulin	Human	36	37
Collagen	Chicken	50	51
Titin	Human	233	234

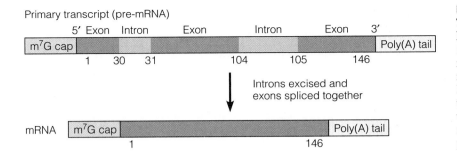

Figure 19-23 An Overview of RNA Splicing.
The capped and tailed primary transcript of the human gene for β-globin contains three exons (dark red) and two introns (light red). The numbers refer to codon positions in the final mRNA. In the mature mRNA that results from RNA splicing, the introns have been excised and the exons joined together to form a molecule with a continuous coding sequence. The cell's ribosomes will translate this message into a polypeptide of 146 amino acids.

remove the introns and splice together the remaining RNA segments (exons). The entire process of removing introns and rejoining the exons is termed **RNA splicing.** As an example, Figure 19-23 shows both the primary transcript (pre-mRNA) of the β-globin gene and the mature end-product (mRNA) that results after removal of the two introns.

RNA splicing must be very precise, because a single nucleotide error would alter the mRNA reading frame and render it useless. Proper splicing can, in fact, be disrupted by simply altering the base sequence of short nucleotide stretches located at either end of an intron, indicating that these sequences determine the exact location of the *5′ and 3′ splice sites*—that is, the points where the two ends of an intron are cleaved during its removal. Analysis of the base sequences of hundreds of different introns has revealed that the 5′ end of an intron typically starts with the sequence GU and the 3′ end terminates with the sequence AG. In addition, a short stretch of bases adjacent to these GU and AG sequences tends to be similar among different introns. The base sequence of the remainder of the intron appears to be largely irrelevant to the splicing process. Though introns vary from a few dozen to thousands of nucleotides in length, most of the intron can be artificially removed without altering the splicing process. One exception is a special sequence located several dozen nucleotides upstream from the 3′ end of the intron and referred to as the *branch-point*. The branch-point plays an important role in the mechanism by which introns are removed.

The process of intron removal is catalyzed by an RNA-protein complex called a **spliceosome,** which is assembled from a group of smaller RNA-protein complexes known as **snRNPs** (small nuclear ribonucleo-proteins) and additional proteins. Each snRNP (pronounced "snurp") contains one or two small molecules of a special type of RNA called **snRNA** (small nuclear RNA). During RNA splicing, a group of several snRNPs bind sequentially to an intron to form the spliceosome (Figure 19-24). The first step in this process is the binding of a snRNP called U1, whose RNA contains a nucleotide sequence that allows it to base-pair with the 5′ splice site. A second snRNP, called U2, then binds to the branch-point sequence. Finally, another group of snRNPs (U4/U6 and U5) brings the two ends of the intron together to form a mature spliceosome. The completed spliceosome, con-

taining five RNAs and more than 50 proteins, is a massive complex almost as big as a ribosome.

At this stage, the pre-mRNA is cleaved at the 5′ splice site and the newly released 5′ end of the intron is covalently joined to an adenine residue located at the branch-point sequence, creating a looped structure called a *lariat*. The 3′ splice site is then cleaved and the two ends of the exon are joined together, releasing the intron for subsequent degradation. Electron micrographs have revealed the presence of snRNPs and mature spliceosomes bound to RNA molecules that are still in the process of being synthesized (Figure 19-25), indicating that introns can be removed before transcription of the pre-mRNA molecule is completed. Although not shown in the figure, both the primary transcript and the mature mRNA are associated along their lengths with non-snRNP proteins, forming **RNP particles** that look something like DNA nucleosomes. The mRNA remains associated with these proteins until it enters the cytoplasm, where it joins with other proteins and the ribosomal subunits.

In addition to the main class of introns characterized by GU and AG sequences located at their 5′ and 3′ boundaries, respectively, a second class of introns containing AU and AC at these two sites has been discovered. These "AU–AC" introns are often excised by a second type of spliceosome that differs in snRNP composition from the spliceosome illustrated in Figure 19-24. But in spite of the complexities provided by the existence of multiple types of introns and spliceosomes, a unifying principle has emerged: The snRNA molecules found in spliceosomes appear to be involved in the catalytic mechanism of splicing, as well as in spliceosome assembly and splice-site recognition. The idea of a catalytic role for snRNAs arose from the discovery of self-splicing RNA introns, our next topic.

Some Introns Are Self-Splicing

Although the participation of spliceosomes is almost always required for intron removal, a few types of genes have *self-splicing RNA introns.* The RNA transcript of such a gene can carry out the entire process of RNA splicing in the absence of any protein (e.g., in a test tube); the intron RNA itself catalyzes the process. As we described in Chapter 6, such RNA molecules that function as catalysts in the absence of protein are called *ribozymes.*

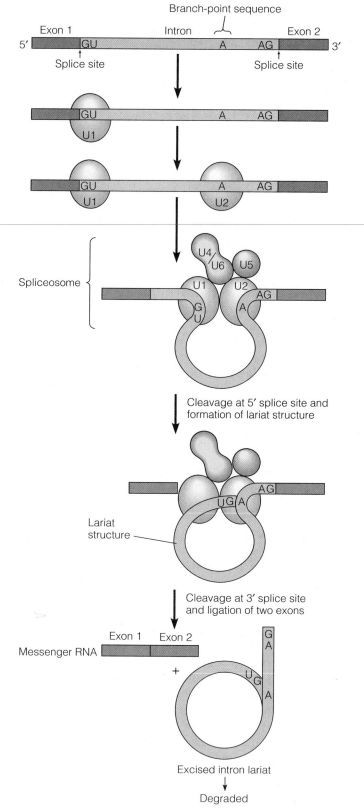

Figure 19-24 Intron Removal by Spliceosomes. The spliceosome is an RNA-protein complex that splices intron-containing pre-mRNA in the eukaryotic nucleus. The substrate here is a molecule of pre-mRNA with two exons and one intron. In a stepwise fashion, the pre-mRNA assembles with the U1 snRNP, U2 snRNP, and U4/U6 and U5 snRNPs (along with some non-snRNP splicing factors), forming a mature spliceosome. The pre-mRNA is then cleaved at the 5′ splice site and the newly released 5′ end is linked to an adenine (A) nucleotide located at the branch-point sequence, creating a looped lariat structure. Next the 3′ splice site is cleaved and the two ends of the exon are joined together, releasing the intron for subsequent degradation.

There are two classes of introns in which the intron RNA functions as a ribozyme to catalyze its own removal. The larger class, called *Group I introns,* are found in the RNA transcripts of certain mitochondrial genes of fungi, of some chloroplast tRNA genes, and of rRNA genes in *Tetrahymena, Physarum* (a slime mold), some fungi, and chloroplasts. In addition, Group I introns have been detected in three pre-mRNAs of bacteriophage T4—one of the few examples of introns in prokaryotic systems. Group I RNA introns are excised as linear pieces of RNA.

Group II introns, on the other hand, excise themselves as lariats, just as in the spliceosome mechanism. Another similarity to spliceosomes is that an adenine within the Group II RNA intron forms the branch-point of the lariat. In present-day organisms, Group II introns have been found only in fungal mitochondria. But biologists think that today's prevailing splicing mechanism probably evolved from such a system, with the intron RNA's catalytic role being taken over by snRNA molecules of the spliceosome.

Why Do Eukaryotic Genes Have Introns?

The burning question about introns is *why* do nearly all genes in multicellular eukaryotes have them? Why do cells have so much DNA that seems to serve no coding function? Why, in generation after generation of cells, is so much energy invested in synthesizing segments of DNA—and of RNA transcripts—that appear to serve no useful function and are destined only for the splicing scrap heap?

In fact, it is not true that introns never perform any functions of their own. In a few cases, at least, intron RNAs are not degraded after being removed from pre-mRNA, but rather are processed to yield other kinds of functional RNA products. For example, some types of snoRNA—whose role in guiding pre-rRNA methylation and cleavage was discussed earlier in the chapter—are derived from introns that have been removed from pre-mRNA and then processed to form the snoRNA.

In spite of such exceptions, in most cases introns are destroyed without serving any apparent function. One possible explanation for this seemingly wasteful arrangement is that it allows for *alternative splicing* of RNAs. Many cases are known in which a given primary transcript can be spliced in several different ways, generating several different mRNAs, and hence polypeptides, from the same gene. This flexibility is possible because the splice-site sequences for different introns are the same, thereby allowing exons to be assembled in different combinations

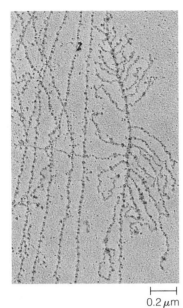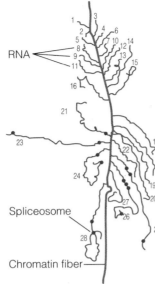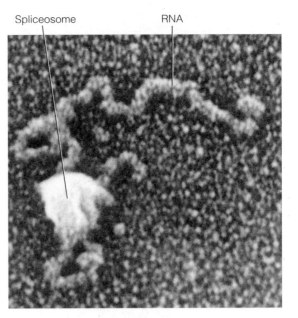

0.2 μm

Figure 19-25 Visualization of Spliceosomes by Electron Microscopy. The electron micrograph on the left shows chromatin fibers in the process of being transcribed. Many newly forming RNA transcripts (each numbered separately in the diagram) protrude from one of the chromatin fibers. The darker granules on the RNA transcripts represent snRNPs that are beginning the process of spliceosome assembly. In the case of RNA transcript number 28, a mature spliceosome has formed. The higher-magnification electron micrograph on the right shows a single RNA molecule with an attached spliceosome (TEMs).

by juxtaposing the splice sites of different introns. Having a single gene encode several different polypeptides may help to explain how the biological complexity of vertebrates is achieved without a major increase in the number of genes compared to simpler organisms. (Recall from Chapter 16 that humans have only about twice the number of genes as do worms or flies.) More than half of all human genes appear to be transcribed into pre-mRNAs that are spliced in more than one way, allowing the roughly 35,000 human genes to produce mRNAs coding for more than 100,000 polypeptides.

Another interesting role proposed for introns is an *evolutionary* one. It is possible that introns hasten the evolution of new and potentially useful proteins. This potential role is based on the discovery that exons often code for different functional regions of polypeptide chains, each of which can independently fold into a separate *domain* (Figure 19-26). For example, the three exons of the β-globin gene correspond to different structural and functional regions of the polypeptide. This kind of arrangement suggests that protein-coding genes with multiple exons may have been assembled during evolution from what were originally separate entities. Introns could be involved in two ways, both of which depend on the fact that introns provide long stretches of DNA where "incorrect" genetic recombination can take place without harming coding sequences. First, crossing over within introns of different genes could lead to the creation of genes containing new combinations of exons—*exon shuffling*. Second, recombi-

nation within other combinations of introns could easily produce duplicates of particular exons within a single gene. These exons might continue as exact duplicates, or one might mutate to a sequence that produces a new activity in the polypeptide.

Many biologists believe that introns are actually relics of ancient unicellular organisms that were the ancestors of all today's organisms, both prokaryotes and eukaryotes. Evidence for this hypothesis includes the presence of introns in modern archaebacteria and in some of the tRNA genes of cyanobacteria. However, over billions of years, evolutionary pressure for a streamlined genome in unicellular organisms may have led to modern bacteria and, to a large extent, unicellular eukaryotes, lacking the introns originally present.

RNA Editing Allows the Coding Sequence of mRNA to Be Altered

About a decade after introns and RNA splicing were first discovered, molecular biologists were surprised by the discovery of yet another type of mRNA processing, called **RNA editing.** During RNA editing, anywhere from a single nucleotide to hundreds of nucleotides may be inserted, removed, or chemically altered within the coding sequence of an mRNA. Such changes often create new initiation and/or stop codons, and can alter the reading frame of the message. The best-studied examples of RNA editing occur in the mitochondrial mRNAs of

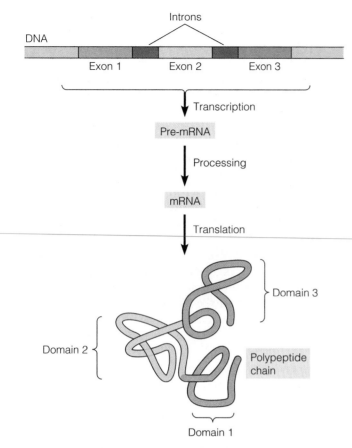

DNA

Introns

Exon 1   Exon 2   Exon 3

↓ Transcription

Pre-mRNA

↓ Processing

mRNA

↓ Translation

Domain 3

Domain 2

Polypeptide chain

Domain 1

Figure 19-26   The Proposed Role of Exons in Coding for Protein Domains. Each domain represents a separate region of a polypeptide chain that is capable of independently folding into a functional unit.

trypanosomes, which are parasitic protozoa. In these mRNAs, editing involves the insertion and deletion of multiple uracil nucleotides at various points in the mRNA. The information for this editing is located in small RNA molecules called *guide RNAs,* which are apparently encoded by mitochondrial genes separate from the mRNA genes. In one proposed editing mechanism, hydrogen bonding causes short complementary regions of the guide RNA and mRNA to come together, and nearby sequences of U's in the guide RNA are then spliced into the mRNA.

A different type of editing occurs in the mitochondrial and chloroplast mRNAs of flowering plants. In these cases, nucleotides are neither inserted nor deleted, but C's are converted to U's (and vice versa) by deamination (and amination) reactions. Similar base conversions have also been discovered in a few mRNAs transcribed from nuclear genes in animal cells. For example, a single codon in the mRNA transcribed from the mammalian apolipoprotein-B gene undergoes a C-to-U conversion during RNA editing. Another type of RNA editing detected in animal cell nuclei converts adenosine (A) to inosine (I), which resembles guanosine (G) in its base-pairing properties. The net result is therefore equivalent to an A-to-G conversion.

In all its manifestations, RNA editing seems relatively rare. However, its existence provides a reason to be cau-

tious in inferring either polypeptide or RNA sequences from genomic DNA sequences. For example, many discrepancies were observed when the amino acid sequences of proteins produced by plant mitochondrial genes were first compared with the amino acid sequences that would be predicted based on the base sequence of mitochondrial DNA. Although some of these discrepancies can be explained by nonstandard codon usage in mitochondria (described earlier in the chapter), most of the unexpected amino acids arise because RNA editing alters the base sequence of various mRNA codons, leading to the incorporation of amino acids that would not have been expected based on the DNA gene sequence.

RNA Proofreading and RNA Surveillance Help to Protect Cells from Producing and Using Defective mRNAs

A discussion of RNA processing would be incomplete without mentioning how cells deal with defective mRNAs. During RNA synthesis, incorrectly base-paired nucleotides are occasionally incorporated into growing RNA chains. Several types of quality-control mechanisms help protect cells from such mistakes. For example, some RNA polymerases possess a $3' \longrightarrow 5'$ exonuclease activity that can remove nucleotides from the 3' end of RNA chains. This activity allows RNA polymerase to remove an improperly base-paired nucleotide from the 3' end of a growing RNA chain immediately after the incorrect base has been incorporated. The result is an **RNA proofreading** mechanism for correcting transcriptional errors that is analogous to the DNA proofreading mechanism for correcting DNA replication errors described in Chapter 17.

Certain mistakes that fail to be corrected by RNA proofreading are subsequently detected by **RNA surveillance**, a quality-control mechanism that monitors RNA molecules for misplaced stop codons that would cause mRNA translation to stop prematurely. During RNA surveillance, ribosomes make an initial attempt to translate newly synthesized mRNAs (as we will learn in Chapter 20, such attempts may even occur while the RNA still resides within the nucleus). If a particular mRNA cannot be successively translated along its entire length, it is shunted into an RNA degradation pathway known as *nonsense-mediated decay*. Besides eliminating defective mRNAs that possess misplaced stop codons generated by base-pairing errors during transcription, surveillance is also useful for eliminating mRNAs that contain incorrect stop codons produced by mutations inherited in an organism's DNA.

Key Aspects of mRNA Metabolism

Before ending this chapter, we should note two key aspects of mRNA metabolism that are important to our overall understanding of how mRNA molecules behave within cells. These are the short life span of most mRNAs and the ability of mRNA to amplify genetic information.

Most mRNA Molecules Have a Relatively Short Life Span

Most mRNA molecules have a high *turnover rate*—that is, the rate at which molecules are degraded and then replaced with newly synthesized versions. In this respect, mRNA contrasts with the other major forms of RNA in the cell, rRNA and tRNA, which are notable for their stability. Because of its short life span, mRNA accounts for most of the transcriptional activity in many cells, even though it represents only a small fraction of the total RNA content. Turnover is usually measured in terms of a molecule's *half-life,* which is the length of time required for degradation of 50% of the molecules present at any given moment. The mRNA molecules of bacterial cells generally have half-lives of only a few minutes, whereas the half-lives of eukaryotic mRNAs range from several hours to a few days.

Since the rate at which a given mRNA is degraded determines the length of time it is available for translation, alterations in mRNA life span can affect the amount of protein a given message will produce. As you will learn in Chapter 21, regulation of mRNA life span is one of the mechanisms by which cells exert control over gene expression.

The Existence of mRNA Allows Amplification of Genetic Information

Because mRNA molecules can be synthesized again and again from the same stretch of template DNA, cells are provided with an important opportunity for *amplification* of the genetic message. If DNA gene sequences were used directly in protein synthesis, the number of protein molecules that could be translated from any gene within a given time period would be strictly limited by the rate of polypeptide synthesis. But in a system using mRNA as an intermediate, multiple copies of a gene's informational content can be made, and each of these can in turn be used to direct the synthesis of the protein product.

As an especially dramatic example of this amplification effect, consider the synthesis of *fibroin,* the major protein of silk. The haploid genome of the silkworm has only one copy of the fibroin gene, but about 10^4 copies of fibroin mRNA are transcribed from the two copies of the gene in each diploid cell of the silk gland. Each of these mRNA molecules, in turn, directs the synthesis of about 10^5 fibroin molecules, resulting in the production of 10^9 molecules of fibroin per cell—all within the 4-day period it takes the worm to make its cocoon! Without mRNA as an intermediate, the genome of the silkworm would need 10^4 copies of the fibroin gene (or about 40,000 days!) to make a cocoon.

Significantly, most genes that code for proteins occur in only one or a few copies per haploid genome. In contrast, genes that code for rRNA and tRNA are always present in multiple copies. It is advantageous for cells to have many copies of genes whose final products are RNA (rather than protein), because in this case there is no opportunity for amplifying each gene's effect by repeated translation.

Perspective

The expression of genetic information is one of the fundamental activities of all cells. Instructions stored in DNA are first transcribed and processed into molecules of mRNA, rRNA, and tRNA. These RNAs then play specific roles in the synthesis of proteins, with the nucleotide sequence of the mRNA providing the information that actually dictates the order of amino acids in the polypeptide product. The mRNA sequence is read as triplet codons. The genetic code specifies which amino acid corresponds to each of the triplet codons. The code is unambiguous, nonoverlapping, degenerate, and nearly universal.

Transcription is the synthesis of RNA by a mechanism that depends on complementary base pairing between incoming nucleotides and the template strand of the DNA. The process of transcription can be divided into four stages: (1) binding of RNA polymerase to the DNA template strand at the promoter region; (2) initiation of RNA synthesis; (3) elongation of the RNA chain; and (4) termination. Prokaryotic cells have only a single kind of RNA polymerase, but eukaryotic cells have several kinds, each used for making particular types of RNA. The DNA promoter sequences used by the three nuclear RNA polymerases are distinctive, and, in contrast to the situation in bacteria, recognition of eukaryotic promoters is primarily the responsibility of transcription factors that are separate from RNA polymerase. Protein-protein interactions are thus more important for eukaryotic transcription than for the prokaryotic process.

Once synthesized, most RNA transcripts must be processed to generate functional RNA molecules. Prokaryotic mRNA is the major exception; it is translated as it is being made. RNA processing can involve cleavage of multigene transcription units, removal of noncoding sequences, addition of special structural features at the 3′ and/or the 5′ end, and insertion, removal, or chemical modification of specific nucleotides. Processing of eukaryotic pre-mRNA is especially elaborate, involving the addition of a 5′ cap and a 3′ poly(A) tail, and removal of introns by an RNA splicing mechanism. Splicing of most eukaryotic pre-mRNAs is carried out by spliceosomes, which consist of RNA and protein; in some other cases, splicing is catalyzed solely by intron RNA. The presence of introns and exons allows each pre-mRNA to be spliced in more than one way, thereby permitting a single gene to produce a series of mRNAs coding for different polypeptides.

Two quality-control mechanisms help protect cells from producing and using defective mRNAs. RNA proofreading is carried out by some forms of RNA polymerase during transcription to remove incorrect nucleotides from growing RNA chains, and RNA surveillance is used to destroy mRNA molecules that contain erroneous stop codons.

Key Terms for Self-Testing

The Directional Flow of Genetic Information
transcription (p. 623)
translation (p. 623)
messenger RNA (mRNA) (p. 623)
ribosomal RNA (rRNA) (p. 623)
transfer RNA (tRNA) (p. 623)

The Genetic Code
genetic code (p. 624)
gene (p. 628)
triplet code (p. 628)
mutagen (p. 628)
frameshift mutation (p. 630)
degenerate code (p. 631)
template strand (p. 632)
coding strand (p. 632)
codon (p. 633)
initiation codon (p. 634)
stop codon (p. 634)

Transcription in Prokaryotic Cells
RNA polymerase (p. 635)
sigma (σ) factor (p. 635)
transcription unit (p. 635)

promoter site (p. 637)
upstream (p. 637)
downstream (p. 637)
DNA footprinting (p. 637)
consensus sequence (p. 637)
termination signal (p. 639)
rho (ρ) factor (p. 639)
hairpin loop (p. 639)

Transcription in Eukaryotic Cells
RNA polymerase I (p. 640)
RNA polymerase II (p. 641)
RNA polymerase III (p. 641)
core promoter (p. 641)
initiator (Inr) (p. 641)
TATA box (p. 641)
TFIIB recognition element (BRE) (p . 641)
downstream promoter element (DPE) (p. 641)
general transcription factor (p. 643)
TATA-binding protein (TBP) (p. 643)

RNA Processing
primary transcript (p. 644)

RNA processing (p. 644)
pre-rRNA (p. 645)
snoRNA (p. 646)
pre-tRNA (p. 646)
pre-mRNA (p. 648)
5′ cap (p. 648)
poly(A) tail (p. 648)
intron (p. 649)
exon (p. 650)
RNA splicing (p. 651)
spliceosome (p. 651)
snRNP (p. 651)
snRNA (p. 651)
RNP particle (p. 651)
RNA editing (p. 653)
RNA proofreading (p. 654)
RNA surveillance (p. 654)

Box 19A: *Reverse Transcription, Retroviruses, and Retrotransposons*
reverse transcriptase (p. 626)

Problem Set

More challenging problems are marked with a •.

19-1.   Codes and Coding. One way to think about the genetic code is to compare it to the International Morse Code, a system of dots, dashes, and spaces used to send messages by telegraph and shortwave radio. In the Morse code, each letter of the alphabet is represented by a unique combination of two, three, or four dots and dashes. For example, • − (dot-dash) represents the letter A, − • • • (dash-dot-dot-dot) represents the letter B, • − • (dot-dash-dot) represents the letter C, and so forth. Based on this information and your knowledge of the genetic code, explain each of the following terms as it relates to coding, and indicate with an M if it is true of the Morse code, with a G if it is true of the genetic code, and with both letters if it is true for both.

(a) Degenerate

(b) Unambiguous

(c) Triplet

(d) Universal

(e) Nonoverlapping

19-2.   The Genetic Code in a T-Even Phage. A portion of a polypeptide produced by bacteriophage T4 was found to have the following sequence of amino acids:

...Lys-Ser-Pro-Ser-Leu-Asn-Ala...

Deletion of a single nucleotide in one location on the T4 DNA template strand with subsequent insertion of a different nucleotide nearby changed the sequence to

...Lys-Val-His-His-Leu-Met-Ala...

(a) What was the nucleotide sequence of the segment of the mRNA that encoded this portion of the original polypeptide?

(b) What was the nucleotide sequence of the mRNA encoding this portion of the mutant polypeptide?

(c) Can you determine which nucleotide was deleted and which was inserted? Explain your answer.

19-3.   Frameshift Mutations. The mutants listed below each have a different mutant form of the gene encoding protein X. Each mutant gene contains one or more nucleotide insertions (+) or deletions (−) of the type caused by acridine dyes. Assume that all the mutations are located very near the beginning of the gene for protein X. In each case, indicate with an "OK" if you would expect the mutant protein to be nearly normal and with a "Not OK" if you would expect it to be obviously abnormal.

(a) −

(b) −/+

(c) −/−

(d) +/−/+

(e) +/−/+/−

(f) +/+/+

(g) +/+/−/+

(h) −/−/+/−/−

(i) −/−/−/−/−

• **19-4.** **Life with an Overlapping Code.** Assume that the genetic code for all forms of life on planet QB9 consists of overlapping triplets of nucleotides, such that the translation apparatus shifts only one nucleotide at a time. Thus, the nucleotide sequence ABCDEF would be read on Earth as two codons (ABC, DEF) but on QB9 as four codons (ABC, BCD, CDE, DEF) and the start of two more. For each of the following kinds of mutations, briefly describe the effect it would have on the amino acid sequence of the coded protein (i) on Earth and (ii) on QB9. (Assume the mutation occurs near the middle of the gene.)

(a) Single-nucleotide substitution

(b) Single-nucleotide deletion

(c) Deletion of three consecutive nucleotides

• **19-5.** **Locating Promoters.** The following table provides data concerning the effects of various deletions in a eukaryotic gene coding for 5S RNA on the ability of this gene to be transcribed by RNA polymerase III.

Nucleotides Deleted	Ability of 5S RNA Gene to Be Transcribed by RNA Polymerase III
−45 through −1	Yes
+1 through +47	Yes
+10 through +47	Yes
+10 through +63	No
+80 through +123	No
+83 through +123	Yes

(a) What do these data tell you about the probable location of the promoter for this particular 5S RNA gene?

(b) If a similar experiment were carried out for a gene transcribed by RNA polymerase I, what kinds of results would you expect?

(c) If a similar experiment were carried out for a gene transcribed by RNA polymerase II, what kinds of results would you expect?

19-6. **RNA Polymerases and Promoters.** For each of the following statements about RNA polymerases, indicate with a B if the statement is true of the bacterial enzyme and with a I, II, or III if it is true of the respective eukaryotic RNA polymerase. A given statement may be true of any, all, or none (N) of these enzymes.

(a) The enzyme is insensitive to α-amanitin.

(b) The enzyme catalyzes an exergonic reaction.

(c) All the primary transcripts must be processed before being used in translation.

(d) The enzyme may sometimes be found attached to an RNA molecule that also has ribosomes bound to it.

(e) The enzyme synthesizes rRNA.

(f) Transcription factors must bind to the promoter before the polymerase can bind.

(g) The enzyme adds a poly(A) sequence to mRNA.

(h) The enzyme moves along the DNA template strand in the 3′ ⟶ 5′ direction.

(i) The enzyme synthesizes a product likely to acquire a 5′ cap.

(j) All promoters used by the enzyme lie mostly upstream of the transcriptional startpoint and are only partially transcribed.

(k) The specificity of transcription by the enzyme is determined by a subunit of the holoenzyme.

19-7. **RNA Processing.** The three major classes of RNA found in the cytoplasm of a typical eukaryotic cell are rRNA, tRNA, and mRNA. For each, indicate the following:

(a) Two or more kinds of processing to which that RNA has almost certainly been subjected

(b) A processing activity unique to that RNA species

(c) A processing activity that you would also expect to find for the same species of RNA from a bacterial cell

19-8. **Spliceosomes.** The RNA processing carried out by spliceosomes in the eukaryotic nucleus involves a number of different kinds of protein and RNA molecules. For each of the following five components of the splicing process, indicate whether it is protein (P), RNA (R), or both (PR). Then briefly explain how each of the five fits into the process.

(a) SnRNA

(b) Spliceosome

(c) SnRNP

(d) Splice sites

(e) Lariat

• **19-9.** **Antibiotic Inhibitors of Transcription.** Rifamycin and actinomycin D are two antibiotics derived from the bacterium *Streptomyces*. Rifamycin binds to the β subunit of *E. coli* RNA polymerase and interferes with the formation of the first phosphodiester bond in the RNA chain. Actinomycin D binds to double-stranded DNA by intercalation (slipping in between neighboring base pairs).

(a) Which of the four stages in transcription would you expect rifamycin to affect primarily?

(b) Which of the four stages in transcription would you expect actinomycin D to affect primarily?

(c) Which of the two inhibitors is more likely to affect RNA synthesis in cultured human liver cells?

(d) Which of the two inhibitors would be more useful for an experiment in which it is necessary to block the initiation of new RNA chains without interfering with the elongation of chains that are already being synthesized?

• 19-10. Copolymer Analysis. In their initial attempts to determine codon assignments, Nirenberg and Matthei first used RNA homopolymers and then used RNA copolymers synthesized by the enzyme polynucleotide phosphorylase. This enzyme adds nucleotides randomly to the growing chain, but in proportion to their presence in the incubation mixture. By varying the ratio of precursor molecules in the synthesis of copolymers, Nirenberg and Matthei were able to deduce base compositions (but usually not actual sequences) of the codons that code for various amino acids. Suppose you carry out two polynucleotide phosphorylase incubations, with UTP and CTP present in both, but in different ratios. In incubation A, the precursors are present in equimolar concentrations; in incubation B, there is three times as much UTP as CTP. The copolymers generated in both incubation mixtures are then used in a cell-free protein-synthesizing system, and the resulting polypeptides are analyzed for amino acid composition.

(a) What are the eight possible codons represented by the nucleotide sequences of the resulting copolymers in both incubation mixtures? What amino acids do these codons code for?

(b) For every 64 codons in the copolymer formed in incubation A, how many of each of the eight possible codons would you expect on the average? How many for incubation B?

(c) What can you say about the expected frequency of occurrence of the possible amino acids in the polypeptides obtained upon translation of the copolymers from incubation A? What about the polypeptides that result from translation of the incubation B copolymers?

(d) Explain what sort of information can be obtained by this technique.

(e) Would it be possible by this technique to determine that codons with 2 U's and 1 C code for phenylalanine, leucine, and serine? Why or why not?

(f) Would it be possible by this technique to decide which of the three codons with 2 U's and 1 C (UUC, UCU, CUU) correspond to each of the three amino acids mentioned in part e? Why or why not?

(g) Suggest a way to assign the three codons of part f to the appropriate amino acids of part e.

• 19-11. Introns. To investigate the possible presence of introns in three newly discovered genes ("X," "Y," and "Z"), you perform an experiment in which the restriction enzyme *Hae*III is used to cleave either the DNA of each gene or the cDNA made by copying its mRNA with reverse transcriptase. The resulting DNA fragments are separated by gel electrophoresis, and the presence of fragments in the gels is detected by hybridizing to a radioactive DNA probe made by copying the intact gene with DNA polymerase in the presence of radioactive substrates. The following results are obtained:

Source of DNA	Number of Fragments After Electrophoresis
Gene "X" DNA	3
cDNA made from mRNA "X"	2
Gene "Y" DNA	4
cDNA made from mRNA "Y"	2
Gene "Z" DNA	2
cDNA made from mRNA "Z"	2

(a) What can you conclude about the number of introns present in gene X?

(b) What can you conclude about the number of introns present in gene Y?

(c) What can you conclude about the number of introns present in gene Z?

Suggested Reading

References of historical importance are marked with a •.

Information Flow and the Genetic Code

Atkins, J. F., and R. F. Gesteland. The twenty-first amino acid. *Nature* 407 (2000): 463.
• Crick, F. H. C. The genetic code. *Sci. Amer.* 207 (October 1962): 66.
• Crick, F. H. C. The genetic code III. *Sci. Amer.* 215 (October 1966): 55.
Hayes, B. The invention of the genetic code. *American Scientist* 86 (1998): 8.
• Khorana, H. G. Nucleic acid synthesis in the study of the genetic code. In *Nobel Lectures: Physiology or Medicine (1963–1970)*, 341. New York: American Elsevier, 1973.
• Nirenberg, M. W. The genetic code II. *Sci. Amer.* 208 (March 1963): 80.
O'Sullivan, J. M., J. B. Davenport, and M. F. Tuite. Codon reassignment and the evolving genetic code: Problems and pitfalls in post-genome analysis. *Trends Genet.* 17 (2001): 20.
• Sarkar, S. Forty years under the central dogma. *Trends Biochem. Sci.* 23 (1998): 312.
Strasser, B. J. Sickle cell anemia, a molecular disease. *Science* 286 (1999): 1488.

Transcription in Prokaryotic Cells

Das, A. Control of transcription termination by RNA-binding proteins. *Annu. Rev. Biochem.* 62 (1993): 893.
Gelles, J., and R. Landick. RNA polymerase as a molecular motor. *Cell* 93 (1998): 13
Landick, R. Shifting RNA polymerase into overdrive. *Science* 284 (1999): 598.
Nudler, E., A. Goldfarb, and M. Kashlev. Discontinuous mechanism of transcription elongation. *Science* 265 (1994): 793.
von Hippel, P. H. An integrated model of the transcription complex in elongation, termination, and editing. *Science* 281 (1998): 660.

Transcription in Eukaryotic Cells

Cramer, P., et al. Architecture of RNA polymerase II and implications for the transcription mechanism. *Science* 288 (2000): 640.
Kutach, A. K., and J. T. Kadonaga. The downstream promoter element DPE appears to be as widely used as the TATA box in *Drosophila* core promoters. *Mol. Cell. Biol.* 20 (2000): 4754.

Orphanides, G., and D. Reinberg. RNA polymerase II elongation through chromatin. *Nature* 407 (2000): 471.

Roeder, R. G. The role of general initiation factors in transcription by RNA polymerase II. *Trends Biochem. Sci.* 21 (1996): 327.

Thomas, M. J., A. A. Platas, and D. K. Hawley. Transcriptional fidelity and proofreading by RNA polymerase II. *Cell* 93 (1998): 627.

Tjian, R. Molecular machines that control genes. *Sci. Amer.* 272 (February 1995): 54.

White, R. J. *Gene Transcription: Mechanisms and Control.* Oxford: Blackwell Science, 2001.

RNA Processing

Culbertson, M. R. RNA surveillance. *Trends Genet.* 15 (1999): 74.

Gott, J. M., and R. B. Emeson. Functions and mechanisms of RNA editing. *Annu. Rev. Genetics* 34 (2000): 499.

Hastings, M. L., and A. R. Krainer. Pre-mRNA splicing in the new millennium. *Curr. Opin. Cell Biol.* 13 (2001): 302.

Lafontaine, D. L. J., and D. Tollervey. Birth of the snoRNPs: The evolution of the modification-guide snoRNAs. *Trends Biochem. Sci.* 23 (1998): 383.

Staley, J. P., and C. Guthrie. Mechanical devices of the spliceosome: Motors, clocks, springs, and things. *Cell* 92 (1998): 315.

Steitz, J. A. Snurps. *Sci. Amer.* 258 (June 1988): 58.

Tarn, W.-Y., and J. A. Steitz. Pre-mRNA splicing: The discovery of a new spliceosome doubles the challenge. *Trends Biochem. Sci.* 22 (1997): 132.

20

Gene Expression: II. Protein Synthesis and Sorting

In the preceding chapter, we took gene expression from DNA to RNA, covering DNA transcription and processing of the resulting RNA transcripts. For genes encoding ribosomal RNA and transfer RNA (and certain other small RNA molecules), RNA is the ultimate expression of a gene. But for the thousands of other genes in an organism's genome, the ultimate gene product is protein. This chapter describes how messenger RNAs produced by protein-coding genes are translated into polypeptides, how polypeptides become functional proteins, and how proteins reach the destinations where they carry out their functions.

Translation, the first and most important phase of protein synthesis, involves a change in language from the nucleotide sequence of an mRNA molecule to the amino acid sequence of a polypeptide chain. In essence, the sequential order of mRNA nucleotides, read as triplet codons, specifies the order in which incoming amino acids are added to a growing polypeptide chain. Ribosomes serve as the intracellular sites of the translation process, while RNA molecules are the agents that ensure insertion of the correct amino acids at each position in the polypeptide. We will start by surveying the cell's cast of characters for performing translation, before examining in detail the steps of the process.

Translation: The Cast of Characters

The cellular machinery for translating mRNAs into polypeptides involves five major components: *ribosomes* that carry out the process of polypeptide synthesis, *tRNA* molecules that align amino acids in the correct order along the mRNA template, *aminoacyl-tRNA synthetases* that attach amino acids to their appropriate tRNA molecules, *mRNA* molecules that encode the amino acid sequence information for the polypeptides being synthesized, and *protein factors* that facilitate several steps in the translation process. In introducing this cast of characters, let us begin with the ribosomes.

The Ribosome Carries Out Polypeptide Synthesis

Ribosomes play a central role in protein synthesis, orienting the mRNA and amino acid–carrying tRNAs in the proper relation to each other so the genetic code is read accurately, and catalyzing formation of the peptide bonds that link the amino acids into a polypeptide. As we saw in Chapter 4, **ribosomes** are particles made of RNA and protein that reside in the cytoplasm of both prokaryotic and eukaryotic cells, as well as in the matrix of mitochondria and the stroma of chloroplasts. In the eukaryotic cytoplasm, they exist both free in the cytosol and bound to membranes of the endoplasmic reticulum and the outer membrane of the nuclear envelope. Ribosomes are often considered to be organelles, but they differ from most other organelles in not being bounded by a membrane. Prokaryotic and eukaryotic ribosomes resemble each other structurally, but they are not identical. Prokaryotic ribosomes are smaller in size (2.5×10^6 instead of 4.2×10^6 Da), contain fewer proteins, have smaller RNA molecules (and one fewer RNA), and are sensitive to different inhibitors of protein synthesis.

The shape of the prokaryotic ribosome as revealed by electron microscopy is shown in Figure 20-1. Like all ribosomes, it consists of two dissociable subunits called the *large* and *small subunits*. A complete prokaryotic ribosome, with a sedimentation coefficient of about 70S, con-

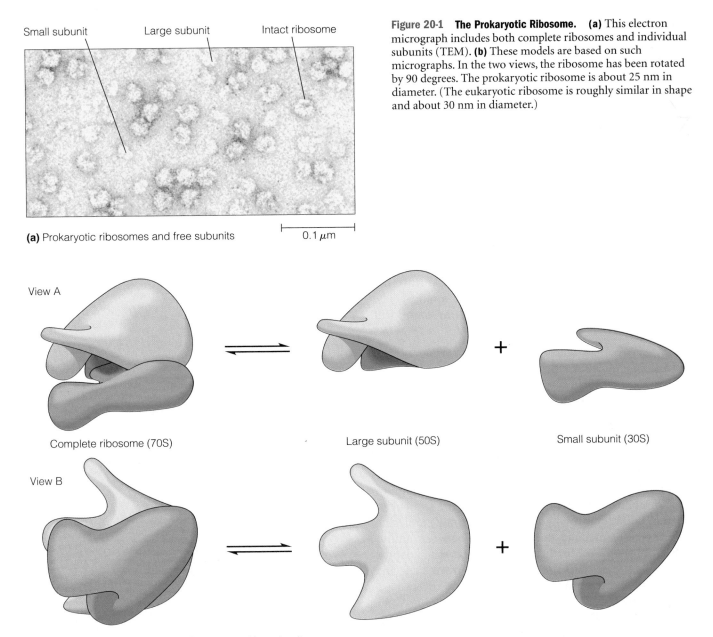

Small subunit   Large subunit   Intact ribosome

Figure 20-1 The Prokaryotic Ribosome. (a) This electron micrograph includes both complete ribosomes and individual subunits (TEM). **(b)** These models are based on such micrographs. In the two views, the ribosome has been rotated by 90 degrees. The prokaryotic ribosome is about 25 nm in diameter. (The eukaryotic ribosome is roughly similar in shape and about 30 nm in diameter.)

(a) Prokaryotic ribosomes and free subunits

0.1 μm

View A

Complete ribosome (70S)   Large subunit (50S)   Small subunit (30S)

View B

(b) Two views of the prokaryotic ribosome and its subunits

sists of a 30S small subunit and a 50S large subunit; its eukaryotic equivalent is an 80S ribosome consisting of a 40S subunit and a 60S subunit. Table 20-1 lists some of the properties of prokaryotic and eukaryotic ribosomes and their subunits. Ribosomal RNAs and protein molecules self-assemble into small and large subunits, but the two types of subunits come together only when bound to mRNA, as we will see shortly. X-ray crystallography has recently allowed the arrangement of all the individual proteins and RNA molecules of the small and large subunits of bacterial ribosomes to be pinpointed down to the atomic level.

Functionally, ribosomes have sometimes been called the "workbenches" of protein synthesis, but their active role in polypeptide synthesis makes "machine" a more apt

label. In essence, the role of the ribosome in polypeptide synthesis resembles that of a large, complicated enzyme constructed from more than 50 different proteins and several kinds of rRNA. For many years it was thought that the rRNA simply provided a structural scaffold for the ribosomal proteins, with the latter actually carrying out the steps in polypeptide synthesis. But today we know that the reverse is closer to the truth—that rRNA performs many of the ribosome's key functions.

Four sites on the ribosome are particularly important for protein synthesis (Figure 20-2). These are an **mRNA-binding site** and three sites where tRNA can bind: an **A (aminoacyl) site** that binds each newly arriving tRNA with its attached amino acid, a **P (peptidyl) site** where the tRNA carrying the growing polypeptide chain

Table 20-1   Properties of Prokaryotic and Eukaryotic Cytoplasmic Ribosomes

| Source | Size of Ribosomes | | Subunit | Subunit Size | | Subunit Proteins | Subunit RNA | |
	S Value*	Mol. Wt.		S Value	Mol. Wt.		S Value	Nucleotides
Prokaryotic cells	70S	2.5×10^6	Large	50S	1.6×10^6	34	23S	2900
							5S	120
			Small	30S	0.9×10^6	21	16S	1540
Eukaryotic cells	80S	4.2×10^6	Large	60S	2.8×10^6	About 45	25–28S	≤4700
							5.8S	160
							5S	120
			Small	40S	1.4×10^6	About 33	18S	1900

*If you are surprised that the S values of the subunits do not add up to that of the whole ribosome, recall that an S value is a measure of the velocity at which a particle sediments upon centrifugation and is only indirectly related to the mass of the particle.

resides, and an **E (exit) site,** from which tRNAs leave the ribosome after they have discharged their amino acids. How these sites function in the process of translation will become clear in a few pages.

Transfer RNA Molecules Bring Amino Acids to the Ribosome

Since the sequence of codons in mRNA ultimately determines the amino acid sequence of polypeptide chains, a mechanism must exist that enables codons to arrange amino acids in the proper order. The general nature of this mechanism was first proposed in 1957 by Francis Crick. With remarkable foresight, Crick postulated that amino acids cannot directly recognize nucleotide base sequences, and that some kind of hypothetical "adaptor" molecule must therefore mediate the interaction between amino acids and mRNA. He further predicted that each adaptor molecule possesses two sites, one that binds to a specific amino acid and the other that recognizes an mRNA base sequence coding for this amino acid.

In the year following Crick's proposal, Mahlon Hoagland discovered a family of adaptor molecules exhibiting these predicted properties. While investigating the process of protein synthesis in cell-free systems, Hoagland found that radioactive amino acids first become covalently attached to small RNA molecules. Adding these amino acid–RNA complexes to ribosomes led to the onset of protein synthesis and the incorporation of radioactive amino acids into new proteins. Hoagland therefore concluded that amino acids are initially bound to small RNA molecules, which then bring the amino acids to the ribosome for subsequent insertion into newly forming polypeptide chains.

The small RNA molecules that Hoagland discovered were named **transfer RNAs (tRNAs).** Appropriate to their role as intermediaries between mRNA and amino acids, tRNA molecules have two kinds of specificity. Each tRNA binds to one specific amino acid, and each recognizes one or more mRNA codons that specify that particular amino acid, as indicated by the genetic code. Transfer RNAs are linked to their corresponding amino acids by an ester bond that joins the amino acid to the 3′ OH group of the

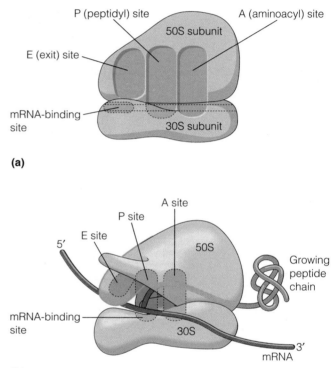

(a)

(b)

Figure 20-2   Important Binding Sites on the Prokaryotic Ribosome. This model of ribosome structure shows the A (aminoacyl) and P (peptidyl) sites as cavities on the ribosome where charged (amino acid–carrying) tRNA molecules bind during polypeptide synthesis. The E (exit) site is the site from which discharged tRNAs leave the ribosome. The mRNA-binding site binds a particular nucleotide sequence near the 5′ end of the mRNA, placing the mRNA in the proper position for the translation of its first codon. **(a)** The diagrammatic representation of a ribosome that is used in this chapter. The pair of horizontal dashed lines indicate where the mRNA molecule lies. **(b)** A more realistic representation. The binding sites are all located at or near the interface between the large and small subunits.

adenine (A) nucleotide located at the 3′ end of all tRNA molecules (Figure 20-3a). Selection of the correct amino acid for each tRNA is the responsibility of the enzymes that catalyze formation of the ester bond, as we will discuss shortly. By convention, the name of the amino acid

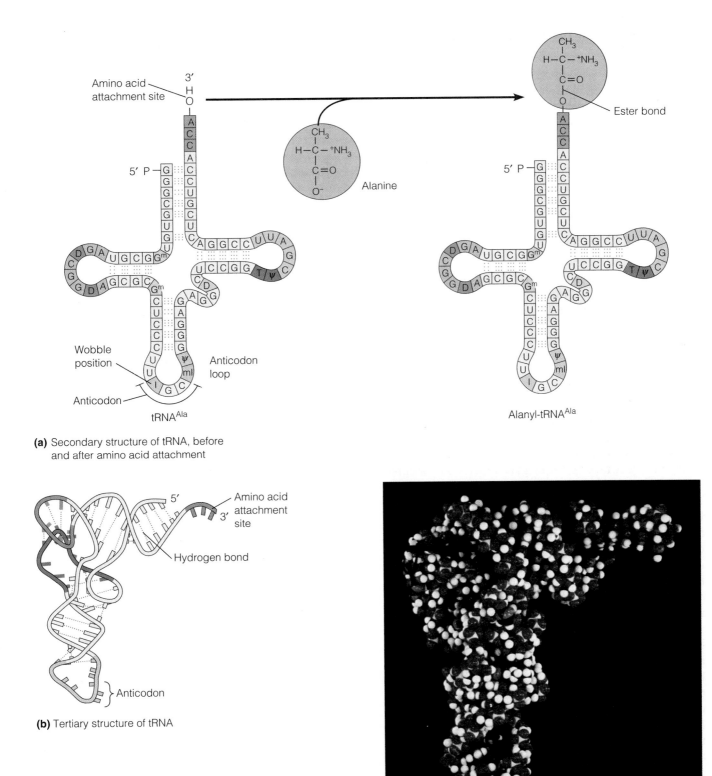

(a) Secondary structure of tRNA, before and after amino acid attachment

(b) Tertiary structure of tRNA

Figure 20-3 Sequence, Structure, and Aminoacylation of a tRNA. **(a)** Yeast alanine tRNA, like all tRNA molecules, contains three major loops, four base-paired regions, an anticodon triplet, and a 3' terminal sequence of CCA, to which the appropriate amino acid can be attached by an ester bond. Modified bases are dark colored, and their names (as nucleosides) are abbreviated I for inosine, mI for methylinosine, D for dihydrouridine, T for ribothymidine, ψ for pseudouridine, and G^m for methylguanosine. (For the significance of the wobble position in the anticodon, see Figure 20-4.) **(b)** In the L-shaped tertiary structure of tRNA, the amino acid attachment site is at one end and the anticodon at the other. Color coding of the tRNA loops correlates the secondary (two-dimensional) structures of part a and the tertiary (three-dimensional) structure on the left in part b. The tRNA model on the right in part b shows the individual atoms. The structure shown here is characteristic of all tRNAs.

that attaches to a given tRNA is indicated by a superscript. For example, tRNA molecules specific for alanine are identified as tRNA^{Ala}. Once the amino acid is attached, the tRNA is called an **aminoacyl tRNA** (e.g., alanyl tRNA^{Ala}). The tRNA is said to be in its *charged* form, and the amino acid is said to be *activated*.

Transfer RNA molecules can recognize codons in mRNA because each tRNA possesses an **anticodon**, a special trinucleotide sequence located within one of the loops of the tRNA molecule (see Figure 20-3a). The anticodon of each tRNA is complementary to one or more mRNA codons that specify the amino acid being carried by that tRNA. Therefore, *anticodons permit tRNA molecules to recognize codons in mRNA by complementary base pairing.* Take careful note of the convention used in representing codons and anticodons: Codons in mRNA are written in the 5′ ⟶ 3′ direction, whereas anticodons in tRNA are usually represented in the 3′ ⟶ 5′ orientation. Thus, one of the codons for alanine is 5′-GCC-3′ and the corresponding anticodon in tRNA is 3′-CGG-5′.

Since the genetic code employs 61 different codons to specify amino acids, you might expect to find 61 different tRNA molecules involved in protein synthesis, each responsible for recognizing a different codon. However, the number of different tRNAs is significantly less than 61, because many tRNA molecules recognize more than one codon. You can see why this is possible by examining the table of the genetic code (see Figure 19-8). Codons differing in the third base often code for the same amino acid. For example, UUU and UUC both code for phenylalanine; UCU, UCC, UCA, and UCG all code for serine; and so forth. In such cases, the same tRNA can bind to more than one codon without introducing mistakes. For example, a single tRNA can recognize the codons UUU and UUC because both code for the same amino acid, phenylalanine.

Such considerations led Francis Crick to propose that mRNA and tRNA line up on the ribosome in a way that permits flexibility or "wobble" in the pairing between the third base of the codon and the corresponding base in the anticodon. According to the **wobble hypothesis,** this flexibility in codon-anticodon binding allows some unexpected base pairs to form, as shown in Figure 20-4. The unusual base inosine (I), which is extremely rare in other RNA molecules, occurs often in the wobble position of tRNA anticodons (see Figure 20-3a). Inosine is the "wobbliest" of all third-position bases, since it can pair with U, C, or A. For example, a tRNA with the anticodon 3′-UAI-5′ can recognize the codons AUU, AUC, and AUA, all of which code for the amino acid isoleucine.

It is because of wobble that fewer tRNA molecules are required for some amino acids than the number of codons that specify those amino acids. In the case of isoleucine, for example, a cell can translate all three codons with a single tRNA molecule containing 3′-UAI-5′ as its anticodon. Similarly, the six codons for the amino acid leucine (UUA, UUG, CUU, CUC, CUA, and CUG) require only three tRNAs because of wobble. Although the existence of wobble means that a single tRNA molecule can recognize more

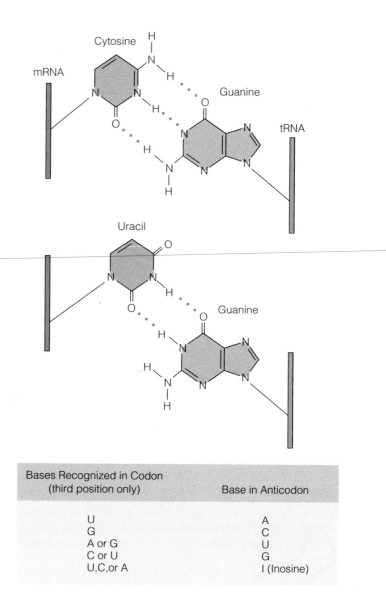

Bases Recognized in Codon (third position only)	Base in Anticodon
U	A
G	C
A or G	U
C or U	G
U, C, or A	I (Inosine)

Figure 20-4 The Wobble Hypothesis. The two diagrams illustrate how a slight shift or "wobble" in the position of the base guanine in a tRNA anticodon would permit it to pair with uracil (bottom) instead of its normal complementary base, cytosine (top). The table summarizes the base pairs permitted at the third position of a codon by the wobble hypothesis.

than one codon, the different codons recognized by a given tRNA always code for the same amino acid, and hence wobble does not cause insertion of incorrect amino acids.

Aminoacyl-tRNA Synthetases Link Amino Acids to the Correct Transfer RNAs

Before a tRNA molecule can bring its appropriate amino acid to the ribosome, that amino acid first must be covalently attached to the tRNA. The enzymes responsible for linking amino acids to their corresponding tRNA molecules are called **aminoacyl-tRNA synthetases.** Cells usually have 20 different aminoacyl-tRNA synthetases, one for each of the 20 amino acids commonly used in protein synthesis. When more than one tRNA exists for a given amino acid, the aminoacyl-tRNA synthetase specific for that particular amino acid recognizes each of the tRNAs. Some cells possess less than 20 aminoacyl-tRNA synthetases. In such cases, the same aminoacyl-tRNA synthetase may catalyze the attachment of two different amino acids to their corresponding tRNAs, or it may attach an incorrect amino acid to a tRNA molecule. These latter "errors" are corrected by a second enzyme that alters the incorrect amino acid after it has been attached to the tRNA.

Aminoacyl-tRNA synthetases catalyze the joining of an amino acid to its corresponding tRNAs via an ester bond, accompanied by the hydrolysis of ATP to AMP and pyrophosphate:

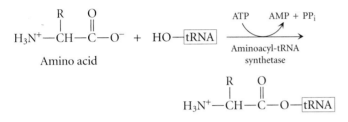

Figure 20-5 outlines the steps by which this reaction occurs. The driving force for the reaction is provided by the hydrolysis of pyrophosphate to $2P_i$.

In the product, aminoacyl tRNA, the ester bond linking the amino acid to the tRNA is said to be a "high-energy" bond. This simply means that hydrolysis of the bond releases sufficient energy to drive formation of the peptide bond that will eventually join the amino acid to a growing polypeptide chain. The process of aminoacylation of a tRNA molecule is therefore also called *amino acid activation,* because it not only links an amino acid to its proper tRNA but also activates it for subsequent peptide bond formation.

How do aminoacyl-tRNA synthetases identify the correct tRNA for each amino acid? Differences in the base sequences of the various tRNA molecules allow them to be distinguished and, surprisingly, the anticodon is not the only feature to be recognized. Changes in the base sequence of either the anticodon triplet or the 3′ end of a tRNA molecule can alter the amino acid to which a tRNA becomes bound. Thus aminoacyl-tRNA synthetases recognize nucleotides located in at least two different regions of tRNA molecules when they pick out the tRNA that is to become linked to a particular amino acid.

Once the correct amino acid has been joined to its tRNA, it is the tRNA itself (and not the amino acid) that recognizes the appropriate codon in mRNA. The first evidence for this was provided by François Chapeville and Fritz Lipmann, who designed an elegant experiment involving the tRNA that carries the amino acid cysteine. They took the tRNA after its cysteine had been attached and treated it with a nickel catalyst, which converted the attached cysteine into the amino acid alanine. The result was therefore alanine covalently linked to a tRNA molecule that normally carries cysteine. When the researchers added this abnormal aminoacyl tRNA to a cell-free pro-

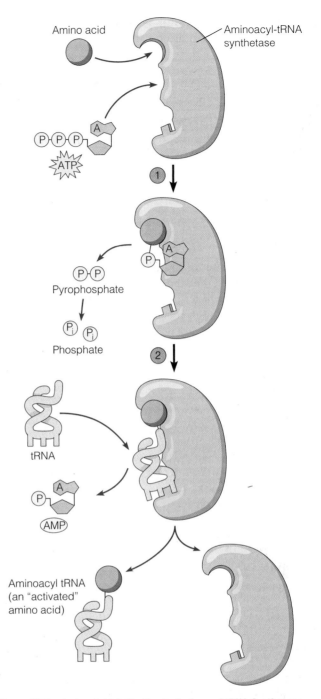

Figure 20-5   Amino Acid Activation by Aminoacyl-tRNA Synthetase. In two chemical steps, this enzyme catalyzes the formation of an ester bond between the carboxyl group of an amino acid and the 3′ OH of the appropriate tRNA. ① The amino acid and a molecule of ATP enter the active site of the enzyme. Simultaneously, ATP loses pyrophosphate, and the resulting AMP bonds covalently to the amino acid. The pyrophosphate is hydrolyzed to $2P_i$. ② The tRNA covalently bonds to the amino acid, displacing the AMP. The aminoacyl tRNA is then released from the enzyme.

tein-synthesizing system, alanine was inserted into poly-peptide chains in locations normally occupied by cysteine. Such results proved that codons in mRNA recognize tRNA molecules rather than their bound amino acids. Hence the specificity of the aminoacyl-tRNA synthetase reaction is crucial to the accuracy of gene expression because it ensures that the proper amino acid is linked to each tRNA.

Messenger RNA Brings Polypeptide-Coding Information to the Ribosome

As we learned in Chapter 19, it is the translation of codons in mRNA that directs the order in which amino acids are linked together during protein synthesis. Hence it is the particular mRNA that binds to a given ribosome that determines which polypeptide the ribosome will manufacture. The heart of messenger RNA is, of course, its message—the sequence of nucleotides that encodes a polypeptide. However, an mRNA molecule also has sequences at either end that are not translated, as shown in Figure 20-6. The nontranslated sequence at the 5′ end of mRNA is called the **leader** because it precedes the **start codon,** the first codon to be translated. The start codon is usually AUG, although it can also be GUG in eukaryotes. The nontranslated sequence at the 3′ end is the **trailer,** and it follows the **stop codon,** which can be UAG, UAA, or UGA. The leader and trailer sequences range from about 30 to several hundred nucleotides in length. While these sequences are not themselves translated, their presence is essential for translation of the message. In addition, eukaryotic mRNAs have 5′ caps and 3′ poly(A) tails, as we saw in Chapter 19.

In eukaryotes, each mRNA molecule typically encodes a single polypeptide. In prokaryotes, however, some mRNAs are *polygenic;* that is, they encode several polypeptides, usually with related functions in the cell. The clusters of genes that give rise to polygenic mRNA molecules are single transcription units called *operons,* which we will describe in Chapter 21 in our study of gene regulation. The protein-coding regions of polygenic mRNAs are often separated from each other by spacer regions.

Protein Factors Are Required for the Initiation, Elongation, and Termination of Polypeptide Chains

In addition to aminoacyl-tRNA synthetases and the protein components of the ribosome, translation also requires the participation of several other kinds of protein molecules. Some of these *protein factors* are required for initiation of the translation process, others for elongation of the growing polypeptide chain, and still others for the termination of polypeptide synthesis. The exact roles played by these factors will become apparent as we now proceed to a discussion of the mechanism of translation.

The Mechanism of Translation

The translation of mRNAs into polypeptides is an ordered, stepwise process that begins the synthesis of a polypeptide chain at its amino-terminal end, or *N-terminus,* and sequentially adds amino acids to the growing chain until the carboxyl-terminal end, or *C-terminus,* is reached. The first experimental evidence for such a mechanism was provided in 1961 by Howard Dintzis, who investigated hemoglobin synthesis in developing red blood cells that had been incubated briefly with radioactive amino acids. Dintzis reasoned that if the time of incubation with radioisotope is kept relatively brief, then the radioactivity present in completed hemoglobin chains should be concentrated at the most recently synthesized end of the molecule. He found that the highest concentration of radioactivity in completed hemoglobin chains was at the C-terminal end, indicating that the C-terminus is the last part of the polypeptide chain to be synthesized. This allowed him to conclude that during mRNA translation, *amino acids are added to the growing polypeptide chain beginning at the N-terminus and proceeding toward the C-terminus.*

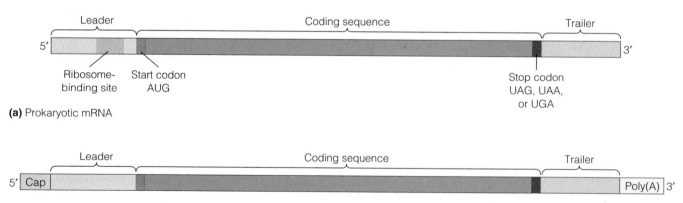

Figure 20-6   Messenger RNA.   (a) A prokaryotic mRNA molecule encoding a single polypeptide has the features shown here. (A polygenic prokaryotic mRNA would generally have a set of these features for each gene.) **(b)** A eukaryotic mRNA molecule has, in addition, a 5′ cap and a 3′ poly(A) tail. It lacks a ribosome-binding site (a nucleotide sequence also called a Shine-Dalgarno sequence, after its discoverers).

In theory, mRNA could be read in either the 5′ → 3′ direction or the 3′ → 5′ direction during this process. The first attempts to determine the direction in which mRNA is actually read involved the use of artificial RNA molecules. A typical example is the synthetic RNA that can be made by adding the base C to the 3′ end of poly(A), yielding the molecule 5′-AAAAAAAAAAA…AAC-3′. When added to a cell-free protein-synthesizing system, this RNA stimulates the synthesis of a polypeptide consisting of a stretch of lysine residues with an asparagine at the C-terminus. Because AAA codes for lysine and AAC codes for asparagine, this means that *mRNA is translated in the 5′ → 3′ direction.* Confirming evidence has come from numerous studies in which the base sequences of naturally occurring mRNAs have been compared with the amino acid sequences of the polypeptide chains they encode. In all cases, the amino acid sequence of the polypeptide chain corresponds to the order of mRNA codons read in the 5′ → 3′ direction.

To understand how translation of mRNA in the 5′ → 3′ direction leads to the synthesis of polypeptides in the N-terminal to C-terminal direction, it is helpful to subdivide the translation process into three stages, as shown in Figure 20-7: ① an *initiation* stage, in which mRNA is bound to the ribosome and positioned for proper translation; ② an *elongation* stage, in which amino acids are sequentially joined together via peptide bonds in an order specified by the arrangement of codons in mRNA; and ③ a *termination* stage, in which the mRNA and the newly formed polypeptide chain are released from the ribosome.

In the following sections, we will examine each of these stages in detail. Our discussion will focus mainly on translation in prokaryotic cells, where the mechanisms are especially well understood, but the comparable events in eukaryotic cells are rather similar. The aspects of translation unique to either prokaryotes or eukaryotes are mostly confined to the initiation stage, as we describe in the next section.

The Initiation of Translation Requires Initiation Factors, Ribosomal Subunits, mRNA, and Initiator tRNA

Prokaryotic Initiation. The initiation of translation in prokaryotes is illustrated in Figure 20-8, which reveals that initiation can be subdivided into three distinct steps. In

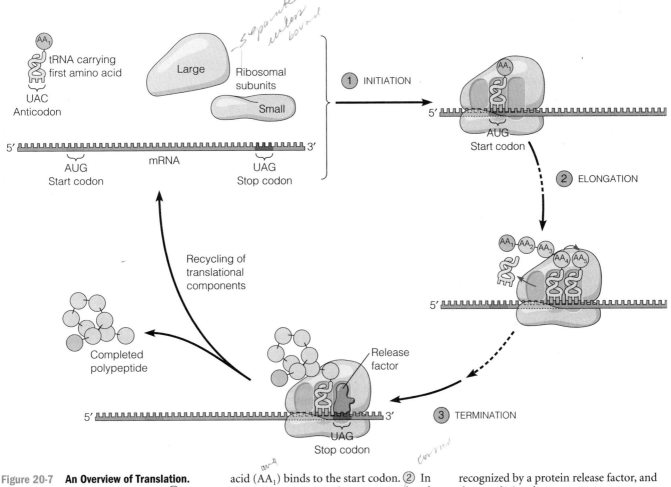

Figure 20-7   An Overview of Translation.
Translation occurs in three stages. ① In initiation, the components of the translational apparatus come together with a molecule of mRNA. A tRNA carrying the first amino acid (AA₁) binds to the start codon. ② In elongation, amino acids are conveyed to the mRNA by tRNAs and are added, one by one, to a growing polypeptide chain. ③ In termination, a stop codon in the mRNA is recognized by a protein release factor, and the translational apparatus comes apart, releasing a completed polypeptide.

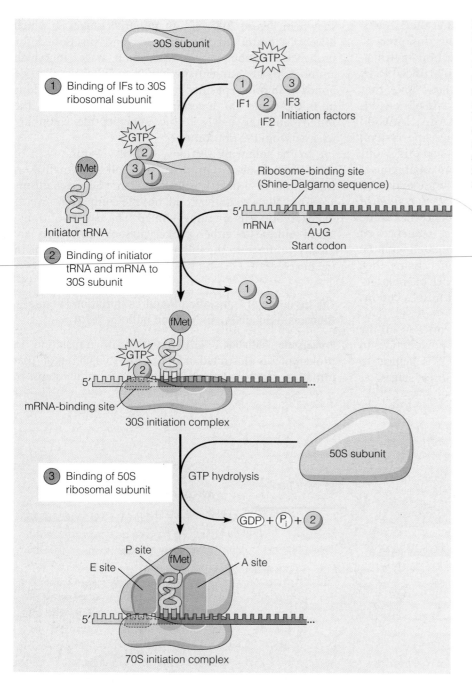

Figure 20-8 Initiation of Translation in Prokaryotes. The formation of the 70S translation initiation complex occurs in three steps. ① Three initiation factors (IF) and GTP bind to the small ribosomal subunit. ② The initiator aminoacyl tRNA and mRNA are attached. The mRNA-binding site is composed, at least in part, of a portion of the 16S rRNA of the small ribosomal subunit. ③ The large ribosomal subunit joins the complex. The resulting 70S initiation complex has fMet-tRNA^{fMet} residing in the ribosome's P site.

NOT IN EUKARYOTES

step ①, three **initiation factors**—called *IF1, IF2,* and *IF3*—bind to the small (30S) ribosomal subunit, with GTP attaching to IF2.

In step ② of Figure 20-8, mRNA and the tRNA carrying the first amino acid (almost always methionine) bind to the 30S ribosomal subunit. The mRNA is bound to the 30S subunit in its proper orientation by means of a special nucleotide sequence called the mRNA's *ribosome-binding site* (also known as the *Shine-Dalgarno sequence,* after its discoverers). This sequence consists of a stretch of 3–9 purine nucleotides (often AGGA) located slightly upstream of the initiation codon. These purines in the mRNA form complementary base pairs with a pyrimidine-rich sequence at the 3′ end of 16S rRNA, which forms the ribosome's

mRNA-binding site. (The importance of this site has been shown by studies involving *colicins,* which are proteins produced by certain strains of *Escherichia coli* that can kill other types of bacteria. One such protein, colicin E3, kills bacteria by destroying their ability to synthesize proteins. Upon entrance into the cytoplasm of susceptible bacteria, colicin E3 catalyzes the removal of a 49-nucleotide fragment from the 3′ end of 16S rRNA, destroying the mRNA-binding site and thereby creating ribosomes that can no longer initiate polypeptide synthesis.)

The binding of mRNA to the mRNA-binding site of the small ribosomal subunit places the mRNA's AUG start codon at the ribosome's P site, where it can then bind to the anticodon of the appropriate tRNA. The first clue that a special kind of tRNA is involved in this step emerged when it was discovered that roughly half the proteins in *E. coli* contain methionine at their N-terminal ends. This was surprising because methionine is a relatively uncommon amino acid, accounting for no more than a few percent of the amino acids in bacterial proteins. The explanation for such a pattern became apparent when it was discovered that bacterial cells contain two different methionine-specific tRNAs. One, designated tRNA^Met, carries a normal methionine destined for insertion into the internal regions of polypeptide chains. The other, called tRNA^fMet, carries a methionine that is converted to the derivative *N-formylmethionine (fMet)* after linkage to the tRNA (Figure 20-9). In *N-formylmethionine*, the amino group of methionine is blocked by the addition of a formyl group and so cannot form a peptide bond with another amino acid; only the carboxyl group is available for bonding to another amino acid. Hence *N-formylmethionine* can be situated only at the N-terminal end of a polypeptide chain, suggesting that tRNA^fMet functions as an **initiator tRNA** that starts the process of translation. This idea was soon confirmed by the discovery that bacterial polypeptide chains in the early stages of synthesis always contain *N-formylmethionine* at their N-terminus. Following completion of the polypeptide chain (and in some cases while it is still being synthesized), the formyl group, and often the methionine itself, is enzymatically removed.

During initiation, the initiator tRNA with its attached *N-formylmethionine* is bound to the P site of the 30S ribosomal subunit by the action of initiation factor IF2 (bound to GTP), which can distinguish initiator tRNA^fMet from other kinds of tRNA. This attribute of IF2 helps to explain why AUG start codons bind to the initiator tRNA^fMet, whereas AUG codons located elsewhere in mRNA bind to the noninitiating tRNA^Met. Once tRNA^fMet enters the P site, its anticodon base-pairs with the AUG start codon in the mRNA, and IF1 and IF3 are released. At this point, the 30S subunit with its associated IF2-GTP, mRNA, and *N-formylmethionyl* tRNA^fMet is referred to as the **30S initiation complex.**

Finally, in step ③ of Figure 20-8, the 30S initiation complex is joined by a free 50S ribosomal subunit, generating the **70S initiation complex.** Binding of the 50S subunit is driven by hydrolysis of the GTP associated with

IF2, which occurs as IF2 leaves the ribosome. At this stage, all three initiation factors have been released.

Eukaryotic Initiation. The initiation process in eukaryotes involves a different set of initiation factors (called *eIFs* in eukaryotes), a somewhat different pathway for assembling the initiation complex, and a special initiator tRNA^Met that—like the normal tRNA for methionine but unlike the initiator tRNA of prokaryotes—carries methionine that does not become formylated. The initiation factor *eIF2*, with GTP already attached, binds to the initiator methionyl tRNA^Met *before* the tRNA then binds to the small ribosomal subunit, along with other initiation factors. The resulting complex next binds to the 5′ end of an mRNA, recognizing the 5′ cap (in some situations the complex may instead bind to an *internal ribosome entry sequence* or *IRES*, which lies directly upstream of the start codon of certain types of mRNA, including viral mRNAs). After binding to an mRNA, the small ribosomal subunit, with the initiator tRNA in tow, scans along the mRNA and usually begins translation at the first AUG triplet it encounters. The nucleotides on either side of the eukaryotic start codon seem to be involved in its recognition; ACCAUGG is a common start sequence, where the underlined triplet is the actual start codon. When the initiator tRNA^Met base-pairs with the start codon, the large ribosomal subunit joins the complex.

Chain Elongation Involves Sequential Cycles of Aminoacyl tRNA Binding, Peptide Bond Formation, and Translocation

Once the initiation complex has been completed, a polypeptide chain is synthesized by the successive addition of amino acids in a sequence specified by codons in mRNA. As summarized in Figure 20-10, this *elongation stage* of polypeptide synthesis involves a repetitive three-step cycle in which ① *binding of an aminoacyl tRNA* to the ribosome brings a new amino acid into position to be joined to the polypeptide chain, ② *peptide bond formation* links this amino acid to the growing polypeptide, and ③ the mRNA is advanced a distance of three nucleotides by the process of *translocation* to bring the next codon into position for translation. Each of these steps is described in more detail below.

Binding of Aminoacyl tRNA. At the onset of the elongation stage, the AUG start codon in the mRNA is located at the ribosomal P site and the second codon (the codon immediately downstream from the start codon) is located at the A site. Elongation begins when an aminoacyl tRNA whose anticodon is complementary to the second codon binds to the ribosomal A site (see Figure 20-10, ①). The binding of this new aminoacyl tRNA requires two protein **elongation factors**, *EF-Tu* and *EF-Ts*, and is driven by the hydrolysis of 2GTP molecules, generating 2GDP and $2P_i$. Note that, from now on, every incoming aminoacyl tRNA binds first to the A (aminoacyl) site—hence the site's name.

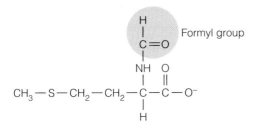

Figure 20-9 The Structure of *N*-Formylmethionine. *N*-formylmethionine (fMet) is the modified amino acid with which every polypeptide is initiated in prokaryotes.

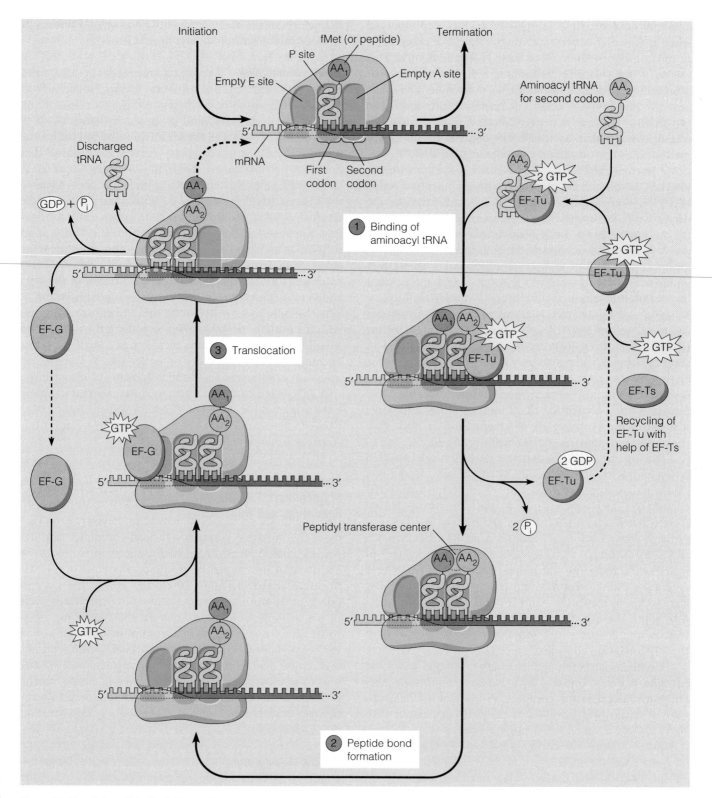

Figure 20-10 Polypeptide Chain Elongation in Prokaryotes. Chain elongation during protein synthesis requires the presence of a peptidyl tRNA or, in the first elongation cycle (as shown here), an fMet-tRNA^fMet at the peptidyl (P) site. ① Elongation begins with the binding of the second aminoacyl tRNA at the ribosomal aminoacyl (A) site. The tRNA is escorted to the A site by the elongation factor EF-Tu, which also carries two bound GTPs. As the tRNA binds, the GTPs are hydrolyzed, and EF-Tu is released. EF-Ts helps recycle the EF-Tu. ② A peptide bond is formed between the carboxyl group of the fMet (or, in a later cycle, of the terminal amino acid) at the P site and the amino group of the newly arrived amino acid at the A site. This reaction is catalyzed by the peptidyl transferase activity of the 23S rRNA molecule in the large ribosomal subunit. ③ After EF-G–GTP binds to the ribosome and GTP is hydrolyzed, the tRNA carrying the elongated polypeptide translocates from the A site to the P site. The discharged tRNA moves from the P site to the E (exit) site and leaves the ribosome. As the peptidyl tRNA translocates, it takes the mRNA along with it. Consequently, the next mRNA codon is moved into the A site, which is open for the next aminoacyl tRNA. This cycle of events is repeated for each amino acid that is added.

The function of EF-Tu, with two bound molecules of GTP, is to convey the aminoacyl tRNA to the A site of the ribosome. The EF-Tu–2GTP complex promotes the binding of all aminoacyl tRNAs to the ribosome *except the initiator tRNA;* hence it ensures that AUG codons located downstream from the start codon do not mistakenly recruit an initiator tRNA to the ribosome. As the aminoacyl tRNA is transferred to the ribosome, the GTPs are hydrolyzed and the EF-Tu–2GDP complex is released. The role of EF-Ts is to regenerate EF-Tu–2GTP from EF-Tu–2GDP for the next round of the elongation cycle.

Elongation factors do not recognize individual anticodons, which means that aminoacyl tRNAs of all types (other than initiator tRNAs) are indiscriminately brought to the A site of the ribosome. Some mechanism must therefore ensure that only the correct aminoacyl tRNA is retained by the ribosome for subsequent use during peptide bond formation. If the anticodon of an incoming aminoacyl tRNA is not complementary to the mRNA codon exposed at the A site, the aminoacyl tRNA does not bind to the ribosome long enough for GTP hydrolysis to take place. When the match is close but not exact, transient binding may occur and GTP is hydrolyzed. However, the mismatch between the anticodon of the aminoacyl tRNA and the codon of the mRNA creates an abnormal structure at the A site that is usually detected by the ribosome, leading to rejection of the bound aminoacyl tRNA. As a result, the final error rate in translation is usually no more than 1 incorrect amino acid per 10,000 incorporated.

Peptide Bond Formation.   After the appropriate aminoacyl tRNA has become bound to the A site of the ribosome, the next step is the formation of a peptide bond between the amino group of the amino acid bound at the A site and the carboxyl group by which the initiating amino acid (or growing polypeptide chain) is attached to the tRNA at the P site. The formation of this peptide bond causes the growing polypeptide chain to be transferred from the tRNA located at the P site to the tRNA located at the A site (see Figure 20-10, ②). Peptide bond formation is the only step in protein synthesis that requires neither nonribosomal protein factors nor an outside source of energy such as GTP or ATP. The necessary energy is provided by cleavage of the high-energy bond that joins the amino acid or peptide chain to the tRNA located at the P site.

For many years, peptide bond formation was thought to be catalyzed by a hypothetical ribosomal protein that was given the name **peptidyl transferase.** However, in 1992 Harry Noller and his colleagues showed that the large subunit of bacterial ribosomes retains peptidyl transferase activity after all ribosomal proteins have been removed. In contrast, peptidyl transferase activity is quickly destroyed when rRNA is degraded by exposing ribosomes to ribonuclease. Such observations suggest that rRNA rather than a ribosomal protein is responsible for catalyzing peptide bond formation. In bacterial ribosomes, peptidyl transferase activity has been localized to the 23S rRNA of the large ribosomal subunit, and high-resolution X-ray data

have pinpointed the catalytic site to a nitrogen atom located in a single adenine nucleotide of the RNA chain. Hence 23S rRNA is an example of a *ribozyme,* an enzyme made entirely of RNA (see Chapter 6, p. 151).

Translocation.   After a peptide bond has formed, the P site contains an empty tRNA and the A site contains a peptidyl tRNA (the tRNA to which the growing polypeptide chain is attached). The mRNA now advances a distance of three nucleotides relative to the small subunit, bringing the next codon into proper position for translation. During this process of **translocation,** the peptidyl tRNA moves from the A site to the P site and the empty tRNA moves from the P site to the E (exit) site, where it is released from the ribosome (see Figure 20-10, ③). The translocation process requires that an elongation factor called *EF-G,* together with a bound molecule of GTP, become transiently associated with the ribosome. The energy that drives translocation comes from the hydrolysis of this GTP.

During translocation, the peptidyl tRNA remains hydrogen-bonded to the mRNA as the mRNA advances by three nucleotides. The central role played by the peptidyl tRNA in the translocation process has been demonstrated using mutant tRNA molecules with *four*-nucleotide anticodons. These tRNAs hydrogen-bond to four nucleotides in the mRNA, and when translocation occurs, the mRNA advances by four nucleotides rather than the usual three. Although this observation indicates that the size of the anticodon loop of the peptidyl tRNA bound to the A site determines how far the mRNA advances during translocation, the physical basis for the mechanism that actually translocates the mRNA over the surface of the ribosome is not well understood.

The net effect of translocation is to bring the next mRNA codon into the A site, so the ribosome is now set to receive the next aminoacyl tRNA and repeat the elongation cycle. The only difference between succeeding elongation cycles and the first cycle is that an initiator tRNA occupies the P site at the beginning of the first elongation cycle, and peptidyl tRNA occupies the P site at the beginning of all subsequent cycles. As each successive amino acid is added, the mRNA is progressively read in the $5' \longrightarrow 3'$ direction. The amino-terminal end of the growing polypeptide is thought to pass out of the ribosome through a tunnel in the 50S subunit, and the chain usually starts to fold while it is being elongated. Polypeptide synthesis is very rapid; in a growing *E. coli* cell, a polypeptide of 400 amino acids can be made in 10 seconds!

Termination of Polypeptide Synthesis Is Triggered by Release Factors That Recognize Stop Codons

The elongation process depicted in Figure 20-10 continues in cyclic fashion, reading one codon after another and adding successive amino acids to the polypeptide chain, until one of the three possible stop codons (UAG, UAA, or UGA) in the mRNA arrives at the ribosome's A site (Figure 20-11). Unlike the situation with other codons,

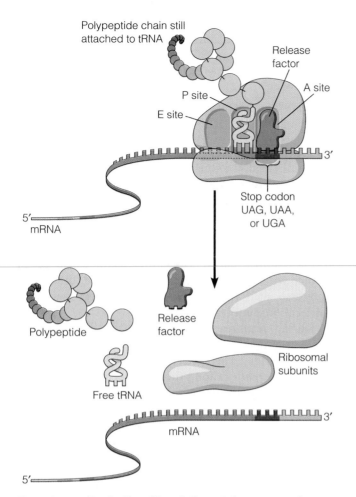

Figure 20-11 **Termination of Translation.** When a stop codon—UAG, UAA, or UGA—arrives at the A site, it is recognized and bound by a protein release factor. This protein causes the polypeptide to be transferred to a molecule of water (not shown), triggering its release from the tRNA and the dissociation of the other components of the elongation complex.

there are no tRNA molecules that recognize stop codons. Instead, the stop codons are recognized by proteins called **release factors,** which mimic the appearance of tRNA molecules and, after binding to the ribosomal A site, terminate translation by triggering the release of the completed polypeptide from the peptidyl tRNA. The cleavage is hydrolytic; in essence, the polypeptide transfers to a water molecule instead of to an activated amino acid, producing a free carboxyl group at the end of the polypeptide, its C-terminus. As the polypeptide is released, the other components of the ribosomal complex come apart. They are now available for reuse in a new initiation complex.

Polypeptide Folding Is Facilitated by Molecular Chaperones

Before newly synthesized polypeptides can perform their normal biological functions, they must be folded into the appropriate three-dimensional conformation. Polypeptides usually begin to fold into their secondary and tertiary structures while they are still being synthesized. As discussed in Chapters 2 and 3, the primary sequence of a protein is sufficient to specify its three-dimensional structure, and indeed many polypeptides can spontaneously fold into their proper conformations in a test tube. However, protein folding inside cells is normally facilitated by proteins called **molecular chaperones** (see p. 33). In fact, the proper folding of some proteins requires the action of several different chaperones acting at different stages of the folding process, beginning when the growing polypeptide chain first begins to emerge from the ribosome. A primary function of chaperones is to bind to polypeptide chains during the early stages of folding, thereby preventing them from interacting with other polypeptides before they have acquired their proper conformation. If the folding process goes awry, chaperones can sometimes rescue improperly folded proteins and help them to fold properly, or the improperly folded proteins may simply be destroyed. However, some kinds of incompletely or incorrectly folded polypeptide chains tend to bind to each other and form insoluble aggregates that become deposited both within and between cells. Such protein deposits trigger disruptions in cell function and may even lead to tissue degeneration and cell death. Box 20A discusses how such events contribute to the development of ailments such as Alzheimer's disease and "mad cow" disease.

Chaperones are found throughout the living world, from bacteria to the various compartments of eukaryotic cells. Two of the most widely occurring chaperone families are called *Hsp70* and *Hsp60*. (The "Hsp" comes from the original designation of these proteins as "heat-shock proteins" because they are produced in bacteria subjected to stressful conditions such as exposure to elevated temperature; under such conditions, chaperones facilitate the refolding of heat-damaged proteins.) The two families of chaperone proteins operate by somewhat different mechanisms, but both have ATPase activity and function by ATP-dependent cycles of binding and releasing their protein substrates. It is the release of the protein that requires ATP hydrolysis. In recent years, it has become clear that chaperone proteins are involved in other activities besides protein folding; we will encounter an example involving protein transport later in this chapter.

Protein Synthesis Typically Utilizes a Substantial Fraction of a Cell's Energy Budget

Polypeptide elongation involves the hydrolysis of at least five "high-energy" phosphoanhydride bonds per amino acid added. Two of these bonds are provided by the ATP that is hydrolyzed to AMP during the aminoacyl-tRNA synthetase reaction. The remainder are supplied by three molecules of GTP: two used in binding the incoming aminoacyl tRNA at the A site, and the other in the

Clinical Applications   PROTEIN-FOLDING DISEASES

Before they can perform their normal functions, polypeptide chains must be properly folded. More than a dozen human diseases have been traced to defects in this folding process. Among the best known is Alzheimer's disease, the memory disorder that affects one in ten Americans over 65 years old. The symptoms of Alzheimer's arise from the degeneration of brain cells associated with at least two abnormalities—intracellular *tangles* of a polymerized form of a microtubule accessory protein called tau, and extracellular *amyloid plaques* containing fibrils built from a small protein called *amyloid-β* (Aβ). Evidence that Aβ might be the primary cause of Alzheimer's emerged in the early 1990s, when it was discovered that some hereditary forms of Alzheimer's are triggered by mutations in a plasma membrane precursor protein called APP, whose cleavage gives rise to Aβ. The mutant APP gives rise to a misfolded form of Aβ that aggregates into long fibrils, creating the amyloid plaques that accumulate in the brain and cause loss of mental function by killing surrounding neurons.

Although this genetic evidence points to the importance of amyloid plaques, most individuals who develop Alzheimer's do not inherit mutations in the precursor APP, and as a result, they produce a normal version of Aβ. Normal Aβ molecules remain soluble and hence harmless in most individuals, and yet in some people, these same Aβ molecules aggregate into fibrils that accumulate and form amyloid plaques. Although the reason for such behavior is not clearly understood, a clue has emerged from the discovery that people inheriting different forms of a protein called *apolipoprotein E* (*apoE*) have differing risks of developing Alzheimer's disease. ApoE functions primarily in cholesterol transport, but tiny amounts are also associated with amyloid plaques in the spaces between nerve cells. The mechanism by which apoE exerts its effect on Alzheimer's disease is not clear, but two general possibilities have been proposed. One is that apoE directly enhances the buildup of amyloid plaques outside nerve cells by disrupting the normal folding of Aβ and/or interfering with its removal from the intercellular space. The other possibility is that apoE delivers cholesterol to nerve cell membranes, producing a cholesterol-rich environment that enhances the cleavage of APP into Aβ. Some support for this idea has come from the discovery that the enzymes involved in converting APP to Aβ reside in cholesterol-rich regions of the plasma membrane.

Our growing understanding of the relationship between amyloid plaques and Alzheimer's disease will hopefully hasten the development of effective treatments. It has already been shown that animals can be protected against amyloid buildup using experimental treatments such as: (a) enzyme inhibitors that block the cleavage of Aβ from its precursor APP, (b) antibiotics that dissolve amyloid plaques or prevent their formation, or (c) Aβ-containing vaccines that stimulate the immune system to clean up plaques and/or prevent Aβ deposits. Such vaccines can even protect mice with Alzheimer's symptoms from further memory loss, providing hope that this devastating illness will be conquered in the not-too-distant future.

Abnormalities in protein folding also lie at the heart of a group of brain-wasting disorders that include *scrapie* in sheep

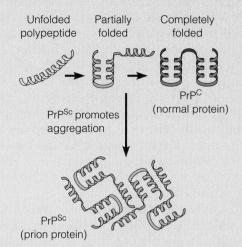

Figure 20A-1   Model Showing How Prions (PrP^Sc) Might Promote Their Own Formation.

and *"mad cow disease"* in cattle. Stanley Prusiner, who received a Nobel Prize in 1997 for his pioneering work in this field, has proposed that these diseases are transmitted by infectious, protein-containing particles called **prions** (briefly discussed in Chapter 4). Because prions do not appear to contain DNA or RNA, Prusiner has formulated a unique theory to explain how prions might transmit disease by causing the infectious spread of abnormal protein folding. According to this theory, a prion protein (designated PrP^Sc) is simply a misfolded version of a normal cellular protein (designated PrP^C). When the misfolded PrP^Sc encounters a normal PrP^C polypeptide chain in the process of folding, it causes the normal polypeptide to fold improperly (Figure 20A-1). The resulting, abnormally folded protein triggers extensive nerve cell damage in the brain, leading to uncontrolled muscle movements and eventual death. Because the presence of even a tiny bit of prion protein can progressively trigger the folding of a cell's normal PrP^C polypeptide chains into more and more improperly folded PrP^Sc prion protein, this scenario allows prion proteins to reproduce themselves without the need for nucleic acid.

Even more surprising has been the discovery of different "strains" of prions that cause slightly different forms of disease. When scientists mix tiny quantities of different PrP^Sc strains in separate test tubes with large amounts of the same, normal PrP^C polypeptide, each tube produces more of the specific PrP^Sc strain than was initially added to that tube. This ability to identify different strains of prions has allowed investigators to show that more than 100 people in Great Britain have become infected with mad cow prions by eating meat derived from diseased cattle, resulting in a fatal, human form of mad cow disease known as *variant Creutzfeldt-Jakob disease*, or *vCJD*. Almost 200,000 infected cattle have been destroyed in the United Kingdom to try to halt the spread of this disease, but thousands more people may eventually die from vCJD as a result of having ingested tainted beef over the past two decades.

translocation step. Assuming each phosphoanhydride bond has a $\Delta G^{o\prime}$ (standard free energy) of 7.3 kcal/mol, the five bonds represent a standard free energy input of 36.5 kcal/mol of amino acid inserted. In addition, one or two additional GTPs are needed per polypeptide chain synthesized, one during formation of the initiation complex and possibly one for termination. Thus, for the synthesis of a polypeptide 100 amino acids long, the $\Delta G^{o\prime}$ value may be as high as 3665 kcal/mol. Clearly, protein synthesis is an expensive process energetically; in fact, it accounts for a substantial fraction of the total energy budget of most cells. When we also consider the energy required to synthesize messenger RNA and the components of the translational apparatus, and the use of ATP by chaperone proteins as well, the cost of protein synthesis becomes even greater.

It is important to note that during translation, GTP does not function as a typical ATP-like energy donor: Its hydrolysis is not directly linked to the formation of a covalent bond. Instead, GTP appears to induce conformational changes in initiation and elongation factors by binding to them and releasing from them. These shape changes, in turn, allow the factors to bind (noncovalently) to, and be released from, the ribosome. In addition, hydrolysis of the GTP attached to EF-Tu apparently contributes to the accuracy of translation by playing a role in the proofreading mechanism that ejects incorrect aminoacyl tRNAs that enter the A site.

A Summary of Translation

We have now seen that translation serves as the mechanism that converts information stored in strings of mRNA codons into a chain of amino acids linked by peptide bonds. For a visual summary of the process, you should refer back to Figure 20-7. As the ribosome reads the mRNA codon by codon in the 5′ ⟶ 3′ direction, successive amino acids are brought into place by complementary base pairing between the codons in the mRNA and the anticodons of aminoacyl-tRNA molecules. When a stop codon is encountered, the completed polypeptide is released, and the mRNA and ribosomal subunits become available for further use.

Most messages are read by many ribosomes simultaneously, each ribosome following closely behind the next on the same mRNA molecule. A cluster of such ribosomes attached to a single mRNA molecule is called a **polyribosome** (see Figure 19-18). By allowing many polypeptides to be synthesized at the same time from a single mRNA molecule, polyribosomes maximize the efficiency of mRNA utilization.

RNA molecules play especially important roles in translation. The mRNA plays a central role, of course, as the carrier of the genetic message. The tRNA molecules serve as the adaptors that bring the amino acids to the appropriate codons. Last but not least, the rRNA molecules have multiple functions. Not only do they serve as structural components of the ribosomes, but one (the 16S rRNA of the small subunit) provides the binding site for incoming mRNA, and another (the 23S rRNA of the large subunit) catalyzes the formation of the peptide bond. The fundamental roles played by RNA may be a vestige of the way in which living organisms first evolved on Earth. As we discussed in Chapter 6, the discovery of RNA catalysts (ribozymes) has fostered the idea that the first catalysts on Earth may have been self-replicating RNA molecules rather than proteins. Hence the present-day ribosome may have evolved from a primitive translational apparatus that was based entirely on RNA molecules.

Nonsense Mutations and Suppressor tRNA

Having described the *normal* process of translation, let us now briefly discuss what happens when mRNAs containing *mutant* codons are translated. Most codon mutations simply alter a single amino acid and hence do not disrupt translation. Moreover, mutations in the third base of a codon frequently do not change the amino acid at all. However, some mutations have a more harmful effect on translation. For example, mutations sometimes convert a codon that codes for an amino acid into a stop codon. Figure 20-12 illustrates such a case, in which mutation of a single base pair in DNA results in the conversion of a lysine codon in the mRNA to a stop signal, UAG. Such mutations usually lead to the production of incomplete, nonfunctional polypeptides that have been prematurely terminated at the mutant stop codon, as Figure 20-12b shows. Mutations that convert amino acid–coding codons into stop codons are referred to as **nonsense mutations.**

Although nonsense mutations in essential genes are often lethal, phages with such mutations can nonetheless grow in certain strains of bacteria. These special bacteria "suppress" the normal chain-terminating effect of nonsense mutations because they have a mutant tRNA that recognizes what would otherwise be a stop codon and inserts an amino acid at that point. In the example shown in Figure 20-12c, a mutant tRNA has an altered anticodon that allows it to read the stop codon UAG as a codon for tyrosine. The inserted amino acid is almost always different from the amino acid that would be present at that position in the wild-type protein, but the crucial feature of suppression is that chain termination is averted and a full-length polypeptide can be made.

A tRNA molecule that somehow negates the effect of a mutation is called a **suppressor tRNA.** As you might imagine, suppressor tRNAs exist that negate the effects of various types of mutations in addition to nonsense mutations (see Problem 20-7 at the end of the chapter). For the cell to survive, suppressor tRNAs must be relatively inefficient; otherwise, the protein-synthesizing

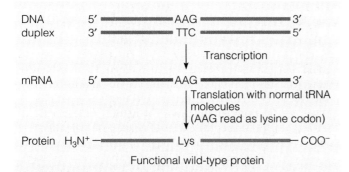

(a) Normal gene, normal tRNA molecules

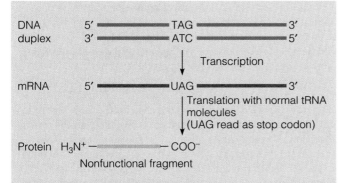

(b) Mutant gene, normal tRNA molecules

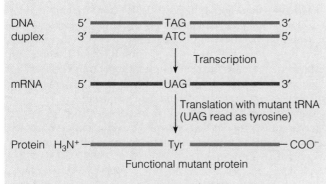

(c) Mutant gene, mutant (suppressor) tRNA molecule

Figure 20-12 Nonsense Mutations and Suppressor tRNAs. **(a)** A wild-type (normal) gene is transcribed into an mRNA molecule that contains the codon AAG at one point. Upon translation, this codon specifies the amino acid lysine (Lys) at one point in the functional, wild-type protein. **(b)** If a mutation occurs in the DNA that changes the AAG codon in the mRNA to UAG, the UAG codon will be read as a stop signal, and the translation product will be a short, nonfunctional polypeptide. Mutations of this sort are called nonsense mutations. **(c)** In the presence of a mutant tRNA molecule that reads UAG as an amino acid codon instead of a stop signal, an amino acid will be inserted, and polypeptide synthesis will continue. In the example shown, UAG is read as a codon for tyrosine because the mutant tyrosine tRNA has as its anticodon 3'-AUC-5' instead of the usual 3'-AUG-5' (which recognizes the tyrosine codon 5'-UAC-3'). The resulting protein will be mutant because a lysine has been replaced by a tyrosine at one point along the chain. The protein may still be functional, however, if its biological activity is not adversely affected by the amino acid substitution.

apparatus would produce too many abnormal proteins. An overly efficient nonsense suppressor, for example, would cause normal stop codons to be read as if they coded for an amino acid, thereby preventing normal termination. In fact, the synthesis of most polypeptides is terminated properly in cells containing nonsense suppressor tRNAs, indicating that a stop codon located in its proper place at the end of an mRNA coding sequence still triggers termination, whereas the same codon in an internal location does not. This suggests that besides requiring the presence of a stop codon, normal termination involves the recognition of a special sequence or three-dimensional configuration located near the end of the mRNA coding sequence.

In addition to helping us understand the behavior of nonsense mutations, knowledge of the genetic code and the mechanism of translation also sheds light on many other kinds of mutations. Box 20B summarizes the major types of mutations and the terminology used to describe them.

Posttranslational Processing

After polypeptide chains have been synthesized, often they must be chemically modified before they can perform their normal functions. In prokaryotes, for example, the N-formyl group located at the N-terminus of polypeptide chains is always removed. Moreover, the methionine to which it was attached is often removed also, as is the methionine that starts eukaryotic polypeptides. As a result, relatively few mature polypeptides have methionine at their N-terminus, even though they all started out that way. Sometimes, whole blocks of amino acids are removed from the polypeptide. Certain enzymes, for example, are synthesized as inactive precursors that must be activated by the removal of a specific sequence at one end or the other. The transport of proteins across membranes also may involve the removal of a terminal *signal sequence,* as we will see shortly, and some polypeptides have internal stretches of amino acids that must be removed to produce an active protein. For instance, insulin is synthesized as a single polypeptide and then processed to remove an internal segment; the two end segments remain linked by disulfide bonds between cysteine residues in the active hormone (Figure 20-13a).

Other common processing events include chemical modifications of individual amino acid groups—by methylation, phosphorylation, or acetylation reactions, for example. In addition, a polypeptide may undergo glycosylation (the addition of carbohydrate side chains; see Chapter 12) or binding to prosthetic groups. Finally, in the case of proteins composed of multiple subunits, individual polypeptide chains must bind to one another to form the appropriate multisubunit proteins or supramolecular complexes.

In its broadest sense, the term *mutation* refers to any change in the nucleotide sequence of a genome. Now that we have examined the processes of transcription and translation, we can understand the effects of a number of different kinds of mutations. Limiting our discussion to protein-coding genes, let's consider some of the main types of mutations and their impact on the polypeptide encoded by the mutant gene.

In this and the previous chapter, we have encountered several types of mutations in which the DNA change involves only one or a few base pairs (Figure 20B-1a). At the beginning of Chapter 19, for instance, we mentioned the genetic allele that, when homozygous, causes sickle-cell anemia. This allele originated from a type of mutation called a *base-pair substitution.* In this case, an AT base pair was substituted for a TA base pair in DNA. As a result, a GUA codon replaces a GAA in the mRNA transcribed from the mutant allele, and in the polypeptide (*β*-globin) a valine replaces a glutamic acid. This single amino acid change, caused by a single base-pair change, is enough to change the conformation of *β*-globin and, in turn, the hemoglobin tetramer, altering the way hemoglobin molecules pack into red cells and producing abnormally shaped cells that become trapped and damaged when they pass through small blood vessels (see Figure 19-3). Such a base-pair substitution is called a *missense mutation,* because the mutated codon continues to code for an amino acid—but the "wrong" one.

Alternatively, a base-pair substitution can create a *nonsense mutation;* that is, it can change an amino acid codon to one of the stop codons. As a result, the translation machinery will terminate the polypeptide prematurely. Unless the nonsense mutation is close to the end of the message or a suppressor tRNA is present, the polypeptide is not likely to be functional. Both nonsense and missense codons can also arise from the *base-pair insertions* and *deletions* that cause *frameshift mutations.*

A single amino acid change (or even a change in several amino acids) does not always affect a protein's function in a major way. As long as the protein's three-dimensional conformation remains relatively unchanged, biological activity may be unaffected. Substitution of one amino acid for another of the same type—for example, valine for isoleucine—is especially unlikely to affect protein function. The nature of the genetic code actually minimizes the effects of single base-pair alterations, because many turn out to be *silent mutations* that change the nucleotide sequence without changing the genetic message. For example, changing the third base of a codon often produces a new codon that still codes for the same amino acid. Here the "mutant" polypeptide is exactly the same as the wild type.

In addition to mutations affecting one or a few base pairs, some alterations involve longer stretches of DNA (Figure 20B-1b). A few affect genome segments so large that the DNA changes can be detected by light microscopic examination of chromosomes. Some of these large-scale mutations are created by *insertions* or *deletions* of long DNA segments, but several other mechanisms also exist. In a *duplication,* a section of DNA is tandemly repeated. In an *inversion,* a chromosome segment is cut out and reinserted in its original position but in the reverse direction. A *translocation* involves the movement of a DNA segment from its normal location in the genome to another place, in the same chromosome or a different one. Because these large-scale mutations may or may not affect the expression of many genes, they have a wide range of phenotypic effects, from no effect at all to lethality.

When we think about the potential effects of mutations, it is useful to remember that genes have important noncoding components and that these, too, can be mutated in ways that seriously affect gene products. A mutation in a promoter, for example, can result in more or less frequent transcription of the gene. Even a mutation in an intron can affect the gene product in a major way if it touches a critical part of a splice-site sequence.

Finally, mutations in genes that encode regulatory proteins—that is, proteins that control the expression of other genes—can have far-reaching effects on many other proteins. We will discuss this topic in Chapter 21.

In addition to the preceding posttranslational events, some proteins undergo a relatively unusual type of processing called **protein splicing,** which is analogous to the phenomenon of *RNA splicing* discussed in Chapter 19. As we saw, intron sequences are removed from RNA molecules during RNA splicing, and the remaining exon sequences are simultaneously spliced together. Likewise, during protein splicing, specific amino acid sequences called *inteins* are removed from a polypeptide chain and the remaining segments, called *exteins,* are spliced together to form the mature protein. Protein splicing is usually intramolecular, involving the excision of an intein from a single polypeptide chain by a self-catalytic mechanism (Figure 20-13b). However, splicing can also take place between two polypeptide chains arising from two different mRNAs. In such cases, the two polypeptides are brought together by binding between their intein segments. An intermolecular splicing reaction then joins the two polypeptides together, accompanied by intein removal (Figure 20-13c). In some cases, the inteins removed during protein splicing turn out to be stable proteins exhibiting their own biological functions. Once considered to be an oddity of nature, protein splicing has now been detected in dozens of different organisms, prokaryotes as well as eukaryotes.

Protein Targeting and Sorting

Now that we have seen how proteins are synthesized, we are ready to explore the mechanisms that route each newly made protein to its correct destination. Think for a moment about a typical eukaryotic cell with its diversity

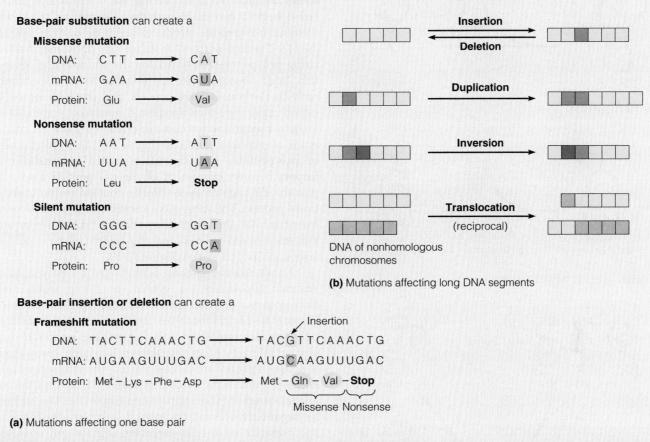

Base-pair substitution can create a

Missense mutation

DNA:	C T T	→	C A T
mRNA:	G A A	→	G U A
Protein:	Glu	→	Val

Nonsense mutation

DNA:	A A T	→	A T T
mRNA:	U U A	→	U A A
Protein:	Leu	→	**Stop**

Silent mutation

DNA:	G G G	→	G G T
mRNA:	C C C	→	C C A
Protein:	Pro	→	Pro

Base-pair insertion or deletion can create a

Frameshift mutation

Insertion

DNA:	T A C T T C A A A C T G	→	T A C G T T C A A A C T G
mRNA:	A U G A A G U U U G A C	→	A U G C A A G U U U G A C
Protein:	Met – Lys – Phe – Asp	→	Met – Gln – Val – **Stop**

Missense   Nonsense

(a) Mutations affecting one base pair

Insertion

Deletion

Duplication

Inversion

Translocation
(reciprocal)

DNA of nonhomologous
chromosomes

(b) Mutations affecting long DNA segments

Figure 20B-1   Types of Mutations. **(a)** Mutations affecting one base pair. **(b)** Mutations affecting long DNA segments.

of organelles, each containing its own unique set of proteins. Such a cell is likely to have billions of protein molecules, representing at least 10,000 different kinds of polypeptides, each of which must find its way either to the appropriate location within the cell or even out of the cell altogether. A limited number of these polypeptides are encoded by the genome of the mitochondrion (and, for plant cells, by the chloroplast genome as well), but most are encoded by nuclear genes and are synthesized by a process that begins in the cytosol. Each of these polypeptides must then be directed to its proper destination and must therefore have some sort of molecular "zip code" that ensures its delivery to the correct place. As our final topic for this chapter, we will consider this process of protein targeting and sorting.

We can begin by grouping the various compartments of eukaryotic cells into three categories: (1) the endomembrane system, an interrelated system of membrane compartments that includes the ER, the Golgi complex, lysosomes, secretory vesicles, the nuclear envelope, and the plasma membrane; (2) the cytosol; and (3) mitochondria, chloroplasts, peroxisomes (and related organelles), and the interior of the nucleus.

Polypeptides encoded by nuclear genes are routed to these compartments using several different mechanisms. The process begins with transcription of the nuclear genes into RNAs that are processed in the nucleus and then transported through nuclear pores for translation in the cytoplasm, where most ribosomes occur. While mRNA translation is largely a cytoplasmic process, recent evidence suggests that as many as 10% of a cell's ribosomes may actually reside in the nucleus, where they can translate newly synthesized RNAs. Nuclear translation appears to function mainly as a quality-control mechanism that

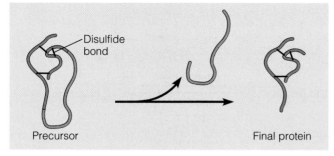

(a) Protein cleavage

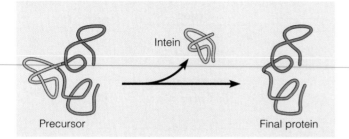

(b) Protein splicing (intramolecular)

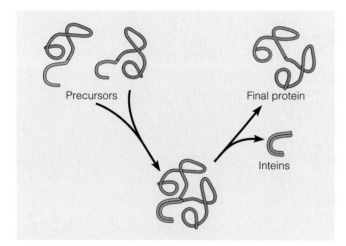

(c) Protein splicing (intermolecular)

Figure 20-13 Processing of Protein Precursors. Some polypeptides are synthesized as inactive precursors that must have amino acids removed to produce an active protein. **(a)** Protein cleavage is one mechanism that can be used to remove a stretch of amino acids from a polypeptide chain. For example, the protein hormone insulin is synthesized in an inactive precursor form (proinsulin) that is converted to the active hormone by enzymatically removing a long internal section of polypeptide. The two chains generated by removing this internal section remain connected by disulfide bonds between cysteine residues. Parts **(b)** and **(c)** illustrate the alternative mechanism of protein splicing, in which amino acid sequences called *inteins* are removed from one or more precursor polypeptides and the resulting polypeptide segments are spliced together to create a continuous polypeptide chain.

checks new mRNAs for the presence of errors (see the discussion of RNA surveillance on p. 654).

In spite of the existence of these functioning nuclear ribosomes, it is clear that most polypeptide synthesis occurs on cytoplasmic ribosomes after mRNAs have been exported through the nuclear pores. Upon arrival in the cytoplasm, these mRNAs become associated with *free ribosomes* (ribosomes not attached to any membrane). Shortly after translation begins, two main pathways for routing the newly forming polypeptide products begin to diverge (Figure 20-14). The first pathway is utilized by ribosomes synthesizing polypeptides destined for the endomembrane system. Such ribosomes become attached to ER membranes early in the translational process, and the growing polypeptide chains are then transferred across (or, in the case of integral membrane proteins, inserted into) the ER membrane as synthesis proceeds (Figure 20-14b). This transfer of polypeptides into the ER is called **cotranslational import** because movement of the polypeptide across or into the ER membrane is directly coupled to the translational process. The subsequent conveyance of such proteins from the ER to their final destinations is carried out by various membrane vesicles and the Golgi complex, as discussed in Chapter 12 (see Figure 12-8).

An alternative pathway is employed for polypeptides destined for either the cytosol or for mitochondria, chloroplasts, peroxisomes, and the nuclear interior (Figure 20-14c). Ribosomes synthesizing these types of polypeptides remain free in the cytosol, unattached to any membrane. After translation has been completed, the polypeptides are released from the ribosomes and either remain in the cytosol as their final destination, or are taken up by the appropriate organelle. The uptake by organelles of such completed polypeptides requires the presence of special targeting signals and is called **posttranslational import.** In the case of the nucleus, polypeptides enter through the nuclear pores, as discussed in Chapter 16 (see Figure 16-30). Polypeptide entrance into mitochondria, chloroplasts, and peroxisomes involves a different kind of mechanism, as we will see shortly.

With this general overview in mind, we are now ready to examine the mechanisms of cotranslational import and posttranslational import in detail.

Cotranslational Import Allows Some Polypeptides to Enter the ER as They Are Being Synthesized

Cotranslational import into the ER is the first step in the pathway by which cells distribute newly synthesized proteins to various locations within the endomembrane system. Proteins destined for distribution via this mechanism are synthesized on ribosomes that become attached to membranes of the rough ER shortly after translation starts. We focus first on the behavior of soluble proteins—a group that includes proteins destined for secretion from the cell as well as the soluble proteins that end up inside the cisternae of endomembrane components such as the Golgi complex, lysosomes, and the ER itself.

Evidence for ER Signal Sequences. The idea that the ER is involved in transporting newly synthesized proteins to

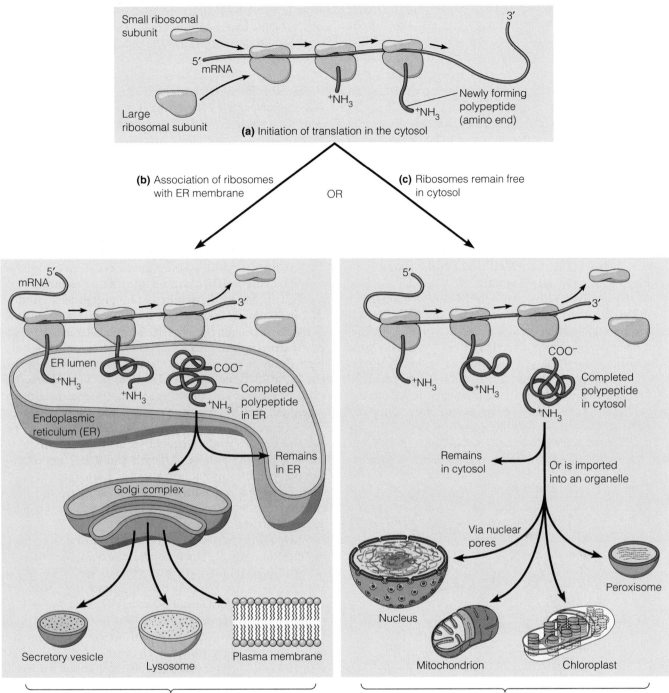

(a) Initiation of translation in the cytosol

(b) Association of ribosomes with ER membrane **OR** **(c)** Ribosomes remain free in cytosol

COTRANSLATIONAL IMPORT
into ER lumen, followed by transport
to final destination

POSTTRANSLATIONAL IMPORT
into various organelles

Figure 20-14 Intracellular Sorting of Proteins. (a) Synthesis of polypeptides encoded by nuclear genes begins in the cytosol. The large and small ribosomal subunits associate with each other and with the 5′ end of an mRNA molecule, forming a functional ribosome that starts making the polypeptide. When the polypeptide is about 30 amino acids long, it enters one of two alternative pathways. **(b)** If destined for any of the compartments of the endomembrane system, the newly forming polypeptide becomes associated with the ER membrane and is transferred across the membrane into the lumen (cisternal space) of the ER as synthesis continues. This process is cotranslational import. The completed polypeptide then either remains in the ER or is transported via various vesicles and the Golgi complex to another final destination. (Integral membrane proteins are inserted into the ER membrane as they are made, rather than into the lumen.) **(c)** If the polypeptide is destined for the cytosol or for import into the nucleus, mitochondria, chloroplasts, or peroxisomes, its synthesis continues in the cytosol. When the polypeptide is complete, it is released from the ribosome and either remains in the cytosol or is transported into the appropriate organelle by posttranslational import. Polypeptide uptake by the nucleus occurs via the nuclear pores, using a mechanism different from that involved in post-translational uptake by other organelles.

various locations within the cell first emerged from studies by Colvin Redman and David Sabatini of protein synthesis in isolated vesicles of rough ER (ER vesicles with attached ribosomes). After briefly incubating rough ER vesicles in the presence of radioactive amino acids and the other components required for protein synthesis, they added the antibiotic *puromycin* to halt the process. In addition to blocking further protein synthesis, puromycin causes the partially completed polypeptide chains to be released from the ribosomes. When the ribosomes and membrane vesicles were then separated and analyzed to see where the newly made, radioactive polypeptide chains were located, a substantial fraction of the radioactivity was found inside the ER lumen (Figure 20-15). Such results suggested that some newly forming polypeptides begin to pass into the lumen of the ER as they are being synthesized, allowing them to be subsequently routed through the channels of the ER to their correct destinations.

If some polypeptides move directly into the lumen of the ER as they are being synthesized, how does the cell determine which polypeptides are to be handled in this way? An answer was first suggested in 1971 by Günter Blobel and David Sabatini, whose model was called the *signal hypothesis* because it proposed that some sort of intrinsic molecular signal distinguishes such polypeptides from the many polypeptides destined to be released into the cytosol. This hypothesis has had such a profound impact on the field of cell biology that Blobel was awarded the Nobel Prize in 1999 for his many years of work in demon-

strating the validity and widespread relevance of the idea that proteins have intrinsic signals that govern their transport and localization within the cell. The signal hypothesis stated that for polypeptides destined for the ER, the first segment of the polypeptide to be synthesized, the N-terminus, contains an **ER signal sequence** that directs the ribosome-mRNA-polypeptide complex to the surface of the rough ER, where the complex anchors at a protein "dock" on the ER surface. Then, as the polypeptide chain elongates during mRNA translation, it progressively crosses the ER membrane and enters the ER lumen.

Evidence for the actual existence of ER signal sequences was obtained shortly thereafter by César Milstein and his associates, who were studying the synthesis of the small subunit, or *light chain*, of the protein *immunoglobulin G*. In cell-free systems containing purified ribosomes and the components required for protein synthesis, the mRNA coding for the immunoglobulin light chain directs the synthesis of a polypeptide product that is 20 amino acids longer at its N-terminal end than the authentic light chain itself. Adding ER membranes to this system leads to the production of an immunoglobulin light chain of the correct size. Such findings suggest that the extra 20–amino acid segment is functioning as an ER signal sequence, and that this signal sequence is removed when the polypeptide moves into the ER. Subsequent studies revealed that other polypeptides destined for the ER also possess an N-terminal sequence that is required for targeting the protein to the ER and that is removed as the polypeptide moves into the ER. Proteins containing such ER signal sequences at their N-terminus are often referred to as *preproteins* (e.g., prelysozyme, preproinsulin, pretrypsinogen, and so forth).

Sequencing studies have revealed that the amino acid compositions of ER signal sequences are surprisingly variable, but several unifying features have been noted. ER signal sequences are typically 15–30 amino acids long and consist of three domains: a positively charged N-terminal region, a central hydrophobic region, and a polar region adjoining the site where cleavage from the mature protein will take place. The positively charged end may promote interaction with the hydrophilic exterior of the ER membrane, and the hydrophobic region may facilitate interaction of the signal sequence with the membrane's lipid interior. In any case, it is now established that only polypeptides with ER signal sequences can be inserted into or across the ER membrane as their synthesis proceeds. In fact, when recombinant DNA methods are used to add ER signal sequences to polypeptides that do not usually have them, the recombinant polypeptides are directed to the ER.

The Roles of SRP and the Translocon. Once the existence of ER signal sequences was established, it quickly became clear that newly forming polypeptides must become attached to the ER membrane before very much of the polypeptide has emerged from the ribosome. If translation were to continue without attachment to the ER, the folding of the growing polypeptide chain might bury the

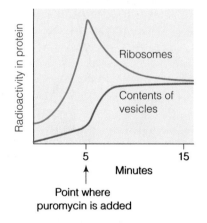

Figure 20-15 Evidence That Proteins Synthesized on Ribosomes Attached to ER Membranes Pass Directly into the ER Lumen. ER vesicles containing attached ribosomes were isolated and incubated with radioactive amino acids to label newly made polypeptide chains. Next, protein synthesis was halted by adding puromycin, which also causes the newly forming polypeptide chains to be released from the ribosomes. The ribosomes were then removed from the membrane vesicles and the amount of radioactive protein associated with the ribosomes and in the membrane vesicles was measured. The graph shows that after the addition of puromycin, radioactivity is lost from the ribosomes and appears inside the vesicles, suggesting that the newly forming polypeptide chains are inserted through the ER membrane as they are being synthesized, and puromycin causes the chains to be prematurely released into the vesicle lumen.

signal sequence. To understand what prevents this from happening, we need to look at the signal mechanism in further detail.

Contrary to the original signal hypothesis, the ER signal sequence does not itself initiate contact with the ER. Instead, the contact is mediated by a **signal-recognition particle (SRP),** which recognizes and binds to the ER signal sequence of the newly forming polypeptide and then binds to the ER membrane. At first, the SRP was thought to be purely protein (the P in its name originally stood for protein). Later, however, the SRP was shown to consist of six different polypeptides complexed with a 300-nucleotide (7S) molecule of RNA. The protein components have three main active sites: one that recognizes and binds to the ER signal sequence, one that interacts with the ribosome to block further translation, and one that binds to the ER membrane.

Figure 20-16 illustrates the role played by the SRP in cotranslational import. The process begins when an mRNA coding for a polypeptide destined for the ER starts to be translated on a free ribosome. Polypeptide synthesis proceeds until the ER signal sequence has been formed and emerges from the surface of the ribosome. At this stage, SRP (shown in orange) binds to the signal sequence and blocks further translation (step ①). The SRP then binds the ribosome to a special structure in the ER membrane called a **translocon** because it carries out the translocation of polypeptides across the ER membrane. (Note that the term *translocation,* which literally means "a change of location," is used to describe both the movement of proteins through membranes and, earlier in the chapter, the movement of mRNA across the ribosome.)

The translocon is a protein complex composed of several components involved in cotranslational import, including an *SRP receptor* to which the SRP binds, a *ribosome receptor* that holds the ribosome in place, a *pore protein* that forms a channel through which the growing polypeptide can enter the ER lumen, and *signal peptidase,* an enzyme that removes the ER signal sequence. As ② shows, SRP (bringing an attached ribosome) first binds to the SRP receptor, allowing the ribosome to become attached to the ribosome receptor. Next, GTP binds to the SRP–SRP-receptor complex, unblocking translation and causing transfer of the signal sequence into the channel of the pore protein (③). GTP is then hydrolyzed, accompanied by release of the SRP (④). With the signal sequence remaining bound at or near the pore protein, the polypeptide elongates into a loop and translocates into the ER lumen (⑤). No energy source beyond the GTP used in polypeptide elongation is needed to drive the translocation. Finally, when polypeptide synthesis is completed, the signal peptidase clips off the signal sequence, releasing the completed polypeptide into the ER lumen (⑥). At the same time, the ribosome detaches from the ER membrane and dissociates into its subunits, releasing the mRNA.

BiP and Protein Disulfide Isomerase. After polypeptides are released into the ER lumen, they fold into their final ter-

tiary form and, in some cases, associate with other polypeptides to form multisubunit proteins exhibiting quaternary structure. As we discussed earlier in the chapter, protein folding is facilitated by molecular chaperones. One of the most abundant chaperones in the ER lumen is a protein called **BiP** (a name derived from the term *binding protein*). BiP, which is a member of the Hsp70 family of chaperones, binds to *hydrophobic regions* of polypeptide chains, especially to regions enriched in the amino acids tryptophan, phenylalanine, and leucine.

In a mature, fully folded protein, such hydrophobic regions are buried in the interior of the protein molecule. However, in an unfolded polypeptide chain, these same hydrophobic regions are exposed to the surrounding aqueous environment, creating an unstable situation in which polypeptides tend to aggregate. By binding to the hydrophobic regions of unfolded polypeptides as they emerge into the ER lumen, BiP stabilizes them and prevents them from aggregating with other unfolded polypeptides. BiP then releases the polypeptide, accompanied by ATP hydrolysis, giving the polypeptide chain a brief opportunity to fold properly (perhaps aided by other chaperones). If the polypeptide folds properly, its hydrophobic regions become buried in the molecule's interior and will no longer bind to BiP. But if the hydrophobic segments fail to fold properly, BiP binds again to the polypeptide and the cycle is repeated. In this way, BiP uses energy released by ATP hydrolysis to promote proper protein folding. Proteins that repeatedly fail to fold properly are eventually shuttled back across the ER membrane to the cytosol, where they are degraded by the ubiquitin-dependent proteasome pathway described in Chapter 21.

Another event associated with proper protein folding in the ER lumen is the formation of disulfide bonds between cysteines located in different regions of a polypeptide chain. In eukaryotes, disulfide bonds are confined to proteins synthesized in the ER. The ER lumen contains the enzyme **protein disulfide isomerase,** which catalyzes the formation and breakage of disulfide bonds between cysteine residues. This process often begins before the synthesis of a newly forming polypeptide has been completed, allowing various disulfide bond combinations to be tested until the most stable arrangement is found.

Sorting of Soluble Proteins. Most of the proteins synthesized on ribosomes attached to the ER are *glycoproteins*—that is, proteins with covalently bound carbohydrate groups. As we learned in Chapter 12, the initial glycosylation reactions that add these carbohydrate side chains take place in the ER, often while the growing polypeptide chain is still in the process of being synthesized.

After polypeptides have been released into the ER lumen, glycosylated, and folded, they are transported by the various membrane channels and vesicles of the endomembrane system to their destinations within the cell. The first stop in this transport pathway is the Golgi complex, where further glycosylation and processing of carbohydrate side chains may occur. The Golgi complex

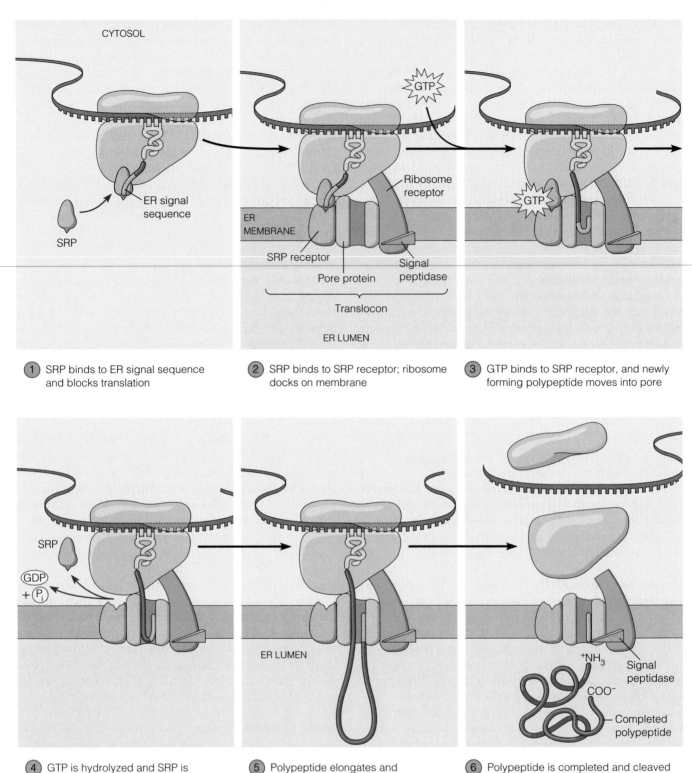

CYTOSOL

ER signal sequence

SRP

ER MEMBRANE

Ribosome receptor

SRP receptor

Pore protein

Signal peptidase

Translocon

ER LUMEN

GTP

① SRP binds to ER signal sequence and blocks translation

② SRP binds to SRP receptor; ribosome docks on membrane

③ GTP binds to SRP receptor, and newly forming polypeptide moves into pore

SRP

GDP + Pᵢ

ER LUMEN

$^+NH_3$

COO⁻

Signal peptidase

Completed polypeptide

④ GTP is hydrolyzed and SRP is released

⑤ Polypeptide elongates and translocates into the ER lumen

⑥ Polypeptide is completed and cleaved by signal peptidase, releasing it from the membrane

Figure 20-16   A Model for the Signal Mechanism of Cotranslational Import.   This figure shows a current model for the signal mechanism. It is now well established that the growing polypeptide translocates through a hydrophilic pore created by one or more membrane proteins. The complex of membrane proteins that carry out translocation is called the *translocon*. The most recent evidence suggests that the ribosome fits tightly across the cytoplasmic side of the pore and that the ER-lumen side is somehow closed off until the polypeptide is about 70 amino acids long (after step ④). In any case, after step ⑥ , the ribosome is released and the pore closes completely (not shown).

then serves as a site for sorting and distributing proteins to other locations.

For soluble proteins, the default pathway takes them from the Golgi complex to secretory vesicles that move to the cell surface and fuse with the plasma membrane, leading to secretion of such proteins from the cell (see Figure 12-8). Those soluble proteins not destined for secretion possess certain carbohydrate side chains and/or short amino acid signal sequences that target them to their appropriate locations within the endomembrane system. For example, we learned in Chapter 12 that many lysosomal enzymes have carbohydrate side chains containing the unusual sugar *mannose-6-phosphate*. This sugar serves as a recognition device that allows the Golgi complex to selectively package such proteins into newly forming lysosomes (see Figure 12-9). Another kind of signal is used for proteins whose final destination is the ER. The C-terminus of these proteins usually contains a *KDEL sequence*, which consists of the amino acids Lys-Asp-Glu-Leu or a closely related sequence. (The letters K, D, E, and L are the one-letter abbreviations for these four amino acids; see Table 3-2.) The Golgi complex employs a receptor protein that binds to the KDEL sequence and delivers the protein back to the ER.

Insertion of Integral Membrane Proteins Using Stop-Transfer Sequences.

So far, we have focused on the cotranslational import and sorting of *soluble proteins* that are destined either for secretion from the cell or for the cisternal spaces of endomembrane components, such as the ER, the Golgi complex, lysosomes, and related vesicles. The other major group of polypeptides synthesized on ER-attached ribosomes are molecules destined to become integral *membrane proteins*. Polypeptides of this type are synthesized by a mechanism similar to the one illustrated in Figure 20-16 for soluble proteins, except that the completed polypeptide chain remains anchored to the ER membrane rather than being released into the ER lumen.

Recall from Chapter 7 that integral membrane proteins are typically anchored to the lipid bilayer by one or more α-helical *transmembrane segments* consisting of 20–30 hydrophobic amino acids. In considering the mechanism that allows such proteins to be retained as part of the ER membrane after synthesis rather than being released into the ER lumen, we focus here on the simplest case: proteins with only a single such transmembrane segment. The principles involved, however, extend to proteins with more complicated configurations. Researchers postulate two main mechanisms by which hydrophobic transmembrane segments anchor newly forming polypeptide chains to the lipid bilayer of the ER membrane.

The first of these mechanisms involves polypeptides with a typical ER signal sequence at their N-terminus, which allows an SRP to bind the ribosome-mRNA complex to the ER membrane. Elongation of the polypeptide chain then continues until the hydrophobic transmembrane segment of the polypeptide is synthesized. As shown in Figure 20-17a, this stretch of amino acids functions as a **stop-transfer sequence** that halts translocation of the polypeptide chain through the ER membrane. Translation continues to completion, but the rest of the polypeptide remains on the cytosolic side of the ER membrane. When the ER signal sequence is subsequently removed by signal peptidase, the result is a transmembrane protein with its N-terminus in the ER lumen and its C-terminus in the cytosol. The hydrophobic stop-transfer signal remains anchored in the membrane, forming the permanent transmembrane segment that anchors the protein to the lipid bilayer.

The second mechanism involves membrane proteins that lack a typical signal sequence at their N-terminus, but instead possess an *internal* **start-transfer sequence** that performs two functions: It first acts as an ER signal sequence that allows an SRP to bind the ribosome-mRNA complex to the ER membrane, and then its hydrophobic region functions as a membrane anchor that permanently attaches the polypeptide to the lipid bilayer (Figure 20-17b). The orientation of the start-transfer sequence at the time of insertion determines which terminus of the polypeptide ends up in the ER lumen and which in the cytosol. Transmembrane proteins with multiple membrane-spanning regions are formed in a similar way, except that an alternating pattern of start-transfer and stop-transfer sequences creates a polypeptide containing multiple transmembrane segments that pass back and forth across the membrane.

Once a newly formed polypeptide has been incorporated into the ER membrane by one of the preceding mechanisms, it can either remain in place to function as an ER membrane protein or be transported to other components of the endomembrane system, such as the Golgi complex, lysosomes, nuclear envelope, or plasma membrane. Transport is carried out by a series of membrane budding and fusing events in which membrane vesicles pinch off from one compartment of the endomembrane system and fuse with another compartment, as we described in Figure 12-8.

Posttranslational Import Allows Some Polypeptides to Enter Organelles After They Have Been Synthesized

In contrast to the cotranslational import of proteins into the ER, proteins destined for the nuclear interior, mitochondrion, chloroplast, or peroxisome are imported into these organelles *after* translation has been completed. Because such proteins are synthesized on free ribosomes and released into the cytosol, each protein must have some sort of signal that targets it to the correct organelle. In Chapter 12, you learned that a serine-lysine-leucine sequence (SKL in single-letter code) located near the C-terminus of a protein targets it for uptake into peroxisomes. And in Chapter 16, you learned that posttranslational import of proteins into the nucleus depends on *nuclear localization signals* that target proteins for transport through nuclear pore complexes (see Figure 16-30). Here we focus on protein import into mitochondria and chloroplasts, which involves signal sequences similar to those used in cotranslational import.

Importing Polypeptides into Mitochondria and Chloroplasts.

Although mitochondria and chloroplasts both contain their own DNA and protein-synthesizing machinery, they synthesize relatively few of the polypeptides they require. Most of the proteins residing in these two organelles, like all proteins found in the nucleus and peroxisomes, are encoded by nuclear genes and synthesized on cytosolic ribosomes. (Those few polypeptides that are synthesized directly within mitochondria are targeted mainly to the inner mito-

chondrial membrane, and polypeptides synthesized within chloroplasts are targeted mainly to thylakoid membranes. Almost without exception, such polypeptides encoded by mitochondrial or chloroplast genes are subunits of multimeric proteins, with one or more of the other subunits encoded by nuclear genes and imported from the cytosol.)

Most mitochondrial and chloroplast polypeptides are synthesized on cytosolic ribosomes, released into the cytosol, and then taken up by the appropriate organelle (mitochondrion or chloroplast) within a few minutes. The targeting signal for such polypeptides is a special sequence called a **transit sequence**. Like the ER signal sequence of ER-targeted polypeptides, the transit sequence is located at the N-terminus of the polypeptide. Once inside the

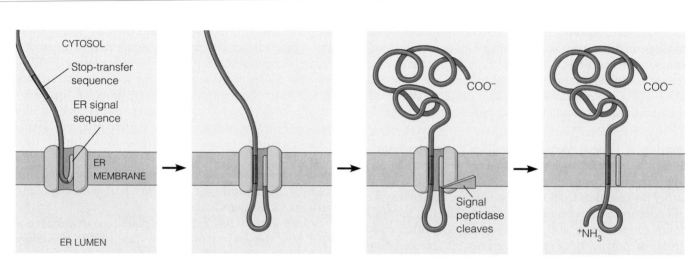

(a) Polypeptide with an internal stop-transfer sequence and a terminal ER signal sequence

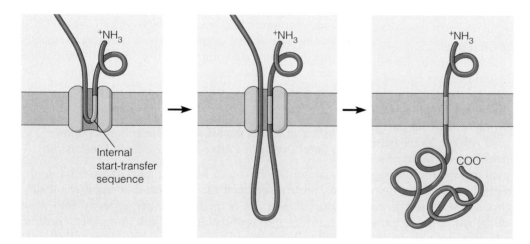

(b) Polypeptide with an internal start-transfer sequence

Figure 20-17 Cotranslational Insertion of Transmembrane Proteins into the ER Membrane. This figure shows two mechanisms postulated for the insertion of integral membrane proteins having a single transmembrane segment. For clarity, the SRP, ribosome, and most other parts of the translocational apparatus have been

omitted. **(a)** Insertion of a polypeptide with both a terminal ER signal sequence and an internal stop-transfer sequence. The terminal peptide is eventually cut off, leaving a transmembrane protein with its N-terminus in the ER lumen and its C-terminus in the cytosol. **(b)** Insertion of a polypeptide with only a single, internal

start-transfer sequence, which both starts polypeptide transfer and anchors itself permanently in the membrane. The amino-carboxyl orientation of the completed protein depends on the orientation of the start-transfer sequence when it first inserts into the translocation apparatus.

mitochondrion or chloroplast, the transit sequence is removed by a *transit peptidase* located within the organelle. Removal of the transit sequence often occurs before the transport process is complete.

The transit sequences of mitochondrial or chloroplast polypeptides typically contain both hydrophobic and hydrophilic amino acids. The presence of positively charged amino acids is critical, although the secondary structure of the sequence may be more important than the specific amino acids. For example, some mitochondrial transit sequences have positively charged amino acids interspersed with hydrophobic amino acids in such a way that, when the sequence is coiled into an α helix, most of the positively charged amino acids are on one side of the helix and the hydrophobic amino acids are on the other.

The uptake of polypeptides possessing transit sequences is mediated by specialized transport complexes located in the outer and inner membranes of mitochondria and chloroplasts. As shown in Figure 20-18, the mitochondrial transport complexes are called **TOM** (translocase of the outer mitochondria membrane) and **TIM** (translocase of the inner mitochondrial membrane); the comparable chloroplast complexes are **TOC** (translocase of the outer chloroplast membrane) and **TIC** (translocase of the inner chloroplast membrane). Polypeptides are initially selected for transport into mitochondria or chloroplasts by components of TOM or TOC known as *transit sequence receptors*. After a transit sequence has bound to its receptor, the polypeptide containing this sequence is translocated across the outer membrane through a *pore* in the TOM or TOC complex. If the polypeptide is destined for the interior of the organelle, movement through the TOM or TOC complex is quickly followed by passage through the TIM or TIC

complex of the inner membrane, presumably at a *contact site* where the outer and inner membranes lie close together.

Evidence supporting this model has come both from electron microscopy, which reveals many sites of close contact between outer and inner membranes, and from biochemical experiments, in which cell-free mitochondrial import systems are incubated on ice to trap polypeptides in the act of being translocated (Figure 20-19). The

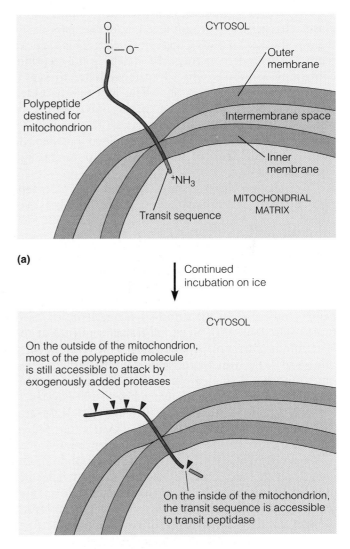

Figure 20-19 Experimental Demonstration That Polypeptides Span Both Mitochondrial Membranes During Import. **(a)** To demonstrate that polypeptides being imported into the mitochondrion span both membranes at the same time, a cell-free import system was incubated on ice, instead of at the usual temperature of 37°C. At low temperature, polypeptides can start to penetrate the mitochondrion, but their translocation then stalls. **(b)** Under these conditions, the transit sequence is cleaved by the transit peptidase present in the matrix, indicating that the N-terminus of the polypeptide is within the mitochondrion. At the same time, most of the polypeptide molecule is readily accessible to attack by exogenously added proteolytic enzymes on the outside of the mitochondrion. Therefore, we conclude that the polypeptide must span both membranes transiently during import, presumably at a contact site between the two membranes.

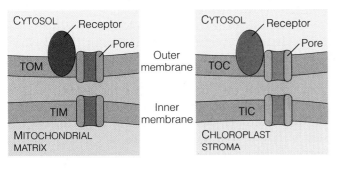

Mitochondrion   **Chloroplast**

Figure 20-18 Main Polypeptide Transport Complexes of the Outer and Inner Membranes of Mitochondria and Chloroplasts. Mitochondrial and chloroplast polypeptides synthesized in the cytosol are transported into these organelles by specialized transport complexes located in their outer and inner membranes. In mitochondria, the outer and inner membrane complexes are called TOM and TIM, respectively; the comparable chloroplast structures are TOC and TIC. The transport complexes of the outer membranes (TOM and TOC) are comprised of two types of components: receptor proteins that recognize and bind to polypeptides targeted for uptake, and pore proteins that form channels through which the polypeptides are translocated.

low temperature causes polypeptide movement across the membranes to halt shortly after it starts. At this point, the polypeptides have already had their transit sequences removed by the transit peptidase enzyme located in the mitochondrial matrix, but they can still be attacked by externally added proteolytic enzymes. Such results indicate that polypeptides can transiently span both membranes during import; in other words, the N-terminus can enter the matrix of the mitochondrion while the rest of the molecule is still outside the organelle.

Polypeptides entering mitochondria and chloroplasts must generally be unfolded before they pass across the membranes bounding these organelles. This requirement has been demonstrated by attaching polypeptides containing a mitochondrial transit sequence to agents that maintain the polypeptide chain in a tightly folded state. Such polypeptides bind to the outer surface of mitochondria but will not move across the membrane, apparently because the size of the folded polypeptide exceeds the diameter of the membrane pore through which it must pass.

To maintain the necessary unfolded state, polypeptides targeted for mitochondria and chloroplasts are bound to *chaperone proteins* similar to those that help newly synthesized polypeptides fold correctly. Figure 20-20 shows a current model for this chaperone-mediated import of polypeptides into the mitochondrial matrix. To start the process, chaperones of the Hsp70 class bind to a newly forming polypeptide that is still in the process of being synthesized in the cytosol, keeping it in a loosely folded state (step ①). Next, the transit sequence at the N-terminus of the polypeptide binds to the receptor component of TOM, which protrudes from the surface of the outer mitochondrial membrane (②). The chaperone proteins then are released, accompanied by ATP hydrolysis, as the polypeptide is translocated through the TOM and TIM pores and into the mitochondrial matrix (③). When the transit sequence emerges into the matrix, it is removed by transit peptidase. As the rest of the polypeptide subsequently enters the matrix, *mitochondrial* Hsp70 molecules bind to it temporarily; their subsequent release also requires ATP hydrolysis (④). This latter ATP hydrolysis is thought to be what actually drives the translocation. Finally, in many cases mitochondrial Hsp60 chaperone molecules bind to the polypeptide and help it achieve its fully folded conformation (⑤).

Both chloroplasts and mitochondria require energy for the import of polypeptides. Mitochondrial import is driven both by ATP hydrolysis and by the electrochemical gradient across the inner membrane. The electrochemical gradient seems to be necessary only for the binding and penetration of the transit sequence. Once this step has occurred, experimental abolition of the membrane potential does not interfere with the rest of the transfer process. Chloroplasts, on the other hand, maintain an electrochemical gradient across the thylakoid membrane, but not across the inner membrane. Presumably, the energy requirement for import into the chloroplast stroma is met by ATP alone.

Targeting Polypeptides to the Proper Compartments Within Mitochondria and Chloroplasts. Because of the structural complexity of mitochondria and chloroplasts, proteins to be imported from the cytosol must be targeted not only to the right organelle but also to the appropriate compartment within the organelle. Mitochondria have four compartments: the outer membrane, the intermembrane space, the inner membrane, and the matrix. Chloroplasts have four similar compartments (with the stroma substituted for matrix), and two additional compartments as well: the thylakoid membrane and the thylakoid lumen. Thus, a polypeptide may have to cross one, two, or even three membranes to reach its final destination.

Given the structural complexity of both organelles, it is perhaps not surprising that many mitochondrial and chloroplast polypeptides require more than one signal to arrive at their proper destinations. For example, targeting of a polypeptide to the outer or inner mitochondrial membrane requires an N-terminal transit sequence to direct the polypeptide to the mitochondrion plus an additional internal sequence, called a *hydrophobic sorting signal*, to target the polypeptide to its final destination. In such cases, the hydrophobic sorting signal acts as a stop-transfer sequence that halts translocation of the polypeptide chain through either the outer or inner membrane. The hydrophobic signal sequence then remains imbedded in the membrane, anchoring the polypeptide to the lipid bilayer, while the N-terminal transit sequence is usually removed. A combination of transit and hydrophobic sorting sequences is also used for targeting polypeptides to the intermembrane space. In this case, the polypeptide passes through the outer membrane and the signal sequences are then removed, leaving the polypeptide in the space between the two membranes.

Multiple signals are also involved in directing some chloroplast polypeptides to their final destination. Polypeptides intended for insertion into (or transport across) the thylakoid membrane, for example, must first be targeted to the chloroplast and transported into the stroma, presumably crossing the inner and outer membranes at a contact site. In the stroma, the transit sequence used for this first step is cleaved from the polypeptide, unmasking a hydrophobic *thylakoid signal sequence* that targets the polypeptide for either the thylakoid membrane or thylakoid lumen. For polypeptides destined for the thylakoid membrane, the hydrophobic signal sequence may spontaneously insert and anchor the polypeptide within the lipid bilayer of the thylakoid membrane. Alternatively, the insertion of some polypeptides into thylakoid membranes requires the participation of a GTP-dependent protein resembling the signal recognition particle (SRP) that directs and binds polypeptides to ER membranes (see Figure 20-16).

Polypeptides destined for the thylakoid lumen are translocated completely across the thylakoid membrane,

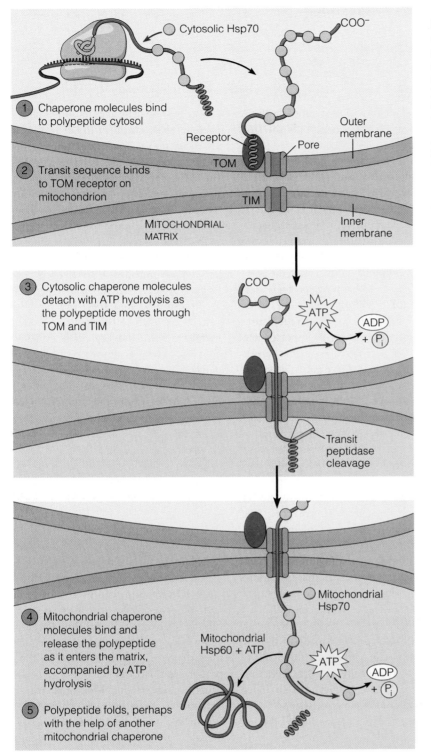

Figure 20-20 A Model for the Posttranslational Import of Polypeptides into the Mitochondrion. Like cotranslational import into the ER, posttranslational import into a mitochondrion involves a signal sequence (called a transit sequence in this case), a membrane receptor, pore-forming membrane proteins, and a peptidase. However, in the mitochondrion, the membrane receptor recognizes the signal sequence directly, without the intervention of a cytosolic SRP. Furthermore, chaperone proteins play several crucial roles in the mitochondrial process: They keep the polypeptide partially unfolded after synthesis in the cytosol so that binding of the transit sequence and translocation can occur (steps ① – ③); they drive the translocation itself by binding to and releasing from the polypeptide *within* the matrix, an ATP-requiring process (step ④); and, in many cases, they help the polypeptide fold into its final conformation (step ⑤). The chaperones included here are cytosolic and mitochondrial versions of Hsp70 (light blue) and a mitochondrial Hsp60 (not illustrated).

accompanied by cleavage of the thylakoid signal sequence as the polypeptide is released into the lumen. Most polypeptides targeted to the lumen are translocated across the thylakoid membrane in an unfolded state by an ATP-dependent process that resembles the mechanism employed for translocating polypeptides across the outer chloroplast or mitochondrial membranes. However, it has recently been discovered that some extensively folded proteins can also be transported across thylakoid membranes using an alternative mechanism driven by energy derived from the proton gradient. Although translocation of extensively folded proteins across membranes is relatively unusual, similar mechanisms have been detected in peroxisomes and in the bacterial plasma membrane.

Perspective

Translation is the process by which polypeptides are synthesized on ribosomes in the cell. The cellular machinery of translation is dominated by RNA molecules of various kinds. Messenger RNA determines the order of amino acids in the polypeptide, tRNA brings the amino acids to the ribosome, and rRNA helps position the mRNA on the ribosome and catalyzes peptide bond formation. In addition, a number of protein factors trigger specific events during the initiation, elongation, and termination stages of the process. GTP binding and hydrolysis drive the necessary conformational changes in the protein factors. The specificity required to link the right amino acids to the right tRNA molecules is a property of the aminoacyl-tRNA synthetases that catalyze these reactions. After the mRNA, ribosomal subunits, and initiator aminoacyl tRNA come together to form the initiation complex, other aminoacyl tRNAs recognize successive codons in the mRNA and add their amino acids to the growing polypeptide chain. Chain termination occurs when one of the stop codons is encountered, and the completed polypeptide is then released from the ribosome. The proper folding of released polypeptides is normally assisted by molecular chaperones. Abnormalities in protein folding can lead to a variety of diseases, such as Alzheimer's and "mad cow" disease.

Knowledge of the genetic code and the details of the translation process enable us to understand how nonsense mutations cause their deleterious effects, and also how they can be suppressed by compensating mutations in tRNA. The phenotypic effect of a mutation that changes an amino acid codon to a stop codon can be largely overcome if a tRNA mutated in its anticodon reads the stop codon as an amino acid.

Proteins reach their final destinations in the cell by two main pathways, both of which involve polypeptide targeting and sorting. The general strategy is that newly made polypeptides have special sequences of amino acids that serve as targeting signals; proteins selectively recognize and bind to these signals, thus sorting the polypeptides. In one pathway, proteins destined for components of the endomembrane system or secretion from the cell are cotranslationally imported into the ER. The signal sequence that targets these polypeptides to the ER is located at the N-terminus of the newly forming polypeptide. An SRP in the cytosol binds to the signal sequence and then to an SRP receptor on the ER membrane, docking the ribosome-mRNA-polypeptide complex to the membrane. As polypeptide synthesis then proceeds, the growing polypeptide is translocated across the ER membrane through a protein pore. The signal sequence is clipped off by a signal peptidase, leaving the remaining polypeptide to fold into its final three-dimensional shape. Polypeptides that insert into the ER membrane have one or more internal stop-transfer sequences, instead of or in addition to a terminal ER signal sequence. Most proteins made in the ER are glycosylated; certain of these oligosaccharide side chains serve as targeting signals that direct the proteins to other parts of the endomembrane system.

In the other sorting pathway, proteins destined for the nuclear interior, mitochondria, chloroplasts, or peroxisomes are synthesized on cytosolic ribosomes (as are proteins that remain in the cytosol) and are then imported posttranslationally into the targeted organelle. Polypeptides destined for peroxisomes contain a special targeting sequence near the C-terminus, whereas those targeted to the nucleus contain nuclear localization signals that promote their entry through the nuclear pores. Targeting to mitochondria and chloroplasts involves a transit sequence located at the N-terminus. Polypeptides are transported into these organelles at contact sites where the inner and outer membranes of the organelle are close together. In this pathway, receptor proteins in the outer membrane recognize the transit sequence directly. The energy needed to transport the unfolded polypeptide into the mitochondrion is provided by ATP hydrolysis associated with chaperone release, and by the electrochemical gradient across the inner membrane. In chloroplasts, transport of unfolded polypeptides into the organelle is driven by ATP hydrolysis alone, but the proton gradient plays a role in driving the transport of some extensively folded proteins into the thylakoid lumen.

Because mitochondria and chloroplasts have multiple compartments (four and six, respectively) to which polypeptides may be targeted, mitochondrial and chloroplast polypeptides often require more than one signal to arrive at their proper destinations. Such polypeptides usually possess an N-terminal transit sequence to direct the polypeptide to the organelle plus a hydrophobic sorting signal to target the polypeptide to its final destination.

Key Terms for Self-Testing

Translation: The Cast of Characters
ribosome (p. 660)
mRNA-binding site (p. 661)
A (aminoacyl) site (p. 661)
P (peptidyl) site (p. 661)
E (exit) site (p. 662)
transfer RNA (tRNA) (p. 662)

aminoacyl tRNA (p. 664)
anticodon (p. 664)
wobble hypothesis (p. 664)
aminoacyl-tRNA synthetase (p. 664)
leader (on mRNA) (p. 666)
start codon (p. 666)
trailer (on mRNA) (p. 666)

stop codon (p. 666)

The Mechanism of Translation
initiation factor (p. 668)
initiator tRNA (p. 669)
30S initiation complex (p. 669)
70S initiation complex (p. 669)

elongation factor (p. 669)
peptidyl transferase (p. 671)
translocation (p. 671)
release factor (p. 672)
molecular chaperone (p. 672)
polyribosome (p. 674)

Nonsense Mutations and Suppressor tRNA
nonsense mutation (p. 674)
suppressor tRNA (p. 674)

Posttranslational Processing
protein splicing (p. 676)

Protein Targeting and Sorting
cotranslational import (p. 678)
posttranslational import (p. 678)
ER signal sequence (p. 680)
signal-recognition particle (SRP) (p. 681)
translocon (p. 681)
BiP (p. 681)

protein disulfide isomerase (p. 681)
stop-transfer sequence (p. 683)
start-transfer sequence (p. 683)
transit sequence (p. 684)
TOM and TIM (p. 685)
TOC and TIC (p. 685)

Box 20A: *Protein-Folding Diseases*
prion (p. 673)

Problem Set

More challenging problems are marked with a •.

20-1.   The Genetic Code and Two Human Hormones. The following is the actual sequence of a small stretch of human DNA:

3′ AATTATACACGATGAAGCTTGTGACAGGGTTTCCAATCATTAA 5′
5′ TTAATATGTGCTACTTCGAACACTGTCCCAAAGGTTAGTAATT 3′

(a) What are the two possible RNA molecules that could be transcribed from this DNA?

(b) Only one of these two RNA molecules can actually be translated. Explain why.

(c) The RNA molecule that can be translated is the mRNA for the hormone vasopressin. What is the apparent amino acid sequence for vasopressin? (The genetic code is given in Figure 19-8.)

(d) In its active form, vasopressin is a nonapeptide (i.e., it has nine amino acids) with cysteine at the N-terminus. How can you explain this in light of your answer to part c?

(e) A related hormone, oxytocin, has the following amino acid sequence:

Cys-Tyr-lle-Glu-Asp-Cys-Pro-Leu-Gly

Where and how would you change the DNA that codes for vasopressin so that it would code for oxytocin instead? Does your answer suggest a possible evolutionary relationship between the genes for vasopressin and oxytocin?

20-2.   Tracking a Series of Mutations. The following diagram shows the amino acids that result from mutations in the codon for a particular amino acid in a bacterial polypeptide:

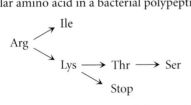

Assume that each arrow denotes a single base-pair substitution in the bacterial DNA.

(a) Referring to the genetic code in Figure 19-8, determine the most likely codons for each of the amino acids and the stop signal in the diagram.

(b) Starting with a population of mutant cells carrying the nonsense mutation, another mutant is isolated in which the premature stop signal is suppressed. Assuming wobble does not occur and assuming a single base change in the tRNA anticodon, what are all the possible amino acids that might be found in this mutant at the amino acid position in question?

20-3.   Initiation of Translation. Figure 20-8 diagrams the initiation of translation in prokaryotic cells. Using the text on page 669 as a guide, draw a similar sketch outlining the steps in *eukaryotic* initiation of translation. What are the main differences between prokaryotic and eukaryotic initiation?

20-4.   Prokaryotic and Eukaryotic Protein Synthesis Compared. For each of the following statements, indicate whether it applies to protein synthesis in prokaryotes (P), in eukaryotes (E), in both (B), or in neither (N).

(a) The mRNA has a ribosome-binding site within its leader region.

(b) AUG is a start codon.

(c) The enzyme that catalyzes peptide bond formation is an RNA molecule.

(d) The mRNA is translated in the 3′ ⟶ 5′ direction.

(e) The C-terminus of the polypeptide is synthesized last.

(f) Translation is terminated by special tRNA molecules that recognize stop codons.

(g) GTP hydrolysis functions to induce conformational changes in various proteins involved in polypeptide elongation.

(h) ATP hydrolysis is required to attach an amino acid to a tRNA molecule.

(i) The specificity required to link the right amino acids to the right tRNA molecules is a property of the enzymes called aminoacyl-tRNA synthetases.

•20-5.   An Antibiotic Inhibitor of Translation. Puromycin is a powerful inhibitor of protein synthesis. It is an analog of the 3′ end of aminoacyl tRNA, as Figure 20-21 reveals. (R represents the functional group of the amino acid; R′ represents the remainder of the tRNA molecule.) When puromycin is added to a cell-free system containing all the necessary machinery for protein synthesis, incomplete polypeptide chains are released from the ribosomes. Each such chain has puromycin covalently attached to one end.

(a) Explain these results.

(b) To which end of the polypeptide chains would you expect the puromycin to be bound? Explain.

(c) Would you expect puromycin to bind to the A or P site on the ribosome, or to both? Explain.

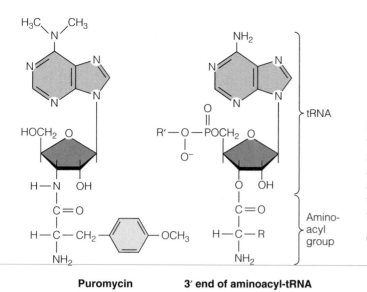

Puromycin   **3′ end of aminoacyl-tRNA**

Figure 20-21   The Structure of Puromycin.   See Problem 20-5.

(d)   Assuming that it can penetrate into the cell equally well in both cases, would you expect puromycin to be a better inhibitor of protein synthesis in a eukaryotic cell or in a prokaryotic cell? Explain.

• **20-6.   A Fictional Antibiotic.** In a study involving a cell-free protein synthesizing system from *E. coli,* the polyribonucleotide AUGUUUUUUUUUUUU directs the synthesis of the oligopeptide fMet-Phe-Phe-Phe-Phe. In the presence of Rambomycin, a new antibiotic just developed by Macho Pharmaceuticals, only the dipeptide fMet-Phe is made.

(a)   What step in polypeptide synthesis does Rambomycin inhibit? Explain your answer.

(b)   Will the oligopeptide product be found attached to tRNA at the end of the uninhibited reaction? Will the dipeptide product be found attached to tRNA at the end of the Rambomycin-inhibited reaction? Explain.

20-7.   Frameshift Suppression. As discussed in the chapter, a nonsense mutation can be suppressed by a mutant tRNA in which one of the three anticodon nucleotides has been changed. Describe a mutant tRNA that could suppress a *frameshift* mutation. (Hint: Such a mutant tRNA was used to investigate the role played by peptidyl tRNA in the translocation of mRNA during protein synthesis.)

• **20-8.   Protein Folding.** The role of BiP in protein folding was briefly described in this chapter. Answer the following questions about observations and situations involving BiP.

(a)   BiP is found in high concentration in the lumen of the ER, but is not present in significant concentrations elsewhere in the cell. How do you think this condition is established and maintained?

(b)   If the gene coding for BiP acquires a mutation that disrupts the protein's binding site for hydrophobic amino acids, what kind of impact might this have on the cell?

(c)   Suppose that recombinant DNA techniques are used to introduce a new gene into a cell, and that this gene codes for polypeptide subunit X of a secreted protein that is constructed from two different kinds of polypeptide subunits, X and Y. If the gene coding for subunit Y is not present in this cell, what do you think will happen to polypeptide X when it is produced?

• **20-9.   Cotranslational Import.** You perform a series of experiments on the synthesis of the pituitary hormone prolactin, which is a single polypeptide chain 199 amino acids long. The mRNA coding for prolactin is translated in a cell-free protein synthesizing system containing ribosomes, amino acids, tRNAs, aminoacyl-tRNA synthetases, ATP, GTP, and the appropriate initiation, elongation, and termination factors. Under these conditions, a polypeptide chain 227 amino acids long is produced.

(a)   How might you explain the discrepancy between the normal length of prolactin (199 amino acids) and the length of the polypeptide synthesized in your experiment (227 amino acids)?

(b)   You perform a second experiment in which you add SRP to your cell-free protein-synthesizing system, and find that translation stops after a polypeptide about 70 amino acids long has been produced. How can you explain this result? Can you think of any purpose this phenomenon might serve for the cell?

(c)   You perform a third experiment in which you add both SRP and ER membrane vesicles to your protein-synthesizing system, and find that translation of the prolactin mRNA now produces a polypeptide 199 amino acids long. How can you explain this result? Where would you expect to find this polypeptide?

20-10.   Two Types of Posttranslational Import. The mechanism by which proteins synthesized in the cytosol are imported into the mitochondrial matrix is different from the mechanism by which proteins enter the nucleus, yet the two mechanisms do share some features. For example, in both cases the protein must cross two membranes. Indicate whether each of the following statements applies to nuclear import (Nu), mitochondrial import (M), both (B), or neither (N). You may want to review the discussion of nuclear import in Chapter 16 before answering this question (see especially Figure 16-30).

(a)   The polypeptide to be transported into the organelle has a specific short stretch of amino acids that targets the polypeptide to the organelle.

(b)   The signal sequence is always at the polypeptide's N-terminus and is cut off by a peptidase within the organelle.

(c)   The signal sequence is recognized and bound by a receptor protein in the organelle's outer membrane.

(d)   ATP hydrolysis is known to be required for the translocation process.

(e)   GTP hydrolysis is known to be required for the translocation process.

(f)   There is strong evidence for the involvement of chaperone proteins during translocation of the protein.

(g)   The imported protein enters the organelle through some sort of protein pore.

(h)   The pore complex consists of many proteins and, with a total mass of over 100 million Da, is large enough to be readily seen with the electron microscope.

Suggested Reading

References of historical importance are marked with a • .

Translation

Cech, T. R. The ribosome is a ribozyme. *Science* 289 (2000): 878.

Dever, T. E. Translation initiation: Adept at adapting. *Trends Biochem. Sci.* 24 (1999): 398.

Frank, J. How the ribosome works. *American Scientist* 86 (1998): 428.

Hentze, M. W. Believe it or not—translation in the nucleus. *Science* 293 (2001): 1058.

Ibba, M., H. D. Becker, C. Stathopoulos, D. L. Tumbula, and D. Söll. The adaptor hypothesis revisited. *Trends Biochem. Sci.* 25 (2000): 311.

Kisselev, L. L., and R. H. Buckingham. Translational termination comes of age. *Trends Biochem. Sci.* 25 (2000): 561.

Lafontaine, D. L. J., and D. Tollervey. The function and synthesis of ribosomes. *Nature Reviews Mol. Cell. Biol.* 2 (2001): 514.

• Nomura, M. Reflections on the days of ribosome reconstitution research. *Trends Biochem. Sci.* 22 (1997): 275.

Rodnina, M. V., and W. Wintermeyer. Ribosome fidelity: tRNA discrimination, proofreading and induced fit. *Trends Biochem. Sci.* 26 (2001): 124.

Weisblum, B. Back to Camelot: Defining the specific role of tRNA in protein synthesis. *Trends Biochem. Sci.* 24 (1999): 247.

Protein Folding and Processing

Hartl, F. U. Molecular chaperones in cellular protein folding. *Nature* 381 (1996): 571.

Paulus, H. Protein splicing and related forms of protein autoprocessing. *Annu. Rev. Biochem.* 69 (2000): 447.

Prusiner, S. B. *Prion Biology and Diseases.* Plainview, NY: Cold Spring Harbor Laboratory Press, 1999.

St George-Hyslop, P. H. Piecing together Alzheimer's. *Sci. Amer.* 283 (December 2000): 76.

Wickner, S., M. R. Maurizi, and S. Gottesman. Posttranslational quality control: Folding, refolding, and degrading proteins. *Science* 286 (1999): 1888.

Cotranslational Protein Import

Anderson, D., and P. Walter. Blobel's Nobel: A vision validated. *Cell* 99 (1999): 557.

Bibi, E. The role of the ribosome-translocon complex in translation and assembly of polytopic membrane proteins. *Trends Biochem. Sci.* 23 (1998): 51.

• Blobel, G., P. Walter, G. N. Chang, B. M. Goldman, A. H. Erickson, and V. R. Lingappa. Translocation of proteins across membranes: The signal hypothesis and beyond. *Symp. Soc. Exp. Biol.* 33 (1979): 9.

Hegde, R. S., and V. R. Lingappa. Regulation of protein biogenesis at the endoplasmic reticulum membrane. *Trends Cell Biol.* 9 (1999): 132.

Johnson, A. E., and N. G. Haigh. The ER translocon and retrotranslocation: Is the shift into reverse manual or automatic? *Cell* 102 (2000): 709.

Matlack, K. E. S., W. Mothes, and T. A. Rapoport. Protein translocation: Tunnel vision. *Cell* 92 (1998): 381.

Powers, T., and P. Walter. A ribosome at the end of the tunnel. *Science* 278 (1997): 2072.

Walter, P., and A. E. Johnson. Signal sequence recognition and protein targeting to the endoplasmic reticulum membrane. *Annu. Rev. Cell Biol.* 10 (1994): 211.

Posttranslational Protein Import

Chen, C., and D. J. Schnell. Protein import into chloroplasts. *Trends Cell Biol.* 9 (1999): 222.

Haucke, V., and G. Schatz. Import of proteins into mitochondria and chloroplasts. *Trends Cell Biol.* 7 (1997): 103.

Pfanner, N., and A. Geissler. Versatility of the mitochondrial protein import machinery. *Nature Reviews Mol. Cell Biol.* 2 (2001): 339.

Pilon, M., and R. Schekman. Protein translocation: How Hsp70 pulls it off. *Cell* 97 (1999): 679.

Teter, S. A., and D. J. Klionsky. How to get a folded protein across a membrane. *Trends Cell Biol.* 9 (1999): 428.

21

The Regulation of Gene Expression

In our exploration of biological information flow thus far, we have identified DNA as the main repository of information in cells, we have seen how DNA is replicated, and we have looked at the steps involved in the expression of DNA's genetic information via transcription and translation. In concluding our consideration of information flow, we will now examine the strategies used by cells to regulate the expression of their genes.

Regulation is an important part of almost every process in nature. Rarely, if ever, is it adequate simply to describe the steps of a process. Instead, we must also ask what turns it on, what turns it off, and what determines its rate. This is especially true of gene expression. Most genes are not expressed all the time. In some cases, selective gene expression enables cells to be metabolically thrifty, synthesizing only those gene products that are of immediate use under the prevailing environmental conditions; this is often the situation with bacteria. In other cases, such as in multicellular organisms, selective gene expression allows cells to fulfill specialized roles.

As you might expect, our first knowledge about the regulation of gene expression came from investigations of prokaryotes. Bacteria are far more amenable to the kinds of genetic and biochemical manipulations that marked the early studies of gene control mechanisms. But in recent years, advances in DNA technology have led to considerable progress with eukaryotes as well. We will look first at the prokaryotes, where mechanisms of gene regulation operate mainly at the level of transcription. Then we will turn to the eukaryotes, to explore both their versions of transcriptional control and some of the other types of genetic control found in these organisms.

Gene Regulation in Prokaryotes

Of the several thousand genes present in a typical bacterial cell, some are so important to the life of the cell that they are active at all times; their expression is not regulated. Such **constitutive genes** include, for example, the genes encoding the enzymes of glycolysis. For many other genes, however, expression is regulated so that the amount of the final gene product—protein or RNA—is carefully tuned to the cell's need for that product. A number of these **regulated genes** encode enzymes for metabolic processes that, unlike glycolysis, are not constantly required. The intracellular concentrations of such enzymes are usually adjusted by starting and stopping gene transcription in response to cellular needs. Because this control of enzyme-coding genes helps bacterial cells adapt to their environment, it is commonly referred to as **adaptive enzyme synthesis.**

Catabolic and Anabolic Pathways Utilize Different Strategies for Adaptive Enzyme Synthesis

Bacteria use two main strategies for regulating enzyme synthesis, depending on whether a given enzyme is involved in a *catabolic* (degradative) or *anabolic* (synthetic) pathway. The enzymes that catalyze such pathways are often regulated coordinately; that is, the synthesis of all the enzymes involved in a particular pathway is turned on and off together. Here we briefly describe two well-understood pathways, one catabolic and one anabolic.

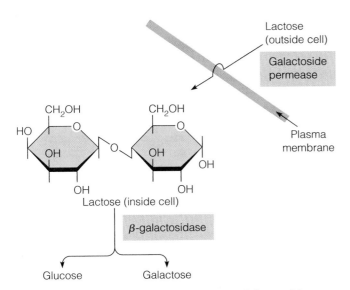

Figure 21-1 A Typical Catabolic Pathway. Breakdown of the disaccharide lactose involves enzymes (green boxes) whose synthesis is regulated coordinately. See Figures 21-3 and 21-4 for the organization and regulation of the genes that code for these enzymes. (The enzymes responsible for the subsequent catabolism of the monosaccharides glucose and galactose are not a part of the same regulatory unit.)

Catabolic Pathways and Substrate Induction.

Catabolic enzymes exist for the primary purpose of degrading specific substrates, often as a means of obtaining energy. Figure 21-1 depicts the steps in the catabolic pathway that degrades the disaccharide lactose into simple sugars that can then be metabolized by glycolysis. The central step in this pathway is the hydrolysis of lactose into the monosaccharides glucose and galactose, a reaction catalyzed by the enzyme β-galactosidase. However, before lactose can be hydrolyzed, it must first be transported into the cell. A protein called galactoside permease is responsible for this transport, and its synthesis is regulated coordinately with β-galactosidase.

Since the function of a catabolic enzyme is to degrade a specific substrate, such enzymes are needed only when the cell is confronted by the relevant substrate. The enzyme β-galactosidase, for example, is useful only when cells have access to lactose; in the absence of lactose, the enzyme is superfluous. Accordingly, it makes sense in terms of cellular economy for the synthesis of β-galactosidase to be turned on, or induced, in the presence of lactose, but to be turned off in its absence. This turning on of enzyme synthesis is called **substrate induction**, and enzymes whose synthesis is regulated in this way are referred to as **inducible enzymes.** Most catabolic pathways in bacterial cells are subject to substrate induction of their enzymes.

Anabolic Pathways and End-Product Repression.

The regulation of anabolic pathways is in a sense just the opposite of that for catabolic pathways. Figure 21-2 summarizes the anabolic pathway for synthesizing the amino acid trypto-

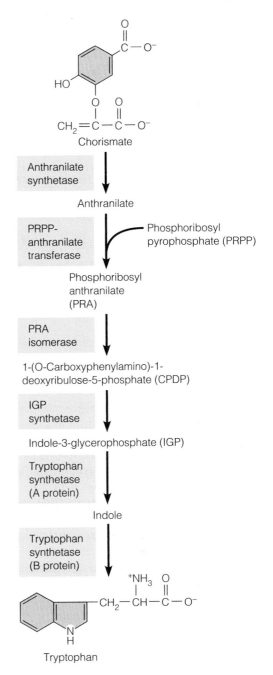

Figure 21-2 A Typical Anabolic Pathway. Synthesis of the amino acid tryptophan from the starting compound chorismate involves a set of enzymes (gold boxes) whose synthesis is regulated in a coordinated way. See Figure 21-6 for the organization and regulation of the genes that code for these enzymes.

phan from the starting compound chorismate. The enzymes that catalyze the six steps of this pathway, like those involved in the first two steps of lactose catabolism, are regulated coordinately at the genetic level.

For anabolic pathways, the amount of enzyme produced by a cell usually correlates inversely with the intracellular concentration of the end-product of the pathway. Such a relationship makes sense. For example, as the concentration of tryptophan rises, it is advantageous for the cell to economize on its metabolic resources by reducing its production of the enzymes involved in synthesizing tryptophan. But it is equally important that the cell be able to turn the production of these enzymes back on when the level of tryptophan decreases again. This kind of control is made possible by the ability of the end-product of an anabolic pathway—in our example, tryptophan—to somehow *repress* (reduce or stop) the further production of the enzymes involved in its formation. Such reduction in the expression of the enzyme-coding genes is called **end-product repression.** Most biosynthetic pathways in bacterial cells are regulated in this way. *Repression* is a general term in molecular genetics, referring to the reduction in expression of any regulated gene.

True genetic repression always has an effect on protein *synthesis,* and not just on protein *activity.* Recall from Chapter 6 that the end-products of biosynthetic pathways often have an inhibitory effect on enzyme activity as well. This *feedback inhibition* differs from repression in both mechanism and result. In feedback inhibition, molecules of enzyme are still present but their catalytic activity is inhibited; in end-product repression, the enzyme molecules are not even made.

Effector Molecules. One feature common to both induction and repression of enzyme synthesis is that control is exerted at the gene level in both cases. Another shared feature is that control is triggered by small organic molecules present within the cell or in the cell's surroundings. Geneticists call small organic molecules that function in this way **effectors.** As we will see shortly, effectors induce shape changes in allosteric proteins that control gene expression. For catabolic pathways, effectors are almost always substrates (lactose in our example), and they function as inducers of gene expression and, thus, of enzyme synthesis. For anabolic pathways, effectors are usually end-products (tryptophan in our example), and they usually lead to the repression of gene expression and thus repression of enzyme synthesis.

The Genes Involved in Lactose Catabolism Are Organized into an Inducible Operon

The classic example of an inducible enzyme system occurs in the bacterium *Escherichia coli* and involves a group of enzymes involved in lactose catabolism—the enzymes that catalyze the steps shown in Figure 21-1. Much of what we know about the regulation of gene expression in bacteria, including the vocabulary used to express that knowledge, is based on the pioneering studies of this system carried out by French molecular geneticists François Jacob and Jacques Monod. In 1961 these investigators published a classic paper that has probably had more influence on our understanding of gene regulation than any other work.

In their paper, Jacob and Monod proposed a general model of gene regulation with far-reaching implications. The cornerstone of this model rested on their discovery

that the control of lactose catabolism involves two types of genes: **structural genes** coding for enzymes involved in lactose uptake and metabolism, and a **regulatory gene** whose product regulates the activity of the structural genes. The structural genes involved in lactose metabolism are (1) the *lacZ* gene, which codes for β-galactosidase, the enzyme that hydrolyzes lactose and other β-galactosides; (2) the *lacY* gene, which codes for galactoside permease, the plasma membrane protein that transports lactose into the cell; and (3) the *lacA* gene, which codes for a transacetylase that adds an acetyl group to lactose as it is taken up by the cell. The *lacZ, lacY,* and *lacA* genes lie next to each other in the bacterial chromosome and are expressed only when an inducer such as lactose is present. Taken together, these observations led Jacob and Monod to suggest that these three genes all belong to a single regulatory unit, or, as they called it, an **operon**—a cluster of genes with related functions, regulated in such a way that all the genes in the cluster are turned on and off together.

The organization of functionally related genes into operons is commonly observed in prokaryotes but not in eukaryotic cells. Nonetheless, the operon model established several basic principles that have shaped our understanding of transcriptional regulation in both prokaryotic and eukaryotic systems.

The *lac* Repressor Is an Allosteric Protein Whose Binding to DNA Is Controlled by Lactose

A key feature of the operon model is the idea that genes with metabolically related functions are clustered together so their transcription can be regulated as a single unit. But how is this regulation accomplished? Jacob and Monod addressed this question by studying the ability of inducers such as lactose to turn on the production of the enzymes involved in lactose metabolism. They found that for induction to occur, an additional gene must be present—a regulatory gene that they named *lacI* (for inducibility). Whereas normal bacteria will produce β-galactosidase, galactoside permease, and transacetylase only when an inducer is present, deletion of the *lacI* gene yielded cells that *always* produce these proteins, even when inducer is absent. Jacob and Monod therefore concluded that the *lacI* gene codes for a product that normally inhibits, and thereby regulates, expression of the *lacZ, lacY,* and *lacA* genes. A regulatory gene *product* that inhibits the expression of other genes is called a **repressor.**

To understand how a repressor can regulate gene expression, we need to take a closer look at the organization of the genes involved in lactose metabolism. As Figure 21-3 shows, the **lac operon** consists of three structural genes (*lacZ, lacY,* and *lacA*) preceded by a **promoter** (P_{lac}) and a special nucleotide sequence called the **operator** (*O*), which actually overlaps the promoter. Transcription of the *lac* operon begins at the promoter, which is the site of RNA polymerase attachment, and then proceeds through the operator and all the structural genes until finally ending at a terminator sequence. The net result is a single

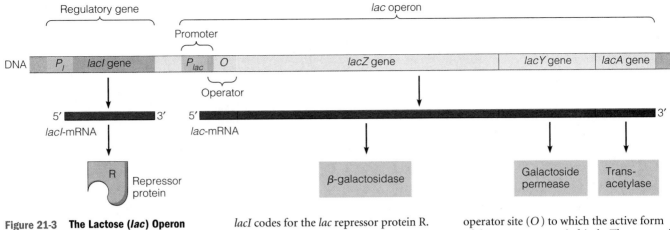

Figure 21-3 The Lactose (*lac*) Operon of *E. coli*. The *lac* operon consists of a segment of DNA that includes three contiguous structural genes (*lacZ*, *lacY*, and *lacA*), which are transcribed and regulated coordinately. The nearby regulatory gene *lacI* codes for the *lac* repressor protein R. Both the regulatory gene and the *lac* operon itself contain promoters (P_I and P_{lac} respectively) at which RNA polymerase binds, and terminators at which transcription halts. P_{lac} overlaps with the operator site (O) to which the active form of the repressor protein binds. The operon is transcribed into a single long molecule of mRNA that codes for all three polypeptides. For details of regulation, see Figure 21-4.

molecule of mRNA coding for the polypeptide products of all three structural genes. Such mRNA molecules, which code for more than one polypeptide, are called **polygenic mRNAs;** they occur only in prokaryotic cells.

The advantage of clustering related genes into an operon for transcription into a single polygenic mRNA is that it allows the synthesis of several polypeptides to be controlled in a single step. The crucial step in this control is the interaction between an operator site in the DNA and a repressor protein. The interaction between these elements for the specific case of the *lac* operon is depicted in Figure 21-4. The repressor protein, called the *lac repressor,* is encoded by the *lacI* regulatory gene, which is located outside the operon (although it happens to be located adjacent to the *lac* operon it regulates). The *lac* repressor is a DNA-binding protein that specifically recognizes and binds to the operator site of the *lac* operon. When the repressor is bound to the operator (Figure 21-4a), RNA polymerase cannot bind to the promoter and thus transcription of the structural genes is not possible. In other words, binding of the repressor to the operator inactivates the operon and keeps its structural genes turned off.

If binding of the repressor to the operator blocks transcription, how do cells turn on transcription of the *lac* operon, as occurs in the presence of inducers such as lactose? The answer is that inducer molecules bind to the *lac* repressor, thereby altering its conformation so that the repressor can no longer bind to the *lac* operator site in the DNA. Without repressor bound to it, the operator site is unoccupied and RNA polymerase can bind to the promoter and proceed down the operon, transcribing the *lacZ*, *lacY*, and *lacA* genes into a single polygenic mRNA molecule (Figure 21-4b).

A crucial feature of a repressor protein, therefore, is its ability to exist in two forms, only one of which binds to the operator. In other words, a repressor is an *allosteric protein.*

As we learned in Chapter 6, an allosteric protein can exist in either of two conformational states, depending on whether or not the appropriate effector molecule is present. In one state the protein is active; in the other state it is inactive, or nearly so. When the effector molecule binds to the protein, it induces a change in the conformational state of the protein and therefore in its activity. The binding is readily reversible, however, and departure of the effector results in the protein's rapid return to the alternative form.

Figure 21-5 shows the reversible interaction of the *lac* repressor with its effector, which is actually not lactose itself but *allolactose,* an isomer of lactose produced after lactose enters the cell. The conformational form assumed by the repressor protein in the absence of allolactose recognizes and binds to the operator (thereby inhibiting transcription), whereas the form with allolactose attached to it does not. The result is that the repressor protein inhibits transcription of the *lac* operon in the absence of allolactose, when there is no need to produce the catabolic enzymes encoded by the *lac* operon. However, in the presence of allolactose, the repressor converts to its inactive form, which does not recognize the operator and hence does not prevent transcription of the *lac* structural genes by RNA polymerase. In this way, lactose triggers the *induction* of the enzymes encoded by the *lac* operon. Because the *lac* operon is turned off unless induced, it is said to be an **inducible operon.** To borrow a computer term, the "default state" of an inducible operon is *off.*

The *lac* Operon Model Is Based Largely on Genetic Analysis

Most of the initial evidence in support of the operon model was based on genetic analyses of mutant bacteria that either produced abnormal amounts of the enzymes of the *lac* operon or showed abnormal responses to the addition or

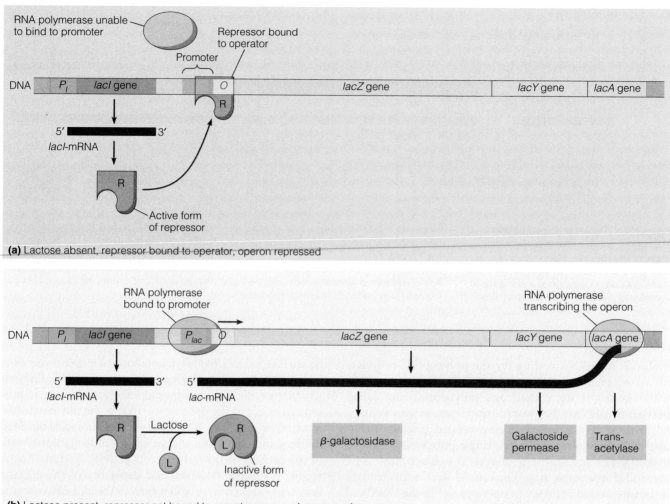

Figure 21-4 Regulation of the *lac* Operon. Transcription of the *lac* operon is regulated by binding of the *lac* repressor (R) to the operator. **(a)** In the absence of lactose, the repressor remains bound to the operator, and RNA polymerase cannot gain access to the promoter. Transcription is therefore blocked, and the operon remains repressed. **(b)** In the presence of lactose, the repressor is converted to the inactive form, which does not bind to the operator. RNA polymerase therefore can bind to the promoter and transcribe the structural genes *lacZ*, *lacY*, and *lacA* into a single polygenic mRNA. The form of lactose that binds to the repressor is an isomer called allolactose (L).

removal of lactose. These mutations were found to map either in the structural genes (*lacZ*, *lacY*, or *lacA*) or in the regulatory elements of the system (O, P_{lac}, or *lacI*). These two classes of mutations can be readily distinguished, because mutations in a structural gene affect only a single protein, whereas mutations in regulatory regions typically affect expression of all the structural genes coordinately.

Table 21-1 summarizes these mutations and their phenotypes, starting in line 1 with the inducible phenotype of the wild-type (nonmutant) cell. The plus signs in the phenotype columns of the table indicate high enzyme levels. The minus signs indicate very low enzyme levels, although not the complete cessation of enzyme production. Even in the absence of lactose, wild-type cells make small amounts of the *lac* enzymes because the binding of repressor to operator is reversible—the active repressor occasionally "falls off" the operator. The resulting low,

background level of transcription is important, because it allows cells to produce enough galactoside permease to facilitate the initial transport of lactose molecules into the cell prior to induction of the *lac* operon.

Experiments such as those represented in Table 21-1 are usually carried out using the synthetic β-galactoside *isopropylthiogalactoside (IPTG)* rather than lactose or allolactose as the inducing molecule. IPTG is a good inducer of the system but cannot be metabolized by cells, so its use avoids possible complications from varying levels of effector caused by its catabolism. Following common usage, we can now formally define the term **inducer** as referring to any effector molecule that turns on the transcription of an inducible operon. But keep in mind that an inducer is not exactly the "opposite" of a repressor, as the words may seem to imply. An inducer is a small molecule that binds to a protein; a repressor is a protein that binds to DNA.

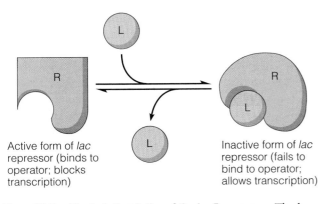

Figure 21-5 **Allosteric Regulation of the *lac* Repressor.** The *lac* repressor (R) is an allosteric protein, capable of reversible conversion between two alternative forms. In the absence of the effector allolactose (L), the protein assumes the form that is active in binding to the *lac* operator. In the presence of the effector, the protein exists preferentially in the alternative conformational state, which does not recognize the operator and is therefore inactive as a repressor of transcription.

Active form of *lac* repressor (binds to operator; blocks transcription)

Inactive form of *lac* repressor (fails to bind to operator; allows transcription)

Examination of Table 21-1 should help clarify how the analysis of genetic phenotypes led Jacob and Monod to formulate the operon model of gene regulation. We will consider each of the six kinds of mutations shown in lines 2–7 of this table in turn; as we do so, you should be able to see how each type of mutation contributed to formulation of the operon model.

Structural Gene Mutations. Mutations in the *lacY* or *lacZ* structural genes can lead to production of altered enzymes with little or no biological activity, even in the presence of inducer (Table 21-1, lines 2 and 3). Such mutants are therefore unable to utilize lactose as a carbon source, either because they cannot transport lactose efficiently into the cell (Y^- mutants) or because they cannot cleave the glycosidic bond between galactose and glucose in the lactose molecule (Z^- mutants). Note that mutations in a bacterial gene or regulatory sequence that render it defec-

tive are usually indicated with a superscript minus sign; for example, the genotype of a bacterium carrying a defective Y gene is written Y^-. The wild-type allele is indicated with a superscript plus sign.

Operator Mutations. Mutations in the operator can lead to a constitutive phenotype; that is, the mutant cells will continually produce the *lac* enzymes, whether inducer is present or not (Table 21-1, line 4). These mutations change the base sequence of the operator DNA so that it is no longer recognized by the repressor. The genotype of such *operator-constitutive mutants* is represented as O^c. As is expected for mutations in a regulatory site, O^c mutations simultaneously affect the synthesis of all three structural genes in the same way.

Promoter Mutations. Promoter mutations can decrease the affinity of RNA polymerase for the promoter. As a result, fewer RNA polymerase molecules bind per unit time to the promoter, and the rate of mRNA production decreases. (However, once an RNA polymerase molecule attaches to the DNA and begins transcription, it elongates the mRNA molecule at a normal rate.) Usually, promoter mutations (P^-) in the *lac* operon decrease both the elevated level of enzyme produced in the presence of inducer *and* the already low, basal level of *lac* enzyme production the cell manages to achieve in the absence of inducer (Table 21-1, line 5).

Regulatory Gene Mutations. Mutations in the *lacI* gene are of two types. Some mutants fail to produce any of the *lac* enzymes, regardless of whether inducer is present, and are therefore called *superrepressor mutants* (I^s in Table 21-1, line 6). Either the repressor molecule in such mutants has lost its ability to recognize and bind the inducer but can still recognize the operator, or else it has a high affinity for the operator regardless of whether inducer is bound to it. In either case, the repressor binds tightly to the operator and represses transcription, and hence enzyme synthesis, under all conditions.

Table 21-1 **Genetic Analysis of Mutations Affecting the *lac* Operon**

Line Number	Genotype of Bacterium*	Phenotype with Inducer Absent		Phenotype with Inducer Present	
		β-galactosidase	Permease	β-galactosidase	Permease
1	$I^+ P^+ O^+ Z^+ Y^+$	−	−	+	+
2	$I^+ P^+ O^+ Z^+ Y^-$	−	−	+	−**
3	$I^+ P^+ O^+ Z^- Y^+$	−	−	−	+
4	$I^+ P^+ O^c Z^+ Y^+$	+	+	+	+
5	$I^+ P^- O^+ Z^+ Y^+$	−	−	−	−
6	$I^s P^+ O^+ Z^+ Y^+$	−	−	−	−
7	$I^- P^+ O^+ Z^+ Y^+$	+	+	+	+

*$P = P_{lac}$

** The defective permease exhibits enough biological activity to transport minimal amounts of lactose into the cell, thereby permitting induction of the *lac* operon.

The other class of *lacI* mutations involves synthesis of a mutant repressor protein that does not recognize the operator (or, in some cases, is not synthesized at all). The *lac* operon in such I^- mutants cannot be turned off, and the enzymes are therefore synthesized constitutively (Table 21-1, line 7).

I^- mutants, along with the O^c and P^- mutants, illustrate the importance of specific recognition of DNA base sequences by regulatory proteins during the control of gene transcription. A small change—either in the DNA sequence of promoter or operator, or in the regulatory protein that binds to the operator—can dramatically affect expression of all the genes in the operon.

The *Cis-Trans* Test Using Partially Diploid Bacteria.
The existence of two different kinds of constitutive mutants, O^c and I^-, raises the question of how one type might be distinguished from the other. A ***cis-trans* test** is often used to differentiate between *cis*-acting mutations, which affect DNA sites (e.g., O^c), and *trans*-acting mutations, which affect proteins (e.g., I^-). The basis of the *cis-trans* test is a cell (or organism) that has two different copies of the DNA segment of interest. As you know, *E. coli* is usually haploid, but Jacob and Monod constructed partially diploid bacteria by inserting a second copy of the *lac* portion of the bacterial genome into the F-factor plasmid of F^+ cells (p. 600). This second copy could be transferred by conjugation into a host bacterium of any desired *lac* genotype to create partial diploids such as the types listed in Table 21-2.

If only one copy of the operon contains an I^- or O^c regulatory mutation, it is possible to determine whether the mutation has an effect on both copies of the operon or only on the copy of the operon where it is located. The mutation is said to act in *cis* if the only structural genes affected are those physically linked to the mutant locus (*cis* means "on this side," in this case referring to a mutation whose influence is restricted to genes located in the same physical copy of the *lac* operon). In contrast, the mutation is said to act in *trans* if the structural genes in both copies of the operon are affected (*trans* means "on the other side," in this case referring to the ability of the mutation to somehow affect the other copy of the *lac* operon).

To determine which copy of the *lac* operon is being expressed in any given cell population, Jacob and Monod used partial diploid cells containing one copy of the *lac* operon with a defective Z gene and the other copy with a defective Y gene. Table 21-2 shows what happens if one of these *lac* operon copies has a defective I^- allele and the other possesses a normal I^+ allele. In such cells, both β-galactosidase and permease are inducible, even though the functional gene for one of these enzymes (Z^+ allele for β-galactosidase) is physically linked to the defective I gene (Table 21-2, line 3). We now know that this occurs because the one functional I gene present in the cell produces active repressor molecules that diffuse through the cytosol and bind to both operator sites in the absence of lactose. The repressor is therefore said to be a ***trans*-acting factor.**

Quite different results are obtained with partial diploids containing both the O^+ and O^c alleles (Table 21-2, line 4). In this case, structural genes linked to the O^c allele are constitutively transcribed (Z^+Y^-), whereas those linked to the wild-type allele are inducible (Z^-Y^+). The O locus, in other words, acts in *cis*; it affects the behavior of structural genes only in the operon of which it is physically a part. Such *cis* specificity is characteristic of mutations that affect binding sites on DNA rather than protein products of genes. Like other noncoding DNA sequences involved in the control of gene expression, the O site is said to be a ***cis*-acting element.**

The Genes Involved in Tryptophan Synthesis Are Organized into a Repressible Operon

Although much of the work leading to the initial formulation of the operon concept involved the *lac* operon of *E. coli*, a number of other bacterial regulatory systems are now known to follow the same general pattern—that is, genes coding for enzymes of a given metabolic pathway are clustered together in a group that serves as a unit of both transcription and regulation. One or more operators, promoters, and regulatory genes are usually involved, although there is sufficient variation from one operon to another to preclude many generalizations.

Operons coding for enzymes involved in catabolic pathways generally resemble the *lac* operon in being

Table 21-2 Diploid Analysis of Mutations Affecting the *lac* Operon

Line Number	Genotype of Diploid Bacterium*	Phenotype with Inducer Absent		Phenotype with Inducer Present	
		β-galactosidase	Permease	β-galactosidase	Permease
1	$I^+ P^+ O^+ Z^+ Y^+/I^+ P^+ O^+ Z^+ Y^+$	−	−	+	+
2	$I^+ P^+ O^+ Z^- Y^+/I^+ P^+ O^+ Z^+ Y^-$	−	−	+	+
3	$I^+ P^+ O^+ Z^- Y^+/I^- P^+ O^+ Z^+ Y^-$	−	−	+	+
4	$I^+ P^+ O^+ Z^- Y^+/I^+ P^+ O^c Z^+ Y^-$	+	−	+	+
5	$I^+ P^+ O^+ Z^- Y^+/I^s P^+ O^+ Z^+ Y^-$	−	−	−	−

*$P = P_{lac}$

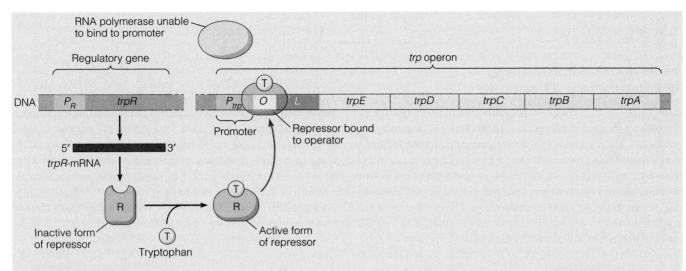

(a) Tryptophan present, repressor bound to operator, operon repressed

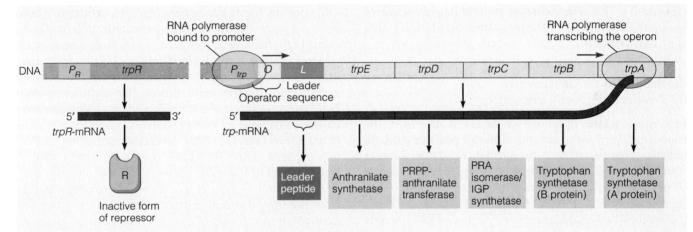

(b) Tryptophan absent, repressor not bound to operator, operon derepressed

Figure 21-6   The Tryptophan (*trp*) Operon of *E. coli.*   (a) The *trp* operon consists of a segment of DNA that includes five contiguous structural genes (*trpE, trpD, trpC, trpB,* and *trpA*) as well as promoter (*P*trp), operator (*O*), and leader (*L*) sequences. The structural genes are transcribed and regulated as a unit. The resulting polygenic message codes for the enzymes of the tryptophan biosynthetic pathway. The repressor protein, encoded by the *trpR* gene, is inactive (cannot recognize the operator site) in the free form. When complexed with tryptophan, the repressor is converted to the active form and binds tightly to the operator, thereby blocking access of RNA polymerase to the promoter and keeping the operon repressed. **(b)** In the absence of tryptophan, the repressor does not bind to the operator site. RNA polymerase is therefore able to bind to the promoter and transcribe the structural genes, giving the cell the capability to synthesize tryptophan. An additional role in regulating expression of this operon is played by the leader segment of the mRNA, as will be explained in Figures 21-8 and 21-9.

inducible; that is, they are turned *on* by a specific allosteric effector, usually the substrate for the pathway involved. In contrast, operons that regulate enzymes involved in anabolic (biosynthetic) pathways are **repressible operons;** they are turned *off* allosterically, usually by an effector that is the end-product of the pathway. The tryptophan (*trp*) operon is a good example of a repressible operon (Figure 21-6). The ***trp* operon** contains the structural genes coding for the enzymes that catalyze the reactions involved in tryptophan biosynthesis, as well as the DNA sequences necessary to regulate the production of these enzymes. The

effector molecule in this case is the end-product of the biosynthetic pathway, the amino acid tryptophan.

Expression of the enzymes produced by the *trp* operon is repressed in the presence of tryptophan (Figure 21-6a) and derepressed in its absence (Figure 21-6b). Thus, unlike the *lac* system, the regulatory gene for this operon, called *trpR,* codes for an allosteric repressor protein that is active (binds to operator DNA) when the effector is attached to it and that is inactive in its free form. The effector in such systems (in this case, tryptophan) is sometimes referred to as a **corepressor** because it is required, along with the repressor protein, to shut off transcription of the operon.

The *lac* and *trp* Operons Illustrate the Negative Control of Transcription

As we have seen, repressor proteins can control the transcription of genes involved in either catabolic or anabolic pathways; for catabolic pathways, the active (DNA-binding) form of the repressor is the *effector-free* repressor protein, whereas for anabolic pathways, the active form is the *effector-bound* repressor protein. But regardless of which is the active form of the repressor, the effect of repressor action is always the same: The active repressor prevents transcription of the operon by blocking the proper attachment of RNA polymerase to the DNA. Repressors, in other words, never turn anything on; their effect is always to turn off the expression of specific genes (or keep it turned off). This is therefore a system of **negative control** in that the active form of the repressor works by inhibiting gene transcription.

Catabolite Repression Illustrates the Positive Control of Transcription

In contrast to the situations we have just described, the transcription of some operons is under **positive control,** meaning that the active form of a key regulatory protein *turns on* expression of the operon. An important example of positive transcriptional control is found in **catabolite repression,** which refers to the ability of glucose to inhibit the synthesis of catabolic enzymes produced by inducible bacterial operons. To understand this phenomenon, we need to recognize that glucose is the preferred energy source for almost all prokaryotic cells (and for most eukaryotic cells, too). This is because the enzymes of the glycolytic and tricarboxylic acid (TCA) pathways are produced constitutively in most cells, so glucose can be catabolized at any time without the synthesis of additional enzymes. Although molecules other than glucose can also be metabolized as energy sources, catabolite repression guarantees that other carbon sources are used only when glucose is not available. For example, *E. coli* cells grown in the presence of both glucose and lactose use the glucose preferentially and have very low levels of the enzymes encoded by the *lac* operon, despite the presence of the inducer for that operon.

Like the other regulatory mechanisms we have encountered, this preferential use of glucose is made possible by a genetic control mechanism that involves an allosteric regulatory protein and a small effector molecule. The actual effector molecule that controls gene expression is not glucose, but a secondary signal that reflects the level of glucose in the cell. This secondary signal is a form of AMP called **cyclic AMP** or **cAMP** (see Figure 10-5). Glucose acts by indirectly inhibiting adenylyl cyclase, the enzyme that catalyzes the synthesis of cAMP from ATP. So the more glucose present, the less cAMP is made.

How does cAMP influence gene expression? Like other effectors, it acts by binding to an allosteric regulatory protein (Figure 21-7a). In this case the regulatory protein, called the **cAMP receptor protein (CRP),** is an *activator* of transcription. By itself, CRP is inactive, but when complexed with cAMP, CRP changes to an active shape that enables it to bind to a particular base sequence within operons that produce catabolic enzymes. This sequence, the *CRP recognition site,* is located upstream of the promoter. Figure 21-7b shows the location of the CRP recognition site (labeled *C*) in the *lac* operon; similar sites are found in a variety of inducible operons. When CRP, in its active form (i.e., the CRP-cAMP complex), attaches to its recognition site in DNA, the binding of RNA polymerase to the promoter is greatly enhanced, thereby stimulating the initiation of transcription. Thus, CRP has a *positive* effect on gene expression.

This phenomenon explains how glucose exerts its influence on the transcription of genes coding for catabolic enzymes. When the glucose concentration inside the cell is high, the cAMP concentration falls and hence CRP is largely in its inactive form. Therefore CRP cannot stimulate the transcription of operons that produce catabolic enzymes. In this way, cells turn off the synthesis of catabolic enzymes that are not needed when glucose is abundantly available as an energy source. Conversely, when the glucose level falls, the cAMP level rises, activating CRP by binding to it. The CRP-cAMP complex greatly enhances transcription of inducible operons, leading to the production of catabolic enzymes that allow cells to obtain energy from the breakdown of nutrients other than glucose. For example, transcription of the *lac* operon can be increased 50-fold in this way—provided, of course, that the repressor for that operon has been inactivated by the presence of its effector (i.e., allolactose).

Inducible Operons Are Often Under Dual Control

As we have just seen, inducible operons such as the *lac* operon are often subject to two types of control, rendering them sensitive to two kinds of signals. Negative control based on repressor-operator interactions allows the presence of an alternative energy source (such as lactose) to turn on a particular operon. And a positive control system, based on the action of CRP, makes transcription of the operon sensitive to the glucose concentration in the cell, as mediated by the cAMP level. An *E. coli* cell might therefore have very low levels of β-galactosidase either because it is growing in the absence of lactose and its *lac* operon is fully repressed, or because it has access to a large supply of glucose. In the latter case, the glucose will suppress the cAMP level, and CRP will therefore be inactive and incapable of stimulating transcription of the *lac* operon, even if the operon is otherwise derepressed by the presence of lactose.

To keep negative and positive types of transcriptional controls clear in your mind, ask about the primary effect of the regulatory protein that binds to the operon DNA. If, in binding to the DNA, the regulatory protein prevents or

Figure 21-7 The cAMP Receptor Protein (CRP) and Its Function.
CRP mediates catabolite repression by activating the transcription of various inducible operons in the presence of cyclic AMP (cAMP), whose concentration is in turn controlled by glucose. **(a)** CRP is an allosteric protein that is inactive in the free form but is converted to the active form by binding to cAMP. **(b)** The resulting CRP-cAMP complex binds at or near the promoter of a variety of inducible operons, including the *lac* operon, increasing the affinity of the promoter for RNA polymerase and thereby stimulating transcription. ① In the *lac* operon, the CRP-cAMP complex binds to the CRP recognition site (*C*) near the promoter region, thereby ② making the promoter more readily bound by RNA polymerase. ③ RNA polymerase binds to the promoter and ④ transcribes the operon.

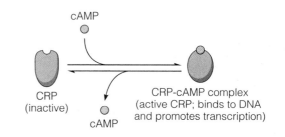

(a) Allosteric activation of CRP

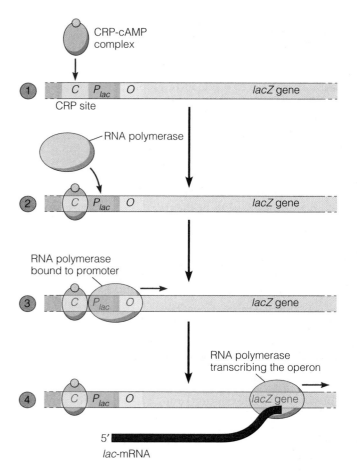

(b) Action of CRP-cAMP complex (active CRP)

turns off transcription, then it is part of a negative control mechanism. If, on the other hand, its binding to DNA results in the activation or enhancement of transcription, then the regulatory protein is part of a positive control mechanism.

Sigma Factors Can Regulate the Initiation of Transcription

In addition to regulatory proteins that activate or repress the transcription of specific operons, bacterial cells also make use of a variety of different sigma (σ) factors in controlling the initiation of transcription. Recall from Chapter 19 that the proper initiation of transcription in bacteria requires the RNA polymerase core enzyme to be combined with a protein called a sigma factor, which is necessary for recognition of the promoter. Each bacterial cell contains several types of sigma factors, each specialized for initiating the transcription of a particular set of genes.

The most common form of bacterial RNA polymerase contains a sigma factor called σ^{70} (molecular weight 70 kDa), which initiates transcription at most prokaryotic promoters. However, changes in a cell's environment, such as an increase in temperature (heat shock), can favor binding of an alternative sigma factor—in this case, σ^{32}—to the RNA polymerase core enzyme. With this sigma factor bound, promoter recognition is altered slightly, initiating transcription of a set of genes encoding proteins that help the cell adapt to the altered environment. Yet another sigma factor, σ^{54}, enables RNA polymerase to preferentially transcribe genes involved in nitrogen utilization. For both σ^{32} and σ^{54}, the promoter DNA sequence is somewhat different from the usual sequence, and its recognition by an alternative sigma factor enables the cell to respond to the changes in its environment. In addition, bacteriophages can take over the transcriptional machinery in cells they infect by coding for specific sigma factors that bind to the bacterial RNA polymerase and cause it to recognize only the viral promoters.

Attenuation Allows Transcription to Be Regulated After the Initiation Step

All the regulatory mechanisms discussed so far control the initiation of transcription. Prokaryotes also employ some

regulatory mechanisms that operate after the initiation step. One of the most important was discovered when Charles Yanofsky and his colleagues found that the *trp* operon has a novel type of negative regulatory site located between the promoter/operator and the first structural gene, *trpE*. This stretch of DNA, called the *leader sequence* (or *L*), is transcribed to produce a leader mRNA segment, 162 nucleotides long, located at the 5′ end of the polygenic *trp* mRNA (see Figure 21-6b).

Analysis of *trp* operon transcripts made under various conditions revealed that, as expected, the full-length, polygenic *trp* mRNA is transcribed when tryptophan is scarce. This allows the enzymes of the tryptophan

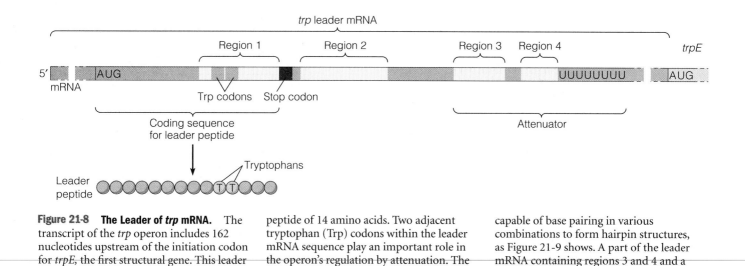

Figure 21-8 **The Leader of *trp* mRNA.** The transcript of the *trp* operon includes 162 nucleotides upstream of the initiation codon for *trpE*, the first structural gene. This leader mRNA includes a section encoding a leader peptide of 14 amino acids. Two adjacent tryptophan (Trp) codons within the leader mRNA sequence play an important role in the operon's regulation by attenuation. The leader mRNA also contains four regions capable of base pairing in various combinations to form hairpin structures, as Figure 21-9 shows. A part of the leader mRNA containing regions 3 and 4 and a string of eight U's is called the attenuator.

biosynthetic pathway to be synthesized, and hence the pathway can produce more tryptophan. On the other hand, when tryptophan is plentiful, the structural genes coding for the enzymes of the tryptophan pathway are not transcribed, also as expected. An unexpected result, however, was that the DNA corresponding to most of the leader sequence *is* transcribed under such conditions. Based on these findings, Yanofsky suggested that the leader sequence contains a control region that is sensitive to tryptophan levels. This control sequence somehow determines not whether *trp* operon transcription can begin, but whether it will continue to completion. The effect of this control element was called **attenuation** because of its role in attenuating, or reducing, the synthesis of mRNA.

The mechanism of attenuation is based on the tight coupling of transcription and translation in prokaryotes—the fact that protein synthesis begins before the mRNA is completed. To understand how attenuation works, we must start with a closer look at the *trp* operon leader mRNA segment (Figure 21-8). This leader has two unusual features that enable it to play a regulatory role. First, in contrast to the nontranslated leader sequences typically encountered at the 5′ end of mRNA molecules (see Figure 20-6), a portion of the *trp* leader sequence *is* translated, forming a *leader peptide* 14 amino acids long. Within the mRNA sequence coding for this peptide are two adjacent codons for the amino acid tryptophan; these will prove important. Second, the *trp* leader mRNA also contains four segments (labeled regions 1, 2, 3, and 4) whose nucleotides can base-pair with each other to form several distinctive hairpin loop structures. The region comprising regions 3 and 4 plus an adjacent string of eight U nucleotides is called the *attenuator*. When base pairing between regions 3 and 4 creates a hairpin, the attenuator acts as a transcription termination signal (Figure 21-9a). Recall that the typical prokaryotic termination signal

shown in Figure 19-12 is also a hairpin followed by a string of U's. The formation of such a structure causes RNA polymerase and the growing RNA chain to detach from the DNA.

As Yanofsky's experiments suggested, translation of the leader RNA plays a crucial role in the attenuation mechanism. A ribosome attaches to its first binding site on the *trp* mRNA as soon as the site appears, and from there it follows close behind the RNA polymerase. When tryptophan levels are low (Figure 21-9b), the concentration of tryptophanyl tRNA (tRNA molecules carrying tryptophan) is also low. Thus, when the ribosome arrives at the tryptophan codons of the leader RNA, it stalls briefly, awaiting the arrival of tryptophanyl tRNA. The stalled ribosome blocks region 1, allowing an alternative hairpin structure to form by pairing of regions 2 and 3. When region 3 is tied up in this way, it cannot pair with region 4 to create a termination structure, and so *the RNA polymerase continues*, eventually producing a complete mRNA transcript of the *trp* operon. Ribosomes use this mRNA to synthesize the tryptophan pathway enzymes, and production of tryptophan therefore increases.

If, however, tryptophan is plentiful and tryptophanyl-tRNA levels are high, the ribosome does not stall at the tryptophan codons (Figure 21-9c). Instead, the ribosome continues to the stop codon at the end of the coding sequence for the leader peptide and pauses there, blocking region 2. This pause permits the formation of the 3-4 hairpin, which is the transcription termination signal. *Transcription by RNA polymerase is therefore terminated* near the end of the leader sequence (after 141 nucleotides), and mRNAs coding for the enzymes involved in tryptophan synthesis are no longer produced.

Following the discovery of attenuation in the *trp* operon, attenuators have been found in numerous other operons, mainly those that code for enzymes involved in

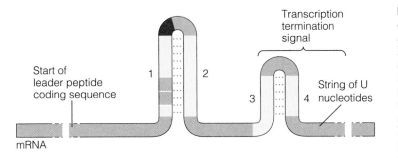

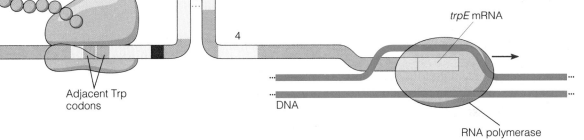

Start of leader peptide coding sequence

Transcription termination signal

String of U nucleotides

mRNA

(a) The most stable secondary structures for *trp* leader mRNA

Figure 21-9 Attenuation in the *trp* Operon. **(a)** Attenuation depends upon the ability of regions 1 and 2 and regions 3 and 4 of the *trp* leader sequence to base-pair, forming hairpin secondary structures. The 3-4 hairpin structure acts as a transcription termination signal; as soon as it forms, the RNA and the RNA polymerase are released from the DNA. **(b)** During periods of tryptophan scarcity, a ribosome translating the coding sequence for the leader peptide may stall when it encounters the two tryptophan (Trp) codons because of the shortage of tryptophan-carrying tRNA molecules. Because a stalled ribosome at this site blocks region 1, a 1-2 hairpin cannot form, and an alternative, 2-3 hairpin is created. The 2-3 base pairing prevents formation of the 3-4 transcription termination hairpin, and therefore RNA polymerase can move on to transcribe the entire operon. **(c)** When tryptophan is readily available, a ribosome can complete translation of the leader peptide without stalling. As it pauses at the stop codon, it blocks region 2, preventing it from base pairing. As a result, the 3-4 structure forms and terminates transcription near the end of the leader sequence.

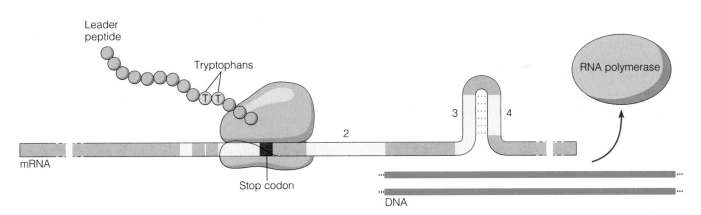

Partial leader peptide

Stalled ribosome

mRNA

Adjacent Trp codons

trpE mRNA

DNA

RNA polymerase

(b) When tryptophan is scarce, ribosome stalls, allowing a 2-3 hairpin to form; RNA polymerase continues transcription

Leader peptide

Tryptophans

mRNA

Stop codon

DNA

RNA polymerase

(c) When tryptophan is plentiful, ribosome continues, and RNA polymerase stops transcribing

amino acid biosynthesis. For some operons, attenuation appears to be the only means of regulation; for others, the attenuator site complements the operator in the regulation of gene expression, as with the *trp* operon. In the latter case, the repressor-effector complex binds to the operator when the intracellular concentration of the relevant amino acid is high, thereby blocking initiation of transcription. As the concentration of effector decreases, the operon becomes derepressed and transcription begins. Fine-tuning of the system then occurs at the attenuator site, allowing greater numbers of RNA polymerase molecules to proceed past the attenuator as the effector becomes scarcer.

While the tight coupling of transcription and translation required for attenuation is the norm in prokaryotic cells, this is not the case in eukaryotes because DNA is transcribed in the nucleus and translation is mainly a cytoplasmic process. It appears, then, that attenuation is restricted to prokaryotes.

Comparison of Gene Regulation in Prokaryotes and Eukaryotes

In the early days of molecular biology, a popular adage claimed that "what is true of *E. coli* is also true of elephants." This maxim expressed the initial conviction of many microbial geneticists that almost everything learned about bacterial function at the molecular level would also be applicable to eukaryotes. Not surprisingly, however, the adage has turned out to be only partly true. In terms of basic metabolic pathways, mechanisms for the transport of solutes across membranes, and such fundamental features as DNA structure, protein synthesis, and enzyme function, there are many similarities between the prokaryotic and eukaryotic worlds, and findings from studies with prokaryotes can be extrapolated to eukaryotes with considerable confidence. But when the discussion turns to the regulation of gene expression, the comparison needs to be scrutinized.

There are certainly basic similarities between prokaryotic and eukaryotic regulation of gene expression. Chief among these is the importance of transcription initiation as a key control point and the general manner in which control at that point is effected. In eukaryotes as well as prokaryotes, many genes are controlled by regulatory proteins (*trans*-acting factors) that interact with specific DNA sequences (*cis*-acting elements) to turn transcription on or off.

However, there are also some dramatic differences between prokaryotic and eukaryotic gene regulation. First, transcriptional regulation is more elaborate in eukaryotes, involving many more regulatory proteins and DNA control elements. Moreover, while some eukaryotic regulatory proteins interact directly with DNA (as do most prokaryotic regulatory proteins), many others exert their effects on DNA indirectly, via interactions with other proteins. In addition, eukaryotes have an elaborate series of mechanisms for *posttranscriptional control*—that is, control at a variety of points in the flow of genetic information from a transcribed gene in the nucleus to an active enzyme (or other gene product) in the cytosol or elsewhere in the cell. Although prokaryotes also exhibit some of these posttranscriptional mechanisms, such control is more prevalent in eukaryotes.

In this section, we will summarize some of the major differences between eukaryotic and prokaryotic gene regulation; in the following sections, we will then examine eukaryotic regulation in detail.

Prokaryotes and Eukaryotes Exhibit Many Differences in Genome Organization and Expression

The greater diversity of regulatory mechanisms in eukaryotic cells reflects basic differences between prokaryotes and eukaryotes in the structure, organization, and compartmentalization of the genome, as well as some fundamental features of the multicellular way of life. We will now review some of these differences, focusing on those that are especially relevant to the regulation of gene expression.

Genome Size and Complexity. As discussed in Chapter 16, eukaryotic genomes are almost always much larger than those of prokaryotes—often by two or three orders of magnitude, or more. For most eukaryotes, genes constitute only a small portion of the genome, leaving the role of the bulk of the DNA, including most of the repeated sequences, unexplained. It is thought that at least some of this "extra" DNA might have regulatory functions, but what these functions might be remains obscure.

In addition to the noncoding DNA located between genes, the noncoding DNA introns located *within* most eukaryotic genes are another genomic feature that may be involved in regulation. Especially intriguing is the finding that the coding sequences of genes—exons—are, in some cases, spliced together in varying combinations to generate alternative mRNAs, which in turn are translated into related, but not identical, proteins. RNA splicing was described in Chapter 19. We will consider an example of alternative splicing later in this chapter.

Genomic Compartmentalization. Another fundamental difference between prokaryotes and eukaryotes is that most of the genetic information of eukaryotes is located in the nucleus, separated from the cytoplasmic sites of protein synthesis by the nuclear envelope. Transcription and translation in eukaryotic cells are therefore separated in both time and space.

This separation has implications for gene regulation because it allows for selectivity in the processing and transport of nuclear RNAs. It is clear, for example, that primary transcripts are extensively modified, cleaved, and spliced in the nucleus, and that most of the RNA sequences produced in the nucleus are degraded there without ever reaching the cytoplasm. Thus, the nuclear envelope may serve as a barrier that screens transcripts for selective passage out of the nucleus. On the other hand, the separation of transcription and translation makes control by attenuation impractical in eukaryotes, since the mechanism of attenuation requires close physical proximity between DNA, RNA polymerase, and ribosomes.

Structural Organization of the Genome. The chromosomes of eukaryotic cells differ from bacterial chromosomes in both chemical composition and structure. As you learned in Chapter 16, the DNA of eukaryotic chromosomes is intimately associated with histone and nonhistone proteins and is highly folded, with several successive levels of structural organization. Bacterial DNA is complexed with small basic proteins that may interact with DNA in much the same way as histones do in eukaryotic chromatin. However, prokaryotic DNA folding is not based on nucleosomes, the basic packaging units of eukaryotic chromosomes, and the successive levels of DNA compaction seen in eukaryotes are not observed in prokaryotes. Because eukaryotic DNA is packaged in a more elaborate way, it is reasonable to hypothesize that eukaryotic transcription is regulated at two stages: the uncoiling of appropriate chromosomal regions, and the binding of

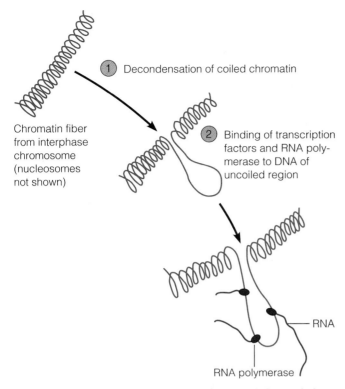

Figure 21-10 Two Stages of Regulation of Eukaryotic Transcription. Transcription in eukaryotes is probably regulated in two stages. ① Localized changes in chromatin structure cause selected regions of the chromatin to decondense. The uncoiled chromatin loops out. ② The DNA is then accessible for the binding of transcription factors and RNA polymerase. Transcription follows.

Within the figure:
① Decondensation of coiled chromatin

Chromatin fiber from interphase chromosome (nucleosomes not shown)

② Binding of transcription factors and RNA polymerase to DNA of uncoiled region

RNA

RNA polymerase

This striking difference can perhaps best be understood in light of what we might call the "lifestyles" of these organisms. Prokaryotes are unicellular organisms with little or no assurance of environmental constancy and must therefore be able to adapt readily to new conditions. Consistent with this need is a short life span for most mRNA molecules and rapid transcriptional regulation via effector molecules and allosteric regulatory proteins. The effector molecules are almost always small molecules such as sugars and amino acids—the very environmental factors to which the cells must respond. Prokaryotic cells can rapidly change their population of mRNAs, and therefore the spectrum of proteins they synthesize, in response to changes in the environmental concentration of these kinds of molecules.

In contrast, many eukaryotic cells are constituents of multicellular organisms and can rely on a much more predictable environment within the organism. Moreover, cells of multicellular eukaryotes are often quite specialized in function, reflecting the division of labor characteristic of the multicellular way of life. Such cells are therefore committed to the synthesis of a relatively small number of gene products required for carrying out that cell's specialized functions. A long-term commitment to the synthesis of a predictable subset of proteins allows eukaryotic cells to use mRNA molecules that are generally longer-lived than their prokaryotic counterparts. Nevertheless, there are significant variations in the stability of different kinds of eukaryotic mRNAs, even within the same cell. And for single-celled eukaryotes such as yeasts, as for bacteria, environmental conditions are of paramount importance, so the mRNAs of these organisms tend to be short-lived.

transcription factors and RNA polymerase to the uncoiled DNA (Figure 21-10).

Genetic studies and DNA sequencing have revealed that eukaryotic genes are almost never clustered into operon-like units of transcription and regulation. Instead, eukaryotes employ a different strategy for coordinating gene transcription. We will see shortly that in addition to its own promoter, each eukaryotic gene in a coordinately transcribed group has its own regulatory DNA elements nearby. These sequences, called *response elements,* allow transcription to be turned on or off in response to various types of signals. By placing the same response element next to genes residing at different chromosomal locations, transcription of the entire group of genes can be turned on and off together by the same regulatory protein even though the genes are not clustered together in an operon.

Stability of Messenger RNA. In Chapter 19, we mentioned that the mRNA molecules of prokaryotes and eukaryotes differ in turnover rate—that is, in their average lifetime in the cell before degradation. Although the values vary among mRNA molecules, the average life span of prokaryotic mRNA molecules is about three minutes, whereas the mRNAs of eukaryotes—especially multicellular eukaryotes—persist much longer, often for several hours or days.

Posttranslational Modification of Proteins. As we discussed in Chapter 20, proteins may be extensively modified after polypeptide synthesis, especially in eukaryotic cells. For example, some eukaryotic proteins must have blocks of amino acids removed before the protein is biologically active; examples include mammalian insulin and many digestive enzymes. Also, signal sequences that target proteins to specific cellular compartments must often be removed. Other important protein modifications include the addition of prosthetic groups, the control of protein activity by phosphorylation and dephosphorylation, and the glycosylation of proteins destined for secretion or integration into the plasma membrane. The wide array of protein modification mechanisms used by eukaryotes reflects the elaborate structure and specialized functions of eukaryotic cells, as compared with prokaryotes, and all offer potential opportunities for regulation.

Protein Turnover. Another difference that arises from the contrasting lifestyles of prokaryotes and eukaryotes is the relative importance of protein degradation as a means of eliminating proteins no longer needed by a cell. Prokaryotes have enzymes that can degrade defective or unneeded proteins, but this is not the only option. Because most prokaryotic cells are continually growing and dividing, they can get rid of unwanted proteins by simply allowing them to be diluted out by successive cell divisions. As a result, prokaryotic cells are not generally dependent on protein degradation as a means of eliminating proteins.

Eukaryotic cells, on the other hand, are more likely to stop dividing and to persist as nondividing, specialized cells for long periods of time thereafter. The nerve cells in your body are an especially good example: Most of them are as old as you are and will never divide again. Clearly, such cells cannot depend on dilution by division to get rid of unnecessary or defective proteins. Instead, they possess specific mechanisms for degrading proteins selectively. The regulated degradation and replacement of proteins, called *protein turnover,* is a more prominent feature of eukaryotic cells than of prokaryotes. Thus, protein degradation is yet another potential level of regulation for eukaryotes.

Stem Cells Give Rise to Specialized Eukaryotic Cell Types by the Process of Cell Differentiation

The nerve cells mentioned in the preceding example illustrate another fundamental difference between eukaryotes and prokaryotes: Multicellular eukaryotes consist of complex mixtures of specialized or "differentiated" cell types—e.g., nerve, muscle, bone, blood, cartilage, and fat—brought together in various combinations to form tissues and organs. Differentiated cells can be distinguished from each other by differences in their microscopic appearances and in the products they manufacture. For instance, red blood cells synthesize hemoglobin, nerve cells produce neurotransmitters, and lymphocytes make antibodies. Such differences indicate that the control of gene expression must play a key role in the mechanism responsible for creating differentiated cells.

Differentiated cells arise from populations of immature, nonspecialized cells called **stem cells** by the process of **cell differentiation**. Stem cells are defined by their capacity for unlimited division and their ability, in the presence of the appropriate signals, to differentiate into a variety of other cell types, usually accompanied by cessation of cell division. The prime example occurs in embryos, whose embryonic stem cells differentiate into all the cell types that make up the adult organism. But stem cells also reside in adult tissues. For instance, stem cells present in bone marrow differentiate into all the cell types found in the blood; more surprising is the recent claim that bone marrow stem cells may also be able to differentiate into other kinds of cells, including bone, cartilage, muscle, and brain.

Our rapidly advancing ability to isolate stem cells and direct their differentiation into other cell types raises the possibility that scientists may eventually be able to grow healthy new cells to replace the damaged tissue found in patients with various diseases. For example, stem cells that can differentiate into nerve cells might be used to repair the brain damage that occurs in patients suffering from stroke, Parkinson's disease, or Alzheimer's disease. Likewise, stem cells might be utilized to replace the defective pancreatic cells of patients with diabetes or the defective skeletal muscle cells present in individuals with muscular dystrophy.

Eukaryotic Gene Expression Is Regulated at Multiple Levels

The differences between prokaryotes and eukaryotes described in the preceding sections underscore the impossibility of explaining eukaryotic regulation of gene expression entirely in terms of known prokaryotic mechanisms—elephants are not just large *E. coli* after all! If we are ever to have a thorough understanding of regulation in eukaryotic cells, we must approach the topic from a eukaryotic perspective.

Gene expression in eukaryotic cells, ultimately measured in terms of the activities of gene products, is the culmination of controls acting at several different levels. Figure 21-11 traces the flow of genetic information from genomic DNA in a eukaryotic nucleus to functional proteins in the cell's cytoplasm, indicating a number of potential control points. As you can see, there are five main levels at which control might be exerted: ① the genome, ② transcription, ③ RNA processing and export from nucleus to cytoplasm, ④ translation, and ⑤ posttranslational events. Regulatory mechanisms in the last three categories are all examples of *posttranscriptional control,* a term that encompasses a wide variety of different processes. In the remainder of the chapter, we will examine each of the five levels in turn.

Eukaryotic Gene Regulation: Genomic Control

Eukaryotic gene regulation almost always involves the selective expression of a specific subset of genes from the same complete genome present in every cell, rather than any cell-specific changes in the genome itself. In this section we begin by examining evidence that all the cells of a multicellular organism usually contain the same set of genes, followed by a discussion of some of the mechanisms used to regulate the structural organization of the genome.

As a General Rule, the Cells of a Multicellular Organism All Contain the Same Set of Genes

In multicellular plants and animals, each specialized cell type expresses only a small fraction of the total number of genes present in the genome. Yet virtually all cells other than the haploid gametes (sperm and eggs) still contain the same exact, complete set of genes. For animals, evidence that even highly specialized cells carry a full complement of genes was first provided by John Gurdon and his colleagues. In studies with *Xenopus laevis,* the African

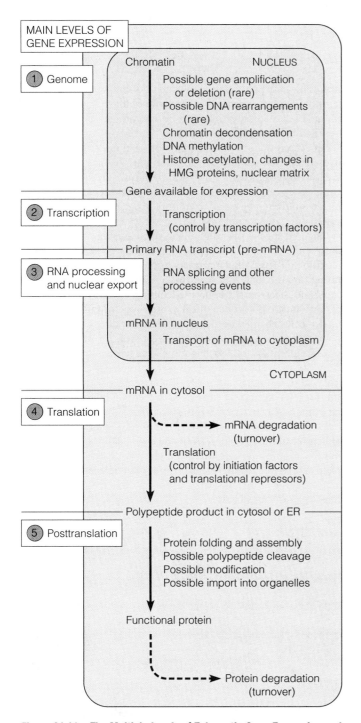

MAIN LEVELS OF
GENE EXPRESSION

NUCLEUS

① Genome

Chromatin

Possible gene amplification
or deletion (rare)
Possible DNA rearrangements
(rare)
Chromatin decondensation
DNA methylation
Histone acetylation, changes in
HMG proteins, nuclear matrix

Gene available for expression

② Transcription

Transcription
(control by transcription factors)

Primary RNA transcript (pre-mRNA)

③ RNA processing
and nuclear export

RNA splicing and other
processing events

mRNA in nucleus

Transport of mRNA to cytoplasm

CYTOPLASM

mRNA in cytosol

④ Translation

mRNA degradation
(turnover)

Translation
(control by initiation factors
and translational repressors)

Polypeptide product in cytosol or ER

⑤ Posttranslation

Protein folding and assembly
Possible polypeptide cleavage
Possible modification
Possible import into organelles

Functional protein

Protein degradation
(turnover)

Figure 21-11   The Multiple Levels of Eukaryotic Gene Expression and Regulation.   Gene expression can be regulated by influencing any of the events that occur within any of these levels. ① The genome level, including rarely occurring amplification or rearrangement of DNA segments, chromatin decondensation (and condensation), and DNA methylation. ② Transcription, where critical control mechanisms determine which genes are active at a given time. ③ Processing of RNA and its export from the nucleus. ④ Translation, the synthesis of polypeptides. This level includes targeting of some newly forming polypeptides to the ER. ⑤ Posttranslational events, including polypeptide folding and assembly, polypeptide cleavage, modifications of polypeptides by the addition of chemical groups, and the import of proteins into organelles (and secretion of some from the cell). Degradation of mRNA and proteins are also subject to regulation.

clawed frog, they transplanted nuclei from differentiated tadpole cells into unfertilized eggs that had been deprived of their own nuclei. Although the frequency of success was low, some eggs containing transplanted nuclei gave rise to viable, swimming tadpoles. A new organism created by this process of nuclear transplantation is said to be a **clone** of the organism from which the original nucleus was taken, since the cells of the new organism all contain nuclear DNA derived from cells of the original organism. The results of these studies indicated that nuclei taken from differentiated cells can direct the development of an entire new organism. Such a nucleus is therefore said to be *totipotent*: It contains the complete set of genes needed to create a new organism of the same type as the organism from which the nucleus was taken.

An especially dramatic example of animal cloning was reported in 1997, when Ian Wilmut and his colleagues in Scotland made newspaper headlines by reporting the birth of a cloned lamb, Dolly—the first animal ever cloned from a cell derived from an adult. Dolly was born from a sheep egg whose original nucleus had been replaced by a nucleus taken from a single cell of an adult sheep. Comparable successes in animal cloning have subsequently been reported by other investigators using mice and cattle. As we discuss in Box 21A, animal cloning is a remarkable feat that raises a series of profound questions about the future applications of such technology.

The ability to produce entire new organisms using genetic information derived from single differentiated cells has also been demonstrated in plants. In an approach pioneered by Frederick Steward, new plants can be created by mitotic division of differentiated cells that have been removed from mature plants. For example, you can take a piece of carrot, cut it into small fragments, and isolate cells that are then placed on a nutrient medium and grown in a test tube. Single cells isolated under these conditions eventually grow into complete carrot plants with normal roots, shoots, and leaves. Thus the nuclei of differentiated plant cells do not require transplantation into an egg to express totipotency. Expression of totipotency is triggered simply by freeing the cells from their normal contacts with neighboring cells. This method for creating plant clones is of great commercial interest because it provides a reliable way of reproducing agricultural plant strains that exhibit desirable genetic traits without the genetic variability inherent in sexual reproduction.

Gene Amplification and Deletion Can Alter the Genome

Although the preceding evidence indicates that the genome tends to be the same in all cells of an adult eukaryotic organism, a few types of gene regulation create exceptions to this rule. One example is **gene amplification,** the selective replication of certain genes. Gene amplification can be regarded as an example of **genomic control**—that is, a regulatory change in the makeup or structural organization of the genome.

Further Insights DOLLY: A LAMB WITH NO FATHER

In January of 1997, who would have expected that the most dramatic scientific breakthrough of the year was comfortably resting in a barn? But the following month, Ian Wilmut made newspaper headlines around the world by introducing us to Dolly, the first animal ever cloned from an adult cell. Dolly was created by removing the nucleus from the cell of an adult sheep and transferring it into a different sheep's egg whose own nucleus had been removed.

Although this *nuclear transfer* technique had been used before to clone animals, it had never been successful with cells taken from an adult. Early studies in frogs revealed that nuclei removed from embryonic cells or tadpoles can program the development of a normal adult frog when transferred into a frog egg. However, when donor nuclei from adult cells were used, development never proceeded beyond the tadpole stage.

Wilmut suspected that these previous failures were caused by the active state of the chromatin in the donor cells. The trick, he said, is to make the DNA of the donor cells behave more like the inactive chromatin of a typical sperm cell that would normally fertilize an egg. His research team accomplished this by taking mammary gland cells from the udder of a 6-year-old female sheep and starving them in culture to force them into the dormant, G0 phase of the cell cycle. This caused many genes to turn off and ensured that the cells remained diploid (unlike cells in S or G2, which have replicated some or all of their DNA). When a nucleus from such donor cells was transplanted into an egg cell lacking a nucleus, it delivered the diploid amount of DNA in a condition that allowed the egg cytoplasm to reprogram the DNA to support normal embryonic development. Implanting such an egg into the uterus of another female sheep led to the birth of a lamb with no father—that is, a lamb whose cells had the same nuclear DNA as the cells of the 6-year-old female sheep that provided the donor nucleus (Figure 21A-1).

Within a few years of the first reports of these pioneering experiments, other investigators had used similar techniques to clone several other kinds of animals, including cattle, mice, goats, and pigs. But what is the practical value of such technology? One possibility is to insert potentially useful genes into the donor cells prior to transferring their nuclei into eggs for cloning. For example, investigators are using recombinant DNA techniques to introduce genes for medically important proteins, such as human blood-clotting factors, that are difficult to produce by other means. This approach has already been used to clone sheep that produce milk containing the blood-clotting factor deficient in people with hemophilia. Thus it is possible to envision in the not-too-distant future a new form of farming in which cloning is used to create herds of identical animals that are used not for milk or meat but as sources of medically important human proteins, and perhaps eventually even tissues and organs.

An even more dramatic application of cloning technology involves current attempts to clone animals that are on the verge of extinction or have recently become extinct. In theory, all it would take is a few cells from an endangered species, whose nuclei could then be transferred into an egg cell of a closely related animal. Attempts are already underway to clone an endangered species of Indian bison (gaur), using normal cows to carry the embryos, and to clone a recently extinct Pyrenees bucardo goat, using normal goats to carry the embryos. Even more dramatic than this, scientific advisory committees, politicians, and the general public are debating the question of whether it would ever be ethical to attempt cloning with human cells. For the moment, producing healthy human clones may prove to be exceedingly difficult (it took 277 attempts to produce Dolly), so society has some time to contemplate this issue. But in the face of nearly universal unease about the prospect of human cloning, Dolly's birth has already sparked calls for laws banning the use of such technology for duplicating human beings. And yet some bioethicists have suggested that society might eventually find cloning acceptable under certain circumstances, such as cloning a dying child or helping an infertile couple have a child. Although such discussions may make people uncomfortable, the rapid pace of scientific developments in this field makes it essential that society not shy away from the debate.

Figure 21A-1   Dolly, the First Animal Cloned from an Adult Cell.

One of the best-studied examples of gene amplification involves the ribosomal RNA genes in *Xenopus laevis*, the organism used by Gurdon in his nuclear transplantation experiments. The haploid genome of *Xenopus* normally contains about 500 copies of the genes that code for 5.8S, 18S, and 28S rRNA. During oogenesis (development of the egg prior to fertilization), the DNA of this entire set of genes is selectively replicated about 4000-fold, so that the mature oocyte contains about 2 million copies of the genes for rRNA. Apparently, this level of amplification is necessary to accommodate the enormous amount of ribosome biosynthesis that must take place during oogenesis, which in turn is required to sustain the high rate of protein synthesis needed for early embryonic development.

The extra rRNA gene copies created by gene amplification are present in extrachromosomal circles of DNA distributed among hundreds of nucleoli that appear in the oocyte nucleus as amplification progresses (Figure 21-12). Note that this example of gene amplification involves genes whose products are *RNA* rather than protein molecules. The expression of genes that encode proteins—even proteins needed in large amounts, such as ribosomal proteins—can usually be increased sufficiently by increasing translation of the mRNA, because each mRNA molecule can be translated numerous times.

In addition to amplifying gene sequences whose products are in great demand, some cells also delete genes whose products are not required. An extreme example of **gene deletion** (also called *DNA diminution*) occurs in mammalian red blood cells, which discard their nuclei entirely after adequate amounts of hemoglobin mRNA have been made. A less extreme example occurs in a group of tiny crustaceans known as copepods. During the embryonic development of copepods, the heterochromatic (transcriptionally inactive) regions of their chromosomes are excised and discarded from all cells except those destined to become gametes. In this way, up to half of the organism's total DNA content is removed from its body cells.

DNA Rearrangements Can Alter the Genome

A few cases are known in which gene regulation is based on the movement of DNA segments from one location to another within the genome, a process known as **DNA rearrangement.** Two particularly interesting examples involve the mechanism used by yeast cells to control mating and the mechanism used by vertebrates to produce millions of different antibodies.

Yeast Mating-Type Rearrangements. In the yeast *Saccharomyces cerevisiae*, mating occurs when haploid cells of two different mating types, called α and *a*, fuse together to form a diploid cell. All haploid cells carry both alleles for

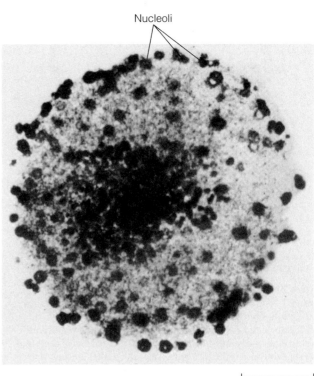

Nucleoli

100 μm

Figure 21-12 Amplification of rRNA Genes in an Amphibian Oocyte. This light micrograph shows a nucleus isolated from a *Xenopus* oocyte, stained to reveal the many nucleoli that are formed during oogenesis by the amplification of genes for ribosomal RNA. Each nucleolus contains multiple copies of the rRNA genes, present as extrachromosomal DNA circles.

mating type; however, a cell's actual mating phenotype depends on which of the two alleles, α or *a*, is present at a special site in the genome called the **MAT locus.** Cells frequently switch mating type, presumably as a means of maximizing opportunities for mating. They do so by moving the alternative allele into the *MAT* locus. This process of DNA rearrangement is called the **cassette mechanism,** because the mating-type locus is like a tape deck into which either the α or the *a* "cassette" (allele) can be inserted and "played" (transcribed).

Figure 21-13 describes the yeast cassette mechanism in more detail. The *MAT* locus, containing either the α or *a* allele, is located on yeast chromosome 3, approximately midway between extra copies of the two alleles. The locus that stores the extra copy of the α allele is called *HMLα*; the locus with the extra copy of the *a* allele is called *HMRa*. In switching mating type, a yeast cell makes a DNA copy of the other mating-type allele, either *HMLα* or *HMRa*, and inserts this new "cassette" into the *MAT* locus. Before the new cassette can be inserted, however, the old DNA cassette at *MAT* must be excised (by a site-specific endonuclease) and discarded. Unlike the DNA

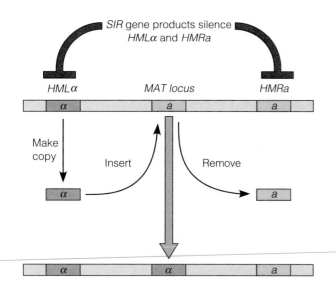

SIR gene products silence
HMLα and HMRa

HMLα   MAT locus   HMRa

Make
copy

Insert   Remove

α   α   a

Figure 21-13   The Cassette Mechanism of the Yeast Mating-Type Switch. Chromosome 3 of *Saccharomyces cerevisiae* contains three copies of the mating-type information. The *HMLα* and *HMRa* loci contain complete copies of the α and *a* forms of the gene, respectively, but the transcription of these loci is inhibited by the products of the *SIR* gene. The cell's actual mating type is determined by the allele present at the *MAT* locus. When a cell switches mating types, the α or *a* DNA at the *MAT* locus is removed and replaced by a DNA copy of the alternative mating-type DNA. As an example, this figure illustrates a switch in mating type from *a* to α.

sequence at the *MAT* locus, the DNA sequences at *HMLα* and *HMRa* never change (except by rare mutations).

The DNA of each mating-type allele actually encodes several different proteins, including secretory proteins and cell-surface receptors. It is these proteins, encoded by the allele inserted at the *MAT* locus, that give the cell either an "α" or an "*a*" mating phenotype. But the presence of extra copies of the alleles at *HMLα* and *HMRa* raises an important question: If the cell contains complete copies of *both* the α and the *a* alleles at these locations, why aren't both sets of proteins made? The answer is that a set of regulatory genes, known as the *silent information regulator (SIR) genes,* act together to prevent expression of the genetic information at *HMLα* and *HMRa*. The proteins encoded by the *SIR* genes block transcription of *HMLα* and *HMRa* by binding to specific DNA sequences that surround the α and *a* DNA cassettes at *HMLα* and *HMRa*.

Antibody Gene Rearrangements. A somewhat different type of DNA rearrangement is used by lymphocytes of the vertebrate immune system for producing antibody molecules. Antibodies are proteins composed of two kinds of polypeptide subunits, called *heavy chains* and *light chains.* Vertebrates make millions of different kinds of antibodies, each produced by a different lymphocyte (and its descendants) and each capable of specifically recognizing and binding to a different foreign molecule. But this enormous diversity of antibody molecules creates a potential problem: If every antibody molecule were to be encoded by a

different gene, virtually all of a person's DNA would be occupied by the millions of required antibody genes.

Lymphocytes get around this problem by starting with a relatively small number of different DNA segments and rearranging them in various combinations to produce millions of unique antibody genes, each one formed in a different, developing lymphocyte. The rearrangement process involves four kinds of DNA sequences, called *V, J, D,* and *C segments.* The C segment codes for a heavy or light chain *constant region* whose amino acid sequence is the same among different antibodies; the V, J, and D segments together code for *variable regions* that differ among antibodies and give each one the ability to recognize and bind to a specific type of foreign molecule.

To see how this works, let's consider human antibody heavy chains, which are constructed from roughly 200 kinds of V segments, more than 20 kinds of D segments, and at least 6 kinds of J segments. As shown in Figure 21-14, the DNA regions containing the various V, D, and J segments are rearranged during lymphocyte development to randomly bring together one V, one D, and one J segment in each lymphocyte. This random rearrangement allows the immune system to create at least $200 \times 20 \times 6 = 24,000$ different kinds of heavy chain variable regions. In a similar fashion, thousands of different kinds of light chain variable regions can also be created (light chains are constructed from their own types of V, J, and C segments; they do not use D segments). Finally, any one of the thousands of different kinds of heavy chains can be assembled with any one of the thousands of different kinds of light chains, creating the possibility of millions of different types of antibodies. The net result is that millions of different antibodies are produced from the human genome by rearranging a few hundred different kinds of V, D, J, and C segments.

The DNA rearrangement process that creates antibody genes also activates transcription of these genes via a mechanism involving special DNA sequences called *enhancers,* which, as we will discuss shortly, increase the rate of transcription initiation. Enhancers are located near DNA sequences coding for C segments, but a promoter sequence is not present in this area and thus transcription does not normally occur. The promoter for gene transcription is located upstream from the DNA coding for V segments, but it is not efficient enough to promote transcription in the absence of an enhancer sequence. Hence prior to DNA rearrangement, the promoter and enhancer sequences of an antibody gene are so far apart that transcription does not occur; only after rearrangement are they close enough for transcription to be activated.

Chromosome Puffs Provide Visual Evidence That Chromatin Decondensation Is Involved in Genomic Control

We encounter another aspect of genome-level control when we consider what is involved in making the eukaryotic genome—that is, chromosomal DNA—accessible to

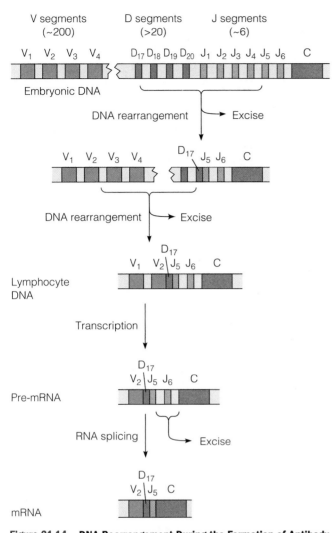

V segments (~200)

D segments (>20)

J segments (~6)

V₁ V₂ V₃ V₄ — D₁₇ D₁₈ D₁₉ D₂₀ J₁ J₂ J₃ J₄ J₅ J₆ C

Embryonic DNA

DNA rearrangement → Excise

V₁ V₂ V₃ V₄ D₁₇ J₅ J₆ C

DNA rearrangement → Excise

D₁₇
V₁ V₂ J₅ J₆ C

Lymphocyte DNA

Transcription

D₁₇
V₂ J₅ J₆ C

Pre-mRNA

RNA splicing → Excise

D₁₇
V₂ J₅ C

mRNA

Figure 21-14   DNA Rearrangement During the Formation of Antibody Heavy Chains. Genes coding for the human antibody heavy chains are created by DNA rearrangements involving multiple types of V, D, and J segments. In this example, an initial DNA excision randomly removes several D and J segments, bringing D_{17} adjacent to J_5. A second random excision removes several V and D segments, bringing V_2 adjacent to D_{17}. After transcription, the sequences separating the $V_2D_{17}J_5$ segment from the C segment are removed by RNA splicing.

the cell's transcription machinery. Recall from Chapter 19 that, to initiate transcription, a eukaryotic RNA polymerase must interact with both DNA and a number of specific proteins (general transcription factors) in the promoter region of a gene. Except when a gene is being transcribed, its promoter region is embedded within a highly folded and ordered chromatin superstructure. Thus, some degree of chromatin decondensation (unfolding) appears to be necessary for the expression of eukaryotic genes.

The earliest evidence that chromatin decondensation is required for gene transcription came from direct microscopic visualization of certain types of insect chromosomes caught in the act of transcription. Because the DNA of most eukaryotic cells is dispersed throughout the nucleus as a mass of intertwined chromatin fibers, it is usually difficult to observe the transcription of individual genes with a microscope. But a way around this obstacle is provided by an unusual type of insect cell. In the fruit fly *Drosophila melanogaster* and related insects, metabolically active tissues such as the salivary glands and intestines grow by an increase in the size of, rather than the number of, their constituent cells. This process generates giant cells whose volumes are thousands of times greater than normal. The development of giant cells is accompanied by successive rounds of DNA replication; but because this replication occurs in cells that are not dividing, the newly synthesized chromatids accumulate in each nucleus and line up in parallel to form multi-stranded structures called **polytene chromosomes.** Each polytene chromosome contains the multiple chromatids generated during replication of both members of each homologous chromosome pair. The four giant polytene chromosomes found in the salivary glands of *Drosophila* larvae, for example, are generated by ten rounds of chromosome replication, and therefore each has 1024 (2^{10}) chromatids aligned in lateral register for each homologue—a total of 2048 chromatids in all!

Polytene chromosomes are thus enormous structures measuring hundreds of micrometers in length and several micrometers in width—roughly ten times longer and a hundred times wider than the metaphase chromosomes of typical eukaryotic cells. The micrograph in Figure 21-15 shows several polytene chromosomes from the nucleus of a *Drosophila* salivary gland cell. Visible in each polytene chromosome is a characteristic pattern of dark bands. Each band represents a chromatin domain that is highly condensed compared with the chromatin in the "interband" regions between the bands. Activation of the genes of a given chromosome band causes the compacted chromatin strands to uncoil and expand outward, resulting in a **chromosome puff.** Figure 21-16 shows a clearer view of a puff-containing region of a polytene chromosome. Such puffs consist of DNA loops that are less condensed than the DNA of bands elsewhere in the chromosome. Though puffs are not the only sites of gene transcription along the polytene chromosome, the extent of chromosome decondensation at puffs correlates well with the enhancement of transcriptional activity at these sites.

As the *Drosophila* larva proceeds through development, each of the polytene chromosomes in salivary gland nuclei undergoes reproducible changes in puffing patterns, under the control of an insect steroid hormone called *ecdysone*. This hormone functions by binding to, and thus activating, a regulatory protein that stimulates the transcription of certain genes. (This is similar to the action of vertebrate steroid hormones, as will be discussed later in the chapter.) It appears, in other words, that the characteristic puffing patterns seen during the development of *Drosophila* larvae are direct visual manifestations of the selective decondensation and transcription of specific DNA segments according to a genetically determined developmental program.

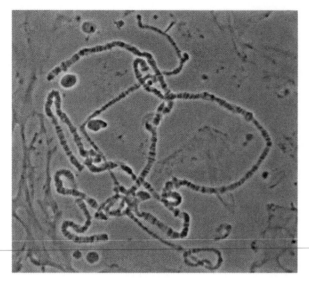

(a)

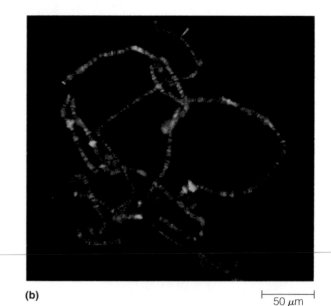

(b)

⊢————⊣
50 μm

Figure 21-15 Transcriptional Activity of Polytene Chromosomes. **(a)** A phase-contrast micrograph showing several polytene chromosomes from the salivary gland of a *Drosophila* larva. The banding pattern is a characteristic property of each chromosome, such that individual bands can be identified. **(b)** The same chromosomes seen with the fluorescence microscope, after incubation with fluorescent antibodies that specifically bind to RNA polymerase II. The chromosomes light up brightly wherever RNA polymerase II molecules are located—that is, where transcription is occurring. The larva from which these chromosomes were obtained had been subjected to an elevated temperature for a short time to activate genes that code for heat-shock proteins.

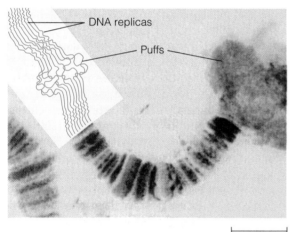

DNA replicas

Puffs

⊢————⊣
25 μm

Figure 21-16 Puffs in Polytene Chromosomes. Puffs are regions in which transcriptionally active chromatin has become less condensed, as indicated diagrammatically. The light micrograph shows part of a polytene chromosome.

DNase I Sensitivity Provides Further Evidence for the Role of Chromatin Decondensation in Genomic Control

The absence of polytene chromosomes in most eukaryotic cells makes it difficult to visualize chromatin decondensation in regions of active genes. Nonetheless, other kinds of evidence support the idea that chromatin decondensation is generally associated with gene transcription. One par-ticularly useful research tool is *DNase I,* an endonuclease isolated from the pancreas. In test-tube experiments, low concentrations of DNase I preferentially degrade transcriptionally active DNA in chromatin. The increased sensitivity of these DNA regions to degradation by DNase I provides evidence that the DNA is uncoiled.

Figure 21-17 illustrates a classic DNase I sensitivity experiment focusing on the chicken gene for a globin polypeptide, one of the polypeptide subunits of hemoglobin. The globin gene is actively expressed in the nuclei of chicken *erythrocytes* (red blood cells). In contrast to the erythrocytes of many other vertebrates, avian erythrocytes retain their nucleus at maturity. If these nuclei are isolated and the chromatin digested with DNase I, the globin gene is completely digested at low DNase I concentrations that do not affect the globin gene in other tissues, such as oviduct tissue. As you might predict, a gene that is not active in erythrocytes (e.g., the gene for ovalbumin, an egg white protein) is not digested by DNase I. The opposite result is obtained when the same procedure is carried out using chromatin isolated from oviduct, where the ovalbumin gene is expressed and the globin gene is inactive. In this case the ovalbumin gene is more sensitive than the globin gene to DNase I digestion. Such data demonstrate that transcription of eukaryotic DNA is correlated with an increased sensitivity to DNase I digestion.

The results of these experiments are compatible with two alternative explanations: Either chromatin uncoiling is necessary to give transcription factors and RNA polymerase access to DNA, or the binding of these proteins to

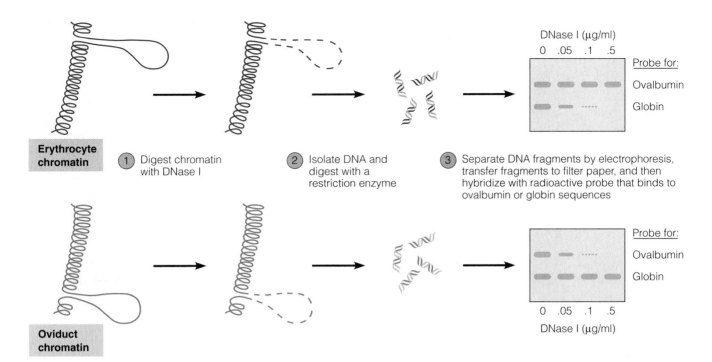

Figure 21-17 Sensitivity of Active Genes in Chromatin to Digestion with DNase I.   The chromatin configuration of active genes can be assayed by exposing cell nuclei to DNase I. As an endonuclease, DNase I digests DNA by repeatedly cutting internal phosphodiester bonds. However, DNA in condensed chromatin is protected from DNase I attack, presumably because it is highly coiled and complexed with proteins. The experiment shown here uses chromatin from two different cell types, chicken erythrocytes and oviduct cells, and focuses on genes for globin and ovalbumin, which are expressed in erythrocytes and oviduct cells, respectively.

In step ① the chromatin is digested with a low concentration of DNase I, which preferentially digests the DNA in uncoiled regions of chromatin. In step ② the chromatin proteins are removed and the DNA is purified and digested with a restriction enzyme that will release a DNA fragment containing an intact globin or ovalbumin gene (if it has not been nicked by DNase I). Finally, the presence of that restriction fragment is detected in step ③ by separating the DNA fragments using electrophoresis, transferring the separated fragments to a filter paper (a Southern blot), and hybridizing with radioactive DNA

probes for the globin and ovalbumin genes. Note that DNA isolated from erythrocyte chromatin treated with increasing amounts of DNase I contains progressively smaller amounts of the intact restriction fragment with the globin gene, and comparable results are obtained for the ovalbumin gene in oviduct chromatin. In contrast, even high concentrations of DNase I have no effect on the globin gene in oviduct chromatin or on the ovalbumin gene in erythrocyte chromatin. In other words, genes are more susceptible to DNase I attack in tissues where they are actively transcribed.

DNA causes the uncoiling. This issue has been resolved by studies showing that sensitivity to DNase I is detected in genes that are being actively transcribed, in genes that have recently been transcribed but are no longer active, and in DNA sequences located adjacent to genes of the preceding two types. Such observations suggest that DNase I sensitivity is not caused by the process of gene transcription itself, but instead reflects an altered chromatin structure in regions associated with active or potentially active genes. Presumably this means that chromatin uncoiling is a prerequisite for—rather than a consequence of—transcriptional activation.

DNase I has also been used in other kinds of experiments. When nuclei are treated with very low concentrations of DNase I, it is possible to detect **DNase I hypersensitive sites,** specific locations in the chromatin that are exceedingly susceptible to digestion. Hypersensitive sites tend to occur up to a few hundred bases upstream from the transcriptional start sites of active genes, and are about ten times more sensitive to DNase I

digestion than the bulk of the DNA associated with these genes. The idea that DNase I hypersensitive sites may represent regions that are free of nucleosomes first emerged from studies involving the eukaryotic virus SV40. When the SV40 virus infects a host cell, its circular DNA molecule becomes associated with histones and forms typical nucleosomes that can be observed with an electron microscope. However, a small region of the viral DNA remains completely uncoiled and free of nucleosomes (Figure 21-18). This region, which includes several DNase I hypersensitive sites, contains DNA sequences that bind regulatory proteins involved in activating transcription.

Although we do not understand the general mechanisms that cause DNA to become uncoiled in preparation for transcription, studies in yeast have provided a few clues. Yeast cells have a regulatory protein complex, known as *SWI/SNF,* that plays an important role in activating many inducible genes. Current evidence suggests that the SWI/SNF complex is involved in **chromatin remodeling**—altering nucleosome structure, packing,

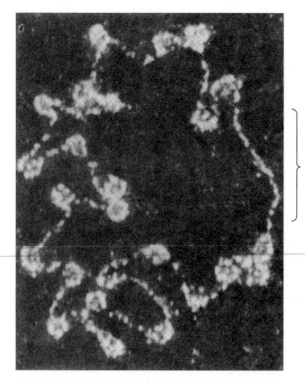

Figure 21-18 The Circular DNA Molecule of an SV40 Virus in an Infected Host Cell. In this high-power electron micrograph of an SV40 DNA molecule, the bracket shows a small region on the right side of the molecule that lacks nucleosomes. This region corresponds to the location of several DNase I hypersensitive sites (TEM).

and/or position to give transcription factors access to their DNA target sites in the promoter region of a gene. The human homologue of the SWI/SNF complex has also been shown to induce an ATP-dependent structural change in nucleosomes. After transcription has been successfully initiated, however, the "normal" nucleosome structure reappears, suggesting that nucleosomes do not interfere with the movement of RNA polymerase along the DNA molecule once transcription has been initiated.

DNA Methylation Is Associated with Inactive Regions of the Genome

Another means of regulating the availability of various regions of the genome is **DNA methylation,** the addition of methyl groups to selected cytosine groups in DNA. The DNA of most vertebrates contains small amounts of methylated cytosine, which tends to cluster in noncoding regions at the 5′ ends of genes. Moreover, methylation patterns are inherited after DNA replication, because the enzyme responsible for DNA methylation is specific for cytosines located in 5′–CG–3′ sequences base-paired to complementary 3′–GC–5′ sequences that are already methylated. This means that if the old strand of a newly formed double helix has a methylated 5′–CG–3′ sequence, then the complementary 3′–GC–5′ sequence in the newly formed strand will become a target for the DNA-methylating enzyme.

Many lines of evidence suggest that DNA methylation influences gene activity. One important example involves **X-chromosome inactivation** in female mammals. The somatic (nongamete) cells of female mammals each contain two copies of the X chromosome, one inherited from each parent. Early in development, one X chromosome in each existing cell is randomly inactivated by condensation into a tight mass of heterochromatin. When interphase cells are examined under a microscope, the inactivated X chromosome is visible as a dark spot called a *Barr body.* The inactivated X chromosome is extensively methylated and does not participate in transcription initiation. Once an embryonic cell has undergone X-chromosome inactivation, all its cellular descendants inactivate the same X chromosome. The benefit to the organism of X inactivation is unclear, but it may function to prevent the buildup of harmfully high levels of the X chromosome's gene products; at any rate, it ensures that female mammals, like males, contain only one active X chromosome per adult cell.

Further evidence suggesting that DNA methylation influences gene activity has come from studies using the restriction enzymes *Msp*I and *Hpa*II. Both of these enzymes cleave the recognition site -CCGG-; however, *Hpa*II works only if the central C is unmethylated, whereas *Msp*I cuts whether the C is methylated or not. Comparing the DNA fragments generated by these two enzymes has confirmed that a number of DNA sites exhibit tissue-specific methylation patterns—that is, the sites are methylated in some tissues but not in others. In general, such sites are unmethylated in tissues where the gene is active or potentially active, and methylated in tissues where the gene is inactive. For example, CG sequences located near the 5′ end of the globin gene are methylated in tissues that do not produce hemoglobin but are unmethylated in red blood cells.

Independent support for the conclusion that decreased DNA methylation is associated with increased gene activity has emerged from experiments employing *5-azacytidine,* a cytosine analog that cannot be methylated because it contains a nitrogen atom in place of carbon at the site where methylation normally occurs. Because methylation patterns tend to be inherited, incorporation of 5-azacytidine into DNA triggers an undermethylated state that is maintained for many cell generations after the drug has been removed. Exposing cells to 5-azacytidine has been found to activate the transcription of several kinds of genes, including some that are located within the heterochromatin of inactivated X chromosomes.

Taken together, the preceding observations suggest that DNA methylation suppresses gene transcription. Although the mechanism for this effect is not well understood, it has been proposed that methylation promotes chromatin condensation, making DNA less accessible to transcription factors and RNA polymerase. We must be careful not to carry such speculations too far, however, as examples are known of methylated DNA sequences that are transcribed and unmethylated sequences that are not. Moreover, the extent of DNA methylation varies significantly among eukaryotes, being prominent in mammals and higher plants but rare in simpler eukaryotes such as *Drosophila* and yeast. So rather than representing an essential component of eukaryotic gene regulation, DNA methylation may simply be one of many factors that contribute to the overall control of gene expression.

Changes in Histones, HMG Proteins, and the Nuclear Matrix Are Associated with Active Regions of the Genome

In addition to alterations in DNase sensitivity and DNA methylation, several other distinctive properties are associated with the structural organization of actively transcribed regions of the genome. For example, given the central role of histones in chromatin organization, changes in histone structure might be expected in the region of active genes. One mechanism for altering histone structure is *acetylation,* the addition of acetyl groups to amino acid side chains in histone molecules. Evidence that histone acetylation is altered in active chromatin has been obtained from experiments in which chromatin was incubated with DNase I to selectively degrade genes that are transcriptionally active. Such treatment causes the release of the acetylated form of histones H3 and H4, suggesting that acetylated histones are preferentially associated with the nucleosomes of active genes.

Further support for this idea has come from studies involving *sodium butyrate,* a fatty acid salt that inhibits the removal of histone acetyl groups and hence causes histones to become excessively acetylated. Chromatin isolated from cells treated with sodium butyrate is more susceptible than normal to digestion with DNase I, suggesting that histone acetylation alters nucleosomal structure in a way that enhances the susceptibility of DNA to enzymatic cleavage. Such acetylation-induced changes in nucleosome structure are thought to loosen the packing of chromatin in ways that facilitate the access of transcription factors to gene promoters.

In addition to acetylation, the amino acid side chains of histone molecules are subject to a variety of other modifications, including phosphorylation and methylation, that may alter chromatin structure and thereby influence gene activity. Another type of histone alteration involves the absence of histone H1 in regions of transcriptionally active chromatin. Since histone H1 is specifically required for folding chromatin into 30-nm chromatin fibers (p. 507), the absence of histone H1 may help to maintain active chromatin in the form of uncoiled 10-nm fibers.

A related feature of transcriptionally active chromatin is its large content of **high-mobility group (HMG) proteins,** a group of nonhistone proteins whose name reflects their rapid mobility during electrophoresis. If HMG proteins are removed from isolated chromatin, the active genes lose their sensitivity to DNase I. Sensitivity to DNase I can be restored by adding back two HMG proteins called HMG 14 and HMG 17. If HMG 14 and 17 are isolated from one tissue and added to HMG-depleted chromatin derived from another tissue, the pattern of DNase I sensitive genes is found to resemble the tissue from which the chromatin, not the HMG proteins, was obtained. For example, if HMG-depleted liver chromatin is mixed with HMG 14 and 17 isolated from brain tissue, the pattern of DNase I sensitive genes in the resulting chromatin resembles that of the liver, not the brain. This means that chromatin must exhibit tissue-specific differences that are recognized by HMG 14 and 17, allowing the HMG proteins to bind selectively to genes that are normally capable of being activated in any given tissue. Binding of HMG 14 and 17 is thought to enhance DNase I sensitivity by helping to uncoil chromatin fibers into a more open configuration, perhaps by displacing histone H1 from the 30-nm fiber.

Finally, actively transcribed genes have a close structural association with the nuclear matrix. This property was first established for the chicken gene coding for ovalbumin, a protein formed only in cells lining the oviduct. In one set of studies, the nuclear matrix was isolated from chick oviduct cells, which actively transcribe the ovalbumin gene, and from chick liver cells, which do not. When the small amount of DNA that remains associated with the nuclear matrix after isolation was hybridized to cloned ovalbumin gene sequences, ovalbumin DNA sequences were found to be enriched in nuclear matrix fractions isolated from oviduct, but not from liver. In other words, DNA sequences coding for ovalbumin are closely associated with the nuclear matrix in tissues where the ovalbumin gene is being actively transcribed. Subsequent studies have revealed that transcriptionally active DNA is bound to the nuclear matrix by special DNA sequences called *matrix attachment regions (MARs).*

Eukaryotic Gene Regulation: Transcriptional Control

We have now described some common regulatory changes in the composition and structure of the genome. However, the existence of structural changes associated with active regions of the genome does not address the underlying question of how the DNA sequences contained in these regions are actually selected for activation. The answer is to be found in the phenomenon of **transcriptional control,** the second main level for controlling eukaryotic gene expression (see Figure 21-11).

Different Sets of Genes Are Transcribed in Different Cell Types

When we consider transcriptional control, we come to a level of gene regulation where knowledge has blossomed in recent years. Direct evidence for the importance of transcriptional regulation in eukaryotes has come from experiments comparing newly synthesized RNA in the nuclei of different mammalian tissues. Liver and brain cells, for example, produce different sets of proteins, although there is considerable overlap between the two sets. How can we determine the source of this difference? If the different proteins produced by the two cell types are a reflection of *differential gene transcription*—that is, transcription of different genes to produce different sets of RNAs—we should see corresponding differences between the populations of nuclear RNAs derived from brain and liver cells. On the other hand, if all genes are equally transcribed in liver and brain, we would find few, if any, differences between the populations of nuclear RNAs from the two tissues, and we would conclude that the tissue-specific differences in protein synthesis were due to posttranscriptional mechanisms that control the ability of various RNAs to be translated.

One way of distinguishing between these alternatives is to use the technique of *nuclear run-on transcription*, which provides a snapshot of the transcriptional activity occurring in a nucleus at a given moment in time (Figure 21-19). Transcriptionally active nuclei are gently isolated from cells and allowed to complete synthesis of RNA molecules in the presence of radioactively labeled nucleoside triphosphates. When such an experiment is performed using liver and brain cells, the newly transcribed (radioactively labeled) RNAs in the liver nuclei contain sequences from liver-specific genes, but these liver-specific

sequences are not detected in the labeled RNAs synthesized by isolated brain nuclei. Likewise, the newly transcribed (radioactively labeled) RNAs in the brain nuclei contain sequences from brain-specific genes, but these brain-specific sequences are not detected in the labeled RNAs synthesized by isolated liver nuclei.

By failing to detect the synthesis of RNA from liver-specific genes in brain cell nuclei and vice versa, these experiments suggest that gene expression is usually regulated in eukaryotes at the transcriptional level. In other words, different cell types transcribe different sets of genes, thereby allowing each cell type to produce those proteins needed for carrying out that cell's specialized functions.

DNA Microarrays Allow the Expression of Thousands of Genes to Be Monitored Simultaneously

Although the preceding kinds of observations reveal that different sets of genes are transcribed in different cell types, they cannot easily determine which of the thousands of different genes in any given cell are turned on or off. However, a recently developed tool, known as a **DNA microarray** (or *gene chip*), makes it relatively straightforward to obtain such information. A DNA microarray is a

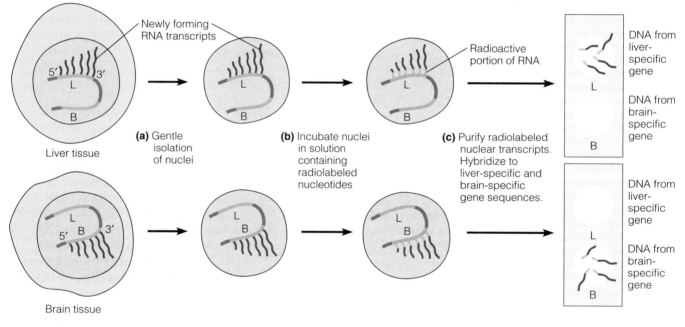

Transcription occurring in intact cell

Transcripts of active genes are completed in isolated nuclei using radiolabeled nucleotides

Specificity of labeled nuclear transcripts is tested

Figure 21-19 Demonstration of Differential Transcription by Nuclear Run-on Transcription Assays. **(a)** Nuclei are gently isolated from brain and liver tissues. B represents a hypothetical gene in the nuclear DNA that is expressed only in brain tissue, and L is a gene expressed only in liver tissue. **(b)** The isolated nuclei are incubated in a solution containing radioactively labeled

ribonucleotides. The labeled nucleotides enter the isolated nuclei through nuclear pores and become incorporated into the mRNA being synthesized by active genes. If different genes are active in liver and brain tissue, some labeled sequences in the liver nuclear transcripts will not be present in brain transcripts, and vice versa. **(c)** The composition of the labeled RNA population

is assayed by allowing the labeled RNA to hybridize with DNA sequences representing different genes that have been attached to a filter paper support. Labeled liver transcripts hybridize with a different set of genes than do labeled brain transcripts, indicating that the identity of the active genes in the two tissues differs.

thin, fingernail-sized chip, usually made of glass, that has been spotted at fixed locations with thousands of single-stranded DNA fragments corresponding to various genes of interest. A single microarray may contain 10,000 or more spots, each representing a different gene. To determine which genes are being expressed in any given cell population, RNA molecules (which represent the products of gene transcription) are isolated from the cells and copied with the enzyme reverse transcriptase into single-stranded cDNA molecules, which are then attached to a fluorescent dye. When the DNA microarray is bathed with the fluorescent cDNA, each cDNA molecule will bind by complementary base-pairing to the spot containing the specific gene from which it was transcribed.

Figure 21-20 illustrates how this approach can be used to compare the pattern of gene expression in two different cell populations. In this particular example, which compares gene expression in cancer cells and normal cells, two different fluorescent dyes are used, a red dye to label cDNAs derived from cancer cells and a green dye to label cDNAs derived from the corresponding normal cells. When the red and green cDNAs are mixed together and placed on a DNA microarray, the red cDNAs will bind to genes expressed in cancer cells and the green cDNAs will bind to genes expressed in normal cells. Red spots therefore represent higher expression of a gene in cancer cells, green spots represent higher expression of a gene in normal cells, yellow spots (caused by mixture of red and green fluorescence) represent genes whose expression is roughly the same, and black spots (absence of fluorescence) represent genes expressed in neither cell type. Thus, the relative expression of thousands of genes in cancer and normal cells can be compared by measuring the intensity and color of the fluorescence of each spot.

Proximal Control Elements Lie Close to the Promoter

In discussing how gene transcription is regulated in different cell types, we will now focus our attention on protein-coding genes, which are transcribed by RNA polymerase II. As we saw in Chapter 19, the specificity of transcription—that is, where on the DNA it initiates—is determined not by RNA polymerase itself, but by a variety of other proteins called *transcription factors*. (Unlike prokaryotic sigma factors, which also determine initiation specificity, none of the eukaryotic transcription factors are considered to be an integral part of an RNA polymerase molecule.) The transcription factors discussed in Chapter 19 were *general transcription factors*, which are essential for the transcription of *all* genes transcribed by a given type of RNA polymerase.

For genes transcribed by RNA polymerase II, the general transcription factors assemble with RNA polymerase at the *core promoter*, a DNA region located in the immediate vicinity of the transcriptional startpoint (see Figure 19-13b). The interaction of general transcription factors and RNA polymerase with the core promoter often initiates transcription at only a low, "basal" rate, so that few transcripts are produced. However, in addition to a core promoter, most protein-coding genes have short DNA sequences farther upstream (and, in some cases, downstream) to which other transcription factors bind, thereby improving the efficiency of the core promoter. When these additional DNA elements are deleted or mutated, the frequency and accuracy of transcription initiation are reduced.

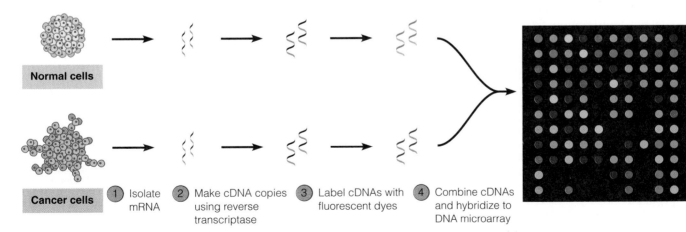

Figure 21-20 Using a DNA Microarray to Profile the Patterns of Gene Expression in Two Different Cell Types. In this example, gene expression in cancer cells and normal cells is compared by ① isolating mRNA from the two populations of cells, ② using reverse transcriptase to make cDNA copies of the mRNA, and ③ attaching a green fluorescent dye to the normal cell cDNAs and a red fluorescent dye to the cancer cell cDNAs. A DNA microarray containing thousands of DNA fragments representing different genes is then ④ bathed with a mixture of the two cDNA populations (only a small section of the DNA microarray is illustrated). Red spots represent genes expressed preferentially in cancer cells, green spots represent genes expressed preferentially in normal cells, yellow spots represent genes whose expression is similar in the two cell populations, and dark regions (missing spots) represent genes that are not expressed in either cell type.

Within the figure:
Normal cells
Cancer cells
① Isolate mRNA
② Make cDNA copies using reverse transcriptase
③ Label cDNAs with fluorescent dyes
④ Combine cDNAs and hybridize to DNA microarray

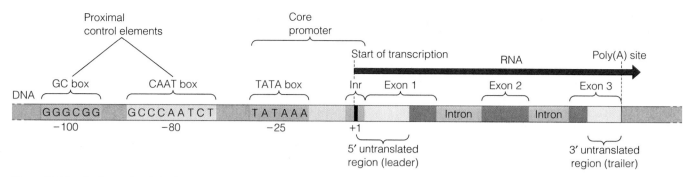

Figure 21-21 **Anatomy of a Typical Eukaryotic Gene, with Its Core Promoter and Proximal Control Region.** This diagram (not to scale) features a typical protein-coding eukaryotic gene, which is transcribed by RNA polymerase II. The promoter—called the core promoter to distinguish it clearly from the proximal control region—is characterized by an initiator (Inr) sequence surrounding the transcriptional startpoint and a sequence called a TATA box located about 25 bp upstream (to the 5′ side) of the startpoint. The core promoter is where the general transcription factors and RNA polymerase assemble for the initiation of transcription. Within about 100 nucleotides upstream from the core promoter lie several proximal control elements, which stimulate transcription of the gene by interacting with regulatory transcription factors. The number, identity, and exact location of the proximal elements vary from gene to gene; here we show a very simple case, with one copy of each of two common elements, the GC box and the CAAT box. The transcription unit includes a 5′ untranslated region (leader) and a 3′ untranslated region (trailer), which are transcribed and included in the mRNA but do not contribute sequence information for the protein product. At the end of the last exon is a site where, in the primary transcript, the RNA will be cleaved and given a poly(A) tail.

In discussing such regulatory DNA sequences, we will use the term **proximal control elements** to refer to sequences located upstream of the core promoter but within about 100–200 base pairs of it. The number, exact location, and identities of these proximal control elements vary with each gene, but three types are especially common: the *CAAT box,* the *GC box,* and the *octamer* (the first two were discussed in Chapter 19 and are illustrated in Figure 21-21). Transcription factors that selectively bind to one of these, or to other DNA control elements located outside the core promoter, we call **regulatory transcription factors.** They increase (or, sometimes, decrease) transcription initiation, apparently by interacting with components of the transcription apparatus. Examples of regulatory transcription factors that selectively bind to particular DNA control elements are given in Table 21-3.

Because many of the same proximal control elements are associated with a variety of different genes, some scientists regard the RNA polymerase II promoter as a region that includes the proximal control elements. However, the particular combinations and locations of the elements are specific to each gene.

Enhancers and Silencers Are Located at Variable Distances from the Promoter

Proximal control elements, like most prokaryotic control elements, lie close to the core promoter on its upstream side. A second class of DNA control sequences are located either upstream or downstream from the genes they regulate and often lie far away from the promoter. This second type of control region is called an **enhancer** if it stimulates gene transcription or a **silencer** if it inhibits transcription. Originally, such control sequences were called *distal control elements* because they can function at distances up to 70,000–80,000 bp from the promoter they regulate (the word *distal* means "away from"). However, the distinctive feature of such sequences is not how far away from the

Table 21-3 Some Eukaryotic DNA Control Elements and Transcription Factors That Bind Them

DNA Control Element	Genes Where Found	DNA Element Consensus Sequence*	Transcription Factor**
TATA box	Many	TATAAAA (in promoter)	TFIID
CAAT box	Many	GGCCAATCT	CTF
GC box	Many	GGGCGG	Sp1
Octamer	Many	ATTTGCAT	Oct-1/Oct-2
Heat shock element	Heat shock genes	CnnGAAnnTTCnnG	Heat-shock transcription factor
Estrogen response element	Ovalbumin, others	AGGTCAnnnTGACCT	Estrogen receptor
κB	Immunoglobulin light chain κ	GGGGACTTTCC	NF-κB

*Some researchers give slightly different consensus sequences. A lowercase n signifies that any nucleotide can be located at that position.

**In some cases, other names have also been assigned to the factors listed here.

promoter they are located, but the fact that their position relative to the promoter can vary significantly, and their orientation can even be reversed, without interfering with their ability to regulate transcription. Thus, enhancers and silencers need not be located at great distances from the promoter. They can be located quite close to the promoter, and are even found occasionally *within* genes; one example is an enhancer found within an intron of certain antibody genes.

Since they are better understood than silencers, we will consider enhancers first. Enhancers, varying in specific sequence but sharing common properties, are associated with many eukaryotic genes. A typical enhancer contains several different control elements within it, each consisting of a short DNA sequence that serves as a binding site for a different regulatory transcription factor. Some of these DNA sequences may be identical to proximal control elements; the octamer and the GC box, for instance, can act both as proximal control elements and as components of enhancers. For an enhancer to function, the regulatory transcription factors that bind to its various control elements must be present; because an enhancer is involved in activating transcription, these regulatory transcription factors are called **activators.**

To investigate the properties of enhancers, researchers have used recombinant DNA techniques to alter enhancer location and orientation with respect to the regulated gene. As shown in Figure 21-22, such studies reveal that enhancers can function properly when relocated at variable distances from the transcription startpoint, as long as the promoter is present (Figure 21-22d and e). Activity is retained even when the orientation of the enhancer relative to the beginning of the gene is reversed (Figure 21-22f), or when the enhancer is relocated *downstream* of the 3′ end of the gene (Figure 21-22g). These properties are often used to distinguish enhancers from proximal control elements, whose precise locations within the DNA tend to be more critical for their function. Nonetheless, as more and more enhancers and proximal control elements have been discovered and investigated, distinctions between these two categories have become less clear-cut, and it now appears that a broad spectrum of transcription control elements exist with overlapping properties. At one extreme are certain proximal control elements that must be precisely located at specific positions near the promoter—in such cases, moving them even 15–20 nucleotides farther from the promoter causes them to lose their influence. At the other extreme are enhancers whose positions can be varied widely, up to tens of thousands of nucleotides away from a promoter, without interfering with their ability to activate transcription.

Although they are not as widespread or as well studied as enhancers, silencers appear to share many of the features of enhancers, except that they inhibit rather than activate transcription. Because the binding of regulatory transcription factors to silencers reduces rather than increases gene transcription rates, such transcription factors are called eukaryotic **repressors.** (While they resemble prokaryotic repressors in turning off transcription, the

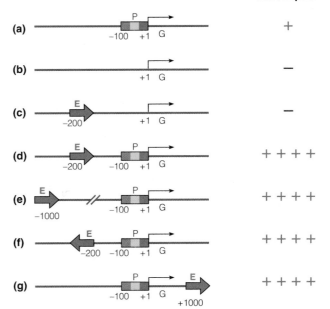

Figure 21-22 Properties of Enhancers. Recombinant DNA techniques can be used to alter the orientation and location of DNA control elements and study the effect of the change on the level of transcription of the gene. The black arrows indicate the direction of transcription of gene G, with the startpoint (first transcribed nucleotide) labeled +1. The other numbers give the positions of nucleotides relative to the startpoint. **(a)** The core promoter (P) alone, in its typical location just upstream of gene G, allows a basal level of transcription to occur. **(b)** When the core promoter is removed from the gene, no transcription occurs. **(c)** An enhancer (E) alone cannot substitute for the promoter region, but **(d)** combining an enhancer with a core promoter results in a significantly higher level of transcription than occurs with the promoter alone. **(e)** This increase in transcription is observed when the enhancer is moved farther upstream, **(f)** when it is inverted in orientation, and **(g)** even when it is moved to the 3′ side of the structural gene.

eukaryotic mechanisms are somewhat different and more varied.) We encountered an example of silencers earlier in the chapter when discussing the yeast *SIR* genes, which produce proteins that inhibit transcription of the *HMLα* and *HMRa* mating-type genes by binding to DNA sequences surrounding these two genes. The proteins produced by the *SIR* genes are examples of repressors, and the DNA sites to which these repressors bind, located near *HMLα* and *HMRa,* are examples of silencers.

Coactivators Mediate the Interaction Between Regulatory Transcription Factors and the RNA Polymerase Complex

Because enhancers and silencers can reside far away from the genes they control, the question arises as to how regulation is achieved over such long distances. In addressing this question, we will again focus on the behavior of enhancers (silencers appear to behave in a basically similar fashion). Two basic principles govern the interaction between enhancers and the genes they regulate. First, looping of the DNA molecule can bring an enhancer into close proximity with a promoter, even though the two lie far apart in terms of linear distance along the DNA double helix. And second, this DNA looping usually involves the participation of **coactivator** proteins that bind both to the regulatory transcription factors (activators) associated with the enhancer and to general transcription factors associated with the promoter, thereby forming a "bridge" between enhancers and promoters.

Figure 21-23 illustrates how such interactions can trigger gene activation. In step ①, a group of activator proteins bind to their respective DNA control elements within the enhancer, forming a multiprotein complex called an *enhanceosome*. Next, one or more of these activator proteins causes the DNA to bend, creating a DNA loop that brings the enhancer close to the core promoter (②). Coactivator proteins then link the enhanceosome to the promoter by binding both to activators associated with the enhanceosome and to general transcription factors in the vicinity of the promoter, most commonly TFIID.

This interaction, illustrated in step ③, either promotes the binding of TFIID to the promoter or assists the binding of other general transcription factors associated with TFIID. In either case, the interaction helps position TFIID properly at the promoter and/or triggers its association with other general transcription factors and RNA

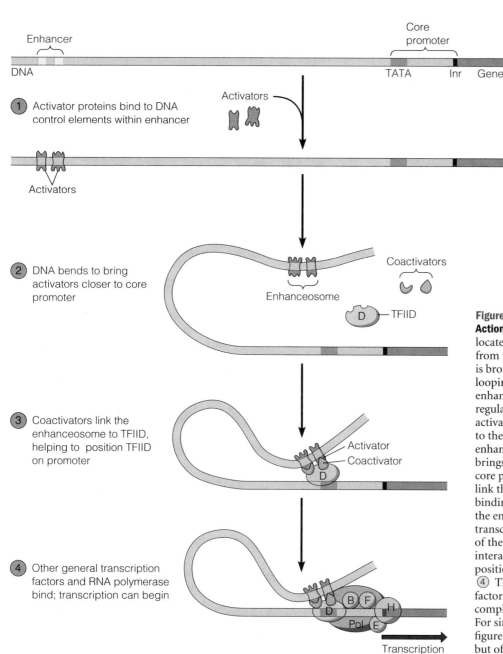

Figure 21-23   A Model for Enhancer Action.   In this model, an enhancer located at a great distance along the DNA from the protein-coding gene it regulates is brought close to the core promoter by a looping of the DNA. The influence of an enhancer on the promoter is mediated by regulatory transcription factors called activators. ① The activator proteins bind to the enhancer elements, forming an enhanceosome. ② Bending of the DNA brings the enhanceosome closer to the core promoter. ③ Coactivator proteins link the enhanceosome to the promoter by binding both to activators associated with the enhanceosome and to general transcription factors located in the vicinity of the promoter, in this case TFIID. This interaction facilitates the correct positioning of TFIID on the promoter. ④ The other general transcription factors and RNA polymerase join the complex, and transcription is initiated. For simplicity, the enhanceosome in this figure is drawn with only two activators, but often half a dozen or more are present.

polymerase, thereby facilitating assembly of the RNA polymerase complex at the promoter and allowing transcription to begin (④). Although the mechanism by which coactivators facilitate this assembly of the RNA polymerase complex may vary, several coactivator proteins exhibit *histone acetyltransferase* activity, which means that they catalyze the acetylation of histones. As mentioned earlier in the chapter, histone acetylation is thought to loosen the packing of nucleosomes in ways that facilitate access of the transcription machinery to gene promoters.

While the binding of coactivator proteins to TFIID plays a central role in the example just described, some coactivators bind instead to other general transcription factors, such as TFIIB and TFIIA. But these details are less important than the main idea: When activator proteins for a particular gene are present in the cell, their binding to an enhancer can stimulate formation of the transcription complex at the promoter, resulting in more efficient initiation of transcription. Thus, differential gene transcription within a cell is determined to a large extent by the activators a cell makes (as well as by factors that control the activity of the activators, a topic we will discuss shortly).

Multiple DNA Control Elements and Transcription Factors Act in Combination

The realization that multiple DNA control elements and their regulatory transcription factors are involved in controlling eukaryotic gene transcription has led to a **combinatorial model for gene regulation.** This model proposes that a relatively small number of different DNA control elements and transcription factors, acting in different combinations, can establish highly specific and precisely controlled patterns of gene expression in different cell types. According to this model, a gene is expressed at a maximum level only when the set of transcription factors produced by a given cell type includes all the regulatory transcription factors that bind to that gene's positive DNA control elements.

The model begins with the assumption that some transcription factors are present in many cell types. These include the general transcription factors, required for transcription in all cells, plus any regulatory factors needed for the transcription of constitutive genes and others that are frequently expressed. In addition, transcription of genes that encode tissue-specific proteins requires the presence of transcription factors or *combinations* of transcription factors that are unique to individual cell types. To illustrate this concept, Figure 21-24 shows how such a model would allow liver cells to produce large amounts of proteins such as albumin but would prevent significant production of these proteins in other tissues, such as brain. The original version of the combinatorial model was "all or none," proposing that transcription of a gene could not be initiated unless the entire set of regulatory factors for the gene was present. It is now considered more likely that there is a continuum of initiation efficiency, ranging from a basal level, occurring when *no* regulatory factors are available, to a maximum level, which can occur only when the full set of regulatory factors is present.

Several Common Structural Motifs Allow Regulatory Transcription Factors to Bind to DNA and Activate Transcription

Although not all transcription factors bind directly to DNA, those that do play critical roles in controlling transcription. Proteins in this category include the general transcription factor TFIID and, more importantly, the wide variety of regulatory transcription factors (activators and repressors) that recognize and bind to specific DNA sequences found in proximal control elements, enhancers, and silencers. What features of these regulatory transcription factors enable them to carry out their functions?

Regulatory transcription factors possess two distinct activities, the ability to bind to a specific DNA sequence and the ability to regulate transcription. The two activities reside in separate protein domains. The domain that recognizes and binds to a specific DNA sequence is called the transcription factor's **DNA-binding domain,** whereas the protein region required for regulating transcription is known as the **transcription regulation domain** (or **activation domain** because most transcription factors activate, rather than inhibit, transcription). The existence of separate DNA-binding and activation domains has been demonstrated by "domain-swap" experiments in which the DNA-binding region of one transcription factor is combined with various regions of a second transcription factor. The resulting hybrid molecule can activate gene transcription only if it contains an activation domain provided by the second transcription factor.

Studies of this type have revealed that activation domains often have a high proportion of acidic amino acids, producing a strong negative charge that is generally clustered on one side of an α helix. Mutations that increase the number of negative charges tend to increase a protein's ability to activate transcription, whereas mutations that decrease the net negative charge or disrupt the clustering on one side of the α helix diminish the ability to activate transcription. In addition to acidic domains, several other kinds of activating domains have been identified in transcription factors. Some are enriched in the amino acid glutamine, and others contain large amounts of proline. Hence several different types of protein structure appear to be capable of creating an activation domain that can stimulate gene transcription.

Several kinds of structures are also found in the DNA-binding domains of transcription factors. In fact, most regulatory transcription factors can be placed into one of a small number of categories based on the secondary structure pattern, or *motif* (p. 50), that makes up the DNA-binding domain. We now briefly describe several of the most common DNA-binding motifs.

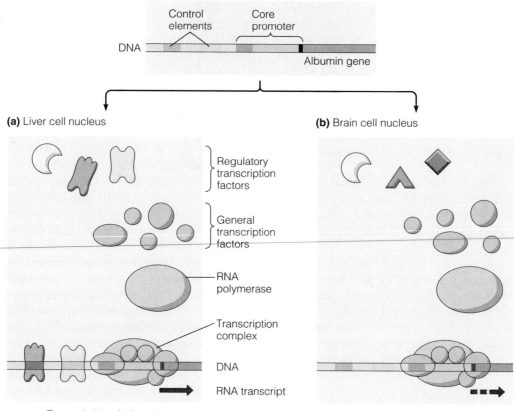

Figure 21-24 A Combinatorial Model for Gene Expression. The gene for the protein albumin, like other genes, is associated with an array of regulatory DNA elements; here we show only two control elements, as well as the core promoter. Cells of all tissues contain RNA polymerase and the general transcription factors, but the set of regulatory transcription factors available varies with the cell type. As shown here, **(a)** liver cells contain a set of regulatory transcription factors that includes the factors for recognizing all the albumin gene control elements. When these factors bind to the DNA, they facilitate transcription of the albumin gene at a high level. **(b)** Brain cells, however, have a different set of regulatory transcription factors, which does not include all the ones for the albumin gene. Consequently, in brain cells, the transcription complex can assemble at the promoter, but not very efficiently. The result is that brain cells transcribe the albumin gene only at a low level.

Helix-Turn-Helix Motif. One of the most common DNA-binding motifs, detected in both eukaryotic and prokaryotic regulatory transcription factors, is the **helix-turn-helix** (Figure 21-25a). This motif consists of two α helices separated by a bend in the polypeptide chain. Although the amino acid sequence of the motif differs among various DNA-binding proteins, the overall pattern is always the same: One α helix, called the *recognition helix,* contains amino acid side chains that recognize and bind to specific DNA sequences by forming hydrogen bonds with bases located in the major groove of the DNA double helix, while the second α helix stabilizes the overall configuration through hydrophobic interactions with the recognition helix. The *lac* and *trp* repressors, the CRP protein, and many phage repressor proteins are examples of prokaryotic proteins exhibiting the helix-turn-helix motif, and transcription factors that regulate embryonic development (the class of factors encoded by homeotic genes, described later) are eukaryotic examples. Figure 21-25b is a model of the phage λ repressor, a helix-turn-helix protein, bound to DNA. Like many DNA-binding regulatory proteins, the phage λ repressor consists of two identical polypeptides, each containing a DNA-binding domain.

Zinc Finger Motif. Initially identified in a transcription factor for the 5S rRNA genes (TFIIIA), the **zinc finger** DNA-binding motif consists of an α helix and a two-segment β sheet, held in place by the interaction of precisely positioned cysteine or histidine residues with a zinc atom. The number of zinc fingers present per protein molecule varies among the transcription factors that possess them, ranging from two fingers to several dozen or more. Figure 21-25c shows a protein with four zinc fingers in a row (TFIIIA has nine). Zinc fingers protrude from the protein surface and serve as the points of contact with specific base sequences in the major groove of the DNA.

Leucine Zipper Motif. The **leucine zipper** motif is formed by an interaction between two polypeptide chains, each

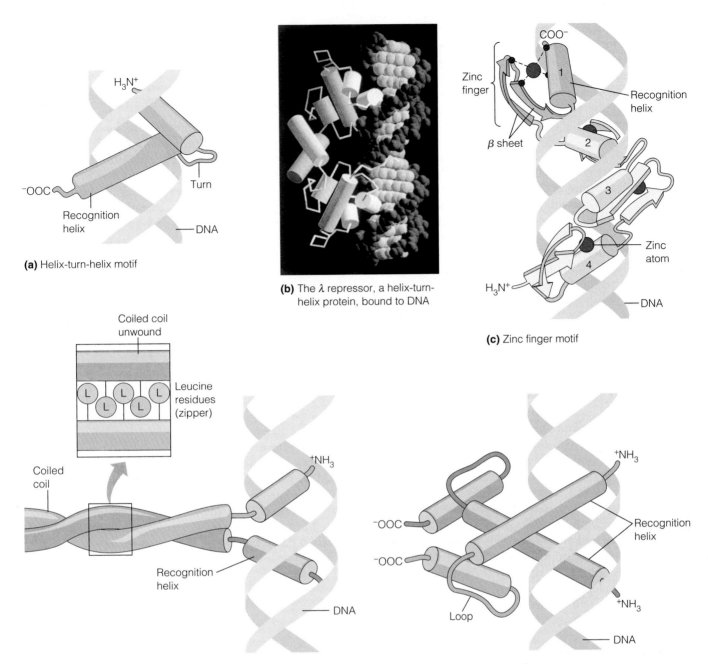

(a) Helix-turn-helix motif

(b) The λ repressor, a helix-turn-helix protein, bound to DNA

(c) Zinc finger motif

(d) Leucine zipper motif

(e) Helix-loop-helix motif

Figure 21-25   Common Structural Motifs in DNA-Binding Transcription Factors.   Several structural motifs are commonly found in the DNA-binding domains of regulatory transcription factors. The parts of these domains that directly interact with specific DNA sequences are usually α helices, called recognition helices, which fit into DNA's major groove. In this figure, all α helices are shown as cylinders. **(a)** The helix-turn-helix motif, in which two α helices are joined by a short flexible turn. **(b)** A computer graphic model showing the λ repressor, a helix-turn-helix protein, bound to DNA. It is a dimer of two identical subunits. The helices of the two DNA-binding domains are light blue. **(c)** The zinc finger motif. Each zinc finger consists of an α helix and a two-segment, antiparallel β sheet (shown as ribbons), all held together by the interaction of four cysteine residues, or two cysteine and two histidine residues, with a zinc atom. At the top of the diagram, these key residues are shown as small purple balls; the zinc atoms are shown as larger red balls. Zinc finger proteins typically have several zinc fingers in a row; here we see four. **(d)** The leucine zipper motif, in which an α helix with regularly arranged leucine residues in one polypeptide (green) interacts with a similar region in a second polypeptide (purple). The two helices coil around each other. **(e)** The helix-loop-helix motif, in which a short α helix connected to a longer α helix by a polypeptide loop interacts with a similar region on another polypeptide to create a dimer.

containing an α helix with regularly spaced leucine residues. Because leucines are hydrophobic amino acids that attract one another, the stretch of leucines exposed on the outer surface of one α helix can interlock with a com-parable stretch of leucines on the other α helix, causing the two helices to wrap around each other into a coil that "zippers" the two α helices together (Figure 21-25d). In some transcription factors, leucine zippers are used to "zip" two identical polypeptides together; in other transcription factors, two kinds of polypeptides are joined together. In either case, DNA binding is made possible by two additional α-helical regions located adjacent to the leucine zipper. These two α-helical segments, one derived from each of the two polypeptides, fit into the DNA's major groove and bind to specific base sequences.

Helix-Loop-Helix Motif. The **helix-loop-helix** motif is composed of a short α helix connected by a loop to another, longer α helix (Figure 21-25e). Like leucine zippers, helix-loop-helix motifs contain hydrophobic regions that usually connect two polypeptides, which may be either similar or different. The formation of the four-helix bundle results in the juxtaposition of a recognition helix derived from one polypeptide with a recognition helix derived from the other polypeptide, creating a two-part DNA-binding domain.

DNA Response Elements Coordinate the Expression of Nonadjacent Genes

So far we have focused our attention on the way in which eukaryotic transcription factors bind to DNA and regulate the transcription of individual genes. But eukaryotic cells, like prokaryotes, often need to activate a group of related genes at the same time. In unicellular eukaryotes, as in bacteria, such coordinate gene regulation may be required to respond to some signal from the external environment. In multicellular eukaryotes, coordinate gene regulation is critical for the development and functioning of specialized tissues. For example, during embryonic development a single fertilized animal egg may give rise to trillions of new cells of hundreds of differing types, each transcribing a different group of genes—nerve cells expressing genes required for nerve function, muscle cells expressing genes required for muscle function, and so forth. How do eukaryotes coordinate the expression of groups of related genes under such conditions?

Unlike the situation in prokaryotes, where genes with related functions often lie next to each other in operons, eukaryotic genes that must be turned on (or off) at the same time are usually scattered throughout the genome. To coordinate the expression of such physically separated genes, eukaryotes employ DNA control sequences called **response elements** to turn transcription on or off *in response to* a particular environmental or developmental signal. Response elements can function either as proximal control elements or as components of enhancers. In either case, placing the same type of response element next to genes residing at different chromosomal locations allows these genes to be controlled together even though they are not located next to one another.

Because they allow groups of genes to be controlled in a coordinate fashion, response elements play important roles in regulating gene expression during embryonic development and during tissue responses to changing environmental and physiological conditions. In the fol-

lowing sections, we will describe several examples of such coordinated gene regulation.

Steroid Hormone Receptors Are Transcription Factors That Bind to Hormone Response Elements

One important group of transcription factors involved in coordinate gene regulation is composed of the **nuclear receptor** proteins that mediate the actions of steroid hormones such as progesterone, estrogen, testosterone, and glucocorticoids. Steroid hormones are lipid signaling molecules synthesized by cells in endocrine tissues. The hormones are released into the bloodstream, where they serve as chemical messengers that transmit signals to target tissues whose cells possess the appropriate receptor proteins. After entering a target cell, a steroid hormone binds to its corresponding receptor protein, triggering a series of events that ultimately activates, or in a few cases inhibits, the transcription of a specific set of genes. (Related receptors for thyroid hormone, vitamin D, and retinoic acid function in a similar way.)

The key to this selective effect on gene expression lies in the ability of steroid hormone receptors to function as transcription factors that bind to DNA control sequences called **hormone response elements.** All of the genes whose transcription is activated by a particular steroid hormone are associated with the same type of response element, allowing them to be regulated together. For example, genes activated by estrogen have a 15-bp *estrogen response element* near the upstream end of their promoters, whereas genes activated by glucocorticoids lie adjacent to a *glucocorticoid response element* exhibiting a slightly different base sequence (Figure 21-26).

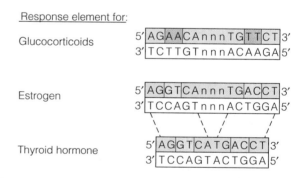

Figure 21-26 Comparison of the DNA Sequence of Several Hormone Response Elements. Note that all three DNA sequences contain inverted repeats (two copies of the same sequence oriented in opposite directions). For example, reading the sequence of the glucocorticoid response element in the 5′ ⟶ 3′ direction from either end yields the same DNA sequence: 5′-AGAACA. The highlighted nucleotides are the only bases of the inverted repeat sequences that vary between the three types of elements. The thyroid hormone element contains the same inverted repeat sequences as the estrogen element, but the three bases that separate the two copies of the sequence in the estrogen element are not present. ("n" signifies that any nucleotide can be located at that position. The dashed lines are included to help you line up the comparable regions of the estrogen and thyroid hormone elements.)

How do steroid hormone receptors activate the transcription of genes associated with the appropriate hormone response element? Steroid hormone receptors belong to the zinc finger category of transcription factors. They typically contain three distinct domains: a domain that binds to a particular steroid hormone, a domain that binds to the appropriate DNA response element, and a domain that activates transcription. Using the glucocorticoid hormone *cortisol* as an example, Figure 21-27 illustrates how the binding of a steroid hormone to its receptor can trigger a change in gene expression. In the absence of cortisol, the glucocorticoid receptor (GR) is located mainly in the cytosol, where it is bound to an inhibitory cytosolic protein that prevents the GR molecule from entering the nucleus and binding to DNA. But the binding of cortisol to GR changes its conformation, causing release of the inhibitory protein. The GR molecule (with its bound cortisol) is then free to move into the nucleus, where it binds to glucocorticoid response elements wherever they may reside in the DNA. The binding of a GR molecule to such a response element in turn facilitates the binding of a second GR molecule to the same response element, creating a GR *dimer* that activates transcription of the adjacent genes.

This binding of *two* GR molecules to the same DNA site is made possible by the fact that the DNA sequence of the glucocorticoid response element contains an **inverted repeat**—that is, *two* copies of the same DNA sequence oriented in opposite directions (see Figure 21-26). Therefore the two GR molecules in a GR dimer each bind to one of the two DNA sequence repeats. Such binding of a protein dimer to an inverted repeat sequence within a DNA control element is a common theme among regulatory transcription factors.

Since glucocorticoid response elements are located near all genes that need to be turned on by cortisol, the presence of cortisol will activate all of these genes simultaneously, regardless of their chromosomal location. This model provides a framework for understanding how steroid hormones, transported through the bloodstream to virtually all cells of the body, can turn on a specific set of genes in the appropriate target tissues. However, some minor differences in detail are observed for various steroid hormones. For example, unlike the glucocorticoid receptor, the receptors for most steroid hormones are already located in the nucleus, even in the absence of hormone. But in the absence of the appropriate hormone, these nuclear receptors do not activate transcription because, like the glucocorticoid receptor, they are bound to inhibitory proteins. Here, too, the binding of the appropriate steroid hormone releases the receptor from its association with its inhibitory protein (or proteins), allowing the receptor to bind to its DNA response element and activate transcription.

Although steroid hormone receptors usually stimulate gene transcription, in a few cases they inhibit it. For example, the glucocorticoid receptor can bind to two different types of DNA response elements—one type associated with genes whose transcription is activated, the other with genes whose transcription is inhibited. After binding to the inhibitory type of response element, the glucocorticoid receptor does not form dimers as it does when bound to activating response elements. Instead, binding of the nondimerized glucocorticoid receptor depresses the initiation of gene transcription by interfering with the binding of other transcription factors to the promoter region. The existence of separate activating and inhibiting response elements for the same hormone receptor allows a single hormone to stimulate transcription of genes coding for one group of proteins while it simultaneously depresses transcription of genes coding for other proteins.

CREBs and STATs Are Examples of Transcription Factors Activated by Phosphorylation

The steroid hormone receptors we have been discussing are allosteric proteins that regulate gene expression by changing their DNA-binding affinity in response to steroid hormones, in much the same way that bacterial repressors change their DNA-binding properties after binding to the appropriate small molecules. An alternative approach for controlling the activity of transcription factors is based on *protein phosphorylation* (addition of phosphate groups). One common example of this type of control involves *cAMP (cyclic AMP)*, the widely employed second messenger whose role in mediating the effects of a

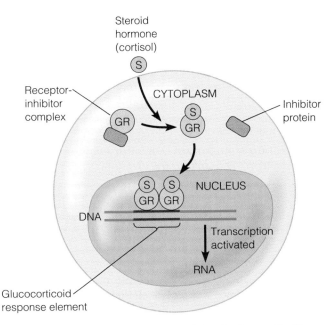

Figure 21-27 Activation of Gene Transcription by Glucocorticoid Receptors. The steroid hormone cortisol (S) diffuses through the plasma membrane and binds to the glucocorticoid receptor (GR), causing the release of an inhibitory protein and activating the GR molecule's DNA-binding site. The GR molecule then enters the nucleus and binds to a glucocorticoid response element in DNA, which in turn causes a second GR molecule to bind to the same response element. The resulting GR dimer activates transcription of the adjacent gene.

variety of extracellular signaling molecules was discussed in Chapter 10. As we saw earlier in this chapter, cAMP also plays a role in controlling catabolite-repressible operons in bacteria. In eukaryotes, cAMP functions by stimulating the activity of protein kinase A, which in turn catalyzes the phosphorylation of a variety of proteins, including a transcription factor called the **CREB protein.** The phosphorylated CREB protein activates the transcription of a specific group of genes by binding to the *cAMP response element (CRE),* which is located adjacent to genes whose transcription is induced by cAMP. Phosphorylation of the CREB protein stimulates its activation domain, thereby allowing it to activate transcription of these cAMP-inducible genes (Figure 21-28).

Protein phosphorylation is also involved in activating a family of transcription factors called **STATs** (for **S**ignal **T**ransducers and **A**ctivators of **T**ranscription). Among the signaling molecules that activate STATs are the *interferons,* a group of glycoproteins produced and secreted by animals cells in response to viral infection. The secreted interferons bind to receptors on the surface of neighboring cells, causing them to produce proteins that make the target cells more resistant to viral infection. The binding of interferon to its cell-surface receptor causes the cytosolic domain of the receptor to associate with and activate an intracellular protein kinase called **Janus activated kinase (JAK).** The activated JAK protein in turn catalyzes the phosphorylation of STAT molecules, which are normally present in an inactive form in the cytoplasm. Phosphorylation causes the STAT molecules to dimerize and move from the cytoplasm to the nucleus, where they bind to appropriate DNA response elements and activate transcription.

The JAK-STAT signaling pathway exhibits considerable specificity. For example, several types of interferon trigger the phosphorylation and activation of different STAT proteins, each of which associates with a particular type of DNA response element and hence activates the transcription of a unique set of genes. In addition to interferons, several other types of signaling molecules bind to cell-surface receptors that trigger the phosphorylation of STATs.

STATs and the CREB protein are only two of the many types of transcription factors known to be activated by phosphorylation. You learned in Chapter 10, for example, that the TGFβ family of growth factors bind to plasma membrane receptors whose activation leads to the phosphorylation of cytoplasmic proteins called *Smads,* which then travel to the nucleus and function as transcription factors that turn on the transcription of specific genes. And we described in Chapter 17 how other growth factors trigger the activation of protein kinases called *MAP kinases,* which phosphorylate and activate transcription factors involved in controlling cell growth and division.

The Heat-Shock Response Element Coordinates the Expression of Genes Activated by Elevated Temperatures

The **heat-shock genes** provide another example of how eukaryotes coordinate the regulation of genes located at different chromosomal sites. Heat-shock genes were initially defined as genes expressed in response to an increase in temperature. But they are now known to respond to other stressful conditions as well, and so they are also called *stress-response genes.* There are a number of different heat-shock genes in both prokaryotic and eukaryotic genomes, and the appropriate environmental trigger activates all of them simultaneously. For example, briefly warming cultured cells by raising the temperature a few degrees activates the transcription of multiple heat-shock genes. Although the functions of these genes are not completely understood, the proteins produced by at least some of them help minimize the damage resulting from thermal denaturation of important cellular proteins.

In cells of *Drosophila,* for example, the most prominent product of the heat-shock genes is Hsp70 (p. 672), a molecular chaperone involved in normal protein folding that can also facilitate the refolding of heat-damaged proteins. The region immediately upstream from the start site of the *hsp70* gene contains several sequences commonly encountered in eukaryotic promoters, such as a TATA box, a CAAT box, and a GC box. In addition, a 14-bp sequence called the **heat-shock response element** is located 62 bases upstream from the transcription start site of the *Drosophila hsp70* gene, and in a comparable location in other genes whose transcription is activated by elevated temperatures.

To investigate the role of this sequence in regulating transcription, Hugh Pelham employed recombinant DNA techniques to transfer the heat-shock response element from the *hsp70* gene to a position adjacent to a viral gene coding for the enzyme thymidine kinase. Transcription of the thymidine kinase gene is not normally induced by

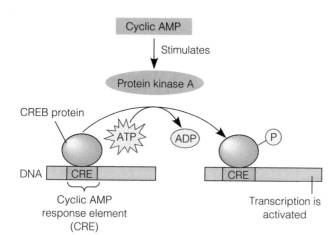

Figure 21-28 Activation of Gene Transcription by Cyclic AMP. Genes activated by cyclic AMP possess an upstream cyclic AMP response element (CRE) that binds a transcription factor called the CREB protein. In the presence of cyclic AMP, cytoplasmic protein kinase A is activated and its activated catalytic subunit then moves into the nucleus, where it catalyzes phosphorylation of the CREB protein. Although the detailed mechanism is not shown, phosphorylated CREB interacts with coactivators that exhibit histone acetyltransferase activity, triggering histone acetylation that loosens the packing of nucleosomes and thereby facilitates access of the transcription machinery to gene promoters.

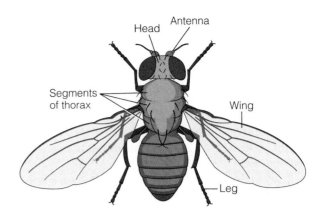

(a) Wild-type *Drosophila*

Figure 21-29 Homeotic Mutants of *Drosophila*. (a) Wild-type *Drosophila* has two wings and six legs extending from its three thoracic segments. **(b)** Mutations in the bithorax gene complex convert the third thoracic segment to a second thoracic segment (the wing-producing segment), and an additional set of wings is formed. **(c)** A mutation in the antennapedia gene complex causes legs to develop where the insect's antennae should be.

heat, but when the hybrid gene was introduced into cells, elevating the temperature triggered the expression of thymidine kinase. In other words, the presence of the heat-shock response element caused the thymidine kinase gene to be activated by heat just like a typical heat-shock gene. Subsequent investigations have revealed that the activation of heat-shock genes is mediated by the binding of a protein called the *heat-shock transcription factor* to the heat-shock response element. The heat-shock transcription factor is present in an inactive form in nonheated cells, but elevated temperatures cause a change in the structure of the protein that allows it to bind to the heat-shock response element in DNA. The protein is then further modified by phosphorylation, which makes it capable of activating gene transcription.

The heat-shock system again illustrates a basic principle of eukaryotic coordinate regulation: Genes located at different chromosomal sites can be activated by the same signal if the same response element is located near each of them.

(b) Bithorax mutant

Homeotic Genes Code for Transcription Factors That Regulate Embryonic Development

Some especially striking examples of coordinate gene regulation in eukaryotes involve an unusual class of genes known as **homeotic genes.** When mutations occur in one of these genes, a strange thing happens during embryonic development—one part of the body is replaced by a structure that normally occurs somewhere else. The discovery of homeotic genes can be traced back to the 1940s, when Edward B. Lewis discovered a cluster of *Drosophila* genes, the *bithorax gene complex,* in which certain mutations cause drastic developmental abnormalities such as the growth of an extra pair of wings (Figure 21-29b). Later, Thomas C. Kaufman and his colleagues discovered a second group of genes that, when mutated, lead to different but equally bizarre developmental changes—for example, causing legs to grow from the fly's head in place of antennae (Figure 21-29c). This group of genes is the *antennapedia gene complex.* The bithorax and antennapedia genes are called homeotic genes because *homeo* means "alike" in Greek, and mutations in these genes change one body segment of *Drosophila* to resemble another.

Although such phenomena might at first appear to be no more than oddities of nature, the growth of an extra pair of wings or the development of legs in the wrong place suggests that homeotic genes play key roles in con-

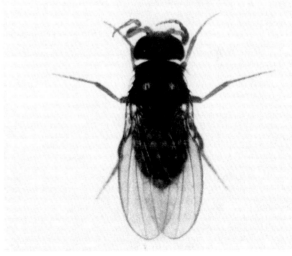

(c) Antennapedia mutant

trolling the formation of the body plan of developing embryos. Homeotic genes exert their influence over embryonic development by coding for a family of regulatory transcription factors that activate (or inhibit) the transcription of developmentally important genes by binding to specific DNA sequences. By binding to all copies of the appropriate DNA element in the genome, each homeotic transcription factor can influence the expression of dozens or even hundreds of genes in the growing embryo. The result of this coordinated gene regulation is the development of fundamental body characteristics such as appendage shape and location. Most homeotic genes control major developmental pathways, and in vertebrates they may also help regulate other important processes, such as histone production and antibody synthesis.

A clue to how homeotic genes work first emerged from the discovery that the bithorax and antennapedia genes each contain a similar, 180-bp sequence near their 3′ end that resembles a comparable sequence found in other homeotic genes. Termed the **homeobox,** this DNA sequence codes for a stretch of 60 amino acids called a **homeodomain** (Figure 21-30). Homeobox sequences have been detected in more than 60 *Drosophila* genes and in a range of other organisms as diverse as sea squirts, frogs, mice, and humans—an evolutionary distance of more than 500 million years. In fact, the C-terminal portion of the typical eukaryotic homeodomain even shows some homology with prokaryotic repressors.

The widespread occurrence of the homeobox suggests that the homeodomain for which it codes must perform an important function that has been highly conserved during evolution. Like the bacterial *lac* repressor, *trp* repressor, and cAMP receptor proteins discussed earlier in the chapter, the homeodomain contains a helix-turn-helix motif, suggesting that it functions in binding to DNA.

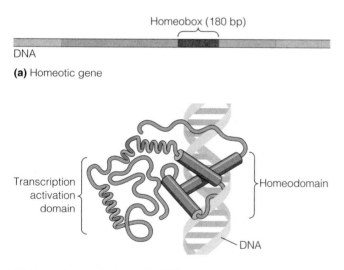

Homeobox (180 bp)

DNA

(a) Homeotic gene

Transcription activation domain

Homeodomain

DNA

(b) Homeotic protein bound to DNA

Figure 21-30 Homeotic Genes and Proteins. (a) Homeotic mutants like the ones in Figure 21-29 have mutations in homeotic genes. These genes all contain a 180-bp segment, called a homeobox, that encodes a homeodomain in the homeotic protein. **(b)** The homeodomain (red) is a helix-turn-helix DNA-binding domain, which, in combination with a transcription activation domain, enables the protein to function as a regulatory transcription factor. Notice that the homeodomain has three α helices (shown as cylinders).

This hypothesis has been confirmed by synthesizing the homeodomain region of the transcription factor encoded by the antennapedia gene and showing that this 60–amino acid stretch binds to DNA in the identical sequence-specific manner as the intact transcription factor.

The role of the homeodomain helix-turn-helix motif in DNA binding has been further investigated in the homeotic gene called *bicoid,* which is required for establishing the proper anterior-posterior polarity during *Drosophila* development. The amino acid lysine is present in position number 9 of the recognition helix of the helix-turn-helix motif in the bicoid protein, whereas glutamine is present in the comparable position in the antennapedia protein. Normally the bicoid and antennapedia proteins function as transcription factors that bind to different DNA response elements. However, if the lysine present at position 9 of the bicoid protein is mutated to glutamine, the altered bicoid protein binds to DNA sequences normally recognized by the antennapedia protein. Thus, not only does the helix-turn-helix motif of the homeodomain mediate DNA binding, but single amino acid differences within this motif can alter the DNA sequence to which the protein binds.

Eukaryotic Gene Regulation: Posttranscriptional Control

We have now examined the first two levels of regulation for eukaryotic gene expression, namely genomic controls and transcriptional controls. After transcription has taken place, the flow of genetic information involves a complex series of *posttranscriptional events,* any or all of which can turn out to be regulatory points (see Figure 21-11, steps ③, ④, and ⑤). Posttranscriptional regulation may be especially useful in providing ways to fine-tune the pattern of gene expression rapidly, allowing cells to respond to transient changes in the intracellular or extracellular environment without changing their overall transcription patterns.

Control of RNA Processing and Nuclear Export Follows Transcription

In eukaryotes, posttranscriptional control begins with the processing of primary RNA transcripts, which provides many opportunities for the regulation of gene expression. As we learned in Chapter 19, virtually all RNA transcripts in eukaryotic nuclei undergo substantial processing, including addition of a 5′ cap and a 3′ poly(A) tail, chemical modifications such as methylation, splicing together of exons accompanied by the removal of introns, and RNA editing. Among these processing events, RNA splicing is an especially important control site because its regulation allows cells to create a variety of different mRNAs from the same pre-mRNA, thereby permitting a gene to generate more than one protein product.

This phenomenon, called **alternative RNA splicing**, is controlled by proteins that bind to pre-mRNA molecules and cause some splice sites to be skipped and others to become activated. By increasing the ways in which a given pre-mRNA can be spliced, such mechanisms make it possible for a single gene to yield dozens or even hundreds of different mRNAs. A remarkable example has been discovered in the inner ear of birds, where 576 alternatively spliced forms of mRNA are produced from a single potassium channel gene. Expression of varying subsets of these mRNAs in different cells of the inner ear facilitates perception of different sound frequencies.

Figure 21-31 illustrates another well-documented example of alternative RNA splicing, in this case involving the mRNA coding for a type of antibody called *immunoglobulin M (IgM)*. The IgM protein exists in two forms, a secreted version and a version that becomes incorporated into the plasma membrane of the cell that makes it. Like all antibodies, IgM consists of four polypeptide subunits, two heavy chains and two light chains. It is the sequence of the heavy chain that determines whether IgM is secreted or membrane bound. The gene coding for the heavy chain has two alternative poly(A) sites where RNA transcripts can terminate, yielding two kinds of pre-mRNA molecules that differ at their 3′ ends. The exons within these pre-mRNAs are then spliced together in two

different ways, producing mRNAs that code for either the secreted version of the heavy chain or the plasma membrane bound form. Only the splicing pattern for the membrane-bound version includes the exons encoding the hydrophobic amino acid domain that anchors the heavy chain to the plasma membrane.

After RNA splicing, the next posttranscriptional step subject to control is the export of mRNA through nuclear pore complexes and into the cytoplasm. We know that export is controlled because RNAs exhibiting defects in capping or splicing have markedly decreased rates of export from the nucleus. Evidence that the export process is closely coupled to RNA splicing has come from experiments in which introns were artificially removed from purified genes prior to the insertion of the genes back into intact cells. Although the altered genes were found to be actively transcribed, their transcripts were not exported from the nucleus unless at least one intron was initially present. Experiments involving HIV, the virus that causes AIDS, have provided even stronger evidence that the export of individual RNAs can be selectively controlled. One type of RNA molecule encoded by HIV is synthesized in the nucleus and remains there until a viral protein called Rev is produced. The amino acid sequence of the

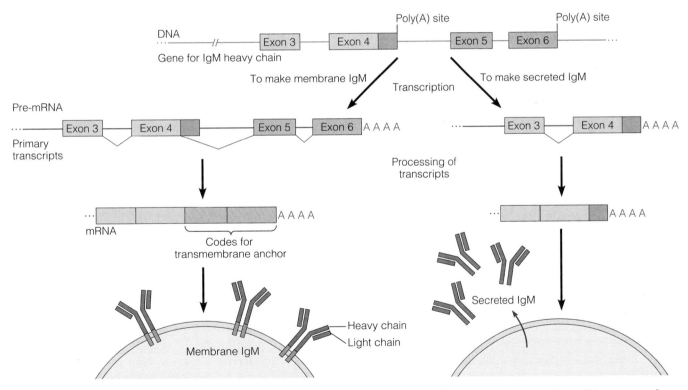

Figure 21-31 Alternative RNA Splicing to Produce Variant Gene Products. The antibody protein immunoglobulin M (IgM) exists in two forms, secreted IgM and membrane-bound IgM, which differ in the carboxyl ends of their heavy chains. A single gene carries the genetic information for both types of IgM heavy chain; the end of the gene corresponding to the polypeptide's C-terminus is shown at the top of the figure. This DNA has two possible poly(A) addition sites, where RNA transcripts can terminate, and four exons that can be used in two alternative configurations. The splices made in the primary transcripts are indicated by V-shaped symbols. A splicing pattern that uses a splice junction within exon 4 and retains exons 5 and 6 results in the synthesis of IgM heavy chains that are held in the plasma membrane by a transmembrane anchor encoded by exons 5 and 6. The alternative product is secreted because a splice within exon 4 is not made and the transcript is terminated after exon 4.

Rev protein includes a *nuclear export signal* (p. 515), which allows the protein to guide the viral RNA out through the nuclear pores and into the cytoplasm.

Translation Rates Can Be Controlled by Initiation Factors and Translational Repressors

Once mRNA molecules have been exported from the nucleus to the cytoplasm, several **translational control** mechanisms are available to regulate the rate at which each mRNA is translated into its polypeptide product. Some translational control mechanisms work by altering ribosomes or protein synthesis factors; others regulate the activity and/or stability of mRNA itself. Translational control takes place in prokaryotes as well as eukaryotes, but we will restrict our discussion to several eukaryotic examples.

One well-studied example of translational control occurs in developing erythrocytes, where *globin* polypeptide chains are the main product of translation. The synthesis of globin depends on the availability of *heme*, the iron-containing prosthetic group that attaches to globin chains to form the final protein product, hemoglobin. Developing erythrocytes normally synthesize globin at a high rate. However, globin synthesis would be wasteful if not enough heme were available to complete the formation of hemoglobin molecules, so erythrocytes have developed a mechanism for adjusting polypeptide synthesis to match heme availability.

This mechanism involves a protein kinase, called **heme-controlled inhibitor (HCI)**, whose activity is regulated by heme. Figure 21-32 shows how HCI works. In the presence of heme, HCI is inactive, but in its absence, HCI specifically phosphorylates and thereby inhibits eIF2 (eukaryotic initiation factor 2), one of several proteins required for the initiation of translation in eukaryotes. Phosphorylated eIF2 cannot form the complex with GTP and methionyl tRNA that is necessary for initiating translation (see Chapter 20). Because eIF2 is involved in the translation of all eukaryotic mRNAs, this mechanism for inhibiting eIF2 in the absence of heme influences all translational activity within the cell. However, globin chains account for more than 90% of the polypeptides produced by developing erythrocytes, so the main effect of HCI in erythrocytes is on globin synthesis. Other translational inhibitors that phosphorylate eIF2 have been detected in a variety of eukaryotic cell types, suggesting that phosphorylation of eIF2 is a widespread means of translational regulation.

The use of protein phosphorylation to control translation rates is not restricted to eIF2. Eukaryotic initiation factor eIF4F, which binds to the 5′ mRNA cap, is also regulated by phosphorylation, although in this case phosphorylation activates rather than inhibits the initiation factor. An example of this type of control is observed in cells infected by adenovirus, which inhibits protein synthesis in the infected cells by triggering the dephosphorylation of eIF4F. Other initiation and elongation factors involved in eukaryotic protein synthesis can also be controlled by

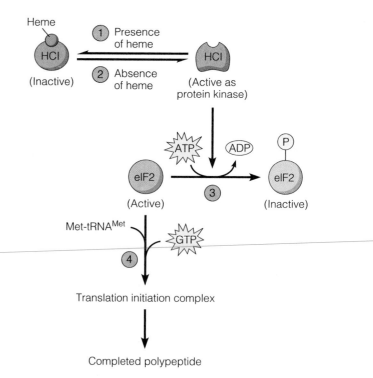

Figure 21-32 Regulation of Translation by Heme in Developing Erythrocytes. The main function of developing erythrocytes is to synthesize hemoglobin, which consists of four globin polypeptides and a heme prosthetic group. These cells contain the protein HCI (heme-controlled inhibitor), which regulates this synthesis in response to the availability of heme. ① When heme is present, it binds to HCI, inactivating it. ② When heme is absent, HCI is active. ③ Active HCI functions as a kinase that catalyzes the phosphorylation of eIF2, a key translation initiation factor. Phosphorylated eIF2 is inactive; it cannot combine with methionyl tRNA and GTP to form the translation initiation complex. Thus, in the absence of heme, translation of all mRNA in the cell is inhibited. The main effect is on globin synthesis, because globin mRNA constitutes most of the cell's mRNA. ④ When heme is present, translation of the mRNA proceeds. Newly made globins combine with heme to form hemoglobin molecules (not shown).

phosphorylation, as can several of the aminoacyl-tRNA synthetases that join amino acids to their respective tRNAs.

The types of translational control we have discussed so far are relatively nonspecific, in that they affect the translation rates of all mRNAs within a cell. An example of a more specific type of translational control is seen with *ferritin*, an iron-storage protein whose synthesis is selectively stimulated in the presence of iron. The key to this selective stimulation lies in the 5′ untranslated leader sequence of ferritin mRNA, which contains a 28-nucleotide segment—the **iron-response element (IRE)**—that forms a hairpin loop required for the stimulation of ferritin synthesis by iron. If the IRE sequence is experimentally inserted into a gene whose expression is not normally regulated by iron, translation of the resulting mRNA becomes iron-sensitive.

Figure 21-33 shows how the IRE sequence works. When the iron concentration is low (left panel), a regula-

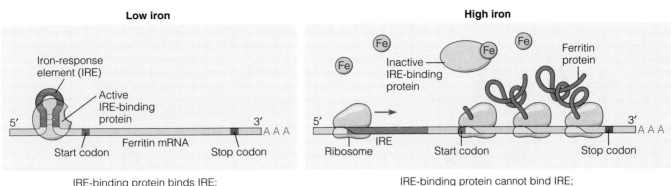

Low iron

Iron-response element (IRE)

Active IRE-binding protein

5'

Start codon

Ferritin mRNA

Stop codon

3' A A A

IRE-binding protein binds IRE; ferritin synthesis low

High iron

Fe

Fe

Fe

Inactive IRE-binding protein

Ferritin protein

5'

Ribosome

IRE

Start codon

Stop codon

3' A A A

IRE-binding protein cannot bind IRE; ferritin synthesis high

Figure 21-33 Translational Control in Response to Iron. The initiation of translation of ferritin mRNA is inhibited by the binding of the IRE-binding protein to the hairpin structure that can be formed by an iron response element (IRE) in the 5′ untranslated leader sequence of the mRNA. The IRE-binding protein is allosteric. When iron binds to it, the IRE-binding protein changes to a conformation that does not recognize the IRE. Therefore, in the presence of iron, ribosomes can assemble on the mRNA and proceed to translate it. Hairpin formation by the IRE does not significantly interfere with ribosome assembly or movement along the mRNA.

tory protein called the *IRE-binding protein* binds to the IRE sequence in ferritin mRNA, preventing the mRNA from forming an initiation complex with ribosomal subunits. But the IRE-binding protein is an allosteric protein whose activity can be controlled by the binding of iron. When more iron is available (right panel), the protein binds an iron atom and undergoes a conformational change that prevents it from binding to the IRE, thereby allowing the ferritin mRNA to be translated. The IRE-binding protein is therefore an example of a **translational repressor** that selectively controls the translation of a particular mRNA. This type of translational control allows cells to respond to specific changes in the environment faster than would be possible by transcriptional control.

Translation Can Also Be Controlled by Regulation of mRNA Half-Life

Translation rates are also subject to control by alterations in mRNA stability—in other words, the more rapidly an mRNA molecule is degraded, the less time available for it to be translated. The *half-life*, or time required for 50% of the initial amount of RNA to be degraded, varies widely among eukaryotic mRNAs, ranging from 30 minutes or less for some growth factor mRNAs to over 10 hours for the mRNA encoding β-globin. The length of the poly(A) tail is one factor that plays a role in controlling mRNA stability. Messenger RNAs with short poly(A) tails tend to be less stable than mRNAs with longer poly(A) tails. In some cases, mRNA stability is also influenced by specific features of the 3′ untranslated region. For example, short-lived mRNAs for several growth factors have a particular AU-rich sequence in this region. The AU-rich sequence triggers removal of the poly(A) tail by degradative enzymes. When the AU-rich sequence is transferred to the 3′ end of a normally stable globin message using recombi-

nant DNA techniques, the hybrid mRNA acquires the short half-life typical of a growth factor mRNA.

An alternative mechanism for regulating mRNA stability is illustrated by the action of iron, which, in addition to participating in translational control as just described, also plays a role in controlling mRNA degradation. The uptake of iron into mammalian cells is mediated by a plasma membrane receptor protein called the *transferrin receptor*. When iron is scarce, synthesis of the transferrin receptor is stimulated by a mechanism that protects transferrin mRNA from degradation, thereby making more mRNA molecules available for translation. As shown in Figure 21-34, this control mechanism involves an IRE (similar to the one in ferritin mRNA) located in the 3′ untranslated region of transferrin mRNA. When intracellular iron levels are low (left panel) and increased uptake of iron is necessary, the IRE-binding protein binds to this IRE and protects the mRNA from degradation. When iron levels in the cell are high (right panel) and additional uptake is not necessary, the IRE-binding protein binds an iron atom and dissociates from the mRNA, allowing the mRNA to be degraded. As a result, synthesis of the transferrin receptor decreases, leading to a decreased rate of iron transport into the cell.

Another way of controlling mRNA degradation is through **RNA interference** (also called *posttranscriptional gene silencing*), a phenomenon that can be demonstrated experimentally by injecting eukaryotic cells with synthetic, double-stranded RNA molecules of known sequence. In response, cells inhibit expression of DNA genes containing the corresponding base sequence by selectively destroying the mRNAs produced by these genes. Normally, RNA interference appears to protect cells from problems involving the presence of double-stranded RNA. For example, in plants infected with viruses that produce double-stranded RNAs, RNA interference helps

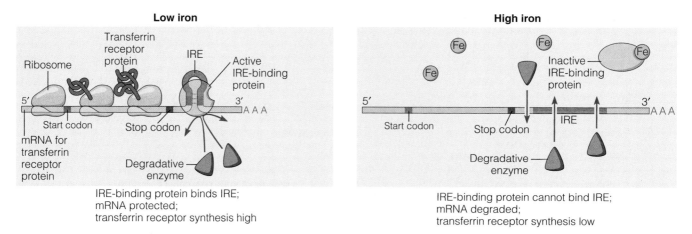

Low iron

Ribosome

Transferrin receptor protein

IRE

Active IRE-binding protein

5′

Start codon

Stop codon

mRNA for transferrin receptor protein

Degradative enzyme

3′ A A A

IRE-binding protein binds IRE;
mRNA protected;
transferrin receptor synthesis high

High iron

Fe

Fe

Fe

Fe

Inactive IRE-binding protein

5′

Start codon

Stop codon

IRE

Degradative enzyme

3′ A A A

IRE-binding protein cannot bind IRE;
mRNA degraded;
transferrin receptor synthesis low

Figure 21-34   Control of mRNA Degradation in Response to Iron. Degradation of the mRNA for the transferrin receptor, a protein required for iron uptake by the cell, is regulated by the same allosteric IRE-binding protein shown in Figure 21-33. The mRNA for the transferrin receptor has an IRE in its 3′ untranslated region. When the intracellular iron concentration is low, the IRE-binding protein remains bound to the IRE, protecting the mRNA from degradation and allowing more transferrin receptor protein to be synthesized. When the iron level is high, iron atoms bind to the IRE-binding protein, causing it to leave the IRE; the mRNA is left vulnerable to attack by degradative enzymes.

defend against the viral infection by shutting down expression of the viral genes.

The mechanism underlying RNA interference involves a ribonuclease that selectively degrades RNA molecules whose base sequences are related to that of the triggering double-stranded RNA. In the first step of this process, the initial double-stranded RNA is bound by the ribonuclease and degraded to small, double-stranded RNA fragments roughly two dozen base pairs in length, which remain bound to the ribonuclease. One strand of these small RNA fragments then binds via complementary base pairing to a target mRNA molecule, which is subsequently destroyed by the attached ribonuclease.

Posttranslational Control Involves Modifications of Protein Structure, Function, and Degradation

After translation of an mRNA molecule has produced a polypeptide chain, there are still many ways of regulating the activity of the polypeptide product. This brings us to the final level of controlling gene expression, namely, the various **posttranslational control** mechanisms that are available for modifying protein structure and function (see Figure 21-11, step ⑤). Included in this category are reversible structural alterations that influence protein function, such as protein phosphorylation and dephosphorylation, as well as permanent alterations, such as proteolytic cleavage. Other posttranslational events subject to regulation include the guiding of protein folding by chaperone proteins, the targeting of proteins to intracellular or extracellular locations, and the interaction of proteins with regulatory molecules or ions, such as cAMP or Ca^{2+}. These kinds of posttranslational events are discussed elsewhere in other

contexts, but we again mention them here to emphasize that the flow of information in the cell is not complete until a properly functioning gene product is available.

So far we have focused on control mechanisms that influence the rate at which a functional protein product for a given gene is produced. But the amount of each protein present in a cell at any particular moment is a function of its rate of degradation as well as its rate of synthesis. Thus, regulating the rate of protein degradation can also be an important means for influencing gene expression.

The relative contributions made by the rates of protein synthesis and degradation in determining the final concentration of any protein in the cell are summarized by the equation

$$P = \frac{K_s}{K_d} \qquad (21\text{-}1)$$

where P is the amount of a given protein present, K_s is the rate order constant for its synthesis, and K_d is the rate order constant for its degradation. Rate constants for degradation are often expressed in the form of a half-life that corresponds to the amount of time required for half the protein molecules existing at any moment to be degraded. The half-lives of cellular proteins range from as short as a few minutes to as long as several weeks. Such differences in half-life can have dramatic effects on the ability of different proteins to respond to changing conditions.

A striking example occurs in liver cells exposed to the steroid hormone cortisone, which triggers a tenfold increase in concentration of the enzyme tryptophan pyrrolase but only a slight increase in concentration of the enzyme arginase. At first glance such data would seem to indicate that cortisone selectively stimulates the synthesis of tryptophan pyrrolase. In fact, cortisone stimulates the synthesis of both tryptophan pyrrolase and arginase to about the same extent. The explanation for the apparent paradox is that tryptophan pyrrolase has a much shorter half-life (larger K_d) than arginase. Using the mathematical relationship summarized in equation 21-1, calculations show that the concentration of a protein exhibiting a short half-life (large K_d) changes more dramatically in response to alterations in synthetic rate than does the concentration of a protein with a longer half-life. For this reason, enzymes that are important in metabolic regulation tend to have short half-lives, allowing their intracellular concentrations to be rapidly increased or decreased in response to changing conditions.

Although the factors that determine the rates at which various proteins are degraded are not completely understood, it is clear that the process is subject to regulation. In rats deprived of food, for example, the amount of the enzyme arginase in the liver doubles within a few days without any corresponding change in the rate of arginase synthesis. The explanation for the observed increase is that the arginase degradation rate slows dramatically in starving rats. This effect is highly selective; most proteins are degraded more rapidly, not more slowly, in starving animals (Figure 21-35).

Ubiquitin Targets Proteins for Degradation by Proteasomes

The discovery that protein breakdown can be preferentially controlled raises the question of how this is accomplished. The most common method for targeting proteins for destruction is to link them to **ubiquitin**, a small protein chain containing 76 amino acids. Ubiquitin is joined to target proteins by a mechanism that involves three components: a *ubiquitin-activating enzyme* (E1), a *ubiquitin-conjugating enzyme* (E2), and a *substrate recognition protein* (E3). As shown in Figure 21-36, ubiquitin is first activated by

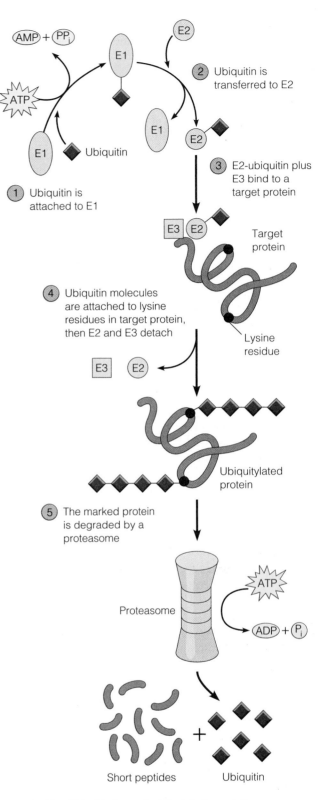

Figure 21-36   Ubiquitin-Dependent Protein Degradation. ① Ubiquitin is first attached to enzyme E1 in an ATP-requiring reaction. ② The ubiquitin molecule is then transferred to enzyme E2, which ③ binds with E3 to the N-terminus of a protein targeted for degradation. ④ Ubiquitin molecules are attached to the protein's lysine residues. ⑤ A proteasome degrades the ubiquitylated protein into short peptides in another ATP-requiring reaction and the ubiquitin is released for reuse.

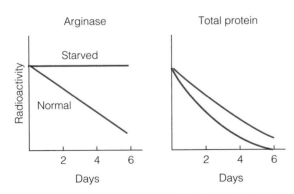

Figure 21-35   Regulation of the Degradation Rate of the Liver Enzyme Arginase. The enzyme degradation rate was measured by injecting rats with a radioactive amino acid to allow proteins to become radioactively labeled, and then measuring the rate at which the radioactivity is lost from the enzyme molecules. The data reveal that the degradation of arginase is inhibited in starved rats, while the degradation rate of total protein is slightly enhanced.

attaching it to E1 in an ATP-dependent reaction. The activated ubiquitin is then transferred to E2 and subsequently linked, in a reaction facilitated by E3, to a lysine residue in a target protein. Additional molecules of ubiquitin are then added in sequence, forming short chains.

These ubiquitin chains serve as targeting signals that are recognized by large, protein-degrading structures called **proteasomes**. Proteasomes are the predominant proteases (protein-degrading enzymes) of the cytosol and are often present in high concentration, accounting for up to 1% of all cellular protein. Each proteasome has a mass of roughly two million daltons and consists of half a dozen proteases associated with several ATPases and a binding site for ubiquitin chains. Shaped like a short cylinder, the proteasome ingests ubiquitylated proteins at one end and hydrolyzes their peptide bonds in an ATP-dependent process, generating small peptide fragments that are released from the other end of the cylinder.

Because most intracellular protein breakdown is carried out by the ubiquitin-proteasome pathway, the key to understanding the control of protein degradation lies in unraveling the mechanisms by which particular proteins are chosen for ubiquitylation. This selectivity is based in part on the existence of multiple forms of E3 that direct the attachment of ubiquitin to different target proteins. One feature recognized by the various forms of E3 is the amino acid present at the N-terminus of a potential target protein. Some N-terminal amino acids cause proteins to be rapidly ubiquitylated and degraded; others make proteins less susceptible.

Another signal that can control ubiquitylation is a sequence of nine amino acids, called a *destruction box*, whose presence allows cells to select particular proteins for destruction at specific times. For example, you saw in Chapter 17 that progression through the final stages of mitosis is controlled by the anaphase-promoting complex, which triggers passage into anaphase and telophase by targeting selected proteins, such as mitotic cyclin, for degradation. The anaphase-promoting complex accomplishes this task by functioning as a substrate recognition protein (E3) that binds to target proteins containing a destruction box, promoting ubiquitylation of these target proteins by E2 and their subsequent degradation by proteasomes. In addition to such selective mechanisms for degrading specific proteins at appropriate times, proteasomes also play a role in general, ongoing mechanisms for eliminating defective proteins from cells. Recent observations suggest that up to 30% of newly synthesized proteins are defective in some way and are immediately tagged with ubiquitin, triggering their destruction by proteasomes.

Although the ubiquitin-proteasome pathway is the primary mechanism used by cells for degrading proteins, it is not the only means available. You learned in Chapter 12 that lysosomes contain digestive enzymes capable of degrading all major classes of macromolecules, including proteins. Lysosomes can take up and degrade cytosolic proteins by an infolding of the lysosomal membrane, creating small vesicles that are internalized within the lysosome and broken down by the organelle's hydrolytic enzymes. This process of **microautophagy** tends to be rather nonselective in the proteins it degrades. The result is the slow, continual recycling of the amino acid building blocks found in most of a cell's protein molecules. Under stressful conditions, however, such nonselective degradation of proteins would be detrimental to the cell. For example, during prolonged fasting the continued nonselective degradation of cellular proteins could lead to depletion of critical enzymes or regulatory proteins. Under these conditions, lysosomes preferentially degrade proteins containing a targeting sequence that consists of glutamine flanked on either side by a tetrapeptide composed of very basic, very acidic, and/or very hydrophobic amino acids. Proteins exhibiting this sequence are targeted for selective degradation, presumably because they are dispensable to the cell.

A Summary of Eukaryotic Gene Regulation

You have now seen that eukaryotic gene regulation encompasses a broad array of control mechanisms operating at five distinct levels (refer again to Figure 21-11 for a visual summary). The first level of regulation, *genomic control*, includes DNA alterations (amplification, deletion, rearrangements, and methylation), chromatin decondensation, histone acetylation, changes in HMG proteins, and altered connections to the nuclear matrix. The second level, *transcriptional control*, involves specific interactions between regulatory transcription factors and various types of DNA control elements, permitting specific genes to be turned on and off in different tissues and in response to changing conditions. The third level involves control of *RNA processing and nuclear export* by mechanisms such as alternative splicing and regulated export of mRNAs through nuclear pores. The fourth level, *translational control*, includes mechanisms for modifying the activity of protein synthesis factors and for controlling the translation of specific mRNAs via translational repressors and/or differential mRNA degradation rates. Finally, the fifth level involves *posttranslational control* mechanisms for reversibly or permanently altering protein structure and function, as well as mechanisms for controlling protein degradation rates.

Thus, when you read that a particular alteration in cell function or behavior is based on a "change in gene expression," you should understand that the underlying explanation might involve mechanisms operating at any one or more of these levels of control. As a final overview, experimental means of distinguishing among these various levels of control are described in Box 21B.

Contemporary Techniques

DISCRIMINATING AMONG MULTIPLE LEVELS OF CONTROL

The control of gene expression is a topic of great interest and excitement, mainly because of a powerful set of techniques that make it possible to dissect and analyze the complex series of events that lead from a specific gene to the functional protein for which that gene codes. In bacteria, gene expression is controlled to a large extent at the level of transcription, by varying the kinds and amounts of mRNA molecules that are synthesized. Although post-transcriptional control mechanisms also exist in bacteria, they are especially elaborate in eukaryotes, where gene expression can be regulated at numerous points along the pathway from DNA to proteins. To investigate these various levels of control, researchers depend on a group of techniques that make it possible to identify and quantify specific proteins and specific mRNA molecules.

Levels of Gene Expression

The expression of a specific gene usually results in an increase in the level of a particular gene product, commonly an enzyme that can be detected because of its activity. Thus, the first indication of the expression of a particular gene is often an increase in the activity of a specific enzyme. Three examples of induction of enzyme activity are illustrated in Figure 21B-1. Figure 21B-1a shows the increase in β-galactosidase activity in *E. coli* cells exposed to lactose, Figure 21B-1b the increase in tyrosine aminotransferase activity in hormone-treated rat hepatoma cells, and Figure 21B-1c the increase in β-amylase activity that occurs in light-grown but not in dark-grown mustard seedlings.

The time scales vary greatly in the three examples, but the results are similar in each case: An enzyme activity that would otherwise remain low (black line) increases many times in response to the appropriate stimulus (solid red line). Notice also that if the stimulus is removed, the amount of enzyme stops increasing, or even decreases (dashed line).

Though an increase in the activity of a specific enzyme reflects the changing expression of a specific gene, it tells us little about the level at which that expression is regulated, particularly in eukaryotic cells. From what we know about the control of gene expression in bacterial cells, we might be tempted to regard any increase in enzyme activity as evidence that transcription of a specific gene has been turned on. But other explanations could be equally consistent with the data and need to be considered. For example, the enzyme we are interested in might already be present in the cell in an inactive form, and the observed increase in activity in response to a stimulus could simply reflect the activation of existing enzyme molecules. Alternatively, perhaps messenger RNA molecules for the desired enzyme are already on hand in the cell but have not been translated for some reason. In this case, the increase in enzyme activity would reflect the translation of preexisting mRNA molecules.

Distinguishing among these several possibilities requires the ability to detect specific proteins and mRNA molecules. We will consider protein and mRNA, in turn.

Detecting and Quantifying Protein Molecules

The detection of specific proteins often depends on the use of *antibodies,* which are blood proteins that are highly specific with

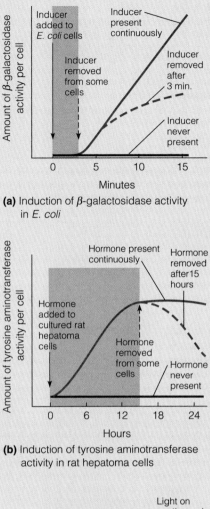

(a) Induction of β-galactosidase activity in *E. coli*

(b) Induction of tyrosine aminotransferase activity in rat hepatoma cells

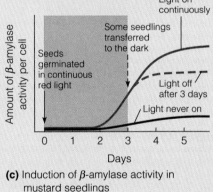

(c) Induction of β-amylase activity in mustard seedlings

Figure 21B-1   Three Examples of Induction of Enzyme Activity.

respect to the particular *antigens* (proteins, usually) to which they will bind. To prepare antibodies against a specific protein, it is necessary to purify the protein and inject it into an experimental animal such as a rabbit. The blood that is subsequently

(continued)

withdrawn from the animal has a high concentration, or *titer*, of antibody capable of reacting specifically with the protein that was injected.

The antibody can then be used to detect and quantify that specific protein by any of a variety of immunological techniques. An especially useful technique is called *immunoblotting*. A homogenate or extract of cells thought to contain the protein is subjected to SDS–polyacrylamide gel electrophoresis to separate molecules based on differences in size (see Figure 7-23). The proteins in the gel are then transferred to a sheet of nitrocellulose paper, and the paper is exposed to the antibody preparation. Because of the specificity of the antibody-antigen reaction, the antibody will bind only to places on the paper where the protein of interest is located. In this way, the protein can be localized to a specific band in the polyacrylamide gel, and the amount of protein in that band can be quantified.

Figure 21B-2 illustrates how we can use the results of an immunological assay such as immunoblotting to analyze gene expression. In addition to assaying for enzyme *activity* at regular time intervals (purple line), we can use the immunoblotting technique to quantify the amount of enzyme *protein* present at the same time points (green and black lines). If the protein level increases concomitantly with enzyme activity (Figure 21B-2a), we can conclude that the increase in enzyme activity is almost certainly due to the accumulation of newly synthesized enzyme molecules.

On the other hand, if the protein is already present at early time points and the amount remains constant as enzyme activity increases (Figure 21B-2b), we can conclude that the increase in enzyme activity is due to the activation of preexisting enzyme molecules. This is an example of posttranslational control, because the regulatory events that lead to activation of the enzyme occur at some point after the synthesis of the enzyme molecule. Common mechanisms of enzyme activation include the removal of a short peptide from one end of an inactive precursor, phosphorylation of specific amino

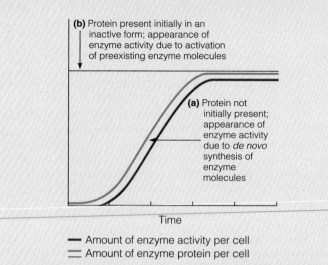

Figure 21B-2   Comparisons of Enzyme Activity Level with Enzyme Protein Level, as Measured by Immunological Assay.

acid side chains, and transport of the protein into the appropriate organelle.

Detecting and Quantifying mRNA Molecules

Many examples of posttranslational control of enzyme activity are known, but for most enzymes, activity increases along with enzyme protein. In such cases, enzyme activity is being regulated at the level of either transcription or translation, depending on whether the mRNA preexists or must be synthesized as needed. To distinguish between these possibilities, we must be able to detect and quantify specific mRNA molecules. The most common techniques take advantage of the sequence complementarity between mRNA and a DNA probe.

Radioactive or dye-labeled complementary DNA (cDNA) probes are first synthesized by reverse transcription of mRNA.

Perspective

Most genes are not expressed all the time. Therefore, in addition to knowing how genetic information is expressed in cells, we need to understand how that expression is regulated. In prokaryotes, coordinately regulated genes are often clustered into operons, and most regulation is effected at the level of transcription. Operons are turned on and off in response to cellular needs. In general, operons that encode enzymes of catabolic pathways are

inducible; their transcription is specifically activated by the presence of substrate. Operons that encode anabolic enzymes, on the other hand, are subject to repression; transcription is specifically turned off in the presence of the endproduct. Both types of regulation are effected by allosteric repressor proteins that exert negative control by binding to the operator and preventing transcription of the associated structural genes.

Some operons coding for catabolic enzymes also have DNA control sites that respond to positive control by the cAMP receptor protein (CRP). This mechanism allows the operon to be shut down in the presence of glucose, ensuring preferential utilization of that sugar. The effect of glucose is mediated by cAMP, the allosteric effector of CRP. Some operons that encode anabolic enzymes, such as those for amino acid biosynthesis, contain an attenuator sequence

There are two main ways to use the resulting cDNAs for quantifying specific mRNAs. In the *Northern blot* procedure, an RNA sample is size-fractionated by gel electrophoresis and transferred to special blotting paper—a step resembling the size-fractionation step used for proteins in the immunoblotting technique. The paper is exposed to a solution of the radioactive cDNA probe, which binds only to places on the paper that contain RNA molecules with a sequence complementary to that of the probe. The amount of mRNA present can be quantified by measuring the amount of radioactive cDNA bound to a given RNA band. Alternatively, if one needs to quantify thousands of different mRNAs at the same time, the *DNA microarray* procedure summarized in Figure 21-20 can be employed.

Regardless of which technique is used to quantify mRNAs, Figure 21B-3 illustrates the results that might be obtained when the same samples used to demonstrate an increase in enzyme activity (purple line) are also assayed for the corresponding mRNA (green and black lines). If the mRNA level is low initially and increases just in advance of the enzyme activity (Figure 21B-3a), we can conclude that either (1) the increase in enzyme activity depends on newly synthesized mRNA molecules that are beginning to be translated as they accumulate in the cell, an example of *transcriptional control;* or (2) the stabilization of mRNA molecules allows increased accumulation within the cell even though transcription rates have not changed, an example of *posttranscriptional control.* Application of the nuclear run-on transcription technique (see Figure 21-19) would allow us to distinguish between these two possibilities.

Alternatively, the mRNA may already be on hand at early time points, such that the amount remains constant as enzyme activity increases (Figure 21B-3b). In this case, the increase in enzyme activity could be due to (1) activation and translation of preexisting mRNA molecules (*translational control*) or (2) stabilization of an enzyme molecule that initially demonstrated a short half-life (*posttranslational control*).

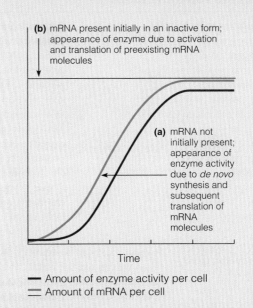

(b) mRNA present initially in an inactive form; appearance of enzyme due to activation and translation of preexisting mRNA molecules

(a) mRNA not initially present; appearance of enzyme activity due to *de novo* synthesis and subsequent translation of mRNA molecules

Time

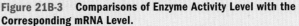

— Amount of enzyme activity per cell
═ Amount of mRNA per cell

Figure 21B-3   Comparisons of Enzyme Activity Level with the Corresponding mRNA Level.

Thus, the expression of a particular gene may be regulated transcriptionally (neither mRNA nor protein present beforehand), translationally (activation or stabilization of preexisting mRNA), or posttranslationally (activation or stabilization of preexisting protein). Techniques that can detect and quantify specific mRNAs and proteins at very low concentrations are currently being used to study many eukaryotic genes to determine the point at which expression is controlled in each case. If present progress is a fair indication, we can look forward to a continued expansion in our understanding of the regulation of gene expression.

that renders expression of the operon sensitive to the intracellular concentration of the corresponding aminoacyl tRNA.

Regulation is more complicated in eukaryotes because of the size and organizational complexity of the eukaryotic genome, as well as the intricate nature of development and cell differentiation in multicellular organisms. As a general rule, the cells of a multicellular organism all contain the same set of genes, although events such as DNA amplification, deletion, rearrangement, and methylation can introduce alterations. Activation of individual genes most likely involves a selective decondensation of chromatin. This structural change can be seen microscopically in the giant polytene chromosomes of insect salivary glands, but it is probably a general phenomenon. Once uncoiled, the DNA is more accessible to RNA polymerases and to the protein factors required for transcriptional initiation. In eukaryotes as well as prokaryotes, transcription initiation is perhaps the most important control point in gene expression. The initiation of gene transcription is controlled by the binding of regulatory transcription factors to various types of DNA control sequences. Especially important among these are DNA response elements, which allow nonadjacent genes to be regulated in a coordinated fashion. For example, hormone response elements coordinate the expression of genes activated by specific hormones, and the heat-shock response element coordinates the expression of genes activated by elevated temperatures and other stresses.

In addition to transcriptional regulation, eukaryotic cells have a variety of posttranscriptional controls, including the generation of variant forms of a gene product by alternative RNA splicing; general, as well as mRNA-specific, mechanisms to regulate translation and mRNA degradation; and posttranslational modulation of protein structure, activity, and degradation. Eukaryotic gene regulation is currently an area of intense research and rapid progress, and we are likely to see more exciting developments in future years.

Key Terms for Self-Testing

Gene Regulation in Prokaryotes
constitutive gene (p. 692)
regulated gene (p. 692)
adaptive enzyme synthesis (p. 692)
substrate induction (p. 693)
inducible enzyme (p. 693)
end-product repression (p. 694)
effector (p. 694)
structural gene (p. 694)
regulatory gene (p. 694)
operon (p. 694)
repressor (p. 694)
lac operon (p. 694)
promoter (p. 694)
operator (p. 694)
polygenic mRNA (p. 695)
inducible operon (p. 695)
inducer (p. 696)
cis-trans test (p. 698)
trans-acting factor (p. 698)
cis-acting element (p. 698)
repressible operon (p. 699)
trp operon (p. 699)
corepressor (p. 699)
negative control (p. 700)
positive control (p. 700)
catabolite repression (p. 700)
cyclic AMP (cAMP) (p. 700)
cAMP receptor protein (CRP) (p. 700)
attenuation (p. 702)

**Comparison of Gene Regulation
in Prokaryotes and Eukaryotes**
stem cells (p. 706)
cell differentiation (p. 706)

**Eukaryotic Gene Regulation:
Genomic Control**
clone (p. 707)
gene amplification (p. 707)
genomic control (p. 707)
gene deletion (p. 709)
DNA rearrangement (p. 709)
MAT locus (p. 709)
cassette mechanism (p. 709)
polytene chromosome (p. 711)
chromosome puff (p. 711)
DNase I hypersensitive site (p. 713)
chromatin remodeling (p. 713)
DNA methylation (p. 714)
X-chromosome inactivation (p. 714)
high-mobility group (HMG) proteins
  (p. 715)

**Eukaryotic Gene Regulation:
Transcriptional Control**
transcriptional control (p. 715)
DNA microarray (p. 716)
proximal control element (p. 718)
regulatory transcription factor (p. 718)
enhancer (p. 718)
silencer (p. 718)
activator (p. 719)
repressor (eukaryotic) (p. 719)
coactivator (p. 720)
combinatorial model for gene regulation
  (p. 721)
DNA-binding domain (p. 721)
transcription regulation domain (p. 721)
activation domain (p. 721)
helix-turn-helix (p. 722)

zinc finger (p. 722)
leucine zipper (p. 722)
helix-loop-helix (p. 724)
response element (p. 724)
nuclear receptor (p. 724)
hormone response element (p. 724)
inverted repeat (p. 725)
CREB protein (p. 726)
STAT (p. 726)
Janus activated kinase (JAK) (p. 726)
heat-shock gene (p. 726)
heat-shock response element (p. 726)
homeotic gene (p. 727)
homeobox (p. 728)
homeodomain (p. 728)

**Eukaryotic Gene Regulation:
Posttranscriptional Control**
alternative RNA splicing (p. 728)
translational control (p. 730)
heme-controlled inhibitor (HCI) (p. 730)
iron-response element (IRE) (p. 730)
translational repressor (p. 731)
RNA interference (p. 731)
posttranslational control (p. 732)
ubiquitin (p. 733)
proteasome (p. 734)
microautophagy (p. 734)

Problem Set

More challenging problems are marked with a •.

21-1.   Laboring with *lac*. Most of what we know about the *lac* operon of *E. coli* has come from genetic analysis of various mutants. In the following list are the genotypes of seven strains of *E. coli*. For each strain, indicate whether the *Z* gene product will be expressed in the presence of lactose and whether it will be expressed in the absence of lactose. Explain your reasoning in each case.

(a) $I^+ P^+ O^+ Z^+$

(b) $I^s P^+ O^+ Z^+$

(c) $I^+ P^+ O^c Z^+$

(d) $I^- P^+ O^+ Z^+$

(e) $I^s P^+ O^c Z^+$

(f) $I^+ P^- O^+ Z^+$

(g) Same as part a, but with glucose present

•21-2.   The Pickled Prokaryote. *Pickelensia hypothetica* is an imaginary prokaryote that converts a wide variety of carbon sources to ethanol when cultured anaerobically in the absence of ethanol. When ethanol is added to the culture medium, however, the organism obligingly shuts off its own production of

ethanol and makes lactate instead. Several mutant strains of *Pickelensia* have been isolated that differ in their ability to synthesize ethanol. Class I mutants cannot synthesize ethanol at all. Mutations of this type map at two loci, *A* and *B*. Class II mutants, on the other hand, are constitutive for ethanol synthesis: They continue to produce ethanol whether it is present in the medium or not. Mutations of this type map at loci *C* and *D*. Strains of *Pickelensia* constructed to be diploid for the ethanol operon have the following genotypes and phenotypes:

$A^+B^-C^+D^+/A^-B^+C^+D^+$ inducible

$A^+B^+C^+D^-/A^+B^+C^+D^+$ inducible

$A^+B^+C^+D^+/A^+B^+C^-D^+$ constitutive

(a) Identify each of the four genetic loci of the ethanol operon.

(b) Indicate the expected phenotype for each of the following partially diploid strains:

(i) $A^-B^+C^+D^+/A^-B^+C^+D^-$

(ii) $A^+B^-C^+D^+/A^-B^+C^+D^-$

(iii) $A^+B^+C^+D^-/A^-B^-C^-D^+$

(iv) $A^-B^+C^-D^+/A^+B^-C^+D^+$

•21-3.   Regulation of Bellicose Catabolism. The enzymes bellicose kinase and bellicose phosphate dehydrogenase are coordinately regulated in the bacterium *Hokus focus*. The genes encoding these proteins, *belA* and *belB*, are contiguous segments on the genetic map of the organism. In her pioneering work on this system, Professor Jean X. Pression established that the bacterium can grow with the monosaccharide bellicose as its only carbon and energy source, that the two enzymes involved in bellicose catabolism are synthesized by the bacterium only when bellicose is present in the medium, and that enzyme production is turned off in the presence of glucose. She identified a number of mutations that reduce or eliminate enzyme production and showed these could be grouped into two classes. Class I mutants are in *cis*-acting elements, whereas those in class II all map at a distance from the structural genes and are *trans*-acting. Thus far, the only constitutive mutations Pression has found are deletions that connect *belA* and *belB* to new DNA at the upstream side of *belA*. The following is a list of conclusions she would like to draw from her observations. Indicate in each case whether the conclusion is consistent with the data (C), inconsistent with the data (I), or irrelevant to the data (X).

(a) Bellicose can be metabolized by *H. focus* cells to yield ATP.

(b) Enzyme production by genes *belA* and *belB* is under positive control.

(c) Phosphorylation of bellicose makes the sugar less permeable to transport across the plasma membrane.

(d) The operator for the bellicose operon is located upstream from the promoter.

(e) Some of the mutations in class I may be in the promoter.

(f) Some of the mutations in class II may be in the operator.

(g) Class II mutations may include mutations in the gene that encodes CRP.

(h) The constitutive deletion mutations connect genes *belA* and *belB* to a new promoter.

(i) Expression of the bellicose operon is subject to regulation by attenuation.

21-4.   Attenuation in 25 Words or Less. Complete each of the following statements about attenuation in 25 words or less.

(a) Attenuators can also be called *conditional terminators* because…

(b) When the gene product of the operon is not needed, the 5′ end of the mRNA forms a secondary structure that…

(c) Implicit in our understanding of the mechanism of attenuation is the assumption that ribosomes follow RNA polymerase very closely. This "tailgating" is essential to the model because…

(d) Measurements of rates of synthesis indicate that translation normally proceeds faster than transcription. This supports the proposed mechanism for attenuation because…

(e) If the operon encoding the enzymes in the biosynthetic pathway for amino acid X is subject to attenuation, the leader sequence probably encodes a polypeptide that…

(f) Attenuation makes sense as a mechanism for regulating operons that code for enzymes involved in amino acid biosynthesis because…

(g) Attenuation is not a likely mechanism for regulating gene expression in eukaryotes because…

21-5.   Positive and Negative Control. Assume you have a culture of *E. coli* cells growing on medium B, which contains both lactose and glucose. At time *t*, 33% of the cells are transferred to medium L, which contains lactose but not glucose; 33% are transferred to medium G, which contains glucose but not lactose; and the remaining cells are left in medium B. For each of the following statements, indicate with an L if it is true of the cells transferred to medium L, with a G if it is true of the cells transferred to medium G, with a B if it is true of the cells left in medium B, and with an N if it is true of none of the cells. In some cases, more than one letter may be appropriate.

(a) The rate of glucose consumption per cell is approximately the same after time *t* as before.

(b) The rate of lactose consumption is higher after time *t* than before.

(c) The intracellular cAMP level is lower after time *t* than before.

(d) Most of the *lac* operator sites have *lac* repressor proteins bound to them.

(e) Most of the cAMP receptor proteins exist as CRP-cAMP complexes.

(f) Most of the *lac* repressor proteins exist as repressor-glucose complexes.

(g) The rate of transcription of the *lac* operon is greater after time *t* than before.

(h) The *lac* operon has both CRP and the repressor protein bound to it.

•21-6.   Polytene Chromosomes. The fruit fly *Drosophila melanogaster* has about 2×10^8 base pairs of DNA per haploid genome, of which about 75% is nonrepeated DNA. The DNA is distributed among four pairs of homologous chromosomes, which have a total of about 5000 visible bands when in polytene form in the salivary gland. The number of genes initially estimated from mutational studies was also about 5000 but recent DNA sequencing studies suggest that the gene number may be somewhat higher.

(a) Why was it tempting to speculate that each band corresponds to a single gene? What does this suggest about the number of different proteins *Drosophila* can make? Does that seem like a reasonable number to you?

(b) Assuming all the nonrepeated DNA is uniformly distributed in the chromosomes, how much nonrepeated DNA (in base pairs) is there in the average band?

(c) How much DNA would it take to code for a single polypeptide with a molecular weight of 50,000? (Assume amino acids have an average molecular weight of 110.) What proportion does that represent of the total nonrepeated DNA in the average band?

(d) How do you account for the discrepancy in part c?

21-7.   Enhancers. An enhancer may increase the frequency of transcription initiation for its associated gene when… (Indicate true or false for each statement, and explain your answer.)

(a) it is located 1000 nucleotides upstream of the gene's core promoter.

(b) it is in the gene's coding region.

(c) no promoter is present.

(d) it causes looping out of the intervening DNA.

(e) it causes alternative splicing of the DNA.

21-8.   Gene Amplification. The best-studied example of gene amplification involves the genes coding for ribosomal RNA during oogenesis in *Xenopus laevis*. The unamplified number of ribosomal RNA genes is about 500 per haploid genome. After amplification, the oocyte contains about 500,000 ribosomal RNA genes per haploid genome. This level of amplification is apparently necessary to allow the egg cell to synthesize the 10^{12} ribosomes that accumulate during the two months of oogenesis in this species. Each ribosomal RNA gene consists of about 13,000 base pairs, and the genome size of *Xenopus* is about 2.7×10^9 base pairs per haploid genome.

(a) What fraction of the total haploid genome do the 500 copies of the ribosomal RNA genes represent?

(b) What is the total size (in base pairs) of the genome after amplification of the ribosomal RNA genes? What proportion of this do the amplified ribosomal RNA genes represent?

(c) Assume that all the ribosomal RNA genes in the amplified oocyte are transcribed continuously to generate the number of ribosomes needed during the two months of oogenesis. How long would oogenesis have to extend if the genes had not been amplified?

(d) Why do you think genes have to be amplified when the gene product needed by the cell is an RNA, but not usually when the desired gene product is a protein?

• 21-9.   Steroid Hormones. Steroid hormones are known to increase the expression of specific genes in selected target cell types. For example, testosterone increases the production of a protein called α2-microglobulin in the liver and hydrocortisone (a type of glucocorticoid) increases the production of the enzyme tyrosine aminotransferase in the liver.

(a) Based on your general knowledge of steroid hormone action, explain how these two steroid hormones are able to selectively influence the expression of two different proteins in the same tissue.

(b) Prior to the administration of testosterone or hydrocortisone, liver cells are exposed to either puromycin (an inhibitor of protein synthesis) or α-amanitin (an inhibitor of RNA polymerase II). How would you expect such treatments to affect the actions of testosterone and hydrocortisone?

(c) Suppose you carry out a domain-swap experiment in which you use recombinant DNA techniques to exchange the zinc finger domains of the testosterone receptor and glucocorticoid receptors with each other. What effects would you now expect testosterone and hydrocortisone to have in cells containing these altered receptors?

(d) If recombinant DNA techniques are used to substitute a testosterone response element for the glucocorticoid response element that is normally located adjacent to the tyrosine aminotransferase gene in liver cells, what effects would you expect testosterone and hydrocortisone to have in such genetically altered cells?

21-10.   Homeotic Genes. Homeotic genes are considered crucial to early development in *Drosophila* because… (Indicate true or false for each statement, and explain your answer.)

(a) they encode proteins containing zinc finger domains.

(b) mutations in homeotic genes are always lethal.

(c) they control the expression of many other genes required for development.

(d) they are identical in all eukaryotes.

(e) homeodomain proteins act by influencing mRNA degradation.

• 21-11.   Protein Degradation. You learned in Chapter 17 that mitotic cyclin is an example of a protein that is selectively degraded at a particular point in the cell cycle, namely the onset of anaphase. Suppose that you use recombinant DNA techniques to create a mutant form of mitotic cyclin that is not degraded at the onset of anaphase.

(a) Sequence analysis of the mutant mitotic cyclin shows that it is missing a stretch of nine amino acids near one end of the molecule. Based on this information, discuss at least three possible explanations of why this form of mitotic cyclin is not degraded at the onset of anaphase.

(b) What kinds of experiments could you carry out to distinguish among these various possible explanations?

(c) Suppose that you discover cells containing a mutation in a second protein, and learn that this mutation also prevents mitotic cyclin from being degraded at the onset of mitosis, even when mitotic cyclin is normal. The degradation of other proteins appears to proceed normally in such cells. A mutation in what kind of protein might explain such results?

21-12.   Levels of Control. Assume liver and kidney tissues from the same mouse each contain about 10,000 species of cytosolic mRNA, but only about 25% of these are common to the two tissues.

(a) Suggest an experimental approach that might be used to establish that some of the mRNA molecules are common to both tissues, but others are not.

(b) One possible explanation for the data is that differential transcription occurs in liver and kidney nuclei. What is another possible explanation? Describe an experiment that would enable you to distinguish between these possibilities.

(c) If all mRNA molecules had been shown to be common to both liver and kidney and yet the two tissues were known to be synthesizing different proteins, what level of control would you have to assume was operating?

• 21-13.   More About Levels of Control. If a frog egg is exposed to an inhibitor of RNA synthesis during the early stages of embryonic development, protein synthesis is not inhibited and development continues. However, if the same inhibitor is added later during embryonic development, protein synthesis is severely depressed and normal embryonic development halts. How might you explain these observations?

Suggested Reading

References of historical importance are marked with a •.

Prokaryotic Gene Regulation

• Dickson, R., J. Abelson, W. Barnes, and W. Reznikoff. Genetic regulation: The *lac* control region. *Science* 187 (1975): 27.

Helmann, J. D., and M. J. Chamberlin. Structure and function of bacterial sigma factors. *Annu. Rev. Biochem.* 57 (1988): 839.

• Jacob, F., and J. Monod. Genetic regulatory mechanisms in the synthesis of proteins. *J. Mol. Biol.* 3 (1961): 318.

Kolb, A., S. Busby, H. Buc, S. Garges, and S. Adhya. Transcriptional regulation by cAMP and its receptor protein. *Annu. Rev. Biochem.* 62 (1993): 749.

Yanofsky, C. Operon specific control by transcription attenuation. *Trends Genet.* 3 (1987): 356.

Eukaryotic Gene Regulation: Genomic Control

Pennisi, E. Opening the way to gene activity. *Science* 275 (1997): 155.

Strahl, B. D., and C. D. Allis. The language of covalent histone modifications. *Nature* 403 (2000): 41.

Wolffe, A. *Chromatin: Structure and Function,* 3d ed. San Diego: Academic Press, 1998.

Workman, J. L., and R. E. Kingston. Alteration of nucleosome structure as a mechanism of transcriptional regulation. *Annu. Rev. Biochem.* 67 (1998): 545.

Eukaryotic Gene Regulation: Transcriptional Control

Brown, C. E., T. Lechner, L. Howe, and J. L. Workman. The many HATs of transcription coactivators. *Trends Biochem. Sci.* 25 (2000): 15.

Carey, M. The enhanceosome and transcriptional synergy. *Cell* 92 (1998): 5.

Darnell, J. E., Jr. STATs and gene regulation. *Science* 277 (1997): 1630.

Friend, S. H., and R. B. Stoughton. The magic of microarrays. *Sci. Amer.* 286 (February 2002): 44.

Hamadeh, H., and C. A. Afshari. Gene chips and functional genomics. *Amer. Scientist* 88 (2000): 508.

Latchman, D. S. *Gene Regulation: A Eukaryotic Perspective,* 3d ed. Cheltenham, UK: Stanley Thornes, 1999.

Lefstin, J. A., and K. R. Yamamoto. Allosteric effects of DNA on transcriptional regulators. *Nature* 392 (1998): 885.

McGinnis, W., and M. Kuziora. The molecular architects of body design. *Sci. Amer.* 270 (February 1994): 58.

McKnight, S. L. Molecular zippers in gene regulation. *Sci. Amer.* 264 (April 1991): 54.

Rhodes, D., and A. Klug. Zinc fingers. *Sci. Amer.* 268 (February 1993): 56.

Tjian, R. Molecular machines that control genes. *Sci. Amer.* 272 (February 1995): 54.

Weatherman, R. V., R. J. Fletterick, and T. S. Scanlan. Nuclear-receptor ligands and ligand-binding domains. *Annu. Rev. Biochem.* 68 (1999): 559.

Welch, W. J. How cells respond to stress. *Sci. Amer.* 268 (May 1993): 56.

Eukaryotic Gene Regulation: Posttranscriptional Control

Goldberg, A. L., S. J. Elledge, and J. W. Harper. The cellular chamber of doom. *Sci. Amer.* 284 (January 2001): 68.

Hammond, S. M., A. A. Caudy, and G. J. Hannon. Post-transcriptional gene silencing by double-stranded RNA. *Nature Reviews Genetics* 2 (2001): 110.

Sach, A. B., and S. Buratowski. Common themes in translational and transcriptional regulation. *Trends Biochem. Sci.* 22 (1997): 189.

Smith, C. W. J., and J. Valcarcel. Alternative pre-mRNA splicing: The logic of combinatorial control. *Trends Biochem. Sci.* 25 (2000): 381.

Sonenberg, N., J. W. B. Hershey, and M. B. Mathews. *Translational Control of Gene Expression.* Plainview, NY: Cold Spring Harbor Laboratory Press, 2000.

Weissman, A. M. Themes and variations on ubiquitylation. *Nature Reviews Molecular Cell Biol.* 2 (2001): 169.

Wilusz, C. J., M. Wormington, and S. W. Peltz. The cap-to-tail guide to mRNA turnover. *Nature Reviews Molecular Cell Biol.* 2 (2001): 237.

Stem Cells and Cloning

Hines, P. J., B. A. Purnell, and J. Marx, eds. Stem cell research and ethics. *Science* 287 (2000): 1417.

Solter, D. Dolly *is* a clone—and no longer alone. *Nature* 394 (1998): 315.

Wilmut, I. Cloning for medicine. *Sci. Amer.* 279 (December 1998): 58.

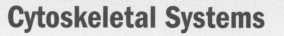

22

Cytoskeletal Systems

In the preceding chapters, we examined a variety of cellular processes and pathways, many of which occur in the organelles of eukaryotic cells. We now come to the cytosol, which is the region of the cytoplasm between and surrounding the organelles. Until a few decades ago, the cytosol of the eukaryotic cell was regarded as the generally uninteresting, gel-like substance in which the nucleus and other organelles were suspended. Cell biologists knew that proteins made up about 20–30% of the cytosol, but these proteins were thought to be soluble and able to move freely within the cytosol. Except for those of known enzymatic activity, little was understood about the structural or functional significance of the cytosolic proteins.

Advances in microscopy and other investigative techniques have revealed that the interior of a eukaryotic cell is highly structured. Part of this structure is provided by the **cytoskeleton:** a complex network of interconnected filaments and tubules that extends throughout the cytosol, from the nucleus to the inner surface of the plasma membrane.

The term *cytoskeleton* expresses well the role of this proteinaceous matrix in providing an architectural framework for eukaryotic cells. It confers a high level of internal organization on cells and enables them to assume and maintain complex shapes that would not otherwise be possible. The name does not, however, convey the dynamic, changeable nature of the cytoskeleton and its critical involvement in a great variety of cellular processes.

The cytoskeleton plays important roles in cell movement and cell division, and positions and actively moves membrane-bounded organelles within the cytosol. It also plays a similar role for messenger RNA and other cellular components. Many enzymes in the cytosol are in fact probably not soluble at all, but are physically clustered and

attached to the cytoskeleton in close proximity to other enzymes involved in the same pathway, thereby facilitating the channeling of intermediates within each pathway. The cytoskeleton is also involved in many forms of cell movement and is intimately related to other processes such as cell signaling and cell-cell adhesion. The cytoskeleton is altered by events at the cell surface and, at the same time, appears to participate in and modulate these events.

In this chapter we will focus on the structure of the cytoskeleton. In Chapter 23 we will look more closely at cytoskeletal function, especially in relation to cellular movement.

The Major Structural Elements of the Cytoskeleton

The three major structural elements of the cytoskeleton are *microtubules, microfilaments,* and *intermediate filaments,* each of which is unique to eukaryotic cells (Figure 22-1). The existence of three distinct systems of filaments and tubules was first revealed by electron microscopy. Biochemical and immunological studies then identified the distinctive proteins of each system, which are also unique to eukaryotic cells. The technique of *immunofluorescence microscopy* (Table 22-2; also the *Guide to Microscopy*) was especially important in localizing specific proteins within the cytoskeleton.

Each of the structural elements of the cytoskeleton has a characteristic size, structure, and intracellular distribution, and each is formed by the polymerization of a different kind of protein monomer (Table 22-1). Microtubules

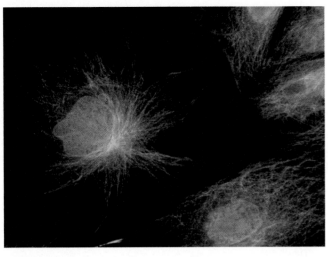

(a) Microtubules

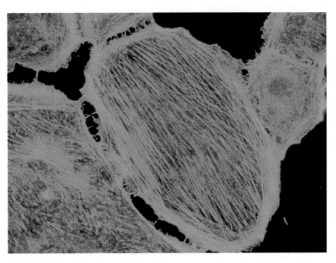

(b) Microfilaments

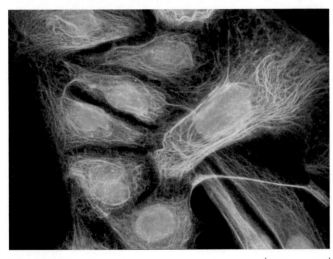

(c) Intermediate filaments

|⊢————⊣|
 5 μm

Figure 22-1 **The Intracellular Distribution of Microtubules, Microfilaments, and Intermediate Filaments.** **(a)** The distribution of microtubules in cells of the kangaroo rat kidney cell line (PtK-1) visualized by immunofluorescence staining for tubulin. For reference, the nucleus has been stained with the red fluorescent DNA stain propidium iodide. **(b)** The distribution of microfilaments in a rat kidney cell line visualized by staining actin with a fluorescent derivative of phalloidin. **(c)** The distribution of intermediate filaments in PtK-1 cells, visualized by immunofluorescence staining for keratin. Here the nucleus has also been stained with propidium iodide.

Table 22-1 **Properties of Microtubules, Microfilaments, and Intermediate Filaments**

	Microtubules	Microfilaments	Intermediate Filaments
Structure	Hollow tube with a wall consisting of 13 protofilaments	Two intertwined chains of F-actin	Eight protofilaments joined end-to-end with staggered overlaps
Diameter	Outer: 25 nm Inner: 15 nm	7 nm	8–12 nm
Monomers	α-tubulin β-tubulin	G-actin	Several proteins; see Table 22–5

Functions	Axonemal: Cell motility
	Cytoplasmic:
	  Organization and maintenance of animal cell shape
	  Chromosome movements
	  Disposition and movement of organelles
		Muscle contraction
		Amoeboid movement
		Cell locomotion
		Cytoplasmic streaming
		Cell division
		Maintenance of animal cell shape
			Structural support
			Maintenance of animal cell shape
			Formation of nuclear lamina and scaffolding
			Strengthening of nerve cell axons (NF protein)
			Keeping muscle fibers in register (desmin)

are composed of the protein *tubulin* and are about 25 nm in diameter. Microfilaments, with a diameter of about 7 nm, are polymers of the protein *actin*. Intermediate filaments have diameters in the range of 8–12 nm. Intermediate fila-

ment protein monomers differ depending on the cell type, but all are similar in size and structure. In addition to its major protein component, each type of cytoskeletal filament has a number of other proteins associated with it.

These *accessory proteins* account for the remarkable structural and functional diversity of the cytoskeletal elements.

Microtubules and microfilaments are best known for their roles in cell motility. Microfilaments are essential components of *muscle fibrils,* and microtubules are the structural elements of *cilia* and *flagella,* appendages that enable certain cells to either propel themselves through a fluid environment or move fluids past the cell. These structures are large enough to be seen by light microscopy and were therefore known and studied long before it became clear that the same structural elements are also integral parts of the cytoskeleton.

We will consider each of the structural elements in detail. In doing so, we will be discussing microtubules, microfilaments, and intermediate filaments as though they were separate entities, each with its own independent functions. Keep in mind, though, that the components of the cytoskeleton are linked together both structurally and functionally, resulting in new architectural properties that are not simply the sum of the parts, as we will see in the last section of this chapter.

Techniques for Studying the Cytoskeleton

Modern Microscopy Techniques Have Revolutionized the Study of the Cytoskeleton

The cytoskeleton is a topic of great current research interest to eukaryotic cell biologists. Much of the recent progress in understanding cytoskeletal structure is due to three powerful microscopy techniques: immunofluorescence microscopy, digital video microscopy, and electron microscopy. Table 22-2 summarizes each of these techniques; they are described in more detail in the *Guide to Microscopy.* In addition, specific drugs and mutations have been used very effectively to analyze cytoskeletal function.

Drugs and Mutations Can Be Used to Disrupt Cytoskeletal Structures

Although microscopy techniques can reveal much about the structure of the cytoskeleton, they do not usually enable us to deduce much about its function. To analyze the function of particular cytoskeletal filaments, we must selectively remove or disrupt the function of the relevant proteins. In the case of tubulin and actin, we can use certain drugs to alter protein function in a known way. By studying the effects of such drugs on specific cellular processes, it is often possible to identify, at least tentatively, functions that depend on microtubules or microfilaments.

For example, *colchicine* (an alkaloid from the Autumn crocus, *Colchicum autumnale*) binds to tubulin monomers, strongly inhibiting their assembly into microtubules and fostering the disassembly of existing ones. In contrast, *taxol* (from the Pacific yew tree, *Taxus brevifolis*)

binds tightly to microtubules and stabilizes them, causing much of the free tubulin in the cell to assemble into microtubules. Sensitivity of a cellular process to colchicine or taxol is therefore a good first indication that microtubules may mediate that process within the cell. In a similar manner, the drug *cytochalasin D,* a fungal metabolite, and *latrunculin A,* a marine toxin isolated from the Red Sea sponge *Latrunculia magnifica,* inhibit the polymerization of actin microfilaments. In contrast, *phalloidin,* a cyclic peptide from the death cap fungus (*Amanita phalloides*), blocks the depolymerization of actin, thereby stabilizing microfilaments. Obviously, processes that are disrupted in cells treated with either of these drugs are likely to depend in some way on microfilaments. The uses and effects of these drugs are discussed in more detail later in the chapter.

In addition to the use of drugs, mutations can be used to study the function of cytoskeletal proteins. Using techniques from genetics and molecular biology, cell biologists have isolated mutant organisms or cell lines in which specific mutations have been introduced in a particular cytoskeletal protein. Such mutations have been useful for identifying what cellular events require particular cytoskeletal proteins, and in determining what portions of a protein are required for its function.

With these techniques in mind, we are now ready to look at each of the three major components of the cytoskeleton. In each case, we will consider the chemistry of the monomer, the structure of the polymer and how it is polymerized, the role of accessory proteins, and some of the structural and functional roles each component plays within the cell. We will look first at microtubules.

Microtubules

Two Types of Microtubules Are Responsible for Many Functions in the Cell

Microtubules (MTs) are the largest of the cytoskeletal elements (see Table 22-1). Microtubules in eukaryotic cells can be classified into two general groups, which differ in both degree of organization and structural stability.

The first group, **axonemal microtubules,** includes the highly organized, stable microtubules found in specific subcellular structures associated with cellular movement, including cilia, flagella, and the basal bodies to which these appendages are attached. The central shaft, or *axoneme,* of a cilium or flagellum consists of a highly ordered bundle of axonemal MTs and associated proteins. Given their order and stability, it is not surprising that the axonemal MTs were the first of the two groups to be recognized and studied. We have already encountered an example of such a structure; the axoneme of the sperm tail shown in Figure 4-12 consists of MTs. We will consider axoneme structure and microtubule-mediated motility further in Chapter 23.

Table 22-2 Techniques for Studying the Cytoskeleton

Technique	Description	Example	
Immunofluorescence microscopy	Primary antibodies bind to cytoskeletal proteins. Secondary antibodies labeled with a fluorescent tag bind to the primary ones, causing the cytoskeletal proteins to glow in the fluorescence microscope.	A fibroblast stained with fluorescent antibodies directed against actin shows bundles of actin filaments.	Figure 1 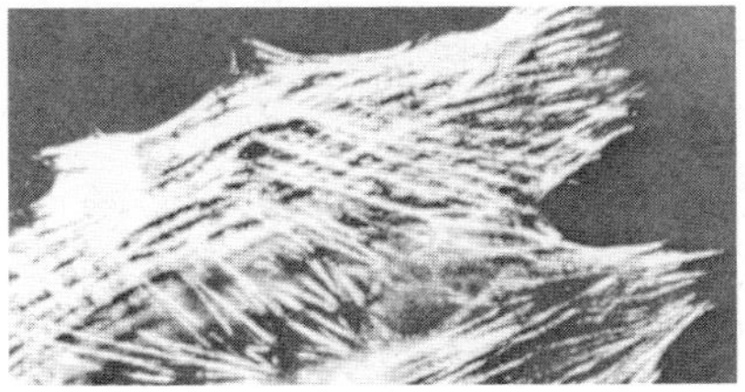
Fluorescence techniques to study the cytoskeleton in living cells	Fluorescent versions of cytoskeletal proteins are made and introduced into living cells. Fluorescence microscopy and video cameras are used to view the proteins as they function in the cells.	Fluorescent tubulin molecules were micro-injected into living fibroblast cells. Inside the cell, the tubulin dimers become incorporated into microtubules, which can be seen easily with a fluorescence microscope.	Figure 2 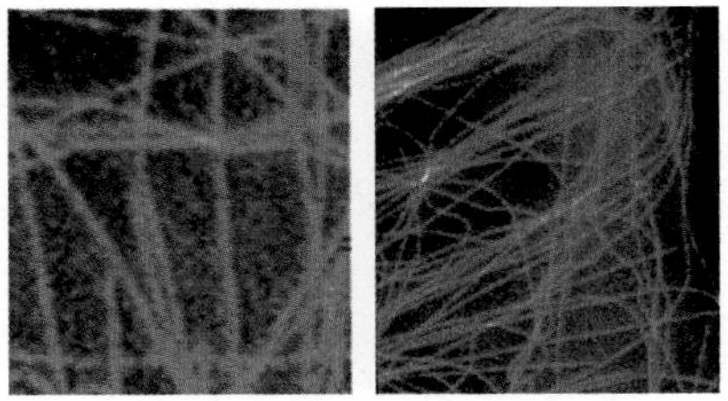
Computer-enhanced digital video microscopy	High-resolution images from a video camera attached to a microscope are computer processed to increase contrast and remove background features that obscure the image.	Two micrographs showing several microtubules processed to make them visible in detail.	Unenhanced / Enhanced Figure 3 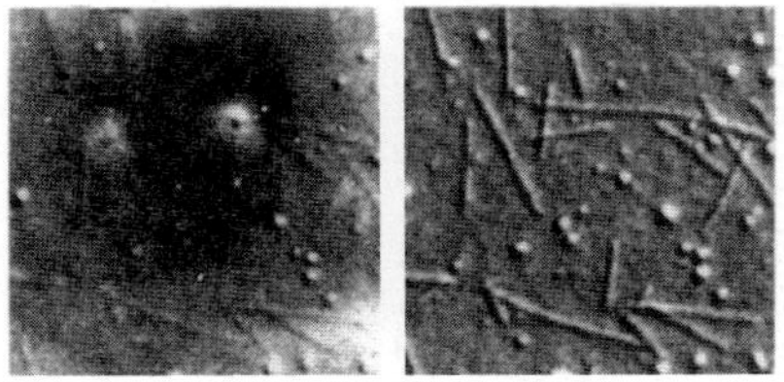
Electron microscopy	Electron microscopy can resolve individual filaments prepared by thin section, quick-freeze deep-etch, or direct-mount techniques.	A fibroblast cell prepared by the quick-freeze deep-etch method. Bundles of actin microfilaments are visible.	Figure 4 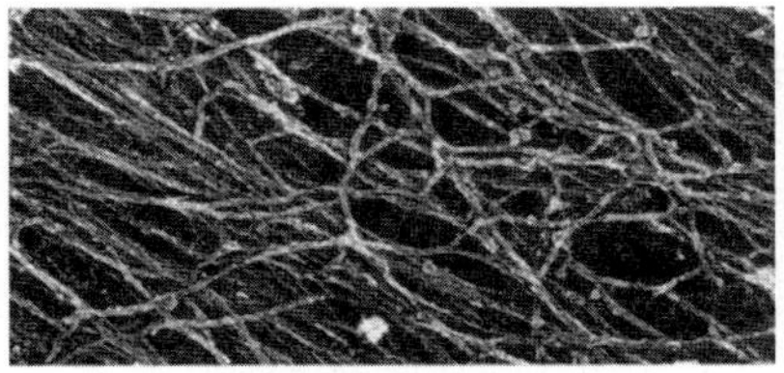

The second group is the more loosely organized, dynamic network of **cytoplasmic microtubules.** The occurrence of cytoplasmic MTs in eukaryotic cells was not recognized until the early 1960s, when better fixation techniques permitted direct visualization of the network of MTs now known to pervade the cytosol of most eukaryotic cells. Since then, immunofluorescence microscopy has revealed the diversity and complexity of MT networks in different cell types.

Cytoplasmic MTs are responsible for a variety of functions (see Table 22-1). For example, in animal cells they are required to maintain axons, nerve cell extensions whose electrical properties we examined in Chapter 9. Some migrating animal cells require cytoplasmic MTs to maintain their polarized shape. In plant cells, cytoplasmic MTs are thought to govern the orientation with which cellulose microfibrils are deposited during the growth of cell walls. In addition, cytoplasmic MTs form the mitotic and meiotic spindles that are essential for the movement of chromosomes during mitosis and meiosis (see Chapter 17).

Cytoplasmic microtubules also contribute to the spatial disposition and directional movement of vesicles and other organelles by providing an organized system of fibers to guide their movement. For example, cytoplasmic MTs help to govern the location of organelles, such as the Golgi complex and the endoplasmic reticulum, and are involved in active movement of vesicles (see Chapter 23).

Tubulin Heterodimers Are the Protein Building Blocks of Microtubules

As mentioned in Chapter 4, MTs are straight, hollow cylinders with an outer diameter of about 25 nm and an inner diameter of about 15 nm (Figure 22-2). Microtubules vary greatly in length. Some are less than 200 nm long; others, such as axonemal MTs, can be many micrometers in length. The MT wall consists of longitudinal arrays of linear polymers called **protofilaments.** There are usually 13 protofilaments arranged side by side around the hollow center, or lumen.

As shown in Figure 22-2, the basic subunit of a protofilament is a heterodimer of the protein **tubulin.** Tubulin is actually a family of similar polypeptides that can be subdivided into five groups based on their structure. The heterodimers that form the bulk of protofilaments are composed of one molecule of α-**tubulin** and one molecule of β-**tubulin.** As soon as individual α- and β-tubulin molecules are synthesized, they bind tightly to each other to produce an $\alpha\beta$-**heterodimer** that does not dissociate under normal conditions.

Individual α- and β-tubulin molecules have diameters of about 4–5 nm and molecular weights of about 50,000. Structural studies show that α- and β-tubulins have nearly identical three-dimensional structures, even though they share only 40% amino acid sequence identity. Each protein folds into three domains: a GTP-binding domain at the N-terminus, a domain in the middle to which the micro-

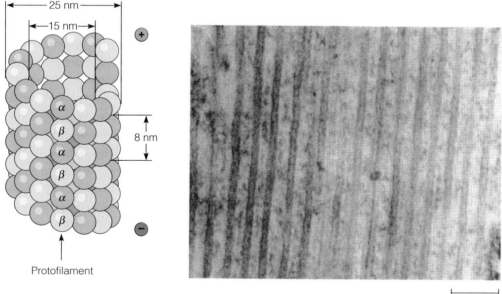

(a) Microtubule structure   **(b)** Microtubules in an axoneme   0.1 μm

Figure 22-2   Microtubule Structure.   (a) A schematic diagram, showing a microtubule as a hollow cylinder enclosing a lumen. The outside diameter is about 25 nm, and the inside diameter is about 15 nm. The wall of the cylinder consists of 13 protofilaments, one of which is indicated by an arrow. A protofilament is a linear polymer of tubulin dimers, each of which consists of two polypeptides, α-tubulin and β-tubulin. All heterodimers in the protofilaments have the same orientation, thereby accounting for the polarity of the microtubule.
(b) Microtubules in a longitudinal section of an axoneme (TEM).

tubule poison colchicine can bind, and a third domain at the C-terminus that interacts with MT-associated proteins (MAPs; we will discuss MAPs later in this chapter).

Within a microtubule, all of the tubulin dimers are oriented in the same direction, such that all of the α-tubulin subunits face the same end. This uniform orientation of tubulin dimers means that one end of the protofilament differs chemically and structurally from the other, giving the protofilament an inherent polarity. Because the orientation of the tubulin dimers is the same for all of the protofilaments in a MT, the MT itself is also a polar structure.

Most organisms have several closely related but nonidentical genes for each of the α- and β-tubulin subunits. These slightly different forms of tubulin are called *tubulin isotypes*. In the mammalian brain, for example, there are five α- and five β-tubulin isotypes. These isotypes differ mainly in the C-terminal domain, the portion of the tubulin molecule that binds to MAPs. This suggests that various tubulin isoforms may have different MAP-binding properties. However, whether or not distinct isoforms have unique functional properties has not been directly investigated in most cases.

Microtubules Form by the Addition of Tubulin Dimers at Their Ends

Microtubules form by the reversible polymerization of tubulin dimers. The polymerization process has been studied extensively in vitro; a schematic representation of MT assembly in vitro is shown in Figure 22-3. When a solution containing a sufficient concentration of tubulin dimers, GTP, and Mg^{2+} is warmed from 0°C to 37°C, the polymerization reaction begins. (MT formation in the solution can be readily measured with a spectrophotometer as an increase in light scattering.) A critical step in the formation of MTs is the aggregation of tubulin dimers into clusters called *oligomers*. These oligomers serve as "nuclei" from which new microtubules can grow, and hence this process is referred to as **nucleation.** Once a MT has been nucleated, it grows by addition of subunits at either end, a process called **elongation.**

Microtubule formation is initially slow, a period referred to as the *lag phase* of MT assembly (Figure 22-4). This period reflects the relatively slow process of MT nucleation. The elongation phase of MT assembly—the addition of tubulin dimers—is relatively fast compared with nucleation. Eventually, the mass of MTs increases to a point where the concentration of free tubulin becomes limiting. This leads to the *plateau phase,* where MT assembly is balanced by disassembly.

Microtubule growth in vitro depends on the concentration of tubulin dimers, in such a way that MTs grow when tubulin concentrations are high and depolymerize when tubulin concentrations are low. Somewhere between these two conditions is a tubulin concentration at which assembly is exactly balanced with disassembly. The concentration of dimers at this point is called the overall **critical concentration.**

Addition of Tubulin Dimers Occurs More Quickly at the Plus Ends of Microtubules

The inherent structural polarity of microtubules means that the two ends differ chemically. Another important

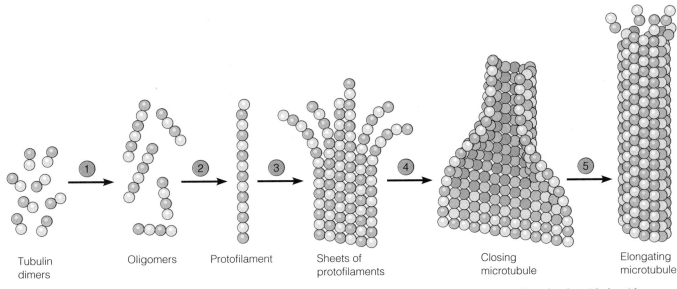

| Tubulin dimers | Oligomers | Protofilament | Sheets of protofilaments | Closing microtubule | Elongating microtubule |

Figure 22-3 A Model for Microtubule Assembly In Vitro. Microtubules are assembled from subunits composed of one molecule of α-tubulin and one molecule of β-tubulin bound together tightly as a dimer, called an $\alpha\beta$-tubulin dimer, or simply a tubulin dimer. ① At the start of the nucleation process, several tubulin dimers can aggregate into clusters called oligomers, ② some of which go on to form linear chains of tubulin dimers called protofilaments. ③ The protofilaments can then associate with each other side-by-side to form sheets. ④ Sheets containing 13 or more protofilaments can close into a tube, forming a microtubule. ⑤ Elongation of the microtubule continues by the addition of tubulin subunits at one or both ends.

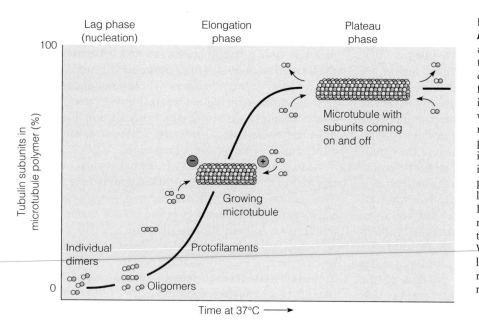

Figure 22-4 The Kinetics of Microtubule Assembly In Vitro.

The kinetics of MT assembly can be monitored by observing the amount of light scattered by a solution containing GTP-tubulin *after* it is warmed from 0°C to 37°C. (Microtubule assembly is inhibited by cold and activated upon warming.) Such light-scattering measurements reflect changes in the MT population as a whole, not the assembly of individual microtubules. When measured in this way, MT assembly exhibits three phases: lag, elongation, and plateau. The lag phase is the period of nucleation. During the elongation phase, MTs grow rapidly, causing the concentration of tubulin subunits in the solution to decline. When this concentration is low enough to limit further assembly, the plateau phase is reached, in which subunits are added and removed from MTs at equal rates.

difference between the two ends of the MT is that one end can inherently grow or shrink much faster than the other. This difference in assembly rate can readily be visualized by mixing MT fragments, such as the MT-associated structures found at the base of cilia, known as *basal bodies*, with tubulin heterodimers. The MT fragments nucleate assembly of the tubulin heterodimers from both of their ends, but the MTs grow much faster from one end of the fragments than the other. (The position of the fragment in the growing MT can be assessed because of its different appearance under the electron microscope; Figure 22-5.) The rapidly growing end of the microtubule is called the **plus end,** and the other end is the **minus end.**

The different growth rates of the plus and minus ends of microtubules reflect the different critical concentrations required for assembly at the two ends of the MT; the critical concentration for the plus end is lower than that for the minus end. If the free tubulin concentration is higher than the critical concentration for the plus end but lower than the critical concentration for the minus end, assembly will occur at the plus end while disassembly takes place at the minus end. This simultaneous assembly and disassembly produces the phenomenon known as **treadmilling** (Figure 22-6). Treadmilling arises when a given tubulin molecule incorporated at the plus end is displaced progressively along the MT and eventually lost by depolymerization at the opposite end. Although treadmilling can be demonstrated in vitro, it is not clear how prevalent treadmilling is in vivo. However, treadmilling has been observed in fragments of fish epidermal cells.

GTP Hydrolysis Contributes to the Dynamic Instability of Microtubules

In the previous section, we saw that tubulin can assemble in vitro in the presence of Mg^{2+} and GTP. In fact, GTP is required for MT assembly. Each tubulin heterodimer binds two GTP molecules. The α-tubulin binds one GTP; the other GTP is bound by β-tubulin, and can be hydrolyzed to GDP sometime after the heterodimer is added to a MT. GTP is apparently needed for MT assembly because the association of GDP-bound tubulin dimers with each other is too weak to support polymerization. However, hydrolysis of GTP is not necessary for assembly, as demonstrated by experiments in which MTs polymerize from tubulin bound to a nonhydrolyzable analogue of GTP.

Studies of MT assembly in vitro using isolated centrosomes (a structure we will discuss in detail below) as nucleation sites show that some microtubules can grow by polymerization at the same time that others shrink by depolymerization. As a result, some MTs effectively enlarge at the expense of others.

To explain how both polymerization and depolymerization might occur simultaneously, Tim Mitchison and Marc Kirschner proposed the **dynamic instability model.** This model presumes two populations of microtubules, one growing in length by continued polymerization at their plus ends, and the other shrinking in length by depolymerization. The distinction between the two populations is that growing MTs have GTP bound to the tubulin at their plus ends, while shrinking MTs have GDP instead. Because bound GTP-tubulin molecules are thought to have a greater affinity for each other than for GDP-tubulin, the presence of a group of such GTP-bound tubulin molecules at the plus end of an MT forms a *GTP cap* that provides a stable MT tip to which further dimers can be added (Figure 22-7a). Loss of GTP is thought to result in an unstable tip, at which depolymerization may occur rapidly.

The concentration of tubulin bound to GTP is crucial to the dynamic instability model. When GTP-tubulin is readily available, it is added to the microtubule quickly, creating a large GTP-tubulin cap. If the concentration of GTP-tubulin falls, however, the rate of tubulin addition decreases. At a sufficiently low concentration of GTP-tubulin, the rate of hydrolysis of GTP on the β-tubulin subunits near the tip of the MT exceeds the rate of addition of new, GTP-bound tubulin. This results in shrinkage of the GTP cap. When the GTP cap disappears, the MT becomes unstable, and loss of GDP-bound subunits from its tip is favored.

Direct evidence for dynamic instability comes from observation of individual microtubules in vitro via light microscopy. An individual MT can undergo alternating periods of growth and shrinkage (Figure 22-7b). When a MT switches from growth to shrinkage, an event called *microtubule catastrophe,* the MT can disappear completely, or it can abruptly switch back to a growth phase, a phenomenon known as *microtubule rescue.* The frequency of catastrophe is inversely related to the free tubulin concentration. High tubulin concentrations make catastrophe less likely, and when catastrophe does occur, higher tubulin concentrations make the rescue of a shrinking MT more likely (Figure 22-7c). At any tubulin concentration, catastrophe is more likely at the plus end of a MT—that is, dynamic instability is more pronounced at the plus end of the MT.

MTs Originate from Microtubule-Organizing Centers Within the Cell

In the previous sections, we primarily discussed the properties that tubulin and microtubules exhibit in vitro, providing a foundation for understanding how MTs function in the cell. However, MT formation in vivo is a more ordered and regulated process, one that produces sets of MTs in specific locations for specific cell functions.

Microtubules commonly originate from a structure in the cell called a **microtubule-organizing center (MTOC).** An MTOC serves as a site at which MT assembly is initiated and acts as an anchor for one end of these MTs. Many cells during interphase have a MTOC called the **centrosome** that is positioned near the nucleus. The centrosome in an animal cell is normally associated with two *centrioles* surrounded by a diffuse granular material known as **pericentriolar material** (Figure 22-8a). In electron micrographs of the centrosome, MTs originate from the pericentriolar material (Figure 22-8b).

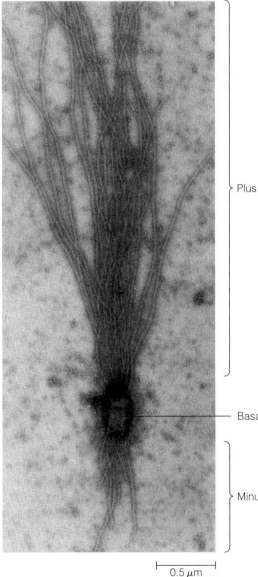

Plus ends

Basal body

Minus ends

0.5 μm

Figure 22-5 Polar Assembly of Microtubules In Vitro. The polarity of MT assembly can be demonstrated by adding basal bodies to a solution of tubulin dimers. The tubulin dimers add to the plus and minus ends of the microtubules in the basal body. However, MTs that grow from the plus end are much longer than those growing from the minus end.

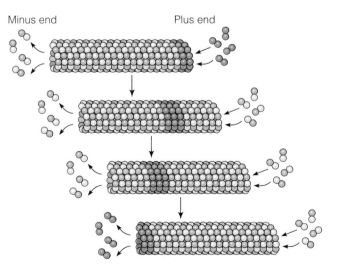

Minus end   Plus end

Figure 22-6 Treadmilling of Microtubules. Microtubule assembly occurs more readily at the plus end of a MT than at the minus end. When the tubulin concentration is higher than the critical concentration for the plus end but lower than the critical concentration for the minus end, the microtubule can add tubulin heterodimers to its plus end while losing them from its minus end.

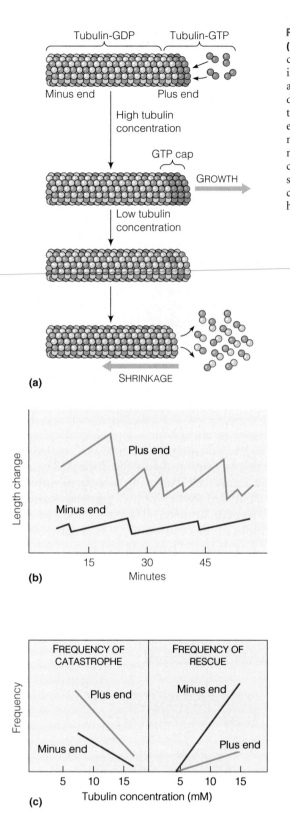

(a)

(b)

(c)

Tubulin-GDP   Tubulin-GTP

Minus end   Plus end

High tubulin concentration

GTP cap

GROWTH

Low tubulin concentration

SHRINKAGE

Length change

Plus end

Minus end

15   30   45

Minutes

FREQUENCY OF CATASTROPHE

FREQUENCY OF RESCUE

Frequency

Plus end

Minus end

Minus end

Plus end

5   10   15   5   10   15

Tubulin concentration (mM)

Figure 22-7 The GTP Cap and Its Role in the Dynamic Instability of Microtubules. **(a)** A model illustrating the postulated role of the GTP cap. When the tubulin concentration is high, tubulin-GTP is added to the microtubule tip faster than the incorporated GTP can be hydrolyzed. The resulting GTP cap stabilizes the MT tip and promotes further growth. At lower tubulin concentrations the rate of growth decreases, thereby allowing GTP hydrolysis to catch up. This creates an unstable tip (no GTP cap) that favors MT depolymerization. Experimental data reveal the existence of dynamic instability. **(b)** In an individual MT observed by light microscopy, growing and shrinking phases alternate with each other. The plus and minus ends grow and shrink independently. **(c)** The frequency of microtubule catastrophe and rescue depends on the tubulin concentration. Catastrophe, the switch from growth to shrinkage, becomes less frequent at high tubulin concentrations. Rescue, the switch from shrinkage to growth, is more frequent at high tubulin concentration.

involved in the formation of *basal bodies*, structures that are important for the formation of cilia and flagella (see Chapter 23). The role of centrioles in non-ciliated cells is less clear. In animal cells, centrioles may serve to recruit pericentriolar material to the centrosome, which then nucleates growth of microtubules. When centrioles are missing from animal cells, microtubule-nucleating material disperses, and the MTOC disappears. Cells lacking centrioles can still divide, probably because chromosomes can organize microtubules to some extent on their own. However, the resulting spindles are poorly organized. In contrast to animal cells, the cells of higher plants lack centrioles; this indicates that centrioles are not essential for the formation of MTOCs.

The pericentriolar material is currently being characterized. Recent studies have identified large, ring-shaped protein complexes in the centrosome that contain another type of tubulin, *γ*-**tubulin**. Other proteins, such as *pericentrin*, are also found within the centrosome. The rings of *γ*-tubulin can be seen at the base of MTs that emerge from the centrosome (Figure 22-9). The *γ*-tubulin ring complexes serve to nucleate the assembly of new MTs away from the centrosome. In addition to the centrosome, some types of cells have other MTOCs. For example, the basal body at the base of each cilium in ciliated cells also serves as a MTOC.

MTOCs Organize and Polarize the Microtubules Within Cells

The centrosome or MTOC plays an important role in controlling the organization of microtubules in cells. The most important aspect of this role is probably the MTOC's ability to nucleate and anchor MTs. As a result of this ability, MTs extend out from a MTOC toward the periphery of the cell. Furthermore, they grow out from a MTOC with a fixed polarity—their minus ends are anchored in the MTOC, and their plus ends extend out toward the cell membrane. The relationship between the MTOC and the distribution and polarity of MTs is shown in Figure 22-10.

The MTOC also influences the number of microtubules in a cell. Each MTOC has a limited number of nucleation and anchorage sites that seem to control how

Embedded within the centrosome is a pair of small cylindrical structures called *centrioles*. The symmetrical structure of centrioles is remarkable: the walls of centrioles are formed by nine pairs of triplet microtubules (Figure 22-8a). In most cases, centrioles are oriented at right angles to one another; the significance of this arrangement is still unknown. Centrioles are known to be

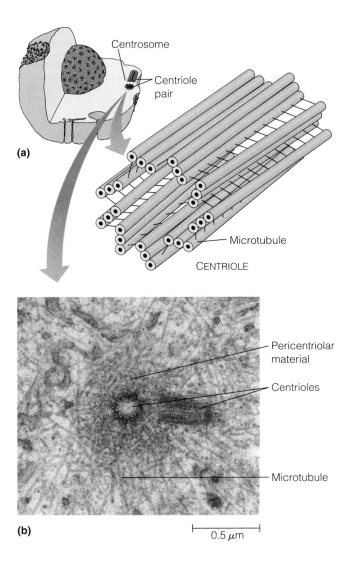

(a)

CENTRIOLE

Centrosome

Centriole pair

Microtubule

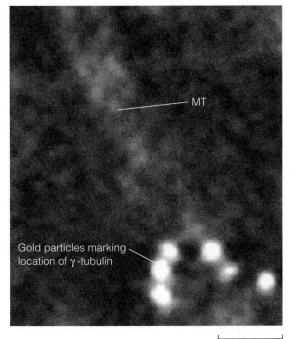

(b)

Pericentriolar material

Centrioles

Microtubule

0.5 μm

Figure 22-8 The Centrosome. **(a)** In animal cells, the centrosome contains two centrioles and associated pericentriolar material. The walls of centrioles are composed of nine sets of triplet microtubules. **(b)** An electron micrograph of a centrosome showing the centrioles and the pericentriolar material. Notice that microtubules originate from the pericentriolar material. **(c)** Nucleation and assembly of MTs at a centrosome in vitro.

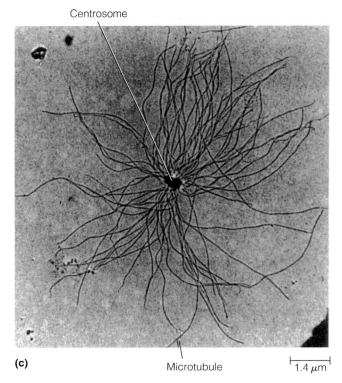

Centrosome

(c)

Microtubule

1.4 μm

MT

Gold particles marking location of γ-tubulin

25 nm

Figure 22-9 γ-Tubulin at the Base of Microtubules Originating from the Centrosome. An electron micrograph of a MT originating from the centrosome. Here, γ-tubulin was labeled with antibodies to which small particles of metal are attached. In the electron micrograph, these antibodies appear as bright spheres (TEM).

many MTs can form. However, the MT-nucleating capacity of the MTOC can be modified during certain processes such as mitosis, in which the number of MTs increases. For example, the abundance of pericentrin fluctuates during mitosis, being highest at prophase and metaphase, when spindle poles show the greatest MT-nucleating activity.

Microtubule Stability Within Cells Is Highly Regulated

The nucleating ability of MTOCs such as the centrosome has an important consequence for microtubule dynamics within cells. Since the minus ends of many MTs are anchored at the centrosome, dynamic growth and shrinkage of these MTs at the plus ends tends to occur at the periphery of cells. Such dynamic shrinkage and growth

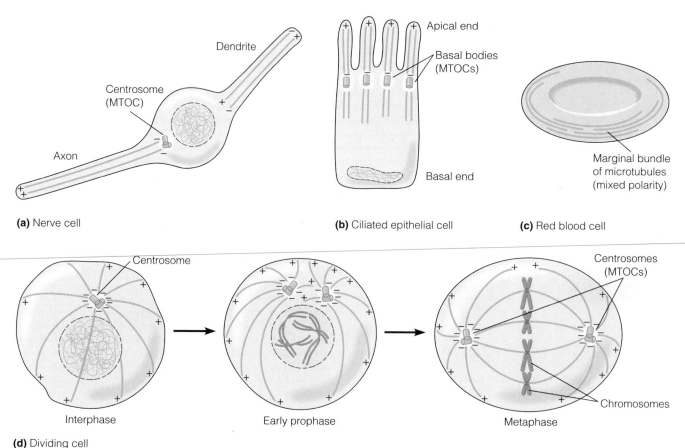

(a) Nerve cell

(b) Ciliated epithelial cell

(c) Red blood cell

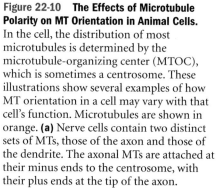

Interphase

Early prophase

Metaphase

(d) Dividing cell

Figure 22-10 **The Effects of Microtubule Polarity on MT Orientation in Animal Cells.** In the cell, the distribution of most microtubules is determined by the microtubule-organizing center (MTOC), which is sometimes a centrosome. These illustrations show several examples of how MT orientation in a cell may vary with that cell's function. Microtubules are shown in orange. **(a)** Nerve cells contain two distinct sets of MTs, those of the axon and those of the dendrite. The axonal MTs are attached at their minus ends to the centrosome, with their plus ends at the tip of the axon. However, dendritic MTs are not associated with the centrosome and are of mixed polarities. **(b)** Ciliated epithelial cells have many MTOCs called basal bodies, one at the base of each cilium. Ciliary MTs originate with their minus ends in the basal bodies and elongate with their plus ends toward the tips of the cilia. **(c)** Mature human red blood cells have no nucleus or MTOC. However, MTs of mixed polarities persist as a circular band at the periphery of the cell. This band helps to maintain the cell's round, disklike shape. **(d)** Throughout the process of mitosis, MTs in a dividing cell are oriented with their minus ends anchored in the centrosome and their plus ends pointing away from the centrosome. Cell division is preceded by the division of the centrosome. The two centrosomes then separate, each forming one of the poles of the mitotic spindle. At metaphase, the centrosomes are at opposite sides of the cell. Each centrosome, or spindle pole, forms half of the spindle MTs, some of which extend from pole to chromosomes, while others extend from one pole to the other pole.

has been demonstrated using video enhanced differential interference contrast (DIC) microscopy to follow the life cycles of individual MTs (Figure 22-11). These studies have shown that dynamic instability phenomena are not restricted to MTs in vitro. In other cases, it has been shown that MTs can detach from the centrosome, freeing their minus ends for subunit addition or loss.

We have seen that cellular microtubules exhibit dynamic instability; they grow out from the centrosome and then disassemble. This process could account for randomly distributed and short-lived MTs, but not for organized and stable arrays of MTs within cells. Such MTs are too unstable to remain intact for long periods of time and will break down unless they are stabilized in some way.

One way to stabilize MTs would be to "capture" and protect their growing plus ends.

An example of how such capture of microtubules may produce a precisely organized set of MTs is seen during mitosis (see Figure 22-10d), as we discussed in Chapter 17. Prior to prophase, the centrosome replicates, giving rise to two daughter centrosomes. These separate during early prophase and move to opposite sides of the cell, where they serve as the poles of the mitotic spindle. When the nuclear envelope breaks down, the chromosomes are connected to the poles by MTs. To establish these connections, the kinetochore on each chromosome is thought to capture the plus ends of MTs. As MTs grow out from the spindle pole, those that encounter a kinetochore are cap-

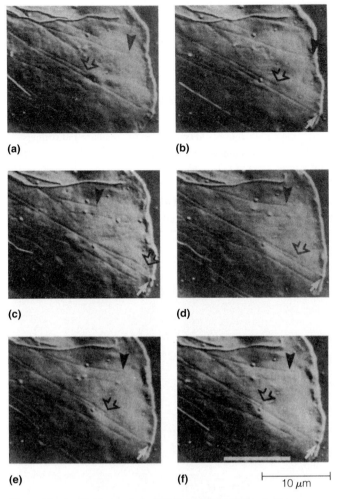

(a) **(b)**

(c) **(d)**

(e) **(f)**

10 μm

Figure 22-11 The Dynamic Instability of Microtubules In Vivo.
Microtubules visualized in a living cell by video-enhanced
differential interference contrast (DIC) microscopy exhibit
dynamic instability in vivo. Here, individual MTs, designated by
several types of arrows, are monitored over time, reading from
(a) to **(f)**. The MTs grow out to the edge of the cell and then rapidly
shorten. For an explanation of DIC, see the *Guide to Microscopy*.

tured and stabilized. Those MTs that miss the kinetochore
will eventually disassemble through dynamic instability
and be replaced by new ones that will go through the same
process. Through repeated cycles of MT growth and
depolymerization, the kinetochore of each chromosome
will eventually capture a MT and become connected to the
spindle poles.

Drugs Can Affect the Assembly of Microtubules

A number of drugs affect microtubule assembly. The best-
known is **colchicine,** introduced earlier in this chapter,
which acts by binding to tubulin dimers. The resulting
tubulin-colchicine complex can still add to the growing
end of a MT, but it then prevents any further addition of
tubulin molecules and destabilizes the structure, thereby
promoting MT disassembly. *Vinblastine* and *vincristine* are

related compounds from the periwinkle plant (*Vinca
minor*) that cause tubulin to aggregate inside the cell.
Nocodazole (a synthetic benzimidazole) is another compound that inhibits MT assembly and is frequently used in
experiments instead of colchicine because its effects are
more readily reversible when the drug is removed.

These compounds are called *antimitotic drugs* because
they disrupt the mitotic spindle of dividing cells, blocking
the further progress of mitosis. The sensitivity of the
mitotic spindle to these drugs is understandable because
the spindle fibers are composed of many microtubules.
Vinblastine and vincristine also find application in medical practice as anticancer drugs. They are useful for this
purpose because cancer cells divide rapidly and are therefore preferentially susceptible to drugs that interfere with
the mitotic spindle.

Taxol, also introduced earlier in this chapter, has the
opposite effect on microtubules: when it binds to MTs, it
stabilizes them. Within cells, it causes free tubulin to
assemble into MTs and arrests dividing cells in mitosis.
Thus, both taxol and colchicine block cells in mitosis, but
they do so by opposing effects on MTs and hence on the
fibers of the mitotic spindle. Taxol is also used in the treatment of some cancers, especially breast cancer.

Microtubules Are Regulated by Microtubule-Associated Proteins (MAPs)

A variety of proteins are known to modulate microtubule
structure, assembly, and function. These **microtubule-
associated proteins (MAPs)** account for 10–15% of the
mass of MTs isolated from cells. MAPs add another level
of regulation to the organization and functions of MTs in
cells. In order to affect MT function, many MAPs bind at
regular intervals along the wall of a microtubule and form
projections from the wall, allowing interaction with other
filaments and cellular structures. MAPs are also important
in regulating MT assembly, most likely by binding to the
growing plus end of a microtubule, thereby stabilizing it
against disassembly. Most MAPs have been shown to
increase MT stability, and some also can stimulate MT
assembly. The many different MAPs vary mainly in how
they link MTs together or to other structures and how
their effects on MTs are regulated.

MAP function has been studied extensively in brain
cells, as they are the most abundant source of these proteins. Two major classes of MAPs are present in MTs from
brain cells: MT-associated motor proteins (motor MAPs)
and nonmotor MAPs. **Motor MAPs,** which include
kinesin and *dynein,* are so named because they use ATP to
drive the transport of vesicles and organelles or to generate sliding forces between MTs. These proteins will be discussed in detail in Chapter 23.

Nonmotor MAPs appear to control microtubule
organization in the cytoplasm. A dramatic example of
such control is seen in neurons. The proper functioning of
the nervous system depends on connections among neurons

Table 22-3 Nonmotor MAPs (Microtubule-Associated Proteins)

MAP	Isoforms	Molecular Mass (kDa)	Tissue	Function
MAP1A (MAP1)	No	350	Nerve dendrites and axons	Induces tubulin polymerization
MAP1B (MAP5)	No	320	Nerve dendrites and axons	Induces neurite elongation
MAP2A	Yes	199	Nerve dendrites	Promotes assembly of tubulin in vitro
MAP2B	Yes	—	Nerve dendrites	Unknown
MAP2C	Yes	70	Nerve dendrites	Bundles MTs; increases MT stability and stiffness
MAP3	Yes	180	Various	Induces tubulin polymerization
MAP4	Yes	200	Various	Induces tubulin polymerization in many cell types during mitosis
Tau	Yes	37–46	Nerve axons	Induces tubulin polymerization, bundles MTs
Radial spoke proteins	—	34–124	Cilia and flagella	Provide structural rigidity?

and with other types of cells. To establish these connections, neurons send out projections called *neurites,* which are reinforced by bundles of MTs. Neurites eventually differentiate into axons, which carry electrical signals away from the cell body of the neuron, and dendrites, which receive signals from neighboring cells and carry them to the cell body. The MT bundles are characteristically denser in axons than they are in dendrites.

These differences arise because dendrites and axons contain different types of MAPs (Table 22-3). For example, an axon-specific MAP called *Tau* causes microtubules to form tight bundles. A family of MAPs called *MAP2,* is present in dendrites and causes the formation of looser bundles of MTs (Figure 22-12).

The importance of MAPs for neurite formation can be demonstrated by introducing the Tau or MAP2C (a specific MAP2) into a nonneuronal cell line that cannot normally make either protein. These cells are normally rounded (Figure 22-13a), but when the *tau* gene is introduced and expressed, these cells extend single long processes that look remarkably similar to nerve axons (Figure 22-13b). In contrast, when the *MAP2C* gene is introduced and expressed, several shorter processes form that resemble dendrites (Figure 22-13c).

The diversity of MAPs may help explain how cells can differ in their microtubule organization. In addition, the function of some MAPs can be altered by phosphorylation, which provides a means for the cell to alter MT organization rapidly. An example of this kind of regulation is seen with MAP4, a MAP that promotes MT assembly when it is not phosphorylated. As the cell goes through mitosis, MAP4 becomes phosphorylated and loses its ability to promote assembly.

Microfilaments

With a diameter of about 7 nm, **microfilaments (MFs)** are the smallest of the cytoskeletal filaments (see Table 22-1).

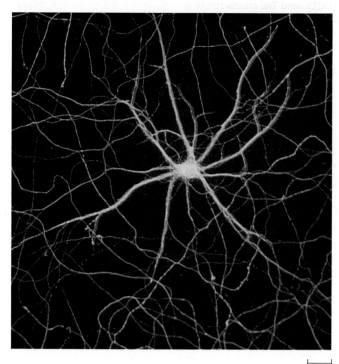

$\vdash\!\!\!\dashv$
10 μm

Figure 22-12 Localization of MAP2 and Tau in a Neuron by Immunofluorescence Staining. The unique subcellular distribution of MAP2 and Tau can be demonstrated by staining a cultured hippocampal neuron for both proteins using two different antibodies, one for MAP2 and one for Tau. When stained in this manner, axons exhibit green fluorescence, indicating the presence of Tau, whereas dendrites show an orange fluorescence, indicating the presence of MAP2.

Microfilaments are best known for their role in the contractile fibrils of muscle cells, where they interact with thicker filaments of myosin to cause the contractions characteristic of muscle. MFs are not confined to muscle cells, however; they occur in almost all eukaryotic cells and are involved in numerous other phenomena, including a variety of locomotory and structural functions.

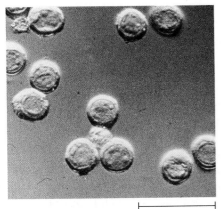

(a) Normal Sf9 cells

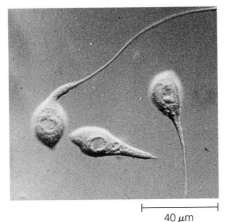

(b) Sf9 cells expressing *tau*

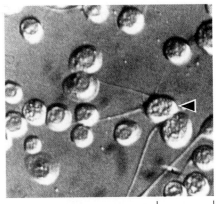

(c) Sf9 cells expressing *MAP2C*

Figure 22-13 **Expression of the Genes for Tau and MAP2C in a Nonneuronal Cell Line.** Both MAP2C and Tau are microtubule-associated proteins whose genes are expressed in neurons and are important for establishing the dendrites and axons of neurons. One way to demonstrate this is to express the *tau* gene or the *MAP2C* gene in a nonneuronal cell line and observe the changes in cell morphology that occur. **(a)** DIC micrograph of cells of the Sf9 tissue culture line (an insect cell line) normally exhibit a round morphology and do not express either *MAP2C* or *tau*. **(b)** When *tau* is expressed in Sf9 cells, a single long process resembling an axon grows out from the cell. **(c)** When *MAP2C* is expressed in Sf9 cells, multiple processes grow out from the cell (arrow), resembling the formation of dendrites.

Examples of cell movements in which microfilaments play a role include *amoeboid movement*, locomotion of cultured cells over a surface to which the cell is attached, and *cytoplasmic streaming*, a regular pattern of cytoplasmic flow in some plant and animal cells. MFs also produce the cleavage furrows that divide the cytoplasm of animal cells during cytokinesis. We will discuss all of these phenomena in detail in Chapter 23. MFs are also found at sites of attachment of cells to one another and to the extracellular matrix (see Chapter 11).

In addition to mediating a variety of cell movements, MFs are important in the development and maintenance of cell shape. Most animal cells, for example, have a dense network of microfilaments called the **cell cortex** just beneath the plasma membrane. The cortex confers structural rigidity on the cell surface and facilitates shape changes and cell movement. Parallel bundles of MFs also make up the structural core of *microvilli*, the fingerlike extensions found on the surface of many animal cells (see Figure 4-2).

Actin Is the Protein Building Block of Microfilaments

Actin is an extremely abundant protein in virtually all eukaryotic cells, including those of plants, algae, and fungi. Actin is synthesized as a single polypeptide consisting of 375 amino acids, with a molecular weight of about 42,000. Once synthesized, it folds into a roughly U-shaped molecule, with a central cavity that binds ATP or ADP (Figure 22-14). Individual actin molecules are referred to as **G-actin** (globular actin). Under the right conditions, G-actin molecules polymerize to form microfilaments; in this form, actin is referred to as **F-actin** (filamentous actin). Actin in the G or F form also binds to a wide variety of other proteins. These

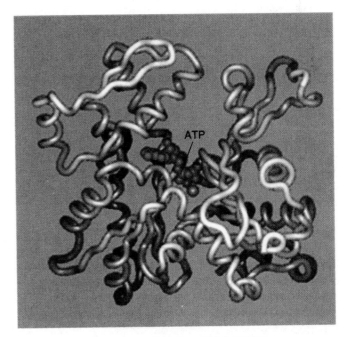

Figure 22-14 **The Molecular Structure of G-Actin Monomer.** X-ray crystallography shows that the G-actin monomer is shaped somewhat like a U. A nucleotide (ATP or ADP) binds reversibly in a groove in the protein. When G-actin monomers polymerize into F-actin, the mouth of the groove is covered by another G-actin monomer, trapping the bound nucleotide inside. In addition, binding of one G-actin to another forces the mouth of the groove to close more tightly on the bound nucleotide, promoting hydrolysis of the ATP.

actin-binding proteins either regulate and modify the function of actin, or are themselves regulated or organized by their association with actin.

Different Types of Actin and Actin-Related Proteins Are Found in Cells

Of the three types of cytoskeletal proteins, actin is the most highly conserved. In functional assays, all actins appear to be identical and actins from diverse organisms will copolymerize into filaments. Despite this high degree of sequence similarity, actins do differ among different organisms and among tissues of the same organism. Based on sequence similarity, actins can be broadly divided into two major groups: the *muscle-specific actins* (α-*actins*) and the *nonmuscle actins* (β- *and* γ-*actins*). In the case of β- and γ-actins, recent studies have shown that these actins localize to different regions of the cell and appear to have different interactions with actin-binding proteins. For example, in epithelial cells, one end of the cell, the apical end, contains microvilli, whereas the opposite side of the cell, known as the basal end, is attached to the extracellular matrix (see Figure 11-22). β-actin is predominantly found at the apical end of epithelial cells, whereas γ-actin is concentrated at the basal end and sides of the cell.

In addition to the various types of actin, another class of proteins known as **actin-related proteins (Arps)** show recognizable, though significantly less, sequence similarity to actin. For example, the actins from yeast and chicken are identical at more than 90% of their amino acids, whereas the Arps show only 50% similarity to a variety of different actins. As we will see later in this chapter, Arp2 and Arp3 are involved in nucleating assembly of new microfilaments in migrating cells.

G-Actin Monomers Polymerize into F-Actin Microfilaments

Like tubulin dimers, G-actin monomers can polymerize reversibly into filaments with a lag phase corresponding to filament nucleation, followed by a more rapid polymer elongation phase. The F-actin filaments that form appear to be composed of two linear strands of polymerized G-actin wound around each other in a helix, with roughly 13.5 actin monomers per turn (Figure 22-15).

Within a microfilament, all the actin monomers are oriented in the same direction, so that an MF, like a microtubule, has an inherent polarity, with one end differing chemically and structurally from the other end. This polarity can be readily demonstrated by incubating MFs with **myosin subfragment 1 (S1)**, a proteolytic fragment of myosin (Figure 22-16). S1 fragments bind to, or "decorate," the actin MFs to give a distinctive arrowhead pattern, with all the S1 molecules pointing in the same direction (Figure 22-16c). Based on this arrowhead pattern, the terms *pointed end* and *barbed end* are commonly used to identify the minus and plus ends of a MF, respectively. The polarity of the MF is important, because it allows for independent regulation of actin assembly or disassembly at each end of the filament.

The polarity of microfilaments is reflected in more rapid addition or loss of G-actin at the plus end, and slower addition or loss of G-actin at the minus end (see Figure 22-15a). If G-actin is polymerized onto short fragments of S1-decorated F-actin, polymerization proceeds much faster at the barbed end, indicating that the barbed end of the filament is also the plus end. Thus even when conditions are favorable for adding monomers to both ends of the filament, the plus end will grow faster than the minus end.

As G-actin monomers assemble onto a microfilament, the ATP bound to them is slowly hydrolyzed to ADP, much the same as the GTP bound to tubulin is hydrolyzed to GDP. Thus the ends of a growing MF tend to have ATP-F-actin, whereas the bulk of the MF is composed of ADP-F-actin. However, ATP hydrolysis is not a strict requirement for MF elongation, since MFs can also assemble from ADP-G-actin or from nonhydrolyzable analogues of ATP-G-actin.

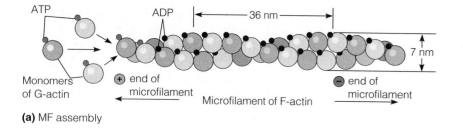

(a) MF assembly

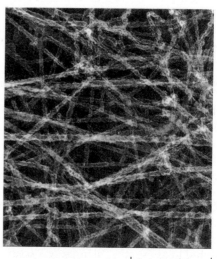

(b) Purified F-actin
0.5 μm

Figure 22-15   A Model for Microfilament Assembly In Vitro.   (a) Monomers of G-actin polymerize into long filaments of F-actin with a diameter of about 7 nm. A full turn of the helix occurs every 36–37 nm, with about 13.5 monomers required for a full turn. The addition of each G-actin monomer is usually accompanied or followed by hydrolysis of the ATP molecule so that it is tightly bound to the monomer, although the energy of ATP hydrolysis is not required to drive the polymerization reaction. **(b)** An electron micrograph of purified F-actin (TEM).

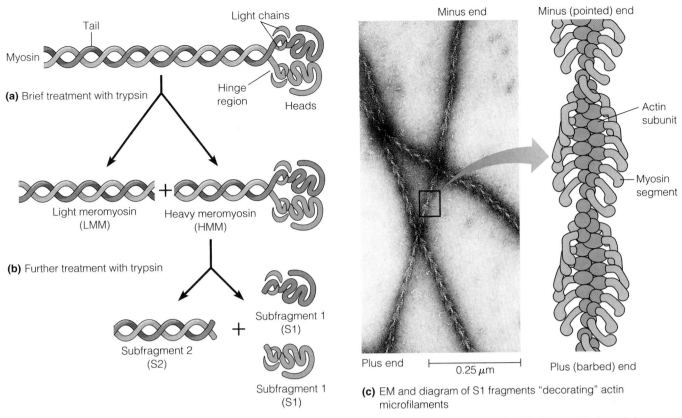

(a) Brief treatment with trypsin

Myosin — Tail — Light chains

Hinge region — Heads

(b) Further treatment with trypsin

Light meromyosin (LMM) + Heavy meromyosin (HMM)

Subfragment 2 (S2) + Subfragment 1 (S1)

Subfragment 1 (S1)

Minus end — Minus (pointed) end

Actin subunit

Myosin segment

Plus end — 0.25 μm — Plus (barbed) end

(c) EM and diagram of S1 fragments "decorating" actin microfilaments

Figure 22-16   The Actin-Binding Protein Myosin II. This protein is part of the contractile machinery found in muscle cells. The globular head of the myosin molecule binds to actin, while the myosin tails can associate with filaments of myosin (the thick myofilaments of muscle cells). **(a)** Myosin II can be cleaved by proteases such as trypsin into two pieces, heavy meromyosin (HMM) and light meromyosin (LMM). **(b)** HMM can be further digested, leaving only the globular head. This fragment, called myosin subfragment 1 (S1), retains its actin-binding properties. **(c)** When actin microfilaments are incubated with myosin S1 and then examined with an electron microscope, the S1 fragments appear to "decorate" the microfilaments like arrowheads. All the S1 arrowheads point toward the minus end, indicating the polarity of the MF.

Actin Polymerization Is Regulated by Small GTP-Binding Proteins

The regulation of microfilament assembly is complex. Cells appear to control the process at several steps, including nucleation of new MFs and elongation of pre-existing MFs; they also control the rate of actin depolymerization. Both proteins and plasma membrane lipids regulate the formation, stability, and breakdown of MFs. We are just beginning to identify some of the molecules that regulate actin assembly. So far, these include the small regulatory GTP-binding proteins known as Rac, Rho, and Cdc42, a variety of actin-binding proteins, and inositol phospholipids, which are found on the inner leaflet of the plasma membrane. We begin with Rac, Rho, and Cdc42.

In addition to the small, monomeric G protein Ras, which is involved in receptor-mediated signal transduction (see Figure 10-18), the cell contains a great number of other small G proteins that regulate the activity of various biological processes, including actin filament assembly. The small G proteins Rac, Rho, and Cdc42 are important modulators of the actin cytoskeleton. Although the effects of these proteins differ, each acts as a biochemical switch that triggers specific changes in the actin cytoskeleton. We will postpone a detailed discussion of how these proteins specifically affect the movement of cells until Chapter 23. For now, recall from Chapter 10 that the state of a small G protein's switch (i.e., whether it is "on" or "off") depends on whether the G protein is bound to GDP or GTP (see Figure 10-4). Normally, the G protein is bound to GDP, corresponding to the off state. A guanine nucleotide exchange factor helps the G protein switch to the on state by facilitating the release of bound GDP in exchange for a molecule of GTP. Once Rac, Rho, or Cdc42 has acquired GTP, it can activate biochemical events in the cell.

One way that these small G proteins appear to regulate actin assembly is by regulating kinases that phosphorylate inositol phospholipids, causing an increase in polyphosphoinositide production. Such regulation could be an important way to modulate the actin cytoskeleton, because a number of proteins that bind to the plus end of microfilaments also bind to polyphosphoinositides, as discussed below.

Specific Proteins and Drugs Affect Polymer Dynamics at the Ends of Microfilaments

In the absence of other factors, the growth of microfilaments depends on the concentration of ATP-bound G-actin. If the concentration of ATP-bound G-actin is high, microfilaments will assemble until the G-actin is limiting. In the cell, however, a large amount of free G-actin is not available for assembly into filaments, because it is bound by the protein *thymosin β4*. A second protein called *profilin* appears to transfer G-actin monomers from the thymosin β4 complex to the end of a growing filament, but this can only happen if there are free filament ends available. Yet another protein, known as *ADF/cofilin,* is known to bind to ADP-G-actin and F-actin; ADF/cofilin is thought to increase the rate of turnover of ADP-actin at the minus ends of MFs. This makes the released ADP-G-actin available for conversion into ATP-G-actin, which can be recycled for addition to the growing plus ends of MFs.

Drugs that result in the depolymerization of microfilaments affect the cell's ability to add G-actin to the plus ends of MFs. For example, the **cytochalasins,** introduced earlier, prevent the addition of new monomers to existing polymerized MFs. As subunits are gradually lost from the minus ends of MFs in cytochalasin-treated cells, they eventually depolymerize. In contrast, latrunculin A acts by sequestering actin monomers, preventing their addition to the plus ends of growing MFs. In either case, the net result is the loss of MFs within the treated cells.

Capping Proteins Stabilize the Ends of Microfilaments

Whether microfilament ends are available for further growth depends on whether or not the filament end is *capped.* Capping occurs when a **capping protein** binds the end of a filament and prevents further addition or loss of subunits, thereby stabilizing it. One such protein that functions as a cap for the plus ends of microfilaments is appropriately named *CapZ.* When CapZ is bound to the end of a filament, further addition of subunits at the plus end is prevented; when CapZ is removed, addition of subunits can resume.

Inositol Phospholipids Regulate Molecules That Affect Actin Polymerization

Inositol phospholipids are one type of membrane phospholipid. One inositol derivative, inositol trisphosphate (InsP$_3$), is a key component of signaling via heterotrimeric G proteins, as we saw in Chapter 10. Several of the hydroxyl groups on the inositol ring of another inositol derivative, phosphatidyl inositol, can be phosphorylated by specific kinases in the cytoplasm to yield a variety of phosphorylated products, which we will simply refer to collectively as **polyphosphoinositides.** Various polyphosphoinositides bind to actin-binding proteins. For example, polyphosphoinositides are known to bind to profilin and CapZ, and are thought to regulate the ability of these proteins to interact with actin. CapZ binds tightly to specific polyphosphoinositides, resulting in their removal from the end of a microfilament, thereby permitting the filament to be disassembled, making its monomers available for assembly into new filaments.

Polyphosphoinositides may also provide a link between Rac and actin polymerization. For example, in blood platelets Rac activation results in increased actin polymerization, and this is accompanied by increased synthesis of a specific polyphosphoinositide. Still other polyphosphoinositides may help recruit guanine nucleotide exchange proteins to specific sites in the cell, which in turn stimulate Rho or other small G proteins. Thus, the formation and breakdown of polyphosphoinositides may represent an important control point for regulating the function of microfilaments.

Actin-Binding Proteins Regulate Interactions Between Microfilaments

As we have seen, microfilaments function in cell motility and cell structure. We have also seen that cytosolic proteins can influence the polymerization of MFs. In addition, as structural components of the cytoskeleton, MFs can form larger arrays of actin with varying degrees of organization. These range from the stiff, parallel bundles of filaments found in microvilli, to the acrosomal processes that extend from the tips of sperm cells of certain kinds of marine organisms to the protrusions used in cell crawling, to the loose network of filaments found in the cell cortex (Figure 22-17). The local structure of the actin cytoskeleton depends on the function of special **actin-binding proteins** and the unique ways in which they interact with microfilaments.

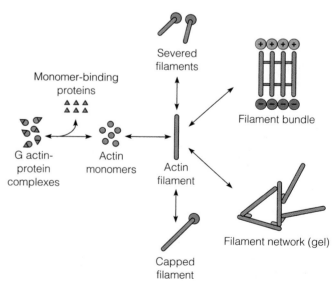

Figure 22-17 Interrelationships Between the Main Structural Forms of Actin. Actin-binding proteins (green, purple, and brown) are responsible for converting actin filaments from one form to another.

Microvilli. Microvilli (singular: **microvillus**) are especially prominent features of intestinal mucosal cells (Figure 22-18a). A single mucosal cell in your small intestine, for example, has several thousand microvilli, each about 1–2 μm long and about 0.1 μm in diameter, which increase the surface area of the cell about 20-fold. This large surface area is essential to intestinal function, because the uptake of digested food depends on an extensive absorptive surface.

As illustrated in Figure 22-18b, the core of the intestinal microvillus consists of a tight bundle of microfilaments. The plus ends point toward the tip, where they are attached to the membrane through an amorphous electron-dense plaque. The MFs in the bundle are also connected to the plasma membrane by lateral crosslinks consisting of the proteins *myosin I* and *calmodulin.* These crosslinks extend outward about 20–30 nm from the bundle to contact electron-dense patches on the inner membrane surface. Adjacent MFs in the bundle are bound tightly together at regular intervals by the crosslinking proteins (also called actin-bundling proteins) *fimbrin* and *villin.*

At the base of the microvillus, the MF bundle extends into a network of filaments called the **terminal web** (Figure 22-19). The filaments of the terminal web are composed mainly of myosin and *spectrin,* which connect the microfilaments to each other, to proteins within the plasma membrane, and perhaps also to the network of intermediate filaments beneath the terminal web. The terminal web apparently gives rigidity to the microvilli by anchoring their MF bundles securely so that they project straight out from the cell surface.

Actin Gels and the Cell Cortex. The cell cortex is a three-dimensional meshwork of microfilaments and associated proteins located just beneath the plasma membrane of most animal cells. The cortex supports the plasma membrane, confers rigidity to the cell surface, and facilitates shape changes and cellular movement. An important function of certain actin-binding proteins in the cortex is to link MFs into a stable network with gel-like properties. One of these crosslinking proteins is *filamin,* a long molecule consisting of two identical polypeptides joined

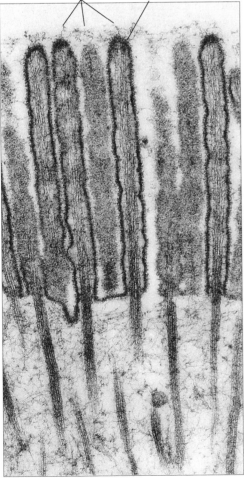

(a) Intestinal microvilli

0.2 μm

Figure 22-18   Microvillus Structure.
(a) An electron micrograph of microvilli from intestinal mucosal cells (TEM). **(b)** A schematic diagram of a single microvillus, showing the core of microfilaments that gives the microvillus its characteristic stiffness. The core consists of several dozen microfilaments oriented with their plus ends facing outward toward the tip and their minus ends facing toward the cell. The plus ends are embedded in an amorphous, electron-dense plaque. The MFs are tightly linked together by actin-bundling (crosslinking) proteins and are connected to the inner surface of the plasma membrane by lateral crosslinks composed of calmodulin and myosin I.

Electron-dense plaque

Actin microfilaments

Lateral cross-links

Actin-bundling proteins

Plasma membrane

(b) Structure of a microvillus

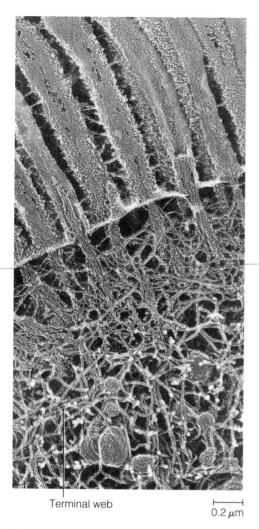

Terminal web

0.2 μm

Figure 22-19 The Terminal Web of an Intestinal Epithelial Cell. The terminal web beneath the plasma membrane is seen in this freeze-etch electron micrograph of an intestinal epithelial cell. Bundles of microfilaments that form the cores of microvilli extend into the terminal web.

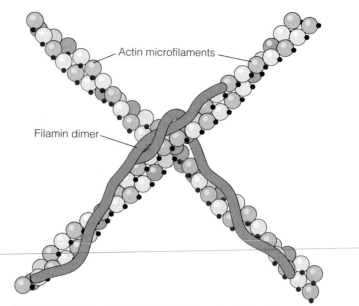

Actin microfilaments

Filamin dimer

Figure 22-20 Crosslinking of Actin Filaments by Filamin. Filamin is one of several crosslinking proteins that link actin filaments into an extensive three-dimensional network. Filamin is a dimer of two identical polypeptides joined head-to-head into a long, flexible molecule. The tail of each polypeptide contains a binding site for actin filaments.

head-to-head, with an actin-binding site at each tail. Molecules of filamin act as "splices," joining two MFs together where they intersect (Figure 22-20). In this way, actin MFs are linked to form large three-dimensional networks.

Other proteins play the opposite role, breaking up the microfilament network and causing the cortical actin gel to soften and liquefy. They do this by severing and/or capping MFs; in some cases such proteins can serve both functions. One of these severing and capping proteins is *gelsolin,* which functions by breaking actin MFs and capping their newly exposed plus ends, thereby preventing further polymerization. Gelsolin is another actin-binding protein that can be regulated by binding to polyphosphoinositides; when gelsolin binds to a specific polyphosphoinositide, it can no longer cap the plus end of a MF, allowing the uncapped end to undergo changes in length.

Dendritic Branched Networks of Actin in Migrating Cells. Cells that crawl have specialized structures called *lamellipodia* that

allow them to move along a surface (we will consider these specialized structures in more detail in Chapter 23). The actin within lamellipodia is more organized than in the cortex but less organized than in microvilli. Lamellipodia contain branched actin filaments that form a tree-like, or *dendritic,* network (Figure 22-21a). As we discussed earlier, the barbed ends of the straight branches form by polymerization of actin monomers via profilin and are probably capped by capping proteins. The difference is that branching is favored in the lamellipodium due to the action of several proteins. A complex of actin-related proteins, the *Arp2/3 complex,* helps branches to form by nucleating new branches on the sides of existing filaments. The branches can be shown to form on the sides of existing microfilaments by adding fluorescent G actin (for example, monomers labeled with a red fluorescent dye) to microfilaments that consist of actin labeled with a different fluorescent marker (e.g., a green dye). New, red branches sprout from existing, green microfilaments when activated Arp2/3 complexes are added (Figure 22-21b). When branches are examined using antibodies that detect Arp2/3, Arp2/3 complexes are found at the branch points (Figure 22-21c). Arp2/3 branching is activated by a family of proteins that includes the *Wiskott Aldrich syndrome protein,* or *WASP.* Human patients who cannot produce functional WASP have defects in the ability of their platelets to undergo changes in shape and so have difficulties in forming blood clots.

A Variety of Proteins Link Actin to Membranes

We have seen that microfilaments are involved in supporting specific structures immediately beneath the plasma

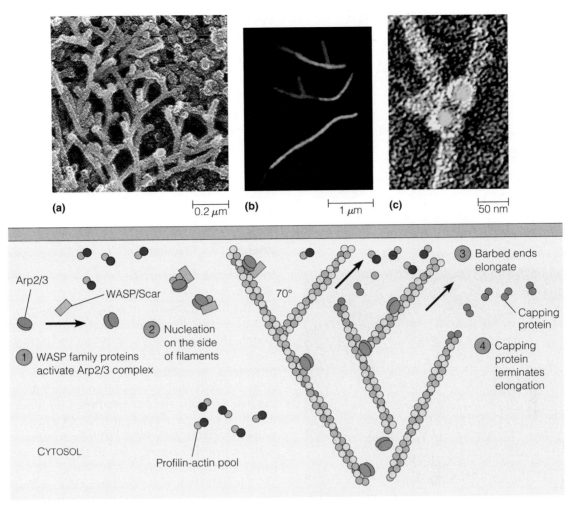

(a) 0.2 μm   **(b)** 1 μm   **(c)** 50 nm

(d)

Figure 22-21 Branched Actin Networks and the Arp2/3 Complex. Actin networks, like those found in migrating cells, have a characteristic pattern of branching. **(a)** Branched actin filaments in a frog keratocyte. Individual branched actin filaments are colored to make them easier to distinguish (deep etch TEM). **(b)** When fluorescently labeled actin monomers (red), WASP, and Arp2/3 protein are added to preexisting, fluorescently labeled actin filaments (green), the actin polymerizes to form new branched structures. Some branches form new actin (left), whereas other branches form on the sides of preexisting actin filaments (right) (fluorescence microscopy). **(c)** Arp2/3 protein (detected by using antibodies that are attached to gold particles, yellow) localizes to branch points (deep etch TEM).

membrane, such as the microvilli of epithelial cells. MFs are also intimately involved in cellular movement and in the pinching in of the cell membrane during cytokinesis. To carry out these different functions, MFs must be connected to the plasma membrane. The connection of MFs to the plasma membrane is indirect, and requires one or more linker proteins that anchor MFs to transmembrane proteins embedded within the plasma membrane.

One group of proteins that appears to function widely in linking microfilaments to membranes is the *band 4.1, ezrin, radixin,* and *moesin (FERM)* family of actin-binding proteins. When these proteins are mutated, a wide variety of cellular processes are affected, including cytokinesis, secretion, and the formation of microvilli. Another example of how actin can be linked to membranes involves the erythrocyte proteins *spectrin* and

ankyrin (mentioned in Chapter 7). As illustrated in Figure 22-22, the plasma membrane of the erythrocyte is supported by a network of spectrin filaments that are crosslinked by very short actin chains. This network is connected to the plasma membrane by molecules of the proteins ankyrin and band 4.1 that link the spectrin filaments to specific transmembrane proteins.

Proteins similar to spectrin, ankyrin, and band 4.1 are found in animal cells other than erythrocytes. For example, members of the family of proteins that includes band 4.1 are found at the apical end of epithelial cells; mutations in these genes in *Drosophila* and in the nematode *Caenorhabditis elegans* disrupt the organization of epithelial cells or their ability to change shape. These and many other experiments demonstrate that the plasma membrane is supported by a cortical network of microfilaments, and that

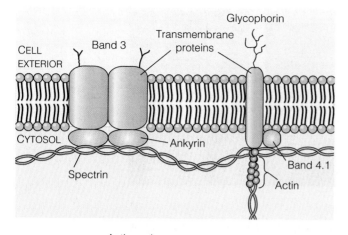

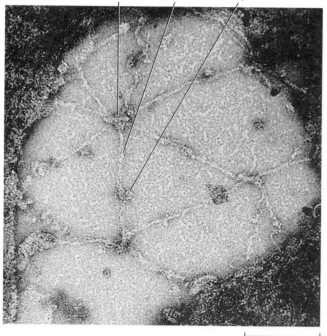

Actin and
Band 4.1   Spectrin   Ankyrin

0.1 μm

Figure 22-22   Support of the Erythrocyte Plasma Membrane by a Spectrin-Ankyrin-Actin Network.   The plasma membrane of a red blood cell is supported on its inner surface by a filamentous network that gives the cell both strength and flexibility. Long filaments of spectrin are crosslinked by short actin filaments. The network is anchored to the band 3 transmembrane protein by molecules of the protein ankyrin.

linkage of this network to the cell surface is required for the normal functioning of animal cells.

Intermediate Filaments

Intermediate filaments (IFs) have a diameter of about 8–12 nm, which makes them intermediate in size between microtubules and microfilaments (see Table 22-1), or between the thin (actin) and thick (myosin) filaments in

muscle cells, where IFs were first discovered. To date, most studies have focused on animal cells, where IFs occur singly or in bundles and appear to play a structural or tension-bearing role. Figure 22-23 is an electron micrograph of IFs from a cultured human fibroblast cell.

Intermediate filaments are the most stable and the least soluble constituents of the cytoskeleton. Treatment of cells with detergents or with solutions of high or low ionic strength removes most of the microtubules, microfilaments, and other proteins of the cytosol, but leaves networks of IFs that retain their original shape. Because of the stability of the IFs, some scientists suggest that they serve as a scaffold to support the entire cytoskeletal framework.

Intermediate Filament Proteins Are Tissue-Specific

Intermediate filaments are only found in multicellular organisms and, in contrast to microtubules and microfilaments, differ markedly in amino acid composition from tissue to tissue. Based on the cell type in which they are found, IFs and their proteins can be grouped into six classes (Table 22-4). Classes I and II comprise the *keratins*, proteins that make up the *tonofilaments* found in the epithelial cells that cover the body surfaces and line its cavities. (The IFs visible beneath the terminal web in the intestinal mucosa cell of Figure 22-19 consist of keratin.) Class I keratins are *acidic keratins*, whereas class II are *basic* or *neutral keratins*; each of these classes contains at least 15 different keratins.

Class III IFs include vimentin, desmin, and glial fibrillary acidic protein. *Vimentin* is present in connective tissue and other cells derived from nonepithelial cells. Vimentin-containing filaments are often prominent features in cultured fibroblast cells, in which they form a network that radiates from the center out to the periphery of the cell. *Desmin* is found in muscle cells, and *glial fibrillary acidic (GFA) protein* is characteristic of the glial cells that surround and insulate nerve cells. Class IV IFs are the *neurofilament (NF) proteins* found in the neurofilaments of nerve cells. Class V IFs are *nuclear lamins* A, B, and C, which form a filamentous scaffold along the inner surface of the nuclear membrane of virtually all eukaryotic cells. Neurofilaments found in cells in the embryonic nervous system are made of *nestin*, which constitutes class VI.

As IF proteins and their genes have been sequenced, it has become clear that these proteins are encoded by a single (though large) family of related genes and can therefore be classified according to amino acid sequence relatedness as well. The six classes of IF proteins have been distinguished on this basis (see Table 22-4).

Because of the tissue specificity of intermediate filaments, animal cells from different tissues can be distinguished on the basis of the IF protein present, as determined by immunofluorescence microscopy. This *intermediate filament typing* serves as a diagnostic tool in medicine. IF typing is especially useful in the diagnosis of cancer, because tumor cells are known to retain the IF proteins characteristic of the tissue of origin, regardless of

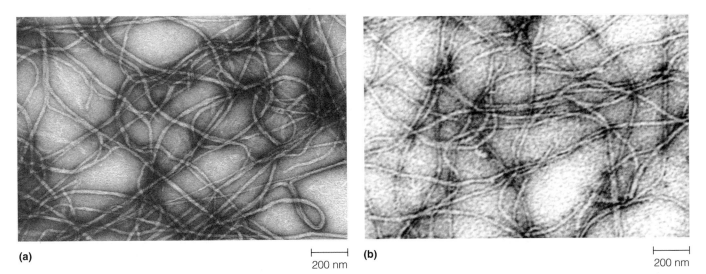

(a)   **(b)**

⊢———⊣   ⊢———⊣
200 nm   200 nm

Figure 22-23   Intermediate Filaments.   Electron micrographs of negatively stained intermediate filaments reconstituted in vitro. **(a)** Filaments formed from keratins 5 and 14; **(b)** vimentin filaments (TEM).

Table 22-4   Classes of Intermediate Filaments

Class	IF Protein	Molecular Mass (kDa)	Tissue	Function
I	Acidic cytokeratins	40–56.5	Epithelial cells	Mechanical strength
II	Basic cytokeratins	53–67	Epithelial cells	Mechanical strength
III	Vimentin	54	Fibroblasts; cells of mesenchymal origin; lens of eye	Maintenance of cell shape
III	Desmin	53–54	Muscle cells, especially smooth muscle	Structural support for contractile machinery
III	GFA protein	50	Glial cells and astrocytes	Maintenance of cell shape
IV	Neurofilament proteins		Central and peripheral nerves	Axon strength; determines axon size
	NF-L (major)	62		
	NF-M (minor)	102		
	NF-H (minor)	110		
V	Nuclear lamins		All cell types	Form a nuclear scaffold to give shape to nucleus
	Lamin A	70		
	Lamin B	67		
	Lamin C	60		
VI	Nestin	240	Neuronal stem cells	Unknown

where the tumor occurs in the body. Because the appropriate treatment often depends on the tissue of origin, IF typing is especially valuable in cases where diagnosis using conventional microscopic techniques is difficult.

Intermediate Filaments Assemble from Fibrous Subunits

As products of a family of related genes, all IF proteins have some common features, although they differ significantly in size and chemical properties. In contrast to actin and tubulin, all IF proteins are fibrous, rather than globular, proteins. All IF proteins have a homologous central rodlike domain of 310–318 amino acids that has been remarkably conserved in size, in secondary structure, and to some extent in sequence. As shown in Figure 22-24, this central domain consists of four segments of coiled helices interspersed with three short linker segments. Flanking the central helical domain are N- and C-terminal domains that differ greatly in size, sequence, and function among IF proteins, presumably accounting for the functional diversity of these proteins.

A possible model for IF assembly is shown in Figure 22-25. The basic structural unit of intermediate filaments consists of two IF polypeptides intertwined into a *coiled coil*. The central helical domains of the two polypeptides are aligned in parallel, with the N- and C-terminal

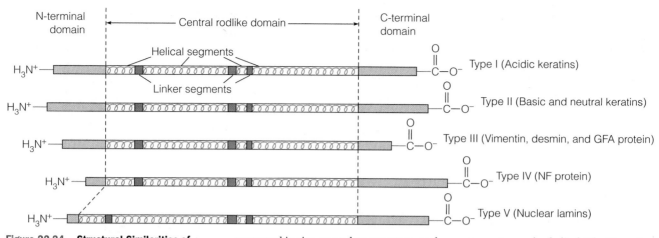

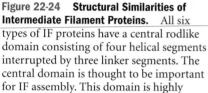

Type I (Acidic keratins)

Type II (Basic and neutral keratins)

Type III (Vimentin, desmin, and GFA protein)

Type IV (NF protein)

Type V (Nuclear lamins)

Figure 22-24 Structural Similarities of Intermediate Filament Proteins. All six types of IF proteins have a central rodlike domain consisting of four helical segments interrupted by three linker segments. The central domain is thought to be important for IF assembly. This domain is highly conserved in size, secondary structure, and sequence, though sequence homologies are confined to the helical regions. In types I–IV, the helical segments contain a total of 276 amino acids and the linker segments are nonhelical. In type V, the helical segments contain 318 amino acids and their linker segments are also helical. The N- and C-terminal domains that flank the central section are nonhelical and are much more variable in size and sequence. (The structure of the sixth type, nestin, is not shown.)

regions protruding as globular domains at each end. Two such dimers then align laterally to form a tetrameric *protofilament*. Protofilaments interact with each other, associating in an overlapping manner to build up a filamentous structure both laterally and longitudinally. When fully assembled, an intermediate filament is thought to be eight protofilaments thick at any point, with protofilaments probably joined end-to-end in staggered overlaps.

Intermediate Filaments Confer Mechanical Strength on Tissues

Intermediate filaments are considered to be important structural determinants in many cells and tissues. Because they often occur in areas of the cell that are subject to mechanical stress, they are thought to have a tension-bearing role. For example, in epithelial cells, *tonofilaments* made of keratin loop through plaques called *desmosomes* that provide strong connecting junctions between two neighboring cells (see Figure 11-19). Another somewhat different structure, the *hemidesmosome,* forms connections between the basal surface of an epithelial cell and the extracellular matrix (see Figure 11-18). The system of desmosomes, hemidesmosomes, and tonofilaments bears most of the mechanical stress when the epithelium is stretched. When keratin filaments are genetically modified in the keratinocytes of transgenic mice, the epidermal cells are fragile and rupture easily. In humans, naturally occurring mutations of keratins give rise to a blistering skin disease called *epidermolysis bullosa simplex (EBS)*. IF defects are also suspected in other pathological conditions, including *amyotrophic lateral sclerosis (ALS)* and certain types of inherited *cardiomyopathies*, which result from defects in the organization of heart muscle.

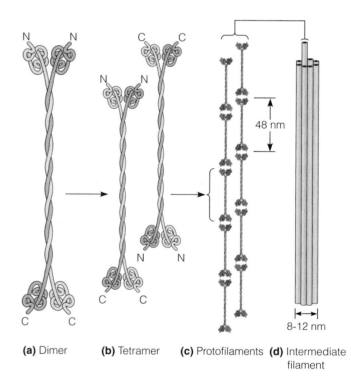

(a) Dimer **(b)** Tetramer **(c)** Protofilaments **(d)** Intermediate filament

Figure 22-25 A Model for Intermediate Filament Assembly In Vitro. (a) The starting point for assembly is a pair of IF polypeptides. The two polypeptides are identical for all IFs except keratin filaments, which are obligate heterodimers with one each of the type I and type II polypeptides. The two polypeptides twist around each other to form a two-chain coiled coil, with their conserved center domain aligned in parallel. **(b)** Two dimers align laterally to form a tetrameric protofilament. **(c)** Protofilaments assemble into larger filaments by end-to-end and side-to-side alignment. **(d)** The fully assembled intermediate filament is thought to be eight protofilaments thick at any point.

Within the cell, intermediate filaments form a structural scaffold called the *nuclear lamina* on the inner surface of the nuclear membrane (discussed in detail in Chapter 17). The nuclear lamina is composed of three separate IF proteins called *nuclear lamins A, B, and C*. These lamins become phosphorylated and disassemble as a part of nuclear envelope breakdown at the onset of mitosis. After mitosis, lamin phosphatases remove the phosphate groups, allowing the nuclear envelope to form again.

The Cytoskeleton Is a Mechanically Integrated Structure

In the preceding sections, we have looked at the individual components of the cytoskeleton as separate entities. When we first look at pictures of the cytoskeleton, it appears to be a tangled web of filaments with little order. In fact, cellular architecture depends on the unique properties of the different cytoskeletal components working together.

In the cytoskeleton, microtubules are generally compression-bearing elements, while microfilaments serve as contractile elements that generate tension. Intermediate filaments are elastic and can withstand tensile forces.

The mechanical integration of intermediate filaments, microfilaments, and microtubules is made possible by specific linker proteins that connect them. Recall that desmosomes and hemidesmosomes are linked to intermediate filaments by members of the *plakin* family of proteins, which includes *plectin, desmoplakin,* and *bullous pemphigoid antigen 1* (BPAG1; see Chapter 11). Such linkages provide mechanical strength to desmosomes and hemidesmosomes. The role of plakins as mechanical linkers is not restricted to desmosomes and hemidesmosomes, however. For example, plectin is a versatile linker protein that is found at sites where intermediate filaments are connected to microfilaments or microtubules (Figure 22-26). Plectin, as well as several other plakins, contains binding sites for intermediate filaments, microfilaments, and microtubules. By linking these major types of polymers, plakins help to integrate them into a mechanically integrated cytoskeletal network. As a result, interconnected cytoskeletal structures can adapt to stretching forces, in such a way that the tension-bearing elements become aligned with the direction of stress. These stress-bearing properties of the cytoskeleton are important in epithelial cells such as those that line the gut. These cells are subjected to stress as smooth muscle within the intestinal wall contracts and puts pressure on the contents of the gut.

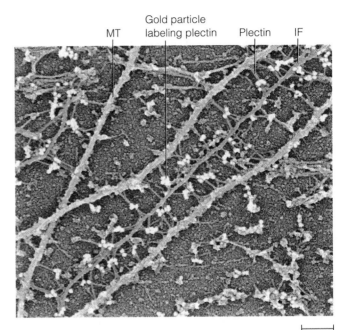

Figure 22-26 Connections Between Intermediate Filaments and Other Components of the Cytoskeleton. Intermediate filaments are linked to both microtubules and actin filaments by a protein called plectin. Plectin (red) links IFs (green) to MTs (orange), and actin filaments (light blue). Myosin is also shown (purple), gold particles (yellow) label plectin (deep etch TEM). Here, IFs serve as strong but elastic connectors between the different cytoskeletal filaments.

Perspective

The cytoskeleton is a structural feature of eukaryotic cells revealed especially well by digital video microscopy, electron microscopy, and immunofluorescence microscopy. It consists of an extensive three-dimensional network of microtubules, microfilaments, and intermediate filaments that determines cell shape and facilitates a variety of cell movements.

Microtubules (MTs) are hollow tubes with walls consisting of heterodimers of α- and β-tubulin polymerized linearly into protofilaments. MTs are polar structures and elongate preferentially from one end, known as the plus end. First identified as components of the axonemal structures of cilia and flagella and the mitotic spindle of dividing cells, microtubules are now recognized as a general cytoplasmic constituent of most eukaryotic cells. Microtubules can undergo cycles of catastrophic shortening or elongation, a phenomenon known as dynamic instability. Within cells, MT dynamics and growth are organized by microtubule-organizing centers (MTOCs). The centrosome is a major MTOC, which contains nucleation sites rich in γ-tubulin and pericentrin that are used to nucleate MT growth. Microtubules are stabilized along their length within cells by microtubule-associated proteins.

Microfilaments (MFs) are double-stranded polymers of actin that were initially discovered because of their role in the contractile fibrils of muscle cells; they are now recognized as a component of virtually all eukaryotic cells. Microfilaments are required for many processes within cells, including cell locomotion and the maintenance of cell shape. Like microtubules, MFs are polar structures, with actin monomers preferentially added to one end and removed from the other. Microfilament assembly within cells is regulated by the small G proteins Rho, Rac, and Cdc42, derivatives of phosphatidyl inositol known as polyphosphoinositides, and capping proteins. Other actin crosslinking, severing, and anchoring proteins regulate the organization of MFs within cells, which range from the parallel arrays of actin in microvilli to branched actin networks.

Intermediate filaments (IFs) are the most stable and least soluble constituents of the cytoskeleton. They appear to play a structural or tension-bearing role. IFs are tissue-specific and can be used to identify cell type. Such typing is useful in the diagnosis of cancer, as tumor cells are known to retain the IF proteins of their tissue of origin. All IF proteins have a highly conserved central domain flanked by terminal regions that differ in size and sequence, presumably accounting for the functional diversity of IF proteins. IFs, MTs, and MFs are interconnected within cells to form cytoskeletal networks that can withstand tension and compression, providing mechanical strength and rigidity to cells.

With this background, we are now ready to proceed to the next chapter, where we will explore in more detail the role of microtubules and microfilaments in cellular motility and contractility.

Key Terms for Self-Testing

cytoskeleton (p. 742)

Microtubules

microtubule (MT) (p. 744)
axonemal microtubule (p. 744)
cytoplasmic microtubule (p. 746)
protofilament (p. 746)
tubulin (p. 746)
α-tubulin (p. 746)
β-tubulin (p. 746)
$\alpha\beta$-heterodimer (p. 746)
nucleation (p. 747)
elongation (p. 747)
critical concentration (p. 747)
plus end (p. 748)
minus end (p. 748)

treadmilling (p. 748)
dynamic instability model (p. 748)
microtubule organizing center (MTOC) (p. 749)
centrosome (p. 749)
pericentriolar material (p. 749)
γ-tubulin (p. 750)
colchicine (p. 753)
nocodazole (p. 753)
microtubule-associated protein (MAP) (p. 753)
motor MAP (p. 753)
nonmotor MAP (p. 753)

Microfilaments

microfilament (MF) (p. 754)

cell cortex (p. 755)
actin (p. 755)
G-actin (p. 755)
F-actin (p. 755)
actin-related protein (Arp) (p. 756)
myosin subfragment 1 (S1) (p. 756)
cytochalasin (p. 758)
capping protein (p. 758)
polyphosphoinositide (p. 758)
actin-binding protein (p. 758)
microvillus (p. 759)
terminal web (p. 759)

Intermediate Filaments

intermediate filament (IF) (p. 762)

Problem Set

More challenging problems are marked with a • .

22-1.   Filaments and Tubules. Indicate whether each of the following descriptions is true of microtubules (MT), microfilaments (MF), intermediate filaments (IF), or none of these (N). More than one response may be appropriate for some statements.

(a) Involved in muscle contraction.

(b) Involved in the movement of cilia and flagella.

(c) More important for chromosome movements than for cytokinesis.

(d) More important for cytokinesis than for chromosome movements in animal cells.

(e) Most likely to remain when cells are treated with solutions of nonionic detergents or solutions of high ionic strength.

(f) Found in bacterial cells.

(g) Differ in composition in muscle cells versus nerve cells.

(h) Can be detected by immunofluorescence microscopy.

(i) Play well-documented roles in cell movement.

(j) Assembled from protofilaments.

22-2.   True, False, or Maybe. Identify each of the following statements as true (T), false (F), or maybe (M), where M indicates a statement that may be either true or false, depending on the circumstances. If you answer F or M, explain why.

(a) The minus end of microtubules and microfilaments is so named because subunits are lost and not added there.

(b) The energy required for tubulin and actin polymerization is provided by hydrolysis of a nucleoside triphosphate.

(c) Microtubules, microfilaments, and intermediate filaments all exist in a typical eukaryotic cell in dynamic equilibrium with a pool of subunit proteins.

(d) Cytochalasin D inhibits chromosome movement but not cytokinesis.

(e) An algal cell contains neither tubulin nor actin.

(f) All of the protein subunits of intermediate filaments are encoded by genes in the same gene family.

(g) All microtubules within animal cells have their minus ends anchored at the centrosome.

(h) As long as actin monomers continue to be added to the plus end of a microfilament, the MF will continue to elongate.

22-3. Cytoskeletal Studies. Described below are the results of several recent studies on the proteins of the cytoskeleton. In each case, state the conclusion(s) that can be drawn from the findings.

(a) Small vesicles containing pigment inside of pigmented fish epidermal cells aggregate or disperse in response to treatment with certain chemicals. When nocodazole is added to cells in which the pigment granules have been induced to aggregate, the granules cannot disperse again.

(b) When an animal cell is treated with colchicine, its microtubules depolymerize and virtually disappear. If the colchicine is then washed away, the MTs appear again, beginning at the centrosome and elongating outward at about the rate (1 μm/min) at which tubulin polymerizes in vitro.

(c) When a gene for heart muscle actin is introduced into and expressed in a cultured cell that normally synthesizes only skeletal muscle actin, the actin produced by the "foreign" gene combines readily with the indigenous actin molecules without any adverse effects on the shape or function of the cell.

(d) Extracts from nondividing frog eggs in the G2 phase of the cell cycle were found to contain structures that could induce the polymerization of tubulin into microtubules in vitro. When examined by immunostaining, these structures were shown to contain pericentrin.

• 22-4. Stabilization and the Critical Concentration. Suppose you have determined the overall critical concentration for a sample of purified tubulin. Then you add a preparation of centrosomes (microtubule-organizing centers), which nucleate microtubules so that the minus end is bound to the centrosome and stabilized against disassembly. When you again determine the overall critical concentration, you find it is different. Explain why the overall critical concentration would change.

• 22-5. Profilin and the Acrosomal Reaction. Not much is known about how the polymerization of G-actin monomers is initiated except in the case of the *acrosomal reaction* in the sperm of some invertebrates. When a sperm cell makes contact with the outer coat of an egg cell, a long, thin, membrane-covered protrusion called the *acrosomal process* shoots out explosively from the sperm cell, puncturing the egg coat and allowing the sperm and egg membranes to fuse. Prior to such contact, the sperm is filled with unpolymerized actin molecules, but with little or no F-actin. Also present in the sperm cell is the protein *profilin*, which is equal in abundance to actin on a molecular basis.

(a) Assuming that the formation of F-actin in the core of the acrosomal process drives its rapid elongation, postulate a mechanism to explain the sudden onset of actin polymerization necessary to accomplish the rapid extension of the acrosomal process upon contact with an egg cell. Suggest a way to test your hypothesis.

(b) The first measurable response of sperm upon contact with an egg cell is a very rapid rise in the intracellular pH of the sperm cell, which occurs even before the polymerization of actin begins. Postulate a possible link between the pH change and the initiation of actin polymerization. Suggest a way to test your hypothesis.

• 22-6. Spongy Actin. You are interested in the detailed effects of cytochalasin D on microfilaments over time. Based on what you know about the mechanism of action of cytochalasins, draw diagrams of what happens to microfilaments within cells treated with the drug. In particular, explain why existing actin polymers eventually depolymerize.

22-7. A New Wrinkle. Fibroblasts can be placed on thin sheets of silicone rubber. Under normal circumstances, fibroblasts exert sufficient tension on the rubber so that it visibly wrinkles. Explain how the ability of fibroblasts to wrinkle rubber would compare with that of normal cells under the following conditions.

(a) The cells are injected with large amounts of purified gelsolin.

(b) The cells are derived from a genetically engineered mouse that lacks vimentin intermediate filaments.

(c) The cells are treated with taxol.

(d) The cells are treated with colchicine, the drug is washed out, and the cells are observed periodically.

Suggested Reading

References of historical importance are marked with a • .

Techniques for Studying the Cytoskeleton

• Bridgman, P. C., and T. S. Reese. The structure of cytoplasm in directly frozen cultured cells. 1. Filamentous meshworks and the cytoplasmic ground substance. *J. Cell Biol.* 99 (1980): 1655.

• Hirokawa, N., and J. E. Heuser. Quick-freeze, deep-etch visualization of the cytoskeleton beneath surface differentiations of intestinal epithelial cells. *J. Cell Biol.* 91 (1981): 399s.

Shotton, D., ed. *Electronic Light Microscopy: Techniques in Modern Biomedical Microscopy.* New York: Wiley-Liss, 1993.

Microtubules

Archer, J., and F. Solomon. Deconstructing the microtubule-organizing center. *Cell* 76 (1994): 589.

Desai, A., and T. Mitchison. Microtubule polymerization dynamics. *Annu. Rev. Cell Develop. Biol.* 13 (1997): 83.

Joshi, H. C. Microtubule dynamics in living cells. *Curr. Opin. Cell Biol.* 10 (1998): 35.

Lee, G. Non-motor microtubule-associated proteins. *Curr. Opin. Cell Biol.* 5 (1993): 88.

Marshall, W. F., and J. L. Rosenbaum. How centrioles work: Lessons from green yeast. *Curr. Opin. Cell Biol.* 12 (2000): 119.

- Mitchison, T., and M. Kirschner. Dynamic instability of microtubule growth. *Nature* 312 (1984): 237.
Moritz, M., and D. A. Agard. Tubulin complexes and microtubule nucleation. *Curr. Opin. Structural Biol.* 11 (2001): 174.
Moritz, M., M. B. Braunfeld, et al. Microtubule nucleation by gamma-tubulin-containing rings in the centrosome. *Nature* 378 (1995): 638.
Nicolaou, K. C., R. K. Guy, et al. Taxoids: New weapons against cancer. *Sci. Amer.* 274 (1996): 94.
Zheng, Y., M. L. Wong, B. M. Alberts, and T. Mitchison. Nucleation of microtubule assembly by a gamma-tubulin-containing ring complex. *Nature* 378 (1995): 578.
Zimmerman, W., C. A. Sparks, and S. J. Doxsey. Amorphous no longer: The centrosome comes into focus. *Curr. Opin. Cell Biol.* 11 (1999): 122.

Microfilaments

Ayscough, K. R. In vivo functions of actin-binding proteins. *Curr. Opin. Cell Biol.* 10 (1998): 102.
Blanchoin, L., K. J. Amann, H. N. Higgs, J.-B. Marchand, D. A. Kaiser, and T. D. Pollard. Direct observation of dendritic actin filament networks nucleated by Arp2/3 complex and WASP/Scar proteins. *Nature* 404 (2000): 1007.
Borisy, G. G., and T. M. Svitkina. Actin machinery: Pushing the envelope. *Curr. Opin. Cell Biol.* 12 (2000): 104.
Heintzelman, M. B., and M. S. Mooseker. Assembly of the intestinal brush border cytoskeleton. *Curr. Topics Develop. Biol.* 26 (1992): 93.
Kwiatikowski, D. J. Functions of gelsolin: Motility, signaling, apoptosis, cancer. *Curr. Opin. Cell Biol.* 11 (1999): 103.
Martin, T. F. J. Phosphoinositide lipids as signaling molecules: Common themes for signal transduction, cytoskeletal regulation, and membrane trafficking. *Annu. Rev. Cell Develop. Biol.* 14 (1998): 231.

Schmidt, A., and M. N. Hall. Signaling to the actin cytoskeleton. *Annu. Rev. Cell Develop. Biol.* 14 (1998): 305.
- Schroder, R. R., D. J. Manstein, W. Jahn, H. Holden, I. Rayment, K. C. Holmes, and J. A. Spudich. Three-dimensional atomic model of F-actin decorated with *Dictyostelium* myosin S1. *Nature* 364 (1993): 171.
Sutherland, J. D., and W. Witke. Molecular genetic approaches to understanding the actin cytoskeleton. *Curr. Opin. Cell Biol.* 11 (1999): 142.
Tapon, N., and A. Hall. Rho, Rac, and Cdc42 GTPases regulate the organization of the actin cytoskeleton. *Curr. Opin. Cell Biol.* 9 (1997): 86.
Welch, M. D., A. Mallavarapu, J. Rosenblatt, and T. J. Mitchison. Actin dynamics in vivo. *Curr. Opin. Cell Biol.* 9 (1997): 54.

Intermediate Filaments

Allen, P. G., and J. V. Shah. Brains and brawn: Plectin as regulator and reinforcer of the cytoskeleton. *Bioessays* 21(1999): 451.
Coulombe, P. A., O. Bousquet, L. Ma., S. Yamada, and D. Wirtz. The "ins" and "outs" of intermediate filament organization. *Trends Cell Biol.* 10 (2000): 420.
Fuchs, E., and D. W. Cleveland. A structural scaffolding of intermediate filaments in health and disease. *Science* 279 (1998): 514.
Herrmann, H., and U. Aebi. Intermediate filaments and their associates: Multi-talented structural elements specifying cytoarchitecture and cytodynamics. *Curr. Opin. Cell Biol.* 12 (2000): 79.
- Osborn, M., and K. Weber. Tumor diagnosis by intermediate filament typing: A novel tool for surgical pathology. *Lab. Invest.* 48 (1983): 372.
Ingber, D. E. The architecture of life. *Sci. Amer.* 278 (1998): 48.

23 Cellular Movement: Motility and Contractility

In the previous chapter, we considered the cytoskeleton of eukaryotic cells and saw that it provides an intracellular scaffolding that organizes structures within the cell and shapes the cell itself. We also noted that this scaffolding plays a dynamic role in cell motility. In this chapter, we will explore the role of these cytoskeletal elements in cellular **motility.** This may involve the movement of a cell (or a whole organism) through its environment, the movement of the environment past or through the cell, the movement of components within the cell, or the shortening of the cell itself. **Contractility,** a related term often used to describe the shortening of muscle cells, is a specialized form of motility.

Motile Systems

Motility occurs at the tissue, cellular, and subcellular levels. The most conspicuous examples of motility, particularly in the animal world, take place at the tissue level. The muscle tissues common to most animals consist of cells specifically adapted for contraction, and the movements produced are often obvious, whether manifested as the bending of a limb, the beating of a heart, or a uterine contraction during childbirth. In humans, about 40% of body weight is due to skeletal muscles, which consume a significant proportion of our total energy budget.

At the cellular level, motility is observed in single cells or in organisms that consist of one or only a few cells. It occurs among cell types as diverse as ciliated protozoa and motile sperm, depending in each case on cilia or flagella, cellular appendages adapted for propulsion. Other examples of motility at the cellular level include cell migration

during animal embryogenesis, amoeboid movement, and the invasiveness of cancer cells in malignant tumors.

Equally important is the movement of intracellular components, which might be regarded as motility at the subcellular level. For example, highly ordered microtubules of the mitotic spindle play a key role in the separation of chromosomes during cell division, as we saw in Chapter 17. In addition, some cells display the phenomenon of *cytoplasmic streaming,* in which the cytoplasm undergoes rhythmic patterns of flow. Other examples of mechanical work at the subcellular level include the characteristic movements of molecular structures that occur during cell growth and differentiation. An example of such a process is the transport of cellulose to the growing wall of a dividing or differentiating plant cell.

Motility is an especially intriguing use of energy by cells because in some cases it involves the conversion of chemical energy directly to mechanical energy. In contrast, most mechanical devices that produce movement from chemicals (such as a steam engine, which depends on the combustion of coal) require an intermediate form of energy, usually heat or electricity. As we will see, the cellular energy for motility typically comes from ATP, whose hydrolysis is often coupled to changes in the shape of specific proteins that mediate movement. The efficiency of such cellular "engines" in transducing energy to produce mechanical work varies, depending on the process. Estimates of such mechanical efficiency suggest that some cellular motors, such as those involved in contracting muscle cells or the movements of intracellular vesicles, are nearly as efficient as—or sometimes more efficient than—automobile engines. In other cases, such as the forces that act during mitosis, the efficiency is many orders of magnitude less than human-made machines.

The microfilaments and microtubules of the cytoskeleton provide a basic scaffolding for specialized **motor proteins,** or **mechanoenzymes,** which interact with the cytoskeleton to produce motion at the molecular level (Table 23-1). The combined effects of these molecular motions produce movement at the cellular level. In cases such as muscle contraction, the combined effects of many cells moving simultaneously produce motion at the tissue level.

In eukaryotes, there are two major motility systems. The first involves interactions between specialized motor molecules and microtubules. Microtubules are abundant in interphase cells and are used for a variety of intracellular movements. A specific example of **microtubule-based movement** is *fast axonal transport,* one of the processes by which a nerve cell transports materials between the central part of the cell and outlying regions. The second type of eukaryotic motility is based on interactions between actin microfilaments and members of the *myosin* family of motor molecules. A familiar example of **microfilament-based movement** is *muscle contraction.* We begin our detailed examination of the two motility systems with two proteins essential for microtubule-based movement.

Intracellular Microtubule-Based Movement: Kinesin and Dynein

Microtubules (MTs) provide a rigid set of tracks for the transport of a variety of membrane-enclosed organelles and vesicles. As we saw in Chapter 22, the centrosome provides organization and orientation to MTs, because the minus ends of most microtubules are embedded in the centrosome. The centrosome is generally located near the center of the cell, so traffic toward the minus ends of MTs might be considered "inbound" traffic. Traffic directed toward the plus ends might likewise be considered "outbound," meaning that it is directed toward the periphery of the cell.

Table 23–1 Selected Motor Proteins of Eukaryotic Cells

Molecules	Typical Function
Microtubule-associated proteins	
Cytoplasmic dynein	Motion toward minus end of microtubule
Axonemal dynein	Activation of sliding in flagellar microtubule
Kinesins	Motion toward plus end of microtubule
Microfilament-associated (actin-binding) proteins	
Myosin I, monomer	Motion along actin filaments
Myosin II, filament	Slides along actin filaments in sarcomere of muscle

While microtubules provide an organized set of tracks along which organelles can move, they do not directly generate the force necessary for movement. The mechanical work needed for movement depends on **microtubule-associated motor proteins (motor MAPs),** which attach to vesicles or organelles and then "walk" along the MT, using ATP to provide the needed energy. Furthermore, motor MAPs recognize the polarity of the MT, with each motor MAP having a preferred direction of movement. At present, we are aware of two major families of motor MAPs: **kinesins** and **dyneins** (Table 23-1).

Motor MAPs Move Organelles Along Microtubules During Axonal Transport

A historically important cell type for studying microtubule-dependent intracellular movement is the neuron. In particular, the squid giant axon has been extremely useful for the biochemical purification of components that interact with MTs (see Figure 9-11 for further details regarding the squid giant axon). A receptor protein or neurotransmitter synthesized in the cell body of the neuron must be transported over distances up to a meter between the cell body and the nerve ending. The need for such transport arises because ribosomes are present only in the cell body, so no protein synthesis occurs in the axons or synaptic knobs. Instead, proteins and membranous vesicles are synthesized in the cell body and transported along the axons to the synaptic knobs. Some form of energy-dependent transport is clearly required, and MT-based movement provides the required mechanism. The process is called **fast axonal transport** and appears to involve the movement of protein-containing vesicles and other organelles along MTs (we will not discuss slow axonal transport, which is less well understood).

The role of microtubules in axonal transport was initially suggested because the process is inhibited by colchicine and other drugs that impair MT function but is insensitive to drugs such as cytochalasins that affect microfilaments. Since then, MTs have been visualized along the axon by the deep-etch technique (a technique described in the *Guide to Microscopy*), and have been shown to be prominent features of the axonal cytoskeleton. Moreover, axonal MTs have small membranous vesicles and mitochondria associated with them (Figure 23-1).

Evidence that a motor MAP drives the movements of organelles was obtained when investigators found that organelles in the presence of ATP could move along fine filamentous structures present in exuded *axoplasm* (the cytoplasm of axons). The organelle movement was visualized by video-enhanced differential interference contrast microscopy (also described in the *Guide to Microscopy*). The rate of organelle movement was shown to be about 2 μm/sec, comparable to the axonal transport rate in intact neurons. The researchers then used a combination of immunofluorescence and electron microscopy to demonstrate that the fine filaments along which the organelles move are single microtubules. They concluded

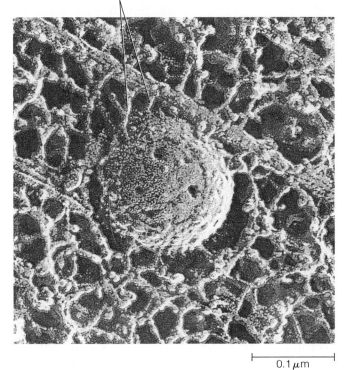

Cross-bridges

0.1 μm

Figure 23-1 Deep-Etch Electron Micrograph Showing a Vesicle Attached to a Microtubule in a Crayfish Axon. Cross-bridges connect the membrane vesicle (round structure at center) to the microtubule.

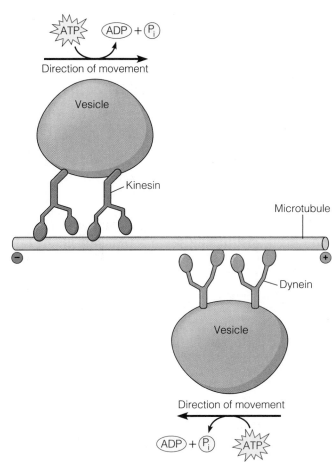

ATP ⟶ ADP + Pᵢ

Direction of movement

Vesicle

Kinesin

Microtubule

Dynein

Vesicle

Direction of movement

ADP + Pᵢ ⟵ ATP

Figure 23-2 Microtubule-Based Motility. Kinesin and dynein are families of molecules that use the energy of ATP hydrolysis to "walk" along microtubules. In the process, they move intracellular structures along MTs. In general, members of the kinesin family move vesicles or organelles toward the plus ends of MTs—that is, from the center of the cell to the periphery. Dynein moves in the opposite direction, toward the minus ends of MTs, and therefore toward the center of the cell where the MTOC is located.

that axonal transport depends on interactions between MTs and organelles.

Since that time, two motor MAPs responsible for fast axonal transport, kinesin and cytoplasmic dynein, have been purified and characterized. To determine the direction of transport by these motor MAPs, purified proteins were used experimentally to drive the transport of polystyrene beads along microtubules of known polarity. To obtain MTs of known polarity, tubulin was polymerized into MTs using purified centrosomes as microtubule-organizing centers, which ensured that the MTs were assembled with their minus ends attached to the centrosome. When polystyrene beads, purified kinesin, and ATP were added to MTs polymerized by centrosomes, the beads moved toward the plus ends (i.e., away from the centrosome). This finding means that in a nerve cell, kinesin mediates transport from the cell body down the axon to the nerve ending (called *anterograde axonal transport*). When similar experiments were carried out with purified dynein, particles were moved in the opposite direction, toward the minus ends of the MTs (called *retrograde axonal transport* when it occurs in neurons). These two motors transport materials in opposite directions within the cytoplasm (Figure 23-2).

Kinesins Move Along Microtubules by Hydrolyzing ATP

The first kinesins were originally identified in the cytoplasm of squid giant axons. Such "classic" kinesins consists of

three parts: a globular head region that attaches to microtubules and is involved in hydrolysis of ATP, a coiled helical region, and a light-chain region that is involved in attaching the kinesin to other proteins or organelles (Figure 23-3a). The movements of single kinesin molecules along MTs have been studied by tracking the movement of beads attached to kinesin, or by measuring the force exerted by single kinesin molecules using calibrated glass fibers or special types of laser beams known as "optical tweezers." It is easiest to visualize this movement as analogous to walking, with the globular heads serving as feet (Figure 23-3b). Classic kinesins move along MTs in 8 nm steps: one of the two globular heads moves forward to make an attachment to a new β-tubulin subunit, followed by detachment of the trailing globular head, which can now make an attachment to a new region of the MT. This movement is coupled to exchange of ATP and ADP at specific sites within the heads. The result is that kinesin moves toward the plus end of a MT in an ATP-dependent fashion. A single kinesin molecule exhibits *processivity*. It can cover long distances before

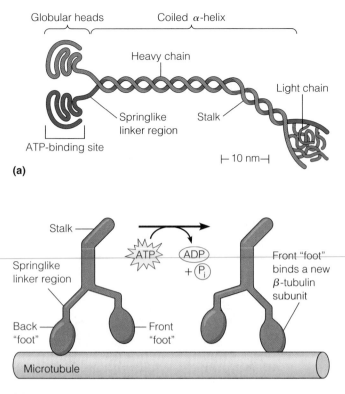

Figure 23-3 Movement of Kinesin. **(a)** The basic structure of a kinesin molecule. **(b)** Kinesins "walk" along microtubules. The front "foot," one of kinesin's two globular heads, detaches from a β-tubulin subunit by hydrolyzing ATP, and moves forward. The rear "foot" then releases by hydrolyzing ATP, and springs forward.

detaching from a MT. A single kinesin molecule can move as far as 1 μm, which is a great distance relative to its size. As a molecular motor, kinesin appears to be quite efficient; estimates of its efficiency in converting the energy of ATP hydrolysis to useful work are on the order of 60–70%.

Kinesins Are a Large Family of Proteins with Varying Structures and Functions

Since the discovery of the original kinesin involved in anterograde transport in neurons, many proteins that share structural similarities with kinesin have been discovered. These **kinesin family members,** or **KIFs,** have a similar motor domain, but it can be found in different areas in different molecules. In some cases, KIFs associate with another identical KIF or with other KIFs. One class of KIFs acts as minus-end directed motor MAPs, rather than as plus-end motors, a property that depends on regions of the protein outside of the motor domain. Biochemical and immuncytochemical studies, as well as analysis of mutant organisms, have shown that KIFs are involved in many different processes within cells. For example, they are involved in moving and localizing substances within cells. Various KIFs localize to the mitotic or meiotic spindle or to kinetochores, where they play roles in the early phases of mitosis and meiosis. Other kinesins

are required for completion of mitosis and cytokinesis. For example, cells in mutant fruit flies or nematode worms lacking one of these KIFs initiate cytokinesis, but the cleavage furrow never deepens and cell division fails.

Dyneins Can Be Grouped in Two Major Classes: Axonemal and Cytoplasmic Dyneins

The dynein family of motor MAPs consists of two basic types: cytoplasmic dynein and axonemal dynein (Table 23-1). In contrast to the kinesins and KIFs, few cytoplasmic dyneins have been identified. **Cytoplasmic dynein** contains two heavy chains that interact with microtubules, three intermediate chains, and four light chains. In contrast to most kinesins, cytoplasmic dynein moves toward the minus ends of MTs. Also, unlike kinesins and KIFs, cytoplasmic dynein cannot efficiently bind to cytoplasmic organelles on its own. Instead, it is associated with the protein complex known as **dynactin** (Figure 23-4). The dynactin complexes includes the actin-related protein *Arp1*, which is capped at one end by a capping protein (see Chapter 22) and at the other by a second protein (*p62*). It also contains a protein known as *p150^Glued*, which may be linked to Arp1 by another protein called *dynamitin*. The dynactin complex probably helps to link cytoplasmic dynein to the cargo (such as a membranous vesicle) it transports along MTs.

At least four types of **axonemal dynein** have been identified. We will discuss the function of axonemal dynein in cilia and flagella in some detail later in this chapter.

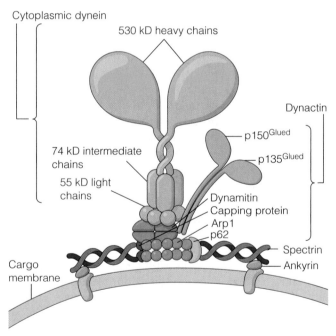

Figure 23-4 Schematic Representation of the Cytoplasmic Dynein/Dynactin Complex. Cytoplasmic dynein is linked with cargo membranes indirectly through the dynactin multiprotein complex, which includes Arp1, dynamitin, and p150^Glued. The dynactin complex binds to spectrin (see Chapter 22, p. 762) in the membrane of the cargo vesicle.

Motor MAPs Are Involved in the Transport of Intracellular Vesicles

We have seen that the cell has an elaborate transportation system of vesicles driven by kinesin and cytoplasmic dynein or similar motor MAPs. A logical question is, why is such a transport system necessary, and how is it used? So far we have seen that such transport systems are critical for fast axonal transport and cell division. Motor MAPs also play a fundamental role in establishing the distribution and structure of the endoplasmic reticulum, the formation of the Golgi complex, the movements of lysosomes and secretory granules, the internalization of vesicles from the plasma membrane, and a variety of vesicle movements within animal cells. We will now look more closely at one of these functions—formation of the Golgi complex.

As we learned in Chapter 12, the Golgi complex is a series of flattened membrane stacks located in the region of the cell centrosome or microtubule-organizing center (MTOC). The function of the Golgi is to receive proteins made in the endoplasmic reticulum (ER) and to process and package those proteins for distribution to the correct cellular destinations. At each step of this process, proteins are transported in vesicles. Thus, there is a continuous flow of vesicles to and from the Golgi. The vesicles are carried by motor MAPs on microtubule tracks (Figure 23-5).

Vesicles traveling from the ER to the Golgi complex appear to be carried by a dyneinlike motor that moves them toward the minus ends of microtubules. The minus ends are anchored in the centrosome, so ER vesicles originating from various parts of the cell will all move toward the centrosome. These vesicles fuse together and become part of the flattened membrane stacks that make up one side of the Golgi complex.

Several experiments demonstrate that microtubule-based transport is crucial for maintenance of the Golgi stacks. For example, if MTs are depolymerized using the drug nocodazole, (see Chapter 22, p. 744) the Golgi complex disperses. When the nocodazole is washed out, the Golgi complex re-forms. Similarly, disruption of the function of the dynactin complex by inducing cells to produce too much dynamitin results in collapse of the Golgi complex and disruption of the transport of intermediates from the ER to the Golgi.

Once in the Golgi complex, proteins are processed as they move through the stacks of Golgi membranes. The finished proteins emerge—still packaged in vesicles—from the other side of the complex. Plus-end directed motor MAPs carry the finished vesicles away from the Golgi complex toward the cell periphery. In organisms in which various KIFs are mutated, specific defects in vesicle transport have been observed, providing direct evidence that these motor MAPs are crucial for plus-end directed vesicular transport. Thus motor MAPs and microtubules allow the two-way flow of vesicle traffic to and from the Golgi complex.

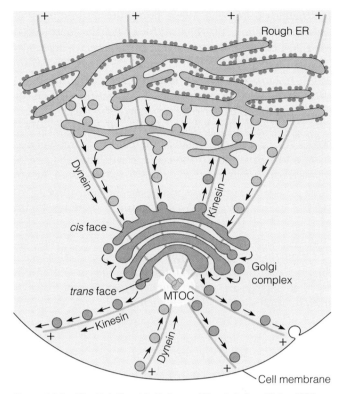

Figure 23-5   The Relationship Between Microtubules, Motor MAPs, and the Golgi Complex.   Vesicles going to and from the Golgi complex are attached to microtubules and are thought to be carried by motor MAPs similar or identical to dynein and kinesin. Dynein is a minus-end directed motor MAP, whereas kinesin is plus-end directed. Thus, vesicles derived from either the ER or the cell membrane are carried toward the Golgi complex and MTOC by dynein, whereas vesicles derived from the Golgi complex are carried toward either the ER or the cell periphery by kinesins.

Microtubule-Based Motility

Cilia and Flagella Are Common Motile Appendages of Eukaryotic Cells

Microtubules are crucial not only for movement within cells, but also for the movements of *flagella* and *cilia,* the motile appendages of eukaryotic cells. The two appendages share a common structural basis and differ only in relative length, number per cell, and mode of beating. **Cilia** (singular: **cilium**) have a diameter of about 0.25 μm, are about 2–10 μm long, and tend to occur in large numbers on the surface of ciliated cells. Each cilium is bounded by an extension of the plasma membrane and is therefore an intracellular structure.

Cilia occur in both unicellular and multicellular eukaryotes. Unicellular organisms, such as protozoa, use cilia for both locomotion and the collection of food particles. In multicellular organisms, cilia serve primarily to move the environment past the cell rather than to propel the cell through the environment. The cells that line the air passages of the human respiratory tract, for example, have several hundred cilia each, which means that every square centimeter of epithelial tissue lining the respiratory tract has about a billion cilia (Figure 23-6a)! It is the coordinated, wavelike beating of these cilia that carries mucus, dust, dead cells, and other foreign matter out of the lungs. One of the health hazards of cigarette smoking lies in the inhibitory effect that smoke has on normal ciliary beating. Certain respiratory ailments can also be traced to defective cilia.

Cilia display an oarlike pattern of beating, with a power stroke perpendicular to the cilium, thereby generating a force parallel to the cell surface. The cilia on the cell surface usually beat in a coordinated manner, ensuring a steady movement of fluid past the cell surface. The cycle of beating for an epithelial cilium is shown in Figure 23-6b. Each cycle requires about 0.1–0.2 sec and involves an active power stroke followed by a recovery stroke.

Flagella (singular: **flagellum**) move cells through a fluid environment. Although they have the same diameter as cilia, flagella are often much longer—from 1 μm to several millimeters, though usually in the range of 10–200 μm—and may be limited to one or a few per cell (Figure 23-6c). Like cilia, flagella are bounded by an extension of the plasma membrane.

Flagella differ from cilia in the nature of their beat. Flagella move with a propagated bending motion that is usually symmetrical and undulatory and may even have a helical pattern. This type of beat generates a force parallel to the flagellum, such that the cell moves in approximately the same direction as the axis of the flagellum. The locomotory pattern of most flagellated cells involves propulsion of the cell by the trailing flagellum, but examples are also known in which the flagellum actually precedes the cell. Figure 23-6d illustrates this type of swimming movement.

Cilia and Flagella Consist of an Axoneme Connected to a Basal Body

Cilia and flagella have a common structure consisting of an **axoneme,** or main cylinder of tubules, about 0.25 μm in

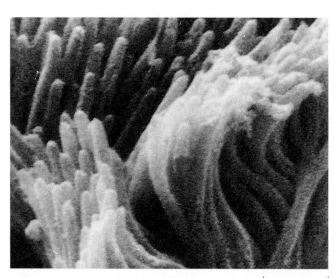

(a) Cilia on a mammalian trachea cell

1 μm

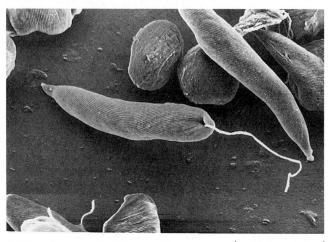

(c) Flagellum on unicellular alga *Euglena*

1 μm

Power stroke   Recovery stroke

1 2 3   4   5   6

(b) Beating of a cilium

(d) Movement of flagellated cell

Figure 23-6 Cilia and Flagella. **(a)** A micrograph of cilia on a mammalian tracheal cell (SEM). **(b)** The beating of a cilium on the surface of an epithelial cell from the human respiratory tract. A beat begins with a power stroke that sweeps fluid over the cell surface. A recovery stroke follows, leaving the cilium poised for the next beat. Each cycle requires about 0.1–0.2 sec. **(c)** A micrograph of the flagellated unicellular alga *Euglena* (SEM). **(d)** Movement of a flagellated cell through an aqueous environment.

diameter. The axoneme is connected to a **basal body** and surrounded by an extension of the cell membrane (Figure 23-7a). Between the axoneme and the basal body is a *transition zone* in which the arrangement of microtubules in the basal body takes on the pattern characteristic of the axoneme. Cross-sectional views of the axoneme, transition zone, and basal body are shown in Figure 23-7b–d.

The basal body is identical in appearance to the centriole. A basal body consists of nine sets of tubular structures arranged around its circumference. Each set is called a *triplet* because it consists of three tubules that share common walls—one complete microtubule and two incomplete tubules. As a cilium or flagellum forms, a centriole migrates to the cell surface and makes contact with

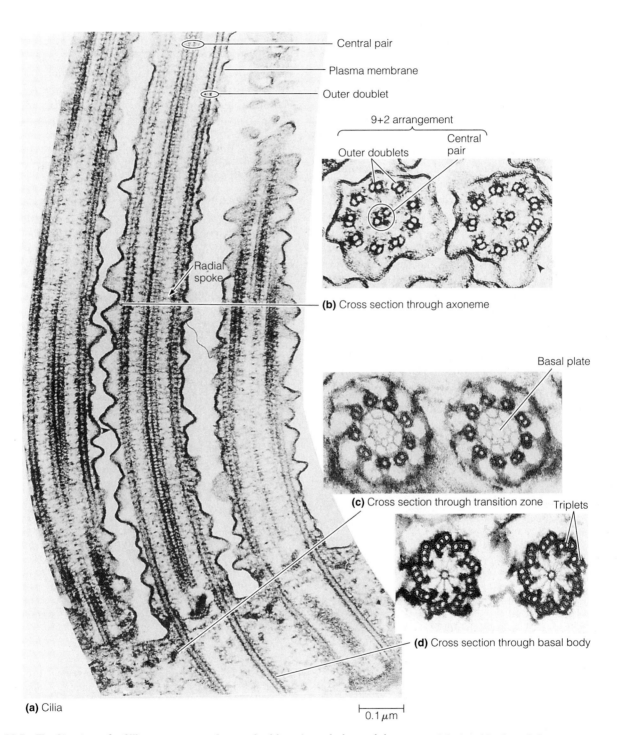

(a) Cilia

0.1 μm

Figure 23-7 The Structure of a Cilium.
(a) These longitudinal sections of three cilia from the protozoan *Tetrahymena thermophila* illustrate several structural features of cilia, including the central pair and outer doublet microtubules, and the radial spokes. Cross-sectional views are shown for **(b)** axonemes, **(c)** the transition zone between cilia and basal bodies, and **(d)** basal bodies. Notice the triplet pattern of the basal body and the 9 + 2 pattern of tubule arrangement in the axoneme of the cilium. (All TEMs.)

the plasma membrane. The centriole then acts as a nucleation site for MT assembly, initiating polymerization of the nine outer doublets of the axoneme. After the process of tubule assembly has begun, the centriole is then referred to as a basal body.

The axoneme has a characteristic "9 + 2" pattern, with nine **outer doublets** of tubules and two additional microtubules in the center, often called the **central pair.** Figure 23-8 illustrates these structural features in greater detail. The nine outer doublets of the axoneme are thought to be extensions of two of the three subfibers from each of the nine triplets of the basal body. Each outer doublet of the axoneme therefore consists of one complete MT, called the **A tubule,** and one incomplete MT, the **B tubule** (Figure 23-8b). The A tubule has 13 protofilaments, whereas the B tubule has only 10 or 11. The tubules of the central pair are both complete, with 13 protofilaments each. All of these structures contain tubulin, together with a second protein called **tektin.** Tektin is related to intermediate filament proteins (see Chapter 22) and is a necessary component of the axoneme. The A and B tubules share a wall that appears to contain tektin as a major component.

In addition to microtubules, axonemes contain several other key structures (Figure 23-8b). The most important of these are the sets of **sidearms** that project out from each of the A tubules of the nine outer doublets. Each sidearm reaches out clockwise toward the B tubules of the adjacent doublet. These arms consist of axonemal dynein, which is responsible for sliding MTs within the axoneme past one

another to bend the axoneme. The dynein arms occur in pairs, one inner arm and one outer arm, spaced along the MT at regular intervals. At less frequent intervals, adjacent doublets are joined by **interdoublet links.** These links are thought to limit the extent to which doublets can move with respect to each other as the axoneme bends.

At regular intervals, **radial spokes** project inward from each of the nine MT doublets, terminating near a set of projections that extend outward from the central pair of microtubules. These spokes are thought to be important in translating the sliding motion of adjacent doublets into the bending motion that characterizes the beating of these appendages. In addition to the radial spoke attachment to the central pair, a protein called **nexin** links adjacent doublets to one another and probably also plays a role in converting sliding into bending motion.

Microtubule Sliding Within the Axoneme Causes Cilia and Flagella to Bend

How does axonemal dynein act on this elaborate structure to generate the characteristic bending of cilia and flagella? The overall length of MTs in cilia and flagella does not change. Instead, the microtubules in adjacent outer doublets slide relative to one another in an ATP-dependent fashion. According to the **sliding-microtubule model** for cilia and flagella, this sliding movement is converted to a localized bending because the doublets of the axoneme are connected radially to the central pair and circumferentially to one another and therefore cannot slide past each

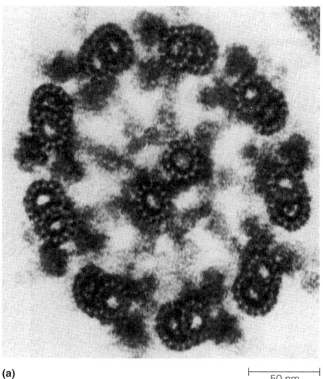

(a)

50 nm

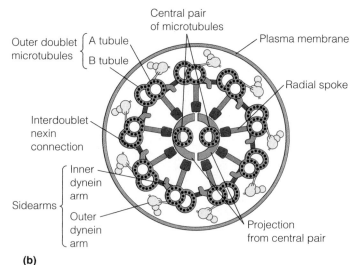

(b)

Figure 23-8   Enlarged Cross Section of an Axoneme.   (a) This micrograph shows an axoneme from a flagellum of *Chlamydomonas* (TEM). **(b)** Diagram of an axoneme in cross section. The microtubules of the central pair have 13 protofilaments each, as do the A tubules of the outer doublets. Each B tubule has 11 protofilaments of its own and shares 5 protofilaments with the A tubule. The dynein side arms have ATPase activity and are thought to be responsible for the sliding of adjacent doublets. The interdoublet links (nexin connections) join adjacent doublets and the radial spokes project inward, terminating near projections that extend outward from the central pair of MTs.

other freely. The resultant bending takes the form of a wave that begins at the base of the organelle and proceeds toward the tip.

Dynein Arms Are Responsible for Sliding. The driving force for MT sliding is provided by ATP hydrolysis, catalyzed by the ATPase activity of the dynein arms. The importance of the dynein arms is indicated by two kinds of evidence. First, when dynein is selectively extracted from isolated axonemes, the arms disappear from the outer doublets, and the axonemes lose both their ability to hydrolyze ATP and their capacity to beat. Furthermore, the effect is reversible: If purified dynein is added to isolated outer doublets, the sidearms reappear and ATP-dependent sliding is restored.

A second kind of evidence comes from studies of nonmotile mutants in such normally motile species as *Chlamydomonas,* a green alga. In some of these mutants, flagella are present but nonfunctional. Depending on the particular mutant their nonmotile flagella lack dynein arms, radial spokes, or the central pair of microtubules. These structures are therefore almost certainly essential to the movement mechanism.

Axonemal dynein is a very large protein. It has multiple subunits, the three largest having ATPase activity and a molecular weight of about 450,000 each. During the sliding process, the stalk of the dynein arm apparently attaches to and detaches from the B tubule in a cyclic manner. Each cycle requires the hydrolysis of ATP and shifts the dynein arm of one doublet relative to the adjacent doublet. In this way, the dynein arms of one doublet move the neighboring doublet, resulting in a relative displacement of the two.

Crosslinks and Spokes Are Responsible for Bending. To convert the dynein-mediated displacement of doublets to a bending motion, the doublets must be restrained in a way that resists sliding but allows deformation. This resistance is provided by the radial spokes that connect the doublets to the central pair of microtubules and possibly also by the nexin crosslinks between doublets (Figure 23-8b). If these crosslinks and spokes are removed (by partial digestion with proteolytic enzymes, for example), the resistance that translates doublet sliding into a bending action is absent, and sliding is uncoupled from bending. Under these conditions, the free doublets move with respect to each other, and the axonemes become longer and thinner as the MTs slide apart.

The importance of the radial spokes is confirmed by electron microscopic observations on bent and straight regions of a cilium. The radial spokes are oriented perpendicular to the doublets in straight regions but are tilted at an angle in bent regions. In other words, the spokes are deformed by the sliding process and provide the resistance that translates the sliding of doublets into the bending of the cilium or flagellum. Eventually, though, the sliding movement of the doublets overcomes the resistance and causes displacement of the radial spokes with respect to the central

pair of tubules. In addition to their role in flagellar and ciliary beating, axonemal dyneins and radial spokes have been implicated in a process that at first glance seems completely unrelated to its function in sliding microtubules: where the internal organs of the body become located (see Box 23A).

Intraflagellar Transport Adds Components to Growing Flagella. The elaborate structure of flagella and cilia raises an interesting question: How are tubulin subunits and other components added to the growing cilia or flagella? Based on the analysis of mutants in *Chlamydomonas* and nematodes, both plus- and minus-end directed microtubule motors are involved in shuttling components to and from the tips of flagella. This process, known as **intraflagellar transport** (in the case of flagella), is in some sense analogous to the processes of axonal transport in nerve cells: a kinesin appears to move material out to the tips of flagella, and a dynein brings material back toward the base.

Actin-Based Cell Movement: The Myosins

Myosins Have Diverse Roles in Cell Motility

Movement of molecules and other cellular components also occurs along another major filament system in the cell, the actin cytoskeleton. As with microtubules, mechanoenzymes act as ATP-dependent motors that exert force on actin microfilaments within cells. These mechanoenzymes are all members of a large superfamily of proteins known as **myosins.** Currently there are at least 17 known families of myosins. All myosins have at least one polypeptide chain, called the *heavy chain,* with a globular head group at one end attached to a tail of varying length (Figure 23-9). The globular head binds to actin and uses the energy of ATP hydrolysis to move along an actin filament. The structure of the tail region varies among the different kinds of myosin, giving myosin molecules the ability to bind to a variety of different molecules or cell structures. The tail structure also determines the ability of myosins to bind to other identical myosins to form dimers or large aggregates. Myosins typically contain small polypeptides bound to the globular head group. These polypeptides, referred to as the *light chains,* often play a role in regulating the activity of the myosin ATPase. Some myosins are unusual in that they have a binding site for actin in their tail region, as well as in the head. In addition, some myosins, such as myosin I and myosin V, appear to bind to membranes, suggesting that these forms of myosin play a role in movement of the plasma membrane or in transporting membrane-enclosed organelles inside the cell.

Several myosins have been shown to have specific functions in events as wide-ranging as muscle contraction, cell movement, and phagocytosis. For example, mutations in the mouse indicate that a myosin V is required for transfer of pigment granules from melanocytes (cells that

Clinical Applications CYTOSKELETAL MOTOR PROTEINS AND HUMAN DISEASE

As it has become easier to isolate genes that are defective in many human genetic disorders, researchers have been able to elucidate many aspects of the mechanisms of human disease, illustrating again how basic research can lead to understanding and treatment of conditions affecting human health. We will discuss two examples of human genetic disorders that are caused by mutations in motors proteins: reversal of body symmetry and genetic deafness.

Axonemal Dyneins, Radial Spokes, and Reversal of the Left-Right Body Axis

In at least 1 in 20,000 live human births, organs in the body cavity are completely reversed left to right. This condition, known as *situs inversum viscerum,* has no medical consequences and is often not recognized until a patient undergoes medical tests for an unrelated condition, at which time the reversed location of organs is discovered. In contrast, when only some organs are reversed (*heterotaxia*) serious health complications often result. In patients suffering from an autosomal recessive condition known as *Kartagener's triad,* there is a 50 percent probability of the complete reversal of the left-right location of internal organs. In addition, such patients suffer from male sterility and bronchial problems. The reason for these abnormalities is a defect in the outer dynein arms of cilia and flagella. Recent studies using mutations in the mouse support the idea that microtubule motor proteins are somehow involved in left-right asymmetry in the developing mammalian embryo. In *inversus viscerum (iv)* mutant mice, the internal organs are reversed in half of the newborn homozygotes. Surprisingly, *iv* encodes a dynein heavy chain, most similar to axonemal dyneins. Current research is aimed at clarifying the mechanisms underlying the randomization of the left-right axis in *iv* mutants.

Myosins and Deafness

Recently, nonmuscle myosins have also been implicated in human genetic disorders. For example, mutations in a myosin VII result in *Usher's syndrome,* an autosomal recessive disorder characterized by congenital profound hearing loss, problems in the vestibular system (i.e., sensing where one is in space), and *retinitis pigmentosum,* which results in blindness. In the inner ear, specialized cells known as *hair cells* contain special actin-

$\vdash\!\!-\!\!2\,\mu m\!-\!\!\dashv$

Figure 23A-1   Scanning Electron Micrograph of Epithelial Cells of the Inner Ear Showing Several Rows of Stereocilia.   Stereocilia are related to microvilli but are much larger. They sense tiny movements caused by sound vibrations or fluid movement in the semicircular canals. (©R. G. Kessel and R. H. Kardon, *Tissues and Organs: A Text-Atlas of Scanning Electron Microscopy,* W. H. Freeman & Co., 1979. All rights reserved.)

rich sensory structures known as *stereocilia* (Figure 23A-1). Hair cells carry out auditory and vestibular transduction. A myosin VII is concentrated in the cell body of hair cells and in stereocilia. A myosin VII is also expressed in the retinal pigmented epithelium and photoreceptor cells in the eye, consistent with its role in functions associated with vision. Ongoing research seeks to clarify the specific role of myosin VII in these processes.

---

produce pigment) to keratinocytes (cells in the hair shaft that normally take up the pigment). A myosin V also appears to be required for normal positioning of the smooth endoplasmic reticulum in nerve cells, suggesting that it functions in vesicle transport or other membrane-associated events. A human disorder called *Griscelli's disease,* which involves partial albinism and neurological defects, has recently been shown to result from a mutation in the same class of myosin. One unexpected function for myosins from several classes is in the maintenance of structures required for hearing in humans (see Box 23A).

The best-understood myosins are the **type II myosins.** They are composed of two heavy chains, each featuring a globular myosin head, a hinge region, a long rodlike tail, and four light chains. These myosins are found in skeletal, cardiac (heart), and smooth muscle cells as well as in non-muscle cells. Type II myosins are distinctive in that they can assemble into long filaments such as the thick filaments of muscle cells. The basic function of myosin II in all cell types is to convert the energy of ATP to mechanical force that can cause actin filaments to slide past the myosin molecule, typically resulting in the contraction of a cell. For example, in *Drosophila* embryos, a nonmuscle type II

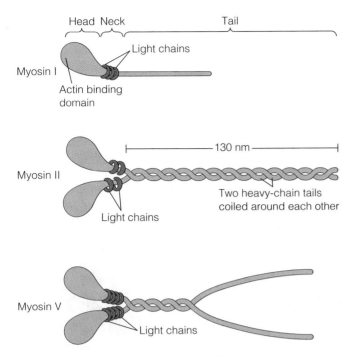

Figure 23-9 Myosin Family Members. All myosins have an actin- and ATP-binding heavy-chain "head," and typically have two or more regulatory light chains. Some myosins, like myosin I, have one head. Others, like myosin II and myosin V, associate via their tails into two-headed proteins.

myosin is found at the free edges of a sheet of epithelial cells that closes rather like a purse string. Analysis of mutants in which this myosin is missing or defective indicates that it is required for the closure of the sheet. The myosin II may allow the free edge of the sheet to contract by the sliding of actin microfilaments at its free edge. There is some evidence in slime mold cells grown in suspension that nonmuscle type II myosins are also involved in the constriction of the contractile ring during cytokinesis.

Myosins Move Along Actin Filaments in Short Steps

Like kinesins, myosins have been studied at the level of single molecules. Several techniques have been used to analyze how myosin heads move along actin filaments and how much force a single myosin head can generate. All rely on the observation that myosins will move along actin filaments in vitro in the presence of ATP. These techniques have shown that the force individual myosin heads exert on actin is similar to that measured for kinesin. The average distance that a myosin II can slide an actin filament is about 12–15 nm. Like kinesin, myosin II is an efficient motor; when myosins must pull against moderate loads, they are about 50% efficient.

It is useful to compare the two best-studied types of cytoskeletal motor proteins, "classic" kinesin and myosin II, because it is now possible to see how they function as biochemical motors at the level of single molecules. Both have two heads that they use to "walk" along a protein filament, and both utilize ATP hydrolysis to change their

shape. Despite these similarities, there are profound differences as well. Conventional kinesins operate alone or in small numbers to transport vesicles over large distances; a single kinesin can move hundreds of nanometers along a single microtubule. In contrast, a single myosin II molecule cannot move an actin filament as far as the next binding site on its own. Instead, myosin II molecules often operate in large arrays. In the case of myosin filaments in muscle, these arrays can contain billions of motors working together. The best-characterized conventional myosin that operates in such huge arrays was the first to be identified, the myosin II involved in contraction of skeletal muscle, the process to which we now turn.

Filament-Based Movement in Muscle

Muscle contraction is the most familiar example of mechanical work mediated by intracellular filaments. Mammals have several different kinds of muscles, including skeletal muscle, cardiac muscle, and smooth muscle. We will first consider skeletal muscle, because much of our knowledge of the contractile process grew out of early investigations of its molecular structure and function.

Skeletal Muscle Cells Are Made of Thin and Thick Filaments

Skeletal muscles are responsible for voluntary movement. The structural organization of skeletal muscle is shown in Figure 23-10. A muscle consists of bundles of parallel **muscle fibers** joined by tendons to the bones that the muscle must move. Each fiber is actually a long, thin multinucleate cell, highly specialized for its contractile function. The multinucleate state arises from the fusion of embryonic cells called *myoblasts* during muscle differentiation. This cell fusion also accounts at least in part for the striking length of muscle cells, which may be many centimeters in length.

At the subcellular level, each muscle fiber (or cell) contains numerous **myofibrils.** Myofibrils are 1–2 μm in diameter and may extend the entire length of the cell. Each myofibril is subdivided along its length into repeating units called **sarcomeres.** The sarcomere is the fundamental contractile unit of the muscle cell. Each sarcomere of the myofibril contains bundles of **thick filaments** and **thin filaments.** Thick filaments consist of myosin, whereas thin filaments consist mainly of actin, although several other important proteins are also present. The thin filaments are arranged around the thick filaments in a hexagonal pattern, as can be seen when the myofibril is viewed in cross section (Figure 23-11).

The filaments in skeletal muscle are aligned in lateral register, giving the myofibrils a pattern of alternating dark and light bands (Figure 23-12a). This pattern of bands, or *striations,* is characteristic of skeletal and cardiac muscle, which are therefore referred to as **striated muscle.** The

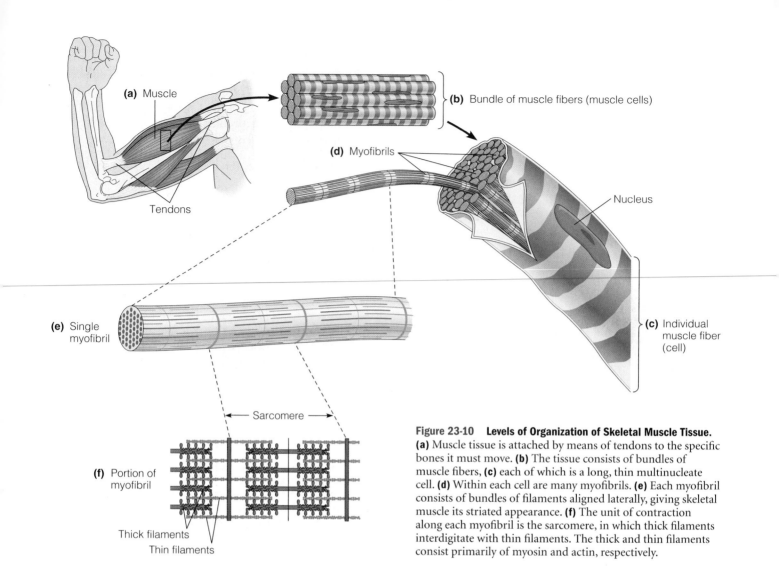

Figure 23-10   Levels of Organization of Skeletal Muscle Tissue. (a) Muscle tissue is attached by means of tendons to the specific bones it must move. (b) The tissue consists of bundles of muscle fibers, (c) each of which is a long, thin multinucleate cell. (d) Within each cell are many myofibrils. (e) Each myofibril consists of bundles of filaments aligned laterally, giving skeletal muscle its striated appearance. (f) The unit of contraction along each myofibril is the sarcomere, in which thick filaments interdigitate with thin filaments. The thick and thin filaments consist primarily of myosin and actin, respectively.

dark bands are called **A bands,** and the light bands are called **I bands.** (The terminology for the structure and appearance of muscle myofibrils was developed from observations originally made with the polarizing light microscope. I stands for *isotropic* and A for *anisotropic,* terms related to the appearance of these bands when illuminated with plane-polarized light.)

As illustrated in Figure 23-12b, the lighter region in the middle of each A band is called the **H zone** (from the German word *hell,* meaning "light"). Running down the center of the H zone is the **M line,** which contains *myomesin,* a protein that links myosin filaments together. In the middle of each I band appears a dense **Z line** (from the German word *zwischen,* meaning "between"). The distance from one Z line to the next defines a single sarcomere. A sarcomere is about 2.5–3.0 μm long in the relaxed state but shortens progressively as the muscle contracts.

Sarcomeres Contain Ordered Arrays of Actin, Myosin, and Accessory Proteins

The striated pattern of skeletal muscle and the observed shortening of the sarcomeres during contraction can be explained in terms of the thick and thin filaments that make up the myofibrils. We will therefore look in some detail at both types of filaments and then return to the contraction process in which they play so vital a role.

Thick Filaments.   The thick filaments of myofibrils are about 15 nm in diameter and about 1.6 μm long. They lie parallel to one another in the middle of the sarcomere (see Figure 23-12). Each thick filament consists of many molecules of myosin, which are oriented in opposite directions in the two halves of the filament. Each myosin molecule is long and thin, with a molecular weight of about 510,000.

Every thick filament consists of hundreds of myosin molecules organized in a staggered array such that the heads of successive molecules protrude from the thick filament in a repeating pattern, facing away from the center (Figure 23-13). Projecting pairs of heads are spaced 14.3 nm apart along the thick filament, with each pair displaced one-third of the way around the filament from the previous pair. These protruding heads can make contact with adjacent thin filaments, forming the cross-bridges between thick and thin filaments that are essential to the mechanism of muscle contraction.

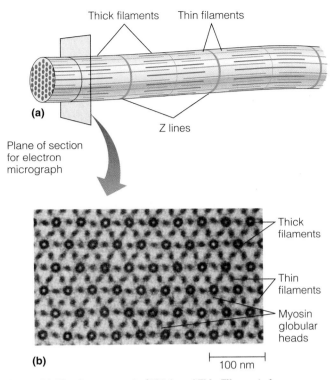

Figure 23-11 Arrangement of Thick and Thin Filaments in a Myofibril. **(a)** A myofibril consists of interdigitated thick and thin filaments. **(b)** The thin filaments are arranged around the thick filaments in a hexagonal pattern, as seen in this cross section of a flight muscle from the fruit fly *Drosophila melanogaster* viewed by high-voltage electron microscopy (HVEM).

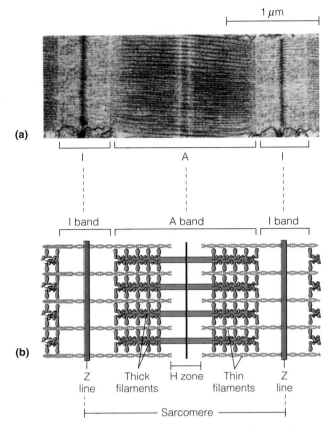

Figure 23-12 Appearance of and Nomenclature for Skeletal Muscle. **(a)** An electron micrograph of a single sarcomere (TEM). **(b)** A schematic diagram that can be used to interpret the repeating pattern of bands in striated muscle in terms of the interdigitation of thick and thin filaments. An A band corresponds to the length of the thick filaments, and an I band represents that portion of the thin filaments that does not overlap with thick filaments. The lighter area in the center of the A band is called the H zone; the line in the middle is known as the M line. The dense zone in the center of each I band is called the Z line. A sarcomere, the basic repeating unit along the myofibril, is the distance between two successive Z lines.

Thin Filaments. The thin filaments of myofibrils interdigitate with the thick filaments. The thin filaments are about 7 nm in diameter and about 1 μm long, and they are the only filaments in the I bands of the myofibril. In fact, each I band consists of *two* sets of thin filaments, one set on either side of the Z line, with each filament attached to the Z line and extending toward and into the A band in the center of the sarcomere. This accounts for the length of almost 2 μm for I bands in extended muscle.

The structure of thin filaments is shown in Figure 23-14. A thin filament consists of at least three proteins. The most important component of thin filaments is actin. Recall from Chapter 22 that actin is synthesized as G-actin, but polymerizes into long, linear strands of F-actin. Thin filaments are strands of F-actin intertwined with the proteins **tropomyosin** and **troponin**. Tropomyosin is a long, rodlike molecule, similar to the myosin tail, that fits in the groove of the actin helix. Each tropomyosin molecule stretches for about 38.5 nm along the filament and associates along its length with seven actin monomers.

Troponin is actually a complex of three polypeptide chains, called **TnT, TnC,** and **TnI.** TnT binds to tropomyosin and is thought to be responsible for positioning the complex on the tropomyosin molecule. TnC binds calcium ions, and TnI binds to actin. (Tn stands for troponin, T for tropomyosin, C for calcium, and I for inhibitory, because TnI inhibits muscle contraction.) One troponin complex is associated with each tropomyosin molecule, so the spacing between successive troponin complexes along the thin filament is 38.5 nm. Troponin and tropomyosin constitute a calcium-sensitive switch that activates contraction in both skeletal and cardiac muscle.

Organization of Muscle Filament Proteins. How can the filamentous proteins of muscle fibers maintain such a precise organization, when the microfilaments in other cells are relatively disorganized? The answer involves structural proteins that play a central role in maintaining the architectural relationships of muscle proteins (Figure 23-15). For instance, *α-actinin* keeps actin filaments bundled into parallel arrays and, along with the capping protein *CapZ*, attaches them to the Z line. *Myomesin* is present at the H zone of the thick filament arrays and performs the same bundling function for the myosin molecules composing the arrays. A third structural protein, *titin*, attaches the

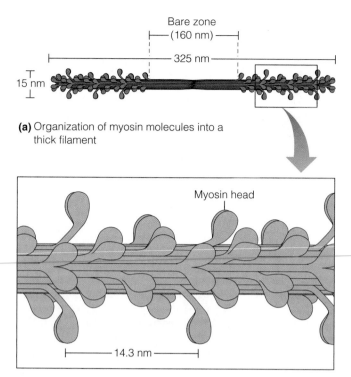

(a) Organization of myosin molecules into a thick filament

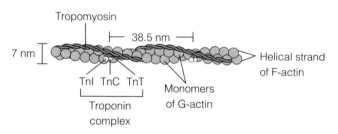

(b) Portion of a thick filament

Figure 23-13 The Thick Filament of Skeletal Muscle. **(a)** The thick filament of the myofibril consists of hundreds of myosin molecules organized in a repeating, staggered array. A typical thick filament is about 1.6 μm long and about 15 nm in diameter. Individual myosin molecules are integrated into the filament longitudinally, with their ATPase-containing heads oriented away from the center of the filament. The central region of the filament is therefore a bare zone containing no heads. **(b)** This enlargement of a portion of the thick filament shows that pairs of myosin heads are spaced 14.3 nm apart.

Figure 23-14 The Thin Filament of Striated Muscle. Each thin filament is a single strand of F-actin in which the G-actin monomers are staggered and give the appearance of a double-stranded helix. One result of this arrangement is that two grooves run along both sides of the filament. Long, ribbonlike molecules of tropomyosin lie in these grooves. Each tropomyosin molecule consists of two α helices wound about each other to form a ribbon about 2 nm in diameter and 38.5 nm long. Associated with each tropomyosin molecule is a troponin complex consisting of the three polypeptides TnT, TnC, and TnI.

thick filaments to the Z lines. Titin is highly flexible; during contraction-relaxation cycles, it can keep thick filaments in the correct position relative to thin filaments. Another protein, *nebulin,* stabilizes the organization of thin filaments. The protein components involved in mus-

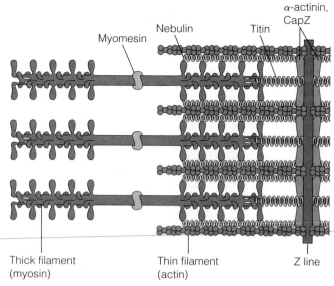

Figure 23-15 Structural Proteins of the Sarcomere. The thick and thin filaments require structural support to maintain their precise organization in the sarcomere. The support is provided by two proteins, α-actinin and myomesin, which bundle actin and myosin filaments, respectively. Titin attaches thick filaments to the Z line, thereby maintaining their position within the thin filament array. Nebulin stabilizes the attachment of thin filaments to the Z line.

cle contraction are summarized in Table 23-2. The contractile process involves the complex interaction of all these proteins.

The Sliding-Filament Model Explains Muscle Contraction

With our understanding of muscle structure, we can now consider what happens during the contraction process. Based on electron microscopic studies, it is clear that the A bands of the myofibrils remain fixed in length during contraction, whereas the I bands shorten progressively and virtually disappear in the fully contracted state. To explain these observations, the **sliding-filament model** illustrated in Figure 23-16 was proposed in 1954 independently by Andrew Huxley and Rolf Niedergerke and by Hugh Huxley and Jean Hanson. According to this model, muscle contraction is due to thin filaments sliding past thick filaments, with no change in the length of either type of filament. The sliding-filament model not only proved to be correct, but was instrumental in focusing attention on the molecular interactions between thick and thin filaments that underlie the sliding process.

As Figure 23-16a indicates, contraction involves sliding of thin filaments such that they are drawn progressively into the spaces between adjacent thick filaments, overlapping more and more with the thick filaments and narrowing the I band in the process. The result is a shortening of the individual sarcomeres and myofibrils, and hence a contracting of the muscle cell and the whole tissue. This, in turn, causes the movement of the body parts attached to the muscle.

Table 23-2 Major Protein Components of Vertebrate Skeletal Muscle

Protein	Molecular Mass (kDa)	Function
Actin	42	Major component of thin filaments
Myosin	510	Major component of thick filaments
Tropomyosin	64	Binds along the length of thin filaments
Troponin	78	Positioned at regular intervals along thin filaments; mediates calcium regulation of contraction
Titin	2500	Links thick filaments to Z line
Nebulin	700	Links thin filaments to Z line
Myomesin	185	Myosin-binding protein present at the M line of thick filaments
α-actinin	190	Bundles actin filaments and attaches them to Z line
Ca^{2+} ATPase	115	Major protein of sarcoplasmic reticulum (SR); transports Ca^{2+} into SR to relax muscle
CapZ	68	Attaches actin filaments to Z line

The interdigitation and sliding of the thin and thick filaments past each other as a means of generating force suggests that there should be a relationship between the force generated during contraction and the degree of shortening of the sarcomere. In fact, when the relationship between shortening and force is measured, it is exactly what the sliding-filament model predicts (Figure 23-16b): The amount of force the muscle can generate during a contraction depends on the number of myosin heads from the thick filament that can make contact with the thin filament. When the sarcomere is stretched, there is relatively little overlap between thin and thick filaments, so the force generated is small. As the sarcomere shortens, the region of overlap increases and the force of contraction gets correspondingly larger. Finally, a point is reached at which continued shortening no longer increases the amount of overlap between thin and thick filaments and the force of contraction stays the same. Any further shortening of the sarcomere results in a dramatic decline in tension as the filaments crowd into one another.

The sliding of thin filaments past thick filaments depends on the elaborate structural features of the sarcomere, and it requires energy. These basic observations raise three questions: (1) What keeps the partially interdigitated thick and thin filaments associated with each other, so that they do not simply fall apart? (2) By what mechanism are the thin filaments drawn or pulled progressively into the spaces between thick filaments to cause the actual contraction? (3) How is the energy of ATP used to drive this process? These questions are answered in the next section.

Cross-Bridges Hold Filaments Together and ATP Powers Their Movement

Cross-Bridge Formation. Regions of overlap between thick and thin filaments, whether extensive (in the contracted muscle) or minimal (in the relaxed muscle), are always characterized by the presence of transient **cross-bridges.** The cross-bridges are formed from links between the F-actin of the thin filaments and the myosin heads of the thick filaments (Figure 23-17). For contraction, the cross-bridges must form and dissociate repeatedly and in such a manner that each cycle of cross-bridge formation causes the thin filaments to interdigitate with the thick filaments more and more, thereby shortening the individual sarcomeres and causing the muscle fiber to contract. A given myosin head on the thick filament repeatedly undergoes a cycle of events in which it binds to specific actin subunits on the thin filament, undergoes an energy-requiring change in shape that pulls the thin filament, then breaks its association with the thin filament and associates with another site farther along the thin filament toward the Z line. Overall contraction, then, is the net result of the repeated making and breaking of many such cross-bridges, with each cycle of cross-bridge formation causing the translocation of a small length of thin filament of a single fibril in a single cell.

ATP and the Contraction Cycle. The driving force for cross-bridge formation is the hydrolysis of ATP, catalyzed by *actin-activated ATPase* located in the myosin head. The requirement for ATP can be demonstrated in vitro because isolated muscle fibers contract in response to added ATP.

The mechanism of muscle contraction is depicted in Figure 23-18 as a four-step cycle. In step ①, a specific myosin head, in a high-energy configuration containing an ADP and a P_i molecule (a hydrolyzed ATP, in effect), binds loosely to the actin filament. The myosin head then proceeds to a more tightly bound configuration that requires the loss of P_i. Step ② is the power stroke. The transition of myosin to the more tightly bound state triggers a conformational change in myosin. This conformational change is associated with a movement of the head, causing the thick filament to pull against the thin filament, which then moves with respect to the thick filament.

Cross-bridge dissociation follows in step ③, as ATP binds to the myosin head in preparation for the next

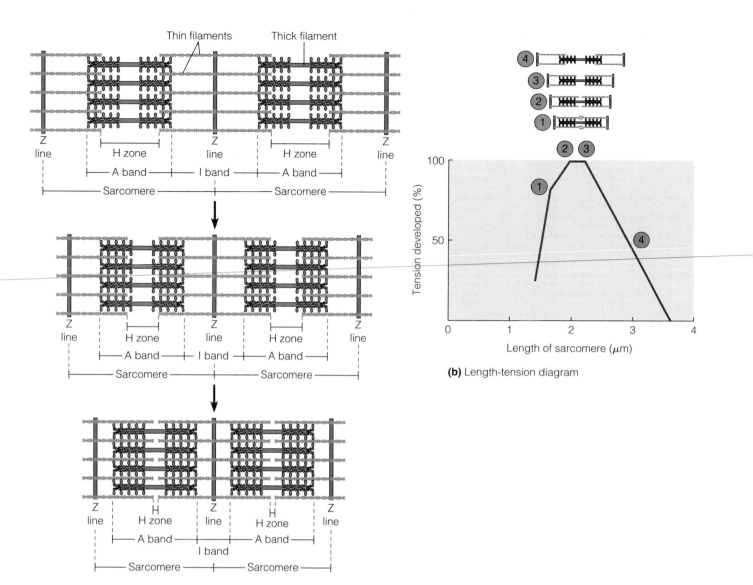

(a) Sliding filament model

(b) Length-tension diagram

Figure 23-16 The Sliding-Filament Model of Muscle Contraction. **(a)** Two sarcomeres of a myofibril during the contraction process. The extended configuration is shown at the top, while the center and bottom views represent progressively more contracted myofibrils. A myofibril shortens by the progressive sliding of thick and thin filaments past each other. The result is a greater interdigitation of filaments without any change in length of individual filaments. The increasing overlap of thick and thin filaments leads to a progressive decrease in the length of the I band as interpenetration continues during contraction. **(b)** This graph shows that the amount of tension developed by the sarcomere is proportional to the amount of overlap between the thin filament and the region of the thick filament containing myosin heads. When the sarcomere begins to shorten, as during a muscle contraction, the Z lines move closer together, increasing the amount of overlap between thin and thick filaments. This overlap allows more of the thick filament to interact with the thin filament. Therefore, the muscle can develop more tension (see 4 to 3). This proportional relationship continues until the ends of the thin filaments move into the H zone. Here they encounter no further myosin heads, so tension remains constant (3 to 2). Any further shortening of the sarcomere results in a dramatic decline in tension (2 to 1) as the filaments crowd into one another.

step. The binding of ATP causes the myosin head to change its conformation in a way that weakens its binding to actin. In the absence of adequate ATP, cross-bridge dissociation does not occur, and the muscle becomes locked in a stiff, rigid state called **rigor**. The *rigor mortis* associated with death results from the depletion of ATP and the progressive accumulation of cross-bridges in the configuration shown at the end of step ②. Note that once detached, the thick and thin filaments would be free to slip back to their previous positions, but they are held together at all times by the many other cross-bridges along their length at any given moment, just as at least some legs of a millipede are always in contact with the surface on which it is walking. In fact, each thick filament has about 350 myosin heads, and each head attaches and detaches about five times per second during rapid contraction, so there are always many cross-bridges intact at any time.

Finally, in step ④, the energy of ATP hydrolysis is used to return the myosin head to the high-energy configuration necessary for the next round of cross-bridge formation and filament sliding. This brings us back to where we started, for the myosin head is now activated and ready to form a bridge to actin again. But the new bridge will be formed with actin farther along the thin filament, because the first cycle resulted in a net displacement of the thin filament with respect to the thick filament. In succeeding cycles, the particular myosin head shown in Figure 23-18 will draw the thin filament in the direction of further contraction. And what about the direction of contraction? Recall from Chapter 22 that actin filaments have plus and minus ends. Myosin always walks toward the *plus* end of the thin filament, thereby establishing the direction of contraction.

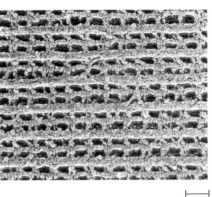

├── 30 nm

Figure 23-17 Cross-Bridges.   The cross-bridges between thick and thin filaments formed by the projecting heads of myosin molecules can be readily seen in this high-resolution electron micrograph (TEM).

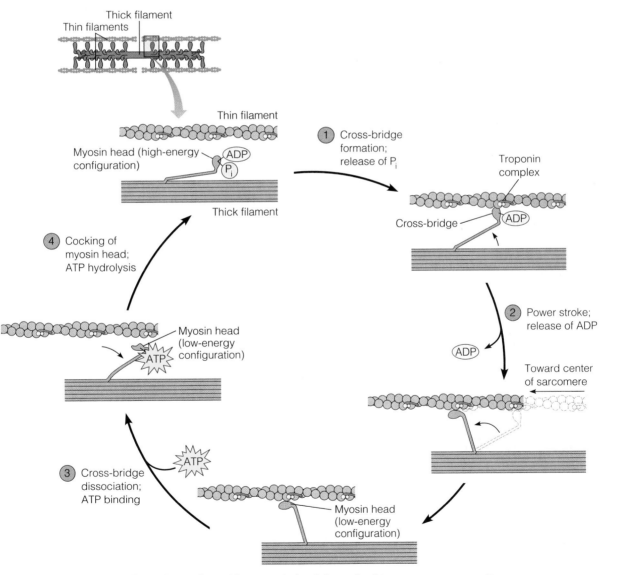

Figure 23-18 The Cyclic Process of Muscle Contraction.   A small segment of adjacent thick and thin filaments (see the orienting inset at upper left) is used to illustrate the series of events in which the cross-bridge formed by a myosin head draws the thin filament toward the center of the sarcomere, thereby causing the myofibril to contract. Step ① shows the cross-bridge configuration of relaxed muscle, whereas the end of step ② shows the configuration of a muscle in rigor. A detailed description of all the steps is given in the text.

The Regulation of Muscle Contraction Depends on Calcium

Our description of muscle contraction so far implies that skeletal muscle ought to contract continuously, as long as there is sufficient ATP. Yet experience tells us that most skeletal muscles spend more time in the relaxed state than in contraction. Contraction and relaxation must therefore be regulated to result in the coordinated movements associated with muscle activity.

The Role of Calcium in Contraction. The regulation of muscle contraction depends on free calcium ions (Ca^{2+}) and on the ability of the muscle cell to raise and lower calcium levels rapidly in the cytosol (called the **sarcoplasm** in muscle cells) around the myofibrils. The regulatory proteins tropomyosin and troponin act in concert to regulate the availability of myosin-binding sites on actin filaments in a way that depends critically on the level of calcium in the sarcoplasm.

To understand how this process works, we must begin by recognizing that the myosin-binding sites on actin are normally blocked by tropomyosin. For myosin to bind to actin and initiate the cross-bridge cycle, the tropomyosin molecule must be moved out of the way. The calcium dependence of muscle contraction is due to troponin C (TnC), which binds calcium ions. When a calcium ion binds to TnC, the TnC molecule undergoes a conformational change that is transmitted to the tropomyosin molecule, causing it to move toward the center of the helical groove of the thin filament, out of the blocking position. The binding sites on actin are then accessible to the myosin heads, allowing contraction to proceed.

Figure 23-19 illustrates how the troponin-tropomyosin complex regulates the interaction between actin and myosin. When the calcium concentration in the sarcoplasm is low ($<10^{-4}$ mM), tropomyosin blocks the binding sites on the actin filament, effectively preventing their interaction with myosin (Figure 23-19a). As a result, cross-bridge formation is inhibited, and the muscle becomes or remains relaxed. At higher calcium concentrations ($>10^{-3}$ mM), calcium binds to TnC, causing tropomyosin molecules to shift their position, which allows myosin heads to make contact with the binding sites on the actin filament and thereby initiate contraction (Figure 23-19b).

When the calcium concentration falls again as it is pumped out of the cytosol (discussed below), the troponin-calcium complex dissociates and the tropomyosin moves back to the blocking position. Myosin binding is therefore inhibited, further cross-bridge formation is prevented, and the contraction cycle ends.

Thus, an increase in the sarcoplasmic calcium concentration stimulates the contraction of skeletal muscle by triggering the following series of events:

1. Calcium binds to troponin and induces a conformational change in the complex.
2. This change in troponin causes a shift in the position of the tropomyosin with which it is complexed.
3. The binding sites on actin become available for interaction with myosin.
4. Cross-bridges form, setting in motion the sequence of events (depicted in Figure 23-18) that leads to contraction.

Regulation of Calcium Levels in Skeletal Muscle Cells. From the previous discussion, you know that muscle contraction is regulated by the concentration of calcium ions in the sarcoplasm. But how is the level of calcium controlled? Think for a moment about what must happen when you

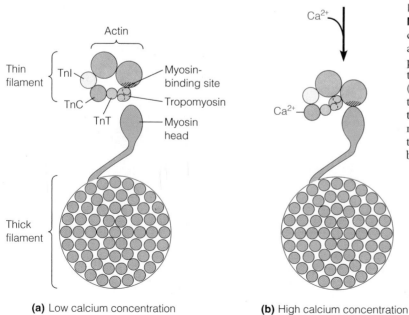

(a) Low calcium concentration   **(b)** High calcium concentration

Figure 23-19 Regulation of Contraction in Striated Muscle. (a) At low concentrations ($<10^{-4}$ mM Ca^{2+}), calcium is not bound to the TnC subunit of troponin, and tropomyosin blocks the binding sites on actin, preventing access by myosin and thereby maintaining the muscle in the relaxed state. **(b)** At high concentrations ($>10^{-3}$ mM Ca^{2+}), calcium binds to the TnC subunit of troponin, inducing a conformational change that is transmitted to tropomyosin. The tropomyosin molecule moves toward the center of the groove in the thin filament, allowing myosin to gain access to the binding sites on actin, thereby triggering contraction.

move any part of your body—when you flex an index finger, for instance. A nerve impulse is generated in the brain and transmitted down the spinal column to the nerve cells, or *motor neurons*, that control a small muscle in your forearm. The motor neurons activate the appropriate muscle cells, which contract and relax, all within about 100 msec. Every time you move, or with every heartbeat, nerve impulses are sent to the appropriate muscles, causing elevation of calcium levels in the sarcoplasm, resulting in contraction. When nerve impulses to the muscle cell cease, calcium levels decline quickly and the muscle relaxes. Therefore, to understand how muscle contraction is regulated, we need to know how nerve impulses cause calcium levels in the sarcoplasm to change and how these changes affect the contractile machinery. Muscle cells have many specialized features that facilitate a rapid change in the sarcoplasmic concentration of calcium ions and a rapid response of the contractile machinery; we discuss these features below.

Events at the Neuromuscular Junction. Recall from Chapter 9 that the signal for a muscle cell to contract is conveyed by a nerve cell in the form of an electrical impulse called an *action potential.* An action potential is carried from the neuron to the muscle cell by nerve axons. The site at which the nerve innervates, or makes contact with, the muscle cell is called the **neuromuscular junction.** At the neuromuscular junction, the axon branches out and forms *axon terminals* that make contact with the muscle cell. These terminals contain the transmitter chemical *acetylcholine,* which is stored in membrane-enclosed vesicles and secreted by axon terminals in response to an action potential. The area of the muscle cell plasma membrane under the axon terminals is called the *motor end plate.* There, in the plasma membrane (called the **sarcolemma** in muscle cells), clusters of acetylcholine receptors are associated with each axon terminal.

Acetylcholine receptors are *ligand-gated channels,* the ligand in this case being acetylcholine (see Chapter 9, p. 247). When the receptor binds acetylcholine, it opens a pore in the plasma membrane through which sodium ions can flow into the muscle cell. The sodium influx in turn causes an electrical impulse to be transmitted away from the sarcolemma at the motor end plate.

Transmission of an Impulse to the Interior of the Muscle. Once an electrical impulse forms at the motor end plate, it spreads out over the sarcolemma via the **transverse (T) tubule system** (Figure 23-20). In contrast to the plasma membrane of most cells, the sarcolemma has a regular pattern of inpocketings, which give rise to a series of tubes called T tubules that penetrate the interior of a muscle cell. The T tubules carry action potentials into the muscle cell, and they are part of the reason that muscle cells can respond so quickly to a nerve impulse.

Inside the muscle cell, the T tubule system comes into contact with the **sarcoplasmic reticulum (SR),** a system of intracellular membranes in the form of flattened sacs or tubes. As the name suggests, the SR is similar to the endoplasmic reticulum found in nonmuscle cells, except that it is highly specialized for accumulating, storing, and releasing calcium ions. The SR runs along the myofibrils, where it is poised to release calcium ions directly into the myofibril, causing contraction, and then to remove calcium from the myofibril, causing relaxation. This close proximity of the SR to the myofibrils is another specialization that facilitates the rapid response of muscle cells to nerve signals.

SR Function in Calcium Release and Uptake. The SR can be functionally divided into two components, referred to as the **medial element** and the **terminal cisternae** (singular: **terminal cisterna;** see Figure 23-20). The terminal cisternae

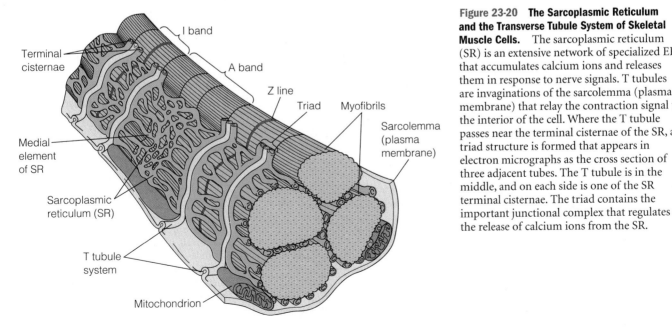

Figure 23-20 The Sarcoplasmic Reticulum and the Transverse Tubule System of Skeletal Muscle Cells. The sarcoplasmic reticulum (SR) is an extensive network of specialized ER that accumulates calcium ions and releases them in response to nerve signals. T tubules are invaginations of the sarcolemma (plasma membrane) that relay the contraction signal to the interior of the cell. Where the T tubule passes near the terminal cisternae of the SR, a triad structure is formed that appears in electron micrographs as the cross section of three adjacent tubes. The T tubule is in the middle, and on each side is one of the SR terminal cisternae. The triad contains the important junctional complex that regulates the release of calcium ions from the SR.

of the SR contain a high concentration of ATP-dependent calcium pumps that continually pump calcium into the lumen of the SR. The ability of the SR to pump calcium ions is crucial for muscle relaxation, but it is also needed for muscle contraction. Calcium pumping produces a high calcium concentration in the lumen of the SR (up to several millimolar). This calcium can then be released from the terminal cisternae of the SR when needed. Figure 23-20 shows how the terminal cisternae of the SR are positioned adjacent to the contractile apparatus of each myofibril. Terminal cisternae are typically found right next to a T tubule, giving rise to a structure called a **triad.** In electron micrographs, a triad appears as three circles in a row. The central circle is the membrane of the T tubule, and the circles on each side are the membranes of the terminal cisternae. Close inspection of the triad reveals that the terminal cisternae appear to be connected to the T tubule by a dense-staining material between the two membranes. This material is referred to as the **junctional complex** (to which we will return shortly).

The proximity of the T tubule, the terminal cisternae of the SR, and the contractile machinery of the myofibril provides the basis for how muscle cells can respond so rapidly to a nerve impulse. The action potential travels from the motor end plate, spreads out over the sarcolemma, and enters the T tubule (Figure 23-21). As the action potential travels down the T tubule, it passes near the terminal cisternae of the SR, causing a special type of calcium channel in the terminal cisternae to open. When this channel opens, calcium rushes into the sarcoplasm, causing contraction.

Letting calcium out of the SR causes a muscle cell to contract, but the cell must also relax in order to be ready for another contraction. For the muscle cell to relax, calcium levels must be brought back down to the resting level. This is accomplished by pumping calcium back into the SR. The membrane of the SR contains an active transport protein, a **calcium pump** or **calcium ATPase,** which can pump calcium ions from the sarcoplasm into the cisternae

of the SR. These pumps are concentrated in the medial element of the SR. The ATP-dependent mechanism by which calcium moves through the pump is similar to that discussed in Chapter 8 for the sodium-potassium pump (see Figure 8-12).

The continued pumping of calcium from the sarcoplasm back into the SR cisternae quickly lowers the sarcoplasmic calcium level to the point at which troponin releases calcium, tropomyosin moves back to the blocking position on actin sites, and further cross-bridge formation is prevented. Cross-bridges therefore disappear rapidly as actin dissociates from myosin and becomes blocked by tropomyosin. This leaves the muscle relaxed and free to be reextended, because the absence of cross-bridge contacts allows the thin filaments to slide out from between the thick filaments.

ATP Generated Under Aerobic or Hypoxic Conditions Meets the Energy Needs Of Muscle Contraction

Because muscle contraction involves the hydrolysis of an ATP molecule for every cycle of attachment and detachment, muscle cells need ways to regenerate ATP continuously. Muscle cells actually have a variety of mechanisms to ensure maximum ATP availability, even during prolonged periods of intense activity.

ATP Generation: The Aerobic Mode. The ATP needs of muscle cells are normally met either by glycolysis or by mitochondrial respiration. The extent to which one or the other of these pathways is favored depends on the kind of muscle involved and whether it is functioning under *aerobic* or *hypoxic* (low-oxygen) conditions. Skeletal muscles that are characterized by frequent use and high activity usually rely on complete respiratory metabolism. The flight muscles of birds are a good example. Such muscles draw both glucose and oxygen from the circulatory system, oxidizing the glucose completely to carbon dioxide

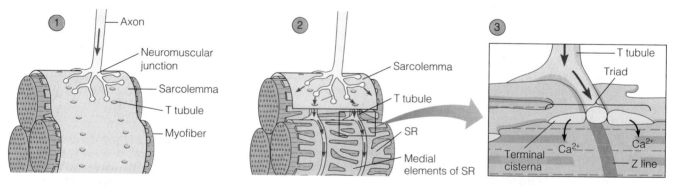

Figure 23-21 Stimulation of a Muscle Cell by a Nerve Impulse. ① An action potential moves down the axon of the neuron until it reaches the end. The ends of the axon branch out over the surface of the muscle cell at the neuromuscular junction to form synapses (contact points) between the neuron and the muscle cell. ② Depolarization of the terminals of the axon causes the release of neurotransmitter molecules, which bind to the acetylcholine receptors on the sarcolemma. Binding of neurotransmitter to the acetylcholine receptors starts an action potential in the muscle cell. As the action potential spreads over the surface of the muscle cell, it travels down into the T tubules. ③ T tubules carry the action potential into the interior of the muscle cell, where it stimulates calcium release from the terminal cisternae of the SR.

and water and generating ATP by oxidative phosphorylation in mitochondria. These muscles are characterized by an abundance of mitochondria and by a red color. Mitochondria occur in close association with the myofibrils in almost all aerobic muscle cells. The red color of such tissue is due to the high degree of vascularization, to the cytochromes and iron-sulfur proteins present in the mitochondria, and to the presence of **myoglobin,** a protein related in structure and function to hemoglobin but localized in muscle cells and used to bind and store oxygen.

ATP Generation: The Hypoxic Mode.

During periods of intense exercise, the demand for ATP regeneration may exceed the rate at which the circulatory system can supply oxygen to the tissue. After depletion of the reserve oxygen available from myoglobin, the tissue begins to function hypoxically, converting glucose to lactate. Because the ATP yield of glucose is greatly reduced under these conditions (from 36 or 38 to 2 molecules of ATP per molecule of glucose; see Chapter 14), much more glucose is required per unit time. The extra glucose is supplied by catabolism of *glycogen,* the storage carbohydrate of muscle cells. The lactate formed under hypoxic conditions is usually released into the blood and eventually reaches the liver, where it is either oxidized fully to carbon dioxide and water or converted back into glucose (see Chapter 14). Intense muscular activity cannot be sustained long under hypoxic conditions because of the rapid depletion of glycogen stores and the accumulation of lactate. However, this method is useful for short bursts of activity when oxygen cannot be supplied fast enough for aerobic respiration.

Other Mechanisms of Energy Storage in Muscle Cells.

ATP clearly serves as the immediate source of energy to drive muscle contraction, and it is also the form in which energy is conserved during glycolysis and respiratory metabolism. However, ATP is not the major form of stored energy for muscle cells. A surprising observation made early in muscle research was that the ATP content of working muscle remains remarkably constant until the muscle is near exhaustion. What decreases instead during prolonged exertion is the cellular level of **creatine phosphate,** a high-energy compound that can be used to recharge ADP, a reaction catalyzed by the enzyme *creatine kinase:*

$$\text{creatine phosphate} + \text{ADP} \longrightarrow \text{creatine} + \text{ATP} \quad \textbf{(23-1)}$$

Because it provides energy to muscles during strenuous activity, creatine has become a popular nutritional supplement among athletes involved in weight training.

As a final backup system, muscle cells also contain an enzyme called *myokinase,* which is capable of phosphorylating one ADP molecule at the expense of another, as follows:

$$2\text{ADP} \longrightarrow \text{ATP} + \text{AMP} \quad \textbf{(23-2)}$$

The myokinase reaction provides a means of extracting energy from the remaining acid anhydride bond of ADP.

The overall picture that emerges of the skeletal muscle cell is one of incredible specialization for its role in contraction—specialization in the design of the contractile elements as well as in the mechanism available to ensure that the ATP needed for contraction can be supplied under virtually any conditions.

The Coordinated Contraction of Cardiac Muscle Cells Involves Electrical Coupling

Cardiac (heart) muscle is responsible for the beating of the heart and the pumping of blood through the body's circulatory system. Cardiac muscle functions continuously; in one year, your heart beats about 40 million times! Cardiac muscle is very similar to skeletal muscle in the organization of actin and myosin filaments and has the same striated appearance (Figure 23-22). However, the two kinds of muscle differ significantly in their metabolism. Cardiac muscle is highly dependent on aerobic respiration; only in emergencies is glycogen used as an energy source. In contrast to skeletal muscle, most of the energy required for the beating of the heart under resting conditions is provided not by blood glucose, but by free fatty acids that are transported from adipose (fat storage) tissue to the heart by serum albumin, a blood protein.

A second difference between cardiac and skeletal muscle is that heart muscle cells are not multinucleate. Instead, cells are joined end-to-end through structures called **intercalated discs.** The discs have gap junctions that electrically couple neighboring cells, providing a way for depolarization waves to spread throughout the heart

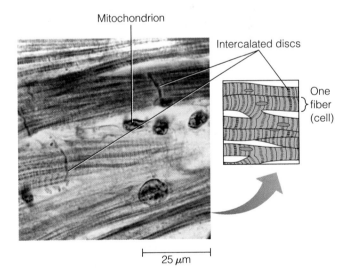

Figure 23-22 Cardiac Muscle Cells. Cardiac muscle cells have a contractile mechanism and sarcomeric structure similar to those of skeletal muscle cells. However, unlike skeletal muscle cells, cardiac muscle cells are joined together end-to-end at the intercalated discs, which allow ions and electrical signals to pass from one cell to the next. This ionic permeability enables a contraction stimulus to spread evenly to all the cells of the heart. In addition, cardiac muscle cells exhibit branches that are not seen in skeletal or smooth muscle cells (LM).

during its contraction cycle. The heart is not activated by nerve impulses, as skeletal muscle is, but contracts spontaneously once every second or so. The heart rate is controlled by a pacemaker region in an upper portion of the heart (right atrium), which contracts 70–80 times per minute, slightly faster than other cardiac tissues. The depolarization wave initiated by the pacemaker then spreads to the rest of the heart to produce the heartbeat.

Smooth Muscle Is More Similar to Nonmuscle Cells than to Skeletal Muscle

Smooth muscle is responsible for involuntary contractions such as those of the stomach, intestines, uterus, and blood vessels. In general, such contractions are slow, taking up to five seconds to reach maximum tension. Smooth muscle contractions are also of greater duration than those of skeletal or cardiac muscle. Though smooth muscle is not able to contract rapidly, it is well adapted to maintain tension for long periods of time, as is required in these organs and tissues.

The Structure of Smooth Muscle. Smooth muscle cells are long and thin, with pointed ends. Unlike skeletal or heart muscle, smooth muscle has no striations (Figure 23-23a). Thick and thin filaments are both present, but they are not organized into myofibrils and are not regularly aligned. Nor is the number of thick filaments always constant. The thin filaments of smooth muscle cells contain actin and tropomyosin, but no troponin.

Smooth muscle cells do not contain Z lines, which are responsible for the periodic organization of the sarcomeres found in skeletal and cardiac muscle cells. Instead, smooth muscle cells contain dense bodies, plaquelike structures in the cytoplasm and on the cell membrane (Figure 23-23b). Bundles of actin filaments are anchored at their ends to these dense bodies. As a result, actin filaments appear in a crisscross pattern, aligned obliquely to the long axis of the cell.

The organization of myosin molecules in the thick filaments of smooth muscle also differs from that of skeletal muscle. Instead of the bare central zone characteristic of thick filaments from skeletal muscle, smooth muscle thick filaments have myosin heads distributed along the entire length of the filament. Cross-bridges connect thick and thin filaments in smooth muscle, but not in the regular, repeating pattern seen in skeletal muscle.

Regulation of Contraction in Smooth Muscle Cells. Smooth muscle cell contraction and nonmuscle cell contraction are regulated in a manner distinct from that of skeletal muscle cells. Although skeletal and smooth muscle cells are both stimulated to contract by an increase in the sarcoplasmic concentration of calcium ions, the mechanisms involved are quite different. When sarcoplasmic calcium concentrations increase in smooth muscle and nonmuscle cells, a cascade of events takes place that includes the activation of **myosin light-chain kinase (MLCK)**. Activated

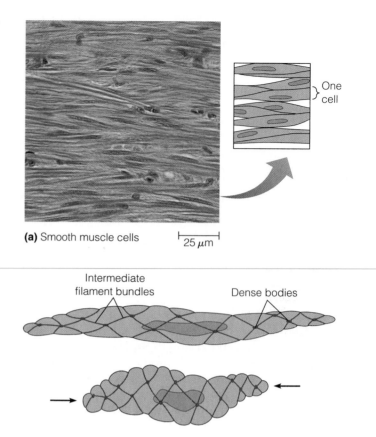

(a) Smooth muscle cells   25 μm

(b) Contraction of smooth muscle cell

Figure 23-23 Smooth Muscle and Its Contraction.   (a) Individual smooth muscle cells are long and spindle-shaped, with no Z lines or sarcomeric structure (LM). **(b)** In the smooth muscle cell, contractile bundles of actin and myosin appear to be anchored at one end of the bundle to a plaquelike structure called a dense body on the plasma membrane, and at the other end to a dense body in the sarcoplasm. The dense bodies are connected to each other by intermediate filaments, thereby orienting the actin and myosin bundles obliquely to the long axis of the cell. When the actin and myosin bundles contract, they pull on the dense bodies and intermediate filaments, producing the cellular contraction shown here.

MLCK then phosphorylates one type of myosin light chain known as a **regulatory light chain** (see Figure 23-9, myosin II).

Myosin light-chain phosphorylation affects myosin in two ways. First, some myosin molecules are curled up so that they cannot assemble into filaments. When the myosin light chain is phosphorylated, the myosin tail uncurls and becomes capable of assembly. Second, the phosphorylation of the light chains activates myosin, enabling it to interact with actin filaments to undergo the cross-bridge cycle.

The cascade of events involved in the activation of smooth muscle and nonmuscle myosin is shown in Figure 23-24a. In response to a nerve impulse or hormonal signal reaching the smooth muscle cell, an influx of extracellular calcium ions occurs, increasing the intracellular calcium concentration and causing contraction. The effect of the increased calcium concentration on con-

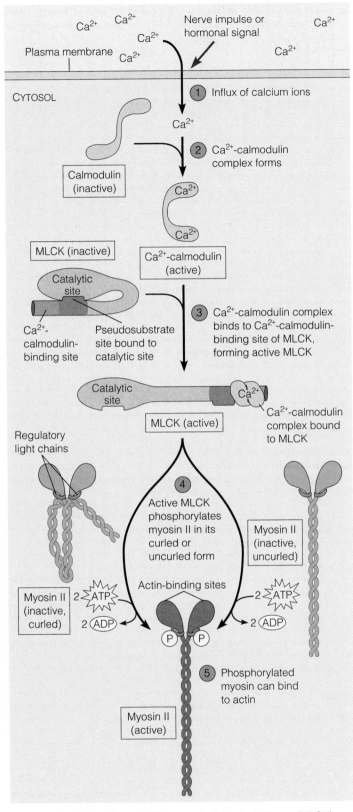

(a) Phosphorylation of myosin II by myosin light-chain kinase (MLCK)

① Influx of calcium ions

② Ca²⁺-calmodulin complex forms

③ Ca²⁺-calmodulin complex binds to Ca²⁺-calmodulin-binding site of MLCK, forming active MLCK

④ Active MLCK phosphorylates myosin II in its curled or uncurled form

⑤ Phosphorylated myosin can bind to actin

Nerve impulse or hormonal signal

Plasma membrane

CYTOSOL

Calmodulin (inactive)

Ca²⁺-calmodulin (active)

MLCK (inactive)

Catalytic site

Ca²⁺-calmodulin-binding site

Pseudosubstrate site bound to catalytic site

Catalytic site

MLCK (active)

Ca²⁺-calmodulin complex bound to MLCK

Regulatory light chains

Myosin II (inactive, curled)

Myosin II (inactive, uncurled)

Actin-binding sites

Myosin II (active)

0.25 μm

(b) Curled and uncurled myosin II molecules

Figure 23-24 Phosphorylation of Smooth Muscle and Nonmuscle Myosin. **(a)** The functions of both smooth muscle and nonmuscle myosin II are regulated by phosphorylation of the regulatory light chains. ① An influx of calcium ions into the cell is triggered by a nerve impulse or a hormonal signal. ② When present at a sufficiently high concentration, calcium ions bind to calmodulin, forming an active calcium-calmodulin complex. ③ The calcium-calmodulin complex in turn binds to a region of myosin light-chain kinase (MLCK) that overlaps the pseudosubstrate site. When this happens, the pseudosubstrate stretch of amino acids is pulled away and prevented from binding to the active site of MLCK, thus activating MLCK. ④ Activated MLCK phosphorylates the myosin light chains, whether the myosin is curled or uncurled. ⑤ The activated (and uncurled) myosin can then bind to actin and undergo the cross-bridge cycle. **(b)** Electron micrographs of curled and uncurled myosin II molecules. (TEMs)

traction is mediated by the binding of calcium to *calmodulin*. This yields the *calcium-calmodulin complex*, which, unlike free calmodulin, can bind to myosin light-chain kinase, thereby activating the enzyme. As a result, myosin light chains become phosphorylated and myosin can interact with actin to cause contractions.

The regulation of MLCK by calcium and calmodulin illustrates a common theme in the regulation of protein kinases. MLCK contains a peptide sequence at one end called a *pseudosubstrate,* a sequence of amino acids similar to the enzyme's normal substrate. In this case, the real substrate is a sequence of amino acids on the myosin light chain that is recognized by MLCK. Within this substrate's sequence of amino acids is the serine that actually accepts the phosphate from ATP. In the pseudosubstrate, this serine is replaced by a different amino acid that cannot accept a phosphate group.

The pseudosubstrate region of MLCK regulates the activity of this enzyme by folding around and binding to the active site of the enzyme and thus inhibiting its activity. Inhibition of enzyme activity in this manner is referred to as *autoinhibition,* because in effect the enzyme inhibits itself. However, there is also a binding site for calcium-calmodulin near the pseudosubstrate site. The pseudosubstrate region cannot bind to the active site of the enzyme at the same time that calcium-calmodulin is also bound. Therefore, when calcium-calmodulin is available to bind, the pseudosubstrate region is prevented from inhibiting the enzyme, and the enzyme becomes active.

Once activated, MLCK can then phosphorylate myosin. Before phosphorylation, smooth muscle or nonmuscle myosin may be simply inactive, or it may be both inactive and in the curled form (Figure 23-24b), preventing assembly. When the myosin light chain is phosphorylated, myosin in both forms becomes active and capable of interacting with actin, and the tail of myosin in the curled form straightens out and can assemble with other myosin molecules into filaments. As the calcium levels within smooth muscle cells drop again, the MLCK is inactivated, and a second enzyme, *myosin light-chain phosphatase,* removes the phosphate group from the myosin light chain. Since the dephosphorylated myosin can no longer bind to actin, the muscle cell relaxes.

Thus, both skeletal muscle and smooth muscle are activated to contract by calcium ions, but from different sources and by different mechanisms. In skeletal muscle, the calcium comes from the sarcoplasmic reticulum. Its effect on actin-myosin interaction is mediated by troponin and is very rapid, because it depends on conformational changes only. In smooth muscle, the calcium comes from outside the cell, and its effect is mediated by calmodulin. The effect is much slower in this case because it involves a covalent modification (phosphorylation) of the myosin molecule.

Actin-Based Motility in Nonmuscle Cells

Actin and myosin are best known as the major components of the thin and thick filaments of muscle cells. In fact, muscle cells represent only one specialized case of cell movements driven by the interactions of actin and myosin. Actins and myosins have now been discovered in almost all eukaryotic cells and are known to play important roles in various types of nonmuscle motility. We saw

one example of actin-dependent, nonmuscle motility during cytokinesis (see Chapter 17). In this section we examine several other examples.

Cell Migration via Lamellipodia Involves Cycles of Protrusion, Attachment, Translocation, and Detachment

Actin microfilaments (MFs) are required for the movement of most nonmuscle cells in animals. Many nonmuscle cells, such as fibroblasts, the growth cones of neurons, and many embryonic cells in animals, are capable of crawling over a substrate. In this section, we will consider a type of crawling involving protrusions known as *lamellipodia.* In a later section, we will consider a specialized form of crawling known as *amoeboid movement.*

Cell crawling involves several distinct events: (1) extension of a protrusion at the cell's leading edge; (2) attachment of the protrusion to the substrate; (3) generation of tension, which pulls the cell forward; and (4) release of attachments and retraction of the "tail" of the cell. These events are summarized in Figure 23-25; a scanning electron micrograph of a crawling cell in vitro is shown in Figure 23-26.

In order to crawl, such cells produce specialized extensions, or *protrusions,* at their front, or *leading edge.* One type of protrusion is a thin sheet of cytoplasm called a **lamellipodium** (plural: **lamellipodia**). Another type of protrusion is a thin, pointed structure known as a **filopodium** (plural: **filopodia**). The form of the protrusion appears to depend on the nature of the cell's movement and on the organization of the actin filaments within the cell. In cells that adhere very tightly to the underlying substratum and that do not move well, organized bundles of actin and myosin, called *stress fibers,* stretch from the tail, or trailing edge, of the cell to the front (Figure 23-27a). Rapidly moving cells typically do not have such striking actin bundles. In such cells, the cell **cortex,** which lies immediately beneath the plasma membrane and is enriched in actin, is crosslinked into a gel or very loosely organized lattice of microfilaments (Figure 23-27b). At the leading edge, and especially in filopodia, microfilaments form highly oriented, polarized cables, with their barbed (plus) ends oriented toward the tip of the protrusion (Figure 23-27c). The actin in lamellipodia is typically less well organized (Figure 23-28). The major events in cell crawling are discussed in the following sections.

Extending Protrusions. Fundamental to the dynamics of protrusions is the phenomenon of **retrograde flow** of F-actin. Retrograde flow has been extensively studied in the crawling appendages of cultured nerve cells known as *growth cones,* and it seems to be a fundamental process in many migrating cells. During normal retrograde flow, there is bulk movement of microfilaments toward the rear of the protrusion as it extends. The filamentous structures disappear gradually when cells are treated with cytochalasin, demonstrating that they are made predominantly of

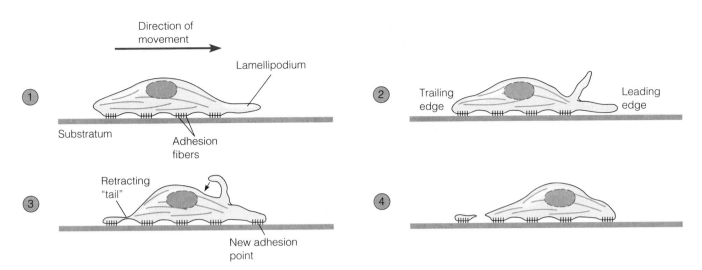

Figure 23-25 **The Steps of Cell Crawling.** Several different processes are involved in cell crawling, including cell protrusion, attachment, and contractile activities. Illustrated schematically here in a macrophage, ① the leading edge of the cell forms protrusions (lamellipodia) that extend in the direction of travel. ② These protrusions can adhere to the substrate, or they can be pulled upward and back toward the cell body. ③ When the protrusions adhere, they provide anchorage points for actin filaments. ④ Tension on the actin filaments can then cause the rest of the cell to pull forward. When the cell adheres very tightly to its substrate, a piece of the tail may break off and be left behind as the cell crawls forward.

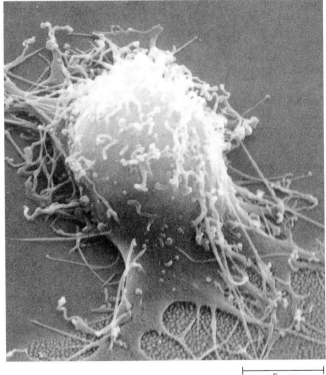

$5 \mu m$

Figure 23-26 **Scanning Electron Micrograph of a Mouse Fibroblast Showing Numerous Filopodia Extending from the Cell Surface.**

actin. Retrograde flow appears to result from two simultaneous processes: *actin assembly* at the tip of the growing lamellipodium or filopodium (see Figure 23-30a), and *rearward translocation* of actin filaments toward the base of the protrusion (see Figure 23-30b). In a typical cell, forward assembly and rearward translocation are thought to balance one another; as one or the other occurs, a protrusion can be extended or retracted.

Actin assembly appears to occur by two major mechanisms. First, subunits can be added to the tips of long, existing filaments; this is likely the predominant mechanism of filament elongation in filopodia. In other cases, new filaments appear to grow as a branched, dendritic network from sites near the leading edge, which involves the recruitment of Arp 2/3 complexes (see Chapter 22).

At the same time that extension of the tip of a protrusion is occurring, the polymerized actin is drawn toward the base of the protrusion, where it is disassembled. Released actin monomers are then available for addition to the plus ends of new or growing microfilaments as the cell continues to crawl forward. The rearward translocation of actin polymers is likely driven by myosins. Nonmuscle myosin II is found to be abundant at the base of lamellipodia in several types of cells (Figure 23-29), and biochemical agents that disrupt myosin motors inhibit retrograde flow.

Cell Attachment. Attachment, or adhesion of the cell to its substrate, is also necessary for cell crawling. New sites of attachment must be formed at the front of a cell, and contacts at the rear must be broken. Attachment sites between the cell and the substrate are complex structures involving the attachment of proteins in the plasma membrane to other proteins both outside and inside the cell. One family of attachment proteins is the *integrins,* which we discussed in Chapter 11. On the outside of the cell, integrins attach to extracellular matrix proteins. Inside the cell, integrins are connected to actin filaments through linker proteins,

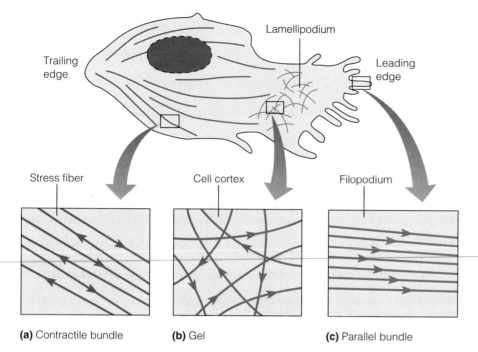

Figure 23-27 The Architecture of Actin in Crawling Cells. Actin is found in a variety of different structures in crawling cells such as this macrophage. **(a)** Running from the trailing edge of the cell to the leading edge are contractile bundles of actin, the stress fibers. **(b)** At the periphery of the cell is the cortex, which contains a three-dimensional meshwork of actin filaments crosslinked into a gel. **(c)** The broad leading edge of lamellipodia can produce thin, fingerlike projections called filopodia. Whereas the bulk of lamellipodia contain an actin meshwork, filopodia contain parallel bundles of actin filaments.

(a) Contractile bundle   **(b)** Gel   **(c)** Parallel bundle

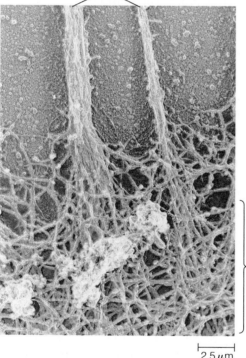

Figure 23-28 Deep-Etch Electron Micrograph Showing Actin Bundles in Filopodia. This view of the periphery of a macrophage shows two prominent actin bundles contained within filopodia that extend from the cell surface. The actin filaments in the filopodia merge with a network of actin filaments lying just beneath the plasma membrane of the lamellipodium.

including *talin, vinculin,* and *α-actinin.* In tightly adhering cells, stress fibers insert at *focal contacts,* sites where integrins are tightly bound to extracellular matrix proteins (see Figure 11-12). Less tightly adhering cells do not have obvious focal contacts, but they do possess similar protein complexes where they attach to the substrate.

How firmly a cell is attached to the underlying substrate probably helps to shift the balance between the processes of retrograde flow and plus-end addition of actin subunits. In this sense, retrograde flow is analogous to a car with its transmission in neutral. Once the car is shifted into a forward gear, it rapidly moves forward. In the same way, firm attachment of the leading edge resists retrograde flow and shifts the balance in favor of protrusion (Figure 23-30).

Translocation and Detachment. Cell crawling coordinates protrusion formation and attachment with forward movement of the entire cell body. Contraction of the rear of the cell squeezes the cell body forward and releases the cell from attachments at its rear. Evidence suggests that contraction is due to interactions between actin and myosins. Nonmuscle myosin II is localized not only at the base of protrusions, but also further toward the rear of the cell. In mutant cells from the cellular slime mold *Dictyostelium* that lack myosin II, the ability of the trailing edge of the cell to retract is reduced, supporting the idea that myosins are involved in contraction.

Contraction of the cell body must be linked to detachment of the trailing edge of the cell. Detachment requires breaking adhesive contacts. Interestingly, contacts at the

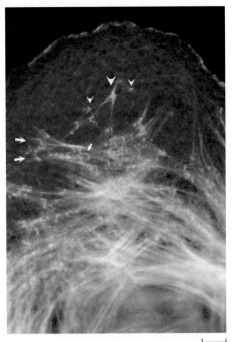

Figure 23-29 Overall Distribution of Myosin II and Actin in a Fibroblast. Close to the leading edge (top), myosin is found in the form of discrete spots (arrowheads) that are not associated with well-defined actin structures. Some spots (large arrowhead) appear where several small actin bundles connect. Farther from the edge, myosin spots align along actin filament bundles (arrows).

rear of the cell are sometimes too tight to be detached, and the tail of the cell actually breaks off as the cell pulls the rear forward (see Figure 23-25). In general, how firmly a cell attaches to a substrate affects how quickly that cell can crawl. Having fewer integrins near the rear of the cell may make detachment of the trailing edge easier; many attachments usually result in poor movement.

Amoeboid Movement Involves Cycles of Gelation and Solation of the Actin Cytoskeleton

Amoebas, slime molds, and leukocytes all exhibit a type of crawling movement referred to as **amoeboid movement** (Figure 23-31). This type of movement is accompanied by protrusions of the cytoplasm called **pseudopodia** (singular: **pseudopodium,** from the Greek for "false foot"). Cells that undergo amoeboid movement have an outer layer of thick, gelatinous cytoplasm called the *ectoplasm* and an inner layer of more fluid cytoplasm called the *endoplasm*. In an amoeba, as a pseudopod is extended from the cell, fluid endoplasm streams forward in the direction of extension and congeals into ectoplasm at the tip of the pseudopod. Meanwhile, at the rear of the moving cell, ectoplasm changes into more fluid endoplasm and streams toward the pseudopod. The alternation between these states of the actin-based cytoskeleton that occurs during such transitions is called **gelation-solation** (Figure

23-31b). Proteins such as gelsolin that are present within these gels may be activated by calcium to convert the gel to the more fluid sol state. Experiments have shown that the forward streaming in the pseudopod does not require squeezing from the rear of the cell: When a pseudopod's cell membrane is removed using detergent, the remaining components can still stream forward if the appropriate mixture of ions and other chemicals is added. Pressure exerted on the endoplasm, possibly due to contraction of an actomyosin network in the trailing edge of the cell, may also squeeze the endoplasm forward, aiding formation of a protrusion at the leading edge.

Cytoplasmic Streaming Moves Components Within the Cytoplasm of Some Cells

Cytoplasmic streaming, an actomyosin-dependent movement of the cytoplasm within a cell, is seen in a variety of organisms that do not display amoeboid movement. In slime molds such as *Physarum polycephalum,* for example, cytoplasm streams back and forth in the branched network of protoplasm that constitutes the cell mass. The flow of cytoplasm reverses direction with predictable periodicity and is therefore called *shuttle streaming.* This periodic streaming seems to provide both nourishment and locomotion, because the streaming process is correlated with the further extension of fingerlike projections by which the slime mold reaches out into its environment in search of nutrients.

Many plant cells display a circular flow of cell contents around a central vacuole. This streaming process, called *cyclosis,* has been studied most extensively in the giant algal cell *Nitella.* In this case, the movement seems to circulate and mix cell contents. Cytoplasmic streaming requires actin filaments, as it is inhibited in cells treated with cytochalasin. In *Nitella,* a dense set of aligned microfilaments are found near sites where cyclosis occurs. Cyclosis probably involves specific myosins that provide the force for movement of components within the cytoplasm. When latex beads coated with various types of myosin are added to *Nitella* cells that have been broken open, the beads move along the actin filaments in an ATP-dependent manner in the same direction as normal organelle movement.

Infectious Microorganisms Can Move Within Cells Using Actin "Tails"

One of the most remarkable findings of modern cell motility research is the discovery that disease-causing microorganisms can co-opt the cell's normal cell adhesion and cell motility systems to penetrate a cell's defenses and enter the cell (see Chapter 11). The best-studied example of such motility is the gram-positive bacterium *Listeria monocytogenes.* One way in which *Listeria* attaches to the host's cells involves the binding of a *Listeria* protein known as *internalin A* to *E-cadherin* on the cell surface. Once bound, *Listeria* enter a cell, move through it at a rate

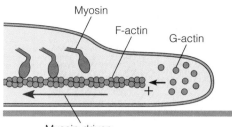

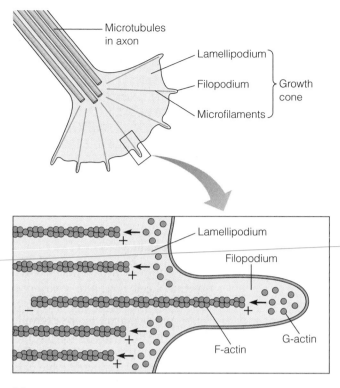

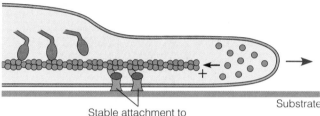

Stationary:
Polymerization and retrograde flow balanced

Myosin F-actin G-actin

Myosin-driven
retrograde flow of actin

Substrate

Moving forward:
Cell attachment resists retrograde flow, resulting in extension

Substrate

Stable attachment to
substrate via integrin

(a)

(b)

Figure 23-30 **Attachment Coupled to Protrusion Formation in a Migrating Nerve Cell.** **(a)** Some nerve cells, such as those from *Aplysia* (a sea slug), have growth cones with a broad lamellipodium and slender filopodia. Microfilaments are oriented with their plus (barbed) ends at the tips of the protrusions. **(b)** A model for how attachment acts as a "clutch" that allows forward movement. In a stationary protrusion, retrograde flow driven by myosin moves microfilaments rearward as new G-actin monomers are added at the tip. When retrograde flow is resisted by attachment to the underlying substrate (i.e., engaging the "clutch"), polymerization results in net elongation of the protrusion and forward movement.

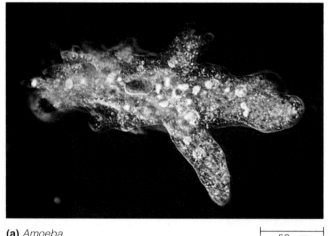

(a) *Amoeba*   50 μm

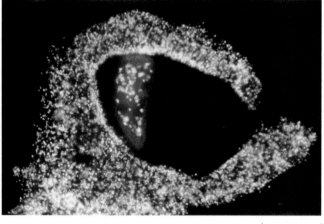

(b) *Amoeba* engulfing prey with pseudopodia   50 μm

Figure 23-31 **Amoeboid Movement.** **(a)** A micrograph of *Amoeba proteus*, a protozoan that moves by extension of pseudopodia. **(b)** This micrograph shows an *Amoeba* cell using its pseudopodia to engulf a smaller ciliated cell on which *Amoeba* feeds by phagocytosis (LMs).

of 11 μm/min, and progress to nearby uninfected cells, where they continue the cycle of infection (Figure 23-32a). Short actin filaments radiate away from the bacteria, forming "comet tails" of branched F-actin (Figure 23-32b). By using fluorescently labeled actin,

investigators have determined that the tails form by polymerization of actin, which is nucleated near the surface of the internalized bacterium. The protein on the surface of *Listeria* that promotes actin polymerization is known as *ActA*. Because the microfilaments nucleated by

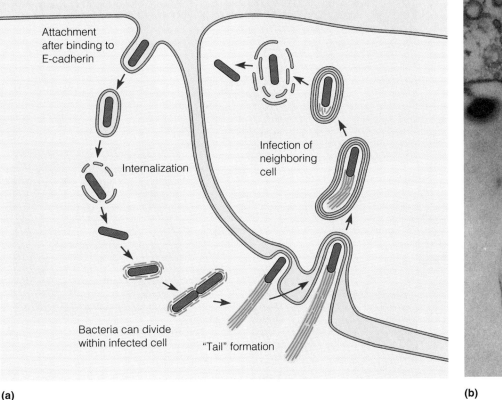

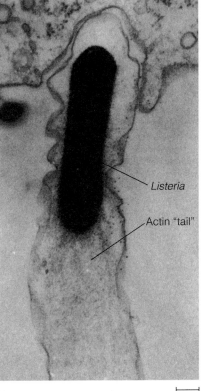

(a)

(b)

$\vdash\!\!\dashv$ 0.1 μm

Listeria

Actin "tail"

Figure 23-32   Infection of a Macrophage by *Listeria monocytogenes*.   **(a)** Life cycle of *Listeria*. A bacterium attaches to the surface of an uninfected cell. The bacterium then moves inside the cell, where it can divide to produce more bacteria in the infected cell, and then spread to a nearby cell by producing a "comet tail" of polymerized actin. **(b)** A transmission electron micrograph showing a *Listeria* within an infected macrophage and the "comet tail" of actin filaments that form behind the bacterium.

Attachment after binding to E-cadherin

Internalization

Infection of neighboring cell

Bacteria can divide within infected cell

"Tail" formation

ActA are strikingly similar to those found at the leading edge of migrating cells, the tails are probably formed using much of the same cellular machinery.

Other pathogens infect host cells in a similar way, using the host's cell-surface proteins as attachment sites and the intracellular machinery of the host cell to move. Some pathogens bind to the cell surface, but are not internalized. For example, the enteropathogenic form of *E. coli*, which causes diarrhea in infants by forming colonies on the surface of intestinal epithelial cells, attaches to the surface of intestinal cells, where it organizes actin-rich "pedestals" that may function like the actin tails induced by *Listeria*.

Rho, Rac, and Cdc42 Regulate the Actin Cytoskeleton

Migrating cells must regulate when and where they form protrusions, assembling actin networks at new sites of protrusion formation and disassembling them where they are no longer needed. One striking case of such regulation is the dramatic change in the cytoskeleton of cells exposed to certain growth factors. For example, in response to stimulation by *platelet-derived growth factor (PDGF)*,

fibroblasts will begin to grow, divide, and form actin-rich membrane extensions that resemble lamellipodia. Other factors, such as *lysophosphatidic acid (LPA)*, induce cells to form stress fibers.

What signals within a cell exposed to growth factors result in such dramatic reorganization of the actin cytoskeleton? Recall from Chapter 22 that the small G proteins *Rho, Rac,* and *Cdc42* regulate the polymerization of actin microfilaments within cells. These proteins are essential for growth factors such as PDGF and LPA to exert their effects. Each of these proteins has profound and different effects on the actin cytoskeleton (Figure 23-33). For example, stimulation of the Rac pathway results in extension of lamellipodia, and inhibition of Rac prevents this normal response to PDGF. Similarly, activation of the Rho pathway results in the formation of focal adhesions and stress fibers, and Rho inactivation prevents the appearance of stress fibers following exposure of fibroblasts to LPA. Finally, activation of Cdc42 results in the formation of filopodia. These results indicate that these small G proteins regulate the formation of different types of protrusions. They do so by regulating the local polymerization of microfilaments, which are then assembled into different types of protrusions.

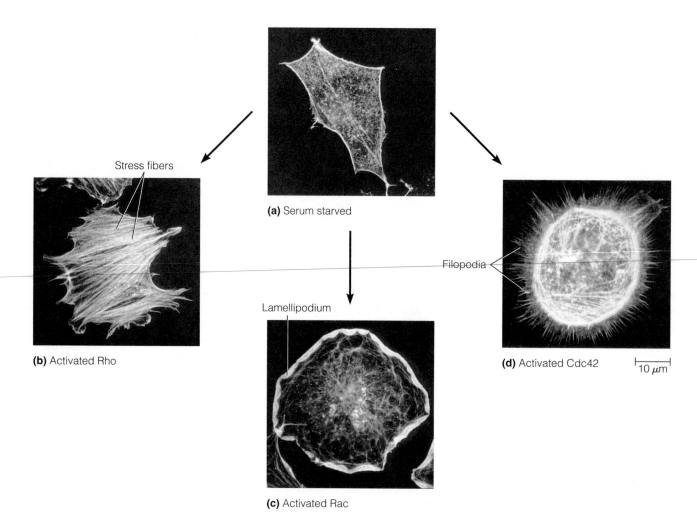

Stress fibers

(a) Serum starved

Lamellipodium

Filopodia

(b) Activated Rho

(c) Activated Rac

(d) Activated Cdc42

10 µm

Figure 23-33   Regulation of Protrusions by Small G Proteins.   (a) When a cultured fibroblast in the absence of growth factors ("serum starved") is stained for actin, it has few actin bundles and shows little protrusive activity. **(b)** Under conditions that activate the Rho signaling pathway (such as addition of lysophosphatidic acid, LPA), stress fibers form. **(c)** When the Rac pathway is activated (in this case by injecting mutated Rac that is always active), lamellipodia form. **(d)** When Cdc42 is activated (e.g., by injecting a guanine nucleotide exchange factor that activates Cdc42), filopodia form.

In addition to their role in regulating protrusions in migratory cells, members of the Rho/Rac/Cdc42 family also affect other actin-mediated processes. For example, a Rac family member is required for cytokinesis in *Dictyostelium*, and Rho and Rac pathways are involved in regulating adhesion and changes in the shape of epithelial cells. The biochemical pathways by which these proteins regulate cell motility is an exciting area of current research.

Chemotaxis Is a Directional Movement in Response to a Graded Chemical Stimulus

A key feature of migrating cells in the body and in embryos is directional migration. One way in which directional migration occurs is through the formation of protrusions predominantly on one side of the cell. In this case, cells must be able to regulate not only whether they form protrusions, but where they form them. Diffusible molecules can act as important cues for such directional migration. When a migrating cell moves toward a greater or lesser concentration of a diffusible chemical, the response is known as **chemotaxis.** The molecule(s) that elicit this response are called *chemoattractants* (when a cell moves toward higher concentrations of the molecule) or *chemorepellants* (when a cell moves away from higher concentrations of the molecule). Chemotaxis has been studied most intensively in white blood cells and in the *Dictyostelium* amoeba. In both cases, increasing the local concentration of a chemoattractant (small peptides in the case of white blood cells and cyclic AMP in the case of *Dictyostelium*) results in dramatic changes in the actin cytoskeleton, including biochemical changes in actin-binding proteins, and migration of the cell toward the source of chemoattractant. These changes occur through local activation of chemoattractant (or repellant) receptors on the cell surface. The receptors in turn stimulate the cell's cytoskeletal machinery to form a protrusion in the direction of migration.

Perspective

Motility is a major theme in cell biology. Our knowledge of the mechanisms underlying eukaryotic motility has increased considerably over the past decade, and we now understand that it is driven at the molecular level by a set of ATP-dependent motor molecules. These molecules use portions of the cytoskeleton as a kind of track to pull subcellular components into position.

Two major eukaryotic motility systems are known, one based on the interaction of dynein or kinesin with microtubules and the other on the interaction of myosin with actin microfilaments. Kinesins and cytoplasmic dynein are motor molecules that move intracellular structures in opposite directions along MT tracks. The axoneme present in both cilia and flagella is a highly specialized example of dynein-tubulin interaction. The nine outer doublets of the axoneme are connected laterally to one another and radially to the central pair of single microtubules. Dynein arms project out from one MT doublet to the next, and are involved in the sliding of one set of microtubules past the next. This sliding is opposed by the radial spokes between the doublets and the central pair of tubules and by the connections between adjacent doublets. As a result, the sliding is converted to a bending motion.

Actin and myosin are found widely distributed in nonmuscle cells, where they are involved in a variety of motility mechanisms, including cell crawling, amoeboid movement, cytoplasmic streaming, and cytokinesis. Skeletal muscle contraction is a specialization of this more general motility process, and it is the best-understood example of cell motility. Muscle contraction involves a progressive sliding of thin actin filaments past thick myosin filaments, driven by the interaction between the ATPase head of the myosin molecules and successive myosin-binding sites on the actin filaments. Contraction is triggered by the release of calcium from the sarcoplasmic reticulum and ceases again as the calcium is actively pumped back into the SR. In skeletal muscle, calcium binds to troponin and causes a conformational change in tropomyosin, which opens myosin-binding sites on the thin filament. In smooth muscle, the effect of calcium is mediated by calmodulin, which activates myosin light-chain kinase, leading to the phosphorylation of myosin.

The crawling of cells using lamellipodia is a striking example of actin-based motility. In crawling, polymerization of actin extends cellular protrusions; attachment of the protrusions to the substrate and contraction of the cell drive forward movement. Specific intracellular signals involving small G proteins regulate this process; particular G proteins are involved in the production of specific types of protrusions. Some disease-causing bacteria use the components of the protrusion-forming machinery to move about within or on top of infected host cells.

Other types of actin-dependent movements include amoeboid movement, in which cycles of gelation and solation occur as pseudopodia are extended by amoeboid cells. Movements of particles within cells also depend on actin. These include cyclosis in plant and algal cells, and cytoplasmic streaming.

Key Terms for Self-Testing

motility (p. 769)
contractility (p. 769)

Motile Systems
motor protein (mechanoenzyme) (p. 770)
microtubule-based movement (p. 770)
microfilament-based movement (p. 770)

Intracellular Microtubule-Based Movement: Kinesin and Dynein
microtubule-associated motor protein (motor MAP) (p. 770)
kinesin (p. 770)
dynein (p. 770)
fast axonal transport (p. 770)
kinesin family member (KIF) (p. 772)
cytoplasmic dynein (p. 772)
dynactin (p. 772)
axonemal dynein (p. 772)

Microtubule-Based Motility
cilium (p. 773)
flagellum (p. 774)
axoneme (p. 774)

basal body (p. 775)
outer doublet (p. 776)
central pair (p. 776)
A tubule (p. 776)
B tubule (p. 776)
tektin (p. 776)
sidearm (p. 776)
interdoublet link (p. 776)
radial spoke (p. 776)
nexin (p. 776)
sliding-microtubule model (p. 776)
intraflagellar transport (p. 777)

Actin-Based Cell Movement: The Myosins
myosin (p. 777)
type II myosin (p. 778)

Filament-Based Movement in Muscle
muscle contraction (p. 779)
skeletal muscle (p. 779)
muscle fiber (p. 779)
myofibril (p. 779)
sarcomere (p. 779)
thick filament (p. 779)

thin filament (p. 779)
striated muscle (p. 779)
A band (p. 780)
I band (p. 780)
H zone (p. 780)
M line (p. 780)
Z line (p. 780)
tropomyosin (p. 781)
troponin (p. 781)
TnT (p. 781)
TnC (p. 781)
TnI (p. 781)
sliding-filament model (p. 782)
cross-bridge (p. 783)
rigor (p. 784)
sarcoplasm (p. 786)
neuromuscular junction (p. 787)
sarcolemma (p. 787)
transverse (T) tubule system (p. 787)
sarcoplasmic reticulum (SR) (p. 787)
medial element (p. 787)
terminal cisterna (p. 787)
triad (p. 788)

junctional complex (p. 788)
calcium pump (calcium ATPase) (p. 788)
myoglobin (p. 789)
creatine phosphate (p. 789)
cardiac (heart) muscle (p. 789)
intercalated disc (p. 789)
smooth muscle (p. 790)

myosin light-chain kinase (MLCK) (p. 790)
regulatory light chain (p. 790)

Actin-Based Motility in Nonmuscle Cells
lamellipodium (p. 792)
filopodium (p. 792)
cortex (p. 792)

retrograde flow (p. 792)
amoeboid movement (p. 795)
pseudopodium (p. 795)
gelation-solation (p. 795)
cytoplasmic streaming (p. 795)
chemotaxis (p. 798)

Problem Set

More challenging problems are marked with a •.

• **23-1. Divide and Conquer.** Kinesin-related proteins known as the *CHO1/MKLP1 kinesin-related proteins* are involved in sliding microtubules in an antiparallel fashion (in other words, these proteins slide aligned MTs whose plus ends are in opposite orientations). Your friend informs you that these KIFs play a role in mitosis. What stage of mitosis would you predict they would be involved in and why?

23-2. Kartegener's Triad. Sterility in human males with Kartegener's triad is due to nonmotile sperm. Upon cytological examination, the sperm of such individuals are found to have tails (i.e., flagella) that lack one or more of the normal structural components. Such individuals are also likely to have histories of respiratory tract disease, especially recurrent bronchitis and sinusitis, caused by an inability to clear mucus from the lungs and sinuses.

(a) What is a likely mechanistic explanation for nonmotility of sperm in such cases of sterility?

(b) Why is respiratory tract disease linked with sterility in affected individuals?

23-3. A Moving Experience. For each of the following statements, indicate whether it is true of the motility system that you use to lift your arm (A), to cause your heart to beat (H), to move ingested food through your intestine (I), or to sweep mucus and debris out of your respiratory tract (R). More than one response may be appropriate in some cases.

(a) It depends on muscles that have a striated appearance when examined with an electron microscope.

(b) It would probably be affected by the same drugs that inhibit motility of a flagellated protozoan.

(c) It requires ATP.

(d) It involves calmodulin-mediated calcium signaling.

(e) It involves interaction between actin and myosin filaments.

(f) It depends heavily on fatty acid oxidation for energy.

(g) It is under the control of the voluntary nervous system.

23-4. Muscle Structure. Frog skeletal muscle consists of thick filaments that are about 1.6 μm long and thin filaments about 1 μm long.

(a) What is the length of the A band and the I band in a muscle with a sarcomere length of 3.2 μm? Describe what happens to the length of both bands as the sarcomere length decreases during contraction from 3.2 to 2.0 μm.

(b) The H zone is a specific portion of the A band. If the H zone of each A band decreases in length from 1.2 to 0 μm

as the sarcomere length contracts from 3.2 to 2.0 μm, what can you deduce about the physical meaning of the H zone?

(c) What can you say about the distance from the Z line to the edge of the H zone during contraction?

23-5. Rigor Mortis and the Contraction Cycle. At death, the muscles of the body become very stiff and inextensible, and the corpse is said to go into *rigor*.

(a) Explain the basis of rigor. Where in the contraction cycle is the muscle arrested? Why?

(b) Would you be likely to go into rigor faster if you were to die while racing to class or while sitting in lecture? Explain.

(c) What effect do you think the addition of ATP might have on muscles in rigor?

• **23-6. AMPPCP and the Contraction Cycle.** AMPPCP is the abbreviation for a structural analogue of ATP in which the third phosphate group is linked to the second by a CH_2 group instead of an oxygen atom. AMPPCP binds to the ATP-binding site of virtually all ATPases, including myosin. It differs from ATP, however, in that its terminal phosphate cannot be removed by hydrolysis. When isolated myofibrils are placed in a flask containing a solution of calcium ions and AMPPCP, contraction is quickly arrested.

(a) Where in the contraction cycle will contraction be arrested by AMPPCP? Draw the arrangement of the thin filament, the thick filament, and a cross-bridge in the arrested configuration.

(b) Do you think contraction would resume if ATP were added to the flask containing the AMPPCP-arrested myofibrils? Explain.

(c) What other processes in a muscle cell do you think are likely to be inhibited by AMPPCP?

23-7. Pulled in Two Directions. In skeletal muscle sarcomeres, the H zone is in the middle and bounded on each side by a Z line. During contraction, the Z lines on either side move in opposite directions toward the H zone. Myosin, however, can only crawl along an actin filament in one direction. How can you reconcile movements of Z lines in opposite directions with the unidirectional movement of myosin along an actin filament?

23-8. Seeing Spots. In a microscopy technique known as *fluorescence photoactivation*, monomers (such as G-actin) labeled with "caged" fluorescein covalently modified so that it cannot fluoresce are injected into a cell. The injected monomers assemble into cytoskeletal elements, and then a local spot can be irradiated with an ultraviolet microbeam to "uncage" the fluorescein, resulting in a fluorescent spot of polymer whose movements can be followed using computer-assisted microscopy. Assume only assembled polymers generate enough signal to see using a fluorescence microscope. If a migrating fibroblast were injected with caged, fluorescent actin, describe the overall movement of actin if the labeled actin were uncaged in the following locations and under the following conditions.

(a) At the leading edge in a migrating fibroblast that has been detached from the substratum

(b) At the leading edge of a migrating fibroblast immediately after treatment with cytochalasin D

(c) At the leading edge of a migrating fibroblast migrating on a fibronectin-rich substrate

23-9.   Motility Potpourri. Answer each of the following questions as concisely as possible.

(a) *Melanophores* are cells of teleost fish that contain pigment granules used to control coloration. When the granules are aggregated, the cell appears light; when the pigment granules are dispersed, the cell appears dark. The movement of the granules is under hormonal control. How could you tell if the distribution and movement of granules is an example of microfilament-based or microtubule-based motility?

(b) Stress fibers of nonmuscle cells contain contractile bundles of actin and myosin II. For stress fibers to contract or

develop tension, how would actin and myosin have to be oriented within the stress fibers?

(c) The beating movements of sperm flagella depend on precise control of where and when dynein ATPase molecules exert force on microtubules. Explain why, and indicate what would happen if all the dynein molecules were to exert force at the same time.

(d) Cultured cells infected with *Listeria* are treated with latrunculin. Describe the effects of latrunculin treatment on the rate of spreading of the *Listeria* infection.

23-10.   Active, Like It or Not. It is possible to make genetically engineered forms of small G proteins that are always active ("constitutively active") or that act as "dominant negatives." The effect of these latter forms of small G proteins is to permanently deactivate the signaling pathway associated with that small G protein.

(a) Explain what happens to a resting fibroblast in normal serum when a dominant negative Rho is introduced into it.

(b) Explain what happens to a similar fibroblast when treated with platelet-derived growth factor (PDGF).

(c) Explain what happens to a fibroblast into which a constitutively activated Rho is introduced when it is treated with lysophosphatidic acid (LPA).

Suggested Reading

References of historical importance are marked with a •.

General References

Bray, D. *Cell Movements: From Molecules to Motility,* 2nd ed. New York: Garland, 2001.

Howard, J. Molecular motors: Structural adaptations to cellular functions. *Nature* 389 (1997): 561.

•Huxley, A. F. *Reflections on Muscle.* Princeton, NJ: Princeton University Press, 1980.

Kreis, T., and R. Vale. *Guidebook to the Cytoskeletal and Motor Proteins,* 2nd ed. New York: Oxford University Press, 1999.

Lauffenburger, D. A., and A. F. Horwitz. Cell migration: A physically integrated molecular process. *Cell* 84 (1996): 359.

•Maruyama, K. Birth of the sliding filament concept in muscle contraction. *J. Biochem.* 117 (1995): 1.

Stossel, T. P. The machinery of cell crawling. *Sci. Amer.* 271 (September 1994): 54–63.

Vale, R. D. Getting a grip on myosin. *Cell* 78 (1994): 733.

Microtubule-Based Motility

Asai, D. J., and M. P. Koonce. The dynein heavy chain: Structure, mechanics and evolution. *Trends Cell Biol.* 11 (2001): 196.

Block, S. M. Leading the procession: New insights into kinesin motors. *J. Cell Biol.* 140 (1998): 1281.

•Brokaw, C. J., D. J. L. Luck, and B. Huang. Analysis of the movement of *Chlamydomonas* flagella: The function of the radial-spoke system is revealed by comparison of wild-type and mutant flagella. *J. Cell Biol.* 92 (1982): 722.

Cross, R. A. Molecular motors: The natural economy of kinesin. *Curr. Biol.* 7 (1997): R631.

•Gibbons, I. R. Cilia and flagella of eukaryotes. *J. Cell Biol.* 91 (1981): 107.

•Goodenough, U. W., and J. E. Heuser. Substructure of the outer dynein arm. *J. Cell Biol.* 95 (1982): 795.

•Grigg, G. Discovery of the 9 + 2 subfibrillar structure of flagella/cilia. *BioEssays* 13 (1991): 363.

Hirokawa, N. Kinesin and dynein superfamily proteins and the mechanism of organelle transport. *Science* (1998) 279: 519.

Karki, S., and E. L. Holzbaur. Cytoplasmic dynein and dynactin in cell division and intracellular transport. *Curr. Opin. Cell Biol.* 11 (1999): 45.

Lippincott-Schwartz, J. Cytoskeletal proteins and Golgi dynamics. *Curr. Opin. Cell Biol.* 10 (1998): 52.

Omoto, C. K. Mechanochemical coupling in cilia. *Internat. Rev. Cytol.* 131 (1991): 255.

Rosenbaum, J. L., D. G. Cole, and D. R. Diener. Intraflagellar transport: The eyes have it. *J. Cell Biol.* 144 (1999): 385.

•Satir, P. How cilia move. *Sci. Amer.* 231 (October 1974): 44.

Witman, G. B. Axonemal dyneins. *Curr. Opin. Cell Biol.* 4 (1992): 74.

Motors and Human Disorders

Mermall, V., P. L. Post, and M. S. Mooseker. Unconventional myosins in cell movement, membrane traffic, and signal transduction. *Science* 279 (1998): 527.

Vogan, K. J., and C. J. Tabin. A new spin on handed asymmetry. *Nature* 397 (1999): 295.

Filament-Based Movement in Muscle

Ashcroft, F. M. Ca^{2+} channels and excitation-contraction coupling. *Curr. Opin. Cell Biol.* 3 (1991): 671.

•Franzini-Armstrong, C., and L. D. Peachey. Striated muscles: Contractile and control mechanisms. *J. Cell Biol.* 91 (1981): 166.

Geeves, M. A., and K. C. Holmes. Structural mechanism of muscle contraction. *Annu. Rev. Biochem.* 68 (1999): 687.

Gulick, A. M., and I. Rayment. Structural studies on myosin II: Communication between distant protein domains. *BioEssays* 9 (1997): 561.

•Huxley, H. E. The mechanism of muscular contraction. *Science* 164 (1969): 1356.

Reedy, M. C. Visualizing myosin's power stroke in muscle contraction. *J. Cell Sci.* 113 (2000): 3551.

Squire, J. M. Architecture and function in the muscle sarcomere. *Curr. Opin. Struct. Biol.* 7 (1997): 247.

•Warrick, H. M., and J. A. Spudich. Myosin structure and function in cell motility. *Annu. Rev. Cell Biol.* 3 (1987): 379.

Nonmuscle Microfilament-Based Movement

Baker, J. P., and M. A. Titus. Myosins: Matching functions with motors. *Curr. Opin. Cell Biol.* 10 (1998): 80.

Goosney, D. L., M. de Ghrado, and B. B. Finlay. Putting *E. coli* on a pedestal: A unique system to study signal transduction and the actin cytoskeleton. *Trends Cell Biol.* 9 (1999): 11.

Grebecki, A. Membrane and cytoskeleton flow in motile cells with emphasis on the contribution of free living amoeba. *Internat. Rev. Cytol.* 148 (1994): 37.

Heidemann S. R., and R. E. Buxbaum. Cell crawling: First the motor, now the transmission. *J. Cell Biol.* 141 (1998): 1.

Kemp, B. E., R. B. Pearson, C. House, P. J. Robinson, and A. R. Means. Regulation of protein kinases by pseudosubstrate prototypes. *Cell Signal* 1 (1989): 303.

Kohama, K., L. H. Ye, K. Hayakawa, and T. Okagaki. Myosin light chain kinase: An actin-binding protein that regulates an ATP-dependent interaction with myosin. *Trends Pharmacol. Sci.* 17 (1996): 284.

• Lin, C. H., E. M. Espreafico, M. S. Mooseker, and P. Forscher. Myosin drives retrograde F-actin flow in neuronal growth cones. *Neuron* 16 (1996): 769.

Means, A. R., M. F. VanBerkum, I. Bagchi, K. P. Lu, and C. D. Rasmussen. Regulatory functions of calmodulin. *Pharmacol. Ther.* 50 (1991): 255.

Pantaloni, D., C. Le Clainche, and M. F. Carlier. Mechanism of actin-based motility. *Science* 292 (2001): 1502.

• Taylor, D. L., and J. S. Condeelis. Cytoplasmic structure and contractility in amoeboid cells. *Internat. Rev. Cytol.* 56 (1979): 57.

• Tilney, L. G., A. D. Portnoy, *et al.* Actin filaments and the growth, movement, and spread of the intracellular bacterial parasite, *Listeria monocytogenes*. *J. Cell Biol.* 109 (1989): 1597.

Welch, M. D., A. Mallavarapu, J. Rosenblatt, and T. J. Mitchison. Actin dynamics in vivo. *Curr. Opin. Cell Biol.* 9 (1997): 54.

Control of Actin-Based Motility

Bear, J. E., M. Krause, and F. B. Gertler. Regulating cellular actin assembly. *Curr. Opin. Cell Biol.* 13 (2001): 158.

• Devreotes, P. N., and S. H. Zigmond. Chemotaxis in eukaryotic cells: A focus on leukocytes and *Dictyostelium*. *Annu. Rev. Cell Biol.* 4 (1988): 649.

Hall, A. Rho GTPases and the actin cytoskeleton. *Science* 279 (1998): 509.

Hall, A., and C. D. Nobes. Rho GTPases: Molecular switches that control the organization and dynamics of the actin cytoskeleton. *Phil. Trans. R. Soc. London Biol. Sci.* 355 (2000): 965.

Glossary

Note: The letters "GM" preceding a page number refer to a page in the *Guide to Microscopy* supplement.

A

A: see *adenine.*

A band: region of a striated muscle myofibril that appears as a dark band when viewed by microscopy; contains thick myosin filaments and those regions of the thin actin filaments that overlap the thick filaments. (p. 780)

A site (aminoacyl site): site on the ribosome that binds each newly arriving tRNA with its attached amino acid. (p. 661)

A tubule: a complete microtubule that is fused to an incomplete microtubule (the B tubule) to make up an outer doublet in the axoneme of a eukaryotic cilium or flagellum. (p. 776)

ABC-type ATPase: type of transport ATPase characterized by an "ATP-binding cassette" (hence the "ABC"), with the term "cassette" used to describe catalytic domains of the protein that bind ATP as an integral part of the transport process; also called ABC transporters. (p. 209) Also see *multidrug resistance transport protein.*

absolute refractory period: brief time during which the sodium channels of a nerve cell are inactivated and cannot be opened by depolarization. (p. 239)

absorption spectrum: relative extent to which light of different wavelengths is absorbed by a pigment. (p. 452)

accelerating voltage: difference in voltage between the cathode and anode of an electron microscope, responsible for accelerating electrons prior to their emission from the electron gun. (p. GM-18)

accessory pigments: molecules such as carotenoids and phycobilins that confer enhanced light-gathering properties on photosynthetic tissue by absorbing light of wavelengths not absorbed by chlorophyll; accessory pigments give distinctive colors to plant tissue,

depending on their specific absorption properties. (p. 453)

acetyl CoA: high-energy two-carbon compound generated by glycolysis and fatty acid oxidation; employed for transferring carbon atoms to the tricarboxylic acid cycle. (p. 406)

acetylcholine: the most common excitatory neurotransmitter used at synapses between neurons outside the central nervous system. (p. 244)

actin: principal protein of the microfilaments found in the cytoskeleton of nonmuscle cells and in the thin filaments of skeletal muscle; synthesized as a globular monomer (G-actin) that polymerizes into long, linear filaments (F-actin). (p. 755)

actin-binding proteins: Proteins that bind to actin microfilaments, thereby regulating the length or assembly of microfilaments or mediating their association with each other or with other cellular structures, such as the plasma membrane. (p. 758)

actin-related protein (Arp): protein whose amino acid sequence is related to that of actin and that may be involved in nucleating assembly of new microfilaments in migrating cells. (p. 756)

action potential: brief change in membrane potential involving an initial depolarization followed by a rapid return to the normal resting potential; caused by the inward movement of Na^+ followed by the subsequent outward movement of K^+; serves as the means of transmission of a nerve impulse. (p. 236)

action spectrum: relative extent to which light of different wavelengths affects a particular light-dependent reaction or process. (p. 454)

activated monomer: a monomer whose free energy has been increased by being linked to a carrier molecule. (p. 30)

activation domain: region of a transcription factor, distinct from the DNA-binding domain, that is responsible for activating transcription. (p. 721)

activation energy (E_A): energy required to initiate a chemical reaction. (p. 131)

activator (transcription): regulatory transcription factor whose binding to DNA control elements leads to an increase in the transcription rate of nearby genes. (p. 719)

active site: region of an enzyme molecule at which the substrate binds and the catalytic event occurs; also called the catalytic site. (p. 133)

active transport: membrane protein-mediated movement of a substance across a membrane against a concentration or electrochemical gradient; an energy-requiring process. (p. 207)

active zone: region of the presynaptic membrane of an axon where neurosecretory vesicles dock. (p. 247)

adaptive enzyme synthesis: regulation of the intracellular concentration of an enzyme by modulating synthesis of that enzyme in response to cellular needs. (p. 692)

adaptor protein (AP): protein found along with clathrin in the coats of clathrin-coated vesicles. (p. 349)

adenine (A): nitrogen-containing aromatic base, chemically designated as a purine, that serves as an informational monomeric unit when present in nucleic acids with other bases in a specific sequence; forms a complementary base pair with thymine (T) or uracil (U) by hydrogen bonding. (p. 55)

adenosine diphosphate (ADP): adenosine with two phosphates linked to each other by a phosphoanhydride bond and to the 5′ carbon of the ribose by a phosphoester bond. (p. 56)

adenosine monophosphate (AMP): adenosine with a phosphate linked to the 5′ carbon of ribose by a phosphoester bond. (p. 56)

adenosine triphosphate (ATP): adenosine with three phosphates linked to each other by phosphoanhydride bonds and to the 5′ carbon of the ribose by a phosphoester bond; principal energy storage compound of most cells, with energy stored in the high-energy phosphoanhydride bonds. (pp. 56, 369)

adenylyl cyclase: enzyme that catalyzes the formation of cyclic AMP from ATP; located on the inner surface of the plasma membrane of many eukaryotic cells and activated by specific ligand-receptor interactions on the outer surface of the membrane. (p. 261)

adherens junction: junction for cell-cell adhesion that is connected to the cytoskeleton by actin microfilaments. (p. 307)

adhesive (anchoring) junction: type of cell junction that links the cytoskeleton of one cell either to the cytoskeleton of neighboring cells or to the extracellular matrix; examples include desmosomes, hemidesmosomes, and adherens junctions. (pp. 98, 306)

ADP: see *adenosine diphosphate*.

ADP ribosylation factor (ARF): a protein associated with COPI (coatomer) in the "fuzzy" coats of COPI-coated vesicles. (p. 351)

adrenergic hormone: epinephrine or norepinephrine. (p. 279)

adrenergic receptor: any of a family of G protein-linked receptors that bind to one or both of the adrenergic hormones, epinephrine and norepinephrine. (p. 279)

adrenergic synapse: a synapse that uses norepinephrine or epinephrine as the neurotransmitter. (p. 244)

aerobic respiration: exergonic process by which cells oxidize glucose to carbon dioxide and water using oxygen as the ultimate electron acceptor, with a significant portion of the released energy conserved as ATP. (pp. 375, 398)

alcoholic fermentation: anaerobic catabolism of carbohydrates with ethanol and carbon dioxide as the end products. (p. 385)

allele: one of two or more alternative forms of a gene. (p. 578)

allergen: a foreign substance that induces the body to make an antibody protein of the class IgE. (p. 280)

allosteric activator: a small molecule whose binding to an enzyme's allosteric site shifts the equilibrium to favor the high-affinity state of the enzyme. (p. 148)

allosteric effector: small molecule that causes a change in the state of an allosteric protein by binding to a site other than the active site. (pp. 148, 694)

allosteric enzyme: an enzyme exhibiting two alternative forms, each with a different biological property; interconversion of the two states is mediated by the reversible binding of a specific small molecule (allosteric effector) to a regulatory site called the allosteric site. (p. 148)

allosteric inhibitor: a small molecule whose binding to an enzyme's allosteric site shifts the equilibrium to favor the low-affinity state of the enzyme. (p. 148)

allosteric regulation: control of a reaction pathway by the effector-mediated reversible interconversion of the two forms of an allosteric enzyme. (pp. 148, 390, 410)

allosteric (regulatory) site: region of a protein molecule that is distinct from the active site at which the catalytic event occurs and that binds selectively to a small molecule, thereby regulating the protein's activity. (p. 148)

alpha beta heterodimer (αβ-heterodimer): protein dimer composed of one α-tubulin molecule and one β-tubulin molecule that forms the basic building block of microtubules. (p. 746)

alpha helix (α helix): spiral-shaped secondary structure of protein molecules, consisting of a backbone of peptide bonds with R groups of amino acids jutting out. (p. 48)

alpha tubulin (α-tubulin): protein that joins with β-tubulin to form a heterodimer that is the basic building block of microtubules. (p. 746)

alternating conformation model: membrane transport model in which a carrier protein alternates between two conformational states, such that the solute-binding site of the protein is open or accessible first to one side of the membrane and then to the other. (p. 203)

alternation of generations: occurrence of alternating haploid and diploid multicellular forms within the life cycle of an organism. (p. 581)

alternative RNA splicing: utilization of different combinations of intron/exon splice junctions in pre-mRNA to produce messenger RNAs that differ in exon composition, thereby allowing production of more than one type of polypeptide from the same gene. (p. 728)

amino acid: monomeric unit of proteins, consisting of a carboxylic acid with an amino group and one of a variety of R groups attached to the α carbon; 20 different kinds of amino acids are normally found in proteins. (p. 41)

amino terminus: see *N-terminus*.

aminoacyl site: see *A site*.

aminoacyl tRNA: a tRNA molecule containing an amino acid attached to its 3′ end. (p. 664)

aminoacyl-tRNA synthetase: enzyme that joins an amino acid to its appropriate tRNA molecule using energy provided by the hydrolysis of ATP. (p. 664)

amoeboid movement: mode of cell locomotion that depends on pseudopodia and involves cycles of gelation and solation of the actin cytoskeleton. (p. 795)

AMP: see *adenosine monophosphate*.

amphibolic pathway: series of reactions that can function both in a catabolic mode and as a source of precursors for anabolic pathways. (p. 414)

amphipathic molecule: molecule having spatially separated hydrophilic and hydrophobic regions. (p. 24)

amylopectin: branched-chain form of starch consisting of glucose repeating subunits linked together by α(1 → 4) glycosidic bonds, with occasional α(1 → 6) linkages creating branches every 12 to 25 units that commonly consist of 20 to 25 glucose units. (p. 63)

amylose: straight-chain form of starch consisting of glucose repeating units linked together by α(1 → 4) glycosidic bonds. (p. 63)

anabolic pathway: series of reactions that results in the synthesis of cellular components. (p. 369)

anaerobic respiration: cellular respiration in which the ultimate electron acceptor is a molecule other than oxygen. (p. 398)

anaphase: stage during mitosis (or meiosis) when the sister chromatids (or homologous chromosomes) separate and move to opposite spindle poles. (p. 545)

anaphase A: movement of sister chromatids toward opposite spindle poles during anaphase. (p. 548)

anaphase B: movement of the spindle poles away from each other during anaphase. (p. 548)

anaphase-promoting complex: large multiprotein complex that targets selected proteins (e.g., securin and mitotic cyclin) for degradation, thereby initiating anaphase and the subsequent completion of mitosis. (p. 561)

anchorage-dependent growth: requirement that cells be attached to a solid surface such as the extracellular matrix before they can grow and divide. (p. 301).

anchoring junction: see *adhesive junction*.

aneuploid: possessing an abnormal number of chromosomes. (p. 570)

angiogenesis: process by which new blood vessels are formed. (p. 571)

angular aperture: half-angle of the cone of light entering the objective lens of a microscope from the specimen. (p. GM-3)

anion exchange protein: antiport carrier protein that facilitates the reciprocal exchange of chloride and bicarbonate ions across the plasma membrane. (p. 206)

annulus: ring of cytoplasm between the desmotubule and the membrane that lines a plasmodesma; thought to provide cytoplasmic continuity between adjacent plant cells. (p. 319)

anoxygenic phototroph: a photosynthetic organism that uses an oxidizable substrate other than water as the electron donor in photosynthetic electron transduction. (p. 447)

antenna pigment: light-absorbing molecule of a photosystem that absorbs photons and passes the energy to a neighboring chlorophyll molecule or accessory pigment by resonance energy transfer. (p. 453)

anterograde transport: movement of material from the ER through the Golgi complex toward the plasma membrane. (p. 335)

anticodon: triplet of nucleotides located in one of the loops of a tRNA molecule that recognizes the appropriate codon in mRNA by complementary base pairing. (p. 664)

antiport: coupled transport of two solutes across a membrane in opposite directions. (p. 204)

AP: see *adaptor protein*.

apoptosis: cell suicide mediated by a group of protein-degrading enzymes called caspases; involves a programmed series of events that leads to the dismantling of the internal contents of the cell. (p. 282)

apoptosome: multiprotein complex involved in triggering apoptosis; assembled from cytochrome *c*, *Apaf-1*, and procaspase-9. (p. 285)

apoptotic bodies: cell fragments produced as a cell is dismantled by the process of apoptosis. (p. 282)

AQP: see *aquaporin*.

aquaporin (AQP): any of a family of membrane channel proteins that facilitate the rapid movement of water molecules into or out of cells in tissues that require this capability, such as the proximal tubules of the kidneys. (p. 207)

archaebacterium (plural, archaebacteria): one of the two main groups of prokaryotes, the other being eubacteria; many thrive under harsh conditions, such as salty, acidic, and hot environments, that would be fatal to most other organisms; also called archaea. (p. 76)

ARF: see *ADP ribosylation factor*.

Arp: see *actin-related protein*.

ascus: small sac enclosing the cells produced by meiosis in fungi such as *Neurospora*. (p. 604)

asexual reproduction: form of reproduction in which a single parent is the only contributor of genetic information to the new organism. (p. 577)

assisted self-assembly: folding and assembly of proteins and protein-containing structures in which the appropriate molecular chaperone is required to ensure that correct assembly will predominate over incorrect assembly. (p. 33)

astral microtubule: type of microtubule that forms asters, which are dense starbursts of microtubules that radiate in all directions from each spindle pole. (p. 545)

asymmetric carbon atom: carbon atom that has four different substituents. Two different stereoisomers are possible for each asymmetric carbon atom in an organic molecule. (p. 20)

ATP: see *adenosine triphosphate.*

ATP synthase: alternative name for an F-type ATPase when it catalyzes the reverse process in which the exergonic flow of protons down their electrochemical gradient is used to drive ATP synthesis; examples include the CF_oCF_1 complex found in chloroplast thylakoid membranes, and the F_oF_1 complex found in mitochondrial inner membranes and bacterial plasma membranes. (pp. 209, 431, 459)

attenuation: mechanism for regulating bacterial gene expression based on the premature termination of transcription. (p. 702)

autonomic nervous system: portion of the nervous system that controls involuntary activities. (p. 225)

autophagic lysosome: mature lysosome containing hydrolytic enzymes involved in the digestion of materials of intracellular origin. (p. 355) Also see *heterophagic lysosome.*

autophagic vacuole: vacuole formed when an old or unwanted organelle or other cellular structure is wrapped in membranes derived from the endoplasmic reticulum prior to digestion by lysosomal enzymes. (p. 356)

autophagy: intracellular digestion of old or unwanted organelles or other cell structures as it occurs within autophagic lysosomes; "self-eating." (p. 356)

autophosphorylation: phosphorylation of a receptor molecule by a receptor molecule of the same type. (p. 271)

autoradiography: procedure for detecting the location of radioactive molecules by overlaying a sample with photographic film, which becomes darkened upon exposure to radioactivity. (p. 494).

axon: extension of nerve cell that conducts impulses away from the cell body. (p. 227)

axon hillock: region at the base of an axon where action potentials are initiated most easily. (p. 240)

axonal transport: see *fast axonal transport.*

axonemal dynein: motor protein in the axonemes of cilia and flagella that generates axonemal motility by moving along the surface of microtubules driven by energy derived from ATP hydrolysis. (p. 772)

axonemal microtubules: microtubules present in highly ordered bundles in the axonemes of eukaryotic cilia and flagella. (p. 744)

axoneme: group of interconnected microtubules that form the backbone of a eukaryotic cilium or flagellum, usually arranged as nine outer doublet microtubules surrounding a pair of central microtubules. (p. 774)

axoplasm: cytoplasm within the axon of a nerve cell. (p. 227)

B

B tubule: an incomplete microtubule that is fused to a complete microtubule (the A tubule) to make up an outer doublet in the axoneme of a eukaryotic cilium or flagellum. (p. 776)

backcrossing: process in a genetic breeding experiment in which a heterozygote is cross-fertilized with one of the original homozygous parental organisms. (p. 592)

bacterial chromosome: circular DNA molecule with bound proteins that contains the main genome of a bacterial cell. (p. 502)

bacteriochlorophyll: type of chlorophyll found in bacteria that is able to extract electrons from donors other than water. (p. 453)

bacteriophage (phage): virus that infects bacterial cells. (pp. 99, 482)

bacteriorhodopsin: transmembrane protein complexed with rhodopsin, capable of transporting protons across the bacterial cell membrane to create a light-dependent electrochemical proton gradient. (p. 215)

bacterium (plural, bacteria): single-celled organism with no nucleus and little or no intracellular compartmentalization; a *prokaryote.* (p. 76).

basal body: microtubule-containing structure located at the base of a eukaryotic flagellum or cilium that consists of nine sets of triplet microtubules; identical in appearance to a centriole. (p. 775)

basal lamina (plural, laminae): thin sheet of specialized extracellular matrix material that separates epithelial cells from underlying connective tissues. (p. 297)

base excision repair: DNA repair mechanism that removes and replaces single damaged bases in DNA. (p. 543)

base pair (bp): pair of nucleotides joined together by complementary hydrogen bonding. (p. 492)

base pairing: complementary relationship between purines and pyrimidines based on hydrogen bonding that provides a mechanism for nucleic acids to recognize and bind to each other; involves the pairing of A with T or U, and the pairing of G with C. (p. 58)

Bcl-2: protein located in the outer mitochondrial membrane that blocks cell death by apoptosis. (p. 285)

benign tumor: tumor that grows only locally, unable to invade neighboring tissues or spread to other parts of the body. (p. 565)

beta oxidation (β oxidation): pathway involving successive cycles of fatty acid oxidation in which the fatty acid chain is shortened each time by two carbon atoms released as acetyl CoA. (pp. 360, 412)

beta sheet (β sheet): extended sheetlike secondary structure of proteins in which adjacent polypeptides are linked by hydrogen bonds between amino and carbonyl groups. (p. 48)

beta tubulin (β-tubulin): protein that joins with α-tubulin to form a heterodimer that is the basic building block of microtubules. (p. 746)

bimetallic iron-copper (Fe-Cu) center: a complex formed between a single copper atom and the iron atom bound to the heme group of an oxygen-binding cytochrome such as cytochrome a_3; important in keeping an O_2 molecule bound to the cytochrome until it picks up four electrons and four protons, resulting in the release of two molecules of water. (p. 419)

binding change model: a mechanism involving the physical rotation of the γ subunit of an F_oF_1 ATP synthase postulated to explain how the exergonic flow of protons through the F_o component of the complex drives the otherwise endergonic phosphorylation of ADP to ATP by the F_1 component. (p. 432)

biochemistry: study of the chemistry of living systems; same as biological chemistry. (p. 5)

bioenergetics: area of science that deals with the application of thermodynamic principles to reactions and processes in the biological world. (p. 112)

bioinformatics: using computers to analyze the vast amounts of data generated by sequencing and expression studies on genomes and proteomes. (p. 498)

biological chemistry: study of the chemistry of living systems; called biochemistry for short. (p. 18)

bioluminescence: production of light by an organism as a result of the reaction of ATP with specific luminescent compounds. (p. 108)

biosynthesis: generation of new molecules through a series of chemical reactions within the cell. (p. 107)

BiP: member of the Hsp70 family of chaperones; present in the ER lumen, where it facilitates protein folding by reversibly binding to the hydrophobic regions of polypeptide chains. (p. 681)

bivalent: pair of homologous chromosomes that have synapsed during the first meiotic division; contains four chromatids, two from each chromosome. (p. 581)

blood platelets: fragments of bone marrow cells that play a central role in blood clotting. (p. 281)

bond energy: amount of energy required to break one mole of a particular chemical bond. (p. 19)

bp: see *base pair.*

BRE (TFIIB recognition element): component of core promoters for RNA polymerase II, located immediately upstream from the TATA box. (p. 641)

brightfield microscopy: light microscopy of specimen that possesses color, has been stained, or has some other property that affects the amount of light that passes through, thereby allowing an image to be formed. (p. GM-6)

bundle sheath cell: internal cell of the leaf of a C_4 plant located in close proximity to the vascular bundle (vein) of the leaf; site of the Calvin cycle in such plants. (p. 471)

buoyant density centrifugation: see *equilibrium density centrifugation.*

C

C: see *cytosine.*

C value: amount of DNA in a single (haploid) set of chromosomes. (p. 588)

C_3 plant: plant that depends solely on the Calvin cycle for carbon dioxide fixation, creating the three-carbon compound 3-phosphoglycerate as the initial product. (p. 471)

C_4 plant: plant that uses the Hatch Slack pathway in mesophyll cells to carry out the initial fixation of carbon dioxide, creating the four-carbon compound oxaloacetate; the assimilated carbon is subsequently released again in bundle sheath cells and recaptured by the Calvin cycle. (p. 471)

cadherin: any of a family of plasma membrane glycoproteins that mediate Ca^{2+}-dependent adhesion between cells. (p. 303)

cal: see *calorie.*

calcium ATPase: see *calcium pump.*

calcium ionophore: a molecule that increases the permeability of membranes to calcium ions. (p. 265)

calcium pump (calcium ATPase): membrane protein that transports calcium ions across a membrane using energy derived from ATP hydrolysis; prominent example occurs in the sarcoplasmic reticulum (SR), where it pumps calcium ions into the SR lumen. (p. 265, 788)

calcium-calmodulin complex: complex formed by the reversible binding to calmodulin of four calcium ions with concomitant changes in the conformation of calmodulin, thereby increasing its affinity for target proteins whose activities are regulated by calmodulin. (p. 265)

calmodulin: calcium-binding protein involved in mediating many of the intracellular effects of calcium ions in eukaryotic cells. (p. 265)

calorie (cal): unit of energy; amount of energy needed to raise the temperature of 1 gram of water 1°C. (pp. 19, 113)

Calvin cycle: cyclic series of reactions used by photosynthetic organisms for the fixation of carbon dioxide and its reduction to form carbohydrates. (p. 462)

CAM: see *crassulacean acid metabolism.*

CAM plant: plant that carries out crassulacean acid metabolism. (p. 473)

cAMP: see *cyclic AMP.*

cAMP receptor protein (CRP): bacterial protein that binds cyclic AMP and then activates the transcription of catabolite-repressible genes. (p. 700)

cancer: uncontrolled, growing mass of cells that is capable of invading neighboring tissues and spreading via body fluids, especially the bloodstream, to other parts of the body; also called a *malignant tumor.* (p. 565)

cap (5'): methylated structure at the 5' end of eukaryotic mRNAs created by adding 7-methylguanosine and methylating the ribose rings of the first, and often the second, nucleotides of the RNA chain. (p. 648)

capping protein: protein that binds to the end of an actin microfilament, thereby preventing the further addition or loss of subunits. (p. 758)

carbohydrate: general name given to molecules that contain carbon, hydrogen, and oxygen in a ratio $C_n(H_2O)_n$; examples include starch, glycogen, and cellulose. (p. 61)

carbon assimilation reactions: portion of the photosynthetic pathway in which fully oxidized carbon atoms from carbon dioxide are fixed (covalently attached) to organic acceptor molecules and then reduced and rearranged to form carbohydrates and other organic compounds required for building a living cell. (p. 445)

carbon atom: the most important atom in biological molecules, capable of forming up to four covalent bonds. (p. 18)

carboxyl terminus: see *C-terminus.*

cardiac (heart) muscle: striated muscle of the heart, highly dependent on aerobic respiration. (p. 789)

carotenoid: any of several accessory pigments found in most plant species that absorb in the blue region of the visible spectrum (420-480 nm) and are therefore yellow or orange in color. (p. 453)

carrier molecule: a molecule that joins to a monomer, thereby activating the monomer for a subsequent reaction. (p. 30)

carrier protein: membrane protein that transports solutes across the membrane by binding to the solute on one side of the membrane and then undergoing a conformational change that transfers the solute to the other side of the membrane. (p. 203)

caspase: any of a family of proteases that degrade other cellular proteins as part of the process of apoptosis. (p. 282)

cassette mechanism: process of DNA rearrangement in which alternative alleles for yeast mating type are inserted into the *MAT* locus for transcription. (p. 709)

catabolic pathway: series of reactions that results in the breakdown of cellular components. (p. 368)

catabolite repression: ability of glucose to inhibit the synthesis of catabolic enzymes produced by inducible bacterial operons. (p. 700)

catalyst: agent that enhances the rate of a reaction by lowering the activation energy without itself being consumed; catalysts change the rate at which a reaction approaches equilibrium, but not the position of equilibrium. (p. 132)

catalytic subunit: a subunit of a multisubunit enzyme that contains the enzyme's catalytic site. (p. 149)

catecholamine: any of several compounds derived from the amino acid tyrosine that functions as a hormone and/or neurotransmitter. (p. 244)

Cdk: see *cyclin-dependent kinase.*

Cdk inhibitor: any of several proteins that restrain cell growth and division by inhibiting Cdk-cyclin complexes. (p. 564)

cDNA: see *complementary DNA.*

cDNA library: collection of recombinant DNA clones produced by copying the entire mRNA population of a particular cell type with reverse transcriptase and then cloning the resulting cDNAs. (p. 612)

cell: the basic structural and functional unit of living organisms; the smallest structure capable of performing the essential functions characteristic of life. (p. 1)

cell body: portion of a nerve cell that contains the nucleus and other organelles and has extensions called axons and dendrites extending from it. (p. 227)

cell cortex: dense network of actin microfilaments and associated proteins located just beneath the plasma membrane of most animal cells; supports the plasma membrane, confers structural rigidity on the cell surface, and facilitates shape changes and cell movement. (pp. 755, 792)

cell cycle: stages involved in preparing for and carrying out cell division; begins when two new cells are formed by the division of a single parental cell and is completed when one of these cells divides again into two cells. (p. 523)

cell differentiation: process by which nonspecialized cells acquire specialized structural and functional traits. (p. 706)

cell division: process by which one cell gives rise to two. (p. 523)

cell junction: specialized connection between the plasma membranes of adjoining cells for the purpose of adhesion, sealing, or communication. (p. 306)

cell membrane: see *plasma membrane.*

cell plate: flattened sac representing a stage in plant cell wall formation, leading to separation of the two daughter nuclei during plant cell division. (p. 554)

cell theory: theory of cellular organization that states that all organisms consist of one or more cells, that the cell is the basic unit of structure for all organisms, and that all cells arise only from preexisting cells. (p. 3)

cell wall: rigid, nonliving structure exterior to the plasma membrane of bacterial, algal, fungal, and plant cells; plant cell walls consist of cellulose microfibrils embedded in a noncellulosic matrix. (pp. 97, 314)

cell-cell adhesion: ability of cells to selectively adhere to one another. (p. 302)

cell-cell recognition: ability of a cell to distinguish one type of cell from another. (p. 302)

cellular respiration: oxidation-driven flow of electrons from reduced coenzymes to an electron acceptor, usually accompanied by the generation of ATP. (p. 398)

cellulose: structural polysaccharide present in plant cell walls, consisting of repeating glucose units linked by $\beta(1 \rightarrow 4)$ bonds. (pp. 63, 315)

central nervous system (CNS): sensory and motor neurons of the brain and spinal cord. (p. 225) Also see *peripheral nervous system (PNS).*

central pair: two parallel microtubules located in the center of the axoneme of a eukaryotic cilium or flagellum. (p. 776)

central vacuole: large membrane-bounded organelle present in many plant cells; helps maintain turgor pressure of the plant cell, plays a limited storage role, and is also capable of a lysosome-like function in intracellular digestion. (p. 93)

centrifugation: process of rapidly spinning a tube containing a fluid to subject its contents to a centrifugal force. (p. 326)

centrifuge: machine for rapidly spinning a tube containing a fluid to subject its contents to a centrifugal force. (p. 326)

centriole: microtubule-containing structure embedded within the centrosome of animal cells, where two centrioles lie at right angles to each other; identical in structure to the basal body of eukaryotic cilia and flagella. (p. 545)

centromere: point along a chromosome where sister chromatids are held together prior to anaphase and where kinetochores are attached; contains simple-sequence, tandemly repeated DNA. (pp. 500, 545)

centrosome: small zone of granular material surrounding two centrioles located adjacent to the nucleus of animal cells; functions as a cell's main microtubule-organizing center. (pp. 545, 749)

cerebroside: an uncharged glycolipid containing the amino alcohol sphingosine. (p. 166)

CF: see *cystic fibrosis.*

CF_1: component of the chloroplast ATP synthase complex that protrudes from the stromal side of thylakoid membranes and contains the catalytic site for ATP synthesis. (p. 459)

CF_o: component of the chloroplast ATP synthase complex that is embedded in thylakoid membranes and serves as the proton translocator. (p. 459)

CF_oCF_1 complex: ATP synthase complex found in chloroplast thylakoid membranes; catalyzes the process by which the exergonic flow of protons down their electrochemical gradient is used to drive ATP synthesis. (p. 459)

CFTR: see *cystic fibrosis transmembrane conductance regulator.*

CGN: see *cis-Golgi network.*

channel gating: closing of a membrane ion channel in such a way that it can reopen immediately in response to an appropriate stimulus. (p. 235)

channel inactivation: closing of a membrane ion channel in such a way that it cannot reopen immediately. (p. 235)

channel protein: membrane protein that forms a hydrophilic channel through which solutes can pass across the membrane without any change in the conformation of the channel protein. (p. 203)

chaperone: see *molecular chaperone.*

Chargaff's rules: observation, first made by Erwin Chargaff, that in DNA the number of adenines is equal to the number of thymines (A = T) and the number of guanines is equal to the number of cytosines (G = C). (p. 486)

charge repulsion: force driving apart two ions, molecules, or regions of molecules of the same electric charge. (p. 370)

chemical messenger: a molecule that transmits a signal from one cell to another. (p. 257)

chemical synapse: junction between two nerve cells where a nerve impulse is transmitted between the cells by neurotransmitters that diffuse across the synaptic cleft from the presynaptic cell to the postsynaptic cell. (p. 243)

chemiosmotic coupling model: model postulating that electron transport pathways establish proton gradients across membranes and that the energy stored in such gradients can then be used to drive ATP synthesis. (p. 426)

chemotaxis: cell movement toward a chemical attractant or away from a chemical repellent. (p. 798)

chemotroph: organism that is dependent on the bond energies of organic molecules such as carbohydrates, fats, and proteins to satisfy energy requirements. (p. 109)

chemotrophic energy metabolism: reactions and pathways by which cells catabolize nutrients such as carbohydrates, fats, and proteins, conserving as ATP some of the free energy that is released in the process. (p. 373)

chiasma (plural, **chiasmata**): connection between homologous chromosomes produced by crossing over during prophase I of meiosis. (p. 583)

chitin: structural polysaccharide found in insect exoskeletons and crustacean cells; consists of *N*-acetylglucosamine units linked by $\beta(1 \rightarrow 4)$ bonds. (p. 65)

chlorophyll: light-absorbing molecule that donates photoenergized electrons to organic molecules, initiating photochemical events that lead to the generation of the NADPH and ATP required for the Calvin cycle; because of its absorption properties, chlorophyll gives plants their characteristic green color. (p. 452)

chlorophyll-binding protein: any of several proteins that bind to and stabilize the arrangement of chlorophyll molecules within a photosystem. (p. 453)

chloroplast: double membrane-enclosed cytoplasmic organelle of plants and algae that contains chlorophyll and the enzymes necessary to carry out photosynthesis. (pp. 87, 447)

cholesterol: lipid constituent of animal cell plasma membrane; serves as a precursor to the steroid hormones. (pp. 70, 166)

cholinergic synapse: a synapse that uses acetylcholine as the neurotransmitter. (p. 244)

chromatid: see *sister chromatid.*

chromatin: DNA-protein fibers that make up chromosomes; constructed from nucleosomes spaced regularly along a DNA chain. (pp. 85, 504)

chromatin fiber (30-nm): fiber formed by packing together the nucleosomes of a 10-nm chromatin fiber. (p. 507)

chromatin remodeling: alterations in nucleosome structure, packing, and/or position designed to give transcription factors access to DNA target sites in the promoter region of a gene. (p. 713)

chromosome: in eukaryotes a single DNA molecule, complexed with histones and other proteins, that becomes condensed into a compact structure at the time of mitosis or meiosis. (pp. 10, 80, 504) Also see *bacterial chromosome.*

chromosome puff: uncoiled region of a polytene chromosome that is undergoing transcription. (p. 711)

chromosome theory of heredity: theory stating that hereditary factors are located on the chromosomes within the nucleus. (p. 10)

cilium (plural, **cilia**): membrane-bounded appendage on the surface of a eukaryotic cell, composed of a specific arrangement of microtubules and responsible for motility of the cell or the environment around the cell; shorter and

more numerous than closely related organelles called flagella. (p. 773) Also see *flagellum.*

cis **face (forming face):** side of the Golgi complex that is oriented toward the transitional elements of the endoplasmic reticulum. (p. 334)

cis-**acting element:** a DNA sequence to which a regulatory protein can bind. (p. 698)

cis-**Golgi network (CGN):** region of the Golgi complex consisting of a network of membrane-bounded tubules that are located closest to the transitional elements. (p. 334)

cisterna (plural, **cisternae**): membrane-bounded flattened sac, such as in the endoplasmic reticulum or Golgi complex. (p. 89)

cisternal maturation model: model postulating that Golgi cisternae are transient compartments that gradually change from *cis*-Golgi network cisternae into medial cisternae and then into *trans*-Golgi network cisternae. (p. 335)

cis-*trans* **test:** analysis used to determine whether a mutation in a bacterial operon affects a regulatory protein or the DNA sequence to which it binds. (p. 698)

clathrin: large protein that forms a "cage" around the coated vesicles and coated pits involved in endocytosis and other intracellular transport processes. (p. 349)

clathrin-dependent endocytosis: see *receptor-mediated endocytosis.*

claudin: transmembrane protein that forms the main structural component of a tight junction. (p. 311)

cleavage: process of cytoplasmic division in animal cells, in which a band of actin microfilaments lying beneath the plasma membrane constricts the cell at the midline and eventually divides it in two. (p. 553)

cleavage furrow: groove formed during the division of an animal cell that encircles the cell and deepens progressively, leading to cytoplasmic division. (p. 553)

clone: organism (or cell or molecule) that is genetically identical to another organism (or cell or molecule) from which it is derived. (p. 707)

cloning vector: DNA molecule to which a selected DNA fragment can be joined prior to rapid replication of the vector in a host cell (usually a bacterium); phage and plasmid DNA are the most common cloning vectors. (p. 607)

CNS: see *central nervous system.*

CoA: see *coenzyme A.*

coactivator: type of protein involved in regulating eukaryotic gene transcription by binding both to regulatory transcription factors associated with an enhancer and to general transcription factors associated with a promoter, thereby serving as a bridge between enhancers and promoters. (p. 720)

coated vesicle: any of several types of membrane vesicles involved in vesicular traffic within the endomembrane system; surrounded by a coat protein such as clathrin, COPI, COPII, or caveolin. (p. 345)

coatomer: see *COPI.*

coding strand: the nontemplate strand of a DNA double helix, which is base-paired to the template strand; identical in sequence to the single-stranded RNA molecules transcribed from the template strand, except that RNA has uracil (U) where the coding strand has thymine (T). (p. 632)

codon: triplet of nucleotides in an mRNA molecule that serves as a coding unit for an amino acid (or a start or stop signal) during protein synthesis. (p. 633)

coenzyme: small organic molecule that functions along with an enzyme by serving as a carrier of electrons or functional groups. (p. 374)

coenzyme A (CoA): organic molecule that serves as a carrier of acyl groups by forming a high-energy thioester bond with an organic acid. (p. 406)

coenzyme Q (CoQ): nonprotein (quinone) component of the mitochondrial electron transport system that serves as the collection point for electrons from both FMN- and FAD-linked dehydrogenases; also called ubiquinone. (p. 419)

coenzyme Q-cytochrome c oxidoreductase: see *complex III.*

cognate receptor: the receptor that is specific for a given ligand. (p. 258)

cohesin: protein that holds sister chromatids together prior to anaphase. (p. 561)

colchicine: plant-derived drug that binds to tubulin and prevents its polymerization into microtubules. (p. 753)

collagen: a family of closely related proteins that form high-strength fibers found in high concentration in the extracellular matrix of animals. (p. 291)

collagen fiber: extremely strong fibers measuring several micrometers in diameter found in the extracellular matrix; constructed from collagen fibrils that are in turn composed of collagen molecules lined up in a staggered array. (p. 391)

combinatorial model for gene regulation: model proposing that complex patterns of tissue-specific gene expression can be achieved by a relatively small number of DNA control elements and their respective transcription factors acting in different combinations. (p. 721)

complementary: in nucleic acids, the ability of guanine (G) to form a hydrogen-bonded base pair with cytosine (C), and adenine (A) to form a hydrogen-bonded base pair with thymine (T) or uracil (U) (p. 487)

complementary DNA (cDNA): DNA molecule copied from an mRNA template by the enzyme reverse transcriptase. (p. 612)

complex I (NADH-coenzyme Q oxidoreductase): multiprotein complex of the electron transport system that catalyzes the transfer of electrons from NADH to coenzyme Q. (p. 423)

complex II (succinate-coenzyme Q oxidoreductase): multiprotein complex of the electron transport system that catalyzes the transfer of electrons from succinate to coenzyme Q. (p. 423)

complex III (coenzyme Q-cytochrome c oxidoreductase): Multiprotein complex of the electron transport system that catalyzes the transfer of electrons from coenzyme Q to cytochrome c. (p. 424)

complex IV (cytochrome c oxidase): multiprotein complex of the electron transport system that catalyzes the transfer of electrons from cytochrome c to oxygen. (p. 424)

compound microscope: light microscope that uses several lenses in combination; usually has a condenser lens, an objective lens, and an ocular lens. (p. GM-5)

concentration work: use of energy to transport ions or molecules across a membrane against an electrochemical or concentration gradient. (p. 108)

condensation reaction: chemical reaction that results in the joining of two molecules by the removal of a water molecule. (p. 30)

condenser lens: lens of a light microscope (or electron microscope) that is the first lens to direct the light rays (or electron beam) from the source toward the specimen. (pp. GM-5, GM-18)

confocal scanning microscope: specialized type of light microscope that employs a laser beam to illuminate a single plane of the specimen at a time. (p. GM-11)

conformation: three-dimensional shape of a polypeptide or other biological macromolecule. (p. 45)

conjugation: cellular mating process by which DNA is transferred from one bacterial cell to another. (p. 600)

connexon: assembly of six protein subunits with a hollow center that forms a channel through the plasma membrane at a gap junction. (p. 313)

consensus sequence: the most common version of a DNA base sequence when that sequence occurs in slightly different forms at different sites. (p. 637)

constitutive gene: gene that is active at all times rather then being regulated. (p. 692)

constitutive heterochromatin: chromosomal regions that are condensed in all cells of an organism at virtually all times and are therefore genetically inactive. (p. 517) Also see *facultative heterochromatin.*

constitutive secretion: continuous fusion of secretory vesicles with the plasma membrane and expulsion of their contents to the cell exterior, independent of specific extracellular signals. (p. 342)

contractile ring: beltlike bundle of actin microfilaments that forms beneath the plasma membrane and acts to constrict the cleavage furrow during the division of an animal cell. (p. 553)

contractility: shortening of muscle cells. (p. 769)

cooperativity: property of enzymes possessing multiple catalytic sites in which the binding of a substrate molecule to one catalytic site causes conformational changes that influence the affinity of the remaining sites for substrate. (p. 149)

COPI (coatomer): protein component of the "fuzzy" coat surrounding vesicles involved in the transport of proteins from the endoplasmic reticulum to the Golgi complex. (p. 351)

CoQ: see *coenzyme Q.*

core glycosylation: attachment of the core oligosaccharide to a newly formed polypeptide chain within the lumen of the endoplasmic reticulum. (p. 336)

core oligosaccharide: initial oligosaccharide segment joined to an asparagine residue during N-glycosylation of a polypeptide chain; consists of two N-acetylglucosamine units, nine mannose units, and three glucose units. (p. 336)

core promoter: minimal set of DNA sequences sufficient to direct the accurate initiation of transcription by RNA polymerase. (p. 641)

core protein: a protein molecule to which numerous glycosaminoglycan chains are attached to form a proteoglycan. (p. 294)

corepressor: effector molecule required along with the repressor to prevent transcription of a bacterial operon. (p. 699)

cortex: see *cell cortex.*

cosmid: cloning vector that can be packaged into phage particles and utilized for cloning large DNA fragments. (p. 613)

cotranslational import: transfer of a growing polypeptide chain across (or, in the case of integral membrane proteins, into) the ER membrane as polypeptide synthesis proceeds. (p. 678)

cotransport: coupled transport of two solutes across a membrane in such a way that transport of either stops if the other is absent. (p. 204)

coupling: relationship between two processes that are mutually dependent on one another, such as the coupling between electron transport and ATP synthesis in respiratory metabolism. (p. 426)

covalent bond: strong chemical bond in which two atoms share two or more electrons. (p. 18)

covalent modification: type of regulation in which the activity of an enzyme (or other protein) is altered by the addition or removal of specific chemical groups. (p. 149)

crassulacean acid metabolism (CAM): pathway in which plants use PEP carboxylase to fix CO_2 at night, generating the four-carbon acid, malate. The malate is then decarboxylated during the day to release CO_2, which is fixed by the Calvin cycle. (p. 473)

creatine phosphate: high-energy compound in muscle cells, used to regenerate the ATP needed for muscle contraction. (p. 788)

CREB protein: transcription factor that activates the transcription of cyclic AMP-inducible genes by binding to cAMP response elements in DNA. (p. 726)

crista (plural, cristae): infolding of inner mitochondrial membrane into the matrix of the mitochondrion, thereby increasing the total surface area of the inner membrane; contains the enzymes of electron transport and oxidative phosphorylation. (pp. 86, 403)

critical concentration: tubulin concentration at which the rate of assembly of tubulin subunits into a polymer is exactly balanced with the rate of disassembly. (p. 747)

critical point dryer: heavy metal canister used to dry a specimen under conditions of controlled temperature and pressure. (p. GM-26)

cross-bridge: structure formed by contact between the myosin heads of thick filaments and the thin filaments in muscle myofibrils. (p. 783)

crossing over: exchange of DNA segments between homologous chromosomes. (p. 583)

CRP: see *cAMP receptor protein.*

cryoprotection: reducing the formation of ice crystals during freezing of a specimen for microscopy. (p. GM-24)

C-terminus (carboxyl terminus): the end of a polypeptide chain that contains the last amino acid to be incorporated during mRNA translation; usually retains a free carboxyl group. (p. 44)

current: movement of positive or negative ions. (p. 229)

cyanobacterium (plural, cyanobacteria): bacteria that have chlorophyll and can carry out photosynthesis; considered by the endosymbiont theory to be the progenitor of eukaryotic chloroplasts. (p. 76).

cyclic AMP (cAMP): adenosine monophosphate with the phosphate group linked to both the 3′ and 5′ carbons by phosphodiester bonds; functions in both prokaryotic and eukaryotic gene regulation; in eukaryotes, acts as a second messenger that mediates the effects of various signaling molecules by activating protein kinase A. (pp. 261, 700)

cyclic electron flow: light-driven transfer of electrons from photosystem I through a sequence of electron carries that returns them to a chlorophyll molecule of the same photosystem, with the released energy used to drive ATP synthesis. (p. 460)

cyclin: any of a group of proteins that activate the cyclin-dependent kinases (Cdks) involved in regulating progression through the eukaryotic cell cycle. (p. 558)

cyclin-dependent kinase (Cdk): any of several protein kinases that are activated by different cyclins and that control progression through the eukaryotic cell cycle by phosphorylating various target proteins. (p. 558)

cyclosis: see *cytoplasmic streaming.*

cystic fibrosis (CF): a disease whose symptoms result from an inability to secrete chloride ions that is in turn caused by a genetic defect in a membrane protein that functions as a chloride ion channel. (p. 212)

cystic fibrosis transmembrane conductance regulator (CFTR): a membrane protein that functions as a chloride ion channel, a mutant form of which can lead to cystic fibrosis. (p. 212)

cytochalasins: family of drugs produced by certain fungi that inhibit a variety of cell movements by preventing actin polymerization. (p. 758)

cytochrome b_6/f complex: multiprotein complex within the thylakoid membrane that transfers electrons from a plastoquinol to plastocyanin as part of the energy transduction reactions of photosynthesis. (p. 457)

cytochrome c: heme-containing protein of the electron transport system that also plays a role in triggering apoptosis when released from mitochondria. (p. 285)

cytochrome c oxidase: see *complex IV.*

cytochromes: heme-containing proteins of the electron transport system, involved in the transfer of electrons from coenzyme Q to oxygen by the oxidation and reduction of the central iron atom of the heme group. (p. 419)

cytokinesis: division of the cytoplasm of a parent cell into two daughter cells; usually follows mitosis. (p. 523)

cytology: study of cellular structure, based primarily on microscopic techniques. (p. 3)

cytoplasm: that portion of the interior of a eukaryotic cell that is not occupied by the nucleus; includes organelles such as mitochondria, as well as the cytosol. (p. 95)

cytoplasmic dynein: cytoplasmic motor protein that moves along the surface of microtubules in the plus-to-minus direction driven by energy derived from ATP hydrolysis; associated with dynactin, which links cytoplasmic dynein to cargo vesicles. (p. 772)

cytoplasmic microtubules: microtubules arranged in loosely organized, dynamic networks in the cytoplasm of eukaryotic cells. (p. 746)

cytoplasmic streaming: movement of the cytoplasm driven by interactions between actin filaments and specific types of myosin; also called cyclosis in plant cells. (pp. 78, 795)

cytosine (C): nitrogen-containing aromatic base, chemically designated as a pyrimidine, that serves as an informational monomeric unit when present in nucleic acids with other bases in a specific sequence; forms a complementary base pair with guanine (G) by hydrogen bonding. (p. 55)

cytoskeleton: three-dimensional, interconnected network of microtubules, microfilaments, and intermediate filaments that provides structure to the cytoplasm of a eukaryotic cell and plays an important role in cell movement. (pp. 95, 742)

cytosol: the semifluid substance in which the organelles of the cytoplasm are suspended. (p. 95)

D

DAG: see *diacylglycerol.*

deep etching: modification of freeze etching technique in which ultrarapid freezing and a volatile cryoprotectant are used to extend the etching period, thereby removing a deeper layer of ice and allowing the interior of a cell to be examined in depth. (p. GM-25)

degenerate code: ability of the genetic code to use more than one triplet code to specify the same amino acid. (p. 631)

dehydrogenation: removal of electrons plus hydrogen ions (protons) from an organic molecule; oxidation. (p. 373)

denaturation: loss of the natural three-dimensional structure of a macromolecule, usually resulting in a loss of its biological activity; caused by agents such as heat, extremes of pH, urea, salt, and other chemicals. (p. 31) Also see *DNA denaturation*.

dendrite: extension of a nerve cell that receives impulses and transmits them inward toward the cell body. (p. 227)

density gradient (rate-zonal) centrifugation: type of centrifugation in which the sample is applied as a thin layer on top of a gradient of solute, and centrifugation is stopped before the particles reach the bottom of the tube; separates organelles and molecules based mainly on differences in size. (p. 328)

deoxyribonucleic acid: see *DNA*.

deoxyribose: five-carbon sugar present in DNA. (p. 54)

dephosphorylation: removal of a phosphate group. (p. 150)

depolarization: change in membrane potential to a less-negative value. (p. 231)

desmosome: junction for cell-cell adhesion that is connected to the cytoskeleton by intermediate filaments; creates buttonlike points of strong adhesion between adjacent animal cells that give tissue structural integrity and allows the cells to function as a unit and resist stress. (p. 309)

desmotubule: tubular structure that lies in the central channel of the plasmodesma between two plant cells. (p. 319)

development (of an organism): series of mitotic divisions and progressive specialization of various groups of cells that leads to the formation of a multicellular embryo and eventually an adult organism. (p. 579)

diacylglycerol (DAG): glycerol esterified to two fatty acids; formed, along with inositol trisphosphate (InsP₃), upon hydrolysis of phosphatidylinositol-4,5-bisphosphate by phospholipase C; remains membrane-bound after hydrolysis and functions as a second messenger by activating protein kinase C, which then phosphorylates specific serine and threonine groups on a variety of target proteins. (p. 263)

diakinesis: final stage of prophase I of meiosis; associated with chromosome condensation, disappearance of nucleoli, breakdown of the nuclear envelope, and initiation of spindle formation. (p. 585)

DIC microscopy: see *differential interference contrast microscopy*.

dictyosome: Golgi complex of plant cells. (p. 357)

differential centrifugation: technique for separating organelles or molecules that differ in size and/or density by subjecting cellular fractions to centrifugation at high speeds and separating particles based on their different rates of sedimentation. (p. 327)

differential interference contrast (DIC) microscopy: technique that resembles phase-contrast microscopy in principle, but is more sensitive because it employs a special prism to split the illuminating light beam into two separate rays. (p. GM-7)

differential scanning calorimetry: technique for determining a membrane's transition temperature by monitoring the uptake of heat during the transition of the membrane from the gel-to-fluid state. (p. 172)

differentiation: see *cell differentiation*.

diffraction: pattern of either additive or canceling interference exhibited by light waves. (p. GM-3)

diffusion: free, unassisted movement of a solute, with direction and rate dictated by the difference in solute concentration between two different regions. (p. 78)

digital deconvolution microscopy: technique in which fluorescence microscopy is used to acquire a series of images through the thickness of a specimen, followed by computer analysis to remove the contribution of out-of-focus light to the image in each focal plane. (p. GM-13)

digital video microscopy: technique in which microscopic images are recorded and stored electronically by placing a video camera in the image plane produced by the ocular lens. (p. GM-14)

diploid: containing two sets of chromosomes and therefore two copies of each gene; can describe a cell, nucleus, or organism composed of such cells. (p. 578)

diplotene: stage during prophase I of meiosis when the two homologous chromosomes of each bivalent begin to separate from each other, revealing the chiasmata that connect them. (p. 583)

direct active transport: membrane transport in which the movement of solute molecules or ions across a membrane is coupled directly to an exergonic chemical reaction, most commonly the hydrolysis of ATP. (p. 208)

directionality: having two ends that are chemically different from each other; used to describe a polymer chain such as a protein, nucleic acid, or carbohydrate; also used to describe membrane transport systems that selectively transport solutes across a membrane in one direction. (pp. 30, 207)

disaccharide: carbohydrate consisting of two covalently linked monosaccharide units. (p. 62)

dissociation constant (K_d): the concentration of free ligand needed to produce a state in which half the receptors are bound to ligand. (p. 258)

disulfide bond: covalent bond formed between two sulfur atoms by oxidation of sulfhydryl groups. The disulfide bond formed between two cysteines is important in stabilizing the tertiary structure of proteins. (p. 45)

DNA (deoxyribonucleic acid): macromolecule that serves as the repository of genetic information in all cells; constructed from nucleotides consisting of deoxyribose phosphate linked to either adenine, thymine, cytosine, or guanine; forms a double helix held together by complementary base-pairing between adenine and thymine, and between cytosine and guanine. (p. 54)

DNA cloning: generating multiple copies of a specific DNA sequence, either by replication of a recombinant plasmid or bacteriophage within bacterial cells or by use of the polymerase chain reaction. (p. 608)

DNA denaturation: separation of the two strands of the DNA double helix caused by disruption of complementary base-pairing. (p. 490)

DNA fingerprinting: technique for identifying individuals based on small differences in DNA fragment patterns detected by electrophoresis. (p. 501)

DNA footprinting: technique for locating the DNA region to which a particular DNA-binding protein has become bound. (p. 637)

DNA gyrase: a type II topoisomerase that can relax positive supercoiling and induce negative supercoiling of DNA; involved in unwinding the DNA double helix during DNA replication. (p. 489)

DNA ligase: enzyme that joins two DNA fragments together by catalyzing the formation of a phosphoester bond between the 3′ end of one fragment and the 5′ end of the other fragment. (p. 532)

DNA melting temperature (T_m): temperature at which the transition from double-stranded to single-stranded DNA is halfway complete when DNA is denatured by increasing the temperature. (p. 490)

DNA methylation: addition of methyl groups to nucleotides in DNA; associated with suppression of gene transcription when selected cytosine groups are methylated in eukaryotic DNA. (p. 714)

DNA microarray: tiny chip that has been spotted at fixed locations with thousands of different DNA fragments for use in gene expression studies. (p. 716)

DNA packing ratio: ratio of the length of a DNA molecule to the length of the chromosome or fiber into which it is packaged; used to quantify the extent of DNA coiling and folding. (p. 507)

DNA polymerase: any of a group of enzymes involved in DNA replication and repair that catalyze the addition of successive nucleotides to the 3′ end of a growing DNA strand, using an existing DNA strand as template. (p. 530)

DNA rearrangement: movement of DNA segments from one location to another within the genome. (p. 709)

DNA renaturation: binding together of the two separated strands of a DNA double helix by complementary base pairing, thereby regenerating the double helix. (p. 490)

DNA-binding domain: region of a transcription factor that recognizes and binds to a specific DNA base sequence. (p. 721)

DNase I hypersensitive site: location near an active gene that shows extreme sensitivity to digestion by the nuclease DNase I; thought to correlate with binding sites for transcriptional factors or other regulatory proteins. (p. 713)

domain: a discrete, locally folded unit of protein tertiary structure, often containing regions of α helices and β sheets packed together compactly. (p. 52)

dominant (allele): allele that determines how the trait will appear in an organism, whether present in the heterozygous or homozygous form. (p. 578)

dominant negative mutation: a loss of function mutation involving proteins consisting of more than one copy of the same polypeptide chain, in which a single mutant polypeptide chain disrupts the function of the protein even though the other polypeptide chains are normal. (p. 273)

double bond: chemical bond formed between two atoms as a result of the sharing of two pairs of electrons. (p. 18)

double helix (model): two intertwined helical chains of a DNA molecule, held together by complementary base-pairing between adenine (A) and thymine (T) and between cytosine (C) and guanine (G). (pp. 10, 58, 487)

double-reciprocal plot: graphic method for analyzing enzyme kinetic data by plotting $1/v$ versus $1/[S]$. (p. 143)

downstream: located toward the 3′ end of the DNA coding strand. (p. 637)

downstream promoter element: see *DPE*.

DPE (downstream promoter element): component of core promoters for RNA polymerase II, located about 30 nucleotides downstream from the transcriptional startpoint. (p. 641)

dynactin: protein complex that helps link cytoplasmic dynein to the cargo (e.g., a vesicle) that it transports along microtubules. (p. 772)

dynamic instability model: model for microtubule behavior that presumes two populations of microtubules, one growing in length by continued polymerization at their plus ends and the other shrinking in length by depolymerization. (p. 748)

dynamin: cytosolic GTPase required for coated pit constriction and the closing of a budding, clathrin-coated vesicle. (p. 350)

dynein: motor protein that moves along the surface of microtubules in the plus-to-minus direction driven by energy derived from ATP hydrolysis; present both in the cytoplasm and in the arms that reach between adjacent microtubule doublets in the axoneme of a flagellum or cilium. (p. 770)

E

E: see *internal energy*.

E face: interior face of the outer monolayer of a membrane as revealed by the technique of freeze-fracturing; called the E face because this monolayer is on the *exterior* side of the membrane. (p. GM-24)

E site (exit site): site on the ribosome to which the empty tRNA is moved during translocation prior to its release from the ribosome. (p. 662)

E′: see *reduction potential*.

E₀′: see *standard reduction potential*.

E2F transcription factor: protein that, when not bound to the Rb protein, activates the transcription of genes coding for proteins required for DNA replication and entrance into the S phase of the cell cycle. (p. 560)

Eₐ: see *activation energy*.

Eadie-Hofstee equation: alternative to the Lineweaver-Burk equation in which enzyme kinetic data are analyzed by plotting $v/[S]$ versus v. (p. 144)

early endosome: vesicles budding off the *trans*-Golgi network that are sites for the sorting and recycling of extracellular material brought into the cell by endocytosis. (p. 347)

ECM: see *extracellular matrix*.

effector: see *allosteric effector*.

EGF: see *epidermal growth factor*.

egg (ovum): haploid female gamete, usually a relatively large cell with many stored nutrients. (p. 579)

elastin: protein subunit of the elastic fibers that impart elasticity and flexibility to the extracellular matrix. (p. 293)

electrical excitability: ability to respond to certain types of stimuli with a rapid series of changes in membrane potential known as an action potential. (p. 228)

electrical synapse: junction between two nerve cells where nerve impulses are transmitted by direct movement of ions through gap junctions without the involvement of chemical neurotransmitters. (p. 243)

electrical work: use of energy to transport ions across a membrane against a potential gradient. (p. 108)

electrochemical equilibrium: condition in which a transmembrane concentration gradient of a specific ion is balanced with an electrical potential across the same membrane, such that there is no net movement of the ion across the membrane. (p. 230)

electrochemical gradient: transmembrane gradient of an ionic species, with both an electrical component due to charge separation and a concentration component. (p. 197)

electrochemical proton gradient: transmembrane gradient of protons, with both an electrical component due to charge separation and a chemical component due to a difference in proton concentration (pH) across the membrane. (p. 426)

electron gun: assembly of several components that generates the electron beam in an electron microscope. (p. GM-18)

electron micrograph: photographic image of a specimen produced by the exposure of a photographic plate to the image-forming electron beam of an electron microscope. (p. GM-19)

electron microscope: instrument that uses a beam of electrons to visualize cellular structures and thereby examine cellular architecture; the resolution is much greater than that of the light microscope, allowing detailed ultrastructural examination. (pp. 6, GM-1)

electron shuttle system: any of several mechanisms whereby electrons from a reduced coenzyme such as NADH are moved across a membrane; consists of one or more electron carriers that can be reversibly reduced, with transport proteins present in the membrane for both the oxidized and reduced forms of the carrier. (p. 436)

electron transport: process of coenzyme reoxidation under aerobic conditions, involving stepwise transfer of electrons to oxygen by means of a series of electron carriers. (p. 418)

electron transport system (ETS): group of membrane-bound electron carriers that transfer electrons from the coenzymes NADH and $FADH_2$ to oxygen. (p. 418)

electronegative: property of an atom that tends to draw electrons toward it. (p. 22)

electrophilic substitution: reaction in which an electron-deficient (electropositive) molecule or region of a molecule attacks an electron-rich (electronegative) molecule or region of a molecule, resulting in the formation of a covalent bond. (p. 139)

electrophoresis: a group of related techniques that utilize an electrical field to separate electrically charged molecules. (p. 180)

elongation (of microtubules): growth of microtubules by addition of tubulin heterodimers to either end. (p. 747)

elongation factors: group of proteins that catalyze steps involved in the elongation phase of protein synthesis; examples include EF-Tu and EF-Ts (p. 669)

Emerson enhancement effect: achievement of greater photosynthetic activity with red light of two slightly different wavelengths than is possible by summing the activities obtained with the individual wavelengths separately. (p. 455)

endergonic: pertaining to an energy-requiring reaction characterized by a positive free energy change ($\Delta G > 0$). (p. 119)

endocrine hormone: a hormone that is released into the circulatory system so that it can act on distant target cells. (p. 276)

endocrine tissue: a tissue that produces and secretes endocrine hormones directly into the bloodstream. (p. 276)

endocytic vesicle: membrane vesicle formed by pinching off of a small segment of plasma membrane during the process of endocytosis. (p. 343)

endocytosis: uptake of extracellular materials by infolding of the plasma membrane, followed by pinching off of a membrane-bound vesicle containing extracellular fluid and materials. (p. 343)

endomembrane system: interconnected system of cytoplasmic membranes in eukaryotic cells composed of the endoplasmic reticulum, Golgi complex, endosomes, lysosomes, and nuclear envelope. (p. 323)

endoplasmic reticulum (ER): network of interconnected membranes distributed throughout the cytoplasm and involved in the synthesis, processing, and transport of proteins in eukaryotic cells. (pp. 88, 323)

endosome: see *early endosome or late endosome*.

endosymbiont theory: theory postulating that mitochondria and chloroplasts arose from ancient bacteria that were ingested by ancestral eukaryotic cells about a billion years ago. (pp. 88, 450)

end-product inhibition: see *feedback inhibition*.

end-product repression: regulation of an anabolic pathway based on the ability of an end product to repress further synthesis of enzymes involved in production of that end product. (p. 694)

energy: capacity to do work; ability to cause specific changes. (p. 106)

energy transduction reactions: portion of the photosynthetic pathway in which light energy is converted to chemical energy in the form of ATP and the coenzyme NADPH, which subsequently provide energy and reducing power for the carbon assimilation reactions. (p. 445)

enhancer: DNA sequence containing a binding site for transcription factors that stimulate transcription and whose position and orientation relative to the promoter can vary significantly without interfering with the ability to regulate transcription. (p. 718)

enthalpy (H): the heat content of a substance, quantified as the sum of its internal energy, E, plus the product of pressure and volume: $H = E + PV$. (p. 114)

entropy (S): measure of the randomness or disorder in a system. (p. 118)

enzyme: biological catalyst; protein (or in certain cases, RNA) molecule that acts on one or more specific substrates, converting them to products with different molecular structures. (pp. 9, 130) Also see *ribozyme*.

enzyme catalysis: involvement of an organic molecule, usually a protein but in some cases RNA, in speeding up the rate of a specific chemical reaction or class of reactions. (p. 130) Also see *catalyst*.

enzyme kinetics: quantitative analysis of enzyme reaction rates and the manner in which they are influenced by a variety of factors. (p. 140)

epidermal growth factor (EGF): protein that stimulates the growth and division of a wide variety of epithelial cell types. (p. 562)

EPSP: see *excitatory postsynaptic potential*.

equilibrium constant (K_{eq}): ratio of product concentrations to reactant concentrations for a given chemical reaction when the reaction has reached equilibrium. (p. 120)

equilibrium density (buoyant density) centrifugation: technique used to separate cellular components by subjecting them to centrifugation in a solution that increases in density from the top to the bottom of the centrifuge tube; during centrifugation, an organelle or molecule sediments to the density layer equal to its own density, at which point movement stops because no further net force acts on the material. (p. 329, 526)

equilibrium membrane potential: membrane potential that exactly offsets the effect of the concentration gradient for a given ion. (p. 230)

ER: see *endoplasmic reticulum.*

ER cisterna (plural, **cisternae**): flattened sac of the endoplasmic reticulum. (p. 323)

ER lumen: the internal space enclosed by membranes of the endoplasmic reticulum. (p. 323)

ER signal sequence: amino acid sequence in a newly forming polypeptide chain that directs the ribosome-mRNA-polypeptide complex to the surface of the rough ER, where the complex becomes anchored. (p. 680)

ETS: see *electron transport system.*

eubacterium (plural, **eubacteria**): member of one of the two main groups of prokaryotes, the other being archaebacteria; includes most present-day bacteria and cyanobacteria. (p. 76)

euchromatin: loosely packed, uncondensed form of chromatin present during interphase; contains DNA that is being actively transcribed. (p. 507) Also see *heterochromatin.*

eukaryote: category of organisms whose cells are characterized by the presence of a true membrane-bounded nucleus and other membrane-bounded organelles. (p. 76)

excision repair: DNA repair mechanism that removes and replaces abnormal nucleotides. (p. 543)

excitatory postsynaptic potential (EPSP): small depolarization of the postsynaptic membrane triggered by binding of an excitatory neurotransmitter to its receptor; if the EPSP exceeds a threshold level, it can trigger an action potential. (p. 250) Also see *inhibitory postsynaptic potential.*

exergonic: pertaining to an energy-releasing reaction characterized by a negative free energy change ($\Delta G < 0$). (p. 119)

exit site: see *E site.*

exocrine tissue: tissue that secretes its product(s) into ducts that then transport the secretion to another region of the body. (p. 276)

exocytosis: fusion of vesicle membranes with the plasma membrane so that contents of the vesicle can be expelled or secreted to the extracellular environment. (p. 342)

exon: nucleotide sequence in a primary RNA transcript that is preserved in the mature, functional RNA molecule. (p. 650) Also see *intron.*

exonuclease: an enzyme that degrades a nucleic acid (usually DNA) from one end, rather than cutting the molecule internally. (p. 535)

expansin: group of proteins that help plant cell walls retain their pliability. (p. 317)

exportin: nuclear receptor protein that binds to the nuclear export signal of proteins in the nucleus and then transports the bound protein out through the nuclear pore complex and into the cytosol. (p. 514)

extensin: group of related glycoproteins that form rigid, rodlike molecules tightly woven into the cell walls of plants and fungi. (p. 316)

extracellular digestion: degradation of components outside a cell, usually by lysosomal enzymes that are released from the cell by exocytosis. (p. 356)

extracellular matrix (ECM): material secreted by animal cells that fills the spaces between neighboring cells; consists of a mixture of structural proteins (e.g. collagen, elastin) and adhesive glycoproteins (e.g. fibronectin, laminin) embedded in a matrix composed of protein-polysaccharide complexes called proteoglycans. (pp. 97, 290)

extracellular structure: structure external to the plasma membrane that is nonetheless an integral part of the cell; consists mainly of macromolecules secreted by the cell (p. 290)

F

F factor: DNA sequence that enables an *E. coli* cell to act as a DNA donor during bacterial conjugation. (p. 600)

F_1 complex: knoblike sphere protruding into the matrix from mitochondrial inner membranes (or into the cytosol from the bacterial plasma membrane) that contains the ATP-synthesizing site of aerobic respiration. (p. 404)

F_1 generation: offspring of the P_1 generation in a genetic breeding experiment. (p. 590)

F_2 generation: offspring of the F_1 generation in a genetic breeding experiment. (p. 590)

F2,6BP: see *fructose-2,6-bisphosphate.*

facilitated diffusion: membrane protein-mediated movement of a substance across a membrane that does not require energy because the ion or molecule being transported is moving down an electrochemical gradient. (p. 203)

F-actin: component of microfilaments consisting of G-actin monomers that have been polymerized into long, linear strands. (p. 755)

facultative heterochromatin: chromosomal regions that have become specifically condensed and inactivated in a particular cell type at a specific time. (p. 517) Also see *constitutive heterochromatin.*

facultative organism: organism that can function in either an anaerobic or aerobic mode. (p. 378)

FAD: see *flavin adenine dinucleotide.*

familial hypercholesterolemia (FH): genetic predisposition to high blood cholesterol levels and heart disease caused by an inherited defect in the gene coding for the LDL receptor. (p. 346)

fast axonal transport: microtubule-mediated movement of vesicles and organelles back and forth along a nerve cell axon. (p. 770)

fatty acid: long, unbranched hydrocarbon chain that has a carboxyl group at one end and is therefore amphipathic; usually contains an even number of carbon atoms and may be of varying degrees of unsaturation. (pp. 68, 169)

fatty acid-anchored membrane protein: protein located on a membrane surface that is covalently bound to a fatty acid embedded with the lipid bilayer. (p. 180)

Fd: see *ferredoxin.*

feedback (end-product) inhibition: ability of the end product of a biosynthetic pathway to inhibit the activity of the first enzyme in the pathway, thereby ensuring that the functioning of the pathway is sensitive to the intracellular level of its product. (p. 147)

fermentation: partial oxidation of carbohydrates by oxygen-independent (anaerobic) pathways, resulting often (but not always) in the production of either ethanol and carbon dioxide or lactate. (p. 375)

ferredoxin (Fd): iron-sulfur protein in the chloroplast stroma involved in the transfer of electrons from photosystem I to $NADP^+$ during the energy transduction reactions of photosynthesis. (p. 458)

ferredoxin-$NADP^+$ reductase (FNR): enzyme located on the stroma side of the thylakoid membrane that catalyzes the transfer of electrons from ferredoxin to $NADP^+$. (p. 458)

fertilization: union of two haploid gametes to form a diploid cell, the zygote, that develops into a new organism. (p. 579)

FGF: see *fibroblast growth factor.*

FGFR: see *fibroblast growth factor receptor.*

FH: see *familial hypercholesterolemia.*

fibroblast growth factor (FGF): any of several related signaling proteins that stimulate growth and cell division in fibroblasts as well as numerous other cell types, both in adults and during embryonic development. (p. 273)

fibroblast growth factor receptor (FGFR): plasma membrane protein that binds to fibroblast growth factor at the outer cell surface, thereby activating the receptor's tyrosine kinase activity and triggering a chain of signal transduction events inside the cell that can lead to cell growth, proliferation, and differentiation. (p. 273)

fibronectin: adhesive glycoprotein found in the extracellular matrix and loosely associated with the cell surface; binds cells to the extracellular matrix and is important in determining cell shape and guiding cell migration. (p. 296)

fibrous protein: protein with extensive α helix or β sheet structure that confers a highly ordered, repetitive structure. (p. 50)

filopodium (plural, **filopodia**): thin, pointed cytoplasmic protrusion that transiently emerges from the surface of eukaryotic cells during cell movements. (p. 792)

first law of thermodynamics: law of conservation of energy; principle that energy can be converted from one form to another but can never be created or destroyed. (p. 113)

Fischer projection: model depicting the chemical structure of a molecule as a chain drawn vertically with the most oxidized atom on top and horizontal projections that are understood to be coming out of the plane of the paper. (p. 62)

5′ cap: methylated structure at the 5′ end of eukaryotic mRNAs created by adding 7-methylguanosine and methylating the ribose rings of the first, and often the second, nucleotides of the RNA chain. (p. 648)

fixative: chemical that kills cells while preserving their structural appearance for microscopic examination. (p. GM-16)

flagellum (plural, **flagella**): membrane-bounded appendage on the surface of a eukaryotic cell composed of a specific arrangement of microtubules and responsible for motility of the cell; longer and less numerous (usually limited to one or a few per cell) than closely related organelles called cilia. (p. 774) Also see *cilium.*

flavin adenine dinucleotide (FAD): coenzyme that accepts two electrons and two protons from an oxidizable organic molecule to generate the reduced form, $FADH_2$; important electron carrier in energy metabolism. (p. 409)

flavoprotein: protein that has a tightly bound flavin coenzyme (FAD or FMN) and that serves as a biological electron donor or acceptor. Several examples occur in the mitochondrial electron transport system. (p. 418)

flippase: see *phospholipid translocator.*

fluid mosaic model: model for membrane structure consisting of a lipid bilayer with proteins associated as discrete globular entities that penetrate the bilayer to various extents and are free to move laterally in the membrane. (p. 164)

fluid-phase endocytosis: nonspecific uptake of extracellular fluid by infolding of the plasma membrane, followed by budding off of a membrane vesicle. (p. 348)

fluorescence: property of molecules that absorb light and then reemit the energy as light of a longer wavelength. (p. GM-8)

fluorescence microscopy: light microscopic technique that focuses ultraviolet rays on the specimen, thereby causing fluorescent compounds in the specimen to emit visible light. (p. GM-8)

fluorescence recovery after photobleaching: technique for measuring the lateral diffusion rate of membrane lipids or proteins in which such molecules are linked to a fluorescent dye and a tiny membrane region is then bleached with a laser. (p. 171)

fluorescent antibody: antibody containing covalently linked fluorescent dye molecules that allow the antibody to be used to locate antigen molecules microscopically. (p. 186)

FNR: see *ferredoxin-NADP+ reductase.*

Fₒ complex: group of hydrophobic membrane proteins that anchor the F₁ complex to either the inner mitochondrial membrane or the bacterial plasma membrane; serves as the proton translocator channel through which protons flow when the electrochemical gradient across the membrane is used to drive ATP synthesis. (p. 404)

FₒF₁ complex: protein complex in the mitochondrial inner membrane and the bacterial plasma membrane that consists of the F₁ complex bound to the Fₒ complex; the flow of protons through the Fₒ component leads to the synthesis of ATP by the F₁ component. (p. 404)

focal adhesion: localized points of attachment between cell surface integrin molecules and the extracellular matrix; contain clustered integrin molecules that interact with bundles of cytoskeletal actin microfilaments via several linker proteins. (p. 300)

focal length: distance between the midline of a lens and the point at which rays passing through the lens converge to a focus. (p. GM-3)

forming face: see *cis face.*

frameshift mutation: insertion or deletion of one or more base pairs in a DNA molecule, causing a change in the reading frame of the mRNA molecule that usually garbles the message. (p. 630)

free energy (G): thermodynamic function that measures the extractable energy content of a molecule; under conditions of constant temperature and pressure, the change in free energy is a measure of the ability of the system to do work. (p. 118)

free energy change (ΔG): thermodynamic parameter used to quantify the net free energy liberated or required by a reaction or process; measure of thermodynamic spontaneity. (p. 118)

freeze etching: cleaving a quick-frozen specimen, as in freeze-fracturing, followed by the subsequent sublimation of ice from the specimen surface to expose small areas of the true cell surface. (p. GM-25)

freeze fracturing: sample preparation technique for electron microscopy in which a frozen specimen is cleaved by a sharp blow followed by examination of the fractured surface, often the interior of a membrane. (pp. 175, GM-23)

fructose-2,6-bisphosphate (F2,6BP): a doubly phosphorylated fructose molecule, formed by the action of phosphofructokinase-2 on fructose-6-phosphate, that plays an important role in regulating both glycolysis and gluconeogenesis. (p. 391)

F-type ATPase: type of transport ATPase found in bacteria, mitochondria, and chloroplasts

that can use the energy of ATP hydrolysis to pump protons against their electrochemical gradient, and can also catalyze the reverse process, in which the exergonic flow of protons down their electrochemical gradient is used to drive ATP synthesis. (p. 208) Also see *ATP synthase.*

functional group: group of chemical elements covalently bonded to each other that confers characteristic chemical properties upon any molecule to which it is covalently linked. (p. 20)

G

G: see *guanine* or *free energy.*

ΔG: see *free energy change.*

ΔG°': see *standard free energy change.*

G protein: any of numerous GTP-binding regulatory proteins located in the plasma membrane that mediate signal transduction pathways, usually by activating a specific target protein such as an enzyme or channel protein. (p. 259)

G protein-linked receptor family: group of plasma membrane receptors that activate a specific G protein upon binding of the appropriate ligand. (p. 259)

G0 state: designation applied to eukaryotic cells that have become arrested in the G1 phase of the cell cycle and thus are no longer proliferating. (p. 525)

G1 checkpoint: control point near the end of G1 phase of the cell cycle where the cycle can be halted until conditions are suitable for progression to the next phase; regulated by factors such as cell size, nutrients, growth factors, and DNA damage. (p. 555)

G1 phase: stage of the eukaryotic cell cycle between the end of the previous division and the onset of chromosomal DNA synthesis. (p. 524)

G2 checkpoint: control point at the junction between the G2 and M phases of the cell cycle where the cycle can be halted until conditions are suitable for the onset of mitosis; regulated by factors such as cell size and DNA damage. (p. 555)

G2 phase: stage of the eukaryotic cell cycle between the completion of chromosomal DNA replication and the onset of cell division. (p. 524)

G-actin: globular monomeric form of actin that polymerizes to form F-actin. (p. 755)

GAG: see *glycosaminoglycan.*

gamete: haploid cells produced by each parent that fuse together to form the diploid offspring; for example, sperm or egg. (p. 579)

gametogenesis: the process that produces gametes. (p. 579)

gametophyte: haploid generation in the life cycle of an organism that alternates between haploid and diploid forms; form that produces gametes. (p. 581)

gamma tubulin (γ-tubulin): form of tubulin located in the centrosome, where it functions in the nucleation of microtubules. (p. 750)

ganglioside: a charged glycolipid containing the amino alcohol sphingosine and negatively charged sialic acid residues. (p. 166)

GAP: see *GTPase activating protein.*

gap junction: type of cell junction that provides a point of intimate contact between two adjacent cells through which ions and small molecules can pass. (pp. 98, 313)

Gb: gigabases; a billion base pairs (p. 492)

GEF: see *guanine-nucleotide exchange factor.*

gel electrophoresis: technique in which proteins or nucleic acids are separated in gels made of

polyacrylamide or agarose by placing the gel in an electric field. (p. 493)

gelation-solation: alternation between a thick, gelatinous state and a more fluid state of the cytoplasm. (p. 795)

gene: "hereditary factor" that specifies an inherited trait; consists of a DNA base sequence that codes for a functional product, usually a polypeptide chain, but in some cases an RNA molecule (e.g., rRNA, tRNA, snRNA). (pp. 10, 628)

gene amplification: mechanism for creating extra copies of individual genes by selectively replicating specific DNA sequences. (p. 707)

gene conversion: phenomenon in which genes undergo nonreciprocal recombination during meiosis, such that one of the recombining genes ends up on both chromosomes of a homologous pair rather than being exchanged from one chromosome to the other. (p. 604)

gene deletion: selective removal within a cell of specific DNA sequences whose products are not required. (p. 709)

gene locus (plural, **loci**): location within a chromosome that contains the DNA sequence for a particular gene. (p. 578)

general transcription factor: a protein that is always required for RNA polymerase to bind to its promoter and initiate RNA synthesis, regardless of the identity of the gene involved. (p. 642)

genetic code: set of rules specifying the relationship between the sequence of bases in a DNA or mRNA molecule and the order of amino acids in the polypeptide chain encoded by that DNA or mRNA. (p. 624)

genetic engineering: application of recombinant DNA technology to practical problems, primarily in medicine and agriculture. (p. 614)

genetic mapping: determining the sequential order and spacing of genes on a chromosome based on recombination frequencies. (p. 598)

genetic recombination: exchange of DNA segments between two different DNA molecules. (p. 582)

genetic transformation: change in the hereditary properties of a cell brought about by the uptake of foreign DNA (pp. 481, 599)

genetics: study of the behavior of genes, which are the chemical units involved in the storage and transmission of hereditary information. (p. 5)

genome: the DNA (or for some viruses, RNA) that contains one complete copy of all the genetic information of an organism or virus (p. 491)

genomic control: a regulatory change in the makeup or structural organization of the genome. (p. 707)

genomic library: collection of recombinant DNA clones produced by cleaving an organism's entire genome into fragments, usually using restriction enzymes, and then cloning all the fragments in an appropriate cloning vector. (p. 612)

genotype: genetic makeup of an organism. (p. 579)

Gᵢ: a G protein that acts as an inhibitor of signal transduction. (p. 259)

glial cell: cell that surrounds, supports, and insulates neurons; category includes microglia, oligodendrocytes, Schwann cells, and astrocytes. (p. 226)

globular protein: protein whose polypeptide chains are folded into compact structures rather the extended filaments. (p. 51)

gluconeogenesis: synthesis of glucose from precursors such as amino acids, glycerol, or lactate; occurs in the liver via a pathway that is essentially the reverse of glycolysis. (p. 386)

glucose: a six-carbon sugar that is widely used as the starting molecule in cellular energy metabolism. (p. 375)

glucose transporter (GluT): membrane carrier protein responsible for the facilitated diffusion of glucose. (p. 205)

GluT: see *glucose transporter.*

glycerol: three-carbon alcohol with a hydroxyl group on each carbon; serves as the backbone for triacylglycerols. (p. 68)

glycerol phosphate shuttle: mechanism for carrying electrons from cytosolic NADH into the mitochondrion, where the electrons are delivered to FAD in the respiratory complexes. (p. 436)

glycocalyx: carbohydrate-rich zone located at the outer boundary of many animal cells. (pp. 186, 302)

glycogen: highly branched storage polysaccharide in animal cells consisting of glucose repeating subunits linked by $\alpha(1 \rightarrow 4)$ bonds and $\alpha(1 \rightarrow 6)$ bonds. (p. 63)

glycolate pathway: light-dependent pathway that decreases the efficiency of photosynthesis by oxidizing reduced carbon compounds without capturing the released energy; occurs when oxygen substitutes for carbon dioxide in the reaction catalyzed by rubisco, thereby generating phosphoglycolate which is then converted to 3-phosphoglycerate in the peroxisome and mitochondrion; also called photorespiration. (p. 469)

glycolipid: lipid molecule containing a bound carbohydrate group. (pp. 70, 166)

glycolysis (glycolytic pathway): series of reactions by which glucose or some other monosaccharide is catabolized to pyruvate without the involvement of oxygen, generating two molecules of ATP per molecule of monosaccharide metabolized. (p. 378)

glycolytic pathway: see *glycolysis.*

glycoprotein: protein with one or more carbohydrate groups linked covalently to amino acid side chains. (pp. 185, 336)

glycosaminoglycan (GAG): polysaccharide constructed from a repeating disaccharide unit containing one sugar with an amino group and one sugar that usually has a negatively charged sulfate or carboxyl group; component of the extracellular matrix. (p. 294)

glycosidic bond: bond linking a sugar to another molecule, which may be another sugar molecule. (p. 62)

glycosylation: addition of carbohydrate side chains to specific amino acid residues of proteins, usually beginning in the lumen of the endoplasmic reticulum and completed in the Golgi complex. (pp. 185, 336)

glyoxylate cycle: modified version of the TCA cycle occurring in plant glyoxysomes; an anabolic pathway that converts two molecules of acetyl CoA to one molecule of succinate, thereby permitting the synthesis of carbohydrates from lipids. (p. 416)

glyoxysome: specialized type of plant peroxisome that contains enzymes responsible for the conversion of stored fat to carbohydrate in germinating seeds. (pp. 91, 361, 416)

Goldman equation: modification of the Nernst equation that calculates the resting membrane potential by adding together the effects of all relevant ions, each weighted for its relative permeability. (p. 232)

Golgi complex: stacks of flattened, disk-shaped membrane cisternae in eukaryotic cells that are important in the processing and packaging of secretory proteins and in the synthesis of complex polysaccharides. (pp. 89, 333)

Golgi stack: a series of flattened Golgi cisternae stacked on top of each other. (p. 333)

GPI-anchored membrane protein: protein bound to the outer surface of the plasma membrane by linkage to glycosylphosphatidylinositol (GPI), a glycolipid found in the external monolayer of the plasma membrane. (p. 180)

granum (plural, grana): stack of thylakoid membranes in a chloroplast. (pp. 87, 448)

group specificity: ability of an enzyme to act on any of a whole group of substrates as long as they possess some common structural feature. (p. 135)

group transfer reaction: chemical reaction that involves the movement of a chemical group from one molecule to another. (p. 372)

growth factor: any of a number of extracellular signaling proteins that stimulate cell division in specific types of target cells; examples include platelet-derived growth factor (PDGF) and epidermal growth factor (EGF). (pp. 273, 562)

G_s: a G protein that acts as a stimulator of signal transduction. (p. 259)

GTPase activating protein (GAP): a protein that speeds up the inactivation of Ras by facilitating the hydrolysis of bound GTP. (p. 271)

guanine (G): nitrogen-containing aromatic base, chemically designated as a purine, which serves as an informational monomeric unit when present in nucleic acids with other bases in a specific sequence; forms a complementary base pair with cytosine (C) by hydrogen bonding. (p. 55)

guanine-nucleotide exchange factor (GEF): a protein that triggers the release of GDP from the Ras protein, thereby permitting Ras to acquire a molecule of GTP. (p. 271)

gyrase: see *DNA gyrase.*

H

H: see *enthalpy.*

H zone: light region located in the middle of the A band of striated muscle myofibrils. (p. 780)

hairpin loop: looped structure formed when two adjacent segments of a nucleic acid chain are folded back on one another and held in that conformation by base pairing between complementary base sequences. (p. 639)

haploid: containing a single set of chromosomes and therefore a single copy of the genome; can describe a cell, nucleus, or organism composed of such cells. (p. 578)

haploid spore: haploid product of meiosis in organisms that display an alternation of generations; gives rise upon spore germination to the haploid form of the organism (the gametophyte, in the case of higher plants). (p. 581)

Hatch-Slack cycle: series of reactions in C_4 plants in which carbon dioxide is fixed in the mesophyll cells and transported as a four-carbon compound to the bundle sheath cells, where subsequent decarboxylation results in a higher concentration of carbon dioxide and therefore a higher rate of carbon fixation by rubisco. (p. 471)

Haworth projection: model that depicts the chemical structure of a molecule in a way that suggests the spatial relationship of different parts of the molecule. (p. 62)

HCI: see *heme-controlled inhibitor.*

heart muscle: see *cardiac muscle.*

heat: transfer of energy as a result of a temperature difference. (p. 108)

heat of vaporization: amount of energy required to convert one gram of a liquid into vapor. (p. 23)

heat-shock gene: gene whose transcription is activated when cells are exposed to elevated temperatures or other stressful conditions. Some heat-shock genes code for molecular chaperones that, in addition to guiding normal protein folding, can also facilitate the refolding of heat-damaged proteins. (p. 726)

heat-shock response element: DNA base sequence located adjacent to heat-shock genes that functions as a binding site for the heat-shock transcription factor. (p. 726)

helicase: any of several enzymes that unwind the DNA double helix, driven by energy derived from ATP hydrolysis. (p. 537)

helix: see *alpha helix* or *double helix.*

helix-loop-helix: DNA-binding motif found in transcription factors that is composed of a short α helix connected by a loop to a longer α helix. (p. 724)

helix-turn-helix: DNA-binding motif found in many regulatory transcription factors; consists of two regions of α helix separated by a bend in the polypeptide chain. (p. 722)

heme-controlled inhibitor (HCI): protein kinase which, in the absence of heme, catalyzes the phosphorylation of eukaryotic initiation factor 2 (eIF2), thereby inhibiting protein synthesis. (p. 730)

hemicellulose: heterogeneous group of polysaccharides deposited along with cellulose in the cell walls of plants and fungi to provide added strength; each consists of a long, linear chain of a single kind of sugar with short side chains. (p. 315)

hemidesmosome: point of attachment between cell surface integrin molecules of epithelial cells and the basal lamina; contains integrin molecules that are anchored via linker proteins to intermediate filaments of the cytoskeleton. (p. 300)

heterochromatin: highly compacted form of chromatin present during interphase; contains DNA that is not being transcribed. (p. 507) Also see *euchromatin.*

heterodimer ($\alpha\beta$-heterodimer): see *alpha beta heterodimer.*

heterokaryon: hybrid cell created by fusing two cells to form a single cell with two nuclei. (p. 556)

heterophagic lysosome: mature lysosome containing hydrolytic enzymes involved in the digestion of materials of extracellular origin. (p. 355) Also see *autophagic lysosome.*

heterophilic interaction: binding of two different molecules to each other. (p. 303) Also see *homophilic interaction.*

heterozygous: having two different alleles for a given gene. (p. 578) Also see *homozygous.*

Hfr cell: bacterial cell in which the F factor has become integrated into the bacterial chromosome, allowing the cell to transfer genomic DNA during conjugation. (p. 601)

hierarchical assembly: synthesis of biological structures from simple starting molecules to progressively more complex structures, usually by self-assembly. (p. 36)

high-energy compound: molecule that possesses one or more bonds whose hydrolysis is highly exergonic. (p. 370)

high-mobility group (HMG) proteins: group of nonhistone proteins associated with eukaryotic chromatin that exhibit high mobility when subjected to electrophoresis; thought to include proteins that regulate gene expression. (p. 715)

high-voltage electron microscope (HVEM): an electron microscope that uses accelerating voltages up to a thousand or more kilovolts, thereby allowing the examination of thicker samples than is possible with a conventional electron microscope. (p. GM-19)

histamine: a paracrine hormone produced by decarboxylation of the amino acid histidine that is responsible for local inflammatory responses. (p. 278)

histone: class of basic proteins found in eukaryotic chromosomes; an octamer of histones forms the core of nucleosomes. (p. 504)

HMG proteins: see *high-mobility group proteins.*

Holliday junction: X-shaped structure produced when two DNA molecules are joined together by single-strand crossovers during genetic recombination. (p. 604)

homeobox: highly conserved DNA sequence found in homeotic genes; codes for a DNA-binding protein domain present in transcription factors that are important regulators of gene expression during development. (p. 728)

homeodomain: amino acid sequence, about 60 amino acids long, found in transcription factors encoded by homeotic genes; contains a helix-turn-helix DNA-binding motif. (p. 728)

homeotic gene: family of genes that control formation of the body plan during embryonic development; code for transcription factors possessing homeodomains. (p. 727)

homeoviscous adaptation: alterations in membrane lipid composition that keep the viscosity of membranes approximately the same despite changes in environmental temperature. (p. 174)

homogenate: suspension of cell organelles, smaller cellular components, and molecules produced by disrupting cells or tissues using techniques such as grinding, ultrasonic vibration, or osmotic shock. (p. 326)

homogenization: disruption of cells or tissues using techniques such as grinding, ultrasonic vibration, or osmotic shock. (p. 326)

homologous chromosomes: two copies of a specific chromosome, one derived from each parent, that pair with each other and exchange genetic information during meiosis. (p. 578)

homologous recombination: exchange of genetic information between two DNA molecules exhibiting extensive sequence similarity. (p. 602)

homophilic interaction: binding of two identical molecules to each other. (p. 303) Also see *heterophilic interaction.*

homozygous: having two identical alleles for a given gene. (p. 578) Also see *heterozygous.*

hopanoid: any of a family of sterol-like molecules that substitute for membrane sterols in some prokaryotes. (p. 166)

hormone: chemical that is synthesized in one organ, secreted into the blood, and able to cause a physiological change in cells or tissues of another organ. (p. 276)

hormone response element: DNA base sequence that selectively binds to a hormone-receptor complex, resulting in the activation (or inhibition) of transcription of nearby genes. (p. 724)

HVEM: see *high-voltage electron microscope.*

hyaluronate: a glycosaminoglycan found in high concentration in the extracellular matrix where cells are actively proliferating or migrating, and in the joints between movable bones. (p. 295)

hybrid: product of the cross of two genetically different parents. (p. 590)

hybrid model: model postulating that late endosomes and lysosomes fuse to form a hybrid organelle, with lipids and proteins from the two organelles not clearly segregated. (p. 354)

hybridization: see *nucleic acid hybridization.*

hydrocarbon: an organic molecule consisting only of carbon and hydrogen atoms; not generally compatible with living cells. (p. 20)

hydrogen bond: weak attractive interaction between an electronegative atom and a hydrogen atom that is covalently linked to a second electronegative atom. (p. 46)

hydrogenation: addition of electrons plus hydrogen ions (protons) to an organic molecule; reduction. (p. 374)

hydropathy index: a value representing the average of the hydrophobicity values for a short stretch of contiguous amino acids within a protein. (p. 179)

hydropathy (hydrophobicity) plot: graph showing the location of hydrophobic amino acid clusters within the primary sequence of a protein molecule; used to determine the likely locations of transmembrane segments within integral membrane proteins. (p. 179)

hydrophilic: describing molecules or regions of molecules that readily associate with or dissolve in water because of a preponderance of polar groups; "water-loving." (p. 23)

hydrophobic: describing molecules or regions of molecules that are poorly soluble in water because of a preponderance of nonpolar groups; "water-hating." (p. 23)

hydrophobic interaction: tendency of hydrophobic groups to be excluded from interactions with water molecules. (p. 46)

hydrophobicity plot: see *hydropathy plot.*

hydroxylation: chemical reaction in which a hydroxyl group is added to an organic molecule. (p. 331)

hyperpolarization (undershoot): state in which the membrane potential is more negative than the normal membrane potential. (p. 238)

hypothesis: a statement or explanation that is consistent with most of the observational and experimental evidence to date. (p. 11)

I

I band: region of a striated muscle myofibril that appears as a light band when viewed by microscopy; contains those regions of the thin actin filaments that do not overlap the thick myosin filaments. (p. 780)

IF: see *intermediate filament.*

IgSF: see *immunoglobulin superfamily.*

immunoelectron microscopy: electron microscopic technique for visualizing antibodies within cells by linking them to substances that are electron dense and therefore visible as opaque dots. (p. GM-22)

immunofluorescence microscopy: technique in which antibodies are labeled with a fluorescent dye to enable them to be identified and localized microscopically based on their fluorescence. (p. GM-9)

immunoglobulin superfamily (IgSF): family of cell surface proteins involved in cell-cell adhesion that are structurally related to the immunoglobulin subunits of antibody molecules. (p. 303)

IMP: see *intramembranous particle.*

importin: receptor protein that binds to the nuclear localization signal of proteins in the cytosol and then transports the bound protein through the nuclear pore complex and into the nucleus. (p. 514)

indirect active transport: membrane transport involving the cotransport of two solutes in which the movement of one solute down its gradient drives the movement of the other solute up its gradient. (p. 208)

indirect immunofluorescence: type of fluorescence microcopy in which a specimen is first exposed to an unlabeled antibody that attaches to specific antigenic sites within the specimen and is then stained with a secondary, fluorescent antibody that binds to the first antibody. (p. GM-9)

induced-fit model: model postulating that the active site of an enzyme is relatively specific for its substrate before it binds, but even more so thereafter because of a conformational change in the enzyme induced by the substrate. (p. 138)

inducer: any effector molecule that activates the transcription of an inducible operon. (p. 696)

inducible enzyme: enzyme whose synthesis is regulated by the presence or absence of its substrate. (p. 693)

inducible operon: group of adjoining genes whose transcription is activated in the presence of an inducer. (p. 695)

informational macromolecule: polymer of non-identical subunits ordered in a nonrandom sequence that carries information important to the function or utilization of the macromolecule. Nucleic acids and proteins are informational macromolecules. (p. 28)

inhibition (of enzyme activity): decreasing the catalytic activity of an enzyme, either by a change in conformation or by chemical modification of one of its functional groups. (p. 146)

inhibitory postsynaptic potential (IPSP): small hyperpolarization of the postsynaptic membrane triggered by binding of an inhibitory neurotransmitter to its receptor, thereby reducing the amplitude of subsequent excitatory postsynaptic potentials and possibly preventing the firing of an action potential. (p. 250) Also see *excitatory postsynaptic potential.*

initial reaction velocity (v): Reaction rate measured over a period of time during which the substrate concentration has not decreased enough to affect the rate and the accumulation of product is still too small to cause any measurable back reaction. (p. 140)

initiation codon (start codon): the codon AUG in mRNA when it functions as the starting point for protein synthesis. (pp. 634, 666)

initiation complex (30S): complex formed by the association of mRNA, the 30S ribosomal subunit, an initiator aminoacyl tRNA molecule, and initiation factor IF2. (p. 669)

initiation complex (70S): complex formed by the association of a 30S initiation complex with a 50S ribosomal subunit; contains an initiator aminoacyl tRNA at the P site and is ready to commence mRNA translation. (p. 669)

initiation factors: group of proteins that promote the binding of ribosomal subunits to mRNA and initiator tRNA, thereby initiating the process of protein synthesis. (p. 668)

initiator (Inr): short DNA sequence surrounding the transcription startpoint that forms part of the promoter for RNA polymerase II. (p. 641)

initiator tRNA: type of transfer RNA molecule that starts the process of translation; recognizes the AUG start codon and carries formylmethionine in prokaryotes or methionine in eukaryotes. (p. 669)

inner membrane: the inner of the two membranes that surround a mitochondrion, chloroplast, or nucleus. (pp. 403, 448)

innervate: to supply a tissue with nerves. (p. 226)

inositol-1,4,5-trisphosphate (InsP₃): triply phosphorylated inositol molecule formed as a product of the cleavage of phosphatidylinositol-4,5-bisphosphate catalyzed by phospholipase C; functions as a second messenger by triggering the release of calcium ions from storage sites within the endoplasmic reticulum. (p. 263)

Inr: see *initiator*.

InsP₃: see *inositol-1,4,5-trisphosphate*.

InsP₃ receptor: ligand-gated calcium channel in the ER membrane that opens when bound to InsP₃, allowing calcium ions to flow from the ER lumen into the cytosol. (p. 264)

integral membrane protein: hydrophobic protein localized within the interior of a membrane but possessing hydrophilic regions that protrude from one or both membrane surfaces. (p. 177)

integral monotopic protein: an integral membrane protein embedded in only one side of the lipid bilayer. (p. 177)

integrin: any of several plasma membrane receptors that bind to extracellular matrix components at the outer membrane surface and interact with cytoskeletal components at the inner membrane surface; includes receptors for fibronectin, laminin, and collagen. (p. 298)

intercalated disc: membrane partition enriched in gap junctions that divides cardiac muscle into separate cells containing single nuclei. (p. 789)

intercellular communication: exchange of cellular components between adjacent cells through direct cytoplasmic connections, either gap junctions in animal cells or plasmodesmata in plant cells. (p. 160)

interdoublet link: link between adjacent doublets in the axoneme of a eukaryotic cilium or flagellum, believed to limit the extent of doublet movement with respect to each other as the axoneme bends. (p. 776)

interference: process by which two or more waves of light combine to reinforce or cancel one another, producing a wave equal to the sum of the two combining waves. (p. GM-3)

intermediate filament (IF): group of protein filaments that are the most stable components of the cytoskeleton of eukaryotic cells; exhibit a diameter of 8-12 nm, which is intermediate between the diameters of actin microfilaments and microtubules. (pp. 97, 762)

intermediate lens (electron microscope): electromagnetic lens positioned between the objective and projector lenses in a transmission electron microscope. (p. GM-18)

intermediate lens (light microscope): lens positioned between the ocular and objective lenses in a light microscope. (p. GM-5)

intermembrane space: region of a mitochondrion or chloroplast between the inner and outer membranes. (pp. 402, 448)

internal energy (E): total energy stored within a system; cannot be measured directly, but the change in internal energy, ΔE, is measurable. (p. 113)

interneuron: nerve cell that processes signals received from other neurons and relays the information to other parts of the nervous system. (p. 226)

interphase: growth phase of the eukaryotic cell cycle situated between successive division phases (M phases); composed of G1, S, and G2 phases. (p. 524)

interspersed repeated DNA: repeated DNA sequences whose multiple copies are scattered around the genome. (p. 501)

intracellular membrane: any cellular membrane internal to the plasma membrane; such membranes serve to compartmentalize functions within eukaryotic cells. (p. 159)

intraflagellar transport: movement of components to and from the tips of flagella driven by both plus- and minus-end directed microtubule motor proteins. (p. 777)

intramembranous particle (IMP): integral membrane protein that is visible as a particle when the interior of a membrane is visualized by freeze-fracture microscopy. (p. GM-24)

intron: nucleotide sequence in an RNA molecule that is part of the primary transcript but not the mature, functional RNA molecule. (p. 649) Also see *exon*.

invasion: direct spread of cancer cells into neighboring tissues. (p. 571)

inverted repeat: DNA segment containing two copies of the same base sequence oriented in opposite directions. (p. 725)

ion channel: membrane protein that allows the passage of specific ions through the membrane; generally regulated by either changes in membrane potential (voltage-gated channels) or binding of a specific ligand (ligand-gated channels). (pp. 206, 233)

ionic bond: attractive force between a positively charged chemical group and a negatively charged chemical group. (p. 46)

IPSP: see *inhibitory postsynaptic potential*.

IRE: see *iron-response element*.

iron-response element (IRE): short base sequence found in mRNAs whose translation or stability is controlled by iron; binding site for an IRE-binding protein. (p. 730)

iron-sulfur (Fe-S) protein: protein that contains iron and sulfur atoms complexed with four cysteine groups and that serves as an electron carrier in the electron transport system. (p. 419)

irreversible inhibitor: molecule that binds to an enzyme covalently, causing an irrevocable loss of catalytic activity. (p. 146)

isoenzyme: any of several, physically distinct proteins that catalyze the same reaction. (p. 467)

J

J: see *joule*.

JAK: see *Janus activated kinase*.

Janus activated kinase (JAK): cytoplasmic protein kinase which, after activation by cell surface receptors, catalyzes the phosphorylation and activation of STAT transcription factors. (p. 726)

joule (J): a unit of energy corresponding to 0.239 calories. (p. 113)

junctional complex: dense-staining material that appears to connect a T tubule to the membranes of the adjoining terminal cisternae in skeletal muscle. (p. 788)

K

karyotype: picture of the complete set of chromosomes for a particular cell type, organized as homologous pairs arranged on the basis of differences in size and shape. (p. 545)

Kb: kilobases; a thousand base pairs (p. 492)

k_cat: see *turnover number*.

K_d: see *dissociation constant*.

K_eq: see *equilibrium constant*.

KIF: see *kinesin family member*.

kinesin: family of motor proteins that generate movement along microtubules using energy derived from ATP hydrolysis. (p. 770)

kinesin family member (KIF): any of numerous proteins with a motor domain resembling that of kinesin, although it can be found in different areas in different molecules. (p. 772)

kinetochore: DNA-protein complex located at the centromere region of a chromosome that provides the attachment site for spindle microtubules during mitosis or meiosis. (p. 545)

kinetochore microtubule: spindle microtubule that attaches to chromosomal kinetochores. (p. 545)

K_m: see *Michaelis constant*.

Krebs cycle: see *tricarboxylic acid cycle*.

L

lac operon: group of adjoining bacterial genes that code for enzymes involved in lactose metabolism and whose transcription is selectively inhibited by the *lac* repressor. (p. 694)

lactate fermentation: anaerobic catabolism of carbohydrates with lactate as the end product. (p. 384)

lagging strand: strand of DNA that grows in the 3′ → 5′ direction during DNA replication by discontinuous synthesis of short fragments in the 5′ → 3′ direction, followed by ligation of adjacent fragments. (p. 534) Also see *leading strand*.

lamellipodium (plural, lamellipodia): thin sheet of flattened cytoplasm that transiently protrudes from the surface of eukaryotic cells during cell crawling; supported by actin filaments. (p. 792)

laminin: adhesive glycoprotein of the extracellular matrix, localized predominantly in the basal lamina of epithelial cells. (p. 297)

large ribosomal subunit: component of a ribosome with a sedimentation coefficient of 60S in eukaryotes and 50S in prokaryotes; associates with a small ribosomal subunit to form a functional ribosome. (p. 94)

late endosome: vesicle containing newly synthesized acid hydrolases plus material fated for digestion; activated either by lowering the pH of the late endosome or transferring its material to an existing lysosome. (p. 354)

lateral diffusion: diffusion of a membrane lipid or protein in the plane of the membrane. (p. 169)

law: a theory that has been so thoroughly tested and confirmed over a long period of time by a large number of investigators that virtually no doubt remains as to its validity. (p. 13)

law of independent assortment: principle stating that the alleles of each gene separate independently of the alleles of other genes during gamete formation. (p. 593)

law of segregation: principle stating that the alleles of each gene separate from each other during gamete formation. (p. 592)

LDL: see *low-density lipoprotein*.

LDL receptor: plasma membrane protein that serves as a receptor for binding extracellular LDL, which is then taken into the cell by receptor-mediated endocytosis. (p. 346)

leader (on mRNA): nontranslated sequence at the 5′ end of an mRNA molecule located prior to the start codon. (p. 666)

leading strand: strand of DNA that grows as a continuous chain in the 5′ → 3′ direction during DNA replication. (p. 532) Also see *lagging strand*.

leaf peroxisome: special type of peroxisome found in the leaves of photosynthetic plant cells that contains the enzymes involved in photorespiration. (pp. 91, 361, 469)

lectin: any of numerous carbohydrate-binding proteins that can be isolated from plant or animal cells and that promote cell-cell adhesion. (p. 304)

leptotene: first stage of prophase I of meiosis, characterized by condensation of chromatin fibers into visible chromosomes. (p. 583)

leucine zipper: DNA-binding motif found in many transcription factors; formed by an interaction between α helices in two polypeptide chains that are "zippered" together by hydrophobic interactions between leucine residues. (p. 722)

LHC: see *light-harvesting complex.*

LHCI: see *light-harvesting complex I.*

LHCII: see *light-harvesting complex II.*

licensing: process of making DNA competent for replication; normally occurs only once per cell cycle. (p. 541)

ligand: substance that binds to a specific receptor, thereby initiating the particular event or series of events for which that receptor is responsible. (p. 257)

ligand-gated ion channel: an integral membrane protein that forms an ion-conducting pore that opens when a specific molecule (ligand) binds to the channel. (p. 233)

light microscope: instrument consisting of a source of visible light and a system of glass lenses that allows an enlarged image of a specimen to be viewed. (pp. 5, GM-1)

light-harvesting complex (LHC): collection of light-absorbing pigments, usually chlorophylls and carotenoids, linked together by proteins; unlike a photosystem, does not contain a reaction center, but absorbs photons of light and funnels the energy to a nearby photosystem. (p. 454)

light-harvesting complex I (LHCI): the light-harvesting complex associated with photosystem I. (p. 458)

light-harvesting complex II (LHCII): the light-harvesting complex associated with photosystem II. (p. 456)

lignin: insoluble polymers of aromatic alcohols that occur mainly in woody plant tissues, where they contribute to the hardening of the cell wall and the structural strength we associate with wood. (p. 316)

limit of resolution: measurement of how far apart adjacent objects must be in order to be distinguished as separate entities. (pp. 5, GM-5)

Lineweaver-Burk equation: linear equation obtained by inverting the Michaelis-Menten equation, useful in determining parameters V_{max} and K_m and in the analysis of enzyme inhibition. (p. 143)

linkage group: group of genes that are transmitted, inherited, and assorted together. (p. 596)

linked genes: genes that are usually inherited together because they are located relatively close to each other on the same chromosome. (p. 596)

lipid: any of a large and chemically diverse class of organic compounds that are poorly soluble or insoluble in water but soluble in organic solvents. (p. 66)

lipid bilayer: unit of membrane structure, consisting of two layers of lipid molecules (mainly phospholipid) arranged so that their hydrophobic tails face toward each other and the polar region of each faces the aqueous environment on one side or the other of the bilayer. (pp. 24, 84, 162)

lipid-anchored membrane protein: protein located on a membrane surface that is covalently bound to one or more lipid molecules residing within the lipid bilayer. (p. 179) Also see *fatty acid-anchored membrane protein, prenylated membrane protein,* and *GPI-anchored membrane protein.*

looped domain: folding of a 30-nm chromatin fiber into loops 50,000-100,000 bp in length by periodic attachment of the DNA to an insoluble network of nonhistone proteins. (p. 507)

low-density lipoprotein (LDL): cholesterol-containing protein-lipid complex that transports cholesterol through the bloodstream and is taken up by cells; exhibits low density because of its high cholesterol content. (p. 346)

lumen: internal space enclosed by a membrane, usually the endoplasmic reticulum or related membrane systems. (p. 89)

lysosomal storage disease: disease resulting from a deficiency of one or more lysosomal enzymes and characterized by the undesirable accumulation of excessive amounts of specific substances that would normally be degraded by the deficient enzymes. (p. 357)

lysosome: membrane-bounded organelle containing digestive enzymes capable of degrading all the major classes of biological macromolecules. (pp. 90, 353)

M

M line: dark line running down the middle of the H zone of a striated muscle myofibril. (p. 780)

M phase: stage of the eukaryotic cell cycle when the nucleus and the rest of the cell divides. (p. 523)

macromolecule: polymer built from small repeating monomer units, with molecular weights ranging from a few thousand to hundreds of millions. (p. 26)

macrophagy: process by which an organelle becomes wrapped in a double membrane derived from the endoplasmic reticulum, creating an autophagic vacuole that acquires lysosomal enzymes which degrade the organelle. (p. 356)

malignant tumor: tumor that can invade neighboring tissues and spread via body fluids, especially the bloodstream, to other parts of the body; also called a *cancer.* (p. 565)

MAP: see *microtubule-associated protein.*

MAP kinase: see *mitogen-activated protein kinase.*

MAPK: see *mitogen-activated protein kinase.*

mast cell: a cell that stores and secretes histamine. (p. 281)

MAT locus: site in the yeast genome where the active allele for mating type resides. (p. 709)

mating bridge: transient cytoplasmic connection through which DNA is transferred from a male bacterial cell to a female cell during conjugation. (p. 600)

mating type: equivalent of sexuality (male or female) in lower organisms, where the molecular properties of a gamete determine the type of gamete with which it can fuse. (p. 579)

matrix: unstructured semifluid substance that fills the interior of a mitochondrion. (pp. 86, 404)

maturing face: see *trans face.*

maximum ATP yield: maximum amount of ATP produced per molecule of glucose oxidized by aerobic respiration; usually 38 molecules of ATP for prokaryotic cells and either 36 or 38 for eukaryotic cells. (p. 435)

maximum velocity (V_{max}): upper limiting reaction rate approached by an enzyme-catalyzed reaction as the substrate concentration approaches infinity. (p. 142)

Mb: megabases; a million base pairs (p. 492)

MDR: see *multidrug resistance transport protein.*

mechanical work: use of energy to bring about a physical change in the position or orientation of a cell or some part of it. (p. 108)

mechanoenzyme: see *motor protein.*

medial cisterna: flattened membrane sac of the Golgi complex located between the membrane tubules of the *cis*-Golgi network and the *trans*-Golgi network. (p. 334)

medial element: region of sarcoplasmic reticulum consisting of an interconnected network of membrane tubules that run between the terminal cisternae in skeletal muscle. (p. 787)

meiosis: series of two cell divisions, preceded by a single round of DNA replication, that converts a single diploid cell into four haploid cells (or haploid nuclei). (p. 580)

meiosis I: the first meiotic division, which produces two haploid cells with chromosomes composed of sister chromatids. (p. 581)

meiosis II: the second meiotic division, which separates the sister chromatids of the haploid cell generated by the first meiotic division. (p. 588)

membrane: permeability barrier surrounding and delineating cells and organelles; consists of a lipid bilayer with associated proteins. (pp. 24, 158)

membrane asymmetry: a membrane property based on differences between the molecular compositions of the two lipid monolayers and the proteins associated with each. (p. 169)

membrane potential: voltage across a membrane created by ion gradients; usually, the inside of a cell is negatively charged with respect to the outside. (pp. 197, 228)

Mendel's laws of inheritance: principles derived by Mendel from his work on the inheritance of traits in pea plants. (p. 592). Also see *law of segregation* and *law of independent assortment.*

mesophyll cell: outer cell in the leaf of a C_4 plant that serves as the site of carbon fixation by the Hatch-Slack cycle. (p. 462)

messenger RNA (mRNA): RNA molecule containing the information that specifies the amino acid sequence of one or more polypeptides. (p. 623)

metabolic pathway: series of cellular enzymatic reactions that convert one molecule to another via a series of intermediates. (p. 368)

metabolism: all chemical reactions occurring within a cell. (p. 368)

metaphase: stage during mitosis or meiosis when the chromosomes become aligned at the spindle equator. (p. 545)

metastable state: condition where potential reactants are thermodynamically unstable but have insufficient energy to exceed the activation energy barrier for the reaction. (p. 131)

metastasis: spread of tumor cells to distant organs via the bloodstream or other body fluids. (p. 571)

MF: see *microfilament.*

Michaelis constant (K_m): substrate concentration at which an enzyme-catalyzed reaction is proceeding at one-half of its maximum velocity. (p. 143)

Michaelis-Menten equation: widely used equation describing the relationship between velocity and substrate concentration for an enzyme-catalyzed reaction: $V = V_{max}[S]/(K_m + [S])$. (p. 142)

microautophagy: process by which lysosomes take up and degrade cytosolic proteins. (p. 734)

microbody: early term for a peroxisome based on its appearance in electron micrographs. (p. 359)

microfibril: aggregate of several dozen cellulose molecules laterally crosslinked by hydrogen bonds; serves as a structural component of plant and fungal cell walls. (p. 315)

microfilament (MF): polymer of actin, with a diameter of about 7 nm, that is an integral part of the cytoskeleton, contributing to the support, shape, and mobility of eukaryotic cells. (pp. 97, 754)

microfilament-based movement: motility based on microfilaments composed of actin and their interaction with myosin; includes muscle contraction, amoeboid movement, cytoplasmic streaming, and cytokinesis in animal cells. (p. 770)

micrometer (μm): unit of measure: 1 micrometer = 10^{-6} meters. (p. 2)

microphagy: process by which small bits of cytoplasm are surrounded by ER membrane to form an autophagic vesicle whose contents are then digested either by accumulation of lysosomal enzymes or by fusion of the vesicle with a late endosome. (p. 356)

microscopic autoradiography: procedure in which specimens being examined by light or electron microscopy are overlaid with a photographic emulsion to permit detection of radioactive molecules. (p. GM-16)

microsome: vesicle formed by fragments of endoplasmic reticulum when tissue is homogenized. (p. 325)

microtome: instrument used to slice an embedded biological specimen into thin sections for light microscopy. (p. GM-16)

microtubule (MT): polymer of the protein tubulin, with a diameter of about 25 nm, that is an integral part of the cytoskeleton and that contributes to the support, shape, and motility of eukaryotic cells; also found in eukaryotic cilia and flagella. (pp. 96, 744)

microtubule-associated motor protein: see *motor MAP*.

microtubule-associated protein (MAP): any of various accessory proteins that bind to microtubules and modulate their assembly, structure, and/or function. (p. 753)

microtubule-based movement: motility based on microtubules; includes motility involving cilia, flagella, and sperm tails, as well as chromosomal movements mediated by spindle microtubules. (p. 770)

microtubule-organizing center (MTOC): structure that initiates the assembly of microtubules, the primary example being the centrosome. (p. 749)

microvillus (plural, microvilli): fingerlike projection from the cell surface that increases membrane surface area; important in cells that have an absorption function, such as those that line the intestine. (p. 759)

middle lamella: first layer of the plant cell wall to be synthesized; ends up farthest away from the plasma membrane, where it functions to hold adjacent cells together. (p. 316)

minus end (of microtubule): slower-growing (or non-growing or shrinking) end of a microtubule. (p. 748)

mismatch repair: DNA repair mechanism that detects and corrects base pairs that are improperly hydrogen bonded. (p. 544)

mitochondrion (plural, mitochondria): double membrane-enclosed cytoplasmic organelle of eukaryotic cells that is the site of aerobic respiration and hence of ATP generation. (pp. 85, 400)

mitogen-activated protein kinase (MAP kinase, or MAPK): a family of protein kinases that are activated when cells receive a signal to grow and divide. (p. 271)

mitosis: process by which two genetically identical daughter nuclei are produced from one nucleus as the duplicated chromosomes of the parent cell segregate into separate nuclei; usually followed by cell division. (p. 523)

mitosis-promoting factor: see *MPF*.

mitotic index: percentage of cells in a population that are in any stage of mitosis at a certain point in time; used to estimate the relative length of the M phase of the cell cycle. (p. 525)

mitotic spindle: microtubular structure responsible for separating chromosomes during mitosis. (p. 545)

MLCK: see *myosin light-chain kinase*.

molecular chaperone: a protein that facilitates the folding of other proteins but is not a component of the final folded structure. (pp. 33, 672)

monomer: small organic molecule that serves as a subunit in the assembly of a macromolecule. (p. 28)

monomeric protein: protein that consists of a single polypeptide chain. (p. 45)

monosaccharide: simple sugar; the repeating unit of polysaccharides. (p. 61)

motif: region of protein secondary structure consisting of small segments of α helix and/or β sheet connected by looped regions of varying length. (p. 50)

motility (cellular): movement or shortening of a cell, movement of components within a cell, or movement of environmental components past or through a cell. (p. 769)

motor MAP (microtubule-associated motor protein): protein, such as kinesin or dynein, that uses energy derived from ATP to drive the transport of vesicles and organelles along microtubules or to generate sliding forces between microtubules. (pp. 753, 770)

motor neuron: nerve cell that transmits impulses from the central nervous system to muscles or glands. (p. 226)

motor protein (mechanoenzyme): protein that uses energy derived from ATP to change shape in a way that exerts force and causes attached structures to move; includes three families of proteins (myosin, dynein, and kinesin) that interact with cytoskeletal elements (microtubules and microfilaments) to produce movements. (pp. 551, 770)

MPF (mitosis-promoting factor): protein kinase consisting of mitotic Cdk bound to mitotic cyclin; controls the G2 checkpoint and spindle assembly checkpoint by phosphorylating proteins involved in key stages of mitosis. (p. 557)

mRNA: see *messenger RNA*.

mRNA-binding site: place on the ribosome where mRNA binds during protein synthesis. (p. 661)

MT: see *microtubule*.

MTOC: see *microtubule-organizing center*.

multidrug resistance (MDR) transport protein: an ABC-type ATPase that uses the energy of ATP hydrolysis to pump hydrophobic drugs out of cells. (p. 210)

multimeric protein: protein that consists of two or more polypeptide chains. (p. 45)

multiphoton excitation microscopy: specialized type of fluorescence light microscope employing a laser beam that emits rapid pulses of light; images are similar in sharpness to confocal microscopy but photodamage is minimized because there is little out-of-focus light. (p. GM-13)

multiprotein complex: two or more proteins (usually enzymes) bound together in a way that allows each protein to play a sequential role in the same multistep process. (p. 54)

muscle contraction: generation of tension in muscle cells by the sliding of thin (actin) filaments past thick (myosin) filaments. (p. 779)

muscle fiber: long, thin, multinucleate cell specialized for contraction. (p. 779)

mutagen: chemical or physical agent capable of inducing mutations. (p. 628)

mutation: change in the base sequence of a DNA molecule. (p. 541)

myelin sheath: concentric layers of membrane that surround an axon and serve as electrical insulation that allows rapid transmission of nerve impulses. (p. 227)

myofibril: cylindrical structure composed of an organized array of thin actin filaments and thick myosin filaments; found in the cytoplasm of skeletal muscle cells. (p. 779)

myoglobin: protein that binds and stores oxygen in muscle cells. (p. 789)

myosin: family of motor proteins that create movements by exerting force on actin microfilaments using energy derived from ATP hydrolysis; makes up the thick filaments that move the actin thin filaments during muscle contraction. (p. 777)

myosin light-chain kinase (MLCK): enzyme that phosphorylates myosin light chains, thereby triggering smooth muscle contraction. (p. 790)

myosin subfragment 1 (S1): proteolytic fragment of myosin that binds to actin microfilaments in a way that yields a distinctive arrowhead pattern, with all the S1 molecules pointing in the same direction. (p. 756)

N

NA: see *numerical aperture*.

Na$^+$/glucose symporter: a membrane transport protein that simultaneously transports glucose and sodium ions into cells, with the movement of sodium ions down their electrochemical gradient driving the transport of glucose against its concentration gradient. (p. 214)

Na$^+$/K$^+$ ATPase: see *Na$^+$/K$^+$ pump*.

Na$^+$/K$^+$ pump: membrane carrier protein that couples ATP hydrolysis to the inward transport of potassium ions and the outward transport of sodium ions to maintain the Na$^+$ and K$^+$ gradients that exist across the plasma membrane of most animal cells. (pp. 211, 230)

NAD$^+$: see *nicotinamide adenine dinucleotide*.

NADH-coenzyme Q oxidoreductase: see *complex I*.

NADP$^+$: see *nicotinamide adenine dinucleotide phosphate*.

nanometer (nm): unit of measure: 1 nanometer = 10^{-9} meters. (p. 3)

native conformation: three-dimensional folding of a polypeptide chain into a shape that represents the most stable state for that particular sequence of amino acids. (p. 50)

N-CAM: see *neural cell adhesion molecule*.

negative control: genetic control in which the key regulatory element acts by inhibiting gene transcription. (p. 700)

negative staining: technique in which an unstained specimen is visualized in a transmission electron microscope against a darkly stained background. (p. GM-22)

NER: see *nucleotide excision repair*.

Nernst equation: equation for calculating the equilibrium membrane potential for a given ion: $E_x = (RT/zF) \ln [X]_{outside}/[X]_{inside}$. (p. 230)

nerve: a tissue composed of bundles of axons. (p. 227)

nerve impulse: signal transmitted along nerve cells by a wave of depolarization-repolarization events propagated along the axonal membrane. (p. 241)

nervous system: group of cells, tissues, and organs that collect, process, and respond to information from the environment and from within the organism by the transmission of electrical impulses and exchange of chemical signals. (p. 225)

NES: see *nuclear export signal.*

N-ethylmaleimide-sensitive factor (NSF): soluble cytoplasmic protein that acts in conjunction with several soluble NSF attachment proteins (SNAPs) to mediate the fusion of membranes brought together by interactions between v-SNAREs and t-SNAREs. (p. 352)

neural cell adhesion molecule (N-CAM): plasma membrane glycoprotein that mediates cell-cell adhesion in neurons. (p. 303)

neuromuscular junction: site where a nerve cell axon makes contact with a skeletal muscle cell for the purpose of transmitting electrical impulses. (p. 787)

neuron: specialized cell directly involved in the conduction and transmission of nerve impulses; nerve cell. (p. 225)

neuropeptide: a molecule consisting of a short chain of amino acids that is involved in transmitting signals from neurons to other cells (neurons as well as other cell types). (p. 244)

neurosecretory vesicle: a small vesicle containing neurotransmitter molecules; located in the terminal bulb of an axon. (p. 245)

neurotoxin: toxic substance that disrupts the transmission of nerve impulses. (p. 247)

neurotransmitter: chemical released by a neuron that transmits nerve impulses across a synapse. (p. 243)

nexin: protein that connects and maintains the spatial relationship of adjacent outer doublets in the axoneme of eukaryotic cilia and flagella. (p. 776)

N-glycosylation: see *N-linked glycosylation.*

nicotinamide adenine dinucleotide (NAD⁺): coenzyme that accepts two electrons and one proton to generate the reduced form, NADH; important electron carrier in energy metabolism. (p. 374)

nicotinamide adenine dinucleotide phosphate (NADP⁺): coenzyme that accepts two electrons and one proton to generate the reduced form, NADPH; important electron carrier in the Calvin cycle and other biosynthetic pathways. (p. 456)

nitric oxide (NO): a gas molecule that transmits signals to neighboring cells by stimulating guanylyl cyclase. (p. 269)

N-linked glycosylation (N-glycosylation): addition of oligosaccharide units to the terminal amino group of asparagine residues in protein molecules. (p. 336)

NLS: see *nuclear localization signal.*

NO: see *nitric oxide.*

nocodazole: synthetic drug that inhibits microtubule assembly; frequently used instead of colchicine because its effects are more readily reversible when the drug is removed. (p. 753)

node of Ranvier: small segment of bare axon between successive segments of myelin sheath. (p. 227)

noncovalent bonds and interactions: binding forces that do not involve the sharing of electrons; examples include ionic bonds, hydrogen bonds, van der Waals interactions, and the hydrophobic effect. (pp. 34, 46)

noncovalent interaction: see *noncovalent bonds and interactions.*

noncyclic electron flow: continuous, unidirectional flow of electrons from water to NADP⁺ during the energy transduction reactions of photosynthesis, with light providing the energy that drives the transfer. (p. 458)

nondisjunction: failure of the two members of a homologous chromosome pair to separate during anaphase I of meiosis, resulting in a joined chromosome pair that moves into one of the two daughter cells. (p. 588)

nonmotor MAP: microtubule-associated protein involved in controlling microtubule organization rather than movement. (p. 753)

nonreceptor tyrosine kinase: a tyrosine kinase that is separate from a receptor, but can bind to the receptor and be activated when the receptor binds its ligand. (p. 271)

nonrepeated DNA: DNA sequences present in single copies within an organism's genome. (p. 500)

nonsense mutation: change in base sequence converting a codon that previously coded for an amino acid into a stop codon. (p. 674)

NOR: see *nucleolus organizer region.*

NPC: see *nuclear pore complex.*

NSF: see *N-ethylmaleimide-sensitive factor.*

N-terminus (amino terminus): the end of a polypeptide chain that contains the first amino acid to be incorporated during mRNA translation; usually retains a free amino group. (p. 44)

nuclear envelope: double membrane around the nucleus that is interrupted by numerous small pores. (pp. 84, 510)

nuclear export signal (NES): amino acid sequence that targets a protein for export from the nucleus (p. 515)

nuclear lamina: thin, dense meshwork of fibers that lines the inner surface of the inner nuclear membrane and helps support the nuclear envelope. (p. 516)

nuclear localization signal (NLS): amino acid sequence that targets a protein for transport into the nucleus. (p. 514)

nuclear matrix: insoluble fibrous network that provides a supporting framework for the nucleus. (p. 515)

nuclear pore: small opening in the nuclear envelope through which molecules enter and exit the nucleus; lined by an intricate protein structure called the nuclear pore complex (NPC). (p. 510)

nuclear pore complex (NPC): intricate protein structure, composed of 100 or more different polypeptide subunits, that lines the nuclear pores through which molecules enter and exit the nucleus, both by simple diffusion of smaller molecules and active transport of larger molecules. (p. 512)

nuclear receptor: nuclear protein that functions as a transcription factor after binding to a specific signaling molecule, usually (but not always) a steroid hormone. (p. 724)

nucleation: act of providing a small aggregate of molecules from which a polymer can grow. (p. 747)

nucleic acid: a linear polymer of nucleotides joined together in a genetically determined order. Each nucleotide is composed of ribose or deoxyribose, a phosphate group, and the nitrogenous base guanine, cytosine, adenosine, or thymine (for DNA) or uracil (for RNA). (p. 54) Also see *DNA* and *RNA.*

nucleic acid hybridization: family of techniques in which single-stranded nucleic acids are allowed to bind to each other by complementary base-pairing; used for assessing whether two nucleic acids contain similar base sequences (p. 491)

nucleic acid probe: see *probe.*

nucleoid: region of cytoplasm in which the genetic material of a prokaryotic cell is located. (p. 502)

nucleolus (plural, nucleoli): large, spherical structure present in the nucleus of a eukaryotic cell; the site of ribosomal RNA synthesis and processing and of the assembly of ribosomal subunits. (pp. 85, 517)

nucleolus organizer region (NOR): stretch of DNA in certain chromosomes where multiple copies of the genes for ribosomal RNA are located and where nucleoli form. (p. 518)

nucleophilic substitution: reaction in which an electron-rich (electronegative) molecule or region of a molecule donates electrons to an electron-deficient (electropositive) molecule or region of a molecule, resulting in the formation of a covalent bond. (p. 138)

nucleoplasm: the interior space of the nucleus, other than that occupied by the nucleolus. (p. 510)

nucleoside: molecule consisting of a nitrogen-containing base (purine or pyrimidine) linked to a five-carbon sugar (ribose or deoxyribose); a nucleotide with the phosphate removed. (p. 56)

nucleoside monophosphate: see *nucleotide.*

nucleosome: basic structural unit of eukaryotic chromosomes, consisting of about 200 base pairs of DNA associated with an octamer of histones. (p. 505)

nucleotide: molecule consisting of a nitrogen-containing base (purine or pyrimidine) linked to a five-carbon sugar (ribose or deoxyribose) attached to a phosphate group; also called a nucleoside monophosphate. (p. 55)

nucleotide excision repair (NER): DNA repair mechanism that recognizes and repairs damage involving major distortions of the DNA double helix, such as that caused by pyrimidine dimers. (p. 543)

nucleus: large, double membrane-enclosed organelle that contains the chromosomal DNA of a eukaryotic cell. (pp. 84, 510)

numerical aperture (NA): property of a microscope corresponding to the quantity $n \sin \alpha$, where n is the refractive index of the medium between the specimen and the objective lens and α is the aperture angle. (p. GM-4)

O

objective lens (electron microscope): electromagnetic lens within which the specimen is placed in a transmission electron microscope. (p. GM-18)

objective lens (light microscope): lens located immediately above the specimen in a light microscope. (p. GM-5)

obligate aerobe: organism that has an absolute requirement for oxygen as an electron acceptor and therefore cannot live under anaerobic conditions. (p. 378)

obligate anaerobe: organism that cannot use oxygen as an electron acceptor and therefore has an absolute requirement for an electron acceptor other than oxygen. (p. 378)

ocular lens: lens through which the observer looks in a light microscope; also called the eyepiece. (p. GM-5)

OEC: see *oxygen evolving complex.*

Okazaki fragments: short fragments of newly synthesized, lagging-strand DNA that are joined together by DNA ligase during DNA replication. (p. 532)

oligodendrocyte: cell type in the central nervous system that forms the myelin sheath around nerve axons. (p. 241)

O-linked glycosylation: addition of oligosaccharide units to hydroxyl groups of serine or threonine residues in protein molecules. (p. 336)

oncogene: any gene whose presence can cause a cell to become malignant; arise by mutation from normal cellular genes called *proto-oncogenes*. (p. 565)

operator: base sequence in an operon to which a repressor protein can bind. (p. 694)

operon: cluster of genes with related functions that is under the control of a single operator and promoter, thereby allowing transcription of these genes to be turned on and off together. (p. 694)

organelle: any discrete intracellular structure that is specialized for carrying out a particular function. Eukaryotic cells contain several kinds of membrane-enclosed organelles, including the nucleus, mitochondria, endoplasmic reticulum, and Golgi complex. Ribosomes, microtubules, and microfilaments are examples of organelles without membranes. (pp. 5, 78)

organic chemistry: the study of carbon-containing compounds. (p. 18)

origin of replication: specific base sequence within a DNA molecule where replication is initiated. (p. 528)

origin of transfer: point on an F factor plasmid at which the transfer of the plasmid from an F⁺ donor bacterial cell to an F⁻ recipient cell begins during conjugation. (p. 600)

osmolarity: solute concentration on one side of a membrane relative to that on the other side of the membrane; drives the osmotic movement of water across the membrane. (p. 200)

osmosis: movement of water through a semipermeable membrane driven by a difference in solute concentration on the two sides of the membrane. (p. 199)

outer doublet: pair of fused microtubules, arranged in groups of nine around the periphery of the axoneme of a eukaryotic cilium or flagellum. (p. 776)

outer membrane: the outer of the two membranes that surround a mitochondrion, chloroplast, or nucleus. (pp. 402, 448)

ovum (plural, **ova**): see *egg*.

oxidation: chemical reaction involving the removal of electrons; oxidation of organic molecules frequently involves the removal of both electrons and hydrogen ions (protons) and is therefore also called a dehydrogenation reaction. (p. 373) Also see *beta oxidation.*

oxidative deamination: release of free ammonia from an amino acid accompanied by oxidation of the molecule's carbon skeleton. (p. 414)

oxidative phosphorylation: formation of ATP from ADP and inorganic phosphate by coupling the exergonic oxidation of reduced coenzyme molecules by oxygen to the phosphorylation of ADP, with an electrochemical proton gradient as the intermediate. (p. 426)

oxygen-evolving complex (OEC): assembly of manganese ions and proteins included within photosystem II that catalyzes the oxidation of water to oxygen. (p. 457)

oxygenic phototroph: organism that utilizes water as the electron donor in photosynthesis, with release of oxygen. (p. 447)

P

P face: interior face of the inner, or cytoplasmic, monolayer of a membrane as revealed by the technique of freeze-fracturing; called the P face because this monolayer is on the *protoplasmic* side of the membrane. (p. GM-24)

P site (peptidyl site): site on the ribosome that contains the growing polypeptide chain at the beginning of each elongation cycle. (p. 661)

P/O ratio: number of molecules of ATP generated as a pair of electrons passes through the electron transport system to reduce a single oxygen atom to water. (p. 427)

P₁ generation: the first parental generation of a genetic breeding experiment. (p. 590)

p53 gene: a tumor suppressor gene that codes for a transcription factor involved in preventing genetically damaged cells from proliferating; most frequently mutated gene in human cancers (p. 568)

P680: the pair of chloroplast molecules that make up the reaction center of photosystem II. (p. 455)

P700: the pair of chloroplast molecules that make up the reaction center of photosystem I. (p. 455)

pachytene: stage during prophase I of meiosis when crossing over between homologous chromosomes takes place. (p. 583)

packing ratio: see *DNA packing ratio.*

paracrine hormone: a hormone that acts only on cells in the immediate environment rather than on cells in distant locations. (p. 276)

passive spread of depolarization: process in which cations (mostly K^+) move away from the site of membrane depolarization to regions of membrane where the potential is more negative (p. 240)

patch clamping: technique in which a tiny micropipette placed on the surface of a cell is used to measure the movement of ions through individual ion channels. (p. 233)

PC: see *plastocyanin.*

PCR: see *polymerase chain reaction.*

PDGF: see *platelet-derived growth factor.*

pectin: branched polysaccharides rich in galacturonic acid and rhamnose found in plant cell walls, where they form a matrix in which cellulose microfibrils are embedded. (p. 316)

pellet: material that sediments to the bottom of a centrifuge tube during centrifugation. (p. 328)

peptide bond: a covalent bond between the amino group of one amino acid and the carboxyl group of a second amino acid. (p. 44)

peptidyl site: see *P site.*

peptidyl transferase: enzymatic activity, exhibited by the rRNA of the large ribosomal subunit, that catalyzes peptide bond formation during protein synthesis. (p. 671)

perfusion: injection of a fixative into the bloodstream of the animal before removing the organs for microscopic examination. (p. GM-16)

pericentriolar material: diffuse granular material surrounding the centrioles of an animal cell centrosome. (p. 749)

perinuclear space: space between the inner and outer nuclear membranes that is continuous with the lumen of the endoplasmic reticulum. (p. 510)

peripheral membrane protein: hydrophilic protein bound through weak ionic interactions and hydrogen bonds to a membrane surface. (p. 179)

peripheral nervous system (PNS): all sensory and motor neurons of the nervous system that are located outside the brain and spinal cord. (p. 225) Also see *central nervous system (CNS).*

peroxisome: single membrane-bounded organelle that contains catalase and one or more hydrogen peroxide-generating oxidases and is therefore involved in the metabolism of hydrogen peroxide. (pp. 90, 358)

PFK-2: see *phosphofructokinase-2.*

phage: see *bacteriophage.*

phagocyte: specialized white blood cell that carries out phagocytosis as a defense mechanism. (p. 344)

phagocytic vacuole: membrane-bounded structure containing ingested particulate matter that fuses with a late endosome or matures directly into a lysosome, forming a large vesicle in which the ingested material is digested. (p. 345)

phagocytosis: type of endocytosis in which particulate matter or even an entire cell is taken up from the environment and incorporated into vesicles for digestion. (p. 344)

phase transition: change in the state of a membrane between a fluid state and a gel state. (p. 172)

phase-contrast microscopy: light microscopic technique that improves contrast without sectioning and staining by exploiting differences in thickness and refractive index; produces an image using an optical material that is capable of bringing undiffracted rays into phase with those that have been diffracted by the specimen. (p. GM-6)

phenotype: observable physical characteristics of an organism attributable to the expression of its genotype. (p. 579)

phosphatidic acid: basic component of phosphoglycerides; consists of two fatty acids and a phosphate group linked by ester bonds to glycerol; key intermediate in the synthesis of other phosphoglycerides. (p. 69)

phosphoanhydride bond: high-energy bond between phosphate groups. (p. 369)

phosphodiester bond: covalent linkage in which two parts of a molecule are joined through oxygen atoms to the same phosphate group. (p. 56)

phosphodiesterase: enzyme that catalyzes the hydrolysis of cyclic AMP to AMP. (p. 262)

phosphoester bond: covalent linkage in which a molecule is joined through an oxygen atom to a phosphate group. (p. 369)

phosphofructokinase-2 (PFK-2): an enzyme that catalyzes the ATP-dependent phosphorylation of fructose-6-phosphate on carbon atom 2 to form fructose-2,6-bisphosphate (F2,6BP), an important regulator of both glycolysis and gluconeogenesis. (p. 392)

phosphoglyceride: predominant phospholipid component of cell membranes, consisting of a glycerol molecule esterified to two fatty acids and a phosphate group. (pp. 69, 166)

phosphoglycolate: two-carbon compound produced by the oxygenase activity of rubisco. Because it cannot be metabolized during the next step of the Calvin cycle, the production of phosphoglycolate decreases photosynthetic efficiency. (p. 469)

phospholipase C: enzyme that catalyzes the hydrolysis of phosphatidylinositol-4,5-bisphosphate into inositol-1,4,5-trisphosphate (InsP₃) and diacylglycerol (DAG). (p. 263)

phospholipid: lipid possessing a covalently attached phosphate group and therefore exhibiting both hydrophilic and hydrophobic properties; main component of the lipid bilayer that forms the structural backbone of all cell membranes. (pp. 69, 166)

phospholipid exchange protein: any of a group of proteins located in the cytosol that transfer specific phospholipid molecules from the ER membrane to the outer mitochondrial, chloroplast, or plasma membranes. (p. 333)

phospholipid translocator (flippase): a membrane protein that catalyzes the flip-flop of membrane phospholipids from one monolayer to the other. (pp. 171, 332)

phosphorylation: addition of a phosphate group. (p. 150)

photoautotroph: organism capable of obtaining energy from the sun and using this energy to drive the synthesis of energy-rich organic molecules, using carbon dioxide as a source of carbon. (p. 445)

photochemical reduction: transfer of photo-excited electrons from one molecule to another. (p. 452)

photoexcitation: excitation of an electron to a higher energy level by the absorption of a photon of light. (p. 452)

photoheterotroph: organism capable of obtaining energy from the sun but dependent on organic compounds, rather than carbon dioxide, for carbon. (p. 445)

photon: fundamental particle of light with an energy content that is inversely proportional to its wavelength. (p. 452)

photophosphorylation: light-dependent generation of ATP driven by an electrochemical proton gradient established and maintained as excited electrons of chlorophyll return to their ground state via an electron transport system. (p. 447)

photoreduction: light-dependent generation of NADPH by the transfer of energized electrons from photoexcited chlorophyll molecules to $NADP^+$ via a series of electron carriers. (p. 447)

photorespiration: light-dependent pathway that decreases the efficiency of photosynthesis by oxidizing reduced carbon compounds without capturing the released energy; occurs when oxygen substitutes for carbon dioxide in the reaction catalyzed by rubisco, thereby generating phosphoglycolate which is then converted to 3-phosphoglycerate in the peroxisome and mitochondrion; also called the glycolate pathway. (p. 469)

photosynthesis: process by which plants and certain bacteria convert light energy to chemical energy that is then used in synthesizing organic molecules. (p. 445)

photosystem: assembly of chlorophyll molecules, accessory pigments, and associated proteins embedded in thylakoid membranes or bacterial photosynthetic membranes; functions in the light-requiring reactions of photosynthesis. (p. 453)

photosystem I (PSI): photosystem containing a pair of chlorophyll molecules (P700) that absorbs 700-nm red light maximally; light of this wavelength can excite electrons derived from plastocyanin to an energy level that allows them to reduce ferredoxin, from which the electrons are then used to reduce $NADP^+$ to NADPH. (p. 455)

photosystem II (PSII): photosystem containing a pair of chlorophyll molecules (P680) that absorb 680-nm red light maximally; light of this wavelength can excite electrons donated by water to an energy level that allows them to reduce plastoquinone. (p. 455)

photosystem complex: a photosystem plus its associated light-harvesting complexes. (p. 454)

phototroph: organism that is capable of utilizing the radiant energy of the sun to satisfy its energy requirements. (p. 109, 445)

phragmoplast: parallel array of microtubules that guides vesicles containing polysaccharides and glycoproteins toward the spindle equator during cell wall formation in dividing plant cells. (p. 553)

phycobilin: accessory pigment found in red algae and cyanobacteria that absorbs visible light in the green-to-orange range of the spectrum, giving these cells their characteristic colors. (p. 453)

phycobilisome: light-harvesting complex found in red algae and cyanobacteria that contains phycobilins rather than chlorophyll and carotenoids. (p. 454)

phytosterol: any of several sterols that are found uniquely, or primarily, in the membranes of plant cells; examples include campesterol, sitosterol, and stigmasterol. (p. 166)

pigment: light-absorbing molecule responsible for the color of a substance. (p. 452)

pilus: see *sex pilus*.

PKA: see *protein kinase A*.

PKC: see *protein kinase C*.

plakin: family of proteins involved in linking the integrin molecules of a hemidesmosome to intermediate filaments of the cytoskeleton. (p. 300)

plaque: dense layer of fibrous material located on the cytoplasmic side of adhesive junctions such as desmosomes, hemidesmosomes, and adherens junctions; composed of intracellular attachment proteins that link the junction to the appropriate type of cytoskeletal filament. (p. 300) The same term can also refer to the clear zone produced when bacterial cells in a small region of a culture dish are destroyed by infection with a bacteriophage. (p. 609)

plasma membrane: bilayer of lipids and proteins that defines the boundary of the cell and regulates the flow of materials into and out of the cell; also called the cell membrane. (pp. 83, 159)

plasmid: small circular DNA molecule in bacteria that can replicate independent of chromosomal DNA; useful as cloning vectors. (p. 504)

plasmodesma (plural, plasmodesmata): cytoplasmic channel through pores in the cell walls of two adjacent plant cells, allowing fusion of the plasma membranes and chemical communication between the cells. (pp. 98, 318)

plasmolysis: outward movement of water that causes the plasma membrane to pull away from the cell wall in cells that have been exposed to a hypertonic solution. (p. 201)

plastid: any of several types of plant cytoplasmic organelles derived from proplastids, including chloroplasts, amyloplasts, chromoplasts, proteinoplasts, and elaioplasts. (pp. 88, 448)

plastocyanin (PC): copper-containing protein that donates electrons to chlorophyll P700 of photosystem I in the light-requiring reactions of photosynthesis. (p. 458)

plastoquinol: fully reduced form of plastoquinone, involved in the light-requiring reactions of photosynthesis; present in the lipid phase of the photosynthetic membrane, where it transfers electrons to the cytochrome b_6/f complex. (p. 456)

plastoquinone: nonprotein (quinone) molecule associated with photosystem II, where it receives electrons from a modified type of chlorophyll called pheophytin during the light-requiring reactions of photosynthesis. (p. 456)

platelet-derived growth factor (PDGF): protein produced by blood platelets that stimulates the proliferation of connective tissue and smooth muscle cells. (p. 562)

plus end (of microtubule): rapidly growing end of a microtubule. (p. 748)

pmf: see *proton motive force*.

PNS: see *peripheral nervous system*.

polar body: tiny haploid cell produced during the meiotic divisions that create egg cells. Polar bodies receive a disproportionately small amount of cytoplasm and eventually degenerate. (p. 588)

polar microtubule: spindle microtubule that interacts with spindle microtubules from the opposite spindle pole. (p. 545)

polarity: property of a molecule that results from part of the molecule having a partial positive charge and another part having a partial negative charge, usually because one region of the molecule possesses one or more electronegative atoms that draw electrons toward that region. (p. 22)

polarized secretion: fusion of secretory vesicles with the plasma membrane and expulsion of their contents to the cell exterior specifically localized at one end of a cell. (p. 343)

poly(A) tail: stretch of about 50-250 adenine nucleotides added to the 3' end of most eukaryotic mRNAs after transcription is completed. (p. 648)

polygenic mRNA: an mRNA molecule that codes for more than one polypeptide. (p. 695)

polymerase chain reaction (PCR): reaction in which a specific segment of DNA is amplified by repeated cycles of (1) heat treatment to separate the two strands of the DNA double helix, (2) incubation with primers that are complementary to sequences located at the two ends of the DNA segment being amplified, and (3) incubation with DNA polymerase to synthesize DNA using the primers as starting points. (p. 530)

polynucleotide: linear chain of nucleotides linked by phosphodiester bonds. (p. 56)

polypeptide: linear chain of amino acids linked by peptide bonds. (pp. 31, 45)

polyphosphoinositide: any of several phosphorylated derivatives of phosphatidyl inositol, some of which regulate actin polymerization. (p. 758)

polyribosome: cluster of two or more ribosomes simultaneously translating a single mRNA molecule. (p. 674)

polysaccharide: polymer consisting of sugars and sugar derivatives linked together by glycosidic bonds. (p. 60)

polytene chromosome: giant chromosome containing multiple copies of the same DNA molecule generated by successive rounds of DNA replication in the absence of cell division. (p. 711)

porin: transmembrane protein that forms pores for the facilitated diffusion of small hydrophilic molecules; found in the outer membranes of mitochondria, chloroplasts, and many bacteria. (pp. 206, 402, 448)

positive control: genetic control in which the key regulatory element acts by activating gene transcription. (p. 700)

postsynaptic neuron: a neuron that receives a signal from another neuron through a synapse. (p. 243)

posttranslational control: mechanisms of gene regulation involving selective alterations in polypeptides that have already been synthesized; includes covalent modifications, proteolytic cleavage, protein folding and assembly, import into organelles, and protein degradation. (p. 732)

posttranslational import: uptake by organelles of completed polypeptide chains after they have been synthesized, mediated by specific targeting signals within the polypeptide. (p. 678)

potential (voltage): tendency of oppositely charged ions to flow toward each other. (p. 229)

pre-mRNA: primary transcript whose processing yields a mature mRNA. (p. 648)

prenylated membrane protein: protein located on a membrane surface that is covalently bound to a prenyl group embedded within the lipid bilayer. (p. 180)

pre-rRNA: primary transcript whose processing yields mature rRNAs. (p. 645)

presynaptic neuron: a neuron that transmits a signal to another neuron through a synapse. (p. 243)

pre-tRNA: primary transcript whose processing yields a mature tRNA. (p. 646)

primary cell wall: flexible portion of the plant cell wall that develops beneath the middle lamella while cell growth is still occurring; contains a loosely organized network of cellulose microfibrils. (p. 316)

primary messenger: a molecule that binds to a receptor, thereby beginning the process of transmitting a signal to the cell. (p. 257)

primary structure: sequence of amino acids in a polypeptide chain. (p. 48)

primary transcript: any RNA molecule newly produced by transcription, before any processing has occurred. (p. 644)

primase: enzyme that uses a single DNA strand as a template to guide the synthesis of the RNA primers that are required for initiation of replication of both the lagging and leading strands of a DNA double helix. (p. 535)

primosome: complex of proteins in bacterial cells that includes primase plus six other proteins involved in unwinding DNA and recognizing base sequences where replication is to be initiated. (p. 536)

prion: infectious, protein-containing particle responsible for neurological diseases such as scrapie in sheep and goats, kuru in humans, and "mad cow" disease in cattle and its human form, vCJD. (pp. 99, 673)

probe (nucleic acid): single-stranded radioactive nucleic acid that is used in hybridization experiments to identify nucleic acids containing sequences that are complementary to the probe. (pp. 491, 611)

procaspase: an inactive precursor form of a caspase. (p. 284)

process: extension or branch from a nerve cell body; an axon or a dendrite. (p. 227)

procollagen: a precursor molecule that is converted to collagen by proteolytic cleavage of sequences at both the N- and C-terminal ends. (p. 292)

projector lens: electromagnetic lens located between the intermediate lens and the viewing screen in a transmission electron microscope. (p. GM-18)

prokaryote: category of organisms characterized by the absence of a true nucleus and other membrane-bounded organelles; includes eubacteria and archaebacteria. (p. 76)

prometaphase: stage of mitosis characterized by nuclear envelope breakdown and attachment of chromosomes to spindle microtubules; also called late prophase. (p. 545)

promoter (site): base sequence in DNA to which RNA polymerase binds when initiating transcription. (pp. 637, 694)

proofreading: removal of mismatched base pairs during DNA replication by the exonuclease activity of DNA polymerase. (p. 535) Also see *RNA proofreading*.

propagation: movement of an action potential along a membrane away from the site of origin. (p. 236)

prophase: initial phase of mitosis, characterized by chromosome condensation and the beginning of spindle assembly. Prophase I of meiosis is more complex, consisting of stages called leptotene, zygotene, pachytene, diplotene, and diakinesis. (p. 545)

proplastid: small, double-membrane-enclosed, plant cytoplasmic organelle that can develop into several kinds of plastids, including chloroplasts. (p. 448)

prostaglandin: any of a series of paracrine hormones that are synthesized from arachidonic acid and function by activating contraction of smooth muscles. (p. 278)

prosthetic group: small organic molecule or metal ion component of an enzyme that plays an indispensable role in the catalytic activity of the enzyme. (p. 134)

proteasome: multiprotein complex that catalyzes the ATP-dependent degradation of proteins linked to ubiquitin. (p. 734)

protein: macromolecule that consists of one or more polypeptides folded into a conformation specified by the linear sequence of amino acids. Proteins play important roles as enzymes, structural proteins, motility proteins, and regulatory proteins. (p. 41)

protein disulfide isomerase: enzyme in the ER lumen that catalyzes the formation and breakage of disulfide bonds between cysteine residues in polypeptide chains. (p. 681)

protein kinase: any of numerous enzymes that catalyze the phosphorylation of protein molecules. (p. 150)

protein kinase A (PKA): a protein kinase activated by the second messenger, cyclic AMP, and which catalyzes the phosphorylation of serine or threonine residues in target proteins. (p. 262)

protein kinase C (PKC): enzyme that phosphorylates serine and threonine groups in a variety of target proteins when activated by diacylglycerol. (p. 264)

protein kinase-associated receptor: receptor exhibiting a protein kinase activity that is activated when the receptor binds to its appropriate ligand. (p. 270)

protein splicing: removal of amino acid sequences called inteins from a polypeptide chain accompanied by splicing together of the remaining polypeptide segments, called exteins. (p. 676)

proteoglycan: complex between proteins and glycosaminoglycans found in the extracellular matrix. (p. 294)

proteolysis: degradation of proteins by hydrolysis of the peptide bonds between amino acids. (p. 413)

proteolytic cleavage: removal of a portion of a polypeptide chain by an enzyme that cleaves peptide bonds. (p. 151)

proteome: the structure and properties of all the proteins produce by a genome. (p. 498)

protoeukaryote: hypothetical evolutionary ancestor of present-day eukaryotic cells whose ability to carry out phagocytosis allowed it to engulf and establish an endosymbiotic relationship with primitive bacteria. (p. 450)

protofilament: linear polymer of tubulin subunits; usually arranged in groups of 13 to form the wall of a microtubule. (p. 746)

proton motive force (pmf): force across a membrane exerted by an electrochemical proton gradient that tends to drive protons back down their concentration gradient. (p. 429)

proton translocator: channel through which protons flow across a membrane driven by an electrochemical gradient; examples include CF_o in thylakoid membranes and F_o in mitochondrial inner membranes. (pp. 431, 459)

proto-oncogene: normal cellular gene that can be converted into an oncogene by point mutation, DNA rearrangement, gene amplification, or chromosomal translocation. (p. 565)

provacuole: vesicle in plant cells comparable to an endosome in animal cells; arises either from the Golgi complex or by autophagy. (p. 357)

proximal control element: DNA regulatory sequence located upstream of the core promoter but within about 100-200 base pairs of it. (p. 718)

pseudopodium (plural, pseudopodia): large, blunt-ended cytoplasmic protrusion involved in cell crawling by amoebas, slime molds, and leukocytes. (p. 795)

PSI: see *photosystem I*.

PSII: see *photosystem II*.

P-type ATPase: type of transport ATPase that is reversibly phosphorylated by ATP as part of the transport mechanism. (p. 208)

purine: two-ringed nitrogen-containing molecule; parent compound of the bases adenine and guanine. (p. 55)

pyrimidine: single-ringed nitrogen-containing molecule; parent compound of the bases cytosine, thymine, and uracil. (p. 55)

Q

quantum: indivisible packet of energy carried by a photon of light. (p. 452)

quaternary structure: level of protein structure involving interactions between two or more polypeptide chains to form a single multimeric protein. (p. 53)

R

Rab GTPase: GTP-hydrolyzing protein involved in locking v-SNAREs and t-SNAREs together during the binding of a transport vesicle to an appropriate target membrane. (p. 352)

radial spokes: inward projections from each of the nine outer doublets to the center pair of microtubules in the axoneme of a eukaryotic cilium or flagellum, believed to be important in converting the sliding of the doublets into a bending of the axoneme. (p. 776)

Ras (protein): a small monomeric G protein bound to the inner surface of the plasma membrane which is a key intermediate in transmitting signals from receptor tyrosine kinases to the cell interior. (p. 271)

rate-zonal centrifugation: see *density gradient centrifugation*.

Rb protein: protein whose phosphorylation controls passage through the G1 checkpoint. (p. 560)

reaction center: portion of a photosystem containing the two chlorophyll molecules that initiate electron transfer, utilizing the energy gathered by other chlorophyll molecules and accessory pigments. (p. 453) Also see *P680* and *P700*.

RecA: bacterial enzyme involved in homologous recombination that catalyzes a "strand invasion" reaction in which a single-stranded DNA segment displaces one of the two strands of a DNA double helix. (p. 606)

receptor: a protein that contains a binding site for a specific signaling molecule. (p. 160)

receptor affinity: a measure of the chemical attraction between a receptor and its ligand. (p. 258)

receptor tyrosine kinase: a receptor whose activation causes it to catalyze the phosphorylation of tyrosine residues in proteins, thereby triggering a chain of signal transduction events inside cells that can lead to cell growth, proliferation, and differentiation. (p. 270)

receptor-mediated (clathrin-dependent) endocytosis: type of endocytosis initiated at coated pits and resulting in coated vesicles; believed to be a major mechanism for selective uptake of macromolecules and peptide hormones. (p. 345)

recessive (allele): allele that is present in the genome but is phenotypically expressed only in the homozygous form; masked by a dominant allele when heterozygous. (p. 578)

recombinant DNA molecule: DNA molecule containing DNA sequences derived from two different sources. (p. 607)

recombinant DNA technology: group of laboratory techniques for joining DNA fragments derived from two or more sources. (p. 606)

redox pair: two molecules or ions that are interconvertible by the loss or gain of electrons; also called a reduction-oxidation pair. (p. 420)

reduction: chemical reaction involving the addition of electrons; reduction of organic molecules frequently involves the addition of both electrons and hydrogen ions (protons) and is therefore also called a hydrogenation reaction. (p. 374)

reduction potential (E'): a measure, in volts, of the electron transfer potential, or relative reducing power, of a redox pair, expressed such that the more highly negative the value, the more readily the reduced form of the pair loses its electron. (p. 420)

reduction-oxidation pair: see *redox pair*.

refractive index: measure of the change in the velocity of light as it passes from one medium to another. (p. GM-4)

regulated gene: gene whose expression is controlled according to the cell's need rather than being continuously active. (p. 692)

regulated secretion: fusion of secretory vesicles with the plasma membrane and expulsion of their contents to the cell exterior in response to specific extracellular signals. (p. 342)

regulatory gene: a gene whose product controls the expression of another gene. (p. 694)

regulatory light chain: type of myosin light chain that is phosphorylated by myosin light-chain kinase in smooth muscle cells, thereby enabling myosin to interact with actin filaments and triggering muscle contraction. (p. 790)

regulatory site: see *allosteric site*.

regulatory subunit: a subunit of a multisubunit enzyme that contains an allosteric site. (p. 149)

regulatory transcription factor: protein that controls the rate at which one or more specific genes is transcribed by binding to DNA control elements located outside the core promoter. (p. 718)

relative refractory period: time during the hyperpolarization phase of an action potential when the sodium channels of a nerve cell are capable of opening again but it is difficult to trigger an action potential because Na^+ currents are opposed by larger K^+ currents. (p. 240)

release factors: group of proteins that terminate translation by triggering the release of a completed polypeptide chain from peptidyl tRNA bound to a ribosome's P site. (p. 672)

renaturation: return of a protein from a denatured state to the native conformation determined by its amino acid sequence, usually accompanied by restoration of physiological function. (p. 31) Also see *DNA renaturation*.

repeated DNA: DNA sequences present in multiple copies within an organism's genome. (p. 500)

replication fork: Y-shaped structure that represents the site at which replication of a DNA double helix is occurring. (p. 527)

replicon: total length of DNA replicated from a single origin of replication. (p. 528)

replisome: large complex of proteins that work together to carry out DNA replication at the replication fork; about the size of a ribosome. (p. 537)

repressible operon: group of adjoining genes that are normally transcribed, but whose transcription is inhibited in the presence of a corepressor. (p. 699)

repressor (eukaryotic): regulatory transcription factor whose binding to DNA control elements leads to a reduction in the transcription rate of nearby genes. (p. 719)

repressor (prokaryotic): protein that binds to the operator site of an operon and prevents transcription of adjacent structural genes. (p. 694)

residual body: mature lysosome in which digestion has ceased and only indigestible material remains. (p. 356)

resolution: minimum distance that can separate two points that still remain identifiable as separate points when viewed through a microscope. (p. GM-3)

resolving power: ability of a microscope to distinguish adjacent objects as separate entities. (p. 5)

resonance energy transfer: mechanism whereby the excitation energy of a photoexcited molecule is transferred to an electron in an adjacent molecule, exciting that electron to a high-energy orbital; important means of passing energy from one pigment molecule to another in photosynthetic energy transduction. (p. 452)

resonance hybrid: the actual structure of functional groups such as carboxylate or phosphate groups that are written formally as two or more structures with one double bond and one or more single bonds to oxygen when the unshared electron pair is in fact delocalized over all of the possible bonds to oxygen; written as single bonds to all possible oxygen atoms and dashed lines indicating delocalization of one electron pair. (p. 370)

resonance stabilization: achievement of the most stable configuration of a molecule by maximal delocalization of an unshared electron pair over all possible bonds. (p. 370)

respiratory complex: subset of carriers of the electron transport system consisting of a distinctive assembly of polypeptides and prosthetic groups organized together to play a specific role in the electron transport process. (p. 423)

respiratory control: regulation of oxidative phosphorylation and electron transport by the availability of ADP. (p. 426)

response element: DNA base sequence located adjacent to physically separate genes whose expression can then be coordinated by binding a regulatory transcription factor to the response element wherever it occurs. (p. 724)

resting membrane potential (V_m): electrical potential (voltage) across the plasma membrane of an unstimulated nerve cell. (p. 228)

restriction enzyme: any of a large family of enzymes isolated from bacteria that cut foreign DNA molecules at or near a palindromic recognition sequence that is usually four or six (but may be eight or more) base pairs long; used in recombinant DNA technology to cleave DNA molecules at specific sites. (p. 492)

restriction fragment length polymorphism (RFLP): difference in restriction maps between

individuals caused by small differences in the base sequences of their DNA. (p. 502)

restriction map: map of a DNA molecule indicating the location of cleavage sites for various restriction enzymes. (p. 496)

restriction site: DNA base sequence, usually four or six (but may be eight or more) base pairs long, that is cleaved by a specific restriction enzyme. (p. 493)

restriction/methylation system: pathway in bacterial cells by which foreign DNA is cleaved by restriction enzymes while the bacterial genome is protected from cleavage by prior methylation. (p. 494)

retrograde flow (of F-actin): bulk movement of actin microfilaments toward the rear of a cell protrusion (e.g., lamellipodium) as the protrusion extends. (p. 792)

retrograde transport: movement of vesicles from Golgi cisternae back toward the endoplasmic reticulum. (p. 335)

reverse transcriptase: enzyme that uses an RNA template to synthesize a complementary molecule of double-stranded DNA. (p. 626)

reversible inhibitor: molecule that causes a reversible loss of catalytic activity when bound to an enzyme; upon dissociation of the inhibitor, the enzyme regains biological function. (p. 146)

RFLP: see *restriction fragment length polymorphism*.

rho (ρ) factor: bacterial protein that binds to the 3′ end of newly forming RNA molecules, triggering the termination of transcription. (p. 639)

ribonucleic acid: see *RNA*.

ribose: five-carbon sugar present in RNA and in important nucleoside triphosphates such as ATP and GTP. (p. 54)

ribosomal RNA (rRNA): any of several types of RNA molecules used in the construction of ribosomes. (p. 623)

ribosome: small particle composed of rRNA and protein that functions as the site of protein synthesis in the cytoplasm of prokaryotes and in the cytoplasm, mitochondria, and chloroplasts of eukaryotes; composed of large and small subunits. (pp. 93, 660)

ribozyme: an RNA molecule with catalytic activity. (p. 151)

ribulose-1,5-bisphosphate carboxylase/oxygenase (rubisco): enzyme that catalyzes the CO_2-capturing step of the Calvin cycle; joins CO_2 to ribulose-1,5-bisphosphate, forming two molecules of 3-phosphoglycerate. (p. 464)

rigor: stiff, rigid state of a muscle that develops when cross-bridges fail to dissociate in the absence of adequate ATP. (p. 784)

RNA (ribonucleic acid): nucleic acid that plays several different roles in the expression of genetic information; constructed from nucleotides consisting of ribose phosphate linked to either adenine, uracil, cytosine, or guanine. (p. 54) Also see *messenger RNA; ribosomal RNA; transfer RNA*.

RNA editing: altering the base sequence of an mRNA molecule by the insertion, removal, or modification of nucleotides. (p. 653)

RNA interference: ability of double-stranded RNA to trigger the selective degradation of mRNA molecules containing a complementary base sequence. (p. 731)

RNA polymerase: any of a group of enzymes that catalyze the synthesis of RNA using DNA as a template; function by adding successive nucleotides to the 3′ end of the growing RNA strand. (p. 635)

RNA polymerase I: type of eukaryotic RNA polymerase present in the nucleolus that synthesizes an RNA precursor for three of the four types of rRNA. (p. 640)

RNA polymerase II: type of eukaryotic RNA polymerase present in the nucleoplasm that synthesizes pre-mRNA and most of the snRNAs. (p. 641)

RNA polymerase III: type of eukaryotic RNA polymerase present in the nucleoplasm that synthesizes a variety of small RNAs, including pre-tRNAs and 5S rRNA. (p. 641)

RNA primer: short RNA fragment, synthesized by DNA primase, that serves as an initiation site for DNA synthesis. (p. 536)

RNA processing: conversion of an initial RNA transcript into a final RNA product by the removal, addition, and/or chemical modification of nucleotide sequences. (p. 644)

RNA proofreading: mechanism for correcting errors in transcription based on the removal of mismatched base pairs by the exonuclease activity of some RNA polymerases. (p. 654)

RNA splicing: excision of introns from a primary RNA transcript to generate the mature, functional form of the RNA molecule. (p. 651)

RNA surveillance: quality-control mechanism that monitors RNA molecules for misplaced stop codons and shunts such aberrant molecules into an RNA degradation pathway. (p. 654)

RNP particle: particle formed by the binding of non-snRNP proteins to pre-mRNA or mature mRNA. (p. 651)

rotation (of lipid molecules): turning of a molecule about its long axis; occurs freely and rapidly in membrane phospholipids. (p. 169)

rough endoplasmic reticulum (rough ER): endoplasmic reticulum that is studded with ribosomes on its cytosolic side because of its involvement in protein synthesis. (pp. 89, 324)

rough ER: see *rough endoplasmic reticulum.*

rRNA: see *ribosomal RNA.*

rubisco: see *ribulose-1,5-bisphosphate carboxylase/oxygenase.*

ryanodine receptor channel: a calcium ion channel whose opening is triggered by the presence of calcium ions; particularly important for calcium release from the sarcoplasmic reticulum of cardiac and skeletal muscle. (p. 265)

S

S: see *entropy.*

S phase: stage of the eukaryotic cell cycle in which DNA is synthesized. (p. 524)

S1: see *myosin subfragment 1.*

saltatory propagation: movement of a wave of membrane depolarization along a myelinated axon in which an action potential is renewed at each node of Ranvier. (p. 241)

sarcolemma: plasma membrane of a skeletal muscle cell. (p. 787)

sarcomere: fundamental contractile unit of striated muscle myofibrils that extends from one Z line to the next and that consists of two sets of thin (actin) and one set of thick (myosin) filaments. (p. 779)

sarcoplasm: cytosol of a muscle cell. (p. 786)

sarcoplasmic reticulum (SR): endoplasmic reticulum of a muscle cell, specialized for accumulating, storing, and releasing calcium ions. (p. 787)

SarI: small GTP-binding protein found along with Sec13/31 and Sec23/24 in the coats of COPII-coated vesicles in yeast. (p. 352)

saturated fatty acid: fatty acid without double or triple bonds such that every carbon atom in the chain has the maximum number of hydrogen atoms bonded to it. (p. 68)

saturation: inability of higher substrate concentrations to increase the velocity of an enzyme-catalyzed reaction beyond a fixed upper limit determined by the finite number of enzyme molecules available. (p. 140)

scanning electron microscope (SEM): microscope in which an electron beam scans across the surface of a specimen and forms an image from electrons that are deflected from the outer surface of the specimen. (pp. 7, GM-19)

scanning probe microscope: instrument that visualizes the surface features of individual molecules by using a tiny probe that moves over the surface of a specimen. (p. GM-26)

Schwann cell: cell type in the peripheral nervous system that forms the myelin sheath around nerve axons. (p. 241)

scientific method: an approach for developing new knowledge based on making observations, formulating hypotheses, designing experiments, collecting data, interpreting results, and drawing conclusions. (p. 11)

Sec13/31: protein complex found along with Sec23/24 and SarI in the coats of COPII-coated vesicles in yeast. (p. 352)

Sec23/24: protein complex found along with Sec13/31 and SarI in the coats of COPII-coated vesicles in yeast. (p. 352)

second law of thermodynamics: the law of thermodynamic spontaneity; principle stating that all physical and chemical changes proceed in a manner such that the entropy of the universe increases. (p. 115)

second messenger: any of several substances, including cyclic AMP, calcium ion, inositol trisphosphate, and diacylglycerol, that transmit signals from extracellular signaling ligands to the cell interior. (p. 257)

secondary cell wall: rigid portion of the plant cell wall that develops beneath the primary cell wall after cell growth has ceased; contains densely packed, highly organized bundles of cellulose microfibrils. (p. 317)

secondary structure: level of protein structure involving hydrogen bonding between atoms in the peptide bonds along the polypeptide backbone, creating two main patterns called the α helix and β sheet conformations. (p. 48)

secretory granule: a large, dense secretory vesicle. (p. 341).

secretory pathway: pathway by which newly synthesized proteins move from the ER through the Golgi complex to secretory vesicles and secretory granules, which then discharge their contents to the exterior of the cell. (p. 341)

secretory vesicle: membrane-bounded compartment of a eukaryotic cell that carries secretory proteins from the Golgi complex to the plasma membrane for exocytosis and that may serve as a storage compartment for such proteins before they are released; large dense vesicles are sometimes referred to as secretory granules. (pp. 90, 341)

sedimentation coefficient: a measure of the rate at which a particle or macromolecule moves in a centrifugal force field; expressed in Svedberg units. (pp. 94, 327)

sedimentation rate: rate of movement of a molecule or particle through a solution when subjected to a centrifugal force. (p. 326)

selectable marker: gene whose expression allows cells to grow under specific conditions that prevent the growth of cells lacking this gene. (p. 610)

selectin: plasma membrane glycoprotein that mediates cell-cell adhesion by binding to specific carbohydrate groups located on the surface of target cells. (p. 306)

self-assembly: principle that the information required to specify the folding of macromolecules and their interactions to form more complicated structures with specific biological functions is inherent in the polymers themselves. (p. 31)

SEM: see *scanning electron microscope.*

semiautonomous organelle: organelle, either a mitochondrion or a chloroplast, that contains DNA and is therefore able to encode some of its polypeptides, although it is dependent on the nuclear genome to encode most of them. (p. 450)

semiconservative replication: mode of DNA replication in which each newly formed DNA molecule consists of one old strand and one newly synthesized strand. (p. 525)

sensory neuron: any of a diverse group of nerve cells specialized for the detection of various types of stimuli. (p. 225)

serine/threonine kinase receptor: a receptor which, upon activation, catalyzes the phosphorylation of serine and threonine residues in target protein molecules. (p. 274)

70S initiation complex: complex formed by the association of a 30S initiation complex with a 50S ribosomal subunit; contains an initiator aminoacyl tRNA at the P site and is ready to commence mRNA translation. (p. 669)

sex chromosome: chromosome involved in determining whether an individual is male or female. (p. 578)

sex pilus (plural, pili): projection emerging from the surface of a bacterial donor cell that binds to the surface of a recipient cell, leading to the formation of a transient cytoplasmic mating bridge through which DNA is transferred from donor cell to recipient cell during bacterial conjugation. (p. 600)

sexual reproduction: form of reproduction in which two parent organisms each contribute genetic information to the new organism; reproduction by the fusion of gametes. (p. 577)

SH2 domain: a region of a protein molecule that recognizes and binds to phosphorylated tyrosines in another protein. (p. 271)

shadowing: deposition of a thin layer of an electron-dense metal on a biological specimen from a heated electrode, such that surfaces facing toward the electrode are coated while surfaces facing away are not. (p. GM-22)

sheet (β sheet): see *beta sheet.*

shuttle vesicle: a vesicle that buds from one cisterna of the Golgi complex and fuses with another cisterna. (p. 335)

sidearm: structure composed of axonemal dynein that projects out from each of the A tubules of the nine outer doublets in the axoneme of a eukaryotic cilium or flagellum. (p. 776)

sigma (σ) factor: subunit of bacterial RNA polymerase that ensures the initiation of RNA synthesis at the correct site on the DNA strand. (p. 635)

signal transduction: detection of specific signals at the cell surface and the mechanisms by which such signals are transmitted into the cell's interior, resulting in changes in cell behavior and/or gene expression. (pp. 160, 257)

signal-recognition particle (SRP): cytoplasmic RNA-protein complex that binds to the ER signal sequence located at the N-terminus of a newly forming polypeptide chain and directs the ribosome-mRNA-polypeptide complex to the surface of the ER membrane. (p. 681)

silencer: DNA sequence containing a binding site for transcription factors that inhibit transcription and whose position and orientation relative to the promoter can vary significantly without interfering with the ability to regulate transcription. (p. 718)

simple diffusion: unassisted net movement of a solute from a region where its concentration is higher to a region where its concentration is lower. (p. 197)

single bond: chemical bond formed between two atoms as a result of sharing a pair of electrons. (p. 18)

single nucleotide polymorphisms (SNPs): variations in DNA base sequence between individuals of the same species. (p. 499)

single-strand binding protein (SSB): protein that binds to single strands of DNA at the replication fork to keep the DNA unwound and therefore accessible to the DNA replication machinery. (p. 537)

sister chromatids: the two replicated copies of each chromosome that remain attached to each other prior to anaphase of mitosis. (p. 523)

site-specific mutagenesis: technique for altering the DNA base sequence at a particular location in the genome, thereby creating a specific mutation whose effects can be studied. (p. 182)

skeletal muscle: type of muscle, striated in microscopic appearance, that is responsible for voluntary movements. (p. 779)

sliding-filament model: model stating that muscle contraction is caused by thin actin filaments sliding past thick myosin filaments, with no change in the length of either type of filament. (p. 782)

sliding-microtubule model: model of motility in eukaryotic cilia and flagella which proposes that microtubule length remains unchanged but adjacent outer doublets slide past each other, thereby causing a localized bending because lateral connections between adjacent doublets and radial links to the center pair prevent free sliding of the microtubules past each other. (p. 776)

Smad (protein): class of proteins involved in the signaling pathway triggered by transforming growth factor β; upon activation, Smads enter the nucleus and regulate gene expression. (p. 274)

small ribosomal subunit: component of a ribosome with a sedimentation coefficient of 40S in eukaryotes and 30S in prokaryotes; associates with a large ribosomal subunit to form a functional ribosome. (p. 94)

smooth endoplasmic reticulum (smooth ER): endoplasmic reticulum that has no attached ribosomes and plays no direct role in protein synthesis; involved in packaging of secretory proteins and synthesis of lipids. (pp. 89, 325)

smooth ER: see *smooth endoplasmic reticulum.*

smooth muscle: muscle lacking striations that is responsible for involuntary contractions such as those of the stomach, intestines, uterus, and blood vessels. (p. 790)

SNAP: see *soluble NSF attachment protein.*

SNAP receptor protein: see *SNARE protein.*

SNARE (SNAP receptor) protein: two families of proteins involved in targeting and sorting membrane vesicles; include the v-SNARES found on transport vesicles and the t-SNARES found on target membranes. (p. 352)

SNARE hypothesis: model explaining how membrane vesicles fuse with the proper target membrane; based on specific interactions between v-SNAREs (vesicle-SNAP receptors) and t-SNAREs (target-SNAP receptors). (p. 352)

snoRNA: group of small nucleolar RNAs that bind to complementary regions of pre-rRNA and target specific sites for methylation or cleavage. (p. 646)

SNPs: see *single nucleotide polymorphisms.*

snRNA: a small nuclear RNA molecule that binds to specific proteins to form a snRNP, which in turn assembles with other snRNPs to form a spliceosome. (p. 651)

snRNP: RNA-protein complex that assembles with other snRNPs to form a spliceosome; pronounced "snurp". (p. 651)

sodium/potassium pump: see Na^+/K^+ *pump.*

soluble NSF attachment protein (SNAP): soluble cytoplasmic protein that acts in conjunction with NSF (N-ethylmaleimide-sensitive factor) to mediate the fusion of membranes brought together by interactions between v-SNAREs and t-SNAREs. (p. 352)

solute: substance that is dissolved in a solvent, forming a solution. (p. 23)

solvent: substance, usually liquid, in which other substances are dissolved, forming a solution. (p. 23)

somatic nervous system: component of the peripheral nervous system that controls voluntary movements of skeletal muscles. (p. 225)

Sos (protein): a guanine-nucleotide exchange factor that activates Ras by triggering the release of GDP, thereby permitting Ras to acquire a molecule of GTP. The Sos protein is activated by interacting with a GRB2 protein molecule that has become bound to phosphorylated tyrosines on an activated tyrosine kinase receptor. (p. 271)

Southern blotting: transfer of DNA fragments separated by gel electrophoresis to a special type of "blotter" paper (nitrocellulose or nylon), which is then hybridized with a radioactive DNA probe. (p. 503)

spatial summation: addition of the effects of depolarizations of a postsynaptic membrane caused by several presynaptic neurons at the same time, bringing the postsynaptic neuron to its threshold and triggering an action potential. (p. 251)

specific heat: amount of heat needed to raise the temperature of 1 gram of a substance 1°C. (p. 22)

sperm: haploid male gamete, usually flagellated. (p. 579)

sphere of hydration: cluster of water molecules around an ion due to the interaction of the ion with the oppositely charged region of the polar water molecule; allows for the dissociation of ion pairs in solution. (p. 24)

sphingolipid: class of lipids containing the amine alcohol sphingosine as a backbone. (pp. 70, 166)

sphingosine: amine alcohol that serves as backbone for sphingolipids; contains an amino group that can form an amide bond with a long-chain fatty acid, and also contains a hydroxyl group that can attach to a phosphate group. (p. 70)

spindle assembly checkpoint: control point located at the junction between metaphase and anaphase where mitosis can be halted if chromosomes are not properly attached to the spindle. (p. 556)

spliceosome: protein-RNA complex that catalyzes the removal of introns from pre-mRNA. (p. 651)

spore: see *haploid spore.*

sporophyte: diploid generation in the life cycle of an organism that alternates between haploid and diploid forms; form that produces spores by meiosis. (p. 581)

sputter coating: vacuum evaporation process used to coat the surface of a specimen with a layer of gold or a mixture of gold and palladium prior to examining the specimen by scanning electron microscopy. (p. GM-26)

squid giant axon: an exceptionally large axon emerging from certain squid nerve cells; its wide diameter (0.5-1.0 mm) makes it relatively easy to insert microelectrodes that can measure and control electrical potentials and ionic currents. (p. 236)

SR: see *sarcoplasmic reticulum.*

SRP: see *signal-recognition particle.*

SSB: see *single-strand binding protein.*

stage: platform on which the specimen is placed in a microscope. (p. GM-5)

staining: incubation of tissue specimens in a solution of dye, heavy metal, or other substance that binds specifically to selected cellular constituents, thereby giving those constituents a distinctive color or electron density. (p. GM-16)

standard free energy change ($\Delta G^{\circ\prime}$): free energy change accompanying the conversion of 1 mole of reactants to 1 mole of products, with the temperature, pressure, pH, and concentration of all relevant species maintained at standard values. (p. 122)

standard reduction potential ($E_0\prime$): convention used to quantify the electron transport potential of oxidation-reduction couples relative to the H^+/H_2 redox pair, which is assigned an $E_0\prime$ value of 0.0 V at pH 7.0. (p. 421)

standard state: set of arbitrary conditions defined for convenience in reporting free energy changes in chemical reactions. For systems consisting of dilute aqueous solutions, these conditions are usually a temperature of 25°C (298 K), a pressure of 1 atmosphere, and reactants other than water present at a concentration of 1 M. (p. 122)

starch: storage polysaccharide in plants consisting of glucose repeating subunits linked together by $\alpha(1 \rightarrow 4)$ bonds and, in some cases, $\alpha(1 \rightarrow 6)$ bonds. The two main forms of starch are the unbranched polysaccharide, amylose, and the branched polysaccharide, amylopectin. (p. 63)

start codon: see *initiation codon.*

start-transfer sequence: amino acid sequence in a newly forming polypeptide that acts as both an ER signal sequence that directs the ribosome-mRNA-polypeptide complex to the ER membrane and as a membrane anchor that permanently attaches the polypeptide to the lipid bilayer. (p. 683)

STAT: type of transcription factor activated by phosphorylation in the cytoplasm catalyzed by Janus activated kinase, followed by migration of the activated STAT molecules to the nucleus. (p. 726)

state: condition of a system defined by various properties, such as temperature, pressure, and volume. (p. 112)

stationary cisternae model: model postulating that each compartment of the Golgi stack is a stable structure, and that traffic between successive cisternae is mediated by shuttle vesicles that bud from one cisterna and fuse with another. (p. 335)

steady state: nonequilibrium condition of an open system through which matter is flowing, such that all components of the system are present at constant, nonequilibrium concentrations. (p. 125)

steady-state ion movements: ion movements across a membrane, whether by leakage or active transport, that maintain steady-state concentrations of these ions on both sides of the membrane. (p. 231)

stem cell: a cell capable of unlimited division that can differentiate into a variety of other cell types. (p. 706)

stereo electron microscopy: microscopic technique for obtaining a three-dimensional view of a specimen by photographing it at two slightly different angles. (p. GM-25)

stereoisomers: two molecules that have the same structural formula but are not superimposable; stereoisomers are mirror images of each other. (p. 20)

steroid: any of numerous lipid molecules that are derived from a four-membered ring compound called phenanthrene. (p. 70)

steroid hormone: any of several steroids derived from cholesterol that function as signaling molecules, moving via the circulatory system to target tissues, where they cross the plasma membrane and interact with intracellular receptors to form hormone-receptor complexes that are capable of activating (or inhibiting) the transcription of specific genes. (p. 71)

sterol: any of numerous compounds consisting of a 17-carbon four-ring system with at least one hydroxyl group and a variety of other possible side groups; includes cholesterol and a variety of other biologically important compounds, such as the male and female sex hormones, that are related to cholesterol. (p. 166)

sticky end: single-stranded end of a DNA fragment generated by cleavage with a restriction enzyme that tends to reassociate with another fragment generated by the same restriction enzyme because of base complementarity. (p. 494)

stoma (plural, **stomata**): pore on the surface of a plant leaf that can be opened or closed to control gas and water exchange between the atmosphere and the interior of the leaf. (p. 462)

stop codon: sequence of three bases in mRNA that instructs the ribosome to terminate protein synthesis. UAG, UAA, and UGA generally function as stop codons. (pp. 634, 666)

stop-transfer sequence: hydrophobic amino acid sequence in a newly forming polypeptide that halts translocation of the chain through the ER membrane, thereby anchoring the polypeptide within the membrane. (p. 683)

storage macromolecule: polymer that consists of one or a few kinds of subunits in no specific order and that serves as a storage form of monosaccharides; examples include starch and glycogen. (p. 29)

striated muscle: muscle whose myofibrils exhibit a pattern of alternating dark and light bands when viewed microscopically; includes both skeletal and cardiac muscle. (p. 779)

stroma: unstructured semifluid matrix that fills the interior of the chloroplast. (pp. 87, 448)

stroma thylakoid: membrane that interconnects stacks of grana thylakoids with each other. (pp. 87, 448)

structural gene: DNA sequence within an operon that codes for a polypeptide that carries out a specific function other than regulating the expression of another gene. (p. 694)

structural macromolecule: polymer that consists of one or a few kinds of subunits in no

specific order and that provides structure and mechanical strength to the cell; examples include cellulose and pectin. (p. 29)

subcellular fractionation: technique for isolating organelles from cell homogenates using various types of centrifugation. (p. 326)

substrate activation: role of an enzyme's active site in making a substrate molecule maximally reactive by subjecting it to the appropriate chemical environment for catalysis. (p. 138)

substrate induction: regulatory mechanism for catabolic pathways in which the synthesis of enzymes involved in the pathway is stimulated in the presence of the substrate and inhibited in the absence of the substrate. (p. 693)

substrate specificity: ability of an enzyme to discriminate between very similar molecules. (p. 134)

substrate-level phosphorylation: formation of ATP by direct transfer to ADP of a high-energy phosphate group derived from a phosphorylated substrate. (p. 381)

substrate-level regulation: enzyme regulation that depends directly on the interactions of substrates and products with the enzyme. (p. 147)

succinate-coenzyme Q oxidoreductase: see *complex II.*

supercoiled DNA: twisting of a DNA double helix upon itself, either in a circular DNA molecule or in a DNA loop anchored at both ends. (p. 489)

supernatant: material that remains in solution after particles of a given size and density are removed as a pellet during centrifugation. (p. 328)

suppressor tRNA: mutant tRNA molecule that inserts an amino acid where a stop codon generated by another mutation would otherwise have caused premature termination of protein synthesis (p. 674)

surface area/volume ratio: mathematical ratio of the surface area of a cell to its volume; decreases with increasing linear dimension of the cell (length or radius), thereby increasing the difficulty of maintaining adequate surface area for import of nutrients and export of waste products as cell size increases. (p. 77)

surroundings: the remainder of the universe when one is studying the distribution of energy within a given system. (p. 112)

Svedberg unit: unit for expressing the sedimentation coefficient of biological macromolecules: One Svedberg unit (S) = 10^{-13} second. In general, the greater the mass of a particle, the greater the sedimentation rate, though the relationship is not linear. (p. 327)

symbiotic relationship: a mutually beneficial association between cells (or organisms) of two different species. (p. 450)

symport: coupled transport of two solutes across a membrane in the same direction. (p. 204)

synapse: tiny gap between a neuron and another cell (neuron, muscle fiber, or gland cell), across which the nerve impulse is transferred by direct electrical connection or by chemicals called neurotransmitters. (p. 227)

synapsis: close pairing between homologous chromosomes during the zygotene phase of prophase I of meiosis. (p. 581)

synaptic cleft: gap between the presynaptic and postsynaptic membranes at the junction between two nerve cells. (p. 243)

synaptonemal complex: zipperlike, protein-containing structure that joins homologous chromosomes together during prophase I of meiosis. (p. 585)

system: the restricted portion of the universe that one decides to study at any given time

when investigating the principles that govern the distribution of energy. (p. 112)

T

T: see *thymine.*

T tubules: see *transverse tubule system.*

tandemly repeated DNA: repeated DNA sequences whose multiple copies are adjacent to one another. (p. 500)

target tissue: tissue that is specifically affected by a particular hormone because it has receptors for that hormone located either in the plasma membrane or within the cell. (p. 276)

target-SNAP receptor: see *t-SNARE.*

TATA box: part of the core promoter for many eukaryotic genes transcribed by RNA polymerase II; consists of a consensus sequence of TATA followed by two or three more A's, located about 25 nucleotides upstream from the transcriptional startpoint. (p. 641)

TATA-binding protein (TBP): component of transcription factor TFIID that confers the ability to recognize and bind the TATA box sequence in DNA; also involved in regulating transcription initiation at promoters lacking a TATA box. (p. 643)

TBP: see *TATA-binding protein.*

TCA cycle: see *tricarboxylic acid cycle.*

TE: see *transitional element.*

tektin: protein component of axonemal microtubules that helps arrange tubulin molecules into the A and B tubules of the doublet. (p. 776)

telomerase: special type of DNA polymerase that catalyzes the formation of additional copies of a telomeric repeat sequence. (p. 539)

telomere: DNA sequence located at either end of a linear chromosome; contains simple-sequence, tandemly repeated DNA. (pp. 500, 539)

telophase: final stage of mitosis or meiosis, when daughter chromosomes arrive at the poles of the spindle accompanied by reappearance of the nuclear envelope. (p. 548)

TEM: see *transmission electron microscope.*

temperature-sensitive mutant: cell that produces a protein that functions properly at normal temperatures but becomes seriously impaired when the temperature is altered slightly. (p. 530)

template: a nucleic acid whose base sequence serves as a pattern for the synthesis of another (complementary) nucleic acid. (p. 57)

template strand: the strand of a DNA double helix that serves as the template for RNA synthesis via complementary base pairing. (p. 632)

temporal summation: rapid addition of the effects of small depolarizations of a postsynaptic membrane over a short period of time, eventually bringing the postsynaptic neuron to its threshold and triggering an action potential. (p. 251)

terminal bulb: region near the end of an axon where neurotransmitter molecules are stored for use in transmitting signals across the synapse. (p. 243)

terminal cisterna (plural, **cisternae**): region of sarcoplasmic reticulum positioned adjacent to a T tubule in skeletal muscle. (p. 787)

terminal glycosylation: modification of glycoproteins in the Golgi complex involving removal and/or addition of sugars to the carbohydrate side chains formed by prior core glycosylation in the endoplasmic reticulum. (p. 336)

terminal oxidase: electron transfer complex that is capable of transferring electrons directly to oxygen. Complex IV (cytochrome *c* oxidase) of the mitochondrial electron transport system is an example. (p. 424)

terminal web: dense meshwork of spectrin and myosin molecules located at the base of a microvillus to which the bundle of actin microfilaments that make up the core of the microvillus is anchored. (p. 759)

termination signal: DNA sequence located near the end of a gene that triggers the termination of transcription. (p. 639)

terpene: a lipid constructed from the five-carbon compound isoprene and its derivatives, joined together in various combinations. (p. 71)

tertiary structure: level of protein structure involving interactions between amino acid side chains of a polypeptide, regardless of where along the primary sequence they happen to be located; results in three-dimensional folding of a polypeptide chain. (p. 50)

tetrahedral (carbon atom): an atom of carbon from which four single bonds extend to other atoms, each bond equidistant from all other bonds, causing the atom to resemble a tetrahedron with its four equal faces. (p. 20)

TFIIB recognition element: see *BRE.*

TGFβ: see *transforming growth factor β.*

TGN: see *trans-Golgi network.*

theory: a hypothesis that has been tested critically under many different conditions—usually by many different investigators using a variety of approaches—and is consistently supported by the evidence. (p. 13)

thermodynamic spontaneity: a measure of whether a reaction can occur, but says nothing about whether the reaction actually will occur. Reactions with a negative free energy change are thermodynamically spontaneous. (p. 114)

thermodynamics: area of science that deals with the laws governing the energy transactions that accompany all physical processes and chemical reactions. (p. 112)

thick filament: myosin-containing filament, found in the myofibrils of striated muscle cells, in which individual myosin molecules are arranged in a staggered array with the heads of the myosin molecules projecting out in a repeating pattern. (p. 779)

thin filament: actin-containing filament, found in the myofibrils of striated muscle cells, in which two F-actin molecules are arranged in a helix associated with tropomyosin and troponin. (p. 779)

thin-layer chromatography (TLC): procedure for separating compounds by chromatography in a medium, such as silicic acid, that is bound as a thin layer to a glass or metal surface. (p. 168)

30-nm chromatin fiber: fiber formed by packing together the nucleosomes of a 10-nm chromatin fiber. (p. 507)

30S initiation complex: complex formed by the association of mRNA, the 30S ribosomal subunit, an initiator aminoacyl tRNA molecule, and initiation factor IF2. (p. 669)

threshold potential: value of the membrane potential that must be reached before an action potential is triggered. (p. 236)

thylakoid: flattened membrane sac suspended in the chloroplast stroma, usually arranged in stacks called grana; contains the pigments, enzymes, and electron carriers involved in the light-requiring reactions of photosynthesis. (pp. 87, 448)

thylakoid lumen: compartment enclosed by an interconnected network of grana and stroma thylakoids. (p. 449)

thymine (T): nitrogen-containing aromatic base, chemically designated as a pyrimidine, which

serves as an informational monomeric unit when present in DNA with other bases in a specific sequence; forms a complementary base pair with adenine (A) by hydrogen bonding. (p. 55)

Ti plasmid: DNA molecule that causes crown gall tumors when transferred into plants by bacteria; used as a cloning vector for introducing foreign genes into plant cells. (p. 614)

TIC: translocase of the inner chloroplast membrane, a transport complex involved in the uptake of specific polypeptides into the chloroplast. (p. 685)

tight junction: type of cell junction in which the adjacent plasma membranes of neighboring animal cells are tightly sealed, thereby preventing molecules from diffusing from one side of an epithelial cell layer to the other by passing through the spaces between adjoining cells. (pp. 98, 310)

TIM: translocase of the inner mitochondrial membrane, a transport complex involved in the uptake of specific polypeptides into the mitochondrion. (p. 685)

TLC: see *thin-layer chromatography.*

T_m: see *transition temperature* or *DNA melting temperature.*

TnC: polypeptide subunit of troponin that binds calcium ions. (p. 781)

TnI: polypeptide subunit of troponin that binds to actin and inhibits muscle contraction. (p. 781)

TnT: polypeptide subunit of troponin thought to be responsible for binding troponin to tropomyosin. (p. 781)

TOC: translocase of the outer chloroplast membrane, a transport complex involved in the uptake of specific polypeptides into the chloroplast. (p. 685)

TOM: translocase of the outer mitochondrial membrane, a transport complex involved in the uptake of specific polypeptides into the mitochondrion. (p. 685)

tonofilament: type of intermediate filament that extends into the cell from the plaque of a desmosome, thereby anchoring the desmosome in the cytoplasm; made primarily of the protein keratin, desmin, or vimentin, depending on the cell type. (p. 310)

topoisomerase: enzyme that catalyzes the interconversion of the relaxed and supercoiled forms of DNA by making transient breaks in one or both DNA strands. (pp. 489, 537)

trailer (on mRNA): nontranslated sequence at the 3′ end of an mRNA molecule located after the stop codon. (p. 666)

***trans* face (maturing face):** side of the Golgi complex that is located opposite from the *cis* (forming) face. (p. 334)

***trans*-acting factor:** regulatory protein that exerts its function by binding to specific DNA sequences. (p. 698)

transamination: transfer of an amino group from an amino acid to an α-keto acid acceptor. (p. 414)

transcription: process by which RNA polymerase utilizes one DNA strand as a template for guiding the synthesis of a complementary RNA molecule. (p. 623)

transcription factor: protein required for the binding of RNA polymerase to a promoter and for the optimal initiation of transcription. (p. 563) Also see *general transcription factor* and *regulatory transcription factor.*

transcription regulation domain: region of a transcription factor, distinct from the DNA-binding domain, that is responsible for regulating transcription (p. 721)

transcription unit: segment of DNA whose transcription gives rise to a single, continuous RNA molecule. (p. 635)

transcriptional control: group of regulatory mechanisms involved in controlling the rates at which specific genes are transcribed. (p. 715)

transcytosis: endocytosis of material into vesicles that move to the opposite side of the cell and fuse with the plasma membrane, releasing the material into the extracellular space. (p. 348)

transduction: transfer of bacterial DNA sequences from one bacterium to another by a bacteriophage. (p. 599)

transfer RNA (tRNA): family of small RNA molecules, each binding a specific amino acid and possessing an anticodon that recognizes a specific codon in mRNA. (pp. 623, 662)

transformation: see *genetic transformation.*

transforming growth factor β (TGFβ): family of growth factors that can exhibit either growth-stimulating or growth-inhibiting properties, depending on the target cell type; regulates a wide range of activities in both embryos and adult animals, including effects on cell growth, division, differentiation, and death. (pp. 274, 564)

transgenic: any organism whose genome contains a gene that has been experimentally introduced from another organism using the techniques of genetic engineering. (p. 614)

***trans*-Golgi network (TGN):** region of the Golgi complex consisting of a network of membrane-bounded tubules that are located on the opposite side of the Golgi complex from the *cis*-Golgi network. (p. 334)

transient fusion model: model postulating that late endosomes form temporary connections with lysosomes for the purpose of transferring material; also called the "kiss-and-run" model. (p. 354)

transit sequence: amino acid sequence that targets a completed polypeptide chain to either mitochondria or chloroplasts. (p. 684)

transition state: intermediate stage in a chemical reaction, of higher free energy than the initial state, through which reactants must pass before giving rise to products. (p. 131)

transition temperature (T_m): temperature at which a membrane will undergo a sharp decrease in fluidity ("freezing") as the temperature is decreased and becomes more fluid again ("melts") when it is then warmed; determined by the kinds of fatty acid side chains present in the membrane. (p. 172)

transition vesicle: membrane vesicle that shuttles lipids and proteins from the endoplasmic reticulum to the Golgi complex. (p. 325)

transitional element (TE): region of the endoplasmic reticulum that is involved in the formation of transition vesicles. (p. 325)

translation: process by which the base sequence of an mRNA molecule guides the sequence of amino acids incorporated into a polypeptide chain; occurs on ribosomes. (p. 623)

translational control: mechanisms that regulate the rate at which mRNA molecules are translated into their polypeptide products; includes control of translation rates by initiation factors, selective inhibition of specific mRNAs by translational repressors, and variations in the rates of mRNA degradation. (p. 730)

translational repressor: regulatory protein that selectively inhibits the translation of a particular mRNA. (p. 731)

translesion synthesis: DNA replication across regions where the DNA template is damaged. (p. 543)

translocation: movement of mRNA across a ribosome by a distance of three nucleotides, bringing the next codon into position for translation. (p. 671) (*Note:* The same term "translocation," which literally means "a change of location," can also refer to the movement of a protein molecule through a membrane channel or to the transfer of a segment of one chromosome to another chromosome.)

translocon: structure in the ER membrane that carries out the translocation of newly forming polypeptides across (or into) the ER membrane. (p. 681)

transmembrane protein: an integral membrane protein possessing one or more hydrophobic regions that span the membrane plus hydrophilic regions that protrude from the membrane on both sides. (p. 177)

transmembrane segment: hydrophobic segment about 20-30 amino acids long that crosses the lipid bilayer in a transmembrane protein. (p. 177)

transmission electron microscope (TEM): type of electron microscope in which an image is formed by electrons that are transmitted through a specimen. (pp. 7, GM-18)

transport: selective movement of substances across membranes, both into and out of cells and into and out of organelles. (pp. 160, 195)

transport protein: membrane protein that recognizes substances with great specificity and assists their movement across a membrane; includes both carrier proteins and channel proteins. (p. 197)

transport vesicle: vesicle that buds off from a membrane in one region of the cell and fuses with other membranes: includes vesicles that convey lipids and proteins from the ER to the Golgi complex, between the Golgi stack cisternae, and from the Golgi complex to various destinations in the cell, including secretory vesicles, endosomes, and lysosomes. (p. 333)

transporter (in the nuclear pore complex): protein granule located in the center of the nuclear pore complex that is thought to move macromolecules across the nuclear envelope. (p. 513)

transverse diffusion: movement of a lipid molecule from one monolayer of a membrane to the other, a thermodynamically unfavorable and therefore infrequent event; also called "flip-flop." (p. 169)

transverse tubule system (T tubules): invaginations of the plasma membrane that penetrate into a muscle cell and conduct electrical impulses into the cell interior, where T tubules make close contact with the sarcoplasmic reticulum and trigger the release of calcium ions. (p. 787)

treadmilling: process by which the addition of tubulin molecules to the plus end of a microtubule is continually balanced by the loss of tubulin molecules from the minus end, creating a situation in which tubulin molecules are continually transferred from the plus end of the microtubule to the minus end, even though the overall length of the microtubule does not change. (p. 748)

triacylglycerol: a glycerol molecule with three fatty acids linked to it; also called a triglyceride (pp. 68, 412)

triad: region where a T tubule passes between the terminal cisternae of the sarcoplasmic reticulum in skeletal muscle. (p. 788)

tricarboxylic acid cycle (TCA cycle): cyclic metabolic pathway that oxidizes acetyl CoA to carbon dioxide in the presence of oxygen,

generating ATP and the reduced coenzymes NADH and $FADH_2$; a component of aerobic respiration; also called the Krebs cycle. (p. 405)

triglyceride: see *triacylglycerol.*

triple bond: chemical bond formed between two atoms as a result of sharing three pairs of electrons. (p. 18)

triplet code: a coding system in which three units of information are read as a unit; a reference to the genetic code, which is read from mRNA in units of three bases called *codons.* (p. 628)

triskelion: structure formed by clathrin molecules consisting of three polypeptides radiating from a central vertex; the basic unit of assembly for clathrin coats. (p. 350)

tRNA: see *transfer RNA.*

tropomyosin: long, rodlike protein associated with the thin actin filaments of muscle cells, functioning as a component of the calcium-sensitive switch that activates muscle contraction; blocks the interaction between actin and myosin in the absence of calcium ions. (p. 781)

troponin: complex of three polypeptides (TnT, TnC, and TnI) that functions as a component of the calcium-sensitive switch that activates muscle contraction; displaces tropomyosin in the presence of calcium ions, thereby activating contraction. (p. 781)

trp operon: group of adjoining bacterial genes that code for enzymes involved in tryptophan biosynthesis and whose transcription is selectively inhibited in the presence of tryptophan. (p. 699)

true-breeding (plant strain): organism that, upon self-fertilization, produces only offspring of the same kind for a given genetic trait. (p. 590)

t-SNARE (target-SNAP receptor): protein associated with the outer surface of a target membrane that binds to a v-SNARE protein associated with the outer surface of an appropriate transport vesicle. (p. 352)

tubulin: family of related proteins that form the main structural component of microtubules. (p. 746) Also see *alpha tubulin, beta tubulin,* and *gamma tubulin.*

tumor: growing mass of cells caused by uncontrolled cell proliferation. (p. 564) Also see *benign tumor* and *malignant tumor.*

tumor suppressor gene: gene that normally functions to restrain cell proliferation and whose loss or inactivation by deletion or mutation can lead to the development of cancer. (p. 567)

turgor pressure: pressure that builds up in a cell as a result of the inward movement of water that occurs because of a higher solute concentration inside the cell than outside; accounts for the firmness, or turgidity, of fully hydrated cells or tissues of plants and other organisms. (p. 201)

turnover number (k_{cat}): rate at which substrate molecules are converted to product by a single enzyme molecule when the enzyme is operating at its maximum velocity. (p. 143)

type II myosin: form of myosin composed of four light chains and two heavy chains, each having a globular myosin head, a hinge region, and a long rodlike tail; found in skeletal, cardiac, and smooth muscle cells, as well as in nonmuscle cells. (p. 778)

U

U: see *uracil.*

ubiquitin: small protein that is linked to other proteins as a way of marking the targeted protein for degradation by proteasomes. (p. 733)

ultracentrifuge: instrument capable of generating centrifugal forces that are large enough to separate subcellular structures and macromolecules on the basis of size, shape, and density. (pp. 9, 326)

ultramicrotome: instrument used to slice an embedded biological specimen into ultrathin sections for electron microscopy. (p. GM-21)

undershoot: see *hyperpolarization.*

unidirectional pumping of protons: the active and directional transport of protons across a membrane such that they accumulate preferentially on one side of the membrane, establishing an electrochemical proton gradient across the membrane; a central component of electron transport and ATP generation in both respiration and photosynthesis. (p. 428)

uniport: membrane protein that transports a single solute from one side of a membrane to the other. (p. 204)

unsaturated fatty acid: fatty acid molecule containing one or more double bonds. (p. 68)

upstream: located toward the 5′ end of the DNA coding strand. (p. 637)

uracil (U): nitrogen-containing aromatic base, chemically designated as a pyrimidine, that serves as an informational monomeric unit when present in RNA with other bases in a specific sequence; forms a complementary base pair with adenine (A) by hydrogen bonding. (p. 56)

useful magnification: measurement of how much an image can be enlarged before additional enlargement provides no additional information. (p. GM-5)

V

v: see *initial reaction velocity.*

vacuole: membrane-bounded organelle in the cytoplasm of a cell, used for temporary storage or transport; acidic membrane-enclosed compartment in plant cells. (pp. 91, 357)

vacuum evaporator: bell jar containing a metal electrode and a carbon electrode in which a vacuum can be created; used in the preparation of metal replicas of the surfaces of biological specimens. (p. GM-22)

valence: a number indicating the number of other atoms with which a given atom can combine. (p. 18)

van der Waals interaction: weak attractive interaction between two atoms caused by transient asymmetries in the distribution of charge in each atom. (p. 46)

variable number tandem repeat (VNTR): short repeated DNA sequences whose variation in length between individuals forms the basis for DNA fingerprinting. (p. 503)

vesicle-SNAP receptor: see *v-SNARE.*

viroid: small, circular RNA molecule that can infect and replicate in host cells even though it does not code for any protein. (p. 99)

virus: subcellular parasite composed of a protein coat and DNA or RNA, incapable of independent existence; invades and infects cells and redirects the host cell's synthetic machinery toward the production of more virus. (p. 98)

V_m: see *resting membrane potential.*

V_{max}: see *maximum velocity.*

VNTR: see *variable number tandem repeat.*

voltage: see *potential.*

voltage sensor: amino acid segment of a voltage-gated ion channel that makes the channel responsive to changes in membrane potential. (p. 235)

voltage-gated ion channel: an integral membrane protein that forms an ion-conducting pore whose permeability is regulated by changes in the membrane potential. (p. 233)

v-SNARE (vesicle-SNAP receptor): protein associated with the outer surface of a transport vesicle that binds to a t-SNARE protein associated with the outer surface of the appropriate target membrane. (p. 352)

V-type ATPase: type of transport ATPase that pumps protons into such organelles as vesicles, vacuoles, lysosomes, endosomes, and the Golgi complex. (p. 208)

W

wavelength: distance between the crests of two successive waves. (p. GM-2)

wild type: normal, nonmutant form of an organism, usually the form found in nature. (p. 596)

wobble hypothesis: flexibility in base-pairing between the third base of a codon and the corresponding base in its anticodon. (p. 664)

work: transfer of energy from one place or form to another place or form by any process other than heat flow. (p. 113)

X

X-chromosome inactivation: random inactivation of one of the two X chromosomes present in the cells of female mammals. (p. 714)

xenobiotic: chemical compound that is foreign to biological organisms. (p. 361)

X-ray diffraction: technique for determining the three-dimensional structure of macromolecules based on the pattern produced when a beam of X-rays is passed through a sample, usually a crystal or fiber. (p. GM-27)

Y

YAC: see *yeast artificial chromosome.*

yeast artificial chromosome (YAC): yeast cloning vector consisting of a "minimalist" chromosome that contains all the DNA sequences needed for normal chromosome replication and segregation to daughter cells, and very little else. (p. 613)

Z

Z line: dark line in the middle of the I band of a striated muscle myofibril; defines the boundary of a sarcomere. (p. 780)

zinc finger: DNA-binding motif found in some transcription factors; consists of an α helix and a two-segment β sheet held in place by the interaction of precisely positioned cysteine or histidine residues with a zinc atom. (p. 722)

zygote: diploid cell formed by the union of two haploid gametes. (p. 579)

zygotene: stage during prophase I of meiosis when homologous chromosomes become closely paired by the process of synapsis. (p. 583)

Photo, Illustration, and Text Credits

Photo Credits

Chapter 1 01-03a, b From Cell Ultrastructure, by William A. Jensen and Roderic B. Park. Copyright ©1967 by Wadsworth Publishing Co., Inc. Used by permission of the authors and publisher. 01-04a, b Dr. Judith Croxdale. Table 01-01(1) ©Biophoto Associates/Photo Researchers, Inc. Table 01-01(2, 4, 5) ©David M. Phillips/Visuals Unlimited. Table 01-01(3) ©Ed Reschke. Table 01-01(6) Courtesy of Noran Instruments.

Chapter 2 02-13a, b Don Fawcett/Photo Researchers, Inc. 02-14a Courtesy of G. F. Bahr, Armed Forces Institute of Pathology. 02-14b Courtesy E. H. Newcomb. 02-14c Courtesy of Eva Frei & R. D. Preston. 02-19 © Graphics Systems Research, IBM UK Scientific Centre.

Chapter 3 03-24a Jeremy Burgess/Photo Researchers, Inc. 03-24b Don Fawcett/Photo Researchers, Inc. 03-25 Courtesy of Eva Frei & R. D. Preston. 3A-1 Barrington Brown/Photo Researchers, Inc.

Chapter 4 04-02 Susumu Ito. 04-03 Courtesy of Gregory J. Brewer, Southern Illinois University. 04-04 Courtesy E. H. Newcomb. 04-05b Courtesy of Richard Rodewald/Biological Photo Service. 04-06b Micrograph by W. P. Wergin; provided by E. H. Newcomb. 04-07 Courtesy of Hans Ris. 04-08 Oscar L. Miller, Jr., University of Virginia. 04-10b Courtesy of Richard Rodewald/Biological Photo Service. 04-10c From J. P. Strafstrom and L. A. Staehelin, *The Journal of Cell Biology* 98 (1984): 699. Reproduced by copyright permission of The Rockefeller University Press. 04-11c Keith Porter/Photo Researchers, Inc. 04-12b From J. B. Rattner and B. R. Brinkley, *J. Ultrastructure Res.* 32(1970): 316. Copyright ©1970 by Academic Press. 04-13 Courtesy of S. M. Wang. 04-13b Courtesy of Clara Franzini-Armstrong. 04-14 Micrograph by W. P. Wergin; provided by E. H. Newcomb. 04-15c Courtesy of H. Stuart Pankratz/Biological Photo Service. 04-15d Courtesy of Barry J. King/Biological Photo Service. 04-16 Courtesy E. H. Newcomb. 04-18b M. Simionescu and N. Simionescu, *J. Cell Biol.* 70(1976):608. Reproduced by copyright permission of The Rockefeller University Press. 04-18c Don Fawcett/Visuals Unlimited. 04-19 ©Barry King/Biological Photo Service. 04-20 From S. E. Frederick and E. H. Newcomb, *Journal of Cell Biology* 43(1969): 343. Reproduced by copyright permission of the Rockefeller University Press; photo provided by E. H. Newcomb. 04-21b Micrograph by P. J. Gruber; provided by E. H. Newcomb. 04-23 From Sigrid Regauer, Werner W. Franke, and Ismo Virtananen, *Journal of Cell Biology* 100 (1988): 997-1009. Reproduced by copyright permission of the Rockefeller University Press. 04-25 From P. H. Raven, R. F. Evert, and H. A. Curtis, *Biology of Plants*, 2nd ed. New York: Worth Publishers, Inc., 1981. Used with the permission of Worth Publishers. 04-26a-c Courtesy of R. C. Williams and H. W. Fisher.

Chapter 5 05-03 From D. A. Cuppels and A. Kelman, *Phytopathology* 70 (1980): 1110. Photo provided by A. Kelman. Copyright ©1980 by the American Phytopathological Society. 05-06 © Krzysztof Kozminski.

Chapter 6 06-07 Richard J. Feldmann, National Institutes of Health.

Chapter 7 07-01a M. Simionescu and N. Simionescu, *J. Cell Biol.* 70(1976): 622. Reproduced by copyright permission of The Rockefeller University Press. 07-01b Courtesy E. H. Newcomb. 07-03e Micrograph courtesy of J. David Robertson. 07-04 Courtesy of Don W. Fawcett, M.D., Harvard Medical School. 07-16b Micrographs of E and P faces courtesy of Philippa Claude. 07-17a Courtesy of Daniel Branton. 07-17b Courtesy of R. B. Park. 07-18a & b Courtesy of David Deamer, University of California, Santa Cruz. 07-20a Ken Eward/Science Source/Photo Researchers, Inc. 07-27 Courtesy of Susumu Ito. 07-29 Courtesy of A. E. Sowers.

Chapter 8 08-14a ©Helen E. Carr/Biological Photo Service.

Chapter 9 09-02 V.I. LAB E.R.I.C./FPG International LLC. 09-03b ©Manfred Kage/Peter Arnold, Inc. 09-18a Courtesy of G. L. Scott, J. A. Feilbach and T. A. Duff. 09-21c Courtesy of S. G. Waxman. 09.24 M. L. Harlow et al., "The architecture of active zone material at the frog's neuromuscular junction," from *Nature* 409: 479-84. Reprinted by permission from *Nature* © 2001, Macmillan Magazines Ltd. Image courtesy Mark L. Harlow, Stanford University. 09-25a From J. Cartaud, E. L. Bendetti, A. Sobel, and J. P. Changeux, *Journal of Cell Science* 29 (1978): 313. Copyright ©1978 The Company of Biologists, Ltd. 09-27b Courtesy of E. R. Lewis, University of California, Berkeley. 09-28 Courtesy of H. Shio and P. B. Lazarow. Reproduced by copyright permission of The Rockefeller University Press; photo provided by P. B. Lazarow.

Chapter 10 10-10 Courtesy of R. D. Burgoyne. 10-13 Courtesy of Y. Hiramoto. 10.15 N. H. Evans and A. Hetherington, "The ups and downs of guard cell signaling," from *Current Biology* 11: R92-94. © 2001. Reprinted with permission from Elsevier Science. Image courtesy Nicola Evans and Alistair Hetherington, Lancaster University, UK. 10-20 Enrique Amaya with Thomas J. Musci and Marc Kirschner 1991, *Cell* 66. Courtesy of Marc Kirschner. 10-27 © Walter Malorni et al., from "Morphological aspects of apoptosis." Image courtesy Walter Malorni, Istituto Superiore de Sanita, Rome.

Chapter 11 11-01 © R. G. Kessel and R. H. Kardon, *Tissues and Organs: A Text-Atlas of Scanning Electron Microscopy* (New York: W. H. Freeman & Co., 1979). All rights reserved. 11-02a Courtesy of Jerome Gross. 11-06a Reproduced from L. Rosenberg, W. Hellmann, and A. K. Kleinschmidt (1975). *Journal of Biological Chemistry* 250:1877-83 by copyright permission of the American Society for Biochemistry and Molecular Biology, Bethesda, MD. 11-08a Reproduced from Dr. Jean Paul Thiery, *The Journal of Cell Biology* 96 (1983): 462-73 by copyright permission of the Rockefeller University Press. 11-08b, c ©Richard Hynes, from Scientific American, June 1986. 11-09 Courtesy of G. W. Willis, M. D., and Biological Photo Service. 11-12b © Don Sakaguchi, Iowa State University. Image courtesy Don Sakaguchi. 11-12d Micrograph courtesy of Douglas E. Kelly. 11-15a-e Courtesy of Dr. Masatoshi Takeichi. 11-16a, b Courtesy of Janet Heasman, University of Minnesota. 11-19a From Douglas E. Kelly, *The Journal of Cell Biology* 28 (1966): 51. Reproduced by copyright permission of The Rockefeller University Press. 11-20c Courtesy of Philippa Claude. 11-21b Courtesy of Daniel S. Friend. 11-23 b From C. Peracchia and A. F. Dulhunty, *The Journal of Cell Biology* 70 (1976): 419. Reproduced by copyright permission of The Rockefeller University Press. 11-23 c Courtesy of Philippa Claude. 11-25 ©Biophoto Associates/ Photo Researchers, Inc. 11-26 ©G. F. Leedale/Photo Researchers, Inc. 11-27a, b Courtesy of K. M̦hlenthaler. 11-28a Micrograph by W. P. Wergin; photo provided by E. H. Newcomb. 11-28c Courtesy E. H. Newcomb. 11-29a, b Courtesy of Werner R. Loewenstein. 11A-1 K. Ireton and P. Cossart, "Interaction of invasive bacteria with host signaling pathways," from *Current Opinions in Cell Biology* 10, no. 2: 276-83. © 1998. Reprinted with permission from Elsevier Science. Image courtesy Pascale Cossart, Institut Pasteur, Paris.

Chapter 12 12-02a Courtesy of Don W. Fawcett, M.D., Harvard Medical School. 12-02b Courtesy of M. Bielinska. 12-03a © Barry King/Biological Photo Service. 12-04b Courtesy Michael J. Wynne. 12-05 Courtesy of William G. Dunphy with Ruud Brands and James E. Rothman, *Cell* 40 (1985): 467, Fig. 6, Panel B. 12-11 From L. Orci and A. Perrelet, *Freeze-Etch Histology*. Heidelberg: Springer-Verlag, 1975. 12-16 From M. M. Perry and A. B. Gilbert, *The Journel of Cell Science* 39 (1979): 257. Copyright © 1979 by The Company of Biologists Ltd. 12-17a Micrograph courtesy of J. Heuser. 12-17b Micrograph courtesy of N. Hirokawa and J. E. Heuser from D. W. Fawcett. 12-18a Micrograph courtesy of J. Heuser. 12-20b Courtesy of Pierre Baudhuin. 12-22 Courtesy of Zdenek Hruban. 12-25 Eldon H. Newcomb/Biological Photo Service. 12B-02a, b R. G. W. Anderson, M. S. Brown, and J. L. Goldstein, *Cell* 10 (1977): 351-64.

Chapter 13 13A-1 Deborah Davis/PhotoEdit.

Chapter 14 14-02 Courtesy of Charles R. Hackenbrock. 14-03 Courtesy of M. L. Vorbeck. 14-04 Keith R. Porter, University of Pennsylvania. 14-05 From L. Packer, *Ann. New York Academy of Sciences* 227 (1974): 166. Copyright ©1974 by the New York Academy of Sciences. Photo provided by H. T. Ngo. 14-06 Courtesy of A. Tzagoloff from *Mitochondria* (New York: Plenum, 1982). 14A-01 From R. N. Trelease, P. J. Gruber, W. M. Becker, and E. H. Newcomb, *Plant Physiology* 48 (1971): 461.

Chapter 15 15-02a Micrograph by M. W. Steer; provided by E. H. Newcomb. 15-02b Courtesy of Professor Linda Graham, Department of Botany, University of Wisconsin, Madison. 15-03a, c Micrographs by W. P. Wergin; provided by E. H. Newcomb. 15-04 N. J. Lang/Biological Photo Service. 15-14 Hartmut Michel, Max Planck Institute of Biophysics, Frankfurt, Germany, and Johann Deisenhofer, University of Texas Southwestern Medical Center at Dallas. Fig. 15-20b ©R. W. Van Norman/Visuals Unlimited.

Chapter 16 16-06bL, R Courtesy James C. Wang. 16-16a Reproduced from H. Kobayashi, K. Kobayashi, and Y. Kobayashi (1977). *Journal of Bacteriology* 132: 262-69 by copyright permission of the American Society for Microbiology. 16-16b Dr. Gopal/SPL/Photo Researchers, Inc. 16-17 Courtesy of Dr. Jack Griffith. 16-19L, R Photographs courtesy of Roger D. Kornberg. 16-21aB Dr. Jack Griffith. 16-21aT Biological Photo Service. 16-21b Courtesy of Barbara Hamkalo. 16-21c Courtesy of J. R. Paulsen and U. K. Laemmli, *Cell*. Copyright Cell Press. 16-21d, e G. F. Bahr/Armed Forces Institute of Pathology. 16-22, Courtesy of Ulrich K. Laemmli. 16-23 Courtesy of D. L. Robberson. 16-25a From L. Orci and A. Perrelet, *Freeze-Etch Histology*. Heidelberg: Springer-Verlag, 1975. 16-25b Micrograph by S. R. Tandon; provided by E. H. Newcomb 16-26a, 16-27 From L. Orci and A. Perrelet, *Freeze-Etch Histology*. Heidelberg: Springer-Verlag, 1975. 16-28 From A. C.

Faberge, *Cell Tiss. Res.* 15 (1974): 403. Heidelberg: Springer-Verlag, 1974. 16-31a Courtesy of Jeffery A. Nickerson, Ph.D., Sheldon Penman, Ph.D., and Gariela Crockmalnic. 16-31b Courtesy of Ueli Aebi. 16-32 D. Phillips/Photo Researchers, Inc. 16-33 From Sasha Koulish and Ruth G. Kleinfeld, *Journal of Cell Biology* 23 (1964): 39. Reproduced by copyright permission of The Rockefeller University Press. 16A-1 ©Lee D. Simon/Science Source/Photo Researchers, Inc. 16A-2 Courtesy of Cellmark Diagnostics, Germantown, MD. 16A-3 Bruce Iverson/Bruce Iverson. 16C-2 Cellmark Diagnostics, Inc

Chapter 17 17-05a Courtesy of *Cold Spring Harbor Symp. Quant. Biol.* (1963) 28: 44. 17-06 From D. J. Burks and P. J. Stambrook, *Journal of Cell Biology* 77 (1978): 762. Reproduced by permission of The Rockefeller University Press. Photos provided by P. J. Stambrook. 17-19a-f Ed Reschke/Ed Reschke. 17-20a-e Phototake/Carolina Biological Supply Company. 17-21b J. Richard McIntosh, University of Colorado. 17-22L J. F. Gennaro/Photo Researchers, Inc. 17-22R CNRI/ SPL/Photo Researchers, Inc. 17-25 Courtesy of Dr. Matthew Schibler, from *Protoplasma* 137 (1987): 29.44. Springer-Verlag. 17-26c, d Courtesy of Jeremy Pickett-Heaps, University of Melbourne. 17-27 ©David M. Phillips/Visuals Unlimited. 17-28 Micrograph by B. A. Palevitz. Courtesy E. H. Newcomb, University of Wisconsin. 17-29a-d ©R. G. Kessel and R. H. Kardon, *Tissues and Organs: A Text-Atlas of Scanning Electron Microscopy*, W. H. Freeman & Co., 1979. All Rights Reserved.

Chapter 18 18-06a-j Courtesy of B. John. 18-08a From P. B. Moens, *Chromosoma* 23 (1968): 418. Copyright © 1968 by Springer-Verlag. 18-20a Courtesy of Charles C. Brinton, Jr., and Ms. Judith Carnahan. 18-20b Omikron/Photo Researchers, Inc. 18-25 Courtesy of Ross B. Inman, University of Wisco nsin, Madison. 18A-01 Courtesy of Dr. Ralph L. Brinster, School of Veterinary Medicine, University of Pennsylvania.

Chapter 19 19-03a Bill Longcore/SS/Photo Researchers, Inc. 19-03b Jackie Lewin/Royal Free Hospital/SS/Photo Researchers, Inc. 19-04a, b Vernon Ingram. 19-15 Courtesy of D. B. Nikolov and S. K. Burley from Nikolov et al., *Nature* (1992) 360: 40-46. 19-18 From O. L. Miller, Jr., B. A. Hamkalo, and C. A. Thomas, Jr. Reprinted with permission from *Science* (1970)169: 392, Fig. 3. Copyright 1970 American Association for the Advancement of Science. 19-25a Courtesy of A. L. Beyer. 19-25b Courtesy of Jack Griffith.

Chapter 20 20-01a Micrograph and figure adapted from J. A. Lake, *Scientific American* (1981) 245: 86. © J. A. Lake. 20-03b Reprinted with permission from Sung-Hou Kim et al., from *Science* (1974) 185:435; ©1974 by the AAAS.

Chapter 21 21-12 From D. D. Brown and I. B. Dawid, reprinted with permission from *Science* (1968) 160: 272, Fig. 1. Copyright 1968 American Association for the Advancement of Science. 21-15a, b Sarah Elgin/Washington University. 21-16 © P. Bryant/Biological Photo Service. 21-18 Reproduced from S. Saragosti, G. Moyne, and M. Yaniv, *Cell* (1980) 20: 65-73 by copyright permission of Cell Press, Cambridge, MA. 21-25b Courtesy of IBM U.K. Ltd. 21-29b, c Courtesy of Edward B. Lewis, California Institute of Technology. 21A-01 AP/Worldwide Photos.

Chapter 22 2-01a-c Courtesy of Mark S. Ladinsky, University of Colorado at Boulder. 22-02b Courtesy of L. E. Roth, Y. Shigenaka, and D. J. Pihlaja/Biological Photo Service. 22-05 Lester Binder and Joel Rosenbaum, *Journal of Cell Biology* 79 (1978): 510. Reproduced by copyright permission of the Rockefeller University Press. 22-08b Courtesy of Kent L. McDonald. 22-08d Mitchison and Kirschner, *Nature* 312 (1984): 235, Fig. 4c. 22-09a Courtesy of Michelle Moritz with Michael B. Braunfield, John W. Sedat, Bruce Alberts and David A. Agard, *Nature* 378: 555, Fig. 1, Panel B.2. 22-11 L. Casimeris, N. Pryer, and E. Salmon, *Journal of Cell Biology* 107 (1988): 2226. Reproduced by copyright permission of the Rockefeller University Press. 22-12 Courtesy of James Mandell and Gary Banker, University of Virginia. 22-13a, b J. Knops et al., *Journal of Cell Biology* 114 (1991): 725-33. Reproduced by copyright permission of the Rockefeller University Press. 22-13c Leclerc at el., © 1990 *Proceedings of the National Academy of Sciences, U.S.A.* 90 (13): 6223-27. 22-14 Adapted from C. E. Schutt et al., *Nature* (1993) 365: 810; courtesy of M. Rozycki. 22-15b Courtesy of R. Niederman and J. Hartwig. 22-16c Courtesy of Roger Craig. 22-18a M. S. Mooseker and L.G. Tilney, *Journal of Cell Biology* (1975) 67: 725-43. Reproduced by copyright permission of the Rockefeller University Press. 22.21a a, b © Borisylab, Northwestern University Medical School. Image courtesy Borisylab. 22.21c L. Blanchoin, K. J. Amann, et al., "Direct observation of dendritic actin filament networks nucleated by Arp2/3 complex and WASP/Scar proteins," from *Nature* 404 (6781):1007-11 (Fig. 1J). Reprinted by permission from *Nature*, © 2000

Macmillan Magazines Ltd. Image courtesy Kurt J. Amann, The Salk Institute. Fig. 22-22 Micrograph courtesy of Daniel Branton. 22.23 P. A. Coulombe et al., "The 'ins' and 'outs' of intermediate filament organization," from Trends in Cell Biology 10: 420-28, Fig. 1. © 2000. Reprinted with permission from Elsevier Science. Image courtesy Pierre A. Coulombe, Johns Hopkins University. 22-24 From M. W. Aynardi, P. M. Steinert and R. D. Goldman, *Journal of Cell Biology* 98(1984): 1407. Reproduced by copyright permission of the Rockefeller University Press. 22-26 Reprinted by permission from E. Fuchs, *Science* 279: 518, Fig. D. Images: T. Svitkina and G. Borisy. Copyright 1998 American Association for the Advancement of Science. T22-01 Fig. 1 Dr. Thomas D. Pollard. T22-02 Fig. 2 Courtesy of Sammak and Borisy, *Nature* (1988) 332: 724-36, Fig. 1b. T22-02 Fig. 3a-d Reprinted from E. D. Salmon, *Trends in Cell Biology* 5: 154-58, Fig. 3, with permission from Elsevier Science. T22-02 Fig. 4 Courtesy of Dr. John Heuser.

Chapter 23 23-01 Courtesy of N. Hirokawa. 23-06a W. L. Dentler/Biological Photo Service. 23-06c Biophoto Associates/Photo Researchers, Inc. 23-07a-d Courtesy of W. L. Dentler. 23-08a Dr. Lewis Tilney, University of Pennsylvania. 23-11 Courtesy of H. Ris. 23-12 Courtesy of Clara Franzini-Armstrong. 23-17 John Heuser, M.D. 23-22 © Allen Bell/University of New England/The Benjamin/Cummings Publishing Co. 23.33 A. Hall,"Rho GTPases and the actin cytoskeleton," from *Science* 279: 509-14, Fig. A, C, E, and G. © 1998. Reprinted with permission from the American Association for the Advancement of Science. Image courtesy Kate Nobes, University College, London. 23-24 Adapted from K. M. Trybus and S. Lowey, *Journal of Biological Chemistry* 259 (1984): 8564-71. 23-26 Courtesy of G. Albrecht-Buehler. 23-28 Courtesy of J. Hartwig. 23-29 © Laboratory of Molecular Biology 1999. A.B. Verkhovsky, T.M. Svitkina, and G.G. Borisy, *Cell Behaviour: Control and Mechanism of Motility*, 1999, 207-22. 23-31a M. Abbey/Visuals Unlimited. 23-31b Animals Animals/©Peter Parks-OSF. 23-32b Courtesy of Lewis G. Tilney with Daniel Portnoy and Pat Connelly, *Journal of Cell Biology* 109:1604, Fig. 18. 23A-01 © R. G. Kessel and R. H. Kardon, *Tissues and Organs: A Text-Atlas of Scanning Electron Microscopy*. W. H. Freeman & Co., 1979. All Rights Reserved.

Illustration and Text Credits

The following illustrations are from L.J. Kleinsmith and V.M. Kish, *Principles of Cell and Molecular Biology,* 2nd ed. (New York, NY: HarperCollins, 1995). Reprinted by permission of Pearson Education, Inc.:

Figs. 3.7, 3.9, 3.13, 3.27, 7.6, 7.7, 7.8, 7.11, 7.19, 7.22, 7.24, 8.4, 8.5, 8.10, 9.18b, 10.1, 10.14, 10.16, 11.3, 11.4, 11.20a, b, 11.21a, 11.24, 11.29, 12.4a, 12.9, 12.14, 12.17c, d, 12.18b, 14.19, 15A.1, 16.8, 16.15, 16.18, 16.19, 17.2, 17.8, 17.10, 17.21a, 17.23, 18.22, 18.23, 18.24, 18.25, 18.29, 19.4, 19.5, 19.12, 19.19, 19.20, 19.22, 19.24, 19.25, 19.26, 20.4, 20.15, 21.13, 21.14, 21.26, 21.28, 21.35, 22.6, 22.18b, and 23.2.

Fig. 3.4: © Irving Geis

Fig. 3.6: © Irving Geis

Fig. 4.22: From Russell, *Genetics* 5th ed. (Menlo Park, CA: Addison Wesley Longman, 1998), Fig. 13.18. Reprinted by permission of Pearson Education, Inc.

Box 6.A: Reprinted with permission from T.R. Cech, *Science* (1987) 236:1532. Copyright ©1987 American Association for the Advancement of Science.

Fig. 7.30: From J. Kyte and R.F. Doolittle, *J. Mol. Biol.* (1982)157:105-32. © Academic Press Ltd.

Fig. 9.3: From Neil Campbell, Jane Reece, and Larry Mitchell, *Biology,* 5th ed. (Menlo Park, CA: Addison Wesley Longman), p. 995. Copyright 1999. Reprinted by permission of Pearson Education, Inc.

Fig. 9.18c: From M.E.T. Boyle et al., *Neuron* 30: 385-97. ©2001.

Fig. 9.24a: From C.G. Garner et al., *Current Opinion Neurobiol.* 10: 321-27, Fig. 1. ©2000.

Fig. 10.18: From M. Berridge, *Nature* (1993) 361:315-325. Copyright 1993 Macmillan Magazines Limited.

Fig. 10.19: From S.F. Gilbert, *Developmental Biology,* 5th ed., (Sunderland, MA: Sinauer Associates, 1997), p. 110. Reprinted by permission of Sinauer Associates, Inc.

Fig. 10.27a: From H. Lodish et al., *Molecular Cell Biology,* 4th ed. (New York: W.H. Freeman), p. 1045. ©2000.

Fig. 10.28: From M.O. Hengartner, *Nature* 407: 770-76. ©2000.

Fig. 11.2: From Neil Campbell, Jane Reece, and Larry Mitchell, *Biology,* 5th ed. (Menlo Park, CA: Addison Wesley Longman), Fig. 36.3. Copyright 1999. Reprinted by permission of Pearson Education, Inc.

Fig. 11.14: From B.D. Angst, C. Marcozzi, Al Magee, *Journal of Cell Science* 114: 629-41. Copyright ©2001. Reprinted by permission of the Company of Biologists, Ltd.

Fig. 11.17: From D. Vestweber, J.E. Blanks, Physiol. Rev. 79: 181-213. © 1999.

Fig. 16.12: From Neil Campbell, Jane Reece, and Larry Mitchell, Biology, 5th ed. (Menlo Park, CA: Addison Wesley Longman, 1999), p. 377. Reprinted by permission of Pearson Education, Inc.

Fig. 18.33: From Neil Campbell, Jane Reece, and Larry Mitchell, Biology, 5th ed. (Menlo Park, CA: Addison Wesley Longman), p. 391. Copyright 1999. Reprinted by permission of Pearson Education, Inc.

Fig. 22.4: From Alberts, et al., Molecular Biology of the Cell, 3rd ed., (New York, NY: Garland Publishing, Inc., 1994), p. 810. © Garland Publishing, Inc.

Fig. 23.3b: From Lodish et al., *Molecular Cell Biology* 4th ed (New York: W.H. Freeman, 2000).

Fig. 23.5: From Gorthesy-Theviaz et al., *J. Cell. Biol.* (1992) 118:1333-45. Copyright permission of The Rockefeller University Press.

Fig. 23.21: From D. Bray, *Cell Movements* (New York, NY: Garland Publishing, Inc. 1992), p. 166. © Garland Publishing, Inc.

Fig. 23.27: From Alberts et al., *Molecular Biology of the Cell*, 3rd ed., (New York, NY: Garland Publishing, Inc. 1994), p. 835. © Garland Publishing, Inc.

Fig. 23.30a: From Lin, et al., Neuron (1996)16:769-782. Courtesy of Paul Forscher.

Index

Note: A *t* following a page number indicates tabular material, an *f* following a page number indicates a figures, and a *b* following a page number indicates a box. Page numbers in **bold** indicate pages on which key terms are defined.

A

AAUAAA signal sequence, 649
A band, muscle, **780**, 781*f*
ABC-type ATPase (ABC transporter), **209**–10
  cystic fibrosis and, 210, 212–13*b*
ABO blood groups, 166, 304–6
Abscisic acid (ABA), opening/closing of plant stomata and, 269, 276
Absolute refractory period, **239**
Absorption spectrum, **452**
  of common plant pigments, 452*f*
  ultraviolet, of DNA, 490*f*
Accessory pigments, **453**
Accessory proteins, 744
  in sarcomeres of muscle myofibrils, 780–82
Acetate, entry of, into TCA cycle, 406, 408*f*
Acetylcholine, **244**
  blood vessel dilation and, 269
  degradation of, at synaptic cleft, 250
  muscle contraction and, 787
  neurotoxin interference with, 146, 248–49, 250*b*
  as neurotransmitter, 244, 246*f*
  receptor for, on postsynaptic neurons, 247–49
Acetylcholine receptor, 247–49
Acetylcholinesterase, 250
Acetyl CoA (acetyl coenzyme A), 400, 405, **406**
  fatty acid catabolism to, 412, 413*f*
  formation of, 406, 407*f*
  in tricarboxylic acid cycle, 406, 408*f*
Achondroplasia, 273
Acid hydrolases, 354
Acidic keratins, 762
Acid phosphatase, cytochemical localization of, 354*f*
Aconitase, 407
Aconitate, 407
Acradine dyes, 628–29
Acrosome, 590
Actin, 79–80, 97, 743, **755**–62

actin-binding proteins and, 755, 758–60
actin-related proteins, 756
architecture of, in crawling cells, 797*f*
branched networks of, and Arp2/3 complex, 760, 761*f*
cell cortex and, 759–60
cell movement based on, 777–79
fibronectins and, 298*f*
G- and F- forms of, 97, 755
GTP-binding proteins and polymerization of, 757
inositol-phospholipid regulation of molecules affecting, 758
interrelationships of main forms of, 758*f*
linkage of membranes to, by proteins, 760–62
movement of myosins along filaments of, 779
muscle contraction and (*see* Muscle contraction)
in myofibril filaments, 780–81
nonmuscle cell motility based in, 792–98
polymerization of G-actin into F-actin microfilaments, 756, 757*f*
Actin-activated ATPase, 783
Actin assembly, 793
Actin-based motility in nonmuscle cells, 792–98
  amoeboid movement, 795, 796*f*
  cell migration, 792–95
  chemotaxis, 798
  cytoplasmic streaming, 795
  movement of infectious microorganisms by actin "tails," 795–97
  role of growth factors in regulating actin cytoskeleton, 797–98
Actin-binding proteins, 755, **758**–60, 770*t*
Actin gels and cell cortex, 759–60
Actin microfilaments (MFs), movement of nonmuscle cells by, 792–95
Actin-related protein (Arps), **756**
Action potential, **236**–43
  defined, 236
  electrical excitability and, 228
  ion concentration changes due to, 240
  ion movement through axonal membrane channels resulting in, 237–40
  membrane-potential changes and, 236–37

muscle contraction and, 787–88
propagation of, 236, 240–41 (*see also* Nerve impulse)
refractory periods after, 239–40
role of axon myelin sheath in, 241–43
transmission of, along nonmyelinated axon, 241*f*
Action spectrum, **454**
Activated monomer, **30**
Activation domain, **721**
Activation energy (E$_A$), **131**
  effect of catalysis on, 131*f*, 132
  metastable state and, 130, 131
  thermal activation, 131*f*, 132
Activation energy barrier, 117, 131
  effect of catalysts for overcoming, 131–32
Activators (regulatory transcription factors), **719**
Active site, enzyme, **133**
Active transport, 160, 197, **207**–10, 400
  direct, and transport ATPases, 208–10
  direct, vs. indirect, 207–8
  directionality of, 207
  examples of, 211–18
  indirect, and ion gradients, 210
  through nuclear pores, 514–15
Active zone, presynaptic neuron membrane, **247**
Adaptive enzyme synthesis, **692**–94
  anabolic pathways and end-response repression, 692, 693–94
  catabolic pathways and substrate induction in, 692, 693, 694
  effector molecules and, 694
Adaptor molecules, 662. *See also* Codon(s)
Adaptor protein (AP), 345, **349**
  lattices composed of clathrin and, 349–50
Adenine (A), **55**, 633
  Chargaff's rules and, 486, 487*t*
  phosphorylated forms of, 56*f*
Adenosine, 369
Adenosine deaminase (ADA), 616
Adenosine diphosphate (ADP), **56**, 370
Adenosine monophosphate (AMP), **56**, 370
Adenosine triphosphate (ATP), 9, **56**, 369–73
  as allosteric regulator, 391
  cellular chemical reactions and role of, 85–86

Adenosine triphosphate (ATP), *continued*
  cellular energy transactions and role of, 371–73
  exergonic nature of hydrolysis of, 370–71
  high-energy phosphoanhydride bonds in, 369–70
  macromolecule synthesis and role of, 30
  muscle contraction and, aerobic and hypoxic modes of, 788–89
  muscle contraction cycle and, 783–84, 785f
  pyruvate formation and generation of, 381–83
  structure and function of, 369f
  synthesis of (*see* Adenosine triphosphate (ATP), synthesis of)
  TCA cycle and formation of, 409
  yields of, in aerobic respiration, 399, 400
Adenosine triphosphate (ATP), synthesis of, 430–39
  aerobic respiration and, 430–34
  electron transport coupled with synthesis of, 426, 427
  glycolysis, fermentation (no oxygen), and, 376–86
  maximum yield in aerobic respiration in, 435–39
  in photosynthesis, 459–61
  TCA cycle and generation of, 408f, 409
Adenovirus, 213
Adenylyl cyclase, **261**, 331
ADF/cofilin, 758
Adherens junction, 306, **307–9**
Adhesion belt, 307
Adhesive glycoproteins, 296
  fibronectins as, 296–97
  laminins as, 297–98
Adhesive (anchoring) junctions, **98, 306–10**
  adherens junctions, 307–9
  desmosomes, 309–10
  types of, 307t
A-DNA, 59, 489
ADP-glucose, 468
ADP-glucose pyrophosphorylase, 468
ADP ribosylation factor (ARF), **351–52**
Adrenal glands, 279
Adrenaline. *See* Epinephrine
Adrenergic hormones, 278, **279**, 280
  coordination of responses to, 280
Adrenergic receptors, 258, 278, **279**, 280
  inositol-phospholipid-calcium pathway and α, 280
  stimulation of G protein-linked signal transduction pathways by, 279f
Adrenergic synapses, **244**
Adrenoleukodystrophy (ALD), 91
Aerobic conditions
  metabolism and, fate of pyruvate under, 383, 384f
  muscle contraction and ATP generation under, 788–89
Aerobic organisms, evolution of, 450
Aerobic respiration, 109, **375**, 377, **398–444**
  ATP synthesis, 430–34
  ATP yield, 399–400
  cellular respiration and, 398–400
  electrochemical proton gradient and oxidative phosphorylation in, 400, 425–30
  electron transport system, 415–25
  equation for, 119
  as exergonic reaction, 119
  glycolysis in (*see* Glycolysis (glycolytic pathway))
  mitochondrion, role of, 400–405
  summary of, 435–39
  tricarboxylic acid (TCA) cycle in, 400, 405–15
Affinity labeling, 182b
African clawed frog (*Xenopus laevis*), genes and cloning of, 706–7, 709

AGA codon, 635
Agarose, 493
Agrobacterium tumefaciens, 614
Alanine, 43f, 44f, 665
  D and L forms of, 21f
  TCA cycle and interconversion of, 414, 415f
Alcohol dehydrogenase, 385
Alcoholic fermentation, 375, **385**
Aldohexose, 375
Aldolase, 380, 464
Aldosterone, 71, 276
Aldosugars, 61, 62f
Algae, photosynthetic pigments in, 453
Alkanes, 361
Allele(s), **578**, 592
  independent assortment of, 592–93
  recessive, 590–92
  segregation of, 592
Allergens, **280–81**
Allolactose, 695
Allosteric activator, **148**, 149f
Allosteric effectors, 146, **148**
Allosteric enzymes, **148**
  allosteric regulation and, 148–49, 694–95
  feedback inhibition and, 147
  subunit cooperativity in, 149
Allosteric inhibitor/inhibition, **148**, 149f
Allosteric protein, repressor as, 694–95
Allosteric regulation, 147, **148–49**, 389, **390**, 410
  of enzymes in glycolytic and gluconeogenic pathways, 390–91
  of enzymes in TCA cycle, 410–12
  of *lac* repressor, 697f
Allosteric (regulatory) site, **148**
α-1,4-glucosidase, 357
α-actinin, 300, 301f, 781, 794
α adarenergic receptors, 258
α-amanitin, 640
α β heterodimer, **746**
α, β-unsaturated acyl CoA, 413
α-bungarotoxin, 248–49, 250
α chains, collagen, 292f
α globin subunits, 45
α helix, protein structure, **48**, 49f, 50, 52, 53f
α-ketoglutarate, transamination and formation of, 414, 415f
α-ketoglutarate dehydrogenase, 409
α-L-iduronidase, 357
α-tubulin, **746**, 747
Alternating conformation model of carrier proteins, **203**, 204f
Alternation of generations, **581**
Alternative RNA splicing, 652–53, **728**, 729f
Altman, Sidney, 152
Alu family of DNA sequences, 501, 627b
Alzheimer's disease, 673b
Amide bond, 44
Amino acid(s), **41–44**
  abbreviations for, 44t
  coding of polypeptide chains by sequences of, 625–28
  linkage of, to tRNAs by aminoacyl-tRNA synthetases, 664–66
  neurotransmitters derived from, 244
  protein structure dependent on, 47–54
  sequence of (*see* Amino acid sequence)
  structure and stereochemistry of, 42f
  structure of twenty, in proteins, 43f
Amino acid activation, 665
D-amino acid oxidase, 359
Amino acid residues, 48
Amino acid sequence, 182
  gene base sequence as code for, in polypeptide chains, 625–28
  genetic code and, 628–31
  protein structure and, 47–54
  triplet codons as determiners of, 628, 630–31

Aminoacyl tRNA, **664**
  binding of, 669–71
Aminoacyl-tRNA synthetase, 660, **664**
  amino acids linked to tRNA by, 664–66
Aminopeptidases, 413
Aminotransferase, 361, 470
Amoeboid movement, **795**, 796f
amp^R gene, 610
Amphibolic pathway, **414**
Amphipathic molecules, **24**, 25f, 67
  of plasma membrane, 83, 84, 161
Amyloid plaques, Alzheimer's disease and, 673
Amylopectin, **63**
Amyloplasts, 88, 448
  starch grains in, 63
Amylose, **63**
Amyotrophic lateral sclerosis (ALS), 764
Anabolic pathway, **369**
  for adaptive enzyme synthesis, 692, 693–94
  TCA cycle as source of precursors for, 414
  typical, 693f
Anaerobic conditions
  early Earth's, 450
  metabolism and fate of pyruvate under, 383, 384f
Anaerobic respiration, 375, **398**. *See also* Fermentation
Anaphase (mitosis), **545**, 547f, 548, 587f
  anaphase A and B, **548**
  chromosome movement and separation during, 549f, 550
Anaphase I (meiosis), 586, 587f, 587f
Anaphase promoting complex, cell-cycle regulation and role of, **561**, 562f
Anchorage-dependent growth, **301**
Anchoring junctions. *See* Adhesive (anchoring) junctions
Anchor proteins, 84
Androgens, 71, 276
Aneuploidy, **570**
Anfinsen, Calvin, 31
Angiogenesis, treating cancer by blocking, 570b, **571**
Angiostatin, 570
Animal(s)
  cloned sheep "Dolly," 707, 708b
  hibernation in, 175
  transgenic, 617b
Animal cell(s)
  adhesive junctions of, 98
  cell-cell recognition and adhesion in, 302–6
  cell junctions in, 306–14
  cytokinesis in, 547f, 553
  egg fertilization and calcium release, 266–68
  extracellular matrix of, 290–96 (*see also* Extracellular matrix (ECM))
  fibronectins and laminas in, 296–302
  gap junctions in, 98
  glycogen in, 29
  meiosis in, 582–83f
  membranes of, 158f
  mitosis in, 546–47f
  nerves and electrical signals (*see* Electrical signals in nerve cells)
  organelles of (*see* Organelle(s))
  osmolarity changes in, 290f
  peroxisomes in, 90–91, 93f, 359f–61
  structure of typical, 81f
  tight junctions of, 98
Animal eggs, calcium release after fertilization of, 266–68. *See also* Egg (ovum)
Animal hormones. *See* Hormone(s)
Anion exchange protein, 178, **206**
Ankyrin, 178f, 179, 184, 761, 762f
Annulus, **319**
Anoxygenic phototrophs, **447**
Antenna-pedia gene complex, 727
Antenna pigment, **453**

Anterograde axonal transport, 771
Anterograde transport, **335**
Anthocyanins, 357
Anthrax, 309*b*
Antibodies
 detection of cell adhesion molecules by, 303
 detection of specific proteins using, 187, 735–36*b*
 gene and DNA rearrangements in, 710
Anticodons, **664**
 wobble hypothesis on binding of codons and, 664*f*
Antidiuretic hormone (vasopressin), 276
Antigens, 735–36*b*
Antimitotic drugs, 753
Antiparallel β sheet, 50
Antiparallel DNA strands, 59
Antiport transport, **204**
 in Calvin cycle, 465
Antisense mRNA, 615
Anucleolar mutants, 518
AP1 transcription factor, 271
Apaf-1 protein, 282, 285
Apical surface of cells, 313
Apoplipoprotein E (apoE), 673
Apoptosis (programmed cell death), **282**–85, 540
 cancer and faulty, 282, 567, 568–69
 major steps in, 284*f*
 triggering of, 284, 285*f*
Apoptotic bodies, **282**
Appressed regions of thylakoids, 448, 461*f*
Aquaporins (AQPs), 206, **207**
Aqueous diffusion channels, 514
Arachidonate, 169
Arachidonic acid, 282
Archaebacteria, **76**
 halophilic, 76
 thermophilic, 137
Arginase, regulation of degradation rate of, 732, 733*f*
Arginine, 43*f*
Arp1 protein, 772
Arp2/3 complex, 760, 760*f*
ARS element (autonomously replicating sequence), 529
Aryl hydrocarbon hydroxylase, 331
Ascus, **604**
Asexual reproduction, **577**
A (aminoacyl) site, **661**
Asparagine, 43*f*, 186*f*
Aspartate, 43*f*
 TCA cycle and interconversion of, 414, 415*f*
Assembly proteins, 350
Assisted self-assembly, **33**
Aster, 545
Astral microtubules, **545**, 551, 552*f*
Astrocytes, 226
Asymmetric carbon atom, **20**–21
Atherosclerosis, 346
Atherosclerotic plaques, 346
ATM protein kinase, 568
Atom(s)
 carbon, 18
 electron configurations of some biologically important, 18*f*
 electronegative, 22
ATP. *See* Adenosine triphosphate (ATP)
ATPase pumps, 208. *See also* Transport ATPases
ATP-dependent calcium pump, 332
ATP-dependent proton pump, 348
ATP synthase, 209, 404, **431**
 F_0F_1 complex as functional, 431–34
 F_1 particles and, 430–31
 polypeptide composition of, 431*t*
ATP synthase complex (CF_0CF_1 complex), 448, **459**, 460
Attached glycocalyx, 302

Attenuation, 701, **702**
 in *trp* operon, 703*f*
A tubule, axoneme, **776**
Autoimmune diseases
 bullous pemphigoid, 300
 myasthenia gravis, 249
Autoinhibition, 792
Autonomic nervous system, **225**
Autophagic lysosomes, **355**
Autophagic vacuole, **356**
Autophagy, 355, **356**
Autophosphorylation, **271**
Autoradiogram, 494
Avers, Charlotte, 402
Avery, Oswald, 10, 481–82
Axon, neuron, **227**
 fast axonal transport, 770–71
 ion concentrations inside and outside, 228–30
 myelinated, 241, 242*f*
 myelinated, action potential transmitted along, 243*f*
 nonmyelinated, action potential transmitted along, 241*f*
Axonemal dynein, 772, 777
 body axis development and, 778*b*
Axonemal microtubules, **744**
Axoneme, 86, 744, **774**–76
 cilia and flagella movement and, 774–75, 776–77
 microtubules sliding in, 96, 746*f*, 776–77
 structure of, 776*f*
Axon hillock, **240**
Axon terminals, 787
Axoplasm, **227**, 770
5-Azacytidine, 714

B

Bacillus anthracis, 309*b*
Back crossing, **592**
Back reaction, 116
Bacteria, 76. *See also* Prokaryote(s); Prokaryotic cell(s)
 ATP synthesis and F_0F_1 complex in, 432*f*
 O. Avery's and F. Griffith's studies on genetic material in, 481–82
 bacteriophage infection of, and potential for genetic recombination, 482–85, 598–99
 cell adhesion systems and enteropathogenic, 308–9*b*
 cell wall, 66*f*
 chromosome, 502–4
 conjugation in, 600, 601*f*
 DNA cloning and, 606–7, 608, 609*f*
 DNA packaging in, 501–4
 DNA replication in, 538*f*
 flagellated, 108*f*
 gene regulation in, 692–741
 genetic recombination in, 598–602
 lactate fermentation in, 385
 lyses of, 133
 plasmids, 504
 purple (*see* Purple bacteria)
 sizes of, 77
 structure of typical, 79*f*
 transformation and transduction in, 599–600, 648*f*
Bacterial chromosomes, **502**–4
Bacteriochlorophyll, **453**, 461–62
Bacteriophage (phage), **99**, **482**–85
 bacterial transduction and, 599–600
 infection of bacteria by, and potential for genetic recombination, 482–85, 598–99
 λ, 485, 598, 639
 λ, as DNA cloning vector, 608, 609, 610, 611*f*
 λ repressor, 722, 723*f*
 as model system for studying genes, 484–86*b*
 plaques of, 484*b*
 prophage, 485*b*, 486*f*

replication of, 485*f*
 T2, 482, 483*f*
 T4, 484*f*, 599, 628–30
 virulent, and temperate, 484*b*
Bacteriorhodopsin, 179, **215**–18
 structure of, 164–65, 178*f*
Bacteriorhodopsin proton pump, 215, 216*f*, 217*f*, 218
Ball-and-stick model of ribonuclease, 51, 52*f*
Baltimore, David, 626
Band 3 protein, 178, 206
Band 4.1 protein, 178*f*, 179, 184, 761, 762*f*
Bangham, Alec, 199
Barbiturate drugs, 331
Barr body, 714
Basal body(ies), 748, 749*f*, 750, **775**
Basal lamina, **297**
 properties of, 297–98
Basal transcription factor, 643–44
Base analogues, 541
Base composition of DNA, 55, 56*t*. *See also* DNA sequence(s)
 Chargaff's rules on, 486
 dependence of DNA melting temperature on, 490, 491*f*
 from select sources, 487*t*
Base excision repair, **543**
Base-modifying agents, 542
Base pairing, **58**
 Chargaff's rules and, 486
 in DNA, 59*f*
 in RNA, 59
Base pairs (bp), **492**
 Chargaff's rules on, 486, 487*t*
 frameshift mutations and addition/deletion of, 628–30, 631*f*, 676*b*, 677*f*
 genome size expressed as number of, 492
 insertions and deletions of, as mutation, 676, 677*f*
 mismatch mutations in, 543–44, 676, 677*f*
 silent mutations in, 676
 substitution of, as mutation, 676, 677*f*
Base sequences. *See* DNA sequence(s); DNA sequencing
Basic fibroblast growth factor (BFGF), 570*b*
Basic keratins, 762
Basolateral surface of cells, 313
Bassham, James, 462
Bcl-2 protein, **285**, 567
B-DNA, 59, 488, 489*f*
Beadle, George, 10, 624–25
Benign tumor, **565**
Benson, Andrew, 462
Berridge, Michael, 263
β-adrenergic receptors, 258
β-α-β motif, 50*f*
β-amylase, 735*f*
β barrel, 178
β carotene, 453
β-catenin, 308, 310
Beta cells, pancreatic, 12
β-galactosidase, 608, 610, 693, 694, 735*f*
β globin subunits, 45
β-glucosamine, 65, 66*f*
β-hydroxyl fatty acyl CoA, 413
β-N-acetylhexosaminidase, 357
β oxidation, **360**, **412**, 413*f*, 416
β sheet, protein structure, **48**, 49*f*, 50, 52, 53*f*, 178
β-tubulin, **746**, 747
Bicarbonate, transport of, across erythrocyte membrane, 198*f*
Bicoid gene, 728
Bifunctional enzymes, 392
Bifunctional proteins, 41
Bimetallic-copper (Fe/Cu) center, **419**
Binding change model, ATP synthesis and, **432**–34

Binding sites (binding pocket), receptor, 257–58
Biochemistry, cell biology and studies in, 4f, **5**, 8–10, **18**. *See also* Chemistry, cellular)
Bioenergetics, 106–29
  of ATP synthesis, 434f
  of carrier protein function, 204
  defined, *112*
  energy in living systems and, 10, 107–12 (*see also* Energy)
  entropy and free energy, 115, 118
  first law of thermodynamics and, 113–14
  free energy and, 115, 116–17b, 118–20
  free energy change (δG), 118–25
  movement toward equilibrium, 125
  second law of thermodynamics and, 114–15
  of simple and facilitated diffusion, 202, 203f
  systems, heat, and work in, 112–13
  of transport across membranes, 218–20
Biogenesis of peroxisomes, 362
Bioinformatics, **498–99**
Biological chemistry, **18**. *See also* Biochemistry, cell biology and studies in; Chemistry, cellular
Bioluminescence, 107f, **108–9**
Biosphere
  energy flows through, 109–11
  matter flows through, 111–12
Biosynthesis, 107
BiP protein, **681**
Birds, skeletal muscles and ATP generation in, 788–89
Bithorax gene complex in fruitflies, 727
Bivalent chromosomes, **581**
  alignment of, at spindle equator, 585–86
Blending theory of inheritance, 592
Blobel, Günter, 680
Blood clotting, 282, 283f
  effect of fibronectin on, 297
Blood clotting factors, 614, 618
Blood glucose levels, 331–32, 376–77
  glucose oxidation and, 375
Blood groups, 166, 304–6
Blood plasma, 273
Blood platelet, **281**
  activation of, by prostaglandins, 281–82, 283f
  effect of elevated cAMP in, 262
Blood serum, 273
Blood types, 304–6
Blood vessels, nitric oxide and relaxation of, 269f, 270
Body axis, axonemal dyneins, radial spokes, and reversal of, 778b
Body fat, blood glucose and synthesis of, 378
Bonds. *See* Chemical bonds
Bone, 290, 291f
Boundaries, membranes and definitions of, 159
Bouqet, chromosome, 585
Boveri, Theodor, 593
Bovine serum albumin (BSA), 12
Box A, B, and C consensus sequences, 642
Brain, oxygen and glucose supplies to, 377
Branch migration, 606
Branch point sequence, 651, 652f
Branton, Daniel, 176, 350
Brassinosteroids, 276
Breakage-and-exchange model of genetic recombination, 602, 603f
Brenner, Sydney, 628, 630, 631
Brightfield microscopy, 5, 7f
5-Bromodeoxyuridine (BrdU), 529–30
Brown, Donald, 518
Brown, Michael, 345, 346
Brown, Robert, 2
B tubule, axoneme, **776**
Buchner, Eduard and Hans, 9, 133
Budding, reproduction by, 577
Bullous pemphigoid, 300

Bullous pemphigoid antigen 1 (BPAG1), 300, 765
Bundle sheath cell, **471**
Burk, Dean, 143
Burkitt's lymphoma, 565, 566, 567
Butylene glycol fermentation, 385

C
C$_3$ plants, **471**
C$_4$ plants
  leaves of, 471f
  photorespiration in, 471–73
CAAT box, 641, 718
Cadherins, role of, in cell-cell recognition and adhesion, 302, **303**, 304f, 305f, 307
Cairns, John, 530
Calcium
  animal egg fertilization, and release of, 266–68
  cadherin requirement for, 303
  cell function regulation and role of, 265, 266f
  neurotransmitter secretion and role of, 245, 246, 247f, 248f
  plant guard cell opening/closing and oscillation of, 268–69
  platelet activation sequence and release of, 282, 283f
  signal transduction process and role of, 264–69
  role of, in regulating muscle contraction, 786–88
  role of, in triggering exocytosis, 343
  smooth ER and storage of, 332
Calcium ATPase, **788**
Calcium-calmodulin complex, **265–66**, 791
  structure and function of, 267f
Calcium indicators, 264, 265f
Calcium ionophore, **265**
Calcium pumps, **265**, **788**
Calmodulin, 264, **265–66**, 759, 791
Calorie (cal), 19, **113**
Calorimeter, 172
Calvin, Melvin, 9, 462
Calvin cycle, 9, 445, **462–66**
  carbon dioxide entry into, and role of rubisco, 463–64
  complete, 465
  formation of G-3-P, 464
  glycolate pathway and, 469–71
  photosynthetic energy transduction and, 466
  regeneration of rubisco, 464–65
  requirements of, for light, 465–66
cAMP. *See* Cyclic AMP (cAMP)
CAM plants, photorespiration in, **473–74**
cAMP receptor protein (CRP), **700**, 701f
Cancer, **565**
  basal lamina and, 298
  defective cell cycle control mechanisms and development of, 564–65
  defective DNA repair and mutations as cause of, 569
  DNA microarray pattern of gene expression in normal cells vs. 717f
  drugs for, 753
  faulty apoptosis and, 282
  fibronectins and, 297
  growth factor signaling, disruption of, and, 275
  hereditary forms of, 544, 567–68
  loss of tumor suppressor genes and development of, 567–69
  oncogenes and development of, 565, 626–27
  oncogenes and growth factor signaling pathway and, 566–67
  proto-oncogenes and, 565–66
  stopping growth of, 570–71
  telomeres and, 541
Capping proteins, **758**
CapZ protein, 758
Carbamoyl esters, 250

Carbohydrate(s), **61–62**. *See also* Polysaccharide(s)
  cell-cell recognition and adhesion and role of, 304–6
  erythrocyte survival and, 304–6
  glycolysis and catabolism of, 386–87
  glyoxylate cycle and conversion of acetyl CoA to, 415
  photosynthesis and synthesis of, 467–68
  smooth ER and metabolism of, 331–32
Carbon, 17, 18–21
  diversity of molecules containing, 19–20
  stability of molecules containing, 18–20
  stereoisomers of molecules containing, 20–21
Carbon assimilation reactions of photosynthesis, **445**, 462–66
Carbon atom (C), **18**
  asymmetric, 20–21
  tetrahedral structure of, 20
  valence, 18
Carbon dioxide (CO$_2$)
  biospheric flow of, 111
  in Calvin cycle, 463–64
  as end product of aerobic respiration
  photorespiration in C$_4$ plants, 471–73
  TCA cycle and release of, 406–9
  transport of, across erythrocyte membranes, 198f
Carbonic anhydrase, 206
Carboxylase, rubisco as, 468
Carboxylation reactions, 388
Carboxypeptidase A, 133
  active site of, 134f
  structure of, 133f
Carboxypeptidases, 41
Cardiac (heart) muscle, **789**
  effect of elevated cAMP in, 262
  electrical coupling and contraction of, 789–90
Cardiomyopathies, 764
Carotenoid pigments, 71, **453**
Carrier molecule, **30**
Carrier protein, 26, **203**
  alternating conformations of, 203
  as analogous to enzymes, 204
  erythrocyte anion exchange as example of, 204–6
  in erythrocyte membrane, 197f
  glucose transporter as example of, 204–6
  kinetics of function in, 204
  uniport, symport, and antiport transport mechanisms of, 204
Cartilage, 290, 291f
Caspase-3, **284**
Caspases, **282–84**, 568
Caspr (contactin-associated protein), 242, 243
Cassette mechanism, **709**, 710f
Catabolic pathway, **369**. *See also* Chemotrophic energy metabolism
  for lactose degradation, 692, 693, 694–98
  typical, 693f
Catabolism
  of fats and proteins, 412–14
  of glucose, 375–78
  as main function of TCA cycle, 414
Catabolite repression, **700**
Catalase, 90, 132, 358, 359, 360f
Catalic mode, 360
Catalysts, **132**
  cell size and need for adequate concentrations of, 78
  enzymes as biological, 132–39
Catalytic event, 139
Catalytic subunit, **149**
Catecholamines, **244**
  as neurotransmitters, 244, 246f
Caveolin, 349
Caveolin-coated vesicles, 349

CD44 protein, 295
CD95/Fas receptors, 284, 285f
cdc2 gene, 558
Cdc42 protein, 757, 797, 798f
Cdk4 oncogene, 567
Cdk-cyclin complex
    cell cycle regulation and, overview of, 562, 563f
    G1, 559–60, 561f
    mitotic, 558–59, 560f, 561
    oncogenes and, 567
    phosphorylation and dephosphorylaton in activation of, 559f
Cdk inhibitor, 564
cDNA as probes to demonstrate transcriptional control of eukaryotic gene expression, 736–37b
cDNA library, 612–13
Cech, Thomas, 151
Celera Genomics, 498
Cell(s), 1–16, 76–83
    animal (see Animal cell(s))
    book chapter cross-references on structures of, 83t
    cell-cell adhesion (see Cell-cell adhesion)
    cell-cell recognition (see Cell-cell recognition)
    cell cycle (see Cell cycle)
    cell signaling in (see Chemical signaling in cells; Electrical signals in nerve cells)
    chemistry of (see Chemistry, cellular)
    development of theory of, 1–3
    energy flow in (see Bioenergetics)
    enzyme catalysis in (see Enzyme(s))
    eukaryotic, 76, 78–98 (see also Eukaryotic cell(s))
    fibronectins and shape of, 296–97, 298f
    growth of, 272–73
    hierarchical nature of structure and assembly of, 27f
    information flow in (see Chromosomes(s); DNA (deoxyribonucleic acid); Gene(s); Gene expression; Genetics)
    intracellular compartments of, 323–67
    lysosomes and digestive processes of, 355–57
    membranes of (see Membrane(s))
    metabolism in (see Chemotrophic energy metabolism; Photosynthesis)
    migration of (see Cell migration)
    modern biology of (see Cell biology)
    motility of (see Motility)
    organelles of, and compartmentalized function, 78 (see also Organelle(s))
    plant (see Plant cell(s))
    prokaryotic, 76, 78–82 (see also Prokaryotic cell(s))
    regulation of function in (see Cell function, regulation of)
    sexual reproduction of (see Sexual reproduction)
    sizes and shapes of, 76–78
    specialization of, and biological unity/diversity, 82–83
    viruses, viroids, and prions as invaders of, 98–101
Cell adhesion, cell crawling and role of, 793–94
Cell adhesion molecules (CAMs), 303
Cell biology, 3–13
    cytological strand of, 5–8
    "facts," scientific method, and, 11–13
    genetic strand of, 10–11
    timeline of, 4f
    units of measurement in, 2–3b
Cell biology techniques
    centrifugation, 92b, 326–30
    differential scanning calorimetry, 172
    discriminating among levels of gene regulation, 735–37b

DNA cloning, 608–14
DNA footprinting, 638b
freeze-fracture, 8
gel electrophoresis of DNA, 493–94
patch clamping, 233–34
polymerase chain reaction (PCR), 503, 530, 533–34b
RFLP analysis and DNA fingerprinting, 502–3b
staining, 7–8
studies of membrane proteins using molecular biology techniques, 182–83b
Cell body, neuron, 227
Cell-cell adhesion, 302–6
    carbohydrate groups and, 304–6
    pathogens and, 308–9b
    transmembrane proteins, 302–4
    types of receptors for, 302f
Cell-cell communication. See Intercellular communication
Cell-cell recognition, 302–6
    blood type determination and, 304–6
    carbohydrates and, 304–6
    cell-cell adhesion and, 302–6
    glycoproteins and, 186
    transmembrane proteins and, 302–6
Cell clone, 608. See also Clone and cloning
Cell cortex, 551, 755
    cell crawling and, 792, 794f
Cell crawling, 792–95
Cell cycle, 523–76
    DNA damage and repair and, 541–44
    DNA replication and, 525–41
    in eukaryotic cell, 524f
    growth control in, and cancer, 562–71
    interphase of, 524
    length of, 554–55
    M phase of, 523, 524f, 544–54
    nuclear and cell division in, 544–54
    overview and phases of, 523–25
Cell cycle regulation, 554–62
    cancer and defective, 564–65
    checkpoints and, 555–56
    control molecules in, 556–57
    cyclin-dependent kinase (Cdk) and, 558
    G1 checkpoint and G1 cdk-cyclin complex, 559–60
    length of cell cycle and, 554–55
    mitosis-promoting factor, 557
    mitotic Cdk-cyclin complex, 558–59, 561
    model of, 562, 563f
    mutants and identification of control molecules in, 557–58
    p53 gene and, 568–69
Cell death signals, apoptosis and, 284, 285f, 567
Cell differentiation, 706
Cell division, 523, 544–54
    cancer and abnormal growth control of (see Cancer)
    cytokinesis and, 523, 551–54
    mitosis and, 544–49
    mitotic spindle and chromosome movement during, 524, 549–51, 552f
    phases of mitosis, 524f
Cell fission, 80
Cell-free systems, 633
Cell function, regulation of
    by calcium, 265, 266f
    by inositol trisphosphate and diacylglycerol, 263t
Cell fusion experiments, 186, 188f
    control molecules in cell cycle and, 556–57
Cell junctions, 306–14
    adhesive, 306–10
    gap, 313–14
    major types of, in animal cells, 307f
    tight, 310–13
Cell migration, 792–95

basal lamina and, 297–98
cell attachment following, 793–94, 796f
contraction detachment, 794–95
extending protrusions in, 792–93, 796f
fibronectins and, 297, 298f
hyaluronates and, 295
lamellipodia and, 760, 792, 794f
steps of cell crawling, 793f
Cell motility. See Motility
Cell plate, 554
Cell theory, development and tenets of, 1–3, 13
Cellular respiration, 109, 398–400
    aerobic (see Aerobic respiration)
    anaerobic (see Anaerobic respiration)
Cellulose, 26–28, 63, 315
    in plant cell walls, 29, 65, 97, 315, 317f, 318
    structure of, 65f, 315f
Cell wall, eukaryotic, 83
Cell wall, plant (see Plant cell wall)
CEN sequences, 545, 613f
Central nervous system (CNS), 225, 226f
Central pair, axoneme, 776
Central vacuole, 93, 94f
Centrifugation, 326–30b
    density gradient, 326, 328–29, 330f
    differential, 92b, 327–28
    equilibrium density, 326, 329–30
    sample preparation for, 326–27
Centrifuge, 326
Centrioles, 545
    microtubules and, 749, 750
Centromere(s), 500, 523, 545
Centrosome, 545, 749, 751f
    microtubules originating from, 749–50, 751f
Ceramide, 70
Cerebrosides, 166
Cervical cancer, 565, 568
CF_0, 459
CF_1, 459
CF_0CF_1 complexes, 459, 460, 466
CFTR gene, 213
CGN (cis-Golgi network), 334
Chain termination method of DNA sequencing, 496–97
Changeux, Jean-Pierre, 148
Channel gating, 235
Channel inactivation, 235
Channel protein, 184, 203, 206–7
    aquaporins and, 206–7
    in erythrocyte membrane, 197f
    ion channels and, 206
    porins and, 206
Chaperone proteins, 686, 687f
Character, genetic, 578
Chargaff, Erwin, 484, 486
Chargaff's rules, 486, 487t
Charge repulsion, 370
Chase, Martha, 10, 482–85
Checkpoints, cell cycle, 555, 556f
Chemical bonds
    bond energy, 19
    covalent, 18, 34
    disulfide, 45
    glycosidic, 65
    hydrogen, 22, 46, 58f (see also Hydrogen bonds)
    ionic, 34, 46
    noncovalent, 34, 46
    peptide, 44
    in protein folding and stability, 45–47
    single, double, and triple, 18
    synthetic work as changes in, 107–8
Chemical equilibrium
    equilibrium constant (K_{eq}), 120–21
    free energy and, 120–23
    movement toward, in diffusion processes, 198–99

Chemical messenger(s), 256
  apoptosis and, 282–85
  calcium as (see Calcium)
  cyclic AMP as (see Cyclic AMP (cAMP))
  G protein-linked receptors for, 259–70
  growth factors as, 257, 272–75
  hormones as, 257, 276–82
  as ligands, 257
  local mediators, 257
  primary and secondary, 257
  protein kinase-associated receptors for, 270–72
  receptor binding to, 257–58
  signal transduction and, 257, 258–59
Chemical method of DNA sequencing, 496–97
Chemical reactions
  changes in enthalpy for, 114f
  depurination and deamination, as cause of DNA damage, 541, 542f
  endergonic and exergonic, 119, 369 (see also Endergonic (energy-requiring) reactions; Exergonic (energy-yielding) reactions)
Chemical signaling in cells, 160, 256–89
  apoptosis and, 282–85
  chemical messengers and receptors in, overview of, 256–59
  endocrine and paracrine hormone system, 276–82
  G protein-linked receptors, 259–70
  growth factors as messengers, 272–75
  growth factor signaling and oncogenes, 566–67
  growth factor signaling via Ras pathway, 564f
  integrins and, 300
  protein kinase-associated receptors, 270–72
Chemical synapse, 243, 245f
  role of neurotransmitters at, 243–44, 245f
Chemiosmotic coupling model of ATP synthesis, 13, 426–27
  experimental evidence for, 427–30
  F_0F_1 complex and, 430–34
  transmembrane proton traffic and, 434, 435f
Chemistry, cellular, 17–40
  carbon and, 17, 18–21
  selectively permeable membranes and, 17, 24–26
  self-assembly of macromolecules and, 17, 31–37
  synthesis by polymerization and, 17, 26–31
  water and, 17, 21–24
Chemoattractant, 798
Chemorepellant, 798
Chemotaxis, 798
Chemotherapy, 571
Chemotrophic energy metabolism, 373–78. See also Aerobic respiration
  ATP as universal energy coupler in, 369–73
  biological oxidations in, 373–74
  coenzymes, role of, 374
  gluconeogenesis, 386–89 (see also Gluconeogenesis)
  glucose as oxidizable substrate in, 375–78
  glycolysis, alternative substrates for, 386
  glycolysis and fermentation (anaerobic respiration), 378–86 (see also Fermentation; Glycolysis (glycolytic pathway))
  metabolic pathways and, 368–69
  oxidation of food molecules as, 375
  oxygen requirements during, as means of classifying living organisms, 378
  regulation of glycolysis and gluconeogenesis, 389–92
  of sugar, in human body, 376–78b
Chemotrophs, 109, 110, 368
  oxidation of organic food molecules by, 375
Chiasma (chiasmata), 583, 584f
Chimeric proteins, 338

Chitin, 29, 65, 66f
Chloride-bicarbonate exchanger, 206
Chloride ions
  calculating free energy change for transport (uptake) of, 219–20
  cystic fibrosis and role of channels for, 210, 212–13b
Chlorophyll, 447, 452–53
  P700 and P680, 455
  types a and b, 452, 453f
Chlorophyll-binding protein(s), 25, 452, 453
Chloroplast(s), 78, 79, 87–88, 92, 447, 448f
  endosymbiont theory of evolution of, 88, 449, 450–51b
  importing polypeptides into, 684–86
  membrane systems of, 448, 449f
  as sight of photosynthesis, 447–49
  starch biosynthesis in stroma of, 468
  starch grains in, 63
  structure of, 87f, 449f
  targeting polypeptides into, 686–87
Cholera, 262
Cholesterol, 70–71, 160, 166, 276
  biosynthesis of, 332–33
  hypercholesterolemia, 345, 346
  LDL receptor, receptor-mediated endocytosis and, 346–47b
  membrane fluidity and effects of, 173–74
  orientation of, in lipid bilayer, 174f
  structure of, 70f, 167f, 168f
Cholinergic synapse, 244
Chondroitin sulfate, 294, 295f
Chromatids
  in bivalent chromosomes, 586, 587f
  crossing over of, genetic recombination and, 597f
  separation of, 550–51
  sister, 523, 581
Chromatin, 78, 85, 504
  decondensation of (see Chromatin decondensation and genomic control)
  levels of DNA packing in, 508f
  nonrandom dispersal of, within nucleus, 516–17
  nucleosomes as basic unit of, 505–6, 507f, 508f
  packaging of nucleosomes and formation of, 506–7, 508f
Chromatin decondensation and genomic control, 710–14
  chromosome puffs as evidence for, 710–11, 712f
  DNase I sensitivity as evidence for, 712–14
Chromatin remodeling, 713–14
Chromatography, lipid analysis using thin-layer, 168–69
Chromophore (light-absorbing pigment), 216
Chromoplasts, 88, 448
Chromosomal DNA replication. See Meiosis
Chromosome(s), 10, 80, 82f, 84–85, 504
  alignment and separation of, during mitosis, 550–51
  attachment of, to mitotic spindle, 548f, 549–50
  bacterial, 591–94
  bivalent, 581, 585–86
  bouquet of, 585
  discovery of genetic information carried on, 593
  DNA packaging in chromatin and, 78–79, 80f, 504–7
  eukaryotic, 504–9
  haploid and diploid, 492, 578
  homologous (see Homologous chromosome(s))
  inversions in, 676
  karyotype of human, 545, 548f
  laws of segregation and independent assortment explained by, 593–94
  linked genes on, 596

  meiosis and, 582–88
  mitosis and, 80
  movement of, and motor proteins, 551, 552f
  nondisjunction of, 588
  polytene, 711, 712f
  prophage propagation within, 486f
  sex, 578
  translocation of, 566, 676
Chromosome puffs, 710, 711, 712f
Chromosome theory of heredity, 10
Cilia (cillium), 744, 773–77
  crosslinks, spokes, and bending of, 776f, 777, 778b
  dynein arms and sliding of, 777
  microtubule sliding within axonemes and bending of, 776–77
  as motile appendages of eukaryotic cells, 773–74
  movement of, 774f
  structure of, 774, 775f, 776
Cimetidine, 258
Circular DNA, 80, 489, 490f
  replication of, 528f
cis-acting element, 698
cis (forming) face, 334
cis-Golgi network (CGN), 334
Cisterna, endoplasmic reticulum, 89, 323
  connection of ER and, by COPI- and COPII-coated vesicles
Cisternae, Golgi complex, 333
Cisternal maturation model, 335
Cis-trans test, 698
Citrate, 407, 414
Citrate synthase, 406
Clathrin, 334, 345, 347, 349
  components of lattices made of, 350
  light and heavy chains of, 350
  triskelions of, 350, 351f
Clathrin cages, 349, 350f
Clathrin-coated vesicles, 349
  lattices of clathrin and adaptor protein surrounding, 349–50
Clathrin-dependent endocytosis, 345–48
Clathrin-independent endocytosis, 344, 348–49
Claude, Albert, 10, 326, 401
Claudins, 311
Clayton, Roderick, 461
Cleavage, 553, 555f
Cleavage furrows, 97, 553
Clone and cloning, 5, 707
  DNA, 606–14
  of sheep "Dolly," 707, 708b
Cloning vector, 606, 608–12
  amplification of, 608–9
  [lambda] bacteriophage as, 608, 609, 610, 611f
  plasmids as, 608, 609f, 610f
Closed system, 112f
Cloverleaf structure, pre-tRNA, 646, 647f
c-myc proto-oncogene, 566, 567
Coactivators, 719, 720, 721
Coated pits, 345, 346, 347, 348f
Coated vesicles, 334, 345, 346, 348f, 349–53
  clathrin-coated, 349–50
  connection of ER and Golgi complex cisternae by COPI and COPII-, 351–52
  in eukaryotic cells, 349t
  SNARE hypothesis connecting target membranes and, 352–53
  vesicle formation and, 350–51
Coatomer. See COPI (coatomer) protein
Coat protein, 35
Cobra toxin, 248–49
Coding strand, 632
Codon(s), 633
  amino acid sequences determined by, 633–34, 662
  anticodons, 664

initiation, 634
nonsense mutations in, 674–75, 676, 677f
start, 666
stop, 634, 666
wobble hypothesis on binding of anticodons
  to, 664f
Coenzyme(s), **374**
  as electron acceptors in biological oxida-
    tion, 374
  electron transport and oxidation of, 418, 427
Coenzyme A, 405, **406**
  structure of, 407f
Coenzyme Q (CoQ), 71, **419**–20
  oxidized and reduced forms of, 420f
Coenzyme Q-cytochrome *c* oxidoreductase com-
    plex, **424**
Cognate receptor, **258**
Cohen, Stanley, 562
Cohesins, **561**
Cohesion of water molecules, 22, 23f
Coiled coil, 763
Colchicine, 343, 545, 744, **753**
Col (colicinogenic) factors, 504
Colicins, 668
Collagen, 51, 83, **291**–92
  assembly of, 293f
  procollagen as precursor to, 292–93
  in proteoglycan matrix, 294
  structure of, 292f
  types, occurrence, and structure of, 293t
Collagen fibers, 97, **291**, 292f
Collins, Francis, 212b
Colony hybridization technique, 611f
Combinatorial model for gene regulation,
    721, 722f
Competitive inhibition
  of carrier proteins, 204
  of enzyme, 146
Complementary DNA (cDNA), **612**
  cloning of, 612f, 613
  quantifying mRNAs using, 736–37b
Complementary DNA strands, 59, **487**
Complex I (NADH dehydrogenase complex),
    423, 424t, 428–29
Complex II (succinate-coenzyme Q oxidoreduc-
    tase), **423**, 424t
Complex III (coenzyme Q-cytochrome c
    oxidoreductase), **424**, 428–29
Complex IV (cytochrome c oxidase), **424**,
    428–29
Complex oligosaccharides, 336
Computerized image processing, 6
Concanavalin, 186
Concentration gradient, rate of simple diffusion
    proportional to, 201–3
Concentration work, 107f, **108**
Condensation of proteins, 342
Condensation reaction, 30, 44
Condensin complex, 559
Condensing vacuoles, 341
Conductance, 234
Confocal scanning microscope, 6, 7f
Conformation, protein, **45**, 50
Conjugation, **600**–602
  F factor and, 600, 601f
  Hfr cells and bacterial chromosome transfer,
    600–602
Connective tissue, 290, 291f
Connexin, 179, 314
Connexons, 184, **313**–14
Consensus sequence, **637**, 642
Constitutive genes, **692**
Constitutive heterochromatin, **517**
Constitutive secretion, 341, **342**
Contractile ring, **553**
Contractility, **769**
Contraction, cell movement and, **794**–95

Control molecules in regulation of cell cycle,
    557–63
Cooperativity in allosteric enzymes, **149**
COPI-coated vesicles, 349
  Golgi complex cisternae and ER connected by,
    351–52
COPII-coated vesicles, 349
  Golgi complex cisternae and ER connected by,
    351–52
COPII proteins, 334, 349
COPI (coatomer) protein, 334, 349, **351**
Copolymers, 633
Copy-choice model of genetic recombination,
    602–3
Core enzyme, 635
Core glycosylation, **336**
Core oligosaccharide, **336**
  assembly of, and transfer to protein, 337f
Core particle, nucleosome, 506
Corepressor, **699**
Core promoter, **641**, 717, 718f
Core protein, **294**, 295f, 296f
Corey, Robert, 48, 49
Cori, Carl and Gerti, 378
Cori cycle, 377f, 378
Correns, Carl, 10
Cortex. *See* Cell cortex
Cortical granules, 266
Corticosteroids, 276
Cortisol, 70f, 71, 276, 725
Cortisone, liver cells and levels of, 732–33
Cosmids, DNA cloning in, **613**–14
Cotransductional mapping, 600
Cotranslational import, **678**, 679f
  BiP and protein disulfide isomerase, 681
  evidence for ER signal sequences, 678–80
  integral-membrane insertion using stop-
    transfer sequences, 683, 684f
  model for signal mechanism of, 682f
  roles of SRP and translocon, 680–81
  sorting of soluble proteins, 681–83
Cotransport (coupled transport), **204**
Counterions, 200b, 229
Coupling, **426**
Covalent bonds, **18**
  in proteins, 34
Covalent modification as enzyme regulation,
    147, **149**–51
Crassulacean acid metabolism (CAM), **473**–74
Creatine kinase, 789
Creatine phosphate, **789**
CREB protein, 725, **726**
Creutzfeldt-Jakob disease (mad cow disease),
    99, 673
Crick, Francis, 10, 58, 60–61b, 487–89, 525, 623,
    628, 630, 631, 662, 664
Cristae of mitochondria, **86**, 403
Critical concentration, microtubule assembly, **747**
Crops, genetically engineered, 615–16
  safety and environmental concerns about, 616
Cross-bridge, muscle fiber, **783**, 784, 785f
Crossing over, genetic recombination due to,
    583, 584f, 585f, 597
Crown gall tumor, 614
CRP recognition site, 700
C-terminus, **44**, 48, 666
Curare, 250
Current (electricity), **229**, 234
C value, **588**
Cyanobacteria, **76**
  evolution of chloroplasts from, 450–51b
  photosynthetic membranes of, 449f
  photosynthetic pigments of, 453
Cyclic AMP (cAMP), 256, **261**, 725–26
  allosteric regulation and role of, 392
  gene expression and role of, 700
  gene transcription activated by, 726f

glycogen degradation and, 280, 281f, 331
  physiological functions regulated by, 279, 280t
  protein kinase A activation by, 262f
  as second messenger for select G proteins,
    261–62
  signal transduction and role of, 261f
  structure and metabolism of, 260f
Cyclic electron flow, 458, **460**
Cyclic GMP (cGMP), 269–70
  binding site for, in rod cells
Cyclic phosphorylation, 460
Cyclin(s), **558**
Cyclin-dependent kinase (Cdk), **558**
  control of cell cycle by, 558
Cyclooxygenase, 282
Cyclosis (cytoplasmic streaming), 78, 795. *See
    also* Cytoplasmic streaming (cyclosis)
Cysteine, 43f, 665
Cystic fibrosis (CF), 210, **212**–13b
  gene therapy for, 616, 618
Cystic fibrosis transmembrane conductance reg-
    ulator (CFTR) protein, 210, **212**, 213f
Cytochalasin(s), **758**
  cytochalasin B, 182, 300
  cytochalasin D, 744
Cytochrome(s), **419**
  copper-containing, 419
  electron transport and role of, 419–20
Cytochrome a, 419
Cytochrome a_3, 419
Cytochrome b, 419
Cytochrome b_6, 457
Cytochrome b/c_1 complex, 424, 461, 462
Cytochrome b_6/f complex, 456, **457**–58, 460, 461
Cytochrome c, **285**, 419, 462
Cytochrome c_1, 419
Cytochrome c oxidase, 419, **424**
Cytochrome f, 457, 458
Cytochrome P-448, 331
Cytochrome P-450, 331
Cytokines as survival factor, 284
Cytokinesis, **523**, 524f, 551–54
  in animal cells, 547f, 553
  in meiosis vs. mitosis, 587f
  in plant cells, 547f, 553–54
Cytology, cell biology and studies in, 3, 4f, 5–8
Cytoplasm, 85, **95**
  division of, by cytokinesis, 523, 524f, 551–54
  RNA components of ribosomes in, 645t
Cytoplasmic dynein, 772
Cytoplasmic microtubules, **746**
Cytoplasmic streaming (cyclosis), **78**, 755,
    769, **795**
Cytosine (C), **55**, 633
  Chargaff's rules and, 486, 487t
Cytoskeleton, 80, 83, **95**–97, **742**–68
  integrins and, 299–300
  intermediate filaments of, 79–80, 96f, 97, 742,
    762–65
  major structural elements of, 742–44
  mechanical integration of structure of, 765
  microfilaments of, 79–80, 95, 96f, 97, 742,
    754–62
  microtubules of, 79–80, 96f, 97, 742, 744–54
  techniques for studying, 744, 745t
Cytosol, 83, **95**, 328
  biosynthesis of sucrose in, 468

D

Danielli, James, 162
Dark reactions of photosynthesis, 447
Davson, Hugh, 162
Davson-Danielli model of membranes, 162
  shortcomings of, 163–64
Deafness, mutations in myosins and
    resulting, 778b
Deamer, David, 176

Deamination reactions, DNA damage and, 541, 542f
Decarboxylase, 470
Decarboxylation reactions, 388, 406
De Duve, Christian, 90, 92b, 326, 353, 358
Degenerate code, **631**–32
Dehydration reaction. *See* Condensation reaction
Dehydrogenases, 373
Dehydrogenation, **373**
Deisenhofer, Johann, 461
ΔG. *See* Free energy change (ΔG)
ΔG$^{0\prime}$. *See* Standard free energy change (ΔG$^{0\prime}$)
DeLucia, Peter, 530
Denaturation
    of DNA, 490, 491
    of polypeptides, **31**
Dendrites, neuron, **227**
Dendritic networks of actin, 760, 761f
Dense tubular system, 282
Density gradient (rate-zonal) centrifugation, 326, **328**–29, 330f
Deoxynucleotides, 496
Deoxyribose, **54**
Dephosphorylation, 149, **150**
Depolarization of membrane, **231**
    action potentials and, 238, 239f, 240–41
Depurination reactions, DNA damage and, 541, 542f
Desmin, 310, 762, 763t
Desmocollins, 310
Desmogleins, 310
Desmoplakin, 310, 765
Desmosome core, 310
Desmosomes, 306, **309**–10
    cadherins in, 304f
    structure of, 310f
Desmotubules, 184, **319**
Destruction box, 734
Development, **579**
Devries, Hugo de, 10
Diabetes, insulin-dependent, 12
Diacylglycerol (DAG), **263**, 279
    cell functions regulated by, 263t
    role of, in signal transduction, 264f
Diakinesis, **585**
Dictyosome, 333, **357**
Dideoxynucleotides, 496
Differential centrifugation, 92b, 326, **327**–28
Differential gene transcription, 715–16
Differential interference contrast (DIC) microscopy, 752, 753f
Differential scanning calorimetry, **172**
Diffusion, **78**
    facilitated (*see* Facilitated diffusion)
    simple (*see* Simple diffusion)
    transverse, and lateral, 169, 171f
Digestive enzymes, 137
Digital video microscopy, 6, 7f
    computer-enhanced, in study of cyto-
        skeleton, 745t
Dihydroquinone (CoQH$_2$, 419
Dihydroxyacetone phosphate (DHAP), 447
    conversion of G-3-P to, 463f, 464
    formation of glucose-1-phosphate from
        G-3-P and, 467
Diisopropyl fluorophosphate, 146
2,4-Dinitrophenol (DNP), 426, 429
Dintzis, Howard, 666
Diploid analysis of *lac* operon mutations, 698t
Diploid state, 492, **578**
    as essential feature of sexual reproduction, 578
    haploid state resulting from, in meiosis, 581–82
    as homozygous or heterozygous, 578–79
    phase of, in life cycle of sexual organisms, 580–81
Diplotene stage of meiosis prophase I, **583**, 584f, 585f

Dipoles, 34, 46
Direct active transport, **208**
    indirect active transport compared to, 208f
    transport ATPases in, 208–10
Directionality, **207**
    of active transport, 207
    of macromolecules, **30**
Disaccharides, **62**
    common, 63f
Disease. *See* Human disease
Disease-causing microorganisms, movement of, using actin "tails," 795–97
Dissociation constant (K$_d$), **258**
Distal control elements, 718
Disulfide bond, **45**
DNA (deoxyribonucleic acid), **54**
    alternative forms of, 59
    apoptosis due to damaged, 285f
    base composition of, from select sources, 487t
    bases, nucleotides and nucleosides of, 56t
    cell nucleus and, 510–18
    Chargaff's rules on base composition of, 486, 487t
    in chloroplasts, 507–9
    cleavage of, by restriction enzymes, 494f
    cloning (*see* DNA cloning)
    crossing over of, 583, 584f, 585f
    damage and repair of, 541–44, 604
    denaturation and renaturation of, 490–91, 499
    discovery of, as genetic material, 10–11, 479–86
    exchange of, during meiosis, 582–88
    flow of genetic information to RNA from, 623–24
    of homologous chromosomes, 594–95
    homologous recombination and breakage-and-exchange of, 602, 603f
    linker, 506, 507f
    mitochondrial, 507–9
    mutations in, 676b, 677f (*see also* Mutation(s))
    organization of, in genomes, 491–501
    packaging (*see* DNA packaging)
    in prokaryotic vs. in eukaryotic cells, 80, 82
    replication (*see* DNA replication)
    sequencing of bases in (*see* DNA sequence(s); DNA sequencing)
    strands of (*see* DNA strands)
    structure (*see* DNA structure)
    transcription of genetic information from (*see* Transcription)
    transfer of, by bacterial conjugation, 600, 601f
DNA-binding domain, **721**
DNA-binding motifs, 721–24
DNA cloning, **608**
    cloning vector use in, 608–12
    in cosmids and yeast artificial chromosomes (YACs), 613–14
    genomic and cDNA libraries useful for, 612–13
    overview of, in bacteria using plasmid vector, 609f
DNA control elements
    combined action of transcription factors and, 721, 722f
    enhancer and silencer, 718–19, 720f
    proximal, 717–18
    response elements, 724–27
    transcription factors binding, 718t
DNA denaturation, **490**–91
DNA fingerprinting, **501**, 534
    by RFLP analysis, 502–3b
DNA footprinting, **637**, 638b
DNA glycosylases, 543
DNA gyrase, **489**, 537
DNA homology, 595
DNA ligase, DNA replication and, **532**–35, 607, 608

DNA melting temperature(T$_m$, **490**, 491f
DNA methylation, 544, **714**
    inactive regions of genome associated with, 714–15
DNA microarray, **716**, 717f, 737b
DNA nucleotide sequences, 48, 54f, 182. *See also* DNA sequence(s); DNA sequencing
DNA packaging, **501**–9
    eukaryotic, in chromosomes and chromatin, 78–79, 504–7
    eukaryotic, in mitochondria and chloroplasts, 507–9
    prokaryotic, in chromosomes and plasmids, 501–4
DNA packing ratio, **507**
DNA polymerase, **530**
    actions and functions of, 532t
    DNA chain elongation catalyzed by, 530, 531f
    proofreading of DNA and exonuclease activity of, 535
DNA primers, 533
DNA probe, **491**
DNA rearrangement, **709**
    antibody production and, 710
    genome alteration by, 709–10
    oncogene formation by, 565–66
DNA renaturation, **490**–91, 499
DNA replication, 525–41
    bidirectionality of, 527–28, 531f
    DNA ligase and joining of DNA segments, 532–35
    DNA polymerases and DNA chain elongation, 530, 531f
    DNA replication proteins, 532t
    licensing of eukaryotic, 541
    multiple replicons in eukaryotes, 528–30
    proofreading of, 535
    RNA primers as initiators of, 535–36
    role of helicases, topoisomerases, and single-strand binding proteins in, 536–37
    semiconservative nature of, 525–27
    summary of, 537, 538f
    telomeres and DNA end-replication, 539–41
    Watson-crick model of, 525f
DNase I, 638b
DNase I hypersensitive sites, **713**
DNase I sensitivity, chromatin decondensation and, 712, 713f
DNA sequence(s), 479
    Alu family of, 501, 627b
    CEN, 545, 613f
    coding of amino acid sequences for polypeptide chains by gene, 625–28
    control elements (*see* DNA control elements)
    end-replication problem, 539–40
    enhancer, 710, 718–19, 720f
    F factor, 600, 601f
    genetic code and, 628–31
    homeobox, 728
    MARs, 715
    oncogenes created by rearrangement of, 565–66
    polymerase chain reaction (PCR) amplification of, 503, 530, 533–34b
    promoter, 635 (*see also* Promoter(s))
    rearrangement of (*see* DNA rearrangement)
    repeated, in eukaryotic genomes, 499–501
    as replicons, 529
    response elements, 705, 724–27
    silencer, 718–19
    similar, in homologous chromosomes, 594–95
    upstream, and downstream, 637
DNA sequencing, 496, 497f
    applied to genomes and genes, 497–98
    applied to study of membrane proteins, 181, 182–83b
    introns and, 650

DNA strands
    antiparallel orientation of, 59, 488
    coding, 632
    complementary, 59, 487
    5′—3′/3′—5′ orientation of, 488
    Holliday junctions between, 604, 605f, 606
    joining of and DNA replication, 532–35 (see also DNA replication)
    leading and lagging, 532, 534
    major and minor grooves of, 488, 489f
    noncomplementary, 604
    separation and rejoining of (denaturation/renaturation), 490, 491f, 499
    template, 487, 632
DNA structure, 57f, 486–91
    alternative, 488, 489f
    denaturation, renaturation, and changes in, 490–91
    double helix, 10, 58, 59f, 60–61b, 487, 488f, 489
    supercoiled and relaxed forms of, 489, 490f
DNA synthesis. See DNA replication
Dolichol phosphate, 336
Dolichols, 71
Domain, globular protein, 52, 53f
    role of exons in coding for, 653, 654f
Dominant allele, 578
Dominant negative mutation, 273, 274f
Dopamine, 244
Double bond, 18
Double helix model of DNA, 10, 58–60, 487. See also DNA strands
    unwinding of, in DNA replication process, 536, 537f
    J. Watson and F. Crick's discovery of, 58, 60–61b, 487–89
Double-reciprocal plot, 143, 144f, 145f
Downstream DNA sequence, 637
Downstream promoter element (DPE), 641
Down syndrome, 588
DPE-driven promoters, 641
Drosophila sp., 351, 498
    heat-shock response elements in, 726
    homeotic genes and mutations in, 727–28
    T. H. Morgan's studies on, 10, 596
    polytene chromosomes in, 711, 712f
Drug(s)
    aspirin, 282
    colchicine, 343, 545, 744, 753
    cytochalasin D, 744
    disruption of cytoskeletal structures by, 744
    effects of, on microtubule assembly, 753
    enzyme inhibition and, 146
    latrunculin A, 744
    neurotransmitter uptake and effects of, 250
    nocodazole, 753
    phalloidin, 744
    resistance to, 210
    smooth ER and detoxification of, 331
    taxol, 744, 753
    tolerance to, 258
    vasodilators, 269–70
Duchenne muscular dystrophy, 650
Duplications, DNA segment, 676
Dynactin, 772
Dynamic instability model of microtubules, 748, 749, 750f
Dynamin, 345, 350–51
Dynamitin, 772
Dyneins, 770
    classes of, 772
    defect in, and development of body axis, 778b
    microtubule sliding and arms of, 771f, 777

E
E2F transcription factor, 560, 561f
Eadie-Hofstee equation, 144

Ear, deafness in, and role of myosin, 778b
Early endosome, 90, 341, 347
E-cadherin, 303, 304f, 795
Ecdysone, 711
Ecology, 111
Ectoplasm, 795
Edema factor (EF), 309b
Edidin, Michael, 186
Effector molecules. See Effectors
Effectors, 694
    allosteric regulation and, 410
    corepressors, 699–700
    as inducers, 696
    substrates as, in enzyme synthesis, 694
EF-G elongation factor, 671
Egg (ovum), 579
    calcium release following fertilization of animal, 266–68
    cleavage of fertilized, 553, 555f
    formation of, during meiosis, 588, 589f, 590
Egg activation, calcium and, 268
Einstein (measurement unit), 19
Elastins, 51, 290, 293–94
    in proteoglycan matrix, 294
    stretching and recoiling of, 294f
Electrical excitability, 228, 233–36
    action potential and, 228, 236–43
    ion channels as gates and, 233
    monitoring of ion channels and, 233–34
    voltage-gated ion channels and, 234–36
Electrical resistance, membranes and, 162
Electrical signals in nerve cells, 160, 225–55
    action potential and, 236–43
    electrical excitability and, 233–36
    integration and processing of, 250–52
    membrane potential and, 228–32
    nervous system and, 225–28
    synaptic transmission, 243–50
Electrical synapse, 243, 244f
Electrical work, energy and, 107f, 108
Electric eel, 108
Electrochemical equilibrium, 229f, 230
Electrochemical gradient, 197
    free energy change of transport and, 218–20
    sodium/potassium pump and maintenance of, 211–13
Electrochemical proton gradient, 400, 425, 426–30
    chemiosmotic model of, 426–30
    electron transport coupled to ATP synthesis and, 426
Electromagnetic radiation, energy and wavelength of, 19f
Electron(s)
    coenzymes as acceptors of, in biological oxidations, 374
    configurations of select atoms and molecules based on, 18f
    transfer of, by cytochrome b_6/f complex, 457–58
    transfer of, by photosystem I, 455f, 458
    transfer of, by photosystem II, 455f, 456–57
Electronegative atoms, 22
Electroneutrality, 229
Electron hole, 454
Electron microscope, 6–8
    high-voltage, 7
    resolution of, 6, 7, 8f
    sample preparation techniques for, 7–8
    scanning electron, 7, 9f
    study of cytoskeleton and, 745t
    transmission, 7
Electron shuttle system, 436
Electron transport, 400, 415–25
    ATP synthesis coupled with, 426
    carriers in system of, 418–22
    defined, 418

of electrons from coenzymes to oxygen, overview of, 418
    respiratory complexes and, 422–25
Electron transport intermediates, 25
Electron transport proteins, 184
Electron transport system (ETS), 400, 418
    carriers in, types of, 418–20
    conveyance of electrons from reduced coenzymes to oxygen by, overview of, 418
    electron carriers in, functions determined by reduction potentials, 420–22
    electron carriers in, types of, 418–20
    respiratory complexes of, 422–25
Electrophilic substitution, 139
Electrophoresis, 180, 625
    gel, 493, 494
    SDS-polyacrylamid, 180, 181f
Electroplaxes, 108, 248, 249f
Electrostatic interactions, 46
Ellis, John, 33
Elongation
    of microtubules, 747, 748f
    of polypeptide chains, 666, 669–71
    RNA synthesis and, 639f
Elongation factors, 669, 671
Embden, Gustav, 9, 380
Embden-Meyerhof pathway, 9, 380
Embolism, 282
Embryonic development
    cadherins and, 303–4, 305f
    growth factors, receptor tyrosine kinases, and, 273–74
    homeotic genes and, 727–28
    stem cells and cell cycle in, 554–55
Emerson, Robert, 454
Emerson enhancement effect, 454f, 455
Endergonic (energy-requiring) reactions, 119, 369. See also Anabolic pathway
Endocrine hormones, 276–80
    adrenergic hormones as examples of, 278–80
    chemical classification and function of, 278t
    target tissues for, 277f
Endocrine tissues, 276, 277t
Endocytic vesicle, 343
Endocytosis, 80, 160, 195, 342, 343–49
    clathrin-independent, 348–49
    neurotransmitter reuptake by, 250
    phagocytosis as, 344–45
    process of, 344f
    receptor-mediated, 258, 345–48
Endomembrane system, 323, 324f, 339f. See also Endoplasmic reticulum (ER); Endosomes; Golgi complex; Lysosomes; Nuclear envelope
Endonucleases, repair, 543
Endopeptidases, 413
Endoplasm, 795
Endoplasmic reticulum (ER), 79, 88–89, 92, 323–33
    composition of, in rat liver cells, 333t
    membrane biosynthesis and role of, 332–33
    protein glycosylation and role of, 336–38
    protein sorting/targeting and, 338–42, 678–80
    rough, 89, 324, 325, 330
    smooth, 89, 325, 330–32
    structure of, 88f
Endoplasmic reticulum-specific proteins, retrieval tags for, 338, 339f
Endosomes, 323
    development of lysosomes from, 354–55
    early, 90, 341, 347
    late, 90, 341, 354
    targeting of lysosomal proteins to, 339–41
Endosperm, 552
Endostatin, 570
Endosymbiont theory, 88, 449, 450–51f
Endothelium of blood vessels, 269
Endothermic reactions, 114
End-product repression, 693, 694

Energy, **106**, 107–12. See also Bioenergetics
  active transport coupled to source of, 207–8
  ATP/ADP system for conserving and releasing, in cells, 372, 373f
  cellular budget of, for protein synthesis, 672–74
  cellular changes caused by, 107–9
  fats as source of, 412–13
  flow of, through biosphere, 109–11
  flow of matter and, 111–12
  internal, 113–14
  mechanisms of storing, in muscle cells, 789
  protein as source of, 413–14
  sunlight and food as sources of, 109
Energy metabolism. See Chemotrophic energy metabolism; Photosynthesis
Energy transduction reactions of photosynthesis, **445**
  accessory pigments and, 453
  Calvin cycle and, 466
  chlorophyll and, 452–53
  complexes of, located on thylakoid, 461f
  model of, 457f, 460
  overview of, 445, 446f
  photophosphorylation (ATP synthesis) and, 459–61
  photoreduction (NADPH synthesis) and, 455–58
  photosystems, light-harvesting complexes, and, 453–55
  in purple bacteria, 461–62
Enhanceosome, 720
Enhancers (DNA sequence), 710, **718**
  model for action of, 720f
  properties of, 719f
Enkephalins, 245
Entactin, 298
Enteropathogenic bacteria, cell adhesions systems and, 308–9b
Enthalpy (H), **114**, 116b
Enthalpy change (δH), 116b
Entropy (S), 111, 115, **118**
Entropy change (δS), 116–17b, 118
  jumping reaction in jumping beans and, 116–17b
  thermodynamic spontaneity measured by, 118
Enveloped viruses, 98
Environmental risks of genetically modified plants, 616
Enzyme(s), **9**, 84, 106, **130–57**
  activation energy, metastable state in cells, and, 130–32
  activation energy barrier lowered by, 117
  active site of, 133–34
  adaptive synthesis of, 692–94
  allosteric regulation of, 390–91, 410–12
  bifunctional, 392
  as biological catalysts, 132–39
  carrier proteins as analogous to, 204
  classes of, 136t
  control of Calvin cycle by, 465
  diversity and nomenclature of, 135, 136t
  endoplasmic-reticulum association, 324
  gene coding for, 624–25
  inducible, 693
  induction of activity in, three examples of, 735f
  inhibition of, 145–47
  isolation of digestive, by lysosomes, 353–54
  kinetics of (see Enzyme kinetics)
  lysosomal, 355–57
  as markers, 160
  membrane-associated, 159–60, 184
  pH sensitivity of, 137
  in plasma membrane, 25
  as proteins, 41, 132–33
  regulation of (see Enzyme regulation)
  restriction (see Restriction enzyme(s))

RNA molecules (ribozymes) as, 151–53
  substrate binding, activation, and reaction in, 137–39
  substrate specificity of, 134–35
  of TCA cycle, 410–12
  temperature sensitivity of, 135–37
Enzyme catalysis, **130**
Enzyme Commission (EC), 135
Enzyme kinetics, **140–47**
  analogy of monkeys shelling peanuts and, 141b
  determining V_{max} and K_m, 144–45
  double-reciprocal plot for linearizing data of, 143–44
  enzyme inhibition and, 145–47
  Michaelis-Menten equation and, 140–42
  significance of, to cell biology, 143
  V_{max}, K_m, and, 142–43
Enzyme regulation, **147–51**
  allosteric regulation, 148–49
  cooperativity and, 149
  covalent modification and, 149–51
  feedback inhibition, 147–48
  proteolytic cleavage, 151
  regulation of, in Calvin cycle
  substrate-level, 147
Epidermal growth factor (EGF), **562**, 563
  receptor-mediated endocytosis of, 345f, 346
  structure and activation of, 270f
Epidermolysis bullosa simplex (EBS), 764
Epinephrine, 244, 276
  effects of, on tissues, 279
  glycogen degradation and role of, 280, 281f
Epithelial cell
  ciliated, microtubules of, 752f
  microvilli in intestinal, 759, 760f
Equilibrium, movement toward, 125
Equilibrium centrifugation, 326, **329**–30
Equilibrium constant (K_{eq}), 116b, **120**–21
  relationship of, to ΔG$^{0\prime}$, 123f
Equilibrium density centrifugation, **526**
  semiconservative DNA replication demonstrated by, 525–27
Equilibrium membrane potential, 229f, **230**
  Nernst equation and, 230
ER. See Endoplasmic reticulum (ER)
ER cisternae, **323**
Ergastoplasm, 324
ER lumen, **323**
ER signal sequences, 678, **680**
Erythroblastosis virus, 567
Erythrocyte(s) (red blood cells)
  anion exchange protein in, 206
  carbohydrates and survival of, 304–6
  chromatin DNase I sensitivity in, 712, 713f
  directions of oxyen, carbon dioxide, and bicarbonate transport in, 198f
  glucose catabolism in, 377
  glucose transporter in, 204–6
  lipid bilayer in plasma membrane of, 161–62
  microtubules of, 752f
  normal and sickled, 628f
  plasma membrane of, structural features of, 178f, 179
  translational control by heme in developing, 730
  transport mechanisms in plasma membrane of, 197
Escherichia coli (E. coli), 174, 195
  ability of, to kill other bacteria, 668–69
  circular DNA in, 80, 528f
  DNA replication in, 527, 528, 536
  genome, 492, 498t
  lactose catabolism and lac operon in, 694–98
  plasmid classes in, 504
  RNA polymerase in, 635
  studies on bacteriophages infecting, 482–85, 600–602
  transcription and translation coupled in, 648f

E-selectin, 306
Eserine, 250
E (exit) site, **662**
Estradiol, 70f, 71
Estrogen(s), 71, 276
Estrogen response element, 724
Ethidium bromide, 494
Ethylene, 276
Ets family of transcription factors, 563
Eubacteria, **76**
Euchromatin, **507**
Eukaryote(s), **76**
  prokaryotes vs., 78–82
Eukaryotes, gene regulation in, 706–28
  genomic control of, 706–15, 734
  multiple levels of, 706, 707f, 735–37b
  posttranscriptional control, 728–34, 737b
  summary of, 734
  transcriptional control, 715–28, 734, 737b
  versus gene regulation in prokaryotes, 704–6
Eukaryotic cell(s), 76, 83–98
  cell cycle of, 524f (see also Cell cycle)
  coated vesicles found within, 349t
  cytoplasm, cytosol, and cytoskeleton of, 95–97
  DNA and chromosomes in, 54f, 78–79, 80, 82f, 504–9 (see also Chromosome(s); DNA (deoxyribonucleic acid))
  DNA expression in, 82
  DNA packaging in chromatin and chromosomes, 504–7
  DNA packaging in mitochondria and chloroplasts, 507–9
  DNA replication in, 528–29, 541
  exocytosis and endocytosis in, 80, 160, 195, 342–49
  extracellular matrix (see Extracellular matrix (ECM))
  extracellular structures of, 291t
  gene expression in (see Gene expression)
  gene regulation in (see Eukaryotes, gene regulation in)
  genome of, compared to prokaryotes, 492 (see also Genome(s))
  internal membranes, 79, 85–96
  major events in evolution of, 451f
  membranes of, 158f (see also Membrane(s))
  messenger RNA in, 666f
  microtubules, microfilaments, and intermediate filaments in, 79–80
  mitochondria of (see Mitochondrion (mitochondria))
  mitosis and meiosis in, and DNA allocation, 80
  nucleus, 78–79, 80f, 84, 85f
  organelles of, 78, 85–95 (see also Organelle(s))
  photosynthetic (see Photosynthesis)
  plasma membrane of, 83–84
  prokaryotic cell versus, 78–82
  repeated DNA sequences in, 500t
  secretion in, 90f
  specialization in, 82–83
  stem cells and specialized types of, 706
  transcription in, 640–55
  translation in, 669–72
  transport processes in, 196f, 197
Evolution, endosymbiont theory, 88, 449, 450–51b
Exchange factor (GEF), **271**, 272f
Excinuclease, 543
Excision repair of DNA, **543**
Excitatory neurotransmitter, 244
Excitatory postsynaptic potential (EPSP), **250**, 251f
Exergonic (energy-yielding) reactions, **119**. See also Catabolic pathway
  biological oxidations as, 373–74
  hydrolysis of ATP as, 370–71
  oxidation of glucose as, 375
  transfer of phosphate groups as, 372f

Exocrine tissues, **276**
Exocytosis, 80, 160, 195, **342**–43
  neurotransmitter secretion by, 246
  polarized secretion and, 343
  process of, 343f
  release of intracellular molecules to extra-
    cellular medium by, 342–43
  role of calcium in triggering, 343
Exons, **650**, 704
  proposed role of, in coding for proteins,
    653, 654f
Exonuclease, DNA proofreading mechanism
  and, **535**
Exopeptidases, 413
Exothermic reactions, 114
Expansins, **317**
Exportins, **515**
Expression vectors, 612
Exteins, 676
Extensins, 315, **316**
Extracellular digestion, 355, 356–57
Extracellular matrix (ECM), 83, **97**–98, **290**–96
  collagens and procollagen in, 291–93
  elastins in, 293–94
  hyaluronate in, 294–95
  proteoglycans and adhesive glycoproteins in
    anchoring of cells to, 295–96
  proteoglycans and glycosaminoglycans of,
    294, 295f, 296f
  types of, 291f
Extracellular structures, **290**
  cell-cell recognition, adhesion, and, 302–6
  cell junctions, 306–14
  cell surface and cell wall in plants, 314–19
  in eukaryotic cells, 291t
  extracellular matrix (ECM) in animal cells,
    290–96
  fibronectins, 296–302
Eye, resolving power of microscopes and
  human, 6f
Ezrin protein, 761

F
F_0 complex, **404**, 405f
F_1 complexes, **404**, 405f, 430–31
F_0F_1 complex, **404**, 405f, 425f, 426–27
  ATP synthesis and, 431–34
  proton translocation and, 431, 432f
F_1 generation, **590**
  backcrossing of, 592f
F_2 generation, **590**, 591f
Facilitated diffusion, 197, **203**–7
  carrier proteins and, 197f, 203–6
  channel proteins, 197f, 203, 206–7
  in erythrocyte, 204–6
  transport through nuclear pores by, 514
F-actin (filamentous actin), 97, **755**, 756f
  G-actin polymerization into microfilaments
    of, 756, 757f
  retrograde flow of, 792
"Facts" and scientific method, 11–12b
Facultative heterochromatin, **517**
Facultative organisms, **378**
FAD. See Flavin adenine dinucleotide (FAD)
$FADH_2$
  electron transport, ATP synthesis, and role
    of, 427
  TCA cycle and generation of, 409–10
Familial hypercholesterolemia (FH), 345, **346**
Famotidine, 258
Farnesyl group, 180
Fast axonal transport, **770**–71
Fast block to polyspermy, 266–68
Fat(s), 68. See also Lipid(s)
  as energy source, 412–13
  in seedlings, 416–17b
Fatty acid(s), 67f, **68**, 169

  catabolism of, to acetyl CoA, 412, 413f
  melting point of, 172, 173f
  membrane fluidity and effects of, 172, 173f
  membrane lipid packing and effects of unsat-
    urated, 173f
  membrane structure/function and essential
    role of, 169
  nomenclature of, 68t
  peroxisomes and oxidation of, 91, 360–61
  saturated, vs. unsaturated, 68, 69f
  structures of select, 170t
Fatty-acid anchored membrane proteins, **180**
Fatty acyl CoA, 412
Fatty acyl dehydrogenase, 413
Feedback (end-product) inhibition, **147**–48
  lactose catabolism and, 694
Fermentation, 109, **375**
  alcoholic, 375, 385
  ATP yield from, 385–86, 400
  lactate, 375, 384–85
  propionate, and butylene glycol, 385
Ferments, 133
Ferredoxin (Fd)
  coordination of photosynthetic energy trans-
    duction and carbon assimilation depend-
    ent on, 465, 466f
  electron transfer from plastocyanin, 457f, **458**
Ferredoxin-NADP+ reductase (FNR), **458**, 460
Ferredoxin-thioredoxin reductase, 465
Ferritin, 186, 730
Fertilization, **579**
Feulgen, Robert, 10
F (fertility) factor, 504, **600**, 601f
FGFR-3 gene and protein, 273–74
Fibrils, collagen, 291, 292f
Fibrin, 282, 296
Fibroblast(s), 273, 291
  distribution of myosin II and actin in, 795f
  filopodia extending from, 793f
  regulation of, 797
Fibroblast growth factor receptors (FGFR), **273**
  dominant negative disruption of, 273, 274f
Fibroblast growth factors (FGF), **273**, 274f
Fibroin, 51
Fibronectin(s), 290, **296**–97
  blood clotting affected by, 297
  cell movement and shape affected by,
    296–97, 298f
  integrins as receptors for, 298–302
  structure of, 297f
Fibrous protein, **50**
Fick's first law of diffusion, 202n
Filaments, eukaryotic, 79–80
Filamin, 759, 760f
Filopodium, **792**, 793f, 794f, 798f
  actin bundles in, 794f
Fimbrin, 759
First law of thermodynamics, **113**–14
Fischer, Emil, 138
Fischer projection, 62
5' cap, 648–49
Fixed-angle roto, 326
Flagella, 744, 773, **774**
  bending and sliding of, 776–77
  intraflagellar transport and, 777
  movement of flagellated cell, 774f
  structure of, 774–76
Flavin adenine dinucleotide (FAD), **409**–10
  electron transport system and role of, 418
  structure of, and oxidation/reduction of, 409f
Flavin mononucleotide (FMN), electron trans-
  port system and role of, 418
Flavoproteins, electron transport and role of,
  418–19
Fleming, Alexander, 12
Flemming, Walther, 10
Flippases (phospholipid translocators), **171**, 332

Fluid mosaic model of membrane structure, 13,
  160, 165f
  lipids component of, 166–75
  model concepts leading up to, 161–64
  protein component of, 175–89
  Singer and Nicolson model, 164–66
  timeline for development of, 161f
Fluid-phase endocytosis, **348**–49
Fluorescein, 187
Fluorescence microscopy, 5, 6, 7f
  study of cytoskeleton using, 745t
Fluorescence recovery after photobleaching, **171**
Fluorescent antibodies, **186**–87
Focal adhesion kinase (FAK), 302
Focal adhesions, **300**, 301f, 302
Focal contacts, 794
Folkman, Judah, 570b
Food(s)
  as energy source, 109, 375
  fermentation and production of, 385
Food crops, genetically modified, 615–16
Fox, C. Fred, 76
Fraenkel-Conrat, Heinz, 35
Fragile X syndrome, 501
Frameshift mutation, 628–29, **630**, 631f, 676b, 677f
Franklin, Rosalind, 487
Free energy (G), 115, **118**–19
  jumping beans reaction and, 116–17b
  release of, from reduced coenzymes, 418,
    420–22
  transport of molecules across membranes
    and, 198
Free energy change (ΔG), 117b, **118**, 120–25
  calculating, 121–22, 124–25
  calculating, for transport of ions, 219–20
  calculating, for transport of molecules, 218
  capacity to do work and, 117
  electrochemical gradient and, for transport of
    charged solutes, 218–20
  equilibrium constant and, 120–21
  meaning of $\Delta G'$ and $\Delta G^{0'}$, 123, 124t
  as measurement of thermodynamic spontane-
    ity, 119–20
  sample calculations, 124–25
  standard, 122–23
Free ribosomes, 678
Freeze-fracture technique, 8
  membrane proteins analyzed by, **175**, 176f
Fructose-1,6-bisphosphatase, 388, 389f
Fructose-1,6-bisphosphate, 467
Fructose-2,6-bisphosphatase, 392
Fructose-2,6-bisphosphate (F2,6BP), **391**, 468
  regulatory role of, 391–92
  synthesis of, 391f
Fructose-6-phosphate, interconversion of glu-
  cose-6-phosphate to, 115f
  calculating ΔG for, 121–22
  equilibrium constant for, 121
Frye, David, 186
F-type ATPase, **208**, 209t
Fumarate, 410
  hydrogenation of, to succinate, 134–35
  stereoisomer of, 135f
Functional groups in biological molecules, **20**
Functions, membranes as site of specific, 159–60
Furchgott, Robert, 269

G
G. See Free energy (G)
G0 state, **525**
G1 checkpoint of cell cycle, **555**, 556f
  control of, by Cdk-cyclin complex, 559–60
G1 cyclins, 558
G1 phase of cell cycle, **524**
G2 checkpoint of cell cycle, **555**, 556f
  control of, by mitotic Cdk-cyclin complex,
    558–59

G2 phase of cell cycle, **524**
GABA (γ-aminobutyric acid), 244
GABA receptor, 249
G-actin (globular actin), 97, **755**
  molecular structure of, 755f
  polymerization of, into F-actin micro-
    filaments, 756, 757f
Galactose, 185, 187f, 294
Galactosemia, 376
Galactose oxidase (GO), 185
Galactoside permease, 693, 694
Gamete(s), **579**
  allele segregation and assortment during
    formation of, 590–95
  effect of nondisjunction on, 588
  meiosis and formation of, 588–90
Gametic meiosis, 581
Gametogenesis, **579**
Gametophyte, **581**
γ-tubulin, **750**, 751f
Gangliosides, **166**, 357
Gap junction, **98**, 160, 184, 306, 307f, **313**–14
  electrical synapses and, 243, 244f
  structure of, 313f
Gastric juice, 208
Gated ion channels, 206
Gb (gigabases), **492**
GC box, 641, 718
GDP, G protein binding to, 259, 261f, 757
Gelation-solation, **795**, 796f
Gel electrophoresis, **493**
  SDS-polyacrylamide, 180, 181f
  separation of DNA restriction fragments by,
    493f, 494
Gelsolin, 760
Geminin, 541
Gene(s), **10**, 479, 501, 592, **628**. See also Chromo-
  some(s); DNA (deoxyribonucleic acid)
  alleles of (see also Allele(s))
  bacteriophages as model systems for study of,
    484–86b
  coding of amino-acid sequences of polypep-
    tide chains by, 625–28
  coding of enzymes by, 624–25
  constitutive, 692
  conversion of, and homologous recombina-
    tion, 603–4
  discovery of chemical nature of, 479–86
  DNA sequences (see DNA sequence(s); DNA
    sequencing)
  examples of, with introns, 650t
  heat-shock, 726–27
  homeotic, 727–28
  linked, 596
  mapping (see Gene mapping)
  mutant, and mutant suppressor tRNA,
    674, 675f
  mutations affecting (see Mutation(s))
  peptide coding and mutations in, 625–28, 629f
  regulation of (see Gene regulation)
  regulatory, 694
  structural, 694
Gene amplification, **707**
  alteration of genome by, 707–9
  oncogene formation by, 566
Gene chip, 716
Gene conversion, 603, **604**
Gene deletion, **709**
Gene expression, 623–91
  direction flow of genetic information and,
    623–24
  genetic code and, 624–35
  levels of eukaryotic, 706, 707f, 735b
  mapping of, 597–98
  nonsense mutations and suppressor tRNA,
    674–75, 676
  posttranslational processing, 675–76

  protein targeting and sorting, 676–87
  regulation of (see Gene regulation)
  transcription, 623
  transcription in eukaryotic cells, 640–55
  transcription in prokaryotic cells, 635–40
  translation and, 623, 660–74
Gene locus, **578**
  mapping of, 597–98
Gene mapping
  gene locus, 597–98
  restriction fragments, 495–96
General transcription factors, **643**–44, 717
  binding of RNA polymerase to promoter site
    and, 643f
Generation time, cell cycle, 524
Gene regulation, 692–741
  combinatorial model of, 721, 722f
  discriminating among multiple levels of,
    735–37b
  in eukaryotes, genomic control, 706–15
  in eukaryotes, posttranscriptional control,
    728–34, 737
  in eukaryotes, transcriptional control,
    715–28, 737
  in prokaryotes, 692–703
  in prokaryotes compared to eukaryotes, 704–6
Gene therapy, 616–18
  for cancer, 571
  for cystic fibrosis, 212–13b, 616, 618
Genetic code, **624**–35
  amino acid codons in mRNA and, 634
  codon dictionary established by synthetic
    RNA and triplets, 633–34
  as degenerate and nonoverlapping, 631–32
  diagram of, 634f
  first look at, 625f
  gene base-sequence coding for amino-acid
    sequence of polypeptide chains, 625–28
  gene coding of enzymes, 624–25
  messenger RNA coding of polypeptide syn-
    thesis, 632–33
  mutations in, 628–30 (see also Mutation(s))
  near universality of, 634–35
  as triplet code, 628–31
Genetic diseases/conditions
  achondroplasia, 273
  body organ heterotaxia, 778b
  cancers, 544, 567, 569–70 (see also Cancer)
  cystic fibrosis, 210, 212–13b, 616, 618
  DNA sequence information and study of, 499
  Down syndrome, 588
  Duchenne muscular dystrophy, 650
  familial hypercholesterolemia, 345, 346
  fragile X syndrome, 501
  galactosemia, 376
  gene therapy for, 212–13b
  Griscelli's disease, 778
  hemophilia, 614, 618
  hereditary retinoblastoma, 567–68
  Huntington's disease, 501
  I-cell disease, 341
  junctional epidermolysis bullosa, 300
  Li-Fraumeni syndrome, 569
  microsatellite DNA abnormalities as cause
    of, 501
  myosins and, 778b
  myotonic dystropohy, 501
  Wiskott Aldrich syndrome (WASP), 760
  X-linked adrenoleukodystrophy, 361
Genetic engineering, **614**–18
  of animals, 617b
  improved food crops using, 615–16
  safety concerns regarding food crops, 616
  treatment of human disease with gene thera-
    py, 616–18
  use of Ti plasmid in plants, 614–15
  valuable proteins produced by, 614

Genetic mapping, 597, **598**
Genetic material, discovery of chemical nature
  of, 479–86
Genetic recombination, **582**, 595–98. See also
  Recombinant DNA technology
  in bacteria and viruses, 598–602
  crossing over and, 583, 584f, 585f, 597
  gene locations mapped by measuring frequen-
    cies of, 597–98
  linked genes and, 596
  during meiosis, 582–88
  molecular mechanisms of homologous, 602–6
Genetics
  cell biology and studies in, 4f, 5, 10–11
  DNA as material of, 479–86 (see also DNA
    (deoxyribonucleic acid))
  Mendelian, 590–95
  terminology of, 578, 579f
Genetic transformation, **481**–82, 599f
Genetic variability, 577–78
  allele segregation and assortment leading to,
    590–95
  meiosis and generation of, 590
  mutations as cause of, 577
  recombination and crossing over leading to,
    585, 595–98
  role of sexual reproduction in, 577–78
Genome(s), 5, **491**–501
  alternation of, by chromatin condensation,
    710–14
  alternation of, by DNA rearrangements,
    709–10
  alternation of, by gene amplification and
    deletion, 707–9
  bioinformatics and study of, 498–99
  compartmentalization of, 704
  DNA base sequencing of, 496–97
  eukaryotic, compared to prokaryotic, 705–6
  gene regulation of (see Genomic control in
    eukaryotes)
  inactive regions of, 714–15
  repeated DNA sequences in eukaryotic,
    499–501
  restriction enzymes and study of DNA in,
    492–96
  sample sequenced, 497–98
  size and complexity of, 492, 704
  structural organization of, 704–5
Genomic control in eukaryotes, 706–15
  chromatin decondensation and, 710–14
  defined, 706, **707**
  DNA methylation and, 714–15
  DNA rearrangements and genome alteration,
    709–10
  gene amplification/deletion and, 707–9
  properties of active regions of genome
    and, 715
  same genes found in organism's cells, 706–7
Genomic DNA, 601
Genomic library, **612**–13
Genotype, 578f, **579**
Geranylgeranyl group, 180
Germ line, 593
Gibbs, Willard, 118
Gilbert, Walter, 496
Gi protein, **259**
Glial cells, **226**
Glial fibrillary acidic (GFA) protein, 762, 763t
Globin
  introns in gene for β, 650
  translational control in synthesis of, 730
Globular proteins **51**–53
  structures of select, 53f
Glucagon, 276
Glucan synthetases, 338
Glucocorticoid, 71
Glucocorticoid receptor, activation of gene

transcription by, 725f
Glucocorticoid response element, 724
Gluconeogenesis, 385, **386–89**
 in fat-storing seedlings, 416–17b
 glycolysis compared to, 387–88, 389f
 in liver, 377–78
 regulation of, 389–92
Glucose, **375**
 D form of, 21f, 63f
 glycogen catabolism and conversion to, 331–32
 glycogen degradation and, 280, 281f
 glycolytic pathway and catabolism of, 378, 379f
 Haworth and Fischer projections of, 62f, 375
 human metabolism of, 376–78b
 oxidation of, as exergonic, 375
 as oxidizable substrate in energy metabolism, 375–78 (see also Aerobic respiration)
 phosphorylation of, 147
 structure of, 62f
 synthesis of (see Gluconeogenesis)
 transcellular transport of, across intestinal epithelium, 312f
 transcription of genes and influence of, 700
 transport of, across membranes, 200, 204–6
Glucose-6-phosphate
 conversion of, to fructose-6-phosphate, 115f, 121–22
 phosphorylation of glucose and production of, 147
Glucose phosphatase, 160, 184, 331, 388
Glucose transporter (GluT), 204, **205**, 206, 313
Glucosidases, 336
Glucuronate (GlcUA), 294
Glutamate, 43f, 51, 244
 TCA cycle and interconversion of, 414, 415f
Glutamic acid, hemoglobin alternations and, 627–28, 629f
Glutamine, 43f
Glyceraldehyde-3-phosphate (G-3-P), 447
 Calvin cycle and formation of, 462, 463f, 464
 carbohydrate synthesis and, 467
Glyceraldehyde-3-phosphate dehydrogenase (GPD), 184, 464
Glycerate, 470
Glycerate kinase, 470
Glycerol, 68, 412
Glycerol phosphate shuttle, **437–38**
Glycine, 43f, 44f, 244, 470
Glycocalyx, **186**, 188f, **302**
 attached and unattached, 302
Glycogen, **63**
 in animal cells, 29
 catabolism of liver, 331, 332f
 degradation of, 262, 280, 281f
 hypoxic conditions, ATP generation and, 789
 storage of blood glucose as, 331, 378b
 structure of, 64f
Glycogen phosphorylase, 280, 331, 386
 regulation of, by phosphorylation, 150f, 151
Glycogen synthase, 150, 280
Glycolate pathway, 361, **469–71**
Glycolipids, 67f, **70**, 166
 structure of, 167f
Glycolysis (glycolytic pathway), 109, 375, **378**
 aerobic respiration and phase of, 399f, 400
 alternatives to glucose as substrate for, 386
 ATP generation by, through catabolism of glucose to pyruvate, 378–83
 Cori cycle and link between gluconeogenesis and, 377f, 378
 fermentation and, 378–86
 gluconeogenesis compared to, 387–88, 389f
 in muscle cells, 377
 pyruvate fate, oxygen and, 383–85
 pyruvate fermentation and NAD+ regeneration, 384–85

 regulation of, 389–92
 summary of, 383
Glycolytic pathway, 378, 400. See also Glycolysis (glycolytic pathway)
Glycophorin, 178, 185, 306
 structure of, 178f
Glycoproteins, 84, **185–86**, 336
 carbohydrates in, 187f
 cells anchored to extracellular matrix by adhesive, 295–98
 receptor, 298–302
 sorting of, 681–83
Glycosaminoglycans (GAGs), **294**
 defective degradation of, 357
 structure of, 295f
Glycosidic bonds, **62–63**
 polysaccharide structure dependent on, 65
Glycosphingolipids, 70, 166
Glycosylation, 89, **336–38**
 of membrane proteins, **185**, 186f
Glycosylphosphatidylinositol (GPI), 180
Glycosyl transferases, 338
Glycylalanine, 44f
Glyoxylate cycle, 361, **416**
 conversion of acetyle CoA to carbohydrates and role of, 415
 glyoxysomes, seed germination, and, 416–17b
Glyoxysomes, 91, 361, **416**
 glyoxylate cycle, seed germination, and, 416–17b
Goldman, David E., 232
Goldman equation, 230, **231**
Goldstein, Joseph, 345, 346
Golgi, Camillo, 89, 333
Golgi complex, 79, **89**, 333–35
 lipid and protein flow through, two models of, 335
 protein glycosylation and role of, 336–38
 protein sorting and role of, 338–42
 relationship of microtubules, motor MAPs, and, 773f
 structure of, 89f, 333, 334f
 transport vesicles surrounding, 333–34
 two faces of stacks of, 334–35
Golgi-complex proteins, sorting of, 339
Golgi stack, **333**
 cis face and trans face, 334–35
Gorter, E., 161–62
GPI-anchored proteins, **180**
Gp protein, 263–64, 279
G protein, **259**
 adrenergic receptors for, 279f, 280
 cyclic AMP as second messenger for one class of, 261–62
 growth factor signaling, oncogenes, and Ras, 567
 human disease caused by disrupted signaling of, 262–63
 nitric oxide coupling with receptors of, effect on human blood vessels, 269–70
 plasma membrane, 567
 regulation of actin cytoskeleton by, 797–98
 signal transduction and, 261f
 structure and activation of, 259, 260f, 261
 structure of receptors for, 259f
G protein-linked receptor family, **259–61**
Gram, Hans Christian, 98
Gram stain, 98
Granum (grana), **87**, **448**, 449f
Grew, Nehemiah, 447
Griffith, Frederick, 481–82
Griscelli's disease, 778
Group I and Group II introns, 652
Group specificity, **135**
Group transfer reaction, **372**
Growth cones, 792

Growth factor(s), 272, **273**–75, **562**
 cancer and disrupted signaling of, 275, 562–64, 566t, 567
 embryonic development and, 273–74
 families of, 273t
 inhibitory, acting through Cdk inhibitors, 555, 564
 Ras pathway activated by, 562–64, 567
 receptor tyrosine kinases and, 270f, 273–74
 serine/threonine kinase receptors and, 274–75
 themes in receptor pathways of, 275
Growth hormone, 614
 creation of transgenic organism using, 616, 617b
Gs protein, **259**, 261, 279
GTP. See Guanosine triphosphate (GTP)
GTPase activating protein (GAP), **271**, 272f
GTP-binding proteins, actin polymerization regulated by, 757
GTP cap, 748, 750f
Guanine (G), **55**, 633
 Chargaff's rules and, 486, 487t
Guanine nucleotide binding protein. See G protein
Guanine nucleotide exchange factor, **271**, 272f, 352
Guanosine triphosphate (GTP)
 dynamic instability of microtubules and hydrolysis of, 748–49, 750f
 gluconeogenesis and role of, 388
 G protein binding to, 259, 261f
 TCA cycle and generation of, 409
Guanylyl cyclase, 269
Guard cells, calcium oscillations and opening and closing of plant, 268–69
Guide to Microscopy (Becker, et al.), 5
Gurdon, John, 518, 706, 709

H

Hair cells (inner ear), 778b
Hair fiber, structure of, 51f
Hairpin loop, 50f
 pre-tRNA, 646, 647f
 RNA, **639**
Half-cell reaction, 420
Half-life of mRNA, 655
 translational control and regulation of, 731–32
Half reactions, 374
Halobacteria (halophiles), 76, 164, 178f
 bacteriorhodopsin proton pump of, 216f
Hanson, Jean, 782
Haploid spore, **581**
Haploid state, 492, **578**
 gametes as haploid cells, 579
 meiosis and conversion of diploid cells to, 581, 582–88
 phase of, in life cycle of sexual organisms, 580–81
Hartwell, Leland, 558
Hatch, Marshall D., 471
Hatch-Slack cycle, **471**, 472f, 473
Haworth projection, **62**
Heart muscle. See Cardiac (heart) muscle
Heart rate, 790
Heat
 energy and, 107f, 108, 113
 enthalpy and, 114
 increasing energy content of systems with, 132
Heat losses, 110
Heat of vaporization, 23f
Heat-shock genes, **726–27**
Heat-shock proteins (Hsp60, Hsp70), 33, 672, 726
Heat-shock response element, **726**
Heat-shock transcription factor, 727
Helicases, **536**
 unwinding of DNA double helix by, 536, 537f

Helix-loop-helix motif, 723*f*, **724**
Helix-turn-helix motif, 50*f*, **722**, 723*f*, 728
Heme, structure of, 419*f*
Heme-controlled inhibitor (HCI), **730**
Hemicellulose, 65, **315**
Hemidesmosomes, **300**, 301*f*, 307*t*, 764
Hemoglobin
    DNase I sensitivity, 712, 713*f*
    sickle-cell anemia and changes in, 53,
      625–28, 629*f*
    structure of, 45*f*
    synthesis of, 666–67
Hemophilia, 614, 618
Henderson, Richard, 164–65
Hereditary nonpolyposis colon cancer
    (HNPCC), 544, 569–70
Hereditary retinoblastoma, 567–68
Heredity, chromosomal theory of, **10**, 593–95
Hershey, Alfred, 10, 482–85
Heterochromatin, **507**, 508*f*
    constitutive, 517
    facultative, 517
αβ-Heterodimer, **746**
Heterogeneous nuclear RNA (hnRNA), 648
Heterokaryon, **556**–57
Heterophagic lysosomes, **355**
Heterophilic interactions, **303**, 306
Heterotaxia, 778*b*
Heterozygous organism, **578**
Hexokinase, 380, 390
    enzyme kinetics of, 144–45
    substrate binding for, 138, 139*f*
    substrate regulation and, 147
Hfr cell, 600, **601**, 602
Hierarchical assembly, 27*f*, **36**–37
    watchmaking example of, 37*b*
High-energy compound, **370**
High-mannose oligosaccharides, 336
High-mobility group (HMG) proteins, **715**
High-voltage electron microscope, 7
Histamine, 244, 276, **278**, 280–82
Histidine, 43*f*, 634
Histone(s), **504**
    acetylation of, 715
    as core of nucleosome, 504–5, 507*f*, 509*f*
    transport of, through nuclear pores, 514
Histone acetyltransferase, 721
HIV (human immunodeficiency virus), 98,
    626, 729
Hoagland, Mahlon, 662
Hodgkin, Alan, 236
Hodgkin cycle, 238*f*
Hoffman, Hans-Peter, 402
Holding electrode, 237
Holliday, Robin, 604
Holliday junction, **604**, 605*f*, 606
Holoenzyme (α₂ββ′σ), **635**
Homeobox, **728**
Homeodomain, **728**
Homeotherms, 108, 135, 174
Homeotic genes, **727**–28
    mutations in, 727*f*
    proteins and, 728*f*
Homeoviscous adaptation, **174**
Homogenate, **327**
Homogenization of tissues for centrifugation, **326**
Homologous chromosome(s), **578**, 582–85. *See
    also* Homologous recombination
    crossing over and DNA exchange between,
      583, 584*f*, 585*f*, 597
    movement of, to opposite spindle poles dur-
      ing meiosis, 586, 587*f*
    separation of, during meiosis anaphase,
      586, 587*f*
    similar DNA base sequences in, 594–95
    spindle equator alignment of, during meiosis,
      585–86

synapsis of, 581, 585*f*
Homologous recombination, **602**–6
    bacterial conjugation as cause of, 600–602
    DNA breakage-and-exchange underlying,
      602–3
    gene conversion resulting from, 603–4
    Holliday junctions as initiators of, 604–6
    infection of bacteria by phages and, 598–99
    meiosis and (*see* Meiosis)
    molecular model for, 602–3
    synaptonemal complex and facilitation of,
      585, 586*f*, 606
    technology involving (*see* Recombinant DNA
      technology)
    transformation/transduction of bacteria by
      phages as cause of, 599–600
Homology searching, 606
Homophilic interactions, **303**
Homopolymer, 633
Homozygous organism, **578**
Hooke, Robert, 1
Hopanoids, **166**–68
    structure of, 168*f*
Hormonal regulation, 277*f*, 280*t*, 389–91
Hormone(s), **276**–82
    adrenergic, 278–80
    chemical classification and function of, 278*t*
    as chemical messengers, 257 (*see also*
      Chemical messenger(s))
    chemical properties of animal, 276–78
    endocrine, 276–80
    histamine and prostaglandin, 276, 278, 280–82
    paracrine, 280–82
    physiological functions controlled by, 276, 277*t*
    steroid (*see* Steroid hormone(s))
    target tissues for, 276, 277*f*
Hormone response elements, **724**
*Hpa*II restriction enzyme, 714
Huber, Robert, 461
Human blood groups, 166, 304–6
Human disease. *See also* Autoimmune diseases;
    Genetic diseases/conditions; Poisons;
    Toxics/toxins
    acetylcholine receptors and, 249
    adrenoleukodystrophy (ALD), 91, 361
    Alzheimer's, 673
    amyotrophic lateral sclerosis (ALS), 764
    anthrax, 309*b*
    cancer (*see* Cancer)
    cardiomyopathies, 764
    cell recognition/adhesion systems and, 308–9*b*
    disrupted G protein signaling as cause of,
      262–63
    dyneins, myosins, and, 778*b*
    epidermolysis bullosa simplex, 764
    faulty glycosphingolipids as cause of, 166
    faulty myelination of axons as cause of, 242
    faulty peroxisomes and, 91, 361
    faulty protein-folding as cause of, 673*b*
    gene therapy for, 616–18
    human papillomavirus (HPV) and, 565,
      568, 569
    hypercholesterolemia, 345, 346
    lysosomal storage diseases, 357
    multiple sclerosis, 242
    myasthenia gravis, 249
    neurological, caused by faulty ion channels, 236
    neurotoxins as cause of, 247, 250*b*
    prions as cause of, 99–101
    rheumatoid arthritis, 356–57
    severe combined immunodeficiency
      (SCID), 616
    sickle-cell anemia, 53, 625–28, 629*f*
    Tay-Sachs disease, 166, 357
    viral, 98
    xeroderma pigmentosum, 543, 570
Human egg cell, 589

Human genome, 498, 499
    mitochondrial, 510*f*
Human Genome Project, 498
Human immunodeficiency virus (HIV), 98, 626
Human papillomavirus (HPV), 565, 568, 569
Hunter syndrome, 357
Huntington's disease, 501
Hurler syndrome, 357
Huxley, Andrew, 236, 782
Huxley, Hugh, 782
Hyaluronate, 294, **295**
Hybrid model for endosome-lysosome material
    transfer, 354–55
Hybrid strains, **590**
Hydrocarbons **20**
Hydrogenation, 134–35, **374**
Hydrogen bonds, 33, **46**
    in nucleic acid structure, 58*f*
    between water molecules, 22
Hydrogen peroxide metabolism, peroxisomes
    and, 360
Hydrolases, 90, 135, 136*t*
Hydrolysis
    of ATP, 370–71
    standard free energies of, for phosphorylated
      compounds involved in energy metabo-
      lism, 371*t*, 372
Hydropathy index, **179**
Hydropathy (hydrophobicity) plot, **179**, 182
Hydrophilic channel, 26
Hydrophilic molecules, **23**
Hydrophobic interactions, 34, **46**–47
Hydrophobic molecules, **23**
    amino acids, 42
Hydrophobic sorting signal, 686
Hydroxylation, **331**
Hydroxylysine, 186*f*, 291
Hydroxymethyl transferase, 470
Hydroxyproline, 186*f*, 291
Hydroxypyruvate, 470
Hypercholesterolemia, 345, 346
Hyperpolarization of membranes (undershoot),
    231, 237, **238**–39
Hypertonic solution, 200*b*
Hypothesis, scientific, **11**–12
Hypotonic solution, 200*b*
Hypoxic conditions, muscle contraction and
    ATP generation under, 788, 789
H zone, muscle, **780**, 781*f*

I

I band, muscle, **780**, 781*f*
ICAM proteins, 306
I-cell disease, 341
IgE antibody protein, 280, 281
Ignarro, Louis, 269
Immunoblotting, 736*b*
Immunofluorescence microscopy, 742, 745*t*
Immunoglobulin(s), 187
    alternative RNA splicing and coding for
      IgM, 729*f*
    G, light chain of, 680
Immunoglobulin superfamily (IgSF), 302, **303**
Immunotherapy, 571
Importin, **514**
Independent assortment. *See* Law of independ-
    ent assortment
Indirect active transport, **208**
    direct active transport compared to, 208*f*
    ion gradients and, 210
Induced-fit model of enzyme specificity, **138**
Inducer(s), **696**
    of transcription in inducible operon, 695, 696
Inducible enzymes, **693**
Inducible operons, **695**
    dual (negative, positive) control of, 700–701
Infectious microorganisms, motility of, 795–97

Information
    amplification of genetic, mRNA and, 655
    cellular requirement for, 106
    flow of genetic, from DNA to RNA, 623–24
    flow of genetic, in cells, 479, 480f
Informational macromolecule, **28–29**
Ingram, Vernon, 626–28
Inheritance
    blending theory of, 592
    chromosomal theory of, 10, 593–95
    Mendel's laws of, 592–93
Inhibition of enzyme activity, 145, **146**, 147
Inhibitory growth factors, 564
Inhibitory neurotransmitter, 244
Inhibitory postsynaptic potential (IPSP),
    250, 251f
Initial reaction velocity, 140
Initiation codon, **634**
Initiation factors, **668**
Initiator (Inr) sequence, **641**, 642f
Initiator tRNA, **669**
Inner membrane, chloroplast, **448**, 449f
Inner membrane, mitochondrion, 85, 86f, **403**
    F_1 and F_0 complexes of, 405f
    functions occurring on, 404t
    glycerol phosphate shuttle in, 437–38
    respiratory complexes in, 422–25
Innervation in nervous system, **226**
Innexins, 314
Inositol-1,4,5-trisphosphate (InsP$_3$), **263–64**,
    279, 280
    calcium levels and, 263, 264, 280
    cell functions regulated by, 263t
    role of, in signal transduction, 264f
Inositol-phospholipid-calcium pathway, 263,
    264, 280
Inositol phospholipids, regulation of molecules
    affecting actin polymerization by, 758
INsP$_3$ receptor, **264**
Insulin, 48, 276
    genetically engineered, 614
    posttranslational processing of, 675, 678f
    primary protein structure of, 48f
    space-filling model of, 34f
Insulin-dependent diabetes, research on, 12
Integral membrane protein, 164, **177–79**
    aquaporins as, 207
    cotranslational insertion of, 683, 684f
    hydropathy analysis of, 179f
    structures of two, 178f
Integral monotopic protein, **177**
Integrin-linked kinase (ILK), 302
Integrins, **298–302**
    cell signaling and, 300–302
    as cell surface receptors, 296, 298–302
    cytoskeleton and, 299–300, 301f
    focal adhesions, hemidesmosomes, and,
        301f, 793
    function of, 300
    structure of, 299, 300f
Inteins, 676, 678f
Interactions
    hydrophobic, 34, 46–47
    noncovalent, 34, 46
    protein folding/stability and, 45–47
    van der Waals, 34, 46
Intercalated discs, **789**
Intercalating agents, 542
Intercellular communication, **160**. *See also*
        Chemical signaling in cells; Electrical signals
        in nerve cells; Signal transduction
    direct, through plant cell wall, 318–19
Interdoublet link, **776**
Interferons, 614
    as activator of STATs, 726
Interleukin, 614
Intermediate energy compound, ATP as, 371

Intermediate filaments (IFs), 79–80, 95, **97**,
    762–65
    assembly of, 763, 764f
    classes of, 762, 763f
    intracellular distribution of, 743f
    mechanical strength in tissues conferred by,
        764–65
    properties of, 743t
    proteins of, 743, 762–63
    structure of, 96f, 764f
Intermediate filament typing, 97, 762
Intermembrane space, chloroplast, **448**, 449f
Intermembrane space, mitochondria, **402**, 403f
    functions occurring in, 404t
Internal energy (E), **113**
    change in (δE), 113–14
Internalin A, 309, 795
International Union of Biochemistry, 135
Interneurons, **226**
Interphase in cell cycle, 85, **524**
Interspersed repeated DNA, **501**
Intestine
    epithelium, glycocalyx of, 188f
    epithelium, transcellular transport of glucose
        across, 312f, 376
    microvilli of mucosal cells in, 78f, 759–60
Intracellular attachment proteins, 307
Intracellular compartments, 323–67
    coated vesicles in cellular transport processes
        and, 349–53
    endoplasmic reticulum, 323–33
    exocytosis, endocytosis, transport across
        plasma membrane, and, 342–49
    golgi complex, 333–35
    lysosomes and cellular digestion, 353–57
    peroxisomes, 358–62
    plant vacuole, 357–58
    protein glycosylation and role of, 336–38
    protein sorting and role of, 338–42
Intracellular membranes, **159**
    organelles and, 85–95
Intracellular protein sorting, 679f
Intracellular signaling, cAMP pathway and,
    280, 281f
Intraflagellar transport, **777**
Intron(s), 612, 646, **649**, 704
    cDNA libraries without, 612–13
    demonstration of, in protein-coding genes, 649f
    detection of, using restriction enzymes, 650f
    discovery of, 649–50
    examples of genes with, 650t
    excision of, 151
    function of, in eukaryotic genes, 652–63
    removal of, from pre-mRNA by spliceosomes,
        650–51, 652f
    self-exision and splicing of, 152f, 651–52
Invasion of tissues, **571**
Inversions, chromosomal, 676
Inverted repeat sequences, 724f, **725**
Ion channels, 203, **206**, **233–34**
    action potentials produced by opening/closing
        of, 237–40
    chloride, and cystic fibrosis disease, 210,
        212–13b
    ligand-gated, 206, 233
    techniques for monitoring, 233–34
    voltage-gated, 206, 234–36
Ion concentrations
    action potential and changes in, 240
    membrane potentials and, 228–32
Ion gradients, indirect active transport driven by,
    and sodium/potassium pump, 210, 211–13
Ionic bonds, 34, **46**
Ion transport across membranes, 201
    action potentials and, 237–40
    calculating free energy change for, 219
IRE-binding protein, 731

Iron
    control of mRNA degradation in response to,
        731, 732f
    translational control in response to, 730, 731f
Iron-protoporphyrin IX, 419
Iron-response element (IRE), **730**, 731f
Iron-sulfur (Fe-S) center, 419, 458
Iron-sulfur proteins, electron transport and role
    of, **419**
Irreversible (enzyme) inhibitor, **146**
Isocitrate, 407
Isocitrate dehydrogenase, 407, 412
Isoenzymes, **467**
Isoleucine, 43f, 51
    synthesis of, from threonine, 148f
Isomerases, 135, 136t, 464
Isoprenoids, 71
Isopropylthiogalactoside (IPTG), 696
Isoproterenol, 258
Isothermal systems, 113
    cells as, 132
Isotonic solution, 200b
Isotope, 9

J
Jacob, François, 10, 148, 694, 698
Janus activated kinase, 275, **726**
Joints, hyaluronate and lubrication of, 294–95
Joule (J), 19n, **113**
Jumping beans and free energy, 116–17b
Jumping reaction, 116
Junctional complex, **788**
Junctional epidermal bullosa, 300
Jun protein, 271
Juxtaparanodal regions, nodes of Ranvier, 242

K
K$_m$. *See* Michaelis constant (K$_m$)
Karjalainen, Jutta, 12
Kartagener's triad, 778b
Karyotype, **545**
    human, 548f
Kb (kilobases), **492**
KDEL sequence, 683
Kennedy, Eugene, 401
Keratan sulfate, 294, 295f
Keratin(s), 51, 300, 310, 762
Keratinocytes, 778
Ketose, 380
Ketosugars, 61, 62f
Khorana, H. Gobind, 633–34
Kinesin, **770**, 771f
    family of, structures and functions of, 772
    movement of, along microtubules, 771, 772f
Kinesin family member (KIF), **772**
Kinetochore, **545**, 550, 586
Kinetochore microtubules, **545**, 550
Kirschner, Marc, 748
Kölliker, Rudolph, 400, 401
Kornberg, Arthur, 530
Kornberg, Roger, 506
Koshland, Daniel, 138
Krebs, Hans, 9, 405
Krebs cycle. *See* Tricarboxylic acid (TCA) cycle
    (Krebs cycle)
Kuru disease, 101

L
lacA gene, 694, 695
lacI gene, 694, 695
lac operon, **694**, 695f
    cis-trans test on, 698
    diploid analysis of mutations affecting, 698t
    genetic analysis of mutations affecting, 697t
    model of, 695–98
    negative control of transcription and, 700
    regulation of, 695, 696f

lac repressor, 694–95, 722
    allosteric regulation of, 695, 697*f*
Lactate, 377–78
Lactate dehydrogenase, 384
Lactate fermentation, 375, **384**–85
Lactoperoxidase (LP), 185
Lactose, 62, 63
    free energy change for transport (uptake) of, 218
    gene regulation in catabolism of, 693, 694–98
    genes involved in prokaryotic catabolism of, 694–98
    metabolism of, 376
Lactose intolerance, 376
lacY gene, 694, 695
lacZ gene, 608, 610*f*, 694, 695
Lagging strand, DNA, **534**
Lamellipodia, 760, **792**, 794*f*, 798*f*
Laminins, 290, 296, **297**–98
    in basal laminae, 297–98, 299*f*
    properties of, 298
    structure of, 299*f*
Langmuir, Irving, 161
Lap-Chee Tsui, 212*b*
Large heterotrimeric G proteins, 259–61
Large ribosomal subunit, **94**
Lariat structure, 652*f*
Late endosomes, 90, 341, **354**
    development of lysosomes from, 354–55
Lateral diffusion, **169**
Latrunculin A, 744, 758
Law, scientific, **13**
Law of independent assortment, 592, **593**
    chromosomal behavior and, 593–94
    meiotic basis for, 595*f*
Law of segregation, **592**
    applied to linked and unlinked genes, 596*f*
    chromosomal behavior and, 593–94
    meiotic basis for, 594*f*
L-cells, 303
LDL receptor, 346*b*, 347*f*
Leader (mRNA sequence), **666**
    trp mRNA, 702, *702*
Leader sequence, pre-tRNA, 646, 647*f*
Leading edge, cell, 792
Leading strand, DNA, **532**–34
Leaf (leaves), C_3 vs. C_4 plants, 471*f*
Leaf peroxisome, **91**, 93*f*, 361, **469**
Lecithin, 161
Lectins, 186, **304**
Leeuwenhoek, Antonie van, 1–2, 447
Lehninger, Albert, 401
Leptotene stage of meiosis prophase I, **583**, 584*f*, 585*f*
Lethal factor (LF), 309*b*
Leucine, 43*f*, 51, 634
Leucine zipper motif, **722**, 723*f*, 724
Leukocyte, selectins and adhesion of, 306*f*
Lewis, Edward B., 727
Licensing in DNA replication, **541**
Life cycles in organisms, 581*f*
Li-Fraumeni syndrome, 569
Ligand, 160, **257**. *See also* Chemical messenger(s)
    binding of receptors to, 257–58
Ligand-binding domain, 271
Ligand-gated ion channel, 206, **233**
    muscle contraction and, 787
    neurotransmitter receptors as, 247–49
Ligases, 135, 136*t*
Light
    Calvin cycle and requirements for, 465–66
    photons and quantum of, 452
    photosynthetic transduction of (*see* Energy transduction reactions of photosynthesis)
Light chain, protein, 680
Light-harvesting complex (LHC), **454**
Light-harvesting complex I (LHCI), **458**

Light-harvesting complex II (LHCII), **456**
Light microscopy, **5**–6, 7*f*
Light reactions. *See* Energy transduction reactions of photosynthesis
Lignin, 98, **316**
Limit of resolution, microscope, **5**
Lineweaver, Hans, 143
Lineweaver-Burk equation, **143**, 144*f*
Linkage group, **596**
Linked genes, **596**
Linker DNA, 506, 507*f*
Linoleate, 169, 170*t*
Lipid(s), **66**–71, **166**–75. *See also* Fat(s)
    chromatography and analysis of, 168–69
    classes of, 68*t*
    endoplasmic reticulum and biosynthesis of, 324
    as energy (ATP) source, 412–13
    fatty acids, 68, 69*f*, 169, 170*t* (*see also* Fatty acid(s))
    flow of, through Golgi complex, 335
    glycolipids, 70, 166, 167*f*
    lipid bilayer (*see* Lipid bilayer)
    in membranes, 26, 166–75
    mobility of, within membranes, 171*f*, 188
    nomenclature of, 68*t*
    phospholipids, 68–70, 166, 167*f*, 168*f*
    steroids, 70–71
    sterols, 166–68
    synthesis of, 28*f*
    terpenes, 71
    triacylglycerols as storage, 68–69, 412
    unequal distribution of, in membranes, 169–71
Lipid-anchored membrane proteins, 164, **179**–80
Lipid bilayer, **24**, 25*f*, **84**, **162**
    assymetrical orientation of membrane proteins across, 184–85
    factors affecting diffusion across, 199*t*
    fluidity of, in membranes, 171–72
    ion permeability across, 201
    membrane function and fluidity of, 172–74
    simple diffusion across, 199–201
    transition temperature of, 172
Lipid bodies, association of glyoxysomes and, in fat-storing plant seedlings, 416*f*
Lipman, Fritz, 9, 405, 665
Lipopolysaccharides, 98
Lipoproteins, 98
    structure of, 347*f*
Liposomes, 182, 199, 213
Listeria monocytogenes, 309*b*
    infection of macrophage by, 795–96, 797*f*
Liver
    composition of ER and plasma membranes of rat, 333*t*
    Cori cycle and gluconeogenesis in, 377*f*, 378
    drug detoxification and, 331
    effect of cortisone on, and arginase degradation, 732–33
    effect of elevated cAMP in, 262
    glucose transporter in, 205
    glycogen catabolism in, 331, 332*f*
Living organisms
    as aerobic, anaerobic, or facultative, 378
    diploid and haploid phases in life cycles of, 580–81
    genome size and type of, 492
    hierarchical nature of, 27*f*
    metabolism and oxygen needs of, 378
    sunlight and food as energy sources for, 109
    transgenic, 614–18
    types of life cycles in, 581*f*
Local mediators, 257
Lock-and-key model of enzyme specificity, 138
Loewenstein, Werner, 314
Looped domain, chromatin, **507**, 508*f*

Low-density lipoprotein (LDL), 345, **346**
    cholesterol, receptor-mediated endocytosis and, 346–37*b*
L-selectin, 306
Luciferase, 362
Lumen, endoplasmic reticulum, **89**, 323
Lumen, microtubule, 96
Lutein, 453
Lyases, 135, 136*t*
Lymphocytes, antibody production by, 710
Lysine, 43*f*
Lysis of bacteria, 133
Lysogenic state, bacteriophage, 485*b*
Lysophosphatidic acid (LPD), 797
Lysosomal proteins, sorting of, 339–41
Lysosomal storage diseases, **357**
Lysosomes, 79, **90**, 323, **353**–57
    development of, from late endosomes, 90, 354–55
    digestive enzymes isolated by, 353–54
    digestive processes and lysosomal enzymes, 355–57, 734
    discovery of, 92*b*
    heterophagic, vs. autophagic, 355
    lysosomal storage diseases, 357
    microautophagy and, **734**
    structure of, 91*f*
    targeting of lysosomal proteins to, 339–41
Lysozyme, 133
    structure of, 133*f*
    substrate binding for, 138, 139*f*
Lytic growth, bacteriophage, 484, 485*f*

M

McCarty, Maclyn, 10, 481
MacLeod, Colin, 10, 481
Macrofibrils, 51, 315
Macromolecule(s), **26**–31, **41**–75
    biologically important, 28, 29*t*
    diffusion rates of, 78
    informational, 28–29
    lipids, 26, 28*f*, 66–71
    nucleic acids, 26, 28*f*, 29*t*, 54–60 (*see also* Nucleic acid(s))
    polysaccharides, 26, 28*f*, 29*t*, 60–66
    proteins, 26, 28*f*, 29*t*, 41–54 (*see also* Protein(s))
    role of, in living systems, 26–28
    self-assembly of, 17, 31–37
    storage, 28, 29
    structural, 28, 29
    synthesis of, 17, 28*f*, 29–31
    transport of, across membranes (*see* Transport across membranes)
    transport of, into and out of nucleus, 513*f*
Macrophages, 344
    infection of, by *Listeria*, 797*f*
Macrophagy, **356**
Mad 2 protein, 561
Mad cow disease, 673
Malate, 473
Malate-aspartate shuttle, 436
Malathion, 250
Maleate, stereoisomer of, 135*f*
Malignant tumor, **565**
Maltase, 376
Maltose, 62
    human metabolism of, 376*b*
Mannose, 185, 187*f*, 386
Mannose-6-phosphate, 683
Mannose-6-phosphate tag, 340*f*
Mannosidases, 336
MAP2 and MAP2C, 754, 755*f*
MAP kinases, 264, **271**, 272*f*, 726
MAPs. *See* Microtubule-associated proteins (MAPs)
Map units, 598

Margulis, Lynn, 450
Markers, enzyme, 160
Mass spectrometry, 498
Mast cells, **281**
Mating bridge, **600**
Mating type, **579**
MAT locus, **709**, 710*f*
Matrix, mitochondrial, **86**, **404**
    functions occurring in, 404*t*
Matrix attachment regions (MARs), 715
Matter flows through biosphere, 111–12
Matthei, J. Heinrich, 633
Maturation-promoting factor (MPF), 557
Maxam, Allan, 496
Maximum ATP yield, **435**–39
Maximum velocity (V_{max}), **142**
    determining, 144–45
    importance of, to cell biologists, 143
    plotting, 143–44
    values of, for select enzymes, 143*t*
Mb (megabases), **492**
MCM complex, 541
Mdm2 protein, 568
Measurement units in cell biology, 2–3*b*
Mechanical work
    energy and, 107*f*, **108**
    muscle tissue and, 108, 109*f*
Mechanoenzymes, **770**
Mechanosensitive channels, 206
Medial cisterna, **334**
Medial element, **787**
Meiosis, 80, **580**–90
    conversion of diploid cell into four haploid
        cells by, 581–82
    genetic diversity generated by, 590
    haploid and diploid phases in life cycle of
        organisms and, 580–81
    laws of segregation and independent assort-
        ment based on, 594*f*, 595*f*
    meiosis I phase of, 581, 582–88
    meiosis II phase of, 588
    mitosis compared to, 587*f*
    phases of, in animal cells, 582–83*f*
    principle of, 580*f*
    sperm and egg generation by, and gamete for-
        mation, 588–90
    synaptonemal complex and homologous
        recombination during, 606
Meiosis I, **581**, 582–88
Meiosis II, 587*f*, **588**
Melanocytes, 777
Membrane(s), **24**, 35, **158**–94
    depolarization of, 231, 238–39
    endoplasmic reticulum and biosynthesis of,
        332–33
    fluidity of, 171–75
    functions of, 159–60
    linkage of actin to, 760–62
    lipids in, 166–75
    phospholipid bilayer structure of, 24–25, 26*f*
    plasma (*see* Plasma (cell) membrane)
    protein, lipid, and carbohydrate content of, 163*t*
    proteins in, 175–89
    as selectively permeable, 25–26
    signal transduction at (*see* Chemical signaling
        in cells; Electrical signals in nerve cells;
        Signal transduction)
    structural models of, 160–66
    transport across (*see* Transport across mem-
        branes)
    trilaminar appearance of, 162*f*, 163
Membrane asymmetry, **169**
Membrane domains, 188, 189
Membrane lipids, 166–75
    classes of, 166–68
    fatty acids essential to structure/function of, 169
    membrane fluidity and role of, 171–75

thin-layer chromatography for analysis of,
    168–69
unequal distribution of, 169–71
Membrane potential (V_m), **197**, 219, **227**–32
    action potentials and rapid changes in,
        236–43
    electrical excitability, action potential, and,
        233–36
    equilibrium, 229*f*, **230**
    ion concentrations and, 228–32
    measuring, 237*f*
    resting, 228–31
Membrane protein(s), 25, 26*f*, 175–89. *See also*
        Enzyme(s); Receptor(s); Transport protein(s)
    asymmetrical orientation of, 184–85
    classes of, 177–79
    contribution of molecular biology to study of,
        181, 182–83*b*
    cotranslational insertion of, into ER, 683, 684*f*
    freeze fracture analysis of, 175–76
    functions of, 184
    glycosylation of, 89, 185–86
    integral, 164, 165*f*, 177–79
    lipid-anchored proteins, 164, 179–80
    mobility of, 186–89
    peripheral, 164, 179
    in plasma membrane, 84
    SDS-polyacrylamide gel electrophoresis of,
        180, 181*f*
    tight junctions and blocked movement of,
        312–13
    transmembrane segments and, 164, 165*f*,
        177–78
Membrane reconstitution, 182
Mendel, Gregor, 5, 10, 479
Mendelian genetics, 590–95
Mendel's laws of inheritance, **592**–93
Menten, Maud, 140
Meristem tissue, 317, 448
Meselson, Matthew, 525–27, 602, 604
Mesoderm, development of embryonic, 273, 274*f*
Mesophyll cells, **462**, 471
Messenger RNA (mRNA), 44, 54, 479, **623**
    alternative RNA splicing and, 729
    amino acids coded by codons of, 633, 634,
        660, 666
    amplification of genetic information by, 655
    antisense, 615
    binding sites on, 661, 662*f*, 668
    detecting and quantifying molecules of,
        736–37*b*
    eukaryotic, 666*f*
    5′ cap and 3′ poly(A) tails on, 648–49
    introns and, 649–53
    life span of, 655
    polygenic, 666
    processing of eukaryotic, 647–51
    prokaryotic, 666*f*
    proofreading and surveilance of, 654
    regulation of half-life of, as translational con-
        trol, 731–32
    RNA editing and alteration in coding
        sequence of, 653–54
    stability of, in eukaryotes vs. prokaryotes, 705
    synthesis of polypeptide chains guided by,
        632–33, 666
    translation of, into polypeptides (*see*
        Translation)
    transport of, through nuclear pores, 514
Metabolic pathway(s), 9, 368–69
Metabolism, **368**
    chemotrophic (*see* Chemotrophic energy
        metabolism)
    mRNA and, 654–55
    phototrophic (*See* Photosynthesis)
Metabolites, 195, 369, 465
Metallothionein (MT), 617

Metaphase (meiosis), 585–86, 587*f*
Metaphase (mitosis), **545**, 547*f*
Metaphase plate, 545, 550
Metastable state, activation energy and, 130, **131**
Metastasis, **571**
Methanobacteria (methanogens), 76
Methionine, 43*f*, 51
Methylation
    DNA, 544
    pre-RNA, 646
Meyerhof, Otto, 9
Michaelis, Leonor, 140
Michaelis constant (K_m), **143**
    determining, 144–45
    values of, for select enzymes, 143*t*
Michaelis-Menten equation, **142**, 147
Michaelis-Menten kinetics, 140–45
Michel, Hartmut, 461
Microautophagy, **734**
Microbodies, **359**
Microfibrils
    of cellulose, 65, **315**, 317*f*
    of hair fiber, 51
Microfilament(s) (MF), 79–80, 95, **97**, **754**–62
    actin as protein building block of, 755
    actin groups and related proteins, 756
    actin linked to membranes by select proteins,
        760–62
    actin polymerization and assembly of, 756, 757*f*
    capping proteins and stabilization of ends
        of, 758
    cell motility based on (*see* Microfilament-
        based movement)
    effect of drugs and specific proteins on, 758
    inositol-phospholipid regulation of molecules
        affecting actin polymerization, 758
    intracellular distribution of, 743*f*
    model for assembly of, 756*f*
    myosin II as actin-binding protein, 757*f*
    properties of, 743*t*
    regulation of actin-polymerization by GTP-
        binding proteins, 757
    regulation of interactions between, by actin-
        binding proteins, 758–60
    structure of, 96*f*
Microfilament-associated (actin-binding)
        proteins, 755, 758–60, 770*t*
Microfilament-based movement, **770**, 779–92.
        See also Muscle contraction
Microglia, 226
Micrometer, **2**–3*b*
Microphagy, **356**
Microsatellite DNA, 501
Microscopy, 5–8
    confocal scanning, 6, 7*f*
    differential interference contrast, 6, 7*f*
    digital video, 6
    early, 1–3
    electron, 6–8
    fluorescence, 5, 6, 7*f*
    light (brightfield), 5, 7*f*
    phase-contrast, 6, 7*f*
    resolving power, 5, 6–7, 8*f*
    study of cytoskeleton and modern, 744, 745*t*
Microsomes, 92*b*, 160, **325**
Microtome, 5
Microtubule(s) (MT), 79–80, 95, **96**–97, **744**–54
    assembly of, 747*f*, 748*f*, 749*f*
    astral, 545, 546*f*, 551, 552*f*
    axonemal, 744, 777
    cell movement based on (*see* Microtubule-
        based movement)
    centrosome and organization of, 749, 750–51
    cytoplasmic, 746
    effect of drugs on assembly of, 753
    GTP hydrolysis and dynamic instability of,
        748–49

Microtubule(s) (MT), *continued*
  intracellular distribution of, 743*f*
  kinetochore, 545
  microtubule-organizing centers and, 749–51
  minus end of, 748, 749*f*
  in mitotic spindle, 545
  plus end of, 748, 749*f*
  polar, 545, 551, 552*f*
  polarity of, 549*f*, 750–51, 752*f*
  properties of, 743*t*
  regulation of, by microtubule-associated
    proteins (MAPs), 753–54
  regulation of stability in, 751–53
  relationship of motor MAPs, Golgi complex,
    and, 773*f*
  sliding of, within axoneme, 776–77
  structure of, 96*f*, 746*f*
  tubulin heterodimers as building blocks of,
    746–47
Microtubule-associated motor proteins (motor
  MAPs), 770–73
Microtubule-associated protein (MAPs),
  753–54, 770*t*
  motor, 753, 770–73
  nonmotor, 753, 754*t*
Microtubule-based movement, **770**–77
  cilia, flagella, and, 773–77
  dyneins and, 772
  human genetic disorders of, 778*b*
  intraflagellar transport, 777
  kinesins and, 771–72
  motor MAPs and axonal transport in, 770–71
  motor MAPs and vesicle transport, 773
Microtubule catastrophe, 749, 750*f*
Microtubule-organizing center (MTOC), **749**–50
  microtubule stability and, 751–53
  organization and polarization of microtubules
    by, 750–51
  origination of microtubules in, 749–50
Microtubule rescue, 749, 750*f*
Microvilli, 78, **759**
  intestinal mucosal cells, 78*f*, 759, 760*f*
  structure of, 759*f*
Middle lamella, **316**
Miescher, Johann Friedrich, 10, 480–81
Minisatellite DNA, 501
Minus end, microtubule, **748**, 749*f*
Mismatch repair of DNA, 543, **544**
Missense mutation, 676*b*, 677*f*
Mitchell, Peter, 426, 428
Mitchell, Robert, 263
Mitchison, Tim, 748
Mitochondrial DNA, 507–8, 509*f*, 510*f*
Mitochondrion (mitochondria), 79, **85**, 86*f*, **400**
  aerobic respiration and role of, 399*f*, 400–405
  apoptosis and, 284–85
  ATP synthesis and, 430–34
  DNA packaging in, 507–9, 510*f*
  electrochemical proton gradient in, 425–30
  endosymbiont theory on evolution of, 88,
    450–51*b*
  genetic code in, 635
  import of polypeptides into, 684–86
  inner membrane of, 402–4 (*see also* Inner
    membrane, mitochondrion)
  as interconnected networks, 401, 402*f*
  localization of metabolic functions within, 404*t*
  location of, and ATP needs, 401
  in muscle cells, 87*f*
  out membrane of, 402–4
  respiratory complexes of, 422–25
  in sperm cells, 86*f*
  structure of, 403*f*
  targeting of polypeptides into, 686–87
Mitogen, 271, 562
Mitogen-activated protein kinase (MAP kinase),
  271, 272*f*

Mitosis, 10, 80, **523**, 524*f*, 544–54, 577
  in animal cell versus plant cell, 546–47*f*
  disappearance of nucleolus during, 518
  drugs affecting, 753
  meiosis compared to, 587*f*
  microtubule orientation and, 752*f*
  mitotic spindle and chromosome movement
    during, 549–51, 552*f*
  phases of, 544–49
Mitosis-promoting factor (MPF), **557**
  evidence for existence of, 557*f*
  fluctuating levels of, 558*f*
  G2 regulation checkpoint and role of, 557
Mitotic Cdk-cyclin complex (MPF), 558–59, 589
  control of G2 checkpoint by, 558–59
  control of spindle assembly checkpoint by,
    561, 562*f*
  mitotic cell cycle and, 560*f*
Mitotic Cdks, 558
Mitotic cyclins, 558
Mitotic index, **525**
Mitotic spindle, 524, **545**
  assembly of, 549–50
  attachment of chromosomes to, 548*f*, 549–50
  chromosome movement during mitosis and,
    549–51
  microtubule polarity in, 549*f*
  spindle assembly checkpoint, 556, 561
  spindle equator (*see* Spindle equator)
Mixed-function oxidases, 331
M line, muscle, **780**, 781*f*
Models, 11
  of membrane structure, 160–66
Moesin (FERM) protein, 761
Molecular biology
  contribution of, to study of membrane
    proteins, 181, 182–83*b*
  genetic information flow from DNA to RNA
    as central dogma of, 623–24
Molecular chaperones, 17, **672**
  polypeptide (protein) folding facilitated by,
    31, 33–34, 50, 672, 673*b*, 681
Molecules. *See also* Macromolecule(s)
  diffusion rates of, in cells, 78
  electron configurations of select, 18*f*
  simple diffusion of small, nonpolar, 199–201
Monod, Jacques, 10, 148, 694, 698
Monofunctional proteins, 41
Monomer(s), **28**
  activated, 30
  amino acid, 41–44
  macromolecule synthesis by polymerization
    of, 29–31
  monosaccharides, 61–63
  most common, in cells, 42*t*
  nucleotide, 55–56
Monomeric protein, **45**, 234
Monooxygenases, 331
Monosaccharides, **61**–63
  structures of, 62*f*
Montgomery, Edward, 593
Morgan, Thomas Hunt, 10, 596, 597
Mos protein kinase, 589
Motifs, protein structural, **50**
Motility, 744, **760**–802
  actin-based, in muscle cells, 780–85
  actin-based, in nonmuscle cells, 792–98
  actin-based, myosins and, 777–79
  of chromosomes during mitosis, 551, 552*f*
  fibronectins and cell migration, 296–97, 298*f*
  filament-based, in muscles, 779–92
  intracellular microtubule-based (kinesin,
    dynein), 770–73
  microfilaments and, 754–55
  microtubule-based, 773–77
  systems of, 769–70
Motility proteins, 41

Motor end plate, 787
Motor MAPs, **753**, 770–73
  dyneins, 770*t*, 771*f*, 772
  kinesins, 770*t*, 771–72
  organelle movement during axonal transport
    and, 770–71
  relationship of microtubules, Golgi complex,
    and, 773*f*
  transport of intracellular vesicles and role
    of, 773
Motor molecule, 770*t*
Motor neurons, **226**, 787
  structure of typical, 227*f*
Motor proteins, **551**, 770
  chromosome movement and role of, 551, 552*f*
Movement. *see* Motility
Moyle, Jennifer, 428
MPF, **557**, 589
M (mitotic) phase of cell division, **523**, 524*f*,
  544–54
M (mitotic) phase promoting factor, **557**
mRNA. *See* Messenger RNA (mRNA)
mRNA-binding site, **661**, 668
*Msp*I restriction enzyme, 714
Mucopolysaccharides. *See* Glycosaminoglycans
  (GAGs)
Mucoproteins. *See* Proteoglycans
Mullis, Kary, 533
Multidrug resistance (MDR) transport
  protein, **210**
Multimeric protein, **45**, 234
Multipass protein, 177, 178–79
Multiple loops, introns and, 649*f*, 650
Multiple sclerosis, 242
Multiprotein complex, **54**, 422–23
Multisubunit proteins, 177
Murad, Ferid, 270
Muscle cell(s)
  actin-based motility in, 777–79
  calcium and function of, 787–88
  energy storage in, 789
  filament-based movement in, 779–92
  microfilaments in contractile fibers of (*see*
    Microfilament(s) (MF))
  mitochondria in, 87*f*
  stimulation of, by nerve impulse, 788*f*
Muscle contraction, 770, **779**–92
  ATP and energy needs of, 783–85, 788–89
  calcium and regulation of, 786–88
  cardiac, 789–90
  cross-bridges formation, 783–85
  cycle of, 783–84, 785*f*
  sarcomeres and, 780–82
  in skeletal muscles, 779–80
  sliding-filament model of, 782–83
  in smooth muscle, 790–92
Muscle fiber, **779**
  actin, myosin, and accessory proteins of,
    780–82
  cross-bridges formation, ATP, and contraction
    of, 783–85
  myofibrils and sarcomeres of, 779–82
  sliding-filament model and contraction of,
    782–83
Muscle fibrils, 744
Muscle-specific actins (α-actins), 756
Muscle tissue. *See also* Muscle fiber
  levels of organization in, 780*f*
  mechanical work and, 108, 109*f*
  rigor in, 784
Mutagens, 541, **628**–30
  DNA damage caused by, 541–42
Mutants
  cell cycle, 557–58
  temperture sensitive, 530, 558
Mutation(s), **541**, 676–77*b*
  cancer-causing, 569–70

DNA damage caused by, 541–42
  frameshift, 628–30, 631f, 676, 677f
  genetic variation due to, 577
  homeotic, 727f
  in *lac* operon, 695–98
  mismatch repair of damage to base pairs
    from, 543–44
  missense, 676, 677g
  nonsense, 674–75, 676, 677f
  oncogenes formed by, 565–66
  point, 565
  sickled red blood cells caused by gene,
    625–28, 629f
  silent, 676
  in structural and regulatory genes, 696–98
  study of cytoskeletal structures using, 744
Mycoplasmas, genetic code of, 635
Myc protein, 566
Myelin sheath, 163, 226, **227**
  around axons, 242f
  role of, in action potential transmission,
    241–42, 243f
Myoblast, 779
Myofibrils, **779**
  arrangement of thick and thin filaments
    in, 781f
  sarcomeres of, 779, 780–82
Myoglobin, **789**
Myokinase, 789
Myomesin, 781
Myosin(s), 553, **777**
  heavy and light chains of, 777, 779f
  human deafness and mutations in, 778b
  movement of, along actin filaments, 779
  in myofibril filaments, 780–82
  regulatory light chain of, 790
  role of, in cell motility, 777–79
  type II, and actin binding, 757f, 778
Myosin I, 759
Myosin II, 757f, 778, 791f
Myosin light-chain kinase (MLCK), **790**, 791f, 792
Myosin light-chain phosphatase, 792
Myosin subfragment I (S1), **756**
Myotonic dystrophy, 501
Myristic acid, 180

N

Na+/glucose symporter, **214**, 215f, 313
Na+/K+ pump, **230**. *See also* Sodium/potassium
  (Na+/K+) pump
N-acetylgalactosamine (GalNAc), 294
N-acetylglucosamine (GlcNAc), 29, 65, 66f, 98,
  133, 185, 187f, 294, 334, 336
N-acetylglucosamine transferase I, 334, 335f
N-acetylmuramic acid (MurNAc), 29, 65, 66f,
  98, 133
NAD+. *See* Nicotinamide adenine dinucleotide
  (NAD+)
NADH (nicotinamide adenine dinucleotide), 331
  electron transport, ATP synthesis and role
    of, 427
  pyruvate fermentation, NAD+ regeneration,
    and reoxidation of, 384–85
  TCA cycle and formation of, 406–9
NADH-coenzyme Q oxidoreductase, **423**
NADH dehydrogenase, 418, 419
NADH dehydrogenase complex, 423
NADPH (nicotinamide adenine dinucleotide
  phosphate), 331
  3-phosphoglycerate reduction, formation of
    G-3-P, and role of, 464
  photosynthesis and synthesis of, 455–60
NADPH-dependent malate dehydrogenase, 473
Nägeli, Karl, 3
Nanometer, 3b
Native conformation, protein, **50**
N-cadherin, 303

Nebulin, 782
Necrosis versus apoptosis, 282
Negative control of transcription, **700**
Negative cooperativity, 149
Negative resting membrane potential, 228
Negative staining, 7–8, 404
Negative supercoil (DNA), 489, 490f
Neher, Erwin, 233
Nernst, Walther, 230
Nernst equation, **230**
Nerve(s), **227**. *See also* Neuron(s)
Nerve cells. 220, 225. *See also* Neuron(s)
  microtubules of, 752f
  migrating, 796f
Nerve gases, 146, 250b
Nerve impulse, 227, **241**
  integration and processing of, 250–52
  myelination of axons and transmission of,
    241–43
  neurotoxins and disruption of, 247, 250b
  stimulation of muscle cell by, 788f
  transmission of, neuron adaptations for, 227–28
Nervous system, **225–28**
  neuron adaptations in, 227–28 (*see also*
    Neuron(s))
  in vertebrates, 226f
Nestin, 762, 763t
N-ethylmaleimide-sensitive fusion protein
  (NSF), **352**, 353f
Neural cell adhesion molecule (N-CAM), **303**
Neuraminidase, 306
Neurexin, 242
Neurites, 754
Neurofilament (NF) proteins, 762, 763t
Neuromuscular junction, **787**
  muscle contraction and, 787, 788f
Neuron(s), **225**, 227–28. *See also* Nerve cells
  action potential in axons of, 228
  electrical excitability of, 228
  integration and processing of nerve signals by,
    250–52
  interneurons, 226
  ion concentrations inside and outside, 228–30
  MAP2 and tau in, 754f
  membrane potential and, 228–32
  motor, 225, 227f
  nonmotor MAPs and neurite function in,
    753–54
  presynaptic, and postsynaptic, 243, 247–49
  sensory, 225, 226f
  shapes of, 228f
  size of, 226f
  structure of, and transmission of electrical
    signals, 227–28
  synaptic transmission between, 243–50
  voltage-gated ion channels and function of,
    234–36
Neuropeptides, **244–45**
Neurosecretory vesicles, **245–46**
  docking and fusion of, with plasma mem-
    brane, 246–47, 248f
  secretion of neurotransmitters and role of,
    245–46
Neurospora sp., 10
  experiments on gene coding of enzymes in,
    624–25
  gene conversion and meiosis in, 604f
Neurotoxins, 146, **247**
  poisons, snake venom, and nerve gases as,
    248–49, 250b
Neurotransmitter, **243**
  calcium and secretion of, at presynaptic
    neurons, 245–46, 247f
  chemical synapses and role of, 243–44, 245f
  detection of, by postsynaptic-neuron recep-
    tors, 247–49
  inactivation of, 249–50

neurosecretory vesicles and secretion of, 245,
  246–47, 248f
  structure and synthesis of, 246f
Neutral keratins, 762
Neutrophils, 344
Nexin, **776**
N-formylmethionine (fMet), 669
Niacin, 374
Nicolson, Garth, 164, 166
Nicotinamide adenine dinucleotide
  (NAD+), **374**
  pyruvate fermentation and regeneration of,
    384–85
Nicotinamide adenine dinucleotide phosphate
  (NADP+), **374, 456**
  photoreduction, NADPH synthesis, and role
    of, 455–58
  structure and oxidation/reduction of, 374f, 456f
Nidogen, 298
Niedergerke, Rolf, 782
Nirenberg, Marshall, 633
Nitric oxide (NO), blood vessel relaxation and,
  269–70
Nitrogen
  biospheric flow of, 111
  photosynthetic assimilation of, 468
Nitroglycerin, 270
N-linked glycosylation (N-glycosylation), 185,
  186f, **336**
Nocodazole, **753**
Nodes of Ranvier, **227**, 241, 242f
Nodules, plant, peroxisomes in, 361–62
Noller, Harry, 153
Nomarski microscopy, 6
Nomenclature
  enzyme, 135, 136t
  fatty acid, 68t
Nonappressed regions of thylakoids, 461
Noncellulosic matrix, 65
Noncompetitive inhibitors of enzymes, 146–47
Noncovalent bonds and interactions, 34, **46**
Noncyclic electron flow in oxygenic phototrophs,
  455f, **458**
Nondisjunction, **588**
Nonheme iron proteins, 419
Nonmotor MAPs, **753**, 754t
Nonmuscle actins (β- and γ-actins), 756
Nonmuscle cells, actin-based motility in, 792–98
Nonoverlapping genetic code, 631, 632f
Nonreceptor tyrosine kinase, **271**
Nonrepeated DNA, **500**
Nonsense-mediated decay, 654
Nonsense mutations, **674–75**, 676b, 677f
Nontranscribed spacers, 645
Norepinephrine, 244, 276, 280
Northern blot procedure, 737b
NO synthase, 269
N-terminus, **44**, 48, 666
Nuclear bodies, 518
Nuclear envelope, 78, 80f, **84**, 323, 510, 511f
  gene regulation and, 704
Nuclear export signal (NLS), **515**, 730
Nuclear lamina, 510, **516**, 765
Nuclear lamins, 762, 763t, 765
Nuclear localization signal (NLS), **514**, 515f, 683
Nuclear matrix, **515**, 516f
Nuclear membranes, 84
Nuclear pore, 510, 511f
  export of mRNA through, 729–30
  structure of, 512f
  transport of molecules through, 513–15
Nuclear pore complex (NPC), **512**
  posttranscriptional control and, 729–30
Nuclear receptor proteins, **724**
Nuclear run-on transcription, 716f
Nuclear transfer, cloning by, 708b
Nucleation, **747**

Nucleic acid(s), 26, **54**–60. *See also* DNA (deoxyribonucleic acid); RNA (ribonucleic acid)
 base composition (*see* Base composition of DNA)
 base pairing (*see* Base pairing; Base pairs (bp))
 DNA and RNA polymers of, 56–58
 DNA structure, 58, 59*f*, 60–61*b*
 hydrogen bonding in, 58*f*
 nucleotide monomers of, 55–56
 repeating units of, 29*t*
 sequences of (*see* DNA sequence(s))
 structure of, 56, 57*f*
 synthesis, 28*f*
Nucleic acid hybridization, **491**
Nucleic acid probe, **611**
Nucleoid, 78, **502**–3, 504*f*
Nucleolus (nucleoli), 78, 80*f*, **85**, **517**
 ribosome formation and, 517–18
Nucleolus organizer region (NOR), **518**
Nucleophilic substitution, **138**–39
Nucleoplasm, 79, 80*f*, **510**
Nucleoplasmins, 33
Nucleoside, **56**
Nucleoside monophosphate, **56**
Nucleosome(s), **505**
 chromatin and chromosomes formed from, 505–7, 508*f*
 evidence for, 506*f*
 histone octamer as core of, 506
 structure of, 507*f*
Nucleotide(s), 10, **55**–56
 bases of (*see* Base composition of DNA; Base pairing; Base pairs (bp))
 changes in genetic code caused by insertion of, into proteins, 631, 632*f*
 coding of amino acid sequence of polypeptide chain by, 625–28
 RNA and DNA structure and, 55, 56*t*
 sequence of (*see* DNA nucleotide sequences)
 structure of, 55*f*, 56
 triplets of, in genetic code (*see* Codon(s))
Nucleotide-binding folds of CFTR protein, 212*b*, 213*f*
Nucleotide excision repair (NER), **543**
Nucleus, 2, **84**, 85*f*, **510**–18
 chromatin fibers in, 516–17
 clones created by transplanting, 707
 division of, and cell division, 544–54
 general transcription factors in gene transcription, 643
 matrix and lamina of, 515–16
 matrix of, associated with active genome regions, 715
 nuclear envelope surrounding, 78, 80*f*, 84, 510–13
 nuclear pores and molecular transport in/out of, 512, 513–15
 nucleolus and ribosome function, 517–18
 prokaryotic versus eukaryotic, 78–79
 structure of, 511*f*
Nurse, Paul, 558

O

Obligate aerobes, **378**
Obligate anaerobes, **378**
Occludin, 311
Octamer, 718
Okazaki, Reiji, 532, 534*f*
Okazaki fragment, **532**, 534*f*, 535
Oleate, 69*f*, 169, 170*t*
Oligodendrocytes, 226, **241**
Oligomers, 747
Oligosaccharyl transferase, 336
O-linked glycosylation, 185, 186*f*
Oncogene, **565**
 categorization of, by protein products, 566*t*
 growth factor signaling pathway components coded by, 566–67
 identifying, 565
 proto-oncogenes, mutations, and development of, 565–66
 viruses and, 565, 626–27*b*
Oncogene transfection assay, 565
One gene-one enzyme hypothesis, 625
One-gene-one polypeptide theory, 628
Oocyte, 557, 588
 gene amplification in, 709
Open system, 112*f*
Operator (O), **694**
 lac operon and, 694–95
 mutations in, 697
Operator-constitutive mutants, 697
Operon(s), 666, **694**
 inducible, 695, 700–701
 lac, **694**, 695–98
 repressible, 699
 trp, **699**, 700–703, 722
Organelle(s), **5**, **78**, 83
 chloroplast, 87–88, 447–49
 coated vesicles in transport processes, 349–53
 compartmentalization of function and, 78, 79
 discovery of, 92*b*
 endoplasmic reticulum, 88–89, 323–33, 336–42
 exocytosis, endocytosis, transport across membranes, and role of, 342–49
 Golgi complex, 89, 333–42
 lysosomes, 90, 92*b*, 353–57
 mitochondria, 85–86, 88 (*see also* Mitochondrion (mitochondria))
 movement of, along microtubules, 770–71
 peroxisomes, 90–91, 358–62
 plant vacuole, 357–58
 posttranslational import of polypeptides into, 683–87
 protein glycosylation and role of, 336–38
 protein sorting and role of, 338–42
 ribosomes, 93–95
 secretory vesicles, 90
 semiautonomous, 450
 vacuoles, 91–93
Organic chemistry, **18**
Organic molecules, 26, 27*f*
Organisms. *See* Living organisms
Origin of replication, **528**, 613*f*
Origin of transfer, **600**
Origin recognition complex (ORC), 528
Osmolarity, **200**–201
Osmosis, **199**, 200–201*b*
 simple diffusion vs., 198*f*
Oubain, 201*b*
Outer doublet, axoneme, **776**
Outer membrane, chloroplast, **448**, 449*f*
Outer membrane, mitochondrion, 85, 86*f*, **402**, 403*f*
 functions occurring on, 404*t*
Ovalbumin chromatin, 712, 713*f*
Overton, Charles, 161, 200
Ovum. *See* Egg (ovum)
Oxaloacetate, 413
 transamination and formation of, 414, 415*f*
Oxalosuccinate, 407
Oxidation(s), **373**, 406
 β, 412, 413*f*
 of coenzymes and electron transport, 418
 coenzymes as electron acceptors in biological electron and proton removal in exergonic biological, 373–74
 of glucose in energy metabolism substrate, 375–80
 of glucose in glycolytic pathway, 378–83
 of glucose in presence of oxygen (*see* Aerobic respiration)
Oxidative deamination, 414
Oxidative decarboxylation, 406, 407*f*
Oxidative phosphorylation, 400, **426**, 429
Oxidizable substrates in energy metabolism, 375
 glucose as, 375–78
Oxidoreductases, 135, 136*t*
Oxygen
 availability of, and fate of pyruvate, 383, 384*f*
 biospheric flow of, 111
 electron transport system and flow of electrons from coenzymes to, 418
 photosynthesis and release of, 445
 transport of, across erythrocyte membrane, 198*f*
Oxygenase, rubisco as, 468–74
Oxygen-evolving complex (OEC), **457**
Oxygenic phototrophs, 445, **447**
 noncyclic electron flow in, 455*f*
 photophosphorylation (ATP synthesis) in, 459–61
 two photosystems of, 454–55

P

P_1 generation, **590**
p53 gene and protein, 560, **568**, 573
 cell apoptosis, cancer, and role of, 285, 568–69
p69 protein, 12
P150^{Glued}, 772
P680 chlorophyll, **455**
P700 chlorophyll, **455**
P960 chlorophyll, 462
Pachytene state of meiosis prophase I, **583**, 584*f*, 585*f*
Palade, George, 6, 326, 341
Palmitate, 69*f*, 169, 170*t*
Palmitic acid, 180
Pancreas
 endocrine and exocrine tissues of, 276
 proteolytic cleavage of protease enzymes of, 151*f*
Pantothenic acid, 406
Paracrine hormones, **276**
 chemical classification and function of, 278*t*
 histamine and prostaglandins as examples of, 280–82
Parallel β sheet, protein structure, 49
Paranodal regions, nodes of Ranvier, 242
Parathion, 250
Parathyroid hormone, 276
Partial DNA digestion, 612
Passive spread of depolarization, **240**
Passive transport, 160. *See also* Facilitated diffusion
 through nuclear pores, 514
Pasteur, Louis, 9
Patch clamping, **233**–34
Pathogens, motility in, 795–97
Pauling, Linus, 48, 49, 625, 626
Paxillin, 302
P-cadherin, 303
PDH kinase, 412
PDH phosphatase, 412
Pectins, 65, 315, **316**
Pellet, **328**
Pemphigus, 310
Penicillin, 12
 as irreversible enzyme inhibitor, 146
PEP. *See* Phosphoenolpyruvate (PEP)
PEP carboxykinase, 388
PEP carboxylase, 471–73
Pepsin, 137
Peptide bond, **44**
 formation of, in translation process, 670*f*, 671
Peptidoglycans, 98
Peptidyl transferase, **671**
Pericentrin, 750
Pericentriolar material, microtubules originating from, **749**, 751*f*

Perinuclear space, 323, **510**, 511*f*
Peripheral membrane protein, 164, **179**
Peripheral nervous system (PNS), **225**, 226*f*
Permeability barrier
  membranes as, 159
  plant cell walls as, 314–15
  tight junctions as, 312*f*
Permeases. *See* Carrier protein
Peroxidatic mode, 360
Peroxisome, 79, **90**–91, 93*f*, 323, **358**–62
  biogenesis of, 362
  catabolism of unusual substances by, 361
  discovery of, 358–59
  disorders linked to, 361
  hydrogen peroxide metabolism and functions
    of, 359–60
  lipid biosynthesis and role of, 91
  metabolism of nitrogen-containing com-
    pounds by, 361
  oxidation of fatty acids by, 360–61
  in plant cells, vs. in animal cells, 361–62
  separation of lysosomes from, 358*f*
Pertussis, 262
pH
  of cytosol, 357
  sensitivity of enzymes to, 137
Phage, **99**, 482. *See also* Bacteriophage (phage)
Phagocytes, **344**
Phagocytic vacuole, **345**
Phagocytosis, 88, **344**–45
  lysosomes and, 355–56
  process of, 344*f*
Phagosomes, 345
Phalloidin, 774
Phase-contrast microscopy, 6, 7*f*
Phenanthrene, 70
Phenobarbital, 331
Phenotype, 578*f*, **579**
  pseudo wild-type, 629
  wild-type, 596
Phenylalanine, 43*f*, 51, 633
Pheophytin (Ph), 456
Phlebitis, 282
Phorbol esters, 264
Phosphatases, 464, 558, 559
Phosphate groups, exergonic transfer of, 372*f*
Phosphate translocator, 465
Phosphatidic acid, **69**
Phosphatidylcholine (PC), 166, 169, 333
Phosphatidylethanolamine (PE), 24, 166, 168
Phosphatidylinositol, 166
Phosphatidylinositol-3-kinase, 272
Phosphatidylserine (PS), 166, 169
Phosphoanhydride bonds, **369**, 380
Phosphodiesterase, **262**
Phosphodiester bond, **56**
Phosphoenol bond, 381
Phosphoenolpyruvate (PEP), 388
  Hatch-Slack cycle and carboxylation of,
    471–73
  thermodynamic instability of, 383*f*
Phosphoester bond, 369, 380
Phosphofructokinase (PFK), 380, 412, 466
Phosphofructokinase-1 (PFK-1), 390
Phosphofructokinase-2 (PFK-2), **392**
  regulatory role of, 392*f*
3-Phosphoglycerate, 462, 470
  reduction of, to form glyceraldehyde-3 phos-
    phate, 464
Phosphoglucomutase, 331, 386
Phosphogluconate pathway, 386
Phosphoglycerides, 24, **69**–70, **166**
  structures of common, 69*f*
Phosphoglycerokinase, 464
Phosphoglycolate, **469**
Phospholipase C, activation of, 272, 279, 282
Phospholipase C₂, 180, **263**

Phospholipases, 163
Phospholipid(s), **69**–70, **166**
  endoplasmic reticulum and biosynthesis of,
    332–33
  in plasma membrane, 83, 166, 167*f*
  rotation and diffusion of, 169, 171*f*
Phospholipid bilayer, 24, 25*f*, 26*f*
  of plasma membrane, 83–84
Phospholipid exchange proteins, **333**
Phospholipid translocator (flippase), **171**, **332**
Phosphoprotein phosphatases, 150
Phosphorolytic cleavage of storage polysaccha-
  rides, 386, 388*f*
Phosphorylase a and b, 150
Phosphorylase kinase, 150, 280
Phosphorylase phosphatase, 150
Phosphorylation, 149, **150**
  of mitotic cdk-cyclin complex, 558, 559*f*
  of myosin II, 791*f*
  oxidative, 400, **426**, 429
  of *p53* gene, 568, 569*f*
  protein, and gene regulation, 725–26, 730
  of Rb protein, and regulation of G1 check-
    point, 559–60
  substrate-level (*see* Substrate-level phosphory-
    lation)
Photoautotrophs, **445**
Photochemical reduction, 452
Photoexcitation, 452
Photoheterotrophs, **445**
Photon, 452
  transfer of energy from, to photosystem
    reaction center, 453, 454*f*
Photophosphorylation, **447**
  ATP synthesis in oxygenic phototrophs as,
    459–61
Photoreduction, **447**, 455–58
Photorespiration, **469**–71
  in C₄ vs. C₃ plants, 471–73
  CAM plants and, 473–74
  role of leaf peroxisomes in, 91, 469
Photorespiration pathway, 361
Photosynthesis, **445**–78
  Calvin cycle and carbon assimilation, 445,
    462–66
  Calvin cycle and energy transduction, 466
  carbohydrate synthesis, 467–68
  chloroplast as sight of, 87, 447–49
  energy transduction in, 452–55
  evolution of chloroplasts and mitochondria,
    450–51*b*
  overview of, 445–47
  photophosphorylation (ATP synthesis),
    459–61
  photoreduction (NADPH synthesis), 455–58
  photosynthetic reaction center from purple
    bacterium, 461–62
  rubisco oxygenase activity and reduction of
    efficiency in, 468–74
Photosynthetic membranes, 449
Photosystem, **453**–54
Photosystem complex, **454**
Photosystem I (PSI) complex, 448, **455**, 460
  electron transfer from plastocyanin to ferre-
    doxin by, 457*f*, 458
Photosystem II (PSII) complex, **455**, 460
  electron transfer from water to plastoquinone
    by, 456–57
Phototrophs, **109**, 110, 368, **445**
  anoxygenic, 447
  noncyclic electron flow in oxygenic, 455*f*
  oxygenic, 445, 447
  photophosphorylation (ATP synthesis) in
    oxygenic, 459–61
  photoreduction (NADPH synthesis) in oxy-
    genic, 455–58
Phragmoplast, **553**

Phycobilins, **453**
Phycobilisome, **454**
Phylloquinone, 458
Physostigmine, 146, 250
Phytosterols, 166, 167*f*
Pigment, **452**
  accessory, 453
  antenna, 453
  chlorophyll, 452–53 (*see also* Chlorophyll)
Pinocytosis, 344
Plakin proteins, **300**, 765
Plakoglobin, 308, 310
Plant(s)
  absorption spectra of common pigments
    in, 452*f*
  C₃ compared to C₄, 471
  CAM, 473–74
  genetically engineered, using Ti plasmid,
    614–15
  glyoxylate cycle, glyoxysomes, and seed germi-
    nation in, 416–17*b*
  photosynthesis in (*see* Photosynthesis)
  transgenic, 614–16
  turgor pressure in, 201*b*
Plant cell(s), 314–19
  cytokinesis in, 547*f*, 553–54
  leaf peroxisomes in, 91, 93*f*
  membranes of, 158*f*
  mitosis in, 546–47*f*
  osmolarity changes in, 201*f*
  peroxisomes in, 360*f*, 361–62
  starch in, 29
  structure of typical, 81*f*
  surface of, 314–19
  vacuoles in, 93, 94*f*, 357–58
Plant cell wall, 35, 65, 97–98, **314**–19
  as permeability barrier, 314–15
  plasmodesmata and cell-cell communication
    through, 98, 318–19
  structure of, 98*f*, 315–16, 317*f*
  synthesis of, 316–18
Plant diseases
  cadang-cadang, 99
  tobacco mosaic virus, 35, 36*f*, 98
  viral and viroid, 98
Plaque, **300**, 301*f*, 307, **609**
  bacteriophage, 484*b*, 485*f*
Plasma, blood, 273
Plasma fibronectin, 297
Plasma (cell) membrane, **83**–84, **159**
  clathrin-coated vesicles and, 350–51
  composition of, in rat liver cells, 333*t*
  endoplasmic reticulum and biosynthesis of,
    332–33
  glycoproteins in, 84
  G protein-linked receptors in, 259–70, 567
  linkage of actin to, 760–62
  lipids in, 166–75
  membrane potential in, 228–32
  membrane proteins of (*see* Membrane
    protein(s))
  models of, 160–66
  neurosecretory vesicle fusion with, and secre-
    tion of neurotransmitters, 246–47, 248*f*
  organization of, 84*f*
  phospholipid composition of select, 168*f*
  protein, lipid, and carbohydrate content of, 163*t*
  protein kinase-associated receptors in, 270–72
  proteins of, 84, 175–89
  transport across (*see* Transport across mem-
    branes)
Plasma membrane receptors
  apoptosis and, 282–85
  G protein-linked, 259–70
  for growth factors, 272–75
  for hormones, 276–82
  protein kinase-associated, 270–72

Plasmid(s), **504**
  as cloning vector, 608, 609*f*, 610*f*, 614–15
  recombinant, 610
  Ti, 614–15
Plasmodesma, **98**, 160, 184, 306, **318**–19
  structure of, 318*f*
Plasmolysis, 201*b*
Plastid, **88**, **448**
Plastocyanin (PC), **458**
  electron transfer from, to ferredoxin, 457*f*, 458
  electron transfer from plastoquinol to, 457–58
Plastoquinol, **456**
  electron transfer from, to plastocyanin, 457–58
Plastoquinone, 71
  photosystem and transfer of electrons to, **456**, 457*f*
Platelet. *See* Blood platelet
Platelet-derived growth factor (PDGF), 273, **562**, 563
  regulation of fibroblasts by, 797
Pleated β sheets, 50
Plectin, 300, 765
Plus end, microtubule, **748**, 749*f*
Poikilotherms, 174–75
Point mutation, 565
Poisons
  enzyme inhibitors as, 146
  neurotoxins as, 146, 247, 250*b*
Polar body, **588**
Polarity
  in Golgi stacks, 334
  of microtubules, 97
  of solutes, 200–201, 202*f*
  of water molecules, **22**
Polarized secretion, **343**
Polar microtubules, **545**, 547*f*, 551, 552*f*
Polar pores, 162
Polyacrylamide gels, 493
Poly(A) polymerase, 648
Poly(A) tail, 644, **648**
  addition of, to pre-mRNA, 648, 649*f*
Polycyclic hydrocarbons, 331
Polygenic mRNa, **695**
Polymer(s), 17
  nucleic acid (DNA, RNA), 56–58
  polypeptides and proteins, 44–45
  polysaccharides, 63–65
Polymerase chain reaction (PCR), 503, **530**, 533–34*b*
Polynucleotide, **56**
Polynucleotide phosphorylase, 633
Polypeptide(s), **31**, **45**. *See also* Protein(s); Protein synthesis
  chain elongation in synthesis of, 666, 669–71
  cotranslational import of, into ER, 678–83
  energy budget for synthesis of, 672–74
  gene coding for amino acid sequence of, 625–28, 629*f*
  initiation of synthesis of, 667–69
  insulin A subunit and B subunit, 48
  light and heavy chains of, 710
  messenger RNA and synthesis of chains of, 632–33, 666
  molecular chaperones and folding of, 31, 33–34, 672, 681
  posttranslational import of, into organelles, 678, 679*f*, 683–87
  ribosomes and synthesis of, 660–62
  role of protein factors in synthesis of, 666
  termination of synthesis of, 671–72
Polyphosphoinositides, **758**
Polyribosome, **674**
Polysaccharide(s), 26, **60**–65. *See also* Carbohydrate(s)
  monosaccharide monomers of, 61–63
  storage and structural polymers, 28–29,

63–65, 66*f*
  structure of, and glycosidic bonds, 65
  as substrates for glycolysis, 386, 388*f*
  subunits, 66*f*
  synthesis, 28*f*
Polyspermy, blocks to, 266–68
Polytene chromosome, **711**
  puffs in, 712*f*
  transcriptional activity of, 712*f*
Poly(U), 633
P/O ratio **427**
Pore, membrane, 203
Pore protein, 681
Porins, 178, 203, **206**, **402**, **448**
Porphyrin, 132, 452
Porter, Keith, 6, 349
Positive control of transcription, **700**
Positive cooperativity, 149
Positive supercoil (DNA), 489, 490*f*
Postsynaptic neuron, **243**, 244*f*
  neurotransmitter detection by receptors on, 247–49
Postsynaptic potentials (PSPs), 250–52
Posttranscriptional control, eukaryotic genes, 728–34, 737*b*
  initiation factors and translational repressors in, 730–31
  of mRNA half-life, 731–32
  protein degradation by proteasomes and, 733–34
  RNA processing and nuclear export in, 728–30
Posttranscriptional gene silencing, 731
Posttranslational control in eukaryotic genes, **732**–33, 737*b*
Posttranslational import, **678**, 679*f*, 683–87
  model for, 687*f*
  polypeptide import into mitochondria and chloroplasts, 684–86
  polypeptide targeting and, 686–87
Posttranslational processing, 675–76, 678*f*
Potassium channels, voltage-gated, 234, 237–38
Potassium ion gradient, 229
Potential (voltage), **229**
Preinitiation complex, 643
Pre-mRNA, **648**
  addition of 5′ cap and poly(A) tail to, 648–49
  removal of introns from, by spliceosomes, 650–51
Prenylated membrane proteins, **180**
Prenyl group, 180
Preproteins, 680
Pre-rRNA, 645–46
  methylation of, 646
Presynaptic neuron, **243**, 244*f*
  active zone, 247
  secretion of neurotransmitters at, 245–47
Pre-tRNA, 646, 647*f*
Pribnow box (TATAAT sequence), 637
Primary cell wall, 35, 98, **316**, 317*f*
Primary messenger, 257
Primary structure of protein, 47*f*, **48**
Primary transcript, **644**, 645*f*. *See also* RNA processing
Primase, **535**
Primosome, **536**
Prions, **99**, **673**
  disease caused by, 99, 101, 673*b*
Probe, DNA, **491**
Procaspase-9, 285
Procaspases, **284**
Processes, neuron, **227**
Processivity, 771
Procollagen, **292**–93
Procollagen peptidase, 292
Profilin, 758
Proflavin, 628–29

Programmed cell death (apoptosis), 282–85
Prokaryote(s), **76**
  eukaryotes vs., 78–82
  genome size, complexity, and organization, 492
Prokaryotes, gene regulation in, 692–703
  adaptive enzyme synthesis and, 692–94
  example of lactose catabolism and, 694–98
  example of tryptophan synthesis and, 693, 698–700
  gene regulation in eukaryotes versus, 704–6
  inducible operons and, 694
  repressible operons and, 698–700
  transcription control and, 700–703
Prokaryotic cell(s), **76**. *See also* Bacteria; Prokaryote(s)
  cell fission in, 80
  DNA organization in, 80
  DNA packaging in, 501–4
  DNA replication in, 528*f*, 538*f*
  eukaryotes vs., 78–82
  gene expression in (*see* Gene expression)
  messenger RNA in, 666*f*
  nucleoid in, 78, 79*f*, 502–4
  photosynthetic membranes in, 449*f*
  plasmids of, 504
  respiratory functions in, 404–5
  transcription in, 635–40
  translation in, 667, 668*f*, 669, 670*f*
Proline, 43*f*
Prometaphase (mitosis), **545**, 546*f*
Promoter(s), 635, **694**
  binding of RNA polymerase to prokaryotic, 635–36
  core, 641
  enhancers, silencers, and, 641, 718–19, 720*f*
  eukaryotic, 640, 641, 642*f*, 643
  in lactose metabolism, 694
  mutations in, 697
  proximal control elements and, 641, 717–18
  typical eukaryotic, 641, 642*f*, 643
  typical prokaryotic, 637*f*
Promoter site, **637**
  binding of RNA polymerase II to TATA-containing, 643*f*
Proofreading, DNA replication and mechanism of, **535**
Propagated action potential, 241. *See also* Nerve impulse
Propagation (action potentials), **236**
  saltatory, 241–42
Prophage, 485*b*, 486*f*
Prophase (meiosis), 582–85, 587*f*
Prophase (mitosis), **545**, 546*f*, 587*f*
Propionate fermentation, 385
Proplastids, **448**
Propranolol, 258
Prostaglandins, 276, **278**, 280–82
  blood platelet activation by, 281–82, 283*f*
Prosthetic groups, **134**
Proteases, 413
Proteasome, **734**
  ubiquitin-dependent protein degradation by, 733–34
Protein(s), 26, **41**–54
  actin-related, 756
  amino acid monomers of, 41–44
  amino acid sequence and interactions in, 47–54
  chimeric, 338
  degradation of, by proteasomes, 733–34
  detecting and quantifying molecules of, 735–36*b*
  DNA replication, 532*t*
  DNA sites for binding of, 637, 638*b*
  as energy (ATP) source, 413–14
  enzymes as, 41, 132–33 (*see also* Enzyme(s))
  flow of, through Golgi complex, 335

folding and stability of, linked to bonds and
    interactions, 45–47
folding of, and molecular chaperones, 672, 681
folding of, faulty, as cause of disease, 673*b*
G (*see* G protein)
glycosylation of, 89
heat-shock, 33, 672
high-mobility group (HMG), 715
histones (*see* Histone(s))
identification of DNA using, 611–12
in membranes (*see* Membrane protein(s))
motor, 551–52, 770
muscle filament, 781–82
phosphorylation of, 725–26
in plasma membrane, 84
polypeptides and, 44–45
posttranslational modification of, 675–76,
    678*f*, 705
repeating units of, 29*t*
self-assembly of, 17, 33–34
splicing, 676, 678*f*
structure of, 47–54
synthesis (*see* Protein synthesis)
targeting and sorting (*see* Protein targeting
    and sorting)
transport of, through nuclear pores, 514, 515*f*
turnover (degradation) of, 705–6
unwinding of DNA by single-strand binding,
    536–37
Protein disulfide isomerase, 330, **681**
Protein factors, translation and effect of, on
    polypeptide chains, 660, 666
Protein kinase, 150, 558
    Janus activated, 275, 726
    MAPs (*see* Microtubule-associated proteins
        (MAPs))
    oncogenes and, 567
    regulation of, 791–92
Protein kinase A (PKA), **262**
    activation of, by cyclic AMP, 262*f*, 726
Protein kinase-associated receptors, **270–72**
Protein kinase C (PKC), **264**
Protein kinase G, 270
Protein/lipid ratio in membranes, 163
Protein microarrays, 499
Protein phosphorylation, 725–26, 730
Protein-protein interactions in eukaryotic tran-
    scription, 640, 643
Protein sorting. *See* Protein targeting and sorting
Protein splicing, **676**, 678*f*
Protein synthesis, 10, 28*f*, 45
    diseases caused by faulty protein folding, 673*b*
    energy budget for, 672–74
    messenger RNA and polypeptide chain syn-
        thesis, 632–33, 666, 669–71
    molecular chaperones and, 33–34, 672, 681
    noncovalent interactions in, 34, 46
    posttranslational protein processing,
        675–76, 678*f*
    protein factors and, 660, 666
    protein targeting and sorting following,
        676–87
    ribosomes as sight of, 33, 660–62
    rough endoplasmic reticulum and, 325, 330
    transfer RNA and, 662–66
    translation mechanisms in, 666–74
Protein targeting and sorting, 676–87
    cotranslational import, 678–83
    posttranslational import, 678, 679*f*, 683–87
Proteoglycans, 83, 290, **294**
    in cartilage, 296*f*
    cells anchored to extracellular matrix by
        adhesive glycoproteins and, 295–96
    collagen and elastin fibers embedded in, 294
Proteolysis, **413**
Proteolytic cleavage, **151**
Proteome, **498–99**

Protoeukaryotes, 88, **450**
Protofibril, 51
Protofilaments, 96, 97, **746**, 764
Proton(s)
    ATP synthesis and transport of, in
        thylakoids, 459
    electron transport and pumping of, from
        mitochondrial matrix, 428
Proton gradient, 429–30
Proton motive force (pmf), **429**
Proton pump
    ATP-dependent, 348
    bacteriorhodopsin, 215–18
    electron transport and, 428
Proton translocator, **459**
Proto-oncogene, **565**
Protrusion, cell, 792–93
    regulation of, by G proteins, 797, 798*f*
Provacuole, **357**
Provirus, 626*b*
Proximal control element, 641, **718**
    location of, near promoter, 717–18
P-selectin, 306
Pseudopodia, **795**, 796*f*
Pseudosubstrate, 792
Pseudo wild-type, 629
P (peptidyl) site, **661–62**
P-type ATPase, **208**, 209*t*
pUC19 plasmid vector, 608, 610*f*
Pumps, 160
    active transport carried out by (*see* Active
        transport)
    bacteriorhodopsin proton, 215–18
    sodium/potassium, 211–13, 214*f*
    transport ATPases as, 208
Punnet square, 592
Purine, 55, 57–58. *See also* Base pairing
Puromycin, 680
Purple bacteria, 216
    evolution of mitochondria from, 450
    photosynthetic reaction center from, 461, 462*f*
Purple membrane, 216
Pyrimidine, 55, 57–58. *See also* Base pairing
Pyrimidine dimer formation, 542
Pyruvate
    conversion of, to acetyl coenzyme A, 406, 407*f*
    fermentation of, and NAD+ regeneration,
        384–85
    formation of, and ATP generation, 378, 379*f*,
        381–83
    oxygen availability and fate of, 383, 384*f*
    transamination and formation of, 414, 415*f*
Pyruvate carboxylase, 388
Pyruvate decarboxylase, 385
Pyruvate dehydrogenase (PDH), 406, 411*f*, 412
Pyruvate dehydrogenase complex, 54
Pyruvate kinase, 383, 390, 412

Q
Quantum, **452**
Quaternary structure of protein, 47*f*, **53–54**

R
Rab4 and Rab5 proteins, 352
Rab GTPase, **352**, 353*f*
Racker, Efraim, 426, 430
Rac protein, 757, 797, 798*f*
Radding, Charles, 604
Radial spokes, axoneme, **776**, 777
    body axis development and role of, 778*b*
Radiation as cancer treatment, 571
Radioactive labeling, 185
Radioisotopes, 9, 482
Radixin protein, 761
Raf protein kinase, 563
Ran protein, 515
Ras gene and protein, 259, **271**, 272*f*, 569, 757

Ras pathway, 271, 272*f*
    growth factor and activation of, 562–63,
        564*f*, 567
Rb gene, cancer and role of, 567–68
Rb protein, **560**
    development of cancer and, 560
    role of, in cell cycle control, 559–60, 561*f*
Reactants, cell size and need for adequate
    concentrations of, 78
Reaction center, photosystem, **453–58**
    from purple bacterium, 461, 462*f*
Reading frame, 629
RecA enzyme, homologous recombination
    and, **606**
Receptor(s), **160**
    affinity of, 258
    apoptosis and, 282–85
    G protein-linked, 259–70
    for growth factors, 272–75, 567
    for hormones, 276–82
    ligand binding to, 257–58
    of neurotransmitters, 243–44, 247–49
    in plasma membrane, 25, 84, 184
    protein kinase-associated, 270–72
    ribosome, 681
    signal transduction and, 257, 258–59
    SRP, 681
Receptor affinity, **258**
Receptor down-regulation, 258
Receptor glycoproteins, 298–302
Receptor-mediated (clathrin-dependent) endo-
    cytosis (RME), 258, 344, **345–48**
    LDL receptor, cholesterol, and, 346–47*b*
    lysosomes and, 355–56
    process of, 345*f*
Receptor tyrosine kinases, **270–71**
    activation of, 270*f*, 271
    as growth-factor receptors, effect on embry-
        onic development, 273–74
    signaling pathways activated by, 272
    signal transduction cascade initiated by,
        271, 272*f*
    structure of, 270*f*, 271
Recessive allele, **578**
Recessive traits, 590–92
Recognition helix, **722**
Recombinant DNA molecule, **607–8**
    generation of, using restriction enzymes, 607*f*
Recombinant DNA technology, 182, 499,
    606–14. *See also* Genetic engineering
    cloning of large DNA segments in cosmids
        and YACs, 613–14
    DNA cloning techniques and, 608–12
    genomic and cDNA libraries used in, 612–13
    promoter sequences and, 637
    restriction enzymes and, 607–8
Recombination. *See* Genetic recombination
Recombination nodules, 585
Red blood cells. *See* Erythrocyte(s) (red blood
    cells)
Redman, Colvin, 680
Redox (reduction-oxidation) pair, **420**
    standard reduction potentials for select, 421*t*
Reductase, 470
Reduction, **374**
Reduction potential (E′), **420–22**
    standard, 420–21
    understanding, 420
Refractory periods after action potentials,
    239–40
Regeneration, reproduction by, 577
Regulated gene, **692**
Regulated secretion, 341, **342**
Regulation. *See* Cell function, regulation of;
    Enzyme regulation; Gene regulation
    of Calvin cycle, 465
    of glycolysis and glyconeogenesis, 389

Regulation, *continued*
  of microtubules, 753–54
Regulatory domain of CFTR protein, 212*b*, 213*f*
Regulatory genes, **694**
  mutations in, 676, 696–98
  repressor as product of, 694–95
Regulatory light chain, myosin, **790**
Regulatory proteins, 41
Regulatory subunit, **149**
Regulatory transcription factors, **718**
  activators, 719
  coactivator mediation of interaction between
    RNA polymerase complex and, 719–21
  repressors, 719
  structural motifs allowing binding of DNA to,
    721–24
Relative refractory period, **240**
Release factors, 671, **672**
Renaturation
  of DNA, 490–91, 499
  of polypeptides, **31**
Repair endonucleases, 543
Repeated DNA, 499, **500**, 501
  interspersed, 501
  tandemly, 500–501
Replication bubble, 528, 529*f*
Replication fork, **527**, 528*f*
  direction of DNA synthesis at, 532, 535*f*
Replicon, **528**–30
Replisome, **537**
Repressible operon, 698, **699**
Repressor(s), **694**
  as allosteric protein, 695
  gene expression control and role of,
    694–95, 719
  lac, in bacteria, 694–95, 697*f*
  as transcription factor, **719**
Residual body, **356**
Resolving power, microscope, **5**
  comparison of, 8*f*
  electron vs. light microscopes, 6–7
  human eye vs. light and electron micro-
    scope, 6*f*
Resonance energy transfer, **452**
Resonance hybrid, **370**
Resonance stabilization, **370**
  decreased, of phosphate and carboxyl
    groups, 371*f*
Respiration. *See* Aerobic respiration
Respiratory complexes, 418, 422, **423**, 424–25, 457
  energetics of, 423*f*
  flow of electrons through, 425*f*
  free movement of, 424–25
  properties of, 423, 424*t*
  role of cytochrome *c* oxidase, 424
Respiratory control, **426**
Response elements, 705, **724**
  heat-shock, 726–27
  hormone, 724–25
  iron, 730, 731*f*
Resting membrane potential (V_m), **228**–32
  effect of ion concentrations on, 228–30
  effect of ions trapped in cell on, 230
  Goldman equation and, 231–32
  Nernst equation and, 230
  steady-state ion concentrations affecting,
    230–31, 232*f*
Restriction enzyme(s), **492**–96
  base sequencing and, 496–97
  cleavage of DNA molecules at specific sites
    by, 494*f*
  closer look at, 494*b*
  common, and their recognition sequences, 495*t*
  detection of introns using, 650*f*
  gel electrophoresis of DNA and, 493*f*, 494
  recombinant DNA technology made possible
    by, 607–8

restriction mapping using, 495–96
Restriction fragment(s), 493, 494–95*b*
  mapping of, 495, 496*f*
  separation of, by gel electrophoresis, 493–94
  sticky ends (cohesive ends), 494*b*
Restriction fragment length polymorphisms
  (RFLPs), **502**
  DNA fingerprinting by analysis of, 502–3*b*
Restriction mapping, 495–96
  introns and, 650
Restriction/methylation system, **494***b*
Restriction point, 555. *See also* G1 checkpoint of
  cell cycle
Restriction site, **493**, 607
  as palindrome, 494*b*
Retinal, 164
  bacteriorhodopsin proton pump and, 216, 217
Retinitis pigmentosum, 778*b*
Retinoids as chemical messengers, 257
Retrieval tages, ER-specific proteins containing,
  338, 339*f*
Retrograde axonal transport, 771
Retrograde flow, **792**
Retrograde transport, **335**
Retrotransposons, 627*b*
Retroviruses, 616–27*b*
  reproductive cycle of, 626*f*
Reverse transcriptase, 612, **626***b*
Reverse transcription, 624, 626–27*b*
Reversible (enzyme) inhibitor, **146**
RGD sequence, 296
Rhamnogalacturonans, 316
Rheumatoid arthritis, 356
Rhodamine, 187
Rho (ρ) factor, **639**, 640, 797, 798*f*
Rho protein, 757
Riboflavin, 409
Ribonuclease
  denaturation and renaturation of, 31, 32*f*
  synthesis and self-assembly of, 32*f*, 33
  tertiary structure of, 51, 52*f*
Ribonuclease P, cleavage of pre-transfer RNAs
  by, 152
Ribose, **54**
Ribosomal RNA (rRNA), 54, 95, 450, 517, **623**
  cleavage of multiple, from common precur-
    sor, 644–46
  gene amplification in, 709
Ribosome(s), 35, **93**–95, 513, **660**
  binding sites on, 152–53, 661, 662*f*, 668
  free, 678
  nucleolus and formation of, 517–18, 644
  prokaryotic, 662*f*
  properties of prokaryotic and eukaryotic, 662*t*
  RNA components of cytoplasmic, 645*t*
  as site of protein synthesis, 660–62 (*see also*
    Protein synthesis)
  structure of, 94*f*
  subunits, large and small, 94, 660, 661*f*
Ribosome-binding site (Shine-Dalgarno
  sequence), 668
Ribosome receptor, 681
Ribozyme(s), 133, **151**–53, 651–52, 671, 674
Ribulose-1,5-bisphosphate carboxylase/oxyge-
  nase (rubisco)
  photosynthetic carbon assimilation and role
    of, 462, 463*f*, **464**, 465
  regeneration of, 464–65
  role of, in C_4 plant photorespiration, 471–73
Ribulose-5-phosphate kinase, 464
Rigor, muscle, **784**
Riordan, John, 212*b*
R looping, 649
RNA (ribonucleic acid), **54**
  base pairing in, 59
  bases, nucleotides, and nucleosides of, 56*t*
  cleavage of, 640, 644

flow of genetic information from DNA to,
  623–24
as intermediates in flow of genetic information, 624*f*
messenger (*see* Messenger RNA (mRNA))
processing of (*see* RNA processing)
structure of, 57*f*
synthesis of, 479, 517*f*, 638–40, 644
transcription of genetic information from
  DNA to (*see* Transcription)
transfer (*see* Transfer RNA (tRNA))
RNA editing, **653**–54
RNA interference, **731**
RNA polymerase(s), **635**
  binding of, to promoter sequence, 635–38
  catalysis of transcription by, 635, 636*f*
  coactivator mediation of interaction between
    regulatory transcription factors and,
    719–21
  general transcription factors and binding of,
    640, 643–44
  initiation of RNA synthesis and role of,
    638–39
  promoters for eukaryotic types of, 641–43
  properties of eukaryotic, 640, 641*t*
  termination of transcription and, 644
RNA polymerase I, **640**, 641*t*, 642, 644
RNA polymerase II, **641**, 642, 717
  binding of, to TATA-containing promoter
    site, 643*f*
RNA polymerase III, **641**, 642
RNA primers, **535**
  DNA replication initiated by, 535, 536*f*
RNA processing, 640, **644**–54
  cleavage of ribosomal RNA, 644–46
  coding sequence of mRNA altered by editing
    and, 653–54
  eukaryotic rRNA genes and, 645*f*
  intron removal from pre-mRNA by spliceo-
    somes, 650–51
  nucleotide modifications of transfer RNA,
    646–47
  PolyA capping and intron removal in mRNA,
    647–51
  posttranscriptional control of, 728–30
  protection against defective mRNA and, 654
  purpose of introns in eukaryotic genes and,
    652–53
  RNA editing, 653–54
  RNA proofreading and surveillance, 654
  self-splicing introns and, 651–52
RNA proofreading, **654**
RNA-protein (ribonucleoprotein), 514
RNA splicing, **651**, 676
  alternative, 652–53, 728–29
  overview of, 651*f*
RNA strand, elongation of, 639*f*
RNA surveillance, **654**
RNA tumor viruses, 626–27*b*
RNP particles, **651**
Robertson, J. David, 162–63
Rosettes, cellulose-synthesizing enzymes, 316, 317*f*
Rotation of phospholipid molecules, **169**, 171*f*
Roth, Thomas, 349
Rough endoplasmic reticulum (rough ER), **89**,
  324, 325*f*
  role of, in protein biosynthesis and process-
    ing, 325, 330, 680
Rous sarcoma virus, 565, 626–27*b*
Roux, Wilhelm, 10
Rubisco. *See* Ribulose-1,5-bisphosphate carboxy-
  lase/oxygenase (rubisco)
Ryanodine receptor channel, **265**

S
Sabatini, David, 680
Sackman, Bert, 233

Saltatory propagation, 241–42
Sanger, Frederick, 48, 496
Sarcolema, **787**
Sarcomas, 627*b*
Sarcomeres, **779**
    actin, myosin, and accessory proteins of, 780–82
    structural proteins of, 782*f*
    thin and thick filaments of, 779, 780*f*
Sarcoplasm, **786**
Sarcoplasmic reticulum (SR), 332, **787**
    muscle contraction and, 787–88
Sarin, 250
SarI protein, 352
Satellite DNA, 500, 501
Saturated fatty acid, **68**, 69*f*
Saturation, **140**
Saturation kinetics, 204
Scanning electron microscope (SEM), **7**, 9*f*
Schiff base, 217
Schimper, Andreas, F. W., 450
Schleiden, Matthias, 2
Schwann, Theodor, 2–3
Schwann cells, 226, **241**, 242*f*
Scientific method, **11**
    biology, "facts," and, 11–12*b*
Scrapie, 99, 673
SDS-polyacrylamide gel electrophoresis of membrane proteins, 180, 181*f*
Sea urchins, 593
    role of calcium in blocking polyspermy in, 266, 268*f*
Sec13/31 protein complex, **352**
Sec23/24 protein complex, **352**
Secondary alcohol, 407
Secondary cell wall, 98, **317–18**
Secondary structure of protein, 47*f*, **48–50**
Second law of thermodynamics, 114, **115**
Second messengers, 160, **257**
    calcium as, 343
    cyclic AMP as, 261–62
    cyclic GMP as, 269–70
    inositol trisphosphate and diacylglycerol as, 263–64
Secretory granules, **341**, 342
Secretory pathways, **341–42**
Secretory proteins, 89
Secretory vesicles, **90**, 341
Securin protein, 561
Sedimentation coefficient, **94**, 327
Sedimentation rate, centrifugation, **326**
Sedoheptulose bisphosphatase, 465
Seed germination, glyoxylate cycle, glyoxysomes, and, 416–17*b*
Segregation, law of. *See* Law of segregation
Selectable marker, **610**
Selectins, 302, **306**
Selectively permeable membranes, 17, 24–26. *See also* Transport across membranes
Selenocysteine, 635
Self-assembly of molecules, 17, **31–37**
    assisted, 33
    hierarchical, 36–37
    limits of, 35–36
    noncovalent interactions in, 34
    of proteins, 31–33
    role of molecular chaperones in protein, 31, 33–34, 672, 681
    tobacco mosaic virus as case study of, 35, 36*f*
Self-splicing RNA introns, 651–52
Semiautonomous organelles, **450**
Semiconservative replication of DNA, **525–27**
Semiquinone (CoQH), 419
Sensory neurons, **225**, 226*f*
Separin protein, 561
Serine, 43*f*, 186*f*
Serine-lysine-leucine (SKL), 362, 683

Serine/threonine kinase receptor, **274–75**
Serotonin, 244
Serum, blood, 273
70S initiation complex, **669**
Severe combined immunodeficiency disease (SCID), 616
Sex chromosome(s), **578**
Sex hormones, 71, 276
Sex pili, **600**
Sexual reproduction, **577–622**
    allele segregation and assortment in, 590–95
    diploid and haploid phases in sexual organisms, 580–81
    diploid state as feature of, 578–79
    gametes produced by, 579
    genetic engineering and, 614–18
    genetic recombination and crossing over in, 595–98
    genetic recombination in bacteria and viruses, 598–602
    genetic variety generated by, 577–78
    homozygosity and heterozygosity in, 578–79
    meiosis and, 580–90
    molecular mechanisms of homologous recombination in, 602–6
    recombinant DNA technology, gene cloning, and, 606–14
SH2 domain, **271**
Shadowing, 8
Shigella flexnerii, 308*b*
Shine-Dalgarno sequence (ribosome-binding site), 668
Shuttle streaming, 795
Shuttle vesicles, **335**
Sialic acid, 185, 187*f*, 304
Sickle-cell hemoglobin and anemia, 53
    gene mutation causing, 625–28
    peptide patterns of normal hemoglobin and, 629*f*
Sidearm, axoneme, **776**
Sigma (σ) factor, **635**
    regulation of transcription initiation by, 701
Signal hypothesis, 680
Signal peptidase, 681
Signal-recognition particle (SRP), 680, **681**, 682*f*
Signal sequence, 675
Signal transduction, 225–89
    action potential and, 236–43
    by adrenergic receptors, 279*f*
    apoptosis and, 282–85
    cotranslational import and, 680–81, 682*f*
    defined, **160**, **257**
    electrical excitability and, 233–36
    endocrine and paracrine hormone systems and, 276–82
    G protein-linked receptors, 259–70
    growth factors as messengers in, 272–75
    integration and processing of nerve signals, 250–52
    integrins and, 300–302
    membrane potential and, 228–32
    nerve impulse, 227
    nervous system and, 225–28
    overview of chemical signals and receptors in, 256–59
    posttranslational import and, 684–86
    protein kinase-associated receptors, 270–72
    receptor binding and initiation of, 258–59
    receptor tyrosine kinases as initiators of, 271, 272*f*
    release of calcium ions in, 264–69
    role of G proteins and cyclic AMP in, 261*f*
    role of InsP$_3$ and DAG in, 264*f*
    synaptic transmission, 243–50
Sildenafil, 270
Silencer (DNA sequence), **718**
Silent information regulator (SIR) genes, 710

Silent mutation, 676*b*, 677*f*
Simple diffusion, 160, 196, **197–203**
    comparison of active transport, facilitated diffusion, and, 202*t*
    concentration gradient and rate of, 201–3
    in erythrocyte, 197*f*
    as limited to small, nonpolar molecules, 199–201
    movement toward equilibrium in, 198–99
    osmosis and, 198*f*, 199, 200–201*b*
Simple-sequence repeated DNA, 500
Singer, S. Jonathan, 164, 166, 203
Single bond, **18**
Single-channel recording, 233
Single nucleotide polymorphisms (SNPs), **499**
Singlepass protein, 177, 178
Single-strand binding protein (SSB), **537**
    unwinding of DNA double helix by, 536–37
sis oncogene, 566
Sister chromatid, **523**, 524*f*
    in meiosis versus in mitosis, 586, 587*f*
Site-specific mutagenesis, **182–83*b***
Situs inversum viscerum, 778*b*
Sjostrand, Fritiof, 6
Skeletal muscle(s), **779–80**
    appearance and nomenclature of, 781*f*
    of birds, 788–89
    effect of elevated cAMP in, 262
    glucose catabolism and glycolysis in, 377–78
    lactate fermentation in, 385
    levels of organization in, 780*f*
    major protein components of, 783*t*
    regulation of calcium levels in, 786–87
    structure of, 779–80, 781*f*
Sliding-filament model of muscle contraction, **782–83**, 784*f*
Sliding-microtubule model, **776–77**
Slow block to polyspermy, 266, 268*f*
Smads protein class, **274–75**, 726
Small intestine, enzyme activity in, 151
Small monomeric G proteins, 259–61
Small ribosomal subunit, **94**
Smooth endoplasmic reticulum (smooth ER), **89**, 325, 330–32
    calcium storage and role of, 332
    carbohydrate metabolism and role of, 331–32
    drug detoxification and role of, 331
Smooth muscle, **790**
    contraction of, 790*f*
    elevation of cAMP in, 262
    nitric oxide and relaxation of, 269–70
    phosphorylation of, 791*f*
    regulation of contraction of, 790–92
    structure of, 790
Snake venom, neurotoxins in, 248–49, 250*b*
SNAPs (soluble NSF attachment proteins), **352**, 353*f*
SNARE hypothesis, 352, 353*f*
SNARE (SNAP receptor) proteins, **352**
snoRNAs (small nucleolar RNAs), **646**
snRNA (small nuclear RNA), 641, **651**
snRNPs (small nuclear ribonucleoproteins), **651**, 652*f*
Sodium butyrate, 715
Sodium channels, voltage-gated, 234–40
Sodium chloride, solubilization of, 23*f*
Sodium dodecyl sulfate (SDS), 180
Sodium/glucose symporter, 214–15
Sodium/potassium (Na+/K+) pump, 201*b*, 208, 210, **211–13**, 228
    model mechanism for, 214*f*
    removal of sodium from cell by, **230**
Solar radiation, 109
Solute(s), **23**
    calculating free energy change for transport of charged and uncharged, 218–20
    carrier protein transport of, 204

Solute(s), *continued*
  membrane permeability and transport
    of, 160, 162 (*see also* Transport across
    membranes)
  simple diffusion and polarity of, 200–201
  simple diffusion and size of, 199–200
Solvent, **23**
  water as, 23–24
Somatic nervous system, **225**
Somatotropin, 276
Sos protein, **271**, 272f
Southern, E. M., 502
Southern blotting, **503**, 611
Spatial summation, **251**
Specific heat, **22**
Specificity
  of carrier proteins, 204
  of enzyme substrates, 134–35
Spectrin, 178f, 179, 184, 759, 761, 762f
Spectrin-ankyrin-actin network, 761, 762f
Sperm, **579**, 744
  formation of, during meiosis, 588, 589f, 590
  mitochondria in, 86f
Spermatocyte, 588
S phase of cell cycle, **524**
Sphere of hydration, **24**
Sphingolipids, **70**, 166
Sphingomyelin, 166, 167f
Sphingosine, **70**
Spindle assembly checkpoint (cell cycle), **556**
  control of, 561, 562f
Spindle equator, 553
  alignment of bivalent chromosomes at,
    during meiosis, 585–86
Spindle fibers, 96
Spindle poles, movement of homologous chro-
  mosomes to, during meiosis, 586
Spiral-and-ribbon model of ribonuclease, 51, 52f
Spliceosomes, 650, **651**
  removal of introns from pre-mRNA by,
    650–51, 652f
  visualization of, by electron microscopy, 653f
Spontaneity, meaning of, 120
Spores, haploid, 581
Sporophyte, **581**
Squid giant axon, **236**
  action potential along, 237f
  calculating resting membrane potential on, 232
  ion channel changes affecting, 239f
  microtubules and, 770
src gene and Src protein, 271, 565
SRP receptor, 681
Stahl, Franklin, 525–27
Staining, negative, 7–8, 404
Standard free energy change ($\delta G^{0\prime}$), 122–24
  for hydrolysis of ATP, 371
  for hydrolysis of phosphorylated compounds
    involved in energy metabolism, 371t, 372
  sample calculations, 124–25
Standard reduction potential ($E_0{}'$), 420, **421**
  determination of, 421f
  for select redox pairs, 421t
Standard state, **122**
Starch, **63**
  in plant cell walls, 29
  structure of, 64f
  synthesis of, from Calvin cycle products,
    467f, 468
Starch grains, 63
Starch phosphorylase, 386
Starch synthase, 468
Start codon, **666**
Startpoint, prokaryotic promoter, 637
Start-transfer sequence, **683**, 684f
STAT (signal transducers and activators of tran-
  scription), 725, **726**
State, **112**

standard, 122
steady, 117, 125
Stationary cisternae model, **335**
Steady state, 117, **125**
Steady-state ion movements, resting membrane
  potential and, **230**–31, 232f
Stearate, 169, 170t
Stem cells, 554, **706**
Stereocilia, 778b
Stereoisomers, 20, 21f
  of amino acids, 42f
Steroid(s), 67f, **70**
Steroid hormone(s), **71**
  as chemical messenger, 257
Steroid hormone receptors, binding of hormone
  response elements to, 724–25
Sterols, **166**–68
  membrane fluidity and effects of, 173–74
Steward, Frederick, 707
Sticky ends, restriction fragment, **494**, 607
Stimulating electrode, 237, 238f
Stomata, **462**
  opening/closing of, 268–69
Stop codon, **634**, **666**
Stop-transfer sequence, **683**, 684f
Storage macromolecules, 28, **29**
  polysaccharides as, 63–65, 66f
Streptococcus pneumoniae, experiments on genet-
  ic transformation of, 481f
Stress fibers, cell migration and, 792, 794f, 798f
Striated muscle, **779**–80, 781f
  regulation of contraction in, 786f
  thin filament of, 781, 782f
Striations
  collagen molecule, 292
  muscle, 779
Stroma, **87**, **448**, 449f
  starch biosynthesis in, 467f, 468
Stroma thykaloids, **87**, 448, 449f
Structural genes, **694**
  mutations in, 696–98
Structural macromolecules, 28, **29**
  polysaccharides as, 63–65, 66f
Structural proteins, 41
Sturtevant, Alfred, 598
Subcellular fractionation, **326**
Submitochondrial particles, 430–31
Subramani, Suresh, 362
Substrate(s), glucose as oxidizable, 375–78
Substrate activation, **138**–39
Substrate analogs, 146
Substrate binding, 137–38, 139f
Substrate concentrations, velocity of enzyme-
  catalyzed reactions related to, 140–43
Substrate induction, **693**, 694
Substrate-level phosphorylation, **381**, 382f,
  400, 426
Substrate-level regulation of enzymes, **147**
Substrate-recognition protein (E3), 733
Substrate specificity, **134**–35
Subthreshold depolarizations, 238
Succinate, 413
  hydrogenation of, from fumarate, 134–35
Succinate-coenzyme Q oxidoreductase complex,
  423, 424t
Succinate dehydrogenase, 134, 418, 419, 423
Succinyl CoA, 409, 414
Sucrose, 62
  synthesis of, from Calvin cycle products,
    467f, 468
Sucrose-6-phosphate, 468
Sucrose-phosphate synthase, 468
Sugars, 61–63
  in glycoproteins, 185, 187f
  metabolism of, in human body, 376–78b
  reactions summarizing synthesis of, 120
Sulfobacteria, 76

Sulfur, photosynthetic assimilation of, 468
Sumner, James B., 133
Sunlight as source of energy, 109
Supercoiled DNA, **489**, 490f
Supernatant, **328**
Superrepressor mutants, 697
Suppressor tRNA, **674**, 675f
Supramolecular structures, 26
Surface area/volume ratio, **77**–78
Surface tension
  membranes and, 162
  of water, 22, 23f
Surroundings, **112**
  heat/work as energy exchange between
    systems and, 112–13
Sutton, Walter, 10, 593
Svedberg, Theodor, 9, 327
Svedberg units (S), 327
Swinging-bucket rotor, 326
Symbiosis, 88
Symbiotic relationship, **450**
Symport transport, **204**, 210
  Na+/glucose, 214–15
Synapse, **227**
  chemical, 243, 245f
  electrical, 227, 243, 244f
  transmission of signal across, 247f
Synapsis, **581**, 585f
Synaptic cleft, **243**
Synaptic knob (terminal bulb), 243
  calcium ions at, 245
Synaptic transmission, 243–50
  calcium levels and secretion of neuro-
    transmitters, 245–47
  detection of neurotransmitters during,
    247–49
  neurotransmission inactivation following,
    249–50
  role of neurotransmitters in relaying signals,
    243–45
Synaptobrevin, 247
Synaptonemal complex, **585**
  homologous recombination facilitated by, 606
Synaptotagmin, 247
Syntaxin, 247
Synthetic work, 107
System(s), **112**
  first law of thermodynamics and, 112–14
  heat, work, and energy exchange between
    surroundings and, 112–13
  open, and closed, 112f
  second law of thermodynamics and, 114–15
  specific state of, 112

T

Tabun, 250
Tags, ER-specific proteins with retrieval, 338,
  339f
Talin, 300, 301f, 794
Tandemly repeated DNA, **500**–501
Tangles, Alzheimer's disease and intracellular, 673
Target tissue for hormones, **276**, 277f
TATAAT sequence (Pribnow box), 637
TATA-binding protein (TBP), **643**, 644f
TATA box, **641**, 642f
TATA-driven promoters, 641
Tatum, Edward L., 10, 624–25
Tau gene, 754, 755f
Tautomerization, 381
Taxol, 744, 753
Taylor, J. Herbert, 602
TCA cycle. See Tricarboxylic acid (TCA) cycle
  (Krebs cycle)
T DNA region, 614
Teichoic acids, 98
Tektin, **776**

Telomerase, **539**, 540*f*
Telomere(s), **500**–**501**, **539**
 DNA end-replication problem solved by, 539–41
 extension of, by telomerase, 540*f*
Telomere capping protein, 540
Telophase (mitosis), 547*f*, **548**–49, 587*f*
Telophase I (meiosis), 586, 587*f*, 588
TEL sequence, 539, 613*f*
Temin, Howard, 626
Temperate phage, 484*b*
Temperature
 enzyme sensitivity to, 135–37
 water and stabilization of, 22–23
Temperature-sensitive mutants, **530**, 558
Template, nucleic acid, **57**
Template strand, **632**
Temporal summation, **251**
-10 sequence, 637
Terminal bulb (synaptic knob), **243**
 calcium ions in, 245
Terminal cisterna, 787
Terminal glycosylation, **336**
Terminal oxidase, **424**
Terminal web, **759**, 760*f*
Termination signal, 635, **639**
Terpene(s), 67*f*, **71**
Tertiary alcohol, 407
Tertiary structure of protein, 47*f*, **50**–**53**
Testosterone, 70*f*, 71, 276
Tetanus, 247
Tetrahedral (carbon atom), **20**
TFIIB recognition element (BRE), **641**
TGN (*trans*-Golgi network), **334**
Thanatophoric dysplasia, 273–74
Theory, scientific, **13**
Thermacidophiles, 76
Thermodynamics, 113
 entropy, free energy, and, 115–25
 first law of, 113–14
 free energy change calculations and, 124–25
 second law of, 114–15
 systems, heat, work, and, 112–13
Thermodynamic spontaneity, **114**
 in electron transport chain, 422
 entropy change as measure of, 118
 free energy change as measure of, 119–20
 meaning of spontaneity, 120
 second law of thermodynamics and, 115
Thermophilic archaebacteria, 137
Thick filament of myofibril sarcomere, **779**, 780, 781*f*, 782*f*
Thin filament of myofibril sarcomere, **779**, 780*f*, 781, 782*f*
Thin-layer chromatography (TLC), **168**, 169*f*
Thioredoxin-mediated activation of Calvin cycle, 465, 466*f*
30-nm chromatin fiber, **507**, 508*f*
30S initiation complex, **669**
-35 sequence, 637
3′, 5′-cyclic AMP (cAMP), **700**
Threonine, 43*f*, 186*f*, 634
 synthesis of isoleucine from, 148*f*
Threonine deaminase, 148
Threshold depolarization, **236**, 238
Threshold potential, **236**
Thrombospondin, 570
Thromboxan A2, 282
Thudicum, Johann, 70
Thylakoid(s), **87**, **448**, 449*f*
 major energy transduction complexes within, 457*f*, 460, 461
Thylakoid lumen, **449**
Thylakoid signal sequence, 686
Thymine (T), **55**
 Chargaff's rules and, 486, 487*t*
 DNA damage repair, uracil, and, 544

Thymosin β4, 758
Thyroxine, 276
TIC (translocase of inner chloroplast membrane), **685**
Tight junctions, **98**, 306, 307*f*, **310**–13
 membrane protein movement, transcellular transport, and role of, 312–13
 permeability barrier created by, 312*f*
 restraints on membrane proteins by, 189
 structure of, 311*f*–12
TIM (translocase of inner mitochondrial membrane), **685**
Ti plasmid, **614**–15
Tissue(s), intermediate filaments and strength of, 764–65
Tissue plasminogen activator (TPA), 614
Titin protein, 781
T lymphocytes, 616
TnC polypeptide chain, **781**
TnI polypeptide chain, **781**
TnT polypeptide chain, **781**
Tobacco mosaic virus (TMV), 98
 self-assembly of, 35, 36*f*
 structure of, 35*f*, 51
TOC (translocase of outer chloroplast membrane), **685**
TOM (translocase of outer mitochondiral membrane), **685**
Tomato, genetically modified, 615*f*
Tonofilaments, **310**, 762, 764
Tonoplast, 93
Topoisomerases, type I and type II, **489**, **536**
 unwinding of DNA strands by, 536–37
Totipotency, 707
Toxics/toxins
 α-amanitin, 640–41
 cholera, 262
 enzyme inhibitors as, 146
 neurotoxins, 146, 247, 250*b*
 pertussis, 262
Trailer (mRNA sequence), **666**
Trait, genetic, 578
Trans-acting factor, **698**
Transamination, 414
Transcellular transport, 312–13
Transcribed spacers, 645
Transcription, 54, 479, **623**
 chromatin decondensation required for, 710–14
 in eukaryotic cells, 640–55
 gene regulation in lactose catabolism and, 695–98
 negative control of, 700
 overview of, 636*f*
 positive control of, 700–701
 in prokaryotic cells, 635–40
 regulation of, after inititation, 701–3
 reverse, 624, 626–27*b*
 sigma factors and regulation of initiation of, 701
 stages of, 635–40
 stages of regulation of eukaryotic, 705*f*
Transcriptional control, **715**–**28**, 737*b*
 activation of CREBs and STATs, 725–26
 coactivator proteins and, 719–21
 combined action of multiple DNA control elements and transcription factors in, 721, 722*f*
 differential gene transcription allowed by, 715–16
 DNA microarray and monitoring, 716–17
 DNA response elements and expression of nonadjacent genes, 724
 general transcription factors and, 644
 heat-shock response element and, 726–27
 homeotic gene-coding of transcripton factors and, 727–28

location of enhancers and silencers from promoter and, 718–19
location of proximal control elements near promoter and, 717–18
steroid hormone receptors as transcription factors, 724–25
structural motifs allowing binding of transcription factors to DNA, 721–24
Transcription factors, **563**, 640, **643**
 activation of, by phosphorylation (CREBs and STATs), 725–26
 AP1, 271
 combined action of DNA control elements and, 721, 722*f*
 DNA control elements bound to, 718*t*
 E2F, 560, 561*f*
 Ets family of, 563
 general, 643–44, 717
 homeotic genes and coding for development-related, 727–28
 oncogenes and, 567
 regulatory (*see* Regulatory transcription factors)
 repressors as, 719
 steroid hormone receptors as, 724–25
Transcription regulation domain, **721**
Transcription unit, **635**
Transcytosis, **348**
Transducing phages, 599
Transduction in bacteria, **599**–600
trans (maturing) face, **334**
Transfection, 565
Transferases, 135, 136*t*
Transferrin receptor, 352, 731
Transfer RNA (tRNA), 31, 44, 54, **623**, 635, 660, **662**–**64**
 amino acids transferred to ribosome by, 662–64
 initiator tRNA, 669
 processing and secondary structure of, 646, 647*f*
 promoters for, 642
 sequence, structure, and aminoacylation of, 663*f*
Transformation, **599**
 in bacteria, 481–82, 599–600
 genetic, 481–82
 oncogene identification and, 565
Transforming growth factor β (TGFβ), 274, 275*f*, **564**
Transgenic organisms, **614**–18
trans-Golgi network (TGN), **334**
 clathrin-coated vesicles and, 350–51
 lysosomal protein sorting in, 339–31
Transient fusion model for endosome-lysosome transfers, 354
Transitional elements (TEs), **325**
Transition state, **131**
Transition temperature (T_m), **172**
Transition vesicles, **325**
Transit peptidase, 685
Transit sequence, **684**
Transit sequence receptors, 685
Transketolases, 464
Translation, 54, 479, **623**, **666**–**74**
 chain elongation in, 669–71
 energy budget of protein synthesis and, 672–74
 initiation of, 667–69
 linkage of amino acids to tRNA by amino-acyl-tRNA synthetases in, 664–66
 molecular chaperones, polypeptide folding, and, 672, 681
 mRNA and polypeptide coding-information transfer during, 666
 nonsense mutations, suppressor tRNA and, 674–75

Translation, *continued*
  overview of, 666, 667f
  posttranslational processing, 675–76
  protein factors and polypeptide chain synthesis in, 666
  rates of, controlled by initiation factors and translational repressors, 730–31
  ribosome as site of polypeptide synthesis and, 660–62
  summary of, 674
  termination of polypeptide synthesis and, 671–72
  tRNA and transfer of amino acids to ribosome in, 662–64
Translational control, **730**–31, 737b
Translational repressor, **731**
Translesion synthesis, **543**
Translocase, 685
Translocation
  chromosomal, 566, 676
  of polypeptides across ER membrane, 681
  of polypeptides into chloroplasts and mitochondria, 685
  polypeptide synthesis and, 670f, **671**
  rearward, of actin filaments, 793, 796f
Translocon, 680, **681**
Transmembrane domains of CFTR protein, 212b, 213f
Transmembrane electrochemical proton gradient, 418
Transmembrane proteins, 84, **177**
  cell-cell recognition and adhesion and, 302–4
  cotranslational insertion of, 683, 684f
  porins as, 448
Transmembrane receptor, phosphorylation of, 275
Transmembrane segments, 164, 165f, **177**–78, 683
Transmission electron microscope (TEM), 7
Transport, 195. *See also* Transport across membranes
  fast axonal, 770–71
  motor MAPs and intracellular-vesicle, 773
  nuclear pores and molecular, 513–15
  tight junctions and transcellular, 312–13
Transport across membranes, 160, 161, 195–224
  by active transport, 160, 197, 207–10
  active transport examples, 211–18
  bioenergetics of, 218–20
  cells and transport processes, 195–97
  coated vesicles and, 349–53
  cystic fibrosis, gene therapy, and, 210, 212–13b, 616, 618
  energetics of, 218–20
  in erythrocyte, 197f
  in eukaryotic cell, 196f, 197
  exocytosis and endocytosis in, 80, 160, 195
  by facilitated diffusion, 160, 197, 203–7
  osmosis, 200–201b
  passive, 160
  simple diffusion, 160, 196, 197–203
  tight junctions and, 312f
Transport ATPases, 184, 208–10
  ABC-type, 209–10
  F-type, 208–9
  main types of, 209t
  P-type, 208, 211f
  V-type, 208
Transporter (nuclear pore complex), **513**
Transport protein(s), 184, **197**
  active transport and role of, 197
  facilitated diffusion and role of, 160, 197, 203–7
  in plasma membrane, 25, 26f, 84
Transport vesicles, 323, **333**–34. *See also* Coated vesicles
Transposable elements, 501
Transposition of retrotranspons, 627b
Transverse diffusion, **169**, 171f
Transverse (T) tubule system, 787

Treadmilling of microtubules, **748**, 749f
Triacylglycerol(s) (triglycerides), 67f, **68**, **412**–13
Triad, muscle cells, **788**
Tricarboxylic acid (TCA) cycle (Krebs cycle), 9, 83, 391, 400, **405**–15
  allosteric regulation of enzymes of, 410–12
  ATP and GTP generation in, 408f, 409
  FADH$_2$ and NADH formation in, 409–10
  fat and protein catabolism and role of, 412–14
  glyoxylate cycle and, 415
  initiation of, with acetyl CoA, 406, 408f
  NADH formation and release of carbon dioxide in, 406–9
  oxidative decarboxylation and formation of acetyl CoA in, 406, 407f
  as source of precursors for anabolic pathways, 414, 415f
  summary of, 410
Trillium, genome of, 492
Triose phosphate isomerase, 464
Triplet code (genetic code), **628**–31
  evidence for, 630–31
  frameshift mutations in, 628–30, 631f, 676, 677f
Triplet repeat amplification, 501
Triskelions, **350**, 351f
trk oncogene, 565
tRNA^{Met} and tRNA^{fMet}, 669
Tropomyosin, **781**
Troponin, **781**
trp operon, **699**, 722
  attenuation of, 701–2, 703f
  leader of *trp* mRNA, 702f
  negative control of transcription and, 700
True-breeding plant strains, **590**
Trypsin, 137, 626, 629f
Trypsinogen, 151
Tryptophan, 43f
  gene regulation in synthesis of, 693–94, 701–3
Tryptophan synthetase, 628, 630f
t-SNAREs (target-SNAP) receptors, **352**, 353f
Tubules, eukaryotic, 79–80
Tubulin, **746**
  α and β forms of, 96–97, 746
  as building blocks of microtubules, 96–97, 551, 743, 746–47
  microtubule assembly by addition of tubulin dimers, 747, 748f, 749f
Tubulin isotypes, 747
Tumor, **564**–65. *See also* Cancer
  attacking blood supply of, 570b
  drug resistance in, 210
Tumor necrosis factor, 284, 614
Tumor suppressor gene, **567**–69
  examples of, 568t
Turgor pressure, 93, 201b, 268
Turnover, protein, 705–6
Turnover number (k$_{cat}$), **143**
  values of, for select enzymes, 143t
Type II glycogenosis, 357
Type II myosin, **778**
Tyrosine, 43f
Tyrosine aminotransferase, 735f
Tyrosine kinase. *See* Receptor tyrosine kinases

U
Ubiquinone, 419
Ubiquitin, 561, 568, **733**
  protein degradation by proteasomes dependent on, 733f, 734
Ubiquitin-activating enzyme (E1), 733
Ubiquitin-conjugating enzyme (E2), 733
UDP-glucose, 468
UGA codon, 635
Ultracentrifuge, **9**–10, **326**
Ultramicrotome, 7
Ultraviolet radiation
  as cause of DNA mutations, 542f

DNA absorption spectrum for, 490f
  hazards of, to biological molecules, 19
Unattached glycocalyx, 302
Uncoating ATPase, 351
Undershoot (hyperpolarization), 237, **238**–39
Ungewickell, Ernst, 350
Unidirectional pumping of protons, **428**
Uniport transport, **204**
  glucose transporter as, 204–6
Unsaturated fatty acid, **68**, 69f
  effects of, on packing of membrane lipids, 173f
Unwin, Nigel, 164–65
Upstream control element, 641, 642f
Upstream DNA sequence, **637**
Uracil (U), **56**, 633
  DNA damage repair, thymine, and, 544
Uracil-DNA glycosylase, 544
Urate oxidase, 358, 359, 361
Urease, 133
Uricase, 361
Uridine triphosphate (UTP), 468
Usher's syndrome, 778b

V
Vacuoles, 79, **91**–93, 94f, **357**–58
Valence, **18**
Valine, 43f, 51
  hemoglobin alterations and, 627–28, 629f
Van der Waals interactions (forces), 34, **46**
Van der Waals radius, 34
Variable loop, pre-tRNA, 646, 647f
Variable number tandem repeat (VNTR), **503**
Variant Creutzfeldt-Jakob disease (vCJD), 673
Vascular endothelial growth factor (VEGF), 570
Vasodilator, 269
Vasopressin (antidiuretic hormone), 276
Velocity of enzyme-catalyzed reactions related to substrate concentration, 140–43
Vesicle, 79
  labeling proteins on surface of membrane, 185f
  motor MAPs and transport of intracellular, 773
  neurosecretory, 245–48
  secretory, 90
Viagra (drug), 270
Villin, 759
Vimentin, 310, 762, 763t
Vinblastine, 753
Vincristine, 753
Vinculin, 300, 301f, 794
Virchow, Rudolf, 3
Viroids, 98
Viron, 35, 36f
Virulent phage, 484b, 485f
Virus(es), **98**–99
  genetic recombination in, 598–602
  A. Hershey and M. Chase's studies on genetic material of, 10, 482–85
  oncogenes and, 565, 566, 567, 626–27b
  retroviruses, 626–27b
  self-assembly of tobacco mosaic, 35, 36f
  sizes and shapes of, 100f
Vitalism, 133
Vitamin(s)
  A$_1$, 71
  pantothenic acid (B), 406
  riboflavin (B), 409
Vitelline envelope, 266
V$_{max}$. *See* Maximum velocity (V$_{max}$)
Voltage (potential), **229**
Voltage-gated ion channel, 206, **233**
  action potential and opening/closing of, 237–40
  domains of, as sensors and inactivators, 234–36
  function of, 235f
  structure of, 234f
Voltage-gated potassium channel, 234, 237–38
Voltage-gated sodium channel, 234–40
Voltage sensor, **235**

Von Tschermak, Ernst, 10
V-SNAREs (vesicle-SNAP) receptors, **352**, 353f
V-type ATPase, **208**, 209t

W

Wald, George, 36
Walker, John, 432
Warburg, Otto, 9, 401
Watchmaking example of hierarchical
    assembly, 37b
Water, 17, 21–24
    biospheric flow of, 111
    cohesion of molecules, 22, 23f
    high temperature-stabilizing capacity of,
        22–23
    hydrogen bonds in molecules of, 22f
    osmosis and transport of, across membranes,
        199, 200–201b
    photosystems and transfer of electrons to
        plastoquinone from, 456–57
    polarity of molecules, 22
    as solvent, 23–24
Water photolysis, 457

Watson, James, 10, 58, 60–61b, 487–89, 525
Watson-Crick model of DNA replication, 525f
Weigle, Jean, 602
Weissman, August, 10
Wheat germ agglutinin, 186
Whooping cough, 262
Wild type cells, **596**
    pseudo, 629
Wilkins, Maurice, 487
Wilmut, Ian, 707, 708b
Wiskott Aldrich syndrome protein (WASP), 760
Wobble hypothesis, **664**
Woese, Carl, 76
Wöhler, Friedrich, 8
Wood, 316
Work, **113**
    energy and types of, in cells, 107–9

X

X-chromosome(s), 578
X-chromosome inactivation, **714**
Xenobiotics, **361**
Xeroderma pigmentosum, 543, 570

X-linked adrenoleukodystrophy, 361

Y

Yanofsky, Charles, 628, 701, 702
Y chromosome(s), 578
Yeast (*Saccharomyces cerevisiae*), 554
    mating-type DNA rearrangements, 709–10
Yeast artificial chromosome (YAC), **613**–14
    construction of, 613f
Yersinia pseudotuberculosis, 308b

Z

Zacharias, Eduard, 480
Z-DNA, 59, 489
Zinc finger motif, **722**, 723f
Z line, muscle, **780**, 781f, 782
Zygote, **579**
Zygotene phase of meiosis prophase I, **583**, 584f,
    585f
Zymogen, activation of, by proteolytic
    cleavage, 151
Zymogen granules, 342f